Volume II

COMPOUNDS OF GERMANIUM, TIN AND LEAD
INCLUDING BIOLOGICAL ACTIVITY
AND COMMERCIAL APPLICATION

FIRST SUPPLEMENT

ORGANOMETALLIC COMPOUNDS

METHODS OF SYNTHESIS
PHYSICAL CONSTANTS AND CHEMICAL REACTIONS

Edited by
MICHAEL DUB
Central Research Department, Monsanto Company

Volume II

COMPOUNDS OF GERMANIUM, TIN AND LEAD
INCLUDING BIOLOGICAL ACTIVITY
AND COMMERCIAL APPLICATION

Second Edition

FIRST SUPPLEMENT

By
RICHARD W. WEISS
Fachhochschule Lippe, Lemgo

Covering the Literature from 1965 to 1968

Springer Science+Business Media, LLC
1973

All rights reserved

No part of this book may be translated or reproduced in
any form without written permission from Springer Science+Business Media, LLC.

© 1972 Springer Science+Business Media New York
Originally published by Springer-Verlag New York Inc. in 1972
Library of Congress Catalog Card Number 66-28249

ISBN 978-3-642-50292-7   ISBN 978-3-642-50290-3 (eBook)
DOI 10.1007/978-3-642-50290-3

PREFACE TO THE FIRST SUPPLEMENT VOLUMES

The output of chemical literature has increased to such an extent that the four-year supplements to Volumes I and II cover more references than the respective main volumes contained for the period from 1937 through 1964. The supplement to the third volume includes almost three fourths as many references as the main volume.

In view of this enormous increase in the publications, the preparation of the supplements was only possible through the cooperation of Dr. Richard W. Weiss, Fachhochschule Lippe, Lemgo, Germany, who further carried out the work on Volume II, and of Drs. Klaus Bauer and Gilbrecht Haller, University of Vienna, Austria, who took over the coverage of Volume I.

This supplement series covers the literature and patents abstracted by Chemical Abstracts, Volumes 62 through 69, and includes some references from Volumes 60 and 61 that were omitted in the main volumes of the series. While the collection of references was based on Chemical Abstracts, for the most part original articles and patent specifications were used in compiling the supplements.

The classification systems of the main volumes have been retained and expanded by inclusion of some new types of compounds unknown through 1964. Each supplement volume was prepared independently; therefore some slight, inadvertent variations in their forms probably occurred.

Numerous patents and several articles from Soviet journals were obtained from the USSR National Public Library for Science and Technology in exchange for our publications. The cooperation of Messrs. V. Orlov and N. Tyshkevich, directors of the Moscow library, in the exchange is very much appreciated.

Support provided by the Monsanto Company Information Center is also acknowledged with thanks.

Michael Dub

St. Louis, Missouri
March 1972

INTRODUCTION TO THE SUPPLEMENT VOLUME II

Presentation and classification follow the previous arrangement. The decimal system is expanded to absorb new compounds and chapters. The enormous increase in publications has delayed completion of this supplement volume considerably.

Literature data are reported for each individual compound. Whenever possible entries are combined in tables for easier reference. Schemes compile synthetic methods common to the entries in every table, regardless of stoichiometry and by-products. Citation to compounds reported in the main volume is given for each compound by a capital M and the page number, e.g. Tetraethyllead (M-521). In the interest of brevity certain numerical data are omitted, but references are always presented. Yields are rounded to two significant figures.

The paragraph Biological Properties includes all tested species for each organic compound of the three metals disregarding activity and systematic order. Entries for groups, e.g. triethyltin are located in Chapter 8.2.

In the paragraph Addition Compounds one will find most of the derivatives where the central metal reaches or exceeds a covalency of five. Due to the classification system and the character of the series no new section for such structures was added.

Question marks following formula or name of a compound usually express the authors uncertainty of a proposed structure. The remark sic indicates the editors suspicion of a misprint in the original paper or the abstract.

I am greatly indebted to my good friend and former colleague Dr. M. W. Dietrich who helped me communicating halfway around the world.

I am very grateful to Mrs. Mary Alice Doiron who again prepared an excellent typescript and who was so friendly assisted by Mrs. Mary Jean Spreitler.

R. Weiss

Fachhochschule Lippe,
Lemgo, December 1972

FROM INTRODUCTION TO VOLUME II

Volume II of the series "Organometallic Compounds" covers organic derivatives of germanium, tin, and lead. The term organometallic compound is used in the strictest sense and includes only compounds with at least one metal-carbon bond. We attempt to present complete information on the organic derivatives of the three metals. Literature data for each compound are given in separate paragraphs on Synthesis, Properties, Chemical Reactions, and Addition Compounds. Two new entries not covered in Volume I for Use and Biological Properties are added. This recognizes the commercial interest reflected in numerous publications connected with the industrial application of organometallic derivatives of tin and lead.

To facilitate the use of this volume and to allow correlation of analogous organometallic derivatives of germanium, tin, and lead, the individual compounds are listed systematically. A decimal system is used to characterize chapters and subchapters. The system and arrangement will be obvious from the Table of Contents. The arrangement follows three simple principles: order with decreasing number of metal-carbon bonds; order with increasing complexity of organic groups; consistent use of a sequence of heteroatoms.

The compounds are listed in order of the organic moieties linked to the central metal in the series aliphatic, aromatic, heterocyclic, unsaturated, and functionally substituted groups; aralkyl groups are arranged according to the location of the carbon-metal bond within the above series: benzyl follows methyl but precedes ethyl; tolyl follows phenyl and precedes naphthyl.

The sequence of the heteroatoms as substituents follows the reverse of their position in the Periodic Table, i.e., halogen, oxygen and homologues, nitrogen and homologues, etc. A group of closely related substituents, with often uncertain linking atoms, is listed as pseudohalides and follows halogen. All salts or esters of oxygen acids, excluding carboxylic acids, are combined under Salts. Polyfunctional groups are listed under the heteroatom next to the central metal or the function first appearing in the system. Cross references are given. The nomenclature is generally in agreement with Chemical Abstracts.

R. Weiss

St. Louis, November 1966

ABBREVIATIONS

The following abbreviations are used in addition to or at variance with those in Chemical Abstracts:

AiBN	azobisisobutyronitrile
BiPy	2,2'-dipyridine
c.	cyclo
cat.	catalyst
corr.	corresponding
dec.	decomposition, decomposes, decomposing
GLC	gas-liquid-phase chromatography
IR	infrared
liq.	liquid
m.	melting point, melts at
M	a metal atom in generic formula
NMR	nuclear (proton) magnetic resonance
NQR	nuclear quadruple resonance
phen	1,10-phenanthroline
prod.	product(s)
PVC	polyvinyl chloride
Py	pyridine
ref.	reference(s)
r.t.	room temperature
rxn.	reaction(s)
sepd.	separated
subl.	sublimes at, sublimation
THF	tetrahydrofuran
TLC	thin-layer chromatography
UV	ultraviolet

CONTENTS

PREFACE TO THE FIRST SUPPLEMENT VOLUMES		V
INTRODUCTION TO THE SUPPLEMENT VOLUME II		VII
FROM INTRODUCTION TO VOLUME II		VIII
ABBREVIATIONS		IX

GERMANIUM

1.	Symmetric Tetraorganogermanes	1
1.1	Nonfunctional Symmetric Tetraorganogermanes	1
1.3	Functional Symmetric Tetraorganogermanes	4
2.	Unsymmetric Tetraorganogermanes	6
2.1	Nonfunctional Unsymmetric Tetraorganogermanes	6
2.1.1	Derivatives with Two Different Organic Groups	6
2.1.2	Tetraorganogermanes with Three Different Organic Groups	16
2.1.3	Nonfunctional Asymmetric Tetraorganogermanes	16
2.3	Heterocyclic Substituted Tetraorganogermanes	16
2.4	Unsymmetric Tetraorganogermanes with Olefinic Substituents	20
2.4.1	Unsymmetric Tetraorganogermanes Containing Halogen Substituted Olefins	29
2.4.3	Unsymmetric Tetraorganogermanes Containing Functionally Substituted Olefins	32
2.5	Unsymmetric Tetraorganogermanes with Acetylenic Substituents	37
2.5.1	Unsymmetric Tetraorganogermanes Containing Halogen Substituted Acetylenes	44
2.5.3	Unsymmetric Tetraorganogermanes Containing Functionally Substituted Acetylenes	45
2.6	Functionally Substituted Unsymmetric Tetraorganogermanes	50
2.6.1	Tetraorganogermanes Containing Halogen	50
2.6.2	Tetraorganogermanes Containing Pseudohalogen Substituents	51
2.6.3	Tetraorganogermanes Containing Oxygen	55
2.6.3.1	Organogermanes with Hydroxy Groups and Their Derivatives	55
2.6.3.3	Tetraorganogermanes Containing Carbonyl Groups	60
2.6.3.4	Tetraorganogermanes Containing Carboxyl Groups and Derivatives	62
2.6.4	Tetraorganogermanes Containing Sulfur	67
2.6.5	Tetraorganogermanes Containing Nitrogen and Phosphorus	67
2.7	Polymetallic Tetraorganogermanes	69
2.7.1	Bridged Polyorganogermanes	69
2.7.1.1	Tetraorganogermanes with Two Germanium Atoms	69
2.7.1.1.1	Nonfunctional Derivatives	69

2.7.1.1.3	Functionally Substituted Bis(tetraorganogermanes)	72
2.7.1.2	Tetraorganogermanes Containing Three to Four Ge-Atoms	73
2.7.1.8	Polymeric Tetraorganogermanes	76
2.7.2	Tetraorganogermanes Containing Organosilicon Substituents	78
2.7.3	Tetraorganogermanes Containing Organotin Substituents	78
2.7.4	Tetraorganogermanes Containing Organolead Substituents	78
2.7.5	Tetraorganogermanes Containing Group III Metals	81
2.7.6	Tetraorganogermanes Containing Magnesium	81
2.7.7	Tetraorganogermanes Containing Alkali Metals	82
2.7.8	Tetraorganogermanes Containing Transition Metals	83
3.	Organogermanium Hydrides	83
3.1	Nonfunctional Organogermanium Hydrides	83
3.2	Unsymmetric Organogermanium Hydrides	92
3.3	Deuterium Substituted Organogermanium Hydrides	94
3.6	Functionally Substituted Organogermanium Hydrides	96
3.8	Organogermanium Hydrides Containing Heteroatom-Germanium Bonds	99
3.8.1	Organogermanium Hydride Halides	99
3.8.2	Organogermanium Hydride Pseudohalides	102
3.8.3	Other Organogermanium Heterohydrides	104
4.	Organogermanium Compounds Containing Heteroatom-Germanium Bonds	106
4.1	Compounds with Halogen	106
4.1.1	Triorganogermanium Halides	106
4.1.1.1	Unsubstituted Symmetric Triorganogermanium Halides	106
4.1.1.2	Unsymmetric Unsubstituted Triorganogermanium Halides	115
4.1.1.6	Functionally Substituted Triorganogermanium Halides	120
4.1.2	Diorganogermanium Dihalides	123
4.1.2.1	Unsubstituted Symmetric Diorganogermanium Dihalides	123
4.1.2.2	Unsymmetric Unsubstituted Diorganogermanium Dihalides	128
4.1.2.6	Functionally Substituted Diorganogermanium Dihalides	129
4.1.3	Organogermanium Trihalides	133
4.1.3.1	Unsubstituted Organogermanium Trihalides	133
4.1.3.4	Unsaturated Organogermanium Trihalides	139
4.1.3.6	Functionally Substituted Organogermanium Trihalides	141
4.1.3.6.1	Halogen Substituted Organogermanium Trihalides	141
4.1.3.6.2	Pseudohalogen Substituted Organogermanium Trihalides	141
4.1.3.6.3	Oxygen Containing Organogermanium Trihalides	141
4.1.3.6.3.1	Organogermanium Trihalides Containing Hydroxy Derivatives	143
4.1.3.6.3.3	Derivatives Containing Carbonyl Groups	143
4.1.3.6.3.4	Compounds Containing Carboxylic Groups	143
4.1.3.6.5	Organogermanium Trihalides Containing Phosphorus	143
4.1.3.6.7	Polymetallic Organogermanium Trihalides	148
4.1.3.6.7.1	Bridged Polyorganogermanium Trihalides	148
4.1.3.6.7.2	Organogermanium Trihalides Containing Silicon and Boron	148
4.1.4	Organogermanium Halides Containing Germanium-Pseudohalogen Bonds	151

4.1.5	Organogermanium Halides Containing Germanium-Oxygen Bonds	151
4.1.6	Organogermanium Halides Containing Germanium-Sulfur Bonds	153
4.1.7	Organogermanium Halides Containing Germanium-Nitrogen Bonds	153
4.1.8	Organogermanium Halides Containing Germanium-Metal Bonds	154
4.2	Organogermanium Salts	154
4.2.1	Organogermanium Pseudohalides	154
4.2.4	Organogermanium Salts or Esters of Oxygen Acids	157
4.3	Organogermanes with Germanium-Oxygen Bonds	159
4.3.1	Organogermanium Oxides and Hydroxides	159
4.3.1.1	Triorganogermanium Oxides and Hydroxides	159
4.3.1.1.1	Unsubstituted Triorganogermanium Oxides and Hydroxides	159
4.3.1.1.6	Substituted Triorganogermanium Oxides and Hydroxides	162
4.3.1.2	Diorganogermanium Oxides	164
4.3.1.3	Monoorganogermanium Oxides	165
4.3.2	Organogermanium Peroxy Compounds	165
4.3.3	Organogermanium Alkoxides and Phenoxides	166
4.3.3.1	Triorganogermanium Alkoxides	166
4.3.3.1.1	Unsubstituted Triorganogermanium Alkoxides	166
4.3.3.1.6	Functionally Substituted Triorganogermanium Alkoxides	172
4.3.3.1.8	Triorganogermanium Alkoxides Derived from Dihydric Alcohols	173
4.3.3.1.9	Triorganogermanium Mixed Oxides	175
4.3.3.2	Diorganogermanium Dialkoxides	175
4.3.3.2.1	Unsubstituted Diorganogermanium Alkoxides	175
4.3.3.2.6	Functionally Substituted Diorganogermanium Dialkoxides	178
4.3.3.2.8	Diorganogermanium Alkoxides Derived from Dihydric Alcohols	178
4.3.3.2.9	Diorganogermanium Mixed Oxides	182
4.3.3.3	Monoorganogermanium Trialkoxides	182
4.3.3.3.1	Unsubstituted Organogermanium Trialkoxides	182
4.3.3.3.6	Functionally Substituted Monoorganogermanium Trialkoxides	184
4.3.3.4	Organogermanium Alkoxides Containing Germanium-Heteroatom Bonds	185
4.3.4	Organogermanium Acetals and Enolates	185
4.3.5	Organogermanium Carboxylates	188
4.3.5.1	Triorganogermanium Carboxylates	188
4.3.5.1.1	Unsubstituted Triorganogermanium Carboxylates	188
4.3.5.1.6	Functionally Substituted Triorganogermanium Carboxylates	189
4.3.5.2	Diorganogermanium Carboxylates	192
4.3.5.3	Monoorganogermanium Carboxylates	192
4.4	Organogermanes with Sulfur, Selenium and Tellurium	194
4.4.1	Organogermanium Sulfides, Selenides and Tellurides	194
4.4.1.4	Organogermanium Thiols and Selenols	197

4.4.2	Organogermanium Mercaptides, Organoselenides and Organotellurides	198
4.4.2.1	Triorganogermanium Mercaptides, Organoselenides and Organotellurides	198
4.4.2.1.1	Unsubstituted Derivatives	198
4.4.2.1.6	Functionally Substituted Triorganogermanium Mercaptides and Organoselenides	202
4.4.2.2	Diorganogermanium Mercaptides	202
4.4.2.2.1	Unsubstituted Diorganogermanium Mercaptides	202
4.4.2.2.6	Functionally Substituted Diorganogermanium Mercaptides	202
4.4.2.3	Monoorganogermanium Trimercaptides	207
4.4.4	Organogermanium Derivatives of Organic Thioacids	207
4.4.7	Metal Substituted Organogermanium Sulfides, Selenides and Tellurides	208
4.5	Organogermanium Derivatives with Nitrogen, Phosphorus, Arsenic, Antimony and Bismuth	210
4.5.1	Organogermanium Amides	210
4.5.1.1	Triorganogermanium Amides	210
4.5.1.2	Diorganogermanium Diamides	213
4.5.1.3	Monoorganogermanium Triamides	215
4.5.2	Organogermanium Compounds with Substituted Amines	215
4.5.2.1	Organogermanium Derivatives of Heterocyclic Amines	215
4.5.2.4	Organogermanium Derivatives of Amidines, Imides and Hydrazine	215
4.5.3	Organogermanium Derivatives of Acid Amides and Acid Imides	218
4.5.4	Organogermanium Metal Imides	219
4.5.5	Organogermanium Amido and Imido Phosphines	221
4.5.6	Organogermanium Phosphides, Arsenides, Antimonides and Bismuthides	223
4.5.6.1	Organogermanium Phosphides	223
4.5.6.4	Organogermanium Arsenides, Antimonides and Bismuthides	224
5.	Organo-Polygermanes	227
5.1	Hexaorganodigermanes	227
5.1.1	Symmetric Derivatives	227
5.1.2	Unsymmetric Hexaorganodigermanes	230
5.2	Organo-Polygermanes Containing Three or More Germanium Atoms	231
5.3	Organo-Polygermanium Hydrides	234
5.4	Organo-Polygermanes Containing Heteroatoms	236
6.	Polymetallic Organogermanes	238
6.1	Organogermanium Derivatives with Group IV Metals	238
6.1.1	Organogermanes with Ge-Si Bonds	238
6.1.2	Organogermanes with Ge-Sn Bonds	241
6.1.2.1	Bimetallic Organogermanes with Ge-Sn Bonds	241
6.1.2.2	Polymetallic Organogermanes with Ge-Sn Bonds	243

6.1.2.6	Functionally Substituted Polymetallic Organogermanes with Ge-Sn Bonds	243
6.1.2.6.3	Derivatives with Oxygen	243
6.1.2.6.5	Nitrogen Substituted Organogermyl Tin Compounds	245
6.1.3	Bimetallic Organogermyl Lead Compounds	246
6.2	Organogermanium Derivatives of Group III Metals	247
6.3	Organogermanium Derivatives of Group II Metals	248
6.4	Organogermyl Alkali Metal Compounds	251
6.5	Polymetallic Organogermanium Derivatives with Coordination Compounds of Transition Metals	255
6.5.1	Bimetallic Organogermanium Derivatives of Groups I, IV and VI Transition Metals	255
6.5.5	Polymetallic Organogermanium Derivatives of Manganese and Rhenium	260
6.5.6	Bimetallic Organogermyl Iron Compounds	263
6.5.7	Bimetallic Organogermanium Derivatives of Cobalt and Iridium	265
6.5.8	Polymetallic Organogermanium Derivatives of Nickel and Homologues	267
7.	Cyclic Organogermanes	273
7.1	Cyclic Tetraorganogermanes	273
7.1.1	Unsubstituted Cyclic Tetraorganogermanes	273
7.1.4	Unsaturated Unsubstituted Cyclic Tetraorganogermanes	275
7.1.6	Functionally Substituted Cyclic Tetraorganogermanes	281
7.1.6.1	Cyclic Functionally Substituted Tetraorganogermanes	281
7.1.6.6	Tetraorganogermanes Containing Heteroatoms in the Cyclic System	285
7.3	Cyclic Organogermanium Hydrides	288
7.4	Cyclic Organogermanes Containing Heteroatom-Germanium Bonds	288
7.4.1	Cyclic Organogermanium Halides	288
7.4.1.1	Cyclic Organogermanium Monohalides	288
7.4.1.2	Cyclic Organogermanium Dihalides	292
7.4.3	Cyclic Organogermanes Containing Germanium Linked to Oxygen, Sulfur and Silicon	296
8.	Miscellaneous Organogermanes	296
8.1	Polymerization Reactions	296
8.2	Unspecified Organogermanes	300

TIN

1.	Symmetric Tetraorganotin Compounds	301
1.1	Symmetric Unsubstituted Tetraorganotin Compounds	301
1.4	Symmetric Unsaturated Tetraorganotin Compounds	311
1.6	Symmetric Functionally Substituted Tetraorganotin Compounds	312

2.	Unsymmetric Tetraorganotin Compounds	315
2.1	Nonfunctional Unsubstituted Tetraorganotin Compounds	315
2.1.1	Derivatives with Two Different Organic Groups	315
2.1.2	Nonfunctional Unsubstituted Tetraorganotin Compounds with Three and Four Different Organic Groups	330
2.3	Heterocyclic Substituted Tetraorganotin Compounds	330
2.4	Unsymmetric Tetraorganotin Compounds with Olefinic Substituents	336
2.4.1	Unsymmetric Tetraorganotin Compounds Containing Halogen Substituted Olefins	350
2.4.2	Unsymmetric Tetraorganotin Compounds Containing Pseudohalogen Substituted Olefins	351
2.4.3	Unsymmetric Tetraorganotin Compounds Containing Functionally Substituted Olefins	351
2.5	Unsymmetric Tetraorganotin Compounds with Acetylenic Substituents	357
2.5.1	Unsymmetric Tetraorganotin Compounds Containing Halogen Substituted Acetylenes	361
2.5.3	Unsymmetric Tetraorganotin Compounds Containing Functionally Substituted Acetylenes	368
2.6	Functionally Substituted Unsymmetric Tetraorganotin Compounds	374
2.6.1	Tetraorganotin Compounds Containing Halogen	374
2.6.2	Tetraorganotin Compounds Containing Pseudohalogen Substituents	383
2.6.3	Tetraorganotin Compounds Containing Oxygen	386
2.6.3.1	Organotin Compounds with Hydroxy Groups and Derivatives	386
2.6.3.3	Tetraorganotin Compounds Containing Carbonyl Groups and Derivatives	389
2.6.3.4	Tetraorganotin Compounds Containing Carboxyl Groups and Derivatives	393
2.6.4	Tetraorganotin Compounds Containing Sulfur	399
2.6.5	Tetraorganotin Compounds Containing Nitrogen and Phosphorus	399
2.6.5.4	Tetraorganotin Compounds Containing Antimony	400
2.7	Polymetallic Tetraorganotin Compounds	402
2.7.1	Bridged Polyorganotin Compounds	402
2.7.1.1	Tetraorganotin Compounds with Two Tin Atoms	402
2.7.1.1.1	Nonfunctional Derivatives	402
2.7.1.1.3	Functionally Substituted Bistetraorganotin Compounds	408
2.7.1.2	Tetraorganotin Compounds with Three or More Tin Atoms	409
2.7.2	Tetraorganotin Compounds Containing Organosilicon Substituents	411
2.7.3	Tetraorganotin Compounds Containing Organogermanium Substituents	412
2.7.4	Tetraorganotin Compounds Containing Organolead Substituents	

2.7.5	Tetraorganotin Compounds with Boron and Aluminum	415
2.7.8	Tetraorganotin Compounds Containing Transition Metals	415
3.	Organotin Hydrides	417
3.1	Nonfunctional Organotin Hydrides	418
3.3	Organotin Deuterides	437
3.6	Functionally Substituted Organotin Hydrides	438
3.8	Organotin Hydrides Containing Heteroatom-Tin Bonds	439
4.	Organotin Compounds Containing Heteroatom-Tin Bonds	441
4.1	Compounds with Halogen	441
4.1.1	Triorganotin Halides	441
4.1.1.1	Unsubstituted Symmetric Triorganotin Halides	441
4.1.1.2	Unsymmetric Unsubstituted Triorganotin Halides	477
4.1.1.4	Unsaturated Triorganotin Halides	478
4.1.1.6	Functionally Substituted Triorganotin Halides	484
4.1.1.6.1	Derivatives with Halogen	484
4.1.1.6.2	Pseudohalogen Substituted Triorganotin Halides	487
4.1.1.6.3	Oxygen Containing Triorganotin Halides	487
4.1.1.6.5	Nitrogen Substituted Triorganotin Halides	491
4.1.1.6.7	Metal Substituted Triorganotin Halides	491
4.1.2	Diorganotin Dihalides	493
4.1.2.1	Unsubstituted Symmetric Diorganotin Dihalides	493
4.1.2.2	Unsymmetric Unsubstituted Diorganotin Dihalides	516
4.1.2.4	Unsaturated Diorganotin Dihalides	521
4.1.2.6	Functionally Substituted Diorganotin Dihalides	521
4.1.2.6.1	Halogen Substituted Diorganotin Dihalides	521
4.1.2.6.2	Pseudohalogen Substituted Diorganotin Dihalides	522
4.1.2.6.3	Oxygen Substituted Diorganotin Dihalides	522
4.1.2.6.3.1	Compounds Containing an Ether Linkage	522
4.1.2.6.3.3	Diorganotin Dihalides Containing Carbonyl Groups	522
4.1.2.6.3.4	Diorganotin Dihalides Containing Carboxyl Groups and Derivatives	527
4.1.2.6.9	Diorganotin Dihalides Containing Nitrogen and Boron	533
4.1.3	Organotin Trihalides	533
4.1.3.1	Unsubstituted Organotin Trihalides	534
4.1.3.4	Unsaturated Organotin Trihalides	543
4.1.3.6	Functionally Substituted Organotin Trihalides	543
4.1.4	Organotin Halides with Tin Linked to Anions	545
4.1.5	Organotin Halides Containing Tin-Oxygen Bonds	545
4.1.5.1	Organotin Halides Containing Oxy- or Hydroxy-Groups	546
4.1.5.3	Organotin Halides Containing Alkoxy Groups	552
4.1.5.3.1	Diorganotin Halides Containing Unsubstituted Alkoxy Groups	552
4.1.5.3.6	Organotin Halides Containing Functionally Substituted Groups	553
4.1.5.4	Organotin Halides Containing Acetal Groups	556
4.1.5.5	Diorganotin Halides Containing Carboxyl Groups	556
4.1.6	Organotin Halides Containing Tin-Sulfur Bonds	556
4.1.7	Organotin Halides Containing Tin-Nitrogen Bonds	560

4.1.9	Organotin Halides with Other Heteroatoms	562
4.2	Organotin Salts	562
4.2.1	Organotin Pseudohalides	562
4.2.1.1	Triorganotin Pseudohalides	562
4.2.1.2	Diorganotin Dipseudohalides	569
4.2.1.3	Monoorganotin Tripseudohalides	570
4.2.1.5	Organotin Pseudohalides Containing Tin-Oxygen Bonds	572
4.2.1.5.1	Organotin Pseudohalides Containing Oxy- or Hydroxy-Groups	572
4.2.1.5.3	Organotin Pseudohalides Containing Alkoxy and Acetal Groups	574
4.2.1.6	Organotin Pseudohalides Containing Tin-Sulfur and Tin-Nitrogen Bonds	574
4.2.4	Organotin Salts or Esters of Oxygen Acids	574
4.2.4.1	Triorganotin Salts or Esters of Oxygen Acids	575
4.2.4.2	Diorganotin Salts or Esters of Oxygen Acids	586
4.2.4.3	Monoorganotin Salts or Esters of Oxygen Acids	587
4.2.4.6	Diorganotin Salts or Esters Containing Tin-Heteroatom Bonds	588
4.3	Organotin Oxygen Compounds	593
4.3.1	Organotin Oxides and Hydroxides	593
4.3.1.1	Triorganotin Oxides and Hydroxides	593
4.3.1.1.1	Symmetric Triorganotin Oxides and Hydroxides	593
4.3.1.1.2	Unsymmetric Triorganotin Oxides and Hydroxides	607
4.3.1.1.6	Functionally Substituted Triorganotin Oxides and Hydroxides	607
4.3.1.2	Diorganotin Oxides and Hydroxides	610
4.3.1.2.1	Unsubstituted Diorganotin Oxides and Hydroxides	610
4.3.1.2.6	Functionally Substituted Diorganotin Oxides and Hydroxides	617
4.3.1.3	Monoorganotin Oxides and Hydroxides	618
4.3.1.9	Organotin Oxides and Hydroxides Containing Tin-Heteroatom Bonds	621
4.3.2	Organotin Peroxy Compounds	621
4.3.3	Organotin Alkoxides and Phenoxides	622
4.3.3.1	Triorganotin Alkoxides	623
4.3.3.1.1	Unsubstituted Triorganotin Alkoxides	623
4.3.3.1.3	Triorganotin Alkoxides Derived from Heterocyclic Alcohols	632
4.3.3.1.4	Triorganotin Alkoxides Derived from Unsaturated Alcohols	632
4.3.3.1.6	Triorganotin Alkoxides Derived from Functionally Substituted Alcohols	636
4.3.3.1.6.1	Triorganotin Alkoxides Derived from Halogen Substituted Alcohols	636
4.3.3.1.6.2	Triorganotin Alkoxides Derived from Pseudohalogen Substituted Alcohols	636
4.3.3.1.6.3	Triorganotin Alkoxides Derived from Alcohols Containing Oxygen Functions	636
4.3.3.1.6.5	Triorganotin Alkoxides Derived from Alcohols Containing Nitrogen and Phosphorus	636

4.3.3.1.6.5.9	O-Triorganotin Derivatives of Oximes and Hydroxylamines	636
4.3.3.1.6.7	Triorganotin Alkoxides Derived from Organometallic Substituted Alcohols	646
4.3.3.1.6.9	Triorganotin Alkoxides Derived from Polyfunctional Alcohols	646
4.3.3.1.9	Triorganotin Mixed Oxides	646
4.3.3.2	Diorganotin Alkoxides	652
4.3.3.2.1	Unsubstituted Diorganotin Alkoxides	652
4.3.3.2.3	Diorganotin Alkoxides Derived from Heterocyclic Alcohols	657
4.3.3.2.6	Diorganotin Alkoxides Derived from Functionally Substituted Alcohols	657
4.3.3.2.6.1	Alkoxides Derived from Halogen Substituted Alcohols	657
4.3.3.2.6.3	Alkoxides Derived from Alcohols Containing Oxygen Functions	657
4.3.3.2.6.5	Alkoxides Derived from Alcohols Containing Nitrogen	657
4.3.3.2.6.8	Diorganotin Alkoxides Derived from Dihydric Alcohols	663
4.3.3.2.9	Miscellaneous Diorganotin Alkoxides	663
4.3.3.2.9.1	Mixed Diorganotin Oxides	663
4.3.3.2.9.3	Diorganotin Alkoxides Containing Acetal Enol and Carboxy Groups Linked to Tin	668
4.3.3.2.9.5	Diorganotin Alkoxides Containing Tin-Heteroatom Bonds	668
4.3.3.3	Monoorganotin Trialkoxides and Phenates	672
4.3.3.3.1	Monoorganotin Trialkoxides from Unsubstituted Alcohols	672
4.3.3.3.6	Monoorganotin Alkoxides Derived from Substituted Alcohols	672
4.3.4	O-Organotin Derivatives of Aldehydes and Ketones	675
4.3.4.1	Triorganotin Derivatives of Aldehydes and Ketones	675
4.3.4.1.1	Triorganotin Acetals and Ketals	675
4.3.4.1.3	Triorganotin Enoxides	677
4.3.4.1.4	Triorganotin Derivatives of 1,2- and 1,3-Diketones	680
4.3.4.2	Diorganotin Derivatives of Aldehydes and Ketones	680
4.3.4.3	Monoorganotin Derivatives of Aldehydes and Ketones	683
4.3.5	Organotin Carboxylates	684
4.3.5.1	Triorganotin Carboxylates	684
4.3.5.1.1	Unsubstituted Triorganotin Carboxylates	685
4.3.5.1.3	Unsubstituted Triorganotin Carboxylates Derived from Heterocyclic Acids	695
4.3.5.1.4	Unsubstituted Triorganotin Carboxylates Derived from Unsaturated Acids	695
4.3.5.1.6	Unsubstituted Triorganotin Carboxylates Derived from Functionally Substituted Acids	700
4.3.5.1.6.1	Unsubstituted Triorganotin Carboxylates Derived from Halogen Substituted Acids	700
4.3.5.1.6.3	Unsubstituted Triorganotin Carboxylates Derived from Oxygen and Sulfur Containing Acids	700
4.3.5.1.6.5	Unsubstituted Triorganotin Carboxylates Derived from Acids Containing Nitrogen	705

4.3.5.1.6.7	Unsubstituted Triorganotin Carboxylates Derived from Metal Substituted Acids	705
4.3.5.1.6.8	Unsubstituted Triorganotin Carboxylates Derived from Polyfunctional Acids	709
4.3.5.1.6.8.1	Bis(triorganotin) Derivatives of Dicarboxylic Acids	709
4.3.5.1.6.8.3	Poly(triorganotin) Derivatives of Polycarboxylic Acids	709
4.3.5.1.6.8.4	Unsubstituted Triorganotin Derivatives of Polyfunctional Carboxylic Acids	709
4.3.5.1.9	Unsymmetric and Functionally Substituted Triorganotin Carboxylates	709
4.3.5.2	Diorganotin Carboxylates	717
4.3.5.2.1	Unsubstituted Diorganotin Dicarboxylates	717
4.3.5.2.2	Unsubstituted Diorganotin Carboxylate Oxides and Hydroxides	723
4.3.5.2.3	Unsubstituted Diorganotin Carboxylates of Heterocyclic and Unsaturated Acids	727
4.3.5.2.6	Unsubstituted Diorganotin Carboxylates Derived from Functionally Substituted Acids	733
4.3.5.2.6.1	Diorganotin Carboxylates of Halogen Substituted Acids	733
4.3.5.2.6.3	Unsubstituted Diorganotin Carboxylates Derived from Acids Containing Oxygen and Sulfur	733
4.3.5.2.6.5	Unsubstituted Diorganotin Carboxylates Derived from Nitrogen Containing Acids	733
4.3.5.2.6.7	Unsubstituted Diorganotin Carboxylates Derived from Polyfunctional and Metal Substituted Acids	739
4.3.5.2.6.8	Unsubstituted Diorganotin Carboxylates with Polycarboxylic Acids	739
4.3.5.2.6.8.1	Unsubstituted Diorganotin Carboxylates Derived from Unsubstituted Dicarboxylic Acids	739
4.3.5.2.6.8.2	Unsubstituted Diorganotin Dicarboxylates Derived from Functionally Substituted Polycarboxylic Acids	739
4.3.5.2.8	Functionally Substituted and Miscellaneous Diorganotin Carboxylates	746
4.3.5.2.9	Diorganotin Carboxylates Containing Tin-Sulfur and Tin-Nitrogen Bonds	747
4.3.5.3	Monoorganotin Carboxylates	747
4.4	Organotin Compounds with Sulfur, Selenium and Tellurium	752
4.4.1	Organotin Sulfides, Selenides and Tellurides	752
4.4.1.1	Triorganotin Sulfides, Selenides and Tellurides	752
4.4.1.2	Diorganotin Sulfides, Selenides and Tellurides	756
4.4.1.3	Monoorganotin Sulfides, Selenides and Tellurides	757
4.4.2	Organotin Mercaptides and Organoselenides	760
4.4.2.1	Triorganotin Mercaptides and Selenides	760
4.4.2.1.1	Unsubstituted Triorganotin Mercaptides and Selenides	760
4.4.2.1.3	Heterocyclic Substituted Triorganotin Mercaptides	764
4.4.2.1.6	Functionally Substituted Triorganotin Mercaptides	764
4.4.2.2	Diorganotin Dimercaptides and Diorganoselenides	769

4.4.2.2.1	Unsubstituted Diorganotin Dimercaptides and Diorganoselenides	770
4.4.2.2.3	Heterocyclic Substituted Diorganotin Dimercaptides	775
4.4.2.2.6	Functionally Substituted Diorganotin Dimercaptides	775
4.4.2.3	Monoorganotin Trimercaptides	776
4.4.2.9	Organotin Sulfur Compounds Containing Tin-Heteroatom Bonds	776
4.4.4	Organotin Thiocarboxylates, Xanthates, Thiamides and Thiocarbamates	781
4.4.4.1	Triorganotin Thiocarboxylates, Xanthates, Thiamides and Thiocarbamates	781
4.4.4.2	Diorganotin Thiocarboxylates, Xanthates, Thiamides and Thiocarbamates	781
4.4.7	Metal Substituted Organotin Sulfides, Selenides and Tellurides	786
4.5	Organotin Compounds with Nitrogen, Phosphorus, Arsenic, Antimony and Bismuth	787
4.5.1	Unsubstituted Organotin Amides, Imides and Nitrides	788
4.5.1.1	Triorganotin Derivatives	788
4.5.1.2	Unsubstituted Diorganotin Amides and Imides	794
4.5.1.3	Unsubstituted Monoorganotin Triamides	795
4.5.2	Substituted Organotin Amides	797
4.5.2.1	Organotin Derivatives of Heterocyclic Amines	797
4.5.2.4	Organotin Derivatives of Amidines, Imides, Hydrazine and Triazene	797
4.5.2.7	Organotin Derivatives of Nitramine	797
4.5.2.8	Organotin Derivatives of Keteneimine	801
4.5.3	Organotin Derivatives of Acid Amides and Acid Imides	804
4.5.3.1	Organotin Derivatives of Amides and Imides of Carboxylic Acids	804
4.5.3.2	N-Organotincarbamates	807
4.5.3.3	Organotin Derivatives of Urea	811
4.5.3.6	Organotin Amides and Imides of Acids of Sulfur	814
4.5.4	Organotin Metal Imides	815
4.5.5	Organotin Phosphorus Imides	817
4.5.6	Organotin Phosphides, Arsenides, Antimonides and Bismuthides	819
4.5.6.1	Organotin Phosphides	819
4.5.6.4	Organotin Arsenides, Antimonides and Bismuthides	823
5.	Organo-Polytin Compounds	826
5.1	Hexaorganoditin Compounds	826
5.1.1	Symmetric Hexaorganoditin Compounds	826
5.1.2	Unsymmetric Hexaorganoditin Compounds	833
5.2	Organo-Polytin Compounds Containing Three or More Tin Atoms	835
5.2.1	Organo-Polytin Compounds with $R_{2n+2}Sn_n$ Structure	835
5.2.1.1	Octaorganotritin Compounds	835

5.2.1.4	Organo-Polytin Compounds Containing Four or More Tin Atoms	838
5.2.2	Organo-Polytin Compounds with $(R_2Sn)_x$ Structure	838
5.2.4	Organo-Polytin Compounds with $(RSn)_x$ Composition	841
5.4	Organo-Polytin Compounds	845
6.	Polymetallic Organotin Compounds	847
6.1	Bimetallic Organotin Compounds with Group IV Metals	847
6.1.1	Organotin Compounds with Sn-Si Bonds	847
6.1.2	Organotin Compounds with Sn-Ge Bonds	847
6.1.3	Organotin Compounds with Sn-Pb Bonds	847
6.2	Bimetallic Organotin Compounds with Group III Metals	849
6.3	Bimetallic Organotin Magnesium Compounds	849
6.4	Bimetallic Organotin Alkali Metal Compounds	849
6.5	Bimetallic Organotin Derivatives of Transition Metal Coordination Compounds	852
6.5.1	Organotin Compounds with Group I-V Metals	852
6.5.4	Organotin Compounds with Group VI Metals	855
6.5.5	Organotin Compounds with Group VII Metals	859
6.5.6	Organotin Compounds with Iron and Ruthenium	863
6.5.7	Organotin Compounds with Cobalt Rhodium and Iridium	868
6.5.8	Organotin Compounds with Nickel and Platinum	873
6.6	Polymetallic Organotin Transition Metal Complexes	873
7.	Cyclic Organotin Compounds	876
7.1	Cyclic Tetraorganotin Compounds	876
7.1.1	Unsubstituted Cyclic Tetraorganotin Compounds	876
7.1.4	Unsaturated Cyclic Tetraorganotin Compounds	878
7.1.6	Functionally Substituted Cyclic Tetraorganotin Compounds	878
7.3	Cyclic Organotin Compounds Containing Heteroatom Tin Bonds	878
8.	Miscellaneous Organotin Compounds	888
8.1	Polymerization Reactions with Organotin Compounds	888
8.2	Unspecified Organotin Compounds	894
9.	Commercial Application of Organotin Compounds	895

LEAD

1.	Symmetric Tetraorganolead Compounds	900
1.1	Unsubstituted Derivatives	900
1.3	Tetraheterocyclic Organolead Compounds	908
1.4	Symmetric Unsaturated Tetraorganolead Compounds	908
1.6	Symmetric Functionally Substituted Tetraorganolead Compounds	908
2.	Unsymmetric Tetraorganolead Compounds	910

2.1	Nonfunctional Derivatives	910
2.3	Heterocyclic Substituted Tetraorganolead Compounds	911
2.4	Tetraorganolead Compounds with Olefinic Substituents	915
2.5	Tetraorganolead Compounds with Acetylenic Substituents	917
2.6	Functionally Substituted Tetraorganolead Compounds	917
2.6.1	Tetraorganolead Compounds Containing Halogen	917
2.6.3	Tetraorganolead Compounds Containing Oxygen and Sulfur	922
2.6.5	Tetraorganolead Compounds Containing Nitrogen and Phosphorus	922
2.7	Polymetallic Tetraorganolead Compounds	922
3.	Organolead Hydrides	927
4.	Organolead Compounds Containing Heteroatom-Lead Bonds	928
4.1	Organolead Halides	928
4.1.1	Triorganolead Halides	928
4.1.2	Diorganolead Dihalides	934
4.1.3	Monoorganolead Trihalides	937
4.2	Organolead Salts	937
4.2.1	Organolead Pseudohalides	937
4.2.4	Organolead Salts or Esters of Oxygen Acids	940
4.3	Organolead Oxygen Compounds	943
4.3.1	Organolead Oxides and Hydroxides	943
4.3.2	Organolead Peroxides	946
4.3.3	Organolead Alkoxides	946
4.3.3.1	Unsubstituted Organolead Alkoxides	946
4.3.3.4	Substituted Organolead Alkoxides	948
4.3.3.8	Organolead Mixed Oxides	948
4.3.3.9	Multifunctional Organolead Alkoxides	948
4.3.4	Organolead Halfacetals and Derivatives of Diketones	953
4.3.5	Organolead Carboxylates	953
4.3.5.1	Triorganolead Carboxylates	953
4.3.5.1.1	Unsubstituted Triorganolead Carboxylates	954
4.3.5.1.3	Triorganolead Carboxylates Derived from Heterocyclic Acids	955
4.3.5.1.4	Unsaturated Triorganolead Carboxylates	955
4.3.5.1.6	Functionally Substituted Triorganolead Carboxylates	955
4.3.5.2	Diorganolead Dicarboxylates	959
4.3.5.2.1	Unsubstituted Diorganolead Dicarboxylates	959
4.3.5.2.2	Substituted Diorganolead Dicarboxylates	961
4.3.5.3	Monoorganolead Tricarboxylates	961
4.3.5.9	Basic Organolead Carboxylates	965
4.4	Organolead Compounds with Sulfur, Selenium and Tellurium	965
4.4.1	Organolead Sulfides, Selenides and Tellurides	965
4.4.2	Organolead Mercaptides, Organoselenides and Organotellurides	966
4.4.2.1	Unsubstituted Organolead Mercaptides and Organotellurides	966
4.4.2.4	Substituted Organolead Mercaptides	968

4.4.4	Organolead Derivatives of Organic Thioacids	969
4.4.7	Metal Substituted Organolead Sulfides, Selenides and Tellurides	969
4.5	Organolead Derivatives of Nitrogen, Phosphorus, Arsenic and Antimony	972
4.5.1	Unsubstituted Organolead Amides	972
4.5.2	Organolead Compounds with Substituted Amines	973
4.5.2.1	Organolead Derivatives of Heterocyclic Amines	973
4.5.2.4	Organolead Derivatives of Amidine, Imidine and Ketimide	973
4.5.3	Organolead Derivatives of Acid Amides and Acid Imides	973
4.5.4	Organolead Metal Imides	977
4.5.5	Organolead Phosphorus Imides	977
4.5.6	Organolead Phosphides, Arsenides and Antimonides	977
5.	Organo-Polylead Compounds	977
5.1	Hexaorganodilead Compounds	977
5.2	Organolead Compounds with Less Than Three Organic Groups per Lead Atom	980
6.	Polymetallic Organolead Compounds	981
6.1	Organolead Derivatives of Group I-IV Metals	981
6.5	Polymetallic Organolead Derivatives of Transition Metal Coordination Compounds	981
6.5.1	Bimetallic Organolead Compounds with Group I-VI Metals	981
6.5.5	Polymetallic Organolead Compounds with Manganese and Rhenium	983
6.5.6	Bimetallic Organolead Compounds with Group VIII Metals	984
7.	Cyclic Organolead Compounds	987
8.	Miscellaneous Organolead Compounds	988
8.1	Polymerization Reactions with Organolead Compounds	988
8.2	Unspecified Organolead Compounds	990

BIBLIOGRAPHY 991

REVIEW PUBLICATIONS AND MONOGRAPHS 1102

Volume II

COMPOUNDS OF GERMANIUM, TIN AND LEAD
INCLUDING BIOLOGICAL ACTIVITY
AND COMMERCIAL APPLICATION

FIRST SUPPLEMENT

GERMANIUM

1. SYMMETRIC TETRAORGANOGERMANES

Tetraorganogermanes are prepared by alkylation of germanium tetrahalides. Organoaluminum compounds are receiving more interest as alkylating agents. This reaction frequently is aided by addition of sodium chloride. New is the generation of alkylaluminum in situ, the use of dimethylcadmium for preparation of highest purity tetramethylgermane and alkylation with organotin and organolead compounds.

The following scheme summarizes common methods for preparation of the symmetric tetraorganogermanes in Table 1.

I Grignard reaction with germanium tetrahalides (IA) in diethyl ether or (IB) in tetrahydrofuran:

$$GeX_4 + RMgX \rightarrow R_4Ge .$$

II Reaction with excess organo derivatives of lithium (IIA) or sodium (IIB):

$$GeX_4 + RM \rightarrow R_4Ge .$$

III Reaction with aluminum alkyls:

$$GeX_4 + R_3Al + NaCl \rightarrow R_4Ge + NaAlCl_4 .$$

IV Alkylation with organotin derivatives:

$$GeCl_4 + Et_3SnCH_2CO_2Me \rightarrow (MeO_2CCH_2)_4Ge + Et_3SnCl .$$

R = organic gooup, M = alkali metal, X = halogen. Tetraorganogermaspiranes are listed in Chapter 7.1.

1.1 NONFUNCTIONAL SYMMETRIC TETRAORGANOGERMANES

Tetramethylgermane (M-1) GeC_4H_{12} Me_4Ge
Synth.: By rxn. of $GeCl_4$ with Me_3Al in N atm., in 78-81% yield (1108), in presence of NaCl, in 73% yield (1108), with $Me_3Al_2Cl_3$ and NaCl in N atm., in 51% yield (1108).
By rxn. of $GeCl_4$ with MeCl and Al in $NaCl-AlCl_3$ - melt at 180°, in 97% yield (2000).
By rxn. of $GeCl_4$ with Me_2Cd in very high purity (2544).
By rxn. of GeI_4 with MeMgI in xylene, in 82% yield (625).
Prop.: Vapor pressure data (2173, 2544), $n_D^{19.6}$ 1.3897, $n_D^{24.9}$ 1.3854, d_{25} 0.9650 - high purity sample (2544), d_{20} 0.9725 (367), spectra: IR (135, 546, 763, 1013, 1108, 1463), (far) IR and assignment (1462), Raman (763, 1423), assignment (1462), UV (424, 1905), NMR (129, 259, 518, 570, 600, 985, 1436, 1821. 1856, 1949, 2296, 2365, 2419, 2723), ^{13}C NMR (2121) and mass (1109, 1756), GLC (431, 432, 2278), charge distribution and chem. shift (1924), cubical expansion coeff. (2544), magnetic susceptibility (1, 699), photolysis (605), self-diffusion and nucl. spin-lattice relaxation time (1307), thermodynamic

data (305, 555).

Rxn. with: R_4Ge + cat. $AlCl_3$ → random R-distribution at 80-120°, by GLC; R = Et (432) Bu (432, 491).
Me_2GeCl_2 (1:1) + cat. $GaCl_3$ → Me_3GeCl (1856).
$MeGeCl_3$ (2:1) + cat. $GaCl_3$ → Me_3GeCl (1856).
$MeGeCl_3$ (1:2) + cat. $GaCl_3$ → Me_3GeCl at 150° or Me_2GeCl_2 at 253° (1856).
HCl + cat. $AlCl_3$ at r.t. → Me_3GeCl + CH_4 (582).
Br + cat. $AlBr_3$ → Me_2GeBr_2 (367).
GeX_4 (3:1) + cat. GaX_3 → Me_3GeX; X = Cl, Br (1856).
$GeCl_4$ (1:1-3) + cat. $GaCl_3$ → Me_2GeCl_2 (1856).
$GeCl_4$ → none random distribution rxn., by NMR (817).
$GaCl_3$ at r.t. → Me_3GeCl + $Me_2Ga_2Cl_4$ (1823).
Me_2CHX → Me_3GeX; X = Cl, Br (367).
AcCl + $AlCl_3$ → Me_3GeCl (1:1:1) or Me_2GeCl_2 (1:2:2) (1781).
SO_3 in CH_2Cl_2 → Me_3GeO_3SMe (2742).

Tetrabenzylgermane (M-5) $GeC_{28}H_{28}$ $(PhCH_2)_4Ge$
Synth.: By rxn. of $(PhCH_2)_3GeH$ with $PhCH_2Li$, in 60% yield (104), with BuLi at -10° in Et_2O, followed by MeI, in 16% yield (104).
Prop.: m. 110° (104), far IR (105), NMR (104) and mass (1109) spectra.
Rxn. with: Li in $(MeOCH_2)_2$ + H_2O → $(PhCH_2)_3GeH$ (104).
Li at 0°, + ROH → $(PhCH_2)_3GeR$ + $(PhCH_2)_2GeR_2$ + $PhCH_3$ + $(PhCH_2)_2$; R = H, D (104).
Br in $(BrCH_2)$, + $LiAlH_4$ → $(PhCH_2)_3GeH$ (104).

Tetraethylgermane (M-2) GeC_8H_{20} Et_4Ge
Synth.: By rxn. of $GeCl_4$ with EtMgBr in Et_2O or MePh in presence of excess Mg at 0-94°, in 71-95% yield (2701), in THF, in 16-58% (2701).
By rxn. of excess $GeCl_4$ with Et_3Al and NaCl in Ar atm. at 140-60° (2603).
By rxn. of $Et_3GeCH_2CH_2Li$ with H_2O, in 68% yield (2849).
By rxn. of Et_3GeLi with EtBr in THF at r.t., in 77% yield (2088), with C_2H_4, followed by H_2O, in 68% yield (2087).
By rxn. of $(Et_3Ge)_2Zn$ with EtBr at 100°, in 49% yield (2854).
By thermal dec. of $(Et_3Ge)_2Zn$ at 100-30°, in 32% yield (2854).
By-prod. in rxn. of EtMgBr with $(EtCO_2)_4Ge$ (2609), with $GeCl_4$ in presence of PrMgBr, in 8% yield (491).
By-prod. in rxn. of Et_3GeH with Et_2Zn at 125°, in 9% yield (584), with Et_4Pb at 165-70°, in 13% yield (2850).
By-prod. in rxn. of $(Et_3Ge)_2Cd$ with Et_3MBr at 80-100°; M = Si, Ge (587a).
Prop.: Vapor pressure equation (2173), spectra: IR (763, 920, 1013, 1108, 1444), from -170 to 20° (588), assignment (105), Raman (763), NMR (1444, 2296, 2373, 2875) and mass (848, 1109, 1756-7), flash photolysis (2299), magnetic susceptibility (1, 699), thermodynamic data (305, 2025, 2856). Anal. detn. by GLC (432, 491, 2278, 2701), by Schoeninger combustion and polarography (2710).
Rxn. with: R_4Si + cat.$AlCl_3$ → random distribution of R, by GLC; R = Me,Pr (432).
R_4Ge + cat. $AlCl_3$ → random distribution, by GLC; R = Me, n-Bu, n-C_5H_{11}, n-C_6H_{13} (432), n-Pr (432, 491).
R_4Ge + $AlCl_3$ → $Et_nR_{3-n}GeCl$; R = n-Pr, n-Bu (491).

Me$_4$Sn + cat. AlCl$_3$ → random distribution, followed by GLC (491).
Br in (CH$_2$Br)$_2$ → Et$_3$GeBr + EtBr (1103).
GeCl$_4$ (3:1) + cat. GaCl$_3$ → Et$_3$GeCl (1856).
SnCl$_4$ → Et$_3$GeCl + EtSnCl$_3$ (2292).
PCl$_3$ + O → Et$_3$GeC$_2$H$_4$POCl$_2$ (995).
BzO$_2$CMe$_3$ (10:1) → Et$_3$GeC$_2$H$_4$OBz + (Et$_3$GeC$_2$H$_4$)$_2$ + organic prod. (2098).
Neutron irradiation → ^{75}GeEt$_4$ (1637) + ^{77}GeEt$_4$ (659, 1637).
Biol. Prop. and Use: Toxicity (845) and metabolism in rats (2659); effect as stabilizer for PVC (439), ^{77}GeEt$_4$ as monitor in hydrocarbon refining (659).

Tetrapropylgermane (M-2)   GeC$_{12}$H$_{28}$ n-Pr$_4$Ge
By rxn of GeCl$_4$ with PrMgBr in Et$_2$O, in 61% yield (2837).
By-prod. in rxn of GeCl$_4$ with EtMgBr and PrMgBr, in 5% yield (491).
Prop.: b$_5$ 86-87°, n$_D^{25}$ 1.4447, d$_{25}$ 0.9491 (2837), vapor pressure equation (2173), far IR (1108) and mass (831, 1757) spectra, dipole moment (2836), magnetic susceptibility (1), thermodynamic data (305). Anal detn. by GLC (432, 491, 2278), by Schoeninger combustion and polarography (2710).
Rxn. with: Et$_4$Si + cat. AlCl$_3$ → Et$_n$GePr$_{4-n}$ + Et$_{4-n}$SiPr$_n$; n = 0 - 4 (491).
R$_4$Ge + cat. AlCl$_3$ at 170° → random distribution of R, by GLC; R = n-Bu, n-C$_5$H$_{11}$, n-C$_6$H$_{13}$ (432), Et (432, 491).
Et$_4$Ge + AlCl$_3$ → Et$_n$(Pr$_{3-n}$)GeCl (491).
n-Bu$_4$Ge + GeBr$_4$, + EtMgX → Et$_x$Pr$_y$GeBu$_z$; x + y + z = 4 (491).
GeCl$_4$ → Pr$_n$GeCl$_{4-n}$; n = 1-3 (2837).
Biol. Prop.: Toxicity (845).

Tetraisopropylgermane   GeC$_{12}$H$_{28}$ (Me$_2$CH)$_4$Ge
Synth.: By-prod. in rxn of GeCl$_4$ with i-PrMgCl in Et$_2$O-C$_6$H$_6$, in 5-20% yield (831).
By-prod. in rxn of i-Pr$_3$GeCl with Na-K alloy in MePh, in 5% yield (831).
Prop.: Far IR (1108), and mass (831) spectra.
Rxn. with: Br in (Ch$_2$Br)$_2$ → i-Pr$_2$GeBr$_2$ (2834).
Biol. Prop.: Toxicity (845).

Tetrabutylgermane (M-3)   GeC$_{16}$H$_{36}$ n-Bu$_4$Ge
Synth.: By rxn of GeCl$_4$ with BuMgCl in dry Et$_2$O in He atm. at -4 and -30°, in 85 and 80% yield, resp. (2591).
By-prod. in rxn of GeCl$_4$ with BuMgBr (1:1) in Et$_2$O, followed by LiAlH$_4$, in 12% yield (2039).
Prop.: b$_{0.3}$ 100°, n$_D^{20}$ 1.4566, d$_{20}$ 0.9367 (2591), vapor pressure equation (2173), far IR (105, 1108) and mass (1757) spectra, magnetic susceptibility (1, 699), thermodynamic data (438). Anal. detn. by GLC (432, 491), by Schoeninger combustion and polarography (2710).
Rxn. with: Et$_4$Si + cat. AlCl$_3$ at 170° → random distribution, by GLC (432).
R$_4$Ge + cat. AlCl$_3$ at 170° → random distribution of R, by GLC; R = Me, Et, n-Pr, n-C$_5$H$_{11}$, n-C$_6$H$_{13}$ (432).
Me$_4$Ge + AlCl$_3$ → Me$_n$Bu$_{3-n}$GeCl (491).

i-C$_5$H$_{11}$X + cat. AlCl$_3$ → Bu$_3$GeX; X = Cl, Br (119).
GeCl$_4$ at 175-250° → Bu$_3$GeCl + BuGeCl$_3$; cat.: GeI$_2$, Ge Bu$_3$GeH, LiAlH$_4$, Bu$_6$Ge$_2$ (1760).
Biol. Prop.: Toxicity (845).

Tetraphenylgermane (M-3)　　　　　　GeC$_{24}$H$_{20}$　　　　　　Ph$_4$Ge
Synth.: By-prod. in rxn. of Ph$_3$GeLi with (R$_3$P)$_2$PtI$_2$ (103).
By-prod. in pyrolysis of (R$_3$P)$_2$Pt(GePh$_3$)$_2$ (103) or its rxn. with PhLi and PhBr (103).
By-prod. in pyrolysis of (Et$_3$P)$_2$Pd(GePh$_3$)$_2$ (795).
Prop.: Spectra: IR (207, 333), far IR and assignment (51, 105, 2784), Raman (2784), UV (335), NMR (1026) and mass (848, 1109, 1519, 2696), effect of ^{60}Co irradiation (422) of neutron irradiation (84, 356), ^{137}Cs irradiation (454), thermal dec. at 421° (547). Anal. detn. by Schoeninger combustion and polarography (2710), by fusion and volumetric titration (265), by x-ray fluorescence (1820), and sepn. from Ph$_3$GeBr by paper chromatography (384), from Ph$_n$GeCl$_{4-n}$; n = 3, 2 by TLC (2635).
Rxn. with: Ph$_4$Si + H-Raney Ni → retardation of ring hydrogenation of Ph$_4$Si (528).
Ph$_2$GeCl$_2$ (1:1) + cat. AlCl$_3$ → Ph$_3$GeCl (1364).
PhGeCl$_3$ (2:1) + cat. AlCl$_3$ → Ph$_3$GeCl (1364).
PhGeCl$_3$ (1:2) + cat. AlCl$_3$ → Ph$_2$GeCl$_2$ (1364).
GeCl$_4$ (1:1) + cat. AlCl$_3$ → PhGeCl$_3$ + Ph$_3$GeCl (1364).
CF$_3$CO$_2$H → Ph$_3$GeO$_2$CCF$_3$ + C$_6$H$_6$ (1789).
KNH$_2$ in liq. NH$_3$ → [K$_3$(Ge$_4$N$_4$)(NH)$_4$]$_x$ (2744).
Addn. Compd.: Ph$_4$Ge·1.75 Ph$_6$Sn$_2$; Ph$_4$Ge + Ph$_6$Sn$_2$ (1:2) in MeNO$_2$ under reflux, in 76% yield; m. 194-96° (1500).
Use: In catalyst for polyester synth. (2945), as antifogging additive for photographic plates (591a), effect on spreading coeff. of solder (1692).

Other nonfunctional symmetric tetraorganogermanes are compiled in Table 1.

1.3 FUNCTIONAL SYMMETRIC TETRAORGANOGERMANES

Tetrakis(perfluorophenyl)germane　　　GeC$_{24}$F$_{20}$　　　　　(C$_6$F$_5$)$_4$Ge
Synth.: By rxn. of GeCl$_4$ with C$_6$F$_5$MgBr in THF, in 73% yield (547), in Et$_2$O-C$_6$H$_6$ at -78° to r.t. (1034).
By rxn. of GeCl$_4$ with C$_6$F$_5$Li (145) in Et$_2$O-C$_6$H$_6$ at -78° to r.t. (1034), in Et$_2$O-C$_6$H$_{14}$, in 88% yield (547).
By rxn. of Ge with C$_6$F$_5$I in sealed ampule at 325° (885).
By rxn. of (C$_6$F$_5$)$_3$GeCl with C$_6$F$_5$Li in Et$_2$O-C$_6$H$_{14}$, in good yield (1034).
Prop.: m. 224-30° (subl.) (145, 1034), 245° (885), 246.5-47.5° (sealed tube) (547), 246-48° (sealed tube) (1034), subl.$_{0.001}$ 170° (1034), stable in air (1034) and in vacuo at 246-48° (1034), thermal dec. at 416° (547), spectra: IR (1034), assignment (547), ^{19}F NMR (2475) and mass (1519); soly. (1034), GLC (547).
Rxn. with: 6 N HCl at reflux → stable for 6 hours (547).

Br (in CH$_2$Br)$_2$ at reflux → no rxn. (145, 1034).
I at 250° → no rxn. (1034).
5 N NaOH at r.t. → no rxn. (1034).
10% aq. KOH → stable at reflux; dec. on addn. of THF (547).
Li in THF → no rxn. (547).

Other functional symmetric tetraorganogermanes are compiled at the bottom of Table 1.

Table 1. Symmetric Tetraorganogermanes   R$_4$Ge

R Groups*	Synth. Method**	Yield	Properties	Ref.
Me$_2$CHCH$_2$ (5)	III	50	b$_{0.01}$ 78°, far IR, NMR spectra (a,b)	1108
n-C$_5$H$_{11}$ (5)	--	--	(c,d,e,f), mass spectrum (1757), GLC (432)	
n-C$_6$H$_{13}$ (5)	--	--	Mass spectrum (1757), GLC (432) (c) (f-1, 699)	
c. C$_6$H$_{11}$CH$_2$	IA	80	m. 66°, b$_{0.04}$ 175°	2591
n-C$_7$H$_{15}$ (6)	--	--	(f)	--
C$_8$H$_{17}$ (6)	--	--	(f-699)	--
o-MeC$_6$H$_4$	--	--	Mass spectrum	1109
m-MeC$_6$H$_4$ (6)	--	--	Far IR (105), mass (1109) spectra	--
p-MeC$_6$H$_4$ (6)	--	--	Far IR (105), mass (848, 1109) spectra	--
R	--	--	Ge-bond refraction D$_{20}$ (g)	81
MeOCH$_2$	IIA (h)	--	b$_{0.05}$ 50°, NMR spectrum	1832
CH$_2$:CH (5)	--	--	NMR spectrum	846
HC⋮C	IIB (i)	34	m. 91-92°, mol. wt., IR, mass spectra (j,k,l)	939
	IB	--	m. 94°, detonates vehemently on shock	1494

* Numbers in parenthesis refer to pages in main volume.
** The characters used here correspond to the synthetic scheme in the introduction to Chapter 1.
(a) Mass spectra (1108, 1109). (b) Rxn. with Br in (CH$_2$Br)$_2$ → i-Bu$_2$GeBr$_2$ (2834).
(c) Rxn. with R$_4$Ge + cat. AlCl$_3$ → random distribution of R, by GLC; R = Et, n-Pr, n-Bu, n-C$_5$H$_{11}$, n-C$_6$H$_{13}$ (432). (d) Rxn. with SnCl$_4$ in MeNO$_2$ → R$_3$GeCl + RSnCl$_3$ (2292). (e) Neutron irradiation → ^{77}GeR$_4$ (659); use of ^{77}GeR$_4$ for monitoring hydrocarbon refining (659). (f) Magnetic susceptibility (1). (g) Use for Ge deposition by microwave discharge (1859); process of purification (516). (h) Synthesis in methylal. (i) By-prod. in synthesis is a red highly explosive solid (939). (j) Sublimes at r.t. at 40-50 torr (939). (k) (Far) IR, Raman spectra and assignment (1780). (l) Rxn. (HC⋮C)$_4$Ge + Br → (BrCH:CBr)$_4$Ge (939, 1494).

Table 1 Continued R_4Ge

R Groups*	Synth. Method**	Yield	Properties	Ref.
BrCH:CBr	(1)	--	m. 123°, mol. wt.	939
	(1)	--	m. 131°	1494
MeO_2CCH_2	IV	77	$b_{0.05}$ 125-26°, n_D^{20} 1.4810, d_{20} 1.3589, IR spectrum (m)	692, 1440
p-BrC_6H_4C⋮C	IIA	49	m. 266°	204
Me_3SiC⋮C	IB	--	m. 160°	1345
	IIB	59	m. 176°	2775
CH_2:$CHCH_2$ (3)	--	--	Mol. magn. rotation (301), GLC	492
cis-MeCH:CH	IIA	28	b_4 77.5-78.5°, n_D^{20} 1.5040, d_{20} 1.0347 (n)	1600
trans-MeCH:CH	IIA	33	b_1 64-66°, n_D^{20} 1.4930, d_{20} 1.0074 (n)	1600
CH_2:CMe	IIA	36	b_2 60-62°, n_D^{20} 1.4935, d_{20} 1.0245 (n)	1600
	IB	60	b_{11} 90-91°, n_D^{18} 1.5110, d_{18} 1.061	2051
MeC⋮C	IB	55	m. 168°, IR spectrum, GLC	2562
C_3H_3	I	73	b_2 115° (o)	1473
Me_2CH:CH	IB	52	b_{11} 139-40°, n_D^{18} 1.5180, d_{18} 1.018	2051
Me_3CC⋮C	IIA	95	m. 190-91°, mol. wt., spectra: IR, mass, NMR, stable to 280°	2664
c. $C_6H_{11}C$⋮C	IIA	32	m. 146°	204

(m) NMR spectrum (1689). (n) IR spectrum (1600). (o) Synth. either from propargyl or allenyl magnesium bromide (1473).

2. UNSYMMETRIC TETRAORGANOGERMANES

A large number of new unsymmetric tetraorganogermanes are recently reported in the literature. The field has benefited from better synthetic methods for preparation of intermediate halides and introduction of modern analytical techniques. Especially useful have been nuclear magnetic resonance (NMR) and mass spectroscopy combined with gas liquid phase chromatography (GLC), the latter both on analytical and preparative scale.

2.1 NONFUNCTIONAL UNSYMMETRIC TETRAORGANOGERMANES

This subchapter comprises derivatives of the type $R_nGeR'_{4-n}$, $R_nR'_mR''_oGe$ and RR'R''R'''Ge (n + m + o = 4). Cyclic tetraorganogermanes are listed in Subchapter 7.1.

2.1.1 DERIVATIVES WITH TWO DIFFERENT ORGANIC GROUPS

The following scheme summarizes methods used in the preparation of the unsymmetric nonfunctional organogermanes listed in Table 2. Synthetic reactions include substition of germanium halides and organogermanium halides with Grignard reagent (I) and alkali metal or mercury derivatives (II). A reversal of the latter method is the use of organic halide and organogermanium alkali metal derivatives (III, IV). Other methods employed are interconversion (V) and redistribution (VI) reactions.

I Grignard synthesis:
   (IA) $GeCl_4 + RMgX + R'MgX \rightarrow R_nGeR'_{4-n}$,
   (IB) $GeCl_4 + RX + R'X + Mg \rightarrow R_nGeR'_{4-n}$,
   (IC) $GeCl_4 + C_6H_{11}MgX + RCl \rightarrow (C_6H_{11})_3GeR$,
   (ID) $R_nGeX_{4-n} + R'MgX \rightarrow R_nGeR'_{4-n}$,
   (IE) $RR'GeX_2 + RMgX \rightarrow R_3GeR'$.

II Synthesis with organolithium derivatives:
   (IIA) $GeCl_4 + RLi + R'Li \rightarrow R_nGeR'_{4-n}$,
   (IIB) $R_nGeX_{4-n} + R'Li \rightarrow R_nGeR'_{4-n}$,
   (IIC) $Me_3GeCR:CHR + EtLi, + H_2O \rightarrow Me_3GeCREtCH_2R + Me_3GeCHRCHREt$.

III Synthesis with triorganogermanium metal derivatives:
   (IIIA) $Me_3GeLi + R'X \rightarrow Me_3GeR'$,
   (IIIB) $Ph_3GeLi + R_3PO_4 \rightarrow Ph_3GeR$,
   (IIIC) $Et_3GeM + R'X \rightarrow Et_3GeR$,
   (IIID) $Et_3GeLi + RCH:CH_2, + H_2O \rightarrow Et_3GeCH_2CH_2R$,
   (IIIE) $(Me_3Ge)_2Hg + RCH:CH_2 \rightarrow Me_3GeCH_2CH_2R$.

IV Synthesis with organogermanium metal derivatives prepared in situ:
   (IVA) $R_4Ge + Li$ in $(MeOCH_2)_2 \rightarrow R_nGeMe_{4-n}$,
   (IVB) $R_nGeH_{4-n} + Li, + R'X \rightarrow R_nGeR'_{4-n}$,
   (IVC) $R_nGeH_{4-n} + BuLi, + R'X \rightarrow R_nGeBu_{4-n} + R_nGeR'_{4-n}$,
   (IVD) Ph_6Ge_2 or $(Ph_2Ge)_n + MC_{10}H_8 + RX \rightarrow Ph_nGeR_{4-n}$.

V Interconversion reactions:
   (VA) $R'_3GeCHRCHRLi + H_2O \rightarrow R'_3GeCHRCH_2R$,
   (VB) $Ph_3GeCHPhOH + MePh + cat. BF_3 \cdot OEt_2 \rightarrow Ph_3GeCHPhC_6H_4Me$.

VI Redistribution reactions:
   (VIA) $R_4Ge + R'_4Ge + cat. AlCl_3 \rightarrow R_nGeR'_{4-n}$,
   (VIB) $R_4Ge + R'_4Ge + GeBr_4, + EtMgBr \rightarrow R_nGeR'_{4-n} + R_nGeEt_{4-n} + \cdots$,
   (VIC) $R_4Ge + R'_4Si + cat. AlCl_3 \rightarrow R_nGeR_{4-n}$.

$R = R'$ = organic group, M = sodium and potassium, X = halogen, n = 1-3.

Benzyltrimethylgermane (M-7)　　　　　　$GeC_{10}H_{16}$　　　　　　Me_3GeCH_2Ph
Synth.: By rxn. of Me_3GeBr with $PhCH_2MgCl$ (2159).
By-prod. in rxn. of $(PhCH_2)_4Ge$ with Li in $(MeOCH_2)_2$ under reflux in N atm. (104).
Prop.: b. 204-206° (104), far IR (105) and UV (203, 1570) spectra, dipole moment (1570, 1572), dielectrical const. (1572).
Rxn. with: Na or K in $(MeOCH_2)_2$ in N atm. → ESR spectrum of rxn. prod. (524). $OCCH_2CH_2CONBr \rightarrow Me_3GeCPhBr_2$ (2159).

Phenyltrimethylgermane (M-4)   GeC$_9$H$_{14}$ Me$_3$GePh
Synth.: By rxn. of PhGeCl$_3$ with excess MeMgBr in Et$_2$O, in 67-80% yield, (1041, 1042), with MeLi (1606).
Prop.: b$_{20}$ 82-85° (1041, 1043), b$_{13}$ 62° (1606), n$_D^{20}$ 1.5080, GLC (1606), UV (1561, 1570) and NMR (2476) spectra, dipole moment (1570), oscilllator strength (2593). Anal. detn. by Schoeninger combustion and polarography (2710).
Rxn. with: MeEtCHCl + cat. AlCl$_3$ → Me$_3$GeCl (367).
BuI + cat. AlCl$_3$ → Me$_3$GeI (367).
Na or K in (MeOCH$_2$)$_2$ → ESR spectrum of rxn. prod. (524).

Methyltriphenylgermane (M-8)   GeC$_{19}$H$_{18}$ MeGePh$_3$
Synth.: By rxn. of Ph$_3$GeX with MeLi in Et$_2$O (1606).
By rxn. of Ph$_6$Ge$_2$ with Na-C$_{10}$H$_8$ in (MeOCH$_2$)$_2$ in N atm., in 95% yield (1606).
By rxn. of Ph$_3$GeLi with Me$_3$PO$_4$ in THF, in 55% yield (1928), with Me$_3$PtI (103).
By rxn. of (n-Pr$_3$P)$_2$Pt(GePh$_3$)$_2$ with MeI, followed by alc. KOH (101, 103).
By rxn. of (Ph$_3$P)AuGePh$_3$ with MeI in sealed tube at 24°, in 84% yield (2420).
Prop.: m. 70-71° (1606), 67-69° (1928), far IR (105) and mass (1109) spectra, GLC (1606).

Butyltriethylgermane (M-9)   GeC$_{10}$H$_{24}$ n-BuGeEt$_3$
Synth.: By hydrogenation of Et$_3$GeCH$_2$CH:CHMe with Raney-Ni in EtOH (467), in 92% yield (1798).
By rxn. of Et$_3$GePPh$_2$ with BuLi at 0°, in 41% yield (50).
By-prod. in redistribution rxn. of n-Pr$_4$Ge with n-Bu$_4$Ge and GeBr$_4$, followed by rxn. with EtMgBr, in 8% yield (491).
Prop.: b$_{14}$ 84° (1798), b$_{0.03}$ 50° (50), n$_D^{20}$ 1.4481 (1798), far IR spectrum (105), GLC (467, 491-2, 1798).
Biol. Prop.: Toxicity (845).

Butyltriphenylgermane (M-11)   GeC$_{22}$H$_{24}$ n-BuGePh$_3$
Synth.: By rxn. of Ph$_3$GeH with BuLi, in low yield (105), with K in (Me$_2$N)$_3$PO-Et$_2$O, followed by BuBr, in 62% yield (1634, 2983).
By rxn. of Ph$_3$GeLi with n-BuBr in (MeOCH$_2$)$_2$ (105), with n-Bu$_3$PO$_4$ in THF, in 67% yield (1928).
By rxn. of Ph$_3$GeSMe with BuLi, in 81% yield (227).
By rxn. of Ph$_3$GeOR with BuLi in THF-hexane or (Me$_2$NCH$_2$)$_2$, in 35-74% yield (2675).
Prop.: m. 87-88° (1928), 82-84° (2675), 80-81° (105), 79-80° (227), b$_{0.01}$ 160°, far IR (105) and mass (1109) spectra.

Other nonfunctional tetraorganogermanes with two different organic substituents are compiled in Table 2.

Table 2. Nonfunctional Tetraorganogermanes With Two Different Substituents $R_nGeR'_{4-n}$

Formula*	Synth. Method**	Yield	Properties	Ref.
Me₃GeR				
R = Et (7)	IE	--	n_D^{20} 1.4080 (a, b, c)	336
n-Pr (7)	IB, IIA	--	GLC (b, c)	493
	VIA (d)	--	--	491
	ID - PrGeX₃	--	b. 87.5°	2175
i-Pr	IB, IIA	--	GLC (b)	493
	ID - i-PrGeI₃	62	b. 87.5°, n_D^{20} 1.4182, d_{20} 0.9816	2877
n-Bu (7)	IA	--	GLC (491, 492) (c, e)	491
	ID - BuGeX₃	--	b. 118°	2175
i-Bu	ID - Me₃GeCl	66	b_{740} 121°, IR, NMR spectra, colorless liquid	475
t-Bu	ID - Me₃CGeCl₃	--	m. 95-95.5°, b. 111°, IR spectrum	364
CH₂CHCH₂CH₂ (7)	--	--	(f)	2629
CHPrPh	ID - Me₃GeI	30	b_3 59-61°, n_D^{20} 1.5153-62	2629
	IIC	71	d_{20} 1.0728-31, IR spectrum	2629
CHEtCH₂Ph	ID - Me₃GeI	21	b_3 62-63°, n_D^{20} 1.5116-23	2629
	IIC	67	d_{20} 1.0713-19, IR spectrum	2629
c. C₅H₉ (7)	--	--	Dipole moment (g)	254
n-C₆H₁₃	IIIE	14	--	2241
p-C₆H₄Ph	IIB - Me₃GeBr	74	m. 55°, UV and NMR spectra (h)	112

* Numbers in parenthesis refer to pages in main volume.
** The characters used here correspond to the synthetic scheme in the introduction to Chapter 2.1.1.

(a) GLC (432, 493). (b) IR spectrum (493). (c) Vapor pressure equation (2173). (d) Synthesis by redistribution rxn. of Et₄Ge with n-Pr₄Ge and cat. AlCl₃, followed by LiAlH₄ and MeMgX before work-up (491). (e) Rxn. with SnCl₄ → Me₂BuGeCl (2292). (f) Rxn. with Hg(OAc)₂ → Me₃GeCH₂CHOHCH₂CH₂HgOAc (127). (g) Anal. detn. by fusion and volumetric titration (265). (h) Rxn. with Na in THF → [Me₃GeR]Na, ESR spectrum at r.t. (112); polarographic reduction (112).

Table 2 Continued

$R_nGeR'_{4-n}$

Formula*	Synth. Method**	Yield	Properties	Ref.
Me₂GeR₂				
R = CH₂Ph	IVA - R₄Ge	--	m. 53-55°, b₀.₀₀₁ 82-100°, spectra: (far) IR (105), UV (203), mass (1109)	104
	IVB - R₃GeH	--		
Et (7)	IB, IIA	--	NMR spectrum (2373),(a, b, c)	493
n-Pr (7)	IB, IIA	--	NMR spectrum (2373), GLC (491-3) (b, c)	493
i-Pr	IB, IIA	--	GLC (b)	493
n-Bu	IA	--	GLC (491-2)	491
Ph (8)	IVD - Ph₂Ge	--	b₀.₃₅ 97°, n_D²⁰ 1.5734, GLC	1606
	ID - Ph₂GeX₂	--	NMR spectrum (1012)	1606
	(i)	5	(g)	1799
p-C₆H₄Ph	ID - Me₂GeCl₂	24	m. 176°	423
	IIB - Me₂GeCl₂	24	IR spectrum	423
MeGeR₃				
R = CH₂Ph	IVA - R₄Ge	27	m. 82-85°, b₀.₀₀₁ 165°	104
	IVC - R₃GeH (j)	50	Far IR spectrum (105)	104
	IVB - R₃GeH	--	Mass spectrum (1109)	104
Et (8)	VIA (d)	26	GLC (491-2), spectra: IR (493, 920), NMR (2373) (a, c)	491
	IB, IIA	--		493
n-Pr (8)	IB, IIA	--	GLC (491-3)	493
	VIA (d)	5	(b, c)	491
i-Pr	IB, IIA	--	GLC, (b)	493
n-Bu (8)	ID - MeGeI₃	--	b. 164-65°, n_D²⁰ 1.3809, d₂₀ 0.9458	2877
	IA	--	GLC (491-2)	491
(PhCH₂)₃GeEt	IIIA	64	m. 34-35°, b₀.₀₀₁ 170-83° (k, l)	104
(PhCH₂)₃GeBu	IA	16	b₀.₀₀₁ 150-63°	104
	IVC	26	(k)	104
(PhCH₂)₂GeBu₂	IA	46	b₀.₀₀₁ 130-33° (k)	104

PhCH$_2$GeR$_3$			
R = Et (8)	--	--	--
	IA	27	b$_{0.001}$ 100-110° (k)	104
n-Bu	IC	39	b$_{0.3}$ 199-200°, n$_D^{20}$ 1.5685, d$_{20}$ 1.1255	2591
c. C$_6$H$_{11}$ (8)	IVC - Ph$_6$Ge$_2$	56	Synth. in THF + PhCH$_2$Cl (k, l)	1634
Ph (9)	--	--	IR and UV spectra	137
R				
p-MeC$_6$H$_4$CHPhGePh$_3$	VB	85	IR and NMR spectra	49
Et$_3$GeR				
R = CH$_2$CH$_2$Ph	IIID	43	--	2087
	VA	43	b$_{1.5}$ 107°, n$_D^{20}$ 1.5078	2849
CHPhCH$_2$Ph	VA	50	b$_{1.5}$ 148-49°, n$_D^{20}$ 1.5550	2849
	VIC	--	GLC	432
n-Pr (9)	IA	30	GLC (491-2)	491
	VIA	27	(g)	491
	VIB	3	--	491
i-Pr	ID - i-PrGeI$_3$	86	b. 176°, n$_D^{20}$ 1.4487, d$_{20}$ 0.9904	2877
	VIA	--	GLC, toxicity (845)	432
n-C$_5$H$_{11}$ (9)	IIID	--	b$_{20}$ 117°, n$_D^{20}$ 1.4580	2087
	VA	9	b$_{20}$ 117°, n$_D^{20}$ 1.4580	2849
n-C$_6$H$_{13}$	VIA	--	GLC	432
Ph (9)	IIB	--	Synth.: X = Ph$_2$P, b$_{0.1}$ 53°	50
	IIIC	94	b$_{10}$ 110-12°, n$_D^{20}$ 1.5145	2088
	IIIA	40	(k, l)	2088
Et$_2$GeR$_2$				
R = n-Pr	VIC	--	GLC	432
	VIA	35	GLC (491-2)	491
	VIB	5	--	491

(i) By-prod. in rxn. of Ph$_2$GeH$_2$ with CH$_2$N$_2$ in Et$_2$O under UV irradiation (1799). (j) BuLi or PhCH$_2$Li are used for metallating R$_3$GeH (104). (k) Far IR spectrum (105). (l) Mass spectrum (1109).

Table 2 Continued

Formula*	Synth. Method**	Yield	Properties	Ref.
Et_2GeR_2 (Cont'd.)				
R = n-Bu (10)	VIB	11	GLC (491-2, 1493)	491
	V (m)	40	b_{18} 111°, n_D^{20} 1.4518, d_{20} 0.9595, IR and mass spectra	1493
n-C_5H_{11}	VIA	--	GLC	432
n-C_6H_{13}	VIA	--	GLC	432
Ph (10)	IVB	2	Synth. from Ph_2GeH_2, GLC (k, l)	104
$EtGeR_3$				
R = n-Pr	VIC	--	GLC	432
	VIA	23	GLC (491-2)	491
	IA	20		491
	VIB	12		491
n-Bu	VIB	8	GLC (491-2)	491
n-C_5H_{11} (10)	VIA	--	GLC	432
n-C_6H_{13}	VIA	--	GLC	432
Ph	--	--	(k, l, n)	
n-Pr_3GeR				
R = i-Pr	ID - Pr_3GeBr	65	b. 227-28°, mass spectrum	831
	ID - i-$PrGeI_3$	62	$b_{1.5}$ 84°, n_D^{20} 1.492, d_{20} 0.9621	2877
n-Bu	ID-$BuGeX_3$	--	$b_{0.35}$ 54.5°, GLC (491-2)	2175
n-C_5H_{11}	VIA	--	GLC	432
n-C_6H_{13}	VIA	--	GLC	432
Ph	ID - Pr_3GeBr	73	b_{20} 101-102°, n_D^{20} 1.4680, d_{20} 0.9784	1788
1-$C_{10}H_7$	ID - Pr_3GeBr	50	b_{10} 101-102°, n_D^{20} 1.4734, d_{20} 0.9622	1735
n-Pr_2GeR_2				
R = i-Pr	ID - n-Pr_2GeBr_2	--	GLC, mass spectrum	831

n-Bu	--	--	GLC	491-2
n-C$_5$H$_{11}$	VIA	--	GLC	432
n-C$_6$H$_{13}$	VIA	--	GLC	432
Ph	ID - Ph$_2$GeBr$_2$	58	b$_6$ 158-60°, n$_D^{20}$ 1.5583, d$_{20}$ 1.1138 (o)	464
	ID - Ph$_2$GeBr$_2$	--	(p)	831

n-PrGeR$_3$

R = i-Pr	ID - i-Pr$_3$GeCl	3	Colorless liq., GLC, mass spectrum	831
	IA	12	Synth. R = R' = Me$_2$CH	831
n-Bu	ID - n-PrX$_3$	--	b$_{0.45}$ 82°, GLC (491-2)	2175
n-C$_5$H$_{11}$	VIA	--	GLC	432
n-C$_6$H$_{13}$	VIA	--	GLC	432

n-Bu$_3$GeR

R = n-C$_5$H$_{11}$	VIA	--	GLC	432
n-C$_6$H$_{13}$	VIA	--	GLC	432
Ph	ID - Bu$_3$GeBr	65	b$_5$ 136-37°, n$_D^{20}$ 1.4661, d$_{20}$ 0.9415	1788
1-C$_{10}$H$_7$	ID - Bu$_3$GeBr	51	b$_4$ 139°, n$_D^{20}$ 1.4700, d$_{20}$ 0.9591	1735

n-Bu$_2$GeR$_2$

R = n-C$_5$H$_{11}$	VIA	--	GLC	432
n-C$_6$H$_{13}$	VIA	--	GLC	432
Ph	ID - Ph$_2$GeBr$_2$	62	b$_7$ 173-74°, n$_D^{20}$ 1.5508, d$_{20}$ 1.0922	464
Ph (11)	IVC - Ph$_2$GeH$_2$	12	b$_{0.001}$ 130-40° (k, p)	104

n-BuGeR$_3$

R = n-C$_5$H$_{11}$	VIA	--	GLC	432
n-C$_6$H$_{13}$	VIA	--	GLC	432

(m) Synth. by rxn. of Et$_2$Ge[(CH$_2$)$_4$Br]$_2$ with sodium in xylene, in 21-40% yield (1493). (n) Rxn. with Br in (CH$_2$Br)$_2$ → EtPh$_2$GeBr (1:2) or EtPhGeBr$_2$ (1:4) (999). (o) Rxn. with Br in (CH$_2$Br)$_2$ → n-Pr$_2$GeBr$_2$ (831). (p) Surface tension (464).

Table 2 Continued

Formula*	Synth. Method**	Yield	Properties	Ref.
i-Bu$_3$GePh	ID - Bu$_3$GeBr	63	b$_8$ 125-26°, n$_D^{20}$ 1.4783, d$_{20}$ 0.9703	1788
i-Bu$_3$GeC$_{10}$H$_7$-1	ID - Bu$_3$GeBr	45	b$_7$ 123-24°, n$_D^{20}$ 1.4674, d$_{20}$ 0.9460	1735
i-Bu$_2$GePh$_2$	ID - Ph$_2$GeBr$_2$	44	b$_4$ 160-62°, n$_D^{20}$ 1.5498, d$_{20}$ 1.0821 (p)	464
s-BuGePh$_3$	IIB - s-BuGeCl$_3$	--	m. 70-71°, (far) IR spectrum, assignment	2695
t-BuGePh$_3$	IIB - Me$_3$CGeCl$_3$	--	m. 160-62°, (far) IR spectrum, assignment	2694
	IIB - Ph$_3$GeX (9)	40	m. 165-66°, NMR spectrum	2675

(n-C$_5$H$_{11}$)$_{4-n}$GeR$_n$

R$_n$ = n-C$_6$H$_{13}$	VIA	--	GLC	432
Ph	ID - RMgBr	75	b$_5$ 154-55°, n$_D^{20}$ 1.4675, d$_{20}$ 0.9273	1788
1-C$_{10}$H$_7$	ID - RMgBr	41	b$_5$ 169-70°, n$_D^{20}$ 1.4706, d$_{20}$ 0.9356	1735
(n-C$_6$H$_{13}$)$_2$	VIA	--	GLC	432
(n-C$_6$H$_{13}$)$_3$	VIA	--	GLC	432

(i-C$_5$H$_{11}$)$_{4-n}$GeR$_n$

R$_n$ = Ph	ID - RMgBr	66	b$_5$ 141-42°, n$_D^{20}$ 1.4697, d$_{20}$ 0.9369	1788
1-C$_{10}$H$_7$	ID - RMgBr	37	b$_5$ 154-55°, n$_D^{20}$ 1.4671, d$_{20}$ 0.9301	1735
Ph$_2$	ID - RMgBr	48	b$_4$ 187-89°, n$_D^{20}$ 1.5392, d$_{20}$ 1.0592 (p)	464

(n-C$_6$H$_{13}$)$_{4-n}$GeR$_n$

R$_n$ = 1-C$_{10}$H$_7$	ID - RMgBr	53	b$_6$ 204-205°, n$_D^{20}$ 1.4740, d$_{20}$ 0.9352 (r)	1735
Ph$_2$	ID - RMgBr	57	b$_4$ 203-205°, n$_D^{20}$ 1.5331, d$_{20}$ 1.0326 (p)	464
Ph$_3$ (11)	(s) - Ph$_3$GeH	(s)	m. 76°, NMR spectrum	1762

(n-C$_7$H$_{15}$)$_{4-n}$GeR$_n$

R$_n$ = Ph	ID - RMgBr	64	b$_{10}$ 242-43°, n$_D^{20}$ 1.4683, d$_{20}$ 0.9143	1788
1-C$_{10}$H$_7$	ID - RMgBr	52	b$_4$ 230-34°, n$_D^{20}$ 1.4705, d$_{20}$ 0.9233	1735
Ph$_2$	ID - RMgBr	55	b$_3$ 215-17°, n$_D^{20}$ 1.5252, d$_{20}$ 1.0148 (p)	464

$(n-C_8H_{17})_{4-n}GeR_n$			
R_n = Ph	ID - RMgBr	61	b_8 270-72°, n_D^{20} 1.4685, d_{20} 0.9053	1788
$1-C_{10}H_7$	ID - RMgBr	40	b_8 251-52°, n_D^{20} 1.4760, d_{20} 0.9276 (r)	1735
Ph_2	ID - RMgBr	62	b_3 225-26°, n_D^{20} 1.5274, d_{20} 0.9980 (p)	464
$(n-C_9H_{19})_{4-n}GeR_n$			
R_n = Ph	ID - RMgBr	61	b_8 290-92°, n_D^{20} 1.4808, d_{20} 0.9197	1788
$1-C_{10}H_7$	ID - RMgBr	37	b_6 285-86°, n_D^{20} 1.4730, d_{20} 0.9105 (r)	1735
Ph_2	ID - RMgBr	69	b_3 229-30°, n_D^{20} 1.5171, d_{20} 0.9913 (p)	464
$(n-C_{10}H_{21})_3GePh$	ID - RMgBr	53	b_8 310-12°, n_D^{20} 1.4740, d_{20} 0.9052	1788
$(n-C_{10}H_{21})_2GePh_2$	ID - RMgBr	84	b_6 234-35°, n_D^{20} 1.5051, d_{20} 0.9761 (p)	464
Ph_2GeR_2			
R = $o-C_6H_4Me$	ID - RMgBr	48	m. 165-66°, IR spectrum	463
$p-C_6H_4Et$	ID - RMgBr	39	m. 111-12°, IR spectrum	463
$p-C_6H_4CHMe_2$	ID - RMgBr	33	m. 127-28°	463
R_3GePh	--	--	IR and UV spectra	137
R_3GeAr	--	--	(t)	--

(q) Synth. from $Ph_3GeOCHPh_2$ with t-BuLi in pentane (2675). (r) Violet fluorescence (1735). (s) Synth. by rxn. of Ph_3GeH with n-BuCH:CH_2 at 80°, with AIBN or under UV irradiation, in 90-95% yield (1762). (t) Discussion of rxn. mechanism of protonolysis and halodemetallation (1379).

2.1.2 TETRAORGANOGERMANES WITH THREE DIFFERENT ORGANIC GROUPS

The nonfunctional organogermanes containing three different organic groups are listed in Table 3. The compounds are prepared by methods summarized in the following scheme.

  I Grignard synthesis:
    (IA) $R_2R'GeCl + R''MgX \rightarrow R_2GeR'R''$,
    (IB) $R'R''GeCl_2 + RMgX \rightarrow R_2GeR'R''$,
    (IC) $\overline{CH_2CH_2CH_2}GeBu_2 + ICl, + EtMgBr \rightarrow EtPrGeBu_2$.

  II Synthesis with organolithium compounds:
    (IIA) $R_3GeCl + R'Li \rightarrow R_3GeR'$,
    (IIB) $R_3GeX + BuLi \rightarrow R_3GeBu$.

  III Synthesis with triorganogermyl lithium derivatives:
    $R_3GeLi + R'X \rightarrow R_3GeR'$.

  IV Synthesis with triorganogermyl compounds formed in situ:
    (IVA) $\overline{CH_2CH_2CH_2}GeEt_2 + BuLi \rightarrow Et_2GePrBu$,
    (IVB) $Ph_2GeH_2 + BuLi, + EtBr \rightarrow EtBuGePh_2$.
    (IVC) $R_3GeH + BuLi, + R'I \rightarrow R_3GeR'$.

  V Redistribution reactions:
    (VA) $R_4Ge + R'_4Ge + cat. AlCl_3, + LiAlH_4, + MeMgCl \rightarrow R_nR'_mGeMe_p$,
    (VB) $R_4Ge + R'_4Ge + GeBr_4, + EtMgBr \rightarrow R_nR'_mGeEt_p$.

(R, R', R'' = organic groups, X = halogen, n = 1-2, n + m + p = 4.)

2.1.3 NONFUNCTIONAL ASYMMETRIC TETRAORGANOGERMANES

Synthesis of asymmetric tetraorganogermanes was only reported quite recently (2292, 2367-9). The new compounds and their stereochemistry are shown in abbreviated form at the bottom of Table 3.

2.3 HETEROCYCLIC SUBSTITUTED TETRAORGANOGERMANES

Heterocyclic substituted tetraorganogermanes are prepared by introduction of the cyclic moiety with organolithium derivatives (I), by forming the heterocycle in a side chain reaction with diazoalkane (II) or by dehydration with hydrogen sulfate (III). The compounds are prepared by methods in the following scheme and are compiled in Table 4 on page 19.

  I $R_3GeBr + R'Li \rightarrow R_3GeR'$.
  (IIA) $Me_3GeC\vdots CR + R_2CN_2 \rightarrow Me_3Ge\overline{C:CRCR_2N:N}$,
  (IIB) $Ph_2GeC\vdots CH + CH_2N_2 \rightarrow Ph_2Ge(\overline{C:CHNHN:CH})_2$.
  III $Et_3GeC(CR_2OH):CHCMeROH + MHSO_4 \rightarrow Et_3Ge\overline{C:CCMeROCR_2}$.

R = hydrogen or organic group, may be different in a molecule, R' = heterocyclic group, M = sodium or potassium. One thiophene derived bis(triorganogermane) and is listed in Chapter 2.7.1.1.3.

Table 3. Nonfunctional Tetraorganogermanes with Three and Four Organic Groups $R'R''R'''GeR$

Formula	Synth. Method*	Yield	Properties	Ref.
Me$_2$EtGeBu-n	--	--	(a)	--
Me$_2$(n-Pr)GeBu-n	IA - PrMgCl	--	b. 170-71°, n_D^{20} 1.4350 (a, b)	2292
MeEt$_2$GePr-n	VA	44	(a)	491
MeEt$_2$GeBu-n	--	--	GLC	492
MeEtGe(Pr-n)$_2$	VA	25	(a)	491
MeEtGe(Bu-n)$_2$	--	--	(a)	--
Me(n-Pr$_2$)GeBu-n	--	--	(a)	--
Me(n-Pr)Ge(Bu-n)$_2$	--	--	GLC	492
Et$_2$(n-Pr)GeBu-n	VB	14	(a)	491
	IVA	80	b$_{10}$ 92°, n_D^{20} 1.4504, d$_{20}$ 0.9629	2575
	--	--	b$_{13}$ 91-92°, n_D^{20} 1.4499 (c)	2292
Et$_2$(n-Pr)GePh	IVA	80	b$_{10}$ 119°, n_D^{20} 1.5108, d$_{20}$ 1.0700	2575
Et(n-Pr$_2$)GeBu-n	VB	18	(a)	491
Et(n-Pr)Ge(Bu-n)$_2$	VB	21	(a)	491
	IC	--	b$_{1.8}$ 90°, n_D^{20} 1.4578, d$_{20}$ 0.9645	1492
Et(i-Pr)GePh$_2$	IA - i-PrMgBr	70	b$_{0.15}$ 104-105°, n_D^{25} 1.5641 (d)	999
Et(n-Bu)GePh$_2$	IVB	21	b$_{0.001}$ 100-110°, (far) IR spectrum (105)	104
EtPhGe(C$_{10}$H$_7$-1)$_2$	IB - EtPhGeBr$_2$	76	m. 166.5-167° (e)	999
MeEt(n-Pr)GeBu-n	IA - EtMgBr	--	b$_{47}$ 104°, n_D^{20} 1.4431 (f, g)	2229

*The characters used here correspond to the synthetic scheme in the introduction to Chapter 2.1.2.
(a) GLC (491-2). (b) Rxn. with SnCl$_4$ → MeSnCl$_3$ + MePrBuGeCl (2292). (c) Rxn. with SnCl$_4$ → EtSnCl$_3$ + EtPrBuGeCl (2292). (d) Rxn. with Br in (CH$_2$Br)$_2$ → Et(i-Pr)PhGeBr (999, 1003). (e) Rxn.with Br in (CH$_2$Br)$_2$ → EtPh(1-C$_{10}$H$_7$)GeBr (999). (f) Rxn. with SnCl$_4$ → EtPrBuGeCl (2292). (g) GLC (492).

Table 3 Continued

Formula	Synth. Method*	Yield	Properties	Ref.
R_3GeR':	$R_3Ge = EtPh(1-C_{10}H_7)Ge$			
$R' =$ Me	IVC - EtI (h)	76	(−)-form, n_D^{25} 1.6245, $b_{0.05}$ 180°, IR spectrum (i,j)	2367
	IVC - EtI (k)	79	(+)-form, n_D^{25} 1.6245 (i,j)	2367
	IVC - (+)-R_3GeH	85	(+)-form, n_D^{25} 1.6228 (i,l,m)	2367, 2369
	IVC - (+)-R_3GeH	50	(+)-form, n_D^{21} 1.6228	2367
	IIA - (−)-R_3GeCl	--	(+)-form (i,j)	2368
PhCH$_2$	III - PhCH$_2$Cl	--	(i,l)	2369
	III - PhCH$_2$X	--	X = Br, I (i,j)	2369
i-Pr	III - Me$_2$CHCl	--	(i,n)	2369
	III - Me$_2$CHBr	--	(i,l)	2369
	III - Me$_2$CHI	--	(i,j)	2369
n-Bu	IIB - (o)	--	(i)	2368
	III - n-BuX	--	X = Cl, Br (i,l)	2369
	III - n-BuI	--	(i,j)	2369
t-Bu	III - Me$_3$CX	--	X = Cl, Br (i,l)	2369
CH$_2$:CHCH$_2$	III - RX	--	X = Cl, Br (i,l)	2369
	III - RI	--	(i,j)	2369

(h) Synth. from (+)-MePh(1-C$_{10}$H$_7$)GeH (2367). (i) Optical rotation. (j) Synth. with inversion of configuration. (k) Synth. from (−)-MePh(1-C$_{10}$H$_7$)GeH (2367). (l) Synth. with retention of configuration. (m) Absolute configuration (2367). (n) Synth. with partial retention of configuration (2369). (o) The stereospecificity of the synthesis is reported as follows: R_3GeX + BuLi → R_3GeBu; inversion of configuration with X = Cl, N-pyrrolyl; retention of configuration with X = MeO, menthoxide; partial inversion and racemization with X = R_3GeS (2368).

Table 4. Heterocyclic Substituted Unsymmetric Tetraorganogermanes R_3GeR'

Formula		Synth. Method*	Yield	Properties	Ref.
Me₃Ge-pyrazole structure		IIA	--	(R = CF₃)	2332
Et₃Ge-dihydrofuran structure					
R_2 = Me₂,	R' = H	III	12	b_6 83-85°, n_D^{20} 1.4703, d_{20} 1.0121	199
(CH₂)₅,	H	III	21	b_8 128-29°, n_D^{20} 1.5008, d_{20} 1.0380	199
Me₂,	CH₂CH₂OMe	III	75	b_5 137-38°, n_D^{20} 1.4730, d_{20} 1.0386	2427
Me + Et	CH₂CH₂OMe	III	83	b_2 137-38°, n_D^{20} 1.4795, d_{20} 1.0396	2427
Me + CH₂CH₂OMe,	CH₂CH₂OMe	III	91	b_2 149-50°, n_D^{20} 1.4765, d_{20} 1.0496	2427
(CH₂)₅,	CH₂CH₂OMe	III	79	b_2 164-65°, n_D^{20} 1.4940, d_{20} 1.0703	2427
Et₃Ge-thiane structure					
R = H		I	50	$b_{3.5}$ 144°, n_D^{22} 1.5440 (a, b)	792
Ph		I	70	$b_{0.3}$ 162°, n_D^{20} 1.5920 (c)	792
Me		I	52	$b_{0.4}$ 104°, n_D^{20} 1.5461	792
		(a)	63	(c)	792

* The characters used here correspond to the synthetic scheme in the introduction of Chapter 2.3. (a) Rxn. of R₃GeCHSCH₂CH₂CH₂S + BuLi in THF-C₆H₁₂ in N atm., followed by MeI → R₃GeCMeSCH₂CH₂CH₂S (792). (b) Rxn. with BuLi, + R'₃MX → Et₃Ge(R'₃M)CSCH₂CH₂CH₂S; R'MX = EtGeBr, MeSiI (792). (c) Rxn. with HgCl₂ + CdCO₃ in aq. THF or Me₂SO → R₃GeCOR (792).

Table 4 Continued

Formula	Synth. Method*	Yield	Properties	R₃GeR' Ref.
 Ph₃Ge-[imidazole]	IIB	--	m. 215°, mol. wt.	939
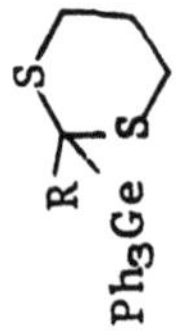 Ph₃Ge-C(R)(S-CH₂CH₂CH₂CH₂-S)				
R = H	I	77	m. 172-73° (a)	792
Me	I	60	m. 182-83°	792
	(a)	64	(c)	792

2.4 UNSYMMETRIC TETRAORGANOGERMANES WITH OLEFINIC SUBSTITUENTS

Olefinic tetraorganogermanes are important intermediates for polymeric organogermanium compounds and for introduction of functional groups in the side chain. Addition and substitution reactions compete especially with vinyl-type derivatives and either or both reaction products are obtained under suitable reaction conditions. Olefinic organogermanes are prepared by reaction of organogermanium halides with Grignard (I) or organoalkali metal (II) reagents. Other methods include hydrogermylation of dienes or acetylenes (III) and interconversion of organogermanes by classical organic methods (IV) in the carbon chain like dehydration or Wittig reaction. The following scheme summarizes methods for preparation of the olefinic tetraorganogermanes in Table 5.

I Grignard synthesis:
   (IA) $R_nGeX_{4-n} + R'MgBr \rightarrow R_nGeR'_{4-n}$,
   (IB) $R_2R'GeX + RMgBr \rightarrow R_3GeR'$,
   (IC) $Et_3GeX + R'Cl + Mg \rightarrow Et_3GeR'$,
   (ID) $HGeCl_3 + diene, + MeMgCl \rightarrow Me_3GeR'$,
   (IE) $i\text{-}Pr_3GeMgCl + R'Cl \rightarrow i\text{-}Pr_3GeR'$,
   (IF) $\overline{CH_2CH:CMeCH_2}GeX_2 + RMgI \rightarrow R_3GeCH_2CH:CMe_2$.

II Synthesis with alkali metal derivatives:
   (IIA) $R_3GeX + R'Li \rightarrow R_3GeR'$,
   (IIB) $R'GeCl_3 + PhLi \rightarrow Ph_3GeR'$,
   (IIC) $Me_2GeCl_2 + NaC_5H_5 \rightarrow Me_2Ge(C_5H_5)_2$.

III Hydrogermylation reactions:
   (IIIA) $R_3GeH + diene \rightarrow R_3GeR'$; cat. H_2PtCl_6 ,
   (IIIB) $Et_3GeH + PhC\vcentcolon CR \rightarrow Et_3GeCR:CHPh$.

IV Interconversion reactions:
   (IVA) $Me_3GeCMe_2CH_2CMe_2OH + KHSO_4 \rightarrow Me_3GeCMe_2CH_2CMe:CH_2$,
   (IVB) $Ph_3GeCMeROH + PBr_3 \rightarrow Ph_3GeCR:CH_2$; R = Me, Ph .
   (IVC) $Ph_3GeCOPh + Ph_3P:CHR \rightarrow Ph_3GeCPh:CHR$; R = H, Me .

R = organic group, may be different in a molecule, R' = olefinic group, X = halogen, n = 1-3. Other unsaturated tetraorganogermanes are listed in Chapters 1.3, 2.1.3, 2.5, 7.1.4 and 7.1.6.6 .

Trimethylvinylgermane (M-13) GeC_5H_{12} $Me_3GeCH:CH_2$
Synth.: By rxn. of Me_3GeI with $CH_2:CHLi$ in Et_2O, in 52% yield (1579); with $CH_2:CHBr$ and Na in Et_2O in presence of cat. EtOAc, in 27% yield (1579); with $CH_2:CHMgBr$ in THF, in 18% yield (1579).
Prop.: b. 70-71°, n_D^{20} 1.4167, d_{20} 0.9980 (1579), IR (763, 1013) and UV (424) spectra, magnetic measurements (984).
Rxn. with: BuLi → relative rate of nucleophilic substitution (2629).
$(SCN)_2$ → relative rate of electrophilic substitution (2629).
Ac_2O_2 → relative reactivity capturing Me-radicals (2630).

α-Styryltrimethylgermane $GeC_{11}H_{16}$ $Me_3GeCPh:CH_2$
Synth.: By rxn. of Me_3GeI with Na and $PhCBr:CH_2$ in Et_2O, in 36% yield (1580).
Prop.: $b_{2.5}$ 57°, n_D^{20} 1.5365, d_{20} 1.1379, mol. cat., IR spectrum and assignment (1580).
Rxn. with: EtLi → $Me_3GeCHPrPh$ (2629).
BuLi → relative rate of nucleophilic substitution (2629).
$(SCN)_2$ → relative rate of electrophilic substitution (2629).

β-Styryltrimethylgermane $GeC_{11}H_{16}$ $Me_3GeCH:CHPh$
Synth.: By rxn. of Me_3GeI with $PhCH:CHMgBr$ in Et_2O, in 46% yield (1580).
Prop.: b_7 93°, b_4 83°, n_D^{20} 1.5425, d_{20} 1.1441, mol. wt., IR spectrum and assignment (1580).
Rxn. with: EtLi → $Me_3GeCHEtCH_2Ph$ (2629).

BuLi → relative rate of nucleophilic substitution (2629).
(SCN)$_2$ → relative rate of electrophilic substitution (2629).
Ac$_2$O$_2$ → relative reactivity capturing Me-radicals (2630).

Allyltrimethylgermane (M-14)   GeC$_6$H$_{14}$ Me$_3$GeCH$_2$CH:CH$_2$
Synth.: By rxn. of MeMgBr with HC⋮CCH$_2$GeCl$_3$ (?) from HGeCl$_3$ and HC⋮CCH$_2$Br (364).
By rxn. of Me$_3$GeH with CH$_2$:C:CH$_2$ in sealed tube in presence of cat. H$_2$PtCl$_6$, in 28% yield (1041-2).
Prop.: b$_{62}$ 30°, n$_D^{20}$ 1.4310 (364), n$_D^{25}$ 1.4278 (1041-2), IR (136, 763, 1041-2), UV (424) and NMR (1041-2) spectra, GLC (492, 1041-2), magnetic measurements (984).
Rxn. with: PhHgCCl$_2$Br → Me$_3$GeCH$_2$$\overline{\text{CHCCl}_2\text{CH}_2}$ (1877), relative rxn. rate and mechanism (1882).

1-Propenyltrimethylgermane (M-14)   GeC$_6$H$_{14}$ Me$_3$GeCH:CHMe
Synth.: By rxn. of Me$_3$GeI with MeCH:CHMgBr in THF, in 53% yield (1580).
Prop.: b$_{752}$ 101°, n$_D^{20}$ 1.4341, d$_{20}$ 1.0045, IR spectrum and assignment (1580).
Rxn. with: Li in THF → isomerization equilibrium; 92% trans-, 8% cis-form (1887).
BuLi → relative rate of nucleophilic substitution (2629).
(SCN)$_2$ → relative rate of electrophilic substitution (2629).
Ac$_2$O$_2$ → relative reactivity capturing Me-radicals (2630).

2-Propenyltrimethylgermane   GeC$_6$H$_{14}$ Me$_3$GeCMe:CH$_2$
Synth.: By rxn. of Me$_3$GeI with CH$_2$CMeLi in Et$_2$O, in 56% yield (1580).
By-prod. in rxn. of Me$_3$GeH with allene in presence of cat. H$_2$PtCl$_6$, in 18% yield (1041-2).
Prop.: b$_{747}$ 97°, n$_D^{20}$ 1.4294 (1580), n$_D^{25}$ 1.4260 (1041-2), d$_{20}$ 1.006 (1580), IR (1041-2), assignment (1580), and NMR (1041-2) spectra, GLC (1041-2).
Rxn. with: BuLi → relative rate of nucleophilic substitution (2629).
(SCN)$_2$ → relative rate of electrophilic substitution (2629).
Ac$_2$O$_2$ → relative reactivity capturing Me-radicals (2630).

Crotyltrimethylgermane (M-16)   GeC$_7$H$_{16}$ Me$_3$GeCH$_2$CH:CHMe
Synth.: By rxn. of MeCH:CHCH$_2$GeCl$_3$ with MeMgCl (364, 1530).
By rxn. of (CH$_2$:CH)$_2$ with HGeBr$_3$, followed by MeMgCl, in 70% yield; the product contains 30% Me$_2$$\overline{\text{GeCH}_2\text{CH:CHCH}_2}$ (161, 365).
By rxn. of (CH$_2$:CH)$_2$ with Me$_3$GeH in the presence of cat. H$_2$PtCl$_6$, in 65% yield (1041-2).
Prop.: b$_{94}$ 68°, n$_D^{20}$ 1.4435 (364), n$_D^{25}$ 1.4430 (cis-form), 1.4375 (trans-form) (1041-2), d$_{20}$ 0.9912 (364), 0.9907 (1530), IR (1041-2, 1530), UV (1530) and NMR (161, 365, 1041-2, 1530) spectra, GLC (1041-2), cis-trans-ratio in product (1041-2).
Rxn. with: CF$_3$CO$_2$H → EtCH:CH$_2$ (1041-2).

Triethylvinylgermane (M-14)   GeC$_8$H$_{18}$ Et$_3$GeCH:CH$_2$
Prop.: Molar magnetic rotation (301).

Rxn. with: BuLi → Ge-contg. polymer (1700).
RCH:CH$_2$ + BuLi → Ge-contg. copolymer; R = Ph, Et$_3$Si, Et$_3$Sn (1700).
Me$_3$SnCF$_3$ + NaI in (MeOCH$_2$)$_2$ → Et$_3$Ge$\overline{\text{CHCH}_2\text{CF}_2}$ + Me$_3$SnI (1874).
PhHgCCl$_2$Br → Et$_3$Ge$\overline{\text{CHCH}_2\text{CCl}_2}$ (1877), relative rxn. rate and mechanism (1882).

Allyltriethylgermane (M-14)   GeC$_9$H$_{18}$ Et$_3$GeCH$_2$CH:CH$_2$
Synth.: By-prod. in rxn. of Et$_3$Ge(CH$_2$)$_3$B(OMe)$_2$ with EtMgBr (357).
Prop.: b$_{10}$ 60-62° (2715), b$_2$ 30-33° (357), n$_D^{20}$ 1.4595 (2715), GLC (492).
Rxn. with: HgBr$_2$ → Et$_3$GeBr + CH$_2$:CHCH$_2$HgBr; kinetic data, influence of solvent and rxn. mechanism (2715).
Biol. Prop.: Toxicity (845).

Crotyltriethylgermane   GeC$_{10}$H$_{22}$ Et$_3$GeCH$_2$CH:CHMe
Synth.: By rxn. of Et$_2$(MeCH:CHCH$_2$)GeX with EtMgBr; X = Cl, Br (467), in 79 and 82% yield, resp. (1798), X = I, in 85% yield (1798).
By-prod. in rxn. of Et$_3$GeH with (CH$_2$:CH)$_2$ in presence of cat. H$_2$PtCl$_6$ at 100°, in 36% yield (1798).
Prop.: b$_{22}$ 92° (1798), b$_{18}$ 88.5° (467, 1798), b$_{14}$ 76-78°, n$_D^{20}$ 1.4639 (2715), 1.4648-57 (1798), 1.4655, d$_{20}$ 0.9971 (467), 0.9940-71 (1798); cis-form: n$_D^{20}$ 1.4667, d$_{20}$ 1.0002, trans-form: n$_D^{20}$ 1.4608, d$_{20}$ 0.9872 (1798), (far) IR spectrum (1798), GLC (467, 1798), cis-trans-ratio in product (467, 1798).
Rxn. with: H + cat. Raney Ni → Et$_3$GeBu (467, 1798).
Br in EtBr → Et$_3$GeBr + MeCH:CHCH$_2$Br (467, 1798).
HgBr$_2$ → Et$_3$GeBr + MeCH:CHCH$_2$HgBr; kinetic data, influence of solvent, rxn. mechanism (2715).

α-Styryltriphenylgermane   GeC$_{20}$H$_{22}$ Ph$_3$GeCPh:CH$_2$
Synth.: By rxn. of Ph$_3$GeBz with MeMgBr, in 48% yield (793); with Ph$_3$P:CH$_2$ in Et$_2$O, in 75% yield (793).
By rxn. of Ph$_3$GeCMePhOH with PBr$_3$ in C$_6$H$_6$ under reflux, in 63% crude yield (1624, 1626).
Prop.: m. 120.5-122.5° (793), 119-120° (1624, 1626), IR (793, 1624, 1626), UV and NMR (793) spectra.
Rxn. with: O$_3$ → Ph$_3$GeBz (1624).

Additional olefinic substituted tetraorganogermanes are summarized in Table 5.

Table 5. Unsymmetric Tetraorganogermanes with Olefinic Substituents $R_nGeR'_{4-n}$

Formula*	Synth. Method**	Yield	Properties	Ref.
Me_3GeR'				
R' = $CH_2CH_2CH:CH_2$ (16)	--	--	IR spectrum (136), magnetic measurements	984
$CH_2CH:CMe_2$	IF	51	b. 146.5°, n_D^{20} 1.4491, d_{20} 0.9737	1495
$CH_2CMe:CHMe$	ID (a)	(b)	IR and NMR spectra (c)	1530
1-cyclo-C_5H_7 (16)	IIIA	25 (b)	n_D^{25} 1.4657, GLC (d, e)	1041-2
2-cyclo-C_5H_7 (16)	IIIA	25 (b)	n_D^{25} 1.4662, GLC (d, e)	1041-2
2-cyclo-C_6H_9	IIIA (f)	60	n_D^{25} 1.4795 (d)	1041-2
	IIIA (g)	49	(h)	1041-2
$CMe_2CH_2CMe:CH_2$	IVA	87	b_{27} 79.5-80°, n_D^{20} 1.4580, d_{20} 0.9891 (i)	1526
C_7H_9	IIIA	80	n_D^{25} 1.4834, GLC (d, j)	1041-2
3-$CH_2CH_2C_6H_9$-cyclo	IIIA (k)	47	n_D^{25} 1.4756, GLC (d, l)	1041-2
2-C_8H_{13}	IIIA (m)	30	n_D^{25} 1.4869, GLC	1041-2
	IIIA (n)	42	(d, h)	1041-2
p-$C_6H_4CH:CH_2$ (16)	--	--	$b_{0.3}$ 58°, n_D^{20} 1.5403, d_{20} 1.109 (o)	2316
$Me_2Ge(CH_2CH:CH_2)_2$ (16)	--	--	GLC	492
$Me_2Ge(C_5H_5)_2$	IIC	61	b_{14} 130°, NMR spectrum, structure (1893)	157
$MeGe(CH_2CH:CH_2)_3$ (17)	--	--	GLC	492
Et_3GeR'				
R' = CD:CHPh	(p)	--	NMR spectrum	1016
CPh:CHPh	(q)	98	b_1 140°, n_D^{20} 1.5750	2849
cis-CH:CHMe	IIA	77	b_{11} 58-60°, n_D^{20} 1.4625, d_{20} 1.0342 (r)	1600
trans-CH:CHMe	IIA	61	b_{10} 60-62°, n_D^{20} 1.4581, d_{20} 1.0108 (r)	1600
$CH_2CH:CHPh$	IC	35	$b_{0.6}$ 105-108°, n_D^{20} 1.5433, IR and UV spectra	2716
$CH_2CH_2CH:CH_2$	IA	--		1494
	IB	--	b_{17} 86°, n_D^{20} 1.4587, d_{20} 0.9888	2578

$CH_2CMe:CH_2$	IA	--	b_{11} 77°, n_D^{20} 1.4648, d_{20} 0.9973	1494
$(CH_2)_3CH:CH_2$	IA	--	b_{12} 93°, n_D^{20} 1.4597, d_{20} 0.9785 (s)	1494
$CH_2CH:CMe_2$	IF	43	(X = I) b_{38} 116.5°, n_D^{20} 1.4678, d_{20} 0.9824, GLC	1495
	IF	(b)	(X = Cl)	1495
$C(:CHPh)CMe:CH_2$	IVA (t)	72	b_2 138-40°, n_D^{20} 1.5360, d_{20} 1.0638	2426
C_5H_5 (15)	(u)	--		832
$CH_2CMe:CMe_2$	IB	75	b_{10} 103°, n_D^{20} 1.4748, d_{20} 0.9859 (v)	467, 1798
$C(:CHPh)CMe:CHMe$	IVA (t)	77	b_2 135-37°, n_D^{20} 1.5280, d_{20} 1.0751	2426
$1-C(:CHPh)C_5H_7c.$	IVA	80	b_2 159-60°, n_D^{20} 1.5300, d_{20} 1.0945	2425
	IIIB	20		2425
$4-CH_2CH_2C_6H_9c.$	IB	90	b_{18} 154°, n_D^{20} 1.4845, d_{20} 1.0234	1798
	IIIA (w)	(b)	b_{20} 154°	1798
$1-C(:CHPh)C_6H_9c.$	IVA	85	b_2 162-63°, n_D^{20} 1.5575, d_{20} 1.1043	2425
$C(CH:CH_2):CHCMe:CHEt$	IVA (x)	46	b_4 175°	2428
$p-C_6H_4CH:CH_2$ (17)	--	--	(o, y)	--

* Numbers in parenthesis refer to page number in main volume.

** The characters used here correspond to the synthetic scheme in the introduction to Chapter 2.4.

(a) $CH_2:CHCMe:CH_2$ used as diene (1550). (b) By-prod. in synth. (c) Rxn. prod. was not sepd. (d) IR and NMR spectra (1041-2). (e) Crude rxn. prod. reacts with $CF_3CO_2H \rightarrow C_5H_8$ (1041-2). (f) Synth. from 1,3-cyclohexadiene. (g) Synth. from 1,4-cyclohexadiene. (h) Rxn. with $CF_3CO_2H \rightarrow Me_3GeO_2CCF_3 + R'H$ (1041-2). (i) NMR spectrum (1526). (j) Mixture of endo-, exo-norborneyl-GeMe$_3$ and nortricyclenyl-GeMe$_3$, not separable by GLC, ratio by NMR = 75:15:10 (1041-2). (k) Synth. from 4-vinylcyclohexene (1041-2). (l) Prod. does not react with CF_3CO_2H (1041-2). (m) Synth. from 1,3-cyclooctadiene. (n) Synth. from 1,5-cyclooctadiene. (o) Rxn. with $(SCN)_2 \rightarrow$ relative reactivity (863); with H and Raney Ni $\rightarrow$ relative rate of hydrogenation (2316). (p) Synth. by rxn. of $Et_3GeC(AlBu_2)$:CHPh with D_2O (1016). (q) Synth. by hydrolysis of Et_3GeCPh:CPhLi in C_6H_6 in inert atm. (2849). (r) IR spectrum. (s) Rxn. with Br $\rightarrow Et_3Ge(CH_2)_3CHBrCH_2Br$ (1494). (t) Synth. from $Et_3GeC(:CHPh)-$ $CMeOHCH_2R$; R = H, Me (2426). (u) By-prod. in thermal dec. of $Et_3GeC_5H_5Mn(CO)_3$. (v) Rxn. with Br $\rightarrow Me_2C:CMe-$ CH_2Br (467, 1798); with O_3 in $HCCl_3 \rightarrow Et_3GeCH_2Ac$ (467, 1798). (w) Diene used in synth. was $(CH_2:CH)_2$ (1798). (x) Synth. from $Et_3GeC(CH:CH_2):CHCMePrOH$. (y) ESR spectrum of rxn. prod. with Na or K in $(MeOCH_2)_2$ in N_2 atm. (524).

Table 5 Continued

Formula*	Synth. Method**	Yield	Properties	Ref. $R_nGeR'_{4-n}$
$Et_2Ge(CH_2CH:CH_2)_2$ (18)	--	--	GLC	492
$EtGe(CH_2CH:CH_2)_3$	--	--	GLC	492
n-Pr_3GeR'				
R' = $CH_2CH:CH_2$	IA	77	b_{23} 132-33°, n_D^{20} 1.4527, d_{20} 0.9426 (z)	2038
CH:CHMe	IA	74	b_3 80-81°, n_D^{18} 1.4562, d_{18} 0.9697 (r)	2051
CMe:CH_2	IA	86	b_5 95-96°, n_D^{18} 1.4536, d_{18} 0.9667 (r)	2051
CH:CHEt	IA	79	b_3 87-88°, n_D^{18} 1.4608, d_{18} 0.9663 (r)	2051
CH:CMe_2	IA	68	b_4 93-96°, n_D^{18} 1.4558, d_{18} 0.9566 (r)	2051
i-Pr_3GeR'				
R' = $CH_2CH:CH_2$	IA	75	b_{23} 90-91°, n_D^{20} 1.4885, d_{20} 1.1988 (z)	2038
	IE	--	b_{26} 112-14°, n_D^{20} 1.4847, d_{20} 0.9952	2591
CMe:CH_2	IA	75	b_{25} 120-21°, n_D^{20} 1.4720, d_{20} 0.9854 (r)	2834
CH:CHEt	IA	75	b_{12} 111-12°, n_D^{20} 1.4780, d_{20} 0.9835 (r)	2834
i-$Pr_2Ge(CMe:CH_2)_2$	IA	71	b_{12} 100-101°, n_D^{20} 1.4785, d_{20} 0.9949 (r)	2834
i-$Pr_2Ge(CH:CHEt)_2$	IA	78	b_4 103-104°, n_D^{20} 1.4810, d_{20} 0.9924 (r)	2834
n-Bu_3GeR'				
R' = CH:CH_2 (15)	--	--	n_D^{20} 1.4508, magnetic data	301
$CH_2CH:CH_2$ (15)	IA	69	b_{23} 159-60°, n_D^{20} 1.4567, d_{20} 0.9316 (aa)	2038
CH:CHMe	IA	87	b_5 127-28°, n_D^{18} 1.4598, d_{18} 0.9443 (r)	2051
CMe:CH_2	IA	77	b_{11} 138-39°, n_D^{18} 1.4588, d_{18} 0.9469 (r)	2051
CH:CHEt	IA	78	b_3 123-24°, n_D^{18} 1.4605, d_{18} 0.9423 (r)	2051
CH:CMe_2	IA	78	b_4 127-28°, n_D^{18} 1.4602, d_{18} 0.9431 (r)	2051
$CH_2CMe:CMe_2$	IB	77	b_{13} 163-64°, n_D^{20} 1.4728, d_{20} 0.9498	467, 1798
	IB (bb)	76	n_D^{20} 1.4730, d_{20} 0.9501 (cc)	467, 1798

$(i-Bu)_{4-n}GeR'_n$					
$R'_n =$ CMe:CH$_2$	IA	67	b$_4$ 111-12°, n_D^{20} 1.4698, d$_{20}$ 0.9786 (r)	2834
CH:CHEt	IA	70	b$_4$ 112-13°, n_D^{20} 1.4684, d$_{20}$ 0.9696 (r)	2834
CH:CMe$_2$	IA	71	b$_5$ 112-13°, n_D^{20} 1.4672, d$_{20}$ 0.9661 (r)	2834
(CMe:CH$_2$)$_2$	IA	70	b$_4$ 110-11°, n_D^{20} 1.4733, d$_{20}$ 0.9809 (r)	2834
(CH:CHEt)$_2$	IA	65	b$_6$ 114-15°, n_D^{20} 1.4770, d$_{20}$ 0.9718 (r)	2834
(CH:CMe$_2$)$_2$	IA	70	b$_5$ 112-13°, n_D^{20} 1.4750, d$_{20}$ 0.9701 (r)	2834

$R_3GeCH_2CH:CH_2$				
R = n-C$_5$H$_{11}$	IA	80	b$_{23}$ 162-63°, n_D^{20} 1.4587, d$_{20}$ 0.9198	2038
n-C$_6$H$_{13}$	IA	57	b$_{10}$ 195-96°, n_D^{20} 1.4627, d$_{20}$ 0.9137	2038
n-C$_7$H$_{15}$	IA	66	b$_{10}$ 225-27°, n_D^{20} 1.4656, d$_{20}$ 0.9109	2038
n-C$_8$H$_{17}$	IA	70	b$_{10}$ 251-52°, n_D^{20} 1.4684, d$_{20}$ 0.9097	2038

Ph_3GeR'				
R' = CH:CH$_2$ (18)	--	--	spectra: IR (1013), NMR (846), Mass (2695) (dd)	--
cis-CPh:CHPh	IIIA (ee)	48	m. 113-14°, UV spectrum, soly.	2283
CH$_2$CH:CH$_2$ (18)	IIB	--	m. 87-89°, (far) IR spectrum and assignment (ff)	2694
CH:CHMe	IA	--	m. 67-68°, b$_6$ 208-10°, IR spectrum (ff 1787)	462
CMe:CH$_2$	IVB	58	m. 96-97°, IR spectrum (gg)	1624, 1626
CPh:CHMe	IVC	81	m. 89-91.5°, UV and NMR spectra, trans-form	793
CH:CMe$_2$	IA	--	m. 119-20°, IR spectrum (ff, 1787)	462
{CMe:CHMe}{CEt:CH$_2$}	IVC	58 (c)	m. 116-20°, IR spectrum	1624, 1626

(z) Thermally stable, does not disproportionate (2038). (aa) Magnetic data (301). (bb) Synth. from Bu(Me$_2$C:CMe-CH$_2$)GeCl$_2$ (467, 1798). (cc) Rxn. with Br → Bu$_2$GeBr + Me$_2$C:CMeCH$_2$Br (467). (dd) Rxn. with Br in HCCl$_3$ → Ph$_3$Ge-CHBrCH$_2$Br (464). (ee) Synth. from (PhC≡)$_2$ and Ph$_3$GeH in presence of cat. Pt on carbon at 130° (2283). (ff) Rxn. with Br in HCCl$_3$ → Ph$_3$GeBr. (gg) Rxn. with O$_3$ → (Ph$_3$Ge)$_2$O (1624, 1626).

Table 5 Continued $R_nGeR'_{4-n}$

Formula*	Synth. Method**	Yield	Properties	Ref.
Ph$_3$GeR' (Cont'd.)				
R' = 9-C$_{13}$H$_9$ (19)	IIA	57	(fluorenyl), m. 166-68°, NMR spectrum (hh)	2675
p-C$_6$H$_4$CH:CH$_2$ (15)	IA	--	m. 91-92°, b$_4$ 250-70° (r)	462
p-C$_6$H$_4$CMe:CH$_2$	IA	--	m. 95-96°, b$_5$ 220-25° (r)	462
p-C$_6$H$_4$CH:CHEt	IA	--	m. 99-100°, IR spectrum	462
p-C$_6$H$_4$CMe:CHMe	IA	--	m. 102-103°, IR spectrum	462
Ph$_2$GeR'$_2$				
R' = CH:CH$_2$ (15)	IA	60	b$_4$ 145-47°, n_D^{20} 1.5867, d$_{20}$ 1.1757 (ii, jj)	464
CH$_2$CH:CH$_2$ (19)	IA	68	b$_3$ 158-60°, n_D^{20} 1.5859, d$_{20}$ 1.1462 (ii, jj)	464
CH:CHMe	IA	--	b$_9$ 179-80°, n_D^{20} 1.5789, d$_{20}$ 1.1363 (r) (ii, 1787)	462
CH:CMe$_2$	IA	--	b$_7$ 180-81°, n_D^{20} 1.5717, d$_{20}$ 1.1126 (r) (ii, 1787)	462
p-C$_6$H$_4$CH$_2$CH:CH$_2$	IA	--	m. > 200°, yellow powder, soly. (r)	462
p-C$_6$H$_4$CMe:CH$_2$	IA	--	m. 117-18°, b$_5$ 240-45° (r, kk, ll)	462
p-C$_6$H$_4$CMe:CHMe	IA	--	m. 130-31°, (r, kk, ll)	462
R$_3$GeCH:CH$_2$	--	--	Ge-bond refractions D$_{20}$	81
R$_3$GeCH$_2$CH:CH$_2$	--	--	Ge-bond refractions D$_{20}$	81

(hh) Rxn. with (CH$_2$CO)$_2$NBr → 9,9-Ph$_3$GeC$_{13}$H$_8$Br (2675). (ii) Monomeric, surface tension (464). (jj) Rxn. with Br in HCCl$_3$ → Ph$_2$GeBr$_2$. (kk) Rxn. with Br in HCCl$_3$ → (p-BrCHRCMeBrC$_6$H$_4$)$_2$GePh$_2$; R = H, Me (1787). (ll) Rxn. with dry HCl in C$_6$H$_6$ → (p-RCH$_2$CMeClC$_6$H$_4$)$_2$GePh$_2$; R = H, Me (1787).

2.4.1 UNSYMMETRIC TETRAORGANOGERMANES CONTAINING HALOGEN SUBSTITUTED OLEFINS

New is a series of derivatives of fluorolefins and synthesis of haloolefins by dehalocarbene insertion reaction. The haloolefin substituted tetraorganogermanes in Table 6 are prepared by methods from the following scheme.

I Catalyzed addition of alkylgermanium hydrides:
  (IA) $R_3GeH + R'C\vdots CH \rightarrow Et_3GeCR':CH_2 + Et_3GeCH:CHR'$,
  (IB) $R_3GeH + (R'C\vdots)_2 \rightarrow Et_3GeCR':CHR'$.

II Condensation with trialkylgermane:
  $Me_3GeH + \overline{CF_2CF_2CCl:CCl} \rightarrow Me_3Ge\overline{C:CXCF_2CF_2} + HCl$; X = H, Cl.

III Halogenation of unsaturated organogermanes:
  $Et_3GeC\vdots CCR:CHR + Br \rightarrow Et_3GeCBr:C:CRCHRBr + Et_3GeCBr:CBrCR:CHR$; R = H, Me.

IV Dehalogenation of fluorovinylgermanes:
  (IVA) with Me_3SnH ;
  (IVB) with Me_2SnH_2 .

V Dihalocarbene addition reactions:
  (VA) $R_3GeC\vdots CCF_3 + Me_3SnCF_3 \rightarrow R_3Ge\overline{C:C(CF_3)CF_2}$,
  (VB) $(R_3Ge)_2Hg + PhHgCCl_2Br \rightarrow R_3GeCCl:CCl_2$.

R = organic group, R' = halogen-substituted organic group. Other tetraorganogermanes containing halogenated olefinic groups are listed in Chapters 1.3, 2.7.1.8, 7.1.6.1 and 7.1.6.6.

Perfluorovinyltrimethylgermane $GeC_5H_9F_3$ $Me_3GeCF:CF_2$
Synth.: By rxn. of Me_3GeCl with $CF_2:CFMgBr$ in THF, in 36% yield (2209).
By-prod. in rxn. with $Me_3GeMn(CO)_5$ with C_2ClF_3 under UV irradiation (872).
Prop.: Colorless oil (872), IR (872, 2209), ^1H and ^{19}F NMR (872) spectra, GLC (2209).
Rxn. with: Me_3SnH at 55° → $Me_3GeCF:CHF + Me_3GeCH:CF_2 + Me_3GeCH:CHF + Me_3SnF$ (2209).
Me_3SnH under UV irradiation → cis-$Me_3GeCF:CHF$ + [$Me_3GeCF(SnMe_3)CF_2H$] + Me_3Ge-$CHFCF_2SnMe_3 + Me_3SnF$ (2209).
Me_2SnH_2 under UV irradiation → $Me_3GeCF:CHF + Me_3GeCH:CF_2 + Me_3GeCH:CHF + Me_2SnF_2$ (2209).

Bis(perflourovinyl)dimethylgermane $GeC_6H_6F_6$ $Me_2Ge(CF:CF_2)_2$
Synth.: By rxn. of Me_2GeCl_2 with $CF_2:CFMgBr$ in THF in 36% yield (2209).
Prop.: IR and NMR spectra, GLC (2209).
Rxn. with: Me_3SnH at 55° → $Me_2Ge(CF:CHF)_2 + Me_2Ge(CH:CF_2)_2 + Me_2Ge(CH:CHF)_2 + Me_3SnF$ (2209).
Me_2SnH_2 at 55° → $Me_2Ge(CF:CHF)_2 + Me_2Ge(CH:CF_2)_2 + Me_2Ge(CH:CHF)_2 + Me_2SnF_2$ (2209).
Me_3SnH under UV irradiation → unstable 1:2 addn. compd., proposed structure from NMR spectra (2209).

Other haloolefin-substituted tetraorganogermanes are listed in Table 6.

Table 6. Unsymmetric Tetraorganogermanes Containing Halogen-Substituted Olefins $R_nGeR'_{4-n}$

Formula*	Synth. Method**	Yield	Properties	Ref.
Me_3GeR'				
R' = CH:CHF	IVA (a)	9	IR and NMR spectra	2209
CH:CF$_2$	IVB (c)	(b)	--	2209
	IVA (a)	40	IR and NMR spectra	2209
	IVB (c)	(b)	--	2209
CF:CHF	IVA (a)	51	IR and NMR spectra	2209
	IVA, B (c)	(b)	cis-trans-ratio = 7:5	2209
CH:CHCl (20)	--	--	Dipole moment (d)	254
CCl:CH$_2$ (20)	--	--	Dipole moment (e)	254
CCl:CCl$_2$	VB	63	n_D^{25} 1.5330, IR spectrum, GLC	1881
C:C(CF$_3$)CF$_2$	VA	--	--	2332
C:CHCF$_2$CF$_2$	II	0.5	IR and NMR spectra	924
C:CClCF$_2$CF$_2$	VA	40	b_{50} 78°, IR, ^1H and ^{19}F NMR spectra	924
Me$_2$Ge(CH:CHF)$_2$	IVA (a)	(f)	IR, ^1H and ^{19}F NMR spectra	2209
	IVB (a)	(g)	--	2209
Me$_2$Ge(CH:CF$_2$)$_2$	IVA (a)	(f)	IR, H and F NMR spectra	2209
	IVB (a)	(g)	--	2209
Me$_2$Ge(CF:CHF)$_2$	IVA (a)	(f)	IR, ^1H and ^{19}F NMR spectra	2209
	IVB (a)	(g)	cis-trans-ratio = 2:5	2209
Et_3GeR'				
R' = {CH:CHCH$_2$Cl / C(CH$_2$Cl):CH$_2$}	IA (h)	65	IR and NMR spectra, GLC (i)	1472
	IA (j)	20	IR and NMR spectra, GLC (k)	1472
C(CF$_3$):CHCF$_3$	IB (l)	--	b_{28} 82°, IR and NMR spectra	923
C:C(CF$_3$)CF$_2$	VA	90	b. 160°, IR and NMR spectra	921
{CBr:CBrCH:CH$_2$ / CBr:C:CHCH$_2$Br}	III (m)	--	$b_{1.5}$ 108-10°, n_D^{20} 1.5542, d_{20} 1.5546, IR and NMR spectra	531
				531

{ CBr:CBrCH:CHMe	III (m)	--	b$_{0.5}$ 117-19°, n$_D^{20}$ 1.5540, d$_{20}$ 1.5500, IR and NMR spectra (n)	531
{ CBr:C:CHCHBrMe				531
CBr:CBrCMe:CH$_2$	III (m)	--	b$_{0.5}$ 103-105°, n$_D^{20}$ 1.5442, d$_{20}$ 1.5470, IR and NMR spectra (n)	531
Bu$_3$GeCCl:CCl$_2$	VB	--	---	1881

* Numbers in parenthesis refer to pages in main volume.

** The characters used here correspond to the synthetic scheme in the Introduction to Chapter 2.4.1.
(a) Synth. in sealed tube at 55° (2209). (b) By-prod. in synth. (2209). (c) Synth. under UV irradiation (2209). (d) Rxn. with Me$_3$GeCl and Na → (Me$_3$GeCH:)$_2$ + Me$_6$Ge$_2$ + trace (Me$_3$GeC:)$_2$ (1533). (e) Rxn. with Me$_3$GeCl + Na → (Me$_3$Ge)$_2$C:CH$_2$ (1533). (f) Unresolved rxn. mixture in 63% conversion (2209). (g) Rxn. mixture separated by GLC (2209). (h) Synth. in presence of H$_2$PtCl$_6$ at 25° without solvent (1472). (i) Isomer mixture contains 30% cis- and 35% trans-form (1472). (j) Synth. in presence of MeCN at 80° (1472). (k) Unresolved rxn. mixture in synth. without solvent contains 35% of geminal form (1472). (l) Synth. under UV irradiation (923). (m) Synth. in HCCl$_3$ at -15 to -10° (531). (n) The original rxn. mixture containing 1,4- and 3,4- adduct rearranges on distillation. Structure postulated from spectra (531).

2.4.3 UNSYMMETRIC TETRAORGANOGERMANES CONTAINING FUNCTIONALLY SUBSTITUTED OLEFINS

Synthetic methods for tetraorganogermanes containing functionally substituted olefinic groups are summarized in the following scheme.

I Addition reaction:
  (IA) $Et_3GeH + RC\vdots CR' \rightarrow Et_3GeCR:CHR'$; catalyzed by H_2PtCl_6 ,
  (IB) $Et_3GePEt_2 + PhC\vdots CH \rightarrow Et_3GeCH:CPhPEt_2 + Et_3GeCPh:CHPEt_2$,
  (IC) $(Et_3Ge)_2Hg + (EtO_2C)_2C:C(CO_2Et)_2 \rightarrow Et_3GeC(CO_2Et)_2C(CO_2Et):C(OEt)OGeEt_3$.

II Interconversion reactions of hydroxy-groups:
  (IIA) Acylation with Ac_2O-KOAc ,
  (IIB) Silylation with Me_3SiCl-pyridine ,
  (IIC) Acetal formation with $BuOCH_2Cl$-Me_2NPh .

Additional tetraorganogermanes containing functionally substituted olefinic groups are listed in Chapters 2.3, 2.5.3, 2.7.1.1, 2.7.1.8, 2.7.2, 2.7.3, 2.7.5 and 2.7.7 .

2,4,3-Methylphenyl(triethylgermyl) $GeC_{17}H_{28}O$ $Et_3GeC(:CHPh)CMe_2OH$
  -3,2-butenol

<u>Synth.</u>: By rxn. of Et_3GeH with $Me_2COHC\vdots CPh$ in presence of cat. H_2PtCl_6 in Me_2CHOH, in 85% yield (2426).
<u>Prop.</u>: b_2 144-45°, n_D^{20} 1.5240, d_{20} 1.0816 (2426).
<u>Rxn. with</u>: $KHSO_4 \rightarrow Et_3GeC(:CHPh)CMe:CH_2$ (2426).
H + Raney-Ni $\rightarrow Et_3GeCH(CH_2Ph)CMe_2OH$ (2426).
Ac_2O + KOAc $\rightarrow Et_3GeC(:CHPh)CMe_2OAc$ (2426).
Me_3SiCl in Py $\rightarrow Et_3GeC(:CHPh)CMe_2OSiMe_3$ (2426).

3,5,4-Methylphenyl(triethylgermyl) $GeC_{18}H_{30}O$ $Et_3GeC(:CHPh)CMeEtOH$
  -4,3-pentenol

<u>Synth.</u>: By rxn. of Et_3GeH with $MeEtCOHC\vdots CPh$ in presence of cat. H_2PtCl_6 in Me_2CHOH, in 80% yield (2426).
<u>Prop.</u>: b_2 152-53°, n_D^{20} 1.5230, d_{20} 1.0759 (2426).
<u>Rxn. with</u>: $KHSO_4 \rightarrow Et_3GeC(:CHPh)CMe:CHMe$ (2426).
H + Raney-Ni $\rightarrow Et_3GeCH(CH_2Ph)CMeEtOH$ (2426).
Ac_2O + KOAc $\rightarrow Et_3GeC(:CHPh)CMeEtOAc$ (2426).
Me_3SiCl + Py $\rightarrow Et_3GeC(:CHPh)CMeEtOSiMe_3$ (2426).

Additional tetraorganogermanes carrying functionally substituted olefinic groups are compiled in Table 7.

Table 7. Unsymmetric Tetraorganogermanes Containing Functionally Substituted Olefins $R_nGeR'_{4-n}$

Formula*	Synth. Method**	Yield	Properties	Ref.
Et_3GeR'				
$R' = \begin{cases} CH:CPhPEt_2 \\ CPh:CHPEt_2 \end{cases}$	IB	47	Mixture, $b_{0.1}$ 108-11°, cis- and trans- forms, NMR spectra	1797 1797
$\underline{CH:CHCH_2OR}$				
R = H (20)	--	--	(a)	670
$CH_2OCH_2CH_2Cl$	IA	55	b_2 127-29°, n_D^{20} 1.4745, d_{20} 1.1324	2559
$CH_2OCH_2CH:CH_2$	IA	28	b_5 117-18°, n_D^{20} 1.4685, d_{20} 1.0432 (b)	2559
$CH_2OC_6H_{13}$	IA	48	b_5 148-49°, n_D^{20} 1.4605, d_{20} 1.0002 (b)	2559
$CHMeOCH_2CH:CH_2$	IA	33	b_5 119-20°, n_D^{20} 1.4679, d_{20} 1.0385 (b)	2559
$CHMeOC_6H_{13}$	IA	50	b_3 150-57°, n_D^{20} 1.5920, d_{20} 0.9865 (b)	2559
Ac	(a)	--	b_4 87-90°, n_D^{20} 1.4655, d_{20} 1.0841	1531
$C(C(OHPh)):CHCH:CH_2$ (20)	--	--	(c)	--
$C(:CHPh)CMe_2OH$	--	--	(d, e)	--
$C(:CHPh)CMe_2OAc$	IIA (d)	11	b_2 135-36°, n_D^{20} 1.5160, d_{20} 1.0906	2426
$C(:CHPh)CMe_2OSiMe_3$	IIB (e)	50	b_2 152-55°, n_D^{20} 1.5020, d_{20} 1.0227	2426
$CH:CHCHOHPr$	IA	--	$b_{0.1}$ 88-89°, n_D^{20} 1.4740, d_{20} 1.0300	1898
$CH:CHCHPrOCH_2OCH_2CH_2Cl$	IA	--	b_1 136°, n_D^{20} 1.4702, d_{20} 1.0789	2766
$CH:CHCHOHPr-i$	IA	--	$b_{0.2}$ 73-74°, n_D^{20} 1.4760, d_{20} 1.0350	1898
$CH:CHCMeEtOCH_2OCH_2CH_2Cl$	IA	--	b_1 129-30°, n_D^{28} 1.4738, d_{20} 1.0966	2766

* Numbers in parenthesis refer to pages in main volume.
** Characters used here correspond to the synthetic scheme in the introduction to Chapter 2.4.3.
(a) Rxn. of $Et_3GeCH:CHCH_2OH + CH_2:C:O \rightarrow Et_3GeCH:CHCH_2OAc$ (1531). (b) IR spectrum (2559). (c) Complete hydrogenation of double bond with cat. Pd on $CaCO_3$ (201). (d) Rxn. with $Ac_2O + KHSO_4$ at 100° $\rightarrow Et_3GeC(:CHPh)CMeOAc-CH_2R$; R = H, Me (2426). (e) Rxn. with Me_3SiCl + Py at r.t. $\rightarrow Et_3GeC(:CHPh)CMeO(SiMe_3)CH_2R$; R = H, Me (2426).

Table 7 Continued

$R_nGeR'_{4-n}$

Formula*	Synth. Method**	Yield	Properties	Ref.
Et_3GeR' (Cont'd.)				
R' = C(:CHPh)CMeEtOH	--	--	(d, e)	--
C(:CHPh)CMeEtOAc	IIA (d)	17	b_2 142-45°, n_D^{20} 1.5195, d_{20} 1.0952	2426
C(:CHPh)CMeEtOSiMe$_3$	IIB (e)	41	b_2 156-58°, n_D^{20} 1.5180, d_{20} 1.0468	2426
C(CH$_2$CH$_2$OH):CHCH:CH$_2$ (20)	--	--	(f, g)	--
CH$_2$(CH$_2$CH$_2$OAc):CHCH:CH$_2$	IIA (g)	17	b_1 106-108°, n_D^{20} 1.4836, d_{20} 1.0763	201
C(CHMeOH):CHCH:CH$_2$ (20)	--	--	(g, h)	--
C(CHMeOAc):CHCH:CH$_2$	IIA (g)	7	b_2 111-12°, n_D^{20} 1.4800, d_{20} 1.0699	201
CX$_2$CX:C(OEt)OGeEt$_3$	IC	95	(X = CO$_2$Et)	1365
CH:CHCMePrOH	IA	--	$b_{0.2}$ 81-83°, n_D^{20} 1.4646, d_{20} 1.0064 (i)	1900
CH:CHCMePrOCHMeOBu	(i)	52	$b_{0.2}$ 114-15°, n_D^{20} 1.4600, d_{20} 0.9782	1900
C(CMe$_2$OH):CHEt	(h)	77	b_1 86°, n_D^{20} 1.4795, d_{20} 1.0415	201
C(CMe$_2$OH):CHCHMeOH	IA (j)	--	b_5 155°, n_D^{20} 1.4870, d_{20} 1.1030 (k)	199
C(CMe$_2$OH):CHCHMeOGeEt$_3$	IA	28	b_1 141°, n_D^{20} 1.4850, d_{20} 1.1130	199
C(CMe$_2$OH):CHCH:CH$_2$ (20)	--	--	(g, h)	--
C(CMe$_2$OAc):CHCH:CH$_2$	IIA (g)	17	$b_{1.5}$ 95-96°, n_D^{20} 1.4820, d_{20} 1.0405	201
C(:CHPh)C$_5$H$_8$OH-1c.	IA	80	b_2 164-66°, n_D^{20} 1.5320, d_{20} 1.0972 (l)	2425
C(CHOHPr):CHEt	(m)	73	b_1 99-100°, n_D^{20} 1.4835, d_{20} 1.0358	201
CH:CHCMeBuOH	IA	--	$b_{0.2}$ 88-90°, n_D^{20} 1.4690, d_{20} 1.0127 (n)	1900
CH:CHCMeBuOCH$_2$CH$_2$CN	(n)	46	$b_{0.2}$ 111-12°, n_D^{20} 1.4740, d_{20} 1.0329	1900
CH:CHCMe$_2$OEt):CMe$_2$OEt	IA	40	b_1 136-37°, n_D^{20} 1.4782, d_{20} 1.1230 (o)	1150
CH:CHCMe(CMe$_3$)OH	IA	66	$b_{0.2}$ 84-86°, n_D^{20} 1.4662, d_{20} 1.0079 (n)	1900
CH:CHCMe(CMe$_3$)OCH$_2$CH$_2$CN	(n)	46	$b_{0.2}$ 101-102°, n_D^{20} 1.4700, d_{20} 1.0270	1900
C(CHOHPr):CHCH:CH$_2$ (20)	--	--	(m, p)	--
C(CHOAcPr):CHCH:CH	IIA (p)	16	b_3 126-27°, n_D^{20} 1.4812, d_{20} 1.0390	201
C(:CHPh)C$_6$H$_{10}$OH-1c.	IA	80	b_2 173-75°, n_D^{20} 1.5350, d_{20} 1.0985 (l)	2425
C(CMe$_2$OH):CHCMeOHCH$_2$CH$_2$OMe	IA	62	b_5 170-72°, n_D^{20} 1.4880, d_{20} 1.0886 (q, r)	2427

Compound				
C(CH:CH$_2$):CHCMePrOH	IA (s)	41	b$_4$ 156-59° (r, t)	2428
C(CH$_2$OMe):CHC$_8$H$_{10}$OH-1c.	IA	49	b$_2$ 147-48°, n$_D^{20}$ 1.4892, d$_{20}$ 1.0746 (u)	2767
C(CH$_2$OPr):CHC$_8$H$_{10}$OH-1c.	IA	49	b$_{0.5}$ 162°, n$_D^{20}$ 1.4902, d$_{20}$ 1.0709 (u)	2767
C(CH$_2$OBu):CHC$_8$H$_{10}$OH-1c.	IA	46	b$_1$ 166-67°, n$_D^{20}$ 1.4866, d$_{20}$ 1.0460 (u)	2767
C(CMeEtOH):CHCMeOHCH$_2$CH$_2$OH	IA	90	b$_{0.5}$ 171-73°, n$_D^{20}$ 1.4920, d$_{20}$ 1.0824	2427
C(CMeOHCH$_2$CH$_2$OMe):CHCMeOHCH$_2$CH$_2$OH	IA	63	b$_2$ 167-69°, n$_D^{20}$ 1.4885, d$_{20}$ 1.0930	2427
C(1-C$_8$H$_{10}$OHc.):CHCHMeOH	IA (v)	15	b$_8$ 153-54°, n$_D^{20}$ 1.5040, d$_{20}$ 1.1225	199
C(1-C$_8$H$_{10}$OHc.):CHCHMeOGeEt$_3$	IA (v)	29	b$_8$ 198°, n$_D^{20}$ 1.4990, d$_{20}$ 1.1256 (w)	199
C(CHOHC$_5$H$_9$c.):CHCH:CH$_2$ (20)	--	--	(c, x)	--
C(CHOAcC$_5$H$_9$c.):CHCH:CH$_2$	IIA (x)	14	b$_3$ 138°, n$_D^{20}$ 1.5108, d$_{20}$ 1.0933	201
C(CHOHC$_6$H$_{11}$c.):CHCH:CH$_2$ (20)	--	--	(c, x)	--
C(CHOAcC$_6$H$_{11}$c.):CHCH:CH$_2$	IIA (x)	21	b$_3$ 146-48°, n$_D^{20}$ 1.5062, d$_{20}$ 1.0872	201
COH(CMe$_2$):CHCPr$_2$OH	IIA	13	b$_3$ 205-207°	2428
C(1-C$_8$H$_{10}$OHc.):CHCMeOHCH$_2$OMe	IA	28	b$_2$ 188-90°, n$_D^{20}$ 1.5010, d$_{20}$ 1.0923 (q, r)	2427
COH(CMeEt):CHCPr$_2$OH	IA	8	b$_3$ 174-75°	2428

(f) Rxn. with H + Raney Ni → Et$_3$GeCHPrCH$_2$CH$_2$OH (201). (g) Rxn. of Et$_3$GeC(YOAc):CHCH:CH$_2$ + Ac$_2$O + NaOAc → Et$_3$GeC(YOH):CHCH:CH$_2$; Y = CH$_2$CH$_2$, CHMe, CMe$_2$ (201). (h) Rxn. of Et$_3$GeC(CRMeOH):CHCH:CH$_2$ with H + cat. Pd on CaCO$_3$ → Et$_3$GeCHPrCRMeOH; R = H, Me, + Et$_3$GeC(CRMeOH):CHEt; R = Me (201). (i) Rxn. of Et$_3$GeCH:CHCMePrOH with CH$_2$:CHOBu in concd. HCl at 85° → Et$_3$GeCH:CHCMePrOCHMeOBu (1900). (j) Catalyst used Pt on carbon (199). (k) Dehydration with KHSO$_4$ → 2,5-dihydro[2,2,5,3-Me$_3$(Et$_3$Ge)]-furan (199). (l) Rxn. with KHSO$_4$ → Et$_3$GeCR:CHPh; R = c.1-C$_5$H$_7$, c.1-C$_6$H$_9$ (2425); with H + Raney Ni → Et$_3$GeCHRCH$_2$Ph; R = c.1-HOC$_5$H$_9$, c.1-HOC$_6$H$_{11}$ (2425). (m) Rxn. of Et$_3$GeC(CHOHPr):CHCH:CH$_2$ + cat. Pd-CaCO$_3$ → Et$_3$GeC(CHOHPr):CHEt + Et$_3$GeCHPrCHPrOH (201). (n) Rxn. of Et$_3$GeCH:CHCMeBuOH with CH$_2$:CHCN + NaOMe → Et$_3$GeCH:CHCMeBuOCH$_2$CH$_2$CN; for n- and t-Bu isomers (1900). (o) Rxn. with H + cat. Pt on carbon → Et$_3$GeCH(CMe$_2$OEt)CH$_2$CMe$_2$OEt (1150). (p) Rxn. of Et$_3$GeC(CHOHPr):CHCH:CH$_2$ with Ac$_2$O + NaOAc → Et$_3$GeC(CHOAcPr):CHCH:CH$_2$ (201). (q) Dehydration with KHSO$_4$ → Et$_3$GeC:CHCMe(CH$_2$CH$_2$OMe)OCR$_2$; R$_2$ = Me$_2$, Et$_2$, (CH$_2$CH$_2$OMe)$_2$, (CH$_2$)$_5$ (2427). (r) Complete hydrogenation of double bond with cat. Raney Ni. (s) Synth. from Et$_3$GeH with CH$_2$:CHC:CCMePrOH and cat. H$_2$PtCl$_6$ at 100° (2428). (t) Rxn. with KHSO$_4$ → Et$_3$GeC(CH:CH$_2$):CHCMe:CHEt (2428). (u) IR spectrum (2767). (v) Synth. from Et$_3$GeH + 1,1-HOC$_6$H$_{10}$C:CCHMeOH + cat. Pt on carbon at 115-30° (199). (w) Dehydration with KHSO$_4$ → Et$_3$GeC(CHCMeOH(CH$_2$)$_4$CH$_2$ (199). (x) Rxn. of Et$_3$GeC(CHOHR):CHCH:CH$_2$ + Ac$_2$O + NaOAc → Et$_3$GeC(CHOAcR):CHCH:CH$_2$; R = c.C$_5$H$_9$, c.C$_6$H$_{11}$ (201).

Table 7 Continued $R_nGeR'_{4-n}$

Formula*	Synth. Method**	Yield	Properties	Ref.
$Pr_3GeCH:CHOEt$	IA	65	b_1 76.5-77°, n_D^{20} 1.4606, d_{20} 0.9968	2483
$Pr_3GeCH:CHOBu$	IA	65	b_1 93.5-94°, n_D^{20} 1.4610, d_{20} 0.9512	2483
Bu_3GeR'				
R' = CH:CHCH$_2$OH (21)	--	--	(y)	1899
CH:CHCH$_2$OCH$_2$OBu	IIC (y)	--	$b_{0.5}$ 149-50°, n_D^{20} 1.4711, d_{20} 0.9925	1899
CH:CHCMe$_2$OH (21)	--	--	(y)	1899
CH:CHCMe$_2$OCH$_2$OBu	IIC (y)	--	b_1 158-60°, n_D^{20} 1.4661, d_{20} 0.9740	1899
CH:CHCHPrOH	IA	58	b_1 146-48°, n_D^{20} 1.4757, d_{20} 0.9864 (y)	1899
CH:CHCHPrOCH$_2$OBu	IIC (y)	53	$b_{0.5}$ 177-80°, n_D^{20} 1.4685, d_{20} 0.9702	1899
CH:CHCMeEtOH	IA	--	b_2 130-31°, n_D^{20} 1.4696, d_{20} 0.9760 (y)	1899
CH:CHCMeEtOCH$_2$OBu	IIC (y)	--	b_1 169-71°, n_D^{20} 1.4610, d_{20} 0.9612	1899

(y) Rxn. of $Bu_3GeCH:CHYOH$ with $BuOCH_2Cl$ and $PhNMe_2$ in $Et_2O \rightarrow Bu_3GeCH:CHYOCH_2OBu$; Y = CH_2, CMe_2, CHPr, CMeEt (1899).

2.5 UNSYMMETRIC TETRAORGANOGERMANES WITH ACETYLENIC SUBSTITUENTS

This chapter reports acetylenic substituted tetraorganogermanes. The tautomeric allene derivatives and enyne-groups are included. The following scheme summarizes methods for preparation of acetylenic substituted tetraorganogermanes listed in Table 8.

I Grignard reactions:
   R_nGeX_{4-n} + $R'MgX$ → $R_nGeR'_{4-n}$.

II Reaction with acetylenic metal compounds:
   (IIA) Ph_3GeBr + $R'Li$ → Ph_3GeR' ,
   (IIB) Bu_2GeCl_2 + $HC⋮CNa$ → $Bu_2Ge(C⋮CH)_2$,
   (IIC) Bu_3GeCl + $NaAl(C⋮CR)_4$ → $Bu_3GeC⋮CR'$.

III Interconversion reactions:
   (IIIA) $Et_3Ge(CH_2)_mBr$ + $HC⋮CCH_2Br$ + Mg → $Et_3Ge(CH_2)_mCH_2C⋮CH$,
   (IIIB) $Et_3Ge(CH_2)_mBr$ + $HC⋮CNa$ → $Et_3Ge(CH_2)_mC⋮CH$,
   (IIIC) $Et_3Ge(CH_2)_mCHBrCH_2Br$ + base → $Et_3Ge(CH_2)_mC⋮CH$,
   (IIID) $Me_2Ge(C⋮CCMeBuOH)_2$ + $KHSO_4$ → $Me_2Ge(C⋮CCBu:CH_2)_2$,
   (IIIE) $R_3GeC⋮CH$ + $RMgBr$, + D_2O → $R_3GeC⋮CD$.

IV Partial catalytic hydrogenation of polyunsaturated organogermanes, leading usually to complex mixtures.

V Substitution of organogermanium amide with phenylacetylene:
   Bu_3GeNMe_2 + $PhC⋮CH$ → $Bu_3GeC⋮CPh$.

R = organic group, R′ = acetylenic organic group, X = halogen, n = 1-3, m = 3-5. Other tetraorganogermanes with acetylenic substituents are listed in Chapters 1.3 and 7.1.4.

Ethynyltrimethylgermane (M-22) GeC_5H_{10} $Me_3GeC⋮CH$
Synth.: By rxn. of C_2H_2 with EtMgBr, followed by Me_3GeCl, in 76% yield (367).
By rxn. of $Me_3GeC⋮CCl$ with PhLi in Et_2O, followed by hydrolysis, in 25% yield (1981).
Prop.: b. 72.5°, n_D^{20} 1.4180, d_{20} 1.0391 (367), IR (1074), (far) IR and assignment (1981), Raman (367) and NMR (1688) spectra, dipole moment (254), GLC (1981).
Rxn. with: EtMgBr, + D_2O → $Me_3GeC⋮CD$ (1981).
$HGeCl_3$ → Me_3GeCl + $Me_3GeCH:CHGeCl_3$ + $Me_3GeCH_2CH(GeCl_3)_2$ (1533).

Phenylethynyltrimethylgermane $GeC_{11}H_{14}$ $Me_3GeC⋮CPh$
Synth.: By rxn. of Me_3GeX with $PhC⋮CMgBr$ in THF (134).
By rxn. of $(Me_3Ge)_2NH$ with $PhC⋮CH$ in presence of cat. $(NH_4)_2SO_4$ at 150°, in 28% yield (1534).
By rxn. of $Me_3GeNHSiMe_3$ with $PhC⋮CH$ at 130-60° in 51% yield (1534).
Prop.: $b_{1.5}$ 70°(134), b_1 64°, n_D^{20} 1.5429 (1534), n_D^{25} 1.5429 (134) d_{20} 1.1474, IR spectrum (1534).
Rxn. with: aq. NaOH in MeOH → Me_3GeOH + $PhC⋮CH$; relative kinetic data of cleavage rxn. (134).

Ethynyltriethylgermane (M-22) GeC_8H_{16} $Et_3GeC\vdots CH$
Prop.: n_D^{20} 1.4450, (1000) IR spectrum and force const. (2144), NMR spectrum (1930-1), mol. magnetic rotation (301).
Rxn. with: EtMgBr, + ^{3}HOH → $Et_3GeC\vdots C^3H$ (1000).

Phenylethynyltriethylgermane (M-22) $GeC_{14}H_{20}$ $Et_3C\vdots CPh$
Synth.: By rxn. of $(Et_3Ge)_2NH$ with $PhC\vdots CH$ and cat. $(NH_4)_2SO_4$ at 180°, in 54% yield (1534).
By-prod. in rxn. of Et_3GePEt_2 with $PhC\vdots CH$ and cat. AIBN (1797).
Prop.: b_7 132°, n_D^{20} 1.5280, d_{20} 1.0981, IR spectrum (1534).
Rxn. with: aq. NaOH in MeOH → $Et_3GeOH + PhC\vdots CH$; kinetic data (134).
i-Bu_2AlH → $Et_3GeC(AlBu_2):CHPh$ (1016).

3,1 Butenynyltriethylgermane $GeC_{10}H_{18}$ $Et_3GeC\vdots CCH:CH_2$
Synth.: By rxn. of Et_3GeBr with $CH_2:CHC\vdots CMgBr$ in Et_2O, in 72% yield (530).
Prop.: b_5 62-63°, n_D^{20} 1.4850, d_{20} 1.0333 (530), IR (530), (far) IR, assignment (1973), NMR (530) and mass (262) spectra, dipole moment, GLC (530).
Rxn. with: H + cat. Pd-C → $Et_3GeC\vdots CEt$ + some $Et_3GeCH:C:CHMe + Et_3GeCH:CHCH:CH_2$ (530).
BuLi, + H_2O → $Et_3GeC\vdots C(CH_2)_5Me + Et_3GeCH:C:CHCH_2BuC$ (1972).
Br in $HCCl_3$ (-15 to -10°) → $Et_3GeCBr:C:CHCH_2Br + Et_3GeCBr:CBrCH:CH_2$ (531).
EtSH + cat. $(Me_3CO)_2$ → $Et_3GeC\vdots CCH_2CH_2SEt + Et_3GeCH:C:CHCH_2SEt$ + some $Et_3GeC(SEt):CHCH:CH_2$ (1998).

3,1- Pentenynyltriethylgermane $GeC_{11}H_{20}$ $Et_3GeC\vdots CCH:CHMe$
Synth.: By rxn. of Et_3GeBr with $MeCH:CHC\vdots CMgBr$ in Et_2O, in 80% yield (530).
By rxn. of $Et_3GeC\vdots CC\vdots CMe$ with H and cat. Pd on $CaCO_3$ at r.t., in 43% yield (2797).
Prop.: b_5 78-79° (530), b_2 64-81° (2797) n_D^{20} 1.4900 (530), 1.4923 (2797), d_{20} 1.0173 (530), 1.0191 (2797), IR (530, 2797), (far) IR, Raman and assignment (1973), NMR (530) and mass (262) spectra, dipole moment (530), GLC (530, 2797).
Rxn. with: H + cat. Pd-C → $Et_3GeC\vdots CPr + Et_3GeCH:C:CHEt + Et_3GeCH:CHCH:CHMe$ (530).
BuLi, + H_2O → $Et_3GeCH:C:CHCH:CH_2 + GeC_{15}H_{30}$ (1972).
Br in $HCCl_3$ at -15 to -10° → $Et_3GeCBr:CBrCH:CHMe + Et_3GeCBr:C:CHCHMeBr$ (531).
EtSH + cat. $(Me_3CO)_2$ → $Et_3GeC\vdots CCH_2CHMeSEt + Et_3GeCH:C:CHCHMeSEt$ (1998).

3,3,1- Methylbutenynyltriethylgermane (M-22) $GeC_{11}H_{20}$ $Et_3GeC\vdots CCMe:CH_2$
Synth.: By rxn. of Et_3GeBr with $CH_2:CMeC\vdots CMgBr$ in Et_2O, in 75% yield (530).
Prop.: b_5 70-71°, n_D^{20} 1.4808, d_{20} 1.0047 (530), IR (530), (far) IR, Raman and assignment (1973), NMR (530) and mass (262) spectra, dipole moment, GLC (530).
Rxn. with: H and cat. Pd-C → $Et_3GeC\vdots CCHMe_2$ + some $Et_3GeCH:C:CMe_2 + Et_3GeCH:CHCMe:CH_2$ (530).
BuLi, + H_2O → $Et_3GeCH:C:CHCHMeBu + Et_3GeC\vdots CCH_2CHMeBu$ (1972).
Br in $HCCl_3$ (at -15 to -10°) → $Et_3GeCBr:CBrCMe:CH_2$ (531).
EtSH + cat. $(Me_3CO)_2$ → $Et_3GeC\vdots CCHMeCH_2SEt$ (1998).

Ethynyltributylgermane (M-23) $GeC_{14}H_{28}$ $n-Bu_3GeC\vdots CH$
Synth.: By rxn. of $n-Bu_3GeI$ with $HC\vdots CMgBr$ in THF in 66% yield (1598).

Prop.: b_8 112°, n_D^{20} 1.4571, d_{20} 0.9374 (1598), ^{13}C NMR spectrum (2719), mol. magnetic rotation (301).
Rxn. with: $Bu_3SnH \rightarrow Bu_3GeCH{:}CHSnBu_3$ (1598).

Bis(ethynyl)diphenylgermane $GeC_{16}H_{12}$ $Ph_2Ge(C{:}CH)_2$
Synth.: By rxn. of Ph_2GeBr_2 with $HC{:}CMgBr$ in THF, in 75% yield (1946).
By rxn. of Ph_2GeCl_2 with C_2HNa in THF in Ar atm., in 56% yield (939).
Prop.: m. 49° (939), 43-45°, b_5 160° (1946), NMR spectrum (1026), mol. wt. (939), soly. (1946).
Rxn. with: Mg in THF $\rightarrow Ph_2Ge(C{:}CMgBr)_2$, sic. (1438).
150-200° in N atm. $\rightarrow HC{:}C(Ph_2GeC{:}CC{:}C)_xPh_2GeC{:}CH + H$ (1438).
$Ph_3GeC{:}CH$ at 150-200° $\rightarrow$ copolymer (1438).
$CH_2N_2 \rightarrow Ph_2Ge(\overline{C{:}CHNHN{:}CH})_2$ (939).
$Ph_2GeH_2 \rightarrow (CH{:}CHPh_2GeCH{:}CH)_x$ (327).

Ethynyltriphenylgermane $GeC_{20}H_{16}$ $Ph_3GeC{:}CH$
Synth.: By rxn. of Ph_3GeBr with $HC{:}CMgBr$ in Et_2O in N atm. (1438), in THF in 60% yield (2562).
By rxn. of Ph_3GeCl with C_2HNa in THF in N atm., in 14% yield (939), in pyridine in good yield (1039).
Prop.: m. 55-70° (1438), $b_{0.5}$ 150° (939), $b_{0.1}$ 147-60° (1438), IR (2562) and NMR (1026, 1474, 1931) spectra, GLC (2562), mol. wt. (939), polymerizes in N atm. at 150-200° (1438).
Rxn. with: $Ph_2Ge(C{:}CH)_2 \rightarrow$ copolymer (1438).
$Ph_3SnNEt_2 \rightarrow Ph_3GeC{:}CSnPh_3$ (1039).

Additional tetraorganogermanes with acetylenic substituents are summarized in Table 8.

Table 8. Unsymmetric Tetraorganogermanes with Acetylenic Substituents $R_nGeR'_{4-n}$

Formula*	Synth. Method**	Yield	Properties	Ref.
$Me_3GeC\!:\!CD$	IIID	35	GLC, (far) IR spectrum, assignment	1981
$Me_3GeC\!:\!CMe$	I	13	GLC, (far) IR spectrum, assignment	1981
$Me_3GeC\!:\!CCH\!:\!CH_2$	I	50	b. 133–34°, n_D^{20} 1.4655, d_{20} 1.0711 (a)	367
$Me_2Ge[C\!:\!CC(CMe_3)\!:\!CH_2]_2$	IIID	--	b_4 118–19°, n_D^{20} 1.4810, d_{20} 0.9814	1901
Et_3GeR'				
$R' = C\!:\!C^3H$	IIIE	--	n_D^{20} 1.4450 (b)	1000
o-$C\!:\!CC_6H_4Me$ (22)	--	--	(c)	--
m-$C\!:\!CC_6H_4Me$ (22)	--	--	(c)	--
p-$C\!:\!CC_6H_4Me$ (22)	--	--	(c)	--
$2,3$-$C\!:\!CC_6H_3Me_2$ (22)	--	--	(c)	--
$2,4,6$-$C\!:\!CC_6H_2Me_3$ (22)	--	--	(c)	--
p-$C\!:\!CC_6H_4CMe_3$ (22)	--	--	(c)	--
$C\!:\!CMe$	I	50	b_{11} 81°, IR spectrum, GLC (d)	2562
$\{CH_2C\!:\!CH \atop CH\!:\!C\!:\!CH_2\}$ (22)	I	70	b_{30} 89°, (e)	1473
$CH_2C\!:\!C^3H$	IIIE (e)	--	b_{60} 100°, n_D^{20} 1.4664 (b)	1000
$\{C\!:\!CEt \;(f) \atop {CH\!:\!C\!:\!CHMe \atop CH\!:\!CHCH\!:\!CH_2}\}$	IV	77	$b_{3.5}$ 56°, n_D^{20} 1.4677, d_{20} 1.0089, isomeric mixture, composition by IR spectrum	530
$\{C\!:\!CPr \;(f) \atop {CH\!:\!C\!:\!CHEt \atop CH\!:\!CHCH\!:\!CHMe}\}$	IV	66	$b_{3.5}$ 63.5–64°, n_D^{20} 1.4642, d_{20} 0.9879, isomeric mixture, composition by IR spectrum (g)	530
$(CH_2)_3C\!:\!CH$	IIIC	--	b_{20} 107°, n_D^{20} 1.4643, d_{20} 1.0060 (h)	1494

Compound		Yield	Properties	Ref.
C:CCHMe$_2$ (f)	IV	85	b$_{3.5}$ 71.5°, n$_D^{20}$ 1.4685, d$_{20}$ 0.9953, isomeric mixture, composition by IR spectrum	530
CH:C:CMe$_2$				
CH:CHCMe:CH$_2$				
C:CCH:CHMe	(i)	--	(j)	2797
CH:C:CHCH:CH$_2$	(j)	--	b$_1$ 70-71°, n$_D^{20}$ 1.4819, d$_{20}$ 0.9937 (k)	1972
C:CC:CMe	I	55	b$_2$ 87-89°, n$_D^{20}$ 1.5120, dec. slowly at r.t., IR and NMR spectra (l)	2797
(CH$_2$)$_4$C:CH	IIIB	60	b$_{15}$ 113°, n$_D^{20}$ 1.4644, d$_{20}$ 0.9943 (l)	1494
C:CCH:CHEt	IV(m)	44	b$_1$ 76-80°, n$_D^{20}$ 1.4915, d$_{20}$ 1.0154 (n)	2797
CH$_2$C:CCH$_2$CH:CH$_2$ (22)	--	--	Mol. magnetic rotation	301
C:CC:CEt (22)	I	36	b$_1$ 91-92°, n$_D^{20}$ 1.5110, d$_{20}$ 1.0330 (m, o)	2797
(CH$_2$)$_5$C:CH	IIIC	--	b$_2$ 98°, n$_D^{20}$ 1.4647, d$_{20}$ 0.9876	1494
	IIIA	--	(By-prod.), GLC	1494
(CH$_2$)$_4$CH:C:CH$_2$	IIIA	--	b$_8$ 122°, n$_D^{20}$ 1.4785, d$_{20}$ 0.9947, GLC	1494
C:CCH:CHPr	(m)	32	b$_1$ 93-95°, n$_D^{20}$ 1.4872, d$_{20}$ 1.4872 (n)	2797
C:CC:CPr	I	60	b$_1$ 104°, n$_D^{20}$ 1.5090, d$_{20}$ 1.0216 (m, o)	2797
C:C(CH$_2$)$_5$Me	(j)	57	Mixture, not separated, b$_1$ 90-91°, n$_D^{21}$ 1.4721, d$_{20}$ 0.9688, IR spectrum (p)	1972
CH:C:CHCH$_2$Bu				

* Numbers in parenthesis refer to page numbers in main volume.
** The characters used here correspond to the synthetic scheme in the introduction to Chapter 2.5.
(a) Raman spectrum, polymerizes on standing (367). (b) Rate of ^{3}H exchange in 20% MeOH at pH 8.05 at 25° (1000). (c) Kinetics of cleavage rxn. by aq. alkali in MeOH (134). (d) NMR spectrum (1930-1). (e) Rxn. of Et$_3$GeCH$_2$C:CH with EtMgBr, followed by ^{3}HOH → Et$_3$Ge(CH$_2$)$_3$Ac (1494). (f) Predominating isomer (530). (g) GLC (2797). (h) Rxn. with aq. HgSO$_4$ → Et$_3$Ge(CH$_2$)$_3$Ac (1494). (i) Rxn. of Et$_3$GeC:CC:CMe with H and Pd-CaCO$_3$ → Et$_3$GeC:CCH:CHMe + by-products (2797). (j) Rxn. of Et$_3$GeC:CCH:CHR with BuLi in Et$_2$O at -20° to r.t. → Et$_3$GeC:CCH$_2$CHRBu + Et$_3$GeCH:C:CH-CHRBu + isomerized starting material; R = H, Me. (1972). (k) Not separated from starting material (1972). (l) IR spectrum, rxn. with Et$_3$GeH → Et$_3$Ge(CH$_2$)$_4$CH:CHGeEt$_3$ (1494). (m) Rxn. of Et$_3$GeC:CC:CR with H and cat. Pd on CaCO$_3$ → Et$_3$GeC:CCH:CHR + other partly hydrogenated species; R = Et, n-Pr (2797). (n) IR spectrum, crude product (2797). (o) IR and NMR spectra, dec. slowly at r.t. (2797), mol. magnetic rotation (301). (p) Rxn. with alc. KOH → n-BuCH$_2$CH$_2$C:CH + BuCH$_2$CH:C:CH$_2$ (1972).

Table 8 Continued $R_nGeR'_{4-n}$

Formula*	Synth. Method**	Yield	Properties	Ref.
Et_3GeR' (Cont'd.)				
R' = (CH$_2$)$_5$C:CMe	III	--	$b_{1.8}$ 116°, n_D^{20} 1.4697, d_{20} 0.9787	1494
(CH$_2$)$_3$C:CCH$_2$CH:CH$_2$	III	--	$b_{1.2}$ 112°, n_D^{20} 1.4787, d_{20} 0.9969	1494
C:CC(CMe$_3$):CH$_2$	IIID	--	b_3 80-82°, n_D^{20} 1.4682, d_{20} 0.9750	1901
$\begin{Bmatrix} C:CCH_2CHMeBu \\ CH:C:CHCHMeBu \end{Bmatrix}$	(j)	22	Mixture, not separated, b_1 95-97°, n_D^{20} 1.4740, d_{20} 0.9619, IR spectrum	1972 1972
C(CMe:CH$_2$):CHC:CCMe:CH$_2$	IIID	7	b_2 121°, n_D^{20} 1.5100, d_{20} 1.0560	198
(CH$_2$)$_5$C:CCH:CMe$_2$	III	--	$b_{1.2}$ 126°, n_D^{20} 1.4848, d_{20} 0.9780	1494
CR:CHC:CR (R = CMe:CHMe)	IIID (q)	--	b_2 125°, n_D^{20} 1.5260, d_{20} 1.0309	198
CR:CHC:CR (R = 1-C$_6$H$_9$c.)	(r)	13	b_7 161°, n_D^{20} 1.5480, d_{20} 1.0860	200
CR:CHC:CR (R = CMe:CMe$_2$)	IIID	31	b_{10} 156°, n_D^{20} 1.5250, d_{20} 0.9874	200
Et$_2$Ge(C:CPh)$_2$ (22)	I (s)	64	$b_{0.01}$ 139-41°, n_D^{21} 1.5981, d_{21} 1.2190, (t)	1221
Et$_2$Ge(CH$_2$C:CH)$_2$	I	--	b_4 71-73°, IR spectrum (u, 1123)	1946
Et$_2$Ge(C:CBu-n)$_2$	I (s)	63	$b_{0.15}$ 98°, n_D^{21} 1.4729, d_{21} 0.9810 (v)	1221
Et$_2$Ge[C:CC(CMe$_3$):CH$_2$]$_2$	IIID	56	b_4 131-32°, n_D^{20} 1.4870, d_{20} 0.9880	1901
EtPhGe(C:CH)$_2$ (23)	I	50	b_{28} 133-36° (w)	1946
EtGe(C:CPh)$_3$ (23)	I	--	m. 72-73°, IR spectrum, soly. (v, 1122)	1946
n-Bu$_3$GeC:CR'				
R' = Ph (23)	V	--	$b_{1.4}$ 150°, n_D^{20} 1.5195, d_{20} 1.0078 (x)	466, 1761
n-Pr (23)	--	--	Mol. magnetic rotation	301
CH$_2$CH:CH$_2$ (23)	--	--	Mol. magnetic rotation	301

Compound	Type	Yield	Properties	Ref.
n-Bu (23)	III	--	$b_{0.7}$ 132°, n_D^{20} 1.4648, d_{20} 0.9381	301
	IIC	81	b_1 130°, n_D^{20} 1.4670, mol. magnetic rotation (301)	1076
CMeEtC⋮CCMe:CH$_2$	IIID	33	b_1 115-16°, n_D^{20} 1.4800, d_{20} 0.9805	502
n-Bu$_2$Ge(C⋮CH)$_2$	IIB	57	$b_{0.5}$ 46°, mol. wt.	939
Ph$_3$GeR'				
R' = C⋮CPh (23)	IIA	--	m. 92.5-93° (c)	134
	I	68	m. 93°, IR and UV spectra	2562
C⋮CMe	I	82	m. 106°, IR spectrum (y)	2562
{CH$_2$C⋮CH / CH:C:CH$_2$}	I	72	m. 88°, mixture, not separated	1473
{CH$_2$C⋮CMe / CH:C:CHMe}	I	60	m. 104°, mixture, not separated	1473
{CHMeC⋮CH / CMe:C:CH$_2$}	I	68	m. 82°, mixture, not separated	1473
C⋮CC⋮CH	I	40	Oil, IR and UV spectra, GLC (y)	2562
C⋮CC⋮CPh	I	52	m. 114°, IR spectrum	2562
Ph$_2$Ge(C⋮CPh)$_2$ (23)	I	58	m. 82.5-83°, soly. (z)	1221
Ph$_2$Ge(C⋮CMe)$_2$	--	--	NMR spectrum	1026
{Ph$_2$Ge(CH$_2$C⋮CH)$_2$ / Ph$_2$Ge(CH:C:CH$_2$)$_2$}	I	--	$b_{0.01}$ 120° (aa), mixture, IR and NMR (1026) spectra	1946
Ph$_2$Ge(C⋮CPr-n)$_2$	I	67	$b_{0.01}$ 136°, n_D^{21} 1.5702, d_{21} 1.188 (z)	1221
PhGe(C⋮CH)$_3$	I	61	Polymerizes at 150-200° in N atm.	1438

(q) Dehydration with POCl$_3$ in pyridine (198). (r) Synth. from Et$_3$GeH with (1-HOC$_6$H$_{10}$C⋮C)$_2$ + cat. H$_2$PtCl$_6$; R = 1-C$_6$H$_9$ (200). (s) Work-up without hydrolysis (1221). (t) Rxn. with Co$_2$(CO)$_8$ → Et$_2$Ge[C⋮CPhCo$_2$(CO)$_6$]$_2$ (236a, 1222). (u) (Far) IR and UV spectra. (v) Decolorizes Br soln. (1221). (w) Rxn. with MePhSiH$_2$ → (EtPhGeCH:CHSi-MePhCH:CH)$_x$ (327). (x) IR spectrum (466, 1761). (y) NMR spectrum (1931). (z) Rxn. with Co$_2$(CO)$_8$ → Ph$_2$Ge[C⋮CR-Co$_2$(CO)$_6$]$_2$ + CO; R = Pr, Ph (236A, 1222). (aa) Thermal dec. at 200° → (Ph$_2$GeCH$_2$C⋮CC⋮CCH$_2$)$_x$ (2546).

2.5.1 UNSYMMETRIC TETRAORGANOGERMANES CONTAINING HALOGEN SUBSTITUTED ACETYLENES

The compounds listed in Table 9 are prepared by methods from the following scheme.

I Grignard reactions:
  (IA) $R_nGeCl_{4-n} + R'MgCl \rightarrow R_nGeR'_{4-n}$,
  (IB) $R''Cl + R'MgCl \rightarrow R''R'$.

II Synthesis with acetylenic lithium compounds:
  $Me_3GeBr + R'Li \rightarrow Me_3GeR'$.

III Interconversion reactions:
  (IIIA) $R''OH + HCl \rightarrow R''Cl$,
  (IIIB) $R''OH + PCl_5 \rightarrow R''Cl$.

R = organic group, R' = acetylenic group containing halogen, R" = tetraorganogermyl group, may contain an acetylenic substituent, n = 1-3. A symmetric bromine substituted acetylenic derivative $(p\text{-}BrC_6H_4C\!\equiv\!C)_4Ge$ is listed in Table 1.

Table 9. Unsymmetric Tetraorganogermanes Containing Halogen-Substituted Acetylenes $R_nGeR'_{4-n}$

Formula*	Synth. Method*	Yield	Properties	Ref.
$Me_3GeC\!\equiv\!CCl$	II	37	b_{65} 51°, n_D^{20} 1.4525 (a, b)	1980
$Me_3GeC\!\equiv\!CCF_3$	IA	--	(c)	2332
$Me_2Ge(C\!\equiv\!CCF_3)_2$	IA	--	--	2332
$Et_3GeC\!\equiv\!CR$				
R = $p\text{-}C_6H_4F$ (25)	--	--	(d)	--
$o\text{-}C_6H_4Cl$ (25)	--	--	(d)	--
$m\text{-}C_6H_4Cl$ (25)	--	--	(d)	--
$p\text{-}C_6H_4Cl$ (25)	--	--	(d)	--
$o\text{-}C_6H_4Br$ (25)	--	--	(d)	--
$m\text{-}C_6H_4Br$ (25)	--	--	(d)	--
$p\text{-}C_6H_4Br$ (25)	--	--	(d)	--
$p\text{-}C_6H_4I$ (25)	--	--	(d)	--
$m\text{-}C_6H_4CF_3$ (25)	--	--	(d)	--

* Numbers in parenthesis refer to pages in main volume.
** The characters used here correspond to the synthetic scheme in the introduction to Chapter 2.5.1.
(a) Characteristic odor, IR spectrum and assignment (1980), ^{35}Cl NQR spectrum (2885). (b) Rxn. with PhLi, + $H_2O \rightarrow Me_3GeC\!\equiv\!CH$ (1981). (c) Rxn. with $(CF_3)_2CN_2 \rightarrow Me_3Ge\overline{C\!:\!C(CF_3)C(CF_3)_2N\!:\!N}$ (2332); with $Me_3SnCF_3 \rightarrow Me_3Ge\overline{C\!:\!C(CF_3)CF_2}$ (2332). (d) Kinetic data for cleavage rxn. with aq. alkali in MeOH (134).

Table 9 Continued $R_nGeR'_{4-n}$

Formula*	Synth. Method**	Yield	Properties	Ref.
$Et_3GeC\vdots CR$ (Cont'd.)				
R = CF_3	I	62	b_{105} 98-100°, IR and NMR spectra (e)	921
CMeEtCl	IIIA	70	b_2 70-71°, n_D^{20} 1.4560, d_{20} 1.0600 (f)	501
$Bu_3GeC\vdots CCMe_2Cl$	IIIB	--	$b_{0.3}$ 95-96°, n_D^{20} 1.4810, d_{20} 1.0295	500
$Bu_3GeC\vdots CCMeEtCl$	IIIB	--	b_{13} 105-106°, n_D^{20} 1.4840, d_{20} 0.9995 (g)	502
$Ph_2Ge(C\vdots Cl)_2$	--	--	NMR spectrum	1026

(e) Rxn. with $Me_3SnCF_3 \rightarrow EtGe\overline{C\vdots C(CF_3)}CF_2 + Me_3SnF$ (921). (f) Rxn. of $Et_3GeC\vdots CCMeEtCl + HOCMeRC\vdots CMgBr \rightarrow Et_3GeC\vdots CCMeEtC\vdots CCMeROH$; R = Me, Et (501).
(g) Rxn. of $Bu_3GeC\vdots CCMeEtCl + HOCMe_2C\vdots CMgBr \rightarrow Bu_3GeC\vdots CCMeEtC\vdots CCMeOH$ (502).

2.5.3 UNSYMMETRIC TETRAORGANOGERMANES CONTAINING FUNCTIONALLY SUBSTITUTED ACETYLENES

Synthetic methods for the tetraorganogermanes compiled in Table 10 are summarized in the following scheme. The methods include introduction of the functionally substituted acetylenic group by Grignard reaction (I) or by synthesis with organolithium reagent (II). Triorganogermane is added to a diacetylenic derivative (III). Interconversion reactions include modification of hydroxyl groups by esterification (ketene), acetalization (vinyl ether, acrylonitrile), silylation, dehydration and deacetalization.

I Grignard synthesis:
  (IA) $R_nGeCl_{4-n} + R'MgBr \rightarrow R_nGeR'_{4-n}$,
  (IB) $R''Cl + R'MgX \rightarrow R''R'$.

II Synthesis with organolithium derivatives:
  $R_3GeBr + R'Li \rightarrow R_3GeR'$.

III Hydrogermylation:
  $Et_3GeH + (R'C\vdots C)_2 \rightarrow Et_3GeCR'\vdots CC\vdots CR'$.

IV Interconversion reactions:
  (IVA) $R''OH + RCH\vdots CO \rightarrow R''O_2CCH_2R$; R = H, Ac ,
  (IVB) $R''OH + CH_2\vdots CHX \rightarrow R''OCHMeX + R''OCH_2CH_2X$; X = BuO, CN ,
  (IVC) $R''OH + ClCH_2OBu + NaOH \rightarrow R''OCH_2OBu$,

(IVD) R″OCHROBu + RMgBr → R″OH ,
(IVE) R″CMeEtOH + POCl$_3$ + Py → R″CMe:CHMe ,
(IVF) R″OH + R$_3$SiCl + Py → R″OSiR$_3$,
(IVG) Et$_3$GeC⋮CCR:CHR + EtSH → Et$_3$GeR′ .

R = organic group, R′ = organic group containing a functionally substituted acetylene, R″ = tetraorganogermyl group containing an acetylene function, n = 1-3. Other tetraorganogermanes with functionally substituted acetylenes are listed in Chapters 1.3, 2.7.1.1.1, 2.7.1.8, 2.7.2, 2.7.3 and 2.7.4.

Table 10. Unsymmetric Tetraorganogermanes Containing Functionally Substituted Acetylenes $R_nGeR′_{4-n}$

Formula*	Synth. Method**	Yield	Properties	Ref.
Me$_3$GeC⋮CPPh$_2$	II	70	m. 35°, b$_{0.05}$ 126-27°, mol. wt. (a, b)	2778
Me$_3$GeC⋮CP(S)Ph$_2$	(b)	65	m. 173°, IR and mass spectra, mol. wt.	2778
Me$_2$Ge(C⋮CCMeOHCMe$_3$)$_2$	IA	--	b$_4$ 140-41°, n$_D^{20}$ 1.4735, d$_{20}$ 1.0011 (c)	1901
Et$_3$GeR′				
R′ = m-C⋮CC$_6$H$_4$OMe (25)	--	--	(d)	--
p-C⋮CC$_6$H$_4$OMe (25)	--	--	(d)	--
C⋮CCH$_2$OH (25)	--	--	Mol. magnetic rotation (e)	301
C⋮CCH$_2$OAc	IVA (e)	--	b$_2$ 87-88°, n$_D^{20}$ 1.4692, d$_{20}$ 1.1052	1531
C⋮CCH$_2$O$_2$CCH$_2$Ac	IVA (e)	77	b$_4$ 136-37°, n$_D^{20}$ 1.4805, d$_{20}$ 1.1397	1532

Compound		Yield	Properties	Ref.
C⋮CCH$_2$CH$_2$SEt	IVG	65	Mixture, b$_2$ 100-102°, n$_D^{20}$ 1.5036, structure	1998
CH:C:CHCH$_2$SEt		8	and composition by (f) and hydrolysis	1998
C(SEt):CHCH:CH$_2$		2		1998
C⋮CCMe$_2$OH (26)	--	--	Mol. magnetic rotation	301
C⋮CCH$_2$CHMeSEt	IVG	17	Mixture, b$_1$ 104-106°, n$_D^{20}$ 1.5130, d$_{20}$	1998
CH:C:CHCHMeSEt		1	1.037, composition by (f) and hydrolysis	1998
C⋮CCHMeCH$_2$SEt	IVG	25	b$_2$ 105-107°, n$_D^{20}$ 1.5040, d$_{20}$ 1.038 (f)	1998
C⋮CCPrHOH (26)	--	--	Mol. magnetic rotation	301
C⋮CCMeEtOH (26)	--	--	(g)	--
C⋮CCMeOHCMe$_3$	IA	52	b$_2$ 90-91°, n$_D^{20}$ 1.4630, d$_{20}$ 1.1100 (h)	1901
(CH$_2$)$_4$C⋮CCHOHPr	--	--	bo.1 113°, n$_D^{20}$ 1.4794, d$_{20}$ 1.0148	1494
(CH$_2$)$_5$C⋮CCMe$_2$OH	--	--	bo.4 113°, n$_D^{20}$ 1.4768, d$_{20}$ 1.0094	1494
C(CMe$_2$OH):CHC⋮CCMe$_2$OH	III	80	b$_2$ 152°, n$_D^{20}$ 1.4990, d$_{20}$ 1.0630 (i, j)	198
C(CMe$_2$OH):CHC⋮CCMe$_2$OSiEt$_3$	IVF (j)	--	b$_3$ 179°, n$_D^{20}$ 1.4940, d$_{20}$ 1.0027	198
C⋮CCMeEtC⋮CCMe$_2$OH	IB (k)	62	b$_3$ 105-107°, n$_D^{20}$ 1.4740, d$_{20}$ 1.0027	501
C⋮CCMeEtC⋮CCMeEtOH	IB (k)	--	b$_3$ 111-12°, n$_D^{20}$ 1.4770, d$_{20}$ 1.0040 (1)	501
C⋮CCMeEtC⋮CCMeEtOCHMeOBu	IVB (1)	41	b$_3$ 158-60°, n$_D^{20}$ 1.4710, d$_{20}$ 1.0020	501
CR:CHC⋮CR; R = CMeEtOH	III	70	b$_2$ 164°, n$_D^{20}$ 1.4964, d$_{20}$ 1.0534 (j, m)	198
C(CMe:CHMe):CHC⋮CCMeEtOH	IVE (m)	--	b$_2$ 148°, n$_D^{20}$ 1.5200, d$_{20}$ 1.0430	198
	III	56	d$_{20}$ 1.0428	198

* Numbers in parenthesis refer to pages in main volume.

**The characters used here correspond to the synthetic scheme in the introduction to Chapter 2.5.3.
(a) Thermal dec. at 280-300°, slowly dec. at r.t., colorless cryst., IR, NMR and mass spectra (2778). (b) Rxn. of R$_3$GeC⋮CPPh$_2$ + S in CS$_2$ → R$_3$GeC⋮CP(S)Ph$_2$; R = Me, Ph (2778). (c) Rxn. with KHSO$_4$ → R$_2$Ge[C⋮CC(Bu-t):CH$_2$]$_2$; R = Me, Et (1901). (d) Kinetic data for cleavage rxn. with aq. alkali in MeOH (134). (e) Rxn. of Et$_3$GeC⋮CCH$_2$OH + RCH:C:O → Et$_3$GeR'; R = H, catalyzed by H$_2$SO$_4$ (1531), Ac, catalyzed by Et$_3$N in C$_6$H$_6$ (1532). (f) IR and NMR spectra (1998). (g) Rxn. with HCl → Et$_3$GeC⋮CCMeEtCl (501). (h) Rxn. with KHSO$_4$ → Et$_3$GeC:C(CMe$_3$):CH$_2$ (1901). (i) Rxn. with KHSO$_4$ → Et$_3$GeC(CMe:CH$_2$):CHC⋮CCMe:CH$_2$ (198). (j) Rxn. of Et$_3$GeC(CMe:CH$_2$):CHC⋮CCMeROH + Et$_3$SiCl + Py at 110° → Et$_3$GeR'; R = Me, Et, i-Pr; MeR = 1-HOC$_5$H$_8$, 1-HOC$_6$H$_{10}$ (198 or 200). (k) Synth. from Et$_3$GeC⋮CMeEtCl + HOCMeRC⋮C-MgBr; R = Me, Et (501). (l) Rxn. of Et$_3$GeCR:CHC⋮CCMeEtOH + BuOCH:CH$_2$ + cat. HCl → Et$_3$GeR' (501). (m) Rxn. of Et$_3$GeC(CMe$_2$OH):CHC⋮CCMe$_2$OH + POCl$_3$ in Py at 60° → Et$_3$GeCR:CHC⋮CCMeEtOH; R = CMe:CHMe (198).

Table 10 Continued

$R_nGeR'_{4-n}$

Formula*	Synth. Method**	Yield	Properties	Ref.
Et$_3$GeR' (Cont'd.)				
R' = C(CMeEtOH):CHC:CCMeEtOSiEt$_3$	IVF (j)	--	b$_1$ 167°, n$_D^{20}$ 1.5100, d$_{20}$ 1.0305	198
CR:CHC:CR; R = CMeOHCHMe$_2$	III	50	b$_1$ 151-53°, n$_D^{20}$ 1.5025, d$_{20}$ 1.0350 (j, n)	200
C(CMeOHCHMe$_2$):CHC:CCMe(CHMe$_2$)OSiEt$_3$	IVF (j)	21	b$_2$ 147-49°, n$_D^{20}$ 1.4886, d$_{20}$ 1.0036	200
CR:CHC:CR; R = 1-HOC$_5$H$_8$c.	III	23	b$_3$ 192°, n$_D^{20}$ 1.5232, d$_{20}$ 1.1232 (j, o)	200
C(1-C$_5$H$_7$):CHC:CC$_5$H$_8$OH-1c.	(o)	9	b$_3$ 184°, n$_D^{20}$ 1.5412, d$_{20}$ 1.1050	200
C(1-HOC$_5$H$_8$):CHC:CC$_5$H$_8$OSiEt$_3$c.	IVF (j)	--	b$_1$ 183°, n$_D^{20}$ 1.4690, d$_{20}$ 1.6363	200
CR:CHC:CR; R = 1-HOC$_6$H$_{10}$c.	III	70	b$_3$ 201-202°, n$_D^{20}$ 1.5440, d$_{20}$ 1.0995 (j)	200
C(1-C$_6$H$_9$):CHC:CC$_6$H$_{10}$OH-1c.	III	7	b$_2$ 170-71°, n$_D^{20}$ 1.5420, d$_{20}$ 1.0917	200
C(1-HOC$_6$H$_{10}$):CHC:CC$_6$H$_{10}$OSiEt$_3$c.	IVF (j)	42	b$_3$ 216-18°, n$_D^{20}$ 1.5315, d$_{20}$ 1.0643	200
Et$_2$Ge(C:CCMeOHCMe$_3$)$_2$	IA	66	b$_2$ 151-52°, n$_D^{20}$ 1.4800, d$_{20}$ 1.0100 (c)	1901
Bu$_3$GeR'				
R' = C:CCMe$_2$OH	IA	37	b$_{0.3}$ 107-108°, n$_D^{20}$ 1.4705, d$_{20}$ 0.9960	500
	IVD (p)	--	(Synth. at 95-135°, R = H, Me) (q, r, s, t, u)	2765
C:CCMe$_2$OCH$_2$OBu	IVC (q)	47	b$_{0.5}$ 150-52°, n$_D^{20}$ 1.4607, d$_{20}$ 0.9712 (p)	2765
C:CCMe$_2$OCH$_2$CH$_2$CN	IVB (t)	52	b$_{0.1}$ 132-33°, n$_D^{20}$ 1.4795, d$_{20}$ 1.0304	500
C:CCMe$_2$OCHMeOBu	IVB (r)	--	b$_1$ 126-27°, n$_D^{20}$ 1.4625, d$_{20}$ 0.9595	500
	IVB (r)	60	b$_{0.6}$ 129-31°, d$_{20}$ 0.9685 (p)	2765
C:CCMe$_2$OSiEt$_3$	IVF (u)	28	b$_2$ 150-51°, n$_D^{20}$ 1.4590, d$_{20}$ 0.9083	500
C:CCMeEtOH	IA	--	b$_3$ 115-16°, n$_D^{20}$ 1.4785, d$_{20}$ 0.9905	500
	IVD (p)	--	(r, s)	2765
C:CCMeEtOCHMeOBu	IVB (r)	--	b$_{0.2}$ 144-45°, n$_D^{20}$ 1.4505, d$_{20}$ 0.9415	500
	IVB (r)	62	b$_1$ 144-45°, n$_D^{20}$ 1.4685, d$_{20}$ 0.9737 (p)	2765
C:CCMeEtC:CCMe$_2$OH	IB	27	b$_5$ 125-26°, n$_D^{20}$ 1.4917, d$_{20}$ 1.0038 (v, w)	502

C≡CCMeEtC≡CMe$_2$OCH$_2$CH$_2$CN	IVB (w)	38	b$_{0.3}$ 103-104°, n$_D^{20}$ 1.4962, d$_{20}$ 1.0267	502
Ph$_3$GeC≡CPPh$_2$	II	55	m. 102-103°, colorless cryst. (b, x, y)	2778
Ph$_3$GeC≡CP(O)Ph$_2$	(x)	81	m. 136-37° (y)	2778
Ph$_3$GeC≡CP(S)Ph$_2$	(b)	55	m. 155° (y)	2778

(n) Rxn. with KHSO$_4$ → Et$_3$GeCR:CHC≡CR; R = CMe:CMe$_2$ (200). (o) Rxn. of Et$_3$GeCR:CHC≡CR + KHSO$_4$ → Et$_3$GeC(1-C$_5$H$_7$): CHC≡CR; R = 1-HOC$_5$H$_8$c. (200). (p) Rxn. of Bu$_3$GeC≡CCMeROCHROBu + R'MgBr → Bu$_3$GeC≡CCMeROH; R = H, Me, Et (2765). (q) Rxn. of Bu$_3$GeC≡CCMe$_2$OH + BuOCH$_2$Cl + NaOH → Bu$_3$GeC≡CCMe$_2$OCH$_2$Bu (2765). (r) Rxn. of Bu$_3$GeC≡CCMeROH + BuOCH:CH$_2$ + cat. HCl at 80° → Bu$_3$GeC≡CCMeROCHMeOBu; R = Me, Et (500, 2765). (s) Rxn. with PCl$_5$ → Bu$_3$GeC≡CCMeRCl; R = Me, Et (500). (t) Rxn. of Bu$_3$GeC≡CCMe$_2$OH + CH$_2$:CHCN + cat. MeONa at 60° → Bu$_3$GeC≡CCMe$_2$OCH$_2$CH$_2$CN (500). (u) Rxn. of Bu$_3$GeC≡CCMe$_2$OH + Et$_3$SiCl + Py → Bu$_3$GeC≡CCMe$_2$OSiEt$_3$ (500). (v) Rxn. with KHSO$_4$ → Bu$_3$GeC≡CCMeEtC≡CCMe:CH$_2$ (502). (w) Rxn. of Bu$_3$GeC≡CCMeEtC≡CCMe$_2$OH + CH$_2$:CHCN = cat. MeONa at 60-80° → Bu$_3$GeC≡CCMeEtC≡CCMe$_2$OCH$_2$CH$_2$CN (502). (x) Rxn. of Ph$_3$GeC≡CPPh$_2$ + H$_2$O$_2$ in Me$_2$CO → Ph$_3$GeC≡CP(O)Ph$_2$ (2778). (y) IR spectrum, mol. wt. (2778).

2.6 FUNCTIONALLY SUBSTITUTED UNSYMMETRIC TETRAORGANOGERMANES

2.6.1 TETRAORGANOGERMANES CONTAINING HALOGEN

Organogermanes containing halogen linked to carbon are prepared from organohalogermanes by Grignard (I) or organolithium (II) reagents. Other methods include hydrogermylation (III) of cyclic or unsaturated organic halides, interconversion (IV) of tetraorganogermanes by conventional organic methods and dihalocarbene insertion reactions (V). The compounds compiled in Table 11 are prepared by methods from the following scheme.

I Grignard synthesis:
- (IA) $R_nGeX_{4-n} + R'MgBr \rightarrow R_nGeR'_{4-n}$,
- (IB) $R'GeCl_3 + RMgCl \rightarrow R_3GeR'$,
- (IC) $Bu_2\overline{GeCH_2CH_2CH_2} + ICl, + EtMgBr \rightarrow EtBu_2Ge(CH_2)_3I$.

II Reactions with organolithium reagents:
- (IIA) $R_nGeX_{4-n} + R'Li \rightarrow R_nGeR'_{4-n}$,
- (IIB) $Ph_3GeLi + R'Cl \rightarrow Ph_3GeR'$.

III Hydrogermylation reactions:
- (IIIA) $Me_3GeH + \overline{CF_2CF_2CX{:}CX} \rightarrow Me_3Ge\overline{CXCHXCF_2CF_2}$; X = H, F, Cl ,
- (IIIB) $Ph_3GeH + CH_2{:}CH(CH_2)_mCN \rightarrow Ph_3Ge(CH_2)_{m+2}CN$; m = 0, 1 .

IV Interconversion reactions:
- (IVA) $Me_3GeCH_2X + NaX' \rightarrow Me_3GeCH_2X'$,
- (IVB) $R''CH{:}CH_2 + Br \rightarrow R''CHBrCH_2Br$,
- (IVC) $R''CR{:}CHR + HCl \rightarrow R''CRClCH_2R$; R = H, Me ,
- (IVD) $Me_3GeCH_2Ph + NBS \rightarrow Me_3GeCBr_2Ph$,
- (IVE) $R''OH + PBr_3 \rightarrow R''Br$.

V Dihalocarbene insertion reactions:
- (VA) $R_3GeH + PhHgBrCl_2 \rightarrow R_3GeCHCl_2$,
- (VB) $R_3Ge(CH_2)_mCH{:}CH_2 + PhHgCBrCl_2 \rightarrow R_3Ge(CH_2)_m\overline{CHCH_2CCl_2}$; m = 0, 1 ,
- (VC) $Et_3GeCH{:}CH_2 + Me_3SnCF_3 \rightarrow Et_3Ge\overline{CHCH_2CF_2}$.

R = organic group, R' = organic group containing halogen or a pseudohalogen group, R" = tetraorganogermyl group, may be functionally sbustituted, X = X' = halogen, NBS = N-bromsuccinimide, n = 1-3. Other tetraorganogermanes containing halogen are listed in Chapters 1.3, 2.4.1, 2.4.3, 2.5.1, 2.6.3.3, 2.6.3.4, 2.6.5, 2.7.1.1.3, 2.7.1.8, 2.7.3, 2.7.6, 2.7.8, 7.1.6.1 and 7.1.6.6.

Trifluoromethyltrimethylgermane $GeC_4H_9F_3$ Me_3GeCF_3
Synth.: By rxn. of CF_3GeCl_3 with Me_2Zn in evacuated tube, in 97% yield; CAUTION: REACTION MIGHT PROCEED EXPLOSIVELY (2446).
Prop.: b. 71°, IR and ^{19}F NMR spectra, mol. wt., stable in sealed tube to 185°, slow dec. above 235° (2446).
Rxn. with: Thermal dec. at 275° $\rightarrow HCF_3 + SiF_4$ (2446).

4-Bromobutyltriethylgermane (M-29) $GeC_{10}H_{23}Br$ $Et_3Ge(CH_2)_4Br$
Synth.: By rxn. of $Et_3Ge(CH_2)_4OH$ with PBr_3 at 15° in 60% yield (343).
Prop.: b_2 104°, IR spectrum (343).
Rxn. with: Mg, + CO_2 → $Et_3Ge(CH_2)_4CO_2H$ (343).
Mg, + CH⋮CCH_2Br → $Et_3Ge(CH_2)_4CH:C:CH_2$ + $Et_3Ge(CH_2)_5C⋮CH$ (1494).
NaC⋮CH in $(Me_2N)_3PO$ → $Et_3Ge(CH_2)_4C⋮CH$ (1494).
KCN → $Et_3Ge(CH_2)_4CN$ (343).

Bis(4-bromobutyl)diethylgermane $GeC_{12}H_{26}Br_2$ $Et_2Ge[(CH_2)_4Br]_2$
Synth.: By rxn. of $Et_2Ge[(CH_2)_4OH]_2$ with PBr_3 in pentane and EtBr at 0°, in 35 (1490) and 53% yield, resp. (343); with 48% aq. HBr in 72% yield (343, 1493).
Prop.: b_1 152° (343), b_1 145° (1490) n_D^{20} 1.5115 (343, 1490), 1.5100 (343), d_{20} 1.4298 (343, 1490), IR spectrum (343).
Rxn. with: Mg, + CO_2 → $Et_2Ge[(CH_2)_4CO_2H]_2$ (343).
Na → $Et_2\overline{Ge(CH_2)_7CH_2}$ + polymer + Et_2GeBu_2 (1493).
KCN + cat. CuCl → $Et_2Ge[(CH_2)_4CN]_2$ (343, 1490).

Additional tetraorganogermanes containing halogen are listed in Table 11.

2.6.2 TETRAORGANOGERMANES CONTAINING PSEUDOHALOGEN SUBSTITUENTS

Tetraorganogermanes containing (iso-)cyano and (iso-)thiocyano groups linked to carbon are listed at the bottom of Table 11. Additional derivatives with pseudohalides not directly linked to Ge are listed in Chapters 2.4.3, 2.5.3, 3.6, 3.8.2 and 6.5.8.

Table 11. Unsymmetric Tetraorganogermanes Containing Halogen $R_nGeR'_{4-n}$

Formula*	Synth. Method**	Yield	Properties	Ref.
Me$_3$GeR'				
R' = CH$_2$Cl (27)	--	--	b. 112° (2131), dipole moment (a, b, c)	254
CH$_2$Br (28)	IVA (d)	--	IR, NMR and mass spectra	2244
CH$_2$I (28)	IVA (c)	--	b. 152°, (a, d)	2131
CPhBr$_2$	IVD	--	(e)	2159
(CH$_2$)$_3$Cl (29)	--	--	NMR spectrum	2373
CH$_2$CH$_2$CHClMe	IB	60	b_{20} 66-67°, n_D^{20} 1.4505, d_{20} 1.1013, IR spectrum	364
CH$_2$CHCH$_2$CCl$_2$	VB	94	n_D^{25} 1.4737	1877
CFCHFCF$_2$CF$_2$	IIIA (f)	86	b. 118°, IR and NMR spectra	924
cis-CHCHClCF$_2$CF$_2$	IIIA (g)	4	IR and NMR spectra	924
trans-CHCHClCF$_2$CF$_2$	IIIA (g)	8	b_{50} 80-82°, NMR spectrum	924
CClCHClCF$_2$CF$_2$	IIIA (g)	27	IR and NMR spectra	924
p-C$_6$H$_4$Br (29)	IIA	--	m. 23°, IR and NMR spectra	2223
p,p'-(C$_6$H$_4$)$_2$Br	IIA	--	b_{10} 80-95° (h)	2223
Me$_2$Ge(CH$_2$Cl)$_2$	--	--	m. 94-98°, crude (h)	
			NMR spectrum	2373
Et$_3$GeR'				
R' = CHCl$_2$	VA (i)	83	n_D^{25} 1.4798, IR and NMR spectra	1876
(CH$_2$)$_3$Cl (29)	--	--	(j) NMR spectrum	2373
CHCH$_2$CF$_2$	VC	39	n_D^{25} 1.4253, IR spectrum	1874
CHCH$_2$CCl$_2$	VB	65	n_D^{25} 1.4868	1877
(CH$_2$)$_5$Br (29)	IVE	70	$b_{0.8}$ 107° (k)	343
(CH$_2$)$_3$CHBrCH$_2$Br	IVB	--	$b_{0.6}$ 123°, n_D^{20} 1.5120, d_{20} 1.4580 (1)	1494
(CH$_2$)$_5$CHBrCH$_2$Br	--	--	$b_{0.15}$ 122°, n_D^{20} 1.5079, d_{20} 1.3928 (1)	1494
Et(i-Pr)PhGeC$_6$H$_4$Br-p	IA	69	$b_{0.01}$ 128-30°, n_D^{25} 1.5856 (m)	1003
EtBu$_2$Ge(CH$_2$)$_3$I	IC	52	$b_{1.7}$ 137°, n_D^{20} 1.5064, d_{20} 1.3059	1492

BuGe(C_6F_5)$_3$	IIA (n)	--	m. 75-76°, IR spectrum (o)	1034

Ph$_3$GeR'

R' = CHCl$_2$	VA	88	m. 154-55°, IR and NMR spectra	1876
CHBrCH$_2$Br	IVB	--	m. 108-109°	464
CPhBrCHMe$_2$	IVE (p)	47	m. 135.5-37°, mol. wt.	1624, 1626
C$_6$F$_5$	IIA	--	m. 114-16°, IR spectrum (o, q, r) mol. wt.	1034
9,9-$C_{13}H_8$Br	(s)	--	m. 166° (t)	2675

Ph$_2$GeR'

R' = (C_6F_5)	IIA	--	m. 125-27°, mol. wt., soly., IR spectrum (r)	1034
o-C_6F_4Br	IIA	76	m. 147-50°, mol. wt., soly. (u, v)	2325
p-C_6H_4Br	IA	--	m. 129-30°	463
p-C_6H_4CCl$_2$Me	IVC	--	m. 125-26°, light yellow powder	1787
p-C_6H_4CBrMeCH$_2$Br	IVB	--	m. 156-57°	1787
p-C_6H_4CClMeEt	IVC	--	m. 109-10°	1787
p-C_6H_4CBrMeCHBrMe	IVB	--	m. 154-55°	1787

---

* Numbers in parenthesis refer to pages in main volume.

**The characters used here correspond to the synthetic scheme in the introduction to Chapter 2.6.1.
(a) NMR spectrum (2131, 2373). (b) Rxn. with I$^-$ or EtO$^-$ → relative reactivity of exchange rxn. and mechanism (766). (c) Rxn. of Me$_3$GeCH$_2$Cl with NaX → Me$_3$GeCH$_2$X; X = I, SCN (2131). (d) Rxn. of Me$_3$GeCH$_2$I with Br → Me$_2$(BrCH$_2$)GeBr + Me$_3$GeCH$_2$Br (?) (2244). (e) Rxn. with AgOAc → Me$_3$GeBz (2159). (f) Synth. from $\overline{CF_2CF_2CF_2CF}$:CF at 230° (924). (g) Synth. from $\overline{CF_2CF_2CCl}$:CCl at 190° (924). (h) Rxn. of p-Me$_3$GeC$_6$H$_4$Br + p,p'-Me$_3$Ge(C$_6$H$_4$)$_2$Br + Mg → p-(p-Me$_3$GeC$_6$H$_4$)$_2$C$_6$H$_4$ + (p,p Me$_3$GeC$_6$H$_4$C$_6$H$_4$)$_2$ (2233). (i) Isolated by GLC (1876). (j) Rxn. with P(OR)$_3$ → Et$_3$Ge(CH$_2$)$_3$PO$_3$R$_2$; R = Et, Bu (2366). (k) Rxn. of Et$_3$Ge(CH$_2$)$_n$Br with KCN + cat. CuCl → Et$_3$Ge(CH$_2$)$_n$CN; n = 4, 5 (343). (l) Rxn. with anhydrous base → Et$_3$Ge(CH$_2$)$_n$C:CH; n = 3, 5 (1494). (m) Rxn. with BuLi, + CO$_2$ → Et(i-Pr)PhGeC$_6$H$_4$-CO$_2$H-p (1003). (n) Synth. with BuLi (1034). (o) Rxn. with NaOH → C$_6$F$_5$H (1034). (p) Synth. from Ph$_3$GeCOHPhCHMe$_2$ (1624, 1626). (q) Subl. in vacuo at 50-70° (1034). (r) ^{19}F NMR spectrum (1032, 2475). (s) Synth. from 9-Ph$_3$Ge-fluorene with N-bromsuccinimide in presence of catalytic amount of Bz$_2$O$_2$ in CCl$_4$ (2675). (t) Rxn. with AgO$_2$CCF$_3$ in aq. Me$_2$CO → 9,9-Ph$_3$GeC$_{13}$H$_8$OH (2675). (u) ^{19}F NMR spectrum (2325). (v) Rxn. with BuLi at -78°, + Ph$_2$GeCl$_2$ → o,o-$\overline{C_6F_4GePh_2C_6F_4GePh_2}$. (2325).

Table 11 Continued

R_nGeR'

Formula*	Synth. Method**	Yield	Properties	Ref.
Me$_3$GeCH$_2$SCN (31)	IVA	--	b$_{11}$ 88°, NMR spectrum	2131
Me$_3$GeCHMeSCN (31)	--	--	NMR spectrum	2373
Et$_3$GeCH(CH$_2$PEt$_2$)CN	IIIB (w)	40	b$_{0.1}$ 100°, n$_D^{20}$ 1.5006, d$_{20}$ 1.0730 (x)	1797
Et$_3$Ge(CH$_2$)$_4$CN	IVA (k)	88	b$_{12}$ 139°, n$_D^{20}$ 1.4650, d$_{20}$ 1.0412 (y)	343
Et$_3$Ge(CH$_2$)$_5$CN	IVA (k)	88	b$_1$ 122°, n$_D^{20}$ 1.4652, d$_{20}$ 1.0281 (z)	343
Et$_2$Ge[(CH$_2$)$_4$CN]$_2$	IVA (k)	96	b$_{0.4}$ 165°, n$_D^{20}$ 1.4777, d$_{20}$ 1.0690 (y, aa)	343, 1490
Ph$_2$Ge(CH$_2$CH$_2$CN)$_2$	IIIB	26	b$_{0.01}$ 203°, IR and NMR spectra (bb)	2577
Ph$_2$Ge[(CH$_2$)$_3$CN]$_2$	IIIB	50	m. 55-56°, b$_{0.04}$ 226° (cc)	2577

(w) Synth. from CH$_2$:CHCN with Et$_3$GePEt$_2$ (1797). (x) IR spectrum, GLC (1797), rxn. with Br → Et$_3$GeBr (1797). (y) Rxn. with H + cat. Raney Ni → Et$_n$Ge[(CH$_2$)$_5$NH$_2$]$_{4-n}$; n = 2, 3 (343). (z) Rxn. with NH$_3$ in aq. (CH$_2$OH)$_2$, followed by HCl → Et$_3$Ge(CH$_2$)$_5$CO$_2$H (343). (aa) Rxn. with KOH in (CH$_2$OH)$_2$, followed by HCl → Et$_2$Ge[(CH$_2$)$_5$CO$_2$H]$_2$ (343, 1490), IR spectrum 343, 1490). (bb) Rxn. with Br in EtBr at -5° → Ph$_2$GeBr$_2$ (2577). (cc) Rxn. with Br in EtBr at -5° → [NC(CH$_2$)$_3$]$_2$GeBr$_2$ + PhBr (2577).

2.6.3 TETRAORGANOGERMANES CONTAINING OXYGEN

2.6.3.1 ORGANOGERMANES WITH HYDROXY GROUPS AND THEIR DERIVATIVES

This chapter contains unsymmetric tetraorganogermanes with a hydroxy group linked through carbon. Compounds derived by substitution of the hydroxy-proton, e.g. ethers and esters are included. Organogermanium alcohols are prepared by typical organometallic methods (I-III) or by common organic synthesis (IV-VII) not involving the germanium atom. The following scheme shows synthetic methods used in preparation of the organogermanes in Table 12.

- I Grignard reactions:
    - (IA) $R_nGeBr_{4-n} + R'MgX \rightarrow R_nGeR'_{4-n}$,
    - (IB) $R'GeX_3 + RMgX \rightarrow R_3GeR'$.

- II Reactions with lithium derivatives:
    $R_3GeLi + R_2CO \rightarrow R_3GeCR_2OH$.

- III Hydrogermylation:
    $Ph_3GeH_3 + R'CH:CH_2 \rightarrow Ph_3GeCHMeR'$.

- IV Introduction of hydroxyl group:
    - (IVA) $R_3Ge(CH_2)_3B(OMe)_2 + H_2O_2 + base \rightarrow R_3Ge(CH_2)_3OH$,
    - (IVB) $R''Br + CF_3CO_2Ag \rightarrow R''OH$,
    - (IVC) $Et_4Ge + BzO_2CMe_3 \rightarrow Et_3GeC_2H_4OBz$.

- V Reduction of carbonyl or carboxyl groups:
    - (VA) $R''COR + LiAlH_4 \rightarrow R''CHOHR$,
    - (VB) $R''CO_2R + LiAlH_4 \rightarrow R''CH_2OH$,
    - (VC) $R''CO_2R + RMgBr \rightarrow R''CR_2OH$.

- VI Hydrogenation of unsaturated germanes:
    $R''CR':CHR + H + cat. \rightarrow R''CHR'CH_2R$.

- VII Derivatization:
    $R_3GeCH_2CH_2OH + RCH:CO \rightarrow R_3GeCH_2CH_2O_2CCH_2R$; R = H, Ac .

R = organic group, R' = organic hydroxo group, R" = tetraorganogermyl group, may contain a hydroxy function, X = halogen, n = 1-3. Other tetraorganogermanes with hydroxy groups are reported in Chapters 1.3, 2.3, 2.4.3, 2.5.3, 2.6.3.3, 2.7.1.1.3, 2.7.1.2, 2.7.5, 2.7.8 and 7.1.6.1.

Table 12. Tetraorganogermanes Containing Hydroxy Groups and Their Derivatives $R_nGeR'_{4-n}$

Formula*	Synth. Method**	Yield	Properties	Ref.
Me_3GeR'				
R' = CH_2OH (33)	IB	--	NMR spectrum	2604
CH_2OMe (33)	IB	76	b_{758} 98°, n_D^{20} 1.4170, d_{20} 1.0126 (a)	1588
CH_2CH_2OH (33)	IB (b)	40	b_{10} 110-12°, n_D^{20} 1.4545, d_{20} 1.0903 (c)	2604
CHMeOH	--	--	NMR spectrum	2373
$(CH_2)_3OH$ (33)	--	--	NMR spectrum	2373
CMe_2OH	IB	85	b_{35} 59°, b. 140°, n_D^{20} 1.4497-9, d_{20} 1.0955-65 (d)	1071
$CHMeCH_2CHMeOH$	IB (e)	80	b_4 69.5-70.5°, n_D^{20} 1.4550, d_{20} 1.0666 (f)	1526
$CMe_2CH_2CMe_2OH$	IB (g)	84	b_7 79-79.5°, n_D^{20} 1.4667, d_{20} 1.0555 (h, i)	1526
	IB (g)	68	b_{25} 106-108°, n_D^{20} 1.4685	2504
p-C_6H_4OMe	--	--	$n_D^{25.5}$ 1.5158, UV spectrum	1561
Et_3GeR'				
R' = CPh_2OH	II	44	n_D^{20} 1.5770	2849
CH_2CH_2OH (33)	VB	77	b_8 98-99°, n_D^{20} 1.4703, d_{20} 1.0980 (j)	24b
CH_2CH_2OAc	VII (j)	72	$b_{6.5}$ 95-98°, n_D^{20} 1.4550, d_{20} 1.0857	24b
C_2H_4OBz	IVC	23	b_3 116-17°, n_D^{20} 1.5010	2098
$(CH_2)_3OH$ (33)	IVA	85	b_2 83-84°, NMR spectrum (k, l)	357
$(CH_3)_3OAc$ (34)	VII (k)	--	b_4 100-102°, n_D^{20} 1.4589, d_{20} 1.0733 (l)	1531
$(CH_3)_3O_2CCH_2Ac$	VII (k)	89	b_2 118-20°, n_D^{20} 1.4690, d_{20} 1.1033	1532
$CH_2CHMeOH$	VB	61	b_8 90-92°, n_D^{28} 1.4702, d_{20} 1.0723 (m)	711
$CHPrCHPhOH$	VI (n)	68	b_2 186°, n_D^{20} 1.5212, d_{20} 1.0925	201
$CH(CH_2Ph)CMe_2OH$	VI (o)	59	b_2 136-37°, n_D^{20} 1.5259, d_{20} 1.0936	2426
$CHMeCH_2CHEtOH$	IB (p)	60	b_{17} 143°, n_D^{20} 1.4737, d_{20} 1.0442,	1496
	IB (q)	--	IR and NMR spectra	1496
$CHPrCH_2CH_2OH$	VI (r)	85	b_2 121°, n_D^{20} 1.4830, d_{20} 1.0508	201
$CHPrCHMeOH$	VI (n)	70	b_1 129-31°, n_D^{20} 1.4880, d_{20} 1.0686	201

Compound				Ref.
CH$_2$CH$_2$CMeEtOH	IB (s)	53	b$_2$ 107°, n$_D^{20}$ 1.4687, d$_{20}$ 1.0321	1496
CH(CH$_2$Ph)CMeEtOH	VI (o)	60	b$_2$ 145-47°, n$_D^{20}$ 1.5280, d$_{20}$ 1.0941	2426
CHPrCMe$_2$OH	VI (n)	50	b$_2$ 121°, n$_D^{20}$ 1.4822, d$_{20}$ 1.0401	201
CH$_2$CH$_2$CEt$_2$OH	IB (t)	39	b$_{14}$ 144°, n$_D^{20}$ 1.4725, d$_{20}$ 1.0328	1496
CH(CH$_2$Ph)C$_5$H$_8$OH-1c.	VI (u)	84	b$_2$ 159-60°, n$_D^{20}$ 1.5340, d$_{20}$ 1.0987	2425
CHPrCHOHPr	VI (n)	85	b$_1$ 116-18°, n$_D^{20}$ 1.4815, d$_{20}$ 1.0296	201
CH(CMe$_2$OEt)CH$_2$CMe$_2$OEt	VI (v)	--	b$_1$ 101-103°, n$_D^{20}$ 1.4769, d$_{20}$ 1.0201	1150
CH$_2$CHMeCEt$_2$OH	IB (t)	38	b$_{14}$ 152°, n$_D^{20}$ 1.4779, d$_{20}$ 1.0362	1496
CH(CH$_2$Ph)C$_6$H$_{10}$OH-1c.	VI (u)	85	b$_2$ 162-63°, n$_D^{20}$ 1.5380, d$_{20}$ 1.0995	2425
CHEtCH$_2$CMePrOH	VI (w)	47	b$_4$ 132-33°	2428

* Numbers in parenthesis refer to pages in main volume.

**The characters used here correspond to the synthetic scheme in the introduction to Section 2.6.3.1.

(a) Dipole moment and dielectric constant (1572), NMR spectrum (1336). (b) Synth. from HGeCl$_3$·2OEt$_2$ with $\overline{CH_2CH_2O}$ followed by MeMgCl (2604). (c) NMR spectrum (2604). (d) IR, Raman and NMR spectra, GLC (1071). (e) Synth. from Cl$_3$GeCHMeCH$_2$CHO with MeMgCl (1525). (f) IR spectrum (1526). (g) Synth. from AcCH$_2$CMe$_2$GeCl$_3$ with MeMgCl (1526, 2504). (h) IR and NMR spectra (1526, 2504). (i) Rxn. with KHSO$_4$ → Me$_3$GeCMe$_2$CH$_2$CMe:CH$_2$ (1526). (j) Rxn. of R$_3$GeCH$_2$CH$_2$OH + CH$_2$:C:O → R$_3$GeCH$_2$CH$_2$OAc (24b). (k) Rxn. with RCH:C:O → Et$_3$Ge(CH$_2$)$_3$O$_2$CCH$_2$R; R = H (1531), Ac (1532). (l) NMR spectrum (2373). (m) Rxn. with H$_2$SO$_4$ at 140-60° → (Et$_3$Ge)$_2$O + C$_3$H$_6$ (711). (n) Synth. by hydrogenation of Et$_3$GeCR:CHCH:CH$_2$ with Pd on CaCO$_3$ cat.; R = PhCHOH, CHMeOH, CMe$_2$OH, CHPrOH, c.C$_5$H$_9$CHOH, c.C$_6$H$_{11}$CHOH (201). (o) Synth. by hydrogenation of Et$_3$GeC(:CHPh)CMeOHCH$_2$R with Raney Ni cat.; R = H, Me (2426). (p) Synth. from I$_3$GeCHMe-CH$_2$CHO with EtMgBr (1496). (q) Synth. from HGeCl$_3$ with MeCH:CHCHO, followed by EtMgBr (1496). (r) Synth. by hydrogenation of Et$_3$GeC(CH$_2$CH$_2$OH):CHCH:CH$_2$ with Raney Ni cat. (201). (s) Synth. from I$_3$GeCH$_2$CH$_2$Ac (1496). (t) Synth. from I$_3$GeCH$_2$CHRCO$_2$Me with EtMgBr; R = H, Me (1496). (u) Synth. by hydrogenation of Et$_3$GeCR:CHPh with Raney Ni cat.; R = c.1-HOC$_5$H$_8$, c.1-HOC$_6$H$_{10}$ (2425). (v) Synth. by hydrogenation of Et$_3$GeC(CMe$_2$OEt):CHCMe$_2$OEt with Pd on carbon cat. (1150). (w) Synth. by hydrogenation of Et$_3$GeC(CH:CH$_2$):CHCMePrOH with Raney Ni cat. (2428).

Table 12 Continued $R_nGeR'_{4-n}$

Formula*	Synth. Method**	Yield	Properties	Ref.
Et_3GeR' (Cont'd.)				
$R' = \underline{CHRCH_2CMeOHCH_2CH_2OMe}$				
R = CMe_2OH	VI (x)	74	b_4 153-55°, n_D^{20} 1.4892, d_{20} 1.0864	2427
$CMeEtOH$	VI (x)	60	b_3 172-74°, n_D^{20} 1.4935, d_{20} 1.0901	2427
$CMeOHCH_2CH_2OH$	VI (x)	65	b_3 162-63°, n_D^{20} 1.4845, d_{20} 1.0746	2427
$C_6H_{10}OH$-1c.	VI (x)	68	b_2 173-74°, n_D^{20} 1.4980, d_{20} 1.0886	2427
$CHPrCHOHC_5H_9c.$	VI (n)	57	b_2 125-26°, n_D^{20} 1.5012, d_{20} 1.0709	201
$CHPrCHOHC_6H_{11}c.$	VI (n)	80	b_2 126-28°, n_D^{20} 1.5000, d_{20} 1.0637	201
$p-C_6H_4OMe$ (34)	--	--	(y)	--
$Et_2Ge[(CH_2)_4OH]_2$	IA (z)	84	b_2 166°, n_D^{20} 1.4872, d_{20} 1.1050 (aa)	343, 1490
R_3GeR' $R_3Ge = EtPh(1-C_{10}H_7)Ge$				
$R' = CH_2OH$	VB	--	Stereochemistry and	2368
	II	--	optical activity	2369
CPh_2OH	II	--	Stereochemistry and optical activity	2369
$2,1-Me(HO)C_8H_9c.$	II	--	Stereochemistry and optical activity	2369
$Pr_3GeCH_2CHOHMe$	VA	60	b_3 86-88°, n_D^{20} 1.4679, d_{20} 1.0226	711
$Pr_3GeCH_2CEt_2OH$	VC (bb)	64	b_2 104-106°, n_D^{20} 1.4718, d_{20} 1.0031	24b
$Pr_3GeCH_2CPr_2OH$	VC (bb)	50	b_1 104-106°, n_D^{20} 1.4686, d_{20} 0.9885	24b
$Pr_2Ge(CH_2CH_2OH)_2$	VB	58	b_5 186-8°, n_D^{20} 1.4865, d_{20} 1.1368	25
$Bu_3GeCH_2CH_2OH$	VB	68	b_3 129-32°, n_D^{20} 1.4700, d_{20} 1.0171 (j)	24b
$Bu_3GeCH_2CH_2OAc$	VII (j)	67	b_2 122-25°, n_D^{20} 1.4600, d_{20} 1.0043	24b
$Bu_3GeCH_2CHOHMe$	VA	74	b_2 120-21°, n_D^{20} 1.4668, d_{20} 0.9954	711
$Bu_2Ge(CH_2CH_2OH)_2$	VB	55	$b_{1.5}$ 147-50°, n_D^{20} 1.4838, d_{20} 1.0966	25

Ph₃GeR'

R' =				
CHOHPh (35)	--	--	(cc)	1624, 1626
CPh₂OH (35)	II	81	m. 150-52°, IR spectrum	2671
CHOHMe	VC (dd)	57	m. 152-55° (ee)	2671
CHOHMe	II	59	m. 107-108°, NMR spectrum	2671
	VA	70	(ff)	
CPhOHMe	II	68	m. 116-18°, IR and NMR spectra (gg)	1624, 1626
(CH₂)₃OH	IVA	95	m. 126-26.5°	357
CMe₂OH	II	40	m. 156-58°, IR spectrum (gg)	1624, 1626
CHMeCH₂CH₂OGePh₃	III	60	m. 183-85°, mol. wt., IR spectrum	806
CMeEtOH	II	37	m. 96-98°, IR spectrum (hh)	1624, 1626
CPhOHCHMe₂	II	84	m. 144-46.5°, IR spectrum (ii)	1624, 1626
CEt₂OH	II	70	m. 88.5-89.5°, IR spectrum (jj)	1624, 1626
CHMeCHMeCH₂OGePh₃	III	--	m. 176°, IR spectrum	806
2-C₆H₁₀OHc.	II	57	m. 137-40°, IR spectrum	2670
CPr₂OH	II	48	m. 77-79°, IR spectrum	1624, 1626
9,9-C₁₃H₈OH	IVB	44	m. 150-51°, NMR spectrum (ll)	2675

Ph₂Ge(C₆H₄OR-p)₂

R =				
Me	IA	44	m. 120-21°, IR spectrum	463
Et	IA	33	m. 110-11°, IR spectrum	463
Ph	IA	24	m. 129-30°, IR spectrum	463

(x) Synth. by hydrogenation of Et₃GeCR:CHCMeOHCH₂CH₂OMe with Raney Ni cat.; R = CMe₂OH, CMeEtOH, CMe(OH)CH₂CH₂OMe, c.1-HOC₆H₁₀ (2427). (y) Rxn. with HCl in dioxane → deuterium effect, mechanism of cleavage reaction (43). (z) Hydroxy group protected during synth. as 2-pyranylether (1490). (aa) Rxn. with PBr₃ → Et₂Ge[(CH₂)₄Br]₂ (343, 1490), same product with HBr (343, 1493). (bb) Synth. from Pr₃GeCH₂CO₂Me + RMgBr; R = Et, Pr (24b). (cc) Rxn. with BF₃ in hexane → Ph₂CH(Ph₂)GeF (49); in MePh → Ph₃GeCHPhC₆H₄Me-p (49). (dd) Synth. from Ph₃GeBz and PhMgBr in N atm. at 0° (2671). (ee) Rxn. with BF₃ → Ph₃C(Ph₂)GeF (49). (ff) Rxn. with CrO₃-H₂SO₄ → Ph₃GeOAc (1677) + (Ph₃Ge)₂O (2671). (gg) Rxn. of Ph₃GeCROHMe with PBr₃ → Ph₃GeCR:CH₂; R = Me, Ph (1624, 1626). (hh) Rxn. with PBr₃ → Ph₃GeCMe:CHMe + (Ph₃Ge)₂O (1624, 1626). (ii) Rxn. with PBr₃ → Ph₃GeCPhBrCHMe₂ (1624, 1626). (jj) Rxn. with PBr₃ → Ph₃GeCEt:CH₂ + (Ph₃Ge)₂O (2670), with (C₆H₁₁N:)₂C + cat. CF₃CO₂H·Py → 2-Ph₃GeC₆H₈O (2670). (ll) Rxn. with NaH → 9-Ph₃GeOC₁₃H₉ + 9-C₁₃H₈OH; C₁₃H₉ = fluorenyl (2675). (kk) Rxn. with CrO₃ in Me₂CO → (Ph₃Ge)₂O (2670).

2.6.3.3 TETRAORGANOGERMANES CONTAINING CARBONYL GROUPS

Unsymmetric tetraorganogermanes containing a carbonyl group are listed in this chapter. Derivatives of the tautomeric enol-form may be found in Chapter 4.3.4. The carbonyl substituted tetraorganogermanes are prepared by organometallic reactions (I-III) and by interconversion reactions (IV) not involving the germanium atom. The compounds summarized in Table 13 are prepared by methods from the following scheme.

I Synthesis with organometallic compounds:
  (IA) $R_3GeX + R_3SnR' \rightarrow R_3GeR'$,
  (IB) $R_3GeI + R'HgX \rightarrow R_3GeR'$,
  (IC) $ClOCCH_2CH_2GeCl_3 + Me_2Cd \rightarrow Me_3GeCH_2CH_2Ac$.

II Acylation of triphenylgermyllithium:
  $Ph_3GeLi + ROCl \rightarrow Ph_3GeC:OR$.

III Hydrogermylation reaction:
  $Et_3GeH + CH_2:CHAc \rightarrow Et_3GeCH_2CH_2Ac$.

IV Interconversion reactions:
  (IVA) $Me_3GeCBr_2Ph + CF_3CO_2Ag \rightarrow Me_3GeBz$,
  (IVB) $Et_3Ge(CH_2)_3C\vdots CH + H_2O + cat.\ HgSO_4 \rightarrow Et_3Ge(CH_2)_3Ac$,
  (IVC) $R_3Ge\ \overline{CRS(CH_2)_3S} + HgCl_2 \rightarrow R_3GeC:OR;\ R = Me, Ph$,
  (IVD) $R''RCHOH + (C_6H_{11}N:)_2C \rightarrow R''RCO$.

V Insertion reactions:
  $Ph_3GeBz + CH_2N_2 \rightarrow Ph_3GeCH_2Bz$.

R = organic group, R' = organic group containing carbonyl, R" = tetraorganogermyl group, X = halogen. Additional tetraorganogermanes with carbonyl functions are listed in Chapters 2.3, 2.4.3, 2.5.3, 2.6.3.1, 2.7.1.1.3, 2.7.2, 2.7.8 and 7.1.6.1.

1-Acetonyltriethylgermane (M-37) $GeC_9H_{20}O$ Et_3GeCH_2Ac
Synth.: By rxn. of Et_3GeX with Bu_3SnCH_2Ac; X = Cl, Br and I in 70, 77 and 92% yield, resp. (711).
By rxn. of Et_3GeI with $(AcCH_2)_2Hg$, in 77% yield (711).
By-prod. in ozonolysis of $Et_3GeCH_2CMe:CMe_2$ (467), in $HCCl_3$ (1798).
Prop.: b_7 85-87° (711), b_{12} 100-110° (1798), n_D^{20} 1.4660 (711), 1.4557 (1798), d_{20} 1.0805 (711), IR (711, 1798) and UV (711) spectra.
Rxn. with: $Me_3SiI \rightarrow Et_3GeI + Me_3SiOCMe:CH_2$ (711).
$LiAlH_4 \rightarrow Et_3GeCH_2CHOHMe$ (711).
$2,4(O_2N)_2C_6H_3NHNH_2 \rightarrow$ dinitrophenylhydrazone $[GeC_{15}H_{24}N_4O_4]$, m. 105° (467, 1798).

Benzoyltriphenylgermane (M-37) $GeC_{25}H_{20}O$ $Ph_3GeCOPh$
Synth.: By rxn. of Ph_3GeLi with BzCl at -78 and -23° in N atm. in 80 and 51% yield, resp. (1624-5).
By-prod. in ozonolysis of $Ph_3GeCPh:CH_2$ in EtOAc at -78°, in 5% yield (1624, 1626).

Prop.: Yellow crystals, m. 103-106° (1624-5), IR (2671), assignment (1624-5), visible (658, 1624-5) and UV (2671) spectra, base strength in aq. H_2SO_4 and relative to PhOH and MeOH (2159), soly. (1624-5).
Rxn. with: MeMgBr → $Ph_3GeCPh:CH_2$ (793).
PhMgBr → Ph_3GeCPh_2OH (2671).
Ph_3GeLi → $(Ph_3Ge)_2CPhOH$ (1624-5).
CH_2N_2 → Ph_3GeCH_2Bz (792a).
$Ph_3P:CHR$ → $Ph_3GeCPh:CHR + Ph_3P:O$; R = H, Me (793).

Acetyltriphenylgermane $GeC_{20}H_{18}O$ $Ph_3GeCOMe$
Synth.: By rxn. of $Ph_3GeCMeS(CH_2)_3S$ with $HgCl_2$ and $CdCO_3$ in aq. THF, in 58% yield (792).
By rxn. of $Ph_3GeCHOHMe$ with $CrO_3-H_2SO_4$ in Me_2CO, in 23% yield (2671).
By rxn. of Ph_3GeLi with AcCl at -80°, followed by oxidation (1617).
Prop.: m. 122-23° (792), 121.5-23° (2671), IR (792, 2671), visible (658) and UV (792, 1677, 2671) spectra, x-ray structural analysis (1160).
Rxn. with: $LiAlH_4$ → $Ph_3GeCHOHMe$ (2671).

Additional tetraorganogermanes with keto groups are compiled in Table 13.

Table 13. Tetraorganogermanes Containing Carbonyl Groups R_3GeR'

Formula*	Synth. Method**	Yield	Properties	Ref.
Me_3GeBz	IVA	--	$b_{0.3}$ 53.5-54°, $n_D^{24.5}$ 1.5364, yellow liq. (a, b)	2159
Me_3GeAc	--	--	(a)	--
Me_3GeCH_2Ac	IA	65	b_{100} 85-95°, n_D^{20} 1.4438, d_{20} 1.1372 (c, d)	2547
$Me_3GeCH_2CH_2Ac$	IC	55	b_{29} 78.5-79.5°, n_D^{20} 1.4410, d_{20} 1.0827 (e)	2602
Et_3GeBz	IVC	63	$b_{0.15}$ 82-83°, yellow oil (f)	792
Et_3GeAc	IVC	54	Liq. (f, g)	792
$Et_3GeCH_2CH_2Ac$ (37)	III	31	b_{17} 118°, n_D^{20} 1.4602, d_{20} 1.0652 (h)	1798

* Numbers in parenthesis refer to page in main volume.
** The characters used here correspond to the synthetic scheme in the introduction to Section 2.6.3.3.
(a) Visible spectrum (658). (b) IR and UV spectra, base strength (2159). (c) Equilibrium mixt. $Me_3GeCH_2Ac \rightleftharpoons Me_3GeOCMe:CH_2$; at r.t. and 170° in 4 and 17% of O-isomer, resp., influence of Me_3GeBr on equilibrium (2547). (d) NMR, IR spectra and assignment (2547). (e) NMR spectrum (2373). (f) IR and UV spectra (792). (g) Isolated as 2,4-dinitrophenylhydrazone [$GeC_{14}H_{22}N_4O_4$] m. 83.5-84.5° (792). (h) IR spectrum, 2,4-dinitrophenylhydrazone [$GeC_{16}H_{26}N_4O_4$] m. 71° (1798).

Table 13 Continued R_3GeR'

Formula*	Method**	Yield	Properties	Ref.
$Et_3Ge(CH_2)_3Ac$	IVB	--	b_{26} 139°, n_D^{20} 1.4606, d_{20} 1.0507	1494
$n-Pr_3GeCH_2Ac$	IB	49	b_3 74-76°, n_D^{20} 1.4634, d_{20} 1.0255 (i)	711
$n-Bu_3GeCH_2Ac$	IA	80	b_1 106-107°, n_D^{20} 1.4664, d_{20} 1.0165 (i)	711
Ph_3GeR'				
R' = CH_2Bz	V	81	m. 81-83° (j)	792a
$COCMe_3$	II	62	m. 103-105°, IR and UV spectra	2671
$2-C_6H_9O-c.$	IVD	--	m. 118-25°, IR spectrum (k)	2670
$p-COC_6H_4F$	II	22	Yellow cryst., m. 118-20°, soly. (l)	1624-5
$p-COC_6H_4OMe$	II	--	Yellow cryst., m. 136-38.5° (l)	1624-5
$p-COC_6H_4CF_3$	II	44	Yellow cryst., m. 116-18°, soly. (l)	1624-5

(i) IR and UV spectra (711), rxn. with $LiAlH_4 \rightarrow R_3GeCH_2CHOHMe$ (711). (j) Stable to 140° in xylene (793a). (k) Decomposes on standing. (l) Visible and IR spectra, assignment (1624-5).

2.6.3.4 TETRAORGANOGERMANES CONTAINING CARBOXYL GROUPS AND DERIVATIVES

This chapter summarizes tetraorganogermanes substituted with a carboxylic acid group. Derivatives as salts, acid chlorides, esters, peroxyesters and amides are included. The germanes listed in Table 14 are prepared by methods from the following scheme.

I   Addition of CO_2 to organometallic derivatives:
    (IA) $R''Br + Mg, + CO_2 \rightarrow R''CO_2H$,
    (IB) $R'Br + BuLi + CO_2 \rightarrow R'CO_2H$,
    (IC) $R_3GeM + CO_2 \rightarrow R_3GeCO_2R'$; $R' = H, R_3Ge$.

II Synthesis with organometallic compounds:
    (IIA) $Et_3GeI + R_3SnR' \rightarrow Et_3GeR'$,
    (IIB) $Pr_2GeI_2 + R'_2Hg \rightarrow Pr_2GeR'_2$,
    (IIC) $(R_3Ge)_2S + R'_2Hg \rightarrow Et_3GeR'$,
    (IID) $R_2GeH_2 + R'_2Hg \rightarrow R_2GeR'_2$.

III Addition reactions:
    (IIIA) $Et_3GeH + CH_2:CHR' \rightarrow Et_3GeCH_2CH_2R'$,
    (IIIB) $R_3GeX + CR_2:CO \rightarrow R_3GeCR_2COX$; $X = OR, NR_2$,

(IIIC) $Me_3GeNMe_2 + (:CCO_2Et)_2 \rightarrow Me_3GeC(CO_2Et):C(CO_2Et)NMe_2$.

IV Modification of carboxyl precursors:
   (IVA) $(Ph_3Ge)_2CO + air \rightarrow Ph_3GeCO_2GePh_3$,
   (IVB) $R''CN + KOH, + HCl \rightarrow R''CO_2H$.

V Interconversion of carboxyl groups:
   (VA) $R''CO_2H + SOCl_2 \rightarrow R''COCl$,
   (VB) $R''CO_2H + ROH \rightleftharpoons R''CO_2R + H_2O$,
   (VC) $R''COCl + RO_2H + base \rightarrow R''CO_3R$.

R = organic group, R' = organic group containing a carboxyl function, R'' = tetraorganogermyl group, M = lithium and potassium. Other tetraorganogermanes with carboxylic substituents are listed in Chapters 1.3, 2.4.3, 2.5.3, 2.6.3.1, 2.7.1.1.3, 2.7.8, 7.1.6.1 and 7.1.6.6.

Methyl Triethylgermylacetate $GeC_9H_{20}O_2$ $Et_3GeCH_2CO_2Me$

Synth.: By rxn. of Et_3GeX with $Bu_3SnCH_2CO_2Me$; X = Cl and I, in 27 and 85% yield, resp. (692, 1440).
By rxn. of Et_3GeI with $(MeO_2CH_2)_2Hg$ in C_6H_6, in 86% yield (24b).
By rxn. of Et_3GeOMe with ketene in presence of HgI_2 and MeOH, in 92% yield (1440).

Prop.: b_9 89-92° (24b, 1440), b_7 80-82° (692, 1440), n_D^{20} 1.4570 (24b, 1440), 1.4570-83 (692, 1440) d_{20} 1.1070 (692, 1440), 1.1052 (24b, 1440), IR (24b, 692, 1440) and NMR (1689) spectra.

Rxn. with: $LiAlH_4 \rightarrow Et_3GeCH_2CH_2OH$ (24b).

Additional tetraorganogermanes with carboxyl functions may be found in Table 14.

Table 14. Tetraorganogermanes Containing Carboxyl Functions $R_nGeR'_{4-n}$

Formula*	Synth. Method**	Yield	Properties	Ref.
$Me_3GeCH_2CONEt_2$	IIIB	70	b_{13} 115-16°, n_D^{20} 1.4679, d_{20} 1.0837 (a, b)	2513
$Me_3GeC(CO_2Et){:}C(CO_2Et)NMe_2$	IIIC	78	$b_{0.02}$ 80°	854
	IIIC	78	Green liq., n_D^{25} 1.4840, IR and NMR spectra	2406
$Me_2PhGeCO_2GeMe_3$	(c)	(d)	White solid, m. 112-13° (e)	2882-3
$(PhCH_2)_3GeCO_2Ge(CH_2Ph)_3$	IC	--	m. 77-80°	104

Et_3GeR'

$R' = CO_2H$ (39)	IC	30	n_D^{20} 1.4132, NMR spectrum (f)	815
CH_2CO_2Bu	IIIB	81	b_1 69-70°, n_D^{20} 1.4578, d_{20} 1.0586	1440
	IIA	74	b_1 70-72°, n_D^{20} 1.4580, d_{20} 1.0579	1440
CH_2CONMe_2	IIIB	61	b_7 126-27°, n_D^{20} 1.4803, d_{20} 1.1017 (b)	2513
	IIIB	84	$b_{0.8}$ 112°, n_D^{20} 1.4811, d_{20} 1.1050 (g)	2736
CH_2CONEt_2	IIIB	63	b_1 97-98°, n_D^{20} 1.4793, d_{20} 1.0629 (b)	2513
CPh_2CONMe_2	IIIB	75	(g, h)	2736
$CH_2CH_2CO_2H$ (39)	VB (i)	--	b_3 125°, n_D^{20} 1.4735, d_{20} 1.1433 (j, k)	364
	(l)	20-31		2095
$CH_2CH_2CO_2Na$	--	--	Spectrographic determination	1358
$CH_2CH_2CO_2K$	--	--	Spectrographic determination	1358
$CH_2CH_2CO_2Et$ (39)	--	--	Relative rate of hydrolysis with KOH	987
$CH_2CH_2CO_3CMe_3$	VC (m)	85	Pale yellow liq., n_D^{20} 1.4550, d_{20} 1.0604 (l, n)	2095
$CH_2CH_2CO_2SiMe_3$	IIIA	58	b_3 93-94°, n_D^{20} 1.4560, d_{20} 1.0461 (i)	364
CH_2CH_2COCl	VA (k)	100	b_3 75-76°, n_D^{20} 1.4725, d_{20} 1.1670 (m)	2095
$(CH_2)_4CO_2H$ (39)	IA	62	$b_{0.8}$ 130°, n_D^{20} 1.4685, d_{20} 1.0980	343
$(CH_2)_5CO_2H$	IVB	89	$b_{0.9}$ 151°, n_D^{20} 1.4688, d_{20} 1.0840	343
$Et_2Ge[(CH_2)_4CO_2H]_2$	IA	64	$b_{0.8}$ 210°, n_D^{20} 1.4832, d_{20} 1.1660	343
	IVB	61	$b_{0.02}$ 163-65°, n_D^{20} 1.4830 (o, p)	343, 1430
	IVB	--	$b_{0.07}$ 170-72°, IR spectrum	1493

$Et_2Ge[(CH_2)_4CO_2Na]_2$	VB	--		1493 (q)
$\{Et_2Ge[(CH_2)_4CO_2]_2\}_2Th$	(q)	--		1493 (r)
$Et_2Ge[(CH_2)_4CO_2Et]_2$	VB	95	$b_{0.8}$ 163°, n_D^{20} 1.4615, d_{20} 1.0660 (s)	1493
R_3GeCO_2R'			$R_3Ge = EtPh(1-C_{10}H_7)Ge$	
$R' = H$ R_3Ge	IC --	-- --	Stereochemistry (t) Stereochemistry (u)	2369 2368
$p-Et(i-Pr)PhGeC_6H_4CO_2H$	.IB	47	m. 115.5-17°, stereochemistry (v)	1003
$n-Pr_3GeCH_2CO_2Me$ (40)	IIC	47	$b_{1.5}$ 79-82°, n_D^{20} 1.4582 d_{20} 1.0542 (w)	2648
$n-Pr_2Ge(CH_2CO_2Me)_2$	IIB	73	b_3 122-24°, n_D^{20} 1.4668, d_{20} 1.1634	25
$n-Bu_3GeCH_2CO_2Me$ (40)	--	--	NMR spectrum (b, 24b)	1689

* Numbers in parenthesis refer to pages in main volume.

**The characters used here correspond to the synthetic scheme in the introduction to Section 2.6.3.4.

(a) IR and NMR spectra (2513). (b) Rxn. with $LiAlH_4 \rightarrow R_3GeCH_2CH_2NR_2$. (c) Synth. by thermal dec. of 2,3,1,4,5,6-dicarbomethoxytetraphenyl-7,7-dimethyl-7-germanorbornadiene in dimethyl acetylenedicarboxylate (2882-3). (d) By-prod. (e) IR, NMR and mass spectra, structure $Me_3GeO_2CGeMe_2Ph$ prefered over $Me_2PhGeO_2CGeMe_3$ from NMR spectrum (2883-3). (f) Thermal dec. at 100° $\rightarrow (Et_3Ge)_2O + CO$ (815), rxn. with $HCl \rightarrow Et_3GeCl + CO + H_2O$ (815). (g) IR and NMR spectra (2736). (h) Rxn. with aq. $NH_3 \rightarrow (Et_3Ge)_2O + Ph_2CHCONMe_2$ (2736), rxn. prod. contains $Et_3GeOC(:CPh_2)NMe_2$ (2736). (i) Rxn. of $Et_3GeCH_2CH_2CO_2SiMe_3 + H_2O \rightarrow Et_3GeCH_2CH_2CO_2H$ (2095). (j) Dissociation const. (364). (k) Rxn. of $Et_3GeCH_2CH_2CO_2H + SOCl_2 \rightarrow Et_3GeCH_2CH_2COCl$ (2095). (l) Thermal dec. of $Et_3GeCH_2CO_3CMe_3$ at 100-105° $\rightarrow Et_3GeCH_2CH_2CO_2H + Et_3GeOCMe_3 + CO_2 + CH_4 + C_2H_4 + Me_2CO + Me_3COH$ (2098). (m) Rxn. of $Et_3Ge-CH_2CH_2COCl + Me_3CO_2H + KOH \rightarrow Et_3GeCH_2CH_2CO_3CMe_3$ (2095). (n) IR spectrum, dec. on distn. (2095). (o) Rxn. with $Et_2Ge[(CH_2)_5NH_2]_2 \rightarrow$ polyamide polymer (343, 1490). (p) Rxn. of $Et_2Ge[(CH_2)_4CO_2H]_2$ with $EtOH \rightarrow Et_2Ge[(CH_2)_4CO_2Et]_2$ (1493). (q) Rxn. of $Et_2Ge[(CH_2)_4CO_2Na]_2$ with $Th(NO_3)_4 \rightarrow \{Et_2Ge[(CH_2)_4CO_2]_2\}_2Th$ (1493). (r) Pyrolysis at 400° $\rightarrow Et_2Ge(CH_2)_4CO(CH_2)_3CH_2$ (1493). (s) Rxn. with $Na \rightarrow Et_2Ge(CH_2)_4COCHOH(CH_2)_3CH_2$ (1493). (t) Rxn. with $BuLi \rightarrow R_3GeLi$ (2369). (u) Rxn. with $LiAlH_4 \rightarrow R_3GeH + R_3GeCH_2OH$ (2368). Quinine salt, m. 147.5-48°, diastereoisomer $[GeC_{38}H_{46}N_2O_4]$, stereochemistry (1003). (w) Rxn. with $RMgBr \rightarrow Pr_3GeCR_2OH$ (24b).

Table 14 Continued

$R_nGeR'_{4-n}$

Formula*	Synth. Method**	Yield	Properties	Ref.
$Bu_2Ge(CH_2CO_2Me)_2$	IIIA	52	b_2 124-28°, n_D^{20} 1.4640, d_{20} 1.1201 (x)	25
	IID (y)	78	b_2 123-26°, n_D^{20} 1.4650, d_{20} 1.1175	25
Ph_3GeCO_2H (41)	--	--	m. 187-90°, ionization const.	2800
$Ph_3GeCO_2GePh_3$ (41)	IVA	19	White solid, IR and UV spectra (z)	794
$(p-MeC_6H_4)_3GeCO_2H$ (41)	--	--	IR (105) and mass (1109) spectra	--

(x) Rxn. $LiAlH_4 \rightarrow R_2Ge(CH_2CH_2OH)_2$ (25). (y) Synth. from $Bu_2R'GeH + R'_2Hg$ in THF (25). (z) Hydrolyzes on silica gel $\rightarrow Ph_3GeOH$ (794).

2.6.4 TETRAORGANOGERMANES CONTAINING SULFUR

Trimethylgermylmethylphenyl Sulfoxide $\quad\quad\quad\quad\quad\quad\quad\quad\quad$ Me$_3$GeCH$_2$S:OPh

<u>Synth.</u>: By rxn. of Me$_3$GeCH$_2$MgCl with PhSO$_2$Me in THF at 0°, in 78% yield (2281).
<u>Prop.</u>: b$_{0.01}$ 120°, n$_D^{22}$ 1.5528, IR and NMR spectra (2281).
<u>Rxn. with</u>: H$_2$O at 60° → (Me$_3$Ge)$_2$O + MeSOPh (2281).

Additional sulfur containing tetraorganogermanes are reported in Chapters 2.3, 2.5.3, 2.6.2 2.7.1.1.3, 2.7.2, 2.7.5 and 7.1.6.6.

2.6.5 TETRAORGANOGERMANES CONTAINING NITROGEN AND PHOSPHORUS

Organogermanes containing amido, imino and imido groups, phosphonates and phosphonyl chlorides are reported in the literature. All compounds listed in Table 15 are prepared by methods from the following scheme.

   I Synthesis with organolithium compounds:
      Et$_3$GeBr + R'Li

  II Hydrogermylation reaction:
      Et$_3$GeH + CH$_2$:CHR' + cat. H$_2$PtCl$_6$ → Et$_3$GeCH$_2$CH$_2$R' .

 III Phosphorylation reaction:
      (IIIA) Et$_4$Ge + PCl$_3$ + air → Et$_3$GeC$_2$H$_4$POCl$_2$,
      (IIIB) R"Cl + PX$_3$ + air → R"POX$_2$; X = Cl, EtO, BuO .

 IV Interconversion reactions:
      (IVA) R"CONR$_2$ + LiAlH$_4$ → R"CH$_2$NR$_2$,
      (IVB) R"CN + H + cat. Raney Ni → R"CH$_2$NH$_2$,
      (IVC) R"N(SiMe$_3$)$_2$ + base → R"NH$_2$,
      (IVD) R"POCl$_2$ + EtOH → R"PO$_3$Et$_2$.

R = organic group, R' = organic group containing nitrogen or phosphorous, R" = tetraorganogermyl group. Other tetraorganogermanes containing nitrogen and phosphorous are reported in Chapters 2.3, 2.4.3, 2.5.3, 2.6.2, 2.6.3.3, 2.6.3.4, 2.7.1.1.3, 2.7.2 and 2.7.5.

Table 15. Tetraorganogermanes Containing Nitrogen or Phosphorus $R_nGeR'_{4-n}$

Formula*	Synth. Method**	Yield	Properties	Ref.
$Me_3GeCH_2NEt_2$ (44)	--	--	Basicity (1715), NMR spectrum (2373)	--
$Me_3GeCH_2CH_2NEt_2$	IVA	60	b_{18} 64-65°, n_D^{20} 1.4450, d_{20} 0.9731	2513
Et_3GeR'				
$R' = CH_2NEt_2$	--	--	Basicity (1350)	--
$CH_2CH_2NMe_2$	IVA	62	b_{23} 106-107°, n_D^{20} 1.4592, d_{20} 0.9889 (a)	2513
$CH_2CH_2NEt_2$	IVA	58	b_3 101-102°, n_D^{20} 1.4620, d_{20} 0.9773	2513
$(CH_2)_3NH_2$	II	80	b_{12} 95.5-96.5°, n_D^{20} 1.4690, d_{20} 1.0364 (b)	364
$(CH_2)_4NH_2$	IVB	80	b_9 112°, n_D^{20} 1.4690, d_{20} 1.0105	343
$(CH_2)_5NH_2$	IVB	88	b_{20} 148°, n_D^{20} 1.4692, d_{20} 0.9985	343
$p-C_6H_4NH_2$	IVC (c)	97	$b_{0.1}$ 94°, n_D^{25} 1.5487	2109
$p-C_6H_4N(SiMe_3)_2$	I	86	$b_{0.35}$ 111°, n_D^{25} 1.4996 (c)	2109
$Et_2Ge[(CH_2)_5NH_2]_2$	IVB	84	$b_{0.3}$ 127°, n_D^{20} 1.4835, d_{20} 1.0153 (d, e)	343, 1490
$\gamma-Ph_3GeC_7H_4(:NEt)NHEt$ (f)	I	--	m. 186°, NMR spectrum (g)	1004
$(\gamma-Ph_3GeC_7H_4NEtNEt)_2Ni$ (f)	(g)	--	m. 304°, NMR spectrum	1004
Et_3GeR'				
$R' = C_2H_4POCl_2$	IIIA	83	$b_{1.5}$ 118-19°, n_D^{20} 1.5021, d_{20} 1.3564 (h, i)	995
$CH_2CH_2POCl_2$	II	17	b_4 136°, n_D^{20} 1.4962, d_{20} 1.2947 (i, j)	2366
$C_2H_4PO_3Et_2$	IVD (1)	38	$b_{3.5}$ 106-107°, n_D^{20} 1.4550, d_{20} 1.1074 (h)	995
$CH_2CH_2PO_3Et$	II	51	b_3 112°, n_D^{20} 1.4572, d_{20} 1.1088	2366
	IVD (1)	38	b_4 120°, n_D^{20} 1.4554, d_{20} 1.1050 (j)	2366
$(CH_2)_3POCl_2$	II	52	b_{20} 77°, n_D^{20} 1.4610, d_{20} 1.1952	2366
$(CH_2)_3PO_3Et_2$	II	60	b_{10} 166°, n_D^{20} 1.4580, d_{20} 1.0950	2366
	IIIB	73	b_2 150°, n_D^{20} 1.4560, d_{20} 1.0976 (j)	2366
$(CH_2)_3PO_3Bu_2$	II	37	b_3 166°, n_D^{20} 1.4585, d_{20} 1.0591	2366
	IIIB	35	b_2 157°, n_D^{20} 1.4577, d_{20} 1.0510 (j)	2366

2.7 POLYMETALLIC TETRAORGANOGERMANES

2.7.1 BRIDGED POLYORGANOGERMANES

2.7.1.1 TETRAORGANOGERMANES WITH TWO GERMANIUM ATOMS

2.7.1.1.1 NONFUNCTIONAL DERIVATIVES

Several nonfunctional tetraorganogermanes containing two germanium atoms linked by carbon bridges are reported in the literature. The compounds listed in Table 16 are prepared by methods in the following scheme.

I Grignard reactions:
(IA) $Cl_3GeYGeCl_3 + RMgX \rightarrow (R_3Ge)_2Y$,
(IB) $Me_3GeYBr + Me_3GeY'Br + Mg \rightarrow Me_3Ge\text{-}YY'GeMe_3$,
(IC) $Me_3GeCl + Y(MgX)_2 \rightarrow (Me_3Ge)_2Y$.

II Wurtz reaction:
$Me_3GeYCl + Me_3GeCl + Na \rightarrow (Me_3Ge)_2Y$.

III Synthesis with alkali metal derivatives:
(IIIA) $Et_3GeK + CH_2Cl_2 \rightarrow (Et_3Ge)_2CH_2$,
(IIIB) $R_3GeBr + YLi_2 \rightarrow (R_3Ge)_2Y$,
(IIIC) $Et_3GeYLi + Et_3GeBr \rightarrow (Et_3Ge)_2Y$,
(IIID) $R_3GeCl + HC\!:\!CNa \rightarrow (R_3Ge C\!:\!)_2$,
(IIIE) $GeCl_4 + HC\!:\!CNa \rightarrow [(HC\!:\!C)_3GeC\!:\!]_2$.

IV Synthesis with organotin derivatives:
$Ph_3GeCl + (Bu_3SnC\!:\!)_2 \rightarrow (Ph_3GeC\!:\!)_2$.

V Hydrogermylation:
$R_3GeH + R_3GeYC\!:\!CH \rightarrow R_3GeYCH\!:\!CHGeR_3$.

VI Miscellaneous reactions:
(VIA) $Et_4Ge + BzO_2CMe_3 \rightarrow (Et_3GeC_2H_4)_2$,
(VIB) $(Ph_3Ge)_2COHMe + PBr_3 \rightarrow (Ph_3Ge)_2C\!:\!CH_2$.

R = organic group, X = halogen, Y = hydrocarbon bridge. Cyclic organogermanes are reported in Chapter 7.1.6.6.

---

* Numbers in parenthesis refer to pages in the main volume.

** The characters used here correspond to the synthetic scheme in the introduction to Chapter 2.6.5. (a) Determination of Ge by combustion and photometry (2654). (b) NMR spectrum (2654). (c) Rxn. of p-Et_3Ge-$C_6H_4N(SiMe_3)_2$ with aq. NaOH in MeOH → p-$Et_3GeC_6H_4NH_2$ (2109). (d) IR spectrum (343, 1490). (e) Rxn. with $Et_2M[(CH_2)_4CO_2H]_2$ → polyamide polymer; Me = Ge (343, 1490), Si (343). (f) Y,N,N'-diethylaminotroponeimine. (g) Rxn. of $Ph_3GeC_7H_4(:NEt)NHEt_2$ with Ni(OAc)$_2$ in aq. EtOH-C_6H_6 → $(Ph_3GeC_7H_4\overline{NEtNEt})_2Ni$ (1004). (h) Mixture of 1- and 2-ethanephosphonyl chloride or the corresponding ethyl ester (2366). (i) Rxn. of $Et_3GeYPOCl_2$ with EtOH ↑ $Et_3GeYPO_3Et_2$, Y = C_2H_4 (995), CH_2CH_2 (2366). (j) IR spectrum (2366).

Bis(trimethylgermyl)ethylene (M-46) $Ge_2C_8H_{20}$ $(Me_3GeCH:)_2$

Synth.: By rxn. of C_2H_2 with $HGeBr_3$, followed by MeMgCl, in 70% yield, not sepd. from by-prod. $(Me_3GeCH_2)_2$ (161, 365).
By rxn. of $HGeCl_3$ with C_2H_2 at r.t., followed by MeMgBr, in 50-60% yield (382a).
By rxn. of $(Cl_3GeCH:)_2$ with MeMgBr, in 60% yield (1529).
Prop.: b_{35} 79-79.5°, n_D^{20} 1.4628, d_{20} 1.1480, dipole moment, GLC (1529), IR (1529), Raman (2079), UV (1529, 1693, 2079) and NMR (161, 365, 1529, 2079) spectra.

1,3-Bis(trimethylgermyl)-1,2,3,4- $Ge_2C_{16}H_{28}$ 1,3-$(Me_3Ge)_2C_{10}H_{10}$
   tetrahydronaphthalene

Synth.: By rxn. of $HGeCl_3$ with $C_{10}H_8$ at 110-30°, followed by MeMgBr, in 53% yield (278), in 55% yield by GLC (1336).
Prop.: $b_{0.6}$ 110-13° (1336), $b_{0.2}$ 100-102° (278), n_D^{20} 1.5379 (278), 1.5372, d_{20} 1.2096 (1336), IR (1336) and NMR (278, 1336) spectra, GLC, TLC (1336), mixt. of 2 main isomers 88:12 (1336).
Rxn. with: Se → $C_{10}H_8$ (278, 1336).
Chloranil → $C_{10}H_8$ (1336).
Pt-carbon → no rxn. up to 300-310° (1336).

Other nonfunctional bis(organogermanes) are listed in Table 16.

Table 16. Tetraorganogermanes Containing Two Carbon-Chain $R_3GeYGeR_3$
           Bridged Germanium Atoms

Formula*	Synth. Method**	Yield	Properties	Ref.
$Me_3GeYGeMe_3$				
Y = CH_2 (46)	IA	83	b_{748} 154°, n_D^{20} 1.4502	363
CH_2CH_2 (46)	IA (a)	30	NMR spectrum, not isolated	161, 365
	IA (b)	50-60		382a
C:CH_2 (46)	II	--	b_{28} 72.3°, IR and Raman spectra (c)	1533
C⋮C (46)	IC	Low	m. 35°, b° 163-64°	367
	II	(d)	GLC, Raman spectrum	1533

* Numbers in parenthesis refer to pages in main volume.
** The characters used here correspond to the synthetic scheme in the introduction to Subsection 2.7.1.1.1.
(a) By-prod. in rxn. of C_2H_2 with $HGeBr_3$ in Et_2O, followed by MeMgCl (161, 365). (b) By-prod. in rxn. of C_2H_4 with $HGeCl_3·2Et_2O$, followed by MeMgBr (382a). (c) UV spectrum (1693). (d) By-prod. in rxn. of Me_3GeCl + Me_3Ge-CH:CHCl with Na in Et_2O (1533).

Table 16 Continued $R_3GeYGeR_3$

Formula*	Synth. Method**	Yield	Properties	Ref.
Me$_3$GeYGeMe$_3$ (Cont'd.)				
Y = CH$_2$CH:CH	IA (e)	--	b$_{16}$ 75.5-76.5, n$_D^{20}$ 1.4680, d$_{20}$ 1.1480	364
	IA (f)	Good	IR spectrum (g)	364
(CH$_2$CH:)$_2$	IA	83	b$_{12}$ 85°, n$_D^{20}$ 1.4713, d$_{20}$ 1.1190 (h)	308
(C:CH$_2$)$_2$	IA	76	b$_{17}$ 85-86°, n$_D^{20}$ 1.4797, d$_{20}$ 1.1480	2403
	IA (i)	62	b$_{15}$ 80-86°, n$_D^{20}$ 1.4813 (j, k)	2403
p-C$_6$H$_4$ (46)	IIIB	50	m. 91-92°, soly., polarography (l, m)	2223
(p-C$_6$H$_4$)$_2$	IIIB	80	m. 112°, polarography (l, n, o)	112
(p-C$_6$H$_4$)$_3$	IB	--	m. 202-205°, polarography (l)	2223
(p-C$_6$H$_4$)$_4$	IB	--	m. 278-80°, polarography (l)	2223
1,4-C$_{10}$H$_6$	IIIB	--	m. 104-105°, polarography (l)	2223
1-MeC$_{10}$H$_9$	IA (p)	--	b$_{0.2}$ 100-104°	278
	IA (p)	62		1336
2-MeC$_{10}$H$_9$	IA	--		1336
Et$_3$GeYGeEt$_3$				
Y = CH$_2$	IIIA	65	b$_{14}$ 138-39°, n$_D^{20}$ 1.4822	815
CH$_2$CH$_2$ (47)	IIIC	69	b$_1$ 110-12°, n$_D^{20}$ 1.4780	2849
CH:CPh	IIIC	34	b$_1$ 91-93°, n$_D^{20}$ 1.5120	2849
CHPhCHPh	IIIC	29	m. 81-82°	2849
CPh:CPh	IIIC	85	m. 64°	2849
(C$_2$H$_4$)$_2$	VIA	11	b$_5$ 152-53°, n$_D^{20}$ 1.4878	2098
(C:CH$_2$)$_2$	IA (q)	--	b$_4$ 125-25.5°, n$_D^{20}$ 1.5015, d$_{20}$ 1.1929 (r)	2403

(e) By-prod. in rxn. of Cl$_3$GeCH$_2$CH$_2$CH(GeCl$_3$)$_2$ with MeMgBr (364). (f) Synth. by rxn. of Cl$_3$GeCH$_2$C⋮CH with MeMgBr in Et$_2$O (364). (g) Detn. by fusion and volumetric titration (265). (h) IR and Raman spectra (308). (i) Synth. from Cl$_3$GeC(CH$_2$Cl):C(CH$_2$Cl)GeCl$_3$ with MeMgCl (2403). (j) IR, UV and NMR spectra, dipole moment (2403). (k) Catalytic hydrogenation → uptake of 2 mols H (2403). (l) UV and ESR spectra of anion formed with alkali metal in THF (2223). (m) Oscillator strength (2593). (n) NMR spectrum (930). (o) ESR spectrum of ion formed with K in THF (112, 930). (p) Synth. by rxn. of n-MeC$_{10}$H$_7$ with HGeCl$_3$, followed by MeMgBr, n = 1, 2 (278, 1336). (q) Synth. from mixt. of Cl$_3$GeC(CH$_2$Cl)C(CH$_2$Cl)GeCl$_3$ and (CH$_2$:CGeCl$_3$)$_2$ with EtMgBr (2403). (r) IR and Raman spectra (2403).

Table 16 Continued $R_3GeYGeR_3$

Formula*	Synth. Method**	Yield	Properties	Ref.
$Et_3GeYGeEt_3$ (Cont'd.)				
Y = $(CH_2)_3C\vdots C$	--	--	$b_{0.8}$ 140°, n_D^{20} 1.4815, d_{20} 1.0780	1494
$(CH_2)_4CH\vdots CH$	V	--	$b_{0.8}$ 148°, n_D^{20} 1.4818, d_{20} 1.0614	1494
$[(HC\vdots C)_3GeC\vdots]_2$	IIIE	10	Pink cryst., m. 143°, mol. wt., (s)	939
$(Bu_3GeC\vdots)_2$	V (t)	--	b_2 150-53°, n_D^{20} 1.4810 (u)	1598
$[Bu_2(HC\vdots C)GeC\vdots]_2$	IIID (v)	--	$b_{0.5}$ 130°, mol. wt.	939
$(Ph_3Ge)_2C\vdots CH_2$	VIB	62	m. 145-46°, IR and NMR spectra (w)	1624, 1626
$(Ph_3GeC\vdots)_2$ (48)	IIID	55	m. 148-49°, mol. wt.	939
	IV	79	m. 127-28°	2264

(s) IR, Raman and mass spectra (939). (t) By-prod. in rxn. of Bu_3GeH with $Et_3SiC\vdots CH$ (1598). (u) IR and NMR spectra (1598). (v) By-prod. in rxn. of Bu_2GeCl_2 with NaC_2H in THF in Ar atm. (939). (w) Rxn. with $O_3 \rightarrow (Ph_3Ge)_2O$ (1624, 1626).

2.7.1.1.3 FUNCTIONALLY SUBSTITUTED BIS(TETRAORGANOGERMANES)

Functionally substituted tetraorganogermanes containing two germanium atoms are summarized in this chapter. The compounds listed in Table 17 are prepared by methods from the following scheme.

I Grignard reaction:
  $Y(GeCl_3)_2 + RMgX \rightarrow (R_3Ge)_2Y$.

II Synthesis with alkali metal derivatives:
  (IIA) $Ph_3GeLi + RCOCl \rightarrow (Ph_3Ge)_2CROH$,
  (IIB) $Et_3GeYLi + Et_3GeX \rightarrow (Et_3Ge)_2Y$,
  (IIC) $Me_2\overline{GeSGeMe_2CH_2SCH_2} + MeLi \rightarrow (Me_3GeCH_2)_2S$.

III Synthesis with other metal derivatives:
  (IIIA) $(Et_3Ge)_2Cd + CCl_4 \rightarrow (Et_3Ge)_2CCl_2$,
  (IIIB) $(Et_3Ge)_2Hg + RC\vdots CCO_2R \rightarrow Et_3GeCR\vdots C(CO_2R)GeEt_3$.

IV Hydrogermylation reactions:
  (IVA) $Et_3GeH + CH\vdots CYCH\vdots CH_2 \rightarrow Et_3GeCH\vdots CHYCH_2CH_2GeEt_3$,
  (IVB) $Et_3GeH + Bu_2\overline{GeCH_2CH_2CH_2} \rightarrow Et_3Ge(CH_2)_3GeBu_2H$,
  (IVC) $HGeCl_3 + Me_3GeC\vdots CH \rightarrow Me_3GeCH\vdots CHGeCl_3$.

V Interconversion reactions:
   (VA) $(Ph_3Ge)_2CO + LiAlH_4 \rightarrow (Ph_3Ge)_2CHOH$,
   (VB) $(Et_3Ge)_2\overline{CS(CH_2)_3S} + HgCl_2 + CdCO_3 \rightarrow (Et_3Ge)_2CO$,
   (VC) $Me_2\overline{GeCH_2GeMe_2CH_2} + XCl \rightarrow Me_2(XCH_2)GeCH_2GeMe_2Cl$; X = H, Cl.

IV Ylide reactions:
   $Me_3GeCl + Ph_3P:CH_2 \rightarrow (Me_3Ge)_2C:PPh_3$.

R = organic group, X = halogen, Y = hydrocarbon bridge. Other functionally substituted bis(organogermanes) are reported in Chapters 2.4.3, 2.6.3.1 and 7.1.6.6.

Bis(triphenylgermyl)ketone $Ge_2C_{37}H_{30}O$ $(Ph_3Ge)_2CO$
Synth.: By rxn. of $(Ph_3Ge)_2CHOH$ with $(C_6H_{11}N:)_2C$ and cat. $CF_3CO_2H \cdot Py$ in Me_2SO, in 77% yield (794).
Prop.: m. 152-54°, orange-pink solid, IR and UV spectra, dec. in soln., stable in solid form (794).
Rxn. with: Thermal dec. $\rightarrow (Ph_3Ge)_2O + Ph_6Ge_2 + CO$ (794).
Air $\rightarrow Ph_3GeCO_2GePh_3$ (794).
$LiAlH_4 \rightarrow (Ph_3Ge)_2CHOH$ (794).
UV irradiation $\rightarrow Ph_6Ge_2 + Ph_3GeCO_2GePh_3$ (only in presence of air) (794).
UV irradiation in $HCCl_3$ or $CCl_4 \rightarrow Ph_3GeCl$ (794).

Additional functionally substituted bistetraorganogermanes are listed in Table 17.

2.7.1.2 TETRAORGANOGERMANES CONTAINING THREE TO FOUR GERMANIUM ATOMS

Polymetallic tetraorganogermanes with three or four germanium atoms linked by carbon bridges are listed in Table 18 on page 76. Additional tetraorganogermanes with three and four germanium atoms are listed in Chapters 5.1.2 and 7.1.6.6.

Table 17. Functionally Substituted Tetraorganogermanes Containing Two Germanium Atoms $R_3GeYGeR'_nR_{3-n}$

Formula*	Synth. Method**	Yield	Properties	Ref.
$Me_3GeCH_2GeMe_2Cl$	VC	--	IR and NMR spectra	2244
$Me_3GeCH_2GeMe_2GeMe_3$	--	--	See Table 73	--
$Me_2(ClCH_2)GeCH_2GeMe_2Cl$	VC	--	IR and NMR spectra	2244
$(Me_3Ge)_2C:PPh_3$	VI	51	Yellow, m. 145-47°, $b_{0.001}$ 158-61° (a)	1826
$(Me_3GeCH_2)_2S$	IIC	--	$b_{0.1}$ 80°, NMR spectrum	2134
$Me_3GeCH:CHGeCl_3$ (46)	IVC	--	b_{12} 100-103° (b)	1533
$(Me_3GeCHMe)_2O$	IA	57	b_8 70°, n_D^{20} 1.4458, d_{20} 1.1166 (c)	1526
$(Me_3Ge)_2C_4H_8S$	IA (d)	75	$b_{0.4}$ 72-75°, n_D^{20} 1.5093, d_{20} 1.2483 (e)	1336
$Et_3GeYGeEt_3$				
$Y = CCl_2$	IIIA	55	b_2 140-42°, n_D^{20} 1.5130	2088
CO	VB (f)	--	Pink soln. IR and UV (794) spectra	792
$CS(CH_2)_3S$	IIB	17	$b_{0.15}$ 144°, n_D^{22} 1.5447 (f)	792
$CH:C(CO_2Me)$	IIIB	55	$b_{0.7}$ 120°	1365
$C(CO_2Et):C(CO_2Et)$	IIIB	--	$b_{0.001}$ 129°	1365
$CH:CHCH_2OCH_2O(CH_2)_3$	IVA	10	b_5 137-38°, n_D^{20} 1.4720, d_{20} 1.0636 (g)	2559
$CH:CHCH_2OCHMeO(CH_2)_3$	IVA	10	b_4 143-44°, n_D^{20} 1.4719, d_{20} 1.0645 (g)	2559
$CR:CHCH:CR; R = CMe_2OH$	IVA (h)	--	b_2 163-65°, n_D^{20} 1.4893, d_{20} 1.0921	198
$Et_3Ge(CH_2)_3GeBu_2H$	IVB	--	$b_{0.9}$ 143°, n_D^{20} 1.4760, d_{20} 1.0672 (i)	1491
$Ph_3GeYGePh_3$				
$Y = CHOH$	VA	71	m. 138-40°, IR and NMR spectra	794
	IIA (j)	68	(k)	794
CPhOH	IIA	49	Light yellow cryst., m. 183-85°	1624-5
	IIA (l)	57	IR spectrum and assignment	1624-5
$C(C_6H_4F-p)OH$	IIA	60	m. 187-90°, solv. (m)	1624-5

CMeOH	IIA (n)	85	m. 208-10°, dec. on standing (m, o)	1624-5
	IIA (p)	61	m. 195-200°	2671
CEtOH	IIA (n)	37	m. 178-81°	1624-5

---

* Numbers in parenthesis refer to pages in main volume.

**The characters used here correspond to the synthetic scheme in the introduction to Subsection 2.7.1.1.3.
(a) Spectra: (far) IR and assignment, NMR, mol. wt. (1826). (b) Rxn. with MeMgCl → (Me₃GeCH:)₂ (1533). (c) GLC and NMR spectrum (1526). (d) Synth. by rxn. of HGeCl₃ with thiophene at -70 to -85°, followed by MeMgBr (1336). (e) Mol. wt., IR and NMR spectra, isomer ratio by GLC 1:2.9:1.2 (1336). (f) Rxn. of (Et₃Ge)₂CS(CH₂)₃S with HgCl₂ and CdCO₃ in aq. THF-MeOH → (Et₃Ge)₂CO (792). (g) IR spectrum (2559). (h) Synth. from Et₃GeH and (Me₂COHC:C)₂ in presence of cat. H₂PtCl₆ in N atm. (198). (i) NMR spectrum (1491). (j) Synth. from Ph₃GeLi and HCO₂Et in THF, followed by LiAlH₄ (794). (k) Rxn. with (cyclo-C₆H₁₁N:)₂C and CF₃CO₂H·Py → (Ph₃Ge)₂CO (794). (l) Synth. from Ph₃GeLi with Ph₃GeCOPh in THF in N atm. (1624-5). (m) IR spectrum and assignment (1624-5). (n) Synth. at -23° to -78° (1624-5). (o) Rxn. with PBr₃ → (Ph₃Ge)₂C:CH₂ (1624-5). (p) Synth. at -80° to r.t. (2671).

Table 18. Tetraorganogermanes Containing Three or Four $R_3GeY'(GeR_3)_n$
Germanium Atoms

Formula	Synth. Method*	Yield	Properties	Ref.
$(Me_3Ge)_3COH$	(a)	--	Mass spectrum	832
$Me_3GeCH_2CH(GeCl_3)_2$	IVC	11	m. 30-50°, b_7 138-39° (b, c)	1533
$Me_3GeCH_2CH_2CH(GeMe_3)_2$	I	--	b_7 104-105°, n_D^{20} 1.4940, d_{20} 1.2368 (d)	364
$Me_3GeCH(CH_2GeMe_3)_2$	--	--	Analytical detn.	265
$1,3,5-(Me_3Ge)_3C_6H_8OMe$	I (e)	32	m. 66°, $b_{0.6}$ 105-107°	278
	I (e)	32	Mol. wt., IR and NMR spectra (f)	1336
$1,3,5-(Me_3Ge)C_6H_8OEt$	I (e)	80	$b_{0.4}$ 92-103°	278
	I (e)	80	IR and NMR spectra	1336
$(Me_3Ge)_4C_{10}H_{12}$ (?)	I (e)	(f)	Isomeric mixture	1336

*The characters used here correspond to the synthetic scheme in the introduction to Subsection 2.7.1.1.3.
(a) Intermediate in oxidation of $Me_3GeC_5H_5Mo(CO)_3$ (832). (b) n_D^{20} 1.5570, d_{20} 1.8441 (1533). (c) Rxn. with MeMgCl → $Me_3GeCH_2CH(GeMe_3)_2$ (1533). (d) IR spectrum (364). (e) Synth. by rxn. of $HGeCl_3$ with anisole, phenetole or naphthalene, followed by MeMgBr (278, 1336). (f) Total yield of all isomers formed in this reaction is 65% by GLC (1336).

2.7.1.8 POLYMERIC TETRAORGANOGERMANES

Oligomeric and polymeric tetraorganogermanes contain more than four germanium atoms linked by carbon bridges. The derivatives listed in Table 19 are prepared by methods described in the following scheme.

I Grignard reactions:
  (IA) $(YGeCl_2)_x + RMgX → (R_2GeY)_x$,
  (IB) $Ph_2GeBr_2 + Y(MgBr)_2 → (Ph_2GeY)_x$.

II Synthesis with alkali metal derivatives:
  $R_2GeBr_2 + YM_2 → (R_2GeY)_x$; M = Li, Na.

III Hydrogermylation reaction:
  $Ph_2GeH_2 + Y(C\!:\!CH)_2 → (Ph_2GeCH\!:\!CHYCH\!:\!CH)_x$; Y = C_6H_4, Ph_2Ge.

IV Interconversion reactions:
  (IVA) $(Ph_2GeC\!:\!C)_x + Br → (Ph_2GeCBr\!:\!CBr)_x$,
  (IVB) $Ph_2Ge(CH_2C\!:\!CH)_2 + heat → (Ph_2GeCH_2C\!:\!CC\!:\!CCH_2)_x$.

R = organic group, Y = hydrocarbon bridge.

Table 19. Carbon-bridged Polyorganogermanes $(R_2GeY)_x$

Formula	Synth. Method*	Yield	Properties	Ref.
(p-Me$_3$GeC$_6$H$_4$CHCH$_2$)$_x$	--	--	Thermal degradation	1410
(Me$_2$GeCH$_2$CH$_2$)$_x$	IA	--	White powder, m. 103-107° (a)	1589
	IA (b)	25		382a
(Me$_2$GeCH:CH)$_x$	IA	--	White powder, m. 120-50° (a)	1589
	IA (b)	30		382a
(Me$_2$GeCH$_2$CMe$_2$)$_x$	IA	--	White powder, m. 160-90° (a)	1589
(Me$_2$GeCH$_2$CH:CHCH$_2$)$_x$	IA	--	Mol. wt. (a)	1589
	IA (b)	30		382a
(Me$_2$GeCH$_2$CH:CMeCH$_2$)$_x$	IA	--	Viscous liq. (a)	1589
(Me$_2$GeCH$_2$CH:CHCHMe)$_x$	IA	--	Viscous liq. (a)	1589
(Me$_2$GeCH$_2$CHBu)$_x$	IA	--	(a)	1589
(9,10-Me$_2$GeC$_{14}$H$_{10}$)$_x$	II	(c)	C$_{14}$H$_{10}$ = anthracene	1726
(Ph$_2$GeCH:CH)$_x$	III	--	Amber, mol. wt. (d)	327
(Ph$_2$GeCBr:CBr)$_x$	IVA (f)	--	m. >200°	1439
(Ph$_2$GeC⋮C)$_x$	IB	--	m. 130-40°, mol. wt. (e, f)	1439
(Ph$_2$GeC⋮CC⋮C)$_x$	--	--	(e)	--
(Ph$_2$GeCH$_2$C⋮CC⋮CCH$_2$)$_x$	IVD	--	Thermal stability (g)	2546
(Ph$_2$GeC⋮CC$_6$H$_4$C⋮C)$_x$	(h)	--	(e)	1437
R(Ph$_2$GeC⋮CC$_6$H$_4$C⋮C)$_n$H	IB	70	(R = HC⋮CC$_6$H$_4$C⋮C), n = 3-4 (h, i)	1437
(Ph$_2$GeCH:CHC$_6$H$_4$-CH:CH-p)$_x$	III	--	Brown solid, soly. (d)	327

*The characters used here correspond to the synthetic scheme in the introduction to Chapter 2.7.1.8.
(a) IR and NMR spectra (1589). (b) By-prod. in rxn. of C$_2$H$_2$ or (CH$_2$:CH)$_2$ with HGeCl$_3$·2Et$_2$O at r.t., followed by MeMgBr (382a). (c) By-prod. (d) Thermogravimetric analysis (327). (e) NMR spectrum (1026). (f) Rxn. with Br in CCl$_4$ → (Ph$_2$GeCBr:CBr)$_x$ (1439). (g) Isomerizes on heating to form allenic structure (2546), hydrogenation, + HCl → C$_6$H$_6$ + C$_6$H$_{14}$ + GeCl$_4$ (2546). (i) Softening point ~220°, IR spectrum (1437). (h) Oxidation of R(Ph$_2$GeC⋮CC$_6$H$_4$C⋮C)$_n$H with cat. CuCl-Py → (Ph$_2$GeC⋮CC$_6$H$_4$C⋮C)$_x$ + Ph$_2$GeC⋮CC$_6$H$_4$C⋮CC⋮CC$_6$H$_4$C⋮C)$_y$ (1437).

2.7.2 TETRAORGANOGERMANES CONTAINING ORGANOSILICON SUBSTITUENTS

This chapter includes tetraorganogermanes with silicon atoms linked through carbon bridges. All compounds compiled in Table 20 are prepared by methods from the following scheme.

I Grignard reaction:
  (IA) $Me_3GeBr + R'MgCl \rightarrow Me_3GeR'$,
  (IB) $R'GeCl_3 + RMgX \rightarrow R_3GeR'$.

II Synthesis with alkali metal derivatives:
  (IIA) $Me_3GeCl + R'Cl + Na \rightarrow Me_3GeR'$,
  (IIB) $R_3GeNEt_2 + HC\vdots CNa, + R_3SiX \rightarrow R_3GeC\vdots CSiR_3$,
  (IIC) $Et_3GeYLi + R_3SiX \rightarrow Et_3GeYSiR_3$.

III Reaction with organogermylamines:
  $(Et_3Ge)_2NH + R'C\vdots CH \rightarrow Et_3GeC\vdots CR'$.

IV Hydrogermylation reactions:
  (IVA) $R_2GeH_2 + R_2Si(C\vdots CH)_2 \rightarrow (R_2GeCH:CHSiR_2CH:CH)_x$,
  (IVB) $Bu_3GeH + Et_3SiC\vdots CH \rightarrow Bu_3GeCH:CHSiEt_3$.

V Miscellaneous reactions:
  (VA) $Et_3Ge\overline{CR'S(CH_2)_3S} + HgCl_2 + CdCO_3 \rightarrow Et_3GeC(O)R'$,
  (VB) $Me_3GeCl + R''CH:PMe_3 \rightarrow Me_3GeCR':PMe_3$.

R = organic group, R' = organosilyl group, X = halogen, Y = hydrocarbon bridge. Other silicon substituted tetraorganogermanes including compounds where silicon is linked via a heteroatom are listed in Chapters 1.3, 2.4.3, 2.5.3, 2.6.3.4, 2.6.5 and 6.5.8.

2.7.3 TETRAORGANOGERMANES CONTAINING ORGANOTIN SUBSTITUENTS

Tetraorganogermanes containing tin atoms linked by carbon bridges are listed in Table 20.

2.7.4 TETRAORGANOGERMANES CONTAINING ORGANOLEAD SUBSTITUENTS

Triphenylgermylethynyltriphenyllead is listed at the bottom of Table 20.

Table 20. Tetraorganogermanes Containing Carbon-Bridged Derivatives of Group IV Metals R_3GeR'

Formula*	Synth. Method**	Yield	Properties	Ref.
Me_3GeR'				
R' = CH_2SiMe_3 (49)	IA	91	b. 139°, n_D^{20} 1.4329, d_{20} 0.9541 (a)	367
CH_2SiHMe_2	IA	51	b. 132-34°, colorless liq. stable to moisture (b)	475
$C(:PMe_3)SiMe_3$	VB	62	m. 14-15°, colorless liq. b_1 60-65° (e)	1826
$CH:CHSiMe_3$ (49)	IIA	56	b. 158°, n_D^{20} 1.4473, d_{20} 0.9557 (d)	367, 1533
	IB	81	b_{755} 159.5°, n_D^{20} 1.4498, d_{20} 0.9642	1533
$C(:CH_2)SiMe_3$ (49)	IIA	62	b_{757} 164.5°, n_D^{20} 1.4523, d_{20} 1.9717 (d)	367, 1533
$C:CSiMe_3$	IA	--	m. 25°, b. 150-51°, n_D^{28} 1.4405 (e)	367
$p-C_6H_4SiMe_3$	IA	89	m. 98°, b_7 104-107°, IR spectrum	367
Et_3GeR'				
R' = $C(O)SiMe_3$	VA (f)	--	Unstable pink soln. (g)	792
$\overline{C(SiMe_3)S(CH_2)_3S}$	IIO	79	$b_{0.5}$ 119°, n_D^{20} 1.5491 (f)	792
$C:CSiMe_3$	III	49	b_{755} 210-11°, n_D^{20} 1.4551, d_{20} 0.9709 (h)	1534
$p-C_6H_4CH_2SiMe_3$	--	--	(i)	--

* Numbers in parenthesis refer to pages in main volume.
**The characters used here correspond to the synthetic scheme in the introduction to Chapter 2.7.2.

(a) IR and Raman spectra (367). (b) IR and NMR spectra (475). (c) NMR, (far) IR spectra and assignment (1826). (d) IR (1533), Raman (367, 1533) and UV (1693) spectra, dipole moment (254). (e) d_{28} 0.9642, IR and Raman spectra (367). (f) Rxn. of $Et_3GeC(SiMe_3)S(CH_2)_3S$ with $HgCl_2$ and $CdCO_3$ in aq. alc. THF → $Et_3GeC(O)SiMe_3$ (792). (g) IR (792) and UV (792, 794) spectra. (h) IR spectrum (1534). (i) Rxn. with HCl in aq. dioxane → mechanism of Et_3Ge cleavage rxn., isotope effect (43).

Table 20 Continued

Formula*	Synth. Method**	Yield	Properties	Ref.
(EtPhGeCH:CHSiMePhCH:CH)$_x$	IVA (j)	--	Brown solid (k)	327
Bu$_3$GeCH:CHSiEt$_3$	IVB	low	b$_2$ 120-22°, n$_D^{20}$ 1.4727, IR and NMR spectra	1598
Ph$_3$GeC:CSiPh$_3$	IIB	low	m. 152-53°, mol. wt.	1039
(Ph$_2$GeCH:CHYCH:CH)$_x$ Y = p-(Me$_2$Si)$_2$C$_6$H$_4$	IVA	--	Yellow solid (k)	327
(Ph$_2$GeCH:CHCH$_2$SiMe$_2$CH$_2$CH:CH)$_x$	IVA	--	Brown solid, mol. wt. (k)	327
Me$_3$GeCHFCF$_2$SnMe$_3$	IV (b)	--	Liq., slightly volatile at -23°	2209
Me$_3$GeCF(CF$_2$H)SnMe$_3$	(l)	--	dec. at r.t. (m)	2209
Me$_3$GeC:CSnMe$_3$	IIB (n)	40	m. 41-44°, b$_1$ 47° (o)	1039
Bu$_3$GeCH:CHSnBu$_3$	IVB (p)	80	b$_2$ 168-70°, n$_D^{20}$ 1.4897, d$_{20}$ 1.0825, IR and NMR spectra	1598
Ph$_3$GeC:CSnPh$_3$	IIB	87	(From Ph$_3$SnNEt$_2$), m. 149°	1039
	III	64	(From Ph$_3$SnNEt$_2$), mol. wt.	1039
(Ph$_2$GeCH:CHCH$_2$SnMe$_2$CH$_2$CH:CH)$_x$	IVA	--	Brown solid, soly. (k)	327
Ph$_3$GeC:CPbPh$_3$	IIB (q)	58	m. 134-35°, mol. wt.	1039

(j) Synth. from EtPhGe(C:CH)$_2$ and MePhSiH$_2$ in i-PrOH with cat. H$_2$PtCl$_6$ (327).   (k) Thermogravimetric analysis (327).   (l) Instable intermediate in rxn. of Me$_3$SnH$_2$ with Me$_3$GeCF:CF$_2$ (2209).   (m) IR, ^{1}H and ^{19}F NMR spectra, stability of isomers (2209), dec. at r.t. → Me$_3$GeCF:CHF + Me$_3$SnF (2209).   (n) Synth. from R$_3$SnNR$_2$ (1039).   (o) IR and NMR spectra (1039).   (p) Synth. from Bu$_3$SnH (1598).   (q) Synth. by successive rxn. of Ph$_3$PbCl with R$_2$NLi, NaC:CH and Ph$_3$GeX in Ar atm. (1039).

2.7.5 TETRAORGANOGERMANES CONTAINING GROUP III METALS

Tetraorganogermanes containing carbon-linked boron and aluminum were reported recently. The compounds listed in Table 21 are prepared by methods from the following scheme.

I $Et_3GeC\vdots CPh + i\text{-}Bu_2AlH \rightarrow Et_3Ge(i\text{-}Bu_2Al)C:CHPh$.

II $R_3GeH + CH_2:CH(CH_2)_nB(OR)_2 \rightarrow R_3Ge(CH_2)_{n+2}B(OR)_2$.

IIIA $Et_3Ge(CH_2)_3B(OMe)_2 + EtMgBr \rightarrow [Et_3Ge(CH_2)_3]_2BOMe$,

IIIB $Ph_3Ge(CH_2)_3B(OMe)_2 + H_2NCH_2CH_2OH \rightarrow Ph_3Ge(CH_2)_3B(OCH_2CH_2NH_2)$,

IIIC $Et_3Ge(CH_2)_3B(OMe)_2 + B(SEt)_3 \rightarrow Et_3Ge(CH_2)_3B(SEt)_2$.

Table 21. Tetraorganogermanes Containing Group III Metals $\quad\quad R_3GeR'$

Formula	Synth. Method*	Yield	Properties	Ref.
$[Et_3Ge(CH_2)_3]_2BOMe$	IIIA (a)	--	b_1 210-20°, n_D^{20} 1.4797, d_{20} 1.0880	357
$Et_3Ge(CH_2)_3B(OMe)_2$	II	78	b_{12} 118-20°, n_D^{20} 1.4499, d_{20} 1.0315 (a, b, c)	357
$Et_3Ge(CH_2)_3B(SEt)_2$	IIIC (b)	90	b_1 135-37°, n_D^{20} 1.5100, d_{20} 1.0562 (d)	357
$Bu_3GeCH_2CH_2\overline{BOCH_2CH_2O}$	II	--	$b_{0.5}$ 123-31°, NMR spectrum	47
$Ph_3Ge(CH_2)_3B(OMe)_2$	II	100	Viscous oil (c, e)	357
$Ph_3Ge(CH_2)_3B(OCH_2CH_2NH_2)_2$	IIIB	87	m. 125-26.5°	357
$Et_3Ge(i\text{-}Bu_2Al)C:CHPh$	I	--	trans-form in heptane (f)	1016
			cis-form in presence of $\overline{CH_2CH_2CH_2CON}Me$	1016

* The characters used here correspond to the synthetic scheme in the introduction to Chapter 2.7.5.
(a) Rxn. of $Et_3Ge(CH_2)_3B(OMe)_2$ with EtMgBr $\rightarrow [Et_3Ge(CH_2)_3]_2BOMe + Et_3GeCH_2$-$CH:CH_2 + Et_3B$ (357). (b) $Et_3Ge(CH_2)_3B(OMe)_2$ with $(EtS)_3B \rightarrow Et_3Ge(CH_2)_3B(SEt)_2$ (357). (c) Rxn. with alc. NaOH and $H_2O_2 \rightarrow R_3Ge(CH_2)_3OH$ (357). (d) Oxidizes in air (357). (e) Rxn. of $Ph_3Ge(CH_2)_3B(OMe)_2$ with $H_2NCH_2CH_2OH$ in MeOH $\rightarrow$ $Ph_3Ge(CH_2)_3B(OCH_2CH_2NH_2)_2$ (357). (f) Not isolated, structure by hydrolysis: $X_2O \rightarrow Et_3GeCX:CHPh$; X = H, D (1016).

2.7.6 TETRAORGANOGERMANES CONTAINING MAGNESIUM

Tetraorganogermanes containing magnesium are prepared from halogen substituted organogermanes with magnesium in ether solvents. A direct synthesis of diethynyldiphenylgermane and magnesium in diethyl ether (2281) is mentioned. The organogermanium Grignard derivatives are not isolated but used as synthetic intermediates. For this reason the compilation is incomplete.

Chloromethyltrimethylgermane Magnesium   GeMgC$_4$H$_{11}$Cl   Me$_3$GeCH$_2$MgCl
Synth.: By rxn. of Me$_3$GeCH$_2$Cl with Mg (2281).
Rxn. with: PhSO$_2$Me → Me$_3$GeCH$_2$S(O)Ph (2281).

Diethynyldiphenylgermane Magnesium   -----   -----
Synth.: By rxn. of Ph$_2$Ge(C⋮CH)$_2$ with Mg in THF (1438).
Rxn. with: Anhyd. CoBr$_2$ → HC⋮C(Ph$_2$GeC⋮CC⋮C)$_x$Ph$_2$GeC⋮CH + Co (1438).

2.7.7 TETRAORGANOGERMANES CONTAINING ALKALI METALS

The organolithium tetraorganogermanes reported in this chapter are prepared by addition of triethylgermyl lithium to unsaturated hydrocarbons under exclusion of air. The compounds are used as synthetic intermediates and are not isolated in substance.

2-Triethylgermylethyl Lithium   GeLiC$_8$H$_{19}$ Et$_3$GeCH$_2$CH$_2$Li
Synth.: By rxn. of Et$_3$GeLi with C$_2$H$_4$ in sealed ampul at r.t. (2849).
Rxn. with: H$_2$O → Et$_4$Ge (2849).
Et$_3$GeBr → (Et$_3$GeCH$_2$)$_2$ + LiBr (2849).

2,1,1-Triethylgermylphenylethyl Lithium   GeLiC$_{14}$H$_{23}$ Et$_3$GeCH$_2$CHPhLi
Synth.: By rxn. of Et$_3$GeLi with PhCH:CH$_2$ in sealed ampul in C$_6$H$_6$ (2849).
Prop.: Raspberry red soln., sensitive to air (2849).
Rxn. with: H$_2$O → Et$_3$CH$_2$CH$_2$Ph + LiOH (2849).

2,1-Triethylgermylstyryl Lithium   GeLiC$_{14}$H$_{21}$ Et$_3$GeCH:CPhLi
Synth.: By rxn. of Et$_3$GeLi with PhC⋮CH in C$_6$H$_6$ (2849).
Rxn. with: Et$_3$GeBr → Et$_3$GeCH:CPhGeEt$_3$ (2849).

2,1,2-Triethylgermyldiphenylethyl Lithium   GeLiC$_{20}$H$_{21}$ Et$_3$GeCHPhCHPhLi
Synth.: By rxn. of Et$_3$GeLi with stilbene in C$_6$H$_6$ (2849).
Prop.: Red-orange soln., sensitive to air (2849).
Rxn. with: H$_2$O → Et$_3$GeCHPhCH$_2$Ph + LiOH (2849).
Et$_3$GeBr → (Et$_3$GeCHPh)$_2$ + LiBr (2849).

2,1,2-Triethylgermyldiphenylethenyl Lithium   GeLiC$_{20}$H$_{25}$ Et$_3$GeCPh:CPhLi
Synth.: By rxn. of Et$_3$GeLi with tolane in C$_6$H$_6$ (2849).
Rxn. with: H$_2$O → Et$_3$GeCPh:CHPh + LiOH (2849).
Et$_3$GeBr → (Et$_3$GeCPh:)$_2$ + LiBr (2849).

2-(1-Triethylgermyl)hexyl Lithium   Et$_3$GeCH$_2$CHBuLi
Synth.: By rxn. of Et$_3$GeLi with BuCH:CH$_2$ in sealed ampul at 90° (2849).
Rxn. with: H$_2$O → n-C$_6$H$_{13}$GeEt$_3$ (2849).

2.7.8 TETRAORGANOGERMANES CONTAINING TRANSITION METALS

Tetraorganogermanes containing a transition metal carbon band in a side chain are listed in this chapter. Transition metal π-complexes linked to germanium through an aromatic or pseudoaromatic system are listed in Chapter 6.5. The bimetallic organogermanes in this chapter are prepared by cleavage of a cyclopropane ring with mercuric acetate or by insertion of fluoroolefins in germanium manganese bonds.

2,4-Hydroxymercurioacetato-butyltrimethylgermane $GeHgC_9H_{20}O_3$ $Me_3GeCH_2CHOHCH_2CH_2HgOAc$

Synth.: By rxn. of $Me_3GeCH_2\overline{CHCH_2CH_2}$ with $Hg(OAc)_2$ at 30°, in 70% yield (127).
Prop.: b_{745} 126°, n_D^{20} 1.4355, d_{20} 1.0290 (127).

2-Pentacarbonylmanganesetetra-fluorethyltrimethylgermane $GeMnC_{10}H_9F_4O_5$ $Me_3GeCF_2CF_2Mn(CO)_5$

Synth.: By rxn. of $Me_3GeMn(CO)_5$ with C_2F_4 under UV irradiation in pentane at 50° (872).
Prop.: m. 32°, IR and NMR spectra (872).

ω-Pentacarbonylmanganesepolytetra-fluorethyltrimethylgermane $Me_3Ge(CF_2CF_2)_xMn(CO)_5$

Synth.: By-prod. in rxn. of $Me_3GeMn(CO)_5$ with C_2F_4 under UV irradiation at 50° (872).
Prop.: IR spectrum, structure from analogy to Sn derivative (872).

4-Pentacarbonylmanganeseperfluor-buten-3-yltrimethylgermane $GeMnC_{12}H_9F_6O_5$ $Me_3GeCF_2CF_2CF{:}CFMn(CO)_5$

Synth.: By rxn. of $Me_3GeMn(CO)_5$ with $(CF_2{:}CF)_2$ in pentane under UV irradiation, in low yield (1141).
Prop.: IR, NMR and mass spectra, proposed structure (1141).

3 ORGANOGERMANIUM HYDRIDES

3.1 NONFUNCTIONAL ORGANOGERMANIUM HYDRIDES

Organogermanium hydrides are prepared by reduction of organogermanium halides or other organogermanium derivatives of heteroatoms. The following reducing or hydrogenating agents are employed: lithium hydride, lithium aluminum hydride and several other complexes, e.g., $LiAlH(OR)_3$, sodium borohydride and water. Organogermanium hydrides are shown as by-products in the synthesis of tetraorganogermanes from germanium tetrahalide with Grignard reagents or with organoaluminum derivatives. They originate from two reactions.

a) Hydrolysis of a postulated intermediate with a germanium metal bond, e.g., $R_3GeMgCl$ requires a proton source to yield a triorganogermane

$$R_3GeMgCl + HX \rightarrow R_3GeH + MgXCl$$

b) Without protonolysis proceeds the cleavage of propene during reaction of germanium tetrachloride with isopropylmagnesium chloride at low temperatures in inert atmosphere, leading to triisopropylgermane in moderate yield.

The following scheme shows reactions producing organogermanes as by-products.

I $GeCl_4$ with excess Grignard reagent.

II R_4Ge with lithium in dimethoxy ethane.

III $GeCl_4$ with triisobutyl aluminum.

IV GeR_4 with $AlCl_3$, followed by treatment with $LiAlH_4$.

Additional organogermanes are reported in Chapters 3.3 and 7.3.

Trimethylgermane (M-53) GeC_3H_{10} Me_3GeH
<u>Synth.</u>: By rxn. of Me_3GeBu with $LiAlH_4$ in $(MeOCH_2)_2$ at 70°, in 70% yield (1041-2).
By rxn. of Me_3GeCl with $LiAlH(OCMe_3)_3$ in Et_2O in 65% yield (582).
By hydrogenation of the following platinum complexes: $Me_3Ge(Et_3P)_2PtCl$ and $Me_3Ge(Et_3P)_2PtI$ in xylene, in 66 and 58% yield, resp. (1106), $Me_3Ge(Ph_3Ge)$-$Pt(PEt_3)_2$ in xylene, in 83% yield (1106), $[Me_3Ge(Et_3P)Pt(CH_2PPh_2)_2]Cl \cdot C_6H_6$ at 500 torr in MeOH (1107).
By rxn. of $Me_3Ge(Ph_3Ge)Pt(PEt_3)$ with anhyd. HCl in C_6H_6 or Et_2O in 60% yield (1107).
By-prod. in rxn. of $Me_3Ge(Et_3P)_2PtPh$ or $[Me_3Ge(Et_3P)Pt(CH_2PPh_2)_2]$ with dry HCl (1107).
By-prod. in rxn. of Me_3GeBr with H_3SiK in diglyme in N atm. (2224).
By-prod. in rxn. of $GeCl_4$ with large excess of Me_3Al in N atm., in < 10% yield (1108).
<u>Prop.</u>: (Far) IR, Raman and assignment (2081), NMR (259, 582, 1821, 2373) and mass (1109, 1436) spectra, dipole moment (582), force constant (138).
<u>Rxn. with</u>: $Et_2Hg \rightarrow (Me_3Ge)_2Hg + C_2H_6$ (996).
$(Et_3P)_2PdCl_2 \rightarrow Me_6Ge_2 + Me_3GeCl + trans\text{-}(Et_3P)_2PdHCl + Pd + H$ (795).
$(Et_3P)_2PtHCl \rightleftharpoons trans\text{-}Me_3Ge(Et_3P)_2PtCl + H$ (1107).
$CH_2C{:}CH_2^* \rightarrow Me_3GeCH_2CH{:}CH_2 + Me_3GeCMe{:}CH_2$ (3:2) (1041-2).
$(CH_2{:}CH)_2^* \rightarrow cis\text{-} + trans\text{-}Me_3GeCH_2CH{:}CHMe$ (1041-2).
Cyclopentadiene* $\rightarrow$ 1- + 2-$Me_3GeC_5H_7$ (1041-2).
1,3- or 1,4-Cyclohexadiene* $\rightarrow$ 2-$Me_3GeC_6H_9$ (1041-2).
1,3- or 1,5-Cyclooctadiene* $\rightarrow$ 2-$Me_3GeC_8H_{13}$ (1041-2).
1,5-Cyclooctadiene* $\rightarrow$ 1,3-C_8H_{12} (1041-2).
4-$CH_2{:}CHC_6H_9^* \rightarrow$ 3-$Me_3GeCH_2CH_2C_6H_9$ (1041-2).
Bicyclo[2.2.1]heptadiene-2,5* $\rightarrow$ endo- + exo-norbornenyl-$GeMe_3$ + nortricyclo-$GeMe_3$ (1041-2).
$\overline{CF_2CF_2CCl{:}CCl}$ at 190° $\rightarrow Me_3Ge\overline{CClCHClCF_2CF_2} + Me_3Ge\overline{CHCHClCF_2CF_2} + Me_3Ge\overline{C{:}CClCF_2CF_2}$ + trace $Me_3Ge\overline{C{:}CHCF_2CF_2} + Me_3GeCl$ (924).
$(CF_3)_2CO \rightarrow Me_3GeOCH(CF_3)_2$ (109).

---

* H_2PtCl_6 is used as catalyst.

Dimethylgermane (M-53) GeC_2H_8 Me_2GeH_2

<u>Synth.:</u> By rxn. of Me_2GeCl_2 with $LiAlH(OCMe_3)_3$ in Et_2O (582), with NaH and BPh_3 in mineral oil at 110° under exclusion of moisture (2242).

<u>Prop.:</u> (Far) IR, Raman and assignment (2081), NMR (259, 582, 735, 1821, 2361, 2373), microwave and NQR (2029) spectra, dipole moment (582), force constant (138).

<u>Rxn. with:</u> $NaNH_2$ in liq. NH_3 → relative acidity (735).

$Fe_3(CO)_{12}$ → $Me_6Ge_3Fe_2(CO)_6$ (2284).

Methylgermane (M-53) $GeCH_6$ $MeGeH_3$

<u>Synth.:</u> By rxn. of $MeGeCl_3$ with $LiAlH(OCMe_3)_3$ in Et_2O (582).

By rxn. of H_3GeK with MeI in $(Me_2N)_3PO$, in 94% yield (910), in $(CH_2OMe)_2$ (2224, 2724).

<u>Prop.:</u> Spectra: IR (957), (far) IR, Raman and assignment (1313, 2081), NMR (259, 582, 735, 1821), microwave (957, 1556), Coriolis coefficient and ground state centrifugal distortions (869), dipole moment (582), deuterium effect on dipole moment (1556), nonbonded electron effect on internal rotation (1161), electrostatic model to treat hindered rotation (1432), force constant (138, 1067), mol. polarizability (1566), thermodynamic properties and data (957, 1067), GLC (910).

<u>Rxn. with:</u> $NaNH_2$ in liq. NH_3 → relative acidity (735).

Tribenzylgermane $GeC_{21}H_{22}$ $(PhCH_2)_3GeH$

<u>Synth.:</u> By rxn. of $(PhCH_2)_4Ge$ or $(PhCH_2)_6Ge_2$ with Li in $(CH_2OMe)_2$, followed by H_2O, in 76% yield (104).

By rxn. of $(PhCH_2)_4Ge$ with Br in $(CH_2Br)_2$, followed by $LiAlH_4$, in 85% yield (104).

<u>Prop.:</u> m. 80-82°, $b_{0.001}$ 164-87° (104), IR (104), (far) IR and assignment (105), NMR (104) and mass (1109) spectra.

<u>Rxn. with:</u> Pyrolysis at 370-90° → $(PhCH_2)_2$ + trans-$(PhCH:)_2$ (104).

$PhCH_2Li$ + MeI → $Me_nGe(CH_2Ph)_{4-n}$; n = 0-2 + $Me(PhCH_2)_2GeH$ (104).

BuLi at -12°, + MeI → $MeGe(CH_2Ph)_3$ + $(PhCH_2)_4Ge$ + $(PhCH_2)_6Ge_2$ + $(PhCH_2)_3GeBu$ (104).

Li in $(CH_2OMe)_2$, + MeI → $Me_2Ge(CH_2Ph)_2$ (104).

Triethylgermane (M-53) GeC_6H_{16} Et_3GeH

<u>Synth.:</u> By rxn. of $Et_3GeOCH:CPh_2$ with $LiAlH_4$ (2714).

By rxn. of Et_3GePPh_2 with $LiAlH_4$ (466).

By rxn. of $(Et_3Ge)_3M$ with Et_3SnH at 170-80° in a sealed ampul; M = Sb and Bi, in 88 and 62% yield, resp. (2086), with AcOH at 170°; M = Bi, in 53% yield (2086).

By rxn. of $(Et_3Ge)_nBiEt_{3-n}$ with AcOH or Et_3SnH; n > 0 (291).

By rxn. of $(Et_3Ge)_2Te$ with Et_3SnH at 70°, in 57% yield (2853).

By rxn. of $Et_3GeMgSeBu$ with BuSeH at liq. N temp., in 80% yield (2855).

By rxn. of $Et_3GeZn(GeEt_2)_xGeEt_3$ (?) with H_2O (584).

By rxn. of $(Et_3Ge)_2Zn$ with BuSeH at liq. N temp., in 82% yield (2855).

By rxn. of $(Et_3Ge)_2Cd$ with AcOH, n-PrOH and Et_3SnH at r.t., in 31, 56 and 95%

yield, resp. (587), with BuSeH at liq. N temp., in 36% yield (2855).
By rxn. of $(Et_3Ge)_2Hg$ with AcOH (291), in a sealed ampul at 130-40°, in 62% yield (2086), with BuSeH at liq. N to r.t., in 74% yield (2855).
By rxn. of $(Et_3Ge)_3Tl$ with H_2O or Et_3SnH, in 73 and 77% yield, resp. (1361).
By-prod. in rxn. of $GeCl_4$ with EtMgBr in Et_2O, THF or MePh in the presence of excess Mg, in up to 4% yield (2701).
By-prod. in rxn. of Ge_2H_6 with C_2H_4 at 154-60° in a sealed tube (2551).
By-prod. in rxn. of Et_4Ge with n-Pr_4Ge and cat. $AlCl_3$, followed by $LiAlH_4$, in 25% yield (491).
Prop.: IR (1444), NMR (1444, 2373) and mass (1109) spectra, force constant (138), GLC (491, 2701).
Rxn. with: $\overline{CH_2CH_2CH_2GeBu_2}$ → $Et_3Ge(CH_2)_3GeBu_2H$ (1491).
Et_4Pb at 165-70° → Et_4Ge + Et_6Pb_2 + C_2H_6 + Pb + organo-GePb compd. (2850).
Et_3SnOMe at 180° → Et_3GeOMe + Et_6Sn_2 + MeOH (2097).
Et_3GeLi → Et_6Ge_2 + LiH (2087, 2852).
$Et_3GeHgEt$ → $(Et_3Ge)_2Hg$ + C_2H_6 (2852).
$Et_3SiHgEt$ → $Et_3GeHgSiEt_3$ + C_2H_6 (583a, 2852).
Et_5Si_2HgEt → $Et_3GeHgSi_2Et_5$ + C_2H_6 (583a).
$(Et_3Si)_3M$ → $(Et_3Ge)_3M$ + Et_3SiH; M = Sb, Bi (2086).
$(Et_3Si)_2Te$ → $(Et_3Ge)_2Te$ + Et_3SiH (2853).
MeMgX → Et_3GeMe (491).
Et_2Zn → $Et_3GeZn(GeEt_2)_xGeEt_3$ (?) + C_2H_6 + Zn + Et_4Ge (584).
Et_2Hg → $(Et_3Ge)_2Hg$ + $Et_3GeHgEt$ + C_2H_6 + Hg (2852).
$PhHgCBrCl_2$ → $Et_3GeCHCl_2$ + PhHgBr (1876).
Et_3Tl → $(Et_3Ge)_3Tl$ + C_2H_6 (1361, 2089).
Et_3Bi → $(Et_3Ge)_nBi_{3-n}$; n = 1-3 (291).
Et_3M → $(Et_3Ge)_3M$ + C_2H_6; M = Sb, Bi (2086).
$(MeS)_2$ → Et_3GeSMe + MeSH (465).
$(CH_2SH)_2$ → $(Et_3GeSCH_2)_2$ (465).
BuSH → Et_3GeSBu (465).
Et_2Se → $(Et_3Ge)_2Se$ (2091).
Et_2Te → Et_3GeTeH (1:1) + $(Et_3Ge)_2Te$ (2:1) + C_2H_6 (2092).
Y (1:1) → Et_3GeYH; Y = S, Se (2091-2).
Y → $(Et_3Ge)_2Y$; Y = S, Se, Te (2091).
CH_2:CHR → $Et_3GeCH_2CH_2R$; R = $CH_2B(OMe)_2$ (357), CH_2NH_2, CO_2SiMe_3 (364), $POCl_2$, PO_3Et_2, CH_2POCl_2, $CH_2PO_3Et_2$, $CH_2PO_3Bu_2$ (2366).
$\overline{(CH_2:CH)_2}$* → $Et_3GeCH_2CH:CHMe$ + $Et_3GeCH_2CH_2C_6H_9$ c. (1798).
CH_2:$\overline{CXCRCH_2O}$* → Et_6Ge_2 + $Et_3GeOCH_2CR:CXMe$; R = H, Me, X = H, Cl (806).
$\overline{MeCH:CCHMeO}$* → $Et_3GeOCHMeCH:CHMe$ (2636).
RC⋮CR'* → Et_3GeCR:CHR' and/or Et_3GeCR':CHR, numerous derivatives (199, 1150, 2425-8, 2767).
HC⋮CR* → Et_3GeCH:CHR, numerous acetylenic derivatives (670, 1898, 1900, 2559, 2766).

---

* Reactions usually catalyzed by H_2PtCl_6.

(CF$_3$C!)$_2$ + UV irradiation → Et$_3$GeC(CF$_3$):CHCF$_3$ (923).
Et$_3$Ge(CH$_2$)$_4$C!CH → Et$_3$Ge(CH$_2$)$_4$CH:CHGeEt$_3$ (1494).
HC!CCH$_2$Cl* → Et$_3$GeCH:CHCH$_2$Cl + Et$_3$GeC(CH$_2$Cl):CH$_2$ + Et$_3$GeCl + HC!CMe (1472).
1-HOC$_5$H$_8$C!CPhc. → Et$_3$GeC(:CHPh)C$_5$H$_8$OH-1 + Et$_3$GeC(:CHPh)C$_5$H$_7$-1 (2425).
HC!CCH$_2$OCHROCH$_2$CH:CH$_2$* → Et$_3$GeCH:CHCH$_2$OCHROCH$_2$CH:CH$_2$ + Et$_3$GeCH$_2$CH:CHOCHRO-(CH$_2$)$_3$GeEt$_3$; R = H, Me (2559).
Diyne → Et$_3$GeCR:CHC!CR + (Et$_3$GeCR:CH)$_2$ (198, 200).
Ph$_2$C:CO → Et$_3$GeOCH:CPh$_2$ (2714).
$\overline{\text{OCCH}_2\text{CH}_2\text{CON}}$Br → Et$_3$GeBr (164).
GeCl$_4$ → GeCl$_{1.5}$ (2591).
GeCl$_4$ → Et$_3$GeCl + HGeCl$_3$ (1471).
Use: In catalyst for olefin polymerization (411), reactivity in catalyst for C$_2$H$_4$ polymerization (1956).

Diethylgermane (M-56)   GeC$_4$H$_{12}$ Et$_2$GeH$_2$
Synth.: By-prod. in rxn. of Ge$_2$H$_6$ with C$_2$H$_4$ at 154-60° in sealed tube (2551).
By-prod. in dissocn. of Et$_2$GeHF (1471).
Prop.: IR (1444) and NMR (735, 1444, 1471, 2373) spectra, force constant (138).
Rxn. with: HgCl$_2$ → Et$_2$GeHCl (1471).
MeOCH$_2$Cl + cat. AlCl$_3$ → Et$_2$GeHCl (1471).
NaNH$_2$ in liq. NH$_3$ → relative acidity (735).

Ethylgermane (M-56)   GeC$_2$H$_8$ EtGeH$_3$
Synth.: By rxn. of H$_3$GeK with EtBr in (Me$_2$N)$_3$PO, in 65% yield (910).
By-prod. in rxn. of G$_2$H$_6$ with C$_2$H$_4$ at 154-60° in sealed tube (2551).
By-prod. in dec. of EtGeH$_2$F (1471).
Prop.: IR, assignment (1444-5) and NMR (735, 1444, 2550) spectra, force constant (138) and GLC (910).
Rxn. with: HgCl$_2$ → EtGeH$_2$Cl + Hg (1471).
$\overline{\text{OCCH}_2\text{CH}_2\text{CON}}$X → EtGeH$_2$X + EtGeHX$_2$; X = Br, I (1471).
NaNH$_2$ in liq. NH$_3$ → relative acidity (735).
Na + MeCl in liq. NH$_3$ → MeEtGeH$_2$ (735).

Triisopropylgermane (M-57)   GeC$_9$H$_{22}$ i-Pr$_3$GeH
Synth.: By-prod. in rxn. of GeCl$_4$ with i-PrMgCl in Et$_2$O, in presence of excess Mg or after work-up with LiAlH$_4$, in 10-39% yield (831).
By-prod. in rxn. of GeCl$_4$ with i-PrMgCl in dry Et$_2$O in He atm. at 4° under evolution of C$_3$H$_6$ and C$_3$H$_8$, in 32% yield, rxn. mechanism is discussed (2591).
By-prod. in rxn. of MeI and i-PrBr with GeCl$_4$ and Mg in Et$_2$O (493).
Prop.: b. 176° (831), b$_{42}$ 88-89°, n$_D^{20}$ 1.4500, d$_{20}$ 0.9750 (2591), IR (2591), far IR and assignment (105) and NMR (831) spectra, GLC (493).
Rxn. with: Se at 210° → i-Pr$_3$GeSeH (2092).

---

* Reactions usually catalyzed by H$_2$PtCl$_6$.

Tributylgermane (M-54) $GeC_{12}H_{28}$ n-Bu$_3$GeH
Synth.: By rxn. of Bu$_3$GeI with LiAlH$_4$ in Et$_2$O in Ar atm., in 86% yield (1598).
By-prod. in rxn. of BuMgBr with GlCl$_4$ (1:1) in Et$_2$O in N atm., followed by LiAlH$_4$, in 23% yield (2032).
Prop.: b$_{20}$ 117-18°, n$_D^{20}$ 1.4502 (1598), IR and NMR spectra (2039), GLC (491).
Rxn. with: Ph$_2$Sn(NEt$_2$)$_2$ → (Bu$_3$Ge)$_2$SnPh$_2$ (913).
Et$_3$SiC⋮CH → Bu$_3$GeCH:CHSiEt$_3$ (1598).
Bu$_2$SbC⋮CH → (Bu$_2$SbCH:)$_2$ (1601).
HC⋮CR + cat. H$_2$PtCl$_6$ → Bu$_3$GeCH:CHR; R = PrCHOH, MeEtCOH (1900).
CH$_2$:CHBOCH$_2$CH$_2$O → Bu$_3$GeCH$_2$CH$_2$BOCH$_2$CH$_2$O (47).
Bu$_4$Ge + GeCl$_4$ → activity as catalyst for scrambling rxn. (1760).

Dibutylgermane (M-55) GeC_8H_{20} n-Bu$_2$GeH$_2$
Synth.: By rxn. of Bu$_2$GeCH$_2$CMe:CMeCH$_2$ with I in EtI, followed by LiAlH$_4$, in 83% yield (1495).
By rxn. of Bu$_2$GeCl$_2$ with K in (Me$_2$N)$_3$PO and hydrolysis, in 22% yield (815).
By-prod. in dec. of Bu$_2$GeHF (1471).
By-prod. in rxn. of GeCl$_4$ with BuMgBr in Et$_2$O, followed by LiAlH$_4$ in N atm., in 33% yield (2039).
Prop.: IR (336), assignment (2039) and NMR (2039) spectra, GLC (1495).
Rxn. with: R$_2$Hg → Bu$_2$RGeH (1:1) + Bu$_2$GeR$_2$ (1:2); R = CH$_2$CO$_2$Me (25).
HgCl$_2$ → Bu$_2$GeHCl (24, 1471).
MCl$_4$ → H/Cl exchange rxn., measured by NMR spectrum (2039).

Butylgermane (M-57) GeC_4H_{12} n-BuGeH$_3$
Prop.: Very sensitive to air, NMR, IR spectra and assignment (2039).
Rxn. with: HgCl$_2$ → BuGeHCl$_2$ + Hg (467).
HgCl$_2$ → BuGeH$_2$Cl (1471) at (2:1) + HCl + Hg (2039).
HgBr$_2$ (1·5:1) → Bu$_2$GeH$_2$Br + Hg + HBr (2039).
HgX$_2$ → BuGeH$_2$X + Hg + HX; X = CN (at 1:1), SCN (at 4:3) (2039).
I (7:5) → BuGeH$_2$I + HI + Hg (sic.) (2039).
MeOCH$_2$Cl → BuGeH$_2$Cl (1471).
MCl$_4$ → H/Cl exchange rxn., kinetic data; M = C, Ge (2039).

Tricyclohexylgermane (M-57) $GeC_{18}H_{34}$ $(C_6H_{11})_3$GeH
Synth.: By rxn. of excess C$_6$H$_{11}$MgCl with GeCl$_4$ in Et$_2$O at 0°, in up to 30% yield (353), at 4° in He atm., in 29% yield (2591), by-prods. are C$_6$H$_{10}$ + C$_6$H$_{12}$ (2591), (C$_6$H$_{11}$)$_3$GeMgCl is discussed as rxn. intermediate (353).
By rxn. of GeCl$_{1.5}$ with C$_6$H$_{11}$MgCl in dry Et$_2$O in He atm. (2591).
Prop.: b$_{0.3}$ 141-43°, n$_D^{25}$ 1.5223, d$_{25}$ 1.0838 (2591), IR spectrum (353, 2591).
Rxn. with: Thermal dec. at 400° → Ge + H + C$_6$H$_{12}$ + C$_6$H$_6$ + C$_6$H$_{10}$ + condensed prod. (1691).
Et$_2$Te at 200° → [(C$_6$H$_{11}$)$_3$Ge]$_2$Te + C$_2$H$_6$ (2092).
Se at 210° → [(C$_6$H$_{11}$)$_3$Ge]$_2$Se (2092).
MgCl$_2$ in Et$_2$O, + D$_2$O → no R$_3$GeD found (2591).

Triphenylgermane (M-55) $GeC_{18}H_{16}$ Ph_3GeH

Synth.: By rxn. of $Ph_3GeOCH:CPh_2$ with $LiAlH_4$ (2714).
By rxn. of Ph_3GeLi with AcR in Et_2O at -20° to r.t., R = t-Bu, Et and Me, in 93, 45 and 43% crude yield, resp., higher in THF as solvent (1624, 1626), with 2-methylcyclohexanone, in 83% yield (1624, 1626).
By rxn. of $(Ph_3Ge)_2M$ with aq. MeOH; M = Sr, Bu, Zn (675).
By rxn. dry HCl with $(Ph_3Ge)_2Pt(CH_2PPh_2)_2$ in C_6H_6 (2285), with $Ph_3Ge(Me_3Ge)Pt(PEt_3)_2$ in C_6H_6 (1107), with $(Ph_3Ge)HPt(PEt_3)_2$ in Et_2O (2285), with $(Ph_3Ge)_2Pt(PR_3)_2$ as by-prod. (101, 103), with $(Ph_3Ge)_2Pd(PEt_3)_2$ (795).
By hydrogenation of $(Ph_3Ge)_2Pt(PR_3)_2$ (101, 103), of $(Ph_3Ge)_2Pd(PEt_3)_2$ at 100 atm. as by-prod (795), of $(Ph_3Ge)_2Pt(CH_2PPh_2)_2$ at 100 atm. (2285).
By rxn. of $(Ph_3Ge)HPt(PEt_3)_2$ with $(Ph_2PCH_2)_2$ in C_6H_6 in N atm., in 76% yield (2285).
By rxn. of $Ph_3GeAuPPh_2$ with $MgBr_2$ in Et_2O at r.t. in N atm. and hydrolysis (2420).
By polarographic reduction of Ph_3GeCl and Bu_4NClO_4 (963), of Ph_6Ge_2, followed by Ph_3MCl; M = Si, Ge, Sn (963), of Ph_6Sn_2, followed by Ph_3GeCl (963).
By-prod. in rxn. of $(Ph_3Ge)_2Pt(PEt_3)_2$ with PhLi, followed by MgI_2 and $LiAlH_4$ (103).
By-prod. in rxn. of PhMgBr with $GeCl_4$ (1:1), followed by $LiAlH_4$, in 7% yield (50).

Prop.: $b_{0.01}$ 128-36° (50), IR (1364), assignment (105), NMR (104, 460, 735) and mass (1109) spectra, force constant (138), effect of neutron irradiation (84), rel. acidity from rxn. with $NaNH_2$ in liq. NH_3 (735).

Rxn. with: $Ph_2GeCl_2 \rightarrow Ph_2GeHCl$ (1364).
$R_3SnNEt_2 \rightarrow Ph_3GeSnR_3$; R = Et, Ph (913), Me, Bu (1618).
$Et_2Sn(NEt_2)_2$ (2:1) $\rightarrow (Ph_3Ge)_2SnEt_2$ (913).
$R_2Sn(NEt_2)_2$ (1:1) $\rightarrow Ph_3GeSnR_2NEt_2$; R = Et, Ph (913).
$EtSn(NEt_2)_3 \rightarrow (Ph_3Ge)_nSnEt(NEt_2)_{3-n}$; n= 0-2, depending on ratios (913).
$R_3PbNEt_2 \rightarrow Ph_3GePbR_3 + Et_2NH$; R = n-Bu, i-Bu, $c.C_6H_{11}$, Ph (1608).
BuLi $\rightarrow Ph_3GeLi$ (1624-5).
BuLi $\rightarrow Ph_3GeBu$ (105).
$Et_2Cd \rightarrow (Ph_3Ge)_2Cd + C_2H_6$ (2295).
$PhHgCBrCl_2 \rightarrow Ph_3GeCHCl_2 + PhHgBr$ (1876).
$(Et_3P)_2PdCl_2 \rightarrow Ph_3GeCl + (Et_3P)_2PdHCl$ (795).
$(Me_2PhP)_2PtCl_2 \rightarrow Ph_3Ge(Me_2PhP)_2PtCl$ (855).
$CH_2:CHCH_2B(OMe)_2 \rightarrow Ph_3Ge(CH_2)_3B(OMe)_2$ (357).
$CH_2:CHCRCH_2O$ + cat. $H_2PtCl_6 \rightarrow Ph_3GeCHMeCHRCH_2OGePh_3$; R = H, Me (806).
$MeCH:CHCH_2OH \rightarrow Ph_3GeOCH_2CH:CHMe$ (806).
1-Hexene $\rightarrow n-C_6H_{13}GePh_3$; relative rxn. rates and mechanism (1762).
$(PhC!)_2$ + cat. Pt on C $\rightarrow$ cis-$Ph_3GeCPh:CHPh$ (2283).
$Ph_2C:CO \rightarrow Ph_3GeOCH:CPh_2$ (2714).
BzH + cat. AiBN $\rightarrow Ph_3GeOCH_2Ph$ (1762).
$c.C_6H_{10}O \rightarrow Ph_3GeOC_6H_{11} + (Ph_3Ge)_2O + C_6H_{11}OH$ (1762).
$AcCH_2CHMe_2$ + cat. AiBN $\rightarrow Ph_3GeOCHMeCH_2CHMe_2$ (1762).
K in $(Me_2N)_3PO$, + BuBr $\rightarrow BuGePh_3$ (1634, 2983).

Use: In catalyst for olefin polymerization (411).
Additive to improve corona resistance of polyimide films (2585).

Diphenylgermane (M-56)　　　　　　　　　$GeC_{12}H_{12}$　　　　　　　　　Ph_2GeH_2

<u>Synth.</u>: By-prod. in rxn. of $GeCl_4$ with PhMgBr (1:1), followed by $LiAlH_4$, in 13% yield (50).

By-prod. in thermal dec. of $PhGeH_3$ (1762) of Ph_2GeHCl (1762).

By polarographic reduction of Ph_2GeCl_2 and Bu_4NClO_4 in $(CH_2OMe)_2$ (963).

<u>Prop.</u>: $b_{0.01}$ 79° (50), IR (1364), (far) IR and assignment (105) and NMR (104, 460) spectra, force constant (138), effect of neutron irradiation (84).

<u>Rxn. with</u>: $Ph_3GeSnEt_2NEt_2$ (1:2) → $Ph_3GeSnEt_2)GePh_2$ (913).

R_3SnNEt_2 → $Ph_2Ge(SnR_3)_2 + Et_2NH$; R = Et, Ph (913).

$R_2Sn(NEt_2)_2$ → $(Ph_2GeSnR_2)_x$; R = Et, Ph (913).

$Ph_2Pb(NEt_2)_2$ → $Ph_2Ge(PbPh_3)_2 + Pb + Et_2NH$ (1608).

Et_2Hg → $(Ph_2Ge)_4 + (Ph_2Ge)_x$ (396).

n-BuLi, + EtBr, + H_2O → $Et_2GePh_2 + EtBuGePh_2 + Bu_2GePh_2 + (EtPh_2Ge)_2$ (104).

n-BuLi + H_2O → $BuPh_2GeH + (Ph_2GeH)_2 + BuPh_2GeGePh_2H$ (105).

n-BuLi + D_2O → $BuPh_2GeD$ (105).

$Fe_3(CO)_{12}$ → $Ph_2GeFe_2(CO)_8$ (2284).

CH_2:CHR + cat. AIBN → $Ph_2Ge(CH_2CH_2R)_2 + Ph_2(RCH_2CH_2)GeH$ + polymer; R = CN, CH_2CN (2577).

CH_2N_2 + UV irradiation or Cu → $MePh_2GeH + Me_2GePh_2$ (1799).

Y(C⋮CH)$_2$ → $(Ph_2GeCH:CHYCH:CH)_x$; Y = C_6H_4, Ph_2Ge, $CH_2SiMe_2CH_2$, $CH_2SnMe_2CH_2$, p-$Me_2SiC_6H_4SiMe_2$ (327).

$ClCH_2OMe$ + cat. $AlCl_3$ → Ph_2GeHCl (1762, 1799).

$\overline{OCCH_2CH_2CONX}$ → $Ph_2GeHX + Ph_2GeX_2$; X = Cl, Br, I (1762), Br (1799).

HgX_2 → Ph_2GeHX; X = Cl, Br (1762).

Phenylgermane (M-58)　　　　　　　　　GeC_6H_8　　　　　　　　　$PhGeH_3$

<u>Synth.</u>: By rxn. of $PhGeCl_3$ with $LiAlH_4$ in Et_2O (1762), in Ar atm. in 75% yield (1364) in N atm., in 29% yield (2593).

By rxn. of $GeCl_4$ with PhMgBr (1:1), followed by $LiAlH_4$ in Et_2O, in 28% yield (50).

By-prod. in thermal dec. of $PhGeH_2Cl$ at 150° (1762).

<u>Prop.</u>: b_{23} 48° (50), b_{22} 40° (1762), b_{12} 41-43° (1364), n_D^{20} 1.5353, d_{20} 1.2371 (1762), IR (1364, 1762), far IR and assignment (105, 2364), Raman (2364), UV (2593) and NMR (460, 1762) spectra, force constant (138), GLC (2364), oscillator strength (2593), thermal dec. at 150-200° → Ph_2GeH_2 + GeH_4 + H; in 6-20% conversion (1762).

<u>Rxn. with</u>: UV irradiation at 30° → 12% dec. (1762).

$ClCH_2OMe$ + cat. $AlCl_3$ → $PhGeH_2Cl$ (1:1) (1762, 1799) + $PhGeHCl_2$ (1:2) (1762).

Br in $HcCl_3$ → $PhGeBr_3$ (50).

$HgCl_2$ → $PhGeH_2Cl$ (1762).

$\overline{OCCH_2CH_2CONX}$ → $PhGeH_2X$; X = Br, I (1762), Br (1799) + $PhGeI_3 + GeI_4$ (1762).

CH_2N_2 + UV irradiation or Cu cat. → $Me_2PhGeH + MePhGeH_2$ (1799).

Additional nonfunctional organogermanes are compiled in Table 22.

Table 22. Nonfunctional Organogermanium Hydrides R_nGeH_{4-n}

Formula*	Synth. Method**	Yield	Properties	Ref.
$(PhCH_2)_2GeH_2$	II	8	$b_{0.001}$ 80-85° (a-105, b)	104
$HC\!\equiv\!CGeH_3$	--	--	(a-1431, c)	--
$n\text{-}Pr_3GeH$ (54)	IV	3	GLC (d, e)	491
$n\text{-}Pr_2GeH_2$ (56)	--	--	(f, g)	--
$i\text{-}Pr_2GeH_2$ (57)	--	--	(h)	--
$i\text{-}BuGeH_3$	III	--	IR and mass spectra	1108
$(cyclo\text{-}C_5H_9)_3GeH$	I	35	$b_{0.4}$ 121-22°, n_D^{20} 1.5162, d_{20} 1.0954, IR spectrum	2591
$(n\text{-}C_6H_{13})_3GeH$ (57)	--	--	(h)	--
$cyclo\text{-}C_6H_{11}GeH_3$	--	--	IR spectrum	2039
$(o\text{-}MeC_6H_4)_3GeH$ (58)	--	--	(a-105)	--
$(m\text{-}MeC_6H_4)_3GeH$ (58)	--	--	(a-105)	--
$(p\text{-}MeC_6H_4)_3GeH$ (58)	--	--	(a-105)	--
$(Me_3C_6H_2)_3GeH$	--	--	Mesityl (a-105)	--
$(1\text{-}C_{10}H_7)_3GeH$ (58)	--	--	(h)	--
$R GeH_{4-n}$	--	--	(i)	--

* Numbers in parenthesis refer to pages in main volume.
** The characters used here correspond to the synthetic scheme in the introduction to Chapter 3.1.
(a) (Far) IR spectrum and assignment. (b) IR and NMR spectra (104). (c) Microwave spectrum (2029-30), force constant and thermodynamic properties (1734). (d) Rxn. with MeMgX → $n\text{-}Pr_3GeMe$ (491). (e) Rxn. with ROC≡CH → $Pr_3GeCH\!:\!CHOR$; R = Et, Bu (2483). (f) Rxn. with HgX_2 → Pr_2GeHX; X = Cl (24), I (25). (g) Rxn. with R_2Hg → Pr_2RGeH; R = CH_2CO_2Me (25). (h) Use in cat. for olefin polymerization (411). (i) IR spectra (135) and force constant calcn. (1014), influence of molecular vicinity on Ge-H by IR and NMR spectra, based on literature data (1476), Ge-bond refractions D_{20} (81).

3.2 UNSYMMETRIC ORGANOGERMANIUM HYDRIDES

Unsymmetric organogermanes containing two and three different organic groups are included in this chapter. The compounds compiled in Table 23 are prepared by methods from the following scheme.

I Hydrogenation with lithium aluminum hydride:
- (IA) $Et_2RGeX + LiAlH_4 \rightarrow Et_2RGeH$,
- (IB) $R_4Ge + R'_4Ge + AlCl_3, + LiAlH_4 \rightarrow R_2R'GeH$,
- (IC) $Bu_2\overline{GeCH_2CH_2CH_2} + LiAlH_4 \rightarrow PrBu_2GeH$.

II Alkylation reactions:
- (IIA) $RGeH_nX_{3-n} + R'MgX \rightarrow RR'_{3-n}GeH_n$
- (IIB) $EtGeH_2Na + MeCl \rightarrow MeEtGeH_2$,
- (IIC) $PhGeH_3 + CH_2N_2 \rightarrow Me_nPhGeH_{3-n}$,
- (IID) $Ph_2GeH_2 + BuLi, + H_2O \rightarrow BuPh_2GeH$.

R = organic group, X = halogen, n = 1, 2. Other unsymmetric organogermanium hydrides are listed in Chapter 7.3.

Methyl-α-Naphthyl Phenyl Germane (M-58) $GeC_{17}H_{16}$ $MePh(1-C_{10}H_7)GeH$
Synth.: By rxn. of $MePh(1-C_{10}H_7)GeLi$ with MeOH under retention of configuration (417).
Prop.: IR and NMR spectra (2674).
Rxn. with: Br → $MePh(1-C_{10}H_7)GeBr$; under racemization (417).
$O\overline{CCH_2CH_2CON}Br$ → $MePh(1-C_{10}H_7)GeBr$; under racemization (417).
(+)-form + BuLi, + EI → (-)-$MeEtPhGe(C_{10}H_7-1)$ (2367).
(-)-form + BuLi, + EI → (+)-$MeEtPhGe(C_{10}H_7-1)$ (2367).
$LiAlD_4$ in THF → slight racemization + D-exchange (-5%) (2674).

Ethyl-α-Naphthyl Phenyl Germane (M-59) $GeC_{18}H_{18}$ $EtPh(1-C_{10}H_7)GeH$
Synth.: By rxn. of R_3GeSPh^* and $N-R_3GeNC_4H_4^*$ with $LiAlH_4$ in Et_2O under inversion of configuration (2368).
By rxn. of $R_3GeOC_{10}H_{19}^*$ with $LiAlH_4$ in Et_2O yielding (+)-form (999, 2368), rxn. of (-)-R_3GeOMe^* with $LiAlH_4$ proceeds under retention (2368), resolved R_3Ge-$OCHMePh^*$ with $LiAlH_4$ yields (-)-form (999).
By rxn. of $(R_3Ge)_2O^*$ with $LiAlH_4$ in Et_2O with retention (2368).
By rxn. of R_3GeLi^* with H_2O under retention (2368).
Stereochemistry in rxn. of $R_3GeCO_2GeR_3^*$ with $LiAlH_4$ (2368).
Prop.: (±)-Form oil, $b_{0.01}$ 140-41°, n_D^{25} 1.6270, IR spectrum (999).
(+)-Form m. 31-32.5°, $[\alpha]_D^{25}$ + 23.6° (c. 5.5 in C_6H_6), fluoresces strongly under UV irradiation, IR spectrum (999).
(-)-Form oil, n_D^{25} 1.6226, $[\alpha]_D^{25}$ -10.9° (c. 13.3 in C_6H_6), IR spectrum (999).
Absolute configuration (2367).

(See page 94 for continuation of rxns. of above compound.)

---

* R_3Ge = $EtPh(1-C_{10}H_7)Ge$.

Table 23. Unsymmetric Organogermanium Hydrides RR'_nGeH_{3-n}

Formula*	Synth. Method**	Yield	Properties	Ref.
Me$_2$EtGeH (60)	IIA - EtGeHCl$_2$	--	b$_1$ 62°, n$_D^{20}$ 1.4078, d$_{20}$ 1.0077 (a)	366
Me$_2$BuGeH	IB	--	GLC (b, c)	491
Me$_2$PhGeH	IIC	14	b$_{50}$ 80°, n$_D^{20}$ 1.5170, d$_{20}$ 1.1583, IR and NMR spectra	1799
MeEtGeH$_2$	IIB	--	NMR spectrum (d)	735
MeBu$_2$GeH	IB	--	GLC (b)	491
MeBuGeH$_2$	IIA - BuGeH$_2$Br	--	b. 90-105°, IR and NMR spectra (e)	2039
MePh$_2$GeH (60)	IIC - Ph$_2$GeH$_2$ (f)	43	b$_{0.5}$ 90°, n$_D^{20}$ 1.5830, d$_{20}$ 1.2128,	1799
	IIC - Ph$_2$GeH$_2$ (g)	25	IR and NMR spectra (h)	1799
MePhGeH$_2$	IIC - PhGeH$_3$ (f)	59	b$_{35}$ 70°, n$_D^{20}$ 1.5292, d$_{20}$ 1.2024 (i)	1799
Et$_2$PrGeH	IB	49	GLC (b, c)	491
Et$_2$BuGeH	IB	--	GLC (b)	491
Et$_2$(CH$_2$:CHCH$_2$CH$_2$)GeH	IA	--	b. 160°, n$_D^{20}$ 1.4562, d$_{20}$ 1.0088	2578
Et$_2$(CH$_2$:CHCHMeCH$_2$)GeH	IA	--	b$_{20}$ 68°, n$_D^{20}$ 1.4568, d$_{20}$ 0.9906	2578
EtPr$_2$GeH	IB	24	GLC (b, c)	491
EtBu$_2$GeH	IB	--	GLC (b)	491
PrBu$_2$GeH	IC	--	b$_{13}$ 96°, n$_D^{20}$ 1.4569, d$_{20}$ 0.9774,	344,1492
		40	IR spectrum, GLC	1492
Bu$_2$PhGeH (60)	--	--	IR spectrum	336
BuPh$_2$GeH	IID	36	b$_{0.01}$ 120° (h)	105
BuPhGeH$_2$	IIA - BuGeH$_2$Br	--	b$_1$ 45-54°, b. 208-13° (j)	2039

* Numbers in parenthesis refer to pages in main volume.

** The characters used here correspond to the synthetic scheme in the introduction to Chapter 3.2.
(a) IR (366) and NMR (2373) spectra. (b) Rxn. with R"MgX → R$_2$R'GeR" (491). (c) Rxn. with SOCl$_2$ → R$_2$R GeCl (491). (d) Relative basicity vs. NaNH$_2$ in liq. NH$_3$ (735). (e) Rxn. with CCl$_4$ → H/Cl exchange reaction followed by NMR (2039). (f) Synth. under UV irradiation. (g) Powd. Cu as cat. (h) (Far) IR spectrum and assignment (105). (i) IR and NMR spectra (1799). (j) NMR, IR spectra and assigment (2039).

(Cont. from Page 92)

<u>Rxn. with:</u> (±)-Form + Cl → (±)-R$_3$GeCl* (999).
(+)-Form + Cl → (−)-R$_3$GeCl* (999).
(±)-Form + BuLi, + MeI → (±)-R$_3$GeMe* (2637).
(+)-Form + BuLi, + MeI → (+)-R$_3$GeMe* (2637).
n-BuLi → R$_3$GeLi* + C$_4$H$_{10}$ (2369).

Listing of unsymmetric organogermanium hydrides concludes in Table 23.

3.3 DEUTERIUM SUBSTITUTED ORGANOGERMANIUM HYDRIDES

This chapter includes deuterides as well as compounds having deuterium-substitution in the carbon moiety. All compounds are listed in Table 24. Cyclic organogermanium deuterides may be found in Chapter 7.3. The following scheme summarizes synthetic methods for deuterium substituted organogermanium hydrides.

  I   Reaction GeCl$_4$ with excess RMgCl, followed by D$_2$O .

 II   Cleavage of R$_4$Ge with Li in dimethoxy ethane, followed by D$_2$O .

III   Reduction of RGeCl$_3$ with LiAlH$_4$ or LiAlD$_4$.

 IV   Metallation of Ph$_2$GeH$_2$ with BuLi, followed by D$_2$O .

---

* R$_3$Ge = EtPh(1-C$_{10}$H$_7$)Ge .

Table 24. Deuterium Substituted Organogermanium Hydrides   D-(R$_n$GeH$_{4-n}$)

Formula*	Synth. Method**	Yield	Properties	Ref.
Me$_2$GeHD	--	--	NMR spectrum	2361
MeGeD$_3$ (61)	--	--	(a-1067, b, c)	--
CD$_3$GeH$_3$ (61)	--	--	(b, c)	--
CD$_3$GeD$_3$	--	--	(b)	--
(PhCH$_2$)$_3$GeD	II	77	m. 81°, b$_{0.01}$ 170-76° (d, e, f)	104
(PhCH$_2$)$_2$GeD$_2$	II	--	(d, e)	104
Et$_3$GeD (61)	--	--	NMR spectrum	1444
EtGeD$_3$ (61)	--	--	IR spectrum and assignment	1445
HC≡CGeD$_3$	--	--	(Far) IR spectrum and assignment (a-1734)	1431
1-Pr$_3$GeD	--	--	(d)	--
BuPh$_2$GeD	IV	30	(d)	105
(cyclo-C$_6$H$_{11}$)$_3$GeD	I	30	R$_3$GeMgCl as synth. intermediate	353
	I	29	b$_{0.1}$ 143-44°, n$_D^{20}$ 1.5250, d$_{20}$ 1.0913, far IR spectrum, D content 94%	2591
Ph$_3$GeD	--	--	(d)	--
Ph$_2$GeD$_2$	--	--	(f)	--
PhGeD$_3$	III	--	Molecular distillation, GLC (g)	2364
(C$_6$D$_5$)GeH$_3$	III	--	b$_{13}$ 33-35°, GLC (g)	2364
(C$_6$D$_5$)GeD$_3$	III	56	b. 136-37°, GLC (g)	2364

* Numbers in parenthesis refer to pages in main volume.
** The characters used here correspond to the synthetic scheme in the introduction to Chapter 3.3.
(a) Force constant and thermodynamic properties. (b) Coriolis coefficient and ground state centrifugal distortion const. (869). (c) Calculation of vibrational spectra (1713). (d) (Far) IR spectrum and assignment (105). (e) IR spectrum (104). (f) Mass spectrum (1109). (g) (Far) IR, Raman spectra and assignment (2364).

3.6 FUNCTIONALLY SUBSTITUTED ORGANOGERMANIUM HYDRIDES

Organogermanes containing functional groups are treated in this chapter. The compounds are prepared by alkylation, reduction and addition reactions as explained in the following scheme.

- I Alkylation reaction of organogermanium halides or hydrides:
    - (IA) With organotin,
    - (IB) With organomercury,
    - (IC) With alkali metal derivatives .

- II Reduction of germanium halides:
    - (IIA) With lithium aluminum hydride,
    - (IIB) With organotin hydride .

- III Addition reactions of metal hydride:
    - (IIIA) To stressed cyclic compounds,
    - (IIIB) To methylene,
    - (IIIC) To olefin .

Potassium 2-Germaacetate $\quad$ GeKCH$_3$O$_2$ $\quad$ KO$_2$CGeH$_3$

<u>Synth.:</u> By rxn. of KGeH$_3$ with CO$_2$ at -196° to r.t. (2525).
<u>Prop.:</u> IR spectrum, hydrolyzes in air, pK of free acid, fairly stable in aq. soln. at r.t., dec. at 85° (2525).
<u>Rxn. with:</u> Thermal dec. at 480° → Ge + K$_2$CO$_3$ + CO + H (2525).
D$_2$O → rate of hydrolysis (2525).
H$_2$O → GeH$_4$ + KHCO$_3$ (2525).
HCl → GeH$_4$ + CO + GeH-contg. solid (2525)

Additional functionally substituted organogermanium hydrides are listed in Table 25 and in Chapters 2.7.1.1.3, 5.3 and 6.5.8.

Table 25. Functionally Substituted Organogermanium Hydrides

R'_nGeH_{4-n}

Formula	Synth. Method*	Yield	Properties	Ref.
$Me_2(ClCH_2)GeH$	IIA - $Me_2(ClCH_2)GeCl$	84	b_{740} 93.5-95° (a, b)	475
$Me_2(ClMgCH_2)GeH$	(b)	--	NMR spectrum, dec. on standing	475
$Et_2R'GeH$				
$R' = MeSiH_2(CH_2)_3$	IIA (c)	--	b_{22} 87°, n_D^{20} 1.4570, d_{20} 0.9793 (d)	1491
$MePh_2Si(CH_2)_3$	IIIA - $Et_2GeCH_2CH_2CH_2$	< 80	$b_{0.9}$ 143°, n_D^{20} 1.4760, d_{20} 1.0627	1491
$MeCl_2Si(CH_2)_3$	IIIA - $Et_2GeCH_2CH_2CH_2$	< 80	$b_{0.2}$ 148°, n_D^{20} 1.5481, d_{20} 1.0752 (c, d)	1491
$n-Pr_2R'GeH$				
$R' = HOCH_2CH_2$	IIA - $Pr_2(MeO_2CCH_2)GeCl$	60	$b_{7.5}$ 98-99°, n_D^{20} 1.4670, d_{20} 1.0866	24
	IIA - $Pr_2(MeO_2CCH_2)GeH$	80	$b_{1.5}$ 75-76°, d_{20} 1.0885 (e, f)	25
MeO_2CCH_2	IB - Pr_2GeHI	74	b_2 64-65.5°, n_D^{20} 1.4539, d_{20} 1.1006 (e, f)	24
	IB - Pr_2GeH_2	59	$b_{1.5}$ 54-54.5°, n_D^{20} 1.4532, d_{20} 1.0980	25
	IA - Pr_2GeHCl	70	b_{10} 110-12°, n_D^{20} 1.4545, d_{20} 1.0903 (g)	2238
$AcCH_2$	IA - Pr_2GeHI	78	$b_{1.5}$ 69-70°, n_D^{20} 1.4642, d_{20} 1.0991	2238
$n-Bu_2R'GeH$				
$R' = HOCH_2CH_2$	IIA - $Bu_2(MeO_2CCH_2)GeH$	72	b_1 82-84°, n_D^{20} 1.4679, d_{20} 1.0446 (e)	25
MeO_2CCH_2	IB - Bu_2GeHI	71	b_6 106-109°, n_D^{20} 1.4560, d_{20} 1.0661 (e, h)	24
	IB - Bu_2GeH_2	62	b_1 73-74.5°, n_D^{20} 1.4552, d_{20} 1.0613	25

* The characters used here correspond to the synthetic scheme in the introduction to Chapter 3.6.
(a) NMR, IR spectra and assignment (475). (b) Rxn. of $Me_2(ClCH_2)GeH$ with Mg in THF → $Me_2(ClMgCH_2)GeH$ (475). (c) Rxn. of $R_2(MeSiCl_2CH_2CH_2CH_2)GeH$ with $LiAlH_4$ → $R_2(MeSiH_2CH_2CH_2CH_2)GeH$; R = Et, Bu (1491). (d) IR and NMR spectra (1491). (e) IR spectrum (24, 25). (f) Rxn. with HgX_2 → $Pr_2R'GeX + Pr_2GeHX$ (25). (g) Rxn. with $LiAlH_4$ → $Pr_2(HOCH_2CH_2)GeH$ (25). (h) Rxn. with $HgCl_2$ → $Bu_2R'GeCl$ (24), with $LiAlH_4$ → $Bu_2R'GeH$ (25), with $(MeO_2CCH_2)_2Hg$ → $Bu_2GeR'_2$ (25).

Table 25 Continued

$R'_n GeH_{4-n}$

Formula	Synth. Method*	Yield	Properties	Ref.
n-Bu$_2$R'GeH (Cont'd.)				
R' = MeSiH$_2$(CH$_2$)$_3$	IIA (c)	--	b$_{20}$ 130°, n$_D^{20}$ 1.4621, d$_{20}$ 0.9453 (d)	1491
MeBu$_2$Si(CH$_2$)$_3$	IIIA - Bu$_2$GeCH$_2$CH$_2$CH$_2$	< 80	b$_{0.8}$ 152°, n$_D^{20}$ 1.4650, d$_{20}$ 0.9246 (i)	1491
MePh$_2$Si(CH$_2$)$_3$	IIIA - Bu$_2$GeCH$_2$CH$_2$CH$_2$	< 80	b$_{0.1}$ 156°, n$_D^{20}$ 1.5353, d$_{20}$ 1.0261 (i)	1491
Et$_3$Si(CH$_2$)$_3$	IIIA - Bu$_2$GeCH$_2$CH$_2$CH$_2$	< 80	b$_{1.5}$ 143°, n$_D^{20}$ 1.4670, d$_{20}$ 0.9406 (i)	1491
Bu$_2$SiH(CH$_2$)$_3$	IIA (j)	--	b$_{1.3}$ 148°, n$_D^{20}$ 1.4660, d$_{20}$ 0.9233 (d)	1491
MeCl$_2$Si(CH$_2$)$_3$	IIIA - Bu$_2$GeCH$_2$CH$_2$CH$_2$	< 80	b$_{18}$ 160°, n$_D^{20}$ 1.4699, d$_{20}$ 1.0912 (c, d)	1491
Bu$_2$ClSi(CH$_2$)$_3$	IIIA - Bu$_2$GeCH$_2$CH$_2$CH$_2$	< 80	b$_{1.3}$ 165°, n$_D^{20}$ 1.4702, d$_{20}$ 0.9814 (d, j)	1491
Ph$_2$(NCCH$_2$CH$_2$)GeH	IIIC - Ph$_2$GeH$_2$ (k)	33	b$_{0.03}$ 136°, n$_D^{20}$ 1.5862, d$_{20}$ 1.2472 (1)	2577
Ph$_2$(NCCH$_2$CH$_2$CH$_2$)GeH	IIIC - Ph$_2$GeH$_2$ (k)	35	b$_{0.05}$ 163°, n$_D^{20}$ 1.5796, d$_{20}$ 1.2105	2577
Ph(Ph$_2$CHCHOH)GeH$_2$	IIA - Ph(Ph$_2$CHCO)GeCl$_2$	--	Dec. at 120-40° (m)	2714
MeOCH$_2$GeH$_3$	IC - H$_3$GeK (n)	41	GLC	910
	IC - H$_3$GeNa	11	m. - 121.6 ± 0.3°, mol. wt. GLC (o)	1095
Ph$_2$CHCHOHGeH$_3$	IIA - Ph$_2$CHCOGeCl$_3$	--	Dec. at 120-40° (m)	2714
(MeO$_2$CCH$_2$)$_3$GeBr	IIB - (MeO$_2$CCH$_2$)$_3$GeBr	70	b$_{0.7}$ 110-11°, n$_D^{20}$ 1.4752, d$_{20}$ 1.3580 (p)	2238
HO(CH$_2$)$_3$GeH$_3$	IIA - HO$_2$CCH$_2$CH$_2$GeI$_3$	46	b$_{758}$ 143°, n$_D^{20}$ 1.4628, d$_{20}$ 1.2585	1496
HOCH$_2$CH$_2$CHMeGeH$_3$	IIA - OHCCH$_2$CHMeGeI$_3$	37	b$_{18}$ 68°, n$_D^{20}$ 1.4682, d$_{20}$ 1.1976	1496
[H$_2$N(CH$_2$)$_4$]$_2$GeH$_2$	IIA - [NC(CH$_2$)$_4$]$_2$GeBr$_2$	--	b$_{20}$ 140°, n$_D^{20}$ 1.4988, d$_{20}$ 1.1078 (l, q)	2577
HOCH$_2$CH$_2$C$_4$H$_6$GeH$_3$ (r)	IIA - HO$_2$CCH$_2$C$_4$H$_6$GeI$_3$	47	b$_{13}$ 89°, n$_D^{20}$ 1.4906, d$_{20}$ 1.152 (s)	1496
p-MeOC$_6$H$_4$GeH$_3$	II (t)	32	b$_{0.5}$ 49°, UV spectrum (u)	2593

(i) NMR spectrum (1491). (j) Rxn. of Bu$_2$(Bu$_2$SiCl CH$_2$CH$_2$CH$_2$)GeH with LiAlH$_4$ → Bu$_2$(Bu$_2$SiHCH$_2$CH$_2$CH$_2$)GeH (1491). (k) Synth. in sealed tube with cat. AIBN (2577). (l) IR and NMR spectra (2577). (m) IR and NMR spectra (2714). (n) Synth. with Li and Na derivative in 9 and 19% yield, resp. (910). (o) Vapor pressure equation, IR, NMR and mass spectra and relative base strength (1095). (p) IR and NMR spectra (2238). (q) Sensitive to air (2577). (r) HOCH$_2$CH$_2$C$_4$H$_6$GeH$_3$ = HOCH$_2$CH$_2$CH(CH:CHMe)GeH$_3$. (s) IR and NMR spectra (1496). (t) Synth. from GeI$_2$ and p-MeOC$_6$H$_4$I in sealed tube at 165°, followed by LiAlH$_4$ (2593). (u) Oscillator strength (2593).

3.8 ORGANOGERMANIUM HYDRIDES CONTAINING HETEROATOM-GERMANIUM BONDS

3.8.1 ORGANOGERMANIUM HYDRIDE HALIDES

Organogermanes containing both hydrogen and halogen bound to the same germanium atom are listed in this chapter. The following scheme summarizes methods for preparation of the compounds compiled in Table 26.

I Alkylation reactions:
$HGeCl_3 + Me_4Sn \rightarrow MeHGeCl_2$.

II Addition reactions:
(IIA) $BuH_2GeCl + CH_2{:}CRCH_2Cl \rightarrow Bu(ClCH_2CHRCH_2)GeHCl$,
(IIB) $HGeCl_3 \cdot 2Et_2O + C_2H_2 \rightarrow (Cl_3GeCH{:}CH)HGeCl_2$.

III Halogen exchange reactions:
(IIIA) $BuGeH_3 + HgX_2 \rightarrow BuH_2GeX$,
(IIIB) $RGeH_3 + (CH_2CO)_2NX \rightarrow RH_2GeX + RHGeX_2$.

IV Neutralization reactions:
(IVA) $(R_2GeH)_2O + HF \rightarrow R_2HGeF$,
(IVB) $(RGeH_2)_2O + HX \rightarrow RH_2GeX$,
(IVC) $(RHGeO)_x + HX \rightarrow RH_2GeX + RHGeX_2$.

Diethylchlorogermane (M-63) $GeC_4H_{11}Cl$ Et_2HGeCl
Synth.: By rxn. of Et_2GeH_2 with $MeOCH_2Cl$ and cat. $AlCl_3$ at 60°, in 90% yield (1471), with $HgCl_2$, in 80% yield (1471).
Prop.: IR (1798) and NMR (1471) spectra.
Rxn. with: $(CH_2{:}CH)_2 \rightarrow Et_2(MeCH{:}CHCH_2)GeCl$ (467) $+ Et_2(c.C_6H_{11}CH_2CH_2)GeCl +$ $c.4\text{-}C_6H_9CH{:}CH_2$ (1798).
$(CH_2{:}CMe)_2 \rightarrow Et_2(Me_2C{:}CMeCH_2)GeCl$ (467, 1798).
$HC{:}CCH_2Cl \rightarrow Et_2(ClCH_2CH{:}CH)GeCl + CH_2{:}C(GeEt_2Cl)CH_2Cl + Et_2GeCl_2 + MeC{:}CH$ (1472).
$CH_2{:}CHCOR \rightarrow Et_2(ROCCH_2CH_2)GeCl$; R = Me, EtO (1798).
$Li_3N \rightarrow (Et_2GeH)_3N$ (1471).
Aq. $NH_3 \rightarrow (Et_2GeH)_2O$ (1471).
Dry NH_3 in $Et_2O \rightarrow (Et_2GeH)_2NH + (Et_2GeH)_3N$ (1471).
$(EtS)_2Pb \rightarrow Et_2HGeSEt$ (465).

Ethylchlorogermane GeC_2H_7Cl EtH_2GeCl
Synth.: By rxn. of $EtGeH_3$ with $HgCl_2$ at -40° without solvent, in 65% yield (1471).
Prop.: b. 90.5°, n_D^{20} 1.4577, d_{20} 1.3702, IR and NMR spectra (1471).
Rxn. with: $MeOCH_2Cl$ + cat. $AlCl_3 \rightarrow EtHGeCl_2$ (1471).
Aq. $NH_3 \rightarrow (EtGeH_2)_2O$ (1471).
Dry NH_3 in $Et_2O \rightarrow (EtGeH_2)_3N$ (1471).

Ethyldichlorogermane (M-64) $GeC_2H_6Cl_2$ $EtHGeCl_2$
Synth.: By rxn. of EtH_2GeCl with $MeOCH_2Cl$ at 60°, in presence of cat. amt. of $AlCl_3$, in 84% yield (1471).

By-prod. in rxn. of Et$_4$Ge with HGeCl$_3$·OEt$_2$ (366).
Prop.: b. 129°, n$_D^{20}$ 1.4747, d$_{20}$ 1.5329 (1471), IR (366, 1471) and NMR (1471, 2373) spectra.
Rxn. with: MeMgCl → Me$_2$EtGeCl (366).
HC⋮CCH$_2$Cl → Et(ClCH$_2$CH:CH)GeCl$_2$ + Et(HC⋮CCH$_2$)GeCl$_2$ + HCl (1472).
Aq. NH$_3$ → (EtGeHO)$_x$ (1471).

Dipropylchlorogermane   GeC$_6$H$_{15}$Cl   Pr$_2$HGeCl
Synth.: By rxn. of Pr$_2$GeH$_2$ with HgCl$_2$ in 65% yield (24).
By rxn. of Pr$_2$(MeO$_2$CCH$_2$)GeH with HgCl$_2$ in Et$_2$O under reflux (24).
Prop.: b$_{20}$ 88-90°, b$_7$ 54-55°, n$_D^{20}$ 1.4604-24, d$_{20}$ 1.1608-22, IR spectrum (24).
Rxn. with: R$_2$Hg → Pr$_2$RGeCl; R = CH$_2$CO$_2$Me (24).
Bu$_3$SnCH$_2$CO$_2$Me → Pr$_2$(MeO$_2$CCH$_2$)GeH (2238).

Dipropyliodogermane   GeC$_6$H$_{15}$I   n-Pr$_2$HGeI
Synth.: By rxn. of Pr$_2$GeH$_2$ with HgI$_2$ in THF, in 68% yield (25).
By rxn. of Pr$_2$(HOCH$_2$CH$_2$)GeH with HgI$_2$ in Et$_2$O, in 32% yield (25).
Prop.: b$_{2.5}$ 64-66°, n$_D^{20}$ 1.5226, d$_{20}$ 1.5725, 1.5898, IR spectrum (25).
Rxn. with: R$_2$Hg → Pr$_2$RGeH; R = CH$_2$CO$_2$Me (24).
Bu$_3$SnOMe → Pr$_2$HGeOMe (2238).
Bu$_3$SnCH$_2$Ac → Pr$_2$(AcCH$_2$)GeH (2238).

Dibutylchlorogermane (M-63)   GeC$_8$H$_{19}$Cl   n-Bu$_2$HGeCl
Prop.: IR, assignment (2039) and NMR (1471, 2039) spectra.
Rxn. with: (CH$_2$:CMe)$_2$ → Bu$_2$(Me$_2$C:CMeCH$_2$)GeCl (467, 1798).
R$_2$Hg → Bu$_2$RGeCl; R = CH$_2$CO$_2$Me (24).
Aq. NH$_3$ → (Bu$_2$GeH)$_2$O (1471).

Dibutyliodogermane (M-64)   GeC$_8$H$_{19}$I   n-Bu$_2$HGeI
Synth.: (M-64).
Prop.: IR (336), assignment (2039) and NMR (1471, 2039) spectra.
Rxn. with: R$_2$Hg → Bu$_2$RGeH; R = CH$_2$CO$_2$Me (24).
I → Bu$_2$GeI$_2$ + HI (2039).
MCl$_4$ → relative activity in Cl exchange rxn., by NMR; M = C, Ge (2039).

Butylchlorogermane (M-64)   GeC$_4$H$_{11}$Cl   n-BuH$_2$GeCl
Synth.: By rxn. of BuGeH$_3$ with MeOCH$_2$Cl at 60° in presence of cat. AlCl$_3$, in 80% yield (1471).
By rxn. of BuGeH$_3$ with HgCl$_2$ in N atm. at -10°, in 89% yield (2039).
Prop.: IR, assignment (2039) and NMR (1471, 2039) spectra.
Rxn. with: CH$_2$:CRCH$_2$Cl → Bu(ClCH$_2$CHRCH$_2$)GeHCl; R = H, Me (2575).
MeOCH$_2$Cl + cat. AlCl$_3$ → BuHGeCl$_2$ (1471).
Aq. NH$_3$ → (BuGeH$_2$)$_2$O (1471).
HgCl$_2$ → BuHGeCl$_2$ + Hg (1471) + BuGeCl$_3$ (1798).
Pb(SEt)$_2$ (2:1) → BuH$_2$GeSEt + PbCl$_2$ (2039).

Bromobutylgermane (M-64) $GeC_4H_{11}Br$ BuH_2GeBr

Prop.: b_{20} 60° (2039), IR (1471), assignment (2039) and NMR (1471, 2039) spectra.

Rxn. with: RMgX → $BuRGeH_2$; R = Me, Ph (2039).

$GeCl_4$ → relative reactivity in Cl-exchange rxn. by NMR (2039).

Butyldichlorogermane $GeC_4H_{10}Cl_2$ $BuHGeCl_2$

Synth.: By rxn. of $BuGeH_3$ with $HgCl_2$, in 35% yield (467), in 33% yield (1471), separated by GLC (1798).

Prop.: b_{28} 78° (467, 1798), b_{25} 74.5° (1471), n_D^{20} 1.4738 (467, 1798), 1.4725 (1471), d_{20} 1.3722 (467, 1798), 1.3708 (1471) IR (1471, 1798) and NMR spectra (1471), GLC (1798).

Rxn. with: $(CH_2:CMe)_2$ → $Bu(Me_2C:CMeCH_2)GeCl_2$ (467, 1798).

Aq. NH_3 → $(BuGeHO)_n$; n = 4, x (1471).

Diphenylfluorogermane $GeC_{12}H_{11}F$ Ph_2HGeF

Synth.: By rxn. of $(Ph_2GeH)_2O$ with 25% aq. HF, in 87% yield (1762).

Prop.: $b_{0.002}$ 66°, n_D^{20} 1.5790, d_{20} 1.3436, IR and NMR spectra (1762).

Rxn. with: Thermal dec. in N atm. at 200° → 2% dec. in 4 hours (1762).

1-Hexene → $n\text{-}C_6H_{13}Ph_2GeF$; rel. rate and rxn. mechanism (1762).

Chlorodiphenylgermane $GeC_{12}H_{11}Cl$ Ph_2HGeCl

Synth.: By rxn. of Ph_2GeCl_2 with Ph_3GeH in presence of cat. $AlCl_3$ in Ar atm. (1364).

By rxn. of Ph_2GeH_2 with $MeOCH_2Cl$ and cat. $AlCl_3$ (1799), at 50-60°, in 86% yield (1762), with $HgCl_2$ at 50-60°, in 67% yield (1762), with $(CH_2CO)_2NCl$ in $MeNO_2$ at 150° (1762).

By rxn. of $(Ph_2GeH)_2O$ with HCl (1799).

By-prod. in thermal dec. of PhH_2GeCl at 150° (1762).

Prop.: $b_{0.4}$ 102°, n_D^{20} 1.6010, d_{20} 1.3514 (1762, 1799) IR (1364, 1762, 1799), and NMR (1762, 1799) spectra.

Rxn. with: Dec. at r.t. → Ph_2GeH_2 + Ph_2GeCl_2 + Ph_3GeCl (1762).

Thermal dec. at 200° → 5% dec. after 4 hours (1762).

1-Hexene $n\text{-}C_6H_{13}(Ph_2)GeCl$; relative rate and rxn. mechanism (1762).

$c.C_6H_{10}O$ → $Ph_2Ge(OC_6H_{11})Cl$ + $Ph_2Ge(OC_6H_{11})_2$ + Ph_2GeCl_2 + trace Ph_2GeH_2 (1762).

Me_2CHCHO + UV irradiation → $Ph_2Ge(OCH_2CHMe_2)Cl$ (1762).

Aq. NH_3 → $(Ph_2GeH)_2O$ (1762, 1799).

Bromodiphenylgermane $GeC_{12}H_{11}Br$ Ph_2HGeBr

Synth.: By rxn. of Ph_2GeH_2 with $HgBr_2$ at 70°, in 35% yield (1762), with $(CH_2CO)_2NBr$ in petr. ether at 60°, in 72% yield (1762).

By rxn. of $(Ph_2GeH)_2O$ with concn. HBr in 75% yield (1762).

Prop.: $b_{0.001}$ 90°, n_D^{20} 1.6186, d_{20} 1.5340, IR and NMR spectra (1762, 1799).

Rxn. with: Thermal dec. at 200° → 50% dec. after 4 hours (1762).

1-Hexene → $n\text{-}C_6H_{13}(Ph_2)GeBr$; relative rate, rxn. mechanism (1762).

Diphenyliodogermane   GeC$_{12}$H$_{11}$Br   Ph$_2$HGeI

Synth.: By rxn. of Ph$_2$GeH$_2$ with (CH$_2$CO)$_2$NI in petr. ether at 60°, in 30% yield (1762).
By rxn. of (Ph$_2$GeH)$_2$O with 25% HI, in 90% yield (1762).
Prop.: m. 27°, IR and NMR spectra (1762, 1799).
Rxn. with: Thermal dec. at 200° → 60% dec. after 4 hours (1762).
1-Hexene → n-C$_6$H$_{13}$(Ph$_2$)GeI; relative rate, rxn. mechanism (1762).

Chlorophenylgermane   GeC$_6$H$_7$Cl   PhH$_2$GeCl

Synth.: By rxn. of PhGeH$_3$ with MeOCH$_2$Cl (1:1) in presence of cat. AlCl$_3$, in 85% yield (1762), with HgCl$_2$ at 25°, in 75% yield (1762).
By rxn. of (PhGeH$_2$)$_2$O with aq. HCl (1762).
Prop.: b$_{13}$ 80°, n$_D^{20}$ 1.5650, d$_{20}$ 1.4364, IR and NMR spectra (1762, 1799).
Rxn. with: Thermal dec. at 150° → PhGeH$_3$ + PhHGeCl$_2$ + PhGeCl$_3$ + Ph$_2$HGeCl$_2$ + Ph$_2$GeCl$_2$ + HGeCl$_3$ + H + C$_6$H$_6$ + GeH$_4$ (1762).
ClCH$_2$OMe + cat. AlCl$_3$ → PhHGeCl$_2$ (1762).
Aq. NH$_3$ → (PhGeH)$_2$O (1762, 1799).

Dichlorophenylgermane   GeC$_6$H$_6$Cl$_2$ PhHGeCl$_2$

Synth.: By rxn. of PhGeH$_3$ with MeOCH$_2$Cl in presence of cat. AlCl$_3$ at 60°, in 84% yield (1762).
By rxn. of PhH$_2$GeCl with MeOCH$_2$Cl in presence of cat. AlCl$_3$ (1799).
By-prod. in thermal dec. of PhH$_2$GeCl at 150° (1762).
Prop.: b$_{11}$ 94°, n$_D^{20}$ 1.5642, d$_{20}$ 1.5435, IR and NMR spectra (1762, 1799).
Rxn. with: 1-Hexene → n-C$_6$H$_{13}$(Ph)GeCl$_2$; relative rate and rxn. mechanism (1762).
PhGe(OR)Cl$_2$ → PhGeCl$_3$ + (PhGeCl$_2$)$_2$ + ROH; R = Me, i-Bu, cyclo-C$_6$H$_{11}$ (1762).
Ph$_2$C:CO → Ph(Ph$_2$C:COH)GeCl$_2$ (2714).
Aq. NH$_3$ → (PhGeHO)$_x$ (1762).
i-PrCHO + UV irradiation → PhGe(OCH$_2$CHMe$_2$)Cl$_2$ + i-BuOH (1762).
Cyclo-C$_6$H$_{10}$O + UV irradiation → PhGe(OC$_6$H$_{11}$)Cl$_2$ + C$_6$H$_6$ + C$_6$H$_{11}$OH + trace (PhGeCl$_2$)$_2$ (1762).

Additional organogermanium hydride halides are listed in Table 26. Derivatives of cyclic organogermanes may be found in Chapter 7.3.

3.8.2 ORGANOGERMANIUM HYDRIDE PSEUDOHALIDES

Organogermanes containing hydrogen and (iso-)cyano or (iso-)thiocyanato groups at the same germanium atom may be found at the bottom of Table 26.

Table 26. Organogermanium Hydride Halides　　　　　　　　　　　　　　　$R_nH_mGeX_o$

Formula*	Synth. Method**	Yield	Properties	Ref.
MeHGeCl$_2$ (64)	I	--	b. 101.5°, n_D^{20} 1.4701, d_{20} 1.6356 (a, b)	366
Et$_2$HGeF	IVA	84	b. 112°, n_D^{20} 1.4132, d_{20} 1.2158 (c, d)	1471
Et$_2$HGeBr (64)	--	--	IR and NMR spectra (e-467, 1798)	1471
Et$_2$HGeI (64)	--	--	(c, e-1798)	--
EtH$_2$GeF	IVB	--	Dec. rapidly to EtGeH$_3$ (c)	1471
EtH$_2$GeBr	IIIB	46	b. 112-15°, n_D^{20} 1.4945, d_{20} 1.7350	1471
	IVB	83	b. 113°, n_D^{20} 1.4962, d_{20} 1.7341	1471
	IVC	--	IR and NMR spectra	1471
EtH$_2$GeI	IIIB	15	b. 137-38°, n_D^{20} 1.5613, IR and NMR spectra	1471
	IVB	82	n_D^{20} 1.5612, d_{20} 2.0277	1471
EtHGeBr$_2$	IVC	85	IR and NMR spectra	1471
EtHGeI	IIIB	--	Dec. upon distn., IR and NMR spectra	1471
(Cl$_3$GeCH$_2$CH$_2$)HGeCl$_2$	IIB	--	Mixt. with (Cl$_3$GeCH$_2$), from C$_2$H$_4$	1588
(Cl$_3$GeCH:CH)HGeCl$_2$	IIB	--	Mixt. with (Cl$_3$GeCH:)$_2$	1588
Bu$_2$HGeF	IVA	80	b$_{60}$ 114°, n_D^{20} 1.4344, d_{20} 1.0848 (c, f)	1471
Bu$_2$HGeBr (64)	--	--	IR spectrum and assignment (c, g)	2039
Bu(ClCH$_2$CH$_2$CH$_2$)HGeCl	IIA	--	b$_{17}$ 137°, n_D^{20} 1.4876, d_{20} 1.2795 (h)	2575
Bu(ClCH$_2$CHMeCH$_2$)HGeCl	IIA	--	b$_{15}$ 137°, n_D^{20} 1.5008, d_{20} 1.2558 (h)	2575
BuH$_2$GeI (64)	IIIA (?)	74	b. 182°, b$_{20}$ 92°, IR spectrum (g)	2039

* Numbers in parenthesis refer to pages in main volume.
** The characters used here correspond to the synthetic scheme in the introduction to Chapter 3.8.1.
(a) IR (366) and NMR (2373) spectra. (b) Rxn. with C$_2$H$_4$ → MeEtGeCl$_2$ (366).
(c) NMR spectrum (1471). (d) Dec. → Et$_2$GeH$_2$ + Et$_2$GeF$_2$ (1471). (e) Rxn. with (CH$_2$:CH)$_2$ at 110° → Et$_2$(MeCH:CHCH$_2$)GeX (467, 1798). (f) Decomposes slowly to Bu$_2$GeF$_2$ + Bu$_2$GeH$_2$ (1471). (g) Rxn. with GeCl$_4$ → relative exchange rate, measured by NMR technique (2039). (h) Rxn. with Na in MePh → $\overline{CH_2CHRCH_2GeHBu}$; R = H, Me (2575).

Table 26 Continued $R_nH_mGeX_o$

Formula*	Synth. Method**	Yield	Properties	Ref.
PhH$_2$GeF	IVB	86	b$_{18}$ 76°, n$_D^{20}$ 1.5329, d$_{20}$ 1.4422 (i)	1762
PhH$_2$GeBr	IIIB	80	b$_{0.2}$ 38°, n$_D^{20}$ 1.5961, d$_{20}$ 1.7322	1762, 1799
	IVB	89	IR and NMR spectra	1762, 1799
PhH$_2$GeI	IIIB	20	m. 29°	1762
	IVB	83	IR and NMR spectra	1762, 1799
PhHGeF$_2$	IVC	90	m. 38°, IR and NMR spectra (j)	1762
PhHGeBr$_2$	IVC	80	b$_{0.001}$ 38°, n$_D^{20}$ 1.6190, d$_{20}$ 2.0469 (i, j)	1762
PhHGeI$_2$	IVC	63	b$_{0.002}$ 80°, d$_{20}$ 2.4 (i, j)	1762
BuH$_2$GeCN	IIIA	--	b. 143°, IR and NMR spectra (g, k)	2039
BuH$_2$GeSCN	IIIA	69	b. 128-31°, IR and NMR spectra (g)	2039

(i) IR and NMR spectra (1762). (j) Rxn. with 1-hexene → Ph(n-C$_6$H$_{13}$)GeX$_2$; relative rate and rxn. mechanism (1762). (k) Equilibrium between cyano- and isocyano-form from IR spectrum (2039).

3.8.3 OTHER ORGANOGERMANIUM HETEROHYDRIDES

This chapter comprises organogermanium hydrides having germanium directly linked to oxygen, sulfur or nitrogen. The compounds listed in Table 27 are prepared by methods from the following scheme from organogermanium hydride halides.

I Hydrolysis with aqueous base:
$R_nH_{3-n}GeCl + NH_3 \rightarrow (R_nGeH_{3-n})_2O$.

II Exchange reactions:
  (IIA) $Pr_2HGeI + Bu_3SnOMe \rightarrow Pr_2HGeOMe$,
  (IIB) $R_nH_{3-n}GeCl + Pb(SEt)_2 \rightarrow R_nH_{3-n}GeSEt$,
  (IIC) $Et_2HGeCl + Li_3N \rightarrow (Et_2HGe)_3N$.

III Soloalysis in anhydrous medium:
$Et_2HGeCl + NH_3 \rightarrow (Et_2GeH)_nNH_{3-n}$.

Cyclic organogermanium heterohydrides are listed in Chapter 7.3, metal-substituted heterohydrides may be found in Sections 6.5.8.

Table 27. Organogermanium Heterohydrides $R_nH_mGeX_o$

Formula	Method*	Yield	Properties	Ref.
$(Et_2GeH)_2O$	I	93	b_{21} 97°, n_D^{20} 1.4545, d_{20} 1.2115 (a, b, c)	1471
$(EtGeH_2)_2O$	I	Quant.	b. 145°, n_D^{20} 1.4532, d_{20} 1.3489 (a, d)	1471
$(EtHGeO)_x$	I	--	(a, e)	1471
$Pr_2HGeOMe$	IIA	95	b_{45} 79-81°, n_D^{20} 1.4355, d_{20} 1.0514	2238
$(Bu_2GeH)_2O$	I	Quant.	$b_{0.1}$ 104°, n_D^{20} 1.4604, d_{20} 1.0727 (a, b)	1471
$(BuGeH_2)_2O$	I	Quant.	b_{16} 104°, n_D^{20} 1.4590, d_{20} 1.2115 (a)	1471
$(BuHGeO)_4$	I	--	$b_{0.2}$ 144°, n_D^{20} 1.4810, d_{20} 1.3621 (a, f)	1471
$(Ph_2GeH)_2O$	I	55	$b_{0.005}$ 195°, n_D^{20} 1.6130, d_{20} 1.3437 (a, g)	1762, 1799
$(PhGeH_2)_2O$	I	82	$b_{0.001}$ 90°, n_D^{20} 1.5830, d_{20} 1.4340 (a, g)	1762, 1799
$(PhHGeO)_4$	I	72	n_D^{20} 1.615, mol. wt. (a, h)	1762
$Et_2HGeSEt$	IIB	81	b_{25} 83°, n_D^{20} 1.4909, d_{20} 1.1214 (i)	465
BuH_2GeSEt	IIB	--	b. 150.5°, colorless liq. (a, j)	2039
$(Et_2GeH)_3N$	III	72	b_{15} 159°, n_D^{20} 1.4918, d_{20} 1.2550 (a)	1471
	IIC	35	b_{19} 163°, n_D^{20} 1.4913, d_{20} 1.260	1471
$(Et_2GeH)_2NH$	III	18	b_2 59°, n_D^{20} 1.4833, d_{20} 1.2491 (a)	1471
$(EtGeH_2)_3N$	III	63	b_{12} 99.5°, n_D^{20} 1.4891, d_{20} 1.3959 (a)	1471

* The characters used here correspond to the synthetic scheme in the introduction to Chapter 3.8.3.
(a) IR and NMR spectra. (b) Rxn. with HF → R_2HGeF (1471). (c) Rxn. with Br → Et_2GeBr_2 (1471), rxn. with oxygen → $(Et_2GeO)_x$ (1471). (d) Rxn. with HX → EtH_2GeX; X = F, Br, I (1471). (e) Rxn. with HBr → EtH_nGeBr_{3-n}; n = 1-3 (1471). (f) Mol. wt. (1471). (g) Rxn. with HX → $Ph_nH_{3-n}GeX$; X = F, Br, I (1762), Cl, Br, I (1799). (h) Rxn. with HX → $PhHGeX_2$; X = F, Br, I (1762). (i) IR spectrum (465). (j) Rxn. with $GeCl_4$ → relative rate of Cl exchange, by NMR technique (2039).

4. ORGANOGERMANIUM COMPOUNDS CONTAINING HETEROATOMGERMANIUM BONDS

4.1 COMPOUNDS WITH HALOGEN

4.1.1 TRIORGANOGERMANIUM HALIDES

4.1.1.1 UNSUBSTITUTED SYMMETRIC TRIORGANOGERMANIUM HALIDES

Triorganogermanium halides are prepared by dealkylation (I) of tetraorganocompounds, alkylation (II) of germanium halides or by interconversion reactions (III) involving triorganogermane heterofunctions. The following scheme gives examples for preparation of triorganogermanium halides listed in Table 28.

I Dealkylation reactions:
  (IA) $(C_5H_{11})_4Ge$ with $SnCl_4$,
  (IB) R_4Ge with $AlCl_3$.

II Alhylation reactions:
  (IIA) GeX_4 with excess $RMgBr$,
  (IIB) $GeCl_4$ with $i-Bu_3Al$.

III Interconversion reactions:
  (IIIA) Ph_3GeBr with PbF_2 ,
  (IIIB) $(Et_3Ge)_2O$ with $MePOF_2$,
  (IIIC) $i-Bu_3GeOR$ with HCl ,
  (IIID) $Ph_3GeO_2CCF_3$ by thermal decomposition.

R = organic group, X = halogen.

Trimethylgermanium Fluoride (M-71) GeC_3H_9F Me_3GeF
Synth.: By rxn. of $(Me_3Ge)_2O$ with KF (2365), with $MePOF_2$ at 90°, in 62% yield (1103).
By rxn. of Me_3GeBr with HF, in 39% yield (367).
By rxn. of Me_3GeCl with KHF_2 (2541).
By-prod. in rxn. of $Me_3GeMn(CO)_5$ with C_2HF_3 or C_2F_4 under UV irradiation (872).
By-prod. in rxn. of GeX_4 with Al_4C_3 and HF at 150-200° (2114).
Prop.: b. 76° (367) b_{751} 77-79° (1103), n_D^{20} 1.3874 (367), 1.3835 (1103), IR (872) far IR, Raman and assignment (2540-1) and NMR (872, 1430, 1436, 1807, 1825) spectra, dipole moment (255).
Rxn. with: $NaGeH_3 \rightarrow Me_3GeGeH_3$ (2365).
$Et_3GeM \rightarrow Me_3GeGeEt_3 + Me_6Ge_2 + Et_6Ge_2$; M = Li, Na, K (815).

Trimethylgermanium Chloride (M-66) GeC_3H_9Cl Me_3GeCl
Synth.: By rxn. of Me_4Ge with HCl and $AlCl_3$ (1:1:1) at r.t. (582), with AcCl and $AlCl_3$ (1:1:1) at r.t. (1781), with $GaCl_3$ at r.t. in inert atm. (1823), with i-PrCl and $AlCl_3$ under reflux, in 95% yield (367).
By $GaCl_3$ catalyzed rxn. of Me_4Ge with Me_2GeCl_2 (1:1) at 50-97°, in 95% yield (1856), with $MeGeCl_3$ (2:1) at 110 and 240° in sealed tube in 27 and 93% yield,

resp. (1856), with MeGeCl₃ (1:2) at 150° in sealed tube, in quant. yield (1856), with GeCl₄ at 150° in sealed tube, in 96% yield (1856).

By rxn. of Me₃GePh with AlCl₃ and MeEtCHCl, in 85% yield (367).

By rxn. of (Me₃Ge)₂O with SOCl₂ (1826).

By-prod. in rxn. of powd. Ge and Cu with MeCl at 400°, in 2.7% yield (1540), in presence of Al at 400-450°, in 6% yield (2192).

By-prod. in rxn. of Me₃GeC⋮CH with HGeCl₃ (1533).

By-prod. in rxn. of Me₃GeH with (Et₃P)₂PdCl₂ (795), with $\overline{CF_2CF_2CCl:CCl}$, at 190° in 20% yield (924).

By-prod. in rxn. of (Me₃Ge)₂O with Et₃SiCl, in 32% yield (1535), in presence of FeCl₃ in 48% yield (1535).

By-prod. in rxn. of Me₃GeOLi with GeCl₃ (685).

By-prod. in rxn. of (Me₃Ge)₂NH with Et₃GeCl, in 43% yield (1535), with Et₃SiCl, in 38% yield (1535), in presence of FeCl₃, in 68% yield (1535).

By-prod. in rxn. of (Me₃Ge)₂Hg with PhHgCBrCl₂ in C₆H₆ at 65-75° in Ar atm. (1881).

By-prod. in rxn. of Me₃GeC₅H₅Mo(CO)₃ with HCl (832).

By-prod. in rxn. of Me₃Ge(Et₃P)₂PtCl with HCl in Et₂O or C₆H₆, in 28% yield (1107), with (CH₂Cl)₂ in sealed tube at r.t. (1106).

By-prod. in rxn. of Me₃Ge(Ph₃Ge)Pt(PEt₃)₂ with anhydr. HCl in C₆H₆ or Et₂O (1107).

By-prod. in rxn. of [Me₃Ge(Et₃P)Pt(CH₂PPh₂)₂]·C₆H₆ with anhydr. HCl in C₆H₆ (1107).

By-prod. in rxn. of GeCl₄ with Al₄C₃ and HCl at 150-200° (2114).

Prop.: b. 97.8°, n_D^{20} 1.4350, d_{20} 1.2435 (367), IR (135, 1013, 1463), (far) IR and assignment (1462, 2080), Raman (367) and assignment (1462, 2080), NMR (582, 1430, 1436, 1807, 1856, 2373), ³⁵Cl NQR (2753), and mass (1109) spectra, GLC (432), dipole moment (255, 582).

Rxn. with: RLi → Me₃GeR; R = Me₃SiO (474a), Me₂(ClCH₂)SiO (2129), Me₂N (1796, 1853), PhNH (2705), N-pyrazolyl (2378), Ph₂C:N (852) Me₃GeNMe (1855), Me₃Si-NCMe₃ (2749), Me₃SiNMeCPh:N (2738) Me₃SiNMeSiMe₂NH (1810), t-Bu₂PNH (1813, 2739), R′₃P:N; R′ = Me, Et, Ph (1824); Me₂As (2201), B₅H₈ (2398).

Me₃SnLi → Me₃GeSnMe₃ + Me₆Ge₂ + Me₆Sn₂ (1844).

Me₃SiNLiP(CMe₃)₂ → Me₃GeP(CMe₃)₂:NSiMe₃ (1813).

Me₃MNHP(CMe₃)₂ + BuLi → Me₃GeP(CMe₃)₂:NMMe₃ ; M = Si, Ge (2739).

Me₂NLi + PhCN → Me₃GeN:CPhNMe₂ (2738).

Li₃N → (Me₃Ge)₃N (1796).

LiAlH(OCMe₃)₃ → Me₃GeH (582).

Et₃GeM → Me₃GeGeEt₃ + Me₆Ge₂ + Et₆Ge₂; M = Li, Na, K (815).

Na₃Y in liq. NH₃ → (Me₃Ge)₃Y; Y = P, As (1853), Bi (1854).

Me₃MCH:CHCl + Na → Me₃GeCH:CHMMe₃; M = Si (367, 1533), Ge (1533).

Me₃MCCl:CH₂ + Na → Me₃GeC(:CH₂)MMe₃; M = Si (367, 1533), Ge (1533).

KH₂F → Me₃GeF (2541).

KX → Me₃GeX; X = NCSe (2825), OAc (367).

AgNO₃ → Me₃GeNO₃ (476).

RMgCl → Me₃GeR; R = i-Bu, Me₂HSiCH₂ (475).

RMgBr → Me₃GeR; R = CF₂:CF (2209).

RMgI → Me₃GeR; R = CF₃C⋮C (2332).
C₂H₂ + EtMgBr → Me₃GeC⋮CH + (Me₃GeC⋮)₂ (367).
Me₃SiCl + NH₃ → Me₃GeNHSiMe₃ (1534).
Me₃SnNEt₂ + NaC⋮CH → Me₃GeC⋮CSnMe₃ (1039).
Me₃SiCH:PMe₃ → Me₃Ge(Me₃Si)C:PMe₃ (1826).
CH₂:PPh₃ → (Me₃Ge)₂C:PPh₃ (1826).
$\overline{CH_2CH_2O}$ → Me₃GeOCH₂CH₂Cl (2196).
GeCl₄ (2:1-1:2) + cat. GaCl₃ → Me₂GeCl₂ (1856).
MeGeCl₃ (1:1) + cat. GaCl₃ → Me₂GeCl₂ (1856).
Cl at 20° in liq. phase → Me₂(ClCH₂)GeCl + Me₂(Cl₂CH)GeCl + C (2133).
Cl at 100° in gas phase and UV irradiation → Me₂(ClCH₂)GeCl + Me₂(Cl₂CH)GeCl + Me₂GeCl₂ + MeGeCl₃ + C (2133).
NH₃ at -60° in Et₂O → (Me₃Ge)₃N + (Me₃Ge)₂NH (1796).

Trimethylgermanium Bromide (M-66)   GeC₃H₉Br   Me₃GeBr

Synth.: By rxn. of Me₄Ge with Br in Pr₂O under reflux, in 98% yield (367), in presence of cat. AlCl₃, in 83% yield (643), with i-PrBr and AlBr₃, in 92% yield (367), with GeBr₄ and cat. GaBr₃ at 165° in a sealed tube, in 83% yield (1856).
By rxn. of Me₃GePh with Br in EtBr in 98% yield (1041-2).
By-prod. in rxn. of powdered Ge and Cu at 400° with MeBr, in 1.2% yield (1540).
By-prod. in rxn. of Me₃GeC₅H₅M(CO)₃ with (CH₂Br)₂; M = Mo, W (832).
Prop.: b. 113.5° (367), n_D^{20} 1.4750 (1103), n_D^{25} 1.4672 (367), 1.4688 (1041-2), d₂₀ 1.5604 (367), IR (367) and assignment (872), Raman (367) and NMR (872, 1430, 1436, 1807, 1856) spectra, dipole moment (255).
Rxn. with: RLi → Me₃GeR; R = p-PhC₆H₄, (p-C₆H₄)₂ (112), p-BrC₆H₄, p,p'-BrC₆H₄C₆H₄, 1,4-BrC₁₀H₆, 1,4-Me₃GeC₁₀H₆ (2223), ClC⋮C (1980), Ph₂PC⋮C (2778).
PhLi + Et₂NH → Me₃GeNEt₂ (643).
YLi₂ → (Me₃Ge)₂Y; Y = p-C₆H₄ (2223).
Y'Li₃ → (Me₃Ge)₃Y'; Y' = N (644), Sb (676).
LiAl(PH₂)₄ → Me₃GePH₂ (2652).
Li in THF → Me₆Ge₂ (367).
RNa → Me₃GeR; R = GeH₃ (2365), PhOCH₂CH₂O (2353), Mn(CO)₅ (872), C₅H₅M(CO)₃; M = Mo, W (832), Cr, W (2303).
EtYNa → Me₃GeYEt; Y = O, S (643).
Na-Hg → (Me₃Ge)₂Hg (996).
K + Me₃SiBr → Me₆Ge₂ + Me₃GeSiMe₃ + Me₆Si₂ (367).
K at 140° → Me_{n+2}Ge_n + Me₃GeCH₂GeMe₂GeMe₃ + (Me₃Ge)₂GeMeGeMe₂Et (2419).
H₃SiK → Me₃GeH + Me₃GeSiH₃ + SiH₄ + Si₂H₆ + MeSiH₃ (2224).
KSCN → Me₃GeSCN (367).
Aq. KOH → (Me₃Ge)₂O (643, 1103).
AgOCN → Me₃GeNCO (2027).
RMgCl → Me₃GeR; R = PhCH₂ (2159).
RMgBr → Me₃GeR; R = MeC⋮C (1981).
Me₃SiYMgX → Me₃GeYSiMe₃; Y = CH₂, C⋮C, p-C₆H₄ (367).
Me₃SiSEt → Me₃GeSEt + Me₃SiBr (643).
Ph₃SnCH₂Ac → Me₃GeCH₂Ac + Me₃GeOCMe:CH₂ (2547).

$\overline{CH_2CH_2O}$ → $Me_3GeOCH_2CH_2Br$ (367).
H_2S + Py → $(Me_3Ge)_2S$ (644).
MeSH + Py → Me_3GeSMe (227).
HX → Me_3GeX; X = F (367), N_3 (556-7).

Trimethylgermanium Iodide (M-67) GeC_3H_9I Me_3GeI

Synth.: By rxn. of Me_3GePh and BuI in the presence of cat. $AlCl_3$, in 79% yield (367).
By-prod. in rxn. of Ge-sponge with MeI at 320° (2176), in the presence of powdered Cu at 400°, in 1.3% yield (1540).
By-prod. in rxn. of $Me_3Ge(C_5H_5)W(CO)_3$ with I in C_6H_6 at r.t. (832).
Prop.: b. 133.5°, n_D^{20} 1.5190, d_{20} 1.7962 (367), IR (367) and NMR (1012, 1430, 1436, 1540, 1807) spectra, dipole moment (255).
Rxn. with: RLi → Me_3GeR; R = $CH_2:CH$ (1579), $CH_2:CMe$ (1580).
$CH_2:CHCl$ + Na → $Me_3GeCH:CH_2$ (1579).
RBr + Na → Me_3GeR; R = $CH_2:CPh$ (1580).
Dry Ag_2S → $(Me_3Ge)_2S$ (927).
RMgBr → Me_3GeR; R = $CH_2:CH$ (1579), MeCH:CH, PhCH:CH (1580), PrPhCH, $PhCH_2$-CHEt (2629).

Triethylgermanium Chloride (M-67) $GeC_6H_{15}Cl$ Et_3GeCl

Synth.: By rxn. of Et_4Ge with $SnCl_4$ at 210°, in 100% conversion (2292), in AcCl or $MeNO_2$ at 20-50°, in 75-100% conversion (2292), with $GeCl_4$ (3:1) in the presence of cat. $GaCl_3$ at 123-75°, in 65% yield (1856).
By rxn. of Et_3GeCO_2H with HCl (815).
By rxn. of Et_3GeH with H_2PtCl_6, in 35% yield (1472), in MeCN in 80% yield (1472).
By rxn. of $(Et_3Ge)_2O$ with $n-Pr_3GeCl$ under reflux, in 48-54% yield (1535).
By rxn. of $(Et_3Ge)_2NH$ with $AlCl_3$, in 47% yield (1535), with $HC\vdots CCH_2Cl$ at 70-200°, in 48% yield (1534).
By rxn. of Et_3GePEt_2 with gaseous HCl (1797).
By rxn. of $Et_3GeZn(GeEt_2)_xGeEt_3$ with $HgCl_2$ (584).
By rxn. of $(Et_3Ge)_2Cd$ with $HgCl_2$ and Et_3SnCl, in 31 and 60% yield, resp. (587), with CCl_4 in Et_2O at -75°, in 41% yield (2088).
By rxn. of $(Et_3Ge)_2Hg$ with $C_5H_5Fe(CO)_2Cl$ in C_6H_6 at 100°, in 62% yield (2414).
By rxn. of $Et_3GeC_5H_5Mo(CO)_3$ with dry HCl (832).
By rxn. of $GeCl_4$ with Et_3Al at 90-100° in Ar atm., in 52% yield (2603), at 110-15°, followed by NaCl, in 85% yield (2603).
By-prod. in redistribution rxn. of Et_4Ge with $(n-C_6H_{13})_4Ge$ cat. by $AlCl_3$ at 170° (437), with Pr_4Ge, in 28% yield (491).
By-prod. in rxn. of Et_4BuGe_2Cl with Cl (2292).
By-prod. in rxn. of $2-XC_6H_{10}OGeEt_3$ with dry HCl in C_6H_6; X = Cl, Br (2532).
By-prod. in rxn. of $GeCl_4$ with Et_2AlCl at 120-50°, followed by NaCl, in Ar atm., in 6% yield (2603).
By-prod. in rxn. of $GeCl_4$ with EtMgCl in Et_2O, THF or MePh in the presence of excess Mg, in up to 11% yield (2701).
Prop.: b. 180° (1856), b_{756} 172° (1535), $b_{770.5}$ 171° (1534), b_{17} 65-67°

(584), n_D^{20} 1.4583 (1534), 1.4579-89 (1535), 1.4575 (584), d_{20} 1.1656 (1534), IR (1013, 1444, 1534), NMR (1444, 2373), ^{35}Cl NQR (2753) and mass (1109) spectra, GLC (432, 491, 2603, 2701).

Rxn. with: LiNH$_2$ in THF → (Et$_3$Ge)$_2$NH + LiCl + NH$_3$ (1796).
RLi → Et$_3$GeR; R = Me$_2$N, Et$_2$N (466), Et$_2$P (1797).
Et$_3$GeK → Et$_6$Ge$_2$ (814).
KSeCN → Et$_3$GeNCSe (2825).
K → Et$_3$GeK (814).
RMgBr → Et$_3$GeR; R = Me$_2$N, Et$_2$N, C$_6$H$_{13}$NH (466).
PhN(MgBr)$_2$ → (Et$_3$Ge)$_2$NPh + MgClBr (1796).
(Me$_3$Ge)$_2$NH → Me$_3$GeCl + (Et$_3$Ge)$_2$NH (1535).
Bu$_3$SnR → Et$_3$GeR; R = CH$_2$CO$_2$Me (692, 1440), CH$_2$Ac (711), BuO (1440).
(EtS)$_2$Pb → Et$_3$GeSEt (465).
Cyclohexene oxide → 2-ClC$_6$H$_{10}$OGeEt$_3$, mechanism of epoxide cleavage rxn. (2532).

Triethylgermanium Bromide (M-67)   GeC$_6$H$_{15}$Br   Et$_3$GeBr

Synth.: By rxn. of Et$_4$Ge with Br in (CH$_2$Br)$_2$ at 60-100°, in 90% yield (1103).
By rxn. of Et$_3$GeCH$_2$CR:CRMe with Br in EtBr; R = H, Me (467), H, in 93% yield (1798).
By rxn. of Et$_3$GeCHCNCH$_2$PEt$_2$ with Br, in 95% yield (1797).
By rxn. of Et$_3$GeCH$_2$CH:CHR with HgBr$_2$; R = H, Me (2715).
By rxn. of Et$_3$GeH with N-bromosuccinimide (164).
By rxn. of Et$_3$GeI with Br, in 76% yield (2851).
By rxn. of (Et$_3$Ge)$_2$M with Br in C$_6$H$_6$ under exclusion of air; M = S, Se, Te, in 67-85% yield (2851).
By rxn. of Et$_3$GeSeH with Br in C$_6$H$_6$ under exclusion of air, in 82% yield (2851).
By rxn. of (Et$_3$Ge)$_3$Sb with Br in C$_7$H$_{16}$ under exclusion of air, in 83% yield (2851), with RBr at 100-160°; R = PhCH$_2$, c.C$_5$H$_9$ and BrCH$_2$CH$_2$, in 54, 80 and 100% yield, resp. (2089), with PhBr at 30° under UV irradiation, in 53% yield (2089).
By rxn. of (Et$_3$Ge)$_2$Zn with (CH$_2$Br)$_2$, in 61% yield (2854).
By rxn. of (Et$_3$Ge)$_2$Hg with (CH$_2$Br)$_2$ at r.t., in 58% yield (584), with C$_5$H$_5$-Fe(CO)$_2$Br in a sealed tube at 100°, in 78% yield (2414).
By rxn. of (Et$_3$Ge)$_3$Tl with Br in C$_7$H$_{16}$, in 80% yield (2851), with (CH$_2$Br)$_2$ at r.t., in 89% yield (1361, 2089).
By rxn. of Et$_3$GeC$_5$H$_5$Mo(CO)$_3$ with EtBr (832).
By rxn. of Et$_3$GeC$_5$H$_5$Fe(CO)$_2$ with Br in C$_6$H$_6$, in 85% yield (2414, 2416).

Prop.: $b_{0.01}$ 32-40° (50), n_D^{20} 1.4881 (1103), 1.4845 (584), IR, assignment (1444), far IR, assignment (105), NMR (1444) and mass (1109) spectra. Anal. detn. of Br (1690).

Rxn. with: RLi → Et$_3$GeR; R = MeCH:CH (1600), p-(Me$_3$Si)$_2$NC$_6$H$_4$ (2109), $\overline{S(CH_2)_3SCR'}$; R' = H, Me, Ph, Et$_3$Ge (792).
Et$_3$GeYLi → Et$_3$GeYGeEt$_3$; Y = CH$_2$CH$_2$, (CHPh)$_2$, CH:CPh, (:CPh)$_2$ (2849).
RLi → Et$_3$GeR; R = (PhCH$_2$)$_3$Ge in Et$_2$O (104), Et$_3$Si (2852), Et$_3$GeSe (2090), (Et$_3$Ge)$_2$N (2502).
(PhCH$_2$)$_3$GeLi in (MeOCH$_2$)$_2$ → (PhCH$_2$)$_6$Ge$_2$ + Et$_6$Ge$_2$ (104).

Na in NH_3 → $(Et_3Ge)_2NH$ (1534).
$NaNH_2$ in C_6H_6 → $(Et_3Ge)_2NH$ (2502).
RNa → Et_3GeR; R = Me_3CO (2095), $C_5H_5M(CO)_3$; M = Mo, W (832), W (2414),
$C_5H_5Fe(CO)_2$ (2414, 2416).
KSCN → Et_3GeNCS (367).
Aq. KOH → $(Et_3Ge)_2O$ (1103).
$(C_5H_9O)O(CH_2)_nMgCl$, + H_3PO_4 → $Et_3Ge(CH_2)_nOH$ + C_5H_6O; n = 4, 5 (343).
RC⋮CMrBr → $Et_3GeC⋮CR$; R = CH_2:CH, CH_2:CMe, MeCH:CH (530), MeC⋮C, EtC⋮C,
n-PrC⋮C (2797).
$Me_3CCMe(OMgBr)C⋮CMgBr$ → $Et_3GeC⋮CCMe(OH)CMe_3$ (1901).
RMgI → Et_3GeR; R = $CF_3C⋮C$ (921), Et_3GeSe (2090).
$(Et_3Ge)_2Cd$ → Et_4Ge + $CdBr_2$ (587a).
Bu_3SnCH_2Ac → Et_3GeCH_2Ac + Bu_3SnBr (711).
$(EtO)_3P$ + EtCHO → $Et_3GeOCHEtPO_3Et_2$ (452).
Cyclohexene oxide → $2\text{-}BrC_6H_{10}OGeEt_3$, mechanism of epoxide cleavage rxn. (2532).

Triethylgermanium Iodide (M-71) $GeC_6H_{15}I$ Et_3GeI
Synth.: By rxn. of Et_3GeCH_2Ac with Me_3SiI, in 81% yield (711).
By rxn. of $(Et_3Ge)_2Te$ with I under exclusion of air, in 74% yield (2851).
By-prod. in rxn. of Et_3GePPh_2 with MeI (1:2), isolated as Et_3GePh, in 31% yield (50).
Prop.: NMR and IR spectra and assignment (1444), flash photolysis (2314).
Anal. detn. by fusion and volumetric titration (265).
Rxn. with: B_5H_8Li → $Et_3GeB_5H_8$ (2398).
Ag_2S → $(Et_3Ge)_2S$ (465).
Bu_3SnR → Et_3GeR; R = MeO (692), CH_2CO_2Me (692, 1440), CH_2CO_2Bu (1440), $AcCH_2$ (711).
R_2Hg → Et_3GeR + HgI_2; R = CH_2CO_2Me (24b), $AcCH_2$ (711).
Br → Et_3GeBr + I (2851).

Tripropylgermanium Chloride (M-71) $GeC_9H_{21}Cl$ n-Pr_3GeCl
Synth.: By rxn. of $2\text{-}XC_6H_{10}OGePr_3$ with dry HCl in C_6H_6; X = Cl, Br (2532).
By rxn. of $GeCl_4$ with Pr_3Al at 110-30°, followed by NaCl, in 85% yield (2603).
By-prod. in rxn. of Et_4Ge and Pr_4Ge with $AlCl_3$, in 3% yield (491).
By-prod. in rxn. of Pr_4Ge with $GeCl_4$ (2837).
Prop.: b_{15} 103-105° (2603), b_4 80-81.5°, n_D^{25} 1.4620, d_{25} 1.0851 (2837),
GLC (491, 2603), dipole moment (2836). Anal. detn. by Schoeninger combustion and polarography (2710).
Rxn. with: $(Et_3Ge)_2O$ → Et_3GeCl + $(Pr_3Ge)_2O$ (1535).
Cyclohexene oxide → $2\text{-}ClC_6H_{10}OGePr_3$; mechanism of epoxide cleavage rxn. (2532).

Tripropylgermanium Bromide (M-71) $GeC_9H_{21}Br$ n-Pr_3GeBr
Rxn. with: $(Pr_3Ge)_2NLi$ → $(Pr_3Ge)_3N$ (2502).
NaR → Pr_3GeR; R = $C_5H_5Mo(CO)_3$ (832).
$NaNH_2$ → $(Pr_3Ge)_2NH$ (2502).
RMgBr → Pr_3GeR; R = i-Pr (831), CH_2:$CHCH_2$ (2038), MeCH:CH, CH_2:CMe, EtCH:CH,

Me$_2$C:CH (2051), Ph (1788), 1-C$_{10}$H$_7$ (1735).
RCHCH$_2$O → Pr$_3$GeOCHRCH$_2$Br; kinetic data on epoxide cleavage rxn., R = H, Me, MeOCH$_2$, ClCH$_2$ (1396).
Cyclohexene oxide → 2-BrC$_6$H$_{10}$OGePr$_3$; mechanism of cleavage rxn. (2532).

Triisopropylgermanium Chloride (M-72)   GeC$_9$H$_{21}$Cl   i-Pr$_3$GeCl
Synth.: By-prod. in rxn. of GeCl$_4$ with i-PrMgCl in Et$_2$O-C$_6$H$_6$ in 31-47% yield (831), in Et$_2$O at 4° in He atm., in 43% yield (2591).
Prop.: b$_{42}$ 120-21° (2591), far IR spectrum and assignment (105).
Rxn. with: Na-K in MePh, + H$_2$O → i-Pr$_3$GeOH + i-Pr$_3$GeH + i-Pr$_3$GePr + i-Pr$_4$Ge + (i-Pr$_2$GeH)$_2$ + i-Pr$_6$Ge$_2$ + (i-Pr$_3$Ge)$_2$O (831).

Tributylgermanium Chloride (M-68)   GeC$_{12}$H$_{27}$Cl   n-Bu$_3$GeCl
Synth.: By rxn. of Bu$_4$Ge with i-C$_5$H$_{11}$Cl and 4% cat. AlCl$_3$ under reflux, in 55% yield (119), with GeCl$_4$ in the presence of cat. Ge, GeI$_2$, Bu$_3$GeH, LiAlH$_4$ or Bu$_6$Ge$_2$, in high yield (1760).
Prop.: b$_{13}$ 142°, n$_D^{20}$ 1.4641 (119), GLC (1760). Elemental analysis (426).
Rxn. with: R$_2$NLi → Bu$_3$GeNR$_2$; R = Me, Et (465, 1761).
Na in liq. NH$_3$ → (Bu$_3$Ge)$_2$NH (1761).
NaAl(C⋮CBu)$_4$ → Bu$_3$GeC⋮CBu (1076).
K in (Me$_2$N)$_3$PO → Bu$_3$GeK (814).
Bu$_3$GeK → Bu$_6$Ge$_2$ (814).
KR → Bu$_3$GeR; R = Bu$_2$PO (1232), (CH$_2$CO)$_2$N, o-C$_6$H$_4$(CO)$_2$N (1761).
RMgBr → Bu$_3$GeR; R = HOCMe$_2$C⋮C, HOCMeEtC⋮C (500), Me$_2$N (465), Et$_2$N (465, 2452).
Pb(SR)$_2$ → Bu$_3$GeSR; R = Et, Bu (465).
H$_2$NCN + NaOEt → (Bu$_3$GeN:)$_2$C (1761).
ROH + NH$_3$ → Bu$_3$GeOR; R = Me, Et, i-Pr, t-Bu (1503).
RSH + NH$_3$ or Et$_3$N → Bu$_3$GeSR; R = PhCH$_2$, Et, n-Pr, i-Pr, n-Bu, i-Bu, t-Bu, n-C$_{12}$H$_{25}$, Ph (1502).
Biol. Prop.: Toxicity (845). Insecticidal activity against Heliothis zea and H. virescens (2145).

Tributylgermanium Bromide (M-72)   GeC$_{12}$H$_{27}$Br   n-Bu$_3$GeBr
Synth.: By rxn. of Bu$_4$Ge with i-C$_5$H$_{11}$Br and 4% cat. AlBr$_3$ under reflux, in 67% yield (119).
By rxn. of Bu$_3$GeCH$_2$CMe:CH$_2$ with Br (467).
Prop.: b$_{10}$ 141°, n$_D^{20}$ 1.4805 (119). Elemental analysis (426).
Rxn. with: (Bu$_3$Ge)$_2$NLi → (Bu$_3$Ge)$_3$N (2502).
NaNH$_2$ in C$_6$H$_6$ → (Bu$_3$Ge)$_2$NH$_2$ (2502).
RMgBr → Bu$_3$GeR; R = CH$_2$:CHCH$_2$ (2038), MeCH:CH, EtCH:CH, Me$_2$C:CH (2051), Ph (1788), 1-C$_{10}$H$_7$ (1735).

Tributylgermanium Iodide (M-72)   GeC$_{12}$H$_{27}$I   n-Bu$_3$GeI
Rxn. with: LiAlH$_4$ → Bu$_3$GeH (1598).
KSeCN → Bu$_3$GeNCSe (2825).
AgCN → Bu$_3$GeCN (2825).
Ag$_2$S → (Bu$_3$Ge)$_2$S (465).

HC⋮CMgBr → Bu$_3$GeC⋮CH (1598).
Et$_3$SnCH$_2$Ac → Bu$_3$GeCH$_2$Ac + Et$_3$GeI (711).

Triphenylgermanium Chloride (M-69)   GeC$_{18}$H$_{15}$Cl   Ph$_3$GeCl
Synth.: By rxn. of Ph$_4$Ge with Ph$_2$GeCl$_2$ (1:1) and cat. AlCl$_3$ at 120°, in 95% yield (1364).
By rxn. of Ph$_3$GeAuPPh$_3$ with (CH$_2$Cl)$_2$ in N atm. in 84% yield (2420), with SnCl$_4$ in C$_6$H$_6$ at r.t. in N atm., in 84% yield (2420).
By-prod. in rxn. of (Ph$_3$Ge)$_2$CO with HCCl$_3$ or CCl$_4$ under UV irradiation (794).
By-prod. in thermal dec. of Ph$_2$HGeCl (1762).
By-prod. in rxn. of (Ph$_3$Ge)$_2$Pd(PEt$_3$)$_2$ with dry HCl (795).
By-prod. in rxn. of (Ph$_3$Ge)$_2$Pt(PR$_3$)$_2$ with HCl (101, 103) or with CCl$_4$ (103).
By-prod. in rxn. of Ph$_3$GePt(CH$_2$PPh$_2$)$_2$ with C$_5$H$_5$W(CO)$_3$HgCl or with HgCl$_2$ (832).
Prop.: m. 116-17° (1364), far IR, assignment (105, 333), (far) IR, assignment (1443), UV (335) and mass (1109) spectra, GLC (1364), TLC, detn. and sepn. from Ph$_n$GeCl$_{4-n}$; n = 4, 2 (2635).
Rxn. with: Polarographic reduction → Ph$_3$GeH (963).
Polargraphic reduction + Ph$_6$M$_2$ → Ph$_3$GeH; M = Ge, Sn (963).
(Ph$_3$Sn)$_3$SnLi → (Ph$_3$Sn)$_3$SnGePh$_2$ (68a, 175).
Ph$_3$GeLi + SnCl$_2$ → (Ph$_3$Ge)$_4$Sn + Ph$_6$Ge$_2$ (1928).
RLi → Ph$_3$GeR; R = fluorenyl (2675), Et$_2$N (1039), 4-N,N-diethylaminotroponeimine, N-N,N'-diethylaminotroponeimine (C$_7$H$_5$N$_2$Et$_2$) (1004), Ph$_2$C:N (852), Ph$_3$SnPPhP-(SnPh$_3$)$_2$ (1850).
RNa → Ph$_3$GeR; R = HC⋮C, Ph$_3$SnC⋮C (1039), Ph$_2$As, Ph$_2$Sb (1847), Ph$_3$PIr(CO)$_3$ (891), OC$_6$H$_{11}$, Me$_2$CHCH$_2$ (1762), 9-OC$_{13}$H$_9$ (2675).
NaC⋮CH → Ph$_3$GeC⋮CH = (Ph$_3$GeC⋮)$_2$ (939).
NaOMe + Ph$_2$CHOH → Ph$_3$GeOCHPh$_2$ (2675).
N-Pyrrylpotassium → N-Ph$_3$GeNC$_4$H$_4$ (1761).
GeCl$_4$ + cat. AlCl$_3$ → Ph$_2$GeCl$_2$ + PhGeCl$_3$ (1364).
PhPH$_2$ + Et$_3$N → (Ph$_3$Ge)$_2$PPh (1849).
PH$_3$ + Et$_3$N → (Ph$_3$Ge)$_3$P (1849).
Ph$_2$AsO$_2$H + Et$_3$N → Ph$_3$GeO$_2$AsPh$_2$ (1847).

Triphenylgermanium Bromide (M-69)   GeC$_{18}$H$_{15}$Br   Ph$_3$GeBr
Synth.: By rxn. of Ph$_3$GeCH$_2$CH:CH$_2$ with Br in HCCl$_3$ (464).
By rxn. of Ph$_3$GeCH:CRMe with Br in HCCl$_3$; R = H, Me (1787).
By rxn. of Ph$_3$GeOCHPh$_2$ with N-bromosuccinimide in presence of cat. Bz$_2$O$_2$ in CCl$_4$ (2675).
By rxn. of Ph$_3$GePPh$_2$ with EtBr, in 25% yield (50).
By rxn. of (Ph$_3$Ge)$_2$Cd with (CH$_2$Br)$_2$ in C$_6$H$_6$, in 87% yield (2295).
By rxn. of Ph$_3$GeSn(OAc)$_3$ with Br in C$_6$H$_6$ (2128).
By rxn. of Ph$_3$GeAuPR$_3$ with (CH$_2$Br)$_2$ in sealed tube at r.t.; R = Et, Ph (2420).
By rxn. of (Ph$_3$Ge)$_2$Pd(PEt$_3$)$_2$ with)CH$_2$Br)$_2$ (795).
By rxn. of (Ph$_3$Ge)$_2$Pt(PR$_3$)$_2$ with (CH$_2$Br)$_2$ (101, 103).
Prop.: (Far) IR, assignment (105, 1443) and mass (1109) spectra.
Anal. detn. by fusion and volumetric titration (265), and sepn. from Ph$_4$Ge by paper chromatography (384).

<u>Rxn. with:</u> RLi → Ph₃GeR; PhC⋮C (134), C₆F₅ (1034), Ph₃SiC⋮C (1039), Ph₂PC⋮C (2778), Ph₃GeY; Y = S, Se, Te (484), $\overline{S(CH_2)_3SCR'}$; R' = H, Me (792).
RNa → Ph₃GeR; R = Mn(CO)₅ (1593), Re(CO)₅ (1594).
Na-K in THF → Ph₃GeK (2128).
Ph₃SiK → Ph₃GeSiPh₃ (2128).
Ph₃SnK → Ph₆Ge₂ + Ph₆Sn₂ + Ph₃GeSnPh₃ (2128).
AgX → Ph₃GeX; X = NCS (1443, 2795), CN, NCO (2795) CNO (28), N(CN)₂ (2501).
Ag₂Y → (Ph₃Ge)₂Y; Y = SO₄, SeO₃, SeO₄ (1970).
RCO₂Ag → Ph₃GeO₂CR; R = Me, PhCHOH, Ph, o-HOC₆H₄ (1970).
RMgBr → Ph₃GeR; R = MeCH:CH, Me₂C:CH (462), HC⋮C (1438), p-R'C₆H₄; R' = CH₂:CH, CH₂:CMe, EtCH:CH, MeCH:CMe (462).
MeSMgI → Ph₃GeSMe + Ph₃GeI (227).
PbX₂ → Ph₃GeX; X = F, NCO (1443).
Co[Co(CO)₄]₂ → Ph₃GeCo(CO)₄ (1672).
H₂O₂(98%) → Ph₃GeO₂H (937).
RSH + Py → Ph₃GeSR; R = Me (227), Ph (121, 227).
Me$\overline{CHCH_2O}$ → kinetic data for epoxide cleavage (1396).
Cyclohexene oxide → 2-BrC₆H₁₀OGePh₃; mechanism of cleavage rxn. (2532).

Triphenylgermanium Iodide (M-73)   GeC₁₈H°₅I   Ph₃GeI
<u>Synth.</u>: By rxn. of Ph₃GeBr with MeSMgI in Et₂O-C₆H₆, in 55% yield (227).
By rxn. of Ph₃GeSMe with MeI, in 65% yield (227).
By rxn. of Ph₃GePPh₂ with MeI in C₆H₆, in 85% yield (50).
By-prod in rxn. of (Ph₃Ge)₂Pt(PR₃)₂ with MeI (101, 103), with MgI₂ or I (103).
By-prod. in rxn. of Ph₃Ge(Pr₃P)₂PtI with MgI₂ (103).
<u>Prop.</u>: m. 155-56° (227), m. 144-47° (103), IR (50), far IR, assignment (105) and mass (1109) spectra.

Additional symmetric unsubstituted triorganogermanium halides are listed in Table 28.

Table 28. Unsubstituted Symmetric Triorganogermanium Halides   R₃GeX

Formula*	Synth. Method**	Yield	Properties	Ref.
(PhCH₂)₃GeCl	--	--	(a)	--
(PhCH₂)₃GeBr	--	--	(b)	--
Et₃GeF (71)	IIIB	74	b₇₄₄ 147-49°, n$_D^{20}$ 1.4221 (c)	1103
n-Pr₃GeF (71)	--	--	(c)	--
n-Pr₃GeI (71)	--	--	(d)	--
i-Pr₃GeBr (72)	--	--	(e, f)	--
i-Pr₃GeI (72)	IIA	--		2877
n-Bu₃GeF (72)	--	--	(c)	--
i-Bu₃GeCl	IIB	67	b₁.₅ 80-81°, n$_D^{20}$ 1.4659, d₂₀ 1.0401, GLC	2603
	IIIC	--		2532

i-Bu$_3$GeBr	--	--	(f, g, h, i)	--
(C$_5$H$_{11}$)$_3$GeCl (72)	IA	88		2292
(n-C$_5$H$_{11}$)$_3$GeBr	--	--	(e, h, i)	--
(i-C$_5$H$_{11}$)$_3$GeBr	--	--	(e, h, i)	--
(c.C$_5$H$_9$)$_3$GeCl	IIA	43	b$_{0.1}$ 142-43°, n$_D^{20}$ 1.5299, d$_{20}$ 1.1965	2591
(n-C$_6$H$_{13}$)$_3$GeCl (72)	IB	(j)	GLC (i)	432
(n-C$_6$H$_{13}$)$_3$GeBr (72)	--	--	(e)	--
(c.C$_6$H$_{11}$)$_3$GeCl (72)	IIA	40	White cryst. m. 100-101°	2591
(n-C$_7$H$_{15}$)$_3$GeBr	--	--	(e, h, i)	--
(n-C$_8$H$_{17}$)$_3$GeBr	--	--	(e, h, i)	--
(n-C$_9$H$_{19}$)$_3$GeBr	--	--	(h, i)	--
(n-C$_{10}$H$_{21}$)$_3$GeBr	--	--	(h)	--
Ph$_3$GeF (73)	IIIA	--	m. 75° (k)	1443
	IIID	--	(Thermal dec. at 300°)	1789
(o-MeC$_6$H$_4$)$_3$GeBr (73)	--	--	(a)	--
(p-MeC$_6$H$_4$)$_3$GeBr (73)	--	--	(l, m)	--
R$_3$GeCl	--	--	(n, o)	--
R$_3$GeBr	--	--	(o, p)	--
R$_3$GeI	--	--	(o)	--

* Numbers in parenthesis refer to pages in main volume
** The characters used here correspond to the synthetic scheme in the introduction to Chapter 4.1.1.1.
(a) Far IR spectrum and assignment (105). (b) Rxn. with Na → R$_6$Ge$_2$ (104).
(c) (Far) IR and Raman spectra, assignments (2540). (d) Rxn. with AcCH$_2$HgI → Pr$_3$GeCH$_2$Ac + HgI$_2$ (711). (e) Rxn. with CH$_2$:CHCH$_2$MgBr → R$_3$GeCH$_2$CH:CH$_2$ (2038).
(f) Rxn. with R'MgBr → R$_3$GeR'; R'= CMe:CH$_2$, CH:CMe$_2$, CH:CHEt (2834). (g) Rxn. with cyclohexene oxide → 2-BrC$_6$H$_{10}$OGeR$_3$; mechanism of epoxide cleavage rxn. (2532). (h) Rxn. with PhMgBr → R$_3$GePh (1788). (i) Rxn. with 1-C$_{10}$H$_7$MgBr → 1-C$_{10}$H$_7$GeR$_3$ (1735). (j) By-prod. in low yield. (k) (Far) IR spectrum and assignment (1443). (l) Rxn. with NaMn(CO)$_5$ → [(p-MeC$_6$H$_4$)$_3$Ge]$_2$O (1592). (m) Rxn. with AgX in C$_6$H$_6$ → (p-MeC$_6$H$_4$)$_3$GeX; X = BzO (1970), CN, NCS, NCO (2795).
(n) ^{35}Cl NQR spectrum (33, 2754). (o) Ge-bond refractions d$_{20}$ (81). (p) ^{81}Br NQR spectrum (2754).

4.1.1.2 UNSYMMETRIC UNSUBSTITUTED TRIORGANOGERMANIUM HALIDES

Triorganogermanium halides containing two or three different organic groups are prepared by dealkylation (I) of tetraorganogermanes, by hydrogermylation (II) and by interconversion reactions (III) of triorganogermanium derivatives. Optically active triorganogermanium halides have been prepared and resolved. Triorganogermanium halides containing olefinic unsaturation are listed on the bottom of Table 29. The following scheme shows reactions for preparation of the unsymmetric unsubstituted triorganogermanium halides listed in Table 29.

I Dealkylation:
    (IA) $R_3R'Ge + SnCl_4 \rightarrow R_3GeCl$,
    (IB) $R_3PhGe + Br \rightarrow R_3GeBr$,
    (IC) $Ph_3GeCRPhOH + BF_3 \rightarrow Ph_2(Ph_2CR)GeF$; R = H, Ph,
    (ID) $Et_2\overline{GeCH_2CHRCHOHCH_2} + PX_3 \rightarrow Et_2(CH_2:CHCHRCH_2)GeX$,
    (IE) $R_2\overline{Ge(CH_2)_nCH_2} + HX \rightarrow R_2Me(CH_2)_nGeX$,
    (IF) $R_4Ge + R'_4Ge + AlCl_3 \rightarrow R_2R'GeCl + RR'_2GeCl$.

II Hydrogermylation:
    (IIA) $R_2HGeX + CH_2:CHR' \rightarrow R_2(R'C_2H_4)GeX$,
    (IIB) $Et_2HGeCl + (CH_2:CR)_2 \rightarrow Et_2R'GeCl$; R = H, Me .

III Interconversion:
    (IIIA) $R_3GeH + SOCl_2 \rightarrow R_3GeCl$,
    (IIIB) $R_3GeOH + HX \rightarrow R_3GeX$.

R = organic group, may be different in one molecule, R' = organic or olefinic unsaturated group, X = halogen.

Ethyl-1-Naphthylphenylgermanium Chloride $GeC_{18}H_{17}Cl$ $EtPh(1-C_{10}H_7)GeCl$
Synth.: By rxn. of (±)-EtPh(1-$C_{10}H_7$)GeH with Cl in CCl_4 at -10°, in 89% yield (999), (-) form (+)-R_3GeH* (999).
Prop.: m. 88-90°, IR spectrum (999), $[\alpha]_D^{25}$ -7 to -10.7° in C_6H_6 (2368).
(-)-form m. 95-98°, $[\alpha]_D^{25}$ -9.04° (c. 6.2 in C_6H_6), IR spectrum (999).
Rxn. with: (-)-isomer + $LiAlH_4$ → (+)-R_3GeH* (999, 2368).
$R'Li \rightarrow R'GeR_3$* ; R' = Me, Bu, N-C_4H_4N, with inversion (2368).
MeOH + i-Pr_2NH → R_3GeOMe*, with inversion (2368).
PhSH + i-Pr_2NH → R_3GeSPh*, with inversion (2368).
$H_2S + Et_3N \rightarrow (R_3Ge)_2S$*, with inversion (2368).

Ethyl-1-Naphthylphenylgermanium Bromide $GeC_{18}H_{17}Br$ $EtPh(1-C_{10}H_7)GeBr$
Synth.: By rxn. of $(1-C_{10}H_7)_2$GeEtPh with Br in $(CH_2Br)_2$, in 64% yield (999).
Prop.: m. 76.5-78°, $b_{0.02}$ 156-58°, n_D^{20} 1.6560 (supercooled) (999).
Rxn. with: Na(-)-Menthoxide → (±)-EtPh(1-$C_{10}H_7$)Ge$OC_{10}H_{19}$-(-) (999).
(-)-MePhCHONa → (±)-EtPh(1-$C_{10}H_7$)GeOCHMePh-(-) (999).

Additional unsymmetric unsubstituted triorganogermanium halides are listed in Table 29 and in Subchapter 7.4.1.1.

---

* R_3Ge = EtPh(1-$C_{10}H_7$)Ge

Table 29. Unsymmetric Unsubstituted Triorganogermanium Halides   RR'$_2$GeX

Formula*	Synth. Method**	Yield	Properties	Ref.
Me$_2$EtGeCl (74)	IF	--	GLC	432
Me$_2$BuGeCl	IF	--	GLC	491
	--	--	b. 168°, n_D^{20} 1.4490 (a)	2292
MeEt$_2$GeCl	IF	--	GLC	432
MePrBuGeCl	IA - Me$_2$PrBuGe	--	b$_{28}$ 104°, n_D^{20} 1.4567 (b)	2292
MeBu$_2$GeCl	IF	--	GLC	491
MePh(1-C$_{10}$H$_7$)GeCl (74)	--	--	(c, d)	--
Ph$_2$(Ph$_2$CH)GeF	IC	95	IR and NMR spectra	49
Ph$_2$(Ph$_3$C)GeF	IC	52	(In Et$_2$O), IR and NMR spectra	49
	IC	81	(In C$_6$H$_{14}$)	49
Et$_2$PrGeCl	IF	49	GLC	491
	IIIA	--	(e)	491
Et$_2$BuGeCl	IF (f)	--	GLC (e)	432,491
Et$_2$(n-C$_6$H$_{13}$)GeCl	IF	--	GLC	432
EtPr$_2$GeCl	IF	20	GLC (e)	491
	IIIA	--		491
EtPrBuGeCl	IA - Et$_2$PrBuGe	--	b$_{12}$ 86-89°, n_D^{20} 1.4611	2292
	IA - MeEtPrBuGe	--	GLC (491)	2292
Et(i-Pr)PhGeBr	IB - Et(i-Pr)GePh$_2$	84	b$_{0.15}$ 74.5-75°, n_D^{25} 1.5473 (g)	999,1003

* Numbers in parenthesis refer to pages in main volume.
** The characters used here correspond to the synthetic scheme in the introduction to Chapter 4.1.1.2.
(a) Rxn. with PrMgCl → Me$_2$GePrBu → MeEtPrBuGe (2292). (b) Rxn. with EtMgBr → MeEtPrBuGe (2292). (c) Relative rate of Li$^+$ catalyzed racemization, solvent effect (2307). (d) MeOH + i-Pr$_2$NH → MePh(1-C$_{10}$H$_7$)GeOMe under racemization (417).
(e) Rxn. with LiAlH$_4$ → Et$_n$R$_{3-n}$GeH; R = n-Pr, n-Bu (491). (f) By-prod. in rxn. of Et$_4$BuGe$_2$Cl with Cl (2292).
(g) Rxn. with p-BrC$_6$H$_4$MgBr → Et(i-Pr)PhGeC$_6$H$_4$Br (1003).

Table 29 Continued

Formula*	Synth. Method**	Yield	Properties	Ref.
EtBu$_2$GeCl	IF (f)	--	GLC (e)	432,491
Et(n-C$_8$H$_{13}$)$_2$GeCl	IF	--	GLC	432
EtPh$_2$GeBr	IB - EtGePh$_3$	78	b$_{0.2}$ 127-29.5°, n$_D^{25}$ 1.6025 (h)	999
PrBu$_2$GeCl	IE - n = 2	86	b$_{13}$ 124°, n$_D^{20}$ 1.4645, d$_{20}$ 1.0593	344,1492
	IIIB (i)	> 90	GLC (491)	1492
PrBu$_2$GeBr	IE - n = 2	85	b$_{15}$ 137°, n$_D^{20}$ 1.4816, d$_{20}$ 1.2243	344
PrBu$_2$GeI	IE - n = 2	88	b$_{14}$ 144°, n$_D^{20}$ 1.5082, d$_{20}$ 1.3609	344
Ph$_2$(n-C$_8$H$_{13}$)GeF	IIA (k)	92	b$_{0.2}$ 124°, n$_D^{20}$ 1.5420, d$_{20}$ 1.1733	1762
	IIA (l)	95	NMR spectrum	1762
	IIA (m)	95		1762
Ph$_2$(n-C$_8$H$_{13}$)GeCl	IIA (j)	81	b$_{0.1}$ 148°, n$_D^{20}$ 1.5586, d$_{20}$ 1.1842	1762
	IIA (m)	85	NMR spectrum	1762
Ph$_2$(n-C$_8$H$_{13}$)GeBr	IIA (j)	91	b$_{0.07}$ 128°, n$_D^{20}$ 1.5740, d$_{20}$ 1.3368,	1762
	IIA (l)	90	NMR spectrum	1762
Ph$_2$(n-C$_8$H$_{13}$)GeI	IIA (j)	70	b$_{0.05}$ 138°, n$_D^{20}$ 1.5998, d$_{20}$ 1.4574	1762
	IIA (k)	85	NMR spectrum	1762
	IIA (l)	70		1762

Et$_2$R'GeX

Formula		Synth. Method	Yield	Properties	Ref.
R' = CH$_2$:CHCH$_2$CH$_2$ X = Cl		IE (n)	--	b$_{19}$ 100°, n$_D^{20}$ 1.4742, d$_{20}$ 1.1406 (o, p)	2578
CH$_2$:CHCH$_2$CH$_2$	Br	IE (n)	--	b$_{21}$ 114°, n$_D^{20}$ 1.4951, d$_{20}$ 1.3384 (o, p)	2578
CH$_2$:CHCH$_2$CH$_2$	I	IE (n)	--	b$_{19}$ 116°, n$_D^{20}$ 1.5288, d$_{20}$ 1.5130 (o, p)	2578
MeCH:CHCH$_2$	Cl	IIB	--	b$_{22}$ 99.5°, n$_D^{20}$ 1.4812 (p, q, r)	467
		IIB	31	b$_{22}$ 99-100°, n$_D^{20}$ 1.4818, d$_{20}$ 1.1524	1798
MeCH:CHCH$_2$	Br	IIB	49	b$_{22}$ 112°, n$_D^{20}$ 1.5033, d$_{20}$ 1.3525 (p, q)	467,1798
MeCH:CHCH$_2$	I	IIB	83	b$_{22}$ 126°, n$_D^{20}$ 1.5405, d$_{20}$ 1.5501 (p)	1798
CH$_2$:CHCHMeCH$_2$	F	IE (n)	--	b$_{18}$ 80°, n$_D^{20}$ 1.4502, d$_{20}$ 1.0951	2578

$CH_2:CHCHMeCH_2$	Cl	ID	--	b_{16} 98°, n_D^{20} 1.4739, d_{20} 1.1138	2578
		IE (n)	--	IR and NMR spectra (e)	2578
$CH_2:CHCHMeCH_2$	Br	ID	--	b_2 84°, n_D^{20} 1.4936, d_{20} 1.3062	2578
		IE (n)	--	IR and NMR spectra (e)	2578
$CH_2:CHCHMeCH_2$	I	IE (n)	--	b_{17} 114°, n_D^{20} 1.5202, d_{20} 1.4278	2578
$CH_2:CMeCHMeCH_2$	Cl	IE (s)	--	b_{24} 119°, n_D^{20} 1.4474, d_{20} 1.1096	2578
$CH_2:CMeCHMeCH_2$	Br	IE (s)	--	b_{23} 130°, n_D^{20} 1.4963, d_{20} 1.2841	2578
$CH_2:CMeCHMeCH_2$	I	IE (s)	--	b_{17} 150°, n_D^{20} 1.5264, d_{20} 1.4456	2578
$Me_2C:CMeCH_2$	Cl	IIB	52	b_{10} 110°, n_D^{20} 1.4878, d_{20} 1.1149 (p, t)	467, 1798
$4-C_6H_9CH_2CH_2$	Cl	IIA	44	b_{18} 162°, n_D^{20} 1.4970, d_{20} 1.1463	1798
		IIB	--	IR spectrum (p)	1798
$Bu_2(Me_2C:CMeCH_2)GeCl$		IIB	--	b_{13} 159°, n_D^{20} 1.4837, d_{20} 1.0545,	467
		IIB	41	IR spectrum (u)	1798

(h) Rxn. with i-PrMgBr → Et(i-Pr)GePh$_2$ (999). (i) By-prod. in rxn. of HSiCl$_3$ with $Bu_2GeCH_2CH_2CH_2$ in autoclave with cat. AIBN (1491). (j) Synth. without catalyst. (k) Synth. catalyzed by AIBN. (l) Synth. catalyzed by H$_2$PtCl$_6$. (m) Synth. under UV irradiation. (n) Synth. from $Et_2GeCH_2CHOHCHRCH_2$ and concn. HX; R = H, Me (2578). (o) Rxn. with LiAlH$_4$ → $Et_2R'GeH$. (p) Rxn. with EtMgBr → Et_3GeR'. (q) Cis- and trans-isomers indicated by GLC (467). (r) Isomers indicated by IR and NMR spectra (1798). (s) Synth. from $Et_2GeCH_2CHMeCMeOHCH_2$ with concn. HX (2578). (t) IR spectrum GLC (1798). (u) Rxn. with BuMgBr → $Bu_3GeCH_2CMe:CMe_2$ (467, 1798).

4.1.1.6 FUNCTIONALLY SUBSTITUTED TRIORGANOGERMANIUM HALIDES

This chapter comprises triorganogermanium halides containing halogen, oxygen and cyano groups in the organic substituent. A few bimetallic derivatives with germanium, boron and iron are listed at the bottom of Table 30. The following scheme summarizes synthetic methods for the compounds listed in Table 30.

I Dealkylation reactions:
- (IA) $Me_3GeCH_2I + Br \rightarrow Me_2(BrCH_2)GeBr$,
- (IB) $R_2\overline{GeCH_2CHRCH_2} + X \rightarrow R_2(XCH_2CHRCH_2)GeX$; R = H, Me ,
- (IC) $Bu_2\overline{GeCH_2CH_2CH_2} + GeCl_4 \rightarrow Bu_2(Cl_3GeCH_2CH_2CH_2)GeX$,
- (ID) $(Me_2GeCHBrCHBr)_2 \rightarrow Me_2(BrCH:CH)GeBr$ by thermal decomposition,
- (IE) $(Me_2GeCPh:CPh)_2 + Br \rightarrow Me_2(BrCPh:CPh)GeBr$.

II Alkylation reactions:
- (IIA) $GeCl_4 + R_3SnR' \rightarrow R'_3GeCl$,
- (IIB) $R'_2GeCl_2 + R_3SnR' \rightarrow R'_3GeCl$,
- (IIC) $GeX_4 + R'Li \rightarrow R'_3GeX$,
- (IID) $Me_2GeCl_2 + YLi_2 \rightarrow Me_2GeCl(YGeMe_2)_nCl$, n = 1, 6,
- (IIE) $R_2HGeX \rightarrow R'_2Hg \rightarrow R_2R'GeX$,
- (IIF) $(Me_2N)_4Ge + R'H + AlCl_3 \rightarrow R'_3GeCl$.

III Hydrogermylation reactions:
- (IIIA) $Et_2HGeCl + HC\vdots CR' \rightarrow Et_2(R'C_2H_2)GeCl$,
- (IIIB) $R_2HGeCl + CH_2:CHR' \rightarrow R_2(R'C_2H_4)GeCl$.

IV Miscellaneous reactions:
- (IVA) $R_2R'GeH + HgX_2 \rightarrow R_2R'GeX$,
- (IVB) $Me_3GeCl + Cl \rightarrow Me_2(ClCH_2)GeCl + Me_2(Cl_2CH)GeCl$.

R = organic group, R' = functionally substituted organic group, X = halogen, Y = carborane group.

Tris(pentafluorophenyl)germanium Chloride $GeC_{18}ClF_{15}$ $(C_6F_5)_3GeCl$
Synth.: By rxn. of $GeCl_4$ with C_6F_5Li (145), in $Et_2O-C_6H_{14}$ at -78° to r.t. in N atm. (1034).
Prop.: m. 103-104°, stable in air (145, 1034), soly. (1034), IR (1034) and ^{19}F NMR spectra (2475).
Rxn. with: RLi $\rightarrow (C_6F_5)_3GeR$; R = n-Bu, C_6F_5 (1034).
$Ph_3SnLi \rightarrow Ph_6Sn_2 + Ph_3SnCl$ (1034).
$H_2O \rightarrow (C_6F_5)_3GeOH$ (1034).
Aq. $NH_3 \rightarrow [(C_6F_5)_3Ge]_2O$ (1034).
Aq. $NaOH \rightarrow C_6F_5OH$ (1034).
Addn. Compd.: Forms hydroscopic pyridine adduct (145).

Chloromethyldimethylgermanium Chloride $GeC_3H_8Cl_2$ $Me_2(ClCH_2)GeCl$
Synth.: By rxn. of Me_3GeCl at 20° in liq. phase, at 100° in daylight and under UV irradiation, in 45, 40 and 84% yield, resp. (2133).
By-prod. in rxn. of $Me(ClCH_2)GeCl_2$ with MeLi at -70° (2131).

Prop.: b. 148-49° (2131), 148° (2133), NMR spectrum (475, 2131, 2373).
Rxn. with: Me$_2$(ClCH$_2$)SiOLi → Me$_2$(ClCH$_2$)GeOSi(CH$_2$Cl)Me$_2$ (2129).
LiAlH$_4$ → Me$_2$(ClCH$_2$)GeH (475).
o-HOC$_6$H$_4$YH + Et$_3$N → o-Me$_2$GeOC$_6$H$_4$YCH$_2$; Y = O, NH (2130).
4,1,2-MeC$_6$H$_3$(SH)$_2$ + Et$_3$N → 1,2,4-Me$_2$GeSC$_6$H$_3$MeSCH$_2$ + 1,2,4-Me$_2$GeCH$_2$SC$_6$H$_3$MeS (2130).
Aq. NaOH in Et$_2$O → [Me$_2$(ClCH$_2$)Ge]$_2$O (2131, 2134).
H$_2$S + Et$_3$N → Me$_2$GeSGeMe$_2$CH$_2$SCH$_2$ (2134).
Dry NH$_3$ in Et$_2$O → [Me$_2$(ClCH$_2$)Ge]$_2$NH (2134).

Additional functionally substituted triorganogermanium halides are listed in Table 30 and Chapters 2.7.1.1.3, 6.5.5, 6.5.7 and 6.5.8.

Table 30. Functionally Substituted Triorganogermanium Halides   RR'R"GeX

Formula*	Synth. Method**	Yield	Properties	Ref.
(MeO$_2$CCH$_2$)$_3$GeCl	IIA	77	b$_1$ 131-33°, n$_D^{20}$ 1.4860, d$_{20}$ 1.4412	692, 1440
	IIB	80	b$_2$ 140-42°, n$_D^{20}$ 1.4855, d$_{20}$ 1.4365 (a, b)	692, 1440
(MeO$_2$CCH$_2$)$_3$GeBr	--	--	(c)	--
(C$_6$F$_5$)$_3$GeBr	IIC	--	m. 105-107°, IR spectrum (d)	1034
Me$_2$(Cl$_2$CH)GeCl	IVB	30	b. 172°	2133
Me$_2$(BrCH$_2$)GeBr	IA	--	IR, NMR and mass spectra (e)	2244
Me$_2$(BrCH:CH)GeBr (75)	ID	--	IR and NMR spectra	2079
Me$_2$(BrCPh:CPh)GeBr	IE	Low	m. 134-35° (f)	248
Et$_2$(ClCH$_2$CH$_2$CH$_2$)GeCl (75)	--	--	(g)	--
Et$_2$(BrCH$_2$CH$_2$CH$_2$)GeBr	IB	97	b$_{1.5}$ 111°, n$_D^{25}$ 1.5250, d$_{20}$ 1.6640	344, 1492
Et$_2$(ClCH$_2$CH:CH)GeCl	IIIA	42	cis-:trans-form 3:2 (h)	1472
Et$_2$(ClCH$_2$C:CH$_2$)GeCl		9	(g)	1472
Et$_2$(ClCH:CMe)GeCl		34	cis-:trans-form 1:1 (h)	1472
Et$_2$(EtO$_2$CCH$_2$CH$_2$)GeCl (75)	--	--	(i)	--

* Numbers in parenthesis refer to pages in main volume.
** The characters used here correspond to the synthetic scheme in the introduction to Chapter 4.1.1.6.
(a) IR (692, 1440) and NMR (1689) spectra. (b) Rxn. with Et$_3$SnOMe → (MeO$_2$CCH$_2$)$_3$GeOMe (692). (c) Rxn. with Et$_3$SnH → (MeO$_2$CCH$_2$)$_3$GeH (2238). (d) ^{19}F NMR spectrum (2475). (e) Rxn. with Mg in THF → (Me$_2$GeCH$_2$)$_n$; n = 2, 3 (2244). (f) IR, UV and mass spectra (248). (g) Rxn. with K-Na in MePh → R$_2$GeCH$_2$CH$_2$CH$_2$ (344, 1492). (h) Mixt. not sepd., IR and NMR spectra, GLC (1472). (i) IR and NMR spectra (1798).

Table 30 Continued   RR'R"GeX

Formula*	Synth. Method**	Yield	Properties	Ref.
$Et_2(AcCH_2CH_2)GeCl$ (75)	--	--	(i, j)	--
$Pr_2(HOCH_2CH_2)GeI$	IVA	52	b_2 116-18°, n_D^{20} 1.5368, d_{20} 1.5800 (k)	25
$Pr_2(MeO_2CCH_2)GeCl$	IIE	45	b_3 106-10°, n_D^{20} 1.4690, d_{20} 1.2030	24
	IVA	47	b_1 82-83°, n_D^{20} 1.4670, d_{20} 1.2003 (l, m)	24
$Bu_2(NCCH_2CH_2)GeCl$ (76)	--	--	(i)	--
$Bu_2(MeO_2CCH_2)GeCl$	IIE	40	b_2 97-101°, n_D^{20} 1.4665, d_{20} 1.1533	24
	IVA	35	$b_{1.5}$ 93-95°, n_D^{20} 1.4678, d_{20} 1.1557 (m)	24
$Bu_2(ClCH_2CH_2CH_2)GeCl$ (76)	--	--	(g, n)	--
$Bu_2(ICH_2CH_2CH_2)GeI$	IB	80	$b_{0.4}$ 136-37°, n_D^{20} 1.5603, d_{20} 1.7127	1492
$Bu_2(MeO_2CCH_2CH_2)GeCl$ (76)	--	--	(i)	--
$Bu_2(ClCH_2CHMeCH_2)GeCl$ (76)	--	--	(o)	--
$Bu_2(ClCH_2CHMeCH_2)GeBr$	IB	88	b_1 140-41°, n_D^{20} 1.5108, d_{20} 1.4217	1492
$(C_5H_5FeC_5H_4)_3GeCl$	IIF	25	Golden platelets, m. 224-26° (dec.), soly.	1952
$o\text{-}(Me_2GeCl)_2C_2B_{10}H_{10}$	IID	42	m. 118°, IR and mass spectra (p)	1835
$m\text{-}(Me_2GeCl)_2C_2B_{10}H_{10}$	IID	20	m. 59-61°, $b_{0.05}$ 130-32°	1835
$Me_2ClGe(m\text{-}CB_{10}H_{10}CGeMe_2)_6Cl$	IID	80	Softening > 95°	1835
$Me_2ClGe(p\text{-}CB_{10}H_{10}CGeMe_2)_6Cl$	IID	12	m. 460-80° (dec.)	1835
$m\text{-}(Me_2GeBr)_2C_6H_4$	--	--	(q)	--
$Bu_2(Cl_3GeCH_2CH_2CH_2)GeCl$	IC	52	$b_{0.4}$ 140-45°, n_D^{20} 1.5041, d_{20} 1.3828	1492

(j) Far IR spectrum, GLC (1798), dinitrophenylhydrazone, m. 82° [$GeC_{14}H_{21}ClN_4O_4$] (1798). (k) IR spectrum (25). (l) Rxn. with $LiAlH_4 \rightarrow Pr_2(HOCH_2CH_2)GeH$ (24). (m) IR spectrum (24). (n) Rxn. with $Na_2S \rightarrow \overline{Bu_2GeCH_2CHRCH_2S}$; R = H, Me (2576). (o) Rxn. with Na in xylene $\rightarrow \overline{Bu_2GeCH_2CHMeCH_2}$ (1492). (p) Rxn. with $H_2Y \rightarrow \overline{Me_2GeCB_{10}H_{10}CGeMe_2Y}$; Y = O, NH (1835). (q) Rxn. with $NaN_3 \rightarrow m\text{-}(Me_2GeN_3)_2C_6H_4$ (2115).

4.1.2 DIORGANOGERMANIUM DIHALIDES

4.1.2.1 UNSUBSTITUTED SYMMETRIC DIORGANOGERMANIUM DIHALIDES

Unsubstituted symmetric organogermanium dihalides contain one organic moiety and one kind of halogen. The following scheme depicts synthetic methods for unsubstituted organogermanium dihalides listed in Table 31.

I Dealkylation reactions:
   (IA) $R_4Ge + Br \rightarrow R_2GeBr_2 + RBr$,
   (IB) $R_2GePh_2 + Br \rightarrow R_2GeBr_2 + PhBr$,
   (IC) $EtGePh_3 + Br \rightarrow EtPhGeBr_2 + PhBr$.

II Alkylation reactions:
   (IIA) $GeCl_4 + i\text{-}Bu_2AlCl + NaCl \rightarrow i\text{-}Bu_2GeCl_2$,
   (IIB) $MeGeCl_3 + RAlCl_2 + NaCl \rightarrow MeRGeCl_2$,
   (IIC) $GeX_4 + Al_4C_3 + HX \rightarrow Me_2GeX_2$.

III Hydrogermylation reactions:
   (IIIA) $RHGeX_2 + CH_2{:}CHR' \rightarrow R(R'C_2H_4)GeX_2$,
   (IIIB) $EtHGeCl_2 + HC{:}CCH_2Cl \rightarrow Et(HC{:}CCH_2)GeCl_2 + HCl$.

IV Other reactions with organogermanium hydrides:
   (IVA) $Bu_2HGeI + I \rightarrow Bu_2GeI_2$,
   (IVB) $R_2HGeF + heat \rightarrow R_2GeF_2$,
   (IVC) $Ph_2GeH_2 + NIS \rightarrow Ph_2GeI_2$.

V Reactions involving organopolygermanes:
   (VA) $(Ph_2Ge)_x + I \rightarrow Ph_2GeI_2$,
   (VB) $R_4BuGe_2Cl + Cl \rightarrow RBuGeCl_2 + others$.

VI Interconversion reactions:
   (VIA) $Bu_2GeI_2 + SbF_3 \rightarrow Bu_2GeF_2 + SbI_3$,
   (VIB) $Bu_2Ge(OR)_2 + AcBr \rightarrow Bu_2GeBr_2$.

VII Redistribution reactions:
   (VIIA) $Me_2GeX_2 + Me_2GeX'_2 \rightarrow Me_2GeXX'$,
   (VIIB) $Me_2GeX_2 + Me_2SiX'_2 \rightarrow MeGeXX'$.

R = organic group, X = halogen, NIS = N-iodosuccinimide.

Dimethylgermanium Dichloride (M-77) $GeC_2H_6Cl_2$ Me_2GeCl_2
Synth.: By rxn. of Me_4Ge in the presence of cat. $GaCl_3$ in a sealed tube with $GeCl_4$ (1:1 to 1:3) at 235-240°, in quant. yield (1856), with $MeGeCl_3$ (1:2), in 94% yield (1856).
By rxn. of Me_4Ge with AcCl and $AlCl_3$ (1:2:2) at r.t. in 70% yield (1781).
By rxn. of Me_3GeCl in the presence of cat. $GaCl_3$ with $GeCl_4$ (2:1) in a sealed tube at 165 and 235°, in 27 and 92% yield, resp. (1856), with $GeCl_4$ (1:2) as sole prod. at 235° (1856), with $MeGeCl_3$ (1:1) at 235° in quant. yield (1856).
By rxn. of $Me_2GeCH_2CH{:}CHCH_2$ with Cl in the dark, in 84% yield (1540).
By rxn. of Ge with MeCl, effect of radio frequency field on formation (2486).

By rxn. of Ge-Cu and MeCl at 400°, in 61% yield (1540), at 480°, in 69% yield (2193), at 350-500°, in 46-58% yield (582), in the presence of Al at 400-50°, in 71-73% yield (2195).

By-prod. in rxn. of Me_3GeCl with Cl at 100°, in daylight and under UV irradiation, in 10 and 4% yield, resp. (2133).

By-prod. in rxn. of $GeCl_4$ with Al_4C_3 and HCl at 150-200° (2114).

Prop.: Spectra: IR (135, 1013, 1463), (far) IR and assignment (1462, 2080), Raman and assignment (1462, 2080), NMR (582, 1544, 1856, 2373), ^{35}Cl NQR (583, 737, 2753), dipole moment (582), magnetic susceptibility (1), effect of radio frequency field on dec. rxn. on Ge surface at 422° and 2 atm. (2486).

Anal. detn. by Schoeninger combustion and polarography (2710).

Rxn. with: RLi → Me_2GeR_2; R = Et_2N (2162), $p-PhC_6H_4$ (423).

RLi → Me_2RGeCl; R = $Me_3SiN(CMe_2)$ (2749).

RLi → RH + R_2; R = $p-PhCH_2C_6H_4$ (423).

YLi_2 → Me_2GeY; Y = $NMeCH_2CH_2NMe$ (2162), CPh:CPhCPh:CPh (2447).

$1,2-Li_2C_2B_{10}H_{10}$ → $o-(Me_2GeCl)_2C_2B_{10}H_{10}$ (1835).

$1,7-LiCB_{10}H_{10}CLi$ → $m-(Me_2GeCl)_2C_2C_{10}H_{10}$ + $m-Me_2GeCl(CB_{10}H_{10}CGeMe_2)_6Cl$ (1835).

$1,12-LiCB_{10}H_{10}CLi$ → $p-Me_2GeCl(CB_{10}H_{10}GeMe_2)_6Cl$ (1835).

$LiAl(PH_2)_4$ → $Me_2Ge(PH_2)_2$ (2625).

Li in THF → $(Me_2Ge)_n$; n = 6 (cyclic), x (1585).

M in THF → $(Me_2Ge)_n$; n = 4-6 (cyclic, ≥ 55), M = Li, Na (1586).

RNa → Me_2GeR_2; R = C_5H_5 (157), H in presence of BPh_3 (2242), OEt (2192, 2837), OPr (2192), OPh, suitable for sepn. from $MeGeCl_3$ (277).

NaX → Me_2GeClX + Me_2GeX_2; X = $Co(CO)_4$ (1672).

$Na_2Fe(CO)_4$ → $[Me_2GeFe(CO)_4]_2$ (1269).

Et_3GeK → $(Et_3Ge)_2GeMe_2$ (815).

CF_3CO_2Ag → $Me_2Ge(O_2CCF_3)_2$ (2446).

RMgBr → Me_2GeR_2; R = $p-PhC_6H_4$ (423), $CF_2:CF$ (2209).

$Me_3CCMe(OMgBr)C \vdots CMgBr$ → $Me_2Ge(C \vdots CCMeOHCMe_3)_2$ (1901).

$CF_3C \vdots CMgI$ → $Me_2Ge(C \vdots CCF_3)_2$ (2332).

Me_4Ge + cat. $GaCl_3$ → Me_3GeCl (1856).

$Me_2GeX_2 \rightleftharpoons Me_2GeClX$; X = Br, I* (1544, 2605).

$Me_2GeX_2 + Me_2SiX'_2 \rightleftharpoons *$; X = Br, I, PhO, X' = Cl, Br, I; nonrandom distribution (2605).

$(Me_2GeY)_n$ → $Me_2GeCl(YMe_2Ge)_nCl*$; Y = O (1541-2), S (368); S, n = 1, 2 (2607); nonrandom distribution (1541-2).

$Me_2Ge(YMe)_2 \rightleftharpoons Me_2GeYMeCl*$; Y = O (1542), S (368).

$MeGeX_3 \rightleftharpoons *$; X = Br, I, PhO (2606).

$Me_2SiX_2 \rightleftharpoons *$; X = Br, I, CN (1544), MeO, MeS (2062).

$(Me_2GeO)_x + Me_2SiCl_2 + (Me_2SiO)_x$ → * (1547).

$(Bu_2SnO)_x$ → $(Me_2GeCl)OBu_2SnCl$ (946).

$Pb(SMe)_2$ → $Me_2Ge(SMe)_2$ (2062).

R_2NH → $Me_2Ge(NR_2)Cl$; R = Me, Et, + $Me_2Ge(NR_2)_2$; R = Me (2163).

$(CH_2NHMe)_2$ → $Me_2ClGeNMeCH_2CH_2NMeGeClMe_2$ (2163).

---

* Equilibrium constants for redistribution rxns., measured by NMR spectroscopy.

MeNH$_2$ (1:3) → c.(Me$_2$GeNMe)$_3$ + (Me$_2$GeNMe)$_x$ (1855).
Abs. ROH + NH$_3$ → Me$_2$Ge(OR)$_2$; R = Et, Pr, Me$_3$C, Ph, CH$_2$:CHCH$_2$ (2192).
ROH + Et$_3$N → Me$_2$Ge(OR)$_2$; R = Me (1542), Ph (2605).
$\overline{CH_2CH_2O}$ → Me$_2$Ge(OCH$_2$CH$_2$Cl)Cl + Me$_2$Ge(OCH$_2$CH$_2$Cl)$_2$ (2196).

Dimethylgermanium Dibromide (M-78)   GeC$_2$H$_6$Br$_2$ Me$_2$GeBr$_2$
Synth.: By rxn. of Me$_4$Ge with Br in the presence of AlBr$_3$, in 79% yield (367).
By rxn. of powdered Ge-Cu at 400° with MeBr, in 38% yield (1540), in presence of Al at 450°, in 73% yield (2195).
Prop.: IR (367), (far) IR and assignment (1462), Raman (367) and assignment (1462) and NMR (1544).
Rxn. with: 9,10-Na$_2$C$_{14}$H$_{10}$ → 9,10-Me$_2$GeC$_{14}$H$_{10}$ + (9,10-Me$_2$GeC$_{14}$H$_{10}$)$_x$ (1762).
Me$_2$GeX$_2$ ⇌ Me$_2$GeBrX*; X =
(Me$_2$GeY)$_n$ → Me$_2$GeBr(GeMe$_2$Y)Br*; Y = O, prod. distribution (1542), S, nonrandom distribution (368).
Me$_2$Ge(YMe)$_2$ ⇌ Me$_2$Ge(YMe)Br*; Y = O (1542), S (368).
MeGeX$_3$ ⇌ *; X = Cl, I, MeO, PhO, MeS (2606), prod. distribution (2606).
Me$_2$SiX$_2$ ⇌ *; X = Cl, I, CN (1544), MeO, MeS (2062).
Me$_2$GeX$_2$ + Me$_2$SiX'$_2$ ⇌ *, nonrandom distribution; X = Cl, I, PhO, X' = Cl, Br, I (2605).
Me$_2$SiBr$_2$ + (Me$_2$SiO)$_x$ + (Me$_2$GeO)$_x$ ⇌ *; metal distribution in chains (2061).

Dimethylgermanium Diiodide (M-80)   GeC$_2$H$_6$I$_2$ Me$_2$GeI$_2$
Synth.: By rxn. of Ge with MeI at 360° (626, 2176) in presence of Cu at 400°, in 16% yield (1540).
Prop.: b. 204° (626, 2176), n_D^{20} 1.70, d$_{20}$ 2.677 (2176), NMR spectrum (1012, 1540, 1544).
Rxn. with: Me$_2$GeX$_2$ ⇌ Me$_2$GeXI*; X = Cl, Br (1544, 2605).
(Me$_2$GeY)$_x$ ⇌ Me$_2$GeI(Me$_2$GeY)I*; Y = O, prod. distribution (1542), S, nonrandom distribution.
Me$_2$Ge(YMe)$_2$ ⇌ Me$_2$Ge(YMe)I*; Y = O (1542), S (368).
MeGeX$_3$ ⇌ *; prod. distribution; X = Br, PhO (2606).
Me$_2$SiX$_2$ ⇌ *; X = Cl, Br (1544), MeO, MeS (2062).
Me$_2$GeX$_2$ + Me$_2$SiX'$_2$ ⇌ *; nonrandom distribution, X = Cl, Br, PhO, X' = Cl, Br, I (2605).

Diethylgermanium Dichloride (M-78)   GeC$_4$H$_{10}$Cl$_2$ Et$_2$GeCl$_2$
Synth.: By rxn. of GeCl$_4$ with Et$_4$Pb at 120-40° in Ar atm., in 83% yield (1364), with Et$_3$Al at 90-100° in Ar atm., in 17% yield (2603), at 110-30° after hydrolytic work-up, in 76% yield (2603), with Et$_2$AlCl at 120-50°, followed by NaCl, in 78% yield (2603).
By rxn. of powdered Ge-Cu on nichrome column packing with EtCl at 400°, in 72% yield (2194).

---

* Equilibrium constants for redistribution rxns., measured by NMR spectroscopy.

By-prod. in rxn. of Et$_2$HGeCl with HC⋮CCH$_2$Cl at 80° in presence of AIBN or under UV irradiation at 150° in sealed tube, in 15 and 20% yield, resp. (1472). By-prod. in rxn. of Et$_4$BuGe$_2$Cl with HCl (2292).

Prop.: IR and assignment (1444), NMR (1444), ^{35}Cl NQR (2753) spectra, GLC (1364, 2603).

Rxn. with: Li or Na in THF → (Et$_2$Ge)$_x$; x = 4, 5 (cyclic), ⩾ 55 (1586).
Na in liq. NH$_3$ → c.(Et$_2$GeNH)$_3$ (1535).
NaOMe → Et$_2$Ge(OMe)$_2$ (465, 2192).
Na$_2$Fe(CO)$_4$ → [Et$_2$GeFe(CO)$_4$]$_2$ (1269).
RMgCl → Et$_2$GeR$_2$; R = CH$_2$C⋮C (1946).
RMgBr → Et$_2$GeR$_2$; R = PhC⋮C, n-BuC⋮C (1221).
Me$_3$CCMe(OMgBr)C⋮CMgBr → Et$_2$Ge(C⋮CCMeOHCMe$_3$)$_2$ (1901).
Et$_2$GeR$_2$ → Et$_2$RGeCl + Et$_2$GeR$_2$; R = polynuclear group (2703).

Diethylgermanium Dibromide (M-80) GeC$_4$H$_{10}$Br$_2$ Et$_2$GeBr$_2$

Synth.: By rxn. of (Et$_2$GeH)$_2$O with Br in EtBr, in 74% yield (1471).
By rxn. of Et$_2$GeCH$_2$CMe:CMeCH$_2$ with Br in EtBr at -80° (1495).

Prop.: n_D^{20} 1.5263, IR and assignment (1444) and NMR (1444) spectra. Anal. detn. by fusion and volumetric titration (265) and spectrographic detn. (1358).

Rxn. with: (C$_5$H$_9$O)O(CH$_2$)$_4$MgCl, + H$_3$PO$_4$ → Et$_2$Ge[(CH$_2$)$_4$OH]$_2$ + C$_5$H$_8$O (343, 1490).

Dipropylgermanium Dichloride (M-80) GeC$_6$H$_{14}$Cl$_2$ n-Pr$_2$GeCl$_2$

Synth.: By rxn. of Pr$_4$Ge with GeCl$_4$, in 54% yield (2837).

Prop.: b$_4$ 66-66.5°, n_D^{25} 1.4703, d$_{25}$ 1.2615 (2837), dipole moment (2836), magnetic susceptibility (1). Germanium analysis by Schoeninger combustion and polarographic detn. (2710).

Rxn. with: NaOEt → Pr$_2$Ge(OEt)$_2$ (2837).

Dibutylgermanium Dichloride (M-81) GeC$_8$H$_{18}$Cl$_2$ n-Bu$_2$GeCl$_2$

Synth.: By rxn. of Bu$_4$Pb with GeCl$_4$ at 140°, in quant. yield (1364), with BuGeCl$_3$ at 140°, in quant. yield (1364).

Prop.: Magnetic susceptibility (1).

Rxn. with: NaC$_2$H → Bu$_2$Ge(C⋮CH)$_2$ + [Bu$_2$(HC⋮C)GeC⋮]$_2$ (939).
Na + CH$_2$(CH$_2$Cl)$_2$ → Bu$_2$GeCH$_2$CH$_2$CH$_2$ (344, 1492).
Na in liq. NH$_3$ → (Bu$_2$GeNH)$_3$ (1761).
C$_4$H$_4$NK in MePh → (N-C$_4$H$_4$N)$_2$GeBu$_2$ (1761).
K, + H$_2$O → Bu$_2$GeH$_2$ (815).
ROH + NH$_3$ in C$_6$H$_6$ → Bu$_2$Ge(OR)$_2$; R = Me, Et, n-, i-Pr, n-, s-, i-Bu, i-C$_5$H$_7$ (337).
ROH + Py·NH$_3$ in C$_6$H$_6$ → Bu$_2$Ge(OR)$_2$; R = Me$_3$C, t-C$_5$H$_{11}$ (327).
Pb(SEt)$_2$ → Bu$_2$Ge(SEt)$_2$ (465).
RSH + Et$_3$N or H$_3$N → Bu$_2$Ge(SR)$_2$; R = PhCH$_2$, Et, n-, i-Pr, n-, i-, t-Bu, n-C$_{12}$H$_{21}$, Ph (1502).

Addn. Compd.: Bu$_2$GeCl$_2$·phen., from components in alc., m. 117° (dec.) [GeC$_{20}$H$_{26}$Cl$_2$N$_2$] (232a, 1212).

Biol. Prop.: Toxicity (845).

Diphenylgermanium Dichloride (M-79) $GeC_{12}H_{10}Cl_2$ Ph_2GeCl_2

Synth.: By rxn. of Ph_4Ge with $PhGeCl_3$ (1:2) at 130-40° and cat. $AlCl_3$, in 74% yield (1364).
By rxn. of Ph_3GeCl with $PhGeCl$ (1:1) and cat. $AlCl_3$, in 82% yield (1364).
By-prod. in rxn. of Ph_3GeCl with $GeCl_4$ (1:1) and cat. $AlCl_3$ at 120° (1364).
By-prod. in rxn. of $HgCl_2$ with Ph_2GeH_2, in 10% yield (1762), with Ph_2HGeCl at r.t. (1762), with PhH_2GeCl at 150° (1762).
By-prod. in rxn. of Ph_2HGeCl with $C_6H_{10}O$, in 14-35% yield (1762).
Prop.: (Far) IR and assignment (333, 1443) and UV (335) spectra, GLC (1364), TLC, detn. and sepn. from Ph_nGeCl_{4-n}; n = 3, 4 (2635).
Rxn. with: Polarographic reduction → Ph_2GeH_2 (963).
RLi → Ph_2GeR_2; R = C_6F_5 (1034), $2-BrC_6F_4$ (2325), Me_2N (2163).
YLi_2 → Ph_2GeY; Y = $(2-C_6F_4)_2$ (883), $(Ph_2Si)_4$ (1177), $(2-C_6F_4)_2S$ (2325), $2,2-C_6F_4GePh_2C_6F_4$ (2325).
$(PhNLi)_2$ → $Ph_2\overline{GeNPhNPh_2GePh_2NPhN}Ph$ + $(PhN:)_2$ + $(PhNH)_2$ (1089).
NaC_2H → $Ph_2Ge(C⋮CH)_2$ (939).
NaR → Ph_2GeRCl; R = $Co(CO)_4$ (1672), $c.C_6H_{11}O$ (1762).
Na in xylene → $(Ph_2Ge)_n$; n = 4, x (396).
Na in $C_{10}H_8-(MeOCH_2)_2$ → $(Ph_2Ge)_n$; n = 5, 6, x (396).
RMgBr → Ph_2GeR_2; R = PhC⋮C, n-PrC⋮C (1221).
Ph_4Ge (1:1) + cat. $AlCl_3$ → Ph_3GeCl (1364).
Ph_3GeH → Ph_2GeHCl (1364).
$(Ph_2GeS)_x$ → $Cl(Ph_2GeS)_xGePh_2Cl$ (2607).
$GeCl_4$ (1:1) + cat. $AlCl_3$ → $PhGeCl_3$ (1364).
$(CH_2NHMe)_2$ → $Ph_2\overline{GeNMeCH_2CH_2N}Me$ (2163).
ROH + NH_3 → $Ph_2Ge(OR)_2$; R = Me, Et, i-Pr, n-, s-, t-Bu (1503).
$NaSCH_2CO_2Na$ or $HSCH_2CO_2H$ + Py → $Ph_2\overline{GeO_2CCH_2S}$ (121).
Addn. Compd.: Ph_2GeCl_2·phen., from components in alc., m 136° (dec.) $[GeC_{24}H_{18}Cl_2N_2]$ (232a, 1212).

Diphenylgermanium Dibromide (M-79) $GeC_{12}H_{10}Br_2$ Ph_2GeBr_2

Synth.: By rxn. of Ph_2GeR_2 with Br in $HCCl_3$; R = CH_2:CH, CH_2:$CHCH_2$ (464), MeCH:CH, Me_2C:CH (1787), $NCCH_2CH_2$ in EtBr at -5° (2577).
By rxn. of Ph_6Ge_2 with Br in $(CH_2Br)_2$ under reflux (831).
By rxn. of $GeBr_4$ with PhI and powdered Cu under reflux, in 28% yield (1528).
By-prod. in rxn. of Ph_2GeH_2 with N-bromosuccinimide, in 8% yield (1762).
Prop.: b_{24} 208-14°, n_D^{20} 1.6380 (1528), spectra: IR (1528), far IR and assignment (105), (far) IR and assignment (1443).
Rxn. with: Ph_2PLi → $Ph_2Ge(PPh_2)$ + P_2Ph_4 (50).
$HC⋮CC_6H_4C⋮CLi$ → $(Ph_2GeC⋮CC_6H_4C⋮C)_x$ (1437).
$NaMn(CO)_5$ → $Ph_2Ge(Br)Mn(CO)_5$ + $Ph_2Ge[Mn(CO)_5]_2$ (1592).
$NaMn(CO)_5$ → $Ph_2Ge[Mn(CO)_5]_2$ + $[Ph_2GeMn(CO)_5]_2$ (1593).
$NaRe(CO)_5$ → $Ph_2Ge[Re(CO)_5]_2$ (1594).
AgSCN → polymeric product (2794).
RMgBr → Ph_2GeR_2; R = Pr, i-Bu, i-C_5H_{11}, C_6H_{13}, C_7H_{15}, C_8H_{17}, C_9H_{19}, $C_{10}H_{21}$ (464), HC⋮C (1946), MeCH:CH, Me_2C:CH, $p-C_6H_4R'$; R' = CH_2:$CHCH_2$, CH_2:CMe, MeCH:CMe (462), R = $p-C_6H_4R'$; R' = Me, i-Pr, MeO, EtO, PhO, Br, o-Me (463).

$(CH_2CH_2MgBr)_2 \rightarrow \overline{CH_2(CH_2)_3GePh_2}$ (1492).
$(CH_2CH_2MgBr)_2$, + Br in $(CH_2Br)_2$, + $LiAlH_4 \rightarrow \overline{CH_2(CH_2)_3GeH_2}$ (2039).
$(\text{!CMgBr})_2 \rightarrow (Ph_2GeC\text{!}C)_x$ (1439).
MeSH + Py → $Ph_2Ge(SMe)_2$ (227).
$HYCH_2CH_2SH$ + Py → $Ph_2Ge\overline{YCH_2CH_2S}$; Y = O, S (121).

Additional unsubstituted symmetric diorganogermanium dihalides are listed in Table 31.

4.1.2.2 UNSYMMETRIC UNSUBSTITUTED DIORGANOGERMANIUM DIHALIDES

Unsymmetric diorganogermanium dihalides contain either two different unsubstituted organic groups or different halogens per molecule. All compounds are listed at the bottom of Table 31. Other derivatives may be found in Sections 3.8.1 and 7.4.1.2.

Table 31. Unsubstituted Diorganogermanium Dihalides R_2GeX_2

Formula*	Synth. Method**	Yield	Properties	Ref.
Me_2GeF_2 (80)	IIC	--		2114
$(PhCH_2)_2GeCl_2$ (80)	--	--	(a)	--
Et_2GeF_2	IVB	--	(By thermal dec.)	1471
Et_2GeI_2 (80)	--	--	IR and NMR spectra (1444)	--
n-Pr_2GeBr_2 (80)	IA	--	(b)	831
n-Pr_2GeI_2 (80)	--	--	(c)	--
i-Pr_2GeBr_2 (80)	IA	--	(d)	2834
n-Bu_2GeF_2 (80)	IVB	--	(By thermal dec.)	1471
	VIA (e)	88	b_{14} 105°	2039
n-Bu_2GeBr_2 (81)	VIB	--	b_9 130°	337
n-Bu_2GeI_2 (81)	IVA	96	$b_{0.25}$ 75° (e)	2039
i-Bu_2GeCl_2	IIA	66	b_1 62-63°, n_D^{20} 1.4736, d_{20} 1.1894, GLC	2603
i-Bu_2GeBr_2	IA	--	(f)	2834
Ph_2GeI_2 (81)	VA	100		396
	IVC	40		1762

* Numbers in parenthesis refer to pages in main volume.
** The characters used here correspond to the synthetic scheme in the introduction to Chapter 4.1.2.1.
(a) Rxn. with RNa → $(PhCH_2)_2GeR_2$; R = $Ph_3PMn(CO)_4$ (185). (b) Rxn. with i-PrMgBr → n-$Pr_2Ge(CHMe_2)_2$ (831). (c) Rxn. with R_2Hg → n-Pr_2GeR_2; R = CH_2CO_2Me (25). (d) Rxn. with RMgBr → i-Pr_2GeR_2; R = $CMe:CH_2$, CH:CHEt (2834). (e) Rxn. of Bu_2GeI_2 with SbF_3 → Bu_2GeF_2 + SbI_3 (2039). (f) Rxn. with RMgBr → i-Bu_2GeR_2; R = $CMe:CH_2$, CH:CHEt, $CH:CMe_2$ (2834).

Table 31 Continued R_2GeY_2

Formula*	Synth. Method**	Yield	Properties	Ref.
$Me_2GeClBr$	VIIA,B	--	NMR spectrum	1544
Me_2GeClI	VIIA,B	--	NMR spectrum	1544
Me_2GeBrI	VIIA,B	--	NMR spectrum	1544
$MeEtGeCl_2$ (80)	IIB	56	b. 148-50°, n_D^{20} 1.4659 (g)	2603
$EtBuGeCl_2$	VB	--		2292
$EtPhGeBr_2$	IC	--	$b_{0.02}$ 78-80°, n_D^{25} 1.5897 (h)	999
$n-C_6H_{13}PhGeF_2$	IIIA	81	$b_{0.2}$ 86°, n_D^{20} 1.4865, d_{20} 1.2568 (i)	1762
$n-C_6H_{13}PhGeCl_2$	IIIA	84	$b_{0.1}$ 94°, n_D^{20} 1.5280, d_{20} 1.2546 (i)	1762
$n-C_6H_{13}PhGeBr_2$	IIIA	89	$b_{0.02}$ 105°, n_D^{20} 1.5600, d_{20} 1.5665 (i)	1762
$n-C_6H_{13}PhGeI_2$	IIIA	92	$b_{0.006}$ 128°, n_D^{20} 1.6260, d_{20} 1.8704 (i)	1762
$Et(HC\!:\!CCH_2)GeCl_2$	IIIB	25	IR and NMR spectra, GLC	1472
$Bu(Me_2C\!:\!CMeCH_2)GeCl_2$	IIIA	77	b_{17} 141°, n_D^{20} 1.4953, d_{20} 1.1968 (j)	467, 1798
R_2GeF_2	--	--	(k)	--
R_2GeCl_2	--	--	(k)	--
R_2GeBr_2	--	--	(k)	--

(g) IR spectrum (366). (h) Rxn. with $1\text{-}C_{10}H_7MgBr \rightarrow EtPhGe(C_{10}H_7\text{-}1)_2$ (999).
(i) NMR spectrum (1762). (j) IR spectrum (1798), rxn. with $BuMgBr \rightarrow Bu_3Ge\text{-}CH_2CMe\!:\!CMe_2$ (467, 1798), synth. from $BuHGeCl_2$ and $(CH_2\!:\!CMe)_2$ (467, 1798).
(k) Ge-bond refractions D_{20} (81).

4.1.2.6 FUNCTIONALLY SUBSTITUTED DIORGANOGERMANIUM DIHALIDES

This chapter comprises diorganogermanes containing halogen, oxygen and cyano groups in the organic substituent. Bimetallic compounds containing germanium of boron atoms in the side chain are also included as are a few oligomeric or polymeric organogermanium dihalides. The following scheme summarizes synthetic methods for the functionally substituted diorganogermanium dihalides listed in Table 32.

I Dealkylation reactions:
$R'_2GePh_2 + Br \rightarrow R'_2GeBr_2 + PhBr$.

II Alkylation reactions:
- (IIA) $GeCl_4 + R'MgBr \rightarrow R'_2GeCl_2$,
- (IIB) $GeCl_4 + R'Li \rightarrow R'_2GeCl_2$,
- (IIC) $GeBr_4 + Et_3SnR' \rightarrow R'_2GeBr_2$,
- (IID) $GeBr_4 + R'I + Cu \rightarrow R'_2GeBr_2$,
- (IIE) $R'GeCl_3 + CH_2N_2 \rightarrow R'(ClCH_2)GeCl_2$.

III Hydrogermylation reactions:
- (IIIA) $EtHGeCl_2 + HC\vdots CR' \rightarrow Et(R'CH:CH)GeCl_2$,
- (IIIB) $HGeCl_3 + C_2H_2 \rightarrow (CH:CHGeCl_2)_x$,
- (IIIC) $HGeCl_3 + CH_2:CR_2 \rightarrow (CR_2CH_2GeCl_2)_x$,
- (IIID) $HGeCl_3 + (RCH:CR)_2 \rightarrow (RCH_2CR:CRCHRGeCl_2)_x$; R = H, Me ,
- (IIIE) $PhHGeCl_2 + Ph_2C:CO \rightarrow Ph(Ph_2C:COH)GeCl_2$.

IV Interconversion reaction:
$Ph(Ph_2C:COH)GeCl_2 + heat \rightarrow Ph(Ph_2CHC:O)GeCl_2$.

R = organic group, R' = functionally substituted organic group, X = halogen.

Methyl(Dichlorogermylene)Diacetate $GeC_6H_{10}Cl_2O_4$ $(MeO_2CCH_2)_2GeCl_2$
Synth.: By rxn. of $Et_3SnCH_2CO_2Me$ with $GeCl_4$ at 100°, in 75% yield (692, 1440), with $MeO_2CCH_2GeCl_3$ at 100°, in 76% yield (692, 1440).
Prop.: b_2 106-12°, $b_{0.5}$ 95-96°, n_D^{20} 1.4870, d_{20} 1.5426-30 (692, 1440), IR (692, 1440) and NMR (1689) spectra.
Rxn. with: $Et_3SnCH_2CO_2Me \rightarrow (MeO_2CCH_2)_3GeCl$ (692, 1440).
$Et_3SnOMe \rightarrow (MeO_2CCH_2)_2Ge(OMe)_2$ (692).

Additional functionally substituted diorganogermanium dihalides are listed in Table 32 and in Chapters 6.5.6 and 6.5.7.

Table 32. Functionally Substituted Diorganogermanium Dihalides R'_2GeX_2

Formula*	Synth. Method**	Yield	Properties	Ref.
(ClCH$_2$)$_2$GeCl$_2$ (82)	--	--	^{35}Cl NQR spectrum	2753
(MeO$_2$CCH$_2$)$_2$GeBr$_2$	IIC	70	b$_2$ 114-16°, n$_D^{20}$ 1.5248, d$_{20}$ 1.9220	1440
(B$_{10}$H$_{10}$C$_2$Ph)$_2$GeCl$_2$	IIB	59	m. 205.5-206°	627
(NCCH$_2$CH$_2$CH$_2$)$_2$GeBr$_2$	I	--	(a)	2577
(C$_6$F$_5$)$_2$GeCl$_2$	--	--	Hydrolysis → (R$_2$GeO)$_x$	145
(C$_6$F$_5$)$_2$GeBr$_3$	IIA	--	b$_{12}$ 180-85° (b)	1034
(2-C$_4$H$_3$S)$_2$GeBr$_2$	IID	10	Colorless liq. b$_{20}$ 212-14° (c)	1528
Me(ClCH$_2$)GeCl$_2$ (82)	--	--	NMR spectrum (d)	2131
ClCH$_2$(p-RC$_6$H$_4$)GeCl$_2$				
R = H (82)	IIE	95	n$_D^{25}$ 1.5640, GLC (e)	1885
Me	IIE	98	n$_D^{25}$ 1.5599, GLC (e)	1885
F	IIE	100	n$_D^{25}$ 1.5473, GLC (e)	1885
Cl	IIE	98	n$_D^{25}$ 1.5777, GLC (e)	1885
MeO	IIE	99	n$_D^{25}$ 1.5707, GLC (e)	1885
Et(ClCH$_2$CH:CH)GeCl$_2$	IIIA	90	cis-:trans-form 3:2, GLC (f)	1472
Ph(Ph$_2$C:COH)GeCl$_2$	IIIE	--	IR spectrum, stable at r.t. (g)	2714

* Numbers in parenthesis refer to pages in main volume.
** The characters used here correspond to the synthetic scheme in the introduction to Chapter 4.1.2.6.
(a) Rxn. with LiAlH$_4$ → [H$_2$N(CH$_2$)$_4$]$_2$GeH$_2$ (2577). (b) Sensitive towards moisture, IR spectrum (1034), rxn. with H$_2$O → [(C$_6$F$_5$)$_2$GeO]$_x$ (1034). (c) IR spectrum (1528). (d) Rxn. with Me$_2$(ClCH$_2$)SiOLi → Me(ClCH$_2$)Ge(OSiMe$_2$CH$_2$Cl)$_2$ (2129), with MeLi → Me$_2$(ClCH$_2$)GeCl + Me$_3$GeCH$_2$Cl (2131). (e) NMR spectrum (1885). (f) IR and NMR spectra.
(g) Ph(Ph$_2$C:COH)GeCl$_2$ rearranges at 90° → Ph(Ph$_2$CHC:O)GeCl$_2$ (2714).

Table 32 Continued

Formula*	Synth. Method**	Yield	Properties	R'$_2$GeX$_2$ Ref.
Ph(Ph$_2$CHC:O)GeCl$_2$	IV (g)	--	Dec. 130° (f, h)	2714
p-(PhGeCl$_2$)$_2$C$_6$H$_4$ (82)	--	--	(i)	--
(CH:CHGeCl$_2$)$_x$	IIIE	45	m. 340-60° (dec.) (f, j, k)	1589
(CR$_2$CH$_2$GeCl$_2$)$_x$				
R$_2$ = H$_2$	IIIC	50	m. 230-300° (dec.), mol. wt. (1)	1589
H + CN	IIIC	--	Glass-like solid, m. 45-55° (f)	1589
Me$_2$	IIIC	97	White powder, m. 225-65° (f, 1)	1589
H + Bu	IIIC	94	Viscous liq. (f, l, m)	1589
(CH$_2$CH:CHCH$_2$GeCl$_2$)$_x$	IIID (n)	--	White powder, m. 62-67°, mol. wt. (f, o)	1589
(MeCHCH:CHCH$_2$GeCl$_2$)$_x$	IIID	--	Viscous liq., mol. wt. (f, o)	1589
(CH$_2$CMe:CHCH$_2$GeCl$_2$)$_x$	IIID	--	White powder, m. 48-55°, mol. wt. (f, o, p)	1589
[$\overline{\text{CH}}$(CH$_2$)$_4$CHGeCl$_2$]$_x$	IIID	--	Glass-like polymer, m. 55-56°	1589

(h) Rxn. with LiAlH$_4$ → Ph(Ph$_2$CHCH$_2$OH)GeH$_2$ (2714). (i) Rxn. with Ph$_3$As + NaN$_3$ (1:4:4) → p-[(Ph$_3$As:N)$_2$GePh]$_2$C$_6$H$_4$ (2116). (j) Glass-like mass or white powder (1589). (k) Rxn. with MeMgBr → (CH:CHGeMe$_2$)$_x$ (1589). (l) Rxn. with MeMgBr → (CR$_2$CH$_2$GeMe$_2$)$_x$ (1589). (m) Rxn. with base → (CR$_2$CH$_2$GeMe$_2$)$_x$ (1589). (n) Synth. from H$_3$GeCl and (CH$_2$:CH)$_2$ (1589), from GeCl$_2$·dioxane and (CH$_2$:CH)$_2$ (1337). (o) Rxn. with MeMgBr → (CHRCR:CHCH$_2$GeMe$_2$)$_x$; R = H, Me (1589). (p) Also obtd. as viscous liquid (1589).

4.1.3   ORGANOGERMANIUM TRIHALIDES

4.1 3.1 UNSUBSTITUTED ORGANOGERMANIUM TRIHALIDES

Synthetic method for organogermanium trihalides have improved and high yields are reported for redistribution reactions of germanium tetrachloride with tetraorganoderivatives of germanium and lead. The following scheme shows methods for preparation of organogermanium trihalides listed in Table 33.

I   Rochow synthesis:
    Ge-Cu + RX at 290-425° → $RGeX_3$.

II Alkylation reactions:
    (IIA) $GeCl_4 + R_2Hg → RGeCl_3$,
    (IIB) $GeBr_4 + RI + Cu → RGeBr_3$,
    (IIC) $GeCl_4 + R_4Pb → RGeCl_3$,
    (IID) $GeX_4 + Al_4C_3 + HX → MeGeX_3$,
    (IIE) $CsGeX_3 + RX → RGeX_3$,
    (IIF) $GeCl_4 + R_3SnR' → R'GeCl_4$.

III Hydrogermylation:
    (IIIA) $HGeCl_3 + RCH:CHR' → RCH_2CHR'GeCl_3$,
    (IIIB) $HGeCl_3 + PhCHCH_2CH_2 → PhEtCHGeCl_3$.

IV Condensation with germachloroform:
    (IVA) $HGeCl_3 + ROH → RGeCl_3 + H_2O$.
    (IVB) $HGeX_3 + RCl → RGeX_3 + HCl$

V   Interconversion reactions:
    (VA) $PhGeH_3 + NIS → PhGeI_3$,
    (VB) $(EtHGeO)_x + HBr → EtGeBr_3$,
    (VC) $RGeCl_3 + HBr → RGeBr_3$.

R = R' = organic group, X = halogen, NIS = N-iodosuccinimide.

Methylgermanium Trichloride (M-83) $GeCH_3Cl_3$ $MeGeCl_3$
Synth.: By rxn. of Ge with MeCl at 460-550° (2132), in presence of Cu at 350-500°, in 13-35% yield (582), at 400°, in 20% yield (1540), effect of Ge-Cu ratio on yield (582), in presence of Al at 400-450° at 370-390°, in 19-21% yield (2195), in presence of Cu, As, Sb and $ZnCl_2$ at 480°, in 69% yield (2193), investigation of optimum conditions (2193).
By rxn. of $HGeCl_3$ with MeOH, in 47% yield (1590)
By rxn. of $CsGeCl_3$ with MeI in THF under reflux, in 47% yield (2694).
By rxn. of Me_4Sn with $HGeCl_3$ (366).
By-prod. in rxn of Ge-Cu with CH_2Cl_2 at 370-90°, in 27% yield (363).
By-prod. in rxn of Me_3GeCl with HCl at 100° in daylight and under UV irradiation, in 15 and 5% yield, resp. (2133).
Prop.: n_D^{20} 1.4660 (363), spectra: IR (135, 363, 1013, 1463), and assignment (1462, 2080, 2082), far IR and assignment (1462, 2080, 2082, 2694), Raman and assignment (1462, 2080, 2082), NMR (363, 582, 1543, 1545, 1856, 2082,

2373, 2694), ^{35}Cl NQR (737, 2753) and mass (1109), dipole moment (582, 2082), GLC (2278), structure (2082). Germanium analysis by Schoeninger combustion and polarographic determination (2710).

Rxn. with: LiAlH(OCMe$_3$)$_3$ → MeGeH$_3$ (582).
RONa → MeGe(OR)$_3$; R = Et (2192, 2837), Pr (2192), Ph, suitable for sepn. from Me$_2$GeCl$_2$ (277).
EtAlCl$_2$ → MeEtGeCl$_2$ (2603).
Me$_4$Ge (1:2) + cat. GaCl$_3$ → Me$_3$GeCl (1856).
Me$_4$Ge (2:1) + cat. GaCl$_3$ → Me$_2$GeCl$_2$ (1856).
Me$_3$GeCl (1:1) + cat. GaCl$_3$ → Me$_2$GeCl$_2$ (1856).
Me$_2$GeX$_2$ ⇌ *, product distribution, X = Br, PhO (2606).
MeGeX$_3$ ⇌ MeGeCl$_n$X$_{3-n}$*, X = Br, I (1543), MeO, MeS; n = 0-3 (1545).
MeSiX$_3$ ⇌ MeGeCl$_n$X$_{3-n}$ + MeSiX$_n$Cl$_{3-n}$*; X = Br (1543, 2060), X = MeO, MeS, Me$_2$N, n = 0-3 (1546).
MeGeBr$_3$ + MeGeI$_3$ + [MeGe(OPh)$_3$] ⇌ *, product distribution (2608).
ROH + NH$_3$ in C$_6$H$_6$ → MeGe(OR)$_3$; R = Ph, CH$_2$:CHCH$_2$ (2192).
Me$_3$COH + NH$_3$ in C$_6$H$_6$ → MeGe(OMe)$_2$Cl + [MeGe(OCMe$_3$)$_2$]$_2$NH (2192).
$\overline{CH_2CH_2O}$ → MeGe(OCH$_2$CH$_2$Cl)$_3$ + MeGe(OCH$_2$CH$_2$Cl)Cl$_2$ (2196).

Methylgermanium Tribromide (M-84)   GeCH$_3$Br$_3$ MeGeBr$_3$
Synth.: By rxn. of Ge-Cu with MeBr at 400°, in 40% yield (1540), in presence of Al at 450°, in 27% yield (2195).
Prop.: (far) IR and assignment (1462), Raman and assignment (523, 1462) and NMR (1012, 1543, 1545) spectra.
Rxn. with: Me$_2$GeX$_2$ ⇌ * product distribution; X = Cl, I, MeO (2606).
MeGeX$_3$ ⇌ MeGeBr$_n$X$_{3-n}$*, X = I (1543), X = MeO, MeS, n = 0-3 (1545).
MeSiX$_3$ ⇌ MeGeBr$_n$X$_{3-n}$ + MeSiX$_n$Br$_{3-n}$*, X = Br (1543), X = MeO, MeS, Me$_2$N, n = 0-3 (1546).
MeGeCl$_3$ + MeGeI$_3$ + [MeGe(OPh)$_3$] ⇌ *, product distribution (2608).
PhOH + Et$_3$N in C$_6$H$_6$ → MeGe(OPh)$_3$ (2608).
$\overline{(HOCH_2CH_2)_3N}$ → $\overline{MeGe(OCH_2CH_2)_3N}$ (2847).

Methylgermanium Triiodide (M-86)   GeCH$_3$I$_3$ MeGeI$_3$
Synth.: By rxn. of Ge sponge with MeI at 320° (2176).
By rxn. of powd. Ge + Cu at 400° with MeI, in 11% yield (1540).
By rxn. of GeI$_2$ with MeI in sealed tube at 110° (1672, 2446).
Prop.: Far IR, assignment (105) and NMR (1540, 1543, 1545) spectra.
Rxn. with: NaCo(CO)$_4$ → MeI$_2$GeCo(CO)$_4$ (1672).
CF$_3$CO$_2$Ag → MeGe(O$_2$CCF$_3$)$_3$ (2446).
Me$_2$CHMgI → i-Pr$_3$GeMe + (MeGeCHMe$_2$)$_x$ (2877).
Me$_2$GeX$_2$ ⇌ *, product distribution; X = Cl, Br, PhO (2606).

---

* Equil. const. for redistribution rxns., measured by NMR spectroscopy.

MeGeX$_3$ → * X = Br (1543), x = MeO, MeS, n = 0-3 (1545).
MeSiX$_3$ → * X = Cl, Br (1543) X = MeO, MeS, Me$_2$N (1546).
MeGeCl$_3$ + MeGeBr$_3$ + [MeGe(OPh)$_3$] → *, product distribution (2608).

Benzylgermanium Trichloride (M-86)   GeC$_7$H$_7$Cl$_3$ PhCH$_2$GeCl$_3$
Synth.: By rxn. of PhCH$_2$Cl with freshly prepared GeCl$_2$ at 50° (381a, 382a), with GeCl$_2$·dioxane (1337), with CsGeCl$_3$ in MeOH under reflux, in 30% yield (2694).
Prop.: m. 33-36° (2694), b$_{10}$ 98° (1588), far IR, assignment and NMR spectra (2694). Anal. detn. of Ge and Cl (580).

Ethylgermanium Trichloride (M-84)   GeC$_2$H$_5$Cl$_3$ EtGeCl$_3$
Synth.: By rxn. of Et$_4$M with HGeCl$_3$·2Et$_2$O; M = Sn, Pb (366).
By rxn. of GeCl$_4$ with Et$_6$Ge$_2$ (813).
By thermal dec. of HGeCl$_3$·2Et$_2$O, in 39% yield (381a, 382a).
By-prod. in rxn. of Ge-Cu on nichrome packed column with EtCl at 400° (2194).
Prop.: IR (1013), assignment (1444), NMR (1444, 2373) and ^{35}Cl NQR (583, 737, 2753) spectra.
Rxn. with: PCl$_3$ + 0 → Cl$_2$OPC$_2$H$_4$GeCl$_3$ (995).
RONa → EtGe(OR)$_3$; R = Me, Et (2192).
Abs. PrOH + NH$_3$ in C$_6$H$_6$ → EtGe(OPr)$_3$ (2192).

Propylgermanium Trichloride (M-84)   GeC$_3$H$_7$Cl$_3$ n-PrGeCl$_3$
Synth.: By rxn. of GeCl$_4$ with Pr$_4$Pb at 100° in Ar atm., in 98% yield (1364).
By rxn. of Ge-Cu-Al with PrCl at 400-450°, in 26% conversion (2195).
By-prod. in rxn. of Pr$_6$Ge$_2$ with GeCl$_4$ at 200° (813).
Prop.: b$_4$ 47.5-48.5°, n$_D^{25}$ 1.4730 (2837), n$_D^{20}$ 1.4745 (1364) d$_{20}$ 1.5092 (2837), IR spectrum and basicity (2837), dipole moment (2836).
Ge analysis by Schoeninger combustion and polarography (2710).
Rxn. with: PCl$_3$ + 0 → Cl$_2$OPC$_3$H$_6$GeCl$_3$ (995).
RONa → PrGe(OR)$_3$; R = Me (2192), Et (2837).

Butylgermanium Trichloride (M-86)   GeC$_4$H$_9$Cl$_3$ n-BuGeCl$_3$
Synth.: By rxn. of GeCl$_4$ with Bu$_4$Pb at 100-140° in Ar atm. in 78% yield (1364), with BuMgCl (1:1) in Et$_2$O, in 25% yield (2039), with Bu$_4$Ge in presence of cat.: Ge, GeI$_2$, Bu$_3$GeH, LiAlH$_4$ or Bu$_6$Ge$_2$ at 175-200°, in high yield (1760).
By rxn. of n-BuI with CsGeCl$_3$ at 160°, in 70% yield (2694).
By rxn. of Ge-Cu with BuCl at 375-425° in N atm., in 38% yield (2064).
By rxn. of HGeCl$_3$ with BuOH, in 47% yield (1590) or by thermal dec. of HGeCl$_3$·2Bu$_2$O , in 35% yield (381a).

---

* Equil. const. for redistribution reaction, measured by NMR spectroscopy.

By-prod. in rxn. of Bu$_6$Ge$_2$ with GeCl$_4$ at 200° (813, 1760).
By-prod. in rxn. of BuH$_2$GeCl with HgCl$_2$, in 20% yield (1798).
By rxn. of GeCl$_4$ with Bu$_4$Sn: method IC (M-86) line 14 from above.
Prop.: b. 183-87° (2694), n$_D^{20}$ 1.4744 (1364), far IR, assignment (2064, 2694) and NMR (2694) spectra, GLC (1760, 1798). Elemental analysis (426).
Rxn. with: Bu$_4$Pb → Bu$_2$GeCl$_2$ (1364).
ROH + NH$_3$ → BuGe(OR)$_3$; R = Me, Et, Me$_2$CH (1506).
Biol. Prop.: Toxicity (845).

Tertiarybutylgermanium Trichloride   GeC$_4$H$_9$Cl$_3$ Me$_3$CGeCl$_3$
Synth.: By rxn. of H$_3$GeCl with Me$_3$CCl in Et$_2$O, in 80% yield (364), with Me$_3$COH, in 70% yield (1590).
By rxn. of CsGeCl$_3$ with Me$_3$CCl in sealed tube at 140°, in 40% yield (2694).
Prop.: m. 95-98° (364), 66-67° (subl.) (2694), b$_{68}$ 92.5-94° (364), IR (364), far IR, assignment and NMR (2694) spectra.
Rxn. with: PhLi → Me$_3$CGePh$_3$ (2694).
MeMgBr → Me$_3$GeCMe$_3$ (364).
HBr → Me$_3$CGeBr$_3$ (364).

Octylgermanium Trichloride (M-86)   GeC$_8$H$_{17}$Cl$_3$ C$_8$H$_{17}$GeCl$_3$
Synth.: (A) By rxn. of n-C$_5$H$_{11}$CHCH$_2$CH$_2$ with HGeCl$_3$ in Et$_2$O, in 35-40% yield (382).
(B) By rxn. of BuCHCHMeCH$_2$ with HGeCl$_3$ in Et$_2$O, in 65% yield (382).
(C) By rxn. of Ge-Cu and C$_8$H$_{17}$Cl at 375-425°, in 48% yield (2064).
Prop.: (A) Mixture of three isomers, b$_{10}$ 112-12.5°, n$_D^{20}$ 1.4838 (382).
(B) Mixture of three isomers, b$_7$ 103-104°, n$_D^{20}$ 1.4813 (382).
(C) Mixture of n- and i- derivatives (2064), (far) IR spectrum and Cl detn. (2064).

Phenylgermanium Trichloride (M-85)   GeC$_6$H$_5$Cl$_3$ PhGeCl$_3$
Synth.: By rxn. of GeCl$_4$ with PhI and powd. Cu, in 80% yield (362), in 63% yield (1528), with Ph$_4$Ge (3:1) and cat. AlCl$_3$ at 120°, in 82% yield (1364), with Ph$_2$GeCl$_2$ and cat. AlCl$_3$ at 120°, in 90% yield (1364), with Ph$_2$Hg in dry o-xylene in N atm. in sealed tube at 160-65°, in 76% yield (1885), with Ph$_2$Hg in HOKE bomb at 130-55°, in 57% yield (2593), with PhCl and Ge-Ag at 450° in N atm. (2064).
By-prod. in rxn. of Ph$_2$GeH$_2$ with HgCl$_2$, in 2% yield (1762).
By-prod. in thermal dec. of PhH$_2$GeCl at 150° (1762).
Prop.: n$_D^{20}$ 1.5540, (362, 1528), n$_D^{20}$ 1.5531 (1364), n$_D^{25}$ 1.5532 (1885), d$_{20}$ 1.5972 (362, 1528) Spectra: IR (2064) and assignment (994, 1528, 2694), far IR (2064) and assignment (333, 2694, 2784), Raman (1528) and assignment (994, 2784), UV (335), GLC (1364, 1885). Detn. of Cl (2064).
Rxn. with: LiAlX$_4$ → PhGeX$_3$; X = H (1364, 2364, 2593), D (2364).
C$_4$Ph$_4$Li$_2$ → PhC:CPhCPh:CPhGePhCl (931).
RNa → PhRGeCl$_2$; R = MeO, c.C$_6$H$_{11}$O (1762), Co(CO)$_4$ (1672).
RNa → PhR$_2$GeCl; R = Mn(CO)$_5$ (1592).

KN$_3$ → PhGe(N$_3$)$_3$ (2115).
MeMgBr → Me$_3$GePh (1041-2).
MeMgI → Me$_3$GePh (2159).
Ph$_4$Ge → Ph$_3$GeCl (1:2) + Ph$_2$GeCl$_2$ (2:1) (1364).
Ph$_3$GeCl (1:1) → Ph$_2$GeCl$_2$ (1364).
BiPy W(CO)$_4$ → PhCl$_2$GeBiPyW(CO)$_3$Cl + CO (1372).
C$_5$H$_5$Co(CO)$_2$ → (PhGeCl$_2$)$_2$C$_5$H$_5$CoCO (1373).
CH$_2$N$_2$ → Ph(ClCH$_2$)GeCl$_2$; relative rxn. rates and mechanism (1885).
MeCHO + (EtO)$_3$P → PhGe(OCHMePO$_3$Et$_2$)$_3$ (452).
(C$_4$H$_3$O)CHO + (EtO)$_3$P + $\overline{\text{MeCHCH}_2\text{O}}$ + EtCHOH → PhGe(OCH$_2$CHMeCl)[OCH(C$_4$H$_3$O)PPhO$_2$Et]-(OCHEtPO$_3$Et$_2$) (?) (452).

Phenylgermanium Tribromide   GeC$_6$H$_5$Br$_3$ PhGeBr$_3$
<u>Synth.</u>: By rxn. of GeBr$_4$ with PhI and powd. Cu, in 51% yield (1528), in 80% yield (362).
By rxn. of PhGeH$_3$ with Br in HCCl$_3$ at 0°-20°, in 68% yield (50).
<u>Prop.</u>: b$_{24}$ 160-61° (362, 1528), b$_{0.01}$ 82-83° (50), n$_D^{20}$ 1.6330, d$_{20}$ 2.2641 (362, 1528), spectra: IR (1528), far IR and assignment (105, 2363), Raman (1528) and assignment (2363).
<u>Rxn. with:</u> Ph$_2$PLi → (PhGePPh$_2$)$_2$ + Ph$_4$P$_2$ (50).
HC⋮CMgBr → PhGe(C⋮CH)$_3$ (1438).
MeSH + Py → PhGe(SMe)$_3$ (227).

Additional unsubstituted organogermanium trihalides are listed in Table 33.

Table 33. Unsubstituted Organogermanium Trihalides   RGeX$_3$

Formula*	Synth. Method**	Yield	Properties	Ref.
MeGeF$_3$ (86)	IID	--	--	2114
Ph$_2$CHGeCl$_3$	IVA (a)	89	m. 78-79°, b$_{0.5}$ 147-52°, soly. (b)	1590
Ph$_3$CGeCl$_3$	IVB	77	m. 215-18°, IR spectrum (c)	364
Ph$_3$CGeBr$_3$	VC (c)	--	--	364
EtGeBr$_3$ (86)	VB	15	NMR spectrum (d)	1471
EtGeI$_3$ (84)	I	--	b. 280°, n$_D^{20}$ 1.70, d$_{20}$ 2.9879 (d)	2714, 2716
n-PrGeBr$_3$ (86)	I	--	(e)	2175

* Numbers in parenthesis refer to pages in main volume.
** The characters used here correspond to the synthetic scheme in the introduction to Chapter 4.1.3.1.
(a) By-prod. in rxn. of HGeCl$_3$ + Ph$_2$CO (1590). (b) NMR spectrum (1590).
(c) Rxn. of RGeCl$_3$ + 48% HBr → RGeBr$_3$ (364). (d) NMR, IR spectra and assignment (1444), rxn. with (HOCH$_2$CH$_2$)$_3$N → EtGe(OCH$_2$CH$_2$)$_3$N (2847). (e) Rxn. with RMgI → PrGeR$_3$; R = Me, Bu (2175).

Table 33 Continued   RGeX$_3$

Formula*	Synth. Method**	Yield	Properties	Ref.
n-PrGeI$_3$	I	--	b$_{0.5}$ 86°, n$_D^{20}$ 1.7, d$_{20}$ 2.8639 (e)	2175, 2176
i-PrGeCl$_3$ (86)	IIE	4	NMR, IR spectra and assignment	2694
	I	87	d$_{20}$ 1.5612	2877
i-PrGeBr$_3$ (86)	I	97	b$_{2.5}$ 63°, n$_D^{20}$ 1.5683, d$_{20}$ 2.287	2877
i-PrGeI$_3$	I	59	b$_{1.5}$ 114°, n$_D^{20}$ > 1.7, d$_{20}$ 2.843 (f)	2877
EtPhCHGeCl$_3$	IIIB	85	b$_8$ 125-26°, n$_D^{20}$ 1.5549	382
n-BuGeBr$_3$ (86)	I	--	Ge and Br detn. (580) (g)	2175
n-BuGeI$_3$ (86)	I	--	b$_{0.35}$ 111°, n$_D^{20}$ 1.70, d$_{20}$ 2.677 (g)	2175, 2176
i-BuGeCl$_3$ (86)	IIC	75	b$_{10}$ 62°, n$_D^{20}$ 1.4719 (h)	1364
	I (i)	28	(Far) IR spectrum, Cl detn.	2064
s-BuGeCl$_3$	IVA	42	(j)	1590
	IIE	26	b. 184-85°, far IR and NMR spectra	2694
Me$_3$CGeBr$_3$	VC	--	--	364
C$_5$H$_{11}$GeCl$_3$ (86)	I (i)	63	(Far) IR spectrum, Cl detn.	2064
c.C$_5$H$_9$GeCl$_3$ (86)	--	--	^{35}Cl NQR spectrum (737)	--
C$_6$H$_{13}$GeCl$_3$ (86)	I (i)	47	(Far) IR spectrum, Cl detn.	2064
c.C$_6$H$_{11}$GeCl$_3$ (86)	IIC	83	b$_{12}$ 103°, n$_D^{20}$ 1.5121 (k)	1364
	I	14	(Far) IR spectrum, Cl detn.	2064
c.C$_6$H$_{11}$GeBr$_3$	IIIA	low	b$_4$ 110°, n$_D^{20}$ 1.5895, d$_{20}$ 2.1178	161
	IIIA (1)	60	NMR spectrum	161, 365
c.MeC$_6$H$_{10}$GeCl$_3$	IIIA (m)	--	b$_9$ 110°, n$_D^{20}$ 1.5153	382
n-C$_7$H$_{15}$GeCl$_3$ (86)	IIIA	68	b$_8$ 100°, n$_D^{20}$ 1.4742, d$_{20}$ 1.3071 (h)	364
C$_7$H$_{15}$GeCl$_3$	I (i)	52	(Far) IR spectrum, Cl detn.	2064
n-C$_9$H$_{19}$GeCl$_3$	IIIA	86	b$_{10}$ 163.5°, n$_D^{20}$ 1.4732, d$_{20}$ 1.2499	995
C$_9$H$_{19}$GeCl$_3$	I (i)	53	(Far) IR spectrum, Cl detn.	2064
C$_{10}$H$_{21}$GeCl$_3$ (87)	I (i)	57	(Far) IR spectrum, Cl detn.	2064
PhGeI$_3$ (87)	VA	5	(o)	1762
C$_6$D$_5$GeCl$_3$	IIE	--	b$_{0.5}$ 88-95° (p-993, q)	993
C$_6$D$_5$GeBr$_3$	--	--	(p-2363)	--
m-MeC$_6$H$_4$GeBr$_3$	IIB	65	b$_{37}$ 186-87°, n$_D^{20}$ 1.6215, d$_{20}$ 2.1440 (r)	1528
p-MeC$_6$H$_4$GeCl$_3$	IIA	45	b$_{24}$ 129-31°, n$_D^{25}$ 1.5502, GLC (s)	1885

1-$C_{10}H_7$GeBr$_3$	IIA	65	m. 52.5-53.5°, $b_{3.5}$ 193-94° (t)	1528
(2-C_4H_3S)GeBr$_3$	IIA	37	b_{23} 152-53°, n_D^{20} 1.6495, d_{20} 2.428 (u)	1528

(f) Rxn. with RMgX → i-PrGeR$_3$; R = Me, Et, Pr (2877). (g) Rxn. with RMgX → BuMgR$_3$; R = Pr (2175). (h) Ge and Cl detn. (580). (i) Synth. yields mixture of n- and i- derivatives (2064). (j) Rxn. with PhLi → s-BuGePh$_3$ (2694). (k) Rxn. with PCl$_3$ and O → Cl$_2$OPC$_6$H$_{10}$GeCl$_3$ (995). (l) Synth. in presence of GeBr$_4$ (161, 365). (m) Synth. from HGeCl$_3$ and norcarane (382). (n) Rxn. with PCl$_3$ and O → Cl$_2$OPC$_9$H$_{18}$GeCl$_3$ (995). (o) Rxn. with C$_5$H$_5$Co(CO)$_2$ → PhI$_2$GeC$_5$H$_5$CoI(CO) (1373). (p) (Far) IR, Raman spectra and assignment (993, 2363). (q) Rxn. with LiAlX$_4$ → C$_6$D$_5$GeX$_3$; X = H, D (2364). (r) Raman spectrum (1528). (s) Rxn. with CH$_2$N$_2$ → p-MeC$_6$H$_4$(ClCH$_2$)GeCl$_2$, relative rxn. rates (1885). (t) Thick colorless liq. n_D^{20} 1.6988, d_{20} 2.1871, IR spectrum (1528). (u) Colorless liq., IR spectrum (1528).

4.1.3.4 UNSATURATED ORGANOGERMANIUM TRIHALIDES

This chapter includes organogermanium trihalides with olefinic and acetylenic unsaturation. Derivatives containing chloroolefinic groups are listed in Tables 35 and 39.

Vinylgermanium Trichloride (M-85)   GeC$_2$H$_3$Cl$_3$ CH$_2$:CHGeCl$_3$
Synth.: By cleavage of Cl$_3$GeCH$_2$CHClGeCl$_4$ with AlCl$_3$ or quinoline, in 75% yield (1529).
By-prod. in thermal dec. of Cl$_2$BCH$_2$CH(BCl$_2$)GeCl$_3$ at 150° (2329).
By-prod. in rxn. of C$_2$H$_2$ with HGeCl$_3$·OEt$_2$ at -30° (1529).
Prop.: b_{50} 40-50°, n_D^{20} 1.4606 (1529), 1.4813 (1533), IR (309) and Raman (1529) spectra.
Rxn. with: RLi → CH$_2$:CHGeR$_3$; R = [(C$_{12}$H$_{25}$)$_3$AsMn(CO)$_4$] (185).
B$_2$Cl$_4$ → BCl$_2$CH$_2$CH(BCl$_2$)GeCl$_3$; relative rxn. rates (2329).

Allylgermanium Trichloride (M-85)   GeC$_3$H$_5$Cl$_3$ CH$_2$:CHCH$_2$GeCl$_3$
Synth.: By rxn. of HGeCl$_3$·2Et$_2$O with CH$_2$:CHCH$_2$COCl under reflux, in 30% yield (2602).
By rxn. of CsGeCl$_3$ with CH$_2$:CHCH$_2$I at 120°, in 60% yield (2694).
Prop.: b_{26} 58-59.5°, n_D^{20} 1.4920 (2602), spectra: IR (2602), (far) IR and assignment (2694) UV and structure (424), GLC (2602).
Rxn. with: PhLi → CH$_2$:CHCH$_2$GePh$_3$ (2694).

Allylgermanium Tribromide (M-88)   GeC$_3$H$_5$Br$_3$ CH$_2$:CHCH$_2$GeBr$_3$
Synth.: By rxn. of CH$_2$:CHCH$_2$Br with HGeBr$_3$ in Et$_2$O, in 15% yield (161), with GeBr$_2$, in 65% yield (161).
By-prod. in rxn. of CH$_2$:CHCH$_2$Cl with HGeBr$_3$ and GeBr$_4$ (161).

Prop.: b_6 66-67°, n_D^{20} 1.5885-95 (161), UV (424) and NMR (161) spectra, structure (424).

Propargylgermanium Trichloride $GeC_3H_3Cl_3$ $HC\vdots CCH_2GeCl_3$
Synth.: By rxn. of $HGeCl_3$ with $HC\vdots CCH_2Cl$ in Et_2O at 25°, in 90% yield (1472), with $HC\vdots CCH_2Cl$ in Et_2O, in low yield (364).
Prop.: b_{16} 53-57°, n_D^{20} 1.5283 (364), IR (364, 1472) and NMR (1472) spectra, GLC (1472).
Rxn. with: $MeMgBr \rightarrow Me_3GeCH_2CH:CH_2 + Me_3GeCH_2CH:CHGeMe_3$ (364).

2-Butenylgermanium Trichloride (M-88) $GeC_4H_7Cl_3$ $MeCH:CHCH_2GeCl_3$
Synth.: By rxn. of $HGeCl_3$ with $MeCHClCH:CH_2$ in Et_2O, in 60% yield (364), without solvent in 20% yield (364).
By-prod. in rxn. of $HGeCl_3 \cdot 2Et_2O$ with $(CH_2:CH)_2$ (1530).
Prop.: b_4 39-40°, b_{32} 83.5-84°, n_D^{20} 1.4990, 1.5004, d_{20} 1.4934-94 (364), IR spectrum (364).
Rxn. with: $MeMgCl \rightarrow Me_3GeCH_2CH:CHMe$ (1530).

Additional unsaturated organogermanium trihalides are listed in Table 34.

Table 34. Unsaturated Organogermanium Trihalides $R'GeX_3$

Formula*	Synth. Method**	Yield	Properties	Ref.
$CH_2:CHGeBr_3$ (88)	--	--	Dipole moment (a)	254
$HC\vdots CGeCl_3$	IIF	95	b. 97-98°, n_D^{20} 1.4587, d_{20} 1.6653 (b)	2881
$CH_2:CHCH_2GeBr_2Cl$ (?)	IVB	--	b_{10} 50-76° (!), NMR spectrum	161
$MeC\vdots CGeCl_3$	IIF	96	b_{20} 54°, n_D^{20} 1.4840, d_{20} 1.5953 (b)	1891, 2881
$CH_2:CHCH:CH_2GeCl_3$ (?)	IIIA (c)	13	b_7 53.5-54.5°, n_D^{20} 1.5250, d_{20} 1.5346 (d)	364, 2403
$\{EtCH:CHCH_2GeCl_3\}$ $(MeCH:CHCHMeGeCl_3)$	IIIA (e)	--	b_{13} 73.5°, n_D^{20} 1.5035, d_{20} 1.4306, IR and NMR spectra (f)	1530
$Me_2C:C:CHgeCl_3$	IVB (g)	19	b_{24} 88-89°, n_D^{20} 1.5135, d_{20} 1.4447 (h)	2403
$2-C_5H_7GeCl_3$	IVB	76	b_5 63.5°, n_D^{20} 1.5280, d_{20} 1.5621 (a, i)	308, 364

* Numbers in parenthesis refer to pages in main volume.
** The characters used here correspond to the synthetic scheme in the introduction to Chapter 4.1.3.1.
(a) UV spectrum and structure (424). (b) IR spectrum (2881). (c) Synth. from $HGeCl_3$ and $HC\vdots CCH:CH_2$ at -60° (364, 2403). (d) Mixture of components, IR spectrum indicates also allenic and acetylenic unsaturation (2403). (e)

Synth. from HGeCl$_3$·OEt$_2$ with MeCH:CHCH:CH$_2$ (1530). (f) Rxn. with MeMgCl →
Me$_3$GeCH$_2$CH:CHEt + Me$_3$GeCHMeCH:CHMe (1530). (g) Synth. from HGeCl$_3$·2Et$_2$O and
HC⋮CCMe$_2$Cl at r.t. (2403). (h) IR and NMR spectra (2403). (i) IR and Raman
spectra (308).

4.1.3.6 FUNCTIONALLY SUBSTITUTED ORGANOGERMANIUM TRIHALIDES

4.1.3.6.1 HALOGEN SUBSTITUTED ORGANOGERMANIUM TRIHALIDES

All compounds are listed in Table 35. Additional organogermanium trihalides with halogen substituents may be found in Subchapters 4.1.3.6.3.1, 4.1.3.6.3.4 and 4.1.3.6.7.1.

4.1.3.6.2 PSEUDOHALOGEN SUBSTITUTED ORGANOGERMANIUM TRIHALIDES

Two cyano group containing derivatives are listed at the bottom of Table 35.

4.1.3.6.3 OXYGEN CONTAINING ORGANOGERMANIUM TRIHALIDES

Organogermanium trihalides containing oxygen in form of hydroxy, carbonyl, carboxyl groups or their derivatives are compiled in this chapter. The compounds are prepared by methods from the following scheme.

I Alkylation reactions:
   (IA) $GeX_4 + R'_2Hg \rightarrow R'GeX_3$,
   (IB) $GeX_4 + R_3SnR' \rightarrow R'GeX_3$,
   (IC) $GeI_2 + HI + RR'CHOH \rightarrow RR'CHGeI_3$.

II Hydrogermylation:
   (IIA) $HGeX_3^* + RR'C:CHR' \rightarrow R'CH_2CRR'GeCl_3$; R = H, Me ,
   (IIB) $HGeX_3 + R_2C:O \rightarrow RRC(OH)GeX_3$,
   (IIC) $HGeCl_3 + 2R_2CO \rightarrow R'GeCl_3$.

III Other reactions with germachloroform:
   $HGeCl_3 + R'X \rightarrow R'GeCl_3$; X = Cl, AcO.

IV Addition of alkoxytrichlorogermane:
   $ROGeCl_3 + CH_2:CO \rightarrow RO_2CCH_2GeCl_3$.

V Interconversion reactions:
   (VA) $R''CO_2H + SOCl_2 \rightarrow R''COCl$,
   (VB) $R''CN + HCl + SOCl_2 \rightarrow R''COCl$,
   (VC) $C_nH_{2n+1}GeCl_3 + PCl_3 + O \rightarrow Cl_2OPC_nH_{2n}GeCl_3$; n = 2, 3, 6, 9 .

---

*HGeCl$_3$, HGeBr$_3$, neat and as etherates, and GeI$_2$ + 57% aq. HI. R = organic group or hydrogen, R' = organic group, may be functionally substituted, R'' = trihalogermane containing organic group, X = halogen.

Table 35. Halogen Substituted Organogermanium Trihalides

Formula*	Synth. Method**	Yield	Properties	R'GeX$_3$ Ref.
CF$_3$GeCl$_3$ (90)	--	--	(a)	--
ClCH$_2$GeCl$_3$ (90)	--	--	Dipole moment (b)	254
Cl$_2$CHGeCl$_3$	--	--	(b)	--
ClCH$_2$CH$_2$GeCl$_3$ (89)	--	--	GLC (b-737, 2753, c)	520
MeCHClGeCl$_3$ (90)	--	--	GLC	520
BrCH$_2$CH$_2$GeCl$_3$	IIIA	39	b$_8$ 73-74°, n$_D^{20}$ 1.5365, d$_{20}$ 2.0211 (b-737)	364
ClCH:CHGeCl$_3$ (90)	--	--	Dipole moment (c)	254
CH$_2$:CClGeCl$_3$ (90)	--	--	(d)	--
CF$_3$CH$_2$CH$_2$GeCl$_3$ (90)	IIIA	55	b$_{748}$ 142°, n$_D^{20}$ 1.4240, d$_{20}$ 1.7115 (e, f, g)	1261
CF$_3$CH$_2$CHClGeCl$_3$	(g)	33	b$_8$ 46°, n$_D^{20}$ 1.4475, d$_{20}$ 1.8047 (e, h)	1261
CF$_3$CH$_2$CCl$_2$GeCl$_3$	(g)	58	b$_8$ 50°, n$_D^{20}$ 1.4600, d$_{20}$ 1.8458 (e, i)	1261
CF$_3$CH:CHGeCl$_3$	(h)	87	b. 124-26°, n$_D^{20}$ 1.4260, d$_{20}$ 1.7188 (e)	1261
CF$_3$CH:CClGeCl$_3$	(i)	88	b. 141-42°, n$_D^{20}$ 1.4430, d$_{20}$ 1.7893 (e)	1261
Cl(CH$_2$)$_3$GeCl$_3$ (90)	--	--	GLC (b-2753)	520
Cl(CH$_2$)$_3$GeBr$_3$	IIIA	60	b$_6$ 109-11°, n$_D^{20}$ 1.5820, d$_{20}$ 2.3383 (j)	161
EtCHClGeCl$_3$	--	--	GLC	520
ClCH$_2$CHMeGeCl$_3$ (91)	--	--	(c)	--
MeCHClCH$_2$CH$_2$GeCl$_3$	IIIA	69	b$_4$ 68°, n$_D^{20}$ 1.4990, d$_{20}$ 1.5667 (k, l)	364
2-ClC$_5$H$_8$GeCl$_3$c.	IIIA	57	b$_7$ 111-12°, n$_D^{20}$ 1.5295, d$_{20}$ 1.6445 (b-737, c)	364
m-FC$_6$H$_4$GeBr$_3$	IIB	49	b$_{23}$ 151-52°, n$_D^{20}$ 1.6120, d$_{20}$ 2.3250 (m)	1528
p-FC$_6$H$_4$GeCl$_3$	IIA	34	b$_{17}$ 106-107°, n$_D^{25}$ 1.5342, GLC (n)	1885
C$_6$F$_5$GeBr$_3$	IIA (o)	--	--	77a
p-ClC$_6$H$_4$GeCl$_3$ (91)	IIA	64	b$_{10}$ 116-17°, n$_D^{25}$ 1.5678, GLC (n)	1885
p-IC$_6$H$_4$GeBr$_3$	IIB	19	m. 73-74°, subl. IR spectrum	1528
NCCH$_2$CH$_2$GeCl$_3$ (91)	--	--	(p)	--
NCCH$_2$CHMeGeCl$_3$	IIIA	24	b$_2$ 81-81.5°, NMR spectrum (p)	2602

4.1.3.6.3.1 ORGANOGERMANIUM TRIHALIDES CONTAINING HYDROXY DERIVATIVES

Organogermanium trihalides containing a free hydroxy group or derivatives like ethers and esters are listed in Table 36 and Subchapter 4.1.3.6.7.1.

4.1.3.6.3.3 DERIVATIVES CONTAINING CARBONYL GROUPS

All derivatives are listed in Table 37.

4.1.3.6.3.4 COMPOUNDS CONTAINING CARBOXYLIC GROUPS

All compounds are listed in Table 38 and Subchapter 4.1.3.6.7.1

4.1.3.6.5 ORGANOGERMANIUM TRIHALIDES CONTAINING PHOSPHORUS

All compounds are listed at the bottom of Table 38.

---

* Numbers in parenthesis refer to pages in main volume.
** The characters used here correspond to the synthetic scheme in the introduction to Chapter 4.1.3.1.
(a) Rxn. with $CF_3CO_2Ag \to CF_3GeO_2CCF_3$ (2446), with $Me_2Zn \to Me_3GeCF_3$, CAUTION! Rxn. might proceed with explosive violence (2446). (b) ^{35}Cl NQR spectra (737, 2753). (c) Ge and Cl detn. (580). (d) Rxn. with $HGeCl \to Cl_3Ge-CH_2CHClGeCl_3$ (1533). (e) ^1H and ^{19}F NMR spectra (1261). (f) IR (1261) and ^{35}Cl NQR (583, 737, 2753) spectra. (g) Rxn. of $CF_3CH_2CH_2GeCl_3$ with Cl under UV irradiation $\to CF_3CH_2CHClGeCl_3 + CF_3CH_2CCl_2GeCl_3$ (1261). (h) Distn. of $CF_3CH_2CHClGeCl_3$ from quinoline $\to CF_3CH:CHGeCl_3$ (1261). (i) Distn. of $CF_3CH_2CCl_2GeCl_3$ from quinoline $\to CF_3CH:CClGeCl_3$ (1261). (j) NMR spectrum (161). (k) IR spectrum (36). (l) Rxn. with $MeMgCl \to Me_3GeCH_2CH_2CHClMe$ (36). (m) IR and Raman spectra (1528). (n) Rxn. of p-$XC_6H_4GeCl_3$ with $CH_2N_2 \to p$-$XC_6H_4(ClCH_2)GeCl_2$; X = F, Cl (1885). (o) Synth. from $GeBr_4$ and $MeHgC_6F_5$ (77a). (p) Rxn. with concn. HCl followed by $SOCl_2 \to ClCOCH_2CHRGeCl_3$; R = H (1527), Me (2602).

Table 36. Organogermanium Trihalides Containing Hydroxy Derivatives

Formula	Synth. Method*	Yield	Properties	R'GeX$_3$ Ref.
HOCH$_2$GeCl$_3$	IIB	59	b$_3$ 62.5-63°, n$_D^{20}$ 1.5241, d$_{20}$ 1.9048 (a, b)	2604
MeOCH$_2$GeCl$_3$	III (c)	55	b$_{15}$ 61°, n$_D^{20}$ 1.4855, d$_{20}$ 1.6154	381a, 1588
	III	77	b$_{19}$ 58.5-59°, n$_D^{20}$ 1.4870 (d)	1526
ClCH$_2$OCH$_2$GeCl$_3$	III	17	b$_8$ 70.5-71°, n$_D^{20}$ 1.5053, d$_{20}$ 1.7705 (e)	1526
EtOCH$_2$GeBr$_3$	IIB (f)	40	b$_2$ 71.5-72.5°, n$_D^{20}$ 1.5690, d$_{20}$ 2.2922 (a)	2604
AcOCH$_2$GeCl$_3$	--	--	^{35}Cl NQR spectrum (737)	--
HOCMe$_3$GeCl$_3$	--	--	IR spectrum (1526)	--
AcOCMe$_2$GeCl$_3$	IIA	50	m. 29.8°, b$_9$ 107°, n$_D^{20}$ 1.4987, d$_{20}$ 1.5752	1071
	IIA (g)	89	m. 31.2°, b$_7$ 95°, n$_D^{20}$ 1.5008, d$_{20}$ 1.5953 (h)	1071
p-MeOC$_6$H$_4$GeCl$_3$	IA	62	b$_5$ 104-105°, n$_D^{25}$ 1.5610, GLC (i)	1885

* The characters used here correspond to the synthetic scheme in the introduction to Chapter 4.1.3.6.3.
(a) NMR spectrum (2604). (b) Rxn. with MeMgCl → Me$_3$GeCH$_2$OH (2604). (c) Synth. with GeCl$_2$ and MeOCH$_2$Cl (381a, 1588). (d) IR spectrum (1526), mol. wt. (1588), rxn. with MeMgBr → Me$_3$GeCH$_2$OMe (1588). (e) IR and NMR spectra (1526). (f) Synth. from HGeBr$_3$·2Et$_2$O and H$_2$CO at r.t. (2604). (g) Synth. from HGeCl$_3$ and Me$_2$CO containing CH$_2$:CO (1071). (h) IR (1071, 1526), Raman and NMR spectra (1071), rxn. with MeMgCl → Me$_3$GeCMe$_2$OH + Me$_3$COH (1071). (i) Rxn. with CH$_2$N$_2$ → p-MeOC$_6$H$_4$(ClCH$_2$)GeCl$_2$; relative reaction rate and mechanism (1885).

Table 37. Organogermanium Trihalides Containing Carbonyl Groups

Formula	Synth. Method*	Yield	Properties	R'GeX$_3$ Ref.
AcGeCl$_3$·Ac$_2$O	III	45	m. 129.5-30.5° (dec.) (a)	2602
Ph$_2$CHC(:O)GeCl$_3$	IIB	--	Instable, IR and NMR spectra (b)	2714
OCHCH$_2$CH$_2$GeCl$_3$	IIA	65	b$_8$ 97.5-98°, n$_D^{20}$ 1.5178, d$_{20}$ 1.7380 (c, d)	1526

Compound	Type	Yield	Properties	Ref.
OHCCH$_2$CHMeGeCl$_3$	IIA	80	b$_{4.5}$ 88.5°, n$_D^{20}$ 1.5130, d$_{20}$ 1.6375,	1526
OHCCH$_2$CHMeGeI$_3$	IIA (e)	--	NMR spectrum (c, f)	1526
OHCCH$_2$CHMeGeI$_3$	IIA	--	Dec. 0.01 140°, n$_D^{20}$ > 1.7, d$_{20}$ 2.807 (g, h)	1496
AcCH$_2$CH$_2$GeI$_3$	IIA	62	b$_{0.01}$ 136-37°, n$_D^{20}$ > 1.7, d$_{20}$ 2.838 (g, i)	1496
BzCH$_2$CMePhGeCl$_3$	IIC (j)	73	m. 172-73°, soly. (c, k)	1526
	IIC (j)	96	m. 172-74°, soly.	1590
	IIC (l)	40	--	2504
AcCH$_2$CMeGeCl$_3$	IIA	67	m. 50-50.5°, b$_2$ 109-11° (subl.), mol. wt.,	1071
	IIC (m)	76	IR, Raman and NMR spectra (c)	1071, 1526
	IIC (l)	46	m. 48-49°, b$_1$ 107-11° (n, o)	2504
AcCH$_2$CMeGeBr$_3$	IIC	25	m. 62-62.5°, b$_4$ 135°, soly. (p)	1526
AcCH$_2$CMeGeI$_3$	IIA	57	m. 65°, soly (g)	1496
1-(2-C$_5$H$_7$O)C$_5$H$_8$GeCl$_3$	IIB (q)	43	m. 74-75°, soly. (c)	1526
(c.C$_6$H$_9$O)C$_6$H$_{10}$GeCl$_3$	IIC (r)	90	m. 173-75°	1335
2-(2-C$_6$H$_9$O)C$_6$H$_{10}$GeCl$_3$	IIC (s)	89	m. 170-70.5°, soly. (c)	1526
1-(2-C$_6$H$_9$O)GeH$_{10}$GeCl$_3$	IIA (t)	85	m. 174-76°, soly.	1590
	III	65	Mol. wt. (2504)	1590
	III (u)	60	NMR spectrum (2504)	1590
	IIA (v)	35	IR spectrum	1590
1-(2-C$_7$H$_{11}$O)C$_7$H$_{12}$GeCl$_3$	IIC (w)	98	m. 112-15°, soly. (x)	1590, 2504

---

* The characters used here correspond to the synthetic scheme in the introduction to Chapter 4.1.3.6.3.
(a) IR and NMR spectra (2602). (b) Dec. at r.t. → (Ph$_2$CHC:O)$_2$ + GeCl$_4$ + GeCl$_2$ (2714), rxn. with LiAlH$_4$ → Ph$_2$CHCHOHGeH$_3$ (2714). (c) IR spectrum (1526). (d) Polymerizes on standing (1526). (e) By-prod. in rxn. of HGeCl$_3$·2Et$_2$O with AcH or paraldehyde (1526). (f) Rxn. with MeMgCl → Me$_3$GeCHMeCH$_2$CHOHMe (1526). (g) NMR spectrum (1496). (h) Orange liq. (1496), rxn. with EtMgBr → Et$_3$GeCHMeCH$_2$CHOHEt (1496), with LiAlH$_4$ → HOCH$_2$CH$_2$CHMeGeH$_3$ (1496). (i) Rxn. with EtMgBr → Et$_3$GeCH$_2$CH$_2$COHMeEt (1496). (j) Synth. from HGeCl$_3$·2Et$_2$O and AcPh under reflux (1526, 1590). (k) IR and NMR spectra (1590). (l) From GeCl$_2$·dioxane and AcR; R = Me, Ph (2504). (m) From HGeCl$_3$ or HGeCl$_3$·2Et$_2$O and Me$_2$CO (1071, 1526). (n) IR and NMR spectra (2504). (o) Rxn. with MeMgCl → Me$_3$GeCMe$_2$CH$_2$-CMe$_2$OH (1526, 2504). (p) IR and NMR spectra (1526). (q) From cyclopentanone (1526). (r) From cyclohexanone or 1-methoxycyclohexene (1335). (s) From cyclohexanone (1526). (t) From 2-(1-cyclohexenyl)cyclohexanone (1590). (u) From GeCl$_2$·dioxone and 2-(1-chlorocyclohexyl)cyclohexanone (1590). (v) From 1-methoxycyclohexene (1590). (w) From cycloheptanone (1590, 2504). (x) IR spectrum (1590, 2504).

Table 38. Organogermanium Trihalides Containing Carboxyl Derivatives   R'GeX₃

Formula	Synth. Method*	Yield	Properties	Ref.
$MeO_2CCH_2GeCl_3$	IB	87	$b_{6.5}$ 70-71°, n_D^{20} 1.4820, d_{20} 1.6760	692, 1440
	IV	62	b_7 68-70°, n_D^{20} 1.4835, d_{20} 1.6765 (a, b, c)	1440
$MeO_2CCH_2GeBr_3$	IB	88	b_2 81-82°, n_D^{20} 1.5613, d_{20} 2.3534	1440
$MeO_2CCH_2GeI_3$	IB	65	b_1 116-17°, d_{20} 2.9	1440
$EtO_2CCH_2GeCl_3$	IV	61	$b_{1.5}$ 49-50°, n_D^{20} 1.4750, d_{20} 1.5756	1440
$HO_2CCH_2CH_2GeCl_3$	IIA	98	m. 83-85°, dec. > 85° (d, e)	1527
$HO_2CCH_2CH_2GeI_3$	IIA	72	m. 112°, soly. (f, g)	1496
$MeO_2CCH_2CH_2GeI_3$	IIA	87	$b_{0.01}$ 120°, n_D^{20} 1.6998, d_{20} 2.772 (f, h, i)	1496
$ClOCCH_2CH_2GeCl_3$	IIA	83	b_7 89-91°, n_D^{20} 1.5115, d_{20} 1.7514	1527
	VB	34	(d)	1527
	VA (e)	--	(j)	1527
$HO_2CCH_2CHPhGeCl_3$	IIA	50	m. 105.5-107.5°, soly. (d)	2602
$ClOCCH_2CHPhGeCl_3$	IIA	36	$b_{1.5}$ 111.5-12°, n_D^{20} 1.5690, d_{20} 1.5927 (d)	2602
$HO_2CCHMeCH_2GeCl_3$	IIA	62	m. 84°, soly (d, k)	2602
$HO_2CCHMeCH_2GeI_3$	IIA	55	m. 63-64° (f)	1496
$MeO_2CCHMeCH_2GeI_3$	IIA	73	$b_{0.01}$ 107-108°, n_D^{20} 1.6848, d_{20} 2.629 (f, h, i)	1496
$ClOCCHMeCH_2GeCl_3$	IIA	91	$b_{3.5}$ 80-81°, n_D^{20} 1.5050, d_{20} 1.6607	2602
	VA (k)	86	n_D^{20} 1.5052	2602
	IIA (l)	--	(d)	2602
$HO_2CCH_2CHMeGeCl_3$	IIA	50	m. 78-79°, soly. (d, k)	2602
$HO_2CCH_2CHMeGeI_3$	IIA	74	m. 78°, soly. (f)	1496
$ClOCCH_2CHMeGeCl_3$	IIA	87	b_5 92-93°, n_D^{20} 1.5080, d_{20} 1.6586	2602
	VA (k)	--	(d)	2602
	IIA (l)	--	--	2602
	IIA (m)	--	--	2602
	VB	--	b_3 86°, n_D^{20} 1.5067	2602
$HO_2CCH_2CH(CO_2H)GeCl_3$	IIA (n)	70	m. 151.5-53°, IR spectrum	2602

HO$_2$CH$_2$CH(CH:CHMe)GeI$_3$	IC (o)	--	Dec. 0.01 190° (h, p)	1496
HO$_2$C(CH$_2$)$_{10}$GeCl$_3$	IIA	16	b$_3$ 181-83°, n$_D^{20}$ 1.4909, d$_{20}$ 1.3280 (d)	2602
Cl$_2$OPC$_2$H$_4$GeCl$_3$	VC	19	b$_{1.5}$ 96°, n$_D^{20}$ 1.5305, d$_{20}$ 1.8067 (q)	995
Cl$_2$OPC$_3$H$_6$GeCl$_3$	VC	88	b$_1$ 123-25°, n$_D^{20}$ 1.5290, d$_{20}$ 1.7700	995
Cl$_2$OPC$_6$H$_{10}$GeCl$_3$	VC	19	b$_5$ 196-98°, n$_D^{20}$ 1.5488, d$_{20}$ 1.6891	995
Cl$_2$OPC$_9$H$_{18}$GeCl$_3$	VC	60	b$_1$ 170°, n$_D^{20}$ 1.5133, d$_{20}$ 1.4536	995

* The characters used here correspond to the synthetic scheme in the introduction to Chapter 4.1.3.6.3. (a) IR (692, 1440) and NMR (1689) spectra. (b) Rxn. with Et$_3$SnCH$_2$CO$_2$Me → (MeO$_2$CCH$_2$)$_2$GeCl$_2$ (692, 1440). (c) Rxn. with Bu$_3$SnOMe → MeO$_2$CCH$_2$Ge(OMe)$_3$ (692). (d) IR and NMR spectra (2602). (e) Heating above 85°, + SOCl$_2$ → ClOCCH$_2$CH$_2$GeCl$_3$ (1527). (f) NMR spectrum (1496). (g) Rxn. with LiAlH$_4$ → HO(CH$_2$)$_3$GeH$_3$ (1496). (h) IR spectrum (1496). (i) Rxn. with EtMgBr → Et$_3$GeCH$_2$CHRCEt$_2$OH; R = H, Me (1496). (j) Rxn. with Me$_2$Cd → Me$_3$GeCH$_2$CH$_2$Ac (2602). (k) Heating HO$_2$CCHRCHRGeCl$_3$ to 180°, + SOCl$_2$ → ClOCCHRCHRGeCl$_3$; R = H, Me (2602). (l) From RCH:CHCO$_2$SiMe$_3$ with HGeCl$_3$, + SOCl$_2$; R = H, Me (2602). (m) From CH$_2$:CHCH$_2$CO$_2$H + HGeCl$_3$, + SOCl$_2$ (2602). (n) From fumaric acid (2602). (o) From sorbic acid (1496). (p) Rxn. with LiAlH$_4$ → HOCH$_2$CH$_2$CH(CH:CHMe)GeH$_3$ (1496). (q) IR spectrum, weak indication of CH$_3$ (995).

4.1.3.6.7 POLYMETALLIC ORGANOGERMANIUM TRIHALIDES

Organogermanium trihalides with two or more germanium atoms, silicon and boron are reported in this chapter. The compounds are prepared by methods from the following scheme.

I Rochow synthesis:
$$Ge-Cu + CH_2Cl_2 \rightarrow CH_2(GeCl_3)_2 .$$

II Reactions with organo metallic intermediates:
(IIA) $GeCl_4 + YLi_2 \rightarrow (Cl_3Ge)_2Y$,
(IIB) $GeBr_4 + YI_2 + Cu \rightarrow (Br_3Ge)_2Y$.

III Hydrogermylation reactions:
(IIIA) $HGeX_3 + R'CH:CH_2 \rightarrow X_3GeCH_2CHR'GeX_3$,
(IIIB) $HGeX_3 + R'C\vdots CH \rightarrow X_3GeCH:CR'GeX_3$,
(IIIC) $HGeCl_3 + C_{10}H_8 \rightarrow (Cl_3Ge)_nC_{10}H_{8+n}$.

IV Condensation with germachloroform:
(IVA) $HGeCl_3 + YCl_2 \rightarrow (Cl_3Ge)_2Y$,
(IVB) $HGeCl_3 + MeCHO \rightarrow (Cl_3GeCHMe)_2O$.

V Addition of diboron tetrachloride:
$$CH_2:CHGeCl_3 + B_2Cl_4 \rightarrow Cl_2BCH_2CH(BCl_2)GeCl_3 .$$

R = organic group, R' = organic group, may be metal-substituted, Y = hydrocarbon bridge, may be metal-substituted, X = halogen.

4.1.3.6.7.1 BRIDGED POLYORGANOGERMANIUM TRIHALIDES

The compounds are listed in Table 39. Additional derivatives are listed in Subchapters 2.7.1.1.3, 2.7.1.2 and 3.8.1.

4.1.3.6.7.2 ORGANOGERMANIUM TRIHALIDES CONTAINING SILICON AND BORON

The compounds are listed at the bottom of Table 39.

Table 39. Polymetallic Organogermanium Trihalides

Formula*	Method**	Yield	Properties	Ref.
$CH_2(GeCl_3)_2$	I	23	b_{12} 98-99°, n_D^{20} 1.5300, d_{20} 2.0015 (a, b)	363
$(CH_2GeCl_3)_2$ (92)	IIIA (c)	(d)	Mixture, concn. Ge-H derivatives	1588
$(CH_2GeBr_2)_2$	IIIA (c)	10	m. 122-24°, NMR spectrum	161, 365
$(:CHGeCl_3)_2$ (92)	IIIB (e)	24	m. 71-72°, dipole moment (f, g)	1529
	IIIB (e)	(d)	Mixture contg. Ge-H derivatives	1588
$(:CHGeBr_3)_2$	IIIB (e)	--	m. 121.5-23°, NMR spectrum (h, i)	161, 365
$Cl_3GeYGeCl_3$				
Y = CH_2CHCl	IIIA (j)	57	b_{10} 126°, n_D^{20} 1.5461, d_{20} 1.9951 (a, k)	1533
CH_2OCH_2	IVA	31	$b_{3.5}$ 110.5-11°, n_D^{20} 1.5302, d_{20} 1.9486 (a)	1526
CH_2CHMe	IIIA (l)	31	b_5 110°, n_D^{20} 1.5400, d_{20} 1.8927 (a)	2403
$CH:CMe$	IIIB	45	$b_{4.5}$ 94-95°, n_D^{20} 1.5455, d_{20} 1.8953 (a, m)	2403
$CH_2CH_2CHGeCl_3$	IIIB (n)	29	b_{10} 166.5-67°, n_D^{20} 1.5660, d_{20} 2.0691 (o, p)	364
$CH_2CH(COCl)$	IIIB (q)	40	m. 36.5-38°, b_2 116-18°	2602
$CH_2CH:CHCH_2$	IIIA (r)	44	m. 56-58°, $b_{0.5}$ 146-47° (m, o, s)	308

* Numbers in parenthesis refer to pages in main volume.
** The characters used here correspond to the synthetic scheme in the introduction to Chapter 4.1.3.6.7.

(a) IR and NMR spectra. (b) Rxn. with MeMgCl → $(Me_3Ge)_2CH_2$ (363). (c) From C_2H_2 (363). (d) By-prod. (e) From C_2H_2. (f) IR (1529, 1533), Raman (1529, 1533), UV (1693) and NMR (1529) spectra (g) Rxn. with MeMgBr → $(Me_3GeCH:)_2$ (1529). (h) UV spectrum (1693). (i) Rxn. with MeMgCl → $(Me_3GeCH:)_2$ + $(Me_3GeCH_2)_2$ (161, 365). (j) From $CH_2:CClGeCl_3$. (k) Rxn. with $AlCl_3$ or quinoline → $GeCl_4$ + $CH_2:CHGeCl_3$ (1533). (l) From $CH_2:C:CH_2$ at -70°. (m) Raman spectrum. (n) From $HC:CCH_2Cl$. (o) IR spectrum. (p) Rxn. with MeMgBr → $Me_3GeCH_2CH_2CH(GeMe_3)_2$ + $Me_3GeCH_2CH:CHGeMe_3$ (364). (q) From $HC:CCO_2H$ at -10°. (r) From $(ClCH_2CH:)_2$. (s) Rxn. with MeMgBr → $:Me_3GeCH_2CH:)_2$ (308).

Table 39 Continued

Formula*	Synth. Method**	Yield	Properties	Ref.
$Cl_3GeYGeCl_3$ (Cont'd.)				
Y = $C(:CH_2)C(:CH_2)$	IIIB (t)	25	m. 34-35°, b_2 98-98.5°, n_D^{20} 1.5518 (a, u)	2403
$C(CH_2Cl):C(CH_2Cl)$	IIIB (t)	15	m. 88-90°, $b_{2.5}$ 150-52°, mol. wt. (a, u)	2403
$CH:C(CHMe_2)$	IIIB (v)	30	b_2 104-106°, n_D^{20} 1.5420, d_{20} 1.7641 (a)	2403
$CH_2C(:CMe_2)$	IIIB (w)	16	$b_{2.5}$ 125-27°, n_D^{20} 1.5560, d_{20} 1.8206 (a)	2403
CHMeOCHMe	IVB	50	b_1 102.5°, n_D^{20} 1.5248, d_{20} 1.7814 (a, x)	1526
$C_{10}H_{10}$	IIIC	--	Dec. o.3-0.5 torr, NMR spectrum (y)	278,1336
$C_{10}H_{15}GeCl_3$ (?)	IIIC (z)	16		278,1336
$C_{10}H_{14}(GeCl_3)_2$ (?)	IIIC	75	(aa)	1336
$C_{10}H_{13}(GeCl_3)_3$ (?)	IIIC	20		278
p-$C_6H_4(GeBr_3)_2$	IIB	23	Colorless cryst. m. 187-88° (subl.)	1528
$Cl_3GeCH_2SiCl_3$	--	--	^{35}Cl NQR spectrum	737
$Cl_3GeCH:CHSiMe_3$	IIIB (bb)	50	b_{11} 84-85°, n_D^{20} 1.4933, d_{20} 1.3490 (o, cc)	1533
$BCl_2CH_2CH(BCl_2)GeCl_3$	V	--	(Far) IR and NMR spectra, mol. wt. (dd)	2329
m-$B_{10}H_{10}(CGeCl_3)_2$	IIA	34	$b_{0.2}$ 124-26°	1835

(t) From $(ClCH_2C:)_2$. (u) Rxn. with RMgX → $(CH_2:CGeR_3)_2$; R = Me, Et (2403). (v) From $Me_2CHC:CH$. (w) From $Me_2CClC:CH$. (x) Rxn. with MeMgCl → $(Me_3GeCHMe)_2O$ (1526). (y) Rxn. with MeMgBr → $(Me_3Ge)_2C_{10}H_{10}$ (278, 1336). (z) From tetrahydronaphthalene. (aa) Rxn. with MeMgBr → $(Me_3Ge)_4C_{10}H_{14}$ (1336). (bb) From $Me_3SiC:CH$. (cc) Rxn. with MeMgBr → $Me_3GeCH:CHSiMe_3$ (1533). (dd) Thermal decomposition: at 150° → $C_2H_3GeCl_3$, at 230° → BCl_3 + $GeCl_4$ + HCl (2329).

4.1.4 ORGANOGERMANIUM HALIDES CONTAINING GERMANIUM-PSEUDOHALOGEN BONDS

Spectroscopic evidence for existence of dimethylcyanogermanium halides is reported in the literature. Data for the two compounds are listed at the top of Table 40.

4.1.5 ORGANOGERMANIUM HALIDES CONTAINING GERMANIUM-OXYGEN BONDS

A number of organogermanium alkoxide halides are reported in Table 40. Evidence for the existence of several derivatives is only by means of spectroscopy. The following scheme shows methods used in preparation of organogermanium halides containing germanium-heteroatom bonds.

I Reaction with organogermanium hydrides:
$$R_nHGeCl_{3-n} + R_2CO \rightarrow R_nGe(OCHR_2)Cl_{3-n} .$$

II Reactions with organogermanium halides:
   (IIA) $R_nGeCl_{4-n} + \overline{CH_2CH_2O} \rightarrow R_nGe(OCH_2CH_2Cl)Cl_{3-n}$,
   (IIB) $R_nGeCl_{4-n} + R'OH + base \rightarrow R_nGe(OR')Cl_{3-n}$,
   (IIC) $Ph_2GeCl_2 + ROH \rightarrow Ph_2Ge(OR)Cl$,
   (IID) $Me_2GeCl_2 + R_2NH \rightarrow Me_2Ge(NR_2)Cl$.

III Reaction with organogermanium alkoxides:
$$Bu_2Ge(OR)_2 + AcBr \rightarrow Bu_2Ge(OR)Br .$$

IV Interconversion reactions:
$$PhGe(OR)Cl_2 + R'OH \rightarrow PhGe(OR')Cl_2 .$$

V Redistribution reactions:
   (VA) $Me_2GeX_2 + Me_2GeX'_2 \rightarrow Me_2GeXX'$,
   (VB) $Me_2GeX_2 + Me_2SiX'_2 \rightarrow Me_2GeXX'$,
   (VC) $MeGeX_3 + MeGeX'_3 \rightarrow MeGeX_nX'_{3-n}$,
   (VD) $MeGeX_3 + MeSiX'_3 \rightarrow MeGeX_nX'_{3-n}$

R = organic group, R' = organic group, X = halogen, X' = halogen, pseudohalogen, alkoxy, thioalkoxy, and alkylamino group, resp.

Table 40. Organogermanium Halides Containing Germanium-Oxygen and Germanium-Pseudohalogen Bonds

$R_nGeX_mX'_o$
$n + m + o = 4$

Formula	Synth. Method*	Yield	Properties	Ref.
$Me_2Ge(CN)Cl$	VB (a)	--	NMR spectrum (b)	1544
$Me_2Ge(CN)Br$	VB (a)	--	NMR spectrum (b)	1544
$Me_2Ge(OMe)Cl$	VA (c)	--	NMR spectrum (b)	1542
$Me_2Ge(OMe)Br$	VA (c)	--	NMR spectrum (b)	1542
$Me_2Ge(OMe)I$	VA (c)	--	NMR spectrum (b)	1542
$Me_2Ge(OCH_2CH_2Cl)Cl$	IIA (d)	--	b_4 103-105°, n_D^{20} 1.4670, d_{20} 1.4276 (e)	2196
$MeGe(OMe)Cl_2$	VC (a)	--	NMR spectrum (b)	1545
	VD	--	NMR spectrum (b)	1546
$MeGe(OMe)_2Cl$	VC (a)	--	NMR spectrum (b)	1545
	VD	--	NMR spectrum (b)	1546
$MeGe(OCH_2CH_2Cl)Cl$	IIA (d)	--	b_{22} 81°, n_D^{20} 1.4756, d_{20} 1.5897 (e)	2196
$MeGe(OCMe_3)_2Cl$	IIB (f)	41	b_{15} 80-82°, n_D^{20} 1.4320, d_{20} 1.1277	2192
$MeGe(OMe)Br_2$	VC (a)	--	NMR spectrum (b)	1545
	VD	--	NMR spectrum (b)	1546
$MeGe(OMe)_2Br$	VC (a)	--	NMR spectrum (b)	1545
	VD	--	NMR spectrum (b)	1546
$MeGe(OMe)I_2$	VC (a)	--	NMR spectrum (b)	1545
	VD	--	NMR spectrum (b)	1546
$MeGe(OMe)_2I$	VC (a)	--	NMR spectrum (b)	1545
	VD	--	NMR spectrum (b)	1546
$ClCH_2Ge(OEt)_2Cl$	--	--	IR and NMR spectra	2244
$Bu_2Ge(OEt)Br$	III	--	b_4 104-10°	337
$Bu_2Ge(OCHMe_2)Br$	III	--	b_5 104-20°	337
$Bu_2Ge(OCMe_3)Br$	III	--	b_4 112-16°	337
$Ph_2Ge(OCH_2CHMe_2)Cl$	I (g)	76	$b_{0.04}$ 106°, n_D^{20} 1.5560, d_{20} 1.2432	1762
$Ph_2Ge(OC_6H_{11})Cl$	I (g)	80	$b_{0.003}$ 136°, n_D^{20} 1.5720, GLC	1762
	I (h)	35	IR and NMR spectra	1762
	IIB (i)	80	Stable at 100°	1762
$Ph_2Ge(OC_9H_6N-8)Cl$	IIC	28	Yellow cryst. dec. > 34°, crude	1215
$PhGe(OMe)Cl_2$	IIB (j)	90	b_{28} 122°, n_D^{20} 1.5310, d_{20} 1.4764 (k, l)	1762
$PhGe(OCH_2CHMe_2)Cl_2$	I (g)	25	$b_{0.2}$ 68°, n_D^{20} 1.5100, d_{20} 1.3179	1762
	IV (k)	85	IR and NMR spectra (l)	1762
$PhGe(OC_6H_{11})Cl_2$	I (g)	--	$b_{0.1}$ 108°, n_D^{20} 1.5350, d_{20} 1.3503	1762
	IIB (i)	43	IR and NMR spectra (l)	1762

* The characters used here correspond to the synthetic scheme in the introduction to Chapter 4.1.5.
(a) Synth. at 120°, (b) Compound not isolated, evidence from spectra.
(c) Synth. at r.t. to 35°. (d) Synth. cat. by H_2PtCl_6 or $AlCl_3$. (e) IR spectrum, sensitive to moisture (2196). (f) From $Me_3COH + NH_3$. (g) Synth. under UV irradiation at 30°. (h) Synth. at 170° in N atm., with AIBN at 80° and in presence of $ZnCl_2$, in 30, 25, and 35° yield, resp. (i) From c.$C_6H_{11}ONa$ in THF. (j) From MeONa (k) Rxn. with i-BuOH → $PhGe(OCH_2CHMe_2)Cl_2$ (1762).
(l) Rxn. with $PhHGeCl_2$ → $(PhGeCl_2)_2 + PhGeCl_3 + ROH$ (1762).

4.1.6 ORGANOGERMANIUM HALIDES CONTAINING GERMANIUM-SULFUR BONDS

Organogermanium halides containing Ge-SR and Ge-S-Ge bonds are compiled in Table 41.

4.1.7 ORGANOGERMANIUM HALIDES CONTAINING GERMANIUM-NITROGEN BONDS

Amidoorganogermanium halides are listed at the bottom of Table 41.

Table 41. Organogermanium Halides Containing Ge-S and Ge-N Bonds $\quad R_nGeX_mX'_o \quad n + m + o = 4$

Formula	Method*	Yield	Properties	Ref.
$Me_2GeCl_2 \cdot Me_2GeS$	VA (a)	73	$b_{0.1}$ 50°, sensitive to moisture (b)	2607
$Me_2GeCl_2 \cdot 2Me_2GeS$	VA (a)	27	$b_{0.01}$ 70°, sensitive to moisture (b)	2607
$Me_2Ge(SMe)Cl$	VA (c)	--	NMR spectrum (d)	368
$Me_2Ge(SMe)Br$	VA (c)	--	NMR spectrum (d)	368
$Me_2Ge(SMe)I$	VA (c)	--	NMR spectrum (d)	368
$MeGe(SMe)Cl_2$	VC (a)	--	NMR spectrum (d)	1545
	VD	--	NMR spectrum (d)	1546
$MeGe(SMe)_2Cl$	VC (a)	--	NMR spectrum (d)	1545
	VD	--	NMR spectrum (d)	1546
$MeGe(SMe)Br_2$	VC (a)	--	NMR spectrum (d)	1545
	VD	--	NMR spectrum (d)	1546

* The characters used here correspond to the synthetic scheme in the introduction to Chapter 4.1.5.
(a) Synth. in N atm. at 120°. (b) Use as insecticides, functional fluids and intermediates (2607). (c) Synth. at r.t. to 35° (368). (d) Compound not isolated.

Table 41 Continued $\qquad$ $R_nGeX_mX'_o$
$n + M + O = 4$

Formula	Synth. Method*	Yield	Properties	Ref.
MeGe(SMe)$_2$Br	VC (a)	--	NMR spectrum (d)	1545
	VD	--	NMR spectrum (d)	1546
MeGe(SMe)I$_2$	VC (a)	--	NMR spectrum (d)	1545
	VD	--	NMR spectrum (d)	1546
MeGe(SMe)$_2$I	VC (a)	--	NMR spectrum (d)	1545
	VC	--	NMR spectrum (d)	1546
Ph$_2$GeCl$_2$·nPh$_2$GeS	--	--	Use as insecticide, functional fluid	2607
Me$_2$Ge(NMe$_2$)Cl	IID	--	b$_{745}$ 144°, GLC (e, f)	2163
Me$_2$Ge(NEt$_2$)Cl	IID	65	b$_{743}$ 177°, mol. wt. (f, g, h)	2163
(Me$_2$ClGeNMeCH$_2$)$_2$	IID	--	b$_9$ 90°, mol. wt.	2163
	(h)	--	IR and NMR spectra	2163
Me$_2$Ge(NCMe$_3$SiMe$_3$)Cl	IID (i)	21	b$_{0.5}$ 72°, mol. wt. (j, k)	2749
MeGe(NMe$_2$)Cl$_2$	VD	--	NMR spectrum (d)	1546
MeGe(NMe$_2$)$_2$Cl	VD	--	NMR spectrum (d)	1546
MeGe(NMe$_2$)Br$_2$	VD	--	NMR spectrum (d)	1546
MeGe(NMe$_2$)$_2$Br	VD	--	NMR spectrum (d)	1546
MeGe(NMe$_2$)I$_2$	VD	--	NMR spectrum (d)	1546
MeGe(NMe$_2$)$_2$I	VD	--	NMR spectrum (d)	1546

(e) Mixture with Me$_2$Ge(NMe$_2$)$_2$, not sepd. (2163). (f) NMR spectrum (2163). (g) Dec. slowly on standing at r.t. (2163). (h) Rxn. with (CH$_2$NHMe)$_2$ → (Me$_2$ClGeNMeCH$_2$)$_2$ (2163). (i) Synth. from Me$_3$SiN(CMe$_3$)Li and Me$_2$GeCl$_2$ in Et$_2$O in N atm. at 0° (2749). (j) NMR and (far) IR spectra and assignment (2749). (k) Rxn. with MeLi → Me$_3$GeN(CMe$_3$)SiMe$_3$ (2749).

4.1.8 ORGANOGERMANIUM HALIDES CONTAINING GERMANIUM-METAL BONDS

Intermetallic organogermanium halides are listed elsewhere. The compounds may be found in Chapters 5.4, 6.1.1, 6.5.1, 6.5.5, 6.5.6, 6.5.7 and 6.5.8.

4.2 ORGANOGERMANIUM SALTS

4.2.1 ORGANOGERMANIUM PSEUDOHALIDES

New organogermanium pseudohalides include dicyanimides and fulminates. Structure of some of the compounds has not been elucidated. From far infrared spectra an "iso-"structure is proposed for cyanates, thiocyanates and selenocyanates. Trimethylgermanium cyanide was shown by x-ray structural analysis to have the cyanide structure. For other cyanides only spectroscopic evidence is cited. The following scheme gives synthetic methods for the compounds in Table 42.

I Metathetic Substitution:
   (IA) $R_3GeX + AgX' \rightarrow R_3GeX'$,
   (IB) $R_nGeX_{4-n} + KX' \rightarrow R_nGeX'_{4-n}$,
   (IC) $Ph_3GeBr + PbX'_2 \rightarrow Ph_3GeX'$.

II Exchange reaction:
   $R_3GeX + aq.HN_3 \rightarrow R_3GeN_3$.

III Neutralization reaction:
   $(R_3Ge)_2O + HSCN \rightarrow R_3GeNCS$.

IV Interconversion reactions:
   (IVA) $R_3GeCN + Se \rightarrow R_3GeNCSe$,
   (IVB) $Me_3Ge(N_3)_2 + PR_3 \rightarrow Me_2Ge(N:PR_3)N_3$.

R = organic group, X = bromide or iodide, X' = pseudohalogen, i.e., cyanide, isocyanate, fulminate, isothiocyanate, isoselenocyanate, dicyanamide and azide. Other organogermanes containing germanium-pseudohalide bonds are listed in Chapters 3.8.2 and 4.1.4.

Table 42. Organogermanium Pseudohalides $\quad\quad\quad\quad\quad\quad\quad\quad R_nGeX'_{4-n}$

Formula*	Synth. Method**	Yield	Properties	Ref.
Me_3GeCN (94)	--	--	(Far) IR spectra, assignment (a, b)	2825
Me_3GeNCO	IA	27	Colorless liq., b.120-22° (c)	2027
Me_3GeNCS (94)	IB	81	b_8 63°, n_D^{20} 1.5145, d_{20} 1.2676 (c, d)	367
$Me_3GeNCSe$	IB	--	m. 13-14°, b_{232} 118-19°	2825
	IVA (a)	--	(c)	2825
Me_3GeN_3 (94)	II	48	b. 135-36°, IR and UV spectra (c, e, f, g)	556, 557

* Numbers in parenthesis refer to pages in main volume.
** The characters used here refer to the synthetic scheme in the introduction to Chapter 4.2.1.
(a) Rxn. with Se → $R_3GeNCSe$ (2825). (b) X-ray structural analysis (1817). (c) (Far) IR spectrum and assignment (2027, 2825). (d) IR and Raman spectra (367). (e) Molecular orbital calculation of electronic spectra (601). (f) Rxn. with PR_3 → $Me_3GeN:PR_3$ + N; R = Me, Et (2743). (g) m. = -65° (2027).

Table 42 Continued $R_nGeX'_{4-n}$

Formula*	Synth. Method**	Yield	Properties	Ref.
m-(Me$_2$GeN$_3$)$_2$C$_6$H$_4$	IB (h)	--	(i)	2115
Me$_2$Ge(CN)$_2$	--	--	NMR spectrum (j)	1544
Me$_2$Ge(N$_3$)$_2$ (94)	--	--	(k)	--
Me$_2$Ge(N:PMe$_3$)N$_3$	IVB (k)	83	m. 87-89°, b$_{0.1}$ 80-85°, mol. wt. (l, m)	2743
Et$_3$GeNCS (94)	IB	84	b$_{13}$ 117.5°, n_m^{20} 1.5152, d$_{20}$ 1.1792	367
Et$_3$GeNCSe	IB	--	b$_{240}$ 175-76°, (far) IR spectrum, assignment	2825
Bu$_3$GeCN	IA	--	b. 292-94°, (far) IR spectrum, assignment (a)	2825
Bu$_3$GeNCSe	IVA (a)	--	b$_{100}$ 221-22°, (far) IR spectrum, assignment	2825
	IB	--		2825
Ph$_3$GeCN	IA	50	m. 140-41°, mol. wt. IR spectrum, assignment	2795
Ph$_3$GeNCO	IC	80	m. 117-18.5°, (far) IR spectrum, assignment	1443
	IA	55	m. 117-18°, (far) IR spectrum, assignment	2795
Ph$_3$GeCNO	IB (n)	--	m. 122-24°, soly., mol. wt. (o)	
Ph$_3$GeNCS (96)	IA	61	m. 105°, (far) IR spectrum,	1443
	IA	60		2795
	III	30	m. 106-107°, (far) IR spectrum, assignment	1443
Ph$_3$GeN(CN)$_2$	IA	--	m. 145°, soly., mol. wt., struct. (p)	2501
Ph$_3$GeN$_3$·BBr$_3$	(q)		Pale yellow solid, IR spectrum (r)	556, 558
Ph$_3$GeN$_3$·SnCl$_4$	(q)	--	White solid, IR spectrum (r)	556, 558
PhGe(N$_3$)$_3$	IB	--	(i)	2115
(p-MeC$_6$H$_4$)$_3$GeCN	IA	40	m. 151-53°, mol. wt. (s)	2795
(p-MeC$_6$H$_4$)$_3$GeNCO	IA	50	m. 222-23°, mol. wt. (t)	2795
(p-MeC$_6$H$_4$)$_3$GeNCS	IA	55	m. 213-14°, mol. wt. (t)	2795
R$_3$GeNCS	--	--	Ge bond refraction	81

(h) From NaN$_3$. (i) Use as fungicide, insecticide, herbicide, oil additive and blowing agent (2115). (j) Rxn. with Me$_2$SiX$_2$ → equilibrium const. for redistribution rxn., X = Cl, Br, measured by NMR spectroscopy (1544). (k) Rxn. with PMe$_3$ → Me$_2$Ge(N:PMe$_3$)N$_3$ (2743). (l) Colorless cryst., subl. in vacuo, IR spectrum (2743). (m) Rxn. with PMe₃ → Me₂Ge(N:PMe₃)₂ (2743), with Me₃SiOLi → Me₂Ge(OSiMe₃)N:PMe₃ + Me₂Ge(OSiMe₃)₂ + LiN₃ (2743). (n) Synth. under exclusion of light and air (28). (o) Colorless prisms, IR spectrum, sensitive to moisture, no conductance in Me₂CO (28). (p) Colorless cryst., hydrolyzes easily, (far) IR spectrum, assignment (2501). (q) From components in CCl₄. (r) Hydrolyzes easily (556, 558). (s) IR spectrum and assignment (2795). (t) (Far) IR spectrum and assignment (2795).

4.2.4 ORGANOGERMANIUM SALTS OR ESTERS OF OXYGEN ACIDS

Organogermanium derivatives of inorganic acids are listed in Table 43. Methods of synthesis are depicted in the following scheme. Cleavage of tetraorganogermanes with sulfur dioxide and sulfur trioxide represents a new preparative way to sulfinates and sulfonates.

I   Cleavage reaction:
   (IA) $Me_4Ge + SO_3 \rightarrow Me_3GeO_3SMe$,
   (IB) $R_2\overline{GeCH_2CHMeCH_2} + SO_2 \rightarrow R_2\overline{GeCH_2CHMeCH_2S(O)O}$.

II Metathetic substitution:
   (IIA) $Me_3GeCl + AgNO_3 \rightarrow Me_3GeNO_3$,
   (IIB) $Bu_3GeCl + Ph_2POK \rightarrow Bu_3GeOPPh_2$,
   (IIC) $Ph_3GeCl + Ph_2AsO_2H + Et_3N \rightarrow Ph_3GeO_2AsPh_2$.

III Neutralization - Esterification:
   (IIIA) $(Et_3Ge)_2O + Ph_2PO_2H \rightarrow Et_3GeO_2PPh_2$,
   (IIIB) $(Me_3Ge)_2O + P_2O_5 \rightarrow (Me_3Ge)_3PO_4$,
   (IIIC) $(R_3Ge)_2O + MePOF_2 \rightarrow R_3GeO_2(F)PMe$.

IV Exchange reaction:
   $Me_3GeSMe + Me_2SO_4 \rightarrow Me_3Ge(Me)SO_4$.

V   Oxidation of phosphines:
   $R_3GeYPh_2 + dry\ O \rightarrow R_3GeO_2YPh_2$.

R = organic group, Y = P or As.

Table 43. Organogermanium Salts or Esters of Oxygen Acids　　　　　　　　　　　　　　　　　$R_nGeX'_{4-n}$

Formula*	Synth. Method**	Yield	Properties	Ref.
Me_3GeNO_3 (97)	IIA	38	m. 5°, b_{11} 62°, n_D^{18} 1.4436 (a)	476
$(Me_3Ge)_3PO_4$ (97)	IIIB	62	(a)	478
$Me_3GeO_2(F)PMe$	IIIC (b)	57	b_5 49-49.5°, n_D^{20} 1.4123, d_{20} 1.3656	522,1103
$Me_3Ge(Me)SO_4$	IV	70	$b_{0.2}$ 55-57°, IR and NMR spectra	227
Me_3GeO_3SMe	I	92	$b_{0.1}$ 56-57°, mol. wt.	2742
	IIIB (c)	70	NMR spectrum	2742
$Et_3GeO_2(F)PMe$	IIIC (b)	64	b_9 111-12°, n_D^{21} 1.4378, d_{20} 1.2610	522,1103
$Et_3GeO_2PPh_2$ (97)	V	--	Stable liq., $b_{0.001}$ 160-62°	50
	IIIA	--	IR spectrum (d, e)	50
$Et_2GeCH_2CH_2CH_2SO_2$	I (f)	--	$b_{0.5}$ 103°, n_D^{20} 1.5111, d_{20} 1.3505 (g)	2362
$Et_2GeCH_2CHMeCH_2SO_2$	I (f)	--	$b_{0.2}$ 104°, n_D^{20} 1.5026, d_{20} 1.2799 (g)	2362
$EtBuGeCH_2CH_2CH_2SO_2$	I (f)	--	$b_{0.1}$ 115°, n_D^{20} 1.5050, d_{20} 1.2664 (g)	2362
$Bu_3GeOPPh_2$	IIIB	72	b_1 136-38°, NMR spectrum	1232
$Bu_2GeCH_2CH_2CH_2SO_2$	I (f)	--	$b_{0.03}$ 111°, n_D^{20} 1.4980 (g)	2632
$Bu(n-C_8H_{17})Ge(CH_2)_3SO_2$	I (f)	--	$b_{0.1}$ 165°, n_D^{28} 1.4902, d_{20} 1.1184 (g)	2362
$Ph_3GeO_2PPh_2$	V	--	Pale viscous oil, IR spectrum	1847
$Ph_3GeO_2AsPh_2$	V (h)	100	m. 178°, soly.	1847
	IIIC	97		1847
$(Ph_3Ge)_2SO_4$	IIA	40	m. 225-26°, soly., IR spectrum	1970
$(Ph_3Ge)_2SeO_3$	IIA	50	m. 159°, soly., IR spectrum	1970
$(Ph_3Ge)_2SeO_4$	IIA	40	m. 172°, soly., IR spectrum	1970
$[(p-MeC_6H_4)_3Ge]_2SO_4$	IIA	--	m. 147-49°, soly., IR spectrum	1970

* Numbers in parenthesis refer to pages in main volume.
** The characters used here correspond to the synthetic scheme in the introduction to Chapter 4.2.4.
(a) Use as additive for engine or rocket fuel. (b) From $MePOF_2$. (c) From $Me_2S_2O_5$. (d) Far IR spectrum and assignment (105). (e) Rxn. with $H_2O \rightarrow$ stable volatile hydrate (1:1); excess $\rightarrow (Et_3Ge)_2O + Ph_2PO_2H$ (50). (f)

4.3 ORGANOGERMANES WITH GERMANIUM-OXYGEN BONDS

4.3.1 ORGANOGERMANIUM OXIDES AND HYDROXIDES

4.3.1.1 TRIORGANOGERMANIUM OXIDES AND HYDROXIDES

The following scheme summarizes synthetic methods for the triorganogermanium oxides and hydroxides listed in Table 44.

I Cleavage reactions of cyclic tetraorganogermanes:
    (IA) with sulfuric acid, followed by aqueous base,
    (IB) with phosphorous oxychloride and pyridine,
    (IC) with alcoholic potassium hydroxide.

II By-product in Grignard reactions with germanium tetrahalide.

III Hydrolysis of organogermanium chlorides:
    (IIIA) with water,
    (IIIB) with basic alkali metal salts,
    (IIIC) with ammonia in ether,
    (IIID) with sodium potassium alloy in toluene.

IV Hydrolysis:
    (IVA) of trimethylgermylsilane,
    (IVB) oxidation of organogermanes.

V Interconversion:
    (VA) triorganogermanium hydroxide to the oxide at 130°,
    (VB) bis(chloromethyldimethylgermanium)oxide with sodium sulfide to a cyclic oxide.

4.3.1.1.1 UNSUBSTITUTED TRIORGANOGERMANIUM OXIDES AND HYDROXIDES

Bistrimethylgermanium Oxide (M-99) $GeC_6H_{18}O$ $(Me_3Ge)_2O$
Synth.: By rxn. of Me_3GeBr with aq. KOH, in 57% yield (643), in 60% yield (1103).
By rxn. of Me_3GeOLi with $GaCl_3$ (685).
By hydrolysis of $(Me_3Ge)_3N$ in THF, in quantitative yield (1976), of $(Me_3Ge)_2NH$, in 84% yield (1535).
By air-oxidation of $(Me_3Ge)_2Hg$ (996), of $(Me_3Ge)_3Bi$ (1854), of $Me_3GeAuPPh_3$ in C_6H_6 (2420).
By oxidation of $Me_3Ge(Et_3P)_2PtCl$ with aq. $NaNO_2$ in Me_2CO (406).
By hydrolysis of $Me_3Ge[(Et_3P)Pt(CH_2PPh_2)_2]Cl \cdot C_6H_6$ in MeOH (1107), of $Me_3Ge(Et_3P)_2PtCl$ in diglyme (1106).
Prop.: b. 138-39° (1535), b_{732} 132-34° (1103), n_D^{20} 1.4396 (1535), 1.4326 (1103), n_D^{21} 1.4290 (693), d_{20} 1.2314 (1103), (far) IR, Raman and assignment (2534), UV (927), NMR (645), and mass (1109) spectra, dipole moment (928), relative

From $R_2GeCH_2CHR'CH_2$; $R' = H$, Me. (g) Stable to 200°, IR and mass spectra (2362), rxn. concn HCl → $R_2(HO_2SCH_2\text{-}CHRCH_2)GeCl$ (2363). (h) Rxn. with dry oxygen or aq. alc. H_2O_2.

basicity (644-5).
Rxn. with: KF = Me$_3$GeF (2365).
Et$_3$SiCl + cat. FeCl$_3$ → Me$_3$GeCl + (Et$_3$Si)$_2$O (1535).
P$_2$O$_5$ → (Me$_3$Ge)$_3$PO$_4$ (478).
(MeSO$_2$)$_2$O → Me$_3$GeO$_3$SMe (2742).
MePOF$_2$ → Me$_3$GeF + MeP(F)O$_2$GeMe$_3$ (522, 1103).
Me$_3$CSH → Me$_3$GeSCMe$_3$ (643).
Biol. Prop.: Toxicity, relative to homologues (45).

Bistriethylgermanium Oxide (M-99)   Ge$_2$C$_{12}$H$_{30}$O   (Et$_3$Ge)$_2$O
Synth.: By rxn. of Et$_3$GeCH$_2$CHOHMe with H$_2$SO$_4$ at 130-60° (466).
By rxn. of Et$_3$GeCPh$_2$CONMe$_2$ with 40% aq. NH$_3$ (2736).
By thermal dec. of Et$_3$GeCO$_2$H (815).
By rxn. of Et$_3$GeBr with aq. KOH at 100°, in 90% yield (1103).
By hydrolysis of Et$_3$GeOCMe$_3$ in air (2095), of Et$_3$GeOCH:CHCRPEt$_2$; R = Me, Ph (2737), of Et$_3$GeO$_2$PPh$_2$ at r.t. (50), of Et$_3$GeOC(:CPh$_2$)NMe$_2$ in 40% aq. NH$_3$ (2736).
By hydrolysis of (Et$_3$Ge)$_2$NH, in 87% yield (1535), of Et$_3$GeNMe$_2$ (466) of Et$_3$GeNPhCOCHPh$_2$ (2736), of Et$_3$GeNPhCYNEt$_2$; Y = O, S (466), of Et$_3$GeNPhAc with 40% aq. NH$_3$ (2736), of Et$_3$GePPh$_2$ in (MeOCH$_2$)$_2$ at r.t., in 45% yield (50).
By rxn. of Et$_3$GePPh$_2$ with PhNCY, followed by H$_2$O; Y = O, S (1797).
By autoxidation of (Et$_3$Ge)$_2$Hg with O in n-octane at -20° to r.t., in 92% yield (2706).
By hydrolysis of (Et$_3$Ge)$_3$Tl with H$_2$O, in 46% yield (1361).
By-prod. in rxn. of Et$_3$GeH with diynes (198).
By-prod. in rxn. of (EtCO$_2$)$_4$Ge with EtMgBr (2609).
Prop.: b. 250-51° (1535), b$_5$ 100-102° (1103), b$_5$ 95-98° (2095), b$_3$ 95-97° (198), b$_{0.02}$ 60° (50), n$_D^{20}$ 1.4624 (1535), 1.4615 (198), 1.4602 (1103), 1.4575 (2095), d$_{20}$ 1.1402 (198), far IR and assignment (105), IR (2095), NMR (1009, 1444) and mass (1109) spectra, dipole moment (928), thermodynamic data (2025), catalytic effect on autoxidation of (Et$_3$Ge)$_2$Hg (2706).
Rxn. with: n-Pr$_3$GeCl → Et$_3$GeCl + (n-Pr$_3$Ge)$_2$O (1535).
MePOF$_2$ → Et$_3$GeF + MeP(F)O$_2$GeEt$_3$ (522, 1103).
C$_3$H$_4$N$_2$ → N-Et$_3$GeC$_3$H$_3$N$_2$; imidazole (1761), pyrazole (2378).
BuSH → Et$_3$GeSBu (465).
HC⋮CCH$_2$OH → Et$_3$GeOCH$_2$C⋮CH (1534).
Biol. Prop.: Toxicity, relative to homologues (45), activity as insecticide against Musca domestica, LD$_{50}$ and ATP inhibition (425).

Bistripropylgermanium Oxide (M-101)   Ge$_2$C$_{18}$H$_{42}$O   (n-Pr$_3$Ge)$_2$O
Synth.: By rxn. of Pr$_3$GeCl with (Et$_3$Ge)$_2$O under reflux, in 50-54% yield (1535).
Prop.: b$_2$ 128-32°, n$_D^{25}$ 1.4638 (1535).
Rxn. with: HCl → Pr$_3$GeCl (2837).
EtOH + cat. p-MeC$_6$H$_4$SO$_3$H → Pr$_3$GeOEt (2837).
Biol. Prop.: Toxicity, comparative to homologues (45).

Bistributylgermanium Oxide (M-101) $GeC_{24}H_{54}O$ $(n-Bu_3Ge)_2O$
Prop.: (Far) IR and Raman spectra, assignments (2534), dipole moment (928).
Rxn. with: ROH + cat. $MeC_6H_4SO_3H$ → Bu_3GeOR; R = Et, i-Pr, t-Bu (1504).
$H_2NCH_2CH_2OH$ → $Bu_3GeOCH_2CH_2NH_2$ (1504).
$Y(OH)_2$ → $(Bu_3GeO)_2Y$; Y = CH_2CH_2, CH_2CHMe, $NH(CH_2CH_2)_2$ (1504).
CH_2Ac_2 → $Bu_3Ge(CHAc_2)$ (1504).
n-BuSH + cat. $MeC_6H_4SO_3H$ → Bu_3GeSBu (1502).
$C_3H_4N_2$ → $N-Bu_3GeC_3H_3N_2$; pyrazole, imidazole (1761).
1,2,4-Triazole → $1-Et_3GeC_2H_2N_3$ (1761).
Biol. Prop.: Toxicity, comparative to homologues (45), activity as insecticide against Musca domestica, LD_{50}, ATP inhibition (425).

Bistriphenylgermanium Oxide (M-100) $Ge_2C_{36}H_{30}O$ $(Ph_3Ge)_2O$
Synth.: By rxn. of $Ph_3GeCR:CH_2$ with O_3 in heptane or AcOEt; R = Me and Ph_3Ge, in 25 and 45% crude yield, resp. (1624, 1626).
By chromic acid oxidation of $Ph_3GeCHOHMe$ in Me_2CO, in 61% yield (2671), of $2-Ph_3GeC_6H_{11}OH$ in N atm. in 87% yield (2670).
By rxn. of Ph_3GeCEt_2OH with PBr_3 in C_6H_6 at reflux and aq. acidic work-up, in 84% yield (1624, 1626).
By rxn. of Ph_3GeH with BzH and $ZnCl_2$, in 70% yield (1762), with c.$C_6H_{10}O$ at 180°, in 17% yield (1762).
By rxn. of Ph_3GeOH with $AlCl_3$ at 225-50°, in 98% yield (958).
By recrystallization of Ph_3GeOH from ligroin-$HCCl_3$ (958).
By rxn. of $9-Ph_3GeOC_{13}H_9$ with BuLi, followed by $PhCH_2Cl$, in 35% yield (2675).
By rxn. of Ph_3GeSMe with anhydr. EtOH in N atm., with NO_2 (227), with H_2O_2 in Me_2CO and with aq. $AgNO_3$, in 70 and 90% yield, resp. (227).
By-prod. in thermal dec. of $(Ph_3Ge)_2CO$ (794).
By-prod. in hydrolysis of $Ph_3GeOCHPh_2$ (2675), of $Ph_3GeAuPPh_3$ with wet diglyme (2420).
By-prod. in rxn. of $Ph_3GeAuPPh_3$ with PPh_3 in air (2420).
By-prod. in rxn. of Ph_3GeLi with BzCl in THF at -23°, in 4% yield (1624-5).
By-prod. in rxn. of $(EtCO_2)_4Ge$ with excess PhMgBr (2609).
Prop.: IR and assignment (1443), far IR and assignment (105, 1443) and mass (1109) spectra, effect of neutron irradiation (84). Thermogravimetric analysis (958).
Rxn. with: HSCN → Ph_3GeNCS (1443).
Biol. Prop.: Insecticidal activity against Musca domestica, LD_{50}, and ATP inhibition (425), against Heliothis zea and H. virescens (2145).

Triphenylgermanium Hydroxide (M-100) $GeC_{18}H_{16}O$ Ph_3GeOH
Synth.: By hydrolysis of $Ph_3GeCO_2GePh_3$ on chromatographic silica gel (794).
By alkaline hydrolysis of $(Ph_3Ge)_3P$ (1849) of $(Ph_3Ge)_2PPh$ (1849), of $Ph_3GePPhSnPh_3$ (1850), of $Ph_3GeP(SnPh_3)_2$ (1850).
By-prod. in rxn. of $(EtCO_2)_4Ge$ with excess PhMgBr (2609).
Prop.: IR spectrum (342, 985), effect of neutron irradiation (84), relative basicity (342). Thermogravimetric analysis (958).
Rxn. with: BuLi → Ph_3GeOLi + C_4H_{10} (685).

$R_3Ga \rightarrow (Ph_3GeOGaR_2)_2 + RH$; R = Me (685, 1822), Ph (685).
Ph_3SiOH + cat. $AlCl_3 \rightarrow (Ph_3GeOSiPh_2)_2$ (958).
$Ph_2Si(OH)_2$ + cat. $AlCl_3 \rightarrow (Ph_3GeO)_3(Ph_5Si_4O_4)$ (958).
$PcSn(OH)_2 \rightarrow PcSn(OGePh_3)_2$; Pc = phthalocyanine, $C_{32}H_{16}N_8$ (1721).
$Et_3SnOH \rightarrow Ph_3GeOSnEt_3$ (2344).
$Ph_3PbOH \rightarrow Ph_3GeOPbPh_3$ (2344).
$AlCl_3 \rightarrow (Ph_3Ge)_2O$ (958).
$HCCl_3$ + ligroin $\rightarrow (Ph_3Ge)_2O$ (958).
H_2S in EtOH $\rightarrow (Ph_3Ge)_2S$ (1970).

Additional unsubstituted triorganogermanium oxides and hydroxides are listed in Table 44.

4.3.1.1.6 FUNCTIONALLY SUBSTITUTED TRIORGANOGERMANIUM OXIDES AND HYDROXIDES

Bischloromethyldimethylgermanium Oxide (M-101) $Ge_2C_6H_{16}Cl_2O$ $(Me_2GeCH_2Cl)_2O$
Synth.: By hydrolysis of $Me_2(ClCH_2)GeCl$ with aq. NaOH in Et_2O (2134), in 88% yield (2134), of $[Me_2(ClCH_2)Ge]_2NH$ (2134).
Prop.: b_{12} 112° (2131), b_{13} 113-15° (2134), NMR spectrum (2131, 2134).
Rxn. with: Aq. $H_2S \rightarrow (Me_2GeCH_2Cl)_2S$ (2134).
NaS in $Et_2O \rightarrow \overline{Me_2GeOGeMe_2CH_2SCH_2}$ (2134).
H_2S in $Et_3N \rightarrow \overline{Me_2GeSGeMe_2CH_2SCH_2}$ (2134).

o-Hydroxyphenyloxymethyldimethyl- $GeC_9H_{14}O_3$ $Me_2(o-HOC_6H_4OCH_2)GeOH$
   germanium Hydroxide
Synth.: By rxn. of $o-\overline{Me_2GeOC_6H_4OCH_2}$ with aq. HCl (2130).
Prop.: b_1 94-95°, mol. wt., IR and NMR spectra (2130).
Rxn. with: $Me_3SiCl + Et_3N \rightarrow o-Me_3SiOC_6H_4OCH_2GeMe_2OSiMe_3$ (2130).

o-Bis(dimethylgermyl)carborane Oxide $Ge_2B_{10}C_6H_{22}O$
Synth.: By rxn. of $1,2-(Me_2GeCl)_2C_2B_{10}H_{10}$ with H_2O in Me_2CO, in quant. yield (1835).
Prop.: m. 200-202°, IR and mass spectra (1835).

2,2,5,5-Tetramethyl-3,4-dicarbomethoxy- $Ge_2C_{10}H_{18}O_5$
   2,5-digerma-2,5-dihydrofurane
Synth.: By-prod. in rxn. of I with excess $(MeO_2CC!)_2$ at 55-60° (2882-3).
Prop.: White solid, m. 53-54°, NMR and mass spectra, GLC (2882-3).

I =

Bistriferrocenylgermanium Oxide $Ge_2Fe_6C_{60}H_{54}O$ $[(C_5H_5FeC_5H_4)_3Ge]_2O$

Synth.: By rxn. of ferrocene with $(Me_2N)_4Ge$ and cat. $AlCl_3$, followed by H_2O, in 25% yield (1952), with $(Me_2N)_2GeCl_2$ in below 1% yield (1952).

Prop.: Golden-tan platelets, m. 205-206°, soly. (1952).

Additional functionally substituted triorganogermanium oxides and hydroxides are listed at the bottom of Table 44.

Table 44. Triorganogermanium Oxides and Hydroxides $(R_3Ge)_2O$
    R_3GeOH

Formula*	Synth. Method**	Yield	Properties	Ref.
Me_3GeOH (lol)	IVA	--	--	2224
$[Me_2(Ph_2C_4H_7)Ge]_2O$	IA (a)	78	$b_{0.4}$ 247-49°, n_D^{20} 1.5712	1587
$[Me_2(MeC_6H_4)_2C_4H_7Ge]_2O$	IA (b)	37	$b_{0.6}$ 250-55°, n_D^{20} 1.5720	1587
$[(PhCH_2)_3Ge]_2O$	--	--	(c)	--
$[Et_2(CH_2:CMeCHMeCH_2)Ge]_2O$	I (d)	--	b_2 147°, n_D^{20} 1.4810, d_{20} 1.0834	2578
$[EtPh(1-C_{10}H_7)Ge]_2O$	--	--	$[\alpha]_D^{25}$ + 1.7° (C_6H_6) (e)	2368
$(PrBu_2Ge)_2O$	IA (f)	85	$b_{0.3}$ 132-33°, n_D^{20} 1.4642, d_{20} 1.0294	1492
	IC (f)	60	(g)	1492
$(i-Pr_3Ge)_2O$ (101)	II	(h)	--	831
	IIID	7	--	831
$i-Pr_3GeOH$ (101)	IIID	16	--	831
$[(n-C_5H_{11})_3Ge]_2O$ (101)	--	--	Elemental anal. (i)	426
$[(i-C_5H_{11})_3Ge]_2O$	--	--	Elemental anal.	426
$[(C_6H_{13})_3Ge]_2O$ (101)	--	--	(i)	
$[(p-MeC_6H_4)_3Ge]_2O$ (101)	IIIB (j)	--	m. 147-48°	1592

* Numbers in parenthesis refer to pages in main volume.
** The characters used here correspond to the synthetic scheme in the introduction to Chapter 4.3.1.1.
(a) From 1,1,X,X-dimethyldiphenylgermacylopentane. (b) From 1,1,X,X-dimethylditolylcyclopentane. (c) Far IR spectrum and assignment (105). (d) From $\overline{Et_2GeCH_2CMeOHCHMeCH_2}$. (e) Rxn. with $LiAlH_4 \rightarrow EtPh(1-C_{10}H_7)GeH$ under retention of configuration (2368). (f) From $\overline{Bu_2GeCH_2CH_2CH_2}$. (g) Rxn. with $HX \rightarrow PrBu_2GeX$; X = Cl, Br, I (1492). (h) By-prod. (i) Toxicity, relative to homologues (45). (j) From $NaMn(CO)_5$.

Table 44 Continued $(R_3Ge)_2O$
R_3GeOH

Formula*	Synth. Method**	Yield	Properties	Ref.
$Me_2GeCH_2SCH_2GeMe_2O$	--	61	b_1 79-81°, NMR spectrum	2134
$[(C_6F_5)_3Ge]_2O$	IIIC	--	m. 269-71°, IR spectrum	1034
	V (k)	100	m. 270-71°	1034
$(C_6F_5)_3GeOH$	IIIA	--	m. 115-17°, IR and NMR spectra (k)	1034

(k) ^{19}F NMR (2475), heating to 130° → $[(C_6F_5)_3Ge]_2O$ (1034).

4.3.1.2 DIORGANOGERMANIUM OXIDES

Dimethylgermanium Oxide (M-102) $(GeC_2H_6O)_n$ $(Me_2GeO)_n$
Prop.: Raman spectrum, n = 4 (2041).
Rxn. with: MeLi → Me_3GeOLi (477).
Me_2GeX_2 → $Me_2GeX(Me_2GeO)_nX$; X = Cl, Br, I, prod. distribution by NMR spectroscopy (1542).
Me_2GeX_2 + Me_2SiX_2 + $(Me_2SiO)_x$ → equil. distribution by NMR; X = Cl (1547), Br (2061).
Biol. Prop.: Toxicity (842), teratogenic effect on chicken embryo (844).

Trisdibutylgermanium Oxide (M-103) $Ge_3C_{24}H_{54}O_3$ $(n-Bu_2GeO)_3$
Synth.: By rxn. of $Bu_2Ge(OR)_2$ with water in presence of $MeC_6H_4SO_3H$; R = Et, i-Pr, t-Bu (337).
Prop.: b_1 180-81°, trimeric (337). Elemental analysis (426).
Rxn. with: ROH * → $Bu_2Ge(OR)_2$; R = Me, Et, n-, i-Pr, n-, s-, t-Bu (1504).
EtOH + Me_3COH * → $Bu_2Ge(OCMe_3)OEt$ (1504).
$Y(OH)_2$* → Bu_2GeOYO; Y = CH_2CH_2, CH_2CHMe, $(CH_2CH_2)_2NH$, $(CH_2CH_2)_2NCH_2CH_2OH$ (1504).
$H_2NCH_2CH_2OH$ → $Bu_2GeOCH_2CH_2NH$ (1504).
n-BuSH* → $Bu_2Ge(SBu)_2$ (1502).
Biol. Prop.: Toxicity (842).

Diphenylgermanium Oxide (M-103) $(GeC_{12}H_{10}O)_x$ $(Ph_2GeO)_x$
Rxn. with: ROH + cat. $MeC_6H_4SO_3N$ → $Ph_2Ge(OR)_2$; R = Et, i-Pr, t-Bu (1504).
$(CH_2OH)_2$ → $Ph_2GeOCH_2CH_2O$ (1504).
$H_2NCH_2CH_2OH$ → $Ph_2GeOCH_2CH_2NH$ (1504).

---

* Rxn. in presence of cat. $MeC_6H_4SO_3H$.

$CH_2Ac_2 \to Ph_2Ge(CHAc_2)_2$ (1504).

Additional diorganogermanium oxides are listed in Table 45 and in Subchapter 7.4.3.

4.3.1.3 MONOORGANOGERMANIUM OXIDES

All compounds are listed at the bottom of Table 45 and in Subchapter 4.1.5.

4.3.2 ORGANOGERMANIUM PEROXY COMPOUNDS

All compounds are listed at the bottom of Table 45.

Table 45. Organogermanium Oxides and Peroxides

Formula*	Synth. Method**	Yield	Properties	Ref.
$(Et_2GeO)_x$ (102)	--	--	NMR spectrum (a)	1444
n = 3-4	IVB	48	b. 140-80°	1471
$(i-Pr_2GeO)_3$ (103)	II	1-6	Colorless liq., mass spectrum	831
	II	10	$b_{0.8}$ 135-37°	2591
$(i-Pr_2GeO)_4$	II	1-6	Colorless liq., mass spectrum	831
$(i-Pr_2GeO)_x$	II	--	(b)	2877
$(CHBuCH_2GeO)_x$	III	91	Nonmelting solid, IR spectrum (c)	1589
$[(C_5H_5FeC_5H_4)_2GeO]_3$	(d)	50	Orange cryst., m. 338-40°, soly.	1952
$[(C_6F_5)_2GeO]_4$	II	--	m. 238-40°, soly.	145
	IIIA	--	IR spectrum, mol. wt., subl. in vacuo	145, 1034
$[(p-ClC_6H_4)_2GeO]_x$ (103)	--	--	Elemental analysis by simultaneous detn.	1088
$(Me_2Ge_2O_3)_x$ (104)	III (e)	--	(f-2085)	2196

* Numbers in parenthesis refer to pages in main volume.
** The characters used here correspond to the synthetic scheme in the introduction to Chapter 4.3.1.1.
(a) Rxn. with $PhSH \to Et_2Ge(SPh)_2$ (465). (b) Far IR spectrum and assignment (105). (c) Gels in C_6H_6, forms colorless well adhering films on glass (1589). (d) From ferrocene and $(Me_2N)_2GeCl_2$ with cat. $AlCl_3 + H_2O$ (1952). (e) From $MeGe(OCH_2CH_2Cl)_3$ in air. (f) Rxn. with $N(CH_2CH_2OH)_3 \to RGe(OCH_2CH_2)_3N$ (2085, 2847).

Table 45 Continued

Formula*	Synth. Method**	Yield	Properties	Ref.
(n-Bu$_2$Ge$_2$O$_3$)$_x$ (104)	--	--	(g)	--
(Ph$_2$Ge$_2$O$_3$)$_x$ (104)	--	--	(f-2085, 2847)	--
[(1-C$_{10}$H$_7$)$_2$Ge$_2$O$_3$]$_x$	--	--	(f-2847)	--
(Et$_3$GeO)$_2$ (105)	--	--	(h)	--
Ph$_3$GeO$_2$H	III (i)	55	m. 135-36°, IR spectrum, (j)	937

(g) Rxn. with n-BuOH → BuGe(OBu)$_3$ (1506). (h) No effect on autoxidation of (Et$_3$Ge)$_2$Hg (2706). (i) From Ph$_3$GeBr with 98% H$_2$O$_2$. (j) Kinetic data for thermal decomposition (937).

4.3.3 ORGANOGERMANIUM ALKOXIDES OR PHENOXIDES

4.3.3.1 TRIORGANOGERMANIUM ALKOXIDES

4.3.3.1.1 UNSUBSTITUTED TRIORGANOGERMANIUM ALKOXIDES

Unsubstituted organogermanium alkoxides are prepared by rearrangement of tetraorganogermanes (I), substitution reactions of organogermanium halides (II), organogermanium oxides, thioalkoxides or imides (V), hydrogermylation (III) of epoxides or carbonyl derivatives, reaction with bimetallic organogermanes (IV) or by interconversion reactions (VI). The following scheme summarizes methods for preparation of the unsubstituted triorganogermanium alkoxides or phenoxides compiled in Table 46.

I Reactions with tetraorganogermanes:
(IA) Et$_2$GeCH$_2$CHMeCHOHCH$_2$ + MeC$_6$H$_4$SO$_2$Cl + Py, + NaOEt → Et$_2$R'GeOEt ,
(IB) Et$_3$GeCH$_2$CH$_2$CO$_3$CMe$_3$ + heat → Et$_3$GeOCMe$_3$,
(IC) 9,9-Ph$_3$GeC$_{13}$H$_8$OH + NaH → 9-Ph$_3$GeOC$_{13}$H$_9$.

II Substitution reactions with organogermanium halides:
(IIA) R$_n$GeX$_{4-n}$ + R'ONa → R$_n$Ge(OR')$_{4-n}$,
(IIB) R$_n$GeCl$_{4-n}$ + R'OH + i-Pr$_2$NH → R$_n$Ge(OR')$_{4-n}$,
(IIC) R$_3$GeCl + Bu$_3$SnOR' → R$_3$GeOR' .

III Reactions with organogermanes:
(IIIA) Et$_3$GeH + RR'CCH$_2$O + cat. H$_2$PtCl$_6$ → Et$_3$GeOCH$_2$CHRR' ,
(IIIB) R$_n$GeH$_{4-n}$ + RR'CO + cat. AiBN → R$_n$Ge(OCHRR')$_{4-n}$; n = 2, 3 ,
(IIIC) Ph$_3$GeH + MeCH:CHCH$_2$OH + Cu → Ph$_3$GeOCH$_2$CH:CHMe .

IV (Et$_3$)$_2$Cd + ROH → Et$_3$GeOR .

V Substitution reactions with organogermanium heterocompounds:
  (VA) $R_{2n}Ge_2O_{4-n} + R'OH + cat.\ MeC_6H_4SO_3H \rightarrow R_nGeOR'_{4-n}$; n = 1-3 ,
  (VB) $Ph_3GeSR + EtOH \rightarrow Ph_3GeOEt$,
  (VC) $(R_3Ge)_2NH + EtOH \rightarrow R_3GeOEt$.

VI $R_nGe(OR'')_{4-n} + R'OH \rightarrow R_nGe(OR')_{4-n} + R''OH$.

R = R' = R'' = organic group, X = halogen, n = 1-3.

Trimethylgermanium Methoxide (M-107) $GeC_4H_{12}O$ Me_3GeOMe
Synth.: By rxn. of $Me_3GeOC_6H_{11}$ with $n-Bu_3SnOMe$ at 190°, in 77% yield (1682).
Prop.: b. 87-9° (1682), b. 87° (928), n_D^{20} 1.4020, d_{20} 1.081 (1682), (far) IR, Raman and assignment (2534) and NMR (926) spectra, dipole moment (928).

Trimethylgermanium Ethoxide $GeC_5H_{14}O$ Me_3GeOEt
Synth.: By rxn. of Me_3GeBr with NaOEt in Et_2O, in 61% yield (643).
By rxn. of Me_3GeCl with NaOEt in petr. ether, in 65% yield (2837).
Prop.: b. 100-101° (643), 100° (2837), n_D^{20} 1.4067 (643), n_D^{25} 1.4048 (2837), d_{20} 1.06 (643), d_{25} 1.0566 (2837), IR (2837), (far) IR spectra and assignment (2534), basicity by IR (644-5, 2837), by NMR (645) spectroscopy, dipole moment (2836). Germanium analysis by Schoeninger combustion and polarographic detn. (2710).
Rxn. with: PhSH → Me_3GeSPh (643).

Triethylgermanium Methoxide (M-107) $GeC_7H_{18}O$ Et_3GeOMe
Synth.: By rxn. of Et_3GeI with Bu_3SnOMe at r.t., in 97% yield (692).
By rxn. of Et_3GeH with Et_3SnOMe at 180° in sealed tube, in 48% yield (2097).
Prop.: b_{16} 56-57° (692), b_{15} 62-65° (2097), n_D^{20} 1.4412 (2097), 1.4376 (692), d_{20} 1.0696 (692). Anal. detn. by fusion and volumetric titration (265).
Rxn. with: $CH_2:CO$ → $Et_3GeOC(:CH_2)CH_2CO_2Me$ (1440).
$CH_2:CO + cat.\ HgI_2 + MeOH$ → $Et_3GeCH_2CO_2Me$ (1440).

Triethylgermanium Cyclohexyloxide (M-108) $GeC_{12}H_{26}O$ $Et_3GeOC_6H_{11}$
Synth.: By rxn. of $(Et_3Ge)_2Hg$ with dicyclohexyl percarbonate in MePh at r.t., in 79% yield, $EtGeOCO_2C_6H_{11}$ is discussed as intermediate (583a).
By rxn. of $(Et_3Ge)_2Te$ with $(c.C_6H_{11}OCO_2)_2$ in C_6H_6 under exclusion of air at r.t., in 94% yield (2093).
Prop.: b_1 70-72°, n_D^{20} 1.4640 (2093), 1.4645 (583a). Spectrographic detn. (1358).
Rxn. with: Et_3GeLi → $Et_6Ge_2 + C_6H_{11}OH$ (2097).

Ethyl-1-naphthylphenylgermanium Methoxide $GeC_{19}H_{20}O$ $EtPh(1-C_{10}H_7)GeOMe$
Synth.: By rxn. of $EtPh(1-C_{10}H_7)GeCl$ with MeOH and $i-Pr_2NH$ under inversion of configuration (2368).
Prop.: $[\alpha]_D^{25}$ -4.9° in C_6H_6 (2368).
Rxn. with: $LiAlH_4$ in Et_2O → $EtPh(1-C_{10}H_7)GeH$ with retention (2368).
BuLi in Et_2O → $EtPh(1-C_{10}H_7)GeBu$ with retention (2368).

Ethyl-1-naphthylphenylgermanium   GeC$_{26}$H$_{26}$O   EtPh(1-C$_{10}$H$_7$)GeOCHMePh
  1-Phenylethoxide (M-110)
Synth.: By rxn. of EtPh(1-C$_{10}$H$_7$)GeBr with (-)-MePhCHONa in Et$_2$O-C$_6$H$_6$ in N atm., in 64% yield (999).
Prop.: Crude b$_{0.01}$ 190-95°, n$_D^{20}$ 1.6199 (999), after recrystn. from pentane at -20 to -10°: m. 66-67°, [α]$_D^{20}$ -52.3° (c. 10.9 in C$_6$H$_6$) (999).
Rxn. with: LiAlH$_4$ → (-)-EtPh(1-C$_{10}$H$_7$)GeH (999).

Ethyl-1-naphthylphenylgermanium   GeC$_{28}$H$_{36}$O   EtPh(C$_{10}$H$_7$)GeOC$_{10}$H$_{19}$
  Menthoxide (M-110)
Synth.: By rxn. of EtPh(1-C$_{10}$H$_7$)GeBr with (-)-C$_{10}$H$_{19}$ONa in C$_6$H$_6$, in 60% yield (999).
Prop.: Crude b$_{0.01}$ 172-74°, n$_D^{20}$ 1.5757, [α]$_D^{25}$ -59.7° (c. 5.1 in C$_6$H$_6$) (999). Recrystd. pentane at -70° in N atm.: m. 79-80.5°, [α]$_D^{25}$ -64.5 (c. 2.6 in (C$_6$H$_6$) (999).
Rxn. with: LiAlH$_4$ in Et$_2$O → (+)-EtPh(1-C$_{10}$H$_7$)GeH (999, 2368).
BuLi in Et$_2$O → EtPh(1-C$_{10}$H$_7$)GeBu with retention (2368).

Tripropylgermanium Ethoxide   GeC$_{11}$H$_{26}$O   n-Pr$_3$GeOEt
Synth.: By rxn. of (Pr$_3$Ge)$_2$O with EtOH in the presence of cat. MeC$_6$H$_4$SO$_3$H, in 66% yield (2837).
Prop.: b$_{10}$ 80°, n$_D^{25}$ 1.4405, d$_{25}$ 0.9916 (2837), IR spectrum (2837), relative basicity (2837), dipole moment (2836). Anal. detn. by photometry after combustion (2654), by polarography after Schoeninger combustion (2710).

Tributylgermanium Methoxide (M-109)   GeC$_{13}$H$_{30}$O   n-Bu$_3$GeOMe
Synth.: By rxn. of Bu$_3$GeCl with MeOH and NH$_3$ in C$_6$H$_6$, in 93% yield (1503).
By rxn. of Bu$_2$GeCH$_2$CH$_2$CH$_2$ with MeONa, followed by MeI in MeOH-THF, in 80% yield (2375).
Prop.: b$_{9.5}$ 128-32°, n$_D^{20}$ 1.4480 mol. wt. (1503), (far) IR and Raman spectra, assignment (2534). Anal. detn. by photometry after combustion (2654).
Rxn. with: (RN:)$_2$C → Bu$_3$GeNRC(:NR)OMe; R = p-MeC$_6$H$_4$ (1228).
Cl$_3$CCHO → Bu$_3$GeOCH(OMe)CCl$_3$ (1228).
PhNCO → Bu$_3$GeNPhCO$_2$Me (1228).
PhNCS in air → PhNHC(S)OMe (1228).
RSH → Bu$_3$GeSR; R = n-C$_8$H$_{17}$, Ph (465).

Tributylgermanium Ethoxide (M-109)   GeC$_{14}$H$_{32}$O   n-Bu$_3$GeOEt
Synth.: By rxn. of Bu$_3$GeCl with EtOH and NH$_3$ in C$_6$H$_6$, in 75% yield (1503).
By rxn. of (Bu$_3$Ge)$_2$O with EtOH in C$_6$H$_6$ in the presence of cat. MeC$_6$H$_4$SO$_3$H, in 82% yield (1504).
By rxn. of Bu$_2$GeCH$_2$CH$_2$CH$_2$ with EtONa, followed by MeI in Et$_2$O, in 80% yield (2575).
Prop.: b$_8$ 127-28° (1504), b$_{2.5}$ 104° (1450), n$_D^{20}$ 1.4444 (1504), 1.4440 (1503), n$_D^{25}$ 1.4450 (1450), GLC (1450), mol. wt. (1504). Anal. detn. by volumetric titration after rxn. with Ac$_2$O and dry HBr (1450).

Rxn. with: HOYCO$_2$H → Bu$_3$GeOYCO$_2$GeBu$_3$; Y = CHPh, o-C$_6$H$_4$ (1477).
Y(OH)$_2$ → (Bu$_3$GeO)$_2$Y; Y = CH$_2$CHMe, CH$_2$CH$_2$CHMe, (CHMe)$_2$, O(CH$_2$CH$_2$)$_2$, (CMe$_2$)$_2$, (CH$_2$)$_n$; n = 2-6 (1505).
RCO$_2$H → Bu$_3$GeO$_2$CR; R = Me, Ph (1477).
PhNCO → Bu$_3$GeNPhCO$_2$Et (1228).
PhNCS in air → PhNHC(S)OMe (1228).

Triphenylgermanium Diphenylmethoxide   GeC$_{31}$H$_{26}$O   Ph$_3$GeOCHPh$_2$
Synth.: By rxn. of Ph$_3$GeCl with MeONa, followed by Ph$_2$CHOH in xylene, in 54% yield (2675).
Prop.: Viscous oil, b$_1$ 220-40°, NMR spectrum (2675).
Rxn. with: x-BuLi → BuGePh$_3$ + Ph$_2$CHOH; x = n-, t- (2675).
Moisture → (Ph$_3$Ge)$_2$O + Ph$_2$CHOH (2675).
(CH$_2$CO)$_2$NBr + cat. Bz$_2$O$_2$ → Ph$_3$GeBr + BzPh (2675).

Additional unsubstituted triorganogermanium alkoxides are listed in Table 46.

Table 46. Unsubstituted Triorganogermanium Alkoxides R_3GeOR

Formula*	Synth. Method**	Yield	Properties	Ref.
Me$_3$GeOCHMe$_2$	--	--	(a)	--
Me$_3$GeOCMe$_3$	--	--	(b)	--
Me$_3$GeOC$_6$H$_{11}$c.	--	--	(b, c)	--
Me$_3$GeOC$_8$H$_{17}$-n	--	--	(b)	--
MePh(1-C$_{10}$H$_7$)GeOMe (110)	IIB	--	Synth. under inversion of configuration	417
Et$_3$GeOR				
R = Pr (108)	IV	84	b. 190-91°, n_D^{20} 1.4370	587
CH$_2$C:CH (108)	VA	61	n_D^{20} 1.4598, IR spectrum	1534
	VC	86	b$_4$ 64-65°, n_D^{20} 1.4600, d$_{20}$ 1.0862	1534
Bu (108)	IIC	90	b$_9$ 81-82°, n_D^{20} 1.4420 (d)	1440
CMe$_3$ (108)	IIA	65	b$_{10}$ 62-63°, n_D^{20} 1.4360, IR spectrum,	2095
	IB	11	Hydrolyzes easily → (Et$_2$Ge)$_2$O	2095
CH$_2$CH:CHMe (108)	IIIA (e)	< 60	b$_{15}$ 101-104°, n_D^{20} 1.4535, d$_{20}$ 1.0357 (f)	806
CH$_2$CMe:CHMe	IIIA (g)	< 60	b$_{15}$ 110-14°, n_D^{20} 1.4558, d$_{20}$ 1.0349 (h)	806
CHMeCH:CHMe	IIIA (f)	--	b$_{10}$ 110-13°, n_D^{20} 1.4620, d$_{20}$ 1.0446 (h)	2636
(−)-C$_{10}$H$_{19}$	--	--	(Menthoxide) (j)	--
Ph (108)	--	--	(k)	--
Et$_2$(CH$_2$:CHCHMeCH$_2$)GeOEt	IA	--	b$_{10}$ 90°, n_D^{20} 1.4544, d$_{20}$ 1.0235 (1)	2578
n-Bu$_3$GeOCHMe$_2$	IIB	88	b$_7$ 127-30°, n_D^{20} 1.4400, mol. wt. (b)	1503
	VA	85	b$_{3.8}$ 113-14°, n_D^{20} 1.4400, mol. wt.	1504
n-Bu$_3$GeOCMe$_3$	IIB	70	b$_8$ 136-40°, n_D^{20} 1.4420, mol. wt.	1503
	VA	30	b$_4$ 116°, n_D^{20} 1.4420, mol. wt.	1504
n-Bu$_3$GeOC$_6$H$_{11}$c. (109)	--	--	(b)	--
n-Bu$_3$GeOC$_8$H$_{17}$-n	--	--	(a)	--

Ph$_3$GeOR				
R = Me (110)	--	--	b$_{2.2}$ 148°, n$_D^{25}$ 1.5564, GLC (m, n)	1450
CH$_2$Ph	IIIB	65	m. 78°, IR and NMR spectra	1762
Et	VB	--	m. 83-85°, IR spectrum	227
CH$_2$CHMe$_2$	IIIB	93	m. 47°, b$_{0.005}$ 146°,	1762
	IIA	86	IR and NMR spectra	1762
CH$_2$CH:CHMe	IIIC	--	NMR spectrum (o)	806
CHMeCH$_2$CHMe$_2$	IIIB	54	b$_{0.03}$ 154°, n$_D^{20}$ 1.5700, d$_{20}$ 1.1462,	1762
	VI (n)	75	IR and NMR spectra	1762
c.C$_6$H$_{11}$	IIIB (p)	92	b$_{0.003}$ 160°, n$_D^{20}$ 1.5945	1762
	IIA	65	IR and NMR spectra	1762
9-C$_{13}$H$_9$	IC	70	m. 124-26°, NMR and mass spectra	2675
	IIA	29	(Fluorenyl) (q)	2675

* Numbers in parenthesis refer to pages in main volume

** The characters used here correspond to the synthetic scheme in the introduction to Chapter 4.3.3.1.1.

(a) (Far) IR, Raman spectra and assignment (2534). (b) (Far) IR spectrum and assignment (2534). (c) Rxn. with Bu$_3$SnOMe at 190° → Me$_3$GeOMe + Bu$_3$SnOC$_6$H$_{11}$ (1682). (d) Rxn. with CH$_2$:CO and cat. HgI$_2$ → Et$_3$GeCH$_2$CO$_2$Bu (1440). (e) From CH$_2$:CHCHCH$_2$O. (f) IR spectrum. (g) From CH$_2$:CHCMeCH$_2$O. (h) IR and NMR spectra. (i) From MeCH:CCHMeO. (j) Rxn. with benzofuran + cat. AlCl$_3$ → optically active (C$_8$H$_6$O)$_x$ (1164). (k) Rxn. with K in (Me$_2$N)$_3$PO, + PhCH$_2$Cl → PhCH$_2$OPh (1635). (l) Rxn. with LiAlH$_4$ → Et$_2$(CH$_2$:CHCHMeCH$_2$)GeH (2578). (m) Anal. detn. by volumetric titration after rxn. with Ac$_2$O or NH$_4$Br (1450). (n) Rxn. with MeCHOHCH$_2$CHMe$_2$ → Ph$_3$GeOCHMeCH$_2$CHMe$_2$ + MeOH (1762). (o) Rxn. with Ph$_3$GeH + cat. H$_2$PtCl$_6$ → Ph$_3$GeCHMeCH$_2$CH$_2$OGePh$_3$ (806). (p) Synth. without catalyst or under UV irradiation, in 70 and 86% yield, resp. (1762). (q) Rxn. with BuLi, + PhCH$_2$Cl → BuGePh$_3$ + (Ph$_3$Ge)$_2$O + C$_{13}$H$_9$OH + 9-PhCH$_2$C$_{13}$H$_9$ (2675).

4.3.3.1.6 FUNCTIONALLY SUBSTITUTED TRIORGANOGERMANIUM ALKOXIDES

Triorganogermanium alkoxides containing halogen, oxygen, sulfur, nitrogen or phosphorous in the alkoxide or the organo group linked directly to germanium are listed in Table 47. The compounds are prepared by methods from the following scheme.

I Metathetic substitution:
  (IA) $R_nGeX_{4-n} + R'ONa \rightarrow R_nGe(OR')_{4-n}$,
  (IB) $Me_2(ClCH_2)GeCl + o\text{-}HOC_6H_4YH + Et_3N \rightarrow Me_2GeCH_2YC_6H_4O$; Y = O, NH ,
  (IC) $R'_nGeCl_{4-n} + R_3GeOMe \rightarrow R'_nGe(OMe)_{4-n}$.

II Hydrogermylation reaction:
  (IIA) $Et_3GeH + CH_2{:}CClCHCH_2O + H_2PtCl_6 \rightarrow Et_3GeOCH_2CH{:}CClMe$,
  (IIB) $Me_3GeH + RR'CO \rightarrow Me_3GeOCHRR'$

III Addition reactions with organogermanium halides or imides:
  (IIIA) $Me_nGeX_{4-n} + CH_2CH_2O + \text{cat. } H_2PtCl_6 \rightarrow Me_nGe(OCH_2CH_2X)_{4-n}$,
  (IIIB) $R_3GeX' + RR'CO \rightarrow R_3GeOCRR'X'$; X' = EtS, Me_2N, Et_2P ,
  (IIIC) $Et_nGeBr_{4-n} + RCHO + (EtO)_3P \rightarrow Et_nGe(OCHRPO_3Et_2)_{4-n}$; n = 1,3 ,
  (IIID) $(Et_3Ge)_2NH + Ph_2C{:}CO \rightarrow [Et_3GeOC({:}CPh_2)]_2NH$.

IV Condensation reactions:
  (IVA) $(Et_3Ge)_2O + R'OH \rightarrow Et_3GeOR' + H_2O$,
  (IVB) $R_{2n}Ge_2O_{4-n} + Y(OH)_2 + \text{cat. } MeC_6H_4SO_3H \rightarrow (R_nGeO)_{n-1}Y + H_2O$;
         n = 3,2
         $R_2Ge_2O_3 + Y(OH)_3 \rightarrow (RGeO_3)Y + H_2O$,
  (IVC) $R_nGe(OEt)_{4-n} + Y(OH)_2 \rightarrow (R_nGeO)_{n-1}Y + EtOH$; n = 3,2 ,
  (IVD) $(R_3Ge)_2NH + Y(OH)_2 \rightarrow (R_3GeO)_2Y + NH_3$.

R = organic group, R' = functionally substituted organic group, X = halogen, Y = hydrocarbon bridge, may carry a functional group, n = 1-3 .

Additional functionally substituted triorganogermanium alkoxides are reported in Subchapters 2.4.3, 2.6.3.1, 4.1.5 and 4.3.3.1.7 .

Table 47. Functionally Substituted Triorganogermanium Alkoxides R_3GeOR'

Formula	Method*	Yield	Properties	Ref.
Me_3GeOR'				
R' = CH_2CH_2Cl	IIIA	--	b_{763} 153°, n_D^{20} 1.4421, d_{20} 1.2304 (a)	2196
CH_2CH_2Br	IIIA	82	b_{13} 64°, n_D^{20} 1.4670, d_{20} 1.4823 (b)	367
CH_2CH_2OPh	IA	50	GLC, mol. wt., mass spectrum	2353
$CH(CF_3)_2$	IIB	100	b_{758} 117°, ^1H, ^{19}F NMR and IR spectra (c)	109

C(CF$_3$)$_2$SEt	IIIB	--	b$_{15}$ 88°, n$_D^{23}$ 1.4088, ^{1}H and ^{19}F NMR spectra	2203
o-Me$_2$GeCH$_2$OC$_6$H$_4$O	IB	--	b$_1$ 95-96°, mol wt, IR and NMR spectra (d)	2130
o-Me$_2$GeCH$_2$NHC$_6$H$_4$O	IB	--	b$_1$ 122°, mol. wt, NMR spectrum	2130

Et$_3$GeOR′

R′ = CHPhPEt$_2$	IIIB	70	b$_{0.09}$ 117°, IR and NMR spectra	2735
CHEtPO$_3$Et$_2$	IIIC	--	b$_{0.1}$ 98-101°, n$_D^{25}$ 1.4475	452
CH$_2$CH:CClMe	IIA	< 60	b$_{15}$ 120-23°, n$_D^{20}$ 1.4655, d$_{20}$ 1.1272 (e)	806
CHPrPEt$_2$	IIIB	70	b$_{0.15}$ 85°, IR and NMR spectra	2735
2-ClC$_6$H$_{10}$c.	IIIB	--	b$_{2.5}$ 121-25°, n$_D^{25}$ 1.4765 (f)	2532
2-BrC$_6$H$_{10}$c.	IIIB	--	b$_{2.5}$ 134-36°, n$_D^{25}$ 1.4890 (f)	2532
(MeO$_2$CCH$_2$)$_3$GeOMe	IC	62	b$_1$ 131-33°, n$_D^{20}$ 1.4750, d$_{20}$ 1.3645 (g)	692
n-Pr$_3$GeOC$_6$H$_{10}$Cl-2c.	IIIB	--	b$_{2.2}$ 135-38°, n$_D^{25}$ 1.4742 (f)	2532
n-Pr$_3$GeOC$_6$H$_{10}$Br-2c.	IIIB	--	b$_{2.8}$ 150-55°, n$_D^{25}$ 1.4880 (f)	2532
n-Bu$_3$GeOCH$_2$CH$_2$NH$_2$	IVA	80	b$_2$ 130°, n$_D^{20}$ 1.4585, d$_{20}$ 1.4585 (h)	1504
n-Bu$_3$GeOC(CF$_3$)$_2$NMe$_2$	IIIB	80	b$_{0.01}$ 110°, n$_D^{20}$ 1.4220, ^{1}H and ^{19}F NMR spectra	2199
i-Bu$_3$GeOC$_6$H$_{10}$Br-2c.	IIIB	--	b$_1$ 105-10°, n$_D^{26}$ 1.4854 (f)	2532
Ph$_3$GeOC$_6$H$_{10}$Br-2c.	IIIB	--	m. 179-80°	2532

* The characters used here correspond to the synthetic scheme in the introduction to Chapter 4.3.3.1.6.
(a) Easily hydrolyzed in air IR spectrum (2196). (b) IR and Raman spectra (367). (c) Addn. compd.: Me$_3$GeOCH(CF$_3$)$_2$·OC(CF$_3$)$_2$, ^{1}H and ^{19}F NMR spectra, dissocn. at r.t. (109). (d) Rxn. with aq HCl → o-HOC$_6$H$_4$CH$_2$GeMe$_2$OH (2130). (e) IR and NMR spectra (806). (f) Rxn. with dry HCl in C$_6$H$_6$ → R$_3$GeCl + 2-XC$_6$H$_{10}$OH; ratio of cis- and trans- forms, X = Cl, Br (2532). (g) NMR spectrum (1685). (h) Mol. wt., struct. (1504).

4.3.3.1.8 TRIORGANOGERMANIUM ALKOXIDES DERIVED FROM DIHYDRIC ALKOHOLS OR PHENOLS

All compounds are listed in Table 48.

Table 48. Bimetallic Triorganogermanium Alkoxides R_3GeOR'

Formula*	Synth. Method**	Yield	Properties	Ref.
[Et$_3$GeOC(:CPh$_2$)]$_2$NH	IIID	47	Dec. $_{0.05}$ 118°, IR spectrum (a)	2736
p-(Et$_3$GeO)$_2$C$_6$H$_4$ (108)	IVD	--	--	1796
Bu$_3$GeOYOGeBu$_3$				
Y = CH$_2$CH$_2$	IVB	85	b$_{0.4}$ 180°, n$_D^{20}$ 1.4565, mol. wt.	1504
	IVC	84	b$_{0.4}$ 180°, n$_D^{20}$ 1.4625, mol. wt.	1505
(CH$_2$)$_3$	IVC	85	b$_{0.3}$ 180°, n$_D^{20}$ 1.4630, mol. wt.	1505
CH$_2$CHMe	IVB	88	b$_{0.2}$ 167-69°, n$_D^{20}$ 1.4580, mol. wt.	1504
	IVC	65	b$_{0.1}$ 165°, n$_D^{20}$ 1.4685, mol. wt.	1505
(CH$_2$)$_4$	IVC	85	b$_{0.25}$ 187°, n$_D^{20}$ 1.4650, mol. wt.	1505
CH$_2$CH$_2$CHMe	IVC	93	b$_{0.2}$ 175°, n$_D^{20}$ 1.4635, mol. wt.	1505
CHMeCHMe	IVC	88	b$_{0.2}$ 165°, n$_D^{20}$ 1.4625	1505
(CH$_2$CH$_2$)$_2$O	IVC	94	b$_{0.1}$ 185°, n$_D^{20}$ 1.4650	1505
(CH$_2$CH$_2$)$_2$NH	IVB	77	b$_{0.8}$ 152-55°, n$_D^{20}$ 1.4600, mol. wt.	1504
(CH$_2$)$_5$	IVC	70	b$_{0.2}$ 186°, n$_D^{20}$ 1.4660	1505
(CH$_2$)$_6$	IVC	81	b$_{0.5}$ 184°, n$_D^{20}$ 1.4605	1505
CMe$_2$CMe$_2$	IVC	60	b$_{0.2}$ 141°, n$_D^{20}$ 1.4625	1505

* Numbers in parenthesis refer to pages in main volume.
** The characters used here correspond to the synthetic scheme in the introduction to Chapter 4.3.3.1.6.
(a) Rxn. with air → (Ph$_2$CHCO)$_2$NH (2736).

4.3.3.1.9 TRIORGANOGERMANIUM MIXED OXIDES

Triorganogermanium derivatives containing lithium, aluminum, gallium and indium linked to germanium by oxygen are very sensitive towards hydrolysis. Mixed oxides with other organo derivatives of group IV metals are usually stable in air. The compounds listed in Table 49 are prepared by the following methods.

I Reaction with organometallic compounds:
   (IA) $Ph_3GeOH + BuLi \rightarrow Ph_3GeOLi + C_4H_{10}$,
   (IB) $R_3GeOH + R_3Ga \rightarrow R_3GeOGaR_2 + RH$,
   (IC) $(Me_2GeO)_x + MeLi \rightarrow Me_3GeOLi$.

II Metathetic substitution reactions:
   (IIA) $Me_3GeOLi + R_2MCl \rightarrow Me_3GeOMR_2$; M = Al, Ga, In ,
   (IIB) $Me_3GeCl + R_3SiOLi \rightarrow Me_3GeOSiR_3$.

III Condensation reactions:
   $Ph_3GeOH + (R_nM)OH \rightarrow Ph_3GeO(MR_n) + H_2O$; M = Si, Sn, Pb .

4.3.3.2   DIORGANOGERMANIUM ALKOXIDES

4.3.3.2.1 UNSUBSTITUTED DIORGANOGERMANIUM ALKOXIDES

Dimethylgermanium Dimethoxide (M-107) $GeC_4H_{12}O_2$ $Me_2Ge(OMe)_2$
Synth.: By rxn. of Me_2GeCl_2 with MeOH and Et_3N, in 63% yield (1542).
Prop.: b. 115-17° (1542), NMR spectrum (926, 1012), GLC (1450), dipole moment (929). Anal. detn. by volumetric titration after rxn. with Ac_2O or NH_4Br (1450).
Rxn. with: $Me_2GeX_2 \rightleftharpoons Me_2Ge(OMe)X$; X = Cl, Br, I, equilibrium const. (1542); X = MeS (2062).
$Me_2SiX_2 \rightleftharpoons$ equilibrium const. for redistribution rxn.; X = Cl, Br, I, MeS (2062).
$MeGeBr_3 \rightleftharpoons$ equilibrium const. for redistribution rxn. by NMR (2606).

Dimethylgermanium Diethoxide (M-107) $GeC_6H_{16}O_2$ $Me_2Ge(OEt)_2$
Synth.: By rxn. of Me_2GeCl_2 with EtONa in C_6H_6, in 36% yield (2192), in petr. ether, in 64% yield (2837).
By rxn. of Me_2GeCl_2 with dry EtOH and NH_3 in C_6H_6, in 66% yield (2192).
Prop.: b. 130° (2837), b_{746} 139° (2192), n_D^{20} 1.4128 (2192), n_D^{25} 1.4119 (2837), d_{20} 1.1129 (2192), d_{25} 1.1142 (2837), IR (2837) and NMR (1012) spectra, dipole moment (2836), basicity (2837). Anal. detn. by Schoeninger combustion and polarography (2710).

(Additional unsubstituted diorganogermanium alkoxides may be found on page 177, following Table 49.)

Table 49. Triorganogermanium Mixed Oxides R_3GeOMR_n

Formula*	Synth. Method**	Yield	Properties	Ref.
Me$_3$GeOLi (111)	IC	--	Not isolated (a-d)	477
(Me$_3$GeOAlMe$_2$)$_2$	IIA (a)	41	m. 64-66°, subl. o.1 50-55°, b$_{20}$ 128-30 (e-g)	685
(Me$_3$GeOGaMe$_2$)$_2$	IIA (a)	77	m. 42-44°, subl. o.1 55-58°, b$_1$ 84-85° (f-j)	685, 1822
(Me$_3$GeOGaPh$_2$)$_2$	IIA (b)	63	m. 186-88° (f-h)	685
(Me$_3$GeOInMe$_2$)$_2$	IIA (a)	69	m. 14-16°, b$_1$ 98-100° (f-i)	685
Me$_3$GeOSiMe$_3$ (106)	IIB	--	m. -68°, b$_{725}$ 117° (k)	474a
Me$_3$GeOSiMe$_2$CH$_2$Cl	IIB	80	b$_{12}$ 62°, NMR spectrum	2129
Me$_2$(ClCH$_2$)GeOSiMe$_2$CH$_2$Cl	IIB	74	b$_{12}$ 96°, NMR spectrum	2129
Ph$_3$GeOLi	IA	--	Not isolated (l)	685
(Ph$_3$GeOGaMe$_2$)$_2$	IIA (l)	60	m. 177-79°	685
(Ph$_3$GeOGaPh$_2$)$_2$	IB	94	m. 175-77° (f-i)	685, 1822
(Ph$_3$GeOSiPh$_2$)$_2$O	IB	59	m. 291-93° (f, h)	685
(Ph$_3$GeO)$_3$(Ph$_5$Si$_4$O) (?)	III (m)	62	White wax, softening 90-97° (n)	958
Ph$_3$GeOSnEt$_3$	III (o)	57	Clear hard resin, softening 50° (n)	958
(Ph$_3$GeO)$_2$SnPc (p)	III	--	Oil, IR spectrum	2344
Ph$_3$GeOPbPh$_3$	III	--	IR spectrum (q)	1721
	III	--	m. 127.5-28.5°, IR spectrum	2344

* Numbers in parenthesis refer to pages in main volume.
** The characters used here correspond to the synthetic scheme in the introduction to Chapter 4.3.3.1.9.
(a) Rxn. of Me$_3$GeOLi with Me$_2$MCl in N atm. → (Me$_3$GeOMMe$_2$)$_2$; M = Al, Ga, In (685), Ga (1822). (b) Rxn. of Me$_3$GeOLi with Ph$_2$GaCl in N atm. → (Me$_3$GeOGaPh$_2$)$_2$ (685). (c) Rxn. with GaCl$_3$ → (Me$_3$Ge)$_2$O + Me$_3$GeCl (685). (d) Use as catalyst and additive for propellants and explosives (477). (e) Pyrophoric (685). (f) Solubility, structure (685). (g) Monomeric, NMR spectrum (685). (h) IR spectrum and assigment (685). (i) Colorless crystalline (685). (j) X-ray structural analysis (685). (k) IR (474a) and NMR (474) spectra. (l) Rxn. of Ph$_3$GeOLi with Me$_2$GaCl in N atm. → (Ph$_3$GeOGaMe$_2$)$_2$ (685). (m) From Ph$_3$GeOH and Ph$_3$SiOH + cat. AlCl$_3$ at 225-50° (958). (n) IR spectrum, thermogravimetric analysis (958). (o) From Ph$_3$GeOH and Ph$_2$Si(OH)$_2$ + cat. AlCl$_3$ at 225-50° (958). (p) Pc = phthalocyanine. (q) Rxn. with aq. HCl → PcSnCl$_2$ (1721).

Dimethylgermanium Diphenoxide $GeC_{14}H_{16}O_2$ $Me_2Ge(OPh)_2$

Synth.: By rxn. of Me_2GeCl_2 with PhONa (277), with dry PhOH in C_6H_6 with NH_3, in 62% yield (2192), with PhOH and Et_3N (2605).

Prop.: b_{14} 176-77° (2192), $b_{0.6}$ 135° (2605), $b_{0.5}$ 126-27° (277), $b_{0.4}$ 130° (1450), n_D^{20} 1.5633 (2192), n_D^{25} 1.5604 (1450), d_{20} 1.2754 (2192), IR (277) and NMR (2605) spectra, GLC (1450). Anal. detn. by volumetric titration after rxn. with Ac_2O or dry HBr (1450).

Rxn. with: $Et_2NLi \rightarrow Me_2Ge(NEt_2)_2$ (277).
$(Me_3Si)_2NNa \rightarrow Me_2Ge(OPh)N(SiMe_3)_2$ (277).
Me_2GeX_2 (+ Me_2SiX_2) $\rightleftharpoons$ equilibrium const. for redistribution rxn., nonrandom distribution; X = Cl, Br, I (2605).
$MeGeX_3 \rightleftharpoons$ equilibrium const. for redistribution rxn., prod. distribution; X = Cl, Br (2606).
HCl $\rightarrow Me_2GeCl_2$ (277).

Diethylgermanium Dimethoxide $GeC_6H_{16}O_2$ $Et_2Ge(OMe)_2$

Synth.: By rxn. of MeONa at 50-70°, in 66% yield (2192), in C_6H_6 (465).
By rxn. of $(Et_2GeNH)_3$ with anhyd. MeOH, in 84% yield (1535)

Prop.: b. 165-66° (1535), 160° (465), b_{755} 159° (2192), n_D^{20} 1.4369 (2192) 1.427 (465), n_D^{23} 1.4321 (1535), d_{20} 1.1417 (465), 1.0692 (2192).

Rxn. with: $(Et_2GeS)_3 \rightarrow MeO(Et_2GeS)_xEt_2GeOMe$ (2607).
PhSH $\rightarrow Et_2Ge(SPh)_2$ (465).

Dibutylgermanium Diethoxide $GeC_{12}H_{28}O_2$ $n-Bu_2Ge(OEt)_2$

Synth.: By rxn. of Bu_2GeCl_2 with EtOH and NH_3 in C_6H_6 under anhyd. conditions, in 84% yield (337).
By rxn. of $(Bu_2GeO)_x$ with EtOH in C_6H_6 and cat. $MeC_6H_4SO_3H$, in 78% yield (1504).
By rxn. of $Bu_2Ge(SPr)_2$ with excess EtOH and azeotropic distn., in 90% yield (1502).

Prop.: b_6 106° (337), b_6 103-106° (1504), n_D^{20} 1.4357 (337), 1.4330 (1504), mol. wt. (337, 1504). Microanal. detn. (426).

Rxn. with: H_2O in EtOH $\rightarrow (Bu_2GeO)_3$ (337).
ROH $\rightarrow Bu_2Ge(OR)_2$; R = MeEtCH, Me_2EtC (338).
$Y(OH)_2 \rightarrow Bu_2GeOYO$; Y = $(CH_2)_n$; n = 2-4, CH_2CH_2CHMe, $(CH_2CH_2)_2O$, $CHMeCH_2CMe_2$ (338).
$HOYCO_2H \rightarrow Bu_2GeO_2CYO$; Y = PhCH, $o-C_6H_4$ (1477).
$HOCHRCH_2SH \rightarrow Bu_2GeSCH_2CHRO$; R = H, CH_2OH (2805).
$RCO_2H \rightarrow Bu_2Ge(O_2CR)OEt + Bu_2Ge(O_2CR)_2$; R = Me, Ph (1477).
H_2S in EtOH $\rightarrow (Bu_2GeS)_2$ (337).
$n-C_{12}H_{25}SH \rightarrow Bu_2Ge(SC_{12}H_{25})_2$ + EtOH (1502).
$HSCH_2CH_2CO_2H \rightarrow (Bu_2GeSCH_2CH_2CO_2)_3$ + EtOH (2805).
$HSCHMeCO_2H \rightarrow Bu_2GeO_2CCHMeS$ + EtOH (2805).
AcBr $\rightarrow Bu_2GeBr_2 + Bu_2Ge(OEt)Br$ + EtOAc (337).

Dibutylgermanium Diisopropoxide   GeC$_{14}$H$_{32}$O$_2$ Bu$_2$Ge(OCHMe$_2$)$_2$

Synth.: By rxn. of Bu$_2$GeCl$_2$ with i-PrOH and NH$_3$ in C$_6$H$_6$, in 84% yield (337). By rxn. of (Bu$_2$GeO)$_x$ and i-PrOH in C$_6$H$_6$ with cat. MeC$_6$H$_4$SO$_3$H, in 70% yield (1504), in 5% yield without catalyst (1504).

Prop.: b$_5$ 103-108° (337), b$_{4.5}$ 106-109°, b$_{0.5}$ 72° (1504), n$_D^{20}$ 1.4350 (337), 1.4340-45 (1504), mol. wt. (337, 1504).

Rxn. with: H$_2$O in i-PrOH → (Bu$_2$GeO)$_3$ (337).
n-BuOH → Bu$_2$Ge(OBu)$_2$ (338).
Y(OH)$_2$ → Bu$_2$GeOYO; Y = (CH$_2$)$_2$, (CHMe)$_2$, CHMeCH$_2$CMe$_2$ (338).
H$_2$S → (Bu$_2$GeS)$_2$ (337).
AcBr → Bu$_2$GeBr$_2$ + Bu$_2$Ge(OCHMe$_2$)Br + i-PrOAc (337).

Dibutylgermanium Ditertiarybutoxide   GeC$_{16}$H$_{36}$O$_2$ Bu$_2$Ge(OCMe$_3$)$_2$

Synth.: By rxn. of Bu$_2$GeCl$_2$ with Me$_3$COH and NH$_3$/Py in C$_6$H$_6$, in 70% yield (337). By rxn. of (Bu$_2$GeO)$_x$ with Me$_3$COH in C$_6$H$_6$ and cat. MeC$_6$H$_4$SO$_3$H, in 38-85% yield (1504).

Prop.: b$_{0.3}$ 105-108°, n$_D^{20}$ 1.4355, monomeric (337, 1504).

Rxn. with: H$_2$O → (Bu$_2$GeO)$_3$ (337).
AcBr → Bu$_2$GeBr$_2$ + Bu$_2$Ge(OCMe$_3$)Br + t-BuOAc (337).
H$_2$S → (Bu$_2$GeS)$_2$ (337).

Diphenylgermanium Diethoxide   GeC$_{16}$H$_{20}$O$_2$ Ph$_2$Ge(OEt)$_2$

Synth.: By rxn. of Ph$_2$GeCl$_2$ with EtOH and NH$_3$ in C$_6$H$_6$, in 86% yield (1503). By rxn. of (Ph$_2$GeO)$_x$ with EtOH in C$_6$H$_6$ and cat. MeC$_6$H$_4$SO$_3$H, in 75% yield (1504).

Prop.: b$_3$ 156-58° (1504), b$_1$ 130-32 (1503), n$_D^{20}$ 1.5400 (1503-4), mol. wt. (1503-4).

Rxn. with: Y(OH)$_2$ → Ph$_2$GeOYO; Y = (CH$_2$)$_n$; n = 2-6, CH$_2$CHMe, CH$_2$CH$_2$CHMe, Y = (CHMe)$_2$, (CMe$_2$)$_2$, O(CH$_2$CH$_2$)$_2$ (1505).
RCO$_2$H → Ph$_2$Ge(O$_2$CR)OEt + Ph$_2$Ge(O$_2$CR)$_2$; R = Me, Ph (1477).
o-HOC$_6$H$_4$CO$_2$H → o-Ph$_2$GeO$_2$CC$_6$H$_4$O (1477).

Additional unsubstituted diorganogermanium alkoxides are listed in Table 50.

4.3.3.2.6 FUNCTIONALLY SUBSTITUTED DIORGANOGERMANIUM DIALKOXIDES

Diorganogermanium alkoxides containing halogen, oxygen and nitrogen in the organo group or the alkoxide moiety are listed at the top of Table 51.

4.3.3.2.8 DIORGANOGERMANIUM DIALKOXIDES DERIVED FROM DIHYDRIC ALCOHOLS

All compounds are listed in Table 51.

Table 50. Unsubstituted Diorganogermanium Alkoxides $R_2Ge(OR)_2$

Formula	Synth. Method*	Yield	Properties	Ref.
$Me_2Ge(OPr)_2$	IIA	43	b_{755} 174°, n_D^{20} 1.4212, d_{20} 1.0658	2192
	IIB (a)	86	Anal. detn. of PrO- (266)	2192
$Me_2Ge(OCHMe_2)_2$	--	--	b_{92} 92-93°, n_D^{25} 1.4097, GLC (b)	1450
$Me_2Ge(OCH_2CH:CH_2)_2$	IIB (a)	75	b_{30} 92°, n_D^{20} 1.4470, d_{20} 1.1278	2192
$Me_2Ge(OCMe_3)_2$	IIB (a)	73	b_{13} 70-72°, n_D^{20} 1.4220, d_{20} 1.0181	2192
$Pr_2Ge(OEt)_2$	IIA	74	b_3 61°, n_D^{25} 1.4315, d_{25} 1.0419 (c-e)	2837
$Bu_2Ge(OR)_2$				
R = Me	IIB (a)	84	b_{13} 113-16°, n_D^{20} 1.4365, mol. wt. (f)	337
	VA	25	$b_{7.5}$ 103-106°, n_D^{20} 1.4365, mol. wt.	1504
Et + CMe_3	IA	80	$b_{5.5}$ 123-25°, n_D^{20} 1.4370, mol. wt.	1504
n-Pr	IIB (a)	80	b_5 127°, n_D^{20} 1.4357, monomeric	337
	VA	78	$b_{3.8}$ 117-20°, n_D^{20} 1.4355, mol. wt.	1504
n-Bu	IIB (a)	75	$b_{0.5}$ 106-108°, n_D^{20} 1.4388, monomeric	337
	VA	80	$b_{0.01}$ 92-95°, n_D^{20} 1.4380, mol. wt.	1504
	VI	88	--	338
CH_2CHMe_2	IIB (a)	62	b_2 114-16°, n_D^{20} 1.4390, monomeric	337
CHMeEt	IIB (a)	50	b_2 110°, n_D^{20} 1.4373, monomeric	337
	VA	76	b_3 116-21°, n_D^{20} 1.4350, mol. wt.	1504
	VI	81	--	338
CHMePr	IIB (a)	97	b_4 142-45°, n_D^{20} 1.4400, monomeric	337
CMe_2Et	IIB (g)	52	b_5 141°, n_D^{20} 1.4452, monomeric	337
	VI	65	b_1 108°	337

* The characters used here correspond to the synthetic scheme in the introduction to Chapter 4.3.3.1.1. (a) Synth. with NH_3. (b) Anal. detn. by volumetric titration after rxn. with Ac_2O or NH_4Br (1450). (c) IR spectrum, basicity (2837). (d) Anal. detn. by Schoeninger combustion and polarography (2710). (e) Dipole moment (2836). (f) Rxn. with $Y(OH)_2 \rightarrow Bu_2GeOYO$; Y = $CHMeCH_2CMe_2$ (338). (g) Synth. with NH_3-Py.

Table 50 Continued $R_2Ge(OR)_2$

Formula	Synth. Method*	Yield	Properties	Ref.
$Ph_2Ge(OR)_2$				
R = Me	IIB (a)	91	$b_{0.5}$ 119-22°, n_D^{20} 1.5495, mol. wt.	1503
CHMe$_2$	IIB (a)	84	$b_{0.7}$ 129-30°, n_D^{20} 1.5247, mol. wt.	1503
	VA	80	$b_{0.6}$ 122-25°, n_D^{20} 1.5230, mol. wt.	1504
n-Bu	IIB (a)	91	$b_{0.6}$ 154-55°, n_D^{20} 1.5240, mol. wt.	1503
CHMeEt	IIB (a)	75	$b_{0.2}$ 124-25°, n_D^{20} 1.5130, mol. wt.	1503
CMe$_3$	IIB (a)	65	$b_{2.5}$ 152-53°, n_D^{20} 1.5210, mol. wt.	1503
	VA	40	$b_{0.4}$ 125°, n_D^{20} 1.5240, mol. wt.	1504
$c.C_6H_{11}$	IIIB	68	m. 57-58° (h)	1762

(h) By-prod. in rxn. of Ph_2HGeCl with $c.C_6H_{10}O$ at 170°, at 80° with cat. AIBN and with $ZnCl_2$ at 130°, in 32, 35 and 21% yield (1762).

Table 51. Functionally Substituted Diorganogermanium Alkoxides $R_2Ge(OR')_2$

Formula*	Method**	Yield	Properties	Ref.
$Me_2Ge(OCH_2CH_2Cl)_2$	IIIA	--	b_{28} 150-53°, n_D^{20} 1.4695, d_{20} 1.3842 (a)	2196
$(MeO_2CCH_2)_2Ge(OMe)_2$	IC	61	b_2 101-102°, n_D^{20} 1.4630, d_{20} 1.3683 (b)	692
$Ph_2Ge(OC_9H_6N-8)_2$	IA (c)	61	Dec. > 140°, yellow cryst. IR spectrum	232a, 1215

Bu$_2$GeOYO

Y =				
CH$_2$CH$_2$	IVC	73	b$_4$ 106°, n$_D^{20}$ 1.4712	338
	IVC (d)	76	--	338
(CH$_2$)$_3$	IVB	60	b$_{25}$ 98-103°, n$_D^{20}$ 1.4653, mol. wt.	1504
CH$_2$CHMe	IVC	81	b$_4$ 108°, n$_D^{20}$ 1.4705, monomeric	338
(CH$_2$)$_4$ (110)	IVB	60	b$_{3.5}$ 106-109°, n$_D^{20}$ 1.4540, mol. wt.	1504
CH$_2$CH$_2$CHMe	IVC	81	b$_4$ 111°, n$_D^{20}$ 1.4725, monomeric	338
CHMeCHMe	IVC	85	b$_{1.5}$ 94°, n$_D^{20}$ 1.4555, monomeric	338
(CH$_2$CH$_2$)$_2$O	IVC (d)	86	b$_3$ 105°, n$_D^{20}$ 1.4550, monomeric	338
	IVC	66	b$_3$ 122°, n$_D^{20}$ 1.4695, monomeric	338
(CH$_2$CH$_2$)$_2$NH	IVB	75	b$_{0.6}$ 127-30°, n$_D^{20}$ 1.4750, mol. wt.	1504
CHMeCH$_2$CMe$_2$	IVC (e)	88	b$_3$ 106-108°, n$_D^{20}$ 1.4500	338
	IVC	85	--	338
	IVC (d)	82	b$_4$ 116-18°, monomeric	338
(CH$_2$CH$_2$)$_2$NCH$_2$CH$_2$OH	IVB	82	b$_{0.1}$ 136°, n$_D^{20}$ 1.4810, mol. wt.	1504

Ph$_2$GeOYO

Y =				
CH$_2$CH$_2$	IVB	50	b$_{0.1}$ 140-42°, mol. wt.	1504
	IVC	80	b$_{0.6}$ 140°, mol. wt.	1505
(CH$_2$)$_3$	IVC	85	b$_{0.6}$ 150°	1505
CH$_2$CHMe	IVC	78	b$_{0.6}$ 127°, mol. wt.	1505
(CH$_2$)$_4$	IVC	85	b$_{0.2}$ 142°, mol. wt.	1505
CH$_2$CH$_2$CHMe	IVC	81	b$_{0.8}$ 135°, n$_D^{20}$ 1.5665, mol. wt.	1505
CHMeCHMe	IVC	85	b$_{0.6}$ 127°, n$_D^{20}$ 1.5680, mol. wt.	1505
(CH$_2$CH$_2$)$_2$O	IVC	70	b$_{0.2}$ 155°, mol. wt.	1505
(CH$_2$)$_5$	IVC	78	b$_{0.2}$ 138°	1505
(CH$_2$)$_6$	IVC	82	b$_{0.2}$ 129°, n$_D^{20}$ 1.5570, mol. wt.	1505
(CMe$_2$)$_2$	IVC	85	b$_{0.3}$ 133°, n$_D^{20}$ 1.5595	1505

* Numbers in parenthesis refer to pages in main volume.
** The characters used here correspond to the synthetic scheme in the introduction to Chapter 4.3.3.1.6.
(a) IR spectrum, hydrolyzes in air (2196). (b) NMR spectrum (1689). (c) From Ph$_2$GeCl$_2$ and 8-hydroxyquinoline in EtOH with NH$_3$ (232a, 1215). (d) From Bu$_2$Ge(OCHMe$_2$)$_2$ (338). (e) From Bu$_2$Ge(OMe)$_2$ (338).

4.3.3.2.9 DIORGANOGERMANIUM MIXED OXIDES

Bis(trimethylsiloxy)dimethylgermane $GeSi_2C_8H_{24}O_2$ $Me_2Ge(OSiMe_3)_2$
(M-111)
Synth.: By-prod. in rxn. of $Me_2Ge(N:PMe_3)N_3$ with Me_3SiOLi in Et_2O (2734).

Bis(chloromethyldimethylsiloxy)chloro- $GeSi_2C_8H_{22}Cl_2O_2$ $Me(ClCH_2)GeO-$
methylmethylgermane $(SiMe_2CH_2Cl)_2$
Synth.: By rxn. of $Me(ClCH_2)GeCl_2$ with $Me_2(ClCH_2)SiOLi$ in 68% yield (2129).
Prop.: b_{12} 147°, NMR spectrum (2129).

Other diorganogermanium mixed oxides are listed in Subchapter 4.3.3.4.

4.3.3.3 MONOORGANOGERMANIUM TRIALKOXIDES

4.3.3.3.1 UNSUBSTITUTED ORGANOGERMANIUM TRIALKOXIDES

Methylgermanium Trimethoxide (M-107) $GeC_4H_{12}O_3$ $MeGe(OMe)_3$
Prop.: NMR spectrum (926, 1012, 1545-6), dipole moment (929).
Rxn. with: $MeGeX_3 \rightleftharpoons MeGe(OMe)_nX_{3-n}$, equilibrium const., X = Cl, Br, I,
n = 0-3 (545).
$MeSiX_3 \rightleftharpoons MeGe(OMe)_nX_{3-n} + MeSi(OMe)_nX_{3-n}$, equilibrium const., X = Cl, Br, I,
n = 0-3 (1546).
$Me_2GeBr_2 \rightleftharpoons$ equilibrium const. for redistribution rxn., prod. distribution
(2606).

Methylgermanium Triethoxide (M-107) $GeC_7H_{18}O_3$ $MeGe(OEt)_3$
Synth.: By rxn. of $MeGeCl_3$ with NaOR at 50-70°, in 74% yield (2192), in petr.
ether, in 70% yield (2837).
Prop.: b_{737} 165° (2192), b_3 43-44° (2837), n_D^{20} 1.4128 (2192), n_D^{25} 1.4102
(2837), d_{20} 1.1280 (2192), d_{25} 1.1317 (2837), IR (2837) and NMR (1012) spectra,
basicity (2837), dipole moment (2836). Anal. detn. by fusion and volumetric
titration (265), by Schoeninger combustion and polarography (2710), of EtO-
group (266).

Methylgermanium Triphenoxide $GeC_{19}H_{18}O_3$ $MeGe(OPh)_3$
Synth.: By rxn. of $MeGeCl_3$ with PhONa (277), with PhOH and NH_3 in C_6H_6, in
86% yield (2192).
By rxn. of $MeGeBr_3$ with PhOH and Et_3N in C_6H_6 (2608).
Prop.: b_3 217° (2192), $b_{0.5}$ 169-70° (277), $b_{0.45}$ 168° (2608), n_D^{20} 1.5835, d_{20}
1.3010 (2192).
Rxn. with: HCl → $MeGeCl_3$ (277).
$Me_2GeX_2 \rightleftharpoons$ * prod. distribution; X = Cl, Br, I (2606).
$MeGeCl_3 + MeGeBr_3 + MeGeI_3 \rightleftharpoons$ * prod. distribution (2608)

Additional unsubstituted monoorganogermanium trialkoxides are listed in Table 52.

---
* Equilibrium constants for redistribution rxn. measured by NMR spectroscopy.

Table 52. Unsubstituted Monoorganogermanium Trialkoxides   RGe(OR')$_3$

Formula*	Synth. Method**	Yield	Properties	Ref.
MeGe(OPr)$_3$	IIA	61	b_{15} 95-96°, n_D^{20} 1.4222, d_{20} 1.7014	2192
MeGe(OCH$_2$CH:CH$_2$)$_3$	IIB (a)	76	b_7 91-94°, n_D^{20} 1.4540, d_{20} 1.1427	2192
EtGe(OMe)$_3$ (109)	IIA	77	b_{761} 154°, n_D^{20} 1.4178, d_{20} 1.2446 (b)	2192
EtGe(OEt)$_3$ (109)	IIA	61	b_{761} 180°, n_D^{20} 1.4178, d_{20} 1.1105	2192
EtGe(OPr)$_3$	IIB (a)	77	b_{25} 118.5-20°, n_D^{20} 1.4258, d_{20} 1.0563	2192
PrGe(OMe)$_3$	IIA	83	b_{737} 169°, n_D^{20} 1.4185, d_{20} 1.2024	2192
PrGe(OEt)$_3$	IIA	68	b_3 55-56°, n_D^{25} 1.4239, d_{25} 1.1225 (c, d)	2837
BuGe(OR)$_3$				
R = Me	IIB (a)	60	b_9 76-78°, n_D^{20} 1.4245, mol. wt. (e)	1506
Et	IIB (a)	90	b_8 90-93°, n_D^{20} 1.4330, mol. wt. (e, f)	1506
n-Pr	VI (f)	80	$b_{2.5}$ 95-97°, n_D^{20} 1.4250, mol. wt.	1506
CHMe$_2$	IIB (a)	85	$b_{7.5}$ 94-95°, n_D^{20} 1.4151, mol. wt. (e)	1506
n-Bu	VA	70	$b_{0.5}$ 111-14°, n_D^{20} 1.4310, mol. wt.	1506
	VI (f)	81	$b_{0.3}$ 109-11°, n_D^{20} 1.4310, mol. wt.	1506
CHMeEt	VI (f)	98	b_1 96-98°, n_D^{20} 1.4260, mol. wt.	1506
CMe$_3$	VI (f)	66	$b_{4.5}$ 99-100°, n_D^{20} 1.4280, mol. wt.	1506

* Numbers in parenthesis refer to pages in main volume.
** The characters used here refer to the synthetic scheme in the introduction to Chapter 4.3.3.1.1.
(a) Synth. with NH$_3$. (b) Anal. detn. by fusion and volumetric titration (265). (c) IR spectrum and basicity (2837), dipole moment (2836). (d) Anal detn. by Schoeninger combustion and polarography (2710). (e) Sensitive to moisture (1506). (f) Rxn. of BuGe(OEt)$_3$ with ROH and cat. MeC$_6$H$_4$SO$_3$H → BuGe(OR)$_3$ + EtOH; R = Pr, n-, s-, t-Bu (1506).

4.3.3.3.6 FUNCTIONALLY SUBSTITUTED MONOORGANOGERMANIUM TRIALKOXIDES

All compounds are listed in Table 53.

Table 53. Functionally Substituted Monoorganogermanium $RGe(OR')_3$ Trialkoxides

Formula	Synth. Method*	Yield	Properties	Ref.
$MeGe(OCH_2CH_2Cl)_3$	IIIA	--	b_2 134-38°, n_D^{20} 1.4795, d_{20} 1.4557 (a)	2196
$MeGe(OCH_2CH_2)_3N$	IVB	--	m. 158-59°, soly.	2085
	IA (b)	85	m. 158-59°	2847
$ClCH_2Ge(OEt)_3$	--	--	IR and NMR spectra (2244)	--
$EtGe(OCH_2CH_2)_3N$	IA (b)	70	m. 146-47° (c)	2847
$MeO_2CCH_2Ge(OMe)_3$	IC	89	b_1 66-67°, n_D^{20} 1.4420, d_{20} 1.3530	692
	IC (d)	67	b_1 67-68°, n_D^{20} 1.4400, d_{20} 1.3596 (e)	692, 1440
$PhGe(OCH_2CH_2)_3N$	IVB	89	m. 232-32.5°, NMR spectrum	2085, 2847
$PhGe(OCHEtPO_3Et_2)_3$	IIIC	--	--	452
$PhGe(OR)(OR')(OR'')$ (?)	IIC	--	R = $CH_2CHClMe$, R' = CH-(C_4H_3O)PPh_2O_2Et, R'' = $CHEtPO_3Et_2$	452
$1-C_{10}H_7Ge(OCH_2CH_2)_3N$	IVB	84	m. 262-63.5°	2847

* The characters used here correspond to the synthetic scheme in the introduction to Chapter 4.3.3.1.6.
(a) IR spectrum (2196), moist air → $(Me_2Ge_2O_3)_x$ (2196). (b) From $RGeBr_3$ and $(HOCH_2CH_2)_3N$ in xylene (2847). (c) X-ray structural analysis, structure: 1-ethylgermatrane: (2254).

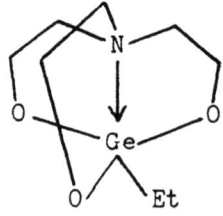

(d) Synth. from $(MeO)_3GeCl$ and $Bu_3SnCH_2CO_2Me$ (692, 1440). (e) NMR spectrum (1689).

4.3.3.4 ORGANOGERMANIUM ALKOXIDES CONTAINING GERMANIUM-HETEROATOM BONDS

Organogermanium alkoxides which contain germanium bonded to oxygen, sulfur and nitrogen are listed in Table 54. The compounds are prepared by methods from the following scheme.

I Reactions with alkali metal derivatives:
   (IA) $Me_2Ge(OPh)_2 + (Me_3Si)_2NNa \rightarrow Me_2Ge(OPh)N(SiMe_3)_2$,
   (IB) $Me_2Ge(N_3)X' + Me_3SiOLi \rightarrow Me_2Ge(OSiMe_3)X'$.

II Alcoholysis of organogermanium halides:
   (IIA) $MeGeCl_3 + ROH + NH_3 \rightarrow [MeGe(OR)_2]_2NH$.
   (IIB) $Ph_2GeBr_2 + HOCH_2CH_2SH \rightarrow Ph_2GeSCH_2CH_2O$.

III Interconversion of organogermanium alkoxides:
   (IIIA) $Bu_3GeOEt + HOYCO_2H \rightarrow Bu_3GeOYCO_2GeBu_3$,
   (IIIB) $R_2Ge(OEt)_2 + RCO_2H \rightarrow R_2Ge(O_2CR)OEt$,
   (IIIC) $R_2Ge(OEt)_2 + HOY'H \rightarrow R_2GeOY'$.

R = organic group, X' = N:PMe$_3$, Y = hydrocarbon bridge, Y' = functionally substituted hydrocarbon bridge.

4.3.4 ORGANOGERMANIUM ACETALS AND ENOLATES

Organogermanium acetals and enolates are listed in Table 55. It is possible indeed that compounds listed in Chapters 2.6.3.3 of the main volume and this supplement exist partly in the tautomeric enolate form. The acetals and enolates of this chapter are prepared by methods from the following scheme.

I Organometallic reactions:
   (IA) $Me_3GeBr + Pr_3SnCH_2Ac \rightarrow Me_3GeOCMe:CH_2$,
   (IB) $Et_3GeCl + 1-Et_3SnOC_6H_9 \rightarrow 1-Et_3GeOC_6H_9$.

II Hydrogermylation reaction:
   $R_3GeH + Ph_2C:CO \rightarrow R_3GeOCH:CPh_2$.

III Addition reactions with organogermanium heterocompounds:
   (IIIA) $Et_3GeOMe + CH_2:CO \rightarrow Et_3GeOC(:CH_2)CH_2CO_2Me$,
   (IIIB) $Bu_3GeOMe + CCl_3CHO \rightarrow Bu_3GeOCH(OMe)CCl_3$,
   (IIIC) $Et_3GeNMe_2 + Ph_2C:CO \rightarrow Et_3GeOC(:CPh_2)NMe_2$,
   (IIID) $Et_3GePEt_2 + RCH:CHCHO \rightarrow Et_3GeOCH:CHCHRPEt_2$.

IV Formation of acetonylacetonates, catalyzed by $MeC_6H_4SO_3H$:
   (IVA) $(Bu_3Ge)_2O + CH_2Ac_2 \rightarrow Bu_3GeOCMe:CHAc$,
   (IVB) $(R_2GeO)_x + CH_2Ac_2 \rightarrow R_2Ge(OCMe:CHAc)_2$.

Table 54. Organogermanium Alkoxides Containing Germanium-Heteroatom Bonds

Formula	Synth. Method*	Yield	Properties	Ref.
$Me_2Ge(OPh)N(SiMe_3)_2$	IA	--	b_1 135°, viscous liq., IR spectrum	277
$Me_2Ge(OSiMe_3)N:PMe_3$	IB	68	m. -12-10°, $b_{0.2}$ 41-42°, mol. wt.	2743
$[MeGe(OCMe_3)_2]_2NH$	IIA	40	b_{17} 152-54°, n_D^{20} 1.4408, d_{20} 1.1086	2192
$MeO(Et_2GeS)_xEt_2GeOMe$	--	--	Insecticide, functional fluid	2607
$Bu_3GeOCHPhCO_2GeBu_3$	IIIA	87	$b_{0.2}$ 160°, n_D^{20} 1.5065, mol. wt.	1477
$o-Bu_3GeOC_6H_4CO_2GeBu_3$	IIIA	70	$b_{0.4}$ 203°, n_D^{20} 1.4875, mol. wt.	1477
$Bu_2Ge(OAc)OEt$	IIIB	70	b_4 110°, n_D^{20} 1.4400, mol. wt.	1477
$Bu_2Ge(OBz)OEt$	IIIB	60	$b_{0.2}$ 140°, n_D^{20} 1.4955, mol. wt.	1477
$Bu_2GeO_2CCHPhO$	IIIC	90	Subl. o.4 150-160°, soly.	1477
$o-Bu_2GeO_2CC_6H_4O$	IIIC	85	$b_{0.5}$ 158°, n_D^{20} 1.5395, mol. wt.	1477
$Bu_2GeSCH_2CH_2O$	IIIC	95	$b_{0.2}$ 80°, mol. wt.	2805
$Bu_2GeSCH_2CHOCH_2OH$	IIIC	94	$b_{1.5}$ 152°, mol. wt., IR spectrum	2805
$Bu_2GeNHCH_2CH_2O$	IIIC (b)	50	b_2 121-22°, n_D^{20} 1.4650, mol. wt.	1504
$Ph_2Ge(OAc)OEt$	IIIB	70	$b_{0.4}$ 140°, n_D^{20} 1.5557, mol. wt.	1477
$Ph_2Ge(OBz)OEt$	IIIB	40	$b_{0.2}$ 180°, n_D^{20} 1.5600	1477
$o-Ph_2GeO_2CC_6H_4O$	IIIC	95	Mol. wt.	1477
$Ph_2GeSCH_2CH_2O$	IIB	17	m. 143-45°, IR spectrum	121
$Ph_2GeNHCH_2CH_2O$	IIIC (b)	--		1504

* The characters used here correspond to the synthetic scheme in the introduction to Chapter 4.3.3.4. (a) IR and NMR spectra (2743). (b) Synth. from $(R_2GeO)_x$ in presence of cat. $MeC_6H_4SO_3H$ (1504).

Table 55. Organogermanium Acetals and Enolates $R_nGe(OR')_{4-n}$

Formula	Synth. Method*	Yield	Properties	Ref.
Bu₃GeOCH(OMe)CCl₃	IIIB	--	--	1228
Me₃GeOCMe:CH₂	IA	--	IR and NMR spectra (a)	2547
Et₃GeOR'				
R' = CH:CPh₂	II	80	$b_{0.09}$ 139°, n_D^{20} 1.5740, d_{20} 1.1411 (b, c)	2714
C(:CPh₂)NMe₂	IIIC	(d)	IR and NMR spectra (e)	2736
CH:CHCHPhEt₂	IIID	75	$b_{0.15}$ 113°, IR and NMR spectra (f, g)	2735
CH:CHCHMeEt₂	IIID	75	$b_{0.15}$ 93-95°, IR and NMR spectra (f, h)	2735
C(:CH₂)CO₂Me	IIIA	69	b_1 81-83°, n_D^{20} 1.4615, d_{20} 1.1272 (i)	1440
1-C₆H₉c.	IB	65	$b_{2.5}$ 93-94°, n_D^{20} 1.4820, d_{20} 1.1084 (j)	1709
Bu₃GeOCMe:CHAc	IVA	73	Yellow liq. $b_{1.8}$ 166-68°, n_D^{20} 1.4670, mol. wt.	1504
Bu₂Ge(OCMe:CHAc)₂	IVB	--	Orange-red liq.	1504
Ph₃GeOCH:CPh₂	II	--	$b_{0.01}$ 240-45°, n_D^{20} 1.6410 (b, c)	2714
Ph₂Ge(OCMe:CHAc)₂	IVB	40	Light yellow low metling solid, $b_{0.6}$ 201° (k)	1504

* The characters used here correspond to the synethetic scheme in the introduction to Chapter 4.3.4.
(a) Equilibrium Me₃GeCH₂Ac ⇌ Me₃GeOCMe:CH₂ at r.t. and 170° yielding 4 and 17% O-isomer, resp. (2547), isolated by vacuum fractionation (2547). (b) IR and NMR spectra (2714). (c) Rxn. with LiAlH₄ → R₃GeH + Ph₂CHCH₂OH (2714). (d) By-prod. (e) Not isolated, by-prod. in Et₃GeCPh₂CONMe₂ (2736). (f) Rxn. with H₂O → (Et₃Ge)₂O + Et₂PCHRCH₂CHO; R = Me, Ph (2735). (g) Isomer ratio cis: trans = 3 (2735). (h) Isomer ratio cis: trans = 2 (2735). (i) IR and NMR spectra (1440). (j) IR and NMR spectra (1709). (k) Molecular weight (1504).

4.3.5 ORGANOGERMANIUM CARBOXYLATES

4.3.5.1 TRIORGANOGERMANIUM CARBOXYLATES

4.3.5.1.1 UNSUBSTITUTED TRIORGANOGERMANIUM CARBOXYLATES

Unsubstituted organogermanium carboxylates listed in Table 56 are prepared by methods from the following scheme.

I Cleavage reaction with tetraorganogermane:
$Et_2GeCH_2CHOHCHMeCH_2 + AcOH \rightarrow Et_2(CH_2:CHCHMeCH_2)GeOAc$.

II Reaction with organometallic derivatives:
 (IIA) $(PhCH_2)_3GeLi + CO_2 \rightarrow (PhCH_2)_3GeO_2CCH_2Ph$,
 (IIB) $(R_3Ge)_2Cd$ with $Bz_2O_2 \rightarrow R_3GeOBz$.

III Metathetic substitution reactions:
 (IIIA) $Me_3GeCl + RCO_2K \rightarrow Me_3GeO_2CR$,
 (IIIB) $Ph_3GeBr + RCO_2Ag \rightarrow Ph_3GeO_2CR$,
 (IIIC) $Bu_3GeOEt + RCO_2H \rightarrow Bu_3GeO_2CR$.

IV Interconversion or exchange reaction:
$R_3GeOAc + RCO_2H \rightarrow R_3GeO_2CR + AcOH$.

Triethylgermanium Acetate (M-113) $GeC_8H_{18}O_2$ Et_3GeOAc
Synth.: By rxn. of AcOH with $(Et_3Ge)_2Cd$ at r.t., in 75% yield (587), with $Et_3GeHgEt$ at 100°, in 83% yield (2858), with $(Et_3Ge)_2Hg$ (291), at 130-40°, in 47% yield (2086).
By rxn. of AcOH with $(Et_3Ge)_nBiEt_{3-n}$ (291), n = 3 at 170°, in 69% yield (2086).
Prop.: b_{60} 99-100°, n_D^{20} 1.4430 (587), empirical equation for calculation of boiling point (680).
Biol. prop.: Toxicity (100), activity as fungicide and bactericide (514).

Triethylgermanium Benzoate (M-113) $GeC_{13}H_{20}O_2$ Et_3GeOBz
Synth.: By rxn. of Bz_2O_2 with $Et_3GeZn(Et_2Ge)_xEt$ (584); x = 0, in 64% yield (2854).
By rxn. of $(Et_3Ge)_3Tl$ with Bz_2O_2 in 63% yield (1361, 2089).
By rxn. of $(Et_3Ge)_nBiEt_{3-n}$ with Bz_2O_2 (291); n = 3 at 15-20°, in 38% yield (2086).
By rxn. of $(Et_3Ge)_2Te$ and $Et_3GeTeEt$ with Bz_2O_2 in C_6H_6 in inert atm., in 68 and 59% yield, resp. (2093).
Prop.: b_3 113-15° (584), $b_{1.5}$ 111-15° (2093), n_D^{20} 1.5111 (2093), 1.5110 (584).

Tripropylgermanium Acetate (M-114) $GeC_{11}H_{24}O_2$ $n-Pr_3GeOAc$
Prop.: b_{14} 115° (2697), empirical equation for boiling point calculation (680).
Rxn. with: $RCO_2H \rightarrow Pr_3GeO_2CR + AcOH$; R = Et, i-Pr, t-Bu (2697).

H_2O-dioxane → relative rate of hydrolysis, rxn. mechanism (1714, 2697).
Biol. prop.: Toxicity (100), activity as fungicide and bactericide (514).

Tributylgermanium Acetate (M-115) $GeC_{14}H_{20}O_2$ Bu_3GeOAc
Synth.: By rxn. of Bu_3GeOEt with AcOH in C_6H_6, in 95% yield (1477).
By rxn. of Bu_3GeNMe_2 with Ac_2O (466).
Prop.: $b_{0.2}$ 80°, n_D^{20} 1.4635, mol. wt. (1477), empirical equation for boiling point calculation (680).
Biol. prop.: Toxicity (100), activity against Mycobacterium phlei, Streptococcus lactis, Leuconostoc mesenteroides (1926).

Other unsubstituted triorganogermanium carboxylates are listed in Table 56.

4.3.5.1.6 FUNCTIONALLY SUBSTITUTED TRIORGANOGERMANIUM CARBOXYLATES

Triorganogermanium carboxylates containing halogen, oxygen or nitrogen are listed in Table 57. The compounds are prepared by methods from the following scheme.

I   Cleavage reactions with tetraorganogermanes:
     (IA) $R_3GeR' + RCO_2H \rightarrow R_3GeO_2CR$,
     (IB) $R_2GeY + R'CO_2H \rightarrow R_2(HY)GeO_2CR$,
     (IC) $Me_2\overline{GeCPh:CPhCPh:C}Ph$ + excess $(:CCO_2Me)_2 \rightarrow Me_2\overline{GeO_2CCH:C}CO_2Me$.

II   Metathetic substitution reactions:
     (IIA) $R_nGeX_{4-n} + AgO_2CR' \rightarrow R_nGe(O_2CR')_{4-n}$,
     (IIB) $(Et_3Ge)_2NH + (CH_2CH_2H)_2 \rightarrow (Et_3GeO_2CCH_2)_2$,
     (IIC) $R_nGe(OEt)_{4-n} + RCO_2H \rightarrow R_nGe(O_2CR)_{4-n}$.

III   Insertion reactions:
     (IIIA) $R_3GeNEt_2 + \overline{OCH_2CH_2C}O \rightarrow R_3GeO_2CCH_2CH_2NEt_2$,
     (IIIB) $Et_3GeNMe_2 + CO_2 \rightarrow Et_3GeO_2CNMe_2$.

R = organic group, may be functionally substituted, R' = functionally substituted organic group, X = halogen, Y = hydrocarbon bridge, n = 1-3. Other functionally substituted triorganogermanium carboxylates are listed in Subchapter 2.6.3.4.

Table 56. Unsubstituted Triorganogermanium Carboxylates R_3GeO_2CR

Formula*	Synth. Method**	Yield	Properties	Ref.
Me_3GeOAc (113)	IIIA	87	b. 129° n_D^{20} 1.4215, d_{20} 1.1961 (a, b)	367
$(PhCH_2)_3GeO_2CCH_2Ph$	IIA	(c)	m. 146-48°, colorless cryst.	104
Et_3GeO_2CH (113)	--	--	(d)	--
Et_3GeO_2CEt (113)	--	--	Anal. detn. (d)	265
Et_3GeO_2CPr-n (113)	--	--	(d)	--
Et_3GeO_2CBu-n (113)	--	--	(d)	--
$Et_2(CH_2:CHCHMeCH_2)GeOAc$	I	--	b_{16} 104°, n_D^{20} 1.4617, d_{20} 1.0885	2578
Et_3GeO_2CR	--	--	(e)	--
n-Pr_3GeO_2CH (114)	--	--	(d)	--
n-Pr_3GeO_2CEt (114)	IV	--	b_{10} 111°, (d, f-2697)	2697
n-$Pr_3GeO_2CCHMe_2$	IV	--	b_{10} 109° (f-2697)	2697
n-$Pr_3GeO_2CCMe_3$	IV	--	b_{10} 120° (f-2697)	2697
i-Pr_3GeOAc (114)	--	--	(d, f-1714)	--
i-Pr_3GeO_2CEt (114)	--	--	(d)	--
i-Pr_3GeO_2CPr-n (115)	--	--	(d)	--
i-Pr_3GeO_2CBu-n (115)	--	--	(d)	--
i-Pr_3GeO_2CR	--	--	(e)	--
n-Bu_3GeO_2CH (115)	--	--	(d)	--
n-$Bu_3GeO_2CC_6H_{13}$-n (115)	--	--	(d)	--
n-Bu_3GeOBz	IIIC	87	$b_{1.5}$ 168°, n_D^{20} 1.4930, mol. wt.	1477
(n-$C_5H_{11})_3GeOAc$ (115)	--	--	Microanal. detn. (426)	--
(i-$C_5H_{11})_3GeOAc$ (115)	--	--	Microanal. detn. (426)	--
(n-$C_6H_{13})_3GeOAc$ (116)	--	--	(f-1714)	--
(c.$C_6H_{11})_3GeOAc$ (116)	--	--	(f-1714)	--
n-Bu_3GeOBz	IIIC	87	$b_{1.5}$ 168°, n_D^{20} 1.4930, mol. wt.	1477
Ph_3GeOAc (116)	IIIB	50	m. 83°, soly., mol. wt. (f-1714, g, h)	1970
Ph_3GeOBz	IIIB	60	m. 112-13°, soly., mol. wt. (h, i)	1970
	IIB	--	m. 107°, soly.	2295
$Ph_3GeO_2CC_6H_4Me$-p	--	--	(i)	--
(p-$MeC_6H_4)_3GeOBz$	IIIB	35	m. 160°, soly., mol. wt. (h)	1970

* Numbers in parenthesis refer to pages in main volume.
** The characters used here correspond to the synthetic scheme in the introduction to Chapter 4.3.5.1.1.

(a) IR and Raman spectra (367). (b) Toxicity (100). (c) By-prod. (d) Empirical equation for boiling point calculation (680). (e) Ge-bond refraction D_{20}.(81). (f) Rxn. with H_2O-dioxane → relative rate of hydrolysis, rxn. mechanism (1714, 2697). (g) Microanalytical detn. (426). (h) IR spectrum (1970). (i) Mass spectrum (2282).

Table 57. Functionally Substituted Triorgano- R_3GeO_2CR'
germanium Carboxylates

Formula*	Synth. Method**	Yield	Properties	Ref.
$Me_3GeO_2CCF_3$	IA (a)	--	n_D^{20} 1.3820, IR spectrum, GLC	1041, 1042
$Me_3GeO_2CCH_2CH_2NEt_2$	IIIA	67	b_3 86.9-87.2°, IR spectrum (b)	1238
$Me_2GeO_2CCH:CCO_2Me$	IC	25	m. 127-28°, white solid, mol. wt. (c)	2882, 2883
Et_3GeO_2CR'				
R' = NMe_2	IIIB	90	b_5 115°, n_D^{20} 1.4529, d_{20} 1.0906	466
CF_3 (117)	--	--	(d)	--
CH_2Cl (117)	--	--	(d)	--
$CHCl_2$ (117)	--	--	(d)	--
CH_2Br (117)	--	--	(d)	--
C_2F_5 (117)	--	--	(d)	--
CH_2CH_2Cl (117)	--	--	(d)	--
$CHClMe$ (117)	--	--	(d)	--
C_3F_7 (117)	--	--	(d)	--
$(Et_3GeO_2CH_2)_2$	IIB	100	$b_{0.25}$ 130-31°, n_D^{20} 1.4679	1796
$(Et_2(CH_2:CHCHMeCH_2GeO_2CCH_2Cl$	IB (e)	--	b_{23} 168°, n_D^{20} 1.4793, d_{20} 1.2460	2578

* Numbers in parenthesis refer to pages in main volume.
** The characters used here correspond to the synthetic scheme in the introduction to Chapter 4.3.5.1.6.
(a) R' = 2-c.C_6H_9 or 2-c.C_8H_{13} (1041-2). (b) Kinetic data and rxn. mechanism for synth. (1236-7), rxn. with HCl in Et_2O → $Et_2NCH_2CH_2CO_2H\cdot HCl$ (1238). (c) IR, UV, NMR and mass spectra, GLC (2882-3). (d) Empirical equation for boiling point calculation (680). (e) From $Et_2GeCH_2CHOHCHMeCH_2$ (2578).

Table 57 Continued   R_3GeO_2CR'

Formula*	Synth. Method**	Yield	Properties	Ref.
n-Pr$_3$GeO$_2$CCF$_3$ (118)	--	--	(d)	--
n-Pr$_3$GeO$_2$CCH$_2$Cl (118)	--	--	(d)	--
n-Pr$_3$GeO$_2$CCHCl$_2$ (118)	--	--	(d)	--
n-Pr$_3$GeO$_2$CCH$_2$Br (118)	--	--	(d)	--
PrBu$_2$GeO$_2$CCH$_2$Cl	IB (f)	85	b$_{10}$ 161-63°, n$_D^{20}$ 1.4664, d$_{20}$ 1.1262 (g)	1492
i-Pr$_3$GeO$_2$CCF$_3$ (118)	--	--	(d)	--
i-Pr$_3$GeO$_2$CCH$_2$Cl (118)	--	--	(d)	--
i-Pr$_3$GeO$_2$CCHCl$_2$ (118)	--	--	(d)	--
i-Pr$_3$GeO$_2$CCH$_2$Br (118)	--	--	(d)	--
i-Pr$_3$GeO$_2$CCHClMe (118)	--	--	(d)	--
n-Bu$_3$GeO$_2$CCF$_3$ (119)	--	--	(d)	--
n-Bu$_3$GeO$_2$CCH$_2$CH$_2$NEt$_2$	IIIA	67	b$_3$ 96.2-97.2°	2452
Ph$_3$GeO$_2$CCF$_3$	IA (h)	95	m. 120° (sealed tube), subl. $_{0.001}$ 120° (i)	1789
Ph$_3$GeO$_2$CCHOHPh	IIA	60	m. 112-13° (j)	1970
p-Ph$_3$GeO$_2$CC$_6$H$_4$Br	--	--	(k)	--
o-Ph$_3$GeO$_2$CC$_6$H$_4$OH	IIA	55	m. 146-47°, soly. (j)	1970
p-Ph$_3$GeO$_2$CC$_6$H$_4$NO$_2$	--	--	(k)	--

(f) From Bu$_2$Ge$\overline{\text{CH}_2\text{CH}_2\text{CH}_2}$ (1492). (g) IR spectrum (1492). (h) From Ph$_4$Ge (2452). (i) Sensitive to moisture, IR spectrum, thermal dec. at 300° → Ph$_3$GeF + CO + CO$_2$ + CF$_3$COF (1789). (j) IR spectrum, mol. wt. (1970). (k) Mass spectrum (2282).

4.3.5.2 DIORGANOGERMANIUM CARBOXYLATES

All compounds are listed in Table 58 and in Subchapter 4.3.3.4.

4.3.5.3 MONOORGANOGERMANIUM CARBOXYLATES

All compounds are listed in Table 59 on page 194.

Table 58. Diorganogermanium Carboxylates $R_2Ge(O_2CR')_2$

Formula*	Synth. Method**	Yield	Properties	Ref.
$Et_2Ge(O_2CH)_2$ (114)	--	--	(a)	--
$Et_2Ge(OAc)_2$ (114)	--	--	(a)	--
$n\text{-}Pr_2Ge(OAc)_2$ (114)	--	--	(a)	--
$i\text{-}Pr_2Ge(OAc)_2$ (115)	--	--	(a)	--
$i\text{-}Pr_2Ge(O_2CEt)_2$ (115)	--	--	(a)	--
$i\text{-}Pr_2Ge(O_2CPr\text{-}n)_2$ (115)	--	--	(a)	--
$Bu_2Ge(OAc)_2$	IIC	60	b_5 127°, n_D^{20} 1.4475, mol. wt.	1477
$Bu_2Ge(O_2CCHEtBu)_2$	--	--	(b)	--
$Bu_2Ge(OBz)_2$	IIC	80	$b_{0.4}$ 198°, mol. wt.	1477
$Ph_2Ge(OAc)_2$	IIC	60	$b_{0.2}$ 150°	1477
$Ph_2Ge(OBz)_2$	IIC	--	--	1477
$Me_2Ge(O_2CCF_3)_2$	IIA	91	b_{755} 167° (c)	2446
$n\text{-}Pr_2Ge(O_2CCH_2Cl)_2$ (118)	--	--	(a)	--
$i\text{-}Pr_2Ge(O_2CCH_2Cl)_2$ (119)	--	--	(a)	--
$(Bu_2GeO_2CCH_2CH_2S)_3$	IIC (d)	99	Dec. on distn., mol. wt.	2805
$\overline{Bu_2GeO_2CCHMeS}$	IIC (e)	97	$b_{1.5}$ 154°, mol. wt.	2805
$\overline{Ph_2GeO_2CCH_2S}$	IIA (f)	11	m. 80-81°, monomeric	121
	IIA (g)	80	IR spectrum	121

* Numbers in parenthesis refer to pages in main volume.

** The characters used here correspond to the synthetic scheme in the introduction to Chapter 4.3.5.1.6.

(a) Empirical equation for boiling point calculation (680). (b) Microanalytical determination (426). (c) Thermal decomposition at 240° → GeO_2 + $(CF_3CO)_2O$ + CF_3COCl + SiF_4 + CO_2 (2446). (d) From $HSCH_2CH_2CO_2H$ (2805). (e) From $HSCHMeCO_2H$ (2805). (f) From Ph_2GeCl_2 with $HSCH_2CO_2H$ and Py in Et_2O (121). (g) From Ph_2GeCl_2 and $NaSCH_2CO_2Na$ in MePh (121).

Table 59. Monoorganogermanium Carboxylates $RGe(O_2CR')_3$

Formula*	Synth. Method**	Yield	Properties	Ref.
$EtGe(O_2CH)_3$ (114)	--	--	(a)	--
$EtGe(OAc)_3$ (114)	--	--	(a)	--
$EtGe(O_2CEt)_3$ (114)	--	--	(a)	--
$EtGe(O_2CPr-n)_3$ (114)	--	--	(a)	--
$EtGe(O_2CBu-n)_3$ (114)	--	--	(a)	--
$EtGe(O_2CR)_3$	--	--	(b)	--
$i-PrGe(O_2CR)_3$	--	--	(b)	--
$MeGe(O_2CCF_3)_3$	IIA	91	b. 167° (c)	2446
$CF_3Ge(O_2CCF_3)_3$	IIA	82	b. 120°, stable at 240° (d)	2446

* Numbers in parenthesis refer to pages in main volume.
** The characters used here correspond to the synthetic scheme in the introduction to Chapter 4.3.5.1.6.
(a) Empirical equation for boiling point calculation (680). (b) Ge-bond refractions D_{20} (81). (c) Thermal decomposition at 240° → GeO_2 + $(CF_3CO)_2O$ + CF_3COCl + SiF_4 + CO_2 (2446). (d) Addn. compd.: $CF_3Ge(O_2CCF_3)_3 \cdot 2AgO_2CCF_3$, from components in C_6H_6, white to pale yellow solid [$GeAg_2C_{11}F_{18}O_{10}$] (2446).

4.4 ORGANOGERMANES WITH SULFUR, SELENIUM AND TELLURIUM

4.4.1 ORGANOGERMANIUM SULFIDES, SELENIDES AND TELLURIDES

Several new organogermanium tellurides are listed in this chapter. The sulfides, selenides and tellurides in Table 60 are prepared by methods from the following scheme.

I Metathetic substitution reactions:
   (IA) $R_3GeCl + H_2S → (R_3Ge)_2S$,
   (IB) $R_3GeCl + H_2S + Et_3N → (R_3Ge)_2S$,
   (IC) $Ph_3GeBr + Ph_3GeYLi → (Ph_3Ge)_2Y$; Y = S, Se, Te .

II Reaction with organogermanium hydride:
   $R_3GeH + Et_2Te → (R_3Ge)_2Te$.

III Reaction with organogermanium alkoxide:
   $Bu_2Ge(OR)_2 + H_2S → (Bu_2GeS)_2$.

Bistrimethylgermanium Sulfide (M-120) $Ge_2C_6H_{18}S$ $(Me_3Ge)_2S$
Synth.: By rxn. of Me_3GeBr with H_2S and Py, in 53% yield (644).
By rxn. of Me_3GeI with anhydr. Ag_2S (927).

Prop.: $b_{7.5}$ 71° (927), b_1 40° (644), n_D^{21} 1.4980, d_{20} 1.278 (644), IR (644-5), UV (927) and NMR (645-6) spectra, relative basicity (644-5), dipole moment (928).

Bis(chloromethyldimethylgermanium) $Ge_2C_6H_{16}Cl_2S$ $[Me_2(ClCH_2)Ge]_2S$
  Sulfide
Synth.: By rxn. of $[Me_2(ClCH_2)Ge]_2O$ with aq. H_2S at 80°, in 75% yield (2134).
Prop.: b_{13} 139-41°, NMR spectrum (2134).
Rxn. with: H_2S + Et_3N → $\overline{Me_2GeSGeMe_2CH_2SCH_2}$ + $Et_3N \cdot HCl$ (2134).
Na_2S in Et_2O → $\overline{Me_2GeSGeMe_2CH_2SCH_2}$ + NaCl (2134).

1,1,3,3-Tetramethyl-2,5,1,3-Dithiadi- $Ge_2C_6H_{16}S_2$ $\overline{Me_2GeSGeMe_2CH_2SCH_2}$
  germacyclohexane
Synth.: By rxn. of $[Me_2(ClCH_2)Ge]_2S$ or $Me_2(ClCH_2)GeCl$ with H_2S and Et_3N in C_6H_6, in 76% yield (2134).
By rxn. of $[Me_2(ClCH_2)Ge]_2S$ with Na_2S in Et_2O under anhydr. cond., in 67% yield (2134).
By rxn. of $[Me_2(ClCH_2)Ge]_2Y$ with H_2S and Et_3N; Y = O, NH, in 90% yield (2134).
Prop.: m. 76°, $b_{0.1}$ 71-72°, NMR spectrum, mol. wt., soly (2134).
Rxn. with: MeLi → $(Me_3GeCH_2)_2S$ (2134).

Dimethylgermanium Sulfide (M-120) $Ge_3C_6H_{18}S_3$ $(Me_2GeS)_3$
Prop.: NMR spectrum (368, 2607).
Rxn. with: MeLi → Me_3GeSLi (477).
Me_2GeX_2 ⇌ equilibrium const. measured by NMR spectroscopy, X = Cl, Br, I (368).
Me_2GeCl_2 → $(Me_2GeCl)_2S$ + $Cl(Me_2GeS)_2GeMe_2Cl$ (2607).

Bistrietylgermanium Sulfide (M-120) $Ge_2C_{12}H_{30}S$ $(Et_3Ge)_2S$
Synth.: By rxn. of Et_3GeI with Ag_2S at 260° in sealed tube, in 74% yield (465).
By rxn. of Et_3GeH with S at 170°, in 20% yield (2091).
By rxn. of Et_3GeSH with Et_2Hg at 100° under exclusion of air, in 74% yield (2090).
By rxn. of $(Et_3Ge)_2Se$ with S at 170° (2853).
Prop.: b_{10} 148.5° (2090), $b_{1.5}$ 95-98° (2091), n_D^{20} 1.5142 (465), 1.5121 (2090), 1.5111 (2091), d_{20} 1.1881 (465), 1.1860 (2091), IR (1010), far IR, assignment (2374) and NMR (1009) spectra.
Rxn. with: Bz_2O_2 → Et_3GeOBz + S (2093).
Br in C_6H_6 → Et_3GeBr + S (2851).

Bistriphenylgermanium Sulfide (M-120) $Ge_2C_{36}H_{30}S$ $(Ph_3Ge)_2S$
Synth.: By rxn. of Ph_3GeBr with Ph_3GeSLi in THF in N atm., in 68% yield (484).
By rxn. of Ph_3GeOH with H_2S in EtOH, in 45% yield (1970).
Prop.: m. 138° (1970), m. 136-38°, monomeric (484), IR (1970), (far) IR spectra and assignment (1848), mol. wt. (1970).

Bistriethylgermanium Selenide $Ge_2C_{12}H_{30}Se$ $(Et_3Ge)_2Se$

<u>Synth.</u>: By rxn. of $Et_3GeSeLi$ with $(CH_2Br)_2$ in THF under exclusion of air at 70°, in 85% yield (2090), with Et_3GeBr at r.t., in 54% yield (2090).
By rxn. of Et_3GeSeH with $HgEt_2$ at r.t., in 82% yield (2090), with Et_2Se at 200°, in 46% yield (2091), with Se at 200°, in 18% yield (2091), with MeMgBr, followed by Et_3GeBr at 100°, in 54% yield (2090).

<u>Prop.</u>: b_{10} 152-53°, b_5 140-41° (2090), b_3 129-32° (2091), n_D^{20} 1.5298 (2090), 1.5287 (2090-1), d_{20} 1.3280 (2091), IR (1010, 2090), far IR, assignment (2374) and NMR (1009) spectra.

<u>Rxn. with</u>: S at 170° → $(Et_3Ge)_2S$ + Se (2851).
Br in C_6H_6 → Et_3GeBr + Se (2853).

Bistriphenylgermanium Selenide $Ge_2C_{36}H_{30}Se$ $(Ph_3Ge)_2Se$

<u>Synth.</u>: By rxn. of Ph_3GeBr with $Ph_3GeSeLi$ in N atm. in THF, in 22% yield (484).

<u>Prop.</u>: m. 150°, very sensitive to moisture (484), (far) IR spectra and assignment (1848).

<u>Rxn. with</u>: H_2O → Se (484).

Bistriethylgermanium Telluride $Ge_2TeC_{12}H_{30}$ $(Et_3Ge)_2Te$

<u>Synth.</u>: By rxn. of Et_3GeH with $(Et_3Si)_2Te$ at 230°, in 79% yield (2858), with Te at 210°, in 60% yield (2091), in 75% yield (2092), with Et_2Te at 140°, in 60% yield (2092).

<u>Prop.</u>: b_1 113-16° (2092), b_1 112-15° (2091-2), n_D^{20} 1.5610 (2091-2), 1.5601 (2091-2), d_{20} 1.448 (2092), spectra: IR (1010-1), far IR and assignment (2374), NMR (1009).

<u>Rxn. with</u>: Et_3SnH → $(Et_3Sn)_2Te$ + Et_3GeH (2853).
$(c.C_6H_{11}OCO_2)_2$ → $Et_3GeOC_6H_{11}$ + CO_2 + Te (2093).
X → Et_3GeX + Te; X = Br, I (2851).

Bistriphenylgermanium Telluride $Ge_2C_{36}H_{30}Te$ $(Ph_3Ge)_2Te$

<u>Synth.</u>: By rxn. of Ph_3GeBr with $Ph_3GeTeLi$ in N atm. in THF, in 11% yield (484).

<u>Prop.</u>: m. 120° (dec.), yellow needles, monomeric, very sensitive to air and moisture (484), (far) IR spectra and assignment (1848).

Additional organogermanium sulfides, selenides and tellurides are listed in Table 60.

Table 60. Organogermanium Sulfides, Selenides and Tellurides $(R_3Ge)_2Y$
$(R_2GeY)_x$

Formula*	Synth. Method**	Yield	Properties	Ref.
[EtPh(1-$C_{10}H_7$)Ge]$_2$S	IB	--	$[\alpha]_D^{25}$ + 29.9° (C_6H_6) (a)	2368
($Pr_3Ge)_2S$	--	--	(b)	--
($Bu_3Ge)_2S$	IA	61	$b_{0.3}$ 190-91°, n_D^{20} 1.4940, d_{20} 1.0448	465
($Et_2GeS)_3$	--	--	(c)	--
[(CH_2:CH)$_2$GeS]$_3$	--	--	(d)	--
n(CH_2:CH)$_2$GeS·(CH_2:CH)$_2$Ge(SMe)$_2$	(d)	--	Insecticide, functional fluid	2607
($Bu_2GeS)_2$	III	--	b_1 193-98°	337
($Ph_2GeS)_3$ (120)	--	--	(e)	--
($Me_2GeSe)_x$ (121)	--	--	(f)	--
[(c.$C_6H_{11})_3Ge]_2Te$	II	78	m. 128-29°, soly., sensitive to air	2092

* Numbers in parenthesis refer to pages in main volume.
** The characters used here correspond to the synthetic scheme in the introcution to Chapter 4.4.1.
(a) Synth. with inversion of configuration (2368), rxn. with BuLi in Et_2O → R_3GeBu, stereochemistry of reaction (2368). (b) Rxn. with Hg(CH_2CO_2Me)$_2$ → $Pr_3GeCH_2CO_2Me$ + HgS (2648). (c) Rxn. with $Et_2Ge(OMe)_2$ → MeO($Et_2GeS)_xEt_2GeOMe$ (2607). (d) Rxn. with (CH_2:CH)$_2$Ge(SMe)$_2$ → MeS[(CH_2:CH)$_2$GeS]$_x$(CH_2:CH)$_2$GeSMe (2607). (e) Rxn. with Ph_2GeCl_2 → Cl($Ph_2GeS)_xPh_2GeCl$ (2607). (f) Rxn. with MeLi → $Me_3GeSeLi$ (477).

4.4.1.4 ORGANOGERMANIUM THIOLS AND SELENOLS

Triethylgermanium Thiol $GeC_6H_{16}S$ Et_3GeSH
<u>Synth.</u>: By rxn. of Et_3GeH with S (1:1) at 140°, in 50% yield (2091-2).
<u>Prop.</u>: b_{35} 86°, n_D^{20} 1.4852, d_{20} 1.1270 (2091-2), IR (1010) and NMR (1009, 2375) spectra. Anal. detn. by Zerewitinoff method (2090).
<u>Rxn. with:</u> Thermal dec. at 130° → ($Et_3Ge)_2S$ (2092).
Li in THF → Et_3GeSLi (2090).
Et_2Hg (2:1) → ($Et_3Ge)_2S$ + HgS (2090).
Et_3SnH → $Et_3GeSSnEt_3$ + H (2092).

Triethylgermanium Selenol $GeC_6H_{16}Se$ Et_3GeSeH
<u>Synth.</u>: By rxn. of Et_3GeH with Se at 140°, in 63% yield (2091-2).
<u>Prop.</u>: b_{35} 95-97°, n_D^{20} 1.5058, d_{20} 1.3700 (2091-2), IR (1010) and NMR (1009, 2375) spectra. Anal detn. by Zerewitinoff method (2090).

Rxn. with: Li in THF → Et$_3$GeSeLi + H + some H$_2$Se + C$_2$H$_6$ (2090).
MeMgI, + Et$_3$GeBr → (Et$_3$Ge)$_2$Se + MgBrI + CH$_4$ (2090).
Et$_2$Hg (2:1) → (Et$_3$Ge)$_2$Se + HgSe (2090).
Et$_3$SnH → Et$_3$GeSeSnEt$_3$ + H (2091-2).
Br in C$_6$H$_6$ → Et$_3$GeBr + Se (2851).

Triisopropylgermanium Selenol   GeC$_9$H$_{22}$Se   i-Pr$_3$GeSeH
Synth: By rxn. of i-Pr$_3$GeH with Se at 210°, in 67% yield (2092).
Prop.: b$_{15}$ 111-13°, n$_D^{20}$ 1.5129 (2092).

Tricyclohexylgermanium Selenol   GeC$_{18}$H$_{24}$Se   (c.C$_6$H$_{11}$)$_3$GeSeH
Synth.: By rxn. of (C$_6$H$_{11}$)$_3$GeH with Se at 210°, in 31% yield (2092).
Prop.: m. 80-83° (2092).

4.4.2 ORGANOGERMANIUM MERCAPTIDES, ORGANOSELENIDES AND ORGANOTELLURIDES

The organogermanium mercaptides and organoselenides listed in Tables 61 and 63 are prepared by methods from the following scheme.

I   Reactions with organometallic derivatives:
    (IA) Et$_3$GeYLi + PhCH$_2$Cl → Et$_3$GeYCH$_2$Ph; Y = S, Se ,
    (IB) (Et$_3$Ge)$_2$Hg + BuSeH → Et$_3$GeSeBu .

II Reactions with organohalogermanes:
    (IIA) R$_n$GeX$_{4-n}$ + (R'S)$_2$Pb → R$_n$Ge(SR')$_{4-n}$,
    (IIB) R$_n$GeX$_{4-n}$ + R'SH + organic base → R$_n$Ge(SR')$_{4-n}$.

III Reactions with organogermanes:
    (IIIA) Et$_3$GeH + BuSH → Et$_3$GeSBu + H; catalyzed by Ni-Kieselguhr ,
    (IIIB) Et$_3$GeH + Me$_2$S$_2$ → Et$_3$GeSMe .

IV Reactions with heteroorganogermanes:
    (IVA) R$_{2n}$Ge$_2$O$_{4-n}$ + BuSH + cat. → R$_n$Ge(SBu)$_{4-n}$,
    (IVB) R$_n$Ge(OR')$_{4-n}$ + RSH → R$_n$Ge(SR)$_{4-n}$; R' = Me, Et ,
    (IVC) R$_3$GeNR$_2$ + R'SH → R$_3$GeSR' .

V   Exchange reaction:
    Me$_n$Ge(SEt)$_{4-n}$ + RSH + cat. → Me$_n$Ge(SR)$_{4-n}$.

R = organic group, may be different in one molecule, R' = organic group, may be functionally substituted, X = halogen, n = 2,3 , cat. = reaction might require MeC$_6$H$_4$SO$_3$H as catalyst.

4.4.2.1 TRIORGANOGERMANIUM MERCAPTIDES ORGANOSELENIDES AND ORGANOSELENIDES

4.4.2.1.1 UNSUBSTITUTED DERIVATIVES

Trimethylgermanium Methylmercaptide   GeC$_4$H$_{12}$S   Me$_3$GeSMe

Synth.: By rxn. of Me$_3$GeBr with MeSH and Py in hexane, in 73% yield (227).
By rxn. of Me$_3$GeX with MeSH and aq. NaOH (728).
Prop.: b$_{734}$ 135° (728), b. 129-31° (227), n$_D^{20}$ 1.4792, d$_{20}$ 1.18 (728), IR (644) and NMR (227, 646, 728) spectra, relative basicity (644).
Rxn. with: MeI → Me$_3$SI (227).
Me$_2$SO$_4$ → Me$_3$Ge(Me)SO$_4$ + Me$_3$S(Me)SO$_4$ (227).

Trimethylgermanium Ethylmercaptide   GeC$_5$H$_{14}$S   Me$_3$GeSEt

Synth.: By rxn. of Me$_3$GeBr with EtSNa in Et$_2$O, in 94% yield (643).
By rxn. of Me$_3$GeBr with Me$_3$SiSEt, in 56% yield (643).
Prop.: b. 148°, n$_D^{21}$ 1.4779-88, d$_{20}$ 1.10 (643), IR (644-5) and NMR (645) spectra, relative basicity (644-5).
Rxn. with: PhSH → Me$_3$GeSPh + EtSH (643).
(CF$_3$)$_2$CO → Me$_3$GeOC(CF$_3$)$_2$SEt (2203).

Triphenylgermanium Methylmercaptide   GeC$_{19}$H$_{18}$S   Ph$_3$GeSMe
(M-123)

Synth.: By rxn. of Ph$_3$GeBr with MeSH and Py in C$_6$H$_6$ at 0°, in 81% yield (227), with MeSMgI in Et$_2$O-C$_6$H$_6$, in 32% yield (227).
Prop.: m. 85°, NMR spectrum (227).
Rxn. with: BuLi → BuGePh$_3$ (227).
H$_2$O$_2$ or NO$_2$ in Me$_2$CO → (Ph$_3$Ge)$_2$O (227).
EtOH → Ph$_3$GeOEt + (Ph$_3$Ge)$_2$O (227).
MeI → Ph$_3$GeI + Me$_3$SI (227).

Triethylgermanium Ethyltelluride   GeTeC$_8$H$_{20}$ Et$_3$GeTeEt

Synth.: By-prod. in rxn. of Et$_3$GeH with Et$_2$Te at 140°, in 28-39% yield (2092).
Prop.: b$_1$ 61°, n$_D^{20}$ 1.5458 (2092), IR spectrum (1011).
Rxn. with: Et$_3$GeH at 140° → (Et$_3$Ge)$_2$Te (2092).
Et$_3$SnH at 20° → Et$_3$GeTeSnEt$_3$ + H (2092).
Bz$_2$O$_2$ → Et$_3$GeOBz (2093).

Additional unsubstituted triorganogermanium mercaptides and organoselenides are listed in Table 61.

Table 61. Unsubstituted Triorganogermanium Mercaptides and Organoselenides

Formula*	Synth. Method**	Yield	Properties	Ref. R_3GeSR R_3GeSeR
Me_3GeSBu-n	IVC	85	b_8 62°, n_D^{22} 1.4736, d_{20} 1.08	643
$Me_3GeSCMe_3$	IVA	23	$b_{0.1}$ 25–26°, n_D^{22} 1.4729, d_{20} 1.08	643
Me_3GeSPh	IVB	72	$b_{0.001}$ 37°, n_D^{23} 1.5560	643
	V	80	n_D^{20} 1.5564	643
Et_3GeSR				
R = Me	IIIB	92	b_{15} 84°, n_D^{20} 1.492, d_{20} 1.1133	465
CH_2Ph (122)	IA	66	$b_{0.5}$ 113–16°, n_D^{20} 1.5435 (a, b)	2090
Et	IIA	66	b_{10} 93°, n_D^{20} 1.49, d_{20} 1.0868	465
$CHMe_2$	IVC	93	b_{17} 116°, n_D^{20} 1.4868, d_{20} 1.059	465
Bu (122)	IIIA	74	--	465
	IVA	85	--	465
n-C_6H_{13} (122)	--	--	(b)	--
n-C_7H_{15} (122)	--	--	(b)	--
n-C_8H_{17}	IVC	89	b_7 161°, n_D^{20} 1.4839, d_{20} 1.005	465
Ph (122)	--	--	(b)	--
o-MeC_6H_4 (122)	--	--	(b)	--
m-MeC_6H_4 (122)	--	--	(b)	--
2-$C_{10}H_7$ (122)	--	--	(b)	--
R			(c)	
$EtPh(1$-$C_{10}H_7)GeSPh$	IIB	--	$[\alpha]_D^{25}$ 21.5° (C_6H_6) (d)	2368
Bu_3GeSR				
R = CH_2Ph	IIB	84	$b_{1.2}$ 171–74°, mol. wt., IR spectrum	1502
Et	IIA	67	b_{17} 164°, n_D^{20} 1.483, d_{20} 1.0082	465
	IIB	90	$b_{0.8}$ 103°, mol. wt., IR spectrum	1502

Pr	IIB	95	$b_{0.5}$ 115-16°, mol. wt., IR spectrum	1502
CHMe$_2$	IIB	90	b_1 120°, mol. wt., IR spectrum	1502
Bu	IIA	61	b_9 169°, n_D^{20} 1.4824, d_{20} 0.9855	465
	IIB	93	b_2 130-32°, mol. wt.,	1502
	IVA	92	IR spectrum	1502
CHMeEt	IIB	98	b_2 136-38°, mol. wt., IR spectrum	1502
CMe$_3$	IIB	92	b_1 125°, mol. wt., IR spectrum	1502
n-C$_8$H$_{17}$	IVB	87	$b_{0.2}$ 143°, n_D^{20} 1.481, d_{20} 0.9648	465
	IIB	90	$b_{0.4}$ 192-93°, mol. wt., IR spectrum	1502
n-C$_{12}$H$_{25}$	IVB	80	$b_{0.12}$ 134°, n_D^{20} 1.528, d_{20} 1.0555	465
Ph	IIB	90	$b_{0.5}$ 143°, mol. wt., IR spectrum	1502
(Me$_3$C)$_3$GeSCMe$_3$	IVC	90	b_{18} 169°, n_D^{25} 1.4835, d_{25} 0.9828	465
Ph$_3$GeSCH$_2$CH:CH$_2$	IIB	>70	m. 62°	121
Ph$_3$GeSC$_{12}$H$_{25}$-n	IIB	>70	m. 56°	121
Ph$_3$GeSPh	IIB	>70	m. 96°	121
	IIB	80	m. 90.5-91.5°	227
Ph$_3$GeSC$_{10}$H$_7$-2	IIB	>70	m. 88-89°	121
Et$_3$GeSeCH$_2$Ph	IA	73	b_1 118-20°, n_D^{20} 1.5652 (a)	2092
Et$_3$GeSeBu	IB	78	b_7 105-108°, n_D^{20} 1.5031	2855
	IB (e)	82	b_6 105-106°, n_D^{20} 1.5041	2855

* Numbers in parenthesis refer to pages in main volume.
** The characters used here correspond to the synthetic scheme in the introduction to Chapter 4.4.2.
(a) (Far) IR spectra and assignment (2090). (b) Empirical equation for boiling point calculation (680). (c) Germanium bond refraction D$_{20}$ (81). (d) Synth. with inversion of configuration (2368), rxn. with LiAlH$_4$ → R$_3$GeH; with inversion (2368). (e) From Et$_3$GeHgEt at 80° (2855).

4.4.2.1.6 FUNCTIONALLY SUBSTITUTED TRIORGANOGERMANIUM MERCAPTIDES AND ORGANO-SELENIDES

The functionally substituted triorganogermanium derivatives with RS- and RSe- groups listed in Table 62 are prepared by methods from the following scheme.

I Cleavage of cyclic tetraorganogermane with sulfur and selenium.
$\overline{CH_2CHRCH_2GeR_2}$ + S at 250° → $R_2\overline{GeCH_2CHRCH_2S}$, R = H, Me .

II Substitution reactions with organohalogermanes:
(IIA) $R_3GeX + R'SH(+Et_3N$ or $Py) \rightarrow R_3GeSR'$,
(IIB) $R_3GeX + Y(SH)_2 + Et_3N \rightarrow R_3GeSYSGeR_3$,
(IIC) $R_2(XY)GeX + Na_2S \rightarrow R_2\overline{GeYS}$.

III Reaction with organogermane, catalyzed by Ni-Kieselguhr.
$Et_3GeH + Y(SH)_2 \rightarrow (Et_3GeS)_2Y$.

R = organic group, R' = functionally substituted organic group, X = halogen, Y = hydrocarbon bridge. Other functionally substituted triorganogermanium mercaptides are reported in Chapters 4.4.1 and 7.4.3.

4.4.2.2 DIORGANOGERMANIUM MERCAPTIDES

4.4.2.2.1 UNSUBSTITUTED DIORGANOGERMANIUM MERCAPTIDES

All compounds are listed in Table 63 on page 204.

4.4.2.2.6 FUNCTIONALLY SUBSTITUTED DIORGANOGERMANIUM MERCAPTIDES

All compounds are listed on the bottom of Table 63 on page 204. Other functionally substituted diorganogermanium mercaptides are listed in Sub-chapters 4.3.3.4 and 4.3.5.2.

Table 62. Functionally Substituted Triorganogermanium Mercaptides and Organoselenides R_3GeSR' R_3GeSeR'

Formula*	Synth. Method**	Yield	Properties	Ref.
![7-Me-benzodithiin-GeMe2] (a)	IIC	--	b_1 131-32°, IR and NMR spectra	2130
2-$Et_3GeSCH_2C_4H_3O$ (122)	--	--	(b)	--
$(Et_3GeSCH_2)_2$	III	46	b_{17} 165°, n_D^{20} 1.5149, d_{20} 1.1899	465
o-$Et_3GeSC_6H_4NH_2$ (122)	--	--	(b)	--
$Et_2GeSCH_2CH_2CH_2$	I	85	b_{23} 107°, n_D^{20} 1.5241, d_{20} 1.2102 (c)	2576
$Bu_2GeSCH_2CH_2CH_2$	I	84	b_4 130°, n_D^{20} 1.5089, d_{20} 1.1043	1492, 2576
	IIC	85	$b_{0.4}$ 94° (c, d, e)	2576
$Bu_2GeSCH_2CHMeCH_2$	I	90	$b_{0.7}$ 103°, n_D^{20} 1.5048, d_{20} 1.0836	2576
	IIC	97	(c, d)	2576
Ph_3GeSR'				
R' = CH_2CH_2OH	IIA	(f)	Oil	121
CH_2CO_2H	IIA	(f)	m. 149-50°	121
CH_2CO_2Me	IIA	(f)	m. 44-45° (g)	121
CH_2CONH_2	IIA	(f)	m. 151°	121
	(g)	90		121

* Numbers in parenthesis refer to pages in main volume.
** The characters used here correspond to the synthetic scheme in the introduction to Chapter 4.4.2.1.6.
(a) Mixture of 6- and 7-Me-isomers, ratio 2 (2130). (b) Empirical equation for boiling point calculation (680).
(c) NMR spectrum (2576). (d) (Far) IR spectrum (2576). (e) Mass spectrum (2576). (f) Yield at 70-90% (121).
(g) Rxn. of $Ph_3GeCH_2CO_2Me$ with dry NH_3 in MeOH → $Ph_3GeSCH_2CONH_2$ (121).

Table 62 Continued

Formula*	Synth. Method**	Yield	Properties	R_3GeSR' R_3GeSeR' Ref.
Ph$_3$GeSR'				
C$_6$F$_5$	IIA	(f)	m. 96-97°	121
p-C$_6$H$_4$Cl	IIA	(f)	m. 101-102°	121
17-testosterone	IIA	(f)	m. 188° [GeC$_{37}$H$_{44}$OS]	121
(Ph$_3$GeSCH$_2$)$_2$	IIB	(f)	m. 152°	121
(Ph$_3$GeSCH$_2$C$_6$H$_4$)$_2$O	IIB	(f)	Syrup	121
Bu$_2$GeSeCH$_2$CH$_2$CH$_2$	I	50	b$_{0.3}$ 98°, n$_D^{20}$ 1.5252, d$_{20}$ 1.2754 (c, e)	2576

Table 63. Diorganogermanium Dimercaptides

Formula	Synth. Method*	Yield	Properties	$R_2Ge(SR')_2$ Ref.
Me$_2$Ge(SMe)$_2$	IIB	--	b$_{15}$ 85°, n$_D^{20}$ 1.5470, d$_{20}$ 1.30	728
	IIA	65	b$_1$ 45° (a, b)	2062
Me$_2$Ge(SBu)$_2$	--	--	b$_{0.05}$ 93°, n$_D^{25}$ 1.5083, GLC (c)	1450
Et$_2$Ge(SPh)$_2$	IVA	94	b$_{0.15}$ 153-54°, n$_D^{20}$ 1.6242, d$_{20}$ 1.247	465
	IVB	91	b$_3$ 162°, n$_D^{20}$ 1.624, d$_{20}$ 1.2483	465
(CH$_2$:CH)$_2$Ge(SMe)$_2$	--	--	(d)	--
Bu$_2$Ge(SR)$_2$				
R = CH$_2$Ph	IIB	88	b$_{17}$ 233-35°, mol. wt., IR spectrum	1502

Et	IIA	58	b_{10} 158°, n_D^{20} 1.5132, d_{20} 1.0782	465
Pr	IIB	84	b_1 120-22°, mol. wt., IR spectrum	1502
CHMe$_2$	IIB	80	$b_{0.2}$ 115°, mol. wt., IR spectrum (e, f)	1502
Bu	IIB	94	$b_{0.5}$ 114-15°, mol. wt., IR spectrum	1502
	IVA	97	$b_{0.5}$ 140-43°, mol. wt., IR spectrum	1502
CH$_2$CHMe$_2$	IIB	80	$b_{1.5}$ 162-63°	1502
CMe$_3$	IIB	89	$b_{0.7}$ 141°, mol. wt., IR spectrum	1502
n-C$_{12}$H$_{25}$	IIB	75	$b_{1.5}$ 145°, mol. wt., IR spectrum	1502
	IVB	99	$b_{0.8}$ 238°, mol. wt., IR spectrum	1502
Ph	IIB	97		1502
	V (f)	88	$b_{1.2}$ 204-206°, mol. wt.	1502
			IR spectrum	

Ph$_2$Ge(SR)$_2$

R = Me	IIB	85	$b_{0.2}$ 132-35°, NMR spectrum	227
CH$_2$Ph	IIB	--	m. 50-51° (g)	121
CH$_2$CH:CH$_2$	IIB	--	$n_D^{28.5}$ 1.6229 (g)	121
n-C$_{12}$H$_{25}$	IIB	--	$n_D^{28.5}$ 1.5343 (g)	121
Ph	IIB	--	n_D^{22} 1.6776 sic. (g)	121
2-C$_{10}$H$_7$	IIB	--	m. 90-91° (g)	121

Ph$_2$Ge(SR')$_2$

R' = CH$_2$CO$_2$Me	IIB	--	Crude, dec. on distn. (h)	121
CH$_2$CONH$_2$	(h)	--	m. 140°	121

* The characters used here corresonpd to the synthetic scheme in the introduction to Chapter 4.4.2.
(a) NMR spectrum (728). (b) Rxn. with Me$_2$GeX$_2$ → equilibrium constants for redistribution reactions by NMR spectroscopy; X = Cl, Br, I (368), X = MeO (2062), same reaction with Me$_2$SiX$_2$; X = Cl, Br, I, MeO (2062). (c) Analytical detn. by volumetric titration after reaction with Ac$_2$O or Et$_4$NBr (1450). (d) Rxn. with (R$_2$GeS)$_3$ → MeS(R$_2$GeS)$_x$R$_2$GeSMe; R = CH$_2$:CH (2606). (e) Rxn. with anhydr. EtOH → Bu$_2$Ge(OEt)$_2$ by azeotropic distillation with excess EtOH (1502). (f) Rxn. of Bu$_2$Ge(SPr)$_2$ with PhSH + cat. MeC$_6$H$_4$SO$_3$H → Bu$_2$Ge(SPh)$_2$ (1502). (g) White colorless, thermally stable (121). (h) Rxn. of Ph$_2$Ge(CH$_2$CO$_2$Me)$_2$ with dry NH$_3$ in MeOH → Ph$_2$Ge(SCH$_2$CONH$_2$)$_2$ (121).

Table 63 Continued $R_2Ge(SR')_2$

Formula	Synth. Method*	Yield	Properties	Ref.
$Ph_2Ge(SR')_2$ (Cont'd.)				
R' = p-C_6H_4Cl	IIB	--	n_D^{22} 1.6751	121
p-$C_6H_4NH_2$	IIB	15	m. 115°	121
p-$C_6H_4NO_2$	IIB	--	m. 114-15°	121
$\overline{Ph_2GeSCH_2CH_2S}$	IIB (i)	--	m. 87-88°, monomeric, IR spectrum	121

(i) From Ph_2GeBr_2 with $(CH_2SH)_2$ and pyridine in C_6H_6 (121).

4.4.2.3 MONOORGANOGERMANIUM TRIMERCAPTIDES

Methylgermanium Trimethylmercaptide $GeC_4H_{12}S_3$ $MeGe(SMe)_3$
Synth.: By rxn. of $MeGeX_3$ with aq. NaOH, followed by MeSH at pH 4-5 (728).
Prop.: b_{15} 130°, n_D^{20} 1.5908, d_{20} 1.39 (728), NMR spectrum (728, 1545-6).
Rxn. with: $Me_2GeBr_2 \rightleftarrows$ *product distribution (2606).
$MeGeX_3 \rightleftarrows$ *X = Cl, Br, I (1545).
$MeSiX_3 \rightleftarrows$ *X = Cl, Br, I (1546).

Phenylgermanium Trimethylmercaptide $GeC_9H_{14}S$ $PhGe(SMe)_3$
Synth.: By rxn. of $PhGeBr_3$ with MeSH in pyridine, in 81% yield (227).
Prop.: $b_{0.2}$ 110°, NMR spectrum (227).

4.4.4 ORGANOGERMANIUM DERIVATIVES OF ORGANIC THIOACIDS

Triethylgermanium Dimethyldithio- $GeC_9H_{21}NS_2$ $Et_3GeS_2CNMe_2$
carbamate
Synth.: By rxn. of Et_3GeNMe_2 with CS_2, in 90% yield (466).
Prop.: b_{16} 179-80°, n_D^{20} 1.5700, d_{20} 1.2156 (466). Microanalytical detn. (426).

Triethylgermanium Diethyldithio- $GeC_{11}H_{25}NS_2$ $Et_3GeS_2CNEt_2$
carbamate
Synth.: By rxn. of Et_3GeNEt_2 with CS_2 (466).
Prop.: $b_{0.4}$ 122°, n_D^{20} 1.5772, d_{20} 1.1618 (466).

Triethylgermanium Diethylphosphino- $GeC_{11}H_{25}PS_2$ $Et_3GeS_2CPEt_2$
dithioformate
Synth.: By rxn. of Et_3GePEt_2 with CS_2 in C_6H_6, in 38% yield (1797).
Prop.: $b_{0.1}$ 98°, n_D^{20} 1.5493, d_{20} 1.1694, IR and NMR spectra (1797).

Triphenylgermanium Thioacetate $GeC_{20}H_{18}OS$ Ph_3GeSAc
Synth.: By rxn. of Ph_3GeBr with AcSH and Et_3N or pyridine in C_6H_6, in 60-90% yield (121).
Prop.: m. 105°, dec. during chromatographic purification on silica or alumina (121).

Triphenylgermanium Thiobenzoate (M-123) $GeC_{25}H_{20}OS$ Ph_3GeSBz
Synth.: By rxn. of Ph_3GeSLi with BzCl in THF, in 59% yield (484).
Prop.: m. 142-43°, monomeric (484).

---

* Equilibrium constants for redistribution rxn, measured by NMR spectroscopy.

Bis(S-Triphenylgermanium) Benzenedithio- $Ge_2C_{44}H_{64}O_2S_2$ $(Ph_3GeSOC)_2C_6H_4$
  dicarboxylate

<u>Synth.</u>: By rxn. of Ph_3GeBr with $C_6H_4(COCl)_2$ and pyridine or NEt_3 in C_6H_6, in 70-90% yield (121).

<u>Prop.</u>: m. 255-57° (121).

4.4.7 METAL SUBSTITUTED ORGANOGERMANIUM SULFIDES, SELENIDES AND TELLURIDES

New in this chapter are a number of mixed tellurides. The new tellurides are less sensitive towards hydrolysis than selenides which in turn are more resistant than sulfides. Stability also increases going from germanium to the higher homologues. All alkali metal derivatives are very sensitive to air and moisture. The compounds are listed in Table 64. Methods of preparation are summarized in the following scheme.

I Alkali metal derivatives:

  (IA) $(Me_2GeY)_x + MeLi \rightarrow Me_3GeYLi$,
  (IB) $Ph_3GeBr + Na_2S \rightarrow Ph_3GeSNa$,
  (IC) $Ph_3GeLi + Y_n \rightarrow Ph_3GeYLi$,
  (ID) $Et_3GeYH + Li \rightarrow Et_3GeYLi$.

II Group IV metal derivatives:

  (IIA) $Ph_3GeYLi + Ph_3MX \rightarrow Ph_3GeYMPh_3$,
  (IIB) $Et_3GeYH + Et_3SnH$ at 130° $\rightarrow Et_3GeYSnEt_3$,
  (IIC) $Et_3GeTeEt + Et_3SnH$ at 20° $\rightarrow Et_3GeTeSnEt_3$.

Y = sulfur, selenium or tellurium, M = Sn, Pb.

Table 64. Metal-Substituted Organogermanium Sulfides, Selenides and Tellurides R_3GeSM' R_3GeSMR_3

Formula*	Synth. Method**	Yield	Properties	Ref.
Me$_3$GeSLi (124)	IA	--	(a)	477
Me$_3$GeSeLi (124)	IA	--	(a)	477
Et$_3$GeSLi	ID	--	Not isolated (b, c)	2090
Et$_3$GeSSnEt$_3$	IIA (c)	63	b$_2$ 121-28°, n$_D^{20}$ 1.5270, IR spectrum	2090
Et$_3$GeSSnEt$_3$	IIB	57	b$_1$ 100-103°, n$_D^{20}$ 1.5279, d$_{20}$ 1.2990 (d)	2092
Et$_3$GeSeLi	ID	--	Not isolated (b, e)	2090
Et$_3$GeSeSnEt$_3$	IIB	37	b$_{0.5}$ 111°, n$_D^{20}$ 1.5470 (f)	2091-2
Et$_3$GeTeSnEt$_3$	IIC	62	b$_1$ 126-28°, n$_D^{20}$ 1.5723 (d, f)	2092
Ph$_3$GeSLi	IC	--	Deep red-brown soln., negative Gilman test (g, h)	484
Ph$_3$GeSNa (124)	IB	--	White cryst., m. 180-97°, soly.	227
Ph$_3$GeSSnPh$_3$ (124)	IIA (g)	72	m. 136°, monomeric (i)	484
Ph$_3$GeSPbPh$_3$	IIA (g)	45	m. 128-29°, yellow needles, monomeric (i, j)	484
Ph$_3$GeSeLi	IC	--	Light grey soln. (k)	484
Ph$_3$GeSeSnPh$_3$ (124)	IIA (k)	52	m. 144-45°, monomeric, stable in H$_2$O (i, l)	484
Ph$_3$GeSePbPh$_3$	IIA (k)	45	m. 119°, colorless cryst., monomeric (i m)	484
Ph$_3$GeTeLi	IC	--	Light grey soln., sensitive to air (n. o)	484
Ph$_3$GeTeSnPh$_3$	IIA (o)	48	m. 142-46°, colorless, cryst., (i, l)	484
Ph$_3$GeTePbPh$_3$	IIA (o)	42	m. 115-17°, yellow cryst. (i, m)	484

* Numbers in parenthesis refer to pages in main volume.

** The characters used here correspond to the synthetic scheme in the introduction to Chapter 4.4.7.

(a) Use as catalyst and additive for explosives and propellants (477). (b) Rxn. with PhCH$_2$Cl → Et$_3$GeYCH$_2$Ph + LiCl; Y = S, Se (2090). (c) Rxn. of Et$_3$GeSLi with Et$_3$SnCl → Et$_3$GeSSnEt$_3$ (2090). (d) NMR spectrum (1009). (e) Rxn. with (CH$_2$Br)$_2$ → (Et$_3$Ge)$_2$Se + LiBr + C$_2$H$_4$ (2090). (f) IR spectrum (1010). (g) Rxn. of Ph$_3$GeSLi with Ph$_3$MK → Ph$_3$GeSMPh$_3$; MX = GeBr, SnCl, PbCl (484). (h) Rxn. with BzCl → Ph$_3$GeSBz (484). (i) (Far) IR spectrum and assignment (1848). (j) Rxn. with aq. HCl → H$_2$S (484). (k) Rxn. of Ph$_3$GeSeLi with Ph$_3$MX → Ph$_3$GeSeMPh$_3$; MX = GeBr, SnCl, PbCl (484). (l) Rxn. with H$_2$O under reflux → Se or Te (484), polarography (963, 969). (m) Rxn. with dilute mineral acid → Se or Te (484). (n) Rxn. with air → Te (484). (o) Rxn. of Ph$_3$GeTeLi with Ph$_3$GeMX → Ph$_3$GeTeMPh$_3$; MX = GeBr, SnCl, PbCl (484).

4.5 ORGANOGERMANIUM DERIVATIVES WITH NITROGEN, PHOSPHORUS, ARSENIC ANTIMONY AND BISMUTH

4.5.1 ORGANOGERMANIUM AMIDES

4.5.1.1 TRIORGANOGERMANIUM AMIDES

Triorganogermanes containing one to three organogermyl groups linked to a nitrogen atom are listed in this chapter. Most compounds are sensitive to protonolysis. They are useful as intermediates for synthesis of substituted organogermanes. The compounds listed in Tables 65 and 67 are prepared by methods from the following scheme.

I Grignard reactions:
(IA) $Et_3GeCl + RNHMgBr \rightarrow Et_3GeNHR$,
(IB) $Et_3GeCl + PhN(MgBr)_2 \rightarrow (Et_3Ge)_2NPh$.

II Reactions with alkalimetal derivatives:
(IIA) $R_3GeCl + RR'NLi \rightarrow R_3GeNRR'$; $R' = H, R$,
(IIB) $R_3GeX + (R_3Ge)_nMe_{2-n}NLi \rightarrow (R_3Ge)_{n+1}NMe_{2-n}$; $n = 1,2$,
(IIC) $Pr_3GeBr + NaNH_2$ in $C_6H_6 \rightarrow (Pr_3Ge)_2NH$.

III Interconversion reactions:
(IIIA) $R_3GeNMe_2 + R'NH_2 \rightarrow (R_3Ge)_nH_{2-n}NR' + Me_2NH$,
(IIIB) $[Me_2(ClCH_2)Ge]_2NH + Na_2S \rightarrow Me_2GeNHGeMe_2CH_2SCH_2$.

IV Miscellaneous reactions:
(IVA) $(Et_3Ge)_2Hg + RN:NR \rightarrow (Et_3GeNR)_2$,
(IVB) $(Et_3Ge)_2O + R_2NH \rightarrow Et_3GeNR_2$.

R = organic group; R' = organic group, may be functionally substituted, X = halogen.

Tris(trimethylgermyl)amine (M-126) $Ge_3C_9H_{27}N$ $(Me_3Ge)_3N$
<u>Synth.:</u> By rxn. of Me_3GeCl with NH_3 in Et_2O at -60°, in 32% yield (1796), with Li_3N in THF, in 55% yield (1796).
By rxn. of Me_3GeBr with Li_3N, in 36% yield (644).
By rxn. of Me_3GeNMe_2 with NH_3 in Et_2O at 0°, in 28% yield (1796).
By rxn. of $(Me_3Ge)_2NLi$ with Me_3GeCl in Et_2O, in 70% yield (1796).
<u>Prop.:</u> Colorless liq. m. 13° (644), b_{13} 103° (1796), $b_{0.05}$ 45° (644), n_D^{20} 1.4788 (1796), n_D^{21} 1.4775, d_{20} 1.29 (644), IR (644) and NMR (1796) spectra, relative basicity (1796).
<u>Rxn. with:</u> $C_8H_{17}NH_2 \rightarrow (Me_3Ge)_2NC_8H_{17} + NH_3$ (1796).
$Ph_2PH \rightarrow Me_3GePPh_2 + NH_3$ (1796).

Bis(trimethylgermyl)amine (M-126) $Ge_2C_6H_{19}N$ $(Me_3Ge)_2NH$
<u>Synth.:</u> By rxn. of Me_3GeCl with NH_3 in Et_2O at -60°, in 50% yield (1796).
By rxn. of Me_3GeNMe_2 with NH_3 in Et_2O at 0°, in 32% yield (1796).
<u>Rxn. with:</u> $Et_3GeCl \rightarrow Me_3GeCl + (Et_3Ge)_2NH$ (1535).
$Et_3SiCl + cat. FeCl_3 \rightarrow Me_3GeCl + (Et_3Ge)_2NH$ (1535).

PhC⋮CH → Me₃GeC⋮CPh (1534).
C₈H₁₇NH₂ → (Me₃Ge)₂NC₈H₁₇ (1796).
Ph₂PH → Me₃GePPh₂ + NH₃ (1796).
H₂O → (Me₃Ge)₂O + NH₃ (1535).

Trimethylgermyldimethylamine GeC₅H₁₅N Me₃GeNMe₂
<u>Synth.:</u> By rxn. of Me₃GeCl with Me₂NLi in inert atm. (1853), in Et₂O (1796).
<u>Prop.:</u> b. 103°, n_D^{20} 1.4246 (1796).
<u>Rxn. with:</u> C₅H₅M(CO)₃H → Me₃GeC₅H₅M(CO)₃; M = Mo, W (2303).
(⋮CCO₂Et)₂ → Me₃GeC(CO₂Et):C(CO₂Et)NMe₂ (854).
NH₃ in Et₂O → (Me₃Ge)₂NH + (Me₃Ge)₃N + Me₂NH (1796).
C₈H₁₇NH₂ → (Me₃Ge)₂NC₈H₁₇ + NH₃ (1796).
YNH → N-Me₃GeNY + Me₂NH; YN = 3-Me-pyrazolyl, 3,5-Me₂-pyrazolyl, imidazolyl (2378).
MH₃ → (Me₃Ge)₂M; M = P, As (1853).

Trimethylgermyldiethylamine GeC₇H₁₉N Me₃GeNEt₂
<u>Synth.:</u> By rxn. of Me₃GeBr with Et₂NLi in Et₂O, in 51% yield (643).
<u>Prop.:</u> b. 138-39°, n_D^{20} 1.4304, d₂₀ 1.01 (643), IR (644-5) and NMR (645), spectra, relative basicity (644-5).
<u>Rxn. with:</u> $\overline{OCH_2CH_2C}$:O → Me₃GeO₂CCH₂CH₂NEt₂ (1238); kinetic data, solvent effect and rxn. mechanism (1236-7).
CH₂:CO → Me₃GeCH₂CONEt₂ (2513).
n-BuSH → Me₃GeSBu + Et₂NH (643).

Bis(chloromethyldimethylgermyl)amine Ge₂C₆H₁₇Cl₂N [Me₂(ClCH₂)Ge]₂NH
<u>Synth.:</u> By rxn. of Me₂(ClCH₂)GeCl with NH₃ in Et₂O at -70° to r.t., in 69% yield (2134).
<u>Prop.:</u> b₀.₁ 71-74°, hydroscopic liq., NMR spectrum (2134).
<u>Rxn. with:</u> H₂O → [Me₂(ClCH₂)Ge]₂O (2134).
Na₂S in Et₂O → $\overline{Me_2GeNHGeMe_2CH_2SCH_2}$ (2134).
H₂S in Et₃N → $\overline{Me_2GeSGeMe_2CH_2SCH_2}$ (2134).

Bis(triethylgermyl)amine Ge₂C₁₂H₃₁N (Et₃Ge)₂NH
<u>Synth.:</u> By rxn. of Et₃GeCl with LiNH₂ in THF under reflux (1796), with (Me₃Ge)₂NH, in 18% yield (1535).
By rxn. of Et₃GeBr with Na in liq. NH₃, in 84% yield (1534), with NaNH₂ in C₆H₆ in N atm., in 56% yield (2502).
By rxn. of Et₃GeNMe₂ in liq. NH₃ in autoclave at 50°, in high yield (1796).
By rxn. of Et₃GePEt₂ with liq. NH₃ in autoclave at 150° (1797).
<u>Prop.:</u> Colorless liq. (2502). b₇₅₇ 257° (1534), b₀.₂ 92°, n_D^{20} 1.4755 (2502), 1.4699 (1534), d₂₀ 1.1361 (1534), IR spectrum (1534, 1796) mol. wt., hydrolyzes in air (2502).
<u>Rxn. with:</u> BuLi → (Et₃Ge)₂NLi (2502).
Et₃Al → (Et₃Ge)₂NAlEt₂ + C₂H₆ (1535).
RC⋮CH → Et₃GeC⋮CR ; R = Ph, Me₃Si (1534).
HC⋮CCH₂OH → Et₃GeOCH₂C⋮CH (1534).

HC⋮CCH$_2$Cl → Et$_3$GeCl + polymer (1534).
H$_2$O → (Et$_3$Ge)$_2$O (1535).
p-(HO)$_2$C$_6$H$_4$ → p-(Et$_3$GeO)$_2$C$_6$H$_4$ + NH$_3$ (1796).
Ph$_2$C:CO → [Et$_3$GeOC(:CPh$_2$)]$_2$NH (2736).
(CH$_2$CO$_2$H)$_2$ → (Et$_3$GeO$_2$CCH$_2$)$_2$ + NH$_3$ (1796).

Triethylgermylphenylamine   GeC$_{12}$H$_{21}$N   Et$_3$GeNHPh
Synth.: By rxn. of Et$_3$GeNR$_2$ with PhNH$_2$; R = Me, Et (466).
Prop.: b$_{11}$ 141°, n$_D^{20}$ 1.5373, d$_{20}$ 1.1282 (466), IR spectrum (466).
Rxn. with: CH$_2$:CO → Et$_3$GeNPhAc (2736).
Ph$_2$C:CO → Et$_3$GeNPhCOCHPh$_2$ (2736).

Triethylgermyldimethylamine   GeC$_8$H$_{21}$N   Et$_3$GeNMe$_2$
Synth.: By rxn. of Et$_3$GeCl with Me$_2$NM in THF or heptane, in 75-85% yield;
M = Li, MgBr (466).
Prop.: b. 176°, n$_D^{20}$ 1.4498, d$_{20}$ 1.0235 (466).
Rxn. with: CH$_2$:CO → Et$_3$GeCH$_2$CONMe$_2$ (2513, 2736).
Ph$_2$C:CO → Et$_3$GeCPh$_2$CONMe$_2$ + Et$_3$GeOC(:CPh$_2$)NMe$_2$ (2736).
NH$_3$ → (Et$_3$Ge)$_2$NH + Me$_2$NH (1796).
RNH$_2$ → Et$_3$GeNHR; R = C$_6$H$_{13}$, Ph Ac (466).
3,5-Me$_2$C$_3$H$_2$N$_2$ → N-Et$_3$GeC$_3$HN$_2$Me$_2$ + Me$_2$NH (2378).
Ph$_2$PH → Et$_3$GePPh$_2$ (466).
CY$_2$ → Et$_3$GeY$_2$CNMe$_2$; Y = O, S (466).
i-PrSH → Et$_3$GeSCHMe$_2$ + Me$_2$NH (465).

Triethylgermyldiethylamine   GeC$_{10}$H$_{25}$N   Et$_3$GeNEt$_2$
Synth.: By rxn. of Et$_3$GeCl with Et$_2$NM in THF or heptane, in 75-85% yield;
M = Li, MgBr (466).
Prop.: b$_{10}$ 86°, n$_D^{20}$ 1.4551, d$_{20}$ 1.0010 (466).
Rxn. with: CH$_2$:CO → Et$_3$GeCH$_2$CONEt$_2$ (2513).
RNH$_2$ → Et$_3$GeNHR; R = C$_6$H$_{13}$, Ph (466).
PhNCY → Et$_3$GeNPhC(:Y)NEt; Y = O, S (466).
CS$_2$ → Et$_3$GeS$_2$CNEt$_2$ (466).
n-C$_8$H$_{17}$SH → n-C$_8$H$_{17}$SGeEt$_3$ + Et$_2$NH (465).

Bis(tributylgermyl)amine (M-126)   Ge$_2$C$_{24}$H$_{55}$N   (Bu$_3$Ge)$_2$NH
Synth.: By rxn. of Bu$_3$GeBr with NaNH$_2$ in C$_6$H$_6$ in N atm., in 92% yield (2502).
Prop.: b$_{0.05}$ 147-50°, colorless liq., easily hydrolyzed, mol. wt. (2502).
Rxn. with: BuLi → (Bu$_3$Ge)$_2$NLi (2502).
(CH$_2$)$_4$NH → N-Bu$_3$GeN(CH$_2$)$_4$ (1761).

Tributylgermyldimethylamine   GeC$_{14}$H$_{33}$N   Bu$_3$GeNMe$_2$
Synth.: By rxn. of Bu$_3$GeCl with Me$_2$NM in THF or heptane, in 75-85% yield;
M = Li, MgBr (466), in Et$_2$O, M = Li, in 60% yield (1761).
Prop.: b$_{19}$ 144-45° (466), b$_{15}$ 134-35° (1761), n$_D^{20}$ 1.4588 (1761), 1.4583
(466), d$_{20}$ 0.9607 (466).
Rxn. with: PhC⋮CH → Bu$_3$GeC⋮CPh (466, 1761).

$(CF_3)_2CO \rightarrow Bu_3GeOC(CF_3)_2NMe_2$ (2199).
$Ac_2O \rightarrow Bu_3GeOAc + Me_2NAc$ (466).

Tributylgermyldiethylamine (M-126)　　　$GeC_{16}H_{37}N$　　　Bu_3GeNEt_2
<u>Synth.</u>: By rxn. of Bu_3GeCl with Et_2NMgBr in THF or heptane, in 75-85% yield (466), in Et_2O, in 80% yield (2452).
<u>Prop.</u>: b_9 150° (466), b_2 105-105.5° (2452) n_D^{20} 1.4609, d_{20} 0.9573 (466), very sensitive to moisture (1761).
<u>Rxn. with</u>: $Ph_3SnH \rightarrow Bu_3GeSnPh_3 + Et_2NH$ (913).
$Bu_2SnH_2 \rightarrow (Bu_3GeSnBu_2)_2$ (913).
$nPhNCO \rightarrow Bu_3Ge(NPhCO)_nNEt_2$; n = 1-3 (2452).
$PhNCS \rightarrow$ adduct, $b_{0.07}$ 145° (2452).
$Cl_3CCHO \rightarrow$ 1:1 adduct (2452).
$\overline{OCH_2CH_2C}$:O $\rightarrow Bu_3GeO_2CCH_2CH_2NEt_2$ (2452); kinetic data and rxn. mechanism of O-alkyl cleavage (1236).
$\overline{OCH_2CH_2C}$:O (excess) $\rightarrow$ polymerization; kinetic data and activation energy (2452).
$(p\text{-}MeC_6H_4N:)_2C$, + HCl in $Et_2O \rightarrow p\text{-}(MeC_6H_4NH)_2C(NEt_2)Cl$ (2452).

Other triorganogermanium amides are listed in Table 65.

4.5.1.2 DIORGANOGERMANIUM DIAMIDES

Diorganogermanes containing germanium linked to two nitrogen atoms exist as monomers, cyclic small ring derivatives and polymers. The compounds listed in Table 66 on page 216 are prepared by the following methods.

I　Reaction with N-lithium derivatives:
　　$Ph_2GeCl_2 + Me_2NLi \rightarrow Ph_2Ge(NMe_2)_2$.

II　Condensation with organohalogermanes:
　　(IIA)　$Me_2GeCl_2 + RNH_2 \rightarrow$ c.$(Me_2GeNR)_3$; R = H, alkyl,
　　(IIB)　$Me_2GeCl_2 + Me_2NH \rightarrow Me_2Ge(NMe_2)_2$,
　　(IIC)　$Ph_2GeCl_2 + (MeNHCH_2)_2 \rightarrow Ph_2\overline{GeNMeCH_2CH_2NMe}$.

III　Interconversion reactions:
　　(IIIA)　$Me_2\overline{GeNMeCH_2CH_2NMe} + (NH_4)_2SO_4 \rightleftharpoons (Me_2GeNMeCH_2CH_2NMe)_x$,
　　(IIIB)　$Me_2Ge(NEt_2)_2 + Y(NH_2)_2 \rightarrow (Me_2GeNHYNH)_x$,
　　(IIIC)　$Me_2Ge(NEt_2)_2 + HN(CH_2CH_2)_2NH \rightarrow Me_2\overline{GeN(CH_2CH_2)_2N}$.

R = organic group, Y = hydrocarbon bridge.

Table 65. Triorganogermanium Amides $(R_3Ge)_nNR_{3-n}$

Formula*	Synth. Method**	Yield	Properties	Ref.
(Me$_3$Ge)$_2$NMe (126)	IIB	80	Mol. wt., NMR spectrum, soly.	1855
(Me$_3$Ge)$_2$NC$_8$H$_{17}$	IIIA (a)	82	b$_{13}$ 147°, n$_D^{20}$ 1.4604	1796
Me$_3$GeNHPh	IIA	--	b$_4$ 67-68°, mol. wt. (b)	2705
Me$_3$GeN(CH$_2$)$_4$CH$_2$	--	--	(c)	--
Me$_2$GeCH$_2$SCH$_2$GeMe$_2$NH	IIIB	74	m. 170-72°, colorless cryst., soly. (d)	2134
(Et$_3$Ge)$_3$N	IIB	40	b$_{0.04}$ 148-51°, n$_D^{20}$ 1.5108 (e)	2502
(Et$_3$Ge)$_2$NPh	IB	30	b$_{0.2}$ 120°, n$_D^{20}$ 1.5290	1796
Et$_3$GeNHC$_6$H$_{13}$	IB-IIIA	--	b$_{20}$ 135°, n$_D^{20}$ 1.4526, d$_{20}$ 0.9863 (f)	466
(Pr$_3$Ge)$_3$N	IIB	61	b$_{0.5}$ 198-200°, n$_D^{20}$ 1.4937 (e)	2502
(Pr$_3$Ge)$_2$NH	IIC	47	b$_{1.5}$ 128-30°, n$_D^{20}$ 1.4720 (e, g)	2502
(Bu$_3$Ge)$_3$N	IIB	39	b$_{0.05}$ 222-23°, n$_D^{20}$ 1.4880 (e)	2502
Bu$_3$GeNHC$_8$H$_{17}$ (126)	--	--	Microanal. detn.	426
Bu$_3$GeN(CH$_2$)$_3$CH$_2$ (127)	--	--	Very sensitive to moisture	1761
Bu$_3$GeN(CH$_2$)$_4$CH$_2$	--	--	(h)	--
(Me$_3$C)$_3$GeNMe$_2$	--	--	(i)	--
Ph$_3$GeNEt$_2$	IIA	--	Not isolated (j)	1039

* Numbers in parenthesis refer to pages in main volume.
** The characters used here correspond to the synthetic scheme in the introduction to Chapter 4.5.1.1.

(a) Synth. from C$_8$H$_{17}$NH$_2$ and (Me$_2$Ge)$_2$NH and (Me$_3$Ge)$_3$N at 50-100 and 120-80°, in 66 and 54% yield, resp. (1796).
(b) Spectra: IR and assignment (2705), ^{14}N (2705) and ^{15}N NMR (1733, 2705). (c) Rxn. with $\overline{OCH_2CH_2CO}$ → kinetic data, solvent effect and rxn. mechanism for CH$_2$-O-cleavage (1236-7). (d) Sensitive to moisture (2134). (e) Very sensitive to moisture, mol. wt. (2502). (f) IR spectrum (466). (g) Rxn. with BuLi → (Pr$_3$Ge)$_2$NLi (2502).
(h) Effect on polymerization of $\overline{OCH_2CH_2C:O}$, kinetic data and activation energy (2452). (i) Rxn. with t-BuSH → (Me$_3$C)$_3$GeSCMe$_3$ + Me$_2$NH (465). (j) Rxn. with NaC$_2$H in THF, + Ph$_3$SiCl → Ph$_3$GeC:CSiPh$_3$ (1039).

Dimethylgermylbis(diethylamine)　　　　$GeC_{10}H_{26}N_2$　　　　　　　$Me_2Ge(NEt_2)_2$

<u>Synth.</u>: By rxn. of Me_2GeCl_2 with Et_2NLi in Et_2O-C_6H_{14} at r.t. in N atm., in 70% yield (2162).
By rxn. of $Me_2Ge(OPh)_2$ with Et_2NLi in petr. ether under reflux (277).
<u>Prop.</u>: b_2 106-108°, n_D^{20} 1.5025 (277), IR (277), assignment (2162) and NMR (1731) spectra, stable in boiling 3NNaOH (277).
<u>Rxn. with</u>: (H_2NCH_2 → polymeric dimethylgermylamine (2163).
$(MeNHCH_2)_2$ at 150° → $\overline{Me_2GeNMeCH_2CH_2NMe}$ + Et_2NH (2162).
Piperazine at 120° ⇌ $Me_2GeN(CH_2CH_2)_2N$ + $[Me_2GeN(CH_2CH_2)_2N]_x$ (2163).

1,2,2,3-Tetramethyl-1,3,2-Diaza-　　$GeC_6H_{16}N_2$　　　　$\overline{Me_2GeNMeCH_2CH_2NMe}$
  germacyclopentane

<u>Synth.</u>: By rxn. of $Me_2Ge(NEt_2)_2$ with $(MeNHCH_2)_2$ at 150° with cat. $(NH_4)_2SO_4$ or at 100° without catalyst, in 58% yield (2162).
By rxn. of Me_2GeCl_2 with $(CH_2NMeLi)_2$ in C_6H_6-C_6H_{14} at 0° in N atm., in 45% yield (2162).
<u>Prop.</u>: b_{742} 152-53°, mol. wt. (2162), IR (1731, 2162), assignment (2161) and NMR (1731, 2162) spectra, basicity (1731).
<u>Rxn. with</u>: $PhPOCl_2$ → $\overline{PhPNMeCH_2CH_2NMe}$ (2162).
Cat. $(NH_4)_2SO_4$ ⇌ $(Me_2GeNMeCH_2CH_2NMe)_x$ (2163).

Other diorganogermanium diamides are listed in Table 66.

4.5.1.3 MONOORGANOGERMANIUM TRIAMIDES

Methylgermaniumtris(dimethylamine) is listed at the bottom of Table 66.

4.5.2 ORGANOGERMANIUM COMPOUNDS WITH SUBSTITUTED AMINES

4.5.2.1 ORGANOGERMANIUM DERIVATIVES OF HETEROCYCLIC AMINES

Triorganogermylamines with heterocyclic bases are listed in Table 67 on page 217. A quinoline derivative is listed in Subsection 4.3.3.2.6.

4.5.2.4 ORGANOGERMANIUM DERIVATIVES OF AMIDINES, IMIDES, AND HYDRAZINE

Organogermanes containing a germanium linkage to nitrogen of ketoneimide amidine and hydrazine are listed at the bottom of Table 67 on page 217.

Table 66. Diorganogermanium Diamides $R_2Ge(NR'_2)_2$

Formula*	Synth. Method**	Yield	Properties	Ref.
c.(Me$_2$GeNMe$_2$)$_3$ (126)	IIA	78	b$_2$ 80°, mol. wt., NMR spectrum (a)	1855
Me$_2$Ge(NMe$_2$)$_2$	IIB	--	Mixture with Me$_2$Ge(NMe$_2$)Cl, b$_{745}$ 144° (b)	2163
(Me$_2$GeNHCH$_2$CH$_2$NH)$_x$	IIIB	--	Paste, struct., NMR spectrum	2163
(Me$_2$GeNMeCH$_2$CH$_2$NMe)$_x$	IIIA	--	Mol. wt., depolymerizes on distn. (c)	2163
Me$_2$GeN(CH$_2$CH$_2$)$_2$N	IIIC	--	NMR spectrum, mol.wt., mixture (d)	2163
c.(Et$_2$GeNH)$_3$	IIA (e)	41	b$_3$ 137°, n$_D^{20}$ 1.4891, d$_{20}$ 1.3235 (f)	1535
c.(Bu$_2$GeNH)$_3$ (127)	--	--	(g)	--
Ph$_2$Ge(NMe$_2$)$_2$	I	41	b$_{13}$ 181-82°	2163
Ph$_2$GeNMeCH$_2$CH$_2$NMe	IIC	--	b$_{13}$ 172°, mol. wt. (h)	2163
MeGe(NMe$_2$)$_3$	--	--	NMR spectrum (i)	1546

* Numbers in parenthesis refer to pages in main volume.
** The characters used here correspond to the synthetic scheme in the introduction to Chapter 4.5.1.2.

(a) Rxn. with MeLi → Me$_3$GeNMeLi (1835). (b) NMR spectrum, GLC (2163). (c) Structure of polymer, polymerization followed by refraction index and viscosity (2163). (d) Not separated from by-prod. polymer (2163). (e) From R$_2$GeCl$_2$ with liq. NH$_3$ and Na (1535). (f) IR spectrum (1535), rxn. with MeOH → Et$_2$Ge(OMe)$_2$ + NH$_3$ (1535), with AlCl$_3$ → Et$_3$GeCl (1535). (g) Very sensitive to moisture, mol. wt (1761). (h) IR spectrum and assignment (2161). (i) Rxn. with MeSiX$_3$ → equilibrium constants for products of redistribution rxn. followed by NMR spectroscopy (1546).

Table 67. Organogermanium Derivatives of Substituted Amines $R_nGe(NRR')_{4-n}$

Formula*	Synth. Method**	Yield	Properties	Ref.
$Me_3GeN_2C_3H_3$ (a)	IIA	--	b_{65} 106°, n_D^{20} 1.4859 (b, c)	2378
$Me_3GeN_2C_3H_2Me$-3 (a)	IIIA	--	b_{25} 95°, n_D^{20} 1.4860 (b)	2378
$Me_3GeN_2C_3HMe_2$-3,5 (a)	IIIA	--	b_{28} 109°, n_D^{20} 1.4900 (b)	2378
$Me_3GeN_2C_7H_5$ (d)	IIIA	--	b_{15} 146°, n_D^{20} 1.5668 (b)	2378
$Et_3GeN_2C_3H_3$ (a)	IVB	--	b_{23} 121°, n_D^{20} 1.4878 (b)	2378
$Et_3GeN_2C_3HMe_2$-3,5 (a)	IIIA	--	b_{10} 121°, n_D^{20} 1.4939 (b)	2378
$EtPh(1-C_{10}H_7)GeNC_4H_4$	IIA	--	$[\alpha]_D^{25}$ -1.6° (C_6H_6) (e)	2368
$Bu_3GeNC_4H_2Me_2$-2,4 (d) (126)	--	--	Microanal. detn. (f)	426
$Bu_3GeNC_8H_6$ (d) (127)	--	--	Microanal. detn. (f)	426
$Bu_3GeN_2C_3H_3$ (a) (127)	--	--	IR spectrum (f)	1761
$Bu_3GeN_2C_3H_3$ (d) (127)	--	--	Microanal. detn. (f)	426
$Me_3GeN:CPh_2$	IIA	--	$b_{0.2}$ 109-10°, yellow oil (g)	852
$Me_3GeN:CPhNMe_2$	II (h)	14	m. 30-32°, $b_{0.1}$ 76-77° (i)	2738
$Me_3GeN:CPhNMeSiMe_3$	IIIA	42	$b_{0.1}$ 79-81°, mol. wt. (i)	2738
$(Et_3GeNPh)_2$	IVA	85	m. 56°	1365
$(Et_3GeNCO_2Et)_2$	IVA	95	$b_{0.0001}$ 120°	1365
$Ph_3GeN:CPh_2$	IIA	--	m. 127-32°, yellow cryst., mol. wt. (g)	852
$Ph_3Ge(EtN)_2C_7H_5$ (j)	IIA	--	Not isolated, NMR spectrum	1004
$(Ph_2GeNPhNPh)_2$	II (k)	23	m. 306-307° (dec.)	1089

* Numbers in parenthesis refer to pages in main volume.

** The characters used here correspond to the synthetic scheme in the introduction to Chapter 4.5.1.1.
(a) $C_3H_3N_2$ = pyrazolyl, 3-$MeC_3H_2N_2$ = 3-methylpyrazolyl, 3,5-$Me_2C_3HN_2$ = 3,5-dimethylpyrazolyl. (b) NMR spectrum (2378). (c) Contains 7-8% $(Me_3Ge)_2O$ (2378). (d) $C_7H_5N_2$ = imidazolyl, C_4H_4N = pyrryl, C_8H_6N = indolyl, $C_3H_3N_2$ = imidazolyl, 2,4-$Me_2C_4H_2N$ = 2,4-dimethylpyrrolyl   (e) Synth. with inversion of configuration (2368), rxn. with $LiAlH_4 \rightarrow R_3GeH$ (2368), rxn. with BuLi → n-$BuGeR_3$; both rxns. with inversion (2368).   (f) Sensitive to moisture (1761). (g) Sensitive to moisture, UV, IR spectra and assignment (852). (h) Synth. from Me_2NH with BuLi and PhCN and Me_3GeCl (2738). (i) IR and NMR spectra (2738). (j) Triphenylgermyl-N,N'-diethylaminotroponeimine. (k) Synth. from Ph_2GeCl_2 and $(PhNLi)_2$ (1089).

4.5.3 ORGANOGERMANIUM DERIVATIVES OF ACID AMIDES AND ACID IMIDES

This chapter comprises derivatives of acid amides and cyclic acid imides. Also included are N-organogermanium derivatives of carbamic acid, urea and similar compounds. The following scheme shows synthetic methods for the compounds listed in Table 68.

I Condensation reaction:
$Et_3GeNMe_2 + AcNH_2 \rightarrow Et_3GeNHAc + Me_2NH$.

II Addition to organogermanium amides and phosphides:
(IIA) $Et_3GeNHPh + R_2C:CO \rightarrow Et_3GeNPhCOCHR_2$,
(IIB) $R_3GeNEt_2 + PhNCY \rightarrow R_3Ge(NPhCY)_nNEt_2$,
(IIC) $Et_3GePEt_2 + PhNCY \rightarrow Et_3GeNPhC(Y)PEt_2$.

III Addition to organogermanium alkoxides:
(IIIA) $Bu_3GeOR + PhNCO \rightarrow Bu_3GeNPhCO_2R$,
(IIIB) $Bu_3GeOMe + (RN:)_2C \rightarrow Bu_3GeNRC(OMe):NR$.

R = organic group or hydrogen, Y = oxygen and sulfur n = 1-3.

Table 68. Organogermanium Derivatives of Acid Amides R_3GeNRR'

Formula*	Synth. Method**	Yield	Properties	Ref.
$Et_3GeNHAc$	I	--	$b_{0.2}$ 82°, n_D^{20} 1.4713, d_{20} 1.1381	466
$Et_3GeNPhAc$	IIA	60	$b_{0.007}$ 85-86°, n_D^{20} 1.5252 (a, b)	2736
$Et_3GeNPhCOCHPh_2$	IIA	76	$b_{0.4}$ 200°, n_D^{20} 1.5781 (a, c)	2736
$Et_3GeNPhCONEt_2$	IIB	80	$b_{0.2}$ 119°, n_D^{20} 1.520, d_{20} 1.1120 (d)	466
$Et_3GeNPhCSNEt_2$	IIB	60	$b_{0.3}$ 143°, n_D^{20} 1.566, d_{20} 1.1288 (d)	466
$Et_3GeNPhCOPEt_2$	IIC	55	$b_{0.1}$ 83°, n_D^{20} 1.5390 (e)	1797
$Et_3GeNPhCSPEt_2$	IIC	66	$b_{0.2}$ 92°, n_D^{20} 1.5745 (e)	1797
$Bu_3GeN(COCH_2)_2$ (127)	--	--	(f)	--
$Bu_3GeN(CO)_2C_6H_4$-o (127)	--	--	(f)	--
$Bu_3GeNPhCONEt_2$	IIB	--	IR spectrum (g)	2452
$Bu_3Ge(NPhCO)_2NEt_2$	IIB	--	IR spectrum (h)	2452
$Bu_3Ge(NPhCO)_3NEt_2$	IIB	--	IR spectrum (i)	2452
$Bu_3GeNRC(OMe):NR$	IIIB	--	R = p-MeC_6H_4	1228
$Bu_3GeNPhCO_2Me$	IIIA	--	$b_{0.08}$ 125-27° (j)	1228
$Bu_3GeNPhCO_2Et$	IIIA	--	$b_{0.1}$ 134-35° (j)	1228

* Numbers in parenthesis refer to pages in main volume.
** The characters used here correspond to the synthetic scheme in the introduction to Chapter 4.5.3.

(a) IR and NMR spectra (2736). (b) Rxn. with aq. $NH_3 \rightarrow (Et_3Ge)_2O$ + AcNHPh (2736). (c) Rxn. with air $\rightarrow$ (Et Ge)$_2$O + $Ph_2CHCONHPh$ (2736). (d) Rxn. with $H_2O \rightarrow (Et_3Ge)_2O$ + $PhNHCYNEt_2$; Y = O, S (466). (e) IR and NMR spectra (1797). (f) Microanalytical detn. (426). (g) Rxn. with $H_2O \rightarrow PhNHCONEt_2$ (2452). (h) Rxn. with $H_2O \rightarrow PhNHCONPhCONEt_2$ (2452). (i) Rxn. with $H_2O \rightarrow PhNH(CONPh)_2$-$CONEt_2$ (2452). (j) Very sensitive to moisture (1228).

4.5.4 ORGANOGERMANIUM METAL IMIDES

A number of organogermanium metal imides containing group I, II and IV metals have recently been reported. The derivatives of lithium and aluminum are very sensitive to air and moisture. Derivatives with boron and other group IV metals are rather stable. The following scheme lists synthetic method for the organogermanium metal imides in Table 69.

I Reactions organolithium compounds:
  (IA) $Me_3GeNRLi + Me_3MCl \rightarrow Me_3GeNRMMe_3$,
  (IB) $R'Me_2SiNRLi + Me_3GeCl \rightarrow Me_3GeNRSiMe_2R'$;
     R' = Me, Me_3SiNMe ,
  (IC) $Me_2Ge(NRSiMe_3)Cl + MeLi \rightarrow Me_3GeNRSiMe_3$.

II Condensation reactions with ammonia:
  (IIA) $Me_3GeCl + Me_3SiCl + NH_3 \rightarrow Me_3GeNHSiMe_3 + NH_4Cl$,
  (IIB) $(Me_2ClGeC)_2B_{10}H_{10} + NH_3 \rightarrow (Me_2GeCB_{10}H_{10}CGeMe_2NH)n + NH_4Cl$;
     n = 1, x .

III Reaction with aluminumalkyl in Ar atm.:
  $(Et_3Ge)_2NH + Et_3Al \rightarrow (Et_3Ge)_2NAlEt_2$.

Lithium Trimethylgermylmethylamide $GeLiC_4H_{12}N$ $Me_3GeNMeLi$
  (M-128)
Synth.: By rxn. of $(Me_2GeNMe)_3$ or the crude rxn. prod. Me_2GeCl_2 with $MeNH_2$ (1:3) with MeLi in Et_2O in N atm. under reflux (1855).
Prop.: NMR spectrum, stable in N atm. in the dark (1855).
Rxn. with: $Me_3MCl \rightarrow Me_3GeNMeMMe_3 + LiCl$; M = Si, Ge, Sn, Pb (1855)

Lithium Bis(triethylgermyl)amide $Ge_2LiC_{12}H_{30}N$ $(Et_3Ge)_2NLi$
Synth.: By rxn. of $(Et_3Ge)_2NH$ with BuLi in hexane in N atm. under reflux (2502).
Prop.: Brownish solid dec. on heating in vacuo, sensitive to air, soly. (2502).
Rxn. with: $Et_3GeBr \rightarrow (Et_3Ge)_3N$ (2502).

Lithium Bis(tripropylgermyl)amide $Ge_2LiC_{18}H_{42}N$ $(Pr_3Ge)_2NLi$
Synth.: By rxn. of $(Pr_3Ge)_2NH$ with BuLi in hexane in N atm. under reflux (2502).
Prop.: Orange-red syrup, sensitive to air soly. (2502).
Rxn. with: $Pr_3GeBr \rightarrow (Pr_3Ge)_3N$ (2502).

Addn. Compd.: $(Pr_3Ge)_2NLi \cdot C_4H_8O_2$, from $(Pr_3Ge)_2NLi$ and 1,4-dioxane in N atm., white solid, dec. on standing $[Ge_2LiC_{22}H_{50}NO_2]$ (2502).

Lithium Bis(tributylgermyl)amide $Ge_2C_{24}H_{54}N$ $(Bu_3Ge)_2NLi$
Synth.: By rxn. of $(Bu_3Ge)_2NH$ with BuLi in hexane in N atm. under reflux (2502).
Prop.: Orange-red syrup, sensitive to air, soly. (2502).
Rxn. with: $Bu_3GeBr \rightarrow (Bu_3Ge)_3N$ (2502).
Addn. Compd.: $(Bu_3Ge)_2NLi \cdot C_4H_8O_2$, from $(Bu_3Ge)_2NLi$ and 1,4-dioxane in N atm., white solid, dec. on standing, soly. $[Ge_2LiC_{28}H_{62}NO_2]$ (2502).

Additional organogermanium metal imides are listed in Table 69 and in Subchapters 4.1.7, 4.3.3.4 and 4.5.5.

Table 69. Organogermanium Metal Imides R_3GeNRM

Formula*	Synth. Method**	Yield	Properties	Ref.
$(Me_3Ge)_2NLi$	--	--	(a)	--
$Me_3GeN(SiMe_3)_2$ (128)	--	--	NMR spectrum (474)	--
$Me_3GeNHSiMe_3$	IIA	38	b_{752} 133-34°, n_D^{20} 1.4183, d_{20} 0.9587 (b)	1534
$Me_3GeNMeSiMe_3$ (128)	IA	74	b_{12} 42°, mol. wt., soly. (c)	1855
$Me_3GeNMeSnMe_3$ (128)	IA	78	b_2 28°, mol. wt., soly. (c)	1855
$Me_3GeNMePbMe_3$ (128)	IA	48	b_2 49°, mol. wt., soly. (c, d)	1855
$Me_3GeNHSiMe_2NMeSiMe_3$	IB	--	$b_{0.2}$ 45-47° (e)	1810
$Me_3GeNCMe_3SiMe_3$	IB	4	b_1 47°, mol. wt., NMR and (far) IR spectra and assignment	2749
	IC	24		2749
$Me_3Ge(Me_3Sn)NSiMe_2NMeSiMe_3$	I (e)	--	$b_{0.8}$ 84-86°	1810
o-$Me_2GeCB_{10}H_{10}CGeMe_2NH$ (f)	IIB	100	m. 166-68°, IR and mass spectra	1835
(m-$Me_2GeCB_{10}H_{10}CGeMe_2NH)_x \cdot NH_3$ (g)	IIB	--	m. 182-88°, IR spectrum, mol. wt.	1835
$(Et_3Ge)_2NAlEt_2$	III	--	b_3 241°	1535

* Numbers in parenthesis refer to page numbers in main volume.
** The characters used here correspond to the synthetic scheme in the introduction to Chapter 4.5.4.
(a) Rxn. with $Me_3GeCl \rightarrow (Me_3Ge)_3N$ + LiCl (1796). (b) Rxn. with PhC⋮CH $\rightarrow$ $Me_3GeC⋮CPh$ + $(Me_3Si)_2NH$ + NH_3 (1534). (c) NMR spectrum (1855). (d) Sensitive

to light (1855). (e) Rxn. of Me$_3$GeNHSiMe$_2$NMeSiMe$_3$ with MeLi at -78°, followed by Me$_3$SnCl → Me$_3$Ge(Me$_3$Sn)NSiMe$_2$NMeSiMe$_3$ (1810). (f) 1,2-carborane. (g) 1,7-carborane.

4.5.5 ORGANOGERMANIUM AMIDO AND IMIDO PHOSPHINES

A series of new organogermanium imino phosphines has been reported in the literature. Two new synthetic routes (I) and (III) leading to imino phosphines are used as shown in the following scheme.

I Reaction with lithium amide:
   Me$_3$GeCl + R$_3$P:NLi → Me$_3$GeN:PR$_3$.

II Azide route:
   Me$_n$Ge(N$_3$)$_{4-n}$ + R$_3$P → Me$_n$Ge(N:PR$_3$)$_{4-n}$; n = 2,3.

III Interconversion reactions:
   (IIIA) Me$_3$GeNHPR$_2$ + BuLi, + MeCl → Me$_3$GeN:PMeR$_2$,
   (IIIB) Me$_3$GeNHPR$_2$ + BuLi, + Me$_3$MCl → Me$_3$GeN:PR$_2$MMe$_3$; M = Si, Ge, Sn ,
   (IIIC) Me$_3$GeNHPR$_2$ + CCl$_4$ → Me$_3$GeN:PR$_2$Cl ,
   (IIID) Me$_3$GeNHPR$_2$ + S → Me$_3$GeNHP(S)R$_2$.

Trimethylgermylaminodi(tertiarybutyl)-   GeC$_{11}$H$_{28}$NP   Me$_3$GeNHP(CMe$_3$)$_2$
   phosphorane

Synth.: By rxn. of Me$_3$GeCl with t-Bu$_2$PNHLi in Et$_2$O-hexane in N atm., in 73% yield (1813 2739).
Prop.: m. -6 -4°, b$_{0.5}$ 46°, IR and NMR spectra, mol. wt. (1813. 2739).
Rxn. with: RLi, + MeCl → Me$_3$GeN:PMe(CMe$_3$)$_2$ (2739).
RLi, + Me$_3$MCl → Me$_3$GeN:P(CMe$_3$)$_2$MMe$_3$; M = Si (1813, 2739), Ge, Sn (2739).
CCl$_4$ → Me$_3$GeN:P(CMe$_3$)$_2$Cl (2739).
S → Me$_3$GeNHPS(CMe$_3$)$_2$ (2739).

Other organogermanium imino phosphoranes are listed in Table 70 and in Chapters 4.2.1 and 4.3.3.4.

Table 70 Organogermanium Imino and Amino Phosphines $R_3GeN:PR_3$

Formula*	Synth. Method**	Yield	Properties	Ref.
$Me_3GeN:Ph_3$	I	--	b_{12} 69-70° (a)	1824
	II	79	Mol. wt., NMR spectrum	2743
$Me_3GeN:PMe(CMe_3)_2$	IIIA	83	$b_{0.1}$ 50°, stable at 200° (b)	2739
$Me_3GeN:PEt_3$	I	48	$b_{0.5}$ 55-57°, mol. wt. (a)	1824
	II	89	$b_{0.8}$ 57-58°, NMR spectrum	2743
$Me_3GeN:PPh_3$	I	36	m. 78-80°, mol. wt. (a)	1824
$Me_3GeN:P(CMe_3)_2Cl$	IIIA	75	m. 4°, $b_{0.05}$ 46°, mol. wt. (b)	2739
$Me_3GeNHPS(CMe_3)_2$	IIID	92	m. 96-98°, mol. wt. (b)	2739
$Me_3GeN:P(CMe_3)_2SiMe_3$	IIIB	75	Subl. 0.1 90°, soly. (b, c)	1813, 2739
$Me_3GeN:P(CMe_3)_2GeMe_3$	IIIB	71	Subl. 0.1 120°, mol. wt. (b)	2739
$Me_3GeN:P(CMe_3)_2SnMe_3$	IIIB	52	Subl. 0.1 120°, mol. wt. (b)	2739
$Me_2Ge(N:PMe_3)_2$	II	33	m. 9-10°, $b_{0.01}$ 72-75°, mol. wt. (d)	2743
$Bu_2Ge[N:P(OMe)_3]_2$ (128)	--	--	(e)	--

* Numbers in parenthesis refer to pages in main volume.

** The characters used here correspond to the synthetic scheme in the introduction to Chapter 4.5.5.
(a) NMR and (far) IR spectra, assignment (1824). (b) IR and NMR spectra (2739), very sensitive to moisture (2739). (c) Thermal rearrangement → $Me_3GeP(CMe_3)_2:NSiMe_3$ (2739). (d) IR and NMR spectra, polymerizes above 75° (2743). (e) The series of tri- and diorganogermanium imino phosphoranes, arsanes and stibanes listed in main volume Table 45, page 128 is claimed as UV stabilizers and antioxidants for synthetic resins (2116).

4.5.6 ORGANOGERMANIUM PHOSPHIDES, ARSENIDES ANTIMONIDES, AND BISMUTHIDES

4.5.6.1 ORGANOGERMANIUM PHOSPHIDES

Organogermanium phosphorous derivatives include phosphines, phosphine imines, tris(trimethylgermyl)phosphine complexes with transition metals and mixed organosilicon and organotin phosphides. The compounds listed in Table 71 are prepared by methods from the following scheme.

I Synthesis with organolithium compounds:
  (IA) $Ph_nGeX_{4-n} + R'Li \rightarrow Ph_nGeR'_{4-n}$,
  (IB) $Me_3GeCl + R_2PN(SiMe_3)Li \rightarrow Me_3GePR_2:NSiMe_3$,
  (IC) $Me_nGeCl_{4-n} + LiAl(PH_2)_4 \rightarrow Me_nGe(PH_2)_{4-n}$,
  (ID) $Et_3GeLi + R'Br \rightarrow Et_3GeR'$.

II Reactions with heteroorganogermanes:
  (IIA) $Ph_3GeCl + Ph_nPH_{3-n} \rightarrow (Ph_3Ge)_{3-n}PPh_n$,
  (IIB) $(Me_3Ge)_nNH_{3-n} + Ph_2PH \rightarrow Me_3GePPh_2$,
  (IIC) $Et_3GeH + Et_3Bi \rightarrow (Et_3Ge)_nBiEt_{3-n}$.

III Miscellaneous reactions:
  (IIIA) $(Me_3Ge)_3P + M(CO)_{m+1} \rightarrow (Me_3Ge)_3PM(CO)_m$,
  (IIIB) $Me_3GeN:PR_2SiMe_3 + \text{heat} \rightarrow Me_3GePR_2:NSiMe_3$,
  (IIIC) [structure: Ph-substituted bicyclic GeMe$_2$ compound with CO$_2$Me groups] + BzH $\rightarrow (Me_2GePPh_3)_x$, + Ph$_3$P
  (IIID) $(Et_3Ge)_3Sb + PhCH_2Br \rightarrow (Et_3Ge)_2SbCH_2Ph$.

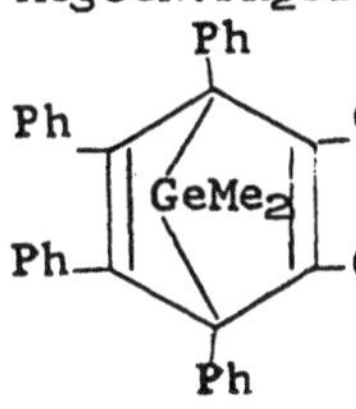

R = organic group, R' = phosphine, arsine or stibine group, X = halogen, M = Cr, Fe, Ni, Co(NO), m = 2-5, n = 1-3.

Tris(trimethylgermyl)phosphine $Ge_3C_9H_{27}P$ $(Me_3Ge)_3P$
Synth.: By rxn. of Me_3GeNMe_2 with PH_3 in Et_2O in inert atm., in 60% yield (1853).
By rxn. of Me_3GeCl with PNa_3 in liq. NH_3 (1853).
Prop.: Colorless liq. $b_{0.1}$ 62-63° (1853), spectra: (far) IR and assignment (1019, 1853), Raman and assignment (1019) and ^{31}P NMR (1019), mol. wt. (1853), sensitive to air and moisture (1853).
Rxn. with: $Cr(CO)_6$ under UV irradiation $\rightarrow (Me_3Ge)_3PCr(CO)_5 + CO$ (1852).
$Fe(CO)_5$ under UV irradiation $\rightarrow (Me_3Ge)_3PFe(CO)_4 + CO$ (2748).
$Fe_2(CO)_9$ under UV irradiation $\rightarrow (Me_3Ge)_3PFe(CO)_4 + Fe(CO)_5$ (2748).
$Co(CO)_3NO \rightarrow (Me_3Ge)_3PCo(CO)_2NO + CO$ (1852).
$Ni(CO)_4 \rightarrow (Me_3Ge)_3PNi(CO)_3 + CO$ (1851).

Triethylgermyldiethylphosphine $GeC_{10}H_{25}P$ Et_3GePEt_2
Synth.: By rxn. of Et_3GeCl with Et_2PLi in THF, in 77% yield (1797).
Prop.: b_{15} 120°, n_D^{20} 1.4845 (1797).

Rxn. with: PhC⋮CH + cat. AIBN → Et$_3$GeCH:CHCPhPEt$_2$ + Et$_3$GeCPh:CHPEt$_2$ + Et$_3$Ge-C⋮CPh + Et$_2$PH (1797).
CH$_2$:CHCN → Et$_3$GeCH(CN)CH$_2$PEt$_2$ (1797).
NH$_3$ → (Et$_3$Ge)$_2$NH + Et$_2$PH (1797).
RCHO → Et$_3$GeOCHRPEt$_2$; R = n-Pr, Ph (2735).
RCH:CHCHO → Et$_3$GeOCH:CHCHRPEt$_2$; R = Me, Ph (2735).
PhNCY → Et$_3$GeNPhC(Y)PEt$_2$; Y = O, S (1797).
CS$_2$ → Et$_3$GeS$_2$CPEt$_2$ (1797).
HX → Et$_3$GeX + Et$_2$PH ; X = Cl, PhNH (1797).

Triethylgermyldiphenylphosphine (M-127)   GeC$_{18}$H$_{25}$P   Et$_3$GePPh$_2$
Synth.: By rxn. of Et$_3$GeBr with Ph$_2$PLi in THF, in 79% yield (50).
By rxn. of Et$_3$GeNMe$_2$ with Ph$_2$PH (466).
Prop.: b$_{0.001}$ 146°, IR (50) and far IR spectra (105).
Rxn. with: PhLi → Et$_3$GePh + Ph$_2$PLi (50).
LiAlH$_4$ → Et$_3$GeH + Ph$_2$PH (466).
AgI → [Et$_3$GePPh$_2$AgI]$_4$ (50).
Dry O$_2$ in Et$_2$O → Et$_3$GeO$_2$PPh$_2$ (50).
PhNCY, + H$_2$O → (Et$_3$Ge)$_2$O + Ph$_2$PC(Y)NHPH; Y = O, S (1797).
MeI → Et$_3$GeI + [Me$_2$PPh$_2$]I (50).

Triphenylgermyldiphenylphosphine   GeC$_{30}$H$_{25}$P   Ph$_3$GePPh$_2$
Synth.: By rxn. of Ph$_3$GeBr with Ph$_2$PLi in THF (50).
Prop.: m. 154-56° (50), far IR spectrum and assignment (105).
Rxn. with: H$_2$O in (MeOCH$_2$)$_2$ → (Ph$_3$Ge)$_2$O + Ph$_2$PH (50).
Dry O$_2$ in C$_6$H$_6$ → Ph$_3$GeO$_2$PPh$_2$ (50).
EtBr → Ph$_3$GeBr + [Et$_2$PPh$_2$]Br (50).
MeI → Ph$_3$GeI + [Me$_2$PPh$_2$]I (50).

Other organogermanium phosphides are listed in Table 71 and in Chapter 6.5.1.

4.5.6.4 ORGANOGERMANIUM ARSENIDES, ANTIMONIDES AND BISMUTHIDES

Tris(trimethylgermyl)arsine   Ge$_3$AsC$_9$H$_{27}$ (Me$_3$Ge)$_3$As
Synth.: By rxn. of Me$_3$GeCl with Na$_3$As in liq. NH$_3$ under exclusion of air (1853).
By rxn. of Me$_3$GeNMe$_2$ with AsH$_3$ in Et$_2$O in inert atm., in 60% yield (1853).
Prop.: Colorless liq., b$_{0.1}$ 67-68°, NMR, far IR spectra and assignment, sensitive to air and moisture (1853).

Tris(trimethylgermyl)stibine   Ge$_3$SbC$_9$H$_{27}$ (Me$_3$Ge)$_3$Sb
Synth.: By rxn. of Me$_3$GeBr with Li$_3$Sb in Et$_2$O in N atm. under reflux, in 85% yield (676).
Prop.: Colorless liq. m. 11-13°, vapor pressure at -25°, monomeric, dec. in air, pyrolysis at 300-400° → Sb-Ge mirror (676).

Tris(trimethylgermyl)bismuthine $Ge_3BiC_9H_{27}$ $(Me_3Ge)_3Bi$

<u>Synth.:</u> By rxn. of Me_3GeCl with Na_3Bi in liq. NH_3 in dry Ar atm., in 5-10% yield (1854).

<u>Prop.:</u> Slightly yellow oil b_1 114-16°, NMR spectrum dec. on standing, sensitive to moisture, air and light (1854).

<u>Rxn. with:</u> Thermal dec. → Me_6Ge_2 + Bi (1854).

Moisture → $(Me_3Ge)_2O$ + Bi (1854).

Tris(triethylgermyl)stibine $Ge_3SbC_{18}H_{45}$ $(Et_3Ge)_3Sb$

<u>Synth.:</u> By rxn. of Et_3GeH with Et_3Sb at 200° in sealed tube, in 75% yield (2086), with $(Et_3Si)_3Sb$ at 230° in sealed tube, in 59% yield (2086).

<u>Prop.:</u> $b_{2.5}$ 164-67°, b_1 157-61°, d_{20} 1.386-1.392 (2086), NMR (1009) and far IR spectra and assignment (2374), thermal dec. at 280° (2086), sensitive to air (2086).

<u>Rxn. with:</u> Et_3SnH → Et_3GeH + $(Et_3Sn)_3Sb$ (2086).

PhBr at 30° under UV irradiation → Et_3GeBr + Ph_3Sb + some Sb (2086).

RBr → Et_3GeBr + R_3Sb ; R = CH_2Ph, C_5H_{11}, + $(Et_3Ge)_2SbCH_2Ph$ (2086).

$(CH_2Br)_2$ → Et_3GeBr + Sb + C_2H_4 (2086).

Br in C_7H_{16} → Et_3GeBr + $SbBr_3$ (2851).

Tris(triethylgermyl)bismuthine $Ge_3BiC_{18}H_{45}$ $(Et_3Ge)_3Bi$

<u>Synth.:</u> By rxn. of Et_3GeH with Et_3Bi at 140-45°, in 62% yield (291), in sealed tube, in 55% yield (2086), with $(Et_3Si)_3Bi$ at 180° in sealed tube, in 73% yield (2086).

<u>Prop.:</u> $b_{2.5}$ 163-68°, d_{20} 1.586 (291, 2086), readily oxidized by air, stable in N atm. to 200° (291).

<u>Rxn. with:</u> Thermal dec. at 270° → Et_6Ge_2 + Bi (291, 2086).

Et_3SnH → Et_3GeH + Et_6Sn_2 + Bi (291, 2086).

AcOH → Et_3GeH + Et_3GeOAc + Bi (291, 2086).

Bz_2O_2 → Et_3GeOBz + Bi (291, 2086).

Other organogermanium arsenides, antimonides and bismuthides are listed at the bottom of Table 71 and in Chapter 6.1.1.

Table 71. Organogermanium Phosphides, Arsenides, Antimonides and Bismuthides $(R_3Ge)_nPR'_{3-n}$

Formula	Synth. Method*	Yield	Properties	Ref.
$(Me_3Ge)_3PCr(CO)_5$	IIIA (a)	--	Yellow cryst., dec. 150°, stable in air (b)	1852
$(Me_3Ge)_3PFe(CO)_4$	IIIA (a, c)	100	Yellow cryst., dec. 120°, air sensitive (d)	2748
$(Me_3Ge)_3PCo(CO)_2NO$	IIIA	--	Red cryst., dec. 132°, stable in air (b)	1852
$(Me_3Ge)_3PNi(CO)_3$	IIIA	100	Dec. 100°, stable in air (e)	1851
Me_3GePH_2	IC	69	--	2652
$Me_3GeP(GMe_3)_2:NSiMe_3$	IB	88	Subl. o.1 90-100°, soly., mol. wt.,	1813, 2739
	IIIB	100	Sensitive to moisture (f)	1813, 2739
Me_3GePPh_2	IIB	--	b_{12} 185-87°, n_D^{20} 1.6089	1796
$Me_2Ge(PH_2)_2$	IC	46	--	2652
$(Me_2GePPh_3)_x$	IIIC	--	Pale yellow solid, m. 210-55° (g)	2882-3
$(Ph_3Ge)_3P$	IIA	--	m. 128° (h, i)	1849
$(Ph_3Ge)_2PPh$	IIA	--	m. 110° (h)	1849
$Ph_3GePPhSnPh_3$	IA	31	m. 115-19° (j)	1850
$Ph_3GeP(SnPh_3)_2$	IA	50	m. 160° (j)	1850
$Ph_2Ge(PPh_2)_2$	IA	--	Colorless needles, m. 182-85° (k)	50
$[PhGePPh_2]_2$	IA	--	m. 108-12°	50
$Me_3GeAsMe_2$	IA	91	b_{100} 85-87°, (far) IR and NMR spectra	2201
$(Et_3Ge)_2SbCH_2Ph$	IIID	57	$b_{1.5}$ 155-59°, n_D^{20} 1.5979, d_{20} 1.379	2086
$Et_3GeSbEt_2$	ID	51	$b_{1.5}$ 75-82°	2415
$(Et_3Ge)_2BiEt$	IIC	86	Yellow oil (1)	291
$Et_3GeBiEt_2$	IIC	9	b_2 132-36° (1)	291
$Ph_3GeAsPh_2$	IA (m)	64	m. 114°, mol. wt. (n, o)	1847
$Ph_3GeSbPh_2$	IA (p)	12	m. 120°, mol. wt. (n)	1847

* The characters used here correspond to the synthetic scheme in the introduction to Chapter 4.5.6.1.
(a) Synth. under UV irradiation. (b) (Far) IR spectrum and assignment (1852), NMR spectrum (1852), subl. 60-100° at 10^{-5} mm (1852). (c) From $Fe(CO)_5$ or $Fe_2(CO)_9$ (2748). (d) NMR, (far) IR spectra and assignment, mol. wt.

(2748). (e) (Far) IR spectrum and assigment (1851). (f) NMR (1813, 2739) spectra, rxn. with CCl$_4$ → Me$_3$SiN:P(CMe$_3$)$_2$Cl (1813). (g) IR spectrum (2882-3). (h) Far IR spectrum and assignment (1849), rxn. with alc. KOH in air → Ph$_3$GeOH + H$_3$PO$_4$ or PhPO$_3$H$_2$ (1849). (i) ^{31}P NMR, (far) IR and Raman spectra and assignment (1019). (j) Far IR spectrum and assignment (1850), rxn. with alc. KOH in air → Ph$_3$GeOH + Ph$_3$SnOH + H$_3$PO$_4$ or PhPO$_3$H$_2$ (1850). (k) Far IR spectrum and assignment (105). (l) Stable in N atm. to 200° (291). (m) From Ph$_2$AsNa (1847). (n) Very sensitive to air (1847). (o) Rxn. with aq. alc. H$_2$O$_2$ → Ph$_3$GeO$_2$AsPh$_2$ (1847). (p) From Ph$_2$SbNa (1847).

5. ORGANO-POLYGERMANES

5.1 HEXAORGANODIGERMANES

5.1.1 SYMMETRIC DERIVATIVES

Hexamethyldigermane   Ge$_2$C$_6$H$_{18}$ Me$_6$Ge$_2$
(M-130)

<u>Synth.:</u> By rxn. of Me$_3$GeBr with Li in THF below 30°, in 53% yield (367), with K and Me$_3$SiBr in N atm., in 70% yield (367).
By-prod. in rxn. of Me$_3$GeX with Et$_3$GeM ; X = Cl, F, M = Li, Na, K (815).
By-prod. in rxn. of Me$_3$GeCl with Na and Me$_3$MCH:CHCl in Et$_2$O ; M = Ge, Si (1533), and Me$_3$SiCCl:CH$_2$ (1533), with Et$_3$SnLi in N atm. below 0° in THF (1844).
By-prod. in rxn. of Me$_3$GeH with (Et$_3$P)$_2$PdCl$_2$ (795).
By-prod. in base catalyzed disproportionation of Me$_3$Ge-GeEt$_3$ (812).
By-prod. in thermal dec. of (Me$_3$Ge)$_2$Hg in 1-hexene at 40-100°, in > 15% yield (2241), of (Me$_3$Ge)$_3$Bi in inert atm. (1854).
By-prod. in oxidation of Me$_3$GeSnPPh$_3$ in C$_6$H$_6$ (2420).
By-prod. in rxn. of GeCl$_4$ with MeI, i-PrBr and Mg in Et$_2$O (493), with Me$_3$Al in N atm., in low yield (1108).
<u>Prop.:</u> b. 136°, n$_D^{20}$ 1.4569, d$_{20}$ 1.1690 (367), IR (367) far IR (1108), Raman (367), UV (1905), NMR (2419), and mass (2713) spectra, GLC (492-3).
<u>Rxn. with:</u> Li in (Me$_2$N)$_3$PO → Me$_3$GeLi (815).
SnCl$_4$ → Me$_3$GeGeMe$_2$Cl + MeSnCl$_3$ (2292).

Hexabenzyldigermane   Ge$_2$C$_{42}$H$_{42}$ (PhCH$_2$)$_6$Ge$_2$
<u>Synth.:</u> By rxn. of (PhCH$_2$)$_3$GeBr with Na in 9% yield (104).
By rxn. of (PhCH$_2$)$_3$GeH with BuLi, followed by MeI or Et$_3$GeBr, in 13% yield (104).
By rxn. of GeI$_2$ with PhCH$_2$MgCl in 11% yield; (PhCH$_2$)$_3$Ge-MgI is discussed as intermediate prod. (104).
<u>Prop.:</u> m. 185° (104), far IR (105) and mass (1109) spectra.
<u>Rxn. with:</u> Li in (MeOCH$_2$)$_2$ → (PhCH$_2$)$_3$GeLi + MePh (104).

Hexaethyldigermane (M-129) Ge$_2$C$_{12}$H$_{30}$ Et$_6$Ge$_2$

Synth.: By rxn. of Et$_3$GeLi in inert atm. with Et$_3$GeH at 60-70°, in 36% yield (2087, 2852), with Et$_3$GeOC$_6$H$_{11}$ in sealed tube in C$_6$H$_6$, in 64% yield (2097), with Ph$_2$SiCl$_2$ and Me$_2$EtSiCl in THF, in 16 and 13% yield, resp. (2852), with Et$_3$GeX ; X = F, Cl (815).
By rxn. of Et$_3$GeNa with Et$_3$GeX ; X = F, Cl (815).
By rxn. of Et$_3$GeK with Et$_3$GeX ; X = Cl (814-5), F (815), with HCCl$_3$ or CCl$_4$ (815).
By thermal dec. of (Et$_3$Ge)$_2$Zn at 100-130° in 45% yield (2854), of (Et$_3$Ge)$_3$Tl at 170° (2089).
By rxn. of Et$_3$GeZn(GeEt$_2$)$_n$GeEt$_3$ with Bz$_2$O$_2$ or H$_2$O (584).
By rxn. of (Et$_3$Ge)$_n$BiEt$_{3-n}$ with Et$_3$SnH (291).
By-prod. in rxn. of Et$_3$GeBr with (PhCH$_2$)$_3$GeLi in (MeOCH$_2$)$_2$ (104).
By-prod. in base catalyzed disproportionation of Me$_3$GeGeEt$_3$ (812).
By-prod. in rxn. of Et$_3$GeH with CH$_2$:CXCRCH$_2$O and H$_2$PtCl$_6$ in i-PrOH; R = H, Me, X = Cl, H (806), with Et$_4$Pb at 165-70° in inert atm., in 18% yield (2850).
By-prod. in rxn. of GeCl$_4$ with EtMgBr in Et$_2$O, MePh and THF in 10-43% yield (2701), with Et$_3$Al and NaCl in inert atm., in 15% yield (1108), at 140-60°, in 6-12% yield (2603).
Prop.: IR (493, 588), far IR (1108) and mass (1109, 2713) spectra, GLC (2701), thermodynamic data (2025).
Rxn. with: Me$_3$GeLi → Et$_3$GeLi + Me$_3$GeGeEt$_3$ (815).
K in (Me$_2$N)$_3$PO → Et$_3$GeK (814).
MCl$_4$ → Et$_3$GeGeEt$_2$Cl + (Et$_2$GeCl)$_2$; M = C, Si (813).
MCl$_4$(1:1) → Et$_3$GeGeEt$_2$Cl + EtMCl$_3$; M = Ge (813), Sn (813, 2292).
MCl$_4$(2:1) → Et$_3$GeGeEt$_2$Cl + (EtGeCl)$_2$ + EtMCl$_3$; M = Ge, Sn (813).

Hexapropyldigermane Ge$_2$C$_{18}$H$_{42}$ n-Pr$_6$Ge$_2$

Synth.: By-prod. in rxn. of GeX$_4$ with excess PrMgX (1332).
Prop.: Colorless to light yellow liq. b$_{0.04}$ 123-27°, n$_D^{20}$ 1.4911, stable in air, mol. wt. (1332), magnetic susceptibility (1).
Rxn. with: GeCl$_4$ → Pr$_3$GeGePr$_2$Cl + PrGeCl$_3$ (813).

Hexabutyldigermane Ge$_2$C$_{24}$H$_{54}$ n-Bu$_6$Ge$_2$

Synth.: By rxn. of Bu$_3$GeK with Bu$_3$GeCl in (Me$_2$N)$_3$PO (814).
By-prod. in rxn. of GeX$_4$ with excess BuMgX (1332).
Prop.: Colorless to light yellow liq. b$_2$ 182-86° (1332), b$_{0.2}$ 131° (814), n$_D^{20}$ 1.4858 (814), 1.4822 (1332), stable in air (1332), elemental analysis (426), magnetic susceptibility (1).
Rxn. with: K in (Me$_2$N)$_3$PO → Bu$_3$GeK (813).
GeCl$_4$ → Bu$_3$GeGeBu$_2$Cl + (Bu$_2$GeCl)$_2$ + BuGeCl$_3$ (813, 1760).
GeCl$_4$ + Bu$_4$Ge → activity as cat. in redistribution rxn. (1760).

Hexaphenyldigermane (M-129) Ge$_2$C$_{36}$H$_{30}$ Ph$_6$Ge$_2$

Synth.: By rxn. of Ph$_3$GeLi in inert atm. with SnCl$_2$, followed by Ph$_3$GeCl, in 16% yield (1928), with GeCl$_4$ in THF, in 43% yield (1928), with COCl$_2$ in C$_6$H$_6$-THF at -78°, in 78% yield (1624-5), with (Et$_3$P)$_2$PtI$_2$ (102-3).

By rxn. of Ph$_3$GeK with Ph$_3$SnCl in up to 50% yield (2128), with Ph$_3$GeBr, in 15% yield (2128).

By thermal dec of (Ph$_3$Ge)$_2$Cd in MePh at 150°, in 97% yield (2295).

By-prod. in thermal dec of (Ph$_3$Ge)$_2$Pt(PR$_3$)$_2$ (103), of (Ph$_3$Ge)$_2$Pd(PEt$_3$)$_2$ (795), in presence of CO in C$_6$H$_6$ at r.t. (2285), of (Ph$_3$Ge)$_2$CO, also under UV irradiation (794).

By-prod. in rxn. of Br$_3$GeMn(CO)$_5$ with PhMgBr in Ar atm. in THF (2227).

By-prod. in rxn. of (Ph$_3$Ge)$_2$Pt(PR$_3$)$_2$ with MgI$_2$ (103).

Prop.: m. 346-47° (1105), spectra: IR and assignment (1443), far IR and assignment (105, 1443), UV (2031) and mass (1109), polarography (968), in presence of azulene, quinone, (O$_2$N)$_3$C$_6$H$_3$, and transition metal complexes (965), of i-PrBr (966), of AgClO$_4$ and Ph$_3$MCl ; M = Si, Ge, Sn (963).

Rxn. with: Li in (MeOCH$_2$)$_2$ → Ph$_3$GeLi (105), in THF (1928).

Na-C$_{10}$H$_8$ in (MeOCH$_2$)$_2$, + Me$_2$SO$_4$ → Me$_3$GePh (1606).

K in (Me$_2$N)$_3$PO, + PhCH$_2$Cl → PhCH$_2$GePh$_3$ (1634).

M in liq. NH$_3$ → (Ph$_3$Ge)$_2$M ; M = Sr, Ba (675).

Se(SeCN)$_2$ → Se + PhSeCN (697).

Listing of symmetric hexaorganodigermanes concludes in Table 72.

Table 72. Symmetric Hexaorganodigermanes   R$_6$Ge$_2$

Formula*	Synth. Method**	Properties	Ref.
R$_6$Ge$_2$			
R = CH$_2$:CH (130)	--	IR spectrum (1013)	--
Me$_2$CH	IA	m. 235-40° (sealed tube)	831
	II (2)	Subl.$_{740}$ mm 125° (a)	831
Me$_2$CHCH$_2$	IB (20)	m. 48-49°, colorless needles (b)	1108
o-MeC$_6$H$_4$ (130)	--	(c)	--
m-MeC$_6$H$_4$ (130)	--	(c)	--
p-MeC$_6$H$_4$ (130)	--	(c)	--

* Numbers in parenthesis refer to pages in main volume.

** Synthesis by rxn. of GeCl$_4$ with - (IA) RMgCl - (IB) with R$_3$Al. (II) By rxn. of i-Pr$_3$GeCl with Na-K in MePh, followed by hydrolysis, (yields).

(a) Far IR (1108) and UV (831) spectra. (b) GLC (1108), far IR (1108) and mass (1108-9) spectra. (c) Mass (1109) and far IR spectra and assignment (105).

5.1.2 UNSYMMETRIC HEXAORGANODIGERMANES

A number of new unsymmetric hexaorganodigermanes of the structure $R_3GeGeR'_3$ or $R_nGe_2R'_{6-n}$ are reported in the newer literature. Several of the compounds were not characterized but identified by GLC retention times and/or IR spectra. The compounds in Table 73 are prepared by the following methods.

I Grignard reactions:
    (IA) $GeCl_4 + MeI + RX + Mg \rightarrow Me_nGe_2R_{6-n}$,
    (IB) $R_nGeI_{4-n} + MeMgI \rightarrow Me_nGe_2R_{6-n}$,
    (IC) $Bu_3GeGeBu_2Cl + MeMgBr \rightarrow MeBu_5Ge_2$.

II Reactions with alkali metal derivatives:
    (IIA) $GeCl_4 + MeLi + EtLi \rightarrow Me_nGe_2Et_{6-n}$,
    (IIB) $R_3GeX + R'_3GeM \rightarrow R_3GeGeR'_3$,
    (IIC) $(Ph_2GeBr)_2 + MeLi \rightarrow (MePh_2Ge)_2$,
    (IID) $Ph_2GeH_2 + BuLi , + EtBr \rightarrow (EtPh_2Ge)_2$.

III Reaction with trimethylaluminum:
    $GeI_2 + Me_3Al \rightarrow Me_3GeGeMe_2CH_2GeMe_3$.

R = R' = organic groups, M = lithium or potassium, X = halogen, n = 1-5.

1,1,1,2,2,2-Trimethyltriethyldigermane $Ge_2C_9H_{24}$ $Me_3GeGeEt_3$
Synth.: By rxn. of Et_3GeK with Me_3GeCl in $(Me_2N)_3PO$ (814).
By-prod. in rxn. of Et_3GeM with Me_3GeX in $(Me_2N)_3PO$; X = F, Cl, M = Li, Na, K (815).
By-prod. in rxn. of Et_6Ge_2 with Me_3GeLi, in low yield (815).
Prop.: b_{14} 80-82°, n_D^{20} 1.4804 (814), IR (814), NMR (814) and mass (2713) spectra.
Rxn. with: $Et_3GeLi \rightarrow Me_3GeLi + Et_6Ge_2$ (815).
$SnCl_4 \rightarrow Et_3GeGeMe_2Cl$ (2292).
Base $\rightarrow Me_6Ge_2 + Et_6Ge_2$; kinetic data (812).

Other unsymmetric hexaorganodigermanes are listed in Table 73.

Table 73. Unsymmetric Hexaorganodigermanes $R_nGe_2R'_{6-n}$

Formula	Synth. Method*	Yield	Properties	Ref.
$Me_nGe_2Et_{6-n}$	IA	7	GLC (492-3), n = 1-5	493
	IIA	9	n = 3-5, IR spectra (a)	493
$[MeGe(CH_2Ph)_2]_2$	--	--	Far IR spectrum	105
$Me_nGe_2(Pr-n)_{6-n}$	IA	10	GLC (492-3) n = 2-5 (b, c)	493
$Me_nGe_2(Pr-i)_{6-n}$	IA	35	GLC, n = 3-5 (c, d)	493
$Me_nGe_2(CH_2CH:CH_2)_{6-n}$	IB	--	n = 4-5, GLC (492)	493
$Me_3GeGeBu_3$	IIB (e)	--	$b_{0.1}$ 78-80°, n_D^{20} 1.4800 (f)	814
$MeGe_2Bu_5$	IC	84	$b_{0.1}$ 125-26°, n_D^{20} 1.4841	813

Me$_3$GeGePh$_3$	--	--	IR and mass spectra (849)	--
(MePh$_2$Ge)$_2$	II (g)	--		396, 1606
	IIC	75	m. 131-32°	1606
(PhCH$_2$)$_3$GeGeEt$_3$	IIB (h)	--	b$_{0.001}$ 220-30° (i, j)	104
Et$_5$Ge$_2$Bu	--	--	b$_{0.03}$ 84-88°, n$_D^{20}$ 1.4947 (k)	2292
(EtPh$_2$Ge)$_2$	IIC	28	m. 125-26°, IR spectrum (j)	104
Me$_3$GeGeMe$_2$CH$_2$GeMe$_3$	II (l)	--	NMR and mass spectra	2419
	III	--	GLC	2419

* The characters used here correspond to the synthetic scheme in the introduction to Chapter 5.1.2.
(a) n = 3,4 Isomer ratio by IR spectrum (493). (b) n = 4,5, identified by IR spectrum (493). (c) n = 4, isomer ratio by IR spectrum (493). (d) n = 4-5, identified by IR spectrum (493). (e) From Bu$_3$GeK (814). (f) IR and NMR spectra (814). (g) By-prod. in rxn. of c.(Ph$_2$Ge)$_n$ with NaC$_{10}$H$_8$ in (MeOCH$_2$)$_2$ (396, 1606). (h) From (PhCH$_2$)$_3$GeLi (104). (i) NMR spectrum (104). (j) Far IR spectrum and assignment (105). (k) Rxn. with SnCl$_4$ → Et$_4$BuGe$_2$Cl (2292). (l) From Me$_3$GeBr and K (2419).

5.2 ORGANO-POLYGERMANES CONTAINING THREE OR MORE GERMANIUM ATOMS

Understanding of the nature or organo-polygermanes as cyclic or chain-type compounds with Ge-Ge bonds has greatly improved, mainly through the continued work of Neumann, et al. and also by Glockling and coworkers. The compounds in Table 74 are prepared by methods from the following scheme.

I Grignard reactions:
   (IA) GeCl$_4$ + MeMgCl → Me$_{2n+2}$Ge$_n$,
   (IB) HGeCl$_3$ + RMgX → R$_{2n+2}$Ge$_n$.

II Synthesis with organic derivatives of alkali metals:
   (IIA) HGeCl$_3$·2Et$_2$O + MeLi → (Me$_2$Ge)$_x$,
   (IIB) GeCl$_4$ + Me$_3$GeLi → (Me$_3$Ge)$_4$Ge ,
   (IIC) GeI$_2$ + Me$_3$GeBr + K → Me$_{2n+2}$Ge$_n$,
   (IID) (Ph$_2$Ge)$_4$I$_2$ + RLi → (RPh$_2$GeGePh$_2$)$_2$,
   (IIE) (Et$_2$GeCl)$_2$ + Et$_3$GeK → Et$_{10}$Ge$_4$,
   (IIF) R$_2$GeCl$_2$ + Li or Na in THF → (R$_2$Ge)$_n$.

III Synthesis with trialkylaluminum:
   (IIIA) GeCl$_4$ + excess R$_3$Al(+NaCl) → R$_{2n+2}$Ge$_n$.

Octamethyltrigermane (M-131)   Ge$_3$C$_8$H$_{24}$ Me$_8$Ge$_3$
Synth.: By rxn. of MeLi and RLi with GeCl$_4$ in Et$_2$O, in < 8% yield (493).
By-prod. in rxn. of GeCl$_4$ with large excess AlCl$_3$ in N atm. (1108).
Prop.: IR (493), UV (1905) and mass (1108) spectra, GLC (493, 1072), mol. wt. (1108), struct. (493).
Rxn. with: SnCl$_4$ in MeNO$_2$ at 100° → Me$_3$GeGeMe$_2$GeMe$_2$Cl + MeSnCl$_3$ (2292).

Octaethyltrigermane $Ge_3C_{16}H_{40}$ Et_8Ge_3

Synth.: By rxn. of $Et_3GeGeEt_2Cl$ with Et_3GeK in $(Me_2N)_3PO$, in 60% yield (815).
By-prod. in rxn. of $GeCl_4$ with excess EtMgX (1332), with EtMgBr and excess Mg in Et_2O, THF or MePh, in 3-22% yield (2701).
By-prod. in rxn. of $GeCl_4$ with Et_3Al and NaCl, in N atm., in 3% yield (1108).
Prop.: Slightly yellow liq. (1332) $b_{0.1}$ 152-56° (815), $b_{0.05}$ 133-37° (1332), n_D^{20} 1.5350 (815), 1.5321 (1332), mass spectrum (1108), GLC (2701), mol. wt. (1332), stable in air (1332).

Octaphenylcyclotetragermane (M-131) $Ge_4C_{48}H_{40}$ $c.(Ph_2Ge)_4$

Prop.: m. 260-70° (capilary), 238-39° (Kofler miscroscope) (396), IR, UV, ESR (396) and mass (2520) spectra, mol. wt., struct., magnetic moment (396).
Rxn. with: Thermal dec. at m.p. → $(Ph_2Ge)_x$ (396).
$Na-C_{10}H_8$, + Me_2SO_4 → Me_2GePh_2 + $(MePh_2Ge)_2$ + $MeNaSO_4$ (396, 1606).
I in C_6H_6 at 20° → Ph_2GeI_2 (1:8); quantitative rxn. + $(Ph_2Ge)_4I_2$ (1:2) (396).
Li, + PhBr → $(Ph_3GeGePh_2)_2$ + $c.(Ph_2Ge)_5$ (396).

Decaphenylcyclopentagermane (M-131) $Ge_5C_{60}H_{50}$ $c.(Ph_2Ge)_5$

Synth.: By rxn. of $c.(Ph_2Ge)_4$ with Li in THF, followed by PhBr, in 58% yield (396).
By rxn. of Ph_2GeCl_2 with $Na-C_{10}H_8$ in $(MeOCH_2)_2$, in 29-37% yield (396).
Prop.: m. > 360° (396), mass spectrum (2520), mol. wt., soly. (396).
Rxn. with: $Na-C_{10}H_8$ + Me_2SO_4 → Me_2GePh_2 + $(MePh_2Ge)_2$ + $MeNaSO_4$ (1606).

Poly(diphenylgermane) (M-131) $(GeC_{12}H_{10})_x$ $(Ph_2Ge)_x$

Synth.: By rxn. of Ph_2GeH_2 with Et_2Hg in xylene under reflux, in 66% yield (396), at low temp. under UV irradiation (396).
By rxn. of Ph_2GeCl_2 with Na in refluxing xylene, in 67% yield (396), with $Na-C_{10}H_8$ in $(MeOCH_2)_2$, in 57% yield (396).
By-prod. in thermal dec. of $c.(Ph_2Ge)_4$ at 230-70° (396).
Prop.: Yellow laquer, IR spectrum, mol. wt., mixt. of cyclic compounds, contn. some Ge-O bonds (396).
Rxn. with: I in C_6H_6 → Ph_2GeI_2 (396).
HCl in Et_2O, + $Na-C_{10}H_8$ → $c.(Ph_2Ge)_5$ + $c.(Ph_2Ge)_6$ + $(Ph_2Ge)_x$ (396).

Other organo-polygermanes are listed in Table 74 and in Subsection 2.7.1.1.3.

Table 74. Organopolygermanes with Three or more Germanium Atoms R_nGe_m

Formula*	Synth. Method**	Yield	Properties	Ref.
$Me_{2n+2}Ge_n$				
n = 4 (131)	IIIA	(a)	GLC (1072), mass spectrum	1108
	IIIB	(a)	GLC, NMR spectrum, proposed	2419
	IIF (b)	(a)	struct. $(Me_3Ge)_3GeMe$	2419
5	IIIA	(a)	GLC (1072), mass spectrum	1108
	IIIB	7	Structure: $(Me_3Ge)_2GeMeGeMe_2$-	1108
			$GeMe_3$	2419
6	IIIA	(a)	Mass spectrum	1108
	IIIB	⩾ 7		1108
7 (131)	IIIB	⩾ 7	Mass spectrum	1108
8	IIB	(a)	Stable in air, struct. $(Me_3Ge)_6$-	
			Ge_2 proposed	2419
10	IIIB	low	GLC, mass spectrum	2419
$(Me_3Ge)_4Ge$	IIB	(a)	Stable in air, NMR spectrum	2419
$(Me_2Ge)_n$				
n = 4	IIF	--	Mixture: n = 4-6, ⩾ 55 (c)	1586
	IB - IIA	--	m. 89-90°, $b_{0.3}$ 60-70°	1588
6	IIF - Li	--	m. 207-9°, GLC	1585
	IB - IIA	--	$b_{1.5}$ 130-55°	1588
$Et_{10}Ge_4$	IIE	40	b. 155-60 × $5 \cdot 10^{-4}$ mm, n_D^{20} 1.5748	815
	IA	10	GLC	2701
$(Et_2Ge)_n$	IIF	--	n = 4-6 (cyclic), ⩾ 55	1586
$n\text{-}Pr_8Ge_3$	IA	(a)	$b_{0.01}$ 138-42°, n_D^{20} 1.5102 (d)	1332
$c.Pr_8Ge_4$	IA	4	Orange semi-solid (e)	831
$n\text{-}Bu_8Ge_3$	IA	(a)	$b_{0.04}$ 194-99°, n_D^{20} 1.5018 (d)	1332
$i\text{-}Bu_8Ge_3$	IIIA	(a)	Mass spectrum	1108
$(Ph_3GeGePh_2)_2$	IID	42	m. 274-76°, mol. wt.	396
	II (f)	(a)		396

* Numbers in parenthesis refer to pages in main volume.
** The characters used here correspond to the synthetic scheme in the introduction to Chapter 5.2.
(a) As by-prod. (b) From Me_3GeBr and K at 140° (2419). (c) Thermal dec. in presence of $C_2H_4 \rightarrow Me_2\overline{Ge(CH_2)_3CH_2} + [Me_2Ge(C_2H_4)_x]_y$ (1586). (d) Slightly yellow liq., stable in air, mol. wt. (1332). (e) IR and mass spectra (831). (f) From $c.(Ph_2Ge)_4$ and Li in THF, followed by PhBr (396).

Table 74 Continued   RnGem

Formula*	Synth. Method**	Yield	Properties	Ref.
c.$(Ph_2Ge)_6$	IIF (g)	13	m. > 360°, soly., mol. wt. (h)	396
$Me_nGe_3Et_{8-n}$				
n = 7	II	(i)	IR spectrum, GLC	493
6	II	(i)	GLC	493
	IIE	50	b_{18} 175-76°, n_D^{20} 1.5256 (j)	815
5	II	(i)	GLC	493
$(Me_3Ge)_2GeMeGeMe_2Et$	IIC	(a)	NMR and mass spectra, GLC	2419
$(Me\ i\text{-}PrGe)_x$	IB (k)	(a)	Red-brown solid, m. > 260°	2877
$(MePh_2GeGePh_2)_2$	IID	--	m. 189-91°, mol. wt., soly. (l)	396
$MeGe(GePh_3)_3$ (131)	--	--	Far IR spectrum, assignment (105)	--

(g) From Ph_2GeCl_2 and $Na\text{-}C_{10}H_8$ in $(MeOCH_2)_2$ and separation from cyclic pentamer by C_6H_6-extraction (396). (h) Mass spectrum (2520), rxn. with Na-$C_{10}H_8$ in $(MeOCH_2)_2$, + Me_2SO_4 → Me_2GePh_2 + $(MePh_2Ge)_2$ + $MeNaSO_4$ (1606). (i) By rxn. of $GeCl_4$ with MeLi and EtLi in Et_2O, in below 8% yield (493). (j) Proposed struct. $Me_2Ge(GeEt_3)_2$ (815). (k) From $MeGeI_3$ and i-PrMgI (2877). (l) IR spectrum (396).

5.3 ORGANO-POLYGERMANIUM HYDRIDES

Organo-polygermanium hydrides are prepared by exchange of halogen with lithium aluminum hydride (I), by hydrolysis of organogermanium metal derivatives (II) and by partial hydrogermylation of olefins (III).

(IA) $R_nGe_2Cl_{6-n}$ + $LiAlH_4$ → $R_nGe_2H_{6-n}$,
(IB) $GeCl_4$ + i-PrMgBr, + H_2O, + $LiAlH_4$ → $(i\text{-}Pr_2GeH)_2$.
(IIA) i-Pr_3GeCl + Na/K , + H_2O → $(i\text{-}Pr_2GeH)_2$,
(IIB) $Et_3GeZn(GeEt_2)_nEt$ + H_2O → Et_5Ge_2H .
(III) Ge_2H_6 + C_2H_4 → $Et_nGe_2H_{6-n}$.

1,1,1-Trimethyldigermane $Ge_2C_3H_{12}$ Me_3GeGeH_3
Synth.: By rxn. of Me_3GeX with excess H_3GeNa at -20 to -4°; X = F and Br, in 36 and 8% yield, resp. (2365).
Prop.: m. -89.6°, b. 74.4°, IR and NMR spectra, vapor pressure from 0-54°, mol. wt., thermodynamic data (2365).

Methyldigermane Ge_2CH_8 $MeGeH_2GeH_3$
Synth.: By rxn. of Ge_2H_5I with MeMgI in Bu_2O at -45°, in 60% yield (2550).
Prop.: b. 54.7°, NMR, mass, (far) IR spectra and assignment, GLC, vapor pressure, mol. wt. and thermodynamic data (2550).

Ethyldigermane   Ge$_2$C$_2$H$_{10}$ EtGe$_2$H$_5$

Synth.: By rxn. of Ge$_2$H$_5$I with excess EtMgX in Bu$_2$O at $-45°$, in 67% yield (2550).
By-prod. in rxn. of Ge$_2$H$_6$ with C$_2$H$_4$ at 154-60° in sealed tube, in 16% yield (2551).
Prop.: b. 88.6°, NMR, mass, (far) IR spectra and assignment, GLC, vapor pressure, mol. wt. and thermodynamic date (2550).

1,1,2,2-Tetraphenyldigermane   Ge$_2$C$_{24}$H$_{22}$ (Ph$_2$GeH)$_2$

Synth.: By rxn. of Ph$_2$GeH$_2$ with BuLi in Et$_2$O, followed by hydrolysis (105).
By rxn. of Ph$_2$GeBr$_2$ with LiAlH$_4$ in Et$_2$O, in above 35% yield (1364).
Prop.: m. 73-75° (dec.) (1364), IR spectrum (105).
Rxn. with: aq. alkali → c.(Ph$_2$GeGePh$_2$O) (1364).

Other organo-polygermanium hydrides are listed in Table 75.

Table 75. Organo-Polygermanium Hydrides   R$_n$Ge$_m$H$_o$

Formula*	Synth. Method**	Yield	Properties	Ref.
Et$_5$Ge$_2$H	IIA	(a)	b$_5$ 115°, n$_D^{20}$ 1.4808	584
Et$_2$GeHGeEtH$_2$	III	(a)	(b)	2551
Et$_2$GeHGeH$_3$	III	(a)	(b)	2551
(EtGeH$_2$)$_2$	III	(a)	(b) NMR spectrum	2551
EtGe$_3$H$_7$ (?)	III	(a)	(b)	2551
(i-Pr$_2$GeH)$_2$	IB	(a)	Colorless liq.	831
	IIA	24	Mass spectrum	831
Bu$_5$Ge$_2$H	IA	50	b$_{0.1}$ 115-16°, n$_D^{20}$ 1.4851	813
BuPh$_2$GeGePh$_2$H	II (c)	--	NMR spectrum only	105
(PhGeH$_2$)$_2$	IA	low	b$_{0.3}$ 113°, n$_D^{20}$ 1.6200 (d)	1762
(Ph$_3$Ge)$_3$GeH (133)	--	--	(e)	--
(Ph$_2$Ge)$_4$H$_2$	IA	11	m. 147-49°, IR spectrum	1364

* Numbers in parenthesis refer to pages in main volume.
** The characters used here correspond to the synthetic scheme in the introduction to Chapter 5.3.
(a) By-prod. (b) (Far) IR, assignment and mass spectra (2551). (c) From Ph$_2$GeH$_2$ and BuLi, followed by hydrolysis (105). (d) IR and NMR spectra (1762). (e) Far IR spectrum and assignment (105).

5.4 ORGANO-POLYGERMANES CONTAINING HETEROATOMS

Organo-polygermanium derivatives of heteroatoms include halides, oxides, hydroxides and carboxylates. The compounds listed in Table 76 are prepared by methods from the following scheme.

I Dealkylation reactions of organo-polygermanes $R_{2n+2}Ge_n$ with:
   (IA) $SnCl_4$ at 50-200° in $MeNO_2$,
   (IB) $GeCl_4$ at 200° .

II Condensation of $PhGe(OR)Cl_2$ with $PhHGeCl_2$.

III Cleavage of $Et_3GeZn(GeEt_2)_nEt$ (?):
   (IIIA) with $HgCl_2$,
   (IIIB) with Bz_2O_2 .

IV Hydrolysis of $R_nGe_2Cl_{6-n}$ or $(Ph_2Ge)_4I_2$:
   (IVA) with H_2O
   (IVB) with $PhNH_2$ in aq. EtOH .

Pentaethyldigermanium Chloride $\qquad$ $Ge_2C_{10}H_{25}Cl$ $\qquad$ Et_5Ge_2Cl
<u>Synth.</u>: By rxn. of Et_6Ge_2 with MCl_4 (1:1) at 200°, in 80-88% yield; M = Ge, Sn (813), with $SnCl_4$ at 50° in AcCl or $MeNO_2$, in 100% conversion (2292), with CCl_4 or $SiCl_4$ at 200° as by-prod. (813).
<u>Prop.</u>: b_{16} 126-27°, n_D^{20} 1.5092 (813).
<u>Rxn. with:</u> $H_2O \rightarrow (Et_3GeGeEt_2)_2O$ (813).
Et_3GeK in $(Me_2N)_3PO \rightarrow Et_8Ge_3$ (813).

1,2-Tetraethyldigermanium Dichloride $\qquad$ $Ge_2C_8H_{20}Cl_2$ $\qquad$ $(Et_2GeCl)_2$
<u>Synth.</u>: By rxn. of Et_6Ge_2 with MCl_4 (1:2) at 200° ; M - Ge, cat. GeI_2, or Sn, in 64-86% yield, with CCl_4 or $SiCl_4$ at 200° as by-prod. (813).
<u>Prop.</u>: b_{16} 130-32°, n_D^{20} 1.5197 (813).
<u>Rxn. with:</u> Et_3GeK in $(Me_2N)_3PO \rightarrow Et_{10}Ge_6$ (813).
$X \rightarrow Et_2GeClX$ (813).
Thermal dec. at 200° $\rightarrow Et_2GeCl_2 + (Et_2Ge)_x$ (?) (813).

Pentabutyldigermanium Chloride $\qquad$ $Ge_2C_{20}H_{45}Cl$ $\qquad$ $n\text{-}Bu_3GeBu_2GeCl$
<u>Synth.</u>: By rxn. of Bu_6Ge_2 with $GeCl_4$ at 200°, in 87% yield (813, 1760).
<u>Prop.</u>: $b_{0.06}$ 130-32°, n_D^{20} 1.4932 (813, 1760).
<u>Rxn. with:</u> $LiAlH_4 \rightarrow Bu_5Ge_2H$ (813).
$MeMgBr \rightarrow Bu_5Ge_2H$ (813).

1,2-Tetraphenyldigermanium Dibromide $\qquad$ $Ge_2C_{24}H_{20}Br_2$ $\qquad$ $(Ph_2GeBr)_2$
   (M-133)
<u>Rxn. with:</u> $MeLi \rightarrow (MePh_2Ge)_2$ (1606).
$LiAlH_4 \rightarrow (Ph_2GeH)_2$ (1364).
$PhNH_2$ in aq. EtOH $\rightarrow$ c.$(Ph_2GePh_2GeO)_2$ (1364).

1,4-Octaphenyltetragermanium Diodide $Ge_4C_{48}H_{40}I_2$ $(Ph_2Ge)_4I_2$
(M-133)

Synth.: By rxn. of c.$(Ph_2Ge)_4$ with I (1:2) at 20° in C_6H_6, in 82% yield (396).
Prop.: m. 210-12°, mol. wt., soly. (396).
Rxn. with: RLi → $RPh_2Ge(GePh_2)_2GePh_2R$; R = Me, Ph (396).
$LiAlH_4$ → $(Ph_2Ge)_4H_2$ (1364).
$PhNH_2$ in aq. EtOH → $Ph_2\overline{Ge(GePh_2)_3O}$ (396).
Aq. NaOH in C_6H_6 → $Ph_2\overline{Ge(GePh_2)_3O}$ (396).

Additional organo-polygermanium compounds with heteroatoms are listed in Table 76.

Table 76. Organo-Polygermanium Compounds with Heteroatoms $R_nGe_mX_o$

Formula*	Synth. Method**	Yield	Properties	Ref.
Me_5Ge_2Cl	IA	--	b_{18} 62-63°, n_D^{20} 1.4911	2292
$Et_3GeGeMe_2Cl$	IA (a)	--	b_{14} 108°, n_D^{20} 1.5011	2292
Et_5Ge_2Cl	IIIA	--	b_{17} 120-27°, n_D^{20} 1.5015	584
$Et_3Ge(GeEt_2)_2Cl$	IA (b)	--	b. 62-64° x 5 x 10^{-4} mm, n_D^{20} 1.5410	2292
$\{Et_2BuGeGeEt_2Cl\}$ $(Et_3GeGeEtBuCl)$	IA (c)	--	$b_{0.5}$ 114-16°, n_D^{20} 1.5042, Mixture, GLC (d)	2292 2292
$n-Pr_5Ge_2Cl$	IB	60	$b_{0.4}$ 110-12°, n_D^{20} 1.5007	813
$(n-Bu_2GeCl)_2$	IB	86	$b_{0.2}$ 133-38°, n_D^{20} 1.5027 (e)	813
$(PhGeCl_2)_2$	II (f)	--	Instable (g)	1762
$(Et_3GeGeEt_2)_2O$	IVA	70	$b_{0.3}$ 162-65°, n_D^{20} 1.5185	813
Et_5Ge_2OBz	IIIB	--	b_2 140-46°, n_D^{20} 1.5287	584
c.$(Bu_2GeGeBu_2O)_2$	IVA	80	b. 130-32° x 10^{-5} mm, n_D^{20} 1.4993	813
$(Ph_2GeOH)_2$	IVB	--	m. 206-208°, soly. (h, i)	1364
c.$(Ph_2GeGePh_2O)_2$	(h)	--	m. 216-17° (i)	1364
$Ph_2\overline{Ge(GePh_2)_3O}$ (133)	IVB	65	m. 206-208°, soly.,	396
	IVB (j)	--	Mol. wt., IR spectrum	396

** Numbers in parenthesis refer to pages in main volume.
** The characters used here correspond to the synthetic scheme in the introduction to Chapter 5.4.
(a) From $Me_3GeGeEt_3$ (2292). (b) In 100% conversion (2292). (c) From Et_5BuGe_2 (2292). (d) Rxn. with Cl → Et_3GeCl + Et_2GeCl_2 + $Et_2BuGeCl$ + $EtBuGeCl_2$ (2292). (e) Rxn. of $(n-Bu_2GeCl)_2$ with H_2O → c.$(Bu_2GeGeBu_2O)_2$ (813). (f) From PhGe-(OR)Cl ; R = Me, i-Bu, c.C_6H_{11} (1762). (g) Rxn. with $LiAlH_4$ → $(PhGeH_2)_2$ (1762). (h) Dehydration in desiccator or reflux with HCO_2H → c.$(Ph_2GeGePh_2O)_2$ (1364). (i) IR spectrum (1364). (j) From $(Ph_2Ge)_4I_2$ with aq. NaOH in C_6H_6 (396).

6. POLYMETALLIC ORGANOGERMANES

6.1 ORGANOGERMANIUM DERIVATIVES OF GROUP IV METALS

6.1.1 ORGANOGERMANES WITH Ge-Si BONDS

This chapter comprises bimetallic organogermanes containing germanium-silicon bonds. Compound with longer metal-metal chains and other metals are also included, providing the other metals, e.g., Li, Hg, Sn, Sb are linked to germanium through silicon. One compound with a germanium chlorine bond $Me_3SiGeEt_2Cl$ is listed on the bottom of Table 77. The compounds in this table are prepared by methods from the following scheme.

I Reactions with lithium derivatives:
- (IA) $Et_3GeLi + R_nSiCl_{4-n} \rightarrow (Et_3Ge)_{4-n}SiR_n$,
- (IB) $Et(Et_3Ge)_2SiLi + Et_2SbBr \rightarrow Et(Et_3Ge)_2SiSbEt_2$,
- (IC) $Ph_2GeCl_2 + \overline{(Ph_2Si)_4Li_2} \rightarrow Ph_2Ge(SiPh_2)_3SiPh_2$.

II Reactions with potassium derivatives:
- (IIA) $R_3GeK + R_3SiCl \rightarrow R_3GeSiR_3$,
- (IIB) $Me_3GeBr + Me_3SiBr + K \rightarrow Me_3GeSiMe_3$.

III Reaction with silane: $Et_3GeSiEt_2H + Et_2Hg \rightarrow (Et_3GeSiEt_2)_2Hg + C_2H_6$.

IV Interconversion reactions:
- (IVA) $[(Et_3Ge)Si]_2Hg + (CH_2Br)_2 \rightarrow (Et_3Ge)_3SiBr + Hg + C_2H_4$,
- (IVB) $[Et(Et_3Ge)_2Si]_2Hg + O_2 \rightarrow [Et(Et_3Ge)_2Si]_2O$,
- (IVC) $[Et_n(Et_3Ge)_{3-n}Si]_2Hg + UV \rightarrow [Et_n(Et_3Ge)_{3-n}Si]_2$.

R = organic group or hydrogen.

Trimethylgermylsilane $GeSiC_3H_{12}$ Me_3GeSiH_3
Synth.: By rxn. of Me_3GeBr with H_3SiK in diglyme in N atm., in 10% yield (2224).
Prop.: Vapor pressure, IR and NMR spectra, mol. wt. (2224).
Rxn. with: Aq. KOH $\rightarrow Me_3GeOH + Si(OH)_4 + H$ (2224).

Trimethylsilyltribenzylgermane $GeSiC_{24}H_{30}$ $(PhCH_2)_3GeSiMe_3$
Synth.: By rxn. of $(PhCH_2)_3GeLi$ with Me_3SiCl (104).
Prop.: White needles, m. 63.5 - 64.5°, $b_{0.001}$ 183° (104), IR (849), far IR, assignment (105) and mass (849) spectra, monomeric (104).

Tris(triethylgermyl)silane $Ge_3SiC_{18}H_{46}$ $(Et_3Ge)_3SiH$
Synth.: By rxn. of Et_3GeLi with $HSiCl_3$ in Ar atm. in THF or C_6H_6, in 32% yield (2087), in inert atm., in 37% yield (2852), with $(EtO)_3SiH$ in C_6H_6 in sealed tube, in 78% yield (2097).
Prop.: b_1 177° (2087, 2852), b_1 160-61° (2097), n_D^{20} 1.5425 (2097), 1.5380 (2087, 2852).
Rxn. with: $Et_2Hg \rightarrow [(Et_3Ge)_3Si]_2Hg + C_2H_6$ (2415).

Bis[tris(triethylgermyl)silyl] Mercury $Ge_6HgSi_2C_{36}H_{90}$ $[(Et_3Ge)_3Si]_2Hg$

Synth.: By rxn. of $(Et_3Ge)_3SiH$ with Et_2Hg in sealed tube in N atm., at 150°, in 60% yield (2415).

Prop.: Viscous yellow-orange liq., nondistillable (2415).

Rxn. with: $(CH_2Br)_2$ at 100° → $(Et_3Ge)_3SiBr$ + Hg + C_2H_4 (2415).

Bis(triethylgermyl)ethylsilane Lithium $Ge_2LiSiC_{14}H_{35}$ $(Et_3Ge)_2SiEtLi$

Synth.: By rxn. of $[Et(Et_3Ge)_2Si]_2Hg$ with Li in N atm. at r.t. in C_6H_6 (2415).

Prop.: Not isolated (2415).

Rxn. with: Et_3SnH → $(Et_3Ge)_2SiEtSnEt_3$ + LiH (2415).
Et_2SbBr → $(Et_3Ge)_2SiEtSbEt_2$ + LiBr (2415).

Bis[bis(triethylgermyl)ethylsilyl]-Mercury $Ge_4HgSi_2C_{28}H_{70}$ $[(Et_3Ge)_2EtSi]_2Hg$

Synth.: By rxn. of $(Et_3Ge)_2SiEtH$ with Et_2Hg in sealed tube in N atm. at 150°, in 66% yield (2415).

Prop.: $b_{0.2}$ 215-16°, sensitive to air (2415).

Rxn. with: UV irradiation → $[Et(Et_3Ge)_2Si]_2$ + Hg (2415).
O_2 in Et_2O → $[Et(Et_3Ge)_2Si]_2O$ + Hg (2415).
Li in C_6H_6 → $Et(Et_3Ge)_2SiLi$ + Hg (2415).

Hexaethylsilylgermane (M-136) $GeSiC_{12}H_{30}$ $Et_3GeSiEt_3$

Synth.: By rxn. of Et_3GeBr with Et_3SiLi in C_6H_6 under exclusion of air, in 52% yield (2852).

By rxn. of Et_3GeLi with Et_3SiH in THF in Ar atm. at 80-90°, in 48% yield (2087, 2852).

By photolysis of $Et_3GeHgSiEt_3$ (533a, 2852).

Prop.: $b_{1.5}$ 92° (583a), 89-91° (2852), b_1 87° (2087, 2852), n_D^{20} 1.4879 (2087, 2852), 1.4858 (583a), 1.4830 (2852).

Other organogermanes with Ge-Si bonds are listed in Table 77 and in Chapters 6.3 and 7.4.3.

Table 77. Organogermane with Germanium-Silicon Bonds R_3GeSiR_3

Formula*	Synth. Method**	Yield	Properties	Ref.
$Me_3GeSiMe_3$	IIB	20	GLC	367
$(Et_3Ge)_3SiBr$	IVA	52	b_1 174-75°, n_D^{20} 1.5590	2415
$(Et_3Ge)_2EtSiH$	IA	52	b_1 110-12°, n_D^{20} 1.5159 (a)	2415
$[(Et_3Ge)_2EtSi]_2O$	IVB	48	b_1 108-10°, n_D^{20} 1.5060	2415
$(Et_3Ge)_2SiSbEt_2$	IB	50	b_1 122-24°	2415
$[(Et_3Ge)_2SiEt]_2$	IVC	100	m. 168-69°	2415
$(Et_3Ge)_2EtSiSnEt_3$	IB (b)	24	b_1 154°, n_D^{20} 1.5590	2415
$(Et_3Ge)_2SiPh_2$	IA	77	$b_{1.5}$ 199-200°, n_D^{23} 1.5853, d_{20} 1.1167	2087, 2852
$Et_3GeSiMe_3$	IIA	--	b_{30} 89-91°, n_D^{20} 1.4670 (c, d)	814
$Et_3GeSiMe_2Et$	IA	46	b_1 65°, n_D^{23} 1.4720, d_{20} 0.9762	2087, 2852
$Et_3GeSiEt_2H$	IA	47	b_1 57-58°, n_D^{20} 1.4840 (e)	2415
$(Et_3GeSiEt_2)_2$	IVC (f)	--	b_1 157-60°, n_D^{20} 1.5440	2415
$(Et_3GeSiEt_2)_2Hg$	III	31	b_1 178-88° (f)	2415
$Ph_3GeSiEt_3$ (136)	--	--	IR and mass spectra	849
$Ph_3GeSiPh_3$ (136)	IIA (g)	84	--	2128
$\overline{Ph_2Ge(SiPh_2)_3SiPh_2}$	IC	--	m. 316°, mol. wt.	1177
$Me_3SiGeEt_2Cl$	(d)	--	b_{65} 126°, n_D^{20} 1.4819, GLC (h)	2292

* Numbers in parentheses refer to pages in main volume.
** The characters used here correspond to the synthetic scheme in the introduction to Chapter 6.1.1.
(a) Rxn. of $(Et_3Ge)_2EtSiH$ with $Et_2Hg \rightarrow [(Et_3Ge)_2EtSi]_2Hg + C_2H_6$ (2415). (b) From Et_3SnH (2415). (c) IR and NMR spectra (814). (d) Rxn. of $Et_3GeSiMe_3$ with $SnCl_4 \rightarrow Me_3SiGeEt_2Cl + EtSnCl_3$ (2292). (e) Rxn. of $Et_3GeSiEt_2H$ with $Et_2Hg \rightarrow (Et_3GeSiEt_2)_3Hg + C_2H_6$ (2415). (f) Photolysis of $(Et_3GeSiEt_2)_2Hg \rightarrow (Et_3GeSiEt_2)_2 + Hg$ (2415). (g) Synth. from Ph_3SiK with Ph_3GeBr in THF, in 81% yield (2128). (h) Prod. contains 2% $EtSnCl_3$ by GLC (2292).

6.1.2 ORGANOGERMANES WITH Ge-Sn BONDS

6.1.2.1 BIMETALLIC ORGANOGERMANES WITH Ge-Sn BONDS

Several new triorganogermanes with a direct germanium-tin linkage are listed in Table 78. The compounds carry identical or different organic groups on both metal atoms. Derivatives with phenylacetylene in the tin moiety are known. New synthetic methods comprise reactions involving lithium derivatives (I) potassium derivatives (II) and condensation of organometal amides with hydrides (III) and with phenylacetylene (IV). The following scheme summarizes preparation of bimetallic organogermanium-tin compounds:

I Synthesis with lithium derivatives:
  (IA) $R_3GeLi + R_3SnX \rightarrow R_3GeSnR_3$,
  (IB) $Me_3SnLi + MeGeCl \rightarrow Me_3GeSnMe_3$.

II Synthesis with potassium derivatives:
  $Et_3GeK + Me_3SnCl \rightarrow Et_3GeSnMe_3$.

III Condensation reactions with amide and hydride:
  (IIIA) $Ph_3GeH + R_3SnNEt_2 \rightarrow Ph_3GeSnR_3 + Et_2NH$,
  (IIIB) $Bu_3GeNEt_2 + Ph_3SnH \rightarrow Bu_3GeSnPh_3$.

IV Interconversion reaction:
  $Ph_3GeSnR_n(NEt_2)_{3-n} + PhC\vdots CH \rightarrow Ph_3GeSn(C\vdots CPh)_{3-n}R_n$.

R = organic group, X = bromine, hydrogen or methoxy group.

Hexaphenylgermyl Tin (M-136) $GeSnC_{36}H_{30}$ $Ph_3GeSnPh_3$
Synth.: By rxn. of Ph_3GeLi with Ph_3SnCl in N atm., in THF, in 71% yield (1928).
By rxn. of Ph_3GeK with Ph_3SnCl in THF, in 86% yield (2128).
By rxn. of $PhGeH$ with Ph_3SnNEt_2 in N atm., in 17% yield (913).
By rxn. of Ph_3GeBr with Ph_3SnK in THF, in 10% yield (2128).
Prop.: m. 293-97° (1928), 288° (2128), 282-94° (913), UV spectrum (913).
Rxn. with: AcOH (excess) → $(AcO)_3GeSn(OAc)_3$ in N atm. (2128).
AcOH in air → $(AcO)_3GeOSn(OAc)_3$ (2128).
AcOH (1:3) → $Ph_3GeSn(OAc)_3$ (2128).

Other bimetallic organogermanes with Ge-Sn bonds are summarized in Table 78.

Table 78. Bimetallic Organogermanes with Ge-Sn Bonds $\qquad$ R_3GeSnR_3

Formula	Synth. Method*	Yield	Properties	Ref.
Me₃GeSnMe₃	IB	50	Colorless liq. b. 154-56°, mol. wt. (a)	1844
Me₃GeSnPh₃	IA (b)	--	m. 110-11° (c)	849
Et₃GeSnMe₃	II	--	b_{14} 90-93°, n_D^{20} 1.5040 (d)	814
Et₃GeSnEt₃	IA (e)	34	$b_{1.5}$ 109°, n_D^{22} 1.5130, d_{20} 1.2870	2087, 2852
	IA (f)	87	$b_{1.5}$ 105°, n_D^{22} 1.5131	2097
Bu₃GeSnPh₃	IIIB	68	White solid, m. 24-25.5° (g, H)	913
Ph₃GeSnMe₃	IIIA	92	m. 89°, soly., stable as solid (c, g)	1618
Ph₃GeSnEt₃	IIIA	52	White solid, m. 45-52° (g, h)	913
Ph₃GeSn(C:CPh)Et₂	IV	100	Yellow oil, n_D^{20} 1.6501 (i)	913
Ph₃GeSn(C:CPh)₂Et	IV	90	Slightly yellow solid, m. 40° (i)	913
Ph₃GeSnBu₃	IIIA	83	m. 23°, $b_{0.001}$ 178-81°, soly.	1618
Ph₃GeSn(C:CPh)Ph₂	IV	77	IR spectrum	913

* The characters used here correspond to the synthetic scheme in the introduction to Chapter 6.1.2.1. (a) (Far) IR spectrum and assigment (1844). (b) From (Me₃Ge)₂Hg with Li in THF, followed by Ph₃SnBr (849). (c) IR and mass spectra (849). (d) IR and NMR spectra (814). (e) From Et₃SnH (2087, 2852). (f) From Et₃SnOMe (2097). (g) Far IR spectrum (828). (h) UV spectrum (913). (i) IR spectrum (913).

6.1.2.2 POLYMETALLIC ORGANOGERMANES WITH Ge-Sn BONDS

Organogermanes with three and more metal atoms are listed in this chapter. The compounds are prepared by methods from the following scheme.

I Reactions with organolithium derivatives:
- (IA) $Ph_3GeLi + SnCl_2, + Ph_3GeCl \rightarrow (Ph_3Ge)_4Sn$,
- (IB) $(Ph_3Sn)_3SnLi + Ph_3GeCl \rightarrow (Ph_3Sn)_3SnGePh_3$.

II Reactions with organogermanium hydrides:
- (IIA) $R_3GeH + R'_nSn(NEt_2)_{4-n} \rightarrow (R_3Ge)_{4-n}SnR'_n$,
- (IIB) $Ph_2GeH_2 + R'_3SnNEt_2 \rightarrow Ph_2Ge(SnR'_3)_2$,
- (IIC) $Ph_2GeH_2 + R_2Sn(NEt_2)_2 \rightarrow (Ph_2GeSnR_2)_x$.

III Reactions with organostannanes:
- (IIIA) $Bu_3GeNEt_2 + Bu_2SnH_2 \rightarrow (Bu_3GeSnBu_2)_2$,
- (IIIB) $R'_3GeNR_2 + R_2SnH_2 \rightarrow (R'_3Ge)_2SnR_2 + (R'_3GeSnR_2)_2$,
- (IIIC) $Ph_3GeSnEt_n(NEt_2)_{3-n} + Ph_3SnH \rightarrow Ph_3GeSnEt_n(SnPh_3)_{3-n}$.

R = organic group or N_i-CHO, R' = organic group or trialkylgermyl, n = 1,2. All compounds are listed in Table 79 and in Chapter 6.1.1.

6.1.2.6 FUNCTIONALLY SUBSTITUTED POLYMETALLIC ORGANOGERMANES WITH Ge-Sn BONDS

6.1.2.6.3 DERIVATIVES WITH OXYGEN

Oxides, hydroxides, phenoxides and carboxylates of organogermanium-tin compounds are listed in this chapter. The derivatives all compiled in Table 80 are prepared by methods from the following scheme.

I Acetylation of hexaphenylgermyltin:
$Ph_3GeSnPh_3 + AcOH \rightarrow Ph_3GeSn(OAc)_3$.

II Protonolysis of organogermyltin amides:
- (IIA) $(Ph_3Ge)_nSnR_{3-n}NEt_2 + H_2O \rightarrow (Ph_3Ge)_nSnR_mOH$,
- (IIB) $Ph_3GeSnEt_2NEt_2 + PhOH \rightarrow Ph_3GeSnEt_2OPh$.

Table 79. Polymetallic Organogermanes with Ge-Sn Bonds $R_nGe_mSn_o$

Formula	Synth. Method*	Yield	Properties	Ref.
$(Bu_3Ge)_2SnPh_2$	IIA	36	Greenish oil, n_D^{20} 1.6360 (a)	913
$(Ph_3Ge)_2SnPh_2$	IIA	70	White, m. 133-37°, mol. wt. (a)	913
$Ph_3GeSnEt_2SnPh_3$	IIIC	85	White, m. 125-29°, mol. wt. (a)	913
$Ph_2Ge(SnEt_3)_2$	IIB	65	Slightly yellow, n_D^{20} 1.6119, mol. wt. (a)	913
$Ph_2Ge(SnPh_3)_2$	IIB	63	White, m. 169-78°, (a)	913
$(Bu_3GeSnBu_2)_2$	IIIA	95	Slightly greenish oil, n_D^{20} 1.5213	913
$(Ph_3Ge)_3SnEt$	IIA	3	White, m. > 330° (a)	913
$Ph_3GeSnEt(SnPh_3)_2$	IIIC	50	Slightly yellow, m. 250°	913
$(Ph_3Ge)_4Sn$	IA	27	m. 407-10° (dec.) (b)	1928
$(Ph_3GeSnEt_2)_2GePh_2$	IIB	75	White solid (a)	913
$(Ph_3GeSnEt_2)_2SnEt_2$	IIIB	35	White gel, n_D^{20} 1.6754 (a)	913
$(Ph_3GeSnEt_2)_2SnPh_2$	IIIB	25	Slightly yellow, m. 40° (a)	913
$Ph_3GeSn(SnPh_3)_3$	IB	18	IR spectrum (c)	68a, 175
$(Ph_3GeSnEt_2SnPh_2)_2$	IIIB	24	Greenish oil (a)	913
$[(Ph_3Ge)_2SnEt]_2SnPh_2$	IIIB	26	Slightly yellow, m. 175°	913
$[(Ph_3Ge)_2SnEtSnPh_2]_2$	IIIB	70	Yellow solid	913
$(Ph_2GeSnEt_2)_x$	IIA	74	Slightly yellow, m. > 260°	913
$(Ph_2GeSnPh_2)_x$	IIA	52	Slightly yellow, m. > 260°	913

* The characters used here correspond to the synthetic scheme in the introduction to Chapter 6.1.2.2. (a) UV spectrum (913). (b) X-ray powder diffraction pattern, isomorphous with $(Ph_3Sn)_4Sn$ (175, 1928). (c) Melting point not depressed by $(Ph_3Sn)_4Sn$ (68a, 175).

Table 80. Oxygen Substituted Organogermyl Tin Compounds   R₃GeSnR₂OH

Formula	Synth. Method*	Yield	Properties	Ref.
(Ph₃Ge)₂SnEtOH	IIA	61	Slightly yellow, m. > 266°	913
(Ph₃GeSnEt₂)₂O	IIA	71	White, m. 85-95°, soly.	913
(Ph₃GeSnEtO)ₓ	IIA	50	Yellow, dec. 200°, soly.	913
Ph₃GeSnPh₂OH	IIA	--	Light green solid, soly.	913
Ph₃GeSnEt₂OPh	IIB	87	Yellow oil, n_D^{20} 1.6443	913
Ph₃GeSn(OAc)₃	I	--	Yellow solid (a)	2128

* The characters used here correspond to the synthetic scheme in the introduction to Chapter 6.1.2.6.3.
(a) Rxn. with Br → Ph₃GeBr + BrSn(OAc)₃ (2128).

6.1.2.6.5 NITROGEN SUBSTITUTED ORGANOGERMYL TIN COMPOUNDS

Triphenylgermyldiethyltin Diethylamide   GeSnC₂₆H₃₅N   Ph₃GeSnEt₂NEt₂
Synth.: By rxn. of Ph₃GeH with Et₂Sn(NEt₂)₂ (1:1) in N atm., in 72% yield (913).
Prop.: Colorless oil, n_D^{20} 1.6085 (913).
Rxn. with: Ph₃SnH → Ph₃GeSnEt₂SnPh₃ + Ph₆Sn₂ (913).
Ph₂GeH₂ (2:1) → (Ph₃GeSnEt₂)₂GePh₂ (913).
PhC⋮CH → Ph₃GeSn(C⋮CPh)Et₂ + Et₂NH (913).
PhNHCHO → Ph₃GeSnEt₂NPhCHO (913).
H₂O → (Ph₃GeSnEt₂)₂O (913).
PhOH → Ph₃GeSnEt₂OPh + Et₂NH (913).

Triphenylgermyldiethyltin N-Phenyl-   GeSnC₂₉H₃₁NO   Ph₃GeSnEt₂NPhCHO
  formanide
Synth.: By rxn. of Ph₃GeSnEt₂NEt₂ with PhNHCHO in N atm., in quant. yield (913).
Prop.: Colorless oil (913).
Rxn. with: R₂SnH₂ (2:1) → (Ph₃GeSnEt₂)₂SnR₂ ; R = Et, Ph (913).
Ph₂SnH₂ (1:1) + Et₂NH → (Ph₃GeSnEt₂SnPh₂)₂ + H₂ (913).

Triphenylgermylethyltin Bis(diethyl-   GeSnC₂₈H₄₀N₂   Ph₃GeSnEt(NEt₂)₂
  amide)
Synth.: By rxn. of Ph₃GeH with EtSn(NEt₂)₃ (1:1) (913).
Rxn. with: PhC⋮CH → Ph₃GeSn(C⋮CPh)₂Et (913).
H₂O → (Ph₃GeSnEtO)ₓ (913).
PhNCHO → Ph₃GeSnEt(NPhCHO)₂ (913).

Triphenylgermylethyltin   GeSnC$_{34}$H$_{32}$N$_2$O$_2$ Ph$_3$GeSnEt(NPhCHO)$_2$
  Bis(N-Phenylformamide)
Synth.: By rxn. of Ph$_3$GeSnEt(NEt$_2$)$_2$ with PhNHCHO in N atm. (913).
Prop.: Not isolated (913).
Rxn. with: Ph$_3$SnH → Ph$_3$GeSnEt(SnPh$_3$)$_2$ (913).

Triethylgermyldiphenyltin   GeSnC$_{34}$H$_{35}$N   Ph$_3$GeSnPh$_2$NEt$_2$
  Diethylamide
Synth.: By rxn. of Ph$_3$GeH with Ph$_2$Sn(NEt$_2$)$_2$ in N atm. (913).
Prop.: Not isolated (913).
Rxn. with: PhC⋮CH → Ph$_3$GeSn(C⋮CPh)Ph$_2$ (913).
H$_2$O → Ph$_3$GeSnPh$_2$OH (913).

Bis(triphenylgermyl)ethyltin   Ge$_2$SnC$_{42}$H$_{45}$N   (Ph$_3$Ge)$_2$SnEtNEt$_2$
  Diethylamide
Synth.: By rxn. of Ph$_3$GeH with EtSn(NEt$_2$)$_3$ (2:1) in N atm. (913).
Prop.: Not isolated (913).
Rxn. with: H$_2$O → (Ph$_3$Ge)$_2$SnEtOH (913).
PhNHCHO → (Ph$_3$Ge)$_2$SnEtNPhCHO (913).

Bis(triphenylgermyl)ethyltin   Ge$_2$SnC$_{45}$H$_{41}$NO   (Ph$_3$Ge)$_2$SnEtNPhCHO
  N-Phenylformamide
Synth.: By rxn. of (Ph$_3$Ge)$_2$SnEtNEt$_2$ with PhNHCHO in N atm. (913).
Prop.: Not isolated (913).
Rxn. with: Ph$_2$SnH$_2$ (2:1) → [(Ph$_3$Ge)$_2$SnEt]$_2$SnPh$_2$ (913).
Ph$_2$SnH$_2$ (1:1) → [(Ph$_3$Ge)$_2$SnEtSnPh$_2$]$_2$ (913).

6.1.3 BIMETALLIC ORGANOGERMYL LEAD COMPOUNDS

Organogermanium lead compounds containing two and three metal atoms linked in one chain are reported in Table 81.

Table 81. Bimetallic Organogermyl Lead Compounds   R$_3$GePbR$_3$

Formula	Synth. Methods	Properties	Ref.
Ph$_3$GePbR$_3$			
R = n-Bu	(a)	Oil, mol. wt. (b)	1608
i-Bu	(a)	m. 86°, mol. wt. (b)	1608
c.C$_6$H$_{11}$	(a)	m. 174° (dec.), mol. wt. (b)	1608
Ph	(a)	m. 227° (dec.), mol. wt. (b)	1608
Ph$_2$Ge(PbPh$_3$)$_2$	(c)	Yellow, dec. 154°, mol. wt.	1608

(a) Synth. from Ph$_3$GeH with R$_3$PbNEt$_2$ in inert atm. in good yield (1608). (b) Analytical determination by I titration (1608). (c) Synth. from Ph$_2$GeH$_2$ and

Ph$_2$Pb(NEt$_2$)$_2$ at -70 to -20° (1608).

6.2 ORGANOGERMANIUM DERIVATIVES OF GROUP III METALS

Organogermanium compounds containing germanium bonds to boron, aluminum, gallium, indium and thallium are included in this chapter. All metal derivatives are sensitive to air and moisture.

μ-Trimethylgermylpentaborane (9)　　　GeB$_5$C$_3$H$_{17}$　　　　　　　μ-Me$_3$GeB$_5$H$_8$
Synth.: By rxn. of Me$_3$GeCl with LiB$_5$H$_8$ in Et$_2$O in N atm. at -45 to -22°, in 88% yield (2398).
Prop.: m. 11.5°, no vapor pressure at r.t., (far) IR, ^{1}H, ^{11}B NMR and mass spectra, struct. (2398).
Rxn. with: Thermal rearrangement at r.t. → 2-Me$_3$GeB$_5$H$_8$ (2398).
Br → 1-BrB$_5$H$_7$GeMe$_3$ (2398).

2-Trimethylgermylpentaborane (9)　　　GeB$_5$C$_3$H$_{17}$　　　　　　　2-Me$_3$GeB$_5$H$_8$
Synth.: By thermal rearrangement of μ-Me$_3$GeB$_5$H$_8$ at r.t., in 50% yield (2398).
Prop.: Vapor pressure 1 mm at 26°, (far) IR, ^{1}H, ^{11}B NMR and mass spectra, struct. (2398).

1-Bromo-μ-Trimethylgermyl-　　　　　GeB$_5$C$_3$H$_{16}$Br　　　　μ-Me$_3$GeB$_5$H$_7$Br (1)
　pentaborane (9)
Synth.: μ-Me$_3$GeB$_5$H$_8$ with Br at -196° to r.t. (2398).
Prop.: Subl. in vacuo at 40°, (far) IR, ^{1}H and ^{11}B NMR spectra (2398).

Triethylgermylpentaborane (9)　　　GeB$_5$C$_6$H$_{23}$　　　　　　　Et$_3$GeB$_5$H$_8$
Synth.: By rxn. of Et$_3$GeI with LiB$_5$H$_8$ in Et$_2$O in N atm. at -22 to -45°, in 80% yield (2398).
Prop.: Exists in two isomers μ- and 2-form, given for both: (far) IR, ^{1}H and ^{11}B NMR spectra, no vapor pressure at r.t., struct. (2398).
Rxn. of: μ-form at r.t. → 2-Et$_3$GeB$_5$H$_8$ (2398).

Tris(triethylgermyl) Thallium　　Ge$_3$TlC$_{18}$H$_{45}$　　　　　　(Et$_3$Ge)$_3$Tl
Synth.: By rxn. of Et$_3$GeH with Et$_3$Tl in inert atm. at 100°, in 91% yield (1361, 2089).
Prop.: Red oil, d$_{20}$ 1.535 (1361, 2089), far IR spectrum and assignment (2374).
Rxn. with: Thermal dec. at 170° → Et$_6$Ge$_2$ + Tl (2089).
Et$_3$SnH → Et$_3$GeH + Et$_6$Sn$_2$ + Tl (1361).
Li in THF → Et$_3$GeLi + Tl (2087, 2852).
Hg → (Et$_3$Ge)$_2$Hg + Tl (1361, 2089).
H$_2$O → Et$_3$GeH + (Et$_3$Ge)$_2$O + Tl (1361).
Bz$_2$O$_2$ → Et$_3$GeOBz + TlOBz (1361, 2089).
Br in C$_7$H$_{16}$ → Et$_3$GeBr + TlBr (2851).
(CH$_2$Br)$_2$ → Et$_3$GeBr + TlBr + C$_2$H$_4$ (1361, 2089).

Far IR spectra and assignments are given without any other information for (Et$_3$Ge)$_3$Ga [Ge$_3$GaC$_{18}$H$_{15}$], (Et$_3$Ge)$_2$GaBr [Ge$_2$GaC$_{12}$H$_{10}$Br] and (Et$_3$Ge)$_3$In [Ge$_3$InC$_{18}$H$_{15}$] (2374).

The following triorganogermyl-Group III metal derivatives are mentioned in the patent literature as agents for metal plating: Me$_3$GeBEt$_2$, Bu$_3$GeBPr$_2$; Et$_3$GeAlEt$_2$, MeEtPrGeAlEtPr, Bu$_3$GeAlBu$_3$; Me$_3$GeGaMe$_2$, Me$_3$GeGaPr$_2$, Pr$_3$GeGa-(C$_{11}$H$_{23}$)$_2$; Me$_3$GeInMe$_2$, Me$_3$GeInBu$_2$ (603).

6.3 ORGANOGERMANIUM DERIVATIVES OF GROUP II METALS

This chapter comprises organogermanes containing germanium bonds to Group II metals. Derivatives of main group elements and transition metals are included. All compounds are very sensitive to air and moisture. The organogermyl metal compounds listed in Table 82 are prepared by methods from the following scheme.

I Reactions with organogermanium hydride:
   Et$_3$GeH + REt$_2$SiHgEt → Et$_3$GeHgSiEt$_2$R ;
   R = Et, Et$_3$Si .

II Interconversion reactions:
   Et$_3$GeMR + BuSeH → Et$_3$GeMSeBu ;
   R = Et, Et$_3$Ge ; M = Cd, Hg .

III Metallation with potassium:
   Bu$_3$GeR + K → Bu$_3$GeK ;
   R = Cl, Bu$_3$Ge .

Triisopropylgermyl Magnesium Chloride   GeMgC$_9$H$_{21}$Cl   i-Pr$_3$GeMgCl

Synth.: Discussed as intermediate in rxn. of GeCl$_4$ with i-PrMgCl in dry Et$_2$O in He atm. at 4° (2591).
Prop.: Not isolated (2591).
Rxn. with: H$_2$O → i-Pr$_3$GeH (2591).
CH$_2$:CHCH$_2$Cl → i-Pr$_3$GeCH$_2$CH:CH$_2$ (2591).

Tricyclohexylgermyl Magnesium Chloride   GeMgC$_{18}$H$_{33}$Cl   (c.C$_6$H$_{11}$)$_3$GeMgCl

Synth.: Discussed as intermediate in rxn. of GeCl$_4$ with excess C$_6$H$_{11}$MgCl in dry Et$_2$O in He atm. at 4° (353, 2591).
Prop.: Not isolated (353, 2591).
Rxn. with: X'$_2$O → (C$_6$H$_{11}$)$_3$GeX' ; X' = H, D (353, 2591).
PhCH$_2$Cl → (C$_6$H$_{11}$)$_3$GeCH$_2$Ph + PhCH$_2$C$_6$H$_{11}$ (2591).

Triphenylgermyl Magnesium Bromide   GeMgC$_{18}$H$_{15}$Br   Ph$_3$GeMgBr
Synth.: By rxn. of Ph$_3$GeAuPPh$_3$ with anhydr. MgBr$_2$ in Et$_2$O in N atm. at r.t. (2420).
Prop.: Orange soln., not isolated (2420).

Rxn. with: $H_2O \rightarrow Ph_3GeH$ (2420).

Bis(triphenylgermyl) Strontium $Ge_2SrC_{36}H_{30}$ $(Ph_3Ge)_2Sr$
Synth.: By rxn. of Ph_6Ge_2 with Sr in liq. NH_3 at $-40°$, isolated as 1:1 adduct with 1,4-dioxane (675).
Prop.: Dec. 40-45°, sensitive to air and light, soly., forms 1:1 adduct with THF (675).
Rxn. with: $AgNO_3 \rightarrow Ag$ (675).
H_2O in MeOH $\rightarrow Ph_3GeH + Sr(OH)_2$ (675).

Bis(triphenylgermyl) Barium $Ge_2BaC_{36}H_{30}$ $(Ph_3Ge)_2Ba$
Synth.: By rxn. of Ph_6Ge_2 with Ba in liq. NH_3 at $-40°$, isolated as 1:1 adduct with 1,4-dioxane (675).
Prop.: Dec. 60-70°, sensitive to air and light, soly., slowly dec. in soln., forms 1:1 adduct with THF (675).
Rxn. with: $AgNO_3 \rightarrow Ag$ (675).
$H_2O \rightarrow Ph_3GeH + Ba(OH)_2$ (675).

Bis(triethylgermyl) Zinc $Ge_2ZnC_{12}H_{30}$ $(Et_3Ge)_2Zn$
Synth.: By rxn. of Et_3GeLi with $ZnCl_2$ in THF at 0-5°, in 73% yield (2854).
Prop.: Greenish-yellow liq., sensitive to air (2854), far IR spectrum and assignment (2374).
Rxn. with: Thermal dec. at 100-130° $\rightarrow Et_4Ge + Et_6Ge_2 + Zn$ (2854).
$Bz_2O_2 \rightarrow Et_3GeOBz + Zn(OBz)_2$ (2854).
$BuSeH \rightarrow Et_3GeH + Zn(SeBu)_2$ (2854).
$EtBr \rightarrow Et_4Ge + ZnBr_2$ (2854).
$(CH_2Br)_2 \rightarrow Et_3GeBr + ZnBr_2 + C_2H_4$ (2854).

--   -- $Et_3GeZn(GeEt_2)_xEt$ (?)
Synth.: By rxn. of Et_3GeH with Et_2Zn at 125° in vacuo, in 55% yield (584).
Prop.: Yellow undistillable solid, dec. above 90° (584).
Rxn. with: $HgCl_2 \rightarrow Et_3GeCl + Et_6Ge_2 + Hg + Zn$ (584).
$H_2O \rightarrow Et_3GeH + Et_6Ge_2 + Et_5Ge_2H + Zn(OH)_2$ (584).
$Bz_2O_2 \rightarrow Et_6Ge_2 + Et_3GeOBz + Et_5Ge_2OBz + Zn(OBz)_2$ (584).
$EtBr \rightarrow Et_4Ge + Et_6Ge_2 + Et_8Ge_3$ (?) $+ ZnBr_2$ (584).
$(CH_2Br)_2 \rightarrow Et_3GeBr + ZnBr_2 + C_2H_4$ (584).

Bis(triphenylgermyl) Zinc $Ge_2ZnC_{36}H_{30}$ $(Ph_3Ge)_2Zn$
Synth.: By rxn. of Ph_3GeM with $ZnCl_2$ in liq. NH_3 at $-40°$; M = Na, K, isolated as 1:1 adduct with 1,4-dioxane (675).
Prop.: Yellow solid, dec. 110-20°, sensitive to air and moisture, soly. (675).
Rxn. with: $AgNO_3 \rightarrow Ag$ (675).
$H_2O \rightarrow Ph_3GeH + Zn(OH)_2$ (675).
Addn. Compd.: $(Ph_3Ge)_2Zr \cdot 1/2$ THF$\cdot 1/2$ NH_3, from Ph_3GeNa and $ZnCl_2$ in NH_3-THF, pale yellow (675).

Bis(triethylgermyl) Cadmium (M-134) $Ge_2CdC_{12}H_{30}$ $(Et_3Ge)_2Cd$
Prop.: d_{20} 1.446 (587), far IR spectrum and assignment (2374).
Rxn. with: UV irradiation → Et_6Ge_2 + Cd (587).
Li → Et_3GeLi + Cd (2088).
Hg → $(Et_3Ge)_2Hg$ (587).
Et_3MBr → Et_4M + $CdBr_2$; M = Ge, Si (587a).
Et_3SnX → Et_3GeX + Et_6Sn_2 + Cd ; X = H, Cl (587).
$HgCl_2$ → Et_3GeCl + $CdCl_2$ + Hg (587).
ROH → Et_3GeOR + Et_3GeH + Cd ; R = n-Pr, Ac (587).
BuSeH → $Et_3GeCdSeBu$(?) + $Cd(SeBu)_2$ + Et_3GeH (587).
CCl_4 → Et_3GeCl + $(Et_3Ge)_2CCl_2$ + $CdCl_2$ (2088).

Bis(triphenylgermyl) Cadmium $Ge_2CdC_{36}H_{30}$ $(Ph_3Ge)_2Cd$
Synth.: By rxn. of Ph_3GeH with Et_2Cd in MePh in N atm. at 100°, in 86% yield (2295).
Prop.: Light orange cryst., dec. 105°, dec. slowly in sealed tube at r.t. or in soln. (2295).
Rxn. with: Thermal dec. at 150° → Ph_6Ge_2 + Cd (2295).
Bz_2O_2 → Ph_3GeOBz + $Cd(OBz)_2$ (2295).
$(CH_2Br)_2$ → Ph_3GeBr + $CdBr_2$ + C_2H_4 (2295).
Addn. Compd.: $(Ph_3Ge)_2Cd \cdot 2C_4H_8O$, Ph_3GeH with Et_2Cd in THF, colorless cryst., dec. 108° [$Ge_2CdC_{44}H_{46}O_2$] (2295).

Bis(trimethylgermyl) Mercury $Ge_2HgC_6H_{18}$ $(Me_3Ge)_2Hg$
Synth.: By rxn. of Me_3GeBr with Na-Hg in cyclohexane in N atm., in 35% yield (996).
By rxn. of Me_3GeH with Et_2Hg at 60° in Ar atm. (996).
Prop.: Yellow cryst., NMR and mass spectra, soly. (996).
Rxn. with: Li in THF → Me_3GeLi (2285).
Li in THF, + Ph_3SnBr → $Me_3GeSnPh_3$ (849).
$(Et_3P)_2PdCl_2$ → Hg + Pd + $(Et_3P)_2PdCl_4$ (2285).
cis-$(Et_3P)_2PtCl_2$ → trans-$Me_3Ge(Et_3P)_2PtCl$ + Hg (1106).
cis-$(Me_2PhP)_2PtCl_2$, + NaI → trans-$Me_3Ge(Me_2PhP)_2PtI$ (1106).
Ph_3PAuCl → $Me_3GeAuPPh_3$ + Hg (2420).
$PhHgCBrCl_2$ → $Me_3GeCCl:CCl_2$ + Me_3GeCl + C_2Cl_4 + PhHgBr ; rxn. mechanism (1881).
1-Hexene → Me_6Ge_2 + 1-$Me_3GeC_6H_{13}$ + isomeric C_6H_{12} (2241).
Air → $(Me_3Ge)_2O$ + Hg (996).

Bis(triethylgermyl) Mercury (M-134) $Ge_2HgC_{12}H_{30}$ $(Et_3Ge)_2Hg$
Synth.: By rxn. of Et_3GeH with $Et_3GeHgEt$ in N atm., in 80% yield (2852).
By rxn. of Hg in inert atm. with $(Et_3Ge)_2Cd$ at 50°, in 79% yield (587), with $(Et_3Ge)_3Tl$ at 20°, in 78% yield (1361, 2089).
Prop.: $b_{1.5}$ 115-18° (1361), b_1 115-18° (587), d_{20} 1.724 (1361, 2852), spectra: (far) IR and assignment (2374), NMR (1009).
Rxn. with: Li → Et_3GeLi + Hg (2087-8, 2852).
Na, + PhBr → Et_3GePh + Hg (2088).

[C$_5$H$_5$Fe(CO)$_2$]$_2$ → Et$_3$GeC$_5$H$_5$Fe(CO)$_2$ + Hg (2414, 2416).
C$_5$H$_5$Fe(CO)$_2$X → Et$_3$GeX + Et$_3$GeC$_5$H$_5$Fe(CO)$_2$ + Hg ; X = Cl, Br (2414).
[C$_5$H$_5$M(CO)$_3$]$_2$ → Et$_3$GeC$_5$H$_5$M(CO)$_3$ + Hg ; M = Mo, W (2414).
[C$_5$H$_5$NiCO]$_2$ → Et$_3$GeC$_5$H$_5$(CO) + Hg (2414).
(:CCO$_2$Et)$_2$ → Et$_3$GeC(CO$_2$Et):C(CO$_2$Et)GeEt$_3$ (1365).
[:C(CO$_2$Et)$_2$]$_2$ → Et$_3$GeC(CO$_2$Et)$_2$C(CO$_2$Et):C(OEt)OGeEt$_3$ (1365).
(:NR)$_2$ → Et$_3$GeNRNRGeEt$_3$; R = Ph, CO$_2$Et (1365).
PhC:NN:CPhN:N → Et$_3$GeNCPh:NN:CPhNGeEt$_3$ (1365).
O in octane at -20° → (Et$_3$Ge)$_2$O, rxn. rates and mechanism, cat. effect of (Et$_3$Ge)$_2$O and (Et$_2$GeO)$_2$ (2706).
AcOH → Et$_3$GeH + Et$_3$GeOAc + Hg (291, 2086).
(c.C$_6$H$_{11}$OCO$_2$)$_2$ → Et$_3$GeOC$_6$H$_{11}$ + CO$_2$ + Hg (583a).
n-BuSeH → Et$_3$GeSeBu + Et$_3$GeH + Hg (2855).
(CH$_2$Br)$_2$ → Et$_3$GeBr + C$_2$H$_4$ + Hg (584).

Triethylgermyl Ethylmercury   GeHgC$_8$H$_{20}$ Et$_3$GeHgEt
Synth.: By-prod. in rxn. of Et$_3$GeH with excess Et$_2$Hg in N atm., in 8% yield (2852).
Prop.: Light yellow liq., dec. in air (2852), (far) IR spectrum and assignment (2374).
Rxn. with: Et$_3$GeH → (Et$_3$Ge)$_2$Hg + C$_2$H$_6$ (2852).
n-BuSeH → Et$_3$GeSeBu + C$_2$H$_6$ + Hg (2855).
AcOH → Et$_3$GeOAc + C$_2$H$_6$ + Hg (2855).

Other organogermanium derivatives of Group II metals are listed in Table 82 and in Chapter 6.1.1.

6.4 ORGANOGERMYL ALKALI METAL COMPOUNDS

Organogermanium derivatives of lithium, sodium and potassium are reported. None of the compounds has been characterized, they are used as synthetic intermediates. All compounds are very sensitive to air and moisture and have to be prepared and handled in inert atmosphere.

Trimethylgermyl Lithium   GeLiC$_3$H$_9$ Me$_3$GeLi
Synth.: By rxn. of Me$_6$Ge$_2$ with Li in (Me$_2$N)$_3$PO at 20° (815).
By rxn. of (Me$_3$Ge)$_2$Hg with Li in THF (2285).
By rxn. of Et$_3$GeLi with Me$_3$GeGeEt$_3$, in high yield (815).
Rxn. with: Me$_3$GeGeEt$_3$ ⇌ Et$_3$GeLi + Me$_6$Ge$_2$; kinetic data (812).
Et$_6$Ge$_2$ ⇌ Et$_3$GeLi + Me$_3$GeGeEt$_3$ (815), kinetic data (812).
Me$_3$GeX → Me$_6$Ge$_2$ (812).
(R$_3$P)$_2$PtHCl → (Me$_3$Ge)HPt(PR$_3$)$_2$; R = Et, Ph (?) (2285).
GeCl$_4$ → (Me$_3$Ge)$_4$Ge + Me$_{18}$G$_8$ (2419).

Tribenzylgermyl Lithium   GeLiC$_{21}$H$_{21}$ (PhCH$_2$)$_3$GeLi
Synth.: By rxn. of Li in (MeOCH$_2$)$_2$ with (PhCH$_2$)$_4$Ge, (PhCH$_2$)$_3$GeH or (PhCH$_2$)$_6$Ge$_2$ (104).

Prop.: Deep brown soln. in $(MeOCH_2)_2$ (104).
Rxn. with: Li in $(MeOCH_2)_2$ → $(PhCH_2)_nGeMe_{4-n}$ + MePh + CH_2:CHOMe + EtPh + $MeOCH_2CH_2OH$; n = 1-3 (104).
Et_3GeBr in $(MeOCH_2)_2$ → $)PhCH_2)_6Ge_2$ + Et_6Ge_2 (104).
RX → $(PhCH_2)_3GeR$; R = Me, Et, Me_3Si, Et_3Ge (104).
$HSiCl_3$ → polymeric matl. (104).
CO_2 → $(PhCH_2)_3GeO_2CCH_2Ph$ + $(PhCH_2)_3GeO_2CGe(CH_2Ph)_2$ (104).
ROH → $(PhCH_2)_3GeR$; R = H, D (104).

Triethylgermyl Lithium (M-137) $GeLiC_6H_{15}$ Et_3GeLi
Synth.: By rxn. of Et_6Ge_2 with Li in $(Me_2N)_3PO$ at 20° (815).
By rxn. of Li in inert atm. in THF or C_6H_6 with $(Et_3Ge)_2Hg$ (2087, 2852), with $(Et_3Ge)_2Cd$ in over 40% yield (2088), with $(Et_3Ge)Tl$ (2087, 2852).
By rxn. of Me_3GeLi with Et_6Ge_2, in low yield (815).
Prop.: Far IR spectrum and assignment (2374).
Rxn. with: Me_6Ge_2 ⇌ Me_3GeLi + $Me_3GeGeEt_3$; kinetic data (812).
$Me_3GeGeEt_3$ ⇌ Me_3GeLi + Et_6Ge_2 (815), kinetic data (812).
Et_3MH → Et_3GeMEt_3 + LiH; M = Si, Ge, Sn (2087, 2852).
$Me_2EtSiCl$ → $Et_3GeSiMe_2Et$ (2087) + Et_6Ge_2 (2852).
Ph_2SiCl_2 → $(Et_3Ge)_2SiPh_2$ (2087) + Et_6Ge_2 (2852).
$HSiX_3$ → $(Et_3Ge)_3SiH$; X = Cl (2087, 2852), EtO (2097).
Et_nHSiCl_{n-3} → $Et_n(Et_3Ge)_{3-n}SiH$ (2415).
Me_3GeX → $Me_3GeGeEt_2$ + Me_6Ge_2 + Et_6Ge_2 ; X = F, Cl (812, 815).
$Et_3GeOC_6H_{11}$ → Et_6Ge_2 + $c.C_6H_{11}OH$ (2097).
Et_3SnOMe → $Et_3GeSnEt_3$ (2097).
Et_2SbBr → $Et_3GeSbEt_2$ (2415).
$ZnCl_2$ → $(Et_3Ge)_2Zn$ (2852).
CH_2:CHR ⇌ Et_3GeCH_2CHRLi , R = H, Bu, Ph (2849).
CH_2:CHR, + H_2O → $Et_3GeCH_2CH_2R$; R = H, Bu, Ph (2087).
PhC⋮CR → Et_3GeCR:CPhLi ; R = H, Ph (2849).
Ph_2CO, + H_2O → Et_3GeCPh_2COH (2849).
RBr → Et_3GeR ; R = Et, Ph (2088).

Triethylgermyl Sodium $GeNaC_6H_{15}$ Et_3GeNa
Synth.: By rxn. of Et_6Ge_2 with Na at r.t. in $(Me_2N)_3PO$ (815).
By rxn. of $(Et_3Ge)_2Hg$ with Na in THF, in inert atm. (2088).
Rxn. with: Me_3GeX → $Me_3GeGeEt_3$ + Me_6Ge_2 + Et_6Ge_2 ; X = F, Cl (815).
PhBr → Et_3GePh (2088).

Triethylgermyl Potassium $GeKC_6H_{15}$ Et_3GeK
Synth.: By rxn. of Et_3GeCl or Et_6Ge_2 with K in $(Me_2N)_3PO$ at 20° (814-5).
Rxn. with: Me_3MCl → Me_3MGeEt_3 ; M = Si, Ge, Sn (814).
REt_2GeCl → $REt_2GeGeEt_3$; R = Et (814), Et_3Ge (815).
YCl_2 → $(Et_3Ge)_2Y$; Y = CH_2, Me_2Ge, $Et_2GeGeEt_2$ (815).
MeCl → $MeGeEt_3$ (815).
$HCCl_3$ or CCl_4 → Et_6Ge_2 + C (815).
CO_2 → Et_3GeCO_2H (815).

Ethyl-1-naphthylphenylgermyl Lithium   GeLiC$_{18}$H$_{17}$ EtPh(1-C$_{10}$H$_7$)GeLi
<u>Synth.</u>: By rxn. of (+)-EtPh(1-C$_{10}$H$_7$)GeH with BuLi in Et$_2$O (2367, 2369).
By rxn. of (-)-EtPh(1-C$_{10}$H$_7$)GeCO$_2$H with BuLi in Et$_2$O (2369).
<u>Rxn. with</u>: R'X → R$_3$GeR'* ; R' = i-Pr, n-Bu, t-Bu, CH$_2$:CHCH$_2$, X = Cl, Br (2369).
2-MeC$_6$H$_9$O → (-)-R$_3$GeCOHCHMe(CH$_2$)$_3$CH$_2$ * (2369).
H$_2$CO → (+)-R$_3$GeCH$_2$OH* (2369).
H$_2$O → (+)-R$_3$GeCH$_2$OH* (2369).
H$_2$O → (+)-R$_3$GeH* (2369).
CO$_2$, + H$_2$O → (-)-R$_3$GeCO$_2$H* (2369).
Ph$_2$CO → (-)-EtPh(1-C$_{10}$H$_7$)GeCPh$_2$OH (2369).
PhCH$_2$X → EtPh(1-C$_{10}$H$_7$)GeCH$_2$Ph ; X = Br, I, with inversion, Cl partly retention (2369).
RI → EtPh(1-C$_{10}$H$_7$)GeR with inversion; R = Me, i-Pr, n-Bu, CH$_2$CH:CH$_2$ (2369).

Triphenylgermyl Lithium (M-134)   GeLiC$_{18}$H$_{15}$ Ph$_3$GeLi
<u>Synth.</u>: By rxn. of Ph$_6$Ge$_2$ with Li in THF, in 90% yield (1928).
By rxn. of Ph$_4$Ge in N atm. in THF, in 75% yield (1928).
By rxn. of Ph$_3$GeH with BuLi in THF in N atm. at -23° (1106, 1624-5).
<u>Prop.</u>: Anal. detn. by double titration with CH$_2$:CHCH$_2$Br (68a, 173), with RBr (1928).
<u>Rxn. with</u>: Dec. in THF → comparative kinetic data (68a, 176).
Ph$_3$GeCOPh → (Ph$_3$Ge)$_2$CPhOH (1624-5).
Ph$_3$SnCl → Ph$_4$Sn + Ph$_3$GeSnPh$_3$ (1928).
GeCl$_4$ → Ph$_6$Ge$_2$ (1928).
SnCl$_2$, + Ph$_3$GeCl → (Ph$_3$Ge)$_4$Sn + Ph$_6$Ge$_2$ (175, 1928).
Ph$_3$PAuCl → Ph$_3$GeAuPPh$_3$ (2420).
(C$_5$H$_5$)$_2$MCl$_2$ → Ph$_3$GeM(C$_5$H$_5$)$_2$Cl ; M = Ti, Zr (907), Zr, Hf (2492).
(C$_5$H$_5$)$_2$TiCl → Ph$_3$GeTi(C$_5$H$_5$)$_2$ (907).
(R$_3$P)$_2$PdBr$_2$ → (Ph$_3$Ge)$_2$Pd(PR$_3$)$_2$ (795).
(R$_3$P)$_2$PtX$_2$ → (Ph$_3$Ge)$_2$Pt(PR$_3$)$_2$ (101, 103, 2285).
(Et$_3$P)$_2$PtI$_2$ (2:1), + H$_2$O → (Ph$_3$Ge)$_2$Pt(PEt$_3$)$_2$ + (Ph$_3$Ge)HPt(PEt$_3$)$_2$ (103) + (Ph$_3$Ge)PtI(PEt$_3$)$_2$ (102).
(Et$_3$P)$_2$PtI$_2$ (1:1), + H$_2$O → (Ph$_3$Ge)HPt(PEt$_3$)$_2$ + Ph$_6$Ge$_2$ + (Et$_3$P)$_2$PtHI (102-3).
(Et$_3$P)$_2$PtI$_2$ (1:1), + EtOH → (Ph$_3$Ge)$_2$Pt(PEt$_3$)$_2$ + Ph$_3$Ge(EtO)Pt(PEt$_3$)$_2$ + Ph$_4$Ge (102-3).
Me$_3$MPtCl(PEt$_3$)$_2$ → Ph$_3$Ge(Me$_3$M)Pt(PEt$_3$)$_2$; M = Si, Ge (1106).
Me$_3$PtI → MeGePh$_3$ + black pyrophic residue (103).
R$_3$PO$_4$ → RGePh$_3$; R = Me, Bu (1928).
Fluorene, + CO$_2$ → 9-C$_{13}$H$_9$CO$_2$H (68a, 174).
1,2-OC$_6$H$_{10}$ → 2-Ph$_3$GeC$_6$H$_{10}$OH (2670).
AcH → Ph$_3$GeCHMeOH (2671).
R$_2$CO → Ph$_3$GeCR$_2$OH + Ph$_3$GeH ; R = Me, Et, Pr, Ph, (Me + Ph), (Me + Et), (Ph + i-Pr) (1624-5).

---

* R$_3$Ge = EtPh(1-C$_{10}$H$_7$)Ge ; all rxns. with retention of configuration.

RCOCl → (Ph$_3$Ge)$_2$CROH + Ph$_3$GeCOR ; R = Me (2671), Ph, p-FC$_6$H$_4$ (1624-5).
RCOCl → (Ph$_3$Ge)$_2$CROH ; R = Me, Et (1624-5).
RCOCl → Ph$_3$GeCOR ; R = Me (1677), Me$_3$C (2671), p-MeOC$_6$H$_4$, p-CF$_3$C$_6$H$_4$ (1624-5).
COCl$_2$ → Ph$_6$Ge$_2$ (1624-5).
HCO$_2$Et, + LiAlH$_4$ → (Ph$_3$Ge)$_2$CHOH (794).
Y → Ph$_3$GeYLi ; Y = S, Se, Te (484).

Triphenylgermyl Potassium (M-135) GeKC$_{18}$H$_{15}$ Ph$_3$GeK

<u>Synth.</u>: By rxn. of Ph$_3$GeBr or Ph$_6$Ge$_2$ with Na-K in THF in N atm. (2128).
By rxn. of Ph$_3$GeH with K in (Me$_2$N)$_3$PO at 60° (1634, 2983).
<u>Rxn. with</u>: Ph$_3$SiCl → Ph$_3$GeSiPh$_3$ (2128).
Ph$_3$SnCl → Ph$_3$GeSnPh$_3$ + Ph$_6$Ge$_2$ + Ph$_6$Sn$_2$ (2128).
ZnCl$_2$ in liq. NH$_3$ → (Ph$_3$Ge)$_2$Zn (675).
BuBr → BuGePh$_3$ + KBr (2983).

Additional organogermanium derivatives of alkali metals may be found at the bottom of Table 82 and in Chapter 6.1.1.

Table 82. Organogermanium Derivatives of Group II R$_3$GeM^I
and Alkali Metals R$_3$GeM^{II}R

Formula*	Synth. Method (Yield)**	Properties	Ref.
Et$_3$GeMgSeBu	--	(a)	--
R$_3$GeMgX	--	(b)	--
Et$_3$GeCdSeBu (?)	II (c)	Yellow nondistillable liq.	2855
Et$_3$GeHgSiEt$_3$ (137)	I (26)	Lemon-yellow liq. (d)	583a, 2852
Et$_3$GeHgSi$_2$Et$_5$ (137)	I (50)	Orange-yellow liq., sensitive to air	583a
Et$_3$GeHgSeBu (?)	II	Not isolated (e)	2855
(Bu$_3$Ge)$_2$Hg	--	(f)	--
MePh$_2$GeLi	--	(g)	--
MePh(1-C$_{10}$H$_7$)GeLi (137)	--	(h)	--
Bu$_3$GeK	III	(i)	814
Ph$_3$GeNa (135)	--	(j)	--

* Numbers in parenthesis refer to pages in main volume.
** The characters used here correspond to the synthetic scheme in the introduction to Chapter 6.3.
(a) Rxn. with BuSeH → Et$_3$GeH + Mg(SeBu)$_2$ (2855). (b) Discussed as intermediate in rxn. of GeCl$_4$ with RMgX, leading to R$_6$Ge$_2$ with R$_3$GeX and R$_3$GeH on hydrolysis (831). (c) From (Et$_3$Ge)$_2$Cd (2855). (d) UV irradiation → Et$_3$GeSiEt$_3$ + Hg (583a, 2852). (e) Rxn. with BuSeH → Et$_3$GeSeBu + Hg (2855). (f) Rxn. with PhHgCBrCl$_2$ → Bu$_3$GeCCl:CCl$_2$ (1881). (g) Rxn. with (Me$_2$PPh)$_2$PtCl$_2$ → (MePh$_2$Ge)$_2$Pt(PMe$_2$Ph)$_2$ (856). (h) Rxn. with Ph$_2$CO → MePh(1-C$_{10}$H$_7$)GeCPh$_2$OH (2369),

with MeOH → MePh(1-C$_{10}$H$_7$)GeH ; stereochemistry (417). (i) Rxn. with R$_3$GeCl →
R$_3$GeGeBu$_3$; R = Me, Bu (814). (j) Rxn. with Zn in liq. NH$_3$, + THF → (Ph$_3$Ge)$_2$
Zn·0.5 NH$_3$·0.5 THF (675).

6.5 POLYMETALLIC ORGANOGERMANIUM DERIVATIVES WITH COORDINATION COMPOUNDS OF TRANSITION METALS

Chemistry of organogermanes with direct metal germanium bonds has received great attention during the past four years; judged by the number of new compounds and published papers. Special emphasis was given to organogermanium compounds with Group VI metals and with platinum, mainly by Glockling and coworkers. Complexes listed in this chapter may not contain a direct Ge-metal σ-bond, but some orbital overlap might occur.

6.5.1 BIMETALLIC ORGANOGERMANIUM DERIVATIVES OF GROUP I, IV AND VI TRANSITION METALS

This chapter comprises organogermanes with bonds to silver, gold, titanium, zirconium, chromium, molybdenum and tungsten. The compounds listed in Table 83 are prepared by methods from the following scheme.

I Reaction with organogermyl lithium derivatives:
   Ph$_3$GeLi + R"Cl → Ph$_3$GeR" .

II Metathetic substitution with alkali metal salts of transition metal
   complexes:
   R$_n$GeX$_{4-n}$ + NaR" → R$_n$GeX$_m$R$_o$" ; m = 0-2 .

III Interconversion reactions:
   (IIIA) R$_3$GeR" + X → R$_n$X$_{3-n}$GeR" ,
   (IIIB) Ph$_2$GeR$_2$" + HCl → Ph$_2$(Cl)GeR" .

IV Ligand exchange reactions:
   (IVA) R$_3$GeR"CO + R'$_3$M → R$_3$GeR"MR'$_3$; M = P, Sb, As ,
   (IVB) Ph$_3$GeAuPR$_3$ + PR'$_3$ → Ph$_3$GeAuPR'$_3$.

V Insertion reaction:
   PhGeCl$_3$ + BiPyW(CO)$_4$ → Ph(Cl$_2$)GeBiPyW(CO)$_3$Cl .

R, R' = organic groups, may be different in one molecule, R" = transition metal coordination group, may contain also an organogermyl moiety, X = halogen. Other organogermanium metal derivatives with groupings like (R$_3$Ge)$_3$PCr- or Ge-Pt-W are listed in Chapters 4.5.6 and 6.5.8, respectively. (Compounds erroneously listed in main volume p. 140, line 9 and lines 7 and 6 from bottom are derivatives of manganese.)

Triethylgermyldiphenylphosphine   Ge$_4$Ag$_4$C$_{72}$H$_{100}$I$_4$P$_4$ [Et$_3$GePPh$_2$·AgI]$_4$
   Silver Iodide (M-127)
<u>Synth.</u>: By rxn. of Et$_3$GePPh$_2$ with AgI in c.MeC$_6$H$_{11}$ in inert atm., in 60% yield

(50).
Prop.: Colorless cryst. m. 183° (dec.), IR (50), far IR spectra and assignment (105), mol. wt., sensitive to air and moisture (50).

Trimethylgermyltriphenyl-phosphine Gold $GeAuC_{21}H_{24}P$ $Me_3GeAuPPh_3$

Synth.: By rxn. of $(Me_3Ge)_2Hg$ with Ph_3PAuCl in N atm. in C_6H_6, in 16% yield (2420).
Prop.: Buff solid, m. 126° (dec.), NMR spectrum, stable as solid in air, slow dec. in refluxing C_6H_6, soly. (2420).
Rxn. with: Air → $(Me_3Ge)_2O + Me_6Ge_2$ (2420).

Triphenylgermyltriphenyl-phosphine Gold $GeAuC_{24}H_{30}P$ $Ph_3GeAuPEt_3$

Synth.: By rxn. of $Ph_3GeAuPPh_3$ with PEt_3 in C_6H_6 in N atm. (2420).
Prop.: m. 159°, mass spectrum, soly. mol. wt. (2420).
Rxn. with: $(CH_2Br)_2$ → $Ph_3GeBr + Ph_3PAuBr$ (sic) (2420).

Triphenylgermyltriphenyl-phosphine Gold (M-139) $GeAuC_{36}H_{30}P$ $Ph_3GeAuPPh_3$

Synth.: By rxn. of Ph_3GeLi with Ph_3PAuCl in Et_2O in N atm., in 40-60% yield (2420).
Prop.: Yellow solid, IR spectrum (105).
Rxn. with: $C_5H_5W(CO)_3HgCl$ → $Ph_3PAu(C_5H_5)W(CO)_3 + Au + Hg$ (832).
$MgBr_2$ (anhydr.) → $Ph_3GeMgBr + Ph_3PAuBr$ (2420).
$SnCl_4$ → $Ph_3GeCl + Au + (Ph_3P)_2AuSnCl_3 + Ph_3PAuCl + SnCl_2$ (2420).
Et_3P → $Ph_3GeAuPEt_3 + Ph_3P$ (2420).
Et_3P in C_6H_6 in air → $(Ph_3Ge)_2O + Ph_3P + (Et_3P)_xAu_y$ (2420).
$(Ph_2PCH_2)_2$ → $Ph_3GeAu(PPh_2CH_2)_2$ (2420).
Wet diglyme in air → $(Ph_3Ge)_2O + Au + Ph_3PO$ (2420).
MeI → $MeGePh_3 + Au + CH_4 + C_2H_6 + Ph_3PAuI$ (2420).
$(CH_2Cl)_2$ → $Ph_3GeCl + Ph_3PAuCl + C_2H_4 + Au + (Ph_3PCH_2)_2Cl_2$ (2420).

π-Cyclopentadienyl(trimethyl-germyl)tricarbonyl Molybdenum $GeMoC_{11}H_{14}O_3$ $Me_3GeC_5H_5Mo(CO)_3$

Synth.: By rxn. of $NaC_5H_5Mo(CO)_3$ with Me_3GeBr in inert atm. in THF, in 65% yield (832).
By rxn. of $C_5H_5Mo(CO)_3H$ with Me_3GeNMe_2 in CO atm. in THF at 70° (2303).
Prop.: Colorless cryst. m. 87-88° (832), pale yellow cryst., m. 84° (2303), subl. 70° x 10^{-3} mm (2303), 50-55° x 10^{-4} mm (832), IR (832), far IR, Raman and assignment (2303), NMR (832, 2303) and mass (2303) spectra, sensitive to air, light (2303), partly dec. at 200° (832).
Rxn. with: $MgBr_2$ → $C_5H_5Mo(CO)_3MgBr \cdot THF$ (832).
Ph_3P in C_6H_6 → $[Me_3GeC_5H_5Mo(CO)_3PPh_3?] + Me_3GeC_5H_5Mo(CO)_2PPh_3 + CO$ (832).
HCl → $Me_3GeCl + C_5H_5Mo(CO)_3H$ (832).
$(CH_2Br)_2$ → $Me_3GeBr + C_5H_5Mo(CO)_3Br$ (832).
O_2 in C_6H_6 → $(Me_3Ge)_2O + CO + CO_2 + (Me_3Ge)_3COH(?)$ (832).

π-Cyclopentadienyl(triethyl- GeMoC$_{14}$H$_{20}$O$_3$ Et$_3$GeC$_5$H$_5$Mo(CO)$_3$
   germyl)tricarbonyl Molybdenum

Synth.: By rxn. of Et$_3$GeBr with NaC$_5$H$_5$Mo(CO)$_3$ in N atm. in THF, in 45% yield (832).
By rxn. of (Et$_3$Ge)$_2$Hg with [C$_5$H$_5$Mo(CO)$_3$]$_2$ in C$_6$H$_6$ at 70°, in 54% yield (2414).
Prop.: Pink needles, m. 26.5° (832), pale liq. b$_{0.1}$ 148-50° (2414).
IR, UV (832) and NMR (832, 2414) spectra, dec. in air (2414).
Rxn. with: Thermal dec. at 150-200° → Et$_3$GeC$_5$H$_5$ + CO + red solid (832).
R$_3$P → Et$_3$GeC$_5$H$_5$Mo(CO)$_2$PR$_3$; R$_3$P = Et$_2$HP, Ph$_3$P (832).
Air → C$_{14}$H$_{20}$GeMoO$_7$ + CO + CO$_2$ (832).
HCl (anhydr.) → Et$_3$GeCl + C$_5$H$_5$Mo(CO)$_3$H (832).
EtBr → Et$_3$GeBr + [C$_5$H$_5$Mo(CO)$_3$]$_2$ + C$_2$H$_4$ + C$_2$H$_6$ (832).

π-Cyclopentadienyl(trimethyl- GeWC$_{13}$H$_{14}$O$_3$ Me$_3$GeC$_5$H$_5$W(CO)$_3$
   germyl)tricarbonyl Tungsten

Synth.: By rxn. of Me$_3$GeBr with NaC$_5$H$_5$W(CO)$_3$ in Bu$_2$O-THF, in 50% yield (832), at r.t. in THF, in 96% yield (2303).
By rxn. of Me$_3$GeNMe$_2$ with C$_5$H$_5$W(CO)$_3$H in THF (2303).
By rxn. of Me$_3$GeC$_5$H$_5$W(CO)$_2$PPh$_2$ with CO (832).
Prop.: Pale yellow, m. 114° (2303), m. 106-107° (832), subl. 90° x 10^{-3}mm (2303), IR (832), far IR, Raman and assignment (2303), NMR (832, 2303) and mass (2303) spectra, mol. wt., stable in air (2303).
Rxn. with: Thermal dec. at 180-90° → [C$_5$H$_5$W(CO)$_3$]$_2$ + CO (832).
HgCl$_2$ → C$_5$H$_5$W(CO)$_3$HgCl (832).
R$_3$P → Me$_3$GeC$_5$H$_5$W(CO)$_2$PR$_3$ + CO ; R$_3$P = Me$_3$P, Et$_2$HP, Ph$_3$P (832).
Air → (Me$_3$Ge)$_2$O + CO + CO$_2$ (832).
I → Me$_3$GeI + C$_5$H$_5$W(CO)$_2$I (832).
EtBr → Me$_3$GeBr + [C$_5$H$_5$W(CO)$_3$]$_2$ + C$_2$H$_4$ + C$_2$H$_6$ (832).

π-Cyclopentadienyl(triethyl- GeWC$_{14}$H$_{20}$O$_3$ Et$_3$GeC$_5$H$_5$W(CO)$_3$
   germyl)tricarbonyl Tungsten

Synth.: By rxn. of Et$_3$GeBr with NaC$_5$H$_5$W(CO)$_3$ in Bu$_2$O-THF, in 48% yield (832), in (MeOCH$_2$)$_2$ at 60°, in 34% yield (2414).
By rxn. of (Et$_3$Ge)$_2$Hg with [C$_5$H$_5$W(CO)$_3$]$_2$ in C$_6$H$_6$ at 55° (2414).
Prop.: m. 37-39° (2414), m. 36° (832), b$_{0.02}$ 158° (2414), IR (832, 2414) and NMR (832, 2414) spectra.
Rxn. with: MgBr$_2$ anhydr. in THF → C$_5$H$_5$W(CO)$_3$·MgBr·THF (832).

Additional organogermanium derivatives with coordination compounds of Group I, IV and VI transition metals are listed in Table 83.

Table 83. Bimetallic Organogermanium Derivatives of Group I-VI Transition Metals

Formula*	Synth. Method**	Yield	Properties	Ref.	R₃GeR' Ref.
Ph₃GeAgPPh₃ (139)	--	--	IR spectrum	105	
Ph₃GeAuPMe₃ (139)	--	--	IR spectrum	105	
[(Ph₃Ge)₂Au]NEt₄ (139)	--	--	IR spectrum	105	
Ph₃GeAu(PPh₂CH₂)₂	IVB	--	m. 227° (dec.), mol. wt.	2420	
Ph₃GeTi(C₅H₅)₂Cl	I	--	Dark green, m. 193-96° (dec.) (a)	907	
Ph₃GeTi(C₅H₅)₂·THF	I	--	Emerald green, m. 110° (dec.) (b)	907	
Ph₃GeZr(C₅H₅)₂Cl	I	--	Pale yellow, impure, NMR spectrum	907	
	I	--	Orange, subl. 190° × 5 × 10⁻⁴ mm (c)	2492	
Ph₃GeHf(C₅H₅)₂Cl	I	--	Yellow, subl. 200° × 10⁻⁴ mm (c)	2492	
Me₃GeC₅H₅Cr(CO)₃	II	--	Yellow green cryst., m. 86-88° (d)	2303	
Me₃GePhCr(CO)₃ (140)	--	--	NMR spectrum	2476	
R₂GeCr(CO)₅					
R = Et	--	--	(e)	--	
Pr	--	--	(e)	--	
n-C₆H₁₃	--	--	(e)	--	
C₁₈H₃₇	--	--	(e)	--	
Bu₂Ge[C₅H₅Cr(CO)₃]₂	--	--	(e)	--	
Ph₃GeC₅H₅Cr(CO)₃	II	24	Yellow green, sensitive to air (f)	1670	
(Me₂C₆H₃)₃GeC₅H₅Cr(CO)₃	--	--	(e)	--	
Me₃GeC₅H₅Mo(CO)₃PPh₃ (?)	IVA (g)	--	Buff, m. 100-200° (dec.)	832	
Me₃GeC₅H₅Mo(CO)₂PPh₃	IVA	--	Pale yellow cryst., m. 215-16° (sealed tube) (h)	832	
Et₃GeC₅H₅Mo(CO)₂PHEt₂	IVA	--	Pale yellow, sensitive to air (i)	832	
Et₃GeC₅H₅Mo(CO)₂PPh₃	IVA	75	Buff cryst., m. 165-70° (dec.) (i)	832	
Pr₃GeC₅H₅Mo(CO)₃	II	10	m. -25°, soly. (i)	832	
Ph₃GeC₅H₅Mo(CO)₃	II	74	Pale green, m. 219-22° (dec.) (j)	1670	

Compound				
(Me$_2$C$_6$H$_3$)$_3$GeC$_5$H$_5$Mo(CO)$_3$	--	--	(e)	--
Me$_3$GeC$_5$H$_5$W(CO)$_2$PMe$_3$	IVA	--	Yellow cryst., m. 80°, IR spectrum	832
Me$_3$GeC$_5$H$_5$W(CO)$_2$PHEt$_2$	IVA	--	Golden yellow air sensitive liq. (i)	832
Me$_3$GeC$_5$H$_5$W(CO)$_2$PPh$_3$	IVA	34	IR spectrum (k)	832
Bu$_2$Ge[C$_5$H$_5$W(CO)$_3$]$_2$	--	--	(e)	--
Ph$_3$GeC$_5$H$_5$W(CO)$_3$	II	90	Pale yellow, m. 240° (dec.) (j)	1670
PhCl$_2$Ge BiPy W(CO)$_3$Cl	V	--	Yellow cryst., dec. 190°, IR spectrum	1372
(Me$_2$C$_6$H$_3$)$_3$GeC$_5$H$_5$W(CO)$_3$	--	--	(e)	--

* Numbers in parenthesis refer to pages in main volume.
** The characters used here correspond to the synthetic scheme in the introduction to Chapter 6.5.1.

(a) IR and NMR spectra, diamagnetic (907). (b) IR spectrum, paramagnetic (907). (c) NMR spectrum (2492). (d) Subl. 50° x 10^{-3}mm, spectra: (far) IR, Raman and assignment, NMR and mass, sensitive to air and light (2303). (e) Use as agent for metal plating on hot surfaces (603). (f) IR and NMR spectra, dec. on standing in inert atm. (1670). (g) Synth. from Me$_3$GeC$_5$H$_5$Mo(CO)$_3$, identified by mass spectrum (832). (h) IR and NMR spectra (832). (i) IR and mass spectra (832). (j) IR, UV and NMR spectrum, stable in air (1670). (k) Rxn. with CO → Me$_3$Ge-C$_5$H$_5$W(CO)$_3$ (832).

6.5.5 POLYMETALLIC ORGANOGERMANIUM DERIVATIVES OF MANGANESE AND RHENIUM

Some organogermanium-manganese compounds in the main volume are listed erroneously as molybdenum derivatives. Correct the following formulae (M-140) line 9: $R_3GeMo(CO)_5$, line 7 f.b.: $Ph_2BrGeMo(CO)_5$ and line 6 f.b.: $R_3GeMo(CO)_3$·R' to read Mn instead of Mo.

Trimethylgermylpentacarbonyl $GeMnC_8H_9O_5$ $Me_3GeMn(CO)_5$
Manganese

Synth.: By rxn. of Me_3GeBr with $NaMn(CO)_5$ in N atm. in THF, in 51% yield (872).
Prop.: m. 33.5-34.5°, subl. 40° x 10^{-3}mm, NMR and IR spectra and assignment, dec. in air (872).
Rxn. with: Thermal dec. at 130° → CO ; 98% recovery after 18 hours (872).
UV irradiation → 95% recovery after 5 hours at 50-60° (872).
C_2H_4* → $Mn_2(CO)_{10}$ (872).
C_2F_4* → Me_3GeF + $Me_3GeCF_2CF_2Mn(CO)_5$ + $Me_3Ge(C_2F_4)_xMn(CO)_5$ + $C_5F_9 \cdot Mn(CO)_3$ + CO (872).
C_2HF_3* → Me_3GeF + $CHF:CFMn(CO)_5$ + CO (872).
C_2F_3Cl* → $Me_3GeCF:CF_2$ + $Mn_2(CO)_{10}$ + $CF_2:CFCOMn(CO)_5$ + $[ClMn(CO)_4]_2$ (872).
C_3F_6* → $Me_3GeC_xF_y$ + $Mn_2(CO)_{10}$ + $CF_3CF:CFMn(CO)_5$ + CO (872).
$(CF_2:CF)_2$* → $Me_3GeCF_2CF_2CF:CFMn(CO)_5$ + $CF_2:CFCF:CFMn(CO)_5$ (1141).

Triphenylgermylpentacarbonyl $GeMnC_{23}H_{15}O_5$ $Ph_3GeMn(CO)_5$
Manganese (M-140)**

Synth.: By rxn. of Ph_3GeBr with $NaMn(CO)_5$ in Ar atm. in THF at r.t., in 61% yield (1593).
By rxn. of $Br_3GeMn(CO)_5$ with PhMgBr in Ar atm. in THF, in 20% yield (2227).
Prop.: White, m. 163-64° (1593), 162° (2057), 161-64° (1252), spectra: IR (1132, 1252, 1592-3, 2057, 2337), (far) IR and assignment (1597), force const. (1132, 1252, 2337), dipole moment (2057), x-ray struct. analysis (1311, 1996).
Rxn. with: Ph_3M → $Ph_3GeMn(CO)_4MPh_3$; M = P, As, Sb (1593).
Br in $(CH_2Br)_2$ → $Br_3GeMn(CO)_5$ (1593).
Cl in CCl_4 → $Cl_3GeMn(CO)_5$ (1593).

Bromodiphenylgermylpentacar- $GeMnC_{17}H_{10}BrO_5$ $Ph_2(Br)GeMn(CO)_5$
bonyl Manganese (M-140)**

Synth.: By rxn. of Ph_2GeBr_2 with $NaMn(CO)_5$ in Ar atm. at r.t. in THF, in 81% yield (1592).
By rxn. of $Ph_2Ge[Mn(CO)_5]_2$ with Br in C_6H_6-C_7H_{16} at r.t. (1593).

---

* Rxn. under UV irradiation.
** Correct formula in main vol. Table 50, page 140, line 9 to $R_3GeMn(CO)_5$, on page 141, line 2 from bottom read (f) $R_3GeMn(CO)_5$ + ... → $R_3GeMn(CO)_3$·diene.

Prop.: White cryst., m. 88.5-89°, subl. in high vacuo (1592-3), IR (1592-3), (far) IR spectra and assignment (1597), soly. (1592-3).
Rxn. with: Br → Ph(Br$_2$)GeMn(CO)$_5$ + Br$_3$GeMn(CO)$_5$ (1592).
Br$_3$GeMn(CO)$_5$ → Ph(Br$_2$)GeMn(CO)$_5$ (1593).

Triphenylgermylpentacarbonyl   GeReC$_{23}$H$_{15}$O$_5$ Ph$_3$GeRe(CO)$_5$
  Rhenium

Synth.: By rxn. of Ph$_3$GeX with NaRe(CO)$_5$ in N atm. (1252), in THF, in 87% yield (1594).
Prop.: White, m. 157.5-58.5° (1252), pale yellow, m. 156° (2057), colorless cryst., m. 155° (1594), IR spectrum (1252, 1594, 2057, 2337), force const. (1252, 2337), soly., stable in air (1594), x-ray struct. analysis (1996).
Rxn. with: Ph$_3$M → Ph$_3$GeReMPh$_3$(CO)$_4$; M = P, As, Sb (1594).
Br → Br$_3$GeRe(CO)$_5$ (1594).

Additional organogermanes containing Mn and Re are listed in Table 84 and Chapter 2.7.8.

Table 84. Polymetallic Organogermanium Derivatives of Group VIII Transition Metals $R_nGeR''_{4-n}$

Formula	Synth. Method*	Yield	Properties	Ref.
$(PhCH_2)_2Ge[Ph_3AsMn(CO)_4]_2$	II	--	(Synth. in p-xylene at 40°)	185
$Et_3Ge(CH_2:CH)_3AsMn(CO)_4$	--	--	(a)	--
$EtGeI_2(CH_2:CH)_3AsMn(CO)_4$	(a)	--	--	187
$CH_2:CHGe[(C_{12}H_{25})_3AsMn(CO)_4]_3$	II	--	(Synth. in cumene from LiR'')	185
$Bu_2Ge[C_6H_{10}Mn(CO)_3]_2$	--	--	$(C_6H_{10} = CH_2:C\equiv tCHCH_2)$ (b)	--
$Bu_2Ge[PhMn(CO)_2]_2$	--	--	(b)	--
$Ph_3Ge(Ph_3P)Mn(CO)_4$	IVA	--	m. 249-51° (c, d)	1593
$Ph_3Ge(Ph_3As)Mn(CO)_4$	IVA	--	Light yellow, m. 229.5-31° (c)	1593
$Ph_3Ge(Ph_3Sb)Mn(CO)_4$	IVA	--	Light yellow, m. 214.5-16.5° (c)	1593
$Ph_2Ge[Mn(CO)_5]_2$	II (e)	40	(d, f)	1592
	II (e)	32	Lemon yellow, m. 145-47°	1593
$[Ph_2GeMn(CO)_5]_2$	II (e)	62	White, m. 205-206°, mol. wt. (c, d)	1593
$Ph(Cl)Ge[Mn(CO)_5]_2$	II	90	m. 101-105°, subl. in high vacuum (d)	1592
$Ph(Br_2)GeMn(CO)_5$	IIIA	39	m. 68-68.5° (c, d, h)	1592
	(g)	29	m. 68-68.5°	1593
$[[(OC)_5Mn]_2PhGe]_2$	II (h)	--	White, m. 155-57°	1592
$(Me_2C_6H_3)_3Ge(C_4H_6)Mn(CO)_3$	--	--	$(C_4H_6 = CH_2:CHCH:CH_2)$ (b)	--
$(Me_2C_6H_3)_3GePhMn(CO)_3$	--	--	(b)	--
$Ph_3Ge(Ph_3P)Re(CO)_4$	IVA	92	White cryst., m. 231-33° (i)	1594
$Ph_3Ge(Ph_3As)Re(CO)_4$	IVA	--	White cryst., m. 225-27° (i)	1594
$Ph_3Ge(Ph_3Sb)Re(CO)_4$	IVA	15	White cryst., m. 216-18° (i)	1594
$Ph_2Ge[Re(CO)_5]_2$	II	60	Colorless cryst., m. 167-68° (i, j)	1594
$Ph_2(Cl)GeRe(CO)_5$	IIIB	--	White cryst., m. 83° (j)	1594

* The characters used here correspond to the synthetic scheme in the introduction to Chapter 6.5.1. (a) Rxn. of $Et_3GeR_3AsMn(CO)_4$ with I in $EtOCH_2Ph$ at 74-86° → $EtGeI_2(CH_2:CH)_3AsMn(CO)_4$ (187). (b) Use as agent for metal plating on hot surfaces (603). (c) IR spectrum, soly. (1593). (d) (Far) IR spectrum and assigment (1597).

(e) From Ph$_2$GeBr$_2$ and NaMn(CO)$_5$. (f) Rxn. with Br in C$_6$H$_6$-C$_7$H$_{16}$ → Ph$_2$(Br)GeMn(CO)$_5$ + BrMn(CO)$_5$ (1593). (g) Synth. from Ph$_2$(Br)GeMn(CO)$_5$ with Br$_3$GeMn(CO)$_5$ in Ar atm. at 160-65° (1593). (h) Rxn. of Ph(Br$_2$)GeMn(CO)$_5$ with NaMn(CO)$_5$ in Ar atm. in THF → [GePh[Mn(CO)$_5$]$_2$]$_2$ (1592). (i) IR spectrum, soly. (1594). (j) Stable in air (1594), rxn. with anhydr. HCl in CCl$_4$ → Ph$_2$(Cl)GeRe(CO)$_5$ (1594).

6.5.6 BIMETALLIC ORGANOGERMYL IRON COMPOUNDS

The bimetallic organogermanes in this chapter may not have a direct germanium-iron σ-bond, but some orbital overlap might occur. Structures for most compounds are not certain. The derivatives listed in Table 85 are prepared by methods from the following scheme.

I   Synthesis with organic derivatives of alkali metals:
   (IA) R″GeI$_2$ + MeLi → Me$_2$GeR″ ,
   (IB) R″$_2$GeCl$_2$ + C$_5$H$_5$Na → (C$_5$H$_5$)$_2$GeR″$_2$.

II Metathetic substitution with sodium salt of iron tetracarbonyl.

III Grignard reaction:
   R″$_2$GeCl$_2$ + RMgBr → R$_2$GeR″$_2$.

IV Synthesis with organogermane:
   R$_2$GeH$_2$ + Fe$_3$(CO)$_{12}$ → (R$_2$Ge)$_n$Fe$_2$(CO)$_{9-n}$; n = 1, 3.

V   Insertion reaction:
   GeCl$_2$·C$_4$H$_8$O$_2$ + MeR″ → Me(Cl$_2$)GeR″ .

R = organic group, R″ = transition metal coordination group.

π-Cyclopentadienyl(tri-   GeFeC$_{13}$H$_{20}$O$_2$ Et$_3$GeC$_5$H$_5$Fe(CO)$_2$
  ethylgermyl)dicarbonyl Iron

Synth.: By rxn. of Et$_3$GeBr with NaC$_5$H$_5$Fe(CO)$_2$ in THF (2414), in 85% yield (2416).
By rxn. of (Et$_3$Ge)$_2$Hg in sealed tube at 100° in C$_6$H$_6$ with [C$_5$H$_5$Fe(CO)$_2$]$_2$, in 49% yield (2414, 2616), with C$_5$H$_5$Fe(CO)$_2$X; X = Cl, Br, in 66% yield (2414).
Prop.: Orange liq. b$_1$ 118-20°, n$_D^{20}$ 1.5868, n$_D^{22}$ 1.5850, IR and NMR spectra, dec. slowly in air, mol. wt, soly. (2414, 2416).
Rxn. with: Br → Et$_3$GeBr + CO (2414, 2416).

Bis(diethylgermyltetra-   Ge$_2$Fe$_2$C$_{16}$H$_{20}$O$_8$ [Et$_2$GeFe(CO)$_4$]$_2$
  carbonyl Iron)

Synth.: By rxn. of Et$_2$GeCl$_2$ with aq. Na$_2$Fe(CO)$_4$ in N atm. (1269).
Prop.: IR (734, 1272, 2339), (far) IR and assignment (1269) and Raman (734) spectra, dipole moment (1272), force const. (734, 2339), struct. (1269, 1272).
Use: Agent for metal plating on heated substrates (603).

Other bimetallic organogermyl iron compounds are listed in Table 85.

Table 85. Bimetallic Organogermyl Iron Compounds $R_nGeR''_{4-n}$

Formula*	Synth. Method**	Yield	Properties	Ref.
$Me_2Ge[C_5H_5Fe(CO)_2]_2$ (141)	IA	74	Orange cryst., m. 129-30°, subl.o.1 100° (a)	1045
$[Me_2GeFe(CO)_4]_2$	II	--	Struct. (b-e)	1269
$Me_8Ge_2Fe_2(CO)_8$	IV	--	Yellow needles, x-ray struct. analysis (f, g)	2284
$Me(Cl_2)GeC_5H_5Fe(CO)_2$	V	78	Yellow, m. 74-75°, struct. (h)	2643
$(Et_3Ge)_2Fe(CO)_4$	--	--	(c, i)	--
$R_2Ge[C_5H_5Fe(CO)_2]_2$				
R = Et	III	76	Yellow, m. 122-24° (j)	2640
Pr	--	--	(e)	--
CH_2:$CHCH_2$	III	30	Yellow, m. 136-37°, NMR spectrum	2640
Bu	III	81	Yellow, m. 117-18°, NMR spectrum (e)	2640
C_5H_5	IB	33	m. 139-41° (dec.) (j)	2640
$EtBuCHCH_2$	--	--	(e)	--
Ph	III	27	Orange-red, m. 98-99°, NMR spectrum	2640
$Me_2C_6H_3$	--	--	(e)	--
$Ph_3GeC_5H_5Fe(CO)_2$ (141)	--	--	Yellow, m. 160° (k)	2057
$Ph_2Ge[Fe(CO)_4]_2$	IV	--	Yellow cryst., struct. (f)	2284

* Numbers in parenthesis refer to pages in main volume.
** The characters used here correspond to the synthetic scheme in the introduction to Chapter 6.5.6.
(a) IR and NMR spectra, sensitive to air (1045). The correct formula for the compound listed in main volume is $Me_2(F)GeC_5H_5Fe(CO)_2$. (b) (Far) IR spectrum and assignment (1269). (c) Raman and IR spectra, force const. (734). (d) IR spectrum and force const. (2339). (e) Use as agent for metal plating on heated substrates (603). (f) IR spectrum (2284). (g) NMR and mass spectra (2284). (h) IR spectrum (2643). (i) IR spectrum, dipole moment, struct. (1272). (j) IR and NMR spectra (2640). (k) IR and NMR spectra (2057). The correct formula for the compound listed in main volume is $Ph_3GeC_5H_5Fe(CO)_2$.

6.5.7 BIMETALLIC ORGANOGERMANIUM DERIVATIVES OF COBALT AND IRIDIUM

The remarks concerning the structure of organogermanium derivatives of metals stated in the preceeding chapter also stand for the cobalt and iridium compounds. The organogermanes listed in Table 86 are prepared by the following methods.

   I Metathetic substitution with sodium salts of metal coordination compounds:
$$R_nGeX_{4-n} + NaR'' \rightarrow R_nGeX_mR''_o \ ; \ n + m + o = 4.$$

   II Condensation reaction:
$$Ph_3GeX + R''H \cdot Py \rightarrow Ph_3GeR'' \cdot Py + HX.$$

   III π-Complex formation with dicobaltoctacarbonyl:
$$R_2GeR'_2 + Co_2(CO)_8 \rightarrow R_2Ge[R'Co_2(CO)_6]_2 \ ; \ R' = PrC\vdots C, PhC\vdots C.$$

   IV Decarbonylation reaction:
      (IVA) $PhGeCl_3 + R''CO \rightarrow (PhGeCl_2)_2R'' + CO$,
      (IVB) $PhGeI_3 + R''CO \rightarrow Ph(I_2)GeR''I + CO$.

R' = organic group, R'' = metal coordination group, X = halogen, n = 1-3.

Triethylgermyltetracarbonyl $GeCoC_{10}H_{15}O_4$ $Et_3GeCo(CO)_4$
  Cobalt

Prop.: IR (734, 1272, 2198, 2338), (far) IR and assignment (1270) and Raman (734) spectra, dipole moment (1270, 1272, 2057), monomeric (1272), force const. (734, 2198, 2338), struct. (1272).

Triphenylgermyltetracarbonyl $GeCoC_{22}H_{15}O_4$ $Ph_3GeCo(CO)_4$
  Cobalt

Synth.: By rxn. of Ph_3GeBr with $Co[Co(CO)_4]_2$ in MeOH-THF at 40° (1672).
Prop.: White cryst., m. 130° (1672), IR spectrum (1672-3, 2198, 2338), force const. (2198, 2338), soly. (1672).

Additional organogermanes containing cobalt and iridium listed in Table 86.

Table 86. Bimetallic Organogermanium Derivatives of Cobalt and Iridium $R_nGeR''_{4-n}$

Formula	Synth. Method*	Yield	Properties	Ref.
$Me_2[GeCo(CO)_4]_2$	I	--	Yellow cryst., m. 39-41° (a)	1672
$Me_2(Cl)GeCo(CO)_4$	I	--	Pale yellow cryst., m. 51-54°, soly. (a, b)	1672
$Me(I_2)GeCo(CO)_4$	I	--	Orange cryst., m. 66-68° (a, b)	1672
$Et_2Ge[C:CPhCo_2(CO)_6]_2$	III	70	Dark orange cryst., m. 87-89° struct. (c)	236a, 1222
$Bu_2Ge[C_4H_6Co(CO)_2]_2$	--	--	$(C_4H_6 = CH_2:CHCH:CH_2)$ (d)	--
$Bu_2Ge(PhGeCO)_2$	--	--		--
$Ph_2Ge[C:CPrCo_2(CO)_6]_2$	III	65	Dark orange cryst., m. 132-34°, soly. (c)	1222
$Ph_2Ge[C:CPhCo_2(CO)_6]_2$	III	66	m. 154°, soly. (c)	1222
$Ph_3GePyCo[H(ON:CMe)_2]_2$	II	--	Orange, dec. 172-80°, stable in air, soly. (e)	486
$Ph_2(Cl)GeCo(CO)_4$	I	--	Pale yellow, m. 59-61° (a, b)	1672
$Ph_2(I)GeCo(CO)_4$	--	--		--
$Ph(Cl_2)GeCo(CO)_4$	I	--	Pale yellow cryst., m. 59-61° (a, b)	1672
$Ph(I_2)GeCo(CO)_4$	--	--		--
$(PhGeCl_2)_2C_5H_5CoCO$	IVA	--	Yellow cryst., stable in air and light (f)	1373
$Ph(I_2)GeC_5H_5CoICO$	IVB	24	Black cryst., stable in air	1373
$(Me_2C_6H_3)_3GeC_4H_6Co(CO)_2$	--	--	$(C_4H_6 = CH_2:CHCH:CH_2)$ (d)	--
$(Me_2C_6H_3)_3GePhCoCO$	--	--		--
$Ph_3GeIr(CO)_3PPh_3$	I (g)	71	White solid, mol. wt. struct. (h)	891

* The characters used here correspond to the synthetic scheme in the introduction to Chapter 6.5.7.
(a) IR spectrum (1672). (b) IR spectrum and force const. (2338). (c) IR spectrum, mol. wt. (1222). (d) Use as agent for metal plating on hot surfaces (603). (e) Triphenylgermylbis(dimethylglyoxime)pyridino cobalt is decomposed by strong acid (486). (f) IR spectrum, soly. (1373). (g) From $[ClIr(PPh_3)_2CO]_2$ and 1% Na-Hg in THF at 4 atm. CO pressure, followed by Ph_3GeCl (891).

6.5.8 POLYMETALLIC ORGANOGERMANIUM DERIVATIVES OF NICKEL AND HOMOLOGUES

The remarks concerning structure of organogermanium complexes with metal coordination compounds in Chapter 6.5.6 also fit the derivatives of nickel and its homologues. The Ni, Pd and Pt derivatives summarized in Table 87 are prepared by the following methods.

I   Reactions with lithium derivatives:
    (IA) $R''Cl + PhLi \rightarrow R''Ph$,
    (IB) $R''X_n + R_3GeLi \rightarrow (R_3Ge)_nR''$.

II Condensation with organogermane:
    $Ph_3GeH + (R_3P)_2PtCl_2 \rightarrow Ph_3Ge(Cl)Pt(PR_3)_2 + HCl$.

III Condensation with organogermanium mercury derivatives:
    (IIIA) $(Et_3Ge)_2Hg + (C_5H_5NiCO)_2 \rightarrow Et_3GeC_5H_5NiCO + Hg$,
    (IIIB) $(Me_3Ge)_2Hg + (R_3P)_2PtCl_2 , + NaI \rightarrow Me_3Ge(I)Pt(PR_3)_2 + Hg$.

IV Exchange reactions:
    (IVA) $R_3GeR'' + M'X \rightarrow R''X$ or R''_2 ; $M' = HgCl, MgI, H$,
    (IVB) $R''Cl + M'X \rightarrow R''X$; $M' = Li, Na$; $X = Br, I, CN, SCN, BPh_4$,
    (IVC) $(R_3Ge)_2M(PR_3)_2 + H \rightarrow R_3Ge(H)M(PR_3)_2$: $M = Pd, Pt$,
    (IVD) $(R_3Ge)_2M(PR_3)_2 + KCN \rightarrow K_2[(R_3Ge)_2M(CN)_2]$; $M = Pd, Pt$,
    (IVE) $R''OEt + ROH \rightarrow R''OR$; $R = H, Me, Pr$.

V   Miscellaneous reactions:
    (VA) $Ph_3GeR'' + C_5H_5WHgCl \rightarrow R''C_5H_5W(CO)_3$,
    (VB) $Ph_3GeY(NEt)_2H + Ni(OAc)_2 \rightarrow [Ph_3GeY(NEt)_2]_2Ni$.

R = organic group, R'' = metal coordination group, may contain an organogermyl moiety, Y = tropone skeleton.

Bis(triphenylgermyl)bis(triethyl- $GePdC_{48}H_{60}P_2$ $(Ph_3Ge)_2Pd(PEt_3)_2$
  phosphine) Palladium
Synth.: By rxn. of Ph_3GeLi with trans-$(Et_3P)_2PdBr_2$ (795).
Prop.: Stable to 97° in solid state, dec. in soln. at -20° (795).
Rxn. with: Thermal dec. at 107° $\rightarrow Ph_4Ge + Ph_6Ge_2 + Pd + Et_3P + C_6H_6 + C_2H_4 +$ H (795).
H_2 at 100 atm. $\rightarrow Ph_3Ge(H)Pd(PEt_3)_2 + Ph_3GeH$ (795).
KCN $\rightarrow K_2[(Ph_3Ge)_2Pd(CN)_2] + Et_3P$ (795).
CO $\rightarrow Ph_6Ge_2$ (2285).
HCl, anhydr. $\rightarrow Ph_3GeH + (Et_3P)_2PdCl_2 + Ph_3GeCl + (Et_3P)_2PdHCl$ (795).
$(CH_2Br)_2 \rightarrow Ph_3GeBr + (Et_3P)_2PdBr_2 + C_2H_4$ (795).

Trimethylgermyltriphenylgermylbis- $Ge_2PtC_{33}H_{54}P_2$ $Me_3Ge(Ph_3Ge)Pt(PEt_3)_2$
  (triethylphosphine) Platinum
Synth.: By rxn. of $Me_3Ge(Cl)Pt(PEt_3)_2$ with Ph_3GeLi in N atm. in Et_2O, in 41% yield (1106).
Prop.: Bright yellow needles, m. 98-100°, far IR spectrum and assignment, soly. (1106).

Rxn. with: H$_2$ in xylene → Me$_3$GeH + Ph$_3$Ge(H)Pt(PEt$_3$)$_2$ + Me$_3$Ge(H)Pt(PEt$_3$)$_2$ (?) (1107) + Ph$_3$GeH + (Et$_3$P)$_2$PtHCl in C$_6$H$_6$ (1106).
HCl, anhydr. → Me$_3$GeH + (Et$_3$P)$_2$Pt(Cl)GePh$_3$ + Me$_3$GeCl + (Et$_3$P)$_2$Pt(H)GePh$_3$ (1107).

Bis(triethylphosphine)tri- GePtC$_{15}$H$_{39}$ClP$_2$ trans-Me$_3$Ge(Cl)Pt(PEt$_3$)$_2$
 methylgermyl Platinum Chloride
Synth.: By rxn. of Me$_3$GeH with (Et$_3$P)$_2$PtHCl in N atm. in xylene, in 6% yield (1107).
By rxn. of (Me$_3$Ge)$_2$Hg with cis-(Et$_3$P)$_2$PtCl$_2$ in N atm. in refluxing C$_6$H$_6$, in 75% yield (1106).
Prop.: Buff cryst., m. 34-36°, spectra: far IR and assignment, NMR, and mass, monomeric, soly. (1106).
Rxn. with: PhLi → Me$_3$Ge(Ph)Pt(PEt$_3$)$_2$ (1106).
Ph$_3$GeLi → Me$_3$Ge(Ph$_3$Ge)Pt(PEt$_3$)$_2$ (1106).
MX in Me$_2$CO → Me$_3$Ge(X)Pt(PEt$_3$)$_2$; MX = LiBr, NaI, NaSCN, KCN (1106).
KCN (excess) → (Et$_3$P)$_2$Pt(CN)$_2$ (1106).
NaNO$_3$ → (Et$_3$P)$_2$Pt(H)NO$_2$ + (Me$_3$Ge)$_2$O (1106).
(Ph$_2$PCH$_2$)$_2$ → [(Ph$_2$PCH$_2$)$_2$Pt(PEt$_3$)GeMe$_3$]Cl + Et$_3$P (1107).
H$_2$ → Me$_3$GeH + trans-(Et$_3$P)$_2$PtHCl (1106).
H$_2$O in diglyme → (Me$_3$Ge)$_2$O + trans-(Et$_3$P)$_2$PtHCl (1106).
HCl in Et$_2$O → Me$_3$GeCl + trans-(Et$_3$P)$_2$PtHCl (1107).
(CH$_2$Cl)$_2$ → Me$_3$GeCl + cis-(Et$_3$P)$_2$PtCl$_2$ + trans-(Et$_3$P)$_2$PtHCl + C$_2$H$_4$ (1106).

1,2-Bis(diphenylphosphino)ethane- GePtC$_{35}$H$_{48}$ClP$_3$
 triethylphosphinetrimethylgermyl
 Platinum(II) Chloride
Synth.: By rxn. of trans-Me$_3$Ge(Cl)Pt(PEt$_3$)$_2$ with Ph$_2$PCH$_2$CH$_2$PPh$_2$ in C$_6$H$_6$, in 61% yield as (1:1) adduct with C$_6$H$_6$ and recrystallization from Me$_2$CO (1107).
Prop.: m. 135-37°, conductivity (1107).
Rxn. with: H$_2$O → (Me$_3$Ge)$_2$O + [(Ph$_2$PCH$_2$)$_2$Pt(H)PEt$_3$]Cl (1107).
Addn. Compd.: [(Ph$_2$PCH$_2$)$_2$Pt(GeMe$_3$)PEt$_3$]Cl·C$_6$H$_6$, see above; colorless cryst., m. 117-21° (dec.), [GePtC$_{41}$H$_{54}$ClP$_3$] (1107). Rxn. with H$_2$ in MeOH → Me$_3$GeH + [(Ph$_2$PCH$_2$)$_2$Pt(H)PEt$_3$]Cl (1107), with HCl in C$_6$H$_6$ → Me$_3$GeH + Me$_3$GeCl + (Et$_3$P)$_2$-PtHCl + [(Ph$_2$PCH$_2$)$_2$Pt(H)PEt$_3$]Cl + (Ph$_2$PCH$_2$)$_2$PtCl$_2$ + [(Ph$_2$PCH$_2$)$_2$Pt(Cl)PEt$_3$]Cl + H (1107), with NaBPh$_4$ in MeOH → [(Ph$_2$PCH$_2$)$_2$Pt(GeMe$_3$)PEt$_3$]·BPh$_4$ (1107).

Bis(triphenylgermyl)bis(tri- Ge$_2$PtC$_{48}$H$_{60}$P$_2$ (Ph$_3$Ge)$_2$Pt(PEt$_3$)$_2$
 ethylphosphine) Platinum
Synth.: By rxn. of Ph$_3$GeLi with trans-(Et$_3$P)$_2$PtCl$_2$ in Et$_2$O, in 91% yield (101-3), with cis-form, in 65% yield (101-3), with cis-(Et$_3$P)$_2$PtBr$_2$ in (MeOCH$_2$)$_2$ in N atm., in 58% yield (2285), with (Et$_3$P)$_2$PtI$_2$ in (MeOCH$_2$)$_2$, in 54% yield (103).
Prop.: Pale yellow needles, m. 160° (dec.) (101, 103), IR (103, 105), far IR (2285) and UV (103) spectra, dipole moment (2285), monomeric, stable in air

and water, soly. (103), exists in cis- and trans- form (2285).
Rxn. with: Thermal dec. at 180-230° → Ph$_4$Ge + Ph$_6$Ge$_2$ + Et$_3$P + C$_6$H$_6$ + C$_2$H$_4$ + black C$_6$H$_6$-soluble residue (103).
HgCl$_2$ → [(Ph$_3$Ge)Pt(PEt$_3$)$_2$]$_2$ + Hg (832).
KCN → K$_2$[(Ph$_3$Ge)$_2$Pt(CN)$_2$] + PEt$_3$ (2285).
KCN (excess) → Ph$_6$Ge$_2$ + K$_2$Pt(CN)$_6$ + K$_2$[Ph$_3$GePt(CN)$_3$] (?) (2285).
H$_2$ in EtOAc → Ph$_3$GeH + Ph$_3$Ge(H)Pt(PEt$_3$)$_2$ (101, 103).
Ph$_2$PCH$_2$CH$_2$PPh$_2$ → (Ph$_3$Ge)$_2$Pt(PPh$_2$CH$_2$)$_2$ (2285).
HCl (anhydr.) → Ph$_3$GeH + Ph$_3$GeCl + trans-(Et$_3$P)$_2$PtHCl + trans-(Et$_3$P)$_2$PtCl$_2$ (101, 103).
CCl$_4$ → Ph$_3$GeCl + (Et$_3$P)$_2$PtCl$_2$ + trace MeGePh$_3$ (103).
(CH$_2$Br)$_2$ → Ph$_3$GeBr + (Et$_3$P)$_2$PtBr$_2$ + C$_2$H$_4$ (101, 103).
MeI → Ph$_3$GeI + MeGePh$_3$ (101) + (Et$_3$P)$_2$PtI$_2$ (103).
I → Ph$_3$GeI + trans-(Et$_3$P)$_2$PtI$_2$ + trace (Ph$_3$Ge)$_2$O (103).

Bis(triethylphosphine)tri-phenylgermyl Platinum Hydride   GePtC$_{30}$H$_{46}$P$_2$ Ph$_3$Ge(H)Pt(PEt$_3$)$_2$

Synth.: By hydrogenation of (Ph$_3$Ge)$_2$Pt(PEt$_3$) in EtOAc at 1 atm. (101, 103), of Ph$_3$Ge(Me$_3$M)Pt(PEt$_3$)$_2$ in xylene ; M = Si and Ge, in 79 and 68% yield, resp. (1106).
By-prod. in rxn. of Me$_3$Ge(Ph$_3$Ge)Pt(PEt$_3$)$_2$ with anhydr. HCl in Et$_2$O or C$_6$H$_6$ (1107), with aq. diglyme (1106).
By-prod. in rxn. of Ph$_3$GeLi with (Et$_3$P)$_2$PtI$_2$ in (MeOCH$_2$)$_2$ (103).
Prop.: Colorless cryst., m. 150° (1106), with dec. (101, 103), IR spectrum (101, 103, 105, 1106), soly. (103), stable in air (101, 103), H$_2$O, EtOH, R$_2$CO and KOH (103).
Rxn. with: Ph$_2$PCH$_2$CH$_2$PPh$_2$ → Ph$_3$GeH + (Ph$_2$PCH$_2$)$_2$Pt (2285).
HCl anhydr. in Et$_2$O → Ph$_3$GeH + (Et$_3$P)$_2$PtHCl + H (2285).

Bis(triethylphosphine)tri-phenylgermyl Platinum Ethoxide   GePtC$_{32}$H$_{50}$OP$_2$ Ph$_3$Ge(EtO)Pt(PEt$_3$)$_2$

Synth.: By rxn. of Ph$_3$GeLi with (Et$_3$P)$_2$PtI$_2$ (1:1), followed by EtOH, in 10% yield (103).
By rxn. of Ph$_3$Ge(HO)Pt(PEt$_3$)$_2$ with EtOH (103).
Prop.: m. 160-70° (dec.), IR and NMR spectra, stable towards air, H$^+$ and R$_3$P, partial x-ray struct. analysis (103).
Rxn. with: ROH → Ph$_3$Ge(RO)Pt(PEt$_3$)$_2$; R = H, Me, i-Pr (103).

Bis(triethylphosphine)tri-methylsilyltriphenylgermyl Platinum   GePtSiC$_{33}$H$_{54}$P$_2$ Ph$_3$Ge(Me$_3$Si)Pt(PEt$_3$)$_2$

Synth.: By rxn. of Ph$_3$GeLi with Me$_3$Si(Cl)Pt(PEt$_3$)$_2$ in N atm. in Et$_2$O, in 37% yield (1106).
Prop.: Yellow cryst., m. 96-98°, far IR spectrum and assignment (1106).
Rxn. with: H$_2$ → Ph$_3$Ge(H)Pt(PEt$_3$)$_2$ + Me$_3$SiH (1106).
Aq. diglyme → Ph$_3$Ge(H)Pt(PEt$_3$)$_2$ + (Me$_3$Si)$_2$O (1106).

Bis(diphenylphosphino)ethane- $Ge_2PtC_{62}H_{54}P_2$ $(Ph_3Ge)_2Pt(PPh_2CH_2)_2$
   bis(triphenylgermyl) Platinum

Synth.: By rxn. of Ph_3GeLi with $(Ph_2PCH_2)_2PtI_2$ in C_6H_6-$(MeOCH_2)_2$-Et_2O in N atm. under reflux, in 98% yield (2285).
By rxn. of $(Ph_3Ge)_2Pt(PEt_3)_2$ with $Ph_2PCH_2CH_2PPh_2$ in N atm. in C_6H_6, in 93% yield (2285).

Prop.: Colorless cryst., m. 260-80°, stable in air, H_2O, dipole moment, soly. (2285).

Rxn. with: $HgCl_2 \rightarrow Ph_3GeCl + Hg$ (832).
$C_5H_5W(CO)_3HgCl \rightarrow Ph_3Ge(C_5H_5)W(CO)_3Pt(PPh_2CH_2)_2 + Ph_3GeCl$ (832).
KCN in aq. Me_2CO, + $Me_4NBr \rightarrow (Me_4N)_2[(Ph_3Ge)_2Pt(CN)_2] + [Ph_2P(O)CH_2]_2$ (2285).
H_2 in $C_6H_6 \rightarrow Ph_3GeH + Ph_3Ge(H)Pt(PPh_2CH_2)_2$ (2285).
HCl in $C_6H_6 \rightarrow (Ph_2PCH_2)_2PtCl_2 + Ph_3GeH$ (2285).

Bis(triphenylgermylbis(tri- $Ge_2PtC_{54}H_{72}P_2$ $(Ph_3Ge)_2Pt(PPr_3)_2$
   propylphosphine) Platinum

Synth.: By rxn. of Ph_3GeLi with trans-$(n-Pr_3P)PtCl_2$ in N atm., in 98% yield (103).

Prop.: Pale yellow cryst., dec. 150° (103), IR (103, 105) and UV (103) spectra, dipole moment (103).

Rxn. with: Thermal dec. at 120-250° in vacuo $\rightarrow Ph_4Ge + Ph_6Ge_2 + C_6H_6 + C_3H_6 + Pr_3P$ (103).
PhLi $\rightarrow Ph_3GeH + Ph_4Ge + (Pr_3P)_2PtPh_2$ (103).
$LiAlH_4 \rightarrow Ph_3GeH$ + pyrophoric Pt-complex (103).
MgI_2, + EtOH $\rightarrow [Ph_3Ge(I)Pt(PPr_3)_2 + Ph_3GeMgI] \rightarrow (Pr_3P)_2PtI_2 + Ph_3GeH$ (103).
MgI_2, + air-free $H_2O \rightarrow Ph_6Ge_2 + Ph_3GeI + Ph_3GeH + (Pr_3P)_2PtI_2$ (103).
KCN $\rightarrow K_2[(Ph_3Ge)_2Pt(CN)_2]$ (2285).
H_2 in MePh $\rightarrow Ph_3GeH + Ph_3Ge(H)Pt(PPr_3)_2$ (103).
HCl $\rightarrow Ph_3GeH + Ph_3GeCl +$ cis-$(Pr_3P)_2PtCl_2 + (Pr_3P)_2PtHCl$ (103).
MeI $\rightarrow Ph_3GeI +$ trans-$(Pr_3P)_2PtI_2 +$ trans-$(Pr_3P)_2Pt_2I_4 + CH_4 + C_2H_6$ (103).
MeI, + alc. KOH $\rightarrow MeGePh_3$ (103).
$(CH_2Br)_2 \rightarrow Ph_3GeBr + (Pr_3P)_2PtBr_2 + C_2H_4$ (103).

Addn. compd: $(Ph_3Ge)_2Pt(PPr_3)_2 \cdot C_6H_6$, from components at r.t., looses C_6H_6 at 45° in vacuo [$Ge_2PtC_{60}H_{78}P_2$] (103).

Other organogermanium derivatives with Ni, Pd and Pt coordination groups are listed in Table 87.

Table 87. Polymetallic Organogermanium Derivatives of Nickel and Homologues $R_nGeR''_{4-n}$

Formula*	Synth. Method**	Yield	Properties	Ref.
$Me_2GeNi(CO)_3$	--	--	(a)	--
$Et_3GeC_5H_5NiCO$	IVA	75	Orange liq., $b_{0.2}$ 79.8°, n_D^{20} 1.5932, d_{20} 1.320 (b)	2414
$Et_2GeNi(CO)_3$ (142)	--	--	(a, c)	--
$Pr_2GeNi(CO)_3$	--	--	(a)	--
$Bu_2Ge[C_5H_5NiCO]$	--	--	(a)	--
$i-Bu_2GeNi(CO)_3$	--	--	(a)	--
$(i-C_5H_{11})_2GeNi(CO)_3$	--	--	(a)	--
$[j-Ph_3GeC_7H_9(NEt)_2]_2Ni$	VB	--	m. 304°, NMR spectrum (d)	1004
$(Me_2C_6H_3)_3GeC_5H_5NiCO$	--	--	(a)	--
$[(Ph_3Ge)_2Pd(CN)_2]K_2$	IVB	96	Dec. 112-20°, IR spectrum (e)	795
$[(Ph_3Ge)_2Pd(CN)_2](NMe_4)_2$	(e)	low	Colorless needles	795
$Ph_3Ge(H)Pd(PEt_3)_2$	IVC	--	IR spectrum	795
$(Ph_3Ge)_2Pd(PPh_2CH_2)_2$	I (f)	--	--	795
$Me_3Ge(I)Pt(PPh_2CH_2)_2$	IIIB	--	Yellow orange, m. 115-17°, trans-form (g)	1106
$Me_3Ge(Ph)Pt(PEt_3)_2$	IA	60	m. 109-11° (dec.), cis-form (h, i)	1106
$Me_3Ge(H)Pt(PEt_3)_2$	IB	--	Oil, IR and NMR spectra	2285
$Me_3Ge(H)Pt(PEt_3)_2$	IVC (j)	--	Oil, IR and NMR spectra	1106

* Numbers in parenthesis refer to pages in main volume.
** The characters used here correspond to the synthetic scheme in the introduction to Chapter 6.5.8.
(a) Use for metal plating on hot substrates (603). (b) Sensitive to air (2414). (c) Use for metal plating on ceramics (603). (d) Bis(j-Triphenylgermyl-N,N',N'-diethylaminotroponeimino)nickolate (II). (e) Rxn. of $(Ph_3Ge)_2$-$Pd(CN)_2$ with aq. alc. $Me_4NBr \rightarrow (Me_4N)_2[(Ph_3Ge)_2Pd(CN)_2]$ (795). (f) Synth. from Ph_3GeLi and $(Et_3P)_2PdCl_2$ in $(MeOCH_2)_2$ at -30° (sic.) (795). (g) Mass spectrum (1106). (h) Spectra: far IR, assignment and NMR (1106). (i) Rxn. with anhydr. HCl in $Et_2O \rightarrow Me_3GeH + (Et_3P)_2PtPhCl$ (1107). (j) From $Me_3Ge(Ph_3Ge)Pt(PEt_3)_2$ (1106).

Table 87 Continued $R_nGeR''_{4-n}$

Formula*	Synth. Method**	Yield	Properties	Ref.
[Me₃Ge(Et₃P)Pt(PPh₂CH₂)₂]BPh₄	IVB	--	White cryst., m. 193° (dec.) (k)	1107
Me₃Ge(Br)Pt(PEt₃)₂	IVB	65	Yellow cryst., m. 53-54°, trans-form (h)	1106
Me₃Ge(I)Pt(PEt₃)₂	IVB	84	Yellow cryst., m. 70-71°, trans-form (h, l)	1106
Me₃Ge(NC)Pt(PEt₃)₂	IVB	--	Colorless, m. < r.t., trans-form (g, h)	1106
	IVB		m. 102-103°, cis-form (g, h)	1106
Me₃Ge(SCN)Pt(PEt₃)₂	IVB	68	m. 71-72°, trans-form (g, h)	1106
(MePh₂Ge)₂Pt(PMe₂Ph)₂	IB	--	m. 169-71° (n)	856
MePh₂Ge(Cl)Pt(PMe₂Ph)₂	IVA (n)	--	m. 156-58°, IR spectrum	856
[(Ph₃Ge)₂Pt(CN)₂]K₂	IVD (o)	56	White cryst., dec. > 300°, soly. (p, q)	2285
[(Ph₃Ge)₂Pt(CN)₂](NMe₄)₂	(q)	--	Colorless cryst., dec. 260° (p)	2285
Ph₃Ge(Cl)Pt(PMe₂Ph)₂	II	--	m. 170-72°, ³¹P NMR spectrum (2437)	855
[Ph₃GePt(PEt₃)₂]₂	IVA	--	Light brown cryst., m. 151-52° (r)	832
Ph₃Ge(Cl)Pt(PEt₃)₂	IVA (s)	42	m. 123-26°, IR spectrum	1107
Ph₃Ge(HO)Pt(PEt₃)₂	IVE	--	m. 153-56° (dec.) (t)	103
Ph₃Ge(MeO)Pt(PEt₃)₂	IVE	--	m. 172-80° (dec.), IR spectrum	103
Ph₃Ge(PrO)Pt(PEt₃)₂	IVE	--	m. 162-72° (dec.), IR spectrum	103
Ph₃Ge(H)Pt(PPh₂CH₂)₂	IVC	--	m. 221-23°, mol. wt. (p, u)	2285
Ph₃Ge(H)Pt(PPr₃)₂	IVC	--	IR and NMR spectra (v)	103
Ph₃Ge(I)Pt(PPr₃)₂	IVA	--	Pale yellow cryst., m. 148-49°	103
	IB	--	IR spectrum (103) (w)	102
Ph₃Ge[C₅H₅W(CO)₃]Pt(PPh₂CH₂)₂	VA	68	Light yellow, m. 200° (dec.) (x)	832

(k) Conductivity in Me₂CO (1107). (l) Rxn. with H → Me₃GeH + trans-(Et₃P)₂PtHI (1106). (n) Rxn. of (MePh₂Ge)₂-Pt(PMe₂Ph)₂ with anhydr. HCl in C₆H₆ → MePh₂Ge(Cl)Pt(PMe₂Ph)₂ (856). (o) From (Ph₃Ge)₂Pt(PR₃)₂ with alc. KCN in N atm.; R = Et, Pr (2285). (p) IR spectrum (2285). (q) Rxn. of K₂[(Ph₃Ge)₂Pt(CN)₂] with alc. Me₄NBr → (Me₄N)₂-[(Ph₃Ge)₂Pt(CN)₂] (2285). (r) IR and mass spectra (832). (s) From Me₃Ge(Ph₃Ge)Pt(PEt₃)₂ with dry HCl in Et₂O

7. CYCLIC ORGANOGERMANES

This chapter includes only such cyclic organogermanes having the metal atom adjacent to two carbon atoms in the heterocyclic system. Other heteroatoms may be located in the same ring, however, only in 3- or 4- position for a five membered ring and in 3- , 4- and 5- position for a six-membered system. Monocyclic derivatives and spirocompounds are known. Heterocyclic organogermanes having oxygen sulfur, nitrogen, metals and other heteroatoms in the ring system connected to germanium are located in the appropriate sections of Chapters 4, 5 and 6.

7.1 CYCLIC TETRAORGANOGERMANES

Organogermanes having all four valencies of germanium linked to carbon are listed in this section. The compounds summarized in Tables 88, 89 and 91 are prepared by methods from the following scheme.

I Grignard synthesis:

(IA) $YGeRX + R'MgX \rightarrow YGeRR'$,

(IB) $GeCl_4 + Y(MgX)_2 \rightarrow Y_2Ge$.

II Reactions with alkali metal derivatives:

(IIA) $R_2GeCl_2 + YLi_2 \rightarrow YGeR_2$,

(IIB) $GeCl_4 + YLi_2 \rightarrow Y_2Ge$,

(IIC) $Et_2Ge[(CH_2)_4Br]_2 + Na \rightarrow Et_2Ge(CH_2)_7CH_2$.

III Hydrogermylation reactions:

(IIIA) $YGeRH + R'CH{:}CH_2 \rightarrow YGe(CH_2CH_2R')R$,

(IIIB) $YGeRH + R'C{\vdots}CH \rightarrow YGe(CH{:}CHR')R$.

IV Alkylation of organopolygermane:

$(Me_2Ge)_x + C_2H_4 \rightarrow CH_2(CH_2)_3GeMe_2$.

V Catalytic hydrogenation of unsaturated cyclic organogermane:

$CH{:}CHCH{:}CHGeR_2 + H + \text{Raney Ni} \rightarrow CH_2(CH_2)_3GeR_2$.

R = R′ = organic group, may be different in one molecule, Y = hydrocarbon bridge, may be functionally substituted X = halogen.

7.1.1 UNSUBSTITUTED CYCLIC TETRAORGANOGERMANES

(1107). (t) Rxn. with EtOH → $Ph_3Ge(EtO)Pt(PEt_3)_2$ (103). (u) Mass spectrum (2285). Rate of synth. at 0 and 25° by H-uptake (103). (w) Rxn. with MgI_2 → $(Pr_3P)_2PtI_2 + Ph_3GeI$ (103). (x) Sensitive to air, mass spectrum (832).

4-Germaspiro[3.4]octane GeC_7H_{14}

Synth.: By rxn. of $\overline{CH_2(CH_2)_3Ge}(CH_2CH_2CH_2Cl)Cl$ with powd. Na in xylene, in 67% yield (1492), in MePh (2576).
Prop.: b_{38} 88-89°, n_D^{20} 1.5185, d_{20} 1.2043, IR spectrum (1492).
Rxn. with: S at 250° → $\overline{CH_2(CH_2)_3Ge(CH_2)_3S}$ (2576).
Br → $\overline{CH_2(CH_2)_3Ge}(CH_2CH_2CH_2Br)Br$ (1492).

1,1-Diethylgermacyclobutane GeC_7H_{16} $\overline{Et_2GeCH_2CH_2CH_2}$

Synth.: By rxn. of $Et_2(ClCH_2CH_2CH_2)GeCl$ with Na-K in MePh under reflux, in 35% yield (344, 1492) with Na in MePh (2576).
Prop.: b_{80} 77-78°, n_D^{20} 1.4738, d_{20} 1.0853 (344, 1492), IR (344, 1492) and mass (1492) spectra.
Rxn. with: RLi → Et_2PrGeR ; R = n-Bu, Ph, rxn. mechanism (2575).
$MeHSiR_2$ + cat. $HClO_4$ → $Et_2(MeR_2SiCH_2CH_2CH_2)GeH$ + polymer; R = Ph, Cl (1491).
S at 250° → $Et_2\overline{Ge(CH_2)_3S}$ (2576).
SO_2 at -10° → $Et_2\overline{Ge(CH_2)_3SO_2}$ (2362).
Br → $Et_2(BrCH_2CH_2CH_2)GeBr$ (344, 1492).

1,1-Dibutylgermacyclobutane $GeC_{11}H_{24}$ $n\text{-}Bu_2\overline{GeCH_2CH_2CH_2}$

Synth.: By rxn. of $Bu_2(ClCH_2CH_2CH_2)GeCl$ with Na in refluxing xylene, in 75% yield (344, 1492).
By rxn. of Bu_2GeCl_2 with $CH_2(CH_2Cl)_2$ and Na in xylene (344), in 13% yield (1492).
Prop.: b_{18} 111-12° (344, 1492), b_{11} 100-108° (1492), n_D^{20} 1.4742, d_{20} 1.0163 (344, 1492), IR (344, 1492) and mass (1492) spectra; reduces $AgNO_3$ in the cold, $HgCl_2$ on warming (344, 1492).
Rxn. with: $LiAlH_4$ → $PrBu_2GeH$ (344, 1492).
RONa, + MeI → Bu_3GeOR ; R = Me, Et, rxn. mechanism (2575).
Et_3MH + cat. $HClO_4$ → $Bu_2(Et_3MCH_2CH_2CH_2)GeH$ + polymer; M = Ge, Si (1491).
Bu_2HSiCl + cat. $HClO_4$ → $Bu_2(Bu_2ClSiCH_2CH_2CH_2)GeH$ + polymer (1491).
MeR_2SiH + cat. $HClO_4$ → $Bu_2(MeR_2SiCH_2CH_2CH_2)GeH$ + polymer; R = Bu, Ph, Cl (1491).
$HSiCl_3$ + cat. AiBN → $PrBu_2GeCl$ (1491).
$GeCl_4$ → $Bu_2(Cl_3GeCH_2CH_2CH_2)GeCl$ (1492).
KOH in EtOH → $(PrBu_2Ge)_2O$ (1492).
93% H_2SO_4, + NaOH → $(PrBu_2Ge)_2O$ (1492).
SO_2 at -10° → $Bu_2\overline{GeCH_2CH_2CH_2SO_2}$ (2362).
Y at 250° → $Bu_2\overline{GeCH_2CH_2CH_2Y}$; Y = S (1492), S, Se (2576).
I → $Bu_2(ICH_2CH_2CH_2)GeI$ (1492).
ICl, + EtMgBr → $EtBu_2GeCH_2CH_2CH_2I$ + $EtPrGeBu_2$ (1492).
HX → $PrBu_2GeX$; X = Cl, Br, I (344, 1492).
$ClCH_2CO_2H$ → $PrBu_2GeO_2CCH_2Cl$ (1492).

1,1,3,1-Dibutylmethylgerma- $GeC_{12}H_{26}$ $n\text{-}Bu_2\overline{GeCH_2CHMeCH_2}$
   cyclobutane

Synth.: By rxn. of $Bu_2(ClCH_2CHMeCH_2)GeCl$ with N in xylene, in 67% yield

(1492).
Prop.: b_9 112-14°, n_D^{20} 1.4690, d_{20} 0.9937; IR spectrum, GLC (1492).
Rxn. with: S at 250° → $Bu_2GeCH_2CHMeCH_2S$ (2576).
Br → $Bu_2(BrCH_2CHMeCH_2)GeBr$ (1492).

1,1,3,4,1-Diethyldimethylgerma- $GeC_{10}H_{22}$
cyclopentane

Synth.: By rxn. of $Et_2GeCH_2CMe:CMeCH_2$ with Raney Ni and
H in EtOH under pressure, in 44% yield (1495), at 100°, prod. contn. 86% cis
form (2579), with BH_3, followed by $EtCO_2H$, for cis-form (2579).
Prop.: b_{47} 90° (2579), b_{26} 100° (1495), n_D^{20} 1.4708, d_{20} 1.0373 (1495), cis-
form: n_D^{20} 1.0433, d_{20} 1.4035, trans-form: n_D^{20} 1.4665, d_{20} 1.0272 (2579), exists
in cis- and trans-form (1495), IR spectrum (1495), GLC (1495, 2579).
Rxn. with: $AlCl_3$ at 140° → cis-form rearranges to trans-form (2579).

Other unsubstituted cyclic tetraorganogermanes are listed in Table 88.

7.1.4 UNSATURATED UNSUBSTITUTED CYCLIC TETRAORGANOGERMANES

The unsaturated cyclic tetraorganogermanes in this section may have olefinic
or acetylenic linkages in a sidechain or single and multiple double bonds in
the heterocyclic system. No aromatic cyclic organogermanes are known.

1,1,1-Dimethylgermacyclopentene-3 GeC_6H_{12}
(M-146)
Synth.: By rxn. of $(CH_2:CH)_2$ with $HGeBr_3$ in Et_2O, followed by MeMgCl (161),
matl. contn. 70% $Me_3GeCH_2CH:CHMe(!)$ (161, 365), with $HGeCl_3 \cdot 2Et_2O$ at r.t.,
followed by MeMgBr, in 60-65% yield (382a, 1530).
Prop.: IR (308, 1530), UV (1530) and NMR (161, 308, 365, 1530) spectra,
dipole moment (1530).
Rxn. with: Cl (in the dark) → Me_2GeCl_2 (1530).

1,1,1-Diethylgermacyclopentene-3 GeC_8H_{16}
Synth.: By rxn. of $CH_2CH:CHCH_2GeI_2$ with EtMgBr, in 82% yield
(2579).
Prop.: b_{17} 65°, n_D^{20} 1.4813, d_{20} 1.1013 (2579).
Rxn. with: H + Raney Ni → $Et_2Ge(CH_2)_3CH_2$ (2579).
BH_3, + H_2O_2-NaOH → $Et_2GeCH_2CHOHCH_2CH_2$ (2579).
N_2CHCO_2Et + cat. Cu at 160° → $Et_2GeCH_2CHCH(CO_2Et)CHCH_2$ (2579).
C_5Cl_6 at 150° → Diels Alder adduct see formula (2579).

Table 88. Unsubstituted Cyclic Tetraorganogermanes

Formula*	Synth. Method**	Yield	Properties	Ref.
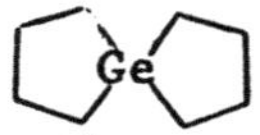 ⬠Ge(144)	IB	30	b_{10} 70-72°, n_D^{25} 1.5035, IR spectrum	17
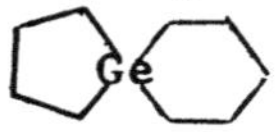 ⬡Ge(144)	IB	50	IR spectrum	17
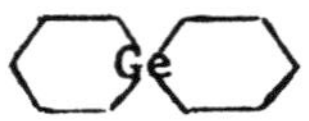 ⬡Ge(144)	IB	30	b_{10} 99-101°, n_D^{25} 1.5005	17

R'⬠GeR$_2$

R' = H	R$_2$ = Et, n-Bu	IA	--	b_{13} 75°, n_D^{20} 1.4720, d_{20} 1.0401 (a)	2575
H	Bu, n-C$_6$H$_{13}$	IIIA	50	--	2575
H	Bu, n-C$_8$H$_{17}$	IA	--	$b_{0.3}$ 103°, n_D^{20} 1.4738, d_{20} 0.9799 (a)	2575
Me	Et$_2$	--	--	--	--
Me	Bu, Ph	IA	--	$b_{0.05}$ 82°, n_D^{20} 1.5355, d_{20} 1.0909 (a)	2572

R'$_2$⬠GeR$_2$
R''$_2$

R' = H, R'' = H, R = Me (145)	IV	--	--	1586
H	H	Et (145)	V	--	--	2579
H	H	Ph (145)	IA	70	m. 33°, $b_{0.4}$ 135-36° (b)	1492
D	D	Bu, Ph	--	--	Mass spectrum	991
D	D	Ph	--	--	Mass spectrum	991
Me	H	Et	V	--	b_{14} 71°, n_D^{20} 1.4658, d_{20} 1.0427	2579
Me	Me	Ph	V	58	$b_{0.4}$ 156°, n_D^{20} 1.5833, d_{20} 1.1710	1495

R′ = R″ = Ph (c) Me (146)	--	$b_{0.3}$ 141-48°, n_D^{20} 1.5910	1587
MeC₆H₄ (c) Me (146)	--	$b_{0.15}$ 146-48°, n_D^{20} 1.5821	1587
$Et_2Ge\overline{(CH_2)_7CH_2}$	IIC	51	b_{20} 114°, n_D^{20} 1.4523, d_{20} 0.9559 (d)	1493

* Numbers in parenthesis refer to pages in main volume.

** The characters used here correspond to the synthetic scheme in the introduction to Chapter 7.1.
(a) Rxn. with SO_2 at -10° → $R_2Ge\overline{CH_2CHR'CH_2SO_2}$; R′ = H, Me (2362). (b) Dielectric const. and dipole moment (1572), rxn. with Br in EtBr → $\overline{CH_2(CH_2)_3GeBr_2}$ (1492). (c) Position of R′ and R″ unknown (1587). (d) IR spectrum, GLC (1493).

277

1,1,3,1-Diethylmethylgerma-   GeC$_9$H$_{18}$
  cyclopentene-3

Synth.: By rxn. of $\overline{CH_2CMe:CHCH_2GeX_2}$ in Et$_2$O with EtMgBr; X = I at -60°, in 40% yield (1495), X = Cl under reflux, in 34% yield (1495).
Prop.: b$_{24}$ 89°, n$_D^{20}$ 1.4805, d$_{20}$ 1.0754 (1495), mass spectrum (991), GLC (1495).
Rxn. with: H + Raney Ni → $\overline{Et_2GeCH_2CHMeCH_2CH_2}$ (2579).
BH$_3$, + H$_2$O$_2$-NaOH → $\overline{Et_2GeCH_2CHOHCHMeCH_2}$ (2579).
N$_2$CHCO$_2$Et → $\overline{Et_2GeCH_2CHCH(CO_2Et)CHCH_2}$ (2579).

1,1,3,4,1-Diethyldimethylgerma-   GeC$_{10}$H$_{20}$
  cyclopentene-3

Synth.: By rxn. of $\overline{CH_2CMe:CMeCH_2GeI_2}$ with EtMgBr in Et$_2$O, in 71% yield (1495).
Prop.: b$_{26}$ 102°, n$_D^{20}$ 1.4850, d$_{20}$ 1.0625 (1495), mass spectrum (991).
Rxn. with: H + Raney Ni → $\overline{CH_2CHMeCHMeCH_2GeEt_2}$ (1495, 2579).
BH$_3$, + H$_2$O$_2$-NaOH → $\overline{Et_2GeCH_2CMeOHCHMeCH_2}$ (2579).
BH$_3$, + EtCO$_2$H → $\overline{Et_2GeCH_2CHMeCHMeCH_2}$ (2579).
N$_2$CHCO$_2$Et → $\overline{Et_2GeCH_2CMeCH(CO_2Et)CMeCH_2}$ (2579).
Br at -80° → Et$_2$GeBr$_2$ (1495).

1,1,3,4,1-Dibutyldimethylgerma-   GeC$_{14}$H$_{28}$
  cyclopentene-3

Synth.: By rxn. of $\overline{CH_2CMe:CMeCH_2GeX_2}$ with BuMgBr in Et$_2$O; X = Cl, in 71% yield (1495), X = I, in 82% yield (1495).
Prop.: b$_{36}$ 161°, n$_D^{20}$ 1.4820, d$_{20}$ 1.0059 (1495).
Rxn. with: I in EtI, + LiAlH$_4$ → Bu$_2$GeH$_2$ (1495).

3,4,1,1,1-Dimethyldiphenylgerma-   GeC$_{18}$H$_{20}$
  cyclopentene-3

Synth.: By rxn. of $\overline{CH_2CMe:CMeCH_2GeI_2}$ with PhMgBr, in 58% yield (1495).
Prop.: m. 40-41°, b$_{0.25}$ 142°, n$_D^{50}$ 1.5832, d$_{50}$ 1.1743 (1495).
Rxn. with: H + Raney Ni in ROH → $\overline{CH_2CHMeCHMeCH_2GePh_2}$ (1495).

1,1-Dimethyltetraphenyl-   GeC$_{30}$H$_{26}$
   germole (M-146)

Synth.: By rxn. of Me$_2$GeCl$_2$ with (CPh:CPhLi)$_2$ in
Et$_2$O (2447).
By-prod. in thermal dec. of (I) in refluxing c.C$_6$H$_{10}$ (2882-3).
Prop.: Bright greenish-yellow solid, m. 179-81°, (far) IR and NMR spectra,
mol. wt. (2447).
Rxn. with: PhC⋮CH → C$_6$HPh$_5$ (2447, 2882-3).
(RC⋮)$_2$ → $\overline{\text{CPh:CRCR:CPhGeMe}_2}$ + o-R$_2$C$_6$Ph$_4$; R = CF$_3$ (2447).
(RC⋮)$_2$ → $\overline{\text{CR:CRCR:CRGeMe}_2}$ + o-R$_2$C$_6$Ph$_4$ (2447) or (II) (2882-3); R = MeO$_2$C .
(:CHCO)$_2$O → (I) (2447).

(I)   (II)

Hexaphenylgermole   GeC$_{40}$H$_{30}$
Synth.: By rxn. of Ph$_2$GeX$_2$ with (CPh:CPhLi)$_2$ in
Et$_2$O in N atm. (2533).
Prop.: m. 198-99° (931), ESR spectrum and polaro-
graphy, forms instable blue radical anion (964).
Rxn. with: BuLi → intense purple soln., with weak ESR signal (931).
Use: Antiknock agent, insecticide (2533).

Other unsaturated cyclic tetraorganogermanes are listed in Table 89.

Table 89. Unsaturated Unsubstituted Cyclic Tetra-
organogermanes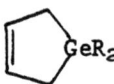

Formula*	Synth. Method**	Yield	Properties	Ref.
[structure: Ge with Bu, Ph, Ph, Bu and fused benzo rings]	IIB	28	NMR spectrum	1736
[structure: spiro Ge with two fluorene units] (144) (a)	IIB	29	m. 244.5-46°, vaporizes at 470° without dec. (b)	(a)
[structure: germacyclobutane with Bu and R]				
R = CH:CH$_2$	IA	--	b$_{18}$ 80°, n$_D^{20}$ 1.4825, d$_{20}$ 1.0560	2575
C⋮CH	IA	--	b$_{10}$ 67°, n$_D^{20}$ 1.4822, d$_{20}$ 1.0795	2575
CH:CHC$_6$H$_{13}$-n	IIIB	66	b$_{0.1}$ 75°, n$_D^{20}$ 1.4828, d$_{20}$ 1.0096 (c)	2575
[structure: germacyclopentane with R and C:CH]				
R = Et	IA	75	b$_{67}$ 80°, n$_D^{20}$ 1.4862, d$_{20}$ 1.1402	1492
Bu	IA	--	b$_{28}$ 101°, n$_D^{20}$ 1.4805, d$_{20}$ 1.0737 (d)	1492
[structure: cyclopentene with Me, R', GeMe$_2$]				
R' = H	IA	--	Crude prod., contn. Me$_3$Ge-CH$_2$CMe:CHMe	1530
Me (146)	IA	56	b$_{27}$ 71°, n$_D^{20}$ 1.4799, d$_{20}$ 1.0902, GLC (d)	1495

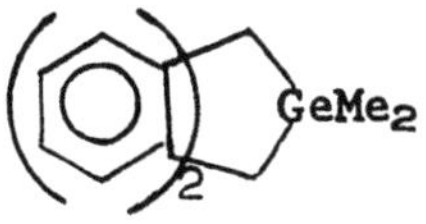 (structure: dibenzo-GeMe₂)	IIA (e)	--	(f)	1726
 (structure: cyclopentane-GeZ₂)	--	--	UV spectrum, struct.	425

---

* Numbers in parenthesis refer to pages in main volume.
** The characters used here correspond to the synthetic scheme in the introduction To Chapter 7.1.
(a) See Formula Index Volume I-III, p. 331. (b) Mass spectrum (908), rxn. with K in THF or $(CH_2OMe)_2$ → paramagnetic anion and dianion (908). (c) Ratio cis- :trans-form = 9 (2572). (d) Mass spectrum (991). (e) From $9,10-Na_2-C_{14}H_{10}$ (1726). (f) Use as fungicide, fuel additive and stabilizer (1726).

7.1.6 FUNCTIONALLY SUBSTITUTED CYCLIC TETRAORGANOGERMANES

7.1.6.1 CYCLIC FUNCTIONALLY SUBSTITUTED TETRAORGANOGERMANES

Only functionally substituted organogermanes with one germanium atom in the cyclic system are listed. The compounds of Table 90 are prepared by methods from the following scheme.

I Synthesis with dilithium derivatives in inert atmosphere:
   (IA) With Ph_2GeCl_2 ,
   (IB) With $GeCl_4$,
   (IC) With Ge at 390° .

II Decarboxylation of thorium salt at 400°.

III Interconversion reactions of germacycloundecanes:
   (IIIA) Dione with $LiAlH_4$ to diole ,
   (IIIB) Reduction of acyloine to ketone with zinc dust and HCl in AcOH.

IV Diels-Alder reactions of tetraphenylgermole:
   (IVA) With $(\vdots CCO_2Me)_2$,
   (IVB) With $(\vdots CCF_3)_2$,
   (IVC) With maleic anhydride; of germacyclopentene:
   (IVD) With hexachorocyclopentadiene
   (IVE) With diazoacetate.

1,1,3-Diethylmethyl-4-hydroxy- $GeC_9H_{20}O$
1-germacyclopentane

(structure: 4-hydroxy-3-methyl-1,1-diethyl-germacyclopentane)

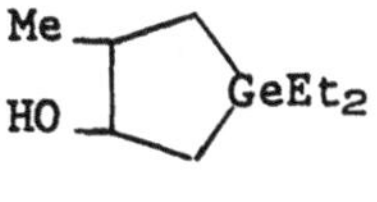

Synth.: By rxn. of $Et_2GeCH_2CMe:CHCH_2$ with BH_3 in THF, followed by H_2O_2-NaOH, in 70% yield (2579).
Prop.: b_{16} 116°, n_D^{20} 1.4855, d_{20} 1.1443, NMR spectrum (2579).
Rxn. with: concd. HX → $Et_2(CH_2:CHCHMeCH_2)GeX$; X = F, Cl, Br, I, AcO,

Cl$_2$CHCO$_2$ (2578).

PX$_3$ → Et$_2$(CH$_2$:CHCHMeCH$_2$)GeX; X = Cl, Br (2578).

MeC$_6$H$_4$SO$_2$Cl, + NaOEt → Et$_2$(CH$_2$:CHCHMeCH$_2$)GeOEt (2578).

1,1,3,4-Diethylmethyl-3-hydroxy-1-germacyclopentane   GeC$_{10}$H$_{22}$O

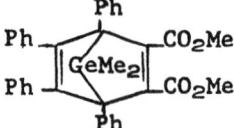

Synth.: By rxn. of Et$_2$GeCH$_2$CMe:CMeCH$_2$ with BH$_3$ in THF, followed by H$_2$O$_2$-NaOH, in 70% yield (2579).

Prop.: b$_{12}$ 112°, n$_D^{20}$ 1.4889, d$_{20}$ 1.1272, NMR spectrum (2579).

Rxn. with: Concd. HX → Et$_2$(CH$_2$:CMeCHMeCH$_2$)GeX; X = Cl, Br, I (2578).

POCl$_3$ in Py → [Et$_2$(CH$_2$:CMeCHMeCH$_2$)Ge]$_2$O (2578).

1,1-Diethylgermacycloundecanolone-6,7   GeC$_{14}$H$_{28}$O$_2$ $\overline{\text{Et}_2\text{Ge(CH}_2)_4\text{CHOHCO(CH}_2)_3\text{CH}_2}$

Synth.: By rxn. of Et$_2$Ge[(CH$_2$)$_4$CO$_2$Et]$_2$ with powd. Na in Ar atm. in xylene, in 58% yield (1493).

Prop.: b$_{0.6}$ 152°, n$_D^{20}$ 1.5084, d$_{20}$ 1.1644, IR spectrum (1493).

Rxn. with: Zn + HCl in AcOH → $\overline{\text{Et}_2\text{Ge(CH}_2)_4\text{CO(CH}_2)_4\text{CH}_2}$ (1493).

Cu(OAc)$_2$ in AcOH-MeOH → $\overline{\text{Et}_2\text{Ge(CH}_2)_4\text{COCO(CH}_2)_3\text{CH}_2}$ (1493).

2,4-(O$_2$N)$_2$C$_6$H$_3$NHNH$_2$ → 2,4-diphenylhydrazone derivative, m. 157° [GeC$_{20}$H$_{32}$N$_4$O$_5$] (1493).

1,1-Diethylgermacycloundecanedione-6,7   GeC$_{14}$H$_{26}$O$_2$ $\overline{\text{Et}_2\text{Ge(CH}_2)_4\text{COCO(CH}_2)_3\text{CH}_2}$

Synth.: By rxn. of $\overline{\text{Et}_2\text{Ge(CH}_2)_4\text{CHOHCO(CH}_2)_3\text{CH}_2}$ with Cu(OAc)$_2$ in AcOH-MeOH, in 76% yield (1493).

Prop.: b$_{0.1}$ 105°, n$_D^{20}$ 1.4983, d$_{20}$ 1.1521, IR spectrum (1493).

Rxn. with: LiAlH$_4$ → $\overline{\text{Et}_2\text{Ge(CH}_2)_4\text{CHOHCHOH(CH}_2)_3\text{CH}_2}$ (1493).

2,4-(O$_2$N)$_2$C$_6$H$_3$NHNH$_2$ → bis(dinitrophenylhydrazone) derivative, m. 248° [GeC$_{26}$H$_{34}$N$_8$O$_8$] (1493).

2,3,7,7,7-(Dicarbomethoxydimethylgerma)tetraphenyl-norbornadiene   GeC$_{36}$H$_{32}$O$_4$

Synth.: By rxn. of $\overline{\text{Me}_2\text{GeCPh:CPhCPh:CPh}}$ with (:CCO$_2$Me)$_2$ in N atm. at 40-48°, in 77% yield (2882-3).

Prop.: Stable at -20°, dec. slowly at r.t., IR and NMR spectra, mol. wt. (2882-3).

Rxn. with: c.C$_6$H$_{10}$ → $\overline{\text{Me}_2\text{GeCPh:CPhCPh:CPh}}$ + o-C$_6$Ph$_4$(CO$_2$Me)$_2$ (2882-3).

(:CCO$_2$Me)$_2$ → $\overline{\text{Me}_2\text{GeO}_2\text{CCH:CCO}_2\text{Me}}$ + $\overline{\text{Me}_2\text{GeOGeMe}_2\text{C(CO}_2\text{Me):CCO}_2\text{Me}}$ + (:CHCO$_2$Me)$_2$ + o-C$_6$Ph$_4$(CO$_2$Me)$_2$ + Me$_3$Ge(O$_2$C)GeMe$_2$Ph (?) (2882-3).

Ph$_3$P + BzH → (Me$_2$GePPh$_3$)$_x$ (2882-3).

C$_6$H$_6$, + Br → Me$_2$GeBr$_2$ + o-C$_6$Ph$_4$(CO$_2$Me)$_2$ (2882-3).

Other functionally substituted cyclic tetraorganogermanes are listed in Table 90.

Table 90. Cyclic Functionally Substituted Tetraorganogermanes

Formula	Synth. Method*	Yield	Properties	Ref.
(spirobis-perfluorocyclohexyl-Ge structure)	IB	48	m. 230-32°, IR spectrum, mol. wt.	883-4
	IC	--	m. 230-32°	885
HO-cyclopentyl-GeEt$_2$	IIIC	70	b$_{15}$ 111°, n$_D^{20}$ 1.4944, d$_{20}$ 1.1806 (a)	2579
R'',R'-substituted germacyclopentadiene GeR$_2$				
R' = Ph, R'' = CF$_3$, R = Me	IVB	--	(Far) IR and NMR spectra, mol. wt.	2447
R' = R'' = CO$_2$Me, R = Me	IVA	--	Dark red resin, NMR spectrum	2447
R' + R'' = (CF)$_4$, R = Ph	IA	--	m. 139-41°, IR spectrum, mol. wt. (b)	883-4
Et$_2$Ge(CH$_2$)$_4$CO(CH$_2$)$_3$CH$_2$	II	10	b$_{0.6}$ 116-17°, n$_D^{20}$ 1.5008, d$_{20}$ 1.0628 (c, d)	1493
Et$_2$Ge(CH$_2$)$_4$(CHOH)$_2$(CH$_2$)$_3$CH$_2$	IIIA	--	b$_{0.1}$ 146° (dec.), n$_D^{20}$ 1.5148, d$_{20}$ 1.1529 (c)	1493
Et$_2$Ge(CH$_2$)$_4$CO(CH$_2$)$_4$CH$_2$	IIIB	--	n$_D^{20}$ 1.500, d$_{20}$ 1.1096, GLC (c, e)	1493

* The characters used here correspond to the synthetic scheme in the introduction to Chapter 7.1.6.1.
(a) NMR spectrum (2579), rxn. with HX → Et$_2$(CH$_2$:CHCH$_2$CH$_2$)GeX; X = Cl, Br, I (2578). (b) Subl. 120-30° x 10^{-4} mm (883-4). (c) IR spectrum (1493). (d) 2,4-Dinitrophenylhydrazone, m. 93° [GeC$_{19}$H$_{30}$N$_4$O$_4$] (1493). (e) 2,4-Dinitrophenylhydrazone, m. 85° [GeC$_{20}$H$_{32}$N$_4$O$_4$] (1493).

Table 90 Continued

Formula: (cyclopentane with GeR'$_2$)

Formula	Synth. Method*	Yield	Properties	Ref.
(Et$_2$Ge cyclopentane with R, R', CO$_2$Et)				
R = H, R' = H	IVE	--	α-(endo-)isomer, b$_{0.7}$ 102°, n$_D^{20}$ 1.4899, d$_{20}$ 1.11637	2579
	IVE	--	β-(exo-)isomer, d$_{20}$ 1.1665 (f, g)	2579
H Me	IVE	--	b$_{0.4}$ 92°, n$_D^{20}$ 1.4850, d$_{20}$ 1.1381 (f, g)	2579
Me Me	IVE	--	α-(endo-)isomer, b$_{0.3}$ 94°, n$_D^{20}$ 1.4844, d$_{20}$ 1.1092	2579
	IVE	--	β-(exo-)isomer, n$_D^{20}$ 1.4862, d$_{20}$ 1.1213 (f, g)	2579
(Et$_2$Ge-CCl$_2$ polycyclic with Cl, Cl, Cl, Cl)	IVD	70	m. 43°, b$_{0.1}$ 180°, n$_D^{20}$ 1.5593, d$_{20}$ 1.5480 (f, h)	2579
(GeMe$_2$ with Ph, Ph, Ph, Ph and anhydride)	IVC	--	White solid, (far) IR and NMR spectra	2447

(f) IR and NMR spectra (2579). (g) GLC (2579). (h) Tends to supercooling (2579).

7.1.6.6 TETRAORGANOGERMANES CONTAINING HETEROATOMS IN THE CYCLIC SYSTEM

Heterocyclic tetraorganogermanes with sulfur and two or more germanium atoms in the same ring are listed.

1,1,3,3-Tetramethyl-1,3-digermacyclobutane $\quad Ge_2C_6H_{16} \quad \overline{Me_2GeCH_2GeMe_2CH_2}$

Synth.: By-prod. in rxn. of $Me_2(BrCH_2)GeBr$ with Mg in THF under reflux (2244).
Prop.: IR, NMR and mass spectra (2244).
Rxn. with: Anhydr. HCl → $Me_3GeCH_2GeMe_2Cl$ (2244).
Cl → $Me_2(ClCH_2)GeCH_2GeMe_2Cl$ (2244).

1,1,4,4,1,4-Tetramethyldigermacyclohexadiene (M-147) $\quad Ge_2C_8H_{16}$

Prop.: Spectra: Raman and assignment (2079), UV (2079), NMR (2079) and mass (578), mol. wt. (578, 2079).
Rxn. with: H + Raney Ni → $\overline{Me_2GeCH_2CH_2GeMe_2CH_2CH_2}$ (2079).
AgOBz + I → $\overline{Me_2GeCH:CHGeMe_2CH(I)CHOBz}$ (2079).
Br in CCl_4 → $\overline{Me_2GeCH:CHGeMe_2CHBrCHBr}$ (1:2) + $(Me_2GeCHBrCHBr)_2$ (1:4) (2079).

1,1,4,4,2,3-Tetramethyldibromo-1,4-digermacyclohexene-5 $\quad Ge_2C_8H_{16}Br_2$

Synth.: By rxn. of $(Me_2GeCH:CH)_2$ with Br (1:2) in CCl_4 at -5°, in 71% yield (2079).
Prop.: m. 72-74°, IR and NMR spectra, mol. wt. (2079).
Rxn. with: Br in CCl_4 (1:2) → $(Me_2GeCHBrCHBr)_2$ (2079).

Other tetraorganogermanes containing heteroatoms in the cyclic system are listed in Table 91.

Table 91. Tetraorganogermanes Containing Heteroatoms in the Cyclic System

$$\begin{array}{c} R_2 \\ Ge \\ Y \end{array}$$

Formula*	Synth. Method**	Yield	Properties	Ref.
[structure with F4, S, Ge, S]	IIB	49	m. 229.5-31.5°, mol. wt.	2325
[structure with F4, S, GePh2]	IIA	46	Colorless cryst., m. 140-43°, mol. wt. (a)	2325
Me₂Ge(HX)(HX)—(HX)(HX)GeMe₂ ring; X = H (147)	V	--	b_{14} 79-80°, n_D^{20} 1.4861, mol. wt. (b)	2079
Br	(c)	--	m. 145-49°, mol. wt. (b)	2079
R₂Ge—R'R''—GeR₂ (148); R = Ph, R' = R'' = H	IA	--	m. 149-50°, soly. (d)	2079
Me, Ph	--	--	m. 305°, UV and mass spectra, struct. (e)	248
Et, Ph	--	--	Mass spectrum	248

Ph, R' + R" = (CF)$_4$	IIA	17	Colorless plates, m. 278.5-82°	2325
Me$_2$Ge(CH$_2$GeMe$_2$)$_2$CH$_2$	IA (f)	90	b$_4$ 71°, n$_D^{20}$ 1.4951, d$_{20}$ 1.3214 (g)	363
	IA (h)	--	IR and NMR spectra	2244

Me$_2$Ge⌐GeMe$_2$
    └OBz
     I
	(i)	--	Syrupy mass	2079

* Numbers in parenthesis refer to pages in main volume.

** The characters used here correspond to the synthetic scheme in the introduction to Chapter 7.1.
(a) Subl. 130-40 x 10^{-3} mm, ^{19}F NMR (2325). (b) IR and NMR spectra (2079). (c) From Me$_2$Ge:(CH:CH)$_2$:GeMe$_2$ or Me$_2$GeCHBrCHBrGeMe$_2$CH:CH with Br in CCl$_4$ at -5° (2079). (d) X-ray struct. analysis, mol. wt. (754a, 755, 2079). (e) Mol. wt. (2079), rxn. with Br → Me$_2$(BrCPh:CPh)GeBr (?) (248). (f) From Cl$_2$Ge(CH$_2$GeCl$_2$)$_2$CH$_2$.
(g) IR and NMR spectra (363). (h) From Me$_2$(BrCH$_2$)GeBr and Mg in THF as by-prod. (2244). (i) From (Me$_2$GeCH:CH)$_2$, AgOBz and I in C$_6$H$_6$ (2079).

7.3 CYCLIC ORGANOGERMANIUM HYDRIDES

The cyclic organogermanes listed in Table 92 are prepared (I) by reaction of the corresponding halides with $LiAlH_4$ or $LiAlD_4$. The cyclic hydride chlorides are obtained by reaction of a dihydride (II) with mercuric chloride.

1,1-Butylgermacyclobutane GeC_7H_{16} n-BuHGeCH$_2$CH$_2$CH$_2$
Synth.: By rxn. of Bu(ClCH$_2$CH$_2$CH$_2$)GeHCl with Na in MePh, in 45% yield (2575).
Prop.: b_{25} 70°, n_D^{20} 1.4781, d_{20} 1.0879, dec. > 150°, IR and NMR spectra (2575).
Rxn. with: n-C$_6$H$_{13}$C⋮CH → C$_6$H$_{13}$CH:CH(Bu)GeCH$_2$CH$_2$CH$_2$ (2575).
n-BuCH:CH$_2$ → Bu(n-C$_6$H$_{13}$)GeCH$_2$CH$_2$CH$_2$ (2575).
RX → CH$_2$CH$_2$CH$_2$GeBuX + RH; RX = CCl$_4$, n-C$_6$H$_{13}$Br, MeI (2575).

1,3,1-Butylmethylgermacyclobutane   n-BuHGeCH$_2$CHMeCH$_2$
Synth.: By rxn. of Bu(ClCH$_2$CHMeCH$_2$)GeHCl with Na in MePh, in 20% yield (2575).
Prop.: b_{15} 61°, n_D^{20} 1.4701, d_{20} 1.0361, dec. > 150°, IR and NMR spectra (2575).
Rxn. with: MeI → CH$_2$CHMeCH$_2$GeBuI (2575).

1-Germacyclopentane (M-149) GeC_4H_{10} CH$_2$(CH$_2$)$_3$GeH$_2$
Synth.: By rxn. of CH$_2$(CH$_2$)$_3$GeBr$_2$ with $LiAlH_4$ in N atm., in 90% yield (1492).
By rxn. of $GeCl_4$ with (CH$_2$CH$_2$MgBr)$_2$ in Et$_2$O, followed by $LiAlH_4$, in 21% yield (2039).
By rxn. of Ph$_2$GeBr$_2$ with (CH$_2$CH$_2$MgBr)$_2$, followed by Br and $LiAlH_4$, in 57% yield (2039).
Prop.: b. 91-92° (1492, 2039), IR (336, 1492, 2039) NMR (2039), microwave (2029) and mass (130, 991) spectra, very sensitive to air (2039).
Rxn. with: $HgCl_2$ → CH$_2$(CH$_2$)$_3$GeHCl + Hg + HCl (1492).
SO_2Cl_2 → CH$_2$(CH$_2$)$_3$GeCl$_2$ (1492).

Other cyclic organogermanium hydrides are listed in Table 92.

7.4 CYCLIC ORGANOGERMANES CONTAINING HETEROATOM-GERMANIUM BONDS

7.4.1 CYCLIC ORGANOGERMANIUM HALIDES

7.4.1.1 CYCLIC ORGANOGERMANIUM MONOHALIDES

Cyclic organogermanium monohalides listed in Table 93 are prepared by reaction of organogermanium trihalide with organodilithium reagent (I), by hydrogermylation of olefin with cyclic hydride halide (II), by exchange of hydride with alkyl halide (III) and by cleavage of germacyclobutane moiety with halogen. Cyclic organogermanium halide hydrides are listed in Chapter 7.3.

Table 92. Cyclic Organogermanium Hydrides   YGeRH

Formula*			Synth. Method**	Yield	Properties	Ref.
R'_2Ge(cyclopentane ring)R'_2 with R, R"						
$R' = H, R'' = D, R = D$			I	97	Mass spectrum, GLC	130
H	H	Pr	I (a)	90	b_{65} 89°, n_D^{20} 1.4754, d_{20} 1.0995 (b, c)	1492
H	H	Ph (149)	--	--	IR spectrum	336
H	H	Cl	II	77	b_{745} 154°, n_D^{20} 1.5078, d_{20} 1.4218 (b-d)	1492
H	H	I (149)	--	--	IR spectrum	336
H	D	Bu	--	--	(b)	--
D	H	Ph	--	--	(b)	--

Me—(ring)—GeH₂ with R		Synth. Method**	Yield	Properties	Ref.
R = H		I	50	b. 117°, n_D^{20} 1.5212, d_{20} 1.2751 (e, f)	1495
Me		I	27	b. 142°, n_D^{20} 1.5025, d_{20} 1.1765, GLC (e)	1495

* Numbers in parenthesis refer to pages in main volume.
** (I) Synth by rxn. of corresponding halide with LiAlH₄ and LiAlD₄, resp., (II) by reaction of dihydride with HgCl₂.
(a) From $\overline{CH_2(CH_2)_3}$Ge(CH₂CH₂CH₂Br)Br with LiAlH₄ (1492). (b) Mass spectrum (991). (c) IR spectrum (1492). (d) Rxn. with CH₂:CHCH₂Cl → $\overline{CH_2(CH_2)_3}$Ge(CH₂CHCH₂Cl)Cl (1492). (e) IR spectrum (1495). (f) Stable in inert atm., polymerizes on standing (1495).

Table 92 Continued   YGeRH

Formula*	Synth. Method**	Yield	Properties	Ref.
Ph₄ [Ge(H)(R) cyclopentadiene]				
R = H	I	--	m. 192-94°, IR spectrum	931
Ph	I	--	m. 187-88°, IR spectrum (g)	931
RH = Z	--		UV spectrum, struct.	424
[Ge(H)(R) cyclohexane]				
R = H (149)	I	35	IR and NMR spectra (h)	2039
Ph (149)	--	--	(h)	--
I (149)	--	--	(h)	--
1/2 O	--	--	(h)	--
$\overline{CH_2(CH_2)_5GeH_2}$	I	--	b. 126-31°, NMR spectrum	2039

(g) Rxn. with BuLi at -78°, + Me₃SiCl → $\overline{CPh{:}CPhCPh{:}CPhGe(SiMe_3)Ph}$ (931). (h) IR spectrum (336).

Table 93. Cyclic Organogermanium Monohalides   YGeRX

Formula*	Synth. Method**	Yield	Properties	Ref.
![cyclobutyl Ge(Bu)(X)] R' = H, X = Cl	III (a)	90	b_{18} 94°, n_D^{20} 1.4900, d_{20} 1.1912 (b)	2575
H, Br	III (c)	90	b_{18} 107°, n_D^{20} 1.5157, d_{20} 1.4520 (b)	2575
H, I	III (d)	90	b_{18} 121°, n_D^{20} 1.5558, d_{20} 1.6358 (b)	2575
Me, I	III (d)	90	b_{15} 109°, n_D^{20} 1.5410, d_{20} 1.5431 (b)	2575
![cyclopentyl Ge(R)(X)] R'$_2$, R'$_2$ = H, R = Et, X = Br (151)	--	--	(e)	--
H, Cl(CH$_2$)$_3$	II	78	b_{12} 136-37°, n_D^{20} 1.5158, d_{20} 1.3693 (f)	1492
H, Br(CH$_2$)$_3$	IV	92	$b_{1.1}$ 118°, n_D^{20} 1.7933, d_{20} 1.5574 (g)	1492
H, Bu Br (151)	--	--	(e)	--
H, Ph Br (151)	--	--	(h)	--
D, Bu Br	--	--	Mass spectrum	991
![cyclopentadienyl] Ph$_4$GePhCl	I	70	m. 210-11° (i)	931

* Numbers in parenthesis refer to pages in main volume.
** The Roman numerals correspond to synthetic methods in the introduction to Chapter 7.4.1.1.

(a) Synth. with CCl$_4$ (2575). (b) Rxn. with RMgX → $\overline{CH_2CHR'CH_2GeBuR}$; R = Et, CH:CH$_2$, C:CH, n-C$_8H_{17}$, Ph (2575). (c) Synth. with n-C$_6$H$_{13}$Br (2575). (d) Synth. with MeI (2575). (e) Rxn. with HC:CMgBr → $\overline{CH_2(CH_2)_3Ge(C:CH)R}$; R = Et, Bu (1492). (f) Rxn. with powd. Na in xylene → $\overline{CH_2CH_2CH_2Ge(CH_2)_3CH_2}$ (1492). (g) Rxn. with LiAlH$_4$ → $\overline{CH_2(CH_2)_3GePrH}$ (1492). (h) Rxn. with CH$_2$:CHCH$_2$MgBr → $\overline{CH_2(CH_2)_3Ge(CH_2CH:CH_2)Ph}$ (1492). (i) Rxn. with LiAlH$_4$ → CPh:CPhCPh:CPhGePhH (931).

7.4.1.2 CYCLIC ORGANOGERMANIUM DIHALIDES

Cyclic organogermanium dihalides listed in Table 94 are prepared by methods from the following scheme.

I Cyclization reactions:
  (IA) $GeCl_4 + YLi_2 \rightarrow YGeCl_2$,
  (IB) $GeX_2 + (CH_2:CH)_2 \rightarrow \overline{CH_2CH:CHCH_2}GeX_2$,
  (IC) $Ge-Cu + CH_2Cl_2 \rightarrow c.(CH_2GeCl_2)_3$.

II Dealkylation reaction:
  $YGePh_2 + Br \rightarrow YGeBr_2$.

III Exchange reaction:
  $YGeH_2 + SO_2Cl_2 \rightarrow YGeCl_2$.

IV Interconversion reactions:
  (IVA) $YGeI_2 + AgX \rightarrow YGeX_2$,
  (IVB) $Cl_2Ge\langle\rangle GeCl_2 + C_5H_6 \rightarrow 1:1 + 1:2$-Diels-Alder adducts.

Y = hydrocarbon bridge, X = halogen.

1-Germacyclopentene-3 Dichloride $GeC_4H_6Cl_2$
  (M-152)

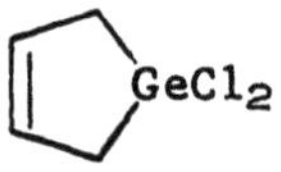

Synth.: By rxn. of $(CH_2:CH)_2$ with $HGeCl_3 \cdot OEt_2$ at -60° (1530) at -30°, in 65-90% yield (1588), with $GeCl_2 \cdot$dioxane (1337).
Prop.: b_{17} 66-67°, crude, assay 70-95% (1588), spectra: IR (308, 1530), Raman (308), ^{35}Cl NQR (737).
Rxn. with: $MeMgCl \rightarrow \overline{CH_2CH:CHCH_2}GeMe_2$ (1530).

3,1-Methylgermacyclopen- $GeC_5H_8Cl_2$
  tene-3 Dichloride

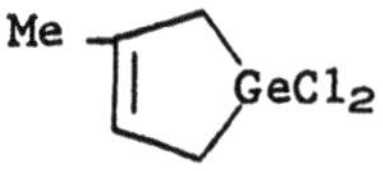

Synth.: By rxn. of $HGeCl_3 \cdot OEt_2$ with isoprene (1530).
By rxn. of $\overline{CH_2CMe:CHCH_2}GeI_2$ with AgCl in heptane, in 69% yield (1495).
Prop.: Crude, contn. linear isomer $Me_2C:CHCH_2GeCl_3$ (1495, 1530), b_{34} 98° n_D^{20} 1.5128, d_{20} 1.4694 (1495), IR and NMR spectra (1530).
Rxn. with: $RMgX \rightarrow R_3GeCH_2CH:CMe_2 + \overline{CH_2CMe:CHCH_2}GeR_2$; R = Me (1495), Et (1530).

3,1-Methylgermacyclopen- $GeC_5H_8I_2$
  tene-3 Diiodide

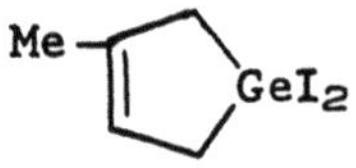

Synth.: By rxn. of GeI_2 with isoprene at 60° in N atm. (1495).
Prop.: Crude n_D^{20} 1.6511, d_{20} 2.319, dec. on distn., sensitive to air (1495).
Rxn. with: $RMgX \rightarrow R_3GeCH_2CH:CMe_2$; R = Me, Et (1495).
EtMgBr at -60° $\rightarrow \overline{CH_2CMe:CHCH_2}GeEt_2$ (1495).
$AgX \rightarrow \overline{CH_2CMe:CHCH_2}GeX_2$; X = Cl, Br (1495).
$LiAlH_4 \rightarrow \overline{CH_2CMe:CHCH_2}GeH_2$ (1495).

3,4,1-Dimethylgermacyclopen-   GeC₆H₁₀I₂
   tene-3 Diiodide

Synth.: By rxn. of GeI₂ with (CH₂:CMe)₂ in N atm. at 110°
(1495).
Prop.: m. 32-34°, dec. on distn., n_D^{20} 1.654, d_{20} 2.224 (1495).
Rxn. with: RMgX → $\overline{CH_2CMe:CMeCH_2GeR_2}$; R = Me, Et, Bu, Ph (1495).
AgCl → $\overline{CH_2CMe:CMeCH_2GeCl_2}$ (1495).
LiAlH₄ → $\overline{CH_2CMe:CMeCH_2GeH_2}$ (1495).
Aq. alc. KOH → $(\overline{CH_2CMe:CMeCH_2GeO})_x$ (1495).

1,4-Digermacyclohexa-   Ge₂C₄H₄Cl₄
   diene Tetrachloride (M-152)

Prop.: IR (1070), Raman (2079), NMR (2079) and mass (1121)
spectra, x-ray structural analysis (2079, 2260), x-ray vapor
phase diffraction anal. (2075), force const. (1070).
Rxn. with: C₅H₆ → Diels-Alder adducts I and II, compare formulae below
(2079).

(I) 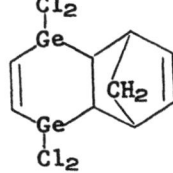   (II)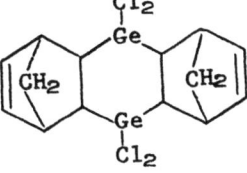

1,4-Digermacyclohexadiene   Ge₂C₄H₄I₂
   Tetraiodide (M-153)

Prop.: m. 302-303° (248), IR, UV (248), NMR (2079) and mass
(248) spectra, x-ray structural analysis (755, 2079).
Rxn. with: PhMgBr → (Ph₂GeCH:CH)₂ (2079).

Other cyclic organogermanium dihalides are listed in Table 94.

Table 94. Cyclic Organogermanium Dihalides

Formula*	Synth. Method**	Yield	Properties	Ref.
GeX$_2$ cyclopentane				
X = Cl (151)	III	90	b_{64} 94°	1492
Br (151)	II	80	b_{35} 112-13° (a)	1492
I (151)	--	--	(b)	--
R', R'' cyclopentene GeX$_2$				
R' = H, R'' = H, X = Br	IB	28	$b_{20.5}$ 95.5°, n_D^{20} 1.5880, d_{20} 2.1742 (c)	161, 365
H H I	IB	--	$b_{0.2}$ 77°, n_D^{30} 1.6890, d_{30} 2.5693 (d, e)	2579
- - Z	--	--	UV spectrum, struct.	424
H Me Br	IVA	57	b_{25} 125°, n_D^{20} 1.5825, d_{20} 2.059 (f)	1495
Me Me Cl (152)	IVA	76	b_{26} 120°, n_D^{20} 1.5178, d_{20} 1.4249 (g)	1492
Ph$_4$ cyclopentadiene GeCl$_2$	IA	<70	m. 197-99° (h)	931

Structure	Group	Yield	Properties	Ref.
$Cl_2Ge(CH_2GeCl_2CH_2)_2CH_2GeCl_2CH_2$ (6-membered ring)	IC	19	m. 91-92°, b_5 150-52° IR and NMR spectra (i)	363
Norbornene-fused $Cl_2Ge-GeCl_2$	IVB	73	White cryst., m. 175-77°, soly., mol. wt.	2079
Bis-norbornene-fused Ge-Ge Cl_2	IVB	--	m. 264-65°, mol. wt.	2079

* Numbers in parenthesis refer to pages in main volume.
** The characters used here correspond to the synthetic scheme in the introduction to Chapter 7.4.1.2.
(a) Rxn. with $LiAlH_4 \rightarrow CH_2(CH_2)_3GeH_2$ (1492). (b) Rxn. with $LiAlD_4 \rightarrow CH_2(CH_2)_3GeD_2$ (130). (c) NMR spectrum (161, 365). (d) IR and NMR spectra (2579). (e) Rxn. with $EtMgBr \rightarrow Et_2GeCH_2CH:CHCH_2$ (2579). (f) Rxn. with aq. alc. $KOH \rightarrow (CH_2CMe:CHCH_2GeO)_x$ (1495). (g) Rxn. with $BuMgBr \rightarrow CH_2CMe:CMeCH_2GeBu_2$ (1492). (h) Rxn. with $LiAlH_4 \rightarrow PhC:CPhCPh:CPhGeH_2$ (931). (i) Rxn. with $MeMgCl \rightarrow$ c.$(Me_2GeCH_2)_3$ (363).

7.4.3 CYCLIC ORGANOGERMANES CONTAINING GERMANIUM LINKED TO OXYGEN, SULFUR AND SILICON

3,1-Methylgermacyclopen- $(GeC_5H_8O)_x$
tene-3 Oxide
Synth.: By rxn. of $\overline{CH_2CMe:CHCH_2GeBr_2}$ with aq. alc. KOH (1495).
Prop.: Polymeric, soly. (1495).

3,4,1-Dimethylgermacyclo- $(GeC_6H_{10}O)_x$
pentene-3 Oxide
Synth.: By rxn. of $\overline{CH_2CMe:CMeCH_2GeI_2}$ with aq. alc. KOH, in 40% yield (1495).
Prop.: IR spectrum, soly. (1495).

5,1-Thiagermaspiro[4.4]- $GeC_7H_{14}S$
nonane
Synth.: By rxn. of $\overline{CH_2(CH_2)_3GeCH_2CH_2CH_2}$ with S at 250°, in 82% yield (2576).
Prop.: b_{18} 120°, n_D^{20} 1.5669, d_{20} 1.3198, NMR spectrum (2576).

1-Trimethylsilylpentaphenyl- $GeSiC_{37}H_{34}$
germole
Synth.: By rxn. of pentaphenylgermole with BuLi, followed by Me_3SiCl, in 90% yield (931).
Prop.: m. 178-80°, IR and NMR spectrum (931).

8 MISCELLANEOUS ORGANOGERMANES

8.1 POLYMERIZATION REACTION

Table 95 gives a very brief summary of polymerization reactions reported for unsaturated or functionally substituted organogermanes. The following polymerization methods and catalysts are used.

I Self polymerization:
   (IA)   at 150-200°,
   (IB)   at -70°.

II Polymerization with metallic catalysts:
   (IIA) With BuLi,
   (IIB) With Ni-H_2
   (IIC) With $CoBr_2$-Py
   (IID) With CuCl-Py-O_2.

III Radical reactions with azodiisobutyronitride (AiBN).

IV Polycondenzation:
   (IVA) With Na ,
   (IVB) With base .

Additional information on organogermanium polymers may be found in Subchapters 2.7.1.8, 2.7.2, 2.7.3, 4.1.2.6 and 4.3.1.3.

Table 95. Polymerization Reactions with Organogermanes

Polymerization System Components	Method*	Remarks	Ref.
I UNSATURATED TETRAORGANOGERMANES			
Me$_3$GeC:CCH:CH$_2$	I	Glassy polymer	367
Et$_3$GeCH:CH$_2$	IIA	Prop. of prod., mol. wt.	1700
Et$_3$GeCH:CH$_2$ + (a, b)	IIA	Prop. of prod., mol. wt.	1700
Ph$_3$GeC:CH	IA	Prop. of prod., spectra	1438
Ph$_3$GeC:CH + Ph$_2$Ge(C:CH)$_2$	IA	Prop. of prod., spectra	1438
Ph$_2$Ge(C:CH)$_2$	IA	Prop. of prod., spectra	1438
(Ph$_2$GeC:CC$_6$H$_4$C:C)$_x$	IID (c)	Prop. of prod.	1437
(Ph$_2$GeCBr:CBr)$_x$	IIB	Saturated polymer, spectra	1439
PhGe(C:CH)$_3$	IA	Prop. of prod., spectra	1438
II ORGANOGERMANIUM HYDRIDES			
Ph$_2$GeH$_2$ + CH$_2$:CHCN	III	--	2577
Ph$_2$GeH$_2$ + CH$_2$:CHCH$_2$CN	III	--	2577
R$_2$GeH$_2$ + alkyne	I	(d)	1349
R$_2$GeH$_2$ + alkene	I	(d)	1349
HGeCl$_3$ + alkene (e)	I	Prop. of prod.	1589
HGeCl$_3$ + diene (f)	I	Prop. of prod. (g)	1589
HGeCl$_3$ + C$_2$H$_2$	I	Prop. of prod. (g)	1589
III TETRAORGANOGERMANES			
Et$_2$Ge[(CH$_2$)$_2$Br]$_2$ + Na	IVA	--	1493
Et$_2$Ge[(CH$_2$)$_5$NH$_2$]$_2$ + Et$_2$M[(CH$_2$)$_4$CO$_2$H]$_2$	IV	Heat resistent polypeptide, M = Ge M = Si	1490 343, 1490
Ph$_2$Ge(C:CMgBr)$_2$	IIC	Prop. of prop., spectra	1438

IV IV GERMANIUM HALIDES

$Et_2GeCl_2 + ArNa_x$ (h)	IVA	Oil	2703
$Ph_2GeBr_2 + C_6H_4(C:CLi)_2$	IV	Prop. of prod. (c)	1437
$Ph_2GeBr_2 + C_6H_4(C:CMgBr)_2$	IV	--	1439
$Ph_2GeX_2 + Y(SH)_2$ (i)	IVB	Prop. of prod. (j)	121
$GeCl_2 \cdot dioxane + (CH_2:CH)_2$	I	Prop. of prod.	1337

V MISCELLANEOUS DERIVATIVES

$(Ph_2GeSYS)_x$ (i)	(j)	Prop. of polymer	121
$(Me_2Ge)_x + C_2H_4$	IA	--	1586

* The characters used here correspond to the scheme in the introduction to Chapter 8.1.
(a) $Et_3MCH:CH_2$; M = Si, Sn. (b) $PhCH:CH_2$. (c) Oxidation with CuCl-Py-O → polymer (1437). (d) Use as sensitizer in photosemiconductor (1349). (e) Alkenes: C_2H_4 , $CH_2:CMe_2$, $c\text{-}C_6H_{10}$, $CH_2:CHCN$ (1589). (f) Diene: butadiene, isoprene, 1,3-pentadiene (1589). (g) Alkylation of polymers by MeMgBr, spectra for prod. (1589). (h) Ar = polynuclear aromatic oil (2703). (i) Y = $(CH_2)_5$, $(CH_2C_6H_4)_2O$ (121). (j) Rxn. with $GeCl_4$ → Ge-contg. polymer (121).

8.2 UNSPECIFIED ORGANOGERMANES

A certain number of incompletely identified organogermanes appears in the literature. Whenever possible these compounds are treated in the appropriate chapter, usually at the bottom of a table, e.g., diorganogermanium difluoride in Table 31. A few compounds that cannot be classified in this manner are listed in the following table.

Table 96. Unspecified Organogermanes

Me_3Ge-	Group electronegativity (234).
Et_3Ge-	Ge-bond refractions D_{20} (81).
R_xGe	Automated C-H-N analysis (2397).
C-Ge-O-	Evidence for organogermanium compounds in coal (1778).

T I N

1. SYMMETRIC TETRAORGANOTIN COMPOUNDS

This chapter comprises organotin derivatives with four identical groups attached to tin. The symmetric tetraorganotin compounds listed in Tables 97 through 99 are prepared by methods from the following scheme.

I Grignard synthesis from tin tetrahalide:
   (IA) In diethyl ether,
   (IB) In hydrocarbon solvent without ethers in presence of catalytical triphenyl phosphite,
   (IC) With alkyl chloride and magnesium in ether with catalytical amount of alkyl bromide.

II Synthesis from tin tetrachloride and alkyl lithium in diethyl ether or other solvents:
   (IIA) With alkyl sodium in hydrocarbon solvent,
   (IIB) With sodium and alkyl chloride in presence of mercury as catalyst.

III Reaction of tin tetrachloride with acetylenes in presence of:
   (IIIA) Sodium and
   (IIIB) Secondary amine.

IV Alkylation of stannous fluoride with ethylene, hydrogen and aluminum in presence of triethylaluminum catalyst at 200° in autoclave.

V Interconversion reactions of a tin-substituted carboxylic acid with:
   (VA) Diazomethane,
   (VB) Thionyl chloride,
   (VC) Alcoholic sodium hydroxide and
   (VD) Hydrogenation with $LiAlH_4$ of a tin-substituted nitrile.

Cyclic symmetric organotin compounds are listed in Chapter 7.1. Additional references to tetraorganotin derivatives are given in Chapters 8.1 and 9.

1.1 SYMMETRIC UNSUBSTITUTED TETRAORGANOTIN COMPOUNDS

Tetramethyltin (M-158) SnC_4H_{12} Me_4Sn

<u>Synth.:</u> By rxn. of $SnCl_4$ with MeMgI in Et_2O and direct distn. without hydrolysis (429), with MeMgCl in THF (2962).
By rxn. of $SnCl_2$ with MeMgBr in ether-hydrocarbon, in > 86% yield (149).
By rxn. of $SnCl_4$ with MeCl and Al in molten $NaCl-AlCl_3$ at 220°, in 95% yield (2000-1).
By-prod. in rxn. of $SnCl_4$ with RMgI; R = mixture Me and Et (430).
By-prod. in rxn. of Me_3SnCl with Cu-Zn in THF (2830), of Me_3SnN_3 with PR_3 at 130-200°; R = Me, Et, Ph (2743).
By-prod. in rxn. of Me_6Sn_2 with R_3SnCl in MeOH; R = Me, Et (2012), with C_2H_4,

$CH_2:CF_2$, C_3F_6, C_2F_3X; X = H, F, Cl, Br under UV irradiation (871), with CF_3COI or $(CF_3CO)_2CO$ (922).

By-prod. in rxn. of c.$(Me_2Sn)_6$ with C_2F_4 under UV irradiation (871).

By-prod. in thermal dec. of $Me_3SnCH_2CHCHRCCl_2$ at 140-50° and in presence of $ZnCl_2$ at 80-125°; R = H, Me (2755), of 2,3,4,5-$Cl_4C_6HSnMe_3$ at 230° (1883), of o-$Me_3SnC_6H_4I$ at 235°, in 7% yield (1021), of $[Me_3SnC(CF_3):]_2$ at 150° (108), of Me_3SnN_2CH at 360° (1393), of $Me_3SnSnEt_3$ at 190° (399) of Me_3SnLi-THF (2127), of $(Me_3Sn)_2Fe(CO)_4$ at 140° in sealed tube, in 20% yield (904).

By-prod. in x-ray radiolysis of Me_2SnEt_2 (1191).

Prop.: IR (135, 546, 763, 1013), far IR and assignment (2322), Raman (1423), UV (424, 1350, 1715, 1905), NMR (55, 129, 267, 447, 518, 570, 581, 932, 985-6, 1220, 1350, 1430, 1436, 1924, 1949, 2267, 2296, 2475, 2733, 2875), ^{13}C NMR (1498, 2121), Moessbauer resonance (902, 1112, 1180, 1309, 1350, 1489, 1960, 1989-91, 2424, 2489, 2665, 2840) and mass (621, 1173, 1191, 1648, 1756, 2268, 2511, 2528, 2869) spectra, GLC (429, 431, 442, 1191-2, 1718, 2278), effect of mol. redistribution on refractive index (1175), parachor, refrachor and surface tension (1717), magnetic susceptibility (1), calcd. and exptl. (699), self-diffusion coeff. and nucl. spin-lattice relaxation time (1307), solvent prop., mixture with CCl_4 (457), thermodynamic data (305, 555, 2025, 2856).

Rxn. with: X-ray irradiation → Me_6Sn_2 + $(Me_3Sn)_2CH_2$ + CH_4 + C_2H_6 + H (1191), investigation by ESR spectroscopy (2443).

γ-irradiation → by ESR spectroscopy (2531).

R_4Sn (+ cat. $AlCl_3$ at 120-70°) → random redistribution by GLC; R = Et, n-Pr and n-Bu (432).

Et_4Ge (+ cat. $AlCl_3$ at 120° → random redistribution by GLC (432).

X → kinetic data of cleavage rxn. and mechanism, solvent effect of Me_2SO, MeOH, AcOH, PhCl, DMF; X = Br (653, 767, 769) under UV irradiation (768), I (653, 767, 1098), Br in presence of Me_3SnBr (767), in presence of n-Pr_4Sn or n-Bu_4Sn under UV irradiation (768).

HCl in C_6H_6 → kinetic data and mechanism of cleavage rxn. (653).

$SnCl_4$ → Me_3SnCl + $MeSnCl_3$ + MeCl, kinetic data in CCl_4 (191), Me_2SnCl_2 (2962), heat of redistribution rxn. (380).

SnX_4 → Me_nSnX_{4-n}, temp. influence*; X = Cl, Br (725).

$GeCl_4$ in CCl_4 → Me_3SnCl + $MeGeCl_3$ + MeCl,* (191).

$SiCl_4$ in CCl_4 → Me_3SnCl + Si_nCl_{2n+2} + MeCl * (191).

$TiCl_4$ → CH_4 + MeCl, influence of H_2O (544).

PF_5 → $MePF_4$ + [Me_3SnPF_6?] → Me_3SnF (567).

Et_3SnCl ⇌ Me_3SnCl + $MeSnEt_3$ (2012).

Me_2SnX_2 → Me_3SnX; kinetic data and mechanism; X = Cl (2686), Cl, Br, I * (725).

$RBuSnCl_2$ → Me_3SnCl + $RMeBuSnCl$; R = Me at 75-85°, Et at 120-30° (1370).

---

* Kinetic data on redistribution rxn. by NMR spectroscopy.

BuSnCl$_3$ (1:1 at 0°) → Me$_3$SnCl + MeBuSnCl$_2$ + Me$_2$BuSnCl + Me$_2$SnCl$_2$ (1370).
BuSnCl$_3$ (1:2 at 0-180°) → Me$_2$SnCl$_2$ + MeBuSnCl$_2$ (1370).
PhSnCl$_3$ (at 0°) → Me$_3$SnCl + MePhSnCl$_2$ (1370).
HGeCl$_3$·OEt$_2$ → Me$_3$SnCl + MeHGeCl$_2$ (366).
AcCl + AlCl$_3$ (1:1:1) → Me$_3$SnCl (1781).
AcCl + AlCl$_3$ (1:2:2) → Me$_2$SnCl$_2$ (1781).
Et$_3$PO$_4$ (at 180°) → Et$_2$O + unidentified prod. (!) (2355).
Et$_4$SiO$_4$ (+ cat. AlCl$_3$ at 120°) → Et$_2$O + unidentified prod. (2355).
Al$_2$O$_3$ → chemisorption followed by Moessbauer spectroscopy (253, 2480).
H$_2$SO$_4$ (100%) → Me$_3$Sn(H)SO$_4$ + CH$_4$ (1100).
SO$_3$ (in CH$_2$Cl$_2$ at -78°) → Me$_3$SnO$_3$SMe (1:1) or Me$_2$Sn(O$_3$SMe)$_2$ (1:2) (2742).
CrO$_3$ (in AcOH) → kinetic data and mechanism of oxidn. (653).
C$_6$Cl$_5$CO$_2$H → Me$_3$SnO$_2$CC$_6$Cl$_5$ + CH$_4$ (956).
Use: As solvent in synth. of Me$_6$Sn$_2$ from Me$_3$SnCl and Na (1548a).
In catalyst for olefin polymerization (184).
In catalyst for (CH$_2$:CH)$_2$ polymerization (1650a).

Tetrabenzyltin (M-160)　　　　　　　　SnC$_{28}$H$_{28}$　　　　　　　　(PhCH$_2$)$_4$Sn
Synth.: By rxn. of SnCl$_4$ and PhCH$_2$MgCl in Et$_2$O-MePh, in 20% yield (841).
By rxn. of Sn with PhCH$_2$Cl in SnCl$_2$-KCl melt at 300° (2000).
By rxn. of (PhCH$_2$)$_2$SnCl$_2$ with PhCH$_2$Cl and Mg in hydrocarbon and Ph$_3$PO$_3$ or with PhCH$_2$MgCl in dry Et$_2$O, in 80% yield (1871a).
Prop.: m. 43.3° (841), b$_{0.04}$ 155° (841), NMR (2069), (far) IR spectra and assignment (841).
Use: In catalyst for (CH$_2$:CH)$_2$ polymerization (1650a).

Tetraethyltin (M-160)　　　　　　　　SnC$_8$H$_{20}$　　　　　　　　Et$_4$Sn
Synth.: By rxn. of EtMgBr in Et$_2$O with (Et$_3$Sn)$_2$O (2:1), in 87% yield (1906), with Et$_3$SnOSiR$_3$; R = Et and Me, in 90 and 96% yield, resp. (1906), with (Et$_2$SnO)$_x$, in 65% yield (505), with (Et$_3$Sn)$_2$S (2813), with Et$_3$SnC⫶CCl in presence of CoCl$_2$, in 70% yield (2180).
By rxn. of Et$_3$SnC⫶CBr with EtLi in Et$_2$O at -50° (2180).
By rxn. of Et$_3$SnH with Et$_2$Zn (584), with Et$_4$Pb at 130° in inert atm., in 21% yield (2850).
By rxn. of Et$_3$SnOH with Et$_3$Al in C$_6$H$_6$ in N atm. (2663).
By-prod. in rxn. of Et$_3$SnCl with Zn in BuCl-Et$_3$N at 160°, in 38% yield (1939), with Na in BuCl at 160°, in 3% yield (1939).
By thermal dec. of Me$_3$SnSnEt$_3$ at 190° (399), of (Et$_2$Sn)$_x$ at 160°, in 13% yield (1941), of (Et$_3$Sn)$_3$Sb at 150° in presence of cat. AlBr$_3$, in 92% yield (2086).
By-prod. in photooxidn. of (Et$_3$Sn)$_2$O (2217).
By-prod. in rxn. of (Et$_2$Sn)$_x$ in presence of Et$_3$N in BuCl at 170°, in 13% yield (1941).
By-prod. in rxn. of Bu$_2$SnCl$_2$ with Zn in EtCl-Et$_3$N at 160°, in 8% yield (1939).
By electrolysis of NaBEt$_4$ on Sn anode and Hg cathode, in 10% current yield (2186).

Prop.: IR (321, 588), far IR and assignment (321, 2322), NMR (267, 1220, 1430, 1938, 2068, 2296) and mass (848, 850, 1173, 1191, 1309, 1648, 1756, 2268, 2410, 2838) spectra, magnetic susceptibility (1), parachor, refrachor and surface tension (1717), preparative GLC (1192), thermodynamic data (305, 2025, 2856). Anal. detn. by GLC (431-2, 442, 1191, 1428, 1718, 1939, 2278), in air (1868), in n-Bu_4Sn from Grignard synth. by chromatography and mass spectroscopy (2838).

Rxn. with: X-ray irradiation → C_2H_6 + C_2H_4 + C_4H_{10} + H (1191), investigation of prod. by ESR spectroscopy (2443).

Me_4Sn(+ cat. $AlCl_3$ at 120° → random distribution by GLC (432).

X → kinetic data and mechanism of cleavage rxn., solvent effect; X = Br (653, 767, 769), under UV irradiation and in presence of Me_3SnR; R = n- , i-Pr (768), X = I (653, 767, 1098).

HCl(in C_6H_6) → kinetic data and mechanism for dealkylation (653).

$SnCl_4$ → heat of redistribution rxn. (380).

$SnCl_4$(+ cat. $AlCl_3$) → Et_3SnCl (705).

$TiCl_4$ → C_2H_6 + EtCl, influence of H_2O (544).

HgI_2(in 96% MeOH) → Et_3SnI + EtHgI, kinetic data and mechanism (654, 2205).

PF_5 → Et_3SnF + POF_3 + $EtPF_4$ (567).

CCl_4(in air at 100°) → Et_3SnCl + $HCCl_3$ + $Cl_3CCH_2CH_2Cl$ + C_2H_4 + C_2H_6 (2094).

Me_3SnBr → complete random redistribution without catalyst (432).

$RSnCl_3$ → $EtRSnCl_2$ + Et_3SnCl; R = n-Bu, Ph (1370).

$HGeCl_3 \cdot OEt_2$ → Et_3SnCl + $EtGeCl_3$ + $EtHGeCl_2$ + C_2H_6 + H (366).

$Et_3SnO_2CMe_3$(at 80-160° in decane) → Et_2SnO + $(Et_3Sn)_2O$ + Et_3SnOH + $Me_3CO\cdot$- degradation prod., rxn. kinetics and mechanism (665).

$Me_3SiCH_2CH_2CO_3CMe_3$ → $Et_3SnO_2CCH_2CH_2SiMe_2$ + Me_3SiEt + C_2H_6 + C_2H_4 + CH_4 (586).

O(under UV irradiation) → Et_3SnO_2Et + Et_2SnO + EtOH + AcH, rxn. mechanism (662), rel. kinetic data (2216).

O_3(at -78°) → AcH + H_2O + oligomer (663).

(At 70° in H_2O) → Et_3SnOH + Et_2SnO + AcH + H_2O_2 (663) discussion of kinetic data and intermediates at temperatures between (2219).

RCO_3CMe_3 → Et_3SnO_2CR + Me_3COH + CO_2 + CH_4 + C_2H_6 ; R = Me, Et (2095), R = Bz, $Et_3SnOCMe_3$ also observed (2098).

CrO_3(in AcOH) → kinetics and mechanism of oxidn. rxn. (653).

PhNCO + urans → catalytic effect and kinetic data (133).

O_2NNRH → $Et_3SnNRNO_2$; R = Me, i-Pr (611).

Biol. Prop.: Toxicity (1944, 2659) and effect on electro encephalogram (2574).

Insecticide activity against Musca domestica, LD_{50} and ATP inhibition (425), Culex pipiens berbericus larvae (73), Culex pipiens pipiens larvae (1137).

Bactericide activity against Staphylococcus aureus and Escherichia coli (702).

Effect on cerebral amino acid transport in vitro (860).

Use: In catalyst for polymerization of α-olefins (688, 2430), of caprolactam (257), of CH_2:CHCl (379), of CH_2:CHOR (2153), of CH_2:CHCN with olefin (2997), of c.C_2H_4O (3000).

Relative reactivity as catalyst for polyurethan prepn. (1603).
Effect on BuLi catalyzed vinyl polymerization (2213).
Effect on degradation of PVC (439).
Additive for flame resistant polyester (2702).
Agent for vacuum tin plating by electron beam (2076).

Tetrapropyltin (M-162) $SnC_{12}H_{28}$ n-Pr_4Sn
Synth.: By rxn. of $SnCl_2$ with PrMgBr in Et_2O-hydrocarbon, in > 86% yield (149).
By rxn. of $SnCl_4$ with PrBr and Mg in C_5H_{12} and cat. Ph_3PO_3 without ether (1871a).
By-prod. in rxn. of $(Pr_2Sn)_x$ with Et_3N in BuCl at 160°, in 4% yield (1941).
Prop.: Spectra: (far) IR and assignment (2058), IR, Raman and conformation (925), NMR (1220, 1321) and mass (1648, 2268, 2410), GLC (442, 1718), magnetic susceptibility (1), parachor, refrachor and surface tension (1717), thermodynamic data (305).
Rxn. with: R_4Sn(+ cat. $AlCl_3$ at 130-60°) → random redistribution by GLC; R = Me, n-Bu (432).
Et_4Pb(+ cat. $AlCl_3$ at 140°) → random redistribution (432).
X → kinetic data and mechanism of cleavage rxn., solvent effect; X = Br (653, 768), I (653).
HCl(in C_6H_6) → kinetic data and rxn. mechanism (653).
BF_3 → $PrBF_2$ (909).
$TiCl_4$ → PrCl + C_3H_8 , influence of H_2O (544).
CCl_4(+ air at 100°) → Pr_3SnCl + $HCCl_3$ + $Cl_3CCH_2CHClMe$ (2094).
PF_5 → $PrPF_4$ (567).
$RSnCl_3$ → Pr_3SnCl + $PrRSnCl_2$; R = n-Bu, Ph (1370).
CrO_3(in AcOH) → kinetic data and mechanism of oxidn. (653).
Biol. prop.: Bactericide activity against Staphylococcus aureus and Escherichia coli (702).
Use: In catalyst for polymerization of styrene (678, 2556) with C_3H_6 (688).
Agent for tin vacuum plating by electron beam (2076).

Tetraisopropyltin (M-163) $SnC_{12}H_{28}$ i-Pr_4Sn
Prop.: NMR (1220, 2070) and mass (2268) spectra, GLC (442, 1718), parachor, refrachor and surface tension (1717), thermodynamic data (305, 886, 2856), rate of speed of sound (1679).
Rxn. with: X → kinetic data and mechanism of cleavage rxn., solvent effect; X = Br (653, 767-9), I (653).
HCl(in C_6H_6) → kinetic data and mechanism of dealkylation (653).
BF_3 → i-$PrBF_2$ (909).
$SnCl_4$(+ cat. $AlCl_3$) → i-Pr_3SnCl (705).
CCl_4(in air at 60°) → i-Pr_3SnCl + $HCCl_3$ + $Cl_3CCH_2CHClMe$ + Me_2CO + C_3H_6 (2094).
CCl_4(+ Bz_2O_2 at 70-85°) → i-Pr_3SnCl + i-Pr_3SnOBz + $Cl_3CCH_2CHClMe$ + BzOH + C_3H_6 (2094).
CrO_3(in AcOH) → kinetic data and mechanism of oxidn. (653).
Use: In catalyst for $(CH_2:CH)_2$ polymerization (1650a).

Tetrabutyltin (M-163) $SnC_{16}H_{36}$ n-Bu_4Sn

<u>Synth.</u>: By rxn. of BuMgCl with $^{113}SnCl_4$ in Et_2O (2049), with $SnCl_4$ in THF-xylene in inert atm. at 30-40°, in 68% yield (1186a), with $(BuSnO_2H)_x$ in THF, in 80% yield (2815), with Bu_3SnCl (2813).

By rxn. of BuMgBr with $SnCl_2$ in MePh-PhCl, in 98% yield (1005), with $Bu_2Sn(OAc)_2$ (2610), with $(Bu_2SnS)_x$ in THF (2813), with Bu_nSnBr_{4-n} (n = 0-3) in MePh and cat. Ph_3PO_3 in absence of ethers (1871a).

By rxn. of Mg and BuCl with $(Bu_3Sn)_2S$ in Et_2O (2813).

By rxn. of Mg and BuBr with Bu_nSnX_{4-n} (n= 0, 2, X = Cl, Br) in MePh or c.C_6H_{12} in presence of cat. Ph_3PO_3 or R_2NCHO in absence of ethers (1871a).

By rxn. of BuLi in Et_2O and/or THF in Ar atm. with $(Bu_3Sn)_2O$, $(Bu_2SnO)_x$, $(Bu_2SnS)_3$ and $Bu_2Sn(OAc)_2$, in 62-94% yield (2709), with $Bu_3SnN:C:C(CN)CHPhR$ (2789).

By rxn. of Bu_3SnOPh with K in $(Me_2N)_3PO$, followed by BuBr, in 77% yield (1635).

By rxn. of Bu_2SnCl_2 with BuCl-Et_3N at 160° in presence of Na, Mg, Al or Zn, in 34-88% yield (1939).

By rxn. of Zn-Et_3N at 160° with Bu_3SnCl in 56% yield, with $(Bu_2SnCl)_2$ and BuCl, in 80% yield, with Bu_2SnCl_2 and EtCl, in 24% yield, with Et_3SnCl and BuCl, in 2% yield (1939).

By rxn. of Bu_3Al with $SnCl_4$ in pentane-hexane in N atm., in 70% yield (2961), with Bu_nSnH_{4-n} in C_6H_{14} in N atm. at 20-100°, (n = 2, 3), in > 90% yield (1830).

By-prod. in rxn. of Bu_3SnCl with CH_2:$CHMgX$ in THF in N atm. (2147), of $(Bu_2Sn)_x$ with BuCl-Et_3N at 160°, in 5% yield (1941), of $(Bu_2Sn)_x$ with $AlBr_3$ (585).

By continuous Grignard process, in 91% yield (2994).

<u>Prop.</u>: Spectra: IR (165), (far) IR and assignment (1509), IR, Raman, and assignment (925, 1078), UV (927), NMR (1309), Moessbauer resonance (1309), and mass (1648, 2207, 2268, 2410, 2838), magnetic susceptibility (1), calcd. and exptl. data (699), effect of molecular redistribution on refractive index (1175), parachor and refrachor (1717), surface tension (1717, 1959), thermodynamic data (305), vapor pressure at 100-300° (2839). Anal. detn. by GLC (442, 536, 1072, 1079, 1718, 1939) by TLC, sepn. from Bu_nSnCl_{4-n} and $(C_8H_{17})_2SnCl_2$ (388), by chromatographic and mass spectroscopy (2838), by chromatography in presence of Bu_nSnX_{4-n}; X = Cl, Br, n = 2-3 (165), by paper chromatography (551), by degradation with HNO_3 and complexometric titrn. (168), in air (1868).

<u>Rxn. with:</u> R_4Sn(+ cat. $AlCl_3$ at 110-70°) → random redistribution, by GLC; R = Me, n-Pr (432).

X → kinetic data and mechanism of redistribution rxn., solvent effect, X = Br, (653, 768-9), I (653).

HgI_2 in 96% MeOH → Bu_3SnI + BuHgI, kinetic data and mechanism (654).

$^{113}SnCl_4$ → $Bu_3^{113}SnCl$ (2824), Bu_2SnCl_2 + $Bu_2^{113}SnCl_2$ (158).

$SnCl_4$(with $^{113}SnBu_4$) → $Bu_3^{113}SnCl$ + $BuSnCl_3$ (2049).

$SnCl_4$ (1:1 at 0°, and flash distn.) → Bu_3SnCl + $BuSnCl_3$ (1646).

$SnCl_4$(+ cat. $AlCl_3$ at 100°) → Bu_3SnCl, Bu_2SnCl_2 or $BuSnCl_3$, depending on

concn. (705).
$SnCl_4$ → heat of redistribution rxn. (380).
$SnCl_2$ → Bu_3SnCl + Bu_2SnCl_2 (1005).
$TiCl_4$ → C_4H_{10} + C_8H_{18} + R'H, effect of H_2O (544).
VCl_4 + $\overline{CH_2CH:CHCHMe}$ → rel. reactivity as polymerization catalyst (2336).
$i-C_5H_{11}X$ + AlX_3 → Bu_3SnX; X = Cl, Br (119).
CrO_3(in AcOH) → kinetics and mechanism of oxidn. (653).
$(MeCH:)_2$ (under UV irradiation) → cis- and trans-$(MeCH:)_2$ 43:57 (1369).
$MeCH:CHCO_2Et$ (+ $t-Bu_2O_2$) → $Bu_3Sn(CH_2)_4CHMeCO_2Et$ (2722).
Biol. prop.: Toxicity (618, 1069).
Activity as insecticide against Musca domestica, LD_{50} and ATP inhibition (425), Culex pipiens herbericus larvae (73), C.p. pipiens larvae (1137).
Activity as bactericide against Staphylococcus aureus and Escherichia coli (702).
Activity as fungicide (1005), against Pityrosporum ovale (2286).
Use: In catalyst for polymerization of C_2H_4 (119a, 122a, 320, 679, 1434, 2990-1, 2993), of C_3H_6 (2969), of $Me_2C:CH_2$ (2154, 2156), of $CH_2:CHCl$ (764, 1686), of $CH_2:CHCN$ (1250), with $CH_2:CHPh$ and $CH_2:CMeCO_2Me$ (1686).
In catalyst for olefin dimerization (2430).
As additive for corona discharge resistant polyimide films (2585), for flame resistant polyester (2702), for oxidation resistant polyphenyl ether (1979).
As stabilizer of PVC (2599, 2832).
Effect on spreading coefficient of solder (1692).
As solvent for synth. of Bu_6Sn_2 from Bu_3SnCl and Na (1548a).

Tetra(2,2-methylphenylpropyl)-tin $SnC_{40}H_{52}$ $(PhCMe_2CH_2)_4Sn$
Synth.: By-prod. in rxn. of $SnCl_4$ with $PhCMe_2CH_2MgCl$ in Et_2O, in 9% (1742), in $Et_2O-C_6H_6$, in 27% (2188).
Prop.: m. 96-97° (2188), m. 90-91.5° (1742), NMR (1742, 2188) and Moessbauer resonance (222, 224, 1742) spectra, mol. wt. and soly. (1742).
Rxn. with: Br in CCl_4 → R_3SnBr (1742).
Br(1:2 in CCl_4, + $NaHSO_3$, + NaOH, + HCl) → R_2SnCl_2 (2188).
HBr(in xylene) → R_3SnBr (1742).
$SnBr_4$(at 200-235°) → R_3SnBr (1742).

Tetraneohexyltin $SnC_{24}H_{52}$ $(Me_3CCH_2CH_2)_4Sn$
Synth.: By rxn. of $Me_3CCH_2CH_2MgCl$ with $SnCl_4$ in dry THF at 80°, in 80% yield (2887).
Prop.: Colorless prisms, m. 76-77°, soly. (2887).
Rxn. with: Br(in CCl_4 in the dark) → R_3SnBr (2887).
$SnCl_4$(at 200°) → R_2SnCl_2 (2887).
Removal of R_6Sn_2 impurity by rxn. with I in CCl_4, + KF as R_3SnF (2887).

Tetraoctyltin (M-166) $SnC_{32}H_{68}$ $(n-C_8H_{17})_4Sn$
Synth.: By rxn. of $SnCl_4$ with $C_8H_{17}MgBr$ in Et_2O-hydrocarbon, in > 86% yield (149), in MePh and cat. Ph_3PO_3 without ether (1871a).

By rxn. of $SnCl_4$ with $C_8H_{17}Cl$ and Mg in Et_2O-heptane and cat. $C_8H_{17}Br$, in 27-88% yield (1583a), in MePh or hexane and cat. Ph_3PO_3 in absence of ether (1871a).

Prop.: TLC (1172a).

Rxn. with: $SnCl_4$ (1:1 at 5-20°) → $C_8H_{17}SnCl_3$ (1435).
$^{113}SnCl_4$ → $(C_8H_{17})_2^{113}SnCl_2$ + $(C_8H_{17})_2SnCl2$ (158).

Biol. prop.: Activity as insecticide against Culex pipiens berbericus larvae (73) C.p. pipiens larvae (1137).

Use: As solvent in synth. of $(C_8H_{17})_6Sn_2$ from $(C_8H_{17})_3SnCl$ and Na (1548a).

Tetraphenyltin (M-166) $SnC_{24}H_{20}$ Ph_4Sn

Synth.: By rxn. of PhMgBr with $SnCl_2$ in PhMe-PhCl (1005), with $Sn(OBz)_4$ in C_6H_6 (1549), with Ph_3SnBz in Et_2O in the dark, in 16% yield (2671).
By rxn. of PhMgCl with $SnCl_4$ in THF-xylene in N atm. at 30-40°, in 64% yield (1186a).
By rxn. of $(Ph_2Sn \cdot PhMgBr)_2$ with Ph_3SnCl, in 35% yield (2330).
By rxn. of PhLi with $SnCl_2$ in Et_2O, in 47% yield (68a, 175), with $(Ph_3Sn)_2O$, Ph_3SnOH, $(Ph_3Sn)_2S$ or $(Ph_2SnO)_x$ in THF in Ar atm., in 85-99%, with $(i-PrO)_4Sn$, in 81% yield (2709), with $Ph_3SnCH_2CH:CHMe$ in Et_2O, in 98% yield (1877), with SnO_2 or SnS_2 in 1-20% yield (2709).
By rxn. of Ph_3SnLi with $SnCl_2$ (68a, 175), in THF in N atm. with BuCl and $CH_2(CH_2Br)_2$ in < 6 and 39% yield, resp. (1928), with $(Ph_2Si)_n$, n = 4 and 5 in 37 and 10% yield, resp. (1928).
By rxn. of Ph_3SnCl with Zn-Cu in THF, in 75-98% yield (2830), with $Ph_3SnZnCl$, in 88% yield (2830).
By-prod. in rxn. of $(Ph_3Sn)_4Sn$ with $(CH_2Cl)_2$ (68a, 175).
By-prod. in rxn. of $(Ph_3Sn)_2O$ with $(MeHSiO)_x$ or Ph_3SiH (1165).
By-prod. in rxn. of $PhCH_2SnPh_3$ with $LiAlH_4$ in THF at r.t. (2673).
By-prod. in thermal dec. of Ph_3SnH (123), of Ph_3SnN_3 (2667), of Ph_3SnNO_3 (568) in presence of $o-Cl_2C_6H_4$ (568, 2388), of Ph_3SnO_2SPh at 255° and 0.2-205 torr., in 33-56% yield (1410).

Prop.: Spectra: IR (207, 333, 1390) and assignment (2592), far IR and assignment (51, 194, 1703, 2784), Raman (2784), UV (335, 1390), NMR (1453), Moessbauer resonance (211, 222-4, 1094, 1114-5, 1309, 1400, 1960, 1989-91, 2621, 2665), mass (848-50, 1648) and luminiscence (2383), effect on plastic scintillator (2625), singlet-triplet transition by phosphorescence exitation (1461), configuration by depolarized Raleigh diffusion (2059), electrostriction in polar solvents (517), solubility product in MeOH and DMF (668), thermodynamic data (2024-5), thermal decomposition 352° (547), vapor pressure at 180-215° (2581). Effect of irradiation with UV (1687), with x-rays (2443), with neutrons (84), with ^{60}Co source (422, 1687) decay of $^{119}SnPh_4$ (223), retainment of radio active isotopes after (p, γ) irradiation (753). Anal. detn. by TLC (2276) and sepn. from Ph_3SnX; X = Cl, OH, OAc (1253), from Ph_nSnCl_{4-n}; n = 1-3 (2635), by Br degradation in presence of Ph_6Sn_2 (1165), by HNO_3 degradation and complexometric titrn. (168), by Na_2O_2 fusion and detn. with cacotheline (155).

Rxn. with: Ph_4Si + Ni-H → inhibition of ring hydrogenation of Ph_4Si (528).

X (in HCCl$_3$) → Ph$_3$SnX; X = I (1719), Br at -10° (1719).
PdCl$_2$ + CO(in AcOH) → BzOH (2439).
SnCl$_2$ (at 220°) → Ph$_3$SnCl (1005).
SnCl$_4$ (+ cat. AlCl$_3$) → Ph$_3$SnCl (705).
SnCl$_4$ (1:1 at 90°) → Ph$_3$SnCl + PhSnCl$_3$ (1643).
CCl$_4$ (+ cat. BiCl$_3$) → Ph$_3$SnCl + PhCl (1716).
VCl$_4$ (+ AlCl$_3$) → organometallic V(II) complex (1508).
MF$_n$ (at 115-40°) → Ph$_3$SnF; M = B, S, Si, P, As (498).
PF$_5$ + HF → C$_6$H$_6$ (498).
RLi ⇌ R$_4$Sn + PhLi; R = MeCH:CH with retention of configuration (496).
RLi ⇌ R$_4$Sn + PhSnR$_3$ + PhLi; R = CH$_2$:CMe (496).
8-HOC$_9$H$_6$N(at 220-30°) → C$_6$H$_6$ (386).
CF$_3$CO$_2$H → Sn(O$_2$CCF$_3$)$_4$ (1789).
H$_2$SO$_4$ (100%) → H$_2$Sn(HSO$_4$)$_6$ + PhSO$_3$H + HSO$_4^-$ + H$_3$O$^+$ (1100).
KNH$_2$ (in NH$_3$) → K$_2$Sn(NH$_2$)$_6$ (480).
Y' (at 280-400°) → Ph$_3$Y'; Y' = P, As, Sb (483a).
M (at 370-475°) → Ph$_2$; M = B, Al, Ga, In, Tl, Bi (483a).

Addn. compd.: Ph$_4$Sn·1.75Ph$_6$Sn$_2$; rxn. of Ph$_3$SnCl with 2-C$_5$H$_4$NMgBr in Et$_2$O-C$_6$H$_6$ or Ph$_4$Sn with Ph$_6$Sn$_2$ (1:2) in MeNO$_2$ under reflux, in 81% yield. Prop.: m. 196-97°, soly., mol. wt., IR, NMR and Moessbauer resonance spectra. Rxn. with I in HCCl$_3$ → Ph$_3$SnI + Ph$_4$Sn (1500).

Biol. prop.: Toxicity (1338).
Activity as insecticide against Musca domestica, effect on reproduction (260), LD$_{50}$ and ATP inhibition (425).
Activity as molluscicide against Australorbis glabratis, Bulinus contortus, effect on insects, crustaceans and aquatic plants, discussion of use in paints (962).

Use: In catalyst for polymerization of C$_2$H$_4$ (1507, 1638, 1994), effect on morphology of polymer (1227), of C$_3$H$_6$ (1638, 2969), of CH$_2$:CHR; R = Ph CN (1686), Cl (764, 1686), CH:CH$_2$ (1650a), of CH$_2$:CMeCO$_2$Me (1686), of olefins (184, 833).
As stabilizer for polycarbonates (264), for flame retardant polymers (1185a), for PVC, relative activity (1573) and mechanism (2832), for mineral oil and synthetic lubricants (981), for polyphenyl ether (1979), for phosphinate hydraulic fluids (1497).
As scintillator (575, 861-2).
As component in electrophotographic developer (2720).
As antifogging additive for photographic plates (591a).
Effect on spreading coeff. of solder (1692).

Additional unsubstituted symmetric tetraorganotin compounds are listed in Table 97.

Table 97. Symmetric Tetraorganotin Compounds R_4Sn

R Group*	Synth. Method**	Yield	Properties	Ref.
$PhCH_2CH_2$	--	--	(a)	--
$\overline{CH_2CH_2CH}$ (163)	II	88	$b_{0.25}$ 69-70°, n_D^{25} 1.5200 (b-d)	494a
$PhCH_2CH_2CH_2$	IA	97	$b_{1.5}$ > 290° (e)	2188
$PhCHMeCH_2$	IA	71	b_3 290°	2188
Me_2CHCH_2 (165)	--	--	GLC (f, g)	442, 1718
$MeEtCH$ (170)	IA	(h)	b_{10} 148-50°, n_D^{20} 1.4977	1618
n-C_5H_{11} (165)	--	--	(g)	--
C_5H_{11} (170)	IA	> 86	Rate of speed of sound (1679)	149
	IC	23		1583a
$MeEtCHCH_2$	--	--	(i)	--
Me_3CCH_2 (170)	--	--	NMR spectrum	2188
$c.C_5H_9$ (170)	--	--	m. 68-68.5°, subl.$_{0.05}$ 60° (j)	2268
n-C_6H_{13} (165)	IA	> 86	(g, k, l)	149
$c.C_6H_{11}$ (166)	--	--	(j, m-o)	--
$Me_2CHCH_2CMe_2CH_2$	IA	23	b_1 183°, n_D^{25} 1.4798, d_{27} 1.4798	2188
$C_{12}H_{25}$ (170)	IC	25	--	1583a
$C_{18}H_{37}$ (170)	IB (p)	--	b. 250-70° sic.	1871a
o-MeC_6H_4 (169)	IIC	45	IR spectrum, assignment (q, r)	1968
m-MeC_6H_4 (169)	--	--	(a)	--
p-MeC_6H_4 (169)	IIC	--	(q, t)	1968
	(s)	--		2388
Mesityl (170)	--	--	(a)	--
p-PhC_6H_4 (170)	--	--	(u)	--
1-$C_{10}H_7$ (171)	--	--	(a)	--
R	IV	--	R = $H(C_2H_4)_{1-4}$	1775
R	--	--	(v, w)	--

* Numbers in parenthesis refer to pages in main volume.
** The characters used here correspond to the synthetic scheme in the introduction to Chapter 1.

(a) Use in catalyst for butadiene polymerization (1650a). (b) IR spectrum (494a). (c) Rxn. with $SnBr_4$ (3:1) → R_4Sn (1873). (d) Use as herbicide and pesticide (1873). (e) Rxn. with $SnCl_4$ (3:1 at 205°) → R_3SnCl (2188). (f) Mass spectrum (2268, 2410), parachor, refrachor and surface tension (1717), TLC (2276). (g) Thermodynamic data (305). (h) As by-prod. (i) Effect of molecular redistribution on refractive index (1175). (j) Mass spectrum (2268). (k) Rxn. with air and $(RO)_2Zn$ → n-$C_6H_{13}OH$ (2149). (l) Rxn. prod. with $SnCl_4$ in DMF → use as catalyst and stabilizer for polyurethan foam (742). (m) Moessbauer resonance spectrum (222, 224). (n) Rxn. with $SnCl_4$ (at 90-100°) → R_nSnCl_{4-n} + $SnCl_2$ + RCl; n = 2,3; (in Me_2SO at > 170°) → R_3SnCl + $R_2SnCl_2 \cdot 2Me_2SO$ + $SnCl_4 \cdot 2Me_2SO$ (1385). (o) Effect on spreading coeff.

of solder (1692). (p) Grignard reagent may be prepared in situ from Mg and RBr in absence of ethers (1871a). (q) Rxn. with I (1:2 in HCCl$_3$ or CCl$_4$) → R$_3$SnI + RI ⎯⎯ (1967-8). (r) Rxn. with I (1:4) → R$_2$SnI$_2$ (1968). (s) By-prod. in thermal decomposition of R$_3$SnNO$_3$. (t) Rxn. with SnBr$_4$ (1:3 at 205°) → R$_3$SnBr (2009). (u) Rxn. with SnCl$_4$ → RSnCl$_3$ (2111). (v) R = alkyl, process of purification (516), method to produce pure tin (2987). (w) Tin deposition by microwave discharge (1859), relative reactivity as polymerization catalyst for ethylene oxide (2572).

1.4 SYMMETRIC UNSATURATED TETRAORGANOTIN COMPOUNDS

Tetravinyltin (M-171)   SnC$_8$H$_{12}$ (CH$_2$:CH)$_4$Sn
Prop.: IR (1013), NMR (743, 846, 1220), ^{13}C NMR (328) and mass (1648) spectra, GLC (442, 1718), parachor, refrachor and surface tension (1717), thermodynamic data (305).
Rxn. with: I$_3^-$ → kinetic data for cleavage rxn., solvent effect (2234).
BCl$_3$ → (CH$_2$:CH)$_2$SnCl$_2$ + CH$_2$:CHBCl$_2$ (52a).
TiCl$_4$, + H$_2$O → C$_2$H$_4$ (544).
PF$_5$ → CH$_2$:CHPF$_4$ (567).
RBuSnCl$_2$ ⇌ RBu(CH$_2$:CH)SnCl + (CH$_2$:CH)$_3$SnCl; R = Me, Et (1370).
X(CF$_2$)$_n$I (under UV irradiation, + SnCl$_4$, + Zn in ROH) → X(CF$_2$)$_n$CH$_2$CH$_2$Sn(OR)$_3$; X = H, F, Cl, n = 8, 10, 12, R = Me, Ac (745a).
BuLi → CH$_2$:CH:i (2099).
Co$_2$(CO)$_8$ → (CH$_2$:CH)$_2$Sn[Co(CO)$_4$]$_2$ (1674).
Anhydr. RCO$_2$H → (RCO$_2$)$_4$Sn (217).
Use: In catalyst for polymerization of C$_3$H$_6$ (2969), of butadiene (571, 1266).

Tetrapropenyltin (M-173)   SnC$_{12}$H$_{20}$ (MeCH:CH)$_4$Sn
Prop.: cis-Form b$_4$ 91°, n$_D^{20}$ 1.5205, IR spectrum (82).
Trans-Form b$_1$ 84°, n$_D^{20}$ 1.5110, IR spectrum (82).
Rxn. with: PhLi ⇌ Ph$_4$Sn + cis- or trans- MeCH:CHLi with retention of configuration (496).
Li(in Et$_2$O) → MeCH:CHLi with isomerization of excess (MeCH:CH)$_4$Sn (1887).
HgBr$_2$ → MeCH:CHHgBr + (MeCH:CH)$_2$SnBr$_2$ with retention (1888).

Tetraallytin (M-172)   SnC$_{12}$H$_{20}$ (CH$_2$:CHCH$_2$)$_4$Sn
Synth.: By rxn. of SnCl$_4$ with Mg and CH$_2$:CHCH$_2$Br or CH$_2$:CHCH$_2$MgBr in c.C$_6$H$_{12}$ and cat. R$_2$SO in absence of ethers (1871a).
Prop.: NMR spectrum (743) and TLC (2276).
Rxn. with: UV irradiation in i-C$_8$H$_{18}$ contn. 0.3% polymer → stable Sn-dispersion (559).
Use: In catalyst for polymerization of C$_3$H$_6$ (2969) of butadiene (2044, 2044a).

Tetraisopropenyltin (M-173)   SnC$_{12}$H$_{20}$ (CH$_2$:CMe)$_4$Sn
Synth.: By rxn. of SnBr$_4$ with CH$_2$:CMeLi in Et$_2$O in Ar atm., in 72% yield (496).

By-prod. in rxn. of Ph_4Sn with $CH_2:CMeLi$ in Et_2O in Ar atm. (490).
Prop.: $b_{0.1}$ 47-50°, n_D^{25} 1.5010, IR and NMR spectra (496).
Rxn. with PhLi $\rightleftharpoons$ $PhSn(CMe:CH_2)_3$ + $CH_2:CMeLi$ (496).

Additional unsaturated derivatives are listed in Table 98 and in Subchapter 7.1.4.

Table 98. Symmetric Unsaturated Tetraorganotin Compounds R'_4Sn

R'-group*	Synth. Method**	Yield	Properties	Ref.
HC⋮C (173)	--	--	(Far) IR spectra, assignment	1780
PhC⋮C (173)	IIIB	26	(a)	2495
MeC⋮C (173)	IIIA	73	m. 150°, IR, NMR spectra	443
HC⋮CCH₂	IB (b)	--	--	1871a
C₃H₃	IA (c)	83	--	1473
MeCH:CHCH₂	--	--	(d)	--
C₅H₅ (173)	II	53	m. 81-82° light yellow cryst. (e)	157
MeC₅H₄	IIA	83	Yellow oil, NMR spectrum, struct.	157
Me₃CC⋮C	II (f)	96	m. 187-88°, mol. wt. (g)	2664
n-PrCH:CH(CH₂)₃	IB (b)	--	--	1871a
c.C₆H₁₁C⋮C	II	44	m. 133°	206
n-C₈H₁₇CH:CH(CH₂)₇ (173)	IB (b)	--	--	1871a
n-C₈H₁₇C⋮C(CH₂)₇	IB (b)	--	--	1871a

* Numbers in parenthesis refer to pages in main volume.
** The characters used here correspond to the synthetic scheme in the introduction to Chapter 1.
(a) Rxn. with AcCl in $Et_2O \rightarrow SnCl_4 \cdot OEt_2$ + PhC⋮CAc (2644). (b) The Grignard reagent may be prepared in situ from RBr and Mg in absence of ether (1871a). (c) Synth. from SnX_4 and Grignard reagent from propyne or allene (1473). (d) Rxn. with RLi → $MeCH:CHCH_2Li$; R = Et, Bu, isomer distribution by NMR spectroscopy (1886). (e) Stable in N atm., no decomposition up to 200° (157), structure (1839). (f) Synth. in THF at -78° (2664). (g) IR, mass and NMR spectra, mol. wt., dec. above m.p. (2664).

1.6 SYMMETRIC FUNCTIONALLY SUBSTITUTED TETRAORGANOTIN COMPOUNDS

Tetrachloromethyltin (M-174) $SnC_4H_8Cl_4$ $(ClCH_2)_4Sn$
Synth.: By rxn. of $SnCl_4$ with excess CH_2N_2 in Et_2O, in 71% yield (2512).
Prop.: $b_{1.5}$ 120-20.5°, n_D^{20} 1.5777, IR, NMR (2512), ^{35}Cl NQR (2752) and mass

spectra (2512).
Rxn. with: Br in CCl$_4$ → (ClCH$_2$)$_2$SnBr$_2$ (655).

Tetra(2-cyanoethyl)tin (M-175) SnC$_{12}$H$_{16}$N$_4$ (NCCH$_2$CH$_2$)$_4$Sn
Synth.: By electrolysis of CH$_2$:CHCN on tin cathode at pH 8.5, in 44% yield (773), at pH 7-9 at -1.85 V in aq. MeCN or DMF < 30°, in 71-80% yield (1947).
Prop.: m. 110-11°, sic. (773), m. 22-24° (1947), n$_D^{22}$ 1.5330 (1747), IR (1745, 1747, 1947) and NMR (1947) spectra, soly. (1947).
Rxn. with: Br → (NCCH$_2$CH$_2$)$_3$SnBr (1745).
LiAlH$_4$ → (H$_2$NCH$_2$CH$_2$CH$_2$)$_4$Sn (1947).
SnX$_4$ (3:1) → (NCCH$_2$CH$_2$)$_3$SnX; X = Cl, Br (1747).
SnBr$_4$ (1:1) → R$_n$SnBr$_{4-n}$; n = 1-3 (1747).
SnBr$_4$ (1:3) → R$_n$SnBr$_{4-n}$; n = 1-3 (1747).
NaOH-MeOH → (NaO$_2$CCH$_2$CH$_2$)$_4$Sn (1948).
NaOH, + HCl at 0° → (HO$_2$CCH$_2$CH$_2$)$_4$Sn (1744, 1947).
NaOH, + HCl at r.t. → (HO$_2$CCH$_2$CH$_2$)$_3$SnCl + EtCO$_2$H + [(HO$_2$CCH$_2$CH$_2$)$_2$SnCH$_2$CH$_2$CO$_2$]$_x$ (1744).
NaOH, + HCl under reflux → (HO$_2$CCH$_2$CH$_2$)$_3$SnCl (2327).
Addn. compd.: (NCCH$_2$CH$_2$)$_4$Sn·SnCl$_2$: from components at r.t. in CH$_2$Cl$_2$ in all ratios, in 99% yield at (1:1), m. 122-24° (sealed tube), IR spectrum (1747) [Sn$_2$C$_{12}$H$_{16}$Cl$_4$N$_4$].
(NCCH$_2$CH$_2$)$_4$Sn·SnBr$_4$: as above, in 98% yield, m. 119-21° (sealed tube), IR spectrum (1747).

Tetra(2-carboxylethyl)tin SnC$_{12}$H$_{20}$O$_8$ (HO$_2$CCH$_2$CH$_2$)$_4$Sn
Synth.: By rxn. of (NCCH$_2$CH$_2$)$_4$Sn with NaOH-MeOH at reflux, followed by concn. HCl at 0°, in 77% yield (1744, 1947-8).
Prop.: Cream solid, m. 112-13° (1947), m. 104-105° (1744), stable at 0-5° for several months, soly. (1744), IR (1744, 1947) and NMR (1947) spectra, thermal gravimetric and differential thermal analysis (1744).
Rxn. with: Thermal dec. (at 60° in aq. soln.) → [(HO$_2$CCH$_2$CH$_2$)$_2$SnCH$_2$CH$_2$CO$_2$]$_x$ + EtCO$_2$H (1744), (at reflux) → [(O$_2$CCH$_2$CH$_2$)$_2$Sn]$_x$ (1744).
SOCl$_2$ → (ClOCCH$_2$CH$_2$)$_4$Sn (1947).
CH$_2$N$_2$ (in THF) → (MeO$_2$CCH$_2$CH$_2$)$_4$Sn (1947).

Tetrakispentafluorphenyltin (M-174) SnC$_{24}$F$_2$ (C$_6$F$_5$)$_4$Sn
Synth.: By rxn. of SnBr$_4$ with C$_6$F$_5$MgBr in Et$_2$O under reflux for three days, in 51% yield (77).
By rxn. of SnCl$_4$ with C$_6$F$_5$Li in Et$_2$O, in 91% yield (547).
By rxn. of Sn with (C$_6$F$_5$)$_2$Hg at elevated temperature, in 60% yield (64), with C$_6$F$_5$I in sealed tube at 240° (885).
By thermal dec. of (C$_6$F$_5$)$_3$SnOH at 150° (1198).
Prop.: m. 221° (77), 220-22° (547), 220° (885), 218-19° (64, 1198), stable for 2 hours at 400°, subl. 0.001 160° (77), IR (77) and assignment (547), UV, NMR (77), ^{19}F NMR (2475) and Moessbauer resonance (211, 1989) spectra, GLC (547), thermal dec. temperature (T$_D$) 399° (547), x-ray powder diagram (1198), stable towards H$_2$O, Cl at 20° and 8 atm., UV irradiation, HBr and Br at 20°,

BF$_3$ at 220°, HgCl$_2$ in MeOH and SnCl$_4$ at 150° (1198).

Rxn. with: Anhydr. HCl → C$_6$F$_5$H (77).
6N aq. HCl (at reflux) → dec. after 6 hours (547).
Br in (CH$_2$Br)$_2$ + cat. AlCl$_3$ → 85% recovery of starting matl.(547).
SnCl$_4$ (at 160°) → (C$_6$F$_5$)$_n$SnCl$_{4-n}$; n = 1, 3 (1198).
SF$_4$ → (C$_6$F$_5$)$_3$SnF (498).
Li (in THF) → no rxn. (547).
10% NaOH → dec. (547).
8-HOC$_9$H$_6$N in EtOH → (C$_6$F$_5$)$_2$Sn(OC$_9$H$_6$N-8)$_2$ (1198).

Other functionally substituted symmetric tetraorganotin compounds are listed in Table 99 and in Chapter 8.1.

Table 99. Symmetric Functionally Substituted Tetraorganotin R'$_4$Sn Compounds

R'-group*	Synth. Method**	Yield	Properties	Ref.
MeOCH$_2$	II	(a)	b$_{0.05}$ 64°, NMR spectrum	1832
p-ClC$_6$H$_4$C⋮C	II	57	m. 161°	204
p-BrC$_6$H$_4$C⋮C	II	53	m. 170°	204
Me$_3$SiC⋮C	IA (b)	--	Dec. 140°	1345
	IIA	38	Dec. 140°	2775
ROCCH$_2$CH$_2$				
R = NaO	VC	--	m. > 300° (c, d)	1948
MeO	VA	--	(e)	1947
EtO	(f)	--	Light yellow liq. (e)	1947
OCH$_2$CHCH$_2$O	(d)	56	(g)	1948
Cl	VB	--	Tan liq., IR spectrum (f, h)	1947
H$_2$N	(h)	--	Hydroscopic solid (e)	1947
H$_2$N(CH$_2$)$_3$	VD	--	Yellow vis. liq., IR spectrum (i)	1947
B$_{10}$H$_{10}$C$_2$HCH$_2$	IA	82	m. > 350° (j)	2878
p-FC$_6$H$_4$ (174)	IA	85	m. 219-21° (k)	428
p-ClC$_6$H$_4$ (174)	--	--	(1-1309)	--
Cl$_5$C$_6$	IB (m)	4	Light brown, m. 446-49° (dec) (n)	1101
m-F$_3$CC$_6$H$_4$ (174)	--	--	(o)	--
p-F$_3$CC$_6$H$_4$	--	--	(o)	--
p-PhOC$_6$H$_4$	II (p)	7	m. 234-35°, mol. wt. (q, r)	1378
p-C$_6$F$_5$OC$_6$F$_4$	II	--	m. 255-57°	2666

* Numbers in parenthesis refer to pages in main volume.
** The characters used here correspond to the synthetic scheme in the introduction to Chapter 1.
(a) Synthesis in methylal, in low yield, from rxn. prod. of Sn with MeOCH$_2$Cl,

followed by MeOCH₂Li, in good yield (1832). (b) Synthesis in THF (1345). (c) Rxn. with HCl → (HO₂CCH₂CH₂)₄Sn (1948). (d) Rxn. of (NaO₂CCH₂CH₂)₄Sn with ŌCH₂CHCH₂Cl → (ŌCH₂CHCH₂O₂CCH₂CH₂)₄Sn (1948). (e) IR and NMR spectra (1947). (f) Rxn. of (ClOCCH₂CH₂)₄Sn with EtOH-Py at r.t. → (EtO₂CCH₂CH₂)₄Sn (1947). (g) Rxn. with hexahydrophthalic anhydride and catalytic PhCH₂NMe₂ → tin containing polymer (1948). (h) Rxn. of (ClOCCH₂CH₂)₄Sn with aq. NH₃ → (H₂NOC-CH₂CH₂)₄Sn (1947). (i) Yielded picrate in EtOH → 4ArOH·R₄Sn, m. 156-57° [SnC₃₆H₄₄N₁₆O₂₈] (1947). (j) R₄Sn·2THF m. > 350°, looses THF at 1 Torr. and 130° [SnB₄₀C₂₀H₆₈O₂] (2874). (k) ¹H, ¹⁹F NMR (1453) and Moessbauer resonance (222) spectra. (l) Moessbauer resonance spectrum (1309), rxn. with SnCl₄ → RSnCl₃ (2111). (m) In THF-C₆H₁₄-xylene. (n) Mol. wt., mass spectrum (1101). (o) Moessbauer resonance spectrum (222, 224). (p) RLi from Ph₂O and BuLi in Et₂O-THF-C₆H₁₄. (q) (Far) IR and UV spectra, soly. (1378). (r) Use as antioxidant for polyphenyl ethers (1979).

2. UNSYMMETRIC TETRAORGANOTIN COMPOUNDS

2.1 NONFUNCTIONAL UNSUBSTITUTED TETRAORGANOTIN COMPOUNDS

The nonfunctional unsymmetric tetraorganotin compounds have two to four alkyl, aryl or aralkyl groups linked to each tin atom. The compounds listed in Tables 100 and 101 are prepared by methods from the ensuing scheme.

I Grignard reaction with (organo)tin halides:
  (IA) R$_n$SnX$_{4-n}$ + R'MgX(in THF) → R$_n$SnR'$_{4-n}$; n = 2, 3 ,
  (IB) SnCl$_4$ + RMgI + R'MgI → R$_n$SnR'$_{4-n}$; n = 0-4 ,
  (IC) (Ph$_2$Sn)$_x$ + RI + MeMgI → Ph$_n$R$_m$SnMe$_p$; n + m + p = 4 ,
  (ID) (Ph$_2$Sn)$_x$ + R$_2$SnI$_2$ + R'MgI → Ph$_n$R$_m$SnR'$_p$; n + m + p = 4 .

II Reactions with lithium derivatives:
  (IIA) R$_n$SnX$_{4-n}$ + R'Li → R$_n$SnR'$_{4-n}$; n = 2, 3 ,
  (IIB) Ph$_3$SnLi + RX → Ph$_3$SnR .

III Other organometallic reactions:
  (IIIA) R$_3$SnBr + Zn + R'Cl → R$_3$SnR' ,
  (IIIB) R$_n$SnH$_{4-n}$ + R'$_3$Al → R$_n$SnR'$_{4-n}$; n = 2, 3 .

IV Hydrostannation reactions:
  (IVA) R$_3$SnH + R$_2$C:CHR $\xrightarrow{UV}$ R$_3$SnCHRCHR$_2$,
  (IVB) R$_3$SnH + R$_2$C:CHR + cat. R$_3$Al, R$_2$AlH → R$_3$SnCHRCHR$_2$.

V Redistribution reactions:
  (VA) R$_4$Sn + R'$_4$Sn + cat. AlCl$_3$ → R$_n$SnR'$_{4-n}$,
  (VB) R$_4$Sn + Et$_4$Pb + cat. AlCl$_3$ → R$_n$SnEt$_{4-n}$.

VI Miscellaneous reactions:
  (VIA) Me$_3$SnC:CPh + PhBr + α-pyrone → o-Ph$_3$SnC$_6$H$_4$Ph .
  (VIB) Ph$_3$SnC:CMe + H + Raney Ni → PrSnPh$_3$.

2.1.1 DERIVATIVES WITH TWO DIFFERENT ORGANIC GROUPS

Benzyltrimethyltin (M-179) $SnC_{10}H_{16}$ Me_3SnCH_2Ph
Synth.: By rxn. of Me_3SnBr with $PhCH_2Cl$ and Mg (1210).
Prop.: UV (1570), NMR (1938, 2267) and mass (621, 2869) spectra, dipole moment (1210, 1570).
Rxn. with: K or Na in $(MeOCH_2)_2$ → ESR spectrum of prod. (524).
Br in PhCl under UV irradiation → relative kinetic data and mechanism of cleavage rxn. (768).

Ethyltrimethyltin (M-179) SnC_5H_{14} Me_3SnEt
Synth.: By rxn. of Me_3SnCl with $(Et_2AlCl)_2$, in 86% yield (1206).
By-prod. in rxn. of $SnCl_4$ with MeMgI and EtMgI (430).
By-prod. in x-ray irradiation of Me_2SnEt_2 (1191).
Prop.: b. 107°, n_D^{25} 1.453 (1206), d_{25} 1.251 (1717), NMR (1220, 1938, 2267), Moessbauer resonance (1309) and mass (621, 1173, 1191, 2268, 2528, 2869) spectra, GLC (431, 442, 1191, 1718), parachor, refrachor and surface tension (1717), thermodynamic data (305).
Rxn. with: X → kinetic data and mechanism of cleavage rxn., solvent effect; X = Br (767-9), I (767), effect of other R_4Sn (768-9), of UV irradiation (768).

Propyltrimethyltin SnC_6H_{16} Me_3SnPr
Synth.: By-prod. in rxn. of $SnCl_4$ with MeMgI and PrMgI (430).
Prop.: d_{25} 1.206 (1717), NMR (1220, 2267) and mass (621, 2268, 2528, 2869) spectra, GLC (429, 442, 1718), parachor, refrachor and surface tension (1717), thermodynamic data (305).
Rxn. with: X → kinetic data and mechanism of cleavage rxn., solvent effect; X = Br (767-9), I (767), effect of R_4Sn (767, 769), of UV irradiation (768).
$PhHgCBrCl_2$ → $Me_3SnCH_2CHMeCHCl_2$ + PhHgBr (1889).

Isopropyltrimethyltin SnC_6H_{16} $Me_3SnCHMe_2$
Synth.: By rxn. of Me_3SnCl with i-PrK in petr. ether, in 53% yield (150).
Prop.: Colorless liq., b. 120° (2267), d_{25} 1.198 (1717), NMR (1220, 2267) and mass (2268, 2869) spectra, GLC (442, 1718), parachor, refrachor and surface tension (1717), thermodynamic data (305).
Rxn. with: X → kinetic data and mechanism of cleavage rxn., solvent effect; X = Br (767-9), I (767), effect of other R_4Sn (768-9), of UV irradiation (768).

Butyltrimethyltin (M-179) SnC_7H_{18} Me_3SnBu
Synth.: By rxn. of BuLi with $(Me_3Sn)_2CCl_2$, in 78% yield (1879), with $Me_3SnCH:CHMe$ in Et_2O and Ar atm. (496), with $Me_3SnCHCl_2$ (1879).
Prop.: NMR (1938, 2267) and mass (2268, 2869) spectra, GLC (429).
Rxn. with: X → kinetic data and mechanism of cleavage rxn., solvent effect; X = Br (767-9), I (767); effect of R_4Sn (768-9) of UV irradiation (768).

Tertiarybutyltrimethyltin SnC_7H_{18} Me_3SnCMe_3
Synth.: By rxn. of Me_3SnX with Me_3CMgCl in Et_2O (2267).

By rxn. of Me$_3$SnCHCl$_2$ with Me$_3$CLi (1879).
Prop.: White solid, m. 42-43°, subl.$_{0.1}$ 25° (2267), NMR (1220, 2267) and mass (2268, 2869) spectra, GLC (442, 1718), thermodynamic data (886).
Rxn. with: X → kinetic data and mechanism of cleavage rxn., solvent effect; X = Br, I (767).

1,2,2-Methyldiphenylcyclopropyl-trimethyltin SnC$_{18}$H$_{24}$ Me$_3$SnC̄MeCPh$_2$C̄H$_2$
Synth.: By rxn. of Me$_3$SnLi with C̄H$_2$CPh$_2$CMeBr in THF in N atm. at 0° from (+)- and (-)-form, in 44 and 61% yield, with retention of configuration, resp. (1942).
Prop.: (+)-(S)-form $[\alpha]_D^{21}$ + 11.6°, n_D^{21} 1.5742, NMR spectrum and GLC, (-)-(R)-form $[\alpha]_D^{19}$ - 16.1°, NMR spectrum and GLC (1942).
Rxn. with: HCl → optically active C̄H$_2$CPh$_2$C̄HMe (1942).
I in CCl$_4$ → IC̄MeCPh$_2$C̄H$_2$ - racemate (1942).

Cyclohexyltrimethyltin SnC$_9$H$_{20}$ c.C$_6$H$_{11}$SnMe$_3$
Synth.: By rxn. of Me$_3$SnH with C$_6$H$_{10}$ in sealed tube in Ar atm. at 10-15° under UV irradiation, in 49% yield (1953).
Prop.: b$_{10}$ 74-76°, n_D^{20} 1.4937 (1953), d$_{25}$ 1.216 (1717), NMR (1220) and mass (2268) spectra, GLC (442, 1718), parachor, refrachor and surface tension (1717).
Rxn. with: Br(in MeOH at 0°) → Me$_2$(c.C$_6$H$_{11}$)SnBr + MeBr (2269).

Phenyltrimethyltin (M-177) SnC$_9$H$_{14}$ Me$_3$SnPh
Synth.: By rxn. of (Ph$_2$Sn·PhMgBr)$_x$ with MeI at 140°, followed by MeMgI, in 60% yield (2330).
By-prod. in rxn. of (Ph$_2$Sn)$_x$ (from Ph$_3$SnZnCl and Ph$_3$SnCl in THF) with I, followed by MeMgI, in 13% yield (2830).
By-prod. in rxn. of SnCl$_4$ with MeMgI and PhMgI (430).
Prop.: b. 198° (997), n_D^{25} 1.5365 (1561), 1.5329 (997), d$_{25}$ 1.326 (1717), UV (1561, 1570) NMR (1220, 1938, 2267, 2476, 2830), Moessbauer resonance (902) and mass (621, 2411, 2869) spectra, dipole moment (1210, 1570), GLC (442, 1718), parachor, refrachor and surface tension (1717).
Rxn. with: HCl in MeOH → rel. rate of Ph-cleavage (62).
I in ROH → rel. kinetic data for Ph-cleavage (59, 61, 2631-2).
Aq. alc. KOH → rel. kinetic data for Ph-cleavage (997).
K or Na in (MeOCH$_2$)$_2$ → ESR spectrum of prod. (524).
SO$_3$(in CH$_2$Cl$_2$ at -78°) → Me$_3$SnO$_3$SPh (2742).

Paratolyltrimethyltin (M-180) SnC$_{10}$H$_{16}$ p-Me$_3$SnC$_6$H$_4$Me
Prop.: b$_{1.2}$ 69° (296), b$_{0.1}$ 58-60° (997), n_D^{25} 1.5294 (997), UV (930), NMR (296, 2267, 2356, 2848) and mass (2411) spectra.
Rxn. with: Alc. I → relative rate of cleavage rxn. (61, 2632).
Aq. alc. KOH → relative rate of cleavage rxn. (997).
Alc. HCl → relative rate of cleavage rxn. (62).

Diethyldimethyltin (M-177) SnC_6H_{16} Me_2SnEt_2

<u>Synth.:</u> By rxn. of $SnCl_4$ with MeMgI and EtMgI, in 40-60% yield (430).

<u>Prop.:</u> NMR (1938) and mass (1173, 1191) spectra, GLC (431, 1191), thermodynamic data (305).

<u>Rxn. with:</u> Me_2SnCl_2 (in MeOH at 25°) → Me_3SnCl + $MeEt_2SnCl$; kinetic data and rxn. mechanism (2686).

X-ray irradiation → Me_nSnEt_{4-n}, + C_nH_{2n+2}, n = 1-4, C_2H_4 + H (1191).

Dimethyldiphenyltin $SnC_{14}H_{16}$ Me_2SnPh_2

<u>Synth.:</u> By rxn. of Ph_2SnCl_2 with MeMgBr in Et_2O (296).

By rxn. of $(Ph_2Sn·PhMgBr)_2$ with MeI at 140°, followed by MeMgI, in 40% yield (2330).

By rxn. of $(Ph_2Sn)_x$, from $Ph_3SnZnCl$ with Ph_3SnCl in THF, with I, followed by MeMgI, in 71% yield (2830).

By-prod. in rxn. of $(Ph_2Sn)_x$ at 140° in sealed tube with RX, followed by MeMgI; R = Et, Pr, Bu, X = Br, I, in 1-5% yield (1938), with n-Bu_2SnI_2, followed by MeMgI, in 5% yield (1938).

<u>Prop.:</u> NMR spectrum (1220, 1938, 2830), GLC (1938).

Methyltriethyltin (M-181) SnC_7H_{18} $MeSnEt_3$

<u>Synth.:</u> By-prod. in rxn. of $SnCl_4$ with MeMgI and EtMgI (430).

By-prod. in rxn. of Me_6Sn_2 with Et_3SnCl (2012).

By-prod. in x-ray cleavage of Me_2SnEt_2 (1191).

<u>Prop.:</u> IR and assignment (1313), NMR (1938), Moessbauer resonance (1309) and mass (1173, 1191, 2268) spectra, GLC (431, 1191), thermodynamic data (305).

<u>Rxn. with:</u> X → kinetic data and mechanism of cleavage rxn., solvent effect; X = Br (767-8, 1022), I (767, 1022).

$Me_3SnCl \rightleftharpoons Et_3SnCl + Me_4Sn$ (2012).

Me_2SnCl_2 → Me_3SnCl + Et_3SnCl, kinetic data and rxn. mechanism (2686).

Methyltriphenyltin (M-181) $SnC_{19}H_{18}$ $MeSnPh_3$

<u>Synth.:</u> By rxn. of Ph_3SnCl with MeMgBr in Et_2O (296), with MeMgI, in 93% yield (2412).

By rxn. of Ph_3SnLi with Me_3PO_4 in THF, in N atm., in 81% yield (1928).

By-prod. in rxn. of $(Ph_2Sn)_x$, from $Ph_3SnZnCl$ and Ph_3SnCl in THF, with I, followed by MeMgI, in 17% yield (2830).

By-prod. in rxn. of $(Ph_2Sn)_x$ in sealed tube at 140° with MeI, followed by BuMgI, in 18% yield (1938), with n-Bu_2SnI_2, followed by MeMgI, in 6% yield (1938).

<u>Prop.:</u> m. 62.5-63° (2412), NMR (1938, 2830) and mass (849, 2412) spectra, GLC (1938).

<u>Rxn. with:</u> I in Et_2O → $MePh_2SnI$ + PhI (521), in $HCCl_3$ (1938).

I in $HCCl_3$ → $MePhSnI_2$ (1719).

I in $HCCl_3$, + NaOH, + HCl → $MePh_2SnCl$ + Ph_3SnCl (68a, 176).

Br in EtOH or EtOH-PhCl → $MePh_2SnBr$ + $MePhSnBr_2$ (2412).

HCl in dry MeOH → $MePh_2SnCl$ (2412).

<u>Use:</u> As antifogging additive for photographic plates (591a).

Butyltriethyltin (M-182)　　　　　SnC$_{10}$H$_{24}$　　　　　Et$_3$SnBu-n

Synth.: By rxn. of (Et$_3$Sn)$_2$S with BuMgX or BuX and Mg (2813).
By rxn. of Et$_3$SnCl with Na in BuCl at 160°, in 35% yield (1939).
By rxn. of Et$_3$SnH with Bu$_3$Al in N atm. at 80°, in C$_6$H$_{12}$ (1830).
By-prod. in rxn. of Et$_3$SnCl with Zn in BuCl-Et$_3$N at 160°, in 20% yield (1939).
By-prod. in rxn. of Bu$_2$SnCl$_2$ with Zn in EtCl-Et$_3$N at 160°, in 5% yield (1939).
By-prod. in rxn. of residue from (Et$_3$Sn)$_2$O and EtMgBr, after distn. of Et$_4$Sn and Et$_3$SnBr, with BuMgBr (1906).
By-prod. in rxn. of (Et$_2$Sn)$_x$ with BuCl at 160° in presence of Et$_3$N, in 7% yield (1941).
Prop.: b$_{3.5}$ 68-71°, n$_D^{20}$ 1.4730 (1906), mass spectrum (2268) GLC (1939).
Rxn. with: X → kinetic data and mechanism of cleavage rxn., solvent effect; X = Br (767-8), I (767).

Phenyltriethyltin (M-178)　　　　　SnC$_{12}$H$_{20}$　　　　　Et$_3$SnPh

Prop.: NMR (1938) and mass (848, 850) spectra, retarding effect on hydrogenation of isoprene with Raney Ni (2316).
Rxn. with: HCl → mechanism of Ph-cleavage, deuterium effect (43).
I → EtPhSnI$_2$ (1410).
Hg(OAc)$_2$ in THF → Et$_3$SnOAc + PhHgOAc; relative rate and rxn. mechanism (1162).

Dibutyldiethyltin (M-183)　　　　　SnC$_{12}$H$_{28}$　　　　　n-Bu$_2$SnEt$_2$

Synth.: By rxn. of (Bu$_2$SnO) with EtBr-EtCl and Mg in THF, in 92% yield (2814).
By rxn. of (Et$_2$SnS)$_x$ with BuMgX or Mg and BuX in Et$_2$O (2813).
By-prod. in rxn. of Et$_3$SnCl with Zn in BuCl-Et$_3$N at 160°, in 4% yield (1939).
By-prod. in rxn. of Bu$_2$SnCl$_2$ with Zn in EtCl-Et$_3$N at 160°, in 4% yield (1939).
Prop.: n$_D^{20}$ 1.4730 (2814), thermodynamic data (305). Anal. detn. by GLC (1939) in Bu$_4$Sn from Grignard synthesis by mass spectrum and chromatographic method (2838).

Ethyltributyltin (M-184)　　　　　SnC$_{14}$H$_{32}$　　　　　EtSnBu$_3$

Synth.: By rxn. of n-Bu$_3$SnH with Et$_3$Al in N atm. in C$_6$H$_{12}$ at 80° (1830).
By-prod. in rxn. of Et$_3$SnCl with Zn in BuCl-Et$_3$N at 160°, in 3% yield (1939).
By-prod. in rxn. of Bu$_3$SnCl$_2$ with Zn in EtCl-Et$_3$N at 160°, in 20% yield (1939).
Prop.: GLC (1939). Anal. detn. in n-Bu$_4$Sn from Grignard rxn. by mass spectroscopy and chromatographic method (2838).
Rxn. with: Br in RCl → EtBr + BuBr, ratio of alkyl bromides; R = Ph, CCl$_3$ (2266).
Br in CCl$_4$ → EtBr; kinetic data and rxn. mechanism (1022).

Ethyltriphenyltin (M-184)　　　　　SnC$_{20}$H$_{20}$　　　　　EtSnPh$_3$

Synth.: By-prod. in rxn. of (Ph$_2$Sn)$_x$ with EtX at 140°, in sealed tube, followed by MeMgI; X = Br and I, in 14 and 29% yield, resp. (1938).
By-prod. in air oxidn. of Ph$_6$Sn$_2$ in presence of MeONa and EtBr (2046).
Prop.: NMR (1938, 2068) and mass (848-50) spectra, GLC (1938).
Rxn. with: I in Et$_2$O → EtPh$_2$SnI + PhI (521).

Phenyltriisopropyltin (M-185) $SnC_{15}H_{26}$ $(Me_2CH)_3SnPh$

Synth.: By rxn. of i-Pr_3SnX with PhLi; X = Cl, in 42% yield (998), Br, in 33% yield (59).

Prop.: b_1 113-14° (59), $b_{0.3}$ 82° (998), n_D^{25} 1.5302 (998), n_D^{20} 1.5377 (59).

Rxn. with: I → kinetic data and mechanism of cleavage rxn. (59), solvent effect (2631-2).

HCl in MeOH → kinetic data for cleavage rxn., relative reactivity (62).

Butyltriphenyltin (M-186) $SnC_{22}H_{24}$ n-$BuSnPh_3$

Synth.: By rxn. of BuLi with $(Ph_3Sn)_2O$ in Et_2O in N atm. (1840), with $(Ph_3Sn)_nPPh_{3-n}$; n = 2, 3 (1850).

By rxn. of Ph_3SnLi with BuCl in THF, in 45% yield (1928), with Bu_3PO_4 in 45% yield (1928).

By-prod. in rxn. of $(Ph_2Sn)_x$ with BuX in sealed tube at 140°, followed by MeMgI; X = Br and I, in 20 and 34% yield, resp. (1938).

By-prod. in air-oxidn. of Ph_6Sn_2 in BuBr-THF in presence of MeONa (2046).

Prop.: m. 56-63° (2046), 56-58° (1928), GLC (1938).

Biol. prop.: Control and effect on reproduction of Musca domestica (260).

Additional unsymmetric tetraorganotin compounds with two different organic groups are listed in Table 100.

Table 100. Nonfunctional Unsymmetric Tetraorganotin Compounds with Two Different Organic Groups $R_nSnR'_{4-n}$

Formula*	Synth. Method**	Yield	Properties	Ref.
Me₃SnR				
R = o-MeC₆H₄CH₂	--	--	NMR (2267, 2356), mass (2268) spectra	--
m-MeC₆H₄CH₂	--	--	NMR (2267), mass (2268) spectra	--
p-MeC₆H₄CH₂	--	--	NMR (2267), mass (2268) spectra	--
$\overline{CH_2CH_2CH}$ (179)	IVA	96	Use as herbicide, pesticide	1873
Me₂CHCH₂ (179)	I, RMgBr	64	b_{37} 58°, n_D^{20} 1.4552 (a, b)	1953
MeEtCH	IVA, (MeCH:)₂	15	b. 145.5-46°, n_D^{20} 1.4614, IR and NMR spectra, GLC (a)	1369
	IVA, (MeCH:)₂	23	b. 146-48°, n_D^{20} 1.4630, NMR spectrum	1369
$\overline{CH_2CHMeCH}$	IA, RMgX	--	Isomers identified by NMR spectrum (c, d)	1953
n-C₅H₁₁ (179)	IV, 1-C₅H₁₀	--	GLC	698
PrCHMe	IVA, MeCH:CHEt	18	b_{25} 69°, n_D^{20} 1.4669 (e)	429
Et₂CH	IVA, MeCH:CHEt	15	(e)	1953
Me₂CHCHMe	IVA, Me₂C:CHMe	51	b_{30} 71-72°, n_D^{20} 1.4669, GLC (f)	1953
Me₂EtC	I, RMgCl	--	m. 46-48°, b. 156-58°, GLC	1953
c.C₅H₉	IVA, c.C₅H₈	76	b_{25} 77°, n_D^{20} 1.4884	1953
n-C₆H₁₃	IV, 1-C₆H₁₂	--	GLC	429
Et₂CHCH₂	IVA, Et₂C:CH₂	97	b. 182-84°, n_D^{20} 1.4654 (f)	1953
Me₂CHCHMeCH₂	IVA, i-PrCMe:CH₂	98	b_{22} 80°, n_D^{20} 1.4649 (f)	1953

* Numbers in parenthesis refer to pages in main volume.
** The characters used here correspond to the synthetic scheme in the introduction to Chapter 2.1.
(a) Mass spectrum (2869). (b) IR (475) and NMR (475, 1953) spectra, colorless liq. unpleasant odor, toxic (?) (475). (c) Rxn. with Br → relative kinetic data and mechanism (768). (d) Rxn. with X → $\overline{CH_2CHMeCHX}$ under retention of configuration, mechanism; X = Br, I (698). (e) Mixture of R = PrCHMe and Et₂CH (1953). (f) NMR spectrum (1953).

Table 100 Continued

$R_nSnR'_{4-n}$

Formula*	Synth. Method**	Yield	Properties	Ref.
Me_3SnR (Cont'd.)				
R = n-C_7H_{15}	IV, 1-C_7H_{14}	--	GLC	429
c.1-MeC$_6$H$_{10}$	IVA, c.1-MeC$_6$H$_8$	61	b_{10} 88-89°, n_D^{20} 1.5003, GLC, struct. by (f)	1953
c.C_7H_{13}	IVA, c.C_7H_{12}	26	b_{10} 95-96°, n_D^{20} 1.5018	1953
c.C_7H_9 (g)	IVA, C_7H_8	(h)	IR spectrum	2052
n-C_8H_{17} (179)	IV, 1-C_8H_{16}	--	GLC	429
n-C_9H_{19}	IV, 1-C_9H_{18}	--	GLC	429
o-MeC$_6$H$_4$	I, RMgBr	48	b_7 90°, n_D^{25} 1.5379, mass (2411) spectrum (i-k)	998
m-MeC$_6$H$_4$ (179)	--	--	$b_{0.9}$ 55°, n_D^{25} 1.5309 (1, j, l)	997
2,3-Me$_2$C$_6$H$_3$	I, RMgBr	32	$b_{0.8}$ 78°, n_D^{25} 1.5419	998
2,4-Me$_2$C$_6$H$_3$	I, RMgBr	56	$b_{2.5}$ 84°, n_D^{25} 1.5363, NMR (2267) spectrum (k)	998
2,6-Me$_2$C$_6$H$_3$	--	--	NMR (2267) and mass (2411) spectra (k)	--
2,4,6-Me$_3$C$_6$H$_2$	IIA, RLi	68	m. 35-36°, $b_{0.1}$ 65-66°, NMR spectrum	296
3,4,5-Me$_3$C$_6$H$_2$	I, RMgBr	47	m. 29-30°	998
p-Me$_3$CC$_6$H$_4$	IIA, RLi	--	m. ~25°, $b_{0.1}$ 80-81°, NMR spectrum	296
p-Me$_3$CCH$_2$C$_6$H$_4$	I, RMgBr	92	$b_{0.1}$ 74°, NMR (2267) spectrum (j)	998
o-PhC$_6$H$_4$	VIA	--	NMR spectrum	2267
m-PhC$_6$H$_4$	IIA, RLi	29	m. 57-58°, $b_{0.06}$ 86-92° (m)	1021
p-PhC$_6$H$_4$ (180)	IIA, RLi	76	$b_{0.2}$ 107°, n_D^{25} 1.5960 (k)	998
1-$C_{10}H_7$ (180)		11	m. 53°, NMR spectrum, polarography (k, n)	112
2-$C_{10}H_7$ (180)	--	--	NMR (2267), mass (2411) spectra (j, l, 59)	--
9-$C_{14}H_9$ (180)	--	--	NMR (2267), mass (2411) spectra (l, 59)	--
	--	--	NMR (2267), mass (2411) spectra (l)	--
Me_2SnR_2				
R = PhCH$_2$	I, MeMgBr	42	m. 26°, b_1 132-35°, NMR (1938) spectrum (o)	841
o-MeC$_6$H$_4$CH$_2$	I, MeMgBr	54	$b_{0.4}$ 150-52°, n_D^{29} 1.5830 (o)	841

Pr	IB	(h)	--	430
Bu (180)	ID	35	NMR (1220, 1938) spectrum	1938
	IIA, Bu$_2$SnO	60	--	2708
n-C$_8$H$_{17}$ (180)	--	--	(p)	--
n-C$_{10}$H$_{21}$	--	--	(p)	--
n-C$_{12}$H$_{25}$	--	--	(p)	--
p-MeC$_6$H$_4$ (180)	IIA, RLi	--	b$_{0.1}$ 110-12°, NMR spectrum	296
2,4,6-Me$_3$C$_6$H$_2$	IIA, RLi	--	m. 101°, b$_{0.1}$ 170-72°, NMR spectrum	296
3,4,5-Me$_3$C$_6$H$_2$	IIA, RLi	--	m. 56°, NMR spectrum	296
p-PhC$_6$H$_4$ (180)	IIA, Me$_3$SnCl	15	m. 173-75°	112
MeSnR$_3$				
R = PhCH$_2$	--	--	NMR spectrum	1938
Pr (181)	IB	(h)	(q, r-2685)	430
Me$_2$CH	--	--	(q, r-2685)	--
Bu (181)	IIA, (Bu$_3$Sn)$_2$O	80	NMR spectrum (1938) (q, r-2680)	2709
p-MeC$_6$H$_4$	I, MeMgBr	--	m. 69°, NMR spectrum	296
2,4,6-Me$_3$C$_6$H$_2$ (181)	IIA, RLi	--	m. 155°, NMR spectrum (s, t)	296
3,4,5-Me$_3$C$_6$H$_2$	IIA, RLi	--	m. 133°, NMR spectrum	296

(g) From bicyclo(2.2.1)heptadiene, proposed structure tricyclo(2.2.1.0^{2,6})hept-3-yltrimethyltin (2052). (h) By-prod. (i) NMR spectrum (2267, 2356). (j) Relative rate of aryl-group cleavage with aq. alc. KOH (997). (k) Relative rate of aryl-group cleavage with I in MeOH at 20° (2632). (l) Relative rate of aryl cleavage with alc. HCl (62), with alc. I (61). (m) IR, UV and NMR spectra (1021), rxn. with I in HCCl$_3$ → Me$_3$SnI + o-IC$_6$H$_4$Ph (1021). (n) Rxn. with Na-K in THF → cleavage of Me$_3$Sn-group, ESR spectrum of prod. (112). (o) (Far) IR spectra and assignment (841). (p) Rxn. with SnCl$_4$ → Me$_2$SnCl$_2$ + R$_2$SnCl$_2$ + MeRSnCl$_2$; prod. depending on rxn. cond. (1482). (q) Rxn. with X → kinetic data and mechanism of cleavage rxn., solvent effect; X = Br, I (1022). (r) Rxn. with Me$_2$SnCl$_2$ → Me$_3$SnCl + R$_3$SnCl; kinetic data and rxn. mechanism. (s) Relative toxicity (1338). (t) Refractometric study (1391).

Table 100 Continued $R_nSnR'_{4-n}$

Formula*	Synth. Method**	Yield	Properties	Ref.
PhCH$_2$SnEt$_3$	--	--	(u)	--
PhCH$_2$SnPh$_3$ (182)	IIB	13	m. 91.5-92.5° (v-w)	1928
Ph$_3$CSnBu$_3$	IIIA	60	--	2382
Et$_3$SnR				
R = Pr (182)	VB	--	Moessbauer resonance spectrum (1309), GLC (y, z)	432
Me$_2$CH (182)	I , RMgBr	70	b$_{732}$ 192-94°, b$_{10}$ 73-74°, n$_D^{20}$ 1.4730 (y, z, aa)	2216
n-C$_8$H$_{17}$ (182)	IVB, 1-C$_8$H$_{16}$	93	b$_{10}$ 142°, b$_{0.7}$ 92°, n$_D^{20}$ 1.4717	1612
Me$_3$CCH$_2$CHMeCH$_2$	IVB, (bb)	33	b$_{10}$ 123°, n$_D^{20}$ 1.4747	1612
m-MeC$_6$H$_4$	--	--	(cc)	--
p-MeC$_6$H$_4$ (183)	--	--	(cc)	--
Et$_2$Sn(CH$_2$Ph)$_2$	I , EtMgBr	30	b$_1$ 150-52° (far) IR spectra, assignment	841
Et$_2$SnPr$_2$	--	--	GLC	432
Et$_2$Sn(CHCH$_2$CH$_2$)$_2$	--	--	(dd), Use: herbicide, pesticide	1873
Et$_2$SnPh$_2$ (183)	--	--	NMR (1938) and mass (848, 850) spectra	--
EtSnPr$_3$ (184)	--	--	GLC	432
EtSn-mesityl$_3$ (184)	--	--	(s, t)	--
PhCH$_2$CH$_2$SnPr$_3$ (184)	IV , PhCH:CH$_2$	--	NMR spectrum	2070
PhCH$_2$CH$_2$SnPh$_3$ (184)	IV , PhCH:CH$_2$	--	Mass (850) and NMR spectrum	2070
PhCHMeSnPh$_3$	--	--	(x)	--
Pr$_3$SnBu (184)	VA	--	GLC	432
	IIIB, Bu$_3$Al	--	(q, ee, ff)	1830
	ID, (Pr$_2$Sn)$_x$	2		1941
Pr$_3$SnPh (184)	--	--	(k)	--
Pr$_3$SnC$_6$H$_4$Me$_2$-2,4	--	--	(k)	--
Pr$_3$SnC$_6$H$_4$Me$_2$-2,6	--	--	(k)	--
Pr$_2$SnBu$_2$	--	--	GLC (ee)	432

Compound				
Pr$_2$Sn(C$_6$H$_{11}$c.-)$_2$	I , PrMgBr	72	b$_8$ 204-205°, n$_D^{18}$ 1.5170, d$_{18}$ 1.1540, parachor (gg)	1772
PrSnBu$_3$	--	--	GLC (q, ee, ff)	432
PrSnPh$_3$ (185)	VIB	--	TLC (2276) (hh)	443
	IC, PrBr	14	GLC, from PrI in 28% yield	1938
PrSn-mesityl$_3$ (185)	--	--	(t)	--
i-Pr$_3$SnCHCH$_2$CH$_2$	--	--	Use: herbicide, pesticide	1873
i-Pr$_3$SnC$_6$H$_4$Me-o	--	--	(k)	--
i-Pr$_3$SnC$_{10}$H$_7$-1 (185)	IIA, ArLi	7	m. 25° (1-59, 62)	59
i-Pr$_3$SnC$_{10}$H$_7$-2 (185)	IIA, ArLi	19	b$_2$ 158-59°, n$_D^{20}$ 1.5827 (1-55, 62)	55
i-Pr$_2$Sn(C$_6$H$_{11}$c.-)$_2$	I , PrMgBr	83	b$_{13}$ 190°, n$_D^{18}$ 1.5245, d$_{18}$ 1.1700, parachor (gg)	1772
i-PrSn(C$_6$H$_{11}$c.-)$_3$	I , PrMgBr	54	b$_3$ 190-91°, n$_D^{20}$ 1.5323, d$_{20}$ 1.2190 (ii)	1773
i-PrSnPh$_3$	I , PrMgBr	60	m. 84°, b$_4$ 219-20°	521
	I , PrMgBr	100	(hh-521, 1719)	1719
i-PrSn-mesityl$_3$ (185)	--	--	(t)	--
(CH$_2$CH$_2$CH)$_3$SnBu (185)	--	--	Use: herbicide, pesticide	1873
(CH$_2$CH$_2$CH)$_3$SnCHCMe$_2$CH$_2$	II , R$_3$SnCl	--	Use: herbicide, pesticide	1873
(CH$_2$CH$_2$CH)$_3$SnC$_8$H$_{17}$	I , RSnCl$_3$	--	Use: herbicide, pesticide	1873
(CH$_2$CH$_2$CH)$_3$SnPh (185)	--	--	Use: herbicide, pesticide	1873
(CH$_2$CH$_2$CH)$_2$SnBu$_2$ (185)	--	--	Use: herbicide, pesticide	1873

(u) Photooxidn. in C$_9$H$_{20}$ → Et$_2$SnO + Et$_2$Sn(OBz)$_2$ + Et$_3$SnOBz + (Et$_3$Sn)$_2$O + PhCHO; rates of formation, quantum yield (2215). (v) Rxn. with LiAlH$_4$ → Ph$_4$Sn (2673). (w) Use in catalyst for butadiene polymerization (2044). (x) Contol and effect on reproduction of Musca domestica (260). (y) Mass spectrum (2268). (z) Rxn. with X → kinetic data and mechanism of cleavage rxn., solvent effect; X = Br (767-9), I (767). (aa) Oxidn. under UV irradiation in C$_9$H$_{20}$ → kinetic data, rxn. mechanism and prod. composition (2216). (bb) Synth. from Me$_3$CCH$_2$CMe:CH$_2$ (1612). (cc) Rxn. with Hg(OAc)$_2$ in THF → Et$_3$SnOAc; relative rate and rxn. mechanism (1162). (dd) Rxn. with HgCl$_2$ → Et$_2$(CH$_2$CH$_2$CH)SnCl (1873). (ee) Anal. detn. in Bu$_4$Sn from Grignard synth. by mass spectra and chromatography (2838). (ff) Rxn. with Br → ratio of alkyl bromides (2266). (gg) Rxn. with X → (n-, i)-Pr$_2$(C$_6$H$_{11}$)SnX; X = Br, I (2721). (hh) Rxn. with I in Et$_2$O → R'Ph$_2$SnI + PhI; R, n-Pr, i-Pr, i-Bu (521). (ii) Parachor, surface tension, stable in air, IR spectrum (1773).

Table 100 Continued

$R_nSnR'_{4-n}$

Formula*	Synth. Method**	Yield	Properties	Ref.
(CH₂CH₂CH)₂Sn(C₈H₁₇-n)₂	I , C₈H₁₇MgBr	--	Use: herbicide, pesticide	1873
CH₂CH₂CHSnBu₃ (186)	--	--	Use: herbicide, pesticide	1873
CH₂CH₂CHSnPh₃ (186)	--	--	Use: herbicide, pesticide (x)	1873
Bu₃SnR				
R = PhCHMeCH₂	IV , PhCMe:CH₂	80	$b_{0.2}$ 128-32°, n_D^{20} 1.511	1612
Me₂CHCH₂ (186)	IIIB, R₃Al	--	--	1830
MeEtCH	--	--	(ee)	--
i-C₇H₁₅	I , Bu₃SnI	76	b_4 178-80°, n_D^{20} 1.5249, d_{20} 1.4040 (jj, kk)	617
n-C₈H₁₇	IVB, 1-C₈H₁₆	100	$b_{0.08}$ 126°, n_D^{20} 1.4734	1612
	IIIB, (11)	--	--	1830
BuEtCHCH₂	IVB, BuEtC:CH₂	90	$b_{0.1}$ 111°, n_D^{20} 1.476	1612
	IIIB, R₃Al	--	--	1830
Me₃CCH₂CHMeCH₂	IVB, (Me₂C:CH₂)₂	74	$b_{0.25}$ 96°, n_D^{20} 1.4730	1612
	IIIB, R₃Al	--	--	1830
Ph (178)	--	--	NMR spectrum (1-62, 59)	1220
1-C₁₀H₇ (186)	--	--	(1-62, 59)	--
2-C₁₀H₇ (186)	--	--	(1-62, 59)	--
Bu₂SnR₂				
R = Me₂CHCH₂	IIIB, R₃Al	--	--	1830
MeEtCH	IIIB, R₃Al	--	(ee)	1830
Me₃CCH₂ (186)	--	--	(mm)	--
c.C₆H₁₁ (186)	I , BuMgBr	92	b_7 214-15°, n_D^{18} 1.5132, d_{18} 1.1310 (nn)	1772
n-C₈H₁₇	IIIB, R₃Al	--	(oo)	1830
n-C₁₀H₂₁	--	--	(oo)	--
n-C₁₂H₂₅	--	--	(oo)	--
Ph (178)	ID	3	NMR spectrum (1220), GLC (pp)	1938

BuSnR$_3$

R =				
Me$_2$CHCH$_2$	IIIB, Bu$_3$Al	--	--	1830
MeEtCH	IIIB, Bu$_3$Al	--	(ee)	1830
c.C$_6$H$_{11}$	I , BuMgBr	63	m. 116-17°, soly., IR spectrum	1773
n-C$_8$H$_{17}$	IIIB, Bu$_3$Al	--	--	1830
mesityl	--	--	(t)	--
i-Bu$_3$SnPh	--	--	TLC	2276
i-Bu$_2$Sn(c.C$_6$H$_{11}$)$_2$	I , i-BuMgBr	83	m. 82°	1772
i-Bu$_2$SnPh$_2$	--	--	TLC	2276
i-BuSn(c.C$_6$H$_{11}$)$_3$	I , i-BuMgBr	62	m. 103-104°, IR spectrum	1773
i-BuSn(n-C$_8$H$_{17}$)$_3$ (187)	IVB, 1-C$_8$H$_{18}$	59	b$_{0.01}$ 180°, n_D^{20} 1.4740	1612
i-BuSnPh$_3$	I , Ph$_3$SnI	70	m. 65°, TLC (2276) (kk)	521
i-BuSn-mesityl$_3$ (187)	--	--	(t)	--
(n-C$_5$H$_{11}$)$_2$Sn(CH$_2$CMe$_3$)$_2$	I , Me$_3$CCH$_2$MgBr	78	b$_{1.6}$ 135°, n_D^{25} 1.4742, d$_{27}$ 1.0118 (qq)	2188
(n-C$_5$H$_{11}$)$_2$Sn(c.C$_6$H$_{11}$)$_2$	I , C$_5$H$_{11}$MgBr	92	b$_7$ 227-28°, n_D^{18} 1.5090, d$_{18}$ 1.1100 (rr, ss)	1772
n-C$_5$H$_{11}$Sn(c.C$_6$H$_{11}$)$_3$	I , C$_5$H$_{11}$MgBr	59	b$_3$ 201-202°, n_D^{20} 1.5135, d$_{20}$ 1.1190 (ii)	1773
n-C$_5$H$_{11}$SnPh$_3$	I , Ph$_3$SnI	76	b$_3$ 222°, n_D^{21} 1.5970, d$_{21}$ 1.2570	2787
	IVA, 1-C$_5$H$_{10}$	100	b$_2$ 209°, n_D^{21} 1.5985 (tt)	123
(1-C$_5$H$_{11}$)$_3$SnC$_9$H$_{19}$-i	I , R$_3$SnI	74	b$_3$ 225-27°, n_D^{20} 1.524, d$_{20}$ 1.3683 (jj, uu)	617
(i-C$_5$H$_{11}$)$_2$Sn(c.C$_6$H$_{11}$)$_2$	I , C$_5$H$_{11}$MgBr	82	m. 45° (ss)	1772
i-C$_5$H$_{11}$Sn(c.C$_6$H$_{11}$)$_3$	I , C$_5$H$_{11}$MgBr	53	m. 57-58°, IR spectrum	1773
i-C$_5$H$_{11}$SnPh$_3$ (187)	I , Ph$_3$SnI	80	m. 27°, b$_3$ 216°, n_D^{27} 1.5930, d$_{27}$ 1.2640 (tt)	2787

(jj) Parachor, surface tension (617). (kk) Rxn. with I in mesitylene → Bu$_2$RSnI + BuI (617). (ll) With R$_3$Al or i-Bu$_2$AlR.(1830). (mm) Rxn. with Br(1:2) in Et$_2$O, + aq. NaOH, + AcOH → BuR$_2$SnOAc (2188). (nn) Thick colorless liquid, parachor (1772). (oo) Rxn. with SnCl$_4$ → BuRSnCl$_2$ (1482). (pp) Use as esterification catalyst (821), antioxidant for polyphenyl ether (1979), stabilizer for phosphinate hydraulic fluids (1497). (qq) Rxn. with Br (1:4 in CCl$_4$) → (Me$_3$CCH$_2$)$_2$SnBr$_2$ (2188). (rr) Parachor, colorless liquid (1772). (ss) Rxn. with I in HCCl$_3$ → (C$_5$H$_{11}$)$_2$(C$_6$H$_{11}$)SnI (1772). (tt) Rxn. with I in Et$_2$O → RPhSnI$_2$ + PhI; R = n-, i-C$_5$H$_{11}$, n-C$_6$H$_{13}$, n-C$_7$H$_{15}$, n-C$_9$H$_{19}$ (2787). (uu) Rxn. with I in mesitylene → (i-C$_5$H$_{11}$)$_2$(i-C$_9$H$_{19}$)SnI (617).

Table 100 Continued

$R_nSnR'_{4-n}$

Formula*	Synth. Method**	Yield	Properties	Ref.
i-C_5H_{11}Sn-mesityl$_3$ (187)	--	--	(s, t)	--
c.C_5H_9SnPh$_3$ (187)	--	--	m. 112°	123
(n-C_6H_{13})$_2$Sn(c.C_6H_{11})$_2$	I , C_6H_{13}MgBr	89	b_5 234-35°, n_D^{18} 1.5070, d_{20} 1.0870 (vv)	1774
n-C_6H_{13}SnPh$_3$ (187)	IVA, 1-C_6H_{12}	100	(tt)	123
(c.C_6H_{11})$_3$SnPh (188)	--	--	NMR spectrum	1220
(c.C_6H_{11})$_2$SnR$_2$				
R = C_7H_{15}	I , RMgBr	90	b_9 253-54°, n_D^{18} 1.5040, d_{20} 1.0640 (vv)	1774
C_8H_{17}	I , RMgBr	90	b_3 255° (part dec.), n_D^{18} 1.5020, d_{18} 1.0530 (vv)	1774
C_9H_{19}	I , RMgBr	60	b_3 262-63° (part dec.), n_D^{18} 1.4985, d_{18} 1.0230 (vv)	1774
$C_{10}H_{20}$	I , RMgBr	68	b_3 282-83° (part dec.), n_D^{18} 1.4942, d_{18} 1.0180 (ww)	1779
Ph (188)	--	--	NMR spectrum	1220
c.C_6H_{11}SnPh$_3$ (188)	--	--	NMR spectrum	1220
n-C_7H_{15}SnPh$_3$ (188)	IVA, 1-C_7H_{14}	100	(tt)	123
n-C_8H_{17}SnPh$_3$ (188)	IVA, 1-C_8H_{16}	100	--	123
	IIB, C_8H_{17}F	39	m. 53-55°	1928
n-C_8H_{17}Sn-mesityl$_3$ (188)	--	--	(s, t)	--
n-C_9H_{19}SnPh$_3$ (189)	IVA, 1-C_9H_{18}	100	(tt)	123
n-$C_{10}H_{21}$SnPh$_3$	I , Ph$_3$SnI	70	m. 39°, b_3 260° (tt)	2787
(PrCHCMe$_2$CH)$_3$SnC$_6H_4$Me-p	I , ArSnCl$_3$	--	Use: herbicide, pesticide	1873
Ph$_3$SnR				
R = o-MeC$_6H_4$ (189)	--	--	Moessbauer resonance spectrum	1309
p-MeC$_6H_4$ (189)	--	--	Moessbauer resonance spectrum	1309
p-EtC$_6H_4$ (189)	--	--	Activity as scintillator in polyvinyltoluene	936
mesityl (189)	I , Ph$_3$SnX	50	m. 157°	303
o-PhC$_6H_4$	I , Ph$_3$SnCl	50	White cryst. m. 95-96°, UV spectrum	1278

1-C₁₀H₇ (189)	--	--	UV spectrum	1278
Ph₂Sn-mesityl₂	I , Ph₂SnX₂	46	m. 158°	303
(o-MeC₆H₄)₃Sn-mesityl	I , Ar₃SnX	46	m. 123°; Ar = o-MeC₆H₄	303
(o-MeC₆H₄)₂Sn-mesityl₂	I , Ar₂SnX₂	28	m. 154°; Ar = o-MeC₆H₄	303
(p-MeC₆H₄)₂Sn-mesityl₂	I , Ar₂SnCl₂	45	m. 211°; Ar = p-MeC₆H₄	303
(p-Me₂C₆H₃)₃Sn-mesityl	I , Ar₃SnX	43	m.1186° (s); Ar = p-Me₂C₆H₃	303
(p-Me₂C₆H₃)₂Sn-mesityl₂	I , Ar₂SnX₂	25	m. 200° (s); Ar = p-Me₂C₆H₃	303
Mesityl₂Sn(1-C₁₀H₇)₂	I , Mesityl-MgX	40	m. 140° (s)	303
Mesityl-Sn(1-C₁₀H₇)₃	I , Mesityl-MgX	27	m. 187° (s)	303
PhCH₂SnR₃	--	--	UV spectrum	2771
R₃SnPh	--	--	UV spectrum	2771
R₃SnAr	--	--	(xx)	--
Ar$_n$SnR$_{4-n}$	--	--	(yy)	--

(vv) Colorless liquid, IR spectrum, rxn. with I in HCCl₃ → (c.C₆H₁₁)R₂SnI + c.C₆H₁₁I (1774). (ww) Rxn. with Br → C₆H₁₁(C₁₀H₂₁)₂SnBr + C₆H₁₁Br (1774), IR spectrum, colorless liq., soly. (1774). (xx) Rxn. with HX or X → discussion of rxn. mechanism of protonolysis and halodemetallation, nucleophilic assistance in electrophilic substitution reaction (1379). (yy) Relative toxicity for R = Me, Et, i-C₅H₁₁ and n-C₈H₁₇ , Ar = methyl-substituted aromatics, x-naphthyl (1338).

2.1.2 NONFUNCTIONAL UNSUBSTITUTED TETRAORGANOTIN COMPOUNDS WITH THREE AND FOUR DIFFERENT ORGANIC GROUPS

All compounds are listed in Table 101.

2.3 HETEROCYCLIC SUBSTITUTED TETRAORGANOTIN COMPOUNDS

The unsymmetric heterocyclic tetraorganotin derivatives in Table 102 are prepared by methods from the following scheme.

I Grignard reaction:
   Bu_3SnCl with $R'MgBr \rightarrow Bu_3SnR'$.

II Synthesis with alkali metal derivatives:
   (IIA) $R_3SnCl + R'Cl + Na \rightarrow R_3SnR'$,
   (IIB) $R_3SnCH_2Cl + R'K \rightarrow R_3SnR'$,
   (IIC) $Ph_3SnLi + R'X \rightarrow Ph_3SnR'$.

III Ring closure reactions:
   (IIIA) $Et_3SnC\vdots CPh + PhNCO \rightarrow Et_3SnCPH:R'$,
   (IIIB) $Me_3SnCHN_2 + PhNCO \rightarrow Me_3SnR'$,
   (IIIC) $R_3SnC\vdots CH + CH_2N_2 \rightarrow R_3SnR'$,
   (IIID) $Ph_3SnC\vdots CPh + $ tetracine $\rightarrow Ph_3SnR' + N_2$.

IV Hydrostannation reaction:
   $Ph_3SnH + CH_2:CHR' \rightarrow Ph_3SnCH_2CH_2R'$.

V Miscellaneous reactions:
   (VA) $R_3SnCH_2Cl + R'H \rightarrow R_3SnCH_2R'$,
   (VB) $R_3SnCH_2CH_2R' + MCl_n \rightarrow MCl_n \cdot mPh_3SnCH_2CH_2R'$.

R = organic group, R' = heterocyclic organic group, M = Ph_3Sn-, Ph_2Sn=, Cu-, Zn=, Co=, Ni≡, m = 1-4, n = 1, 2. Cyclic organotin compounds are summarized in Chapter 7.

Table 101. Nonfunctional Unsymmetric Tetraorganotin Compounds with Three and Four Different Organic Groups   RR'R''R'''Sn

Formula	Synth. Method*	Yield	Properties	Ref.
$Me_2EtSnPh$	IC, EtI	5	NMR spectrum, GLC, with EtBr in 4% yield	1938
	IB	--	--	430
$Me_2PrSnPh$	IC, PrI	5	With PrBr in 4% yield, GLC	1938
$Me_2(i-Pr)SnC_6H_{11}c.$	I , i-PrMgX	--	$b_{1.7}$ 76.5° (a)	2269
$Me_2BuSnPh$	IC, BuMgI	4	GLC, NMR spectrum	1938
	IC, BuI	6	With BuBr in 5% yield	1938
	ID, MeMgI	5	--	1938
$MeEt_2SnPh$	IC, EtI	10	NMR spectrum, GLC	1938
$MeEtSnPh_2$	IC, EtI	17	NMR spectrum, GLC, with EtBr in 5% yield	1938
$MePr_2SnPh$	IC, PrI	15	NMR spectrum, GLC, with PrBr in 3% yield	1938
$MePrSnPh_2$	IC, PrI	11	NMR spectrum, GLC, with PrBr, in 13% yield	1938
$Me(i-Pr)SnPh_2$	I , MeMgI	70	$b_{0.4}$ 118°, NMR spectrum (b, c)	1719
	I , i-PrMgX	--	$b_{0.15}$ 118°, n_D^{20} 1.580, mass spectrum	2412
$MeBu_2SnPh$	IC, BuMgI	14	NMR spectrum, GLC	1938
	IC, BuI	12	With BuBr in 5% yield	1938
	ID, MeMgI	42	--	1938
$MeBuSnPh_2$	I , BuMgI	--	b_3 144–45°, n_D^{27} 1.5620	1938
	IC, BuMgI	27	NMR spectrum, GLC	1938
	IC, BuI	8	With BuBr in 29% yield	1938

* The characters used here correspond to the synthetic scheme in the introduction to Chapter 2.1.
(a) Rxn. with Br in MeOH at 0° → Me(i-Pr)c.C$_6$H$_{11}$SnBr + MeBr (2269). (b) Rxn. with I in HCCl$_3$ at -40° → Me(iPr)-PhSnI (1719). (c) Rxn. with Br in EtOH at -65° → Me(i-Pr)PhSnBr (2412).

Table 101 Continued

Formula	Synth. Method*	Yield	Properties	Ref.
MePhSn(1-C$_{10}$H$_7$)$_2$	I , MePhSnI$_2$	--	m. 127°, NMR spectrum (d)	1719
MeEt(i-Pr)SnC$_6$H$_{11}$c.	I , EtMgBr	97	b$_{0.7}$ 73.5°, GLC (e)	2269
Me(i-Pr)RSnPh				
R = Et	I , EtMgBr	--	b$_{0.05}$ 45-46°, mass spectrum	2412
Pr	I , PrMgBr	--	b$_{0.03}$ 49-50°, mass spectrum	2412
Bu	I , BuMgBr	--	b$_{0.03}$ 56-60°, mass spectrum	2412
Me$_2$CHCH$_2$	I , i-BuMgBr	--	b$_{0.03}$ 51.5-54°, mass spectrum	2412
MeEtCH	I , s-BuMgBr	--	b$_{0.03}$ 55-59°, mass spectrum	2412

(d) Rxn. with I in HCCl$_3$ at -40° → MePh(1-C$_{10}$H$_7$)SnI (1719). (e) Rxn. with Br in PhCl → MeBr + EtBr + i-PrBr + c.C$_6$H$_{11}$Br (2269), with I in MeOH → MeI (2269).

Table 102. Heterocyclic Substituted Tetraorganotin Compounds R$_n$SnR'$_{4-n}$

Formula*	Method**	Yield	Properties	Ref.
Me$_3$SnCH$_2$-N⟨pyrazole⟩	IIB	74	b$_{0.8}$ 55.7-56°, n$_D^{20}$ 1.5279, d$_{20}$ 1.3511, (a-c)	1309,1350
Me$_3$Sn⟨oxazolidinone⟩	IIIB	--	--	1393

Compound		Yield	Properties	Ref.
Me$_3$SnCH$_2$—N(morpholine)	VA	81	b$_{1.2}$ 66-67°, n$_D^{20}$ 1.4998, d$_{20}$ 1.3207 (a-f)	1309, 1350
Me$_3$SnCH$_2$—N(morpholine)·MeI	(e)	94	m. 192.8-93° (b, c)	1309
Me$_3$SnCH$_2$—N(morpholine)·HCl	(f)	--	m. 106.6-107.8°	1350
Et$_3$SnCH$_2$N(pyrrole)	IIB	93	b$_{0.2}$ 86°, n$_D^{20}$ 1.5259, d$_{20}$ 1.2584 (g)	283
Et$_3$Sn(imidazole)	IIIC	71	b$_{0.1}$ 86-88°, n$_D^{20}$ 1.5190, d$_{20}$ 1.3429 (h)	2760
Et$_3$SnCPh:N—C(:O)—N(Ph)—O (hydantoin-type)	IIIA	73	Colorless cryst., m. 102-103° (i)	2644
Et$_3$SnCH$_2$N(morpholine)	VA	--	b$_{0.1}$ 83.5-84.5°, n$_D^{20}$ 1.5039, d$_{20}$ 1.2427 (g)	283

* Numbers in parentesis refer to pages in main volume.

** The characters used here correspond to the synthetic scheme in the introduction to Chapter 2.3.

(a) (Far) IR spectrum (1309). (b) NMR and Moessbauer resonance spectra (1309, 1350). (c) Mass spectrum (1350). (d) UV spectrum (1350) and basicity (1350, 1715). (e) Rxn. of Me$_3$SnCH$_2$NC$_4$H$_8$O with MeI in Et$_2$O → Me$_3$SnCH$_2$NC$_4$H$_8$O·MeI (1309). (f) Rxn. of Me$_3$SnCH$_2$NC$_4$H$_8$O with HCl → Me$_3$SnCH$_2$NC$_4$H$_8$O·HCl (1350). (g) Moessbauer resonance and mass spectra (1309). (h) Colorless liq. NMR and IR spectra, assignment (2760). (i) Rxn. with alc. HCl → Et$_3$SnCl + PhCH:CCONPhCONPh (2644).

Table 102 Continued

Formula*	Synth. Method**	Yield	Properties	Ref.
Pr$_3$Sn-[imidazole-NH]	IIIC	68	b$_{0.01}$ 111-12°, n$_D^{20}$ 1.5192, d$_{20}$ 1.2443 (h)	2760
Bu$_3$Sn-[pyrrole]	I	--	b$_{0.3}$ 140-45°	1500
Bu$_3$Sn-[triazine(NEt)$_2$]	IIA	--	Slightly yellow, b$_{0.005}$ 148-52°, n$_D^{20}$ 1.5137 (j)	1322a
Ph$_3$SnCH$_2$CH$_2$-[pyridine] (192)	IV	--	(k, l)	2689

MX·nPh$_3$SnCH$_2$CH$_2$-[pyridine] (1)

MX = Ph$_3$SnCl, n = 1	VB	--	m. 166-68°, far IR spectrum and assignment	2689
Ph$_2$SnCl$_2$ 2	VB	--	m. 175-78°, far IR spectrum and assignment	2689
CuCl 1 (m)	VB	--	m. 106° (dec.), far IR spectrum	2689
ZnCl$_2$ 2	VB	--	m. 64°, far IR spectrum and assignment (n)	2689
	(n)	--	m. 159-62°, far IR spectrum and assignment	2689
CoCl$_2$ 4	VB	--	m. 150-52°, far IR spectrum and assignment (o)	2689
NiCl$_2$ 4	VB	--	m. 184-85°, far IR spectrum and assignment (p)	2689

Structure	Type			Ref.
Ph₃Sn-[3,5-diphenyl pyridazine]	IIID	80	m. 251° (q)	2644
Ph₃Sn-[triazine]-NHMe, X = Cl	IIC	--	(r)	1322a
X = Br	IIC	--	(r)	1322a

(j) Use as biocide (1322a). (k) Rxn. with HgCl₂ → PhHgCl (2689). (l) Rxn. with metal halides → addition compounds (2689). (m) Synth. from CuCl₂ or CuCl (2689). (n) Resolidifies on heating (2689). (o) Purple solid (2689). (p) Pale green solid (2689). (r) Rxn. with AcOH → 3,4,6-triphenylpyridazine + Ph₃SnOAc (2644). (r) Use as bactericide, fungicide and herbicide (1322a).

2.4 UNSYMMETRIC TETRAORGANOTIN COMPOUNDS WITH OLEFINIC SUBSTITUENTS

This chapter includes organotin derivatives of unsubstituted olefinic hydrocarbons. Compounds with several double bonds are reported. Compounds containing olefinic and acetylenic unsaturation in one hydrocarbon chain are included here when the olefinic bond is nearer to the tin atom. Otherwise, the compounds will be listed in Chapter 2.5. The compounds compiled in Table 103 are prepared by methods from the following scheme.

I Grignard reaction:
- (IA) R_nSnX_{4-n} + R'MgX in THF → $R_nSnR'_{4-n}$; n = 2, 3 ,
- (IB) R_3SnX + R'X + Mg → R_3SnR' ,
- (IC) Et_3SnCH_2Cl + R'MgX → Et_3SnCH_2R' ,
- (ID) $(Bu_3Sn)_2O$ + R'MgX → Bu_3SnR' .

II Reactions with lithium derivatives:
- (IIA) R_3SnCl + R'Li → R_3SnR' ,
- (IIB) R_3SnLi + R'Br → R_3SnR' ,
- (IIC) R'_4Sn + PhLi → R'_3SnPh .

III Synthesis with organotin sodium compounds:
R_3SnNa + R'X in NH_3-Et_2O → R_3SnR' .

IV Hydrogermylation reactions: R_nSnH_{4-n} with dienes, allenes and acetylenes - initiated by:
- (IVA) azobisisobutyronitrile (AiBN)
- (IVB) ultraviolet irradiation and
- (IVC) by diisobutylaluminum hydride.

V Reaction of organotin amides with acidic hydrocarbon:
R_3SnNR_2 + C_5H_6 → $R_3SnC_5H_5$.

R = organic group, R' = unsaturated organic group, X = halogen. Additional tetraorganotin compounds with olefinic unsaturation may be found in Chapter 7.1.4 and 8.1.

Trimethylvinyltin (M-194) SnC_5H_{12} $Me_3SnCH:CH_2$
Synth.: By-prod. in photolysis of $Me_3Sn(CH_2)_3COR$, in 6-24% yield; R = Me, Ph (1488).
Prop.: d_{25} 1.277 (1717), IR (763, 1013), UV (424), NMR (1220) and mass (621, 2869) spectra, GLC (442, 1718), dielectric const. and dipole moment (1572), magnetic measurement (984), parachor, refrachor and surface tension (1717), struct. (424), thermodynamic data (305).
Rxn. with: I_3^- in MeOH → kinetic data and rxn. mechanism (2234).
$PhHgCBrCl_2$ → $Me_3SnCHCH_2CCl_2$ + PhHgBr + some $(CH_2:CH)_2Hg$ (1877).
$PhCH:CH_2$ and $CH_2:CMeCO_2R$ → radical initiated polymerization (1522).
$RCH:CH_2$ + H + Raney Ni → inhibiting effect on hydrogenation; R = $CH_2:CMe$, Ph, p-$Me_3CC_6H_4$, p-$Me_3SnC_6H_4$ (2316).

Allyltrimethyltin (M-194) SnC_6H_{14} $Me_3SnCH_2CH:CH_2$
Synth.: By rxn. of Me_3SnCl with $CH_2:CHCH_2MgCl$ in Et_2O, in 73% yield (1367).

By rxn. of Me$_3$SnH at 100° in presence of AiBN with allene, in 35% yield (295), with MeC⋮CH, in 0.3% yield (295).
Prop.: b. 126-28° (1367), IR (136, 295, 763), UV (424), NMR (295) and mass (2869) spectra, GLC (295, 1367) and struct. (424).
Rxn. with: HCl in aq. MeOH → kinetic and thermodynamic data for olefin cleavage, isotope effect (1367).
Me$_3$SnCF$_3$ + NaI in (MeOCH$_2$)$_2$ → Me$_3$SnCH$_2$CHCF$_2$CH$_2$ + Me$_3$SnI (1874).
HCCl$_3$ + KOCMe$_3$ → Me$_3$SnCH$_2$CHCCl$_2$CH$_2$ (1877).

Crotyltrimethyltin (M-199)   SnC$_7$H$_{16}$ Me$_3$SnCH$_2$CH:CHMe
Synth.: By rxn. of MeCH:CHCH$_2$MgCl with Me$_3$SnCl in Et$_2$O, in 81% yield (1367).
By rxn. of Me$_3$SnH with MeCH:C:CH$_2$ at 100° in presence of AiBN, cis- and trans form in 2.5 and 7% yield, resp. (295), in 33 and 22% yield by GLC, resp. (1041, 1043).
By rxn. of Me$_3$SnNa with MeCH:CHCH$_2$Cl in li2. NH$_3$, in 79% yield (1877).
Prop.: b. 148-54°, cis-trans-ratio 0.3 (1877), b$_{0.25}$ 69-72° (1041, 1043), IR (295, 1041, 1043, 1877), UV (1041, 1043) and NMR (295, 1877) spectra, GLC (295, 1041, 1043, 1367, 1877). Pure trans-form b. 147-48°, n$_D^{25}$ 1.4762, contn. < 5% cis (1877). Pure cis-form b. 150-51°, n$_D^{25}$ 1.4824 (1877).
Rxn. with: HCl in aq MeOH → kinetic and thermocynamic data for olefin cleavage, isomer distribution in olefin (1367).
HCCl$_3$ + KOCMe$_3$ → Me$_3$SnCH$_2$CH:CCl$_2$CHMe without isomerization (1887).
BuLi (cat. amount) → isomerization to 0.66 cis-: trans-ratio (1886).
BuLi → MeCH:CHCH$_2$Li, isomer distribution by NMR spectroscopy and Me$_3$SiCl-quenching (1886).
Li in Et$_2$O → rapid isomerization to 0.66 cis-: trans-ratio (1886); 21:79 (1887).
Li in THF, + Et$_3$SiCl → Me$_3$SnSiEt$_3$ + Et$_3$SiCH$_2$CH:CHMe + MeSiEt$_3$ (1887).
CF$_3$CO$_2$H → Me$_3$SnO$_2$CCF$_3$ + 1-C$_4$H$_8$ (1041, 1043).

3-Butenyltrimethyltin   SnC$_7$H$_{16}$ Me$_3$SnCH$_2$CH$_2$CH:CH$_2$
Synth.: By rxn. of CH$_2$:CHCH$_2$CH$_2$MgBr in Et$_2$O with Me$_3$SnBr, in 78% yield (1877), with Me$_3$SnCl, in 35% yield (1041, 1043).
By-prod. in rxn. of Me$_3$SnH with butadiene in presence of AiBN at 100°, in 4% yield (1041, 1043).
Prop.: b$_{15}$ 48-50°, n$_D^{25}$ 1.4685 (1041, 1043), IR (136, 1041, 1043, 1877) and NMR (1041, 1043, 1877) spectra, GLC (1041, 1043).
Rxn. with: HCCl$_3$ + KOCMe$_3$ → Me$_3$SnCH$_2$CH$_2$CHCCl$_2$CH$_2$ (1877).
CF$_3$CO$_2$H → Me$_3$SnO$_2$CCF$_3$ + 1-C$_4$H$_8$ (1041, 1043).

1,1-Methylisopropenyltri-   SnC$_7$H$_{16}$ Me$_3$SnCMe:CHMe
methyltin
Synth.: By rxn. of Me$_3$SnH with CH$_2$:C:CHMe at 100° in presence of AiBN, cis- and trans-form in 28 and 12% yield, resp. (295).
Prop.: cis-form n$_D^{25}$ 1.4738, trans-form n$_D^{25}$ 1.4753. IR (295) and NMR (295, 2234) spectra, GLC (295, 2234).
Rxn. with: I$_3^-$ in MeOH → relative kinetic data, mechanism and stereochamistry

of obtained C_4H_7I (2234).

$CF_3CO_2H \rightarrow Me_3SnO_2CCF_3$ + 94% cis-2-butene + 6% trans-2-butene, from cis-form (295).

2-Cyclohexenyltrimethyl-tin SnC_9H_{18} $c.2\text{-}C_6H_9SnMe_3$

Synth.: By rxn. of Me_3SnH with $1,3\text{-}C_6H_8$ in presence of AiBN at 100°, in 36% yield (1041, 1043).
By rxn. of Me_3SnNa with $3\text{-}BrC_6H_9$ in liq. $NH_3\text{-}Et_2O$, in 22% yield (1367).
Prop.: b_{15} 97-101° (1367), $b_{0.08}$ 30-30.5°, IR spectrum (1041, 1043), GLC (1041, 1043, 1367).
Rxn. with: HCl (in aq. MeOH) → kinetic and thermodynamic data on olefin cleavage (1367).
$CF_3CO_2H \rightarrow Me_3SnO_2CCF_3$ (1041, 1043).

3-Cyclohexenyltrimethyl-tin SnC_9H_{18} $c.3\text{-}C_6H_9SnMe_3$

Synth.: By rxn. of Me_3SnH with $1,4\text{-}c.C_6H_8$ in Ar under UV irradiation, in 37% yield (1953), with $1,3\text{-}c.C_6H_8$ in presence of AiBN at 100° in 19% yield (1041, 1043).
By rxn. of $4\text{-}C_6H_9MgBr$ with Me_3SnCl, in 24% yield (1041, 1043).
Prop.: b_{12} 80-82°, n_D^{20} 1.5067 (1953), 1.5032 IR and NMR spectra, GLC (1041, 1043).
Rxn. with: $CF_3CO_2H \rightarrow$ sepn. from $2\text{-}C_6H_9SnMe_3$ (1041, 1043).

2-Cyclooctenyltrimethyl-tin $SnC_{11}H_{22}$ $c.2\text{-}C_8H_{13}SnMe_3$

Synth.: By rxn. of Me_3SnH with $1,3\text{-}C_8H_{12}$ in presence of AiBN at 100°, in 24% yield (1043, 2052) and under UV irradiation (1043, 2052).
By rxn. of Me_3SnCl with $3\text{-}C_8H_{13}MgBr$ in Et_2O in 32% yield (1043, 2052).
By rxn. of Me_3SnNa with $3\text{-}BrC_8H_{13}$ in liq. $NH_3\text{-}Et_2O$, in 46% yield (1367).
Prop.: $b_{0.5}$ 68-70° (1367), $b_{0.25}$ 63.5-68.5°, n_D^{25} 1.5172 (1043 2052), IR and NMR spectra (1043, 2052), GLC (1043, 1367, 2052).
Rxn. with: HCl in aq. MeOH → kinetic and thermodynamic data of olefin cleavage rxn., deuterium effect (1367).
$CF_3CO_2H \rightarrow Me_3SnO_2CCF_3$ (1043, 2052).

p-Styryltrimethyltin (M-199) $SnC_{11}H_{16}$ $p\text{-}Me_3SnC_6H_4CH:CH_2$
Prop.: $b_{0.4}$ 78°, n_D^{20} 1.5645, d_{20} 1.299 (1569, 2316), dipole moment (1333).
Rxn. with: H + Raney Ni → relative rate of hydrogenation, inhibiting effect of $Me_3SnCH:CH_2$ (2316).
$(CH_2:CH)_2$ → tin-contn. rubber (1569).
$(SCN)_2$ → relative reactivity (863).

Triethylvinyltin (M-195) SnC_8H_{18} $Et_3SnCH:CH_2$
Prop.: Moessbauer resonance spectrum (2213), thermodynamic data (305).
Rxn. with: ^{60}Co irradiation → polymerization (210).

I_3^- in MeOH → relative kinetic data and mechanism of cleavage rxn. (2234).
BuLi → polymerization (1700).
R_3MCH:CH_2 + BuLi → copolymerization R_3M = Et_3Si, Et_3Ge, Ph (1700).
RCH:CH_2 + BuLi → inhibition of polymerization by complex formation (2213).
CH_2:CHCl → Et_3Sn-contn. polymer (2514).
Me_3SnCF_3 + NaI in $(MeOCH_2)_2$ → $Et_3SnCHCH_2CF_2$ + Me_3SnI (1874).
Addn. compd.: $Et_3SnCH:CH_2 \cdot 2BuLi$, Moessbauer resonance spectrum (2213).

Allyltriethyltin (M-195) SnC_9H_{20} $Et_3SnCH_2CH:CH_2$
Synth.: By rxn. of Et_3SnCl with $CH_2:CHCH_2$MgCl in Et_2O, in 76% yield (1367).
Prop.: Anal. detn. by GLC (1376), by iodometric titration (1619).
Rxn. with: HCl in aq. MeOH → kinetic and thermodynamic data on olefin cleavage rxn. (1367).
Et_3SnH (+ cat. i-Bu_2AlH) → $(Et_3SnCH_2)_2CH_2$ (1612).
$H_2S_{5.2}$ in CCl_4 → $(Et_3SnCH_2CHMe)_2S_6$ (1857).
BzCl → Et_3SnCl (1331).
RCHO → $Et_3SnOCHRCH_2CH:CH_2$ (1331).
$(RCO)_2O$ → Et_3SnO_2CR; R = Me, Ph, 1/2 CH_2CH_2 + $CH_2:CHCH_2$Ac; R = Me (1331).

Crotyltriethyltin $SnC_{10}H_{22}$ $Et_3SnCH_2CH:CHMe$
Synth.: By rxn. of Et_3SnCl with $MeCH:CHCH_2$MgX; X = Cl, in 70% yield (1367), Br, in 77% yield (1619).
By rxn. of Et_3SnH with butadiene in presence of AiBN in autoclave in inert atm. in 16% (1619), in presence of i-Bu_2AlH at 85° as by-prod. (1612).
Prop.: b_{20} 99-105° (1367), b_{12} 95-98° (1619), b_{10} 87-90° (1612), n_D^{20} 1.4931 (1619), 1.483 (1612). Anal. detn. by iodometric titration (1619) and sepn. of cis- and trans-isomers by GLC (1367).

3-Butenyltriethyltin $SnC_{10}H_{22}$ $Et_3SnCH_2CH_2CH:CH_2$
Synth.: By rxn. of Et_3SnCl with $CH_2:CHCH_2CH_2$MgBr in Et_2O, in 84% yield (1619).
By rxn. of Et_3SnH with butadiene and AiBN in autoclave in inert atm in 60% yield (1619).
By rxn. of Et_3SnCH_2Cl and $CH_2:CHCH_2$Br with Mg, in 64% yield (1619).
Prop.: b_{41} 121-22°, b_{14} 96-98°, n_D^{20} 1.4839-44, IR spectrum, mol. wt. no rxn. with I at 20° (1619).
Rxn. with: Et_3SnH + AiBN → $(Et_3SnCH_2CH_2)_2$ (1619).

Diallyldibutyltin (M-203) $SnC_{14}H_{28}$ $Bu_2Sn(CH_2CH:CH_2)_2$
Prop.: Colorless liq., b_4 121-23° (1857), anal. detn. by I titration (1619).
Rxn. with: Br (in $HCCl_3$ at -60°) → $Bu_2(CH_2:CHCH_2)$SnBr + $Bu_2(CH_2:CHCH_2)$SnOH (4).
i-Bu_2SnH_2 + cat. R_2AlH → polymer (1617).
$RSnH_3$ + cat. R_2AlH → crosslinked polymer; R = i-Bu, C_8H_{17} (1617).
$(Et_2HSnCH_2CH_2CH_2)_2$ + cat. R_2AlH → polymer (1617).

Triphenylvinyltin (M-196) $SnC_{20}H_{18}$ $Ph_3SnCH:CH_2$
Prop.: IR (1013), NMR (846), Moessbauer resonance (2665) and mass (2695)

spectra.
Rxn. with: ^{60}Co irradiation → polymerization (210).
Biol. prop.: Control and effect on reproduction of Musca domestica (260).

Allyltriphenyltin (M-196) $SnC_{21}H_{20}$ $Ph_3SnCH_2CH:CH_2$
Synth.: By rxn. of Ph_3SnLi with $CH_2:CHCH_2X$ in THF in N atm., X = Cl and Br, in 70 and 28% yield, resp. (1928).
Prop.: IR spectrum (1857), GLC (1367), TLC (2276).
Rxn. with: HCl (in aq. MeOH) → kinetic and thermodynamic data for C_3H_6 cleavage, deuterium effect (1367).
Biol. prop.: Control and effect on reproduction of Musca domestica (260, 1064) and Tribolum confusum (260). Activity as chemosterilant against azuki-bean weevil (2624).
Use: In catalyst for butadiene polymerization (2044).
Trade name: Dowco 187.

Other unsymmetric tetraorganotin compounds with olefinic substituents are listed in Table 102.

Table 103. Unsymmetric Tetraorganotin Compounds with Olefinic Substituents $R_nSnR'_{4-n}$

Formula*	Synth. Method**	Yield	Properties	Ref.
Me_3SnR				
R = PhCH:CH (194)	IV, PhC!CH	71	cis/trans-ratio 2.5 by NMR spectrum	1414
CH$_2$:CPh	IA	21	b_2 110°, n_D^{20} 1.5570, d_{20} 1.3050	16
	IV, PhC!CH	5	NMR spectrum	1414
MeCH:CH (195)	IVA, MeC!CH	99	b. 125-26°, cis-:trans-ratio 3:7 (a)	295
CH$_2$:CMe (199)	IVA, (CH$_2$:)$_2$C	32	b. 123.5-24.5°, n_D^{25} 1.4608, GLC (b)	295
Me$_2$C:CH	--	--	NMR spectrum, GLC (c)	2234
CH$_2$:CEt	IVA, CH$_2$:C:CHMe	23	b_{25} 55-57°, n_D^{25} 1.4656, GLC (b-d)	295
CH$_2$:CMeCH$_2$ (199)	I	81	b. 147-48°, GLC (e)	1367
CH$_2$:CHC(:CH$_2$)	--	--	(f)	--
PrCH:CH	IV, 1-C$_5$H$_8$	--	IR spectrum, GLC (g)	429
EtCH:CHCH$_2$ } (199)	IV,	--	GLC, not separated	429
MeCH:CHCH$_2$CH$_2$ }	MeCH:CHCH:CH$_2$	--	--	429
EtCH:CMe	--	--	IR spectrum	429
CH$_2$:CPr	IV, 1-C$_5$H$_8$	--'	GLC (g)	429
Me$_2$C:CHCH$_2$	IVA, isoprene	30	n_D^{25} 1.4766, IR, NMR spectra, GLC (h)	1041, 1043

* Numbers in parenthesis refer to pages in main volume.
** The characters used here correspond to the scheme in the introduction to Chapter 2.4.

(a) Rxn. with n-BuLi ⇌ MeCH:CHLi + Me$_3$SnBu with isomerization (496). (b) IR and NMR spectra (295). (c) Rxn. with I_3^- in MeOH → relative kinetic data, stereochemistry of C_4H_7I and rxn. mechanism (2234). (d) NMR spectrum, GLC (2234). (e) Rxn. with HCl in aq. MeOH → kinetic and thermodynamic data for olefin cleavage (1367). (f) Rxn. with AIBN → tin-contn. polymer (2933). (g) Yield 15-fold exceeding that for 1:2 adduct (429). (h) Rxn. with CF_3CO_2H → $Me_3SnO_2CCF_3$ + CH_2:CHCHMe$_2$ (1041, 1043).

Table 103 Continued

$R_nSnR'_{4-n}$

Formula*	Synth. Method**	Yield	Properties	Ref.
Me$_3$SnR (Cont'd.)				
R = MeCH:CMeCH$_2$	IVA, isoprene	29	n_D^{25} 1.4823, IR, NMR spectra, GLC (i)	1041, 1043
CH$_2$CMe:CHCH$_2$	IVA, isoprene	4	IR, NMR spectrum (i)	1041, 1043
CH$_2$:CHCHMeCH$_2$	IVA, isoprene	4	IR spectrum (h)	1041, 1043
MeCH:CEt	IVA, (MeCH:)$_2$C	65	b$_{1.4}$ 39-41°, n_D^{25} 1.4750, GLC, isomer ratio (b)	295
Me$_2$CHC(:CH$_2$)	IVA, CH$_2$:C:CMe$_2$	19	b$_{2.5}$ 37.5-39°, n_D^{25} 1.4678, GLC (b)	295
Me$_2$C:CMe	IVA, CH$_2$:C:CMe$_2$	52	n_D^{25} 1.4847, GLC (b)	295
c-2-C$_5$H$_7$	IVA, C$_5$H$_6$	52	At 175°, b$_{0.3}$ 30°, n_D^{25} 1.5031, GLC	1041, 1043
c-3-C$_5$H$_7$	IVA, C$_5$H$_6$	37	At 100°, IR, NMR spectrum (e, j)	1041, 1043
	IVA, C$_5$H$_6$	> 6	At 175°, b$_{1.2}$ 35-37°, n_D^{25} 1.4935, GLC	1041, 1043
	IVA, C$_5$H$_6$	29	At 100°, IR, NMR spectrum	1041, 1043
	I	28	(j-No rxn.)	1041, 1043
o-$\overline{CH_2C_6H_4CH_2CH}$	IVB, indene	98	b$_{10}$ 128°, n_D^{20} 1.5508, NMR spectrum	1953
C$_5$H$_5$ (199)	V	50	b$_1$ 56-60°, yellow liq. dec. on standing	250
BuCH:CH	IV , 1-C$_6$H$_{10}$	--	GLC	429
	IV , 1-C$_6$H$_{10}$	84	NMR spectrum, cis-:trans-ratio = 0.2	1414
CH$_2$:CHBu	IV , 1-C$_6$H$_{10}$	2	NMR spectrum	1414
Me$_2$C:CMeCH$_2$	IVA, (CH$_2$:CMe)$_2$	51	b$_{1.2}$ 46-48°, n_D^{25} 1.4850, GLC (k)	1041, 1043
CH$_2$:CMeCHMeCH$_2$	IVA, (CH$_2$:CMe)$_2$	15	b$_{1.2}$ 46-48°, n_D^{25} 1.4748, GLC (k)	1041, 1043
Me$_2$C:CEt	IVA, Me$_2$C:C:CHMe	37	n_D^{25} 1.4837, GLC (b)	295
MeCH:CPr-i	IVA, Me$_2$C:C:CMe	45	b$_{1.2}$ 39.5-40°, n_D^{25} 1.4779, GLC (b)	295
Me(CH$_2$)$_4$CH:CH	IV , 1-C$_7$H$_{12}$	--	IR spectrum, GLC	429
CH$_2$:C(C$_5$H$_{11}$)	IV , 1-C$_7$H$_{12}$	--	IR spectrum, GLC	429
BuCH:CMe	IV , 2-C$_7$H$_{12}$	--	IR spectrum, GLC	429

PrCH:CEt	IV , 3-C_7H_{12}	--	IR spectrum, GLC	429
c.-2-C_7H_{11}	III	26	b_2 74-75°, GLC (e)	1367
C_7H_9 (1)	IV (1)	--	Isomer distribution (2677), IR spectrum	2052
Me$(CH_2)_5$CH:CH	IV , 1-C_8H_{14}	--	IR spectrum, GLC	429
CH_2:C(C_6H_{13})	IV , 1-C_8H_{14}	--	GLC	429
PrCH:CPr	IV , 3-C_8H_{14}	--	IR spectrum, GLC	429
c.-3-$C_6H_9CH_2CH_2$	IVA, (m)	40	$b_{0.85}$ 70°, IR, NMR spectra	1043, 2052
c.-3-C_8H_{13}	IVA, 1,3-C_8H_{12}	4	n_D^{25} 1.5125, IR, NMR spectra, GLC	1043, 2052
c.-4-C_8H_{13}	IVB, 1,4-C_8H_{12}	48	$n_D^{24.5}$ 1.5127, IR, NMR spectra, GLC	1043, 2052
[bicyclic structure]	IVA, 1,5-C_8H_{12}	17	n_D^{26} 1.5028, IR, NMR spectra	1043, 2052
[bicyclic structure]	IVA, 1,5-C_8H_{12}	4	IR spectrum	1043, 2052
Me$(CH_2)_6$CH:CH	IV , 1-C_9H_{16}	--	IR spectrum, GLC	429
CH_2:C(C_7H_{15})	IV , 1-C_9H_{16}	--	GLC	429
BuCH:CPr	IV , 4-C_9H_{16}	--	IR spectrum, GLC	429
HC:C$(CH_2)_5$CH:CH	IV , 1,8-C_9H_{12}	low	GLC	429
Me$_2$Sn(CH:CH$_2$)$_2$ (200)	--	--	NMR spectrum	1220
Me$_2$Sn(CH$_2$CH:CH$_2$)$_2$	--	--	(n)	--

Et$_3$SnR

R = PhCH:CH (200)	IV , PhC:CH	75	NMR spectrum, cis/trans-ratio 1.5	1414
PhCH:CHCH$_2$	I	20	$b_{0.5}$ 108°, n_D^{20} 1.5645 (o)	2716
CH_2:CMeCH$_2$ (200)	--	--	Anal. det. by I titration	1619
CH_2:CHC(:CH$_2$)	--	--	(f)	--

(i) Rxn. with CF_3CO_2H → Me$_3$SnO$_2$CCF$_3$ + CH_2:CMeEt (1041, 1043). (j) Rxn. with CF_3CO_2H → Me$_3$SnO$_2$CCF$_3$ (1041, 1043). (k) IR and NMR spectra (1041, 1043). (l) 5,2-bicyclo[2.2.1]heptenyl, from norbornadiene (2052). (m) From 4-vinylcyclohexene. (n) Use in catalyst for propylene polymerization (2969). (o) Rxn. with aq. alc. NaOH → kinetic data and mechanism for cleavage rxn. (2716).

Table 103 Continued $R_nSnR'_{4-n}$

Formula*	Synth. Method**	Yield	Properties	Ref.
Et_3SnR (Cont'd.)				
R = HC≡CCH:CH	IV , (HC≡C)$_2$	--	b_2 50-51°, n_D^{20} 1.5062, d_{20} 1.2366 (p)	631
EtCH:CHCH$_2$	I	71	Anal. detn. by I titration	1619
			IR spectrum (q, r)	1619
MeCH:CHCH$_2$CH$_2$	IVA, CH$_2$:CHCH:CHMe	53	b_{12} 100-104°, n_D^{20} 1.4896 (r)	1619
CH$_2$:CH(CH$_2$)$_3$	IVC, (CH$_2$:CH)$_2$CH$_2$	17	b_{10} 95°, n_D^{20} 1.4755	1612
Me$_2$C:CHCH$_2$	IIB	39	b_{12} 103-108°, n_D^{20} 1.5030	1619
	IVA, CH$_2$:CHCMe:CH$_2$	62	(t)	1619
MeCH:CMeCH$_2$	IVA, CH$_2$:CHCMe:CH$_2$	(s)	(t)	1619
CH$_2$:CMeCH$_2$CH$_2$	IVA, CH$_2$:CHCMe:CH$_2$	14	(t)	1619
	IC	30	b_{12} 100-104°, n_D^{20} 1.4857, IR spectrum	1619
CH$_2$:CHCHMeCH$_2$	IVA, CH$_2$:CHCMe:CH$_2$	(s)	(t)	1619
CH$_2$:CHCHEt	IVA, CH$_2$:CHCH:CHMe	(s)	(q)	1619
Me(CH:CH)$_2$	IV , HC≡CCH:CHMe	63	$b_{0.2}$ 68°, n_D^{20} 1.5070, d_{20} 1.1758 (u)	2558
CH$_2$:CMeCH:CH	IV , HC≡CCMe:CH$_2$	--	$b_{0.2}$ 48-51°, n_D^{20} 1.5021	1410
{c.2-C_5H_7} {c.3-C_5H_7}	IVA, C_5H_6	48 9	b_{12} 110-13°, n_D^{20} 1.5111, mol. wt. not separated, IR spectrum	1619
BuCH:CH	IV , 1-C_6H_{10}	82	NMR spectrum, GLC, cis/trans-ratio 0.5	1414
CH$_2$:CBu	IV , 1-C_6H_{10}	3	--	1411, 1414
	IVA, 1-C_6H_{10}	--	Effect of AIBN on isomer ratio	1413
{Me$_2$C:CMeCH$_2$} {CH$_2$:CMeCMeCH$_2$}	IVA, (CH$_2$:CMe)$_2$	63 31	Mixture, b_{11} 109-11°, n_D^{20} 1.4950, IR spectrum, mol. wt.	1619 1619
EtC≡CCH:CH	IV , HC≡CC≡CEt	--	b_2 88-89°, n_D^{20} 1.5093, d_{20} 1.1757	631
c.3-$C_6H_9CH_2CH_2$	IVC, 4-CH$_2$:CHC$_6H_9$	80	$b_{0.15}$ 82°, n_D^{20} 1.5045	1612

Compound			Ref
(cyclopentadiene dimer structure)	IVA, (C$_5$H$_6$)$_2$	48	1619
	IVA, C$_5$H$_8$	--	1619
p-CH$_2$:CHC$_6$H$_4$ (201)	I	40	b$_{0.6}$ 101°, n$_D^{20}$ 1.5540 (v)	79
Et$_2$RSnC(:CH$_2$)CH:CH$_2$	--	--	R = C$_5$H$_{11}$, C$_8$H$_{17}$ (f)	--
Pr$_3$SnCH:CH$_2$ (201)	--	--	(c)	--
Pr$_3$SnCH:CHPh (201)	--	--	IR spectrum, assignment	1410
i-Pr$_3$SnCH:CH$_2$	--	--	(c)	--
i-Pr$_3$SnCH$_2$CH:CH$_2$ (202)	I	82	b$_7$ 100-101°, GLC (e)	1367
(CH$_2$CH$_2$CH)$_2$Sn(CH:CH$_2$)$_2$	I, C$_3$H$_5$MgBr	--	Use: herbicide, pesticide	1873
(CH$_2$CH$_2$CH)$_2$PhSnCH:CH$_2$	I, C$_2$H$_3$MgBr	--	Use: herbicide, pesticide	1873
Bu$_3$SnR				
R = CH$_2$:CH (195)	--	--	(w)	--
CH$_2$:CHCH$_2$ (202)	--	--	(x)	--
PhCH:CHCH$_2$	IB	10	b$_{0.5}$ 138-42°, n$_D^{20}$ 1.5185 (o, y)	2716
CH$_2$:CHC(:CH$_2$)	ID	76	b$_{0.4}$ 80-90°, IR, UV, NMR spectra (z)	13
C$_5$H$_5$ (202)	V	99	b$_{0.8}$ 134°, n$_D^{20}$ 1.5047	250
{Me$_2$C:CMeCH$_2$ CH$_2$:CMeCHMeCH$_2$}	IVA, (CH$_2$:CMe)$_2$	65	Mixture b$_{0.1}$ 100-102°, n$_D^{20}$ 1.4890, IR spectrum	1619
		29		1619
p-CH$_2$:CHC$_6$H$_4$ (202)	--	--	b$_3$ 130-31°, IR spectrum (aa)	617

(p) NMR spectrum (1688). Et$_3$SnCH$_2$CH:CHEt containing some Et$_3$SnCHEtCH:CH$_2$ from IR spectrum (1619). (r) Mixture of R = EtCH:CHCH$_2$ and MeCH:CHCH$_2$CH$_2$ (1619). (s) By-prod. (t) Four component system, structure by IR spectra, not sepd. (1619). (u) IR spectrum, impure matl. (2558). (v) Rxn. with K or Na in (MeOCH$_2$)$_2$ → ESR spectra of prod. (524). (SCN)$_2$ → relative reactivity; prod. SnCl$_6$H$_{22}$N$_2$S$_2$, m. 84° (863). (w) Rxn. with PhCH:CH$_2$ or CH$_2$:CMeCO$_2$Me → radical initiated polymerization (1522). (x) Rxn. with PrCOSH → Bu$_3$Sn(CH$_2$)$_3$SCOR (534), with Cl$_3$CCO$_2$Na in (MeOCH$_2$)$_2$ → Bu$_3$SnCH$_2$CHCH$_2$CCl$_2$ (2047). (y) IR and UV spectra (2716). (z) Rxn. with AIBN → cis[CH$_2$C(SnBu$_3$):CHCH$_2$]$_x$ (14, 2933). (aa) Decolorizes Br in HCCl$_3$ and KMnO$_4$ soln. (617).

Table 103 Continued

Formula*	Synth. Method**	Yield	Properties	Ref.
Bu$_3$SnR (Cont'd.)				
R = p-CH$_2$:CMeC$_6$H$_4$	IA	39	b$_3$ 170-72°, n$_D^{20}$ 1.5469, IR spectrum (aa, bb)	617
p-EtCH:CHC$_6$H$_4$	IA	36	b$_5$ 191-93°, n$_D^{20}$ 1.5338, d$_{20}$ 1.097 (bb)	617
Bu$_2$SnR$_2$				
R = CH$_2$:CH (196)	--	--	NMR (1220) and mass (1648) spectra (cc)	--
MeCH:CH	IA, C$_3$H$_5$MgBr	--	b$_1$ 106-108°, IR, NMR spectra (dd)	499
CH$_2$:CMeCH$_2$ (203)	--	--	IR spectrum	1857
PrCH:CH	IV , PrC:CH	--	Colorless oil, IR, NMR spectra	499
(Me$_2$C:CMeCH$_2$) (CH$_2$:CMeCHMeCH$_2$)	IVA, (CH$_2$:CH)$_2$	30 / 68	Mixture, b$_{12}$ 181-85°, n$_D^{20}$ 1.5065, IR spectra	1619 / 1619
CH$_2$:CH(CH$_2$)$_6$ (203)	IVC, 1,7-C$_8$H$_{14}$	73	b$_{0.01}$ 135°, n$_D^{20}$ 1.4826 (ee)	1612
R + MeCH:CHCH$_2$	--	--	R = CH$_2$:CHCH$_2$, IR spectrum	4
R + CH$_2$:CHCH$_2$CH$_2$	I , Bu$_2$RSnBr	63	R = CH$_2$:CHCH$_2$, b$_{0.6}$ 98-102°, IR spectrum	4
R + CH$_2$:CH(CH$_2$)$_3$	I , Bu$_2$RSnBr	62	R = CH$_2$:CHCH$_2$, b$_{0.3}$ 122-25°, n$_D^{25}$ 1.4869-73	4
Bu(CH$_2$CHMeCH$_2$)$_2$SnCH:CH$_2$	--	--	Use: herbicide, pesticide (ff)	1873
i-Bu$_3$SnCH$_2$CH:CH$_2$	I	47	b$_3$ 123-24°, n$_D^{20}$ 1.4972, d$_{20}$ 1.3631 (gg)	562
{i-Bu$_3$SnCH$_2$CMe:CMe$_2$ / i-Bu$_3$SnCH$_2$CHMeCMe:CH$_2$}	IVA, (CH$_2$:CMe)$_2$	74 / 17	Mixture, not sepd , b$_{0.15}$ 88-91°, n$_D^{20}$ 1.4919, IR spectra	1619 / 1619
i-Bu$_3$SnC$_6$H$_4$CH$_2$CH:CH$_2$	I	52	b$_3$ 173-75°, n$_D^{20}$ 1.5332, d$_{20}$ 1.1802 (gg)	562
i-Bu$_2$Sn(CH$_2$CH:CH$_2$)$_2$	I	44	m. 245-46°	562

Compound				
1-Bu$_2$Sn[(CH$_2$)$_6$CH:CH$_2$]$_2$	IVC, 1,7-C$_8$H$_{14}$	62	b$_{0.001}$ 135°, n$_D^{20}$ 1.4829, d$_{20}$ 1.003 (hh)	1612
1-Bu$_2$Sn(C$_6$H$_4$CH$_2$CH:CH$_2$)$_2$	I	33	--	562
(i-C$_5$H$_{11}$)$_3$SnC$_6$H$_4$CH:CH$_2$	IA	37	b$_3$ 150-52°, IR spectrum (aa)	617
(i-C$_5$H$_{11}$)$_3$SnC$_6$H$_4$CMe:CH$_2$	IA	50	b$_4$ 190-92°, n$_D^{20}$ 1.5256, d$_{20}$ 1.1219 (aa, bb)	617
(EtCHCH$_2$CH)$_3$SnCH:CH$_2$	--	--	Use: herbicide, pesticide (ii)	1873
(n-C$_6$H$_{13}$)$_3$SnCH$_2$CH:CH$_2$	I	51	b$_3$ 189-91°, n$_D^{20}$ 1.5030, d$_{20}$ 1.0915 (gg)	562
(n-C$_6$H$_{13}$)$_3$SnC(:CH$_2$)CH:CH$_2$	--	--	(f)	--
(n-C$_6$H$_{13}$)$_3$SnC$_6$H$_4$CH$_2$CH:CH$_2$	I	50	b$_3$ 200-201°, n$_D^{20}$ 1.5490, d$_{20}$ 1.1431 (gg)	562

(c.C$_6$H$_{11}$)$_3$SnR

R = CH$_2$:CHCH$_2$				
	I	41	GLC (e)	1367
{EtCH:CHCH$_2$ / MeCH:CHCH$_2$CH$_2$}	IVA, Me(CH:CH)$_2$H	47 41	Mixture, not sepd., b$_{0.001}$ 148-51°, n$_D^{20}$ 1.5411, IR spectra	1619 1619
{Me$_2$C:CMeCH$_2$ / CH$_2$:CMeCHMeCH$_2$}	IVA, (CH$_2$:CMe)$_2$	56 25	Mixture, not sepd., b$_{0.001}$ 166-69° n$_D^{20}$ 1.5458, IR spectra	1619 1619
C$_6$H$_4$CH:CH$_2$	--	--	m. 104-105° (jj)	284
(n-C$_7$H$_{15}$)$_3$SnCH$_2$CH:CH$_2$	I	47	b$_3$ 220-21°, n$_D^{20}$ 1.5230, d$_{20}$ 1.1017 (gg)	562
(n-C$_7$H$_{15}$)$_3$SnC$_6$H$_4$CH$_2$CH:CH$_2$	I	53	b$_3$ 228-29°, n$_D^{20}$ 1.5558, d$_{20}$ 1.1385	562
(n-C$_8$H$_{17}$)$_3$SnCH$_2$CH:CH$_2$	I	66	b$_3$ 235-37°, n$_D^{20}$ 1.5460, d$_{20}$ 1.1049 (gg)	562
(n-C$_8$H$_{17}$)$_3$SnC(:CH$_2$)CH:CH$_2$	--	--	(f)	--
(n-C$_8$H$_{17}$)$_3$SnC$_6$H$_4$CH$_2$CH:CH$_2$	I	43	b$_3$ 257-59°, n$_D^{20}$ 1.5570, d$_{20}$ 1.1031 (gg)	562
[CH$_2$CH$_2$CH(CH$_2$)$_5$]$_3$SnCH$_2$CH:CHPr	--	--	Use: herbicide, pesticide	1873

(bb) IR spectrum (617). (cc) Use as corrosion inhibitor for iron, zinc and their alloys (2101). (dd) Rxn. with (CH$_2$CO)$_2$NBr in CCl$_4$ → MeCH:CHBr (499). (ee) Correct formula main volume p. 203, line 10 from CH$_2$:CHC$_6$H$_{13}$-n to CH$_2$:CH(CH$_2$)$_6$. (ff) Rxn. with AcOH → Bu(CH$_2$CHMeCH$_2$)$_2$SnOAc (1873). (gg) Surface tension (562). (hh) Rxn. with Bu$_2$SnH$_2$ + cat. R$_2$AlH → polymer (1617). (ii) Rxn. with (MeCHCMe$_2$CH)$_3$SnH → R$_3$SnCH$_2$CH$_2$SnR'$_3$ (1873).

Table 103 Continued

Formula*	Synth. Method**	Yield	Properties	Ref.
$(n\text{-}C_9H_{19})_3SnCH_2CH:CH_2$	I	49	m. 289-90°	562
Ph_3SnR				
R = PhCH:CH (204)	IV , PhC:CH	90	cis/trans-ratio 0.26	1414
MeCH:CH	IIA	low	m. 100-101°, IR spectrum	1928
	IIB	--	--	1928
PhCH:CHCH$_2$	IB	44	m. 69.5°, IR, UV spectra	2716
EtCH:CH	--	--	(kk)	--
MeCH:CHCH$_2$ (205)	III	65	White, m. 57-59° (e, 11)	1877
	III	--	m. 51-52°, GLC	1367
CH$_2$:CMeCH$_2$ (205)	--	--	(kk)	--
CH$_2$:CHCHMe	--	--	(kk)	--
CH$_2$:CHCH:CH	--	--	(kk)	--
$\begin{Bmatrix}\text{EtCH:CHCH}_2\\\text{MeCH:CHCH}_2\text{CH}_2\end{Bmatrix}$	IVA, Me(CH:CH)$_2$H	53 40	Mixture, not sepd., b$_{0.001}$ 171-74°, n_D^{20} 1.6162	1619 1619
$\begin{Bmatrix}\text{c.2-C}_5\text{H}_7\\\text{c.3-C}_5\text{H}_7\end{Bmatrix}$	IVA, C$_5$H$_8$	>83 <9	Mixture, not sepd., m. 84°, soly., mol. wt. (m. 82-84° 1416)	1619 1619
C$_5$H$_5$ (197)	V	71	m. 129° (kk, mm)	250
C$_9$H$_7$ (205)	V	74	m. 129°	250
BuCH:CH	IV , BuC:CH	80	NMR spectrum, cis/trans-ratio 5	1414
$\begin{Bmatrix}\text{Me}_2\text{C:CMeCH}_2\\\text{CH}_2\text{:CMeCHMeCH}_2\end{Bmatrix}$ (205)	IVA, (CH$_2$:CMe)$_2$	<9 76	Mixture, not sepd., m. 65-67° --	1619 1619
$\begin{Bmatrix}\text{c.2-C}_6\text{H}_9\\\text{c.3-C}_6\text{H}_9\end{Bmatrix}$	IVA, 1,3-C$_6$H$_8$	36 52	Mixture, not sepd., m. 127-30°, soly.	1619 1619

![cyclopentadiene structure]	IVA, (C$_5$H$_6$)$_2$	--	Colorless cryst., m. 86°, soly.	1619
	IVA, C$_5$H$_6$	--	--	1619
CH$_2$:CHC$_6$H$_4$ (197)	--	--	m. 109-10° (jj-284, 461, nn, oo)	284
CH$_2$:CH(C$_6$H$_4$)$_2$ (205)	--	--	m. 142-43° (jj-284, 461)	284
1,8-PhC!CC$_{10}$H$_6$	IA	28	White cryst., m. 219-20°, mol. wt. (pp)	1278
9-C$_{13}$H$_8$:CPh	IA, (gg)	--	Bright yellow cryst., m. 286-89° (pp)	1278
Ph$_2$Sn(CH:CH$_2$)$_3$ (197)	--	--	(cc)	2276
Ph$_2$Sn(CH$_2$CH:CH$_2$)$_2$ (205)	--	--	TLC (rr)	2665
PhSn(CH:CH$_2$)$_3$ (206)	--	--	Moessbauer resonance spectrum	2665
PhSn(CMe:CH$_2$)$_3$	IIA	--	b$_{0.1}$ 70-72°, IR spectrum, GLC	496
PhSn(CH$_2$CH:CH$_2$)$_3$	--	--	TLC	2276
(C$_6$F$_5$)$_2$Sn(CH:CH$_2$)$_2$	I , ArMgX	75	b$_{0.3}$ 107-109°, n$_D^{20}$ 1.5014	428

(jj) Rxn. with PhCH:CH$_2$ and cat. AIBN → copolymerization, relative reactivity (284). (kk) Control and effect on reproduction of Musca domestica (260). (ll) Rxn. with PhLi → Ph$_4$Sn + MeCH:CHCH$_2$Li (1877). (mm) Rxn. with O → formation of endoperoxide (276). (nn) Rxn. with MeC$_6$H$_4$CH:CH$_2$ → copolymer (461, 569), activity as scintillator (936). (oo) Moessbauer resonance spectrum (1309). (pp) UV spectrum (1278). (qq) Synth. from Ph$_3$SnCl and 2,2'-PhC!CC$_6$H$_4$C$_6$H$_4$MgI (1278). (rr) Use in catalyst for butadiene polymerization (2044).

2.4.1 UNSYMMETRIC TETRAORGANOTIN COMPOUNDS CONTAINING HALOGEN SUBSTITUTED OLEFINS

The unsymmetric tetraorganotin compounds listed in Table 104 are prepared by methods from the ensuing scheme.

I Grignard reaction:
$Et_3SnCl + R'MgBr \rightarrow Et_3SnR'$.

II Synthesis with organolithium derivatives:
$R_3SnCl + R'Li \rightarrow R_3SnR'$.

III Hydrostannation reactions:
(IIIA) $R_nSnH_{4-n} + RC\vdots CR' \rightarrow R_nSn(CR\!:\!CHR')_{4-n}$,
(IIIB) $R_2R'SnH + C_2F_4 \rightarrow R_2R'SnCF_2CF_2H$.

IV Dehydrohalogenation reactions:
(IVA) With Me_3SnH ,
(IVB) With Me_2SnH_2 .

V Reaction of triorganotin amides:
(VA) $R_3SnNR_2 + H'R \rightarrow R_3SnR'$,
(VB) $R_3SnNR_2 + RC\vdots CR \rightarrow R_3SnCR\!:\!CRNR_2$.

VI Miscellaneous reactions:
(VIA) $Me_nSn(C\vdots CCF_3)_{4-n} + Me_3SnCF_3 \rightarrow Me_nSn[\overline{C\!:\!C(CF_3)}CF_2]_{4-n}$,
(VIB) $R_3SnC\vdots CR' + C_5Cl_6 \rightarrow 2,3\text{-}R_3SnC_7Cl_6R'$; $R' = Ph, H$,
(VIC) $Et_3SnOMe + RCH\!:\!CHMe \rightarrow Et_3SnCMe\!:\!CHR$.

R = organic group, R' = olefinic functionally substituted organic group, n = 2, 3.

Perfluorovinyltrimethyl-tin $SnC_5H_9F_3$ $Me_3SnCF\!:\!CF_2$

<u>Synth.</u>: By rxn. of Me_3SnBr with $CF_2\!:\!CFMgBr$ in dry THF (731).
By-prod. in rxn. of Me_3SnH with $Me_2Sn(CF\!:\!CF_2)_2$ at 70° or under UV irradiation (731).
By-prod. in rxn. of $Me_3SnMn(CO)_5$ with C_2HF_3 under UV irradiation in pentane (874).
<u>Prop.</u>: b. 110.5°, IR and NMR spectra, GLC (874).
<u>Rxn. with</u>: Me_3SnH (at 50-70° in the dark) $\rightarrow Me_3SnCF\!:\!CHF$ (731).
Me_3SnH (at 25° under UV irradiation) $\rightarrow Me_3SnF$ + cis- + trans-$Me_3SnCF\!:\!CHF$ + $Me_3SnC_2H_2F$ (731, 2209), + $(Me_3Sn)_2CFCHF_2$(?) + $Me_3SnCHFCF_2SnMe_3$ (?) (2209).

1,2-Difluorovinyltrimethyl-tin $SnC_5H_{10}F_2$ $Me_3SnCF\!:\!CHF$

<u>Synth.</u>: By rxn. of Me_3SnH with $Me_3SnCF\!:\!CF_2$ at 25° under UV irradiation, in < 60% yield by GLC, predominantly in cis- form (731, 2209).
By-prod. in rxn. of $Me_2Sn(CF\!:\!CF_2)_2$ at 70° or at 25° under UV irradiation (731, 2209).
By-prod. in rxn. of Me_3SnH with cis-$Me_3SnCF\!:\!CHF$, in 10% yield of trans-form by GLC (731).

Prop.: IR and NMR spectra (731, 2209), GLC, sepn. of cis-form (731).
Rxn. with: Me$_3$SnH → Me$_3$SnC$_2$H$_2$F + trans-Me$_3$SnCF:CHF (731).
CF$_3$CO$_2$H → CHF:CHF (731).

2,3-Methyl(perchlorobicyclo-   SnC$_{11}$H$_{12}$Cl$_6$
[2.2.1]heptadienyl)trimethyltin
Synth.: By rxn. of Me$_3$SnC⋮CMe with C$_5$Cl$_6$ in xylene in inert atm. in presence of hydroquinone (1883).
Prop.: Colorless prisms, m. 83-85°, IR, UV and NMR spectra, stable in inert atm. at 0°, slow dec. at 20° → Me$_3$SnCl (1883).
Rxn. with: Thermal dec. → Me$_3$SnCl + 2,3MeC$_7$HCl$_6$ + 2-MeC$_7$Cl$_7$ (1883).
Cl in Et$_2$O-MeOH → Me$_3$SnCl + MeC$_7$Cl$_7$ (1883).
Br in MeOH → MeC$_7$Cl$_6$Br (1883).
UV irradiation in Et$_2$O → 2,3-Me(Me$_3$Sn)C$_7$Cl$_6$ (1883).

Dimethyldiperfluorovinyltin (M-207)   SnC$_6$H$_6$F$_6$ Me$_2$Sn(CF:CF$_2$)$_2$
Prop.: IR and NMR spectra (731).
Rxn. with: Me$_3$SnH (at 70°) → Me$_2$SnF$_2$ + Me$_3$SnCF:CF$_2$ + cis-Me$_3$SnCF:CHF + H + C$_2$F$_3$H (731).
Me$_3$SnH (at 25° under UV irradiation) → Me$_2$SnF$_2$ + Me$_3$SnF + cis- + trans-Me$_3$Sn-CF:CHF + Me$_3$SnCF:CF$_2$ + (cis-CHF:CF)$_2$SnMe$_2$ + Me$_3$SnC$_2$H$_2$F + Me$_3$SnCH:CF$_2$ (731).
Me$_2$SnH$_2$ (at 60° in the dark) → Me$_2$(CF$_2$:CF)SnH + cis- + trans-(CHF:CF)$_2$SnMe$_2$ + C$_2$HF$_3$ (731).
Me$_2$SnH$_2$ (at 25° under UV irradiation) → Me$_2$SnF$_2$ + cis- + trans-(CHF:CF)$_2$SnMe$_2$+ Me$_2$Sn(CH:CF$_2$)$_2$ + Me$_2$Sn(C$_2$H$_2$F)$_2$ + H (731).

Other tetraorganotin compounds with halogen substituted olefins are listed in Table 104 and 105.

2.4.2 UNSYMMETRIC TETRAORGANOTIN COMPOUNDS CONTAINING PSEUDOHALOGEN SUBSTITUTED OLEFINS

All compounds are listed at the bottom of Table 104.

2.4.3 UNSYMMETRIC TETRAORGANOTIN COMPOUNDS CONTAINING FUNCTIONALLY SUBSTITUTED OLEFINS

The olefinic tetraorganotin compounds in this chapter contain oxygen, sulfur, nitrogen, phosphorus, and antimony in the olefinic moiety. A deplorably large number of the reported compounds are only characterized by means of spectroscopy due to difficulty in separation and handling. All derivatives are summarized in Table 105.

Additional tetraorganotin compounds with olefinic unsaturation may be found in Chapters 2.5.3, 2.6.3.4, 2.7.1.1.1, 2.7.1.1.3, 2.7.1.2, 2.7.2, 2.7.8 and 7.

Table 104. Unsymmetric Tetraorganotin Compounds Containing Halogen and Pseudohalogen Substituted Olefins $R_nSnR'_{4-n}$

Formula*	Synth. Method**	Yield	Properties	Ref.
Me₃SnR				
R = C_2H_2F	IVA (a)	30	IR and NMR spectra, GLC	731
	IVA (b)	15	(c)	731
$CF_2:CH$	IVA (d)	2	IR and NMR spectra, GLC	731, 2209
	IVA (c)	2	(e)	731, 2209
$CCl_2:CCl$	VA	80	b_{20} 111°, IR spectrum, GLC	865
$\{CF_3CH:CH \atop CH_2:C(CF_3)\}$	IIIA	--	Mixture, not sepd., b. 129°, IR NMR spectra	923
$CF_3CH:C(CF_3)$	IIIA	100	b. 124°, IR, NMR spectra	923
	(f)	--	IR, NMR spectra	108
$\overline{CF_2C(CF_3):C}$	VIA	39	(g)	2332
$Cl_2CHCH_2CH:CHCH_2$	IIIA (h)	35	$b_{0.7}$ 64–66°, n_D^{25} 1.5131 (i)	1877
$2,3\text{-}MeC_7Cl_6$	VIB	--	Colorless cryst., m. 101.5–102.5° (j)	1883
$2,3\text{-}BrC_7Cl_6$	II	36	m. 85–86.5° (k)	1884
Me₂SnR₂				
R = C_2H_2F	IVA (b)	low	IR and NMR spectra	731
	IVB (b)	low	--	731
$CF_2:CH$	IVA (b)	low	IR and NMR spectra	731
	IVB (b)	low	--	731
$CHF:CF$	IVA (b)	6	IR and NMR spectra, cis-form	731
	IVB (b)	--	Cis- and trans-form	731
$\overline{CF_2C(CF_3):C}$	VIA	--	(g)	2332
$Me_2(CHF_2CF_2)SnCF:CF_2$	IIIB	--	IR spectrum, mol. wt.	731

Compound	Type	Yield	Properties	Ref.
$Et_3SnC(CF_3):CHCF_3$	IIIA	100	b_{24} 86°, IR NMR spectra	923
2,3-$Et_3SnC_7Cl_6H$	VIB	44	$b_{0.05}$ 112°, n_D^{20} 1.5595, d_{20} 1.5919	2507
p-$Et_3SnC_6H_4CF:CFCl$ (207)	I	40	b_4 125°, n_D^{20} 1.5450	527
	II	31	--	527
$Bu_3SnC(CH_3):CHCF_3$	IIIA	100	$b_{0.001}$ 72°, IR spectrum	923
$Bu_3SnCH_2CH:CHCH_2CHCl_2$	IIIA (h)	17	$b_{0.1}$ 108-10°, n_D^{25} 1.4999 (i)	1877
p-$Bu_3SnC_6H_4CF:CFCl$	I	40	b_2 162-65°, n_D^{20} 1.5130	527
	II	45	^{19}F NMR spectrum (1872)	527
$Bu_2Sn[C(CF_3):CHCF_3]_2$	IIIA	100	$b_{0.001}$ 63°, IR, NMR spectra	108, 923
$Me_3SnCH:CHCN$	IIIA (1)	--	NMR spectrum, cis- and trans-form	1414
$Me_3SnC(CN):CH_2$	IIIA	90	b_{18} 78-80°, n_D^{20} 1.4982 (m)	313
	IIIA	100	NMR spectrum	1414
	IIIA	90	IR, NMR spectra, GLC (n)	1412
$Me_3SnC(CN):CHD$	IIIA	--	IR, NMR spectra, cis/trans-ratio 20	1412, 1414
$Me_3SnC(CN):CHCN$	IIIA	80	NMR spectrum, struct.	313, 1414
$Et_3SnC(CN):CH_2$	IIIA	100	$b_{0.1}$ 53-54°, GLC (m)	1412
	IIIA	--	Solvent effect, IR, NMR spectra	1412
$Et_3SnC(CN):CHD$	IIIA	95	NMR spectrum, cis/trans-ratio	1412
$Et_3SnC(CN):CHCN$	IIIA	80	NMR spectrum, struct.	1414
	IIIA	--	Solvent effect	1412
$Bu_3SnC(CN):CH_2$	IIIA	90	$b_{0.1}$ 100°, n_D^{20} 1.4928, GLC (m)	313
	IIIA	100	NMR spectrum	1412, 1414
$Bu_3SnC(CN):CHD$	IIIA	--	NMR spectrum, cis/trans-ratio 20	1412

* Numbers in parenthesis refer to pages in main volume.
** The characters used here correspond to the synthetic scheme in the introduction to Chapter 2.4.1.
(a) Synth. from $Me_3SnCF:CHF$. (b) Synth. from $Me_2Sn(CF:CF_2)_2$. (c) Rxn. with $CF_3CO_2H \rightarrow C_2H_3F$ (731). (d) Synth. From $Me_3SnCF:CF_2$. (e) Rxn. with $CF_3CO_2H \rightarrow CF_2:CH_2$ (731). (f) By thermal dec. of $[Me_3SnC(CF_3):]_2$ (108). (g) Rxn. with $H_2O \rightarrow CF_3C:CHCF_2$ (2332). (h) Synth. from $CH_2:CHCHCl_2CH_2$ (1877). (i) IR and NMR spectra (1877). (j) IR, UV and NMR spectra, stable at 0° in inert atm. slow dec. at r.t. $\rightarrow Me_3SnCl$, struct. 2,3-phenylperchlorobi-cyclo[2.2.1]heptadienyltrimethyltin (1883). (k) (Far) IR spectrum, column chromatography, soly., struct. 2,3-bromoperchlorobicyclo[2.2.1]heptadienyltrimethyltin (1884). (l) Synth. in polar solvent (1414). (m) IR and NMR spectra (313). (n) Rxn. with R_3SnH + AIBN $\rightarrow$ isomerization (1413).

Table 105. Unsymmetric Tetraorganotin Compounds Containing Functionally Substituted Olefins $R_nSnR'_{4-n}$

Formula*	Synth. Method**	Yield	Properties	Ref.
Me₃SnR				
R = Me₂NCPh:CCl	VB	40	$b_{0.04}$ 60-89° (a)	854
EtOCH:CH	(b)	--	NMR spectrum, GLC, not isolated	1413
Bu₂SbCH:CH	IIIA	--	$b_{0.5}$ 102-104°, n_D^{20} 1.5400	1601
MeO₂CCH:CH	IIIA	42	b_{27} 83-88°, n_D^{20} 1.4784	1410
	IIIA	8	GLC (c)	311, 1414
MeO₂CC(:CH₂)	IIIA	32	GLC (c, d)	1412, 1414
EtO₂CCH:CH	IIIA	35	b_{12} 74-76°, n_D^{20} 1.4700	1410
	IIIA	7	NMR spectrum (311) (c)	311, 1414
EtO₂CC(:CH₂)	IIIA	13-32	GLC (311) (c)	1412, 1414
{MeCH:C(CO₂Et)} {EtO₂CCH:CMe}	IIIA	65	NMR spectrum (e) NMR spectrum (e)	313, 1411 313, 1411
EtO₂CCH:C(CO₂Et)	IIIA	100	$b_{0.07}$ 71-72°, n_D^{20} 1.4845, GLC (f)	313, 1412
EtO₂C(NMe₂):C(CO₂Et)	VB	98	$b_{0.02}$ 100° (g)	854
Et₃SnR				
R = EtOCH:CH₂	IIIA	96	NMR spectrum, cis/trans-ratio 18 (b, h, i)	1414
	IIIA	98	b_9 86-88°, n_D^{20} 1.4825, d_{20} 1.1956	2483
CH₂:C(OEt)	IIIA	4	Not isolated, NMR spectrum	1414
BuOCH:CH	IIIA	69	b_9 115.5°, n_D^{20} 1.4805, d_{20} 1.1501 (i)	2483
BuSCH:CH	IIIA	70	NMR spectrum, cis/trans-ratio 3.4	1414
{HOCH₂CH:CH} {HOCH₂C(:CH₂)}	IIIA	75	Not sepd., cis/trans-ratio 1.5, NMR spectrum	313, 1414
BuOCHMeCH₂CH:CH	IIIA	85	$b_{0.5}$ 106-107°, n_D^{20} 1.4748, d_{20} 1.1318	1912, 2078

MeO$_2$CCH:CH	IIIA	48	b$_4$ 73-75°, n$_D^{20}$ 1.4900	1410
MeO$_2$CC(:CH$_2$)	IIIA	--	GLC (c)	311, 1414
MeO$_2$CC(:CHD)	IIIA (k)	32	NMR spectrum, cis/trans-ratio 0.08 (1)	311, 1412, 1414
EtO$_2$CCH:CH	IIIA	44	b$_1$ 85-87°, n$_D^{20}$ 1.4860	1412
EtO$_2$CCH:CH	IIIA	--	GLC (c)	1410
EtO$_2$CC(:CH$_2$)	IIIA	--	GLC (c)	311, 1414
HOCH$_2$CH$_2$CH:CH	IIIA	81	b$_{1.5}$ 86-87°, n$_D^{20}$ 1.4930, d$_{20}$ 1.1827	311
CH$_2$:CHOCH$_2$CH$_2$CH:CH	IIIA	79	b$_2$ 106-107°, n$_D^{20}$ 1.5031, d$_{20}$ 1.2511	2078
BuOCHMeOCH$_2$CH$_2$CH:CH	IIIA	85	b$_{0.4}$ 116°, n$_D^{20}$ 1.4738, d$_{20}$ 1.1129 (j)	2078
Et$_3$SnOCH$_2$CH$_2$CH:CH	IIIA	96	b$_1$ 128-32°, n$_D^{20}$ 1.5100, d$_{20}$ 1.3302	1912, 2078
BuOCHMeOCHMeCH:CH	IIIA	--	b$_2$ 126°, n$_D^{20}$ 1.4702, d$_{20}$ 1.1067	1538
MeOCH:CHCH:CH	IIIA (m)	53	b$_{0.03}$ 100-102°, n$_D^{20}$ 1.5102, d$_{20}$ 1.2656	2078
			b$_{0.007}$ 79-81°, n$_D^{20}$ 1.5222, impure (n)	2558
EtSCH:CHCH:CH	IIIA (m)	50	b$_{0.03}$ 72-73°, n$_D^{20}$ 1.5230, d$_{20}$ 1.2502 (n)	1410
Et$_2$NCH:CHCH:CH	IIIA (m)	46	b$_1$ 110-11°, n$_D^{20}$ 1.5272, d$_{20}$ 1.2610 (n)	2558
EtO$_2$CCH:C(CO$_2$Et)	IIIA (208)	--	b$_{0.04}$ 90-93°, GLC (o)	2558
EtO$_2$CCD:C(CO$_2$Et)	IIIA (k)	--	--	313, 1412, 1414
HOCMe$_2$CH:CH	IIIA	--	b$_{15}$ 118-24°, n$_D^{20}$ 1.4950	1412
				1410

* Numbers in parenthesis refer to pages in main volume.
** The characters used here correspond to the scheme in the introduction to Chapter 2.4.1.

(a) Decomposing during distillation, IR and NMR spectrum (2406). (b) Rxn. of Et$_3$SnCH:CHOEt with Me$_3$SnH ⇌ Me$_3$SnCH:CHOEt + Et$_3$SnH, ratio of cis- and trans-isomers, equil. accallerated by AIBN, rxn. inhibited by phenoxyl (1413). (c) By-prod., isomeric mixture of R = R'O$_2$CCH:CH and R'O$_2$CC(:CH$_2$), not sepd., structure by IR and NMR spectra, cis/trans-ratio (311, 313, 1414). (d) Rxn. with Br → CH$_2$:CBrCO$_2$Me (313), effect of R$_3$Sn* on isomer distribution (1413). (e) GLC (1411), struct. and isomer ratio from NMR spectra (1411, 1414), effect of AIBN on isomer ratio (1413). (f) IR and NMR spectra (313), cis/trans-ratio 0.11-0.13 (313, 1414), cis/trans-ratio 9 (2483). (i) 1.4700, IR and NMR spectra (2406). (h) NMR spectrum (1413, 2483), GLC (1413), cis/trans-ratio 9 (2483). (i) Colorless liquid, stable in air (2483). (j) IR spectrum (1912, 2078). (k) Synth. with Et$_3$SnD (1412). (1) Rxn. with Et$_3$SnD + cat. AIBN → isomerization (1413). (m) From XCH:CHC:CH; X = MeO, EtS, Et$_2$N (2558). (n) IR spectrum (2558). (o) IR (313) and NMR (313, 1412, 1414) spectra, cis/trans-ratio 1.9 (313, 1414).

Table 105 Continued

Formula*	Synth. Method**	Yield	Properties	Ref.
Et_3SnR (Cont'd.)				
R = $BuOCHMeCMe_2CH:CH$	IIIA	--	b_2 125-26°, n_D^{20} 1.4731, d_{20} 1.1034	2078
$Et_3SnOCMe_2CH:CH$	IIIA	70	b_1 125-27°, n_D^{20} 1.5032, d_{20} 1.2950	1538
$Et_3SnOCHMeCH:CMe$	VIC	60	b_1 126-27°, n_D^{20} 1.5030, d_{20} 1.2942	1538
$Pr_3SnCH:CHOEt$	IIIA	82	b_1 89-90.5°, n_D^{20} 1.4792, d_{20} 1.1275 (i)	2483
$Pr_3SnCH:CHOBu$	IIIA	61	$b_{1.5}$ 101-102°, n_D^{20} 1.4791, d_{20} 1.0990	2483
$\{Pr_3SnCH:CHCO_2Me$	IIIA	--	Not isolated, IR, NMR spectra (f)	311, 1414
$\{Pr_3SnC(:CH_2)CO_2Me\}$	IIIA	22	--	311, 1414
$Bu_3SnCH:CHOEt$	IIIA	93	b_2 123-24°, n_D^{20} 1.4787, d_{20} 1.0879 (i)	2483
$Bu_3SnCH:CHOBu$	IIIA	78	b_1 136-37°, n_D^{20} 1.4790, d_{20} 1.0669 (i)	2483
$\{Bu_3SnCMe:CHCO_2Et$ $\{Bu_3SnC(CO_2Et):CHMe\}$	IIIA	--	Not sepd., NMR spectrum cis/trans-ratio 4 (1414)	313
$Bu_3SnC(CO_2Et):CHCO_3Et$	IIIA	--	$b_{0.003}$ 130°, cis/trans-ratio 0.1 (f)	313, 1412, 1414
$Bu_2Sn(CH:CHCH_2NEt_2)$	IIIA	72	$b_{0.05}$ 148-52°, IR, NMR spectra, trans-form	499
Ph_3SnR				
R = $EtOCH:CH$	IIIA	90	NMR spectrum, cis/trans-ratio 29	1414
$Bu_2SbCH:CH$	IIIA	--	b_1 169-72°, n_D^{20} 1.5910	1601
$Ph_2SbCH:CH$	IIIA	--	$b_{0.001}$ 206-210°, n_D^{20} 1.6900	1601
$Ph_2PCPh:CH$	VB (p)	78	m. 45°	1841
$\{HOCH_2CH:CH$ (209) $\{HOCH_2C(:CH_2)\}$	IIIA	76 4	NMR spectrum, GLC, cis/trans-ratio 0.9 NMR spectrum GLC	1414 1414
$EtO_2CCH:C(CO_2Et)$	IIIA	85	Not isolated, IR, NMR spectra	1412, 1414

2276	$CH_2{:}CHCH_2CH(CH_2OH)CH_2$	--	--	TLC

(p) Synth. from Ph_3SnPPh_2 and $PhC{\vdots}CH$ (1841).

2.5 UNSYMMETRIC TETRAORGANOTIN COMPOUNDS WITH ACETYLENIC SUBSTITUENTS

This chapter comprises organotin derivatives of unsubstituted acetylenic hydrocarbons and their tautomeric allenic forms. Compounds with olefinic and acetylenic groups follow the principle of the next neighbors, i.e., compounds with the olefinic group nearer to tin shall be found in Chapter 2.4. The derivatives listed in Tables 106 and 107 are prepared by methods from the following scheme.

I Grignard synthesis:

(IA) $R_nSnX_{4-n} + R'MgX \rightarrow R_nSnR'_{4-n}$; $n = 3\text{-}1$,

(IB) $R_2SnO + R'MgX \rightarrow R_2SnR'_2$.

II Synthesis with organolithium derivatives:

(IIA) $R_nSnX_{4-n} + R'Li \rightarrow R_nSnR'_{4-n}$; $n = 3,2$,

(IIB) $Bu_2SnO + R'Li \rightarrow Bu_2SnR'_2$,

(IIC) $Ph_3SnLi + R'Br \rightarrow Ph_3SnR'$.

III Synthesis with organosodium derivatives:

(IIIA) $R_3SnBr + R'C{\vdots}CH$ + Na in hydrocarbon(-ether) $\rightarrow R_3SnC{\vdots}CR'$,

(IIIB) $Ph_nSnI_{4-n} + R'C{\vdots}CH$ + Na in liq. $NH_3 \rightarrow Ph_nSn(C{\vdots}CR')_{4-n}$,

(IIIC) $Et_3SnX + R'Na$ in liq. $NH_3 \rightarrow Et_3SnR'$,

(IIID) $R_3SnNa + R'X \rightarrow R_3SnR'$.

IV Metathetic substitution reactions:

(IVA) $R_{2n}Sn_2O_{4-n} + HR' \rightarrow R_nSnR'_{4-n}$; $n = 3, 2$,

(IVB) $R_3SnOH + HR' \rightarrow R_3SnR'$,

(IVC) $R_3SnOR + HR' \rightarrow R_3SnR'$,

(IVD) $Et_2SnCl_2 + HR' + (c.C_6H_{11})_2NH \rightarrow Et_2SnR_2'$.

V Substitution reaction with organotin amide:

$R_3SnNR_2 + R'C{\vdots}CH \rightarrow R_3SnC{\vdots}CR'$.

VI Miscellaneous reactions:

(VIA) $Me_3SiC{\vdots}CC_6H_4SnR_3$ + alkali $\rightarrow R_3SnC_6H_4C{\vdots}CH$,

(VIB) $Et_3SnC{\vdots}CH + NaC_2H + RCl \rightarrow Et_3SnC{\vdots}CR$,

(VIC) $R_3SnC{\vdots}CCH{:}CH_2 + C_5Cl_6 \rightarrow R_3SnC{\vdots}CR'$.

R = organic group, R' = organic group, may contain halogen and/or an acetylenic group, X = halogen.

Ethynyltrimethyltin (M-212) SnC_5H_{10} $Me_3SnC\vdots CH$

<u>Synth.</u>: By rxn. of $(Me_3Sn)_2O$ with C_2H_2 in autoclave at 11.5 atm., in 66% yield (506, 1339).
By rxn. of Me_3SnBr with $HC\vdots CMgBr$ in THF, in 15% yield (443), in 11% yield (1981).
By rxn. of Me_3SnCl with $HC\vdots CNa$ in liq. NH_3, in 18% yield (2181).
<u>Prop.</u>: b_{745} 98° (506, 1339), b. 95° (443), b_{190} 60-75° (1981), b_{80} 40° (2181), n_D^{20} 1.4626 (506, 1339), 1.4537 (2181), d_{20} 1.3602 (506, 1339) 1.2618 (2181), IR (443, 1074, 2181), (far) IR, assignment (1981) and NMR (1688) spectra, GLC (1981).

Phenylethynyltrimethyltin (M-212) $SnC_{11}H_{14}$ $Me_3SnC\vdots CPh$

<u>Synth.</u>: By rxn. of Me_3SnBr with $PhC\vdots CMgBr$ in Et_2O, in 40% yield (443).
By rxn. of Me_3SnOH with excess $PhC\vdots CH$ at 100-110°, in 79% yield (1921).
By rxn. of Me_3SnNMe_2 with $PhC\vdots CH$, in 92% yield (250).
<u>Prop.</u>: $b_{1.5}$ 107° (443), b_1 84-86°, n_D^{20} 1.5689, d_{20} 1.3408 (1921), IR (443, 2433), UV (443), NMR (443) and Moessbauer resonance (2665 2718) spectra.
<u>Rxn. with:</u> $Ph_3SnCl \rightarrow Ph_3SnC\vdots CPh + Me_3SnCl$ (2433).
$MCl_3 \rightarrow (PhC\vdots C)_3M$; M = P, As, relative reactivity, solvent effect (2433).
$PhPCl_2 \rightarrow (PhC\vdots C)_2PPh$, relative reactivity (2433).
α-pyrone → $o\text{-}Me_3SnC_6H_4Ph$ (1021).
$C_5Cl_6 \rightarrow 2,3\text{-}(Me_3Sn)Ph$-perchlorobicyclo[2.2.1]heptadiene (1883).

Ethynyltriethyltin (M-212) SnC_8H_{16} $Et_3SnC\vdots CH$

<u>Synth.</u>: By rxn. of Et_3SnCl with HC_2Na in $Et_2O\text{-}NH_3$ in 70% yield, in liq. NH_3, in 14% yield, rxn. mechanism and intermediates (2181).
By rxn. of Et_3SnOH with HC_2Na in Et_2O, in 33% yield (1342), with C_2H_2 at 100-120° in autoclave, in 22% yield (504).
By rxn. of $(Et_3Sn)_2O$ with C_2H_2 at 10 atm., in 65% yield (1339), in N-methylpyrrolidone in presence of CaC_2 at 20° in Ar atm. (2495).
By-prod. in rxn. of Et_3SnOMe with HC_2Na in Et_2O, in 7% yield (1342).
By-prod. in rxn. of $Et_3SnC\vdots CBr$ with EtLi at -50° in Et_2O (2180).
<u>Prop.</u>: b_{15} 60° (504), b_{14} 68° (1342), b_{12} 61° (2495), b_7 49-50° (506, 1339), b_5 45° (2181), n_D^{20} 1.4783 (2181), 1.4770 (504, 506, 1339, 1342), d_{20} 1.2486 (2181), 1.2458 (506, 1339), 1.2455 (504), IR (1909, 2144, 2763), NMR (1688) and Moessbauer resonance (2683, 2718) spectra, force const. (2144), dipole moment (2597). Anal. detn. by iodometric titration (2495, 2644).
<u>Rxn. with:</u> $MCl_4 \rightarrow Et_3SnCl + HC\vdots CMCl_3$; M = Ge, Si (2881).
$C_5Cl_6 \rightarrow 3,2\text{-}H(Et_3Sn)$-perchlorobicyclo[2.2.1]heptadiene (2507).
$Et_3SnOH \rightarrow (Et_3Sn)_2O + C_2H_2$ (506).
$(R_3Sn)_2O \rightarrow Et_3SnC\vdots CSnR_3$; R = Et (506, 2495), Bu (2495).
$(Et_3SnONa + PhC\vdots CH)$, $+ (Et_3Sn)_2O \rightarrow (Et_3SnC\vdots)_2$ (1342).
$NaC_2H + RCl \rightarrow (Et_3SnC\vdots)_2 + Et_3SnC\vdots CR$; R = Et, Pr, Et_3Sn (2181).
$NaNH_2$, $+ NH_4Cl \rightarrow (Et_3SnC\vdots)_2$ (2181).
$CH_2N_2 \rightarrow 3\text{-}Et_3SnC_3H_3N_2$ (2760).
CCl_3CHO, $+ CH_2(CO_2H)_2 \rightarrow (Et_3SnO_2C)CH_2 + HC\vdots CCHOHCCl_3$ (2644).
PhNCO (1:2), + HCl-MeOH → $Et_3SnCl + CH_2\vdots \overline{CNPhCONPhCO}$ (2644).

Phenylethynyltriethyltin (M-211) SnC$_{14}$H$_{20}$ Et$_3$SnC⋮CPh

Synth.: By rxn. of PhC⋮CH with (Et$_3$Sn)$_2$O at 100-140° in 53% yield (509), in Et$_2$O at r.t., in 42% yield (1342), in C$_6$H$_6$ by azeotropic distillation, in 82% yield (391), in 88% yield (2495).

By rxn. of PhC⋮CH with Et$_3$SnOH at 100-110°, in 87% yield (509, 1921), with Et$_3$SnCl and alkali metal hydroxide, in 80% yield (512), with Et$_3$SnCl in C$_6$H$_6$ in presence of (C$_6$H$_{11}$)$_2$NH, in 62% yield (2495), with Et$_3$SnOCH$_2$CH$_2$C⋮CH at 100°, in 84% yield (508).

By rxn. of PhC⋮CH with Et$_3$SnNMe$_2$, in 92% yield (250).

By rxn. of PhC⋮CH with Et$_3$SnCH$_2$OAc at 100-110°, in 70% yield (437).

By rxn. of PhC⋮CMgBr with Et$_3$SnOSiR$_3$ in Et$_2$O; R = Me and Et, in 78 and 80% yield, resp. (1906).

By rxn. of PhC⋮CNa in Et$_2$O at r.t. with (Et$_3$Sn)$_2$O, in 55% yield (1342), with Et$_3$SnOH and CH$_2$:CC⋮CH, in 67% yield (1342).

By rxn. of Et$_3$SnCl with LiAl(C⋮CPh)$_4$ in THF, in 88% yield (1076).

Prop.: b$_1$ 108-9° (508), 105° (1076), 102-104° (1342), 93-94° (1921), b$_{0.15}$ 88-90° (509, 512), b$_{0.1}$ 79° (2495), n$_D^{20}$ 1.5580 (509, 512, 2495), 1.5574 (1921), 1.5525 (250), d$_{20}$ 1.2470 (508, 1921), 1.2450 (437, 509, 512), IR (1908, 2433, 2763) and Moessbauer resonance (2683, 2718) spectra. Anal detn. by iodometric titration (2495, 2644).

Rxn. with: Ph$_3$SnCl → Ph$_3$SnC⋮CPh (2433).

MCl$_3$ → (PhC⋮C)$_3$M; M = P, As, relative reactivity, solvent effect (2433).

PhPCl$_2$ → (PhC⋮C)$_2$PPh; relative reactivity (2433).

BzCl → Et$_3$SnCl + BzC⋮CPh (2644).

H$_2$O ⇌ (Et$_3$Sn)$_2$O + 2PhC:CH, rxn. mechanism (2495).

c.C$_6$H$_{10}$O → 1,1-PhC⋮CC$_6$H$_{10}$OSnEt$_3$ (1331).

c.C$_6$H$_{10}$O (+ cat. ZnCl$_2$), + CH$_2$(CO$_2$H)$_2$ → (Et$_3$SnO$_2$C)$_2$CH$_2$ + 1,1-PhC⋮CC$_6$H$_{10}$OH (2644).

CCl$_3$CHO → Et$_3$SnOCH(CCl$_3$)C⋮CPh (1537, 2644).

PhNCO(1:2) → Et$_3$SnCPh:CCONPhCONPh + (PhNCO)$_x$ (2644).

Triethyl-2-vinylethynyltin (M-211) SnC$_{10}$H$_{18}$ Et$_3$SnC⋮CCH:CH$_2$

Synth.: By rxn. of HC⋮CCH:CH$_2$ with Et$_3$SnOH at r.t., in 63% yield (504), in presence of PhC⋮CNa in Et$_2$O, in 31% yield (1342).

By rxn. of HC⋮CCH:CH$_2$ with (Et$_3$Sn)$_2$O at r.t., in 82% yield (1341).

By rxn. of HC⋮CCH:CH$_2$ with Et$_3$SnOMe (1920).

By rxn. of Et$_3$SnOSiR$_3$ with CH$_2$:CHC⋮CMgBr in Et$_2$O; R = Me and Et, in 94 and 96% yield, resp. (1906).

By-prod. in rxn. of Et$_3$SnONa with PhC⋮CH, (Et$_3$Sn)$_2$O and excess HC⋮CCH:CH$_2$, in 39% yield (1342).

Prop.: b$_3$ 78-79° (504), b$_2$ 62-63° (1341), n$_D^{20}$ 1.5082 (1920), 1.5085 (504), 1.5095 (1341), d$_{20}$ 1.2173 (1920), 1.2176 (1341), 1.2180 (504), IR spectrum (1908, 2763).

Rxn. with: (BuO)$_2$BCl → Et$_3$SnCl + CH$_2$:CHC⋮CB(OBu)$_2$ (1891).

C$_5$Cl$_6$ → 2-(Et$_3$SnC⋮C)-hexachlorobicyclo[2.2.1]heptene-5 (1343, 1909).

1-Hexynyltriethyltin $SnC_{12}H_{24}$ $Et_3SnC\vdots CBu$

Synth.: By rxn. of Et_3SnCl with $NaAl(C\vdots CBu)_4$ in THF, in 75% yield (1076), with $HC\vdots CBu$ in presence of $(C_6H_{11})_2NH$ in Ar atm. in 51% yield (2495).
By rxn. of $(Et_3Sn)_2O$ with $BuC\vdots CH$ in C_6H_6 under reflux in presence of CaH_2 in vapor phase in Ar atm., in 76% yield (2495).
Prop.: b_3 105° (1076), $b_{0.2}$ 60° (2495), n_D^{20} 1.4820 (1076).
Anal. detn. by iodometric titration (2495, 2644).
Rxn. with: $H_2O \rightleftharpoons (Et_3Sn)_2O + 2BuC\vdots CH$; rxn. mechanism (2495).

Ethynyltripropyltin $SnC_{11}H_{22}$ $Pr_3SnC\vdots CH$

Synth.: By rxn. of C_2H_2 with $(Pr_3Sn)_2O$ under pressure, in 50% yield (1339), with Pr_3SnOH at 100-120° and 15-18 atm., in 37% yield (504).
By rxn. of Pr_3SnCl with NaC_2H in Et_2O-NH_3, in 68% yield (1281).
By-prod. in rxn. of $(Pr_3Sn)_2O$ with NaC_2H in Et_2O (506).
Prop.: b_{10} 91° (504), b_2 67° (506, 1339), $b_{0.3}$ 46° (2181), n_D^{20} 1.4770 (1339), 1.4772 (2181), 1.4780 (504, 506), d_{20} 1.1545 (506), 1.1555 (504), 1.1574 (1339), 1.1595 (2181), dipole moment (2597), Moessbauer resonance spectrum (2718).
Rxn. with: $Pr_3SnH \rightarrow (Pr_3SnCH\vdots)_2$ (1602).
$CH_2N_2 \rightarrow 3\text{-}(Pr_3Sn)C_3H_3N_2$ (2760).

Ethynyltributyltin (M-213) $SnC_{14}H_{28}$ $Bu_3SnC\vdots CH$

Synth.: By rxn. of C_2H_2 with Bu_3SnOH at 100-120° in autoclave (504), with $(Bu_3Sn)_2O$ under pressure, in 47% yield (1339), in N-methylpyrrolidone in presence of CaH_2 in Ar atm., in 50% yield (2495).
By rxn. of Bu_3SnCl with NaC_2H in Et_2O-NH_3, in 69% yield (2189).
Prop.: b_3 100-101° (504), 98-99° (506, 1339), $b_{0.25}$ 86° (2495), $b_{0.2}$ 76° (2181), n_D^{20} 1.4765 (506, 1339, 2495), 1.4767 (504), in 1.4770 (2181), d_{20} 1.1034 (504), 1.1038 (506, 1339), 1.1113 (2181), IR (763) and Moessbauer resonance (2683, 2718) spectra, dipole moment (2597). Anal. detn. by iodometric titration (2495).
Rxn. with: $Bu_3SnH \rightarrow (Bu_3SnCH\vdots)_2$ (1598).
$Bu_2SnH_2 \rightarrow (Bu_3SnCH\vdots CH)_2SnBu_2 + (Bu_3SnCH\vdots)_2$ (1599).

Phenylethynyltriphenyltin (M-214) $SnC_{26}H_{20}$ $Ph_3SnC\vdots CPh$

Synth.: By rxn. of Ph_3SnCl with $PhC\vdots CH$ and $(C_6H_{11})_2NH$ in C_6H_6, in 59% yield (2495), with $R_3SnC\vdots CPh$ in C_6H_6 at 100°; R = Me, Et, Bu (2433).
By rxn. of $(Ph_3Sn)_2O$ with $PhC\vdots CH$ in C_6H_6 and azeotropic distillation, in 79% yield (391, 2495).
By rxn. of Ph_3SnNMe_2 with $PhC\vdots CH$ without solvent, in 75% yield (250).
Prop.: m. 62° (250, 391, 2495), IR and UV spectra (443). Anal. detn. by iodometric titration (2495, 2644).
Rxn. with: $I \rightarrow Ph_3SnI + PhC\vdots CI$ (2644).
$HClO_4 \rightarrow$ kinetic data for cleavage rxn. (443), mechanism (2562).
$AcCl \rightarrow Ph_3SnCl + AcC\vdots CPh$ (2466).
$Ph\overline{C\vdots NN\vdots CPhN\vdots N} \rightarrow Ph_3Sn\overline{C\vdots CPhN\vdots NCPh\vdots CPh} + N_2$ (2466).
$C_5H_5CoC_8H_{12} \rightarrow C_5H_5Co\overline{CPh\vdots C(SnPh_3)CPh\vdots CSnPh_3}$ (1174).

Listing of unsymmetric tetraorganotin compounds with acetylenic substituents continues in Table 106.

2.5.1 UNSYMMETRIC TETRAORGANOTIN COMPOUNDS CONTAINING HALOGEN SUBSTITUTED ACETYLENES

Chloroethynyltriethyltin $\quad\quad SnC_8H_{15}Cl \quad\quad Et_3SnC⋮CCl$

Synth.: By rxn. of Et_3SnCl with $ClC⋮CNa$ in liq. NH_3, in 76% yield (2180).
Prop.: b_1 50-51°, n_D^{20} 1.5010, d_{20} 1.3543 (2180), IR (2180) and Moessbauer resonance (2683, 2781) spectra, dipole moment (2597).
Rxn. with: Et_3SnNa in liq. $NH_3 \rightarrow Et_6Sn_2 + ClC_2Na$ (2180).
$EtMgBr$ in presence of $CoCl_2 \rightarrow Et_4Sn$ (2180).
Et_3PO_3 at 150° $\rightarrow Et_3SnC⋮CPO_3Et_2 + EtCl$ (2179).

Bromoethynyltriethyltin $\quad\quad SnC_8H_{15}Br \quad\quad Et_3SnC⋮CBr$

Synth.: By rxn. of Et_3SnCl with BrC_2Na in liq. NH_3, in 69% yield (2180).
By rxn. of Et_3SnNa with HC_2Br in liq. NH_3 in 17% yield (2180).
Prop.: b_1 62-63°, n_D^{20} 1.5196-5202, d_{20} 1.5556-76 (2180), IR (2180) and Moessbauer resonance (2718) spectra, dipole moment (2597).
Rxn. with: Et_3SnNa in liq. $NH_3 \rightarrow (Et_3SnC⋮)_2$ (2180).
$EtLi$ in Et_2O at -50° $\rightarrow Et_4Sn + Et_3SnC⋮CH$ (2180).
$EtSNa$ in liq. $NH_3 \rightarrow Et_3SnC⋮CSEt + (Et_3SnC⋮)_2$ (2180).
Et_2NLi in Et_2O at -50° $\rightarrow (Et_3SnC⋮)_2$ (2180).
Et_3PO_3 at 170° $\rightarrow Et_3SnC⋮CPO_3Et_2 + EtBr$ (2179).

Additional unsymmetric tetraorganotin compounds containing halogen substituted acetylenic groups are listed in Table 107 on page 367.

Table 106. Unsymmetric Tetraorganotin Compounds with Acetylenic Substituents $R_nSnR'_{4-n}$

Formula*	Synth. Method**	Yield	Properties	Ref.
Me₃SnR				
R = MeC⋮C (212)	IIIA	50	$b_{0.2}$ 25-26°, IR spectrum (a)	443
	IA	4	b_{44} 67-68°, n_D^{20} 1.4836, sensitive to moisture	1981
{HC⋮CCH₂ CH₂:C:CH}	IA	12 28	Mixture, not sepd., b_{755} 135°, IR spectra	444
PhCH:C:CMe	IA, IIID	--	(b)	881
MeCH:C:CPh	IA, IIID	--	(b)	881
PrC⋮C (212)	V	86	b. 172°, n_D^{20} 1.4716	250
Me₂CHC⋮C (212)	V	90	Moessbauer resonance spectrum (2718)	250
MeCH:C:CMe	IA, IIID	--	(b, c)	881
CH₂:CMeC⋮C (212)	--	--	Moessbauer resonance spectrum	2718
BuC⋮C (212)	V	--	b_{12} 82°, n_D^{20} 1.4714	250
EtC⋮CC⋮C	--	--	Moessbauer resonance spectrum	2718
c.1-C₆H₉C⋮C	IA	15	b_3 68-70°, IR spectrum (d)	443
m-HC⋮CC₆H₄	VIA	55	b_1 72-73°, n_D^{25} 1.5595 (e)	1001
p-HC⋮CC₆H₄	VIA	51	m. 32-33°, $b_{0.35}$ 56.5 (e)	1001
Me₂Sn(C⋮CPh)₂ (212)	IIIA	50	m. 66.7°, IR, UV, NMR spectra	443
Me₂Sn(CH₂C⋮CH)₂	IA	70	b_5 62°, IR spectrum (f)	1946
Et₃SnR				
R = MeC⋮C (212)	--	--	IR spectrum (g, h, i)	1908, 2763
HC⋮CCH₂	--	--	NMR spectrum	1688
EtC⋮C	VIB (j)	--	b_1 52°, n_D^{20} 1.4903, d_{20} 1.1953 (g, h)	2181
HC⋮C⋮C	IVA	50	b_1 60°, n_D^{20} 1.5262, d_{20} 1.2626	506
	IIIC	--	b_2 67-68°, n_D^{20} 1.5273, d_{20} 1.5273 (k)	629

PrC⋮C	VIB	--	b_1 64°, n_D^{20} 1.4812, d_{20} 1.1625	2181
Me$_2$CHC⋮C (212)	--	--	(h-2718)	--
CH$_2$CH$_2$CHC⋮C (213)	--	--	(h-2718)	--
MeCH:C:CMe	IA, IIID	--	(b)	881
Me$_3$CC⋮C	--	--	$b_{1.5}$ 59°, n_D^{20} 1.4710, d_{20} 1.1162	506
HC⋮CCH$_2$CH$_2$C⋮C	IIIC	70	b_1 100-101°, n_D^{20} 1.5293, d_{20} 1.2127, IR, NMR	629
	IIID (1)	70	spectra	629
PrCH:C:CEt	IIA	70	$b_{0.5}$ 82-83°, n_D^{20} 1.4909, d_{20} 1.0891 (m)	2315
BuCH:C:CEt	IIA	--	b_2 108-109°, n_D^{20} 1.4930, d_{20} 1.0905 (m)	2315
PrCH:C:CPr	IIA	70	b_2 107-108°, n_D^{20} 1.4992, d_{20} 1.1080 (m)	2315
Et$_2$SnR$_2$				
R = Ph	IA (n)	--	Oil, dec. on distn., n_D^{20} 1.5919 impure $b_{0.001}$ 165°, anal. detn. by I titrn. (o, p)	1212, 2495
	IVC	62	b_1 62°, n_D^{20} 1.4877, d_{20} 1.2332 (q)	505
MeC⋮C	IB	66	b_1 96°, n_D^{20} 1.5428, d_{20} 1.2267	505, 1919
CH$_2$:CHC⋮C	IB	70	$b_{0.1}$ 66°, n_D^{20} 1.4969, d_{20} 1.221 (o, p)	1221
PrC⋮C	IA (n)	65	$b_{0.15}$ 101°, n_D^{20} 1.4888, d_{20} 1.118 (o, p)	1221
BuC⋮C	IA (n)	63		
	IVC	--	$b_{0.07}$ 100°, anal. detn. by I titrn.	2495

* Numbers in parenthesis refer to pages in main volume.

** The characters used here correspond to the synthetic scheme in the introduction to Chapter 2.5.

(a) NMR (443, 1930-1), (far) IR spectra and assigment (1981), rxn. with C$_5$Cl$_6$ → 2,3-Me$_3$Sn(Me)-hexachlorobicyclo-[2.2.1]heptadiene (1883). (b) Rxn. with HCl → R′CH:C:CMe + R′CH$_2$C:CMe; R′ = Me, Ph (881, 1368). (c) Rxn. with X → MeC:CCHMeX; X = Cl, Br, with 2,4-(O$_2$N)$_2$C$_6$H$_3$SO$_2$Cl → MeC:CCHMeO$_2$SC$_8$H$_3$(NO$_2$)$_2$ (881, 1368). (d) Extremely instable (443). (e) Relative kinetic data on aryl-cleavage by aq. alkali (997). (f) Rxn. with Ph$_2$GeH$_2$ → (Ph$_2$GeCH:CHCH$_2$-SnMe$_2$CH$_2$CH:CH)$_x$ (327). (g) Dipole moment (2597). (h) Moessbauer resonance spectra (2683, 2718). (i) Rxn. with MCl$_4$ → MeC:CMCl$_3$; M = Si, Ge (1891, 2881), with PCl$_3$ → MeC:CPCl$_2$ (1891). (j) Synth. also from (Et$_3$SnC⋮)$_2$ with excess NaC$_2$H in Et$_2$O-NH$_3$, followed by EtCl, in 33% yield (2181). (k) IR (629) and NMR (629, 1688) spectra. (1) Synth. from Et$_3$SnNa and (BrC⋮CCH$_2$)$_2$ (629), also from (Et$_2$Sn)$_x$ with (HC⋮CCH$_2$) and Na in liq. NH$_3$, followed by EtCl (629). (m) IR spectrum (2315). (n) Work-up without hydrolysis (1221). (o) Easily hydrolyzed (1221). (p) Rxn. with Fe$_3$(CO)$_{12}$ → [R$_2$′SnFe(CO)$_4$] ; R′ = Et, Pr, Bu (1222). (q) IR spectrum (1909, 2763).

Table 106 Continued

Formula*	Synth. Method**	Yield	Properties	Ref.
$Pr_3SnC\colon CPh$	--	--	b_1 150°, n_D^{20} 1.5388, d_{20} 1.1811 (r)	506
$Pr_3SnC\colon CCH\colon CH_2$	IVB	60	b_3 100-101°, n_D^{20} 1.4938, d_{20} 1.1391 (s)	504
	IVC	--	b_3 97°, n_D^{20} 1.4910, d_{20} 1.1293	1920
	IVA	--	b_5 108-109°, n_D^{20} 1.4920, d_{20} 1.1395	1341
$Pr_3SnC\colon CC\colon CH$	--	--	b_1 103°, n_D^{20} 1.5110, d_{20} 1.1343	506

Pr_2SnR_2

Formula*	Synth. Method**	Yield	Properties	Ref.
R = HC:C	IB	35	b_2 70°, n_D^{20} 1.4785, d_{20} 1.1819	505
PhC:C	IB	--	$b_{0.9}$ 135° (polymerization) IR spectrum	505
	IA (n)	--	n_D^{20} 1.5991, dec. on distn. (o, p)	1221
$CH_2\colon CHC\colon C$	IB	58	b_1 102°, n_D^{20} 1.5343, d_{20} 1.1840	505,1919
PrC:C	IA (n)	66	$b_{0.15}$ 96°, n_D^{20} 1.4881, d_{20} 1.143 (o, p)	1221
BuC:C	IA (n)	65	$b_{0.15}$ 109-10°, n_D^{20} 1.4899, d_{20} 1.137 (o, p)	1221
$(\overline{CH_2CH_2CH})_3SnCH_2C\colon CEt$	IIA	--	Use: herbicide, pesticide	1873
$Bu_3SnC\colon CPh$ (213)	IVA	90	$b_{0.001}$ 129°, n_D^{20} 1.5329 (t)	391,2495
$Bu_3SnC\colon CCH\colon CH_2$ (213)	IVA	high	b_4 135°, n_D^{20} 1.4955, d_{20} 1.0928	1341
	IVB	50	b_4 135°, n_D^{20} 1.4955, d_{20} 1.0928	504
	IVC	--	b_3 135°, n_D^{20} 1.4950, d_{20} 1.0923	1920

Bu_2SnR_2

Formula*	Synth. Method**	Yield	Properties	Ref.
R = PhC:C (214)	IA (n)	--	Oil, n_D^{20} 1.5878, dec. on distn. (p)	1221
	IVA	31	$b_{0.008}$ 170°, n_D^{20} 1.5881, IR spectrum	2495
MeC:C	IIB (v)	47	$b_{0.3}$ 103°, n_D^{27} 1.4920, mol. wt.	2709
$HC\colon CCH_2$	--	--	(Far) IR and UV spectra	1123
$CH_2\colon CHC\colon C$	IB	65	b_1 108° n_D^{20} 1.5188, d_{20} 1.1352	505,1919
	IIB (v)	60	$b_{0.2}$ 145°, n_D^{25} 1.5444, mol. wt.	2709
$PhCH_2CH_2C\colon C$	IIA	--	IR and NMR spectra	2709

Compound	Method	Yield	Properties	Ref.
PrC⋮C	IA (n)	68	b$_{0.1}$ 99°, n$_D^{20}$ 1.4884, d$_{20}$ 1.249 (p)	1221
BuC⋮C	IA (n)	72	b$_{0.15}$ 114°, n$_D^{20}$ 1.4827, d$_{20}$ 1.063 (p, w)	1221
BuSn(C⋮CPh)$_3$ (214)	IA (n)	67	m. 70-71°	1221
(c-C$_6$H$_{11}$)$_3$SnCMe⋮C⋮CHMe	IA	--	--	881
(i-BuCHCH$_2$CH)$_3$SnCH$_2$C⋮CEt	IIA	--	Use: herbicide, pesticide	1873
(n-C$_8$H$_{17}$)$_3$SnC⋮CPh	--	--	(x)	--
Ph$_3$SnR				
R = HC⋮C (214)	IIIB	70	m. 37-38°, IR, NMR (1474) spectra (a, y)	443
MeC⋮C (214)	IA	85	m. 74-75°, IR, NMR (1931) spectra (z)	443
HC⋮CCH$_2$ (214)	IA	75	m. 81-83°, IR, NMR spectra (aa)	444
CH$_2$:C:CH (214)	IIC	30	m. 61-62°, IR spectrum (aa)	444
	IA	10	(By-prod. with R = HC⋮CCH$_2$)	444
9-C$_{14}$H$_9$CH$_2$C⋮C	IA	70	m. 150°, IR spectrum (anthranyl)	443
MeC⋮CCH$_2$	IA	60	m. 68-69°, IR, NMR spectra	444
HC⋮CCHMe (214)	IA	60	m. 94°, IR, NMR spectra (bb)	444
MeCH:C:CH	IIC	32	Liq., IR, NMR spectra (bb)	444
	IA	20	(By-prod. with R = HC⋮CCHMe)	444
CH$_2$:CHC⋮C (214)	IVA	--	m. 57-58°	1341
HC⋮CC⋮C (214)	--	--	UV (443) and NMR spectra	443, 1931
PhC⋮CC⋮C (214)	IIIB	55	m. 88°, IR, UV spectra (cc)	443

(r) Rxn. with PCl$_3$ in THF, + KF → Pr$_3$SnF + (PhC⋮C)$_3$P (2433), with MCl$_3$ at r.t. → (PhC⋮C)$_3$M; M = P, As; relative reactivity, solvent effect (2433). (s) Rxn. with C$_5$Cl$_6$ → 2-(Pr$_3$SnC⋮C)-hexachlorobicyclo[2.2.1]heptene-5 (1343, 1909). (t) Anal. detn. by iodometric titration (2495, 2644). (u) Rxn. with I in Et$_2$O → Bu$_3$SnI + PhC⋮CI (2644), with Ph$_3$SnCl → Ph$_3$SnC⋮CPh (2433). (v) Synth. with (Bu$_2$SnS)$_3$ and RLi, in 60-70% yield (2709). (w) Rxn. with H$_2$O or aq. base → (Bu$_2$SnO)$_2$ + RC⋮CH, with concn. HCl → Bu$_2$SnCl$_2$ + RC⋮CH, with EtSH → Bu$_2$Sn(SEt)$_2$ (1221). (x) Rxn. with Ph$_2$AsCl → Ph$_2$AsC⋮CPh (2433). (y) Rxn. with NaNH$_2$ in liq. NH$_3$ → (Ph$_3$Sn)$_3$N (443). (z) Rxn. with H and Raney Ni → PrSnPh$_3$ (443). (aa) Rxn. of Ph$_3$SnCH$_2$C⋮CH with EtOH under reflux → Ph$_3$SnCH:C:CH$_2$ (444). (bb) Rxn. of Ph$_3$SnCHMeC⋮CH with EtOH under reflux → Ph$_3$SnCH:C:CHMe (444). (cc) Kinetic data for cleavage rxn. with aq. alc HClO$_4$ (443, 2562), with alc. NaOH (2565).

Table 106 Continued

$R_nSnR'_{4-n}$

Formula*	Synth. Method**	Yield	Properties	Ref.
Ph_3SnR (Cont'd.)				
R = MeCH:C:CMe	IA	--	(dd)	--
CH_2:CMeC:C (215)	IA	77	m. 64°, IR, UV spectra	443
MeC:CC:C (215)	--	--	UV (443), NMR spectra	443, 1931
c.1-C_6H_9C:C (214)	IA	66	m. 106°, IR spectrum	443
$Ph_2Sn(C:CPh)_2$ (215)	IA (n)	71	m. 71-72° (o)	1221
$Ph_2Sn(C:CBu)_2$	IA (n)	--	Oil, dec. on dist., n_D^{20} 1.6048 (o)	1221

(dd) Rxn. with HCl → (MeCH:)$_2$C + MeC:CCH$_2$Me (881, 1368).

Table 107. Tetraorganotin Compounds Containing Halogen Substituted Acetylenes $R_nSnR'_{4-n}$

Formula	Synth. Method*	Yield	Properties	Ref.
$Me_3SnC\!:\!CCl$	IIIA	72	b_{44} 67–68°, n_D^{20} 1.4915 (a)	1980
$Me_3SnC\!:\!CBr$	IIID	15	b_{14} 67–68°, n_D^{20} 1.5175, d_{20} 1.7396 (b)	630, 2180
$Me_3SnC\!:\!CCF_3$	IA	64	b. 125° (c)	2332
$Me_2Sn(C\!:\!CCF_3)_2$	IA	--	(c)	2332

$R_3SnC\!:\!C$— (tetrachlorobicyclic structure with CCl_2 and Cl substituents)

R = Et	VIC	60	b_2 190°, n_D^{20} 1.5503, d_{20} 1.5408	1343, 1909
	IVA	62	Same (d)	1918
Pr	VIC	49	$b_{2.5}$ 209°, n_D^{20} 1.5410, d_{20} 4195	1343, 1909
$Ph_3SnC\!:\!CBr$	IIIB (e)	75	m. 104°, IR spectrum (f)	443
$p\text{-}Ph_3SnC\!:\!CC_6H_4Br$	IA	36	m. 128°, IR, UV spectra	443

* The characters used here correspond to the synthetic scheme in the introduction to Chapter 2.5.
(a) Very sensitive to moisture, stable in inert atm. (1980), ^{35}Cl NQR (2885) and (far) IR spectra and assignment (1980). (b) IR spectrum, easily hydrolyzed in air (630). (c) Rxn. with $H_2O \rightarrow CF_3C\!:\!CH$ (2332), with Me_3SnCF_3 → $Me_3SnF + Me_nSn[C\!:\!C(CF_3)\overline{C(CF_3)}CF_2]_{4-n}$; n = 3, 2 (2332). (d) Rxn. with concd. HCl → 2,4,5,6,7,7-(HC:C)Cl$_6$-bicyclo-[2.2.1]heptene-5 (1909). (e) Synth. from Ph_3SnBr, Na and $(CH_2Br)_2$ in liq. NH_3 (443). (f) Kinetic data for cleavage rxn. with aq. alc. $HClO_4$ (443).

2.5.3 UNSYMMETRIC TETRAORGANOTIN COMPOUNDS CONTAINING FUNCTIONALLY SUBSTITUTED ACETYLENES

The functionally substituted acetylenic groups linked to tin may contain various heteratoms. These include oxygen, sulfur, nitrogen, phosphorous, tin and silicon or combinations of these elements. One compound with a cyano group is included. The derivatives summarized in Table 108 are prepared by methods from the following scheme. Additional tetraorganotin compounds with acetylenic unsaturation may be found in Subchapters 2.7.1.1.1, 2.7.1.1.3, 2.7.7, 2.7.8.

I   Grignard reaction:
    $Ph_3SnBr + R'MgBr$ in THF → Ph_3SnR'.

II Synthesis with organolithium derivatives:
    R_3SnCl with $Ph_2PC⋮CLi$ in THF-hexane → $R_3SnC⋮CPPh_2$.

III Synthesis with organosodium derivatives:
    (IIIA) $R_3SnX + R'Na$ in Et_2O → R_3SnR',
    (IIIB) $R_3SnX + R'Na$ in liq. NH_3 → R_3SnR',
    (IIIC) $Et_3SnC⋮CBr + EtSNa$ in liq. NH_3 → $Et_3SnC⋮CSEt$.

IV Synthesis with acetylenes and organotin oxygen derivatives:
    (IVA) With organotin oxides,
    (IVB) With organotin hydroxides,
    (IVC) With organotin alhoxides.

V   Synthesis with acetylenes and other organotin-hetero derivatives:
    (VA) With organotin amides,
    (VB) With organotin hydrides as by-product,
    (VC) With 2-dimethylacetamidotriethyltin,
    (VD) With organotin chloride and dicyclohexyl amine as hydrogen chloride acceptor.

VI Reactions of triethylstannoxyalkynyltriethyltin derivatives:
    (VIA) With methyl iodide at 190-200°,
    (VIB) With R'Cl; R' = acetyl, benzoyl, methoxymethyl, trimethylsilyl.

Diethyl 2-Triethyltinethynyl- $SnC_{12}H_{25}O_3P$ $Et_3SnC⋮CPO_3Et_2$
  phosphonate

<u>Synth.:</u> By rxn. of $Et_3SnC⋮CX$ with $(EtO)_3P$; X = Cl and Br at 150 and 170° in N atm., in 82 and 24% yield, resp. (2179).
<u>Prop.:</u> b_2 157-58°, b_1 150-51°, n_D^{20} 1.4899-908, d_{20} 1.2797-.2809 (2179), IR, NMR (2179) and Moessbauer resonance (2683, 2718) spectra, dipole moment (2597).
<u>Rxn. with:</u> Br in $HCCl_3$ → $Et_3SnBr + BrC⋮CPO_3Et$ (2179).
AcOH → $Et_3SnOAc + HC⋮CPO_3Et_2$ (2179).

3,1-Triethylstannoxypropynyl- $Sn_2C_{15}H_{32}O$ $Et_3SnC⋮CCH_2OSnEt_3$
  triethyltin (M-215)

<u>Synth.:</u> By rxn. of $HC⋮CCH_2OH$ at 100-110° with Et_3SnOH and $(Et_3Sn)_2O$, in 91

and 97% yield, resp. (1921).
By rxn. of Et$_3$SnOMe with HC⋮CCH$_2$OH and EtMgBr, in 79% yield (2506).
Prop.: b$_5$ 160°, n$_D^{20}$ 1.5145, d$_{20}$ 1.3541 (506, 2506), IR spectrum (1911).
Rxn. with: Me$_3$SiCl → Et$_3$SnCl + Et$_3$SnC⋮CCH$_2$OSiMe$_3$ (1536).
MeI → Et$_3$SnI + Et$_3$SnC⋮CCH$_2$OMe (1910).
MeOCH$_2$Cl → Et$_3$SnCl + Et$_3$SnC⋮CCH$_2$OCH$_2$OMe (1910).
AcCl → Et$_3$SnCl + Et$_3$SnC⋮CCH$_2$OAc (1910).

| 4,1-Triethylstannoxybutynyl-triethyltin (M-215) | Sn$_2$C$_{16}$H$_{34}$O | Et$_3$SnC⋮CCH$_2$CH$_2$OSnEt$_3$ |

Synth.: By rxn. of Et$_3$SnOMe with HC⋮CCH$_2$CH$_2$OH at -10° to r.t., in 75% yield, influence of temp. on product distribution (508).
By rxn. of Et$_3$SnCl with NaOCH$_2$CH$_2$C⋮CH, in 40% yield (508).
By thermal dec. of Et$_3$SnOCH$_2$CH$_2$C⋮CH at 120°, in 92% yield (508).
Prop.: b$_1$ 134-35°, n$_D^{20}$ 1.5086, d$_{20}$ 1.3412 (508, 1911), IR spectrum (1911).
Rxn. with: Dry HCl → Et$_3$SnCl + HC⋮CCH$_2$CH$_2$OH (508).
Me$_3$SiCl → Et$_3$SnCl + Et$_3$SnC⋮CCH$_2$CH$_2$OSiMe$_3$ (1910).
MeOCH$_2$Cl → Et$_3$SnCl + Et$_3$SnC⋮CCH$_2$CH$_2$OCH$_2$OMe (1536, 1922).
AcCl → Et$_3$SnCl + Et$_3$SnC⋮CCH$_2$CH$_2$OAc (1910).

| 3,3,1-Triethylstannoxymethyl-propynyltriethyltin (M-216) | Sn$_2$C$_{16}$H$_{34}$O | Et$_3$SnC⋮CCHMeOSnEt$_3$ |

Synth.: By rxn. of (Et$_3$Sn)$_2$O with HC⋮CCHMeOSnEt$_3$, in 43% yield (508).
By rxn. of Et$_3$SnOH and HC⋮CCHMeOH in C$_6$H$_6$ (2506).
Prop.: b$_1$ 130-32°, n$_D^{20}$ 1.5060, d$_{20}$ 1.3242 (506, 2506), IR spectrum (1911).
Rxn. with: Me$_3$SiCl → Et$_3$SnC⋮CCHMeOSiMe$_3$ (1910).
MeI → Et$_3$SnI + Et$_3$SnC⋮CCHMeOMe (1910).
MeOCH$_2$Cl → Et$_3$SnCl + Et$_3$SnC⋮CCHMeOCH$_2$OMe (1910).
AcCl → Et$_3$SnCl + Et$_3$SnC⋮CCHMeOAc (1910).

| 3,3,3,1-Triethylstannoxydimethylpropynyltriethyltin (M-216) | Sn$_2$C$_{17}$H$_{36}$O | Et$_3$SnC⋮CCMe$_2$OSnEt$_3$ |

Synth.: By rxn. of (Et$_3$Sn)$_2$O with HC⋮CCMe$_2$OH in 23% yield (508, 2506), in C$_6$H$_6$ and azeotropic distillation, in 78% yield (508).
Prop.: b$_2$ 138° (506, 2506), n$_D^{20}$ 1.4987 (508, 1911), 1.5005 (506, 2506), d$_{20}$ 1.2969 (506, 2506), 1.3006 (508, 1911), IR spectrum (1911).
Rxn. with: Thermal dec. at 200-250° → (Et$_3$SnC⋮)$_2$ (2776).
Gaseous HCl at 0° → Et$_3$SnCl + HC⋮CCMe$_2$OH (1910).
MeI at 190-200° → Et$_3$SnI + Et$_3$SnC⋮CCMe$_2$OMe (513).
RCl → Et$_3$SnCl + Et$_3$SnC⋮CCMe$_2$OR; R = Me$_3$Si, MeOCH$_2$, Ac, Bz (1910).

Listing of unsymmetric tetraorganotin compounds containing functionally substituted acetylenes continues in Table 108.

Table 108. Unsymmetric Tetraorganotin Compounds Containing Functionally Substituted Acetylenes $R_nSnR'_{4-n}$

Formula*	Synth. Method**	Yield	Properties	Ref.
Me₃SnC⋮CR				
R = Ph₂P	VA	90	Colorless cryst. m. 35-36°, b₀.₀₅ 146-47°,	2778
	II	62	Dec. 190-210°, NMR, mass spectra (a, b)	2778
Ph₂(S)P	(b)	49	m. 168-70°, mass spectrum (a)	2778
MeO₂C	VB	2-7	IR, NMR spectra, GLC, not isolated (c)	311
EtO₂C	VB	2	IR, NMR spectra, CLG, not isolated (c)	311
BuSCH:CH	IVA	73	b₃ 125.5°, n_D^{20} 1.5482, d_{20} 1.2441	2772
Et₃SnC⋮CR				
R = p-MeOC₆H₄	IVA	90	b₀.₂ 119°, anal. detn. by I titration	391, 2495
EtO	VC	69	b₂ 80-81°, b₀.₁ 69-70°, n_D^{20} 1.4825-30	1710
	IVC	52	d_{20} 1.2310-49, IR spectrum	1710
EtS	IIIC	15	b₁ 96-98°, n_D^{20} 1.5212, d_{20} 1.2706 (d)	2180
NC	IVA	83	m. 47°, b₁₃ 126°, dec. in Ar atm. (e)	391, 2495
MeOCH₂	VIA	44	b₀.₅ 60°, n_D^{20} 1.4878, d_{20} 1.2404	1910
	IVA	27	b₀.₆ 62-63°, IR spectrum (1911)	511, 1921
	IVC	87	d_{20} 1.2428	511
MeOCH₂OCH₂	VIB	73	b₂ 114-15°, n_D^{20} 1.4834, d_{20} 1.2457 (f)	1910
CH₂:CHOCH₂CH₂OCH₂	IIIA	35	b₁ 105°, n_D^{20} 1.4922 (g, h)	1907
	IVA	56	b₁ 100-102°, n_D^{20} 1.4347, d_{20} 0.8615	1921
CH₂:CH(OCH₂CH₂)₂OCH₂	IIIA	33	b₁ 143-45°, n_D^{20} 1.4900, d_{20} 1.2072	1907
BuOCHMeOCH₂ (215)	IVB	87	b₅ 142°, n_D^{20} 1.4775, d_{20} 1.1363	279, 504
	IVC	--	b₀.₃₅ 105-106°, n_D^{20} 1.4742, d_{20} 1.1460	507
	IVA	53	b₁ 103-104°, n_D^{20} 1.4750, d_{20} 1.1450 (i)	507, 1921
Me₃SiOCH₂	VIB	88	b₂.₅ 100-101°, n_D^{20} 1.4740, d_{20} 1.1452	1536
	VIB	44	b₂.₅ 100-101°, n_D^{20} 1.4740, d_{20} 1.1452	1913

AcOCH$_2$	VIB	71	b$_{0.5}$ 84°, n$_D^{20}$ 1.4887, d$_{20}$ 1.2735	1910
MeO$_2$C	IVC	82	b$_{0.4}$ 82-84°, n$_D^{20}$ 1.4985, GLC	311
	VD	41	b$_{0.001}$ 73° (j, k)	2469
	VB	1-4	b$_{0.1}$ 83-86°, n$_D^{20}$ 1.4996	311, 1410, 1414
EtO$_2$C	VB	4	Not isolated, GLC (j)	311, 1414
H$_2$NCO	IVA	76	b$_{0.35}$ 139°, slowly dec. in Ar atm on standing (l)	391, 2495
MeOCH$_2$CH$_2$	--	--	b$_{0.7}$ 67°, n$_D^{20}$ 1.4848, d$_{20}$ 1.2069 (i)	1911
MeOCH$_2$OCH$_2$CH$_2$	VIB	87	b$_1$ 101-102°, n$_D^{20}$ 1.4823, d$_{20}$ 1.2166 (f)	1536, 1922
BuOCHMeOCH$_2$CH$_2$ (215)	IVC	--	b$_{0.3}$ 111-12°	507
	IVA	<70	(i)	511
CH$_2$:CHOCH$_2$CH$_2$	IVA	56	b$_2$ 107-108°, n$_D^{20}$ 1.4925, d$_{20}$ 1.1993	511
	IVC	75	d$_{20}$ 1.2016 (1911) (i, m)	511
Me$_3$SiOCH$_2$CH$_2$	VIB	82	b$_{2.5}$ 108-10°, n$_D^{20}$ 1.4723, d$_{20}$ 1.1160	1910
	IVC	52	n$_D^{20}$ 1.4725, d$_{20}$ 1.1173	1910
AcOCH$_2$CH$_2$	VIB	81	b$_{0.5}$ 92°, n$_D^{20}$ 1.4858, d$_{20}$ 1.2412	1910
	--	--	b$_1$ 84°, n$_D^{20}$ 1.5002, d$_{20}$ 1.2714 (i)	1911
MeOCH:CH$_2$	IVA	82	b$_{13}$ 138°, dec. on standing in Ar atm.	391, 2495
	VD	75	(f)	2495
BuOCH:CH	IVA	98	b$_{0.5}$ 107-108°, n$_D^{20}$ 1.5098, d$_{20}$ 1.1760	506, 2772

* Numbers in parenthesis refer to pages in main volume.

** The characters used here correspond to the synthetic scheme in the introduction to Chapter 2.5.3.

(a) IR spectrum, mol. wt. (2778). (b) Rxn. of R'$_3$SnC:CPPh$_2$ with S in CS$_2$ → R'$_3$SnC:CPSPh$_2$; R' = Me, Ph (2778).
(c) Hydrolysis → HC:CCO$_2$SnMe$_3$ (311). (d) IR spectrum (2180). (e) Anal. detn. by iodometric titration (2495).
(f) IR spectrum (1894). (g) Rxn. with HCl in aq. dioxane → Et$_3$SnCl + HC:CCH$_2$OCH$_2$CH$_2$OH (1907). (g) Rxn. with AcOH → Et$_3$SnOAc + HC:CCH$_2$OCH$_2$CH$_2$OCH:CH$_2$ (1907). (i) IR spectrum (1911). (j) IR and NMR spectra (311). (k) Partly dec. on GLC (311), hydrolysis in air → Et$_3$SnO$_2$CC:CH (311). (l) Rxn. with H$_2$O ⇌ (Et$_3$Sn)$_2$O + H$_2$N(O)CC:CH; mechanism of hydrolysis (2495). (m) Rxn. with BuOH at 80-100° → R'$_3$SnOBu + HC:CCH$_2$CH$_2$OCH:CH$_2$; R = Et, Bu (2078).

Table 108 Continued $R_nSnR'_{4-n}$

Formula*	Synth. Method**	Yield	Properties	Ref.
$Et_3SnC\vdots CR$ (Cont'd.)				
R = BuSCH:CH	IVA	98	b_2 153-54°, n_D^{20} 1.5436, d_{20} 1.1891 (n)	2772
$Me_2NCH:CH$	VB	35	$b_{0.03}$ 79-80°, n_D^{20} 1.5427, d_{20} 1.1732 (o, p)	2558
	IIIA, B	47	$b_{0.02}$ 87°, b_2 125-34°, n_D^{20} 1.5408, d_{20} 1.2200	2687
$Et_2NCH:CH$	IVA	80	$b_{1.5}$ 125-26°, n_D^{20} 1.5470, d_{20} 1.1648	506, 2772
	IIIA, B	69	$b_{0.01}$ 97-99°, n_D^{20} 1.5375, d_{20} 1.1162 (p)	2687
MeOCHMe	VIA	41	$b_{0.8}$ 57-57.5°, n_D^{20} 1.4790, d_{20} 1.1969	1910
$MeOCH_2OCHMe$	VIB	45	b_1 95°, n_D^{20} 1.4756, d_{20} 1.1967	1910
BuOCHMeOCHMe (215)	IVB	89	b_2 120°, n_D^{20} 1.4722, d_{20} 1.1075 (i)	279, 504
$Me_3SiOCHMe$	VIB	93	$b_{1.5}$ 86°, n_D^{20} 1.4672, d_{20} 1.1089	1910
AcOCHMe	VIB	76	$b_{0.7}$ 78°, n_D^{20} 1.4818, d_{20} 1.2226	1910
$MeOCMe_2$	VIA	31	$b_{0.4}$ 58-59°, n_D^{20} 1.4759, d_{20} 1.1669	513
$MeOCH_2OCMe_2$	VIB	92	$b_{0.5}$ 73-74°, n_D^{20} 1.4746, d_{20} 1.1814	1910
	IVC	79	IR spectrum (f)	1910
$BuOCHMeOCMe_2$ (216)	IVB	--	$b_{1.5}$ 112°, n_D^{20} 1.4670, d_{20} 1.0727	279, 504
	IVC	--	(i)	507
$MeSiOCMe_2$	VIB	77	$b_{0.8}$ 70°, n_D^{20} 1.4646, d_{20} 1.0898 (f)	1910
$AcOCMe_2$	VIB	67	$b_{0.6}$ 67-68°, n_D^{20} 1.4793, d_{20} 1.2044	1910
$BzOCMe_2$	VIB	85	b_1 135°, n_D^{20} 1.5228, d_{20} 1.2124 (f)	1910
$Et_3SnOCMeEt$	IVB	47	b_2 131°, n_D^{20} 1.4979, d_{20} 1.2767 (q)	1149
$c.1-Et_3SnOC_5H_8$	IVB	15	b_2 157°, n_D^{20} 1.5041, d_{20} 1.2966	1149
$c.1-Et_3SnOC_6H_{10}$	IVB	60	b_1 152°, n_D^{20} 1.5118, d_{20} 1.2906	1149
$Pr_3SnC\vdots CR$				
R = MeO_2C	VB	0.4	Struct. by GLC, (j)	311
BuOCH:CH	IVA	82	b_1 148-49°, n_D^{20} 1.5038, d_{20} 1.1467	2772
BuSCH:CH	IVA	72	b_2 171.5°, n_D^{20} 1.5321, d_{20} 1.1370	2772

Et$_2$NCH:CH	IVA	50	b$_3$ 156-57°, n$_D^{20}$ 1.5105, d$_{20}$ 1.1590	2772
Pr$_3$SnOCHMe	IVA, B	--	b$_1$ 165°, n$_D^{20}$ 1.4970, d$_{20}$ 1.2139	506, 2506
Pr$_3$SnOCMe$_2$	IVA, B	--	b$_4$ 172°, n$_D^{20}$ 1.4925, d$_{20}$ 1.2073	506, 2506
Pr$_3$SnOCMeEt	IVB	58	b$_2$ 168°, n$_D^{20}$ 1.4930, d$_{20}$ 1.1826	1149
c.1-Pr$_3$SnOC$_6$H$_{10}$	IVB	65	b$_2$ 200°, n$_D^{20}$ 1.5051, d$_{20}$ 1.2112	1149

Bu$_3$SnC:CR

R = CH$_2$:CHOCH$_2$CH$_2$	--	--	b$_1$ 114-16°, n$_D^{20}$ 1.4875, d$_{20}$ 1.1186 (m)	2078
BuOCH:CH	IVA	70	b$_1$ 165-65.5°, n$_D^{20}$ 1.4987, d$_{20}$ 0.0902 (sic)	2772
Et$_2$NCH:CH	IVA	51	b$_2$ 170°, n$_D^{20}$ 1.5010, d$_{20}$ 1.1414	2772
Bu$_3$SnOCHMe	IVA, B	--	b$_3$ 200°, n$_D^{20}$ 1.4875, d$_{20}$ 1.1558	506, 2506
Ac	--	--	(f)	--
Bu$_3$SnOCMe$_2$	IVA, B	--	b$_2$ 198°, n$_D^{20}$ 1.4890, d$_{20}$ 1.1277 (r)	506, 2506

Ph$_3$SnC:CR

R = Ph$_2$P	II	55	Colorless cryst., m. 88-90°, stable at r.t.,	2778
	VA	68	IR spectrum, mol. wt., soly. (s)	2778
	(s)	43	m. 133-34°, IR spectrum, mol. wt.	2778
Ph$_2$(S)P	I	80	m. 90°, IR, NMR spectra	443
HOCH$_2$	VD	16	m. 59-60° (e)	2495
MeO$_2$C	VD	16	m. 59-60° (e)	2495
MeOCH:CH				

(n) Rxn. with H$_2$O → (Et$_3$Sn)$_2$O + BuSCH:CHC:CH (2772), with C$_5$Cl$_6$ → Et$_3$SnCl (2772). (o) IR spectrum (2558). (p) IR and NMR spectra, dipole moment, stable in sealed tube at -5° (2687). (q) Rxn. with H and Raney Ni → Et$_3$SnH + HC:CCHMeEt (1149). (r) Thermal dec. at 200-250° → (Bu$_3$SnC:)$_2$ (2776). (s) Rxn. of Ph$_3$SnC:CPPh$_2$ with S in CS$_2$ → Ph$_3$SnC:CPPh$_2$S (2778).

2.6 FUNCTIONALLY SUBSTITUTED UNSYMMETRIC TETRAORGANOTIN COMPOUNDS

2.6.1 TETRAORGANOTIN COMPOUNDS CONTAINING HALOGEN

The unsymmetric halogen containing tetraorganotin derivatives listed in Table 109 are prepared by methods from the ensuing scheme.

I Grignard reactions:
- (IA) $R_nSnX_{4-n} + R'MgX \rightarrow R_nSnR'_{4-n}$; n = 1-3 ,
- (IB) $R_3SnX + R'I + Mg \rightarrow R_3SnR'$,
- (IC) $(ClCH_2)_3SnCl + MeMgBr \rightarrow Me_xEt_nSn(CH_2Cl)_m$; m + n + x = 4.

II Synthesis with organolithium derivatives:
- (IIA) $R_nSnX_{4-n} + R'Li \rightarrow R_nSnR'_{4-n}$; n = 2, 3 ,
- (IIB) $R'SnCl_3 + RLi \rightarrow R_3SnR'$,
- (IIC) $Me_3SnCCl_2Li + MeI \rightarrow Me_3SnCCl_2Me$.

III Addition reactions to perfluoroalkenes:
- (IIIA) $Me_6Sn_2 + C_2F_4$ under UV irradiation $\rightarrow Me_3Sn(CF_2)_nH$; n = 2, 4 ,
- (IIIB) $Me_6Sn_2 + CF_2{:}CFCF_3$ under UV irradiation $\rightarrow Me_3SnCF_2CHFCF_3$,
- (IIIC) $R_3SnH + \overline{CF_2CF_2CF{:}CF} \rightarrow R_3Sn\overline{CFCHFCF_2CF_2}$.

IV Rearrangement reactions:
- (IVA) $Me_3SnR' \rightarrow Me_nSnR'_{4-n}$,
- (IVB) $(Bu_3SnO)_2CHCBr_3 \rightarrow Bu_3SnCBr_3$,
- (IVC) $(Bu_3SnO)_2C(CCl_3)_2 \rightarrow Bu_3SnCCl_3$.

V Miscellaneous organometallic reactions:
- (VA) $R_3SnNEt_2 + HCX_3 \rightarrow R_3SnCX_3$; X = Cl, Br .
- (VB) $(Me_3Sn)_2Ar + X \rightarrow Me_3SnArX$; X = Br, I,
- (VC) $R_nSnBr_{4-n} + $ alkenyl bromide in Py $\rightarrow R_nSn(R'Py)_{4-n}$; n = 2, 3 ,
- (VD) $R_2(R'Py)SnBr + $ alkenyl bromide in Py $\rightarrow R_2Sn(R'Py)(R''Py)$.

VI Dihalocarbene addition to olefins:
- (VIA) With $PhHgCBrCl_2$ in refluxing C_6H_6 ,
- (VIB) With Me_3SnCF_3 and NaI in $(MeOCH_2)_2$,
- (VIC) With $HCCl_3$ and Me_3COK at -30° ,
- (VID) With CCl_3CO_2Na in $(MeOCH_2)_2$.

VII Dichlorocarbene insertion in Sn-H and C-H bonds:
- (VIIA) $R_3SnH + CCl_3CO_2Na \rightarrow R_3SnCHCl_2$,
- (VIIB) $R_3SnCH_2CH_2Me + PhHgCBrCl_2 \rightarrow R_3SnCH_2CHMeCHCl_2$.

VIII Miscellaneous organic reactions:
- (VIIIA) $Et_2Sn(CH_2OEt)_2 + AcCl \rightarrow Et_2Sn(CH_2Cl)_2$,
- (VIIIB) $Me_3SnCH_2Cl + KF \rightarrow Me_3SnCH_2F$.

R = organic group, R' = halogen-substituted organic group, Ar = aryl group, X = halogen. Additional tetraorganotin compounds containing halogen may be found in Chapters 2.3, 2.4.1, 2.4.3, 2.5.1, 2.6.5, 2.7.1.1.3, 2.7.2, 2.7.7, 2.7.8, and 7.1.6.

Trifluoromethyltrimethyltin (M-217) $SnC_4H_9F_3$ Me_3SnCF_3

Synth.: By rxn. of Me_6Sn_2 with CF_3COI, in 74% yield (922).
By-prod. in rxn. of $(Me_2Sn)_x$ with CF_3I under UV irradiation (871).
Prop.: Moessbauer resonance spectra (2665).
Rxn. with: $(Me_2C:)_2$ + NaI in $(MeOCH_2)_2$ → Me_3SnI + $Me_2\overline{CCF_2C}Me_2$ (494), numerous other gemminal difluorocyclopropane derivatives, relative reactivity of other dihalocarbene sources (1882).
c.C_6H_{10} + NaI → Me_3SnI + 7,7-difluorobicyclo[4.1.0]heptane (494).
$CH_2:CHR$ + NaI → Me_3SnI + $\overline{CH_2CF_2C}HR$; R = Et_3Ge, Et_3Sn, Me_3SnCH_2 (1874).
$CF_3C \vdotseq CR$ → Me_3SnI + $\overline{CF_2CR:C}CF_3$; R = H (2332), Et_3Ge (921, 2332), Me_3Si, Me_2, As (921).
$Me_nSn(C \vdotseq CCF_3)_{4-n}$ at 145° → $Me_nSn[\overline{C:C(CF_3)C}F_2]_{4-n}$; n = 2, 3 (2332).
PF_5(excess) → CF_3PF_4 + (trace) $(CF_3)_nPF_{5-n}$ + $Me_3SnCF_3 \cdot PF_5$ (853).
$(CF_3)_2PF_3$ at 65° → CF_4 + C_2F_6 + HCF_3 (853).
Me_2CO + NaI in $(MeOCH_2)_2$ → Me_3SnI + HCF_3 (1874).
BzOH in C_6H_6 → Me_3SnF + $BzOCF_2H$ (1874).
Addn. compd.: $Me_3SnCF_3 \cdot PF_5$, from components. dissocn. at 120° → PF_5 + POF_3 + SiF_4 + Me_3SnF, IR spectrum, struct. $Me_3Sn[CF_3PF_5]$, $[SnC_4H_9F_8P]$ (853).
$Me_3SnCF_3 \cdot (CF_3)_2PF_3$, subl. in vacuo at 135-40°, IR spectrum $[SnC_6H_9F_{12}P]$ (853).
$Me_3SnCF_3 \cdot (CF_3)_3PF_2$, IR spectrum, soly. $[SnC_7H_9F_{14}P]$ (853).

Chloromethyltrimethyltin (M-219) $SnC_4H_{11}Cl$ Me_3SnCH_2Cl

Synth.: By-prod. in rxn. of $(ClCH_2)_3SnCl$ with MeMgBr in Et_2O (2512).
Prop.: b_{57} 69.5-71° (2512), b_{18} 46.5-48° (1309, 1350), n_D^{20} 1.4890 (2512), 1.4893 (1309, 1350), d_{20} 1.5071 (1309, 1350), 1.5121 (2512), IR (2512), (far) IR (1309), NMR (1309, 1350, 2512), ^{35}Cl NQR (2752), Moessbauer resonance (1309, 1350, 2489) and mass (1350, 2512) spectra.
Rxn. with: I^- and EtO^- → relative reactivity in cleavage rxn. and mechanism (776).
Morpholine → $Me_3SnCH_2\overline{NCH_2CH_2OCH_2C}H_2$ (1309, 1350).
$MeNH_2$ → Me_3SnCH_2NHMe + $(Me_3SnCH_2)_2NMe$ (1715).
Me_2NH → $Me_3SnCH_2NMe_2$ (1715).
KNY → Me_3SnCH_2NY; Y = CH_2CH_2 , CH:CHCH:CH (1309, 1350).
NaOMe → Me_3SnCH_2OMe (1309, 1350).
NaOAc → Me_3SnCH_2OAc (1350).
KF → Me_3SnCH_2F (1309, 1350).

Dichloromethyltrimethyltin $SnC_4H_{10}Cl_2$ $Me_3SnCHCl_2$

Synth.: By rxn. of Me_3SnCl with $CHCl_2Li$ in 66% yield (1879).
By hydrolysis of Me_3SnCCl_2Li at -115° with dilute HCl, in 51% yield (1879).
By-prod. in rxn. of Me_3SnCl with Me_3SiCCl_2Li, in 10% yield (1879).
Prop.: b_{38} 86-87°, NMR spectrum (1879).
Rxn. with: RLi → Me_3SnR; R = Bu, Me_3C (1879).

Trichloromethyltrimethyltin $SnC_4H_9Cl_3$ Me_3SnCCl_3

Synth.: By rxn. of Me_3SnNR_2 with $HCCl_3$; R = Me at 0° in 72% yield (865), Et (948), with CCl_3CHO at 0°; R = Me, Et (2454).

By rxn. of Me$_3$SnCl with Cl$_3$CLi at -110°, in 52% yield (1875).
By-prod. in rxn. of Me$_3$SnCl with PhHgCBrCl$_2$ (1875).
Prop.: m. 54-59° (dec.) (948), 43-45° (865), > r.t. (1875), IR (865, 948) and NMR (948, 1875, 2454) spectra, GLC (1875).
Rxn. with: Br in CCl$_4$ → Me$_3$SnBr (948).
BCl$_3$ at 25° → Me$_2$(Cl$_3$C)SnCl + MeBCl$_2$ (865).
H$_2$O → HCCl$_3$ (948) + Me$_3$SnOH (1875).
Cycloalkene → Me$_3$SnCl + dichlorobicyclo[n.1.0]alkane; n = 6, C$_9$H$_{14}$Cl$_2$ (1875), n = 4, C$_7$H$_{10}$Cl$_2$ (2454).

Bromodichloromethyltrimethyltin   SnC$_4$H$_9$BrCl$_2$ Me$_3$SnCBrCl$_2$
Synth.: By rxn. of Me$_3$SnBr with PhHgCBrCl$_2$ at 80°, in 63% yield (1875).
By-prod. in rxn. of Me$_3$SnCl with PhHgCBrCl$_2$ (1875).
Prop.: Solid, m. > r.t., NMR spectrum, GLC (1875).
Rxn. with: H$_2$O → HCBrCl$_2$ + Me$_3$SnOH (1875).
c.C$_6$H$_{10}$ → Me$_3$SnBr + 7,7-dichlorobicyclo[4.1.0]heptane + (trace)Me$_3$SnCl + 7,7-bromochloronorcarane (1875).
c.C$_8$H$_{14}$ → Me$_3$SnBr + 9,9-dichlorobicyclo[6.1.0]nonane + some Me$_3$SnCl + some BrCl-bicyclo derivative (1875).

Trimethyl-pentafluorophenyltin   SnC$_9$H$_9$F$_5$ Me$_3$SnC$_6$F$_5$
(M-219)
Synth.: By rxn. of Me$_3$SnX with C$_6$F$_5$MgBr in Et$_2$O; X = Br in 60% yield (77), Cl in Et$_2$O-Bu$_2$O, in 60% yield (428).
Prop.: b$_{50}$ 118-19° (428), b$_{0.02}$ 34-36° (77), n$_D^{20}$ 1.4744 (428), dec. 350° (77), IR, UV (77), NMR (39, 77), ^{19}F NMR (2475) and Moessbauer resonance (2665) spectra.
Rxn. with: Anhydr. HCl(1:1) → Me$_3$SnCl + C$_6$HF$_5$ (77).
HX → relative rates of aryl-cleavage (1002).
KF in anhydr. EtOH → Me$_3$SnF + C$_6$HF$_5$ (77).
NaI → Me$_3$SnI + C$_6$HF$_5$ + tar (851).

2,3,4,5-Tetrachlorophenyltrimethyltin   SnC$_9$H$_{10}$Cl$_4$ 2,3,4,5-Cl$_4$C$_6$HSnMe$_3$
Synth.: By rxn. of (Me$_3$SnC!)$_2$ with 5,5-(MeO)$_2$C$_5$Cl$_4$ in xylene in inert atm. under reflux in the dark in presence of hydroquinone as by-prod. (1883).
Prop.: m. 65.5-68°, IR, UV and NMR spectra, mol. wt., TLC (1883).
Rxn. with: Thermal dec. at 230° → Me$_4$Sn + Me$_3$SnCl + Ar$_2$SnMe$_2$* (1883).
HCl in MeOH-Et$_2$O → ArH* (1883).
p-MeC$_6$H$_4$I → p-MeC$_6$H$_4$Ar* (1883).

---

*Ar = 2,3,4,5-Cl$_4$C$_6$H .

Pentachlorophenyltrimethyltin   SnC$_9$H$_9$Cl$_5$ Me$_3$SnC$_6$Cl$_5$

__Synth.__: By rxn. of Me$_3$SnCl with C$_6$Cl$_5$Li in Et$_2$O and Cl$_5$C$_6$MgCl in THF, in 46 and 56% yield, resp. (865), with C$_6$Cl$_5$Li in THF at -78°, in 73% yield (2764).
__Prop.__: White needles (865), m. 119-20° (865, 2764), slow dec. at r.t. (865), IR, NMR (865) and Moessbauer resonance (2665) spectra, mol. wt. (2764).
__Rxn. with__: Thermal dec. at 300° → Me$_3$SnCl + C$_6$HCl$_5$ + perchloropolyphenyl (865).
BCl$_3$ at 100° → Me$_2$SnCl$_2$ + Cl$_5$C$_6$BCl$_2$ + MeBCl$_2$ (865).
Alc. KOH or KF → C$_6$HCl$_5$ (865).

Di(pentafluorophenyl)dimethyl-   SnC$_{14}$H$_6$F$_{10}$ Me$_2$Sn(C$_6$F$_5$)$_2$
   tin (M-220)

__Synth.__: By rxn. of Me$_2$SnX$_2$ with C$_6$F$_5$MgBr in N atm. (1198), in 58% yield (77).
By-prod. in rxn. of Me$_2$(Cl)SnMn(CO)$_5$ with C$_6$F$_5$MgBr and C$_6$F$_5$Li in N atm. in Et$_2$O at -78° (2035).
__Prop.__: b$_{1.7}$ 94-96° (1198), b$_{0.02}$ 74-76° (77), n$_D^{20}$ 1.4970 (1198), IR, UV (77), NMR (63, 77, 2035) ^{19}F NMR (63, 1032, 2475) and Moessbauer resonance (1989) spectra.
__Rxn. with__: Anhydr. HCl → C$_6$HF$_5$ (77).
BCl$_3$ (1:2) → Me$_2$SnCl$_2$ + C$_6$F$_5$BCl$_2$ (77).
SnX$_4$ → Me$_2$SnX$_2$ + (C$_6$F$_5$)$_2$SnX$_2$; X = Cl, Br (1198).

Chloromethyltriethyltin (M-220)   SnC$_7$H$_{17}$Cl   Et$_3$SnCH$_2$Cl

__Synth.__: By rxn. of Et$_2$(ClCH$_2$)SnCl with EtMgBr, in 55% yield (283).
__Prop.__: b$_{10}$ 84.5-85.5°, n$_D^{20}$ 1.4947, d$_{20}$ 1.3383 (283), Moessbauer resonance and mass spectra (1309).
__Rxn. with__: YNH → Et$_3$SnCH$_2$NY$_2$; Y = Me$_2$, CH$_2$CH$_2$OCH$_2$CH$_2$ (283).
YNK → Et$_3$SnCH$_2$NY; Y = CH$_2$CH$_2$, (CH:CH)$_2$ (283).
KCN → Et$_3$SnCH$_2$CN (1309, 1350).
NaOR → Et$_3$SnCH$_2$OR; R = Me (1309, 1350), Ac (1350)
Mg + CH$_2$:CRCH$_2$X → Et$_3$SnCH$_2$CH$_2$CR:CH$_2$; R = H, Me (1619).

Listing of halogen substituted tetraorganotin compounds continues in Table 109.

Table 109. Halogen Substituted Tetraorganotin Compounds $R_nSnR'_{4-n}$

Formula*	Synth. Method**	Yield	Properties	Ref.
Me₃SnR				
R = FCH₂	VIIIB	33	b_{745} 97-101°, n_D^{20} 1.4443, d_{20} 1.4328 (a)	1309, 1350
Br₃C	VA	--	m. 89-94°, IR, NMR spectra	948
p-BrC₆H₄CH₂ (219)	--	--	NMR spectrum	2267
HCF₂CF₂	IIIA	(b)	Mol. wt. (c)	871
MeCCl₂	IIA	47	m. 58-59°	1879
	IIC	26	NMR spectrum	1879
CF₃CHFCF₂	IIIB	(b)	(c)	87, 871
CCl₂CH₂CH	VIA (d)	55	n_D^{25} 1.5059, IR, NMR spectra, GLC	1877
H(CF₂)₄	IIIA	(b)	(c)	871
CH₂CF₂CHCH₂	VIB (e)	54	n_D^{25} 1.4450, NMR spectrum	1874
CF₂CF₂CHFCF	IIIC	94	IR, NMR spectra (f)	924
Cl₂CHCHMeCH₂	VIIB	70	n_D^{25} 1.5005, NMR spectrum	1889
CH₂CCl₂CHCH₂	VIC (g)	68	$b_{1.4}$ 57.5-58°, n_D^{25} 1.5041 (h)	1877
CH₂CCl₂CHCH₂CH₂	VIC (i)	53	b_3 81-83°, n_D^{25} 1.5004	1877
MeCHCCl₂CHCH₂	VIC (j)	81	Trans-form $b_{2.2}$ 73.5-74°, n_D^{25} 1.5018	1877
	VIC (k)	62	$b_{4.5}$ 83-85°, n_D^{25} 1.5004 (l)	1877
(Cl-Me-CCl₂-Cl cyclohexane structure)	(m)	--	m. 64-67.5°, TLC, IR, UV, NMR spectra (n)	1883
o-FC₆H₄ (219)	--	--	$b_{0.5}$ 32°, n_D^{25} 1.5227 (o)	997
m-FC₆H₄	--	--	(o)	--
p-FC₆H₄	IA	--	b_{10} 88°, n_D^{20} 1.5176, dipole moment	1210
	--	--	b_1 45°, n_D^{25} 1.5149 (o)	997
o-BrC₆F₄	IIA	82	Mass, ¹⁹F NMR spectra (p)	851

Compound	Yield	Properties	Ref.
o-ClC$_6$H$_4$	--	b$_{0.5}$ 47°, n$_D^{25}$ 1.5977 (o)	997
m-ClC$_6$H$_4$ (219)	--	b$_{3.5}$ 84°, n$_D^{25}$ 1.5491 (o)	997
p-ClC$_6$H$_4$ (219)	--	Dipole moment (o, q)	1210
o-BrC$_6$H$_4$	VB	b$_{0.03}$ 47°, n$_D^{24}$ 1.5734 (r)	1021
m-BrC$_6$H$_4$	IA	b$_{0.3}$ 63°, n$_D^{25}$ 1.5672 (o)	998
p-BrC$_6$H$_4$ (218)	IA	b$_{18}$ 132-34°, n$_D^{20}$ 1.5704	1410
3,5-Br$_2$C$_6$H$_3$	--	b$_{0.2}$ 64°, n$_D^{25}$ 1.5669 (o, s)	997
o-IC$_6$H$_4$	IA	b$_{1.5}$ 109°, n$_D^{25}$ 1.6000	998
p-IC$_6$H$_4$	VB	b$_{0.04}$ 70°, n$_D^{25}$ 1.6031, TLC (r, t)	1021
o-CF$_3$C$_6$H$_4$	VB	--	60
m-CF$_3$C$_6$H$_4$	IA	b$_{3.8}$ 67°, n$_D^{25}$ 1.4920 (o)	998
p-CF$_3$C$_6$H$_4$	IA	b$_{1.4}$ 53°, n$_D^{25}$ 1.4790 (o)	998
3,5-(CF$_3$)$_2$C$_6$H$_3$	--	(o)	998
2,2'-I(C$_6$H$_4$)$_2$	IA	b$_{8.5}$ 81-82°, n$_D^{25}$ 1.4445	998
	(t)	b$_{0.02}$ 100°, TLC (r, u)	1021

* Numbers in parenthesis refer to pages in main volume.

** The characters used here correspond to the synthetic scheme in the introduction to Chapter 2.6.1.

(a) (Far) IR (1309), UV (1350, 1715), NMR (1309, 1350) and Moessbauer resonance (1309, 1350) spectra. (b) As by-prod. (c) IR, NMR (871) and ^{19}F NMR (870) spectra. (d) From Me$_3$SnCH:CH$_2$ (1877). (e) From Me$_3$SnCH$_2$CH:CH$_2$ (1874). (f) Dec. on distn. → Me$_3$SnF + CF$_2$CF$_2$CF$_2$CF:CH (924). (g) From Me$_3$SnCH$_2$CH:CH$_2$ (1877). (h) Thermal dec. at 145-50° in N atm. → Me$_4$Sn + Me$_2$SnCl$_2$ + CH$_2$:CClCH:CH$_2$ + Me$_3$SnCl (2755), with ZnCl$_2$ at 70° → Me$_4$Sn + CH$_2$:CClCH:CH$_2$ (2755). (i) From Me$_3$SnCH$_2$CH:CH$_2$ (1877). (j) From trans-Me$_3$SnCH$_2$CH:CHMe cis/trans-ratio 0.4 (1877). (k) From Me$_3$SnCH$_2$CH:CHMe cis/trans-ratio 3 (2755). (m) By rearrangement of 2,3-Me(Me$_3$Sn)heptachlorobicyclo[2.2.1]heptadiene under UV irradiation in inert atm. (1883). (n) Rxn. with Br in Et$_2$O-MeOH → RBr (1883), with HCl in aq. MeOH → RH (1883). (o) Kinetic data for aryl cleavage with aq. alc. KOH (997). (p) Rxn. with NaI → Me$_3$SnI + tar (851). (q) NMR (2848) and Moessbauer resonance (902) spectra, rxn. with Me$_2$SnCl$_2$ and Na in MePh → p-(Me$_3$Sn)$_2$C$_6$H$_4$ + (p-Me$_3$SnC$_6$H$_4$)$_2$SnMe$_3$ (1410), same prod. with Mg and Me$_2$SnCl$_2$ in THF (1410). (r) IR, UV and NMR spectra (1021). (s) with alc. HCl and mass (2411) spectra, dipole moment (1210), kinetic data for cleavage rxn. with alc. I$^-$ (61), with alc. HCl (62). (t) Thermal dec. of o-Me$_3$SnC$_6$H$_4$I at 235° → Me$_4$Sn + Me$_3$SnI + 2,2'-Me$_3$SnC$_6$H$_4$C$_6$H$_4$I + Me$_2$Sn(C$_6$H$_4$I-o)$_2$ (1021). (u) Rxn. with alc. HCl → o-IC$_6$H$_4$Ph (1021).

Table 109 Continued

Formula*	Synth. Method**	Yield	Properties	$RnSnR'_{4-n}$ Ref.
$Me_2EtSnCH_2Cl$	IC	(b)	b_{27} 76-77.5°, n_D^{20} 1.4917, d_{20} 1.4781	2512
Me_2SnR_2				
R = $ClCH_2$	IC	(b)	$b_{3.7}$ 66-67°, n_D^{20} 1.5221, d_{20} 1.6559 (v)	2512
o-$ClC_6H_4CH_2$	IA	34	$b_{0.8}$ 163°, n_D^{19} 1.6100 (w)	841
p-ClC_6H_4	IA	--	$b_{0.15}$ 135-40°, n_D^{20} 1.6041 (x)	1410
o-Cl_4C_6H	IVA	--	Colorless cryst., m. 140-41.5° (y)	1883
o-$C_6F_5C_6F_4$	IIA	12	m. 100-102°	1033
o-$I(C_6H_4)_2$	(t)	(b)	NMR spectrum, TLC, mixt.	1021
$MeEtSn(CH_2Cl)_2$	IC	(b)	b_4 88-94°, n_D^{20} 1.5287, d_{20} 1.6012	2512
$MeSn(C_6F_5)_3$ (220)	IA	69	m. 72-73°, subl.0.001 60-65° (aa)	77
Et_3SnR				
R = Cl_2CH	VIIA	55	b_{92} 133-35°, n_D^{20} 1.5078, d_{20} 1.4420	2047
CH_2CF_2CH	VIB	52	n_D^{25} 1.4576, IR spectrum	1874
CF_2CF_2CHFCF	IIIC	3	IR, NMR spectra (cc)	924
m-ClC_6H_4	--	--	(dd)	--
p-ClC_6H_4	--	--	(dd)	--
m-$CF_3C_6H_4$	I	71	$b_{0.3}$ 63°, n_D^{25} 1.4869	998
$Et_2Sn(CH_2Cl)_2$	VIIIA	80	b_{20} 119-20°, d_{20} 1.517	307
Pr_3SnR				
R = o-FC_6H_4	IB	57	b_5 120-25°, n_D^{30} 1.707	1247
p-FC_6H_4	IB	78	b_4 155°, $n_D^{32.5}$ 1.5020, GLC	1247a
o-ClC_6H_4	IB	53	b_8 155-56°, n_D^{30} 1.710	1247
p-ClC_6H_4	IB	79	b_9 93-97°, $n_D^{32.5}$ 1.5000, GLC	1247a

o-BrC$_6$H$_4$	IB	55	b$_5$ 165°, n$_D^{30}$ 1.548	1247
p-BrC$_6$H$_4$ (221)	IB	80	b$_5$ 135-40°, n$_D^{32.5}$ 1.5250, GLC	1247a
o-IC$_6$H$_4$	IB	44	b$_8$ 189-92°, n$_D^{30}$ 1.665	1247
p-IC$_6$H$_4$	IB	81	b$_5$ 115-20°, n$_D^{32.5}$ 1.5290, GLC	1247a
2,4-(CH$_2$CH$_2$CH)$_3$SnC$_6$H$_3$Cl$_2$	IIB	--	Use: herbicide, pesticide	1873

Bu$_3$SnR

R = Cl$_2$CH	VIIA	45	b$_{1.3}$ 114-16°, n$_D^{20}$ 1.4965, d$_{20}$ 1.2590	2047
CCl$_3$	VA	--	Yellow oil, IR spectrum	948
CBr$_3$ (?)	IVC	--	(ee)	954
CH$_2$CCl$_2$CHCH$_2$	IVB	(b)	(ff)	954
	VID (gg)	47	b$_{17}$ 164-65°, n$_D^{20}$ 1.4900, d$_{20}$ 1.2040	2047
o-FC$_6$H$_4$	IB	75	b$_{12}$ 127-30°, n$_D^{25}$ 1.491	1247
p-FC$_6$H$_4$	IB	78	b$_{22}$ 100°, n$_D^{25}$ 1.5060, GLC	1247a
o-ClC$_6$H$_4$ (222)	IB	62	b$_8$ 130-35°, n$_D^{25}$ 1.522-25	1247
p-ClC$_6$H$_4$ (222)	IB	65	b$_4$ 140°, n$_D^{25}$ 1.5235, GLC	1247
o-BrC$_6$H$_4$	IB	66	b$_8$ 115-20°, n$_D^{25}$ 1.495-96	1247
p-BrC$_6$H$_4$	IB	58	b$_{12}$ 160°, n$_D^{25}$ 1.5139, GLC	1247a
o-IC$_6$H$_4$	IB	67	b$_8$ 127-32°, n$_D^{25}$ 1.519	1247
p-IC$_6$H$_4$	IB	50	b$_4$ 120°, n$_D^{25}$ 1.5009, GLC	1247a
Bu$_2$Sn(C$_6$F$_5$)$_2$ (222)	--	--	(hh)	--

(v) (Far) Ir, NMR, mass (2512) and ^{35}Cl NQR (2752) spectra. (w) (Far) IR spectra and assignment (841). (x) Rxn. with Me$_3$SnCl and Na in MePh → p-(Me$_3$Sn)$_2$C$_6$H$_4$ + (p-Me$_3$SnC$_6$H$_4$)$_2$SnMe$_2$ (1410). (y) IR, UV and NMR spectra (1883). (z) Rxn. with alc. HCl → 2-IC$_6$H$_4$Ph + PhI (1021). (aa) IR, UV, NMR (77), ^{19}F NMR (2475) and Moessbauer resonance (1989) spectra, rxn. with anhydr. HCl → C$_6$F$_5$H (77). (bb) From Et$_3$SnCH:CH$_2$ (1874). (cc) Thermal dec. at 20° → Et$_3$SnF (924). (dd) Rxn. with Hg(OAc)$_2$ in THF → Et$_3$SnOAc + RHgOAc; relative rxn. rates and mechanism (1162). (ee) Rxn. with HCl in Et$_2$O → HCCl$_3$ (954). (ff) Rxn. with MeOH → Bu$_3$SnOMe + HCBr$_3$ (954). (gg) From Bu$_3$SnCH$_2$CH:CH$_2$ (2047). (hh) Rxn. with BBr$_3$ → BuBBr$_2$ + C$_6$F$_5$BBr$_2$ (428).

Table 109 Continued

$R_nSnR'_{4-n}$

Formula*	Synth. Method**	Yield	Properties	Ref.
Ph₃SnR				
R = Br₂C₃H₅·Py	VC	56	m. 129-30°, IR spectrum, proposed struct.	2592
Br₂C₆H₉·Py	VC	58	m. 131-32° (dec.), proposed struct. (ii)	2592
C₆F₅ (223)	IA	68	m. 86° (aa, jj)	77
p-ClC₆H₄ (222)	--	--	Dipole moment (kk-1309)	1210
C₆Cl₅	IA	67	m. 170-72°, UV spectrum (kk-2665)	1101
p-BrC₆H₄ (222)	--	--	(kk-1309)	--
m-CF₃C₆H₄ (223)	IA	70	m. 110°	998
Ph₂Sn(C₃H₅Br₂·Py)₂	VC	60	m. 170-71°, IR spectrum	2592
	VC	68	In EtOH under reflux, m. 166-67°	2592
Ph₂Sn(C₃H₅Br₂)C₆H₉Br₂·2Py	VD	76	Synth. with CH₂:CHCH₂Br	2592
	VD	78	Synth. with 3-C₆H₉Br, m. 144-49° (dec.)	2592
Ph₂Sn(C₆H₉Br₂·Py)₂	VC	76	m. 166° (dec.), proposed struct. (ii)	2592
Ph₂Sn(C₆F₅)₂ (223)	IA	54	m. 85°, rapid dec. at 400° (kk-1989), m.78° (1197)	77
Ph₂Sn(C₆Cl₅)₂	IA	40	m. 237-40°, UV spectra, mol. wt.	1101
PhSn(C₆F₅)₃ (223)	IA	85	m. 95-96°, subl.o.o1 80-90°	77
	IA	--	m. 100-102° (aa, nn)	1198
(o-MeC₆H₄)₂Sn(C₆F₅)₂	IA	--	m. 174° (nn)	1198
(p-MeC₆H₄)₃SnC₆H₄CF₃-m	IA	76	m. 133-34°	998
(p-MeC₆H₄)₂Sn(C₆F₅)₂ (223)	IA	--	m. 73-75° (kk-1989, oo)	77
p-MeC₆H₄Sn(C₆F₅)₃ (223)	IA	80	m. 107°, subl.o.o1 140° (kk, 1989, pp)	77
Me₃SnC₆H₃(I)CO₂Me	VB	89	IR and NMR spectra, struct.	1021
Ph₃SnCH₂CH(PPh₂)CH₂Cl	V (qq)	65	m. 39°	1841

(ii) IR and NMR spectra (2592). (jj) Moessbauer resonance spectrum (211. 2665), rxn. with anhydr. HCl(1:1) → Ph₂(C₆F₅)SnCl + C₆H₆ (77). (kk) Moessbauer resonance spectrum. (ll) IR, UV, NMR (77) and ¹⁹F NMR (1032, 2475)

2.6.2 TETRAORGANOTIN COMPOUNDS CONTAINING PSEUDOHALOGEN SUBSTITUENTS

The tetraorganotin derivatives containing a cyano group linked through carbon to tin listed in Table 110 are prepared by methods from the following scheme.

I   Synthesis with organolithium derivatives:
    $R_3SnCl + CNCMe_2Li \rightarrow R_3SnCMe_2CN$.

II Hydrostannation reaction of unsaturated nitriles with organotin hydrides R_3SnH and R_2SnH_2 :
    (IIA) At elevated temperatures ,
    (IIB) Under UV irradiation .

III Reactions with organotin heterocompounds:
    (IIIA) $R_3SnNMe_2 + CH_2:CMeCN \rightarrow R_3SnCMe(CN)CH_2NMe_2$,
    (IIIB) $Bu_3SnOMe + R_3SiCHR'CN \rightarrow Bu_3SnCHR'CN$; R' = H, Et.

IV Exchange reaction:
    $Et_3SnCH_2Cl + KCN \rightarrow Et_3SnCH_2CN$.

1-Cyano-3-methylpropyl-tributyltin $SnC_{17}H_{35}N$ $Bu_3SnCH(CN)CHMe_2$

<u>Synth.</u>: By rxn. of $Me_2C:CHCN$ with Bu_3SnH under UV irradiation in Ar atm. in 68% yield (2680).
<u>Prop.</u>: b_1 150°, n_D^{20} 1.4910, d_{20} 1.114, IR and NMR spectra (2680).
<u>Rxn. with</u>: Bu_3SnH under UV irradiation $\rightarrow Bu_6Sn_2 +$ Me_2CHCH_2CN (2680).
Ph_3SnH at 150° $\rightarrow Bu_3SnH + Me_2CHCH_2CN$ (2680).
$MeOH \rightarrow Bu_3SnOMe + Me_2CHCH_2CN$ (2680).

Additional tetraorganotin derivatives containing a cyano group are listed in Table 110 and 108.

spectra, rxn. with anhydr. HCl $\rightarrow C_6H_6$ (77). (mm) Rxn. with anhydr. HCl(1:3) $\rightarrow C_6H_6 + C_6HF_5$ (77). (nn) Rxn. with HCl $\rightarrow (C_6F_5)_2SnCl_2 + MePh$ (1198). (oo) Rxn. with anhydr. HCl(1:2) $\rightarrow MePh + C_6HF_5$ (77). (pp) Rxn. with anhydr. HCl $\rightarrow MePh + C_6HF_5 + (C_6F_5)_3SnF$ (77). (qq) Synth. from Ph_3SnPPh_2 and $CH_2:CHCH_2Cl$ in presence of AIBN in C_6H_6 , without catalyst in 26% yield (1841).

Table 110. Pseudohalogen Substituted Tetraorganotin Compounds $R_nSnR'_{4-n}$

Formula*	Synth. Method**	Yield	Properties	Ref.
Me$_3$SnR				
R = NCCH$_2$CH$_2$	IIA	67	NMR spectrum, GLC	1417
NCCHMe	IIA	23	NMR spectrum, GLC (a)	1417
Me$_2$NCH$_2$CMeCN	IIIA	54	b$_{0.02}$ 78-80°, n$_D^{30}$ 1.5871, mol. wt.	854, 2406
{c.3-NCC$_8$H$_{10}$}	IIB	52	b$_{13}$ 138-40°, n$_D^{20}$ 1.5061, GLC (b)	1953
{c.4-NCC$_8$H$_{10}$}	IIB	--	Mixture, not sepd.	1953
Et$_3$SnR				
R = NCCH$_2$	IV	70	b$_1$ 49.8-50°, n$_D^{20}$ 1.4760, d$_{20}$ 1.2730 (c)	1309, 1350
NCCH$_2$CH$_2$ (224)	IIA	70	b$_{1.5}$ 91-92.5°, n$_D^{20}$ 1.4931, d$_{20}$ 1.2739, IR spectrum	1715
	IIA	20-100	NMR spectrum, GLC (d, e)	1417
NCCHMe	IIA	70-80	NMR spectrum, GLC (e)	1417
Me$_2$NCH$_2$CHCN	IIIA	69	b$_{0.02}$ 97°, n$_D^{28}$ 1.5853, mol. wt.	854, 2406
NCCMe$_2$	I	53	b$_{12}$ 125°, n$_D^{20}$ 1.4938, stable to 250° (f, g)	1620
c.1-C$_6$H$_{10}$CN	I	34	b$_{0.001}$ 98°, n$_D^{20}$ 1.5144, IR spectrum	1620
Pr$_3$SnCH$_2$CH$_2$CN (224)	IIA	--	NMR spectrum	2070
Pr$_2$Sn(CH$_2$CH$_2$CN)$_2$ (224)	IIA	--	NMR spectrum	2070
i-Pr$_3$SnCH$_2$CH$_2$CN	IIA	77	NMR spectrum, GLC	1417
i-Pr$_3$SnCHMeCN	IIA	13	NMR spectrum, GLC	1417
Bu$_3$SnR				
R = NCCH$_2$ (224)	IIIB	78	b$_{0.2}$ 119°, n$_D^{20}$ 1.4869, d$_{20}$ 1.151 (h)	1682
NCCH$_2$CH$_2$ (224)	IIB	56	b$_{0.3}$ 118-19°, IR spectrum	2680
	IIA	40-67	NMR spectrum, GLC (i, j)	1417, 2680
NCCHMe	IIA	27-60	NMR spectrum, GLC (e, i)	1417, 2680

Compound			Properties	Ref.
NCCHEt	IIIB	81	$b_{0.5}$ 132-34°, n_D^{20} 1.4912, d_{20} 1.149	1682, 2680
	IIB	29	IR, NMR spectra (k, l, m)	2680
	IIA	64	$b_{0.3}$ 128-30°, n_D^{20} 1.487, d_{20} 1.118	2680
NCCH$_2$CHMe	IIB	54	$b_{1.2}$ 139°, n_D^{20} 1.4871, d_{20} 1.113 (k, m)	2680
	(m)	--	IR, NMR spectra	2680
Ph$_3$SnCH$_2$CH$_2$CN (224)	IIA	100	NMR spectrum, GLC (j)	1417
	--	--	m. 95°, IR, NMR spectra (n)	2680
Ph$_3$SnCMe$_2$CN	I	27	m. 135°, soly., IR spectrum	1620
Ph$_2$Sn(CH$_2$CH$_2$CN)$_2$	IIA	--	Not characterized (o)	1745

* Numbers in parenthesis refer to pages in main volume.

** The characters used here correspond to the synthetic scheme in the introduction to Chapter 2.6.2.
(a) Rxn. with Br → MeCHBrCN (1417). (b) GLC shows three isomers, none of which is assigned (1953). (c) Moessbauer resonance spectrum (1309, 1350). (d) Rxn. with MeMgI, + H$_2$O → Et$_3$SnCH$_2$CH$_2$Ac (2489), with LiAlH$_4$ → Et$_3$Sn(CH$_2$)$_3$NH$_2$ (1715). (e) Product distribution depending on rxn. conditions (1417). (f) IR spectrum (1620). (g) Rxn. with I → Me$_2$C(CN)I (1620). (h) Rxn. with RCHO → Bu$_3$SnOCHRCH$_2$CN; R = CCl$_3$, C$_6$F$_5$ (1631). (i) Mixt. of Bu$_3$SnCHCH$_2$CH$_2$CN and Bu$_3$SnCHMeCN not sepd. (1417). The latter product absent in rxn. initiated by UV irradiation (2680). (j) NMR spectrum (2070). (k) UV irradiation suppresses formation of Bu$_3$SnCHEtCN in favor of Bu$_3$SnCH-MeCH$_2$CN (2680). (l) Rxn. with Br → EtCHBrCN (2680). (m) Rxn. of component mixt. Bu$_3$SnR(NCCHEt/NCCH$_2$CHMe 1:3) + Me$_3$SiCl → Bu$_3$SnCl + Me$_3$SiCHEtCN + Bu$_3$SnCHMeCH$_2$CN (2680). (n) Rxn. with Br in CCl$_4$ → (NCCH$_2$CH$_2$)SnBr$_3$ (1745). (o) Rxn. of crude product with Br in HCCl$_3$ → (NCCH$_2$CH$_2$)$_2$SnBr$_2$ (1745).

2.6.3 TETRAORGANOTIN COMPOUNDS CONTAINING OXYGEN

2.6.3.1 ORGANOTIN COMPOUNDS WITH HYDROXY GROUPS AND DERIVATIVES

This chapter comprises unsymmetric tetraorganotin compounds containing a hydroxy group in a carbon side chain. Derivatives issuing from substitution of the hydroxylic proton are included. All compounds are listed in Table 111 and are prepared by methods from the following scheme.

I Grignard reaction:
- (IA) $R_nSnX_{4-n} + R'MgX \rightarrow R_nSnR'_{4-n}$; n = 2, 3 ,
- (IB) $R'_nSnX_{4-n} + RMgX \rightarrow R_{4-n}SnR'_n$; n = 1, 2 .

II Synthesis with organotin lithium compound:
$Ph_3SnLi + R'I \rightarrow Ph_3SnR'$.

III Hydrostannation reaction:
$Ph_3SnH + CH_2:CHR' \rightarrow Ph_3SnCH_2CH_2R'$.

IV Interconversion reaction with tetraorganotin derivatives:
- (IVA) $R_3SnCH_2Cl + RONa \rightarrow R_3SnCH_2OR$; R = Me, Ac ,
- (IVB) $Bu_2Sn(CH_2CH_2CO_2Me)_2 + LiAlH_4 \rightarrow Bu_2Sn[(CH_2)_3OH]_2$,
- (IVC) $Ph_3SnCOMe + LiAlH_4 \rightarrow Ph_3SnCHMeOH$,
- (IVD) $Ph_3SnCOPh + PhMgBr \rightarrow Ph_3SnCPh_2OH$.

R = organic group, may be different in one molecule, R' - oxygen substituted organic group, X = halogen. Additional tetraorganotin compounds containing hydroxy groups and derivatives may be found in Chapters 2.3, 2.4.3, 2.5.3, 2.7.1.1.3, 2.7.2 and 7.1.6.

Table 111. Tetraorganotin Compounds with Hydroxy Groups and Derivatives $R_nSnR'_{4-n}$

Formula*	Synth. Method**	Yield	Properties	Ref.
Me₃SnR				
R = MeOCH₂	IVA	46	b₆₅ 59.2-59.5°, n_D^{20} 1.4570, d_{20} 1.3190 (a)	1309, 1350
AcOCH₂	IVA	53	b₁₈ 61.2-62.6°, n_D^{20} 1.4662, d_{20} 1.3991	1350
	IVA	--	Struct. (b)	2489
o-MeOC₆H₄	IA	35	b₁,₂ 70°, n_D^{25} 1.5369 (c)	998
m-MeOC₆H₄	IA	78	b₀.₄ 61°, n_D^{25} 1.5389 (c, d)	998
p-MeOC₆H₄ (226)	--	--	b₄ 102°, n_D^{25} 1.5392 (c-f)	997
	--	--	$n_D^{25.5}$ 1.5385, dipole moment (1210), UV spectrum	1561
Me(i-Pr)PhSnC₆H₄OMe-p	IA	--	b₀.₀₃ 123.5°, n_D^{20} 1.581, mass spectrum	2412
Et₃SnR				
R = MeOCH₂	IVA	60	b₁₅ 68-69°, n_D^{20} 1.4754, d_{20} 1.2462 (g)	1309, 1350
AcOCH₂	IVA	72	b₁₀ 48.5-50°, n_D^{20} 1.4874, d_{20} 1.3299	1350
	IVA	72	b₁ 48.5-50°, IR, mass spectra (h)	2489
HOCEt₂CH₂CH₂	IB	--	b₁.₅ 107-109°, n_D^{20} 1.4908, mol. wt.	2649
HOCEt₂CHMeCH₂	IB	--	b₄.₅ 110-13°, n_D^{20} 1.4937, mol. wt.	2649
m-MeOC₆H₄	--	--	(i)	--
p-MeOC₆H₄ (226)	--	--	(i)	--

* Numbers in parenthesis refer to pages in main volume.
** The characters used here correspond to the synthetic scheme in the introduction to Chapter 2.6.3.1.
(a) (Far) IR (1309), NMR (1309, 1350), Moessbauer resonance (1309, 1350) and mass (1350) spectra. (b) IR (2489), NMR (1350), Moessbauer resonance (1350, 2489) and mass (1350, 2489) spectra. (c) Relative kinetic data for aryl-cleavage reaction (997). (d) NMR (2267) and mass (2411) spectra. (e) NMR (2848) and Moessbauer resonance (902) spectra. (f) Relative kinetic data for aryl cleavage with alcoholic I⁻ (61), with alcoholic HCl (62). (g) (Far) IR, mass (1309) and Moessbauer resonance (1309, 1350) spectra. (h) Moessbauer resonance spectra (1350, 2489). (i) Rxn. with Hg(OAc)₂ in THF → Et₃SnOAc + ArHgOAc; relative rate and mechanism of cleavage rxn. (1162).

Table 111 Continued $R_nSnR'_{4-n}$

Formula*	Synth. Method**	Yield	Properties	Ref.
$Et_2Sn(CH_2OEt)_2$	IA	67	b_{1e} 111°, n_D^{20} 1.4688, d_{20} 1.2114 (j, k)	307
$Et_2Sn(CH_2CH_2CEt_2OH)_2$	IB	<70	$b_{0.001}$ 152-57°, n_D^{20} 1.5020, mol. wt.	2649
$Et_2Sn(CH_2CHMeCEt_2OH)_2$	IB	<70	$b_{0.001}$ 156-65°, n_D^{20} 1.5042, mol. wt.	2649
$EtSn(CH_2OEt)_3$	IA	54	b_{15} 87-89°, n_D^{20} 1.4720, d_{20} 1.2194	307
Bu_3SnCH_2OEt	IA	29	b_1 116-18°, n_D^{20} 1.4775, d_{20} 1.503	307
$Bu_2Sn(CH_2OMe)_2$	IA	21	b_{21} 147-48°, n_D^{20} 1.4775, d_{20} 1.1881 (k)	307
$Bu_2Sn(CH_2CH_2CH_2OH)_2$	IVB	83	Clear, visc. liq., n_D^{24} 1.5005, dec. on standing (l, m)	1383
$Bu_2Sn[(CH_2)_3OAc]_2$ (226)	(m)	--	n_D^{29} 1.4748, mol. wt. TLC	1383
$Bu_2Sn[(CH_2)_4\overline{CHCHEtO}]_2$	--	--	Use: stabilizer for PVC	1442
$(c.C_6H_{11})_3SnC_6H_4OMe-p$ (227)	--	--	m. 75°, b_3 207°, IR spectrum	1773
$(c.C_6H_{11})_3SnC_6H_4OEt-p$	IA	--	m. 70°, b_3 225°, IR spectrum	1773
Ph_3SnR				
R = Ph_2COH	IVD	19	m. 256-59°	2671
$PhOCH_2CH_2$ (227)	III	--	NMR spectrum	2070
$AcOCH_2CH_2$ (227)	III	--	NMR spectrum	2070
$HOCHMe$	IVC	72	m. 93-95°, dec. in air, IR, NMR spectra	1677, 2671
$4,3,5-HO(Me_3C)_2C_6H_2$	II	--	(n)	535
$Ph_2Sn(C_6H_4OPh-p)_2$ (228)	--	--	(o)	--
$(C_6H_{13})_3SnC(CO_2Et)_2CH_2\overline{CHCH_2O}$	(p)	--	Use: stabilizer of PVC	1442

(j) Anal. detn. by Schoeninger combustion (448), rxn. with AcCl → $Et_2Sn(CH_2Cl)_2$ (307). (k) Use as catalyst for polymerization of $(Me_2SiO)_2$ (307). (l) Thermal dec. at 140-60° in vacuo → $Bu_2SnOCH_2CH_2CH_2$ (1383). (m) Rxn. of $Bu_2Sn(CH_2CH_2CH_2OH)_2$ with Ac_2O-NaOAc → $Bu_2Sn(CH_2CH_2CH_2OAc)_2$ (1383). (n) Dehydrogenation in C_6H_6 → $4,2,6-Ph_3Sn-(Me_3C)_2C_6H_2O\cdot$; mesomeric radical, green solution, ESR spectrum (535). (o) Use: stabilizer for mineral oil and synthetic lubricants (981), antioxidant for polyphenyl ethers (1979). (p) Synth. from $\overline{OCH_2CHCH(CO_2Et)_2}$ with NaOMe and R_3SnCl (1442).

2.6.3.3 TETRAORGANOTIN COMPOUNDS CONTAINING CARBONYL GROUPS AND DERIVATIVES

Previously most of the carbonyl derivatives of the organotin moiety have been regarded as Sn-C derivatives of the tautomeric ketone. It is now generally accepted that derivatives of 1,2- and 1,3-dicarbonyl compounds are Sn-O chelates with the diketone acting as bidentate ligand. The corresponding ketone derivatives may exist in an equilibrium of both Sn-C and Sn-O isomer with steric and electronic reasons for prevalence of either form. The reader is advised to check Chapter 4.3.4 and Tables 192-194 for any given compound. As for the main volume caution should be exercised regarding the formulae used there. The tetraorganotin compounds containing carbonyl groups and derivatives listed in Table 112 are prepared by methods from the following scheme.

I Grignard reaction:
   (IA) $R_3SnCl + MeO(PhO)CHMgBr \quad R_3SnCH(OMe)OPh$,
   (IB) $Et_3SnCH_2CH_2CN + MeMgI \rightarrow Et_3SnCH_2CH_2Ac$.

II Reaction with organotin lithium derivatives:
   $Ph_3SnLi + Me_3CCOCl \rightarrow Ph_3SnCOCMe_3$.

III Reaction with organomercury compounds:
   $(Bu_3Sn)_2S + (EtCOCH_2)_2Hg \rightarrow Bu_3SnCH_2COEt$.

IV Addition reactions of organotin heterocompounds to olefins:
   (IVA) $Ph_3SnH + CH_2:CHAc \rightarrow Ph_3SnCH_2CH_2Ac$,
          $Bu_3SnH + MeCH:CHAc \rightarrow Bu_3SnCEtAc$,
   (IVB) $Me_3SnH + AcCHN_2 \rightarrow Me_3SnCH_2Ac$,
   (IVC) $Et_3SnNMe + RCH:CHCHO \rightarrow Et_3SnCH(CHO)CHRNMe_2$.
   (IVD) $Bu_3SnOMe + R_2C:CROAc \rightarrow Bu_3SnCR_2COR$.

R = organic group or hydrogen. Additional tetraorganotin compounds containing carbonyl groups and derivatives may be found in Chapters 2.3, 2.4.3, 2.5.3, and 2.6.4.

Acetonyltriethyltin (M-230) $SnC_9H_{20}O$ Et_3SnCH_2Ac
Synth.: By rxn. of Et_3SnNMe_2 with $AcOCH:CHMe$, in 50% yield (1710).
By rxn. of Et_3SnOMe with $AcOCMe:CH_2$, in 88% yield (2489).
Prop.: b_3 89.5-90° (2489), $b_{0.1}$ 78-80° (1710), n_D^{20} 1.4960 (1710), 1.4981 (2489), d_{20} 1.2858 (2489), 1.2864 (1710), IR (1710, 2489), NMR, Moessbauer resonance and mass (2489) spectra, mol. wt. (2489).
Rxn. with: $MeOH \rightarrow Et_3SnOMe + Me_2CO$ (437).
$\overline{CH_2CH_2CH_2C(CO_2Et):COH} \rightarrow \overline{CH_2CH_2CH_2C(CO_2Et):COSnEt_3} + Me_2CO$ (1711).
Me_2NH (at -78°) $\rightarrow (Et_3Sn)_2O + Me_2CO + AcCH:CMe_2$ (250).
$PhC\vdots CH \rightarrow Et_3SnC\vdots CPh + Me_2CO$ (437).
$RCHO \rightarrow Et_3SnOCHRCH_2Ac$; R = Ph (437, 1631), Pr, C_6F_5 , (C_4H_3O), o-, p-ClC_6H_4 , m-, p-$O_2NC_6H_4$ (1631).
$R_2CO \rightarrow Et_3SnOCR_2CH_2Ac$; R_2 = $(ClCH_2)_2$, (Ph, CF_3), $(CH_2)_5$ (1631).
$MeCOCH_2CO_2R \rightarrow Et_3SnOCMe:CHCO_2R + Me_2CO$; R = Me, Et, Pr (1711).
$Ac_2CH_2 \rightarrow Et_3SnOCMe:CHAc$ (1711).
$AcCl \rightarrow Et_3SnCl + AcOCMe:CH_2$ (437).

Ph$_2$PX → Ph$_2$PCH$_2$Ac; X = Cl, I (2653).
BuP(OEt)Cl, + S → Bu(AcCH$_2$)P(S)OEt (2653).

Acetonyltripropyltin (M-230)　　　　　　SnC$_{12}$H$_{26}$O　　　　　　Pr$_3$SnCH$_2$Ac
Rxn. with: Me$_3$SiCl → Pr$_3$SnCl + Me$_3$SiOCMe:CH$_2$ (24a).
R$_3$SiX → Pr$_3$SnX + Me$_3$SiOCMe:CH$_2$; R = Me, Et, X = Cl, I (711).
Me$_3$GeBr → Pr$_3$SnBr + Me$_3$GeCH$_2$Ac + Me$_3$GeOCMe:CH$_2$ (2547).
AcCH$_2$COR → Pr$_3$SnOCMe:CHCOR + Me$_2$CO; R = Me, MEO (1711).

Acetonyltributyltin (M-230)　　　　　　SnC$_{15}$H$_{32}$O　　　　　　Bu$_3$SnCH$_2$Ac
Synth.: By rxn. of Bu$_3$SnOMe with AcOCMe:CH$_2$ and exclusion of moisture, in 79%
yield (1681).
Prop.: b$_{1.9}$ 132°, n$_D^{20}$ 1.4863, d$_{20}$ 1.1316, IR and NMR spectra (1681), isomeric
with enoxide Bu$_3$SnOCMe:CH$_2$ (719a).
Rxn. with: R$_3$SnH → Me$_2$CO; R = Bu, Ph (2680).
R$_3$SiX → Bu$_3$SnX + R$_3$SiOCMe:CH$_2$; R = Et, Pr, X = Cl, Br, I (711).
Et$_3$GeX → Bu$_3$SnX + Et$_3$GeCH$_2$Ac; X = Cl, Br, I (711).
Pr$_2$GeHI → Pr$_2$(AcCH$_2$)GeH (2238).
(CH$_2$:CMeO)$_3$SiH → Bu$_3$SnH + (CH$_2$:CMeO)$_4$Si (719).
HSiCl$_3$ → HSi(OCMe:CH$_2$)$_3$ (719a).
SiCl$_4$ → Si(OCMe:CH$_2$)$_4$ (719a).

Acetyltriphenyltin　　　　　　SnC$_{20}$H$_{18}$O　　　　　　Ph$_3$SnCOMe
Synth.: By rxn. of Ph$_3$SnLi with AcCl in THF at -70° (1677), in N atm. in the
dark, in 88) yield (2671).
Prop.: Waxy solid, easily dec. (1677, 2671), IR (2671), UV (1677, 2671)
spectra, sensitive to light (2671).
Rxn. with: Air → Ph$_3$SnOAc (1677, 2671).
LiAlH$_4$ → Ph$_3$SnCHMeOH (1677, 2671).

Benzoyltriphenyltin　　　　　　SnC$_{25}$H$_{20}$O　　　　　　Ph$_3$SnCOPh
Synth.: By rxn. of Ph$_3$SnLi with BzCl at -70° in N atm. in the dark, in 33%
yield (2671).
Prop.: m. 214-17°, IR, UV and mass spectra (2671).
Rxn. with: Moisture → BzH (2671).
O(under UV irradiation) → BzOH (2671).
PhMgBr → Ph$_4$Sn + Ph$_3$SnCPh$_2$OH + (Ph$_3$Sn)$_2$O + Ph$_2$CHOH (2671).

Additional tetraorganotin compounds containing carbonyl groups and derivatives
are listed in Table 112.

Table 112. Tetraorganotin Compounds Containing Carbonyl Groups and Derivatives $R_nSnR'_{4-n}$

Formula	Synth. Method*	Yield	Properties	Ref.
Me$_3$SnR				
R = MeO(PhO)CH	IA	36	b$_1$ 88-89°, n$_D^{20}$ 1.4931, d$_{20}$ 0.9817	1303
AcCH$_2$	IVB	15	Red oil, n$_D^{20}$ 1.5060, struct. (a)	2489
BzCH$_2$CH$_2$	--	--	Photolysis: Me$_2$Sn + Me$_3$SnCH:CH$_2$ + CH$_4$	1488
AcCH$_2$CH$_2$	--	--	Photolysis: Me$_2$Sn + CH$_4$	1488
Bz(CH$_2$)$_3$	--	--	(b)	--
Ac(CH$_2$)$_3$	--	--	Photolysis: Me$_2$Sn + Me$_3$SnCH:CH$_2$ + CH$_4$	1488
Et$_3$SnCH(OMe)OPh	IA	53	b$_{1.5}$ 122-23°, n$_D^{20}$ 1.5024, d$_{20}$ 0.9793	1303
Et$_3$SnCH$_2$CH$_2$Ac	IB	58	b$_1$ 88-89°, n$_D^{20}$ 1.4880, d$_{20}$ 1.2449, struct. (c)	2489
Bu$_3$SnR'				
R' = PhO(MeO)CH	IA	17	b$_{1.5}$ 186-88°, n$_D^{20}$ 1.5140, d$_{20}$ 1.1642	1303
BzCH$_2$ (d)	IVD	51	b$_{0.7}$ 138°, n$_D^{20}$ 1.5249, d$_{20}$ 1.1514, IR and NMR spectra, contains 22% Sn-O isomer	1681
⇌ CH$_2$:CPhO				1681
EtCOCH	III	56	b$_2$ 136-38°, n$_D^{20}$ 1.4899, d$_{20}$ 1.1577	2648
AcCHMe (d)	IVD	44	b$_{0.8}$ 115°, n$_D^{20}$ 1.4820, d$_{20}$ 1.1152, IR and NMR spectra, contains 23% Sn-O isomer	1681
⇌ MeCH:CMeO				1681
EtCOCHMe (d)	IVD	57	b$_{0.8}$ 121-23°, n$_D^{20}$ 1.4808, d$_{20}$ 1.1084, IR and NMR spectra, contains 70% Sn-O isomer	1681
⇌ MeCH:CEtO				1681
AcCEt (d)	IVA	47	b$_1$ 126-28°, n$_D^{20}$ 1.4812, d$_{20}$ 1.0894, IR and NMR spectra, contains 75% Sn-O isomer	1684
⇌ EtCH:CMeO	IVD	65		1681, 1684

* The characters used here correspond to the synthetic scheme in the introduction to Chapter 2.6.3.3. (a) Moessbauer resonance spectrum, dec. on repeated distillation (2489). (b) Photolysis → Me$_2$Sn + Me$_3$SnCH:CH$_2$ + CH$_4$ + BzMe + (BzCH$_2$)$_2$ + (PhCMeOH)$_2$ (1488). (c) IR, Moessbauer resonance and mass spectra (2489). (d) Compare Table 193, Chapter 4.3.4.1.3 for correspond Sn-O isomers.

Table 112 Continued

Formula	Synth. Method*	Yield	Properties	$R_nSnR'_{4-n}$ Ref.
Bu_3SnR' (Cont'd.)				
R' = c.2-OC$_5$H$_7$ (d) ⇌ c.1-C$_5$H$_7$O	IVD	37	$b_{1.4}$ 144-46°, n_D^{20} 1.4930, d_{20} 1.1397, IR and NMR spectra, contains 43% Sn-O isomer	1681 1681
Me$_3$CCOCH$_2$ (d) ⇌ CH$_2$:C(CMe$_3$)O	IVD	52	$b_{0.3}$ 118°, n_D^{20} 1.4811, d_{20} 1.0840, IR and NMR spectra, contains 25% Sn-O isomer	1681 1681
Ph$_3$SnCH$_2$CH$_2$Ac	IVA	52	NMR spectrum of crude prod.	1418
Ph$_3$SnCOCMe$_3$	II	63	m. 90-94° (dec.) (e)	2671
Et$_3$SnCH(CHO)CH$_2$NMe$_2$	IVC	99	$b_{0.03}$ 86°, yellow liq.	854, 2406
Et$_3$SnCH(CHO)CHMeNMe$_2$	IVC	95	$b_{0.01}$ 80°, orange liq., n_D^{25} 1.5840, mol. wt.	854, 2406
Et$_3$SnCH(CHO)CHPhNMe$_2$	IVC	88	$b_{0.02}$ 69-71°, yellow liq., n_D^{28} 1.5881, mol. wt.	854

(e) IR, UV and NMR spectra (2671), rxn. with air and light → (Ph$_3$Sn)$_2$O + Ph$_3$SnO$_2$CCMe$_3$ + (Me$_3$CCO)$_2$O (2671).

2.6.3.4 TETRAORGANOTIN COMPOUNDS CONTAINING CARBOXYL GROUPS AND DERIVATIVES

This chapter comprises unsymmetric tetraorganotin compounds containing a carboxylate group. Derivatives of this carboxylic acid group like esters, amides and organometallic esters are included. A group of derivatives of diazocarboxylates proposed to having a carbon-tin linkage are also added. The compounds listed in Table 113 are prepared by methods from the following scheme.

I Grignard reaction:
$$MeR'SnX_2 + MeMgI \rightarrow Me_3SnR'.$$

II Reaction with organomercury compounds:
$$(Ph_3Sn)_2S + R'_2Hg \rightarrow Ph_3SnR'.$$

III Hydrostannation reaction:
  (IIIA) $R_nSnH_{4-n} + CH_2{:}CHR' \rightarrow R_nSn(CHRCH_2R')_{4-n}$; n = 2, 3 ,
  (IIIB) $R_3SnH + HC{:}CR' \rightarrow R_3SnCH_2CH_2R'$,
  (IIIC) $Bu_3SnH + CH_2{:}CHCO_2Et$ in MeOH $\rightarrow Bu_3SnCH_2CH_2CO_2R$; R = Me, Et.

IV Ketene addition reactions:
  (IVA) $R_3SnOR'' + CH_2{:}C{:}O \rightarrow R_3SnCH_2CH_2CO_2R''$; $R'' =$ alkyl, R_3Sn,
  (IVB) $R_3SnNR_2 + CH_2{:}C{:}O \rightarrow R_3SnCH_2CONR_2$.

V Interconversion reactions:
  (VA) $R_2Sn(CH_2CHMeCO_2Me)_2 + ROH \rightarrow R_2Sn(CH_2CHMeCO_2R)_2 + MeOH,$
  (VB) $Ph_3SnCN_2CO_2Et + Me_2C{:}CH_2$ under UV irradiation $\rightarrow Ph_3Sn\overline{C(CO_2Et)CH_2C}$-$Me_2$.

VI Miscellaneous reactions:
  (VIA) $Me_3SnNEt_2 + \overline{OCH_2CH_2CO} \rightarrow Me_3SnCH_2CH_2CONEt_2$,
  (VIB) $Et_3SnNMe_2 + CH_2{:}CHCO_2Me \rightarrow Et_3SnCH(CO_2Et)CH_2NMe_2$,
  (VIC) $R_nSn(NR_2)_{4-n} + N_2CHCO_2Et \rightarrow R_nSn(CN_2CO_2Et)_{4-n}$; n = 2, 3 ,
  (VID) $Bu_4Sn + MeCH{:}CHCO_2Et \rightarrow Bu_3Sn(CH_2)_4CHCHMeCH_2CO_2Et$; catalyzed by Bz_2O_2 .

R = organic group or hydrogen, R' = organic group containing a carboxy function, X = halogen. Additional tetraorganotin compounds containing carboxyl groups and derivatives may be found in Chapters. 2.4.3, 2.5.3, 2.6.1 2.6.3.1 and 2.7.1.1.3.

Methyl Triethyltinacetate (M-234)　　　　$SnC_9H_{20}O$　　　　$Et_3SnCH_2CO_2Me$
<u>Synth.</u>: By rxn. of Et_3SnOMe with ketene, in 87% yield (2489).
<u>Prop.</u>: b_2 83-83.5°, n_D^{20} 1.4832, d_{20} 1.2944, IR, Moessbauer resonance and mass spectra, struct. (2489).
<u>Rxn. with</u>: $(MeO_2CCH_2)_nGeCl_{4-n} \rightarrow (MeO_2CCH_2)_mGeCl_{4-n}$; n = o-2, m = 2-4 (692, 1440).
$GeBr_4 \rightarrow (MeO_2CCH_2)_2GeBr_2$ (1410).
$GeI_4 \rightarrow MeO_2CCH_2GeI_3$ (1410).
$SiCl_4 \rightarrow Cl_3SiOC(OMe){:}CH_2 + MeO_2CCH_2SiCl_3$ (818).
$MCl_3 \rightarrow Et_3SnCl + (MeO_2CCH_2)_3M$; M = P (440), Sb (2246).

Ethyl Triethyltinacetate (M-234) $SnC_{10}H_{22}O_2$ $Et_3SnCH_2CO_2Et$
<u>Prop.</u>: NMR spectrum (1689).
<u>Rxn. with</u>: RCHO → $Et_3SnOCHRCH_2CO_2Et$; R = Ph, CCl_3, C_6F_5 (1631).
$BzCF_3$ → $Et_3SnOCPh(CF_3)CH_2CO_2Et$ (1631).
PCl_3(2:1) → Et_3SnCl + $(EtO_2CCH_2)_2PCl$ (440).
$EtOPCl_2$(2:1) → Et_3SnCl + $(EtO_2CCH_2)_2POEt$ (440).

Methyl Tripropyltinacetate (M-235) $SnC_{12}H_{26}O_2$ $Pr_3SnCH_2CO_2Me$
<u>Rxn. with</u>: Et_3SiI → Pr_3SnI + $Et_3SiOC(OMe):CH_2$ (24a).
R_2SiCl_2 → Pr_3SnCl + $R_2(Cl)SiOC(OMe):CH_2$ + $R_2(MeO_2CCH_2)SiCl$; R = Me and/or Cl (818).
AcCl → Pr_3SnCl + AcOMe + $CH_2:C(OMe)OAc$ + $AcCH_2CO_2Me$ (437).

Methyl Tributyltinacetate $SnC_{15}H_{32}O_2$ $Bu_3SnCH_2CO_2Me$
<u>Synth.</u>: By rxn. of Bu_3SnOMe with $(MeO_2CCH_2)_2Hg$ in 75% yield (2648).
By rxn. of Bu_3SnOMe with ketene (36).
<u>Prop.</u>: $b_{1.5}$ 119-22°, n_D^{20} 1.4811, d_{20} 1.1560 (2648).
<u>Rxn. with</u>: R_3GeCl → $R_3GeCH_2CO_2Me$; R = Et, MeO, Cl (692, 1440).
Pr_2HGeCl → $Pr_2(MeO_2CCH_2)GeH$ (2238).
Et_3GeI → $Et_3GeCH_2CO_2Me$ (692).
$GeBr_4$ → $MeO_2CCH_2GeBr_3$ (1440).
R_2SiCl_2 → $R_2(Cl)SiOC(OMe):CH_2$ + $R_2(MeO_2CCH_2)SiCl$; R = Me and/or Cl (818).
PCl_3(1:1) → $MeO_2CCH_2PCl_2$ + Bu_3SnCl (440).

Ethyl Tributyltinacetate (M-236) $SnC_{16}H_{34}O_2$ $Bu_3SnCH_2CO_2Et$
<u>Synth.</u>: By rxn. of Bu_3SnOEt with ketene, in 82% yield (440).
By rxn. of Bu_3SnOMe with $Me_3SiCH_2CO_2Et$ at 170° in $(MeOCH_2)_2$, in 82% yield (1682).
<u>Prop.</u>: b_1 129° (440), $b_{0.6}$ 120°, n_D^{20} 1.4741 (1682), 1.4750 (440), d_{20} 1,121 (1682), 1.1271 (440).
<u>Rxn. with</u>: Ph_3SnH → Bu_3SnH + AcOEt (440).
$(EtO)_2PCl$ → $EtO_2CCH_2P(OEt)_2$ + Bu_3SnCl (2680).

Ethyl 4,2-Methyltributyltinbutyrate $SnC_{19}H_{40}O_2$ $Bu_3SnCH(CO_2Et)CHMe_2$
<u>Synth.</u>: By rxn. of Bu_3SnH with $Me_2C:CHCO_2Et$ under UV irradiation in 57% yield (2680).
<u>Prop.</u>: $b_{0.2}$ 120°, n_D^{20} 1.4787, d_{20} 1.095, very sensitive to moisture, IR and NMR spectra (2680).
<u>Rxn. with</u>: Bu_3SnH under UV irradiation → Bu_6Sn_2 + $Me_2CHCH_2CO_2Et$ (2680).
Ph_3SnH under UV irradiation → Bu_3SnH + $Me_2CHCH_2CO_2Et$ (2680).
MeOH → Bu_3SnOMe + $Me_2CHCH_2CO_2R$ + EtOH; R = Me, Et (2680).

Listing tetraorganotin compounds containing carboxylate groups continues in Table 113.

Table 113. Tetorganotin Compounds Containing Carboxyl Groups and Derivatives $R_nSnR'_{4-n}$

Formula*	Synth. Method**	Yield	Properties	Ref.
Me_3SnR				
R = Me_2NOCCH_2	IVB	99	Yellow solid, subl. > 160°, monomeric (a, b)	167
Et_2NOCCH_2	IVB	40	Colorless liq., b_1 68-69°, n_D^{20} 1.5075, d_{20} 1.4134	1710
Et_2NOCCH_2	IVB	60	Colorless liq. b_2 82-84°, n_D^{20} 1.4961, d_{20} 1.2878 (c)	1710
$MeO_2CCH_2CH_2$	I	70	b_{20} 92-95°, n_D^{20} 1.4690	2651
$EtO_2CCH_2CH_2$	IIIB	8	Not sepd., IR, NMR spectra, GLC	311, 1414
$EtO_2CCH_2CH_2$	I	53	b_{20} 99-100°, n_D^{20} 1.4660	2651
$EtO_2CCH_2CH_2$	IIIB	2	Not sepd., IR, NMR spectra, GLC	311, 1414
$Et_2NOCCH_2CH_2$	VIA	50	$b_{0.04}$ 85-87° (d)	1238
Et_3SnR				
R = Me_2NOCCH_2	IVB	74	$b_{1.5}$ 98-100°, n_D^{20} 1.5050, d_{20} 1.2801 (e)	1710
Et_2NOCCH_2	IVB	69	b_1 101-103°, n_D^{20} 1.4995, d_{20} 1.2250 (f)	1710
$MeO_2CCH_2CH_2$ (234)	IIIB	5	IR, NMR spectra, GLC	311, 1414
$EtO_2CCH_2CH_2$ (234)	IIIB	6	IR, NMR spectra, GLC (g)	311, 1414
$Et_2Sn(CH_2CH_2CO_2CH_2CH:CH_2)_2$	IIIA	83	$b_{0.005}$ 122°, n_D^{20} 1.4940, d_{20} 1.2439	2845
$Et_2Sn(CH_2CHMeCO_2CH_2CH:CH_2)_2$	IIIA	87	$b_{0.007}$ 121°, n_D^{20} 1.4908, d_{20} 1.2049	2845

* Numbers in parenthesis refer to pages in main volume.
** The characters used here correspond to the synthetic scheme in the introduction to Chapter 2.6.3.4.
(a) IR spectrum and asignment, struct. (167). (b) IR and NMR spectra, struct., very sensitive to moisture (1710). (c) IR spectrum (1710). (d) Rxn. with HCl → $HOCH_2CH_2CONEt_2$ (1238). (e) Rxn. with HC⋮COEt → $Et_3SnC⋮COEt$ + $AcNMe_2$ (1710), with $AcCH_2CO_2Et$ → $Et_3SnOCMe:CHCO_2Et$ + $AcNMe_2$ (1710). (f) Rxn. with RCHO → $Et_3SnOCHRCH_2CO-NEt_2$; R = $2-C_4H_3O$, CCl_3 , C_6F_5 (1631). (g) Relative rate of cleavage rxn. with KOH (987).

Table 113 Continued

Formula*	Synth. Method**	Yield	Properties	Ref.
Pr$_3$SnR				
R = EtO$_2$CCH$_2$ (235)	--	--	(h, i)	--
PrO$_2$CCH$_2$ (235)	--	--	(i)	--
Me$_2$NOCCH$_2$	IVB	46	b$_{2.5}$ 114-16°, n$_D^{20}$ 1.5014, d$_{20}$ 1.1960	1710
MeO$_2$CCH$_2$CH$_2$ (235)	IIIB	0.4	IR, NMR spectra (j)	311, 1414
MeO$_2$C(CH$_2$)$_{10}$	IIIA	--	b$_{0.01}$ 160-61°, n$_D^{20}$ 1.4768 (k)	598
Pr$_2$Sn(CH$_2$CH$_2$CO$_2$Me)$_2$ (236)	--	--	NMR spectrum (l)	2070
Pr$_2$Sn(CH$_2$CHMeCO$_2$Me)$_2$	--	--	(m)	--
Pr$_2$Sn(CH$_2$CHMeCO$_2$CH$_2$CH:CH$_2$)$_2$	VA (m)	74	b$_{0.02}$ 136°, n$_D^{20}$ 1.4889, d$_{20}$ 1.1722	2844
Bu$_3$SnR				
R = BuO$_2$CCH$_2$	IVA	80	b$_1$ 122-23°, n$_D^{20}$ 1.4772, d$_{20}$ 1.1003 (n)	1440
Me$_2$NOCCH$_2$	IVB	55	b$_{0.03}$ 130-32°, n$_D^{20}$ 1.4925, d$_{20}$ 1.1316 (b)	1710
MeO$_2$CCH$_2$CH$_2$ (236)	IIIA	--	NMR spectrum	2070
	IIIC	30	(o)	2680
EtO$_2$CCH$_2$CH$_2$	IIIA	83	b$_{0.8}$ 130°, n$_D^{20}$ 1.4752, d$_{20}$ 1.117	2680
	IIIC	66	IR, NMR spectra	2680
EtO$_2$CCH$_2$CHMe	IIIA (p)	33	b$_{0.8}$ 128°, mixt. with R = EtO$_2$CCHEt (q)	2680
EtO$_2$CCHEt	IIIA (p)	89	b$_{0.8}$ 128-30°, n$_D^{20}$ 1.4761, d$_{20}$ 1.104 (q)	2680
Bu$_3$SnO$_2$CCMe$_2$	IVA	--	--	36
MeO$_2$CCH$_2$CH(CO$_2$Me)	IIIA	--	b$_{0.001}$ 122-27°, n$_D^{20}$ 1.4799 (r)	1654
EtO$_2$CCH$_2$CH(CO$_2$Et)	IIIA	--	b$_{0.001}$ 124-28°, n$_D^{20}$ 1.4777 (r)	1654
PrO$_2$CCH$_2$CH(CO$_2$Pr)	IIIA	--	b$_{0.001}$ 125-28°, n$_D^{20}$ 1.4599 (r)	1654
BuO$_2$CCH$_2$CH(CO$_2$Bu)	IIIA	--	b$_{0.001}$ 128-35°, n$_D^{20}$ 1.4582 (r)	1654
MeO$_2$CCH(CHMe$_2$)	IIIA (s)	60	b$_{0.2}$ 132-33°, n$_D^{20}$ 1.4810, d$_{20}$ 1.118 (u)	2680
MeO$_2$CCH$_2$CH(CO$_2$Me)CH$_2$	IIIA (t)	< 80	b$_{0.005}$ 85-95°, n$_D^{20}$ 1.4788 (u)	1652
EtO$_2$CCH$_2$CH(CO$_2$Et)CH$_2$	IIIA (t)	< 80	b$_{0.007}$ 125-35°, n$_D^{20}$ 1.4738 (u)	1652

Compound	Type	Yield (%)	Properties	Ref.
PrO$_2$CCH$_2$CH(CO$_2$Pr)CH$_2$	IIIA (t)	<80	b$_{0.003}$ 140-50°, n$_D^{20}$ 1.4738	1652
(EtO$_2$C)$_2$CHCHMe	IIIA (v)	--	b$_{0.001}$ 140-45°, n$_D^{20}$ 1.4765 (r, w)	1654
EtO$_2$CCH$_2$CHMe(CH$_2$)$_4$	VID	--	Brown viscous liq.	2722
HO$_2$C(CH$_2$)$_{10}$	IIIA	--	n$_D^{20}$ 1.4818 (k)	598
MeO$_2$C(CH$_2$)$_{10}$	IIIA	--	b$_{0.001}$ 168-69°, n$_D^{20}$ 1.4758 (k)	598
Pr$_3$SnO$_2$C(CH$_2$)$_{10}$	IIIA	--	n$_D^{20}$ 1.4960 (k)	598
Bu$_2$Sn(CH$_2$CH$_2$CO$_2$Me)$_2$	IIIA	92	b$_{0.075}$ 126°, n$_D^{24}$ 1.4808, mol. wt. (x)	1383
Bu$_2$Sn(CH$_2$CH$_2$CO$_2$CH$_2$CH$_2$CH:CH$_2$)$_2$	VA (m)	79	b$_{0.025}$ 152-53°, n$_D^{20}$ 1.4895, d$_{20}$ 1.1702, mol. wt.	2844
Bu$_2$Sn(CH$_2$CH$_2$CONH$_2$)$_2$	IIIA	--	m. 88-89.5°, IR spectrum (y)	1168, 2434
Ph$_3$SnR				
R = MeO$_2$CCH$_2$	II	53	m. 54°	2648
MeO$_2$CCH$_2$CH$_2$ (237)	--	--	NMR spectrum	2070
EtO$_2$CCH$_2$CH$_2$ (237)	IIIA	98	Viscous liq., IR, NMR spectra	2680
Me$_2$CCH$_2$C(CO$_2$Et)	VB	35	IR, NMR spectra	1831
MeO$_2$C(CH$_2$)$_{10}$	IIIA	--	m. 55.5-57° (k)	598
Me$_3$SnCN$_2$CO$_2$Et	VIC	100	Light yellow cryst., m. 28-30° (z, aa)	2545
Me$_2$Sn(CN$_2$CO$_2$Et)$_2$	VIC	70	Light brown liq., b$_{0.1}$ 135-40° (z)	2545
Et$_3$SnCN$_2$CO$_2$Et	VIC	85	Light yellow liq., b$_{0.1}$ 90-95° (z)	2545

(h) Rxn. with AcCl → Pr$_3$SnCl + AcOR + AcCH$_2$CO$_2$R; R = Et, Pr (437). (1) Rxn. with AcCH$_2$CO$_2$Et → Pr$_3$SnOCMe:CHCO$_2$Et + EtOAc (437). (j) NMR spectrum (2070). (k) Use as pesticide, bactericide (598). (l) Rxn. with Br → Pr(MeO$_2$CCH$_2$CH$_2$)$_2$SnBr + (MeO$_2$CCH$_2$CH$_2$)$_2$SnBr$_2$ (2070). (m) Rxn. of R'$_2$Sn(CH$_2$CHRCO$_2$Me)$_2$ with CH$_2$:CHCH$_2$OH → R'$_2$Sn(CH$_2$CHRCO$_2$CH$_2$CH:CH$_2$)$_2$; R = H, Me, R' = Pr, Bu (2844). (n) Rxn. with Et$_3$GeI → Et$_3$GeCH$_2$CO$_2$Bu (1440). (o) Rxn. with Br → Bu$_2$-(MeO$_2$CCH$_2$CH$_2$)SnBr (2070). (p) From MeCH:CHCO$_2$Et (2680). (q) IR and NMR spectra (2680), synth. under UV irradiation suppresses formation of Bu$_3$SnCHMeCH$_2$CO$_2$Et (2680). (r) IR and NMR spectra (1654). (s) From Me$_2$C:CHCO$_2$Me under UV irradiation (2680). (t) From R'O$_2$CC(:CH$_2$)CH$_2$CO$_2$R'; R' = Me, Et, Pr (1652). (u) Rxn. with Br → Bu$_2$RO$_2$-CCH$_2$CH(CO$_2$R)CH$_2$SnBr; R = Me, Pr (1654), Et (1652). (v) From MeCH:C(CO$_2$Et)$_2$ (1654). (w) Rxn. with Br → Bu$_2$RSnBr (1654). (x) Rxn. with LiAlH$_4$ → Bu$_2$Sn(CH$_2$CH$_2$CH$_2$OH)$_2$ (1383). (y) Rxn. with HBr → Bu$_2$RSnBr (1168). (z) IR and NMR spectra (2545), mol. wt., slightly sensitive to air and moisture, soly. (2545). (aa) Struct., subl. in high vacuo (2545).

Table 113 Continued

Formula*	Synth. Method**	Yield	Properties	$RnSnR'_{4-n}$ Ref.
$Et_3SnCH(CO_2Me)CH_2NMe_2$	VIB	93	$b_{0.02}$ 88°, n_D^{25} 1.5836, mol. wt. (bb)	854, 2406
$Et_2Sn(CN_2CO_2Et)_2$	VIC	60	Light yellow liq., $b_{0.1}$ 140-42° (z)	2545
$Bu_3SnCN_2CO_2Et$	VIC	80	Light yellow liq., $b_{0.1}$ 130-35° (z)	2545
$Bu_2Sn(CN_2CO_2Et)_2$	VIC	35	Light yellow liq., $b_{0.1}$ 145-50° (z)	2545
$Ph_3Sn(CN_2CO_2Et)$	II	80	IR spectrum	1831
	VIC	100	Yellow-brownish cryst., m. 49-51° (z, cc)	2545

(bb) NMR spectrum (2406). (cc) Rxm. with $Me_2C:CH_2$ under UV irradiation → $Ph_3Sn(EtO_2C)\overline{CCH_2CMe_2}$ (1831).

2.6.4 TETRAORGANOTIN COMPOUNDS CONTAINING SULFUR

This chapter includes unsymmetric tetraorganotin derivatives with sulfide, thioacetal, thioketal and thiocarboxylate groups. The compounds summarized in Table 114 are prepared by the following methods:

I Grignard reaction:
   $Me_3SnCl + p\text{-}MeSC_6H_4MgBr \rightarrow p\text{-}Me_3SnC_6H_4SMe$, in 73% yield.

II Synthesis with organolithium derivatives:
   $Ph_3SnX + \overline{SCH_2CH_2CH_2SCRLi} \rightarrow Ph_3Sn\overline{CRSCH_2CH_2CH_2S}$; R = H, Me, Ph, in 79-80% yield.

III Addition reaction of thiocarboxylic acid under UV irradiation:
   $Bu_3Sn(CH_2)_nCH{:}CH_2 + RCOSH \rightarrow Bu_3Sn(CH_2)_{n+2}SCOR$, in 80-85% yield.

R = organic group, X = halogen. Additional tetraorganotin compounds containing sulfur may be found in Chapters 2.4.3, 2.5.3, 2.7.2 and 7.1.6.

Table 114. Tetraorganotin Compounds Containing Sulfur R_3SnR'

Formula*	Synth. Method**	Properties	Ref.
p-$Me_3SnC_6H_4SMe$	I	$b_{0.4}$ 85°, n_D^{25} 1.5832 (a)	998
$Bu_3SnCH_2CH_2SAc$ (239)	III	$b_{0.2}$ 115°, n_D^{23} 1.524 (b)	534
$Bu_3Sn(CH_2)_3SAc$ (239)	III	$b_{0.2}$ 120°, n_D^{23} 1.4986 (b)	534
$Bu_3Sn(CH_2)_3SCOPr$	III	$b_{0.4}$ 118° (b)	534
$Ph_3Sn\overline{CHS(CH_2)_3S}$	II	m. 149-50°	792
$Ph_3Sn\overline{CMeS(CH_2)_3S}$	II	m. 149.5-51°	792
$Ph_3Sn\overline{CPhS(CH_2)_3S}$	II	m. 139-41°	792

* Numbers in parenthesis refer to pages in main volume.
** The characters used here correspond to the synthetic scheme in the introduction to Chapter 2.6.4.
(a) Relative kinetic data for aryl-cleavage reaction (997). (b) Use as herbicide (534).

2.6.5 TETRAORGANOTIN COMPOUNDS CONTAINING NITROGEN AND PHOSPHORUS

This chapter comprises unsymmetric tetraorganotin compounds containing nitrogen and phosphorous. Quarternary ammonium salts and a nickel complex of N,N'-diethylaminotroponeimine are included. Some unsaturated derivatives with phosphorous and antimony may be found in Chapters 2.4.3, 2.5.1 and 2.6.1.

The tetraorganotin compounds with nitrogen and phosphorous summarized in Table 115 are prepared by methods from the ensuing scheme.

I   Reaction with organolithium derivatives:
$R_3SnX + R'Li \rightarrow R_3SnR'$.

II Reaction with organomercury compounds:
$(Et_3Sn)_2S + R'_2Hg \rightarrow Et_3SnR'$.

III Addition of organotin amide and phosphide:
(IIIA) $Me_3SnNMe_2 + MeC_6H_4NC \rightarrow p\text{-}Me_3SnC(NMe_2):NC_6H_4Me$,
(IIIB) $Ph_3SnPPh_2 + PhCH:CH_2 \rightarrow Ph_3SnCH_2CHPhPPh_2$.

IV Interconversion reactions:
(IVA) $R_3SnCH_2Cl + \overline{MeNRH} \rightarrow R_3SnCH_2NMeR$; R = H, Me ,
(IVB) $R_3SnCH_2Cl + \overline{CH_2CH_2NK} \rightarrow R_3SnCH_2\overline{NCH_2CH_2}$,
(IVC) $Et_3SnCH_2CH_2CN + LiAlH_4 \rightarrow R_3Sn(CH_2)_3NH_2$.

R = organic group or hydrogen, R' = organic group containing a nitrogen function, X = halogen. Additional tetraorganotin compounds containing nitrogen and phosphorus may be found in Chapters 2.3, 2.4.2, 2.4.3, 2.5.3, 2.6.1, 2.6.2, 2.6.3.3, 2.6.3.4, 2.6.5, 2.7.1.1.3, 2.7.2 and 7.1.6.

Diazomethyltrimethyltin $SnC_4H_{10}N_2$ Me_3SnCHN_2
<u>Synth.</u>: By rxn. of Me_3SnNMe_2 with CH_2N_2 (1393).
By rxn. of Me_3SnCl with $LiCHN_2$ (1393).
<u>Prop.</u>: $b_{0.1}$ 64-70°, NMR spectrum, mol. wt. preferred struct. $Me_3SnN^-N:CH^+$ over $Me_3SnCH^-N:N^+$ (1393).
<u>Rxn. with</u>: Aq. base $\rightarrow (Me_3Sn)_2O + CH_4 + C_2H_6 + C_3H_8 + N$ (1393).
HX $\rightarrow Me_3SnX + N$; X = Cl, OH (1393).
Br $\rightarrow Me_3SnBr + N + C + HBr$ (1393).
PhNCO $\rightarrow Me_3Sn\overline{CHCONPhN:N}$ (1393).

Additional unsymmetric tetraorganotin derivatives containing nitrogen and phosphorus are listed in Table 115.

2.6.5.4 TETRAORGANOTIN COMPOUNDS CONTAINING ANTIMONY

Three 1,2-triorganotin diorganostibine ethene derivatives $R_3SnCH:CHSbR_2$ are listed in Table 105.

Table 115. Tetraorganotin Compounds Containing Nitrogen and Phosphorous $R_nSn R'_{4-n}$

Formula*	Synth. Method**	Yield	Properties	Ref.
Me_3SnR				
R = $MeNHCH_2$	IVA	31	b_{31} 59.5-60.5°, n_D^{20} 1.4817 (a-c)	1715
Me_2NCH_2	IVA	89	b_{47} 67-68.5°, n_D^{20} 1.4719, d_{20} 1.2426 (a-d)	1715
$HCl \cdot Me_2NCH_2$	--	--	m. 51-52.5° (a)	1715
$I \cdot Me_3NCH_2$	--	--	m. 179-80°	1715
$p\text{-}MeC_6H_4N:C(NMe_2)$	IIIA	37	m. 40° (sealed tube), IR spectrum	2406
$CH_2CH_2NCH_2$	IVB	59	b_{20} 54-55.5°, n_D^{20} 1.4820, d_{20} 1.3104 (a,e,f)	1309, 1350
$H_2N(CH_2)_3$	--	--	(c)	--
$m\text{-}H_2NC_6H_4$	--	--	(g)	--
$p\text{-}Me_2NC_6H_4$ (241)	I	37	m. 38°, $b_{0.85}$ 97° (g-i)	998
	I	--	b_6 102-104°, dipole moment	1210
$p\text{-}I \cdot Me_3NC_6H_4$	(i)	92	m. 186-87° (g)	998
Et_3SnR				
R = Me_2NCH_2	IVA	87	b_1 56.5-57.5°, n_D^{20} 1.4843, d_{20} 1.1829 (e,j)	283
$CH_2CH_2NCH_2$	IVB	89	$b_{0.2}$ 43.5-44.5°, n_D^{20} 1.4912, d_{20} 1.2325 (e)	283
$H_2N(CH_2)_3$	IVC	59	b_2 87.5-88.5°, n_D^{20} 1.4949, d_{20} 1.2211 (b,c)	1715
$p\text{-}Me_2NC_6H_4$ (241)	II	81	b_1 134-36°, n_D^{20} 1.5562, d_{20} 1.2670	2648
$Ph_3SnC_7H_4(NEt)_2H$ (k)	I	59	Yellow cryst., m. 151°, NMR spectrum (1)	1004
$[Ph_3SnC_7H_4(NEt)_2]_2 \cdot Ni$	(1)	--	m. 298°, NMR spectrum	1004
$Ph_3SnCH_2CHPhPPh_2$	IIIA	68	m. 59°	1841

* Numbers in parenthesis refer to pages in main volume.

** The characters used here correspond to the scheme in the introduction to Chapter 2.6.5.

(a) NMR spectrum (1715). (b) IR and mass spectra (1715). (c) Basicity (1715). (d) NMR spectrum (1351). (e) Moessbauer resonance and mass spectra (1309). (f) (Far) IR (1309), UV (1350, 1715), NMR (1309, 1350), Moessbauer resonance (1350) and mass (1350) spectra. (g) Relative kinetic data for aryl-cleavage rxn. with aq. alc. KOH (997). (h) Moessbauer resonance spectrum (902). (i) Rxn. of $p\text{-}Me_3SnC_6H_4NMe_2$ with MeI in EtOH → $p\text{-}Me_3SnC_6H_4NMe_3 \cdot I$ [$SnC_{12}H_{22}IN$] (998). (j) Basicity (1350). (k) γ-N,N'-diethylaminotroponeimine triphenyltin (1004). (1) Rxn. of $\gamma\text{-}Ph_3SnC_7H_4(NEt)_2H$ with $Ni(OAc)_2$ in aq. EtOH → $(\gamma\text{-}Ph_3SnC_7H_4NEtNEt)_2Ni$ (1004).

2.7 POLYMETALLIC TETRAORGANOTIN COMPOUNDS

2.7.1 BRIDGED POLYORGANOTIN COMPOUNDS

2.7.1.1 TETRAORGANOTIN COMPOUNDS WITH TWO TIN ATOMS

2.7.1.1.1 NONFUNCTIONAL DERIVATIVES

This chapter comprises organic derivatives having two tin atoms linked by hydrocarbon chains. The compounds listed in Table 116 are prepared by methods from the following scheme.

I Grignard synthesis:
  (IA) $R_3SnX + Y(MgBr)_2 \rightarrow (R_3Sn)_2Y$,
  (IB) $(R_2SnCl)_2Y + R'MgBr \rightarrow (R_2R'Sn)_2Y$,
  (IC) $R_3SnX + YI_2 + Mg \rightarrow (R_3Sn)_2Y$.

II Synthesis with organolithium derivatives:
  (IIA) $R_3SnCl + YLi_2 \rightarrow (R_3Sn)_2Y$,
  (IIB) $Ph_3SnLi + YX_2 \rightarrow (Ph_3Sn)_2Y$.

III Reactions with organosodium derivatives:
  (IIIA) $R_3SnCl + YNa_2 \rightarrow (R_3Sn)_2Y$,
  (IIIB) $R_3SnX + NaC_2H \rightarrow (R_3Sn)_2Y$,
  (IIIC) $R_3SnNa + YX_2 \rightarrow (R_3Sn)_2Y$,
  (IIID) $Me_3SnCl + (p\text{-}ClC_6H_4)_2SnMe_2 + Na \rightarrow p\text{-}(Me_3Sn)_2C_6H_4 + (p\text{-}Me_3SnC_6H_4)_2SnMe_2$.

IV Hydrostannation reactions:
  (IVA) $R_3SnH + CH_2{:}CHR'' \rightarrow R_3SnCH_2CH_2R''$,
  (IVB) $R_3SnH + (CH_2{:}CH)_2Y \rightarrow (R_3SnCH_2CH_2)_2Y$,
  (IVC) $Me_3SnH + HC{\vdots}CR \rightarrow Me_3SnCH_2CHRSnMe_3$,
  (IVD) $Me_3SnH + (HC{\vdots}C)_2Y \rightarrow (Me_3SnCH{:}CH)_2Y$,
  (IVE) $R_3SnH + HC{\vdots}CSnR_3 \rightarrow (R_3SnCH{:})_2 + (Bu_3SnCH{:}CH)_2SnBu_2$.

V Condensation with acetylenes:
  (VA) $(Bu_3Sn)_2O + Et_3SnC{\vdots}CH \rightarrow Et_3SnC{\vdots}CSnBu_3 + H_2O$,
  (VB) $(Et_3Sn)_2O + (HC{\vdots}C)_2 \rightarrow (Et_3SnC{\vdots}C)_2 + H_2O$.

VI Miscellaneous reactions:
  (VIA) $(Me_3SnC{\vdots})_2 + 5\text{-Me-}\alpha\text{-pyrone} \rightarrow 3,4\text{-}(Me_3Sn)_2C_6H_3Me$,
  (VIB) $(Me_3Sn)_2Y + R_3SnCl \rightarrow (R_3Sn)_2Y$,
  (VIC) $Me_4Sn + \text{x-ray irradiation} \rightarrow (Me_3Sn)_2CH_2$.

R = organic group, may be different in one molecule, R' = organic group, R" = organic group containing a tetraorganotin substituent, X = halogen, Y = hydrocarbon bridge.

Bis(trimethyltin)acetylene (M-245) $Sn_2C_8H_{18}$ $(Me_3SnC{\vdots})_2$
<u>Synth.</u>: By rxn. of NaC_2H with Me_3SnCl in $Et_2O\text{-}NH_2$, in 61% yield (2181), with $(Me_3Sn)_2O$ in Et_2O, in 96% yield (1342).

By rxn. of Me$_3$SnNa with HC⋮CBr in liq. NH$_3$ (630), with Me$_3$SnC⋮CBr in liq. NH$_3$, in 40-50% yield (2180).
By-prod. in rxn. of Me$_3$SnCl with Ph$_2$PC⋮CLi in hexane (2778).
Prop.: b$_{23}$ 102° (506, 1342), b$_3$ 70-72° (2181), b$_1$ 63° (630), n$_D^{20}$ 1.5202, d$_{20}$ 1.5556 (630), stable to hydrolysis and oxidation (630).
Rxn. with: O:$\overline{\text{CCH:CHCR:CHO}}$ → 3,4-(Me$_3$Sn)$_2$C$_6$H$_3$R + CO$_2$; R = H, Me, CO$_2$Me (1021).
5-MeO$_2$C-α-pyrone + α-pyrone → 3,4-(Me$_3$Sn)$_2$C$_6$H$_3$CO$_2$Me (1021).
5-Me-α-pyrone + α-pyrone → 3,4-(Me$_3$Sn)$_2$C$_6$H$_3$Me + o-(Me$_3$Sn)$_2$C$_6$H$_4$ (1.2:1) (1021).
C$_5$Cl$_6$ in Bu$_2$O → 2,3-(Me$_3$Sn)$_2$-hexachlorobicyclo[2.2.1]heptadiene (1883).
5,5-(MeO)$_2$C$_5$Cl$_4$ → o-(Me$_3$Sn)$_2$C$_6$Cl$_4$ + o-Me$_3$SnC$_6$HCl$_4$ + 1,2,3,4-C$_6$H$_2$Cl$_4$ (1883).
Ph$_3$SnCl → (Ph$_3$SnC⋮)$_2$ (2433).
Ph$_2$PCl → Me$_3$SnCl + (Ph$_2$PC⋮)$_2$ (2778).
Ph$_2$PCl, + KF → Me$_3$SnF + (Ph$_2$PC⋮)$_2$ (2433).
Ph$_2$MCl → (Ph$_2$MC⋮)$_2$; relative reactivity, solvent effect, M = As, Sb (2433).
(1-C$_{10}$H$_7$)$_2$AsCl → [(1-C$_{10}$H$_7$)$_2$AsC⋮]$_2$, relative reactivity (2433).

Bis(trimethyltin)diacetylene (M-245)   Sn$_2$C$_{10}$H$_{18}$ (Me$_3$SnC⋮C)$_2$
Prop.: Hydrolyzes in moist air (2433).
Rxn. with: Ph$_3$SnCl → (Ph$_3$SnC⋮C)$_2$ (2433).
R$_3$SnCl → (R$_3$SnC⋮C)$_2$; R = c.C$_6$H$_{11}$, p-MeC$_6$H$_4$, p-ClC$_6$H$_4$ (205, 2433).
R$_2$AsCl → (R$_2$AsC⋮C)$_2$; R = Ph, 1-C$_{10}$H$_7$ (205, 2433).

o-Bis(trimethyltin)benzene   Sn$_2$C$_{12}$H$_{22}$ o-(Me$_3$Sn)$_2$C$_6$H$_4$
Synth.: By rxn. of (Me$_3$SnC⋮)$_2$ with α-pyrone in PhBr under reflux in N atm., in 51% yield (1021).
Prop.: b$_{0.15}$ 92-94°, n$_D^{25}$ 1.5640, IR, UV and NMR spectra, GLC (1021).
Rxn. with: Br (1:4 in HCCl$_3$) → o-Br$_2$C$_6$H$_4$ + Me$_3$SnBr (1021).
X (1:2) → Me$_3$SnX + o-Me$_3$SnC$_6$H$_4$X; X = Br, I (1021).

Bis(triethyltin)acetylene (M-245)   Sn$_2$C$_{14}$H$_{30}$ (Et$_3$SnC⋮)$_2$
Synth.: By rxn. of NaC$_2$H with Et$_3$SnCl in liq. NH$_3$, in 81% yield (2181), in 10-25% yield in presence of Et$_2$O (2181), with Et$_3$SnOH, in Et$_2$O, in 50% yield (1342), with Et$_3$SnOMe in Et$_2$O, in 63% yield (1342), with Et$_3$SnNa in liq. NH$_3$, followed by NH$_4$Cl, in 63% yield (2181), with Et$_3$SnC⋮CH, followed by RX (2181).
By rxn. of Na$_2$C$_2$ with Et$_3$SnOH in Et$_2$O under reflux, in 62% yield (504).
By rxn. of Et$_3$SnC⋮CH with (Et$_3$Sn)$_2$O (506), in THF in Ar atm., in 53% yield (2495), with NaNH$_2$, followed by RX (2181).
By rxn. of BrC⋮CH with Et$_3$SnNa in liq. NH$_3$, in 40-50% yield (2180).
By rxn. of Et$_3$SnC⋮CBr with Et$_3$SnNa, EtSNa and Et$_2$NLi in liq. NH$_3$ or Et$_2$O, in 92, 64 and 60% yield, resp. (2180).
By thermal dec. of Et$_3$SnC⋮CCMe$_2$OSnEt$_3$ at 200-250°, in 78% yield (2776) or by rxn. of Et$_3$SnOMe with HC⋮CCMe$_2$OH at 200-250°, in 63% yield (2776).
Prop.: b$_{13}$ 155-56° (506), b$_9$ 109-10° (2776), b$_5$ 139° (1342), 136-37° (504), b$_1$ 114-18° (2180), b$_{0.5}$ 96° (2495), n$_D^{20}$ 1.5085 (1342, 2495), 1.5083-96 (2180), 1.5099 (2776), d$_{20}$ 1.3386-425 (2180), 1.3421 (2776), 1.3430 (504), IR (2495) and Moessbauer resonance (2718) spectra. Anal. detn. by iodometric titration (2495, 2644).
Rxn. with: Excess NaC⋮CH, + EtBr → Et$_3$SnC⋮CH + Et$_3$SnC⋮CEt (2181).

Bis(tributyltin)acetylene (M-246) $Sn_2C_{26}H_{54}$ $(Bu_3SnC\vdots)_2$

<u>Synth.</u>: By rxn. of Bu_3SnNMe_2 and C_2H_2 under pressure in petr. ether, in 95% yield (250).

By rxn. of Bu_3SnOH with Na_2C_2 in Et_2O under reflux, in 59% yield (504).

By thermal dec. of $Bu_3SnC\vdots CCMe_2OSnBu_3$ at 200-250°, in 64% yield (2776).

By-prod. in rxn. of $(Bu_3Sn)_2O$ with C_2H_2 under pressure (1339).

<u>Prop.</u>: b_3 204° (504, 1339), b_1 193-95° (2776), n_D^{20} 1.4868 (504, 1339), 1.4925 (504, 1339), d_{20} 1.1401 (504, 1339), 1.1496 (2776).

<u>Rxn. with:</u> $HgCl_2$ → $Cl(HgC\vdots C)_5HgCl$ + $Cl(HgC\vdots C)_5SnBu_3$ + $(ClHgC\vdots)_2$ + Bu_3SnCl (2264).

$RHgCl$ → $(RHgC\vdots)_2$ + Bu_3SnCl; R = $PhCH_2$, Ph (2264).

Ph_3MX → $(Ph_3MC\vdots)_2$ + Bu_3SnX; MX = SiCl, GeBr (2264).

Ph_2MCl → $(Ph_2MC\vdots)_2$ + Bu_3SnCl; M = P, As, Sb (2264).

Listing of compounds with two tetraorganotin moieties bridged by a carbon chain continues in Table 116. Cyclic tetraorganotin derivatives with two tin atoms in one ring system are listed in Chapter 2.6.4.

Table 116. Compounds with Two Tetraorganotin Moieties Bridged by a Carbon Chain $(R_3Sn)_2Y$

Formula*	Synth. Method**	Yield	Properties	Ref.
$(Me_3Sn)_2Y$				
Y = CH_2 (245)	VIC	--	Mass spectrum, GLC	1191
CH_2CHPr	IVC	--	GLC	429
CH_2CHBu	IVC	--	IR spectrum, GLC	429
$CH_2CH(C_5H_{11})$	IVC	--	GLC	429
$CH:CH(CH_2)_3CH:CH$	IVD	--	IR spectrum, GLC, struct.	429
$CH_2CH(C_6H_{13})$	IVC	--	GLC	429
$CH_2CH(C_7H_{15})$	IVC	--	GLC	429
$CH:CH(CH_2)_5CH:CH$	IVD	--	IR spectrum, GLC, struct.	429
$p-C_6H_4$ (245)	IIID (a)	40	m. 123-26°, NMR spectrum (2207) (b)	1410
$3,4-C_6H_3Me$	VIA	39	$b_{0.05}$ 90°, n_D^{27} 1.5602, GLC (c)	1021
$(p-C_6H_4)_2$	IIA	70	m. 125-26° UV (930), NMR spectra (d)	112
$Me_3SnC{:}CSnEt_3$	--	--	$b_{1.5}$ 59°, n_D^{20} 1.4710, d_{20} 1.1162	506
$p-(Me_2CH_2{:}CHSn)_2C_6H_4$	IB	37	$b_{0.003}$ 97°, n_D^{20} 1.5633 (e)	1410
$[(PhCH_2)_3Sn]_2C_{14}H_{10}$ (f)	IIIA	--	Synth. in THF	1726
$(Et_3Sn)_2Y$				
Y = $(CH_2)_3$ (246)	IVA	68	$b_{0.2}$ 108°, n_D^{20} 1.5088, mol. wt.	1612
$(CH_2)_4$	IVB	(g)	$b_{0.001}$ 93-97°, n_D^{20} 1.516	1612
	IVA	67	$b_{0.001}$ 101-104°, n_D^{20} 1.5076	1619
$(C{:}C)_2$	VB	46	$b_{0.5}$ 156°, n_D^{20} 1.5483, d_{20} 1.3646 (h)	506

* Numbers in parenthesis refer to pages in main volume.
** The characters used here correspond to the scheme in the introduction to Chapter 2.7.1.1.1.

(a) Synth. also from Me_2SnCl_2 and $p-Me_3SnC_6H_4Cl$ with Na in xylene and Mg in THF, in 40% yield (1410). (b) Rxn. with alc. I → $p-Me_3SnC_6H_4I + Me_3SnI$, comparison of reactivity with Si-analog (60). (c) IR, UV and NMR spectra (1021). (d) Rxn. with Na or Na/K → cleavage of Me_3Sn-groups, ESR spectrum (112). (e) Rxn. with Ph_3SnH → $p-(Ph_3-SnCH_2CH_2SnMe_2)_2C_6H_4$ (1410). (f) $C_{14}H_{10}$ = 9,10-dihydro-9,10-anthranyl (1726). (g) By-prod. in synthesis. (h) Rxn. with C_5Cl_6 → Et_3SnCl (2507).

Table 116 Continued

Formula*	Synth. Method**	Yield	Properties	$(R_3Sn)_2Y$ Ref.
$(Et_3Sn)_2Y$ (Cont'd.)				
Y = $(CH_2)_5$ (246)	IVB	42	$b_{0.001}$ 108°, n_D^{20} 1.5048	1612
$CH_2CHMeCH:CH$	IVC (1)	(g)	$b_{0.015}$ 105-12°, n_D^{20} 1.5210	1410
$Et_3SnC:CSnBu_3$	VA	51	$b_{0.2}$ 164-66°, $n_D^{22.5}$ 1.4930 (j)	2495
$[(CH_2:CH)_3Sn]_2C_{14}H_{10}$ (f)	IIIA	--	Synth. in THF	1726
$(Pr_3Sn)_2Y$				
Y = CH:CH	IVE	42	$b_{1.5}$ 164°, n_D^{20} 1.5090, d_{20} 1.2190 (k)	1602
C:C (246)	IIIB (1)	92	b_6 188-89°, n_D^{20} 1.5040, d_{20} 1.2461	506
	IIIA (m)	54	b_3 178-79°	504
	IIIB	20	--	2181
o-C_6H_4	IC	25	b_4 145°, n_D^{30} 1.713	1247
p-C_6H_4	IC	52	b_5 160°, $n_D^{32.5}$ 1.5230, GLC	1247a
$[\overline{(CH_2CH_2CH)}_3Sn]_2Y$				
Y = CH_2	IIIC	--	Use: herbicide, pesticide	1873
$(CH_2)_{10}$	IC	--	Use; herbicide, pesticide	1873
1,4-$C_{10}H_6$	IIIC	--	Use: herbicide, pesticide	1873
$(Bu_3Sn)_2Y$				
Y = CH:CH	IVE	74	b_2 177-78°, n_D^{20} 1.5030, d_{20} 1.1503	1598
	IVE (n)	(g)	b_2 180-82°, n_D^{20} 1.5041, d_{20} 1.1574 (o)	1599
o-C_6H_4	IC	35	b_5 166-70°, n_D^{25} 1.520	1247
p-C_6H_4	IC	46	b_5 155°, n_D^{25} 1.5169	1247a
$C_{14}H_{10}$ (f)	IIIC	--	Yellow oil, blue fluorescence (p)	1726
$[\overline{(CH_2CMe_2CH)}_3SnCPh:]_2$	IIA	--	Use: herbicide, pesticide	1873

R$_3$SnCH$_2$CH$_2$SnR'$_3$ (q)				
[(C$_6$H$_{13}$)$_3$SnC!C]$_2$	IVA	--	Use: herbicide, pesticide	1873
[(c.C$_6$H$_{11}$)$_3$SnC!C]$_2$	IIIA	>76	Oil, n_D^{20} 1.4946	205
[(c.C$_6$H$_{11}$)$_3$Sn]$_2$C$_{14}$H$_{10}$ (f)	VIB	84	m. 210°, IR spectrum	205, 2433
(c.C$_6$H$_{11}$Ph$_2$SnCH$_2$CH$_2$)$_2$	IIIA	--	Synth. in THF	1726
[(C$_8$H$_{17}$)$_3$SnC!C]$_2$	IB	--	m. 105-106°, dipole moment	1518, 2189
1,4-(R$_3$Sn)$_2$C$_6$H$_2$Me$_2$-2,5 (r)	IIIA	>76	Oil, n_D^{20} 1.4905	205
[(C$_{12}$H$_{25}$)$_3$Sn]$_2$C$_{14}$H$_{10}$ (f)	IIIC	--	Use: herbicide, pesticide	1873
	IIIA	--	Synth. in THF	1726

(Ph$_3$Sn)$_2$Y

Y = CH$_2$	IIA	--	m. 104-106° (s)	1410
CH$_2$CH$_2$ (246)	--	--	(t)	--
C!C (246)	IIIB (u)	--	m. 152°	443
	IVB	80	m. 152°	2433
(CH$_2$)$_4$ (247)	IIB	61	White, cryst., m. 154°	1518
	IIIC	85	Dipole moment (v)	1518
(CH:CH)$_2$ (247)	IVC (w)	(g)	m. 201-203.5°, soly. (t)	1601
(C!C)$_2$ (247)	VIB	--	Stable towards dil. KOH	2433
(CH$_2$)$_5$	IIA	--	White cryst., m. 74°, dipole moment (v, x)	1518
(CH$_2$)$_6$	IIA	--	White cryst., m. 90°, dipole moment (v)	1518
o-CH$_2$CH$_2$C$_6$H$_4$CH$_2$CH$_2$	IVB	80	m. 111-14°, IR spectrum	1410
o-CH:CHC$_6$H$_4$CH:CH	IVE	50	m. 140-45°, IR spectrum	1410
9,10-C:CC$_{14}$H$_8$C!C	IA	57	m. 225° (dec.), IR spectrum, anthranyl	443
C$_{14}$H$_{10}$ (f)	IIIA	--	Synth. in THF	1726
[(p-MeC$_6$H$_4$)$_3$SnC!C]$_2$	VIB	94	m. 208°, IR spectrum	205, 2433

(i) From HC!CCMe:CH$_2$ (1410). (j) Anal. detn. by iodometric titration (2495). (k) NMR spectrum, 96.5% in transform (1602), rxn. with HgCl$_2$ → Pr$_3$SnCl + (:CHHgCl)$_2$ (1602). (l) From (Pr$_3$Sn)$_2$O (506). (m) From Pr$_3$SnOH (504). (n) From Bu$_2$SnH$_2$ (1599). (o) IR and NMR spectra (1598), rxn. with Cl in CCl$_4$ → Bu$_3$SnCl (1598). (p) Reduces AgNO$_3$ soln. (1726), rxn. with HCl → 9,10-C$_{14}$H$_{12}$ + Bu$_3$SnCl (1726), use as fuel additive and stabilizer (1726). (q) R = EtCHCH$_2$CH, R' = MeCHCMe$_2$CH (1873). (r) R = $\overline{CH_2CH_2CH(CH_2)_5}$ (1873). (s) Rxn. with I → Ph$_3$SnCH$_2$SnPh$_2$I + PhI (1410). (t) Control and effect on reproduction of Musca domestica (260). (u) Synth. from Ph$_3$SnI with C$_2$H$_2$ and Na in liq. NH$_3$/Et$_2$O (443). (v) Rxn. with I (1:4) in HCCl$_3$ → (Ph$_2$SnI)$_2$(CH$_2$)$_n$; n = 4, 5, 6 (1518, 2189). (w) Synth. from Ph$_3$SnH and SbC!CH (1601). (x) TLC (2276).

2.7.1.1.3 FUNCTIONALLY SUBSTITUTED BISTETRAORGANOTIN COMPOUNDS

This chapter includes organotin derivatives having two tin atoms linked by a hydrocarbon chain. Heteroatoms like halogen, oxygen, nitrogen and phosphorus may be substituted at or in the chain. The compounds listed in Table 117 are prepared by methods from the following scheme.

I Reactions with organolithium derivatives:
  (IA) $Me_3SnCl + YLi_2 \rightarrow (Me_3Sn)_2Y$,
  (IB) $Me_3SnCCl_2Li + MeI \rightarrow (Me_3Sn)_2CCl_2$.

II Hydrostannation reactions:
  (IIA) $Ph_3SnH + (CH_2:CH)_2CHOH \rightarrow (Ph_3SnCH_2CH_2)_2CHOH$,
  (IIB) $Me_3SnH + HC⋮CCF_3 \rightarrow (Me_3Sn)_2CHCH_2CF_3$.

III Condensation reactions:
  (IIIA) $Et_3SnOMe + (HC⋮C)_2Y \rightarrow (Et_3SnC⋮C)_2Y + MeOH$,
  (IIIB) $Me_3SnCl + Ph_3P:CH_2 \rightarrow (Me_3Sn)_2C:PPh_3 + HCl$.

IV Insertion reactions:
  (IVA) $Me_6Sn_2 + PhHgCBrClX \rightarrow (Me_3Sn)_2CClX$; X = Cl, Br ,
  (IVB) Me_6Sn_2 + fluorolefin + UV irradiation $\rightarrow (Me_3Sn)_2Y$,
  (IVC) $Me_6Sn_2 + (CF_3C⋮)_2$ + UV irradiation $\rightarrow [Me_3SnC(CF_3):]_2$.

V Diels-Alder reactions:
  (VA) $(Me_3SnC⋮C)_2 + 5,5-(MeO)_2C_5Cl_4 \rightarrow o-(Me_3Sn)_2C_6Cl_4$,
  (VB) $(Me_3SnC⋮)_2 + 5-MeO_2C-\alpha-pyrone \rightarrow 3,4-(Me_3Sn)_2C_6H_3CO_2Me$.

VI Exchange reactions:
  (VIA) $(Me_3SnC⋮C)_2 + R'_3SnCl \rightarrow (R'_3SnC⋮C)_2$,
  (VIB) $(Me_3Sn)_2CBrCl + NaI \rightarrow (Me_3Sn)_2CClI$,
  (VIC) $(Ph_3Sn)_2CH_2 + I \rightarrow Ph_3SnCH_2SnPh_2I$,
  (VID) $Me_3SnCH_2Cl + MeNH_2 \rightarrow (Me_3SnCH_2)_2NMe$.

R = organic group, R' = functionally substituted organic group, X = halogen, Y = functionally substituted hydrocarbon bridge.

2,3-Bis(trimethyltin)hexachlorobi- $Sn_2C_{13}H_{18}Cl_6$ $2,3-(Me_3Sn)_2C_7Cl_6$
  cyclo[2.2.1]heptadiene-2,5

Synth.: By rxn. of $(Me_3SnC⋮)_2$ with C_5Cl_6 in Bu_2O in inert atm., in 66% yield (1883), in xylene and toluene, in 53 and 16% yield resp. (1883).
Prop.: Colorless cryst., m. 89-90°, IR, UV and NMR spectra, soly., stable in inert atm. at 0°, mol. wt. (1883).
Rxn. with: Thermal dec. at r.t. $\rightarrow Me_3SnCl$ (1883).
UV irradiation $\rightarrow 2,3-(Me_3Sn)_2-1,4,5,6,7,7$-hexachloroquadricyclo$[2.2.1.0^{2,6}.0^{3,5}]$-heptane (1883), compare structural formula on page 409.
HCl in Et_2O-MeOH, + KF $\rightarrow Me_3SnF + 2,3-C_7H_2Cl_6$ (1883).
Br in $HCCl_3$-MeOH $\rightarrow 2,3-C_7Br_2Cl_6$ (1883).

2,3-Bis(trimethyltin)-1,4,5,6,7,7-hexachloroquadricyclo [2.2.1.0^{2,6}.0^{3,5}]heptane*   Sn$_2$C$_{13}$H$_{18}$Cl$_6$

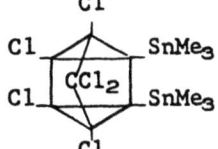

Synth.: By UV irradiation of 2,3-(Me$_3$Sn)$_2$C$_7$Cl$_6$ (preceeding compound) in Et$_2$O in inert atm. with external cooling, in 45% yield (1883).

Prop.: m. 139-41° (dec.), IR, UV and NMR spectra, TLC, mol. wt. (1883).

Rxn. with: HCl (excess) → YH$_2$* (1883).

Listing of functionally substituted bistetraorganotin compounds continues in Table 117.

2.7.1.2 TETRAORGANOTIN COMPOUNDS WITH THREE AND MORE TIN ATOMS

All compounds are listed in Table 118.

---

* (Me$_3$Sn)$_2$Y = (Me$_3$Sn)$_2$C$_7$Cl$_6$.

Table 117. Functionally Substituted Bistetraorganotin Compounds $R_3SnYSnR'_3$

Formula	Synth. Method*	Yield	Properties	Ref.
$(Me_3Sn)_2Y$				
$Y = CCl_2$	IVA	53	$b_{0.0002}$ 48–50°, n_D^{25} 1.5326	1878
	IB	37	--	1879
	IB (a)	35	(b)	1879
CBrCl	IVA	39	$b_{0.0002}$ 61°, n_D^{25} 1.5502 (c)	1878
CClI	(c)	63	Unstable in air	1878
$(Ph_3P\colon)C$	IIIB	55	Yellow, m. 129–30°, b_1 183°, mol. wt. (d)	1826
CH_2NMeCH_2	VID	31	b_1 74–76.5°, n_D^{20} 1.5111, d_{20} 1.4478 (e)	1715
$(CF_3CH_2)CH$	IIB	--	b_{29} 108°, IR, NMR spectra	923
$CF_2CF(CF_3)$	IVB (f)	--	(^{19}F) NMR spectra	870-1
$(CF_2CHF)_2$	IVB (g)	--	^{19}F NMR spectrum, mol. wt.	871
$CF(CF_3)\colon CF(CF_3)$	IVC	--	$b_{0.001}$ 53° (h)	108
$o\text{-}C_6Cl_4$	VA	35	m. 118–20°, TLC (i)	1883
$p\text{-}C_6Cl_4$	IA	32	m. 148–49.5°, mol. wt.	2764
$3,4\text{-}C_6H_3CO_2Me$	VB	65	$b_{0.09}$ 120°, n_D^{25} 1.5613 (j)	1021
$(Et_3SnC\colon CCHMeO)_2CHMe$	IIIA	--	b_1 156°, n_D^{20} 1.4962, d_{20} 1.2449 (k)	2078
$(Ph_3SnCH_2CH_2)_2CHOH$	IIA	--	Viscous oil	1410
$[(p\text{-}ClC_6H_4)_3SnC\colon Cl]_2$	VIA	94	m. 204°, IR spectrum	205, 2433
$Ph_3SnCH_2SnPh_2I$	VIC	--	--	1410

* The characters used here correspond to the synthetic scheme in the introduction to Chapter 2.7.1.1.3.
(a) Synth. from Me_3SnCCl_2Li and Me_3SiCl (1879). (b) Rxn. with BuLi at -130° → $Me_3SnCCl_2Li + Me_3SnBu$ (1879). (c) Rxn. of $(Me_3Sn)_2CBrCl$ with NaI → $(Me_3Sn)_2CClI$ (1878). (d) NMR, (far) IR spectra and assignment (1826). (e) IR and NMR spectra (1715). (f) From $CF_2\colon CFCF_3$ (870-1). (g) From C_2HF_3 (871). (h) IR, NMR and ^{19}F NMR spectra (108), thermal dec. at 150° → $Me_4Sn + Me_8Sn_2 + Me_3SnC(CF_3)\colon CHCF_3$ (108). (i) IR, UV and NMR spectra (1883), rxn. with absorption silica → $C_6H_2Cl_4$ (1883). (j) IR, UV and NMR spectra (1021), rxn. with I (1:2) in $HCCl_3$ → Me_3SnI + $3,4\text{-} + 4,3\text{-}(Me_3Sn)IC_6H_3CO_2Me$ (1021). (k) IR spectrum (1894).

Table 118. Tetraorganotin Compounds with Three and More Tin Atoms

Formula	Synth. Method*	Properties	Ref.
(p-Me$_3$SnC$_6$H$_4$)$_2$SnMe$_2$	IIID (a)	m. 94-96°	1410
(p-Me$_3$SnC$_6$H$_4$CHCH$_2$)$_x$	--	Thermal degradation	1410
(Bu$_3$SnCH:CH)$_2$SnBu$_2$	IVE	b$_{0.02}$ 155-57°, n$_D^{20}$ 1.5176	1599
(Bu$_2$SnCH:CHCH$_2$CH$_2$CH:CH)$_x$	IVB (b)	IR spectrum (c)	499
(Ph$_3$SnCH$_2$CH$_2$SnMe$_2$)$_2$C$_6$H$_4$	IVB (d)	m. 189-98°	1410
Ph$_2$Sn(CH$_2$SnPh$_2$Cl)$_2$	(e)	m. 180-81° (f)	1352
Ph$_2$Sn(CH$_2$SnPh$_2$OH)$_2$	(f)	Colorless solid	1352

* The characters used here correspond to the synthetic scheme in the introduction to Chapter 2.7.1.1.1.
(a) Synth. also from Me$_2$SnCl$_2$ and p-Me$_3$SnC$_6$H$_4$Cl with Na in xylene and Mg in THF, in 5-10% yield (1410). (b) Synth. with Bu$_2$SnH$_2$ and 1,5-hexadiyne, by thermal degradation of Bu$_2$SnCH:CHCH$_2$CH$_2$CH:CH above 90° (499). (c) Thermal decomposition at 220-40° → Bu$_2$SnCH:CHCH$_2$CH$_2$CH:CH (499), rxn. with DX-D$_2$O → (DCH:CHCH$_2$)$_2$ + Bu$_2$SnX$_2$ (499). (d) Synth. from Ph$_3$SnH and p-(CH$_2$:CHSnMe$_2$)$_2$C$_6$H$_4$, in 27% yield (1410). (e) Synth. from Ph$_2$SnCl$_2$ and CH$_2$(MgBr)$_2$ in Et$_2$O (1352). (f) Rxn. of Ph$_2$Sn(CH$_2$SnPh$_2$Cl)$_2$ with NaOH → Ph$_2$Sn(CH$_2$SnPh$_2$OH)$_2$ (1352).

2.7.2 TETRAORGANOTIN COMPOUNDS CONTAINING ORGANOSILICON SUBSTITUENTS

The tetraorganotin derivatives containing silicon moieties in the carbon chain listed in Table 119 are prepared by the following methods.

I Grignard reactions:
  (IA) R$_3$SnX + R″MgBr → R$_3$SnR″ ,
  (IB) R$_2$SnCl$_2$ + (Me$_2$SiCH$_2$Cl)$_2$O + Mg → (R$_2$SnCH$_2$SiMe$_2$)$_2$O .

II Synthesis with organolithium derivatives:
  (IIA) Me$_3$SnCl + R″Li → Me$_3$SnR″ ,
  (IIB) Me$_3$SnCCl$_2$Li + Me$_3$SiCl → Me$_3$SnCCl$_2$SiMe$_3$.

III Synthesis with organosodium derivatives:
  Ph$_3$SnC⋮CNa + R$_3$SiCl → Ph$_3$SnC⋮CSiR$_3$.

IV Hydrostannation reactions:
  (IVA) Me$_3$SnH + CF$_2$:CFR″ → Me$_3$SnYR″ ,
  (IVB) R$_3$SnH + R$_3$SiC⋮CH → R$_3$SnCH:CHSiR$_3$.

V Condensation reactions:
  (VA) Me$_3$SnNEt$_2$ + Me$_3$SiC⋮CH → Me$_3$SnC⋮CSiMe$_3$ + Et$_2$NH ,
  (VB) (Et$_3$Sn)$_2$O + Me$_3$SiC⋮CH → Et$_3$SnC⋮CSiMe$_3$ + H$_2$O ,
  (VC) Me$_3$SnCl + Me$_3$SiCH:PMe$_3$ → Me$_3$SnC(:PMe$_3$)SiMe$_3$ + HCl .

VI Thermal decomposition:
$Et_3SnOCMe_2C$⋮$CSiMe_3$ at 200-250° → Et_3SnC⋮$CSiMe_3$.

R = organic group or hydrogen, R″ = organic group containing an organometallic substituent, X = halogen. Additional tetraorganotin derivatives containing silicon may be found in Chapters 2.5.3 and 8.1.

2-Triethylsilyltripropyltinethylene $SnSiC_{17}H_{38}$ $Pr_3SnCH{:}CHSiEt_3$
Synth.: By rxn. of Pr_3SnH with Et_3SiC⋮CH in Ar atm. at 60-65°, in 64% yield (1602).
Prop.: b_2 126°, n_D^{20} 1.4860, d_{20} 1.0585, IR and NMR spectra (1602).
Rxn. with: Cl in CCl_4 → $Et_3SiCH{:}CHCl$ (1602).
HCl in MeOH → Pr_3SnCl + $Et_3SiCH{:}CH_2$ (1602).
$HgCl_2$ → $Et_3SiCH{:}CHHgCl$ (1602).

Additional tetraorganotin compounds with organosilicon substituents are summarized in Table 119.

2.7.3 TETRAORGANOTIN COMPOUNDS CONTAINING ORGANOGERMANIUM SUBSTITUENTS

All compounds are listed in Part I Germanium on pages 79 and 80 in Table 20.

2.7.4 TETRAORGANOTIN COMPOUNDS CONTAINING ORGANOLEAD SUBSTITUENTS

Triphenyltintriphenyllead acetylene is listed at the bottom of Table 119.

Table 119. Tetraorganotin Compounds Containing Silicon and Lead R_3SnYMR_3

Formula*	Synth. Method**	Yield	Properties	Ref.
Me_3SnR'				
R = Me_2HSiCH_2	IA (a)	77	Colorless liq. b. 153-54°, unpleasant odor	475
$Me_2(KO)SiCH_2$	--	--	(c)	--
$CH_2SiMe_2(OSiMe_2)_xCH_2$	(c)	--	$b_{1.5} > 250°$, n_D^{20} 1.414, mol. wt. 2280	1066
Me_3SiCCl_2	IIA	70	b_{10} 84°, n_D^{25} 1.4992	1879
	IIB	36	--	1879
$Me_3SiC(:PMe_3)$	VC	--	m. 18-21°, b_1 63°, dec. > 120° (d, e)	1475
	VC	52	Colorless liq., m. 11-13° (impure), b_1 51-53°	1826
$\{Me_3SiCHFCF_2\}$ $\{Me_3SiCF(CHF_2)\}$	IVA	77	Mixt., ratio 3:2, not sepd., dec. in vacuo at 25°, (far) IR spectrum, assignment (f)	2208
$Me_3SiC:C$	VA	68	m. 41-44°, $b_1 \sim 25°$, NMR spectrum (g)	1039
m-$Me_3SiC_6H_4$	IIA	57	$b_{1.5}$ 87°, n_D^{25} 1.5171 (h)	998
p-$Me_3SiC_6H_4$ (249)	--	--	NMR (2267) and mass (2411) spectra (h, i)	--
p-$Me_3SiC_6H_4CH_2$ (249)	--	--	Mass spectrum	2411
m-$Me_3SiC:CC_6H_4$	IA	57	$b_{0.2}$ 85.7-88.5°, n_D^{25} 1.5428 (j)	1001
p-$Me_3SiC:CC_6H_4$	IA	57	m. 31-32.5°, $b_{0.7}$ 112-12°, n_D^{25} 1.5466 (j)	1001
$[Me_2Sn(CH_2SiMe)_2O]_x$ (249)	IB	--	Yellowish liq., $b_3 > 85°$, n_D^{20} 1.4861, soly. (k)	1066

* Numbers in parenthesis refer to pages in main volume.
** The characters used here correspond to the scheme in the introduction to Chapter 2.7.2.
(a) Synth. from $Me_2(ClMg)SiH$ (475). (b) IR and NMR spectra, stable to air and moisture, toxic (475). (c) Rxn. of $Me_3SnCH_2SiMe_2OK$ with $(Me_2SiO)_xCl_2 \rightarrow Me_3SnCH_2SiMe_2(OSiMe_2)_xCH_2SnMe_3$ (1066). (d) NMR, (far) IR spectra and assignment (1826). (e) Rxn. with MeI, + $HPF_6 \rightarrow [Me_3Sn(Me_3Si)CMePMe_3]PF_6$ (1475), with EtOH, followed by HCl gas $\rightarrow Me_3SnCl$ (1475). (f) NMR and ^{19}F NMR spectra (2208-9). (g) IR spectrum (1039, 2763). (h) Relative rate of aryl cleavage by aq. alc. KOH (997). (i) Rxn. with I in MeOH $\rightarrow Me_3SnI$ + p-$Me_3SiC_6H_4I$, comparison with Sn analog (60). (j) Rxn. with aq. alc. NaOH $\rightarrow Me_3SnC_6H_4C:CH$ (1001). (k) Cyclic polymer, mol. wt., viscosity (1066), rxn. with alkali, followed by Me_2SiCl_2 and Me_3SiCl (1:2) $\rightarrow [Me_2Si(OSiMe_2CH_2SnMe_2CH_2SiMe_2OSiMe_3)_2]_x$ (159, 1066).

Table 119 Continued R_3SnYMR_3

Formula*	Synth. Method**	Yield	Properties	Ref.
$Et_3SnC\!:\!CSiMe_3$	VB	44	b_9 107°, n_D^{20} 1.4805, d_{20} 1.1450 (l)	506, 1344
	VI	51	b_4 70-71°, n_D^{20} 1.4770, d_{20} 1.1049	2776
$[Et_2Sn(CH_2SiMe_2)_2O]_x$	IB	--	Yellowish liq., soly., cyclic polymer, mol. wt.	1066
$Bu_3SnCH\!:\!CHSiEt_3$	IVB	41	b_2 144-46°, n_D^{20} 1.4851 d_{20} 1.0282 (m)	1598
Ph_3SnR				
R = $Me_3SiCH\!:\!CH$	IVB	70	m. 72.5-74.5° (n)	1354
$Me_2PhSiCH\!:\!CH$	IVB	60	m. 48.5-50.5° (n)	1354
$Me_3SiC\!:\!C$	IA	61	m. 74-76° (n)	1354
$Me_2PhSiC\!:\!C$	IA	48	m. 41-42.5° (n)	1354
$Ph_3SiC\!:\!C$	III (o)	--	m. 124-26°, mol. wt.	1039
$(Ph_3Sn)_2Y$				
Y = $CH\!:\!CHSiMePhCH\!:\!CH$	IVB (p)	36	m. 95.5 (n)	1354
$CH\!:\!CHSiPh_2CH\!:\!CH$	IVB (p)	72	m. 142-44° (n)	1354
$C\!:\!CSiMePhC\!:\!C$	IA	33	m. 100.5-102° (n)	1354
$C\!:\!CSiPh_2C\!:\!C$	IA	38	m. 124-25° (n)	1354
$p-(CH_2CH_2SiMe_2)_2C_6H_4$	IVA (q)	74	m. 130-35°	1410
$p-(CH_2CH_2SiPh_2)_2C_6H_4$	IVA (q)	42	m. 162-72°	1410
$Ph_3SnC\!:\!CPbPh_3$	III (r)	56	m. 143-45°, mol. wt.	1039

(l) IR spectrum (1908). (m) Rxn. with Cl in $CCl_4 \rightarrow Bu_3SnCl + Et_3SiCH\!:\!CHCl$ (1598). (n) NMR, IR spectra and assignment (1354). (o) Synth. also from Ph_3SnNEt_2 and HC_2Na in THF, followed by Ph_3SiCl in Ar atm., in 68% yield (1039). (p) Synth. from $RPhSi(C\!:\!CH)_2$; R = Me, Ph (1354). (q) Synth. from $p-(CH_2\!:\!CHSiR_2)_2C_6H_4$; R = Me, Ph (1410). (r) Synth. from Ph_3PbCl with R_2NLi, followed by HC_2Na and Ph_3SnCl in Ar atm. (1039).

2.7.5 TETRAORGANOTIN COMPOUNDS WITH BORON AND ALUMINUM

The tetraorganotin derivatives listed in Table 120 are prepared by the following methods.

I Synthesis with organolithium derivatives:
    (IA) $R_3SnCl + B_{10}H_{10}C_2Li_2 \rightarrow (R_3SnC)_2B_{10}H_{10}$,
    (IB) $R_nSnCl_{4-n} + R'CB_{10}H_{10}CLi \rightarrow R_nSn(CB_{10}H_{10}CR')_{4-n}$; $R' = H$, CMe_3 , Ph,
    (IC) $R_2SnCl_2 + B_{10}H_{10}C_2Li \rightarrow (R_2SnCB_{10}H_{10}C)_x \cdot R_2SnCl_2$.

II Hydrostannation catalyzed by H_2PtCl_6 :
    $Et_3SnH + MeCH{:}CH\overline{BOCH_2CH_2O} \rightarrow Et_3SnCHMeCH_2\overline{BOCH_2CH_2O}$.

Cyclic tetraorganotin carborane derivatives may be found in Table 265.

2.7.7 TETRAORGANOTIN COMPOUNDS WITH ALKALI METALS

Dichloromethyltrimethyltin Lithium $SnLiC_4H_9Cl_2$ Me_3SnCCl_2Li
<u>Synth.</u>: By rxn. of BuLi with $(Me_3Sn)_2CCl_2$ in $THF-Et_2O-CH_2(OMe)_2-C_5H_{12}$ at $-130°$, in high yield (1879).
<u>Prop.</u>: Not isolated (1879).
<u>Rxn. with:</u> Dil. HCl (at $-120°$) $\rightarrow Me_3SnCHCl_2$ (1879).
MeI $\rightarrow Me_3SnBu + (Me_3Sn)_2CCl_2 + Me_3SnCCl_2Me$ (1879).
$Me_3SiCl \rightarrow (Me_3Si)_2CCl_2 + Me_3SnCCl_2SiMe_3 + (Me_3Sn)_2CCl_2$ (1879).

Triphenyltinacetylene Sodium $SnNaC_{20}H_{15}$ $Ph_3SnC{:}CNa$
<u>Synth.</u>: By rxn. of Ph_3SnNEt_2 with $HC{:}CNa$ in THF in Ar atm. (1039).
<u>Prop.</u>: Stable in inert atm., neutralization equivalent (1039).
<u>Rxn. with:</u> $Ph_3SiCl \rightarrow Ph_3SnC{:}CSiPh_3$ (1039).

2.7.8 TETRAORGANOTIN COMPOUNDS CONTAINING TRANSITION METALS

Structure of tetraorganotin compounds containing transition metals linked to tin by a hydrocarbon bridge is not always certain. Compounds reported to have a transition metal tin bond are listed in Chapter 6.5. The compounds summarized in Table 121 are prepared by methods from the following scheme.

I Reaction of $Bu_3SnC{:}CSnBu_3$ with $HgCl_2$.

II Reaction of triorganotin metalcarbonyl complexes under UV irradiation with:
    (IIA) Tetrafluoroethylene,
    (IIB) Bistrifluoromethylacetylene.

Additional tetraorganotin derivatives containing transition metals may be found in Chapters 2.3 and 2.6.5.

Table 120. Tetraorganotin Compounds with Boron $R_nSnR'_{4-n}$

Formula	Synth. Method*	Yield	Properties	Ref.
$(Me_3SnC)_2B_{10}H_{10}$ (a)	IA	--	m. 68-68.5°	2279
$Me_2Sn(CB_{10}H_{10}CCMe_3)_2$ (a)	IB	--	m. 107.7-108°	2279
$Me_2Sn(CB_{10}H_{10}CPh)_2$ (a)	IB	--	m. 142.3°	2279
$(Me_2SnCB_{10}H_{10}C)_x$ (a)	IC	--	m. 205° (dec.)	785
$(m-Me_2SnCB_{10}H_{10}C)_n \cdot Me_2SnCl_2$ (b)	IC	42	n = 8, m. 217-21°, mol. wt.	1835
	IC	--	n = 9,	1835
	IC	45	n = 24, m. 236-43°	1835
	IC	57	n = 28, m. 244-47°	1835
	IC	21	n = 30, m. 250-55°	1835
$(p-Me_2SnCB_{10}H_{10}C)_x$ (c)	IC	50	m. 420-30°	1835
$Et_3SnCHMeCH_2BOCH_2CH_2O$	II	--	b_2 125°, NMR spectrum	47
$(Pr_3SnC)_2B_{10}H_{10}$ (d)	IA	30	m. 42-43°, $b_{0.002}$ 140-43° (e)	627
$Pr_3SnCB_{10}H_{10}CH$ (d)	IB	35	b_1 148-50°, n_D^{20} 1.5470 (e, f)	627
$Pr_3SnCB_{10}H_{10}CPh$ (d)	IB	45	b_1 128-30° (e, f)	784
$(Bu_3SnC)_2B_{10}H_{10}$ (a)	IA	--	$b_{0.004}$ 134-35°, n_D^{20} 1.5314	2279
$(Bu_2SnCB_{10}H_{10}C)_x$ (a)	IC	--	m. 190-230°, mol. wt. 1210-1815	785
$Bu_2Sn(CB_{10}H_{10}CPh)_2$ (d)	IB	65	m. 149-50° (e, g)	784
$(Ph_2SnC)_2B_{10}H_{10}$ (d)	--	--	(e)	--
$(Ph_3SnC)_2B_{10}H_{10}$ (a)	IA	--	m. 191-92.5°	2279
$Ph_3SnCB_{10}H_{10}CH$ (d)	IB	56	m. 161-62° (e, f)	784
$Ph_3SnCB_{10}H_{10}CPh$ (d)	IB	84	m. 186-87° (e, h)	627
$(Ph_2SnCB_{10}H_{10}C)_x$ (a)	IC	<90	m. 52-250°, soly. (1)	785
$(m-Ph_2SnCB_{10}H_{10}C)_8 \cdot Ph_2SnCl_2$ (b)	IC	30	m. 235-40°, mol. wt.	1835
$(p-Ph_2SnCB_{10}H_{10}C)_{11} \cdot Ph_2SnCl_2$ (c)	IC	13	m. 404-12°, mol. wt.	1835

* The characters used here correspond to the synthetic scheme in the introduction to Chapter 2.7.5. (a) Neocarborane. (b) 1,7-Carborane or m-carborane. (c) 1,12-Carborane or p-Carborane. (d) Barene or carborane. (e) Moessbauer resonance spectrum (660). (f) Rxn. with aq. alc. KOH → $B_{10}H_{10}(CH)_2$ (784). (g) Rxn. with aq. KOH in C_6H_6 → $(Bu_2SnO)_x$ + $B_{10}H_{10}C_2PhH$ (784). (h) Rxn. with aq. alc. KOH → Ph_3SnOH + $B_{10}H_{10}C_2PhH$ (784). (f) Stable in air, mol. wt. 515-5750 (785).

Table 121. Tetraorganotin Compounds Containing Transition Metals R_3SnYM

Formula	Synth. Method*	Properties	Ref.
$Bu_3Sn(C\!:\!CHg)_5Cl$	I	Infusible solid	2264
$Me_3SnCF_2CF_2Mn(CO)_5$	IIA	Pale yellow solid	87
	IIA	White cryst., m. 57.5° (a)	874

$$R_3Sn\begin{array}{c}CF_3\\|\\|\\CF_3\end{array}\!\!-\!\!\!=\!\!\!-\!\!\begin{array}{c}CF_3\\|\\|\\CF_3\end{array}\!\!Mn(CO)_5$$

Formula	Synth. Method*	Properties	Ref.
R = Me	IIB	m. 109.5°, monomeric (b, c)	733
Ph	IIB	IR spectrum	733
$Me_3SnC(CF_3)\!:\!C(CF_3)FeC_5H_5(CO)_2$	IIB	m. 86-88°, monomeric (b)	733
$Me_3SnCF_2CF_2Co(CO)_4$	IIA	IR, NMR and ^{19}F NMR spectra	730

* The characters used here correspond to the synthetic methods explained in the introduction to Chapter 2.7.8.
(a) (Far) IR spectrum and assignment (874). (b) IR and mass spectra (733).
(c) ^{19}F NMR spectrum (733).

3. ORGANOTIN HYDRIDES

The increase in activity in the field of organotin hydrides is mostly due to the facility of addition of the tin-hydrogen bond to unsaturated systems and to the multiple mechanistic investigations of the "hydrostannation" reaction. New is a number of organometallic reactions, i.e., addition to transition metal complexes, condensation with metal alkyls yielding tin-metal derivatives and with metal amides for synthesis of compounds with longer tin chains.

The organotin hydrides in Tables 122 and 124 are prepared by reduction of organotin halides with:

I Lithiumaluminum hydride,

II Diethylaluminum hydride and

III Tributyltin hydride.

Other synthetic methods include:

IV Reaction of dimethyltin dihydride with bis(trifluoroethylene)dimethyltin under exclusion of light.

Polymerization reactions involving organotin hydrides are listed in Table 267.

3.1 NONFUNCTIONAL ORGANOTIN HYDRIDES

Trimethyltin Hydride (M-255) SnC_3H_{10} Me_3SnH

<u>Synth.</u>: By rxn. of Me_3SnCl with $LiAlH_4$ in $(MeOCH_2)_2$, in 87% yield (1043), with $NaBH_4$ in diglyme, in 92% yield (736).
By rxn. of Me_3SnCl with Bu_3SnH and distillation at 50°, in 90-95% yield (1953).
By-prod. in rxn. of Me_3SnCl with H_3SiK in diglyme in N atm. (2224).

<u>Prop.</u>: b. 59-61° (258), IR (258), NMR (258-9, 331, 447, 581, 1436, 1489) and Moessbauer resonance (902, 1180 1489, 1666, 1960, 2424) spectra, force const. (138), thermodynamic data (305, 2856).

<u>Rxn. with</u>: $R_3SnNEt_2 \rightarrow Me_3SnSnR_3$; R = Me, Et (1618).

$Me_2Sn(NEt_2)_2 \rightarrow (Me_3Sn)_2SnMe_2$ (1955).

$Et_3SnX \rightarrow Me_3SnX + Et_3SnH$, solvent effect; X = OMe, OPh, NPhCHO, NEt_2, NPh_2, PPh_2 (915).

cis-$Et_3SnCH:CHOEt \rightleftharpoons Me_3SnCH:CHOEt + Et_3SnH$, isomerization, effect of free radical inhibitor and accelerator (1413).

$Me_3SnCF:CF_2 \rightarrow Me_3SnF$, $Me_3SnCF:CHF + Me_3SnC_2H_2F$ (731) + $(Me_3Sn)_2CFCF_2H + Me_3Sn$-$CHFCF_2SnMe_3$ (?) (2209).

cis-$Me_3SnCF:CHF \rightarrow Me_3SnC_2H_2F$ (731).

$Me_2Sn(CF:CF_2)_2 \rightarrow Me_3SnF + Me_2SnF_2 + Me_3SnCF:CHF + Me_3SnCF:CF_2 + Me_3SnC_2H_2F + Me_3SnCH:CF_2 + Me_2Sn(CF:CHF)_2 + [Me_2Sn(C_2H_2F)_2 + Me_2Sn(CH:CF_2)_2]$ (731).

$Me_3GeCF:CF_2 \rightarrow Me_3SnF + Me_3GeCF:CHF + Me_3GeCH:CF_2 + Me_3GeCH:CHF + Me_3GeCHFCF_2$-$SnMe_3 + Me_3Sn(Me_3Ge)CFCF_2H$ (2209).

$Me_2M(CF:CF_2)_2 \rightarrow Me_3SnF + Me_2M(C_2HF_2)_2 + Me_2M(C_2H_2F)_2$; M = Ge (2209), Si (2208).

$Me_3SiCF:CF_2 \rightarrow Me_3SnF + Me_3SiC_2HF_2 + Me_3SiC_2H_2F + Me_3Sn(Me_3Si)CFCHF_2 + Me_3SnCF_2$-$CHFSiMe_3$ (2208).

$Bu_2SbC\vdotdbl CH \rightarrow Me_3SnCH:CHSbBu_2 + Me_3SnSbBu_2$ (1601).

$Fe(CO)_5 \rightarrow (Me_3Sn)_2Fe(CO)_4 + [Me_2SnFe(CO)_4]_2 + Me_4Sn_3[Fe(CO)_4]_4 + Sn[Fe(CO)_4]_4$ (904).

$Ru_3(CO)_{12} \rightarrow (Me_3Sn)_2Ru(CO)_4$ (905, 2328) + $Me_{10}Sn_4Ru_2(CO)_6$ (905).

$(RPh_2P)_2Ir(CO)X \rightarrow Me_3Sn(RPh_2P)_2Ir(CO)HX$; R = Me, Ph, X = Cl, Br (2530).

$(Ph_3P)_2Pt(NH_2)Cl \rightarrow Me_3Sn(Ph_3P)_2PtCl$ (825).

RX → kinetic data and rate const. for halogen exchange; RX = t-BuCl, t-BuBr, c.$C_6H_{11}Cl$ (2306).

RBr → RH; R = $\overline{CH_2CPh_2C}Me$ (1942).

$CH_2:CH(CH_2)_nH \rightarrow Me_3Sn(CH_2)_{n+2}H$; n = 1-7 (429).

$CH_2:CR_2$ (under UV irradiation) → $Me_3SnCH_2CHR_2$; R = Me, Et, (Me + Me_2CH) (1953).

Internal and cyclic olefins (under UV irradiation) → adducts from $(MeCH:)_2$, MeCH:CHEt, MeCH:CMe_2, indene, c.C_5H_8, c.C_6H_{10}, 1-MeC_6H_9, C_7H_{12} (1953), 1,3-C_8H_9CN (1953).

$(MeCH:)_2 \rightarrow Me_3SnCHMeEt$ + cis- and trans $(MeCH:)_2$ (1:3) (1369).

1,2-diene → 1,2- and 1,3- addition with: C_3H_4, C_4H_6, n-2,3-C_5H_8, 3,1,2-Me-C_4H_5, 2,2,3-MeC_5H_7 (295).

1,3-diene (+ AiBN) → 1,2- + 1,4- addition with: $(CH_2:CH)_2$, $(CH_2:CMe)_2$, c.C_5H_6 (1041, 1043), c.C_6H_8 (1041, 1043, 1953), c.C_8H_{12} (1043, 2052), C_5H_8 (429, 1605).

4-CH_2:CHC_6H_9 (+ AIBN) → 3-$Me_3SnCH_2CH_2C_6H_9$ (1043, 2052).
1,5-C_8H_{12} (+ AIBN) → 3-$C_8H_3SnMe_3$ + bicyclic systems (1043, 2052).
Norbornadiene → 2-(Me_3Sn)C_7H_9 (2052), isomer distribution in UV catalyzed rxn. (2677).
CH_2:CHCN → $Me_3SnCHMeCN$ + $Me_3SnCH_2CH_2CN$; prod. ratio and rxn. mechanism (1417).
$\overline{CH_2CCl_2CHCH}$:CH_2 → Me_3SnCH:$CHCH_2CHCl_2$ + $\overline{CH_2CHClCHCH}$:CH_2 (1877).
1,3-C_6H_9CHO → 3-$Me_3SnOCH_2C_6H_9$ (1953).
[O_2CCH:$CHCO_2(CH_2)_4$]$_x$ + AIBN → Me_3Sn-containing polyester (599).
HC⋮CR → Me_3SnCH:CHR + Me_3SnCR:CH_2 + $Me_3SnCH_2CHRSnMe_3$; R = n -Pr, -Bu, -C_5H_{11}, -C_6H_{13}, -C_7H_{15} (429).
HC⋮CR → Me_3SnCH:CHR + Me_3SnCR:CH_2 ; R = Bu, Ph (1414).
Heptyne → GLC retention ratio for 1-, 2- and 3- isomer (429).
HC⋮C(CH_2)$_n$C⋮CH → prop. and struct. of adducts, n = 3, 5 (429).
HC⋮CCF_3 → (Me_3Sn)$_2CHCH_2CF_3$ + Me_3SnCH:$CHCF_3$ + $Me_3SnC(CF_3)$:CH_2 (923).
(⋮CCF_3)$_2$ → $Me_3SnC(CF_3)$:$CHCF_3$ (923).
HC⋮CCN → $Me_3SnC(CN)$:CH_2 (313, 1414), rate of rxn. and mechanism (312), solvent and deuterium effect (1412).
(⋮CCN)$_2$ → $Me_3SnC(CN)$:CHCN, rxn. mechanism and solvent effect (1412, 1414).
HC⋮CCO_2R → Me_3SnCH:$CHCO_2R$ + $Me_3SnC($:$CH_2)CO_2R$ + $Me_3SnCH_2CH_2CO_2R$ + Me_3SnC⋮CCO_2R; R = Me, Et (311, 1414), Me (1412), kinetic data and rxn. mechanism (312).
HC⋮CCO_2R → Me_3SnCH:$CHCO_2R$ + HC⋮CCO_2SnMe_3 + ROH; R = Me, Et (1410).
RC⋮CCO_2Et → $Me_3SnC(CO_2Et)$:CHR; R = Me, CO_2Et (313, 1414) + Me_3SnCR:$CHCO_2Et$; R = Me, prod. ratio and rxn. mechanism (1411), R = CO_2Et, kinetic data and solvent effect (1412).
RCHO → Me_3SnOCH_2R (+ $Me_3SnOCHROCH_2R$?); R = CCl_3, C_6F_5 (2538), kinetic data (2537).
(CF_3)$_2$CO → $Me_3SnOCH(CF_3)_2$ (109), kinetic data (2537).
$CF_3COC_6F_5$ → $Me_3SnOCH(CF_3)C_6F_5$ (2538).
Ph_2CO → relative kinetic data (2537).
PHNCO → kinetic data, solvent effect and rxn. mechanism (314), relative rates (1415).
RNCS → Me_3SnSCH:NR; R = Ph, p-$EtOC_6H_4$, rxn. mechanism and solvent effect (429).
Use: In catalyst for olefin polymerization (411).

Dimethyltin Dihydride (M-256) SnC_2H_8 Me_2SnH_2
Synth.: By rxn. of Me_2SnCl_2 with $NaBH_4$ in diglyme, in 96% yield (736).
By rxn. of Me_2SnCl_2 with Bu_3SnH (1:2) in N atm. (1805).
By rxn. of (Me_2SnO)$_x$ with ($MeHSiO$)$_x$ in N atm. at r.t. in presence of cat I (1165).
Prop.: IR (1013, 1805), NMR (259, 331, 447, 581, 1180, 1489, 1805, 2361) and Moessbauer resonance (1180, 1489, 1666, 1960, 2424) spectra, force const. (138) and thermodynamic data (305, 3856).
Rxn. with: Me_3SnNEt_2 → (Me_3Sn)$_2SnMe_2$ (1955).
Me_2SnCl_2 + HC⋮CCN → Me_2(NCCH:CH)SnCl (1613).
Me_2SnX_2 → Me_2SnHX; X = Cl, BrI (1292).
$Me_2Sn(CF$:$CF_2)_2$ → Me_2SnF_2 + $Me_2Sn(CF$:$CHF)_2$ + $Me_2Sn(CH$:$CF_2)_2$ + $Me_2Sn(C_2H_2F)_2$ +

$Me_2(CF_2:CF)SnH + C_2F_3H + H$ (731).
$Me_3GeCF:CF_2 \rightarrow Me_2SnF_2 + Me_3GeCF:CHF + Me_3GeCH:CF_2 + Me_3GeCH:CHF$ (2209).
$Me_2M(CF:CF_2)_2 \rightarrow Me_2SnF_2 + Me_2M(C_2HF_2)_2 + Me_2M(C_2H_2F)_2$; M = Si (2208), Ge (2209).
$Me_3SiCF:CF_2 \rightarrow Me_2SnF_2 + Me_3SiC_2HF_2 + Me_3SiC_2H_2F$ (2208).
$Fe(CO)_5 \rightarrow [Me_2SnFe(CO)_4]_2$ (904).
$Ru_3(CO)_{12} \rightarrow Me_{10}Sn_4Ru_2(CO)_8$ (905).
$[O_2CCH:CHCO_2(CH_2)_4]_x + AIBN \rightarrow Me_2Sn$-contn. polyester (599).
$o\text{-}(HC\vdots C)_2C_6H_4 \rightarrow o\text{-}\overline{CH:CHC_6H_4CH:CH}SnMe_2 + (Me_2SnCH:CHC_6H_4CH:CH)_x$ (310).
$(CF_3)_2CO \rightarrow Me_2Sn[OCH(CF_3)_2]_2$ (109).
Use: In catalyst for olefin polymerization (411).

Methyltin Trihydride (M-263)　　　　　　$SnCH_6$　　　　　　$MeSnH_3$
Synth.: By rxn. of H_3SnK with MeI in $(MeOCH_2)_2$ in N atm. in the dark at -30°, in 63% yield (2225).
Prop.: IR (95, 2400), (far) IR and assignment (1313), NMR (331, 447, 581, 1489), Moessbauer resonance (1180, 1489, 1666, 1960, 2424) and microwave (957) spectra, calculation of vibrational spectra (1713), force const. (138), mathematical treatment of hindered rotation (1432), molar polarizabicity (1566), thermodynamical data (305, 957, 2400).

Tribenzyltin Hydride　　　　　　$SnC_{21}H_{22}$　　　　　　$(PhCH_2)_3SnH$
Synth.: By rxn. of $(PhCH_2)_3SnCl$ with $LiAlH_4$ in Et_2O (2069).
Prop.: m. 52-53°, NMR spectrum, soly. (2069).
Rxn. with: $Ru_3(CO)_{12} \rightarrow [(PhCH_2)_3Sn]_2Ru(CO)_4$ (905).
$RCO_2H \rightarrow (PhCH_2)_3SnO_2CCR$; R = 2-bicyclo[2.2.1]heptene-5, 1-adamantane (2391).

Dibenzyltin Dihydride　　　　　　$SnC_{14}H_{16}$　　　　　　$(PhCH_2)_2SnH_2$
Synth.: By rxn. of $(PhCH_2)_2SnCl_2$ with $LiAlH_4$ at 0° in Et_2O (394).
Rxn. with: Me_2NCHO (+ cat. R_2SnCl_2) $\rightarrow$ c.$[(PhCH_2)_2Sn]_4$ (394).
$(PhCH_2)_2Sn(NEt_2)_2 \rightarrow$ c.$[(PhCH_2)_2Sn]_4$ (1614).

Triethyltin Hydride (M-256)　　　　　　SnC_6H_{16}　　　　　　Et_3SnH
Synth.: By rxn. of Et_3SnOR with $(MeHSiO)_x$ and distillation (238).
By rxn. of Et_3SnCl with NaH and BPh_3 in mineral oil at 110° under exclusion of moisture (2242), with Bu_3PbH at 0° (397).
By rxn. of $(Et_3Sn)_2O$ with Et_2SnH_2 at 20° (1955), with $Ph_2P(O)H$, in 74-86% yield (1233).
By rxn. of Et_3SnOH with $R_3SnOPPh_3$, in 91% yield (1233).
By-prod. in rxn. of $Et_3SnC\vdots CCMeEtOSnEt_3$ with H and Raney Ni (1149).
Prop.: IR (258), NMR (258, 297, 331, 990, 1309) and Moessbauer resonance (1309) spectra, force const. (138) and thermodynamic data (305).
Rxn. with: $R_3SnNEt_2 \rightarrow Et_3SnSnR_3$; R = Me (1618), Et (525, 1618), relative rxn. rates and mechanism (915).
$Et_2Sn(NEt_2)_2 \rightarrow (Et_3Sn)_2(SnEt_2)_n$; n = 0, 1, 2 (1955).
$R_2Sn(NEt_2)_2 \rightarrow (Et_3Sn)_2SnR_2$; R = Me, Bu (525, 1955).
$Et_3SnX \rightarrow Et_6Sn_2 + HX$; X = $C_6H_{13}NCHO$, relative rxn. rates (915).

$Et_3SnN_3 \rightarrow Et_6Sn_2 + HN_3$ (1429).

$(Et_3Sn)_2O \rightarrow Et_3SnSnR_3$; R = Et (400, 1618), i-Bu (1618), Ph (400).

$(Et_2SnO)_x \rightarrow (Et_3Sn)_2SnEt_2$ (400, 1955) + Et_6Sn_2 + $Et_{2n+2}Sn_n$; n = >3 (1955).

$Et_3SnOPh \rightarrow Et_6Sn_2 + PhOH$ (914).

$Sn(NEt_2)_4$ (4:1) $\rightarrow Et_6Sn_2 + Sn$ (1955).

Et_4Pb (at 100-130°) $\rightarrow Et_6Sn_2 + Et_4Sn + Pb + C_2H_6 + H + C_2H_4$ (2850).

$R_3PbX + HC\vdots CCN \rightarrow Et_3SnX + R_3PbCH:CHCN + R_3PbC(CN):CH_2$; X = AcO, imidazolyl, R = Et, Bu (912).

$Bu_3PbOAc + HC\vdots CCO_2Me \rightarrow Et_3SnOAc + Bu_3PbCH:CHCO_2Me$ (912).

$(MeO_2CCH_2)_3GeBr \rightarrow (MeO_2CCH_2)_3GeH$ (2238).

$Et_3GeLi \rightarrow Et_3SnGeEt_3$ (2087).

$(Et_3Ge)_2EtSiLi \rightarrow (Et_3Ge)_2EtSiSnEt_3 + LiH$ (2415).

$(Et_3Ge)_3Sb \rightarrow (Et_3Sn)_3Sb + Et_3GeH$ (2086).

$(Et_3Ge)_nBiEt_{3-n} \rightarrow Et_6Sn_2 + Et_3GeH + Bi$ (291); n = 3 (2086).

$Et_3GeYH \rightarrow Et_3SnYGeEt_3$; Y = S (2092), Se (2091-2).

$Et_3MTeEt \rightarrow Et_3SnTeMEt_3$; M = Si, Ge (2092).

$(Et_3Ge)_2Te \rightarrow (Et_3Sn)_2Te + Et_3GeH$ (2853).

$(Et_3Ge)_2Cd \rightarrow Et_6Sn_2 + Et_3GeH + Cd$ (587).

$(Et_3Ge)_3Tl \rightarrow Et_6Sn_2 + Et_3GeH + Tl$ (1361).

$Et_2Zn \rightarrow Et_6Sn_2 + Zn + C_2H_6 + Et_4Sn$ (584).

$Bu_3Al \rightarrow Et_3SnBu + Bu_2AlH$; relative rxn. rates (1830).

$Ru_3(CO)_{12} \rightarrow (Et_3Sn)_2Ru(CO)_4$ (905).

$Y \rightarrow (Et_3Sn)_2Y$; Y = S, Se (2091), Te (2091-2).

$Et_2M \rightarrow (Et_3Sn)_2M$; M = Se (2091) + $Et_6Sn_2 + C_2H_6 + H$; M = Te (2092)

$Et_3M \rightarrow (Et_3Sn)_3M$; M = Sb, Bi (2092).

$HBr \rightarrow Et_3SnBr + H$ (156).

$Ph_3CCl \rightarrow Et_3SnCl + Ph_3CH$ (585).

Ph_3C*, + AcOH $\rightarrow Ph_3CH + H + Et_3SnOAc + Et_6Sn_2$ (585).

Cl_3CCO_2Na in $(MeOCH_2)_2 \rightarrow Et_3SnCHCl_2$ (2047).

$CH_2:CHR* \rightarrow Et_3SnCH_2CH_2R$; R = C_6H_{13}, 3-C_6H_9, Et_3SnCH_2 (1612).

t-$BuCH_2CMe:CH_2* \rightarrow Et_3SnCH_2CHMeCH_2Bu$-$t$ + H (1612).

$\overline{CF_2(CF_2)_nCCl:CCl}$ at 100° $\rightarrow Et_3SnCl$; n = 1, 2 (924).

$CH_2:CHCN \rightarrow Et_3SnCH_2CH_2CN$ (1715) + $Et_3SnCHMeCN$; prod. ratio and rxn. mechanism (1417).

$RR'C:C(CN)_2 \rightarrow Et_3SnN:C:C(CN)CHRR'$; RR' = H, Me, EtO, substd. C_6H_4 (1621, 2615).

$CH_2:CH\overline{CRCH_2O} \rightarrow Et_6Sn_2 + CH_2:CHCHRCH_2OH + MeCH:CRCH_2OH$; R = H, Me (807).

1,3-Diene $\rightarrow$ 1,2- and 1,4- adduct with $(CH_2:CH)_2$, $(CH_2:CMe)_2$, $CH_2:CHCMe:CH_2$, $CH_2:CHCH:CHMe$ (1619).

$(CH_2:CH)_2* \rightarrow Et_3SnCH_2CH:CHMe + (Et_3SnCH_2CH_2)_2 + Et_6Sn_2$ (1612).

---

* i-Bu_2AlH is used as catalyst.

$C_5H_6 \rightarrow$ 1,2- and 2,1- adduct (1619).

$(C_5H_6)_2 \rightarrow$ 1,2- and 2,1 adduct (1619).

$(CH_2:CH)_2CH_2* \rightarrow Et_3Sn(CH_2)_3CH:CH_2 + (Et_3SnCH_2CH_2)_2CH_2$ (1612).

$HC \vdots CCMe:CH_2 \rightarrow Et_3SnCH:CHCMe:CH_2 + Et_3SnCH:CHCHMeCH_2SnEt_3$ (?) (1410).

$HC \vdots CCH:CHR \rightarrow Et_3Sn(CH:CH)_2R$; R = Me, EtS, Et_2N (2558), MeO (1410, 2558).

$HC \vdots CCH:CHNMe_2 \rightarrow Et_3SnC \vdots CCH:CHNMe_2$ (2558).

$HC \vdots CBu \rightarrow Et_3SnCBu:CH_2 + Et_3SnCH:CHBu$; prod. ratio and rxn. mechanism (1411, 1413-4).

$HC \vdots CPh \rightarrow Et_3SnCH:CHPh$ (1414).

$HC \vdots CC \vdots CR \rightarrow Et_3SnCH:CHC \vdots CR$; R = H, Et (631).

$HC \vdots CX \rightarrow Et_3SnCH:CHX$; X = EtO, BuO (2483), BuS (1414).

$HC \vdots CR \rightarrow Et_3SnCH:CHR$; R = $HOCH_2CH_2$, $CH_2:CHOCH_2CH_2$ (2078) $HOCMe_2$ (1410), BuOCH-MeOY; Y = CH_2, $(CH_2)_2$, CHMe, CMe_2 (2078).

$HC \vdots CCN \rightarrow Et_3SnCH(CN):CH_2$ (313, 1414), kinetic data and rxn. mechanism (312), solvent and deuterium effect (1412).

$HC \vdots CR \rightarrow Et_3SnCHR:CH_2 + Et_3SnCH:CHR$; R = CH_2OH (313, 1414), EtO (1414).

$HC \vdots CCO_2R \rightarrow Et_3SnCH:CHCO_2R + HC \vdots CCO_2SnEt_3$; R = Me, Et, + $Et_3SnC \vdots CCO_2Me$ (1410).

$HC \vdots CCO_2R \rightarrow Et_3SnC(:CH_2)CO_2R + Et_3SnCH:CHCO_2R + Et_3SnCH_2CH_2CO_2R + Et_3SnC \vdots CCO_2R$; R = Et (311), Me (311, 1414); R = Me, solvent effect and kinetic data (1414).

$(\vdots CX)_2 \rightarrow Et_3SnCX:CHX$; X = CF_3 (923), CN (1414), CO_2Et (313, 1414), solvent effect and rxn. mechanism for X = CN, CO_2Et (1412).

$PhCH:NR \rightarrow Et_3SnNRCH_2Ph$; R = Ph, $p-MeC_6H_4$ (390).

$p-H_2NC_6H_4N:NPh \rightarrow p-(H_2N)_2C_6H_4$ (390).

$(RN:C)_2C \rightarrow Et_3SnNRCH:NR$; R = $c.C_6H_{11}$ (390).

$PhN:CCMe_2CN \rightarrow Et_3SnCN + PhNHN:CMe_2$ (1620).

$(NCYN:)_2 \rightarrow Et_3SnCN + Et_6Sn_2 + HYCN + (Y:N)_2 + (NCY)_2 + N + H + C_2H_6$; Y = CMe_2, $1,1-C_6H_{10}$, + $C_6H_{10}:NNHC_6H_{10}CN$; kinetic data and rxn. mechanism (1620).

$(Me_3CON:)_2 \rightarrow Et_3SnOCMe_3 + Me_3COH$ (2645).

$(PhCH_2ON:)_2 \rightarrow Et_3SnOCH_2Ph$; cat. effect of Et_3Sn* on dec. of oxydimide to N_2 + $(PhCHO)_4$ + PhCHO + $PhCH_2OH$ (1611), rxn. mechanism (2645).

$PhN:NSO_2Ph \rightarrow Et_3SnO_2SPh + Et_3SnN(NHPh)SO_2Ph + N_2 + C_6H_6$ (1611), cat. effect of Et_3SnH on dec. of azoderivative (2646).

$p-Me_3CC_6H_4SN:NPh$ (2:1) $\rightarrow Et_6Sn_2 + p-Me_3CC_6H_4SSnEt_3 + N + C_6H_6 + MePh + p-Me_3C-C_6H_4SNHNHPh$ (2646).

PhNO (1:1) $\rightarrow (PhN:)_2 + [PhNOSnEt_3 \rightarrow]$ PhNHOH (390).

RNO (1:2) $\rightarrow Et_6Sn_2 + RN:N(O)R$; R = Ph, $p-Me_2NC_6H_4$ (390).

$PhN(NO)Ac \rightarrow Et_3SnOAc + N + C_6H_6$ (585).

RCHO $\rightarrow Et_3SnOCH_2R$; R = Pr, i-Pr, Ph, o-, $m-HOC_6H_4$, $p-MeOC_6H_4$, C_4H_3O (389).

RCHO $\rightarrow Et_3SnOCH_2R$ ($+ Et_3SnOCHROCH_2R$?) ; R = CCl_3, C_6F_5 (2538), rxn. mechanism (2537).

$PhCOCF_3 \rightarrow$ kinetic data, solvent effect and mechanism of addition rxn. (2537).

---

* $i-Bu_2AlH$ is used as catalyst.

RR'CO → Et$_3$SnOCHRR'; R, R' = Me$_2$, Et$_2$, (CH$_2$)$_5$; R = Me, R' = Et, p-MeOC$_6$H$_4$ (389), R = CF$_3$, R' = C$_6$F$_5$ (2538).
RCO$_2$H → Et$_3$SnO$_2$CR + H; R = p-MeC$_6$H$_4$ (1615), 2-C$_7$H$_9$, 2-bicyclo[2.2.1]heptenyl-5, 1-C$_{10}$H$_{15}$, 1-adamantanyl (2391).
Me$_3$CO$_2$H → (Et$_3$Sn)$_2$O + Me$_3$COH (585).
(RCO$_2$)$_2$ → Et$_3$SnO$_2$CR + CO$_2$ + H + Et$_2$Sn(OBz)$_2$; R = Me, Ph, rxn. inhibited by galvinoxyl, kinetic data and rxn. mechanism (1616).
(C$_{11}$H$_{23}$CO$_2$)$_2$ → Et$_3$SnO$_2$CC$_{11}$H$_{23}$ + C$_{12}$H$_{26}$ (1616).
RNCO → kinetic data, solvent effect and mechanism of addition rxn., R = Ph, c.C$_6$H$_{11}$ (1415).
RNCS → Et$_3$SnSCH$_2$:NR; R = Ph, p-EtOC$_6$H$_4$, rxn. mechanism, solvent effect (1415).
PhSO$_2$H → Et$_3$SnO$_2$SPh + H (2646).
MePO$_3$RH → Et$_3$Sn(R)O$_3$PMe (2761).
MePS(OBu-i)OH → Et$_3$SnO(i-BuO)SPMe (2761).
Use: In catalyst for polymerization of C$_2$H$_4$ (1956) with PhCH:CH$_2$ (2466), of C$_2$H$_4$, C$_4$H$_6$ and isoprene (2657).

Diethyltin Dihydride (M-257)   SnC$_4$H$_{12}$ Et$_2$SnH$_2$
Synth.: By rxn. of Et$_2$Sn(OR)$_2$ or (Et$_2$SnO)$_x$ with (MeHSiO)$_x$ (238), in 61% yield (1242), R = Me (2461).
By rxn. of Et$_2$SnCl$_2$ with Bu$_3$SnH in N atm. (1805).
Prop.: b. 95-97° (238), IR (1805) and NMR (331, 990, 1805) spectra, force const. (138) and thermodynamic data (305, 2856).
Rxn. with: Et$_3$SnNEt$_2$ → (Et$_3$Sn)$_2$SnEt$_2$ (525, 1955).
Et$_2$Sn(NEt$_2$)$_2$ → c.(Et$_2$Sn)$_6$ (525, 1614).
Et$_3$SnNPhCHO → Et$_3$SnSnEt$_2$H (1:1) and (Et$_3$Sn)$_2$SnEt$_2$ (2:1) (95).
Et$_3$SnSnEt$_2$NPhCHO → (Et$_3$SnSnEt$_2$)$_2$SnEt$_2$ (95).
Ph$_3$GeSnEt$_2$NPhCHO → (Ph$_3$GeSnEt$_2$)$_2$SnEt$_2$ (913).
Et$_2$SnX$_2$ → Et$_2$SnHX; X = Cl, Br, I (1292).
Et$_2$SnX$_2$ + CH$_2$:CHR → Et$_2$(RCH$_2$CH$_2$)SnX; X + R = Br + CN, I + CH$_2$CN (1613).
Et$_2$SnCl$_2$ + RMeCO → Et$_2$Sn(OCHMeR)Cl; R = Et, p-NCC$_6$H$_4$; MeR = (CH$_2$)$_5$ (1613).
Et$_2$SnCl$_2$ + R$_2$Hg → Et$_2$RSnCl; R = MeO$_2$CCH$_2$ (24).
(Et$_3$Sn)$_2$O (at 20°) → Et$_3$SnH + (Et$_2$SnO)$_x$ (1955).
CH$_2$:CHCH$_2$O$_2$CCR:CH$_2$ → Et$_2$Sn(CH$_2$CHRCO$_2$CH$_2$CH:CH$_2$)$_2$; R = H, Me (2845).
C$_{10}$F$_{21}$C$_4$H$_8$CH:CH$_2$ + AiBN, + SnCl$_4$ → fluororganotin derivative (745a).
o-(CH$_2$:CH)$_2$C$_6$H$_4$ → o-CH$_2$CH$_2$C$_6$H$_4$CH$_2$CH$_2$SnEt$_2$ + (Et$_2$SnCH$_2$CH$_2$C$_6$H$_4$CH$_2$CH$_2$)$_x$ (310).
o-(HC⋮C)$_2$C$_6$H$_4$ → o-CH:CHC$_6$H$_4$CH:CHSnEt$_2$ + (Et$_2$SnCH:CHC$_6$H$_4$CH:CH)$_x$ (310).
Et$_2$NH(+ cat. ZnCl$_2$) → c.(Et$_2$Sn)$_6$ (1614).

Ethyltin Trihydride (M-258)   SnC$_2$H$_8$ EtSnH$_3$
Prop.: NMR spectrum (331, 990), force const. (138) and thermodynamic data (305).
Rxn. with: EtSnBr$_3$ + PhC⋮CH → Et(PhCH:CH)SnBr$_2$ (1613).
HBr (at -78°) → EtSnH$_2$Br + H + (EtSnBr)$_x$ (156).

Tripropyltin Hydride (M-258)   SnC$_9$H$_{22}$ Pr$_3$SnH
Synth.: By thermal dec. of Pr$_3$SnO$_2$CH at 10-12 torr. and 160-70° (408), at

100°, in 92% yield (1650).
By rxn. of $(Pr_3Sn)_2O$ with $(MeHSiO)_x$ and distillation in N atm., in 66% yield (1165), with Ph_3SiH, in 49% yield (1165).

Prop.: b_{37} 108-109° (258), b_{12} 79° (297), b_4 59-64° (1165), n_D^{25} 1.4698, d_{25} 1.1571 (1165), IR (258) and NMR (258, 297, 331) spectra and thermodynamic data (1971).

Rxn. with: $Pr_3SnC⋮CH → (Pr_3SnCH:)_2$ (1602).
$Et_3SiC⋮CH → Pr_3SnCH:CHSiEt_3$ (1602).
$Bu_3Al → Pr_3SnBu + Bu_2AlH$, relative rxn. rates (1830).
$Ru_3(CO)_{12} → (Pr_3Sn)_2Ru(CO)_4$ (905).
$CH_2:CHR → Pr_3SnCH_2CH_2R$; $R = MeO_2C(CH_2)_8$ (598), MeO_2C, Ph, CN (2070).
$HC⋮CCO_2Me → Pr_3C(:CH_2)CO_2Me + Pr_3SnCH:CHCO_2Me + Pr_3SnCH_2CH_2CO_2Me + Pr_3SnC⋮CO_2Me$ (311, 1414).
$HC⋮COR → Pr_3SnCH:CHOR$; R = Et, Bu (2483).
$RCO_2H → Pr_3SnO_2CR$; R = c.Pr, c.Bu, $c.C_5H_9$, $c.C_6H_{11}$, bicycloheptene, adamantane (2391).
$HCl → Pr_3SnCl + H$; heat of rxn. (1971).

Use: In catalyst for olefin polymerization (1650).

Dipropyltin Dihydride (M-258) $\quad\quad SnC_6H_{16} \quad\quad Pr_2SnH_2$
Synth.: By rxn. of Pr_2SnCl_2 with Bu_3SnH (2:1) in N atm. in pentane (1805).
Prop.: b_{35} 46.5-47.5° (258), IR (258, 1805), NMR (258, 331, 1309, 1805) and Moessbauer resonance (1309) spectra.
Rxn. with: $Pr_2SnX_2 → Pr_2SnHX$; X = Cl, Br, I (1292).
$Pr_2SnCl_2 + R_2Hg → Pr_2RSnCl$; $R = MeO_2CCH_2$ (24).
$CH_2:CHR → Pr_2Sn(CH_2CH_2R)_2$; $R = MeO_2C$, CN (2070).

Triisopropyltin Hydride (M-263) $\quad\quad SnC_9H_{22} \quad\quad (Me_2CH)_3SnH$
Prop.: b_{16} 68-69° (258), b_{13} 69-70° (297), IR (258) and NMR (258, 297, 313) spectra.
Rxn. with: $CH_2:CHCN → i-Pr_3SnCHMeCN + i-Pr_3SnCH_2CH_2CN$; prod. ratio and rxn. mechanism (1417).
Use: In catalyst for olefin copolymerization (1446).

Tributyltin Hydride (M-258) $\quad\quad SnC_{12}H_{28} \quad\quad Bu_3SnH$
Synth.: By rxn. of Bu_3SnCl with $NaBH_4$ in glycol ethers, in 96% yield (736).
By rxn. of $(Bu_3Sn)_2O$ with $(MeHSiO)_x$ and distillation (241, 1239, 2457), in 97% yield (238), in 87% yield (1165, 1242), in presence of Bu_3SnOMe (2459), with $(Ph_2SiH)_2O$, in > 65% yield (1165), with Ph_3SiH in 41-69% yield (1165), with $(Me_2SiH)_2O·(MeHSiO)_x$ in low yield (1165), with $(H_2Si_2O_3)_x(Me_2SiO)_y$ 1:2, in 68% yield (1165), with $R_2P(O)H$ in 64-87% yield (1233), with Bu_2SnH_2 at r.t., in 71% yield (468), with R_3SiH (719, 719a).
By rxn. of Bu_3SnOR with $(MeHSiO)_x$; R = Et (1242), in 86% yield (1165), R = Me, Pr, i-Pr, Bu, t-Bu, Ph, Me_3Si and relative reactivity (1242), Pr (2458), Bu_3Si and Ph_3Si, in 88 and 76% yield, resp. (1165).
By rxn. of $(Bu_3Sn)_2S$ with $i-Bu_2SnH_2$ at 110°, in 54% yield (1955), with Ph_3SnH under exclusion of air at r.t., in 11% yield (1618).

By rxn. of Bu_3SnCH_2Ac with R_3SiH (719, 719a).
By thermal dec. of Bu_3SnO_2CH at 1 torr. and 170-80° (408), at 100° (1650).
By-prod. in rxn. of $(Bu_2SnO)_x$ with $(MeHSiO)_x$ in N atm., in 3% yield (1165).
By-prod. in rxn. of Ph_3SiH in Ar atm. with $Bu_3SnCH(CN)CHMe_2$ at 150°, in 16% yield (2680), under UV irradiation with $Bu_3SnCH(CO_2Et)CHMe_2$ and $Bu_3SnCH_2CO_2Et$ in 20 and 8% yield, resp. (2680).

Prop.: b_8 112.5-13.5° (258), b_4 84° (238), $b_{0.1}$ 70-71° (297), n_D^{24} 1.4694, d_{25} 1.0976 (1165), IR (258, 1804), NMR (258, 297, 331, 990, 1180, 1804) and Moessbauer resonance (1180, 1666) spectra and thermodynamic data (1971). Anal. detn. with isatin or ninhydrin (1048), by iodometric titration in THF (1165).

Rxn. with: $Bu_3SnNEt_2 \rightarrow Bu_6Sn_2$ (1618).
$Bu_2Sn(NEt_2)_2 \rightarrow (Bu_3Sn)_2SnBu_2$ (525) + Bu_6Sn_2 (1955).
$Bu_3SnPBu_2 \rightarrow Bu_6Sn_2$ (1618).
$(Bu_3Sn)_2O \rightarrow Bu_6Sn_2$ (400, 468, 1618, 1802).
$Me_3SnOH \rightarrow Me_3SnSnBu_3$ (1802).
$Bu_3SnOCMe_3 \rightarrow Bu_6Sn_2$ (1618).
$Bu_3SnOCMe:CHCHMe_2 \rightarrow Bu_6Sn_2$ + i-BuAc + $(Bu_3Sn)_2O$ (1684).
$(Bu_2SnO)_x \rightarrow (Bu_3Sn)_2SnBu_2$ (468).
$Me_3SnCl \rightarrow Me_3SnH + Bu_3SnCl$ (1953).
$R_2SnCl_2 \rightarrow Bu_3SnCl + R_2HSnCl$ (1:1) and R_2SnH_2 (2:1); R = Me, Et, Pr, i-Bu, $c.C_6H_{11}$, $n-C_8H_{17}$, Ph (1805).
$Bu_2SnX_2 \rightarrow Bu_3SnX + Bu_2HSnX$ (1:1) and Bu_2SnH_2 (2:1); X = Cl, Br, I (1804).
$Bu_2SnF_2 \rightarrow Bu_2HSnF + Bu_2SnH_2$ (1804).
$BuSnCl_3 \rightarrow BuSnH_3 + BuHSnCl_2$? (1804).
Bu_3SnCH_2Ac under UV irradiation $\rightarrow Me_2CO$ (2680).
$Bu_3SnCHRCHMe_2 \rightarrow Bu_6Sn_2 + Me_2CHCH_2R$; R = EtO_2C, CN (2680).
$Bu_3MC\vdots CH \rightarrow Bu_3SnCH:CHMBu_3$; M = Sn, Ge (1598).
$Et_3SiC\vdots CH \rightarrow Bu_3SnCH:CHSiEt_3$ (1598).
$(Me_2N)_2BCl \rightarrow Bu_3SnCl + (Me_2N)_2BH$; kinetic data and rxn. mechanism (1623).
$o-\overline{OC_6H_4O}BCl \rightarrow o-\overline{OC_6H_4O}BH$ (1623).
$R_3Al \rightarrow Bu_3SnR + R_2AlH$; R = Et, n-, i-Bu, C_8H_{17}, $EtBuCHCH_2$, $Me_3CCH_2CHMeCH_2$; relative rxn. rates and mechanism (1830).
$i-Bu_2Al(C_8H_{17}-n) \rightarrow n-C_8H_{17}SnBu_3 + i-Bu_2AlH$ (1830).
$PhHgCClBrX$, + KF $\rightarrow Bu_3SnF + PhHgCHClX$; X = Cl, Br (1876).
$PhHgCBr_3$, + KF $\rightarrow Bu_3SnF + PhHgCHBr_2 + PhHgBr$ (1876).
$Ru_3(CO)_{12} \rightarrow (Bu_3Sn)_2Ru(CO)_4 + Bu_{10}Sn_4Ru_2(CO)_8$ (905).
HCl $\rightarrow Bu_3SnCl$; heat of rxn. (1971).
RX $\rightarrow$ RH, kinetic data for dehalogenation and rxn. mechanism for Me_3CCl and Me_3CBr (2305-6), rate const. and thermodynamic data for MeI, $n-C_5H_{11}Cl$, $n-C_6H_{13}Br$, $c.C_6H_{11}Br$ (2306).
$YHX \rightarrow YH_2$; Y = saturated and unsaturated cyclic and polycyclic compd., X = Cl, Br, relative rxn. rates (2113).
Propargyl chlorides $\rightarrow$ acetylene and allene derivatives, kinetic data and mechanism (2384).
cis- and trans-9-$ClC_{10}H_{17} \rightarrow$ mixed cis- and trans-decalin, independent of temp. and starting matl. (1143).
YFCl $\rightarrow$ YHF; Y = bicyclopropanes, sterospecific dechlorination (682).

$CH_2:CH(CH_2)_4Br \rightarrow \overline{CH_2(CH_2)_3CHMe}$ + $CH_2:CHBu$ (2106).

1-Br$(CH_2)_4C_6H_9 \rightarrow$ cis- + trans-decalin + spiro[4.5]decane + 4-(1-cyclohexnyl)-butyl radical (2804).

RCOCl $\rightarrow$ Bu$_3$SnCl + RCHO + RCO$_2$CH$_2$R; relative ratio of prod., effect of ketones and R'Br, rxn. mechanism (1371).

CCl$_3$CO$_2$Na in (MeOCH$_2$)$_2 \rightarrow$ Bu$_3$SnCHCl$_2$ (2047).

$CH_2:CHCHCCl_2CH_2$* $\rightarrow$ Bu$_3$SnCH$_2$CH:CHCH$_2$CHCl$_2$ + $\overline{CH_2CHClCHCH:CH_2}$ (1877).

$CH_2:CRR' \rightarrow$ Bu$_3$SnCH$_2$CHRR'; R = H, R' = n-C$_6$H$_{13}$, R = Me, R' = Me$_3$CCH$_2$, Ph, R = Et, R' = Bu (1612).

CHD:CHR* $\rightarrow$ adduct + isomerization of excess olefin, R = Bu, Ph (1369).

(MeCH:)$_2$* $\rightarrow$ isomerization, cis/trans-form 0.3 + some C$_4$H$_{10}$ (1369).

(CH$_2$:CMe)$_2$ + cat. AiBN $\rightarrow$ Bu$_3$SnCH$_2$CMe:CMe$_2$ + Bu$_3$SnCH$_2$CHMeCMe:CH$_2$ (1619).

CH$_2$:CHCN $\rightarrow$ Bu$_3$SnCHMeCN + Bu$_3$SnCH$_2$CH$_2$CN (1417, 2680), prod. ratio and rxn. mechanism (1417).

RCH:CHCN (at 150°) or * $\rightarrow$ RCH$_2$CH$_2$CN; R = H, Me and CH$_2$:CMeCN (1683).

MeCH:CHCN (at 150°) $\rightarrow$ Bu$_3$SnCHEtCN + PrCN + Bu$_6$Sn$_2$ (2680).

MeCH:CHCN * $\rightarrow$ Bu$_3$SnCHEtCN + Bu$_3$SnCHMeCH$_2$CN (2680).

Me$_2$C:CHCN at 150° $\rightarrow$ Bu$_6$Sn$_2$ + i-BuCN (2680).

Me$_2$C:CHCN * $\rightarrow$ Bu$_3$SnCH(CN)CHMe$_2$ (2680).

R$_2$C:CHCN + MeOH under reflux $\rightarrow$ R$_2$CHCH$_2$CN + Bu$_3$SnOMe + Bu$_6$Sn$_2$; R = H, Me (2680).

MeCH:CHAc * $\rightarrow$ Bu$_3$SnCHEtAc + Bu$_3$SnOCMe:CHEt (1684).

Me$_2$C:CHCOR $\rightarrow$ Bu$_3$SnOCR:CHCHMe$_2$; R = Me (421) and i-BuAc + (Bu$_3$Sn)$_2$O + Bu$_6$Sn$_2$; R = i-Bu, Me$_2$C:CH (1684).

RCH:CHBz $\rightarrow$ Bu$_3$SnOCPh:CHCH$_2$R + BzCH$_2$CH$_2$R; R = H, Ph, prod. ratio (1418).

CH$_2$:CHCO$_2$Me $\rightarrow$ Bu$_3$SnCH$_2$CH$_2$CO$_2$Me (2070) and EtCO$_2$Me + Bu$_3$SnOMe in presence of MeOH (2680).

CH$_2$:CHCO$_2$Et (at 150° or *) $\rightarrow$ Bu$_3$SnCH$_2$CH$_2$CO$_2$Et + EtCO$_2$Et + (CH$_2$CHCO$_2$Et)$_x$ (2680).

CH$_2$:CHCO$_2$Et (at 150° or *) $\rightarrow$ EtCO$_2$Et (1683).

CH$_2$:CHCO$_2$Et + MeOH $\rightarrow$ EtCO$_2$R + Bu$_3$SnOR + Bu$_6$Sn$_2$; R = Me, Et (2680).

MeCH:CHCO$_2$R $\rightarrow$ PrCO$_2$R; R = Et (1683) + Bu$_3$SnOMe in presence of MeOH; R = Me (2680).

MeCH:CHCO$_2$Et (at 150°) $\rightarrow$ Bu$_3$SnCHEtCO$_2$Et + Bu$_3$SnCHMeCH$_2$CO$_2$Et + PrCO$_2$Et + Bu$_3$Sn-OEt + Bu$_6$Sn$_2$ (2680).

MeCH:CHCO$_2$Et * $\rightarrow$ Bu$_3$SnCHEtCO$_2$Et + PrCO$_2$Et (2680).

Me$_2$C:CHCO$_2$R (at 70°*) $\rightarrow$ Me$_2$CHCH$_2$CO$_2$R; R = Me, Et (1683) + Bu$_3$SnOMe in presence of MeOH; R = Me (2680).

Me$_2$C:CHCO$_2$R * $\rightarrow$ Bu$_3$SnCH(CO$_2$R)CHMe$_2$ + i-BuCO$_2$R; R = Me, Et (2680).

MeCH:C(CO$_2$Et)$_2 \rightarrow$ Bu$_3$SnCHMeCH(CO$_2$Et)$_2$ (1654).

(:CCO$_2$R)$_2 \rightarrow$ Bu$_3$SnCH(CO$_2$R)CH$_2$CO$_2$R (1654).

CH$_2$:C(CO$_2$R)CH$_2$CO$_2$R $\rightarrow$ Bu$_3$SnCH$_2$CH(CO$_2$R)CH$_2$CO$_2$R; R = Me, Et, Pr (1652).

---

* Under UV irradiation

$CH_2:CH(CH_2)_8CO_2R \rightarrow Bu_3Sn(CH_2)_{10}CO_2R$; R = H, Me, Pr_3Sn (598).
$[O_2CCH:CHCO_2(CH_2)_4]_x$ + cat. AIBN $\rightarrow Bu_3Sn$-contn. polyester (599).
$RR'C:C(CN)_2 \rightarrow Bu_3SnN:C:C(CN)RR'$; R,R' = H, Me, EtO, subst. aryl, CN (1621, 2615).
$HC\vdots CCN \rightarrow Bu_3SnC(CN):CH_2$ (313, 1414), kinetic data for rxn. in PrCn, mechanism, (312), deuterium effect (1412).
$(RC\vdots)_2 \rightarrow Bu_3SnCR:CHR$; R = CF_3 (923), CO_2Et, relative rxn. rates (1412).
$HC\vdots COR \rightarrow Bu_3SnCH:CHOR$; R = Et, Bu (2483).
$HC\vdots CCO_2Et \rightarrow Bu_3SnC(CO_2Et):CHMe + Bu_3SnCMe:CHCO_2Et$ (313, 1414).
$PhCH:NPh \rightarrow Bu_3SnNPhCH_2Ph$ (390).
$(1,1-NCC_6H_{10}N:)_2 \rightarrow Bu_3SnCN + c.C_6H_{11}CN + (C_6H_{11}N:)_2 + N + H$ + trace C_4H_{10} + C_4H_8, kinetic data and rxn. mechanism (1620).
RNO_2 * $\rightarrow$ kinetic data and mechanism of photoreduction; R = Ph, $1-C_{10}H_7$ (2833).
PHNCO $\rightarrow$ kinetic data for addition rxn., solvent effect (314), relative rxn. rates (1415).
$PhNCS \rightarrow Bu_3SnSCH:NPh$, rxn. mechanism and solvent effect (1415).
$PhCH_2N:C$ (+ cat. $Me_3CO)_2 \rightarrow Bu_3SnCN + MePh$ (2725).
$c.C_6H_{11}N:C$ (+ AIBN) $\rightarrow Bu_3SnCN + c.C_6H_{12}$ (2725).
MeOH $\rightarrow Bu_3SnOMe$ (2680).
$(RO)_2 \rightarrow Bu_6Sn_2 + ROH$; R = Me_3C, Me_2PhC (1616).
$(Me_3CO)_2$ (1:1) $\rightarrow Bu_3SnOCMe_3 + Me_3COH$ + same Bu_4Sn (1616).
$(AcO)_2 \rightarrow Bu_3SnOAc + CO_2 + H + CH_4 + C_4H_{10}$ (1616).
$(C_{11}H_{23}CO_2)_2 \rightarrow Bu_3SnOR + CO_2 + C_{12}H_{26}$ (1616).
$[RC(^{18}O)O]_2 \rightarrow Bu_3SnOC(^{18}O)R + H + CO_2$; R = Me, Ph, mechanism of peroxide cleavage rxn. (1615).
$RCHO \rightarrow Bu_3SnOCH_2R$; R = $m-HOC_6H_4$ (389).
RCHO + RR'CO $\rightarrow$ competitive ester formation and rxn. mechanism (2107).
RCHO + R'Br $\rightarrow$ relative reactivity of competing aldehydes (2107).
2-acetonaphthone $\rightarrow$ kinetic study of reduction by flash photolysis (1657).
$CCl_3CHO \rightleftharpoons Bu_3SnOCH_2CCl_3 + [Bu_3SnOCH(CCl_3)OCH_2CCl_3 + Bu_3SnCl?]$ (2538).
$C_6F_5CHO \rightarrow Bu_3SnOCH_2C_6F_5$ (+ $Bu_3SnOCHC_6F_5OCH_2C_6F_5$) (2538), relative kinetic data (2537).
$Me_2CO \rightarrow$ photo reduction, preferentially in triplet state (2104).
$RR'CO \rightarrow Bu_3SnOCHRR'$; R = Me, R' = Et (389), R = CF_3, R' = C_6F_5 (2538), R = Ph, R' = CF_3, relative kinetic data (2537).
$x-MeC_6H_9:O \rightarrow Bu_3SnOC_6H_{10}Me + Bu_6Sn_2 + x-MeC_6H_{10}OH$; x = 1-, 2-, 3- (436).
$R'CO_2H \rightarrow Bu_3SnO_2CR'$; R' = C_3-C_8 cyclic and C_6-C_9 polycyclic compd. (2391).
BzOR at 80-130° or * $\rightarrow$ RH; R = $PhCH_2$, Ph_2CH, Ph_3C, PhCHMe, i-Pr, $CH_2:CHCH_2$, $PhCH:CHCH_2$, n-, t-Bu, $c.C_6H_{11}$, $c.C_6H_9$, effect of radical initiation, rxn. mechanism (2488).
$PhCON_3 \rightarrow BzNH_2 + PhNHCONHBz$ (1050).

---

* Under UV irradiation

PhCON$_3$ (+ cat. AiBN) → Bu$_3$SnNHBz + N (1050).
S (at 75°) → (Bu$_3$Sn)$_2$S (2091).
Use: In catalyst for polymerization of olefins (241, 1650), of C$_2$H$_4$ (1238a, 1241, 2463-4, 2467, 2469) in presence of C$_2$H$_2$ (2468), PhCH:CH$_2$ (2466) and CH$_2$:CMeCO$_2$Me (2467), of CH$_2$:CMeCO$_2$Me (1240).
Effect on spreading of solder (1692).
As active agent in antifouling coatings (2302).

Dibutyltin Dihydride (M-259)   SnC$_8$H$_{20}$ Bu$_2$SnH$_2$

Synth.: By rxn. of Bu$_2$SnCl$_2$ with NaBH$_4$ in (MeOCH$_2$)$_2$, in 56% yield (736).
By rxn. of (MeHSiO)$_x$ with (Bu$_2$SnO)$_x$ in N atm., in 51% yield (1165), in 24% yield (1242), with Bu$_2$Sn(OR)$_2$, in 24% yield (1242); R = Et, in 66% yield (1165), c.C$_6$H$_{11}$, in 97% yield (2461), with (Bu$_2$SnOPh)$_2$O and Bu$_2$Sn(OCH$_2$CH:CH$_2$)$_2$ (2460).
By rxn. of (Bu$_2$SnO)$_x$ with (Ph$_2$SiH)$_2$O and Ph$_3$SiH, in 78 and 4% yield, resp. (1165).
By-prod. in rxn. of (Bu$_2$SnCl)$_2$ with LiAlH$_4$ (1955).
Prop.: b$_7$ 61-62° (258), IR (258, 1804), NMR (259, 331, 990, 1804) and Moessbauer resonance (1180, 1666) spectra. Anal. detn. with isatin and ninhydrin (1048), by iodometric titration (1165).
Rxn. with: UV irradiation → (Bu$_2$Sn)$_x$ + H + C$_4$H$_{10}$ + C$_4$H$_8$ + some C$_2$H$_6$ (585).
R$_3$SnNEt$_2$ → (R$_3$Sn)$_2$SnBu$_2$; R = Et (400, 1955), Bu (1955).
Bu$_3$SnNPhCHO → Bu$_3$SnSnBu$_2$H (95).
Bu$_2$Sn(NEt$_2$)$_2$ → c.(Bu$_2$Sn)$_6$ (525).
(Ph$_3$Ge)$_2$SnEtNPhCHO → [(Ph$_3$Ge)$_2$SnEt]$_2$SnBu$_2$ (913).
(Bu$_3$Sn)$_2$O (at r.t.) → Bu$_3$SnH + (Bu$_2$SnO)$_x$ (468).
(Bu$_3$Sn)$_2$O at 100° → (Bu$_3$Sn)$_2$SnBu$_2$ + H$_2$O (468, 1803).
(Bu$_2$SnO)$_x$ → (Bu$_2$Sn)$_x$ + H$_2$O (468, 1802).
Bu$_2$Sn(OMe)$_2$ → (Bu$_2$Sn)$_n$ (400, 468, 1802), n = 6 (1614).
Bu$_2$SnX$_2$ at (r.t.) → Bu$_2$SnHX; X = F, Cl, Br, I (469).
Bu$_2$SnX$_2$ ⇌ Bu$_2$SnHX; X = Cl, BrI (2737).
R$_2$SnCl$_2$ → Bu$_2$SnHCl + R$_2$SnHCl; R = Me, Et, Pr, i-Bu, c.C$_6$H$_{11}$, n-C$_8$H$_{17}$, Ph (2737).
BuSnCl$_3$ → BuSnH$_3$ + BuSnH$_2$Cl ?(1804).
Bu$_3$SnC⋮CH → (Bu$_3$SnCH:CH)$_2$SnBu$_2$ + (Bu$_3$SnCH:CH)$_2$ (1599).
i-Bu$_2$Sn[(CH$_2$)$_6$CH:CH$_2$]$_2$ + (cat. R$_2$AlH) → Sn-contn. polymer (1617).
Bu$_3$GeNEt$_2$ → (Bu$_3$GeSnBu$_2$)$_2$ (913).
Bu$_3$Al → Bu$_4$Sn + Bu$_2$AlH, relative rxn. rate (1830).
HCl → Bu$_2$SnHCl (469).
t-BuCl → kinetic data for dechlorination (2306).
cis-, or trans- 9-ClC$_{10}$H$_{17}$ → cis- and trans-decalin, mixt. independent of temp. and starting matl. (1143).
PVC (at 140°) → Bu$_3$Sn-contn. PVC, rxn. rate and solvent effect (2770).
PhCH:CHD (excess) under UV irradiation → adduct, no isomerization of excess olefin (1369).
(CH$_2$:CMe)$_2$ (1:2 + cat. AiBN) → Bu$_2$Sn(CH$_2$CMe:CMe$_2$)$_2$ + Bu$_2$Sn(CH$_2$CHMeCMe:CH$_2$)$_2$ (1619).

1,7-Octadiene (+ cat. i-Bu$_2$AlH) → Bu$_2$Sn[(CH$_2$)$_6$CH:CH$_2$]$_2$ (1612) + polymer (1612, 1617).
1,7-Octadiene (+ i-BuSnH$_3$) → polymer (1617).
CH$_2$:CHCH$_2$OH (+ cat. AiBN) → Bu$_2$SnOCH$_2$CH$_2$CH$_2$ + Bu$_2$Sn(OCH$_2$CH:CH$_2$)$_2$ (1383).
CH$_2$:CHCO$_2$Me (+ cat. AiBN) → Bu$_2$Sn(CH$_2$CH$_2$CO$_2$Me)$_2$ (1383).
CH$_2$:CHCONH$_2$, + HBr → Bu$_2$Sn(CH$_2$CH$_2$CONH$_2$)$_2$ (1168) + Bu$_2$(H$_2$NOCCH$_2$CH$_2$)SnBr (2434).
CH$_2$:CHCH$_2$SH (+ cat. AiBN) → Bu$_2$Sn(SCH$_2$CH:CH$_2$)$_2$ + polymer (1383).
PhCH:C(CN)$_2$ (at 40°) → PhCH$_2$CH(CN)$_2$ (2789).
[O$_2$CCH:CHCO$_2$(CH$_2$)$_4$] (+ cat. AiBN) → Bu$_2$Sn-contn. polymer (599).
HC⋮CR → Bu$_2$Sn(CH:CHR)$_2$; R = Pr, Et$_2$NCH$_2$ (499).
HC⋮CCHRCH$_2$C⋮CH .(at 220-300°) → Bu$_2$SnCH:CHCHRCH$_2$CH:CH; R = H, MeO (499).
p-(HC⋮C)$_2$C$_6$H$_4$ → (p-Bu$_2$SnCH:CHC$_6$H$_4$CH:CH)$_x$ (327).
HC⋮CCHXCH$_2$C:CH → Bu$_3$SnX; X = Br, AcO (499).
(CF$_3$C⋮)$_2$ → Bu$_2$Sn[C(CF$_3$):CHCF$_3$]$_2$ (923).
CH$_2$CH$_2$O → (CH$_2$CH$_2$O)$_x$; relative reactivity as polymerization catalyst (2572).
(CF$_3$)$_2$CO → Bu$_2$Sn[OCH(CF$_3$)$_2$]$_2$ (109).
Bz$_2$O$_2$ → Bu$_2$Sn(OBz)$_2$ + (Bu$_2$SnOBz)$_2$ + H (585).
Py in MePh or MeONa in THF → c.(Bu$_2$Sn)$_6$ (1617).

Butyltin Trihydride (M-263)   SnC$_4$H$_{12}$ BuSnH$_3$
<u>Synth.:</u> By rxn. of BuSnCl$_3$ with NaBH$_4$ in diglyme, in 16% yield (736).
By-prod. in rxn. of BuSnCl$_3$ with Bu$_3$SnH or Bu$_2$SnH$_2$ in N atm. at r.t. (1804).
<u>Prop.:</u> IR (258, 1180, 1804), NMR (258, 331, 990, 1804) and Moessbauer resonance (1180 1666) spectra.
<u>Rxn. with:</u> 1-C$_{10}$F$_{21}$CH:CH$_2$ (+ cat. AiBN), + SnCl$_4$ → C$_{10}$F$_{21}$CH$_2$CH$_2$SnCl$_3$ + Bu-(C$_{10}$F$_{21}$CH$_2$CH$_2$)SnCl$_2$ (745a).
(CH$_2$:CH)$_2$Y + SnCl$_4$ + 1-C$_{10}$F$_{21}$I (+ cat. AiBN) → 1-C$_{10}$F$_{21}$CH$_2$CH$_2$YCH$_2$CH$_2$SnCl$_3$; Y = CH$_2$CH$_2$, CH$_2$OCH$_2$ (745a).

Triisobutyltin Hydride (M-260)   SnC$_{12}$H$_{28}$ (Me$_2$CHCH$_2$)$_3$SnH
<u>Prop.:</u> b$_{12}$ 103° (297), NMR spectrum (297, 990).
<u>Rxn. with:</u> UV irradiation → i-Bu$_3$Sn* , ESR spectrum of prod. (1827).
R$_3$SnNEt$_2$ → i-Bu$_3$SnSnR$_3$; R = Et, i-Bu (525).
Ph$_3$SnCl + Et$_3$N (1:1:1) → Ph$_6$Sn$_2$ + i-Bu$_6$Sn$_2$ + i-Bu$_3$SnSnPh$_3$ + Et$_3$N·HCl (1618).
(R$_3$Sn)$_2$O → i-Bu$_3$SnSnR$_3$; R = Et (400, 1618), i-Bu (1618).
Bu$_3$Al → i-Bu$_3$SnBu + Bu$_2$AlH; relative rxn. rates (1830).
(CH$_2$:CMe)$_2$ + (cat. AiBN) → i-Bu$_3$SnCH$_2$CMe:CMe$_2$ + i-Bu$_3$SnCH$_2$CHMeCMe:CH$_2$ (1619).
RR'C:C(CN)$_2$ → i-Bu$_3$SnN:C:C(CN)CHRR' (2615).
(NCCMe$_2$N:)$_2$ → i-Bu$_3$SnCN + Me$_2$CHCN + (Me$_2$CCN)$_2$ + N; kinetic data and rxn. mechanism (1620).
RCHO → i-Bu$_3$SnOCH$_2$R; R = i-Pr, o-HOC$_6$H$_4$ (389).
MeEtCO → i-Bu$_3$SnOCHMeEt (389).

Diisobutyltin Dihydride (M-260)   SnC$_8$H$_{20}$ (Me$_2$CHCH$_2$)$_2$SnH$_2$
<u>Synth.:</u> By rxn. of i-Bu$_2$SnCl$_2$ with Bu$_3$SnH (1:2) in N atm. (1805).
By-prod. in rxn. of (i-Bu$_2$SnCl)$_2$ with LiAlH$_4$ (1955).

Prop.: IR (1805) and NMR (990, 1805) spectra.
Rxn. with: $R_3SnNEt_2 \rightarrow (R_3Sn)_2SnBu\text{-}i_2$; R = Et, i-Bu (525, 1955), c.C_6H_{11}, Ph (1955).
$i\text{-}Bu_2Sn(NEt_2)_2 \rightarrow c.(i\text{-}Bu_2Sn)_9$ (525, 1614).
$(Ph_3Sn)_2O$ (at 85°) $\rightarrow Ph_6Sn_2 + (i\text{-}Bu_2Sn)_xH_2$ (1955).
$(Bu_3Sn)_2S$ (at 110°) $\rightarrow Bu_6Sn_2 + Bu_3SnH + (i\text{-}Bu_2SnS)_x + H + H_2S + SnS$ (1955).
$i\text{-}Bu_2SnCl_2$ (+ Et_2NH) $\rightarrow (i\text{-}Bu_2SnCl)_2$ (1955).
$Bu_2Sn(CH_2CH:CH_2)_2 \rightarrow (i\text{-}Bu_2SnCH_2CH_2CH_2SnBu_2CH_2CH_2CH_2)_x$ (1617).
$R_2SnCl_2 \rightleftharpoons i\text{-}Bu_2SnHCl + R_2SnHCl$; R = Me, Et, Pr, n-, i-Bu, c.C_6H_{11}, n-C_8H_{17}, Ph (2737).
$i\text{-}Bu_2SnCl_2 + CH_2:CHCO_2Me \rightarrow i\text{-}Bu_2(MeO_2CCH_2CH_2)SnCl$ (1618).
$i\text{-}Bu_2SnCl_2 + PhCH:CHR \rightarrow i\text{-}Bu_2(PhCH_2CHR)SnCl$; R = Me, Bu, Ph (1613).
$i\text{-}Bu_2SnX_2 + CH_2:CHR \rightarrow i\text{-}Bu_2(RCH_2CH_2)SnX$; RX = CH_2OH + F, Cl (1613), MeO_2C + Cl, Br (1613).
$i\text{-}Bu_2SnCl_2 + HC\vdots CR \rightarrow i\text{-}Bu_2(RCH:CH)SnCl$; R = CN, Ph, p-$MeOC_6H_4$ (1613).
$i\text{-}Bu_2SnCl_2 + RCHO \rightarrow i\text{-}Bu_2Sn(OCH_2R)Cl$; R = Et, n-, i-Pr, n-$C_7H_{15}$ (1613).
$i\text{-}Bu_2SnCl_2 + c.C_6H_{10}O \rightarrow i\text{-}Bu_2Sn(OC_6H_{11})Cl$; (1613).
$Bu_3Al \rightarrow i\text{-}Bu_2SnBu_2 + Bu_2AlH$; relative rxn. rate (1830).
1,7-Octadiene (+ cat. $i\text{-}Bu_2AlH$) $\rightarrow i\text{-}Bu_2Sn[(CH_2)_6CH:CH_2]_2$ (1612).
Polybutadiene (+ cat. R_2O_2) $\rightarrow$ Sn-contg. polymer (2900).
RCHO $\rightarrow i\text{-}Bu_2Sn(OCH_2R)_2$; R = i-Pr, o-HOC_6H_4 (389).
Py $\rightarrow c.(i\text{-}Bu_2Sn)_9$ (1614).

Isobutyltin Trihydride (M-263) SnC_4H_{12} $i\text{-}BuSnH_3$
Prop.: NMR spectrum (990).
Rxn. with: Bu_2SnH_2 + 1,7-octadiene $\rightarrow$ polymerization, effect of $i\text{-}BuSnH_3$ (1617).
1-Octene (+ cat. $i\text{-}Bu_2AlH$) $\rightarrow (n\text{-}C_8H_{17})_3SnCH_2CHMe_2$ (1612).
$Bu_2Sn(CH_2CH:CH_2)_2$ (+ cat. R_2AlH) $\rightarrow$ crosslinked polymer (1617).

Trisecondarybutyltin Hydride $SnC_{12}H_{28}$ $(MeEtCH)_3SnH$
Synth.: By rxn. of $s.Bu_3SnI$ with $LiAlH_4$ under exclusion of air, in 66% yield (1618).
Prop.: $b_{0.4}$ 65°, n_D^{20} 1.4863, d_{20} 1.12, IR spectrum (1618).
Rxn. with: $R_3SnNEt_2 \rightarrow i\text{-}Bu_3SnSnR_3$; R = Et, i-Bu (1618).
$(i\cdot Bu_3Sn)_2O \rightarrow$ no exchange rxn. (1618).
$Bu_3Al \rightarrow s\text{-}Bu_3SnBu + Bu_2AlH$; relative rxn. rates (1830).

Ditertiarybutyltin Dihydride SnC_8H_{20} $(Me_3C)_2SnH_2$
Synth.: By rxn. of $t\text{-}Bu_2SnCl_2$ with $LiAlH_4$ in Ar atm., in 83% yield (1614).
Prop.: b_{11} 38°, n_D^{20} 1.4634, d_{20} 1.17 (1614), IR (1614) and NMR (990, 1454) spectra.
Rxn. with: $t\text{-}Bu_2Sn(NEt_2)_2 \rightarrow c.(t\text{-}Bu_2Sn)_4$ (1614).
$Bu_3Al \rightarrow t\text{-}Bu_2SnBu_2 + Bu_2AlH$; relative rxn. rate (1830).

Tricyclohexyltin Hydride SnC_8H_{34} $(c.C_6H_{11})_3SnH$
Synth.: By rxn. of $(C_6H_{11})_3SnBr$ with $LiAlH_4$ in inert atm., in 89% yield (1618).
Prop.: $b_{0.001}$ 147-50°, n_D^{20} 1.5411, d_{20} 1.40, IR spectrum (1618).

Rxn. with: R$_3$SnNEt$_2$ → (C$_6$H$_{11}$)$_3$SnSnR$_3$; R = Me, i-Bu, C$_6$H$_{11}$ (1618).
(CH$_2$:CMe)$_2$ (+ cat. AiBN) → (C$_6$H$_{11}$)$_3$SnCH$_2$CMe:CMe$_2$ + (C$_6$H$_{11}$)$_3$SnCH$_2$CHMeCMe:CH$_2$ (1617).
CH$_2$:CHCH:CHMe (+ cat. AiBN) → (C$_6$H$_{11}$)$_3$SnCH$_2$CH:CHEt + (C$_6$H$_{11}$)$_3$SnCH$_2$CH$_2$CH:CHMe (1619).
[O$_2$CCH:CHCO$_2$(CH$_2$)$_4$]$_x$ (+ cat. AiBN) → (C$_6$H$_{11}$)$_3$Sn-contn. polyester (599).
Use: In catalyst for copolymerization of C$_2$H$_4$ and PhCH:CH$_2$ (2466).

Dioctyltin Dihydride (M-264)　　　　　SnC$_{16}$H$_{36}$　　　　　(n-C$_8$H$_{17}$)$_2$SnH$_2$
Synth.: By rxn. of (C$_8$H$_{17}$)$_2$SnCl$_2$ with Bu$_3$SnH (1:2) in N atm. (1805).
Prop.: IR (1805) and NMR (990, 1805) spectra.
Rxn. with: R$_2$SnCl$_2$ ⇌ (C$_8$H$_{17}$)$_2$SnHCl + R$_2$SnHCl; R = Me, Et, Pr, n-, i-Bu, c.C$_6$H$_{11}$, n-C$_8$H$_{17}$, Ph (2737).
(C$_8$H$_{17}$)$_2$SnCl$_2$ + CH$_2$:CHR → RCH$_2$CH$_2$(C$_8$H$_{17}$)$_2$SnCl; R = HOCH$_2$CH$_2$OCH$_2$ (1613).
Bu$_3$Al → Bu$_2$Sn(C$_8$H$_{17}$)$_2$ + Bu$_2$AlH, relative rxn. rate (1830).

Triphenyltin Hydride (M-260)　　　　　SnC$_{18}$H$_{16}$　　　　　Ph$_3$SnH
Synth.: By rxn. of Ph$_3$SnCl with NaBH$_4$ in (MeOCH$_2$)$_2$, in 82% yield (736).
By rxn. of (Ph$_2$Sn)$_2$O with (MeHSiO)$_x$ in N atm., in 33% yield (1165), with Ph$_2$SnH$_2$ in C$_6$H$_6$ at 20° (1955).
By polarographic reduction of Ph$_6$Sn$_2$ in (MeOCH$_2$)$_2$-Bu$_4$NClO$_4$, followed by HCl (963).
By rxn. of Ph$_3$SnMgBr·NEt$_3$ with H$_2$O (2330).
Prop.: b$_2$ 177.5-78.5° (258), IR (258), NMR (258, 331, 460) and Moessbauer resonance (1180, 1666, 1991) spectra, force const. (138), polarography (967). Anal. detn. with isatin or ninhydrin (1048).
Rxn. with: Thermal dec. at 80-120° → free radical rxn.; in C$_6$H$_6$, EtOH and octane (123).
UV irradiation → Ph$_3$Sn* ; ESR spectrum of prod. (1827).
R$_3$SnNPhCHO → R$_3$SnSnPh$_3$; R = Me, Et, Bu, n-C$_8$H$_{17}$ (97).
Et$_3$SnNRCHO → Et$_3$SnSnPh$_3$ + RNHCHO; solvent effect, R = C$_6$H$_{13}$, Ph, p-ClC$_6$H$_4$ (96, 915), n-, t-Bu, O$_2$NC$_6$H$_4$ (915).
R$_2$Sn(NPhCHO)$_2$ → (Ph$_3$Sn)$_2$SnR$_2$; R = Et, Bu, Ph (97).
Bu$_3$SnSnBu$_2$H + PhNCHO → Bu$_3$SnSnBu$_2$SnPh$_3$ (95).
R$_3$SnNEt$_2$ → R$_3$SnSnPh$_3$; R = n-, i-Bu (525), Me, i-Bu, c.C$_6$H$_{11}$ (1618).
Et$_3$SnR → Et$_3$SnSnPh$_3$ + RH; R = Et$_2$N, PhNHNPh, solvent effect (96, 915), Ph$_2$N (915).
Ph$_3$GeSnEt$_2$NEt$_2$ → Ph$_3$GeSnEt$_2$SnPh$_3$ (913).
Ph$_3$GeSnEt(NPhCHO)$_2$ → Ph$_3$GeSnEt(SnPh$_3$)$_2$ (913).
R$_3$SnCl + Et$_3$N → R$_3$SnSnPh$_3$ + Et$_3$N·HCl ; R = Me (1618), Et (400).
i-Bu$_3$SnI + Et$_3$N (1:1:1) → i-Bu$_3$SnSnPh$_3$ + Et$_3$N·HI (1618).
Et$_3$SnCN + Et$_3$N (1:1:1) → Et$_3$SnSnPh$_3$ + Ph$_3$SnCN (1618).
Bu$_2$SnBr$_2$ + Et$_3$N (2:1:2) → Bu$_2$Sn(SnPh$_3$)$_2$ + Et$_3$N·HBr + Ph$_6$Sn$_2$ (1955).
Et$_3$SnMPh$_2$ → Et$_3$SnSnPh$_3$ + Ph$_2$MH; M = P, As (915).
Et$_3$SnPPh$_2$ → Ph$_3$SnPPh$_2$ + Et$_3$SnH, rxn. followed by NMR spectroscopy (915).
Bu$_3$SnPBu$_2$ → Bu$_3$SnSnPh$_3$ + Bu$_2$PH + some Ph$_6$Sn$_2$ (1618).
(R$_3$Sn)$_2$O → R$_3$SnSnPh$_3$; R = Et (400, 1618), Ph (400).

$(Ph_2SnO)_x$ (at 100°) → Ph_6Sn_2 + $(Ph_3Sn)_2(SnPh_2)_n$; n = 1, 2, x (1955).
Et_3SnOR → $Et_3SnSnPh_3$ + ROH; rxn. mechanism, R = Me_3C, $p-XC_6H_4$; X = H, Cl, MeO, O_2N (914-5), R = Me, $PhCH_2$, Et, EtMeCH, solvent effect, competitive exchange rxn. (915).
Et_3SnOPh → Ph_3SnOPh + Et_3SnH; rxn. followed by NMR spectroscopy (915).
Bu_3SnOMe → $Bu_3SnSnPh_3$ (1618).
Ph_3SnOR → Ph_6Sn_2 ; R = H, Me (1618).
Bu_3SnOAc + Et_3N → Ph_6Sn_2 + $Bu_3SnSnPh_3$ (1618).
$(Bu_3Sn)_2S$ → $Bu_3SnSnPh_3$ + Bu_3SnH + Ph_6Sn_2 + SnS (1618).
Et_3SnSBu → $Et_3SnSnPh_3$ + BuSH; relative rxn. rates, solvent effect and mechanism (915).
Bu_3SnR → Bu_3SnH + RH; R = CH_2Ac, CH_2CO_2Et, i-PrCHCN, i-PrCH(CO_2Et) (2680).
$Me_2RSiC\!:\!CH$ → $Ph_3SnCH\!:\!CHSiMe_2R$; R = Me, Ph (1354).
$p-(CH_2\!:\!CHMR_2)_2C_6H_4$ → $(Ph_3SnCH_2CH_2MR_2)_2C_6H_4$; RM = MeSn, MeSi, PhSi (1410).
$PhRSi(C\!:\!CH)_2$ → $(Ph_3SiCH\!:\!CH)_2SiPhR$; R = Me, Ph (1354).
Bu_3GeNEt_2 → $Ph_3SnGeBu_3$ + Et_2NH (913).
$N-Bu_3PbC_3H_3N_2$ + $HC\!:\!CCO_2Me$ → $N-Ph_3SnC_3H_3N_2$ + $Bu_3PbCH\!:\!CHCO_2Me$ (912).
R_3PbX + $HC\!:\!CCN$ → Ph_3SnX + $R_3PbCH\!:\!CHCN$ + $R_3PbC(CN)\!:\!CH_2$; R = Et, Bu; X = AcO, imidazolyl (912).
$R_2SbC\!:\!CH$ → $Ph_3SnCH\!:\!CHSbR_2$; R = Ph (1601) + $(Ph_3SnCH\!:)_2$; R = Bu (1601).
$EtMgBr \cdot NEt_3$ → $Ph_3SnMgBr \cdot NEt_3$ + C_2H_6 (2330).
$Et_2Zn \cdot L$ → $(Ph_3Sn)_2Zn \cdot L$; L = $(Me_2NCH_2)_2$, $(MeOCH_2)_2$, BiPy (2043).
$EtZnCl \cdot L$ → $Ph_3SnZnCl \cdot L$; L = $(Me_2NCH_2)_2$, $(MeOCH_2)_2$ (2043).
$Me_2Cd \cdot L$ → $(Ph_3Sn)_2Cd \cdot L$; L = $(Me_2NCH_2)_2$, $(MeOCH_2)_2$, BiPy (2043).
$MeCdCl \cdot L$ → $Ph_3SnCdCl \cdot L$; L = $(Me_2NCH_2)_2$ (2043).
$[(Me_3Si)_2N]_2Hg$ → $(Ph_3Sn)_2Hg$ + $(Me_3Si)_2NH$ (2370).
$M(NR_2)_4$ (4:1) → $(Ph_3Sn)_4M$ + Me_2NH; MR = TiMe, ZrEt (2331).
$Ti(NMe_2)_4$ (2:1), + MeOH → $(Ph_3Sn)_2Ti(OMe)_2$ + Me_2NH (2331).
$Ti(NMe_2)_4$ (2:1), + H_2O → $[(Ph_3Sn)_2TiO]_x$ (2331).
$Zr(NEt_2)_4$ (3:1), + PhNHCHO, + Ph_2SnH_2 → $[(Ph_3Sn)_3Zr]_2SnPh_2$ (2331).
$Zr(NEt_2)_4$ (2:1), + PhNHCHO, + Ph_2SnH_2 → $[(Ph_3Sn)_2ZrSnPh_2]_x$ (2331).
$C_5H_5FeC_5H_4CO_2H$ → $Ph_3SnO_2CC_5H_4FeC_5H_5$ + $(Ph_2SnO_2CC_5H_4FeC_5H_5)_2O$ (1310, 1374).
$C_5H_5FeC_5H_4CO_2H$ (1:2) $Ph_2Sn(O_2CC_5H_4FeC_5H_5)_2$ (1310, 1374).
$Ru_3(CO)_{12}$ → $(Ph_3Sn)_2Ru(CO)_4$ (905).
$(Ph_2RP)_2Ir(CO)X$ → $Ph_3Sn(Ph_2RP)Ir(CO)HX$; R = Ph, X = Cl, Br, I, RX = MeCl (2530).
HBr (at -78°) → Ph_2SnHBr + C_6H_6 (1:1) and $PhSnHBr_2$ (1:2) (156).
HBr in Et_2O → $PhSnBr_3 \cdot OEt_2$ (156).
RCl → RH (1251).
Me_3CX → kinetic data and rxn. mechanism X = Cl, Br (2305-6), heat of reduction X = Cl (2306).
cis- and trans-$ClC_{10}H_{17}$ → cis- and trans-decalin (1143).
RX → RH; R = norbornenyl, nortricyclenyl, X = Cl, Br, prod. ratio (2113).
BzCl → Ph_3SnCl + $PhCH_2OBz$ (299).
RC_6H_4COCl → Ph_3SnCl; R = Me, MeO (299), in presence of AiBN, R = F, CF_3 (299).
RCOCl → Ph_3SnCl + RCO_2CH_2R + < 16% RCHO; R = alkyl, aryl, bridgehead groups (1310 1374).

RCOCl → Ph$_3$SnCl + RCHO; R = triptycenyl (1310, 1374).

RCOCl + AcPh → MePhCHO$_2$CR; several acid chlorides and ketones (1280).

Ph$_2$PCl → Ph$_3$SnCl + Ph$_2$PH (495).

RCN → Ph$_6$Sn$_2$; R = Pr, C$_5$H$_{11}$ (2680).

CH$_2$:CHPh → Ph$_3$SnCH$_2$CH$_2$Ph (2070).

CHD:CHPh → no isomerization of excess olefin (1369).

CHD:CHBu → complete isomerization of excess olefin (1369).

1-C$_8$H$_{16}$ (at 120°) → Ph$_6$Sn$_2$ + n-C$_8$H$_{17}$SnPh$_3$ (123).

1-Olefin under UV irradiation or with AiBN → Ph$_3$Sn(CH$_2$)$_n$Me; n = 4-8 (123).

Cycloolefin in air or under UV irradiation → Ph$_6$Sn$_2$ (123).

(MeCH:)$_2$ under UV irradiation → partial isomerization of olefin (1369).

CH$_2$:CHCH:CHMe (+ AiBN) → Ph$_3$SnCH$_2$CH$_2$CH:CHMe + Ph$_3$SnCH$_2$CH:CHEt (1619); rxn. mechanism (1605).

Diene (+ AiBN) → 1,2- and 1,4-adduct, with C$_5$H$_6$ (1410, 1619), (C$_5$H$_6$)$_2$, c.C$_6$H$_8$ (1619), kinetic data and effect of galvinoxyl with (CH$_2$:CMe)$_2$ (1619).

(CH$_2$:CH)$_2$Y → (Ph$_3$SnCH$_2$CH$_2$)$_2$Y; Y = CHOH; o-C$_6$H$_4$ (1410).

CH$_2$:CHCN → Ph$_3$SnCH$_2$CH$_2$CN (1417, 2070, 2680).

CH$_2$:CHCN (at 150°) → EtCN (1683).

MeCR:CHCN (at 150°) → MeCHRCH$_2$CN + Ph$_6$Sn$_2$; R = H, Me (2680).

PhCH:C(CN)$_2$ → Ph$_3$SnN:C:C(CN)CH$_2$Ph (1:1) and Ph$_6$Sn$_2$ + PhCH$_2$CH(CN)$_2$ (2:1) (2789).

CH$_2$:CHR → Ph$_3$SnCH$_2$CH$_2$R; R = PhO, AcO, MeO$_2$C (2070), EtO$_2$C (2680), (CH$_2$)$_8$CO$_2$Et (598).

CH$_2$:CHAc → Ph$_3$SnCH$_2$CH$_2$Ac + EtAc (1418).

RCH:CHBz → Ph$_3$SnOCPh:CHCH$_2$R + RCH$_2$CH$_2$Bz; R = H, Ph (1418), at (2-3:1) only RCH$_2$CH$_2$Bz (1418).

MeCR:CHCO$_2$Et → MeCHRCH$_2$CO$_2$Et, R = H, Me (1683) and Ph$_6$Sn$_2$ (2680).

Me$_2$C:CHAc → Ph$_6$Sn$_2$ + Me$_2$CHCH$_2$Ac (421, 1684), other unsaturated ketones reduced in similar manner: phorone, dihydrophorone, AcCH:CH$_2$, AcCH:CHMe, (Me$_2$C:CH)$_2$CO, i-BuCOCH:CMe$_2$ (1684).

CH$_2$:CHCH$_2$NH$_2$ → Ph$_6$Sn$_2$; no NH$_3$ or CH$_2$:CHMe, cf.(M-261, ref. 731) (537).

[O$_2$CCH:CHCO(CH$_2$)$_4$]$_x$ + AiBN → Ph$_3$Sn-contn. polyester (599).

HC⋮CR → Ph$_3$SnCH:CHR; R = Ph, Bu, EtO (1414) and PhCHR:CH$_2$; R = CH$_2$OH (1414).

HC⋮CCN in PrCN → kinetic data and rxn. mechanism (312).

o-(HC⋮C)$_2$C$_6$H$_4$ → o-(Ph$_3$SnCH:CH)$_2$C$_6$H$_4$ (1410).

(EtO$_2$CC⋮)$_2$ → Ph$_3$SnC(CO$_2$Et):CHCO$_2$Et; relative rxn. rates (1412, 1414).

PhNCO → rate of rxn. and mechanism, solvent effect (314), relative reactivity (1415).

RNCS → Ph$_3$SnSCH:NR; R = Ph, p-EtOC$_6$H$_4$, rxn. mechansim, solvent effect (1415).

RCHO → Ph$_3$SnOCH$_2$R; R = o-HOC$_6$H$_4$ (389).

BzH → Ph$_6$Sn$_2$ + PhCH$_2$OH (389).

RR'CO → relative reaction kinetics; RR' = CF$_3$Ph, HC$_6$F$_5$ (2537).

RCO$_2$H → Ph$_3$SnCO$_2$R with 1-apocamphane, 1-norborane, 1-tryptycene, 2-pyrrole acids (1310 1374).

RCO$_2$H → Ph$_2$Sn(O$_2$CR)$_2$; with 2-pyridine (1310, 1374), 2-bicycloheptene, 1-adamantane acids (2391).

Bz$_2$O$_2$ → (Ph$_2$SnOBz)$_2$O + Ph$_3$SnOBz (123).

Use: In catalyst for polymerization of olefins (411, 656, 1656) of C$_2$H$_4$ with

PhCH:CH$_2$ (2466).
As additive for flame resistant polester (2702).

Diphenyltin Dihydride (M-262)   SnC$_{12}$H$_{12}$ Ph$_2$SnH$_2$
Synth.: By rxn. of Ph$_2$SnCl$_2$ with Bu$_3$SnH (1:2) in Et$_2$O in N atm. (1805).
Prop.: m. -17° (392), IR (1805), NMR (259, 460, 1805) and Moessbauer resonance (1180, 1666, 1991) spectra, force const. (138).
Rxn. with: R$_3$SnPhCHO (1:2) → (R$_3$Sn)$_2$SnPh$_2$; R = Me, Et, Bu, n-C$_8$H$_{17}$ (97).
Bu$_3$SnSnBu$_2$H + Bu$_3$SnNPhCHO → Bu$_3$SnSnBu$_2$SnPh$_2$SnBu$_3$ (95).
Et$_2$Sn(NPhCHO)$_2$ → (Et$_2$SnSnPh$_2$)$_x$ + PhNHCHO (97).
(Ph$_3$Ge)$_n$SnEt$_{3-n}$PhCHO (1:2) → [(Ph$_3$Ge)$_n$SnEt$_{3-n}$]$_2$SnPh$_2$; n = 1,2 (913).
(Ph$_3$Ge)$_n$SnEt$_{3-n}$NPhCHO (1:1) → [(Ph$_3$Ge)$_n$SnEt$_{3-n}$SnPh$_2$]$_2$; n = 1, 2 (913).
Ph$_2$Sn(NEt$_2$)$_2$ → c.(Ph$_2$Sn)$_6$ (1610).
Ph$_3$SnH + Zr(NEt$_2$)$_4$, + PhNHCHO → [(Ph$_3$Sn)$_3$Zr]$_2$SnPh$_2$ + [(Ph$_3$Sn)$_2$ZrSnPh$_2$]$_x$ (2331).
(Bu$_3$Sn)$_2$O → (Bu$_3$Sn)$_2$SnPh$_2$ + H$_2$O; n-Bu (1803), i-Bu (1955).
(Ph$_3$Sn)$_2$O (at 20°) → Ph$_3$SnH + (Ph$_2$SnO)$_x$ (1955).
Ph$_3$SnOMe (1:2 at 20°) → Ph$_6$Sn$_2$ + Ph$_3$Sn(SnPh$_2$)$_n$SnPh$_3$; n = 1-3 (1955).
R$_2$SnCl$_2$ ⇌ Ph$_2$SnHCl + R$_2$SnHCl; R = Me, Et, Pr, n-, i-Bu, c.C$_6$H$_{11}$, n-C$_8$H$_{17}$, Ph (2737).
Ph$_2$SnCl$_2$ + PhMeC:CH$_2$ → Ph$_2$(PhMeCHCH$_2$)SnCl (1613).
Ph$_2$SnCl$_2$ + (CH$_2$:CHCH$_2$)$_2$ → Ph$_2$[CH$_2$:CH(CH$_2$)$_4$]SnCl (1613).
o-(CH$_2$:CH)$_2$C$_6$H$_4$ → (o-Ph$_2$SnCH$_2$CH$_2$C$_6$H$_4$CH$_2$CH$_2$)$_n$; n = 1, 2, x (310).
CH$_2$:CHCN → Ph$_2$Sn(CH$_2$CH$_2$CN)$_2$ (1745).
CH$_2$:CHCOR → EtCOR; R = Me, Ph (1418).
MeCH:CHBz → PrBz (1418).
[O$_2$CCH:CHCO$_2$(CH$_2$)$_4$]$_x$ + AIBN → Ph$_2$Sn-contn. polyester (599).
o-(HC⋮C)$_2$C$_6$H$_4$ → (o-Ph$_2$SnCH:CHC$_6$H$_4$CH:CH)$_n$; n = 1, x (310).
RCO$_2$H → Ph$_2$Sn(O$_2$CR)$_2$; R = 2-pyridyl, ferrocenyl (1310, 1374).
Py → H + c.(Ph$_2$Sn)$_6$ (392).

Phenyltin Trihydride (M-264)   SnC$_6$H$_8$ PhSnH$_3$
Prop.: NMR (460) and Moessbauer resonance (1180, 1666) spectra, force const. (138).
Rxn. with: Bu$_3$SnNPhCHO → (Bu$_3$Sn)$_2$SnPhH + PhNHCHO (95).
R$_3$SnNPhCHO → (R$_3$Sn)$_3$SnPh + PhNHCHO; R = Et, n-C$_8$H$_{17}$ (97).

Diparatolytin Dihydride   SnC$_{14}$H$_{16}$ (p-MeC$_6$H$_4$)$_2$SnH$_2$
Synth.: By rxn. of (p-MeC$_6$H$_4$)$_2$SnCl$_2$ with Et$_2$AlH in inert atm. in Et$_2$O at -40 to +20°, in 94% yield (393).
Prop.: Colorless cryst., m. 24-25° IR spectrum, mol. wt. (393).
Rxn. with: Me$_2$NCHO → H + c.[p-MeC$_6$H$_4$)$_2$Sn]$_6$ (393).
(p-MeC$_6$H$_4$)$_2$SnCl$_2$ + PhMeC:CH$_2$ → PhCHMeCH$_2$(p-MeC$_6$H$_4$)$_2$SnCl (1613).

Other nonfunctional organotin hydrides are listed in Table 122.

Table 122. Nonfunctional Organotin Hydrides R_nSnH_{4-n}

Formula*	Synth. Method**	Yield	Properties	Ref.
$Et_2[CH_2:CH(CH_2)_4]SnH$	--	--	(a)	--
$EtPhSnH_2$	I (b)	--	$b_{0.7}$ 59-64°, n_D^{20} 1.5527 (c)	1410
$PrSnH_3$ (263)	I	--	Stable at -78°, dec. at r.t., NMR spectrum	331
$i-Pr_2SnH_2$	--	--	NMR spectrum (d)	331
$i-PrSnH_3$	--	--	NMR spectrum (e)	331
$(MeEtCH_2)_2SnH_2$	--	--	(f)	--
$(PhCMe_2CH_2)_3SnH$	--	--	Moessbauer resonance spectrum	224
$(c.C_6H_{11})_2SnH_2$	I	86	$b_{0.001}$ 94-96°, n_D^{20} 1.5295 (g)	1614
	III	--	IR (1614) and NMR spectra	1805
$(Me\overline{CHCMe_2CH})_3SnH$	--	--	Use as herbicide and pesticide (h)	1873
$(C_8H_{17})_3SnH$ (263)	I	--	IR and NMR (990) spectra (i)	297
$C_8H_{17}SnH_3$	--	--	NMR spectrum (j)	990
$(1-C_8H_{17})_2SnH_2$	--	--	(k)	--
$(p-MeC_6H_4)_3SnH$ (264)	--	--	IR and NMR spectra (l)	297
$Mesityl_3SnH$ (264)	--	--	(m)	--

* Numbers in parenthesis refer to pages in main volume.
** The characters used here correspond to the synthetic scheme in the introduction to Chapter 3.

(a) Polymerizes with R_2AlH (1617). (b) Synth. also by reduction of reaction product of Et_3SnPh and $SnCl_4$ with $LiAlH_4$ (1410). (c) Rxn. with $o-(HC:C)_2C_6H_4 \rightarrow (o-EtPhSnCH:CHC_6H_4CH:CH)_n$; n = 1, 2, x (310, 1410). (d) Rxn. with $i-Pr_2SnX_2 \rightarrow i-Pr_2SnHX$; X = Cl, Br, I (1292). (e) Moessbauer resonance spectrum (1180 1666). (f) Rxn. with $Bu_3Al \rightarrow s-Bu_2SnBu_2 + Bu_2AlH$; relative reactivity (1830). (g) Rxn. with $(c.C_6H_{11})_2Sn(NEt_2)_2 \rightarrow c.[(c.C_6H_{11})_2SnI]_5$ (1614). (h) Rxn. with $(Et\overline{CHCH_2CH})_3SnCH:CH_2 \rightarrow R_3SnCH_2CH_2SnR'_3$ (1873). (i) Rxn. with $Bu_3Al \rightarrow (C_8H_{17})_3SnBu + Bu_2$-AlH; relative reactivity (1830). (j) Rxn. with $Bu_2Sn(CH_2CH:CH_2)_2 + $ cat. $R_2AlH \rightarrow$ crosslinked polymer (1617). (k) Rxn. with $(1-C_8H_{17})_2SnCl_2 + CH_2:CHR \rightarrow RCH_2CH_2(1-C_8H_{17})_2SnCl$; R = $HOCH_2CH_2OCH_2$ (1613). (l) Rxn. with $CCl_4 \rightarrow$ attempted kinetic study (537). (m) Rxn. with PhI $\rightarrow$ solvent effect on yield (537).

Table 122 Continued R_nSnH_{4-n}

Formula*	Synth. Method**	Yield	Properties	Ref.
$(p-PhC_6H_4)_2SnH_2$	II	52	Colorless plates, m. 140-41° (n, o)	393
$(1-C_{10}H_7)_2SnH_2$	II	84	Colorless cryst., m. 83°, soly. (n, o)	393
$(2-C_{10}H_7)_2SnH_2$	II	45	Colorless needles, dec. 270°, (n, p)	393
R_3SnH	II	--	IR spectrum (135) and force const. (1014) (q-s)	398
R_2SnH_2	II	--	IR spectrum (135) and force const. (1014) (q-s)	398
$RSnH_3$	II	--	IR spectrum (135) and force const. (1014) (r, s)	398

(n) IR spectrum and mol. wt. (393). (o) Rxn. with Py → H + $(R_2Sn)_6$; R = $p-PhC_6H_4$, $1-C_{10}H_7$ (388a, 393). (p) Rxn. with $HCONMe_2$ + cat. R_2SnCl_2 → $[(2-C_{10}H_7)_2Sn]_6$ (388a, 393). (q) Anal. detn. with isatin and ninhydrin (1048). (r) Use for preparation of pure tin (2987). (s) Use in catalyst for olefin polymerization (241).

3.3 ORGANOTIN DEUTERIDES

Organotin deuterides may contain deuterium linked to tin or also in the organic moiety. The only novel synthetic method for an organotin deuteride listed in Table 123 is solvolysis of Bu_2SnNa_2 with D_2O leading to Bu_2SnD_2.

Triethyltin Deuteride (M-265) $SnC_6H_{15}D$ Et_3SnD
Synth.: By rxn. of Et_3SnNa with D_2O (2521).
Rxn. with: $Et_3SnC(CO_2Me):CHD$ + AiBN → isomerization of olefin (1413).
RBr → RD; R = MeCH(CN), $PhCH_2CH_2$, $p-O_2NC_6H_4COCH_2$ (2521).
PhI → PhD (2521).
RCOCl → RCOD + RCD_2O_2CR; R = Me, Ph, $p-O_2NC_6H_4$, $p-NCC_6H_4$ (2521).
HC⋮CR → cis- and trans-$Et_3SnCR:CHD$; kinetic data and rxn. mechanism, solvent effect R = CN, CO_2Me (1412).
$(EtO_2C⋮)_2$ → kinetic data for addn. rxn. (1412).
PhNCO → $Et_3SnNPhCDO$ (1415).

Tributyltin Deuteride $SnC_{12}H_{27}D$ Bu_3SnD
Synth.: By rxn. of Bu_3SnNa with D_2O (2521).
Rxn. with: RX → RD (2105), kinetic data for t-BuBr (2306).
cis- or trans-$9-ClC_{10}H_{17}$ → cis- and trans-decalin (1143).
syn- or anti-$7-C_7H_9Br$ → 7-deuteronorbornene (2112).
1,3,5,7-Tetrabromoadamantane → 1,3,5,7-D_4-adamantane (1553a).
RCHO → RCHDOD; R = MeCH:CH, Ph (2521).
Ph_2CO + Ni → Ph_2CDOD (2521).
Bz_2 + Pt → PhCDODCOPh (2521).
Optically active BzOCHMePh (+ cat. AiBN) → MeCHDPh (racemate) (2488).

Triphenyltin Deuteride (M-265) $SnC_{18}H_{15}D$ Ph_3SnD
Synth.: By rxn. of Ph_3SnNa with D_2O (2521).
Rxn. with: $(Ph_2RP)_2Ir(CO)X$ → $Ph_3Sn(Ph_2RP)Ir(CO)XD$; R = Ph, X = Cl, Br, RX = MeCl (2530).
RX → RD (2105); RX = EtI, $BzCH_2Br$ (2521).
1,3,5,7-Tetrabromoadamantane → 1,3,5,7-D_4-adamantane (1553a).

Other organotin deuterides are listed in Table 123.

Table 123. Organotin Deuterides R_nSnD_{4-n}

Formula*	Properties	Ref.
Me$_3$SnD (265)	(a)	--
Me$_2$SnHD	NMR spectrum	2361
MeSnD$_3$	(Far) IR spectrum, assignment	1313
CD$_3$SnH$_3$	(Far) IR spectrum, assignment	1313
Bu$_2$SnD$_2$	(b)	2521
Ph$_2$SnHD	NMR spectrum	2361

* Numbers in parenthesis refer to pages in main volume.
(a) Rxn. with (Ph$_3$P)$_2$Ir(CO)Cl → Me$_3$Sn(Ph$_3$P)Ir(CO)DCl (2530). (b) Rxn. with Bz$_2$O$_2$ (2:1) → (PhCDOH)$_2$ (2521).

3.6 FUNCTIONALLY SUBSTITUTED ORGANOTIN HYDRIDES

All compounds are listed in Table 124.

Table 124. FUNCTIONALLY SUBSTITUTED ORGANOTIN HYDRIDES R'_nSnH_{4-n}

Formula*	Synth. Method**	Properties	Ref.
Me$_2$(CF$_2$:CF)SnH	IV	Mol. wt., IR and NMR spectra (a)	731
Et$_2$SnH(CH$_2$)$_5$SnEt$_2$H (266)	--	(b)	--
Et$_2$SnH(CH$_2$)$_6$SnEt$_2$H (266)	--	(c)	--
(p-FC$_6$H$_4$)$_3$SnH	--	Moessbauer resonance spectrum	222, 224
(p-MeOC$_6$H$_4$)$_3$SnH	I	m. 105-106° (d)	914
(p-EtOC$_6$H$_4$)$_2$SnH$_2$	II	Colorless plates, m. 53-54°, (e)	393

* Numbers in parenthesis refer to pages in main volume.
** The characters used here correspond to the synthetic scheme in the introduction to Chapter 3.
(a) Rxn. with C$_2$F$_4$ at 60° in the dark → Me$_2$(CF$_2$:CF)SnCF$_2$CHF$_2$ (731). (b) Rxn. with Me$_3$SnNEt$_2$ (1:2) → Me$_3$SnSnEt$_2$(CH$_2$)$_5$SnEt$_2$SnMe$_3$ (1618). (c) Rxn. with Bu$_2$Sn(CH$_2$CH:CH$_2$)$_2$ + R$_2$AlH cat. → polymer (1617). (d) Rxn. with Et$_3$SnX → (p-MeOC$_6$H$_4$)$_3$SnSnEt$_3$ + HX; relative rxn. rates, R = OPh (914), N(C$_6$H$_{13}$)CHO, NPhCHO (915). (e) IR spectrum, mol. wt. and soly. (393), rxn. with Me$_2$NCHO → c.[(p-EtOC$_6$H$_4$)$_2$Sn]$_6$ (393).

3.8 ORGANOTIN HYDRIDES CONTAINING HETEROATOM-TIN BONDS

Existence of organotin hydride halides in free form is doubtful in several cases. The compounds may exist only in equilibrium mixtures. The derivatives listed in Table 125 are prepared by methods from the following scheme. Organotin hydrides containing a tin-tin linkage are now listed in Chapter 5.4.

I Hydrogenation of organotin halides and alkoxides:
- (IA) $R_2SnX_2 + Bu_3SnH \rightarrow R_2SnHX$,
- (IB) $BuSnCl_3 + Bu_3SnH \rightarrow BuSnHCl_2$,
- (IC) $BuSnCl_3 + Bu_2SnH_nCl_{2-n} \rightarrow BuSnH_2Cl$; n = 1, 2 ,
- (ID) $Et_2Sn(OEt)_2 + (MeHSiO)_x \rightarrow Et_2Sn(OEt)H$.

II Cleavage reaction with anhydric hydrobromic acid at -78°:
- (IIA) $Ph_3SnH + {}_nHBr \rightarrow Ph_{3-n}SnHBr$; n = 1, 2 ,
- (IIB) $EtSnH_3 + HBr \rightarrow EtSnH_2Br$.

III Comproportionation reaction:
$R_2SnH_2 + R_2SnX_2 \rightleftharpoons 2R_2SnHX$.

Dibutyltin Hydride Chloride (M-267) $SnC_8H_{19}Cl$ Bu_2SnHCl
Synth.: By rxn. of Bu_2SnH_2 with Bu_2SnCl_2 at r.t. in N atm. in equilibrium rxn. (469), in 93% yield (2737).
By rxn. of Bu_2SnH_2 with gaseous HCl in dioxane (469).
By rxn. of Bu_2SnCl_2 with Bu_3SnH (1:1) in N atm. at r.t. (469).
Prop.: Dec. at 100° (469), IR (469, 1292), NMR (469, 1292, 2737) and Moessbauer resonance (1180, 1666) spectra.
Rxn. with: Thermal dec. at 100° → $(Bu_2SnCl)_2$ (469).
$Bu_3SnH \rightarrow Bu_2SnH_2 + Bu_3SnCl$ (1804).
$BuSnCl_3 \rightarrow BuSnH_2Cl$ (?) (1804).
BuCH:CHD (excess under UV irradiation) → complete isomerization of excess olefin (1369).
PhCH:CHD (excess under UV irradiation) → no isomerization of excess olefin (1369).
$CH_2:CHCMe:CH_2 + AiBN \rightarrow Bu_2(Me_2C:CHCH_2)SnCl + Bu_2(CH_2:CMeCH_2CH_2)SnCl + Bu_2(CH_2:CHCMeCH_2)SnCl + Bu_2(MeCH:CMeCH_2)SnCl$ (1619).
Diene + AiBN → 1,2- and 1,4-adduct with $CH_2:CHCH:CHMe$, $(CH_2:CMe)_2$, C_5H_6 (1619).
i-BuCl → kinetic and thermodynamic data, rate const. (2306).

Diisobutyltin Hydride Chloride $SnC_8H_{19}Cl$ $(Me_2CHCH_2)_2SnHCl$
Synth.: By rxn. of $i-Bu_2SnCl_2$ with Bu_2SnH_2 in N atm., in 92% yield (2737).
By rxn. of $i-Bu_2SnCl_2$ with Bu_3SnH (1:1) in N atm. (1805).
Prop.: IR (1805) and NMR (1805, 2737) spectra.
Rxn. with: $CH_2:CHCMe:CH_2 + AiBN \rightarrow i-Bu_2(Me_2C:CHCH_2)SnCl + i-Bu_2(CH_2:CMeCH_2CH_2)SnCl + i-Bu_2(CH_2:CHCHMeCH_2)SnCl + i-Bu_2(MeCH:CMeCH_2)SnCl$ (1619).
Diene + AiBN → 1,2- and 1,4-adduct with $CH_2:CHCH:CHMe$, $(CH_2:CMe)_2$, $c.C_5H_6$ (1619).

Additional organotin hydrides containing tin-halogen and tin-oxygen bonds are listed in Table 125.

Table 125. Organotin Hydrides Containing Heteroatom-Tin Bonds $R_nSnH_mX_o$

Formula*	Synth. Method**	Properties	Ref.
Me_2SnHCl	III	IR, NMR spectra	1292
	IA	IR, NMR spectra	1805
Me_2SnHBr	III	IR, NMR spectra	1292
Me_2SnHI	III	IR, NMR spectra	1292
Et_2SnHCl	III	IR, NMR spectra (a)	1292
	IA	IR, NMR spectra	1805
Et_2SnHBr	III	IR, NMR spectra	1292
Et_2SnHI	III	IR, NMR spectra	1292
$EtSnH_2Br$ (267)	IIB	White cryst., dec. at -65° (b)	156
Pr_2SnHCl	III	IR, NMR spectra	1292
	IA	IR, NMR spectra	1805
Pr_2SnHBr	III	IR, NMR spectra	1292
Pr_2SnHI	III	IR, NMR spectra	1292
$(Me_2CH)_2SnHCl$	III	IR, NMR spectra	1292
$(Me_2CH)_2SnHBr$	III	IR, NMR spectra	1292
$(Me_2CH)_2SnHI$	III	IR, NMR spectra	1292
Bu_2SnHF	III (c)	IR, NMR spectra, dec. 100°, sticky solid	469
	IA	IR, NMR spectra (1292)	1804
Bu_2SnHBr	III (d)	IR, NMR spectra, dec. 100° → H	469
	IA	IR (1292), NMR (1292, 2737) spectra	1804
Bu_2SnHI	III (d)	IR, NMR spectra, dec. 100° → H	469
	IA	IR (1292), NMR (1292, 2737) spectra	1804
$BuSnH_2Cl$	IC	IR, NMR spectra, very unstable	1804
$BuSnHCl_2$	IB	IR, NMR spectra, very unstable	1804
$(c.C_6H_{11})_2SnHCl$	IA	IR, NMR spectra	1805
$(n-C_8H_{17})_2SnHCl$	IA	IR, NMR spectra	1805
	III	In 94% yield, NMR spectrum	2737
Ph_2SnHCl	IA	IR, NMR spectra	1805
	III	In 92% yield, NMR spectrum	2737
Ph_2SnHBr	IIA	IR spectrum (e)	156
$PhSnHBr_2$ (267)	IIA	Stable at -78°, IR spectrum (e)	156
$(Me_2SnH)_2O$	--	(f)	--
$Bu_2Sn(OEt)H$	ID	b_6 105-11°, n_D^{25} 1.4713, d_{25} 1.1040 (g)	1165

* Numbers in parenthesis refer to pages in main volume.
** The characters used here correspond to the synthetic scheme in the introduction to Chapter 3.8.
(a) Use in catalyst for olefin polymerization (657). (b) Dec. at r.t. → (EtSnBr)$_n$ + H (156), rxn. with BuLi at -78° → EtBuSnH$_2$ (156). (c) In 70% conversion (469). (d) In 87% yield (2737). (e) Dec. at r.t. → C_6H_6 + polymer (156). (f) Use in catalyst for olefin polymerization (411). (g) Synth. in 76% yield, detn. by iodometric titration (1165).

4. ORGANOTIN COMPOUNDS CONTAINING HETERATOM-TIN BONDS

4.1 COMPOUNDS WITH HALOGEN

4.1.1 TRIORGANOTIN HALIDES

4.1.1.1 SYMMETRIC UNSUBSTITUTED TRIORGANOTIN HALIDES

The symmetric unsubstituted triorganohalides summarized in Table 130 are prepared by methods from the ensuing scheme.

I Dealkylation reactions:
- (IA) $R_4Sn + Br \rightarrow R_3SnBr$,
- (IB) $R_4Sn + HBr \rightarrow R_3SnBr$,
- (IC) $Pr_3SnCH_2Ac + R_3SiI \rightarrow Pr_3SnI$,
- (ID) $R_6Sn_2 + I \rightarrow R_3SnI$.

II Redistribution reaction at above 200°:
- $3R_4Sn + SnX_4 \rightarrow 4R_3SnX$; X = Cl, Br.

III Partial alkylation of tin or tin halides:
- (IIIA) $SnX_4 + Mg + RX \rightarrow R_3SnX$; X = Cl, Br ,
- (IIIB) $SnF_2 + R_2Hg \rightarrow R_3SnF$,
- (IIIC) $Sn + RX \rightarrow R_3SnX$.

IV Interconversion reactions:
- (IVA) $R_3SnX + KF \rightarrow R_3SnF$; X = Cl, Br ,
- (IVB) $R_3SnI + KOH, + HX \rightarrow R_3SnX$; X = F, Cl ,
- (IVC) $(R_3Sn)_2O + HI \rightarrow R_3SnI$,
- (IVD) $R_3SnOH + HX \rightarrow R_3SnX$; X = F, Br, I ,
- (IVE) $R_3SnOH + NH_4I \rightarrow R_3SnI$.

R = organic group, X = halogen. Additional references to triorganotin halides may be found in Chapters 7.3 and 8.1.

Trimethyltin Fluoride (M-269) SnC_3H_9F Me_3SnF
Synth.: By rxn. of Me_3SnCF_3 with $HC\vdots CCF_3$ (2332), with $RC\vdots CCF_3$; R = Me_3Si, Et_3Ge, Me_2As (921), with $Me_nM(C\vdots CCF_3)_{4-n}$; M = Sn, n = 2,3, Ge, n = 3 (2332), with BzOH in C_6H_6 at 150° (1874).
By thermal dec. of $Me_3SnCF_3 \cdot PF_5$ at 120° (853), of $Me_3Sn\overline{CFCHFCF_2CF_2}$ (924).
By rxn. of $Me_nSn(CF\vdots CF_2)_{4-n}$ and Me_3SnH under UV irradiation; n = 2, 3 (731).
By photolysis of $Me_3SnCF_2CF_2Mn(CO)_5$ (87, 874).
By rxn. of $Me_3SnC_6F_5$ with KF in alc. under reflux, in 31% yield (77).
By rxn. of Me_4Sn with PF_5 after hydrolysis of nonvolatile rxn. residue (567).
By rxn. of $(Me_3SnC\vdots)_2$ with Ph_2PCl without solvent, followed by aq. KF (2433).
By rxn. of $(Me_3Sn)_2C_7Cl_6$ with HCl, followed by KF (1883).
By rxn. of Me_3SnH with $Me_nGe(CF\vdots CF_2)_{4-n}$ in 40% yield; n = 2, 3 (2209), with $Me_nSi(CF\vdots CF_2)_{4-n}$; n = 2, 3 (2208), with $(PF_3)_4RuH$ (825).
By rxn. of $(^{14}CH_3)_3SnI$ with fluoride in aq. alkaline soln. (71).
By rxn. of $Me_3{}^{113}SnClO_4$ with fluoride (836, 838).

By rxn. of Me$_3$SnNMe$_2$ with (CF$_3$)$_2$CHF at 25°, in quant. yield (865), with BF$_3$·OEt$_2$ (1090).
By rxn. of Me$_6$Sn$_2$ with (CF$_3$C!)$_2$ at 100° (108), with C$_2$F$_4$, C$_3$F$_6$, C$_2$ClF$_3$ or C$_2$BrF$_3$ under UV irradiation (871).
By rxn. of Me$_3$SnLi with C$_5$F$_5$N at -78° in THF (2423).
By rxn. of Me$_3$SnMn(CO)$_5$ with C$_2$ClF$_3$ (874), with (CF$_2$:CF)$_2$ under UV irradiation (1141).
By rxn. of Me$_3$SnCo(CO)$_4$ with C$_2$F$_4$, C$_2$HF$_3$ or CF$_3$I upon heating or under UV irradiation (730).
By rxn. of SnX$_4$ with Al$_4$C$_3$ and HF at 150-200° (2114).
Prop.: IR (619), Raman (619), NMR (1430, 1436) and Moessbauer resonance (224 902, 1400. 1960, 1988, 2424, 2801, 2840) spectra, x-ray diffraction pattern, struct. (619). Anal. detn. and complex formation, stability const. (836-8), formation of [Me$_3$SnF$_n$]$^{-(n-1)}$ by anionic paper chromatography (71).
Addn. Compd.: Me$_3$SnBF$_4$ (M-269), from Me$_3$SnC$_6$F$_5$ and BF$_3$ in CCl$_4$ (77).

Trimethyltin Chloride (M-269)   SnC$_3$H$_9$Cl   Me$_3$SnCl
Synth.: By rxn. of tin with MeCl in SnCl$_2$-KCl melt at 300°, in 55% yield (2000).
By rxn. of Me$_4$Sn with RSnCl$_3$ at 0-85°, in 54-99% yield; R = n-, i-Bu, Ph (1370), with MeBuSnCl$_2$ at 75-85°, in 96% yield (1370), with EtBuSnCl$_2$ at 120-30°, in 91% yield (1370), with SnCl$_4$ study of effect of temp. by NMR spectroscopy (725), at 25-78°, in high yield, effect of hexane as solvent (2524), with AcCl and AlCl$_3$ (1:1:1) at r.t., in 70% yield (1781).
By thermal dec. of Me$_3$SnC$_7$Cl$_6$Me, in 51% yield (1883), of Me$_3$SnC$_6$Cl$_5$ at 300°, in 87% yield (865).
By rxn. of (Me$_3$Sn)$_2$C$_7$Cl$_6$ or Me$_3$SnC$_7$Cl$_6$Me with Cl (1883).
By rxn. of Me$_3$SnC!CPh with Ph$_3$SnCl in C$_6$H$_5$ at 100° (2433).
By rxn. of (Me$_3$SnC!)$_2$ with Ph$_2$PCl in C$_6$H$_6$ or HCCl$_3$ (2433), at r.t., in 85% yield (2778), with Ph$_3$SnCl in C$_6$H$_6$ or HCCl$_3$ (2433).
By thermal dec. of Me$_3$SnCH$_2$CHCHRCCl$_2$ at 145° or at 70-125° in presence of cat. ZnCl$_2$; R = H, Me (2755).
By rxn. of Me$_3$SnAr with HCl in MeOH (62), under anhydrous cond., Ar = C$_6$F$_5$ (77).
By rxn. of Me$_3$SnNMe$_2$ with HCl in Et$_2$O, in 90% yield (250).
By rxn. of Me$_3$SnNR$_2$ with Cl$_3$CCHO in refluxing c.C$_6$H$_{10}$; R = Me and Et, in 60 and 78% yield, resp. (2454).
By rxn. of Me$_3$SnCHN$_2$ with HCl (1393).
(Me$_3$Sn)$_3$Y or (Me$_3$Sn)$_2$YPh with Ph$_2$YCl in Et$_2$O in Ar atm., in high yield; Y = P, As (1846).
By thermal dec. of Me$_3$SnOCH(SMe)CCl$_3$ above 80° (2456), of (Me$_2$SnCl)$_2$O at 220-300°, in 71% yield (2611).
By rxn. of (Me$_3$Sn)$_2$S with YCl$_3$; Y = B, P, As, Sb (647).
By rxn. of Me$_3$SnSR with YCl$_3$; Y = B, P, As, Sb (647), with PCl$_3$; R = Ph (2), with Cl$_3$CCHO (1:3) > 80°, R = Me and Et, in 80 and 91% yield, resp. (2456).
By rxn. of Me$_3$SnM(CO)$_n$ with RCl in THF or (MeOCH$_2$)$_2$ at -40 to +50°; Mn = Mn$_5$, Co$_4$; R = Ph$_2$P, (Ph$_3$P)Au (2202).
By-prod. in rxn. of Me$_3$SnCXCl$_2$ with olefin; X = Cl, Br (1875).

By-prod. in thermal dec. of o-Me$_3$SnC$_6$HCl$_4$ at 230° (1883).
By-prod. in rxn. of Me$_3$Sn(Me$_3$Si)C:PMe$_3$ in EtOH with anhydr. HCl, in 25% yield (1475).
By-prod. in rxn. of Me$_4$Sn with Et$_3$SnCl (2012).
By-prod. in rxn. of Me$_3$SnO$_2$CCCl$_3$ with cyclooctene (1871).
By-prod. in rxn. of Me$_2$Sn(OC$_7$H$_5$)$_2$ with Me$_2$SnCl$_2$ in refluxing MeCN (2510).
By-prod. in rxn. of Me$_6$Sn$_2$ with Me$_2$SnCl$_2$ (2012), with C$_2$ClF$_3$ under UV irradiation (871).
By rxn. of SnCl$_4$ with Al$_4$C$_3$ and HCl at 150-200° (2114).
Formation of chlorocomplexes from Me$_3$SnClO$_4$ (838, 840).

Prop.: b. 163-65° (2000), IR (135, 796), far IR (534), assignment (2322), NMR (31, 55, 581, 724, 796, 1220, 1430, 1436, 1478, 2071) and Moessbauer resonance (211, 224, 902, 1400, 1583, 1960, 1989, 2347, 2424, 2840) spectra, dipole moment (325, 724, 1211), polarography (1249), thermodynamic data (380, 2856). Anal. detn. by GLC (1370, 2278), by electrometric titration, isohydric point and hydroxo complex formation (2232), by potentiometric titration (1701).

Rxn. with: X-ray irradiation → radiolysis followed by ESR spectroscopy (2443).
(p-ClC$_6$H$_4$)$_2$SnMe$_2$ + Na in MePH → p-(Me$_3$Sn)$_2$C$_6$H$_4$ + some (p-Me$_3$SnC$_6$H$_4$)$_2$SnMe$_2$ (1410).
Ph$_3$SnH + Et$_3$N (1:1:1) → Me$_3$SnSnPh$_3$ + Et$_3$N·HCl (1618).
Me$_3$MNHP(CMe$_3$), + BuLi → Me$_3$SnP(CMe$_3$)$_3$:NMMe$_3$; M = Si, Ge (2739).
Me$_3$SnX → rate of halogen exchange, X = Br, I (727).
MeBuSnCl$_2$ ⇌ Me$_2$BuSnCl + Me$_2$SnCl$_2$ (1370).
RSnCl$_3$ → Me$_2$SnCl$_2$ + MeRSnCl$_2$; R = Bu, MeEtCH (1370).
MeSnCl$_3$ → Me$_2$SnCl$_2$; influence of temp. on prod. distribution (725).
Me$_6$Sn$_2$ → Me$_4$Sn + (Me$_2$Sn)$_x$; kinetic data (2012).
Me$_3$SiCH:PMe$_3$ → Me$_3$Sn(Me$_3$Si)C:PMe$_3$ (1475, 1826).
Me$_3$SiCCl$_2$Li → Me$_3$SnCCl$_2$SiMe$_3$ + (Me$_3$Si)$_2$CCl$_2$ + Me$_3$SnCHCl$_2$ (1879).
Me$_3$SiSCHMe$_2$ → Me$_3$SnSCHMe$_2$ + Me$_3$SiCl (643).
Me$_3$SiNHNa → (Me$_3$Sn)$_2$NSiMe$_3$ (1814).
H$_3$SiK in monoglyme → (Me$_3$Sn)$_4$Si + (Me$_3$Sn)$_3$SiH + Me$_3$SnH + SiH$_4$ + Si$_2$H$_6$ (2224).
H$_3$SiK in diglyme → Me$_3$SnSiH$_3$ + Me$_3$SnH + SiH$_4$ + Si$_2$H$_6$ (2224).
RLi → Me$_3$SnR; R = aryl (296, 998), subst. aryl (998), C$_6$Cl$_5$ (865, 2764), CCl$_3$ (1875), CHCl$_2$, MeCCl$_2$ (1879), ClC⋮C (1980).
RLi → Me$_3$SnR; R = RR'N (251, 324, 1729, 2148, 2705), MeN:CPhNMe (2738), t-Bu$_2$PNH (2739), R'$_3$P:N (1824), Me$_3$SiNMe (472, 1812), Me$_3$Si(Me$_3$C)N (2749), Me$_3$SiNMeSiMe$_2$NMe (1811), Me$_3$SiNMeSiMe$_2$N(GeMe$_3$) (1810), Me$_3$SiN:CPhNMe (2738).
RLi → Me$_3$SnR + Me$_2$SnR$_2$; R = p-PhC$_6$H$_4$ (112), o-BrC$_6$F$_4$ (851), Ph$_2$C:N (852), (Me$_3$As)Mn(CO)$_4$ (185).
Ph$_2$PC⋮CLi → Me$_3$SnC⋮CPPh$_2$ + (Me$_3$SnC⋮)$_2$ (2778).
RMeNLi, + PhCN → Me$_3$SnN:CPhNMeR; R = Me, Me$_3$Si (2738).
YLi$_2$ → (Me$_3$Sn)$_2$Y; Y = (p-C$_6$H$_4$)$_2$ (112), p-C$_6$Cl$_4$ (2764), MeN, EtN (251), B$_{10}$H$_{10}$C$_2$ (2279).
SbLi$_3$ → (Me$_3$Sn)$_3$Sb (676).
Na in Me$_4$Sn → Me$_6$Sn$_2$ (1548a).
HC⋮CNa → (Me$_3$SnC⋮)$_2$ + Me$_3$SnC⋮CH (2181).
NaX → Me$_3$SnX; X = Mn(CO)$_5$ (1252), Cr(CO)$_5$CN, Mo(CO)$_5$CN, W(CO)$_5$CN (1314),

$C_5H_5Mo(CO)_3$, $C_5H_5W(CO)_3$ (1670), $C_5H_5Mo(CO)_2P(OPh)_3$ (2491), $C_5H_5Fe(CO)_2$ (2490), $Co(CO)_4$ (730), $(Ph_3P)_2Rh(CO)_2$, $Ph_3PIr(CO)_3$ (891).

NaX → Me_3SnX; X = PhO (471), $8\text{-}OC_9H_6N$ (1291), MeSe (651), PhSe (643).

$NaBH_4$ in diglyme → $Me_3SnH = B_2H_6 + NaCl$ (736).

$NaBPh_4$ + L → $[Me_3SnL_2]BPh_4$ (938).

$Na_2Ru(CO)_4$ → $(Me_3Sn)_2Ru(CO)_4$ (905).

$Na_2B_{10}H_{13}CN$, + H_2O_2 → $B_{10}H_{12}CNH_3$ (2745).

RK → Me_3SnR; R = Me_2CH (150), H_3Si (2224).

KX → Me_3SnX; X = Me_3CO (2707), NCS (72).

$M[Me_3SiRu(CO)_4]$ → $Me_3Sn(Me_3Si)Ru(CO)_4$ (905).

$M_2Fe(CO)_4$ → $(Me_3Sn)_2Fe(CO)_4$ (1268).

AgX → Me_3SnX; X = O_2NNMe (611), NCO (2027).

$CF_3C{\vdots}CMgI$ → $Me_3SnC{\vdots}CCF_3$ (2332).

RMgBr → Me_3SnR; R = aryl (998), subst. aryl (998, 1410), C_6F_5 (428), alkyl (1369), cycloalkyl (1873), cycloalkenyl (1041, 1043, 2052), $CH_2{:}CHCH_2CH_2$ (1041, 1043), $PhO(MeO)CH$ (1303), Me_2HSiCH_2 (457).

RMgCl → Me_3SnR; R = Me_2EtC (1953), alkenyl (1367), C_6Cl_5 (865).

$PhHgCCl_2Br$ → Me_3SnCCl_3 + Me_3SnCCl_2Br (1875).

$(Et_2AlCl)_2$ → Me_3SnEt + $(EtAlCl_2)_2$ (1206).

$Pb(SMe)_2$ → Me_3SnSMe (643).

Zn-Cn in THF → Me_4Sn + Sn + $ZnCl_2$ (2830).

Zn-Cn + $(Me_2NCH_2)_2$ or BiPy → Me_4Sn + Me_6Sn_2 (2830).

Zn-Cn + MeOH → CH_4 + Sn + $ZnCl_2$ (2830).

$Co[Co(CO)_4]_2$ → $Me_3SnCo(CO)_4$ (1672).

HCl concn. → Me_2SnCl_2 (31).

HN_3 → Me_3SnN_3 (556-7).

$p\text{-}BrC_6H_4CH{:}CH_2$ → $p\text{-}Me_3SnC_6H_4CH{:}CH_2$ (sic) (1569).

$(FSO_2)_2NCl$ → $Me_3SnN(SO_2F)_2$ + Cl (1776).

SO_3 in CH_2Cl_2 at $-78°$ → $Me_2Sn(O_3SMe)_2$ (2742).

$PhPH_2$ + Et_3N → $(Me_3Sn)_2PPh$ (1845).

$Ph_3P{:}CH_2$ → $(Me_3Sn)_2C{:}PPh_3$ (1826).

Me_2AsH + aq. NaOH → $Me_3SnAsMe_2$ (2201).

<u>Addn. compd.</u>: $Me_3SnCl \cdot Py$ (M-270), IR (757, 757b), far IR, assignment (2322), NMR (726, 757, 757b, 1220) and Moessbauer resonance (224, 902, 1742, 1991, 2801) spectra, thermodynamic data (757, 757b) [$SnC_6H_{14}ClN$].

$Me_3SnCl \cdot BiPy$, white needles, m. 70° (147), m. 69-70° (879), dissocn. in all solvents (879), instable (147), far IR spectrum (879), assignment (2322), conductivity (147) [$SnC_{13}H_{17}ClN_2$].

$Me_3SnCl \cdot ONC_5H_5$, from Me_3SnCl and pyridine oxide in C_6H_6, m. 86-87° (dec.), (far) IR spectrum and assignment, stability const. (1296) [$SnC_8H_{14}ClNO$].

Additional addition compounds of Me_3SnCl are listed in Table 126.

<u>Biol. prop.</u>: Activity as insecticide against Musca domestica, LD_{50} and ATP inhibition (425), toxicity against insecticide resistant Musca d. (1091).

<u>Use:</u> In catalyst for olefin polymerization (32).

As catalyst for making polyurethan (229).

Table 126. Addition Compounds of Trimethyltin Chloride $Me_3SnCl \cdot L$

Ligand*	Synth. Method**	Properties	Ref.
2 NH_2 (270)	IA	IR spectrum	86
NH_3	IA	IR spectrum	86
Me_2NH	II	NMR spectrum	865
Et_4NCl	--	Moessbauer resonance spectrum	2347
2,6-$Me_2C_5H_3N$	--	NMR spectrum	757, 757c
o-phen	--	Moessbauer resonance spectrum	1583
4-azaphen (a)	--	Moessbauer resonance spectrum	1583
p-ONC_5H_4R (b)			
R = Me	IB	UV spectrum, stability const.	1296
Et	IB	Stability const.	1296
Cl	IB	Stability const.	1296
$PhCH_2O$	IB	Stability const.	1296
EtO_2C	IB	Stability const.	1296
O_2N	IB	Stability const.	1296
MeCN	--	NMR spectrum (c)	757, 757b
Me_2NCN	--	NMR spectrum, thermodynamic data	757c
$MeCONMe_2$ (270)	--	IR, NMR spectra (c-e)	757, 757b
$MeCSNMe_2$	--	NMR spectrum (c)	757
Bu_3P	--	NMR spectrum	757, 757c
$(Me_2N)_3PO$	--	IR, NMR spectra (c, d)	757, 757b
$(HPrN)_3PO$	IC (f)	m. 123-26° (g)	1838a
$(Me_2N)_3PS$	--	NMR spectrum	757, 757c
Et_2O	--	NMR spectrum (h)	757, 757c
$(CH_2)_4O$	--	NMR spectrum (h)	757, 757c
Me_2CO	--	IR, NMR spectra (d, h)	757, 757b
Et_2S	--	NMR spectrum (h)	757, 757c
Me_2SO	I (i)	IR spectrum	1388
	--	IR, NMR spectra (c, d)	757, 757b

* Numbers in parenthesis refer to pages in main volume.
** Synthetic methods comprise reaction I from components - (IA) under exclusion of moisture - (IB) in C_6H_6 - (IC) in alcohol. II By reaction of Me_3SnNMe_2 with C_2HCl_5 at 0°
(a) 4-Azaphenantkrene. (b) Derivatives of N-pyridine oxide. (c) Thermodynamic data (757b). (d) Thermodynamic data (757). (e) Thermodynamic data (757a). (f) In 72% yield (1838a). (g) Use as bactericide, herbicide, insecticide, acaricide (1838a). (h) Thermodynamic data (757c). (i) In 94% yield (1388).

Trimethyltin Bromide (M-270) SnC_3H_9Br Me_3SnBr

<u>Synth.</u>: By rxn. of tin foil with $Br(CH_2)_3CO_2Me$ at 160° as by-prod. (2650).
By rxn. of Me_3SnCCl_2Br with olefin as by-prod. (1875).
By rxn. of Me_4Sn with Me_2SnBr_2 or $SnBr_4$ (3:1); influence of temp. on prod. distribution by NMR spectroscopy (725).
By rxn. of Br with $o\text{-}(Me_3Sn)_2C_6H_4$ in $HCCl_3$ or Et_2O-MeOH (1021), with Me_3SnCCl_3 in CCl_4 (948), with Me_3SnCHN_2 (1393).
By rxn. of Br with Me_3SnSPh in CCl_4 (2), with $Me_3SnMn(CO)_5$ in CCl_4 at 0° (2202), with $(Me_3Sn)_3SnLi \cdot 3THF$ in THF at -78° (2127).
By thermal dec. of $1,2\text{-}Bu_3SnOC_6H_{10}Br$, $1,1,2\text{-}Bu_3SnOC_6H_9MeBr$ and $1,2\text{-}(Bu_3SnO)\text{-}C_5H_8Br$, from cis- and trans-forms at 180-200° and 80-100°, resp. (2348).

<u>Prop.</u>: IR (1013), far IR and assignment (2322), NMR (724, 1012, 1098, 1220, 1430, 1436, 2071) ^{13}C NMR (1498), Moessbauer resonance (902, 1989, 2290, 2424) and mass (621, 2869) spectra, dipole moment (724) polarography (2352) and thermodynamic data (305, 2856).

<u>Rxn. with:</u> $Et_4Sn \rightarrow$ complete random redistribution without cat. (432).
$Me_3SnX \rightarrow$ rate of halogen exchange by NMR spectroscopy; X = Cl, I (727).
$MeSnBr_3$ (at 100°) $\rightarrow Me_2SnBr_2$, effect of temp. on prod. distribution (725).
RLi $\rightarrow Me_3SnR$; R = $3,5\text{-}Br_2C_6H_4$ (998), $p\text{-}Me_2NC_6H_4$ (1210), B_5H_8 (2398).
$MePH_2$ + LiPh $\rightarrow (Me_3Sn)_2PMe$ (57, 703).
NaR $\rightarrow Me_3SnR$; R = MeC⋮C (443), $C_5H_5Cr(CO)_3$ (2303), $Mn(CO)_5$ (874), CF_3CO_2 (2756).
$NaBPh_4$ + L $\rightarrow (Me_3SnL_2)BPh_4$; L = $HCONMe_2$, $AcNMe_2$, Me_2SO (938).
AgX $\rightarrow Me_3SnX$; X = NO_3 (86), $o\text{-}BrC_6H_4CO_2$ (1880).
$Ag_2Y \rightarrow (Me_3Sn)_2Y$; Y = SO_4, CrO_4 (85).
RMgCl $\rightarrow Me_3SnR$; R = CH_2:CPh (16), HC⋮CCH_2 (444).
RMgBr $\rightarrow Me_3SnR$; R = CH_2:$CHCH_2CH_2$ (1877), HC⋮C, PhC⋮C, c.$1\text{-}C_6H_9$C⋮C (443), $p\text{-}FC_6H_4$ (1210), C_6F_5 (77), m-, $p\text{-}Me_3SiC$⋮CC_6H_4 (1001).
$PhHgCCl_2Br \rightarrow Me_3SnCCl_2Br + PhHgBr + C_2Cl_4$ (1875).
Br in PhCl $\rightarrow$ relative kinetic data for cleavage rxn. in presence of Me_4Sn or $i\text{-}Pr_4Sn$ (767), of Et_3SnPr and $i\text{-}PrSnEt_3$ (767, 769).
Br in AcOH $\rightarrow$ discussion of rxn. mechanism (653).
PhNCO + ureas $\rightarrow$ cat. effect, kinetic data (133).

<u>Addn. compd.</u>: $Me_3SnBr \cdot 2NH_3$, from components in C_6H_6, subl. in vacuo at 25°, IR spectrum and assignment, x-ray diffraction pattern (86), thermal dec. at 100° $\rightarrow Me_3SnBr \cdot NH_3$ (86) [$SnC_3H_{12}BrN_2$].
$Me_3SnBr \cdot NH_3$, s. ab., IR spectrum, x-ray diffraction pattern (86) [$SnC_3H_{12}BrN$].
$Me_3SnBr \cdot Py$ (M-271), from components in petr. ether (86), IR (86), far IR, assignment (2322), NMR (443, 726) and Moessbauer resonance (1583) spectra [$SnC_8H_{14}BrN$].
$Me_3SnBr \cdot BiPy$, needles, m. 65°, instable, conductivity (147), far IR spectrum and assignment (2322) [$SnC_{13}H_{17}BrN_2$].

Other addition compounds with Me_3SnBr are listed in Table 127.

<u>Use:</u> In catalyst for polymerization of olefins (184).

Table 127. Addition Compounds of Trimethyltin Bromide $Me_3SnBr \cdot L$

Ligand	Synth. Method	Properties	Ref.
$PhNH_2$	--	NMR spectrum	1220
Quinoline	--	Moessbauer resonance spectrum	1583
Isoquinoline	--	Moessbauer resonance spectrum	1583
Acridine	--	Moessbauer resonance spectrum	1583
1-Azaphen (a)	--	Moessbauer resonance spectrum	1583
4-Azaphen (a)	--	Moessbauer resonance spectrum	1583
9-Azaphen (a)	--	Moessbauer resonance spectrum	1583
C_5H_5NO (b)	(c)	Dec. 86-88°, NMR spectrum, struct. (d, e)	2340
$4\text{-}MeC_5H_4NO$ (b)	(c)	Dec. 92-94°, IR and NMR spectra	2494
$2,4,6\text{-}Me_3C_5H_2NO$ (b)	(c)	Dec. 92°, IR and NMR spectra (f)	2494
$4\text{-}ClC_5H_4NO$ (b)	(c)	Dec. 91-95°, IR and NMR spectra (f)	2494
$4\text{-}MeOC_5H_4NO$ (b)	(c)	Dec. 89-92°, IR and NMR spectra (f)	2494
Ph_3PO	(c)	Dec. 150-52°, NMR spectrum, struct. (d)	2340
$(CH_2(CH_2)_3N)_3PO$	(g)	m. 106-108° (h)	1838a
Ph_3AsO	(c)	Dec. 156-58°, NMR spectrum, struct. (d)	2340
Me_2SO	--	Far IR and NMR spectra, struct.	2340

(a) x-Azaphenanthrene. (b) N-Pyridine oxide. (c) From components in $HCCl_3$ or H_2CCl_2. (d) (Far) IR spectra and assignment (2340). (e) IR and NMR spectra (2494). (f) White cryst., nonhydroscopic (2494). (g) From components in alc., in 87% yield (1838a). (h) Use as acaricide, bactericide, herbicide, insecticide (1838a).

Trimethyltin Iodide (M-271) SnC_3H_9I Me_3SnI

Synth.: By rxn. of SnI_4 with $^{14}CH_3MgI$ (71).
By rxn. of o-$Me_3SnC_6H_4I$ at 235°, in 28% yield (1021).
By rxn. of I with p-$(Me_3Sn)_2C_6H_4$ or p-$Me_3SnC_6H_4SiMe_3$ in MeOH (60), with $Me_3SnCo(CO)_4$ in hexane at -10° (2202).
By rxn. of Me_4Sn with Me_2SnI_2 at 120° (725).
By rxn. of NaI with olefin and Me_3SnCF_3, in 90% yield (494, 1874), with $Me_3SnC_6F_5$ or o-$Me_3SnC_6F_4Br$ (851).
By rxn. of Me_6Sn_2 with CF_3COI (922).
By rxn. of c.$(Me_2Sn)_6$ with CF_3I under UV irradiation (871).
By rxn. of $Me_3SnCo(CO)_4$ with Me_2AsI in THF, at 0° (2202), with CF_3I under heating or UV irradiation (730).

Prop.: Far IR assignment (2322), NMR (724, 727, 1430, 1436), Moessbauer resonance (902, 2424) and mass (621, 2869) spectra, dipole moment (724) and thermodynamic data (305).

Rxn. with: Me_3SnX → rate of halogen exchange; X = Cl, Br (727).
$MeSnI_3$ (at 140°) → Me_2SnI_2 (725).
F^- (in aq. alc. soln.) → $(^{14}CH_3)_3SnF$ (71).

Addn. compd.: $Me_3SnI\cdot Py$ (M-271), far IR, assignment (2322) and NMR (726) spectra, struct. (726) [$SnC_8H_{14}IN$].
$Me_3SnI\cdot BiPy$, white needles, m. 71° (147), far IR spectrum and assignment (2322), very instable, conductivity (147) [$SnC_{13}H_{17}IN_2$].
$Me_3SnI\cdot MeCONMe_2$, NMR spectrum and thermodynamic data (757, 757c).

Tribenzyltin Chloride (M-271) $SnC_{21}H_{21}Cl$ $(PhCH_2)_3SnCl$

Synth.: By rxn. of powd. tin with $PhCH_2Cl$ and cat. $HgCl_2$; influence of solvent polarity and time, rxn. mechanism, in 10-53% yield (1940).
By rxn. of $SnCl_4$ with $PhCH_2Cl$ and Mg in Et_2O-MePh, in 20% yield (841).
By rxn. of $(PhCH_2)_2SnCl_2$ and $[(PhCH_2)_2SnCl]_2O$ with tin under reflux in 92 and 82% yield, resp. (1940).

Prop.: m. 108° (841), (far) IR, assignment (841), UV (203) and NMR (2069) spectra.

Rxn. with: $LiAlH_4$ → $(PhCH_2)_3SnH$ (2069).
NaX → $(PhCH_2)_3SnX$; X = I in Me_2CO (841), N_3 in Et_2O (1965).
$Na_2Ru(CO)_4$ → $[(PhCH_2)_3Sn]_2Ru(CO)_4$ (905).
9,10-$Na_2C_{14}H_{10}$ → 9,10-$(Ph_3Sn)_2C_{14}H_{10}$ (anthranyl) (1726).
KOH, + HI → $(PhCH_2)_3SnI$ (1967).
KSCN in PrOH → $(PhCH_2)_3SnSCN$ (841).
$(CH_2O_2CCH_2SH)_2$ → $(PhCH_2)_3SnSCH_2CO_2CH_2CH_2O_2CCH_2SH$ (2903).
$C_2N_2S(SH)_2$ + alc. NaOH → $[(PhCH_2)_3SnS]_2C_2N_2S$, 2.5-dimercapto-1,3,4-thiadiazole (1262).

Addn. compd.: $(PhCH_2)_3SnCl\cdot OP(\overline{NCH_2CH_2OCH_2CH_2})_3$, from components in ROH, in 75% yield, m. 119-20°, use as acaricide, bactericide, insecticide, herbicide (1838a) [$SnC_{25}H_{29}ClN_3O_4P$].

Use: Effect on degradation of PVC (371).

Tribenzyltin Iodide SnC$_{21}$H$_{21}$I (PhCH$_2$)$_3$SnI

<u>Synth.</u>: By rxn. of (PhCH$_2$)$_3$SnCl with NaI in Me$_2$CO, in 45% yield (841), with 30% aq. KOH, followed by HI, in 58% yield (1967).

<u>Prop.</u>: m. 101° (1967), m. 98-100°, (far) IR spectra and assignment (841).

<u>Rxn. with:</u> OH$^-$ → [(PhCH$_2$)$_3$Sn]$_2$O (1964).

AgX in C$_6$H$_6$ → (PhCH$_2$)$_3$SnX + AgI; X = NO$_3$ (1964), AcO (2793).

Pb(NCO)$_2$ in C$_6$H$_6$ → (PhCH$_2$)$_3$SnNCO (1967).

Triethyltin Fluoride (M-272) SnC$_6$H$_{15}$F Et$_3$SnF

<u>Synth.</u>: By rxn. of Et$_4$Sn with PF$_5$ as by-prod. (567).

By thermal dec. of Et$_3$SnCFCHFCF$_2$CF$_2$ (924).

<u>Prop.</u>: (Far) IR, assignment (321), NMR (1430) and mass (1648, 2410) spectra, polarographic reduction (972).

<u>Rxn. with:</u> PhNCO + ureas → cat. effect of, kinetic data (133).

<u>Biol. prop.</u>: Activity as bactericide against Staphylococcus aureus and Escherichia coli (702).

Triethyltin Chloride (M-272) SnC$_6$H$_{15}$Cl Et$_3$SnCl

<u>Synth.</u>: By rxn. of Et$_4$Sn with SnCl$_4$ and cat. AlCl$_3$, in 89% yield (705), with RSnCl$_3$; R = Bu at 25-100°, in 83% yield (1370), Ph at 20° (1370), with HGeCl$_3$·OEt$_2$ as by-prod. (336).

By rxn. of Et$_3$SnR with gaseous HCl; R = HC⋮C (in MeOH in presence of PhNCO) (2644), CH$_2$:CHOCH$_2$CH$_2$OCH$_2$C⋮C in dioxane (1907), OCNPhCONPhC:CPh in MeOH (2644), Et$_3$SnOCH$_2$CH$_2$C⋮C (508), Et$_3$SnOCMe$_2$C⋮C at 0 and 50°, in 49 and 90% yield, resp. (1910).

By rxn. of MeOCH$_2$Cl with Et$_3$SnC⋮CYOSnEt$_3$ at 90°, Y = CH$_2$CH$_2$, in 87% yield (1922), CMe$_2$, in 95% yield (1910), Y = CH$_2$, CH$_2$CH$_2$ in 87-100% yield (1536).

By rxn. of Et$_3$SnR with AcCl; R = AcCH$_2$, in 94% yield (437), Et$_3$SnOYC⋮C with Y = CH$_2$, CH$_2$CH$_2$, CHMe, CMe$_2$, in > 95% yield (1910).

By rxn. of Et$_3$SnR with BzCl; R = PhC⋮C, in 75% yield (2644), CH$_2$:CHCH$_2$ (1331), Et$_3$SnOCMe$_2$C⋮C (1910).

By rxn. of Et$_4$Sn with CCl$_4$ and air at 100°, in 67% yield (2094).

By rxn. of Et$_3$SnR with C$_5$Cl$_6$; R = Et$_3$SnC⋮C (2507), BuSCH:CHC⋮C (2772).

By rxn. of Et$_3$SnC⋮CYOSnEt$_3$ with Me$_3$SiCl at 0-60°, in > 90% yield; Y = CH$_2$ (1536, 1910, 1913), CH$_2$CH$_2$ (1536, 1910), CHMe, CH$_2$CH$_2$ (1910).

By rxn. of Et$_3$SnR with MCl$_4$; R, M = HC⋮C, MeC⋮C, Ge, Si (2881), CH$_2$:CHC⋮C, Ge (1891), RO$_2$CCH$_2$, Ge, in 98% yield (692), RO$_2$CCH$_2$ with RGeCl$_3$ in 98% yield (692).

By rxn. of Et$_3$SnR with MCl$_3$; R, M = MeO$_2$CCH$_2$, Sb, in 80% yield (2246), RO$_2$CCH$_2$, P, in 98% yield; also RO$_2$CCH$_2$ with EtOPCl$_2$ (440).

By rxn. of Et$_3$SnH with Ph$_3$CCl, in 70% yield (585), with CF$_2$(CF$_2$)$_n$CCl:CCl, in quant. yield (924).

By rxn. of Et$_3$SnN$_3$ or Et$_3$SnN:PPh$_3$ with HCl in dry Et$_2$O (1429).

By rxn. of Et$_3$SnN:C:C(CN)CHRR' with AcCl or BzCl (1954, 2789), with Ph$_3$CCl (2789).

By rxn. of Et$_3$SnN(NHPh)S(O$_2$)P with AcCl at 60°, in C$_6$H$_6$ (2646).

By rxn. of (Et$_3$Sn)$_2$O with excess CCl$_4$ and pptd. Cu, in 83% yield (1737), with EtMgBr as by-prod. (1906).

By rxn. of Et$_3$SnOR with gaseous HCl; R = HC⋮CCH$_2$OCH(CCl$_3$), in 95% yield (1910), PhC⋮CCH(CCl$_3$) in Et$_2$O (1537), AcCH$_2$CHPh in alc., in 80% yield (437), HC⋮CCMe$_2$ at r.t., in 95% yield (1910).
By rxn. of Et$_3$SnOYC⋮CH with MeOCH$_2$Cl; Y = CH$_2$ (1536, 1910), CH$_2$CH$_2$ (1536), CHMe (1910, 1922), CMe$_2$ (1910).
By rxn. of Et$_3$SnOYC⋮CH with Me$_3$SiCl in high yield; Y - CH$_2$ (1536, 1913), CH$_2$CH$_2$ (1536, 1910), CHMe, CMe$_2$ (1910).
By rxn. of (Et$_3$Sn)$_2$S with excess CCl$_4$ and pptd. Cu in N atm. at r.t., in 70-90% yield (1737).
By rxn. of (Et$_3$Sn)$_2$Te with Cl in C$_6$H$_6$ in inert atm., in 52% yield (2851).
By rxn. of Et$_2$SnCl$_2$ with powd. Sn in H$_2$O at 160° in autoclave, in 76% yield (1939), with Fe and BuCl-NEt$_3$ at 160°, in 36% yield (1939).
By rxn. of (Et$_3$SnCl)$_2$O with powd. Sn in H$_2$O at 160°, in 79% yield (1939).
By rxn. of Bu$_2$SnCl$_2$ with Fe in EtCl-Et$_3$N, in 2% yield (1939).
By rxn. of Et$_6$Sn$_2$ with CCl$_4$ and pptd. Cu, in 22% yield (1737), with excess CuCl$_2$ quant. (1737), with CF$_3$COCl under UV irradiation (922).
By rxn. of (Et$_2$Sn)$_x$ with BuCl and Et$_4$NI at 160°, in 25% yield (1941).
By-prod. in rxn. of powd. Sn with EtCl and Et$_4$NI at 160°, in 8% yield (1943).
Prop.: Far IR (543), assignment (321), NMR (573, 1220, 1430, 2068, 2071) and mass (2410) spectra, GLC (1370, 1939, 1941), dipole moment (325, 1211), magnetic susceptibility (1), polarography (972) and thermodynamic data (305, 380).
Rxn. with: Me$_4$Sn ⇌ MeSnEt$_3$ + Me$_3$SnCl (2012).
Prod. of rxn. Et$_3$SnC⋮CH + NaC$_2$H → (Et$_3$SnC⋮)$_2$ (2181).
Ph$_3$SnH in Et$_3$N → Et$_3$SnSnPh$_3$ (400).
(Bu$_2$SnO)$_x$ → Et$_3$SnOSnBu$_2$Cl (944).
Me$_6$Sn$_2$ → MeSnEt$_3$ + Me$_4$Sn + (Me$_2$Sn)$_x$ (2012).
(Et$_3$Ge)$_2$Cd → Et$_6$Sn$_2$ + Et$_3$GeCl + Cd (587).
Et$_3$GeSLi → Et$_3$SnSeSnEt$_3$ (2090).
Bu$_3$PbH → Et$_3$SnH + Bu$_3$PbCl (397).
Li in THF → Et$_3$SnLi (1928, 2179).
RLi → Et$_3$SnR; R = PrCH:C:CEt, PrCH:C:CPr, BuCH:C:CEt (2315), 1-C$_6$H$_{10}$CN (1620), p-FClC:CFC$_6$H$_4$ (527), Et$_2$N (251, 323), Me$_2$CCN (1620).
LiAl(C⋮CPh)$_4$ → Et$_3$SnC⋮CPh (1076).
LiAl(C$_3$F$_7$)$_2$I$_2$ → C$_3$F$_7$-contn. tin compd. (977).
PhLi + MePH$_2$ → (Et$_3$Sn)$_2$PMe (57, 703).
Na + BuCl → Et$_3$SnBu + Et$_4$Sn + Et$_6$Sn$_2$ (1939).
NaR → Et$_3$SnR; R = Me$_2$NCH:CHC⋮C, Et$_2$NCH:CHC⋮C (2687), BrC⋮C, ClC⋮C (2180), CH$_2$:CH(OCH$_2$CH$_2$)$_n$OCH$_2$C⋮C; n = 1, 2 (1907).
NaC$_2$H → Et$_3$SnC⋮CH + (Et$_3$SnC⋮)$_2$ (2181).
NaAl(C⋮CBu)$_4$ → Et$_3$SnC⋮CBu (1076).
NaH + BPh$_3$ → Et$_3$SnH (2242).
NaX → Et$_3$SnX; X = Mn(CO)$_5$, C$_5$H$_5$Fe(CO)$_2$ (2057), N$_3$ (2115), Et$_2$NCS$_2$ (761).
Na$_2$Ru(CO)$_4$ → (Et$_3$Sn)$_2$Ru(CO)$_4$ (905).
NaOR → Et$_3$SnOR; R = Et, (350), HC⋮CCH$_2$CH$_2$ (508), p-O$_2$NC$_6$H$_4$ (791), 8-C$_9$H$_6$N (1291), CH$_2$:CH(OCH$_2$CH$_2$)n ; n = 1-3 (1914), $\overline{CH_2OCMeR'OCHCH_2}$; R ' = Me, Et, Pr (690).
PhC⋮CH + MOH → Et$_3$SnC⋮CPh; M = Na, K (511).
RMgCl → Et$_3$SnR, R = EtOCH$_2$ (307), CH$_2$:CHCH$_2$, MeCH:CHCH$_2$ (1367), CH$_2$:CMeCH$_2$

(1857).

RMgBr → Et$_3$SnR; R = MeO(PhO)CH (1303), Me$_2$CH (2216), CH$_2$:CHCH$_2$ (1857), MeCH:CH-CH$_2$, CH$_2$:CHCH$_2$CH$_2$, EtCH:CHCH$_2$ (1619), m-CF$_3$C$_6$H$_4$ (998), p-FClC:CFC$_6$H$_4$ (527).

Y(MgX)$_2$ → (Et$_3$Sn)$_2$Y; Y = CMeEtC⋮C, $\overline{CH_2(CH_2)_nCC}$⋮C, n = 3, 4 (1149).

Zn + BuCl + Et$_3$N at 160° → Et$_n$SnBu$_{4-n}$; n = 0-4 (1939).

Fe(CO)$_5$ → [Et$_2$SnFe(CO)$_4$]$_2$ + Et$_4$Sn$_3$[Fe(CO)$_4$]$_4$ + Sn[Fe(CO)$_4$]$_4$ (904).

ROH + base → Et$_3$SnOR; R = CH$_2$:CH(OCH$_2$CH$_2$)$_n$; n = 1-3 (510).

(EtO)$_2$PS$_2$NH$_4$ → Et$_3$SnS$_2$P(OEt)$_2$ (292).

RC⋮CH + (c.C$_6$H$_{11}$)$_2$NH → Et$_3$SnC⋮CR + (C$_6$H$_{11}$)$_2$NH$_2$Cl; R = MeO$_2$C, MeOCH:CH, Bu, Ph (2495).

O$_3$ in C$_6$H$_{14}$ → EtSnOCl·2Et$_3$SnCl·O$_3$ + AcH (664).

PhNCO + ureas → catalytic effect of, kinetic data (133).

<u>Addn. compd.</u>: Et$_3$SnCl·AlCl$_3$ (M-274), catalyzes polymerization of benzofurane, optically active polymer in presence of (-)-menthoxy-SnEt$_3$ (545).

2Et$_3$SnCl·EtSnOCl, dec. of EtSnOCl·2Et$_3$SnCl·O$_3$ at r.t. and 1 torr., m. 169-71°, soly. (664) [Sn$_3$C$_{14}$H$_{35}$Cl$_3$O].

2Et$_3$SnCl·EtSnOCl·O$_3$, Et$_3$SnCl with O$_3$ at 0° in hexane, m. 117-19°, soly. dec. r.t. at 1 torr   AcH + 2Et$_3$SnCl·EtSnOCl (664) [Sn$_3$C$_{14}$H$_{35}$Cl$_3$O$_4$].

Et$_3$SnCl·N(CH$_2$CH$_2$OH)$_3$, from components in MeOH, in 96% yield, m. 50° (1226). Activity as fungicide against Aspergillus niger, Penicillium citricum, as bactericide against Humicola grisea, Flavobacter peregrinum and Bacillus subtilis (1226) [SnC$_{12}$H$_{30}$ClNO$_3$].

Et$_3$SnCl·8-OC$_9$H$_6$N, from components in MeOH, in 33-38% yield (1047). Activity as fungicide against Aspergillus niger, A. arnstelodami, Penicillium cyclopium, P. brevi-compactum, Chaetomium globosum. Paecilomyces varioti and Stachybotris atra (1047) [SnC$_{15}$H$_{21}$ClNO].

<u>Biol. prop.</u>: Distribution in animal body (1770).

Activity as insecticide against Culex pipiens berbericus larvae (73), C.p. pipiens larvae (1137), Musca domestica, LD$_{50}$ and ATP inhibition (425).

Activity as bactericide against Bacillus subtilis, Micrococcus roseus (623), Staphylococcus aureus, Escherichia coli (702).

Activity as fungicide against Aspergillus niger, Torulaspora spp. (623).

<u>Use</u>: In catalyst for polymerization of CH$_2$:CMe$_2$ (2155), (CH$_2$:CH)$_2$O (2868), CH$_2$:CHCN with olefins (2997).

As catalyst for polymerization of ethylene oxides (2727), for polyurethane, relative reactivity (1603).

As fungicide for slime control (2668).

Triethyltin Bromide (M-274)   SnC$_6$H$_{15}$Br   Et$_3$SnBr

<u>Synth.</u>: By rxn. of Et$_3$SnC⋮CPO$_3$Et$_2$ with Br in HCCl$_3$ at -30°, in 77% yield (2179).

By rxn. of Et$_3$SnCH$_2$Ac with Bu$_3$GeBr (711).

By rxn. of Et$_3$SnH with HBr (156).

By rxn. of Et$_3$SnN:C:C(CN)CHRR' with PhCH$_2$Br (2789), with CH$_2$:CHCH$_2$Br (1954, 2789) with Br (2789).

By rxn. of Et$_3$SnN(NHPh)SO$_2$Ph with PhCH$_2$Br at 85° in Ar atm., in 97% yield (2646).

By rxn. of (Et$_3$Sn)$_3$Sb with Br in heptane, in 82% yield (2851), with PhCH$_2$Br and c.C$_5$H$_9$Br at 100-150°, in 86 and 82% yield, resp. (2086).

By rxn. of (Et$_3$Sn)$_2$Y with Br in C$_6$H$_6$ in inert atm.; Y = Se and Te, in 83 and 67%

yield, resp. (2851).
By rxn. of $(Et_2Sn)_x$ with RBr in sealed tube at 140°; R = Pr and Bu in 14 and 18% yield, resp. (1941).
By rxn. of powd. tin with EtBr in Et_3N and cat. I at 140°, in 10% yield (1943).
Prop.: IR (321), far IR, assignment (321, 2322), NMR (1430, 2068, 2071), Moessbauer resonance, NQR and (2290) and mass (2410) spectra, GLC (1941) and polarography (2352).
Rxn. with: $Et_3SiLi \rightarrow Et_3SnSiEt_3 + Et_4Si + Et_6Sn_2$ (2852).
$R_2NLi \rightarrow Et_3SnNR_2$; R = Me, (H + Ph) (251).
$Ph_2YNa \rightarrow Et_3SnYPh_2$; Y = As, Sb (68).
$AgX \rightarrow Et_3SnX$; X = $o\text{-}BrC_6H_4CO_2$ (1880).
Br → kinetic data of cleavage rxn. in AcOH (653), in PhCl and influence of n-, i-$PrSnMe_3$ (767, 769).
$Et_3PO_3 \rightarrow Et_3Sn(Et)O_3PEt$; rxn. mechanism (441).
Addn. compd.: Moessbauer resonance spectra are reported for 1:1 adducts of Et_3SnBr with pyridine quinoline, isoquinoline, acridine and 1-, 4-, 9- azaphenanthrene (1583).
Biol. prop.: Activity as bactericide against Staphylococcus aureus and Escherichia coli (702).

Triethyltin Iodide (M-275) $SnC_6H_{15}I$ Et_3SnI
Synth.: By rxn. of Et_4Sn with HgI_2 in aq. MeOH (654, 2205).
By rxn. of $Et_3SnC\vdots CYOSnEt_3$ with MeI; Y = CH_2, CHMe at 100-120°, in 99.5% yield (1910), CMe_2 at 190-200°, in 61% yield (513).
By rxn. of $Et_3SnOCH_2CH_2C\vdots CH$ with MeI at 100-110°, in 39% yield (513).
By rxn. of $Et_3SnN{:}PPh_3$ with MeI (1429).
By rxn. of $Et_3SnN{:}C{:}C(CN)CHRR'$ with I (2789).
By rxn. of $(Et_2Sn)_x$ with EtI (1:4) at 140° in sealed tube, in 54% yield (1941).
By rxn. of distillation residue of thermal dec. of $(Et_2Sn)_x$ with I in THF, in 13% yield (1941).
Prop.: IR (321), far IR, assignment (321, 2322), NMR (1430, 2068) and mass (2410) spectra, GLC (1941).
Rxn. with: $O_2NNRAg \rightarrow Et_3SnNRNO_2$; R = Me, Et, i-Pr, t-Bu, Ph (611).
Et_3PO_3 (at 150°) → EtI (441).
Biol. prop.: Activity against Staphylococcus aureus and Escherich a coli (702).

Tripropyltin Fluoride (M-282) $SnC_9H_{21}F$ Pr_3SnF
Synth.: By rxn. of $Pr_3SnC\vdots CPh$ with PCl_3 at r.t., followed by aq. KF (2433).
By rxn. of $(Pr_3Sn)_2O$ with aq. H_2SiF_6 (1468).
By rxn. of SnF_2 with Pr_2Hg in ligroine in inert atm. (2557).
Prop.: m. 324-25° (2557), 269° (2433), > 250° (1468), colorless needles (2433), mass spectrum (2410).
Biol. prop.: Toxicity (933).
Activity as bactericide (1468) against Staphylococcus aureus, Escherichia coli (702).
Activity as fungicide against Fusarium culmorum, Alternaria tenuis, Rhizoctonia solani (933).
Activity as insecticide (1468).

Use: As fungicide in PVC (2334).

Tripropyltin Chloride (M-275) $SnC_9H_{21}Cl$ Pr_3SnCl

Synth.: By rxn. of Pr_4Sn with $RSnCl_3$; R = Ph at 60°, in 98% yield (1370), Bu at 180-90°, in high yield (1370), with CCl_4 in air at 100°, in 55% yield (2094).
By rxn. of $Pr_3SnC\vdots CCSiEt_3$ with dry HCl in MeOH, in 90% yield (1602).
By rxn. of Pr_3SnR with AcCl; R = CH_2CO_2R' at 60-80° in 88-95% yield (437), OCMe:CHAc, in 85% yield (1711).
By rxn. of Pr_3SnCH_2Ac with R_3SiCl; R = Me (24a) and Et, in 88 and 74% yield, resp. (711).
By rxn. of $Pr_3SnCH_2CO_2R$ with Me_nSiCl_{4-n} ; n = 0-2 (818).
By rxn. of $(Pr_3SnCH:)_2$ with $HgCl_2$ in Me_2CO, in 83% yield (1602).
By rxn. of Pr_3SnH with dry HCl (1971).
By rxn. of $(Pr_3Sn)_2O$ with NH_4Cl in c.MeC_6H_{11} under reflux, in 62% yield (2660).
By rxn. of powd. Sn with Pr_2SnCl_2 and $(Pr_2SnCl)_2O$ in H_2O at 160°, in 62 and 84% yield, resp. (1939).
By rxn. of Pr_2SnI_2 with PrI, Zn dust and BuOH at 125-30°, followed by NaOH and HCl (2811).
By rxn. of $(Pr_2Sn)_x$ with BuCl and Et_3N at 160°, in 32% yield (1941).

Prop.: b_4 98-100° (2660), IR (925), (far) IR, assignment (2058), far IR (543), Raman (925), NMR (1309), Moessbauer resonance (1309) and mass (2410) spectra, GLC (1370) and magnetic susceptibility (1).

Rxn. with: RLi → Pr_3SnR; R = $B_{10}H_{10}C_2H$barene, (627), $B_{10}H_{10}C_2$Phcarborane (784).
YLi_2 → $(Pr_3Sn)_2Y$; Y = $B_{10}H_{10}C_2$ (627).
NaC_2H → $Pr_3SnC\vdots CH + (Pr_3SnC\vdots)_2$ (2181).
$NaS_2P(OEt)_2$ → $Pr_3SnS_2P(OEt)_2$ (745).
$Na_2S_2O_3$ → $(Pr_3Sn)_2S$ (1791).
$Na_2Ru(CO)_4$ → $(Pr_3Sn)_2Ru(CO)_4$ (905).
Mg + $n-XC_6H_4I$ → $n-Pr_3SnC_6H_4X$; X = F, Cl, Br, I, n = -o (1247), p- (1247a).
Mg + $n-I_2C_6H_4$ → $(Pr_3Sn)_2C_6H_4$; o- (1247), p- (1247a).
$Y(MgX)_2$ → $(Pr_3Sn)_2Y$; Y = $C\vdots CC(CH_2)_4CH_2$ (1149).

Addn. compd.: $Pr_3SnCl \cdot N(CH_2CH_2OH)_3$, from components in MeOH, in 90% yield, m. 75° (1226). Activity as bactericide against Humicola grisea, Flavobacter peregrinum and Bacillus subtilis (1226). Activity as fungicide against Aspergillus niger and Penicillium citricum (1226) [$SnC_{15}H_{36}ClNO_3$].

$Pr_3SnCl \cdot C_{18}H_{37}N[(CH_2CH_2O)_xH]_2$; 2x = 10, from components at 121°, amber liq. d_{25}^{25} 1.0577, in 97% yield * (1427) [$SnC_{47}H_{100}ClNO_{10}$].

$Pr_3SnCl \cdot C_{18}H_{37}N[(CH_2CH_2O)_xH]_2$, 2x = 50, in 94% yield, waxy solid* (1427) [$SnC_{127}H_{260}ClNO_{50}$].

$Pr_3SnCl \cdot OP(OC_6H_3MeCl-2,4)-(morpholine)_2$, from components in ROH, in 98% m. 60-62°, use as acaricide, bactericide, herbicide, fungicide (1838a) [$SnC_{24}H_{45}Cl_2N_2O_4P$].

---

* Use as fungicide, germicide and wood preservative (1427).

$2Pr_3SnCl \cdot OP(OC_6H_3MeCl-2,4)-(morpholine)_2$ from components at 120°, in 97% yield, d_{25}^{25} 1.1093, reddish brown liq.* (1427) $[Sn_2C_{33}H_{66}Cl_3N_2O_4P]$.
Biol. prop.: Effect on Mg-ATP of erythrocyte (361), on myosin A ATP (1273).
Physiological effect on humans (933).
Activity as bactericide against Staphylococcus aureus and Escherichia coli (702), Pseudomonas aeruginosa, Aerobacter aerogenes (2183).
Activity as fungicide against Alternaria tenuis, Fusarium culmorum, Rhizoctonia solani (933), Penicillium funiculosum, Aspergillus niger, Candida albicans (2183).
Activity as insecticide against Musca domestica, LD_{50} and ATP inhibition (425), Heliothis zea, H. virescence (2145).
Use: For slime control in pulp and paper industry (2668).
In antifouling coatings (2203).

Tripropyltin Bromide (M-282) $SnC_9H_{21}Br$ Pr_3SnBr
Synth.: By rxn. of Pr_3SnCH_2Ac with Me_3GeBr at r.t. (2547).
By rxn. of $(Pr_3Sn)_2O$ with NH_4Br under reflux, in 80% yield (2660).
By rxn. of $(Pr_2Sn)_x$ with BuBr, in sealed tube at 140°, in 4% yield (1941).
Prop.: NMR (1309), Moessbauer resonance (1304, 2290) and mass (2410) spectra.
Rxn. with: $Ph_2YNa \rightarrow Pr_3SnYPh_2$; Y = As, Sb (68).
Br in AcOH → mechanism of cleavage rxn. (653).
Biol. prop.: Activity as bactericide against Staphylococcus aureus and Escherichia coli (702).

Triisopropyltin Chloride (M-275) $SnC_9H_{21}Cl$ $(Me_2CH)_3SnCl$
Synth.: By rxn. of $i-Pr_4Sn$ with $SnCl_4$ and cat. $AlCl_3$, in 84% yield (705), with CCl_4 in air at 60°, in 43% yield (2094), with CCl_4 and cat. Bz_2O_2 in N atm., in 45% yield (2094).
Prop.: b. 134-37.5° (705).
Rxn. with: PhLi → $i-Pr_3SnPh$ (998).
RMgCl → $i-Pr_3SnR$; R = $CH_2:CHCH_2$ (1367), $\overline{CH_2CH_2CH}$ (1873).
Use: In antifouling coatings (2302).

Tricyclopropyltin Chloride (M-282) $SnC_9H_{15}Cl$ $(\overline{CH_2CH_2CH})_3SnCl$
Synth.: By rxn. of $(\overline{CH_2CH_2CH})_4Sn$ with $HgCl_2$ (1873).
Rxn. with: $\overline{CH_2CMe_2CHLi} \rightarrow (\overline{CH_2CH_2CH})_3Sn\overline{CHCMe_2CH_2}$ (1873).
Na → $(\overline{CH_2CH_2CH})_6Sn_2$ (1873).
NaX → $(\overline{CH_2CH_2CH})_3SnX$; X = I, $C_7H_{15}CO_2$ (1873).
$Y(MgBr)_2 \rightarrow [(\overline{CH_2CH_2CH})_3Sn]_2Y$; Y = $(CH_2)_{10}$ (1873).
Use: As herbicide, pesticide (1873).

---

* Use as fungicide, germicide and wood preservative (1427).

Tributyltin Fluoride (M-282) $SnC_{12}H_{27}F$ Bu_3SnF

Synth.: By rxn. of Bu_4Sn with SF_4, as quoted in (52).
By rxn. of Bu_3SnH with $PhHgCBrX_2$ in N atm., followed by KF; X = Cl, Br (1876).
By rxn. of $(Bu_3Sn)_2O$ with aq. H_2SiF_6 (1468).
By rxn. of Bu_6Sn_2 with $(CF_3C!)_2$ as by-prod. (108).
By rxn. of SnF_2 with Bu_2Hg in ligroine in inert atm. (2557).
Prop.: m. 341-42° (2557), 250° (1468), far IR, assignment (52) and mass (2410) spectrum.
Addn. compd.: $Bu_3SnF \cdot BF_3$, from $(Bu_3Sn)_2O$ and HBF_4 in H_2O, m. 216° (595), use as fungicide to control mold (242) [$SnBC_{12}H_{27}F_4$].
Biol. prop.: Activity as bactericide against Staphylococcus aureus and Escherichia coli (702).
Activity as fungicide against Fusarium culmorum, Alternaria tenuis and Rhizoctonia solani (933).
Biological method for detn. of 5ppb in seawater (456).
Use: As bactericide (1468).
Fungicide in paint (2045), in PVC (2334) to control mold (242).
As insecticide (1468).
As molluscicide in antifouling coatings (2302), in marine paints (2682).

Tributyltin Chloride (M-276) $SnC_{12}H_{27}Cl$ Bu_3SnCl

Synth.: By rxn. of Bu_4Sn with $^{113}SnCl_4$ (2824), from $Bu_4^{113}Sn$ and $SnCl_4$ at 0° in pentane (2049), with $SnCl_4$ and cat. $AlCl_3$ at 100°, in 90% yield (705), with $SnCl_2$ at 220° (1005), with $i-C_5H_{11}Cl$ and 4% $AlCl_3$ under reflux, in 48% yield (119).
By rxn. of 9,10-$(Bu_3Sn)_2C_{14}H_{10}$ with concn. HCl under reflux (1726).
By rxn. of $Bu_3SnCH:CHMR_3$ with Cl in CCl_4, in 93% yield; RM = BuSn, EtSi (1598).
By rxn. of $(Bu_3SnC!)_2$ with MCl in 63-90% yield, M = HgCl, $PhCH_2Hg$, PhHg, Ph_3Si, Ph_2P, Ph_2As, Ph_2Sb (2264).
By rxn. of $Bu_3SnCHEtCN$ with Me_3SiCl under reflux (2680).
By rxn. of $Bu_3SnCH_2CO_2R$ with PCl_3 or $(EtO)_2PCl$, in 96% yield (440).
By rxn. of Bu_3SnH with Me_3SnCl and distillation at 50° (1953), with R_2SnCl_2 in N atm. (1805), with Bu_2SnCl_2 (2:1) or Bu_2SnHCl (1:1) in N atm. at r.t. (1804), with dry HCl (1971), with CCl_3CHO at 0° as by-prod. (2538), with RCOCl (1371), with $(Me_2N)_2BCl$ (1623).
By rxn. of Bu_3SnNEt_2 with RCl at 45-100°, R = $PhCH_2$, Bu, $CH_2:CHCH_2$ (2691).
By rxn. of $Bu_3SnN:C:C(CN)CHRR'$ with $ClCO_2Et$ (2789).
By thermal dec. of Bu_3SnX at 100-115°; X = $Bu_3SnOCH(CCl_3)NMeCONBu$ (2258), $Bu_3SnOC(CCl_3):N$, in 50% yield (750).
By rxn. of $(Bu_3Sn)_2O$ with Me_3SiCl (2465), with NH_4Cl in c.MeC_6H_{11} under reflux, in 72% yield (2660), with CCl_4 and pptd. Cu, in 63% yield (1737).
By rxn. of Bu_3SnOMe with $Cl(CH_2)_nOH$ on heating; n = 2, 4 (2690).
By rxn. of Bu_3SnOR with $PhCH_2Cl$ (435), with R_3SiCl (434).
By rxn. of $Bu_3SnOCH(C_6F_5)CH_2CN$ with dry HCl in Et_2O (1631).
By rxn. of Bu_3SnOR with AcCl; R = $Me_2CHCH:CMe$ (1684), $Br_3CCH(OMe)$ (953), Ac (953).

By rxn. of $Bu_3SnOC^{18}OR$ with PCl_5; R = Me (1615), Ph, in 100% yield (1615).
By rxn. of $Bu_3SnO_2CCCl_3$ with Ph_3P, also in presence of ArCHO (2656).
By rxn. of Bu_3SnO_2CNHEt with PhNCS in CCl_4, followed by HCl (2258).
By rxn. of $(Bu_3Sn)_2S$ with excess CCl_4 and pptd. Cu in N atm., in 70-90% yield (1737).
By thermal dec. of Bu_3Sn-contn. PVC (2598).
By rxn. of Bu_2SnCl_2 with powd. M in H_2O or MeOH at 160°; M = Sn, Fe, Al, in 32-54% yield (1939), Mg, Ni, Co, in 4-15% yield (1939), in $BuCl-Et_3N$, M = Sn, Fe in 44-117% yield, based on Bu in Bu_2SnCl_2 (1939), M = Zn, Al, Mg, Na, in 22-66% yield (1939).
By rxn. of $(Bu_2SnCl)_2O$ with powd. Sn at 160° in H_2O, in 32-49% yield (1939), in $BuCl-Et_3N$, in 73% yield (1939), with powd. Zn in 35% yield (1939).
By rxn. of Bu_6Sn_2 with CCl_4 and pptd. Cu, in 43% yield (1737).
By rxn. of $(Bu_2Sn)_x$ with Et_2SnCl_2 and Et_3N at 160°, in 5% yield, with Bu_2SnCl_2, in 30% yield (1941), with BuCl at 160°, in 50% yield (66), in presence of Et_3N, in 60% yield (1941).
By rxn. of $(R_2Sn)_x$ with $BuCl/Et_3N$ and Et_4NX at 160°; R = Et, Pr, in 3% yield (1941).
By rxn. of BuCl with powd. Sn in presence of I and Et_3N at 160° in 54% yield, effect of N-bases or ammonium halides (1943).
By rxn. of tin foil with BuCl in presence of $SnCl_2$ and LiI in polyglycol (880b).
By electrolysis of $SnCl_4$ and/or $SnCl_2$ with BuBr in presence of $ZnBr_2$ on Mg cathode and graphite anode, in 48-52% yield (684).
Continual manufacturing process based on Grignard rxn. (2994).
Continual manufacturing process with Bu_4Sn and $SnCl_4$, purification by Bu_2SnCl_2 removal with aq. amine treatment (2280), at 0° and purification by $BuSnCl_3$ removal through flash distillation at 8 torr and 94°, followed by fractionation (1646).
Prop.: $b_{1.2}$ 118° (1005), IR (165, 905), assignment (1078, 1509), far IR, assignment (1509, 2322), Raman (905), assignment (1078), NMR (1309), Moessbauer resonance (1309) and mass (2410) spectra, dipole moment (325, 1211), light resistance (1063), magnetic susceptibility (1), polarography (1249, 2588-9) and surface tension (1959). Anal. detn. by GLC (1079, 1939), TLC (388), by paper chromatography (551), in presence of Bu_4Sn by mass spectrography and chromatography (2838), of Bu_2SnCl_2 and Bu_4Sn by chromatography (165), of Bu_2SnCl_2 by polarography (247) of Bu_2SnCl_2 and $BuSnCl_3$ by polarography (2453), Cl detn. by titration with NaOMe in Py (1144).
Rxn. with: $(Ph_3Sn)_3SnLi \rightarrow (Ph_3Sn)_4Sn$ (68a, 175).
Ph_3SnCl (2:1) $\rightarrow Bu_2PhSnCl$ (1577).
$(Bu_2SnO)_x \rightarrow Bu_3SnOSnBu_2Cl$ (944).
$Bu_2Sn(OMe)_2 \rightarrow Bu_2SnCl_2 \cdot Bu_2Sn(OMe)_2 + Bu_3SnOMe$ (2184).
RLi $\rightarrow Bu_3SnR$; R = $p-FClC:CFC_6H_4$ (527), Me_2N (251, 1392), Et_2N (323, 2053).
$B_{10}H_{10}(CLi)_2 \rightarrow (Bu_3SnC)_2B_{10}H_{10}$ (2279).
$Et_2PLi + \overline{CH_2CH_2O} \rightarrow Bu_3SnOCH_2CH_2PEt_2$ (2835).
Na $\rightarrow Bu_6Sn_2$, in C_6H_6 or Et_3N (1939), in Bu_4Sn (1548a).
$NaC_2H \rightarrow Bu_3SnC\colon CH + (Bu_3SnC\colon)_2$ (2181).

$Na_2Y \to (Bu_3Sn)_2Y$; $Y = 9,10-C_{14}H_{10}$ = dihydroanthracene (1726).

$RONa \to Bu_3SnOR$; R = Me, n-, i-Pr (334), Et (334, 692), MeCH:CH (1051), Me_3CO (1618), Ph (36), Et_2NCH_2CHPh, Et_2NCH_2CHEt, $1,2-Et_2NC_6H_{10}$ (2053), o-$Et_2NC_6H_4$ (2835), 8-C_9H_6N (1291).

$RCO_2Na \to Bu_3SnO_2CR$; $R = PhCH_2$ (1051), $BzNHCH(CH_2Ph)$ (151), $Me(CH_2)_5CMe_2$ (1169a).

$Y(CO_2Na)_2 \to (Bu_3SnO_2C)_2Y$; Y = CH:CH, C_6H_4, $(CH_2)_n$, n = 2, 4, 8 (1051).

$NaBH_4$ in glycol ether $\to Bu_3SnH + NaCl + B_2H_6$ (736).

$NaX \to Bu_3SnX$; $X = Ta(CO)_6$ (1304), $Me_2C:NO$ (2661), $2,4,6-(Et_3N)C_3N_3$ = sym. triazine (1322a), $(RO)_2PS_2$, R = Me (1565), Et, n-, i-Pr (745), $(EtO)_2SPO$ (745), N-benzoxazolinone-X; X = 6-Cl, 6-Br, 5,6-Cl_2, N-benzoxazolinethione-5-chloro (2590).

$Et_3PO_4 + NaOH \to Bu_3Sn(Et)_2PO_4$ (292).

$(RO)_2POH + Na$ in alc. $C_6H_6 \to Bu_3SnPO_3R_2$; R = Me, Et (1564).

$PhCH_2C(CN)_2Na \to Bu_3SnN:C:C(CN)CH_2Ph$ (2615).

$(H_2C:NONa)_3 \to Bu_3SnON:CH_2$ (2125).

$EtO_2CNHNa \to Bu_3SnNCO + EtOH$ (373).

$Na_2S_2O_3$ in MeOH $\to (Bu_3Sn)_2S$ (1791).

$Et_2NCH_2CH_2SH + Na$ in THF $\to Bu_3SnSCH_2CH_2NEt_2$ (2835).

$Na_2Y \to (Bu_3Sn)_2Y$; Y = S (1399, 1514), Se (1618).

$KX \to Bu_3SnX$; X = SCN (2751), Bu_2PO (1232), $MeO_2CCH:CHCO_2$ (1557), $MeO_2CCHCH_2CMe:CMeCH_2CHCO_2$ (1558), N-benzoxazolinone-X'; X' = 6,Cl, 6-Br, 5,6-Cl_2, N-benzoxazolinethione-5-chloro (2590).

$(R_2N)_2POCl + KOH \to Bu_3SnO_2P(NR_2)_2$; R = Me, Et (292).

$(KS)_2C:NCN \to (Bu_3Sn)_2S + Bu_3SnNCS$ (2751).

Base $\to (Bu_3^{113}Sn)_2O$ (2824).

$C_{12}H_{25}N(CH_2CH_2OH)_2$ or $p-C_9H_{19}C_6H_4(OCH_2CH_2)_xOH$ + alkali $\to Bu_3Sn$-alkoxides (1958).

RSH + base $\to Bu_3SnSR$; R = N-heterocycles, with Et_3N in Me_3CO and NaOH in Et_2O (2204).

R_4NCl or LiCl in Me_2CO or MeCN $\to$ potentiometric titration and complex formation; R = Me, Et (2015).

$M_2Ru(CO)_4 \to (Bu_3Sn)_2Ru(CO)_4$ (905).

AgNCO $\to Bu_3SnNCO$ (373).

$(EtO)_2PS_2H + Et_3N \to Bu_3SnS_2P(OEt)_2$ (745).

$(EtO)_2PS_2NH_4 \to Bu_3SnS_2P(OEt)_2$ (292).

$RMgCl \to Bu_3SnR$; $R = EtOCH_2$ (307), $CH_2:CMeCH_2$ (1857).

$RMgBr \to Bu_3SnR$; $R = CH_2CH_2CH$ (1873), $CH_2:CHCH_2$ (1857), $MeO(PhO)CH$ (1303), C_6F_5 (428), $p-FClC:CFC_6H_4$ (527), $2-C_5H_4N$ (1500).

Mg + $n-I_2C_6H_4 \to (Bu_3Sn)_2C_6H_4$; n = o- (1247), p- (1247a).

Mg + $n-IC_6H_4X \to n-Bu_3SnC_6H_4X$; X = F, Cl, Br, I, n = o- (1247), p- (1247a).

$CH_2:CHMgX \to Bu_3SnCH:CH_2 + Bu_4Sn + Bu_2Sn(CH:CH_2)_2$; method for assay of vinyl-Grignard reagent by GLC (2147).

$CaS_2C:NCN \to (Bu_3SnS)_2C:NCN$ (1976).

Zn + Et_3N at 160° $\to Bu_4Sn + Bu_6Sn_2$ (1939).

$Fe(CO)_5 \to [Bu_2SnFe(CO)_4]_2 + Bu_4Sn_3[Fe(CO)_4]_4 + (Bu_2SnCl)_2Fe(CO)_4 + Sn[Fe(CO)_4]_4$ (904).

Br in PhCl → kinetic data for cleavage rxn., mechanism (767, 769).
Olefin → comparative stability of π-complexes (182).
PhNCO + ureas → catalytic effect of, kinetic data (133).
$(H_2N)_2CO$ → Bu_3SnNCO + NH_4Cl (532b).
Polyhydroxy compd. → Bu_3Sn-contn. polymer (1515).
$(CH_2CH_2CH_2NCO)_2 + O(CH_2CH_2OH)_2$ → catalytic effect of, kinetic data (2542-3).
S at 180-90° → Bu_2SnCl_2 + SnS, cf. (M-277), ref. 2531 (1705).
$Y(SH)_2$ → $Bu_3SnSYSH$; Y = 2,5(1,3,4-thiadiazole) (1567), $CH_2CO_2(CH_2)_4O_2CCH_2$ (2903).

Addn. compd.: $Bu_3SnCl·AlCl_3$, from components, soly., activity as polymerization catalyst for 1-olefins (32) [$SnACC_{12}H_{27}Cl_4$].

$Bu_3SnCl·8-OC_9H_6N$, from components in 28-30% yield (1047).
Activity as fungicide against Aspergillus niger, A. arnstelodami, Penicillium cyclopium, P. brevi-compactum, Chaetomium globosum, Paecilomyces variota and Stachybotrys atra (1047).
Use as catalyst for foaming polyurethan (450) [$SnC_{21}H_{33}ClNO$].

$Bu_3SnCl·C_{18}H_{37}N[(CH_2CH_2O)_xH]_2$; 2x = 10, from components in i-PrOH, in 84% yield, use as germicide, fungicide and wood preservative (1427).
2x = 50, from components in i-PrOH, in 94% yield, use as germicide, fungicide, and wood preservative (1427).

$Bu_3SnCl·OP(OMe_2)R$; R = N-morpholine, from components in alc., in 89% yield, $b_{0.2}$ 73°, use as acaricide, bactericide, herbicide, insecticide (1838a) [$SnC_{18}H_{42}ClNO_4P$].

$Bu_3SnCl·OP(NMe_2)_2OBu$, from components in ROH, in 85% yield, $b_{0.2}$ 80°, use as acaricide, bactericide, herbicide, insecticide (1838a). [$SnC_{20}H_{48}ClN_2O_2P$].

$Bu_3SnCl·OP(NMe_2)_2SBu$, from components in ROH, in 90% yield, $b_{0.05}$ 80°, use as acaricide, bactericide, herbicide, insecticide (1838a) [$SnC_{20}H_{48}ClN_2OPS$].

Moessbauer resonance spectra are reported for 1:1 adducts of Bu_3SnCl with pyridine, quinoline, isoquinoline, acridine, 1-, 4-, 9- azaphenanthrene (1583).

Biol. prop.: Toxicity (1099, 2678), inhibition of oxidative phosphorylation in mitochondria (526), of photophosphorylation of chloroplasts (1267), phytotoxicity towards barley (376), rice (548), biol. method for detn. of 5 ppb. in seawater (456).
Activity as bactericide (1099), against Escherichia coli, Staphylococcus aureus (702), Bacillus subtilis, Micrococcus roseus (623), Aerobacter aerogenes, B. cereus var. mycoides (894).
Activity as fungicide (1099), against Aspergillus niger (623, 894), Penicillium expansum (894), P. funiculosum (1099), Colletotrichum (1057), Coriolellus pallustris (1063), Ceratostomella (Ceratocystis) coerulescens, Speira heptaspora, Fusarium culmorum (1099), F. nivale, Rhizoctonia solani (1057), Piricularia oryzae (rice blast disease) (548), Torulaspora spp. (623).
Activity as growth inhibitor of bacteria, fungi, yeast (622).
Activity as insecticide against Musca domestica, LD_{50} and ATP inhibition (425), investigation of resistant strain (2448).

Use: As catalyst for foaming polyurethan (1029).
In catalyst for polymerization of C_2H_4 (2462), $CH_2:CMe_2$ (2155).
As stabilizer in polyethylene (263), mechanism of PVC stabilization (2599,

2832).

As rodent repellent (30), in marine paints (1023), antifouling coatings (2302), wood preservative (451).

Tributyltin Bromide (M-277) $\quad\quad\quad\quad SnC_{12}H_{27}Br \quad\quad\quad\quad Bu_3SnBr$
Synth.: By rxn. of Bu_4Sn with $i-C_5H_{11}Br$ and 4% $AlCl_3$ under reflux, in 64% yield (119).
By rxn. of $(Bu_3Sn)_2$ with Ph_3GeBr, in 80% yield (2264).
By rxn. of Bu_3SnH with Bu_2SnBr_2 in N atm. at r.t. (1804).
By rxn. of Bu_2SnH_2 with $HC⋮CCHBrCH_2C⋮CH$ in hexane (499).
By rxn. of Bu_3SnOMe with $BuBr$ or $CH_2:CHCH_2Br$ (435).
By rxn. of Bu_3SnNEt_2 with RBr at 25-100°; $R = PhCH_2$, Bu, $CH_2:CHCH_2$ (2691).
By rxn. of $Bu_3SnN:C:C(CN)CHRR'$ with $CH_2:CHCH_2Br$ (1954, 2789).
By rxn. of $(Bu_2Sn)_x$ with $BuBr$ in sealed tube at 140°, in 70% yield (1941), in 45% yield (66).
By rxn. of Bu_2SnCl_2 with $BuBr$ in presence of Mg and Ph_3PO_3 in absence of ethers (1871a).
By rxn. of Sn with $BuBr$ in $(MeOCH_2CH_2)_2O$ under reflux in presence of cat. $LiBrBu_2SnBr_2$ in 84% yield (880b), with cat. Bu_4NBr, in 53% yield (1209a), under ^{60}Co irradiation (616a), in 1-11% yield (1031), in presence of HBr in 13% yield (2634), in presence of BuOH, H_2O, in 4-7% yield (2601).
By electrolysis of BuBr in presence of $ZnBr_2$ and Br on Mg cathode and Sn anode, in 50% yield (683), with $SnCl_2$ and/or $SnCl_4$ on graphite anode, in 45-69% yield (684).
Prop.: Far IR, assignment (2322), Moessbauer resonance (2290) and mass (2410) spectra, polarography (1249). Anal. detn. by GLC (530), in presence of Bu_4Sn and Bu_2SnBr_2 by IR spectroscopy (165).
Rxn. with: $(Bu_2SnO)_x \rightarrow Bu_3SnOSnBu_2Br$ (946).
$RNa \rightarrow Bu_3SnR$; R = polynuclea group (2703).
$Ph_2MNa \rightarrow Bu_3SnMPh_2$; M = As, Sb (68).
$Zn + Ph_3CCl$ in DMF $\rightarrow Bu_3SnCPh_3$ (2382).
Biol. prop.: Toxicity (618), activity as bactericide against Staphylococcus aureus and Escherichia coli (702).

Tributyltin Iodide (M-283) $\quad\quad\quad\quad SnC_{12}H_{27}I \quad\quad\quad\quad Bu_3SnI$
Synth.: By rxn. of Bu_4Sn with HgI_2 in aq. MeOH (654).
By rxn. of Bu_3SnCH_2Ac with Et_3MI; M = Si and Ge, in 83 and 75% yield, resp. (711).
By rxn. of $Bu_3SnC⋮CPh$ with I in Et_2O in Ar atm., in 98% yield (2644).
By rxn. of Bu_3SnH with Bu_2SnI_2 in N atm. at r.t. (1804).
By rxn. of Bu_3SnX with RI; R = Bu, $CH_2:CHCH_2$, X = NEt_2 (2691), OMe (435).
By rxn. of $(Bu_3Sn)_2O$ or Bu_3SnOMe with NH_4I in $c.MeC_6H_{11}$ or MePh, in 91 and 94% yield, resp. (2660).
By rxn. of Bu_3SnSR with I in CCl_4, in quant. yield; R = Me, Ph (1676).
By rxn. of Bu_3SnMPh_2 with MeI, M = As, Sb (68).
By rxn. of Bu_2SnI_2 with powd. Cu at 125-30° (2731), with Cu-Zn alloy and BuI at 80-90° (2808), with Zn dust and BuI in BuOH (2811).

By rxn. of $(Bu_2Sn)_x$ with BuI in sealed tube at 140°, in 65% yield (1941).
By thermal dec. of $(Bu_2Sn)_x$ at 160°, followed by I in THF, in 7% yield (1941).
By-prod. in rxn. of tin with BuI and cat. BuOH + Mg (2955).
Prop.: b_{12} 173° (2644), $b_{0.9}$ 118-20° (1941), $b_{0.3}$ 130-35° (2660), far IR, assignment (2322) and mass (2410) spectra.
Rxn. with: RMgX → Bu_3SnR; R = i-C_7H_{15} (517), p-R'C_6H_4 ; R' = CH_2:CH, CH_2:CMe, EtCH:CH (517).
Biol. prop.: Activity as bactericide against Escherichia coli and Staphylococcus aureus (702).
Use: As slime control agent in paper and paper mill industry (2810).

Triisobutyltin Chloride (M-283) $SnC_{12}H_{27}Cl$ $(Me_2CHCH_2)_3SnCl$
Synth.: By rxn. of $SnCl_4$ with i-Bu_3Al in pentane-hexane in N atm. (2961).
Prop.: Mass spectrum (2410).
Rxn. with: $LiNEt_2$ → i-Bu_3SnNEt_2 (1618).
Aq. NaOH in C_6H_6 → (i-$Bu_3Sn)_2O$ (1618).
Use: In antifouling coatings (2302).

Tris(2,2-methylphenylpropane)tin $SnC_{30}H_{39}Cl$ $(PhCMe_2CH_2)_3SnCl$
  Chloride
Synth.: By rxn. of $SnCl_4$ with $PhCMe_2CH_2MgCl$ in Et_2O, in 44% yield (1742), in Et_2O-C_6H_6, in 48% yield (2188).
Prop.: m. 117.5-18.5° (1742), 116.5-17.5° (2188), NMR (1742, 2188) and Moessbauer resonance (222, 224, 1742) spectra.
Rxn. with: Na in liq. NH_3 → $(PhCMe_2CH_2)_6Sn_2$ (2188).
NaOH → $(PhCMe_2CH_2)_3SnOH$ in aq. alc. (1741) or $[(PhCMe_2CH_2)_3Sn]_2O$ in aq. Et_2O (2188).
Na_2S → $[(PhCMe_2CH_2)_3Sn]_2S$ (1742).
KF in MeOH → $(PhCMe_2CH_2)_3SnF$ (2188).

Trihexyltin Chloride (M-284) $SnC_{18}H_{39}Cl$ $(C_6H_{13})_3SnCl$
Synth.: By rxn. of $SnCl_4$ with $(C_6H_{13})_3Al$ in CH_2Cl_2 in N atm. (90).
Rxn. with: $SnCl_4$ at 180° → $(C_6H_{13})_2SnCl_2$ (90).
$(NaC\vdots C)_2$ in liq. NH_3 → $[(C_6H_{13})_3SnC\vdots C]_2$ (205).
NaR → $(C_6H_{13})_3SnR$; R = $(EtO_2C)_2C(CH_2\overline{CHCH_2O})$ (1442).
Use: As rodent repellent (2484).

Tricyclohexyltin Chloride (M-284) $SnC_{18}H_{33}Cl$ $(c.C_6H_{11})_3SnCl$
Synth.: By rxn. of $(C_6H_{11})_4Sn$ at 100-170° in Me_2SO, yield depending on rxn. conditions (1385).
By rxn. of $SnCl_4$ with $C_6H_{11}MgCl$ (1:3) in THF-xylene in N atm. at 75-80°, in 63-87% yield (1186a).
By rxn. of $(C_6H_{11})_2SnCl_2$ with $C_6H_{11}MgCl$ in THF-xylene in N atm. at 70-90°, in 81% yield (1583b).
Prop.: IR spectrum (1385), differential thermal analysis (1385).
Rxn. with: $(Me_3SnC\vdots C)_2$ → $[(C_6H_{11})_3SnC\vdots C]_2$ (205, 2433).
$9,10-Na_2C_{14}H_{10}$ → $9,10-[(C_6H_{11})_3Sn]_2C_{14}H_{10}$, 9,10-dihydroanthracene (1726).

CH$_2$:CHCH$_2$MgCl → (C$_6$H$_{11}$)$_3$SnCH$_2$CH:CH$_2$ (1367).
Biol. prop.: Activity controling arachnids (1306).

Tricyclohexyltin Bromide (M-277)   SnC$_{18}$H$_{33}$Br   (c.C$_6$H$_{11}$)$_3$SnBr
Synth.: By rxn. of (C$_6$H$_{11}$)$_3$SnOH with NH$_4$Br in c.MeC$_6$H$_{11}$ under reflux in 85% yield (2660).
Prop.: m. 76-77° (2660).
Rxn. with: LiAlH$_4$ → (C$_6$H$_{11}$)$_3$SnH (1618).
LiNEt$_2$ → (C$_6$H$_{11}$)$_3$SnNEt$_2$ (1618).
RCH$_2$ONa → (C$_6$H$_{11}$)$_3$SnOCH$_2$R; R = CH$_2$OH, CH$_2$OHCHOH (2721).
RMgBr → (C$_6$H$_{11}$)$_3$SnR; R = i-Pr, n-, i-Bu, n-, i-C$_5$H$_{11}$, p-MeOC$_6$H$_4$, p-EtOC$_6$H$_4$ (1773).

Trioctyltin Chloride (M-284)   SnC$_{24}$H$_{51}$Cl   (C$_8$H$_{17}$)$_3$SnCl
Synth.: By rxn. of SnCl$_4$ with C$_8$H$_{17}$Cl, Mg and cat. C$_8$H$_{17}$Br in Et$_2$O-heptane, in 65% yield (1583a).
By rxn. of n-C$_8$H$_{17}$Cl with powd. tin and cat. Et$_3$N and I, in 20% yield (1943).
By rxn. of tin sheet in glycol ether or polyglycol with C$_8$H$_{17}$Cl and cat. LiBr, SnCl$_2$ or R$_2$SnCl$_2$ and I (880b).
Rxn. with: Na in (C$_8$H$_{17}$)$_4$Sn → (C$_8$H$_{17}$)$_6$Sn$_2$ (1548a).
(NaC⋮C)$_2$ in liq. NH$_3$ → [(C$_8$H$_{17}$)$_3$SnC⋮C]$_2$ (205).
Na$_2$S$_2$O$_3$ in MeOH → [(C$_8$H$_{17}$)$_3$Sn]$_2$S (1791).
CaS$_2$C:NCN → [(n-C$_8$H$_{17}$)$_3$SnS]$_2$C:NCN (1976).
Biol. prop.: Activity as insecticide against Musca domestica, LD$_{50}$ and ATP inhibition (425), Heliothis zea, H. virescens (2145).
Activity as molluscicide against Australorbis glabratus, Bulinus contortus; effect on insects, crustaceans and aquatic plants, use in paint (962).

Triphenyltin Fluoride (M-284)   SnC$_{18}$H$_{15}$F   Ph$_3$SnF
Synth.: By rxn. of Ph$_4$Sn and MF$_n$ in steel bomb, with BF$_3$, SiF$_4$ at 140°, PF$_5$ at 135°, AsF$_5$ at 120° and SF$_4$ at 115° (498).
By rxn. of Ph$_3$SnCl with aq. HF (52).
By rxn. of (Ph$_3$Sn)$_2$O with aq. H$_2$SiF$_6$ (1468).
By rxn. of Ph$_3$SnNO$_3$ with excess NaF in EtOH, in quant. yield (1969).
By rxn. of Ph$_3$SnCo(CO)$_4$ with CF$_3$I, C$_2$F$_4$ and C$_2$F$_3$H under UV irradiation or on heating (730).
Prop.: m. > 250° (1468), far IR, assignment (52, 1703), Moessbauer resonance (211, 224, 2665) and mass (850) spectra, polarographic reduction (2842).
Addn. compd.: Ph$_3$SnF·BF$_3$, from (Ph$_3$Sn)$_2$O and HBF$_4$ in H$_2$O, m. 312° (595), use as fungicide for mold control (242) [SnBC$_{18}$H$_{15}$F$_4$].
Biol. prop.: Activity as insecticide and effect on reproduction of Musca domestica (260).
Use: As bactericide and insecticide (260), as fungicide in PVC (2334), for mold control (242).

Triphenyltin Chloride (M-278)   SnC$_{18}$H$_{15}$Cl   Ph$_3$SnCl
Synth.: By rxn. of Ph$_4$Sn with SnCl$_4$ and cat. AlCl$_3$ at 100°, in 87% yield

(705), with CCl_4 and cat. $BiCl_3$ under reflux, in 99% yield (1716).
By rxn. of $Ph_3SnC\vdots CPh$ with AcCl at 100° in Ar atm. (2644).
By rxn. of Ph_3SnH with BzCl in Et_2O or C_6H_6 (299), with RCOCl, in 63-90% yield (1310, 1374), with Ph_2PCl (495).
By rxn. of Ph_3SnN_3 with $CaCl_2$ during drying (1402), with CH_2Cl_2 under UV irradiation (2471).
By rxn. of Ph_3SnOH with NH_4Cl in $c.MeC_6H_{11}$ under reflux, in 67% yield (2660), with aq. NaCl and solvent extraction (487).

By dec. of $Ph_3SnO_2CCCl_3$ in refluxing octadiene (1875), of Ph_3SnNO_3 in $o-Cl_2C_6H_4$ (2388).
By rxn. of $(Ph_3Sn)_2S$ with Ph_2SnCl_2 in refluxing C_6H_6 (1705).
By rxn. of Ph_3SnSPh with $HgCl_2$ in Et_2O or $SnCl_2$ in C_6H_6 (1704).
By rxn. of $Ph_3SnNBuCO_3CMe_3$ with HCl in Et_2O, in 76% yield (2256).
By rxn. of Ph_6Sn_2 with Ph_3CCl (2046), with halides of H, Cu, Hg, Fe, in 15-99% yield (2046).
By rxn. of Ph_3SnLi with $(C_6F_5)_3GeCl$ (1034).
By-prod. in rxn. of PhMgCl with $SnCl_4$ in THF-xylene in N atm. at 30-40°, in 20% yield (1186a).
Two step manufacturing process, reacting $PhSnCl_3$ with NaPh in petr. ether at 100-120° and converting Ph_4Sn with $SnCl_4$ (1:1) at 90° to $Ph_3SnCl + PhSnCl$, stripping the latter at 10 torr and 120° (1643).

Prop.: IR (194, 333), assignment (2449), far IR (194), assignment (1703, 1706, 1962, 2449, 2784) Raman (2784), UV (335), NMR (1220, 1453, 2067), Moessbauer resonance (211, 222, 224, 503, 1114, 1309, 1960, 1988-9, 1991, 2665), ^{35}Cl NQR (1142, 1962) and mass (850) spectra, dipole moment (194, 325, 1211, 1518), polarography (963, 967, 1249, 2352), x-ray struct. analysis (754). Anal. detn. by spectrography (1358) by TLC (1253, 2635), by reverse potentiometric titration (2013), of Cl (1690).

Rxn. with: ^{60}Co irradiation → evidence of Ph_3Sn* formation by ESR spectroscopy (693).
$(Me_3Sn)_2Y → (Ph_3Sn)_2Y + Me_3SnCl$; $Y = C\vdots C$, $C\vdots CC\vdots C$ (2433).
$R_3SnC\vdots CPh → Ph_3SnC\vdots CPh + R_3SnCl$; R = Me, Et, Bu (2433).
$i-Bu_3SnH + Et_3N$ (1:1:1) → $Ph_6Sn_2 + i-Bu_6Sn_2 + i-Bu_3SnSnPh_3 + Et_3N\cdot HCl$ (1618).
Bu_3SnCl (1:2) → $Bu_2PhSnCl$ (1577).
$(Me_3Sn)_3SnLi$ in THF → Ph_6Sn_2 (2127).
$Ph_3SnLi + SnCl_2$ in THF → $(Ph_3Sn)_4Sn + Ph_6Sn_2$ (68a, 175).
$Ph_3SnMgBr\cdot NEt_3 → Ph_6Sn_2$ (2330).
$(Ph_2Sn\cdot PhMgBr)_2 → Ph_4Sn$ (2330).
$Ph_3SnZnCl$ in THF → $Ph_4Sn + (Ph_2Sn)_x$ (2830).
$Ph_3GeLi → Ph_4Sn + Ph_3SnGePh_3$ (1928).
Ph_3GeK in THF → $Ph_3SnGePh_3 + Ph_6Sn_2 + Ph_6Ge_2$ (2128).
Ph_3SiK in THF → $Ph_3SnSiPh_3$ (2128).
Ph_6Ge_2 and polarographic reduction → $Ph_6Sn_2 + Ph_3GeH$ (963).
$Ph_3GeYLi → Ph_3SnYGePh_3$; Y = S, Se, Te (484).
$Ph_3PbYLi → Ph_3SnYPbPh_3$; Y = S, Se (485).
Li in THF → Ph_6Sn_2 (1928).
Excess Li → Ph_3SnLi (1928).

Li in THF, + SnCl$_2$ → (Ph$_3$Sn)$_4$Sn + Ph$_6$Sn$_2$ (68a, 175).
PhLi → Ph$_4$Sn; rxn. suitable for quant. anal. detn. of PhLi (68a, 173).
RLi → Ph$_3$SnR; R = MeCH:CH (1928), Ph$_2$PC⋮C (2778), Me$_2$CCN (1620) B$_{10}$H$_{10}$C$_2$H (784), B$_{10}$H$_{10}$C$_2$Ph (627), $\overline{C:CHCH:C}$(NHEt)C(:NEt)CH:CH (1004).
B$_{10}$H$_{10}$(CLi)$_2$ → (Ph$_3$SnC)$_2$B$_{10}$H$_{10}$ (2279).
LiX → Ph$_3$SnX; X = Me$_2$N (251), Ph$_2$C:N (852).
6-LiSC$_5$H$_3$N$_4$ → 6-(N-Ph$_3$SnC$_5$H$_2$N$_4$)SSnPh$_3$·C$_5$H$_3$N$_4$SSnPh$_3$ (1375).
Na-C$_{10}$H$_8$ in (MeOCH$_2$)$_2$ → Ph$_3$SnNa (395).
9,10-Na$_2$C$_{14}$H$_{10}$ → 9,10-(Ph$_3$Sn)$_2$C$_{14}$H$_{10}$; 9,10-dihydroanthracene (1726).
NaR → Ph$_3$SnR; R = C$_5$H$_5$M(CO)$_3$, M = Cr, W (1670), Mo (967, 1459, 1670), C$_5$H$_5$Mo(CO)$_2$PX$_3$; X = Ph, MeO, PhO (1459), C$_5$H$_5$Mo(CO)$_2$SbPh$_3$ (1459).
NaBH$_4$ in (MeOCH$_2$)$_2$ → Ph$_3$SnH + NaCl + B$_2$H$_6$ (736).
EtO$_2$CNHNa → Ph$_3$SnNCO + EtOH (373).
NaX → Ph$_3$SnX; X = Ph$_2$As (1842), Ph$_2$Sb (1843), Mn(CO)$_5$ (967), Co(CO)$_4$ (216, 730), Ph$_3$PIr(CO)$_3$ (891).
Na$_2$M(CO)$_4$ → (Ph$_3$Sn)$_2$M(CO)$_4$; M = Fe, Ru (905).
NaX → Ph$_3$SnX; X = p-RC$_6$H$_4$SO$_2$; R = H, Me, Br, AcNH (1410), 1/2(SiO$_2$)$_n$O, n = 1, 2 (2826), Ph$_2$PYS, Y = O, S (2108), (RO)$_2$PS$_2$, R = Et (745, 1565), n-, i-Pr (745).
NaOR → Ph$_3$SnOR; R = Ph, p-O$_2$NC$_6$H$_4$ (791), ethoxylated oleic acid (911), 8-OC$_9$H$_6$N (1291), C$_7$H$_5$O tropolone (2510).
NaO$_2$CR (at r.t.) → Ph$_3$SnO$_2$CR; R = Me, ClCH$_2$, Et, Ph (1051), C$_5$H$_5$FeC$_5$H$_4$ in EtOH under reflux (1310, 1374).
(NaO$_2$C)$_2$Y (at r.t.) → (Ph$_3$SnO$_2$C)$_2$Y; Y = (CH$_2$)$_2$, (CH$_2$)$_4$ (1051).
NaX → Ph$_3$SnX; X = C$_6$F$_5$S (1676), R$_2$NCS$_2$, R = Me (298), Et, Ph (298, 761), 2-C$_2$N$_2$S(SH), 5-mercapto-1,3,4-thiadiazole (1262).
(RO)$_2$POH + Na in alc. C$_6$H$_6$ → Ph$_3$SnPO$_3$R$_2$; R = Me, Et (1564).
Aq. alc. Na$_2$S$_2$O$_3$ → (Ph$_3$Sn)$_2$S (1791).
(HNCS)$_3$ + alc. NaOH → 2,4,6-(Ph$_3$SnS)$_3$C$_3$H$_3$-sym. triazine (1322a).
K in liq. NH$_3$ → Ph$_3$SnK (2128).
KX → Ph$_3$SnX; X = SCN (2751), SeCN (697, 2259), O:NC in liq. NH$_3$ (28).
K$_2$Ru(CO)$_4$ → (Ph$_3$Sn)$_2$Ru(CO)$_4$ (2328).
Et$_3$PO$_4$ + KOH → Ph$_3$Sn(Et$_2$)PO$_4$ (292).
(R$_2$N)$_2$POCl + KOH → Ph$_3$SnO$_2$P(NR$_2$)$_2$; R = Me, Et, H + Et, Pr, Ph (292).
(KS)$_2$C:NCN in aq. THF → Ph$_3$SnNCS + (Ph$_3$Sn)$_2$S (2751).
KO$_2$CR → Ph$_3$SnO$_2$CR; R = MeO$_2$CCH:CH (1557), MeO$_2$CCHCH$_2$CMe:CMeCH$_2$CH (1558).
p-C$_9$H$_{19}$C$_6$H$_4$(OCH$_2$CH$_2$)$_{9-10}$OH + alkali → Ph$_3$Sn-alkoxide (1958).
LiCl and R$_4$NCl in MeCN and Me$_2$CO, resp. → potentiometric titration and complex formation, R = Me, Et (2015).
AgX → Ph$_3$SnX; X = Ph$_2$N$_3$ (48a), NCO (373), NO$_3$ (85, 1030, 1934), Me$_3$CCO$_2$ (2671), O$_2$NNMe, O$_2$NNPh (611).
RMgX → Ph$_3$SnR; R = Me (296, 2412), $\overline{CH_2CH_2CH}$ (1873), o-PhC$_6$H$_4$ (1278), 1,8-PhC⋮C-C$_{10}$H$_6$ (1278), C$_6$F$_5$ (428), m-F$_3$CC$_6$H$_4$ (998), C$_6$Cl$_5$ (1101), Me$_3$SiC⋮C, Me$_2$PhSiC⋮C (1354).
Y(MgBr)$_2$ → (Ph$_3$Sn)$_2$Y; Y = MePhSi(C⋮C)$_2$, Ph$_2$Si(C⋮C)$_2$ (1354).
EtMgBr in Et$_2$O-C$_6$H$_6$ → Ph$_6$Sn$_2$ (1500).
2-(C$_5$H$_4$N)MgBr → Ph$_4$Sn·Ph$_6$Sn$_2$ (1:1.75) (1500).
2,2'-PhC⋮C(C$_6$H$_4$)$_2$MgI → Ph$_3$SnCPh:$\overline{CC_6H_4C_6H_4}$-o (1278).

Zn-Cu in THF → Ph_6Sn_2 (2830).
Zn-Cu + $(Me_2NCH_2)_2$ in THF → Ph_4Sn + Sn + $ZnCl_2$ (2830).
Zn-Cu in alc. THF → C_6H_6 + Sn + MeOZnCl (2830).
Olefin → comparative stability of π complexes (182).
NH_4X → Ph_3SnX; X = Ph_2PO_4 (292), $(RO)_2PS_2$; R = Me, Et, Pr (292).
Ph_3-$_nPH_n$ + Et_3N → $(Ph_3Sn)_nPPh_{3-n}$; n = 1, 2 (1849).
Ph_2AsO_2H + Et_3N → $Ph_3SnO_2AsPh_2$ (1842).
R_3SbS → $(Ph_3Sn)_2S$ + R_3SbCl_2 ; R = Me, c.C_6H_{11} (1903).
8-HOC_9H_6N (at 140°) → $Ph_2Sn(OC_9H_6N)_2$ + $(8-C_9H_6NO)_2SnCl_2$ (386).
Carboxymethyl cellulose → polymeric Ph_3Sn-compd. (1515).
Tropolone → $PhSn(O_2C_7H_5)_2Cl$ (372).
RCO_3H → $[Ph_3SnO_3CR]$ → $Ph_2Sn(OPh)O_2CR$ + PhOH, kinetic data and mechanism of rearrangement (790).
X_2 and polarographic reduction → Ph_3SnX; X = $C_5H_5Mo(CO)_3$, $Mn(CO)_5$, $C_5H_5Fe(CO)_2$ (967).
$Co[Co(CO)_4]_2$ in MeOH → $Ph_3SnCo(CO)_4$ (1672).
$[Ni(NH_3)_6]Cl_2$ + Me_2CO → $(C_6H_{14}N_2)_2NiCl_2$ (1248).
$(Ph_3P)_2PtL$ → $Ph_3Sn(PPh_3)PtCl$ + L; L = C_2H_4 , Ph_3P, kinetic data and relative reactivity (2248).
$Pt(PPh_3)_4$ → $Ph_3SnPt(PPh_3)_2Cl$ (1397).
trans-$(Ph_3P)_2PtHCl$ → $(Ph_3P)_2PtPhCl$ (700).
Addn. compd.: All compounds are listed in Table 128.
Biol. prop.: Toxicity (1099, 2073), effect on ovarian tissue of rats (2647). Phytotoxicity towards barley (376), citrus varieties (2291), cotton, lima beans and squash (1738).
Activity against Arachnids: Phyllocoptruta oleivora (2291) and Aculus pelekassi (1739), Tetranychus urtica (1738), Eriophyidae (2023).
Activity as bactericide (1099).
Activity as fungicide (1099), against Ceratostomella (Ceratocystis) coerulescens, Fusarium culmorum, Penicillium funiculosum, Speira hepatospora (1099), Cereospora beticola-sugar beets, Phytophthora infestans-on tomatoes, Septoria apri-celery (2741). Pericularia oryzae - rice blast (548).
Activity as insecticide against Culex quinque fasciatus, Hippelatus colluscor (2616), C. pipiens larvae (73), Musca domestica, LD_{50} and ATP inhibition (425), control and effect on reproduction (260, 1064), effect on reproduction (1315, 2436), effect as chemosterilant (172a), Heliothis zea, H. vivescens (2145), Oryzaephilus surinamensis, Sitophilus granarius, Sitotroga cerealella, Tribolium confusum, Trogoderma parabile larvae (2803), Popilla japonica as chemosterilant (1382), Tineola bisselliella larvae (162).
Activity as molluscicide against Australorbis glabratus (1860).
Use: As additive for flame resistant polyester (2702), for corona resistence in polyimide films (2585).
As antioxidant for polyphenyl ethers (1979).
In antifouling coatings (2302), in fungicidal compositions with decreased phytotoxocity (132, 1809a).

Table 128. Addition Compounds of Triphenyltin Chloride   Ph$_3$SnCl·L

Ligand	Synth. Method*	Yield	Properties	Ref.
Me$_4$NCl	--	--	(a-1963)	--
PhCH$_2$NH$_2$	--	--	(b)	--
4-Ph$_3$SnCH$_2$CH$_2$C$_5$H$_4$N	IC	--	m. 166-68°, soly. (a)	2689
4-PhC$_5$H$_4$N	IA	24	m. 113-15° (dec.), soly., struct. (a)	1706
4,4'- BiPy	IA	85	m. 132-34°, soly., struct. (a)	1706
PyO (c)	IA	--	m. 133-35° (dec.)	1296
	ID	--	m. 135-36°, IR spectrum, struct.	2340
2-MeC$_5$H$_4$NO (c)	IC	84	m. 131-32° (d)	2907
2-(n-C$_9$H$_{19}$)C$_5$H$_4$NO (c)	IC	99	m. 136-38° (d)	2907
4-(O$_2$N)C$_5$H$_4$NO (c)	IC	56	m. 115-17° (d)	2907
2,4-Me(O$_2$N)C$_5$H$_3$NO (c)	IC	89	m. 94-96° (d)	2907
	IC	94	m. 97-99° (d)	2907
C$_9$H$_7$NO	IC	89	m. 147-49°, quinoline (d, e)	2907
2-MeC$_9$H$_6$NO	IC	93	m. 96-98°, quinoline (d)	2907
1/2 (CH$_2$R·X)$_2$				
R, X = Me$_3$N, I	IE	29	m. 138-40° (f)	1837
Me$_2$EtN, Br	IE	46	m. 201-203° (f)	1837
Me$_2$PrN, Br	IE	35	m. 171°, soly. (f)	1837
Py, Br	IE	38	m. 257° (f)	1837
4-MeC$_5$H$_4$N, Br	IE	30	m. 247° (f)	1837

* Synthesis of triphenyltin chloride adducts is carried out from components I in C$_6$H$_6$-(IA), Me$_2$CO-(IB), alcohol-(IC), HCCl$_3$ or CH$_2$Cl$_2$-(ID), at 80-150°-(IE) or by fusion (IF). II by rxn. of Ph$_3$SnCl with R$_3$P and RX or R$_4$PX in MeOH.

(a) Far IR spectrum and assignment. (b) Rxn. with 6-HS-purine → N-Ph$_3$SnC$_5$H$_2$N$_4$SSnPh$_3$·C$_5$H$_3$SSnPh$_3$-6 (1375). (c) N-oxides. (d) Use as bactericide, fungicide and molluscicide with low toxicity (2907). (e) Activity as fungicide against Cercospora beticola - sugar beets, Phytophthora infestans - tomato, Septoria apii - celery (2741). (f) Use as algicide, molluscicide, bacteriostatic and fungistatic agent (1837).

Table 128 Continued   Ph$_3$SnCl·L

Ligand	Synth. Method*	Yield	Properties	Ref.
1/2 (PyCH$_2$)$_2$CH$_2$·BrCl	IE	47	m. 190° (f)	1837
1/2 (CH$_2$CH$_2$R·X)$_2$				
R, X = Me$_2$(C$_{12}$H$_{25}$)N, Br	IE	34	m. 130° (f)	1837
C$_5$H$_{10}$NH, Br	IE	53	m. 239-40°, 2-piperidine (f)	1837
MeNC$_4$H$_8$O, Br	IE	60	m. 220°, N-methylmorpholine (f)	1837
Py, Br	IE	80	m. 225°, soly. (f)	1837
Py, SCN	IE	79	m. 168-70° (f)	1837
2-MeC$_5$H$_4$N, Br	IE	62	m. 260° (f)	1837
4-MeC$_5$H$_4$N, Br	IE	65	m. 236-38° (f)	1837
2-H$_2$NC$_5$H$_4$N, Br	IE	62	m. 283° (f)	1837
C$_9$H$_7$N, Br	IE	69	m. 239-40°, quinoline (f)	1837
1/2 Me$_2$RN(CH$_2$)$_8$NRMe$_2$·X$_2$				
R, X = PhCH$_2$, Cl	IE	61	m. 214-15°, soly. (f)	1837
p-ClC$_6$H$_4$CH$_2$, Cl	IE	48	m. 224-25° (f)	1837
Et, Br	IE	73	m. 175-77° (f)	1837
Pr, Br	IE	85	m. 163° (f)	1837
C$_6$H$_{13}$, Br	IE	37	m. 70° (f)	1837
C$_8$H$_{17}$, Br	IE	29	m. 116-17° (f)	1837
C$_{12}$H$_{25}$, Br	IE	58	m. 94-95° (f)	1837
MeO$_2$CCH$_2$, Br	IE	45	m. 156-58° (f)	1837
PhCH$_2$O$_2$CCH$_2$, Cl	IE	7	m. 206-207° (f)	1837
EtO$_2$CCH$_2$, Cl	IE	43	m. 160° (f)	1837
PrO$_2$CCH$_2$, Cl	IE	18	m. 151-53° (f)	1837
BuO$_2$CCH$_2$, Cl	IE	40	m. 140-41° (f)	1837
C$_5$H$_{11}$O$_2$CCH$_2$, Cl	IE	41	m. 178-81° (f)	1837
C$_6$H$_{13}$O$_2$CCH$_2$, Cl	IE	41	m. 185-87° (f)	1837

$C_7H_{15}O_2CCH_2$, Cl	IE	61	m. 133° (f)	1837
$C_{10}H_{21}O_2CCH_2$, Cl	IE	50	m. 128-31° (f)	1837
$C_{12}H_{25}O_2CCH_2$, Cl	IE	23	m. 108-11° (f)	1837
$(PhCH_2)_2NOCCH_2$, Cl	IE	51	m. 225° (f)	1837
$PhNHOCCH_2$, Cl	IE	52	m. 195° (f)	1837
p-$MeC_6H_4NHOCCH_2$, Cl	IE	70	m. 193-95° (f)	1837
o-$ClC_6H_4NHOCCH_2$, Cl	IE	57	m. 175-76° (f)	1837
$2,4$-$Cl_2C_6H_3NHOCCH_2$, Cl	IE	73	m. 74-75° (f)	1837
p-$MeOC_6H_4NHOCCH_2$, Cl	IE	63	m. 170-72° (f)	1837
p-$EtOC_6H_4NHOCCH_2$, Cl	IE	40	m. 182° (f)	1837
2-$C_{10}H_7NHOCCH_2$, Cl	IE	77	m. 90-95° (f)	1837
$1/2$ $Py(CH_2)_{10}Py \cdot Br_2$	IE	45	m. 112° (f)	1837
$(ClCH_2)_4PCl$	II	74	m. 124-26° (g)	2908
$Et_3(PhCH_2)PBr$	II	71	m. 129° (g)	2908
Ph_3RPX				
R, X = H, Cl	II	33	m. 152-53° (g)	2908
Me, Cl	II	76	m. 146-47° (g)	2908
$ClCH_2$, Cl	II	82	m. 171-72° (g)	2908
$HOCH_2$, Cl	II	49-80	m. 138-42° (g)	2908
$MeOCH_2$, Cl	II	55	m. 136-38° (g)	2908
$EtOCH_2$, Cl	II	54	m. 117-19° (g)	2908
$EtSCH_2$, Cl	II	83	m. 160-61° (g)	2908
$PhCH_2$, Cl	II	67	m. 105-107° (g)	2908
p-$ClC_6H_4CH_2$, Cl	II	68	m. 152-54° (g)	2908
Ph_2CH , Cl	II	54	m. 205-206° (g)	2908
EtO_2CCH_2 , Cl	II	77	m. 177-78° (g)	2908
$C_{10}H_{21}$, Br	II	84	m. 119° (e, g)	2908
Ph, Cl	II	78	m. 203-205° (g)	2908
p-HOC_6H_4 , Cl	II	54	m. 201-203° (g)	2908

(g) Use as bacteriostatic and fungistatic agent (2908).

Table 128 Continued							Ph₃SnCl·L

Ligand		Synth. Method*	Yield	Properties	Ref.
R₂R'PO					
R' = H	R = Ph	IB, C	43	m. 139-40° (h)	1838
Ph	PhCH₂	IB, C	81	m. 160° (h)	1838
Et	Et	IB, C	40	m. 177° (h)	1838
Ph	Et	IB, C	80	m. 133-34 (h)	1838
Et	Ph	IB, C	87	m. 149° (h)	1838
Ph	Pr	IB, C	90	m. 158-60° (h)	1838
Ph	Bu	IB, C	90	m. 128-29° (h)	1838
Bu	Ph	IB, C	86	m. 143-45° (h)	1838
Me₂CHCH₂CH₂	Ph	IB, C	80	m. 155-57° (h)	1838
C₆H₁₃	Ph	IB, C	95	m. 93-95° (h)	1838
C₁₂H₂₅	Ph	IB, C	78	m. 76-78° (h)	1838
Ph	Ph	ID	--	Dec. 163-64°, IR spectrum	2340
		IB, C	77	m. 167-68.5, soly. (h)	1838
PhCH₂	Pr + EtO	IC	72	m. 105-107° (i)	1838a
Me₂N	Ph	IC	77	m. 156-57° (i)	1838a
PrNH	Ph	IC	72	m. 123-26° (i)	1838a
Me₂N	Bu	IC	44	m. 108-11° (i)	1838a
Ph	BuNH + C₅H₁₀N	IC	44	m. 89-99° (i)	1838a
MeO	o-MeC₆H₄O	IC	72	m. 68° (i)	1838a
OC₄H₈N	p-ClC₆H₄O	IC	97	Morpholine, m. 116° (i)	1838a
MeO	Me₂N	IC	88	m. 97-102° (i)	1838a
MeO	C₅H₁₀N	IC	73	Piperidine, m. 132-35° (i)	1838a
MeO	C₆H₁₀N (?)	IC	79	m. 132-35° (i)	1838a
PhO	Me₂N	IC	65	m. 78-80° (i)	1838a
PhO	Me₂N + RNH	IC	56	R = CH₂:CHCH₂ , m. 80-87° (i)	1838a
PhO	MePhN	IC	49	m. 91-92° (i)	1838a

468

PhO	PhCH$_2$NH	IC	90	m. 89-90° (i)	1838a
PhO	BuNH	IC	77	m. 85-86° (i)	1838a
Me$_2$N	PhS	IC	87	m. 100° (i)	1838a
(C$_7$H$_4$NS)S	OC$_4$H$_8$N	IC	55	m. 105-108° (i, j)	1838a
Me$_2$N	Me$_2$N	IC	77	m. 158-60° (i)	1838a
$\overline{CH_2CH_2N}$	$\overline{CH_2CH_2N}$	IC	78	m. 130-31° (i)	1838a
C$_5$H$_{10}$N	C$_5$H$_{10}$N	IC	89	Piperidine m. 161-63° (i)	1838a
c.C$_6$H$_{11}$NH	c.C$_6$H$_{11}$NH	IC	68	m. 178-79° (i)	1838a
PhC$_2$HN$_3$NH	Me$_2$N	IC	70	m. 99-102° (i, k)	1838a
Me$_2$N	PhNHNH + R	IC	57	R = morpholine, m. 135-38° (i)	1838a
2-C$_{10}$H$_7$NH	OC$_4$H$_8$N	IC	72	Morpholine, m. 183° (i)	1838a
C$_7$H$_5$N$_2$	OC$_4$H$_8$N	IC	60	Benzoimidazole, morpholine, m. 156-58° (i)	1838a
Et$_2$N(OC$_4$H$_8$N)ClPO		IC	52	Morpholine, m. 72-80° (i)	1838a
PhS(OC$_4$H$_8$N)$_2$PS		IC	82	Morpholine, m. 50-55° (i)	1838a
(OC$_4$H$_8$N)$_3$PS		IC	60	Morpholine, m. 95-97° (i)	1838a
Ph$_3$AsO		ID	--	Dec. 200-202°, IR spectrum	2340
Ph$_3$SBr		IB	--	m. 165° (i)	1182
R$_2$SO					
R = Me		I	91	--	1388
		IC	87	m. 114-15°, soly. (e, m)	1836
PhCH$_2$		IC	32	m. 91° (m)	1836
Et		IC	65	m. 115-17° (m)	1836
R$_2$ = (CH$_2$)$_4$		IC	72	m. 113-114° (m)	1836
O(CH$_2$CH$_2$)$_2$		IC	64	m. 116-117° (m)	1836
S(CH$_2$CH$_2$)$_2$		IF	> 50	m. 127-28.5°, soly. IR spectrum	1707

(h) Use as bactericide and fungicide with low phytoxicity (1838). (i) Activity as bactericide against Staphylococcus aureus (1838a), as fungicide against Aspergillus niger, Candida albicans (1838a), use as acaricide, herbicide, insecticide (1838a). (j) 2-benzothiazole, morpholine. (k) PhC$_2$HN$_3$NH$_2$ = 3,5-phenyl(amino)-1,2,4-triazole. (1) Use as esterification catalyst (1182). (m) Activity against Staphylococcus aureus, Aspergillus niger and Candida albicans, Phytophthora infestans, Cercospora beticola (1836), use as algicide and molluscicide with low toxicity (1836).

Table 128 Continued   Ph$_3$SnCl·L

Ligand	Synth. Method*	Yield	Properties	Ref.
R$_2$SO (Cont'd.)				
R$_2$ = PhCH$_2$ + Ph	IC	64	m. 102° (m)	1836
Et + Ph	IC	67	m. 90-93° (m)	1836
Pr + Ph	IC	68	m. 81-83° (m)	1836
i-Pr + Ph	IC	55	m. 88-95° (m)	1836
CH$_2$:CHCH$_2$ + Ph	IC	70	m. 89-90° (m)	1836
Bu + p-ClC$_6$H$_4$	IC	68	m. 81-82° (m)	1836
C$_6$H$_{13}$ + Ph	IC	71	m. 93-97° (m)	1836
R = Ph	IC	87	m. 102-104° (m)	1836
p-MeC$_6$H$_4$	IC	82	m. 106-108° (m)	1836
2,4-Me$_2$C$_6$H$_3$	IC	69	m. 97-100° (m)	1836
2,5,4-Me$_2$ClC$_6$H$_2$	IC	84	m. 139-42° (m)	1836
p-MeOC$_6$H$_4$	IC	71	m. 105-107° (m)	1836
p-AcOC$_6$H$_4$	IC	65	m. 130-33° (m)	1836
p-MeHNC$_6$H$_4$	IC	77	m. 130-33° (m)	1836
p-Me$_2$NC$_6$H$_4$	IC	84	m. 149-50° (m)	1836
p-Et$_2$NC$_6$H$_4$	IC	55	m. 93-96° (m)	1836

Triphenyltin Bromide (M-280) $SnC_{18}H_{15}Br$ Ph_3SnBr

Synth.: By rxn. of Ph_4Sn with Br in $HCCl_3$ at $-10°$ (1719).
By rxn. of $(Ph_3Sn)_4Sn$ with $(CH_2Br)_2$, in 64% yield (68a, 175).
By rxn. of $Ph_3SnO_2CCBr_3$ under reflux in C_8H_{14} (1875).
By solvent extraction of Ph_3SnOH with aq. $NaBr$ (487).
By-prod. in rxn. of Ph_3SnLi with C_6F_5Br (1034).
By-prod. in rxn. of $(Ph_3Sn)_2Hg$ with Br in C_6H_6 (2370).

Prop.: IR, assignment (2592), far IR, assignment (1703), Moessbauer resonance (211, 2290, 2665), NQR (2290) and mass (850) spectra.

Rxn. with: $(Me_3Ge)_2Hg$ + Li in THF → $Ph_3SnGeMe_3$ (849).
Na + $(CH_2Br)_2$ in liq. NH_3 → $Ph_3SnC{:}CBr$ (443).
RCO_2Na → Ph_3SnO_2CR; R = $1\text{-}C_9H_{15}$, 1-apocamphane (1310, 1374).
RCO_2Na → $Ph_2Sn(O_2CR)_2$; R = 2-pyridyl (1310, 1374).
MSeCN → $Ph_3SnNCSe$; M = Na, K (2259).
AgX → Ph_3SnX; X = ClO_4, NO_3 (85).
RMgBr → Ph_3SnR; R = C_6F_5 (77), mesityl (303).
$RC{:}CMgBr$ → $Ph_3SnC{:}CR$; R = Me, $CH_2{:}CH$, $CH_2{:}CMe$, $1\text{-}c.C_6H_9$, $9\text{-}C_{14}H_9CH_2$, $HOCH_2$, $p\text{-}BrC_6H_4$ (443).
$9,10\text{-}(BrMg)_2C_{14}H_{10}$ → $(Ph_3Sn)_2C_{14}H_{10}$ 9,10-dihydroanthracene (443).
$CH_2{:}CHCH_2Br$ + Py → $Ph_3SnC_3H_5Br_2 \cdot Py$ (2592).
$3\text{-}C_6H_9Br$ + Py → $Ph_3SnC_6H_9Br_2 \cdot Py$ (2592).
R_3SbS → $(Ph_3Sn)_2S$ + R_3SbBr_2 ; R = Me, $c.C_6H_{11}$ (1903).
RCO_2H + $PhCH_2NH_2$ → Ph_3SnO_2CR + $PhCH_2NH_3Br$; R = $1\text{-}C_7H_{11}$, $1\text{-}C_9H_{15}$, $1\text{-}C_{20}H_{13}$, $C_5H_5FeC_5H_4$ (1310, 1374).
$Ni(NH_3)_6Br_2$ in Me_2CO → $(C_6H_{14}N_2)_2NiBr_2$ (1248).

Addn. compd.: See following Table 129.

Biol. prop.: Activity as insecticide against Culex pipiens berbericus larvae (73), C.p. pipiens larvae (1137), Tineola bisselliella (162).

Use: In catalyst for C_2H_4 polymerization (1507).
As agent to control pecan bunch disease (2335).

Table 129. Addition Compounds of Triphenyltin Bromide* $Ph_3SnBr \cdot L$

Ligand	Yield	Properties	Ref.
[(PhCH$_2$)$_3$PCH$_2$Cl]Cl	86	m. 156-57°	2908
[Et$_3$PCH$_2$Ph]Cl	71	m. 129°	2908
[Et$_2$PhPCH$_2$Ph]Cl	79	m. 114-15°	2908
[EtPh$_2$PCH$_2$Ph]Cl	100	m. 178°	2908
[Bu$_3$PC$_{12}$H$_{25}$-n]Cl	16	m. 118°	2908
[Ph$_3$PR]Cl			
R = PhCH$_2$	78	m. 195-96°	2908
Et	61	m. 139-40°	2908
Pr	50	m. 114-15°	2908
CH$_2$:CHCH$_2$	67	m. 117-18°	2908
Ph(CH$_2$)$_3$	57	m. 125-26°	2908
Bu	65	m. 127-29°	2908
MeCHCH$_2$CH$_2$	73	m. 150-53°	2908
n-C$_6$H$_{13}$	80	m. 93-94°	2908
n-C$_{10}$H$_{21}$	84	m. 119°	2908
p-ClC$_6$H$_4$	68	m. 161-63°	2908
p-O$_2$NC$_6$H$_4$	81	m. 201-202°	2908

* The compounds are prepared from Ph$_3$SnBr, R$_3$P and R'X or R$_3$R'PX in MeOH (2908). All complexes are claimed to have use as bacteriostatic and fungistatic agents (2908).

Triphenyltin Iodide (M-280)   SnC$_{18}$H$_{15}$I   Ph$_3$SnI

Synth.: By rxn. of Ph$_3$SnR with I; R = Ph in HCCl$_3$, in 80% yield (1719), PhC⋮C in Et$_2$O in Ar atm., in 84% yield (2644).
By rxn. of Ph$_4$Sn·1.75Ph$_6$Sn$_2$ with I in HCCl$_3$ (1500).
By rxn. of Ph$_3$SnOH with NH$_4$I in c.MeC$_6$H$_{11}$ under reflux, in 90% yield (2660), with aq. Na^{131}I and solvent extraction (487).
By rxn. of Ph$_6$Sn$_2$ with MeI at 140° in C$_6$H$_6$ in sealed bottle, in 6% yield (1938).
By rxn. of (Ph$_2$Sn)$_x$ with Ph$_2$SnI$_2$ at 140° in C$_6$H$_6$ in sealed bottle, in 53% yield (1938).
By rxn. of (Ph$_3$Sn)$_4$Sn with I in HCCl$_3$, in 82% yield (68a, 175).
By rxn. of Ph$_3$SnCo(CO)$_4$ with CF$_3$I under UV irradiation (730).
By rxn. of Ph$_3$Sn(Ph$_3$P)Ir(CO)$_3$ with I in THF (891).
Prop.: Far IR, assignment (1703), Moessbauer resonance (1309, 2290, 2665), NQR (2290) and mass (849-50) spectra, dipole moment (1518).
Rxn. with: RNa → Ph$_3$SnR; R = HC⋮C, PhC⋮CC⋮C (443).
Na$_2$C$_2$ → (Ph$_3$SnC⋮)$_2$ (443).
AgX → Ph$_3$SnX; X = NO$_3$ (1969), NCC(CN)$_2$ (717), N(CN)$_2$ (2501).

$Ag_2N_2O_2 \rightarrow (Ph_3Sn)_2N_2O_2$ (716).

$RMgBr \rightarrow Ph_3SnR$; R = i-Pr (521, 1719), i-Bu (521), n-, i-C_5H_{11}, n-$C_{10}H_{21}$ (2787), HC⋮CCH_2, HC⋮CCHMe, MeC⋮CCH_2 (444).

$Ni(NH_3)_6I_2 + Me_2CO \rightarrow (C_6H_{14}N_2)_2NiI_2$ (1248).

Addn. compd.: $Ph_3SnI \cdot Me(PhCH_2)_2PCl$, from components in MeOH, in 22% yield, m. 121°, use as bacteriostatic and fungistatic (2908).

Biol. prop.: Activity as insecticide, effect on reproduction of Musca domestica (260), against Tineola bisselliella larvae (162).

Use: As bactericide (529), agent to control pecan bunch disease (2335).

Triorthotolyltin Iodide (M-285) $SnC_{21}H_{21}I$ $(o\text{-}MeC_6H_4)_3SnI$

Synth.: By rxn. of $(o\text{-}MeC_6H_4)_4Sn$ with I in $HCCl_3$ or CCl_4, in 55% yield (1967), in 65% yield (1968).

Prop.: m. 119°, soly. (1967-8).

Rxn. with: Alkali $\rightarrow [(o\text{-}MeC_6H_4)_3Sn]_2O$ (1964).

Aq. KOH in EtOH $\rightarrow (o\text{-}MeC_6H_4)_3SnOH$ (1968).

AgX in $C_6H_6 \rightarrow (o\text{-}MeC_6H_4)_3SnX$; X = NO_3 (1964), SCN (1966), AcO (2793).

$Pb(NCO)_2$ in $C_6H_6 \rightarrow (o\text{-}MeC_6H_4)_3SnNCO$ (1967).

Triparatolyltin Chloride (M-281) $SnC_{21}H_{21}Cl$ $(p\text{-}MeC_6H_4)_3SnCl$

Synth.: By-prod. in thermal dec. of $(p\text{-}MeC_6H_4)_3SnNO_3$ in $o\text{-}Cl_2C_6H_4$ (2388).

Rxn. with: $(Me_3Sn⋮C)_2 \rightarrow [(p\text{-}MeC_6H_4)_3Sn⋮C]_2$ (205, 2433).

Aq. NaOH in $Et_2O \rightarrow [(p\text{-}MeC_6H_4)_3Sn]_2O$ (1375).

NaN_3 in $Et_2O \rightarrow (p\text{-}MeC_6H_4)_3SnN_3$ (1965).

$CH_2{:}CMeCH_2MgCl \rightarrow (p\text{-}MeC_6H_4)_3SnCH_2CMe{:}CH_2$ (1857).

$RMgBr \rightarrow (p\text{-}MeC_6H_4)_3SnR$; R = Me (296), $CH_2{:}CHCH_2$ (1857), $m\text{-}F_3CC_6H_4$ (998).

Triparatolyltin Iodide $SnC_{21}H_{21}I$ $(p\text{-}MeC_6H_4)_3SnI$

Synth.: By rxn. of $(p\text{-}MeC_6H_4)_4Sn$ with I in $HCCl_3$ (1966), in 61% yield (1967).

Prop.: m. 120° (1967).

Rxn. with: Alkali $\rightarrow [(p\text{-}MeC_6H_4)_3Sn]_2O$ (1964).

AgX in $C_6H_6 \rightarrow (p\text{-}MeC_6H_4)_3SnX$; X = NO_3 (1964), SCN (1966), AcO (2793).

$Pb(NCO)_2$ in $C_6H_6 \rightarrow (p\text{-}MeC_6H_4)_3SnNCO$ (1967).

Additional symmetric unsubstituted triorganotin halides are listed in Table 130.

Table 130. Symmetric Unsubstituted Triorganotin Halides

Formula*	Method**	Yield	Properties	Ref. R_3SnX
(PhCH$_2$)$_3$SnBr	III	11	m. 108-10° (a)	841
Pr$_3$SnI (282)	IC	92	--	711
	ID	--	--	24a
i-Pr$_3$SnBr (282)	IVE	88	b$_2$ 128-32°	2660
(c.C$_3$H$_5$)$_3$SnF (282)	--	--	NMR spectrum (d)	2070
(c.C$_3$H$_5$)$_3$SnBr (282)	--	--	Use: herbicide, pesticide	1873
(c.C$_3$H$_5$)$_3$SnI (282)	II	--	Use: herbicide, pesticide (e)	1873
(PhCH$_2$CH$_2$)$_3$SnF	--	--	Use: herbicide, pesticide	1873
(PhCH$_2$CH$_2$)$_3$SnCl	(f)	84	m. 209-211°	2188
i-Bu$_3$SnF (283)	II	56	m. 61-61.5° (f)	2188
i-Bu$_3$SnBr	--	--	(b)	--
1-Bu$_3$SnI	IIIC	(g)	(b, c)	616a
s-Bu$_3$SnI	ID	100	b$_0.1$ 76-78°, n$_D^{20}$ 1.5269, d$_{25}$ 1.44 (b, k)	1618
(CH$_2$CHMeCH)$_3$SnCl	ID	--	b$_0.1$ 89-92°, n$_D^{20}$ 1.5332 (i)	1618
(PhMe$_2$CH$_2$)$_3$SnF	--	--	Use: herbicide, pesticide (j)	1873
(PhMe$_2$CH$_2$)$_3$SnBr	IVA	87	m. 98.5-99.5° (k)	2188
	IVD	--	m. 98-100°, soly., NMR spectrum	1742
	IA	--	m. 109.5-10°, soly., NMR spectrum	1742
	II	23	Moessbauer resonance spectrum (2290)	1742
	IVD	--	(k)	1742
(PhCMe$_2$CH$_2$)$_3$SnI	IVD	--	m. 116.5-17.5°, soly., NMR spectrum (k)	1742
(n-C$_5$H$_{11}$)$_3$SnF	IIIB	--	m. 376-78°	2557
(n-C$_5$H$_{11}$)$_3$SnCl (283)	IIIA	68	--	1583a
(1-C$_5$H$_{11}$)$_3$SnI (283)	--	--	(l)	--
(Me$_3$CCH$_2$)$_3$SnCl (284)	--	--	NMR spectrum (m)	2188
(CH$_2$CMe$_2$CH)$_3$SnCl	--	--	Herbicide, pesticide (n)	1873
(n-C$_6$H$_{13}$)$_3$SnF	IIIB	--	m. 398-99° (dec.)	2557
(n-C$_6$H$_{13}$)$_3$SnI	--	--	(o)	--

Compound				
(Me₃CCH₂CH₂)₃SnF	ID	--	m. 333-36°, subl.o.₁ 200°	2887
(Me₃CCH₂CH₂)₃SnBr	IVA	--	--	2887
(c.C₆H₁₁)₃SnI	IA	48	Needles 128-30°, soly. (p)	2887
(n-C₇H₁₅)₃SnI (284)	IVE	94	m. 65°	2660
(1-BuCHCH₂CH)₃SnI	--	--	(o)	1873
(n-C₈H₁₇)₃SnI	--	--	Use: herbicide, pesticide (q)	--
[CH₂CH₂CH(CH₂)₅]₃SnCl	--	--	(o)	1873
(n-C₉H₁₉)₃SnI	--	--	Use: herbicide, pesticide (r)	--
(C₁₀H₂₁)₃SnCl	--	--	(t)	--
(C₁₂H₂₅)₃SnCl (284)	IIIA	47	(u)	1583a
(o-MeC₆H₄)₃SnF	IVB	93	m. 245°, NMR spectrum (2356) (v)	1968
(o-MeC₆H₄)₃SnCl	II	10	m. 103°, soly.	1968
(o-MeC₆H₄)₃SnCl (280)	IVB	65	(v, w)	1968
(o-MeC₆H₄)₃SnBr (284)	--	--	(x, y)	--

* Numbers in parenthesis refer to pages in main volume.
** The characters used here correspond to the synthetic scheme in the introduction to Chapter 4.1.1.1.

(a) (Far) IR spectra and asigment (841). (b) Mass spectrum (2410). (c) Activity as bactericide against Escherichia coli and Staphylococcus aureus (702). (d) Rxn. with RLi → 1-Pr₃SnR; R = Ph, 1-, 2-C₁₀H₇ (59). (e) Rxn. with LiCH₂C:CEt → (c.C₃H₅)₃SnCH₂C:CEt (1873), with aq. KOH → [(c.C₃H₅)₃Sn]₂O (1873). (f) Rxn. of R₃SnCl with aq. KF in MeOH → R₃SnF; R = Ph(CH₂)₃ (2188). (g) As by-prod. in synth. under ⁶⁰Co irradiation (616a). (h) Rxn. with Ph₃SnH and Et₃N (1:1:1) → 1-Bu₃SnSnPh₃ (1618), with RMgBr → 1-Bu₃SnR; R = CH₂:CHCH₂ , CH₂:CHCH₂C₆H₄ (562). (i) Rxn. with LiAlH₄ → s-Bu₃SnH (1618), with aq. NaOH → (s-Bu₃Sn)₂O (1618). (j) Rxn. with Na → (CH₂CHMeCH₂)₆-Sn₂ (1873). (k) Moessbauer resonance spectrum (222, 224, 1742). (l) Rxn. with RMgX → (1-C₅H₁₁)₃SnR; R = p-CH₂: CHC₆H₄ , p-CH₂:CMeC₆H₄ , 1-C₉H₁₉ (617). (m) Rxn. with aq. NaOAc in C₆H₆ → (Me₃CCH₂)₃SnOAc (2188). (n) Rxn. with (CPh:CPhLi)₂ → [(CH₂CMe₂CH)₃SnCPh:CPh]₂ (1873). (o) Rxn. with RMgBr → R'₃SnR; R = CH₂:CHCH₂ , CH₂:CHCH₂C₆H₄ ; R' = n-C₆H₁₃, n-C₇H₁₅, n-C₈H₁₇ (562). (p) Rxn. with Li in THF → (Me₃CCH₂CH₂)₆Sn₂ (2887). (q) Rxn. with EtC:CCH₂Li → Me₂CHCHCH₂CHCH₂CHSnCH₂C:CEt (1873). (r) Rxn. with PrCH:CHCH₂Li → [CH₂CH₂CH(CH₂)₅]₃SnCH₂CH:CHPr (1873). (t) Rxn. with CH₂:CHCH₂MgBr → (n-C₉H₁₉)₃SnCH₂CH:CH₂ (562). (u) Rxn. with 9,10-Na₂C₁₄H₁₀ → (R₃Sn)₂C₁₄H₁₀; R = C₁₀H₂₁, C₁₂H₂₅ (1726). (v) IR spectrum and assignment (1968). (w) Rxn. with NaN₃ in Et₂O → (o-MeC₆H₄)₃SnN₃ (1965). (x) Rxn. with RMgBr → R'₃SnR; R' = o-tolyl, p-xylyl, 1-C₁₀H₇ ; R = mesityl (303). (y) Rxn. with C₂N₂S(SH)₂ in alc. NaOH → [(o-MeC₆H₄)₃SnS]₂C₂N₂S, 1,3,4-thiadiazole (1262).

Table 130 Continued

Formula*	Synth. Method**	Yield	Properties	Ref.
(p-MeC$_6$H$_4$)$_3$SnBr (285)	II	--	m. 98°, soly. (z)	2009
(p-Me$_2$C$_6$H$_3$)$_3$SnCl	--	--	Toxicity (aa)	1338
(p-Me$_2$C$_6$H$_3$)$_3$SnBr	--	--	Toxicity (x)	1338
(p-Me$_2$C$_6$H$_3$)$_3$SnI	--	--	Toxicity	1338
Mesityl$_3$SnBr (285)	--	--	Toxicity (bb)	1338
Mesityl$_3$SnI (285)	--	--	Toxicity (bb)	1338
(1-C$_{10}$H$_7$)$_3$SnBr (285)	--	--	Toxicity (x)	1338
(1-C$_{10}$H$_7$)$_3$SnI	--	--	Toxicity	1338
R$_3$SnX	IIIC (cc)	--	(dd) X = Cl, Br; R = alkyl (ff)	1348
Ar$_3$SnX	IIIC (ee)	--	X = Br, I, toxicity	2970
				1338

(z) Rxn. with Na in xylene → (p-MeC$_6$H$_4$)$_6$Sn$_2$ (2009). (aa) Rxn. with C$_2$N$_2$S(SH)$_2$ in alc. NaOH → (p-Me$_2$C$_6$H$_4$)$_3$SnSC$_2$N$_2$S-(SH), 1,3,4-thiadiazole (1262). (bb) Refractometric data (1391). (cc) Synth. by electrolysis in ether alc. with ZnBr$_2$ catalyst and Mg cathode (1348). (dd) Use in catalyst for epoxide resin hardening (2916), with Et$_2$POEt → R$_3$SnX·OPEt$_3$ addn. compd., IR spectrum (441). (ee) Synth. by heating in presence of RI and Et$_3$N as catalyst (2970). (ff) ^{35}Cl and ^{81}Br NQR spectra, resp. (2754).

4.1.1.2 UNSYMMETRIC UNSUBSTITUTED TRIORGANOTIN HALIDES

Unsymmetric triorganotin halides listed in this chapter may contain two or three different unsubstituted organic groups linked to a tin atom.

The triorganotin halides summarized in Tables 131 through 135 are prepared by the following methods.

I Dealkylation reactions:
  (IA) $R_3SnR' + X \rightarrow R_2R'SnX$; X = Br, I ,
  (IB) $Ph_3SnR + HCl \rightarrow Ph_2RSnCl$,
  (IC) $(C_6F_5)_4Sn + SF_4 \rightarrow (C_6F_5)_3SnF$.

II Redistribution reactions:
  (IIA) $R_4Sn + R'SnX_3 \rightarrow R_2R'SnX$,
  (IIB) $R_4Sn + R'R''SnX_2 \rightarrow R_2R'SnX + R_2R''SnX + RR'R''SnX$,
  (IIC) $R_3SnCl + R'_3SnCl \rightarrow R_nR'_{3-n}SnCl$.

III Alkylation reactions:
  (IIIA) $R_2SnX_2 + R'MgX \rightarrow R_2R'SnX$,
  $(R_2Sn)_x + R'MgX, + HX \rightarrow R_2R'SnX$,
  (IIIB) $R_2SnI_2 + Zn$ or Zn-Cu alloy $+ R'I \rightarrow R_2R'SnI$,
  (IIIC) $(R_2Sn)_x + R'X \rightarrow R_2R'SnX$; X = Br, I ,
  (IIID) $R_2SnBr_2 + PhLi \rightarrow R_2PhSnX$,
  (IIIE) $R_nSnX_{4-n} + CH_2N_2 \rightarrow R_n(XCH_2)_{3-n}SnX$; n = 0-2 .

IV Addition reactions with organotin hydrides and halides:
  (IVA) $R_2SnH_2 + R_2SnX_2 + R'RC:CH_2 \rightarrow R_2(R'RCHCH_2)SnX$,
  (IVB) $Bu_2SnHCl + diene \rightarrow Bu_2R'SnCl$,
  (IVC) $RSnH_3 + RSnBr_3 + diene \rightarrow RR'_2SnBr$,
  (IVD) $R_2SnH_2 + R_2SnCl_2 + R'C\vdots CH \rightarrow R_2(R'CH:CH)SnCl$.

V Interconversion reaction:
  $R_3SnOH + HX \rightarrow R_3SnX$.

R = organic group, may be different in one molecule; R' = R'' = organic group, may be unsaturated or substituted; X = halogen.

Diphenylmethyltin Chloride $SnC_{13}H_{13}Cl$ MePh$_2$SnCl
Synth.: By rxn. of MeSnPh$_3$ with I in HCCl$_3$, followed by extraction with NaOH and HCl, in 44% yield (68a, 176).
By rxn. of Ph$_2$SnI$_2$ with MeI and Zn dust, followed by aq. NaOH and HCl or anhydr. HCl in MeOH (2811).
Prop.: $b_{0.15}$ 129-30° (68a, 176).
Rxn. with: Li in THF $\rightarrow$ MePh$_2$SnLi (68a, 176).
Me$_2$CHMgBr $\rightarrow$ Me(i-Pr)SnPh$_2$ (2412).

Diphenylmethyltin Iodide $SnC_{13}H_{13}I$ MePh$_2$SnI
Synth.: By rxn. of MeSnPh$_3$ with I in Et$_2$O (521), in HCCl$_3$ under reflux (1938).
Prop.: b_3 185°, n_D^{20} 1.6492, d_{20} 1.7800 (521).
Rxn. with: 15% aq. KOH in Et$_2$O $\rightarrow$ MePh$_2$SnOH (2787).

Aq. KOH in Et$_2$O, + AcOH → MePh$_2$SnOAc (521).
I in Et$_2$O → MePhSnI$_2$ + PhI (2787).
MeI at 140°, + n-BuMgI → MeBuSnPh$_2$ (1938).

Butyldiethyltin Chloride   SnC$_8$H$_{19}$Cl   Et$_2$BuSnCl
<u>Synth.:</u> By rxn. of Et$_2$SnCl$_2$ with powd. Fe in BuCl-Et$_3$N at 160°, in 17% yield
(1939), with (Bu$_2$Sn)$_x$ and Et$_3$N, in 7% yield (1941).
By rxn. of (Et$_2$Sn)$_x$ with BuCl and Et$_4$NI at 160°, in 31% yield (1941).
By rxn. of Bu$_2$SnCl$_2$ with powd. Fe in EtCl-Et$_3$N at 160°, in 5% yield (1939).
<u>Prop.:</u> n$_D^{20}$ 1.4998, GLC (1939, 1941).

Dibutylethyltin Chloride   SnC$_{10}$H$_{23}$Cl   EtBu$_2$SnCl
<u>Synth.:</u> By rxn. of Bu$_2$SnCl$_2$ with powd. Fe in EtCl-Et$_3$N at 160°, in 26% yield
(1939), with EtI and Cu-Zn alloy at 80-90° (2808).
By rxn. of Et$_2$SnCl$_2$ with powd. Fe in BuCl-Et$_3$N at 160°, in 3% yield (1939),
with (Bu$_2$Sn)$_x$ and Et$_3$N at 160°, in 12% yield (1941).
By rxn. of (Et$_2$Sn)$_x$ with BuCl and Et$_4$NI at 160°, in 6% yield (1941).
<u>Prop.:</u> n$_D^{20}$ 1.4940 (1941), GLC (1939, 1941).

Listing of unsymmetric unsubstituted triorganotin halides continues in Table 131.

4.1.1.4 UNSATURATED TRIORGANOTIN HALIDES

Trivinyltin Chloride (M-292)   SnC$_6$H$_9$Cl   (CH$_2$:CH)$_3$SnCl
<u>Prop.:</u> Dipole moment (1211).
Rxn. with: 9,10-Na$_2$C$_{14}$H$_{10}$ → 9,10-[(CH$_2$:CH)$_3$Sn]$_2$C$_{14}$H$_{10}$, 9,10-dihydroanthracene
(1726).
<u>Addn. compd.:</u> (CH$_2$:CH)$_3$SnCl·OP(NMe$_2$)$_3$, from components in ROH, in 79% yield,
m. 81-82°, activity as acaricide, bactericide, herbicide, insecticide (1838a).
<u>Biol. prop.:</u> Activity as bactericide against Pseudomonas aeruginosa, Aerobacter aerogenes (2183).
Activity as fungicide against Penicillium funiculosum, Aspergillus flavus
(2183), Candida albicans (2183).

Other unsaturated triphenyltin halides are compiled in Table 132. Functionally
substituted unsaturated triorganotin halides are listed in Subsections 4.1.1.6.1
and 4.1.1.6.2. Table 132 is on page 482.

Table 131. Unsymmetric Unsubstituted Triorganotin Halides RR'_2SnX

Formula*	Synth. Method**	Yield	Properties	Ref.
$Me_2EtSnCl$ (287)	IIIA, Me_2SnO	18-29	--	149
$Me_2(CH_2CH_2CH)SnCl$	--	--	Use: herbicide, pesticide (a)	1873
$Me_2BuSnCl$	IIA, $BuSnCl_3$	high	b_{30} 108°, n_D^{20} 1.4971	1370
	IIA, $MeBuSnCl_2$	91	GLC	1370
$Me_2(i-Bu)SnCl$	IIA, $i-BuSnCl_3$	low	--	1370
$Me_2(c.C_6H_{11})SnBr$	IA, $Me_3SnC_6H_{11}$	99	(b)	2269
$Me_2(c.C_6H_{11})SnI$ (287)	IIIB, Me_2SnI_2	--	b_1 93-94°	2809
$MeEt_2SnI$	IIIC, Et_2Sn	49	b_{35} 90-92°, n_D^{20} 1.5607, GLC	1941
$MeEtBuSnCl$	IA, $EtBuSnCl_2$	76	b_{35} 127-28°, GLC	1370
$MePr_2SnI$	IIIC, Pr_2Sn	63	b_4 85-87°, n_D^{20} 1.5468, GLC	1941
$Me(i-Pr)c.C_6H_{11}SnBr$	IA, $Me_2PrC_6H_{11}$	98	GLC (c)	2269
$Me(i-Pr)PhSnBr$	IA, $MePrSnPh_2$	high	$b_{0.15}$ 81.3-81.6°, n_D^{20} 1.578, mass spectrum (d)	2412
$Me(i-Pr)PhSnI$	IA, $MePrSnPh_2$	90	$b_{0.001}$ 90°, NMR spectrum (e)	1719
$MeBu_2SnI$ (288)	IIIC, Bu_2Sn	58-74	$b_{2.5}$ 93-96°	1941
	IIIB, Bu_2SnI_2	--	b_1 98-99°	2809
$Me(PhCMe_2CH_2)PhSnCl$	--	--	NMR spectrum, sterochemistry	2672
$MePh_2SnBr$ (288)	IA, $MeSnPh_3$	70	$b_{0.2}$ 125.5°, n_D^{20} 1.629, mass spectrum (f)	2412
$PhCH_2Bu_2SnBr$	IIIC, Bu_2Sn	49	b_1 140-55°	66
$Et_2PrSnBr$	IIIC, Et_2Sn	12	n_D^{20} 1.5182, GLC	1941
Et_2PrSnI (288)	IIIC, Et_2Sn	39	b_5 85-86°, n_D^{20} 1.5470, GLC	1941

* Numbers in parenthesis refer to pages in main volume.
** The characters used here correspond to the synthetic scheme in the Introduction of Chapter 4.1.1.2.
(a) Rxn. with NaOH → $Me_2CH_2CH_2CHSnOH$ (1873). (b) Rxn. with i-PrMgBr → $Me_2(i-Pr)SnC_6H_{11}$ (2269). (c) Rxn. with EtMgBr → $MeEt(i-Pr)C_6H_{11}Sn$ (2269). (d) Rxn. with RMgBr → $Me(i-Pr)RSnPh$; R = n-Pr, n-, i-, s-Bu, p-$MeOC_6H_4$ (2411). (e) Rxn. with $KOC_{10}H_{17}$ → $Me(i-Pr)PhSnOC_{10}H_{17}$-bornyl (1719). (f) Rxn. with i-PrMgBr → $Me(i-Pr)SnPh_2$ (2412).

Table 131 Continued

Formula*	Synth. Method**	Yield	Properties	Ref.
Et$_2$BuSnBr	IIIC, Et$_2$Sn	28	n_D^{20} 1.5150, GLC	1941
Et$_2$BuSnI (288)	IIIC, Et$_2$Sn	41	b$_5$ 100-102°, n_D^{20} 1.5411, GLC	1941
Et$_2$(C$_5$H$_{11}$)SnCl	IIIA, Et$_2$SnO	18-29	(g)	149
Et$_2$(C$_6$H$_{13}$)SnCl (288)	--	--	(h)	--
EtPr$_2$SnI	IIIC, Pr$_2$Sn	36	b$_6$ 105-.10°, n_D^{20} 1.5491, GLC	1941
Et(CH$_2$CH$_2$CH)$_2$SnCl	--	--	Use: herbicide, pesticide (i)	1873
EtBu$_2$SnBr	IIIC, Bu$_2$Sn	26	n_D^{20} 1.5103, GLC	1941
EtBu$_2$SnI	IIIC, Bu$_2$Sn	56	b$_{1.5}$ 102-104°	1941
EtPh$_2$SnI	IA, EtSnPh$_3$	81-98	b$_4$ 198°, n_D^{20} 1.6464, d$_{20}$ 1.7348	521
	IIIB, Ph$_2$SnI$_2$ (j)	--	(k)	2808
	IIIB, Ph$_2$SnI$_2$	--	--	2809
Pr$_2$BuSnCl	IIIC, Pr$_2$Sn	33	n_D^{20} 1.4938, GLC (g)	1941
Pr$_2$BuSnBr	IIIC, Pr$_2$Sn	25	n_D^{20} 1.5092, GLC	1941
Pr$_2$BuSnI (289)	IIIC, Pr$_2$Sn	48	b$_3$ 105-107°, n_D^{20} 1.5309, GLC	1941
Pr$_2$(C$_5$H$_{11}$)SnI	--	--	(l)	--
Pr$_2$(c.C$_6$H$_{11}$)SnBr	IA, Pr$_2$Sn(C$_6$H$_{11}$)$_2$	96	b$_{2.5}$ 165-66°, n_D^{15} 1.5412, d$_{15}$ 1.4640 (m)	2721
Pr$_2$(c.C$_6$H$_{11}$)SnI	IA, Pr$_2$Sn(C$_6$H$_{11}$)$_2$	96	b$_4$ 208-10°, n_D^{15} 1.5708, d$_{15}$ 1.5640	2721
PrBu$_2$SnCl	IIIC, Pr$_2$Sn	7	n_D^{20} 1.4912, GLC	1941
PrBu$_2$SnBr	IIIC, Bu$_2$Sn	23	n_D^{20} 1.5058, GLC	1941
PrBu$_2$SnI	IIIC, Bu$_2$Sn	64	b$_{0.8}$ 110-15°	1941
PrPh$_2$SnI	IA, PrSnPh$_3$	81-98	b$_3$ 190°, n_D^{25} 1.6355, d$_{22}$ 1.6790 (n)	521
1-Pr$_2$(c.C$_6$H$_{11}$)SnBr	IA, 1-Pr$_2$Sn(C$_6$H$_{11}$)$_2$	87	b$_3$ 163°, n_D^{15} 1.5320, d$_{15}$ 1.5585 (o)	2721
1-Pr$_2$(c.C$_6$H$_{11}$)SnI	IA, 1-Pr$_2$Sn(C$_6$H$_{11}$)$_2$	93	b$_4$ 177-78°, n_D^{15} 1.5810, d$_{15}$ 1.6230	2721
1-PrPh$_2$SnI	IA, 1-PrSnPh$_3$	--	b$_3$ 192°, n_D^{24} 1.6420, d$_{24}$ 1.7340	521
1-PrPh$_2$SnI	IA, 1-PrSnPh$_3$	75	b$_{0.4}$ 150°, NMR spectrum (n, p)	1719
(CH$_2$CH$_2$CH)$_2$PhSnCl	--	--	Use: herbicide, pesticide (q)	1873
PhCHMeCH$_2$Ph$_2$SnCl (289)	IVA	87	--	1613
PhCHMeCH$_2$Ar$_2$SnCl	IVA	--	Ar = p-MeC$_6$H$_4$	1613

Compound	Method	Yield (%)	Properties	Ref.
$Bu_2(1-C_5H_{11})SnI$ (289)	IA, Bu_3SnR	--	b_4 178-80°, n_D^{20} 1.5294, d_{20} 1.4040 (r, s)	617
$Bu_2PhSnCl$ (289)	IIC	--	$b_{0.2}$ 130-40° (t)	66
$Bu_2PhSnBr$ (289)	IIIC, Bu_2Sn	22	$b_{0.5}$ 110-20°	1577
i-$BuPh_2SnI$	IA, 1-$BuSnPh_3$	--	b_3 188°, n_D^{29} 1.6450 (n)	521
$(n-C_5H_{11})_2c$-$C_6H_{11}SnI$	IA, $R_2SnR'_2$	82	b_3 216°, n_D^{18} 1.5325, d_{18} 1.2870 (u)	1772
n-$C_5H_{11}Ph_2SnI$	IA, $RSnPh_3$	87	b_3 216°, n_D^{23} 1.6158, d_{23} 1.5942 (v)	2787
$(1-C_5H_{11})_2c$-$C_6H_{11}SnI$	IA, $R_2SnR'_2$	78	b_6 216°, n_D^{18} 1.5288, d_{18} 1.2650 (u)	1772
$(1-C_5H_{11})_2$$1$-$C_9H_{19}SnI$	IA, R_3SnR'	--	b_3 225-27°, n_D^{20} 1.3683, d_{20} 1.3683 (s)	617
i-$C_5H_{11}Ph_2SnI$	IA, $RSnPh_3$	92	b_3 205°, n_D^{27} 1.6112, d_{27} 1.6053 (v)	2787

c-$C_6H_{11}R_2SnX$

Compound	Method	Yield (%)	Properties	Ref.
$R = C_6H_{13}$, $X = I$	IA, $(C_6H_{11})_2SnR_2$	90	b_3 228-29°, n_D^{18} 1.5482, d_{18} 1.4070 (w)	1774
C_7H_{15} I	IA, $(C_6H_{11})_2SnR_2$	90	b_3 233°, n_D^{18} 1.5462, d_{18} 1.3950 (w)	1774
C_8H_{17} I	IA, $(C_6H_{11})_2SnR_2$	87	b_{10} 250-51°, n_D^{18} 1.5340, d_{18} 1.3340 (w, x)	1774
C_9H_{19} I	IA, $(C_6H_{11})_2SnR_2$	53	b_{10} 255°, n_D^{18} 1.5160, d_{18} 1.2250 (w, x)	1774
$C_{10}H_{21}$ Br	IA, $(C_6H_{11})_2SnR_2$	41	b_{13} 231°, n_D^{18} 1.5035, d_{18} 1.1740 (w, x)	1774

RPh_2SnI

Compound	Method	Yield (%)	Properties	Ref.
$R = n$-C_6H_{13}	IA, $RSnPh_3$	95	b_3 223° (dec.), n_D^{25} 1.6000, d_{25} 1.5361	2787
n-C_7H_{15}	IA, $RSnPh_3$	89	b_3 225° (dec.), n_D^{24} 1.5922, d_{24} 1.5090 (v)	2787
n-C_9H_{19}	IA, $RSnPh_3$	80	b_3 232° (dec.), n_D^{20} 1.5758, d_{20} 1.4650 (v)	2787
n-$C_{10}H_{21}$	IA, $RSnPh_3$	93	b_3 (dec.), n_D^{20} 1.5695, d_{20} 1.4362 (v)	2787

(g) Use in antifouling coatings (2302). (h) Rxn. with NaOR → $Et_2(C_6H_{13})SnOR$, R = ethoxylated tall oil fatty acid (911). (i) Rxn. with $Me_3COMgBr$ → $Et(c$-$C_3H_5)_2SnOCMe_3$ (1873). (j) Synth. with EtI and Cu-Zn alloy (2808). (k) Rxn. with aq. NaOH → $(EtPh_2Sn)_2O$ (2808). (l) Rxn. with aq. NaOH → $(Pr_2SnC_5H_{11})_2O$ (2808). (m) Rxn. with NaOR → $Pr_2(c$-$C_6H_{11})SnOR$; R = $H(OCH_2CH_2)_n$; n = 1, 2 (2721). (n) Rxn. with aq. KOH in Et_2O, + AcOH → RPh_2SnOAc; R = n-, i-Pr, i-Bu (521). (o) Rxn. with $NaOCH_2CH_2OH$ → 1-$Pr_2(c$-$C_6H_{11})SnOCH_2CH_2OH$ (2721). (p) Rxn. with MeMgI → $Me(1-Pr)$-$SnPh_2$ (1719). (q) Rxn. with CH_2:$CHMgBr$ → $(CH_2CH_2CH)_2PhSnCH$:CH_2 (1873). (r) Rxn. with KOH, + AcOH → $Bu_2(1-C_5H_{11})$-$SnOAc$ (617). (s) Surface tension, parachor (617). (t) Rxn. with NaOH → $Bu_2PhSnOH$ (1577). (u) Parachor, colorless thick liq. (1772). (v) Rxn. with 15% aq. KOH in Et_2O → RPh_2SnOH; R = n-, 1-C_5H_{11}, n-C_7H_{15}, n-C_9H_{19}, n-$C_{10}H_{21}$ (2787). (w) slightly yellow liq., IR spectrum, soly. (1774). (x) Slight dec. on distillation (1774).

Table 132. Unsaturated Triorganotin Halides $R_2R'SnX$

Formula*	Synth. Method**	Yield	Properties	Ref.
MeBu(CH$_2$:CH)SnCl	IIB, MeBuSnCl$_2$	--	b$_{16}$ 96°, GLC	1370
EtBu(CH$_2$:CH)SnCl	IIB, EtBuSnCl$_2$	--	GLC	1370
Bu$_2$RSnX				
R = CH$_2$:CHCH$_2$, X = Br (29%)	IIIC, Bu$_2$Sn	70	b$_1$ 108-12°, n$_D^{20}$ 1.5225 (a)	66
{MeCH:CHCH$_2$CH$_2$, Cl} {EtCH:CHCH$_2$, Cl}	IVB, MeCH:CHCH:CH$_2$	39 / 49	b$_{0.2}$ 92-94°, n$_D^{20}$ 1.5039, IR spectrum (b)	1619
{CH$_2$:CMeCH$_2$CH$_2$, Cl Me$_2$C:CHCH$_2$, Cl CH$_2$:CHCMeCH$_2$, Cl MeCH:CMeCH$_2$, Cl}	IVB, CH$_2$:CHCMe:CH$_2$	26 / 54 / low / low	b$_{0.2}$ 98-102°, n$_D^{20}$ 1.5068, IR spectrum (b)	1619
{c.2-C$_5$H$_7$, Cl} {c.3-C$_5$H$_7$, Cl}	IVB, C$_5$H$_6$	5 / 84	b$_{0.2}$ 90-94°, n$_D^{20}$ 1.5225, IR spectrum (b)	1619
{CH$_2$:CMeCHMeCH$_2$, Cl Me$_2$C:CMeCH$_2$, Cl}	IVB, (CH$_2$:CMe)$_2$	24 / 54	b$_{0.3}$ 102-104°, n$_D^{20}$ 1.5071, IR spectrum (b)	1619
1-Bu$_2$RSnCl				
R = PhCH:CH	IVD, PhC⋮CH	92	b$_{0.04}$ 120°	1613
{MeCH:CHCH$_2$ CH$_2$:CHCH$_2$CH$_2$}	IVB, (CH$_2$:CH)$_2$	68 / 16	b$_{0.2}$ 79-81°, n$_D^{20}$ 1.5043, IR spectrum, struct. (b)	1619
{MeCH:CHCH$_2$CH$_2$ EtCH:CHCH$_2$}	IVB, MeCH:CHCH:CH$_2$	39 / 50	b$_{0.2}$ 94-96°, n$_D^{20}$ 1.5018, IR spectrum, struct. (b)	1619

$\begin{Bmatrix} CH_2:CMeCH_2CH_2 \\ Me_2C:CHCH_2 \\ CH_2:CHCMeCH_2 \\ MeCH:CMeCH_2 \end{Bmatrix}$	22 69 low low	IVB, $CH_2:CHCMe:CH_2$	$b_{0.1}$ 87–90°, n_D^{20} 1.5052, IR spectrum, struct. (b)	1619
$\begin{Bmatrix} CH_2:CMeCHMeCH_2 \\ Me_2C:CMeCH_2 \end{Bmatrix}$	27 70	IVB, $(CH_2:CMe)_2$	$b_{0.2}$ 99–102°, n_D^{20} 1.5061, IR spectrum, (b)	1619
$\begin{Bmatrix} c.\,2\text{-}C_5H_7 \\ c.\,3\text{-}C_5H_7 \end{Bmatrix}$	6 85	IVB, C_5H_6	$b_{0.2}$ 89–92°, n_D^{20} 1.5214, IR spectrum (b)	1619
$Ph_2[CH_2:CH(CH_2)_4]SnCl$	94	IVA, $1,5\text{-}C_6H_{10}$	$b_{0.5}$ 85°, n_D^{20} 1.5091	1613

* Numbers in parenthesis refer to pages in main volume.
** The characters used here correspond to the synthetic scheme in the introduction to Chapter 4.1.1.2.
(a) Rxn. with $CH_2:CH(CH_2)_n MgBr \rightarrow Bu_2(CH_2:CHCH_2)Sn(CH_2)_n CH:CH_2$; n = 2, 3 (4). (b) Mixture, not sepd. (1619).

4.1.1.6 FUNCTIONALLY SUBSTITUTED TRIORGANOTIN HALIDES

The functionally substituted triorganotin halides of this subchapter may contain halogen, cyano groups, oxygen, nitrogen and metal atoms substituted in a carbon chain.

4.1.1.6.1 DERIVATIVES WITH HALOGEN

Tri(chloromethyl)tin Chloride (M-295) $SnC_3H_6Cl_4$ $(ClCH_2)_3SnCl$
Synth.: By rxn. of $SnCl_4$ with CH_2N_2 in C_6H_6 at 2° (2071), at 3-6°, in 14% yield (2512), in presence of powd. Cu at r.t. in 95% yield (2512).
Prop.: Viscous liquid (2071), b_2 119.5-20° (2512), n_D^{20} 1.5920-53, d_{20} 1.9922 (2512), (far) IR (2512), NMR (2071, 2512), ^{35}Cl NQR (2752) and mass (2512) spectra, mol. wt. (2512).
Rxn. with: MeMgBr → Me_3SnCH_2Cl + $Me_2EtSnCH_2Cl$ + $Me_2Sn(CH_2Cl)_2$ + $MeEtSn(CH_2Cl)_2$ (2512).

Tris(pentafluorophenyl)tin Chloride (M-296) $SnC_{18}ClF_{15}$ $(C_6F_5)_3SnCl$
Synth.: By rxn. of $p-MeC_6H_4Sn(C_6F_5)_3$ with HCl at 100°, in 75% yield (77).
By rxn. of $(C_6F_5)_4Sn$ with $SnCl_4$ at 160° (1198).
By rxn. of $SnCl_4$ and C_6F_5MgBr (1:2.5) and vacuum sublimation without hydrolysis as by-prod. (1198).
Prop.: m. 108-109° (77), 103-104° (1198), IR, UV, NMR (77) and Moessbauer resonance (211) spectra.
Rxn. with: $SnCl_4$ (at 150°) → $(C_6F_5)_2SnCl_2$ (2:1) or $C_6F_5SnCl_3$-excess (1198).
Aq. NH_3 → $[(C_6F_5)_3Sn]_2O$ (77).
NaOH → $(C_6F_5)_3SnOH$ (1198).
$8-HOC_9H_6N$ → $(C_6F_5)_2Sn(OC_9H_6N)_2$, quinoline (1198).
Addn. compd.: $(C_6F_5)_3SnCl \cdot 2NH_3$, from components in anhydr. Et_2O, m. 155-70° (77) $[SnC_{18}H_6ClF_{15}N_2]$.

Tri(parachlorophenyl)tin Chloride (M-296) $SnC_{18}H_{21}Cl_4$ $(p-ClC_6H_4)_3SnCl$
Prop.: Moessbauer resonance spectrum (222, 224).
Rxn. with: $(Me_3SnC \vdots C)_2$ → $[(p-ClC_6H_4)_3SnC \vdots C]_2$ (205, 2433).
Aq. KOH in Et_2O → $[(p-ClC_6H_4)_3Sn]_2O$ (1375).
$HSC \vdots NN \vdots C(SH)S$ in alc. NaOH → $R_3SnSC \vdots NN \vdots CS(SnR_3)S$; R = $p-ClC_6H_4$ (1262).
Addn. compd.: $(p-ClC_6H_4)_3SnCl \cdot OP(NMe_2)_3$, from components in ROH, in 78% yield, m. 168-69° (1838a). Activity as bactericide and fungicide against Staphylococcus aureus, Aspergillus niger and Candida albicans (1838a). Use as acaricide, herbicide, insecticide (1838a).
$(p-ClC_6H_4)_3SnCl \cdot OSMe_2$, from components in MeOH, in 50% yield, m. 121-22° (1836). Activity as bactericide and fungicide with low phytotoxicity against S. aureus A. niger, C. albicans, Phytophthora infestans and Cercospora beticola (1836). Use as algicide, fungicide and molluscicide (1836).
Use: As moth proofing agent for keratin-containing matl. (473).

Other symmetric halogen substituted triorganotin halides are listed in Table 133.

Dimethyltrichloromethyltin Chloride $SnC_3H_6Cl_4$ $Me_2(Cl_3C)SnCl$

Synth.: By rxn. of Me_3SnCCl_3 with BCl_3 at 25°, in 75% yield (865).
Prop.: m. 74-76°, subl.$_{0.5}$ 40°, IR and NMR spectra (865).
Rxn. with: Thermal dec. at 120° → Me_2SnCl_2 + C_2Cl_4 (865).
BCl_3 at 60° → Me_2SnCl_2 + C_2Cl_4 (865).
H_2O → Me_2SnCl_2 + $HCCl_3$ + $(Me_2SnCl)_2O$ (865).

Listing of unsymmetric halogen substituted triorganotin halides concludes in Table 134.

Table 133. Symmetric Halogen Substituted Triorganotin Halides R'_3SnX

Formula*	Synth. Method**	Properties	Ref.
$(BrCH_2)_3SnBr$ (295)	--	NMR spectrum	2071
(trans-ClCH:CH)$_3$SnCl (296)	--	IR, NMR spectra (a)	2638
$(p-FC_6H_4)_3SnCl$ (296)	--	^1H, ^{19}F NMR spectra (b)	1453
$(p-FC_6H_4)_3SnI$	--	Moessbauer resonance spectrum	224
$(C_6F_5)_3SnF$	IC	Dec. at 300° (c)	498
$(C_6F_5)_3SnBr$	V	m. 107-108°, IR spectrum (d)	1198
$(p-ClC_6H_4)_3SnF$	--	(e)	--
$(p-ClC_6H_4)_3SnBr$ (296)	--	(e)	--
$(p-ClC_6H_4)_3SnI$ (296)	--	(e)	--
$(m-F_3CC_6H_4)_3SnBr$ (296)	--	Moessbauer resonance spectrum	224
$(F_3CC_6H_4)_3SnBr$	--	Moessbauer resonance spectrum	2290

* Numbers in parenthesis refer to pages in main volume.
** The characters used here correspond to the synthetic scheme in the introduction to Chapter 4.1.1.2.
(a) Rxn. with $HgCl_2$ → trans-ClCH:CHHgCl (2638). (b) Rxn. with aq. NaOH in Et_2O → $(p-FC_6H_4)_3SnOH$ (1375). (c) Far IR spectrum and assignment (52), no rxn. with alc. KOH under reflux (498). (d) Moessbauer resonance spectrum (211, 2290). (e) Use as moth proofing agent for keratin-contn. matl. (473).

Table 134. Unsymmetric Halogen Substituted Triorganotin Halides RR'₂SnX

Formula*	Synth. Method**	Yield	Properties	Ref.
Me₂RSnX				
R, X = ClCH₂ , Cl (297)	--	--	³⁵Cl NQR spectrum	2752
Br(CPh:CPh)₂ , Br (297)	--	--	(a)	--
Br(CPh:CPh)₂ , Br, d₂-	I (b)	--	m. 141.5-43°, cis,cis-form (c)	2728
Br(CPh:CPh)₂ , Br, d₄-	I (b)	--	cis,cis-form (d)	2728
C₆F₅ , Cl (297)	--	--	NMR spectrum	63
p-ClC₆H₄ , Cl	IIIA	--	m. ~115°, b₀.₀₅ 88-92°	1410
	IIB (e)	--	--	1410
Et₂(ClCH₂)SnCl	IIIE	77	b₁ 67°, n_D²⁰ 1.5350, d₂₀ 1.6323 (f)	283
Et₂[Br(CH₂)₅]SnBr	IIIA	18-29	From Et₂SnO	149
Bu₂RSnBr				
R = Br(CH₂)₄	IIIC	71	b₀.₅ 120-40°	66
BrCH₂CH₂CMe₂CH₂CH₂	--	--	(g)	--
Ph₂RSnX				
R, X = Br₂C₃H₅ , Br·Py	IV (h)	45	m. 152-59° (dec.) (i)	2592
Br₂C₃H₅ , Br·2Py	IV (h)	43	m. 128-31°, IR, NMR spectra (i)	2592
Br₂C₆H₉ , Br·2Py	IV (h)	57	m. 128-31°, IR, NMR spectra (j)	2592
C₆F₅ , Cl	IB	95	Rearranges on distn.	77
Ph(BrCH₂)₂SnBr	IIIA, D	--	b₀.₄ 160-66°	1410

* Numbers in parenthesis refer to pages in main volume.
** The characters used here correspond to the synthetic scheme in the introduction to Chapter 4.1.1.2.
(a) Single crystal x-ray diffraction analysis and configuration (752). (b) From d₂- and d₄-dimethyltetraphenylstannole, resp. (2728). (c) Rxn. with O at 140° → cis-(PhCOCPh:)₂ , rxn. mechanism (2728). (d) Formation of d₄-C₄Ph₄Br radical → (CPh:CPh)₄ , deuterium distribution under varying rxn. conditions (2728). (e) From Me₂SnCl₂ and (p-ClC₆H₄)₂SnMe₂ (1410). (f) Rxn. with EtMgBr → Et₃SnCH₂Cl (283). (g) Activity as Acaricide against Tetra-

nychus althaeae (433). (h) From Ph_2SnBr_2 and allyl and 3-cyclohexenyl bromide in Py (2592). (i) Rxn. with 3-C_6H_9Br in Py → $Ph_2(Br_2C_3H_5)SnC_6H_9Br_2 \cdot 2Py$ (2592). (j) Rxn. with CH_2:$CHCH_2Br$ in Py → $Ph_2(Br_2C_3H_5)SnC_6H_9Br_2 \cdot 2Py$ (2592).

4.1.1.6.2 PSEUDOHALOGEN SUBSTITUTED TRIORGANOTIN HALIDES

Tris(2-Cyanoethyl)tin Bromide (M-298) $SnC_9H_{12}BrN_3$ $(NCCH_2CH_2)_3SnBr$

Synth.: By rxn. of $(NCCH_2CH_2)_4Sn$ with Br in 46-74% yield (1746).
Prop.: m. 103-104°, monomeric in THF, associated in solid state (1745), IR (1745) and Moessbauer resonance (2290) spectra.
Rxn. with: AgOAc → $(NCCH_2CH_2)_3SnOAc$ (895b).
TlOAc → $(NCCH_2CH_2)_3SnOAc$ (895b, 1746).
Na_2S → $[(NCCH_2CH_2)_3Sn]_2S$ (2326).
Tl_2O → $(NCCH_2CH_2)_3SnOH \cdot H_2O$ + TlBr (1746).

Additional pseudohalogen substituted triorganotin halides are listed in Table 135.

Table 135. Pseudohalogen Substituted Triorganotin Halides $R_2R'SnX$

Formula*	Synth. Method**	Yield	Properties	Ref.
$(NCCH_2CH_2)_3SnF$ (298)	--	--	(a)	503
Me_2(NCCH:CH)SnCl	IVD	--	--	1613
$Et_2(NCCH_2CH_2)SnBr$ (298)	IVA	89	$b_{0.7}$ 119° (b)	1613
$Et_2(NCCH_2CH_2CH_2)SnI$	IVA	95	$b_{0.02}$ 102° (c)	1613
i-$Bu_2(NCCH_2CH_2)SnCl$ (298)	IVA	89	$b_{0.7}$ 107° n_D^{20} 1.5157 (b)	1613
i-Bu_2(NCCH:CH)SnCl	IVD	--	--	1613

* Numbers in parenthesis refer to pages in main volume.
** The characters used here correspond to the synthetic scheme in the introduction to Chapter 4.1.1.2.
(a) Moessbauer resonance spectrum. (b) From CH_2:CHCN. (c) From CH_2:$CHCH_2CN$.

4.1.1.6.3 OXYGEN CONTAINING TRIORGANOTIN HALIDES

The triorganotin halides containing oxygen in the organic moiety as hydroxy, ether and carboxy groups in Tables 136 and 137 are prepared by methods explained in the following scheme.

I Dealkylation reactions:
(IA) R_4-SnR'_n + Br → $R_{3-n}R'_nSnBr$; n = 1, 2 ,
(IB) $R_2SnR'_2$ + HBr → $R_2R'SnBr$,
(IC) $(NCCH_2CH_2)_4Sn$ + alc. NaOH, + HCl → $(HO_2CCH_2CH_2)_3SnCl$.

II Alkylation reactions:
   (IIA) $R'MgBr + SnBr_4 \rightarrow R'_3SnBr$,
   (IIB) $MeR'SnX_2 + MeMgBr \rightarrow Me_2R'SnX$,
   (IIC) $R_2SnH_2 + R_2SnCl_2 + R'_2Hg \rightarrow R_2R'SnCl$.

III Hydrostannation reactions:
   (IIIA) $R_2SnH_2 + R_2SnCl_2 + CH_2{:}CHR' \rightarrow R_2(R'CH_2CH_2)SnCl$,
   (IIIB) $R_2SnH_2 + R_2SnCl_2 + HC{:}CR' \rightarrow R_2(R'CH{:}CH)SnCl$.

R = organic group; R' = organic group carrying an oxygen function; X = halogen. Symmetric derivatives are listed in Table 136. Unsymmetric compounds may be found in Table 137.

Table 136. Symmetrical Oxygen Substituted Triorganotin Halides R'_3SnX

Formula*	Synth. Method**	Yield	Properties	Ref.
$(HO_2CCH_2CH_2)_3SnCl$	IC	37	m. 96.5-98° (a)	1744, 2327
$(p\text{-}MeOC_6H_4)_3SnCl$ (299)	--	--	(b)	--
$(p\text{-}MeOC_6H_4)_3SnBr$	--	--	(c)	--
$(1\text{-}C_{10}H_6R)_3SnBr$				
R = 2-MeO (299)	--	--	(d)	--
2-EtO (299)	--	--	(d)	--
2-PrO	IIA	32	m. 169°	303
2-Me$_2$CHO	IIA	24	m. 236°	303
2-BuO	IIA	28	m. 159°	303
2-i-BuO	IIA	15	m. 170°	303
2-n-C$_5$H$_{11}$O	IIA	23	m. 142°	303
2-n-C$_6$H$_{13}$O	IIA	33	m. 110°	303

* Numbers in parenthesis refer to pages in main volume.
** The characters used here correspond to the synthetic scheme in the introduction to Chapter 4.1.1.6.3.
(a) Stable at r.t. (1744, 2327). (b) Addn. compound $(p\text{-}MeOC_6H_4)_3SnCl \cdot OP(NMe_2)_3$, from components in ROH, in 69% yield, m. 146-48° (1838a). Activity against Staphylococcus aureus, Aspergillus niger, Candida albicans (1838a). Use as acaricide, herbicide, insecticide (1838a) [$SnC_{27}H_{39}ClN_3O_3P$]. (c) Rxn. with $LiAlH_4 \rightarrow (p\text{-}MeOC_6H_4)_3SnH$ (914). (d) Refractometric data (1391).

Table 137. Unsymmetrical Oxygen Substituted Triorganotin Halides $R_2R'SnX$

Formula*	Synth. Method**	Yield	Properties	Ref.
$Me_2(MeO_2CCH_2)SnCl$	IIB	--	m. 35.5-37.5°, b_2 107-109°	2651
$Me_2(MeO_2CCH_2CH_2)SnBr$	IIB	--	m. 58-59°	2651
$Me_2(EtO_2CCH_2CH_2)SnBr$	IIB	--	$b_{1.5}$ 107-11°, n_D^{20} 1.5215	2651
$Et_2(MeO_2CCH_2)SnCl$	IIC	73	b_3 104-107°, n_D^{20} 1.5264, d_{20} 1.5498 (a)	24
$Pr_2(MeO_2CCH_2)SnCl$	IIC	77	$b_{1.5}$ 110-14°, n_D^{20} 1.5165, d_{20} 1.4343 (a, b)	24
$Pr(MeO_2CCH_2CH_2)_2SnBr$	--	--	NMR spectrum	2070

Bu_2RSnX

Formula*	Synth. Method**	Yield	Properties	Ref.
R, X = $HOCH_2CH_2O(CH_2)_3$, Cl (300)	IIIA	93	$b_{0.3}$ 151°, n_D^{20} 1.5103	1613
$MeO_2CCH_2CH_2$, Br (300)	--	--	NMR spectrum	2070
$H_2NOCCH_2CH_2$, Br	IB	--	m. 99.5-101°, IR spectrum	1168
	IIIA	--	--	2344
$(EtO_2C)_2CHCHMe$, Br	IA	--	$b_{0.001}$ 153-57°, n_D^{20} 1.5004, NMR spectrum	1654
$MeO_2CCH_2CH(CO_2Me)CH_2$, Br	IA	--	$b_{0.001}$ 151-55°, n_D^{20} 1.4787 (d)	1654
$EtO_2CCH_2CH(CO_2Et)CH_2$, Br	IA	67	$b_{0.04}$ 155-57°, n_D^{20} 1.5013 (d)	1652
$PrO_2CCH_2CH(CO_2)r)CH_2$, Br	IA	--	$b_{0.001}$ 155-64°, n_D^{20} 1.5006 (d)	1654

i-Bu_2RSnX

Formula*	Synth. Method**	Yield	Properties	Ref.
R, X = p-$MeOC_6H_4$, Cl	IIIB	--	--	1613
$HO(CH_2)_3$, F	IIIA	90	Dec. 220°	1613
$HO(CH_2)_3$, Cl (300)	IIIA	95	$b_{0.4}$ 152°, n_D^{20} 1.5162	1613
$MeO_2CCH_2CH_2$, Cl (300)	IIIA	89	$b_{0.3}$ 111°, n_D^{20} 1.4982	1613

* Numbers in parenthesis refer to pages in main volume.
** The characters used here correspond to the synthetic scheme in the introduction to Chapter 4.1.1.6.3.
(a) IR spectrum (24). (b) Rxn. with air → $(Pr_2SnCl)_2O$ (24). (c) Found after work-up of rxn. of Bu_2SnH_2 with CH_2:$CHCONH_2$ (2434). (d) IR and NMR spectra.

Table 137 Continued $R_2R'SnX$

1-Bu$_2$RSnX (Cont'd.)

Formula*	Synth. Method**	Yield	Properties	Ref.
R, X = MeO$_2$CCH$_2$CH$_2$, Br	IIIA	--	--	1613
EtO$_2$CCH$_2$CHMe, Cl (300)	IIIA	90	n_D^{20} 1.4942	1613
(n-C$_8$H$_{17}$)$_2$RSnCl	IIIA	--	R = HOCH$_2$CH$_2$O(CH$_2$)$_3$	1613
(i-C$_8$H$_{17}$)$_2$RSnCl	IIIA	--	R = HOCH$_2$CH$_2$O(CH$_3$)$_3$	1613
Ph(MeO$_2$CCH$_2$CH$_2$)SnI (300)	--	--	NMR spectrum	2070

4.1.1.6.5 NITROGEN SUBSTITUTED TRIORGANOTIN HALIDES

Tri(nitromethyl)tin Chloride $SnC_3H_6ClN_3O_6$ $(O_2NCH_2)_3SnCl$
Synth.: By rxn. of $SnCl_4$ with $MeNO_3$ in $i-PrNH_2$ (318).
By rxn. of $(O_2NCH_2)_2SnCl_2$ with $EtNH_2$ (318).
Prop.: Colorless cryst., no proof of structure aside from poor elemental anal. (318).

Additional nitrogen substituted triorganotin halides may be found in Subsections 4.1.1.6.2 and 4.1.1.6.3.

4.1.1.6.7 METAL SUBSTITUTED TRIORGANOTIN HALIDES

This chapter contains triorganotin halides with boron or two tin atoms linked by carbon bridges. The compounds are prepared by the following methods.

I Dealkylation di(organotin) compounds:
$(Ph_3Sn)_2Y + I \rightarrow (Ph_2SnI)_2Y + PhI$.

II Alkylation of tin halide with organolithium derivatives:
(IIA) $SnCl_4 + R'Li \rightarrow R'_3SnCl$,
(IIB) $Ph_2SnCl_2 + YLi_2 \rightarrow (Ph_2SnCl)_2Y + (Ph_2SnCl)YH$.

III Hydrostannation reaction:
$Et_2SnH_2 + Et_2SnCl_2 + diene \rightarrow (Et_2SnCl)_2Y$.

R = organic group, R' = boron substituted organic group, Y = hydrocarbon bridge. Additional metal substituted triorganotin halides are listed in Subsections 2.7.1.1.3 and 2.7.1.2.

Bis-1,4-(diphenyliodotin)butane $Sn_2C_{28}H_{28}I_2$ $(Ph_2ISnCH_2CH_2)_2$
Synth.: By rxn. of $(Ph_3SnCH_2CH_2)_2$ with I (1:4) in $HCCl_3$ at $-20/-10°$ (1518, 2189).
Prop.: White cryst., m. 96-97°, x-ray struct. analysis (1518, 2189). NMR spectrum, dipole moment (1518).
Rxn. with: $RCO_2Ag \rightarrow [Ph_2Sn(O_2CR)CH_2CH_2]_2$; R = Me, Ph (1518, 2189).
$NaOH \rightarrow [Ph_2Sn(OH)CH_2CH_2]_2$ (1518, 2189).
$RMgBr \rightarrow (Ph_2RSnCH_2CH_2)_2$; $R = c.C_6H_{11}$ (1518, 2189).

Other metal substituted triorganotin halides are listed in Table 138.

Table 138. Metal Substituted Triorganotin Halides $R_2R'SnX$

Formula*	Synth. Method**	Yield	Properties	Ref.
$(B_{10}H_{10}C_2Ph)_3SnCl$ (a)	IIA	25	m. 287-89° (b)	627
o-$(Ph_2ClSnC)_2B_{10}H_{10}$ (a)	IIB	18	m. 143-44°	1835
m-$(Ph_2ClSnC)_2B_{10}H_{10}$ (c)	IIB	--	--	2279
m-$Ph_2ClSnCB_{10}H_{10}CH$ (c)	IIB	57	m. 72-73° (d)	1835
p-$(Me_2SnCl)_2C_6H_4$ (310)	--	--	(e)	--
$(Me_2SnClC_6H_4)_2CMe_2$	--	--	(f)	--
$(Ph_2ClSnCH_2CH_2)_2CH_2$	I	--	Colorless oil (g)	1518, 2189
$(Ph_2ClSnCH_2CH_2CH_2)_2$	I	--	Colorless oil, n_D^{20} 1.6656 (g)	1518, 2189
$(p-PhF_2SnC_6H_4)_2O$	--	--	(h)	--

* Numbers in parenthesis refer to pages in main volume.
** The characters used here correspond to the synthetic scheme in the introduction to Chapter 4.1.1.6.7.
(a) o- or 1,2-Carborane. (b) Moessbauer resonance spectrum (660). (c) Neocarborane, equal to m- or 1,7-carborane. (d) Distillable in vacuo (1835). (e) Rxn. with CH_2:CHMgCl → p-$(CH_2$:CHSnMe$_2)_2C_6H_4$ (1410). (f) Rxn. with LiN$_3$ (1:2) in Py → [Me$_2$Sn(N$_3$)C$_6$H$_4$]$_2$CMe$_2$ (2115). (g) Purified by column chromatography, dipole moment (1518). (h) Rxn. with NaN$_3$ (1:4) → [p-Ph(N$_3)_2$SnC$_6$H$_4$]$_2$O (2115).

4.1.2 DIORGANOTIN DIHALIDES

The diorganotin dihalides summarized in Tables 142 through 146 are prepared by methods from the following scheme.

I Dealkylation of organotin compounds:
- (IA) $R_4Sn + X \rightarrow R_2SnX_2$; X = Br, I ,
- (IB) $Pr_4Sn + SF_4 \rightarrow Pr_2SnF_2$,
- (IC) $R_nSnPh_{4-n} + X \rightarrow RPhSnX_2$,
- (ID) $RPh_2SnX + I \rightarrow RPhSnI$.

II Cleavage reaction of organo(polytin) compounds:
- $(R_2Sn)_x + I \rightarrow R_2SnI_2$.

III Redistribution reactions:
- (IIIA) $R_4Sn + SnCl_4 \rightarrow R_2SnCl_2$,
- (IIIB) $R_2SnR'_2 + SnCl_4 \rightarrow R_2SnCl_2 + R'_2SnCl_2 = RR'SnCl_2$,
- (IIIC) $R_4Sn + R'SnCl_3 \rightarrow RR'SnCl_2$,
- (IIID) $R_3SnCl + R'SnCl_3 \rightarrow RR'SnCl_2$.

IV Partial alkylation of tin or tin halides:
- (IVA) $SnX_4 + RMgX \rightarrow R_2SnX_2$,
  $SnX_4 + RX + Mg \rightarrow R_2SnX_2$,
- (IVB) $SnCl_2 + RMgX, + I \rightarrow R_2SnI_2$,
- (IVC) $Sn + RX + cat. \rightarrow R_2SnX_2$; X = Cl, BrI ,
- (IVD) $SnX_4 + CH_2N_2 \rightarrow (XCH_2)_2SnX_2$,
- (IVE) $SnCl_2 + R'_2Hg \rightarrow R'_2SnCl_2$.

V Interconversion reactions:
- (VA) $R_2SnX_2 + NaI \rightarrow R_2SnI_2$,
- (VB) $(R_2SnO)_x + HX \rightarrow R_2SnX_2$,
- (VC) $R'_2SnX_2 + alkali, + HX' \rightarrow R'_2SnX'_2$; X = Br, I; X' = Cl, Br.

R = R' = organic group may be different in one molecule, may be functionally substituted. X = X' = halogen, n = 1-3. Additional references to diorganotin dihalides are listed in Chapters 7.3, 8.1 and 9.

4.1.2.1 UNSUBSTITUTED SYMMETRIC DIORGANOTIN DIHALIDES

Dimethyltin Difluoride $SnC_2H_6F_2$ Me_2SnF_2
<u>Synth.:</u> By rxn. of Me_2SnCl_2 with AgF in H_2O (873), with Ag_nMF_x ; MF_x^{n-} = BF_4^-, PF_6^-, SiF_6^{--} (873).
By rxn. of Me_2SnH_2 with $Me_2Sn(CF:CF_2)_2$ under UV irradiation (731), with Me_nGe-$(CF:CF_2)_{4-n}$ at 55° or under UV irradiation; n = 2, 3, in 50% yield (2209), with $Me_nSi(CF:CF_2)_{4-n}$; n = 2, 3 (2208).
By-prod. in rxn. of Me_3SnH with $Me_2Sn(CF:CF_2)_2$ at 70° or under UV irradiation (731).
By rxn. of SnX_2 with Al_4C_3 and HF at 150-200° (2114).
<u>Prop.:</u> NMR spectrum (1430), x-ray struct. analysis (1818), free rotation of Me- groups by neutron spectrography (1777).

<u>Rxn. with:</u> $F^- \rightarrow [Me_2SnF_n]^{2-n}$; n = 1-4, complex formation, stability const. (838-9).

<u>Addn. compd.:</u> $Me_2SnF_2 \cdot 2KF$, far IR spectrum, struct. (605b) [$SnK_2C_2H_6F_4$].

$Me_2SnF_2 \cdot 2KF \cdot 2H_2O$, from components in H_2O, far IR spectrum, struct. (605b). [$SnK_2C_2H_{10}F_4O_2$].

$Me_2SnF_2 \cdot NH_4F$, from components in H_2O, far IR spectrum, struct., x-ray powder diagram (605b) [$SnC_2H_{10}F_3N$].

$Me_2SnF_2 \cdot 2NH_4F$, from components in H_2O, far IR spectrum, struct., x-ray powder diagram (605b) [$SnC_2H_{14}F_4N_2$].

$Me_2SnF_2 \cdot 2BF_3 \rightleftarrows Me_2Sn(BF_4)_2$, from Me_2SnCl_2 and $AgBF_4$, obtained in crude form (873) [$SnB_2C_2H_6F_8$].

Dimethyltin Dichloride (M-303) $SnC_2H_6Cl_2$ Me_2SnCl_2

<u>Synth.:</u> By thermal dec. of $Me_2(Cl_3C)SnCl$, in 55% yield (865), by hydrolysis at 60° as by-prod (865).

By thermal dec. of $Me_3SnCH_2\overline{CHCHRCCl_2}$ at 140-45° and with cat. $ZnCl_2$ at 78-125°; R = H, Me (2755).

By rxn. of Me_4Sn with $SnCl_4$; influence of temp. on prod. distribution (725), in 91% yield (2962), with $RSnCl_3$ (1:2) at 0°-180°, in quant. yield (1370), R = Bu (1:1) at 0° as by-prod. (1370), with AcCl and $AlCl_3$ (1:2:2) at r.t., in 90% yield (1781).

By rxn. of Me_2SnR_2 with $SnCl_4$; R = $n-C_8H_{17}$, $n-C_{10}H_{21}$, $n-C_{12}H_{25}$, in > 90% yield (1482), R = $\underline{C_6F_5 \text{ at } 150°}$ (1198).

By rxn. of $2,2'-Me_2SnC_6H_4NMeC_6H_4$ with $PhPCl_2$ (1376), 2,8-dibromo-derivative in refluxing xylene with $SnCl_4$ and $PhPCl_2$, in 46 and 61% yield, resp. (1376), 1,9-dimethyl-derivative with $PhPCl_2$ (1376).

By rxn. of BCl_3 with $Me_2(Cl_3C)SnCl$ in 54% yield (865), with $Me_2Sn(C_6F_5)_2$ (1:2) (77), with $Me_3SnC_6Cl_5$, in 93% yield (865).

By rxn. of Me_3SnCl with $RSnCl_3$; R = Me, influence of temp. on prod. distribution (725), R = Bu, i-Bu at 180-90°, in 97% yield (1370), with concn. HCl (31).

By rxn. of $(Me_2SnPPh)_3$ with anhydr. HCl in Et_2O (2746).

By rxn. of $SnCl_4$ with Al_4C_3 and HCl at 150-200° (2114).

<u>Prop.:</u> IR (135, 796), far IR (543), assignment (2322), NMR (31, 55, 347, 581, 602, 724-5, 727, 796, 1024, 1220, 1317, 1430, 1478), ^{13}C NMR (1498), Moessbauer resonance (211, 224, 902, 1960, 1989, 2347, 2424) and ^{35}Cl NQR (729, 1142, 2752) spectra, dipole moment (325, 724, 1211), mol. polarizability (1566), polarography (1249, 1701, 2352), thermodynamic data (305, 380, 1479, 2856), x-ray struct. analysis (729).

Anal. detn. by GLC (1370), with alizarin S at pH 4.2 (70), with 4-(2-pyridylazo)-2-resorcinol (2686), by potentiometric titration (1701, 2013), by electrometric titration and formation of hydroxo complexes (2232).

<u>Rxn. with:</u> $MeSnR_3$ (in MeOH at 30°) $\rightarrow Me_3SnCl + R_3SnCl$; kinetic data and rxn. mechanism, effect of NaCl, $NaClO_4$, relative reactivity; R = Me, Et, n-, i-Pr Bu, (Et_2 + Me) (2686); R = Me, prod. distribution and effect of temp. (725).

$(p-ClC_6H_4)_2SnMe_2 \rightarrow Me_2(p-ClC_6H_4)SnCl$ (1410).

$Bu_3SnH \rightarrow Bu_3SnCl + Me_2SnHCl$ (1:1) and Me_2SnH_2 (1:2) (1805).

$R_2SnH_2 \rightleftarrows Me_2SnHCl + R_2SnHCl$; R = Me (1292), n-, i-Bu, $n-C_8H_{17}$, Ph (2737).

Me_2SnH_2 + HC⋮CCN → Me_2(NCCH:CH)SnCl (1613).

p-$Me_3SnC_6H_4Cl$ + Na in MePh (or Mg in THF) → p-$(Me_3Sn)_2C_6H_4$ + some $(p-Me_3SnC_6H_4)$-$SnMe_2$ (1410).

Me_2SnX_2 → rate of halogen exchange by NMR spectroscopy; X = Br, I (727).

$(PhCH_2)_2SnCl_2 \cdot 2Me_2SO$ → ligand exchange by NMR spectroscopy (1317).

Me_2SnO → $Me_2SnCl \cdot 2MeSnO$ (2431).

Aq. Et_2SnO → $Me_2SnCl_2 \cdot 3Et_2SnO \cdot H_2O$ (2431).

R_2SnO → $(Me_2SnCl)O(R_2SnCl)$; R = Et, Bu (2431).

$Me_2Sn(OR)_2$ → $Me_2Sn(OR)Cl$; R = Me (942), 8-C_9H_6N (602, 1291).

$Me_2Sn(O_2C_7H_5)_2$ (at r.t.) → $Me_2Sn(O_2C_7H_5)Cl$ (2510)
 (in MeCN under reflux) → Me_3SnCl + $MeSn(O_2C_7H_5)_2SnCl$ (2510).

$(Me_2SnS)_2$ → $(Me_2SnCl)_2S$ (945).

Me_6Sn_2 → Me_3SnCl + $(Me_2Sn)_x$ (2012).

Me_3SiCl, + aq. NH_3 → $(Me_2SnOSiMe_3)_2O$ (409a).

$Me_3Si(OSiMe_2)_xOH$ + aq. NH_3 → $Me_2Sn[(OSiMe_2)_xSiMe_3]_2$ (12).

$(Me_2SiCH_2Cl)_2O$ + Mg → $[Me_2Sn(CH_2SiMe_3)_2]_xO$ (1066).

R_2SiCl_2 + aq. KOH → HO-terminated poly(stannasiloxanes); R = Me, (Me + Ph),
$CH_2:CH$, (Ph + $CH_2:CH$) (1008).

Li in THF → $Me_3SnLi \cdot THF$ (2127).

RLi → Me_2SnR_2 ; R = $p-MeC_6H_4$, $2,4,6-Me_3C_6H_2$, $3,4,5-Me_3C_6H_2$ (296), $2-C_6H_4-C_6F_5$ (1033), Me_2N, Et_2N, $i-Pr_2N$ (251), Me_3SiNMe (472, 1812), $Me_3SiNCMe_3$ (2749).

RLi → Me_3SnR + $MeSnR_3$; R = Me_2N (324).

$Me_3SiNCMe_3Li$ (1:1) → $Me_2Sn(NCMe_3SiMe_3)Cl$ (2749).

YLi → $Me_2\overline{SnY}$; Y = PhC:CHCH:CPh (691), $(2-C_6F_4)_2$ (883), $(2-C_6H_4)_2O$, $(2-C_6H_4)_2SO_2$ (1378), $(2-C_6H_3R)_2NMe$; R = H, 4-Me, 4-Br (1376).

YLi_2 → $(Me_2SnY)_n$; Y_n = $(o-C_2B_{10}H_{10})_2$ (1835), $(m-CB_{10}H_{10}C)_x$ (785), $(p-CB_{10}H_{10}C)_x$ (1835).

$m-LiCB_{10}H_{10}CLi$ (1:1-2) → $Me_2ClSn(CB_{10}H_{10}CSnMe_2)_nCl$; n = 8-30, depending on solvent (1835).

$(CH_2NMeLi)_2$ → polymeric organostannylamine (2163).

$PhCH:CH_2$ + Li → glasslike polymer (283).

Na in liq. NH_3, + $i-BrC_{10}H_{15}$ → $[Me_2(1-C_{10}H_{15})Sn]_2$, adamantane (1748).

$9,10-Na_2C_{14}H_{10}$ → $9,10-Me_2SnC_{14}H_{10}$ + $(9,10-Me_2SnC_{14}H_{10})_x$, 9,10-dihydroanthracene (1726).

$NaCH_2CO_2Na$ → $(Me_2SnCH_2CO_2)_x$ (124).

$NaBH_4$ in diglyme → Me_2SnH_2 + NaCl + B_2H_6 (736).

$NaBPh_4$ + L in H_2O → $(Me_2SnL_4)(BPh_4)_2$; L = Me_2SO, $HCONMe_2$, $MeCONMe_2$ (938).

NaX → Me_2SnX_2 ; X = $C_5H_5W(CO)_3$ (1670), $Re(CO)_5$ (2034), $Ph_3PIr(CO)_3$ (891).

NaX → Me_2SnX_2 + Me_2ClSnX; X = $Mn(CO)_5$ (1252), $C_5H_5Mo(CO)_3$ (1670).

$NaO_2C_7H_5$ in EtOH → $Me_2Sn(O_2C_7H_5)_2$, tropolone (2510).

$o,o'-HOC_6H_4N:CHC_6H_4OH$ + NaOMe → $Me_2\overline{SnOC_6H_4N:CHC_6H_4O}$ (2553).

CH_2Ac_2 + NaOMe → $Me_2Sn(OCMe:CHAc)_2$ (1649, 2055).

NaX → Me_2SnX_2 ; X = SCN (77), NCS (591), Ph_2PS_2 (2108), PhSe (651).

$(NaSCR:)_2$ → $Me_2\overline{SnSCR:CRS}$; R = H, CN (652).

Aq alkali → $[Me_2Sn \cdot aq]^{++}$ and other complexes, equil. const. and assocn. (689).

$M_2Fe(CO)_4$ → $[Me_2SnFe(CO)_4]_2$ (1268).

Aq. KX → Me$_2$SnX$_2$; X = (NC)$_2$C:C:CN, (NC)$_2$N (1330).

Me$_2$NCS$_2$K → Me$_2$Sn(S$_2$CNMe$_2$)$_2$ (1:2) and Me$_2$Sn(S$_2$CNMe$_2$)Cl (1:1) (2445).

AgX → Me$_2$SnX$_2$; X = NO$_3$ (1120), Ph$_2$N$_3$ (48a), F (873), ClO$_4$ (70).

Ag$_2$Y → Me$_2$SnY; Y = CO$_3$, SO$_4$, B$_{12}$Cl$_{12}$ (873).

Ag$_n$MF$_x$ → Me$_2$SnF$_2$; MF$_x$ = BF$_4^-$, PF$_6^-$, SiF$_6^{--}$ (873).

RMgCl → Me$_2$SnR$_2$; R = HC:CH$_2$ (1946), Me$_2$N$_3$ (48b).

p-ClC$_6$H$_4$MgCl → Me$_2$Sn(C$_6$H$_4$Cl-p)$_2$ + some Me$_2$(p-ClC$_6$H$_4$)SnCl (1410).

RMgX → Me$_2$SnR$_2$; R = CF$_3$C:C (2332), C$_6$F$_5$ (428).

Me$_2$N$_3$MgX → Me$_2$Sn(N$_3$Me$_2$)X; X = Br, I (48b).

SnCl$_4$ in Me$_2$SO → MeSnCl$_3$·2Me$_2$SO (1385, 1388).

Co[Co(CO)$_4$]$_2$ in MeOH → MeSn[Co(CO)$_4$]$_2$ (1672).

Co$_2$(CO)$_8$ in THF → Me$_2$Sn[Co(CO)$_4$]Cl (1673).

Et$_3$N in EtOH → (Me$_2$SnCl)O(Me$_2$SnOH) (10).

PhPH$_2$ + Et$_3$N → c.(Me$_2$SnPPh)$_3$ + polymer + Et$_3$N·HCl (2746).

Koji acid + aq. NH$_3$ → Me$_2$Sn(C$_6$H$_5$O$_4$)$_2$ (1658).

ArOH + aq. NH$_3$ → (Me$_2$SnOAr)$_2$O; Ar = α-(ON)-βC$_{10}$H$_6$ (550).

8-HOC$_9$H$_6$N in EtOH → Me$_2$Sn(OC$_9$H$_6$N)$_2$ (602), in aq. MeOH (1216).

8-HOC$_9$H$_6$N in MeOH → Me$_2$Sn(OC$_9$H$_6$N)Cl (1216).

5-(8-HOC$_9$H$_5$N)$_2$CH$_2$ + NH$_3$ → [Me$_2$Sn(OC$_9$H$_5$N)$_2$CH$_2$·H$_2$O]$_x$ (1871).

o-HOC$_6$H$_4$N:CMeCH$_2$COR → Me$_2$$\overline{\text{SnOC}_6\text{H}_4\text{N:CMeCH:CRO}}$; R = Me, Ph (2553).

RCO$_2$H + aq. alc. NH$_3$ → (Me$_2$SnO$_2$CR)$_2$O; R = o-BrC$_6$H$_4$ (1880).

Pb(SMe)$_2$ → Me$_2$Sn(SMe)$_2$ (643).

(C$_6$H$_{11}$)$_3$SbS → (Me$_2$SnS)$_3$ + (C$_6$H$_{11}$)$_3$SbCl$_2$ (1903).

(CH$_2$SH)$_2$ + aq. alkali → Me$_2$$\overline{\text{SnSCH}_2\text{CH}_2\text{S}}$ (2, 141).

H$_2$N(CH$_2$)$_6$NH$_2$ + Y(COCl)$_2$ in aq. alkali → organotin-polyamide; Y = (CH$_2$)$_4$, p-C$_6$H$_4$ (1245).

(p-HOC$_6$H$_4$)$_2$CMe$_2$ + p-C$_6$H$_4$(COCl)$_2$ in aq. alkali → organotin-polyester (1246).

<u>Addn. compd.:</u> Me$_2$SnCl$_2$·2Py (M-304), by rxn. of Me$_2$Sn(OC$_9$H$_6$N-8)Cl with Py, in 40% yield (1291), from components in EtOH (876), m. 158-61° (1291), IR, assignment (2819), far IR, assignment (876, 2819), NMR (726) and Moessbauer resonance (902) spectra, struct. (726, 2819), isomorphous with corresponding dibromide (876) [SnC$_{12}$H$_{16}$Cl$_2$N$_2$].

Me$_2$SnCl$_2$·BiPy (M-304), from components (147), in alc. (1212), white cryst., m. 238-39° (147), 235° (dec.) (1212), transition point 160° (1212), far IR, assignment (876, 878-9, 2322), UV (280) and Moessbauer resonance (902, 1044, 1991) spectra, conductivity (147), stability const. (280, 1479), struct. (876, 878-9), thermodynamic data (1479) [SnC$_{12}$H$_{14}$Cl$_2$N$_2$].

2Me$_2$SnCl$_2$·2,2',2"-terpyridyl, from components, m. 182-86° (147), Moessbauer resonance spectrum (2347), conductivity (147), struct. (147), x-ray struct. analysis (1015, 2376) [Sn$_2$C$_{19}$H$_{23}$Cl$_4$N$_3$].

Me$_2$SnCl$_2$·(1,10-phenathroline) (M-304), from components in alc., m. 260° (dec.) (1212), transition point 190° (1212), far IR, assignment (879, 2322) and Moessbauer resonance (1044) spectra, struct. (1044) [SnC$_{14}$H$_{14}$Cl$_2$N$_2$].

Me$_2$SnCl$_2$·2C$_5$H$_5$NO, from components in HCCl$_3$ or CH$_2$Cl$_2$ (2340), in C$_6$H$_6$ (1296), m. 135-36.5° (dec.) (1296), 132-33° (dec.) (2340), IR (2494), assignment (1296), far IR, assignment (876, 1296, 2340) and NMR (2340, 2494) spectra, isomorphous with corresponding dibromide (876), stability const. (1296), struct. (876, 2340) [SnC$_{12}$H$_{16}$Cl$_2$N$_2$O$_2$].

Me$_2$SnCl$_2$·2Ph$_3$PO, from components in HCCl$_3$ or CH$_2$Cl$_2$, dec. 135-36° (2340), IR, assignment (875, 2340), far IR, assignment (875-6, 2340) and NMR (2340) spectra, struct. (2340) [SnC$_{38}$H$_{36}$Cl$_2$O$_2$P$_2$].

Me$_2$SnCl$_2$·2Ph$_3$AsO, from components in EtOH (875), in HCCl$_3$ or CH$_2$Cl$_2$ (2340), dec. 206-208° (2340), (far) IR, assignment (875 2340) and NMR (2340) spectra, struct. (2340) [SnAs$_2$C$_{38}$H$_{36}$Cl$_2$O$_2$].

Me$_2$SnCl·2SSbMe$_3$, from components in MeOH (1902-3), from (Me$_2$SnS)$_3$ + Me$_3$SbS + Me$_3$SbCl$_2$ (1:3:3) in HCCl$_3$ in 98% yield in HCCl$_3$ (1903), colorless cryst., m. 147-47.5° (1903), dissocn. in HCCl$_3$ (1902) in polar solvents (1903), IR (1902), assignment (1903) and NMR (1903) spectra, x-ray diffraction pattern (1903) [SnSb$_2$C$_8$H$_{24}$Cl$_2$S$_2$].

Me$_2$SnCl$_2$·2Me$_2$SO; from components, in > 64% yield (1388), in Et$_2$O (2021), in HCCl$_3$ (873), from Me$_2$SnCl$_2$ and (PhCH$_2$)$_2$SnCl$_2$·2Me$_2$SO (1317), white cryst. (2021), m. 113° (1388), 110-10.5° (2021), IR (1388), assignment (873), far IR, assignment (2021, 2340, 2451) and NMR (1317, 2340) spectra, soly. (1388), struct. (2021, 2340), x-ray struct. analysis (2451).

Rxn. with SnCl$_4$ in Me$_2$SO → MeSnCl$_3$·2MeSO (1385, 1388) [SnC$_6$H$_{18}$Cl$_2$O$_2$S$_2$].

Additional addition compounds of Me$_2$SnCl$_2$ are listed in Table 139 and 120.

<u>Use:</u> In catalyst of polymerization of olefins (184).
As catalyst for polyurethan synthesis (1209).

Dimethyltin Dibromide (M-304)   SnC$_2$H$_6$Br$_2$ Me$_2$SnBr$_2$

<u>Synth.:</u> By rxn. of Me$_4$Sn with SnBr$_4$ at 100-120°, effect of temp. on product distribution (725).

By rxn. of Me$_2$Sn(C$_6$F$_5$)$_2$ with SnBr$_4$ at 150° as by-prod. (1198).

By rxn. of Me$_3$SnBr with MeSnBr$_3$ at 100-120°, effect of temp. on prod. distribution (725).

By rxn. of Sn with Br(CH$_2$)$_3$CO$_2$Me at 160°, in 16% yield (2650).

<u>Prop.:</u> Far IR, assignment (2322), NMR (443, 724-5, 727, 1220, 1430) and Moessbauer resonance (1989) spectra, dipole moment (724), molar polarizability (1566). Anal. detn. with alizarin S at pH 4.2 (70).

<u>Rxn. with:</u> Me$_4$Sn(at 120°) → Me$_3$SnBr; effect of temp. on prod. distribution (725).

Me$_2$SnH$_2$ → Me$_2$SnHBr (1292).

Me$_2$SnX$_2$ → rate of halogen exchange; X = Cl, I (727).

Me$_2$Sn(O$_2$C$_7$H$_5$)$_2$ (at r.t.) → Me$_2$Sn(O$_2$C$_7$H$_5$)Br, tropolone (2510).

Me$_2$SnS → Me$_2$SnBr)$_2$S (945).

Br → MeSnBr$_3$ (2).

PhC⋮CNa → Me$_2$Sn(C⋮CPh)$_2$ (443).

Me$_2$NCS$_2$K → Me$_2$Sn(S$_2$CNMe$_2$)Br (2445).

AgSCN → Me$_2$Sn(NCS)$_2$ (591).

C$_6$F$_5$MgBr → Me$_2$Sn(C$_6$F$_5$)$_2$ (77).

BrMg(CH$_2$)$_5$MgBr → Me$_2$Sn(CH$_2$)$_5$CH$_2$ (2888).

(C$_6$H$_{11}$)$_3$SbS → (Me$_2$SnS)$_3$ + (C$_6$H$_{11}$)$_3$SbBr$_2$ (1903).

<u>Addn. compd.:</u> All derivatives are listed in Table 140 on page 499.

<u>Use:</u> In catalyst for olefin polymerization (184).

Table 139. Addition Compounds of Dimethyltin Dichloride $\qquad$ $Me_2SnCl_2 \cdot nL$

Ligand*	Synth. Method**	Properties	Ref.
Ph_3CCl	--	Conductivity (a)	2880
2CsCl	--	Far IR spectrum (b)	876
NH_4Cl	--	Far IR spectrum (c)	876
Me_4NCl	--	Far IR spectrum	876
$2Me_4NCl$ (304)	--	Far IR spectrum	876
Et_4NCl	--	Far IR spectrum	876
$2Et_4NCl$	--	Far IR spectrum	876
2(Py·HCl) (304)	--	Struct. (d)	1044
$(CH_2NH_2)_2$	ID	(e)	1702
MeCN	IA	Thermodynamic data	1479
2 $4-MeC_5H_4NO$ (f)	IB, C	Dec. 171-73° (g)	2494
2 $2,4-Me_2C_5H_3NO$ (f)	IB, C	Dec. 152-55° (g)	2494
2 $2,4,6-Me_3C_5H_2NO$ (f)	IB, C	Dec. 150-52° (g)	2494
2 $4-ClC_5H_4NO$ (f)	IB, C	Dec. 162-65° (g)	2494
2 $4-MeOC_5H_4NO$ (f)	IB, C	Dec. 201-202° (g)	2494
2 $Me_2N:\overline{CSCH_2CH_2S}Cl$	(h)	Colorless cryst., struct. (i)	2820
$HCONMe_2$	IA	m. 56-58.5° (j)	2564
2 $HCONMe_2$	IA	m. 84-85.5°, stable in vacuo (j)	2564
$p-Me_2NC_6H_4CHO$	IA	m. 86.5-88.5°, UV spectrum (j, k)	2564
$p-MeOC_6H_4Ac$	IA	m. 49-51° (j, k)	2564
$p-Me_2NC_6H_4Ac$	IA	Dec. > 60° (j, k)	2564
$(p-Me_2NC_6H_4)_2CO$	IA	m. 140-42.5° (j, k)	2564
Ph_4AsCl	--	Mol. wt. conductivity	2880
Ph_3SBr	IE	m. 102° (l)	1182
$2Me_2SeO$	IC	White cryst., m. 188-89° (m)	2821

* Numbers in parenthesis refer to pages in main volume.
** The compounds are prepared from components I in CCl_4-(IA), in $HCCl_3$-(IB), in CH_2Cl_2-(IC), in Bu_2O-(ID) and in Me_2CO-(IE).
(a) Equil. const. by photometric measurements (2880). (b) Isomorphous with corresponding dibromide (876). (c) Moessbauer resonance spectrum (2347). (d) Moessbauer resonance spectrum (1044). (e) Rxn. with $HSCH_2CO_2R \rightarrow Me_2Sn(SCH_2CO_2-R)$; $R = i-C_8H_{17}$ (1702). (f) Pyridine oxides. (g) White nonhydroscopic cryst., IR and NMR spectra (2494). (h) By rxn. of $Me_2Sn(S_2CNMe_2)_2$ with $(CH_2Cl)_2$ under reflux, in 85% yield (2820). (i) IR and NMR spectra (2820). (j) (Far) IR spectrum, assignment, struct. (2564). (k) Stability const. (2564). (l) Use as catalyst for polyurethan and polyester synth. (1182). (m) (Far) IR spectrum, assignment, struct., soly. (2821).

Table 140. Addition Compounds of Dimethyltin Dibromide $Me_2SnBr_2 \cdot nL$

Ligand*	Synth. Method**	Properties	Ref.
2CsBr	--	Far IR spectrum (a)	876
Me₄NBr	--	Far IR spectrum	876
Et₄NBr	--	Far IR spectrum	876
2Et₄NBr (305)	--	Far IR spectrum	876
2Py (305)	--	Far IR spectrum, struct. (a, b)	876
BiPy	I	m. 224°, conductivity	147
	--	Far IR spectrum, struct. (c)	876
1/2 2,2',2"-terpyridyl	I	m. 178-80°, conductivity, struct.	147
1,10-phenanthroline	IF	m. 232-35° Dec. (c)	2322
$2C_5H_5NO$ (d)	--	Far IR spectrum (a)	876
$Me_2N:\overline{CSCH_2CH_2S}Br$	(e)	Colorless cryst., m. 176° (f)	2820
HCONMe₂	IA	m. 72.5-74° (g)	2564
2HCONMe₂	IA	m. 68-70.5°, loses HCONMe₂ in vacuo (g)	2564
p-Me₂NC₆H₄CHO	IA	m. 86-88.5° (g)	2564
2Ph₃PO	--	Far IR spectrum, struct. (h)	876
2Ph₃AsO	IE	(h)	875
2Me₃SbS	IB,D,E	m. 123-24°, mol. wt., IR spectrum,	1903
	IB (i)	assignment, dissocn. in polar solvent	1903
2Me₂SO	IA	White cryst., m. 119-20° (j)	2021
2Me₂SeO	IC	White cryst., m. 119-20°, struct (k)	2821

* Numbers in parenthesis refer to pages in main volume.
** The adducts are prepared from components I, in CCl_4-(IA), in $HCCl_3$-(IB), in CH_2Cl_2-(IC), in Me_2CO-(ID), in alcohols-(IE) and in C_6H_6-(IF).
(a) Isomorphous with corresponding dichloride (876). (b) (Far) IR, assignment (2819) and NMR (443, 726) spectra, struct. (876, 2819). (c) Far IR spectrum and assignment (2322). (d) Pyridine oxide (876). (e) By rxn. of $Me_2Sn(S_2CNME_2)_2$ and $Me_2Sn(S_2CNMe_2)Br$ with $(CH_2Br)_2$ under reflux, in 90% and low yield, resp. (2820). (f) IR and NMR spectra, struct. (2820). (g) (Far) IR spectrum, struct. (2564). (h) (Far) IR spectrum and assignment (875). (i) By rxn. of $(Me_2SnS)_3 + Me_3SbS + Me_3SbBr_2$ (1:3:3) in $HCCl_3$ (1903). (j) Far IR spectrum and assignment, struct. (2021). (k) (Far) IR spectrum and assignment (2821).

Dimethyltin Diiodide (M-305) $SnC_2H_6I_2$ Me_2SnI_2

Synth.: By rxn. of Me₃SnI with MeSnI₃ at 140°, effect of temp. on prod. distribution (725).
By rxn. of $(Me_2SnO)_x$ with NH₄I in c.MeC₆H₁₁ under reflux, in 60% yield (2660).
By-prod. in rxn. of tin foil with $I(CH_2)_3CO_2Me$ at 160° (2650).
Prop.: m. 43-44° (2660), far IR, assignment (2322) and NMR (724-5, 727, 1430) spectra, dipole moment (724), molar polarizability (1566).
Rxn. with: Me₂SnH₂ → Me₂SnHI (1292).

Me$_4$Sn (at 120°) → Me$_n$SnI$_{4-n}$; effect of temp. on prod. distribution (725).

Me$_2$SnX$_2$ → rate of halogen exchange; X = Cl, Br (727).

Aq. Et$_2$SnO → Me$_2$SnI$_2$·3Et$_2$SnO·H$_2$O (2431).

Me$_2$Sn(OMe)$_2$ → Me$_2$Sn(OMe)I (942).

Me$_2$NCS$_2$K → Me$_2$Sn(S$_2$CNMe$_2$)I (2445).

c.C$_6$H$_{11}$I + Zn in DMF → Me$_2$(C$_6$H$_{11}$)SnI (2809).

<u>Addn. compd.</u>: Me$_2$SnI$_2$·2Py (M-305), IR, assignment (2819), far IR, assignment (876, 2819) and NMR (726) spectra, struct. (726, 2819) [SnC$_{12}$H$_{16}$I$_2$N$_2$].

Me$_2$SnI$_2$·BiPy, cryst., m. 214°, conductivity (147), far IR spectrum (876) and assignment (2322) [SnC$_{12}$H$_{14}$I$_2$N$_2$].

Me$_2$SnI$_2$·2,2',2"-terpyridyl, yellow powder, m. 202° (dec.), conductivity, struct. (147) [SnC$_{17}$H$_{17}$I$_2$N$_3$].

Me$_2$SnI$_2$·(1,10-phenanthroline), from components in C$_6$H$_6$, m. 235-45° (dec.), far IR spectrum and assignment (2322) [SnC$_{14}$H$_{14}$I$_2$N$_2$].

Me$_2$SnI$_2$·C$_5$H$_5$NO, far IR spectrum (876) [SnC$_{12}$H$_{16}$I$_2$N$_2$O$_2$].

Me$_2$SnI$_2$·2Ph$_3$PO, IR, assignment (875), far IR spectra (876) and assignment (875) [SnC$_{38}$H$_{36}$I$_2$O$_2$P$_2$].

Me$_2$SnI$_2$·2Ph$_3$AsO, from components in EtOH, yellow cryst., (far) IR spectrum and assignment (875) [SnAs$_2$C$_{38}$H$_{36}$I$_2$O$_2$].

Dibenzyltin Dichloride (M-305)   SnC$_{14}$H$_{14}$Cl$_2$ (PhCH$_2$)$_2$SnCl$_2$

<u>Synth.</u>: By rxn. of metallic tin with PhCH$_2$Cl in presence of cat. BuI and BuNH$_2$ (565), in presence of cat. HgCl$_2$ at 108-14°, in 3-62% yield, influence of solvent polarity and time (1940).

By rxn. of (PhCH$_2$)$_2$SnO with HCl (2069).

<u>Prop.</u>: (Far) IR, assignment (841) and NMR (2069) spectra.

<u>Rxn. with</u>: Sn in H$_2$O → (PhCH$_2$)$_3$SnCl + [(PhCH$_2$)$_2$SnCl]$_2$O + SnO$_2$ + SnCl$_2$ + (PhCH$_2$)$_3$SnOH (1940).

M in H$_2$O → (PhCH$_2$)$_3$SnCl + [(PhCH$_2$)$_2$SnCl]$_2$O + (PhCH$_2$)$_3$SnOH + (PhCH$_2$)$_2$SnO; M = Mg, Zn, Cd, Al, Fe (1940).

LiAlH$_4$ → (PhCH$_2$)$_2$SnH$_2$ (394).

9,10-Na$_2$C$_{14}$H$_{10}$ → 9,10-(PhCH$_2$)$_2$SnC$_{14}$H$_{10}$ + (9,10-Ph$_2$SnC$_{14}$H$_{10}$)$_x$, 9,10-dihydroanthracene (1726).

RNa → (PhCH$_2$)$_2$SnR$_2$; R = ethoxylated soybean fatty acids (911).

RCO$_2$Na → (PhCH$_2$)$_2$Sn(O$_2$CR)$_2$; R = C$_7$H$_{15}$, C$_{11}$H$_{23}$ (1357).

Y(CO$_2$Na)$_2$ → (PhCH$_2$)$_2$Sn(O$_2$C)$_2$Y ; Y = o-C$_6$H$_4$, CH:CH, (CH$_2$)$_n$; n = 3, 4, 8 (1357).

KSCN in PrOH → (PhCH$_2$)$_2$Sn(SCN)$_2$ (841).

RMgX → (PhCH$_2$)$_2$SnR$_2$; R = Me, Et (841), PhCH$_2$, in absence of ether (1871a).

PhCH$_2$Cl + Mg + cat. Ph$_3$PO$_3$ → (PhCH$_2$)$_4$Sn, in absence of ether (1871a).

H$_2$O → [(PhCH$_2$)$_2$SnCl]$_2$O + HCl (1940).

<u>Addn. compd.</u>: (PhCH$_2$)$_2$SnCl$_2$·BiPy, from components in i-PrOH, soly. (1940) [SnC$_{24}$H$_{22}$Cl$_2$N$_2$].

(PhCH$_2$)$_2$SnCl$_2$·2C$_5$H$_5$NO, from components in MeOH, dec. 104-107°, IR and NMR spectra, struct. (2340) [SnC$_{24}$H$_{24}$Cl$_2$N$_2$O$_2$].

(PhCH$_2$)$_2$SnCl$_2$·2Ph$_3$PO, from components in MeOH, dec. 94-95°, IR and NMR spectra, struct. (2340) [SnC$_{50}$H$_{44}$Cl$_2$O$_2$P$_2$].

(PhCH$_2$)$_2$SnCl$_2$·2Me$_2$SO, NMR spectrum, ligand exchange with Me$_2$SnCl$_2$ and Ph$_2$SnCl$_2$ (1317) [SnC$_{18}$H$_{26}$Cl$_2$O$_2$S$_2$].

Use: As antioxidant for polyphenyl ethers (1979).

Dibenzyltin Dibromide (M-316) $SnC_{14}H_{14}Br_2$ $(PhCH_2)_2SnBr_2$
Synth.: By rxn. of metallic tin with $PhCH_2Br$ by heating in Bu_2O, in 46% yield (841), by ^{60}Co irradiation (616a).
By rxn. of $(PhCH_2)_2SnO$ with concn. HBr (2069).
Prop.: m. 123-24°, NMR (2069), (far) IR spectra and assignment (841).
Rxn. with: KCNO in EtOH → $(PhCH_2)_2SnCN:O$ (841).
NaI in Me_2CO → $(PhCH_2)_2SnI_2$ (841).

Diethyltin Dichloride (M-306) $SnC_4H_{10}Cl_2$ Et_2SnCl_2
Synth.: By rxn. of EtCl and powd. tin in presence of Et_3N and I at 160°, in 34% yield, influence of cat. iodides (1943).
By rxn. of Et_6Sn_2 with CCl_4 and pptd. Cu, followed by HCl cleavage of rxn. residue, as by-prod. (1737), with CF_3COCl under UV irradiation, as by-prod. (922).
By rxn. of $Bu_2Sn(OMe)_2 \cdot Et_2SnCl_2$ with $SnCl_4$, in 80% yield (2184).
Prop.: IR (589), far IR, assignment (321, 2322), NMR (573, 1430, 2068), Moessbauer resonance (1960) and NQR (1142) spectra, dipole moment (325, 1211), polarography (971, 1701, 2352) and thermodynamic data (305, 1479).
Anal. detn. by GLC (1939), by colorimetric method with diphenylcarbazone (490), with cacotheline after Na_2O_2 fusion (155), by electrometric titration, investigation of hydroxocomplexes (2232), by potentiometric titration (1701, 2013), by polarographic method (2265).
Rxn. with: Polarographic reduction → $(Et_2Sn)_x$ (1550).
Bu_3SnH → $Bu_3SnCl + Et_2SnHCl$ (1:1) and Et_2SnH_2 (2:1) (1805).
Et_2SnH_2 → Et_2SnHCl (1292).
$R_2SnH_2 \rightleftharpoons R_2SnHCl + Et_2SnHCl$; R = n-, i-Bu, n-$C_8H_{17}$, Ph (2737).
$Et_2SnH_2 + (MeO_2CCH_2)_2Hg$ → $Et_2(MeO_2CCH_2)SnCl$ (24).
$Et_2SnH_2 + (CH_2:CHCH_2)_2$ → $Et_2ClSn(CH_2)_6SnEt_2Cl$ (1613).
$Et_2SnH_2 + RMeCO$ → $Et_2Sn(OCHMeR)Cl$; R = Et, p-NCC_6H_4, RMe = $(CH_2)_5$ (1613).
$Me_2Sn(OR)_2$ → $Me_2Sn(OR)Cl$; R = Me (942), 8-C_9H_6N (1291).
$(Bu_2Sn)_x + Et_3N$ at 160° → $Et_nBu_{3-n}SnCl + Sn$; n = 0-3 (1941).
Me_3SiCl + aq. NH_3 → $(Me_3SiOSnEt_2)_2O$ (409a).
$(Me_2SiCH_2Cl)_2O + Mg$ → $[Et_2Sn(CH_2SiMe_2)_2O]_x$ (1066).
Et_2NLi → $Et_2Sn(NEt_2)_2$ (323, 1955).
$(2-C_6H_4Li)_2O$ → $Et_2\overline{SnC_6H_4OC_6H_4}$ (1378).
NaX → Et_2SnX_2; X = $Pr_3AsMn(CO)_4$ (185), NCS (591), X = RCO_2; R = Ph, Cl_2CH, o-ClC_6H_4 (1051).
$C_7H_5O_2Na$ → $Et_2Sn(O_2C_7H_5)_2$, tropolone (2510).
$Y(CO_2Na)_2$ → $(Et_2\overline{SnO_2CYCO_2})_x$; Y = (o-, (p-$C_6H_4)_2SiMe_2$ (1516).
$(:CRSNa)_2$ → $Et_2\overline{SnSCR:CRS}$; R = H, CN (652).
Na_2Y → $(Et_2SnS)_3$; Y = S (1514), S_2O_3 (1790).
Na_2S_2 (2:1) → $(Et_2SnCl)_2S_2$ (1514).
CH_2Ac_2 + NaOMe → $Et_2Sn(OCMe:CHAc)_2$ (1649, 2055).
$M_2Fe(CO)_4$ → $[Et_2SnFe(CO)_4]_2$ (1268).
RMgX → Et_2SnR_2; R = $EtOCH_2$ (307), $CH_2:CHCH_2$ (1857), $PrC \vdots C$, $BuC \vdots C$, $PhC \vdots C$ (1221).

$CH_2N_2 \rightarrow Et_2(ClCH_2)SnCl$ (283).
$RC\vdots CH + (c.C_6H_{11})_2NH$ in $C_6H_6 \rightarrow Et_2Sn(C\vdots CR)_2 + (C_6H_{11})_2NH_2Cl$; R = Bu, Ph (2495).
$Sn + H_2O \rightarrow Et_3SnCl + (Et_2SnCl)_2O$ (1939).
Fe in $BuCl-Et_3N \rightarrow Et_nBu_{3-n}SnCl$; n = 3, 2, 1 (1939).
5,1,10-X-phenanthroline (in buffered AcOH) → detn. of complex formation and equil. const. by polarography, ion exchange and potentiometry; X = H, Me, Cl, NO_2 (655).
$H_2O + Et_3N \rightarrow [Et_4Sn_2(OH)ClO]_2$ (10).
ArOH + aq. $NH_3 \rightarrow (Et_2SnOAr)_2O$; Ar = α-ON-β-$C_{10}H_6$ (550).
$(c.C_6H_{11})_3SbS \rightarrow (Et_2SnS)_3 + (C_6H_{11})_3SbCl_2$ (1903).
$Fe(CO)_5 \rightarrow (Et_2SnCl)_2Fe(CO)_2 + Et_4Sn_3[Fe(CO)]_4$ (904).

<u>Addn. compd.</u>: $Et_2SnCl_2 \cdot Ph_3CCl$, equil. const. conductivity (2880) [$SnC_{23}H_{25}Cl_3$].

$Et_2SnCl_2 \cdot 2Me_3(PhCH_2)NCl$, from components, m. 130°, conductivity (126), use for electrolytical tin plating and metal refining (126) [$SnC_{24}H_{42}Cl_4N_2$].

$Et_2SnCl_2 \cdot 2Py$, (far) IR spectrum and assignment (2819) [$SnC_{14}H_{20}Cl_2N_2$].

$Et_2SnCl_2 \cdot BiPy$ (M-307), from components in CCl_4 (1479), from $Et_2Sn(OMe)Cl$ and BiPy in EtOH, m. 198° (942), far IR spectrum and assignment (879, 2322), formation const. and thermodynamic data (1479) [$SnC_{14}H_{18}Cl_2N_2$].

$Et_2SnCl_2 \cdot$ (1,10-phenanthroline) (M-307), far IR spectrum and assignment (879, 2322) [$SnC_{16}H_{18}Cl_2N_2$].

$Et_2SnCl_2 \cdot MeCN$, from components in CCl_4, thermodynamic data (1479) [$SnC_6H_{13}Cl_2N$].

$Et_2SnCl_2 \cdot 2Me_2N\overline{:CSCH_2CH_2S}Cl$, from $Et_2Sn(S_2CNMe_2)_2$ with $(CH_2Cl)_2$ under reflux, in 80-90% yield, colorless cryst., m. 86-89°, IR and NMR spectra, struct. (2820) [$SnC_{16}H_{30}Cl_4N_2S_2$].

$Et_2SnCl_2 \cdot Ph_4AsCl$, from components in aq. HCl-NaCl, m. 165-66° (2014), mol. wt. and conductivity (2880) [$SnAsC_{28}H_{25}Cl_3$].

$Et_2SnCl_2 \cdot 2Me_3SbS$, from components in MeOH, $HCCl_3$ or Me_2CO, from $(Et_2SnS)_3 + Me_3SbS + Me_3SbCl_2$ (1:3:3) in $HCCl_3$, m. 122-23°, dissocn. in polar solvents, IR spectrum and assignment (1903) [$SnSb_2C_{10}H_{28}Cl_2S_2$].

$Et_2SnCl_2 \cdot 2Me_2SO$, from components in Et_2O, white cryst., m. 64°, for IR spectra assignment and struct. (2021) [$SnC_8H_{22}Cl_2O_2S_2$].

$Et_2SnCl_2 \cdot 2Me_2SeO$, from components in CH_2Cl_2, white cryst., m. 159-60°, (far) IR spectra, assignment, soly., struct. (2821) [$SnC_8H_{22}Cl_2O_2Se_2$].

<u>Biol. prop.</u>: Effect on metabolism of rats, mode of excretion studied with $(Me^{14}CH_2)_2SnCl_2$ (787).

Activity as bactericide against Escherichia coli (1263).

Activity as fungicide against Aspergillus niger, Penicillium citricum and Serratia marcescens (677).

Activity as insecticide against Musca domestica, LD_{50} and ATP inhibition (425), Culex pipiens berbericus larvae (73), C.p. pipiens larvae (1137).

<u>Use:</u> As catalyst for polyurethan synthesis (1603), for polymerization of $R\overline{CHCH_2O}$ with $(CH_2CO)_2O$; R = Me, Ph (2727).

In catalyst for polymerization of $CH_2:CMe_2$ (2155), of $CH_2:CHCN$ with olefins (2997).

Effect on spreading coeff. of solder (1692).

Mechanism of PVC stabilization (2832).

Diethyltin Dibromide (M-307) $SnC_4H_{10}Br_2$ Et_2SnBr_2

<u>Synth.</u>: By rxn. of $(Et_2SnO)_x$ with $HC\!\equiv\!CMgBr$ in THF, in 71% yield (505).
By rxn. of tin foil with $Br(CH_2)_3CO_2Et$ at 160°, as by-prod. (2650).
By rxn. of powd. Sn with EtBr and Et_3N at 140°, in 60% yield (1943).

<u>Prop.</u>: IR, assignment (321), far IR, assignment (321, 2322), NMR (1430, 2068) and NQR (2290) spectra.

<u>Rxn. with</u>: $Et_2SnH_2 \rightarrow Et_2SnHBr$ (1292).
$Et_2SnH_2 + CH_2\!:\!CHCN \rightarrow Et_2(NCCH_2CH_2)SnBr$ (1613).
$Pr_2SnO \rightarrow (Et_2SnBr)O(Pr_2SnBr)$ (2431).
$Me_2Sn(OMe)_2 \rightarrow Me_2Sn(OMe)Br$ (942).
$BrMg(CH_2)_5MgBr \rightarrow Et_2\overline{Sn(CH_2)_5CH_2}$ (2888).
$(c.C_6H_{11})_3SbS \rightarrow (Et_2SnS)_2 + (C_6H_{11})_3SbBr_2$ (1903).

<u>Addn. compd.</u>: $Et_2SnBr_2 \cdot BiPy$, from components in C_6H_6 (2322), from $Et_2Sn(OMe)Br$ and BiPy in Et_2O (942), m. 205-206° (942), 201-202° (2322), far IR spectrum and assignment (2322) [$SnC_{14}H_{18}Br_2N_2$].

$Et_2SnBr_2 \cdot$(1,10-phenanthroline), from components in C_6H_6, m. 225-30° (dec.), far IR spectrum and assignment (2322) [$SnC_{16}H_{18}Br_2N_2$].

$Et_2SnBr_2 \cdot 2Me_2N\!:\!\overline{CSCH_2CH_2S}Br$, from $Et_2Sn(S_2CNMe_2)_2$ with $(CH_2Br)_2$ under reflux, in 80-90% yield, colorless cryst., m. 76-78°, IR and NMR spectra (2820) [$SnC_{16}H_{30}Br_4N_2S_2$].

$Et_2SnBr_2 \cdot 2Me_3SbS$, from components in MeOH, Me_2CO or $HCCl_3$, from $(Et_2SnS)_3$, Me_3SbS and Me_3SbBr_2 (1:3:3) in $HCCl_3$, m. 121°, IR spectrum and assignment (1903) [$SnSb_2C_{10}H_{28}Br_2S_2$].

Diethyltin Diiodide (M-308) $SnC_4H_{10}I_2$ Et_2SnI_2

<u>Synth.</u>: By rxn. of distillation residue from thermal dec. of $(Et_2Sn)_x$ with I in THF, in 29% yield (1941).
By-prod. in rxn. of $o\text{-}Et_2\overline{SnCH\!:\!CHC_6H_4CH\!:\!CH}$ with I (1410).

<u>Prop.</u>: IR, assignment (321), far IR, assignment (321, 2322), NMR (1430, 2068), Moessbauer resonance (2290) and NQR (2290) spectra.

<u>Rxn. with</u>: $Et_2SnH_2 \rightarrow Et_2SnHI$ (1292).
$Et_2SnH_2 + CH_2\!:\!CHCH_2CN \rightarrow Et_2(NCCH_2CH_2CH_2)SnI$ (1613).
$Et_3PO_3 \rightarrow EtI$ (441).

<u>Addn. compd.</u>: $Et_2SnI_2 \cdot BiPy$, from components in C_6H_6, m. 177-79°, far IR spectrum and assignment (2322) [$SnC_{14}H_{18}I_2N_2$].

$Et_2SnI_2 \cdot$(1,10-phenanthroline), from components in C_6H_6, m. 202-203°, far IR spectrum and assignment (2322) [$SnC_{16}H_{18}I_2N_2$].

Dipropyltin Dichloride (M-308) $SnC_6H_{14}Cl_2$ Pr_2SnCl_2

<u>Prop.</u>: IR (589, 925), assignment (2058), far IR, assignment (2058), Raman (925), Moessbauer resonance (1960) and NQR (1142) spectra. Anal. detn. by reverse potentiometric titration (2013).

<u>Rxn. with</u>: $Bu_3SnH \rightarrow Bu_3SnCl + Pr_2SnHCl$ (1:1) and Pr_2SnH_2 (1:2) (1805).
$Pr_2SnH_2 \rightarrow Pr_2SnHCl$ (1292).
$R_2SnH_2 \rightleftharpoons R_2SnHCl + Pr_2SnHCl$; R = n-, i-Bu, C_8H_{17}, Ph (2737).
$Pr_2SnH_2 + (MeO_2CCH_2)_2Hg \rightarrow Pr_2(MeO_2CCH_2)SnCl$ (24).
$Bu_2SnO \rightarrow (Pr_2SnCl)O(Bu_2SnCl)$ (243).

$Pr_2Sn(OR)_2 \rightarrow Pr_2Sn(OR)Cl$; R = $8\text{-}C_9H_6N$ (1291).
Me_3SiCl + aq. $NH_3 \rightarrow (Pr_2SnOSiMe_3)_2O$ (409a).
$NaSCN \rightarrow Pr_2Sn(NCS)_2$ (591).
$C_7H_5O_2Na \rightarrow Pr_2Sn(O_2C_7H_5)_2$, tropolone (2510)
$AgNO_3$ in MeOH $\rightarrow Pr_2Sn(NO_3)_2$ (1120).
$RMgBr \rightarrow Pr_2SnR_2$; R = $PrC\vdots C, BuC\vdots C, PhC\vdots C$ (1221).
Sn + $H_2O \rightarrow Pr_3SnCl + (Pr_2SnCl)_2O$ (1939).
$R_3SbS \rightarrow (Pr_2SnS)_3 + R_3SbCl_2$; R = $c.C_6H_{11}$ (1903).
ArOH + aq. $NH_3 \rightarrow (Pr_2SnOAr)O(Pr_2SnCl)$; Ar = $\alpha\text{-}ON\text{-}\beta\text{-}C_{10}H_6$ (550).
Addn. compd.: $Pr_2SnCl_2 \cdot Ph_3CCl$, equil. const. (2880) [$SnC_{25}H_{29}Cl_3$].
$Pr_2SnCl_2 \cdot BiPy$ (M-309), from components in CCl_4, formation const. and thermodynamic data (1479) [$SnC_{18}H_{22}Cl_2N_2$].

Dipropyltin Dibromide (M-309) $SnC_6H_{14}Br_2$ Pr_2SnBr_2
Synth.: By rxn. of $(Pr_2SnO)_x$ with $HC\vdots CMgBr$ in THF, in 53% yield (505). By rxn. of metallic tin and PrBr under ^{60}Co irradiation (616a), by heating to 160° in presence of Et_3N, in 55% yield (1943).
Prop.: b_{15} 86° (505), NMR spectrum (2070).
Rxn. with: $Pr_2SnH_2 \rightarrow Pr_2SnHBr$ (1292).
$Et_2SnO \rightarrow (Et_2SnBr)O(Pr_2SnBr)$ (2431).
$R_3SbS \rightarrow (Pr_2SnS)_3 + R_3SbBr_2$; R = $c.C_6H_{11}$ (1903).

Diisopropyltin Diiodide (M-316) $SnC_6H_{14}I_2$ $(Me_2CH)_2SnI_2$
Synth.: By rxn. of metallic tin with i-PrI under reflux in 60% yield (1292), at 80° (2070), in $(MeOCH_2CH_2)_2O$ in presence of cat. HgI_2 and Et_3N, in 80% yield (1904).
Prop.: b. 265-68° (2070), b_{17} 145-47° (1292), b_4 100-102° (1904), NMR spectrum (2070).
Rxn. with: Aq. NH_3, + HX $\rightarrow$ i-Pr_2SnX_2; X = Cl, Br (1292).
i-$Pr_2SnH_2 \rightarrow$ i-Pr_2SnHI (1292).

Dibutyltin Difluoride (M-317) $SnC_8H_{18}F_2$ Bu_2SnF_2
Synth.: By rxn. of $(Bu_2SnO)_x$ with aq. H_2SiF_6 (1468).
By rxn. of SnF_2 with C_2H_4 + H + Al + cat. Et_3Al in C_6H_{12} at 200° (1775).
Rxn. with: $Bu_3SnH \rightarrow Bu_2SnHF + Bu_2SnH_2$ (1804).
$Bu_2Sn(OMe)_2 \rightarrow Bu_2Sn(OMe)F$ (942).
$Bu_2SnS \rightarrow (Bu_2SnF)_2S$ (945).
PhNCO + ureas $\rightarrow$ cat. effect of, kinetic data (133).
Use: As stabilizer for PVC in combination with Bu_2SnS- or Bu_2SnO-derivatives (2953).
As bactericide, insecticide (1468).

Dibutyltin Dichloride (M-309) $SnC_8H_{18}Cl_2$ Bu_2SnCl_2
Synth.: By rxn. of Bu_4Sn with excess $SnCl_4$ at 203° in autoclave on pilot plant scale (2909), with $^{113}SnCl_4$ (158), with $SnCl_2$ (1005), with $SnCl_4$ and $AlCl_3$ at 100°, in 72% yield (705).
By rxn. of $Bu_2Sn(C\vdots CBu)_2$ with HCl (1221).

By rxn. of (Bu$_2$SnCH:CHCH$_2$CHRCH:CH)$_n$; n = 1 with HCl, BCl$_3$ (499); n = X with DCl (499).
By rxn. of Bu$_3$SnCl with S at 180-90° as by-prod (1705).
By rxn. of Bu$_2$SnCl$_2$·Bu$_2$Sn(OMe)$_2$ with SnCl$_4$ in hexane, in 90% yield (2184).
By thermal dec. of (Bu$_2$SnCl)$_2$Y at 1 torr. and 110-30°; Y = O, S (1514).
By rxn. of (Bu$_2$SnO)$_x$ with NH$_4$Cl in c.MeC$_6$H$_{11}$ under reflux, in 70% yield (2660), with PNCl$_2$)$_3$ in MePh under reflux in Ar (2539).
By rxn. of Bu$_2$Sn(OBu)$_2$ with NH$_4$Cl in MePh, in 73% yield (2660).
By rxn. of (Bu$_2$SnPPh)$_3$ with anhydr. Et$_2$O-HCl (2746).
By rxn. of Bu$_6$Sn$_2$ with CCl$_4$ and pptd. Cu, followed by HCl cleavage of rxn. residue (1737).
By rxn. of (Bu$_2$Sn)$_x$ with Ph$_3$CCl in C$_6$H$_6$ at 100°, in 93% yield (66), with Cl, in 54% yield (585).
By rxn. of SnCl$_4$ with BuMgCl in THF in N atm. at 30-40°, in 7% yield (1186a).
By rxn. of metallic tin and BuCl at 120-170° in presence of additives and catalysts: Et$_3$N + I, in 42% yield, influence of N-bases (1943), ZnI$_2$ or HgI$_2$, with Na activated Sn, in 80% yield (1139), BuI and BuNH$_2$, in 94% yield (565), Mg, I and ROH or RSH (880a), LiBr, SnBr$_2$ and I in polyglycol (880b), Bu$_4$NBr under reflux, in 36% yield (1209a), Bu$_4$NBr and SnCl$_4$ under reflux, in 78% yield (1209a).
By rxn. of Bu$_2$SnBr$_2$ with aq. HCl, recovery of Bu$_2$Sn derivatives from Grignard production wastes (2479).
Purification by removal of (up to 20%) Bu$_3$SnCl by aq. amine treatment (2280) by NH$_3$, in 99% yield (2954).
Prop.: n_D^{20} 1.5065 (2479), IR (165, 925, 1038), assignment (1079, 1142, 1509), far IR, assignment (1509, 2322), Raman (925), assignment (1079, 1142), UV (1116-7), NMR (1309, 1514, 2070), Moessbauer resonance (1113, 1309, 1960) and ^{35}Cl NQR (524) spectra, dipole moment (325, 1211), polarography (1249, 2393, 2453, 2588-9) surface tension (1959), thermodynamic data (1479).
Anal. detn. by GLC (1078, 1939), TLC in presence of Bu$_n$SnCl$_{4-n}$, (C$_8$H$_{17}$)$_2$SnCl$_2$ (388), paper chromatography (551), by colorimetry with diphenylcarbazone (490), by polarography (2265, 2823), in presence of Bu$_3$SnCl (247) in presence of BuSnCl$_3$ (2453), by chromatography and IR spectroscopy in presence of Bu$_4$Sn and Bu$_3$SnCl (165), by complexometric titration in presence of BuSnCl$_3$ (1007), after HNO$_3$ degradation (168), by reverse potentiometric titration (2013) calorimetric titration with Bu$_2$Sn(YBu)$_2$; Y = O, S (1116-7), of chloride by titration with NaOMe in Py (1144).
Rxn. with: Bu$_3$SnH → Bu$_3$SnCl + Bu$_2$SnHCl (1:1) and Bu$_2$SnH$_2$ (2:1) (1804).
R$_2$SnH$_2$ ⇌ R$_2$SnHCl + Bu$_2$SnHCl; R = n-, i-Bu, n-C$_8$H$_{17}$, Ph (2737).
i-Bu$_2$SnH$_2$ + Py → c.(i-Bu$_2$Sn)$_9$, cat. effect of (1614).
Bu$_2$SnH$_2$ + CH$_2$:CHCH$_2$OCH$_2$CH$_2$OH → Bu$_2$(HOCH$_2$CH$_2$OCH$_2$CH$_2$CH$_2$)SnCl (1613).
Bu$_2$SnX$_2$ → Bu$_2$SnXCl; X = MeO, AcO (942), MeO$_2$CNEt (943), BuO (846a), BuS (942, 1116-7).
Me$_2$SnO → (Me$_2$SnCl)O(Bu$_2$SnCl) (946).
Bu$_2$SnO → Bu$_2$(Cl)Sn(OSnBu$_2$)$_n$Cl; n = 2, 3 (947, 2857).
Bu$_2$SnO + H$_2$O → Bu$_2$SnCl$_2$·3Bu$_2$SnO·H$_2$O (2431).
Bu$_2$SnO + EtOH → Bu$_4$Sn$_2$Cl(OEt)O (10).
Bu$_2$SnO + HSCH$_2$CH$_2$CO$_2$H (1:10:10) → Bu$_2$(Cl)Sn(SCH$_2$CH$_2$CO$_2$SnBu$_2$)$_x$Cl (178).

$(C_8H_{17})_2SnO \rightarrow Bu_2(Cl)SnO(C_8H_{17})_2SnCl$ (944).

$Bu_2SnS \rightarrow (Bu_2SnCl)_2S$ (945, 1514).

$(Bu_2Sn)_x + Et_3N$ at 160° $\rightarrow Bu_3SnCl + Sn$ (1941).

$Me_3SiCl + $ aq. $NH_3 \rightarrow (Bu_2SnOSiMe_3)_2O$ (409a).

$R_2SiCl_2 + $ aq. KOH $\rightarrow$ OH-terminated poly(stannasiloxane); R = Me, CH_2:CH, Ph (1008).

M in H_2O at 160° $\rightarrow Bu_3SnCl + (Bu_2SnCl)_2O$, M = Al, Fe, Ni, Sn also in MeOH (1939).

M in H_2O at 160° $\rightarrow Bu_3SnCl + (Bu_2SnO)_x$; M = Co, Mg, Zn (1939).

M in Et_3N-BuCl at 160° $\rightarrow Bu_4Sn + Bu_3SnCl + (Bu_2SnCl)_2O$; M = Al, Mg, Na, Zn (1939).

Fe in Et_3N-EtCl at 160°, + HCl $\rightarrow Bu_nEt_{3-n}SnCl$; n = 0-3 (1939).

Zn in Et_3N-EtCl at 160° $\rightarrow Et_nSnBu_{4-n}$; n = 0-4 (1939).

RLi $\rightarrow Bu_2SnR_2$; R = $PhCH_2CH_2C\vdots C$ (2709), $B_{10}H_{10}C_2Ph$ (784), Me_2N (251), Et_2N (323).

$YLi_2 \rightarrow \overline{Bu_2SnYSnBu_2Y}$; Y = $B_{10}H_{10}C_2$ (627), $(2-C_6H_4)_2O$ (1378).

$YLi_2 \rightarrow (Bu_2SnY)_x$; Y = $CB_{10}H_{10}C$ (785).

NaR $\rightarrow Bu_2RSnCl + Bu_2SnR_2$; R = polynuclear organic radical (2703).

$9,10-Na_2C_{14}H_{10} \rightarrow (9,10-Bu_2SnC_{14}H_{10})_n$; n = 1, X 9,10-dihydroanthracene (1726, 2172).

$NaBH_4 \rightarrow Bu_2SnH_2 + NaCl + B_2H_6$ (736).

10% aq. NaOH $\rightarrow (Bu_2SnO)_x$ (565).

NaSCN $\rightarrow Bu_2Sn(NCS)_2$ (591).

NaOR $\rightarrow Bu_2Sn(OR)_2$; R = Et (349), t-Bu (2096), Me, $\overline{OCH_2CHCH_2}$ (1442), 2,6,4-t-Bu_2-$(Me_2NCH_2)C_6H_2O$ (554), PhCH:N (2661).

NaOR $\rightarrow Bu_2Sn(OR)Cl + Bu_2Sn(OR)_2$; R = Me_2C:N (2661).

$p-C_9H_{19}C_6H_4(OCH_2CH_2)_{9-10}OH$ + alkali $\rightarrow Bu_2Sn$-alkoxide (1958).

NaOH + $(p-\overline{OCH_2CHCH_2}C_6H_4NH)_2CO \rightarrow$ butyltin epoxide (1442).

$C_7H_5O_2Na \rightarrow Bu_2Sn(O_2C_7H_5)_2$ tropolone (2510).

$RCO_2Na \rightarrow Bu_2Sn(O_2CR)_2$; R = Cl_2CH, PhCHOH, $o-ClC_6H_4$, $p-H_2NSO_2C_6H_4$, m-, $p-O_2NC_6H_4$, $3,5-(O_2N)_2C_6H_3$ in aq. petr. ether (1051), C_7H_{15}, $C_{11}H_{23}$ in alc. (1357).

$Y(CO_2Na)_2 \rightarrow (Bu_2SnO_2CYCO_2)_x$ in aq. petr. ether; various acids and acid mixtures (1049), $(p-CH_2C_6H_4)_2SiMe_2$, $(o-, (p-C_6H_4)_2SiMe_2$ in aq. petr. ether (1516), CH:CH, C_6H_4, $(CH_2)_n$, n = 3,4,8 in aq. MeOH (1357).

$Na_2S \rightarrow (Bu_2SnS)_x$ (1:1) and $(Bu_2SnCl)_2S$ (2:1) (1514).

$Na_2S_2O_3$ in aq. MeOH $\rightarrow (Bu_2SnS)_3$ (1790).

Na_2S_x (1:1) $\rightarrow (Bu_2SnS)_n + S$; x = 2, 4 (1514).

Na_2S_x (2:1) $\rightarrow (Bu_2SnCl)_2S_n$; n = 2.5, 4 (1514).

NaX $\rightarrow Bu_2SnX_2$; X = BuS (1117), Et_2PS_2 (1363), Ph_2PO_2, Ph_2POS, Ph_2PS_2 in Me_2CO (2108), $(EtO)_2PS$ (2529).

$(:\overline{CRSNa})_2 \rightarrow Bu_2SnSCR:CRS$; R = CN (1975, 2613), NO_2 (2613).

$Y(CH_2SH)_2$ + NaOH $\rightarrow (Bu_2SnSCH_2YCH_2S)_x$; Y = C_6H_4 , $m-Me_2C_6H_2$, 1,2,4-, 1,3,5-Me_3C_6H (445).

$MeO_2CCH:CHCO_2K \rightarrow Bu_2Sn(O_2CCH:CHCO_2Me)Cl + Bu_2Sn(O_2CCH:CHCO_2Me)_2 + Sn(O_2CCH:)_2$? (1557).

$(KS)_2C$:NCN $\rightarrow Bu_2SnS_2C$:NCN (2751), also from Ca salt (1976).

$M_2Fe(CO)_4 \rightarrow [Bu_2SnFe(CO)_4]_2$ (1268).

AgX → Bu_2SnX_2 ; X = NCO (373), NO_3 (1120).
RMgX → Bu_2SnR_2 ; R = $MeOCH_2$ (307), $\overline{CH_2CH_2CH}$ (1873), MeCH:CH (499), CH_2:$CHCH_2$, CH_2:$CMeCH_2$ (1857), PhC⋮C, PrC⋮C, BuC⋮C (1221).
BuMgBr or BuBr + Mg → Bu_4Sn; rxn. in absence of ethers, cat. R_3PO_3 or $HCONR_2$ (1871a).
9,10-$MgC_{14}H_{10}$ → (9,10-$Bu_2SnC_{14}H_{10})_n$; n = 1, X 9,10-dihydroanthracene (1727).
Mg + arene-olefin → Bu_2Sn-derivatives; arene-olefin = anthracene-isoprene (1727), anthracene-styrene, styrene-isoprene, styrene (1728).
$R_2O_3PSNEt_4$ → $Bu_2Sn(SPO_3R_2)_2$; R = Et, Bu (2529).
$\overline{OCH_2CHCH_2OH}$ + Et_3N → $Bu_2Sn(OCH_2\overline{CHCH_2O})_2$ (1442).
$PhPH_2$ + Et_3N → $(Bu_2SnPhP)_n$ + $Et_3N·HCl$; n = c.3, X (2746).
RH_2PO_4 → $Bu_2Sn(PO_4HR)_2$; R = $EtBuCHCH_2$ (1183).
H_2O in EtOH → $[Bu_4Sn_2Cl_2O]_2$ (10).
EtOH in NEt_3 → $Bu_4SnCl(OEt)O$ (10).
ArOH + aq. NH_3 → $(Bu_2SnOAr)O(Bu_2SnCl)$; Ar = α-ON-8-$C_{10}H_6$ (550).
8-HOC_9H_6N in MeOH → $Bu_2Sn(OC_9H_6N)Cl$; obtained in aq. MeOH as monohydrate (1216).
RCO_2H → $Bu_2Sn(O_2CR)_2$-mixture; R = o-ClC_6H_4 + p-$O_2NC_6H_4$ (2395).
$(HSCH_2CO_2)_2Y$ → $Bu_2Sn(SCH_2CO_2YCCH_2SH)_2$; Y = CHEt, CH_2CHPr, $(CH_2)_n$; n = 4, 5, 8 (2903).
Olefin → comparative stability of π-complexes formed (182).
ArNCO + MeOH → kinetic data and mechansim of urethan formation, cat. effect of (1604).
PhNCO + ureas → cat. effect of, kinetic data (133).
R_4NCl and LiCl in Me_2CO and MeCN → potentiometric titration and complex formation (2015).

<u>Addn. compd.:</u> $Bu_2SnCl_2·2NH_3$ (M-311), crude Bu_2SnCl_2 + NH_3 in Me_2CO, method for purification of Bu_2SnCl_2 (2954), rxn. with HX → Bu_2SnX_2 + NH_4Cl ; X = AcCH:CMeO, $C_{11}H_{23}CO_2$, naphthenate, $C_{12}H_{25}S$ (2954) [$SnC_8H_{24}Cl_2N_2$].

$Bu_2SnCl_2·(H_2NCH_2)_2$, from components in Bu_2O (1702), rxn. with $HSCH_2CO_2R$ → $Bu_2Sn(SCH_2CO_2R)_2$; R = $C_{12}H_{25}$ (1702) [$SnC_{10}H_{26}Cl_2N_2$]$_3$.

$Bu_2SnCl_2·2PhNH_2$, from components in Bu_2O (1702), rxn. with $HSCH_2CO_2R$ → $Bu_2Sn(SCH_2CO_2R)_2$; R = i-C_8H_{17} (1702) [$SnC_{20}H_{32}Cl_2N_2$].

$Bu_2SnCl_2·4$-PhC_5H_4N, from components in C_6H_6 , in 77% yield, m. 133-34°, far IR spectrum, assignment, struct. (1706) [$SnC_{30}H_{36}Cl_2N$].

$Bu_2SnCl_2·(2,2'$-BiPy) (M-311), from $Bu_2Sn(O_2CCH:CHCO_2Me)Cl$ and BiPy, m. 184-86° (1557), from components (232a), m. 179-80° (2660), far IR, assignment (879, 2322), UV (280) and Moessbauer resonance (1044) spectra, const. of formation (1479), stability const. (280), struct. (1044), thermodynamic data (1479), use for separation of Bu_2SnCl_2 and $(Bu_2SnS)_3$ (1705) [$SnC_{18}H_{26}Cl_2N_2$].

$Bu_2SnCl_2·(4,4'$-BiPy), from components in C_6H_6 , in 85% yield, m. 173-75°, far IR spectrum, assignment, soly., struct. (1706) [$SnC_{18}H_{26}Cl_2N_2$].

$Bu_2SnCl_2·(1,10$-phenanthroline) (M-311), from components in alc., m. 230°, transition point (232a, 1212), far IR, assignment (879, 2322) and Moessbauer resonance (1044, 1559) spectra, dipole moment (1559), struct. (1044) [$SnC_{20}H_{26}Cl_2N_2$].

$Bu_2SnCl_2·MeCN$, from components in CCl_4, thermodynamic data (1479) [$SnC_{10}H_{21}Cl_2N$].

$Bu_2SnCl_2·OC(NH_2)_2$, IR spectrum (286) [$SnC_9H_{22}Cl_2NO$].

$Bu_2SnCl_2 \cdot 2\ OC(NH_2)_2$, IR spectrum (286) [$SnC_{10}H_{26}Cl_2N_2O_2$].

$Bu_2SnCl_2 \cdot AlCl_3$ (M-311), from components, soly., active as polymerization catalyst for α-olefins (31) [$SnAlC_8H_{18}Cl_5$].

$Bu_2SnCl_2 \cdot ClAsPh_4$, from components in aq. HCl-NaCl, m. 94-96° (2014). [$SnAsC_{32}H_{38}Cl_3$].

Biol. prop.: Toxicity (2404), effect on encephalogram (2574).
Activity as anthelmintic, relative reactivity (2022), for chickens against Davainea proglottina (609).
Activity as fungicide against Rhizoctonia solani (190).
Activity as insecticide against Culex pipiens berbericus larvae (73), C.p. pipiens larvae (1137), Musca domestica, LD_{50} and ATP inhibition (425).

Use: In catalyst for polymerization of $CH_2:CMe_2$ (2155), terpene (1961), for olefin dimerization (261), for vulcanization of chlorine containing polymers (704) for polyurethan synthesis (252, 1027, 1480) for esterification (2298), for crosslinking OH-containing polymers (1751).
As curing agent for unsaturated amides (143).
As fungicide (1005).
As stabilizer for PVC, effect on degradation (371), mechanism of stabilization (2832).

Dibutyltin Dibromide (M-312) $SnC_8H_{18}Br_2$ Bu_2SnBr_2

Synth.: By rxn. of $Bu_2SnCH:CHCH_2CH_2CH:CH$ with $PhBBr_2$ (449).
By rxn. of $(Bu_2SnO)_x$ with NH_4Br in $c.MeC_6H_{11}$ under reflux in 80% yield (2660).
By rxn. of $Bu_2Sn(OR)_2$ with R'Br (435).
By rxn. of $(Bu_2Sn)_x$ with 40% HBr at 120°, in 32% yield (2096), with RBr (1:2); R = allyl and benzyl, in 68 and 36% yield, resp. (66).
By rxn. of metallic tin with BuBr in presence of various additives and catalysts at 120-160°: in $(MeOCH_2CH_2)_2O$, THF or BuOH, Mg, I, Bu_2SnBr_2 and MBr; M = Li, Na, in 16-87% yield (880b), Mg, I and RSH (880a), in Et_3N in 85% yield (1943), R_4NBr or Bu_4PBr, $SnBr_2$ and/or Bu_2SnBr_2, in 86-98% yield (1209a), Bu_4NBr under reflux, in 43% yield (1209a), ZnI_2 or HgI_2 with Na-activated tin, in 85% yield (1139), BuI and $BuNH_2$ (565).
By rxn. of metallic tin with BuBr under ^{60}Co irradiation (144, 616a), in 90% yield (2072), in presence of cat. BuOH and H_2O (2601), optimum conditions (2493), conversion and dose rate, in 84-96% yield (1031), optimum dose yield at 20-40 megarad 63-66% yield (15).
By electrolysis of BuBr on Mg cathode and tin anode in presence of cat. Br and $ZnBr_2$, in 50% yield (683), in 100% current efficiency in BuOAc in presence of Br at 50-75° (2229).
By electrolysis of $SnCl_2$ or $SnCl_4$ on Mg cathode and graphite or Zn anode at 50-70° in presence of Br and $ZnBr_2$, in 16-20% yield (684).
Refining of crude prod. by addition of Br and $SnCl_4$ and heating at 190° (2240).

Prop.: IR (165), far IR, assignment (2322) and Moessbauer resonance (7, 1960) spectra, dipole moment (2187), soly. (2096), calorimetric titration with $Bu_2Sn(OMe)_2$ (2184) and $Bu_2Sn(SBu)_2$ (1116-7).
Anal. detn. by GLC (536), in presence of Bu_4Sn and Bu_3SnBr by chromatography and IR spectroscopy (165).

Rxn. with: $Bu_3SnH \rightarrow Bu_3SnBr + Bu_2SnHBr$ (1:1) and Bu_2SnH_2 (2:1) (1804).
$Bu_2SnH_2 \rightleftharpoons Bu_2SnHBr$ (2737).
$Bu_2SnX_2 \rightarrow Bu_2SnBrX$; X = MeO (942), BuS (1117).
BuBr + Mg or BuMgBr $\rightarrow Bu_4Sn$, rxn. in absence of ether, cat. R_3PO_3 or $HCONR_2$ (1871a).
$BrMg(CH_2)_{n+1}MgBr \rightarrow Bu_2\overline{Sn(CH_2)_nCH_2}$; n = 3, 4 (2888).
Aq. HCl $\rightarrow Bu_2SnCl_2$ + HBr (2479).
$Et_3N + Ph_3SnH$ (1:2:2) $\rightarrow (Ph_3Sn)_2SnBu_2 + Ph_6Sn_2 + Et_3N \cdot HBr$ (1955).
Addn. compd.: $Bu_2SnBr_2 \cdot BiPy$ (M-312), from components in EtOH, m. 176-77° (2660), from $Bu_2Sn(OMe)Br$ and BiPy in Et_2O, m. 180° (942), far IR spectrum and assignment (879, 2322) [$SnC_{18}H_{26}Br_2N_2$].
$Bu_2SnBr_2 \cdot (4,4'-BiPy)$, from components in C_6H_6, in 85% yield, m. 153-55° (1706) [$SnC_{18}H_{26}Br_2N_2$].
$Bu_2SnBr_2 \cdot (1,10\text{-phenanthroline})$ (M-312), far IR, assignment (879, 2322) and Moessbauer resonance (1559) spectra, dipole moment (1559) [$SnC_{20}H_{26}Br_2N_2$].
Biol. prop.: Toxicity (618).
Use: In catalyst for synthesis of Bu_nSnBr_{4-n} from Sn and BuBr; n = 1-3 (880b, 1209a).

Dibutyltin Diiodide (M-312) $SnC_8H_{18}I_2$ Bu_2SnI_2

Synth.: By rxn. of (Bu_2SnO) and $Bu_2Sn(OBu)_2$ with NH_4I in $c.MeC_6H_{11}$ and MePh under reflux, in 92-93% yield (2660).
By rxn. of $(Bu_2Sn)_x$ after thermal dec. at 160° with I in THF, in 86% yield (1941).
By rxn. of $(Bu_2Sn)_x$ with 45% HI, in 65% yield (2096).
By rxn. of metallic tin with BuI in presence of Li or LiX in O-contn. solvents (406), of Bu_2Se or $Zn(S_2CNR_2)_2$ (880b), of Bu_2NH_2 (565), of HgI_2 and Et_3N (1904), of BuOH and Mg (2955).
Prop.: Far IR, assignment (2322) and Moessbauer resonance (1960) spectra.
Rxn. with: $Bu_3SnH \rightarrow Bu_3SnI + Bu_2SnHI$ (1:1) and Bu_2SnH_2 (2:1) (1804).
$Bu_2SnH_2 \rightleftharpoons Bu_2SnHI$ (2737).
$Bu_2SnX_2 \rightarrow Bu_2SnIX$; X = MeO, AcO (942).
$(Ph_2Sn)_x$, + MeMgI $\rightarrow Ph_6Sn_2 + Me_2SnPh_2 + Me_2BuSnPh + MeBu_2SnPh + MeSnPh_3 + Me_2SnBu_2$, rxn. mechanism (1938).
BuI + powd. Cu, + NaOH $\rightarrow Bu_3SnI + (Bu_3Sn)_2O$ (2731).
RI + powd. Zn, + NaOH $\rightarrow (Bu_2RSn)_2O$; R = Et (2811), Bu (2809, 2811).
MeI + Zn in DMF $\rightarrow MeBu_2SnI$ (2809).
RI + Cu-Zn alloy $\rightarrow Bu_2RSnI$; R = Et, Bu (2808).
Addn. compd.: $Bu_2SnI_2 \cdot BiPy$ (M-312), from components in EtOH, m. 162-63° (2660), from $Bu_2\overline{SnSCH_2CH_2S}$ with I and BiPy in $HCCl_3$, m. 167° (1704), far IR spectrum, assignment (879, 2322) [$SnC_{18}H_{22}I_2N_2$].
$Bu_2SnI_2 \cdot (4,4'-BiPy)$, from components in C_6H_6, in 85% yield, m. 100-102° (1706) [$SnC_{18}H_{22}I_2N_2$].
$Bu_2SnI_2 \cdot (1,10\text{-phenanthroline})$ (M-312), far IR, assignment (879, 2322) and Moessbauer resonance (1559) spectra, dipole moment (1559) [$SnC_{20}H_{22}I_2N_2$].

Diisobutyltin Dichloride (M-312) $SnC_8H_{18}Cl_2$ $(MeCHCH_2)_2SnCl_2$

<u>Synth.:</u> By-prod. in rxn. of $SnCl_4$ with $i-Bu_3Al$ in N atm. in pentane-hexane (2961).

Purification of technical matl. by rxn. with NH_3 in Me_2CO (2954).

<u>Rxn. with:</u> $Bu_3SnH \rightarrow Bu_3SnCl + i-Bu_2SnHCl$ (1:1) and $i-Bu_2SnH_2$ (2:1) (1805).
$R_2SnH_2 \rightleftharpoons i-Bu_2SnHCl + R_2SnHCl$; R = n-, i-Bu, $n-C_8H_{17}$, Ph (2737).
$i-Bu_2SnH_2 + Et_2NH \rightarrow (i-Bu_2SnCl)_2$ (1955).
$i-Bu_2SnH_2 + CH_2:CHR$ + cat. AIBN $\rightarrow i-Bu_2(RCH_2CH_2)SnCl$; R = $HOCH_2$, MeO_2C, CN (1613).
$i-Bu_2SnH_2 + MeCH:CHCO_2Et \rightarrow i-Bu_2(EtO_2CCH_2CHMe)SnCl$ (1613).
$i-Bu_2SnH_2 + RC:CH \rightarrow i-Bu_2(RCH:CH)SnCl$; R = Ph, $p-MeOC_6H_4$, CN (1613).
$i-Bu_2SnH_2 + RCHO \rightarrow i-Bu_2Sn(OCH_2R)Cl$; R = Et, n-, i-Pr, $n-C_7H_{15}$ (1613).
$i-Bu_2SnH_2 + c.C_6H_{10}O \rightarrow i-Bu_2Sn(OC_6H_{11})Cl$ (1613).
$i-Bu_2SnH_2 + PhCH:NR \rightarrow i-Bu_2Sn(NRCH_2Ph)$; R = Me, Bu, Ph (1613).
$Et_2NLi \rightarrow i-Bu_2Sn(NEt_2)_2$ (1955).
$NaOMe \rightarrow i-Bu_2Sn(OMe)_2$ (2555).
$Y(CO_2Na)_2 \rightarrow (i-Bu_2SnO_2CYCO_2)_x$; adipate, citrate (1049).
$YNa_2 \rightarrow (i-Bu_2SnY)_x$; Y = $p-O_3SNHC_6H_4CO_2$, $p-SCH_2C_6H_4CH_2S$ (1049).
Mg in THF $\rightarrow c.(i-Bu_2Sn)_9$ (1614).
$i-Bu_2SnH_2 + Py \rightarrow c.(i-Bu_2Sn)_9$, cat. of ~ (1614).

<u>Addn. compd.:</u> $i-Bu_2SnCl_2 \cdot 2NH_3$, from components in Me_2CO (2954), rxn. with $RCO_2H \rightarrow i-Bu_2Sn(O_2CR)_2 + NH_4Cl$ (2954), use for refining crude $i-Bu_2SnCl_2$ (2954) [$SnC_8H_{24}Cl_2N_2$].

<u>Use:</u> As additive for flame resistant polyester (2702).

Dineopentyltin Dibromide (M-318) $SnC_{10}H_{22}Br_2$ $(Me_3CCH_2)_2SnBr_2$

<u>Synth.:</u> By rxn. of $(n-C_5H_{11})_2Sn(CH_2CMe_3)_2$ with Br in refluxing CCl_4, in 90% yield (2188).

<u>Prop.:</u> $b_{0.1}$ 86.7°, n_D^{25} 1.5292 (2188).

<u>Rxn. with:</u> NaOH in aq. Et_2O, + AcOH $\rightarrow (Me_3CCH_2)_2Sn(OAc)_2$ (2188).
NaOH in aq. Et_2O, + $Y(CO_2H)_2 \rightarrow (Me_3CCH_2)_2Sn(O_2C)_2Y$; Y = $Me_2C(SCH_2CH_2)_2$ (1170).
$BrMg(CH_2)_{n+1}MgBr \rightarrow (Me_3CCH_2)_2\overline{Sn(CH_2)_nCH_2}$; n = 3, 4, 5 (2888).

Dihexyltin Dichloride (M-318) $SnC_{12}H_{26}Cl_2$ $(n-C_6H_{13})_2SnCl_2$

<u>Synth.:</u> By rxn. of $(C_6H_{13})_3SnCl$ with $SnCl_4$ at 180° (90).
By rxn. of powd. tin with $C_6H_{13}Cl$ in presence of cat. RI and SbI_3 at 180° (1244a).

<u>Prop.:</u> Detn. and sepn. by paper chromatography (551).

<u>Rxn. with:</u> NaOH in aq. $CH_2Cl_2 \rightarrow [(C_6H_{13})_2SnO]_x$ (90).
$\overline{CH_2CH_2CH}MgBr \rightarrow (C_6H_{13})_2Sn(\overline{CHCH_2CH_2})_2$ (1873).

Dicyclohexyltin Dichloride (M-318) $SnC_{12}H_{22}Cl_2$ $(c.C_6H_{11})_2SnCl_2$

<u>Synth.:</u> By rxn. of $(C_6H_{11})_4Sn$ with $SnCl_4$ at 100-190°, yield depending on rxn. conditions (1385).

<u>Prop.:</u> IR spectrum, differential thermal analysis (1385).

<u>Rxn. with:</u> $Bu_3SnH \rightarrow Bu_3SnCl + (C_6H_{11})_2SnHCl$ (1:1) and $(C_6H_{11})_2SnH_2$ (2:1) (1805).

$R_2SnH_2 \rightleftharpoons (C_6H_{11})_2SnHCl + R_2SnHCl$; R = n-, i-Bu, n-$C_8H_{17}$, Ph (2737).
c.$C_6H_{11}MgCl \rightarrow (C_6H_{11})_3SnCl$ (1583b).
Addn. compd.: (c.$C_6H_{11})_2SnCl_2 \cdot 2Me_2SO$, from $(C_6H_{11})_4Sn$ and $SnCl_4$ in Me_2SO above 170°, IR spectrum (1385) [$SnC_{16}H_{34}Cl_2O_2S_2$].

Dicyclohexyltin Dibromide (M-318) $SnC_{12}H_{22}Br_2$ $(c.C_6H_{11})_2SnBr_2$
Rxn. with: $LiAlH_4 \rightarrow (C_6H_{11})_2SnH_2$ (1614).
$NaOCH_2R \rightarrow (C_6H_{11})_2Sn(OCH_2R)_2$; R = CH_2OH, $CH_2OHCHOH$ (2721).
$RMgBr \rightarrow (C_6H_{11})_2SnR_2$; R = n-, i-Pr, n-, i-Bu, n-, i-C_5H_{11} (1772); C_6H_{13}, C_7H_{15}, C_8H_{17}, C_9H_{19}, $C_{10}H_{21}$ (1774).

Dioctyltin Dichloride (M-313) $SnC_{16}H_{34}Cl_2$ $(n-C_8H_{17})_2SnCl_2$
Synth.: By rxn. of $(C_8H_{17})_4Sn$ with $^{113}SnCl_4$ (158).
By rxn. of $Me_2Sn(C_8H_{17})_2$ with $SnCl_4$, in > 85% yield (1482).
By rxn. of metallic tin with n-$C_8H_{17}Cl$ at 120-80° in presence of Et_3N and I, in 18% yield, effect of iodides (1943), of RSH as by-prod. (880a), of LiBr and $SnCl_2$ or $(C_8H_{17})_2SnCl_2$ in glycol ether or polyglycol (880b), of RI and SbI_3, in 66% yield (1244a).
By rxn. of $[(C_8H_{17})_2SnO]_x$ with aq. HCl, method of purification (2481).
Purification of tech. grade matl. by conversion to R_2SnCl_2 ammonia complex (2954).
Prop.: NQR spectrum (1142), polarography (1249), thermal stability at 180° (1052). Anal. detn. by TLC (388, 1052) and sepn. by paper chromatography (551).
Rxn. with: $Bu_3SnH \rightarrow Bu_3SnCl + (C_8H_{17})_2SnHCl$ (1:1) and $(C_8H_{17})_2SnH_2$ (2:1) (1805).
$R_2SnH_2 \rightleftharpoons (C_8H_{17})_2SnHCl + R_2SnHCl$; R = n-, i-Bu, n-$C_8H_{17}$, Ph (2737).
$(C_8H_{17})_2SnCl_2 + CH_2:CHR \rightarrow (C_8H_{17})_2(RCH_2CH_2)SnCl$; R = $HOCH_2CH_2OCH_2$ (1613).
$RCO_2Na \rightarrow (C_8H_{17})_2Sn(O_2CR)_2$; R = C_7H_{15}, $C_{11}H_{23}$ (1357).
$Y(CO_2Na) \rightarrow (C_8H_{17})_2Sn(O_2C)_2Y$; Y = CH:CH, o-$C_6H_4$, $(CH_2)_n$; n = 3, 4, 8 (1357).
NaOH in aq. MePh $\rightarrow [(C_8H_{17})_2SnO]_x$ (2481).
$RSH \rightarrow C_8H_{17})_2Sn(SR)_2$; R = $HSCH_2CO_2(CH_2)_4O_2CCH_2$ (2903).
$Na_2S_2O_3$ in aq. MeOH $\rightarrow [(C_8H_{17})_2SnS]_3$ (1790).
$[:C(CN)SNa]_2 \rightarrow (C_8H_{17})_2SnSC(CN):C(CN)S$ (1975).
$Ph_2PYNa \rightarrow (C_8H_{17})_2Sn(YPPh_2)_2$; Y = S_2, SO (2108).
$(KS)_2C:NCN \rightarrow (C_8H_{17})_2SnS_2C:NCN$ (2751), from Ca derivative (1976).
Addn. compd.: $(n-C_8H_{17})_2SnCl_2 \cdot 2NH_3$, from components in Me_2CO, use for purification of R_2SnCl_2 (2954), rxn. with HX $\rightarrow (C_8H_{17})_2SnX_2$; X = $RO_2CCH:CHCO_2$, $EtBuCHCO_2$, $RO_2CCH_2CH(CO_2R)S$, PhS (2954) [$SnC_{16}H_{40}Cl_2N_2$].
$(n-C_8H_{17})_2SnCl_2 \cdot BiPy$ (M-313), far IR spectrum and assignment (879) [$SnC_{26}H_{42}Cl_2N_2$].
$(n-C_8H_{17})_2SnCl_2 \cdot (4,4'-BiPy)$, from components in C_6H_6 in 85% yield, m. 153-55°, far IR spectrum and assignment, struct. (1706) [$SnC_{26}H_{42}Cl_2N_2$].
$(n-C_8H_{17})_2SnCl_2 \cdot (1,10$-phenanthroline) (M-313), far IR spectrum and assignment (879) [$SnC_{28}H_{42}Cl_2N_2$].
Biol. prop.: Toxicity (1130-1)
Activity as anthelmintic for chicken against nematodes and cestodes (1130-1).
Activity as insecticide against larvae of Culex pipiens berbericus (73), C.p.

pipiens (1137).
Use: As catalyst in synthesis of $(C_8H_{17})_nSnCl_{4-n}$ from Sn and $C_8H_{17}Cl$ (880b). As additive for flame resistant polyester (2702).

Dioctyltin Dibromide (M-318) $SnC_{16}H_{34}Br_2$ $(n-C_8H_{17})_2SnBr_2$
Synth.: By rxn. of metallic tin with $C_8H_{17}Br$ in presence of ZnI_2 or HgI_2, in 61% yield (1139), at 180° in presence of Bu_4NBr and $SnBr_2$, in high yield (1209a), in presence of AsI_3 or SbI_3, in 62% yield (1244a).
Refining by treatment with small amt. of Br and heating with $SnCl_4$ to 190° (2240).
Prop.: Polarographic detn. (2265).
Rxn. with: NaOH in aq. MePh → $[(C_8H_{17})_2SnO]_x$ (2481).

Dilauryltin Dichloride (M-319) $SnC_{24}H_{50}Cl_2$ $(n-C_{12}H_{25})_2SnCl_2$
Synth.: By rxn. of $Me_2Sn(C_{12}H_{25})_2$ with $SnCl_4$, in > 85% yield (1482).
By rxn. of $C_{12}H_{25}Cl$ and $SnCl_4$ with Mg and cat. $C_{12}H_{25}Br$ in Et_2O-heptane, in 3% yield (1583a).
By rxn. of $C_{12}H_{25}Cl$ and powd. tin and cat. $C_{12}H_{25}I$ and SbI_3 at 180° (1244a).
Prop.: $b_{1.5}$ 235° (1482).
Rxn. with: Ph_2PYNa → $(C_{12}H_{25})_2Sn(YPPh_2)_2$; Y = S_2, SO, O_2 (2108).
Addn. compd.: $(n-C_{12}H_{25})_2SnCl_2 \cdot 2NH_3$, from components in Me_2CO (2954). Use for refining tech. grade R_2SnCl_2 (2954), rxn. with HX → $(C_{12}H_{25})_2SnX_2$; X = BzO, BzS (2954) $[SnC_{24}H_{56}Cl_2N_2]$.

Diphenyltin Dichloride (M-313) $SnC_{12}H_{10}Cl_2$ Ph_2SnCl_2
Synth.: By rxn. of Na-activated tin with PhCl in presence of ZnI_2 or HgI_2, in 53% yield (1139).
By rxn. of $(Ph_2SnPPh)_3$ with HCl in Et_2O (2746).
By-prod. in rxn. of $SnCl_4$ with PhMgCl in THF-xylene at 30-40° in N atm., in 18% yield (1186a).
Prop.: IR (333), far IR (194), assignment (1703, 1706, 2784) Raman (2784), UV (335), NMR (1453, 2067), Moessbauer resonance (7, 67, 211, 224, 503, 1114, 1960, 1989, 1991, 2235, 2347, 2617) and ^{35}Cl NQR (542, 1142, 1962) spectra, dipole moment (194, 325, 1271, 1518), magnetic susceptibility (7). Anal detn. by TLC and sepn. from Ph_3SnCl (1253), from Ph_nSnCl_{4-n} ; n = 1, 3, 4 (2635), with alizarin S at pH 4.2 (70), by reverse potentiometric titration (2013).
Rxn. with: ^{60}Co irradiation → formation of Ph_2SnCl radical by ESR evidence (693).
Polarographic reduction → $(Ph_2SnCl)_2$ → $(Ph_2Sn)_x$ (963).
Olefin → comparative stability of π-complexes formed (182).
Bu_3SnH → Bu_3SnCl + Ph_2SnHCl (1:1) and Ph_2SnH_2 (2:1) (1805).
$R_2SnH_2 \rightleftharpoons Ph_2SnHCl + R_2SnHCl$; R = n-, i-Bu, $n-C_8H_{17}$, Ph (2737).
$Ph_2SnH_2 + RR'C:CH_2$ → $Ph_2(RR'CHCH_2)SnCl$; RR' = H, $CH_2:CHCH_2CH_2$; Me, Ph (1613).
$(PhCH_2)_2SnCl_2 \cdot 2Me_2SO$ → ligand exchange (1317).
$Ph_2Sn(OC_9H_6N-8)_2$ → $Ph_2Sn(OC_9H_6N)Cl$ (602).
$(Ph_3Sn)_2S$ → $Ph_3SnCl + (Ph_2SnS)_3$ (1705).
Li in THF → $(Ph_2Sn)_x$ (4).

$Me_2NLi \rightarrow Ph_2Sn(NMe_2)_2$ (251).

$YLi_2 \rightarrow Ph_2SnY$; $Y = (2-C_6H_4)_2O$, $(2-C_6H_4)_2SO_2$ (4, 1378), $(2-C_6H_4)_2S$ (4), $(2-C_6F_4)_2$ (883), $(2-C_6H_3R)_2NMe$; $R = 4-Me$, $4-Br$, X (1376), $(SiPh_2)_4$ (1177).

$o-(LiC)_2B_{10}H_{10} \rightarrow o-Ph_2ClSnCB_{10}H_{10}CSnClPh_2$ (1835).

$m-(LiC)_2B_{10}H_{10} \rightarrow m-Ph_2ClSnCB_{10}H_{10}CH + m-Ph_2ClSn(CB_{10}H_{10}CSnPh_2)_6Cl$ (1835).

$m-(LiC)_2B_{10}H_{10} \rightarrow (m-Ph_2SnCB_{10}H_{10}C)_x$ (785).

$p-(LiC)_2B_{10}H_{10} \rightarrow p-Ph_2ClSn(CB_{10}H_{10}CSnPh_2)_{11}Cl$ (1835).

$(2-LiC_6H_4)_2C_6H_4 \rightarrow (o-C_6H_4)_3$ (1410).

$(PhNLi)_2 \rightarrow (Ph_2Sn)_x + (PhN:)_2$ (1089).

Na in $C_{10}H_8 \rightarrow$ c.$(Ph_2Sn)_6$ (392).

$NaR \rightarrow Ph_2RSnCl$; R = polynuclear organic radical (2703).

$9,10-Na_2C_{14}H_{10} \rightarrow (9,10-Ph_2SnC_{14}H_{10})_n$; n = 1, X 9,10-dihydroanthracene (1726).

$NaX \rightarrow Ph_2SnX_2$; $X = Ph_2As$ (1842), Ph_2Sb (1843), $Co(CO)_4$ (216), $C_5H_5Fe(CO)_2$ (1595), $AcCH:CMeO$ (347), SCN (2617), Et_2PS_2 (1363).

$NaX \rightarrow Ph_2SnClX + Ph_2SnX_2$; $X = Re(CO)_5$ (2034).

$Ph_2PYNa \rightarrow Ph_2Sn(YPPh_2)_2$; $Y = O_2$, SO, S_2 (2108).

$PhSO_2Na$ in $H_2O \rightarrow Ph_2Sn(O_2SPh)_2$ (1:2) and $Ph_2Sn(Cl)O_2SPh \cdot Ph_2SnO$ (1410).

$Ac_2CH_2 + NaOMe \rightarrow Ph_2Sn(OMe:CHAc)_2$ (2055).

$C_7H_5O_2Na \rightarrow Ph_2Sn(O_2C_7H_5)_2$ tropolone (2510).

$RCO_2Na \rightarrow Ph_2Sn(O_2CR)_2$; R = 2-pyridine, ferrocene (1310, 1374).

$Y(CO_2Na)_2 \rightarrow (Ph_2SnO_2CYCO_2)_n$ n, Y = 1, $(CH_2CH_2)_2S$ (214), X, $(CH_2)_4$ (1049).

$NaS \rightarrow (Ph_2SnS)_x$ (1514, 1705).

$NaX \rightarrow Ph_2SnX_2$; $X = R_2NCS_2$; R = Me, Et, Ph ($PhCH_2$ + H) (298).

$NaX \rightarrow Ph_2SnClX$; $X = R_2NCS_2$; R = Me, ($PhCH_2$ + H) (298), Et, Ph (298, 761).

$(:CHRSNa)_2 \rightarrow Ph_2\overline{SnSCR:CRS}$; R = CN (1975, 2613) CO_2Et (2613).

$(KS)_2C:NCN \rightarrow Ph_2SnS_2C:NCN$ (2751).

$Y(SH)_2$ + aq. $KOH \rightarrow Ph_2\overline{SnSYS}$; $Y = CH_2CH_2$, $1,3,4-MeC_6H_3$ (141).

$YH_2 + KOH \rightarrow (Ph_2SnY)_x$; $Y = p-OC_6H_4O$, $p-SCH_2C_6H_4CH_2S$ (610).

$AgX \rightarrow Ph_2SnX_2$; $X = NO_3$ (1030), ClO_4 (70).

$RMgBr \rightarrow Ph_2SnR_2$; R = Me (296), mesityl (303), BuC⋮C, PhC⋮C (1221) C_6F_5 (428).

$C_6Cl_5MgCl \rightarrow Ph_2Sn(C_6Cl_5)_2 + C_6Cl_6$ (1101).

$CH_2Br_2 + Mg \rightarrow Ph_2Sn(CH_2SnClPh_2)_2$ (1352).

$CaY \rightarrow (Ph_2SnY)_2$; $Y = Fe(CO)_4$ (216).

$PHC⋮CH + RNH_2$ in $C_6H_6 \rightarrow Ph_2Sn(C⋮CPh)_2 + RNH_3Cl$ (2495).

$PhPH_2 + Et_3N \rightarrow (Ph_2SnPPh)_n + Et_3N \cdot HCl$; n = c.3, x (2746).

$Ph_2AsO_2H + Et_3N \rightarrow Ph_2Sn(O_2AsPh_2)_2$ (1842).

$R_3SbS \rightarrow (Ph_2SnS)_3 + R_3SbCl_2$; R = Me, c.$C_6H_{11}$ (1903).

$o,o'-HOC_6H_4N:CHC_6H_4OH \rightarrow Ph_2\overline{SnOC_6H_4N:CHC_6H_4O}$ (2553).

$o'-HOC_6H_4N:CMeCH_2COR$; $Ph_2SnOC_6H_4N:CMeCH:CRO$; R = MePh (2553).

$8-HOC_9H_6N$ in aq. $MeOH \rightarrow Ph_2Sn(OC_9H_6N)Cl$ (385, 602) and $Ph_2Sn(OC_9H_6N)_2$ (385).

$8-HOC_9H_6N + NH_3 \rightarrow Ph_2Sn(OC_9H_6N)_2$ (1216).

$C_7H_5O_2 \rightarrow PhSn(O_2C_7H_5)_2Cl$ tropolone (372).

$H \rightarrow {}_2SnCl_2$; = Ac_2CH (385-6), Bz_2CH, AcCHBz (386), $o-OC_6H_4CHO$, $8-OC_9H_6N$ (386), kinetic data and mechanism (1465).

$AcCH_2Bz \rightarrow Ph_2Sn(CHAcBz)_2$ (1591).

$HSCH_2CO_2H + Py \rightarrow (SnC_{14}H_{12}O_2S)_x$ (121).

$CS_2 + NH_3 \rightarrow (Ph_2SnS)_3$ (298).

(RCOCHCOR)Tl → Ph$_2$Sn(OCR:CHCOR)$_2$; R = Me, Ph (Me + Ph) (385).
BiPyM(CO)$_4$ → Ph$_2$ClSnBiPyM(CO)$_3$Cl; M = Mo, W (1373).
[C$_5$H$_5$Fe(CO)$_2$]$_2$ → PhCl$_2$SnC$_5$H$_5$Fe(CO)$_2$ + [C$_5$H$_5$Fe(CO)$_2$]$_2$SnCl$_2$ + (C$_5$H$_5$)$_2$Fe (1006).
Co$_2$(CO)$_8$ + Py → Ph$_2$Sn[Co(CO)$_4$]$_2$ (759).
Co$_2$(CO)$_8$ in THF → Ph$_2$ClSnCo(CO)$_4$ (1673).
Co[Co(CO)$_4$]$_2$ in MeOH → Ph$_2$Sn[Co(CO)$_4$]$_2$ (1672).
H$_2$N(CH$_2$)$_6$NH$_2$ + ClOC(CH$_2$)$_4$COCl in aq. alkali → Sn-contn. polyamide (1245).
(p-HOC$_6$H$_4$)$_2$CMe$_2$ + p-C$_6$H$_4$(COCl)$_2$ in aq. alkali → Sn-contn. polyester (1246).
R$_4$NCl and LiCl in Me$_2$CO and MeCN → potentiometric titration, complex formation (2015).

Addn. compd.: All derivatives are listed in Table 141 and 120.
Biol. prop.: Toxicity (1129, 1131).
Activity as anthelmintic for chicken against nematodes and cestodes (1129, 1131).
Activity as insecticide against Culex pipiens berbericus larvae (73), C.p. pipiens larvae (1137), Musca domestica, LD$_{50}$ and ATP inhibition (425), Tineola bisselliella larvae (162).
Use: In catalyst for polymerization of C$_2$H$_4$ (1507), (CH$_2$:CH)$_2$ (1560), α-olefins (1265).
As catalyst for polyurethan synthesis (2166).
As antioxidant for polyphenyl ethers (1979).

Table 141. Addition Compounds of Diphenyltin Dichloride   Ph$_2$SnCl$_2$·L

Ligand*	Synth. Method**	Properties	Ref.
nH$_2$O	--	M.R.S. (a)	503
Ph$_3$CCl	--	Conductivity (b)	2880
Et$_4$NCl	--	M.R.S. (a)	2347
2Py·HCl	--	M.R.S. (a)	1044
2 4-Ph$_3$SnCH$_2$CH$_2$C$_5$H$_4$N	ID	m. 175-78°, far IR spectrum, assignment	2689
Pyrazine	IA (23)	m. 108-14°, soly. (c)	1706
BiPy (315)	I	m. 242-45° (1705) (d)	232a
1/2 BiPy	ID (86)	m. 195°, soly. mol. wt. (e, f)	2100
4,4'-BiPy	IA (85)	m. 209-10°, soly. (c)	1706
o-Phen (315) (g)	I	(e)	232a
1/2 TerPy (h)	I	m. 104°, M.R.S. (a-2347) (i, j)	147
2PyO (k)	IB, C	m. 162-63°, (dec.) (l)	234a
	IA	m. 166-68° (dec.)	1296
2C$_5$H$_4$RN	ID	m. 175-78°, soly. (n)	2689
1/2 (BzCH$_2$CMe:NCH$_2$)$_2$	IA	m. 128-30°	386
2Ph$_3$PO	IB, C	Dec. 135-36° (l)	2430
Ph$_4$AsCl	IE	m. 225-27° (o)	2014
2Ph$_3$AsO	IB, C	Dec. 183° (l)	2430
2Me$_2$SO	I	m. 135°, IR spectrum	1388
	ID	White cryst. (p)	2021

		(q)	NMR spectrum	1317
2(CH$_2$)$_4$SO		I (95)	IR spectrum	1388
2C$_4$H$_8$S$_2$O (r)		ID (> 50)	m. 171-75° (dec.) (s)	1707
cis-C$_4$H$_8$S$_2$O$_2$ (t)		ID	m. 140-43° (dec.) (s, u)	1707
		ID	m. 184-88° (dec.)	1707
trans-C$_4$H$_8$S$_2$O$_2$ (t)		ID (> 50)	m. 238° (dec.) (s)	1707
2Me$_2$SeO		IC	White cryst., m. 158-60° (v)	2821

* Numbers in parenthesis refer to pages in main volume.
** Synthesis from components I, in C$_6$H$_6$-(IA), in HCCl$_3$-(IB), in CH$_2$Cl$_2$-(IC), in alcohol-(ID) and in aq. NaCl-HCl-(IE), yield given in parenthesis.
(a) M.R.S. = Moessbauer resonance spectrum. (b) Stability const. (2880). (c) Far IR spectrum and assignment (1706). (d) Used for separating (Ph$_2$SnS)$_x$ and Ph$_2$SnCl$_2$ (1705). (e) Moessbauer resonance spectrum (2617). (f) Forms clear yellow film after melting, useful for films and protective coatings (2100). (g) o-Phen = 1,10-Phenanthroline (232a). (h) TerPy = 2,2',2"-terpyridine (147). (i) Conductivity, proposed struct. [Ph$_2$SnCl·TerPy] [Ph$_2$SnCl$_3$] (147). (j) Rxn. with NaBPh$_4$ in ROH → [Ph$_2$SnCl·TerPy]BPh$_4$ (147). (k) PyO = N-pyridine oxide (2340). (l) IR spectrum, struct. (2340). (m) R = 4-Ph$_3$SnCH$_2$CH$_2$ (2689). (n) Far IR spectrum and assignment, struct. (2689). (o) Conductivity (2882). (p) Far IR spectrum and assignment, struct. (2021). (q) From Ph$_2$SnCl$_2$ and (PhCH$_2$)$_2$SnCl$_2$·2Me$_2$SO (1317). (r) 1,4-Dithiane sulfoxide (1707). (s) IR spectrum, struct., soly (1707). (t) 1,4-Dithiane disulfoxide (1707). (u) Exists in two not interconvertible forms (1707). (v) (Far) IR spectra and assignment, struct. (2821).

Diphenyltin Dibromide (M-319) SnC$_{12}$H$_{10}$Br$_2$ Ph$_2$SnBr$_2$
Synth.: By rxn. of (Ph$_3$Sn)$_2$Hg with Br in C$_6$H$_6$ as by-prod. (2370).
Prop.: IR spectrum and assignment (2592). Anal. detn. with alizarin S at pH 4.2 (70).
Rxn. with: AgNCO → [Ph$_2$Sn(NCO)OH]$_2$ (373).
AgNCO in air → [(Ph$_2$SnNCO)$_2$O]$_2$ + [Ph$_8$Sn$_4$(NCO)$_2$(OH)$_2$O$_2$] (373).
RMgBr → Ph$_2$SnR$_2$; R = C$_6$F$_5$ (77).
BrMg(CH$_2$)$_{n+1}$MgBr → Ph$_2$Sn(CH$_2$)$_n$CH$_2$ (2888).
R$_3$SbS → (Ph$_2$SnS)$_3$ + R$_3$SbBr$_2$; R = Me, c.C$_6$H$_{11}$ (1903).
Co$_2$(CO)$_8$ → Ph$_2$Sn(Br)Co(CO)$_4$ (1673).
8-HOC$_9$H$_6$N (at 170-202°) → (C$_9$H$_6$NO)$_2$SnBr$_2$ + C$_6$H$_6$ (1465).
RBr in Py → Ph$_2$(Br$_2$R)SnBr·Py (1:1) and Ph$_2$Sn(RBr$_2$)·2Py; R = CH$_2$:CHCH$_2$, c.3-C$_6$H$_9$ (2592).
Addn. compd.: Ph$_2$SnBr$_2$·2Py, IR and NMR spectra (2592) [SnC$_{12}$H$_{16}$Br$_2$N$_2$].
Ph$_2$SnBr$_2$·BiPy, from components in 88% yield, m. 245-48° (373) [SnC$_{22}$H$_{18}$Br$_2$N$_2$].
2Ph$_2$SnBr$_2$·2,2',2"-Terpyridine, white, m. 120°, conductivity, proposed struct.
[Ph$_2$SnBr·TerPy] [Ph$_2$SnBr$_3$] (147), rxn. with NaBPh$_4$ → [Ph$_2$SnBr·TerPy]BPh$_4$ (147) [Sn$_2$C$_{39}$H$_{31}$Br$_4$N$_3$].
Use: In catalyst for C$_2$H$_4$ polymerization (1507).

Diphenyltin Diiodide (M-319) $SnC_{12}H_{10}I_2$ Ph_2SnI_2

Synth.: By rxn. of c.$(Ph_2Sn)_n$ with I in C_6H_6 at 20°, in quant. yield, n = 5, 6 (392).
Prop.: Far IR spectrum and assignment (1703), dipole moment (1518).
Rxn. with: $(Ph_2Sn)_x$ at 140° → Ph_3SnI + Sn (1938).
AgX → Ph_2SnX_2 ; X = NCS (2794), N(CN)$_2$ (1330), NO$_3$ (1030).
$AgNO_3$ in air → $(Ph_2SnNO_3)_2O$ (1030).
MeI + Zn dust, + NaOH, + HCl → $MePh_2SnCl$ (2811).
EtI + M → $EtPh_2SnI$, M = Cu-Zn alloy (2808), Zn (2809).
PhI + Zn, + NaOH → Ph_3SnOH (2809).

Additional unsbustituted symmetric diorganotin dihalides are listed in Table 142.

4.1.2.2 UNSYMMETRIC UNSUBSTITUTED DIORGANOTIN DIHALIDES

Butylmethyltin Dichloride $SnC_5H_{12}Cl_2$ $MeBuSnCl_2$
Synth.: By rxn. of Me_3SnCl with $BuSnCl_3$ (1:1) at 180-90°, in 88% yield (1370).
By rxn. of Me_4Sn with $BuSnCl_3$ (1:2) at 0 and 180°, in 37 and 99% yield, resp. (1370).
Prop.: m. 42-43°, b_{16} 117-19°, GLC (1370).
Rxn. with: Me_4Sn at 75-80° → Me_3SnCl + $Me_2BuSnCl$ (1370).
$(CH_2:CH)_4Sn$ at 20-60° ⇌ $(CH_2:CH)_3SnCl$ + $MeBu(CH_2:CH)SnCl$ (1370).
Me_3SnCl at 150° ⇌ Me_2SnCl_2 + $Me_2BuSnCl$ (1370).

Methylphenyltin Diiodide $SnC_7H_8I_2$ $MePhSnI_2$
Synth.: By rxn. of $MeSnPh_3$ with I in $HCCl_3$ (1719).
By rxn. of $MePh_2SnI$ with I in Et_2O at 30°, in 83% yield (2787).
Prop.: m. 40° (1719), yellow liq. b_3 163°, n_D^{20} 1.6622, d_{20} 2.3575 (2787), NMR spectrum (1719).
Rxn. with: $1-C_{10}H_7MgX$ → $MePhSn(C_{10}H_7-1)_2$ (1719).
15% aq. KOH in Et_2O → $(MePhSnO)_x$ (2787).

Listing of unsymmetric unsubstituted diorganotin dihalides concludes in Table 143 on page 520.

Table 142. Unsubstituted Symmetric Diorganotin Dihalides R_2SnX_2

Formula*	Synth. Method**	Yield	Properties	Ref.
(PhCH$_2$)$_2$SnI$_2$ (316)	II	100	m. 88°	394
	VA	37	(Far) IR spectra, assignment	841
	VB	--	NMR spectrum	2069
(o-MeC$_6$H$_4$CH$_2$)$_2$SnCl$_2$ (316)	--	--	(a)	--
(o-MeC$_6$H$_4$CH$_2$)$_2$SnBr$_2$	IVC	35	m. 121-22° (far) IR spectra, assignment	841
(m-MeC$_6$H$_4$CH$_2$)$_2$SnBr$_2$	IVC	65	m. 118-19°, (far) IR spectra, assignment	841
(p-MeC$_6$H$_4$CH$_2$)$_2$SnBr$_2$	IVC	8	m. 189-90° (far) IR spectra, assignment	841
Et$_2$SnF$_2$ (316)	--	--	(Far) IR spectra, assignment (b)	321
Pr$_2$SnF$_2$	IB	--	Far IR spectrum, assignment (c)	52
Pr$_2$SnI$_2$ (309)	--	--	(d, e)	--
i-Pr$_2$SnCl$_2$ (316)	VB (f)	--	(d)	1292
i-Pr$_2$SnBr$_2$ (316)	VB (f)	--	(d)	1292
(CH$_2$CH$_2$CH)$_2$SnCl$_2$ (316)	--	--	Use: herbicide, pesticide	1873
(CH$_2$CH$_2$CH)$_2$SnBr$_2$ (316)	--	--	Use: herbicide, pesticide (g)	1873
(CH$_2$CH$_2$CH)$_2$SnI$_2$ (316)	--	--	Use: herbicide, pesticide	1873
i-Bu$_2$SnF$_2$	--	--	(h)	--
i-Bu$_2$SnBr$_2$ (317)	--	--	(i)	--
i-Bu$_2$SnI$_2$ (317)	--	--	(j)	--
s-Bu$_2$SnBr$_2$ (317)	IVC	--	Under ^{60}Co irradiation	616a

* Numbers in parenthesis refer to pages in main volume.
** The characters used here correspond to the synthetic scheme in the introduction to Chapter 4.1.2.

(a) Rxm. with MeMgBr → (o-MeC$_6$H$_4$CH$_2$)$_2$SnMe$_2$ (841). (b) NMR spectrum (1430). (c) Rxn. with Pr$_2$Sn(OMe)$_2$ → Pr$_2$Sn(OMe)F (942). (d) Rxn. with R$_2$SnH$_2$ → R$_2$SnHX; R = n-, i-Pr; X = Cl, Br, I (1292). (e) Rxn. with C$_5$H$_{11}$I + Zn in DMF, + NaOH → (Pr$_2$SnC$_5$H$_{11}$)$_2$O (2809), with PrI and Zn dust in BuOH, + NaOH, + HCl → (Pr$_3$Sn)$_2$O (2811). (f) Rxm. of i-Pr$_2$SnI$_2$ with aq. NH$_3$, followed by HX → i-Pr$_2$SnX$_2$; X = Cl, Br (1292). (g) Rxn. with NaX → (c.C$_3$H$_5$)$_2$SnX$_2$; X = NO$_3$, CH$_3$:CHCO$_2$ (1873). (h) Rxn. with i-Bu$_2$SnH$_2$ + CH$_2$:CHCH$_2$OH + cat. AIBN → i-Bu$_2$(HOCH$_2$CH$_2$CH$_2$)SnF (1613). (i) Rxn. with i-Bu$_2$SnH$_2$ + CH$_2$:CHCO$_2$Me + cat. AIBN → i-Bu$_2$(Me$_2$O$_2$CCH$_2$CH$_2$)SnBr (1613). (j) Rxn. with RMgBr → i-Bu$_2$SnR$_2$; R = CH$_2$:CHCH$_2$, CH$_2$:CHCH$_2$C$_6$H$_4$ (562).

Table 142 Continued

Formula*	Synth. Method**	Yield	Properties	Ref. R$_2$SnX$_2$
(Me$_3$C)$_2$SnCl$_2$ (317)	--	--	Dipole moment (k)	1211
(Me$_3$C)$_2$SnBr$_2$	IVC	--	Under ^{60}Co irradiation	616a
(Me$_3$C)$_2$SnI$_2$ (317)	II	--	m. 80-84°	438
(PhCMe$_2$CH$_2$)$_2$SnCl$_2$	IA (1)	71	m. 50.5-51.5°, soly. NMR spectrum	2188
(C$_5$H$_{11}$)$_2$SnCl$_2$ (317)	IVA	4	(m)	1583a
(Me$_3$CCH$_2$)$_2$SnCl$_2$ (318)	--	--	NMR spectrum	2188
(C$_6$H$_{13}$)$_2$SnBr$_2$ (318)	IVC (n)	--	Polarographic detn. (2265)	1244a
(Me$_3$CCH$_2$CH$_2$)$_2$SnCl$_2$	IIIA	--	m. 162-63°	2887
(c.C$_6$H$_{11}$)$_2$SnI$_2$ (318)	--	--	(o)	--
(Me$_2$CHCHCH$_2$CH)$_2$SnBr$_2$	--	--	Use: herbicide, pesticide (p)	1873
(C$_7$H$_{15}$)$_2$SnBr$_2$ (318)	--	--	Polarographic detn.	2265
(C$_8$H$_{17}$)$_2$SnF$_2$	--	--	Use: stabilizer for PVC	2953
(C$_8$H$_{17}$)$_2$SnI$_2$ (318)	IVC	50	HgI$_2$ as catalyst, b$_{0.4}$ 168-72° (q)	1904
(1-C$_8$H$_{17}$)$_2$SnCl$_2$	(r)	--	Detn. by TLC (r, s)	388
(1-C$_8$H$_{17}$)$_2$SnCl$_2$·(H$_2$NCH$_2$)$_2$	--	--	(t)	1702
(EtBuCHCH$_2$)$_2$SnCl$_2$ (318)	IIIB	>85	(u)	1482
(C$_{10}$H$_{21}$)$_2$SnCl$_2$ (319)	IVC (n)	--	m. 46-48°, b$_{1.5}$ 215-20°	1244a
(C$_{12}$H$_{25}$)$_2$SnF$_2$	--	--	Use: stabilizer for PVC	2953
(C$_{18}$H$_{37}$)$_2$SnCl$_2$	--	--	(v)	--
(o-MeC$_6$H$_4$)$_2$SnCl$_2$ (319)	--	--	NMR spectrum (w)	2356
(o-MeC$_6$H$_4$)$_2$SnBr$_2$	--	--	NMR spectrum	--
(o-MeC$_6$H$_4$)$_2$SnI$_2$	IA	55	m. 83°, IR spectrum, assignment (x)	2356
(p-MeC$_6$H$_4$)$_2$SnCl$_2$ (320)	--	--	(w, y, z)	1968
(p-MeC$_6$H$_4$)$_2$SnI$_2$	II	100	--	--
(p-Me$_2$C$_6$H$_4$)$_2$SnCl$_2$	IIIA	--	m. 72° (w)	393
(2,4,6-Me$_3$C$_6$H$_2$)$_2$SnBr$_2$	--	--	Refractometric study	305
(p-PhC$_6$H$_4$)$_2$SnCl$_2$ (320)	--	--	(z)	1391
				--

(p-PhC$_6$H$_4$)$_2$SnI$_2$	IVB	74	Colorless needles, m. 128°, soly.	393
	II	--	Soly. (aa)	393
(1-C$_{10}$H$_7$)$_2$SnCl$_2$ (315, 320)	II	--	(w, z)	--
(1-C$_{10}$H$_7$)$_2$SnI$_2$ (320)	II	60	--	393
(2-C$_{10}$H$_7$)$_2$SnCl$_2$	--	--	(z)	--
(2-C$_{10}$H$_7$)$_2$SnI$_2$	II	100	--	393
R$_2$SnI$_2$	IVC	--	R = alkyl	2970
R$_2$SnX$_2$	IVC	--	Cat. RI + NEt$_3$, (bb)	2970
	IVC	--	Electrolytical on Sn anode with ZnBr$_2$	1348

(k) Rxn. with Me$_3$CMgCl in THF → (t-Bu$_2$Sn)$_4$ (438), with LiAlH$_4$ → t-Bu$_2$SnH$_2$ (1614). (l) Synth. from R$_4$Sn with Br (1:2) in CCl$_4$, followed by NaHSO$_3$, NaOH and HCl in sequence (2188). (m) Moessbauer resonance spectrum (224), detn. and sepn. by paper chromatography (551), rxn. with Me$_3$CCH$_2$MgCl → (C$_5$H$_{11}$)$_2$Sn(CH$_2$CMe$_3$)$_2$ (2188). (n) RI and SbI$_3$ are used as catalysts in synth. (1244a). (o) Rxn. with MeI and Zn dust in BuOH, + NaOH → [Me(C$_8$H$_{17}$)$_2$-Sn]$_2$O (2811). (p) Rxn. with KOH → [Me$_2$CHCHCH$_2$CH)$_2$SnO]$_x$ (1873). (q) Rxn. with NaOH in aq. MePh → [(C$_8$H$_{17}$)$_2$SnO]$_x$ (2481). (r) Rxn. of (i-C$_8$H$_{17}$)$_2$SnCl$_2$ with (CH$_2$NH$_2$)$_2$ in BuOH (1702). (s) Rxn. with (i-C$_8$H$_{17}$)$_2$SnH$_2$ + CH$_2$:CHR + cat. AiBN → (i-C$_8$H$_{17}$)$_2$RCH$_2$CH$_2$SnCl; R = HOCH$_2$CH$_2$OCH$_2$ (1613). (t) Rxn. with HSCH$_2$CO$_2$R → (i-C$_8$H$_{17}$)$_2$Sn(SCH$_2$CO$_2$R)$_2$; R = 1-C$_8$H$_{17}$ (1702). (u) Rxn. with RSH → (EtBuCHCH$_2$)$_2$Sn(SR)$_2$; R = HSCH$_2$CO$_2$CH$_2$CH$_2$O$_2$CCH$_2$ (2903). (v) Rxn. with 9,10-C$_{14}$H$_{10}$ → [9,10-(C$_{18}$H$_{37}$)$_2$SnC$_{14}$H$_{10}$]$_n$; n = 1, X 9,10-dihydroanthracene (1726). (w) Rxn. with mesityl-MgBr → R$_2$Sn-mesityl$_2$; R = o-, p-tolyl, p-xylyl, 1-naphthyl (303). (x) NMR spectrum (2356). (y) Rxn. with (p-MeC$_6$H$_4$)$_2$-SnH$_2$ + PhMeC:CH$_2$ → (p-MeC$_6$H$_4$)$_2$PhMeCHCH$_2$SnCl (1613), with C$_6$F$_5$MgBr → (p-MeC$_6$H$_4$)$_2$Sn(C$_6$F$_5$)$_2$ (77). (z) Rxn. with Et$_2$AlH → R$_2$SnH$_2$; R = p-tolyl, p-biphenylyl, 1-, 2-naphthyl (393). (aa) Rxn. with Na-C$_8$H$_{10}$ in THF → [(p-PhC$_6$H$_4$)$_2$-Sn] (393). (bb) Relative reactivity as catalyst for $\overline{CH_2CH_2O}$ polymerization (2572, 2916).

Table 143. Unsymmetric Unsubstituted Diorganotin Dihalides   RR′SnX$_2$

Formula*	Synth. Method**	Yield	Properties	Ref.
Me$_2$SnClI·TerPy	--	--	TerPy = 2,2′,2″-terpyridine, Moessbauer resonance spectrum, struct.	2347
MeRSnCl$_2$				
R = 1-Bu	IIID	80	m. 41°, b$_{27}$ 124-25°, GLC	1370
	IIIC	32	Synth. at 0°	1370
C$_8$H$_{17}$	IIIB	--	m. 32-32.5°, b$_1$ 121-22°	1482
C$_{10}$H$_{21}$	IIIB	--	m. 39-41°	1482
C$_{12}$H$_{25}$	IIIB	--	m. 46-47.5°	1482
Ph	IIIC	85	m. 43°, GLC	1370
MePhSnBr$_2$	ID	30	b$_{0.15}$ 97-98°, n$_D^{20}$ 1.633 (a)	2412
EtBuSnCl$_2$	IIIC	--	m. 39°, b$_{14}$ 126-27°, GLC (b)	1370
EtPhSnCl$_2$	IIIC	98	m. 66°, GLC	1370
EtPhSnI$_2$	ID	--	(c)	1410
PrBuSnCl$_2$ (316)	IIIC	63	m. 46-47°, b$_{15}$ 140-42°, GLC	1370
PrPhSnCl$_2$	IIIC	69	m. 40°, GLC	1370
Bu(C$_8$H$_{17}$)SnCl$_2$	IIIB	--	b$_1$ 144-65°	1482
Bu(C$_{10}$H$_{21}$)SnCl$_2$	IIIB	--	m. 26.5-28°, b$_3$ 178-81°	1482
Bu(C$_{12}$H$_{25}$)SnCl$_2$	IIIB	--	m. 33-35°, b$_1$ 182-85°	1482
(1-C$_5$H$_{11}$)PhSnI$_2$	ID	84	b$_4$ 190°, n$_D^{23}$ 1.6319, d$_{23}$ 2.0070 (d)	2787

* Numbers in parenthesis refer to pages in main volume.
** The characters used here correspond to the synthetic scheme in the introduction to Chapter 4.1.2.
(a) Mass spectrum (2412). (b) Rxn. with Me$_4$Sn at 120-30° → Me$_3$SnCl + MeEtBuSnCl (1370), with (CH$_2$:CH)$_4$Sn at 20-60° ⇌ (CH$_2$:CH)$_3$SnCl + EtBu(CH$_2$:CH)SnCl (1370). (c) Rxn. with LiAlH$_4$ → EtPhSnH$_2$ (1410). (d) Rxn. with 15% aq. KOH in Et$_2$O → (1-C$_5$H$_{11}$PhSnO)$_x$ (2787).

4.1.2.4 UNSATURATED DIORGANOTIN DIHALIDES

Divinyltin Dichloride (M-321) $SnC_4H_6Cl_2$ $(CH_2:CH)_2SnCl_2$
Synth.: By rxn. of $(CH_2:CH)_4Sn$ with BCl_3 in petr. ether (52a).
Prop.: NMR (1220), ^{13}C NMR (328) and NQR (1142) spectra, dipole moment (1211).
Rxn. with: $2\text{-}LiC_6H_4CBu:CPhLi \rightarrow (o\text{-}C_6H_4CBu:CPh)_2Sn$ (1736).
$9,10\text{-}Na_2C_{14}H_{10} \rightarrow [(CH_2:CH)_2SnC_{14}H_{10}]_n$; n = 1, X 9,10-dihydroanthracene (1726).
Na-oleate $\rightarrow (CH_2:CH)_2Sn(oleate)_2$ (1442).
$Ac_2CH_2 + NaOMe \rightarrow (CH_2:CH)_2Sn(OCMe:CHAc)_2$ (2055).
$(:CRSNa)_2 \rightarrow (CH_2:CH)_2SnSCR:CRS$; R = H, CN (652).
$RMgBr \rightarrow (CH_2:CH)_2SnR_2$; R = CH_2CH_2CH (1873), C_6F_5 (428).
$B_2Cl_4 \rightarrow SnCl_2$ (2329).
$ClNO_3 \rightarrow (O_3NCH_2CHCl)_2SnCl_2$ (148).
5,1,10-X-phenanthroline in buffered AcOH → detn. of complex formation and equil. const. by polarography, spectrophotometry and ion exchange; X = H, Me, Cl, NO_2 (655).

Diallyltin Dichloride $SnC_6H_{10}Cl_2$ $(CH_2:CHCH_2)_2SnCl_2$
Synth.: By rxn. of $CH_2:CHCH_2Cl$ with tin in presence of RSH, Mg and I (880a).
Rxn. with: $1\text{-}C_{10}F_{21}I$ at 300°, + Zn + MeOH → $(1\text{-}C_{10}F_{21}CH_2CH_2CH_2)_2Sn(OMe)_2$ (745a).

Dipropenyltin Dibromide (M-323) $SnC_6H_{10}Br_2$ $(MeCH:CH)_2SnBr_2$
Synth.: By rxn. of $(MeCH:CH)_4Sn$ with $HgBr_2$ in Et_2O in Ar atm. without isomerization (1888).

Diparastyryltin Dichloride $SnC_{16}H_{14}Cl_2$ $(p\text{-}CH_2:CHC_6H_4)_2SnCl_2$
Rxn. with: Na-oleate → $(p\text{-}CH_2:CHC_6H_4)_2Sn(oleate)_2$ (1442).

Ethyl-(β-styryl)tin Dibromide $SnC_{10}H_{12}Br_2$ $Et(PhCH:CH)SnBr_2$
Synth.: By rxn. of $EtSnH_3$ and $EtSnBr_3$ with PhC⋮CH in presence of cat. AIBN, in 73% yield (1613).
Prop.: m. 21°, $b_{0.07}$ 115° (1613).

Unsaturated functionally substituted diorganotin dihalides are listed in Table 144. Additional references to diorganotin dihalides may be found in Chapter 8.1.

4.1.2.6 FUNCTIONALLY SUBSTITUTED DIORGANOTIN DIHALIDES

4.1.2.6.1 HALOGEN SUBSTITUTED DIORGANOTIN DIHALIDES

Bis(pentafluorophenyl)tin Dichloride (M-325) $SnC_{12}Cl_2F_{10}$ $(C_6F_5)_2SnCl_2$
Synth.: By rxn. of $SnCl_4$ with C_6F_5MgBr in N atm. and subl. without hydrolysis (1198), with $(C_6F_5)_3SnCl$ (1:2) (1198), with $R_2Sn(C_6F_5)_2$ at 150°; R = Me, o-Me-C_6H_4 (1198).
Prop.: b_3 113°, $b_{0.15}$ 61°, d_{20} 2.4 (1198).
Rxn. with: Aq. $Me_2CO \rightarrow [(C_6F_5)_2SnCl]_2O$ (1198).

8-HOC$_9$H$_6$N → (C$_6$F$_5$)$_2$Sn(OC$_9$H$_6$N)$_2$, quinoline (1198).

Other halogen substituted diorganotin dihalides are listed in Table 144 and 145.

4.1.2.6.2 PSEUDOHALOGEN SUBSTITUTED DIORGANOTIN DIHALIDES

Bis(2-cyanoethyl)tin Dibromide   SnC$_6$H$_8$Br$_2$N$_2$ (NCCH$_2$CH$_2$)$_2$SnBr$_2$
Synth.: By rxn. of [(NCCH$_2$CH$_2$)$_2$SnO]$_x$ with HBr (1483).
By rxn. of Ph$_2$SnCl$_2$ with LiAlH$_4$, followed by CH$_2$:CHCN and Br in HCCl$_3$, in 39% yield (1745).
Prop.: m. 186-88° (1483), 178-82° (dec.) (1745), IR spectrum, mol. wt. (1483, 1745), conductivity, soly. (1483). Anal. detn. of Br by alkalimetric titration (1483).
Rxn. with: Aq. Na$_2$S → [(NCCH$_2$CH$_2$)$_2$SnS]$_3$ (1746, 2326).
Aq. NH$_3$ → [(NCCH$_2$CH$_2$)$_2$SnOH]$_2$O (1746).
i-C$_8$H$_{17}$O$_2$CCH$_2$SH + Et$_3$N → (NCCH$_2$CH$_2$)$_2$Sn(SCH$_2$CO$_2$C$_8$H$_{17}$)$_2$ (895b).

Additional pseudohalogen substituted diorganotin dihalides are listed at the bottom of Table 144.

4.1.2.6.3 OXYGEN SUBSTITUTED DIORGANOTIN DIHALIDES

4.1.2.6.3.1 COMPOUNDS CONTINAING AN ETHER LINKAGE

All compounds are listed in Table 145 on page 525.

4.1.2.6.3.3 DIORGANOTIN DIHALIDES CONTAINING CARBONYL GROUPS

All compounds are listed in Table 146 on page 526.

Table 144. Halogen Substituted Diorganotin Dihalides and Dipseudohalides R'_2SnX_2

Formula*	Synth. Method**	Yield	Properties	Ref.
$(ClCH_2)_2SnCl_2$ (324)	IVD	22	Acicular cryst., m. 88-89°, mol. wt.	2512
	IVD	--	Colorless needles, m. 89.5-90°, soly. (a)	2071
$(ClCH_2)_2SnBr_2$	IA	--	m. 95-96°, subl. in vacuo 100° (b)	655
$(BrCH_2)_2SnBr_2$	IVD	--	m. 87°, soly., NMR spectrum (c)	2071
$(o\text{-}ClC_6H_4CH_2)_2SnCl_2$	IVC	53	m. 110° (far) IR spectra, assignment (d)	841
$(o\text{-}ClC_6H_4CH_2)_2SnI_2$	--	--	(Far) IR spectra, assignment	841
$(p\text{-}ClC_6H_4CH_2)_2SnCl_2$ (324)	IVC	38	(Far) IR spectra, assignment (e)	841
$(p\text{-}ClC_6H_4CH_2)_2SnI_2$	VA (e)	70	m. 115° (Far) IR spectrum, assignment	841
$(C_6F_5CH_2)_2SnBr_2$	IVC	21	m. 112°	841
$(ClCH:CH)_2SnCl_2$ (325)	IVE	--	cis-Form liq., IR and NMR spectra	2638
	--	--	trans-Form m. 78_1 (f)	2638
$(O_3NCH_2CHCl)_2SnCl_2$	V (g)	100	Dec. on heating	148
$(MeCCl:CHCH_2)_2SnCl_2$	IVC	--	Heavy brown liq., b_1 103-106°, fumes in air (h)	2408
$(MeCCl:CHCH_2)_2SnCl_2 \cdot 2Py$	(h)	--	m. 170° (Dec.)	2408
$[Br(CH_2)_5]_2SnBr_2$	IA (i)	--	Isolated as complex (j)	17
$[Br(CH_2)_5]_2SnBr_2 \cdot BiPy$	(j)	--	m. 157-59°	17
$[Br(CH_2)_5]_2SnBr_2 \cdot o\text{-phen}$ (k)	(j)	--	m. 149-51°	17
$[I(CH_2)_5]_2SnI_2$	IA (i)	--	Isolated as complex (j)	17

* Numbers in parenthesis refer to pages in main volume.
** The characters used here correspond to the synthetic scheme in the introduction to Chapter 4.1.2.

(a) (Far) IR (2512), NMR (2071, 2512) and ^{35}Cl NQR (2752) spectra. (b) Rxn. with 5,1,10-X-phenanthroline in buffered AcOH → detn. of complex formation by polarography, ion exchange and spectrophotometry; X = H, Me, Cl, NO$_2$ (655). (c) Rxn. with PhM (1:2) → Ph(BrCH$_2$)$_2$SnBr; M = Li, MgBr (1410). (d) Rxn. with MeMgBr → Me$_2$Sn(CH$_2$-C$_6$H$_4$Cl)$_2$ (841). (e) Rxn. of (p-ClC$_6$H$_4$CH$_2$)$_2$SnCl$_2$ with NaI in Me$_2$CO → (p-ClC$_6$H$_4$CH$_2$)$_2$SnI$_2$ (841). (f) Synth. without isomerization, rxn. with HgCl$_2$ → ClCH:CHHgCl without isomerization (2638). (g) Synth. by rxn. of (CH$_2$:CH)$_2$-SnCl$_2$ with ClNO$_3$ at -100° to 0° (148). (h) Rxn. with (MeCCl:CHCH$_2$)$_2$SnCl$_2$ with Py → (MeCCl:CHCH$_2$)$_2$SnCl$_2$·2Py (2408). (i) Synth. from 6-stannaspiro[5.5]undecane with Br or I (17). (j) Cleavage product of the spirane is isolated as BiPy and o-phen complex (17). (k) o-phen = 1,10-phenanthroline.

Table 144 Continued

R'_2SnX_2

Formula*	Synth. Method**	Yield	Properties	Ref.
[I(CH$_2$)$_5$]$_2$SnI$_2$·BiPy	(j)	--	m. 145-47°	17
[I(CH$_2$)$_5$]$_2$SnI$_2$·o-phen (k)	(j)	--	m. 172-74°	17
(BrC$_7$H$_{14}$)$_2$SnBr$_2$ (325)	--	--	BrC$_7$H$_{14}$ = BrCH$_2$CH$_2$CMe$_2$CH$_2$CH$_2$ (l)	--
(p-FC$_6$H$_4$)$_2$SnCl$_2$	--	--	^{1}H and ^{19}F NMR spectra	1453
(C$_6$F$_5$)$_2$SnBr$_2$	IIIB	--	Crude, b$_{0.2}$ 68° contn. Me$_2$SnBr$_2$	1198
Bu(C$_{10}$F$_{21}$CH$_2$CH$_2$)SnCl$_2$	IIIB	--	Use as dirt, water and oil repellent	745a
(NCCH$_2$CH$_2$)$_2$SnCl$_2$ (322)	VB	--	m. 186-88°, IR spectrum, mol. wt., soly (m)	1483
(NCCH$_2$CH$_2$)$_2$SnI$_2$	IVC (n)	--	m. 128-29°, IR spectrum, mol. wt. (m, o)	1483
[NC(CH$_2$)$_3$]$_2$SnI$_2$	IVC (p)	--	m. 84.5-85.5°, IR spectrum, mol. wt., soly. (m, q)	1483
(NCCH$_2$CHMe)$_2$SnI$_2$	IVC (p)	--	m. 82-89°, IR spectrum, mol. wt.	1483
[NC(CH$_2$)$_4$]$_2$SnBr$_2$	VB	--	m. 61-63°, IR spectrum, mol. wt. (m)	1483
[NC(CH$_2$)$_4$]$_2$SnI$_2$	IVC (p)	--	m. 74.5-76°, IR spectrum, mol. wt. (m, r)	1483

(k) o-phen = 1,10-phenanthroline. (l) Activity as acaricide against Tetranychus althaeae (433). (m) Halogen detn. by alkalimetric titration (1483). (n) Mg, BuI and BuOH are used as catalyst in synth. (1483). (o) Rxn. with alkali [(NCCH$_2$CH$_2$)$_2$SnO]$_x$ (1483). (p) Mg is used as catalyst in synthesis (1483). (q) Rxn. with NaHCO$_3$ → [(NCCH$_2$CH$_2$CH$_2$)$_2$SnO]$_x$ (1483). (r) Rxn. with NaOH → {[NC(CH$_2$)$_4$]$_2$SnO}$_x$ (1483).

Table 145. Diorganotin Dihalides Containing an Ether Linkage R'_2SnX_2

Formula*	Synth. Method**	Yield	Properties	Ref.
[PhO(CH$_2$)$_3$]$_2$SnCl$_2$	VC	--	IR spectrum, struct.	2435
[PhO(CH$_2$)$_3$]$_2$SnBr$_2$	VC	--	IR spectrum, struct. (a)	2435
[PhO(CH$_2$)$_3$]$_2$SnI$_2$	IVC (b)	80	IR spectrum, struct. (a)	2435
[PhO(CH$_2$)$_4$]$_2$SnCl$_2$	VC	--	IR spectrum, struct.	2435
[PhO(CH$_2$)$_4$]$_2$SnBr$_2$	IVC (b)	80	IR spectrum, struct. (a)	2435
[PhO(CH$_2$)$_4$]$_2$SnI$_2$	IVC (b)	80	IR spectrum, struct. (a)	2435
[p-MeC$_6$H$_4$O(CH$_2$)$_4$]$_2$SnCl$_2$	VC	--	IR spectrum, struct.	2435
[p-MeC$_6$H$_4$O(CH$_2$)$_4$]$_2$SnBr$_2$	IVC (b)	--	IR spectrum, struct. (a)	2435
[p-MeC$_6$H$_4$O(CH$_2$)$_4$]$_2$SnI$_2$	IVC (b)	80	IR spectrum, struct. (a)	2435
[p-ClC$_6$H$_4$O(CH$_2$)$_4$]$_2$SnCl$_2$	VC	--	IR spectrum, struct.	2435
[p-ClC$_6$H$_4$O(CH$_2$)$_4$]$_2$SnBr$_2$	VC	--	IR spectrum, struct.	2435
[p-ClC$_6$H$_4$O(CH$_2$)$_4$]$_2$SnI$_2$	IVC (b)	--	(a)	2435
[p-BrC$_6$H$_4$O(CH$_2$)$_4$]$_2$SnCl$_2$	IVC (b)	80	IR spectrum, struct. (a)	2435
[p-BrC$_6$H$_4$O(CH$_2$)$_4$]$_2$SnBr$_2$	VC	--	IR spectrum, struct.	2435
[p-BrC$_6$H$_4$O(CH$_2$)$_4$]$_2$SnI$_2$	VC	--	IR spectrum, struct.	2435
]p-BrC$_6$H$_4$O(CH$_2$)$_4$]$_2$SnI$_2$	IVC (b)	--	(a)	2435
[2-C$_{10}$H$_7$O(CH$_2$)$_4$]$_2$SnCl$_2$	IVC (b)	80	IR spectrum, struct. (a)	2435
[2-C$_{10}$H$_7$O(CH$_2$)$_4$]$_2$SnBr$_2$	VC	--	IR spectrum, struct.	2435
[2-C$_{10}$H$_7$O(CH$_2$)$_4$]$_2$SnI$_2$	VC	--	IR spectrum, struct.	2435
[2-C$_{10}$H$_7$O(CH$_2$)$_4$]$_2$SnI$_2$	IVC (b)	80	IR spectrum, struct. (a)	2435
(p-EtOC$_6$H$_4$)$_2$SnCl$_2$ (326)	--	--	(c)	--
(p-EtOC$_6$H$_4$)$_2$SnI$_2$	II	100	--	393

* Numbers in parenthesis refer to pages in main volume.
** The characters used here correspond to the synthetic scheme in the Introduction to Chapter 4.1.2.
(a) Rxn. with alkali, followed by HX → R$_2$SnX$_2$; X = Cl, Br (2435). (b) Mg, BuI and THF are used as catalyst in synthesis (2435). (c) Rxn. with Et$_2$AlH in Et$_2$O → (p-EtOC$_6$H$_4$)$_2$SnH$_2$ (393), with Na and C$_{10}$H$_8$ in THF → [(p-EtOC$_6$H$_4$)$_2$Sn]$_6$ (393).

Table 146. Diorganotin Dihalides Containing Carbonyl Groups   R'$_2$SnX$_2$

Formula	Synth. Method*	Properties	Ref.
(BzCH$_2$CH$_2$)$_2$SnCl$_2$	IVC (a)	m. 140-41°, struct. (b, c)	1481
	IVC (d)	UV spectrum, stable in moist air	1484
(BzCH$_2$CH$_2$)$_2$SnBr$_2$	IVC	m. 171-73°, struct. (b, c)	1481
	IVC, VB	UV spectrum, stable in moist air	1484
(BzCH$_2$CH$_2$)$_2$SnI$_2$	IVC (a)	m. 184-86°, struct. (b)	1481
	IVC (d)	UV spectrum, stable in moist air	1484
(AcCH$_2$CH$_2$)$_2$SnCl$_2$	IVC (a)	m. 112-13°, struct. (b)	1481
	IVC (d)	UV spectrum, stable in moist air	1484
(AcCH$_2$CH$_2$)$_2$SnBr$_2$	IVC (a)	m. 122-24°, struct. (b)	1481
	IVC (d)	UV spectrum, stable in moist air	1484
(AcCH$_2$CH$_2$)$_2$SnI$_2$	IVC (a)	m. 116-18°, struct. (b)	1481
	IVC (d)	UV spectrum, stable in moist air	1484
(EtCOCH$_2$CH$_2$)$_2$SnCl$_2$	IVC (a)	m. 91-93°, struct. (b)	1481
	IVC (e)	UV spectrum, stable in moist air	1484
(EtCOCH$_2$CH$_2$)$_2$SnBr$_2$	IVC (a)	m. 108-10°, struct. (b)	1481
	IVC (e)	UV spectrum, stable in moist air	1484
(EtCOCH$_2$CH$_2$)$_2$SnI$_2$	IVC (a)	m. 107-108°, struct. (b)	1481
	IVC (e)	UV spectrum, stable in moist air	1484
(PrCOCH$_2$CH$_2$)$_2$SnCl$_2$	IVC (a)	m. 31-32°	1481
	IVC (e)	UV spectrum	1484
(PrCOCH$_2$CH$_2$)$_2$SnBr$_2$	IVC (a)	--	1481
	IVC (e)	UV spectrum, stable in moist air	1484
(PrCOCH$_2$CH$_2$)$_2$SnI$_2$	IVC (a)	--	1481
	IVC (e)	UV spectrum, stable in moist air	1484
(AcCH$_2$CMe$_2$)$_2$SnCl$_2$	IVC (a)	m. 155-56°, struct. (b)	1481
	IVC (e)	UV spectrum, stable in moist air	1484
(AcCH$_2$CMe$_2$)$_2$SnBr$_2$	IVC (a)	m. 156-57°, struct. (b)	1481
	IVC (d)	UV spectrum, stable in moist air	1484
(AcCH$_2$CMe$_2$)$_2$SnI$_2$	IVC (a)	m. 160° (dec.), struct. (b)	1481
	IVC (d)	UV spectrum, stable in moist air	1484

* The characters used here correspond to the synthetic scheme in the introduction to Chapter 4.1.2.
(a) Mg, BuI and BuOH are used as catalyst in synthesis (1481). (b) IR spectrum, proposed structure with hexacoordinated chelated tin, having halogen and carbonyl groups in trans-position (1481, 1484). (c) Rxn. with alc. NaOH → (BzCH$_2$CH$_2$)$_2$SnO (1484). (d) MgO, BuI and BuOH are used as catalysts in synthesis (1484).

4.1.2.6.3.4 DIORGANOTIN DIHALIDES CONTAINING CARBOXYL GROUPS AND DERIVATIVES

The majority of the iodides and bromides in this chapter is synthesized by reaction of metallic tin with the corresponding halides of the carboxylic acid esters or amides at 120 to 170° in presence of catalytical amounts of magnesium, alkyl iodides and polar solvents. The chlorides and some bromides issue from interconversion reactions of the heavier halides via oxides or hydroxides which are usually not isolated. Yields for the heavy halides prepared by the direct synthesis with metallic tin vary from below 20 to 70% and are greatly enhanced by the above mentioned catalysts. Structure as octahedral coordination compounds under inclusion of the carbonyl group; but not the amide group is stated and spectroscopic evidence for this structure is forewarded. The compounds compiled in Table 147 are prepared by the following methods.

I Dealkylation of tetraorganotin compounds:
  (IA) $Ph_3SnCH_2CH_2CO_2Me + Br \rightarrow Ph(MeO_2CCH_2CH_2)SnBr_2$,
  (IB) $R_2SnR'_2 + Br \rightarrow R'_2SnBr_2$.

II Reaction of metallic tin with functionally substituted organic halides in presence of:
  (IIA) Mg and THF ,
  (IIB) Mg, THF and glycol ,
  (IIC) Mg, BuI, and THF ,
  (IID) Mg, BuI, THF and glycol.

III Interconversion reactions:
  (IIIA) R'_2SnX_2 + satd. $NaHCO_3$, + HX' → R'_2SnX' ; X = I, Br; X' = Br, Cl,
  (IIIB) R'_2SnI_2 + alkali, + HX' → $R'_2SnX'_2$; X' = Br, Cl.

Conventional organic reaction:

IV of the free carboxylic acid:
  $(HO_2CY)_2SnX_2 + ROH \rightarrow (RO_2CY)_2SnX_2$.

V of a carboxylic acid ester:
  (VA) $(RO_2CCHMeCH_2)_2SnI_2 + KOH, + HX \rightarrow (HO_2CCHMeCH_2)_2SnX_2$,
  (VB) $\overline{Sn(YCO_2)_2} + SOCl_2$ or $HBr \rightarrow (HO_2CY)_2SnX'_2$.

VI of an amide group:
  (VIA) $(H_2NCOCHMeCH_2)_2SnI_2 + NaOH, + HBr, + MeOH \rightarrow (MeO_2CCHMeCH_2)_2SnBr_2$,
  (VIB) $(R_2NOCCH_2CH_2)_2SnX_2 + KOH, + HX' \rightarrow (HO_2CCH_2CH_2)_2SnX'_2$.

VII of a nitrile:
  (VIIA) $(NCY)_2SnI_2$ + aq. NaOH, + MeOH-HCl → $(MeO_2CY)_2SnCl_2$,
  (VIIB) $[NC(CH_2)_4]_2SnO$ + alc. NaOH, + HX' → $[HO_2C(CH_2)_4]_2SnX'_2$.

R = organic group, R' = functionally substituted organic group, Y = hydrocarbon bridge, X = halogen, X' = chlorine, bromine.

Table 147. Diorganotin Dihalides Containing Carboxyl Groups and Derivatives $RR'SnX_2$

Formula*	Synth. Method**	Properties	Ref.
Me(MeO$_2$CCH$_2$CH$_2$)SnCl$_2$	--	m. 84-85°	2651
Me(MeO$_2$CCH$_2$CH$_2$)SnI$_2$	II (a)	b$_{0.001}$ 130-40°, n_D^{20} 1.6553 (b)	2651
Me(EtO$_2$CCH$_2$CH$_2$)SnCl$_2$	--	m. 62-63°	2651
Me(EtO$_2$CCH$_2$CH$_2$)SnBr$_2$	II (a)	m. 52-54° (b)	2651
Me(EtO$_2$CCH$_2$CH$_2$)SnI$_2$	II (a)	n$_{0.001}$ 143-46°, n_D^{20} 1.6383 (b)	2651
Me[MeO$_2$C(CH$_2$)$_3$]SnBr$_2$	IIC	m. 125-35°, IR spectrum, struct.	2650
Me[MeO$_2$C(CH$_2$)$_3$]SnI$_2$	IIC	m. 140-47°, IR spectrum, struct.	2650
Et[EtO$_2$C(CH$_2$)$_3$]SnBr$_2$	IIC	IR spectrum, struct.	2650
Ph(MeO$_2$CCH$_2$CH$_2$)SnBr$_2$ (326)	IA	NMR spectrum	2070
(HO$_2$CCH$_2$CH$_2$)$_2$SnCl$_2$	VB	m. 153-61°, IR spectrum (c)	1485
(HO$_2$CCH$_2$CH$_2$)$_2$SnBr$_2$	VB	m. 157-60°, IR spectrum (c)	1485
(MeO$_2$CCH$_2$CH$_2$)$_2$SnCl$_2$	IIIA (d)	m. 131-32°, IR spectrum	1485
	VIA	m. 131-32°, IR spectrum	1483
(MeO$_2$CCH$_2$CH$_2$)$_2$SnBr$_2$ (326)	IIC	m. 139-40° (d, e)	1481, 1485
	IB	NMR spectrum	2070
(MeO$_2$CCH$_2$CH$_2$)$_2$SnI$_2$	IIC	m. 120.5-22°, IR spectrum, struct. (f)	1481, 1485
(PhCH$_2$O$_2$CCH$_2$CH$_2$)$_2$SnI$_2$	IIC	m. 145-46.5°, IR spectrum, struct.	1481, 1485
(EtO$_2$CCH$_2$CH$_2$)$_2$SnBr$_2$	IIC	m. 99-101.5°, IR spectrum, struct. (d)	1481, 1485
(EtO$_2$CCH$_2$CH$_2$)$_2$SnI$_2$	IIC	m. 91.5-92.5°, IR spectrum, struct. (f)	1481, 1485
(PrO$_2$CCH$_2$CH$_2$)$_2$SnBr$_2$	IIC	m. 98.5-99.5°, IR spectrum, struct. (d)	1481, 1485
(PrO$_2$CCH$_2$CH$_2$)$_2$SnI$_2$	IIC	m. 91-92°, IR spectrum, struct. (f)	1481, 1485
(BuO$_2$CCH$_2$CH$_2$)$_2$SnBr$_2$	IIC	m. 42-44°, IR spectrum, struct. (d)	1481, 1485
(BuO$_2$CCH$_2$CH$_2$)$_2$SnI$_2$	IIC	m. 56.5-58°, IR spectrum, struct.	1481, 1485
(C$_8$H$_{17}$O$_2$CCH$_2$CH$_2$)$_2$SnCl$_2$	--	Liq. IR spectrum	1485
(C$_8$H$_{17}$O$_2$CCH$_2$CH$_2$)$_2$SnBr$_2$	IIC	Liq. IR spectrum	1485
(H$_2$NOCCH$_2$CH$_2$)$_2$SnBr$_2$	ID	m. 258° (dec.), IR spectrum, struct. (g)	1166, 2434
(H$_2$NOCCH$_2$CH$_2$)$_2$SnI$_2$	ID	m. 234-35°, IR spectrum, struct. (g)	1166, 2434

528

Compound	Class	Properties	Ref.
(MeHNOCCH$_2$CH$_2$)$_2$SnCl$_2$	IIIB	m. 206-208°	2434
(MeHNOCCH$_2$CH$_2$)$_2$SnBr$_2$	IID	m. 218-19°, IR spectrum, struct. (g)	1166, 2434
(MeHNOCCH$_2$CH$_2$)$_2$SnI$_2$	IID	m. 225.5-27°, IR spectrum, struct. (g)	1166, 2434
(EtHNOCCH$_2$CH$_2$)$_2$SnCl$_2$	IIIB	m. 153-54°	2434
(EtHNOCCH$_2$CH$_2$)$_2$SnBr$_2$	IID	m. 160-61°, IR spectrum, struct. (g)	1166, 2434
(EtHNOCCH$_2$CH$_2$)$_2$SnI$_2$	IID	m. 191-92°, IR spectrum, struct. (g)	1166, 2434
(PhHNOCCH$_2$CH$_2$)$_2$SnCl$_2$	IIIB	m. 233-36°	2434
(PhHNOCCH$_2$CH$_2$)$_2$SnBr$_2$	IID	m. 243-44°, IR spectrum, struct. (g)	1166, 2434
(PhHNOCCH$_2$CH$_2$)$_2$SnI$_2$	IID	m. 215-16°, IR spectrum, struct. (g)	1166, 2434
(p-MeC$_6$H$_4$NHOCCH$_2$CH$_2$)$_2$SnCl$_2$	IIIB	m. 243-44°, 237-44°	2434
(p-MeC$_6$H$_4$NHOCCH$_2$CH$_2$)$_2$SnBr$_2$	IID	m. 222-23°, IR spectrum, struct. (g)	1166, 2434
(Me$_2$NOCCH$_2$CH$_2$)$_2$SnCl$_2$	IIIB	m. 208-10°	2434
(Me$_2$NOCCH$_2$CH$_2$)$_2$SnBr$_2$	IID	m. 201-202°, IR spectrum, struct. (g)	1166, 2434
(Et$_2$NOCCH$_2$CH$_2$)$_2$SnCl$_2$	IIIB	m. 135-36°	2434
(Et$_2$NOCCH$_2$CH$_2$)$_2$SnBr$_2$	IIB	m. 133-34°, IR spectrum, struct. (g)	1166, 2434
(EtO$_2$CCH$_2$NHOCCH$_2$CH$_2$)$_2$SnBr$_2$	IIIB	In 37% yield, m. 212.5-14° (h)	1167
(EtO$_2$CCH$_2$NHOCCH$_2$CH$_2$)$_2$SnI$_2$	IIA	In 68% yield, m. 207-208° (h-j)	1167
[(EtO$_2$C)$_2$CH]$_2$SnBr$_2$	II	Cryst., dec. instantly in air	1652
(HO$_2$CCHMeCH$_2$)$_2$SnCl$_2$	VA	m. 180-83° (k)	1632
(HO$_2$CCHMeCH$_2$)$_2$SnBr$_2$	VA	m. 180.5-82° (k)	1632
(MeO$_2$CCHMeCH$_2$)$_2$SnCl$_2$	IV, IIIA	m. 122-24°	1632
(MeO$_2$CCHMeCH$_2$)$_2$SnBr$_2$	VIIA	m. 104-105°	1168
(MeO$_2$CCHMeCHMeCH$_2$)$_2$SnBr$_2$	IV, IIIA	m. 106.5-107.5°	1632

* Numbers in parenthesis refer to pages in main volume.

** The characters used here correspond to the synthetic scheme in the introduction to Chapter 4.1.2.6.3.4.

(a) Synth. in presence of MeI (2651). (b) Rxn. with MeMgI → Me$_3$SnCH$_2$CH$_2$CO$_2$R; R = Me, Et (2651), with MeMgI, + Br → Me$_2$(RO$_2$CCH$_2$CH$_2$)SnBr; R = Me, Et (2651). (c) Rxn. with ROH → (RO$_2$CCH$_2$CH$_2$)$_2$SnCl$_2$ (1485). (d) Rxn. with aq. NaHCO$_3$, + HCl → (MeO$_2$CCH$_2$CH$_2$)$_2$SnCl$_2$ (1485). (e) Rxn. with 50% aq. alkali → Sn(CH$_2$CH$_2$CO$_2$)$_2$ (1485). (f) Rxn. with EtMgBr → Et$_2$Sn(CH$_2$CHRCEt$_2$OH)$_2$; R = H, Me (2649). (g) Rxn. with KOH, + HX → (R$_2$NOCCH$_2$CH$_2$)$_2$SnX$_2$; X = Cl, Br (1166, 2434). (h) IR spectrum, soly (1167). (i) Rxn. with aq. alc. NaOH → Sn(CH$_2$CH$_2$CO$_2$)$_2$; also with aq. MeC$_6$H$_4$SO$_3$H (1167), with ROH + cat. MeC$_6$H$_4$SO$_3$H → ester exchange (1167). (j) Rxn. with KOH, + HBr → (EtO$_2$CCH$_2$NH-OCCHRCHR)$_2$SnBr$_2$; R = H, Me (1167). (k) Rxn. with ROH → (RO$_2$CCHMeCH$_2$)$_2$SnX$_2$; X = Cl, Br (1632).

Table 147 Continued

Formula*	Synth. Method**	Properties	RR'SnX$_2$ Ref.
(MeO$_2$CCHMeCH$_2$)$_2$SnI$_2$	IIC	In 23-30% yield, m. 97.5-99.5° (f, l)	1632
(EtO$_2$CCHMeCH$_2$)$_2$SnCl$_2$	IV	m. 79-82°	1632
(EtO$_2$CCHMeCH$_2$)$_2$SnBr$_2$	IV	m. 76.5-78.5°	1632
(EtO$_2$CCHMeCH$_2$)$_2$SnI$_2$	IIC	In 23-30% yield, m. 75.5-77° (f)	1632
(PrO$_2$CCHMeCH$_2$)$_2$SnCl$_2$	IV	m. 83-85.5°	1632
(PrO$_2$CCHMeCH$_2$)$_2$SnBr$_2$	IV	m. 85.5-87.5°	1632
(PrO$_2$CCHMeCH$_2$)$_2$SnI$_2$	IIC	m. 90-92.5° (f)	1632
(i-PrO$_2$CCHMeCH$_2$)$_2$SnI$_2$	IIC	m. 80-82.5°	1632
(i-BuO$_2$CCHMeCH$_2$)$_2$SnI$_2$	IIC	m. 118.5-21°	1632
(^{n}H$_2$NOCCHMeCH$_2$)$_2$SnCl$_2$	IIIA	m. 243-44°, IR spectrum	1168, 2434
(H$_2$NOCCHMeCH$_2$)$_2$SnBr$_2$	IIIA	m. 234-36°, IR spectrum	1168
(H$_2$NOCCHMeCH$_2$)$_2$SnI$_2$	IIB	m. 243° (dec.), IR spectrum (m)	1168, 2434
(MeHNOCCHMeCH$_2$)$_2$SnCl$_2$	IIIB	m. 214-16°	1168, 2434
(MeHNOCCHMeCH$_2$)$_2$SnBr$_2$	IIIB	m. 207-209°, IR spectrum	1168, 2434
(MeHNOCCHMeCH$_2$)$_2$SnI$_2$	IIB	m. 210.5-11.5° (n)	1168, 2434
(EtHNOCCHMeCH$_2$)$_2$SnCl$_2$	IIIB	m. 177-79°	1168, 2434
(EtHNOCCHMeCH$_2$)$_2$SnBr$_2$	IIIB	m. 156-57°	1168, 2434
(EtHNOCCHMeCH$_2$)$_2$SnI$_2$	IIB	m. 147.5-49° (n)	1168, 2434
(Me$_2$NOCCHMeCH$_2$)$_2$SnCl$_2$	IIID	m. 215-17°	1168, 2434
(Me$_2$NOCCHMeCH$_2$)$_2$SnBr$_2$	IIIB	m. 193-94.5°	1168, 2434
(Me$_2$NOCCHMeCH$_2$)$_2$SnI$_2$	IIB	m. 200.5-202.5° (n)	1168, 2434
(EtO$_2$CCH$_2$NHOCCHMeCH$_2$)$_2$SnCl$_2$	IIIB	m. 195-97°, struct. (h)	1167
(EtO$_2$CCH$_2$NHOCCHMeCH$_2$)$_2$SnBr$_2$	IIIB	m. 208.5-10.5° (h)	1167
(EtO$_2$CCH$_2$NHOCCHMeCH$_2$)$_2$SnI$_2$	IIA	In 46% yield, m. 190-91.5°, struct. (h, j)	1167
(MeO$_2$CCH$_2$CHMe)$_2$SnCl$_2$	IIIB	m. 83.5-85°, struct.	2649
(MeO$_2$CCH$_2$CHMe)$_2$SnBr$_2$	IIIB	m. 102-104°, struct.	2649
(MeO$_2$CCH$_2$CHMe)$_2$SnI$_2$	II	m. 97.5-99.5°, mol. wt., struct. (o)	2649
(EtO$_2$CCH$_2$CHMe)$_2$SnCl$_2$	IIIB	m. 68-70°, struct.	2649

Compound	Method	Properties	Ref.
(EtO$_2$CCH$_2$CHMe)$_2$SnBr$_2$	IIIB	m. 71.5-73°, struct.	2649
(EtO$_2$CCH$_2$CHMe)$_2$SnI$_2$	II	m. 88.5-90.5°, mol. wt., struct. (o)	2649
(PrO$_2$CCH$_2$CHMe)$_2$SnCl$_2$	IIIB	m. 65.5-67.5°, struct.	2649
(PrO$_2$CCH$_2$CHMe)$_2$SnBr$_2$	IIIB	m. 94.0-94.5°, struct.	2649
(PrO$_2$CCH$_2$CHMe)$_2$SnI$_2$	II	m. 98.5-100.5°, mol. wt., struct. (o)	2649
(H$_2$NOCCH$_2$CHMe)$_2$SnCl$_2$	IIIA	Dec. 284°, IR spectrum	1168, 2434
(H$_2$NOCCH$_2$CHMe)$_2$SnBr$_2$	IID	Dec. 272°, IR spectrum	1168, 2434
(H$_2$NOCCH$_2$CHMe)$_2$SnI$_2$	IIB	Dec. 263°, IR spectrum	1168, 2434
(MeHNOCCH$_2$CHMe)$_2$SnCl$_2$	IIIB	m. 218.5-20°	1168, 2434
(MeHNOCCH$_2$CHMe)$_2$SnBr$_2$	IIIB	m. 212-14°	1168, 2434
(MeHNOCCH$_2$CHMe)$_2$SnI$_2$	IIB	m. 220-22° (n)	1168, 2434
(PhHNOCCH$_2$CHMe)$_2$SnCl$_2$	IIIB	m. 227-28°	1168, 2434
(PhHNOCCH$_2$CHMe)$_2$SnBr$_2$	IIB	m. 232-33° (n)	1168, 2434
(PhHNOCCH$_2$CHMe)$_2$SnI$_2$	IIB	m. 199.5-201°	1168, 2434
(p-MeC$_6$H$_4$NHOCCH$_2$CHMe)$_2$SnI$_2$	IIIB	m. 213-15°	1168, 2434
(Me$_2$NOCCH$_2$CHMe)$_2$SnCl$_2$	IID	m. 229-31°, IR spectrum (n)	1168, 2434
(Me$_2$NOCCH$_2$CHMe)$_2$SnBr$_2$	IIIB	m. 198-99° (h)	1168, 2434
(EtO$_2$CCH$_2$NHOCCH$_2$CHMe)$_2$SnBr$_2$	IIA	In 41% yield, m. 218-19°, struct. (k, j)	1167
(EtO$_2$CCH$_2$NHOCCH$_2$CHMe)$_2$SnI$_2$	II	In 23-30% yield, m. 94-95°	1167
(MeO$_2$CCHEt)$_2$SnI$_2$			1632

[RO$_2$CCH$_2$CH(CO$_2$R)]$_2$SnX$_2$

R = Me,	X = Br	IIC	cis-Form m. 130-31°, NMR spectrum	1652
			trans-Form m. 157.5-58°, NMR spectra	1652
Me,	I	IIC	m. 164.5-65.5°	1652

(l) Rxn. with aq. KOH, + HCl → $\overline{\text{Sn(CH}_2\text{CHMeCO}_2\text{)}_2}$ (1632), with aq. NaHCO$_3$, + HX → (MeO$_2$CCHMeCH$_2$)$_2$SnX$_2$; X = Cl, Br (1632). (m) Rxn. with aq. NaOH, + HBr, + MeOH → (MeO$_2$CCHMeCH$_2$)$_2$SnBr$_2$ (1168), with NaHCO$_3$, + HX → (H$_2$NOCCHMeCH$_2$)$_2$-SnX$_2$; X = Cl, Br (1168, 2434). (n) Rxn. with aq. NaOH, + HX → (R$_2$NOCCHRCHR)$_2$SnX$_2$; X = Cl, Br; R = H, Me (1168, 2434). (o) Rxn. with alkali, + HX → (RO$_2$CCH$_2$CHMe)$_2$SnX$_2$; X = Cl, Br; R = Me, Et, Pr (2649).

Table 147 Continued

RR'SnX$_2$

Formula*		Synth. Method**	Properties	Ref.
[RO$_2$CCH$_2$CH(CO$_2$R)]$_2$SnX$_2$ (Cont'd.)				
R = Et	X = Br	IIC	cis-Form m. 114-15° (p, q)	1486
			trans-Form m. 119-21°, IR spectrum, mol. wt.	1486
			cis-Form m. 114-15°, cryst. struct. (r)	2873
Et	I	IIC	m. 99-101°	1486
			m. 118.5-19.5°, IR spectrum	1486
Pr	Br	IIC	Not cryst., mol. wt. (q, s)	1652
Pr	I	IIC	m. 77-78°	1652
Bu	Br	IIC	Not cryst., mol. wt. (q, s)	1652
C$_8$H$_{17}$	Br	IIC	(s)	1652
[HO$_2$C(CH$_2$)$_4$]$_2$SnCl$_2$		VIIB	m. 146-48°, IR spectrum, mol. wt. (t)	1483
[HO$_2$C(CH$_2$)$_4$]$_2$SnBr$_2$		VIIB	m. 134.5-36.5°, IR spectrum, mol. wt. (t)	1483

(p) Proposed struct. for trans-form: octahedral tin with coordination bonds from carbonyl groups in 2-position and halogens in trans position, no model for cis-form advanced (1486). (q) Reflux in H$_2$O → (CH$_2$CO$_2$R)$_2$ + (CH$_2$CO$_2$H)$_2$ (1653). (r) cis-Form with adjacent Br atoms (2873). (s) IR spectrum, struct. (1652). (t) Neutralization value (1483).

4.1.2.6.9 DIORGANOTIN DIHALIDES CONTAINING NITROGEN AND BORON

Di(nitromethyl)tin Dichloride $SnC_2H_4Cl_2N_2O_4$ $(O_2NCH_2)_2SnCl_2$
Synth.: By rxn. of $SnCl_4$ with $MeNO_2$ in i-$PrNH_2$ (318).
Prop.: Colorless powder (318). No proof for struct. but rather poor elemental analysis (Ed.).

Di(phenylcarboranyl)tin Dichloride $SnB_{20}H_{16}H_{30}Cl_2$ $(PhCB_{10}H_{10}C)_2SnCl_2$
Synth.: By rxn. of $SnCl_4$ with $(B_{10}H_{10}C_2Ph)Li$ in Et_2O-C_6H_6, in 37% yield (627).
Prop.: m. 216.5-17.5° (627), Moessbauer resonance spectrum (660).
Rxn. with: KOH in aq. $C_6H_6 \rightarrow [(B_{10}H_{10}C_2Ph)_2SnO]_x + B_{10}H_{10}C_2PhH$ (784).
KOH in aq. alc. $\rightarrow B_{10}H_{10}C_2PhH$ (784).

Di(phenylcarboranyl)tin Dibromide $SnB_{20}C_{16}H_{30}Br_2$ $(PhCB_{10}H_{10}C)_2SnBr_2$
Synth.: By rxn. of $SnBr_4$ with $B_{10}H_{10}C_2PhLi$ in Et_2O-C_6H_6, in 61% yield (627).
By rxn. of $(B_{10}H_{10}C_2Ph)_2SnO$ with HBr (784).
Prop.: m. 203-204° (627), Moessbauer resonance spectrum (660).

Additional nitrogen substituted diorganotin dihalides may be found in Subsections 4.1.2.6.2 and 4.1.2.6.3.4.

4.1.3 ORGANOTIN TRIHALIDES

The following scheme summarizes methods for preparation of organotin trihalides listed in Tables 148 and 149.

I   Dealkylation reactions:
    (IA) $Ph_3SnR + HgX_2 \rightarrow RSnX_3$,
    (IB) $Ph_3SnR + Br \rightarrow RSnBr_3$.

II   Alkylation of metallic tin:
    (IIA) $Sn + RBr$ in presence of cat. $\rightarrow RSnBr_3$,
    (IIB) $Sn + RBr$ under ionizing irradiation $\rightarrow RSnBr_3$,
    (IIC) $Sn + RX$ by electrolysis $\rightarrow RSnX_3$.

III   Alkylation of tin halides:
    (IIIA) $SnX_2 + RX + Ph_3PO_3 \rightarrow RSnX_3$,
    (IIIB) $SnCl_4 + R_4Sn \rightarrow RSnCl_3$,
    (IIIC) $SnCl_4 + R_4Ge \rightarrow RSnCl_3$,
    (IIID) $SnBr_4 + MeHgR \rightarrow RSnBr_3$,
    (IIIE) $SnCl_4 + CH_2:N_2 \rightarrow ClCH_2SnCl_3$.

IV   Miscellaneous reactions:
    (IVA) $CH_2:CHSnCl_3 + B_2Cl_4 \rightarrow BCl_2CH_2CH(BCl_2)SnCl_3$,
    (IVB) $BuSnH_3$ + diene, + $C_{10}F_{21}I$, + NaHg $\rightarrow C_{10}F_{21}YSnCl_3$,
    (IVC) $BuSnH_3 + C_{10}F_{21}CH:CH_2 + AIBN$, + $SnCl_4 \rightarrow C_{10}F_{21}CH_2CH_2SnCl_3$.

R = organic group, may be functionally substituted, X = halogen, Y = hydrocarbon bridge.

4.1.3.1 UNSUBSTITUTED ORGANOTIN TRIHALIDES

Methyltin Trifluoride $SnCH_3F_3$ $MeSnF_3$
Synth.: By rxn. of SnX_4 with Al_4C_3 at 150-200° in presence of HF as by-prod. (2114).
Prop.: Anal. detn. and complex formation $[MeSnF_{3+n}]^{-n}$ by anionic paper chromatography and paper electrophoresis; n = 1-2 (71).
Rxn. with: F^- + Fe^{3+} → competitive fluoride complex formation, kinetic data and stability const. (835, 838).

Methyltin Trichloride (M-328) $SnCH_3Cl_3$ $MeSnCl_3$
Synth.: By rxn. of $SnCl_4$ with $(Me_3Si)_2O$ at 270° in sealed tube (2157), with Me_3GeBu, $Me_2GePrBu$ and $MeEtGePrBu$ in > 98% yield (2292), with $Me_3GeGeEt_3$ or Me_6Ge_2 (2292).
By rxn. of $SnCl_2$ with MeCl at 360° (2004).
By rxn. of $(MeSnO_2H)_x$ with anhyd. HCl (289).
By rxn. of SnX_4 with Al_4C_3 and HCl at 150-200° as by-prod. (2114).
Formation of weak chlorocomplexes from $MeSn(ClO_4)_3$ (838, 840).
Prop.: m. 53° (289), 48° (2157), IR (135, 796, 2400), assignment (289), far IR (543, 878), assignment (289, 879, 2322), Raman, assignment (289) and NMR (31, 55, 581, 723-4, 796, 1220, 1430), dipole moment (325, 724) force const. (304), molar polarizability (1566), thermodynamic data (380, 2400). Anal. detn. by GLC (2004), with alizarin S at pH 4.2, indication for 1:2 adduct (70), by polarography (973), by potentiometry, detn. of dissocn. (2554), by reverse potentiometry (2013), by polarography (973).
Rxn. with: Me_3SnCl (at 75-120°) → Me_2SnCl_2, effect of temp. on product distribution (725).
$MeSnX_3$ → rate of halogen exchange; X = Br, I (727).
RLi → $MeSnR_3$; R = mesityl, $3,4,5-Me_3C_6H_2$ (296), Me_2N (324).
$NaM(CO)_5$ → $MeSn[M(CO)_5]_3$; M = Mn, Re (2034).
$Na_2Fe(CO)_4$ → $[Me_2SnFe(CO)_4]_2$ + $Me_4Sn_3Fe_4(CO)_{16}$ (2004).
Aq. $NaCO_3$ → $(Me_2Sn_2O_3)_x$ (2372).
Me_2NCS_2K → $MeSn(S_2CNMe_2)_2Cl$ (2445).
Concn. KOH → $[MeSn(OH)_5]^{2-}$ (289).
Aq. KOH → $[MeSnCl_3(OH)_2]^{2-}$ (289).
$AgClO_4$ → $MeSn(ClO_4)_3$ (70).
Ac_2CH_2 → $MeSn(OCMe:CHAc)_2Cl$ (2055).
Concn. HCl → instable $H_2(MeSnCl_5)$, by NMR spectroscopy (31).
$LM(CO)_4$ → $MeSnCl_2LMCl(CO)_3$ + CO; M = Mo, W, L = BiPy, 1,10-phenanthroline (1372).
$Co_2(CO)_8$ → $MeSn[Co(CO)_4]_3$ (1671, 1674).
$Co[Co(CO)_4]_2$ → $MeSnCl[Co(CO)_4]_2$ (1672).
$C_5H_5Co(CO)_2$ → $(MeSnCl_2)_2C_5H_5CoCO$ (1373).
Addn. compd.: $MeSnCl_3 \cdot BiPy$ (M-328), far IR spectrum and assignment (878-9, 2322) $[SnC_{11}H_{11}Cl_3N_2]$.
$MeSnCl_3 \cdot (1,10-phenanthroline)$ (M-328), far IR spectrum and assignment (879, 2322) $[SnC_{13}H_{11}Cl_3N_2]$.

MeSnCl$_3$·2Me$_2$SO, from Me$_2$SnCl$_2$ and SnCl$_4$ in Me$_2$SO at reflux, in 90% yield, m. 188-90°, IR spectrum and differential thermal analysis (1385, 1388) [SnC$_5$H$_{15}$Cl$_3$O$_2$S$_2$].

Visible spectra and dissocn. const. of 1:1 adducts which were not isolated are reported for 2,4-, 3,4-, 2,5-diaminonitrobenzene, m-nitroaniline, 2,4- and 4,2-aminonitrotoluene (2110).

Methyltin Tribromide (M-331)   SnCH$_3$Br$_3$ MeSnBr$_3$

Synth.: By rxn. of Me$_4$Sn with SnBr$_4$ at 100° (725).
By rxn. of Me$_2$SnBr$_2$ with Br (2).
By rxn. of MeSnO$_2$H with anhydr. HBr (289).
Prop.: m. 55° (289), 50-52° (2), b$_{12}$ 82-83° (289), IR, assignment (289), far IR, assignment (289, 2322) Raman, assignment (289), NMR (551, 724-5, 1430), ^{13}C NMR (1489) and Moessbauer resonance (1989) spectra, dipole moment (55, 724), molar polarizability (1566). Anal. detn. by potentiometric Br titration (2554), with alizarin S at pH 4.2, indication of (1:2) adduct (70).
Rxn. with: Me$_3$SnBr → Me$_2$SnBr$_2$; effect of temp. on prod. distribution (725).
MeSnX$_3$ → rate of halogen exchange; X = Cl, I (727).
Me$_3$SiNMeLi → MeSn(NMeSiMe$_3$)$_3$ (1812).
Me$_2$NCS$_2$K → MeSn(S$_2$CNMe$_2$)$_3$ (2445).
C$_6$F$_5$MgBr → MeSn(C$_6$F$_5$)$_3$ (77).
C$_5$H$_5$Co(CO)$_2$ → (MeSnBr$_2$)$_2$(C$_5$H$_5$)CoCO + (MeSnBr$_2$)C$_5$H$_5$CoBrCO (1373).
CH$_2$Ac$_2$ → MeSn(OCMe:CHAc)$_2$Br (2055).
RSH + aq. NaOH → MeSn(SR)$_3$; R = Me (643), Et, Pr (2).
Addn. compd.: MeSnBr$_3$·BiPy, from components, m. 245° (dec.), conductivity (147), far IR spectrum and assignment (2322) [SnC$_{11}$H$_{11}$Br$_3$N$_2$].
MeSnBr$_3$·(1,10-phenanthroline), from components in C$_6$H$_6$, m. 220-30°, far IR spectrum and assignment (2322) [SnC$_{13}$H$_{11}$Br$_3$N$_2$].
2MeSnBr$_3$·2,2',2"-terpyridine, from components, dec. > 210°, conductivity, struct. (147) [Sn$_2$C$_{17}$H$_{17}$Br$_6$N$_2$].
MeSnBr$_3$·2pyridine oxide, from components in C$_6$H$_6$, m. 138-42° (dec.), far IR spectrum and assignment (1296) [SnC$_{11}$H$_{13}$Br$_3$N$_2$O$_2$].

Methyltin Triiodide (M-328)   SnCH$_3$I$_3$ MeSnI$_3$

Prop.: IR (796), far IR, assignment (2322) and NMR (55, 724-5, 796, 1430) spectra, dipole moment (55, 724), molar polarizability (1566).
Rxn. with: Me$_3$SnI at 140° → Me$_2$SnI$_2$, effect of temp. on prod. distribution (725).
MeSnX$_3$ → rate of halogen exchange; X = Cl, Br (727).
AcCH:CMeONa → MeSn(OCMe:CHAc)$_2$I (2055).
Me$_2$NCS$_2$K → MeSn(S$_2$CNMe$_2$)$_2$I (2445).
BiPyM(CO)$_4$ → MeSnI$_2$BiPyM(CO)$_3$I + CO; M = Mo, W (1372).
C$_5$H$_5$Co(CO)$_2$ → MeSnI$_2$C$_5$H$_5$Co(CO)I (1373).
Addn. compd.: MeSnI$_3$·BiPy, from components, yellow needles, m. 260° (dec.), conductivity (147), far IR spectrum and assignment (2322) [SnC$_{11}$H$_{11}$I$_3$N$_2$].
MeSnI$_3$·(1,10-phenanthroline), from components in C$_6$H$_6$, m. 230° (dec.), far IR spectrum and assignment (2322) [SnC$_{13}$H$_{11}$I$_3$N$_2$].

MeSnI$_3$·(2,2',2"-terpyridine), orange brown powder, dec. > 170°, conductivity, struct. (147) [SnC$_{16}$H$_{14}$I$_3$N$_3$].

MeSnI$_3$·2Ph$_3$PO, yellow brown solid, (far) IR spectra and assignment (875) [SnC$_{37}$H$_{33}$I$_3$O$_2$P$_2$].

MeSnI$_3$·2Ph$_3$AsO, (far) IR spectrum and assignment (875) [SnAs$_2$C$_{37}$H$_{33}$I$_3$O$_2$].

Ethyltin Trichloride (M-328)   SnC$_2$H$_4$Cl$_3$ EtSnCl$_3$

Synth.: By rxn. of SnCl$_4$ with Et$_4$Ge at 210° or in AcCl and MeCN, in 75-100% conversion (2292), with Et$_6$Ge$_2$ (813), in AcCl and MeCN at 50°, in 100% conversion (2292), with Et$_8$Ge$_3$ in MeNO$_2$, in 100% conversion (2292).

By rxn. of SnCl$_4$ with Et$_2$PrBuGe, Me$_3$SiGeEt$_3$ or Et$_5$BuGe$_2$, the latter in 80% yield (2292).

Prop.: IR, assignment (321), far IR, assignment (321, 879, 2322) and NMR (323, 573, 1430, 2068) spectra, dipole moment (325) magnetic susceptibility (1), polarography (970, 973-4, 2352) reverse potentiometric titration (2013), thermodynamic data (305, 380).

Rxn. with: Polarographic reduction → (EtSn)$_x$ (970), in alc. NaOH (973-4), at pH 5 → (EtSnO$_2$H)$_x$ (970).

(Me$_2$SnO)$_x$ → (Me$_2$SnCl)O(EtSnCl$_2$) (944).

Et$_2$NLi → EtSn(NEt$_2$)$_3$ (323).

LiAlH$_4$ + Et$_2$NH → (EtSn)$_x$ (973).

Aq. NaOH → EtSn(OH)$_2$Cl·H$_2$O (1435).

OH$^-$ loaded ion exchange resin → (EtSnO$_2$H)$_3$ (1435).

NaOH at pH 4-5, + AgClO$_4$, + K$_2$CO$_2$ → (EtSnO$_2$H)$_x$ (2351).

KF → EtSnF$_3$·2KF (2351).

(CHOHCO$_2$K)$_2$ → EtSn(OH)$_2$OCH(CO$_2$K)CHOHCO$_2$K (2351).

CH$_2$OH(CHOH)$_4$CH$_2$OH + KOH → Et(KO)$\overline{\text{SnOEtSn(OH)OCH}_2\text{(CHOH)}_4\text{CH}_2\text{O}}$ (2351).

AgClO$_4$ → EtSn(ClO$_4$)$_3$ (2351).

Moist air → EtSn(OH)Cl$_2$·H$_2$O (1435).

Addn. compd.: EtSnCl$_3$·Ph$_3$CCl, photometric detn. of equil. const., conductivity (2880) [SnC$_{21}$H$_{20}$Cl$_4$].

EtSnCl$_3$·BiPy, far IR spectrum and assignment (879, 2322) [SnC$_{12}$H$_{13}$Cl$_3$N$_2$].

EtSnCl$_3$·(1,10-phenanthroline), far IR spectrum and assignment (879, 2322) [SnC$_{14}$H$_{13}$Cl$_3$N$_2$].

EtSnCl$_3$·Ph$_4$AsCl, from components in aq. NaCl-HCl, m. 180-85° (2014), mol. wt., conductivity (2880) [SnAsC$_{26}$H$_{25}$Cl$_4$].

Biol. prop.: Metabolism and modes of excretion by rats (787).

Activity as insecticide against Culex pipiens berbericus larvae (73), C.p. pipiens (1137), Heliothis zea, H. virescens (2145), Musca domestica, LD$_{50}$ and ATP inhibition (425).

Use: As catalyst for polyurethan synth., relative reactivity (1603).

Ethyltin Tribromide (M-331)   SnC$_2$H$_5$Br$_3$ EtSnBr$_3$

Prop.: IR, assignment (321), far IR, assignment (321, 2322) and NMR (1430, 2068) spectra.

Rxn. with: EtSnH$_3$ + PhC⋮CH → Et(PhCH:CH)SnBr$_2$ (1613).

Aq. CH$_2$Ac$_2$ → EtSn(OCMe:CHAc)$_2$Br (2055).

MeSH + aq. NaOH → EtSn(SMe)$_3$ (2).

Addn. compd.: EtSnBr$_3$·BiPy, from components in C$_6$H$_6$, m. 240-41°, far IR spectrum and assignment (2322) [SnC$_{12}$H$_{13}$Br$_3$N$_2$].

EtSnBr$_3$·(1,10-phenanthroline), from components in C$_6$H$_6$, m. 270-80°, far IR spectrum, assignment (2322) [SnC$_{14}$H$_{13}$Br$_3$N$_2$].

EtSnBr$_3$·HCONMe$_2$, use as catalyst for improved polyurethan foam (742) [SnC$_5$H$_{12}$Br$_3$NO].

Butyltin Trichloride (M-329)   SnC$_4$H$_9$Cl$_3$ BuSnCl$_3$

Synth.: By rxn. of Bu$_4$Sn with excess SnCl$_4$ in autoclave at 203° on pilot plant scale (2909), with SnCl$_4$ in presence of cat. AlCl$_3$, in 16% yield (705), with SnCl$_4$ 1:1 at 0° and flash-distn. at 8 torr. and 94°, contin. process (1646).

By rxn. of SnCl$_4$ with BuMgCl in THF-xylene at 30-40° in N atm., in 20% yield (1186a).

By rxn. of metallic tin with BuCl at 120-170°, in presence of cat. and additives: RSH, ROH or (RO)$_2$PS$_2$H and Mg + I (880a), Bu$_4$NBr and SnCl$_2$, in 12% yield (1209a), Bu$_4$NBr in autoclave, in 64% yield (1209), Bu$_4$NBr, LiBr and SnCl$_2$ in glycol ether (880b).

By rxn. of SnCl$_2$ with BuCl in autoclave in presence of Mg, I, Bu$_2$S$_2$ or C$_{12}$H$_{25}$SH at 175° (1209b), in presence of RNH$_2$ or MePh at 200°, in good yield (1110a).

Prop.: IR (925, 2405), assignment (1078), far IR assignment (879, 2322), Raman (925), assignment (1078), NMR (1309, 2070, 2405), ^{35}Cl NQR (542, 1142) and Moessbauer resonance (1309, 2347) spectra, GLC (1079), dipole moment (325, 1211), magnetic susceptibility (1), polarography (1249, 2588-9), surface tension (1959), extraction from H$_2$O with Bu$_3$PO$_4$-BuOH or Bu$_3$PO$_4$-C$_6$H$_6$ (2142). Anal. detn. by paper chromatography (551), by complexometry in presence of Bu$_n$SnCl$_{4-n}$, n = 2, 3 (2453) by EDTA-titration in technical Bu$_2$SnCl$_2$ (1007), by reverse potentiometric titration (2013).

Rxn. with: R$_4$Sn (1:1) → R$_3$SnCl + RBuSnCl$_2$; Me, Et, Pr (1370).

Me$_4$Sn (1:2) → Me$_2$SnCl$_2$ + MeBuSnCl$_2$ at 0-180° (1370).

Bu$_3$SnH → BuSnH$_3$ + BuSnHCl$_2$ (?) (1804).

Bu$_2$SnHCl → BuSnH$_2$Cl (?) (1804).

Me$_3$SnCl at 180-90° → Me$_2$SnCl$_2$ + MeBuSnCl$_2$ (1370).

Bu$_2$SnY → (Bu$_2$SnCl)Y(BuSnCl$_2$) ; Y = O (944), S (945).

Bu$_2$SnO → BuSn[(OSnBuCl)$_x$Cl][(OSnBu$_2$)$_y$Cl]$_2$; X = 0-4, Y = 1-4 (947).

R$_3$SbS → Bu$_2$Sn$_2$S$_3$ + R$_3$SbCl$_2$; R = Me, c.C$_6$H$_{11}$ (1903).

Co$_2$(CO)$_8$ → BuSn[Co(CO)$_3$]$_3$ (236a) or BuSnCl$_2$Co(CO)$_4$ (1672) or BuSn[Co(CO)$_4$]$_3$ (1671, 1674).

Co[Co(CO)$_4$]$_2$ → BuSnCl[Co(CO)$_4$]$_2$ (1672).

BiPyM(CO)$_4$ → BuSnCl$_2$BiPyMCl(CO)$_3$ + CO; M = Mo, W (1372).

RLi → BuSnR$_3$; R = Me$_2$N (251), Et$_2$N (323).

9,10-C$_{14}$H$_{10}$Na$_2$ → Bu(9,10-C$_{14}$H$_{10}$)SnCl (1726).

RNa → BuR$_3$Sn; R = Co(CO)$_4$ (1671, 1674), Mn(CO)$_5$, Re(CO)$_5$ (2034).

NaX → BuSnX$_3$; X = Me$_2$C:NO (2661), HCO$_2$ (1319).

NaBH$_4$ → BuSnH$_3$ + NaCl + B$_2$H$_6$ (736).

NaCHAc$_2$ in MeOH → [BuSn(CHAc$_2$)Cl(OMe)]$_2$ (1300).

$Na_2S \rightarrow$ polymeric organotin compound (980).
Aq. NaOH $\rightarrow$ BuSn(OH)$_2$Cl (1435) or (Bu$_2$Sn$_2$O$_3$)$_x$ (2372).
Aq. NaOH (50%) $\rightarrow$ BuSnO$_2$H (1170a).
Aq. alkali (1:3) $\rightarrow$ BuSnO$_2$H (1435).
RMgBr $\rightarrow$ BuSnR$_3$; R = PhC⋮C (1221), $\overline{CH_2CH_2CH}$ (1873), C$_6$F$_5$ (428).
Moist air $\rightarrow$ BuSn(OH)Cl$_2\cdot$H$_2$O (1435).
H$_2$O $\rightarrow$ dissocn., complex formation and electrochemistry (2405).
H$_2$O $\rightarrow$ (BuSnCl$_2$)$_2$O (?) (947).
8-HOC$_9$H$_6$N $\rightarrow$ BuSn(OC$_9$H$_6$N-8)$_2$Cl (1291).
PhNCO + ureas $\rightarrow$ cat. effect on rxn., kinetic data (133).
RSH $\rightarrow$ BuSn(SR)$_3$; R = HSCHMeCH$_2$CO$_2$C$_3$H$_6$O$_2$CCH$_2$CHMe (2903).
n-C$_8$H$_{17}$SH + Et$_3$N $\rightarrow$ BuSn(SC$_8$H$_{17}$)$_2$Cl (2892).

<u>Addn. compd.:</u> BuSnCl$_3\cdot$BiPy (M-329), far IR, assignment (879, 2322) and UV (1479) spectra, thermodynamic data (1479) [SnC$_{14}$H$_{17}$Cl$_3$N$_2$].
3BuSnCl$_3\cdot$2(2,2',2''-terpyridyl), m. 223° (dec.) (147), Moessbauer resonance spectrum (2347), conductance, suggested struct. (147) [Sn$_3$C$_{42}$H$_{49}$Cl$_9$N$_6$].
BuSnCl$_3\cdot$1,10-phenanthroline, far IR spectrum, assignment (879, 2322) [SnC$_{16}$H$_{17}$Cl$_3$N$_2$].
BuSnCl$_3\cdot$2NEt$_4$Cl, Moessbauer resonance spectrum (2347) [SnC$_{20}$H$_{49}$Cl$_5$N$_2$].
BuSnCl$_3\cdot$Ph$_4$AsCl, from components in aq. NaCl-HCl, m. 142-43° (2014), mol. wt., conductance (2880) [SnAsC$_{28}$H$_{29}$Cl$_4$].
BuSnCl$_3\cdot$Ph$_3$CCl, photometric detn. of equil. const., conductance (2880) [SnC$_{23}$H$_{24}$Cl$_4$].
BuSnCl$_3\cdot$OC(NH$_2$)$_2$, IR spectrum, stability (286) [SnC$_5$H$_{13}$Cl$_3$N$_2$O].
BuSnCl$_3\cdot$ 2 OC(NH$_2$)$_2$, IR spectrum, stability (286) [SnC$_6$H$_{17}$Cl$_3$N$_4$O$_2$].
BuSnCl$_3\cdot$HCONMe$_2$, use in cat. for improved polyurethan foam (742) [SnC$_7$H$_{16}$Cl$_3$NO].
BuSnCl$_3\cdot$OC(NHPh)$\overline{NCH_2CH_2NHCH_2CH_2}$, from components, m. 63-64°, dissocn. 80° (133) [SnC$_{15}$H$_{24}$Cl$_3$N$_3$O].
BuSnCl$_3\cdot$EtOAc, IR spectrum (2288) [SnC$_8$H$_{17}$Cl$_3$O$_2$].
BuSnCl$_3\cdot$MeCN, from components in CCl$_4$, thermodynamic data (1479)[SnC$_6$H$_{12}$Cl$_3$N].
Visible spectra and dissocn. const. for 1:1 adducts of BuSnCl$_3$ with 2,4-, 3,4-, 2,5-diaminonitrobenzene, m-nitroaniline, 2,4-nitroaminotoluene (2110).

<u>Biol. prop.:</u> Activity against Musca domestica, LD$_{50}$ and ATP inhibition (425).
<u>Use:</u> In catalyst for polymerization of (CH$_2$:CH)$_2$ (2044a), for vulcanizing chlorine-contn. polymers (704), for polyurethan synthesis (3005).
As additive for flame resistant polyester (2702), for surface coating of dye pigments and carbon black (34).

Butyltin Tribromide (M-331)   SnC$_4$H$_9$Br$_3$ BuSnBr$_3$
<u>Synth.:</u> By rxn. of SnBr$_2$ with BuBr in presence of LiBr and (EtOCH$_2$CH$_2$)$_2$O (2893), in presence of R$_4$NBr under reflux, in 2-15% yield (1209a).
By rxn. of metallic tin with BuBr in presence of Mg, I and RSH (880a), in presence of LiBr, SnCl$_2$ in THF, BuOH or (MeOCH$_2$CH$_2$)$_2$O (880b).
By-prod. in rxn. of metallic tin with BuBr under ionizing irradiation (616a), in 1-9% yield (1031), in presence of aq. BuOH, in 1-2% yield (2601), in presence of 10% Br, in 2% yield (2634).

Prop.: IR (165), (far) IR spectra and assignments (2322), GLC (536).
Rxn. with: $Bu_2SnS \rightarrow (Bu_2SnBr)S(BuSnBr_2)$ (945).
$BuMgBr \rightarrow Bu_4Sn$, rxn. in abs. of ethers in MePh and cat. Ph_3PO_3 (1871a).
$R_3SbS \rightarrow Bu_2Sn_2S_3 + R_3SbBr_2$; R = Me, c.C_6H_{11} (1903).
Addn. compd.: Bu...·BiPy (M-331), from components in C_6H_6, m. 217-18° (dec.), far IR spectrum and assignment (2322) [$SnC_{14}H_{17}Br_3N_2$].
$BuSnBr_3$·1,10-phenanthroline, from components in C_6H_6, m. 225-30° (dec.), far IR spectrum and assignment (2322) [$SnC_{16}H_{17}Br_3N_2$].

Butyltin Triiodide (M-331)　　　　$SnC_4H_9I_3$　　　　　　$BuSnI_3$
Synth.: By rxn. of metallic tin with BuI in presence of cat. Mg in $(EtOCH_2)_2$ (2734), in presence of cat. Bu_2Se or $Zn(S_2CNR_2)_2$ (880a).
Rxn. with: $BuMgCl \rightarrow (Bu_2SnO)_x + (Bu_3Sn)_2O$ (2734).
$NaOH \rightarrow (BuSnO_2H)_x$ (2815).

Octyltin Trichloride (M-332)　　　　$SnC_8H_{17}Cl_3$　　　　　n-$C_8H_{17}SnCl_3$
Synth.: By rxn. of $(C_8H_{17})_4Sn$ with $SnCl_4$ (1:1) at 5-20°, in 78% yield (1435), (1:4) at 210-15° (2909).
By rxn. of $SnCl_2$ with $C_8H_{17}Cl$ in presence of Bu_4NBr in $(EtOCH_2CH_2)_2O$ at reflux (2893), of cat. Mg, I and RSH or R_2Se_2 (1209b).
By rxn. of metallic tin and $C_8H_{17}Cl$ in presence of cat. RSH, Mg and I (880a), of glycol ethers and cat. LiBr, I and $SnCl_2$ (880b), of cat. RI, SbI_3 at 180°, in 20% yield (1244a).
Prop.: $b_{0.1}$ 102-110°, n_D^{20} 1.5037 (1435), magnetic susceptibility (1).
Rxn. with: $RMgBr \rightarrow C_8H_{17}SnR_3$; R = $\overset{.}{C}H_2CH_2\overset{.}{C}H$ (1873).
Aq. $Na_2CO_3 \rightarrow [(C_8H_{17}Sn)_2O_3]_n$; n = 1-5 (2372).
Aq. KOH (1:3) $\rightarrow [C_8H_{17}Sn(OH)_2]O[C_8H_{17}Sn(OH)Cl]$ (1435).
$H_2O \rightarrow C_8H_{17}Sn(OH)_2Cl$ (1435).
Moist air $\rightarrow C_8H_{17}Sn(OH)Cl_2·H_2O$ (1435).
Addn. compd.: $C_8H_{17}SnCl_3·HCONMe_2$, use as cat. for improved polyurethan foam (742) [$SnC_{11}H_{24}Cl_3NO$].
Use: For surface coatings of dye pigments (34).

Phenyltin Trichloride (M-330)　　　　$SnC_6H_5Cl_3$　　　　　　$PhSnCl_3$
Synth.: By rxn. of Ph_4Sn and $SnCl_4$ at 90° and fract. distn. from Ph_3SnCl in manufacturing process (1643).
Prop.: IR (333), assignment (994), far IR (194), assignment (1703, 1706, 2784), Raman, assignment (994, 2784) UV (335), NMR (1453, 1962, 2067), NQR (542, 1142) and Moessbauer resonance (1114, 1989) spectra, dipole moment (325, 1211). Anal. detn. by polarography (973, 2352) by TLC and sepn. from Ph_nSnCl_{4-n}; n = 4, 3, 2 (2635), by TLC in Ph_3SnCl (1253), with alizarin S (70), by reverse potentiometric titration (2013).
Polarographic redn. at pH > 12 $\rightarrow Ph_2Hg$ (973), in alc. NaOH $\rightarrow (PhSn)_x$ (973), in alc. $HClO_4 \rightarrow (Ph_2Sn)_x$ (973).
Rxn. with: R_4Sn at 0-60° $\rightarrow R_3SnCl + PhRSnCl_2$; R = Me, Et, n-Pr (1370).
$Bu_2SnS \rightarrow (Bu_2SnCl)S(PhSnCl_2)$ (945).
$RLi \rightarrow PhSnR_3$; R = Me_2N (251), $[(2-MeC_6H_4)_3AsMn(CO)_4]$ (185).

LiAlH$_4$ in Et$_2$O-Et$_2$NH → (PhSn)$_x$ (973).

NaR → PhSnR$_3$; R = Ph (1643), Ph$_2$As (1842), Ph$_2$Sb (1843), Mn(CO)$_5$, Re(CO)$_5$ (1596, 2034), C$_5$H$_5$Fe(CO)$_2$ (1595), Co(CO)$_4$ (1596, 1671, 1674).

Na[C$_5$H$_5$M(CO)$_3$] → PhClSn[C$_5$H$_5$M(CO)$_3$]$_2$; M = Mo, W (1596).

NaO$_2$CH → PhSn(O$_2$CH)$_3$ (1319).

AgN$_3$Ph$_2$ → PhSn(N$_3$Ph$_2$)$_2$Cl (48a).

AgClO$_4$ → PhSn(ClO$_4$)$_3$ (70).

RMgBr → PhSnR$_3$; R = Me (296), $\overline{CH_2CH_2CH}$ (1873), C$_6$F$_5$ (428).

BiPyM(CO)$_4$ → (PhSnCl$_2$)BiPyMCl(CO)$_3$ + CO; M = Mo, W (1372).

Fe(CO)$_5$ → trans-(SnCl$_3$)$_2$Fe(CO)$_4$ (2523).

Co$_2$(CO)$_8$ in THF → PhSnCl$_2$Co(CO)$_4$ (1673).

Co[Co(CO)$_4$]$_2$ in MeOH → PhSnCl[Co(CO)$_4$]$_2$ (1672).

C$_5$H$_5$Co(CO)$_2$ → (PhSnCl$_2$)C$_5$H$_5$Co(SnCl$_3$)CO (1373).

Aq. CH$_2$Ac$_2$ → PhSn(CHAc$_2$)$_2$Cl (2055).

Tropolone → PhSn(O$_2$C$_7$H$_5$)$_2$Cl (372).

PH$_3$ + Et$_3$N → (PhSnP)$_4$ + Et$_3$NHCl + polymeric matl. (2747).

Ph$_2$AsO$_2$H + Et$_3$N → PhSn(O$_2$AsPh$_2$)$_3$ (1842).

Olefin → comparative stability of π-complexes (182).

<u>Addn. compd.:</u> PhSnCl$_3$·2,2'-BiPy from components in C$_6$H$_6$, in 96% yield, m. 276-80° (373) [SnC$_{16}$H$_{13}$Cl$_3$N$_2$].

PhSnCl$_3$·4,4'-BiPy, from components in C$_6$H$_6$, in 85% yield, m. 220-24°, far IR spectrum, assignment sensitive to moisture, soly., struct. (1706) [SnC$_{16}$H$_{13}$Cl$_3$N$_2$].

PhSnCl$_3$·Ph$_3$CCl (M-330), photometric detn., equil. const., conductance (2880) [SnC$_{25}$H$_{20}$Cl$_4$].

PhSnCl$_3$·Ph$_4$AsCl, from components in aq. NaCl-HCl, m. 161-62° (2014), conductance (2880) [SnAsC$_{30}$H$_{25}$Cl$_4$].

PhSnCl$_3$·2Me$_2$SO, from components. in 95% yield, IR spectrum, soly. (1388) [SnC$_{10}$H$_{17}$O$_2$S$_2$].

PhSnCl$_3$·cis-O$_2$S$_2$C$_4$H$_8$(1,4-dithianedioxide), from components in EtOH, in 19% yield, m. 172°, IR spectrum, struct. (1707) [SnC$_{10}$H$_{13}$Cl$_3$O$_2$S$_2$].

PhSnCl$_3$·trans-O$_2$S$_2$C$_4$H$_8$(1,4-dithianedioxide), from components in EtOH, in > 50% yield, m. 220° (dec.), IR spectrum, struct. (1707) [SnC$_{10}$H$_{13}$Cl$_2$O$_2$S$_2$].

PhSnCl$_3$·2BrSPh$_2$, in Me$_2$CO, m. 255°, use in cat. for polyurethan-polyester synth. (1182) [SnC$_{30}$H$_{25}$Br$_2$Cl$_3$S$_2$].

PhSnCl$_3$·EtOAc, IR spectrum (2288) [SnC$_{10}$H$_{13}$Cl$_3$O$_2$].

PhSnCl$_3$·perinaphthenone *), IR spectrum, equil. const. (1548) [SnC$_{16}$H$_{13}$Cl$_3$O].

Visible spectra and dissocn. const. are reported for 1:1 adducts of PhSnCl$_3$ with 2,4-, 3,4-, 2,5-diaminonitrobenzene, m-, p-nitroaniline, 2,4-, 2,5-, 4,2-, 5,2-aminonitrotoluene and 2,5,4-dimethylnitroaniline (2110).

---

* Struct. for perinaphthenone:

Biol. prop.: Activity as insecticide against Culex pipiens larvae (73, 1137), against Musca domestica, LD_{50} and ATP inhibition (425).
Use: In catalyst for ethylene polymerization (1507).

Phenyltin Tribromide (M-332) $SnC_6H_5Br_3$ $PhSnBr_3$
Prop.: Anal. detn. with alizarin S (70).
Rxn. with: $RMgBr \rightarrow PhSnR_3$; $R = C_6F_5$ (77).
$Co_2(CO)_8 \rightarrow (PhSnBr_2)Co(CO)_4$ (1673).
$C_5H_5Co(CO)_2 \rightarrow (PhSnBr_2)C_5H_5Co(CO)Br$ (1373).
Addn. compd.: $PhSnBr_3 \cdot 2Et_2O$, from Ph_3SnH and HBr (1:2) in Et_2O (156) $[SnC_{14}H_{25}Br_3O_2]$.
Use: In catalyst for ethylene polymerization (1507).

p-Tolyltin Trichloride (M-332) $SnC_7H_7SnCl_3$ $p\text{-}MeC_6H_4SnCl_3$
Synth.: By rxn. of $(p\text{-}MeC_6H_4)_4Sn$ with $SnCl_4$ (2111).
Prop.: $\overline{b_{10}\ 120°}$ (2111).
Rxn. with: $RMgBr \rightarrow p\text{-}MeC_6H_4SnR_3$; $R = n\text{-}Pr\overline{CHCMe_2C}H$ (1873), C_6F_5 (77).
Addn. compd.: Visible spectra and equil. const. of 1:1 adducts of $p\text{-}MeC_6H_4SnCl_3$ with m-nitroaniline, 2,4-, 4,2-aminonitrotoluene, 1,4,3- and 1,3,4-diaminonitrobenzene (2111).

p-Biphenylyltin Trichloride $SnC_{12}H_9Cl_3$ $p\text{-}PhC_6H_4SnCl_3$
Synth.: By rxn. of $(p\text{-}PhC_6H_4)Sn$ with $SnCl_4$ (2111).
Prop.: m. 64-66° (2111).
Addn. compd.: Visible spectra and dissocn. const. of 1:1 adducts of $p\text{-}PhC_6H_4SnCl_3$ with 1,4,3-diaminonitrobenzene and 4,2-aminonitrotoluene (2111).

Listing of unsubstituted organotin trihalides concludes in Table 148.

Table 148. Unsubstituted Organotin Trihalides $RSnX_3$

Formula*	Synth. Method**	Properties	Ref.
$EtSnF_3 \cdot 2KF$	IVC	m. > 250°, IR spectrum	2351
$EtSnI_3$	--	(Far) IR, assignment (321, NMR (1430) spectra	--
$PrSnCl_3$	--	(Far) IR, assignment (2058), NMR (2070) spectra	--
$PrSnBr_3$	IIB	--	616a
$CH_2CH_2CHSnCl_3$ (331)	IA	n_D^{25} 1.5447 (a)	1873
$CH_2CH_2CHSnBr_3$ (331)	IA	n_D^{25} 1.6282 (a)	1873
$Me_2CHCH_2SnCl_3$ (331)	--	(b)	--
$MeEtCHSnBr_3$	IIB	--	616a
$n\text{-}C_5H_{11}SnCl_3$	IIIC	In 100% conversion	2292
$C_8H_{17}SnBr_3$	IIIA	In excellent yield	1642
$C_{12}H_{25}SnCl_3$	IIA	In low yield	1209a
$C_{18}H_{37}SnCl_3$	--	(c)	--
$PhSnF_3$	--	(c)	--
$PhSnF_3 \cdot 2KF$	IVC	(d)	--
$PhSnI_3$ (332)	--	m. > 250°, IR spectrum	2351
$RSnX_3$	IIA	Far IR spectrum, assignment (e)	1703
	IIIA	R = alkyl, X = Cl, Br	2970
	IIC (h)	(f, g)	1642
		Synth. in presence of cat. $ZnBr_2$	1348

* Numbers in parenthesis refer to pages in main volume.
** The characters used here correspond to the synthetic scheme in the introduction to Chapter 4.1.3.1.
(a) Use as herbicide and pesticide (1873). (b) Rxn. with Me_4Sn at 0° → Me_3SnCl + 1-$BuMeSnCl_2$ + some Me_2SnCl_2 + 1-$BuMe_2SnCl$ (1370), with Me_3SnCl at 180-90° → Me_2SnCl_2 + 1-$BuMe_2SnCl$ (1370). (c) Rxn. with aq. Na_2CO_3 → $(R_2Sn_2O_3)_x$; X = 1-15; R = $C_{12}H_{25}$, $C_{18}H_{37}$ (2372). (d) Anal. detn. and complex formation $[PhSnF_n]^{(3-n)-}$ by anionic paperchromatography and paper electrophoresis (71). (e) Rxn. with $C_5H_5Co(CO)_2$ → $PhSnI_2(C_5H_5)CoICO$ (1373). (f) Relative reactivity as catalyst for CH_2CH_2O polymerization; X = Cl, Br, I (2572). (g) Use as catalyst for epoxide resin hardening (2916). (h) Electrolysis on tin anode and magnesium cathode (1348).

4.1.3.4 UNSATURATED ORGANOTIN TRIHALIDES

Vinyltin Trichloride (M-334) $SnC_2H_3Cl_3$ $CH_2:CHSnCl_3$
Prop.: IR spectrum (309), dipole moment (1211).
Rxn. with: $NaM(CO)_5 \rightarrow CH_2:CHSn[M(CO)_5]_3$; M = Mn, Re (2034).
$NaCo(CO)_4 \rightarrow CH_2:CHSn[Co(CO)_4]_3 + CH_2:CHSnCl[Co(CO)_4]_2$ (1672, 1674).
$BiPyM(CO)_4 \rightarrow CH_2:CHSnCl_2BiPyM(CO)_3Cl + CO$; M = Mo, W. (1372).
$LW(CO)_4 \rightarrow CH_2:CHSnCl_2LW(CO)_3Cl + CO$; L = 1,10-phenanthroline (1372).
$C_5H_5Co(CO)_2 \rightarrow CH_2:CHSnCl_2(C_5H_5)Co(SnCl_3)CO + (Cl_3Sn)_2C_5H_5CoCO$ (1373).
$B_2Cl_2 \rightarrow BCl_2CH_2CH(BCl_2)SnCl_3$; relative rxn. rates (2329).
$1-C_8F_{17}I$, + Zn in EtOH $\rightarrow 1-C_8F_{17}CH_2CH_2Sn(OEt)_3$ (745).
$X(CF_2)_{10}I$, + Zn in ROH $\rightarrow X(CF_2)_{10}CH_2CH_2Sn(OR)_3$; R = Me, Et; X = F, Cl (745a).

Allyltin Trichloride $SnC_3H_5Cl_3$ $CH_2:CHCH_2SnCl_3$
Rxn. with: $RK \rightarrow CH_2:CHCH_2SnR_3$; R = $[Ph_3SbMn(CO)_4]$ (185).
$Cl(CF_2)_{10}I$, + Zn in MeOH $\rightarrow Cl(CF_2)_{10}(CH_2)_3Sn(OMe)_3$ (745a).

4.1.3.6 FUNCTIONALLY SUBSTITUTED ORGANOTIN TRIHALIDES

Pentafluorophenyltin Trichloride $SnC_6Cl_3F_5$ $C_6F_5SnCl_3$
Synth.: By rxn. of $SnCl_4$ with $MeHgC_6F_5$ at r.t., in 92% yield (77), with $(C_6F_4)_4Sn$ at 140° (1198).
By rxn. of $(C_6F_5)_3SnCl$ with excess $SnCl_4$ (1198).
Prop.: b_{16} 120-22° (1198), $b_{0.02}$ 54-56° (77), fumes in moist air (77), IR and NMR spectra (77).
Rxn. with: $H_2O \rightarrow C_6F_5SnO_2H$ (1198).

Additional functionally substituted organotin trihalides are compiled in Table 149.

Table 149. Functionally Substituted Organotin Trihalides   R'SnX$_3$

Formula*	Synth. Method**	Yield	Properties	Ref.
ClCH$_2$SnCl$_3$ (335)	IIIE	--	b$_5$ 72-73°, Colorless liq. (a)	2071
	IIIE	24	b$_2$ 67-70°, n$_D^{20}$ 1.5730, d$_{20}$ 2.2092	2512
NCCH$_2$CH$_2$SnBr$_3$	IB	76	m. 75-77°, mol. wt., IR spectrum (b)	1745
BCl$_2$CH$_2$CH(BCl$_2$)SnCl$_3$	IVA	--	(Far) IR and NMR spectra (c)	2329
C$_{10}$F$_{21}$(CH$_2$)$_3$O(CH$_2$)$_3$SnCl$_3$	IVB	--	(d, e)	745a
H(CF$_2$)$_8$CH$_2$CH$_2$SnCl$_3$	IIIB	--	(e)	745a
Cl(CF$_2$)$_8$CH$_2$CH$_2$SnCl$_3$	IIIB	--	(e)	745a
F(CF$_2$)$_{10}$CH$_2$CH$_2$SnCl$_3$	IVC	--	(e)	745a
C$_{10}$F$_{21}$(CH$_2$)$_6$SnCl$_3$	IVB	--	(d, e)	745a
p-FC$_6$H$_4$SnCl$_3$	--	--	^{1}H and ^{19}NMR spectra	1453
C$_6$F$_5$SnBr$_3$	IIID	--	--	77
p-ClC$_6$H$_4$SnCl$_3$ (335)	IIIB	--	m. 38-40° (f, g)	2111
2,4-Cl$_2$C$_6$H$_3$SnCl$_3$	--	--	(h)	--
m-(Cl$_3$Sn)$_2$C$_6$H$_4$ (335)	--	--	(i)	--

* Numbers in parenthesis refer to pages in main volume.

** The characters used here correspond to the synthetic scheme in the introduction to Chapter 4.1.3.1.
(a) Fuming in air, dec. on standing (2071), mol. wt. (2512), far IR, mass (2512), NMR (2071, 2512) and ^{35}Cl NQR (2752) spectra. (b) Rxn. with Na$_2$S → (NCCH$_2$CH$_2$Sn)$_2$S$_3$ (2326). (c) Thermal dec. at 100° → SnCl$_2$ + BCl$_3$ + C$_2$H$_3$BCl$_3$ + HCl (2329). (d) Rxn. with NaOMe → R$_2$Sn$_2$O$_3$ (745a). (e) Useful as water, dirt, oil repellent, anti-adhesive and release agent (745a). (f) Visible spectra and dissocn. const. of 1:1 adducts with m-nitroaniline, 2,4-, 4,2-aminonitroaniline, 1,2,4-, 1,3,4-, 1,4,3-diaminonitrobenzene (2111). (g) Use for reducing oil absorption of carbon black (34). (h) Rxn. with $\overline{CH_2CH_2CHL1}$ → 2,4-Cl$_2$C$_6$H$_3$Sn($\overline{CHCH_2CH_2}$)$_3$ (1873). (i) Rxn. with Me$_3$PO$_3$ + NaN$_3$ (1:6:6) in Py → m-{(MeO)$_3$P:N}$_3$Sn}$_2$C$_6$H$_4$ (2116).

4.1.4 ORGANOTIN HALIDES WITH TIN LINKED TO ANIONS

This chapter comprises a number of organotin halides containing anions like tetraphenyloborate, benzenesulfinate and perchlorate. The terpyridine complexes are proposed to contain six-coordinated tin. The compounds are prepared by the following methods.

I Partial dehalogenation:
$Ph_2SnCl_2 + NaX' \rightarrow Ph_2SnX'Cl$.

II Metathetic substitution of terpyridine complexes:
(IIA) $2 \cdot Ph_2SnX_2 \cdot TerPy + NaBPh_4 \rightarrow [Ph_2SnCl \cdot TerPy]BPh_4$,
(IIB) $2RSnX_3 \cdot TerPy + NaBPh_4 \rightarrow [RSnX_2 \cdot TerPy]BPh_4$.

R = organic group, X = halogen, X' = benzenesulfinate. All compounds are listed in Table 150.

Table 150. Organotin Halides Containing Tin Linked to Anions $\quad R_nSnX_{3-n}X'$

Formula	Synth. Method*	Properties	Ref.
$[Me_2SnClTerPy]BPh_4$ (a)	--	Struct. (b)	2347
$[Me_2SnClTerPy]ClO_4$ (a)	--	Struct. (b)	2347
$[Ph_2SnClTerPy]BPh_4$ (a)	IIA	Pale yellow cryst., m. 82°, conductance (b)	147
$Ph_2SnCl(O_2SPh) \cdot Ph_2SnO$	I	m. 200° (dec.)	1410
$[Ph_2SnClTerPy]ClO_4$ (a)	--	Struct. (b)	2347
$[Ph_2SnBrTerPy]BPh_4$ (a)	IIA	m. 186°, conductance	147
$[MeSnBr_2TerPy]BPh_4$ (a)	IIB	Dec. > 210°, conductance	147
$[BuSnCl_2TerPy]BPh_4$ (a)	--	m. 223° (dec.), conductance (b)	147
$[BuSnCl_2TerPy]ClO_4$ (a)	--	Struct. (b)	2347

* The numbers correspond to the synthetic scheme in the introduction to Chapter 4.1.4.
(a) TerPy = 2,2',2''-terpyridine. (b) Moessbauer resonance spectra (2347).

4.1.5 ORGANOTIN HALIDES CONTAINING TIN-OXYGEN BOND

This chapter, numbered 4.1.4 in the main volume, comprises organotin halide derivatives with a tin-oxygen linkage. There is considerable evidence in the recent literature for the four-membered "stannoxane" ring for compounds of the type $(R_2SnX-O-R_2SnX)_2$, compare Main Volume page 336. However, this has not been accepted by all authors.

For easier reference some compounds again have been rearranged, e.g., $(Me_2SnCl)_2O$ to $Me_2SnCl_2 \cdot Me_2SnO$ and $(R_2SnCl)O(R_2SnOR')$ to $(R_4Sn_2ClOR')O$ or similar formulations; R' may stand for hydrogen, an alkyl or acyl group.

Of special interest are a number of new organotin halide chelates comprising hydroxyquinoline, nitrosonaphthol, koji acid, acetylacetone and tropolone as ligand that present tin with a covalence of five and possible six.

4.1.5.1 ORGANOTIN HALIDES CONTAINING OXY- OR HYDROXY-GROUPS

The compounds listed in Table 151 are prepared by methods from the following scheme.

I Partial hydrolysis of alkyltin halides, alkoxides or dithiocarbamates with:
   (IA) Et_3N, for R_2SnCl_2,
   (IB) With H_2O, for $RSnX_3$ or $R_2Sn(OR)X$,
   (IC) With moist air, for $RSnX_3$ or $(R_4Sn_2XOR)O$,
   (ID) With aq. alkali, for $RSnCl_3$ or $(R_2SnCl)_2O$.

II Reaction between diorganotin oxides and:
   (IIA) R_nSnX_{4-n},
   (IIB) MX; M = Cl_3Sn, $R_nX_{3-n}Si$, R_2XGe, R_2XPb, PhHg, Ph_2Tl.

III Partial halogenation of organotin oxides with hydrochloric acid involving:
   (IIIA) R_2SnO,
   (IIIB) $EtSnO_2H$.

IV Alkylation of metallic tin with alkyl halide in aqueous medium.

Bis(dimethylchlorotin) Oxide (M-338) $Sn_4C_8H_{24}Cl_4O_2$ $[(Me_2SnCl)_2O]_2$
Synth.: By hydrolysis of $Me_2Sn(OAc)Cl$ (409a), of $Me_2(Cl_3C)SnCl$ in H_2O at 60° (865).
Prop.: m. > 200° (dec.) (2611).
Rxn. with: Thermal dec. at 220-300° → Me_3SnCl (2611).
Me_3SiCl + aq. NH_3 → $(Me_3SiOSnMe_2)_2O$ (409a).

Bis(dibenzylchlorotin) Oxide $Sn_4C_{56}H_{56}Cl_4O_2$ $\{[(PhCH_2)_2SnCl]_2O\}_2$
Synth.: By rxn. of $(PhCH_2)_2SnCl_2$ with H_2O under reflux, in 82% yield (1940), in presence of powd. tin in 4-97% yield (1940), in presence of Al, Cu, $FeCl_2$, Ti, Zn, in 17-98% yield (1940).
By-prod. in rxn. of $PhCH_2Cl$ with powd. tin and cat. $HgCl_2$ at 108-14° in H_2O, in 11% yield (1940).
Rxn. with: Sn in H_2O → $(PhCH_2)_3SnCl + SnO_2 + SnCl_2$ (1940).

Bis(diethylchlorotin) Oxide (M-339) $Sn_4C_{16}H_{40}Cl_4O_2$ $[(Et_2SnCl)_2O]_2$
Synth.: By hydrolysis of $Et_2Sn(OMe)Cl$ (942), of $Et_2Sn(NPhCO_2Me)Cl$ (943).
By-prod. in rxn. of Et_2SnCl_2 with powd. Sn in H_2O at 160° (1939).
Prop.: m. 174-76° (409a, 942).
Rxn. with: Sn in H_2O at 160° → Et_3SnCl (1939).
Me_3SiCl + aq. NH_3 → $(Me_3SiOSnEt_2)_2O$ (409a).
NaSCN in EtOH → $(Et_2SnNCS)O(Et_2SnOR)$; R = H, Et (591).
NH_4SCN in alc. C_6H_6 → $(Et_2SnNCS)_2O$ (2165).
RCO_2Na in aq. H_2CCl_2 → $(Et_2SnO_2CR)_2O$; R = $HCCl_2$, Ph, o-ClC_6H_4 (1051).

Aq. Y(CO$_2$Na)$_2$ → (Et$_2$SnOSnEt$_2$O$_2$CYCO$_2$)$_x$; Y = (o-C$_6$H$_4$)$_2$SiMe$_2$ (1516).

Bis(dipropylchlorotin) Oxide (M-339)   Sn$_4$C$_{24}$H$_{56}$Cl$_4$O$_2$ [(Pr$_2$SnCl)$_2$O]$_2$
<u>Synth.:</u> By hydrolysis of Pr$_2$(MeO$_2$CCH$_2$)SnCl in air (24).
By rxn. of Pr$_2$SnCl$_2$ with powd. tin H$_2$O at 160°, in 14% yield (1939).
<u>Prop.:</u> m. 121-22° (2611), 121° (409a), 110-12° (24), IR spectrum (2158). Anal. detn. of hydrate water by K. Fischer titration (2158).
<u>Rxn. with:</u> Thermal dec. 0.3 torr- 85° → (Pr$_3$Sn)$_2$O (261).
Sn in H$_2$O at 160° → Pr$_3$SnCl (1939).
Me$_3$SiCl + aq. NH$_3$ → (Me$_3$SiOSnPr$_2$)$_2$O (409a).
NaSCN in EtOH → (Pr$_2$SnNCS)O(Pr$_2$SnOR); R = H, Et (591).
<u>Use:</u> As catalyst for polyurethan synthesis (2166).

Bis(dibutylchlorotin) Oxide (M-337)   Sn$_4$C$_{32}$H$_{72}$Cl$_4$O$_2$ [(Bu$_2$SnCl)$_2$O]$_2$
<u>Synth.:</u> By rxn. of (Bu$_2$SnCl)$_2$ with air (469).
By rxn. of Bu$_2$SnCl$_2$ with H$_2$O in org. solvent at 0°-r.t., in good yield (1051), with H$_2$O at 160° in Et$_3$N-BuCl in presence of various metals at 2-98% yield (1939), with H$_2$O at 160° in presence of various metals in 14-97% yield (1939).
By hydrolysis of Bu$_2$SnXCl; X = MeO (942), MeO$_2$CNEt, MeO$_2$CNPh (943).
By rxn. of (Bu$_2$SnO)$_x$ with HCl in C$_6$H$_6$ or MePh under reflux, in 48% yield (2595).
<u>Prop.:</u> m. 112° (409a), 110-12° (469), 110° (942, 2595), dimeric (632), IR (469) and mass (1514) spectra, dipole moment, configuration (632).
<u>Rxn. with:</u> Thermal dec. on distn. → Bu$_2$SnCl$_2$ + (Bu$_2$SnO)$_x$ (1514).
Thermal dec. at 0.3 torr-115° → (Bu$_3$Sn)$_2$O (2611).
Sn in H$_2$O at 160° → Bu$_3$SnCl (1939).
Zn in H$_2$O at 160° → Bu$_4$Sn + Bu$_3$SnCl (1939).
Me$_3$SiCl + aq. NH$_3$ → (Me$_3$SiOSnBu$_2$)$_2$O (409a).
NaSCN → (Bu$_2$SnNCS)$_2$O (591).
NaO$_2$CR in aq. petr. ether → (Bu$_2$SnO$_2$CR)$_2$O; R = HCCl$_2$, PhCHOH, o-ClC$_6$H$_4$, p-H$_2$NSO$_2$C$_6$H$_4$, m-, p-O$_2$NC$_6$H$_4$, 3,5-(O$_2$N)$_2$C$_6$H$_3$ (1051).
Aq. Y(CO$_2$Na)$_2$ → (Bu$_2$SnOSnBu$_2$O$_2$CYCO$_2$)$_x$; Y = (p-CH$_2$C$_6$H$_4$)$_2$SiMe, (o-C$_6$H$_4$)$_2$SiMe$_2$ (1516).
NaOH in aq. H$_2$CCl$_2$ → Bu$_2$SnCl$_2$·nBu$_2$SnO; n = 2, 3, 5 (2594).
NH$_4$OCN in alc. C$_6$H$_6$ → (Bu$_2$SnNCY)$_2$O; Y = O, S (2165).
ArOH + aq. alc. NH$_3$ → (Bu$_2$SnCl)O(Bu$_2$SnOAr); Ar = 1,2-ONC$_{10}$H$_6$ (550).
Me$_3$CO$_2$H + Et$_3$N → (Bu$_2$SnCl)O(Bu$_2$SnO$_2$CMe$_3$) (2257).
Py in H$_2$CCl$_2$ → Bu$_2$SnCl$_2$·3Bu$_2$SnO (2594).
o-ClC$_6$H$_4$CO$_2$H → Bu$_2$Sn(O$_2$CC$_6$H$_4$Cl-o)$_2$ (2395).
(CH$_2$CH$_2$CO$_2$H)$_2$ → [Bu$_2$Sn(Cl)O$_2$CCH$_2$CH$_2$]$_2$ (2395).
H$_2$S → (Bu$_2$SnCl)$_2$S (1514).
H$_2$S, + BiPy → Bu$_2$SnCl$_2$·BiPy + (Bu$_2$SnS)$_3$ (1705).
BuSH → Bu$_2$Sn(SBu)Cl (1116-7).
$\overline{CH_2CH_2O}$ → relative reactivity as polymerization catalyst (2572).
<u>Use:</u> As catalyst for polyurethan synthesis (2166-7, 2485).

Dibutylchlorotin(dibutylhydroxytin) $Sn_4C_{32}H_{74}Cl_2O_4$ $[(Bu_2SnCl)O(Bu_2SnOH)]_2$
Oxide (M-340)

Synth.: By rxn. of Bu_2SnCl_2 with Et_3N in EtOH, in 69% yield (2594).
By rxn. of $(Bu_2SnO)_x$ with aq. HCl in Me_2CO (2169).
By hydrolysis of $(Bu_2SnCl)O(Bu_2SnOEt)$ in air, (10).
Prop.: m. 102° (2594), IR spectrum (10, 2594), proposed struct. cyclic (10), $Cl(Bu_2SnO)_4Cl$ (2594).
Rxn. with: NaSCN → $(Bu_2SnNCS)O(Bu_2SnOH)$ (591).
Me_3CO_2H → $(Bu_2SnCl)O(Bu_2SnO_2CMe_3)$ (2257).
Use: As polymerization catalyst (2169, 2485).
As additive and curing agent for polyurethan-polyester resins (2573).

Bis(dibutyltinoxy)dibutyltin $Sn_3C_{24}H_{54}Cl_2O_2$ $Bu_2SnCl(OSnBu_2)_2Cl$
Dichloride

Synth.: By rxn. of $(Bu_2SnO)_x$ with Bu_2SnCl_2 (947), in MePh, in 73% yield (2857).
By rxn. of $(Bu_2SnCl)_2O$ with aq. NaOH at 0°, in 88% yield (2594).
Prop.: m. 104° (2857), 97° (2594), 89-90° (947), mol. wt., soly. (2594).
Anal. Cl-detn. by alkalimetric titration (2594).
Rxn. with: RCO_2Na in aq. H_2CCl_2 → $(Bu_2SnO_2CR)(OSnBu_2)_2O_2CR$; R = m-, p-$O_2NC_6H_4$ (2594).

Tris(dibutyltinoxy)dibutyltin $Sn_4C_{32}H_{72}Cl_2O_3$ $Bu_2SnCl(OSnBu_2)_3Cl$
Dichloride

Synth.: By rxn. of $(Bu_2SnO)_x$ with Bu_2SnCl_2 (947), in 86% yield (2857).
By rxn. of $(Bu_2SnCl)_2O$ with aq. NaOH at 0°, in 98% yield (2594), with aq. Py (1:1) in H_2CCl_2 (2594).
By rxn. of $(Bu_2SnO)_x$ with aq. HCl in C_6H_6 or MePh (2595), with $PNCl_3$ in MePh in Ar atm. (2539).
Prop.: m. 102° (2594-5, 2857), m. 94-95° (947), mol. wt., soly. (2594). Anal. Cl-detn. by alkalimetric titration (2594).
Rxn. with: RCO_2Na in aq. H_2CCl_2 → $Bu_2SnO_2CR(OSnBu_2)_3O_2CR$; R = Me, CH_2Cl, $HO_2CC\colon C$, m-$O_2NC_6H_4$ (2594), p-$O_2NC_6H_4$ (2594-5).
$Y(CO_2Na)$ → $[Bu_2Sn(OSnBu_2)_3O_2CYCO_2]_x$; Y = CH_2CH_2, $C\colon C$, p-C_6H_4 (2594), $(CH_2)_4$ (2594-5).
Addn. compd.: $Bu_2SnCl(OSnBu_2)_3Cl\cdot H_2O$, from components by fusion, (far) IR spectra, mol. wt., soly., struct. by x-ray diffraction data (2431).

Listing of organotin halide oxides and organotin halide hydroxides concludes in Table 151.

Table 151. Organotin Halides Containing Oxy- or Hydroxy-Groups $(R_2SnX)_2O$

Formula*	Synth. Method**	Properties	Ref.
$(Et_3Sn)O(Bu_2SnCl)$	IIA	Dissocn. > 150°, IR spectrum	944
$(Bu_3Sn)O(Bu_2SnCl)$	IIA	Dissocn. > 150°, IR spectrum	944
$(Bu_3Sn)O(Bu_2SnBr)$	IIA	m. 140° (dec.), IR spectrum	946
$(Me_2SnCl)O(Me_2SnOH)$ (338)	IA	m. > 300°, soly., struct.	10
$Me_2SnCl_2 \cdot 2Me_2SnO$ (338)	IIA	m. 95°, far IR spectrum, soly. (a, b)	2431
$(Me_2SnCl)O(Et_2SnCl)$	IIA	m. 191°, (far) IR spectra, struct., soly.	2431
$Me_2SnCl_2 \cdot 3Et_2SnO \cdot H_2O$	IIA	(far) IR spectra, struct. soly., mol. wt.	2431
$(Me_2SnCl)O(Bu_2SnCl)$	IIA	m. 92–94°	946
	IIA	m. 95°, (far) IR spectra, struct. soly.	2431
$Me_2SnI_2 \cdot 3Et_2SnO \cdot H_2O$ (338)	IIA	(Far) IR spectra, struct. soly. (a)	2431
$(Et_2SnCl)O(Et_2SnOH)$ (339)	IA	m. 236°, mol. wt., struct. (c, d)	10
$(Et_2SnBr)_2O$ (339)	IB	m. 173°	942
$(Et_2SnBr)O(Pr_2SnBr)$ (339)	IIA (e)	m. 102–103° (far) IR spectra, struct.	2431
	IIA (f)	m. 104–105° (far) IR spectra, soly.	2431
$(Pr_2SnCl)O(Pr_2SnOH)$ (340)	IIIA	m. 174–94° (c, g)	2169
$(Pr_2SnBr)_2O$ (340)	--	Use as cat. for polyurethan	2485
$(Pr_2SnCl)O(Bu_2SnCl)$	IIA	m. 103° (far) IR spectra, struct., soly.	2431

* Numbers in parenthesis refer to pages in main volume.
** The characters used here correspond to the synthetic scheme in the introduction to Chapter 4.1.5.1.
(a) X-ray diffraction data, struct. (2431). (b) Conductivity (2431). (c) Rxn. with NaSCN in EtOH → $(R_2SnNCS)O(R_2SnOH)$; R = Et, Pr (591). (d) Rxn. with Me_3CO_2H → $(Et_2SnCl)O(Et_2SnO_2CMe_3)$ (2257). (e) Synth. from Et_2SnBr_2 and $(Pr_2SnO)_x$ at 110° by fusion (2431). (f) Synth. from Pr_2SnBr_2 and $(Et_2SnO)_x$ at 110° by fusion (2431). (g) Use as polymerization catalyst (2169).

Table 151 Continued   (R$_2$SnX)$_2$O

Formula*	Synth. Method**	Properties	Ref.
Bu$_2$SnCl(OSnBu$_2$)$_n$Cl			
n = 4	IIA	m. 90-92°	947
5	IIA	m. 100-102°	947
6	ID	(78% yield), m. 110°, mol. wt. (h)	2594
9	ID	m. 178-80°	2594
12	ID	m. ~140°	2594
	ID	m. ~178°	2594
(Bu$_2$SnI)$_2$O	--	(i)	--
(Bu$_2$SnI)O(Bu$_2$SnOH) (340)	IC	Mol. wt.	10
(Ph$_2$SnCl)$_2$O (340)	IC (j)	(63% yield), dimeric	298
(Me$_2$SnCl)O(EtSnCl$_2$)	IIA	m. 89-91°, IR spectrum	944
(Bu$_2$SnCl)O(BuSnCl$_2$)	IIA	m. 34-35°, IR spectrum	944
BuSn⟨(OSnBuCl)$_x$Cl / [(OSnBu$_2$)$_y$Cl]$_2$			
X = 0, Y = 1	IIA (k)	m. 100-102°	947
1, 1	IIA (k)	m. 109-110°	947
2, 1	IIA (k)	m. 85-86°	947
1, 2	IIA (k)	m. 92-93°	947
2, 2	IIA (k)	m. 89-90°	947
3, 3	IIA (k)	m. 100-102°	947
4, 4	IIA (k)	m. 109-110°	947
[(C$_8$H$_{17}$)$_2$SnCl]O(BuSnCl$_2$)	IIA	m. 42-43°, IR spectrum	944
(Bu$_2$SnCl)OSn(OH)Cl$_2$	IIB	m. 46-47°, IR spectrum	944

[(C$_8$H$_{17}$)$_2$SnCl]OSn(OH)Cl$_2$	IIB	m. 42-44°, IR spectrum	944
EtSn(OH)Cl$_2$·H$_2$O	IC	m. 94-96°	1435
	(i)	IR spectrum	2351
EtSn(OH)$_2$Cl	IIIB	IR spectrum (i)	2351
EtSn(OH)$_2$Cl·H$_2$O	ID	(37-48% yield), m. > 255°	1435
(BuSnCl$_2$)$_2$O	IB	m. 34-35°	947
BuSn(OH)Cl$_2$	IC	(44% yield), m. 80-87°	1435
BuSn(OH)$_2$Cl	ID	(65-98% yield), m. 107-112° (m)	1435
(BuSnClO)$_4$	(m)	Glasslike solid, mol. wt.	1435
C$_8$H$_{17}$Sn(OH)Cl$_2$·H$_2$O	IC	(60% yield), m. 45-56°	1435
C$_8$H$_{17}$Sn(OH)$_2$Cl	IB	(100% yield), m. 122-27° (n)	1435
[C$_8$H$_{17}$Sn(OH)Cl]O[C$_8$H$_{17}$Sn(OH)$_2$]	ID	(100% yield), m. 119-20°, mol. wt.	1435
(C$_8$H$_{17}$SnClO)$_3$	(n)	Mol. wt.	1435
RO$_2$CCH$_2$(RO$_2$C)CHSn(OH)Br$_2$			
R = Et	IV	IR spectrum, mol. wt., struct.	1653
Pr	IV	IR spectrum, mol. wt., struct.	1653
Bu	IV	IR spectrum, mol. wt., struct.	1653
(Me$_2$SnCl)O(SiMe$_2$Cl)	IIB	m. 180° (dec.)	946
(Bu$_2$SnCl)O(SiMe$_3$)	IIB	Use as stabilizer for polyolefins, PVC	1096
(Bu$_2$SnCl)O(SiMe$_2$Cl)	IIB	m. 38.5-40°, IR spectrum	946
(Bu$_2$SnCl)O(GeMe$_2$Cl)	IIB	m. ~25°, IR spectrum	946
(Bu$_2$SnCl)O(PbBu$_2$Cl)	IIB	m. 120-30° (dec.)	946
(Bu$_2$SnCl)O(TlPh$_2$)	IIB	m. 160° (dec.)	946
(Bu$_2$SnCl)O(HgPh)	IIB	m. 206-208°	946
[(C$_8$H$_{17}$)$_2$SnCl]O(SiMePhCl)	IIB	m. 81-84°, IR spectrum	946
(Ph$_2$SnCl)O(SiMePhCl)	IIB	Oil, IR spectrum	946

(h) Rxn. with (CH$_2$CH$_2$CO$_2$Na)$_2$ in aq. H$_2$CCl$_2$ → [(Bu$_2$SnO)$_5$SnBu$_2$O$_2$C(CH$_2$)$_4$CO$_2$]$_x$ (2594). (i) Rxn. with EtOH → (Bu$_2$SnI)O(Bu$_2$SnOEt) (10). (j) From Ph$_2$Sn(S$_2$CNPh$_2$)Cl at 250° (298). (k) From (Bu$_2$SnO)$_x$ and BuSnCl$_3$ in verying ratio (947). (l) Rxn. of EtSn(OH)$_2$Cl with EtSnCl$_3$ → EtSn(OH)Cl$_2$ (2351). (m) Dehydration of BuSn(OH)$_2$Cl at 140° in vacuo → (BuSnClO)$_4$ (1435). (n) Dehydration of C$_8$H$_{17}$Sn(OH)$_2$Cl at 125° in vacuo → (C$_8$H$_{17}$SnOCl)$_3$ (1435).

4.1.5.3 ORGANOTIN HALIDES CONTAINING ALKOXY GROUPS

Organotin compounds proposed to carry halogen and alkoxide groups on tin are compiled in this chapter. The compounds are prepared by the following methods.

I   Partial dehalogenation of organotin halides with:
   (IA) ROH and amines, from R_2SnCl_2,
   (IB) ROH, from R_nSnCl_{4-n} and $(R_2SnX)_2O$,
   (IC) RONa, from R_nSnX_{4-n}.

II Reaction of diorganotin dihalide with diorganotin dialkoxide.

III Partial halogenation of organotin oxides with hydrochloric or hydrobromic acid and ROH.

IV Miscellaneous reactions:
   (IVA) $Me_2Sn(OR)Cl + NaI \rightarrow Me_2Sn(OR)I$,
   (IVB) $R_2SnCl_2 + R_2SnH_2 + RR'CO \rightarrow R_2Sn(OCHRR')Cl$,
   (IVC) $Et_2Sn(OMe)Br + RCHO \rightarrow Et_2Sn(OCHROMe)Br$,
   (IVD) $Ph_2SnCl_2 + ROH + Py \rightarrow PhSn(OR)_2Cl$,
   (IVE) $Me_2Sn(OR)Cl + Py \rightarrow MeSn(OR)_2Cl$.

R = organic group, may be functionally substituted, R' = R or H, X = halogen, n = 1, 2.

4.1.5.3.1 DIORGANOTIN HALIDES CONTAINING UNSUBSTITUTED ALKOXY GROUPS

Dibutyltin Chloride Methoxide $SnC_9H_{21}ClO$ $Bu_2Sn(OMe)Cl$
<u>Synth.</u>: By rxn. of Bu_2SnCl_2 with $Bu_2Sn(OMe)_2$ (942, 2184), in hexane (181), with Bu_3SnOMe (2184).
<u>Prop.</u>: m. 25-26° (181, 2184), 5° (942), $b_{1.5}$ 166-67° (181, 2184), n_D^{30} 1.5094, d_{30} 1.3775 (2184), IR (181, 942, 2184) and NMR (942) spectra, dipole moment (181, 2184), monomeric (942), thermodynamic data (181, 1116-7, 2184).
<u>Rxn. with</u>: $SnCl_4 \rightarrow Bu_2SnCl_2 + Bu_2Sn(OMe)_2 \cdot SnCl_4$ (2184).
$H_2O \rightarrow (Bu_2SnCl)_2O$ (942).
$RNCO \rightarrow Bu_2Sn(NRCO_2Me)Cl$; R = Et, Ph (943).
$PhNCS \rightarrow Bu_2Sn[SC(OMe):NPh]Cl$ (943).
$CS_2 \rightarrow Bu_2Sn(S_2COMe)Cl$ (943).

Dibutyltin Chloride Butoxide $SnC_{12}H_{27}ClO$ $Bu_2Sn(OBu)Cl$
<u>Synth.</u>: By rxn. of $Bu_2Sn(OBu)_2$ with Bu_2SnCl_2 (846a), in hexane (181).
<u>Prop.</u>: IR spectrum (181), dipole moment (181) thermodynamic data (181, 1116-7, 2184).
<u>Rxn. with</u>: $SnCl_4 \rightarrow Bu_2Sn(OBu)_2 \cdot SnCl_4$ (181).
<u>Use</u>: As catalyst for curing silicone rubber (846a).

Dibutyltin Bromide Methoxide $SnC_9H_{21}BrO$ $Bu_2Sn(OMe)Br$
<u>Synth.</u>: By rxn. of Bu_2SnBr_2 with $Bu_2Sn(OMe)_2$ (942, 2184), in hexane (181).
By-prod. in rxn. of $Bu_2Sn(OMe)_2$ with RBr, in 35-65% yield (435).
<u>Prop.</u>: m. 49-50° (181, 435), 45-50° (2184), 35.5-37.5° (942), b_3 169-70° (181,

2184), $b_{1.5}$ 159° (435), IR (181, 942) and NMR (942) spectra, dipole moment (181, 2184), thermodynamic data (181, 1116-7, 2184).
Rxn. with: AcOH → $Bu_2Sn(OAc)Br$ + MeOH (435).
EtNCO → $Bu_2Sn(NEtCO_2Me)Br$ (943).
BiPy → $Bu_2SnBr_2 \cdot BiPy$ (942).

Additional organotin halides containing unsubstituted alkoxy groups are listed in Table 152.

4.1.5.3.6 ORGANOTIN HALIDES CONTAINING FUNCTIONALLY SUBSTITUTED ALKOXY GROUPS

Dimethyltin 8-Oxyquinolate Chloride $SnC_{11}H_{12}ClNO$ $Me_2Sn(OC_9H_6N)Cl$
Synth.: By rxn. of Me_2SnCl_2 with $Me_2Sn(OC_9H_6N)_2$ in C_6H_6, in 70% yield (1291), in 85% yield (602), with $8-HOC_9H_6N$ in MeOH, in 30% yield (1216), in abs. alc., in 63% yield (1291).
Prop.: m. 147-48° (602), 137° (1291), dec. > 140° (1216), yellow cryst. (602), IR (386, 602, 1216), far IR (1291), UV (602, 1216, 1291) and NMR (602, 1291) spectra, conductivity (1216), monomeric in CCl_4 (602), mol. wt. (1216, 1291).
Rxn. with: BuLi → $Me_2BuSn(OC_9H_6N)$ (602).
NaI → $Me_2Sn(OC_9H_6N)I$ (602).
Bz_2CHTl → $Me_2Sn(OC_9H_6N)CHBz_2$ (602).
TlOEt → $Me_2Sn(OC_9H_6N)OEt$ (?) (602).
H_2O → $Me_2Sn(OC_9H_6N)_2$ (1291).
Py → $Me_2SnCl_2 \cdot 2Py$ + $Me_2Sn(OC_9H_6N)_2$ (1291)
Py at 160° → Me_3SnCl + $MeSn(OC_9H_6N)_2Cl$ (1291).

Diphenyltin 8-Oxyquinolate Chloride $SnC_{21}H_{16}ClNO$ $Ph_2Sn(OC_9H_6N)Cl$
Synth.: By rxn. of Ph_2SnCl_2 with $Ph_2Sn(OC_9H_6N)_2$ in C_6H_6, in 70% yield (602), with $8-HOC_9H_6N$ in MeOH, in 87% yield (1216).
Prop.: Orange cryst. m. 167-68° (602), 156° (1216), IR (602, 1216), UV (602, 1216), NMR (602) and Moessbauer resonance (2617) spectra, no dec. after 3 hours refluxing in C_6H_6 (602), conductivity (1216), dielectrical const. (2617), mol. wt. (1216), monomeric in C_6H_6 (602), struct. (602, 2617).
Rxn. with: BuLi → $BuPh_2Sn(OC_9H_6N)$ (602).
NaI → $Ph_2Sn(OC_9H_6N)_2$ + Ph_2SnI_2 (?) (602).
KSCN → $Ph_2Sn(OC_9H_6N)NCS$ (2617).
Ag_2SO_4 → $(Ph_2SnOC_9H_6N)_2SO_4$ (1216).
RCOCHBzTl → $Ph_2Sn(OC_9H_6N)RCOCHBz$; R = Me, Ph (602).
TlOEt → $Ph_2Sn(OC_9H_6N)OEt$ (?) + TlCl (602).

Other organotin halides containing functionally substituted alkoxy groups are summarized in Table 153.

Table 152. Diorganotin Halides Containing Unsubstituted Alkoxy Groups $R_2Sn(OR)X$

Formula*	Synth. Method**	Yield	Properties	Ref.
$Me_2Sn(OMe)Cl$	II	--	m. 178-80° (sealed tube), hydroscopic (a)	942
$(Me_2SnCl)O(Me_2SnOMe)$ (341)	III	--	m. 200° (b)	2169
$Me_2Sn(OMe)I$	II	--	m. 183-84° (sealed tube), IR spectrum	942
$Et_2Sn(OMe)Cl$	II	--	m. 135-36° (sealed tube) (a, c, d, e)	942
$Et_2Sn(OCHMeEt)Cl$	IVB	--	--	1613
$Et_2Sn(c.OC_6H_{11})Cl$	IVB	90	m. 42°	1613
$Et_2Sn(OMe)Br$	II	--	m. 138-40° (sealed tube) (a, c, d, f)	942
$Et_2SnCl_2 \cdot Bu_2Sn(OMe)_2$	--	--	Rxn. with $SnCl_4 \rightarrow Et_2SnCl_2$	2184
$Pr_2Sn(OMe)F$	II	--	m. 180-85°, IR spectrum, dimeric	942
$Bu_2Sn(OMe)F$	II	--	m. 95-97°, IR spectrum, mol. wt.	942
$Bu_2Sn(OMe)I$	II	--	m. 73° (a, g)	942
$(Bu_2SnCl)O(Bu_2SnOEt)$	IA	45	White waxy, m. 85-140°	10
	II	81	(h, i)	10
$(Bu_2SnI)O(Bu_2SnOEt)$	IC	--	m. 71.5-85° (i)	10
$1-Bu_2Sn(OR)Cl$				
R = Et	IVB	--	--	1613
Pr	IVB	--	--	1613
Me_2CH	IVB	--	m. 65°, $b_{0.5}$ 120°	1613
$c.C_6H_{11}$	IVB	--	--	1613
C_7H_{15}	IVB	--	--	1613

* Numbers in parenthesis refer to pages in main volume.
** The characters used here correspond to the synthetic scheme in the introduction to Chapter 4.1.5.3.
(a) IR and NMR spectra (942). (b) Use as polymerization catalyst (2169), for polyurethan (2166). (c) Rxn. with $H_2O \rightarrow (R_2SnX)_2O$; R = Et; X = Cl, Br (942). (d) Rxn. with BiPy $\rightarrow R_2SnX_2 \cdot BiPy$; R = Et; X = Cl, Br (942). (e) Rxn. with PhNCO $\rightarrow Et_2Sn(NPhCO_2Me)Cl$ (943). (f) Rxn. with $Cl_3CCHO \rightarrow Et_2Sn(OCHCCl_3OMe)Br$ (943). (g) Rxn. with EtNCO $\rightarrow Bu_2Sn(NEtCO_2Me)I$ (943). (h) Rxn. with 8-hydroxyquinoline $\rightarrow Bu_2Sn(OC_9H_6N)_2$ (10). (i) Rxn. with air $\rightarrow (Bu_2SnX)O-Bu_2SnOH$); X = Cl, I (10).

Table 153. Organotin Halides Containing Functionally Substituted Alkoxide Groups $R_nSn(OR')_{3-n}X$

Formula	Method*	Yield	Properties	Ref.
$Me_2Sn(OC_9H_6N)I$ (a)	IVA	36	Yellow cryst., m. 134-35°, monomeric (b)	602
$Et_2Sn(OCHMeC_6H_4CN$-p$)Cl$	IVB	--	--	1613
$Et_2Sn(OC_9H_6N)Cl$ (a)	II	--	Yellow cryst., m. 119-20.5°	1291
$Pr_2Sn(OC_9H_6N)Cl$ (a)	II	--	Yellow cryst., m. 90-91°	1291
$(Pr_2SnCl)O(Pr_2SnOC_{10}H_6NO)$ (c)	IA	40	m. 159-60°, IR spectrum	550
$Bu_2Sn(ON:CMe_2)Cl$	IC	--	bo.02 85° (d)	2661
$Bu_2Sn(OC_9H_6N)Cl$ (a)	IB	25	Dec. > 48°, mol. wt. (e, f)	1216
$Bu_2Sn(OC_9H_6N)Cl \cdot H_2O$ (a)	IB	26	Dec. > 130°, mol. wt. (f)	1216
$(Bu_2SnCl)O(Bu_2SnOC_{10}H_6NO)$ (c)	IA	50	m. 140-41°, IR spectrum	550
$MeSn(C_6H_5O_4)_2Cl$ (g)	III	--	m. 205-208° (dec.) (h)	1658
$MeSn(OC_9H_6N)_2Cl$ (a)	IVE	55	m. 246-47°, far IR, UV and NMR spectra	1291
	III	92	--	1291
$MeSn(C_6H_5O_4)_2Br$ (g)	III	--	m. 113-14° (dec.) (h)	1658
$MeSn(OC_9H_6N)_2Br$ (a)	IC	--	Yellow, m. 258-60°, UV spectrum	142
$BuSn(OC_9H_6N)_2Cl$ (a)	IB	--	m. 182-83°, mol. wt.	1291
$PhSn(OC_9H_6N)_2Cl$ (a)	IVD	30	Orange cryst., m. 150°, monomeric	385
	IC	--	Yellow, m. 218-19°, UV spectrum	142

* The characters used here correspond to the synthetic scheme in the introduction to Chapter 4.1.5.3.
(a) HOC_9H_6N = 8-hydroxyquinoline. (b) IR, UV and NMR spectra (602). (c) $HOC_{10}H_6NO$ = α-nitroso-β-naphthol. (d) Use as catalyst for polyurethan and curing silicone rubber (2661). (e) IR and UV spectra (1216). (f) Dipole moment and conductance (1216). (g) $HC_6H_5O_4$ = koji acid. (h) IR and NMR spectra, struct. (1658).

4.1.5.4 ORGANOTIN HALIDES CONTAINING ACETAL GROUPS

Methyltin Bis(acetylacetonate) Chloride $SnC_{11}H_{17}ClO_4$ $MeSn(O_2C_5H_7)_2Cl$
Synth.: By rxn. of $MeSnCl_3$ with CH_2Ac_2 in H_2O, in 60% yield (2055).
By rxn. of $(MeSnO_2H)_x$ with CH_2Ac_2 in aq. NH_3, followed by aq. HCl at pH 5 (1418).
Prop.: Colorless cryst. (2055), m. 135-36° (1649, 2055), 133-35° (2418), IR (1301, 1649) and NMR (1293, 1298-9, 2418) spectra, magnetic anisotropy in C_6H_6 (1295), struct. (1649).
Rxn. with: $CH_2Ac_2 \rightarrow$ rate of ligand exchange (2418).
MeOH $\rightarrow$ $MeSn(O_2C_5H_7)Cl(OMe)$ (1300).

Additional organotin halides containing acetal groups are listed in Table 154.

4.1.5.5 DIORGANOTIN HALIDES CONTAINING CARBOXY GROUPS

All compounds are listed in Table 155 on page 558.

4.1.6 ORGANOTIN HALIDES CONTAINING TIN-SULFUR BONDS

The chapter, numbered 4.1.5 in the Main Volume, includes derivatives with sulfide polysulfide, mercaptide and dithiocarbamate groupings. The compounds are prepared by the following methods.

I Partial dehalogenation of organotin halides with:
  (IA) RSH in presence of base,
  (IB) Aq. Na_2S_2,
  (IC) R_2NCS_2K.

II Reaction of organotin halides with:
  (IIA) $(R_2SnS)_3$,
  (IIB) $Bu_2Sn(SR)_2$.

III Thioalkylation of $(Bu_2SnCl)_2O$ with BuSH.

IV Addition of a tin-methoxide linkage to the carbon-sulfur double bond in:
  (IVA) Carbondisulfide,
  (IVB) Phenylthioisocyanate.

Bis(dibutylchlorotin)oxide (M-342) $Sn_2C_{16}H_{36}Cl_2S$ $(Bu_2SnCl)_2S$
Synth.: By rxn. of Bu_2SnCl_2 with $(Bu_2SnS)_3$ without solvent (945), with aq. Na_2S (2:1) in petr. ether or H_2CCl_2 at 0°, in 89% yield (1514).
By rxn. of $(Bu_2SnCl)_2O$ with H_2S in Me_2CO, in 85% yield (1514).
Prop.: m. 35.5-37° (945), 35° (1514), NMR and mass spectra (1514), dissocn. on distn. (1514).
Rxn. with: $RCO_2H + NaOH \rightarrow Bu_2Sn(O_2CR)_2$; R = PhCHOH, m-, p-$O_2NC_6H_4$ (1514).
$Y(CO_2H)_2 + NaOH \rightarrow (Bu_2SnO_2CYCO_2)_x + Bu_2SnS$; Y = $(CH_2)_2$, $(CH_2)_4$, o-, p-C_6H_4 (1514).

Listing of organotin halide sulfides concludes in Table 156.

Table 154. Organotin Halides Containing Acetal Groups $R_nSn(OR')_{3-n}X$

Formula	Synth. Method*	Yield	Properties	Ref.
Me$_2$Sn(O$_2$C$_7$H$_5$)Cl (a)	II	60	Yellow cryst., m. 111-13° (b-d)	2510
Me$_2$Sn(O$_2$C$_7$H$_5$)Br (a)	II	--	m. 85-88° (b, c)	2510
Et$_2$Sn[OCH(OMe)CCl$_3$]Br	IVC	--	IR and NMR spectra	943
MeSn(O$_2$C$_7$H$_5$)$_2$Cl (a)	III	83	Pale yellow cryst., m. 228-30°, mol. wt. (e)	2510
	II	50	--	2510
MeSn(O$_2$C$_5$H$_7$)$_2$Br (f)	IB	60	Colorless cryst., m. 129°, soly. (g, h, i)	2055
MeSn(O$_2$C$_7$H$_5$)$_2$Br (a)	III	--	m. 228-30° (e)	2510
MeSn(O$_2$C$_5$H$_7$)$_2$I (f)	IC	--	m. 115-16°, soly. (g, h)	2055
EtSn(O$_2$C$_5$H$_7$)$_2$Cl (f)	--	--	NMR spectrum	1292
EtSn(O$_2$C$_5$H$_7$)$_2$Br (f)	IB	--	m. 94.5°, soly. (g, h)	2055
BuSn(O$_2$C$_7$H$_5$)$_2$Cl (a)	III	--	m. 160-62°, mol. wt. (b)	2510
BuSn(O$_2$C$_7$H$_5$)$_2$Br (a)	III	--	m. 176-77.5° (b)	2510
PhSn(O$_2$C$_5$H$_7$)$_2$Cl (f)	IB	--	m. 149-52°, soly. (g)	2510
PhSn(O$_2$C$_7$H$_5$)$_2$Cl (a)	IB (j)	--	m. 248-52°, struct.	372
	--	--	m. 247.5-52° (b, k)	2510
PhSn(O$_2$C$_5$H$_7$)$_2$Br (f)	--	--	NMR spectrum (1293, 1297-8)	--
[RSn(O$_2$C$_5$H$_7$)X(OMe)]$_2$ (f)				
R = Me, X = Cl	(h)		m. 188° (dec.) (far) IR spectra, struct.	1300
Me Br	(h)		m. 187°, (dec.) (far) IR spectra, struct.	1300
Et Br	(h)		m. 175-76° (dec.) (far) IR spectra, struct.	1300
Bu Cl	IC	--	m. 99-100°, (far) IR spectra, struct.	1300

* Numbers used here correspond to the synthetic scheme in the introduction to Chapter 4.1.5.3.
(a) HO$_2$C$_7$H$_5$ = tropolone. (b) (Far) IR spectra and assigment, struct. (2510). (c) NMR spectrum (2510). (d) Refluxing Me$_2$Sn(O$_2$C$_7$H$_5$)Cl (a) in MeCN → Me$_3$SnCl + MeSn(O$_2$C$_7$H$_5$)$_2$Cl (2510). (e) (Far) IR spectra, assigment, struct. (2510). (f) HC$_5$H$_7$O$_2$ = acetylacetone. (g) IR (1301) and NMR (1293, 1298-9) spectra. (h) Rxn. of RSn(O$_2$C$_5$H$_7$)$_2$X (f) with MeOH → RSn(O$_2$C$_5$H$_7$)X(OMe); R = Me, Et; X = Cl, Br (1300). (i) Magnetic anisotropy in C$_6$H$_6$ (1295). (j) Synth. from Ph$_n$SnCl$_{4-n}$ and tropolone in C$_6$H$_6$; n = 1-3! (372). (k) Rxn with NaC$_7$H$_5$O$_2$ → PhSn(O$_2$C$_7$H$_5$)$_3$ (a). (372).

Table 155. Diorganotin Halides Containing Carboxyl Groups $R_2Sn(O_2CR')X$

Formula*	Synth. Method**	Yield	Properties	Ref.
$Me_2Sn(OAc)Cl$ (343)	--	--	(a)	--
$Bu_2Sn(OAc)Cl$ (343)	II	--	Dipole moment, IR spectrum	181
$Bu_2Sn(OAc)Cl$ (343)	II	--	White cryst., m. 56.5-57.5° (b)	942
$Bu_2Sn(O_2CCH{:}CHCO_2Me)Cl$	IC	--	m. 41-43°, IR spectrum, soly., mol. wt. (c)	1557
$(Bu_2SnClO_2CCH_2CH_2)_2$	IB	85	m. 66-67° (d, e)	2395
$Bu_2Sn(O_2CC_6H_4Cl\text{-}o)Cl$	IB	--	(f)	2395
$Bu_2Sn(OAc)Br$ (343)	IB	--	Synth. from $Bu_2Sn(OMe)Br$	435
$Bu_2Sn(OAc)I$	II	--	White cryst., m. 59-60° (sealed tube) (g)	942

* Numbers in parenthesis refer to pages in main volume.
** The characters used here correspond to the synthetic scheme in the introduction to Chapter 4.1.5.3, acyl groups are to be substituted for R in the appropriate places.

(a) Rxn. with $H_2O \rightarrow (Me_2SnCl)_2O$ (409a). (b) IR (942, 2184), UV (1116-7) and NMR (942) spectra, thermodynamic data (181, 2184), no rxn. with Ag_2S (1514). (c) Rxn. with $H_2O \rightarrow Bu_2SnCl_2 \cdot BiPy$ (1557). (d) Rxn. with aq. Na_2S $[Bu_2SnO_2C(CH_2)_4CO_2SnBu_2S]_x$ (1514). (e) Rxn. with $p\text{-}O_2NC_6H_4CO_2Na \rightarrow Bu_2Sn(O_2CC_6H_4NO_2\text{-}p)_2$ (2395). (f) Rxn. with $p\text{-}O_2NC_6H_4CO_2H \rightarrow Bu_2Sn(O_2CC_6H_4NO_2\text{-}p)_2 + Bu_2Sn(O_2CC_6H_4Cl\text{-}o)_2$ (2395). (g) IR and NMR spectra (942).

Table 156. Organotin Halides Containing Tin-Sulfur Bonds $(R_2SnX)_2S$
$R_2Sn(SR)X$

Formula*	Synth. Method**	Yield	Properties	Ref.
$(Me_2SnCl)_2S$	IIA	--	m. 59-61°	945
$Me_2Sn(S_2CNMe_2)Cl$	IC	--	m. 135-37°, mol. wt., struct. (a, b)	2445
$(Me_2SnBr)_2S$	IIA	--	m. 39-42°	945
$Me_2Sn(S_2CNMe_2)Br$	IC	--	m. 128-30°, mol. wt., struct. (a)	2445
$Me_2Sn(S_2CNMe_2)I$	IC	--	m. 141-44°, mol. wt., struct. (a)	2445
$(Et_2SnCl)_2S_2$	IB	--	m. 50°, instable	1514
$(Bu_2SnCl)_2S_{2.5}$	IB	--	Yellow, m. 26-27°, instable (c)	1514
$(Bu_2SnCl)_2S_{3.5}$	--	--	(d)	--
$(Bu_2SnCl)_2S_4$	IB	79	Yellow, m. 29-30°, instable (e)	1514
$Bu_2Sn(SBu)Cl$ (342)	IIB	65	m. 116°, IR spectrum	942
	IIB	--	n_D^{20} 1.5209, d_{20} 1.225, far IR and	1116-7
	III	--	UV spectra, mol. wt. (f)	1116-7
$Bu_2Sn(S_2COMe)Cl$	IVA	--	IR and NMR spectra	943
$Bu_2Sn[SC(:NPh)OMe]Cl$	IVB	--	(?) NMR spectrum	943
$Bu_2Sn(SBu)Br$	IIB	--	Thermodynamic data	1116-7
$Ph_2SnCl(Ph_2SnS_4)_xCl$	IB	--	Yellow solid (g)	1514
$Ph_2Sn(S_2CNR_2)Cl$				
R_2 = H + $PhCH_2$	IC	83	m. 139-40° (h, i)	298
2Me	IC	85	m. 139-41° (h)	298
2Et	IC	88	m. 143-45° (h)	298
2Ph	IC	88	m. 203-204° (h, j)	298
$(Bu_2SnCl)S(BuSnCl_2)$	IIA	--	m. 28-30°	945
$(Bu_2SnCl)S(BuSnBr_2)$	IIA	--	Oil	945
$(Bu_2SnCl)S(PhSnCl_2)$	IIA	--	m. 38-38.5°	945
$MeSn(S_2CNMe_2)_2Cl$	IC	--	m. 201-203°, mol. wt. soly. struct. (a)	2445

* Numbers in parenthesis refer to pages in main volume.
** The characters used here correspond to the synthetic scheme in the introduction ot Chapter 4.1.6.
(a) NMR, (far) IR spectra and assignment (2445). (b) Rxn. with $(CH_2Br)_2$ → [$Me_2N:\overline{CSCH_2CH_2S}$] [$Me_2SnBr_3$] (2820). (c) Rxn. with $ArCO_2Na$ → $(Bu_2SnO_2CAr)_2S_{2.5}$; Ar = $m\text{-}O_2NC_6H_4$ (1514). (d) Rxn. with RCO_2Na → $(Bu_2SnO_2CR)_2S_{3.5}$; R = $2\text{-}C_4H_3O$, $m\text{-}O_2NC_6H_4$ (1514). (e) Rxn. with $Y(COH)_2$ and aq. NaOH → $(O_2CYCO_2SnBu_2S_4)_x$; Y = $(CH_2)_4$, $(CH_2)_8$, $p\text{-}C_6H_4$, trans-CH:CH (1514), with $HO_2CC\vdots CCO_2Na$ → $(Bu_2SnO_2CC\vdots CCO_2H)S_4$ (1514). (f) Thermodynamic data (1116-7), dipole moment (1116). (g) Rxn. with RCO_2Na → $Ph_2SnO_2CR(Ph_2SnS_4)_xO_2CR$; R = $m\text{-}O_2NC_6H_4$ (1514). (h) IR and UV spectra (298). (i) Decomposes in refluxing C_6H_6 → $PhCH_2NCS$ + HCl + $(Ph_2SnS)_3$ (298). (j) Thermal dec. at 250° → $(Ph_2SnCl)_2O$ + Ph_2NH (298).

Table 156 Continued

(R$_2$SnX)$_2$S
R$_2$Sn(SR)X

Formula*	Synth. Method**	Yield	Properties	Ref.
MeSn(S$_2$CNMe$_2$)$_2$Br	IC	--	m. 192-93°, mol. wt., soly. struct. (a, k)	2445
MeSn(S$_2$CNMe$_2$)$_2$I	IC	--	m. 155° (dec.), soly., struct. (a)	2445
BuSn(SC$_8$H$_{17}$-n)$_2$Cl	IA	--	Colorless liq., stabilizer for PVC	2892

(k) NMR spectrum (1201).

4.1.7 ORGANOTIN HALIDES CONTAINING TIN-NITROGEN BONDS

No compounds of this structure have been reported in the older literature, though there is only a formal difference between a dialkyl(dialkyltriazeno)tin halide reported in this chapter to dialkyl(bipyridino)tin dihalide listed in Chapter 4.1.2 as addition compound of dialkyltin dihalide. The compounds listed in Table 157 are prepared by methods from the following scheme.

I Organometallic synthesis:
(IA) Me$_2$SnCl$_2$ + R'MgBr → Me$_2$SnR'Cl ,
(IB) Me$_2$SnCl$_2$ + R'Li → Me$_2$SnR'Cl ,
(IC) PhSnCl$_3$ + R'Ag → PhSnR'$_2$Cl .

II Addition reactions:
(IIA) R$_2$SnH$_2$ + R$_2$SnCl$_2$ + PhCH:NR → R$_2$Sn(NRCH$_2$Ph)Cl,
(IIB) R$_2$Sn(OMe)X + RNCO → R$_2$Sn(NRCO$_2$R)X .

III Conproportionation reaction:
Bu$_2$SnCl$_2$ + Bu$_2$SnR'$_2$ → Bu$_2$SnR'Cl.

R = organic group, R' = nitrogen containing group, X = halogen.

Dimethyl(1,3-dimethyltriazeno)tin Iodide   SnC$_4$H$_{12}$IN$_3$ Me$_2$Sn(N$_3$Me$_2$)I

Synth.: By rxn. of Me$_2$SnCl$_2$ with Me$_2$N$_3$MgI, in 85% yield (48b).
Prop.: b$_{0.2}$ 50°, pale yellow liq., NMR spectrum, monomeric, struct. (48b).
Rxn. with: Me$_2$NLi → Me$_2$Sn(N$_3$Me$_2$)NMe$_2$·LiI + Me$_2$Sn(N$_3$Me$_2$)$_2$ (48b).
Ph$_2$N$_3$Ag → Me$_2$Sn(N$_3$Ph$_2$)$_2$ + Me$_2$Sn(N$_3$Me$_2$)$_2$ (48b).
Addn. compd.: Me$_2$Sn(N$_3$Me$_2$)NMe$_2$·LiI, see above, in 8% yield, m. 110°, white solid [SnLiC$_6$H$_{18}$IN$_4$] (48b).

Additional organotin halides containing tin-nitrogen bonds are listed in Table 157.

Table 157. Organotin Halides Containing Tin-Nitrogen Bonds $R_2Sn(NR_2)X$

Formula	Synth. Method*	Yield	Properties	Ref.
$Me_2Sn(NCMe_3SiMe_3)Cl$	IB	66	b_1 74°, mol. wt. (a)	2749
$Me_2Sn(N_3Me_2)Br$ (b)	IA	45	$b_{0.2}$ 38°, colorless liq., monomeric (c)	48b
$Et_2Sn(NPhCO_2Me)Cl$	IIB	--	IR and NMR spectra	943
$Bu_2Sn(NEtCO_2Me)Cl$	IIB	--	IR and NMR spectra	943
	III	--	(d)	943
$Bu_2Sn(NPhCO_2Me)Cl$	IIB	--	IR and NMR spectra (d)	943
$Bu_2Sn(NEtCO_2Me)Br$	IIB	--	IR and NMR spectra	943
$Bu_2Sn(NEtCO_2Me)I$	IIB	--	IR and NMR spectra	943
$i\text{-}Bu_2Sn(NMeCH_2Ph)Cl$	IIA	--	--	1613
$i\text{-}Bu_2Sn(NBuCH_2Ph)Cl$	IIA	--	--	1613
$i\text{-}Bu_2Sn(NPhCH_2Ph)Cl$	IIA	--	--	1613
$PhSn(N_3Ph_2)_2Cl$ (b)	IB	--	Orange cryst., m. 203°, monomeric	48a

* The characters used here correspond to the synthetic scheme in the introduction to Chapter 4.1.7.
(a) NMR, IR spectra and assignment (2749). (b) N_3R_2H = dicarbyltriazene. (c) NMR spectrum, struct. (48b).
(d) Hydrolysis → $(Bu_2SnCl)_2O$ (943).

4.1.9 ORGANOTIN HALIDES WITH OTHER HETEROATOMS

Organotin halide hydrides are reported in Chapter 3.8. Organotin halides containing tin-metal bonds may be found in Chapters 5.4, 6.5.4, 6.5.5, 6.5.6, 6.5.7 and 6.6.

4.2 ORGANOTIN SALTS

4.2.1 ORGANOTIN PSEUDOHALIDES

This chapter comprises organotin compounds having a pseudohalogen group linked to tin. Aside from (iso)cyanides, (iso)cyanates (iso)thiocyanates and azides reported in the Main Volume, a number of novel derivatives are added. These include (iso)selenocyanate-NCSe, fulminate-CN:O and N-derivatives of dicyanamide-$N(CN)_2$ and tricyanomethane-$N:C:C(CN)_2$. The compounds listed in Tables 158 and 159 are prepared by the following methods.

I Metathetic substitution reactions:
- (IA) $R_nSnX_{4-n} + MX' \rightarrow R_nSnX'_{4-n}$,
- (IB) $R_nSnX_{4-n} + AgX' \rightarrow R_nSnX'_{4-n}$,
- (IC) $R_3SnI + PbX'_2 \rightarrow R_3SnX'$.

II Reactions with organotin oxides, hydroxides and alkoxides:
- (IIA) $R_3SnOH + NH_4X' \rightarrow R_3SnX'$,
- (IIB) $R_3SnOH + HX' \rightarrow R_3SnX'$,
- (IIC) $R_{2n}Sn_2O_{4-n} + HX' \rightarrow R_nSnX'_{4-n}$,
- (IID) $(R_3Sn)_2O + (NH_2)_2CO \rightarrow R_3SnNCO$,
- (IIE) $Pr_3SnOMe + H_2NCO_2Et \rightarrow Pr_3SnNCO$.

III Reactions with organotin amides:
$R_nSn(NR_2)_{4-n} + HX' \rightarrow R_nSnX'_{4-n}$.

IV Interconversion reactions:
- (IVA) $R_nSnX'_{4-n} + nL \rightarrow R_nSnX'_{4-n} \cdot nL$,
- (IVB) $Et_3SnCN + Ph_3SnH + Et_3N \rightarrow Ph_3SnCN$

V Miscellaneous reactions:
- (VA) $i\text{-}Bu_3SnH + AIBN \rightarrow i\text{-}Bu_3SnCN$,
- (VB) $Ph_6Sn_2 + Se(SeCN)_2 \rightarrow Ph_3SnNCSe$,
- (VC) Me_3SnCHN_2 at $360° \rightarrow Me_3SnCN$.

R = organic group, M = Li, Na, K; X = halogen, X' = pseudohalogen group, L = ligand, n = 1, 2, 3.

4.2.1.1 TRIORGANOTIN PSEUDOHALIDES

Trimethyltin Isothiocyanate (M-345) SnC_4H_9NS Me_3SnNCS
Synth.: By rxn. of Me_3SnCl with KSCN in Me_2CO (72).
By rxn. of Me_3SnOH with NH_4SCN in $c.MeC_6H_{11}$ under reflux, in 75% yield (2660).
Prop.: m. 107-110° (2027), 105-108° (far) IR spectra and assignment (2027,

2103), struct. (2103).

Rxn. with: MSCN → Complex formation and dissocn. by electrophoresis and anion exchange paper chromatography; M = Na, K (2385).

Addn. compd.: Me$_3$SnNCS·Py, m. 112-14° (2660), 110-13° (2103), (far) IR spectra and assignment, struct. (2103) [SnC$_9$H$_{14}$N$_2$S].

Me$_3$SnNCS·NMe$_4$SCN, m. 157-58°, (far) IR spectra and assignment, struct. (2103) [SnC$_9$H$_{21}$N$_3$S$_2$].

Me$_3$SnNCS·Et$_4$NSCN, from components in EtOH, m. 120-121° (dec.), complex formation by paper chromatography and electrophoresis (72) [SnC$_{13}$H$_{29}$N$_3$S$_2$].

Trimethyltin Azide (M-345)   SnC$_3$H$_9$N$_3$ Me$_3$SnN$_3$

Synth.: By rxn. of Me$_3$SnCl with aq. HN$_3$ and extraction with Et$_2$O, in 58% yield (556-7).

By rxn. of Me$_3$SnNR$_2$ with HN$_3$-Et$_2$O in N atm. at r.t.; R = Me, Et (1429).

By rxn. of Me$_3$SnNHP(CMe$_3$)$_2$ with Me$_3$SiN$_3$ in Et$_2$O in N atm. at 0°, in 85% yield (2739).

Prop.: Colorless cryst. (1429), m. 120-21° (556-7), 115-17° (1429), stable to 120° (1429), IR (556-7), assignment (2027), far IR, assignment (2027), UV (556-7), NMR (1429) and Moessbauer resonance (224) spectra.

Rxn. with: PR$_3$ → Me$_4$Sn + Me$_2$Sn(N:PR$_3$)N$_3$ + N; R = Me, Et, Ph (2743).

Addn. compd.: Me$_3$SnN$_3$·NEt$_3$, from components in CH$_2$Cl$_2$, m. 125-28°, yellowish solid, IR spectrum, stable in air, hydrolyzes rapidly (2027) [SnC$_9$H$_{24}$N$_4$].

Me$_3$SnN$_3$·Py, from components in CH$_2$Cl$_2$, m. 136-37°, yellow solid, IR spectrum, stable in air, hydrolyzes rapidly (2027) [SnC$_8$H$_{14}$N$_4$].

Me$_3$SnN$_3$·BBr$_3$, from components in CCl$_4$, IR spectrum, very sensitive to moisture (556, 558) [SnBC$_3$H$_9$Br$_3$N$_3$].

Triethyltin Cyanide (M-345)   SnC$_7$H$_{15}$N   Et$_3$SnCN

Synth.: By rxn. of Et$_3$SnNEt$_2$ and HCN at -30° in pentane, in 94% yield (322).

By rxn. of Et$_3$SnH with 1,1'-(NCC$_6$H$_{10}$N:)$_2$ in presence of AIBN or PhN:NCMe$_2$CN at 80-100° (1620).

By thermal dec. of Me$_2$C:C:NSnEt$_3$ (1620).

Prop.: m. 165°, IR spectrum (322). Anal. detn. with cacotheline after Na$_2$O$_2$ fusion (155).

Rxn. with: Ph$_3$SnH + Et$_3$N (1:1:1) → Et$_3$SnSnPh$_3$ + Ph$_3$SnCN (1618).

Triethyltin Isocyanate (M-345)   SnC$_7$H$_{15}$NO   Et$_3$SnNCO

Synth.: By rxn. of (Et$_3$Sn)$_2$O with HCNO in C$_6$H$_6$ (2799), with (H$_2$N)$_2$CO at 135°, in 92% yield (532).

By rxn. of Et$_3$SnOH with (H$_2$N)$_2$CO at 135°, in 91% yield (532b).

By rxn. of Et$_3$SnOMe with H$_2$NCO$_2$Et, in 90% yield (857).

Prop.: m. 53-54° (857), 51-53° (532), 34° (532b), b$_8$ 113-14° (857), b$_{0.4}$ 70° (532, 532b, 2799), IR spectrum (532, 532b).

Use: As pesticide (532b).

As catalyst for foaming polyurethan (532b, 2799).

Triethyltin Azide $SnC_6H_{15}N_3$ Et_3SnN_3

Synth.: By rxn. of Et_3SnCl with NaN_3 in MePh (2115).
By rxn. of Et_3SnNEt_2 with HN_3 in dry Et_2O at r.t. in N atm., in 65% yield (1429).

Prop.: Colorless cryst., m. 37-39°, b_1 63-65°, NMR spectrum (1429).

Rxn. with: $Et_3SnH \rightarrow Et_6Sn_2 + HN_3$ (1429).
$R_3P \rightarrow Et_3SnN{:}PR_3 + N$; R = Bu, Ph, Me_2N (1429).
HCl-$Et_2O \rightarrow Et_3SnCl$ (1429).
$H_2O \rightarrow (Et_3Sn)_2O$ (1429).
$EtOH \rightarrow Et_3SnOEt$ (1429).

Use: As fungicide, herbicide, insecticide, oil additive and blowing agent (2115).

Tripropyltin Isothiocyanate $SnC_{10}H_{21}NS$ Pr_3SnNCS

Synth.: By rxn. of $(Pr_3Sn)_2O$ with NH_4SCN at r.t. in C_6H_6, in 30% yield (2660).

Prop.: Colorless liq. $b_{0.2}$ 126-28°, IR spectrum (2660).

Addn. compd.: $Pr_3SnNCS \cdot OP(NHC_6H_{13})_2OMe$, from components in ROH, in 34% yield, $b_{0.06}$ 165°, use as acaricide, bactericide, herbicide and insecticide (1838a) $[SnC_{23}H_{52}N_3O_2PS]$.

$Pr_3SnNCS \cdot OP(NMe_2)_2Ph$, from components in ROH, in 79% yield, $b_{0.03}$ 125°, use as acaricide, bactericide, herbicide and insecticide (1838a) $[SnC_{20}H_{38}N_3OPS_2]$.

Tributyltin Cyanide (M-346) $SnC_{13}H_{27}N$ Bu_3SnCN

Synth.: By rxn. of Bu_3SnNEt_2 with HCN in pentane at -30°, in 95% yield (322).
By rxn. of Bu_3SnH with $(1\text{-}NCC_6H_{11}N{:})_2$ in presence of AIBN at 80° (1620).
By rxn. of Bu_3SnH with RN:C in presence of AIBN or $(Me_3CO)_2$ in N atm. under reflux; R = $c.C_6H_{11}$ and $PhCH_2$, in 52 and 82% yield, resp. (2725).

Prop.: m. 88.5° (322), 88-89° (2725), IR spectrum (322, 2725), GLC (2725).

Tributyltin Isocyanate $SnC_{13}H_{27}NO$ Bu_3SnNCO

Synth.: By rxn. of Bu_3SnCl with EtO_2CNHNa in xylene, in 81% yield (373), with $(H_2N)_2CO$ at 150° (532b), with AgNCO, in Et_2O, in 63% yield (373).
By rxn. of $(Bu_3Sn)_2O$ with $(H_2N)_2CO$ at 125-40° in N atm., in 82% yield (532, 532b), with HNCO, in C_6H_6, in 90% yield (532, 2799), with H_2NCONR_2 in xylene under reflux; R = H, Ph, (H + Ph) or with EtO_2CNH_2 (949).

Prop.: $b_{1.3}$ 144-47° (373), $b_{0.6}$ 126-27° (949), $b_{0.5}$ 112-15° (532, 532b), $b_{0.3}$ 104-106° (2799), n_D^{20} 1.4885 (373), n_D^{22} 1.490 (532, 532b), IR spectrum (532, 532b, 949), monomeric, stable in air (373).

Rxn. with: NH_3 in alc. $\rightarrow Bu_3SnOH + CO(NH_2)_2$ (373).
$H_2O \rightarrow (Bu_3Sn)_2O$ (532).

Use: As catalyst for foaming polyurethan (450, 532b, 2799).
As pest control agent (532b).

1,3,5-Tris(tributyltin)-s-triazine- $Sn_3C_{39}H_{81}N_3O_3$ $(Bu_3SnNCO)_3$
2,4,6-trione

Synth.: By rxn. of $(Bu_3Sn)_2O$ with cyanuric acid, in almost quantitative yield (532), in vacuo (532a), at 100° (949).

Prop.: Liq. at -30° (532, 532a), $b_{0.04}$ > 300° (949), $b_{0.03}$ > 240° (532, 532a), stable to 280° (532), to 320° (532a), n_D^{22} 1.5099 (532, 532a), IR spectrum (532, 532a, 949), stable in MeOH (949).
Use: As pesticide (532a).

Tributyltin Isothiocyanate (M-346) $SnC_{13}H_{27}NS$ Bu_3SnNCS
Synth.: By rxn. of $(Bu_3Sn)_2O$ with NH_4SCN in $c.C_6H_{11}Me$ under reflux in 88% yield (2660), with $(H_2N)_2CS$ at 170-95°, in 50% yield (532b).
By rxn. of Bu_3SnOBu with NH_4SCN in MePh under reflux, in 75% yield (2660).
By rxn. of Bu_3SnCl with KSCN in 92% yield (2751), with $(KS)_2C:NCN$ in aq. THF, in 23% yield (2751).
Prop.: $b_{0.5}$ 152-53° (2751), $b_{0.3}$ 150-52° (2660), IR (532a) and ^{15}N NMR (2473) spectra.
Addn. compd.: $Bu_3SnNCS \cdot OPOEt(NMe_2)NHBu$, from components in ROH, in 69% yield, $b_{0.03}$ 123°, use as acaricide, bactericide, herbicide, insecticide (1838a) [$SnC_{21}H_{48}N_3O_2PS$].
$Bu_3SnNCS \cdot OPPh(N\text{-morpholine})_2$, from components in ROH, in 61% yield, $b_{0.008}$ 150-54°, use as acaricide, bactericide, herbicide, insecticide (1838a) [$SnC_{27}H_{48}N_3O_3PS$].
Use: As pesticide (532b).
As catalyst for polyurethan synth. (532b)

Triphenyltin Isocyanate (M-346) $SnC_{19}H_{15}NO$ Ph_3SnNCO
Synth.: By rxn. of Ph_3SnCl with AgNCO in Et_2O, in 89% crude yield (373), with $NaNHCO_2Et$ in xylene, in 83% yield (373).
Prop.: m. 100-103°, monomeric, stable in air (373).
Biol. prop.: Activity as bactericide (529) against Bacillus mycoides, Staphylococcus aureus, Streptococcus lactis (2796).
Activity as fungicide against Candida albicans, Cryptococcus neoformans, Glomerella cingulata and Trychophyton mentagrophytes (2796).
Activity as insecticide against Tineola bisselliella larvae (162).

Triphenyltin Isothiocyanate (M-346) $SnC_{19}H_{15}NS$ Ph_3SnNCS
Synth.: By rxn. of Ph_3SnCl with KSCN in EtOH, in 96% yield (2751), with $(KS)_2\text{-}C:NCN$ in aq. THF, in 12% yield (2751).
By rxn. of Ph_3SnOH with NH_4SCN in $c.MeC_6H_{11}$ under reflux, in 90% yield (2660).
Prop.: m. 171-72° (2660), 166-68° (2751), IR spectrum (2751).
Addn. compd.: $Ph_3SnNCS \cdot Ph_3(HOCH_2CH_2)PSCN$, from components in MeOH, in 88% yield, m. 124-25°, activity as bacteriostat, fungistat (2908) [$SnC_{40}H_{35}N_2OPS_2$].
$Ph_3SnNCS \cdot OP(CH_2Ph)Pr(OEt)$, from components in ROH, in 84% yield, m. 141-43°* (1838a) [$SnC_{31}H_{34}NO_2PS$].

---

* Activity as acaricide, as bactericide against Staphylococcus aureus, as fungicide against Aspergillus niger, Candida albicans, as herbicide and as insecticide (1838a).

$Ph_3SnNCS \cdot OPC_6H_4Me-p(OEt)\overline{N(CH_2)_4CH_2}$, from components in ROH, in 90% yield, m. 71-76°* (1838a) [$SnC_{33}H_{37}N_2O_2PS$].

$Ph_3SnNCS \cdot OP(OC_6H_4Me-p)_3$, from components in ROH, in 63% yield, m. 107°* (1838a) [$SnC_{40}H_{36}NO_4PS$].

$Ph_3SnNCS \cdot O\overline{P(NHC_6H_4NH-o)}\overline{N(CH_2)_4CH_2}$, from components in ROH, in 82% yield, m. 248-50°* (1838a) [$SnC_{30}H_{31}N_4OPS$].

$Ph_3SnNCS \cdot OP(\overline{NCH_2CH_2OCH_2CH_2})_2NHC_2H_2N_3$ (1-amino-1,3,4-triazol), from components in ROH, in 83% yield, m. 70-76°* (1838a) [$SnC_{29}H_{34}N_7O_3PS$].

$Ph_3SnNCS \cdot OSMe_2$, from components in MeOH, in 63% yield, m. 147-50°** (1836) [$SnC_{21}H_{21}NOS_2$].

$Ph_3SnNCS \cdot OSEt_2$, from components in MeOH, in 75% yield, m. 111-12°** (1836) [$SnC_{23}H_{25}NOS_2$].

$Ph_3SnNCS \cdot OSPh_2$, from components in MeOH, in 67% yield, m. 117-19** (1836) [$SnC_{31}H_{25}NOS_2$].

Biol. prop.: Activity as bactericide (529) against Bacillus mycoides, Staphylococcus aureus and Streptococcus lactis (2796).

Activity as fungicide against Alternaria tenuis (2297), Candida albicans, Cryptococcus neoformans, Glomerella cingulata (2796), Phytophthora infestans (2297), Trichophyton mentagrophytes (2796) and Ventura inaequalis (2297).

Activity as insecticide against Leptinotarsa decemlineata and antifeeding effect, Musca domestica, chemosterilant effect (805, 2297), Tinoela bisselliella larvae (162).

Use: As antiwear addtivie for polyphenyl ether lubricants (1950).

Triphenyltin Azide (M-346) $SnC_{18}H_{15}N_3$ Ph_3SnN_3

Synth.: By rxn. of Ph_3SnNR_2 with HN_3 at r.t. in N atm. (1429).

Prop.: Moessbauer resonance spectrum (224), differential thermal analysis and thermal gravimetrical analysis (2667).

Rxn. with: Thermal dec. → Ph_4Sn (2667).

UV irradiation → Ph_3SnCl in H_2CCl_2 or $(Ph_3Sn)_2O$ + Ph_3SnOH in THF (2471).

$CaCl_2$-during drying → Ph_3SnCl (1402).

R_3P → $Ph_3SnN:PR_3$ + Ph_4Sn; R = Bu, C_8H_{17} , Ph (1402).

Addn. compd.: $Ph_3SnN_3 \cdot BBr_3$, from components in CCl_4 , IR spectrum, very sensitive to moisture (556, 558) [$SnBC_{18}H_{15}Br_3N_3$].

$Ph_3SnN_3 \cdot SnCl_4$, from components in CCl_4 , white solid, IR spectrum, very sensitive to moisture (556, 558) [$Sn_2C_{18}H_{15}Cl_4N_3$].

Listing of triorganotin pseudohalides concludes in Table 158.

---

* Activity as acaricide as bactericide against Staphylococcus aureus, as fungicide against Aspergillus niger, Candida albicans as herbicide and as insecticide (1838a).

** Activity as algicide, as bacteriostat against Staphylococcus aureus, as fungistat against Aspergillus niger, Candida albicans, as fungicide against Phytophthora infestans and Cercospora beticola, as molluscide (1836).

Table 158. Triorganotin Pseudohalides R_3SnX'

Formula*	Synth. Method**	Yield	Properties	Ref.
Me_3SnCN (345)	III	92	m. 182°, IR spectrum	322
	VC	--	X-ray struct. analysis (1816) (a)	1393
Me_3SnNCO	IB	48	m. 105-107° (b)	2027
$Me_2C(C_6H_4SnMe_2N_3)_2$	IA	--	(c)	2115
$(PhCH_2)_3SnNCO$	IC	49	m. 119°, mol. wt. soly. (d)	1967
$(PhCH_2)_3SnNCS$	IA	49	m. 109-10°, (far) IR spectra, assignment	841
$(PhCH_2)_3SnN_3$	IA	43	m. 119°, soly., IR spectrum, struct.	1965
$Et_3SnNCS \cdot OPRR'_2$ (e)	IVA	78	m. 98° (f)	1838a
Pr_3SnNCO	IIE	87	$b._{0.03}$ 78-78.5°, n_D^{20} 1.4938, d_{20} 1.2768 (g)	857
$i-Pr_3SnNCS \cdot OP(NEt_2)_2OMe$	IVA	58	$b._{0.2}$ 106° (f)	1838a
Bu_3SnN_3 (346)	III	--	Colorless liq., b_1 126° (h)	1429
$i-Bu_3SnNCS$	VA	low	--	1620
$i-Bu_3SnNCO$	IIC	--	$b._{0.3}$ 102°	2799
	IID	81	$b._{0.3}$ 103°, n_D^{21} 1.488 (i)	532, 532b
$(PhCMe_2CH_2)_3SnN_3$	IIB	--	m. 96-96.5°, IR spectrum (j)	1742
$(c.C_6H_{11})_3SnNCS$	IIA	78	m. 123°	2660
$[(n-C_8H_{17})_3SnNCO]_3$ (k)	IIC	--	Use as pesticide	532a
Ph_3SnCN (346)	III	98	m. 266°, IR spectrum	322
	IVB	40	m. 267°, soly.	1618

* Numbers in parenthesis refer to pages in main volume.
** The characters used here correspond to the synthetic scheme in the introduction to Chapter 4.2.1.
(a) Moessbauer resonance spectrum (1987). (b) (Far) IR spectra and assigment (2027). (c) Use as fungicide, herbicide, insecticide, oil additive and blowing agent (2115). (d) (Far) IR spectra and assigment (1967). (e) $R = \alpha-C_{10}H_7O$, $HR' = N$-morpholine. (f) Activity as acaricide, bactericide herbicide and insecticide (1838a). (g) Rxn. with $Me_3SiCl \to Me_3SiNCO$ (857). (h) Rxn. with R_3P at 160° $\to Bu_3SnN:PR_3 + N$; $R = Bu$, NMe_2 (1429). (i) Use as pesticide (532), as catalyst for foaming polyurethan (2799). (j) Moessbauer resonance spectrum (222, 224, 1742). (k) N-Derivative of cyanuric acid.

Table 158 Continued

Formula*	Synth. Method**	Yield	Properties	Ref.
$(Ph_3SnNCO)_3$ (k)	IIB	--	m. > 360°, colorless cryst. stable to 340° (l)	532, 532a
Ph_3SnCN:O (m)	IA	--	m. 146-48°, colorless cryst., soly. (n)	28
$Ph_3SnN(CN)_2$ (o)	IB	80	m. 169-70°, colorless cryst. soly. (p)	2501
$Ph_3SnN:C:C(CN)_2$ (q)	IC	48	m. 91° (dec.), pale yellow cryst., soly., mol. wt.,	717
	IIB	86	(Far) IR spectra, assignment, struct.	717
$Ph_3SnNCSe$	IA	--	White cryst., soly.	697
	IA	--	IR, assignment, UV spectra	2259
	VB	71	IR spectrum	697
$(o-MeC_6H_4)_3SnNCO$	IC	54	m. 95°, soly. (r, s)	1967
$(o-MeC_6H_4)_3SnNCS$	IB	49	m. 120°, soly. struct. (s, t)	1966
$(o-MeC_6H_4)_3SnN_3$	IA	50	m. 81°, IR spectrum, soly. struct. (s)	1965
$(p-MeC_6H_4)_3SnNCO$	IC	51	m. 114°, soly. (r)	1967
$(p-MeC_6H_4)_3SnNCS$	IB	53	m. 128°, soly. struct. (t)	1966
$(p-MeC_6H_4)_3SnN_3$	IA	48	m. 115°, IR spectrum, soly. struct.	1965
$(p-ClC_6H_4)_3SnCN$	--	--	Use for moth control	473
$(p-ClC_6H_4)_3SnSCN$	--	--	Use for moth control	473
R_3SnCN	--	--	Use as catalyst for epoxide resin hardening	2916

(l) Use as pesticide (532a). (m) Fulminate. (n) IR spectrum, stable in H_2O, no conductivity in Me_2CO, monomeric (28). (o) N-Derivative of dicyanamide. (p) (Far) IR spectra, assignment, mol. wt., struct. hydrolyzes easily (2501). (q) N-Derivative of tricyanomethane. (r) (Far) IR spectra, assignment (1967). (s) NMR spectrum (2356). (t) IR spectrum, stable to hydrolysis (1966).

4.2.1.2 DIORGANOTIN DISPSEUDOHALIDES

Dimethyltin Diisothiocyanate (M-345) $SnC_4H_6N_2S_2$ $Me_2Sn(NCS)_2$
Synth.: By rxn. of Me_2SnCl_2 with NaSCN in EtOH (72).
By rxn. of Me_2SnBr_2 with AgSCN in C_6H_6-MeOH (591), with NaSCN in EtOH, in 84% yield (591).
Prop.: m. 194-96° (591), (far) IR spectrum, assignment (2103), soly. (591), struct. (2103).
Addn. compd.: $Me_2Sn(NCS)_2 \cdot 2Me_4NSCN$, m. 211-12°, (far) IR spectra, assignment, struct. (2103) $[SnC_{14}H_{30}N_6S_4]$.
$Me_2Sn(NCS)_2 \cdot 2Et_4NSCN$, from components in EtOH, m. 113-15°, struct. complex formation followed by paper chromatography and paper electrophoresis (72) $[SnC_{22}H_{46}N_6S_4]$.
$Me_2Sn(NCS)_2 \cdot nMSCN$, M = Na, K, complex formation and dissocn. by paper chromatography, anion exchange and paper electrophoresis (2385).
$Me_2Sn(NCS)_2 \cdot BiPy$, from components in Et_2O-MeOH, in 95% yield, m. 219-220.5° (951), (far) IR spectra, assignment, struct. (2103) $[SnC_{14}H_{14}N_4S_2]$.

Diethyltin Diisothiocyanate $SnC_6H_{10}N_2S_2$ $Et_2Sn(NCS)_2$
Synth.: By rxn. of Et_2SnCl_2 with NaSCN in EtOH, in 95% yield (591).
Prop.: m. 188.5-90° (591), (far) IR spectrum, assignment (2103), soly. (591), struct. (2103).
Rxn. with: $(Et_2SnO)_x \rightarrow (Et_2SnNCS)_2O$ (591).
ROH $\rightarrow (Et_2SnNCS)O(Et_2SnOR)$; R = H (591, 2168), Me (2168).
Addn. compd.: $Et_2Sn(NCS)_2 \cdot BiPy$, from components in Et_2O-MeOH, in 92% yield, m. 220-22° (591), (far) IR spectra, assignment, struct. (2103) $[SnC_{16}H_{18}N_4S_2]$.

Dipropyltin Diisothiocyanate $SnC_8H_{14}N_2S_2$ $Pr_2Sn(NCS)_2$
Synth.: By rxn. of Pr_2SnCl_2 with NaSCN in EtOH, in 95% yield (591).
Prop.: m. 135-36° (591), (far) IR spectra, assignment (2103), soly. (591), struct. (2103).
Rxn. with: $(Pr_2SnO)_x \rightarrow (Pr_2SnNCS)_2O$ (591).
Addn. compd.: $Pr_2Sn(NCS)_2 \cdot BiPy$, from components in Et_2O, in 80% yield, m. 158-59° (591) (far) IR spectra, assignment, struct. (2103) $[SnC_{18}H_{22}N_4S_2]$.

Dibutyltin Diisothiocyanate (M-346) $SnC_{10}H_{18}N_2S_2$ $Bu_2Sn(NCS)_2$
Synth.: By rxn. of Bu_2SnCl_2 with NaSCN in EtOH, in 86% yield (591).
By rxn. of $Bu_2Sn(OMe)_2$ or $(Bu_2SnO)_x$ with NH_4SCN in $c.MeC_6H_{11}$ or MePh under reflux, in 81 and 75% yield, resp. (2660).
By-prod. in rxn. of $[(Bu_2SnNCS)_2O]_2$ with H_2S in Me_2CO (1705).
Prop.: m. 143-45° (2660), 142-42.5° (591), 138-40° (1705), (far) IR, assignment (2103) and Moessbauer resonance (2617) spectra, dielectrical const. (2617), soly. (591), struct. (2103, 2617).
Rxn. with: $Bu_2SnY \rightarrow (Bu_2SnNCS)_2Y$; Y = O (591), S (945).
$Bu_2SnX_2 \rightarrow Bu_2Sn(NCS)X$; X = OMe (942), $8-OC_9H_6N$ (2617).
Addn. compd.: $Bu_2Sn(NCS)_2 \cdot BiPy$ (M-346), from components in Et_2O, in 88% yield (591), in EtOH, m. 152-53° (2660), 150-50.5° (591), far IR, assignment (2103),

Moessbauer resonance (2617) spectra, dielectrical const. (2617), struct. (2103, 2617) [$SnC_{20}H_{26}N_4S_2$].

$Bu_2Sn(NCS)_2 \cdot 1,10$-phenanthroline, from components in EtOH, m. 208-10°, Moessbauer resonance spectrum, dielectrical const., struct, (2617) [$SnC_{22}H_{26}N_4S_2$].

Diphenyltin Diisothiocyanate $SnC_{14}H_{10}N_2S_2$ $Ph_2Sn(NCS)_2$

Synth.: By rxn. of Ph_2SnCl_2 with NaSCN in EtOH (2617).
By rxn. of Ph_2SnI_2 with AgSCN in C_6H_6, in 60% yield (2794).

Prop.: White cryst., m. 176-77° (2794), dec. > 155° (2794), IR (2794) and Moessbauer resonance (2617) spectra, dielectrical const. (2617), soly., stable in air (2794), struct. (2617).

Addn. compd.: $Ph_2Sn(NCS)_2 \cdot BiPy$, from components in EtOH, m. > 210°, Moessbauer resonance spectrum, dielectrical const., struct. (2617) [$SnC_{24}H_{18}N_4S_2$].

$Ph_2Sn(NCS)_2 \cdot 1,10$-phenanthroline, from components in EtOH, m. 195-97°, Moessbauer resonance spectrum, dielectrical const., struct. (2617)[$SnC_{26}H_{18}N_4S_2$].

Additional diorganotin dipseudohalides are listed in Table 159.

4.2.1.3 MONOORGANOTIN TRIPSEUDOHALIDES

All compounds are listed at the bottom of Table 159.

Table 159. Organotin Pseudohalides $R_nSnX'_{4-n}$

Formula*	Synth. Method**	Yield	Properties	Ref.
$Me_2Sn(CN)_2$	III	96	m. > 400° ( ), IR spectrum (a)	322
$Me_2Sn[N(CN)_2]_2$ (b)	IA	--	Dec. > 220°, struct. (c)	1330
$Me_2Sn[N:C:C(CN)_2]_2$ (d)	IA	--	m. 179°, struct. (c)	1330
$Me_2Sn(N_3)_2$	III	--	Colorless, m. 151-53° (dec.), soly. (e)	1429
$(PhCH_2)_2Sn(NCO)_2$	IA	69	m. 140°	841
$(PhCH_2)_2Sn(NCS)_2$	IA	22	m. 204-6°	841
$Et_2Sn(CN)_2$	III	94	m. 226°, IR spectrum	322
$Et_2Sn(N_3)_2$	III	--	Colorless, m. 134-37° (dec.), soly. (e)	1429
$(CH_2CH_2CH)_2Sn(NCS)_2$ (346)	IA	--	Use: herbicide, pesticide	1873
$Bu_2Sn(CN)_2$	III	93	m. 220°, IR spectrum	322
$Bu_2Sn(NCO)_2$	IB	94	m. 48-51°, struct., dec. by moisture (f)	373
$Bu_2Sn(N_3)_2$	III	--	Colorless cryst., m. 63-65°, bo.1 170-80° (e)	1429
$Ph_2Sn(CN)_2$	III	93	m. 265°, IR spectrum	322
$Ph_2Sn[N(CN)_2]_2$ (b)	IB	--	Dec. > 180°, struct. (c)	1330
$Ph_2Sn[N:C:C(CN)_2]_2$ (d)	IIC	--	Dec. 160°, dec. in air, struct. (c)	1330
$Ph_2Sn(NCO)_2 \cdot Py \cdot 1.5C_6H_6$	IB (g)	--	m. 129-30°, IR spectrum	373
$2Ph_2Sn(NCO)_2 \cdot BiPy$	IB (g)	75	m. 294-206°	373
$O[C_6H_4Sn(N_3)_2Ph]_2$	IA	--	(h)	2115
$MeSn(NCS)_3$	IA	--	--	72
$MeSn(NCS)_3 \cdot 2Et_4SCN$	IVA	--	m. 163-69° (dec.), anal. detn.	72

* Numbers in parenthesis refer to pages in main volume.
** The characters used here correspond to the synthetic scheme in the introduction to Chapter 4.2.1.

(a) Use as antiwear additive for polyphenyl ether lubricants (1950). (b) N-Derivative of dicyanamide. (c) IR spectrum, assignment (1330). (d) N-Derivative of tricyanomethane. (e) NMR spectrum, detonates in flame (1429). (f) Rxn. with H_2O → $[Bu_2SnNCO)_2O]_2$ + $[(Bu_2SnNCO)OBu_2SnOH]_2$ (373). (g) Synth. in presence of Py and BiPy, resp. (373). (h) Use as fungicide, herbicide, insecticide, oil additive, blowing agent (2115).

4.2.1.5 ORGANOTIN PSEUDOHALIDES CONTAINING TIN-OXYGEN BONDS

Organotin pseudohalides containing tin-oxygen bonds are only quite recently known. Several of the new derivatives may contain a four-membered "stannoxane"-ring, cf. the structural formula in the Main Volume, page 336. The following scheme gives synthetic methods for the compounds summarized in Tables 160 and 161.

- I Metathetic substitution of organotin halides with:
  - (IA) NaX' with $(R_2SnCl)O(R_2SnOR')$,
  - (IB) AgX' with R_2SnX_2,
  - (IC) NH_4X' with $(R_2SnCl)_2O$,
  - (ID) KX' with $Ph_2SnX''Cl$.
- II Conproportionation reactions of diorganotin dipseudohalide:
  - (IIA) With $(R_2SnO)_x$ or $(R_2SnS)_3$,
  - (IIB) With $R_2SnX''_2$.
- III Condensation reactions of diorganotin oxide with:
  - (IIIA) HX',
  - (IIIB) $(H_2N)_2CO$.
- IV Solvolysis with R'OH of:
  - (IVA) $R_2SnX'_2$,
  - (IVB) $(R_2SnX')_2O$.
- V Miscellaneous reactions:
  - (VA) Me_3SnN_3 with PR_3,
  - (VB) $Bu_2Sn(OMe)X'$ with EtN:CO and $Cl_3CCH:O$.

R = organic group; R' = H, R; X = halogen; X' = pseudohalogen, X" = heteroatomic group like OR, OH, functionally substituted alkoxy group.

4.2.1.5.1 ORGANOTIN PSEUDOHALIDES CONTAINING OXY- OR HYDROXY-GROUPS

Bis(dibutylisothiocyanatotin) Oxide (M-346) $Sn_2C_{18}H_{36}N_2OS_2$ $(Bu_2SnNCS)_2O$
<u>Synth.</u>: By rxn. of $(Bu_2SnO)_x$ with HSCN in Me_2CO-Et_2O, in 100% yield (2164), with $Bu_2Sn(NCS)_2$ in C_6H_6, in 88% yield (591).
By rxn. of $(Bu_2SnCl)_2O$ with NaSCN in EtOH (591), with NH_4SCN in C_6H_6-EtOH at 100° (2165).
<u>Prop.</u>: m. 83.5-84.5° (591, 2164-5), IR spectrum (591, 2103), soly. (591).
<u>Rxn. with:</u> $H_2S \rightarrow Bu_2Sn(NCS)_2 + (Bu_2SnS)_3$ (1705).
<u>Use:</u> As catalyst for polymerization of lactone (1999), of urethan (2164-5, 2485).

Additional diorganotin pseudohalides containing oxy- or hydroxy-groups are listed in Table 160.

Table 160. Organotin Pseudohalides Containing Oxy- and Hydroxy-Groups $(R_2SnX')_2O$

Formula	Synth. Method*	Yield	Properties	Ref.
$(Me_2SnNCS)O(Me_2SnOH)$	--	--	IR spectrum	2103
$(Et_2SnNCS)_2O$	IIA	99	m. 178-79°, soly.	591
	IC	--	IR spectrum (2103) (a, b)	2165
$(Et_2SnNCS)O(Et_2SnOH)$	IVB (a)	--	IR spectrum (2103) (c)	591, 2168
	IA	60	Cubic cryst., m. 170-76° (dec.)	591
$Pr_2SnNCS)_2O$	IIA	93	m. 108-108.5°, soly., mol. wt. (a, d)	591
	IIIA	100	m. 108°, IR spectrum (2103)	2164
$(Pr_2SnNCS)O(Pr_2SnOH)$	IVB (a)	--	m. 162-67° (dec.)	591
	IA	55	IR spectrum (2103)	591
$[(Bu_2SnNCO)_2O]_2$	IVA	82	m. 100-102°, struct. (e)	373
$[(Bu_2SnNCO)_2O]_x$	IIIA	--	m. 163-66°	2164
	IIIB	--	Wax-like matl., m. 195-215°	532, 532b
	IC	--	m. 163-66° (b, d, f)	2165
$(Bu_2SnNCO)O(Bu_2SnOH)$	IVA	66	m. 162-64°	373
$(Bu_2SnNCS)O(Bu_2SnOH)$	IA	70	m. 123-24° (dec.), IR spectrum (2103)	591
$[(Ph_2SnNCO)_2O]_2$	IB (g)	78	m. 158-60°, dec. by moisture (h)	373
$[Ph_2Sn(OH)NCO]_2$	IB	89	m. 99-100°, IR spectrum, struct. (g)	373
$[(Ph_2SnNCO)O(Ph_2SnOH)]_2$	IB	91	m. 300-301°, IR spectrum	373
	IV (g, h)	--	Struct.	373

* The characters used here correspond to the synthetic scheme in the introduction to Chapter 4.2.1.5.
(a) Rxn. of $(R_2SnNCS)_2O$ with $H_2O \rightarrow (R_2SnNCS)O(R_2SnOH)$ (591). (b) Use as catalyst for polyurethan synth. (2165).
(c) Use as catalyst for polyurethan synth. (2168). (d) Use as catalyst for polyurethan synth. (2164). (e) Use as catalyst for lactone polymerization (1999). (f) IR spectrum, soly. (532b), use as pesticide (532b), as catalyst for polyurethan synth. (532b, 2166). (g) Rxn. of $[Ph_2Sn(OH)NCO]_2$ with air $\rightarrow [(Ph_2SnNCO)_2O]_2$ + $[(Ph_2SnNCO)O(Ph_2SnOH)]_2$ + $(Ph_2SnO)_x$ (373). (h) Rxn. of $(Ph_2SnNCO)_2O$ with air $\rightarrow [(Ph_2SnNCO)O(Ph_2SnOH)]_2$ (373).

4.2.1.5.3 ORGANOTIN PSEUDOHALIDES CONTAINING ALKOXY AND ACETAL GROUPS

All compounds are listed in Table 161.

4.2.1.6 ORGANOTIN PSEUDOHALIDES CONTAINING TIN-SULFUR AND TIN-NITROGEN BONDS

All compounds may be found at the bottom of Table 161.

Table 161. Diorganotin Pseudohalides Containing Tin-Heteroatom Bonds $R_2SnX'X''$

Formula	Synth. Method*	Yield	Properties	Ref.
$Me_2Sn(OC_9H_6N)NCS$ (a)	IIB	--	m. 124° (dec.) (b)	1291
$Et_2Sn(OC_9H_6N)NCS$ (a)	--	--	(b)	--
$(Et_2SnNCS)O(Et_2SnOMe)$	IVA	--	m. 140-41° (c)	2168
$Pr_2Sn(OC_9H_6N)NCS$ (a)	IIB	--	Yellow cryst., m. 134° (b)	1291
$Bu_2Sn(OMe)NCS$	IIB	--	m. 81-84°, hydrolyzes easily (d-f)	942
$Bu_2Sn(OC_9H_6N)NCS$ (a)	IIB	--	Dec. > 74°, soly, struct. (g)	2617
$Bu_2Sn(NCS)OCHOMeCCl_3$	IVB (f)	--	IR spectrum	943
$(Bu_2SnNCS)O(Bu_2SnOEt)$	--	--	(h)	--
$Ph_2Sn(OC_9H_6N)NCS$ (a)	ID	--	Yellow cryst., dec. > 76°, struct. (g)	2617
$Me_2Sn(N:PMe)N_3$	VA	63	m. 238-42° (dec.), IR spectrum	2743
$Me_2Sn(N:PEt_3)N_3$	VA	58	m. 218-24° (dec.), IR spectrum	2743
$Me_2Sn(N:PPh_3)N_3$	VA	63	m. 225-30° (dec.), IR spectrum	2743
$(Bu_2SnNCS)_2S$	IIA	--	m. 115-17°	945
$Bu_2Sn(NEtCO_2Me)NCS$	VB (e)	--	IR, NMR spectra	943

* The characters used here correspond to the synthetic scheme in the introduction to Chapter 4.2.1.5.
(a) HOC_9H_6N = 8-hydroxyquinoline. (b) IR spectrum (2103). (c) Use as catalyst for polyurethan synth. (2168). (d) IR and NMR spectra (942). (e) Rxn. of $Bu_2Sn(OMe)NCS$ with $EtN:CO \rightarrow Bu_2Sn(NEtCO_2Me)NCS$ (943). (f) Rxn. of $Bu_2Sn(OMe)NCS$ with $CCl_3CCH:O \rightarrow Bu_2Sn(NCS)OCHOMeCCl_3$ (943). (g) Moessbauer resonance spectrum, dielectrical const. (2617). (h) Use as catalyst for polyurethan foaming (2485).

4.2.4 ORGANOTIN SALTS OR ESTERS OF OXYGEN ACIDS

Organotin compounds with inorganic acids do not behave like true salts or esters. The term oxygen acid is understood to include such derivatives of the acid H_nMO_m where H-, HO- or O- groups are partially or totally substituted with organic groups, halogen, HS- S= or R_2N-groups.

4.2.4.1 TRIORGANOTIN SALTS OR ESTERS OF OXYGEN ACIDS

The organotin compounds summarized in Table 162 are prepared by the following methods.

I   Dealkylation reactions of R_4Sn with:
  (IA) H_2SO_4 ,
  (IB) SO_2 or PF_5 .

II Cleavage reaction of R_6Sn_2 with SO_2 .

III Metathetic substitution of R_3SnX with:
  (IIIA) MX' ,
  (IIIB) AgX' ,
  (IIIC) NH_4X' ,
  (IIID) $(R_2N)_2POCl$ and KOH ,
  (IIIE) HX' and Et_3N .

IV Substitution reaction of $(R_3Sn)_2O$ or R_3SnOH with:
  (IVA) HX' ,
  (IVB) $NaHSO_3$ or $NaHSO_4$,
  (IVC) Acid anhydrides $P_2O_3F_4$, SO_2 and $P_2S_5 + ROH$,
  (IVD) $R_2SnPOCl$.

V   Substitution reactions of triorganotin heterocompounds:
  (VA) R_3SnOR or R_3SnNR_2 with SO_2 ,
  (VB) Et_3SnNEt_2 with Ph_2POH ,
  (VC) Et_3SnH with HX' .

VI Transesterification reactions of triorganotin heterocompounds:
  (VIA) Bu_3SnOAc with Bu_3BO_3 ,
  (VIB) R_3SnCl with R_3PO_4 and $NaOH$,
  (VIC) Et_3SnBr with R_3PO_3 .

VII Oxidation reactions of:
  (VIIA) R_3SnYR_2 with air or sulfur,
  (VIIB) Et_3SnH with $PhN:NSO_2Ph$,
  (VIIC) $R_3SnOPPh_2$ with Et_3SnOH ,
  (VIID) Ph_2POH with $(R_3Sn)_2O$.

M = sodium or potassium, R = organic group, may be different in one molecule, X = halogen, X' = anion of oxygen acid HX' , Y = phosphorous or arsenic.

Trimethyltin Tetraphenyloborat $SnBC_{27}H_{29}$ [Me_3SnBPh_4]
Addn. compd.: $Me_3SnBPh_4 \cdot 2H_2O$, Me_3SnCl with $NaBPh_4$ in H_2O, m. 80-84° (2102),
  82° (938), IR (2102), NMR (938) and Moessbauer resonance (2347) spectra,
  stable in H_2O, dec. slowly in air (2102), soly., struct. (938) [$SnBC_{27}H_{33}O_2$].
$Me_3SnBPh_4 \cdot 2Me_2NCHO$, Me_3SnBr with $NaBPh_4$ in $DMF-H_2O$, m. 144°, IR and NMR
  spectra, soly. (938) [$SnBC_{33}H_{43}N_2O_2$].
$Me_3SnBPh_4 \cdot 2Me_2NCOMe$, Me_3SnBr with $NaBPh_4$ in $MeCONMe_2-H_2O$, white needles, m.
  80°, IR and NMR spectra, hygroscopic, soly., struct. (938) [$SnBC_{37}H_{47}N_2O_2$].

Me$_3$SnBPh$_4 \cdot$2Me$_2$SO, Me$_3$SnBr with NaBPh$_4$ and Me$_2$SO-H$_2$O, white cryst., m. 107°, IR and NMR spectra, soly., struct. (938) [SnBC$_{31}$H$_{41}$O$_2$S$_2$].

Trimethyltin Nitrate (M-348)   SnC$_3$H$_9$NO$_3$ Me$_3$SnNO$_3$
Synth.: By rxn. of Me$_3$SnOH with dil. HNO$_3$ at r.t. (620).
By vacuum subl. of Me$_3$SnNO$_3 \cdot$H$_2$O (86, 620).
Prop.: m. 140° (in sealed tube, dec.) (620), IR (86), assignment (620) and Moessbauer resonance (902) spectra, hygroscopic (86, 620).
Rxn. with: Moist air $\rightleftharpoons$ Me$_3$SnNO$_3 \cdot$H$_2$O (86, 620).
Addn. compd.: Me$_3$SnNO$_3 \cdot$H$_2$O, m. 147° (86), 98-99° (620), IR (86), assignment (620) and Raman (620) spectra [SnC$_3$H$_{11}$NO$_4$].

Bis(trimethyltin) Sulfate (M-348)   Sn$_2$C$_6$H$_{18}$O$_4$S   (Me$_3$Sn)$_2$SO$_4$
Synth.: By rxn. of Me$_3$SnBr with Ag$_2$SO$_4$ in dry MeOH (85).
By rxn. of Me$_4$Sn with 100% H$_2$SO$_4$ and ppt. with little H$_2$O (1100).
Prop.: White cryst., IR spectrum, hygroscopic (85), cryoscopic and conductiometric measurements in 100% H$_2$SO$_4$ (1100).
Addn. compd.: (Me$_3$Sn)$_2$SO$_4 \cdot$2MeOH, from components, hydrolyzes in moist air, soly. (85) [Sn$_2$C$_8$H$_{26}$O$_6$S].

Trimethyltin Perchlorate (M-348)   SnC$_3$H$_9$ClO$_4$ Me$_3$SnClO$_4$
Synth.: By rxn. of Me$_3$SnOH with aq. HClO$_4$ (2040).
Prop.: IR spectrum (86), anal. detn. by ion exchange and alkalimetric titration, dissocn. const. and hydration (2040).
Rxn. with: F$^- \rightarrow$ Me$_3^{113}$SnF + [Me$_3^{113}$SnF$_2$]$^-$, stability const. (836, 838).
Cl$^- \rightarrow$ [Me$_3$SnCl$_{n+1}$]$^{n-}$, stability const. (838, 840).
Addn. compd.: Me$_3$SnClO$_4 \cdot$2Py, from components, IR spectrum, struct. (86) [SnC$_{13}$H$_{19}$ClN$_2$O$_4$].

Bis(triethyltin) Sulfate (M-350)   Sn$_2$C$_{12}$H$_{30}$O$_4$S   (Et$_3$Sn)$_2$SO$_4$
Prop.: (Far) IR spectra and assignment (321).
Biol. prop.: Toxicity and symptoms (2441), effect on optical and peripheral nerve response (1809), on liver and brain mitrochondria in vitro (1663), on ^{32}P phosphate incorporation into brain phospholipids (1771), formation of cerebral edema and prevention by corticosteroids (553) by glycerol (1456).

Tris(tripropyltin) Borate   Sn$_3$BC$_{27}$H$_{63}$O$_3$ pPr$_3$Sn)$_3$BO$_3$
Synth.: By rxn. of (Pr$_3$Sn)$_2$O with B(OH)$_3$ (1426), in 75% yield (2126), with B$_2$O$_3$ in N atm. (2063).
Prop.: b$_{0.001}$ 174-78° (2063), b$_{0.001}$ 138-40° (2126) n$_D^{20}$ 1.4880 (2126), d$_{25}$ 1.2698 (1426).
Biol. prop.: Activity as bactericide against Staphylococcus aureus, Salmonella typhosa (2126).
Activity as fungicide against Aspergillus niger, Rhizoctonia solani, Phytium ultimum, Venturia inaequalis (2126).
Activity against Aedes mosquito (2063).
Use: As fungicide (1426), soil sterilant (2126), wood preservative (1426).

Tris(tributyltin) Borate $Sn_3BC_{36}H_{81}O_3$ $(Bu_3Sn)_3BO_3$

<u>Synth.:</u> By rxn. of $(Bu_3Sn)_2O$ with $B(OH)_3$, in 99% yield (1426), in N atm. as by-prod. (2063), in 75% yield (2126).
<u>Prop.:</u> $b_{0.001}$ 195-200° (2126), $b_{0.001}$ 192-96° (2063), n_D^{20} 1.4848 (2126), d_{25} 1.1844 (1426).
<u>Biol. prop.:</u> Activity as bactericide against Staphylococcus aureus (2126).
Activity as fungicide against Aspergillus niger (2126).
Activity as insecticide against Aedes mosquito (2063), Contrachelus nenuphar, Prodenia eridania (2126).
<u>Use:</u> As fungicide (1426), soil sterilant (2126), wood preservative (1426).

Tributyltin Borate $SnBC_{12}H_{27}O_2$ Bu_3SnBO_2

<u>Synth.:</u> By rxn. of $(Bu_3Sn)_2O$ with $B(OH)_3$, in 99% yield (1426).
<u>Prop.:</u> d_{20} 1.2266 (1426).
<u>Biol. prop.:</u> Toxicity (1099).
Relative reactivity as bactericide (1099).
Activity as fungicide against Ceratostomella coerulescens, Fusarium culmorum, Penicillium funiculosum, Speira heptaspora (1099).
<u>Use:</u> As fungicide and wood preservative (1426).

Bis(tributyltin) Sulfite $Sn_2C_{24}H_{54}O_3S$ $(Bu_3Sn)_2SO_3$

<u>Synth.:</u> By rxn. of SO_2 with $(Bu_3Sn)_2O$ (36, 750), with $(Bu_3SnNPh)_2SO$ or with $(Bu_3SnO)_2CHCCl_3$ in dry CCl_4 (2345), with $Bu_3SnNPhCO_2SnBu_3$ in pentane (2345).
By rxn. of $(Bu_3Sn)_2O$ with aq. $NaHSO_3$ in i-PrOH, in 82% yield (2146a).
By hydrolysis of $Bu_3SnOSO(OR)$; R = Me (750), Et (2345), of $(Bu_3SnNPh)_2SO$ (2345).
<u>Prop.:</u> IR spectrum (750, 2345).
<u>Rxn. with:</u> Thermal dissocn. at 0.1 torr. 120° → $(Bu_3Sn)_2O$ + SO_2 (750).
PhNCO → $Bu_3SnNPhCO_2SnBu_3$ + PhNSO + CO_2 (2345).

Ethyl Tributyltin Sulfite $SnC_{14}H_{32}O_3S$ $Bu_3SnOS(O)OEt$

<u>Synth.:</u> By rxn. of Bu_3SnOEt with dry SO_2 in CCl_4, in 97% yield (2345).
By rxn. of $Bu_3SnNPhCO_2Et$ with SO_2 in CCl_4 (2345).
<u>Prop.:</u> IR and NMR spectra, viscous oil, dec. on distn. (2345).
<u>Rxn. with:</u> Moist air → $(Bu_3Sn)_2SO_3$ (2345).
EtNCO → $Bu_3SnNEtCO_2Et$ + SO_2 (2345).
PhNCO → $Bu_3SnNPhCO_2Et$ + SO_2 + PhNSO (2345).
$Cl_3CCHO \rightleftharpoons Bu_3SnOCH(OEt)CCl_3$ + SO_2 (2345).

Tris(triphenyltin) Borate $Sn_3BC_{54}H_{45}O_3$ $(Ph_3Sn)_3BO_3$

<u>Synth.:</u> By rxn. of Ph_3SnOH with $B(OH)_3$ in C_6H_6, in 80% yield (2126).
<u>Prop.:</u> m. 131-32° (2126).
<u>Biol. prop.:</u> Activity as bactericide against Staphylococcus aureus (2126).
Activity as fungicide against Aspergillus niger, Venturia inaequalis (2126).
Activity as insecticide against yellow fever mosquito larvae (2126), Musca domestica, effect on reproduction (260), Prodenia eridania larvae (2126).
Phytotoxicity to foliage of woody plants (2126).

Triphenyltin Nitrate (M-355)　　　　　$SnC_{18}H_{15}NO_3$　　　　　Ph_3SnNO_3

<u>Synth.</u>: By rxn. of Ph_3SnCl with $AgNO_3$ in MeCN, in 81% yield (1030).
By rxn. of $(Ph_3Sn)_2O$ with aq. HNO_3 (1969).
By rxn. of Ph_6Sn_2 with $AgNO_3$ in Me_2CO at 0°, in 90% yield (568).
<u>Prop.</u>: m. 184-86° (1030), 181-82° (568), 170-70.5° (1934), IR (85, 568, 1030, 1934) and UV (1934) spectra, stable in dry form (85, 586), dec. at 186° (1030).
<u>Rxn. with:</u> Thermal dec. → $PhNO_2$ + Ph_4Sn + $(Ph_2SnO)_x$ (568).
$o-Cl_2C_6H_4$ at reflux → Ph_4Sn + $Ph_2Sn(NO_3)_2$ (?) (568), + Ph_3SnCl + $(Ph_2SnO)_x$ (2388).
Air → $(Ph_2SnNO_3)_2O$ (1030).
NaF → Ph_3SnF (1969).
trans-$H(Ph_3P)_2PtCl$ → $Ph_3Sn(Ph_3P)_2PtCl$ (700).
Ph_3CH at 150° → Ph_3COH + Ph_3SnOH + $(Ph_3Sn)_2O$ + O + N_xO_y (568).
Ph_2NH at 185° → Ph_3SnOH + C_6H_6 + N_2O_4 (568).
Anhydr. NH_3 → $(Ph_3Sn)_2O$ + NH_4NO_3 sic. (85).
O in $HCCl_3$, anhydrous → $(Ph_2SnNO_3)_2O$ + Ph_2 (1300).
<u>Addn. compd.</u>: $Ph_3SnNO_3 \cdot 2Me_2SO$, from components at 100°, IR and UV spectra, dec. > 110° (1934) [$SnC_{22}H_{27}NO_5S_2$].
<u>Biol. prop.</u>: Activity as bactericide (529), against Bacillus mycoides (2796), B. subtilis, Mycobacterium phelei (1969), Staphylococcus aureus, Streptococcus lactis (2796).
Activity as fungicide against Aspergillus niger (1969), Candida albicans, Cryptococcus neoformans, Glomerella cingulata (2796), Rhizopus nigricans (1969), Trichophyton mentagrophytes (2796).

Triphenyltin Diphenylphosphinate (M-355)　　$SnC_{30}H_{25}O_2P$　　　$Ph_3SnO_2PPh_2$

<u>Synth.</u>: By rxn. of $(Ph_3Sn)_2O$ with Ph_2PO_2H in C_6H_6, in 98% yield (292), with Ph_2PHO in C_6H_6, in 98% yield (1233).
By air oxidation of $Ph_3SnOPPh_2$ in C_6H_6 (1840).
By rxn. of Ph_3SnPPh_2 with $PhCH:N(I)Ph$ in C_6H_6 or other oxidizing agents in inert atm. (1840).
<u>Prop.</u>: m. 360° (1840), > 250° (292), 250° (1233), (far) IR spectrum and assignment (1840).
<u>Biol. prop.</u>: Activity as fungicide and insecticide, phytotoxicity (292).

Triphenyltin Diphenylphosphenate　　$SnC_{30}H_{25}OP$　　　$Ph_3SnOPPh_2$

<u>Synth.</u>: By rxn. of Ph_3SnOLi with Ph_2PCl in C_6H_6 in N atm., in 80% yield (1840).
By rxn. of Ph_3SnLi with Ph_2POCl in THF, in N atm., in 98% yield (1840).
<u>Prop.</u>: m. 120° (dec.), (far) IR spectrum and assignment, mol. wt., sensitive to air, soly. (1840).
<u>Rxn. with:</u> Air in C_6H_6 → $Ph_3SnO_2PPh_2$ (1840).

Triorthotolyltin Nitrate　　$SnC_{21}H_{21}NO_3$　　　$(o-MeC_6H_4)_3SnNO_3$

<u>Synth.</u>: By rxn. of $(o-MeC_6H_4)_3SnI$ with $AgNO_3$ in C_6H_6 at reflux (1964).
By rxn. of $(o-MeC_6H_4)_6Sn_2O$ with HNO_3 in CCl_4 or $HCCl_3$ (1964).
<u>Prop.</u>: m. 110°, IR spectrum and assignment (1964).

Biol. prop.: Activity as bactericide against Bacillus subtilis, Escherichia coli, Salmonella typhi, Staphylococcus aureus (1964).
Activity as fungicide against Aspergillus niger, Candida albicans, Trichophyton mentagrophytes (1964).

Triparatolyltin Nitrate $SnC_{21}H_{21}NO_3$ $(p-MeC_6H_4)_3SnNO_3$
Synth.: By rxn. of $(p-MeC_6H_4)_3SnI$ with $AgNO_3$ in C_6H_6 at r.t. (1964).
By rxn. of $(p-MeC_6H_4)_6Sn_2O$ with HNO_3 in CCl_4 or $HCCl_3$ at r.t. (1964).
Prop.: m. 208°, IR spectrum, assignment (1964).
Rxn. with: Thermal dec. in $o-Cl_2C_6H_4$ → $(p-MeC_6H_4)_4Sn$ + $(p-MeC_6H_4)_3SnCl$ + $[(p-MeC_6H_4)_2SnO]_2$ + other prod. (2388).
Biol. prop.: See the preceeding o-derivative.

Additional triorganotin esters and salts may be found in Table 162.

Table 162. Triorganotin Salts or Esters of Oxygen Acids $(R_3Sn)_nX'$

Formula*	Synth. Method**	Yield	Properties	Ref.
Me_3SnBF_4 (269)	--	--	From $Me_3SnC_6F_5$ and BF_3	77
$(Me_3Sn)_2PO_4$	--	--	NMR spectrum	1220
Me_3SnPF_6 (?)	IB	--	White solid, dec. in air, IR spectrum (a, b)	567
$Me_3SnPF_6 \cdot 2NH_3$	(b)	--	White solid, soly.	567
$Me_3SnO_2AsPh_2$	IVA	59	m. 207-208°, IR spectrum, tetramer	68
Me_3SnSO_4H	IA	--	Conductivity, acidity in H_2SO_4	1100
Me_3SnO_3SMe	IB	72	m. 143-44°, subl.o.1 170°, mol. wt. (c)	2742
Me_3SnO_3SPh	IB	87	m. 130°, soly. (c)	2742
$Me_3SnO_2SNMe_2$	VA	99	Dec. on distn., IR spectrum	167
$Me_3Sn \cdot SO_2$	II	--	White solid, dec. > 110°, soly. struct. (d)	827
$(Me_3Sn)_2CrO_4$	IIIB	--	Yellow cryst., stable in air, dec. H_2O (e)	85
$(PhCH_2)_3SnNO_3$	IIIB	--	m. 161°, IR spectrum, assignment, for biol. prop. see the o-tolyl-derivation page 579.	1964
	IVA	--		1964
$Et_3SnS_2P(OEt)_2$	IIIC	75	bo.1 115-18°	292
	IVC	--	bo.1 117° (f, g)	1576
$Et_3Sn(Et)O_3PMe$	VC	62	bo.001 215°, n_D^{20} 1.4660, d_{20} 1.3534	2761
$Et_3Sn(i-Pr)O_3PMe$	VC	65	bo.002 216°, n_D^{20} 1.4530, d_{20} 1.3039	2761
$Et_3Sn(i-Bu)O_2(s)PMe$	VC	82	bo.003 115°, d_{20} 1.2255	2761
$Et_3Sn(Et)O_3PEt$ (349)	VIC	--	--	441
$Et_3SnS_2PEt_2$	IIIA	--	Colorless liq., b_2 190°, monomeric, soly. (h)	760
$Et_3SnO_2PPh_2$ (350)	VIIC, D	99	m. 246-47°	1233
$Et_3Sn(O)PPh_2$	VB	--	m. 54-57°, bo.01 145-47°, soly. (i)	1233
$Et_3SnO_2AsPh_2$	IVA	74	m. 170° (slow heating at 220-23°) dimeric,	68
	VIIA	54	m. 166-67°, IR spectrum	68
Et_3SnO_2SPh (350)	VC	--	IR spectrum	2646
	VIIB	--	m. 67°	1611, 2646
$(Pr_3Sn)_4B_2O_5$	IVA	99	d_{20} 1.2824	1426
	IVA	--	bo.001 184-92° (j-1)	2063

Compound	Group	Yield (%)	Properties	Ref.
Pr_3SnBO_2	IVA	--	d_{20} 1.3030 (1)	1426
$(Pr_3Sn)_3PO_4$	--	--	Use as stabilizer for polyolefins	1864
$Pr_3SnS_2P(OEt)_2$	IIIA	90	n_D^{20} 1.5238, d_{20} 1.2279	745
$Pr_3SnO_2PPh_2$ (352)	IVA	83	m. 209° (f)	292
$Pr_3SnO_2AsPh_2$	IVA	59	m. 204-205°, dimeric in C_6H_8	68
	VIIA	51	--	68
$(Pr_3Sn)_2SO_4$	IVB	89	m. 183-86°	2146a
$Pr_3SnO_3SNH_2$	IVA	--	Waxy solid, soly. (m, n)	1578, 1629
$(Pr_3Sn)_2SO_3$	IVB	89	Viscous oil	2146a
$(i-Pr_3Sn)_2SO_4$	IVB	100	Cryst. solid	2146a
$(i-Pr_3Sn)_2SO_3$	IVB	96	White powder	2146a
$(Bu_3Sn)_4B_2O_5$	IVA	99	d_{20} 1.1923 (k, l)	1426
	IVA	--	$b_{0.001}$ 230-35°	2063
$(Bu_3Sn)_2BuBO_3$	VIA	--	$b_{2.5}$ 128-29° (o)	2620
$Bu_3Sn(Bu)_2BO_3$	VIA	--	m. 66-69° (o)	2620
$Bu_3Sn(i-Bu)_2BO_3$	VIA	--	m. 75-78° (o)	2620
Bu_3SnBF_4	--	--	See Bu_3SnF, Chapter 4.2.4.1	242, 595
$(Bu_3Sn)_2SiF_6$	--	--	Use: fungicide, for mold control	242
$(Bu_3Sn)_3PO_4$ (351)	--	--	Use: light stabilizer for $(MeCHCH_2)_x$	1234
$Bu_3SnO_2PF_2$	IVC	--	Dec. on distn., mol. wt. (p)	1767
$(Bu_3Sn)_2MeO_3P:S$	IVC	--	n_D^{15} 1.5056 (g)	1576

* Numbers in parenthesis refer to pages in main volume.
** The characters used here correspond to the synthetic scheme in the introduction to Chapter 4.2.4.1.
(a) Rxn. with $H_2O \rightarrow Me_3SnF$ (567). (b) Rxn. of Me_3SnPF_6 with dry $NH_3 \rightarrow Me_3SnF + Me_3SnPF_6 \cdot 2NH_3$ (567). (c) IR and NMR spectra (2742). (d) (Far) IR and NMR spectra (827). (e) IR spectrum, soly. (85). (f) Activity as fungicide and insecticide, phytotoxicity (292). (g) Activity as bactericide, insecticide, germination inhibitor for plant seed (1576). (h) Stable in air, H_2O, conductivity in org. solvent (760). (i) Rxn. with $Et_3SnOH \rightarrow R_3SnO_2PPh_2 + Et_3SnH$; R = Et, Bu (1233). (j) Activity as insecticide against Aedes mosquito (2063). (k) Activity as bactericide, fungicide and herbicide (2063). (l) Use as fungicide, wood preservative (1426). (m) Reacts antiseptic, activity as bactericide (1578). (n) Use as fungicide, fungistat, wood preservative (1629). (o) Use as bactericide, as additive for polymers (2620). (p) IR and ^{19}F NMR spectra (1767).

Table 162 Continued

Formula*	Synth. Method**	Yield	Properties	$(R_3Sn)_nX'$ Ref.
$Bu_3Sn(Et_2)PO_4$ (351)	VIB	--	Liq. (f)	292
$Bu_3SnSPO_3Et_2$	IIIC	92	n_D^{20} 1.4988, d_{20} 1.2197	745
$Bu_3SnS_2P(OR)_2$				
R = Me	IIIA	--	n_D^{21} 1.4965, dec. on distn. (g)	1565
	IVC	--	Pale yellow oil, n_D^{15} 1.5298	1576
Et (351)	IIIC	86	$b_{0.0001}$ 132-36° (f)	292
	IIIA, E	96	$b_{0.03}$ 115-16°, n_D^{20} 1.5183, d_{20} 1.2005	745
Pr	IIIA	93	n_D^{20} 1.5105, d_{20} 1.1924	745
i-Pr	IIIA	91	$b_{0.02}$ 110-12°, n_D^{20} 1.5094, d_{20} 1.1813	745
EtBuCHCH$_2$	IVC	--	n_D^{15} 1.5018 (g)	1576
$(RO)_2 = OCH_2CH_2S$	--	--	(q)	--
$(Bu_3Sn)_2O_3PCH_2OH$	IVA	--	n_D^{20} 1.5122 (r)	538
$(Bu_3Sn)_2O_3PCCl_3$	IVA	--	n_D^{30} 1.5175 (r)	538
$(Bu_3Sn)_2O_3PPh$	IVA	--	n_D^{20} 1.5292 (r)	538
$Bu_3SnO_2PR_2$				
R = CH_2OH	IVA	--	m. 60.5° (r)	538
Bu	IVA	74	$b_{0.01}$ 200-231°, very viscous oil, IR spectrum, tetrameric	1233
	VIID	--		1233
C_6H_{13}	IVA	--	n_D^{20} 1.4901 (r)	538
$C_7H_{14}OH$	IVA	--	n_D^{22} 1.4938 (r)	538
Ph (352)	IVA	85	m. 218° (f)	292
	VIIC, D	95	m. 218°, IR spectrum	1233
Me_2N	IIID	36	m. 158° (f)	292
Et_2N	IIID	50	m. 151-3° (f)	292
$(Bu_3Sn)_2HPO_3$	--	--	Use as stabilizer for PVC copolymers	1866

Compound	Group	Yield	Properties	Ref
Bu$_3$SnOPBu$_2$	IIIA	77	b$_1$ 165-67°, NMR spectrum	1232
Bu$_3$SnOPPh$_2$	VB	--	b$_{0.01}$ 160-62° (i)	1233
Bu$_3$SnO$_2$AsPh$_2$	IVA	54	m. 196-98°, dimeric in C$_6$H$_6$	68
	VIIA	41	IR spectrum (s)	68
(Bu$_3$Sn)$_2$SO$_4$	IVB	89	White cryst., m. 148-49.5°	2146a
Bu$_3$SnO$_3$SMe	--	--	(s)	--
Bu$_3$SnO$_3$SC$_6$H$_4$Me	IVA	--	m. 85-89°(q, t)	572
Bu$_3$SnO$_3$SC$_6$H$_4$NH$_2$-p	IVA	--	m. 161-62° (u)	2966
5,2-Bu$_3$SnO$_3$S(HO)C$_6$H$_4$CO$_2$SnBu$_3$	IVA	--	m. 149-53° (u) (M-352)*	2966
(Bu$_3$Sn)$_2$O$_3$S)$_2$NH	IVA (v)	--	Waxy, m. 115° (m)	1629
Bu$_3$SnO$_3$SNH$_2$	IVA (v)	--	Yellow oil, soly.	1578
	IVA	--	(m, n)	1629
Bu$_3$SnO$_3$SNHCONH$_2$	IVA	--	Waxy solid (m)	1578
Bu$_3$SnO$_2$SOMe (352)	VA (w)	--	Viscous oil, b$_{0.08}$ 125° (dec.) (x)	36,750
Bu$_3$SnO$_2$SOPh	VA	--	--	750
(Bu$_3$Sn)$_2$CrO$_4$		--	Use in antifouling coatings	2302
Bu$_2$PhSnO$_3$SNH$_2$		--	Soly., use as antiseptic for wood	1577
1-Bu$_3$SnO$_3$SMe (354)		--	(s)	--
(PhCMe$_2$CH$_2$)$_3$SnNO$_3$	IVA	--	m. 122-23°, mol. wt., soly. (y)	1742
(PhCMe$_2$CH$_2$)$_3$SnClO$_4$	IVA	--	m. 162-63°, mol. wt., soly. (y)	1742
[(c.C$_6$H$_{11}$)$_3$Sn]$_2$SO$_4$	IVB	80	Cryst., m. 154-57.5°	2146a
[(c.C$_6$H$_{11}$)$_3$Sn]$_2$SO$_3$	IVB	90	White solid	2146a
(Ph$_3$Sn)$_2$SiO$_3$	IIIA	80	m. 270° (z)	2826
(Ph$_3$Sn)$_2$Si$_2$O$_5$	IIIA	--	(z)	2826

(q) Activity as light resistant fungicide against Coriolellus palustris (1063). (r) Use as bactericide, bacteriostat, fungicide and fungistat (538). (s) Activity as bactericide (1099), as fungicide against Ceratostomella coerulescens, Fusarium culmorum, Penicillium funiculosum, Speira heptaspora (1099). (t) Use as antifouling agent (572). (u) Use as bacteriostatic agent for textiles, activity as bactericide against Staphylococcus aureus, as fungicide (2966). (v) Synth. from (R$_3$Sn)$_2$O and H$_2$NSO$_3$H (1578). (w) Synth. also from Bu$_3$SnOMe and MeNSO (2345). (x) IR spectrum (750), rxn. with air → (Bu$_3$Sn)$_2$SO$_3$ (750). (y) IR (1742), NMR (1742) and Moessbauer resonance (224, 1742) spectra. (z) Relative reactivity as fungicide, phytotoxicity (2826).

Table 162 Continued

Formula*	Synth. Method**	Yield	Properties	$(R_3Sn)_3X'$ Ref.
$(Ph_3Sn)_2SiF_6$	--	--	Use as fungicide, for mold control	242
$(Ph_3Sn)_2N_2O_2$	IIIC	50	m. 128° (dec.), monomeric, soly. (aa)	716
$(Ph_3Sn)_3PO_4$ (355)	--	--	Use as bactericide (bb-1863)	529
$Ph_3Sn(Et_2)PO_4$	VIB	82	m. 193° (f)	292
$Ph_3Sn(Ph_2)PO_4$	IIIC	41	m. 170° (f)	292
$Ph_3SnS_2P(OR)_2$				
R = Me	IVA	81	m. 86°	292
	IIIC	50	(f, g)	292
	IVA	--	m. 83°	1576
Et	IVA	79	m. 105°	292
	IIIC	74	(f)	292
	IIIA	87	m. 97-98°	745
	IIIA	--	m. 124-25°, soly.	1565
Pr	IIIC	76	m. 63° (f)	292
	IIIA	76	m. 54-56°	745
i-Pr	IIIA	74	m. 47-49°	745
Bu	IVA	69	m. 69.5° (f)	292
$Ph_3SnS_2PEt_2$	IIIA	--	White, m. 88-89°, mol. wt. (h)	760
$Ph_3SnS(O)PPh_2$	IIIA	--	(bb-2108)	2108
$Ph_3SnS_2PPh_2$	VIIA	27	m. 128-30°, (far) IR spectra, assignment	1840
	IIIA	--	(bb-2108)	2108
Ph_3SnO_2PNRR'				
R = H, R' = Et	IIID	75	m. 218-21° (f)	292
H Pr	IIID	--	m. 153° (f)	292
H Ph	IVD	73	m. 187° (f)	292

Me	Me	IIID	60	m. > 250° (f)	292
Et	Et	IIID	58	m. 250° (f)	292
Ph₃SnH₂PO₂		--	--	Use: stabilizer for PVC-copolymers	1866
Ph₃SnO₂AsPh₂		IVA	90	m. 323-24°, IR spectrum, mol. wt.	68
		IIIE	98	m. 323°, soly. (s)	1842
		VIIA	80	Synth. with H₂O₂	1842
(Ph₃Sn)₂SO₄ (355)		--	--	Use as bactericide	529
Ph₃SnO₃SMe (356)		--	--	(s)	--
Ph₃SnO₃SNH₂		IVA	--	Soly. use as antiseptic for wood	1577
(Ph₃Sn)₂SO₃		IVC (cc)	100	White cryst., m. 82-85°, IR spectrum	2345
		IVB	94	m. 150-54°	2146a
Ph₃SnO₂SPh (356)		IIIA	--	m. 240° (dd)	1410
Ph₃SnO₂SC₆H₄Me-p		IIIA	--	m. 225° (dec.) (dd)	1410
Ph₃SnO₂SC₆H₄Br-p		IIIA	--	m. 245° (dec.)	1410
Ph₃SnO₂SC₆H₄NHAc-p		IIIA	--	m. 175° (dec.)	1410
Ph₃Sn·SO₂		II	--	White cryst., m. 153°, IR spectrum (ee, ff)	827
Ph₃Sn·2SO₂		II (ff)	--	White, dec. 201°, soly., struct.	827
Ph₃SnHCrO₄		IVA	--	Stability const.	487
Ph₃SnClO₄		IIIB	--	IR spectrum, soly., very hygroscopic (gg)	85
[(o-MeC₆H₄)₃Sn]₂SO₄		--	--	NMR spectrum	2356
R₃SnNO₂		IIIA	--	R = alkyl, aryl	2487

(aa) Dec. in high vacuum at 80-90° → (Ph₃Sn)₂O + N₂O, (far) IR spectra and assigment (716). (bb) Use as stabilizer for polyolefins (1863 or 2108). (cc) Synth. with SO₂ or PhNSO. (dd) Thermal dec. at 0.2 torr - 235-55° → Ph₄Sn + RSnPh₃ ; R = Ph, p-MeC₆H₅ (1410). (ee) Soly., struct., dimeric in C₆H₆ (827). (ff) Rxn. of Ph₃Sn·SO₂ with SO₂ → Ph₃Sn·2SO₂ (827). (gg) Rxn. with dry NH₃ in MeOH → (Ph₃Sn)₂O + NH₄ClO₄ (85).

4.2.4.2 DIORGANOTIN SALTS OR ESTERS OF OXYGEN ACIDS

The diorganotin derivatives of oxygen acids and their derivatives summarized in Table 163 are prepared by the following methods.

- I Dealkylation reactions of:
    - (IA) Me_4Sn by SO_3,
    - (IB) Me_3SnCl by SO_3,
    - (IC) Et_3SnLi by $PhSO_2Cl$.
- II Metathetic substitution reaction of R_2SnX_2 with:
    - (IIA) NaX',
    - (IIB) *AgX',
    - (IIC) NH_4X',
    - (IID) *HX'.
- III Substitution reaction of $(R_2SnO)_x$ with:
    - (IIIA) HX',
    - (IIIB) *P_2S_5 + ROH.
- IV Miscellaneous reactions:
    - (IVA) R_2SnCl_2 with P_2O_5 and ROH,
    - (IVB) $R_2Sn(AsR_2)_2$* and H_2O_2,
    - (IVC) R_2SnX_2 and ligand.

* The corresponding monoorganotin derivatives are to be used for synthesis of monoorganotin salts. R = organic group, X = halogen, X' = anion of oxygen acid HX'

Dimethyltin Bistetraphenyloborat $SnB_2C_{50}H_{46}$ $Me_2Sn(BPh_4)_2$
Addn. compd.: $Me_2Sn(BPh_4)_2 \cdot 4Me_2NCHO$, Me_2SnCl_2 with $NaBPh_4$ in $DMF-H_2O$, white cryst., m. 116-18°, IR and NMR spectra, soly., struct. (938) $[SnB_2C_{62}H_{74}N_4O_4]$.
$Me_2Sn(BPh_4)_2 \cdot 4Me_2NCOMe$, Me_2SnCl_2 with $NaBPh_4$ and $MeCONMe_2-H_2O$, m. 80-85°, IR and NMR spectra, hygroscopic, soly., struct. (938) $[SnB_2C_{66}H_{82}N_4O_4]$.
$Me_2Sn(BPh_4)_2 \cdot 4Me_2SO$, Me_2SnCl_2 with $NaBPh_4$ and Me_2SO-H_2O, white cryst., m. 92-93°, IR and NMR spectra, soly., stable in vacuo, struct. (938) $[SnB_2C_{58}H_{70}O_4S_4]$.

Dimethyltin Diperchlorate (M-349) $SnC_2H_6Cl_2O_8$ $Me_2Sn(ClO_4)_2$
Synth.: By rxn. of Me_2SnCl_2 with $AgClO_4$ (70).
By rxn. of $(Me_2SnO)_x$ with aq. $HClO_4$ (2417).
Prop.: (Far) IR, Raman (2596) and NMR (347, 1024) spectra, hydrolysis const. (1695), formation and stability of $(Me_2SnOH)_2^{2+}$ in dependance on hydrogen ion concn. (2792). Anal. detn. with alizarin S (70), by amperometric titration (1695), by ion exchange and alkalimetric titration, dissocn. const. and hydration (2040).
Rxn. with: $LiX \rightleftharpoons$ evidence of complex formation at pH 1; X = Cl, Br, I (69), $[Me_2SnX_{2+n}]^{-n}$; X = Cl, Br (69).
$Cl^- \rightarrow [Me_2SnCl_{2+n}]^{-n}$, stability const. (838, 840).
$H_2O \rightarrow$ hydration and exchange by ^{17}O NMR spectroscopy (2417).
NaSCN → complex formation and stability const. (2699).

1-(2-Pyridylazo)-2-naphthol → spectrophotometrical evidence for 1:1 chelate (1695).

Diethyltin Diperchlorate $SnC_4H_{10}Cl_2O_8$ $Et_2Sn(ClO_4)_2$
Synth.: By rxn. of $(Et_2SnO)_2$ with aq. $HClO_4$ (2040).
Prop.: Hydrolysis const. (1695), dissocn. const. and hydration (1695). Anal. detn. by amperometric titration (1695), by ion exchange and alkalimetric titration (2040).
Rxn. with: 1-(2-Pyridylazo)-2-naphthol → spectrophotometric evidence for 1:1 chelate (1695).

Dibutyltin Dinitrate $SnC_8H_{18}N_2O_6$ $Bu_2Sn(NO_3)_2$
Synth.: By rxn. of Bu_2SnCl_2 with $AgNO_3$ in MeOH; in 66% yield (1120).
By vacuum dehydration of dihydrate below (2158).
Prop.: Deliquescent cryst., m. 103-104.5° (1120), m. 83-87° (2158), dec. slowly at r.t., stable at -18°, soly. (1120).
Addn. compd.: $Bu_2Sn(NO_3)_2 \cdot 2H_2O$, $(Bu_2SnO)_x$ with HNO_3 in Me_2CO, cryst. m. 83-87°, X-ray diffraction pattern, H_2O detn. by K. Fischer titration, dehydrates in vacuo to $Bu_2Sn(NO_3)_2$ (2158) [$SnC_8H_{22}N_2O_8$].
$Bu_2Sn(NO_3)_2 \cdot 1,10$-phenanthroline, from components in EtOH, in 96% yield, white cryst., m. 209-12° (dec.) (1120) [$SnC_{20}H_{26}N_4O_6$].

Diphenyltin Dinitrate $SnC_{12}H_{10}N_2O_6$ $Ph_2Sn(NO_3)_2$
Synth.: By rxn. of Ph_2SnX_2 with $AgNO_3$ in inert atm.; X = Cl in MeCN, I in Me_2CO, in 91% yield (1030).
By oxidation of Ph_3SnNO_3 in $HCCl_3$, in 20-94% yield (1030).
By-prod. in thermal dec. of Ph_3SnNO_3 in $o\text{-}Cl_2C_6H_4$ under reflux (568).
Prop.: m. 195-97° (dec.), IR spectrum, mol. wt., soly. (1030).
Rxn. with: H_2O → $(Ph_2SnNO_3)_2O$ (1030).
Addn. compd.: $Ph_2Sn(NO_3)_2 \cdot 1,10$-phenanthroline, from components in alc. Me_2CO, in 95% yield, m. 276-78° (dec.), IR spectrum (1030) [$SnC_{24}H_{18}N_4O_6$].

Diphenyltin Diperchlorate $SnC_{12}H_{10}Cl_2O_8$ $Ph_2Sn(ClO_4)_2$
Synth.: By rxn. of Ph_2SnCl_2 with $AgClO_4$ (70).
Prop.: Hydrolysis const. (1695). Anal. detn. with alizarin S (70), by amperometric titration (1695).
Rxn. with: 1-(2-Pyridylazo)-2-naphthol → spectrophotometric evidence for 1:1 chelate, stability const. (1095).

Listing of salts or esters of the diorganotin moiety concludes in Table 163.

4.2.4.3 MONOORGANOTIN SALTS OR ESTERS OF OXYGEN ACIDS

Methyltin Triperchlorate $SnCH_3Cl_3O_{12}$ $MeSn(ClO_4)_3$
Synth.: By rxn. of $MeSnCl_3$ with $AgClO_4$ (70).
Prop.: Anal. detn. with alizarin S (70), by potentiometry in presence of $HClO_4$ and NaCl or NaBr (2554).

<u>Rxn. with:</u> LiX ⇌ evidence for complex formation; X = Cl, BrI (69), $[MeSnX_{3+n}]^{n-}$; X = Cl, Br (69).
Cl$^-$ ⇌ $[MeSnCl_{3+n}]^{n-}$, stability const. (840).
NaSCN → complex formation and stability const. (2699).

Phenyltin Triperchlorate $SnC_6H_5Cl_3O_{12}$ $PhSn(ClO_4)_3$
<u>Synth.:</u> By rxn. of PhSnCl$_3$ with AgClO$_4$ (70).
<u>Prop.:</u> Anal. detn. with alizarin S (70).
<u>Rxn. with:</u> LiX ⇌ evidence for complex formation; X = Cl, Br, I (69),
$[PhSnX_{3+n}]^{-n}$; X = Cl, Br (69).

Additional monoorganotin salts or esters are listed at the bottom of Table 163.

4.2.4.6 DIORGANOTIN SALTS OR ESTERS CONTAINING TIN-HETEROATOM BONDS

Diorganotin salts or esters of oxygen acids containing tin-heteroatom bonds are listed in Table 164. The heterofunctions linked to tin comprise oxy, hydroxy, deuterooxy, alkoxy and thioalkoxy groups. Derivatives of diorganotin salts and esters containing a tin-halogen bond are listed in Chapter 4.1.4. The compounds are prepared by methods from the following scheme.

I Substitution reactions of AgX' with:
  (IA) Ph$_2$SnI$_2$,
  (IB) Ph$_2$Sn(OAr)Cl ,
  (IC) of Na$_2$X' with R$_2$SnX$_2$.

II Substitution reactions of (R$_2$SnO)$_x$ with:
  (IIA) HX' ,
  (IIB) RSH and B(OH)$_3$.

III Substitution reactions of Bu$_2$Sn(OMe)$_2$ or Ph$_2$Sn(OAr)$_2$ with:
  (IIIA) SO$_2$,
  (IIIB) Bu$_2$SnX'$_2$,
  (IIIC) AgX' .

R = organic group, ArOH = 8-hydroxyquinoline, X' = anion of oxygen acid HX' or H$_2$X' .

Table 163. Organotin Salts or Esters of Oxygen Acids $(R_2Sn)_nX'$

Formula*	Synth. Method**	Yield	Properties	Ref.
$Me_2Sn(NO_3)_2$ (348)	IIB	92	NMR spectrum (347, 1024), stable at -18°	1120
$Me_2Sn(S_2PPh_2)_2$	IIA	--	White, m. 161-62°, NMR spectrum, soly. (b, c)	760, 2108
Me_2SnSO_4 (349)	IIB	--	IR spectrum, assignment, struct. soly.	873
$Me_2SnSO_4 \cdot AgCl \cdot MeOH$	IIB	--	Synth. in dry MeOH, IR spectrum	873
$Me_2SnSO_4 \cdot Py$	IVC	--	IR spectrum	873
$Me_2SnSO_4 \cdot Me_2SO$	IVC	--	IR spectrum, assignment	873
$Me_2Sn(O_3SMe)_2$ (349)	IA	67	Dec. 200°, IR spectrum	2742
	IB	90	--	2742
Me_2SnCrO_4	IIB	--	Dec. in aq. soln., IR spectrum (d)	85
Me_2SnMoO_4 (349)	--	--	(d)	--
Me_2SnWO_4 (349)	--	--	(d)	--
$(PhCH_2)_2Sn(i-PrHPO_4)_2$	IVA	--	--	2711
$Et_2Sn(S_2PPh_2)_2$	IIA	--	White, m. 162-63° (b)	760
$Et_2Sn(O_2SPh)_2$	IC	50	Colorless cryst., m. 264°, IR spectrum	2646
$Pr_2Sn(NO_3)_2$	IIB	63	m. 137-38°, stable at -18°, soly. (a)	1120
$Pr_2Sn(NO_3)_2 \cdot C_{12}H_8N_2$ (e)	IVC	94	White cryst. m. 205-206° (dec.)	1120
$Pr_2Sn(ClO_4)_2$	IIIA	--	Dissocn. const. hydration (f)	2040
$(\overline{CH_2CH_2CH})_2Sn(NO_3)_2$	IIA	--	Activity as herbicide, fungicide	1873
$Bu_2Sn(EtBuCHCH_2)_2(HPO_4)_2$	IID	--	Use as lubrication additive	1183
$S-Bu_2Sn(Et_2PO_3S)_2$	IIC	84	m. 197-98°	2529

* Numbers in parenthesis refer to pages in main volume.
** The characters used here correspond to the synthetic scheme in the introduction to Chapter 4.2.4.2.
(a) Deliquescent, dec. slowly at r.t. (1120), Moessbauer resonance spectrum (1143a). (b) Monomeric in $HCCl_3$, C_6H_6, stable in air and H_2O, no conductivity in org. solvents (760). (c) Use as stabilizer for polyolefins (2108). (d) NMR and Moessbauer resonance (2848) spectra. (e) $C_{12}H_8N_2$ = 1,10-phenanthroline. (f) Anal. detn. by ion exchange and alkalimetric titration (2040).

Table 163 Continued

Formula*	Synth. Method**	Yield	Properties	$(R_2Sn)_nX'$ Ref.
S-Bu$_2$Sn(Bu$_2$PO$_3$S)$_2$	IIC	--	m. 183-84°, supercooled oil, n_D^{20} 1.4961	2529
S-Bu$_2$Sn[S$_2$P(OEt)$_2$]$_2$	IIA	--	Oil, n_D^{20} 1.5553, d_{20} 1.3310	2529
S-Bu$_2$Sn[S$_2$P(OCH$_2$CHEtBu)$_2$]$_2$	IIIB	--	n_D^{15} 1.5120 (g)	1576
S-Bu$_2$Sn[S$_2$P(OC$_{10}$H$_{21}$)$_2$]$_2$	--	--	Use as lubrication additive for drilling mud	2858
Bu$_2$Sn(S$_2$PEt$_2$)$_2$	IIA	96	Colorless cryst., m. 40.5°, stable to 180° (h)	1363
Bu$_2$Sn(O$_2$PPh$_2$)$_2$ (353)	IIA	--	(c)	2108
S-Bu$_2$Sn(OSPPh$_2$)$_2$	IIA	--	(c)	2108
Bu$_2$Sn(S$_2$PPh$_2$)$_2$	IIA	--	m. 138-40° (c)	2108
Bu$_2$SnHPO$_3$ (353)	--	--	Stabilizer for PVC-copolymers (i)	1866
Bu$_2$SnSn(O$_3$SC$_{10}$H$_{15}$O)$_2$	--	--	Camphorsulfonate (j)	942
Bu$_2$Sn(ClO$_4$)$_2$	--	--	(k)	--
S-(C$_8$H$_{17}$)$_2$Sn(SOPPh$_2$)$_2$	IIA	--	(c)	--
(C$_8$H$_{17}$)$_2$Sn(S$_2$PPh$_2$)$_2$	IIA	--	(c)	--
(C$_8$H$_{17}$)$_2$SnHPO$_3$	--	--	Stabilizer for PVC-copolymers	1866
(C$_{12}$H$_{25}$)$_2$Sn(O$_2$PPh$_2$)$_2$	IIA	--	(c)	2108
S-(C$_{12}$H$_{25}$)$_2$Sn(SOPPh$_2$)$_2$	IIA	--	(c)	2108
(C$_{12}$H$_{25}$)$_2$Sn(S$_2$PPh$_2$)$_2$	IIA	--	(c)	2108
Ph$_2$Sn(Et$_2$PO$_4$)$_2$	--	--	Moessbauer resonance spectrum, struct.	1044
Ph$_2$Sn(S$_2$PEt$_2$)$_2$	IIA	83	m. 149.5°, stable to 180° (h)	1363
Ph$_2$Sn(O$_2$PPh$_2$)$_2$ (356)	IIA	--	(c)	2108
S-Ph$_2$Sn(SOPPh$_2$)$_2$	IIA	--	(c)	2108
Ph$_2$Sn(S$_2$PPh$_2$)$_2$	IIA	--	(c)	2108
Ph$_2$Sn(O$_2$AsPh$_2$)$_2$	IVB	80	m. 330°, soly.	1842
Ph$_2$Sn(O$_2$AsPh$_2$)$_2$	IID	68	Synth. in presence of Et$_3$N	1842
Ph$_2$Sn(O$_2$SPh)$_2$	IIA	--	m. 240° (dec.)	1410
EtSn(ClO$_4$)$_3$	IIB	90	m. ~ 50° (violent dec.) IR spectrum	2351
BuSn[S$_2$P(OMe)$_2$]$_3$	IIIB	--	n_D^{15} 1.5967 (g)	1576

BuSn[S$_2$P(OCH$_2$CHEtBu)$_2$]$_3$	IIIB	--	n_D^{15} 1.5202 (g)	1576
PhSn(O$_2$AsPh$_2$)$_3$	IVB	70	m. 178-80° (dec.) soly.	1842
	ID	85	Synth. in presence of Et$_3$N	1842

(g) Use as bactericide, insecticide and seed germination inhibitor (1576). (h) Dipoly moment, cis-struct. (1363). (i) Stabilizer for ABS copolymers - CH$_2$:CHCN-(CH$_2$:CH)$_2$-PhCH:CH$_2$ (1867). (j) Rxn. with Bu$_2$Sn(OMe)$_2$ → Bu$_2$Sn(OMe)-O$_3$SC$_{10}$H$_{15}$O (942). (k) Anal. detn. by amperometric titration (1695), rxn. with 1-(2-pyridazo)-2-naphthol → spectrophotometric evidence for 1:1 adduct, stability const. (1695).

Table 164. Diorganotin Salts or Esters Containing Tin-Heteroatom Bonds   R$_2$Sn(OR)X'

Formula*	Synth. Method**	Yield	Properties	Ref.
Me$_2$Sn(OH)NO$_3$	IIA	--	m. > 250°, struct. (a-c)	620
	IIA	--	Moessbauer resonance spectrum (902)	2158
Me$_2$Sn(OD)NO$_3$	(c)	--	IR spectrum	620
(Me$_2$SnNO$_3$)$_2$O	(b)	--	m. > 250°, dimeric, struct. (d)	2158
Et$_2$Sn(OH)NO$_3$	IIA	--	m. 214° (dec.)	620
	IIA	--	(a, b, e)	2158
(Et$_2$SnNO$_3$)$_2$O	(b)	--	m. 214° (dec.), dimeric, struct. (d)	2158
Pr$_2$Sn(OH)NO$_3$	IIA (f)	--	m. 183° (dec.) (b, d, g)	2158

* Numbers in parenthesis refer to pages in main volume.
** The characters used here correspond to the synthetic scheme in the introduction to Chapter 4.2.4.6.
(a) (Far) IR spectrum, assignment, recryst. from dil. HNO$_3$ or MeOH (620). (b) Dehydration of R$_2$Sn(OH)NO$_3$ at 10 torr and 80-200° → (R$_2$SnNO$_3$)$_2$O; R = Me, Et (200°), Pr (100°), Bu (80°) (2158). (c) Rxn. of Me$_2$Sn(OH)NO$_3$ with D$_2$O (repeated recryst.) → Me$_2$Sn(OD)NO$_3$ (620). (d) (Far) IR spectrum and assignment (2158). (e) Anal. detn. of -OH by K. Fischer titration (2158). (f) Rxn. prod. Pr$_2$Sn(OH)NO$_3$·H$_2$O is dehydrated in vacuo over anhydr. CaCl$_2$ (2158). (g) Rxn. of R$_2$Sn(OH)NO$_3$ with aq. NaOH (2:1) in Me$_2$CO → (R$_2$SnNO$_3$)O(R$_2$SnOH); R = Pr, Bu (2158).

Table 164 Continued $R_2Sn(OR)X'$

Formula*	Synth. Method**	Yield	Properties	Ref.
$Pr_2Sn(OH)NO_3 \cdot H_2O$	IIA	--	m. 183° (dec.), IR spectrum (e, f)	2158
$(Pr_2SnNO_3)_2O$	(b)	--	m. 183° (dec.), dimeric (d)	2158
$(Pr_2SnNO_3)O(Pr_2SnOH)$	(g)	--	m. 221-22.5° (dec.)	2158
$[Bu_2Sn(SR)]_3BO_3$ (h)	IIA	--	Use as stabilizer for PVC	2971
$Bu_2Sn(OH)NO_3$	IIA	--	m. 92.5-95°, x-ray diffraction anal. (b, d, g)	2158
$(Bu_2SnNO_3)_2O$ (352)	(b)	--	m. 92.5-95° (d)	2158
$(Bu_2SnNO_3)O(Bu_2SnOH)$	(g)	--	m. 210-13°	2158
$Bu_2Sn(O_3SR)_2 \cdot 3Bu_2SnO$	IIA	89	$R = p\text{-}MeC_6H_4$, m. 190°, mol. wt.	2595
$Bu_2Sn(OMe)O_3SC_{10}H_{15}O$ (i)	IIIB	--	Viscous golden-brown oil, IR spectrum	942
$Bu_2Sn(OMe)OS(O)OMe$	IIIA	--	m. 95.5-96.5°, IR, NMR spectra	943
$[1\text{-}Bu_2SnO_3SNHC_8H_4CO_2\text{-}p]_x$	IC	69	m. 170-80°	1049
$(Ph_2SnNO_3)_2O$	(j)	87	m. 288-90° (dec.), IR spectrum	1030
$Ph_2Sn(OC_9H_6N)NO_3$ (k)	IIIC	72	Dec. > 175°, IR, UV spectra, mol. wt.	1216
$(Ph_2SnOC_9H_6N)_2SO_4$ (k)	IB	35	Dec. > 300°, IR, UV spectra, mol. wt. (1)	1216

(h) $R = CH_2CO_2CHMe(-)CH_2O_2CCH_2SH$. (i) $OC_{15}H_{10}SO_3H$ = Camphorsulfonic acid. (j) Synth. by hydrolysis of $Ph_2Sn(NO_3)_2$ in air (1030). (k) HOC_9H_6N = 8-hydroxyquinoline. (1) Conductivity (1216).

4.3   ORGANOTIN OXYGEN COMPOUNDS

4.3.1   ORGANOTIN OXIDES AND HYDROXIDES

4.3.1.1   TRIORGANOTIN OXIDES AND HYDROXIDES

Organotin hydroxides and oxides are easily convertible. The compounds are not easily distinguished for melting points, are often identical due to dehydration of the hydroxides. Spectroscopic evidence is required. The compounds compiled in the following Tables 165-167 are prepared by the ensuing methods.

   I   Hydrolysis of triorganotin halides.
   II   Alkylation of diorganotin diiodides with alkyl iodide in presence of zinc dust.
   III Dealkylation of $R_2R'_2Sn$ with bromine.
   IV   Interconversion reactions of triorganotin hydroxides:
        (IVA) By dehydration ,
        (IVB) By deuterium exchange with D_2O .

Additional references to triorganotin oxides and hydroxides may be found in Chapters 7.3, 8.1 and 9.

4.3.1.1.1 SYMMETRIC TRIORGANOTIN OXIDES AND HYDROXIDES

Bis(trimethyltin) Oxide (M-358) $Sn_2C_6H_{18}O$ $(Me_3Sn)_2O$
Synth.: By rxn. of Me_3SnNMe_2 with Me_2CO, in 82% yield (250).
By rxn. of Me_3SnN_2CH with aq. base (1393).
Prop.: IR (644-5), assignment (1360), Raman, assignment (1360) and NMR (645) spectra, relative basicity (644-5).
Rxn. with: $NaC_2H \rightarrow (Me_3SnC\vdots)_2$ (1342).
$NaNH_2$ in $Et_2O \rightarrow (Me_3Sn)_3N$ (471).
$RC\vdots CH \rightarrow Me_3SnC\vdots CR + Me_3SnOH$; R = H (506, 1339) $BuSCH\vdots CH$ (2772).
8-Hydroxyquinoline $\rightarrow$ 8-$Me_3SnOC_9H_6N$ (404, 1046).
i-C_8H_{17}-9,10-expoxystearate $\rightarrow$ methyltin epoxystearate (1442).

Trimethyltin Hydroxide (M-358) $SnC_3H_{10}O$ Me_3SnOH
Synth.: By rxn. of $(Me_3Sn)_2O$ with C_2H_2 (506, 1339), with $BuSCH\vdots CHC\vdots CH$ at r.t. (2772).
By hydrolysis of $Me_3SnC_6F_5$ (851) with aq. alc. KOH, in 87% yield (77), of $Me_3SnC_6H_4Br$-o (851), of Me_3SnCCl_3 or $Me_3SnCBrCl_2$ (1875), of Me_3SnNMe_2 in Et_2O, in 97% yield (250), of Me_3SnN_2CH (1393).
By dec. of Me_3SnO_2H in dioxane and MeCN, in 48 and 69% yield, resp. (118).
Manufacturing process (541).
Prop.: m. 118° with subl. (250), IR, assignment (342), NMR (1220) and Moessbauer resonance (224, 902, 1988) spectra, x-ray struct. analysis (256).

Rxn. with: $Bu_3SnH \rightarrow Me_3SnSnBu_3 + H_2O$ (1802).
$NH_4SCN \rightarrow Me_3SnNCS + NH_3 + H_2O$ (2660).
$PhC\!:\!CH \rightarrow Me_3SnC\!:\!CPh$ (1921).
$RCO_2H \rightarrow Me_3SnO_2CR$; R = $ClCH_2$, Cl_2CH, Cl_3C (1933, 2243), FCH_2, F_2CH, CF_3 (2243), $HC\!:\!C$ (311).
$HNO_3 \rightarrow Me_3SnNO_3 \cdot H_2O$ (620).
$HClO_4 \rightarrow Me_3SnClO_4$ (2040).
Dry $CO_2 \rightarrow (Me_3Sn)_2CO_3$ (1800).
$(:CHCO)_2O \rightarrow (Me_3SnO_2CCH:)_2$ (1779).
(Styrene-maleic anhydride)$_x \rightarrow$ tin-containing polymer (1328).
$CH_2:\overline{CCH_2COCH_2} \rightarrow Me_3SnO_2CCH_2Ac$ (2310).
PhOH + molecular sieves $\rightarrow Me_3SnOPh$, useful method for titration, error $\pm$ 4% (342).
$\overline{CH_2CH_2O} \rightarrow$ relative activity as polymerization catalyst (2572).
$H_2O_2 \rightarrow Me_3SnO_2H$ (118).
$Me_3CO_2H \rightarrow Me_3SnO_2CMe_3 + Me_3SnO_2CMe_3 \cdot Me_3CO_2H$ (2257).
Biol. prop.: Activity as bactericide (1328), against Staphylococcus aureus and intestinal bacillus (275).
Activity as insecticide (1328), against Heliothis zea, H. virescens (192, 2145).
Activity in plankton control (1995).
Use: As pesticide (1328), in algaecidal and water repellent coatings (1754). As catalyst for polyurethan synth. (1209).

Tribenzyltin Hydroxide (M-364) $SnC_{21}H_{22}O$ $(PhCH_2)_3SnOH$
Synth.: By rxn. of $PhCH_2Cl$ with powd. tin and cat. $HgCl_2$ in H_2O at 108-114°, in 21% yield (1940).
By rxn. of $(PhCH_2)_2SnCl_2$ with powd. tin in H_2O under reflux, in 4-34% yield (1940), with powd. Al, Mg, Zn, in 22-90% yield (1940).
Prop.: IR spectrum (1940).
Rxn. with: $CH_2:\overline{CCH_2COCH_2} \rightarrow (PhCH_2)_3SnO_2CCH_2Ac$ (2310).
$(:CHCO)_2O \rightarrow (PhCH_2)_2\overline{SnO_2CCH:CHCO_2}$ (1779).
(Styrene-maleic anhydride)$_x \rightarrow$ tin-containing polyester (1328).

Bis(triethyltin) Oxide (M-359) $Sn_2C_{12}H_{30}O$ $(Et_3Sn)_2O$
Synth.: By rxn. of Et_4Sn with $Et_3SnO_2CMe_3$ (665).
By oxidation of Et_6Sn_2 with air at r.t., in 55% yield (1737), in presence of freshly ppt. Cu, in 22% yield (1737), under UV initiation (661), with $Et_3SnO_2CMe_3$ (665).
By rxn. of Et_3SnCH_2Ac with Me_2NH at -78°, in 98% yield (250).
By hydrolysis of $Et_3SnC\!:\!CCH\!:\!CHSBu$ with H_2O at 100°, in 46% yield (2772), of Et_3SnN_3 with H_2O (1429).
By rxn. of Et_3SnH with Me_3CO_2H, in 49% yield (585).
By rxn. of Et_3SnOH with Et_3SnC_2H, in 98% yield (506).
By thermal dec. of $Et_3SnO_2CMe_2Ph$ (666).
By rxn. of $(Et_3Sn)_2S$ with freshly ppt. Cu in air at r.t., in 94% yield (1737).
By-prod. in photooxidation of $PhCH_2SnEt_3$ in n-C_9H_{20}; rate of formation (2215).

Prop.: IR, assignment (321, 1360), far IR, assignment (321), Raman, assignment (1360) and NMR (1009) spectra, molar polarization, dipole moment, configuration (632).

Rxn. with: $Et_3SnC\vdots CH \rightarrow (Et_3SnC\vdots)_2$ (506).
$R_3SnH \rightarrow Et_3SnSnR_3$; R = Et, i-Bu, Ph (400, 1618).
$Et_2SnH_2 \rightarrow (Et_2SnO)_x + Et_3SnH$ (1955).
$Et_3SnONa + PhC\vdots CH$, + excess $CH_2\colon CHC\vdots CH \rightarrow Et_3SnC\vdots CCH\colon CH_2$ (1342).
$Et_3SnONa + PhC\vdots CH$, + $Et_3SnC\vdots CH \rightarrow (Et_3SnC\vdots)_2$ (1342).
$PhC\vdots CNa$ in $Et_2O \rightarrow Et_3SnC\vdots CPh$ (1342).
RMgBr (excess) $\rightarrow Et_3SnR$; R = Et, $CH_2\colon CHC\vdots C$ (1906).
EtMgBr, and distn. $\rightarrow Et_4Sn + Et_3SnBr$; distn. residue with BuMgBr $\rightarrow Et_3SnBu$ (1906).
Cu in $CCl_4 \rightarrow Et_3SnCl + C_5H_8Cl_4 + HCCl_3 + C_2Cl_6 + C_4H_{10} + C_4H_8 + CuCl$ (1737).
$RC\vdots CH \rightarrow Et_3SnC\vdots CR + H_2O$; R = numerous subst. and unsaturated alkyl and aryl groups (391, 506-7, 511, 1918, 1921, 2495), CN (391).

$RC\vdots CH + CaH_2 \rightarrow Et_3SnC\vdots CR$ in high yield; R = Bu, MeOCH:CH, p-MeOC$_6$H$_4$, H$_2$NCO, CN, Et$_3$Sn (2495).
$(HC\vdots C)_2 \rightarrow Et_3SnC\vdots CC\vdots CH + (Et_3SnC\vdots C)_2$ (506).
$RC\vdots CH \rightarrow Et_3SnOH + Et_3SnC\vdots CR$; R = H (1339), Me$_3$Si (1344).
$RCH\colon CHC\vdots CH \rightarrow Et_3SnOH + Et_3SnC\vdots CCH\colon CHR$; R = H (1341), BuO, BuS, Et$_2$N (2772).
$HC\vdots CCH_2OH \rightarrow Et_3SnOCH_2C\vdots CH$ (391) + $Et_3SnOCH_2C\vdots CSnEt_3$ (1921).
$HC\vdots CYOH \rightarrow Et_3SnOYC\vdots CH + Et_3SnOYC\vdots CSnEt_3$; Y = CHMe (508), CMe (508, 2506).
$RC\vdots CYOH \rightarrow Et_3SnOYC\vdots CR$; R = $CH_2\colon CH$, Et$_3$Si; Y = CH$_2$, CHMe, CMe$_2$ (2506).
$HC\vdots CCH_2OAc \rightarrow Et_3SnOAc + Et_3SnOCH_2C\vdots CH$ (391).
$Ph_2P(O)H \rightarrow Et_3SnO_2PPh_2 + Et_3SnH$ (1233).
$H_2O \rightarrow Et_3SnOH$ (342).
$HNCO \rightarrow Et_3SnNCO$ (2799).
$(H_2N)_2CO \rightarrow Et_3SnNCO + H_2O + NH_3$ (532).
$P_2S_5 + EtOH \rightarrow Et_3SnS_2P(OEt)_2$ (1576).
$(\colon CHCO)_2O \rightarrow (Et_3SnO_2CCH\colon)_2$ (1779, 2191).
$(\colon CHCO)_2O$, + $PhCH\colon CH_2 \rightarrow$ tin-contn. polyester (1327).
AcOEt $\rightarrow Et_3SnOAc + Et_3SnOEt$ (?) (949).
AcH $\rightarrow$ solid, m. 123°, 43% Sn (661), readily absorbs $CO_2 \rightarrow (Et_3Sn)_2CO_3$ (321).
PhNCO $\rightarrow Et_3SnNPhCO_2SnEt_3$ (35, 748).
o-HOCH$_2$C$_6$H$_4$OH $\rightarrow$ o-Et$_3$SnOC$_6$H$_4$CH$_2$OH (389).
H_2O_2 at -60° $\rightarrow Et_3SnOH$ (8).
H_2O_2 (2:1) at -60° in presence of dry $MgSO_4 \rightarrow 2Et_3SnO_2H \cdot H_2O_2$ (8).
$Me_2RCO_2H \rightarrow Et_3SnO_2CMe_2R$; R = Me, Ph (667).
O under UV irradiation $\rightarrow (Et_2SnO)_2 + Et_4Sn + C_2H_6 + C_4H_{12}$; rate of photolysis, quantum yield (2217), rxn. intermediates, influence of inhibitors (2218).

Use: In catalyst for C_2H_4 polymerization (240).

Triethyltin Hydroxide (M-359) $SnC_6H_{16}O$ Et_3SnOH

Synth.: By rxn. of Et_4Sn with O_3 at 70° (663), with $Et_3SnO_2CMe_3$ at 80-160° (665).
By rxn. of Et_6Sn_2 with $Et_3SnO_2CMe_3$ at 80-160° (665).
By rxn. of $(Et_3Sn)_2O$ with RC_2H; R = H (1339), $CH_2\colon CH$, in 98% (1341), R =

BuOCH:CH, Et$_2$NCH:CH, BuSCH:CH (2772), Me$_3$Si, in 44% yield (1344), with H$_2$O$_2$ at -60° in Et$_2$O (8), with H$_2$O (342).
By dec. of Et$_3$SnO$_2$CMe$_2$Ph at 140-70° (666).
By rxn. of (Et$_3$SnO)$_x$ with EtMgBr, in 6% yield (505).
Prop.: IR, assignment (321, 342), far IR, assignment (321) and Moessbauer resonance (503) spectra. Anal. detn. by acidimetric titration (151), with PhOH (342), by spectrographic method (1358).
Rxn. with: Ph$_3$GeOH → Et$_3$SnOGePh$_3$ (2344).
Et$_3$SnC⋮CH → (Et$_3$Sn)$_2$O + C$_2$H$_2$ (506).
R$_3$SnOPPh$_2$ → R$_3$SnO$_2$PPh$_2$ + Et$_3$SnH; R = Et, Bu (1233).
Et$_3$Al → Et$_4$Sn + Et$_2$AlOEt + C$_2$H$_6$ (2663).
Et$_2$Cd → Et$_3$SnOCdEt + C$_2$H$_6$ (2663).
Na$_2$C$_2$ → (Et$_3$SnC⋮)$_2$ (504).
NaC$_2$H → Et$_3$SnC⋮CH + (Et$_3$SnC⋮)$_2$ (1342).
PhC⋮CNa + CH$_2$:CHC⋮CH → Et$_3$SnC⋮CPh + Et$_3$SnC⋮CCH:CH$_2$ (1342).
RC⋮CH → Et$_3$SnC⋮CR; R = H, CH$_2$:CH (504), Ph (509, 1921), BuOCHMeOY; Y = CH$_2$, CHMe, CMe$_2$ (279, 504).
HC⋮CYOH → Et$_3$SnC⋮CYOSnEt$_3$; Y = CH$_2$ (1921), CMeEt, $\overline{C(CH_2)_nCH_2}$, n = 3, 4 (1149).
HC⋮CYOH → Et$_3$SnC⋮CYOSnEt$_3$ + HC⋮CYOSnEt$_3$; Y = CH$_2$ in presence of EtMgBr, CHMe (2506).
HClO$_4$ → dissocn. const. of free acid, deuterium effect (564).
RCO$_2$H → Et$_3$SnO$_2$R; deriv. of amino acids (151), copolymer of (:CHCO)$_2$O, PhCH:CH$_2$ + MeOH (868), HC⋮C (311).
CO$_2$ → (Et$_3$Sn)$_2$CO$_3$ (321, 1800).
(:CHCO)$_2$O → Et$_3$SnO$_2$CCH:CHCO$_2$H (1779).
PhCH:CH$_2$-(:CHCO)$_2$O polymer → tin-contg. polyester (1327, 2191), rxn. with excess (:CHCO)$_2$O (1779).
CO(NH$_2$)$_2$ → Et$_3$SnNCO (532, 532b).
PhNHCHO → Et$_3$SnNPhCHO (949).
CS$_2$ + Et$_2$NH → Et$_3$SnS$_2$CNEt$_2$ (761).
ROH → Et$_3$SnOR; R = HC⋮CCMeEt, $\overline{HC⋮CC(CH_2)_nCH_2}$, n = 3, 4 (1149), Ph (342), p-ClC$_6$H$_4$, p-MeOC$_6$H$_4$ (914).
DMF, Et$_2$O or (CH$_4$)$_2$SO ⇌ reversible dehydration to (Et$_3$Sn)$_2$O (342).
Biol. prop.: Activity as insecticide against Culex pipiens larvae (1137).

Bis(tripropyltin) Oxide (M-360)   Sn$_2$C$_{18}$H$_{42}$O   (Pr$_3$Sn)$_2$O
Synth.: By rxn. of Pr$_3$SnOH with C$_2$H$_2$ in autoclave (504).
By thermal dec. of (Pr$_2$SnX)$_2$O at 0.3 torr and 85-150°, in 53-70% yield; X = Cl, OPh, OAc (2611).
Rxn. with: Ph$_3$SiH → Pr$_3$SnH + Pr$_3$SnOSiPh$_3$ (1165).
(MeHSiO)$_x$ → Pr$_3$SnH (1165).
NaC$_2$H → Pr$_3$SnC⋮CH + Pr$_3$SnOH (506).
NaHSO$_n$ → (Pr$_3$Sn)$_2$SO$_n$ + Na$_2$SO$_n$; n = 3, 4 (2146a).
NH$_4$X → Pr$_3$SnX + H$_2$O + NH$_3$; X = Cl, Br, I, NCS (2660).
C$_2$H$_2$ → Pr$_3$SnC⋮CH + Pr$_3$SnOH (1339).
RCH:CH⋮CH → Pr$_3$SnC⋮CCH:CHR + Pr$_3$SnOH; R = H (1341), BuO, BuS, Et$_2$N (2772).
B(OH)$_3$ → tripropyltin borates (1426) (Pr$_3$SnO)$_3$B (2126).

$B_2O_3 \to (Pr_3SnO)_3B + (Pr_3SnO)_4B_2O_5$ (2063).
$H_2SiF_6 \to Pr_3SnF$ (1468).
$Ph_2PO_2H \to Pr_3SnO_2PPh_2$ (292).
$H_2NSO_3H \to Pr_3SnO_3SNH_2$ (1578, 1629).
$RCO_2H \to Pr_3SnO_2CR$; $R = CH_2{:}CHCO_2C_3H_6O_2CCH{:}CH$ (2974), $6\text{-}H_2NC_7H_{10}OS$ penicillic acid (2236).
$C_6H_{6-n}(CO_2H)_n \to (Pr_3SnO_2C)_xC_6H_{6-n}(CO_2H)_{n-x}$; $n = 3, 4$; $x = 1, 2, 3$ (2619).
$C_7H_8(CO)_2O \to 2,3\text{-}(Pr_3SnO_2C)_2C_7H_8 = 5$-bicycloheptene dicarboxylate (1890).
$({:}CHCO)_2O \to (Pr_3SnO_2CCH{:})_2$ (1779).
$PhNCO \to Pr_3SnNPhCO_2SnPr_3$ (748).
8-Hydroxy quinoline $\to 8\text{-}Pr_3SnOC_9H_6N$ (404).
Biol. prop.: Toxicity (933).
Activity as anthelmintic against Hymenolepsis nana and H. diminuta (1892).
Activity as bactericide against Aerobacter aerogenes, Pseudomonas aeruginosa (2183).
Activity as fungicide against Alternaria tenuis (933), Aspergillus flavus, Candida albicans (2183), Fusarium culmorum (933), Penicillium funiculosum (2183), Rhizoctonia solani (190, 933).
Activity as molluscicide (455), against Australorbis glabratus (154, 823, 1860).
Use: As fungicide in PVC (2334), in antifouling coatings (2302).

Tripropyltin Hydroxide (M-360) $SnC_9H_{22}O$ Pr_3SnOH
Synth.: By rxn. of $(Pr_3Sn)_2O$ with C_2H_2 under pressure (1339), with $CH_2{:}CHC{:}CH$ at r.t. (1341), with $RCH{:}CHC{:}CH$; $R = BuO, BuS, Et_2N$ (2772).
Rxn. with: $Na_2C_2 \to (Pr_3SnC{:})_2$ (504).
$RC{:}CH \to Pr_3SnC{:}CR + (Pr_3Sn)_2O$; $R = H, CH_2{:}CH$ (504).
$ROCS_2CH_2CO_2H \to Pr_3SnO_2CCH_2S_2COR$; $R = Et, Pr, Bu$ (1308).
$ROH \to Pr_3SnOR$; $R = HC{:}CCMeEt, HC{:}C\overline{C(CH_2)_nCH_2}$, $n = 3, 4$ (1149).

Bis(triisopropyltin) Oxide (M-364) $SnC_{18}H_{42}O$ $(i\text{-}Pr_3Sn)_2O$
Rxn. with: $NaHSO_x \to (i\text{-}Pr_3Sn)_2SO_n + Na_2SO_n$; $n = 3, 4$ (2146a).
$C_7H_8(CO)_2O \to 2,3\text{-}(i\text{-}Pr_3SnO_2C)C_7H_8 = 5$-bicycloheptene dicarboxylate (1890).
$HRNNO_2 \to i\text{-}Pr_3SnNRNO_2$; $R = Me, t\text{-}Bu$ (611).
Biol. prop.: Activity as bactericide against Aerobacter aerogenes, Pseudomonas aeruginosa (2183).
Activity as fungicide against Aspergillus flavius, Candida albicans, Penicillium funiculosum (2183).

Bis(tributyltin) Oxide (M-361) $Sn_2C_{24}H_{54}O$ $(Bu_3Sn)_2O$
Synth.: By rxn. of Bu_6Sn_2 with air at r.t., in 19% yield (1737), in presence of freshly ppt. Cu, in 7% (1737).
By rxn. of Bu_3SnH with $Me_2C{:}CHAc$ (1684).
By rxn. of $Bu_3{}^{113}SnCl$ with base (2824).
By rxn. of $Bu_3SnOCHEtCH_2NEt_2$ with aq. KOH in dioxane, in 84% yield (2053).
By thermal dec. of Bu_3SnO_2CNHPh (748), of $Bu_3SnOTi(OCH_2CH_2)_3N$ (882).
By UV irradiation of $Bu_3SnOCMe{:}CHCHMe_2$ (1684).
By rxn. of $(Bu_3Sn)_2S$ with freshly ppt. Cu in air at r.t., in 71% yield (1737).

By rxn. of Bu$_3$SnNMe$_3$ with Me$_2$CO, in 97% yield (250).
By hydrolysis of (Bu$_3$SnNPh)$_2$CO (748).
By rxn. of Bu$_2$SnI$_2$ with BuI and powd. Cu at 125-30° (2731) and Zn dust followed by aq. NaOH (2809).
By rxn. of (Bu$_2$SnX)$_2$O at 0.3 torr and 115-52°, in 56-76% yield; X = Cl, PhO, AcO (2611).
By rxn. of (Bu$_2$SnO)$_x$ with ROH in vacuo at 130-215°; R = Ph, Ac, in 57-84% yield (2611), with PhAc at 250°, in 83% yield (1487).
By rxn. of BuSnI$_3$ with BuMgCl (2734).
Prop.: IR (2824), (far) IR, assignment (110, 1360, 1509), Raman, assignment (1360), UV (927, 1117) and NMR (1676) spectra, dipole moment (632, 928), light resistance (1063), molar polarization and configuration (632), surface tension (1959). Anal. detn. by GLC (1801), TLC (2309, 2824), acidimetric titration (151), column chromatography (2824), polarography (2589).
Rxn. with: Bu$_3$SnH → Bu$_6$Sn$_2$ (400, 468, 1618, 1802).
R$_2$SnH$_2$ at 100° → (Bu$_3$Sn)$_2$SnR$_2$; R = Bu (468, 1803), Ph (1803).
R$_2$SnH$_2$ at r.t. → Bu$_3$SnH + (Bu$_2$SnO)$_x$ (468).
R$_3$SiH → Bu$_6$Sn$_2$ + (R$_3$Si)$_2$O; R = Me (1239), Bu (1165).
R$_3$SiH → Bu$_3$SnH + Bu$_3$SnOSiR$_3$; R = Et (719), Ph (1165).
(Me$_2$HSi)$_2$O·(MeHSiO)$_x$ → Bu$_3$SnH + volatile silane (1165).
(MeHSiO)$_x$ → Bu$_3$SnH + Bu$_3$SnO(MeSiO)$_x$ (238, 1239, 1242, 2457).
(MeHSiO)$_x$ (1:2) → Bu$_3$SnH (1165).
(H$_2$Si$_2$O$_3$)$_x$·(Me$_2$SiO)$_y$ → Bu$_3$SnH (1165).
(Ph$_2$SiH)$_2$O → Bu$_3$SnH + c.(Ph$_2$SiO)$_4$ (1165).
Me$_3$SiCl → (Me$_3$Si)$_2$O + Bu$_3$SnOSiMe$_3$ + Bu$_3$SnCl (2465).
Me$_2$SiCl$_2$ → (Bu$_3$SnO)$_2$SiMe$_2$ (2465).
Ph$_2$Si(OH)$_2$ → (Bu$_3$SnO)$_2$SiPh$_2$ (2344).
(EtO)$_4$Si + HSCH$_2$CH$_2$OH → (Bu$_3$SnSCH$_2$CH$_2$O)$_4$Si (2536).
RLi → Bu$_3$SnR; R = Me, Bu (2709).
NaHSO$_n$ → (Bu$_3$Sn)$_2$SO$_n$ + NaSO$_n$; n = 3, 4 (2146a).
NH$_4$X → Bu$_3$SnX + H$_2$O + NH$_3$; X = Cl, I, NCS (2660).
CH$_2$:CHC(:CH$_2$)MgCl → Bu$_3$SnC(:CH$_2$)CH:CH$_2$ (13, 14, 2933).
Cu + CCl$_4$ → Bu$_3$SnCl + C$_5$H$_8$Cl$_4$ + HCCl$_3$ + C$_2$Cl$_6$ + C$_4$H$_{10}$ + C$_4$H$_8$ + CuCl (1737).
(i-PrO)$_4$Ti + N(CH$_2$CH$_2$OH)$_3$ → i-PrO(Bu$_2$SnO)$_n$Ti(OCH$_2$CH$_2$)$_3$N; n = 1, 2 (2324).
RC⋮CH + CaH$_2$ → Bu$_3$SnC⋮CR; R = H, Et$_3$Sn (2495).
C$_2$H$_2$ → Bu$_3$SnC⋮CH + (Bu$_3$SnC⋮)$_2$ + Bu$_3$SnOH (1339).
RC⋮CH → Bu$_3$SnC⋮CR + Bu$_3$SnOH; R = CH$_2$:CH (1341), BuOCH:CH, Et$_2$NCH:CH (2772).
PhC⋮CH → Bu$_3$SnC⋮CPh + H$_2$O (391, 2495).
B(OH)$_3$ → Bu$_3$Sn-borates (1426), (Bu$_3$Sn)$_3$BO$_3$ (2063, 2126) + (Bu$_3$Sn)$_4$B$_2$O$_5$ (2063).
HBF$_4$ → Bu$_3$SnBF$_4$ (595).
H$_2$SiF$_6$ → Bu$_3$SnF (1468).
RPO$_3$H$_2$ → (Bu$_3$Sn)$_2$O$_3$PR; R = Ph, CH$_2$OH, CCl$_3$ (538).
R$_2$PO$_2$H → Bu$_3$SnO$_2$PR$_2$; R = Ph (292), CH$_2$OH, C$_6$H$_{13}$, C$_7$H$_{14}$OH (538).
R$_2$P(O)H → Bu$_3$SnO$_2$PR$_2$ + Bu$_3$SnH; R = Bu, Ph (1233).
P$_2$O$_3$F$_4$ → Bu$_3$SnO$_2$PF$_2$ (1767).
P$_2$S$_5$ + MeOH → Bu$_3$SnS$_2$P(OMe)$_2$ + (Bu$_3$SnO)$_2$MeOPS (1576).
P$_2$S$_5$ + ROH → Bu$_3$SnS$_2$P(OR)$_2$; R = EtBuCHCH$_2$ (1576).

$RSO_3H \rightarrow Bu_3SnO_3SR$; $R = MeC_6H_4$ (572), $p\text{-}H_2NC_6H_4$ (2966), H_2NCONH (1578).

$H_2NSO_3H \rightarrow Bu_3SnO_3SNH_2$ (1578, 1629) + $(Bu_3SnO_3S)_2NH$ (1578).

$2,5\text{-}HO(HO_3S)C_6H_3CO_2H \rightarrow 2,5\text{-}HO(Bu_3SnO_3S)C_6H_3CO_2SnBu_3$ (2966).

$RCO_2H \rightarrow Bu_3SnO_2CR$; $R = HC{\equiv}C$ (1520), Cl_3C, Ph (2966), alkenyl (2626, 2974), aryl (1795, 2257, 2966), adamantyl (2391), heterocyclo (378, 572, 2236) aminoalkyl (151, 2236, 2959) mercaptoalkyl (213, 1170).

Phenylalanine + $PhCH_2O_2CNHCH_2CO_2CO_2Et \rightarrow PhCH_2O_2CNHCH_2CONHCH(CH_2Ph)CO_2H$ (151).

Tryptophane + $PhCH_2O_2CNHCH_2CO_2C_6H_4NO_2 \rightarrow PhCH_2O_2CNHCH_2CONHCH(CH_2C_8H_6N)CO_2H$ (151).

$CH_2{:}CMeCO_2H + AcCH_2CMe_2OH + Bz_2O_2 \rightarrow$ tin-contg. polymer (2949).

$R'(C_4H_6)_xCO_2H + (Me\overline{CHCH_2N})_3PO \rightarrow$ cat. effect, polymerization mechanism (2206).

$Y(CO_2H)_n \rightarrow (Bu_3SnO_2C)_nY$; aliphatic dicarboxylic acids (1515, 1765, 1793, 2966), benzene polycarboxylic acids (572, 1795, 2619).

$Y(CO)_2O \rightarrow (Bu_3SnO_2C)_2Y$; $Y = (CH_2)_n$, $n = 1, 4$ (1765), 1, 2, 3, (2966), $CH{:}CH$ (1779), 2-bicycloheptene (1890), epoxyalkyl (1305).

$({:}CHCO)_2O\text{-}PhCH{:}CH_2$-polymer $\rightarrow$ tin-contg. polyester (1328).

2-Bicycloheptene dicarboxylic anhydride + $(CH_2CH_2OH)_2 \rightarrow Bu_3Sn$-contg. polymer (2020).

$R_2CO_3 \rightarrow Bu_3SnOR + CO_2$; $R = Me, Et$; $R_2 = CH_2CH_2$ (950).

$RCO_2R' \rightarrow Bu_3SnO_2CR$; R = epoxyalkyl (1442).

$AcOEt \rightarrow Bu_3SnOAc + Bu_3SnOEt$ (?) (949).

$ArCH{:}NO(CH_2)_nCO_2Me \rightarrow Bu_3SnO_2C(CH_2)_nON{:}CHAr$; $n = 1, 2$; Ar = substituted phenyl groups (1574).

$Me_2C{:}C{:}O \rightarrow Bu_3SnCMe_2CO_2SnBu_3$ (36).

$YO_2 \rightarrow (Bu_3Sn)_2YO_3$; $Y = C, S$ (36, 750).

$RNHCHO \rightarrow Bu_3SnNRCHO + H_2O$; $R = Me, C_6H_{13}, Ph$ (949).

$RNHAc \rightarrow Bu_3SnNRAc + H_2O$; $R = H, Ph$ (949).

$RCN \rightarrow Bu_3SnN{:}CROSnBu_3$; $R = CCl_3$ (36, 750, 940).

$H_2NCO_2Et \rightarrow Bu_3SnNOC + EtOH$ (949).

$RNHCO_2Me \rightarrow Bu_3SnNRCO_2Me + H_2O$; $R = Et, Ph$, + $(Bu_3SnNPh)_2CO$ (949).

$PhNHC({:}S)OMe \rightarrow Bu_3SnSC(OMe){:}NPh$ (750).

$(NHCO_2Et)_2 \rightarrow [Bu_3SnN(CO_2Et)]_2 + H_2O$ (949).

$(H_2N)_2CY \rightarrow Bu_3SnNCY$; $Y = O$ (532, 532b, 949), S (532b).

$H_2NCONR_2 \rightarrow Bu_3SnNCO + R_2NH$; $R = H + Ph, Ph$ (949).

$(RNH)_2CO \rightarrow (Bu_3SnNR)_2O + H_2O$; $R = Ph, 1\text{-}C_{10}H_7$, + $Bu_3SnNHPh$ (?) (949).

$PhNHCONHC_{10}H_7 \rightarrow (Bu_3SnNPh)_2CO$ (949).

$Ph_2NCONHPh + MeOH \rightarrow (PhNH)_2CO + Ph_2NH$ (949).

$PhNHCONMeCHO \rightarrow (Bu_3SnNPh)_2O + Bu_3SnNMeCHO$ (949).

$(PhNH)_2CS \rightarrow (Bu_3Sn)_2S + (PhN{:})_2C$ (2258).

$HNCO \rightarrow Bu_3SnNCO$ (532, 2799).

$(HCNY)_3 \rightarrow c.(Bu_3SnY)_3C_3N_3$ sym-triazine; $Y = O$ (532, 532a, 949) S (1322a).

$PhNCO + PhOH \rightarrow PhNHCO_2Ph$; catalytic effect (36).

$MeNCO + Me_2COH \rightarrow MeNHCO_2CMe_3$ (2256).

$RNCO \rightarrow Bu_3SnNRCO_2SnBu_3$; $R = Me$ (748), Et (35-6, 748), Ph, $1\text{-}C_{10}H_7$ (35, 748).

$RNCO$ (1:2) $\rightarrow (Bu_3SnNR)_2CO + CO_2$; $R = Me, Bu$ (2258), $Ph, 1\text{-}C_{10}H_7$ (35, 748).

$RNCO$ (excess) $\rightarrow (RNCO)_3$; $R = Me, Et, Ph, 1\text{-}C_{10}H_7$ (746, 2342).

$(PhNCO)_2 \rightarrow (Bu_3SnNH)_2CO + (Bu_3Sn)_2CO_3$ (746).

$RNCS \rightarrow (Bu_3Sn)_2S$ (36) + $Bu_3SnNPhCO_2SnBu_3$ (750).

$RNCS \rightarrow (RN{:})_2C$; $R = CH_2{:}CHCH_2$ (2258).

RNCO + PhNCS → PhN:C:NR + $(Bu_3Sn)_2S$; R = Et, Ph (2258).
EtNCO + CH_2:$CHCH_2$NCS → EtN:C:NCH_2CH:CH_2 + $(Bu_3Sn)_2S$ (2258).
RNCO + CH_2:$CHCH_2$NCS → $(Bu_3SnNR)_2CO$; R = 1-$C_{10}H_7$ (2258).
C(:NPh)$_2$ → $Bu_3SnNPhC$(:NR)$OSnBu_3$ (35-6, 940).
C(:NR)$_2$ → $(Bu_3SnNR)_2CO$; R = i-Pr, c.C_6H_{11}, Ph, 1-$C_{10}H_7$ (750).
C(:NR)$_2$ excess → $(RNH)_2CO$; R = 1-$C_{10}H_7$ (750).
RCHO ⇌ $(Bu_3SnO)_2CHR$ + $Bu_3Sn(OCHR)_xOSnBu_3$; R = CCl_3 (941, 953), CBr_3, C_6F_5 (953).
RR'CO ⇌ $(Bu_3SnO)_2CRR'$; R = R' = CCl_3, R = CCl_2F, R' = $CClF_2$ (953).
$(CCl_3)_2CO$ + MeOH → $HCCl_3$ + CCl_3CO_2Me (953).
ROH → Bu_3SnOR; R = alkyl, aralkyl, CCl_3CH_2 (950), Et_2NCH_2CHPh (2835), substituted phenyl (1792, 1794, 1958), epoxy-ol, epoxyamine (1794, 1958), expoxy carboxyl (911), 8-C_9H_6N (1046).
$\overline{CH_2CH_2O}$ → relative activity as polymerization catalyst (2572).
RR'C:NOH → Bu_3SnON:CRR' ; R,R' = alkyl, alkenyl, aryl (2627), substituted aryl, Ph (2628).
Dioxime → $(Bu_3SnON$:C$)_2Y$; derivatives of 1,4-cyclohexanedione, Ac_2, Bz_2 (2125),
Me_3CO_2H → $Bu_3SnO_2CMe_3$ + H_2O (2257).
RSH → Bu_3SnSR (859); R = Bu (1117), Me, Ph, C_6F_5 (1676), $PhHPCH_2CHMe$ (2835), 2,4,6-(R'RN)$_2$-sym-triazine (1322a).
R_2PH → Bu_3SnPR_2 + H_2O; R = Ph (1233), Bu (1618).
RPH_2 → $(Bu_3Sn)_2PR$ + H_2O; R = Ph (1233).
PrNSO → $Bu_3SnNPrSO_2SnBu_3$ (2345).
ArNSO → $(Bu_3Sn)_2SO_3$ + $(Bu_3SnNAr)_2SO$; Ar = Ph, p-MeC_6H_4 (2345).
EtNSO, + CCl_3CHO → $(Bu_3SnO)_2CHCCl_3$ + EtNSO + EtN:$CHCCl_3$ (2345).
(ArN:)$_2$S → $(Bu_3SnNAr)_2SO$; Ar = Ph, p-MeC_6H_4 (2345).
<u>Biol. prop.</u>: Toxicity (451, 1099).
Toxicity towards Spirogyra longata, Ulthrix zonata, flagellata-Chlamydomonas, infusoria-Chilodon, nematoda-Plectus, rotifera-Euchlanis, hydracariens-Hydrachna (2350).
Phytotoxicity towards barley (376) rice (548).
Activity as bactericide (91, 622) against Aerobacter aerogenes (893), Achromobacter, Brevibacterium, Flavobacterium, Micrococcus, Pseudomonas, Sarcina (1275), intestinal bacillus (275), Staphylococcus aureus (91, 275). Bacillus proteus, Escherichia coli. Pseudomonas pyocyanea (1145).
Activity as fungicide (451, 622, 1320, 1675) against Alternaria tenuis (1275, 1319), Aspergillus niger (1243), Botrytis cinerea (1319), Candida albicans (1145) Ceratostomella coerulescens (1099), Chaetomium (1275), Coriollelus palustris (1063), Corticium centrifugum (340), Epidermophyton (1145), Fusarium culmorum, Penicillium funiculosum (1099), Phialophorae fastigiatea (1275), Phytophthora infestans (1319), Piricularia oryzae (548), Rhizoctonia solani (190), Rhizopus nigricans (1243), Rosellinia necatrix (340), Speira heptaspora (1099), Trichoderma viride (1275).
Activity as insecticide (1328), against Anobium punctatum (701), Culex pipiens larvae (73, 1137), Heliothis zea, H. virescens (192, 2145), Lyctus brunneus (701).
Activity as molluscicide (125, 455, 961), against Australorbis glabratus (1860), Biomphalaria glabrata (2245).

Activity as marine antifoulant (360).
Biological test method for detn. of 5 ppb. in seawater (456).
Use: US limits for use in preservatives and adhesives (2941).
In antifouling coatings (2301-2), as antifouling agent (438, 720), method of dispersion (2333, 2910), in seawater resistant coatings (2981), in fungicidal and molluscicidal paints (78), as marine wood preservative (1023, 2182) leaching rate from paints (901, 2320).
As fungicide (367a), in PVC (2334), for mold control in PVC (30) in PVC based paints (30).
As bactericide and fungicide in paint (720) in paper coatings (2693), in polyurethan (30), preservative in floor polish (1152), as wood preservative (451, 1753), as slime control agent (2768) in paper and paper mill industry (2810).
As rodent repellent (2036).
As additive for flame resistant polyester (2702), PVC (2599), as dopant in liquid for image charge transfer (2247).
In catalyst for depolymerizing polycaprolactone esters (2587), for esterification (458, 1766), for modifying polypropylene (269), for curing polyurethan rubber (539a), for synthesis of polyurethan (979).
As corrosion inhibitor for magnesium (959).

Tributyltin Hydroxide (M-361)　　　　$SnC_{12}H_{28}O$　　　　Bu_3SnOH
Synth.: By rxn. of Bu_3SnNCO with alc. NH_3 (373).
By rxn. of $(Bu_3Sn)_2O$ with H_2C_2 (1339), with $RCH:CHC\vdots CH$; R = H (1341), BuO, BuS, Et_2N (2772).
Rxn. with: $Na_2C_2 \rightarrow (Bu_3SnC\vdots)_2$ (504).
$RC\vdots CH \rightarrow Bu_3C\vdots CR$; R = H, $CH_2:CH$ (504).
$RCO_2H \rightarrow Bu_3SnO_2CR$; R = Me, CH_2F, CHF_2, CF_3, CH_2Cl, $CHCl_2$, CCl_3 (2074).
$MeO_2CCH:CH$ (1557), m-$AcNHC_6H_4$ (281).
$(CH_2CMeCO_2H)_x \rightarrow (Bu_3SnO_2CCMeCH_2)_x$ (285).
$Y(CO_2H)_2 \rightarrow (Bu_3SnO_2C)_2Y$; Y = cis-, trans $CH:CH$ (1557).
$ROCS_2CH_2CO_2H \rightarrow Bu_3SnO_2CCH_2S_2COR$; R = Et, Pr, Bu (1308).

Bis(triisobutyltin) Oxide (M-362)　　　$Sn_2C_{24}H_{54}O$　　　$(i$-$Bu_3Sn)_2O$
Synth.: By rxn. of i-Bu_3SnCl with dil. NaOH in C_6H_6, in quantitative yield (1618).
Rxn. with: $R_3SnH \rightarrow i$-Bu_3SnSnR_3; R = Et, i-Bu (1618).
$Ph_2SnH_2 \rightarrow (i$-$Bu_3Sn)_2SnPh_2$ (1955).
$(H_2N)_2CO \rightarrow i$-Bu_3SnNCO (532, 532b).
$HNCO \rightarrow i$-Bu_3SnNCO (2799).
Use: As slime control agent in paper and pulp industry (2668).

Tris(2,2-Methylphenylpropyltin)　　　$SnC_{30}H_{40}O$　　　$(PhCMe_2CH_2)_3SnOH$
　Hydroxide
Synth.: By rxn. of $(PhCMe_2CH_2)_3SnCl$ with aq. alc. NaOH under reflux, in 73% yield (1742).
Prop.: m. 145-46° (1742), NMR (1742) and Moessbauer resonance (222, 224, 1742) spectra, thermal dec. 340-50° (1742).

Rxn. with: HX → (PhCMe$_2$CH$_2$)$_3$SnX; X = F, Br, I, N$_3$, ClO$_4$, NO$_3$ (1742).
D$_2$O → (PhCMe$_2$CH$_2$)$_3$SnOD (1742).
RCO$_2$H → (PhCMe$_2$CH$_2$)$_3$SnO$_2$CR; R = H, Me, CF$_3$ (1742).
Na$_2$S → [(PhCMe$_2$CH$_2$)$_3$Sn]$_2$S (1742).

Tricyclohexyltin Hydroxide   SnC$_{18}$H$_{34}$O   (c.C$_6$H$_{11}$)$_3$SnOH
Synth.: By rxn. of SnCl$_4$ with c.C$_6$H$_{11}$MgCl in THF-xylene in N atm. at 75-85°, in 67% yield (2633).
Prop.: m. 221.5-23° (2633).
Rxn. with: NaHSO$_4$ → [(c.C$_6$H$_{11}$)$_3$Sn]$_2$SO$_4$ + Na$_2$SO$_4$ (2146a).
NH$_4$X → (c.C$_6$H$_{11}$)$_3$SnX + NH$_3$ + H$_2$O; X = Br, I, NCS (2660).
EDTA → [[(c.C$_6$H$_{11}$)$_3$SnO$_2$CCH$_2$]$_2$NCH$_2$]$_2$ (1387).
Biol. prop.: Phytotoxicity to beans, cotton (2222), apples, pears, peaches (2879).
Activity as acaricide (1306) against Tetranychus urticae (2222, 2231, 2879), T. pacificus, Panonychus citri (2474), P. ulmi (2231), Entetranychus carpini, Stethorus punctillum (2879).
Use: Tradename and synonyms Plictran, Dowco-213.
Activity and use as rodent repellent (2484).

Bis(trioctyltin) Oxide (M-365)   Sn$_2$C$_{48}$H$_{102}$O   [(C$_8$H$_{17}$)$_3$Sn]$_2$O
Prop.: IR (110) and NMR (1220) spectra.
Rxn. with: (HNCO)$_3$ → [(C$_8$H$_{17}$)$_3$SnNCO]$_3$ cyanurate (532a).
R'CO$_2$H → (C$_8$H$_{17}$)$_3$SnO$_2$CR; R = (HO$_2$CCH$_2$S)$_2$CHCMe(SCH$_2$CO$_2$H) (1170), CH$_2$:CMeCO$_2$-CH$_2$CH$_2$O$_2$CCH:CH (2974).
Me$_2$CO$_3$ → (C$_8$H$_{17}$)$_3$SnOMe (950).
Octyl 9,10-epoxystearate → trioctyltin epoxystearate (1442).

Bis(triphenyltin) Oxide (M-362)   Sn$_2$C$_{36}$H$_{30}$O   (Ph$_3$Sn)$_2$O
Synth.: By rxn. of Ph$_6$Sn$_2$ with air in presence of MeONa in THF or with H$_2$O$_2$-NaOH in aq. Me$_2$CO, in 92 and 77% yield, resp. (2046).
By air oxidation of Ph$_3$SnCOCMe$_3$ (2671).
By rxn. of Ph$_3$SnBz with PhMgBr in Et$_2$O at 0°, in N atm. in the dark, in 47% yield (2671).
By dehydration of Ph$_3$SnOH with AlCl$_3$ at 225-50°, in 56% yield (958).
By hydrolysis of Ph$_3$SnI in aq. EtOH, in 95% yield (1969), of Ph$_3$SnO$_2$CR with wet alumina; R = ferrocenyl, pyrryl, in 70% yield (1310, 1374), of Ph$_3$SnNMeCO$_3$CMe$_3$ with aq. CCl$_4$ (2256).
By thermal dec. of (Ph$_3$Sn)$_2$N$_2$O$_2$ in high vacuum at 80-90° (716), of Ph$_3$SnS$_2$NPh$_2$ at 250°, in 84% yield (298), of Ph$_3$SnNO$_3$ in presence of Ph$_3$CH or Ph$_2$NH (568).
By photolysis of Ph$_3$SnN$_3$ in THF (2471).
By rxn. of Ph$_3$SnLi with ClCH$_2$$\overline{\text{CHCH}_2\text{O}}$ in THF, in 15% yield (1928).
Prop.: IR (110), assignment (1360), Raman, assignment (1360) and Moessbauer resonance (224) spectra, thermogravimetric analysis (959), soly. (2046).
Rxn. with: Cold aq. alkali → Ph$_3$SnOH (110).
Ph$_3$SnH → Ph$_6$Sn$_2$ (400).
Ph$_2$SnH$_2$ → Ph$_3$SnH + (Ph$_2$SnO)$_x$ (1955).

i-Bu$_2$SnH$_2$ → Ph$_6$Sn$_2$ + (i-Bu$_2$Sn)$_x$H$_2$ (1955).
Ph$_3$SiH → Ph$_6$Sn$_2$ + Ph$_3$SnOSiPh$_3$ + Ph$_4$Sn (1165).
(MeHSiO)$_x$ → Ph$_3$SnH + Ph$_4$Sn + Ph$_6$Sn$_2$ (1165).
R$_3$SiOH → Ph$_3$SnOSiR$_3$; R = Me, Ph (2344).
Ph$_2$Si(OH)$_2$ → (Ph$_3$SnO)$_2$SiPh$_2$ (2344).
PhLi → Ph$_4$Sn (2709).
BuLi → BuSnPh$_3$ + Ph$_3$SnOLi (1840).
NaHSO$_3$ → (Ph$_3$Sn)$_2$SO$_3$ + Na$_2$SO$_3$ (2146a).
CH$_2$:CHC⋮CH → Ph$_3$SnC⋮CCH:CH$_2$ + Ph$_3$SnOH (1341).
PhC⋮CH → Ph$_3$SnC⋮CPh + H$_2$O (391, 2495).
HBF$_4$ → Ph$_3$SnBF$_4$ (595).
H$_2$SiF$_6$ → Ph$_3$SnF (1468).
HNO$_3$ → Ph$_3$SnNO$_3$ (1969).
Ph$_2$PO$_2$H → Ph$_3$SnO$_2$PPh$_2$ (292).
Ph$_2$P(O)H → Ph$_3$SnO$_2$PPh$_2$ + Ph$_6$Sn$_2$ (1233).
(RO)$_2$PS$_2$H → Ph$_3$SnS$_2$P(OR)$_2$; R = Me, Et, Bu (292).
(PhNH)$_2$POCl → Ph$_3$SnO$_2$P(NHPh)$_2$ (292).
P$_2$S$_5$ + MeOH → Ph$_3$SnS$_2$P(OMe)$_2$ (1576).
SO$_2$ → (Ph$_3$Sn)$_2$SO$_3$ (2345).
RCO$_2$H → Ph$_3$SnO$_2$CR; R = CH$_2$:CHCO$_2$C$_3$H$_6$ (2974). Oxime-derivatives of oxyacetic acid (2626), 6-H$_2$N-penicillic acid (2236).
ArCH:NO(CH$_2$)$_n$CO$_2$Me → Ph$_3$SnO$_2$C(CH$_2$)$_2$ON:CHAr; Ar = Ph, p-ClC$_6$H$_4$, n = 1, 2 (1574).
Me$_2$C:NOH → Ph$_3$SnON:CMe$_2$ (2627).
PhCH$_2$NH$_3$S$_2$CNHCH$_2$Ph → (Ph$_3$Sn)$_2$S + (PhCH$_2$NH)$_2$CS (1377).
PhNSO → (Ph$_3$Sn)$_2$SO$_3$ (2345).
6-R-purine → N-Ph$_3$SnC$_5$H$_2$N$_4$R; R = H, Cl (1375).
6-MeS-purine → N-Ph$_3$SnC$_5$H$_2$N$_4$SMe·C$_5$H$_3$N$_4$SMe (1375).
6-HS-purine → N,S-Ph$_3$SnC$_5$H$_2$N$_4$SSnPh$_3$·C$_5$H$_3$N$_4$SSnPh$_3$ (1375).
Biol. prop.: Phytotoxicity (306, 1809a).
Activity as fungicide (1809a), against Alternaria tenuis (933, 2297), Fusarium culmorum (933), Phytophthora infestans (2297), Rhizoctonia solani (933), Venturia inaequalis (2297).
Activity as insecticide against Leptinotarsu decemlineata, Musca domestica, chemosterilant (805, 2297), Tineola bisselliella larvae (162).
Use: As catalyst for polyurethan synthesis (113).

Triphenyltin Hydroxide (M-363)   SnC$_{18}$H$_{16}$O   Ph$_3$SnOH
Synth.: By rxn. of Ph$_6$Sn$_2$ with excess SnCl$_4$ in C$_6$H$_6$ followed by aq. NaOH, in 32% yield (2046).
By polarographic reduction of Ph$_6$Sn$_2$ (963).
By hydrolysis of Ph$_3$SnC$_2$PhB$_{10}$H$_{10}$ (784), of Ph$_3$SnNMe$_2$, in 99% yield (250), of Ph$_3$SnN:PPh$_3$ (1402) of (Ph$_3$Sn)$_3$P and (Ph$_3$Sn)$_2$PPh (1849), of (Ph$_3$Sn)$_2$PMPh$_3$ and Ph$_3$SnPPhMPh$_3$; M = Ge, Pb (1850).
By photolysis of Ph$_3$SnN$_3$ in THF (2471).
By rxn. of (Ph$_3$Sn)$_2$O with CH$_2$:CHC⋮CH at r.t. (1341).
By rxn. of Ph$_3$SnO$_2$CMe$_3$ with PhNCO, in 73% yield (2256).

By rxn. of Ph_2SnI_2 with PhI and Zn dust in DMF at 60-100°, followed by hydrolysis (2809).

Prop.: IR, assignment (110, 342), far IR, assignment (342), Moessbauer resonance (224, 1987) and mass (302) spectra, soly. (2046). Anal. detn. by TLC (1253), by differential thermal anal. (302), by thermogravimetric anal. (958).

Rxn. with: Dehydration to $(Ph_3Sn)_2O$ by heating to 120°, refluxing in H_2O, CS_2 (110), C_6H_6 (342).

$Ph_3SnH \to Ph_6Sn_2$ (1618).

$Ph_3SnLi \to Ph_4Sn$ (2709).

Ph_3SiOH + cat. $AlCl_3 \to Ph_3SnOSiPh_3 + Ph(Ph_2SnO)_2 \cdot 5Ph_2SiOH$ (958).

$Ph_2Si(OH)_2$ + cat. $AlCl_3 \to (Ph_2SnOH)_2Ph_6Si_4O_6 + Ph_4Sn$ (958).

$AlCl_3$ at 225-50° $\to (Ph_3Sn)_2O$ (958).

$(BuO)_4Zr \to Ph_3SnOZr(OBu)_3$ (1197).

$(BuO)_4Zr + (HOCH_2CH_2)_3N \to Ph_3SnOZr(OCH_2CH_2)_3N$ (2324).

$(i\text{-}PrO)_4Ti + N(YOH)_3 \to Ph_3SnOTi(OY)_3N$; Y = CH_2CH_2, CH_2CHMe (2324).

$N(CH_2CH_2O)_3TiOCHMe_2 \to Ph_3SnOTi(OCH_2CH_2)_3N$ (882).

$B(OH)_3 \to (Ph_3Sn)_3BO_3$ (2126).

H_2SO_4 (100%) $\to H_2Sn(HSO_4)_6 + PhSO_3H + H_2O^+ + HSO_4^-$ (1100).

$H_2NSO_3H \to Ph_3SnO_3SNH_2$ (1577).

$HN:C:C(CN)_2 \to Ph_3SnNC:C(CN)_2$ (716).

A^- in aq. $C_6H_6 \to$ solvent extraction of anion, followed by tracer detn. (487).

$NH_4X \to Ph_3SnX + H_2O + NH_3$; X = Cl, I, NCS (2660).

$RCO_2H \to Ph_3SnO_2CR$; R = $ClCH_2$ (791), $t\text{-}C_8H_{17}$ (2905), $MeO_2CCH:CH$ (1537), $BuOCS_2CH_2$ (1308), numerous subst. aryl and aklyl groups (866).

$Y(CO_2H)_2 \to (Ph_3SnO_2C)_2Y$; Y = cis-, trans-$CH:CH$ (1557).

$Y(CO_2H)_n \to Y(CO_2SnPh_3)_n$; deriv. of EDTA (1387), polymetacrylic acid (285).

$Y(CO)_2O \to (Ph_3SnO_2C)_2Y$; Y = epoxy-$(C_6H_{10})CHCH_2$ (1305).

$(:CHCO)_2O\text{-}PhCH:CH_2$-polymer $\to$ tin-contn. polyester (1575).

$CH_2:\overline{CCH_2COCH_2} \to Ph_3SnO_2CCH_2Ac$ (2310).

$(HNCO)_3 \to (Ph_3SnNCO)_3$-cyanurate (532, 532a).

$ArOH \to Ph_3SnOAr$; Ar = 2,3,5- 4,3,5-$Cl_2C_5H_2N$, 2,3,5-$Br_2C_5H_2N$ (2516), 2,6-, 2,4-$Cl_2C_6H_3$, (2642), 4-ONC_6H_4, 3,5,4-$Me_2(ON)C_6H_2$ (2641).

$RMe_2CO_2H \to Ph_3SnO_2CMe_2R$; R = Me, Ph (2257).

$C_9H_{19}CO_3H \to Ph_2Sn(OPh)O_2CC_9H_{19} + PhOH$; kinetic data and rxn. mechanism (781-90).

H_2O_2 98% in $Et_2O \to Ph_3SnO_2H$ (118).

$X\text{-}HSC_5H_4N \to x\text{-}Ph_3SnSC_5H_4N$; x = 4 (2516), 2 (2642).

2-$HSC_5H_4NO \to Ph_3SnSC_5H_4NO$-derivative of N-pyridine oxide (1836a).

Biol. prop.: Toxicity (2073, 2399), phytotoxicity (132, 306, 978, 1128, 1190, 1256, 1809a, 2291, 2518).

Activity as acaricide against Aculus pelekassi (1739), Eriophyidae (2023), Phyllocoptruta (2291), P. oleivora (1739).

Relative reactivity as biocide (2905).

Activity as fungicide (132, 306, 1809a) against Alternaria tenuis (2297), Aspergillus flavus, A. niger, A. parasiticus, A. terreus (1344), Capnodium, Cephalothecium roseum (978), Cercospora (150a, 1040), C. beticola (2741, 2798), C. fusca (978), C. personata (2843), Colletotrichum graminicolum (2518), C.

lagenarium (1256), Curvularia maculans (1244), Diplodia zeae (2518), Dothistroma pini (1187), Fusarium (332), F. oxysporum (1244), Fusicladium effusum (978, 2237), Glomerella gossypii (2978), Microphaera alni, Mycosphaerella caryigena (978), Penicillium citrinum, P. puberulum, P. rubrum (1244), Peronospora parasitica (2313), Phaeocytostroma ambiguum (2518), Phytophthora infestans (1053, 2297, 2741, 2976), P. palmivora (1622). Piricularia oryzae (548), Pseudoperonospora cubensis (1256), Pythium ultimum (2978), Rhizoctonia (1040), R. solani (1256, 2978), Rhizopus arrhizus, R. oryzae, R. stolonifer, Sclerotium bataticola (1244), Septoria apii (2741), Sphacelotheca reiliana (2518), Tilletia caries, T. foetida (1366), Venturia inaequalis (2297).

Activity as herbicide against Cynodon dactylon, Cyperus rotundus, Sorghum halepense (2975).

Activity as insecticide including chemisterilant and antifeeding effect against Acyrthosiphon pisum (732), Aphis gossypii (2978), Blattella germanica (260), Callosobruchus chinensis (374, 1568), Chilo agamemnon (354), Culex quinquefasciatus (2616), Epitrix hirtipennis (552), Gnorimoschema opercucella (354), Heliothis virescens (192, 2145), H. zea (2145), Hippelatus collusor (2616), Hylema cilicura (346), Leptinotarsa decemlineata (805, 2297), Musca domestica (172a, 260, 686, 805, 1064, 2230, 2297), Pericallia ricini (2806), Popilla japonica (1382), Spodoptera littoralis (2806), Tineola bisselliella larvae (162), Tribolium confusum (260), Trichoplusia ni (2145).

Activity as mulluscicide (445).

Activity controlling anthracnose, scab and soil rot of cucumbers (249), leaf spot on celery (2704) and sugar beet (829), bunch disease of pecan (2335).

Use: Tradenames and synonyms Dowco 186, DuTer (W20) Fentin (hydroxide), Thompson Hayword: TPTH.

U.S. limits as fungicide residue on potatoes (2943), as fungicide in potatoe and sugar beet cultures (2976), in PVC (2334).

As additive for flame retardant polymers (1185a).

As catalyst for curing polyurethan rubber (539a).

Listing of symmetric triorganotin hydroxides and oxides continues in Table 165.

Table 165. Triorganotin Oxides and Hydroxides

Formula*	Synth. Method**	Yield	Properties	Ref.
[(PhCH$_2$)$_3$Sn]$_2$O (364)	I	--	Use: fungicide in PVC (2334) (a-c)	1964
[(CH$_2$CH$_2$CH)$_3$Sn]$_2$O	I	--	Oily solid (d, e)	1873
[(BuCHCH$_2$CH)$_3$Sn]$_2$O	--	--	(d, e)	--
(s-Bu$_3$Sn)$_2$O	I	75	b$_{0.001}$ 138-42°, n$_D^{20}$ 1.5080, d$_{20}$ 1.22	1618
[(PhMe$_2$CH$_2$)$_3$Sn]$_2$O	I	76	m. 143-44°, IR, NMR spectra	2188
(PhCMe$_2$CH$_2$)$_3$SnOD	IVB	--	m. 143-44°, IR spectrum	1742
[(Me$_3$CCH$_2$)$_3$Sn]$_2$O (365)	--	--	NMR spectrum	2188
[(C$_6$H$_{13}$)$_3$Sn]$_2$O (365)	--	--	(f)	--
[(c.C$_6$H$_{11}$)$_3$Sn]$_2$O (365)	--	--	(g, h)	--
[(C$_{12}$H$_{25}$)$_3$Sn]$_2$O	--	--	(f)	--
[(o-MeC$_6$H$_4$)$_3$Sn]$_2$O	I	--	NMR spectrum (c)	1964
(o-MeC$_6$H$_4$)$_3$SnOH (366)	I	--	IR spectrum (j)	1968
[(p-MeC$_6$H$_4$)$_3$Sn]$_2$O	I	39	White, m. 106-107.5°, soly.	1375
[(p-EtC$_6$H$_4$)$_3$Sn]$_2$O	I	--	(c, i, k)	1964
(R$_3$Sn)$_2$O	--	--	(a)	--
(R$_3$Sn)$_2$O	--	--	Stabilized by phenols	2859
R$_3$SnOH	--	--	Cat. for epoxide hardening	2916

* Numbers in parenthesis refer to pages in main volume.
** The characters used here correspond to the synthetic scheme in the introduction to Chapter 4.3.1.1.
(a) Rxn. with 8-hydroxyquinoline → R$_3$SnOC$_9$H$_6$N; R = PhCH$_2$, p-EtC$_6$H$_4$ (1046). (b) Rxn. with ROH → (PhCH$_2$)$_3$SnOR; R = C$_{17}$H$_{33}$CO(OCH$_2$CH$_2$)$_{5-8}$ (911). (c) Rxn. with HNO$_3$ → R$_3$SnNO$_3$; R = PhCH$_2$, o-, p-C$_6$H$_4$ (1964). (d) Rxn. with R'CO$_2$H → R$_3$SnO$_2$CR'; R = CH$_2$CH$_2$CH, BuCHCH$_2$CH; R' = Me, C$_{11}$H$_{23}$, C$_{17}$H$_{35}$, C$_{17}$H$_{33}$ (1873). (e) Use as herbicide, pesticide (1873). (f) Rxn. with Y:NOH → R$_3$SnON:Y; R = C$_6$H$_{13}$, C$_{12}$H$_{25}$; Y = Me$_2$C, Ph$_2$C (2125). (g) Rxn. with NaHSO$_3$ → (c.C$_6$H$_{11}$)$_3$Sn]$_2$SO$_3$ + Na$_2$SO$_3$ (2146a). (h) Use as acaricide (1306), as rodent repellent (2484). (i) (Far) IR and UV spectra (1375). (j) Rxn. with HX → (o-MeC$_6$H$_4$)$_3$SnX; X = F, Cl (1968). (k) Rxn. with 6-HS-purine → R$_3$SnC$_5$H$_2$N$_4$SSnR$_3$·C$_5$H$_3$N$_4$SSnR$_3$ (1375).

4.3.1.1.2 UNSYMMETRIC TRIORGANOTIN OXIDES AND HYDROXIDES

All compounds are summarized in Table 166.

4.3.1.1.6 FUNCTIONALLY SUBSTITUTED TRIORGANOTIN OXIDES AND HYDROXIDES

Tris(pentafluorophenyl)tin Hydroxide $SnC_{18}HF_{15}O$ $(C_6F_5)_3SnOH$

<u>Synth.:</u> By hydrolysis of $(C_6F_5)_3SnCl$ with aq. NaOH in Et_2O (1198).

<u>Prop.:</u> m. 150° (on rapid heating), (far) IR spectrum, mol. wt., soly., x-ray powder diagram (1198).

<u>Rxn. with:</u> Thermal dec. at 150° → $(C_6F_5)_4Sn$ (1198).
Dehydration at 2 torr. 100° → $[(C_6F_5)_3Sn]_2O$ (1198).
Dehydration at 100° x 10^{-3} torr. → $[(C_6F_5)_2SnO]_x$ + $(C_6F_5)_4Sn$ + C_6F_5H + H_2O (1198).
Aq. HBr → $(C_6F_5)_3SnBr$ (1198).
8-Hydroxyquinoline → $(C_6F_5)_2Sn(OC_9H_6N)_2$ (1198).

Listing of functionally substituted triorganotin oxides and hydroxides continues in Table 167.

Table 166. Unsymmetric Triorganotin Oxides and Hydroxides $(RR'_2Sn)_2O$, RR''_2SnOH

Formula*	Synth. Method**	Yield	Properties	Ref.
$Me_2(\overline{CHCH_2CH})_2SnOH$	I	--	White cryst. (a)	1873
$MePh_2SnOH$	I	82	Dec. without melting	2787
$(PhCH_2)Et_2SnOH$ (364)	--	--	IR spectrum, GLC (b)	1582
$(Et_2SnC_6H_{13})_2O$	--	--	(c)	--
$(Et_2SnC_8H_{17})_2O$ (364)	--	--	Fungicide for Rhizoctonia solani (c)	190
$(Et_2SnC_{12}H_{25})_2O$	--	--	Fungicide for Rhizoctonia solani	190
$(EtBu_2Sn)_2O$	II	--	--	2811
$[Et(c.C_6H_{11})_2Sn]_2O$	II	--	--	2811
$(EtPh_2Sn)_2O$	I	--	--	2808
$(Pr_2SnC_5H_{11})_2O$	I	--	--	2808
	II	--	--	2809
$Bu_2PhSnOH$	I	--	(d)	1577
$n-C_5H_{11}Ph_2SnOH$	I	98	m. 68° (e)	2787
$i-C_5H_{11}Ph_2SnOH$	I	95	m. 86° (e)	2787
$n-C_7H_{15}Ph_2SnOH$	I	95	(e)	2787
$n-C_9H_{19}Ph_2SnOH$	I	72	m. 90° (e)	2787
$n-C_{10}H_{21}Ph_2SnOH$	I	65	m. 93° (e)	2787
$R_3SnO(R_2SnO)_2SnR_3$	--	--	$R = C_8H_{17}$ (f)	269

* Numbers in parenthesis refer to pages in main volume.
** The characters used here correspond to the synthetic scheme in the introduction to Chapter 4.3.1.1
(a) Use as herbicide, pesticide (1873). (b) Anal. detn. by TLC in starch (1581-2). (c) Rxn. with R'OH → Et_2RSnOR'; R = C_6H_{13}, C_8H_{17}; R' = ethoxylated soybean fatty acid (911). (d) Rxn. with $H_2NSO_3H → Bu_2PhSnO_3SNH_2$ (1577). (e) Rxn. with AcOH → RPh_2SnOAc (2787). (f) Use as catalyst for modifying polypropylene (269).

Table 167. Functionally Substituted Triorganotin Oxides and Hydroxides $(R'_3Sn)_2O$, R'_3SnOH

Formula*	Synth. Method**	Yield	Properties	Ref.
$[(p-ClC_6H_4CH_2)_3Sn]_2O$	--	--	(a)	--
$(CH_2:CH)_3SnOH$ (364)	--	--	Dec. 20-100° → $(CH_2:CH)_2SnO + C_2H_4$ (b, c)	1386
$(NCCH_2CH_2)_3SnOH \cdot H_2O$	I	92	m. 103-104°, soly. (d)	1746
$(CH_2:CHCH_2)_3SnOH$	--	--	(b)	--
$Bu_2(CH_2:CHCH_2)SnOH$	III	--	White cryst., m. 105-106°, IR spectrum	4
$[[Cl(CH_2)_4]_3Sn]_2O$	--	--	(e)	--
$[Ph_2(HO)SnCH_2CH_2]_2$	I	--	m. 114-16°, dipole moment	1518, 2189
$(p-FC_6H_4)_3SnOH$	I	58	m. 135-36°, soly. (f, g)	1375
$[(p-ClC_6H_4)_3Sn]_2O$	I	61	m. 121-23°, mol. wt., soly. (f, g)	1375
$(p-ClC_6H_4)_3SnOH$ (366)	--	--	(h)	--
$[(C_6F_5)_3Sn]_2O$	IVA	--	X-ray powder diagram (i)	1198
	I	48	White solid	77
$Bu_3SnSnBu_2OAc$	--	--	See Table 209	946

* Numbers in parenthesis refer to pages in main volume.

** The characters used here correspond to the synthetic scheme in the introduction to Chapter 4.3.1.1.

(a) Rxn. with 8-hydroxyquinoline → $(p-ClC_6H_4CH_2)_3SnOC_9H_6N$ (1046). (b) Rxn. with EDTA → $[(R_3SnO_2CCH_2)_2NCH_2]_2$; R = $CH_2:CH$, $CH_2:CHCH_2$ (1387). (c) Use for control of ripening of fruit (1386). (d) Rxn. with AcOH → $(NCCH_2CH_2)_3SnOAc$ (895b). (e) Rxn. with butyl epoxystearate → tris(4-chlorobutyl)tin epoxystearate (1442). (f) (Far) IR and UV spectra (1375). (g) Rxn. with 6-HS-purine → $R_3SnC_5H_2N_4S \cdot C_5H_3N_4SSnR_3$; R = p-FC_6H_4, p-ClC_6H_4 (1375). (h) Use as pesticide for moth control (473). (i) IR (1198) and Moessbauer resonance (211) spectra.

4.3.1.2 DIORGANOTIN OXIDES AND HYDROXIDES

The diorganotin oxides and hydroxides summarized in Tables 168 and 169 are prepared by the following methods.

I Hydrolysis of diorganotin dihalide by aqueous base in organic solvent.

II Oxidation of diorganotin compounds $(R_2Sn)_x$.

III Thermal decomposition of:
   (IIIA) R_3SnNO_3 ,
   (IIIB) R_3SnOH .

Additional references to diorganotin oxides and hydroxides may be found in Chapters 7.3, 8.1 and 9.

4.3.1.2.1 UNSUBSTITUTED DIORGANOTIN OXIDES AND HYDROXIDES

Dimethyltin Oxide (M-367)　　　　　$(SnC_2H_6O)_x$　　　　　Me_2SnO
Synth.: By hydrolysis of $Me_2Sn(C_6F_5)_2$ in aq. alc. catalyzed by F^-, Cl^- and CN^- ions, in quant. yield (77).
By dec. of Me_3SnO_2H in MeCN, in 31% yield (118).
Prop.: (Far) IR, assignment (110) and Moessbauer resonance (224) spectra, cat. effect on cleavage of Me_6Sn_2 with I (44).
Rxn. with: $EtSnCl_3 \rightarrow (Me_2SnCl)O(EtSnCl_2)$ (944).
$Me_2SnCl_2 \rightarrow Me_2SnCl_2 \cdot 2Me_2SnO$ (2431).
MCl $\rightarrow (Me_2SnCl)OM$; MCl = Bu_2SnCl_2 , Me_2SiCl_2 (946).
$(MeHSiO)_x \rightarrow Me_2SnH_2$ (1165).
$Ph_3SiOH \rightarrow Me_2Sn(OSiPh_3)_2$ (2344).
EtMgBr, + HCl $\rightarrow Me_2EtSnCl$ (149).
$NH_4I \rightarrow Me_2SnI_2 + H_2O + NH_3$ (2660).
$X^- \rightarrow [Me_2Sn \cdot aq]^{++}2X^-$, formation of hydrates by Raman, NMR spectroscopy and EMF measurements; X = Cl, Br (1024).
HCl in MeOH $\rightarrow (Me_2SnCl)O(Me_2SnOMe)$ (2169).
HNO_3 (1:1) $\rightarrow Me_2Sn(OH)NO_3$ (620, 2158).
Aq. $HClO_4 \rightarrow Me_2Sn(ClO_4)_2$ (2147).
$B(OH)_3 + RCOSH + RCO_2H + RSH \rightarrow$ composite derivative (2003).
$RCO_2H \rightarrow Me_3SnO_2CR$ (2611).
$RCO_2H \rightarrow Me_2Sn(O_2CR)_2$; even C_8-C_{18} , oleic, ricinoleic (339) and picolinic (347) acids.
$Y(CH_2CH_2CO_2H)_2 \rightarrow (Me_2SnO_2CCH_2CH_2YCH_2CH_2CO_2)_x$; Y = $Me_2SiOMeRSi(OSiMe_2)$, R = H, $CH_2:CH$ (2185).
$RCO_2Ph \rightarrow Me_3SnO_2CR$; R = Me, Et (1487).
$Ac_2O \rightarrow Me_2Sn(OAc)_2$ (1448).
R_2O or ROH $\rightarrow Me_3SnOR + (Me_2SnOR)_2O$; R = Ac, EtCO (2611).
$(EtCO)_2NH \rightarrow Me_2Sn(O_2CEt)NHOCEt$ (1181).
Epoxyoctadecyl 9,10-epoxystearate $\rightarrow$ methyltin-contg. epoxystearate (1442).
Tropolone in $C_6H_6 \rightarrow Me_2Sn(O_2C_7H_5)_2$ (2726).
$(RCO)_2CH_2 \rightarrow Me_2Sn(OCR:CHCOR)_2$; R = Me, Ph, (Me + Ph) (347).
PhOH + AcOH $\rightarrow Me_3SnOAc$ (2611).

PhOH → Me$_3$SnOPh (2611).
Addn. compd.: Me$_2$SnO·2KOH·OH$_2$, from components (347), struct. K$_2$[Me$_2$Sn(OH)$_4$], NMR spectrum (347), Raman spectrum, EMF measurements (563), dissocn. const. (563-4) [SnK$_2$C$_2$H$_{10}$O$_4$].
Use: As catalyst for polyurethan synth. (229).

Diethyltin Oxide (M-368)　　　　(SnC$_4$H$_{20}$O)$_x$　　　　Et$_2$SnO
Synth.: By rxn. of Et$_4$Sn with Et$_3$SnO$_2$CMe$_3$ at 80-160° (665), with O$_3$ (663).
By rxn. of Et$_6$Sn$_2$ with AcOH at 135° (2096), with Et$_3$SnO$_2$CMe$_3$ at 80-160° (665).
By thermal dec. of Et$_3$SnO$_2$CMe$_2$Ph at 140-70° (666) of Et$_3$SnO$_2$H or Et$_2$Sn(OH)O$_2$H at r.t. in vacuo (8).
By rxn. of Et$_2$SnH$_2$ with (Et$_3$Sn)$_2$O at 20° (1955).
By rxn. of SnCl$_2$ with Et$_3$Al in (CH$_2$OMe)$_2$-MePh at 50° (2498).
By-prod. in photooxidation of Et$_4$Sn (662), Et$_6$Sn$_2$ (661), Et$_3$SnCH$_2$Ph (2215), (Et$_3$Sn)$_2$O (2217).
Prop.: (Far) IR spectra and assignment (321), anal. detn. by β-ray reflection technique (687), by polarographic method (2265).
Rxn. with: Et$_3$SnH → (Et$_3$Sn)$_2$(SnEt$_2$)$_n$; n = 1 (400), 0,1 > 2 (1955).
Me$_2$SnCl$_2$ → (Me$_2$SnCl)O(Et$_2$SnCl) (2431).
Me$_2$SnX$_2$ + H$_2$O → Me$_2$SnX$_2$·3Et$_2$SnO·H$_2$O; X = Cl, I (2431).
Pr$_2$SnBr$_2$ → (Et$_2$SnBr)O(Pr$_2$SnBr) (2431).
Et$_2$Sn(NCS)$_2$ → (Et$_2$SnNCS)$_2$O (591).
Et$_2$Sn(OBz)$_2$ or BzOH → BzO(Et$_2$SnO)$_n$Bz; n = 2, 3 (1720).
(MeHSiO)$_x$ → Et$_2$SnH$_2$ (238, 1242).
RMgX, + HX → Et$_2$RSnX; R, X = C$_5$H$_{11}$, Cl, Br(CH$_2$)$_5$, Br, p-BrC$_6$H$_4$, OAc (149).
RMgBr → Et$_2$SnR$_2$; R = Et, HC⋮C, MeC⋮C, CH$_2$:CHC⋮C (505), CH$_2$:CHC⋮C (1919).
HNO$_3$ (1:1) → Et$_2$Sn(OH)NO$_3$ (120, 2158).
Aq. HClO$_4$ → Et$_2$Sn(ClO$_4$)$_2$ (2040).
Ac$_2$O → Et$_2$Sn(OAc)$_2$ (1449).
(ClCH$_2$CH$_2$) 9,10-epoxystearate (2:1) → ethyltin epoxystearate (1442).
(HOCH$_2$)$_2$CH$_2$ → Et$_2$SnO(CH$_2$)$_3$O (2084).
(HOCH$_2$CH$_2$)$_2$O → Et$_2$SnOCH$_2$CH$_2$OCH$_2$CH$_2$O (2084).
(HOCHR)$_2$ → (Et$_2$SnOCHRCHRO)$_2$; R = H, Me (2084).
Me$_3$CO$_2$H → (Et$_2$SnO$_2$CMe$_3$)$_2$O (2257).
Biol. prop.: Metabolism in rats, modes of excretion (787).
Use: In antifouling paints (1147).
In catalyst for modifying polypropylene (269).

Dipropyltin Oxide (M-368)　　　　(SnC$_6$H$_{14}$O)$_x$　　　　Pr$_2$SnO
Rxn. with: Et$_2$SnBr$_2$ → (Et$_2$SnBr)O(Pr$_2$SnBr) (2431).
Pr$_2$Sn(NCS)$_2$ → (Pr$_2$SnNCS)$_2$O (591).
RMgBr → Pr$_2$SnR$_2$ + Pr$_2$SnBr$_2$; R = HC⋮C (505).
RMgBr → Pr$_2$SnR$_2$; R = CH$_2$:CHC⋮C (505, 1919), PhC⋮C (505).
HNO$_3$ (1:1) → Pr$_2$Sn(OH)NO$_3$·H$_2$O (2158).
HCl in Me$_2$CO → (Pr$_2$SnCl)O(Pr$_2$SnOH) (2169).
Aq. HClO$_4$ → Pr$_2$Sn(ClO$_4$)$_2$ (2040).
RCO$_2$H → (Pr$_2$SnO$_2$CR)$_2$O; R = m-, p-MeC$_6$H$_4$, p-Me$_3$CC$_6$H$_4$ (1289), [4,3,5-HO(Me$_3$C)$_2$-C$_6$H$_2$]$_2$CMeCH$_2$CH$_2$ (1575).

$Ac_2O \rightarrow Pr_2Sn(OAc)_2$ (1449).
$HSCN \rightarrow (Pr_2SnNCS)_2O$ (2164).

Dibutyltin Oxide (M-369) $(SnC_8H_{18}O)_x$ Bu_2SnO

<u>Synth.</u>: By air-oxidation of $Bu_3Sn(Bu_2Sn)_nSnBu_3$ (54), of $(Bu_2Sn)_x$ (468, 1802).
By oxidation of $(Bu_2Sn)_x$ with Me_3CO_2H in C_6H_6 (2096).
By hydrolysis of $Bu_2Sn(C_2PhB_{10}H_{10})_2$ with aq. alc. KOH, in 83% yield (784), of $Bu_2Sn(C\vdots CBu)_2$ (1221), $Bu_2Sn(OAc)_2$ (540).
By rxn. of Bu_2SnH_2 with $(Bu_3Sn)_2O$ at r.t., in 97% yield (468).
By thermal dec. of $(Bu_2SnCl)_2O$ (1514).
By rxn. of Bu_2SnCl_2 with powd. Co, Mg, Zn, in 10-94% yield (1939).
By rxn. of $BuSnI_3$ with BuMgCl after hydrolysis (2734).
By rxn. of Sn with BuI, in 32% yield (403), in presence of cat. Li and $EtOCH_2CH_2OH$, in 45-70% yield (1645), in presence of $(i-PrO)_3Al$, in 92% yield (2733).
By rxn. of Sn with Bu_3PO_4 in presence of cat. BuI, Mg and BuOH (2891).
By rxn. of $SnCl_2$ with Bu_3Al in $(CH_2OMe)_2$-MePh at 50°, in 77% yield (2498).

<u>Prop.</u>: (Far) IR, assignment (110) and Moessbauer resonance (2230, 2509) spectra. Anal. detn. by β-ray reflection technique (687), by polarography (2265), by photometric method (1806), detn. of $BuSnO_2H$ in tech. Bu_2SnO, by EDTA titration (1007).

<u>Rxn. with</u>: $Bu_2SnH_2 \rightarrow (Bu_2Sn)_x + H_2O$ (468, 1802).
$R_3SnX \rightarrow (R_3Sn)O(Bu_2SnX)$; R, X = Et, Cl; Bu, Cl (944), Bu, Br; Bu, AcO (946).
$R_2SnCl_2 \rightarrow (R_2SnCl)O(Bu_2SnCl)$; R = Me, Pr (2431).
$Bu_2SnCl_2 \rightarrow Bu_2SnCl(OSnBu_2)_nCl$; n = 1-12 (947), 2, 3 (2857).
Bu_2SnCl_2, $+ H_2O \rightarrow Bu_2SnCl_2 \cdot 3Bu_2SnO \cdot H_2O$ (2431).
Bu_2SnCl_2, $+ EtOH \rightarrow (Bu_2SnCl)O(Bu_2SnOEt)$ (10).
$Bu_2SnX_2 \rightarrow (Bu_2SnX)_2O$; X = NCS (591), OAc (944).
$Bu_2Sn(OR)_2 \rightarrow (Bu_2SnOR)O(Bu_2SnO)_2R$; R = o-, $p-O_2NC_6H_4CO$, $p-ClC_6H_4CO$, $BzNHCH_2CO$, $ClCH_2CO$, $p-O_2NC_6H_4$, $\overline{OCH_2CH_2CON}$ (2857).
$BuSnCl_3 \rightarrow (Bu_2SnCl)O(BuSnCl_2)$ (944, 947).
$BuSnCl_3 \rightarrow BuSn[(OSnBuCl)_xCl][OSnBu_2]_{2y}Cl_2$; X = 0-4, Y = 0-4 (947).
$Bu_6Sn_2 + I \rightarrow$ catalytic effect of Bu_2SnO on cleavage rxn. (44).
$SnCl_4 + H_2O$ (1:1:1) $\rightarrow (Bu_2SnCl)O(SnCl_2OH)$ (944).
$MCl \rightarrow (Bu_2SnCl)OM$; M = Me_3Si (1096), Me_2SiCl, Me_2GeCl, PhHg (946).
$Ph_3SiH \rightarrow Bu_2SnH_2 + Bu_3SnOSiPh_3$ (1165).
$Bu_3SiH \rightarrow Bu_3SnOSiBu_3$ (1165).
$(Ph_2SiH)_2O \rightarrow Bu_2SnH_2$ (1165).
$(MeHSiO)_x \rightarrow Bu_2SnH_2$ (1242, 2461) + some Bu_3SnH (1165).
$Ph_3SiOH \rightarrow Bu_2Sn(OSiPh_3)_2$ (2344).
$Ph_2Si(OH)_2 \rightarrow Bu_2SnOSiPh_2OSiPh_2$ sic. (2344).
$Me_2Si(OAc)_2 \rightarrow (Bu_2SnOAc)O(Me_2SiOAc)$ (946).
$CH_2:CHSi(OEt)_3 \rightarrow (Bu_2SnOEt)OSi(OEt)_2CH:CH_2$ (1096).
$Ph_2TlBr \rightarrow (Bu_2SnBr)OTlPh_2$ (946).
$Ti(OR)_4 \rightarrow [Bu_2Sn(OR)O]_nTi(OR)_{4-n}$; R = i-Pr; n = 1-4; R = $EtBuCHCH_2$; n = 1 (1096).

$(i-PrO)_4Zr \rightarrow (Bu_2SnOPr-i)OZr(OPr-i)_3$ (1096).

$RLi \rightarrow Bu_2SnR_2$; R = Me, Bu, MeC⋮C (2709).

$CH_2:CHC⋮CMgBr \rightarrow Bu_2Sn(C⋮CCH:CH_2)_2$ (505, 1919).

$EtCl + EtBr + Mg \rightarrow Et_2SnBu_2$ (2814).

$NH_4X \rightarrow Bu_2SnX_2 + NH_3 + H_2O$; X = Cl, Br, I, NCS (2660).

$(PNCl_2)_3 \rightarrow Bu_2SnCl_2 + 3Bu_3SnO \cdot Bu_2SnCl_2$ (2539).

$PhC⋮CH \rightarrow Bu_2Sn(C⋮CPh)_2$ (2495).

$B(OH)_3 + RSH \rightarrow B(OSnBu_2SR)_3$; R = $HSCH_2CO_2CHMeCH_2CH_2O_2CCH_2$ (2971).

$B(OH)_3 + RCO_2H + R'SH \rightarrow$ composite ester (2003).

$H_2SiF_6 \rightarrow Bu_2SnF_2$ (1468).

HNO_3 (1:1) $\rightarrow Bu_2Sn(OH)NO_3$ (2158).

HNO_3 (1:2) $\rightarrow Bu_2Sn(NO_3)_2$ (2158).

$HCl \rightarrow (Bu_2SnCl)O(Bu_2SnOH)$ (2169).

$HX \rightarrow Bu_2SnX(OSnBu_2)_nX$; X = Cl, $p-MeC_6H_4SO_3$, n = 1-3 (2595).

$P_2S_5 + EtBuCHCH_2OH \rightarrow Bu_2Sn[S_2P(OCH_2CHEtBu)_2]_2$ (1576).

$RCO_2H \rightarrow Bu_2Sn(O_2CR)_2$; R = straight chain aliphatic (290), ω-Cl-straight chain (293), substituted alkyl (2974), substituted aryl (1558, 2395, 2732); contg. hindred phenol (1576, 2567), mercapto groups (1170), oxime groups (2626), heterocyclic group (1284).

$RCO_2H \rightarrow (Bu_2SnO_2CR)_2O$; R = Me (2169), $NCCH_2$ (2566), $Ph_2C:C(CN)$ (2568), $\omega-Cl(CH_2)_n$ (293).

$RCO_2H \rightarrow (Bu_2SnO)_x(O_2CR)_y$; R = subst. aryl (1289, 2595), Me, $ClCH_2$, $C_{10}H_{19}$ (2595).

$AcOH \rightarrow (Bu_2SnOAc)O(Bu_2SnOH)$ (2169).

$AcOH \rightarrow (Bu_3Sn)_2O$ (2611).

$ArCH:C(CN)CO_2H + C_{11}H_{23}CO_2H \rightarrow Bu_2Sn(O_2CC_{11}H_{23})O_2CC(CN):CHAr$; Ar = $4,3,5-HO-(Me_3C)_2C_6H_2$ (2571).

$ArCH:C(CN)CO_2H + RCO_2H \rightarrow (Bu_2SnO_2CR)OBu_2SnO_2CC(CN):CHAr$; Ar = $4,3,5-HO(Me_3C)_2-C_6H_2$; R = $C_{11}H_{21}$, $BuO_2CCH:CH$ (2569).

$HS(CH_2)_nCO_2H \rightarrow \overline{Bu_2SnO_2C(CH_2)_nS}$; n = 1 (2802), 2 (1035, 1380a).

$HSCH_2CH_2CO_2H \rightarrow (Bu_2SnO_2CCH_2CH_2S)_n$; n = 2 (1035), n = x (1347).

$HSCH_2CH_2CO_2H + Bu_2SnCl_2$ (5:5:1) $\rightarrow Bu_2SnCl(SCH_2CH_2CO_2SnBu_2)_nSCH_2CH_2CO_2SnBu_2Cl$ (178).

$NCCH_2CO_2H + RR'CO \rightarrow Bu_2Sn[O_2CC(CN):CRR']_2$; R,R' = H, $4,3,5-HO(Me_3C)_2C_6H_2$; Me, i-Bu (2565).

$RCO_2H + i-C_8H_{17}O_2CCH_2SH \rightarrow Bu_2Sn(SCH_2CO_2C_8H_{17})_2 + Bu_2Sn(O_2CR)_2$; R = C_8H_{17}, C_9H_{19} (1758).

$MeO_2CCH:CHCO_2H \rightarrow Bu_2Sn(O_2CCH:CHCO_2Me)_2 + [(:CHCO_2)]_2Sn$ (?) (1557).

$Y(CO_2H)_2 \rightarrow \overline{Bu_2SnO_2CYCO_2}$; Y = numerous alkylene and O-substituted groups (1170, 2917), thioacetal and thioketal groups (1170, 1819, 2028, 2917).

$(:CHCO_2H)_2 \rightarrow \overline{Bu_2SnO_2CCH:CHCO_2} + (Bu_2SnO_2CCH:CHCO_2)_n$; n = 3-4 (1557).

$(CH_2CH_2CO_2H)_2 \rightarrow (Bu_2SnO)_3SnBu_2O_2C(CH_2)_4CO_2$ (2595).

$Ac_2O \rightarrow Bu_2Sn(OAc)_2$ (1449).

$(:CHCO)_2O \rightarrow \overline{Bu_2SnO_2CH:CHCO_2}$ (1641, 2122a).

$(:CHCO)_2O + ROH \rightarrow RO(Bu_2SnO)_2O_2CCH:CHCO_2R$; R = H, Me, Et, n-, i-Pr, n-, i-Bu, $C_{12}H_{25}$ (2872).

$C_7H_8(CO)_2O + (CH_2OH)_2 \rightarrow Bu_2Sn$-contg. polymer; C_7H_8 = 2,3-bicycloheptene-5 (2020).

$\overline{OCH_2CH_2OCO}$ → $(Bu_2SnOCH_2CH_2O)_2$ (950).

RCO_2Me → $Bu_2Sn(O_2CR)_2$; R = p-MeOC$_6$H$_4$CH:NOCH$_2$ (1574).

AcOPh (8:3) → $(Bu_3Sn)_2O$ (1487).

EtCO$_2$Ph → Bu_3SnO_2CEt (1487).

$R'CO_2R$ → Bu$_2$Sn-contg. epoxyester (1442).

HSYCO$_2$R → $Bu_2Sn[SYCO_2(SnBu_2O)_nR]_2$; n = 0, 1; Y = (CH$_2$)$_2$, (CH$_2$)$_3$, CH$_2CMe_2$; R = Bu, i-C$_8H_{17}$ (2861).

AcSR → $Bu_2Sn(SR)OAc$; R = Bu, i-C$_6$H$_{13}$, C$_8$H$_{17}$, Ph (1055).

$(H_2N)_2CO$ → $(Bu_2SnNCO)_2O$ (532) + NH$_3$ + (NH$_4$)$_2$CO$_3$ (532b).

$(RCO)_2NH$ → $Bu_2Sn(O_2CR)NHOCR$; R = PhCH$_2$, Et (1181).

$Y(CO)_2NH$ → $(Bu_2SnO_2CYCONH)_x$; Y = CH$_2$CH$_2$, o-C$_6$H$_4$ (1181).

Hexaneamido epoxystearate → butyltinamido epoxystearate (1442).

Ac$_2$CH$_2$ → $Bu_2Sn(OCMe:CHAc)_2$ (391).

HYCN → $(Bu_2SnNCY)_2O$; Y = O, S (2164).

PhOH → $(Bu_3Sn)_2O$ (2611).

ArOH → $Bu_2Sn(OAr)_2$; Ar = Ph, p-MeC$_6$H$_4$, p-ClC$_6$H$_4$ (1740), substituted nitroaryl (533).

C$_6$Cl$_5$OH → $(Bu_2SnOC_6Cl_5)_2O$ (2138).

ArOH → ArO(Bu$_2$SnO)$_n$OAr; n = 3-6; Ar = Ph, p-H$_2$NC$_6$H$_4$, p-O$_2$NC$_6$H$_4$ (2595).

ROH → $Bu_2Sn(OR)_2$; R = Bu (846a), ethoxylated tall oil fatty acid (911).

$Y(OH)_2$ → $Bu_2\overline{SnOYO}$; Y = cis- or trans-c.C$_6$H$_{10}$, (CHMe)$_2$ (895), (CH$_2$)$_4$, (CH$_2$)$_5$, (CH$_2$)$_6$, $\overline{CH_2CH_2OCH_2CH_2}$ (2084).

$Y(OH)_2$ → $Bu_2\overline{SnOYOSnBu_2OYO}$; Y = CH$_2CH_2$ (895, 2084).

$Y(OH)_2$ → $(Bu_2SnOYO)_x$; several bisphenol derivatives (2123).

Me$_3$CO$_2$H → $(Bu_2SnO_2CMe_3)_2O$ (2257).

RSH → $Bu_2Sn(SR)_2$; R = alkyl (1676, 1997, 2802), O-substituted alkyl (1992, 2536, 2802, 2903 2959), Ph (1676, 1997), C$_6$F$_5$ (1676).

$Y(SH)_2$ → $(Bu_2SnSYS)_x$ (330, 445, 2386).

p-MeC$_6$H$_4$SO$_2$NHR → $Bu_2Sn(NRO_2SC_6H_4Me-p)_2$; R = CH$_2$$\overline{CHCH_2O}$ (1442).

p-MeC$_6$H$_4$SO$_2$NHR + R'SH (1:1:1) → $Bu_2Sn(SR')NRO_2SC_6H_4Me-p$; R = CH$_2$$\overline{CHCH_2O}$, R' = i-C$_8H_{17}O_2CH_2$, C$_{12}H_{25}$ (1442).

Addn. compd.: Bu$_2$SnO·x(PNCl$_3$)$_3$, from components in MePh, IR spectrum (2539).

Use: Safety precautions for handling Bu$_2$SnO (1829).

In antifouling paints (1147), as fungistat (1680).

As catalyst for esterification and transesterification (131, 163, 458, 694, 820-1 1467, 1667, 2429, 2586, 2870, 2984), for modifying polypropylene (269), polyethylene (1897), for synthesis of polyurethan (979, 1209).

As stabilizer in chlorine contg. resins and PVC (405, 1170) in flame resistant polyesters (2702).

As curing agent for polythiopolymercaptans (2083).

In thermal coatings of tin oxide on glass (1019a).

As additive to improve adhesion of polychloroprene (2432).

Dioctyltin Oxide (M-370) $(SnC_{16}H_{34}O)_x$ $(C_8H_{17})_2SnO$

Synth.: By rxn. of $(C_8H_{17})_2SnX_2$ with aq. NaOH in MePh at 70-80°; X = Br, I (2841).

By rxn. of metallic tin with C$_8$H$_{17}$X; X = I, in 28% yield (403), in presence of

Zn, in 40% yield (403); X = Br, I at 130-60° in presence of cat. Li + EtO-
CH$_2$CH$_2$OH, in 32-71% yield (1645), X = I in presence of (i-PrO)$_3$Al at 115-30°,
in 88% yield (2733).
Prop.: (Far) IR spectrum and assignment (110), detn. by photometric method
(1806).
Rxn. with: Bu$_2$SnCl$_2$ → (Bu$_2$SnCl)O[(C$_8$H$_{17}$)$_2$SnCl] (944).
SnCl$_4$ + H$_2$O → [(C$_8$H$_{17}$)$_2$SnCl]OSnCl$_2$OH (944).
Bu$_3$SiH → C$_8$H$_{16}$ + SnSiC$_{30}$H$_{68}$ (1165).
MePhSiCl$_2$ → [(C$_8$H$_{17}$)$_2$SnCl]O(MePhSiCl) (946).
(i-PrO)$_n$M → (C$_8$H$_{17}$)$_2$Sn(OPr-i)OM(OPr-i)$_{n-1}$; M$_n$ = Ti$_4$, Al$_3$ (1096).
Bu$_3$VO$_4$ → [(C$_8$H$_{17}$)$_2$SnOBu]OVO$_3$Bu$_2$ (1096).
Aq. HCl → (C$_8$H$_{17}$)$_2$SnCl$_2$ (2481).
B(OH)$_3$ + C$_{12}$H$_{25}$SH + RSH → Bu[OSn(C$_8$H$_{17}$)$_2$SC$_{12}$H$_{25}$]OSn(C$_8$H$_{17}$)$_2$SR; R = HSCH$_2$CO$_2$CH$_2$-
CHMeO$_2$CCH$_2$ (2003).
RCO$_2$H → (C$_8$H$_{17}$)$_2$Sn(O$_2$CR)$_2$; R = CH$_2$:CHCO$_2$C$_3$H$_6$, CH$_2$:CMeCO$_2$CH$_2$CH$_2$ (2974),
several alkylmercapto groups (1170).
PhCH:C(CN)CO$_2$H + RCO$_2$N → [(C$_8$H$_{17}$)$_2$SnO$_2$CR]OSn(C$_8$H$_{17}$)$_2$O$_2$CC(CN):CHPh ; R =
n-C$_{17}$H$_{35}$, EtOCH$_2$CH$_2$O$_2$CCH:CH (2569).
PhCH:C(CN)CO$_2$H + n-C$_{17}$H$_{35}$CO$_2$H → (C$_8$H$_{17}$)$_2$Sn(O$_2$CC$_{17}$H$_{35}$)O$_2$CC(CN):CHPh (2570).
NCCH$_2$CO$_2$H + p-HOC$_6$H$_4$CHO → [(C$_8$H$_{17}$)$_2$SnO$_2$CC(CN):CHC$_6$H$_4$OH]$_2$O (2566).
NCCH$_2$CO$_2$H + PrCHO + 4,3,5-HO(Me$_3$C)$_2$C$_6$H$_2$CHO → tin-contg. condensation prod.
(2565).
C$_8$H$_{17}$CO$_2$H + RSH → (C$_8$H$_{17}$)$_2$Sn(SR)$_2$ + (C$_8$H$_{17}$)$_2$Sn(O$_2$CC$_8$H$_{17}$)$_2$; R = i-C$_8$H$_{17}$O$_2$CCH$_2$
(1758).
i-C$_8$H$_{17}$O$_2$CCH$_2$SH → (C$_8$H$_{17}$)$_2$Sn(SCH$_2$CO$_2$C$_8$H$_{17}$)$_2$ (1992).
p-(HSCH$_2$)$_2$C$_6$H$_4$ → -(C$_8$H$_{17}$)$_2$SnSCH$_2$C$_6$H$_4$CH$_2$S- (445).
Use: As stabilizer for flame resistant polyester (1170), for PVC and halogen
contg. resins (2702).
As catalyst for polyurethan synthesis (1209).

Diphenyltin Oxide (M-370)   (SnC$_{12}$H$_{10}$O)$_x$ Ph$_2$SnO
Synth.: By hydrolysis of Ph$_2$Sn(C$_6$F$_5$)$_2$ in aq. alc. cat. by Cl$^-$, F$^-$ and CN$^-$ions,
in quant. yield (77).
By thermal dec. of Ph$_3$SnO$_2$H (118), of Ph$_3$SnNO$_3$ in refluxing o-Cl$_2$C$_6$H$_4$ (2388),
of Ph$_2$Sn(S$_2$CNPh$_2$)$_2$ at 250°, in 96% yield (298).
By rxn. of Ph$_2$SnH$_2$ with (Ph$_3$Sn)$_2$O at 20° in C$_6$H$_6$ (1955).
Prop.: IR (110) and Moessbauer resonance (224, 1309) spectra, detn. and sepn.
from Ph$_3$SnOH by TLC (1253).
Rxn. with: Ph$_3$SnH → Ph$_6$Sn$_2$ + Ph$_3$Sn(SnPh$_2$)$_n$SnPh$_3$; n = 1, 2...x (1955).
Ph$_3$SiOH → Ph$_2$Sn(OSiPh$_3$)$_2$ (2344).
MePhSiCl$_2$ → (Ph$_2$SnCl)O(MePhSiCl) (946).
R$_2$Si(O$_2$CR′)$_2$ → (Ph$_2$SnO$_2$CR′); R, R′ = Me, Et (1326, 2005).
PhLi → Ph$_4$Sn (2709).
B(OH)$_3$ + HSCH$_2$CH$_2$OH + (HO$_2$CCH:)$_2$ → composite thiol-ester (2003).
(NC)$_3$CH → Ph$_2$Sn[N:C:C(CN)$_2$]$_2$ (1330).
AcOH → (Ph$_2$SnOAc)$_2$O (1704).
RCO$_2$H → Ph$_2$Sn(O$_2$CR)$_2$; R = PhCH$_2$S(MeS)CH (1170).

RCO_2H at r.t. → $Ph_2S(O_2CR)_2$; R = Me, Et (1326, 2005).
Epoxypropyl 9,10-epoxystearate (2:1) → phenyltin epoxyester (1442).
Ac_2CH_2 → $Ph_2Sn(CHAc_2)_2$ (2418).
C_6Cl_5OH → $(Ph_2SnOC_6Cl_5)_2O$ (2138).
$CS_2 + NH_3$ → $(Ph_2SnS)_3$ (298).
$CS_2 + PhCH_2NH_2$ → $Ph_2Sn(S_2CNHCH_2Ph)_2$ (298).
$PhCH_2NH_3S_2CNHCH_2Ph$ → $(Ph_2SnS)_3 + (PhCH_2NH)_2CS$ (946).

Biol. prop.: Toxicity (189, 1131) and activity as anthelmintic in chicken against Raillietina cesticillus, R. echinobothrida, R. tetragona and Hymenolepsis carioca (189), against cestodes (1131).

Use: As modifying agent for polypropylene-copolymers (2899). As stabilizer for polyolefins (2898).

More unsubstituted diorganotin oxides are listed in Table 168.

Table 168. Unsubstituted Diorganotin Oxides $(R_2SnO)_x$

Formula*	Synth. Method**	Yield	Properties	Ref.
MePhSnO	I	73	Dec. 300-305°, soly.	2787
$(PhCH_2)_2SnO$ (368)	--	--	(a, b)	--
$(CH_2:CH)_2SnO$ (372)	IIIB	--	(c)	4, 1386
$HO[(Me_3C)_2SnO]_nH$	II	--	n = 1.2-1.44 (d, e)	714
$(Me_3C)_2SnO$ (372)	--	--	(f)	--
$(C_5H_{11})_2SnO$ (372)	--	--	(Far) IR spectra, assignment	110
$HO[(Me_2EtC)_2SnO]_nH$	II	--	n = 1-2 (d, e)	714
$HO[(Me_3CCH_2)_2SnO]_nH$	II	--	n = 1-2 (d, e)	714
$(i-C_5H_{11})PhSnO$ (372)	I	81	m. 275° (dec.)	2787
$(C_6H_{13})_2SnO$ (372)	I	--	(g)	90
$(c.C_6H_{11})_2SnO$	--	--	(a)	--
$(Me_2CHCHCH_2CH)_2SnO$	I	--	Use: herbicide, pesticide	1873
$(EtBuCHCH_2)_2SnO$ (372)	--	--	(h, i)	--
$(C_{12}H_{25})_2SnO$ (370)	--	--	(Far) IR spectra, assignment (a, j)	110
$(C_{18}H_{37})_2SnO$	--	--	(a)	--
$(p-MeC_6H_4)_2SnO$ (373)	IIIA	--	--	2388
$(p-EtC_6H_4)_2SnO$	--	--	(k)	--
$Mesityl_2SnO$ (373)	--	--	Refractometric data	1391
$(C_{10}H_7)_2SnO$	--	--	(l)	--
R_2SnO	--	--	(m)	--

* Numbers in parenthesis refer to pages in main volume.
** The characters used here correspond to the synthetic scheme in the introduction to Chapter 4.3.1.2.
(a) Rxn. with R'9,10-epoxystearate → organotin epoxystearate (1442). (b) Rxn. with RSH → $(PhCH_2)_2Sn(SR)_2$; R = $i-C_8H_{17}O_2CCH_2$ (751), with HX → $(PhCH_2)_2SnX_2$; X = Cl,Br,I,OAc (2069). (c) Rxn. with $C_6H_{13}CH(C_6H_4OH-p)$ → polymeric bisphenol

derivative (2123). (d) Biol. activity as bactericide Aereobacter aerogenes, Bacillus subtilis, Escherichia coli, Pseudomonas aeruginosa, Staphylococcus aureus (714). (e) Forms water soluble K salt (714). (f) Use as stabilizer for polyolefins (2898), as modifying catalyst for propylene copolymers (2899). (g) Rxn. with (:CHCO)$_2$O → (C$_6$H$_{13}$)$_2$SnO$_2$CCH:CHCO$_2$ (90), with C$_7$H$_8$(CO)$_2$O + HO(CH$_2$)$_6$OH → (C$_6$H$_{13}$)$_2$Sn-contg. polyester; C$_7$H$_8$ = 2,3-bicycloheptene-5 (2020). (h) Rxn. with $\overline{\text{OCHRCH}_2\text{CO}_2}$ → (EtBuCHCH$_2$)$_2$SnO$_2$CCH$_2$CHRCO$_2$; R = epoxy-C$_4$H$_7$O, -C$_6$H$_{11}$O, -C$_8$H$_{15}$O (1305). (i) Use as catalyst for polyurethan synth. (3005). (j) Rxn. with (p-HOC$_6$H$_4$)$_2$CMe$_2$ → polymeric lauryl-tin bisphenol derivative (2123). (k) Rxn. with 8-hydroxyquinoline → (p-EtC$_6$H$_4$)$_2$Sn(OC$_9$H$_6$N)$_2$ (1046). (l) Rxn. with R' epoxytallate (2:1) → naphthyltin epoxytallate (1442). (m) Use as catalyst for water-proofing fabric with polymethylsiloxane (2948), for hardening of epoxide resins (2916).

4.3.1.2.6 FUNCTIONALLY SUBSTITUTED DIORGANOTIN OXIDES AND HYDROXIDES

All compounds are listed in Table 169.

Table 169. Functionally Substituted Diorganotin Oxides and Hydroxides   (R'$_2$SnO)$_x$, (R'$_2$SnOH)$_2$O

Formula*	Synth. Method**	Yield	Properties	Ref.
(p-ClC$_6$H$_4$CH$_2$)$_2$SnO	--	--	(a)	--
(NCCH$_2$CH$_2$)$_2$SnO	I	--	IR spectrum (b, c)	1483
[(NCCH$_2$CH$_2$)$_2$SnOH]$_2$O	II	83	m. 195° (dec.) (d, e)	1746
(NCCH$_2$CH$_2$CH$_2$)$_2$SnO	I	--	IR spectrum	1483
(BzCH$_2$CH$_2$)$_2$SnO	I	--	(f)	1484
[NC(CH$_2$)$_4$]$_2$SnO	I	--	IR spectrum (g, h)	1483
(p-ClC$_6$H$_4$)$_2$SnO (374)	--	--	(a, i)	--
(p-BrC$_6$H$_4$)$_2$SnO (374)	--	--	(i)	--
(p-IC$_6$H$_4$)$_2$SnO (374)	--	--	(i)	--
(C$_6$F$_5$)$_2$SnO	IIIB	--	m. > 300°, IR spectrum	1198
[(o-PhOC$_6$H$_4$)$_2$SnOH]$_2$O (374)	--	--	Dimeric, mol. wt.	10
(C$_4$H$_3$S)$_2$SnO	--	--	(j)	--
(B$_{10}$H$_{10}$C$_2$Ph)$_2$SnO	I	62	m. 218-20° (dec.) (h)	--

* Numbers in parenthesis refer to pages in main volume.
** The characters used here correspond to the synthetic scheme in the introduction to Chapter 4.3.1.2.
(a) Rxn. with 8-hydroxyquinoline → R$_2$Sn(OC$_9$H$_6$N)$_2$; R = p-ClC$_6$H$_4$CH$_2$, p-ClC$_6$H$_4$ (1046). (b) Rxn. with HX → (NCCH$_2$CH$_2$)$_2$SnX$_2$; X = Cl, Br (1483). (c) Rxn. with alc. NaOH, + HCl in MeOH → (MeO$_2$CCH$_2$CH$_2$)$_2$SnCl$_2$ (1483). (d) Rxn. with RSH → (NCCH$_2$CH$_2$)$_2$Sn(SR)$_2$; R = i-C$_8$H$_{17}$O$_2$CCH$_2$ (895b, 1746). (e) Rxn. with RCO$_2$H → [(NCCH$_2$CH$_2$)$_2$SnO$_2$CR]$_2$O; R = EtBuCH (895b, 1746), C$_6$H$_{11}$O$_2$CCH:CH (895b). (f) Rxn. with HBr → (BzCH$_2$CH$_2$)$_2$SnBr$_2$ (1483). (g) Rxn. with HBr → [NC(CH$_2$)$_4$]$_2$SnBr$_2$

(1483). (h) Rxn. with alc. NaOH, + HX → [HO$_2$C(CH$_2$)$_4$]SnX$_2$; X = Cl, Br (1483).
(i) Moessbauer resonance spectrum (1309). (j) Rxn. with (C$_4$H$_8$O)CH$_2$
9,10-epoxystearate (2:1) → thienyltin epoxystearate (1442). (k) Rxn. with
dil. HBr in C$_6$H$_6$ → (B$_{10}$H$_{10}$C$_2$Ph)$_2$SnBr$_2$ (784).

4.3.1.3 MONOORGANOTIN OXIDES AND HYDROXIDES

The following scheme summarizes synthetic methods for the compounds listed in Table 170.

I Hydrolysis of:
   (IA) organotin trihalides ,
   (IB) organotin trialkoxides ,
   (IC) organotin halide during polarography.

II Alkylation of SnX$_4$ with CH$_2$Br$_2$ and activated Mg.

III Reaction of metal hydroxides with ethylstannoic acid.

Additional references to monoorganotin oxides and hydroxides may be found in Chapter 9.

Bis(methyltin) Trioxide (M-376)   (Sn$_2$C$_2$H$_6$O$_3$)$_x$ Me$_2$Sn$_2$O$_3$
Synth.: By rxn. of MeSnCl$_3$ with aq. Na$_2$CO$_3$ (2372).
Prop.: x = 1-15 (2372).
Rxn. with: Koji acid in dil. HX → MeSn(C$_6$H$_5$O$_4$)$_2$X; X = Cl, Br (1658).
8-Hydroxyquinoline in dil. HCl → MeSn(OC$_9$H$_6$N)$_2$Cl (1291).
Tropolone in dil. HX → MeSn(O$_2$C$_7$H$_5$)$_2$X; X = Cl, Br (2510).
Aq. Na$_2$S in dil. HCl → (Me$_2$Sn$_2$S$_3$)$_x$ (1347).
RSH → polymeric methyltin mercaptides (2372).

Methylstannoic Acid (M-376)   (SnCH$_4$O$_2$)$_x$ MeSnO$_2$H
Synth.: By polarography of MeSnCl$_3$ at pH 4-7 (973).
Prop.: NMR spectrum in concd. HCl (31).
Rxn. with: Concd. HCl → H$_2$[MeSnCl$_5$] (31).
Anhyd. HX → MeSnX$_3$; X = Cl, Br (289).
Concd. KOH → [MeSn(OH)$_5$]$^=$ (289).
Ac$_2$CH$_2$ → MeSn(CHAc$_2$)$_2$Cl (2418).

Ethylstannoic Acid (M-376)   (SnC$_2$H$_6$O$_2$)$_x$ EtSnO$_2$H
Synth.: By hydrolysis of EtSnCl$_3$ (2351).
By rxn. of EtSnCl$_3$ with OH$^-$ loaded ion exchange resin, in 59% yield (1435).
By polarography of EtSnCl$_3$ at pH 5 (970).
Prop.: IR spectrum (2351), x = 3, mol. wt., infusible solid (1435).
Rxn. with: HCl → EtSn(OH)$_2$Cl (2351).
MOH → EtSn(OM)$_3$·nH$_2$O; M n = Na 3, K 9 (2351).
Ba(OH)$_2$ → EtSnO$_3$HBa (2351).

Bis(butyltin) Trioxide (M-376)   (Sn$_2$C$_8$H$_{18}$O$_3$)$_x$ Bu$_2$Sn$_2$O$_3$
Synth.: By rxn. of BuSnCl$_3$ with aq. NaOH at 90° (2372).

By thermal dec. of $(Bu_2Sn)_x$ at 160°, followed by I in THF and hydrolysis, in 5% yield (1941).
By azeotropic dehydration of $BuSnO_2H$ in C_6H_6 (1170a).
Prop.: IR spectrum (110), n = 1-15 (2372), anal. detn. by degradation with HNO_3 and complexometric titration (168).
Rxn. with: $RCO_2H \to BuSn(O_2CR)_3$; R = $H_2C:CHCO_2C_3H_6O_2CCH:CH$ (2974).
RCO_2H (1:2.3) → polymeric butyltin carboxylates; R = $HOCH_2$, MeO_2CCH_2 (1170a).
8-Hydroxyquinoline → $BuSn(OC_9H_6N)_3$ (1046).
Tropolone in dil. HX → $BuSn(O_2C_7H_5)_2X$; X = Cl, Br (2510).
HCl, + aq. Na_2S → $(Bu_2Sn_2S_3)_2$ (1347).
$HS(CH_2)_nCO_2C_8H_{17}$-i (1:< 3) → polymeric mercaptide; n = 1 (1170a, 2372), n = 2 (1170a).
Biol. Prop.: Activity as molluscicide (445).
Use: As non-toxic stabilizer for PVC and chlorine contg. polymers (980).

Butylstannoic Acid (M-376) $(SnC_4H_{10}O_2)_x$ $BuSnO_2H$
Synth.: By hydrolysis of $BuSnCl_3$, in 83% yield (1435), of $BuSnI_3$ (1170a, 2815).
By rxn. of metallic tin with Bu_3PO_4 in BuOH in the presence of cat. BuI + Mg (2891).
Prop.: Mol. wt., x = 11 (1435), detn. in tech. Bu_2SnO by titration with EDTA (1007).
Rxn. with: $Ph_3SiOH \to Bu(OSiPh_3)_3$ (2344).
$(EtO)_4Si + HSCH_2CH_2OH \to [BuSn(SCH_2CH_2O)_3]_4Si_3$ (2536).
$BuMgCl \to Bu_4Sn$ (2815).
$RCO_2H \to BuSn(O_2CR)_3$; R = H (1319), derivatives of various substituted mercapto acids (1170).
RCO_2H (1:1) → polymeric butyltin carboxylate; R = Me, CH_2OH (1170a).
$CH_2:CMeCO_2H \to (BuSn)_3O_2(O_2CCMe:CH_2)_5$ (637).
$\overline{CH_2CH_2O}$ → relative reactivity as polymerization catalyst (2572).
$i-C_8H_{17}O_2CCH_2SH$ (1:1) → polymeric butyltin mercaptide (1170a).
C_6H_6 under reflux → $(Bu_2Sn_2O_3)_x$ (1170a).
Use: For surface coatings of dye pigments (37). As stabilizer for PVC (1170).

Listing of monoorganotin oxides and hydroxides concludes in Table 170.

Table 170. Monoorganotin Oxides and Hydroxides $R_2Sn_2O_3$, $RSnO_2H$

Formula*	Synth. Method**	Yield	Properties	Ref.
$Et_2Sn_2O_3$	--	--	(a)	--
$EtSn(ONa)_3 \cdot 3\ H_2O$	III	--	White infusible solid, IR spectrum (b)	2351
$(EtSnO_2)_6K_5H \cdot 9\ H_2O$	III	--	White infusible solid, IR spectrum (b)	2351
$EtSnO_3HBa \cdot 1.5\ H_2O$	III	--	White infusible solid, IR spectrum (b)	2351
$Pr_2Sn_2O_3$	--	--	(c)	--
$PrSnO_2H$	--	--	(d, e)	--
$BuSn(OH)_3$	--	--	(f)	--
$(C_6H_{13}Sn)_2O_3$ (376)	--	--	(g)	--
$(C_8H_{17}Sn)_2O_3$	IA	--	x = 1-15 (g, h)	2372
$C_8H_{17}SnO_2H$	IA	95	(i)	1435
$(C_{12}H_{25}Sn)_2O_3$	IA	--	x = 1-15 (h)	2372
$(C_{18}H_{37}Sn)_2O_3$	IA	--	x = 1-15 (h)	2372
$Ph_2Sn_2O_3$	--	--	IR spectrum (a)	110
$PhSnO_2H$ (376)	IC	--	Detn. in Ph_3SnOH by TLC (1235) (j, k)	973
$HO_2Sn(CH_2SnO)_xOH$	II	--	Colorless solid	1352
$[C_{10}F_{21}(CH_2)_3O(CH_2)_3Sn]_2O_3$	IA	--	(1)	745a
$C_{10}F_{21}CH_2CH_2SnO_2H$	IB	--	(1)	745a
$[Cl(CF_2)_{10}CH_2CH_2Sn]_2O_3$	IB	--	(1)	745a
$[C_{10}F_{21}(CH_2)_6Sn]_2O_3$	IA	--	(1)	745a
$C_6F_5SnO_2H$	IA	--	IR spectrum	1198

* Numbers in parenthesis refer to pages in main volume.
** The characters used here correspond to the synthetic scheme in the introduction to Chapter 4.3.1.3.

(a) Rxn. with 8-hydroxyquinoline → $RSn(OC_9H_6N)_3$; R = Et, Ph (1046). (b) Decomposes on heating, soly. (2351). (c) Rxn. with HCl, + Na_2S → $(Pr_2Sn_2S_3)_x$ (1347). (d) Rxn. with $CH_2:CHCO_2H$ → $(PrSn)_3O_2(O_2CCH:CH_2)_5$ (637). (e) Use for surface coatings of dye pigments (34). (f) Rxn. with RCO_2H → $BuSn(O_2CR)_3$; R = derivs. of thioacetal or thioketal acids (213). (g) Use as non-toxic stabilizer for PVC and chlorine containing polymers (980). (h) Rxn. with RSH → polymeric organotin mercaptides (2372). (i) Rxn. with H_2YCO_2H → $C_8H_{17}SnO_2CY$; Y = $(O_2CCH_2CH_2S)_2CMeCH_2CH_2$ (1170), with RCO_2H (1:1) → polymeric octyltin carboxylate; R = $BuO_2CCH:CH$ (1170a). (j) Rxn. with HCO_2H → $PhSn(O_2CH)_3$ (1319). (k) Rxn. with i-$C_8H_{17}O_2CCH_2SH$ (1:1) → polymeric PhSn-mercaptide (1170a). (1) Use as water, oil and dirt repellent, antiadhesive and release agent (745a).

4.3.1.9 ORGANOTIN OXIDES AND HYDROXIDES CONTAINING TIN-HETEROATOM BONDS

Organotin compounds containing a tin-oxygen and a tin-heteroatom bond, supposedly in the same molecule, are listed elsewhere. The following heteroatoms or groups and sections conform: hydrogen - 3.8, halogen - 4.1.5.1, pseudohalogen - 4.2.1.5.1, anions - 4.2.4.6, alkoxide - Tables 182, 184, 185 and 186, carboxylate - 4.3.5.2.2, 4.3.5.2.8 and Tables 209, 212, 214 through 218 and 221.

4.3.2 ORGANOTIN PEROXY COMPOUNDS

The organotin peroxides and hydroperoxides in Table 171 are prepared by methods from the following scheme.

I Photooxidation of:
    (IA) R_4Sn ,
    (IB) R_6Sn_2.
II Reaction of alkylhydroperoxides and hydrogen peroxide with:
    (IIA) $(Et_3Sn)_2O$,
    (IIB) R_3SnOH ,
    (IIC) Et_3SnOMe ,
    (IID) Et_2SnO.
III Addition of t-BuO_2H to $Me_3SnO_2CMe_3$.

Trimethyltin Hydroperoxide $SnC_3H_{10}O_2$ Me_3SnO_2H
Synth.: By rxn. of Me_3SnOH with 98% H_2O_2 in Et_2O and MePh at 0°, in 81 and 90% yield, resp. (118).
Prop.: m. 97° (dec.), IR spectrum, stable for 1 day at r.t. (118).
Rxn. with: Thermal dec. in MeCN → Me_3SnOH + Me_2SnO + $MeCONH_2$ + MeOH and O; kinetic data (118).
Thermal dec. in dioxane → Me_3SnOH + Me_2SnO + MeOH + O, kinetic data (118).
AcOH → Me_3SnOAc (118).

Triethyl-t-Butylperoxide (M-379) $SnC_{10}H_{24}O_2$ $Et_3SnO_2CMe_3$
Synth.: By rxn. of Et_3SnOMe or $(Et_3Sn)_2O$ with Me_3CO_2H in hexane, in 99% yield (667).
Prop.: b_1 62-62.5° (667).
Rxn. with: Et_4Sn at 80-160° → $(Et_3Sn)_2O$ + Et_2SnO + Me_3COH + C_2H_4 + C_2H_6 + CH_4 + Me_2CO; kinetic data and rxn. mechanism (665).
Et_6Sn_2 → $Et_3SnOCMe_3$ + prod. identical to preceding rxn. (665).

Tributyl-t-Butylperoxide (M-378) $SnC_{16}H_{36}O_2$ $Bu_3SnO_2CMe_3$
Synth.: By rxn. of $(Bu_3Sn)_2O$ with Me_3CO_2H at 0.1 torr., in quant. yield (2257).
Prop.: IR and NMR spectra, stable at r.t., readily hydrolyzed (2257).
Rxn. with: o-$C_6H_4(CO)_2O$ → o-$Bu_3SnO_2CC_6H_4CO_3CMe_3$ (2257).
MeNCO in CCl_4 → Bu_2SnCl_2 + BuCl + $Cl_3CCH_2CHClEt$ + $HCCl_3$ + $MeNHCO_2CMe_3$ (2256).
Me_3CO_2H + RNCO → $RNHCO_3CMe_3$; R = Me, Et (2256).

Triphenyltin-t-Butylperoxide (M-380) $SnC_{22}H_{24}O_2$ $Ph_3SnO_2CMe_3$

<u>Synth.</u>: By rxn. of Ph_3SnOH with Me_3CO_2H and azeotropic distn. with C_6H_6, in 90% yield (2257).

<u>Prop.</u>: m. 61-66°, IR spectrum, hydrolyzes rapidly (2257).

<u>Rxn. with</u>: $o-C_6H_4(CO)_2O \rightarrow o-Ph_3SnO_2CC_6H_4CO_3CMe_3$ (2257).
RNCO → $Ph_3SnNRCO_3CMe_3$; R = Me, Et, Bu (2256).
PhNCO → $Ph_3SnOH + PhNHCO_2CMe_3$ (2256).

Listing of organotin peroxy compounds continues in Table 171.

Table 171. Organotin Peroxy Compounds $(R_3SnO)_2$, $R_{4-n}Sn(O_2R)_n$

Formula*	Synth. Method**	Yield	Properties	Ref.
$Me_3SnO_2CMe_3$ (379)	IIB	61	IR and NMR spectra, mol. wt. (a)	2257
$Me_3SnO_2CMe_3 \cdot Me_3CO_2H$	III	--	m. 28°, soly.	2257
	II	--	IR and NMR spectra	2257
$(Et_3SnO)_2$ (378)	IB	--	IR spectrum (b, c)	661
$2Et_3SnO_2H \cdot H_2O_2$	IIA (c)	--	m. 35-36° (d)	8
Et_3SnO_2Et (379)	IA	16	Mechanism for dec. and oxidation	662
$Et_3SnO_2CMe_2Ph$ (379)	IIA, C	92	b_2 105-10° (e)	667
$Et_2Sn(OH)O_2H$	(d)	--	Explodes at 150-58° (f)	8
$(Et_2SnO_2CMe_3)_2O$ (379)	IID	--	m. 135-43°	2257
Ph_3SnO_2H	IIB	45	Explodes at 75° (g)	118
$Ph_3SnO_2CMe_2Ph$ (380)	IIB	88	m. 105-10°, IR spectrum, stable at r.t.	2257
$Ph_3SnO_2C(O)C_9H_{19}$	IIB	--	(h)	789

* Numbers in parenthesis refer to pages in the main volume.
** The characters used here correspond to the synthetic scheme in the introduction to Chapter 4.3.2.

(a) Slow decomposition on standing (2257). (b) Thermal decomposition → $Et_2SnO + Et_3SnOEt$ (661). (c) Rxn. of $(Et_3SnO)_2$ with H_2O_2 (1:2) in Et_2O, + anhyd. $MgSO_4$ → $2 Et_3SnO_2H \cdot H_2O_2$ (8). (d) Rxn. of $2 Et_3SnO_2H \cdot H_2O_2$ in vacuo at r.t. → $Et_2Sn(OH)O_2H$ (8). (e) Thermal decomposition at 140-70° → $Et_3SnOH + Et_2SnO + (Et_3Sn)_2O + Et_3SnOCMe_2Ph + Me_2PhCOH + AcPh$ (666). (f) Decomposition at r.t. in vacuo → Et_2SnO (8). (g) Decomposition in MeCN → $Ph_2SnO + PhOH$ (118).

4.3.3 ORGANOTIN ALKOXIDES AND PHENOXIDES

There is considerable new activity with organotin alkoxides due to commercial interest and novel synthetic reactions. The latter compose addition of organotin alkoxide and organotin amide to epoxides, carbonyl compounds and lactones and condensation of organotin alkoxide with protic agents.

4.3.3.1 TRIORGANOTIN ALKOXIDES

Triorganotin alkoxides are reported from hydrocarbon alcohols and phenols and a large variety of functionally substituted hydroxy compounds. Derivatives listed in Tables 172 through 180 are prepared by methods from the following scheme.

- I Insertion reactions into labile carbon-tin bonds:
  - (IA) $Et_3SnCH_2R' + RR^xCO \rightarrow Et_3SnOCRR^xCH_2R'$; $R^x = H, R,$
  - (IB) $Et_3SnC \vdots CPh + R_2CO \rightarrow Et_3SnOCR_2C \vdots CR$.

- II Hydrostannation reactions:
  - (IIA) $R_3SnH + RR^xCO \rightarrow R_3SnOCHRR^x$; $R^x = H, R,$
  - (IIB) $R_3SnH + RNO \rightarrow R_3SnONHR$.

- III Reaction with organotin halides:
  $R_3SnX + R'OM \rightarrow R_3SnOR'$.

- IV Reactions with organotin oxides, hydroxides and alkoxides:
  - (IVA) $R_3SnOR^x + R'OH \rightarrow R_3SnOR'$; $R^x = H, R_3Sn,$
  - (IVB) $R_3SnOR + R'OH \rightarrow R_3SnOR'$.

- V Reactions with organotin amides:
  - (VA) $R_3SnNR_2 + R'OH \rightarrow R_3SnOR'$,
  - (VB) $R_3SnNR_2 + R'_2CO \rightarrow R_3SnOCR'_2NR_2$,
  - (VC) $R_3SnNR_2 + \overline{OCH_2CHR'} \rightarrow R_3SnOCH_2CHR'NR_2$,
  - (VD) $R_3SnNR_2 + RNO \rightarrow R_3SnONRNR_2$.

- VI Miscellaneous reactions:
  - (VIA) $R_3SnOR' + R'_2CO \rightarrow R_3SnOR' \cdot OCR'_2$,
  - (VIB) $(Me_2SnOR)_2O + heat \rightarrow Me_3SnOR'$,
  - (VIC) $R_3SnOPh + Hg(OAc)_2 \rightarrow R_3SnOC_6H_4AgOAc$,
  - (VID) $R_3SnOCHMeC \vdots CH + R_3PbOH \rightarrow R_3SnOCHMeC \vdots CPbR_3$.

R = organic group, may be different in one molecule; R' = R, may be functionally substituted; M = Li, Na, K, MgX; X = halogen. Additional references to triorganotin alkoxides may be found in Chapter 8.1.

4.3.3.1.1 UNSUBSTITUTED TRIORGANOTIN ALKOXIDES

Trimethyltin Methoxide (M-384) $SnC_4H_{12}O$ Me_3SnOMe
Synth.: By rxn. of Me_3SnNMe_2 with $CH_2:CMeCO_2Me$ (854).
By rxn. of Me_3SnNEt_2 with MeOH in pentane in N atm. (323).
By rxn. of Me_3SnCl with NaOMe in C_6H_6, in 63% yield (350).
Prop.: m. 75° (323), (far) IR, NMR (323) and Moessbauer resonance (224) spectra, assocn. in Et_2O, C_6H_6 (350).
Rxn. with: $\overline{OCH_2CH_2C:O} \rightarrow Me_3SnOCH_2CH_2CO_2Me + Me_3SnO_2CCH_2CH_2OMe$; solvent effect and prod. ratio (1235).

Trimethyltin Phenate   SnC$_9$H$_{14}$O   Me$_3$SnOPh

Synth.: By rxn. of Me$_3$SnCl with PhONa (471).
By titration of Me$_3$SnOH with PhOH in presence of molecular sieves in CCl$_4$ (342).
By rxn. of Me$_3$SnNMe$_2$ with AcOPh (854).
By thermal dec. of (Me$_2$SnOPh)$_2$O at 220-300° at 8 torr, in 81% yield (2611).
Prop.: b$_{17}$ 122-25° (471), IR (342, 985), UV and NMR (342) spectra.

Triethyltin Methoxide (M-382)   SnC$_7$H$_{18}$O   Et$_3$SnOMe

Synth.: By rxn. of MeOH with Et$_3$SnCH$_2$Ac under reflux, in 92% yield (437), with Et$_3$SnOCH:CMe$_2$ (1712), with Et$_3$SnNMe$_2$ in Et$_2$O, in 98% yield (250), with Et$_3$SnNEt$_2$ in pentane in N atm. (323), with Et$_3$SnN:PPh$_2$ (1429).
Prop.: b$_{15}$ 78-79° (437), b$_{12}$ 70° (250), b$_{0.1}$ 37° (323) n$_D^{20}$ 1.4750 (250, 437), d$_{20}$ 1.3059 (437), NMR (323) and Moessbauer resonance (2489) spectra.
Rxn. with: R$_3$SnH → Et$_3$SnH + R$_3$SnOMe; R = Me, Ph, solvent effect (915).
Et$_3$SnOCHMeCH:CHMe → Et$_3$SnCMe:CHCHMeOSnEt$_3$ (1538).
R$_3$SiH → Et$_6$Sn$_2$ + R$_3$SiOMe + MeOH; R = Et, Pr, Ph (2097).
Me$_3$SiOCH$_2$CH$_2$C⋮CH → Et$_3$SnC⋮CCH$_2$CH$_2$OSiMe$_3$ (1910).
Et$_3$GeH → Et$_6$Sn$_2$ + Et$_3$GeOMe + MeOH (2097).
Et$_3$GeLi → Et$_3$SnGeEt$_3$ (2097).
R$_n$GeCl$_{4-n}$ → R$_n$Ge(OMe)$_{4-n}$; R = MeO$_2$CCH$_2$; n = 2, 3 (692).
NaC$_2$H → Et$_3$SnC⋮CH + (Et$_3$SnC⋮)$_2$ (1342).
RC⋮CH → Et$_3$SnC⋮CR; R = CH$_2$:CH (1920), MeOCH$_2$, CH$_2$:CHOCH$_2$CH$_2$ (511), MeOCH$_2$OCMe$_2$ (1910), MeO$_2$C (311), EtO (1710), BuOCHMeY; Y = CH$_2$, CH$_2$CH$_2$, CHMe, CMe$_2$ (507).
HC⋮CCMe$_2$OH at 200-250° → (Et$_3$SnC⋮)$_2$ (2776).
MeCH(OCHMeC⋮CH)$_2$ → (Et$_3$SnC⋮CCHMeO)$_2$CHMe (2078).
(CH$_2$:C:O)$_2$ → Et$_3$SnOCMe:CHCO$_2$Me (1709).
CH$_2$:CMeOAc → Et$_3$SnCH$_2$Ac + MeOAc (2489).
Y:CHOAc → Et$_3$SnOCH:Y + MeOAc; Y = Me$_2$C, Me$_2$CHCH (1712).
1-C$_6$H$_9$OAc → 1-Et$_3$SnOC$_6$H$_9$c. (1709).
Cellulose acetate or cellulose → tin-contg. polymer (1916).
R'CONHR → Et$_3$SnNRCOR'; R = c.C$_6$H$_{11}$, Ph; R' = H, MeO (402).
HCSNHPh → (Et$_3$Sn)$_2$S (402).
MeCSNHPh → Et$_3$SnSCMe:NPh (402).
H$_2$NCO$_2$Et → Et$_3$SnNCO (857).
PhNHC(S)OEt → Et$_3$SnS(O)CEt (402).
ArOH → Et$_3$SnOAr; Ar = 2,3,5- 4,3,5-Cl$_2$C$_5$H$_2$N, 2,3,5-Br$_2$C$_5$H$_2$N (2516), 2,4-BrFC$_6$H$_3$, 2,6,4-Br$_2$FC$_6$H$_2$, 4-FC$_6$H$_4$ (2639).
Me$_2$RCO$_2$H → Et$_3$SnO$_2$CMe$_2$R; R = Me, Ph (667).
4-HSC$_5$H$_4$N → 4-Et$_3$SnSC$_5$H$_4$N (2516).
Use: As modifying agent for PVC (2774).

Triethyltin Benzyloxide (M-384)   SnC$_{13}$H$_{22}$O   Et$_3$SnOCH$_2$Ph

Synth.: By rxn. of Et$_3$SnH with PhCHO in presence of anhydr. ZnCl$_2$, AIBN or MeOH, in 88% yield (389), without catalyst with five fold excess of aldehyde (389).
By rxn. of Et$_3$SnH with (PhCH$_2$OH:)$_2$ (1611), in ratio 2:1 in C$_6$H$_6$ in Ar atm., in 86% yield (2645).

Prop.: $b_{0.5}$ 99-100°, n_D^{20} 1.5272 (389), IR spectrum, GLC (2645).
Rxn. with: $Ph_3SnH \rightarrow Et_3SnSnPh_3 + PhCH_2OH + Et_3SnH + Ph_3SnOCH_2Ph$; solvent effect (915).

Triethyltin Ethoxide (M-382) $SnC_8H_{20}O$ Et_3SnOEt
Synth.: By rxn. of Et_3SnX with EtONa in EtOH, in 85% yield (350).
By rxn. of EtOH with Et_3SnN_3 (1429), with Et_3SnNEt_2 in pentane in N atm. (323).
Prop.: $b_{0.2}$ 30° (36), $b_{0.1}$ 45° (323) n_D^{20} 1.4680 (1117), NMR spectrum (323), monomeric in C_6H_6 (350), mol. wt. (1117), molar polarization, dipole moment and configuration (632).
Rxn. with: $Ph_3SnH \rightarrow Et_3SnSnPh_3 + Et_3SnH + Ph_3SnNEt_2 + MeOH$, solvent effect (2642).
$(CH_2:C:O)_2 \rightarrow Et_3SnOCMe:CO_2Et$ (1711).
$RNCO \rightarrow Et_3SnNRCO_2Et$; $R = 1-C_{10}H_7$ (36).
$CH_2Ac_2 \rightarrow Et_3SnCHAc_2$ (350).
$ArOH \rightarrow Et_3SnOAr$; $Ar = 2,4-, 2,6-Cl_2C_6H_3$ (2642).
$RSH \rightarrow Et_3SnSR$; $R = Bu, 2-C_5H_4N$ (2642).

Triethyltin t-Butoxide (M-384) $SnC_{10}H_{24}O$ $Et_3SnOCMe_3$
Synth.: By rxn. of Et_4Sn with excess Me_3CO_2Bz in sealed tube at 100-130°, in 32% yield (2098).
By rxn. of Et_6Sn_2 with Me_3CO_2Bz in sealed tube at 100-130°, in 88% yield (2098), with $Et_3SnO_2CME_3$ at 80-160° (665), with $Me_3SiCH_2CH_2CO_3CMe_3$, in 35% yield (586).
By rxn. of Et_3SnH with $(Me_3CON:)_2$ in C_6H_6 at 40° in Ar atm., in 93% yield (2645).
Prop.: $b_{0.03}$ 52-59°, n_D^{20} 1.4780 (914), n_D^{20} 1.4623 (2645), Moessbauer resonance spectrum (2489).
Rxn. with: $Ph_3SnH \rightarrow Et_3SnSnPh_3 + Me_3COH$ (914), relative rxn. rates (915).

Tripropyltin Methoxide (M-384) $SnC_{10}H_{24}O$ Pr_3SnOMe
Rxn. with: $HC:CCH:CH_2 \rightarrow Pr_2SnC:CCH:CH_2$ (1920).
$(CH_2:C:O)_2 \rightarrow Pr_3SnOCMe:CHCO_2Me$ (1711).
$1-C_6H_9OAc \rightarrow 1-Pr_3SnOC_6H_9c$. (1709).
$Me_2C:CHOAc \rightarrow Pr_3SnOCH:CMe_2$ (1712).
$H_2NCO_2Et \rightarrow Pr_3SnNCO$ (857).
Biol. prop.: Toxicity (933).
Activity as fungicide against Alternaria tenuis, Fusarium culmorum, Rhizoctonia solani (933).

Tributyltin Methoxide (M-382) $SnC_{13}H_{30}O$ Bu_3SnOMe
Synth.: By rxn. of MeOH with $Bu_3SnCHRCHMe_2$ under reflux; $R = CN, CO_2Et$ (2680), with $Bu_3SnOC(OMe)RR'$; $R = H$; $R' = CBr_2$, $R = R' = CCl_3$ (953) with $(Bu_3SnO)_2CHCX_3$; $X = Cl$ (953), Br (953-4), with Bu_3SnNR_2; $R = Me$ in Et_2O, in 97% yield (250), $R = Et$, in pentane in N atm. (323), with $Bu_3SnNEtCO_2SnBu_3$ (748), with Bu_3SnH in presence of RCN or RCO_2Me under reflux or UV irradiation; $R = CH_2:CH, MeCH:CH$, in 26-75% yield (2680).
By rxn. of $(Bu_3Sn)_2O$ with Me_2CO_3, in 86% yield (950).

By rxn. of $Bu_2Sn(OMe)_2$ with Bu_3SnCl (2184).
By thermal dec. of Bu_3SnOCO_2Me (750), of $Bu_3SnSC(OMe)NCH_2CH:CH_2$ in vacuo, in 39% yield (750).
By rxn. of $Bu_3SnNPhCO_2Me$ with $EtNCO$ (746).
Prop.: b_2 108° (1117), $b_{0.8}$ 99° (250), $b_{0.5}$ 75° (30), $b_{0.1}$ 90° (950), 87° (323), n_D^{20} 1.4746 (1117), n_D^{19} 1.4773 (36), d_{20} 1.129 (1117), far IR, assignment (334, 1509), IR, assignment (110, 334, 1509) and NMR (323) spectra, dipole moment, molar polarization and configuration (632), mol. wt. (1117).
Rxn. with: $Ph_3SnH \rightarrow Bu_3SnSnPh_3$ (1618).
$Bu_2SnCl_2 \rightarrow Bu_3SnCl + Bu_2SnCl_2 \cdot Bu_2Sn(OMe)_2$ (2184).
$Me_3SiR \rightarrow Bu_3SnR + Me_3SiOMe$; $R = NCCH_2$, $NC(CH_2)_3$, EtO_2CCH_2 (1682), $NCCHEt$ (2680).
$R_3SiH \rightarrow R_3SiOMe$; $R = Me$ (1239), Ph (719).
$(MeHSiO)_x \rightarrow Bu_3SnH$ (1242).
Me_2SiHCl (2:1) $\rightarrow Me_2Si(OMe)_2$ (719).
$R_3SiCl \rightarrow Bu_3SnCl + R_3SiOMe$ (434).
$Me_3SiCl + Cl(CH_2)_4OH \rightarrow Bu_3SnCl + Me_3SiO(CH_2)_4Cl + MeOH$ (2690).
$RMCl_3 \rightarrow RM(OMe)_3$; $R = MeO_2CCH_2$, $M = Si$ (818), $M = Ge$ (692).
$Me_3SiOR \rightarrow Bu_3SnOR + Me_3SiOMe$; $R = c.C_6H_{11}$, Ph, $Me(CH_2)_4CH:CH$ (1628).
$Bu_3SiOBu \rightarrow Bu_3SnOBu + Bu_3SiOMe$ (1682).
$Pr_2GeHI \rightarrow Pr_2HGeOMe$ (2238).
$Et_3GeI \rightarrow Et_3GeOMe$ (692).
$Me_3GeOC_6H_{11} \rightarrow Bu_3SnOC_6H_{11} + Me_3GeOMe$ (1682).
$GeCl_4 \rightarrow (MeO)_4Ge$ (692).
$HC\vdots CCH:CH_2 \rightarrow Bu_3SnC\vdots CCH:CH_2$ (1920).
$RX \rightarrow Bu_3SnX + MeOR$; $X = Cl$; $R = PhCH_2$; $X = Br, I$; $R = Bu, CH_2:CHCH_2$ (435).
$ClCH_2CH_2OH \rightarrow Bu_3SnCl + Bu_3SnOCH_2CH_2Cl + MeOH + MeOCH_2CH_2OH$ (2690).
$X(CH_2)_3OH \rightarrow Bu_3SnO(CH_2)_3X + MeOH$; $R = Cl, Br$ (2690).
$ClCH_2CH_2CHMeOH \rightarrow Bu_3SnOCHMeCH_2CH_2Cl + MeOH$ (2690).
$Cl(CH_2)_4OH \rightarrow Bu_3SnCl + MeOH + C_4H_8O$ (2690).
$CH_2:C:O \rightarrow Bu_3SnCH_2CO_2Me$ (36).
Phthalic anhydride $\rightarrow$ o-$Bu_3SnO_2CC_6H_4CO_2Me$ (36).
$Y:CHOAc \rightarrow Bu_3SnOCH:Y + MeOAc$; $Y = EtCH$ (719a, 1681), Me_2C (1681, 1684), Me_2CHCH, $Me(CH_2)_4CH$ (1681).
$CH_2:CROAc \rightarrow MeOAc + Bu_3SnCH_2OR$; $R = Me, Me_3C, Ph + Bu_3SnOCR:CH_2$; $R = Me_3C, Ph$ (1681).
$RCH:CR'OAc \rightarrow MeOAc + Bu_3SnCHRCOR' + Bu_3SnOCR':CHR$; $R + R' = Me_2$, $(CH_2)_3$, $(Me + Et)$, 1681) $R + R' = Et + Me$ (1681, 1684).
$RCH:CR'OAc \rightarrow MeOAc + Bu_3SnOCR':CHR$; $R + R' = i-Pr_2$, $(Et + Me)$, $(i-Pr + Me)$, $(CH_2)_4$ (1681), $i-Pr + Me$ (1684).
$(EtO_2CNH)_2 \rightarrow [Bu_3SnN(CO_2Et)]_2$ (949).
$RNCO \rightarrow Bu_3SnNRCO_2Me$; $R = Me$ (36), Et (36, 746, 949).
$RNCO$ (excess) $\rightarrow (RNCO)_3$; $Me, Et, 1-C_{10}H_7$ (2342), Ph (746, 2342).
$(PhNCO)_2 \rightarrow Bu_3SnNPhCO_2Me$ (746).
$(PhNCO)_2 + MeOH \rightarrow PhNHCONPhCO_2Me$ (746).
$RNCS \rightarrow Bu_3SnSC(OMe):NR$; $R = Ph$ (36, 750), $CH_2:CHCH_2$ (750).
$C(:NR)_2 \rightarrow Bu_3SnNRC(OMe):NR$; $R = 1-C_{10}H_7$ (36, 750, 940), p-MeC_6H_4 (750).

C(:NR)$_2$ (excess) + MeOH → RNHC(OMe):NR; R = 1-C$_{10}$H$_7$ (750).
RCHO ⇌ Bu$_3$SnOCHROMe; R = CCl$_3$ (36) + Bu$_3$Sn[OC(CCl$_3$)H]$_n$OMe (941, 953); R = Me, Pr, Br$_3$C, C$_6$F$_5$ (953).
R$_2$CO ⇌ Bu$_3$SnOCR$_2$OMe; R = CCl$_3$, (CCl$_2$F + CClF$_2$) (953).
ROH → Bu$_3$SnOR + MeOH; R = n-Bu (334, 1440), t-Bu (334), c.C$_6$H$_{11}$, 1-, 2-, 3-MeC$_6$H$_{10}$ (436).
ArOH → Bu$_3$SnOAr; derivs. of Ph$_2$CO, o-HOC$_6$H$_4$CO$_2$H, benzotriazolylphenol (2972).
RNSO → Bu$_3$SnOS(O)OMe + (RN:)$_2$SO; R = Ph, o-O$_2$NC$_6$H$_4$ (2345).
Addn. compd.: 2Bu$_3$SnOMe·SnCl$_4$, from components, thermodynamic data (2184) [Sn$_3$C$_{26}$H$_{60}$Cl$_4$O$_2$].
Use: As fungicide (1005).
As catalyst for making polyurethan (450, 2982).

Tributyltin Ethoxide (M-385)   SnC$_{14}$H$_{32}$O   Bu$_3$SnOEt
Synth.: By rxn. of EtOH with Bu$_3$SnNEt$_2$ in pentane in N atm. (323), with Bu$_3$SnN(1-C$_{10}$H$_7$)CO$_2$Me (36).
By rxn. of Bu$_3$SnCl with EtONa (334), in 68% yield (692).
By rxn. of (Bu$_3$Sn)$_2$O with Et$_2$CO$_3$ under reflux, in 86% yield (950).
By-prod. in rxn. of Bu$_3$SnH with RCH:CHCO$_2$Et at 150° in Ar atm. in 15% yield; R = H, Me (2680).
Prop.: b$_1$ 110-11° (692), 96° (440), b$_{0.1}$ 80° (323), b$_{0.01}$ 98° (36), n$_D^{20}$ 1.4686 (440), 1.4656 (692), d$_{20}$ 1.1097 (692), 1.0977 (440), (far) IR, assignment (334, 1509) and NMR (323) spectra.
Rxn. with: (MeHSiO)$_x$ → Bu$_3$SnH (1165, 1242).
MePhSiHX → MePhSi(OEt)$_2$; X = Cl, OEt (719).
GeCl$_4$ → (EtO)$_4$Ge (692).
SO$_2$ → Bu$_3$Sn(Et)SO$_3$ (2345).
CH$_2$:C:O → Bu$_3$SnCH$_2$CO$_2$Et (440).
EtNCO → Bu$_3$SnNEtCO$_2$Et (2345).
Cl$_3$CCHO → Bu$_3$SnOCH(OEt)CCl$_3$ (2345).
ROH → Bu$_3$SnOR; R = CH$_2$(CH$_2$)$_n$CHXCR′; X = Cl, Br; n = 2,3; R′ = H, Me; isomer distribution (2348).

Tributyltin Butoxide (M-385)   SnC$_{16}$H$_{36}$O   Bu$_3$SnOBu
Synth.: By rxn. of BuOH with Bu$_3$SnC:CCH$_2$CH$_2$OCH:CH$_2$ at 80-100° in presence of HCl (2078), with Bu$_3$SnOMe under reflux (334), in 84% yield (1440).
By rxn. of Bu$_3$SnOMe with Bu$_3$SiOBu at 145°, in 83% yield (1682).
Prop.: b$_{1.5}$ 112-13°, d$_{20}$ 1.0775 (1440), 1.077 (1682), (far) IR spectra and assignment (334, 1509).
Rxn. with: (MeHSiO)$_x$ → Bu$_3$SnH (1242, 2459).
Et$_3$GeCl → Et$_3$GeOBu (1440).
NH$_4$X → Bu$_3$SnX + BuOH + NH$_3$; X = I, NCS (2660).
CH$_2$:C:O → Bu$_3$SnCH$_2$CO$_2$Bu (1440).
Use: In catalyst for modifying polypropylene (269).

Tributyltin t-Butoxide (M-385)   SnC$_{16}$H$_{36}$O   Bu$_3$SnOCMe$_3$
Synth.: By rxn. of Me$_3$COH with Bu$_3$SnOMe (334), with Bu$_3$SnNPhCO$_2$SnBu$_3$ (748).

By rxn. of Bu_3SnH or Bu_6Sn_2 with $(Me_3CO)_2$ (1616).
By rxn. of Bu_3SnCl with Me_3CONa under exclusion of moisture, in 41% yield (1618).
Prop.: $b_{0.15}$ 82-83°, n_D^{20} 1.4657, d_{20} 1.07 (1618), IR (1616), assignment (334, 1509), far IR spectra and assignment (334, 1509).
Rxn. with: $Bu_3SnH \rightarrow Bu_6Sn_2$ (1618).
$(MeHSiO)_x \rightarrow Bu_3SnH$ (1242).

Tributyltin Cyclohexaneoxide (M-385) $SnC_{18}H_{38}O$ $Bu_3SnOC_6H_{11}$
Synth.: By rxn. of Bu_3SnOMe with $c.C_6H_{11}OH$ (436), with $Me_3SiOC_6H_{11}$ and $Me_3GeOC_6H_{11}$ at 170° and 190°, in 76 and 86% yield, resp. (1682).
Prop.: $b_{0.5}$ 137° (436), $b_{0.2}$ 117° (1682), n_D^{20} 1.4831 (436, 1682) d_{20} 1.110 (436), 1.106 (1682), (far) IR spectra and assignment (1509).
Rxn. with: $Et_3SiCl \rightarrow Bu_3SnCl + Et_3SiOC_6H_{11}$ (434).
$CH_2:CHCH_2Br \rightarrow Bu_3SnBr + c.C_6H_{11}OCH_2CH:CH_2$ (435).

Tributyltin Phenoxide (M-385) $SnC_{18}H_{32}O$ Bu_3SnOPh
Synth.: By rxn. of Bu_3SnOMe with Me_3SiOPh at 180°, in 85% yield (1682).
By rxn. of Bu_3SnCl with $PhONa$ in C_6H_6 (36).
Prop.: $b_{0.01}$ 124° (36), $b_{0.6}$ 137-39°, n_D^{20} 1.5169 (1682).
Rxn. with: $K + (Me_2N)_3PO$, $+ BuBr \rightarrow Bu_4Sn + BuOPh$ (1635).
$(MeHSiO)_x \rightarrow Bu_3SnH$ (1242).
$Hg(OAc)_2 \rightarrow Bu_3SnOC_6H_4HgOAc$ (1225).
$SO_2 \rightarrow Bu_3Sn(Ph)SO_3$ (750).
$PhNCO \rightarrow Bu_3SnNPhCO_2Ph$ (36).
$CX_3CHO \rightarrow Bu_3SnOCH(CX_3)OMe$; $X = Cl$, Br (953).
$PhNCO + PhOH \rightarrow PhNHCO_2Ph$, catalytic effect of Bu_3SnOPh (36).
Use: In catalyst for ethylene polymerization (1238a) for foaming polyurethan (450).

Triphenyltin Methoxide (M-386) $SnC_{19}H_{18}O$ Ph_3SnOMe
Synth.: By rxn. of Ph_3SnX with $MeONa$ in N atm. in MePh, in 73% yield (791).
By rxn. of Ph_3SnNEt_2 with MeOH in pentane in N atm. (323).
Prop.: m. 120° (323), 114-16° (1005), 69-70° (791), NMR spectrum (323).
Rxn. with: $Ph_3SnH \rightarrow Ph_6Sn_2$ (1618).
$Ph_2SnH_2 \rightarrow Ph_6Sn_2 + Ph_3Sn(Ph_2Sn)_nSnPh_3$; $n = 1-3$ (1955).
$RCO_3H \rightarrow Ph_2Sn(OPh)O_2CR + PhOH$; kinetic data and rxn. mechanism (790).
$ArOH \rightarrow Ph_3SnOAr$; $Ar = 2,5-Bz(C_8H_7O)C_6H_3$, $2-PhO_2CC_6H_4$ (2972).
$ArSH \rightarrow Ph_3SnSAr$; deriv. of 2,4,6-bis(alkylamino)-s-triazine (1322a).
Biol. prop.: Activity as fungicide (1005) against Alternaria tenuis, Phytophthora infestans, Venturia inaequalis (2297).
Activity as insecticide and chemosterilant against Leptinotarsa decemlineata Musca domestica (805).
Use: Control of pecan bunch disease (2335).

Listing of unsubstituted triorganotin alkoxides concludes in Table 172.

Table 172. Unsubstituted Triorganotin Alkoxides R_3SnOR

Formula*	Synth. Method**	Yield	Properties	Ref.
Me_3SnOEt	VA	--	$b_{0.1}$ 81°, IR, assignment, NMR spectra (a)	323
$Me_3SnOCMe_3$	III	--	Colorless liq., b_{750} 140-42°	2707
$2\text{-}Me_3SnOC_{10}H_7$	VIB	86	--	2611
$Me(i\text{-}Pr)PhSnOC_{10}H_{17}$	III	--	Bornyloxide, liq. easily hydrolyzed	1719
$(PhCH_2)_3SnOMe$	--	--	(b)	--
Et_3SnOR				
$R = Pr$ (384)	--	--	(c)	--
Me_2CH	IIA	60	b_{12} 83°, n_D^{20} 1.491	389
Bu	IIA	74	b_{11} 98-101°, n_D^{20} 1.4708, IR spectrum	389
	I (d)	30	b_1 69°, n_D^{20} 1.4652, d_{20} 1.1885 (e)	2078
Me_2CHCH_2 (384)	IIA	90	b_{11} 94-96°, n_D^{20} 1.4678, d_{20} 1.165 (f)	389
MeEtCH (384)	IIA	96	b_{12} 93-96°, n_D^{20} 1.479 (g)	389
Et_2CH	IIA	58	b_{13} 110°, n_D^{20} 1.474	389
$c\text{-}C_6H_{11}$ (384)	IIA	61	b_{11} 129°, n_D^{20} 1.495	389
$(-)\text{-}C_{10}H_{19}$	--	--	Menthoxide (h)	--
Ph (384)	IVA	96	IR (985), UV and NMR spectra (i)	342

* Numbers in parenthesis refer to pages in the main volume.
** The characters used here correspond to the synthetic scheme in the introduction to Chapter 4.3.3.1.
(a) Relative basicity (644). (b) Rxn. with ArOH → $(PhCH_2)_3SnOAr$; Ar = 2,6,4-$(4\text{-}ClN_3C_6H_3)C_6H_2(CMe_3)Me$ - benzotriazol deriv. (2972). (c) Rxn. with $(CH_2:CO)_2$ → $R_3SnOCMe:CHCO_2Pr$; R = Et, Pr (1711). (d) Synth. from BuOH and $Et_3SnC:CCH_2CH_2OCH:CH_2$ at 80-100° (2078). (e) Rxn. with $(EtCO)_2O$ → $Et_3SnO_2CEt + EtCO_2Bu$ (389). (f) Rxn. with $PhCH_2SeH$ → $Et_3SnSeCH_2Ph + 1\text{-}BuOH$ (389), with PhNCO → $Et_3SnNPhCO_2Bu\text{-}i$ (389). (g) Rxn. with Ph_3SnH → $Et_3SnSnPh_3 + s\text{-}BuOH + Ph_3SnOBu\text{-}s + Et_3SnH$; solvent effect and kinetic data (915). (h) Rxn. with $Et_3SnCl\cdot AlCl_3$ + benzofuran → optically active $(C_8H_6O)_x$ (545, 1164). (i) Moessbauer resonance spectrum (2637), rxn. with R_3SnH → Et_3SnR_3 + ROH; R = Me, Ph, $p\text{-}MeOC_6H_4$, kinetic data, solvent effect (915).

Table 172 Continued

Formula*	Synth. Method**	Yield	Properties	R_3SnOR Ref.
Et($\overline{CH_2CH_2CH}$)$_2$SnOCMe$_3$	III	--	Use: herbicide, pesticide	1873
Pr$_3$SnOCHMe$_2$	--	--	(c)	--
Pr$_3$SnOPh (385)	--	--	(j)	--
i-Pr$_3$SnOMe (385)	--	--	(far) IR spectra, assignment	111

Bu$_3$SnOR

Formula*	Synth. Method**	Yield	Properties	Ref.
R = Ph$_2$CH	IVA	90	b$_{0.3}$ 140°	950
Ph$_3$C	IVA	90	b$_{0.3}$ 180°	950
MePhCH (385)	IVA	90	b$_{0.3}$ 138°	950
Pr	III	--	(Far) IR spectra, assignment (334, 1509) (k)	334
Me$_2$CH	III	--	(Far) IR spectra, assignment (334, 1509) (k)	334
MeEtCH	IIA	85	b$_{0.4}$ 102°, n$_D^{20}$ 1.476	389
Me$_3$CCH$_2$	IVA	90	b$_{0.3}$ 122°	950
Me$_2$CHCH$_2$CHMe	--	--	(Far) IR spectra, assignment	1509
n-C$_7$H$_{15}$	IVA	90	b$_{0.3}$ 132°	950
c.1-MeC$_6$H$_{10}$	III, IVB	35	b$_{0.5}$ 157°, n$_D^{20}$ 1.4821, d$_{20}$ 1.0947	436
c.2-MeC$_6$H$_{10}$	III, IVB	55	b$_{0.5}$ 155°, n$_D^{20}$ 1.4815, d$_{20}$ 1.0910, GLC	436
c.3-MeC$_6$H$_{10}$	III, IVB	64	b$_{0.5}$ 160°, n$_D^{20}$ 1.4818, d$_{20}$ 1.0928	436
n-C$_8$H$_{17}$	--	--	(Far) IR spectra, assignment	1509
EtBuCHCH$_2$	--	--	Use as cat. foaming polyurethan	450
MeC$_6$H$_4$	--	--	Use as bactericide, fungicide	1984
o-Me$_3$CC$_6$H$_4$	--	--	Use as cat. foaming polyurethan	450
2,6-1-Pr$_2$C$_6$H$_3$	--	--	(l)	--
p-C$_8$H$_{17}$C$_6$H$_4$	--	--	Use as cat. foaming polyurethan	450
C$_9$H$_{19}$C$_6$H$_4$	--	--	Use as bactericide, fungicide	1984
o-PhC$_6$H$_4$	IVA	--	b$_4$ 204-208°, foaming cat. for polyurethan (450) (m)	1792
p-PhC$_6$H$_4$	--	--	Use as cat. foaming polyurethan	450

i-Bu₃SnOCH₂CHMe₂	IIA	89	b_2 122-24°, n_D^{20} 1.4675	389
i-Bu₃SnOCHMeEt	IIA	90	$b_{0.4}$ 88°, n_D^{20} 1.474	389
(n-C₈H₁₇)₃SnOMe (386)	IV (n)	86	$b_{0.1}$ 160° (o)	950
(n-C₈H₁₇)₃SnOPh	--	--	(j)	--
Ph₃SnOEt	VA	--	m. 112°, NMR spectrum (p, q)	323
Ph₃SnOPh	III	62	m. 83-89° (r)	791
Ph₃SnOR	--	--	R = organic group C < 12 (q)	--
R₃SnOPh	--	--	UV spectrum	2771

(j) Rxn. with Hg(OAc)₂ → R₃SnOC₆H₄OHgOAc; R = Pr, C₈H₁₇ (1225). (k) Rxn. with (MeHSiO)ₓ → Bu₃SnH (1242, 2458). (l) Biological activity as molluscicide against Helisoma tenue, H. subcrenatum (1054). (m) Use as bactericide, fungicide and antiseptic agent (1792). (n) Synth. from (R₃Sn)₂O and Me₂CO₃ (950). (o) Rxn. with ArOH → (C₈H₁₇)₃SnOAr; Ar = 2,2'-HOC₆H₄OC₆H₄, 2,4'-PhC₆H₄O₂CC₆H₄ (2972). (p) Biol. activity as fungicide against Alternaria tenuis, Phytophthora infestans, Venturia inaequalis (2297), as insecticide against Musco domestica (805). (q) Use for controlling pecan bunch disease (2335). (r) Use in catalyst for ethylene polymerization (2464).

4.3.3.1.3 TRIORGANOTIN ALKOXIDES DERIVED FROM HETEROCYCLIC ALCOHOLS

Trimethyltin 8-Oxyquinolate $SnC_{12}H_{15}NO$ $8-Me_3SnOC_9H_6N$
<u>Synth.</u>: By rxn. of Me_3SnX with $8-C_9H_6NONa$, in 48% yield (1046, 1291).
By rxn. of $(Me_3Sn)_2O$ with $8-C_9H_6OH$ at 115° (404), in 55% yield (1046).
<u>Prop.</u>: Yellow liq. $b_{0.2}$ 108.5-109.5°, far IR, UV and NMR spectra, slowly decomposing in air, mol. wt. (1291).
<u>Rxn. with</u>: Thermal dec. → Me_4Sn + $Me_2Sn(OC_9H_6N)_2$ (1291).
<u>Biol. Prop.</u>: Activity as fungicide against Aspergillus arnstelodami, A. niger, Chaetomium globosum, Paecilomyces varioti, Penicillium brevi-compactum, P. cyclopium (1046), P. roquefortii (404), Stachybotrys atra (1046).

Additional triorganotin alkoxides derived from heterocyclic alcohols are listed in Table 173.

4.3.3.1.4 TRIORGANOTIN ALKOXIDES DERIVED FROM UNSATURATED ALCOHOLS

Triethyltin Propargyloxide (M-388) $SnC_9H_{18}O$ $Et_3SnOCH_2C\equiv CH$
<u>Synth.</u>: By rxn. of $(Et_3Sn)_2O$ with $HC\equiv CCH_2OR$ in C_6H_6; R = H, Ac, in 80-90% yield (391).
By rxn. of Et_3SnOH with $HC\equiv CCH_2OH$ and $EtMgBr$, in 15% yield (2506).
<u>Prop.</u>: b_{10} 102° (391), b_5 80-81°, n_D^{20} 1.4660, d_{20} 1.2870 (506, 2506), IR spectrum (1911).
<u>Rxn. with</u>: Me_3SiCl → Et_3SnCl + $HC\equiv CCH_2OSiMe_3$ (1536).
$MeOCH_2Cl$ → Et_3SnCl + $HC\equiv CCH_2OCH_2OMe$ (1910).
Cl_3CCHO → $Et_3SnOCH(CCl_3)OCH_2C\equiv CH$ (1910).

Triethyltin 3-Butynoxide (M-388) $SnC_{10}H_{20}O$ $Et_3SnOCH_2CH_2C\equiv CH$
<u>Prop.</u>: IR spectrum (1911).
<u>Rxn. with</u>: Et_3SnH → $Et_3SnCH:CHCH_2CH_2OSnEt_3$ (1538).
Me_3SiCl → Et_3SnCl + $Me_3SiOCH_2CH_2C\equiv CH$ (1910).
$PhC\equiv CH$ → $Et_3SnC\equiv CPh$ + $HC\equiv CCH_2CH_2OH$ (508)
MeI → Et_3SnI + $MeOCH_2CH_2C\equiv CH$ (513).
$MeOCH_2Cl$ → Et_3SnCl + $MeOCH_2OCH_2CH_2C\equiv CH$ (1536).

Triethyltin 1,1,2-Dimethylpropynoxide (M-388) $SnC_{11}H_{22}O$ $Et_3SnOCMe_2C\equiv CH$
<u>Synth.</u>: By rxn. of $(Et_3Sn)_2O$ with $HC\equiv CCMe_2OH$, in 22% yield (508, 2506)
<u>Prop.</u>: b_2 83°, n_D^{20} 1.4755, d_{20} 1.1851 (506, 2506), IR spectrum (1911).
<u>Rxn. with</u>: Et_3SnH → $Et_3SnCH:CHCMe_2OSnEt_3$ (1538).
Me_3SiCl → Et_3SnCl + $HC\equiv CCMe_2OSiMe_3$ (1910).
HCl → Et_3SnCl + $HC\equiv CCMe_2OH$ (1910).
$MeOCH_2Cl$ → Et_3SnCl + $HC\equiv CCMe_2OCH_2OMe$ (1910).

Listing of unsaturated alkoxides of the triorganotin moiety concludes in Table 174. Additional triorganotin alkoxides containing unsaturated substituted groups may be found in Tables 178-180 and in Chapters 2.4.3 and 2.5.3. Triorganotin enoxides are listed in Chapter 4.3.4.1.3.

Table 173. Triorganotin Alkoxides Derived from Heterocyclic Alcohols R_3SnOR'

Formula*	Synth. Method**	Yield	Properties	Ref.
$Me_2BuSnOC_9H_6N$ (a) (?)	VIB	--	m. 165-90°, UV spectrum (b)	602
$(PhCH_2)_3SnOR$	IVB	--	R = 2,6,4-(5-$ClN_3C_6H_3$)$Me_3C(Me)C_6H_2$ (c)	2972
$(PhCH_2)_3SnOC_9H_6N$ (a)	III	56	(d)	1046
	IVA	64	--	1046
Et_3SnOR				
R = 2-(C_4H_3O)CH_2 (e) (387)	IIA	63	b_{12} 130°, n_D^{20} 1.5072	389
2-(C_4H_3O)CH(CH_2CONEt_2) (e)	IA	--	NMR spectrum	1631
2-(C_4H_3O)CH(CH_2Ac) (e)	IA	--	NMR spectrum	1631
3,5,2-$Cl_2C_5H_2N$ (f)	IVB	--	b_2 145-46°, n_D^{20} 1.5550, UV spectrum, struct. (g)	2516
3,5,4-$Cl_2C_5H_2N$ (f)	IVB	--	m. 179-81°, soly., struct. (g)	2516
3,5,2-$Br_2C_5H_2N$ (f)	IVB	--	b_6 180-82°, n_D^{20} 1.5841, struct.	2516
8-C_9H_6N (a) (387)	III	--	$b_{0.05}$ 132-34°	1291
	IVA	--	(h)	404
$Pr_3SnOC_9H_6N$ (a)	IVA	--	(h)	404
Bu_3SnOR	IVB	--	R = 4,2-MeC_6H_3($N_3C_6H_4$) (c)	2972

* Numbers in parenthesis refer to pages in the main volume.
** The characters used here correspond to the synthetic scheme in the introduction to Chapter 4.3.3.1.
(a) HOC_9H_6N = 8-hydroxyquinoline. (b) Recryst. petr. ether → $R_2Sn(OC_9H_6N)_2$, UV indicates component mixture; R = Me, Ph (602). (c) 2-Benzotriazole derivative, use as UV stabilizer for polypropylene and synthetic resins (2972). (d) Biol. activity as fungicide against Aspergillus arnstelodami, A. niger, Chaetomium globosum, Paecilomyces varioti, Penicillium brevi-compactum, P-cyclopium, Stachybotrys atra (1046). (e) C_4H_4O = furan. (f) Dihalopyridine derivative. (g) IR (2516) and Moessbauer resonance (2642) spectra. (h) Biol. active as fungicide against Penicillium roquefortii (404).

Table 173 Continued

R_3SnOR'

Formula*	Synth. Method**	Yield	Properties	Ref.
$Bu_3SnOC_9H_6N$ (a)	III	53	m. 126° (d, i)	1046
	IVA	58	--	1046
	III	--	Yellow liq., b0.007 149-51°, UV spectrum, mol. wt.	1291
$Bu_3SnOC_9H_4NCl_2$ (j)	--	--	(k)	--
$BuPh_2OSnC_9H_6N$ (a) (?)	VIB	--	Yellow cryst., m. 206-21°, IR, NMR spectra (b)	602
Ph_3SnOR				
R = 3,5,2-$Cl_2C_5H_2N$ (f)	IVA	--	m. 112-13°, UV spectrum, soly., struct. (g)	2516
3,5,4-$Cl_2C_5H_2N$ (f)	IVA	--	m. 254-55°, soly., struct. (g)	2516
3,5,2-$Br_2C_5H_2N$ (f)	IVA	--	m. 127-30°, soly., struct.	2516
C_9H_6N (a) (387)	III	--	m. 145-46°, UV spectrum (l)	1291
(p-$ClC_6H_4)_3SnOC_9H_6N$ (a)	III	68	m. 260°	1046
	IVA	75	(d)	1046
(p-$EtC_6H_4)_3SnOC_9H_6N$ (a)	III	59	Liq.	1046
	IVA	64	(d)	1046

(i) Hydrolysis → $BuSn(OC_9H_6N)_3$ (a) (1290). (j) $HOC_9H_4NCl_2$ = 5,7-dichloro-8-hydroxyquinoline. (k) Biol. activity as molluscicide against Heliosoma tenue, H. subcrenatum (1054). (l) IR (1768) and UV (590) spectra, mol. wt. (590), struct. (590, 1768).

Table 174. Triorganotin Alkoxides Derived From Unsaturated Alcohols

Formula*	Synth. Method**	Yield	Properties	R_3SnOR' Ref.
3-$Me_3SnOCH_2C_6H_9$ c.	IIA	73	b. 104-106°, IR spectrum	1953
Et_3SnOR'				
R' = HC⋮CCHMe (388)	IVA	21	b_1 57°, n_D^{20} 1.4817, d_{20} 1.2475 (a)	508
	IVA	--	b_3 73°, n_D^{20} 1.4830, d_{20} 1.2375	2506
CH_2:$CHCH_2CH(Ph)$ (b)	IA	86	$b_{0.003}$ 96°, hydrolyzes in H_2O (c)	1331
MeCH:CHCHMe	--	--	(d)	--
CH_2:CHC⋮CCH_2	IVA	--	b_1 84°, n_D^{20} 1.5110, d_{20} 1.2503	2506
CH_2:CHC⋮CCHMe	IVA	--	b_1 112°, n_D^{20} 1.4920, d_{20} 1.04	2506
HC⋮CCMeEt	IVA	42	b_1 70°, n_D^{20} 1.4788, d_{20} 2106	1149
CH_2:$CHCH_2CH$(CH:CHPh) (b)	IA	84	$b_{0.001}$ 179°, hydrolyzes in H_2O (c)	1331
CH_2:CHC⋮$CCMe_2$	IVA	--	$b_{1.5}$ 81°, n_D^{20} 1.4910, d_{20} 1.1550	2506
1,1-c.-C_5H_8(C⋮CH)	IVA	38	b_2 99°, n_D^{20} 1.4961, d_{20} 1.2231	1149
1,1-c.-C_6H_{10}(C⋮CH)	IVA	63	b_5 110°, n_D^{20} 1.4980, d_{20} 1.2008	1149
	IB	--		1331
CH_2:$CHCH_2CH(C_7H_{15})$ (b)	IA	85	m. 27°, $b_{0.35}$ 122°, hydrolyzes in H_2O (c)	1331
$Pr_3SnOCHMeC$⋮CH	IVA	--	b_1 92°, n_D^{20} 1.4785, d_{20} 1.1544	506, 2506
$Pr_3SnOCMe_2C$⋮CH	IVA	--	b_4 109°, n_D^{20} 1.4735, d_{20} 1.1169	506, 2506
$Pr_3SnOCMeEtC$⋮CH	IVA	45	b_2 101°, n_D^{20} 1.4752, d_{20} 1.1263	1149
1,1-$Pr_3SnOC_6H_{10}C$⋮CHc.	IVA	73	b_2 144°, n_D^{20} 1.4954, d_{20} 1.1617	1149
$Bu_3SnOCHMeC$⋮CH	IVA	--	b_3 117°, n_D^{20} 1.4785, d_{20} 1.1109	506, 2506
$Bu_3SnOCMe_2C$⋮CH	IVA	--	b_2 118°, n_D^{20} 1.4740, d_{20} 1.0867	506, 2506
1-$Bu_3SnOC_8H_9$c.	--	--	(e)	--

* Numbers in parenthesis refer to pages in the main volume.
** The characters used here correspond to the synthetic scheme in the introduction to Chapter 4.3.3.1.
(a) Rxn. with RCl → Et_3SnCl + HC⋮CCHMeOR; R = Me_3Si (1910), $MeOCH_2$ (1910, 1922). (b) Synth. from Et_3SnCH_2CH:CH_2 and RCHO. (c) Rxn. with $CH_2(CH_2CO_2H)_2$ or BzOH → Et_3Sn-carboxylate + R'OH (1331). (d) Rxn. with Et_3SnOMe → $Et_3SnOCHMeCH$:$CMeSnEt_3$ (1538). (e) Rxn. with Et_3SiCl → Bu_3SnCl + $Et_3SiOC_8H_9$ (434).

4.3.3.1.6 TRIORGANOTIN ALKOXIDES DERIVED FROM FUNCTIONALLY SUBSTITUTED ALCOHOLS

4.3.3.1.6.1 TRIORGANOTIN ALKOXIDES DERIVED FROM HALOGEN SUBSTITUTED ALCOHOLS

The compounds are listed in Table 175. Additional triorganotin alkoxides derived from substituted alcohols may be found in Tables 173, 177, 179 and 180.

4.3.3.1.6.2 TRIORGANOTIN ALKOXIDES DERIVED FROM PSEUDOHALOGEN SUBSTITUTED ALCOHOLS

All compounds are listed at the bottom of Table 175.

4.3.3.1.6.3 TRIORGANOTIN ALKOXIDES DERIVED FROM ALCOHOLS CONTAINING OXYGEN FUNCTIONS

The compounds are summarized in Table 176. Additional derivatives containing substituted alcohols may be found in Tables 173, 179 and 180.

4.3.3.1.6.5 TRIORGANOTIN ALKOXIDES DERIVED FROM ALCOHOLS CONTAINING NITROGEN AND PHOSPHORUS

Triorganotin alkoxides with nitrogen or phosphorus containing alcohols are listed in Table 177. Additional substituted derivatives may be found in Subsections 4.3.3.1.3, 4.3.3.1.6.2 and Tables 176 and 180.

4.3.3.1.6.5.9 O-TRIORGANOTIN DERIVATIVES OF OXIMES AND HYDROXYLAMINES

All compounds are listed in Table 178.

Table 175. Triorganotin Alkoxides Derived from Halogen Substituted Alcohols R_3SnOR'

Formula*	Synth. Method**	Yield	Properties	Ref.
Me_3SnOR'				
$R' = C_6F_5CH_2$	IIA	--	$b_{0.3}$ 59-60°, n_D^{20} 1.4723, NMR spectrum (a)	2538
CCl_3CH_2	IIA	--	b_2 71-72°, n_D^{20} 1.5080, NMR spectrum (a)	2538
$C_6F_5(CF_3)CH$	IIA	--	$b_{0.3}$ 62-64°, n_D^{20} 1.4852, NMR spectrum	2538
$(CF_3)_2CH$	IIA	100	b_{58} 76°, IR, H and ^{19}F NMR spectra	109
	II (b)	--	---	109
$(CF_3)_2CH \cdot (CF_3)_2CO$	VIA	--	m. 23° (sealed tube), H, ^{19}F NMR spectra (b)	109
Et_3SnOR'				
$R' = C_6F_5CH_2$	IIA	--	$b_{0.1}$ 86-87°, n_D^{20} 1.4748, NMR spectrum	2538
CCl_3CH_2	IIA	76	$b_{0.3}$ 79°, n_D^{20} 1.4923, NMR spectrum (c)	2538
$C_6F_5(CF_3)CH$	IIA	--	$b_{0.1}$ 80-82°, n_D^{20} 1.4878, NMR spectrum	2538
$p-FC_6H_4$	--	--	$b_{1.5}$ 110-11°, n_D^{20} 1.5215, ^{19}F NMR (2639) spectrum (d)	2637
$p-ClC_6H_4$ (389)	IVA	--	$b_{0.1}$ 115-18°, n_D^{20} 1.5550, (d)	914
	--	--	$b_{1.5}$ 134-35°, n_D^{20} 1.5489 (e)	2637
$p-BrC_6H_4$ (389)	--	--	b_1 141-43°, n_D^{20} 1.5637 (d)	2637
$p-IC_6H_4$	--	--	b_1 146°, n_D^{20} 1.5919	2637
$2,4-BrFC_6H_3$	IVB	--	b_2 152°, n_D^{20} 1.5472, ^{19}F NMR spectrum	2639
$2,4-Cl_2C_6H_3$	IVB	--	b_8 150-55°, n_D^{20} 1.5581, (f)	2642

* Numbers in parenthesis refer to pages in the main volume.
** The characters used here correspond to the synthetic scheme in the introduction to Chapter 4.3.3.1.
(a) Impure, contains 1:2 adducts $Me_3SnOCHR'OCH_2R'$ (2538). (b) Rxn. of $Me_3SnOCH(CF_3)_2 \cdot OC(CF_3)_2$ with $Me_3SnH \rightarrow Me_3SnOCH(CF_3)_2$ (109). (c) Rxn. with $CCl_3CHO \rightleftharpoons Et_3SnOCH(CCl_3)OCH_2CF_3$ (2538). (d) Moessbauer resonance spectrum (2637). (e) Rxn. with $Ph_3SnH \rightarrow Et_3SnSnPh_3 + p-ClC_6H_4OH$ (914), relative rxn. rates (915). (f) Moessbauer resonance spectrum, struct. (2642).

Table 175 Continued

Formula*	Synth. Method**	Yield	Properties	R_3SnOR' Ref.
Et_3SnOR' (Cont'd.)				
$R' = 2,6-Cl_2C_6H_3$	IVB	--	b_6 167-69°, n_D^{20} 1.5575, ^{35}Cl NQR spectrum (2515) (f)	2642
$2,6,4-Br_2FC_6H_2$	--	--	b_1 160-62°, n_D^{20} 1.5702, ^{19}F NMR spectrum	2639
Bu_3SnOR'				
$R' = C_6F_5CH_2$	IIA	--	$b_{0.1}$ 109°, n_D^{20} 1.4736, NMR spectrum	2538
$ClCH_2CH_2$	IVB	--	(g)	2690
Cl_3CCH_2	IVA	90	$b_{0.1}$ 116°	950
$C_6F_5CH(CF_3)$	IIA	--	$b_{0.1}$ 113-15°, n_D^{20} 1.4821, NMR spectrum	2538
$Cl(CH_2)_3$	IVB	--	---	2690
$Br(CH_2)_3$	IVB	--	---	2690
$ClCH_2CH_2CHMe$	IVB	--	---	2690
$2-BrC_5H_8c.$	IVB	--	cis-form (h)	2348
	IVB	--	trans-form (i)	2348
$2-ClC_6H_{10}c.$	IVB	--	cis-form	2348
	IVB	--	trans-form	2348
$2-BrC_6H_{10}c.$	IVB	--	cis-form (h)	2348
	IVB	--	trans-form (i)	2348
$1-Me-2-BrC_6H_9c.$	IVB	--	cis-form (j)	2348
	IVB	--	trans-form (k)	2348
C_6Cl_5 (389)	--	--	Use as foaming cat. for polyurethan (l, m)	450
C_6Br_5	--	--	(l)	--
$4,3-ClMeC_6H_3$	--	--	(l)	--
$2,4-PhClC_6H_3$	IVA	--	b_1 215-20° (n)	1792
$2,6-PhClC_6H_3$	IVA	--	b_1 221-24° (n)	1792
$4,2-PhClC_6H_3$	IVA	--	b_1 200-204° (n)	1792

i-Bu₃SnOC₆H₂Cl₃-2,4,5	--	(1)	--
Ph₃SnOC₆H₃Cl₂-2,4	IVA	m. 92-93°, soly. (f)	2642
Ph₃SnOC₆H₃Cl₂-2,6	IVA	m. 68-70°, ³⁵Cl NQR spectrum, soly. (f)	2642
p-Et₃SnC₆H₄CN	--	m. 59-61°, b₁ 157-59°, n_D^{20} 1.5812	2637
p-Et₃SnC₆H₄SCN	--	b₁ 169°, n_D^{20} 1.5787 (d)	2637
Bu₃SnOCH(C₆F₅)CH₂CN	IA (o)	b₀.₁ 146-56°, n_D^{20} 1.4813, NMR spectrum (p)	1631
Bu₃SnOCH(CCl₃)CH₂CN	IA (o)	b₀.₀₈ 130-33°, n_D^{20} 1.4995, NMR spectrum	1631

(g) Decomposition at 50° → Bu₃SnCl + 1,4-dioxane (2690). (h) Thermal decomposition at 180-200° → Bu₃SnBr + $\overline{CH_2(CH_2)_n}$C:O; n = 3, 4 (2348). (i) Thermal decomposition at 80-100° → Bu₃SnBr + 1,2-cyclopentyl and 1,2-cyclohexyl oxide (2348). (j) Thermal decomposition at 180° → Bu₃SnBr + c.C₅H₉Ac (2348). (k) Thermal decomposition at 100° → 1,1,2-MeC₆H₉O (epoxide) (2348). (l) Biol. activity as molluscicide against Heliosoma tenue, H. subcrenatum (1054). (m) Biol. activity as fungistat (1680). (n) Use as antiseptic agent, bactericide and fungicide (1792). (o) Synth. with Bu₃SnCH₂CN. (p) Rxn. with Et₂O-HCl → Bu₃SnCl + C₆F₅CHOHCH₂CN (1631).

Table 176. Triorganotin Alkoxides Derived from Alcohols Containing Oxygen Functions R_3SnOR'

Formula*	Synth. Method**	Yield	Properties	Ref.
$Me_3SnOCH_2CH_2CO_2Me$	VA	--	IR, NMR spectra	1235
	IVB (a)	--	--	1235
$Me_3SnOCH_2CH_2CONEt_2$	V (a)	95	$b_{0.04}$ 85-87°, IR, NMR spectra	2452
$(PhCH_2)_3Sn(OCH_2CH_2)_{6-8}OR$ (390) (b)	IVA	--	RO = oleate, water miscible	911
$(PhCH_2)_3Sn(OCH_2CH_2)_{20}OR$ (c)	--	--	Use as antistatic for polymers	2818
Et_3SnOR'				
$R' = p-MeOC_6H_4CH_2$ (390)	IIA	95	$b_{0.2}$ 126°, n_D^{20} 1.5327	389
$Me_2\overline{COCH_2CH(O)}CH_2$	III	71	b_2 116°, n_D^{20} 1.4735, d_{20} 1.2618	690
$MeEt\overline{COCH_2CH(O)}CH_2$	III	68	b_3 130°, n_D^{20} 1.4746, d_{20} 1.2408	690
$MePr\overline{COCH_2CH(O)}CH_2$	III	72	b_4 143°, n_D^{20} 1.4730, d_{20} 1.2184	690
EtO_2CCH_2CHPh	IA (d)	98	NMR spectrum, dec. on distn.	1631
$AcCH_2CHPh$	IA (e)	73	b_2 124-26°, n_D^{20} 1.5202, d_{20} 1.2582	437
$AcCH_2CHPr$	IA (e)	--	NMR spectrum (f)	1631
$1,1-AcCH_2C_6H_{10}$	IA (e)	--	NMR spectrum	1631
$p-MeOC_6H_4$	IVB	--	$b_{0.09}$ 111-13°, n_D^{20} 1.5409	914
	--	--	b_1 156°, n_D^{20} 1.5377 (g, h)	2636
$o-HOCH_2C_6H_4$	IIA	78	m. 62-64°	389
$m-HOCH_2C_6H_4$	IVA	82	--	389
	IIA	89	m. 63°	389
$p-OHCC_6H_4$	--	--	m. 76-77°, b_1 170-72°, IR spectrum (1)	2637
$p-MeO_2CC_6H_4$	--	--	m. 56-57°, b_1 167°, n_D^{20} 1.5612, IR spectrum	2637
$Et_2(C_6H_{13})Sn(OCH_2CH_2)_{6-8}OR$ (390) (b)	IVA	--	Water miscible, ROH = tallic acid	911
	III	--	--	911
$Pr_2(c.C_6H_{11})SnOCH_2CH_2OH$	III	62	m. 150-51°, IR spectrum	2721

$Pr_2(c-C_6H_{11})SnOCH_2CHOHCH_2OH$	III	62	m. 61-62°	2721
$i-Pr_2(c-C_6H_{11})SnOCH_2CH_2OH$	III	61	m. 96-97°, IR spectrum	2721

Bu_3SnOR'

R' = $EtOCH_2CH_2$	IVA	--	b_3 146-50° (j)	1794
$BuOCH_2CH_2$	IVA	--	b_1 153-56° (j)	1794
$HO_2CCO(OCH_2CH_2)_x$ (390) (b)	IVA	--	Water miscible prod.	911
$C_{12}H_{25}(OCH_2CH_2)_2$	IVA	--	Water miscible prod.	911
$C_{12}H_{25}(OCH_2CH_2)_3$	IVA	--	Water miscible prod.	911
$C_{17}H_{35}CO(OCH_2CH_2)_4$	IVA	--	Water miscible prod.	911
$C_8H_{17}C_6H_4(OCH_2CH_2)_9$	IVA	--	Water miscible prod.	911
HOC_6H_4	--	--	Use as foaming cat. for polyurethan	450
$p-MeOC_6H_4$	IVA	--	$bo.8$ 177-79° (j)	1794
	--	--	Use as foaming cat. for polyurethan	450
$m-HOCH_2C_6H_4$	IIA	--	IR spectrum	389
$5,2-MeO(Bz)C_6H_3$	IVB	--	(k)	2972
$2-(2,4-MePhC_6H_4)O_2CC_6H_4$	IVB	--	(k)	2972
R (l)	III,IVA	--	(m)	1958
Ar (n)	--	--	Use as stabilizer, biocide for polyamides	2618

* Numbers in parenthesis refer to pages in the main volume.

** The characters used here correspond to the synthetic scheme in the introduction to Chapter 4.3.3.1.
(a) Synth. with $\overline{OCH_2CH_2}C:O$. (b) Correct formulae in main volume, p. 390 line 4 to: $(PhCH_2)_3Sn(OCH_2CH_2)_8-8R$, line 9 to: $Et_2(C_8H_{17})Sn(OCH_2CH_2)_8-8R$, line 10 to: $Et_2(C_6H_{13})Sn(OCH_2CH_2)_8-8R$, line 12 to: R = $(CH_2CH_2O)_nH$, line 13 to: $(CH_2CH_2O)_nR$, line 10 f.b. to: R = $(CH_2CH_2O)_nR$. (c) RO = sorbitan laurate monosuccinate. (d) Synth. with $Et_3SnCH_2CO_2Et$. (e) Synth. with Et_3SnCH_2Ac. (f) Rxn. with alc. HCl → Et_3SnCl (437), with S → $(Et_3Sn)_2S$ (437). (g) Moessbauer resonance spectrum (2636). (h) Rxn. with Ph_3SnH → $Et_3SnSnPh_3$ + $p-MeOC_6H_4OH$ (914), relative rxn. rates (915). (i) Moessbauer resonance spectrum (2637). (j) Use as antiseptic agent, bactericide, fungicide (1794). (k) Use as UV stabilizer for polypropylene and synth. resins (2972). (l) R = various ethoxylated alcohols, phenols and carboxylic acids (1958). (m) Use as biocide, fungicide and stabilizer for PVC (1958). (n) Ar = $2-\{1-[2,3,5-HO(Me_3C)MeC_6H_2CH_2]\}-3,5-(Me_3C)MeC_6H_2$.

Table 176 Continued

Formula*	Synth. Method**	Yield	Properties	R_3SnOR' Ref.
i-$Bu_3SnOC_6H_4CH_2OH$-o	IIA	63	m. 33-34°	389
$(c.C_6H_{11})_3SnOCH_2CH_2OH$	III	87	m. 165-66°	2712
$(c.C_6H_{11})_3SnOCH_2CHOHCH_2OH$	III	63	m. 151-52°, IR spectrum	2712
$(C_8H_{17})_3SnOC_6H_4R$	IVB	--	R = 2,2'-HOC_6H_4CO (k)	2972
$(C_8H_{17})_3SnOC_6H_4R$	IVB	--	R = 2,4'-$PhC_6H_4O_2C$ (k)	2972
$Ph_3SnOCH_2C_6H_4OH$-o	IIA	65	m. 195-96°	389
$Ph_3Sn(OCH_2CH_2)_8$-$8OR$ (390) (b)	III	--	RO = oleate, water miscible prod.	911
$Ph_3Sn(OCH_2CH_2)_{9-10}OC_6H_4C_9H_{19}$-$p$	III	--	(m)	1958
$Ph_3SnOC_6H_3(OC_8H_{17})Bz$-$2,5$	IVB	--	(k)	2972
$Ph_3SnOC_6H_4CO_2Ph$-o	IVB	--	(k)	2972

Table 177. Triorganotin Alkoxides Derived from Alcohols Containing Nitrogen and Phosphorus

Formula*	Synth. Method**	Yield	Properties	R_3SnOR' Ref.
p-$Et_3SnC_6H_4NMe_2$	--	--	b_1 159-60°, n_D^{20} 1.5502	2637
p-$Et_3SnC_6H_4NO_2$ (391)	III	60	m. 31-32°, dec. at 0.5 torr 190°, soly.	791
	--	--	b_1 190-91°, n_D^{20} 1.6136 (a, b)	2637

Bu_3SnOR'

| $R' = Et_2NCH_2CH_2$ | IVA | -- | b_2 162-64° (c) | 1794 |
| $c.C_6H_{11}NHCH_2CH_2$ | IVA | -- | b_2 230-33° (c) | 1794 |

Compound	Type	Yield	Properties	Ref.
Et$_2$NCH$_2$CHPh	VC	70	Colorless liq., b$_{0.2}$ 154°, n$_D^{25}$ 1.4974 (d)	2053
	III	50	b$_{0.05}$ 132°, n$_D^{25}$ 1.4971, IR, NMR spectra	2053
	IVA	--	b$_{0.1}$ 145°, use as fungicide	2835
H$_2$N(CH$_2$)$_3$	IVA	--	Dec. on distn. (c)	1754
Me$_2$N(CH$_2$)$_3$	IVA	--	b$_2$ 155-65° (c)	1754
Et$_2$NCH$_2$CHEt	VC	76	Colorless liq., b$_{0.1}$ 122°, n$_D^{25}$ 1.4705 (d-f)	2053
	III	48	b$_{1.5}$ 160-62°, n$_D^{25}$ 1.4710, IR, NMR spectra	2053
2-Et$_2$NC$_8$H$_{10}$c.	VC	65	Colorless liq., b$_{0.1}$ 144°, n$_D^{25}$ 1.4852 (d-f)	2053
	III	60	b$_{0.15}$ 147°, n$_D^{25}$ 1.4851, IR, NMR spectra	2053
o-Et$_2$NC$_6$H$_4$	III	--	b$_4$ 105-108°, use as fungicide	2835
p-O$_2$NC$_6$H$_4$	--	--	Use as foaming cat. for polyurethan (g)	450
2,4-O$_2$N(Me)C$_6$H$_3$	--	--	(g)	--
p-Ph$_3$SnC$_6$H$_4$NO	IVA	--	m. 165°, visible, IR spectra, soly. (h)	2641
p-Ph$_3$SnOC$_6$H$_4$NO$_2$ (391)	III	85	m. 95-96°, monomeric, soly.	791
Ph$_3$SnOC$_6$H$_2$(NO)Me$_2$-3,5,4	IVA	--	m. 192° (dec.), visible, IR spectra, soly. (h, i)	2641
Bu$_3$SnOCH$_2$CH$_2$PEt$_2$	III	--	b$_4$ 165°, use as fungicide	2835

* Numbers in parenthesis refer to pages in the main volume.
** The characters used here correspond to the synthetic scheme in the introduction to Chapter 4.3.3.1.
(a) Moessbauer resonance spectrum (2637). (b) Rxn. with Ph$_3$SnH → Et$_3$SnSnPh$_3$ + p-O$_2$NC$_6$H$_4$OH (914), relative rxn. rates, mechanism (915). (c) Use as antiseptic agent, bactericide, fungicide (1794). (d) Struct. as penta-coordinate chelate (2053). (e) Rxn. with aq. KOH in dioxane → (Bu$_3$Sn)$_2$O + Et$_2$NCH$_2$CHEtOH (2053). (f) Use as fungicide (2835). (g) Biol. activity as molluscicide against Heliosoma tenue, H. subcrenatum (1054). (h) Structure with isomeric nitrosophenate ⇌ quinoneoxime derivative (2641). (i) Elemental analysis by simultaneous detn. (1088).

Table 178. Triorganotin Derivatives of Oximes and Hydroxylamine $R_3SnON:CR_2$, R_3SnONR_2

Formula	Synth. Method*	Yield	Properties	Ref.
$Et_3SnONHPh$	IIB	--	m. 25°, IR spectrum (a)	390
$Et_3SnONPhNMe_2$	VD	63	$b_{0.02}$ 98-99°	2406
$Bu_3SnON:CHR$				
R = H	III	--	$b_{0.008}$ 75-76°, n_D^{20} 1.4837 (b, c)	2125
PrCH:C(Et)	IVA	--	n_D^{20} 1.4936 (d)	2627
Ph	IVA	--	n_D^{30} 1.5150 (e)	2628
3,5-$(Me_3C)_2C_6H_3$	IVA	--	n_D^{30} 1.5270 (e)	2628
p-ClC_6H_4	IVA	--	n_D^{30} 1.5378 (e)	2628
2,4-$Cl_2C_6H_3$	IVA	--	n_D^{30} 1.5457 (e)	2628
o-HOC_6H_4	IVA	--	n_D^{30} 1.5698 (e)	2628
p-$MeOC_6H_4$	IVA	--	n_D^{30} 1.5325 (e)	2628
3,4-$CH_2O_2C_6H_3$	IVA	--	n_D^{28} 1.5468 (d)	2627
m-$O_2NC_6H_4$	IVA	--	n_D^{30} 1.5426 (e)	2628
2-C_4H_3O	IVA	84	n_D^{20} 1.5180 (d)	2627
			$b_{0.001}$ 113-15°, n_D^{20} 1.5181 (b, c)	2125
$Bu_3SnO:NCRR'$				
R = Me, R' = Me	IVA	100	n_D^{30} 1.4751 (d)	2627
	IVA	86	$b_{0.005}$ 83-85°, n_D^{20} 1.4768 (b, c)	2125
	III	--	$b_{0.01}$ 79° (f)	2661
Me Et	IVA	--	n_D^{28} 1.4728 (d)	2627
Me i-Bu	IVA	--	n_D^{28} 1.4728 (d)	2627
Me Ph	IVA	--	n_D^{30} 1.5239 (d)	2627
Ph Ph	IVA	95	$b_{0.25}$ 215°, n_D^{20} 1.4768 (b, c)	2125
	IVA	--	n_D^{30} 1.5518 (d)	2627
R, R' = $(CH_2)_5$	IVA	--	n_D^{30} 1.4896 (d)	2627

(Bu₃SnON:CMe)₂	IVA	100	Crude yellow oil, n_D^{20} 1.5065 (b, c)	2125
1,4-(Bu₃SnON:)₂C₆H₈c.	IVA	95	Dec. on distn., n_D^{20} 1.5065 (b)	2125
(Bu₃SnON:CPh)₂	IVA	100	Crude, n_D^{20} 1.5382 (b)	2125
(C₆H₁₃)₃SnON:CMe₂	IVA	91	b0.005 118°, n_D^{20} 1.4750 (b)	2125
(C₆H₁₃)₃SnON:CPh₂	IVA	96	b0.01 192-93°, n_D^{20} 1.5388 (b)	2125
(C₁₂H₂₅)₃SnON:CMe₂	IVA	92	m. 45-49° (b)	2125
(C₁₂H₂₅)₃SnON:CPh₂	IVA	98	m. 133-37° (b)	2125
Ph₃SnON:CMe₂	IVA	--	m. 110-15° (d)	2627

* The characters used here correspond to the synthetic scheme in the introduction to Chapter 4.3.3.1. (a) Rxn. with dil. HCl → PhNHOH (390), with PhNO → PhN:N(O)Ph (390). (b) Use as disinfectant, herbicide, insecticide (2125). (c) Biol. activity bacteriostat against Staphylococcus aureus (2125), as fungistat against Aspergillus niger (2125). (d) Use as bactericide, fungicide, herbicide (2627). (e) Use as bactericide, fungicide, herbicide, insecticide (2628). (f) Use as catalyst for hardening silicone rubber (2661), for synth. of polyurethan (2661).

4.3.3.1.6.7 TRIORGANOTIN ALKOXIDES DERIVED FROM ORGANOMETALLIC SUBSTITUTED ALCOHOLS

New compounds in this chapter are mercury containing phenates, a manganese complex previously reported with a manganese-tin bond and the ensuing oxastannacyclopentane. All bis(organotin)alkoxides are included in this chapter.

Additional substituted triorganotin alkoxides containing metallic substituents are listed in Subchapters 2.4.3, 2.5.3 and 2.7.1.2.

2,2,1,2-Dibutyloxastannacyclopentane $SnC_{11}H_{24}O$ $\overline{Bu_2SnCH_2CH_2CH_2O}$
Synth.: By thermal dec. of $Bu_2Sn(CH_2CH_2CH_2OH)_2$ at 25-160° (1383).
By rxn. of Bu_2SnH_2 with $CH_2:CHCH_2OH$ in the presence of AIBN in Ar atm., in 20% yield (1383).
Prop.: m. 219-26° (dec.), subl$_{0.0075}$ 190° with some dec. (1383).
Rxn. with: Thermal dec. 190° → $(Bu_2SnO)_x$ (1383).
AcOH → $[Bu_2(HOCH_2CH_2CH_2)SnOAc]_2$ (1383).

Additional triorganotin alkoxides derived from organometallic alcohols are listed in Table 179.

4.3.3.1.6.9 TRIORGANOTIN ALKOXIDES DERIVED FROM POLYFUNCTIONAL ALCOHOLS

Triorganotin alkoxides derived from polyfunctional alcohols are listed in Table 180. Other substituted derivatives may be found in Subsections 4.3.3.1.6.2 and 4.3.3.1.6.5.9 and Table 176.

4.3.3.1.9 TRIORGANOTIN MIXED OXIDES

This chapter includes new mixed oxides containing cadmium and zirconium. The triorganotin mixed oxides summarized in Table 181 are prepared by methods from the following scheme.

I   Reaction of organotin halides with R_3SiONa.

II Reactions of organotin oxides and hydroxides with:
    (IIA) RLi and $CdEt_2$,
    (IIB) R_3SiH ,
    (IIC) $R_nSi(OH)_{4-n}$, Ph_3GeOH, $(R'O)_3TiOR$ and $Zr(OBu)_4$,
    (IID) Me_nSiCl_{4-n}.

III Reactions of R_3SnOAc with:
    (IIIA) $(RO)_4M$,
    (IIIB) $(RO)_4M$ in presence of $(HOCH_2CH_2)_3N$ and $(HOC_3H_6)_3N$.

R = organic group, $(R'OH)_3$ = triethanolamine, M = Ti, Zr.

Additional references to triorganotin mixed oxides may be found in Chapter 8.1.

Table 179. Triorganotin Alkoxides Derived from Organometallic Alcohols R_3SnOR'

Formula*	Synth. Method**	Yield	Properties	Ref.
$Et_3SnOCH_2C\!:\!CSiEt_3$	IVA	--	b_2 82°, n_D^{20} 1.5000, d_{20} 1.2102	2506
$Et_3SnOCHMeC\!:\!CSiEt_3$	IVA	--	$b_{7.5}$ 154°, n_D^{20} 1.4820, d_{20} 1.0817	2506
$Et_3SnOCHMeC\!:\!CPbMe_3$	VID	--	$b_{0.3}$ 120°, n_D^{20} 1.5400, d_{20} 1.7387, IR spectrum	2713
$Et_3SnOCMe_2C\!:\!CSiMe_3$	--	--	Dec. at 250° → $Et_3SnC\!:\!CSiMe_3$	2776
$Et_3SnOCMe_2C\!:\!CSiEt_3$	IVA	--	$b_{3.5}$ 102°, n_D^{20} 1.4735, d_{20} 1.0577	2506
$p\text{-}(Et_3Sn)_2C_6Cl_4$ (393)	--	--	Use as herbicide (2124a), microbiocide	2142b
$Pr_3SnOC_6H_4HgOAc$	VIC	--	Oil (a)	1225
$Bu_3SnOC_6H_4HgOAc$	VIC	--	Pale brown liq., d_{20} 1.30 (a)	1225
$(Bu_3SnO)_2Y$				
Y = CH_2CH_2	IVA (b)	86	$b_{0.1}$ 170°	950
C_6Cl_4 (393)	--	--	Use as herbicide (2124a), microbiocide	2142b
$2,2'\text{-}(C_6H_4)_2CH_2$	--	--	Use as foaming cat. for polyurethan	450
$2,2'\text{-}(5\text{-}ClC_6H_3)_2CH_2$ (393)	--	--	Use as herbicide (2124a), microbiocide	2124b
$2,2'\text{-}(3,5,6\text{-}Cl_3C_6H)_2CH_2$ (393)	--	--	Use as herbicide (2124a), microbiocide	2124b
$4\text{-}[2,5\text{-}Me_3C(Me)C_6H_2]_2S$	--	--	Use as foaming cat. for polyurethan	450
$(C_8H_{17})_3SnOC_6H_4HgOAc$	VIC	--	Oil (a)	1225
$Ph_3SnOC_6H_4HgOAc$	VIC	--	Oil (a)	1225
$p\text{-}(Ph_3Sn)_2C_6Cl_4$ (393)	--	--	Use as herbicide (2124a), microbiocide	2124b
$Ph_3SnOC_5Ph_4Mn(CO)_3$ (501)	III	53	m. 206-207°, IR spectrum, soly. (c)	1124

* Numbers in parenthesis refer to pages in the main volume.
** The characters used here correspond to the synthetic scheme in the introduction to Chapter 4.3.3.1.
(a) Biol. activity as bactericide against Flavobacter peregrinum (1225), as fungicide against Aspergillus niger, Candida utilis (1225). (b) Synth. with $(CH_2)_2CO_3$. (c) Polarography (964, 967), UV spectrum (967), x-ray struct. analysis, formerly incorrectly formulated as $Ph_3SnC_5Ph_4OMn(CO)_3$ (M-501) (804).

Table 180. Triorganotin Alkoxides Derived from Polyfunctional Alcohols R_3SnOR'

Formula*	Synth. Method**	Yield	Properties	Ref.
$Me_3SnOC(CF_3)_2NMe_2$	VB	60	$b_{0.03}$ 42°, n_D^{20} 1.4025, H, ^{19}F NMR spectra	2199
Et_3SnOR'				
$R' = CH_2:CHOCH_2CH_2$	III	97	b_3 90-90.5°, n_D^{20} 1.4810, d_{20} 1.2318-1.2560	510, 1914
$CH_2:CH(OCH_2CH_2)_2$	III	--	b_2 119-119.5°, n_D^{20} 1.4809, d_{20} 1.2453	510, 1914
$CH_2:CH(OCH_2CH_2)_3$	III	--	b_1 136-136.5°, n_D^{20} 1.4772-80, d_{20} 1.2145	510, 1914
$CH_2:CHCH_2CH(C_6H_4Cl-p)$	IA (a)	81	$b_{0.02}$ 128°, hydrolyzes in H_2O (b)	1331
$CH_2:CHCH_2CH(C_6H_4NO_2-p)$	IA (a)	75	$b_{0.01}$ 166°, hydrolyzes in H_2O (b)	1331
$CH_2:CHCH_2CH(C_6H_4CCl_3-p)$	IA (a)	95	$b_{0.2}$ 98°, hydrolyzes in H_2O (b)	1331
$PhC!CCH(CCl_3)$	IB	70	b_1 160-61°, n_D^{20} 1.5623, d_{20} 1.4090	1537
	IB	50	$b_{0.001}$ 157°, IR (1894) spectrum (c)	2644
$AcCH_2CH(C_6H_4Cl-o)$	IA (d)	--	NMR spectrum	1631
$AcCH_2CH(C_6H_4Cl-p)$	IA (d)	--	NMR spectrum	1631
$AcCH_2CH(C_6F_5)$	IA (d)	--	NMR spectrum	1631
$AcCH_2CH(C_6H_4NO_2-m)$	IA (d)	--	NMR spectrum	1631
$AcCH_2CH(C_6H_4NO_2-p)$	IA (d)	--	NMR spectrum	1631
$EtO_2CCH_2CH(CCl_3)$	IA (c)	--	$b_{0.25}$ 108-114°, n_D^{20} 1.4973, NMR spectrum (f)	1631
$Et_2NOCCH_2CH(CCl_3)$	IA (g)	--	NMR spectrum	1631
$EtO_2CCH_2CPh(CF_3)$	IA (e)	--	Dec. on distn., NMR spectrum	1631
$EtO_2CCH_2CH(C_6F_5)$	IA (e)	--	$b_{0.07}$ 107-108°, n_D^{20} 1.4751	1631
$Et_2NOCCH_2CH(C_6F_5)$	IA (g)	--	NMR spectrum	1631
$AcCH_2CPh(CF_3)$	IA (d)	--	NMR spectrum	1631
$AcCH_2C(CH_2Cl)_2$	IA (d)	--	NMR spectrum (h)	1631
$BuC!CCH(CCl_3)$	---	--	IR spectrum	1894
Bu_3SnOR'				
$R' = C_{17}H_{33}CO(OCH_2CH_2)_{6-8}$ (390)	IVA	--	Water miscible prod. (i)	911

648

$C_{17}H_{33}CO(OCH_2CH_2)_8$	IVA	--	Water miscible prod.	911
$C_{17}H_{33}CO(OCH_2CH_2)_{400}$	IVA	--	Water miscible prod.	911
R″	III, IVA	--	R″ = ethoxy amines (j)	1958
2,4-Cl(O_2N)C_6H_3	--	--	(k)	--
4,2-Cl(O_2N)C_6H_3	--	--	(k)	--

* Numbers in parenthesis refer to pages in the main volume.

** The characters used here correspond to the synthetic scheme in the introduction to Chapter 4.3.3.1.
(a) Synth. from $Et_3SnCH_2CH:CH_2$ and RCHO.   (b) Rxn. with $CH_2(CH_2CO_2H)_2$ or BzOH → Et_3Sn-carboxylates + R′OH (1331).
(c) Rxn. with HCl → Et_3SnCl + PhC!CCHOHCCl$_3$ (1537), with $CH_2(CH_2CO_2H)$ → $(Et_3SnO_2CCH_2)_2CH_2$ + PhC!CCHOHCCl$_3$ (264).
(d) Synth. from Et_3SnCH_2Ac.   (e) Synth. from $Et_3SnCH_2CO_2Et$.   (f) Rxn. with Cl_3CCHO → $Et_3Sn(OCHCCl_3)_2CH_2CO_2Et$
(1631).   (g) Synth. from $Et_3SnCH_2CONEt_2$.   (h) Rxn. with $(CO_2H)_2$ → $(Et_3SnO_2C)_2$ + $(ClCH_2)_2C(OH)CH_2Ac$ (1631).
(i) Correct formula in main volume, p. 390, line 14 to $(CH_2CH_2O)_nR$.   (j) Use as biocide, fungicide and stabilizer
for PVC (1958).   (k) Biol. activity as molluscicide against Heliosoma tenue and H. subcrenatum (1054).

Table 181. Triorganotin Mixed Oxides

R_3SnOMR_n

Formula*	Synth. Method**	Yield	Properties	Ref.
Et_3SnONa	--	--	(a)	1840
Ph_3SnOLi	IIA	91	Dec. 200°, soly. (b)	1906
$Et_3SnOSiMe_3$ (394)	--	--	Struct. (c)	1906
$Et_3SnOSiEt_3$ (394)	--	--	Struct. (c, d)	1165
$Pr_3SnOSiPh_3$	IIB	--	b_3 210-15°, n_D^{25} 1.5689, d_{25} 1.185, IR spectrum	2465
$(Bu_3SnO)_2SiMe_2$	IID	--	b_2 94-96° (e)	2465
$Bu_3SnOSiMe_3$ (394)	IID	--	b_{10} 142-45° (e)	719
$Bu_3SnOSiEt_3$	IIB	High	--	1165
$Bu_3SnOSiBu_3$	IIB (f)	95	b_5 188-93°, n_D^{25} 1.4650, d_{25} 1.009, IR spectrum (g, h)	2344
$(Bu_3SnO)_2SiPh_2$	IIC	98	Oil, IR spectrum	1165
$Bu_3SnOSiPh_3$	I	54	b_4 228-36°, n_D^{25} 1.5507, d_{25} 1.156	1165
	IIB	69	b_5 233-35°, n_D^{25} 1.5547, d_{25} 1.147	1165
	IIB (f)	8	IR spectrum (g, h, i)	2344
$Ph_3SnOSiMe_3$	IIC	100	Oil, IR spectrum	2344
$(Ph_3SnO)_2SiPh_2$ (394)	IIC	100	m. 94-95°, IR spectrum	958
$Ph_3SnOSiPh_3$ (394)	IIC	44	m. 136-38°, IR spectrum	1165
	IIB	--	Thermogravimetric anal. (958)	2344
	IIC	100	m. 138-40°, IR spectrum	958
$Ph(Ph_2SnO)_{2.5}SiPh_2OH$	IIC	--	Waxy, softening 88-94°, IR spectrum (j)	2344
$Et_3SnOGePh_3$	IIC	--	Oil, IR spectrum	2663
$Et_3SnOCdEt$	IIA	--	Viscous mass (k)	--
$(Bu_3SnO)_2Ti(OPr-i)_2$	--	--	(l)	--
$(Bu_3SnO)_2Ti(CHAc_2)_2$	--	--	(l)	882
$Bu_3SnOTi(OCH_2CH_2)_3N$	IIIB	2	$b_{0.05}$ 74°, n_D^{28} 1.4592 (m)	2324
	IIIB	--	IR spectrum, use as fungicide	882
$Bu_3SnOTi(OCH_2CH_2)_3N \cdot L$ (m)	--	--	$L = [N(CH_2CH_2O)_3Ti]_2O$, m. 177° (dec.)	882, 2324
$Ph_3SnOTi(OCH_2CH_2)_3N$	IIC	97	m. 205-208° (dec.), mol. wt.	2324
	IIIB	--	m. 205-208° (dec.) (n, o)	

Ph$_3$SnOTi(OC$_3$H$_6$)$_3$N	IIC	--	m. 208-10° (o)	2324
Me$_3$SnOZr(OBu)$_3$	IIIB	--		1097
Bu$_3$SnOZr(OPr)$_3$	IIIB	--	d$_{20}$ 1.3375, IR spectrum (p-s)	1097
Bu$_3$SnOZr(OPr)$_2$OR	IIIA (t)	--	R = C$_7$F$_{15}$CO	1097
	(s)	--	(p)	1097
Bu$_3$SnOZr(OBu)$_3$	IIIB	--	n$_D^{24}$ 1.4892, d$_{24}$ 1.2020, IR spectrum (p, q, u, v)	1097
Bu$_3$SnOZr(OBu)$_2$OR	(u)	--	R = EtBuCHCH$_2$ (p)	1097
Ph$_3$SnOZr(OPr-i)$_3$	IIIA	--	(p)	1097
Ph$_3$SnOZr(OBu)$_3$	IIC	--		1097
Ph$_3$SnOZr(OCH$_2$CH$_2$)$_3$N	IIIA (w)	--	Use as fungicide	2324

* Numbers in parenthesis refer to pages in main volume.

** The characters used here correspond to the synthetic scheme in the introduction to Chapter 4.3.3.1.9.

(a) Rxn. with PhC⋮CH, + (Et$_3$Sn)$_2$O + CH$_2$:CHC⋮CH (excess) → Et$_3$SnC⋮CCH:CH$_2$ (1342), with PhC⋮CH, + (Et$_3$Sn)$_2$O, + Et$_3$SnC⋮CH → (Et$_3$SnC⋮)$_2$ (1342). (b) Rxn. with Ph$_2$PCl → Ph$_3$SnOPPh$_2$ (1840). (c) Rxn. with R'MgBr → Et$_3$SnR' + R$_3$SiOH; R = Me, Et; R' = Et, CH$_2$:CH, Ph (1906). (d) Rxn. with (:CHCO)$_2$O-PhCH:CH$_2$-polymer → tin-contg. polyester (1228), biol. activity as insecticide (1328). (e) Use as stabilizer for PVC (2465), as water repellent (2465). (f) Synth. from (Bu$_2$SnO)$_x$ → Bu$_3$SnH (1165, 1242). (h) Rxn. with Bu$_3$SnH → Bu$_6$Sn$_2$ (1165). (g) Rxn. with (MeHSiO)$_x$ → Bu$_3$SnH (1165, 1242). (h) Rxn. with Bu$_3$SnH → Bu$_6$Sn$_2$ (1165). (i) Rxn. with Ph$_3$SiH → Bu$_6$Sn$_2$ + (Ph$_3$Si)$_2$O (1165). (j) Thermogravimetric analysis, proposed structure (958). (k) Rxn. with MeOH → Et$_3$SnOCdOMe (?) + C$_2$H$_6$ (2663), with AcOH → Et$_3$SnOAc + Cd(OAc)$_2$ + C$_2$H$_6$ (2663), polymerizes CH$_2$:CMeCO$_2$Me (2663). (l) Use as stabilizer for polyamide fibers (2757), biol. activity as bactericide against Staphylococcus aureus (2757), as fungicide against Aspergillus niger, Trichophyton gypseum (2757). (m) Thermal dec. of Bu$_3$SnOTi(OCH$_2$CH$_2$)$_3$N → (Bu$_3$Sn)$_2$O + Bu$_3$SnOTi(OCH$_2$CH$_2$)$_3$N·[N(CH$_2$CH$_2$O)$_3$Ti]$_2$O (882). (n) Biol. activity as fungicide against Aspergillus niger, Botrytis allii, Pullularia pullulans (2324). (o) Biol. activity as fungicide against Chaetomium globosum, Glomerella cingulata, Penicillium italicum (2324). (p) Use as waterproofing agent (1097). (q) Biol. activity as fungicide against Aspergillus niger, Botrytis allii, Chaetomium globosum, Glomerella cingulata, Pullularia pullulans, Penicillium italicum (1097). (r) Use as catalyst for foaming polyurethan (1097). (s) Rxn. of Bu$_3$SnOZr(OPr)$_3$ with C$_7$F$_{15}$CO$_2$H → Bu$_3$SnOZr(OPr)$_2$O$_2$CC$_7$F$_{15}$ (1097). (t) Synth. with Zr(OR)$_3$O$_2$CC$_7$F$_{15}$. (u) Rxn. of Bu$_3$SnOZr(OBu)$_3$ with EtBuCHCH$_2$OH → Bu$_3$SnOZr·(OBu)$_2$OCH$_2$CHEtBu (1097). (v) Rxn. with HN(CH$_2$CH$_2$OH)$_2$ → polymeric diethanolamine zirconate (1097).

4.3.3.2 DIORGANOTIN ALKOXIDES

Diorganotin dialkoxides $R_2Sn(OR')_2$ and oligomers of the formulae $R'O(R_2SnO)_nR'$ or $R_2Sn(OR')_2 \cdot n-1\ R_2SnO$ are treated in this chapter. The compounds in the ensuing tables are prepared by methods from the following scheme.

I Reactions of diorganotin hydrides with:
    (IA) alcohols,
    (IB) aldehydes and ketones.

II Reactions of triorganotin halides with:
    (IIA) alcohols,
    (IIB) in presence of base,
    (IIC) alkali metal alkoxides.

III Reactions of diorganotin oxides with:
    (IIIA) alcohols,
    (IIIB) diorganotin diphenoxide,
    (IIIC) diorganotin dialkoxide with alcohol.

IV Reaction of diorganotin diamide with alcohol.

V Miscellaneous reactions:
    (VA) complex formation of diorganotin dialkoxide with a suitable ligand,
    (VB) dealkylation of R_4Sn with an acidic alcohol.

Additional references to diorganotin dialkoxides may be found in Chapters 7.3 and 9.

4.3.3.2.1 UNSUBSTITUTED DIORGANOTIN ALKOXIDES

Dimethyltin Dimethoxide (M-396) $SnC_4H_{12}O_2$ $Me_2Sn(OMe)_2$
<u>Synth.</u>: By rxn. of $Me_2Sn(NEt_2)_2$ with MeOH in pentane in N atm. (323).
<u>Prop.</u>: m. 86° (323), 85-86° (942), (far) IR, assignment (323, 2555) and NMR (323, 347, 942) spectra.
<u>Rxn. with</u>: $Me_2SnX_2 \rightarrow Me_2Sn(OMe)X$; X = Cl, I (942).
<u>Use</u>: As water repellent for fabric, wood, brick, leather (1397a).

Diethyltin Dimethoxide (M-396) $SnC_6H_{16}O_2$ $Et_2Sn(OMe)_2$
<u>Synth.</u>: By rxn. of $Et_2Sn(NEt_2)_2$ with MeOH in pentane in N atm. (323).
<u>Prop.</u>: $b_{0.1}$ 105° (323), 70-71° (942), (far) IR, assignment (2555) and NMR (323,942) spectra.
<u>Rxn. with</u>: $Et_2SnX_2 \rightarrow Et_2Sn(OMe)X$; X = Cl, Br (942).
$(MeHSiO)_x \rightarrow Et_2SnH_2$ (2461).
<u>Use</u>: In catalyst for ethylene polymerization (239).

Dibutyltin Dimethoxide (M-397) $SnC_{10}H_{24}O_2$ $Bu_2Sn(OMe)_2$
<u>Synth.</u>: By rxn. of $Bu_2Sn(NEt_2)_2$ with MeOH in N atm. in pentane (323).

Prop.: b_3 150-52° (1117), $b_{1.5}$ 119° (36), $b_{0.3}$ 124° (942), $b_{0.1}$ 110° (323), n_D^{20} 1.4876, d_{20} 1.291 (1117), IR (352, 2184, 2692), assignment (1509, 2555), far IR, assignment (1505, 2555), UV (1116-7) and NMR (942) spectra, dipole moment (2184), mol. wt. (1117, 2184), struct. (1510), calorimetric titration with Bu_2SnCl_2 (1117).

Rxn. with: $Bu_2SnH_2 \rightarrow (Bu_2Sn)_x$ + MeOH (400, 468, 1802); x = 6 (1614).

$Bu_3SnCl \rightarrow Bu_2Sn(OMe)_2 \cdot Bu_2SnCl_2 + Bu_3SnOMe$ (2184).

$Bu_2SnX_2 \rightarrow Bu_2Sn(OMe)X$; X = F, I, NCS, camphor sulfonate (942), Cl, Br, OAc, $C_{11}H_{23}CO_2$, $ClCH_2CO_2$ (942, 2184).

$Bu_2Sn(O_2CR)_2 \rightarrow (Bu_2SnOMe)O(Bu_2SnO_2CR) + MeO_2CR$; R = Me, $ClCH_2$ (634).

$Bu_2SnS \rightarrow (Bu_2SnOMe)_2S$ (945).

$MePhSiHCl \rightarrow MePhSi(OMe)_2$ (719).

$CO_2 \rightarrow Bu_2Sn(OCO_2Me)_2$ (943).

$SO_2 \rightarrow Bu_2Sn(OMe)O_2SOMe$ (943).

$CS_2 \rightarrow Bu_2Sn(OMe)S_2COMe + Bu_2Sn(S_2COMe)_2$ (943).

$RBr \rightarrow Bu_2SnBr_2 + Bu_2Sn(OMe)Br + ROMe$; R = $CH_2:CHCH_2$ (435).

$(CH_2Br)_2 \rightarrow Bu_2SnBr_2 + Bu_2Sn(OMe)Br + (CH_2OMe)_2 + MeOCH_2CH_2Br$ (435).

$RCO_2H \rightarrow Bu_2Sn(O_2CR)_2$; R = $H(CF_2)_{10}$ (1277), $C_8H_{17}O_2CCH:CH$, $C_4H_7OCH_2O_2CCH:CH$ (1283).

$RCN \rightarrow Bu_2Sn(OMe)N:CROMe$; R = CCl_3 (943).

$RNCO \rightarrow Bu_2Sn(OMe)NRCO_2Me$ (2662) + $Bu_2Sn(NRCO_2Me)_2$; R = Et, Ph (943).

EtNCO (excess) $\rightarrow (EtNCO)_3$ (943).

$(PhNCO)_2 \rightarrow Bu_2Sn(NPhCO_2Me)_2$ (943).

2,3-$(ONC)_2C_6H_3Me \rightarrow$ 2,3-$MeC_6H_4N(CO_2Me)SnBu_2NCO_2Me$ (2662).

$RNCS \rightarrow Bu_2Sn(OMe)SC(OMe)NR + Bu_2Sn[SC(OMe)NR]_2$; R = $CH_2:CHCH_2$, Ph (943).

$(RN:)_2C \rightarrow Bu_2Sn(OMe)NRC(OMe):NR + Bu_2[NRC(OMe):NR]_2$; R = 1-$C_{10}H_7$ (943).

$RCHO \rightarrow Bu_2Sn(OMe)OCHROMe + Bu_2Sn(CHROMe)_2$; R = Me, CCl_3 (943).

$ROH \rightarrow Bu_2Sn(OR)_2 + MeOH$; R = 5,6-epoxyoctyl (1442).

2-$HOC_5H_4N \rightarrow N-Bu_2Sn[N(CH)_4CO]_2$ (1230).

$ArOH \rightarrow Bu_2Sn(OAr)_2$; Ar = derivatives of benzophenone, salicylic acid and benzotriazalylphenone (2972).

$(CR_2)_n(OH)_2 \rightarrow Bu_2SnO(CR_2)_nOSnBu_2O(CR_2)_nO + MeOH$; n = 2, 3, R = H, Me (2692).

$(CH_2CH_2OH)_2 \rightarrow Bu_2SnO(CH_2)_4O + MeOH$ (2692).

Addn. compd.: $Bu_2Sn(OMe)_2 \cdot Bu_2Sn(OBu)_2$ (?) from components only at high concentrations, mol. wt., IR spectrum (181) [$Sn_2C_{26}H_{60}O_4$].

$Bu_2Sn(OMe)_2 \cdot SnCl_4$, from $Bu_2Sn(OMe)_2 \cdot Bu_2SnCl_2$ with $SnCl_4$ in hexane, dec. > 200°, IR spectrum, mol. wt., thermodynamic data (2184) [$Sn_2C_{10}H_{24}Cl_4O_2$].

$Bu_2Sn(OMe)_2 \cdot TiCl_4$, from components, dec. > 200°, mol. wt., thermodynamic data (2184) [$SnTiC_{10}H_{24}Cl_4O_2$].

Use: In catalyst for polymerization of butadiene (2044a), for modifying polypropylene (269), polypropylene copolymers (2899), for foaming polyurethan (355a).

As additive for flame resistant polyester (2702).

As water repellent for fabric, wood, brick, leather (1397a).

Dibutyltin Diethoxide (M-397) $SnC_{12}H_{28}O_2$ $Bu_2Sn(OEt)_2$

Synth.: By rxn. of Bu_2SnCl_2 with EtONa in alc. or C_6H_6, in 88% yield (349).

By rxn. of EtOH with $Bu_2Sn(NEt_2)_2$ in N atm. in pentane (323), with $Bu_2Sn(SPr)_2$ in presence of cat. $p\text{-}MeC_6H_4SO_3H$, in 90% yield (1997).
Prop.: Colorless liq., $b_{0.1}$ 117° (1997), $b_{0.1}$ 107° (323), NMR spectrum (323).
Rxn. with: $(MeHSiO)_x$ (1:1) → $Bu_2HSnOEt$ (1165).
$(MeHS:O)_x$ (1:2) → Bu_2SnH_2 (1165).
$HSYCO_2H$ → $Bu_2\overline{SnO_2CYS}$ + EtOH; Y = CH_2CH_2, CHMe (2805).
$AcCH_2COR$ → $Bu_2Sn(CHAcCOR)_2$; R = Me, MeO, EtO (350).
$Y(OH)_2$ → $Bu_2\overline{SnOYO}$, several C_2-C_6 glycols (349).
$RN(CH_2CH_2OH)_2$ → $Bu_2\overline{SnOCH_2CH_2NRCH_2CH_2O}$ + EtOH; R = H, CH_2CH_2OH (1501).
$H_2NCH_2CH_2OH$ → $Bu_2\overline{SnNHCH_2CH_2O}$ (1501).
$HSCH_2CHROH$ → $Bu_2\overline{SnSCHRCH_2O}$ (2805).
RSH → $Bu_2Sn(SR)_2$ + EtOH; R = $PhCH_2$, Pr, n-, i-, t-Bu, $C_{12}H_{25}$, Ph (1997).
Use: As catalyst for curing silicone rubber (375).
As stabilizer for halogen containing polymers (707).

Dibutyltin Dibutoxide (M-398) $SnC_{16}H_{36}O_2$ $Bu_2Sn(OBu)_2$
Synth.: By rxn. of Bu_2SnO with BuOH (1:4) in MePh, in 60% yield (846a).
Prop.: b_4 130-35° (846a), (far) IR spectra and assignment (352, 1509), mol. wt. (2184), struct. (1510), thermodynamic data (2184), calorimetric titration with Bu_2SnCl_2 and $SnCl_2$ (1116-7).
Rxn. with: Bu_2SnCl_2 → $Bu_2Sn(OBu)Cl$ (846a, 2184).
$Bu_2Sn(OAc)_2$ → $Bu_2Sn(OMe)OAc$ (2184).
NH_4X → Bu_2SnX_2 + BuOH + NH_3; X = Cl, Br, I, NCS (2660).
PhNCO → $Bu_2Sn(NPhCO_2Bu)OBu$ (2662).
Addn. Compd.: $Bu_2Sn(OBu)_2 \cdot SnCl_4$, from components in hexane (181, 2184), from $Bu_2Sn(OBu)Cl$ with $SnCl_4$ (181, 2184), IR spectrum, dipole moment (181, 2184), thermodynamic data (118, 1116-7, 2184) [$Sn_2C_{16}H_{36}Cl_4O_2$].
$Bu_2Sn(OBu)_2 \cdot Bu_2Sn(OMe)_2$ (?), from components only in high concn., IR spectrum, mol. wt. (181) [$Sn_2C_{26}H_{60}O_4$].
Use: As catalyst for polyurethan synth. (3005).

Dibutyltin Di-t-Butoxide $SnC_{16}H_{36}O_2$ $Bu_2Sn(OCMe_3)_2$
Synth.: By rxn. of Bu_2SnCl_2 with Me_3CONa in N atm. in hexane under reflux, in 50% yield (2096).
By rxn. of $(Bu_2Sn)_x$ with $(Me_3CO)_2$ in N atm. at 130-35°, in 36% yield (2096).
Prop.: $b_{0.5}$ 107-109°, n_D^{20} 1.4620 (2096), (far) IR spectra and assignment (352, 1509), struct. (1510).
Rxn. with: BzOH → $Bu_2Sn(OBz)_2$ + Me_3COH (2096).

Listing of unsubstituted diorganotin alkoxides concludes in Table 182.

Table 182. Unsubstituted Diorganotin Alkoxides $R_2Sn(OR)_2 \cdot nR_2SnO$

Formula*	Synth. Method**	Yield	Properties	Ref.
$Me_2Sn(OEt)_2$ (396)	IV	--	b$_{0.1}$ 82°, IR, assignment, NMR spectra	323
$(Me_2SnOPh)_2O$ (396)	IIB	>70	m. 202° (409), m. 201-203° (2611) (a)	409, 2611
$(Me_2SnOC_{10}H_7-2)_2O$	IIB	>70	m. 233-35° (a)	2611
$Et_2Sn(OEt)_2$ (396)	IV	--	b$_{0.1}$ 75°, NMR spectrum	323
$Pr_2Sn(OMe)_2$ (397)	--	--	b$_{0.1}$ 125°, NMR spectrum (b, c)	942
$(Pr_2SnOPh)_2O$	--	--	m. 131-34° (d)	2611
$1-Pr_2Sn(OMe)_2$	--	--	(c)	--
$Bu_2Sn(OPr)_2$ (397)	--	--	(Far) (352) IR spectra, assignment (1509), struct.	1510
$Bu_2Sn(OCH_2CH:CH_2)_2$	IA	--	(e)	1383
$(Bu_2SnOBu)_2O$ (398)	--	--	Use as catalyst for modifying polypropylene	269
$Bu_2Sn(OC_5H_{11}-n)_2$	--	--	(Far) (352), IR spectra, assignment	1509
$Bu_2Sn(OC_6H_{11}c.)_2$	--	--	(Far) (352), IR spectra, assignment (e)	1509
$Bu_2Sn(OC_8H_{17}-n)_2$	--	--	(Far) (352), IR spectra, assignment (1509), struct.	1510
$(Bu_2SnOCH_2CHEtBu)_2O$ (398)	--	--	Use as catalyst for polyurethan synth.	2166
$Bu_2Sn(OPh)_2$ (398)	IIIA	84	m. 51-53°, b$_{0.1}$ 146-48°, NMR spectrum	1740
$(Bu_2SnOPh)_2O$ (398)	--	--	m. 138-39° (d, e)	2611
$Bu_2Sn(OPh)_2 \cdot 2Bu_2SnO$	IIIA	100	m. 91°, mol. wt.	2595
$Bu_2Sn(OC_6H_4Me-p)_2$	IIIA	73	m. 65-71°, b$_{0.05}$ 166.5-68.5°, NMR spectrum	1740
$i-Bu_2Sn(OMe)_2$	IIC	48	m. 36-38°, (far) IR spectra, assignment	2555
$i-Bu_2Sn(OBu-i)_2$	IB	47	m. ~20°, b$_{0.002}$ 96°	389
$(C_5H_{11})_2Sn(OMe)_2$	--	--	(f)	--

* Numbers in parenthesis refer to pages in main volume.
** The characters used here correspond to the synthetic scheme in the introduction to Chapter 4.3.3.2.
(a) Thermal decomposition at 220-300° in vacuo → Me_3SnOAr; Ar = Ph, $2-C_{10}H_7$ (2611). (b) Rxn. with Pr_2SnF_2 → $Pr_2Sn(OMe)F$ (942). (c) (Far) IR spectrum and assignment (2555). (d) Thermal decomposition at 0.3 mm 150° → $(R_3Sn)_2O$; R = Pr, Bu (2611). (e) Rxn. with $(MeHSiO)_x$ → Bu_2SnH_2 (2460-1). (f) Rxn. with ArOH → $(C_5H_{11})_2Sn(OAr)_2$; derivatives of PhBz, salicylic acid, benzotriazolylphenol (2972).

Table 182 Continued

Formula*	Synth. Method**	Yield	Properties	Ref.
$(C_8H_{17})_2Sn(OMe)_2$ (399)	--	--	(f-h)	--
$(EtBuCHCH_2)_2Sn(OMe)_2$	--	--	(h)	--
$(C_{12}H_{25})_2Sn(OMe)_2$	--	--	(h)	--
$(C_{10}F_{21}CH_2CH_2CH_2)_2Sn(OMe)_2$	II (i)	--	(j, k)	745a
$(C_{18}H_{37})_2Sn(OMe)_2$	--	--	(h)	--
$Ph_2Sn(OMe)_2$ (399)	IV	--	m. 186°, NMR spectrum (h)	323
$Ph_2Sn(OEt)_2$ (399)	IV	--	m. 95°, NMR spectrum	323
$R_2Sn(OR)_2$	--	--	Use as stabilizer for PVC	2895

(g) Rxn. with $C_8H_{17}\overline{CHCHO}(CH_2)_7CONHC_6H_{13} \rightarrow$ MeOH + octyltin amide (1442). (h) Use as water repellent for fabric, wood, brick, leather (1397a). (i) Synth. from $(CH_2{:}CHCH_2)_2SnCl_2$ with $1\text{-}C_{10}F_{21}I$ at 300°, + Zn in MeOH (745a). (j) Rxn. with $MeSiCl_3$ + MeONa $\rightarrow$ Sn-contg. polymer (745a). (k) Use as water, dirt, oil repellent, antiadhesive, release agent (745a).

4.3.3.2.3 DIORGANOTIN ALKOXIDES DERIVED FROM HETEROCYCLIC ALCOHOLS

Dimethyltin Bis(8-oxyquinolate) (M-401) SnC$_{20}$H$_{18}$N$_2$O$_2$ Me$_2$Sn(OC$_9$H$_6$N)$_2$
Synth.: By rxn. of Me$_2$SnCl$_2$ with 8-HOC$_9$H$_6$N in EtOH, followed by AcONa and NH$_3$, in 87% yield (602), in presence of alc. NH$_3$, in 96% yield (1216). By disproportionation of Me$_2$Sn(OC$_9$H$_6$N)Cl in presence of Py, in 47-55% yield (1291), in presence of BuLi (602), of Me$_2$Sn(OC$_9$H$_6$N)CHBz$_2$ (602).
Prop.: Yellow cryst., m. 236-37° (602), far IR, UV (602, 1291), NMR (347, 602, 1291, 1316) and Moessbauer resonance (902, 1044, 1143a) spectra, struct. (602, 1316), x-ray struct. anal. (1815).
Rxn. with: Me$_2$SnX$_2$ → Me$_2$Sn(OC$_9$H$_6$N)X; X = Cl, NCS (1291).
Me$_2$Sn(O$_2$C$_7$H$_5$)$_2$ Me$_2$Sn(O$_2$C$_7$H$_5$)OC$_9$H$_6$N (1346).

Diphenyltin Bis(8-oxyquinolate) (M-401) SnC$_{30}$H$_{22}$N$_2$O$_2$ Ph$_2$Sn(OC$_9$H$_6$N)$_2$
Synth.: By rxn. of Ph$_2$SnCl$_2$ with 8-HOC$_9$H$_6$N in C$_6$H$_6$, in 74% yield (385), in alc. with aq. NH$_3$, in nearly quant. yield (602, 1216).
By rxn. of Ph$_3$SnCl with 8-HOC$_9$H$_6$N at 140° (386).
By disproportionation of Ph$_2$Sn(OC$_9$H$_6$N)Cl in presence of BuLi (602), of Ph$_2$Sn(OC$_9$H$_6$N)CHAcBz (602).
Prop.: m. 252-53° (602), 251-52° (385), IR (385), far IR (602), UV (385, 602), NMR (602) and Moessbauer resonance (1044, 2617) spectra, conductivity (602), dielectrical const. (2617), dipole moment (385), monomeric (385), stability (602), struct. (385, 602), no resolution to active isomers on D-lactose (385).
Rxn. with: Ph$_2$SnCl$_2$ → Ph$_2$Sn(OC$_9$H$_6$N)Cl (602).
Alc. AgNO$_3$, + H$_2$O → Ph$_2$Sn(OC$_9$H$_6$N)NO$_3$ + AgOC$_9$H$_6$N (1216).
8-Hydroxyquinoline at 300° → (C$_9$H$_6$NO)$_4$Sn (1723).

Additional diorganotin alkoxides derived from heterocyclic alcohols are compiled in Table 183.

4.3.3.2.6 DIORGANOTIN ALKOXIDES DERIVED FROM FUNCTIONALLY SUBSTITUTED ALCOHOLS

4.3.3.2.6.1 ALKOXIDES DERIVED FROM HALOGEN SUBSTITUTED ALCOHOLS

The compounds are listed in Table 184. Additional substituted derivatives may be found in Tables 183 and 187.

4.3.3.2.6.3 ALKOXIDES DERIVED FROM ALCOHOLS CONTAINING OXYGEN FUNCTIONS

All compounds are compiled in Table 185. Additional substituted derivatives may be found in Table 183.

4.3.3.2.6.5 ALKOXIDES DERIVED FROM ALCOHOLS CONTAINING NITROGEN

All compounds are listed in Table 186. Aside from compounds with heterocyclic nitrogen in Chapter 4.3.3.2.3 other derivatives may be found in Tables 184 and 187 and in Subsection 4.3.3.2.9.3.

Table 183. Diorganotin Dialkoxides Derived from Heterocyclic Alcohols $R_2Sn(OR')_2$

Formula*	Synth. Method**	Yield	Properties	Ref.
$Me_2Sn(OC_8H_5O_3)_2$ (a)	IIB	--	m. 194-95.5° (dec.), IR, NMR spectra, struct.	1658
$[Me_2Sn(OC_9H_5N)_2CH_2 \cdot H_2O]_x$ (b)	IIB	53	Bright yellow solid (c)	1871
$R_2Sn(OC_9H_6N)_2$ (d)				
R = p-$ClC_6H_4CH_2$	IIC	82	m. 155°	1046
	IIIA	85	(e)	1046
Et (401)	--	--	(f)	--
Pr (401)	--	--	(g)	--
Bu (401)	IIB (h)	--	m. 154.5-55.5° (i)	10
$Bu_2Sn(OAr)_2$	IIIC	--	Ar = 2-$(N_3C_6H_4)C_6H_4$ (j)	2972
$Bu_2Sn(OAr)_2$	IIIC	--	Ar = 2,4-$(N_3C_6H_4)C_6H_3Me$ (j)	2972
$Bu_2Sn(OAr)_2$	IIIC	--	Ar = 2,6-$(N_3C_6H_4)C_6H_3Me$ (j)	2972
$Bu_2Sn(OAr)_2$	IIIC	--	Ar = 2-$(4-ClN_3C_6H_3)C_6H_4$ (j)	2972
$(C_8H_{17})_2Sn(OAr)_2$	IIIC	--	Ar = 2-$(N_3C_6H_4)C_6H_4$ (j)	2972
$(p-ClC_6H_4)_2Sn(OC_9H_6N)_2$ (d)	IIC	71	m. 255.7-56.3°	1046
	IIIA	75	(e)	1046
$(C_6F_5)_2Sn(OC_9H_6N)_2$ (d)	VB	--	Yellow cryst., m. 277° (dec.), IR spectrum	1198
	IIA (k)	--	X-ray powder diagram	1198
$(p-EtC_6H_4)_2Sn(OC_9H_6N)_2$ (d)	IIC	72	m. 158°	1046
	IVA	78	(e)	1046

* Numbers in parenthesis refer to pages in main volume.
** The characters used here correspond to the synthetic scheme in the introduction to Chapter 4.3.3.2.

(a) $HOC_8H_5O_3$ = koji acid. (b) (8-$HOC_9H_5N)_2CH_2$ = bis(8-hydroxy-5-quinolyl)methane. (c) Thermogravimetric analysis (1871). (d) 8-HOC_9H_6N = 8-hydroxyquinoline. (e) Biol. activity as fungicide against Aspergillus arnstelodami A. niger, Chaetomium globosum, Paecilomyces variota, Penicillium brevi-compactum, P-cyclopium, Stachybotrys atra (1046). (f) Rxn. with $Et_2SnCl_2 \rightarrow Et_2Sn(OC_9H_6N)Cl$ (1291). (g) Rxn. with $Pr_2SnX_2 \rightarrow Pr_2Sn(OC_9H_6N)X$; X = Cl, NCS (1291). (h) Synth. with $(Bu_2SnOEt)O(Bu_2SnCl)$. (i) Moessbauer resonance spectrum, dielectric const. (2617), rxn. with $Bu_2Sn(NCS)_2 \rightarrow Bu_2Sn(OC_9H_6N)NCS$ (2617). (j) Benzotriazol derivative. Use as UV stabilizer for polypropylene and synth. resins (2972). (k) Synth. also from $(C_6F_5)_3SnCl$ or $(C_6F_5)_3SnOH$ (1198).

Table 184. Diorganotin Alkoxides Derived from Halogen Substituted Alcohols $R_2Sn(OR')_2 \cdot nR_2SnO$

Formula*	Synth. Method**	Yield	Properties	Ref.
$Me_2Sn[OCH(CF_3)_2]_2$	IB	--	b_{25} 92°, H, ^{19}F NMR spectra	109
$Me_2Sn[OCH(CF_3)_2]_2 \cdot L$	VA	--	$L = (CF_3)_2CO$, NMR spectrum	109
$Me_2Sn[OCH(CF_3)_2]_2 \cdot 2L$	VA	--	$L = (CF_3)_2CO$, H, ^{19}F NMR spectra	109
$Bu_2Sn[OCH(CF_3)_2]_2$	IB	--	$b_{0.001}$ 75°	109
$Bu_2Sn(OC_6H_4Cl-p)_2$	IIIA	75	m. 80-85°, $b_{0.0007}$ 178-80°	1740
$Bu_2Sn(OC_6H_2Cl_3-2,4,6)OH$ (402)	--	--	(a)	--
$(Bu_2SnOC_6Cl_5)_2O$	IIIA	--	m. 145-47° (b)	2138
$Bu_2Sn(OAr)_2$	IIIA	--	$Ar = 2,6,4-Br_2(O_2N)C_6H_2$, yellow oil (c)	533
$(Ph_2SnOC_6Cl_5)_2O$ (402)	IIIA	--	m. 157-70° (b)	2138

* Numbers in parenthesis refer to pages in main volume.
** The characters used here correspond to the synthetic scheme in the introduction to Chapter 4.3.3.2.
(a) Use as bacteriostat on polyamide (1433). (b) Use as stabilizer for PVC, as fungicide and insecticide, as EP additive for polyphenyl ether (2138). (c) Stable to hydrolysis (533), biol. activity as fungicide, insecticide (533).

Table 185. Diorganotin Alkoxides Derived from Alcohols Containing Oxygen Functions $R_2Sn(OR')_2$

Formula*	Synth. Method**	Yield	Properties	Ref.
$(Me_2SnOC_6H_4OMe-o)_2O$	IIB	>70	m. 235-36°, IR spectrum	409
$(PhCH_2)_2Sn(OR')_2$				
R'O = $RCO_2(CH_2CH_2O)_{6-8}$ (40%) (a)	IIC	--	RCO_2H = epoxylated soy bean fatty acid (b)	911
$C_{17}H_{35}CO_2[(CH_2)_4O]_{25}$	--	--	Use as antistatic agent for thermoplastics	2017
$o\text{-}BuO_2CC_6H_4$	--	--	Use as catalyst for modifying polypropylene	269
$Bu_2Sn(OR')_2$				
R' = $C_{17}H_{35}CO_2CH_2CH_2$	--	--	Use as antistatic agent for thermoplastics	2017
$p\text{-}C_9H_{19}C_6H_4(OCH_2CH_2)_{9-10}$	IIB	--	Use as biocide, stabilizer for PVC	1958
$RCO(OCH_2CH_2)_{6-8}$ (40%) (a)	IIIA	--	RCO_2H = epoxylated tall oil acid (b)	911
$RCO_2C_6H_8O_3H_2(OCH_2CH_2)_{15}$	--	--	R = $C_{11}H_{23}$, use as antistatic agent for polymers	2818
	--	--	Use as antistatic agent for thermoplastics	2018
$RCO_2C_6H_8O_3H_2(OCH_2CH_2)_{15}$	--	--	R = $C_{17}H_{35}$, use as antistatic agent for polymers	2818
$\overline{OCH_2CHCH_2}$	IIB	--	Use as stabilizer for PVC	1442
	IIC	--	--	1442
$Et\overline{OCHCH(CH_2)_4}$	IIIC	--	Liq., use as stabilizer for PVC	1442
$2,4\text{-}Bz(MeO)C_6H_3$	IIIC	--	m. 53-55°, bo.2 155-60° (c, d)	1231
$2,5\text{-}Bz(MeO)C_6H_3$	IIIC	--	(e)	2972
$2,5\text{-}Bz(C_8H_{17})C_6H_3$	IIIC	--	(e)	2972
$2,4\text{-}(o\text{-}HOC_6H_4CO)MeOC_6H_3$	IIIC	--	Solid (d)	1231
$o\text{-}RO_2CC_6H_4$	IIIC	--	R = $p\text{-}C_8H_{17}C_6H_4$ (e)	2972
$2,4\text{-}RO_2CC_6H_3C_8H_{17}$	IIIC	--	R = $o\text{-}C_8H_{17}C_6H_4$ (e)	2972
$i\text{-}Bu_2Sn(OC_6H_4CH_2OH\text{-}o)_2$	IB	94	m. 146°, stable in air	389
$(C_5H_{11})_2Sn(OC_6H_3ROC_{12}H_{25}\text{-}2,5)_2$	IIIC	--	R = $o\text{-}Me_3CC_6H_4O_2C$ (e)	2972
$(c.C_6H_{11})_2Sn(OCH_2CH_2OH)_2$	IIC	70	m. 214-15°, IR spectrum	2721
$(c.C_6H_{11})_2Sn(OCH_2CHOHCH_2OH)_2$	IIC	60	m. 200° (dec.)	2721

$(C_8H_{17})_2Sn(OAr)_2$	IIIC	Ar = 2,5-Bz$(C_{12}H_{25}O)C_6H_3$ (e)	2972
$(C_8H_{17})_2Sn(OAr)_2$	IIIC	Ar = 2-(2,4-MeC$_{12}$H$_{25}$C$_6$H$_3$)O$_2$CC$_6$H$_4$ (e)	2972

* Numbers in parenthesis refer to pages in main volume.
** The characters used here correspond to the synthetic scheme in the introduction to Chapter 4.3.3.2.
(a) Correct in main volume, page 403 formulae line 3 to (PhCH$_2$)$_2$Sn[(OCH$_2$CH$_2$)$_{8-8}$R]$_2$, line 4 to (PhCH$_2$)$_2$Sn-[(OCH$_2$CH$_2$)$_{8-8}$R], line 7 R = (CH$_2$CH$_2$O)$_{8-8}$. (b) Water miscible prod. (911). (c) Use as thermostable UV absorber (1231). (d) Use as stabilizer for polypropylene (554). (e) Use as UV stabilizer for polypropylene and synthetic resins (2972).

Table 185. Diorganotin Alkoxides Derived from Alcohols Containing Nitrogen $R_2Sn(OR')_2 \cdot nR_2SnO$

Formula	Synth. Method*	Yield	Properties	Ref.
$(Me_2SnOC_{10}H_8NO-1,2)_2O$	IIB	90	m. > 230°, IR spectrum	550
$Et_2Sn(OC_{10}H_8NO-1,2)$	IIB	70	m. 163°, IR spectrum	550
$Bu_2SnON:CHPh$	IIC	--	(a)	2661
$Bu_2SnON:CMe_2$	IIC	--	bo.08 100° (a)	2661
$Bu_2Sn(ONCOCH_2CH_2CO)_2$	--	--	(b)	--
$(Bu_2SnONCOCH_2CH_2CO)_2O$	(b)	92	m. 209-210°	2857
$Bu_2Sn(OC_6H_4NH_2-p)_2 \cdot 3\ Bu_2SnO$	IIIA	86	m. 108-15°, mol. wt.	2595
$Bu_2Sn(OC_6H_4NH_2-p)_2 \cdot 5\ Bu_2SnO$	IIIA	73	m. 105-15°, mol. wt.	2595
$Bu_2Sn(OC_6H_4NO_2-p)_2$	--	--	(c)	--
$Bu_2Sn(OC_6H_4NO_2-p)_2 \cdot 2\ Bu_2SnO$	(c)	81	m. 171°	2857
	IIIA	86	m. 171°	2595
$Bu_2Sn(OC_6H_3MeNO_2-4,2)_2$	IIIA	--	m. 46° (d)	533
$Bu_2Sn[OC_6H_2CMe_3(NO_2)_2-2,4,6]_2$	IIIA	--	Yellow oil, dec. on distn. (d)	533
$Bu_2Sn[OC_6H_2(CMe_3)_2CH_2NMe_2-2,6,4]_2$	IIC	--	Yellow solid (e)	554

* The characters used here correspond to the synthetic scheme in the introduction to Chapter 4.3.3.2. (a) Use as catalyst for foaming polyurethan (2661), for curing silicone rubber (2661). (b) Rxn. of $Bu_2SnONCOCH_2CH_2CO$ with $Bu_2SnO \rightarrow (Bu_2SnONCOCH_2CH_2CO)_2O$ (2857). (c) Rxn. with $Bu_2SnO \rightarrow Bu_2Sn(OC_6H_4NO_2-p)_2 \cdot 2\ Bu_2SnO$ (2857). (d) Biol. activity as fungicide, herbicide, insecticide (533), stable in air (533). (e) Use as stabilizer for polypropylene (554).

4.3.3.2.6.8 DIORGANOTIN ALKOXIDES DERIVED FROM DIHYDRIC ALCOHOLS

Bis(dibutyltin glycolate) (M-405) $Sn_2C_{20}H_{44}O_4$ $(Bu_2SnOCH_2CH_2O)_2$

Synth.: By rxn. of Bu_2SnO with $(CH_2OH)_2$ in C_6H_6, in 91% yield (895), in xylene, in 86% yield (2084), with $\overline{OCH_2CH_2OCO}$ under reflux, in 86% yield (950).
By rxn. of $Bu_2Sn(OR)_2$ with $(CH_2OH)_2$ in C_6H_6 under exclusion of moisture; R = Et in 98% yield (349), R = Me in THF (2692).
Prop.: m. 229-30° (2084), 223-26.5° (895), 218-34° (950), 215-20° (2692), cryst. sub$l_{0.3}$ 220-40° (349), IR (895) and NMR (2692) spectra, mol. wt. (895, 2692), dimeric in Py, trimeric in C_6H_6 (349), stable in air overnight (895), easily hydrolyzed (349).
Rxn. with: $(CH_2Br)_2 \rightarrow Bu_2SnBr_2$ + 1,4-dioxane + CH_2:CHBr (435).

Additional diorganotin alkoxides derived from dihydric alcohols are listed in Table 187. An alkoxide from a heterocyclic phenol may be found in Table 183.

4.3.3.2.9 MISCELLANEOUS DIORGANOTIN ALKOXIDES

4.3.3.2.9.1 MIXED DIORGANOTIN OXIDES

Some new mixed oxides with aluminum, titanium, vanadium and zirconium appear in this chapter. Polymeric mixed oxides are listed in Section 8.1. The compounds summarized in Table 188 are prepared by the following methods.

I Cohydrolysis of diorganotin dihalides in presence of ammonia with:
   (IA) R_nSiX_{4-n} ,
   (IB) $(R_2SnCl)_2O$ and R_nSiX_{4-n} ;
   (IC) $R_nSi(OH)_{4-n}$.

II Reactions of diorganotin oxides with:
   (IIA) $R_nSi(OH)_{4-n}$ and $Ti(OR)_4$ in presence of triethanolamine,
   (IIB) $R_nSi(OR)_{4-n}$, $Ti(OR)_4$ and $Zr(OR)_4$.

III Reaction of diorganotin diacetate with $R_2Si(OR)_2$ and $Ti(OR)_4$.

R = organic group, X = halogen, n = 2, 3.

Table 187. Diorganotin Alkoxides Derived from Dihydric Alcohols R_2SnOYO

Formula*	Synth. Method**	Yield	Properties	Ref.
Me_2SnOYO				
Y = o-C_6H_4	IIA	60	White solid, subl. 220-60° in vacuo (a)	409
o-C_5H_4N:CMeCH:CPh	IIC	--	Orange, m. 190-92°, mol. wt., struct.	2553
o-C_5H_4N:CMeCH:CMe	IIC	--	Lt. green, m. 197-98°, struct.	2553
o,o'-C_5H_4N:CHC$_5H_4$	IIC	--	Dark brown, m. 171-73°, mol. wt., struct.	2553
Et_2SnOYO				
Y = $(CH_2)_2$	IIIA	86	m. 280°, cyclic dimer	2084
$CH_2CH_2OCH_2CH_2$	IIIA	87	m. 75-77°, b$_{0.5}$ 106-107°	2084
$(CH_2)_3$	IIIA	80	m. 169-73°, soly.	2084
CH_2CHMe	IIIA	95	m. 255°, cyclic dimer	2084
CHMeCHMe	IIIA	87	m. 246°, cyclic dimer	2084
Bu_2SnOYO				
Y = $CH_2CH_2OCH_2CH_2$ (405)	IIIA	76	m. 35-47°, b$_3$ 145-47°	2084
$CH_2CH_2NHCH_2CH_2$	IIIA	98	White solid, mol. wt., struct.	1501
$CH_2CH_2NRCH_2CH_2$	IIIA	98	R = CH_2CH_2OH, colorless liq., mol. wt., struct.	1501
$(CH_2)_3$	IIIC	78	White cryst., b$_{0.3}$ 184-85°, soly. (b, c)	349
	IIIC	--	m. 88°, b$_{1.5}$ 200°, IR spectrum	2692
CH_2CHMe (405)	IIIC	92	White cryst., sublo.1 210-30°, soly. (b, d)	349
	IIIC	--	m. 185°, cyclic dimer	2692
$CH_2CH(CH_2Cl)$	IIIC	--	White cryst., soly. (b)	349
$(CH_2)_4$	IIIC	78	Colorless viscous liq., b$_{0.4}$ 175-78° (b)	349
	IIIC	--	b$_{0.2}$ 166°, monomeric	2692
CH_2CH_2CHMe (405)	IIIA	92	m. 195° (dec.), polymeric	2084
	IIIC	83	Cryst., b$_{0.3}$ 165°, soly. (b)	349
	IIIC	--	m. 134°, b$_{1.5}$ 185°, dimeric	2692

CHMeCHMe	IIIC	93	Cryst., b$_{0.2}$ 171-73°, dimeric, soly. (b)	349
	IIIA	98	m. 124-26°, IR spectrum, monomeric	895
	IIIC	--	m. 120°, dimeric	2692
(CH$_2$)$_5$	IIIC	72	White solid, b$_{0.15}$ 185-87°, soly. (b)	349
	IIIA	89	m. 174° (dec.), polymeric	2084
CHMeCH$_2$CHMe	IIIC	--	m. 155°, b$_{0.7}$ 159°, mol. wt.	2692
(CH$_2$)$_6$	IIIC	--	White solid, dec. on distn., soly. (b)	349
	IIIA	93	m. 148° (dec.)	2084
hexylene	IIIC	77	Colorless viscous liq., b$_{0.8}$ 163-65° (b)	349
CMe$_2$CH$_2$CHMe	IIIC	--	m. 80°, b$_{1.5}$ 143°, mol. wt.	2692
CMe$_2$CMe$_2$	IIIC	92	White low melting, b$_{1.6}$ 204-205°, dimeric (b)	349
	IIIC	--	m. 45°, b$_{0.4}$ 195°, dimeric	2692
c.1,2-C$_6$H$_{10}$	IIIA	96	cis-form, m. 164-65°, IR spectrum (e)	895
	IIIA	96	trans-form, m. 234-36°, IR spectrum (e)	895
CMe$_2$CH$_2$CMe$_2$	IIIC	--	m. 70°, b$_{0.8}$ 119°, mol. wt.	2692
(C$_5$H$_{11}$)$_2$SnOYO	IIIC	--	Y = 2,2'-C$_6$H$_4$COC$_6$H$_3$OC$_8$H$_{17}$-5 (f)	2972
Ph$_2$SnOYO				
Y = p-C$_6$H$_4$	IIC	--	m. 172-74°, polymeric (g)	610
o-C$_6$H$_4$N:CMeCH:CPh	IIC	--	Orange, m. 191-92°, mol. wt., struct.	2553
o-C$_6$H$_4$N:CMeCH:CMe	IIC	--	Light green, m. 142-45°, mol. wt., struct.	2553
2,2'-C$_6$H$_4$N:CHC$_6$H$_4$	IIC	--	Yellow, m. 215-16°, mol. wt., struct.	2553

* Numbers in parenthesis refer to pages in main volume.
** The characters used here correspond to the synthetic scheme in the Introduction to Chapter 4.3.3.2.
(a) IR spectrum (409). (b) Easily hydrolyzed (349). (c) Rxn. with Ph$_2$SiCl$_2$ → Ph$_2$SiO(CH$_2$)$_3$O (434). (d) Use in catalyst for butadiene polymerization (2044a), as rodent repellent (2036). (e) Monomeric, struct. (895). (f) Use as UV stabilizer for polypropylene and synthetic resins (2972). (g) Use as stabilizer for synthetic polymers (610).

Table 188. Diorganotin Mixed Oxides $R_2Sn(OMR_n)_2$

Formula*	Synth. Method**	Yield	Properties	Ref.
$(C_8H_{17})_2Sn(OCHMe_2)OAl(OPr)_2$	--	--	(a)	--
$[(Me_2SnOSiMe_3)_2O Al(OPr)]_2$	IA	30-95	m. 167-68°, soly. (b, c)	409a
	IB	--	--	409a
$Me_2Sn[(OSiMe_2)_nOSiMe_3]_2$				
n = 30	IC	--	fp. -120°, n_D^{20} 1.4080, d_{20} 0.9910	12
50	IC	--	fp. -120°, n_D^{20} 1.4061, d_{20} 0.9834	12
150	IC	--	fp. -100°, n_D^{20} 1.4057, d_{20} 0.9681	12
x	IC	--	fp. -115°, n_D^{20} 1.4069, d_{20} 0.9900	12
$Me_2Sn(OSiPh_3)_2$ (406)	IIA	--	m. 153-54°, IR spectrum	2344
$[(Et_2SnOSiMe_3)_2O]_2$ (406)	IA	90	m. 126-30° (b)	409a
	IB	--	Mechanism of formation	409a
$[(Pr_2SnOSiMe_3)_2O]_2$ (406)	IA	85	m. 107-108°, soly.	409a
	IB	--	(b)	409a
$[(Bu_2SnOSiMe_3)_2O]_2$ (406)	IA	95	m. 108-109°, soly.	409a
	IB	--	(b)	409a
$Bu_2Sn(OEt)OSi(OEt)_2CH:CH_2$	IIB	--	(a, d)	1096
$Bu_2Sn(OSiPh_3)_2$ (406)	IIA	100	m. 71°, IR spectrum	2344
$Bu_2SnOSiPh_2OSiPh_2$	IIA	100	m. 190-200°, IR spectrum	2344
$AcO(1-Bu_2SnOSiMe_2)_2OEt$	III	--	m. 148.5-153°, soly., mol. wt.	219a
$Ph_2Sn(OSiPh_3)_2$ (406)	IIA	100	m. 146-48°, IR spectrum	2344
$(Ph_2SnOH)_2Si_4Ph_8O_6$	IIA (e)	--	Softening 70°, brittle resin, IR spectrum (f)	958
$[Bu_2SnO_2Ti(OBu)_2]_x$	III	--	Liq., viscosity (g)	195a
$[Bu_2Sn(OCHMe_2)O]_4Ti$	IIB	--	(a, h)	1096
$[Bu_2Sn(OCHMe_2)O]_3TiOCHMe_2$	IIB	--	(a, h)	1096
$[Bu_2Sn(OCHMe_2)O]_2Ti(OCHMe_2)_2$	IIB	--	Liq., n_D^{23} 1.5050 (a, h, i)	1096

[Bu$_2$Sn(OCHMe$_2$)O]Ti(OCHMe$_2$)$_3$	IIB	Yellow liq., n_D^{24} 1.4995-1.5022 (a, h, i)	1096
Bu$_2$Sn(OCHMe$_2$)OTi(OCH$_2$CH$_2$)$_3$N	IIA	(j)	2324
Bu$_4$Sn$_2$(OCHMe$_2$)O$_2$Ti(OCH$_2$CH$_2$)$_3$N	IIA	(j)	2224
Bu$_2$Sn(OBu)OTi(OBu)$_3$	IIB	(a, d)	1096
Bu$_2$Sn(OR)OTi(OR)$_3$	IIB	R = EtBuCHCH$_2$ (a, d)	1096
(C$_8$H$_{17}$)$_2$Sn(OCHMe$_2$)OTi(OCHMe$_2$)$_3$	IIB	(a, i)	1096
Bu$_2$Sn(OCHMe$_2$)OZr(OCHMe$_2$)$_3$	IIB	(a, i)	1096
(C$_8$H$_{17}$)$_2$Sn(OBu)OVO$_3$Bu$_2$	IIB	(a, i)	1096

* Numbers in parenthesis refer to pages in main volume.
** The characters used here correspond to the synthetic scheme in the introduction to Chapter 4.3.3.2.9.1.
(a) Use as stabilizer for PVC and for polyolefins (1096). (b) Decomposing in oxygen containing solvents (409a).
(c) NMR spectrum (92). (d) Use as catalyst for foaming polyurethan (1096). (e) Synth. with Ph$_3$SnOH. (f)
Thermogravimetric analysis (958). (g) Use as vulcanization accelerator for polysiloxanes (195a). (h) Biol.
activity as fungicide against Aspergillus niger, Botrytis allii, Glomerella cingulata, Penicillium italicum
(1096). (i) Use as waterproofing agent (1096). (j) Biol. activity as fungicide against Aspergillus niger,
Botrytis allii, Pullularis pullulans (2324).

4.3.3.2.9.3 DIORGANOTIN ALKOXIDES CONTAINING ACETAL ENOL AND CARBOXY GROUPS LINKED TO TIN

New in this chapter are a number of organotin alkoxide acetals and derivatives of organotin alkoxides with chelating 1,2- and 1,3-diketones. It is doubtful in most cases that the two different moieties are linked to the same tin atom. The compounds compiled in Tables 189 and 190 are prepared by methods from the following scheme.

I   Reaction of diorganotin halide alkoxide with thalium alkoxide.

II   Reaction of dibutyltin oxide with maleic anhydride and alcohol.

III   Reactions of diorganotin dialkoxide with:
    (IIIA)   diorganotin diacetal or dicarboxylate,
    (IIIB)   alcohol or aldehyde,
    (IIIC)   β-amino- or β-thioalcohol.

IV   Reaction of diorganotin dicarboxylate with alcohol.

Additional references to diorganotin alkoxides containing acetal, enol and carboxy groups linked to tin may be found in Chapter 9.

Dibutyltin Methoxide Acetate (M-407)* $SnC_{11}H_{24}O_3$ $Bu_2Sn(OAc)OMe$
Synth.: By rxn. of $Bu_2Sn(OAc)_2$ with $Bu_2Sn(OMe)_2$ (942).
Prop.: White cryst., m. 94-96° (sealed tube), hydrolyzes readily (942).
Rxn. with: H_2O → $[Bu_4Sn_2(OAc)_2]_2O$ (942).
EtNCO → $Bu_2Sn(NEtCO_2Me)OAc$ (943).
PhNCS → $Bu_2Sn[SC(OMe):NPh]OAc$ (943).
Cl_3CCHO → $Bu_2Sn(OAc)OCH(OMe)CCl_3$ (943).
CS_2 → $Bu_2Sn(S_2COMe)OAc$ (943).
Use: As catalyst for modifying polypropylene (269).

Additional diorganotin alkoxides containing acetal, enol or carboxy groups are listed in Table 189.

4.3.3.2.9.5 DIORGANOTIN ALKOXIDES CONTAINING TIN-HETEROATOM BONDS

This chapter comprises a number of new compounds having sulfur or nitrogen linked supossedly to the tin atom of a diorganotin alkoxide. All compounds are listed in Table 190. Additional references to diorganotin hetero derivatives may be found in Chapter 9.

Diorganotin alkoxides containing other heteroatoms linked to tin may be found in the following sections: 3.8 for hydrogen, 4.1.5.3.1 and 4.1.5.3.6 for halogen, 4.2.1.5.3 for pseudohalogen groups and 4.2.4.6 for oxygen containing anions.

---

* Compare also entry 7 in Table 189.

Table 189. Diorganotin Alkoxides Containing Acetal, Enol and Carboxy Groups $R_2Sn(OR)OR'$

Formula*	Synth. Method**	Yield	Properties	Ref.
$Me_2Sn(CHBz_2)OC_9H_8N$ (a, b) (?)	I	96	Yellow cryst., m. 152-230°, dec. on recryst.!	602
$Me_2Sn(C_7H_5O_2)OC_9H_8N$ (b, c)	IIIA	--	Yellow solid, m. 168-70.5°, stable in air (d)	1346
$Bu_2Sn(OCHMeOMe)OMe$	IIIB, MeCHO	--	NMR spectrum, dissocn. at 0.1 mm.	943
$Bu_2Sn(OCHROMe)OMe$	IIIB, RCHO	--	$R = CCl_3$, IR, NMR spectra	943
$Ph_2Sn(CHAcBz)OC_9H_8N$ (b, e) (?)	I	99	Yellow cryst., m. 182-200°, dec. on recryst.! (f)	602
$Ph_2Sn(CHBz_2)OC_9H_8N$ (a, b) (?)	I	97	Yellow cryst., m. 182-86°	602
$Bu_2Sn(OAc)OMe$ (407)	IIIA	--	m. 104-105°, IR spectrum, dipole moment (g)	2184
$(Bu_2SnOMe)O(Bu_2SnOAc)$	IIIA	100	Viscous oil, n_D^{20} 1.3623 (h)	634
$Bu_2Sn(O_2CCH_2Cl)OMe$	IIIA	--	m. 95-96° (g)	2184
$(Bu_2SnOMe)O(Bu_2SnO_2CCH_2Cl)$	IIIA	--	Slightly yellow viscous oil (h)	634
$Bu_2Sn(O_2CC_{11}H_{23})OMe$ (407)	IIIA	--	Oil, IR, NMR spectra (i)	942
$Bu_2Sn(O_2CR)OMe$ (407)	IIIA	--	(g)	2184
$Bu_2Sn(O_2CR)OMe$ (407)	--	--	$R = MeO_2CCH:CH$ (k)	269
$(Bu_2SnOMe)O(Bu_2SnO_2CR)$	II	--	$R = MeO_2CCH:CH$, m. 102-104° (j)	2872
$Bu_2Sn(O_2CR)OMe$ (407)	--	--	$R = o\text{-}MeO_2CC_6H_4$ (k)	270
$Bu_2Sn(O_2CC_7H_{15})OEt$	--	--	(k - 269)	--
$(Bu_2SnOEt)O(Bu_2SnO_2CR)$	II	--	$R = EtO_2CCH:CH$, m. 96-98° (j)	2872

* Numbers in parenthesis refer to pages in main volume.
** The characters used here correspond to the synthetic scheme in the introduction to Chapter 4.3.3.2.9.3.
(a) CH_2Bz_2 = dibenzoylmethane. (b) HOC_9H_8N = 8-hydroxyquinoline. (c) $C_7H_5O_2H$ = tropolone. (d) (Far) IR, NMR spectra, x-ray powder pattern, dissocn. in $HCCl_3$ (1346). (e) CH_2AcBz = benzoylacetone. (f) IR, UV and NMR spectra (602). (g) Thermodynamic data, struct. (2184). (h) Rxn. with aq. MeOH → $RCO_2(Bu_2Sn)_4COR$; R = Me, CH_2Cl (634). (i) Rxn. with EtNCO → $Bu_2Sn(NEtCO_2Me)O_2CC_{11}H_{23}$ (943). (j) Use as stabilizer for PVC (2872), as catalyst for manufacturing polyurethane (2872). (k) Use as catalyst for modifying polypropylene (269, 270).

Table 189 Continued

Formula*	Synth. Method**	Yield	Properties	Ref.
(Bu$_2$SnOPr)O(Bu$_2$SnO$_2$CR)	II	--	R = PrO$_2$CCH:CH, m. 84-87° (j)	2872
(Bu$_2$SnOPr-i)O(Bu$_2$SnO$_2$CR)	II	--	R = Me$_2$CHO$_2$CCH:CH, m. 98-100° (j)	2872
Bu$_2$Sn(OAc)OBu	II	--	(g)	2184
Bu$_2$Sn(O$_2$CR)OBu	--	--	RCO = epoxystearoyl (l)	1381
(Bu$_2$SnOBu)O(Bu$_2$SnO$_2$CR)	II	--	R = BuO$_2$CCH:CH, m. 84-86° (j)	2872
(Bu$_2$SnOBu-i)O(Bu$_2$SnO$_2$CR)	II	--	R = Me$_2$CHCH$_2$O$_2$CCH:CH, m. 94-96° (j)	2872
(Bu$_2$SnOC$_{12}$H$_{25}$)O(Bu$_2$SnO$_2$CR)	II	--	R = C$_{12}$H$_{25}$O$_2$CCH:CH, m. 40-42° (j)	2872
Bu$_2$Sn(OAc)OC$_8$H$_8$N (b)	IV	52	Dec. >142°, IR, UV spectra, mol. wt.	1216
(1-Bu$_2$SnOEt)O(1-Bu$_2$SnOAc)	IV (m)	--	Colorless cryst., m. 194-99° (dec.), mol. wt. (n)	219a

(l) Use as stabilizer for PVC (1381). (m) Synth. with Me$_2$Si(OEt)$_2$. (n) Decomposes in concd. HCl (219a).

Table 190. Diorganotin Alkoxides Containing Tin Heteroatom Bonds $R_2Sn(OR)SR$, $R_2Sn(OR)NR_2$

Formula	Synth. Method*	Yield	Properties	Ref.
$(Bu_2SnOMe)_2S$	III, Bu_2SnS	--	m. 86-88°	945
$Bu_2Sn(SC_{12}H_{25})OMe$	--	--	(a)	--
$Bu_2Sn(S_2COMe)OMe$	III, CS_2	--	m. 64-67°, IR, NMR spectra (b)	943
$Bu_2Sn[SC(OMe):NR]OMe$	III, RNCS	--	R = allyl, IR, NMR spectra	943
$Bu_2Sn[SC(OMe):NPh]OMe$	III, PhNCS	--	IR, NMR spectra, dissocn. on distn. (c)	943
$Bu_2SnOCH_2CH_2S$	IIIC	96	$b_{1.5}$ 194°, dimeric	2805
$Bu_2SnOCH_2CH(CH_2OH)S$	IIIC	--	Dec. on heating, trimeric	2805
$Bu_2Sn(NEtCO_2Me)OMe$	III, EtNCO	--	IR, NMR spectra	943
$Bu_2Sn(NPhCO_2Me)OMe$	III, PhNCO	--	IR, NMR spectra	943
$Bu_2Sn(NPhCO_2Me)OMe$	III, PhNCO	--	Viscous liq. (d)	2662
$Bu_2Sn[NRC(OMe):NR]OMe$	III, $(RN:)_2$	--	R = 1-$C_{10}H_7$, IR, NMR spectra (e)	943
$Bu_2Sn[N:C(OMe)R]OMe$	III, RCN	--	R = CCl_3, white, waxy, m. 93-94° (f)	943
$Bu_2Sn(NPhCO_2Bu)OBu$	III, PhNCO	--	(d)	2662
$Bu_2SnOCH_2CH_2NH$	IIIC	35	Yellow liq., $bo._4$ 174-78°, mol. wt.	1501

* The characters used here correspond to the synthetic scheme in the introduction to Chapter 4.3.3.2.9.3.
(a) Use as catalyst for modifying polypropylene (271), as stabilizer for PVC (2548). (b) Hydrolyzes rapidly in air, dissocn. in CCl_4 (943). (c) Rxn. with $H_2O \rightarrow Bu_2SnO$ + PhNCSOMe (943). (d) Use as curing catalyst for polysilicone elastomers (2662). (e) Very sensitive to moisture (943). (f) IR, NMR spectra, dec. at r.t. in sealed tube (943).

4.3.3.3 MONOORGANOTIN TRIALKOXIDES AND PHENATES

4.3.3.3.1 MONOORGANOTIN TRIALKOXIDES FROM UNSUBSTITUTED ALCOHOLS

The compounds compiled in Table 191 are prepared by methods from the following scheme.

I Reactions of organotin trihalides with:
   (IA) NaOR',
   (IB) zinc in methanol,
   (IC) polyols and aqueous base.

II Reaction of alcohol and phenol with:
   (IIA) monoorganotin oxide,
   (IIB) organotin trialkoxide.

III Reaction of alcohol and phenol with monoorganotin sulfide.

IV Reaction of alcohol with monoorganotin triamide.

4.3.3.3.6 MONOORGANOTIN ALKOXIDES DERIVED FROM SUBSTITUTED ALCOHOLS

The monoorganotin alkoxides derived from functionally substituted alcohols, from oximes and mixed oxides are listed in Table 192. Monoorganotin alkoxides containing tin linked to halogen and pseudohalogen groups may be found in Subsections 4.1.5.4 and 4.2.1.5.3, respectively. Monoorganotin mixed oxides that may be regarded as salts of the hypothetical organostannoic acids are listed in Table 170.

Table 191. Monoorganotin Alkoxides from Unsubstituted Alcohols   RSn(OR')$_3$

Formula	Synth. Method*	Properties	Ref.
MeSn(OMe)$_3$	IV	m. 195°, IR, assignment, NMR spectra	323
MeSn(OEt)$_3$	IV	b.$_{0.1}$ 110°, IR, assignment, NMR spectra	323
EtSn(OMe)$_3$	IV	m. 165°, NMR spectrum, mol. wt.	323
EtSn(OEt)$_3$	IV	b.$_{0.1}$ 127-30°, NMR spectrum	323
BuSn(OMe)$_3$	IV	White powder, dec. > 250°, NMR spectrum (a)	323
BuSn(OEt)$_3$	IV	b.$_{0.1}$ 130°, NMR spectrum	323
BuSn(OPr)$_3$	--	(b)	--
PhSn(OMe)$_3$	IV	m. 212°, NMR spectrum	323
PhSn(OEt)$_3$	IV	m. 119°, NMR spectrum	323
C$_8$F$_{17}$C$_2$H$_4$Sn(OMe)$_3$	IB	(c)	745a
F(CF$_2$)$_{10}$C$_2$H$_4$Sn(OMe)$_3$	IB	(c, d)	745a
F(CF$_2$)$_{10}$C$_2$H$_4$Sn(OEt)$_3$	IB	(c)	745a
Cl(CF$_2$)$_{10}$C$_2$H$_4$Sn(OEt)$_3$	IB	(c, e)	745a
F(CF$_2$)$_{10}$C$_3$H$_8$Sn(OMe)$_3$	IB	(c)	745a

* The characters used here correspond to the synthetic scheme in the introduction to Chapter 4.3.3.3.1.
(a) Rxn. of BuSn(OMe)$_3$ with ROH → BuSn(OR)$_3$; R = 2,5-Bz(MeO)C$_6$H$_3$, 2-(benzotriazolyl)phenyl, o-biphenylylsalicylate (2972). (b) Rxn. with Y(CO$_2$H)$_2$ → Bu$_2$Sn$_2$(O$_2$CYCO$_2$)$_3$; Y = Me$_2$C(SCH$_2$CH$_2$)$_2$ (1170). (c) Use as water, oil and dirt repellent, antiadhesive and release agent (745a). (d) Rxn. with H$_2$O → C$_{10}$F$_{21}$C$_2$H$_4$SnO$_2$H (745a). (e) Rxn. with air → R$_2$Sn$_2$O$_3$, with Na$_2$S → R$_2$Sn$_2$S$_3$; R = Cl(CF$_2$)$_{10}$C$_2$H$_4$ (745a).

Table 192. Monoorganotin Alkoxides Derived from Substituted Alcohols RSn(OR')$_3$

Formula*	Synth. Method**	Yield	Properties	Ref.
MeSn(OC$_9$H$_6$N)$_3$ (a)	III	--	Yellow cryst., m. 280°, soly.	1347
EtSn(OH)$_2$OCH(CO$_2$K)CHOHCO$_2$K	IC	--	IR spectrum, struct.	2351
Et(KO)SnOEtSnOHOCH$_2$(CHOH)$_4$CH$_2$O	IC	--	IR spectrum, struct.	2351
EtSn(OC$_9$H$_6$N)$_3$ (a)	IIA	81	(b)	1046
	IA	85	--	1046
BuSn(ON:CMe$_2$)$_3$	IA	--	(c)	2661
BuSn(OC$_6$H$_3$BzOH-2,5)$_3$	IIB	--	(d)	2972
BuSn(OC$_6$H$_4$N$_3$C$_6$H$_4$-2)$_3$ (e)	IIB	--	(d)	2972
BuSn(OC$_6$H$_4$CO$_2$C$_6$H$_4$Ph-2,2')$_3$	IIB	--	(d)	2972
BuSn(OC$_9$H$_6$N)$_3$ (a) (408)	IA	81	m. 223° (b, f, g)	1046
	IIA	86	IR, UV, NMR spectra (1290)	1046
BuSn(OC$_9$H$_6$N)$_2$OH (a)	(g)	--	Monomeric	1290
[BuSn(OC$_9$H$_6$N)$_2$]$_2$S (a)	III	--	Yellow cryst., m. 225-26°, soly.	1347
PhSn(OC$_9$H$_6$N)$_3$ (a)	IA	99	(b)	1046
	IIA	100	(b)	1046
BuSn(OSiPh$_3$)$_3$	IIA	100	m. 144-46°, IR spectrum	2344

* Numbers in parenthesis refer to pages in main volume.
** The characters used here correspond to the synthetic scheme in the introduction to Chapter 4.3.3.3.1.
(a) HOC$_9$H$_6$N = 8-hydroxyquinoline. (b) Biol. activity as fungicide against Aspergillus arnstelodami, A. niger, Chaetomium globosum, Paecilomyces varioti, Penicillium brevi-compactum, P. cyclopium, Stachybotrys atra (1046). (c) Use as catalyst for curing polysilicone rubber (2661), for synthesis of polyurethan (2661). (d) Use as stabilizer for polypropylene and synthetic resins (2972). (e) 2-Benzotriazole derivative. (f) Biol. activity as molluscicide against Heliosoma subcrenatum, H. tenue (1054). (g) Easily hydrolyzed → BuSn(OC$_9$H$_6$N)$_2$OH

4.3.4 O-ORGANOTIN DERIVATIVES OF ALDEHYDES AND KETONES

Organotin derivatives of the carbonyl group comprise carbon analogs of acetals and ketals. A considerable number of enol derivatives is reported which show the common isomerism between Sn-O and Sn-C linkages. Finally there exists a growing group of triorganotin chelates derived from 1,2- and 1,3-dicarbonyl compounds involving the tin atom with a covalency of five to possibly seven.

4.3.4.1 TRIORGANOTIN DERIVATIVES OF ALDEHYDES AND KETONES

4.3.4.1.1 TRIORGANOTIN ACETALS AND KETALS

Triorganotin acetals and ketals are not very stable compounds. They are sensitive to air, protic agents and other reactive molecules. Several derivatives dissociate easily. The compounds listed in Table 193 are prepared by methods from the following scheme.

I Addition reactions to the carbonyl group of:
   (IA) $(R_3Sn)_2O$,
   (IB) R_3SnOR and R_3SnSR,
   (IC) $R_3Sn(Et)SO_3$.

II Insertion of aldehyde into the Sn-O bond of a previously formed acetal.

Bis(tributyltin) Chloralacetal (M-408) $Sn_2C_{26}H_{55}Cl_3O_2$ $(Bu_3SnO)_2CHCCl_3$
<u>Synth.</u>: By rxn. of $(Bu_3Sn)_2O$ with Cl_3CCHO (36, 941), in 100% yield (953).
<u>Prop.</u>: Dissocn. on distn. at 0.2 torr and 90-120°, IR, NMR spectra, sensitive to H_2O, CO_2 (953).
<u>Rxn. with</u>: MeOH → Bu_3SnOMe + $HCCl_3$ + Bu_3SnO_2CH (953).
AcCl → Bu_3SnCl + Ac_2O + Cl_3CCHO (953).
SO_2 → $(Bu_3Sn)_2SO_3$ + Cl_3CCHO (2345).
Cl_3CCHO → $Bu_3SnO(CHCl_3O)_nSnBu_3$ (941).

Methyl Tributyltin Bromalacetal $SnC_{15}H_{31}Br_3O_2$ $Bu_3SnOCH(OMe)CBr_3$
<u>Synth.</u>: By rxn. of Bu_3SnOMe with Br_3CCHO, in 100% yield (953).
<u>Prop.</u>: Dissocn. on distn., NMR spectrum, sensitive in H_2O, CO_2 (953).
<u>Rxn. with</u>: Dec. in CCl_4 at r.t. → Bu_3SnO_2CH + Bu_3SnCBr_3 (954).
MeOH → Bu_3SnOMe + $HCBr_3$ + HCO_2Me (953).
AcCl → Bu_3SnCl + Br_3CCHO + AcOMe (953).

Additional triorganotin acetals and ketals are listed in Table 193.

Table 193. Triorganotin Acetals and Ketals

Formula*	Synth. Method**	Yield	Properties	R_3SnOR' Ref.
$Me_3SnOCH(SMe)CCl_3$	IB	--	Dissocn. on distn., NMR spectrum (a)	2456
$Et_3SnOCH(CCl_3)OCH_2C{:}CH$	IB	87	b_2 85°, n_D^{20} 1.5081, d_{20} 1.4341 (b)	1910, 1915
$Et_3SnOCH(CCl_3)OCH_2CH_2C{:}CH$	IB	91	b_1 69°, n_D^{20} 1.5059, d_{20} 1.4043	1910
$Et_3Sn(OCHCCl_3)_2CH_2CO_2Et$ (c)	IB	--	NMR spectrum	1631
$(Bu_3SnO)_2CHC_6F_5$	IA	90 (d)	H, ^{19}F NMR spectra, dec. on distn. (e)	953
$Bu_3SnOCH(OMe)C_6F_5$	IB	100 (d)	H, ^{19}F NMR spectra, dec. on distn. (e)	953
$Bu_3SnOCH(OMe)Me$	IB	100 (d)	NMR spectrum, dec. on distn. (e)	953
$Bu_3Sn(OCHCCl_3)_xOSnBu_3$	IA	--	Highly cryst. polymer	941
$Bu_3SnOCH(OMe)CCl_3$ (408)	IB	100 (d)	$b_{0.8}$ 91-92°, IR, NMR spectra (e)	36, 941, 953
$Bu_3Sn(OCHCCl_3)_xOMe$	IB	--	Highly cryst. polymer	941
$Bu_3SnOCH(OEt)CCl_3$	IB	--	Equilibrium mixture in CCl_4, IR and NMR spectra	2345
	IC			2345
$Bu_3SnOCH(OPh)CCl_3$	IB	75 (d)	NMR spectrum, dec. on distn. (e)	953
$(Bu_3SnO)_2CHCBr_3$	IA	100 (d)	NMR spectrum, dec. on distn. (e, f)	953
$Bu_3SnOCH(OPh)CBr_3$	IB	45 (d)	NMR spectrum, dec. on distn. (e)	953
$Bu_3SnOCH(OMe)Pr$	IB	85 (d)	NMR spectrum, dec. on distn. (e)	953
$(Bu_3SnO)_2C(CCl_3)_2$	IA	100 (d)	IR, NMR spectra, dec. on distn. (e, g)	953
$Bu_3SnOC(CCl_3)_2OMe$	IB	100 (d)	IR, NMR spectra, dec. on distn. (e, h)	953
$(Bu_3SnO)_2C(CF_2Cl)CFCl_2$	IA	100 (d)	H, ^{19}F NMR spectra, dec. on distn. (e)	953
$Bu_3SnOC(CF_2Cl)(CFCl_2)OMe$	IB	100 (d)	H, ^{19}F NMR spectra, dec. on distn. (e)	953

* Numbers in parenthesis refer to pages in main volume.
** The characters used here correspond to the synthetic scheme in the introduction to Chapter 4.3.4.1.1.
(a) Thermal dec. > 80° → Me_3SnCl (2456). (b) IR spectrum (1894), rxn. with gaseous HCl → Et_3SnCl + $HC{:}CCH_2OCH(CCl_3)OH$ (1910). (c) Synth. with $Et_3SnOCH(CCl_3)CH_2CO_2Et$. (d) Spectroscopic yield. (e) Sensitive to H_2O, CO_2 (953). (f) Rxn. with MeOH → Bu_3SnOMe + Bu_3SnO_2CH + $HCBr_3$ (953-4). (g) Thermal dec. > 55° → Bu_3SnCCl_3 + Cl_3CCO_2Me (954). (h) Rxn. with MeOH → Bu_3SnOMe + $HCCl_3$ + Cl_3CCO_2Me (953).

4.3.4.1.3 TRIORGANOTIN ENOXIDES

Previously triorganotin enoxides have been usually regarded as derivatives of the tautomeric ketone:

$$R_3Sn-CH_2C(:O)CH_3 \rightleftharpoons R_3Sn-OC(CH_3):CH_2$$

There is recent spectroscopic evidence that compounds prepared via the enol-ester route (II) or the addition of organotin hydride to α,β-unsaturated ketones (I) exist as a mixture of both isomers. In search for a certain compound the reader should always consult Chapter 2.6.3.3 and Table 112.

The compounds listed in Table 194 are prepared by methods from the following scheme.

I Hydrostannation of α,β-unsaturated ketones:
   $R_3SnH + RCH:CHCOR \rightarrow R_3SnOCR:CHCH_2R$.

II Addition of triorganotin alkoxide to a vinyl ether:
   $R_3SnOR + R_2C:CHOAc \rightarrow R_3SnOCH:CR_2$.

III Exchange reaction:
   $R_3SnOR + R_3SiOC:CHR \rightarrow R_3SnOCH:CHR$.

R = hydrogen or organic group.

1-Triethyltin Cyclohexeneoxide (M-230) $SnC_{12}H_{24}O$ $1\text{-}Et_3SnOC_6H_9$
<u>Synth.</u>: By rxn. of Et_3SnOMe with $1\text{-}C_6H_9OAc$ at 25°, in 82% yield (1709).
By rxn. of Et_3SnNEt_2 with $1\text{-}C_6H_9OAc$ at r.t. to 80°, in 72% yield (1709).
<u>Prop.</u>: $b_{0.5}$ 84-84°, $b_{0.01}$ 80-81°, n_D^{20} 1.5080-87, d_{20} 1.2607-1.2621, IR and NMR spectra (1709).
<u>Rxn. with</u>: $Et_3GeCl \rightarrow 1\text{-}Et_3GeOC_6H_9$ (1709).
$R_3SiCl \rightarrow 1\text{-}R_3SiOC_6H_9$; R = Cl, Me (1709).
$Bu_2BCl \rightarrow 1\text{-}Bu_2BOC_6H_9$ (1709).

Tributyltin 1,3-Dimethylbutene-1-oxide $SnC_{18}H_{38}O$ $Bu_3SnOCMe:CHCHMe_2$
<u>Synth.</u>: By rxn. of Bu_3SnH with $Me_2C:CHAc$ (421), under UV irradiation, in 53% yield (1684).
By rxn. of Bu_3SnOMe with cis- and trans-$Me_2CHCH:CMeOAc$ under exclusion of moisture, in 60 and 73% yield, resp. (1681, 1684).
<u>Prop.</u>: $b_{5.5}$ 154-55° (1681, 1684), $b_{0.6}$ 120-22° (421, 1684), n_D^{20} 1.4758 (421, 1684), 1.4749 (1681, 1684), d_{20} 1.0803 (421, 1684), 1.0786-801 (1681, 1684), IR and NMR spectra (1681, 1684), very sensitive to moisture (1684), trans/cis-ratio - 4 (1681, 1684).
<u>Rxn. with</u>: $Bu_3SnH \rightarrow Bu_6Sn_2 + i\text{-}BuAc + (Bu_3Sn)_2O$ (1684).
$R_3SiCl \rightarrow Bu_3SnCl + R_3SiOCMe:CHCHMe_2$ (434).
$AcCl \rightarrow Bu_3SnCl + Me_2CHCH:CMeOAc$ (1684).
$BzOH \rightarrow Bu_3SnCl + i\text{-}BuAc$ (1684).

Additional triorganotin enoxides are summarized in Table 194.

Table 194. Triorganotin Enoxides R_3SnOR'

Formula*	Synth. Method**	Yield	Properties	Ref.
$Et_3SnOCH:CMe_2$	II	77	b_1 61-63°, n_D^{20} 1.4882, d_{20} 1.2259 (a, b)	1712
$Et_3SnOCH:CHCHMe_2$	II	73	$b_{0.5}$ 63-64°, n_D^{20} 1.4838, d_{20} 1.1929, IR spectrum	1712
$Pr_3SnOCH:CMe_2$	II	92	b_1 84-85°, n_D^{20} 1.4830, d_{20} 1.1478, IR spectrum	1712
$c.1-Pr_3SnOC_6H_9$	II	78	$b_{0.01}$ 98-100°, n_D^{20} 1.4989, d_{20} 1.8139 (c)	1709
Bu_3SnR'				
$R' = CH_2:CPhO$ (d)	II	51	$b_{0.7}$ 138°, n_D^{20} 1.5249, d_{20} 1.1514, IR, NMR spectra, contains 78% Sn-C isomer	1681
$\rightleftharpoons BzCH_2$ (230)	--	--		1681
$CH_2:CMeO$ (d) (230)	--	--	$R' \rightleftharpoons AcCH_2$	719a
$MeCH:CPhO$	I	57	NMR spectrum	1418
$PhCH_2CH:CPhO$	I	80	NMR spectrum	1418
$EtCH:CHO$	II	49	$b_{0.8}$ 119-21°, n_D^{20} 1.4802, d_{20} 1.0950 (e, f)	719a, 1618
$Me_2C:CHO$	II	40	$b_{0.7}$ 104-106°, n_D^{20} 1.4812, d_{20} 1.0960 (e, g, h)	1681
	II	38	trans-form, b_3 143-44°, n_D^{20} 1.4790, d_{20} 1.0929	1684
$MeCH:CMeO$ (d)	II	44	$b_{0.8}$ 115°, n_D^{20} 1.4820, d_{20} 1.1152 (e)	1681
$\rightleftharpoons AcCHMe$	--	--	contains 77% Sn-C isomer	1681
$Me_2CHCH:CHO$	II	38	b_3 143-44°, n_D^{20} 1.4790, d_{20} 1.0929 (e, g, i)	1681
	I	40	b_1 125-28°, n_D^{20} 1.4797, d_{20} 1.0930	1684
$EtCH:CMeO$	I	47	b_1 126-28°, n_D^{20} 1.4812, d_{20} 1.0894 (g, j, k)	1684
$\rightleftharpoons AcCEt$	II	65	cis-form, $b_{0.3}$ 117-18°, n_D^{20} 1.4808, d_{20} 1.0936 (e, j, l)	1681, 1684
	II	56	trans-form, $b_{0.4}$ 122-23°, n_D^{20} 1.4818, d_{20} 1.0918 (e, j, m)	1681, 1684
$MeCH:CEtO$	II	57	$b_{0.8}$ 121-23°, n_D^{20} 1.4808, d_{20} 1.1084 (e)	1681
$\rightleftharpoons EtCOCHMe$	--	--	contains 30% Sn-C isomer, cis/trans ratio 1/3	1681
$c.1-C_5H_7$	II	37	$b_{1.4}$ 144-46°, n_D^{20} 1.4930, d_{20} 1.1397 (e)	1681
$\rightleftharpoons c.2-OC_5H_7$	--	--	contains 57% Sn-C isomer	1681

CH$_2$:C(CMe$_3$)O	II	52	b$_{0.3}$ 118°, n$_D^{20}$ 1.4811, d$_{20}$ 1.0840 (e)	1681
⇌ Me$_2$CCOCH$_2$	--	--	contains 75% Sn-C isomer	1681
c.1-C$_6$H$_9$O (230)	II	35	b$_{1.3}$ 146-48°, n$_D^{20}$ 1.4942, d$_{20}$ 1.1324 (e)	1681
Me(CH$_2$)$_4$CH:CHO	II	63	b$_{0.8}$ 142°, n$_D^{20}$ 1.4797, d$_{20}$ 1.0710 (e, n)	1681
	III	86	b$_{1.2}$ 148°, n$_D^{20}$ 1.4792, d$_{20}$ 1.071	1682
Me$_2$CHCH:C(CHMe$_2$)O	II	48	b$_2$ 145-46°, n$_D^{20}$ 1.4732, d$_{20}$ 1.0441 (e, o)	1681
Me$_2$CHCH:C(Bu-i)O	II	62	b$_{0.8}$ 134-35°, n$_D^{20}$ 1.4738, d$_{20}$ 1.0424 (g, p)	1684
Me$_2$CHCH:C(CH:CMe$_2$)O	I	--	b$_{0.6}$ 133-34°, n$_D^{20}$ 1.4860, d$_{20}$ 1.0651	1684
Ph$_3$SnOCPh:CHMe	I	42	NMR spectrum	1418
Ph$_3$SnOCPh:CHCH$_2$Ph	I	38	NMR spectrum	1418

* Numbers in parenthesis refer to pages in main volume.

** The characters used here correspond to the synthetic scheme in the introduction to Chapter 4.3.4.1.3.

(a) IR, UV and NMR spectra (1712). (b) Rxn. with MeOH → Et$_3$SnOMe + Me$_2$CHCHO (1712). (c) IR and NMR spectra (1709). (d) cf. Table 112 for C-Sn isomer. (e) IR and NMR spectra (1681). (f) Rxn. with MeRSiCl$_2$ → MeRSi(OCH:CHEt)$_2$; R = H, Me + MeRSi(OCH:CHEt)Cl; R = Me (719a). (g) IR and NMR spectra (1684). (h) Rxn. with R$_3$SiH → R$_3$SiOCH:CMe$_2$; R = Et, Ph (719, 719a), with MeSiCl$_3$ → MeSi(OCH:CMe$_2$)$_3$ + MeSi(OCH:CMe$_2$)Cl$_2$ (719a). (i) Prod. ratio cis/trans = 43/57 (1684). (j) Prod. ratio: 25% Sn-C, 75% Sn-O isomers (1684). (k) Prod. ratio cis/trans = 1/9 (1684). (l) Prod. contains 9% trans-isomer (1681, 1684). (m) Prod. contains 4% cis-isomer (1681, 1684). (n) Prod. ratio cis/trans = 3/2 (1681). (o) Prod. ratio cis/trans = 1 (1681). (p) Rxn. with Me$_3$SiCl + Bu$_3$SnCl + Me$_3$SiOC(Bu-i):CHCHMe$_2$ (434). (q) IR spectrum, very sensitive to H$_2$O → i-BuCOCH:CH$_2$ (1684).

4.3.4.1.4 TRIORGANOTIN DERIVATIVES OF 1,2- AND 1,3-DIKETONES

Organotin derivatives of diketones were regarded previously as having a Sn-C bond. Compounds of this type were listed in the main volume in Chapter 2.6.3.3. The recent literature leaves no doubt that a chelate structure with pentacovalent tin prevails. The compounds listed in Table 195 are prepared by methods from the following scheme.

- I Replacement of a substituent with a β-keto group in R_3SnCH_2COR' by a 1,3-dicarbonyl compound; $R' = Me, OR, NR_2$.
- II Reaction of triorganotin halide with sodium tropolonate.
- III Reaction of bis(triorganotin)oxide with 1,3-dicarbonyl compounds.
- IV Reaction of triorganotin alkoxides with:
  - (IVA) 1,3-dicarbonyl derivatives,
  - (IVB) diketene.
- V Reaction of triorganotin amide with diketene.

4.3.4.2 DIORGANOTIN DERIVATIVES OF ALDEHYDES AND KETONES

This chapter reports O-diorganotin derivatives of the tautomeric enol-form of aldehydes and ketones. The reader should always consult Chapter 2.6.3.3 to check for the corresponding Sn-C isomer in search for a certain compound.

The diorganotin derivatives of aldehydes, ketones and diketones listed in Table 196 are prepared by methods from the following scheme.

- I Substitution reactions of diorganotin halide with sodium or thallium enoxide.
- II Condensation of diorganotin oxide with a ketone.
- III Reaction of diorganotin alkoxide with:
  - (IIIA) a diketone,
  - (IIIB) an aldehyde.
- IV Reaction of diorganotin sulfide with a diketone.

Additional diorganotin derivatives containing acetal and alkoxide groups are listed in Subsection 4.3.3.2.9.3.

Four individual diorganotin derivatives of diketones are shown following Table 196 on page 683.

Table 195. Triorganotin Derivatives of 1,2- and 1,3-Diketones

Formula	Synth. Method*	Yield	Properties	R_3SnOR' Ref.
$Me_3SnOCMe:CHCONMe_2$	V	--	m. 68-70°, spectroscopic evidence	1205
Et_3SnOR'				
R' = BzCH:CMe	III	>80	--	391
$MeO_2CCH:CMe$	I	71	b_3 94-96°, n_D^{20} 1.5026, d_{20} 1.3065, IR spectrum	1711
	IVB	75	b_3 92-94°, n_D^{20} 1.5001, d_{20} 1.3062	1711
$EtO_2CCH:CMe$	I	51	b_2 92-93°, n_D^{20} 1.4955, d_{20} 1.2598, IR spectrum	1710
	I	85	b_2 91-93°, n_D^{20} 1.4930, d_{20} 1.2625, IR spectrum	1711
	IVB	67	b_2 92-94°, n_D^{20} 1.4920, d_{20} 1.2632 (a)	1711
$PrO_2CCH:CMe$	I	68	b_3 120-23°, n_D^{20} 1.4878	1711
	IVB	70	b_5 126-28°, n_D^{20} 1.4867, d_{20} 1.2160	1711
AcCH:CMe	I	61	m. 41-43°, b_1 94-95°	1711
	III	>80	m. 40°, $b_{0.8}$ 92°	391
	IVA	70	White solid, subl$_{0.15}$ 80-100°, monomeric	350
$c.1-C_5H_8(CO_2Et-5)$ (b)	I	77	$b_{1.5}$ 123-26°, n_D^{20} 1.5123, d_{20} 1.2529	1711
Pr_3SnOR'				
R' = $MeO_2CCH:CMe$	I	63	b_4 128-31°, n_D^{20} 1.4932, d_{20} 1.2154, IR spectrum	1711
	IVB	52	b_3 121-23°, n_D^{20} 1.4918, d_{20} 1.2140	1711
$EtO_2CCH:CMe$	I	50	b_2 114-16°, n_D^{20} 1.4849, d_{20} 1.1766	437
$i-PrO_2CCH:CMe$	IVB	61	b_5 134-36°, n_D^{20} 1.4845, d_{20} 1.2092, IR spectrum	1711
AcCH:CMe	I	48	m. 51-53°, $b_{1.5}$ 106-109° (c)	1711
$Ph_3SnO_2C_7H_5$ (d)	II	85	Pale yellow needles, m. 142-43°, mol. wt.	2510

* The characters used here correspond to the synthetic scheme in the introduction to Chapter 4.3.4.1.4. (a) Rxn. with AcOH → Et_3SnOAc + $AcCH_2CO_2Et$ (1711). (b) 1-Triethyltin-(5-EtO_2C)-cyclopentene-1-oxide. (c) Rxn. with AcCl → Pr_3SnCl + AcCH:CMeOAc (1711). (d) $C_7H_6O_2$ = tropolone. (e) (Far) IR spectra, assignment, struct. (2510).

Table 196. Diorganotin Derivatives of Aldehydes and Ketones $R_2Sn(OR')_2$

Formula	Synth. Method*	Yield	Properties	Ref.
$Me_2Sn(OCPh:CHBz)_2$	II	--	Yellow solid, m. 189-91°, IR, NMR spectra	347
$Me_2Sn(OCMe:CHBz)_2$	II	--	m. 134-35°, IR spectrum	347
$Me_2Sn[OC(CF_3):CHCOCF_3]_2$	II	--	IR, NMR spectra	347
$Et_2Sn(OCMe:CHAc)_2$	I	--	m. 86.5-87.5°, IR spectrum, soly., struct.	1649
	I	--	m. 86.5-87°, soly. (a)	2055
$Et_2Sn(O_2C_7H_5)_2$ (b)	I	--	m. 175-76°, (far) IR spectra, assignment, mol. wt. struct.	2510
$(CH_2:CH)_2Sn(OCMe:CHAc)_2$	IV	--	m. 87-88°, soly. (a)	2055
$Pr_2Sn(O_2C_7H_5)_2$ (b)	I	--	m. 115-17°, (far) IR spectra, assignment, struct.	2510
	IV	--	--	2510

$Bu_2Sn(OR')_2$

R' = MeOCHMe	IIIB	--	NMR spectrum, dissocn. at 0.1 mm	943
$MeOCH(CCl_3)$	IIIB	--	NMR spectrum, viscous oil (c)	943
$MeOCH(CCl_3)$ + Ac	III (c)	--	IR and NMR spectra	943
	IIIB (d)	--	--	943
$MeO_2CCH:CMe$	IIIA	83	White solid, b$_{0.3}$ 136-38°, monomeric	350
$EtO_2CCH:CMe$	IIIA	62	Light yellow liq., b$_{0.3}$ 154°, monomeric	350
C_7H_5O (b)	I	--	m. 110-11.5°, (far) IR spectra, assignment struct.	2510
	IV	--		2510
$Ph_2Sn(OCPh:CHBz)_2$	I	95	Pale yellow, m. 239-41°, mol. wt. (e)	385
$Ph_2Sn(OCMe:CHBz)_2$	I	95	White, m. 181-82°, mol. wt. (e)	385
	I	95	IR (602, 1591) and UV spectra, soly., struct.	1591
$Ph_2Sn(O_2C_7H_5)_2$ (b)	I	--	m. 224-27°, (far) IR spectra, assignment struct.	2510
	II	--		2510

* The characters used here correspond to the synthetic scheme in the introduction to Chapter 4.3.4.2.
(a) IR (1301) and NMR (1298) spectra. (b) $C_7H_6O_2$ = tropolone. (c) Rxn. of $Bu_2Sn[OCH(CCl_3)OMe]_2$ with $Bu_2Sn(OAc)_2$ → $Bu_2Sn(OAc)OCHOMeCCl_3$ (943). (d) Synth. with $Bu_2Sn(OAc)OMe$. (e) IR, UV (385) and NMR (602) spectra, compound not resolved by chromatography on D-lactose, struct. (385).

Dimethyltin Diacetylacetonate (M-230) $SnC_{12}H_{20}O_4$ $Me_2Sn[(OCMe)_2CH]_2$

<u>Synth.</u>: By rxn. of Me_2SnO with CH_2Ac_2 under reflux (347), in H_2O at pH 8 (2418).
By rxn. of Me_2SnCl_2 with MeONa (1:2) and excess CH_2Ac_2 (1649), in MeOH, in 80% yield (2055).
<u>Prop.</u>: Colorless cryst., m. 181-83° (sealed tube) (2418), 177-78° (1649, 2055), dec. > 300° (347), far IR, assignment (347, 1301, 1649), IR, assignment (347, 1301), Raman, assignment (347), NMR (347, 519, 2418) and Moessbauer resonance (1044) spectra, monomeric (347), magnetic anisotropy (1295), stable in air (2055), struct. (347, 1044, 1649).
<u>Rxn. with</u>: CH_2Ac_2 → kinetic data for ligand exchange (2418).

Dimethyltin Ditropolonate $SnC_{16}H_{16}O_4$ $Me_2Sn(O_2C_7H_5)_2$

<u>Synth.</u>: By rxn. of Me_2SnCl_2 with $C_7H_5O_2Na$ in EtOH at r.t., in 84% yield (2510).
By rxn. of Me_2SnO with $C_7H_6O_2$ in C_6H_6 under reflux (2726).
By rxn. of Me_2SnS with $C_7H_6O_2$ in EtOH, in 97% yield (2510).
<u>Prop.</u>: Yellow cryst., m. 181-83° (2510), 181° (1346), (far) IR, assignment (2510) and NMR (2510, 2726) spectra, mol. wt. (2510), struct. (2510, 2726).
<u>Rxn. with</u>: $Me_2Sn(OC_9H_6N)_2 \rightleftharpoons Me_2Sn(OC_9H_6N)O_2C_7H_5$ (1346).

Dibutyltin Diacetylacetone (M-231) $SnC_{18}H_{32}O_4$ $Bu_2Sn[(OCMe)_2CH]_2$

<u>Synth.</u>: By rxn. of $Bu_2SnCl_2 \cdot 2 NH_3$ with CH_2Ac_2 in C_6H_6 (2954).
By rxn. of $Bu_2Sn(OEt)_2$ with CH_2Ac_2 in C_6H_6 under exclusion of moisture, in 75% yield (350).
By rxn. of Bu_2SnO with CH_2Ac_2 and H_2O removal by C_6H_6 or CaH_2 in 80-90% yield (391).
<u>Prop.</u>: Light yellow liq. (350), m. 30° (391), $b_{0.4}$ 132° (350), $b_{0.015}$ 144° (391), n_D^{20} 1.5243 (2954), monomeric, soly. (350).
<u>Use</u>: As stabilizer in PVC (2954).

Diphenyltin Diacetylacetonate $SnC_{22}H_{24}O_4$ $Ph_2Sn[(OCMe)_2CH]_2$

<u>Synth.</u>: By rxn. of Ph_2SnCl_2 with AcCH:CMeONa in abs. EtOH (347), with CH_2Ac_2 and MeONa in MeOH (2055), with AcCH:CMeOTl in C_6H_6 in 90% yield (385).
By rxn. of Ph_2SnO with anhyd. CH_2Ac_2 (2418).
<u>Prop.</u>: m. 125-26° (2055), 123-25° (385), dec. 125° (347), IR (385), (far) IR, assignment (347, 1301), UV (385), NMR (1298, 2418) and Moessbauer resonance (1044) spectra, dipole moment (385), mol. wt. (2055), not resolved by chromatography on D-lactose (385), soly. (2055), struct. (347, 385, 1044).
<u>Rxn. with</u>: CH_2Ac_2 → kinetic data on ligand exchange (2418).

4.3.4.3 MONOORGANOTIN DERIVATIVES OF ALDEHYDES AND KETONES

Butyltin Tritropolonate $SnC_{25}H_{24}O_6$ $BuSn(O_2C_7H_5)_3$

<u>Synth.</u>: By rxn. of $Bu_2Sn_2S_3$ with tropolone in dioxane under reflux, in 80% yield (2510).

<u>Prop.</u>: Pale yellow cryst., m. 240-41.5°, (far) IR spectra and assignment, mol. wt., struct. (2510).

Phenyltin Tritropolonate $SnC_{27}H_{20}O_6$ $PhSn(O_2C_7H_5)_3$
<u>Synth.</u>: By rxn. of $PhSn(O_2C_7H_5)_2Cl$ with $C_7H_5O_2Na$ in MeCN (372).
<u>Prop.</u>: m. 298-302°, mol. wt., struct. (372).
<u>Addn. Compd.</u>: $PhSn(O_2C_7H_5)_3 \cdot 1.5\ CH_2Cl_2$, from components (372).

Monoorganotin derivatives containing halogen and acetal groups are listed in Chapter 4.1.5.4.

4.3.5 ORGANOTIN CARBOXYLATES

4.3.5.1 TRIORGANOTIN CARBOXYLATES

The triorganotin carboxylates summarized in the following tables are prepared by methods from the ensuing scheme.

- I Dealkylation of tetraorganotin compounds with:
  - (IA) peroxy esters,
  - (IB) carboxylic acids.

- II Metathetic substitution of triorganotin halides with:
  - (IIA) AgO_2CR',
  - (IIB) NaO_2CR',
  - (IIC) $R'CO_2H$ and base.

- III Reaction of triorganotin hydrides with:
  - (IIIA) RCO_2H,
  - (IIIB) $(RCO_2)_2$.

- IV Reactions of carboxylic acids and acid anhydrides with:
  - (IVA) R_3SnOH or $(R_3Sn)_2O$, for free carboxylic acid or lactone,
  - (IVB) for CO_2 or $(RCO)_2O$,
  - (IVC) R_3SnOR' for carboxylic acid anhydride or dicarboxylic acid,
  - (IVD) R_3SnO_2CR by transesterification.

- V Rearrangement reactions of dialkyltin oxide with:
  - (VA) RCO_2R,
  - (VB) RCO_2H or $(RCO)_2O$.

- VI Reaction of triorganotin amide with CO_2 or RCO_2H.

- VII Miscellaneous reactions:
  - (VIIA) thermal decomposition of $(Me_2SnO_2CR)_2O$ or $(HO_2CCH_2CH_2)_4Sn$,
  - (VIIB) methanolysis of $(Bu_3SnO)_2CHCX_3$,
  - (VIIC) diene addition to $R_3SnO_2CCH:CHCO_2R$.

R = organic group, may be different in one molecule, R' = R or functionally substituted organic group, X = Cl, Br. Additional references to triorganotin carboxylates may be found in Chapter 8.1.

4.3.5.1.1 UNSUBSTITUTED TRIORGANOTIN CARBOXYLATES

Trimethyltin Acetate (M-411) $SnC_5H_{12}O_2$ Me_3SnOAc

<u>Synth.</u>: By rxn. of Me_3SnCl with AcONa in $c.C_6H_{12}$ and xylene, in 75 and 92% yield, resp. (1197).
By rxn. of Me_3SnO_2H with AcOH at r.t. in dioxane and MeCN, in 79 and 87% yield, resp. (118).
By rxn. of Me_3SnNMe_2 with CH_2:CHOAc (854).
By rxn. of Me_3SnNEt_2 with CH_2:CMeOAc, in 50% yield (250).
By rxn. of Me_2SnO with AcOH and Ac_2O, in 65 and 70% yield, resp. (2611), with PhOH, followed by AcOH at 180-300°, in 83-27% yield (2611), with PhOAc (1:1) at 220°, in 76% yield (1487).
By thermal dec. of $(Me_2SnOAc)_2O$ at 220°, in 78% yield (2611).
<u>Prop.</u>: m. 196° (2611), 191-92° (1487, 1932), (far) IR, NMR (1932) and Moessbauer resonance (224) spectra, soly., struct. (1932). Anal. detn. by alkalimetric titration in Py (1144).
<u>Rxn. with</u>: $(BuO)_4Zr \rightarrow Me_3SnOZr(OBu)_3$ (1097).
$\overline{CH_2CH_2O} \rightarrow$ relative reactivity as polymerization catalyst (2572).
(:CHCO)$_2$O-PhCH:CH$_2$-polymer $\rightarrow$ tin containing polyester (1328).
<u>Biol. Prop.</u>: Toxicity (30), effect on oxidative phosphorylation in rat liver mitochondria (1668).
Activity as bactericide against intestinal bacillus and Staphylococcus aureus (275).
Activity as insecticide (1328), against Culex pipiens larvae (73, 1137), Heliothis virescens, H. zea (192, 2145).
Activity in plankton control (1995).

Triethyltin Acetate (M-409) $SnC_8H_{18}O_2$ Et_3SnOAc

<u>Synth.</u>: By rxn. of Et_4Sn with AcO_2CMe_3 at 120-25° in N atm., in 80% yield (2095).
By rxn. of Et_6Sn_2 with AcOH at 135°, in 63% yield (2096).
By rxn. of AcOH with $Et_3SnC\vdots CCH_2OCH_2CH_2OCH:CH_2$, in 97% yield (1907), with $Et_3SnC\vdots CPO_3Et_2$ in Et_2O, in 98% yield (2179).
By rxn. of $Et_3SnCH_2CH:CH_2$ with Ac_2O (1331).
By rxn. of $Et_3SnC_6H_4R$ with $Hg(OAc)_2$ in THF; R = H, m-, p-Me, m-, p-Cl, m-, p-OMe (1162).
By rxn. of Et_3SnH with Ac_2O_2 in C_6H_6 in inert atm. at 35-45°, in 78-58% yield (1616), with PhN(NO)Ac, in 86% yield (585), with R_3PbOAc and $RC\vdots CH$ in THF at -20-80° (912), with Ph_3C, followed by AcOH (585).
By rxn. of $(Et_3Sn)_2O$ with $HC\vdots CCH_2OAc$ (391).
By rxn. of AcOH with $(Et_3Sn)_2CO_3$ in MeOH (390), with $Et_3SnOCMe:CHCO_2Et$, in 97% yield (1711), with $Et_3SnOCdEt$ (2663).
By rxn. of $Et_3SnSeCH_2Ph$ with Ac_2O (389).
By rxn. of Et_3SnNMe_2 with CH_2:CMeOAc, in 56% yield (250).
By rxn. of $Et_3SnN(NHPh)S(O_2)Ph$ with Ac_2O in C_6H_6 in Ar atm. at 60° (2646).
<u>Prop.</u>: (Far) IR spectrum (321), dipole moment, molar polarization and configuration (632).

<u>Biol. Prop.</u>: Toxicity (30) and symptoms (2441).
Activity as insecticide against Culex pipiens larvae (73, 1137).
<u>Use</u>: As catalyst for polymerization of (:CHCO)$_2$O with RCHCH$_2$O; R = Me, Ph (2727). As marine wood preservative (1023).

Triethyltin Benzoate (M-409) SnC$_{13}$H$_{20}$O$_2$ Et$_3$SnOBz
<u>Synth.</u>: By rxn. of Me$_3$CO$_2$Bz with excess Et$_4$Sn and Et$_6$Sn$_2$ at 100-130° in sealed tube, in 84 and 80% yield, resp. (2098).
By-prod. in photooxidation of PhCH$_2$SnEt$_3$ in nonane, rate of formation (2215).
By rxn. of Et$_3$SnH with Bz$_2$O$_2$ in C$_6$H$_6$ in inert atm., in 89% yield (1616).
By rxn. of Et$_3$SnOCHRCH$_2$CH:CH$_2$ with BzOH in Et$_2$O (1331).
By rxn. of (Et$_3$Sn)$_2$Y with Bz$_2$O$_2$ in C$_6$H$_6$ at r.t. in inert atm.; Y = S, Se, Te, in 83-94% yield (2093).
By rxn. of (Et$_3$Sn)$_3$Y with Bz$_2$O$_2$ in C$_6$H$_6$ at 5-70°; Y = Sb and Bi, in 100 and 87% yield, resp. (2086).
<u>Prop.</u>: (Far) IR spectra (321), thermodynamic data (2025).

Tributyltin Acetate (M-410) SnC$_{14}$H$_{30}$O$_2$ Bu$_3$SnOAc
<u>Synth.</u>: By rxn. of Bu$_3$SnH with Ac$_2$O$_2$ at 60-80°, in 72% yield (1616), with [MeC(^{18}O)O]$_2$ in C$_6$H$_6$ in inert atm., in 62% yield (1615).
By rxn. of Bu$_2$SnH$_2$ with HC:CCH$_2$CHOAcC:CH at 220-270° as by-prod. (499).
By rxn. of (Bu$_3$Sn)$_2$O with Ac$_2$O, in 99% yield (1197), with EtOAc (949).
By rxn. of Bu$_3$SnOH with AcOH, in 98% yield (2074).
By rxn. of Bu$_3$SnO$_2$CCH$_2$NHAc with AcOH in EtOH, in 91% yield (151).
By rxn. of Bu$_3$SnNRCO$_2$Me, with AcOH; R = 1-C$_{10}$H$_7$ (36).
By rxn. of Bu$_3$SnNRCO$_2$SnBu$_3$ with Ac$_2$O; R = Et, Ph (748).
By rxn. of (Bu$_3$SnNPh)$_2$SO with AcOH (2345).
<u>Prop.</u>: IR spectrum (2074), dipole moment, molar polarization and configuration, struct. (632), light resistance (1063). Anal. detn. by TLC (1172a).
<u>Rxn. with</u>: Ph$_3$SnH + Et$_3$N (1:1:1) → Ph$_6$Sn$_2$ + Bu$_3$SnSnPh$_3$ + Et$_3$NHOAc (1618).
Et$_3$SnO$_2$CEt → no exchange below 80° (1615).
Bu$_2$SnO → Bu$_3$SnOBu$_2$SnOAc (946).
BuSn(OAc)$_3$ at 180-85° → Bu$_2$Sn(OAc)$_2$ (540).
B(OR)$_3$ → (Bu$_3$SnO)$_n$B(OR)$_{3-n}$; n = 1, 2; R = i-Pr, Bu (2620).
Zr(OR)$_4$ → Bu$_3$SnOZr(OR)$_3$; R = Pr, Bu (1097).
(PrO)$_3$ZrO$_2$CCF$_3$ → Bu$_3$SnOZr(OPr)$_2$O$_2$CCF$_3$ (1097).
Ti(OBu)$_4$, + N(CH$_2$CH$_2$OH)$_3$ → Bu$_3$SnOTi(OCH$_2$CH$_2$)$_3$N + Bu$_3$SnOTi(OCH$_2$CH$_2$)$_3$N·[N(CH$_2$-CH$_2$O)$_3$Ti]$_2$O + (Bu$_3$Sn)$_2$O (882).
PCl$_5$ with Bu$_3$SnOC(^{18}O)Me → Bu$_3$SnCl + MeC(^{18}O)Cl + POCl$_3$ (1615).
AcCl → Bu$_3$SnCl + Ac$_2$O (953).
AcSH → S-Bu$_3$SnSAc (2406).
(:CHCO)$_2$O-PhCH:CH$_2$-polymer → tin containing polyester (1328).
<u>Addn. Compd.</u>: Bu$_3$SnOAc·BiPy, from components in alc., m. 50° (dec.) (1212) [SnC$_{24}$H$_{38}$N$_2$O$_2$].
Bu$_3$SnOAc·phen, from components in alc., m. 118° (dec.) (1212) [SnC$_{26}$H$_{38}$N$_2$O$_2$].
Bu$_3$SnOAc·N(CH$_2$CH$_2$OH)$_3$, from components in MeOH, in 92% yield, m. 65° (1226).
   Biol. activity as bactericide against Bacillus subtilis, Flavobacter

peregrinum, Humicola grisea (1226), as fungicide against Aspergillus niger, Penicillium citricum (1226) [$SnC_{20}H_{45}NO_5$].

Biol. Prop.: Toxicity (30, 1069) and effects (2678).
Phytotoxicity to barley (376), rice (548).
Activity as bactericide against intestinal bacillus (275), Staphylococcus aureus (275, 1145).
Activity as fungicide (451, 1145), against Alternaria tenuis (833), Coriolellus palustris (1063), Fusarium culmorum (833), Hypochnus sasaki (858), Piricularia oryzae (548), Rhizoctonia solani (190, 933).
Activity as insecticide (451, 1328), against Culex pipiens larvae (73, 1137).
Activity as moluscicide (125, 455), against Australorbis glabratus (833, 1860), Biomphalaria sudanica, Bulinus nasutus (596).
Biol. method for detn. 5 ppm in seawater (456), detn. of leaching rate from paints (2320).
Use: U.S. limits for use as preservative in adhesives (2941).
In antifouling coatings (2302, 3002).
As flame retardant for epoxy resins (1466).
As preserving agent for cured hide and skin (2396), for wood (451).
As rodent repellent (2036).
As slime control agent in paper and paper mill industry (2810).

Tributyltin Laurate (M-414) $SnC_{24}H_{50}O_2$ $Bu_3SnO_2CC_{11}H_{23}$
Synth.: By rxn. of Bu_3SnH with $(C_{11}H_{23}CO_2)_2$ in C_6H_6 in inert atm. (1616).
Prop.: m. 10°, $b_{0.01}$ 156°, n_D^{20} 1.482 (1616).
Biol. Prop.: Toxicity and effects (2678).
Activity as bactericide against Staphylococcus (1145), as fungicide (1145).
Biological method for detn. in seawater (456).
Use: In antifouling coatings (2302).
As catalyst for foaming polyurethan (450).
As stabilizer for PVC; mechanisms and relative reactivity (2599).

Tributyltin Benzoate (M-414) $SnC_{19}H_{32}O_2$ Bu_3SnOBz
Synth.: By rxn. of Bu_3SnH with $[Ph(^{18}O:)CO]_2$ in C_6H_6, in 86% yield (1615).
By rxn. of $Bu_3SnOCMe:CHCHMe_2$ with BzOH (1684).
Synth. of $Bu_3{}^{113}SnOBz$ in three steps from ^{113}Sn, in 37% yield (1659).
By rxn. of $(Bu_3Sn)_2O$ with BzOH at 100-110°, in 98% yield (2966).
Prop.: $b_{0.2}$ 135°, n_D^{20} 1.5161 (1615), d_{20} 1.171 (2966).
Rxn. with: $p\text{-}Et_3SnO_2CC_6H_4Me \rightarrow$ no exchange below 130° (1615).
PCl_5 with $Bu_3SnOC(:^{18}O)Ph \rightarrow Bu_3SnCl + POCl_3 + PhC^{18}OCl$ (1615).
Biol. Prop.: Toxicity and effects (2678).
Activity as bactericide against Bacillus proteus, Escherichia coli, Pseudomonas pyocyanea (1145), Staphylococcus aureus (2966).
Activity as fungicide (2966), against Candida albicaus, Epidermophyton (1145), Fusarium culmorum (281), Hypochnus sasaki (858).
Effect of inactivation of foot and mouth disease virus (1028).
Use: As bacteriostat for textiles (2966).
As disinfectant (11, 482).

As feed additive (2918).
As rodent repellent (2036).
As preservative for hide and skin (2396).

Triphenyltin Acetate (M-410) $SnC_{20}H_{18}O_2$ Ph_3SnOAc

Synth.: By rxn. of Ph_3SnAc with air (1677, 2671).
By rxn. of Ph_3SnH with R_3PbOAc and HC⋮CCN in THF at -80 to -20°, R = Et, Bu (912).
By rxn. of Ph_3SnCl with aq. NaOAc in org. solvent at 0° to r.t., in 92% yield (1051).
By rxn. of $Ph_3Sn-3,4,6-Ph_3C_4N_2$ (pyridazine) with AcOH in Ar atm. (2644).
By rxn. of Ph_3SnSPh with $Hg(OAc)_2$ in EtOH as by-prod. (1704).
By solvent extraction of aq. Ph_3SnOH with ^{14}C tagged NaOAc (487).

Prop.: IR (2394), assignment (2793) and Moessbauer resonance (2394) spectra, adsorption and soly. (1512). Anal. detn. by TLC (1172a), detn. and sepn. from $Ph_nSn(OAc)_{4-n}$ by TLC (1253) in plant matl. by colorimetric method (2170), in fodder by colorimetry and chromatography (2497).

Rxn. with: $(i-PrO)_4Ti + N(CH_2CH_2OH)_2 \rightarrow Ph_3SnOTi(OCH_2CH_2)_3N$ (2324).
$(i-PrO)_4Zr \rightarrow Ph_3SnOZr(OPr-i)_3$ (1097).
$Cl_5C_6CO_2H \rightarrow Ph_3SnO_2CC_6Cl_5 + AcOH$ (956).
$AcO_2H \rightarrow Ph_2Sn(OPh)OAc + PhOH$; kinetic data and rxn. mechanism (790).
$(:CHCO)_2O-PhCH:CH_2$-polymer → tin contg. polyester (1328).
$Ph_2NCS_2Na \rightarrow Ph_3SnS_2CNPh_2$ (298).
$HSC_5H_4NO \rightarrow 2-(Ph_3SnS)C_5H_4NO$, pyridine oxide (1836a).
Extraction with several solvent systems (771).

Addn. Compd.: $Ph_3SnOAc \cdot N(CH_2CH_2OH)_3$, from components in MeOH, in 95% yield, m. 80° (1226). Biol. activity as bactericide against Bacillus subtilis, Flavobacter peregrinum, Humicola grisea (1266), as fungicide against Aspergillus niger, Penicillium citricum (1226) [$SnC_{26}H_{33}NO_5$].

Biol. Prop.: Toxicity (1993, 2073), effects on ovarian tissue of rats (2646). Phytotoxicity (288) to barley (376), citrus varieties (2291), potato (306), swamp rice (1202), tomato (1809a).
Activity as acaricide against Eriophyidae (2023), Phyllocoptruta oleivora (2291).
Activity as bactericide (529) against Bacillus mycoides, Staphylococcus aureus, Streptococcus lactis (2796).
Activity as fungicide (1005), against Alternaria tenuis (2297), Candida albicans (2796), Copnodium, Cephalothecium roseum (978), Cercospora (150a, 1040, 1329), C. beticola (1525, 1694, 1722, 2741), C. fusca (978), Colletotrichum coffeanum (1188), C. lindemuthianum (479), Cryptococcus neoformans (2796), Dothistroma pini (1187), Erisphe betae (2863), E. communis (1359), Fusicladium effurum (22, 978), Glomerella cingulata (140, 2796), Guignardia laricina (2816), Hemileia vastatrix (1189, 2442), Microsphaera alni, Mycosphaerella caryigena (978), M. musicola (287-8), Phytophthora infestans (30, 348, 1053, 1219, 2297, 2741, 2976), P. palmivora (1622), Piricularia oryzae (548), Pyrenophora avena (2762), Rhizoctonia (1040), R. solami (190, 574), Septoria apii (2741), S. apiicola (1895), S. lycopersici (669), Trichophyton

mentagrophytes (2796), Venturia inaequalis (2297). Effect controlling potato blight (2472), rice blast and rice sheath blight (549).
Activity as insecticide (1040, 1328), against Culex pipiens larvae (73, 1137), and chemosterilant effect (2616), Chilo agamemnon, Gnorimoschema opercucella (354), Heliothis virescens, H. zea (192, 2145), Hippelatus collusor, also chemisterilant effect (2616), Leptinotarsa decemlineata, also as chemisterilant effect (805), Musca domestica, also chemisterilant effect (172a, 260, 686, 805, 1064, 1315, 2230, 2436), Popillia japonica (1382), Oryzaphilus surinamensis, Sitophilus granarius, Sitotroga cereallela (2803), Tineola bisselliella (162), Tribolium confusium, Trogoderma parabile larvae (2803), effect on army worm infestation (1040).
Activity as molluscicide (125, 455) against Australorbis glabratus (154, 823, 1860), Biomphalaria glabratus (1202), B. sudanica, Bulinus nasutus (596), anal. method for detn. in seawater at 5 ppm level (1202).
Use: Trade names and synonyms: Brestan, Fentin acetate, Lirostanol.
As fumigant for sugar beets (2178).
In fungicide formulations for potato and sugar beet cultures (2976), with decreased phytotoxicity (132, 1809a).
As marine wood preservative (1023).

Triphenyltin Benzoate (M-415) $SnC_{25}H_{20}O_2$ Ph_3SnOBz
Synth.: By rxn. of Ph_3SnCl with aq. NaOBz in org. solvent at 0° to r.t., in 89% yield (1051).
By solvent extraction of aq. Ph_3SnOH with ^{14}C tagged BzONa (487).
Biol. Prop.: Activity as bactericide (529), against Bacillus mycoides, Staphylococcus aureus, Streptococcus lactis (2796).
Activity as fungicide (1005), against Alternaria tenuis (933, 2297), Candida albicans, Cryptococcus neoformans (2796), Fusarium culmorum (933), Glomerella cingulata (2796), Rhizoctonia solani (933), Trichophyton mentagrophytes (2796), Tropaeoleum majus (281), Venturia inaequalis (2297).
Activity as insecticide against Leptinotarsa decemlineata, including chemosterilant effect (2297), Musca domestica including chemosterilant effect (805), Tineola bisselliella larvae (162).

Listing of unsubstituted triorganotin carboxylates continues in Table 197.

Table 197. Unsubstituted Triorganotin Carboxylates R_3SnO_2CR

Formula*	Synth. Method**	Yield	Properties	Ref.
(Me$_3$Sn)$_2$CO$_3$ (411)	IVB	--	(Far) IR spectra, struct.	1800
Me$_3$SnO$_2$CR				
R = H (411)	--	--	m. 138°, (far) IR, NMR spectra, soly., struct. (a)	1932
Et (411)	VA	--	m. 134-34.5°	1487
	VB	64	Synth. with EtCO$_2$H	2611
	VB	72	Synth. with (EtCO)$_2$O	2611
	VIIA	80	m. 134-34.5°	2611
C$_{11}$H$_{23}$ (411)	--	--	Use as cat. for foaming polyurethan	450
Ph (411)	VIIA	50	Thermodynamic data (2025)	2611
[(PhCH$_2$)$_3$Sn]$_2$CO$_3$	IVB	--	Liq., IR spectrum	1940
(PhCH$_2$)$_3$SnO$_2$CR				
R = H	--	--	Use as depolymerization cat. for (C$_3$H$_6$)$_x$	2750
Me (411)	IIA	71	White, m. 119°, IR spectrum, mol. wt., struct. (b)	2793
1-C$_{10}$H$_{15}$	IIIA	--	R = 1-adamantane, m. 90.5-94° (c)	2391
C$_{11}$H$_{23}$ (411)	--	--	Use as foaming cat. for polyurethane (d)	180
(Et$_3$Sn)$_2$CO$_3$ (412)	VI	95	m. 120° (e)	250
	VI (f)	--	(Far) IR spectra (321, 1800), struct. (1800)	390
Et$_3$SnO$_2$CR				
R = H (412)	--	--	(Far) IR spectra	321
Et (412)	IVC	71	m. 109°, (far) IR spectra (321)	389
	I	58	m. 110-11° (g)	2095
1-C$_{10}$H$_{15}$	IIIA	--	R = 1-adamantane, m. 116.5-17° (c)	2391
C$_{11}$H$_{23}$	IIIB	68	Colorless, waxy needles, m. 73°	1616
p-MeC$_6$H$_4$	IIIA	--	Colorless cryst., m. 54°, b$_{0.2}$ 128° (h)	1615

Compound			
(PhCH$_2$CH$_2$)$_3$SnOAc (412)	--	Phytotoxicity to swamp rice (l)	1202
Pr$_3$SnO$_2$CR			
R = H (412)	--	n_D^{20} 1.5038, thermal dec. → Pr$_3$SnH	408, 1650
Me (412)	--	TLC (1172a), toxicity (j-l)	30
Et (412)	--	Physiological effect on humans (j)	933
c.C$_3$H$_5$	IIIA	m. 115.5-16° (c)	2391
c.C$_4$H$_7$	IIIA	m. 87-90° (c)	2391
c.C$_5$H$_9$	IIIA	m. 87.5-88° (c)	2391
c.C$_6$H$_{11}$	IIIA	m. 78-80° (c)	2391
1-C$_{10}$H$_{15}$	IIIA	R = 1-adamantane, m. 36-38°, bo.1 148-50° (c)	2391
i-Pr$_3$SnOAc·H$_2$O (413)	--	NMR spectrum (2070), toxicity	30
i-Pr$_3$SnOBz (413)	I	--	2094
(c.C$_3$H$_5$)$_3$SnOAc (413)	IVA	White cryst., m. 157-58°, use as herbicide, pesticide	1873
(c.C$_3$H$_5$)$_3$SnO$_2$CC$_7$H$_{15}$	IIB	Use as herbicide, pesticide	1873
(c.C$_3$H$_5$)$_3$SnO$_2$CC$_{11}$H$_{23}$	IVA	Use as herbicide, pesticide	1873
(Bu$_3$Sn)$_2$CO$_3$ (413)	IV	Low	Synth. from (Bu$_3$Sn)$_2$O + (PhNCO)$_2$	746
	IVC	--	Synth. from Bu$_3$SnO$_2$CNHEt + H$_2$O (m, n)	748

* Numbers in parenthesis refer to pages in main volume.

** The characters used here correspond to the synthetic scheme in the introduction to Chapter 4.3.5.1.

(a) Moessbauer resonance spectrum (2801). (b) NMR spectrum (2069), x-ray struct. analysis (2210). (c) Activity as pre- and post-emergence herbicide (2391). (d) Effect on degradation of PVC (371). (e) Rxn. with HCl → CO$_2$ (390), with AcOH in MeOH → Et$_3$SnOAc (390). (f) Synth. with Et$_3$SnPhCH$_2$C$_6$H$_4$Me and air (390). (g) No exchange with Bu$_3$SnOAc < 80° (1615). (h) No exchange with Bu$_3$SnOBz < 130° (1615). (i) Activity as molluscide against Biomphalaria glabrata (1202). (j) Activity as fungicide against Alternaria tenuis, Fusarium culmorum and Rhizoctonia solani (933). (k) Activity as fungicide against Rhizoctonia solani (190). (l) Activity as molluscicide against Australorbis glabratus (154). (m) IR spectrum, viscous oil, dec.o.oi 100° (750), rxn. with air → white needles, m. 78-80° (750). (n) Activity as bactericide, fungicide (1984).

Table 197 Continued R_3SnO_2CR

Formula*	Synth. Method**	Yield	Properties	Ref.
Bu_3SnO_2CR				
R = MeO (413)	IVC	--	Viscous oil, b_{0.05} 60° (dec.), IR spectrum (o)	750
H (413)	VIIB	--	Thermal dec. → Bu_3SnH (408, 1650)	953
	VIIB	--	Synth. from $Bu_3SnOCH(OMe)CBr_3$ (p, q)	954
$PhCH_2$ (413)	IIB	100	m. 61-62°	1051
Et (413)	VA	82	m. 71.5-72° (p, q)	1487
Pr (413)	--	--	(r)	--
Me_2CH (413)	--	--	(q)	--
c-C_3H_5	IIIA	--	m. 91.5-92.5° (c)	2391
$CH_2CHPhCH$	IIIA	--	cis-form: m. 107.5° (c) trans-form: m. 94.5° (c)	2391
c-C_4H_7	IIIA	--	m. 88-89° (c)	2391
C_5H_{11} (413)	--	--	(q, r)	--
c-C_5H_9	IIIA	--	m. 83-84° (c)	2391
RCO_2 = naphthenate	--	--	Use as antifouling coatings (s)	2302
R = c-C_6H_{11}	IIIA	--	m. 66-68°, use as cat. for foaming polyurethan (450)(c)	2391
C_6H_9	IIIA	--	R = 6-exo-bicyclo[3.1.0]hexane, m. 104-104.5° (c)	2391
C_7H_{15} (413)	--	--	(r)	--
EtBuCH	--	--	Use as curing cat. for polyurethan rubber (n, p)	539a
2-C_7H_{11}	IIIA	--	R = 2-bicyclo[2.2.1]heptane, m. 72-72.5° (c)	2391
7-C_7H_{11}	IIIA	--	R = 7-exo-bicyclo[4.1.0]heptane, m. 108-108.5° (c)	2391
C_8H_{17} (413)	--	--	Use as cat. for foaming polyurethan (t)	450
c-C_8H_{15}	IIIA	--	m. 71-72° (c)	2391
8-C_8H_{13}	IIIA	--	R = 8-exo-bicyclo[8.1.0]octane, m. 113-15° (c)	2391
3-C_8H_{11}	IIIA	--	R = 3-nortricyclene, m. 72-73° (c)	2391
C_9H_{19} (414)	--	--	(q, r)	--

R or compound	Group	n or mp	Notes	Ref
Me(CH$_2$)$_5$CMe$_2$	IIB	--	Use as pesticide (p, u, v)	1169a
2-C$_9$H$_{15}$	IIIA	--	R = 9-exo-bicyclo[6.1.0]nonane, m. 125.5-26° (c)	2391
C$_{10}$H$_{21}$ (414)	--	--	(r)	--
1-C$_{10}$H$_{15}$	IVA	--	R = 1-adamantane, m. 32-35°, b$_{0.3}$ 175° (c)	2391
C$_{13}$H$_{27}$ (414)	--	--	(n)	--
C$_{15}$H$_{31}$ (414)	--	--	(q)	--
RCO$_2$ = resinate	--	--	Use for antifouling coatings	2302
(PhCMe$_2$CH$_2$)$_3$SnO$_2$CH	IVA	--	m. 81.5-82.5°, IR, NMR spectra, soly.	1742
(PhCMe$_2$CH$_2$)$_3$SnOAc	IVA	--	m. 86.5-87°, IR, NMR spectra, soly. (w)	1742
(C$_5$H$_{11}$)$_3$SnOAc (414)	--	--	(x)	--
(C$_5$H$_{11}$)$_3$SnO$_2$CC$_{11}$H$_{25}$	--	--	Use as foaming cat. for polyurethan	450
(Me$_3$CCH$_2$)$_3$SnOAc	IIB	47	b$_{0.1}$ 104°, n$_D^{25}$ 1.4747, NMR spectrum	2188
(c.MeC$_5$H$_8$)$_3$SnOAc	--	--	Phytotoxicity to swamp rice (1)	1202
(C$_6$H$_{13}$)$_3$SnOAc (414)	--	--	Toxicity (x-z)	30
(c.C$_6$H$_{11}$)$_3$SnOAc (414)	--	--	X-ray struct. analysis (aa, bb)	2211
(c.C$_6$H$_{11}$)$_3$SnO$_2$CPr	--	--	(aa)	--
(C$_7$H$_{15}$)$_3$SnOAc (414)	--	--	(x)	--
(BuCHCH$_2$CH)$_3$SnO$_2$CC$_{17}$H$_{35}$	IVA	--	Use as herbicide, pesticide	1873
(C$_8$H$_{17}$)$_3$SnOAc (414)	IIA	41	TLC (1172a), toxicity (30) (x)	403
(C$_8$H$_{17}$)$_3$SnO$_2$CC$_{11}$H$_{23}$	--	--	Use as foaming cat. for polyurethan (cc)	450

(o) Rxn. with air → (Bu$_3$Sn)$_2$CO$_3$ (750), thermal dec. → Bu$_3$SnOMe + CO$_2$ (750). (p) Use as rodent repellent (2036). (q) Activity as fungicide against Hypochnus sasaki (858). (r) Activity as bactericide against Staphylococcus aureus (1145), as fungicide (1145). (s) Use as fungicide to prevent mold growth (243). (t) Use for antifouling coatings (2302). (u) U. S. limits for use as preservative in adhesives (2941). (v) Activity as bactericide against Escherichia coli, Staphylococcus aureus (1169), as fungicide against Aspergillus flavus, A. niger (1169). (w) Moessbauer resonance spectrum (224, 1742). (x) Activity as insecticide against Culex pipiens larvae (73, 1137). (y) Activity as insecticide against Musca domestica, LD$_{50}$ and ATP-inhibition (425). (z) Activity as insecticide against Heliothis virescens, H. zea (2145). (aa) Activity as acaricide (1306). (bb) Use as rodent repellent (2484). (cc) Activity as molluscicide against Australorbis glabratus, Bulinus contortus (962), effect on insects, crustaceans and aquatic plants (962), use in antifouling paints (962).

Table 197 Continued

Formula*	Synth. Method**	Yield	Properties	R_3SnO_2CR Ref.
Ph_3SnO_2CR				
R = 1-$C_{10}H_7CH_2$	IVA	89	m. 136-38° (dd)	866
Et (415)	IIA	95	m. 122-23°	1051
	IV (ee)	--	(ff)	487
Me_3C	IIA	51	m. 112-15°, NMR spectrum	2671
	I	--	Synth. from $Ph_3SnCOCMe_3$ and air	2671
RCO_2 = naphthenate	--	--	Use as fumigant for sugar beets	2178
R = EtBuCH	--	--	IR, Moessbauer resonance spectra, struct. (gg)	2394
C_7H_{11}	IIIA	78	R = 1-norbornane, m. 80-82°, IR spectrum,	1310,1374
	IIC	74	mol. wt., soly.	1310,1374
t-C_8H_{17}	IVA	--	Liq., activity as biocide (hh)	2905
1-C_9H_{15}	IIIA	78	R = 1-apocamphane, m. 91-92°, IR spectrum,	1310,1374
	IIB	54	mol. wt., soly.	1310,1374
	IIC	80	--	1310,1374
1-$C_{10}H_{15}$	IIIA	--	R = 1-adamantane, m. 156-59° (c)	2391
$C_{11}H_{23}$	--	--	Use as foaming cat. for polyurethan (ii)	450
$C_{17}H_{35}$	--	--	IR, Moessbauer resonance spectra (jj)	2394
1-$C_{20}H_{13}$	IIIA	84	R = 1-triptycene, m. 215-17°, IR spectrum,	1310,1374
	IIC	79	mol. wt., soly.	1310,1374
(o-$MeC_6H_4)_3SnOAc$	IIA	57	White, m. 95°, IR spectrum, assignment, struct. (kk)	2793
(o-$MeC_6H_4)_3SnOBz$	--	--	NMR spectrum	2356
(m-$MeC_6H_4)_3SnOAc$ (415)	--	--	Phytotoxicity to swamp rice (i)	1202
(p-$MeC_6H_4)_3SnOAc$ (415)	IIA	68	m. 113°, IR spectrum, assignment, struct. (i)	1202

(dd) Activity as fungicide against Aspergillus niger, Penicillium italicum, Sclerotinia sclerotiorum, Xanthomonas malvacearum (866). (ee) Synth. by solvent extraction with aq. ^{14}C tagged $EtCO_2Na$ (487). (ff) Activity as fungicide against Phytophthora infestans (2976), use in potato and sugar beet cultures (2976). (gg) NMR spectrum (1220). (hh) Composition 56% 2,2,4,4-tetramethylvaleric, 27% 2,2,3-isopropyldimethylbutyric, 17% other acids (2905). (ii) Alkalimetric acylate detn. in Py (1144). (jj) Activity as biocide (2905), as insecticide in control of and chemosterilant effect on Musca domestica (260). (kk) NMR spectrum (2356).

4.3.5.1.3 UNSUBSTITUTED TRIORGANOTIN CARBOXYLATES DERIVED FROM HETEROCYCLIC ACIDS

All compounds are listed in Table 198.

4.3.5.1.4 UNSUBSTITUTED TRIORGANOTIN CARBOXYLATES DERIVED FROM UNSATURATED ACIDS

Triorganotin derivatives of unsaturated monocarboxylic acids are listed in Table 199. Triorganotin half esters of unsaturated dicarboxylic acids are summarized in Table 200; Tables 205 and 206 include bis(triorganotin)carboxylates of unsaturated dicarboxylic acids and of polyfunctional acids, respectively. Additional references to triorganotin carboxylates of unsaturated acids may be found in Chapter 8.1.

Tributyltin oleate $\quad Bu_3SnO_2C(CH_2)_7CH:CHC_8H_{17}$
(M-419) $\quad SnC_{30}H_{60}O_2$
Biol. Prop.: Toxicity and effects (2678).
Phytotoxicity to cotton, lima bean, squash (1738).
Activity as acaricide against Tetranychus urtica (1738).
Activity as bactericide against Bacillus proteus, Escherichia coli, Pseudomonas pyocyanea (1145).
Activity as fungicide against Candida albicans, Epidermophyton (1145).
Use: As catalyst for epoxide polymerization (2915), for foaming polyurethan (450).

Additional unsubstituted triorganotin carboxylates derived from unsaturated acids are listed in Tables 199 and 200.

Table 198. Unsubstituted Triorganotin Carboxylates Derived from Heterocyclic Acids R_3SnO_2CR'

Formula*	Synth. Method**	Yield	Properties	Ref.
$Me_3SnO_2CC_4H_3N_2O_2$ (a) (416)	IVA	--	Biol. prop. resembling vitamin B_{13}	2500
$Pr_3SnO_2CC_7H_{11}N_2OS$ (b)	IVA	--	--	2236
Bu_3SnO_2CR'				
$R' = 2-C_4H_3O$ (c) (416)	IVA	--	m. 89.5°, biol. activity as fungicide (d)	378
2,5-BrC_4H_2O (c)	IVA	--	m. 62°, biol. activity as fungicide	378
$C_7H_{11}N_2OS$ (b)	IVA	88	m. 84-85°, IR spectrum, soly. (e)	2236
$3-C_5H_4N$ (f) (416)	IVA	78	m. 136-137° (g)	572
Ph_3SnO_2CR'				
$R' = 2-C_4H_3N$ (h)	IIIA	47	m. 141-43°, IR spectrum, soly.	1310,1374
	IIC	63	Rxn. with alumina → $(Ph_3Sn)_2O + R'CO_2H$	1310,1374
$C_7H_{11}N_2OS$ (b)	IVA	--	--	2236
$3-C_5H_4N$ (f)	IVA	84	m. 182-183° (i)	866
$3-C_8H_6NCH_2$ (j)	IVA	82	m. 72-73° (i)	866
$3-C_8H_6NCH_2CH_2$ (k)	IVA	78	m. 79-81° (i)	866

* Numbers in parenthesis refer to pages in main volume.
** The characters used here correspond to the synthetic scheme in the introduction to Chapter 4.3.5.1.
(a) $C_4H_3N_2O_2CO_2H$ = orotic acid. (b) 6-$(H_2N)C_7H_8NOS$ = 6-aminopenicillic acid. (c) Furane carboxylate. (d) Biol. activity as fungicide against Coriolellus palustris, light resistance (1063). (e) Rxn. with $(RN:)_2C + RCO_2H + PhSK → Bu_3SnSPh + (RNH)_2CO + K(6-RCO)HN-penicillinate$; R = 4,5-phenylthiazolylmethyl, several thizaolyl derivatives (2236). (f) $C_5H_4NCO_2H$ = pyridine carboxylic acid. (g) Use as antifouling agent (572). (h) $C_4H_3NCO_2H$ = pyrrole carboxylic acid. (i) Biol. activity as fungicide against Aspergillus niger, Penicillium italicum, Sclerotina sclerotiorum, Xanthomonas malvacearum (866). (j) α-(3-indolyl)acetate. (k) β-(3-indolyl)-propionate.

Table 199. Unsubstituted Triorganotin Carboxylates Derived from Unsaturated Acids R_3SnO_2CR'

Formula*	Synth. Method**	Yield	Properties	Ref.
$Me_3SnO_2CC!CH$	IVA	74	m. 184-185°, IR spectrum	311
	VI	--	Synth. from $Me_3SnC!CCO_2R + H_2O$	311
	IIIA	--	Synth. with $RO_2CC!CH$; R = Me, Et (a)	1410
$Me_3SnO_2CCMe:CH_2$ (418)	--	--	m. 98-100° (c)	2391
$2-(PhCH_2)_3SnO_2CC_7H_9$ (b)	IIIA	--		311
$Et_3SnO_2CC!CH$	IVA	60	m. 158-159°, IR spectrum	311
	VI	--	Synth. from $Et_3SnC!CCO_2Me + H_2O$	1410
	IIIA	--	Synth. with $RO_2CC!CH$, R = Me, Et	
$2-Et_3SnO_2CC_7H_9$ (b)	IIIA	--	m. 94.5-96° (c)	2391
$Pr_3SnO_2CCH:CH_2$ (418)	--	--	Polymerizes with Bz_2O_2	306
$Pr_3SnO_2CCMe:CH_2$ (418)	--	--	Polymerizes with Bz_2O_2 and RSH	306
$2-Pr_3SnO_2CC_7H_9$ (b)	IIIA	--	m. 56-57° (c)	2391
$(c.C_3H_5)_3SnO_2CC_{17}H_{33}$ (f)	IVA	--	Use as herbicide, pesticide	1373
Bu_3SnO_2CR'				
R' = HC!C (418)	IVA	--	Undistillable yellow oil, IR spectrum (g)	1520
$m-PhCH_2C_6H_4CH:CH$	--	--	Activity as fungicide against Hypochnus sasakii	858
MeCH:CH (419)	IIB	89	m. 81-82°	1051

* Numbers in parenthesis refer to pages in main volume.
** The characters used here correspond to the synthetic scheme in the introduction to Chapter 4.3.5.1.
(a) Rxn. with $(:CHCO)_2O + PhCH:CH_2$ at 65-70° → tin contg. copolymer (1327). (b) $2-C_7H_9CO_2H$ = 2-bicyclo[2.2.1]-heptene-5-carboxylic acid. (c) Biol. activity as pre- and post-emergence herbicide (2391). (d) Rxn. with $(:CHCO)_2O$-$PhCH:CH_2$-polymer → tin contg. polymer (1328). (e) Biol. activity as bactericide against intestinal bacillus, Staphylococcus aureus (275), as insecticide (1328). (f) Oleate. (g) Use as fungicide, herbicide, insecticide (1520), as stabilizer for PVC and copolymers (1511), as stabilizer for Cl-contg. polymers (1520).

Table 199 Continued

Formula*	Synth. Method**	Yield	Properties	R_3SnO_2CR' Ref.
Bu_3SnO_2CR' (Cont'd.)				
R' = CH_2:CMe (417)	--	--	Polymerizes with Bz_2O_2 and RSH (h)	306
3-C_8H_9	IIIA	--	Exo-tricyclo[3.2.1.0.^{2,4}]octene-6, m. 99-100° (c)	2391
p-CH_2:CHC$_6H_4$ (418) (i)	--	--	Polymerizes with $K_2S_2O_8$ + RSH	306
Ph_3SnO_2CR'				
R' = CH_2:CMe (417)	--	--	IR, Moessbauer resonance spectra, struct. (j)	2394
2-C_7H_9 (b)	IIIA	--	m. 101-103° (c)	2391
CH_2:CH(CH_2)$_8$	--	--	IR, Moessbauer resonance spectra, struct. (k)	2394
$C_{17}H_{33}$ (f)	--	--	Acylate detn. by alkalimetry in Py (l)	1144

(h) Rxn. with CH_2:CMeCO$_2$Me + Bz_2O_2 → tin contg. polymer (2949), use as cat. for foaming polyurethan (450). (i) Add in main volume page 418, line 1 from bottom: p-CH_2:CHC$_6H_4$ under PhCH:CH. (j) Polymerizes with Bz_2O_2 (306), biol. activity as biocide (2905). (k) Biol. activity in control and effect on reproduction of Musca domestica (260). (l) Use as fumigant for sugar beets (2178).

Table 200. Triorganotin Carboxylates Derived from Unsaturated Dicarboxylic Acid Half Esters

Formula	Synth. Method*	Yield	Properties	R_3SnO_2CR' Ref.
Et_3SnO_2CCH:CHCO$_2$H	IVB	93	cis-form, m. 126.5-129°	1779
Pr_3SnO_2CCH:CHCO$_2$R	IVA	--	R = CH_2:CHCO$_2C_3H_6$ (a, b)	2974

Bu$_3$SnO$_2$CR'

R' = MeO$_2$CCH:CH	IIB	--	m. 68-70°, IR spectrum, mol. wt.	1357
	IVA	--	cis-form, soly. (c)	1357
RO$_2$CCH:CH	--	--	R = C$_4$H$_7$OCH$_2$, use as stabilizer for PVC	1283
RO$_2$CCH:CH	IVA	--	R = CH$_2$:CMeCO$_2$CH$_2$CH$_2$ (a, b)	2974
RO$_2$CCH:CH	--	--	R = HOC$_3$H$_6$, use as rodent repellent	2036
RO$_2$CCH:CH	IVA	--	R = CH$_2$:CHCO$_2$C$_3$H$_6$, straw colored liq. (b, d)	2974
c.4,5,2-Me$_2$(MeO$_2$C)C$_6$H$_8$	IIB	--	m. 29-31°	1558
	VIC (a)	--	Cyclohexene dicarboxylate	1558
3,5-C$_7$H$_8$(CO$_2$Me)-2 (e)	VIC (a)	82	m. 75-77°	1558
(C$_6$H$_{13}$)$_3$SnO$_2$CCH:CHCO$_2$R	--	--	R = C$_3$H$_6$OH, use as stabilizer for PVC	2897
(C$_8$H$_{17}$)$_3$SnO$_2$CCH:CHCO$_2$R	IVA	--	R = CH$_2$:CMeCO$_2$CH$_2$CH$_2$, cis-form (a, b)	2974
(C$_8$H$_{17}$)$_3$SnO$_2$CCH:CHCO$_2$R	--	--	R = 1-C$_8$H$_{17}$, cis-form, use as stabilizer for PVC	2897

Ph$_3$SnO$_2$CR'

R' = cis-OHCCCl:CCl	IVA	52	m. 154-155° (f)	866
MeO$_2$CCH:CH	IIA	73	m. 113-114°, IR spectrum, soly.	1557
	IVA	92	cis-form (c)	1557
RO$_2$CCH:CH	IVA	--	R = C$_3$H$_6$OH$_4$ (a, b)	2974
RO$_2$CCH:CH	--	--	R = C$_{11}$H$_{23}$CO$_2$C$_6$H$_8$O(OH)$_2$ (g)	--
c.4,5,2-Me$_2$(MeO$_2$C)C$_6$H$_8$	IIB	--	m. 82-84°	1558
	VIC (a)	--	Cyclohexene dicarboxylate	1558
3,5-C$_7$H$_8$(CO$_2$Me)-2 (e)	VIC (a)	--	m. 115-117°, mol. wt., soly.	1558

* The characters used here correspond to the synthetic scheme in the introduction to Chapter 4.3.5.1. (a) Use as polymerization intermediate (2974). (b) Use as microbiocide (2974). (c) Rxn. of R$_3$SnO$_2$CCH:CHCO$_2$Me with diene → Diels Alder adducts; R = Bu, Ph, diene = C$_5$H$_6$, (CH$_2$:CMe)$_2$ (1558). (d) Polymerizes with Bz$_2$O$_2$, copolymers formed with PhCH:CH$_2$ and CH$_2$:CMeCO$_2$Me (2974). (e) 2,3-Bicyclo[2.2.1]heptene-5-dicarboxylate. (f) Biol. activity as fungicide against Aspergillus niger, Penicillium italicum, Sclerotina sclerotiorum, Xanthomonas malvacearum (866). (g) Use as antistatic agent for thermoplastics (2018).

4.3.5.1.6 UNSUBSTITUTED TRIORGANOTIN CARBOXYLATES DERIVED FROM FUNCTIONALLY SUBSTITUTED ACIDS

4.3.5.1.6.1 UNSUBSTITUTED TRIORGANOTIN CARBOXYLATES DERIVED FROM HALOGEN SUBSTITUTED ACIDS

Trimethyltin Trifluoroacetate $SnC_5H_9F_3O_2$ $Me_3SnO_2CCF_3$
Synth.: By rxn. of CF_3CO_2H with $Me_3SnCMe:CHMe$ (295), with $Me_3SnCH_2CH:CHR$ at r.t. (1041, 1043).
By rxn. of Me_6Sn_2 with $(CF_3CO)_2O$ at 60°, as by-prod. (922).
By rxn. of Me_3SnBr with CF_3CO_2Na in THF under reflux, in 61% yield (2756).
By rxn. of Me_3SnOH with CF_3CO_2H, in 85% yield (2243).
Prop.: m. 129° (2243), 122-24° (295), 84-85° (sealed tube) (2756), subl. (2243), (far) IR, assignment and NMR spectra (2243), mol. wt., struct. (2243).

Tributyltin Trichloroacetate (M-422) $SnC_{14}H_{27}Cl_3O_2$ $Bu_3SnO_2CCCl_3$
Synth.: By rxn. of Bu_3SnOH with Cl_3CCO_2H, in 95% yield (2074).
By rxn. of $(Bu_3Sn)_2O$ with Cl_3CCO_2H at 100-110°, in 92% yield (2966).
Prop.: m. 79.8-80° (2966), IR spectrum, struct. (2074).
Rxn. with: $PPh_3 \rightarrow Bu_3SnCl + Ph_3PO$ (2656).
$PPh_3 + BzH \rightarrow Bu_3SnCl + Ph_3PO + PhCH:CCl_2$ (2656).
Biol. Prop.: Activity as bactericide (2466).
Activity as fungicide (2966), against Hypochnus sasakii (858), controlling pea root rot (202).
Use: Trade name or synonym: Stauffer 3446.

Listing of triorganotin carboxylates of halogen substituted carboxylic acids concludes in Table 201.

Additional triorganotin carboxylates of substituted halogen containing acids may be found in Subsections 4.3.5.1.3, 4.3.5.1.6.7 and 4.3.5.1.6.8.4.

4.3.5.1.6.3 UNSUBSTITUTED TRIORGANOTIN CARBOXYLATES DERIVED FROM OXYGEN AND SULFUR CONTAINING ACIDS

The compounds are listed in Table 202. Additional triorganotin carboxylates of substituted oxygen and sulfur containing acids are listed in Subsection 4.3.5.1.3 and in Tables 200, 205 and 206. References to such carboxylates may be found in Chapter 8.1.

Table 201. Unsubstituted Triorganotin Carboxylates Derived from Halogen Substituted Acids

Formula*	Synth. Method**	Yield	Properties	Ref.
Me_3SnO_2CR'				
$R' = CH_2F$	IVA	60	m. 198°, subl., struct. (a)	2243
CHF_2	IVA	70	m. 150°, subl., struct. (a)	2243
CH_2Cl (421)	IVA	--	m. 143-145°, subl. high vacuum, mol. wt., struct.	1933
	IVA	--	Mol. wt., struct. (a-d)	2243
$CHCl_2$ (421)	IVA	--	m. 138°, subl. high vacuum, mol. wt., struct. (a, b)	1933
	IVA	--		2243
CCl_3 (421)	IVA	--	m. 175-176°, subl. high vacuum, mol. wt., struct.	1933
	IVA	--	m. 174-175° (2756) (a, b, e)	2243
C_6Cl_5	IB	52	m. >270°, (far) IR spectra, assignment, mol. wt.	956
o-BrC_6H_4	IIA	48	m. 209-210°, subl. 120° x 10^{-3} mm, IR spectrum (f)	1880
o-$Et_3SnO_2CC_6H_4Br$	IIA	70	m. 101-102.5°, subl. 90° x 10^{-3} mm, IR spectrum	1880
Bu_3SnO_2CR'				
$R' = CH_2F$	IVA	89	IR spectrum, struct. (g)	2074
CHF_2	IVA	94	IR spectrum, struct.	2074
CF_3	IVA	96	IR spectrum, struct.	2074
CH_2Cl (422)	IVA	91	IR spectrum, struct. (g-i)	2074

* Numbers in parenthesis refer to pages in main volume.
** The characters used here correspond to the synthetic scheme in the introduction to Chapter 4.3.5.1.
(a) (Far) IR, assignment and NMR spectra (2243). (b) (Far) IR, assignment and NMR spectra (1933). (c) Rxn. with (:CHCO)$_2$O-PhCH:CH$_2$-polymer → tin contg. polyester (1328). (d) Biol. activity as insecticide (1328). (e) Rxn. with c.C$_8$H$_{14}$ → R$_3$SnX + 9,9-X$_2$-bicyclo[6.1.0]nonane + CO$_2$ + C$_2$Cl$_4$; R, X = Me, Cl; Ph, Cl; Ph, Br (1875). (f) Thermal decomposition at 230° → Me$_4$Sn + (Me$_2$SnO$_2$CC$_6$H$_4$Br-o)$_2$O + o-BrC$_6$H$_4$CO$_2$H (1880). (g) Biol. activity as fungicide against Hypochnus sasakii (858). (h) Rxn. with (EtO)$_2$Ph$_2$Na → Bu$_3$SnO$_2$CCH$_2$S$_2$P(OEt)$_2$, with (RO)$_2$P(S)ONH$_4$ → Bu$_3$SnO$_2$-CCH$_2$SPO$_3$Et$_2$ (744). (i) Use as rodent repellent (2036).

Table 201 Continued R_3SnO_2CR'

Formula*	Synth. Method**	Yield	Properties	Ref.
Bu_3SnO_2CR' (Cont'd.)				
R' = CHCl$_2$ (422)	IVA	92	IR spectrum, struct.	2074
3-C$_9$Cl$_9$	--	--	R = Cl$_9$-pentacyclo[5.2.0^{2,5}.0^{3,8}.0^{4,8}]nonane (j)	2391
(PhCMe$_2$CH$_2$)$_3$SnO$_2$CCF$_3$	IVA	--	m. 108.5-109.5°, IR spectrum, soly.	1742
Ph_3SnO_2CR'				
R' = CF$_3$	--	--	m. 119-121°	2756
CF$_2$Cl	--	--	m. 136-137.5°	2756
CH$_2$Cl (422)	IVA	100	m. 150-151°, activity as bactericide (529)	791
	IIB	99	m. 160-161° (m. 158-159° 2756)	1051
CHCl$_2$	--	--	m. 176-178°	2756
CCl$_3$ (422)	--	--	m. 88-89° (e, k)	2756
CBr$_3$	--	--	m. 135-137° (e)	2756
MeCHCl	IVA	81	m. 139-140° (l)	866
C$_6$Cl$_5$	IVD	--	m. 252°, (far) IR spectra, assignment, struct.	956

(j) n_D^{25} 1.5340, biol. activity as pre- and post-emergence herbicide (2391). (k) Rxn. with PhMeC:CH$_2$ at 145° → PhMeCH$_2$CCl$_2$ + C$_6$H$_6$ (2756). (l) Biol. activity as fungicide against Aspergillus niger, Penicillium italicum, Sclerotina sclerotiorum, Xanthomonas malvacearum (866).

Table 202. Unsubstituted Triorganotin Carboxylates Derived from Oxygen and Sulfur Containing Acids R_3SnO_2CR'

Formula*	Synth. Method**	Yield	Properties	Ref.
$Me_3SnO_2CCH_2OAc$	IVB (a)	78	m. 62-64°, IR and NMR spectra	2310
$Me_3SnO_2CCH_2CH_2OMe$	VI	--	White cryst., m. 84°, IR and NMR spectra	1235
	IVC	--	Synth. with propiolactone	1235
$(PhCH_2)_3SnO_2CCH_2OAc$	IVB (a)	86	m. 53-55°, IR and NMR spectra	2310
Pr_3SnO_2CR'				
$R' = EtOCS_2CH_2$	IVA	--	m. 53-54°	1308
$PrOCS_2CH_2$	IVA	--	n_D^{20} 1.5092, d_{20} 1.2258	1308
$BuOCS_2CH_2$	IVA	--	n_D^{20} 1.5114, d_{20} 1.2175	1308
$(HO_2C)_3C_6H_2$	IVA	--	(b)	2619
Bu_3SnO_2CR'				
$R' = CH_2OH$ (423)	--	--	Use as rodent repellent (c)	2036
$MeOCH_2$	--	--	(d)	--
$EtOCH_2$	--	--	(d)	--
$C_8H_{17}SCH_2$	--	--	Use in antifouling coatings	2302
$1-C_8H_{17}SCH_2$	--	--	Use as rodent repellent	2036
$EtOCS_2CH_2$	IVA	--	m. 54-56°, soly.	1308
$PrOCS_2CH_2$	IVA	--	n_D^{20} 1.5068, d_{20} 1.1883	1308
$BuOCS_2CH_2$	IVA	--	m. 34-36°	1308
$C_{12}H_{25}S(MeS)CH$	IVA	--	Use as stabilizer for PVC	213, 1170

* Numbers in parenthesis refer to pages in main volume.
** The characters used here correspond to the synthetic scheme in the introduction to Chapter 4.3.5.1.
(a) Synth. with diketene. (b) Use as bactericide and crosslinking agent (2619). (c) Rxn. with $(:CHCO)_2O$, + PhHgOH → solid microbiocidal composition (2535). (d) Biol. activity as fungicide against Hypochnus sasakii (858).

Table 202 Continued R_3SnO_2CR'

Formula*	Synth. Method**	Yield	Properties	Ref.
Bu_3SnO_2CR' (Cont'd)				
R' = $Bu_3SnSCH_2CH_2$	--	--	Use as rodent repellent	2036
$(C_{12}H_{25}S)_2CMeCH_2CH_2$	--	--	Use as stabilizer for PVC	213, 1170
o-HOC$_6$H$_4$ (423)	--	--	Liq., d_{20} 1.210 (d, e)	2966
p-HOC$_6$H$_4$	IVA	--	m. 71-73° (f)	1795
o-MeO$_2$CC$_6$H$_4$ (423)	IVC	85	--	36
o-Me$_3$CO$_3$CC$_6$H$_4$	IVC	88	From $Bu_3SnO_2CMe_3$, m. 40-41°, IR spectrum	2257
	IVA		From o-HO$_2$CC$_6$H$_4$CO$_3$CMe$_3$	2257
$(C_8H_{17})_3Sn[H_3(O_2CCH_2S)_4CHCMe]$	IVA	--	Use as stabilizer for PVC	1170
Ph_3SnO_2CR'				
R' = 2-C$_{10}$H$_7$CH$_2$	IVA	58	m. 148.5-150° (g)	866
BuOCS$_2$CH$_2$	IVA	--	m. 134-135°	1308
AcCH$_2$	IVA (a)	80	m. 104-106°, IR and NMR spectra	2310
AcCH$_2$CH$_2$	--	--	IR and Moessbauer resonance spectra, struct.	2394
RO$_2$C(CH$_2$)$_4$	--	--	R = $C_{18}H_{37}$(OCH$_2$CH$_2$)$_{45}$ (h)	--
o-HO$_2$CC$_6$H$_4$	--	--	(i)	--
o-Me$_3$CO$_3$CC$_6$H$_4$	IVC	--	From $Ph_3SnO_2CMe_3$, m. 18-22°, IR spectrum	2257

(e) Biol. activity as bactericide against Staphylococcus aureus (1145, 2966), as bacteriostatic agent for textiles (2966), as fungicide (1145). (f) Biol. activity as bactericide and insecticide (1795). (g) Biol. activity as fungicide against Aspergillus niger, Penicillium italicum, Sclerotina sclerotiorum, Xanthomonas malvacearum (866). (h) Use as antistatic agent (2817). (i) Biol. activity as insecticide and chemosterilant on Musca domestica (805).

4.3.5.1.6.5 UNSUBSTITUTED TRIORGANOTIN CARBOXYLATES DERIVED FROM ACIDS CONTAINING NITROGEN

Tributyltin α-Aminobutyrate $SnC_{16}H_{35}NO_2$ DL-$Bu_3SnO_2CCHEtNH_2$
Synth.: By rxn. of $(Bu_3Sn)_2O$ with $H_2NCHEtCO_2H$ in MePh, in 90% yield (151).
Prop.: m. 110°, $b_{0.07}$ 138-140°, IR spectrum (151). Anal. detn. by alkalimetric titration against thymol blue (151).
Rxn. with: $H_2O \rightarrow H_2NCHEtCO_2H$ (151).
BzOH → $H_2NCHEtCO_2H$ (151).
$PhCH_2O_2CNHCH_2CO_2R \rightarrow PhCH_2O_2CNHCH_2CONHCHEtCO_2H$; R = i-$BuO_2C$, p-$O_2NC_6H_4$ (151).
Cresol → $H_2NCHEtCO_2H$ (151).
$PhCH_2NH_2$ → no rxn. under anhydrous conditions (151).

Additional unsubstituted triorganotin carboxylates derived from nitrogen containing acids may be found in Table 203. Additional functional derivatives are listed in Tables 198 and 205. O-N-Bistriorganotin derivatives of carbamic acids are reported in Subchapter 4.5.3.2.

4.3.5.1.6.7 UNSUBSTITUTED TRIORGANOTIN CARBOXYLATES DERIVED FROM METAL SUBSTITUTED ACIDS

Triorganotin carboxylates carrying metallic substituents in the carboxylic acid moiety are listed in this subchapter. The metallic substituents reported here comprise mercury, silicon and iron. For triorganotin substituted acids the reader should check Subchapter 2.6.3.4.

Triphenyltin Ferrocenecarboxylate $SnFeC_{29}H_{24}O_2$ $Ph_3SnO_2CC_5H_4FeC_5H_5$
Synth.: By rxn. of Ph_3SnH with $C_5H_5FeC_5H_4CO_2H$ in heptane under reflux in N atm., in 28% yield (1310, 1374).
By rxn. of Ph_3SnCl with $C_5H_5FeC_5H_4CO_2Na$ in EtOH, in 50% yield (1310, 1374).
By rxn. of Ph_3SnBr with $C_5H_5FeC_5H_4CO_2H$ and $PhCH_2NH_2$ in $HCCl_3$, in 62% yield (1310, 1374).
Prop.: m. 121-124°, IR spectrum, mol. wt. (1310, 1374).
Rxn. with: $C_5H_5FeC_5H_4CO_2H \rightarrow Ph_2Sn(O_2CC_5H_4FeC_5H_5)_2$ (1310, 1374).
$SiO_2 \cdot$ aq. → $Ph_2Sn(O_2CC_5H_4FeC_5H_5)_2$ (1310, 1374).
Wet alumina → $(Ph_3Sn)_2O + C_5H_5FeC_5H_4CO_2H$ (1310, 1374).

Listing of unsubstituted triorganotin carboxylates derived from metal substituted carboxylic acids concludes in Table 204.

Table 203. Unsubstituted Triorganotin Carboxylates Derived from Acids Containing Nitrogen

Formula*	Synth. Method**	Yield	Properties	Ref.
Me₃SnO₂CNMe₂ (424)	VI	98	White, m. 165°, IR spectrum, assignment	167
	VI	98	m. 156° (dimeric 167) (a)	1392
Et₃SnO₂CR'				
R' = AcNHCH₂	IVA	89	m. 110°, soly., stable in alc., H₂O	151
BzNHCH₂	IVA	92	m. 60°, soly., stable in alc., H₂O	151
PhCH₂O₂CNHCH₂	IVA	82	m. 60°, soly., stable in alc., H₂O (b, c)	151
H₂NCH₂CH₂ (424)	IVA	--	m. 151.2°, monomeric, dec. with H₂O (d)	151
DL-BzNHCH(CH₂Ph)	IVA	97	m. 119°, soly., stable in alc., H₂O (c)	151
DL-AcNHCH(Pr-i)	IVA	86	m. 157°, soly., stable in alc., H₂O (c, e)	151
Bu₃SnO₂CR'				
R' = NHMe	VI (f)	--	m. 64.5-67°, IR and NMR spectra (g)	2256
NHEt	VI (f)	--	m. 72-74°, IR and NMR spectra (h)	35,748
NHPh	VI (f)	--	m. 41-43°, IR and NMR spectra (i)	35,748
NEt₂	--	--	(j)	--
AcNHCH₂ (424)	IVA	96	m. 122°, IR spectrum, soly., stable in H₂O (b, d, k)	151
BzNHCH₂	IVA	90	m. 52°, soly., stable in alc., H₂O (b, d)	151
PhCH₂O₂CNHCH₂	IVA	93	m. 78°, IR spectrum, soly., stable in alc., H₂O (d)	151
H₂NCH₂CH₂	IVA	--	Rxn. with H₂O → H₂NCH₂CH₂CO₂H (l)	151
H₂NCHMe	IVA	75	m. 130-132°, b₀.₀₇ 133-136° (d, m)	151
DL-BzNHCH(CH₂Ph)	IVA	93	m. 35°, IR spectrum, soly.	151
	IIB	53	Stable in H₂O, alc. (b, e)	151
C₁₂H₂₅NHCHMeCH₂	IVA	--	Viscous liq. (n)	2959
H₂NCHEt	IVA	--	m. 109-112°, b₀.₀₅ 135-138°	151
DL-H₂NCH(CHMe₂)	IVA	89	m. 60°, b₀.₀₇ 143-146°, IR spectrum (b, o)	151
DL-AcNHCH(CHMe₂)	IVA	100	m. 121°, IR spectrum, soly.	151
	(o)	93	Stable in H₂O, alc. (b-d)	151

DL-H$_2$NCHBu-i	IVA	72	m. 93-95°, bo.o7 148-150°, IR spectrum (d, p)	151
DL-AcNHCHBu-i	IVA	88	m. 35°, IR spectrum, soly., stable in H$_2$O, alc. (c)	151
o-H$_2$NC$_6$H$_4$ (424)	--	--	(q)	--
m-AcNHC$_6$H$_4$	IVA	--	(r)	281
m-BzNHCH$_2$C$_6$H$_4$NHC$_6$H$_4$	--	--	(q)	--
p-H$_2$NC$_6$H$_4$ (424)	IVA	--	m. 160° (s)	2966
2,x-(O$_2$N)$_2$C$_6$H$_3$	--	--	(q)	--
m-Ph$_3$SnO$_2$CC$_6$H$_4$NHAc	--	--	(t)	--
p-Ph$_3$SnO$_2$CC$_6$H$_4$NHAc (424)	--	--	(u)	--

* Numbers in parenthesis refer to pages in main volume.

** The characters used here correspond to the synthetic scheme in the introduction to Chapter 4.3.5.1.

(a) Rxn. with CS$_2$ → Me$_3$SnS$_2$CNMe$_2$ (167), with PhNCO → Me$_3$SnPhCONMe$_2$ (167). (b) Anal. detn. by titration with MeONa in alc. C$_6$H$_6$ (151). (c) Anal. detn. by titration with anhyd. HClO$_4$ in dioxane (151). (d) Anal. detn. by alkalimetry against thymol blue (151). (e) Anal. detn. by titration with aq. HCl (115). (f) Synth. with Bu$_3$SnNRCO$_2$SnBu$_3$ in presence of MeNH$_2$. (g) Rxn. with HCl in Et$_2$O → CO$_2$ + MeNH$_2$Cl (2256). (h) Rxn. with air → (Bu$_3$Sn)$_2$O (748), with PhNCS, + HCl → Bu$_3$SnCl + H$_2$S + EtNHCSNHPh (2258), thermal decomposition → Bu$_3$SnNEtCO$_2$SnBu$_3$ + CO$_2$ + EtNH$_2$ (748). (i) Thermal decomposition → (Bu$_3$SnNPh)$_2$CO + (Bu$_3$Sn)$_2$O + PhNH$_2$ (748), with H$_2$O → PhNH$_2$ + CO$_2$ (748). (j) Rxn. with EtNCO → Bu$_3$SnNEtCONEt$_2$ (2258), with PhNCS → Bu$_3$SnPhCSNEt$_2$ + CO$_2$ (2258). (k) Rxn. with AcOH → Bu$_3$SnOAc (151), with PhCH$_2$NH$_2$ → AcNHCH$_2$CO$_2$H·H$_2$NCH$_2$Ph (151). (l) Rxn. with RCO$_2$CO$_2$Bu-i → RCONHCH$_2$CH$_2$CO$_2$H; R = PhCH$_2$O$_2$CNHCH$_2$ (151). (m) Rxn. with RCO$_2$CO$_2$Bu-i → RCONHCHMeCO$_2$H; R = PhCH$_2$O$_2$CNHCH$_2$ (151). (n) Rxn. with 1,2,4-MeC$_6$H$_4$(NCO)$_2$ → 1,3,4-[Bu$_3$SnO$_2$CCH$_2$CHMeN(C$_{12}$H$_{25}$)CONH]$_2$C$_6$H$_8$Me (2959). (o) Rxn. with Ac$_2$O and Et$_3$N → DL-Bu$_3$Sn-O$_2$CCH(CHMe$_2$)NHAc (151). (p) Rxn. of H$_2$O with L-ester → L-leucine (151). (q) Biol. activity as fungicide against Hypochnus sasakii (858). (r) Biol. activity as fungicide against Fusarium culmorum, Phytophthora (281). (s) Biol. activity as bactericide, fungicide (2966). (t) Biol. activity as fungicide against Phytophthora and Tropaeoleum majus (281). (u) Biol. activity on reproduction of Musca domestica (172a, 260).

Table 204. Unsubstituted Triorganotin Carboxylates Derived from Metal Substituted Carboxylic Acids R_3SnO_2CR'

Formula*	Synth. Method**	Yield	Properties	Ref.
$(p-Me_3SnO_2CCH_2I \cdot Me_2NC_6H_4)_2Hg$ (425)	--	--	(a)	--
$Et_3SnO_2CCH_2CH_2SiMe_3$	IA	42	m. 89-90°	586
	I	47	Synth. from Et_6Sn_2	586
$Pr_3SnO_2CCH_2I \cdot (p-Me_2NC_6H_4)_2Hg$ (425)	--	--	(a)	--
$Pr_3SnO_2CCH_2I \cdot (3-C_5H_4N)_2Hg$ (425)	--	--	(a)	--
$Bu_3SnO_2CCH_2I \cdot (p-Me_2NC_6H_4)_2Hg$ (425)	--	--	(a)	--
$Bu_3SnO_2CCH_2I \cdot (3-C_5H_4N)_2Hg$ (425)	--	--	(a)	--
$Bu_3SnO_2CCH_2I \cdot (3-C_5H_4N)HgPh$ (425)	--	--	(a)	--
$Bu_3SnO_2CCH_2CH_2I \cdot (3-C_5H_4N)_2Hg$ (425)	--	--	(a)	--

* Numbers in parenthesis refer to pages in main volume.
** The characters used here correspond to the synthetic scheme in the introduction to Chapter 4.3.5.1.
(a) Biol. activity as bactericide against Escherichia coli, Pseudomonas aeruginosa, Staphylococcus aureus (1521).

4.3.5.1.6.8 UNSUBSTITUTED TRIORGANOTIN CARBOXYLATES DERIVED FROM POLY-FUNCTIONAL ACIDS

4.3.5.1.6.8.1 BIS(TRIORGANOTIN) DERIVATIVES OF DICARBOXYLIC ACIDS

All compounds are listed in Table 205. Additional references to bisorganotin dicarboxylates are given in Chapter 8.1.

4.3.5.1.6.8.3 POLY(TRIORGANOTIN) DERIVATIVES OF POLYCARBOXYLIC ACIDS

All compounds are listed at the bottom of Table 205.

4.3.5.1.6.8.4 UNSUBSTITUTED TRIORGANOTIN DERIVATIVES OF POLYFUNCTIONAL CARBOXYLIC ACIDS

Triorganotin carboxylates derived from polyfunctional carboxylic acids are compiled in this subsection. Functional groups comprise any combination of nitrogen, phosphorus, oxygen and halogen. The compounds listed in Table 206 are prepared by methods from the following scheme.

I Reactions of triorganotin oxides and hydroxides with:
   (IA) RCO_2Me,
   (IB) RCO_2H.

II Reaction of tributyltin chloroacetate with:
   (IIA) $(RO)_2PS_2Na$,
   (IIB) $(RO)_2P(S)ONH_4$.

III Reaction of triphenyltin diphenyl phosphine with carbonoxysulfide.

Additional references to triorganotin carboxylates of polyfunctional acids may be found in Chapter 9.

4.3.5.1.9 UNSYMMETRIC AND FUNCTIONALLY SUBSTITUTED TRIORGANOTIN CARBOXYLATES

Tris(2-cyanoethyl)tin Acetate $SnC_{11}H_{15}N_3O_2$ $(NCCH_2CH_2)_3SnOAc$
Synth.: By rxn. of $(NCCH_2CH_2)_4Sn$ with SnX_4 at 104-147°, followed by AgOAc; X = Cl, Br, in 31-57% yield (1747).
By rxn. of $(NCCH_2CH_2)_3SnBr$ with TlOAc in aq. Me_2CO, in quant. yield (1746), with AgOAc, in 90% yield (895b).
By rxn. of $(NCCH_2CH_2)_3SnOH \cdot H_2O$ with AcOH in THF under reflux, in 90% yield (895b).
Prop.: White cryst., m. 150-151° (895b, 1746), m. 146-148° (895b).
Use: As biocide (895b), as stabilizer for plastics (895b).

Listing of unsymmetric and functionally substituted triorganotin carboxylates concludes in Table 207.

Table 205. Polytriorganotin Derivatives of Polycarboxylic Acids $(R_3SnO_2C)_2Y$

Formula*	Synth. Method**	Yield	Properties	Ref.
cis-$(Me_3SnO_2CCH:)_2$	IVB	82	m. 202.5-204°, soly.	1779
$(Et_3SnO_2C)_2CH_2$	IVC	--	Synth. from $Et_3SnOCHRCH_2CH:CH_2$	1331
	IVC	--	Synth. from $Et_3SnOCH(CCl_3)C:CPh$	2644
cis-$(Et_3SnO_2CCH:)_2$	IVB	80	White needles, m. 129-132°	1779, 2191
$(Pr_3SnO_2C)_2Y$				
Y = cis-CH:CH	IVB	70	m. 148-151.5°	1779
2,3-C_7H_8-5 (b)	IVB	--	m. 61-62°, soly. (c - 1890)	1890
$C_6H_3CO_2H$	IVA	--	m. 85-86.5° (d)	2619
$C_6H_2(CO_2H)_2$	IVA	--	m. 182-183° (d)	2619
$C_6H_2(CO)_2O-1,2,4,5$	IVA	--	m. 186-187° (d)	2619
2,3-$(1-Pr_3SnO_2C)_2C_7H_8$-5 (b)	IVB	--	Pale yellow, m. 100-102°, soly. (c - 1890)	1890
$(Bu_3SnO_2C)_2Y$				
Y = --- (427)	IVA	--	Use as antifouling agent	1765
	IVA	88	m. 151-156° (c - 2966)	2966
CH_2 (427)	IVA	--	Use as antifouling agent	1765
	IVA	87	m. 87-87.5° (c - 2966)	2966
CH_2CH_2 (427)	IIB	89	m. 91-93° (e - 1793, f, g)	1051
	IVA	--	m. 96°, soly.	1793
	IVA	93	m. 98.5° (c - 2966)	2966
cis-CH:CH (427)	IVA	79	m. 44-46°, IR spectrum, soly.	1557
	IVA	75	m. 249-252.5° (e - 1793, h)	1779
trans-CH:CH (427)	IIA	98	m. 128°	1051
	IVA	79	m. 128-130°, IR spectrum, soly.	1557
	IVA	--	m. 122-123°, soly. (e - 1793, i)	1793
$(CH_2)_3$	IVA	94	m. 86-87° (c - 2966)	2966

CH$_2$CH$_2$CH(NH$_2$)	IVA	60	m. 80-82°, DL-form	151
(CH$_2$)$_4$ (427)	IIB	93	m. 104°	1051
2,3-C$_7$H$_8$-5 (b)	IVA	--	m. 101-102°, soly. (j - 1765)	1765
(CH$_2$)$_8$ (428)	IVB	--	m. 45-46°, soly. (c - 1890)	1890
CH$_2$CH(C$_6$H$_{11}$O)	IIB	84	m. 100-101°	1051
CH$_2$CH(C$_8$H$_{15}$O)	IVB	100	Epoxide (k)	1305
o-C$_6$H$_4$ (428)	IVB	100	Epoxide, n$_D^{20}$ 1.4934, d$_{29}$ 1.111 (k)	1305
	IVA	--	m. 35-36° (e - 1795, g)	1795
	IVA	--	Liquid (c - 2966)	2966
m-C$_6$H$_4$	IVA	--	m. 46-47° (e - 1795)	1795
p-C$_6$H$_4$ (428)	IIB	94	m. 78°	1051
	IVA	--	m. 80-82° (j - 572)	572
	IVA	--	m. 75-76° (e - 1795, i, l)	1795
C$_6$H$_3$(CO$_2$H)-1,2,4	IVA	--	m. 84.5-90° (d)	2619
C$_6$H$_2$(CO$_2$H)$_2$	IVA	--	m. 180.5° (d)	2619
C$_6$H$_2$(CO)$_2$O-1,2,4,5	IVA	--	m. 170-172° (d)	2619
o-C$_6$H$_4$SC$_6$H$_4$	IVA	--	Use as stabilizer for PVC (m)	213, 1170

(CH$_2$CH$_2$S)$_2$R″

R″ = CH$_2$	IVA	--	Use as stabilizer for PVC	213
CHPh (428)	IVA	--	m. 95-96°, use as stabilizer for PVC (m-o)	213, 1170
CMe$_2$ (428)	IVA	--	m. 98-99°, use as stabilizer for PVC (m, n)	213, 1170
C$_3$H$_6$	--	--	(o)	--

* Numbers in parenthesis refer to pages in main volume.

** The characters used here correspond to the synthetic scheme in the Introduction to Chapter 4.3.5.1.

(a) Rxn. with PhCH:CH$_2$ + cat. Bz$_2$O$_2$ → tin contg. polymer (1779, 2191). (b) Derivative of bicyclo[2.2.1]heptene-5. (c) Biol. activity as bactericide and fungicide (1890, 2966). (d) Use as bactericide, as crosslinking agent (2619). (e) Biol. activity as bactericide and insecticide, resistant to direct sunlight (1793, 1795). (f) Biol. activity as bactericide against Staphylococcus aureus (2966). (g) Use as bactericide for textiles (2966). (h) Use for antifouling coatings (2302), as foaming catalyst for polyurethan (450). (i) Biol. activity as fungicide against Coriolellus palustris, resistant to light (1063). (j) Use as antifouling agent (572, 1765). (k) Use as stabilizer for PVC (1305). (l) Anal. detn. by x-ray spectroscopy (1229). (m) Use as stabilizer for halogen contg. resins (213, 1170). (n) Use as stabilizer for polypropylene (212, 1170). (o) Use as rodent repellent (2036)

Table 205 Continued $(R_3SnO_2C)_2Y$

Formula*	Synth. Method**	Yield	Properties	Ref.
$(Bu_3SnO_2C)_2Y$ (Cont'd.)				
Y = $MeC_6H_3(NHCONRCHMeCH_2)_2$	IV (p)	--	R = $C_{12}H_{25}$, (q)	2959
$(Ph_3SnO_2C)_2Y$				
Y = --- (428)	--	--	(r)	--
CH_2CH_2 (428)	IIB	88	m. 154-155° (s)	1051
cis-CH:CH	IVA	69	m. 118-120°, IR spectrum, soly.	1557
trans-CH:CH	IVA	92	m. 203-205° (t)	866
	IVA	75	m. 208-210°, IR spectrum, soly.	1557
$(CH_2)_4$	IIB	79	m. 166-167° (s)	1051
$CH_2CH(C_6H_{11}O)$	IVB	--	Epoxide, tan, m. 51-52° (u)	1305
$CH_2CH(C_{12}H_{25})$	--	--	(v)	--
$o-C_6H_4$ (428)	--	--	(s, ẅ)	--
$[(R_3SnO_2CCH_2)_2NCH_2]_2$				
R = CH_2:CH	IVA	--	m. 295° (dec.) (x)	1387
CH_2:CHCH$_2$	IVA	--	m. 205° (dec.) (x)	1387
Bu	IVA	60	m. 117-119° (c - 2966)	2966
$c.C_6H_{11}$	IVA	--	m. 188-195° (x)	1387
Ph	IVA	99	m. 133° (x)	1387
$(Pr_3SnO_2C)_3C_6H_2CO_2H$	IVA	--	m. 166-166.5° (d)	2619
$(Bu_3SnO_2C)_3C_6H_3$	IVA	--	m. 101-105° (d)	2619
$(Bu_3SnO_2C)_3C_6H_3-1,3,5$	IVA	--	n_D^{20} 1.5280 (d)	2619
$(Bu_3SnO_2C)_4C_6H_2-1,2,4,5$	IVA	--	m. 93-95° (j - 572)	572
$(Bu_3SnO_2C)_3C_6(CO_2H)_3$	IVA	--	m. 181-181.5° (d)	2619
$[(Bu_3SnO_2CCH_2CH_2S)_2CH]_2$	IVA	--	Use as stabilizer for PVC	213, 1170

$[Bu_3SnO_2CCHS\text{-}CMe_2 \atop Bu_3SnO_2CCH_2]_2$	IVA --	Use as stabilizer for PVC (m) 213, 1170
$[(C_8H_{17})_3SnO_2CCH_2S]_2CHCMe$	IVA --	Use as stabilizer for PVC (m) 213

(p) Synth. from $Bu_3SnO_2CCH_2CHMeNH_2$ with 1,2,4-$MeC_6H_3(NCO)_2$ (2959). (q) Biol. activity as fungicide against Aspergillus flavus, Candia albicans (2959), use as stabilizer for PVC (2959), for synth. of elastomers and adhesives (2959). (r) Use as fungicide in potato and sugar beet cultures, activity against Phytophthora infestans (2976). (s) Biol. activity as bactericide (529). (t) Biol. activity as fungicide against Aspergillus niger, Penicillium italicum, Sclerotina sclerotiorum, Xanthomonas malvacearum (866). (u) Use as stabilizer for PVC (1305). (v) Biol. activity in control and effect on reproduction of Musca domestica (260). (w) Biol. activity as fungicide against Alternaria tenuis, Fusarium culmorum, Rhizoctonia solani (933), as insecticide and chemosterilant for Musca domestica (805). (x) Use as biol. chelating agent (1387).

Table 206. Unsubstituted Triorganotin Derivatives of Polyfunctional Carboxylic Acids

Formula	Synth. Method*	Yield	Properties	Ref.
Bu_3SnO_2CR'				
R' = 2,4-$Cl_2C_6H_3OCH_2$	--	--	m. 75-77°, use as fungicide	1005
2,4,5-$Cl_3C_6H_2OCH_2$	--	--	m. 80-82°, use as fungicide (a)	1005
$C_6Cl_5OCH_2$	--	--	m. 68-70°, use as fungicide	1005
PhCH:$NOCH_2$	IB	--	m. 67-69°, soly. (b)	1574
p-ClC_6H_4CH:$NOCH_2$	IB	--	m. 72-74°, soly. (b)	1574
2,4-$Cl_2C_6H_3$CH:$NOCH_2$	IB	--	m. 62.5°, soly. (b)	1574
p-$MeOC_6H_4$CH:$NOCH_2$	IB	--	m. 69-71°, soly. (b)	1574
m-$O_2NC_6H_4$CH:$NOCH_2$	IB	--	m. 53-55°, soly. (b)	1574
3,5-$(Me_3C)_2C_6H_3$CH:$NOCH_2$	IB	--	m. 64-68°, soly. (b)	1574
Me_2C:$NOCH_2$	IA	--	m. 40-43° (c)	2626
MeEtC:$NOCH_2$	IA	--	m. 62-65° (c)	2626
c.C_6H_{10}:$NOCH_2$	IA	--	m. 53-57° (c)	2626
PrCH:CEtC:$NOCH_2$	IA	--	m. 64-67° (c)	2626
$C_6Cl_5SCH_2$	--	--	m. 85-86°, use as fungicide (a)	1005
$Et_2O_3PSCH_2$	IIB	87	n_D^{20} 1.4982, d_{20} 1.2388	744
$Pr_2O_3PSCH_2$	IIB	85	n_D^{20} 1.4868, d_{20} 1.1931	744
$Bu_2O_3PSCH_2$	IIB	86	n_D^{20} 1.4848, d_{20} 1.1741	744
$(MeO)_2PS_2CH_2$	--	--	Light resistance (d)	1063
$(EtO)_2PS_2CH_2$	IIA	100	n_D^{20} 1.5283, d_{20} 1.2767	744
2,4-$Cl_2C_6H_3$CH:$NOCH_2CH_2$	IB	--	(b)	1574
$MeSCH_2CH_2CH(NH_2)$	--	--	(e)	--
$MeSCH_2CH_2CH(NHAc)$	IA	91	m. 25°, IR spectrum, stable in alc., H_2O (f)	151
2,3,5-$HO(O_2N)_2C_6H_2$	--	--	(g)	--
$Ph_3SnOC(:S)PPh_2$	III	63	m. 97°, IR spectrum, mol. wt.	483

Ph₃SnO₂CR'

R' =				
o-ClC₆H₄OCH₂	IA	67	m. 172-73.5° (h)	866
p-ClC₆H₄OCH₂	IA	88	m. 146-46.5° (h)	866
2,4-Cl₂C₆H₃OCH₂	IA	63	m. 173.5-75.5° (h)	866
	--	--	m. 168-70°, use as fungicide	1005
2,4,5-Cl₃C₆H₂OCH₂	IA	90	m. 185-86.5° (h)	866
	--	--	m. 182-84°, use as fungicide (a)	1005
2,4-MeClC₆H₃OCH₂	--	--	m. 162-64°, use as fungicide	1005
Me₂NC(:S)OCH₂	IA	51	m. 135-35.5° (h)	866
PhCH:NOCH₂	IB	--	m. 105-107°, soly. (b)	1574
p-ClC₆H₄CH:NOCH₂	IB	--	m. 119-20°, soly. (b)	1574
Ph₂C:NOCH₂	IA	--	m. 150-52° (c)	2626
PhMeC:NOCH₂	IA	--	m. 135-36° (c)	2626
CH₂:CHCH:NOCH₂	IA	--	m. 113-14° (c)	2626
Me₂C:NOCH₂	IA	--	m. 123-25° (c)	2626
MeEtC:NOCH₂	IA	--	m. 118-19° (c)	2626
Me₂CHCH₂CMe:NOCH₂	IA	--	m. 102-103° (c)	2626
c.C₆H₁₀:NOCH₂	IA	--	m. 97-99° (c)	2626
PhCH:NOCH₂CH₅	IB	--	m. 78-80° (b)	1574
cis-OHCCCl:CCl	IVA	52	m. 154-55° (h)	866
5,2-Cl(HO)C₆H₃	IA	61	m. 140-41° (h)	866

* The characters used here correspond to the synthetic scheme in the introduction to Chapter 4.3.5.1.6.8.4. (a) Use as fungicide in PVC (2334). (b) Use as bactericide, fungicide and herbicide (1574), as stabilizer for polymers (1574). (c) Use as bactericide, fungicide and herbicide (2626). (d) Biol. activity as fungicide against Coriolellus palustris (1063). (e) Rxn. with ROH → MeSCH₂CH₂CH(NH₂)CO₂H; R = H, PhCH₂ (151), with RCO₂Et → RONHCH(CH₂CH₂SMe)CO₂H; R = PhCH₂O₂CNHCH₂ (151). (f) Anal. detn. by alkalimetric titration (151). (g) Biol. activity as fungicide against Hypochnus sasakii (858). (h) Biol. activity as fungicide against Aspergillus niger, Penicillium italicum, Sclerotina sclerotiorum, Xanthomonas malvacearum (866).

Table 207. Unsymmetric and Functionally Substituted Triorganotin Carboxylates RR'_2SnO_2CR

Formula*	Synth. Method**	Yield	Properties	Ref.
MeBu$_2$SnOAc	--	--	(a)	--
MePh$_2$SnOAc	IVA	84	m. 125°	521
Et$_2$(C$_8$H$_{17}$)SnOAc (430)	--	--	(b)	--
PrPh$_2$SnOAc	IVA	--	m. 106°	521
i-PrPh$_2$SnOAc	IVA	--	m. 83°	521
Bu$_2$(1-C$_7$H$_{15}$)SnOAc	IVA	--	m. 44-45°, b$_5$ 120-122°	617
Bu(CH$_2$CHMeCH)$_2$SnOAc	IB (c)	--	(d)	1873
Bu(Me$_3$CCH$_2$)$_2$SnOAc	IVA	46	m. 44.5-45.5°	2188
i-BuPh$_2$SnOAc	IVA	--	m. 77°	521
RPh$_2$SnOAc				
R = n-C$_5$H$_{11}$	IVA	92	m. 99°	2787
i-C$_5$H$_{11}$	IVA	95	m. 94°	2787
n-C$_7$H$_{15}$	IVA	89	m. 81°	2787
n-C$_8$H$_{18}$	IVA	84	m. 69°	2787
n-C$_{10}$H$_{21}$	IVA	79	m. 62°	2787
Et$_2$(p-BrC$_6$H$_4$)SnOAc (432)	IVA	29	--	149
[Bu$_2$(MeOCH$_2$)SnO$_2$CCH:]$_2$	--	--	Use as stabilizer for PVC	1381
[Bu$_2$HO(CH$_2$)$_3$SnOAc]$_x$	IV (e)	--	n$_D^{29}$ 1.4870, dec. on distn. (f)	1383
Bu$_2$AcO(CH$_2$)$_3$SnOAc	(f)	--	m. 41-42°, subl.o.008 120°, TLC	1383
[Ph$_2$(AcO)SnCH$_2$CH$_2$]$_2$	IIA	--	m. 154-155°, IR spectrum (g)	1518, 2189
[Ph$_2$(BzO)SnCH$_2$CH$_2$]$_2$	IIA	--	m. 110-111°, IR spectrum (g)	1518, 2189
(p-ClC$_6$H$_4$)$_3$SnOAc (432)	--	--	Phytotoxicity (h)	1202
[(HO$_2$CCH$_2$CH$_2$)$_2$SnCH$_2$CH$_2$CO$_2$]$_x$	VIIA	90	m. 180°, IR spectrum (i)	1744

* Numbers in parenthesis refer to pages in main volume.
** The characters used here correspond to the synthetic scheme in the introduction to Chapter 4.3.5.1.

4.3.5.2 DIORGANOTIN CARBOXYLATES

4.3.5.2.1 UNSUBSTITUTED DIORGANOTIN DICARBOXYLATES

The diorganotin dicarboxylates and carboxylates in the ensuing tables are prepared by methods from the following scheme.

I Substitution reactions with diorganotin dihalide or diorganohalotin oxide with:
    (IA)   silver carboxylate for R_2SnX_2,
    (IB)   sodium carboxylate for R_2SnX_2 and $(R_2SnX)_2O$,
    (IC)   carboxylic acid and ammonia for R_2SnX_2.

II Reaction of diorganotin hydrides with acyl-peroxides.

III Reactions of diorganotin oxide with:
    (IIIA) carboxylic acid,
    (IIIB) carboxylic acid anhydride,
    (IIIC) diorganotin dicarboxylate.

IV Reaction of diorganotin alkoxides:
    (IVA) $R_2Sn(OR)_2$ with carbon dioxide and carboxylic acid,
    (IVB) $(R_2SnOR)O(R_2SnO_2CR)$ with water.

V Reaction of diorganotin dicarboxylate with:
    (VA)   carboxylic acid,
    (VB)   water,
    (VC)   diene, for $R_2Sn(O_2CCH:CHCO_2Me)_2$.

Additional references to diorganotin dicarboxylates are given in Chapters 8.1 and 9.

(a) Biol. activity as bactericide (622), as fungicide (622), against Rhizoctonia solani (190), as insecticide against Culex pipiens larvae (73). (b) Biol. activity as insecticide against Heliothis virescens, H. zea (192, 2145). (c) Synth. from $Bu(CH_2CHMeCH)_2SnCH:CH_2$. (d) Use as herbicide and pesticide (1873). (e) Synth. from $Bu_2SnCH_2CH_2CH_2O$ with AcOH. (f) Rxn. of $Bu_2HO(CH_2)_3SnOAc$ with Ac_2O-NaOAc → $Bu_2AcO(CH_2)_3SnOAc$ (1383). (g) Dipole moment (1518). (h) Biol. activity as molluscicide against Biomphalaria glabrata (1202). (i) Resolidifies after melting at 185°, differential thermal analysis, thermal gravimetric analysis, thermal decomposition → $[(O_2CCH_2CH_2)_2Sn]_x$ (1744).

Diethyltin Dicaprylate $SnC_{20}H_{40}O_4$ $Et_2Sn(O_2CC_7H_{15}-n)_2$
Prop.: Anal. detn. by β-ray reflection method (687), by polarography (2265), with diphenylcarbazone (490).
Use: As catalyst for curing polysiloxanes (579), for preparation of polyurethan (252, 2221), kinetic data and polymerization mechanism (2622, 2886), effect on thermal stability of polyurethan (1655, 2822).

Diethyltin Dibenzoate (M-436) $SnC_{18}H_{20}O_4$ $Et_2Sn(OBz)_2$
Synth.: By rxn. of Et_2SnCl_2 with aq. BzONa in CH_2Cl_2 at 20°, in 90% yield (1051).
By rxn. of Et_2SnO with BzOH in hexane at r.t., in 96% yield (1720).
By-prod. in rxn. of Et_3SnH with Bz_2O_2 in inert atm. (1616).
By-prod. in photooxidation of $PhCH_2SnEt_3$ in nonane, kinetic data (2215).
Prop.: m. 123° (1051), 118-120° (1720), IR spectrum, mol. wt. (1720), spectrographic detn. (1358).
Rxn. with: $Et_2SnO \rightarrow Et_2Sn(OBz)_2 \cdot nEt_2SnO$; n = 1, 2 (1720).
Use: As stabilizer for PVC, relative reactivity (2037).

Dibutyltin Diacetate (M-433) $SnC_{12}H_{24}O_4$ $Bu_2Sn(OAc)_2$
Synth.: By rxn. of $(Bu_2SnCH:CHCH_2CH_2CH:CH)$ with AcOD (499).
By rxn. of Bu_3SnOAc with $BuSn(OAc)_3$ at 180-85° (540).
By rxn. of Bu_2SnO with Ac_2O in N atm. (1449).
By rxn. of $Bu_2 \cdot \overline{SnCH_2CH_2S}$ with $Hg(OAc)_2$ (1704).
Prop.: IR (2184), assignment (1449), far IR, assignment (1449) and UV (1116-7) spectra, conductivity in 100% H_2SO_4 (1100), dipole moment, molar polarization and configuration (632), mol. wt. (1449), TLC (1172a,b).
Rxn. with: $Bu_2SnX_2 \rightarrow Bu_2Sn(OAc)X$; X = Cl (181, 942), I (942), MeO (942, 2184), BuO (2184), $MeOCH(CCl_3)O$ (943).
$Bu_2Sn(OMe)_2 \rightarrow (Bu_2SnOMe)O(Bu_2SnOAc) + MeOAc$ (634).
BuLi → Bu_4Sn (2709).
BuMgBr → $Bu_4Sn + Bu_2CMeOH + BuAc$ (2610).
$(Bz_2CH)_nTi(OPr-i)_{4-n}$ → i-PrOAc + Bu_2Sn-titanoxane polymer (2019).
$Ti(OPr-i)_4 + Ac_2CH_2$ → tin contg. polymer (2019).
100% H_2SO_4 → $[Bu_2Sn(HSO_4)_2] + AcOH_2^+ + HSO_4^-$ (1100).
$RCO_2H \rightarrow Bu_2Sn(O_2CR)_2$; R = CH_2Cl (634), 2,5-, 2,4-$(i-C_8H_{17}O_2C)_2C_6H_3$ (1204).
$(:CHCO)_2O$-$PhCH:CH_2$-polymer → tin contg. polymer (1328).
PhNCO + ureas → cat. effect on polymerization, kinetic data (133).
ArNCO + MeOH → cat. effect, kinetic data and addition mechanism (1607).
8-Hydroxyquinoline → $Bu_2Sn(OC_9H_6N)OAc$ (1216).
$p-C_9H_{19}C_6H_4(OCH_2CH_2)_{9.5}OH$ + alkali → Bu_2Sn-alkoxide (1958).
$\overline{CH_2CH_2O}$ → relative activity as polymerization catalyst (2572).
NaOH → Bu_2SnO (540).
Addn. Compd.: $Bu_2Sn(OAc)_2 \cdot BiPy$, cited in (232a).
$Bu_2Sn(OAc)_2 \cdot 1,10$-phenanthraline, cited in (232a).
Biol. Prop.: Toxicity (1069 and histopathology (822).
Activity as insecticide (1328).
Effect on complement fixation test for virus (1151).

Use: As catalyst for curing silicones (120, 375, 1322, 1571, 1739a, 2822a), for esterification (458, 1766), for polyurethan preparation (1027, 2065, 2860, 2982), for polyurethan-siloxane foam preparation (355), for water-proofing fabric with $Me_nSi(OMe)_{4-n}$; n = 1, 2 (2920).
As processing agent for polypropylene (1323).
As rodent repellent (2036).
As stabilizer for flame retardant polymers (1185a), for PVC (1573, 1651).
As tin depositing agent by microwave discharge (1859).

Dibutyltin Dicaprylate (M-437) $SnC_{24}H_{48}O_4$ $Bu_2Sn(O_2CC_7H_{15}-n)_2$
Synth.: By rxn. of Bu_2SnCl_2 with $C_7H_{15}CO_2Na$ in aq. MeOH, in 88% yield (1357).
Prop.: Cryst. m. -22°, viscous liq. (1357).
Rxn. with: $Bu_2SnS \rightarrow (Bu_2SnO_2CC_7H_{15})_2S$ (945).
Use: As anthelmintic and fungicide (1357).
As catalyst for curing silicone rubber (1357), polysiloxane (1739a, 2662), for preparation of polyurethan (1027), kinetic data and polymerization mechanism (2519, 2622, 2886).
As stabilizer for fluorine contg. polymers (1073), for PVC (1082, 1758).

Dibutyltin Di-2-ethylhexoate (M-437) $SnC_{24}H_{48}O_4$ $Bu_2Sn(O_2CCHEtBu)_2$
Biol. Prop.: Toxicity (1069).
Use: As catalyst for curing polysiloxane (2951), polyurethan coatings (1725) and rubber (539a), for preparation of polyurethan (979, 2318); effect on stability and performance of product (329).
As rodent repellent (2036).
As stabilizer for PVC, mechanistic investigation (2618) with tagged compounds (158).

Dibutyltin Dilaurate (M-434) $SnC_{32}H_{64}O_4$ $Bu_2Sn(O_2CC_{11}H_{23})_2$
Synth.: By rxn. of Bu_2SnCl_2 with $C_{11}H_{23}CO_2Na$ in aq. MeOH (1357).
By rxn. of $Bu_2SnCl_2 \cdot 2\ NH_3$ with $C_{11}H_{23}CO_2H$ in C_6H_6, in 99% yield (2954).
Prop.: Moessbauer resonance spectrum (290), dipole moment molar polarization and configuration (632), struct. (290). Anal. detn. with diphenylcarbazone (490) by alkalimetric titration in Py (1144), by ß-ray reflection method (687), by polarography (2265), by TLC (1136), in plastic flakes after extraction by chromatography, polarography and colorimetry (1136).
Rxn. with: $Bu_2Sn(OMe)_2 \rightleftharpoons Bu_2Sn(OMe)O_2CC_{11}H_{23}$ (942, 2184).
$(EtO)_4Si \rightarrow Bu_2Sn$-silicate (1356).
$(BuO)_4Ti \rightarrow [Bu_2SnOTi(OBu)_2O]_x + BuO_2CC_{11}H_{23}$ (195a).
Kinetic data and/or proposed reaction mechanism for the catalytic effect of $Bu_2Sn(O_2CC_{11}H_{23})_2$ are reported for the following systems: $ArNCO + MeOH$ (1604), $PhNCO + ROH$ (446, 1020), $PhNCO$ + ureas (133), $PhNCO + MeCH(OMe)CH_2OH$ and $MeOCH_2CHMeOH$ (1060), $MeC_6H_3(NCO)_2$ + polyester (9), $Y(NCO)_2$ + polyester-polyol (612), $OCN(CH_2)_6NCO + (HOCH_2CH_2)_2O$ (2543).
Addn. Compd.: $Bu_2Sn(O_2CC_{11}H_{23})_2 \cdot HOCHMeCH_2OMe$, IR, UV and NMR spectra (1059) $[SnC_{36}H_{74}O_6]$.
Biol. Prop.: Toxicity (189, 1132), effect on human red blood cells (1151).

Mottling effect on egg yolk after feeding (1061).
Detn. by implanting in tissue and measuring inflammation (401).
Activity as anthelmintic (1357, 2022) for chicken (30, 116) against cestodes and nematodes (1131), against Davainea proglottina (609), Hymenolepis fraterna (1135).
Activity as fungicide (1357).
Use: As additive for drying oils (517a), for thermoplastic copolymers with inorganic fillers (1714a).
As anthelmintic for livestock (2827).
As antioxidant for mineral oil lubricants (2172).
As catalyst for curing cellulose polycarbamate (117), unsaturated amides (143), organosilyl polyethers (2929), polysiloxane (367a, 640, 654a, 1143b, 1356-7, 1384, 1420, 1563, 1630, 1701a, 1808, 1858, 1983, 2031, 2042, 2275, 2662, 2902, 2922, 2944, 2951-2), for esterification (628, 1531, 2064a, 2298, 2308), for gelling hydrocarbon fuel (2141), for hardening polysilicone-SiO_2 (2828), for modifying polyethylene (1897), for preparation of polyurethan (113, 122, 179, 229, 329, 539a, 955, 976, 979, 1027, 1153, 1193, 1209, 1394, 1451, 1460, 1633, 1724-5, 1929, 2066, 2118, 2135, 2167, 2287, 2600, 2717, 2779, 2791, 2807, 2829, 2886, 2894, 2947, 2960, 2967-8, 2988, 3005).
As inhibitor of corrosion in fire extinguishers (245) of radiation induced graft polymerization of $PhCH:CH_2$ and $CH_2:CHCN$ on PVC (1389).
As rodent repellent (2036).
As stabilizer for chloroprene rubber (2884) for flame retardant polymers (1185a), for fluorine contg. polymers (1073), for phenol-formaldehyde resins (1279), for polyethylene (2212), for polysulfide sealant (1102), for rayon tire cord (377), for PVC (597, 681, 782, 1062, 1381, 1464, 1511, 1523, 1573, 1651, 1786, 2016, 2033, 2120, 2549, 2912, 2921, 2953-4); relative effect on degradation of PVC (371, 1035, 1080-2, 1170), for vinyl copolymers (1511, 2921).

Dibutyltin Distearate (M-437) $SnC_{44}H_{88}O_4$ $Bu_2Sn(O_2CC_{17}H_{35})_2$
Prop.: Anal. detn. by alkalimetric titration in Py (1144).
Biol. Prop.: Activity as anthelmintic (2022).
Use: As catalyst for foaming polyurethan (1027).
As rodent repellent (2036).
As stabilizer for chloroprene rubber (2884), for polyethylene (2212), for PVC (2832) and PVC foam (2911).

Diphenyltin Diacetate (M-438) $SnC_{16}H_{16}O_4$ $Ph_2Sn(OAc)_2$
Synth.: By rxn. of Ph_2SnO with AcOH at r.t., in 57% yield (2005), with $R_2Si(OAc)_2$ or AcOH at r.t., in quant. yield (1326, 2005).
Prop.: m. 116-17° (326, 2005), detn. and sepn. by TLC (1253).
Rxn. with: $EtCO_2H \rightarrow Ph_2Sn(O_2CEt)_2$ (2005).
Addn. Compd.: $Ph_2Sn(OAc)_2 \cdot BiPy$, as cited in (232a).
$Ph_2Sn(OAc)_2 \cdot 1,10$-phenanthraline, as cited in (232a).

Additional diorganotin dicarboxylates are listed in Table 208.

Table 208. Unsubstituted Diorganotin Dicarboxylates $R_2Sn(O_2CR')_2$

Formula*	Synth. Method**	Yield	Properties	Ref.
Me_2SnCO_3	IA	--	IR spectrum, x-ray powder pattern	873
$Me_2Sn(O_2CR')_2$				
R' = H (435)	--	--	Moessbauer resonance spectrum	1991
Me	IIIB	--	m. 67°, b_5 93-94°, struct. (a, b)	1448-9
C_7H_{15}	IIIA	--	Use as stabilizer for PVC (c)	339
C_9H_{19}	IIIA	--	Use as stabilizer for PVC	339
$C_{11}H_{23}$ (435)	IIIA	--	Use as stabilizer for PVC (d)	339
$C_{13}H_{27}$	IIIA	--	Use as stabilizer for PVC	339
$C_{15}H_{31}$	IIIA	--	Use as stabilizer for PVC	339
$C_{17}H_{25}$	IIIA	--	Use as stabilizer for PVC	339
Ph	IC	50	Colorless cryst., m. 162-64° (NMR spectrum 2482)	409
$(PhCH_2)_2Sn(O_2CR')_2$				
R' = Me (435)	IIIB	--	NMR spectrum (e)	2069
C_7H_{15}	IB	--	(f)	1357
BuEtCH (435) (g)	--	--	(e)	--
$C_{11}H_{23}$ (435)	IB	--	(e, f)	1357
$Et_2Sn(OAc)_2$ (435)	IIIB	--	m. 44°, b_5 97°, (far) IR spectra, assignment (h)	1449
$Et_2Sn(O_2CC_{11}H_{23})_2$	--	--	(i)	--

* Numbers in parenthesis refer to pages in main volume.
** The characters used here correspond to the synthetic scheme in the introduction to Chapter 4.3.5.2.1.
(a) (Far) IR, assignment and NMR spectra (1448-9). (b) Rxn. with $H_2O \rightarrow (Me_2SnOAc)_2O$ (1448-9). (c) Activity as foaming catalyst for polyurethan (1027). (d) Reactivity as foam gelling catalyst for polyurethan (2906). (e) Use as catalyst for polyurethan preparation (180). (f) Use as anthelmintic, fungicide and curing agent for silicone rubber (1357). (g) Correct formula main volume p. 435, line 13 to: $(PhCH_2)_2Sn(O_2CCHEtBu)_2$. (h) Relative reactivity as stabilizer for PVC (1573, 2037). (i) Activity as insecticide against Culex pipiens larvae (1137).

Table 208 Continued $R_2Sn(O_2CR')_2$

Formula*	Synth. Method**	Yield	Properties	Ref.
$Pr_2Sn(OAc)_2$	IIIB	--	m. 36°, b_5 115°, (far) IR spectra, assignment (j)	1449
$Bu_2Sn(O_2CR')_2$				
R' = MeO	IVA	--	IR, NMR spectra, hydrolyzes in H_2O	943
$EtC_6H_4CH_2$	--	--	Activity as anthelmintic	2022
Et	IIIA	--	Moessbauer resonance spectrum, struct.	290
Bu	IIIA	--	Moessbauer resonance spectrum, struct.	290
C_5H_{11} (437)	--	--	(k)	--
$R'CO_2$ = naphthenate (437)	IIIA	--	n_D^{20} 1.4960, use as stabilizer for PVC	2954
R' = C_6H_{13}	IIIA	--	Moessbauer resonance spectrum, struct.	290
C_9H_{19} (437)	IIIA	--	Moessbauer resonance spectrum, struct. (l)	290
$C_6H_{13}CMe_2$	--	--	Use as rodent repellent	2036
$C_{15}H_{31}$	--	--	Reactivity as anthelmintic	2022
Ph	II	85	m. 66-68° (use as esterification cat. 2870)	585
	IIIA, IVA	85	m. 68-71° (use as rodent repellent 2036)	2096
$p-Me_3CC_6H_4$	IIIA	--	m. 95-96°	2732
$i-Bu_2Sn(OAc)_2$ (438)	IB	76	b_3 123-24° (m)	219a
$i-Bu_2Sn(O_2CC_{11}H_{23})_2$ (438)	IC	--	n_D^{20} 1.4689, use as stabilizer for PVC	2954
$(Me_3CCH_2)_2Sn(OAc)_2$	IIIA	84	m. 66-67°	2188
$(C_8H_{17})_2Sn(OAc)_2$ (438)	--	--	TLC, sepn. from $EtBuCHCH_2$-deriv. (n)	1172a,b
$(C_8H_{17})_2Sn(O_2CC_7H_{15})_2$	IB	--	Use as stabilizer for PVC (1758) (f)	1357
$(C_8H_{17})_2Sn(O_2CC_{11}H_{23})_2$ (438)	IB	--	Use as stabilizer for PVC (152, 1464) (f, n, o)	1357
$(i-C_8H_{17})_2Sn(O_2CC_{11}H_{23})_2$	--	--	(n)	--
$(EtBuCHCH_2)_2Sn(OAc)_2$	--	--	TLC, sepn. from $n-C_8H_{17}$-deriv.	1172a,b
$(Me_3CCH_2CMe_2)_2Sn(O_2CC_{11}H_{23})_2$	--	--	Use as stabilizer for polypropylene	1062a
$(C_{12}H_{25})_2Sn(OBz)_2$ (438)	IC	--	m. 59-61°, use as stabilizer for PVC	2954

Ph$_2$Sn(O$_2$CEt)$_2$	III (p)	1326, 2005
	VA	2005
R$_2$Sn(O$_2$CR')$_2$	52 m. 75-76° Use as stabilizer for PVC (q)	2895

(j) TLC (1172a), rxn. with (:CHCO)$_2$O-PhCH:CH$_2$-polymer → tin contg. polyester (1328), activity as insecticide (1328). (k) Use as stabilizer for phenol-formaldehyde resin (1279), for chlorinated PVC (2889). (1) Use as stabilizer for PVC (1082, 1758). (m) Rxn. with Me$_2$Si(OEt)$_2$ → (i-Bu$_2$SnOEt)O(i-Bu$_2$SnOAc) + [i-Bu$_2$SnOSiMe$_2$O]$_x$ + AcOEt (219a). (n) Use as catalyst for polyurethan preparation (2982). (o) Use as stabilizer for halogen contg. polymers (153). (p) Synth. with Me$_2$Si(O$_2$CEt)$_2$ at r.t. (q) Use in additive to improve dyeability of polypropylene (2156a), as catalyst for curing silicone rubber (772).

4.3.5.2.2 UNSUBSTITUTED DIORGANOTIN CARBOXYLATE OXIDES AND HYDROXIDES

The diorganotin carboxylic oxides and hydroxides are summarized in the new chapter for easier reference. Many compounds probably form a four-membered "stannoxane" ring as set forth in the main volume, page 336.

Bis(dimethylacetatotin) Oxide (M-435)
   Sn$_2$C$_8$H$_{18}$O$_5$ (Me$_2$SnOAc)$_2$O
Synth.: By rxn. of Me$_2$Sn(OAc)$_2$ with moist air (1448), with H$_2$O in quant. yield (1449).
By rxn. of Me$_2$SnO with AcOH (1:1) (1449), with AcOH and Ac$_2$O as by-prod. (2611).
Prop.: m. 236° (1449), 235° (dec.) (2611), (far) IR, assignment and NMR spectra (1449), conductance, mol. wt. and soly (1449).
Rxn. with: Thermal dec. at 220° → Me$_3$SnOAc (2611).
Use: As stabilizer for PVC (1833a).

Bis(dipropylpropionatotin) Oxide
   Sn$_2$C$_{16}$H$_{34}$O$_5$ (Pr$_2$SnOAc)$_2$O
Synth.: By rxn. of Pr$_2$Sn(OAc)$_2$ with H$_2$O (1449).
By rxn. of Pr$_2$SnO with AcOH (1:1) (1449).
Prop.: m. 111-13° (1449, 2611), (far) IR spectrum, assignment and soly. (1449).
Rxn. with: Thermal dec. at 125°/0.3 mm. → (Pr$_3$Sn)$_2$O (2611).
Py in aq. MeOH → (Pr$_2$SnOAc)O(Pr$_2$SnOH) (1449).
Use: As catalyst for preparation of polyurethan (2166-7).

Bis(dibutylacetatotin) Oxide (M-436)
   Sn$_4$C$_{40}$H$_{84}$O$_{10}$ [(Bu$_2$SnOAc)$_2$O]$_2$
Synth.: By rxn. of Bu$_2$SnO with AcOH in MeOH (2169).
By hydrolysis of Bu$_2$Sn(OAc)$_2$ (1449), of Bu$_2$Sn(OMe)OAc (942), of Bu$_2$Sn(SR)OAc; RS = PhN:C(OMe)S, MeOCS$_2$ (943).
Prop.: m. 305-310° (944), 58-60° (1449, 2611), 57-58° (2169), 54-57° (10), 54° (942), IR (944), (far) IR spectra and assignment (1449), dimeric (632), conductance, mol. wt. and soly. (1449), dipole moment, configuration and molar polarization (632).

Rxn. with: $H_2O \rightarrow [(Bu_2SnOH)O(Bu_2SnOAc)]_2$ (10).
Aq. Py in MeOH $\rightarrow [(Bu_2SnOH)O(Bu_2SnOAc)]_2$ (1449).
$Me_3CO_2H + Et_2N \rightarrow (Bu_2SnO_2CMe_3)_2O + Et_3NHOAc$ (2257).
Thermal dec. at 125°/0.3 mm. $\rightarrow (Bu_3Sn)_2O$ (2611).
Use: As catalyst for lactone polymerization (1999), for polyurethan synthesis (2166-7, 2169, 2485, 2998-9).
As stabilizer for PVC (1080-1), for storing polyurethan-polyester resins (2573).

Tetrabutyloxyacetatoditin Oxide* (M-436-7)
$Sn_4C_{40}H_{80}O_8$ $[(Bu_2SnOAc)O(Bu_2SnOH)]_2$

Synth.: By hydrolysis of $Bu_2SnCl(OSnBu_2)_3Cl$ with aq. AcONa in H_2CCl_2 at 0°, in 57-76% yield (2594), of $(Bu_2SnOAc)OBu_2SnOMe$ at r.t. in MeOH, in 78% yield (634), of $(Bu_2SnOAc)_2O$ in air (10), with aq. Py in Me_2CO, in 80% yield (1449). By rxn. of Bu_2SnO with AcOH and azeotropic distn. of C_6H_6 or MePh, in 65% yield (2595), with AcOH in Me_2CO (2169).
Prop.: m. 137-93° (turbid melt) (634), m. 129° (10, 1449), m. 119-22° (2594), m. 107-19° (2169), m. 104° (2595), IR (10), (far) IR spectra and assignment (1449), mol. wt. (10, 1449), soly. (1449).
Use: As catalyst for polyurethan preparation (2166).
As stabilizer for PVC (1081).

Additional diorganotin carboxylate oxides and hydroxides are listed in Table 209.

---

* This compound (10, 1449, 2166, 2169) may be regarded as a monohydrate of and is probably identical to $Bu_2SnOAc(Bu_2SnO)_3OAc$ (2594-5) and $AcO(Bu_2SnO)_4Ac$ (634); all data are compiled here (Ed.).

Table 209. Diorganotin Carboxylate Oxides and Hydroxides $R_2Sn(O_2CR') \cdot R_2SnO$

Formula*	Synth. Method**	Yield	Properties	Ref.
$(Me_2SnO_2CR')_2O$				
R' = H	IC	>70	m. 185° (dec.), IR spectrum	409
Et	IC	>70	m. 185°, IR spectrum	409
	IIIA, B	--	m. 185°, thermal dec. → Me_3SnO_2CEt	2611
Pr	IC	>70	m. 161-62°, IR spectrum	409
Ph (435)	--	--	m. 234.5-35.5°, thermal dec. > Me_3SnOBz	2611
	IC	>70	Rxn. with BzOH → $Me_2Sn(OBz)_2$	409
$(Et_2SnOAc)_2O$ (436)	IIIA	--	m. 105-106°, soly. (a, b)	1449
$(Et_2SnOAc)O(Et_2SnOH)$	IVB (a)	80	m. 200° (dec.), dimeric, soly. (b)	1449
$(Et_2SnOBz)_2O$ (436)	IB	88	m. 213-14°	1051
	IIIC	77	m. 160-65°, IR spectrum, mol. wt.	1720
	IIIA	--	m. 165-68°	1720
$BzO(Et_2SnO)_3Bz$	IIIA, C	--	m. 220°, IR spectrum	1720
$(Pr_2SnOAc)O(Pr_2SnOH)$ (436)	IVB (a)	80	m. 206-208°, dimeric, soly. (b)	1449
$(Pr_2SnO_2CC_6H_4Me-m)_2O$	IIIA	--	m. 149-52°, use as stabilizer for PVC (c)	1289
$Pr_2SnO_2CC_6H_4CMe_3-m)_2O$	IIIA	--	m. 161-64°, use as stabilizer for PVC	1289
$(Bu_2SnOAc)OSnBu_3$	IIIC (d)	--	m. 68-70°, IR spectrum	946
$(Bu_2SnO_2CH)_2O$	--	--	Use as cat. for lactone polymerization	1999
$(Bu_2SnO_2CBu)_2O$	IIIA	--	Moessbauer resonance spectrum, struct.	290
$Bu_2Sn(O_2CC_{10}H_{19})_2 \cdot 7\ Bu_2SnO$	IIIA	65	m. 97-104°	2595

* Numbers in parenthesis refer to pages in main volume.
** The characters used here correspond to the synthetic scheme in the introduction to Chapter 4.3.5.2.1.
(a) Rxn. of $(R_2SnOAc)_2O$ with aq. Py in MeOH → $(R_2SnOAc)O(R_2SnOH)$; R = Et, Pr (1449). (b) (Far) IR spectra and assignment (1449). (c) Use as stabilizer for PVC and vinyl polymers (1289). (d) Synth. from Bu_2SnO and Bu_3SnOAc (946)

Table 209 Continued $R_2Sn(O_2CR')_2 \cdot R_2SnO$

Formula*	Synth. Method**	Yield	Properties	Ref.
$(Bu_2SnO_2CR)_2O$				
R = $C_{11}H_{23}$ (441)	--	--	Use as stabilizer for PVC (e, f)	1080-1
$C_{14}H_{29}$	IIIA	--	Moessbauer resonance spectrum, struct.	290
$C_{17}H_{35}$	IIIA	--	Moessbauer resonance spectrum, struct. (f)	290
p-MeC_6H_4	IIIA	--	m. 144-46° (c)	1289
p-$Me_3CC_6H_4$	IIIA	--	m. 169-72° (c)	1289
p-PhC_6H_4	IIIA	--	123-25° (c)	1289
2-$C_{10}H_7$	IIIA	--	m. 135-37° (c)	1289
$(Ph_2SnOAc)_2O$	IIIA (g)	--	m. > 250°, IR spectrum	1704
$(Ph_2SnOBz)_2O$	II	--	Synth. from Ph_3SnH, m. 184-85°	123

(e) Dipole moment, molar polarization, configuration (632), use as catalyst for polyurethan preparation (2485). (f) Use as catalyst for polyurethan polymerization (2166). (g) Synth. also from Ph_3SnSPh with $Hg(OAc)_2$ in hot EtOH as by-prod. (1704).

4.3.5.2.3 UNSUBSTITUTED DIORGANOTIN CARBOXYLATES OF HETEROCYCLIC AND UNSATURATED ACIDS

Diorganotin dicarboxylates derived from heterocyclic carboxylic acids are listed at the top of Table 210. Diorganotin carboxylates derived from unsaturated acids are compiled in Tables 210 through 212 and in Subsections 4.3.5.2.6.8, 4.3.5.2.8 and 4.3.5.2.9. Tables 210 and 211 contain diorganotin dicarboxylates of unsubstituted and functionally substituted unsaturated acids, respectively. Table 212 comprises "basic" diorganotin carboxylates of unsaturated acids. Subsections 4.3.5.2.6.8.1 and 4.3.5.2.6.8.2 include derivatives of unsaturated dicarboxylic acids. Subsection 4.3.5.2.8 includes a functionally substituted diorganotin derivative of maleic acid. Table 220 summarizes diorganotin carboxylates with tin-heteroatom bonds. Additional references to unsaturated diorganotin carboxylates may be found in Chapters 8.1 and 9.2.

Dibutyltin Di(methyl maleate) (M-440) $SnC_{18}H_{28}O_8$ $Bu_2Sn(O_2CCH:CHCO_2Me)_2$
Synth.: By rxn. of Bu_2SnCl_2 with $MeO_2CCH:CHCO_2K$ in Me_2CO (1557).
By rxn. of Bu_2SnO with $MeO_2CCH:CHCO_2H$ in C_6H_6 (1557).
Prop.: Nondistillable oil, n_D^{23} 1.4930, IR spectrum (1557).
Rxn. with: BiPy → $[(Bu_2SnO_2CCH:CHCO_2Me)_2O]_2$ (1557).
C_5H_6 → Bu_2Sn-di(methyl 2,3-bicyclo[2.2.1]heptene-5-dicarboxylate) (1558).
$(CH_2:CMe)_2$ → Bu_2Sn-di(methyl 1,2-hexene-4-dicarboxylate) (1558).
Use: As stabilizer for PVC and PVC-vinyl-copolymers (729a), tagged species: $Bu_2^{113}Sn(O_2CCH:CHCO_2Me)_2$, $(Pr^{14}CH_2)_2Sn(O_2CCH:CHCO_2Me)_2$ and $Bu_2Sn(O_2CCH:CHCO_2-^{14}CH_3)_2$ are used to study degradation of PVC (158).

Table 210. Unsubstituted Diorganotin Dicarboxylates of Heterocyclic and Unsaturated Acids $R_2Sn(O_2CR')_2$

Formula*	Synth. Method**	Yield	Properties	Ref.
$Me_2Sn(O_2CC_5H_4N-2)_2$ (a)	IIIA	--	m. 267-68°, IR and NMR spectra, mol. wt.	347
$Bu_2Sn(O_2CC_4H_3O)_2$ (b)	IIIA	--	m. 118°, use as fungicide	378
$Bu_2Sn(O_2CC_4H_2OBr-5)_2$ (b)	IIIA	--	m. 120°, use as fungicide	378
$Ph_2Sn(O_2CC_5H_4N-2)_2$ (a)	II (c)	95	m. 281-83°, IR spectrum	1310, 1374
	IB (d)	85	Mol. wt.	1310, 1374
$Me_2Sn(O_2CCMe:CH_2)_2$ (439)	--	--	Use as insecticide (e)	1328
$Me_2Sn(O_2CC_{17}H_{33})_2$	IIIA	--	Oleate, use as stabilizer for PVC	339
$(PhCH_2)_2Sn(O_2CC_{17}H_{33})_2$ (439)	--	--	Oleate, use as polyurethan foaming cat.	180
$(CH_2:CH)_2Sn(O_2CC_{17}H_{33})_2$	IB	--	Oleate (f)	1442
$(\overline{CH_2CH_2CH})_2Sn(O_2CCH:CH_2)_2$	IB	--	Use as herbicide, pesticide	1873
$Bu_2Sn(O_2CCMe:CH_2)_2$ (439)	--	--	Use as stabilizer for PVC (g)	2832
$Bu_2Sn(O_2CC_{17}H_{33})_2$ (439)	--	--	Oleate, use as anthelmintic	2022
$(p-CH_2:CHC_6H_4)_2Sn(O_2CC_{17}H_{33})_2$	--	--	Oleate (h)	1442

* Numbers in parenthesis refer to pages in main volume.
** The characters used here correspond to the synthetic scheme in the introduction to Chapter 4.3.5.2.1.
(a) Pyridinecarboxylate. (b) 2-Furanecarboxylate. (c) Synth. with Ph_3SnH in 91% yield (1310, 1374). (d) Synth. with Ph_3SnBr in 68% yield (1310, 1374). (e) Rxn. with $(:CHCO)_2O-PhCH:CH_2$-polymer → tin contg. polyester (1328). (f) Rxn. with $CF_3CO_3H → (\overline{OCH_2CH})_2Sn[O_2C(CH_2)_7\overline{CHCHOC_8H_{17}}]_2$ (1442). (g) Use as additive for flame resistant polyester (2702). (h) Rxn. with $AcO_2H → (p-\overline{OCH_2CHC_6H_4})_2Sn[O_2C(CH_2)_7\overline{CHCHOC_8H_{17}}]_2$ (1442).

Table 211. Diorganotin Dicarboxylates Derived from Functionally Substituted Unsaturated Acids $R_2Sn(O_2CR')_2$

Formula*	Synth. Method**	Properties	Ref.
$Me_2Sn(O_2CC_{17}H_{32}OH)_2$	IIIA	Ricinolate, use as stabilizer for PVC	339
$Bu_2Sn[O_2CC(CN):CHAr]_2$	IIIA	Ar = 3,5,4-t-$Bu_2HOC_6H_2$, m. 181°	2565
	IIIA (a)	Use as stabilizer for PVC	2567
$Bu_2Sn(O_2CCH:CHCO_2R')_2$		Maleic Acid Half-Ester	
R' = $PhCH_2$	--	Use in stabilizer for PVC (2002, 2956, 2964)	1862
$2-(C_4H_7O)CH_2$	--	Use as stabilizer for PVC	1283
Et (440)	--	Use as esterification catalyst (b, c)	1156
$HOCH_2CH_2$	--	(c)	--
$PhCH_2OCH_2CH_2$	--	Use as stabilizer for PVC	1287
$EtOCH_2CH_2$	--	Use as stabilizer for PVC	1287
$BuOCH_2CH_2$	--	Use as stabilizer for PVC	1287
$C_{12}H_{25}OCH_2CH_2$	--	Use as stabilizer for PVC	1287
$C_{16}H_{33}OCH_2CH_2$	--	Use as stabilizer for PVC	1287
$C_{18}H_{37}OCH_2CH_2$	--	Use as stabilizer for PVC	1287
$C_{18}H_{35}OCH_2CH_2$	--	Use as stabilizer for PVC	1287
$PhOCH_2CH_2$	--	Use as stabilizer for PVC	1287
$CH_2:CMeCO_2CH_2CH_2$	IIIA	Use as microbiocide (d)	2974
$p-C_9H_{19}C_6H_4(OCH_2CH_2)_{35}$	--	Use as antistatic for polymers	2817
$CH_2:CHCH_2$	--	Use as stabilizer for PVC and copolymers	729a
HOC_3H_6	--	Use as stabilizer for PVC	2897
$CH_2:CHCO_2C_3H_6$	IIIA	Use as microbiocide (d)	2974

* Numbers in parenthesis refer to pages in main volume.
** The characters used here correspond to the synthetic scheme in the introduction to Chapter 4.3.5.2.1.
(a) Synth. from Bu_2SnO with $NCCH_2CO_2H$ + ArCHO or 1-BuAr (2565). (b) Use as stabilizer for polyethylene, rxn. mechanism (2212). (c) Use as catalyst for modifying polypropylene (269). (d) Use as intermediate for homo- and co-polymerization (2974).

Table 211 Continued $R_2Sn(O_2CR')_2$

Formula*	Synth. Method**	Properties	Ref.
$Bu_2Sn(O_2CCH:CHCO_2R')_2$ (Cont'd.)			
R' = Bu (440)	--	Use as stabilizer for PVC (c)	1862
c.C_6H_{11}	IIIA	m. 71-73°, use as stabilizer in PVC (2957, 2965)	1110
c.2-MeC_6H_{10}	IIIA	Use as stabilizer for PVC	1110
$EtBuCHCH_2$ (441)	--	Detn. by alkalimetric titration in Py (e)	1144
1-C_8H_{17}	--	Effect on human red blood cells (f-j)	1151
1-C_8H_{17}	--	Fumarate (h)	2897
C_9H_{19}	--	Use as stabilizer for PVC	1464
$C_{12}H_{25}$	--	Use as stabilizer for polyethylene (f, i)	1062a, 2769
$C_{18}H_{33}$	--	Use as stabilizer for chloroprene rubber	2884
R	--	Use as esterification catalyst (k)	2064a
$Bu_2Sn(O_2CR')_2$			
R' = c.4,5,2-$Me_2(MeO_2C)C_6H_8$	IIIA	Cyclohexene dicarboxylate	1558
3,5-$C_7H_8(CO_2Me)$-2 (1)	VC	Viscous liq., dec. on distn.	1558
$C_{17}H_{32}OH$ (441)	VC	m. 56-58°, soly.	1558
$Me_2CHCH_2CHMeC:C(CN)$	IIIA (a)	Ricinolate (m)	--
		m. 123°, use as stabilizer for PVC	2565
$(C_8H_{17})_2Sn(O_2CCH:CHCO_2R')_2$			
R' = $CH_2:CMeCO_2CH_2CH_2$	IIIA	Use as microbiocide (d)	2974
C_3H_6OH	--	Use as stabilizer for PVC	2897
$CH_2:CHCO_2C_3H_6$	IIIA	Use as microbiocide (d)	2974
1-Bu	--	TLC, thermally stable to 180° (n)	1052
$C_{17}H_{35}CO_2C_6H_8O(OH)_2$	--	Use as antistatic for thermoplastics	2018
n-C_8H_{17}	IB	Use as anthelmintic, fungicide (o)	1357
1-C_8H_{17}	--	Use as stabilizer for PVC	2897
$EtBuCHCH_2$	IC	n_D^{20} 1.4812, use as stabilizer for PVC (p)	2954
$C_{12}H_{25}$	--	Use as stabilizer for PVC	2897

Bu$_2$Sn(O$_2$CC$_{11}$H$_{23}$)O$_2$CR	IIIA	R = 3,5,4-t-Bu$_2$HOC$_6$H$_2$CH$_2$CH:CH(CN), d$_{20}$ 1.2237	2571
[Bu$_2$Sn(O$_2$CR)O$_2$CCH:]$_2$			
R = C$_7$H$_{15}$	--	(c)	--
C$_{11}$H$_{23}$ (441)	--	Use as stabilizer for PVC (c, q)	1381
Bu$_2$Sn(O$_2$CCH:CHCO$_2$R)O$_2$CR'			
R = PhCH$_2$ R' = C$_{11}$H$_{23}$	--	Use as stabilizer for PVC	1862
C$_8$H$_{17}$ C$_{11}$H$_{23}$	--	Use as stabilizer for polypropylene	1062a
C$_{11}$H$_{23}$ C$_5$H$_{11}$	--	Use as stabilizer for PVC	635
C$_{12}$H$_{25}$ C$_{11}$H$_{23}$ (441)	--	Use as stabilizer for polypropylene	2769
2-(C$_4$H$_7$O)CH$_2$ C$_8$H$_{17}$O$_2$CCH:CH	IVB	Use as stabilizer for PVC	1283
EtOCH$_2$CH$_2$ PhCH$_2$O$_2$CCH:CH	--	Use as stabilizer for PVC	1287
(C$_8$H$_{17}$)$_2$Sn(O$_2$CR')O$_2$CC$_{17}$H$_{35}$	IIIA	White waxy solid	2570
R' = PhCH:C(CN)	--	Use as antistatic for polymers	2817
C$_{12}$H$_{25}$(OCH$_2$CH$_2$)$_{30}$O$_2$CCH:CH			

(e) Use as stabilizer for chloroprene rubber (2884), for PVC (2953) and copolymers (729a). (f) Effect on complement fixation test for virus (1151). (g) Use as stabilizer for PVC, flame retardant polymers (1185a) for PVC and copolymers (2862). (h) Use as stabilizer for PVC, relative reactivity of maleate and fumarate (2897). (1) Use as stabilizer for PVC, relative activity (1651). (j) Rxn. with Bu^{n}Sn epoxystearate + HSCH$_2$CO$_2$C$_8$H$_{17}$ → rxn. prod. stabilizer for PVC (1442). (k) Use as stabilizer for PVC (713, 1302). (1) 2,3-bicyclo[2.2.1]heptene-5-dicarboxylate. (m) Rxn. with 1,2,4-Me(OCN)$_2$C$_6$H$_3$ in MePh → solid tin contg. polyurethan (2959), use as foaming catalyst for polyurethan (1027). (n) Decomposes in PVC at 180° (1052). (o) U. S. limits for use as stabilizer for PVC (2942), use as curing agent for siloxane rubber (1357). (p) U. S. limits for use in food-packaging cellophane foil (2940). (q) Use as antioxidant for mineral oil lubricant (2172), as catalyst for polyurethan synthesis (2960), as stabilizer for ABS resins (2117), for polypropylene (2769).

Table 212. "Basic" Diorganotin Carboxylates Derived from Unsaturated Acids $R_2Sn(O_2CR') \cdot nR_2SnO$

Formula*	Synth. Method**	Properties	Ref.
$(Bu_2SnO_2CR')_2O$			
R' = CH_2:CH	--	(a)	2566
ArCH:C(CN)	V (b)	Ar = 3,5,4-t-$Bu_2HOC_6H_2$, m. 125° (b)	2566
MePhC:C(CN)	V (b)	(c)	2566
Ph_2C:C(CN)	IIIA	--	2568
MeO_2CCH:CH (441) (d)	VB	m. 91-94°, IR spectrum, dimeric, soly.	1557
$C_{18}H_{37}O_2CCH$:CH	--	Use as stabilizer for polypropylene	2769
$HO(Bu_2SnO)_2OCCH$:$CHCO_2H$	IIIB	m. 280° (dec.) (e)	2872
$3\ Bu_2SnO \cdot Bu_2Sn(O_2CC\textbardbl CCO_2H)_2$	IB (f)	In 84% yield, m. 250°	2594
$(1-Bu_2SnO_2CCH$:$CHCO_2R)_2O$	--	R = $EtBuCHCH_2$, use as stabilizer for PVC	1833a
$[(C_8H_{17})_2SnO_2CR']_2O$			
R' = p-HOC_6H_4CH:C(CN)	IIIA (g)	m. 225° (dec.), use as stabilizer for PVC	2566
i-BuO_2CCH:CH	--	Use as foaming cat. for polyurethan	2876
$(Bu_2SnO_2CR')O(Bu_2SnO_2CAr)$		Ar = 3,5,4-t-Bu_2(HO)C_6H_2CH:C(CN)	
R' = $C_{11}H_{23}$	IIIA	--	2569
BuO_2CCH:CH	IIIA	--	2569
$[(C_8H_{17})_2SnO_2CR']O[(C_8H_{17})_2SnO_2CAr]$		Ar = PhCH:C(CN)	
R' = $EtOCH_2CH_2O_2CCH$:CH	IIIA	--	2569
$C_{17}H_{35}$	IIIA	--	2569

* Numbers in parenthesis refer to pages in main volume.
** The characters used here correspond to the synthetic scheme in the introduction to Chapter 4.3.5.2.1.
(a) Use as stabilizer and hardening agent for polyurethan-polyester resins (2573). (b) Synth. from $(Bu_2SnO_2C$-$CH_2CN)_2O$ with $RR'CO \rightarrow Bu_2SnO_2CC(CN):CRR'$; R = H, R' = Ar; R = Me, R' = Ph (2566). (c) Use as stabilizer for PVC (2566). (d) Correct in main volume page 441, line 12 f.b., third column, R = Me to R = MeO_2CCH:CH, last line R = R' = Me to R = R' = MeO_2CCH:CH. (e) Use as catalyst for polyurethan synthesis (2872), as stabilizer for PVC (2871-2). (f) Synth. from $Bu_2SnCl(Bu_2SnO)_2Cl$ (2594). (g) Synth. from $(C_8H_{17})_2SnO$ with $NCCH_2CO_2H$ and p-HOC_6H_4CHO (2566).

4.3.5.2.6 UNSUBSTITUTED DIORGANOTIN CARBOXYLATES DERIVED FROM FUNCTIONALLY SUBSTITUTED ACIDS

4.3.5.2.6.1 DIORGANOTIN CARBOXYLATES OF HALOGEN SUBSTITUTED ACIDS

Diorganotin carboxylates derived from halogenated carboxylic acids are compiled in Tables 213 and 214. Diorganotin dicarboxylates are listed in Table 213. Table 214 comprises "basic" diorganotin halocarboxylates. Additional diorganotin carboxylates of substituted halogen containing acids may be found in Subsections 4.3.5.2.3 and 4.3.5.2.6.7.

4.3.5.2.6.3 UNSUBSTITUTED DIORGANOTIN CARBOXYLATES DERIVED FROM ACIDS CONTAINING OXYGEN AND SULFUR

All compounds are listed in Table 215. Additional references to oxygen and sulfur substituted diorganotin carboxylates may be found in Chapters 4.3.5.2.3, 4.3.5.2.8 and 8.1 and in Tables 211, 212, 217, 219 and 220.

4.3.5.2.6.5 UNSUBSTITUTED DIORGANOTIN CARBOXYLATES DERIVED FROM NITROGEN CONTAINING ACIDS

Dibutyltin Di-p-nitrobenzoate $SnC_{22}H_{26}N_2O_8$ $Bu_2Sn(O_2CC_6H_4NO_2-p)_2$
Synth.: By rxn. of Bu_2SnCl_2 with aq. $p-O_2NC_6H_4CO_2Na$ in petr. ether at 0°, in 87% yield (1051).
By rxn. of Bu_2SnO with $p-O_2NC_6H_4CO_2H$ under reflux in C_6H_6 or MePh (2732), in 83% yield (2595), rxn. in presence of $o-ClC_6H_4CO_2H$, in 98% yield (2395).
By rxn. of Bu_2SnCl_2 with $p-O_2NC_6H_4CO_2H$ in MePh (2395).
By rxn. of $Bu_2Sn(O_2CC_6H_4Cl-o)Cl$ with $p-O_2NC_6H_4CO_2H$, in 85% yield (2395).
By rxn. of $(Bu_2SnO_2CC_6H_4NO_2-p)_2O$ with $o-ClC_6H_4CO_2H$ in MePh, in 96% yield (2395).
By rxn. of $(Bu_2SnClO_2CCH_2CH_2)_2$ with $p-O_2NC_6H_4CO_2Na$ (2395).
By rxn. of $(Bu_2SnCl)_2S$ with $p-O_2NC_6H_4CO_2H$ and aq. NaOH in petr. ether at 0° (1514).
Prop.: m. 218-19° (1051, 2395), 218° (2595), 205-206° (2732), soly. (2595).
Rxn. with: $Bu_2SnO \rightarrow p-O_2NC_6H_4CO_2Bu_2Sn(Bu_2SnO)_2OCC_6H_4NO_2$ (2857).

Additional diorganotin carboxylates with nitrogen substituted acids are listed in Table 216. Other references to nitrogen substituted diorganotin carboxylates may be found in Subsections 4.3.5.2.3 and 4.3.5.2.8 and in Tables 211, 212, 217, 219 and 220.

Table 213. Diorganotin Dihalocarboxylates

Formula*	Synth. Method**	Yield	Properties	Ref.
$Et_2Sn(O_2CCHCl_2)_2$	IB	88	m. 123-24°	1051
$Et_2Sn(O_2CCHMeCl)_2$	--	--	Use as stabilizer for polyurethan	1660
$Et_2Sn(O_2CC_6H_4Cl$-$o)_2$	IB	100	m. 90-91°	1051
$Bu_2Sn(O_2CR')_2$				
R' = CH_2Cl (443)	VA	76	m. 87-88°, b_5 165-66° (dec.) (a, b)	634
$CHCl_2$ (443)	IB	88	m. 112-14° (b, d)	1051
	VA (c)	--	m. 112-14°	2395
CCl_3 (443)	--	--	Moessbauer resonance spectrum (b)	7
$ClCH_2CH_2$	--	--	Use as stabilizer for polyurethan	1660
$MeCHCl$	--	--	Use as stabilizer for polyurethan	1660
$Cl(CH_2)_3$	--	--	Use as stabilizer for polyurethan	1660
$Cl(CH_2)_4$	IIIA	--	Moessbauer resonance spectrum, struct.	290
$Cl(CH_2)_6$	IIIA	--	Moessbauer resonance spectrum, struct.	290
$Cl(CH_2)_8$	IIIA	--	Moessbauer resonance spectrum, struct.	290
$H(CF_2)_{10}$	IVA	--	(e)	1277
$Br(CH_2)_{10}$	--	--	Use as stabilizer for polyurethan	1660
$Cl(CH_2)_{14}$	IIIA	--	Moessbauer resonance spectrum, struct.	290
o-ClC_6H_4 (443)	IB	83	m. 84-85°, x-ray powder diffraction anal.	1051
	IIIA	92	m. 84-85°	2395
	VA (f, g)	74-92	--	2395
p-ClC_6H_4	--	--	Use as stabilizer for polyurethan (d)	1660
$(C_8H_{17})_2Sn(O_2CCHMeCl)_2$	--	--	Use as stabilizer for polyurethan	1660
$Bu_2Sn(O_2CR')O_2CC_{11}H_{23}$	--	--	R' = $Br(CH_2)_{10}$ (b)	1660

* Numbers in parenthesis refer to pages in main volume.
** The characters used here correspond to the synthetic scheme in the introduction to Chapter 4.3.5.2.1.
(a) Rxn. with $Bu_2Sn(OMe)_2 \rightarrow (Bu_2SnOMe)O(Bu_2SnO_2CCH_2Cl)$ (1660). (c) Synth. from $(Bu_2SnO_2CCH_2Cl) + MeO_2CCH_2Cl$ (634), $Bu_2Sn(OMe)O_2CCH_2Cl$ (2184). (b) Use as stabilizer for polyurethan foam (1660). (c) Synth. from $(Bu_2SnO_2CCH_2Cl) + MeO_2CCH_2Cl$ (634), $Bu_2Sn(OMe)O_2CCH_2Cl$ (2184). (d) Rxn. with $Bu_2SnO \rightarrow Bu_2Sn(O_2CR')_2 \cdot 2$ Bu_2SnO (2857). (e) Use as antiweathering agent for halogenated vinyl compds. (1277). (f) Synth. from $(Bu_2SnO_2CR')_2O$ and $R'CO_2H$ (2395). (g) Synth. from $(Bu_2SnO_2CC_6H_4NO_2$-$p)_2O$ (2395).

Table 214. "Basic" Diorganotin Halocarboxylates $R_2Sn(O_2CR')_2 \cdot nR_2SnO$

Formula	Synth. Method*	Yield	Properties	Ref.
$(Me_2SnO_2CR')_2O$				
R' = CH_2Cl	IC	>70	m. 226-27°, IR spectrum	409
$CHCl_2$	IC	>70	m. 232-33°, IR spectrum	409
CCl_3	IC	>70	m. 221-22°, IR spectrum	409
CH_2Br	IC	>70	m. 172-74°, IR spectrum	409
o-BrC_6H_4	IC (a)	--	m. 253-56°, IR spectrum	1880
$(Et_2SnO_2CCHCl_2)_2O$	IB	87	m. 164-65°, IR spectrum	1051
$(Et_2SnO_2CC_6H_4Cl$-$o)_2O$	IB	100	m. 171-72°, IR spectrum	1051
$Bu_2Sn(O_2CCH_2Cl)_2 \cdot 3Bu_2SnO$	IVB	83	m. 148-49°, soly.	634
$(Bu_2SnO_2CCHCl_2)_2O$	IB	82	m. 192°, IR spectrum	1051
	VA	(b)	m. 192°	2395
$Bu_2Sn(O_2CCHCl_2)_2 \cdot 2Bu_2SnO$	IIIC	46	m. 124-25°	2857
$Bu_2Sn(O_2CCHCl_2)_2 \cdot 3Bu_2SnO$	IIIA	86	m. 131°	2595
	IB (c)	64	m. 131°	2594
$[Bu_2SnO_2C(CH_2)_6Cl]_2O$	IIIA	--	Moessbauer resonance spectrum, struct.	290
$[Bu_2SnO_2C(CH_2)_8Cl]_2O$	IIIA	--	Moessbauer resonance spectrum, struct.	290
$(Bu_2SnO_2CC_6H_4Cl$-$o)_2O$	IB	91	m. 81-82°, x-ray diffraction pattern	1051
	IIIA	56	m. 81°, mol. wt., soly.	2595
$Bu_2Sn(O_2CC_6H_4Cl$-$o)_2 \cdot 2Bu_2SnO$	IIIA	80	m. 119°, mol. wt., soly.	2595
$(Bu_2SnO_2CC_6H_4Cl$-$p)_2O$	IIIA	--	m. 145-49° (d)	1289
$Bu_2Sn(O_2CC_6H_4Cl$-$p)_2 \cdot 2Bu_2SnO$	IIIA	53	m. 144°, mol. wt., soly.	2595
	IIIC	94	m. 144-45°	2857

* The characters used here correspond to the synthetic scheme in the introduction to Chapter 4.3.5.2.1. (a) Synth. also by thermal decomposition of o-$Me_3SnO_2CC_6H_4Br$ at 235° (1880). (b) By-prod. in rxn. of $(Bu_2SnO_2C$-$C_6H_4NO_2$-$p)_2O$ with Cl_2CHCO_2H (2395). (c) Synth. from $Bu_2SnCl_2 \cdot 3Bu_2SnO$ (2594). (d) Use as stabilizer for PVC, polyvinyl acetate (1289).

Table 215. Diorganotin Carboxylates Derived from Acids Containing Oxygen and Sulfur $R_2Sn(O_2CR')_2$

Formula*	Synth. Method**	Properties	Ref.
$Me_2Sn(O_2CCH_2SC_8H_{17}\text{-}i)_2$	--	(a)	--
$Pr_2Sn(O_2CCH_2CH_2CMeAr_2)_2$	IIIA	Ar = 2,5,4-t-$Bu_2HOC_6H_2$, m. 80-85° (b)	1575
$Bu_2Sn(O_2CR')_2$			
R' = PhCHOH	IB	In 100% yield, m. 165-66°	1051
	IB (c)	--	1515
i-$C_8H_{17}SCH_2$	--	Use as stabilizer for PVC (d, e)	1058, 1170
$C_9H_{19}SCH_2$	--	(e)	--
Ar_2CMeCH_2	IIIA	Ar = 2,5,4-t-$Bu_2HOC_6H_2$, m. 60-70° (b)	1575
$BuO_2CCH_2CH_2SCH_2CH_2$	--	Use as stabilizer for polyolefin	214
$Ar_2CMeCH_2CH_2$	IIIA	Ar = 2,5,4-t-$Bu_2HOC_6H_2$, m. 56-62° (b)	1575
$Ar_2CMeCH_2CH_2$	--	Ar = 3,5,4-t-$Bu_2HOC_6H_2$ (f)	1469
$MeO_2C(CH_2)_4$	--	Use as esterification cat.	1156
o-HOC_6H_4 (444)	--	(g, h)	--
o-$HOCH_2CH_2O_2CC_6H_4$	--	(h)	--
o-$MeOCH_2CH_2O_2CC_6H_4$	--	(h)	--
o-$BuOCH_2CH_2O_2CC_6H_4$	--	(h)	--
o-$C_{12}H_{25}O_2CC_6H_4$	--	(h)	--
p-$MeO_2CC_6H_4$	--	Use as esterification catalyst	1156
p-EtBuCHCH$_2O_2CC_6H_4$	IC	n_D^{20} 1.5184, use as stabilizer for PVC	2954
(1-$C_8H_{17}O_2C)_2C_6H_3$	VA	--	1204
[$Bu_2Sn(O_2CCH_2CH_2S)_2CH]_2$	IIIA	Use as stabilizer in PVC	1170
($C_8H_{17})_2Sn(O_2CCH_2SC_8H_{17})_2$	--	Use as stabilizer in PVC (i)	2143
($C_8H_{17})_2Sn(O_2CC_6H_4OH\text{-}p)_2$	--	(j)	--
($C_8H_{17})_2Sn(O_2CC_6H_4CO_2H\text{-}o)_2$	--	Use as stabilizer for polyolefin	1062a
$Ph_2Sn[O_2CCH(SMe)SCH_2Ph]_2$	IIIA	Use as stabilizer for PVC	1170
$R_2Sn[O_2C(CHR')_nOAr]_2$	IIIA	R = alkyl, R' = H, alkyl, n = 1-3 (k)	2255

| (Bu$_2$SnO$_2$CCHOHPh)$_2$O | IB | In 99% yield, m. 190°, IR spectrum | 1051 |

* Numbers in parenthesis refer to pages in main volume.
** The characters used here correspond to the synthetic scheme in the Introduction to Chapter 4.5.3.2.1.
(a) Reactivity as polyurethan foam gelling catalyst (2906). (b) Use as antioxidant for polyolefins (1575). (c) Synth. from (Bu$_2$SnCl)$_2$S (1514). (d) Use as antioxidant for mineral lubricants (2172). (e) Use as stabilizer for oligomeric allyl chlorides (2992). (f) Use as stabilizer for polyolefins (1469). (g) Biol. activity as fungicide against Phytophthora infestans (1562). (h) Use as catalyst for modifying polypropylene (269). (i) Effect of epoxidized soy bean oil on migration of stabilizer (2143). (j) Rxn. with 1,2,4-MeC$_6$H$_3$(NCO)$_2$ → tin contg. polyurethan (2959). (k) Use as fungicide and herbicide (2255).

Table 216. Diorganotin Carboxylates Derived from Nitrogen Substituted Acids   R$_2$Sn(O$_2$CR')$_2$·nR$_2$SnO

Formula	Synth. Method*	Yield	Properties	Ref.
Bu$_2$Sn(O$_2$CR')$_2$				
R' = BzNHCH$_2$	--	--	(a)	--
p-H$_2$NC$_6$H$_4$	IIIA	--	m. 118-19°	2732
o-O$_2$NC$_6$H$_4$	--	--	(a)	--
m-O$_2$NC$_6$H$_4$	IB (b)	--	--	1514
	IB	84	m. 120-22°	1051
3,5-(O$_2$N)$_2$C$_6$H$_3$	IB	89	m. 187-88°	1051

* The characters used here correspond to the synthetic scheme in the Introduction to Chapter 4.3.5.2.1.
(a) Rxn. with Bu$_2$SnO → Bu$_2$SnO$_2$CR'(Bu$_2$SnO)$_2$OCR' (2857). (b) Synth. from (Bu$_2$SnCl)$_2$S, R'CO$_2$H and aq. NaOH in petr. ether (1514).

Table 216 Continued $R_2Sn(O_2CR')_2 \cdot nR_2SnO$

Formula	Synth. Method*	Yield	Properties	Ref.
$(Me_2SnO_2CCH_2CN)_2O$	IC	>70	m. 300°, IR spectrum	409
$(Bu_2SnO_2CCH_2CN)_2O$	IIIA	--	(c)	2566
$Bu_2Sn(O_2CCH_2NHBz)_2 \cdot 2Bu_2SnO$	IIIA	89	m. 140-41°	2857
$Bu_2Sn(O_2CC_6H_4NO_2\text{-}o)_2 \cdot 2Bu_2SnO$	IIIC	96	m. 120°	2857
$(Bu_2SnO_2CC_6H_4NO_2\text{-}m)_2O$	IB	97	m. 148-49°, IR spectrum	1051
$Bu_2Sn(O_2CC_6H_4NO_2\text{-}m)_2 \cdot 2Bu_2SnO$	IB (d)	67	m. 83°	2594
$Bu_2Sn(O_2CC_6H_4NO_2\text{-}m)_2 \cdot 3Bu_2SnO$	IB (e)	98	m. 78°	2594
$(Bu_2SnO_2CC_6H_4NO_2\text{-}p)_2O$	IB	100	m. 187°, IR spectrum	1051
$Bu_2Sn(O_2CC_6H_4NO_2\text{-}p)_2 \cdot 2Bu_2SnO$	IIIA	35	m. 187°, soly. (f, g)	2595
$Bu_2Sn(O_2CC_6H_4NO_2\text{-}p)_2 \cdot 2Bu_2SnO$	IB (d)	91	m. 85°, soly.	2594
	IIIA	79	m. 85°, mol. wt.	2595
	IIIC	90	m. 85°	2857
$Bu_2Sn(O_2CC_6H_4NO_2\text{-}p)_2 \cdot 3Bu_2SnO$	IB (e)	>70	m. 83°	2594
	IB (e)	63	m. 83°, mol. wt.	2595
$Bu_2Sn(O_2CC_6H_4NO_2\text{-}p)_2 \cdot 5Bu_2SnO$	IIIA	54	m. 71°, mol. wt., soly.	2595
$Bu_2Sn(O_2CC_6H_4NO_2\text{-}p)_2 \cdot 112Bu_2SnO$	IIIA	--	Soly.	2595
$[Bu_2SnO_2CC_6H_3(NO_2)_2\text{-}3,5]_2O$	IB	97	m. 204-205°, IR spectrum	1051

(c) Rxn. with $RR'CO \to [Bu_2SnO_2CC(CN):CRR']_2O$; $R + R' = 3,5,4\text{-}t\text{-}Bu_2(HO)C_6H_2 + H$, $Me + Ph$ (2566). (d) Synth. with $Bu_2SnCl_2 \cdot 2Bu_2SnO$ (2594). (e) Synth. with $Bu_2SnCl_2 \cdot 3Bu_2SnO$ (2594). (f) Rxn. with $ClCH_2CO_2H \to Bu_2Sn(O_2CCH_2Cl)_2$ + $(Bu_2SnO_2CCH_2Cl)_2O$ (2395). (g) Rxn. with $o\text{-}ClC_6H_4CO_2H \to Bu_2Sn(O_2CC_6H_4Cl\text{-}o)_2 + Bu_2Sn(O_2CC_6H_4NO_2\text{-}p)_2$ (2395).

4.3.5.2.6.7 UNSUBSTITUTED DIORGANOTIN CARBOXYLATES DERIVED FROM POLYFUNCTIONAL AND METAL SUBSTITUTED ACIDS

All compounds are listed in Table 217.

4.3.5.2.6.8 UNSUBSTITUTED DIORGANOTIN CARBOXYLATES WITH POLYCARBOXYLIC ACIDS

This subchapter comprises diorganotin derivatives of di- or polycarboxylic acids having two or more carboxylate groups bound to tin. Derivatives of half-esters of unsaturated dicarboxylic acids are listed in Subchapter 4.3.5.2.3.

4.3.5.2.6.8.1 UNSUBSTITUTED DIORGANOTIN CARBOXYLATES DERIVED FROM UNSUBSTITUTED DICARBOXYLIC ACIDS

Dibutyltin Maleate (M-442) $SnC_{12}H_{20}O_4$ $\overline{Bu_2SnO_2CCH:CHCO_2}$
Synth.: By rxn. of Bu_2SnCl_2 with $(:CHCO_2Na)_2$ in aq. MeOH (1357).
By rxn. of Bu_2SnO with $(:CHCO_2H)_2$ in C_6H_6 (1557).
Prop.: m. 134° (1641), IR (139) and Moessbauer resonance spectrum (6).
Addn. Compd.: $\overline{Bu_2SnO_2CCH:CHCO_2} \cdot H_2O$, from components, m. 69-71°, IR spectrum, soly. (1557) [$SnC_{12}H_{22}O_5$].
$(\overline{Bu_2SnO_2CCH:CHCO_2} \cdot H_2O)_3$, from components, m. 110-13° (1557) [$Sn_3C_{36}H_{66}O_{15}$].
$(\overline{Bu_2SnO_2CCH:CHCO_2} \cdot H_2O)_4$, from components, m. 134-36°, mol. wt. (1557) [$Sn_4C_{48}H_{88}O_{20}$].
Rxn. with: $C_5H_6 \rightarrow (Bu_2SnO_2C_7H_8CO_2)_x$ 2,3-bicyclo[2.2.1]heptene-5-dicarboxylate (1558).
Biol. Prop.: Toxicity (188-9).
Activity as anthelmintic (1357), against cestodes and nematodes (1131), against Hymenolepsis fraterna (1134).
Activity as fungicide (1357).
Use: As additive for thermoplastic copolymers with inorganic fillers (1714a).
As catalyst for curing silicone rubber (1357, 1384).
As corrosion inhibitor for spinning nozzles (459).
As stabilizer for ABS-copolymers (1285-6), for flame retardant polymers (1185a), for halogen contg. polymers (707), Cl (420), F (1073), for polyethylene (6, 139), for polypropylene (293, 1062a, 1063), for PVC (597, 782, 1082, 1208, 1274, 1285-6, 1381, 1523-4, 1572, 1651, 1862, 2016, 2037, 2817-8, 2832), for PVC copolymers (115, 1662, 2552).

Additional diorganotin dicarboxylates are listed in Tables 218 and 219.

4.3.5.2.6.8.2 UNSUBSTITUTED DIORGANOTIN DICARBOXYLATES DERIVED FROM FUNCTIONALLY SUBSTITUTED POLYCARBOXYLIC ACIDS

All compounds are listed in Table 219.

Table 217. Diorganotin Carboxylates Derived from Polyfunctional and Metal Substituted Acids $R_2SnO_2CR' \cdot nR_2SnO$

Formula	Synth. Method*	Yield	Properties	Ref.
$Bu_2Sn(O_2CR')_2$				
R' = PhCMe:NOCH$_2$	IIIA (a)		m. 95-105° (b)	2626
p-MeOC$_6$H$_4$CH:NOCH$_2$	IIIA (a)		m. 84-87°, soly. (c)	1574
p-H$_2$NSO$_2$C$_6$H$_4$	IB	100	m. 225°	1051
2,4-Cl(O$_2$N)C$_6$H$_4$	IIIA	--	m. 101-102°	2732
$(Bu_2SnO_2CC_6H_4SO_2NH_2$-p$)_2O$	IB	98	m. 250-52°, IR spectrum	1051
$Me_2Si[C_6H_4(CH_2)_nCO_2]_2R'$				
n = 0, R' = o-Et$_2$Sn	IB	26	m. 210°, polymeric	1516
0 o-Et$_2$SnEt$_2$SnO	IB	100	m. 119°, viscosity in C$_6$H$_6$, polymeric	1516
0 p-Et$_2$Sn	IB	54	m. 258°, mol. wt., polymeric	1516
0 o-Bu$_2$Sn	IB	87	m. 61°, viscosity in C$_6$H$_6$, polymeric	1516
0 o-Bu$_2$Sn·Bu$_2$SnO	IB	81	m. 145°, mol. wt., polymeric	1516
0 p-Bu$_2$Sn	IB (d)	55	m. 100°, soly., polymeric	1516
0	IB (e)	--	m. 80°, mol. wt., viscosity in C$_6$H$_6$	1516
1 p-Bu$_2$Sn	IB (d)	28	m. 112°, viscosity in C$_6$H$_6$, polymeric	1516
	IB (e)	54	m. 100°, viscosity in C$_6$H$_6$	1516
1 p-Bu$_2$Sn·Bu$_2$SnO	IB	58	Polymeric	1516
$Ph_2Sn(O_2CC_6H_4FeC_5H_5)_2$	IB	32	m. 240-43°, IR spectrum	1310, 1374
	V (f)	69	Soly.	1310, 1374
$(Ph_2SnO_2CC_6H_4FeC_5H_5)_2O$	II	40	Bronze colored cryst., m. 178-82°	1310, 1374
	II (g)	58	IR spectrum, mol. wt.	1310, 1374

* The characters used here correspond to the synthetic scheme in the introduction to Chapter 4.3.5.2.1. (a) Synth. with Bu$_2$SnO and R'CO$_2$Me. (b) Use as bactericide, fungicide and herbicide (2626). (c) Use as bactericide, fungicide, weed killer (1574), use as stabilizer for polymers (1574). (d) Synth. in petr. ether. (e) Synth. in CCl$_4$. (f) Synth. from Ph$_3$SnO$_2$CC$_5$H$_4$FeC$_5$H$_5$ with C$_5$H$_5$FeC$_5$H$_4$CO$_2$H in alc. under reflux (1310, 1374). (g) Synth. from Ph$_3$SnH (1310, 1374).

Table 218. Diorganotin Dicarboxylates Derived from Unsubstituted Dicarboxylic Acids $R_2SnO_2CYCO_2$

Formula*	Synth. Method**	Yield	Properties	Ref.
$Me_2Sn(O_2C)_2 \cdot H_2O$ (445)	I (a)	--	Recryst. from H_2O	409
$[(PhCH_2)_2SnO_2CYCO_2]_x$				
Y = CH:CH (445)	IB	--	Maleate (b)	1357
	IIIB	26	m. 205.5-207.5°	1779
$(CH_2)_3$	IB	--	(b)	1357
$(CH_2)_4$	IB	--	(b)	1357
$(CH_2)_8$	IB	--	(b)	1357
$o\text{-}C_6H_4$	IB	--	(b)	1357
$(Bu_2SnO_2CYCO_2)_x$				
Y = CH_2	--	--	TLC, paper chromatography, polarography (c)	1136
CH_2CH_2 (445)	IB	78	m. 235°, use as stabilizer for PVC (2832),	1049
	IB (d)	--	for polyethylene (1062a)	1514
CH:CH (445)	IB	91	Fumarate, m. > 300° (e)	1049
C:C	IB	81	m. > 300°	1049
$(CH_2)_3$	IB	--	(b)	1357
$(CH_2)_4$ (445)	IB	High	m. 220°	1049
	IB	--	Use as stabilizer for PVC (2037)	1357
	IB (d)	--	(b, e)	1514

* Numbers in parenthesis refer to pages in main volume.
** The characters used here correspond to the synthetic scheme in the introduction to Chapter 4.3.5.2.1.
(a) Synth. with oxalic acid in H_2O (409). (b) Use as anthelmintic, fungicide and curing agent for silicone rubber (1357). (c) Detn. after extraction from plastic flakes (1136). (d) Synth. from $(Bu_2SnCl)_2S$ (1514). (e) Thermal stability and glass transition (798), thermal degradation (799).

Table 218 Continued $R_2SnO_2CYCO_2$

Formula*	Synth. Method**	Yield	Properties	Ref.
$(Bu_2SnO_2CYCO_2)_x$ (Cont.)				
Y = 2,3-C_7H_8 (f)	VC	--	m. 239-42°, soly.	1558
$(CH_2)_8$ (445)	IB	--	Use as stabilizer for polyethylene (1062a)(b)	1357
o-C_6H_4	IB	--	(b)	1357
	IB (d)	--	--	1514
p-C_6H_4 (445)	IB	70	m. > 300° (e)	1049
	IB (d)	--	m. > 300°	1514
$(CH_2)_4$ + p-C_6H_4	IB	52	Ratio 1:1, m. > 300°	1049
$[1\text{-}Bu_2SnO_2C(CH_2)_4CO_2]_x$	IB	85	m. 210°	1049
$[(C_6H_{13})_2SnO_2CCH:CHCO_2]_x$ (446)	IIIB	--	Use as stabilizer for PVC	90
$[(C_8H_{17})_2SnO_2CYCO_2]_x$				
Y = $(CH_2)_3$	IB	--	(b)	1357
$(CH_2)_4$	IB	--	Use as polypropylene stabilizer (1062a)(b)	1357
$(CH_2)_8$	IB	--	(b)	1357
o-C_6H_4	IB	--	(b)	1357
$Ph_2SnO_2C(CH_2)_4CO_2$	IB	96	m. 185-87°	1049
$Bu_2Sn(O_2C)_2Y \cdot nBu_2SnO$				
n = 3, Y = CH_2CH_2	IB (g)	97	m. 203-20°	2594
	IIIA	76	m. 210-25°, mol. wt.	2595
3, C:C	IB (g)	25	m. 145-50°	2594
3, $(CH_2)_4$	IB (g)	96	m. 158-60°, mol. wt.	2594-5
	IIIA	82	m. 153-60°, mol. wt.	2595
5, $(CH_2)_4$	IB (g)	94	m. < 300°	2594
3, p-C_6H_4	IB (g)	74	m. 160-70°	2594

(f) Deriv. of 2,3-bicyclo[2.2.1]heptene-5 carboxylic acid. (g) Synth. from $Bu_2SnCl_2 \cdot nBu_2SnO$ (2594).

Table 219. Diorganotin Dicarboxylates Derived from Functionally Substituted Polycarboxylic Acids $R_2SnO_2CY'CO_2$

Formula*	Synth. Method**	Yield	Properties	Ref.
[$Bu_2Sn(O_2CCH_2S)_2CH$]$_2$	IIIA	--	(a - 1170)	1170
$Bu_2Sn(O_2CCH_2S)_2Y$				
Y = CH_2	IIIA	--	m. 146-48° (a - 1264, 2028)	2028
	IIIA	--	(b - 1170)	1170
CS (445)	IIIA	93	m. 192-96° (c - 214)	214
PhCH (445)	IIIA	--	m. 157-62°	1170
	IIIA	--	m. 157-62°	2917
	IIIA	--	(a - 1170, 1819, 1834, 2917) (b - 1170)	1819
o-HOC_6H_4CH	IIIA	--	(a - 1170, 2028) (b - 1170)	1170, 2028
3,4-Me(HO)C_6H_3CH	IIIA	--	(a - 2028)	2028
$H_2NC_6H_4CH$	IIIA	--	(a - 2028)	2028
$O_2NC_6H_4CH$	IIIA	--	(a - 2028)	2028
Me_2C (445)	IIIA	--	(a - 1170) (b - 1170)	1170
CH_2:CHCH	IIIA	--	(a - 1819)	1819
PhCH:CHCH	IIIA	--	(a - 1819)	1819
PrCH	IIIA	--	(a - 1058, 1170, 1819)	1170, 1819
MeEtCH	IIIA	--	(a - 1170) (b - 1170)	1170
CH_2CH:$CHCH_2$	IIIA	52	m. 138°	1049
c.1,1-C_6H_{10}	IIIA	--	(a - 1170) (b - 1170)	1170
C_8H_{16}	--	--	(d)	--
1-$C_9H_{19}CH$	IIIA	--	(a - 1170) (b - 1170)	1170

* Numbers in parenthesis refer to pages in main volume.
** The characters used here correspond to the synthetic scheme in the introduction to Chapter 4.3.5.2.1.
(a) Use as stabilizer for PVC. (b) Use as stabilizer for polypropylene. (c) Use as stabilizer for polyolefins.
(d) Effect on radiation induced graft polymerization of PhCH:CH_2 and CH_2:CHCN to PVC (1389).

Table 219 Continued

Formula*	Synth. Method**	Yield	Properties	Ref.
$Bu_2Sn(O_2CCH_2CH_2)_2S$ (445)	IIIA	--	m. 84-92° (b - 214) (c - 1470)	214
$[Bu_2Sn(O_2CH_2CH_2S)_2CH]_2$	IIIA	--	(a - 1170)	1170
$Bu_2Sn(O_2CCH_2CH_2S)_2Y$				
Y = CH_2	IIIA	--	(a - 1170) (b - 1170)	1170
PhCH (446)	IIIA	--	m. 79-89° (a - 1170, 2917) (b - 1170)	1170, 2917
o-HOC_6H_4CH	IIIA	--	m. 131-34° (a - 1170, 2917) (b - 1170)	1170, 2917
CMe_2 (446)	IIIA	--	m. 58-60° (a - 1170, 2917) (b - 1170)	1170, 2917
CMeEt	IIIA	--	(a - 1170) (b - 1170)	1170
EtO_2CCH_2CMe	IIIA	--	(a - 1170) (b - 1170)	1170
Et_2CHCH	IIIA	--	(a - 1170) (b - 1170)	1170
c.1,1-C_6H_{10}	IIIA	--	(a - 1170) (b - 1170)	1170
1-$C_{10}H_{19}CH$	IIIA	--	(a - 1170) (b - 1170)	1170
$Bu_2SnO_2CYCO_2$				
Y = $CH_2CH_2CH(NHO_2CCH_2Ph)$	IB	82	m. 163-65°, polymeric	1049
$Me_2C[SCH(CH_2CO_2H)]_2$	IIIA	--	(a - 1170)	1170
$O(CH_2CH_2O_2CCH:CH)_2$	--	--	Maleate (e)	2876
$O(CHMeCH_2O_2CCH:CH)_2$	--	--	Maleate (f)	1185a
1,2,4,5-$HN(CO)_2C_6H_2$	IB	62	Pyromellitic imide	1049
o-$HOC_6H_4CH(SC_6H_4)_2$	IIIA	--	(a - 1170)	1170
i-Bu_2Sn-citrate	IB	66	m. > 300°	1049
$(Me_3CCH_2)_2Sn(O_2CCH_2CH_2)_2S$	IB	--	m. 161-66° (a - 1170) (c - 214)	214, 1170
$(C_8H_{17})_2Sn(O_2CCH_2)_2Y$				
Y = o-$HOC_6H_4CH(S-)_2$	--	--	(a - 1170) (b - 1170)	--
1-$C_{10}H_{19}CH(S-)_2$	--	--	(a - 1170) (b - 1170)	--
CH_2SCH_2 (446)	IIIA	--	m. 49-56° (c - 214)	214
$PhCH(SCH_2-)_2$	--	--	(a - 1170) (b - 1170)	--
o-$HOC_6H_4CH(SCH_2-)_2$	--	--	(a - 1170) (b - 1170)	--

Me$_2$C(SCH$_2$-)$_2$	IIIA		High boiling liq. (a - 1170) (b - 1170)	1170
1-C$_8$H$_{18}$CH(SCH$_2$-)$_2$	--	--	(a - 1170) (b - 1170)	--
(EtBuCHCH$_2$)$_2$SnO$_2$CYCO$_2$				
Y = CH$_2$CH(C$_4$H$_7$O) (g)	IIIA	100	Viscous liq. (a - 1305)	1305
CH$_2$CH(C$_6$H$_{11}$O) (517) (g,h)	IIIA	100	Dark viscous liq. (a - 1305)	1305
CH$_2$CH(C$_8$H$_{15}$O) (g)	IIIA	100	Light brown viscous liq. (a - 1305)	1305
Ph$_2$Sn(O$_2$CCH$_2$CH$_2$)$_2$S	IB	91	(c - 214)	214

(e) Use as catalyst for polyurethan preparation (2876). (f) Use as stabilizer for flame retardant polymers (1185a). (g) Epoxide derivatives. (h) In main volume page 517 line 5 f.b., third column correct 1/2R = CH$_2$OCH$_2$ to 1/2R = CH$_2$CH(C$_8$H$_{15}$O).

4.3.5.2.8 FUNCTIONALLY SUBSTITUTED AND MISCELLANEOUS DIORGANOTIN CARBOXYLATES

Bis(2-cyanoethyl)tin Di-2-ethylhexanoate

$SnC_{22}H_{38}N_2O_4$ $(NCCH_2CH_2)_2Sn(O_2CCHEtBu)_2$

<u>Synth.</u>: By rxn. of $[(NCCH_2CH_2)_2SnOH]_2O$ with $EtBuCHCO_2H$ in C_6H_6, in 63% yield (895b), in 73% yield (1746).

<u>Prop.</u>: White cryst., m. 154-55° (dec.), mol. wt. (895b, 1746).

<u>Use</u>: As biocide (895b), as stabilizer for plastics (895b).

Bis[di(2-cyanoethyl)cyclohexylmaleatooxytin] Oxide

$Sn_2C_{32}H_{42}N_4O_9$ $[(NCCH_2CH_2)_2SnO_2CCH:CHCO_2C_6H_{11}]_2O$

<u>Synth.</u>: By rxn. of $[(NCCH_2CH_2)_2SnOH]_2O$ with $c.C_6H_{11}O_2CCH:CHCO_2H$ in C_6H_6, in 54% yield (895b).

<u>Prop.</u>: m. 161-63° (dec.) (895b).

<u>Use</u>: As biocide (895b), as stabilizer for plastics (895b).

Di(epoxyethyl)tin Di(epoxystearate)

$SnC_{40}H_{72}O_8$ $(\overline{OCH_2CH})_2Sn[O_2C(CH_2)_7\overline{CHCHO}C_8H_{17}]_2$

<u>Synth.</u>: By rxn. of $(CH_2:CH)_2Sn$-oleate with CF_3CO_3H (1442).

<u>Use</u>: As stabilizer for PVC (1442).

Polybis(2-carboxylatoethyl)tin $(SnC_6H_8O_4)_x$ $[Sn(CH_2CH_2CO_2)_2]_x$

<u>Synth.</u>: By rxn. of $(RO_2CCH_2CH_2)_2SnBr_2$ with 50% aq. alkali, in 90% yield (1485).

By thermal dec. of $(HO_2CCH_2CH_2)_4Sn$ at 100°, in 5% yield (1744).

<u>Prop.</u>: Cryst. (1485), m. 280° (dec.) (1744), IR spectrum (1485, 1744), differential thermal analysis, thermogravimetric analysis (1744), struct. (1485, 1744).

<u>Rxn. with</u>: Thermal dec. at 280° → SnO_2 (1744).

$HBr → (HO_2CCH_2CH_2)_2SnBr_2$ (1485).

$SOCl_2 → (HO_2CCH_2CH_2)_2SnCl_2$ (1485).

Di(p-epoxystyryl)tin Di(epoxystearate)

$SnC_{52}H_{80}O_8$ $(p-\overline{OCH_2CH}C_6H_4)_2Sn[O_2C(CH_2)_7\overline{CHCHO}C_8H_{17}]_2$

<u>Synth.</u>: By rxn. of $(p-CH_2:CHC_6H_4)_2Sn$-oleate with peroxyacetic acid (1442).

<u>Use</u>: As stabilizer for PVC (1442).

Dimethylacetatosiloxane Dibutyltin Acetate

$SnSiC_{14}H_{30}O_5$ $(Bu_2SnOAc)O(SiMe_2OAc)$

<u>Synth.</u>: By rxn. of Bu_2SnO with $Me_2Si(OAc)_2$ under reflux in C_6H_6 (946).

<u>Prop.</u>: Oil, IR spectrum (946).

4.3.5.2.9 DIORGANOTIN CARBOXYLATES CONTAINING TIN-SULFUR AND TIN-NITROGEN BONDS

The diorganotin carboxylates containing heteroatom-tin bonds are summarized in Table 220. The compounds are prepared by methods from the following scheme.

I Reactions with organotin halides:
(IA) Me_2SnCl_2 with $HSCH_2CO_2H$,
(IB) $(R_2SnCl)_2Sn$ with RCO_2Na.

II Reactions of diorganotin oxide with:
(IIA) $HS(CH_2)_nCO_2H$ or $HN(COR)_2$,
(IIB) RSAc.

III Reaction of diorganotin dialkoxide with mercapto acids.

IV Addition reaction between diorganotin dicarboxylate and diorganotin sulfide.

V Insertion reaction of CS_2, PhNCS or EtNCO in the tin-alkoxide bond of $Bu_2Sn(OMe)OAc$.

R = organic group, n = 1-4.

4.3.5.3 MONOORGANOTIN CARBOXYLATES

Monoorganotin carboxylates are reported as organotin tricarboxylates and recently as "basic" tris(monoalkyltin)dioxopentacarboxylate $(RSn)_3O_2(O_2CR')_5$. The compounds listed in Table 221 are prepared by methods from the following scheme.

I Metathetic substitution of organotin trihalides:
$RSnX_3 + R'CO_2Na \to RSn(O_2CR')_3$.

II Esterification of alkylstannoic acids or bis(allyltin)trioxide with carboxylic acids:
$R_2Sn_2O + R'CO_2H \to RSn(O_2CR')_2$,
$RSnO_2H + R'CO_2H \to (RSn)_3O_2(O_2CR')_5$.

III Reaction of organotin trialkoxide with carboxylic acid:
$RSn(OR')_3 + Y(CO_2H)_2 \to (RSn)_2(O_2CYCO_2)_3$.

R = organic group, R' = organic group, may be functionally substituted, X = halogen, Y = divalent hydrocarbon radical.

Table 220. Diorganotin Carboxylates Containing Tin-Sulfur and Tin-Nitrogen Bonds $R_2Sn(X)O_2CR$

Formula*	Synth. Method**	Yield	Properties	Ref.
$Me_2SnO_2CCH_2S$	IA	80	m. 257-59°, IR spectrum	409
$Me(C_6H_{13})Sn(SC_6H_{13})O_2CCH:CHMe$	--	--	(a)	--
$(Bu_2SnO_2CR)_2S$				
R = Me	--	--	(b - 1833a)	--
Et_2CH	--	--	(b - 1833a)	--
EtBuCH	--	--	(b - 1833a)	--
$1-C_7H_{15}$	IV	--	Oil	945
$Me_3CCH_2CHMeCH_2$	--	--	(b - 1833a)	--
$C_{11}H_{23}$	--	--	(b - 1833a)	--
$EtBuCHCH_2O_2CCH:CH$	--	--	(b - 1833a)	--
$1-C_8H_{17}O_2CCH:CH$	--	--	(c)	--
$(Bu_2SnO_2CR')_2S_x$				
x = 3.5 R' = $2-C_4H_3O$	IB	91	m. 27-28°	1514
4 $HO_2CC!C$	IB	26	m. 26°	1514
2.5 $m-O_2NC_6H_4$	IB	--		1514
3.5 $m-O_2NC_6H_4$	IB	100	m. 35°	1514
$(Bu_2SnO_2CCH_2CH_2-)_2S$	I (b)	--	m. 50-52°	1514
$(Bu_2SnO_2CYCO_2SnBu_2S_4)_x$				
Y = trans-CH:CH	IB	61	Paste	1514
$(CH_2)_4$	IB	--	Pasty solid, dec. on standing	1514
$(CH_2)_8$	IB	75	Paste	1514
$p-C_6H_4$	IB	70	Paste	1514

Bu$_2$Sn(SR)OAc

R =			
Bu	IIB	b$_{0.05}$ 140° (d)	1055
1-C$_6$H$_{13}$	IIB	(d)	1055
C$_8$H$_{17}$	IIB	(d)	1055
1-C$_8$H$_{17}$	--	(e - 597)	--
Ph	IIB	(d)	1055
PhN:C(OMe)	V	IR and NMR spectra (f)	943
MeOCS	V	(f)	943

Bu$_2$Sn(SR)O$_2$CR'

R = Bu	R' = MeO$_2$CCH:CH	--	(a)	--
C$_{12}$H$_{25}$	C$_{11}$H$_{23}$ (447)	--	(a) (e - 1035, 2548) (g)	--
C$_{12}$H$_{25}$	C$_{17}$H$_{35}$	--	(a)	--
C$_{12}$H$_{25}$	EtO$_2$CCH:CH	--	(a)	--
C$_{12}$H$_{25}$	C$_{18}$H$_{35}$O$_2$CCH:CH (447)	--	Oleyl (e - 2548)	--
C$_8$H$_{17}$O$_2$CCH$_2$	C$_8$H$_{17}$O$_2$CCH:CH	--	(c)	--

Bu$_2$Sn(SC$_7$H$_4$NS)O$_2$CR   HS(C$_7$H$_4$NS) = 2-mercaptobenzothiazole

R = C$_7$H$_{15}$	--	(h - 707) (i)	--
1-C$_7$H$_{15}$	--	(i)	--
EtBuCH	--	(h - 707) (i)	--

* Numbers in parenthesis refer to pages in main volume.
** The characters used here correspond to the synthetic scheme in the introduction to Chapter 4.3.5.2.9.
(a) Use as transesterification catalyst (1155). (b) Synth. from (Bu$_2$ClSnO$_2$CCH$_2$CH$_2$)$_2$ with aq. Na$_2$S in EtOAc (1514). (c) Use as stabilizer for PVC and CH$_2$:CHCl-CH$_2$:CHOH-copolymers (2862). (d) Biol. activity as bactericide against Escherichia coli (1055), as fungicide against Aspergillus niger, Fusarium solani, Penicillium species and Rhizoctonia solani (1055). (e) Use as stabilizer for PVC. (f) Rxn. with H$_2$O → (Bu$_2$SnOAc)$_2$O (943). (g) Use as esterification catalyst (639). (h) Use as stabilizer for halogen contg. polymers. (i) Use as catalyst for esterification (21), for vulcanization (21), as stabilizer for PVC (21).

Table 220 Continued

$R_2Sn(X)O_2CR$

Formula*	Synth. Method**	Yield	Properties	Ref.
$Bu_2Sn(SC_7H_4NS)O_2CR$ (Cont'd.)			HSC_7H_4NS = 2-mercaptobenzothiazole	
R = $C_{11}H_{23}$	--	--	(h - 707) (i)	--
$C_{17}H_{35}$	--	--	(h - 707) (i)	--
$Bu_2SnO_2CCH_2S$ (447)	IIB	--	m. 176° (j)	2802
$Bu_2SnO_2CCHMeS$	III	98	(j)	2805
$Bu_2SnO_2CCH_2CH_2S$ (447)	IIB	--	White solid (h - 1035)	1035
	IIB	96	(e - 1651, 2963)	1380a
	III	98	(k)	2805
$[(C_8H_{17})_2SnO_2CC_{11}H_{23}]_2S$	--	--	(e - 1833a)	--
$(C_8H_{17})_2SnO_2CCH_2CH_2S$ (447)	--	--	(e - 1651, 2548) (l)	--
$[(C_8H_{17})_2SnO_2CCH_2CH_2S]_x$	IIA	95	m. 90-91°, x = 4-8 (e - 172)	172
$(C_8H_{17})_2Sn(SBz)O_2CCH_2SC_8H_{17}$	--	--	(e - 2548)	--
$Ph_2Sn(O_2CC_6H_4NO_2-m)_2 \cdot 3Ph_2SnS_4$	IB	--	m. 95-100°	1514
$Ph_2Sn(SC_{12}H_{25})O_2CCH:CHCO_2R$	--	--	R = $(CH_2CH_2O)_{45}OCCMe_3$ (m)	2017
$Me_2Sn(NHOCEt)O_2CEt$	IIA	--	m. 190-92°	1181
$Bu_2Sn(NEtCO_2Me)OAc$	V	--	IR and NMR spectra	943
$Bu_2Sn(NHOCCH_2Ph)O_2CCH_2Ph$	IIA	83	m. 160°, soly.	1181
$Bu_2Sn(NHOCEt)O_2CEt$	IIA	--	m. 30°	1181
$Bu_2Sn(NEtCO_2Me)O_2CC_{11}H_{23}$	V	--	IR and NMR spectra	943
$(Bu_2SnNHCOCH_2CH_2CO_2)_x$	IIA	--	m. 250°, x = 5-25	1181
$(o-Bu_2SnNHCOC_6H_4CO_2)_x$	IIA	--	m. > 300°, x = 5-25	1181

(j) Use as stabilizer for ABS-resins (1286). (k) Use as stabilizer for PVC and PVC-copolymers (1866), investigation of PVC degradation with tagged isomers (158). (l) Investigation of PVC degradation with ^{113}Sn tagged compound (158). (m) Use as antistatic agent for thermoplastics (2017).

Table 221. Monoorganotin Carboxylates

Formula*	Synth. Method**	Properties	RSn(O$_2$CR')$_3$ Ref.	(RSn)$_3$O$_2$(O$_2$CR')$_5$ Ref.
MeSn(:O)OCO$_2$K	--	(a)	--	--
(PrSn)$_3$O$_2$(O$_2$CCH:CH$_2$)$_5$	II	Mol. wt.		637
BuSn(O$_2$CR')$_3$				
R' = H	I, II	(b)	1319	
Me (449)	--	(c, d)	--	
EtBuCH (449)	--	(e)	--	
R	--	(f)	--	
C$_4$H$_7$CH$_2$O$_2$CCH:CH	--	(g)	--	
CH$_2$:CHCO$_2$C$_3$H$_6$O$_2$CCH:CH	II	(h)	2974	
C$_{12}$H$_{25}$SCH(SMe)	II	(i)	213, 1170	
(BuSn)$_2$[(O$_2$CCH$_2$CH$_2$S)$_2$CHPh]$_3$ (449)	II	m. 67-68°, polymeric (i, j)	213, 1170	
(BuSn)$_2$[(O$_2$CCH$_2$CH$_2$S)$_2$CMe$_2$]$_3$	II, III	m. 50-80° (i, j)	213, 1170	
(BuSn)$_2$[(o-O$_2$CC$_6$H$_4$S)$_2$CHPh]	II	(i)	213, 1170	
(BuSn)$_3$O$_2$(O$_2$CCMe:CH$_2$)$_5$	II	Mol. wt.		637

* Numbers in parenthesis refer to pages in main volume.
** The characters used here correspond to the synthetic scheme in the introduction to Chapter 4.3.5.3.

(a) Rxn. with MeSH → MeSn(SMe)$_3$ (728). (b) Toxicity and phytotoxicity (1319), activity as fungicide against Alternaria tenuis, Botrytis cinera, Coniphora cerebella and Phytophthora infestans (1319). (c) Rxn. with Bu$_3$SnOAc at 180-85° → Bu$_2$Sn(OAc)$_2$ (540). (d) Use as catalyst for polyurethan preparation (229), relative activity as fungicide (1320). (e) Use as curing catalyst for polyurethan rubber (539a). (f) Effect on complement fixation test for virus (1151). (g) Use as stzbilizer for PVC (1283). (h) Use as microbiocide (2974), as intermediate for homo- and copolymerization (2974). (i) Use as stabilizer for halogen containing resins (213, 1170). (j) Use as stabilizer for polypropylene (1170).

Table 221 Continued

Formula*	Synth. Method**	Properties	Ref.
$C_8H_{17}Sn(OAc)_3$	--	TLC	1172a
$C_8H_{17}Sn[O_2CCH_2CH_2CH_2CMe(SR)_2]_3$	II	R = $HO_2CCH_2CH_2$ (i)	1170
$n\text{-}C_6F_{13}CH_2CH_2Sn(OAc)_3$	I	(k)	745a
$n\text{-}C_{12}F_{25}CH_2CH_2Sn(OAc)_3$	I	(k)	745a
$PhSn(O_2CH)_3$	I, II	(b)	1319
$PhSn(OAc)_3$ (450)	--	TLC	1253

$RSn(O_2CR)_3$ $(RSn)_3O_2(O_2CR')_5$

(k) Use as dirt, oil and water repellent, antiadhesive and release agent (745a).

4.4 ORGANOTIN COMPOUNDS WITH SULFUR SELENIUM AND TELLURIUM

4.4.1 ORGANOTIN SULFIDES, SELENIDES AND TELLURIDES

The organotin sulfides, selenides and tellurides compiled in Tables 222 through 224 are prepared by methods from the following scheme.

I Reaction of diorganotin dihydride with bis(tributyltin)sulfide.

II Reaction of organotin halides with:
 (IIA) Na_2S,
 (IIB) $Na_2S_2O_3$,
 (IIC) R_3SbS.

III Reaction of disodium sulfide with:
 (IIIA) R_3SnOH,
 (IIIB) the reaction product of $R_2Sn_2O_3$ and HCl.

4.4.1.1 TRIORGANOTIN SULFIDES, SELENIDES AND TELLURIDES

Bistrimethyltin Sulfide (M-452)
$Sn_2C_6H_{18}S$ $(Me_3Sn)_2S$
<u>Synth.</u>: By rxn. of Me_3SnSMe with sulfur at 120°, in 53% yield (2).
By rxn. of $(Me_3Sn)_2NMe$ with CS_2 at or below r.t., in 71% yield (1236a), with COS at r.t., in 80% yield (2455), with PhNCS at or below r.t., in 83% yield (1236a), with Ph_2CS at r.t., in 57% yield (2455), with Me_2NCPhS at 130°, in 85% yield (2455).
<u>Prop.</u>: $b_{0.05}$ 50°, n_D^{20} 1.5600 (2), IR (2, 644-5), assignment (1360), Raman, assignment (1360) and NMR (2, 645-6) spectra, relative basicity (644-5).
<u>Rxn. with</u>: $HgCl_2 \rightarrow HgS$ (647).
$SbCl_3 \rightarrow Me_3SnCl + Sb_2S_3$ (647).
$Na_2S + BzCl \rightarrow Me_3SnSBz$ (1747a).
$PhNCO \rightarrow (PhNCO)_3$ (2456).

Bistriethyltin Sulfide (M-452) $Sn_2C_{12}H_{30}S$ $(Et_3Sn)_2S$

<u>Synth.</u>: By rxn. of Et_6Sn_2 with sulfur at 85°, in 65% yield (2091).
By rxn. of Et_3SnH with sulfur at 100°, in 55% yield (2091).
By rxn. of Et_3SnOMe with $PhNHCHS$, in 60% yield (402).
By rxn. of $Et_3SnOCHPhCH_2Ac$ with sulfur in C_6H_6, in 51% yield (437).
By rxn. of $(Et_3Sn)_2Te$ with sulfur at 20° (2853).
<u>Prop.</u>: $b_{0.2}$ 118° (402), $b_{0.03}$ 98-99°, n_D^{20} 1.5481 (437), 1.5488 (2091),
1.5508 (402), d_{20} 1.4130 (437), 1.4160 (2091), (far) IR (1010), assignment
(321, 1360), Raman, assignment (1360) and NMR (1009) spectra.
<u>Rxn. with</u>: $RMgX \rightarrow Et_3SnR$; R = Et, Bu (2813).
$(p-Me_2NC_6H_4)_2Hg \rightarrow p-Et_3SnC_6H_4NMe_2 + HgS$ (2648).
Cu in N atm. $\rightarrow Et_6Sn_2 + CuS$ (1737).
Cu in air $\rightarrow (Et_3Sn)_2O + CuS$; rxn. mechanism (1737).
Cu in $CCl_4 \rightarrow Et_3SnCl + CuS + C_2Cl_4$; rxn. mechanism (1737).
$Bz_2O_2 \rightarrow Et_3SnOBz + S$ (2093).

Bistributyltin Sulfide (M-452) $Sn_2C_{24}H_{54}S$ $(Bu_3Sn)_2S$

<u>Synth.</u>: By rxn. of Bu_3SnH with sulfur at 75°, in 68% yield (2091).
By rxn. of Bu_3SnCl with aq. Na_2S at 0° in petr. ether, in 86% yield (1514),
with aq. $Na_2S_2O_3$ at r.t. (1791), with $(KS)_2C:NCN$ in THF, in 32% yield (2751).
By rxn. of $(Bu_3Sn)_2O$ with RNCS (36, 750); R = Et, $CH_2:CHCH_2$, Ph (2258), with
$(PhNH)_2CS$ (2258).
By rxn. of $(Bu_3Sn)_2NPh$ with MeNCS (2258).
<u>Prop.</u>: $b_{1.5}$ 194-98°, n_D^{20} 1.5151, d_{20} 1.1970 (2091), (far) IR, Raman, assignment (1360) and UV (927) spectra, dipole moment (928), light resistance (1063).
<u>Rxn. with</u>: $BuMgCl \rightarrow Bu_4Sn$ (2813).
Ph_3SnH (1:2) in $Et_3N \rightarrow Bu_3SnH + Ph_6Sn_2 + Bu_3SnSnPh_3 + SnS$ (1618).
$Bu_3SnH \rightarrow$ no exchange observed (1618).
$i-Bu_2SnH_2 \rightarrow Bu_3SnH + (i-BuSnS)_3 + Bu_6Sn_2$ (1955).
$HgR_2 \rightarrow Bu_3SnR + HgS$; R = $EtCOCH_2$, MeO_2CCH_2 (2648), $NPhCHO$ (949).
Cu in N atm. $\rightarrow Bu_6Sn_2 + CuS$; rxn. mechanism (1737).
Cu in air $\rightarrow (Bu_3Sn)_2O + CuS$ (1737).
Cu in $CCl_4 \rightarrow Bu_3SnCl + CuS + C_2Cl_4$; rxn. mechanism (1737).
<u>Biol. Prop.</u>: Toxicity (1099).
Activity as bactericide (720, 1099).
Activity as fungicide (720) against Ceratostomella coerulescens (1099),
Coriolellis palustris (1063), Fusarium culmorum, Penicillium funiculosum and
Speira heptaspora (1099).
Biol. detn. method in sea water to 5 ppb. (456).
<u>Use</u>: For antifouling coatings and paint (360, 720, 1399, 2302).

Bistriphenyltin Sulfide (M-451) $Sn_2C_{36}H_{30}S$ $(Ph_3Sn)_2S$

<u>Synth.</u>: By rxn. of Ph_3SnCl with $(KS)_2C:NCN$ in aq. THF, in 44% yield (2751),
with aq. alc. $Na_2S_2O_3$ at r.t. (1791).
By rxn. of Ph_3SnX with H_2S in presence of org. bases, in quant. yield (121),
with R_3SbS in alc. Me_2CO, in 96% yield; R = Me, $c.C_6H_{11}$, X = Cl, Br (1903),

with PhNHCS$_2$NH$_4$ in HCCl$_3$ at r.t., in high yield (1377).
By rxn. of (Ph$_3$Sn)$_2$O with PhCH$_2$NHCS$_2$NH$_3$CH$_2$Ph at r.t., in 90% yield (1377).
By rxn. of (Ph$_3$Sn)$_2$PPh and (Ph$_3$Sn)$_3$P with sulfur at r.t. in the dark in N atm., in 71 and 50% yield, resp. (1840).
Prop.: m. 96-97° (1791), (far) IR, assignment (1360, 1848), Raman (1360) and NMR (1220) spectra, soly. (1791, 2751). Anal. S detn. (2321).
Rxn. with: Ph$_2$SnCl$_2$ → Ph$_3$SnCl + (Ph$_2$SnS)$_3$ (1705).
HgR$_2$ → Ph$_3$SnR + HgS; R = EtO$_2$CC(N$_2$) (1831), MeO$_2$CCH$_2$ (2648).
PhLi → Ph$_4$Sn (2709).
Biol. Prop.: Toxicity (1099), comparative phytotoxicity (306).
Activity as bactericide (1099).
Activity as fungicide against Ceratostomella coerulescens, Fusarium culmorum, Penicillium funiculosum and Speira heptaspora (1099).
Activity as insecticide against Culex pipiens larvae (73, 1137), Musca domestica, effect on reproduction (260, 1064), Tineola bisselliella larvae (162).
Use: Trade name and synonyms Dowco 188.

Bistriethyltin Selenide   Sn$_2$C$_{12}$H$_{30}$Se   (Et$_3$Sn)$_2$Se
Synth.: By rxn. of Et$_6$Sn$_2$ with Se at 190°, in 23% yield (2091).
By rxn. of Et$_3$SnH with Se at 120°, in 69% yield (2091), with Et$_2$Se at 170°, in 45% yield (2091).
By rxn. of (Et$_3$Sn)$_2$Te with Se at 150° (2853).
Prop.: b$_{1.5}$ 138-40°, n$_D^{20}$ 1.5652, d$_{20}$ 1.5710 (2091), IR (1010) and NMR (1009) spectra.
Rxn. with: Bz$_2$O$_2$ → Et$_3$SnOBz + Se (2093).
Br → Et$_3$SnBr + Se (2851).
S → (Et$_3$Sn)$_2$S + Se (2853).

Bistriethyltin Telluride   Sn$_2$TeC$_{12}$H$_{30}$ (Et$_3$Sn)$_2$Te
Synth.: By rxn. of Et$_3$SnH with Te at 130°, in 64% yield (2091-2), with Et$_2$Te at 35-60° under exclusion of air, in 91% yield (2092), with (Et$_3$Ge)$_2$Te at 70°, in 73% yield (2852).
Prop.: b$_{1.5}$ 144-45° (2091-2), b$_1$ 119-20° (2092), n$_D^{20}$ 1.5950 (2091-2), 1.5972 (2092), d$_{20}$ 1.668 (2091-2), IR (1010) and NMR (1009) spectra.
Rxn. with: Bz$_2$O$_2$ → Et$_3$SnOBz + Te (2093).
X → Et$_3$SnX + Te, X = Cl, Br (2852).
Y → (Et$_3$Sn)$_2$Y + Te; X = S, Se (2853).

Listing of triorganotin sulfides, selenides and tellurides concludes in Table 222.

Table 222. Triorganotin Sulfides, Selenides and Tellurides

Formula*		Synth. Method**	Yield	Properties	Ref.
$(R_3Sn)_2S_n$					
n = 1	R' = $NCCH_2CH_2$	IIA	89	n_D^{28} 1.5745 (a)	2326
1	Pr (452)	IIB	--	b_3 160-65°	1791
1	$PhCMe_2CH_2$	IIA, IIIA	--	m. 97.5-98°, mol. wt., soly.	1742
1	C_8H_{17}	IIA	--	Liquid	1791
2	Ph (452)	--	--	Use as bactericide	529
1	$o\text{-}MeC_6H_4$	--	--	NMR spectrum	2356
$(Me_3Sn)_2Se$ (452)		--	--	IR and Raman spectra, assignments	1360
$(Bu_2Sn)_2Se$		IIA	18	$b_{0.0001}$ 160°, n_D^{20} 1.526 (b)	1618
$(Ph_3Sn)_2Se$ (452)		--	--	(Far) IR spectra and assignment (c)	1848
$(Ph_3Sn)_2Te$ (452)		--	--	(Far) IR spectra and assignment	1848

* Numbers in parenthesis refer to pages in main volume.
** The characters used here correspond to the synthetic scheme in the introduction to Chapter 4.4.1.
(a) Use as biocide and as stabilizer for plastics (2326). (b) No exchange reaction with Bu_3SnH (1618).
(c) Polarographic reduction (963, 969).

4.4.1.2 DIORGANOTIN SULFIDES, SELENIDES AND TELLURIDES

Dimethyltin Sulfide (M-453) $Sn_3C_6H_{18}S_3$ $(Me_2SnS)_3$

<u>Synth.</u>: By rxn. of Me_2SnX_2 with $(c.C_6H_{11})_3SbS$ in $MeOH-Me_2CO$ or $HCCl_3$; X = Cl, Br, in high yield (1903).

<u>Prop.</u>: (Far) IR, assignment (1903, 2517), Raman, assignment (2517), UV (1116-7) and NMR (1220) spectra, dipole moment (1117), mol. wt. (1116-7), struct. (1117), x-ray powder diffraction pattern (1903).

<u>Rxn. with</u>: $Me_2SnX_2 \rightarrow (Me_2SnX)_2S$; X = Cl, Br (945).

$Me_3SbX_3 + Me_3SbS$ (1:3:3) in $HCCl_3 \rightleftharpoons Me_2SnX_2 \cdot 2Me_3SbS$; X = Cl (1902-3), Br (1903).

$Na_2S \rightarrow Me_2Sn(SNa)_2$ (1747a).

Tropolone $\rightarrow Me_2Sn(O_2C_7H_5)_2 + H_2S$ (2510).

<u>Use</u>: As stabilizer for PVC and chlorinated polyethylene (2939).

Diethyltin Sulfide (M-454) $Sn_3C_{12}H_{30}S_3$ $(Et_2SnS)_3$

<u>Synth.</u>: By rxn. of Et_2SnCl_2 with aq. Na_2S in petr. ether at 0°, in 88% yield (1514), with aq. alc. $Na_2S_2O_3$ at r.t. (1790).

By rxn. of Me_2SnX_2 with $(c.C_6H_{11})_3SbS$ in alc. Me_2CO in high yield; X = Cl, Br (1903).

<u>Prop.</u>: (Far) IR (321, 2517) and Raman (2517) spectra with assignments.

<u>Rxn. with</u>: $BuMgX \rightarrow Et_2SnBu_2$ (2813).

$Me_3SbS + Me_3SbX_3$ (1:3:3) in $HCCl_3 \rightleftharpoons Et_2SnX_2 \cdot 2Me_3SbS$ (1903).

$Na_2S \rightarrow Et_2Sn(SNa)_2$ (1747a).

Tropolone $\rightarrow Et_2Sn(O_2C_7H_5)_2 + H_2S$ (2510).

<u>Use</u>: As catalyst for preparation of polyurethan (1790), as stabilizer for polymers (1790).

Dibutyltin Sulfide (M-453) $Sn_3C_{24}H_{54}S_3$ $(Bu_2SnS)_3$

<u>Synth.</u>: By rxn. of Bu_2SnCl_2 with aq. Na_2S at 0° in petr. ether, in 90% yield (1514), with aq. Na_2S_x (1:1) in CH_2Cl_2 (1514), with aq. $Na_2S_2O_3$ in MeOH at r.t. (1790).

By rxn. of $(Bu_2SnX)_2O$ with H_2S and BiPy in Me_2CO (1705).

By rxn. of $(Bu_2SnCl)_2S$ with $p-C_6H_4(CO_2Na)_2$ (1514).

By thermal dec. of $(Bu_2SnCl)_2O$ (1514).

By-prod. in rxn. of $Bu_3SnCH_2CMe:CH_2$ with $H_2S_{5.2}$ in CCl_4 (1857).

<u>Prop.</u>: IR (1705), (far) IR, assignment (2517), Raman, assignment (2517) and NMR (1514) spectra. Anal. detn. of S (232).

<u>Rxn. with</u>: $Bu_2SnX_2 \rightarrow (Bu_2SnX)_2S$; X = Cl (945, 1514), F, SCN, MeO, $C_7H_{15}CO_2$ (945).

$RSnX_3 \rightarrow (Bu_2SnX)S(RSnX_2)$; RX = PhCl, BuCl, BuBr (945).

$RLi \rightarrow Bu_2SnR_2 + Li_2S$; R = Bu, MeC⋮C, $PhCH_2CH_2C⋮C$ (2709).

$Na_2S \rightarrow Bu_2Sn(SNa)_2$ (1747a).

$BuMgX \rightarrow Bu_4Sn$ (2813).

Tropolone $\rightarrow Bu_2Sn(O_2C_7H_5)_2$ (2510).

<u>Use</u>: As additive for two cycle engine lubricants (1749).

As antioxidant for mineral oil lubricants (2149).

As catalyst for preparation of polyurethan (1790), for vulcanization of chlorine containing polymers (704).
As rodent repellent (2036).
As stabilizer for polymers (1790), for PVC (489, 2862, 2953), for PVC copolymers (2862).

Diphenyltin Sulfide (M-453) $Sn_3C_{36}H_{30}S_3$ $(Ph_2SnS)_3$

<u>Synth.</u>: By rxn. of Ph_2SnCl_2 with aq. Na_2S at 0° in petr. ether, in 86% yield (1514), in C_6H_6 at reflux, in 54% yield (1705), with CS_2 and NH_3, in 87% yield (298), with $(Ph_3Sn)_2S$ in C_6H_6 under reflux (1705).
By rxn. of Ph_2SnX_2 with R_3SbS in alc. Me_2CO in high yield; X = Cl, Br, R = Me, c.C_6H_{11} (1903).
By rxn. of Ph_2SnO with CS_2 and NH_3, in 64% yield (298), with $PhCH_2NHCS_2NH_3\text{-}CH_2Ph$ in $HCCl_3$ at r.t., in 79% yield (1377).
By thermal dec. of $Ph_2Sn(S_2CNHCH_2Ph)$ in C_6H_6 at reflux, in 93% yield (298), in 72-86% yield (1377), of $Ph_2Sn(S_2CNHCH_2Ph)Cl$, in 93% yield (298).
<u>Prop.</u>: IR spectrum (1857), soly. (1514).
<u>Rxn. with</u>: $Na_2S \rightarrow Ph_2Sn(SNa)_2$ (1747a).
HCl + BiPy $\rightarrow Ph_2SnCl_2 \cdot BiPy$ (1705).
Tropolane $\rightarrow Ph_2Sn(O_2C_7H_5)_2 + H_2S$ (2510).

Additional diorganotin sulfides and selenides are listed in Table 223.

4.4.1.3 MONOORGANOTIN SULFIDES, SELENIDES AND TELLURIDES

Bis(Butyltin) Trisulfide $Sn_4C_{16}H_{36}S_6$ $(Bu_2Sn_2S_3)_2$

<u>Synth.</u>: By rxn. of Bu_2SnO_3 with HCl, followed by aq. Na_2S (1347).
<u>Prop.</u>: Dec. 150° (1347), mass spectrum (2354), mol. wt. (1347, 2354), soly. (1347), x-ray struct. analysis (2354).
<u>Rxn. with</u>: Na_2S, + $i\text{-}C_8H_{17}O_2CCH_2Cl \rightarrow BuSn(SCH_2CO_2C_8H_{17})$ (1747a).
8-Hydroxyquinoline (1:2) $\rightarrow [BuSn(OC_9H_6N)_2]_2S$ (1347).
Tropolone $\rightarrow BuSn(O_2C_7H_5)_3 + H_2S$ (2510).
<u>Use</u>: As stabilizer for PVC, relative reactivity (1127a).

Listing of monoorganotin sulfides continues in Table 224.

Table 223. Diorganotin Sulfides and Selenides

Formula*	Synth. Method**	Yield	Properties	Ref.
$(R_2SnS)_3$				
R = $NCCH_2CH_2$	IIA	86	m. 173-75°, mol. wt. (a)	1746
Pr (454)	IIB (b)	8	--	1747
Me_2CHCH_2	IIC	High	(c)	2510
C_8H_{17} (454)	I	--	--	1955
$C_{12}H_{25}$ (454)	IIB	--	m. 70-71°, soly. (d, e)	1790
p-MeC_6H_4 (454)	--	--	Use as stabilizer for PVC	2953
R	--	--	White cryst., IR spectrum	1857
	--	--	(f)	--
$(Me_2SnSe)_3$ (455)	--	--	(Far) IR, Raman spectra, assignments	2517
$(Et_2SnSe)_3$	--	--	(Far) IR, Raman spectra, assignments	2517

* Numbers in parenthesis refer to pages in main volume.
** The characters used here correspond to the synthetic scheme in the introduction to Chapter 4.4.1.
(a) Use as biocide and stabilizer for plastics (2326). (b) Synth. by rxn. of $(NCCH_2CH_2)_4Sn$ with $SnBr_4$ at 104-20° in N atm., followed by Na_2S (1747). (c) Rxn. with tropolone → $Pr_2Sn(O_2C_7H_5)_2 + H_2S$ (2510). (d) Rxn. with $Na_2S → (C_8H_{17})_2Sn(SNa)_2$ (1747a). (e) Use as catalyst for polyurethan preparation (1790), as stabilizer for polymers (1790), for PVC (2953). (f) Use in catalyst for hardening of epoxide resins (2916).

Table 224. Monoorganotin Sulfides $(R_2Sn_2S_3)_x$

Formula*	Synth. Method**	Yield	Properties	Ref.
$(Me_2Sn_2S_3)_2$ (457)	IIIB	--	White, insoluble powder, dec. 250° (a, b)	1347
$(NCCH_2CH_2)_2Sn_2S_3$	IIA	--	m. 135° (dec.) (c)	2326
$Pr_2Sn_2S_3$	IIIB	--	White, insoluble powder, dec. 160°	1347
Bu_2SnS_3	IIA	96-99	Polymeric	980
	IIC	High	(d, e)	1903
$(C_6H_{13})_2Sn_2S_3$	IIA	High	Polymeric (d, f)	980
$(C_8H_{17})_2S_2S_3$	IIA	High	Polymeric (d, f)	980
$(C_{11}F_{23}CH_2CH_2)_2Sn_2S_3$	II (g)	--	(h)	745a
$[Cl(CF_2)_{10}CH_2CH_2]_2Sn_2S_3$	IIA	--	(h)	745a

* Numbers in parenthesis refer to pages in main volume.
** The characters used here correspond to the synthetic scheme in the introduction to Chapter 4.4.1.
(a) Mass spectrum, mol. wt., x = 2, x-ray struct. analysis (2354). (b) Rxn. with 8-hydroxyquinoline → MeSn(OC$_9$H$_6$N)$_3$ (1347). (c) Use as biocide and as stabilizer for plastics (2326). (d) Use as non-toxic stabilizer for PVC and chlorine containing polymers (980). (e) Use as stabilizer for PVC (980a, 2963). (f) Use as stabilizer for PVC and PVC-vinyl acetate copolymers (1127a). (g) Synth. by rxn. of CH$_2$:CHSnCl$_3$ with 1-C$_{10}$F$_{21}$I under UV irradiation in i-Pr$_2$O, followed by H$_2$S (745a). (h) Use as dirt, oil and water repellent, antiadhesive and release agent (745a).

4.4.2 ORGANOTIN MERCAPTIDES AND ORGANOSELENIDES

4.4.2.1 TRIORGANOTIN MERCAPTIDES AND SELENIDES

The compounds compiled in the ensuing Tables are prepared by methods from the following scheme.

- I   Reaction of Et_3SnH with $RSN:NAr$.
- II   Reaction of triorganotin halides with:
  - (IIA)   RSH or RSeH and aqueous NaOH or Na_2CO_3 sometimes in organic solvent,
  - (IIB)   RSH in presence of triethylamine or pyridine,
  - (IIC)   mercaptan.
- III   Reaction of triorganotin oxide or hydroxide with HSR or $(HNCS)_3$.
- IV   Reaction of triorganotin alkoxide with RSH or RSeH.
- V   Reaction of triorganotin acetate with RSH.
- VI   Reaction of:
  - (VIA) R_3SnSR with $R'SH$,
  - (VIB) $(Et_3Sn)_2PMe$ with Et_2S_2.

4.4.2.1.1 UNSUBSTITUTED TRIORGANOTIN MERCAPTIDES AND SELENIDES

Trimethyltin Methylsulfide $SnC_4H_{12}S$ Me_3SnSMe
<u>Synth.</u>: By rxn. of Me_3SnCl with $Pb(SMe)_2$ in C_6H_6, in 26% yield (643).
By rxn. of Me_3SnX with MeSH and NaOH, in 60-70% yield (2, 728).
<u>Prop.</u>: b. 163° (2), b. 161-63° (643), n_D^{20} 1.5282 (643) 1.5302, d_{20} 1.453 (2), IR (2, 644-5) and NMR (645-6, 728) spectra, relative basicity (644-5).
<u>Rxn. with</u>: $HgCl_2 \rightarrow Hg(SMe)_2$ (647).
$Mn(CO)_5Br \rightarrow [Mn(CO)_3SMe]_3$ (649).
$RNCO \rightarrow (RNCO)_3$; R = Me, Ph (2456).
Cl_3CCHO (1:1) $\rightleftharpoons Me_3SnOCH(CCl_3)SMe$ (2456).
Cl_3CCHO (1:3) $\rightarrow Me_3SnCl + (MeS)_3CCHO$ (2456).
S at 120° $\rightarrow (Me_3Sn)_2S$ (2).
<u>Addn. Compd.</u>: $Me_3SnSMe \cdot MeI$, from components, in 62% yield, m. 223-26° (dec.), conductivity (2) $[SnC_5H_{15}IS]$.

Trimethyltin Ethylsulfide $SnC_5H_{14}S$ Me_3SnSEt
<u>Synth.</u>: By rxn. of Me_3SnX with EtSH and aq. NaOH, in 60-70% yield (2).
By rxn. of $(Me_3Sn)_3P$ with Et_2S_2 at 40°, in 76% yield (800).
By rxn. of Me_3SnNa with Et_2S_2 in liq. NH_3, in 72% yield (800).
<u>Prop.</u>: b. 177° (2), b_{101} 103-105° (800), n_D^{20} 1.5205 (2), 1.5215-225 (800), d_{20} 1.3943 (800), 1.394 (2), IR (2, 644-5) and NMR (645) spectra, magnetic susceptibility (1), relative basicity (644-5).
<u>Rxn. with</u>: $HgCl_2 \rightarrow Hg(SEt)_2$ (647).
Cl_3CCHO (1:3) $\rightarrow Me_3SnCl + (EtS)_3CCHO$ (2456).

Trimethyltin i-Propylsulfide $SnC_6H_{16}S$ $Me_3SnSCHMe_2$

__Synth.__: By rxn. of Me_3SnCl with $Me_3SiSCHMe_2$, in 84% yield (643).
By rxn. of Me_3SnX with Me_2CHSH and aq. NaOH, in 60-70% yield (2).
__Prop.__: b. 182° (2), $b_{0.01}$ 24-25° (643), n_D^{20} 1.5108 (643), 1.5123, d_{20} 1.318 (2), IR (2, 644-5) and NMR (645) spectra, relative basicity (644-5).

Trimethyltin Butylsulfide $SnC_7H_{18}S$ Me_3SnSBu

__Synth.__: By rxn. of Me_3SnX with BuSH and aq. NaOH, in 60-70% yield (2).
By rxn. of Me_3SnNEt_2 with BuSH, in 92% yield (643).
__Prop.__: $b_{0.05}$ 44° (2), $b_{0.01}$ 40-42° (643), n_D^{20} 1.5090 (643), 1.5098, d_{20} 1.281 (2), IR spectrum (2), magnetic susceptibility (1).
__Rxn. with__: $BCl_3 \rightarrow Me_3SnCl + B(SBu)_3$ (647).
$K_2PtCl_4 \rightarrow Pt(SBu)_2$ (647).
$C_6F_5SH \rightarrow Me_3SnSC_6F_5 + BuSH$ (643).

Triphenyltin Phenylsulfide (M-459) $SnC_{24}H_{20}S$ Ph_3SnSPh

__Synth.__: By rxn. of Ph_3SnX with PhSH in C_6H_6 in presence of Py or Et_3N, in 70-90% yield (121).
__Prop.__: m. 102-103° (121, 2642), Moessbauer resonance (2642) and mass (849) spectra.
__Rxn. with__: $HgCl_2 \rightarrow Ph_3SnCl + PhSHgCl$ (1704).
$Hg(OAc)_2 \rightarrow Ph_3SnOAc + (Ph_2SnOAc)_2O + (PhS)_2Hg + PhHgSPh$ (1704).
$SnCl_4 \rightarrow Ph_3SnCl + (PhS)_4Sn$ (1704).
__Biol. Prop.__: Activity as insecticide against Tineola bisselliella larvae (162).

Listing of unsubstituted triorganotin mercaptides and selenides concludes in Table 225.

Table 225. Unsubstituted Triorganotin Monoxides and Organoselenides R_3SnSR

Formula*	Synth. Method**	Yield	Properties	Ref.
Me$_3$SnSR				
R = Pr	IIA	60-70	b_3 54°, n_D^{20} 1.5178, d_{20} 1.352, (a, b)	2
Me$_3$C	IIA	60-70	$b_{0.1}$ 42°, n_D^{20} 1.5083, d_{20} 1.267 (b-d)	2
C$_8$H$_{17}$	IIA	60-70	$b_{0.1}$ 94°, n_D^{20} 1.5000, d_{20} 1.175 (a, b)	2
Ph	IIA	--	$b_{0.01}$ 69°, n_D^{20} 1.5934, d_{20} 1.418 (b, e)	2
Et$_3$SnSR				
R = Me (458)	IIA	--	b_2 94°, n_D^{20} 1.5274, d_{20} 1.375 (b)	2
Et (458)	IIA	--	$b_{0.7}$ 68°, n_D^{20} 1.5153, d_{20} 1.359 (b, d)	2
Bu (458)	VIB	--	b_3 78°, n_D^{20} 1.5222, d_{20} 1.2689	800
Me$_3$C (458)	IV	--	b_3 107-108°, n_D^{20} 1.5127, struct. (f, g)	2642
Ph (458)	IIA	--	$b_{0.02}$ 47°, n_D^{20} 1.5130, d_{20} 1.240 (b)	2
	IIA	--	$b_{1.7}$ 150°, n_D^{20} 1.5130, d_{20} 1.240 (b)	2
p-Me$_3$CC$_6$H$_4$	--	--	b_2 138°, n_D^{20} 1.5781 (f)	2637, 2642
	I	--	Colorless oil, $b_{0.0001}$ 140°, IR spectrum	2646
Bu$_3$SnSR				
R = Me	III	--	(h-j)	859
	III	--	$b_{0.2}$ 104°, n_D^{22} 1.5110	1676
Bu	III	--	b_2 160-62°, n_D^{20} 1.4981, d_{20} 1.102 (k)	1116-7
C$_{12}$H$_{25}$ (459)	--	--	Use as stabilizer for PVC (1)	2549
Ph (459)	III	--	Colorless liq., $b_{0.3}$ 147°, n_D^{22} 1.5479	1676
	V (m)	--	(h, i)	2236
p-MeC$_6$H$_4$	--	--	(n)	--
C$_{10}$H$_7$	--	--	(o)	--
(C$_8$H$_{17}$)$_3$SnSR	--	--	Use as stabilizer for PVC	2712
Ph$_3$SnSCH$_2$Ph	IIB	70-90	m. 84°	121
Ph$_3$SnSC$_{12}$H$_{25}$ (459)	--	--	Use as stabilizer for PVC and copolymers	1866

Me$_3$SnSeMe	II (p)	62	b$_{0.8}$ 26°, n$_D^{25}$ 1.483, d$_{20}$ 1.67 (q, r)	651
Me$_3$SnSeEt	--	--	(r)	--
Me$_3$SnSePh	IIA	50	b$_{0.001}$ 67-69°, n$_D^{20}$ 1.619, d$_{20}$ 1.65 (r, s)	643
Et$_3$SnSeCH$_2$Ph	IV	~100	b$_{0.0001}$ 115-17°, n$_D^{20}$ 1.5888 (t)	389

* Numbers in parenthesis refer to pages in main volume.

** The characters used here correspond to the synthetic scheme in the introduction to Chapter 4.4.2.1.

(a) Magnetic susceptibility (1). (b) IR spectrum (2). (c) IR and NMR spectra, relative basicity (644-5). (d) Rxn. with PhNCO → (PhNCO)$_3$ (2456). (e) Rxn. with MCl$_3$ → Me$_3$SnCl + (PhS)$_3$M; M = P (2, 647), As (647), with Br → Me$_3$SnBr + (PhS)$_2$ (2). (f) Moessbauer resonance spectrum (2637, 2642). (g) Rxn. with Ph$_3$SnH → Et$_3$SnSnPh$_3$ + BuSH + Et$_3$SnH + Ph$_3$SnSBu, rxn. rates and solvent effect (915). (h) (Far) IR, assignment and NMR spectra (1676). (i) Rxn. with I → Bu$_3$SnI + R$_2$S$_2$; R = Me, Ph (1676). (j) Activity as fungicide against Hypochnus sasakii (859). (k) Far IR and UV spectra (1116-7), dipole moment (1116). (1) Use as catalyst for foaming polyurethan (450). (m) Synth. from Bu$_3$Sn-6-Acylaminopencillate and KSPh (2236). (n) Rxn. with M(CO)$_5$X → [M(CO)$_3$SC$_6$H$_4$Me]$_3$; MX = MnBr, ReCl (649). (o) Activity as molluscicide against Heliosoma subcrenata and H. tenue (1054). (p) Synth. with Me$_2$Se$_2$ and Na in liquid NH$_3$ (651). (q) IR (651) and NMR (646, 651) spectra. (r) Rxn. with M(CO)$_5$X → [M(CO)$_4$SeR]$_2$ + Me$_3$SnX; M = Mn, Re; R = Me, Et, Ph; X = Cl, Br (651). (s) Pale yellow oil, darkens on standing (643). (t) Rxn. with Ac$_2$O → Et$_3$SnOAc + PhCH$_2$SeAc (389).

4.4.2.1.3 HETEROCYCLIC SUBSTITUTED TRIORGANOTIN MERCAPTIDES

$Sn_3C_{64}H_{50}N_8S_2$ $6,9-(Ph_3Sn)_2SC_5H_2N_4 \cdot 6-(Ph_3SnS)C_5H_3N_4$

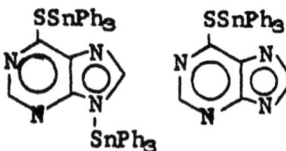

Synth.: By rxn. of $(Ph_3Sn)_2O$ with 6-HS-purine hydrate in Me_2CO, in 97% yield (1375).
By rxn. of $Ph_3SnCl \cdot PhCH_2NH_2$ with $6-HSC_5H_3N_4$, in 60% yield (1375).
By rxn. of Ph_3SnCl with $6-LiSC_5H_3N_4$, in 77% yield (1375).
Prop.: m. 214-15°, (far) IR, UV and NMR spectra, stable in moist air (1375).
Rxn. with: D_2O → D-exchange followed by NMR spectroscopy (1375).

Additional triorganotin mercaptides derived from heterocyclic mercaptans are listed in Tables 226 and 227.

Table 226. S-Triorganotin Purine Derivatives $6,9-(R_3Sn)_2SC_5H_2N_4 \cdot 6-(R_3SnS)C_5H_3N_4$

$R^{(a)}$	Synth. Method	Yield	Properties	Ref.
$p-MeC_6H_4$	(b)	81	m. 203-204° (c)	1375
$p-FC_6H_4$ (d)	(d)	80	m. 226-27° (c)	1375
$p-ClC_6H_4$ (b)	(b)	14	m. 210-14° (c)	1375

(a) For structural formula compare preceding phenyl derivative; R = Ph.
(b) Synth. from 6-mercaptopurine hydrate with $(R_3Sn)_2O$ in Me_2CO. (c) (Far) IR and UV spectra, stable in moist air (1375). (d) Synth. with R_3SnOH.

4.4.2.1.6 FUNCTIONALLY SUBSTITUTED TRIORGANOTIN MERCAPTIDES

All compounds are summarized in Table 228. O,S-bis(tributyltin) β-mercaptopropionate, $Bu_3SnSCH_2CH_2CO_2SnBu_3$, is listed in Table 202. Other references to functionally substituted triorganotin mercaptides may be found in Chapter 8.1.

Table 227. Heterocyclic Substituted Triorganotin Mercaptides

Formula*	Synth. Method**	Properties	R_3SnSR' Ref.
$(PhCH_2)_3SnSC_2N_2S(SH)$ (a)	IIA	m. 96-100° (b)	1262
$[(PhCH_2)_3SnS]_2HSC_3N_3$ (c)	III	m. 110-12° (d-f)	1322a
$2-Et_3SnSC_5H_4N$ (g)	IV	b_8 150-55°, n_D^{20} 1.5775 (h)	2642
$4-Et_3SnSC_5H_4N$ (g)	IV	m. 88-89°, UV spectrum, soly. (h)	2516
$(Pr_3SnS)_3C_3N_3$ (c)	III	n_D^{20} 1.5655 (d-f, i-k)	1322a
$Pr_3Sn(i-PrHN)_2C_3N_3$ (c)	III	(d, f)	1322a
$Pr_3SnS(i-PrHN)C_3N_3NHMe$ (c)	III	(1)	1322a
$2-(CH_2:CHCH_2)_3SnSC_5H_4N$ (g)	IIB	$b_{0.08}$ 67-68° (1)	2204

Bu_3SnSR'

$R' = 2-C_3H_4NS$ (m)	--	$b_{0.01}$ 100°, n_D^{25} 1.5473 (1)	2204
$2-C_7H_4N_2$ (n)	IIB	n_D^{25} 1.5588 (1)	2204
$2-C_7H_4NO$ (o)	IIB	$b_{1.4}$ 204-205°, n_D^{25} 1.5608 (1)	2204
$2-(C_7H_3NS)NO_2-6$ (p)	IIB	n_D^{25} 1.5598 (1)	2204
$2-(C_2N_2S)SH$ (a) (460)	IIC	Yellow cryst., m. 52-53°, soly. (q)	1567
$2-C_5H_4N$ (g)	IIB	b_1 183-86°, n_D^{25} 1.5480 (1)	2204
$4-C_5H_4N$ (g)	IIB	$b_{0.01}$ 120°, n_D^{25} 1.5275 (1)	2204
$4-(3,5-I_2C_5H_2N)$ (g)	IIB	m. 65-66° (1)	2204
$2-(C_4H_3N_2)$ (r)	IIB	$b_{1.6}$ 197°, n_D^{25} 1.5444 (1)	2204

* Numbers in parenthesis refer to pages in main volume.
** The characters used here correspond to the synthetic scheme in the introduction to Chapter 4.4.2.1.
(a) Deriv. of 2,5-dimercapto-1,3,4-thiadiazole. (b) Activity as herbicide and insecticide (1262). (c) Deriv. of sym-triazine. (d) Use as textile preservative (1322a). (e) Activity as bactericide against Staphylococcus aureus (1322a), as fungicide against Aspergillus niger (1322a). (f) Activity as herbicide (1322a). (g) Deriv. of pyridine. (h) Moessbauer resonance spectrum (2642). (i) Activity as molluscicide against Agriolimax reticulatus (1322a). (j) Use as fumigant (1322a). (k) Activity as insecticide (1322a). (l) Use as fungicide, herbicide, insecticide, molluscicide, parasiticide (2204). (m) Deriv. of 1-thiazoline. (n) Deriv. of 2-mercaptobenzimidazole. (o) Deriv. of 2-mercaptobenzoxazole. (p) Deriv. of 2-mercaptobenzothiazole. (q) Use in antifouling paint (1567). (r) Deriv. of pyrimidine.

Table 227 Continued

Formula*		Synth. Method**	Properties	R₃SnSR' Ref.
$(Bu_3SnS)_3C_3N_3$ (c)		III	Clear yellow oil, n_D^{20} 1.5521 (d, e, s, t)	1322a
$Bu_3SnSC_3N_3(NHR)NHR'$ (c)				
R = H	R' = H	III	m. 106-108° (d, e, s-u)	1322a
Me	Et	III	m. 78-79° (d, e, s-u)	1322a
Et	Et	III	m. 95° (d, e, j, s-u)	1322a
Et	Me_2CH	III	m. 81° (d, e, l, t)	1322a
Et	MeEtCH	III	n_D^{20} 1.5520	1322a
Et	Me_3C	III	m. 46-48°	1322a
Me_2CH	Me_2CH	III	Yellow oil, n_D^{20} 1.5488 (d, e, t, v)	1322a
Me_2CH	$MeO(CH_2)_3$	III	n_D^{20} 1.5472 (w)	1322a
Me_2CH	$CH_2:CHCH_2$	III	n_D^{20} 1.5550	1322a
$MeOC_2H_4$	$CH_2:CHCH_2$	III	n_D^{20} 1.5575 (w)	1322a
$C_{18}H_{37}$	$C_{18}H_{37}$	III	m. 39-41°	1322a
$C_{18}H_{37}$	C_4H_8ON (x)	III	m. 25°	1322a
$C_{18}H_{35}$	$C_{18}H_{35}$	III	n_D^{20} 1.5191	1322a
C_4H_8ON	C_4H_8ON (x)	III	n_D^{20} 1.5574	1322a
3-$Bu_3SnS(5,6-Ph_2C_3N_3)$ (y)		--	n_D^{25} 1.5897 (l)	2204
Ph_3SnSR'				
R' = 2-C_7H_4NO (o)		--	(z)	--
2-C_7H_4NS (p) (460)		--	(z)	--
2-$C_2N_2S(SH)$ (a)		--	Yellow, m. 114° (b)	1262
2-$C_2N_2SSSn(C_6H_5Me-o)_3$ (a)		--	Yellow, m. 41-44° (b)	1262
2-C_5H_4N (g)		III	m. 102-103°, soly., struct. (h)	2642
4-C_5H_4N (g)		III	m. 129-30°, UV spectrum, struct. (h)	2516
2-C_5H_4NO (aa)		V	In 92% yield, m. 110-12°, struct., use as bactericide, fungicide	1836a

$(Ph_3SnS)_3C_3N_3$ (c)			IIA	m. 200-202° (d, e, k, u, bb)	1322a

$Ph_3SnSC_3N_3(NHR)NHR'$ (c)

R = Me	R' = Me_2CH		--	Activity as herbicide (bb)	1322a
Et	Me_2CH		IV	m. 160-62°	1322a
Et	MeEtCH		IV	m. 143-45°	1322a
Et	Me_3C		IV	m. 174-75°	1322a
Me_2CH	Me_2CH		IV	m. 68-70° (d, e, k, bb)	1322a
Me_2CH	$CH_2{:}CHCH_2$		IV	m. 153-55°	1322a
Me_2CH	$MeO(CH_2)_3$		IV	m. 52-56°	1322a
$CH_2{:}CHCH_2$	$MeOC_2H_4$		IV	m. 125-27° (bb)	1322a
Bu	$CH_2{:}CHCH_2$		IV	m. 124-26° (bb)	1322a
$C_{18}H_{37}$	$C_{18}H_{37}$		IV	m. 58-61°	1322a
$C_{18}H_{37}$	C_4H_8ON (x)		IV	m. 88-92° (bb)	1322a
$C_{18}H_{35}$	$C_{18}H_{35}$		IV	n_D^{20} 1.5632	1322a
C_4H_8ON	C_4H_8ON (x)		IV	m. 210-12°	1322a
$[(p\text{-}ClC_6H_4)_3SnS]_2C_2N_2S$ (a)			IIA	Yellow, m. 120-23° (b)	1262
$[(o\text{-}MeC_6H_4)_3SnS]_2C_2N_2S$ (a)			IIA	Pale yellow, m. 67-68.5° (b)	1262
$(p\text{-}Me_2C_6H_3)_3SnSC_2N_2S(SH)$ (a)			IIA	Pale yellow, m. 181-86° (dec.) (b)	1262

(s) Activity as fungicide against Pullularia pullulans (1322a). (t) Activity as insecticide against Aedes aegypti and Bruchidius obtectus (1322a). (u) Use as wood preservative (1322a). (v) Use as disinfectant (1322a). (w) Use as fungicide (1322a). (x) N-Morpholine derivative. (y) Derivative of 1,2,4-triazine. (z) Activity as fungicide against Alternaria tenuis, Fusarium culmorum and Rhizoctonia solani (933). (aa) Derivative of pyridine N-oxide. (bb) Activity as anthelmintic against Fasciola hepatica and Hymenolopis nana (1322a).

Table 228. Functionally Substituted Triorganotin Mercaptides R_3SnSR'

Formula*	Synth. Method**	Yield	Properties	Ref.
$Me_3SnSC_6F_5$	VIA	75	$b_{0.01}$ 62°, n_D^{20} 1.5244, d_{20} 1.71	643
$(PhCH_2)_3SnSCH_2CO_2CH_2CH_2O_2CCH_2SH$	--	--	Use as stabilizer for PVC	2903
Et_3SnSR'				
R' = p-FC$_6$H$_4$	--	--	b_1 100-105° (a)	2637
p-ClC$_6$H$_4$	--	--	b_1 132-33°	2637
p-MeOC$_6$H$_4$	--	--	b_1 136-39° (a)	2637
p-Me$_2$NC$_6$H$_4$	--	--	b_1 175-77° (a)	2637
p-O$_2$NC$_6$H$_4$	--	--	b_1 169-71° (a)	2637
$(Bu_3SnSCH_2CH_2O)_4Si$	VI (b)	--	Clear prod., stabilizer for PVC	2536
Bu_3SnSR'				
R' = HOCH$_2$CH$_2$	III	--	Light resistance (1063) (b, c)	2536
(C$_4$H$_7$O)CH$_2$O$_2$CCH$_2$	--	--	Use as stabilizer for PVC	1284
HSCH$_2$CO$_2$(CH$_2$)$_4$O$_2$CCH$_2$	IIC	--	Use as stabilizer for PVC	2903
1-C$_8$H$_{17}$O$_2$CCH$_2$ (461)	--	--	Use as stabilizer for PVC	1404
Et$_2$NCH$_2$CH$_2$	IIC	--	$b_{0.3}$ 165-70° (d)	2835
RCO$_2$CH$_2$CH(O$_2$CR)CH$_2$ (461)	--	--	R = C$_{11}$H$_{23}$ (e)	--
C$_5$H$_{11}$O$_2$CCH$_2$CH$_2$	--	--	(e)	--
PhPHCH$_2$CHMe	III	--	$b_{0.8}$ 210-20° (d)	2835
C$_6$F$_5$	III	--	Colorless liq., $b_{0.03}$ 118°, n_D^{22} 1.5109 (f)	1676
C$_6$Cl$_5$	--	--	(g)	--
1-Bu$_3$SnSC$_6$H$_4$Cl-p	--	--	(g)	--
1-Bu$_3$SnSC$_6$Cl$_5$	--	--	(g)	--
Ph$_3$SnSCH$_2$CO$_2$CH$_2$CH$_2$O$_2$CCH$_2$SH	IIC	--	Use as stabilizer for PVC	2903
Ph$_3$SnSC$_6$F$_5$	IIA	--	m. 76°, subl. in vacuo (f)	1676

Ph$_3$SnSC$_6$H$_4$Cl-p	IIB	--	70-90% yield, m. 96-97°

* Numbers in parenthesis refer to pages in main volume.
** The characters used here correspond to the synthetic scheme in the introduction to Chapter 4.4.2.1.
(a) Moessbauer resonance spectrum (2637). (b) Rxn. of Bu$_3$SnSCH$_2$CH$_2$OH with Si(OEt)$_4$ → (Bu$_3$SnSCH$_2$CH$_2$O)$_4$Si (2536).
(c) Activity as fungicide against coriolellus palustris (1063). (d) Use as fungicide (2835). (e) Use as stabilizer for PVC and copolymers (1404). (f) (Far) IR, assignment and NMR spectra (1676). (g) Use as molluscicide against Heliosoma subcrenatum and H. tenue (1054).

4.4.2.2 DIORGANOTIN DIMERCAPTIDES AND DIORGANO-SELENIDES

The diorganotin dimercaptides in the ensuing tables are prepared by methods from the following scheme.

I Alkylation of diorganotin disodium mercaptide R$_2$Sn(SNa)$_2$ with alkylhalides.

II Reaction of diorganotin dihalide with:
  (IIA) mercaptan RSH and dimercaptan Y(SH)$_2$ and alkali in water or preformed sodium mercaptide RSNa in alcohol,
  (IIB) mercaptans RSH and Y(SH)$_2$ in presence of pyridine or triethylamine,
  (IIC) mercaptan RSH with diorganotin dihalide amine complex,
  (IID) mercaptan RSH.

III Reaction of diorganotin oxide (R$_2$SnO)$_x$ with mercaptans RSH and Y(SH)$_2$.

IV Reaction of diorganotin dialkoxide R$_2$Sn(OR)$_2$ with mercaptan RSH.

V Interconversion reaction of diorganotin dimercaptide R$_2$Sn(SR)$_2$ with:
  (VA) mercaptan RSH,
  (VB) orthosilicate (EtO$_4$Si for R$_2$Sn(SYOH)$_2$,
  (VC) a suitable ligand like pyridine and 1,10-phenanthroline.

R = hydrogen or an organic group, Y = hydrocarbon bridge, may be functionally substituted. Additional references to diorganotin mercaptides may be found in Chapters 8.1 and 9.

4.4.2.2.1 UNSUBSTITUTED DIORGANOTIN DIMERCAPTIDES AND DIORGANOSELENIDES

Dimethyltin Dimethylsulfide $SnC_4H_{12}S_2$ $Me_2Sn(SMe)_2$
Synth.: By rxn. of Me_2SnX_2 with MeSH and aq. NaOH, in 65-70% yield (2, 728).
By rxn. of Me_2SnCl_2 with $Pb(SMe)_2$ in C_6H_6, in 70% crude yield (643).
Prop.: $b_{0.05}$ 44° (2), $b_{0.03}$ 40° (643), n_D^{20} 1.5953 (643), 1.6003, d_{20} 1.547 (2), IR (2) and NMR (646, 728) spectra, magnetic susceptibility (1).
Rxn. with: $MX_2 \to MeMX$; MX = HgBr, HgI, CdCl, CdBr (647).
$HgCl_2 \to Hg(SMe)_2$ (647).
$NiCl_2 \cdot 6H_2O \to [Ni(SMe)_2]_6$ (648).
$M(CO)_5X \to [M(CO)_3SMe]_3$; MX = MnBr, ReCl (648, 650).

Dimethyltin Diethylsulfide $SnC_6H_{16}S_2$ $Me_2Sn(SEt)_2$
Synth.: By rxn. of Me_2SnX_2 with EtSH and aq. NaOH, in 60-75% yield (2).
Prop.: $b_{0.07}$ 58°, n_D^{20} 1.5713, d_{20} 1.440, IR spectrum (2), magnetic susceptibility (1).
Rxn. with: $NiCl_2 \cdot 6H_2O \to [Ni(SEt)_2]_6$ (648).
$K_2PtCl_6 \to Pt(SEt)_2$ (647).
$Mn(CO)_5Br \to [Mn(CO)_3SEt]_3$ (649).

Dimethyltin Dibutylsulfide (M-462) $SnC_{10}H_{24}S_2$ $Me_2Sn(SBu)_2$
Synth.: By rxn. of Me_2SnX_2 with BuSH and aq. NaOH, in 65-70% yield (2).
By rxn. of $Me_2Sn(SNa)_2$ with BuCl, in 64% yield (1747a).
Prop.: $b_{0.7}$ 110° (1747a), $b_{0.1}$ 81° (2), n_D^{25} 1.537 (1747a), n_D^{20} 1.5400, d_{20} 1.280 (2), IR spectrum (2), magnetic susceptibility (1).
Use: As catalyst for polyurethan foaming (1747a), as microbiocide (1747a), as stabilizer for PVC (1747a).

1,3-Dithia-2-stannacyclopentene $SnC_4H_8S_2$ $\overline{Me_2SnSCH{:}CHS}$
Synth.: By rxn. of Me_2SnCl_2 with $(:CHSNa)_2$ in H_2O, followed by sublimation (652)
Prop.: Buff cryst., subl., NMR spectrum, soly. (652).
Rxn. with: $MCl_2 \to (MSCH{:}CHS)_x$; M = Cd, Hg (652).
$CoBr_2, + NEt_4Br \to Et_4N[Co(SCH{:}CHS)_2]$ (652).
Addn. Compd.: $\overline{Me_2SnSCH{:}CHS} \cdot Py$, from components, cream cryst., NMR spectrum, dissocn. in vacuo (652)]$SnC_9H_{13}NS_2$].
$\overline{Me_2SnSCH{:}CHS} \cdot 1,10$-phenanthroline, from components in CH_2Cl_2, orange cryst., stable in air (652) [$SnC_{16}H_{16}N_2S_2$].

Dibutyltin Dibenzylsulfide (M-462) $SnC_{22}H_{32}S_2$ $Bu_2Sn(SCH_2Ph)_2$
Synth.: By rxn. of Bu_2SnO and $Bu_2Sn(OEt)_2$ with $PhCH_2SH$ under reflux in C_6H_6, in 98 and 97% yield, resp. (1997).
By rxn. of $Bu_2Sn(SPr)_2$ with $PhCH_2SH$ in cyclohexane under reflux, in 93% yield (1997).
By rxn. of $Bu_2Sn(SNa)_2$ with $PhCH_2Cl$ at 100° (1640).
Prop.: $b_{1.0}$ 220°, $b_{0.5}$ 210°, mol. wt. (1997).
Use: As stabilizer for PVC (1640).

Dibutyltin Dipropylsulfide $SnC_{14}H_{32}S_2$ $Bu_2Sn(SPr)_2$

<u>Synth.</u>: By rxn. of Bu_2SnO with PrSH under reflux in C_6H_6, 99% yield (1997).
By rxn. of $Bu_2Sn(OEt)_2$ with PrSH under reflux in C_6H_6, in 99% yield (1997).
<u>Prop.</u>: $b_{0.5}$ 140-42°, $b_{0.15}$ 121-23°, mol. wt. (1997).
Rxn. with RSH → $Bu_2Sn(SR)_2$ + PrSH; R = $PhCH_2$, $C_{12}H_{25}$ (1997).
EtOH (excess) + cat. $MeC_6H_4SO_3H$ → $Bu_2Sn(OEt)_2$ + PrSH-EtOH azeotrope (1997).

Dibutyltin Dibutylsulfide $SnC_{16}H_{36}S_2$ $Bu_2Sn(SBu)_2$

<u>Synth.</u>: By rxn. of Bu_2SnCl_2 with BuSNa in alc. C_6H_6 (1117).
By rxn. of Bu_2SnO and $Bu_2Sn(OEt)_2$ with BuSH in C_6H_6 under reflux, in 98 and 99% yield, resp. (1997).
By rxn. of $Bu_2Sn(SEt)_2$ with BuSH (1:2) in hexane, in 85% yield (1997).
<u>Prop.</u>: $b_{2.5}$ 158-60° (1116-7), $b_{1.5}$ 170°, $b_{0.7}$ 157-60° (1997), n_D^{20} 1.5238, d_{20} 1.145 (1116-7), far IR and UV spectra (1116-7), calorimetric titration with Bu_2SnCl_2, Bu_2SnBr_2 (1116-7) and $SnCl_4$ (1116), dipole moment (1116), mol. wt. (1116-7, 1997), polarographic detn. (2265).
<u>Rxn. with</u>: Bu_2SnCl_2 → $Bu_2Sn(SBu)Cl$ (942, 1116).

Dibutyltin Dilaurylsulfide (M-462) $SnC_{32}H_{68}S_2$ $Bu_2Sn(SC_{12}H_{25})_2$

<u>Synth.</u>: By rxn. of $Bu_2SnCl_2 \cdot 2NH_3$ with $C_{12}H_{25}SH$ in C_6H_6 (2954).
By rxn. of Bu_2SnO or $Bu_2Sn(SEt)_2$ with $C_{12}H_{25}SH$ in C_6H_6 under reflux, in 99% yield (1997).
By rxn. of $Bu_2Sn(SPr)_2$ with $C_{12}H_{25}SH$ in cyclohexane, in 98% yield (1997).
<u>Prop.</u>: Colorless liq., dec. on vacuum distn. (1997), mol. wt. (1997). Anal. detn. by GLC (2032) of S (2321).
<u>Biol. Prop.</u>: Effect on complement fixation test for virus, effect on human red blood cells (1151).
<u>Use</u>: In catalyst for condensation of PVC with vinyl monomers and polyester (2006), for polymerization of butadiene (2044a).
As stabilizer for polymers (453), for PVC (597, 1127a, 1464, 1651, 2549, 2953-4, 2963), for PVC-copolymers (1866, 2862).

Additional unsubstituted diorganotin dimercaptides are listed in Table 229.

Table 229. Unsubstituted Diorganotin Dimercaptides and Diselenides $R_2Sn(SR)_2$

Formula*	Synth. Method**	Yield	Properties	Ref.
$Me_2Sn(SR)_2$				
R = $PhCH_2$	I	96	Light yellow liq. (a)	1747a
Pr	IIA	65-70	b_1 74°, n_D^{20} 1.5498, d_{20} 1.323 (b, c)	2
Me_2CH	--	--	(d)	--
C_8H_{17}	IIA	65-70	$b_{0.2}$ 166°, n_D^{20} 1.5129, d_{20} 1.092 (b, c)	2
Ph	VA (d)	65	m. 38-39°, $b_{0.001}$ 130-35°	643
$Et_2Sn(SMe)_2$	IIA	65-70	$b_{0.1}$ 61°, n_D^{20} 1.5793, d_{20} 1.440 (b, e)	2
$Et_2Sn(SEt)_2$	IIA (f)	65-70	$b_{0.5}$ 94°, n_D^{20} 1.5572, d_{20} 1.319 (b)	2
$Bu_2Sn(SR)_2$				
R = Me	III	--	$b_{0.02}$ 94°, n_D^{22} 1.5538 (g, h)	1676
Et (462)	III	98	$b_{0.4}$ 97-100°, mol. wt. (i)	1997
CH_2:$CHCH_2$	IIA (j)	12	$b_{0.075}$ 134-35°, mol. wt., TLC	1383
MeEtCH	III	98	$b_{0.8}$ 147-48°, mol. wt.	1997
	IV	99	--	1997
Me_3C	III	90	$b_{0.1}$ 118-19°, mol. wt.	1997
	IV	92	--	1997
C_8H_{17} (462)	III	--	Polarographic detn. (2265)	2802
Ph (462)	III	--	$b_{0.05}$ 162° (g)	1676
	III	96	$b_{0.1}$ 179°, mol. wt.	1997
	IV	99	$b_{0.8}$ 200°	1997
R	--	--	Use as stabilizer for PVC (k, l)	713, 3004
$(C_8H_{17})_2Sn(SC_{12}H_{25})_2$ (463)	--	--	Anal. S detn. (m)	2321
$(C_8H_{17})_2Sn(SPh)_2$ (463)	IIC	--	n_D^{20} 1.5688, stabilizer for PVC	2954
$(C_8H_{17})_2Sn(SR)_2$	--	--	Use as stabilizer for PVC (l)	2895
$Ph_2Sn(SPh)_2$ (463)	IIB	70-90	m. 65°	121

R_2SnSYS

R =	Y =			
Me	CH_2CH_2 (463)	IIA	52	m. 82°, IR spectrum (n)	2
Me	$CH_2CH_2 \cdot Py$	IIA	--	m. 81.5-82°, Moessbauer resonance spectrum	141
Me	1,2,4-C_6H_3Me	VC	--	Moessbauer resonance spectrum	141
Me	1,2,4-$C_6H_3Me \cdot Py$	IIA	--	m. 137°, Moessbauer resonance spectrum	141
Me	1,2,4-$C_6H_3Me \cdot L$	VC	--	White cryst., m. 106-108° (o)	141
Et	CH:CH	VC	--	L = 1,10-phenanthroline, yellow cryst. (o)	141
CH_2CH	CH:CH	IIA	--	Buff cryst., subl., soly.	652
Bu	CH_2CH_2 (463)	IIA	--	Buff cryst., subl., soly.	652
Bu	$(CH_2)_3$ (463)	--	--	(Far) IR spectra, assignment (p)	1038
Bu	p-$CH_2C_6H_4CH_2$	--	--	(Far) IR spectra, assignment	1038
Bu	m-$CH_2C_6H_3MeCH_2$	III	--	Polymeric, stabilizer for PVC	330,445
Bu	1,2,4-$CH_2C_6HMe_3CH_2$	IIA	--	Viscous oil, solidifies on standing	445
Bu	1,3,5-$CH_2C_6HMe_3CH_2$	IIA	--	Polymeric, viscous oil (q)	445
Bu		IIA	--	Polymeric, viscous oil (q)	445
C_8H_{17}	$CH_2C_6H_4CH_2$	III	--	Polymeric, stabilizer for PVC	445

* Numbers in parenthesis refer to pages in main volume.

** The characters used here correspond to the synthetic scheme in the introduction to Chapter 4.4.2.2
(a) Use as microbiocide, catalyst for polyurethan foaming, as stabilizer for PVC (1747a). (b) IR spectrum (2). (c) Magnetic susceptibility (1). (d) Rxn. of $Me_2Sn(SPr-i)_2$ with PhSH → $Me_2Sn(SPh)_2$ (643). (e) Rxn. with $HgCl_2$ → $Hg(SMe)_2$ (647), with K_2PtCl_4 → $Pt(SMe)_2$ (647). (f) Synth. from $Bu_2Sn(ClCBu)_2$ with EtSH (1221); rxn. with $NiCl_2$ → $[Ni(SEt)_2]_6$ (2146). (g) (Far) IR, assignment and NMR spectra (1676). (h) Anal. detn. by I-titration (1676). (i) Rxn. with BuSH (1:2) and cat. $MeC_6H_4SO_3H$ → $Bu_2Sn(SBu)_2$ + EtSH (1997). (j) By-prod. in rxn. of Bu_2SnH_2 with $CH_2:CHCH_2SH$ in presence of AIBN at 60°, in 15% yield (1383). (k) Use as curing agent for PVC copolymer adhesives (3001). (l) Use as stabilizer for chloroprene rubber (2884). (m) Use in catalyst for condensation of polyesters, vinyl monomers and PVC (2006). (n) Rxn. with $HgCl_2$ → $Bu_2Sn(OAc)_2$ + $(HgSCH_2CH_2S)_x$ (647). (o) Moessbauer resonance spectrum (141). (p) Rxn. with $Hg(OAc)_2$ → $Bu_2Sn(OAc)_2$ + $(HgSCH_2CH_2S)_x$ (1704), with I in $HCCl_3$, + BiPy → $Bu_2SnI_2 \cdot BiPy$ + $(SCH_2CH_2S)_x$ (1704). (q) Use as stabilizer for PVC (445).

Table 229 Continued $R_2Sn(SR)_2$

Formula*		Synth. Method**	Yield	Properties	Ref.
R_2SnSYS (Cont'd.)					
R = Ph	Y = CH_2CH_2 (463)	IIA	--	m. 109-10° (o)	141
		IIB	--	m. 108°, IR spectrum	121
		--	--	(Far) IR spectra, assignment	1038
Ph	$CH_2CH_2 \cdot L$	VC	--	L = 1,10-phenanthroline, yellow cryst. (o)	141
Ph	$(CH_2)_3$ (463)	--	--	(Far) IR spectra, assignment	1038
Ph	1,2,4-C_6H_3Me	IIA	--	m. 151-55°	141
Ph	1,2,4-$C_6H_3Me \cdot Py$	VC	--	m. 105-106° (o)	141
Ph	p-$CH_2C_6H_4CH_2$	IID	--	(r)	610
$R_2Sn(SR)_2$		--	--	Use as stabilizer for PVC (s)	2895
$Me_2Sn(SePh)_2$		IIA	--	m. 32-33°	651

(r) Use as stabilizer for synthetic polymers (610). (s) Use as stabilizer for PVC and copolymers with $CH_2:CMeCO_2Me$ (489).

4.4.2.2.3 HETEROCYCLIC SUBSTITUTED DIORGANOTIN DIMERCAPTIDES

2,5-Dibutyltin 1,3,4-Thiadiazoledimercaptide (M-464)
 $(SnC_{10}H_{18}N_2S_3)_x$
Synth.: By rxn. of Bu_2SnO with $(HS)_2C_2N_2S$ in MePh (193).
Prop.: m. 120° (193).

5-Dibutyltin Bis(3,2-methylthione-1,3,4-thiadiazolemercaptide)
 $SnC_{11}H_{21}N_2S_3$
Synth.: By rxn. of Bu_2SnO with $HS(C_2N_2SMeS)$ in MePh (193).
Prop.: m. 120° (193).

2,5-Dioctyltin 1,3,4-Thiadiazoledimercaptide
 $(SnC_{18}H_{34}N_2S_3)_x$ $(C_8H_{17})_2SnS(C_2N_2S)S*$
Synth.: By rxn. of $(C_8H_{17})_2SnO$ with $(HS)_2C_2N_2S$ in MePh (193).
Prop.: m. 52° (193).

5-Dioctyltin Bis(3,2-methylthione-1,3,4-thiadiazolemercaptide)
 $SnC_{19}H_{37}N_2S_3$ $(C_8H_{17})_2Sn(SC_2N_2S_2Me)_2*$
Synth.: By rxn. of $(C_8H_{17})_2SnO$ with $HS(C_2N_2SMeS)$ in MePh (193).
Prop.: m. 95° (193).

4.4.2.2.6 FUNCTIONALLY SUBSTITUTED DIORGANOTIN DIMERCAPTIDES

1,3-Dithia-2-stanna-4,5-dicyanocyclopentene $SnC_6H_6N_2S_2$ $Me_2\overline{SnSC(CN):C(CN)S}$
Synth.: By rxn. of Me_2SnCl_2 with $[:C(CN)SNa]_2$ in H_2O, in 70-90% yield (652).
Prop.: Buff cryst., NMR spectrum, soly. (652).
Rxn. with: $AgNO_3 \rightarrow [AgSC(CN):]_2$ (652).
$HgX_2 \rightarrow [HgSC(CN):C(CN)S]_x$; X = Cl, Br (652).
$NiBr_2$, + $Et_4NBr \rightarrow \overline{(Et_4N)_2Ni[SC(CN):C(CN)S]_2}$ (652).
Addn. Compd.: $Me_2\overline{SnSC(CN):C(CN)S}\cdot Py$, from components, cream cryst., NMR spectrum, dissocn. in vacuo (652) $[SnC_{11}H_{11}N_3S_2]$.
$Me_2\overline{SnSC(CN):C(CN)S}\cdot 1,10$-phenanthroline, from components without solvent, yellow cryst., stable in air (652) $[SnC_{18}H_{14}N_4S_2]$.

Dibutyltin Bis(i-octyl thioglycolate) $SnC_{28}H_{56}O_4S_2$ $Bu_2Sn(SCH_2CO_2C_8H_{17}-i)_2$
Synth.: By rxn. of $Bu_2SnCl_2\cdot 2H_2NCH_2Ph$ with $i-C_8H_{17}O_2CCH_2SH$ in hydrocarbon (1702).
By rxn. of Bu_2SnO with $i-C_8H_{17}O_2CCH_2SH$ in H_2O at 40-60° (2802), in 98% yield (1992).
By rxn. of $Bu_2Sn(SNa)_2$ with $i-C_8H_{17}O_2CCH_2SH$, in 89% yield (1747a).
Prop.: Slightly yellow liq. (1747a), storage stability, soly. (1992).
Biol. Prop.: Effect on complement fixation test for virus (1151).

---

* For structural formula compare the preceding butyl-analogs.

Use: As antioxidant for polypropylene (2171).
In catalyst for polymerization of vinyl monomers (2958), with unsaturated polyester (1398), for polyurethan foaming (1747a).
As microbiocide (1747a).
As stabilizer for PVC (1464, 1651, 1747a, 2372, 2549, 2963), for PVC-copolymers (1866) with vinyl acetate (1127a, 2862), mechanism of PVC stabilization by degradation with tagged stabilizer (158).

Listing of functionally substituted diorganotin dimercaptides continues in Table 230.

4.4.2.3 MONOORGANOTIN TRIMERCAPTIDES

All compounds are listed in Table 231.

4.4.2.9 ORGANOTIN-SULFUR COMPOUNDS CONTAINING TIN-HETEROATOM BONDS

Organotin sulfides or mercaptides containing tin linked to halogen, pseudohalogen, oxygen anions, alkoxide and carboxylate groups are summarized in Subsections 4.1.6, 4.2.1.6, 4.2.4.6, 4.3.3.2.9.5 and 4.3.5.2.9, respectively. Compounds containing a tin-nitrogen linkage are listed at the bottom of Table 230. Additional references to organotin sulfur compounds containing tin-heteroatom bonds may be found in Chapter 9.

Table 230. Functionally Substituted Diorganotin Dimercaptides $R_2Sn(SR')_2$

Formula*	Synth. Method**	Yield	Properties	Ref.
$Me_2Sn(SCH_2CO_2C_8H_{17}-i)_2$	IIC	--	(a - 1702) (b, c)	1702
$(PhCH_2)_2Sn(SCH_2CO_2C_6H_{11})_2$	--	--	Toxicity	1200
$(PhCH_2)_2Sn(SCH_2CO_2C_8H_{17}-i)_2$	III	--	Thick oil, toxicity (1200) (a - 751)	751
$(NCCH_2CH_2)_2Sn(SCH_2CO_2C_8H_{17}-i)_2$	III	94	Yellow liq., n_D^{25} 1.5215	1746
	IIB		n_D^{28} 1.5212, d_{25} 1.2084 (d)	895b
$Bu_2Sn(SCH_2CH_2OH)_2$ (465)	III	--	(e, f)	2536, 2959
$Bu_2Sn(SCH_2CH_2O_2CC_8H_{17})_2$ (465) (g)	--	--	(a - 1404) (h - 1404)	--
$[Bu_2Sn(SCH_2CH_2O)_2]_2Si$	VB (e)	--	Clear glassy prod. (a - 2536)	2536
$[Bu_2Sn(SCH_2CH_2O)_2]_3(SiOEt)_2$	VB (e)	--	Clear glassy prod. (a - 2536)	2536
$Bu_2Sn(SCH_2COR)_2$				
R = $(C_4H_7O)CH_2O$	III	--	C_4H_7O = tetrahydrofuranyl (a - 1284)	1284
$HOCH_2CH_2O$	III	--	(i)	2536
$C_8H_{17}O$ (466)	--	--	(a - 1758)	--
$EtBuCHCH_2O$ (466)	--	--	(a - 1208)	--
$C_9H_{19}O$ (466)	--	--	(a - 2901) (j)	--

* Numbers in parenthesis refer to pages in main volume.
** The characters used here correspond to the synthetic scheme in the introduction to Chapter 4.4.2.2.
(a) Use as stabilizer for PVC. (b) Use as stabilizer for vinyl chloride vinyl acetate copolymers (1127a). (c) Use as accelerator for polymerization of vinyl monomers with unsaturated polyesters (1398). (d) Use as biocide (895b), as stabilizer for plastics (895b). (e) Rxn. of $Bu_2Sn(SCH_2CH_2OH)_2$ with $(EtO)_4Si \rightarrow [Bu_2Sn(SCH_2CH_2O)_2]_2Si$ + $[Bu_2Sn(SCH_2CH_2O)_2]_3(SiOEt)_2$ (2536). (f) Rxn. with 1,2,4-$MeC_6H_3(NCO)_2 \rightarrow$ tin contg. polyurethan (2959). (g) Correct in main volume p. 465, line 17 formula $C_8H_{11}CO_2CH_2CH_2$ to $C_8H_{17}CO_2CH_2CH_2$. (h) Use as stabilizer for PVC copolymers, halogenated vinyl polymers and chlorinated paraffins. (i) Rxn. of $Bu_2Sn(SCH_2CO_2CH_2CH_2OH)_2$ with $(EtO)_4Si \rightarrow [Bu_2Sn(SCH_2CO_2CH_2CH_2O)_2]_2Si$ (2536). (j) Use as stabilizer for vinyl polymers (2946).

Table 230 Continued $R_2Sn(SR')_2$

Formula*	Synth. Method**	Yield	Properties	Ref.
$Bu_2Sn(SCH_2COR)_2$ (Cont'd.)				
R = $C_{12}H_{25}O$	IIC	--	--	1702
$HSCH_2CO_2CH_2CH_2O$	III	--	(a - 2903)	2903
$HSCH_2CO_2(CH_2)_4O$	IID	--	(a - 2903)	2903
$HSCH_2CO_2(CH_2)_5O$	IID	--	(a - 2903)	2903
$HSCH_2CO_2(CH_2)_8O$	IID	--	(a - 2903)	2903
$HSCH_2CO_2CHEtCH_2CHPrO$	IID	--	(a - 2903)	2903
$C_5H_{11}NH$ (466)	--	--	(h - 1403)	--
$o(CH_2CH_2)_2N$ (k) (466)	--	--	(h - 1403)	--
$[Bu_2Sn(SCH_2CO_2CH_2CH_2O)_2]_2Si$	VB (i)	--	Clear prod. (a - 2536)	2536
$Bu_2Sn[SCH_2CH_2CON(CH_2CH_2)_2O]_2$ (k)	--	--	(h - 1407)	--
$Bu_2Sn[SCH(CO_2Bu)CH_2CO_2Bu]_2$ (466)	--	--	(h - 1407)	--
$Bu_2Sn(SC_6F_5)_2$	III	--	Yellow liq., m. 30°, b0.02 139° (l)	1676
$(C_8H_{17})_2Sn(SR)_2$				
R = $n-C_8H_{17}O_2CCH_2$ (467)	--	--	TLC (a - 1758)	1172a
$i-C_8H_{17}O_2CCH_2$	III	96	Storage stability (a - 722, 980a, 2963) (b, m)	1992, 2802
$EtBuCHCH_2O_2CCH_2$	--	--	Thermal stability, TLC	1052
$HSCH_2CO_2(CH_2)_4O_2CCH_2$	IID	--	(a - 2903)	2903
$RO_2CCH_2CH(CO_2R)$	IIC	--	R = $EtBuCHCH_2$, n_D^{20} 1.4892 (a - 2954)	2954
$(i-C_8H_{17})_2SnSCH_2CO_2C_8H_{17}$	IIC	--	--	1702
$(EtBuCHCH_2)_2Sn(SCH_2CO_2R)_2$	IID	--	R = $HSCH_2CO_2CH_2CH_2$ (a - 2903)	2903
$(C_{12}H_{25})_2Sn(SCH_2CO_2C_8H_{17}-i)_2$	--	--	(a - 2572)	--
R_2SnSYS				
R = Et Y = C(CN):C(CN)	IIA	70-90	Buff cryst., soly.	652

$CH_2:CH$	$C(CN):C(CN)$	70-90	Buff cryst., subl., soly.	652
Bu	$C(CN):C(CN)$	--	m. 165-67° (dec.)	2613
Bu	$C(NO_2):C(NO_2)$	55	m. 155° (dec.) (a - 1975)	1975
Bu	$CH_2Y'CH_2$	--	--	2613
Bu	$CH_2Y'CH_2$ (465)	--	$Y' = CO_2CH_2CH_2CO_2$, polymeric (h - 1405)	--
C_8H_{17}	$C(CN):C(CN)$	51	$Y' = CO_2(CH_2)_4CO_2$, polymeric (h - 1407)	--
Ph	$C(CN):C(CN)$	--	m. 110° (a - 1975)	1975
Ph	$C(NO_2):C(NO_2)$	40	m. 210-20° (dec.)	2613
		--	m. > 200° (dec.) (a - 1975)	1975
$Bu_2Sn(SC_{12}H_{25})NRCH_2\overline{CHCH_2O}$	III	--	R = p-$MeC_6H_4SO_2$, tan solid (a - 1442)	1442
$Bu_2Sn(SCH_2CO_2C_8H_{17})NRCH_2\overline{CHCH_2O}$	III	--	R = p-$MeC_6H_4SO_2$ (a - 1442)	1442

(k) N-morpholine derivative. (l) (Far) IR, assignment and NMR spectra (1676). (m) U. S. limits for use as PVC-stabilizer (2942).

Table 231. Monoorganotin Trimercaptides RSn(SR)$_3$

Formula	Synth. Method*	Yield	Properties	Ref.
MeSn(SMe)$_3$	IIA	51	b$_{0.01}$ 75°, n$_D^{25}$ 1.6352, d$_{20}$ 1.63	643
	(a)	--	NMR spectrum (646, 728)	728
MeSn(SEt)$_3$	IIA	30-50	b$_{0.05}$ 90°, n$_D^{20}$ 1.5972, d$_{20}$ 1.469 (b)	2
MeSn(SPr)$_3$	IIA	30-50	b$_{0.001}$ 95°, n$_D^{20}$ 1.5684, d$_{20}$ 1.337 (b)	2
EtSn(SMe)$_3$	IIA	30-50	b$_{0.001}$ 66°, n$_D^{20}$ 1.6232, d$_{20}$ 1.548 (b)	2
[BuSn(SCH$_2$CH$_2$O)$_3$]$_4$Si$_3$	VB (c)	--	Clear prod. (d - 2536)	2536
BuSn(SCH$_2$CO$_2$CH$_2$C$_4$H$_7$O)$_3$	--	--	(d - 1284)	--
BuSn(SCH$_2$CO$_2$C$_8$H$_{17}$-i)$_3$	I (e)	94	Pink liq., n$_D^{30}$ 1.5020, (d - 1747a) (f)	1747a
BuSn(SCH$_2$CO$_2$C$_{12}$H$_{25}$)$_3$	--	--	(d - 980a)	--
BuSn(SCHMeCH$_2$CO$_2$R)$_3$	IID	--	R = HSCHMeCH$_2$CO$_2$C$_3$H$_6$ (d - 2903)	2903

* The characters used here correspond to the synthetic scheme in the introduction to Chapter 4.4.2.2, monoorganotin species are substituted for the diorganotin entities in the appropriate places.
(a) Synth. from MeSn(O)OCO$_2$K and MeSH (728). (b) IR spectrum (2). (c) Synth. from BuSnO$_2$H with HSCH$_2$CH$_2$OH at 120°, followed by (EtO)$_4$Si (2536). (d) Use as stabilizer for PVC. (e) Synth. from Bu$_2$Sn$_2$S$_3$ with aq. alc. Na$_2$S, followed by i-C$_8$H$_{17}$O$_2$CCH$_2$Cl (1747a). (f) Use as catalyst for foaming polyurethan, as microbiocide (1747a).

4.4.4 ORGANOTIN THIOCARBOXYLATES, XANTHATES, THIAMIDES AND THIOCARBAMATES

The organotin derivatives of organic thioacids listed in Tables 232 and 233 are prepared by methods from the ensuing scheme. Derivatives of thiono-phosphates, -phosphonates and -phosphinates are listed in Tables 162 and 163.

I   Addition of triorganotin hydride R_3SnH to thiocyanates RNCS.

II Acylation of organotin sodium mercaptide $R_nSn(SNa)_{4-n}$ with benzoyl chloride.

III Reaction of organotin halide R_nSnX_{4-n} with:
   (IIIA) thiocarbamates R_2NCS_2M,
   (IIIB) cyanoimidothiocarbamate $NCN:CS_2M'$; $M' = 2Na, Ca$,
   (IIIC) thiocarboxylic acids RCOSH, RCS_2H in presence of organic bases.

IV Reaction of organotin oxides or hydroxides R_3SnOH, R_2SnO with:
   (IVA) thiocarbamate PhNHCSOMe,
   (IVB) carbon disulfide and secondary amine,
   (IVC) thiocarboxylic acid RCOSH.

V   Reaction of organotin alkoxide $R_nSn(OR)_{4-n}$ with:
   (VA) thiocarbamate PhNHCSOEt and thiamide PhNHCSMe,
   (VB) carbon disulfide or thiocyanate RNCS,
   and of triorganotin carboxylate R_3SnO_2CR with:
   (VC) thiocarboxylic acid RCOSH or thiocarbamate R_2NCS_2Na,
   (VD) carbon disulfide for $Me_3SnO_2CNMe_2$.

VI Reactions of:
   (VIA) Me_3SnNR_2 with CS_2 or EtNCS,
   (VIB) Ph_3SnPPh_2 with YCS; $Y = S, 2Cl, 2NH_2$.

R = organic group, R' = acyl or thioacyl group, M = sodium or potassium, n = 2, 3, X = halogen.

4.4.4.1 TRIORGANOTIN THIOCARBOXYLATES, XANTHATES, THIAMIDES AND THIOCARBAMATES

All compounds are listed in Table 232.

4.4.4.2 DIORGANOTIN THIOCARBOXYLATES, XANTHATES, THIAMIDES AND THIOCARBAMATES

All compounds are listed in Table 233. Diorganotin thioacid derivatives containing tin-heteroatom bonds may be found in Subchapters 4.1.6 and 4.3.3.2.9.5.

Table 232. Triorganotin Thiocarboxylates, Xanthates, Thiamides and Thiocarbamates

Formula*	Synth. Method**	Yield	Properties	R_3SnSR' Ref.
Me_3SnSR'				
R' = Bz	II	18	n_D^{27} 1.5906 (a - 1747a) (b)	1747a
PhN:CH	I	--	IR spectrum	1415
p-EtOC$_6$H$_4$N:CH	I	--	IR spectrum	1415
Me$_2$NCS (471)	VIA	95	White, m. 61.5-63°, monomeric	167
	VD	--	IR spectrum, assignment, struct.	167
EtN:C(NMe$_2$)	VIA	96	Dec. on vacuum detn., IR spectrum	2406
Et_3SnSR'				
R' = PhN:CH (471)	I	--	IR spectrum	1415
p-EtOC$_6$H$_4$N:CH	I	--	IR spectrum	1415
PhN:CMe	VA	81	b$_{0.2}$ 108-109°, n_D^{20} 1.5577 (c)	402
Et$_2$NCS	IIIA	91	b$_2$ 165°, IR and UV spectra, dielectric const., dipole moment, struct.	761
	IVB	--		761
PhN:C(OEt)	VA	78	b$_{0.2}$ 62-63°, n_D^{20} 1.5740 (c)	402
Bu_3SnSR'				
R' = Ac (471)	VC	94	b$_{0.02}$ 86-88°	2406
MeOCS (471)	VB	--	Dissocn. at 0.1 mm. 100°, NMR spectrum (750)	36,750
PhN:CH	I	--	IR spectrum	1415
Me$_2$NCS	--	--	Use as fungistat (d)	1680
Et$_2$NCS (471)	--	--	Use as rodent repellent	2036
CH$_2$:CHCH$_2$N:C(OMe)	VB	--	b$_{0.02}$ 84-85° (partly dec.), impure (e)	750
PhN:C(OMe)	VB	--	b$_{0.2}$ 68° (partly dec.),	36,750
	IVA	--	IR spectrum, stable in air	750
(Bu$_3$SnS)$_2$C:NCN	IIIA	--	b$_{0.3}$ 145°, n_D^{23} 1.5276 (f)	1976
[(C$_8$H$_{17}$)$_3$SnS]$_2$C:NCN	IIIA	--	(f)	1976

Ph₃SnSR'

R' = Bz (472)				
PhCS	IIIC	70-90	m. 108-109° (g)	121
PhCS	IIIC	--	m. 97-99°	1410
PhN:CH	I	--	IR spectrum	1415
p-EtOC₆H₄N:CH	I	--	IR spectrum	1415
Me₂NCS	IIIA	96	m. 136-37°, IR, UV spectra (h, i)	298
Me₂NCS	IIIA	65	m. 135-35.5°, NMR spectrum (1220)	866
PhCH₂NHCS (472)	--	--	m. 124-26°, IR, UV spectra (j)	298
EtNHCS	IIIA	91	m. 133-34°, IR, UV spectra	298
EtNHCS	IIIA	88	m. 132° (h)	866
	IIIA	--	UV, NMR spectra, dipole moment	761
PhNHCS	IIIA	81	m. 182-83°, IR, UV spectra, stable in refluxing C₆H₆, (k, l)	298
	VC	70		761
	IIIA	--	m. 194-95°, dipole moment, struct.	
Ph₂PCS	VIB	64	m. 68°, IR spectrum, mol. wt.	483
Ph₂PCCl₂	VIB	87	m. 93°, IR spectrum, mol. wt.	483
Ph₂PC(NH₂)₂	VIB	22	m. 115°, IR spectrum, mol. wt.	483
(Ph₃SnS₂CNHCH₂)₂ (472)	--	--	(i - 260)	--

* Numbers in parenthesis refer to pages in main volume.
** The characters used here correspond to the synthetic scheme in the introduction to Chapter 4.4.4.
(a) Use as stabilizer for PVC. (b) Use as microbiocide, as foaming catalyst for polyurethan (1747a). (c) IR and NMR spectra (402). (d) Use in antifouling coatings (2302). (e) IR spectrum (750), thermal dec. at 85° in vacuo → Bu₃SnOMe (750). (f) Use as antioxidant, polymerization catalyst, stabilizer (1976). (g) Decomposed by silica or alumina (121). (h) Activity as fungicide against Aspergillus niger, Penicillium italicum, Sclerotina sclerotiorum and Xanthomonas malvacearum (866). (i) Activity as insecticide and effect on reproduction of Musca domestica (172a, 260). (j) Thermal decomposition at 130° → (Ph₃Sn)₂S + (PhCH₂NH)₂CS (1377). (k) Thermal decomposition at 250° → (Ph₃Sn)₂O + Ph₂NH (298). (l) White solid, dielectric constant (761).

Table 233. Diorganotin Thiocarboxylates, Xanthates, Thiamides and Thiocarbamates $R_2Sn(SR')_2$

Formula*	Synth. Method**	Yield	Properties	Ref.
$Me_2Sn(SR')_2$				
R' = Bz	II	83	White cryst., m. 134-35° (a, b - 1747a)	1747a
Me_2NCS	IIIA	--	m. 198-200°, mol. wt., soly., struct. (c-e)	2445
Et_2NCS	--	--	NMR spectrum (f)	1201
$(CH_2)_4NCS$	--	--	(f)	--
Ph_2NCS	--	--	(f)	--
$(PhCH_2)_2Sn(SOCC_{17}H_{33})_2$	IVC	--	Oleate, yellow oil (g)	37
$Et_2Sn(SBz)_2$	II	75	White cryst., m. 86-87° (a, b - 1747a)	1747a
$Et_2Sn(S_2CNMe_2)_2$	--	--	(c)	--
$Et_2Sn(S_2CNEt_2)_2$	IIIA	--	White cryst., m. 84°, subl.o.1 180° (h)	761
$Bu_2Sn(SR')_2$				
R' = Bz (471)	II	94	Cryst., m. 51-52° (a, b - 1747a)	1747a
MeOCS	VB	--	NMR spectrum	943
EtOCS	--	--	Anal. detn. by complexometry	168
Me_2CHOCS (471)	--	--	(i)	--
$(PhCH_2)_2NCS$	--	--	(f)	--
$(CH_2)_4NCS$	--	--	(f)	--
Ph_2NCS	--	--	(f)	--
$CH_2:CHCH_2N:C(OMe)$	VB	--	Pale yellow oil, easily hydrolyzed (j)	943
PhN:C(OMe)	VB	--	(j, k)	943
$Bu_2SnS_2C:NCN$	IIIB	70	From Ca salt, m. 145° (l, m)	1976
$(C_8H_{17})_2Sn(SBz)_2$	IIIB	96	From Na salt, m. 150-54° (dec.), stable in H_2O	2751
$(C_8H_{17})_2Sn(SBz)_2$	II	69	Pink-violet oil, n_D^{26} 1.5787 (a, b - 1747a)	1747a
$(C_8H_{17})_2SnS_2C:NCN$	IIIB	--	From Ca salt, m. 135° (l, m)	1976
$(C_8H_{17})_2SnS_2C:NCN$	IIIB	94	From K salt, m. 135-38° (dec.)	2751
$(C_{12}H_{25})_2Sn(SBz)_2$	IIIC (n)	--	n_D^{20} 1.5496 (b - 2954)	2954

$Ph_2Sn(SR')_2$

R' = Bz (472)	II	82	Cryst., m. 152-53°, NMR spectrum (1220)	1747a
$PhCH_2NHCS$	IIIA	83	m. 92-94°, IR and UV spectra	298
	IVB	80	(o)	298
Me_2NCS	IIIA	74	m. 203-205°, IR and UV spectra	298
$(PhCH_2)_2NCS$	--	--	(f)	--
Et_2NCS	IIIA	83	m. 152-54°, IR and UV spectra	298
	IIIA	--	Yellow, m. 145-46°, monomeric (f, h)	761
$(CH_2)_4NCS$	--	--	(f)	--
Ph_2NCS	IIIA	73	m. 216-18°, IR and UV spectra	298
	IIIA	--	Pale yellow, m. 117-18° (f, h, p)	761
$Ph_2SnS_2C:NCN$	IIIA	95	m. 173-75° (dec.), soly. (l)	2751

* Numbers in parenthesis refer to pages in main volume.
** The characters used here correspond to the synthetic scheme in the introduction to Chapter 4.4.4.

(a) Use as foaming catalyst for polyurethan, as microbiocide (1747a). (b) Use as stabilizer for PVC. (c) Rxn. with $(CH_2Cl)_2 \rightarrow [R_2SnCl_4][Me_2N:CSCH_2CH_2S]_2$; R = Me, Et (2880). (d) Rxn. with $(CH_2Br)_2 \rightarrow [Me_2SnBr_3][Me_2N:CSCH_2CH_2S]$ (2880). (e) (Far) IR, assignment (2445) and NMR (1201, 2445) spectra. (f) Moessbauer resonance spectrum, struct. (2392). (g) Use as stabilizer for chlorine containing polymers (37). (h) IR, UV and NMR spectra, dielectric const., dipole moment, struct. (761). (i) Use as stabilizer for halogenated vinyl polymers, copolymers and for chlorinated paraffins (1406). (j) IR and NMR spectra (943). (k) Rxn. with $H_2O \rightarrow Bu_2SnO + PhNCSOMe$ (943). (l) IR spectrum, mol. wt., struct. (2751). (m) Use as antioxidant, as polymerization catalyst, as stabilizer (1976). (n) Synth. from $(C_{12}H_{25})_2SnCl_2 \cdot 2NH_3$ and BzSH (2954). (o) Decomposition in refluxing $C_6H_6 \rightarrow (Ph_2SnS)_3 +$ $PhCH_2NCS$ (298, 1377). (p) Thermal decomposition at 250° $\rightarrow Ph_2SnO + Ph_2NH$ (298).

4.4.7 METAL SUBSTITUTED ORGANOTIN SULFIDES, SELENIDES AND TELLURIDES

The mixed chalcogenides containing organotin groups compiled in Table 233 are prepared by methods from the ensuing scheme.

I   Reaction of diorganotin sulfides with sodium sulfide in isopropanol.

II   Reaction of organotin hydride R_3SnH with:
  (IIA) Et_3GeSH or Et_3GeSeH,
  (IIB) Et_3MTeEt; M = Ge, Si.

III   Reaction of organotin halides R_3SnCl with R_3MYLi; M = Ge, Pb, Y = S, Se, Te.

Additional references to dibutyltin disodium mercaptide may be found in Chapter 9.

Table 234. Metal Substituted Organotin Sulfides, Selenides and Tellurides $R_nSn(SM)_{4-n}$

Formula*	Synth. Method**	Yield	Properties	Ref.
$R_2Sn(SNa)_2$				
R = Me	I	--	(a, b)	1747a
Et	I	--	(a)	1747a
Bu	I	--	Hygroscopic solid, m. 106-10° (a, c)	1747a
C_8H_{17}	I	--	(a)	1747a
Ph	I	--	(a)	1747a
$Et_3SnSGeEt_3$	III	63	b_2 121-28°, n_D^{20} 1.5270, IR spectrum	2090
	IIA	57	b_1 100-103°, n_D^{20} 1.5279, d_{20} 1.2990 (d)	2090
$Et_3SnSeGeEt_3$	IIA	37	$b_{0.5}$ 111°, n_D^{20} 1.5470 (e)	2091-2
$Et_3SnTeSiEt_3$	IIA	91	b_1 109-112°, n_D^{20} 1.5680 (d-f)	2092
$Et_3SnTeGeEt_3$	IIB	62	b_1 126-28°, n_D^{20} 1.5723 (d, e)	2092
$Ph_3SnSGePh_3$ (473)	III	72	m. 136°, monomeric (g)	484

* Numbers in parenthesis refer to pages in the main volume.
** The characters used here correspond to the synthetic scheme in the introduction to Chapter 4.4.7.

(a) Rxn. with BzCl → $R_2Sn(SBz)_2$ (1747a). (b) Rxn. with RCl → $Me_2Sn(SR)_2$; R = $PhCH_2$, Bu (1747a). (c) Rxn. with RCl → $Bu_2Sn(SR)_2$; R = $PhCH_2$ (1640), $C_8H_{17}O_2CCH_2$ (1747a). (d) NMR spectrum (1009). (e) IR spectrum (1010). (f) IR spectrum (1011). (g) (Far) IR spectra, assignment (1848).

Table 234 Continued $R_nSn(SM)_{4-n}$

Formula*	Synth. Method**	Yield	Properties	Ref.
Ph₃SnSPbPh₃ (473)	III	47	m. 137° (g)	485
Ph₃SnSeGePh₃ (473)	III	52	m. 144-45°, monomeric (g-i)	484
Ph₃SnSePbPh₃	III	35	Colorless cryst., m. 138° (g)	485
Ph₃SnTeGePh₃	III	48	Colorless cryst., m. 142-46° (g, i)	484
Ph₃SnTePbPh₃ (473)	--	--	(g)	--

(h) Stable to H₂O at r.t. (484), polarography (963, 969). (i) Rxn. with H₂O at reflux → Se and Te, resp. (484).

4.5 ORGANOTIN COMPOUNDS WITH NITROGEN, PHOSPHORUS, ARSENIC, ANTIMONY AND BISMUTH

This chapter comprises organotin compounds with elements of the fifth Main Group. Derivatives containing tin-nitrogen bonds with a pseudohalogen group are listed in Chapter 4.2.1. The compounds compiled in the ensuing tables are prepared by methods from the following scheme.

I Dealkylation of Ph₃SnC⋮CH with sodium amide.

II Hydrostannation of Schiff bases RCH:NPh with triorganotin hydride in presence of azobisisobutyranitrile.

III Substitution reaction of triorganotin halide R₃SnX with:
   (IIIA) RR'NLi, also for R₂SnX₂ and RSnX₃,
   (IIIB) RHNLi in inert atmosphere.

IV Transamination reactions of triorganotin amides R₃SnNR₂ with:
   (IVA) alkylamine or arylamine, also for R₂Sn(NR₂)₂,
   (IVB) with dialkylamine,
   (IVC) of (R₃Sn)₂PPh with azidobenzene PhN₃.

V Thermal decomposition of (R₃SnNPh)₂SO.

R = organic group, may be different in one molecule. Additional references to organotin-nitrogen compounds may be found in Chapter 8.1. Organotin-nitrogen compounds containing tin-heteroatom bonds are listed elsewhere. Derivatives containing halogen may be found in Chapters 4.1.4 and 4.1.7; compounds containing pseudohalogen in Subchapters 4.2.1.5.1 and 4.2.1.6; compounds containing alkoxide and mercaptide groups are listed in Subsections 4.3.3.2.9.5 and 4.4.2.2.6, respectively.

4.5.1 UNSUBSTITUTED ORGANOTIN AMIDES, IMIDES AND NITRIDES

4.5.1.1 TRIORGANOTIN DERIVATIVES

Tris(trimethyltin) Nitride (M-475) $Sn_3C_9H_{27}N$ $(Me_3Sn)_3N$
<u>Synth.</u>: By rxn. of $(Me_3Sn)_2O$ with $NaNH_2$ in Et_2O, in 70-80% yield (471).
By rxn. of Me_3SnNMe_2 with excess NH_3 (1184) in petr. ether, in 83% yield (251).
<u>Prop.</u>: m. 26-28° (471), $b_{0.2}$ 70°, n_D^{20} 1.5331, d_{20} 1.5084 (251), IR (644),
(far) IR, Raman, assignment and NMR spectra (1184), relative basicity (644),
dec. by moisture (251).
<u>Rxn. with</u>: MeLi → $(Me_3Sn)_2NLi + Me_3SnNLi_2 + Me_4Sn$ (471).

Bis(trimethyltin) Methylamide (M-475) $Sn_2C_7H_{21}N$ $(Me_3Sn)_2NMe$
<u>Synth.</u>: By rxn. of Me_3SnCl with $MeNHLi$ in Et_2O-petr. ether, in 65% yield (251).
By rxn. of Me_3SnNMe_2 with $MeNH_2$, in 83% yield (251, 1392).
<u>Prop.</u>: b_3 64° (251, 1392), n_D^{20} 1.4901, d_{20} 1.4794 (251), mol. wt., dec. by moisture (251).
<u>Rxn. with</u>: Y:CS → $(Me_3Sn)_2S$ + MeNCY; Y = S, PhN (1236a), O (2455).
PhC(:S)R → $(Me_3Sn)_2S$ + PhC(:NMe)R; R = Ph, Me_2N, relative reactivity (2455).
NCNSO → $(Me_3SnN:)_2C$ (2740).

Trimethyltin Anilide (M-475) $SnC_9H_{15}N$ $Me_3SnNHPh$
<u>Synth.</u>: By rxn. of Me_3SnCl with $PhNHLi$ in Et_2O-petr. ether, in 82% yield (251). ^{14}N and ^{15}N derivative (1729).
By transamination of Me_3SnNEt_2 with $PhNH_2$, in 90% yield (251), ^{14}N and ^{15}N derivatives, in 78% yield (2705).
<u>Prop.</u>: $b_{0.05}$ 77°, n_D^{20} 1.5721, d_{20} 1.4255 (251), ^{14}N and ^{15}N IR spectra and assignment (1729, 2705), ^{14}N and ^{15}N NMR spectra (1733, 2705), mol. wt., dec. by moisture (251).
<u>Rxn. with</u>: Moist air → $Me_3SnOH + (Me_3Sn)_2CO_3$ (1729)

Trimethyltin Dimethylamide (M-474) $SnC_5H_{15}N$ Me_3SnNMe_2
<u>Synth.</u>: By rxn. of Me_3SnCl with Me_2NLi in Et_2O-petr. ether, in 91% yield (251), in > 70% yield (324).
By-prod. in rxn. of Me_2SnCl_2 with Me_2NLi in Et_2O-petr. ether by NMR spectroscopic evidence (324, 1732).
<u>Prop.</u>: b. 126° (251), b. 128° (324), n_D^{20} 1.4572, d_{20} 1.2173 (251), NMR spectrum (324, 1732), mol. wt., dec. by moisture (251), relative S-N bond strength (250).
<u>Rxn. with</u>: $(Me_3Si)_2NH$ → $Me_3SnN(SiMe_3)_2$ (324).
R_3B → R_2BNMe_2; R = Bu, Ph (1090).
Ph_2BOMe → Ph_2BNMe_2 (1090).
$B(OMe)_3$ → $B(NMe_2)_3$ (1090).
$(Et_3Al)_2$ → $(Et_2AlNMe_2)_2$ (1090).
$H(C_5H_5)M(CO)_3$ → $Me_3Sn(C_5H_5)M(CO)_3$; M = Mo (2303), W (825).
$Mn(CO)_5Br$ → $Mn(CO)_5NMe_2$ (1090).

$(Ph_3P)_2HIrXCl \rightarrow (Ph_3P)_2IrX$; X = H, Cl (826).
$(PF_3)_4RhH \rightarrow Me_3SnF$ (625).
$Et_2NH \rightarrow Me_3SnNEt_2$ (251).
$\overline{CH_2CH_2NH} \rightarrow Me_3Sn\overline{NCH_2CH_2}$ (1351).
$RNH_2 \rightarrow (Me_3Sn)_2NR$; R = Me (251, 1392), Et (251).
$EtNH_2 \rightarrow Me_3SnNHEt$ (1392).
$ZH_3 \rightarrow (Me_3Sn)_3Z$; Z = N (251, 1184), P, As, Sb (1184).
$Me_2AsH \rightarrow Me_3SnAsMe_2 + Me_2NH$ (2201).
$Ph_2ZH \rightarrow Me_3SnZPh_2$; Z = P, As (250).
$HCl \rightarrow Me_3SnCl + Me_2NH_2Cl$ (250).
$H_2O \rightarrow Me_3SnOH + Me_2NH$ (250).
$HR \rightarrow Me_3SnR$; R = C_5H_5 (250), CCl_3, $CCl_2:CCl$ (865).
$CHRN_2 \rightarrow Me_3SnCRN_2$; R = H (1392), CO_2Et (2545).
$PhCHRCN \rightarrow MeNH$; R = H, Ph (250).
$HC_2Cl_5 \rightarrow Me_3SnCl + Me_2NH$ (865).
$(CF_3)_2CFH \rightarrow Me_3SnF$ (865).
$RC\vdots CH \rightarrow Me_3SnC\vdots CR$; R = Pr, Bu, Ph (250).
$MF_n \rightarrow Me_3SnF + M(NMe_2)_n$; M^{3+} = B, P, As, Sb, M^{4+} = Ti (1090).
$RF \rightarrow RNMe_2$; R = C_6F_5, $CClF:CF$, CF_3NHCF_2 (1090).
$RCl \rightarrow RNMe_2$; R = BCl_2, Me_3Si, Cl (1090).
$BuCl \rightarrow (MeCH:)_2$ (826).
$CCl_3CHO \rightarrow Me_3SnCCl_3 + Me_2NCHO$ (2454).
$Ac_2O \rightarrow AcNMe_2$ (1090).
$AcOPh \rightarrow Me_3SnOPh + AcNMe_2$ (854).
$CH_2:CMeCO_2Me \rightarrow Me_3SnOMe + CH_2:CMeCONMe_2$ (854).
$AcOCH:CH_2 \rightarrow Me_3SnOAc$ + polymer (854).
$RCO_2Et \rightarrow RCONMe_2$; R = Me, $AcCH_2$, $(EtO_2C)_2C:C(CO_2Et)$ (1090).
$Y(CO_2Et)_n \rightarrow Y(CONMe_2)_2$; n, Y = 2-$CH_2$, 4-C:C (1090).
$SO_2 \rightarrow Me_3SnO_2SNMe_2$ (167).
$RC\vdots CR' \rightarrow Me_3SnCR:CR'NMe_2$; R = R' = EtO_2C, R = Cl, R' = Ph (854).
$CH_2:CMeCN \rightarrow Me_3SnCMe(CN)CH_2NMe_2$ (854).
$CH_2:CO \rightarrow Me_3SnCH_2CONMe_2$ (167, 1710).
$CH_2:\overline{CCH_2CO_2} \rightarrow Me_3SnOCMe:CHCONMe_2$ (1205).
$Me_2CO \rightarrow (Me_3Sn)_2O + Me_2NH + Me_2C:CHAc$ (250).
$(CF_3)_2CO \rightarrow Me_3SnOC(CF_3)_2NMe_2$ (2199).
$(NC)_2C:C(CN)_2 \rightarrow Me_3SnN:C(NMe_2)C(CN):C(CN)_2$ (2406).
p-$MeC_6H_4NC \rightarrow p$-$Me_3SnC(NMe_2):NC_6H_4Me$ (2406).
$PhNCO \rightarrow Me_3SnNPhCONMe_2$ (1392).
$EtNCS \rightarrow Me_3SnSC(NMe_2):NEt$ (2406).
$PhNCS \rightarrow Me_3SnNPhC(S)NMe_2$ (167).
$PhNSO \rightarrow Me_3SnNPhS(O)NMe_2$ (?) (167).

<u>Addn. Compd.</u>: $Me_3SnNMe_2 \cdot B_{10}H_{14}$, from components in c.$C_6H_{12}$, orange cryst., m. 49-55° (2311) [$SnB_{10}C_5H_{29}N$].

Trimethyltin Diethylamide (M-475) $SnC_7H_{19}N$ Me_3SnNEt_2
<u>Synth.</u>: By rxn. of Me_3SnCl with Et_2NLi in Et_2O-petr. ether, in 85% yield (251), in 53% yield (2148).

By rxn. of Me$_3$SnNMe$_2$ with Et$_2$NH, in 95% yield (251).
Prop.: b$_8$ 43° (251), b$_6$ 36° (2148), n$_D^{20}$ 1.4618 (251), IR (644-5) and NMR (645) spectra, dec. in moist air, mol. wt. (251), relative basicity (644-5).
Rxn. with: R$_3$SnH → Me$_3$SnSnR$_3$; R = Me, Et, c.C$_6$H$_{11}$, Ph (1618).
Et$_2$SnH(CH$_2$)$_5$SnHEt$_2$ → (Me$_3$SnSnEt$_2$CH$_2$CH$_2$)$_2$CH$_2$ (1618).
Me$_2$SnH → (Me$_3$Sn)$_2$SnMe$_2$ (1955).
(x-Bu$_2$SnH)$_2$ → (Me$_3$SnSnBu-x)$_2$; x = n, i (1955).
Ph$_3$GeH → Me$_3$SnGePh$_3$ (1618).
R$_2$NH → Me$_3$SnNR$_2$; R$_2$ = Bu$_2$, (CH$_2$)$_5$, Ph + H (251).
Me$_2$CHNH$_2$ → (Me$_3$Sn)$_2$NCHMe$_2$ (1618).
Anhyd. HCN → Me$_3$SnCN (322).
MeOCH$_2$CH$_2$CO$_2$H → Me$_3$SnO$_2$CCH$_2$CH$_2$OMe (1235).
MeO$_2$CCH$_2$CH$_2$OH → Me$_3$SnOCH$_2$CH$_2$CO$_2$Me (1235).
BuSH → Me$_3$SnSBu + Et$_2$NH (643).
HR → Me$_3$SnR; R = CCl$_3$, CBr$_3$ (948).
RC⋮CH → Me$_3$SnC⋮CR; R = Me$_3$Si (1039), Ph$_2$P (2778).
NaC$_2$H, + Me$_3$GeCl → Me$_3$SnC⋮CGeMe$_3$ (1039).
CCl$_3$CHO → Me$_3$SnCl + Et$_2$NCHO (2454).
AcOCMe:CH$_2$ → Me$_3$SnOAc + AcNEt$_2$ (250).
CH$_2$:CO → Me$_3$SnCH$_2$CONEt$_2$ (1710).
OCH$_2$CH$_2$C:O → Me$_3$SnCH$_2$CH$_2$CONEt$_2$ (1238, 2452).
(RN:)$_2$C → Me$_3$SnNRC(:NR)NEt$_2$; R = p-MeC$_6$H$_4$ (167).

Triethyltin Dimethylamide (M-475)   SnC$_8$H$_{21}$N   Et$_3$SnNMe$_2$
Synth.: By rxn. of Et$_3$SnBr with Me$_2$NLi in Et$_2$O-petr. ether, in 75% yield (251).
Prop.: b$_9$ 76°, n$_D^{20}$ 1.4783 (251), NMR spectrum (1732), dec. by moisture (251).
Rxn. with: Air → (Et$_3$Sn)$_2$CO$_3$ + Me$_2$NH (250).
AcOCMe:CH$_2$ → Et$_3$SnOAc + AcNMe$_2$ (250).
CH$_2$:CO → Et$_3$SnCH$_2$CONMe$_2$ (1710).
AcOCH:CHMe → Et$_3$SnCH$_2$Ac + AcNMe$_2$ (1710).
CH$_2$:CHR → Et$_3$SnCHRCH$_2$NMe$_2$; R = CN, MeO$_2$C (854).
CH$_2$:CHCHO → Et$_3$Sn[CH(CHO)CH$_2$]$_n$NMe$_2$ (2406).
RCH:CHCHO → Et$_3$SnCH(CHO)CH$_2$R; R = H, Me, Ph (854).
PhNO → Et$_3$SnONPhNMe$_2$ (2406).

Triethyltin Diethylamide (M-475)   SnC$_{10}$H$_{25}$N   Et$_3$SnNEt$_2$
Synth.: By rxn. of Et$_3$SnCl with Et$_2$NLi in Et$_2$O-petr. ether, in 87% yield (251, 323).
Prop.: b. 225° (323), b$_2$ 72° (251), b$_{0.1}$ 40° (323), n$_D^{20}$ 1.4724, d$_{20}$ 1.1692 (251), NMR spectrum (323, 1429), dec. by moisture (251).
Rxn. with: R$_3$SnH → Et$_3$SnSnR$_3$ + Et$_2$NH; R = Me, Et (1618), i-Bu (525, 1618), R = Ph; solvent effect (96), Et, Ph, relative rxn. rates (915).
Me$_3$SnH → Et$_3$SnH + Me$_3$SnNEt$_2$; solvent effect (915).
R$_2$SnH$_2$ → (Et$_3$Sn)$_2$SnR$_2$; R = Et, i-Bu (525, 1955), Bu (400, 1955).
(i-Bu$_2$SnH)$_2$ → (Et$_3$SnSnBu-i$_2$)$_2$ (1955).
Ph$_3$GeH → Et$_3$SnGePh$_3$ (913).

$Ph_2GeH_2 \rightarrow (Et_3Sn)_2GePh_2$ (913).
$RNHCHO \rightarrow Et_3SnNRCHO$; R = n-Bu, t-Bu (915).
$Ph_3P:NH \rightarrow Et_3SnN:PPh_3$ (1429).
HX (anhyd.) $\rightarrow Et_3SnX$; X = CN (322), N_3 (1429).
ROH $\rightarrow Et_3SnOR$; R = Me (250, 323), Et (323).
$Ph_2P(O)H \rightarrow Et_3SnOPPh_2$ (1233).
c.1-$C_6H_9OAc \rightarrow$ 1-$Et_3SnOC_6H_9$ (1709).
$CH_2:CO \rightarrow Et_3SnCH_2CONEt_2$ (1710).
RCN $\rightarrow Et_2NH$; R = Me, Ph_2CH (250).

Bistributyltin Phenylamide $Sn_2C_{30}H_{59}N$ $(Bu_3Sn)_2NPh$
Synth.: By thermal dec. of $(Bu_3SnNPh)_2SO$ at 170°, in 80% yield (2345).
By-prod. in thermal dec. of $(Bu_3SnNPh)SO(NC_6H_4MeSnBu_3)$ (2345).
Prop.: $b_{0.03}$ 168°, IR and NMR spectra (2345).
Rxn. with: MeNCS $\rightarrow (PhN:)_2C + (Bu_3Sn)_2S$ (2345).
p-$MeC_6H_4NSO \rightarrow (Bu_3SnNPh)SO(NC_6H_4MeSnBu_3)$ (2345).

Tributyltin Dimethylamide (M-476) $SnC_{14}H_{33}N$ Bu_3SnNMe_2
Synth.: By rxn. of Bu_3SnCl with Me_2NLi in Et_2O-petr. ether, in 86% yield (215, 1392).
Prop.: $b_{0.1}$ 86° (251, 1392), n_D^{20} 1.4737 (251), NMR spectrum (1732), dec. by moisture, mol. wt. (251).
Rxn. with: $Me_3CNH_2 \rightarrow Bu_3SnNHCMe_3$ (251).
MeOH $\rightarrow Bu_3SnOMe + Me_2NH$ (250).
$C_2H_2 \rightarrow (Bu_3SnC!)_2 + Me_2NH$ (250).
$C_5H_6 \rightarrow Bu_3SnC_5H_5 + Me_2NH$ (250).
$CH_2:CO \rightarrow Bu_3SnCH_2CONMe_2$ (1710).
$Me_2CO \rightarrow Me_2C:CHAc + (Bu_3Sn)_2O + Me_2NH$ (250).
MeCN $\rightarrow Me_2NH$ + unidentified prod. (250).

Tributyltin Diethylamide (M-476) $SnC_{16}H_{37}N$ Bu_3SnNEt_2
Synth.: By rxn. of Bu_3SnCl with Et_2NLi in pentane (323), in 73% yield (2053).
Prop.: Colorless liq., $b_{0.2}$ 95-100° (2053), $b_{0.1}$ 95°, NMR spectrum (323).
Rxn. with: $R_3SnH \rightarrow Bu_3SnSnR_3$; R = Et, Bu (525, 1618), Ph (525).
$Bu_2SnH_2 \rightarrow (Bu_3Sn)_2SnBu_2$ (1955).
$Ph_3GeH \rightarrow Bu_3SnGePh_3$ (1618).
HCN (anhyd.) $\rightarrow Bu_3SnCN$ (322).
$CHCl_3 \rightarrow Bu_3SnCCl_3$ (948).
ROH $\rightarrow Bu_3SnOR$; R = Me, Et (323).
$Ph_2P(O)H \rightarrow Bu_3SnOPPh_2$ (1233).
RX $\rightarrow Bu_3SnX + RNEt_2$; R = $PhCH_2$, $CH_2:CHCH_2$, Bu, X = Cl, Br, I (2691).
Cyclohexane oxide + cat. $LiNEt_2 \rightarrow$ 1,2-$Bu_3SnO(Et_2N)C_6H_{10}$ (2035, 2835).
$\overline{RCHCH_2O}$ + cat. $LiNEt_2 \rightarrow Bu_3SnOCHRCH_2NEt_2$; R = Et (2053, 2835), Ph (2053).
$PhCH_2CHRCN \rightarrow Bu_3SnN:C:CRCH_2Ph$; R = p-$NCC_6H_4$, CN (2615).
PhNCS $\rightarrow Bu_3SnNPhCSNEt_2$ (2258).

Tricyclohexyltin Diethylamide $SnC_{22}H_{43}N$ $(c.C_6H_{11})_3SnNEt_2$

<u>Synth.</u>: By rxn. of $(C_6H_{11})_3SnBr$ with Et_2NLi in inert atm., in 76% yield (1618).

<u>Prop.</u>: m. 79-80° (sealed tube), $b_{0.001}$ 158-61° (1618).

<u>Rxn. with</u>: $R_3SnH \rightarrow (C_6H_{11})_3SnSnR_3$; R = $c.C_6H_{11}$, Ph (1618).

$i\text{-}Bu_2SnH_2 \rightarrow [(C_6H_{11})_3Sn]_2Sn(i\text{-}Bu)_2$ (1955).

$(i\text{-}Bu_2SnH)_2 \rightarrow [(C_6H_{11})_3SnSn(i\text{-}Bu)_2]_2$ (1955).

Triphenyltin Diethylamide (M-476) $SnC_{22}H_{25}N$ Ph_3SnNEt_2

<u>Rxn. with</u>: $i\text{-}Bu_2SnH_2 \rightarrow (Ph_3Sn)_2Sn(i\text{-}Bu)_2$ (1955).

$(i\text{-}Bu_2SnH)_2 \rightarrow (i\text{-}Bu_2SnSnPh_3)_2$ (1955).

$Ph_3GeH \rightarrow Ph_3SnGePh_3$ (913).

$Ph_2GeH_2 \rightarrow (Ph_3Sn)_2GePh_2$ (913).

HCN (anhyd.) $\rightarrow Ph_3SnCN$ (322).

ROH $\rightarrow Ph_3SnOR$; R = Me, Et (323).

RC⋮CH $\rightarrow Ph_3SnC⋮CR$; R = Ph_3Ge (1039), Ph_2P (2778).

NaC_2H, + $Ph_3MCl \rightarrow Ph_3SnC⋮CMPh_3$; M = Si, Ge (1039).

Listing of triorganotin amides, imides and nitriles continues in Table 235.

Table 235. Unsubstituted Triorganotin Amides, Imides and Nitrides $(R_3Sn)_nNR_{3-n}$

Formula*	Synth. Method**	Yield	Properties	Ref.
$(Me_3Sn)_2NEt$ (475)	IVA	90	b_{15} 93°, n_D^{20} 1.4968, d_{20} 1.4805 (a)	251
$(Me_3Sn)_2NCHMe_2$	IVA	76	b_{11} 95-97°	1618
$(Me_3Sn)_2NPh$	IVC	44	b_1 99-100°, mol. wt.	1845
$Me_3SnNHEt$ (475)	IVA	--	b_{15} 95°	1392
$Me_3SnNH(C_6H_4Me-p)$	IIIB	76	$b_{0.5}$ 106°, n_D^{20} 1.5622, d_{20} 1.3536 (a)	251
$Me_3SnNMePh$ (475)	IIIA	83	$b_{0.1}$ 82°, n_D^{20} 1.5757, d_{20} 1.3645 (a)	251
$Me_3SnN(CHMe_2)_2$ (475)	IIIA	80	b_8 63°, n_D^{20} 1.4645, d_{20} 1.539 (a)	251
Me_3SnNBu_2 (475)	IIIA	81	$b_{2.5}$ 74°, n_D^{20} 1.4559, d_{20} 1.1068	251
	IVB	92	(a)	251
$Me_3SnN(c.C_6H_{11})_2$	IIIA	70	$b_{0.2}$ 96°, n_D^{20} 1.5055, d_{20} 1.1972 (a)	251
Me_3SnNPh_2 (475)	IIIA	85	$b_{0.1}$ 108°, n_D^{20} 1.6096, d_{20} 1.3176 (a)	251
$Et_3SnNHPh$	IIIA	82	$b_{0.2}$ 100° (a)	251
$Et_3SnN(CH_2Ph)Ph$ (475)	II	72	$b_{0.002}$ 149-51°, n_D^{20} 1.5907	390
$Et_3SnN(CH_2Ph)C_6H_4Me-p$ (475)	II	63	$b_{0.01}$ 142-43°, n_D^{20} 1.5843 (b)	390
Et_3SnNPh_2	--	--	(c)	--
Pr_3SnNMe_2	IIIA	63	b_1 56-57°, n_D^{20} 1.4778, d_{20} 1.1233 (d)	1710
$(Bu_3Sn)_2NC_6H_4Me-p$	V (e)	--	Straw colored oil, $b_{0.05}$ 172-74° (f)	2345
$Bu_3SnNHCMe_3$	IVA	92	b_1 124°, n_D^{20} 1.4773, d_{20} 1.0577 (a)	251
$Bu_3SnN(CH_2Ph)Ph$	II	~100	Thick oil, $b_{0.002}$ >160°, IR spectrum (g)	390
$(Me_2CHCH_2)_3SnNEt_2$	IIIA	70	$b_{0.2}$ 75-77°, d_{20} 1.08 (h, i)	1618
$(Ph_3Sn)_3N$	I	--	m. 115-17°, IR spectrum	443
Ph_3SnNMe_2 (476)	IIIA	81	m. 62°, $b_{0.1}$ 166° (a, j)	251

* Numbers in parenthesis refer to pages in main volume.
** The characters used here correspond to the synthetic scheme in the introduction to Chapter 4.5.
(a) Decomposed by moisture (251). (b) Rxn. with air → $(Et_3Sn)_2CO_3$ + p-$MeC_6H_4NHCH_2Ph$ (390). (c) Rxn. with Me_3SnH → Me_3SnNH_2 + Et_3SnH; solvent effect (915), with Ph_3SnH → $Et_3SnSnPh_3$ + Ph_2NH; relative rxn. rates (915). (d) Rxn. with CH_2:CO → $Pr_3SnCH_2CONMe_2$ (1710). (e) Synth. also from $(Bu_3SnPh)SO(NC_6H_4MeSnBu_3)$ (2345). (f) IR and NMR spectra (2345). (g) Rxn. with HCl → $PhCH_2NHPh$ (390). (h) Rxn. with R_3SnH → i-Bu_3SnSnR_3; R = i-Bu, Ph (525, 1618), c.C_6H_{11} (1618). (i) Rxn. with i-Bu_2SnH_2 → (i-$Bu_3Sn)_2Sn$(i-Bu)$_2$ (525, 1955). (j) Rxn.with H_2O → Ph_3SnOH + Me_2NH(250).

4.5.1.2 UNSUBSTITUTED DIORGANOTIN AMIDES AND IMIDES

Dimethyltin Bisdimethylamide (M-475) $SnC_6H_{18}N_2$ $Me_2Sn(NMe_2)_2$
Synth.: By rxn. of Me_2SnCl_2 with Me_2NLi in Et_2O-petr. ether, in 68% yield (251), prod. shown to be complex mixture of $Me_nSn(NMe_2)_{4-n}$ by NMR spectroscopy (324, 1732).
By rxn. of $Me_2Sn(NEt_2)_2$ with excess Me_2NH, in 50% yield (324), in sealed tube (323, 1732).
Prop.: b. 138° (251), b_1 45° (324), n_D^{20} 1.4463, d_{20} 1.1482 (251), NMR spectrum (324, 1732).
Rxn. with: $(Me_3Si)_2NH \rightarrow Me_2Sn(NMe_2)N(SiMe_3)_2$ (324).
$RNH_2 \rightarrow c.(Me_2SnNR)_3$; R = Me (251), Et (251, 1392).
$HCCl_3 \rightarrow$ mild explosions in NMR tubes (1730).

Dimethyltin Bisdiethylamide (M-475) $SnC_{10}H_{26}N_2$ $Me_2Sn(NEt_2)_2$
Synth.: By rxn. of Me_2SnCl_2 with Et_2NLi in Et_2O-petr. ether, in 67% yield (251).
Prop.: b_4 78° (251), IR, assignment (2161) and NMR (1731-2) spectra, dec. by moisture (251).
Rxn. with: $R_3SnH \rightarrow R_3SnMe_2SnSnR_3$; R = Me (1955), Et (525, 1955).
Me_2NH (excess) $\rightarrow Me_2Sn(NMe_2)_2$ (323-4, 1732).
$Y(NRH)_2 \rightarrow Me_2\overline{SnNYNR}$; R = Me, Y = $(CH_2)_3$, R = Me_3Si, Y = $(CH_2)_2$, $(CH_2)_3$, $(CH_2)_4$, $o-C_6H_4$ (470).
$(CH_2NHMe)_2 \rightarrow$ polymeric stannylamine (2163).
HCN (anhyd.) $\rightarrow Me_2Sn(CN)_2$ (322).
ROH $\rightarrow Me_2Sn(OR)_2$; R = Me, Et (323).

Diethyltin Bisdiethylamide (M-476) $SnC_{12}H_{30}N_2$ $Et_2Sn(NEt_2)_2$
Synth.: By rxn. of Et_2SnCl_2 with Et_2NLi in pentane (323), in 72% yield (1955).
Prop.: $b_{0.2}$ 58-60° (1955), $b_{0.1}$ 77°, NMR spectrum (323).
Rxn. with: $Et_3SnH \rightarrow Et_3Sn(Et_2Sn)_nSnEt_3$; n = 0, 1, 2 (1955).
$Et_2SnH_2 \rightarrow c.(Et_2Sn)_6$ (525, 1614).
Ph_3GeH (1:1) $\rightarrow Ph_3GeSnEt_2NEt_2$ (913).
Ph_3GeH (1:2) $\rightarrow (Ph_3Ge)_2SnEt_2$ (913).
$Ph_2GeH_2 \rightarrow (Et_2SnGePh_2)_x$ (913).
HCN (anhyd.) $\rightarrow Et_2Sn(CN)_2$ (322).
ROH $\rightarrow Et_2Sn(OR)_2 + Et_2NH$; R = Me, Et (323).

Dibutyltin Bisdiethylamide $SnC_{16}H_{38}N_2$ $Bu_2Sn(NEt_2)_2$
Synth.: By rxn. of Bu_2SnCl_2 with Et_2NLi in pentane (323).
Prop.: $b_{0.1}$ 98°, NMR spectrum (323).
Rxn. with: $R_3SnH \rightarrow (R_3Sn)_2SnBu_2$; R = Et (525, 1955), Bu (525) + Bu_6Sn_2, R = Bu (1955).
$Bu_2SnH_2 \rightarrow c.(Bu_2Sn)_6$ (525, 1614).
HCN (anhyd.) $\rightarrow Bu_2Sn(CN)_2$ (322).
ROH $\rightarrow Bu_2Sn(OR)_2$; R = Me, Et (323).

Diphenyltin Bisdiethylamide (M-476) $SnC_{20}H_{30}N_2$ $Ph_2Sn(NEt_2)_2$

<u>Rxn. with</u>: $Ph_2SnH_2 \to$ c.$(Ph_2Sn)_6$ (1610).
$Bu_3GeH \to (Bu_3Ge)_2SnPh_2$ (913).
$Ph_3GeH \to Ph_3GeSnPh_2NEt_2$ (913).
$Ph_2GeH_2 \to (Ph_2SnGePh_2)_x$ (913).
HCN (anhyd.) $\to Ph_2Sn(CN)_2$ (322).
ROH $\to Ph_2Sn(OR)_2$; R = Me, Et (323).

Additional unsubstituted diorganotin diamides and imides are listed in Table 236.

4.5.1.3 UNSUBSTITUTED MONOORGANOTIN TRIAMIDES

Ethyltin Trisdiethylamide $SnC_{12}H_{35}N_3$ $EtSn(NEt_2)_3$

<u>Synth.</u>: By rxn. of $EtSnCl_3$ with Et_2NLi in pentane (323).
<u>Prop.</u>: $b_{0.1}$ 76°, NMR spectrum (323).
<u>Rxn. with</u>: Ph_3GeH (1:3) $\to (Ph_3Ge)_3SnEt$ (913).
Ph_3GeH (1:2) $\to (Ph_3Ge)_2SnEtNEt_2$ (913).
$Ph_3GeH \to Ph_3GeSnEt(NEt_2)_2$ (913).
ROH $\to EtSn(OR)_3$; R = Me, Et (323).

Listing of organotin triamides continues at the bottom of Table 236.

Table 236. Di- and Monoorganotin Amides

Formula*	Synth. Method**	Yield	Properties	R_nSnNR_m Ref.
c.(Me$_2$SnNMe)$_3$	IVA	63	b$_{0.2}$ 114° (a)	251
Me$_2$Sn(NMeEt)$_2$	IIIA	--	NMR spectrum	1732
c.(Me$_2$SnNEt)$_3$	IVA	64	b$_{0.05}$ 104°, mol. wt. (a)	251, 1392
Me$_2$Sn(Ni-Pr$_2$)$_2$	IIIA	72	b$_{0.05}$ 66°, n$_D^{20}$ 1.4685, d$_{20}$ 1.1060 (a)	251
Me$_2$SnNMe(CH$_2$)$_3$NMe	IVA	77	m. 1-3°, b$_{18}$ 88-90°, NMR spectrum	470
(PhCH$_2$)$_2$Sn(NEt$_2$)$_2$	IIIA	--	Crude, dec. on distn. (b)	1614
Bu$_2$Sn(NMe$_2$)$_2$ (476)	IIIA	78	b$_{0.05}$ 72°, n$_D^{20}$ 1.4747, d$_{20}$ 1.1247 (a, c)	251
(Me$_2$CHCH$_2$)$_2$Sn(NEt$_2$)$_2$	IIIA	65	b$_{0.001}$ 83-86° (d)	1955
(Me$_3$C)$_2$Sn(NEt$_2$)$_2$	IIIA	--	Crude, dec. on distn. (e)	1614
(c.C$_6$H$_{11}$)$_2$Sn(NEt$_2$)$_2$	IIIA	--	Crude, dec. on distn. (f)	1614
Ph$_2$Sn(NMe$_2$)$_2$ (476)	IIIA	83	b$_{0.2}$ 128° (a)	251
R$_2$Sn(NR$_2$)$_2$	--	--	(g)	--
MeSn(NMe$_2$)$_3$ (475)	IIIA	>70	b$_{0.1}$ 50°, NMR spectrum (h)	324
BuSn(NMe$_2$)$_3$ (476)	IIIA	71	b$_{0.1}$ 67° (a)	251
BuSn(NEt$_2$)$_3$	IIIA	--	b$_{0.1}$ 96°, NMR spectrum (i)	323
PhSn(NEt$_2$)$_3$ (476)	--	--	(i)	--

* Numbers in parenthesis refer to pages in main volume.
** The characters used here correspond to the synthetic scheme in the introduction to Chapter 4.5.
(a) Decomposed by moisture (251). (b) Rxn. with (PhCH$_2$)$_2$SnH$_2$ → c.[(PhCH$_2$)$_2$Sn]$_4$ (1614). (c) Rxn. with C$_2$H$_2$ → (Bu$_2$SnC$\vdots$C)$_x$ + Me$_2$NH (250). (d) Rxn. with i-Bu$_2$SnH$_2$ → c.(i-Bu$_2$Sn)$_9$ (525, 1614), with (i-Bu$_2$Sn)$_2$ → c.(i-Bu$_2$Sn)$_9$ (1614). (e) Rxn. with t-Bu$_2$SnH$_2$ → c.(t-Bu$_2$Sn)$_4$ (1614). (f) Rxn. with (c.C$_6$H$_{11}$)$_2$SnH$_2$ → c.[(c.C$_6$H$_{11}$)$_2$Sn]$_5$ (1614). (g) Mild explosive in contact with HCCl$_3$ or on heating (1730). (h) Rxn. with (Me$_3$Si)$_2$NH → MeSn(NMe$_2$)$_2$N(SiMe$_3$)$_2$ (324). (i) Rxn. with ROH → R'Sn(OR)$_3$; R = Me, Et, R' = Bu, Ph (323).

4.5.2 SUBSTITUTED ORGANOTIN AMIDES

This chapter comprises a wide variety of N-substituted organotin derivatives. The amino moiety includes hydrazine, 1,3-diorganotriazene, ketimine, amidine, N,N'-dialkylaminotroponeimine, carbodiimine, nitramine, keteneimine and heterocyclic amines. The compounds summarized in Tables 237 and 238 are prepared by the following methods

I   Dealkylation of tetraethyltin with nitramine.

II   Hydrostannation with triorganotin hydride R_3SnH of:
- (IIA)   RC:N:CR and,
- (IIB) R_3PbNY in presence of a suitable acetylenic acceptor.

III   Substitution reactions of triorganotin halide R_3SnX with:
- (IIIA)   lithium amide which may be prepared in situ,
- (IIIB)   sodium and potassium amide,
- (IIIC)   silver amide and,
- (IIID)   N-Grignard reagent.

IV   Reactions of organotin oxygen compounds:
- (IVA) $(R_3Sn)_2O$ with amine,
- (IVB) $R_2Sn(OR)_2$ with 2-hydroxypyridine.

V   Reactions with organotin amides R_3SnNR_2:
- (VA)   transamination with a less volatile amine,
- (VB)   addition to a cyano compound and to carbodiimide,
- (VC)   to thionylcyanimide NCNSO.

R = organic group, may be different in one molecule, Y = bridging group.

4.5.2.1 ORGANOTIN DERIVATIVES OF HETEROCYCLIC AMINES

All compounds are listed in Table 237.

4.5.2.4 ORGANOTIN DERIVATIVES OF AMIDINES, IMIDES, HYDRAZINE AND TRIAZENE

N-organotin derivatives of amidine, ketimide, carbodiimide, troponeimide, hydrazine and 1,3-diorganotriazene are listed in Table 238. Metal derivatives with such entities may be found in Chapter 4.5.4.

4.5.2.7 ORGANOTIN DERIVATIVES OF NITRAMINE

All organotin nitramides are listed at the bottom of Table 238.

Table 237. N-Organotin Derivatives of Heterocyclic Amines $R_nSn(NY)_{4-n}$

Formula*	Synth. Method**	Yield	Properties	Ref.
Me₃SnNY				
YN = $\overline{CH_2CH_2N}$	VA	60	b_{16} 53-55°, n_D^{20} 1.4950, NMR spectrum	1351
C₄H₄N (a) (478)	--	--	NMR spectrum	1351
C₃H₃N₂ (a) (478)	--	--	Moessbauer resonance spectrum	224
C₇H₅N₂ (a) (478)	--	--	Moessbauer resonance spectrum	224
C₂H₂N₃ (b) (478)	--	--	Moessbauer resonance spectrum	224
C₆H₄N₃ (b) (478)	--	--	Moessbauer resonance spectrum	224
$\overline{CH_2(CH_2)_4N}$	IIIA	76	b_1 48°	251
	VA	97	Dec. in moist air	251
Et₃SnC₃H₃N₂ (a)	IIB	--	---	912
3-Bu₃SnC₇H₃NO₂Cl-6 (c)	IIIB	--	---	2590
3-Bu₃SnC₇H₃NO₂Br-6 (c)	IIIB	--	---	2590
3-Bu₃SnC₇H₂NO₂Cl₂-5,6 (c)	IIIB	--	---	2590
3-Bu₃SnC₇H₃NOSCl-5 (c)	IIIB	--	---	2590
Bu₂Sn(C₅H₄NO)₂ (d)	IVB	--	m. 80°, straw colored liq., n_D^{20} 1.5663 (e)	1230
Ph₃SnNY				
YN = C₃H₃N₂ (a) (479)	IIB	--	---	912
C₇H₅N₂ (a) (479)	--	--	Moessbauer resonance spectrum	224
9-C₅H₃N₄ (f)	IVA	82	m. 304-307°, stable in air (g)	1375
9-C₅H₂N₄-Cl-6 (f)	IVA	77	m. 100-101°, dec. at 200°, stable in air (g)	1375
9-C₅H₂N₄-SMe-6·HNY (f, h)	IVA	94	m. 228-229°, NMR spectrum (g)	1375
C₂H₂N₃ (b) (479)	--	--	Moessbauer resonance spectrum	224
C₆H₄N₃ (b) (479)	--	--	Moessbauer resonance spectrum	224

* Numbers in parenthesis refer to pages in main volume.
** The characters used here correspond to the synthetic scheme in the introduction to Chapter 4.5.2.
(a) C₄H₄NH = pyrrole, C₃H₃N₂H = imidazole, C₇H₅N₂H = benzimidazole. (b) C₂H₂N₃H = 1,2,4-triazole, C₆H₄N₃H = 1,2,3-benzotriazole. (c) C₇H₃NO₂H = 2-benzoxazolinone, 5-ClC₇H₃NOSH = 5-chloro-2-benzoxazolinethione. (d) N-Deriv. of 2-pyridone. (e) Tends to supercooling, IR spectrum, hydrolyzes easily in air, struct. with hexacovalent tin (1230). (f) C₅H₃N₄H = purine. (g) (Far) IR and UV spectra (1375). (h) Adduct 9-Ph₃SnC₅H₂N₄SMe-6·C₅H₃N₄SMe-6·[SnC₃₀H₂₈N₈S₂].

Table 238. N-Organotin Derivatives of Substituted Amines $R_nSn(NRR')_{4-n}$

Formula*	Synth. Method**	Yield	Properties	Ref.
(Me₃SnN:)₂C	VA	82	m. 80°, bo.6 116°, mol. wt.	2740
	VC (a)	--	IR and NMR spectra (b)	2740
Me₃SnNR'₂				
R'₂N = Ph₂C:N	IIIA	--	Yellow solid, bo.15 121°, mol. wt. (c)	852
Me₂NCPh:N (480)	IIIA	43	m. 38-40°, bo.1 84-85°, mol. wt. (d)	2738
	VB	--	Colorless viscous liq., bo.2 79° (e)	167
(NC)₂C:C(CN)C(NMe₂):N	VB	93	Red-brown, m. 110-112° (sealed tube) (f)	2406
MeN:CPhNMe	IIIA	86	Colorless oil, m. 20-22°, bo.1 70°, mol. wt. (d)	2738
Et₂NC(:NR)NR	VB	72	R = p-MeC₆H₄, pale yellow liq., bo.1 168°	167
Me₂Sn(N₃Me₂)₂ (g)	IIID	26	Colorless liq., bo.2 50-52°, monomeric, NMR spectrum, struct.	48b
	IIIC (h)	61	Synth. from Me₂Sn(N₃Me₂)I and Me₂NLi	48b
	IIIA	68		48b
Me₂Sn(N₃Ph₂)₂ (g)	IIIC	--	Orange cryst., m. 164°, monomeric	48a
	IIIC (h)	96	Sensitive to moisture (i)	48b
Et₃SnNPhNHPh (475)	--	--	(j)	--
Et₃SnN(c.C₆H₁₁)CH:NC₆H₄	IIA	73	bo.0001 126°, n_D^{20} 1.5233	390
Ph₃SnN₃Ph₂ (g)	IIIC	--	Yellow cryst., m. 85°, monomeric	48a

* Numbers in parenthesis refer to pages in main volume.
** The characters used here correspond to the synthetic scheme in the introduction to Chapter 4.5.2.
(a) Synth. from Me₃SnNMeSiMe₃ (2740). (b) Polymerizes at 130° (2740). (c) UV, IR spectra and assignment, sensitive to moisture (852). (d) IR and NMR spectra (2738). (e) IR spectrum and assignment, struct. (167). (f) IR spectrum (2406). (g) Deriv. of 1,3-triazene. (h) Synth. from Me₂Sn(N₃Me₂)I with AgN₃Ph₂ (48b). (i) Microanal. C, H and N detn. (604). (j) Rxn. with Ph₃SnH → Et₃SnNPh₃ + (PhNH)₂; solvent effect (96), relative rxn. rates (915).

Table 238 Continued $R_nSn(NRR')_{4-n}$

Formula*	Synth. Method**	Yield	Properties	Ref.
$Ph_3SnN:CPh_2$	IIIA	--	Yellow cryst., m. 78-80° (c)	852
$Ph_3SnC_7H_5(NEt)_2$ (k)	IIIA	55	Yellow, m. 130°, subl.o.5 200° (l)	1004
$Me_3SnNMeNO_2$	IIIC	71	m. 155-56°, IR spectrum, assignment (m)	611
$Et_3SnNRNO_2$				
R = Me	I	10	m. 109.5-10.5°, subl., struct. (m, n)	611
	IIIC	66	IR and Moessbauer resonance spectra	611
Et	IIIC	63	m. 108-109.5°, subl., IR spectrum, (m, o)	611
Me_2CH	I	10	m. 144-44.5°, subl.	611
	IIIC	89	IR spectrum (m)	611
Me_3C	IIIC	80	m. 179-80.5°, IR spectrum (m)	611
Ph	IIIC	48	m. 138-39°, IR spectrum (m)	611
$i-Pr_3SnNMeNO_2$	IVA	68	m. 69-72°, subl., IR spectrum (m)	611
$i-Pr_3SnN(CMe_3)NO_2$	IVA	91	m. 77-81°, subl., IR spectrum (m, o)	611
$Ph_3SnNMeNO_2$	IIIC	63	m. 142-44°, IR spectrum (m, o)	611
$Ph_3SnNPhNO_2$	IIIC	10	m. 138-39° (dec.), IR spectrum (m)	611

(k) Deriv. of N,N'-diethylaminotroponeimine. (l) NMR spectrum, struct., sensitive to moisture (1004). (m) Anal. detn. by titration with alc. NaOH, stable in air (611). (n) Conductance in $PhNO_2$ (611). (o) Monomeric in C_6H_6 (611).

4.5.2.8 ORGANOTIN DERIVATIVES OF KETENEIMINE

N-organotin keteneimides are only reported in the literature very recently. All derivatives known decompose on heating above 50°. In common with most organotin amides, hydrolysis with moist air is facile. The compounds summarized in Table 239 are prepared by the following methods.

I   Reaction with triethyltin lithium:
     $Et_3SnLi + Me_2C(CN)I \rightarrow Et_3SnN:C:CMe_2$.

II Hydrostannation of unsaturated nitriles:
     $R_3SnH + R'_2C:C(CN)_2 \rightarrow R_3SnN:C:C(CN)CHR'_2$.

III Hydrogen abstraction with triethyltin amide.
     $R_3SnNEt_2 + PhCH_2CHRCN \rightarrow R_3SnN:C:CRCH_2Ph$.

Triethyltin Benzylcyanoketeneimide $SnC_{16}H_{22}N_2$ $Et_3SnN:C:C(CN)CH_2Ph$
<u>Synth.</u>: By rxn. of Et_3SnH with $PhCH:C(CN)_2$ in C_6H_6 in Ar atm. (1621, 2615).
<u>Prop.</u>: Viscous yellow oil, dec. on distn. > 50°, IR spectrum (1621, 2615).
<u>Rxn. with</u>: EtOH → $PhCH_2CH(CN)_2$ (1954, 2789).
MeI in DMF → $PhCH_2CMe(CN)_2$ (2789).
RBr → $Et_3SnBr + PhCH_2CR(CN)_2$; R = Et, $PhCH_2$, $p\text{-}O_2NC_6H_4$ (2789), $CH_2:CH$, MeCH:CH, $CH_2:CMe$ (1954, 2789).
RCOCl → $Et_3SnCl + PhCH_2C(CN)_2COR$; R = Me, Ph (1954, 2789).
X → $Et_3SnX + PhCH_2CX(CN)_2$; X = Br (1954, 2789), I (2789).

Triethyltin Furfurylcyanoketeneimide $SnC_{14}H_{20}N_2O$ $Et_3SnN:C:C(CN)CH_2(C_4H_3O)\text{-}2$
<u>Synth.</u>: By rxn. of Et_3SnH with $2\text{-}C_4H_3OCH:C(CN)_2$ in C_6H_6 at 50° in Ar atm. (2615).
<u>Prop.</u>: Viscous oil, dec. > 50°, IR spectrum (2615).
<u>Rxn. with</u>: EtOH → $(Et_3Sn)_2O + 2\text{-}(C_4H_3O)CH_2CH(CN)_2$ (1954, 2789).
$RCH_2Br \rightarrow Et_3SnBr + 2\text{-}(C_4H_3O)CH_2C(CN)_2CH_2R$; R = Ph, $p\text{-}O_2NC_6H_4$ (2789), $CH_2:CH$ (1954, 2789).
AcCl → $Et_3SnCl + 2\text{-}(C_4H_3O)CH_2C(CN)_2Ac$ (1959, 2789).

Tributyltin Benzylcyanoketeneimide $SnC_{22}H_{34}N_2$ $Bu_3SnN:C:C(CN)CH_2Ph$
<u>Synth.</u>: By rxn. of Bu_3SnH with $PhCH:C(CN)_2$ in i-BuCN in Ar atm. (1621), relative rxn. rates (2615).
By rxn. of Bu_3SnCl with $PhCH_2C(CN)_2Na$ in i-BuCN at 40° in Ar atm. (2615).
By rxn. of Bu_3SnNEt_2 with $PhCH_2CH(CN)_2$ in i-BuCN at 40° in Ar atm. (2615).
By rxn. of Bu_3SnLi with $PhCH_2C(CN)_2Br$ in THF in Ar atm. (2615).
<u>Prop.</u>: Oil, dec. above 50°, IR spectrum, mol. wt., proposed struct.(1621,2615).
<u>Rxn. with</u>: EtOH → $(Bu_3Sn)_2O + PhCH_2CH(CN)_2$ (1954, 2789).
BuLi → $Bu_4Sn + PhCH_2C(CN)_2Li$ (2789).
RNa → $PhCH_2C(CN)_2Na$; R = Ph_3C, EtO (2789).
$CH_2:CHCH_2Br \rightarrow Bu_3SnBr + PhCH_2C(CN)_2CH_2CH:CH_2$ (1954, 2789).
$ClCO_2Et \rightarrow Bu_3SnCl + PhCH_2C(CN)_2CO_2Et$ (2789).
$Et_3OBF_4 \rightarrow PhCH_2CEt(CN)_2$ (2789).

Listing of triorganotin keteneimides concludes in Table 239.

Table 239. $R_3SnN:C:CR'R''$

Formula	Synth. Method*	Properties	Ref.
$Et_3SnN:C:CMe_2$	I	$b_{0.6}$ 30-35°, IR spectrum (a)	1620
$Et_3SnN:C:C(CN)CHRR'$			
R = H R' = Me	II	IR spectrum, mol. wt. (b)	2615
H PhCH:CHCH$_2$	II	Yellow oil, IR spectrum (b, c)	1621, 2615
H p-MeC$_6$H$_4$	II	Yellow oil, IR spectrum (b, d, e)	1621, 2615
H p-ClC$_6$H$_4$	II	Yellow oil, IR spectrum (b, d-f)	1621, 2615
H p-BrC$_6$H$_4$	II	Yellow oil, IR spectrum (b, g)	1621, 2615
H p-MeOC$_6$H$_4$	II	Yellow oil, IR spectrum (b, e, h)	1621, 2615
H m-O$_2$NC$_6$H$_4$	II	Yellow oil, IR spectrum (b, i, j)	1621, 2615
H p-O$_2$NC$_6$H$_4$	II	Yellow oil, IR spectrum (b, e)	1621, 2615
Me Me	II	IR spectrum, mol. wt. (b, k)	2615
Me Ph	II	Yellow oil, IR spectrum (b, d)	1621, 2615
Ph Ph	II	Yellow oil, IR spectrum (b, i)	1621, 2615
RR' = (CH$_2$)$_5$	II	Yellow oil, IR spectrum (b, i)	1621, 2615
$Bu_3SnN:C:CRCH_2Ph$	II	R = p-NCC$_6$H$_4$, viscous liq.	2615
	III	Easily hydrolyzed (2789)	2615
$Bu_3SnN:C:C(CN)CH_2R$			
R = EtO	II	Crude, IR spectrum (l)	2615
Me	II	Crude, IR spectrum, mol. wt. (l, m)	2615
2-C$_4$H$_3$O	II	Crude, IR spectrum, mol. wt. (l, m)	2615
PhCH:CH	II	Crude, IR spectrum (l, n)	2615
p-MeC$_6$H$_4$	II	Crude, IR spectrum (l)	2615
p-ClC$_6$H$_4$	II	Crude, IR spectrum (l)	2615
p-Me$_2$NC$_6$H$_4$	II	Crude, IR spectrum (l)	2615
p-O$_2$NC$_6$H$_4$	II	Crude, IR spectrum (l)	1621, 2615

Ph$_3$SnN:C:C(CN)CH$_2$Ph   II   (o)

* The characters used here correspond to the synthetic scheme in the introduction to Chapter 4.5.2.9. (a) Decomposes rapidly at 75° (1620). (b) Viscous liq., dec. > 50° (1621, 2615), rxn. with EtOH → RR'CHCH(CN)$_2$ (1954, 2789). (c) Rxn. with AcCl → Et$_3$SnCl + PhCH:CHCH$_2$CH$_2$(CN)$_2$Ac (1954, 2789). (d) Rxn. with BzCl → Et$_3$SnCl + RR'CHC(CN)$_2$Bz (1954, 2789). (e) Rxn. with X → Et$_3$SnX + R'CH$_2$CX(CN)$_2$; X = Br (1954, 2789), I (2789). (f) Rxn. with MeCH:CHCH$_2$X → Et$_3$SnX + R'CH$_2$C(CN)$_2$CH$_2$CH:CHMe (1954, 2789). (g) Rxn. with Ph$_3$CCl → Et$_3$SnCl + p-BrC$_6$H$_4$CH$_2$C-(CN)$_2$CPh$_3$ (2789). (h) Rxn. with CH$_2$:CMeCH$_2$X → Et$_3$SnX + p-MeOC$_6$H$_4$CH$_2$C(CN)$_2$CH$_2$CMe:CH$_2$ (1954, 2789). (i) Rxn. with PhCH$_2$Br → Et$_3$SnBr + RR'CHC(CN)$_2$CH$_2$Ph (2789). (j) Rxn. with CH$_2$:CHCH$_2$X → Et$_3$SnX + m-O$_2$NC$_6$H$_4$CH$_2$C(CN)$_2$CH$_2$CH:CH$_2$ (1954, 2789), with Ph$_3$CCl → Et$_3$SnCl + Me$_2$CHC(CN)$_2$CPh$_3$ (2789). (k) Rxn. with CH$_2$:CHCH$_2$Br → Et$_3$SnBr + Me$_2$CHC(CN)$_2$CH$_2$CH:CH$_2$ (1954, 2789). (l) Relative rate of formation (2615), viscous yellow oil, proposed struct., soly., easily hydrolyzed (2615), rxn. with EtOH → (Bu$_3$Sn)$_2$O + RCH$_2$CH(CN)$_2$ (1954, 2789). (m) Rxn. with CH$_2$:CHCH$_2$Br → Bu$_3$SnBr + RCH$_2$C(CN)$_2$CH$_2$CH:CH$_2$ (1954, 2789). (n) Rxn. with AcCl, followed by HCO$_2$H + H$_2$O$_2$, Pb(OAc)$_4$ → PhCHO (2615). (o) Rxn. with Ph$_3$SnH → Ph$_6$Sn$_2$ + PhCH$_2$CH(CN)$_2$ (2789).

4.5.3 ORGANOTIN DERIVATIVES OF ACID AMIDES AND IMIDES

Some oxygen free compounds originally listed in this chapter have been moved to Subchapter 4.5.2.4. This includes derivatives of amidines, ketimines and carbodiimide.

4.5.3.1 ORGANOTIN DERIVATIVES OF AMIDES AND IMIDES OF CARBOXYLIC ACIDS

The N-organotin derivatives listed here include also two compounds derived from the isomeric imidoester form of the amide. The compounds summarized in Table 240 are prepared by methods from the following scheme.

I   Reaction of triorganotin hydride R_3SnH with:
   (IA) PhNCO,
   (IB) BzN_3.

II Reaction of bistriorganotin oxide with $RNHCR'O$ or CCl_3CN.

III Reaction of triorganotin methoxide R_3SnOMe with $RNHCR'O$ or CCl_3CN.

IV Reaction of triethyltin amide Et_3SnNR_2 with $BuNHCHO$.

R = organic group, R' = H or organic group.

Triethyltin Phenylformamide (M-480) $SnC_{13}H_{21}NO$ $Et_3SnNPhCHO$
Synth.: By rxn. of Et_3SnOH with $PhNHCHO$ in xylene under reflux (949).
By rxn. of Et_3SnOMe with $PhNHCHO$ at 160°, in 83% yield (402).
By rxn. of Et_3SnH with $PhNCO$, rxn. mechanism (314), relative rxn. rates (1415).
Prop.: m. 49.5-52.5° (949), $b_{0.3}$ 110° (402), $b_{0.1}$ 97° (949), IR and NMR spectra (402, 949), GLC (1415).
Rxn. with: $R_3SnH \rightarrow Et_3SnSnR_3$, relative rxn. rates; R = Ph, p-$MeOC_6H_4$ (915).
$R_3SnH \rightarrow Et_3SnH + R_3SnNPhCHO$; solvent effect, R = Me, Ph (915).
$Ph_3SnH \rightarrow Et_3SnSnPh_3 + PhNHCHO$ (97), solvent effect (96).
$R_2SnH_2 \rightarrow Et_3SnR_2SnH$; R = Et, Ph (95).
R_2SnH_2 (2:1) $\rightarrow (Et_3Sn)_2SnR_2$; R = Et (95), Ph (97).
$Et_3SnEt_2SnH \rightarrow (Et_3Sn)_2SnEt_2$ (95).

Tributyltin Phenylformamide $SnC_{19}H_{33}NO$ $Bu_3SnNPhCHO$
Synth.: By rxn. of Bu_3SnH with $PhNCO$, rxn. mechanism and solvent effect (314), relative rxn. rates (1415).
By rxn. of $(Bu_3Sn)_2O$ with $PhNHCHO$ in xylene under reflux (949).
By rxn. of $(Bu_3Sn)_2S$ with $Hg(NPhCHO)_2$ (949).
By rxn. of $(Bu_3SnNPh)_2SO$ with Cl_3CCHO in CCl_4, in 49% yield (2345).
Prop.: m. 64-67°, IR and NMR spectrum, hydrolyzes rapidly in air (949).
Rxn. with: $Ph_3SnH \rightarrow Bu_3SnSnPh_3$ (97).
$Ph_2SnH_2 \rightarrow (Bu_3Sn)_2SnPh_2$ (97).
$PhSnH_3$ (2:1) $\rightarrow (Bu_3Sn)_2SnPhH$ (95).
$Bu_3SnSnBu_2H \rightarrow (Bu_3Sn)_2SnBu_2 + PhNHCHO$ (95).

Listing of N-organotin amides, imidoesters and imides of carboxylic acids concludes in Table 240.

Table 240. Organotin Derivatives of Amides and Imides of Carboxylic Acids $R_nSn(NRCOR')_{4-n}$

Formula*	Synth. Method**	Properties	Ref.
Me$_3$SnNPhCHO	IA (a)	(b)	314, 1415
Et$_3$SnNY			
NY = NBuCHO	IV	(c)	915
N(CMe$_3$)CHO	IV	(c)	915
N(c.C$_6$H$_{11}$)CHO	III	Yield 86%, b$_{0.3}$ 105-109°, n$_D^{20}$ 1.4918	402
	IA	Mechanism of synth., solvent effect (d)	314
NPhCDO	IA	GLC	1415
N(p-ClC$_6$H$_4$)CHO	--	(e)	--
N(p-O$_2$NC$_6$H$_4$)CHO	--	(c)	--
NPhAc	III	b$_{0.15}$ 111-12°, IR and NMR spectra	402
N(COCH$_2$)$_2$ (f) (480)	--	Spectrographic detn.	1358
Et$_2$Sn(NPhCHO)$_2$	--	(g, h)	--

* Numbers in parenthesis refer to pages in main volume.
** The characters used here correspond to the synthetic scheme in the introduction to Chapter 4.5.3.1.
(a) Rxn. mechanism and solvent effect (314), relative rxn. rates (1415). (b) Rxn. with Ph$_3$SnH → R$_3$SnSnPh$_3$ + PhNHCHO; R = Me, C$_8$H$_{17}$ (97), with Ph$_2$SnH$_2$ → (R$_3$Sn)$_2$SnPh$_2$ + PhNHCHO; R = Me, C$_8$H$_{17}$ (97). (c) Rxn. with Ph$_3$SnH → Et$_3$SnSnPh$_3$ + RNHCHO, relative rxn. rates; R = Bu, Me$_3$C, p-O$_2$NC$_6$H$_4$ (915). (d) IR and NMR spectra (402), rxn. with R$_3$SnH → Et$_3$SnSnR$_3$ + c.C$_6$H$_{11}$NHCHO; solvent effect, R = Ph (96), relative rxn. rates, R = Et, Ph (915). (e) Rxn. with Ph$_3$SnH → Et$_3$SnSnPh$_3$ + p-ClC$_6$H$_4$NHCHO, solvent effect (96) and relative rxn. rates (915). (f) (CH$_2$CO)$_2$NH = succinimide, o-C$_6$H$_4$(CO)$_2$NH = phthalimide. (g) Rxn. with Ph$_3$SnH → (Ph$_3$Sn)$_2$SnR$_2$ + PhNHCHO; R = Et, Bu, Ph (97). (h) Rxn. with Ph$_2$SnH$_2$ → (R$_2$SnSnPh$_2$)$_x$ + PhNHCHO (97).

Table 240 Continued $R_nSn(NRCOR')_{4-n}$

Formula*	Synth. Method**	Properties	Ref.
Bu$_3$SnNY			
NY = NMeCHO	II	m. 37-37.5°, b$_{0.4}$ 136-40°, IR and NMR spectra, sensitive to moisture	949
	II (i)		949
N(C$_6$H$_{13}$)CHO	II	m. 35-37°, b$_4$ 167-70°, IR and NMR spectra (j)	949
NHAc	II	b$_{0.1}$ 142-44°, IR and NMR spectra	949
NHBz	IB	m. 97-100°, b$_{0.001}$ 140-45° (k)	1050
NPhAc	II	b$_{0.03}$ 104-106°, IR and NMR spectra (j)	949
N(CO)$_2$C$_6$H$_4$-o (f) (480)	--	(l)	--
N:C(CCl$_3$)OMe (480)	III	Crude, IR spectrum (m, n)	36,750
N:C(CCl$_3$)OSnBu$_3$	II	Yellow oil, b$_{0.05}$ 115° (dec.), IR spectrum (m, o)	36,750
Bu$_2$Sn(NPhCHO)$_2$	--	(g)	--
(C$_8$H$_{17}$)$_3$SnNPhCHO	--	(b, p)	--
Ph$_3$SnNPhCHO	IA (a)	--	314,1415
Ph$_2$Sn(NPhCHO)$_2$	--	(g)	--

(i) Synth. with PhNHCONMeCHO. (j) Sensitive to moisture (949). (k) Rxn. with EtOH, + AcOH → Bu$_3$SnOAc + BzNH$_2$ (1050). (l) Activity against Fusarium nivale, Rhizoctonia solani (1057), use as seed disinfectant (1057). (m) Hydrolyzes rapidly in air (750), use as biocide (940). (n) Dissocn. on distn. → Bu$_3$SnOMe + CCl$_3$CN (750). (o) Thermal decomposition at 115° → Bu$_3$SnCl (750). (p) Rxn. with PhSnH$_3$ → [(C$_8$H$_{17}$)$_3$Sn]$_3$SnPh + PhNHCHO (97).

4.5.3.2 N-ORGANOTINCARBAMATES

This subchapter reports N-organotin derivatives of carbamic acid esters ($R_2SnNHCO_2R$). A few bistriorganotin compounds with tin bound to the amino and carboxy group of carbamic acid ($R_3SnNRCO_2SnR_3$) are also included. Organotin esters of carbamic acid $H_2NCO_2SnR_3$ may be found in Table 203. The N-organotincarbamates in Table 240 are prepared by methods from the following scheme.

I   Reactions of bistriorganotin oxide with isocyanate and carbamate:
$$(R_3Sn)_2O + R'NCO \rightarrow R_3SnNR'CO_2SnR_3,$$
$$(R_3Sn)_2O + R'NHCO_2R \rightarrow R_3SnNR'CO_2R.$$

II Reactions of organotin alkoxides with isocyanate and carbamate:
(IIA) $R_3SnOMe + RNHCO_2Me \rightarrow R_3SnNRCO_2Me,$
(IIB) $R_nSn(OR')_{4-n} + RNCO \rightarrow R_nSn(NRCO_2R')_{4-n},$
(IIC) $Ph_3SnO_2R' + RNCO \rightarrow Ph_3SnNRCO_3R'.$

III Reaction of organotin sulfite with carbamate:
$$Bu_3Sn(Et)SO_3 + RNCO \rightarrow Bu_3SnNRCO_2Et.$$

$R = R'$ = organic group, $n = 2, 3$.

Methyl N,N-Tributyltinethylcarbamate $SnC_{16}H_{35}NO_2$ $Bu_3SnNEtCO_2Me$
Synth.: By rxn. of $(Bu_3Sn)_2O$ with $EtNHCO_2Me$ in vacuo at 100°, in 85% yield (949), with EtNCO as by-prod. (746).
By rxn. of Bu_3SnOMe with EtNCO, in high yield (36, 949).
By rxn. of $Bu_3SnNEtCO_2SnBu_3$ with MeOH (748).
Prop.: $b_{0.2}$ 90-95° (949), $b_{0.05}$ 101.5° (36), IR spectrum (36, 949).
Rxn. with: RNCO → $R\overline{NCONRCONEtC}O$; R = Ph, $1-C_{10}H_7$ (746).

N,O-Bistributyltin N-Ethylcarbamate $Sn_2C_{27}H_{59}NO_2$ $Bu_3SnNEtCO_2SnBu_3$
Synth.: By rxn. of $(Bu_3Sn)_2O$ with EtNCO (35-6, 748).
By thermal dec. of Bu_3SnO_2CNHEt (748).
Prop.: Oil, dec. on distn., IR spectrum (35, 748).
Rxn. with: MeOH → $Bu_3SnOMe + Bu_3SnNEtCO_2Me$ (748).
Ac_2O → $Bu_3SnOAc + EtNCO$ (748).
$PhNH_2$, + dry CO_2 → Bu_3SnO_2CNHPh (35).
PhNCO → $Bu_3SnNEtCONPhSnBu_3 + CO_2$ (747).
RNCO (1:2) → $Bu_3SnNRCONEtCONRSnBu_3 + CO_2$; R = Ph, $1-C_{10}H_7$ (749).
PhNCO (1:3) → $Ph\overline{NCONPhCONEtC}O$ (749).

Methyl N,N-Tributylphenylcarbamate (M-480) $SnC_{20}H_{35}NO_2$ $Bu_3SnNPhCO_2Me$
Synth.: By rxn. of $(Bu_3Sn)_2O$ with $PhNHCO_2Me$ in vacuo at 100°, in 85% yield (949).
By rxn. of Bu_3SnOMe with PhNCO, in high yield (36, 949), with $(PhNCO)_2$ in C_6H_6 (746).
Prop.: $b_{0.1}$ 118-20° (949), $b_{0.01}$ 98-100° (36), IR spectrum (36, 949).
Rxn. with: Ac_2O → $Bu_3SnOAc + AcNPhCO_2Me$ (36).

$H_2S \rightarrow (Bu_3Sn)_2 + Bu_3SnSH$ (?) + $PhNHCO_2Me$ (36).
$EtNH_2 \rightarrow Bu_3SnOMe + EtNHCONHPh$ (36).
$EtNCO \rightarrow EtNCONEtCONPhCO + Bu_3SnOMe$ (746).

N,O-Bistributyltin N-Phenylcarbamate $Sn_2C_{31}H_{59}NO_2$ $Bu_3SnNPhCO_2SnBu_3$
Synth.: By rxn. of $(Bu_3Sn)_2SO_3$ with PhNCO in CCl_4 (2345).
By rxn. of $(Bu_3Sn)_2O$ with PhNCO (35, 748, 1171), with PhNCS (2:1) (750).
Prop.: Amber liq. (1171), oil, dec. on distn. (35, 748), IR spectrum (748).
Rxn. with: Thermal dec. $\rightarrow (Bu_3Sn)_2O + CO_2 + (Bu_3SnNPh)_2CO$ (35, 748).
$Me_3COH \rightarrow Bu_3SnOCMe_3 + Bu_3SnO_2CNHPh$ (748).
$Ac_2O \rightarrow Bu_3SnOAc + PhNCO + (PhNCO)_3$ (748).
$SO_2 \rightarrow (Bu_3Sn)_2SO_3 + PhNCO + CO_2$ (2345).
$PhNH_2$, + dry $CO_2 \rightarrow Bu_3SnO_2CNHPh$ (35, 748).
$RNCO \rightarrow RNCONRCONPhCO$; R = Et (746), Ph (1171) + $Bu_3SnNRCONPhSnBu_3$; R = 1-$C_{10}H_7$ (749).
Use: As fungicide, as cat. for polyurethan synth., as stabilizer and plasticizer (1171).

Listing of organotin carbamates concludes in Table 241.

Table 241. N-Organotincarbamates $R_nSn(NRCO_2R)_{4-n}$

Formula*	Synth. Method**	Yield	Properties	Ref.
Et₃SnNY				
NY = NBuCO₂Et (480)	IIB	High	$b_{0.01}$ 57.5°, IR spectrum	36
N(C₆H₁₃)CO₂Me	IIA	67	$b_{0.2}$ 92-93°, n_D^{20} 1.4784 (a)	402
NPhCO₂Me	IIA	90	$b_{0.4}$ 105-108°, n_D^{20} 1.5388 (a)	402
NPhCO₂Bu-i	IIB	--	m. 45-47°, hygroscopic, IR spectrum (b)	389
NPhCO₂SnEt₃	I	--	m. 58-60°, IR, NMR spectra (c)	35,748
N(1-C₁₀H₇)CO₂Et (480)	IIB	High	m. 86-93° (sealed tube), IR spectrum	36
Bu₃SnNY				
NY = NMeCO₂Me	IIB	High	$b_{0.005}$ 89-90°, IR spectrum	36
NMeCO₂SnBu₃	I	--	Oil, dec. on distn., IR spectrum (d)	748
NEtCO₂Et	IIB	--	IR and NMR spectra	2345
	III	100	--	2345
	III	100	(e)	2345
NPhCO₂Et	IIB	High	Solid, dissocn. on distn., IR spectrum	36
NPhCO₂Ph	IIB	90	$b_{0.01}$ 120°, IR spectrum (f)	36
N(1-C₁₀H₇)CO₂Me (480)	IIB	--	Oil, dec. on distn., IR spectrum (g)	35,748
N(1-C₁₀H₇)CO₂SnBu₃	I	--		
(Bu₃SnNCO₂Et)₂	I (h)	--	$b_{0.07}$ 125-35°, IR spectrum	949
Bu₂Sn(NEtCO₂Me)₂	IIB	--	IR and NMR spectra (i)	943

* Numbers in parenthesis refer to pages in main volume.
** The characters used here correspond to the synthetic scheme in the introduction to Chapter 4.5.3.2.
(a) IR and NMR spectra (402). (b) Rxn. with moisture → PhNHCO₂Bu-i (389). (c) Decomposes on distn. (748).
(d) Rxn. with MeNH₂ + CO₂ → MeNHCO₂SnBu₃ (2256), with PhNCO → Bu₃SnNMeCONPhSnBu₃ + CO₂ (747, 749). (e) Rxn. with SO₂ → Bu₃Sn(Et)SO₃ + PhNSO (35). (f) Rxn. with HX → Bu₃SnX + 1-C₁₀H₇NHCO₂Me; X = OH, OEt; OAc (36). (g) Thermal decomposition → (Bu₃SnNC₁₀H₇-1)₂CO (35). (h) Synth. from (EtO₂CNH)₂ (949). (i) Rxn. with Bu₂SnCl₂ → Bu₂Sn(NEtCO₂Me)Cl (943), with EtNCO → (EtNCO)₃ (943).

Table 241 Continued $R_nSn(NRCO_2R)_{4-n}$

Formula*	Synth. Method**	Yield	Properties	Ref.
$Bu_2Sn(NPhCO_2Me)_2$	IIB	82	$b_{0.2}$ 154°, NMR spectrum	943
	IIB	--	Synth. with $(PhNCO)_2$ (j)	943, 2662
$Bu_2Sn(NCO_2Me)_2C_6H_3Me-1,2,3$	IIB	--	Synth. with 1,2,3-$MeC_6H_3(NCO)_2$ (j)	2662
$Ph_3SnNMeCO_3CMe_3$	IIC	92	m. 88-90° (sealed tube, dec.) (k, l)	2256
$Ph_3SnNEtCO_3CMe_3$	IIC	77	m. 99-100° (sealed tube, dec.) (k)	2256
$Ph_3SnNBuCO_3CMe_3$	IIC	93	m. 60-68° (sealed tube, dec.) (k, m)	2256
$R_3SnNPhCO_2SnR_3$	--	--	(n)	--

(j) Use as curing catalyst for polysiloxane polymers (2662). (k) IR and NMR spectra, dec. on standing (2256). (l) Rxn. with H_2O in CCl_4 → $(Ph_3Sn)_2O$ + $MeNHCO_3CMe_3$ (2256). (m) Rxn. with HCl in Et_2O → Ph_3SnCl + $BuNHCO_3CMe_3$ (2256). (n) Use as fungicide, as catalyst for synth. of polyurethan, as plasticizer and stabilizer (1171).

4.5.3.3 ORGANOTIN DERIVATIVES OF UREA

This subchapter comprises organotin derivatives of urea, pseudourea, thiourea and biuret. Also reported are two derivatives of the phosphine analog $PhNHCOPPh_2$. The compounds listed in Table 241 are prepared by the following methods.

I Reactions of bistriorganotin oxide with:
   (IA) RNCO,
   (IB) $(RN:)_2C$.

II Reaction of organotin alkoxide with $(RN:)_2C$.

III Reaction of O-organotin carbamate with RNCO or RNCS.

IV Reaction of isocyanate or thioisocyanate with:
   (IVA) R_3SnNR_2,
   (IVB) $R_3SnNRCO_2SnR_3$.

V Reaction of isocyanate or isothiocyanate with Ph_3SnPPh_2.

N,N'-Bis(tributyltin)diphenylurea $Sn_2C_{37}H_{64}N_2O$ $(Bu_3SnNPh)_2CO$
Synth.: By rxn. of $(Bu_3Sn)_2O$ with PhNCO (1:2) (35), with $(PhNCO)_2$ as byprod. (746), with $(PhN:)_2C$ (750), with $(PhNH)_2CO$ under reflux in MePh (949), with $PhNCO_2Me$, PhNHCONMeCHO or $1-C_{10}H_7NHCONHPh$ (949).
By rxn. of $Bu_3SnNPhCO_2SnBu_3$ (1171) or $(Bu_3SnNPh)_2SO$ with PhNCO (2345).
By thermal dec. of $Bu_3SnNPhCO_2SnBu_3$ (35, 748), of Bu_3SnO_2CNHPh (748).
Prop.: Pale yellow oil (949), $b_{0.05}$ 130° (35, 748, 949), IR spectrum (949).
Rxn. with: Air → $(Bu_3Sn)_2O$ + $(PhNH)_2CO$ (748).
Use: As fungicide, as cat. for polyurethan synth., as plasticizer and stabilizer (1171).

N,N'-Bis(tributyltin)di-1-naphthylurea $Sn_2C_{45}H_{68}N_2O$ $(Bu_3SnNC_{10}H_7-1)_2CO$
Synth.: By rxn. of $(Bu_3Sn)_2O$ with $1-C_{10}H_7NCO$ (1:2) (35, 748, 2258), with $(1-C_{10}H_7N:)_2C$ (750), with $(1-C_{10}H_7NH)_2CO$ in refluxing xylene (949).
By thermal dec. of $Bu_3SnN(1-C_{10}H_7)CO_2SnBu_3$ (35, 748).
Prop.: Oil (949), $b_{0.05}$ > 220° (35, 750), IR spectrum (748, 949).
Rxn. with: Moisture → $(1-C_{10}H_7NH)_2CO$ (748).
Use: As biocide (940).

Additional organotin derivatives of urea and biuret are listed in Table 242.

Table 242. Organotin Derivatives of Urea   R₃SnNHCONR₂

Formula*	Synth. Method**	Yield	Properties	Ref.
Me₃SnNPhCONMe₂ (480)	IV	--	Colorless viscous liq., b₀.₅ 103°	1392
	III	95	IR spectrum, assignment, struct.	167
Me₃SnNPhCSNMe₂ (480)	III	--	Liq., n_D^22 1.6024, dec. on distn. (a)	167
(Bu₃SnNMe)₂CO	IA	67	b₀.₀₅ 195° (bath), IR, NMR spectra (b)	2258
Bu₃SnNMeCONBuSnBu₃	IA (c)	--	b₀.₁ 150° (bath), IR, NMR spectra (d)	2258
Bu₃SnNMeCONPhSnBu₃	IVB	--	b₀.₀₅ 149-50°, IR, NMR spectra (e-g)	747,749
Bu₃SnNEtCONEt₂	III	90	b₀.₂ 149°, IR spectrum	2258
Bu₃SnNEtCONPhSnBu₃	IVB	--	b₀.₀₅ 167° (e, f, h)	747
(Bu₃SiNPr-i)₂CO	IB	--	Oil, b₀.₀₅ 150° (bath), IR spectrum (i)	750
Bu₃SnBuCONMeCHO	(j)	--	Not sepd., NMR spectrum	2258
(Bu₃SnNBu)₂CO	IA	--	b₀.₁ 170° (bath), IR spectrum, dec. at 275°	2258
Bu₃SnNBuCONMeCH(CCl₃)OSnBu₃	(d)	--	b₀.₁ 105°, IR, NMR spectra (j)	2258
(Bu₃SnNc.C₆H₁₁)₂CO	IA	--	Oil, b₀.₀₀₁ 155°, IR spectrum (i)	750
Bu₃SnPhCONHEt	(k)	--	IR, NMR spectra	749
Bu₃SnNPhCSNEt₂	III	--	b₀.₁ 160°, IR spectrum	2258
	IVA	--	--	2258

* Numbers in parenthesis refer to pages in main volume.
** The characters used here correspond to the synthetic scheme in the Introduction to Chapter 4.5.3.3.

(a) IR spectrum, assignment and struct. (167). (b) Rxn. with PhNCO (1:2) → (MeNH)₂CO + (PhNH)₂CO + MeNHCONPh + MeNCONMeCONPhCO (2258). (c) Synth. with mixture of MeNCO and BuNCO (2258). (d) Rxn. of Bu₃SnNMeCONBuSnBu₃ with Cl₃CCHO → Bu₃SnNBuCONMeCH(CCl₃)OSnBu₃ (2258). (e) Rxn. of Bu₃SnNRCONPhSnBu₃ with R'NCO → Bu₃SnNR'CONRCONPhSnBu₃; R' = 1-C₁₀H₇, R = Me, Et (749). (f) Rxn. with H₂O → PhNHCONHR; R = Me, Et (747). (g) Rxn. with MeOH → PhNHCONHMe (749). (h) Rxn. with RNCO, + H₂O → (RNHCO)₂NEt; R = Ph, 1-C₁₀H₇ (747). (i) Dissociates partly during distillation (750), rxn. with MeOH → (RNH)₂CO; R = i-Pr, c.C₆H₁₁ (750). (j) Thermal dec. of Bu₃SnNBuCONMeCH(CCl₃)-OSnBu₃ at 100° → Bu₃SnCl + Bu₃SnBuCONMeCHO (2258). (k) Thermal dec. of (Bu₃SnPhCO)₂NEt at 250-290° in vacuo → Bu₃SnNPhCONHEt (749).

Table 242 Continued

R_3SnNHCONR_2

Formula*	Synth. Method**	Yield	Properties	Ref.
Bu_3SnNPhCON(1-C_10H_7)SnBu_3	IVB	--	Not isolated (l)	749
Bu_3SnNPhC(:NPh)OSnBu_3	IB	--	b0.05 200°, IR spectrum (m, n)	35-6,940
Bu_3SnNRC(:NR)OMe	II	--	R = p-MeC_6H_4, crude oil, IR spectrum	750
Bu_3SnNRC(:NR)OMe (480)	II	--	R = 1-C_10H_7, crude, dec. on distn. (n, o)	36,750
Bu_3SnNRC(:NR)OSnBu_3 (480)	--	--	R = 1-C_10H_7 (m)	--
Bu_2Sn[NRC(:NR)OMe]_2	II	--	R = 1-C_10H_7, IR, NMR spectra, dec. on distn.	943
(R_3SnNPh)_2CO	--	--	(p)	--
(Bu_3SnNArCO)_2NR				
R = Me Ar = Ph + 1-C_10H_7	(e)	--	IR, NMR spectra (q)	749
Et Ph	IVB	--	IR spectrum (k, r, s)	749
Et Ph + 1-C_10H_7	(e)	--	IR spectrum	749
Et 1-C_10H_7	IVB	--	Viscous oil, IR spectrum (r)	749
Ph_3SnNPhCOPPh_2	V	64	m. 58° (dec.), IR spectrum	483
Ph_3SnNPhCSPPh_2	V	72	m. 87°, IR spectrum, mol. wt.	483

(l) Synth. also from (Bu_3SnNPh)_2SO with 1-C_10H_7NCO (2345), rxn. with MeOH → 1-C_10H_7NHCONHPh (749). (m) Isomeric with/identical to (Bu_3SnNR)_2CO; R = Ph, 1-C_10H_7, by IR spectrum (35). (n) Use as biocide (940). (o) Rxn. with moisture → 1-C_10H_7NHC(OMe):NC_10H_7-1 (750). (p) Use as fungicide, as catalyst for polyurethan synth., as plasticizer and stabilizer (1171). (q) Rxn. with AcOH → 1-C_10H_7NHCONMeCONHPh (749). (r) Rxn. with MeOH → (ArNHCO)_2-NEt; Ar = Ph, 1-C_10H_7 (749). (s) Rxn. with PhNCO → EtNCONPhCONPhCO (746).

4.5.3.6 ORGANOTIN AMIDES AND IMIDES OF ACIDS OF SULFUR

This subchapter comprises N-organotin derivatives of amides and imides of various acids of sulfur. The compounds listed in Table 243 are prepared by the following methods.

I   Reaction with triorganotin halide:
    $Me_3SnCl + (FO_2S)_2NCl \rightarrow Me_3SnN(SO_2F)_2$.

II Reactions with organotin oxides:
    (IIA) $(Bu_3Sn)_2O + RNSO \rightarrow Bu_3SnNPrSO_2SnBu_3$,
    (IIB) $(Bu_3Sn)_2O + (RN:)_2S \rightarrow (Bu_3SnNR)_2SO$,
    (IIC) $Bu_2SnO + ArSO_2NHR \rightarrow Bu_2Sn(NRSO_2Ar)_2$.

III Reactions with triorganotin amides:
    $Me_3SnNMe_2 + PhNSO \rightarrow Me_3SnNPhSONMe_2$,
    $(Bu_3Sn)_2NPh + RNSO \rightarrow (Bu_3SnNPh)SO(NRSnBu_3)$.

N,N-Triethyltinphenylsulfonyl N'-Phenylhydrazide

$SnC_{18}H_{26}N_2O_2S$ $Et_3SnN(NHPh)O_2SPh$

Synth.: By rxn. of Et_3SnH with $PhN:NSO_2Ph$ in Ar atm. in MePh, in 99% yield (2646).
By-prod. in dec. of $PhN:NSO_2Ph$, catalyzed by Et_3SnH (1611).
Prop.: m. 116° (1611, 2646).
Rxn. with: HCl in aq. $Et_2O \rightarrow PhSO_2NHNHPh$ (2646).
$AcCl \rightarrow Et_3SnCl + AcNPhNHSO_2Ph + AcNPhN(Ac)SO_2Ph$ (2646).
$Ac_2O \rightarrow Et_3SnOAc + AcNPhNHSO_2Ph$ (2646).
$RBr \rightarrow Et_3SnBr + PhNHNHSO_2Ph + PhNRNRSO_2Ph + PhRSO_2 + C_6H_6 + N; R = PhCH_2$ (2646).

N,N'-Bistributyltindiphenyl Thionylimide $Sn_2C_{36}H_{64}N_2SO$ $(Bu_3SnNPh)_2SO$

Synth.: By rxn. of $(Bu_3Sn)_2O$ with $(PhN:)_2S$ (2345), with PhNSO as by-prod. (2345).
Prop.: Red viscous oil, IR and NMR spectra (2345).
Rxn. with: Thermal dec. $\rightarrow$ PhNSO (at 90°) + $(Bu_3Sn)_2NPh$ (at 170°) (2345).
Air $\rightarrow (Bu_3Sn)_2SO_3$ (2345).
$SO_2 \rightarrow (Bu_3Sn)_2SO_3 + PhNSO$ (2345).
$AcOH \rightarrow Bu_3SnOAc + PhNH_2 + PhNSO$ (2345).
$Cl_3CCHO \rightarrow Bu_3SnNPhCHO + Cl_3CCH:NPh$ (2345).
$RNCO \rightarrow Bu_3SnNPhCONRSnBu_3 + PhNSO; R = Ph, 1-C_{10}H_7$ (2345).

Additional N-organotin amides of acids of sulfur are listed in Table 243.

Table 243. Organotin Amides and Imides of Acids of Sulfur $R_nSn(NSY)_{4-n}$

Formula*	Synth. Method**	Properties	Ref.
$Me_3SnN(SO_2F)_2$	I	Dec. on distn., spectra (a)	1776
$Me_3SnNPhSONMe_2$	III	n_D^{22} 1.5642, IR spectrum	167
$Bu_3SnNHSO_2C_6H_4Me-p$ (b) (480)	--	(c)	--
$Bu_3SnNPrSO_2SnBu_3$	IIA	Viscous oil (d, e)	2345
$(Bu_3SnNAr)_2SO$			
Ar = Ph + p-MeC$_6$H$_4$	III	Orange viscous oil (e, f)	2345
p-MeC$_6$H$_4$	IIB	Orange-red viscous oil	2345
	IIA	(e, g)	2345
$Bu_2Sn(NRSO_2C_6H_4Me-p)_2$	IIC	R = $CH_2\overline{CHCH_2O}$, tan powder	1442
p-$Ph_3SnNHSO_2C_6H_4Me$ (b) (481)	--	(h)	--

* Numbers in parenthesis refer to pages in main volume.
** The characters used here correspond to the synthetic scheme in the introduction to Chapter 4.5.3.6.
(a) IR, UV, ^1H and ^{19}F NMR spectra (1776). (b) Sulfonamide derivative.
(c) Activity as fungicide against Fusarium nivale, Rhizoctonia solani, Tilletia tritici (1057). (d) Dissociates at 75° in vacuo (2345), rxn. with Cl_3CCHO → $PrN:CHCl_3$ (2345). (e) IR and NMR spectra (2345). (f) Thermal dec. → $(Bu_3Sn)NPh + (Bu_3Sn)_2NC_6H_4Me-p + PhNSO + p-MeC_6H_4NSO$ (2345). (g) Thermal dec. at 95° → $(Bu_3Sn)_2NC_6H_4Me-p + p-MeC_6H_4NSO$ (2345). (h) Activity as insecticide against Tineola bisselliella larvae (162), use as herbicide (2977).

4.5.4 ORGANOTIN METAL IMIDES

The organotin metal imides are all listed in Table 244. The compounds contain lithium, germanium and silicon bound to nitrogen. A scheme for preparation of the organotin metal imides follows.

I Cleavage of tristrimethyltin nitride with butyllithium.

II Reactions of organotin halides Me_nSnX_{4-n} with alkali metal derivatives:
  (IIA) with metal substituted lithium imide,
  (IIB) in presence of benzonitrile,
  (IIC) with metal substituted sodium imide.

III Condensation of organotin dialkylamides $R_nSn(NR_2)_{4-n}$ with an acidic metal amide:
  (IIIA) R'_2NH,
  (IIIB) $(R'NH)_2Y$.

R = organic group, R' = organometallic group, e.g. Me_3Si, $Me_3SiSiMe_2$, n = 0-3, X = Cl for n = 2, 3, X = Br for n = 0, 1.

Table 244. Organotin Metal Imides $R_nSn(NMR)_{4-n}$

Formula*	Synth. Method**	Yield	Properties	Ref.
(Me$_3$Sn)$_2$NLi	I	--	--	471
Me$_3$SnNLi$_2$	I	--	--	471
(Me$_3$Sn)$_2$NSiMe$_3$	IIC	--	--	1814
Me$_3$SnNMeSiMe$_3$	IIA	68	b$_{30}$ 79-81°, sensitive to air	472
	IIA	62	b$_{11}$ 59-61° (a)	1812, 1814
Me$_3$SnNMeGeMe$_3$ (476)	IIA	78	b$_2$ 28°, NMR spectrum, mol. wt., soly.	1811
Me$_3$SnN(CMe$_3$)SiMe$_3$	IIA	51	b$_1$ 63-66°, NMR spectrum	2749
Me$_3$SnN(SiMe$_3$)$_2$ (476)	IIA	67	b$_{2.5}$ 50°, spectra, mol. wt. (b)	324
Me$_3$SnN(GeMe$_3$)NSiMe$_2$	IIIA	80	b$_{0.1}$ 55°, NMR spectrum (324, 474)	1810
Me$_2$Sn(NMeSiMe$_3$)$_2$	IIA	--	b$_{0.5}$ 61-63°, NMR spectrum	472, 1812
	IIA	60	Sensitive to air and moisture	1814
Me$_2$Sn(NCMe$_3$SiMe$_3$)$_2$	IIA	60	b$_{1.5}$ 135°, spectra, mol. wt. (b)	2749
Me$_2$Sn(NMe$_2$)N(SiMe$_3$)$_2$	IIA	36	b$_{0.1}$ 58°, NMR spectrum	324
Me$_2$SnNSiMe$_3$(CH$_2$)$_2$NSiMe$_3$	IIIB	80	m. 14-16°, b$_{18}$ 121-23°, NMR spectrum	470
Me$_2$SnNSiMe$_3$(CH$_2$)$_3$NSiMe$_3$	IIIB	77	m. -2°, b$_{18}$ 136-38°, NMR spectrum	470
Me$_2$SnNSiMe$_3$(CH$_2$)$_4$NSiMe$_3$	IIIB	65	b$_{0.1}$ 70-71°, NMR spectrum	470
o-Me$_2$SnNSiMe$_3$C$_6$H$_4$NSiMe$_3$	IIIB	63	m. 75°, b$_{0.1}$ 114-17°, NMR spectrum (c)	470
MeSn(NMe$_2$)$_2$N(SiMe$_3$)$_2$	IIIA	--	b$_{0.1}$ 78°, solidifies at r.t., NMR spectrum	324
MeSn(NMeSiMe$_3$)$_3$	IIA	80	m. 22-23°, b$_{0.5}$ 83-85°, NMR spectrum	1812
Sn(NMeSiMe$_3$)$_4$	IIA	55	m. 90-92°, NMR spectrum	1812
Me$_3$SnN:CPhNMeSiMe$_3$	IIB	57	Colorless oil, b$_{0.1}$ 90-91°, mol. wt.	2738
	IIA	67	IR, NMR spectra	2738
		59		

* Numbers in parenthesis refer to pages in main volume.
** The characters used here correspond to the synthetic scheme in the introduction to Chapter 4.5.4.
(a) NMR spectrum (472, 646, 1812), rxn. with NCNSO → (Me$_3$SnN:)$_2$C + (Me$_3$SiN:)$_2$C (2740). (b) (Far) IR, assignment and NMR spectra (2749). (c) Very sensitive to air (470).

4.5.5 ORGANOTIN PHOSPHORUS IMIDES

A glance at the ensuing synthetic scheme will show the advance in chemistry of organotin imides of various phosphorus moieties; all but method (IIA) are new.

I Reaction with triorganotin chloride:
$Me_3SnCl + R_3P:NLi \rightarrow Me_3SnN:PR_3$.

II Reactions with triorganotin azides:
(IIA) $R_3SnN_3 + R_3P \rightarrow R_3SnN:PR_3$,
(IIB) $R_3SnN_3 + (RO)_3P \rightarrow R_3SnNRPO_3R_2$.

III Reactions with organotin amides and phosphides:
(IIIA) $Et_3SnNR_2 + Ph_2P:NH \rightarrow Et_3SnN:PPh_3$,
(IIIB) $(Ph_3Sn)_nPPh_{3-n} + PhN_3 \rightarrow (Ph_3SnNPh)_nPPh_{3-n}:NPh$.

IV Interconversion reactions:
(IVA) $Me_3SnNHPR_2 + S \rightarrow Me_3SnNHP(S)R_2$,
(IVB) $Me_3SnNHPR_2 + CCl_4 \rightarrow Me_3SnN:PClR_2$.

Trimethyltin Di-t-butylphosphineamide $\quad SnC_{11}H_{28}NP \quad Me_3SnNHP(CMe_3)_2$
Synth.: By rxn. of Me_3SnCl with $t\text{-}Bu_2PNHLi$ in Et_2O-hexane in N atm., in 87% yield (2739).
Prop.: m. 27-29°, $b_{0.1}$ 60°, IR and NMR spectra, mol. wt., sensitive to moisture, dec. at 200° (2739).
Rxn. with: $Me_3SiN_3 \rightarrow Me_3SnN_3 + Me_3SiNHP(CMe_3)_2$ (2739).
$CCl_4 \rightarrow Me_3SnN:PCl(CMe_3)_2$ (2739).
$S \rightarrow Me_3SnNHPS(CMe_3)_2$ (2739).

N-Triethyltin P,P,P-Triphenylphosphineimide $\quad SnC_{24}H_{30}NP \quad Et_3SnN:PPh_3$
Synth.: By rxn. of Et_3SnN_3 with Ph_3P at 160° in evacuated sealed tube, in 70% yield (1429).
By rxn. of Et_3SnNEt_2 with $Ph_3P:NH$ in N atm. at reflux, in 50% yield (1429).
Prop.: Colorless cryst., m. 49-52°, $b_{0.1}$ 180-82°, NMR spectrum, mol. wt. (1429).
Rxn. with: HCl anhyd. in $Et_2O \rightarrow Et_3SnCl + Ph_3P:NH_2Cl$ (1429).
$MeOH \rightarrow Et_3SnOMe + Ph_3P:NH$ (1429).
$MeI \rightarrow Et_3SnI + Ph_3P:NMe_2I$ (1429).

Listing of organotin phosphorus imides concludes in Table 245.

Table 245. Organotin Phosphorus Imides $R_3SnNRPR_2$ / $R_3SnN:PR_3$

Formula*	Synth. Method**	Yield	Properties	Ref.
Me_3SnR'				
$R' = Me_3P:N$	I	52	b_{11} 88–89°, spectra, mol. wt. (a)	1824
$Et_3P:N$	I	49	$b_{0.5}$ 73–74°, spectra, mol. wt. (a)	1824
$t\text{-}Bu_2PCl:N$	IVB	90	m. 32–34°, $b_{0.05}$ 70°, spectra, mol. wt. (b)	2739
$t\text{-}Bu_2P(S)NH$	IVA	98	m. 123–26°, spectra, mol. wt.	2739
$Ph_3P:N$	I	41	m. 85–86°, $b_{0.5}$ 172–76°, spectra, mol. wt. (a)	1824
$Me_2Sn[N(PPh_2O)_2]_2$	--	--	Moessbauer resonance spectrum	1143a
$Et_3SnN:PBu_3$	IIA	80	Colorless liq., $b_{0.1}$ 101–105°, mol. wt.	1429
$Et_3SnN:P(NMe_2)_3$	IIA	80	Colorless liq., $b_{0.1}$ 81–84°, NMR spectrum	1429
$Et_3SnNMePO_3Me_2$	IIB	--	NMR spectrum, mol. wt., struct.	1429
$Bu_3SnN:PBu_3$	IIA	70	Colorless liq., $b_{0.1}$ 162–65°, mol. wt.	1429
$Bu_3SnN:P(NMe_2)_3$	IIA	70	Colorless liq., $b_{0.1}$ 138–40°, mol. wt.	1429
Ph_3SnR'				
$R' = Bu_3P:N$	IIA	--	IR spectrum, very sensitive to moisture	1402
$(C_8H_{17})_3P:N$	IIA	--	Viscous oil, IR spectrum, mol. wt.	1402
$Ph_3P:N$	IIA	55	m. 132.5–34°, IR spectrum, mol. wt.	1402
$(Me_2N)_3P:N$ (481) (c)	--	--	(c)	--
$PhN:PPh_2NPh$	IIIB	49	Colorless cryst., dec. 170°, mol. wt. (d)	1845
$(Ph_3SnNPh)_2PPh:NPh$	IIIB	46	Slightly yellow cryst., m. 160°, mol. wt. (d)	1845

* Numbers in parenthesis refer to pages in main volume.
** The characters used here correspond to the synthetic scheme in the introduction to Chapter 4.5.5.
(a) (Far) IR, assignment and NMR spectra (1824). (b) IR add NMR spectra, sensitive to moisture (2739). (c) The series of organotin phosphorane and stibane imines listed in the main volume page 481 in Table 165 is chained as UV stabilizers and antioxidants for synthetic resins (2116). (d) (Far) IR spectra and assignment, sensitive to air, slowly dec. on standing in N atm., soly. (1845).

4.5.6 ORGANOTIN PHOSPHIDES, ARSENIDES, ANTIMONIDES AND BISMUTHIDES

4.5.6.1 ORGANOTIN PHOSPHIDES

This chapter contains organotin compounds with a tin-phosphorous bond. New in the field are a number of derivatives with a metal atom linked to phosphorus, i.e., lithium, silicon, germanium and lead. Interesting is the introduction of organometallic substituted phosphines, e.g., $(Ph_3Sn)_3P$, as ligand in metal carbonyl phosphine complexes. The organotin phosphides listed in Table 246 are prepared by methods from the following scheme.

I Reaction with organotin sodium compounds:
   $R_3SnNa + RPCl_2 \rightarrow (R_3Sn)_2PR$.

II Reactions with organotin halides:
   (IIA) $R_3SnX + RR'PLi \rightarrow R_3SnPRR'$,
   (IIB) $Ph_3MX + R''_2PLi \rightarrow Ph_3MPR''_2$; M = Ge, Pb ,
   (IIC) $R_nSnX_{4-n} + Ph_mPH_{3-m} + Et_3N \rightarrow (R_2SnPPh)_3$ and $(RSnP)_4$,
   (IID) $R_3SnCl + (RO)_3P \rightarrow R_3Sn(O)P(OR)_2$ (?).

III Reaction with organotin oxide:
   $(Bu_3Sn)_2O + R_mPH_{3-m} \rightarrow (Bu_3Sn)_{3-m}PR_m$.

IV Reaction with organotin amide:
   $Me_3SnNMe_2 + Ph_2PH \rightarrow Me_3SnPPh_2$.

V Interconversion reactions:
   (VA) $(Me_3Sn)_3P + M(CO)_{n+1}(NO)_m \rightarrow (Me_3Sn)_3PM(CO)_n(NO)_m$; n = 2-5, m = 0, 1; M = Cr, Fe, Co, Ni ,
   (VB) $(R_3Sn)_{3-n}PPh_n + BuLi \rightarrow (R_3Sn)_{2-n}PPh_nLi$; n = 0, 1 ,
   (VC) $Et_3SnPPh_2 + R_3SnH \rightarrow R_3SnPPh_2$,
   (VD) $(Et_3Sn)_2PMe + Et_2S_2 \rightarrow Et_3SnPMeSEt$.

R = organic group. R' = organosilyl or organic group, R'' = organotin or organic group, n = 1, 2; m = 0, 1; X = halogen.

Tristrimethyltin Phosphide (M-482) $Sn_3C_9H_{27}P$ $(Me_3Sn)_3P$
<u>Synth.:</u> By rxn. of Me_3SnNMe_2 with PH_3 (1184).
<u>Prop.:</u> (Far) IR, assignment, Raman, assignment, NMR (1019, 1184) and ^{31}P NMR (1019) spectra, struct. (1184).
<u>Rxn. with:</u> $Cr(CO)_6$ under UV irradiation $\rightarrow (Me_3Sn)_3PCr(CO)_5 + CO$ (1852).
$Fe_2(CO)_9$ under UV irradiation $\rightarrow (Me_3Sn)_3PFe(CO)_4 + Fe(CO)_5$ (2748).
$Fe(CO)_5$ under UV irradiation $\rightarrow (Me_3Sn)_3PFe(CO)_4 + CO$ (2748).
$Co(CO)_3NO \rightarrow (Me_3Sn)_3PCo(CO)_2NO + CO$ (1852).
$Ni(CO)_4 \rightarrow (Me_3Sn)_3PNi(CO)_3 + CO$ (1851).
$Ph_2YCl \rightarrow Me_3SnCl + (Ph_2Y)_3P$; Y = P, As (1846).
$Et_2S_2 \rightarrow Me_3SnSEt + P$ (800).

Bistrimethyltin Phenylphosphide $Sn_2C_{12}H_{23}P$ $(Me_3Sn)_2PPh$
<u>Synth.:</u> By rxn. of Me_3SnCl with $PhPH_2$ and Et_3N in C_6H_6 in dry N atm. under reflux, in 63% yield (1845).

By-prod. in thermal dec. of c.$(Me_2SnPPh)_3$ at 175° (2746).
Prop.: m. 35°, $b_{0.01}$ 132-36°, mol. wt. (1845).
Rxn. with: $Ph_2AsCl \rightarrow Me_3SnCl + (Ph_2As)_2PPh$ (1846).
PhN_3 (1:3) $\rightarrow (Me_3Sn)_2NPh$ (1845).

Tristriphenyltin Phosphide (M-483) $Sn_3C_{54}H_{45}P$ $(Ph_3Sn)_3P$
Synth.: By rxn. of Ph_3SnCl with PH_3 and Et_3N in C_6H_6 in inert atm. (1847).
Prop.: m. 201°, far IR spectrum and assignment (1849).
Rxn. with: BuLi $\rightarrow (Ph_3Sn)_2PLi + BuSnPh_3$ (1850).
S in $C_6H_6 \rightarrow (Ph_3Sn)_2S + P_2S_5$ (1840).
Alc. KOH in air $\rightarrow Ph_3SnOH + H_3PO_4$ (1849).

Bistriphenyltin Phenylphosphide (M-483) $Sn_2C_{42}H_{35}P$ $(Ph_3Sn)_2PPh$
Synth.: By rxn. of Ph_3SnCl with $PhPH_2$ and Et_3N in C_6H_6 in inert atm. (1849).
Prop.: m. 150° (1849), (far) IR, assignment (1019, 1849), Raman, assignment (1019) and ^{31}P NMR (1019) spectra.
Rxn. with: BuLi $\rightarrow Ph_3SnPPhLi + BuSnPh_3$ (1850).
PhN_3 (1:3) $\rightarrow (Ph_3SnNPh)_2PPh:NPh + N$ (1845).
Alc. KOH in air $\rightarrow Ph_3SnOH + PhPO_3H$ (1849).
S in $C_6H_6 \rightarrow (Ph_3Sn)_2S + (PhPS_3)_2$ (1840).

Triphenyltin Diphenylphosphide (M-483) Ph_3SnPPh_2
Synth.: By exchange rxn. of Et_3SnPPh_2 with Ph_3SnH (915).
Prop.: (Far) IR, assignment (1019, 1840), Raman, assignment and ^{31}P NMR (1019) spectra. Anal. detn. by x-ray fluorescence (1452).
Rxn. with: PhN_3 (1:2) $\rightarrow Ph_3SnNPhPPh_2:NPh + N$ (1845).
$PhC\vdots CH \rightarrow Ph_3SnCH:CPhPPh_2$ (1841).
$CH_2:CHR \rightarrow Ph_3SnCH_2CHRPPh_2$; R = Ph, CH_2Cl (1841).
COS $\rightarrow Ph_3SnOC(S)PPh_2$ (483).
CSY $\rightarrow Ph_3SnSC(Y)PPh_2$; Y = S, Cl_2, $(NH_2)_2$ (483).
PhNC:Y $\rightarrow Ph_3SnNPhC(:Y)PPh_2$; Y = S, O (483).
PhCH:N(O)Ph $\rightarrow Ph_3SnO_2PPh_2 + PhCH:NPh$ (1840).
S in $C_6H_6 \rightarrow Ph_3SnS_2PPh_2$ (1840).

Additional organotin phosphides are listed in Table 246.

Table 246. Organotin Phosphides $R_nSn(PR_2)_{4-n}$

Formula*	Synth. Method**	Yield	Properties	Ref.
$(Me_3Sn)_2PMe$	IIA	89	b_3 89-90°, n_D^{20} 1.5778, d_{20} 1.5601	57, 703
	I	15	b_4 95-97°, n_D^{20} 1.5768, d_{20} 1.5599	58, 703
Me_3SnPPh_2 (482)	IV	80	$b_{0.8}$ 150°	250
	VC	--	--	915
c.$(Me_2SnPPh)_3$	IIC	73	m. 134°, NMR spectrum, TLC (a, b)	2746
$(Et_3Sn)_2PMe$	IIA	73	b_1 143-45°, n_D^{20} 1.5621, d_{20} 1.3725	57, 703
	I	10	b_1 135-40°, n_D^{20} 1.5649 (c)	58, 703
$Et_3SnPMeSEt$	VD (c)	20	b_2 86-88°, n_D^{20} 1.5485, d_{20} 1.2498	800
Et_3SnPPh_2 (482)	IV	70	$b_{0.7}$ 170° (d)	250
$(Bu_3Sn)_2PPh$	III	80	$b_{0.03}$ 178-81°, IR spectrum, mol. wt.	1233
Bu_3SnPBu_2	III	75	$b_{0.15}$ 122-24°, n_D^{20} 1.5045, d_{20} 1.04 (e)	1618
Bu_3SnPPh_2 (483)	III	71	$b_{0.01}$ 160-67°, IR spectrum	1233
$Bu_3Sn(O:)P(OMe)_2$ (f)	IID	--	n_D^{29} 1.4834	1564
$Bu_3Sn(O:)P(OEt)_2$ (f)	IID	--	n_D^{30} 1.4820, dec. on distn.	1564
$Bu_3Sn(O:)PPh_2$ (483)	--	--	Wrong structure (g)	1840
c.$(Bu_2SnPPh)_3$	IIC	61	Oil, NMR spectrum, TLC (a)	2746
$Ph_3Sn(O:)P(OMe)_2$ (f)	IID	--	m. 126-27°	1564
$Ph_3Sn(O:)P(OEt)_2$ (f)	IID	--	m. 128.5-29°	1564

* Numbers in parenthesis refer to pages in main volume.
** The characters used here correspond to the synthetic scheme in the introduction to Chapter 4.5.6.1.
(a) (Far) IR spectrum and assigment (2746), rxn. with $HCl-Et_2O \rightarrow R_2SnCl_2 + PhPH_2$; R = Me, Bu, Ph (2746). (b) Thermal decomposition at 175° $\rightarrow (Me_3Sn)_2PPh + (PhP)_5 + Sn$ (2746). (c) Rxn. of $(Et_3Sn)_2PMe$ with Et_2S_2 at -3° (1:1) $\rightarrow Et_3SnSEt + Et_3SnPMeSEt$ (800), in ratio 1:2 $\rightarrow Et_3SnSEt + MeP(SEt)_2$ (800). (d) Rxn. with $R_3SnH \rightarrow Et_3SnH + R_3SnPPh_2$; R = Me, Ph (915), rxn. with $Ph_3SnH \rightarrow Et_3SnPPh_3 + Ph_2PH$ (1618); relative rxn. rates and solvent effect (915). (e) Rxn. with $Ph_3SnH \rightarrow Ph_6Sn_2 + Bu_3SnPPh_3 + Bu_2PH$ (1618), with $Bu_3SnH \rightarrow Bu_6Sn_2$ (1618). (f) No proof for Sn-P bond, ed. (g) Correct structure $Bu_3SnO_2PPh_2$ (1840).

Table 246 Continued

$R_nSn(PR_2)_{4-n}$

Formula*	Synth. Method**	Yield	Properties	Ref.
c.$(Ph_2SnPPh)_3$ (484)	IIC	53	m. 155-60°, NMR spectrum, TLC (a)	2746
$Ph_{12}Sn_8P_3$	--	--	X-ray fluorescence anal.	1452
$(PhSnP)_4$ (484)	IIC	10	Yellow solid, dec. 160°, TLC (h)	2747
$(Me_3Sn)_3PCr(CO)_5$	VA (i)	--	Yellow cryst., dec. 190°, subl. 60-100° x 10^{-5} mm. (j)	1852
$(Me_3Sn)_3PFe(CO)_4$	VA (i)	100	Yellow cryst., dec. 142°, sensitive to air (k)	2748
$(Me_3Sn)_3PCo(CO)_2NO$	VA	--	Ruby red cryst., dec. 190°, subl. 80-100° x 10^{-5} mm. (j)	1852
$(Me_3Sn)_3PNi(CO)_3$	VA	100	Dec. 90°, stable in air (l)	1851
$Me_3SnP(CMe_3):NSiMe_3$	IIA	70	Subl. 0.1 120°, IR, NMR spectra (m)	2739
$Me_3SnP(CMe_3):NGeMe_3$	IIA	52	Subl. 0.1 120°, NMR spectrum (m)	2739
$(Me_3SnPPh_2)_2Ni(CO)_2$	--	--	IR spectrum	2200
$Me_2Sn(PPh_2)_2Ni(CO)_2$	--	--	IR spectrum	2200
$(Ph_3Sn)_2PLi$	VB	--	(n)	1850
$(Ph_3Sn)_2PGePh_3$	IIB (n)	50	m. 160° (o)	1850
$(Ph_3Sn)_2PPbPh_3$	IIB (n)	40	Dec. 171-72° (o)	1850
$Ph_3SnPPhLi$	VB	--	Yellow solid, very sensitive to air (p)	1850
$Ph_3SnPPhGePh_3$	IIB (p)	31	m. 115-19°, (o)	1850
$Ph_3SnPPhPbPh_3$	IIB (p)	25	Dec. 110°, mol. wt. (o)	1850

(h) Decomposes slowly on standing, (far) IR spectra "cubane" structure proposed (2747). (i) Synth. under UV irradiation. (j) (Far) IR, assignment and NMR spectra, stable in air (1852). (k) (Far) IR, assignment and NMR spectra, mol. wt. (2748). (l) (Far) IR, assignment and NMR spectra (1851). (m) Sensitive to moisture, mol. wt. (2739). (n) Rxn. of $(Ph_3Sn)_2PLi$ with $Ph_3MCl \rightarrow (Ph_3Sn)_2PMPh_3$; M = Ge, Pb (1850). (o) (Far) IR spectra and assignment (1850), rxn. with alc. KOH $\rightarrow Ph_3SnOH + Ph_3MOH + H_3PO_4$ and $PhPO_3H_2$, resp., M = Ge, Pb (1850). (p) Rxn. with $Ph_3MCl \rightarrow Ph_3SnPPhMPh_3$; M = Ge, Pb (1850).

4.5.6.4 ORGANOTIN ARSENIDES, ANTIMONIDES AND BISMUTHIDES

The organotin derivatives with a bond between tin and arsenic, antimony and bithmuth, respectively, are prepared by the following methods.

I Reaction with triphenyltin lithium with halides of antimony and arsenic.

II Reaction of organotin halides with:
  (IIA) Ph_nSbLi_{3-n} ; n = 0-3 and
  (IIB) R_2AsNa and R_2SbNa in liquid ammonia and organic solvent.

III Reaction of triorganotin hydride with $Bu_2SbC\!:\!CH$ and Et_3Bi.

IV Reaction of trimethyltin dimethylamide with arsine and stibine.

V Reaction of trimethyltin dimethylarsenide with carbon disulfide.

Tristriethyltin Antimonide $Sn_3SbC_{18}H_{45}$ $(Et_3Sn)_3Sb$
<u>Synth.</u>: By rxn. of Et_3SnH with Et_3Sb in sealed tube at 170°, in 84% yield (2086), with $(Et_3Ge)_3Sb$ at 180°, in 81% yield (2086).
<u>Prop.</u>: b_2 174-76°, d_{20} 1.615 (2086), NMR spectrum (1009), sensitive to air (2086), thermally stable at 220° (2086).
<u>Rxn. with:</u> Thermal dec. at 150° + cat. $AlBr_3$ → Et_4Sn + Sn + Sb (2086).
Br → Et_3SnBr + $SbBr_3$ (2851).
RBr → Et_3SnBr + R_3Sb ; R = $PhCH_2$, C_5H_9 (2086).
Bz_2O_2 → Et_3SnOBz + Sb (2086).

Triphenyltin Diphenylarsenide $SnAsC_{30}H_{25}$ $Ph_3SnAsPh_2$
<u>Synth.</u>: By rxn. of Ph_3SnLi with Ph_2AsCl in THF in N atm. at r.t., in 81% yield (1842).
By rxn. of Ph_3SnCl with Ph_2AsNa in liq. NH_3 at -60° to r.t., in N atm. in C_6H_6, in 42% yield (1842).
<u>Prop.</u>: Colorless cryst., m. 117-19°, mol. wt., sensitive to air, soly. (1842). Anal. detn. of As and Sn by x-ray fluorescence (1842).
<u>Rxn. with:</u> Alc. H_2O_2 → $Ph_3SnO_2AsPh_2$ (1842).

Additional organotin arsenides, antimonides and bismuthides are listed in Table 247. Polymetallic organotin metal complexes containing trialkylarsine and trialkylstibine groups may be found in Chapter 6.6.

Table 247. Organotin Arsenides, Antimonides and Bismuthides

Formula*	Synth. Method**	Yield	Properties	Ref.
$(Me_3Sn)_3As$	IV	--	IR, Raman, NMR spectra, struct. (a)	1184
$(Me_3Sn)_2AsPh$	--	--	(b)	--
$Me_3SnAsMe_2$	IV	80	b_{10} 57°, n_D^{18} 1.5483, d_{20} 1.57, stable in H_2O,	2201
	IIB (c)	46	(far) IR and NMR (646, 2201) spectra	2201
$Me_3SnAsMe_2 \cdot CS_2$	V	--	IR spectrum, stable at r.t. to 200 torr.	2201
$Me_3SnAsMe_2Ni(CO)_3$	--	--	IR spectrum	2200
$Me_3SnAsPh_2$ (484)	IV	77	$b_{0.05}$ 136°, n_D^{20} 1.6438, d_{20} 1.4682 (d)	250
$Me_2Sn(AsMe_2)_2Cr(CO)_4$	--	--	IR spectrum	2200
$Et_3SnAsPh_2$	IIB	61	$b_{0.15}$ 140-43°, sensitive to air (e-g)	68
$Pr_3SnAsPh_2$	IIB	62	$b_{0.2}$ 159-61°, sensitive to air (e)	68
$Bu_3SnAsPh_2$	IIB	60	$b_{0.08}$ 163-64°, sensitive to air, dipole moment (e, f)	68
$(Ph_3Sn)_3As$ (484)	I	7	Cryst., m. 212-16°, stable in air (h)	1842
$(Ph_3Sn)_2AsPh$	I	45	Cryst., m. 112-15°, stable in air (h)	1842
$Ph_2Sn(AsPh_2)_2$	IIB	33	Cryst., m. 80° (dec.), sensitive to air (h, i)	1842
$PhSn(AsPh_2)_3$	IIB	80	Cryst., m. 85° (dec.), sensitive to air (h, j)	1842
$Sn(AsPh_2)_4$	IIB	37	Cryst., m. 68-70°, sensitive to air (h)	1842
$(Me_3Sn)_3Sb$	IV	--	IR, Raman, NMR spectra, struct.	1184
	IIA	80	Colorless cryst., m. 39°, subl., dipole moment	676
$Me_3SnSbBu_2$	III	--	$b_{0.5}$ 126-30°, n_D^{20} 1.5500	1601
$Et_3SnSbPh_2$	IIB	27	$b_{0.2}$ 144-46°, sensitive to air	68
$Pr_3SnSbPh_2$	IIB	35	$b_{0.1}$ 168-70°, sensitive to air	68
$Bu_3SnSbPh_2$	IIB	26	$b_{0.15}$ 179-80°, sensitive to air (k)	68
$(Ph_3Sn)_3Sb$ (484)	I	3	Cryst., m. 215°, stable in air (l)	1843
$(Ph_3Sn)_2SbPh$	I	40	Cryst., m. 120° (dec.), stable in air (l)	1843
$Ph_3SnSbPh_2$	I	64	Cryst., m. 116°, sensitive to air	1843
	IIA	34	(l)	1843
$Ph_2Sn(SbPh_2)_2$	IIA	29	Cryst., m. 150°, sensitive to air (l)	1843

PhSn(SbPh$_2$)$_3$	IIA	35	Cryst., m. 90°, sensitive to air (1)	1843
Sn(SbPh$_2$)$_3$	IIA	69	Cryst., m. 75°, sensitive to air (1)	1843
(Et$_3$Sn)$_3$Bi	III	70	d_{20} 1.743, sensitive to air (m)	2086

* Numbers in parenthesis refer to pages in main volume.

** The characters used here correspond to the synthetic scheme in the introduction to Chapter 4.5.6.4.
(a) Rxn. with Ph$_2$YCl → Me$_3$SnCl + (Ph$_2$Y)$_3$As; Y = P, As (1846). (b) Rxn. with Ph$_2$YCl → Me$_3$SnCl + (Ph$_2$Y)$_2$AsPh; Y = P, As (1846). (c) Synth. in H$_2$O. (d) Rxn. with Ph$_2$PCl → Me$_3$SnCl + Ph$_2$PAsPh$_2$ (1846). (e) Rxn. with O → R$_3$SnO$_2$AsPh$_2$; R = Et, Pr, Bu (68). (f) Rxn. with MeI → Me$_2$Ph$_2$AsI + R$_3$SnI; R = Et, Bu (68). (g) Rxn. with Ph$_3$SnH → Et$_3$SnSnPh$_3$ + Ph$_2$AsH + some exchange; solvent effect, relative rxn. rates (915). (h) Mol. wt., soly., anal. detn. of Sn, As by x-ray fluorescence (1842). (i) Rxn. with alc. H$_2$O$_2$ → Ph$_2$Sn(O$_2$AsPh$_2$)$_2$ (1842). (j) Rxn. with alc. H$_2$O$_2$ → PhSn(O$_2$AsPh$_2$)$_3$ (1842). (k) Rxn. with MeI → Bu$_3$SnI (68). (l) Mol. wt., soly., anal. detn. of Sn and Sb by x-ray fluorescence (1843). (m) NMR spectrum (1009), thermal decomposition at 160-70° → Et$_6$Sn$_2$ + Bi (2086), with Bz$_2$O$_2$ → Et$_3$SnOBz + Bi (2086).

5 ORGANO-POLYTIN COMPOUNDS

5.1 HEXAORGANODITIN COMPOUNDS

The symmetric and unsymmetric hexaorganoditin derivatives in Tables 248 and 249 are prepared by the following methods.

- I Condensation reactions of triorganotin hydrides with:
  - (IA) triorganotin oxides and hydroxides,
  - (IB) triorganotin alkoxides,
  - (IC) triorganotin amides,
  - (ID) triorganotin halides in presence of triethylamine.
- II Reaction of triorganotin halides with:
  - (IIA) lithium,
  - (IIB) sodium metal.
- III Alkylation reactions:
  - (IIIA) Grignard synthesis with tin tetrachloride, electrolytic alkylation of tin with:
  - (IIIB) ICH_2CH_2CN,
  - (IIIC) $CH_2{:}CHCN$.
- IV Interconversion reaction of functionally substituted hexaorganoditin compound by typical organic methods:
  - (IVA) saponification of a nitrile with strong base,
  - (IVB) reduction of a nitrile with $LiAlH_4$,
  - (IVC) esterification with diazomethane.

5.1.1 SYMMETRIC HEXAORGANODITIN COMPOUNDS

This chapter reports compounds of the formula R_6Sn_2. A few functionally substituted derivatives are listed at the bottom of Table 248.

Hexamethylditin (M-485)　　　　　　　　$Sn_2C_6H_{18}$　　　　　　　　Me_6Sn_2
Synth.: By x-ray irradiation of Me_4Sn (1191).
By rxn. of Me_3SnH with Me_3SnNEt_2 in inert atm., in 95% yield (1618).
By rxn. of Me_3SnCl with Na in Me_4Sn under reflux, in 93-99% yield (1548a), with Cu/Zn in THF in presence of $(Me_2NCH_2)_2$ (2830).
By rxn. of Me_3SnLi with Me_3MCl in THF in N atm. below 0°; M = Si, Ge (1844).
By-prod. in thermal dec. of $[Me_3SnC(CF_3){:}]_2$ at 150° (108), of $Me_3SnSnEt_3$ at 190° (399).
Prop.: m. 22°, b_{12} 62-63° (1618), (far) IR, assignment (1844), UV (1905), NMR (1220, 2267, 2582); Moessbauer resonance (2665) and mass (621, 1191, 2528, 2869) spectra, sensitive to air (1618), electrochemistry and standard potential (983, 2010), thermodynamic data (305, 2856). Anal. detn. by GLC (1191), by coulometric titration (2007), by I titration (1618).
Rxn. with: X-ray irradiation → investigation of radiolysis by ESR spectroscopy (2443).
Me_3SnCl → Me_4Sn + $(Me_2Sn)_x$; kinetic data (2012).

$Et_3SnCl \rightarrow Et_3SnMe + Me_4Sn + (Me_2Sn)_x$ (2012).

$Me_2SnCl_2 \rightarrow Me_3SnCl + (Me_2Sn)_x$ (2012).

Li in THF → $Me_3SnLi \cdot THF$ (2127).

Alc. $AgNO_3 \rightarrow Me_3SnNO_3 + Ag$, rxn. mechanism (2011).

$PhHgCClXBr \rightarrow (Me_3Sn)_2CClX$; X = Cl, Br (1878).

$Ru_3(CO)_{12} \rightarrow (Me_3Sn)_2Ru(CO)_4$ (905).

$PhCBrCPh:CPhCPhBr + Hg \rightarrow (PhC:CPh)_4$ (2728).

I → mechanism of cleavage rxn. (2009), cat. effect of Me_2SnO (44).

$C_2H_4* \rightarrow Me_4Sn$ (871).

$CH_2:CF_2* \rightarrow Me_4Sn + Me_3SnF + C_2H_3F + C_4H_5F_3$ (871).

$C_2HF_3* \rightarrow Me_4Sn + Me_3SnCF_2CHFCF_2CHFSnMe_3$ (871).

$C_2F_4 \rightarrow Me_4Sn + Me_3SnCF_2CF_2H + Me_3Sn(CF_2)_4H + Me_3SnF + 1,3-C_4F_6 + (C_2F_4)_x$ (871).

$C_2ClF_3* \rightarrow Me_4Sn + Me_3SnCl + Me_3SnF$ (871).

$C_2F_3Br* \rightarrow Me_4Sn + Me_3SnF + Me_3SnCF:CF_2$ (871).

$CF_3CF:CF_2* \rightarrow Me_3SnCF_2CF(CF_3)SnMe_3 + Me_3SnCF_2CHFCF_3$ (87) $+ Me_4Sn + Me_3SnF + (C_3F_6)_2 + C_3F_7H$ (871).

$(CF_3C!)_2$ at 100° → Me_3SnF (108).

$(CF_3C!)_2* \rightarrow [Me_3SnC(CF_3):]_2$ (108).

$CF_3COI \rightarrow Me_3SnCF_3 + Me_3SnI + Me_4Sn + CO$ (922).

$(CF_3CO)_2O \rightarrow Me_3SnO_2CCF_3 + CO + Me_4Sn$ (922).

Liq. $SO_2 \rightarrow Me_3Sn \cdot SO_2$ (827).

Biol. Prop.: Activity as insecticide against Agrotis ipsilon, Heliothis zea (1159), mosquito larvae (2786), Peridroma saucia, Prodenia ornithogalli, Pseudaletia unipuncta, Spodoptera frugiperda (1159).

Use: Trade name or synonym Pennsalt TD-5032.

As gasoline additive (2831).

Hexaethylditin (M-486) $Sn_2C_{12}H_{30}$ Et_6Sn_2

Synth.: By rxn. of Et_3SnH with $(Et_3Sn)_2O$ at 130-40°, in 86-88% yield (400, 1618), with Et_2SnO at 100° in C_6H_6, in 25% yield (1955), with Et_3SnOPh (914).

By rxn. of Et_3SnH with Et_3SnNEt_2 (915) in inert atm., in 88% yield (1618), with $Et_3SnN(C_6H_{13})CHO$ (915), with $Et_2Sn(NEt_2)_2$ and $Sn(NEt_2)_4$, in 42 and 77% yield, resp. (1955), with Et_3SnN_3 in Et_2O, in 77% yield (1429).

By rxn. of Et_3SnH with $(Et_3Ge)_nBiEt_{3-n}$ (291), with $(Et_3Ge)_3M$ at 170°; M = Sb and Bi, in 84 and 43% yield, resp. (2086), with $(Et_3Ge)_3Tl$, in 64% yield (1361), with $(Et_3Ge)_2Cd$ at r.t., in 59% yield (587), with Et_4Pb at 100-120° in inert atm., in 67-79% yield (2850), with Et_2Zn (584), with Et_2Te at 35-40° in inert atm., in 8% yield (2092).

By rxn. of Et_3SnH with $p-ONC_6H_4NMe_2$ (390), with aldehydes and ketones (389), with epoxides (807), with azo compounds (1620, 2646), with Ph_3C-radical, followed by AcOH, in 64% yield (385), with butadiene and cat. $i-Bu_2AlH$ at 85° (1612).

By polarographic reduction of Et_3SnF and Et_3SnCl (972).

---

* Reaction under UV irradiation.

By rxn. of Et$_3$SnCl with (Et$_3$Ge)$_2$Cd at r.t. in 42% yield (587), with Na and BuCl at 160°, in 2% yield (1939).

By rxn. of Et$_3$SnBr with Et$_3$SiLi, from Et$_3$SiHgEt and Li in THF (2852).

By rxn. of Et$_3$SnOMe with Et$_3$GeH or R$_3$SiH at 180° in sealed tube; R = Et, Pr, Ph, in 39-53% yield (2097).

By rxn. of (Et$_3$Sn)$_2$S with pptd. Cu in N atm., in quant. yield (1739).

By rxn. of Et$_3$SnLi with Me$_2$SiCl$_2$ at -20° in Et$_2$O, in 25-30% yield (11a), with EtBr in THF, in 25% yield (1928), with ClC!CPO$_3$Et$_2$ in THF (2179).

By rxn. of Et$_3$SnNa with Et$_3$SnC!CCl in liq. NH$_3$, in 86% yield (2180).

By thermal dec. of Me$_3$SnSnEt$_3$ at 190° (399), of Et$_3$SnN:C:CMe$_2$ (1620), of (Et$_2$Sn)$_2$ at 160°, in 5% yield (1941).

Prop.: b$_4$ 143-45° (400, 1955), n$_D^{20}$ 1.5380 (1618, 1955), IR (588), far IR and assignment (321), NMR, Moessbauer resonance (1309) and mass (850) spectra, electrochemistry and standard potential (983), sensitive to air (1618), thermodynamic data (2025). Anal. detn. by coulometric titration (2007), by iodometric method (1618, 1955).

Rxn. with: Et$_3$SnO$_2$CMe$_3$ at 80-160° → Et$_2$SnO + (Et$_3$Sn)$_2$O + Et$_3$SnOH + Et$_3$SnOCMe$_3$ + Me$_3$CO*-degradation prod., kinetic data and rxn. mechanism (665).

Me$_3$SiCH$_2$CH$_2$CO$_3$CMe$_3$ → Et$_3$SnOCMe$_3$ + Et$_3$SnO$_2$CCH$_2$CH$_2$SiMe$_3$ + Me$_3$COH + Me$_2$CO + CH$_4$ (586).

AgNO$_3$ in alc. → Et$_3$SnNO$_3$ + Ag; kinetic data and rxn. mechanism (2011).

Na in C$_{10}$H$_8$ → Et$_3$SnNa (2521).

CuCl$_2$ → Et$_3$SnCl + CuCl (1737).

Cu + CCl$_4$ → Et$_3$SnCl + HCCl$_3$ + C$_2$Cl$_6$ + CuCl (1737).

I in ROH → Et$_3$SnI; kinetic data and rxn. mechanism (2009).

Air → (Et$_3$Sn)$_2$O (1737).

O → kinetic data and rxn. mechanism, effect of inhibitors (2214).

O under UV irradiation → Et$_2$SnO + Et$_3$SnOEt + (Et$_3$Sn)$_2$O (661).

Y → (Et$_3$Sn)$_2$Y; Y = S, Se (2091).

AcOH → Et$_3$SnOAc + H + C$_2$H$_6$ + Et$_2$SnO (2096).

CF$_3$COCl under UV irradiation → Et$_3$SnCl + Et$_2$SnCl$_2$ (922).

Bz$_2$O$_2$ → Et$_3$SnOBz; heat of rxn. (2025).

BzO$_2$CMe$_3$ → Et$_3$SnOBz + Et$_3$SnOCMe$_3$ (2098).

Use: Effect as stabilizer for PVC (439).

Hexabutylditin (M-488)　　　　　　　　Sn$_2$C$_{24}$H$_{54}$　　　　　　　　Bu$_6$Sn$_2$

Synth.: By rxn. of Bu$_3$SnH with Bu$_3$SnCHRCHMe$_2$ under UV irradiation in Ar atm.; R = CN, CO$_2$Et (2680).

By rxn. of Bu$_3$SnH with (Bu$_3$Sn)$_2$O at 110°, in 90% yield (400), at 100-150° (468, 1618, 1802), with Bu$_3$SnOCMe$_3$ at 110°, in 98% yield (1618), with Bu$_3$SnO-SiR$_3$; R = Bu and Ph, in 15 and 23% yield, resp. (1165), with Bu$_3$SnOCMe:CHCHMe$_2$, in 8% yield (1684).

By rxn. of Bu$_3$SnH with Bu$_3$SnNEt$_2$ in inert atm., in 91% yield (1618), with Bu$_2$Sn(NEt$_2$)$_2$ ratio 2:1 (1955), with Bu$_3$SnPBu$_2$ at 110°, in 98% yield (1618).

By rxn. of Bu$_3$SnH with MeOH under reflux (2680), with Me$_2$C:CHAc under UV irradiation (1684), with MeCR:CHCN at 150° in Ar atm.; R = H, Me, in up to 35% yield (2680), with CH$_2$:CHCN and MeOH under reflux (2680), with MeCH:CH-CO$_2$Et at 150° in Ar atm. (2680), with (Me$_2$RCO)$_2$; R = Me, Ph (1616).

By rxn. of Bu$_3$SnCl with Na in Bu$_4$Sn at 100-110°, in 93-95% yield (1548a), with Na in C$_6$H$_6$ or Et$_3$N in 86-92% yield (1939), with Zn and Et$_3$N at 90 and 160°, in 32 and 6% yield, resp. (1939).

By rxn. of (Bu$_3$Sn)$_2$O with Me$_3$SiH (1239), with Bu$_3$SiH, in 82% yield (1165), with Ph$_3$SiH in N atm., in 3-28% yield (1165).

By rxn. of (Bu$_3$Sn)$_2$S with pptd. Cu in inert atm. at r.t., in quant. yield (1737), with i-Bu$_2$SnH$_2$ as by-prod. (1955).

By thermal dec. of (Bu$_2$Sn)$_x$ in sealed tube at 220°, in 48% yield (2096).

Prop.: b$_{0.4}$ 166-67° (468, 1802), b$_{0.2}$ 147-50° (1618), b$_{0.09}$ 149° (400, 1618), n$_D^{20}$ 1.5132 (1618), 1.5010 (400), NMR (1220, 1309) and Moessbauer resonance (1309) spectra, electrochemistry and standard potential (983), mol. wt. (468, 1802), sensitive to air (1618). Anal. detn. by coulometric titration (2002), by titration with Br (468), I (1618).

Rxn. with: AgNO$_3$ in alc. → Bu$_3$SnNO$_3$ + Ag; kinetic data and rxn. mechanism (2011).

Na in C$_{10}$H$_8$ → Bu$_3$SnNa (2521).

Cu + CCl$_4$ → Bu$_3$SnCl + HCCl$_3$ + C$_4$H$_{10}$ + CuCl (1737).

Fe(CO)$_5$ → [Bu$_2$SnFe(CO)$_4$]$_2$ + Sn[Fe(CO)$_4$]$_4$ (904).

I → Et$_3$SnI, cat. effect of Bu$_2$SnO (44), kinetic data and rxn. mechanism (2099).

(CF$_3$C!)$_2$ under UV irradiation → Bu$_2$Sn[C(CF$_3$):CHCF$_3$]$_2$ + Bu$_3$SnF (108).

Air → (Bu$_3$Sn)$_2$O (1737).

(Me$_3$CO)$_2$ → Bu$_3$SnOCMe$_3$ + C$_4$H$_{10}$ + C$_4$H$_8$ + CH$_4$ + Me$_3$COH + polystannane (1616).

Biol. Prop.: Activity as bactericide against Staphylococcus aureus (1548a).

Use: As additive for gasoline (2831), for flame resistant polyester (2702). In antifouling coatings (2302).

Hexaisobutylditin   Sn$_2$C$_{24}$H$_{54}$ (Me$_2$CHCH$_2$)$_6$Sn$_2$

Synth.: By rxn. of i-Bu$_3$SnH under exclusion of air with Ph$_3$SnCl and Et$_3$N (1:1:1), in 53% yield (1618), with (i-Bu$_3$Sn)$_2$O at 160°, in 83% yield (1618), with i-Bu$_3$SnNEt$_2$, in 87% yield (525, 1618).

Prop.: m. 56°, b$_{0.15}$ 130° (525, 1618), stable in air in solid state, air sensitive in soln. (1618). Anal. detn. by titration with I (1618).

Rxn. with: I → i-Bu$_3$SnI (1618).

Biol. Prop.: Activity as herbicide against amaranthus spec. (2785).

Hexaphenylditin (M-489)   Sn$_2$C$_{36}$H$_{30}$ Ph$_6$Sn$_2$

Synth.: By rxn. of Ph$_3$SnH with (Ph$_3$Sn)$_2$O at 20°, in 89% yield (400), with Ph$_2$SnO in C$_6$H$_6$ at 100°, in 30% yield (1955), with Ph$_3$SnOR under exclusion of air; R = H, Me, at 80 and 21°, in 87-89% yield (1618), with Bu$_3$SnOAc at r.t. in inert atm., in 35% yield (1618), with (Ph$_3$Sn)$_2$S and Ph$_3$SnPBu$_2$ under exclusion of air (1618), with Bu$_2$SnBr$_2$ and NEt$_3$ (1:2:2) in Et$_2$O, in 23% yield (1955), with Ph$_3$GeSnEt$_2$NEt$_2$ in N atm. (913).

By rxn. of Ph$_3$SnH with RCN at 150°; R = Pr, C$_5$H$_{11}$, in quant. yield (2680), with MeCR:CHCN or with MeCR:CHCO$_2$Et at 150° in Ar atm.; R = H, Me (2680), with PhCH:C(CN)$_2$ in Ar atm. at 60°, in 96% yield (2789), with Me$_2$C:CHAc (421, 1684), with (Me$_2$C:C)$_2$CO or i-BuCOC:CMe$_2$ (1684), with BzH (389), with RCH:CHAc; R = H, Me (1684).

As by-prod., thermal dec. of Ph$_3$SnH at 80-120° in octane, C$_6$H$_6$ and EtOH (123).

By polarographic reduction of Ph$_3$SnF (2842), Ph$_3$SnCl (963, 967).

By rxn. of Ph$_3$SnCl with i-Bu$_3$SnH and Et$_3$N (1:1:1) in Et$_2$O at r.t. in inert atm., in 53% yield (1618), with Li in THF in N atm. (1928), with EtMgBr in Et$_2$O-C$_6$H$_6$ (1500), with Ph$_3$GeK in THF, in 3% yield (2128), with Ph$_3$SnMgBr·NEt$_3$, in 40% yield (2330), with (Me$_3$Sn)$_3$SnLi in THF at 0° (2127), with Ph$_3$SnZnCl in THF in presence of (Me$_2$NCH$_2$)$_2$ (2830), with Zn/Cu in THF in presence of (Me$_2$NCH$_2$)$_2$, in 60-80% yield (2830).

By rxn. of (Ph$_3$Sn)$_2$O with i-Bu$_2$SnH$_2$ in C$_6$H$_6$ at 85°, in 77% yield (1955), with Ph$_3$SiH or (MeHSiO)$_x$ (1165), with Ph$_2$P(O)H in 99% yield (1233).

By rxn. of Ph$_3$SnOMe with Ph$_2$SnH$_2$ at 20°, in 51% yield (1955).

By rxn. of (Ph$_2$Sn)$_x$ with Bu$_2$SnI$_2$ in C$_6$H$_6$ at 140° in sealed bottle, in 7% yield (1938).

By rxn. of (Ph$_3$Sn)$_4$Si with MeLi in THF-Et$_2$O, followed by PhCH$_2$Cl, in 50% yield (1928).

By rxn. of Ph$_3$SnLi with RCl in THF in N atm. (1928), with Br(CH$_2$)$_n$Br in THF in N atm.; n = 2 and 3, in 56 and 45% yield, resp. (1928), with MeCH:CHX in THF; X = Br and Cl, in 37 and 59% yield (1928), with aq. NH$_4$Br in THF, in 35-44% yield (1928), with SiCl$_4$ in 42-65% yield (1928).

By rxn. of Ph$_3$SnLi with (C$_6$F$_5$)$_3$GeCl (1034), with c.(Ph$_2$Si)$_2$ (1928), with SnCl$_2$ and Ph$_3$SnCl in THF, in 18% yield (68a, 175), with SnCl$_2$ and Li in THF, in 41% yield (68a, 175), with (Ph$_3$Sn)$_4$Sn in THF, in 74% yield (68a, 175), with R$_2$SiCl$_2$; R = Me, Ph, in 15-20% yield (11a).

By rxn. of Ph$_3$SnLi with R$_3$PO$_4$; R = Me, Bu, in 13% yield (1928).

By rxn. of (Ph$_3$Sn)$_3$SnLi with Me$_3$SiCl in Et$_2$O at 24°, in 54% yield (11a), with Me$_3$PO$_4$ (68a, 175).

By rxn. of Ph$_3$SnK with Ph$_3$GeBr in THF, in 40-75% yield (2128).

By thermal dec. of (Ph$_3$Sn)$_2$Hg at 100° in THF at 15° as by-prod. (2370).

By photolysis of Ph$_3$SnCo(CO)$_4$ (730).

By rxn. of PhMgBr with SnCl$_2$ in Et$_2$O, in 22% yield (68a, 175).

Prop.: Far IR, assignment (1065, 1703), UV (203, 392, 967), Moessbauer resonance (222, 224, 1094, 1990-1) and mass (849-50) spectra, no evidence for Ph$_3$Sn-radical formation by ESR spectra (399, 1154), by flash photolysis, I titration and other methods (1154), electrochemistry and standard potential (983), polarography (963, 965, 967-9), soly. (2046), thermodynamic data (2025). Anal. detn. by TLC (2276), by coulometric titration (2007), by iodometry (1165, 1618), with cacotheline after fusion with Na$_2$O$_2$ (155).

Rxn. with: ^{60}Co irradiation → nature of radicals by ESR spectroscopy (693). Polarographic reduction, followed by rxn. with numerous inorganic, organic and organometallic entities (963, 966-8).

Na in C$_{10}$H$_8$ → Ph$_3$SnNa (2521).

SnCl$_4$, + NaOH → Ph$_3$SnOH (2046).

MCl$_n$ → Ph$_3$SnCl; M = Cu, Hg, Fe (2046).
AgNO$_3$ in alc. → Ph$_3$SnNO$_3$ + Ag, kinetic data and rxn. mechanism (2011).
HCl concn. → Ph$_3$SnCl + H (2046).
Ph$_3$CCl → Ph$_3$SnCl + Ph$_2$CO (in presence of air) (2046).
MeONa and RBr → (Ph$_3$Sn)$_2$O + Ph$_3$SnR; R = Et, Bu (2046).
I → kinetic data of cleavage rxn. (1154).
I in ROH → Ph$_3$SnI, kinetic data and rxn. mechanism (2009).
MeI at 140° → Ph$_3$SnI (1938).
Air and MeONa in THF → (Ph$_3$Sn)$_2$O (2046).
H$_2$O$_2$-NaOH in aq. Me$_2$CO → (Ph$_3$Sn)$_2$O (2046).
Se(SeCN)$_2$ → Ph$_3$SnSeCN + Se + PhSeCN (697).
SO$_2$ liq. at r.t. → Ph$_3$Sn·SO$_2$ + Ph$_3$Sn·2SO$_2$ (827).

Addn. Compd.: 7Ph$_6$Sn$_2$·4Ph$_4$Sn, from Ph$_3$SnCl with 2-C$_5$H$_4$NMgBr in Et$_2$O-C$_6$H$_6$ (1500), from components (2:1) in MeNO$_2$, in 81% yield (1500), m. 196-97°, IR, NMR and Moessbauer resonance spectr. (1500), rxn. with I → Ph$_4$Sn + Ph$_3$SnI (1500).

7Ph$_6$Sn$_2$·Ph$_4$Si, from components, in 81% yield, m. 195-97°, Moessbauer resonance spectrum (1500).

7Ph$_6$Sn$_2$·4Ph$_4$Ge, from components, in 76% yield, m. 194-95° (1500).

Biol. Prop.: Activity as insecticide and effect on reproduction of Musca domestica (260).

Use: As antifogging additive for photographic plates (591a).
As antioxidant for polyphenyl ether (1979).
As stabilizer for phosphinate hydraulic fluids (1497, 1861).

Other symmetric hexaorganoditin derivatives are listed in Table 248. Additional references may be found in Chapter 8.1.

Table 248. Symmetric Hexaorganoditin Compounds R_6Sn_2

Formula*	Synth. Method**	Yield	Properties	Ref.
R_6Sn_2				
R = PhCH$_2$	--	--	UV spectrum	2031
Pr (488)	--	--	b$_{15}$ 143° (a-c)	2007
Me$_2$CH	--	--	(c)	--
CH$_2$CH$_2$CH	IIB	--	n_D^{25} 1.5627, use as herbicide, pesticide	1873
MeEtCH	IIIA	--	m. 73-76°, b$_{0.4}$ 160-65° (d)	1618
CH$_2$CHMeCH	IIB	--	Use as herbicide, pesticide	1873
PhCMe$_2$CH$_2$	IIB	57	m. 153.5-54°, NMR spectrum	2188
Me$_3$CCH$_2$ (488)	--	--	NMR spectrum	2188
Me$_3$CCH$_2$CH$_2$	IIA	--	Plates, m. 93-94.5°, mol. wt. (e)	2887
c.C$_6$H$_{11}$ (488)	IC	78	Dec. 300°, mol. wt., dec. in soln. (d, f)	1618
C$_8$H$_{17}$	IIB	96	--	1548a
p-MeC$_6$H$_4$ (488)	IIB	--	m. 253-54°, UV spectrum (203) (b)	2009
R'_6Sn_2				
R' = NCCH$_2$CH$_2$	IIIB	12	m. 111-12°	566
	IIIC	25	(g, h)	773
	IIIC	94	m. 111.5°, IR and NMR spectra	1947
HO$_2$CCH$_2$CH$_2$	VA (g)	--	Crude, m. 284-93°, IR and NMR spectra (i)	1947
MeO$_2$CCH$_2$CH$_2$	VC (i)	--	Light yellow liq., IR and NMR spectra	1947
OCH$_2$CHCH$_2$O$_2$CCH$_2$CH$_2$	--	--	(j)	--
H$_2$NCH$_2$CH$_2$CH$_2$	VB (h)	--	Clear viscous oil, IR spectrum	1947
p-FC$_6$H$_4$	--	--	Moessbauer resonance spectrum	224
p-ClC$_6$H$_4$ (488)	--	--	Moessbauer resonance spectrum	222, 224
m-CF$_3$C$_6$H$_4$ (488)	--	--	Moessbauer resonance spectrum	222, 224

* Numbers in parenthesis refer to pages in main volume.
** The characters used here correspond to the synthetic scheme in the introduction to Chapter 5.1.

5.1.2 UNSYMMETRIC HEXAORGANODITIN COMPOUNDS

1,1,1-Trimethyltriphenylditin (M-488) Me₃SnSnPh₃
Sn₂C₂₁H₂₄

<u>Synth.</u>: By rxn. of Ph₃SnH with Me₃SnNPhCHO (97), with Me₃SnNEt₂ in inert atm., in 92% yield (1618), with Me₃SnH and Et₃N (1:1:1) in inert atm. in Et₂O, in 86% yield (1618).

<u>Prop.</u>: White solid (97), colorless cryst., m. 108° (1618), 106-108.5° (97), 104° (2009), far IR (828), NMR (2267) and mass (849) spectra, soly., stable in solid form, dec. in soln. (1618). Anal. detn. by iodometric titration (1618).

<u>Rxn. with</u>: AgNO₃ in EtOH → Me₃SnNO₃ + Ph₃SnNO₃ + Ag, kinetic data and rxn. mechanism (2011).

I in ROH → Me₃SnI + Ph₃SnI; kinetic data and rxn. mechanism (2009).

<u>Use</u>: As antifogging additive for photographic plates (591a).

1,1,1-Triethyltriphenylditin (M-488) Et₃SnSnPh₃
Sn₂C₂₄H₃₀

<u>Synth.</u>: By rxn. of Ph₃SnH with Et₃SnNRCHO (96, 97, 915, 917), with Et₃SnNEt₂ or Et₃SnNPhNHPh in various solvents (96, 915), with Et₃SnCN and Et₃N (1:1:1) in inert atm., in 51% yield (1618).

By rxn. of Ph₃SnH with Et₃SnCl and Et₃N (400), with (Et₃Sn)₂O at 25°, in 80% yield (400), in 90% yield (1618), with Et₃SnOR; R = Me₃C, p-XC₆H₄; X = H, Cl, MeO, O₂N (914).

By rxn. of Et₃SnH with (Ph₃Sn)₂O at 25°, in 73% yield (400).

<u>Prop.</u>: m. 16.5° (1618), 16° (400), b₀.₀₈ 172-74° (1618), b₀.₀₀₀₄ 186° (400), dec. on distn. (97), n_D^20 1.6327 (97), 1.6247-312 (1618), far IR spectrum (828), sensitive to air (97). Anal. detn. by coulometric titration (2007).

<u>Rxn. with</u>: Alc. AgNO₃ → Et₃SnNO₃ + Ph₃SnNO₃ + Ag; kinetic data and rxn. mechanism (2011).

I in ROH → Et₃SnI + Ph₃SnI; kinetic data and rxn. mechanism (2009).

Listing of unsymmetric hexaorganoditin compounds continues in Table 249 and on page 835.

(a) Detn. by coulometric titration (2007). (b) Rxn. with I in ROH → kinetic data and mechanism of cleavage rxn. (2009). (c) Rxn. with alc. AgNO₃ → Pr₃SnNO₃ + Ag (2011). (d) Rxn. with I → R₃SnI; R = s-Bu, c.C₆H₁₁, rxn. suitable for quantitative detn. (1618). (e) Rxn. with I in CCl₄, + KF → (Me₃CCH₂CH₂)₃SnF (2887). (f) Anal. detn. with cacotheline after Na₂O₂-fusion (155). (g) Rxn. of (NCCH₂CH₂)₆Sn₂ with aq. alc. NaOH, followed by HCl → (HO₂CCH₂CH₂)₆Sn₂ (1947). (h) Rxn. of (NCCH₂CH₂)₆Sn₂ with LiAlH₄ → (H₂NCH₂CH₂CH₂)₆Sn₂ (1947). (i) Rxn. of (HO₂CCH₂CH₂)₆Sn₂ with CH₂N₂ → (MeO₂CCH₂CH₂)₆Sn₂ (1947). (j) Rxn. with acid anhydride → tin contg. polymer (1948).

Table 249. Unsymmetric Hexaorganoditin Compounds $R_nR'_{6-n}Sn_2$

Formula*	Synth. Method**	Yield	Properties	Ref.
$Me_3SnSnEt_3$ (488)	IC, Me_3SnH	80	Stable at 11 mm. at 107° (399)	1618
	IC, Et_3SnH	69	b_{12} 106-108°, n_D^{20} 1.5353 (a, b)	1618
$Me_3SnSnBu_3$	IA, Bu_3SnOH	100	Colorless liq., mol. wt.	1802
$Me_3SnSn(c.C_6H_{11})_3$	IC, Me_3SnNEt_2	67	m. 72°, stable as solid (c)	1618
$(Me_2SnC_{10}H_{15}-1)_2$ (d)	IIC	--	m. 154-56°, IR spectrum	1748
$Et_3SnSnBu_3$	IC, Et_3SnH	78	$b_{0.2}$ 108-110°, n_D^{20} 1.5237 (b, c)	525,1618
$Et_3SnSn(CH_2CHMe_2)_3$	IA, $1-Bu_3SnH$	42	$b_{0.1}$ 75°	400,1618
	IA, Et_3SnH	47	n_D^{20} 1.5250-71 (b)	1618
	IC, $1-Bu_3SnH$	93	$b_{0.3}$ 117°	525,1618
$(Bu_2SnPh)_2$	--	--	Use as stabilizer for polyolefins	1865
$1-Bu_3SnSn(c.C_6H_{11})_3$	IC, $1-Bu_3SnNEt_2$	72	m. 172°, stable as solid (b, c)	1618
$R_3SnSnPh_3$				
R = Me_2CHCH_2	IC, Ph_3SnH	87	m. 79°	525,1618
	ID, Ph_3SnH	68	(b, c)	1618
	ID, Ph_3SnCl	26	Colorless cryst., m. 84°	1618
c.C_6H_{11}	IC, Ph_3SnH	85	Plates, m. 202-204°, stable as solid (b, c)	1618
C_8H_{17}	IC, Ph_3SnH	--	n_D^{20} 1.5599, sensitive to air	97
$(Me_3SnEt_2CH_2CH_2)_2CH_2$	IC, Me_3SnNEt_2	--	$b_{0.0001}$ 164-68°, n_D^{25} 1.5645 (b)	1618
$Et_3SnSn(C_6H_4OMe-p)_3$	IB, Et_3SnOPh	--	Rate of synth. by IR spectroscopy	914
	IC, $Et_3SnNRCHO$	--	--	915
$(c.C_6H_{11})_2RSnSnR_3$ (f)	--	--	Use as stabilizer for polyolefins	1865

* Numbers in parenthesis refer to pages in main volume.
** The characters used here correspond to the synthetic scheme in the introduction to Chapter 5.1.
(a) Thermal dec. at 190° → $Me_4Sn + Et_4Sn + Sn + Me_6Sn_2 + Et_6Sn_2$ (399), sensitive to air (1618). (b) Anal. detn. by Iodometric titration (1618). (c) Decomposition in soln. in air, soly. (1618). (d) Deriv. of adamantane. (e) Far IR spectrum (828). (f) R = $Me_2NC_6H_4$.

1,1,1-Tributyltriphenylditin $SnC_{30}H_{42}$ $Bu_3SnSnPh_3$

Synth.: By rxn. of Ph_3SnH with Bu_3SnNEt_2, in 83% yield (525), with Bu_3Sn-$NPhCHO$ (97), with Bu_3SnPBu_2 in inert atm., in 88% crude yield (1618).
By rxn. of Ph_3SnH in inert atm. with Bu_3SnOMe, in 85% yield (1618), with Bu_3SnOAc and $(Bu_3Sn)_2S$, in 24 and 94% yield, resp. (1618).

Prop.: m. 23°, $b_{0.001}$ 179° (525), n_D^{20} 1.5988, 1.6023 (1618). IR spectrum (1618). Anal. detn. by iodometric titration (1618).

5.2 ORGANO-POLYTIN COMPOUNDS CONTAINING THREE OR MORE TIN ATOMS

The organo-polytin derivatives in this chapter may be divided in two groups with the compositions $R_{2n+2}Sn$ and $(R_2Sn)_x$. In structure the former type is similar to the paraffin series in carbon chemistry; the latter conforms to cycloalkanes. Oligomeric or polymeric derivatives reported as $(R_2Sn)_x$ must carry end groups and belong either in the paraffin-analog series or to the organo-polytin hetero compounds of Chapter 5.4. The solid polymeric $(RSn)_x$ derivatives are of unknown structure.

5.2.1 ORGANO-POLYTIN COMPOUNDS WITH $R_{2n+2}Sn_n$ STRUCTURE

The compounds listed in Tables 250 and 251 are prepared by methods from the following scheme.

I   Reaction of organotin oxides with:
   (IA) R_3SnH for $(R'_2SnO)_x$,
   (IB) R_2SnH_2 for $(R'_3Sn)_2O$.

II Reaction of organotin amides with triorganotin hydrides:
   (IIA) for $R'_2Sn(NR_2)_2$,
   (IIB) for $R'_2Sn(NPhCHO)_2$, with diorganotin dihydride,
   (IIC) for R'_3SnNR_2,
   (IID) for $R'_3SnNPhCHO$, with phenyltin trihydride,
   (IIE) for $R'_3SnNPhCHO$.

III Reaction of organotin hydrides with amines:
   (IIIA) $R_3Sn(SnR'_2)_nH$ with amine,
   (IIIB) Ph_3SnH with R_2SnBr_2 and Et_3N.

IV Reaction of organotin lithium compounds $(R_3Sn)_3SnLi$ with MeX or Me_3PO_4.

V   Alkylation of Mg_2Sn with butyl chloride.

R = R' = organic group, n = 1 and 2.

5.2.1.1 OCTAORGANOTRITIN COMPOUNDS

Octaethyltritin (M-491) $Sn_3C_{16}H_{40}$ $(Et_3Sn)_2SnEt_2$

Synth.: By rxn. of Et_3SnH with Et_2SnO (2:1) at 110°, in 47% yield (400, 1953), with $Et_2Sn(NEt_2)_2$, in 10% yield (1955).

By rxn. of Et_2SnH_2 with Et_3SnNEt_2 (1:2), in 90% yield (525), in 93% yield (1955), with $Et_3SnNPhCHO$ (1:2) (95).
By rxn. of $Et_3SnSnEt_2H$ with $Et_3SnNPhCHO$ (95).
Prop.: $b_{0.001}$ 133° (400), $b_{0.001}$ 118° (525), b. 124-27° x 10^{-4} torr. (1955), n_D^{20} 1.5894 (95), 1.5802 (1955), colorless liq. (95), IR spectrum (95). Anal. detn. by iodometric titration (1955).

Octabutyltritin $\qquad Sn_3C_{32}H_{72} \qquad (Bu_3Sn)_2SnBu_2$
Synth.: By rxn. of Bu_3SnH with $Bu_2Sn(NEt_2)_2$ (2:1), in 49% yield (525, 1955), with Bu_2SnO at r.t. (408).
By rxn. of Bu_2SnH_2 with Bu_3SnNEt_2 (1:2), in 76% yield (1955), with $(Bu_3Sn)_2O$ at 100° (468, 1803).
By rxn. of $Bu_3SnSnBu_2H$ with $Bu_3SnNPhCHO$ (95).
Prop.: Colorless liq. (95), $b_{0.001}$ 165° (partly dec.) (525), $b_{0.0001}$ 175-77° (1955), n_D^{20} 1.5380 (1955), 1.5342 (95), IR spectrum (95), mol. wt. (468, 1955). Anal. detn. by titration with Br (468), I (1955).
Rxn. with: Thermal dec. → Bu_6Sn + $(Bu_2Sn)_x$ (468).

Octaphenyltritin $\qquad Sn_3C_{48}H_{40} \qquad (Ph_3Sn)_2SnPh_2$
Synth.: By rxn. of Ph_3SnH with $Ph_2Sn(NPhCHO)_2$ (97), with Ph_2SnO in C_6H_6 at 100° (1955).
By rxn. of Ph_3SnM with Ph_2SnCl_2; M = Li, Na, 1/2 Mg, in 1-5% yield (395).
By rxn. of Ph_2SnCl_2 with Ph_3SnNa, from Ph_3SnCl and $Na-C_{10}H_8$, in $(MeOCH_2)_2$ and THF, in 70 and 4% yield, resp. (395), with Li in THF as by-prod. (395).
Prop.: White solid, m. 150-53° (97), m. 140-50° (1955), colorless cryst., dec. 280° without melting (395), UV (392, 395) and Moessbauer resonance (1094) spectra, mol. wt. (395), probably mixture of compounds (1955).
Rxn. with: I in C_6H_6 at 30° → Ph_3SnI + Ph_2SnI_2; suitable for quant. detn. (395, 1955).

Listing of octaorganotritin compounds concludes in Table 250.

Table 250. Octaorganotritin Compounds — R_8Sn_3

Formula	Synth. Method*	Yield	Properties	Ref.
Me_8Sn_3	IIC	64	b$_{0.0001}$ 66-68°, n_D^{20} 1.5898	1955
	IIA	low	UV spectrum (1905) (a)	1955
$(Me_3Sn)_2SnPh_2$	IID	--	Colorless liq., n_D^{20} 1.6624 (b)	97
$(Et_3Sn)_2SnMe_2$	IIA	79	b$_{0.0001}$ 113-17°, n_D^{20} 1.5821 (a)	525, 1955
	IID	65	b$_{0.0001}$ 139-40°, n_D^{20} 1.5641	400, 1955
$(Et_3Sn)_2SnBu_2$	IIC	65	b$_{0.0001}$ 133° (a)	525, 1955
	IIA	81	b$_{0.0001}$ 134-38°, n_D^{20} 1.5673 (a)	525, 1955
$(Et_3Sn)_2Sn(CH_2CHMe_2)_2$	IIC	--	Colorless liq., n_D^{20} 1.6202 (b)	97
$(Et_3Sn)_2SnPh_2$	IID	--	Colorless liq., n_D^{20} 1.6516 (b)	97
$(Ph_3Sn)_2SnEt_2$	IIB	--	Light yellow liq., n_D^{20} 1.5886 (b)	97
$(Bu_3Sn)_2SnPh_2$	IID	--	(c)	1803
	IB	--	Slightly yellow liq., n_D^{20} 1.6002 (e)	95
$Bu_3SnSnBu_2SnPh_3$	IIB (d)	--	Colorless liq., n_D^{20} 1.6452 (b)	97
$(Ph_3Sn)_2SnBu_2$	IIB	31	m. 83° (a)	1955
$i-Bu_8Sn_3$	IIIB	83	m. 7°, b$_{0.0001}$ 169-71°, n_D^{20} 1.5431, mol. wt. (a)	525, 1955
	IIC	60	Prism, m. 28°, n_D^{20} 1.5840, soly. (a)	1955
$(i-Bu_3Sn)_2SnPh_2$	IB	88	Colorless cryst., m. 168°, mol. wt., soly. (a)	1955
$[(c.C_6H_{11})_3Sn]_2Sn(i-Bu)_2$	IIC	90	Colorless plates, m. 111°, soly. (a)	1955
$(Ph_3Sn)_2Sn(i-Bu)_2$	IIC	--	Light yellow liq., n_D^{20} 1.5420 (b)	97
$[(C_8H_{17})_3Sn]_2SnPh_2$	IIB			

* The characters used here correspond to the synthetic scheme in the introduction to Chapter 5.2.1. (a) Anal. detn. by iodometric titration (1955). (b) Sensitive to air, decomposes on distillation (97). (c) Useful as biocide, fungicide and catalyst (1803). (d) Synth. with Ph_3SnH and rxn. from $Bu_3SnSnBu_2H$ and $PhNCO$ (95). (e) IR spectrum (95).

5.2.1.4 ORGANO-POLYTIN COMPOUNDS WITH FOUR OR MORE TIN ATOMS

Tetrakis(triphenylstannyl)tin (M-491) $Sn_5C_{72}H_{60}$ $(Ph_3Sn)_4Sn$
<u>Synth.:</u> By rxn. of Ph_3SnLi with $SnCl_2$, followed by Ph_3SnCl (3:1:1), in 69% yield (68a, 175).
By rxn. of Ph_3SnCl with $SnCl_2$ and Li in THF, in 21% yield (68a, 175).
By rxn. of $(Ph_3Sn)_3SnLi$ with Ph_3SiCl, Bu_3SnCl and aq. NH_4Br, in 57, 67 and 34% yield (68a, 175).
By-prod. in rxn. of $SnCl_2$ with PhMgBr in Et_2O, in 2% yield (68a, 175).
<u>Prop.:</u> White solid, m. 315-20° (dec.), IR (68a, 175), far IR (1065), UV (68a, 175) and Moessbauer resonance (1094, 1990) spectra, x-ray powder pattern and single crystal analysis (68a, 175).
<u>Rxn. with:</u> $Ph_3SnLi \rightarrow Ph_6Sn_2$ (68a, 175).
MeLi $\rightarrow Ph_4Sn + Ph_6Sn_2$ (?) (68a, 175).
I $\rightarrow Ph_3SnI + SnI_2$ (?) (68a, 175).
$(CH_2X)_2 \rightarrow Ph_4Sn + Ph_3SnX$; X = Cl, Br (68a, 175).

Listing of organo-polytin compounds with four or more tin atoms concludes in Table 251.

5.2.2 ORGANO-POLYTIN COMPOUNDS WITH $(R_2Sn)_x$ STRUCTURE

This chapter contains organo-cyclopolytin derivatives and some "diorganotin" derivatives of unknown structure. The compounds listed in Table 252 are prepared by methods from the following scheme.

I Dehydrogenation of diorganotin dihydrides R_2SnH_2 with:
(IA) dimethylformamide,
(IB) pyridine,
(IC) diorganotin diamide.

II Disproportionation of hexamethylditin.

III Dealkylation of diorganotin dihalides R_2SnX_2 with:
(IIIA) magnesium in tetrahydrofuran,
(IIIB) with lithium aluminum hydride also for $RSnCl_3$,
(IIIC) with naphthaline-sodium,
(IIID) by polarography of $RSnCl_3$.

IV Alkylation reaction of stannous chloride with:
(IVA) alkyllithium,
(IVB) dialkylmagnesium,
(IVC) Grignard reagent,
(IVD) triisobutyl aluminum.

R = organic group, X = halogen.

Dodecamethyl Cyclohexatin (M-491) $Sn_6C_{12}H_{36}$ $c.(Me_2Sn)_6$
<u>Synth.:</u> By rxn. of Me_2SnCl_2 with $LiAlH_4$ and Et_3N in Et_2O in N atm. (871).
By rxn. of Me_2SnCl_2 with Me_2SnH_2 and Py (871).

Prop.: Yellow solid, NMR spectrum (871).
Rxn. with: C_2F_4 under UV irradiation → Me_4Sn + Sn + $(C_2F_4)_x$ + $[Me_2SnC_2F_4]_8$ (?) (871).
CF_3I under UV irradiation → Me_3SnCF_3 + Me_3SnI + $Me_2Sn(CF_3)_2$ + $Me_2(CF_3)SnI$ (?) (871).

Dodecaethyl Cyclohexatin (M-489) $Sn_6C_{24}H_{60}$ c.$(Et_2Sn)_6$
Synth.: By rxn. of Et_2SnH_2 with $Et_2Sn(NEt_2)_2$, in 83% yield (525, 1614), with Et_2NH in C_6H_6 and cat. $ZnCl_2$ at 60°, in 64% yield (1614).
Prop.: m. ~200° (dec.) (525), mol. wt., fragmentation anal. with Cl (1614).
Rxn. with: X → Me_2SnX_2 ; X = Cl, I quant. rxn. (1614).

Poly(diethyltin) (M-489) $(SnC_4H_{10})_x$ $(Et_2Sn)_x$
Synth.: By rxn. of Et_2SnH_2 with basic cat. (x = 9) (388a).
By polarographic reduction of Et_2SnCl_2 (1550).
Rxn. with: Thermal dec. at 160° → Et_4Sn + Et_6Sn_2 + Sn + residue contn. Et_3Sn- and Et_2Sn- groups (1941).
$(HC!CCH_2)_2$ + Na in liq. NH_3, + EtX → $Et_3SnC!CCH_2CH_2C!CH$ (629).
RI at 140° → Et_2RSnI ; R = Me, Et, Pr Bu (1941).
RBr at 140° → Et_3SnBr + Et_2RSnBr + EtR_2SnBr ; R = Pr, Bu (1941).
BuCl at 140° → Et_4Sn + Et_3SnBu + Et_3SnCl + $Et_2BuSnCl$ + Bu_3SnCl ; rxn. mechanism (1941).

Dodecabutyl Cyclohexatin (M-491) $Sn_6C_{48}H_{108}$ c.$(Bu_2Sn)_6$
Synth.: By rxn. of Bu_2SnH_2 with $Bu_2Sn(NEt_2)_2$ in inert atm. at 50°, in 60% yield (525, 1614), with MeONa in THF and $Bu_2Sn(OMe)_2$ at 50°, in 94 and 52% yield, resp. (1614).
By rxn. of Bu_2SnH_2 in MePh with Py at 60-100°, in presence of cat. Bu_2SnCl_2, in 88% yield (1614).
Prop.: Waxy cryst. (1614), m. 210° (dec.) (525, 1614), mol. wt. (1614).
Rxn. with: I → Bu_2SnI_2, in quant. yield (1614).

Polydibutyltin (M-491) $(SnC_8H_{18})_x$ $(Bu_2Sn)_x$
Synth.: By rxn. of Bu_2SnH_2 with $Bu_2Sn(OMe)_2$ at r.t., in N atm. (400, 1802), with Bu_2SnO at 100° in N atm. (468, 1802).
By photolysis of Bu_2SnH_2 (585).
By rxn. of $BuAlCl_2$ with $SnCl_2$ in THF in N atm., in 75% yield (2904).
Prop.: Viscous yellow-green liq. (468, 1802), IR spectrum (468). Anal. detn. by titration with Br (468).
Rxn. with: Thermal dec. at 220° → Bu_6Sn_2 + Sn (2096).
Thermal dec. at 160°, + I in THF → Bu_3SnI + Bu_2SnI_2 + $BuSnI_3$ (1941).
Et_2SnCl_2 + Et_3N → $EtBu_2SnCl$ + $Et_2BuSnCl$ + Bu_3SnCl + Et_3SnCl + Sn (1941).
Bu_2SnCl_2 + Et_3N → Bu_3SnCl + Sn (1941).
HX → Bu_2SnX_2 + H ; X = Br, I (2096).
Cl → Bu_2SnCl_2 (585).
$AlBr_3$ → Bu_4Sn + trace Bu_3SnBr + Sn (585).
RI → Bu_2RSnI ; R = Me, Et, Pr, Bu (1941).

$CH_2:CHCH_2Br \to Bu_2(CH_2:CHCH_2)SnBr + Bu_2SnBr_2 + (CH_2:CHCH_2)_2$ (66).
$RBr \to Bu_2RSnBr$; $R = PhCH_2$, Bu, Ph, $Br(CH_2)_4$ (66).
$RBr \to Bu_2RSnBr + Bu_3SnBr$; R = Et, Pr, Bu (1941).
$BuCl \to Bu_3SnCl$ (66, 1941) + Bu_4Sn (1941).
PVC at 160° → Bu_2Sn- contg. PVC; rate of rxn., solvent effect (2770).
$RCl \to Bu_2SnCl_2 + R_2$; $R = PhCH_2$, Ph_3C (66).
$BzOH \to (Bu_2SnOBz)_2 + H$ (2096).
Air → $(Bu_2SnO)_x$ (468, 1802).
$Bz_2O_2 \to (Bu_2SnOBz)_2$ (2096).
$(Me_3CO)_2 \to Bu_2Sn(OCMe_3)_2$ (2096).
$Me_3CO_2H \to (Bu_2SnO)_x + Me_3COH$ (2096).

Octadecaisobutyl Cycloenneatin $Sn_9C_{72}H_{162}$ c.$(i-Bu_2Sn)_9$
<u>Synth.</u>: By rxn. of $i-Bu_2SnH_2$ with $i-Bu_2Sn(NEt_2)_2$ in inert atm., in 79% yield (525, 1614), with Py in MePh, in presence of cat. Bu_2SnCl_2 in 33% yield (388a, 1614).
By rxn. of $i-Bu_2Sn(NEt_2)_2$ with $(i-Bu_2SnH)_2$ at 70°, in 57% yield (1614).
<u>Prop.</u>: Yellow cryst. (1614), m. 205° (dec.) (525), 204-206°, UV spectrum, mol. wt. (1614). Fragmentation analysis → 100% $i-Bu_2SnCl_2$ (1614).

Dodecaphenyl Cyclohexatin (M-490) $Sn_6C_{72}H_{60}$ c.$(Ph_2Sn)_6$
<u>Synth.</u>: By rxn. of Ph_2SnH_2 with Py or DMF in presence of cat. Ph_2SnCl_2, in 44% yield (392), with $Ph_2Sn(NEt_2)_2$ (1610).
By rxn. of Ph_2SnCl_2 with $Na-C_{10}H_8$ in THF, in 50% yield (392).
By rxn. of $(Ph_2Sn)_xH_2$ with Py, in 14-28% yield (392).
<u>Prop.</u>: m. 270-74° (dec.) (1990), 270° (dec.) (392), cryst. with 13% MePh (392), IR, UV (392), Moessbauer resonance (1990) and mass (2520) spectra, mol. wt., contains no Ph_3Sn-endgroups by detn. with I (392), soly. (392, 1990), stable in air (392), struct. (1610).
<u>Rxn. with:</u> I in C_6H_6 at 20° → Ph_2SnI_2, in quant. yield (392).

Polydiphenyltin (M-490) $(SnC_{12}H_{10})_x$ $(Ph_2Sn)_x$
<u>Synth.</u>: By rxn. of Ph_2SnH_2 with Ph_3SnNEt_2 as by-prod. (1955).
By rxn. of Ph_2SnCl_2 with Li in THF, in 94% yield (4), with $(PhNLi)_2$ in Et_2O (1089).
By rxn. of $Ph_3SnZnCl$ with Ph_3SnCl (2830), with MeOH or H_2O in THF (2830).
By rxn. of $SnCl_2$ with PhMgBr in THF (x = 13-14) (392), in high dilution x = 6-7 (392), if $SnCl_2$ is run into PhMgBr $x \approx 8$ (392).
By polarographic reduction of $PhSnCl_3$ in alc. $HClO_4$ (973).
<u>Prop.</u>: Light yellow solid, m. 271-73° (1089), ochre powder (392), IR (4) and UV (392) spectra, soly. (392).
<u>Rxn. with:</u> Ph_2SnI_2 (5:3) → $Ph_3SnI + Sn$ (1938).
Bu_2SnI_2 (1:2), + MeMgI → $MeBu_2SnPh + Me_2SnBu_2 + MeSnPh_3 + Me_2BuSnPh + Me_2SnPh_2 + Ph_6Sn_2$ (*) (1938).

---

* Rxn. mechanism is discussed (1938).

I in C$_6$H$_6$ at 20° → Ph$_2$SnI$_2$ + Ph$_3$SnI + PhSnI$_3$ (392).
I, + MeMgI → Me$_n$SnPh$_{4-n}$; n = 1-3 (2830).
MeI (1:4), + BuMgI → BuSnPh$_3$ + MeBuSnPh$_2$ + MeBu$_2$SnPh + Me$_2$BuSnPh + Bu$_2$SnPh$_2$* (1938).
RX (1:4), + MeMgI → MeSnPh$_3$ + MeRSnPh$_2$ + Me$_2$RSnPh + MeR$_2$SnPh + Me$_2$SnPh$_2$; R = Et, Pr, Bu; X = Br, I * (1938).

Listing of poly-organotin compounds with (R$_2$Sn)$_x$ structure concludes in Table 252.

5.2.4 ORGANO-POLYTIN COMPOUNDS WITH (RSn)$_x$ COMPOSITION

Two organo-polytin compounds with a (RSn)$_x$ composition and uncertain structure are listed at the bottom of Table 252.

---

* Rxn. mechanism is discussed (1938).

Table 251. Organo-Polytin Compounds with Four or More Tin Atoms $R_{2n+2}Sn_n$

Formula*	Method**	Yield	Properties	Ref.
$(Me_3Sn)_3SnMe$	IV	--	Synth. with MeI, NMR spectrum	2127
$(Me_3SnSnBu_2)_2$	IIC	60	$b_{0.0001}$ 161-64°, n_D^{20} 1.5785 (a)	1955
$(Me_3SnSni\text{-}Bu_2)_2$	IIC	81	m. 90° $b_{0.0001}$ 150-55°, mol. wt. (a)	1955
$(Ph_3Sn)_3SnMe$	IV	15	Synth. with Me_3PO_4, white, m. 180-85°	68a,175
$(Ph_3Sn)_3SnMe$	IV	5	Synth. with MeI, mol. wt., dec. in soln.	68a,175
$(Et_3SnSnEt_2)_2$ (491)	IIIA	--	Colorless liq., n_D^{20} 1.6283, IR spectrum	95
	IIA	9	$b_{0.0001}$ 165-72°, n_D^{20} 1.6061 (a)	1955
$Et_3Sn(SnEt_2)_3SnEt_3$	IID (b)	--	Yellow liq., n_D^{20} 1.6433, IR spectrum (c)	95
$Et_3Sn(SnEt_2)_4SnEt_3$	IIIA	--	Viscous yellow liq., n_D^{20} 1.678, IR spectrum (c)	95
$Et_3Sn(SnEt_2)_xSnEt_3$	IA	28	Yellow oil	1955
$(Et_3SnSni\text{-}Bu_2)_2$	IIC	86	$b_{0.0001}$ 172-75°, n_D^{20} 1.5842, mol. wt. (a)	1955
$(Et_3SnSnPh_2)_2$	IIIA	--	Light yellow liq., n_D^{20} 1.6877	95
$(Et_3Sn)_3SnPh$	IIE	--	Red liq., n_D^{20} 1.6648, dec. on distn. (d)	97
$(Bu_3SnSnBu_2)_2$	IIIA	--	Colorless liq., n_D^{20} 1.5543, IR spectrum	95
$Bu_3Sn(SnBu_2)_xSnBu_3$	V	--	Pale green, viscous liq. (e)	54
$Bu_3SnSnBu_2SnPh_2SnBu_3$	(f)	--	Orange liq., n_D^{20} 1.6085	95
$[(c.C_6H_{11})_3SnSni\text{-}Bu_2]_2$	IIC	87	m. 207° (a)	1955
$(Ph_3SnSni\text{-}Bu_2)_2$	IIC	80	m. 114°, mol. wt. (a)	1955
$[(C_8H_{17})_3Sn]_3SnPh$	IIE	--	Yellow liq., n_D^{20} 1.5345	97
$Ph_3Sn(SnPh_2)_nSnPh_3$	IB (g)	--	n = 1-3, m. 115-55°	1955
$Ph_3Sn(SnPh_2)_xSnPh_3$	IA	23	Viscous oil, n_D^{20} 1.702	1955

* Numbers in parenthesis refer to pages in main volume.
** The characters used here correspond to the synthetic scheme in the introduction to Chapter 5.2.1.
(a) Anal. detn. by iodometric titration (1955). (b) Synth. from Et_2SnH_2 and $Et_3SnSnEt_3NPhCHO$ (1:2) (95). (c) UV spectrum (913). (d) Sensitive to air (97). (e) Rxn. with air → $(Bu_2SnO)_x$ (54). (f) Synth. from $Bu_3SnSnBu_2H$ and Ph_2SnH_2, followed by $Bu_3SnNPhCHO$ (95). (g) Synth. from Ph_2SnH_2 and Ph_3SnOMe (1955).

Table 252. Organo-Polytin Compounds with $(R_2Sn)_x$ Structure

Formula*	Synth. Method**	Yield	Properties	Ref.
$(Me_2Sn)_x$ (491)	II (a)	--	(b)	2012
c.$[(PhCH_2)_2Sn]_4$	IA	--	Yellow cryst., m. 226-28° (dec.) sensitive to air	394
	IC	53	m. 226-28°, soly. (c)	1614
$(Et_2SnPh_2)_x$	I (d)	--	Yellow, m. 230-233°, sensitive to air	97
c.$(Pr_2Sn)_x$	--	--	(e)	--
c.$(i-Bu_2Sn)_6$	IIIA	66	Waxy solid, m. 200-210°, mol. wt. (f)	1614
	IIIB	--	--	1955
$(i-Bu_2Sn)_x$	IVD	38	--	2904
$[(Me_3C)_2Sn]_4$ (491)	IC	10	--	1614
$(C_5H_5)_2Sn$ (490)	IVB	90	m. 102-105°, subl. 150-80°	1743
c.$[(c.C_6H_{11})_2Sn]_5$	IC	71	Slightly yellow cryst., m. 270-75°	1614
c.$(Ph_2Sn)_5$	IA	--	Pale yellow cryst., mol. wt. (g, h)	392
c.$[(p-MeC_6H_4)_2Sn]_6$	IA	97	Dec. 260-70°, IR spectrum, sensitive to air (i)	393
c.$[(p-PhC_6H_4)_2Sn]_6$	IB	86	Colorless cryst., dec. 265°	393
	IIIC	--	(i)	393
$[(p-PhC_6H_4)_2Sn]_x$	IB	--	--	388a, 393
	IVC	--	(i)	393

* Numbers in parenthesis refer to pages in main volume.
** The characters used here correspond to the synthetic scheme in the introduction to Chapter 5.2.2.
(a) Synth. also by photolysis of $Me_3Sn(CH_2)_nCOR$; R = Me, Ph; n = 2 and 3, in 15-43% yield (1488). (b) Rxn. with air → $(Me_2SnO)_x$ (1488). (c) Rxn. with I at 20° → R_2SnI_2; R = $PhCH_2$ (394), rxn. useful for quant. detn. (1614).
(d) Synth. from Ph_2SnH_2 and $Et_2Sn(NPhCHO)_2$ (97). (e) Rxn. with RI → Pr_2RSnI, R = Me, Et, Pr, Bu (1941), with BuBr → $Pr_3SnBr + Pr_2BuSnBr$ (1941), with BuCl → $Pr_4Sn + Pr_3SnCl + Pr_2BuSnCl + PrBu_2SnCl + Bu_3SnCl$ (1941).
(f) Rxn. with X = $i-Bu_2SnX_2$; X = Cl, I, in quant. yield (1614). (g) Contains no Ph_3Sn-endgroups, prod. contains 10.6% MePh (392). Rxn. with I at 20° → Ph_2SnI_2 (392). (i) Mol. wt., soly., rxn. with I at 20° → R_2SnI_2; R = p-MeC_6H_4, p-PhC_6H_4 (393).

Table 252 Continued

Formula*	Synth. Method**	Yield	Properties	Ref.
[(1-$C_{10}H_7$)$_2$Sn]$_6$	IB	--	Yellow powder dec. 305°, IR spectrum (i)	388a, 393
[(2-$C_{10}H_7$)$_2$Sn]$_6$	IA	42	Colorless needles, dec. 270°	393
(9-$C_{14}H_9$)$_2$Sn (491)	IB	--	IR spectrum (i)	388a, 393
($B_{10}H_{10}C_2Ph$)$_2$Sn	IVA	--	Moessbauer resonance spectrum	706
c.[(p-EtOC_6H_4)$_2$Sn]$_6$	IA	67	Moessbauer resonance spectrum (j)	660
	IIIC	--	Pale yellow cryst., m. 235-40° (dec.)	393
(p-PhOC_6H_4)$_2$Sn (491)	IVC	--	IR spectrum (i)	393
		--	Polymerizes on standing (k)	1982
(EtSn)$_x$	IIID	--	--	970, 973-4
	IIIB	--	--	973
(PhSn)$_x$	IIIB, D	--	--	973

(j) Oxidizes slowly in air (660). (k) Use as antioxidant for polyphenyl ethers (1982).

5.4 ORGANO-POLYTIN HETERO COMPOUNDS

This chapter reports organotin compounds with tin-tin bonds having heteroatoms attached to tin. The heteroatoms comprise hydrogen, halogen, oxygen and nitrogen. Several polymeric diorganotin compounds $(R_2Sn)_2$ will find place here when terminal group analysis is better developed. The compounds listed in Table 253 are prepared by methods from the following scheme.

I Dehydrogenation of diorganotin dihydrides R_2SnH_2 with:
- (IA) MeOH,
- (IB) $R_3Sn(SnR_2)_m NPhCHO$, also for $PhSnH_3$,
- (IC) benzoylperoxide,
- (ID) PhNCO for $Et_3SnSnEt_2H$.

II Synthesis with diorganotin hydride halides (R_2SnHX) by:
- (IIA) thermal decomposition, also for $EtSnH_2Br$,
- (IIB) decomposition with diethylamine.

III Reduction of organotin halides:
- (IIIA) $(R_2SnCl)_2$ with $LiAlH_4$,
- (IIIB) Ph_2SnCl_2 by polarography.

IV Photooxidation of hexaorganoditin Et_6Sn_2.

V Reaction of polymeric diorganotin $(R_2Sn)_n$ with:
- (VA) iodine,
- (VB) benzoic acid,
- (VC) benzoyl peroxide.

R = organic group, m = 0 and 1, n = 4 and x, X = halogen.

Pentaethylditin Hydride (M-267) $Sn_2C_{10}H_{26}$ $Et_3SnSnEt_2H$
Synth.: By rxn. of Et_2SnH_2 with $Et_3SnNPhCHO$ (1:1) in inert atm. (95).
Prop.: IR spectrum (95).
Rxn. with: $Et_3SnNPhCHO \rightarrow (Et_3Sn)_2SnEt_2 + PhNHCHO$ (95).
PhNCO $\rightarrow Et_3SnSnEt_2NPhCHO$ (95).
Py or $Et_2NH \rightarrow (Et_3SnSnEt_2)_2 + H$ (95).

Pentabutylditin Hydride $Sn_2C_{20}H_{46}$ $Bu_3SnSnBu_2H$
Synth.: By rxn. of Bu_2SnH_2 with $Bu_3SnNPhCHO$ (1:1) in inert atm. (95).
Prop.: IR spectrum (95).
Rxn. with: $Bu_3SnNPhCHO \rightarrow (Bu_3Sn)_2SnBu_2 + PhNHCHO$ (95).
Ph_2SnH_2, + $Bu_3SnNPhCHO \rightarrow Bu_3SnSnBu_2SnPh_2SnBu_3$ (95).
PhNCO, + $Ph_3SnH \rightarrow Bu_3SnSnBu_2SnPh_3$ (95).
Py or $Et_2NH \rightarrow (Bu_3SnSnBu_2)_2 + H$ (95).

Listing of organo-polytin hetero compounds is concluded in Table 253.

Table 253. Organo-Polytin Hetero Compounds

Formula*	Synth. Method**	Yield	Properties	$R'R_2SnX$ Ref.
$Et_3SnSnPh_2H$	IB	--	IR spectrum (a)	95
$Et_3SnSnEt_2SnEt_2H$	I (b)	--	IR spectrum (c)	95
$(Bu_2SnH)_2$ (267)	IIIA	45	NMR spectrum (259) (d)	1955
$(Bu_2Sn)_xH_2$	IIIA	Low	--	1955
$(Bu_3Sn)_2SnPhH$	IB	--	n_D^{20} 1.5712, IR spectrum	95
$(1-Bu_2SnH)_2$	IIIA	45	$b_{0.001}$ 109-12°, n_D^{20} 1.518, IR spectrum (e)	1955
$(i-Bu_2Sn)_xH_2$	IIIA	Low	$x > 2$, pale yellow oil	1955
$(t-Bu_2SnH)_2$	--	--	NMR spectrum	1454
$(Ph_2Sn)_6H_2$	IA	82	Pale yellow, IR spectrum, mol. wt. (f)	392
$Et_3SnSnEt_2OEt$	IV	--	(g)	661
$(EtSnBr)_x$	IIA	--	Contains Sn-Sn bonds	156
$(Bu_2SnCl)_2$ (492)	IIA	--	Anal. detn. by Br-titration (h, i)	469
$(Bu_2SnOBz)_2$ (493)	IC	80	m. 29-30°	585
	VB	90	m. 31-32°	2096
	VC	84	--	2096
$(1-Bu_2SnCl)_2$	IIB	98	Colorless oil, n_D^{20} 1.550, dec. on distn. (i)	1955
$(t-Bu_2Sn)_4I_2$	VA	--	m. 160°, mol. wt., dec. in hot solvent	438
$(Ph_2SnCl)_2$	IIIB	--	--	963
$Et_3SnSnEt_2NPhCHO$	ID	--	IR spectrum (b, j)	95

* Numbers in parenthesis refer to pages in main volume.
** The characters used here correspond to the synthetic scheme in the introduction to Chapter 5.4.
(a) Rxn. with Py or $Et_2NH \rightarrow (Et_3SnSnPh_2)_2$ (95). (b) Rxn. of Et_2SnH_2 with $Et_3SnSnEt_2NPhCHO$ (1:1) $\rightarrow Et_3SnSnEt_2$-$SnEt_2H$ (95). (c) Rxn. with Py or $Et_2NH \rightarrow Et_3Sn(SnEt_2)_4SnEt_3$ (95). (d) Rxn. with $Me_3SnNEt_2 \rightarrow (Me_3SnSnBu_2)_2$ (1955). (e) Rxn. with $R_3SnNEt_2 \rightarrow (R_3SnI-Bu_2Sn)_2$; R = Me, Et, $c.C_6H_{11}$, Ph (1955), with $1-Bu_2Sn(NEt_2)_2 \rightarrow$ $c.(1-Bu_2Sn)_8$ (1614). (f) Rxn. with Py $\rightarrow c.(Ph_2Sn)_6$ + H (392). (g) Rxn. with O $\rightarrow Et_2SnO + Et_3SnOEt$ (661). (h) Rxn. with air $\rightarrow (Bu_2SnCl)_2O$ (469). (i) Rxn. with $LiAlH_4 \rightarrow (R_2SnH)_2 + R_2SnH_2 + (R_2Sn)_xH_2$; R = Bu, 1-Bu (1955). (j) Rxn. with Et_2SnH_2 (2:1) $\rightarrow Et_3Sn(SnEt_2)_3SnEt_3$ + PhNHCHO (95).

6. POLYMETALLIC ORGANOTIN COMPOUNDS

This chapter comprises organotin compounds with tin-metal bonds. All transition metal derivatives containing an organotin moiety are listed here unless the author specifically states that no metal-tin bond exists. In these cases the compounds are to be found in Chapters 2.7.2 through 2.7.8. Polymetallic derivatives with tellurium are listed in Chapter 4.4.1; with antimony and bismuth in Subchapter 4.5.6.4.

6.1 BIMETALLIC ORGANOTIN COMPOUNDS WITH GROUP IV METALS

The organotin compounds containing a covalent tin-metal bond compiled in this chapter are prepared by methods from the following scheme.

I Reaction of triorganotin lithium R_3SnLi with R_nSiCl_{4-n}.

II Reaction of triorganotin halide R_3SnX with:
- (IIA) Et_3SiLi,
- (IIB) R_3SiK,
- (IIC) H_3SiK.

III Reaction of triethyltin bromide with $Et_3SiHgEt$.

R = organic group, n = 0-3, X = halogen.

6.1.1 ORGANOTIN COMPOUNDS WITH Sn-Si BONDS

All compounds are listed in Table 254.

6.1.2 ORGANOTIN COMPOUNDS WITH Sn-Ge BONDS

The organogermanes with Ge-Sn bonds are listed in Chapter 6.1.2 in the "Germanium Section" and in Tables 78-80. Tetrakis(triphenyltin) germanium may be found at the bottom of Table 254.

6.1.3 ORGANOTIN COMPOUNDS WITH Sn-Pb BONDS

Tetrakis(triphenyltin) lead may be found at the bottom of Table 254.

Table 254. Organotin Compounds with Group IV Metal-Tin Bonds $(R_3Sn)_nMR_{4-n}$

Formula*	Synth. Method**	Yield	Properties	Ref.
$(Me_3Sn)_4Si$	IIC	--	NMR spectrum	2224
$(Me_3Sn)_3SiH$	IIC	--	IR and NMR spectra	2224
Me_3SnSiH_3	IIC	Low	NMR spectrum	2224
$Me_3SnSiMe_3$	I	40	Colorless liq., b. 144-46°, mol. wt. (a)	1884
$Me_3SnSiEt_3$	I	45	--	1887
$(Et_3Sn)_2SiMe_2$	I	15	b_1 135°, monomeric, soly. (b)	11a
$Et_3SnSiEt_3$	IIA	25	b_1 61-63°, n_D^{20} 1.4831	2087
	III	25	b_1 61-63°, n_D^{20} 1.4835, d_{20} 1.1200	2852
$(Bu_3Sn)_2SiMe_2$	I	15	Colorless oil, b(vacuo) 154°, mol. wt.	11a
$Bu_3SnSiMe_3$ (495)	I	60-70	Orange oil, b_1 94°, mol. wt.	11a
$(Ph_3Sn)_4Si$	I	30	m. 390-95° (dec.), mol. wt. (c, d)	1928
$(Ph_3Sn)_2SiMe_2$	I	35	Colorless cryst., m. 172°, monomeric (e)	11a
$(Ph_3Sn)_2SiPh_2$	I	35	Colorless cryst., m. 199°, monomeric (e)	11a
$Ph_3SnSiMe_3$	I	60-70	Colorless cryst., m. 119°, monomeric (e)	11a
$Ph_3SnSiPh_3$ (495)	IIB	84	Polarography → Ph_3SiH (963, 967) (f)	2128
$Ph_2Sn(SiPh_2)_3SiPh_2$	II (g)	10	Very sensitive to air and moisture	1177
$(Me_3Si)_4Sn$	II (h)	20	m. 235-36° (sealed tube), struct. (i)	810
$(Ph_3Sn)_4Ge$ (495)	--	--	Moessbauer resonance spectrum	1094
$(Ph_3Sn)_4Pb$ (495)	--	--	Moessbauer resonance spectrum	1094

* Numbers in parenthesis refer to pages in main volume.
** The characters used here correspond to the synthetic scheme in the introduction to Chapter 6.1.

(a) (Far) IR spectra and assignment (1844), anal. detn. of Sn and Si by x-ray fluorescence (1844). (b) Rxn. with $TiCl_4 → Et_3SnCl + Me_2SiCl_2 + TiCl_3$ (11a). (c) X-ray powder diffraction pattern (1928). (d) Rxn. with MeLi, + $PhCH_2Cl → Ph_6Sn_2$ (1928). (e) Stable in air, soly. (11a). (f) Rxn. with excess AcOH → $(AcO)_3SnSi(OAc)_3$ (2128). (g) Synth. from Ph_2SnCl_2 and $Li(SiPh_2)_4Li$ in N atm. in THF (1177). (h) Synth. from $SnCl_4$ with Li and Me_3SiCl in THF (810). (i) IR, Raman and NMR spectra (810).

6.2 BIMETALLIC ORGANOTIN COMPOUNDS WITH GROUP III METALS

Trimethyltin Pentaborane(9) $SnB_5C_3H_{17}$ $Me_3SnB_5H_8$
Synth.: By rxn. of Me_3SnBr with LiB_5H_8 at -45 to -22° in N atm., in 60% yield (2398).
Prop.: (Far) IR, H and ^{11}B NMR and mass spectra, no vapor pressure at r.t. (2398).

A number of triorganotin derivatives of boron, aluminum, gallium and indium are claimed as agents for metal plating (603). No other data for the following compounds are reported: Me_3SnBMe_2, Pr_3SnBPr_2, Bu_3SnBEt_2, $(C_5H_{11})_3SnBEt_2$, $Me_3SnAlMe_2$, $Me_3SnAlEt_2$, $Et_3SnAlEt_2$, $EtPr_2SnAlMeEt$, $Bu_3SnAl(i-C_5H_{11})_2$, $Me_3SnGaMe_2$, $Pr_3SnGaPr_2$, $Bu_3SnGa(i-C_5H_{11})_2$, $MeEt_2SnInEt_2$, $Bu_3SnInMe_2$ and $(C_7H_{15})_3SnIn(C_{11}H_{23})_2$ (603).

6.3 BIMETALLIC ORGANOTIN MAGNESIUM COMPOUNDS

Triphenyltin Magnesium Bromide $Sn_2Mg_2C_{36}H_{30}Br_2$ $(Ph_3SnMgBr)_2$
Synth.: By rxn. of Ph_3SnH with $EtMgBr \cdot NEt_3$ at -70°, followed by vacuum treatment at 0.01 torr. and 50° (2330).
Prop.: Slightly yellow solid, sensitive to air, struct. (2330).
Rxn. with: $Ph_3SnCl \rightarrow Ph_4Sn$ (2330).
$Co_2(CO)_8$, + $PPh_3 \rightarrow Ph_2Sn[Co(CO)_3PPh_3]_2$ (2330).
MeI at 140°, + $MeMgI \rightarrow MeSnPh_3 + Me_2SnPh_2$ (2330).
$H_2O \rightarrow C_6H_6$ (2330).
Addn. Compd.: $Ph_3SnMgBr \cdot NEt_3$, from Ph_3SnH and $EtMgBr \cdot NEt_3$, in 94% yield, yellow oil, looses NEt_3 in varuo, rxn. with $Ph_3SnCl \rightarrow Ph_6Sn_2$, with $H_2O \rightarrow Ph_3SnH + C_6H_6$ (2330) [$SnMgC_{24}H_{30}BrN$].

6.4 BIMETALLIC ORGANOTIN ALKALI METAL COMPOUNDS

The organotin alkali metal compounds listed in Table 255 are prepared by methods from the following scheme.

I Reaction of triorganotin chloride R_3SnCl with:
   (IA) Li,
   (IB) K.

II Reaction of hexamethylditin with naphthalene-sodium.

Trimethyltin Lithium (M-496) $SnLiC_3H_9$ Me_3SnLi
Synth.: By rxn. of Me_3SnBr with Li in THF (2423).
By rxn. of Me_2SnCl_2 with Li in THF in Ar atm. (2127).
By rxn. of Me_6Sn_2 with Li in THF (2127).
By rxn. of MeLi with $SnCl_2$ in THF at -60° (2127).
Prop.: Green soln. (2423), deep red soln., stable at r.t. in Ar atm. (2127), H and ^7Li NMR spectra (2127), dec. on removal of THF (2127).
Rxn. with: Vacuum stripping $\rightarrow (Me_3Sn)_3SnLi \cdot 3THF + Me_4Sn$ (2127).

Me$_3$MCl → Me$_3$SnMMe$_3$ + Me$_6$Sn$_2$ + Me$_6$M$_2$; M = Si, Ge (1844).
SnCl$_2$ → (Me$_3$Sn)$_3$SnLi·3THF (2127).
RBr → Me$_3$SnR + RH; R = CH$_2$CPh$_2$CMe; rxn. mechanism (1942).
C$_5$F$_5$N at -78° → Me$_3$SnF + (4-C$_5$F$_4$N)$_2$ (2423).

Trimethyltin Sodium (M-498)   SnNaC$_3$H$_9$ Me$_3$SnNa
Prop.: Moessbauer resonance spectrum (902).
Rxn. with: MeCH:CHCH$_2$Cl in liq. NH$_3$ → Me$_3$SnCH$_2$CH:CHMe (1877).
CH:CH(CH$_2$)$_n$CHBr in liq. NH$_3$-Et$_2$O → Me$_3$SnCH(CH$_2$)$_n$CH:CH; n = 3, 4, 5 (1367).
BrC⋮CH in liq. NH$_3$ → Me$_3$SnC⋮CBr + (Me$_3$SnC⋮)$_2$ (630, 2180).
PhC⋮CCHMeX → Me$_3$SnCMeC:C:CHPh (881).
MeC⋮CCHRX → Me$_3$SnCR:C:CHMe; R = Me, Ph (881).
Et$_2$S$_2$ in liq. NH$_3$ → Me$_3$SnSEt (800).
MePCl$_2$ → (Me$_3$Sn)$_2$PMe (58, 703).

Tris(trimethyltin)tin Lithium Tristetrahydrofuran   (Me$_3$Sn)$_3$SnLi·3THF
Sn$_4$LiC$_{21}$H$_{51}$O$_3$
Synth.: By rxn. of SnCl$_2$ with Me$_3$SnLi in THF in Ar atm. (2127).
By decomposition of Me$_3$SnLi in THF during THF removal (2127).
Prop.: White suspension in THF, very sensitive to air, dec. on standing, in org. solvents. IR, H, ^{7}Li NMR and mass spectra (2127).
Rxn. with: Thermal dec. at 60° → Me$_6$Sn$_2$ (2127).
Ph$_3$SnCl → Ph$_6$Sn$_2$ (2127).
Br → Me$_3$SnBr (2127).
MeBr → (Me$_3$Sn)$_3$SnMe (?) (2127).
Mn(CO)$_5$Br → Me$_3$SnMn(CO)$_5$ (2127).

Triethyltin Lithium (M-498)   SnLiC$_6$H$_{15}$ Et$_3$SnLi
Synth.: By rxn. of Et$_3$SnCl with excess Li in THF in N atm. (1928, 2179).
Prop.: Brown THF soln. (1928).
Rxn. with: Me$_2$SiCl$_2$ → (Et$_3$Sn)$_2$SiMe$_2$ + Et$_6$Sn$_2$ (11a).
EtBr → Et$_6$Sn$_2$ (1928).
Me$_2$C:CHCH$_2$Br → Et$_3$SnCH$_2$CH:CMe$_2$ (1619).
Me$_2$C(CN)I → Et$_3$SnN:C:CMe$_2$ + Et$_3$SnCN + Et$_6$Sn$_2$ (1620).
PhSO$_2$Cl → Et$_2$Sn(O$_2$SPh)$_2$ (2646).
ClC⋮CPO$_3$Et$_2$ → Et$_6$Sn$_2$ + HC⋮CPO$_3$Et$_2$ (2179).

Triethyltin Sodium (M-498)   SnNaC$_6$H$_{15}$ Et$_3$SnNa
Synth.: By rxn. of Et$_3$SnCl with Na in liq. NH$_3$ (2180-1).
By rxn. of Et$_6$Sn$_2$ with C$_{10}$H$_8$-Na (2521).
Rxn. with: Et$_3$SnC⋮CCl → Et$_6$Sn$_2$ + NaC⋮CCl (2180).
Et$_3$SnC⋮CBr → (Et$_3$SnC⋮)$_2$ (2180).
BrC⋮CNa, + NH$_4$Cl → Et$_3$SnC⋮CH + (Et$_3$SnC⋮)$_2$ (2181).
BrC⋮CH → Et$_3$SnC⋮CBr + (Et$_3$SnC⋮)$_2$ (2180).
MeC⋮CCHMeX → Et$_3$SnCMe:C:CHMe (881).
(BrC⋮CCH$_2$)$_2$ → Et$_3$SnC⋮CCH$_2$CH$_2$C⋮CH (629).
MePCl$_2$ → (Et$_3$Sn)$_2$PMe (58, 703).
D$_2$O → Et$_3$SnD (2521).

Tributyltin Lithium (M-496) SnLiC$_{12}$H$_{27}$ Bu$_3$SnLi

Rxn. with: Thermal dec. in THF → comparative kinetic data (68a, 176).
Me$_3$SiCl at -20° → Bu$_3$SnSiMe$_3$ (11a).
(CH$_2$CHCl)$_x$ → Bu$_3$Sn-contg. PVC (2598, 2700).
PhCH$_2$C(CN)$_2$Br in THF → Bu$_3$SnN:C:C(CN)CH$_2$Ph (2615).
Use: As catalyst for diene copolymerization (1639a).

Triphenyltin Lithium (M-497) SnLiC$_{18}$H$_{15}$ Ph$_3$SiLi

Prop.: Anal. detn. by double titration with allylbromide (68a, 173), with RBr (1928).
Rxn. with: Thermal dec. in THF → comparative kinetic data (68a, 176).
(Ph$_3$Sn)$_4$Sn → Ph$_6$Sn$_2$ (68a, 175).
SnCl$_2$ → (Ph$_3$Sn)$_3$SnLi (68a, 175).
SnCl$_2$ → Ph$_4$Sn + (Ph$_3$Sn)$_4$Sn + Ph$_4$Sn (68a, 175).
SnCl$_2$, + Ph$_3$SnCl (3:1:1) → (Ph$_3$Sn)$_4$Sn + Ph$_6$Sn$_2$ (68a, 175).
Me$_3$SiCl at -20° → Ph$_3$SnSiMe$_3$ (11a).
Me$_3$SiCl at 24° → Ph$_3$SnSiMe$_3$ + Ph$_6$Sn$_2$ (11a).
R$_2$SiCl$_2$ → (Ph$_3$Sn)$_2$SiR$_2$ + Ph$_6$Sn$_2$; R = Me, Ph (11a).
c.(Ph$_2$Si)$_n$ → Ph$_6$Sn$_2$ + Ph$_4$Sn; n = 4, 5 (1928).
SiCl$_4$ → Ph$_6$Sn$_2$ + (Ph$_3$Sn)$_4$Si (1928).
(C$_6$F$_5$)$_3$GeCl → Ph$_6$Sn$_2$ + Ph$_3$SnCl (1034).
PhLi, + PhCH$_2$Cl → Ph$_4$Sn + PhCH$_2$SnPh$_3$ (1928).
C$_8$H$_{17}$F → n-C$_8$H$_{17}$SnPh$_3$ + Ph$_6$Sn$_2$ (1928).
BuCl → BuSnPh$_3$ + Ph$_4$Sn + Ph$_6$Sn$_2$ (1928).
CH$_2$:CHCH$_2$X → CH$_2$:CHCH$_2$SnPh$_3$ + Ph$_6$Sn$_2$; X = Cl, Br (1928).
MeCH:CHX → Ph$_6$Sn$_2$; X = Cl + MeCH:CHSnPh$_3$; X = Br (1928).
ClCH$_2$CHCH$_2$O → Ph$_6$Sn$_2$ + (Ph$_3$Sn)$_2$O (1928).
(CH$_2$CHCl)$_x$ → Ph$_3$Sn-contg. PVC (2598, 2700).
s-2,4-X$_2$-6-MeNHC$_3$N$_3$ → 2-Ph$_3$Sn-4,6-X(MeNH)C$_3$H$_3$; X = Cl, Br (1322).
RBr → Ph$_3$SnR; R = HC!CCH$_2$, HC!CCHMe; prod. contg. isomeric allylic deriv. (444).
Br(CH$_2$)$_n$Br → Ph$_6$Sn$_2$; n = 2 + Ph$_4$Sn, n = 3 (1928).
Br(CH$_2$)$_n$Br → Ph$_3$Sn(CH$_2$)$_n$SnPh$_3$; n = 4, 5, 6 (1518, 2189).
C$_6$F$_5$Br → Ph$_3$SnBr + C$_6$F$_5$Li (1034).
Aq. NH$_4$Br → Ph$_6$Sn$_2$ (1928).
4,2,6-I(Me$_3$C)$_2$C$_6$H$_2$OH → 4,3,5-Ph$_3$SnC$_6$H$_2$OH(CMe$_3$)$_2$ (535).
Ph$_2$POCl → Ph$_3$SnOPPh$_2$ (1840).
Ph$_n$MCl$_{3-n}$ → (Ph$_3$Sn)$_{3-n}$MPh$_n$; n = 0-2, M = As (1842), Sb (1843).
RCOCl at -78° → Ph$_3$SnCOR; R = Me (1677, 2671), Me$_3$C, Ph (2671).
R$_3$PO$_4$ → RSnPh$_3$ + Ph$_6$Sn$_2$; R = Me, Bu (1928).

Triphenyltin Sodium (M-498) SnNaC$_{18}$H$_{15}$ Ph$_3$SnNa

Synth.: By rxn. of Ph$_3$SnCl with Na-C$_{10}$H$_8$ in (MeOCH$_2$)$_2$ at -15° (395).
By rxn. of Ph$_6$Sn$_2$ with Na-C$_{10}$H$_8$ (2521).
Prop.: Dark purple soln. (395).
Rxn. with: Ph$_2$SnCl$_2$ (2:1) → (Ph$_3$Sn)$_2$SnPh$_2$ (395).
(C$_5$H$_5$)$_2$TiCl$_2$ → (Ph$_3$Sn)$_2$Ti(C$_5$H$_5$)$_2$ (907).

$(C_5H_5)_2MCl_2 \to Ph_3SnM(C_5H_5)_2Cl$; M = Ti, Zr (907), Zr, Hf (2492).
$(C_5H_5)_2TiCl$ in THF $\to Ph_3SnTi(C_5H_5)_2 \cdot THF$ (907).
$MeCH:CHCH_2Cl \to MeCH:CHCH_2SnPh_3$ (1367, 1877).
$(CH_2CH_2Br)_2 \to (Ph_3SnCH_2CH_2)_2$ (1518).
$D_2O \to Ph_3SnD$ (2521).

Tris(triphenyltin)tin Lithium $Sn_4LiC_{54}H_{45}$ $(Ph_3Sn)_3SnLi$
<u>Synth.</u>: By rxn. of Ph_3SnLi with $SnCl_2$ in THF (68a, 175).
<u>Rxn. with</u>: $Bu_3SnCl \to (Ph_3Sn)_4Sn$ (68a, 175).
$Ph_3GeCl \to (Ph_3Sn)_3SnGePh_3$ (?) (68a, 175).
$MeI \to Ph_6Sn_2 + (Ph_3Sn)_3SnMe$ (68a, 175).
$Me_3PO_4 \to Ph_6Sn_2 + (Ph_3Sn)_3SnMe$ (68a, 175).

Listing of organotin alkali metal compounds concludes in Table 255.

Table 255. Organotin Alkali Metal Compounds R_nSnM_{4-n}

Formula	Synth. Method*	Yield	Properties	Ref.
$MePh_2SnLi$	IA	94	Green-brown THF-soln. (a)	68a, 176
$(\overline{CH_2CH_2CH})_3SnNa$	--	--	(b)	--
Bu_3SnNa	II	--	(c)	2521
Bu_2SnNa_2	II	--	(c)	2521
$[\overline{CH_2CH_2CH(CH_2)_5}]_3SnNa$	--	--	(d)	--
Ph_3SnK	IB	--	(e)	2128

* The characters used here correspond to the synthetic scheme in the introduction to Chapter 6.4.
(a) Relative kinetics of decomposition in THF (68a, 176). (b) Rxn. with $YCl_2 \to [(c.C_3H_5)_3Sn]_2Y$; Y = CH_2, $1,4-C_{10}H_6$ (1873). (c) Rxn. with $D_2O \to Bu_nSnD_{4-n}$; n = 2, 3 (2521). (d) Rxn. with $1,4,2,5-Cl_2C_6H_2Me_2 \to 1,4,2,5-(R_3Sn)_2C_6H_2Me_2$; R = $\overline{CH_2CH_2CH(CH_2)_5}$ (1873). (e) Rxn. with $Ph_3GeBr \to Ph_6Ge_2 + Ph_6Sn_2 + Ph_3SnGePh_3$ (2128).

6.5 BIMETALLIC ORGANOTIN DERIVATIVES OF TRANSITION METAL COORDINATION COMPOUNDS

6.5.1 ORGANOTIN COMPOUNDS WITH GROUP I-V METALS

The compounds listed in Table 256 are prepared by methods from the following scheme.

I Reactions of triphenyltin hydride with:
   (IA) $RMCl \cdot L$; M = Zn, Cd,
   (IB) Me_2Cd,
   (IC) $M(NMe_2)_4$, M = Ti, Zr.

II Reaction of Bu$_3$SnCl with NaTa(CO)$_6 \cdot 3$(MeOCH$_2$)$_2$.

III Reaction of triphenyltin sodium with (C$_5$H$_5$)$_2$MCl$_n$; M = Ti, Zr and Hf, n = 2 and 1.

R = organic group, L = ligand.

Bistriphenyltin Mercury   Sn$_2$HgC$_{36}$H$_{30}$ (Ph$_3$Sn)$_2$Hg
<u>Synth.</u>: By rxn. of Ph$_3$SnH with [(Me$_3$Si)$_2$N]$_2$Hg (2:1) at r.t. (2370).
<u>Prop.</u>: Bright yellow solid, dec. 100°, stable at r.t., sensitive to light, soly. (2370).
<u>Rxn. with</u>: THF at 15° → Ph$_6$Sn$_2$ + Hg (2370).
Br in C$_6$H$_6$ → Ph$_3$SnBr + Ph$_2$SnBr$_2$ + Ph$_6$Sn$_2$ (2370).
(CH$_2$Br)$_2$ → Ph$_6$Sn$_2$ + Hg (2370).

Listing of bimetallic organotin compounds containing Group I-V transition metals concludes in Table 255; additional references may be found in Chapter 8.1.

Table 256. Organotin Derivatives of Group I-V Transition Metals   R₃SnML

Formula	Synth. Method*	Yield	Properties	Ref.
$(Ph_3SnAg)_x$	(a)	--	Stable red soln., polarography (b)	963
$(Ph_3Sn)_2Zn\cdot(MeOCH_2)_2$	IA	48	m. 103-104°	2043
$(Ph_3Sn)_2Zn\cdot(Me_2NCH_2)_2$	IA	84	m. 172-74°	2043
$(Ph_3Sn)_2Zn\cdot BiPy$	IA	85	m. 141-44°	2043
$Ph_3SnZnCl$	IA	--	Yellow solid, dec. 100°, mol. wt., struct. (c)	2043, 2830
$Ph_3SnZnCl\cdot(MeOCH_2)_2$	IA	61	m. 73-76°, struct. (d)	2043
$Ph_3SnZnCl\cdot(Me_2NCH_2)_2$	IA	87	m. 158-59°, struct. (d)	2043
$(Ph_3Sn)_2Cd\cdot(MeOCH_2)_2$	IB	30	m. 110° (dec.)	2043
$(Ph_3Sn)_2Cd\cdot(Me_2NCH_2)_2$	IA	85	m. 175° (dec.)	2043
$(Ph_3Sn)_2Cd\cdot BiPy$	IA	87	m. 154° (dec.)	2043
$Ph_3SnCdCl\cdot(Me_2NCH_2)_2$	IA	82	m. 175° (dec.)	2043
$(Ph_3Sn)_4Ti$	IC	30	Ochre, m. 130°, sensitive to air	2331
$(Ph_3Sn)_2Ti(C_5H_5)_2$	III	--	Green, m. 80° (dec.), diamagnetic (e)	907
$[(Ph_3Sn)_2TiO]_x$	IC (f)	--	White, m. > 260° (dec.), sensitive to air	2331
$(Ph_3Sn)_2Ti(OMe)_2$	IC (g)	--	Yellow, m. > 260° (dec.), sensitive to air	2331
$Ph_3SnTi(C_5H_5)_2Cl$	III	--	Dark green, m. 177-80°, diamagnetic (e)	907
$Ph_3SnTi(C_5H_5)_2\cdot THF$	III	--	Emerald green, m. 80° (dec.) (h)	907
$(Ph_3Sn)_4Zr$	IC	--	Yellow, m. 70-73°, mol. wt., sensitive to air	2331
$[(Ph_3Sn)_3Zr]_2SnPh_2$	IC (i)	22	Yellow, m. 100° (dec.), sensitive to air	2331
$[(Ph_3Sn)_2ZrSnPh_2]_x$	IC (i)	--	Orange, m. 160° (dec.), mol. wt., sensitive to air	2331
$Ph_3SnZr(C_5H_5)_2Cl$	III	--	Pale yellow, NMR spectrum, crude	907
	III	--	Orange, NMR spectrum, crude	2492
$Ph_3SnHf(C_5H_5)_2Cl$	III	--	Yellow solid, NMR spectrum, crude	2492
$Bu_3SnTa(CO)_6$	II	--	Red soln., IR spectrum	1304

* The characters used here correspond to the synthetic scheme in the introduction to Chapter 6.5.1.
(a) Synth. by polarographic reduction of Ph₆Sn₂ in (MeOCH₂)₂-Bu₄NClO₄, followed by AgClO₄ (963). (b) Plates silver on electrolysis (963). (c) Rxn. with Ph₃SnCl in THF → Ph₄Sn + (Ph₂Sn)ₓ (2830), with MeOH or H₂O → C₆H₆ + (Ph₂Sn)ₓ (2830). (d) Rxn. with Ph₃SnCl in THF → Ph₄Sn (2830). (e) IR and NMR spectra (907). (f) Synth. followed by H₂O in N atm. (2331). (g) Synth. followed by MeOH in N atm. (2331). (h) IR spectrum, paramagnetic (907). (i) Synth. followed by PhNHCHO, and by Ph₂SnH₂ (2331).

6.5.4 ORGANOTIN COMPOUNDS WITH GROUP VI METALS

This chapter comprises organotin compounds containing Group VI transition metal complexes. Existence of a tin-metal bond has recently been proven for complexes containing carbonyl groups as well as carbonyl and cyclopentadienyl groups. The compounds listed in Tables 257 and 258 are prepared by methods from the following scheme.

I   Reaction of organotin halides R_nSnX_{4-n} with sodium salts of transition metal complex NaX'.

II Reaction of organochlorotin transition metal complex $R_nSnCl_{3-n}X'$ with LiC_6F_5.

III Addition of organotin halides R_nSnX_{4-n} to transition metal carbonyl amine complex $LM(CO)_4$; M = Mo, W.

IV Carbonyl exchange reaction of an organotin transition metal carbonyl R_3SnX' with:
    (IVA) PR'_3,
    (IVB) C_2H_4.

R = organic group, R' = R or OR, M' = Cr, Mo, W, M^2 = Mn, Re, X = halogen, X' = NCM'(CO)₅, $C_5H_5Cr(CO)_3$, $C_5H_5Mo(CO)_2PR'_3$, $M^2(CO)_5$; n = 2, 3, L = 2,2'-bipyridyl and 1,10-phenanthroline.

Triphenyltin-π-cyclopentadienylchromiumtricarbonyl (M-500)
    $SnCrC_{26}H_{20}O_3$ $Ph_3Sn(C_5H_5)Cr(CO)_3$
Synth.: By rxn. of Ph₃SnCl with NaC₅H₅Cr(CO)₃ in diglyme in N atm., in 46% yield (1670).
Prop.: Yellow, m. 219-21°, IR and NMR spectra, dec. on standing in inert atm., sensitive to air (1670), x-ray struct. analysis (1996).
Use: As metal plating agent (603).

Trimethyltin-π-cyclopentadienylmolybdenumtricarbonyl
    $SnMoC_{11}H_{14}O_3$ $Me_3Sn(C_5H_5)Mo(CO)_3$
Synth.: By rxn. of Me₃SnCl with NaC₅H₅Mo(CO)₃ in THF in N atm., in 48% yield (1670).
By rxn. of Me₃SnNMe₂ with HC₅H₅Mo(CO)₃ in THF at 70° (2203).
Prop.: Pale yellow (1670, 2303), m. 98-99° (2303), 97-98.5° (1670), subl.$_{0.001}$ 100° (2303), IR (415, 1670, 2303), far IR (828, 2303), Raman (2303), UV (1760), NMR (415, 1670, 2303) and mass (2303) spectra, polarography (969), stable in air (1670) for days (2303).

<u>Rxn. with</u>: P(OPh)$_3$ under UV irradiation → Me$_3$Sn(C$_5$H$_5$)Mo(CO)$_2$P(OPh)$_3$ (2491).

Triphenyltin-π-cyclopentadienylmolybdenumtricarbonyl (M-500)

                SnMoC$_{26}$H$_{20}$O$_3$ Ph$_3$Sn(C$_5$H$_5$)Mo(CO)$_3$

<u>Synth.</u>: By rxn. of Ph$_3$SnCl with NaC$_5$H$_5$Mo(CO)$_3$ in EtOH-petr. ether (967), in THF (1459), in N atm., in 39% yield (1670).

By-prod. in polarographic reduction of Ph$_3$SnCl with [C$_5$H$_5$Mo(CO)$_3$]$_2$ (967).

<u>Prop.</u>: White cryst., m. 211-12° (1459), pale green, m. 211-14° (1670), m. 207° (967), IR (1459, 1670), far IR (828), UV (967, 1670) and NMR (1459, 1670, 2057) spectra, polarography (967-9), stable in air (1459, 1670), x-ray struct. anal. (1996).

<u>Rxn. with</u>: Polarographic reduction with [C$_5$H$_5$Fe(CO)$_2$]$_2$ → Ph$_3$Sn(C$_5$H$_5$)Fe(CO)$_2$ (968).

<u>Use</u>: As metal plating agent (603).

Trimethyltin-π-cyclopentadienyltungstentricarbonyl

                SnWC$_{11}$H$_{14}$O$_3$ Me$_3$Sn(C$_5$H$_5$)W(CO)$_3$

<u>Synth.</u>: By rxn. of Me$_3$SnCl with NaC$_5$H$_5$W(CO)$_3$ in THF-diglyme in N atm., in 90% yield (1670).

By rxn. of HC$_5$H$_5$W(CO)$_3$ with Me$_3$SnNMe$_2$ in THF (825).

<u>Prop.</u>: Colorless cryst., m. 120° (825), pale yellow, m. 119-20° (1670), IR (828, 1670, 2303), far IR (828, 2303), Raman (2303), UV (1670) and NMR (1670, 2303) spectra, stable in air (825), for days (2303), dec. in soln. (825).

Listing of organotin compounds containing Group VI transition metal substituents concludes in Table 257; additional derivatives may be found in Chapters 4.5.6 and 6.6.

Table 257. Organotin Compounds with Group VI Transition Metals $R_nSnM(CO)_mL_o$

Formula*	Synth. Method**	Yield	Properties	Ref.
$Me_3SnNCCr(CO)_5$	I	43	White, m. 148-50° (dec.), subl.o.1 135° (a)	1314
$Me_3SnPhCr(CO)_3$ (500)	--	--	NMR spectrum	2476
$Me_3Sn(C_5H_5)Cr(CO)_3$	I	--	Yellow cryst., m. 109°, subl.o.001 80° (b)	2303
$Me_2SnCr(CO)_5$	--	--	(c)	--
$Et_2SnCr(CO)_5$	--	--	(c)	--
$Bu_2SnCr(CO)_5$	--	--	(c)	--
$1-Bu_2SnCr(CO)_5$	--	--	(c)	--
$Me_3SnNCMo(CO)_5$	I	10	White, m. 152-53°, subl.o.1 130-50° (a)	1314
$Me_3SnPhMo(CO)_3$	--	--	NMR and Moessbauer resonance spectra, mol. wt.	2389
$Me_3Sn(C_5H_5)Mo(CO)_2P(OPh)_3$	I	18	"Insoluble" isomer, pale yellow cryst., m. 209-10° (d)	2491
	IVA	75	"Soluble" isomer, white cryst., m. 137-38° (e)	2491
$Me_2Sn[C_5H_5Mo(CO)_3]_2$ (500)	I	57	Yellow, m. 156-59° (dec.), stable in air (f, g)	1670
$Me_2SnCl(C_5H_5)Mo(CO)_3$	I	50	Pale yellow, m. 80-90° (f, h)	1670
$MeSnCl_2(C_5H_5)Mo(CO)_3$	--	--	(Far) IR spectra	828
$MeSnCl_2BiPyMoCl(CO)_3$	III	--	Orange cryst., dec. 190°, soly., struct. (i)	1372
$MeSnCl_2LMoCl(CO)_3$ (j)	III	--	Yellow-orange cryst., dec. 180°, IR spectrum	1372
$MeSnI_2BiPyMoI(CO)_3$	III	--	IR spectrum	1372
$CH_2:CHSnCl_2BiPyMoCl(CO)_3$	III	--	Orange cryst., dec. 161°, IR spectrum	1372
$BuSnCl_2BiPyMoCl(CO)_3$	III	--	Yellow cryst., m. 166-68° (dec.), IR spectrum	1372

* Numbers in parenthesis refer to pages in main volume.
** The characters used here correspond to the synthetic scheme in the introduction to Chapter 6.5.4.
(a) IR, UV and NMR spectra, force const., stable in air (1314). (b) (Far) IR, Raman and assignment, NMR and mass spectra, stable in air (2303). (c) Use as metal plating agent (603). (d) IR spectrum (2491). (e) IR and NMR spectra (2491). (f) IR and NMR spectra (1670). (g) Polarography (969). (h) (Far) IR spectra (828). (i) IR and NMR spectra (1372), x-ray struct. analysis (2377). (j) L = 1,10-Phenanthroline.

Table 257 Continued $R_nSnM(CO)_mL_o$

Formula*	Method**	Yield	Properties	Ref.
$Ph_3Sn(C_5H_5)Mo(CO)_2P(OMe)_3$	I	--	White cryst. m. 181-83°, (k, l)	1459
$Ph_3Sn(C_5H_5)Mo(CO)_2PPh_3$	I	--	White cryst., m. 266-68° (dec.) (k, l)	1459
$Ph_3Sn(C_5H_5)Mo(CO)_2P(OPh)_3$	I	--	White cryst., m. 170-71° (k)	1459
$Ph_2Sn[C_5H_5Mo(CO)_3]_2$ (500)	--	--	IR spectrum (1670), x-ray struct. anal.	1996
$Ph_2SnClBiPyMoCl(CO)_3$	III	--	Yellow cryst., m. 212° (dec.), conductance (m)	1372
$PhSnCl[C_5H_5Mo(CO)_3]_2$	I	--	Orange, m. 142-44°	1596
$PhSnCl_2BiPyMoCl(CO)_3$	--	--	Yellow-orange, dec. 168°, IR spectrum	1372
$Me_3SnNCW(CO)_5$	I	22	White cryst., m. 177-78°, subl.o.1 130° (a, n)	1314
$Me_2SnCl(C_5H_5)W(CO)_3$	--	--	(Far) IR spectra	828
$MeSnCl_2(C_5H_5)W(CO)_3$	--	--	(Far) IR spectra	828
$MeSnCl_2BiPyWCl(CO)_3$	III	--	Golden yellow cryst., dec. 215°, conductance (m, o)	1372
$MeSnCl_2LWCl(CO)_3$ (j)	III	--	Yellow-orange cryst., dec. 195° (m)	1372
$MeSnI_2BiPyWI(CO)_3$	(o)	--	Red cryst.	1372
$CH_2:CHSnCl_2BiPyWCl(CO)_3$	III	--	Orange cryst., dec. 192°, IR spectrum	1372
$CH_2:CHSnCl_2LWCl(CO)_3$ (j)	III	--	Orange cryst., dec. 218°, IR spectrum	1372
$BuSnCl_2BiPyWCl(CO)_3$	III	--	Orange cryst., m. 153-55° (dec.) (m)	1372
$Ph_3Sn(C_5H_5)W(CO)_3$ (500)	I	57	Pale green, m. 227-28°, stable in air (f, p)	1670
$Ph_2Sn[C_5H_5W(CO)_3]_2$ (500)	--	65	X-ray struct. anal.	1996
$Ph_2SnClBiPyWCl(CO)_3$	I	--	Red cryst., m. 225° (dec.) (m)	1372
$PhSnCl[C_5H_5W(CO)_3]_2$	I	--	Orange, m. 183-84°	1596
$PhSnCl_2BiPyWCl(CO)_3$	III	--	Orange, dec. 184°, IR spectrum	1372
$Ph_nSn[M(CO)_m]_o$	--	--	Moessbauer resonance spectrum	1115

(k) IR and NMR spectra (1459). (l) (Far) IR spectra and assignment (2560). (m) IR spectrum, soly (1372). (n) Mass spectrum (1314). (o) Rxn. of $MeSnCl_2BiPyWCl(CO)_3$ with NaI → $MeSnI_2BiPyWI(CO)_3$ (1372). (p) UV spectrum (1670).

6.5.5 ORGANOTIN COMPOUNDS WITH GROUP VII METALS

Trimethyltin Manganesepentacarbonyl $SnMnC_8H_9O_5$ $Me_3SnMn(CO)_5$
Synth.: By rxn. of $NaMn(CO)_5$ in THF in N atm. with Me_3SnCl, in 80% yield (1252), with Me_3SnBr (874).
By rxn. of $(Me_3Sn)_3SnLi \cdot 3THF$ with $Mn(CO)_5Br$ in THF (2127).
Prop.: White cryst., m. 29.5° (874), m. 26-27° (1252), $b_{0.001}$ 47° (874), subl. in vacuo 0° (1252), d_{30} 1.62 (874), far IR (828), assignment (874, 1271), IR (1252, 2337), assignment (874) and NMR (1220, 1252, 2034) spectra, force const. (1132, 1252, 2337), polarography (969), stable at 130° and under UV irradiation (874), x-ray struct. analysis (801).
Rxn. with: $Ph_3PAuCl \rightarrow Me_3SnCl + Ph_3PAuMn(CO)_5$ (2202).
$Ph_2PCl \rightarrow Me_3SnCl + [Ph_2PMn(CO)_4]_2 + CO$ (2202).
$Me_2AsI \rightarrow [Me_2AsI \cdot 2Mn(CO)_3]_2 + CO$ (2202).
$Br \rightarrow Me_3SnBr + BrMn(CO)_5$ (2202).
$C_2H_4{}^* \rightarrow Me_3SnMn(CO)_4C_2H_4 + CO$ (874).
$C_2F_4{}^* \rightarrow Me_3SnF + Me_3SnCF_2CF_2Mn(CO)_5 + [CF_2:CFMn(CO)_4]_2$ (87) $+ C_5F_9Mn(CO)_5 + CF_2:CFCOMn(CO)_5 + CO$ (874).
$C_2HF_3{}^* \rightarrow Me_3SnCF:CF_2 + CHF:CFMn(CO)_5 + CO$ (874).
$C_2ClF_3{}^* \rightarrow Me_3SnF + CF_2:CFCOMn(CO)_5 + ClCF:CFMn(CO)_5$ (?) $+ CO$ (874).
$(CF_2:CF)_2{}^* \rightarrow \underline{Me_3SnF + C_5F_5Mn(CO)_5}$ (1141).
$(CF_3C!)_2{}^* \rightarrow Me_3SnC(CF_3)CMn(CO)_5(CF_3)C(CF_3):CCF_3$ (733).
Addn. compd.: $Me_3SnMn(CO)_5 \cdot 2.5SO_2$, from components in 30-40% yield, pale yellow solid; dec. 150° $\rightarrow Me_4Sn + SO_2 + CO_2$, struct., IR spectrum (827).
$Me_3SnMn(CO)_5 \cdot 1.5SO_2$, from components, in 60-70% yield, red solid, (far) IR and NMR spectra, dec. > 200°, dimeric in $HCCl_3$ (827) $[Sn_2Mn_2C_{16}H_{18}O_{16}S_3]$.

Dimethylchlorotin Manganesepentacarbonyl $SnMnC_7H_6ClO_5$ $Me_2SnClMn(CO)_5$
Synth.: By rxn. of Me_2SnCl_2 with $NaMn(CO)_5$ (1:1) in THF in N atm. (1252).
Prop.: m. 94.5-96.5° (1252), far IR (828), IR (828, 1252, 2337) and NMR (1252) spectra, force const. (1132, 1252, 2337).
Rxn. with: $C_6F_5M \rightarrow Me_2Sn[Mn(CO)_5]_2 + Me_2Sn(C_6F_5)_2$; M = Li, MgBr (2035).
$MC_5H_5Mo(CO)_3 \rightarrow C_5H_5Mo(CO)_3SnMe_2Mn(CO)_5$ (415).

Triphenyltin Manganesepentacarbonyl (M-499) $SnMnC_{23}H_{15}O_5$ $Ph_3SnMn(CO)_5$
Synth.: By polarographic reduction of Ph_3SnCl with $Mn_2(CO)_{10}$ (967).
Prop.: White m. 151° (2057), m. 150-52° (1252), m. 140-42° (967), far IR (788, 828, 1065), assignment (2449), IR (1252, 2056-7, 2337), assignment (770, 2449), Raman, assignment (770), UV (967), Moessbauer resonance (1282) and mass (1419) spectra, dipole moment (2057), force const. (1132, 1252, 2337), polarography (967, 969), x-ray struct. analysis (1458, 2119).

---

* Rxn. under UV irradiation.

Rxn. with: $R_3P \rightarrow Ph_3SnMn(CO)_4PR_3$; R = Bu, PhO (770).
$(CF_3C!)_2$ under UV irradiation $\rightarrow Ph_3Sn\overline{C(CF_3)CMn(CO)_5(CF_3)C(CF_3)}:CCF_3$ (733).
Addn. compd.: $Ph_3SnMn(CO)_5 \cdot 2.5SO_2$, from components, yellow cryst., dec. > 200° dimeric in $HCCl_3$, IR spectrum, soly. (827) $[Sn_2Mn_2C_{46}H_{30}O_{20}S_5]$.
Use: As metal plating agent (603).

Triphenyltinoxytetraphenylcyclopenta- $SnMnC_{50}H_{35}O_4$ $Ph_3SnOC_5Ph_4Mn(CO)_3$
dienyl Manganesetricarbonyl (M-501)

Synth.: By rxn. of Ph_3SnCl with $LiOC_5Ph_4Mn(CO)_3$ in C_6H_6 at 10°, in 53% yield (1124).
By polarographic reduction of Ph_3SnCl and $[C_5Ph_4Mn(CO)_3]_2$ (967).
Prop.: m. 206-207° (1124), IR (1124) and UV (967) spectra, polarography (964, 967), soly. (1124), x-ray struct. analysis, originally formulated with Sn-Mn bond (804).

Diphenylchlorotin Manganesepentacarbonyl $SnMnC_{17}H_{10}ClO_5$ $Ph_2SnClMn(CO)_5$
(M-501)

Prop.: White solid (2057), m. 102-103.5° (2035), 94-95° (2057), IR (2053, 2057) and Moessbauer resonance (1282) spectra, force const. (1132, 2035).
Rxn. with: $C_6F_5Li \rightarrow [Ph_2SnMn(CO)_5]_2 + Ph_2(C_6F_5)SnMn(CO)_5$ (2035).
$NaX \rightarrow Ph_2XSnMn(CO)_5$; X = $C_5H_5Mo(CO)_3$, $C_5H_5W(CO)_3$, $Re(CO)_5$, $Co(CO)_4$ (387).

Triphenyltin Rheniumpentacarbonyl (M-501) $SnReC_{23}H_{15}O_5$ $Ph_3SnRe(CO)_5$
Prop.: Pale yellow, m. 145-46° (2057), white, m. 144.5-46° (1252), IR (1252, 2057, 2337), assignment (770), and Moessbauer resonance (1282) spectra, force const. (1252, 2337), polarography (969), x-ray struct. analysis (1996).
Rxn. with: $MPh_3 \rightarrow Ph_3SnRe(CO)_4MPh_3$; M = P, As, Sb (1594).

Additional organotin compounds containing groups VII transition metal complex substituents are listed in Table 258. Polymetallic organotin complexes are summarized in Chapter 6.6.

Table 258. Organotin Compounds with Group VII Transition Metals $R_nSnM(CO)_nL_o$

Formula*	Synth. Method**	Yield	Properties	Ref.
$Me_3Sn(C_2H_4)Mn(CO)_4$	IVB	43	Yellow oil, subl.o.oo1 25-30°, mol. wt.	874
$Me_2(CF_3)SnMn(CO)_5$	--	--	(Far) IR spectra	828
$Me_2Sn[Mn(CO)_5]_2$ (501) (b)	II	--	White cryst., m. 104-105° (c-f)	2035
$Me_2SnBrMn(CO)_5$	--	--	Far IR spectrum	828
$Me_2SnIMn(CO)_5$	--	--	Force const. (1132), far IR spectrum	828
$MeSn[Mn(CO)_5]_3$	I	> 50	Yellow cryst., m. 120-55° (dec.), subl. in vacuo (d, g)	2034
$MeSnCl_2Mn(CO)_5$	--	--	Force const. (1132), (far) IR spectra	828
$PhCH_2Sn[Mn(CO)_5]_3$	--	--	(h)	--
$Et_3SnMn(CO)_5$	I	--	Colorless liq. bo.3 130°, IR spectrum	2057
$CH_2{:}CHSn[Mn(CO)_5]_3$	I	> 50	Yellow cryst., m. 120-40°, subl. in vacuo (g, i)	2034
$BuSn[Mn(CO)_5]_3$	I	> 50	Yellow cryst., m. 143-45°, subl. in vacuo (g, i)	2034
$Ph_3SnPhMn(CO)_2$	--	--	(j)	--
$Ph_3SnRMn(CO)_3$	--	--	R = $(CH_2{:}CH)_2$ (j)	--
$Ph_3SnMn(CO)_4PBu_3$	IVA	--	IR spectrum and assignment	770
$Ph_3SnMn(CO)_4PPh_3$ (501)	--	--	White, m. 228°, trans-form, dipole moment (j, k)	2057
$Ph_3SnMo(CO)_4P(OPh)_3$	IVA	--	IR spectrum and assignment	770

* Numbers in parenthesis refer to pages in main volume.
** The characters used here correspond to the synthetic scheme in the introduction to Chapter 6.5.4.

(a) (Far) IR spectra and assignment (874). (b) Correct formula (M-501) line 1 to $Me_2Sn[Mn(CO)_5]_2$. (c) NMR spectrum (1252). (d) IR and NMR spectra (2034). (e) IR spectrum (2035). (f) Mass spectrum (2034). (g) Stable in air, light, H_2O, decomposes slowly in hydrocarbon solvent (2034). (h) Rxn. with R_3Sb in $PhCH_2OBu$ at 203-215° → $PhCH_2Sn[R_3SbMn(CO)_4]_3$; R = Me_2CHCH_2 (186). (i) IR spectrum, mol. wt. (2034). (j) Use as metal plating agent (603). (k) IR spectrum (2056-7) and assignment (770), x-ray struct. analysis (802, 1996).

Table 258 Continued $R_nSnM(CO)_nL_o$

Formula*	Synth. Method**	Yield	Properties	Ref.
$Ph_2Sn[Mn(CO)_5]_2$ (501)	--	--	White, m. 137°, IR spectrum (e, l)	2057
$[Ph_2SnMn(CO)_5]_2$	II	--	White cryst., m. 158-61° (dec.), mol. wt. (e)	2035
$Ph_2(C_6F_5)SnMn(CO)_5$	II	--	White cryst., m. 94-95°, stable in air (e, m)	2035
$Ph(C_6F_5)_2SnMn(CO)_5$	II (n)	--	White needles, m. 100-102°, stable in air (e, m, o)	2035
$PhSn[Mn(CO)_5]_3$	I	> 50	Orange-yellow cryst., m. 130-40°	2034
	I	--	Orange, m. 166-67° (g, i)	1596
$PhSnCl_2Mn(CO)_5$ (501)	--	--	White, m. 84-85°, IR spectrum (e, m, n)	2035
$(C_6F_5)_3SnMn(CO)_5$	II	--	Pale yellow cryst., m. 151-54°, stable in air (e, m)	2035
$[(C_6F_5)_2SnMn(CO)_5]_2$	II	--	m. 151-52°, stable in air, light, soly. (e)	2035
$(p-Me_2C_6H_3)_2Sn[Mn(CO)_5]_2$	--	--	(p)	--
$Me_3SnRe(CO)_5$	I	--	White, m. 51-58°, IR spectrum (c, d, q)	1252
$Me_2Sn[Re(CO)_5]_2$	I	--	Pale yellow cryst., mol. wt., soly. (c, d, g)	2034
$MeSn[Re(CO)_5]_3$	I	> 50	Yellow cryst., m. 160-65° (dec.), subl. in vacuo (d, g)	2034
$CH_2:CHSn[Re(CO)_5]_3$	I	> 50	Yellow cryst., m. 175-200° (dec.), subl. in vacuo (g, i)	2034
$BuSn[Re(CO)_5]_3$	I	> 50	Yellow cryst., m. 165-66°, subl. in vacuo (g, i)	2034
$Ph_3SnRe(CO)_4PPh_3$	IVA	--	White cryst., m. 218-20°, soly. (r, s)	1594
$Ph_2Sn[Re(CO)_5]_2$ (502)	I	--	White cryst., m. 135-38°, IR spectrum (s, t, u)	2034
$PhSn[Re(CO)_5]_3$ (502)	I	--	Yellow, m. 189° (g, l, s)	1596
	I	> 50	Yellow cryst., dec. > 195°, subl. in vacuo	2034
$Ph_nSn[M(CO)_m]_o$	--	--	Moessbauer resonance spectrum	1115

(l) Pale yellow cryst., x-ray struct. analysis (738, 1312). (m) Force const. (1132, 2035), soly. (2035). (n) Rxn. of $PhSnCl_2Mn(CO)_5$ with $C_6F_5MgBr \rightarrow Ph(C_6F_5)_2SnMn(CO)_5$ (2035). (o) Mass spectrum (2035). (p) Rxn. with AsR_3 at 190° $\rightarrow (p-Me_2C_6H_3)_2Sn[R_3AsMn(CO)_4]_2$; R = mesityl (186). (q) IR spectrum (2337), force const. (1252, 2337). (r) IR spectrum (1594). (s) Moessbauer resonance spectrum (1282). (t) Stable in air, light slowly oxidizing in soln. (2034). (u) Rxn. with HCl $\rightarrow Cl_2SnRe(CO)_5$ (2034).

6.5.6 ORGANOTIN COMPOUNDS WITH IRON AND RUTHENIUM

The organotin derivatives containing a direct tin-iron and tin-ruthenium bond listed in Table 259 are prepared by methods from the ensuing scheme.

I Reaction of trimetaldodecacarbonyl $M_3(CO)_{12}$ with:
    (IA) $Pr_2Sn(C\vdots CR)_2$,
    (IB) R_3SnH.

II Substitution reaction of tin halide:
    (IIA) R_nSnX_{4-n} with MX',
    (IIB) $X'SnCl_2$ with $MeMgX$,
    (IIC) $Ph_nSnX_{4-n} + Ca[HFe(CO)_4]_2$.

III Decarbonylation of $Fe(CO)_5$ or $[C_5H_5Fe(CO)_2]_2$ with:
    (IIIA) R_3SnCl,
    (IIIB) R_2SnCl_2.

IV Exchange reaction of phosphine with:
    (IVA) $R_3Sn(C_5H_5)Fe(CO)_2$,
    (IVB) With $[Bu_2SnFe(CO)_4]_2$.

R = organic group, M = alkali metal, n = 1-3, X = halogen, $X' = C_5H_5M'(CO)_2$, 1/2 $M'(CO)_4$; M' = Fe, Ru.

Bistrimethyltin Irontetracarbonyl $Sn_2FeC_{10}H_{18}O_4$ $(Me_3Sn)_2Fe(CO)_4$
<u>Synth.</u>: By rxn. of Me_3SnH with $Fe(CO)_5$ in N atm. at 70°, in 47-70% yield (904). By rxn. of Me_3SnCl with aq. $M_2Fe(CO)_4$ in N atm. at r.t. (1268).
<u>Prop.</u>: Colorless liq., m. 8° (904), IR spectrum (904, 1268, 1272, 2339), dipole moment (1272), force const. (2339), struct. (1268, 1272).
<u>Rxn. with</u>: Thermal dec. at 140° in sealed tube → $Me_4Sn + [Me_2SnFe(CO)_4]_2$ (904).

Trimethyltin-π-cyclopentadienyl- $SnFeC_{10}H_{14}O_2$ $Me_3Sn(C_5H_5)Fe(CO)_2$
   irondicarbonyl
<u>Synth.</u>: By rxn. of Me_3SnCl with $NaC_5H_5Fe(CO)_2$ in THF, in 54% yield (2490).
<u>Prop.</u>: d_{25} 1.71 (2490), far IR (828) and NMR (415, 2057, 2490) spectra.
<u>Rxn. with</u>: R_3P^* → $Me_3Sn(C_5H_5)Fe(CO)PR_3$; R = Ph, PhO (2490).
$(Ph_2P)_2Y^*$ → $Me_3Sn(C_5H_5)Fe(PPh_2)_2Y$; Y = CH_2CH_2, cis-CH:CH (2490).
$(CF_3C\vdots)_2^*$ → $Me_3SnC(CF_3):C(CF_3)C_5H_5Fe(CO)_2$ (733).

Dimethyltin Bis(π-cyclopentadienyl- $SnFe_2C_{16}H_{16}O_4$ $Me_2Sn[C_5H_5Fe(CO)_2]_2$
   irondicarbonyl) (M-502)
<u>Synth.</u>: By rxn. of $Cl_2Sn[C_5H_5Fe(CO)_2]_2$ with MeLi in THF at -80°, in 86% yield (1045).
<u>Prop.</u>: Orange plates (739, 741), m. 105-107° (2226), IR (1045, 2226) and NMR (1045, 2226) spectra, polarography (969), sensitive to air, struct. (1045), x-ray struct. analysis (739, 741, 2252).
<u>Use</u>: As metal plating agent (603).

---

* Rxn. under UV irradiation

Bis(dimethyltinirontetracarbonyl) (M-502) $Sn_2Fe_2C_{12}H_{12}O_8$ $[Me_2SnFe(CO)_4]_2$
Synth.: By rxn. of $Fe(CO)_5$ in N atm. at 100-110° with Me_2SnH_2 and Me_3SnH, in 64 and 10% yield, resp. (904).
By rxn. of Me_2SnCl_2 with $M_2Fe(CO)_4$ at r.t. in N atm. in H_2O (1268), in THF under reflux as by-prod. (2004).
By thermal dec. of $(Me_3Sn)_2Fe(CO)_4$ at 140° in sealed tube (904).
Prop.: Yellow cryst. (2004), IR (904, 1268, 2004, 2339) and Moessbauer resonance (1257) spectra, force const. (2339), struct. (1268).
Use: As metal plating agent (603).

Tetramethyltritin Tetrairondodeca- $Sn_3Fe_4C_{20}H_{12}O_{16}$ $Me_4Sn_3[Fe(CO)_4]_4$
  carbonyl*
Synth.: By rxn. of $MeSnCl_3$ with $Na_2Fe(CO)_4$ in N atm. in THF (2004).
By-prod. in rxn. of Me_3SnH with $Fe(CO)_5$ at 100-110° in N atm (904).
Prop.: Dark red cryst m 198-200° (dec.) (2004), IR (904, 2004, 2339), NMR (2004) and Moessbauer resonance (1257) spectra, dec. in air (2004), force const (2339), mol. wt. (2004), x-ray struct. analysis (2004).

Bis(diethyltinirontetracarbonyl) (M-502) $Sn_2Fe_2C_{16}H_{20}O_8$ $[Et_2SnFe(CO)_4]_2$
Synth.: By rxn. of $Et_2Sn(C\vdots CR)_2$ with $Fe_3(CO)_{12}$ in petr. ether at 100°; R = Pr, Bu and Ph, in 34, 32 and 12% yield resp. (1222).
By rxn. of Et_3SnCl with $Fe(CO)_5$ in N atm. under reflux, in 10% yield (904).
By rxn. of Et_2SnCl_2 with aq. $M_2Fe(CO)_4$ at r.t. in N atm. (1268).
Prop.: Pale yellow cryst. (904), m. 138° (dec.) (1222), IR (734, 904, 1222, 1268, 1272, 2339), assignment (1269), far IR, assignment (1269), Raman (734) and NMR (1222) spectra, dipole moment (1272), force const. (734 2339), mol. wt. (904, 1222), stable in air as solid, dec. in soln. in air (1222), struct. (1268, 1272).
Use: As metal plating agent (603).

Bis(dibutyltinirontetracarbonyl) (M-502) $Sn_2Fe_2C_{24}H_{18}O_8$ $[Bu_2SnFe(CO)_4]_2$
Synth.: By rxn. of $Fe(CO)_5$ with Bu_3SnCl in N atm. under reflux, in 32% yield (904), with Bu_6Sn_2, in low yield (904).
By rxn. of $Fe_3(CO)_{12}$ with $Bu_2Sn(C\vdots CR)_2$ in petr. ether at 100°; R = Pr and Bu, in 28 and 32% yield, resp. (1222).
By rxn. of Bu_2SnCl_2 with aq. $M_2Fe(CO)_4$ in N atm. at r.t. (1268).

I $R_4Sn_3[Fe(CO)_4]_3$

II $R_{10}Sn_4Ru(CO)_6$

Prop.: Pale yellow cryst., m. 112° (904), m. 112-13° (1222), IR spectrum (904, 1222, 1268, 2339), dipole moment (1268), force const. (2339), stable as solid, dec. in soln. in air (1222), struct. (1268).
Rxn. with: $Fe(CO)_5 \rightarrow Sn[Fe(CO)]_4 + Fe_3(CO)_{12}$ (904).
$R_3P \rightarrow [Bu_2SnFe(CO)_3PR_3]_2$; R = Et (1268), Ph (1222) + $(Bu_2Sn)_2Fe_2(CO)_7PR_3$; R = Et, Ph, MeO (1268).

Triphenyltin-π-cyclopentadienyliron- $SnFeC_{25}H_{20}O_2$ $Ph_3Sn(C_5H_5)Fe(CO)_2$
    dicarbonyl (M-502)
Synth.: By polarographic reduction of Ph_6Sn_2 in presence of $C_5H_5Fe(CO)_2$ derivatives (967-8), of $[C_5H_5Fe(CO)_2]_2$ in presence of Ph_3SnCl (968).
Prop.: m. 141-43° (967), yellow cryst., m. 140° (2057), IR (2057), assignment (2560), far IR (828), assignment (2560), UV (967), NMR (2057), Moessbauer resonance (1114-5, 2389, 2440) and mass (1419) spectra, polarography (967-969), x-ray struct. analysis (803).
Use: As metal plating agent (603).

Diphenyltin Bis(π-cyclopentadienyl- $SnFe_2C_{26}H_{20}O_4$ $Ph_2Sn[C_5H_5Fe(CO)_2]_2$
    irondicarbonyl)
Synth.: By rxn. of Ph_2SnCl_2 with $NaC_5H_5Fe(CO)_2$ in THF in Ar atm., in 60% yield (1595).
Prop.: Orange m. 148-50° (1595, 2226), IR, NMR (2226) and Moessbauer resonance (1114-5, 2389) spectra, x-ray struct. analysis (741).
Rxn. with: HCl in $CCl_4 \rightarrow Cl_2Sn[C_5H_5Fe(CO)_2]_2$ (1595).
$SO_2 \rightarrow (PhSO_2)_2Sn[C_5H_5Fe(CO)_2]_2$ (2289, 2371).

Bisdimethyltin Rutheniumtetracarbonyl $Sn_2RuC_{10}H_{18}O_4$ $(Me_3Sn)_2Ru(CO)_4$
Synth.: By rxn. of $Ru_3(CO)_{12}$ with Me_3SnH in hexane at 80°, in 68% yield (905, 2328), with Me_6Sn_2 at 90° in sealed tube (905).
By rxn. of Me_3SnCl with $Na_2Ru(CO)_4$ in THF, in 33% yield (905).
Prop.: Yellow liq., $b_{0.01}$ 80°, IR and mass spectra, dec. in air mol wt., cis-struct., column chromatography (905).
Rxn. with: Thermal dec. at 80° → $Me_{10}Sn_4Ru_2(CO)_8$ (905).

Decamethyltetratin Dirutheniumhexa- $Sn_4Ru_2C_{16}H_{30}O_6$ $Me_{10}Sn_4Ru_2(CO)_6$
    carbonyl*
Synth.: By rxn. of $Ru_3(CO)_{12}$ in hexane at 80° with Me_2SnH_2 or Me_3SnH, in 3% yield (905).
By thermal dec. of $(Me_3Sn)_2Ru(CO)_4$ at 80°, in 1% yield (905).
Prop.: Yellow solid, m. 190-92°, IR, NMR and mass spectra, diamagnetic, proposed struct. (905).

Listing of bimetallic organotin compounds with iron and ruthenium concludes in Table 259 with the exception of one compound which appears on page 868. Other bimetallic and polymetallic derivatives may be found in Chapters 4.5.6.1 and 6.6, respectively.

---

* For proposed structure, see formula II on page 864.

Table 259. Organotin Compounds with Iron and Ruthenium $R_nSnM(CO)_nL_o$

Formula*	Synth. Method**	Yield	Properties	Ref.
$Me_3Sn(C_5H_5)FeCOPPh_3$	IVA	40	Orange, m. 127-28°, soly. (a)	2490
$Me_3Sn(C_5H_5)FeCOP(OPh)_3$	IVA	28	Orange, m. 99-100°, soly. (a)	2490
$Me_3Sn(C_5H_5)Fe(PPh_2CH_2)_2$	IVA	49	Orange, m. 190-200°, soly. (a)	2490
$Me_3Sn(C_5H_5)Fe(PPh_2CH:)_2$	IVA	30	cis-form, red, m. 175° (dec.), soly. (a)	2490
$(Et_3Sn)_2Fe(CO)_4$	--	--	IR and Raman spectra, force const.	734
$Et_3Sn(C_5H_5)Fe(CO)_2$	IIA	--	Red-orange liq., $b_{0.4}$ 150-80° (b)	2057
$Et_2Sn[C_5H_5Fe(CO)_2]_2$ (502)	II (c)	60	Dark red, m. 87-89°, IR spectrum (1045) (d)	1595
$(Et_2SnCl)_2Fe(CO)_4$	IIIB	--	Pale yellow, m. ~20°, subl.o.oo1 60-80° (e, f)	904
$Et_4Sn_3Fe_4(CO)_{12}$ (g)	IIIA, B	--	IR spectrum (f)	904
$[Pr_2SnFe(CO)_4]_2$	IA	30	Yellow cryst., m. 128-30°, IR spectrum (h)	1222
$(CH_2:CHCH_2)_2Sn[C_5H_5Fe(CO)_2]_2$	--	--	m. 77-80°, IR and NMR spectra	2226
$(Bu_2Sn)_2Fe_2(CO)_7PEt_3$	IVB	--	IR spectrum and assignment, struct.	1269
$(Bu_2Sn)_2Fe_2(CO)_7PPh_3$	IVB	--	IR spectrum and assignment, struct.	1269
$(Bu_2Sn)_2Fe_2(CO)_7P(OMe)_3$	IVB	--	IR spectrum and assignment, struct.	1269
$[Bu_2SnFe(CO)_3PEt_3]_2$	IVB	--	IR spectrum and assignment	1269
$[Bu_2SnFe(CO)_3PPh_3]_2$ (502)	IVB	69	Yellow cryst., m. 144° (dec.) (i)	1222
$(Bu_2SnCl)_2Fe(CO)_4$	IIIA	--	IR spectrum (f)	904
$Bu_2Sn[C_5H_5Fe(CO)_2]_2$	--	--	m. 90-92°, IR and NMR spectra	2226
$Bu_4Sn_3Fe_4(CO)_{12}$ (g)	IIIA	6	Dark red cryst., m. 168-70° (dec.), IR spectrum (f)	904
$(C_5H_5)_2Sn[C_5H_5Fe(CO)_2]_2$	II (c)	90	Orange, m. 149-50° (d, k)	1595
$(C_6H_{13})_2SnFe(CO)_4$	--	--	(j)	--
$(C_{10}H_{21})_2SnFe(CO)_4$	--	--	(j)	--
$(Ph_3Sn)_2Fe(CO)_4$ (502)	IIA	--	cis-form, IR spectrum, struct.	905
$[Ph_2SnFe(CO)_4]_2$	IIC (1)	32	Dec. 145-50°, dec. in air	216
$PhSn[C_5H_5Fe(CO)_2]_3$	IIC	46	Yellow, dec. in air, IR spectrum (904) (f)	216
$PhSnCl_2(C_5H_5)Fe(CO)_2$	IIA	50	Red, m. 243-45° (m)	1595
$PhSnCl_2(C_5H_5)Fe(CO)_2$	IIIB	10	Pale yellow, m. 108°, IR and NMR spectra (n)	1006

Compound	Type	Yield	Properties	Ref
Me$_2$Sn[C$_5$H$_5$Ru(CO)$_2$]$_2$	IIB	44	Pale cream, m. 85-93°, IR spectrum	2253
[(PhCH$_2$)$_3$Sn]$_2$Ru(CO)$_4$	IIA	9	Pale yellow, m. 99-101°, IR spectrum, trans-form, stable in air (o)	905
	IB	70		905
(Et$_3$Sn)$_2$Ru(CO)$_4$	IIA	49	Yellow liq., b.0.01 100-110°, IR spectrum, struct., dec. in air, mol. wt. (o)	905
	IB	70		905
(Pr$_3$Sn)$_2$Ru(CO)$_4$	IIA	--	Yellow liq., dec. > 150° in vacuo, IR spectrum, struct., dec. in air	905
	IB	70		905
(Bu$_3$Sn)$_2$Ru(CO)$_4$	IIA	--	Yellow liq., dec. > 150° in vacuo, IR spectrum, struct., dec. in air	905
	IB	65		905
Bu$_{10}$Sn$_4$Ru$_2$(CO)$_6$ (p)	IB	--	Yellow, IR spectrum	905
Ph$_n$Sn[M(CO)$_m$]$_o$	--	--	Moessbauer resonance spectrum	1115

* Numbers in parenthesis refer to pages in main volume.

** The characters used here correspond to the synthetic scheme in the introduction to Chapter 6.5.6.
(a) IR, assignment and NMR spectra, column chromatography (2490). (b) IR and NMR spectra (2057). (c) Synth. from Cl$_2$Sn[C$_5$H$_5$Fe(CO)$_2$]$_2$ (1595). (d) IR and NMR spectra (2226). (e) IR spectrum, sensitive to air and moisture (904). (f) IR spectrum, force const. (2339). (g) See structural formula I, page 864. (h) Stable as solid, dec. in soln. in air, mol. wt., soly. (1222). (i) IR spectrum (1222). (j) Use as metal plating agent (603). (k) X-ray struct. analysis (741, 2249). (l) Synth. from Ph$_3$SnOH. (m) Moessbauer resonance spectrum (1114-5, 2389), x-ray struct. analysis (741). (n) IR and NMR spectra (2057). (o) Column chromatography (905). (p) See structural formula II, page 864.

Bistriphenyltin Ruthenlumtetracarbonyl $Sn_2RuC_{40}H_{30}O_4$ $(Ph_3Sn)_2Ru(CO)_4$

<u>Synth.</u>: By rxn. of Ph_3SnH with $Ru_3(CO)_{12}$ in hexane in sealed tube at 80° (905).
By rxn. of Ph_3SnCl with $Na_2Ru(CO)_4$ in THF at 0°, in 6% yield (905), with $H_2Ru(CO)_4$ in MeOH and KOH (2328).
<u>Prop.</u>: Pale yellow (905, 2328), m. 180-82° (905), IR spectrum (905, 2328), column chromatography, trans-struct., stable in air (905).

6.5.7 ORGANOTIN COMPOUNDS WITH COBALT RHODIUM AND IRIDIUM

The organic derivatives containing a direct tin metal bond listed in Table 260 are prepared by the following methods.

I   Reactions of tetraorganotin compounds:
    (IA) R_4Sn with $Co_2(CO)_8$,
    (IB) $Ph_3SnC\colon CPh$ with $C_5H_5CoC_8H_8$.

II Addition of triorganotin hydride or deuteride to $(RPh_3P)_2IrCOX$.

III Substitution reactions of tin halides with:
    (IIIA) MX' for R_nSnX_{4-n},
    (IIIB) RMgX for X'_2SnX_2.

IV Reaction of organotin halides R_nSnX_{4-n} with:
    (IVA) $Co_2(CO)_8$ or $C_5H_5Co(CO)_2$,
    (IVB) $Co_3(CO)_8$.

V   Phosphine exchange reaction with:
    (VA) $R_nSnX'_{4-n}$,
    (VB) $R_nSnX_{3-n}X'$.

R = organic group, M = alkali metal, n = 0-3, X = halogen, X' = $Co(CO)_4$, $Co(CO)_3PR_3$, $(Ph_3P)_2Rh(CO)_2$, $Ph_3PIr(CO)_3$, $(Ph_3P)_2Ir(CO)_2$.

Trimethyltin Cobalttetracarbonyl $SnCoC_7H_9O_7$ $Me_3SnCo(CO)_4$
<u>Synth.</u>: By rxn. of Me_3SnCl in N atm. with $NaCo(CO)_4$ in anhyd. Et_2O, in 95% yield (730), with $Co_3(CO)_8$ in MeOH (1672).
<u>Prop.</u>: White cryst., m. 73-75° (1672), m. 73-74° (730), subl. in vacuo (730, 1672), IR (730, 1672, 2338), far IR (828), assignment (1271) and NMR (730, 1672, 1674, 2034), spectra, force const. (2338), polarography (966), stable in soln., in inert atm. sensitive to air, dec. slowly in MeOH at 90° (730).
<u>Rxn. with</u>: UV irradiation at 60° → 28% dec. after 450 hours (730).
^{14}CO → kinetic data for exchange rxn. (786).
Ph_3PAuCl → Me_3SnCl + $Ph_3PAuCo(CO)_4$ (2202).
Ph_2PCl → Me_3SnCl + $[Ph_2PCo(CO)_3]_2$ + CO (2202).
Me_2AsI → Me_3SnI + $[Me_2AsCo(CO)_3]_x$ + CO (2202).
I at -60° → Me_3SnI + $ICo(CO)_4$ (2202).

$CF_3I^* \rightarrow Me_3SnF + Me_3SnI + Co_3(CO)_9CF + Co(CO)_4CF_3$ (730).
$C_2F_4^* \rightarrow Me_3SnF + Me_3SnCF_2CF_2Co(CO)_4 + Co_3(CO)_9CCF_3 + CO$ (730).
$C_2F_3H^* \rightarrow Me_3SnF + Co_3(CO)_9CCF_3 + Co_3(CO)_9CCH_2CO_2Me$ (730).
Air → oxidizes to purple solid (730).

Triphenyltin Cobalttetracarbonyl $SnCoC_{22}H_{15}O_4$ $Ph_3SnCo(CO)_4$

<u>Synth.</u>: By rxn. of Ph_3SnCl in N atm. with $NaCo(CO)_4$ in petr. ether, in 80% yield (216), in Et_2O, in 99% yield (730), with $Co_3(CO)_8$ in MeOH (1672).

<u>Prop.</u>: m. 123° (216), 120° (730), 110-20° (dec.) (1672), IR (730, 1672-3, 2198, 2338), assignment (770), NMR (730) and Moessbauer resonance (1282) spectra, force const. (2198, 2338), dec. slowly in air (216, 730), magnetic moment (216), polarography (969), soly. (730).

<u>Rxn. with</u>: UV irradiation → Ph_6Sn_2 (730).
^{14}CO → kinetic data for exchange rxn. (786).
$R_3P \rightarrow Ph_3SnCo(CO)_3PR_3$; R = Bu, Ph, PhO (770).
$CF_3I^* \rightarrow Ph_3SnF + Ph_3SnI + Co_3(CO)_9CF$ (730).
$C_2F_4^* \rightarrow Ph_6Sn_2 + Ph_3SnF + Co_3(CO)_9CCF_3 + CO$ (730).
$C_2F_3H \rightarrow Ph_3SnF$ (730).

Diphenyltin Dicobaltoctacarbonyl $SnCo_2C_{20}H_{10}O_8$ $Ph_2Sn[Co(CO)_4]_2$

<u>Synth.</u>: By rxn. of Ph_2SnCl_2 with $NaCo(CO)_4$ in petr. ether, in 43% yield (216), with $Co_3(CO)_8$ in MeOH in N atm. (1672), with $Co_2(CO)_8$ and Py in Et_2O (759).
By rxn. of $Cl_2Sn[Co(CO)_4]_2$ and PhMgBr in Et_2O (759).

<u>Prop.</u>: Straw yellow cryst., dec. 140° (759), dec. 139-41° (216), m. 128-31° (dec.) (1672), IR (759, 1672) and Moessbauer resonance (1282, 2389) spectra, dec. slowly in soln., nonconducting (749), soly. (759, 1672).

<u>Rxn. with</u>: $Bu_3P \rightarrow Ph_2Sn[Co(CO)_3PBu_3]_2$ (759).

Listing of organotin derivatives containing cobalt, rhodium and iridium concludes in Table 260. Other bimetallic and polymetallic organotin-cobalt compounds are summarized in Chapters 4.5.6.1 and 6.6, respectively.

---

* Rxn. at 90° or at 50-60° under UV irradiation, effect of UV initiation on product distribution.

Table 260. Organotin Compounds with Cobalt, Rhodium and Iridium $R_nSnM(CO)_mL_o$

Formula*	Synth. Method**	Yield	Properties	Ref.
$Me_2Sn[Co(CO)_4]_2$ (502)	IVB	--	Yellow cryst., m. 35-37°, subl. in vacuo (a, b)	1672
$Me_2ClSnCo(CO)_4$	IVA	--	White cryst., m. 70-73°, IR spectrum	1673
$MeSn[Co(CO)_4]_3$	IVA	53	Red, m. 69-71°, mass spectrum, dec. >110° (b, c)	1671, 1674
$MeClSn[Co(CO)_4]_2$	IVB	--	Yellow orange, IR and NMR spectra, soly.	1672
$(MeSnCl_2)_2C_5H_5CoCO$	IVA	--	Yellow cryst., IR spectrum, conductance (d)	1373
$(MeSnBr_2)_2C_5H_5CoCO$	IVA	--	Yellow cryst., IR and NMR spectra (d)	1373
$MeBr_2Sn(C_5H_5)CoBrCO$	IVA	--	Black cryst., IR and NMR spectra (e)	1373
$MeI_2Sn(C_5H_5)CoICO$	IVA	--	Black cryst., IR spectrum, conductance (e)	1373
$Et_3SnCo(CO)_4$	--	--	IR and Raman spectra, force const. (f)	734
$Et_2Sn[Co(CO)_3PBu_3]_2$	IIIB	--	Yellow green cryst., m. 77°, soly. (g)	759
$(CH_2:CH)_2Sn[Co(CO)_4]_2$	IA	--	IR spectrum (1672)	1674
$CH_2:CHSn[Co(CO)_4]_3$	IIIA	--	Red cryst., m. 57-60°, IR spectrum	1671
$CH_2:CHSnCl[Co(CO)_4]_2$	IIIA	--	Yellow orange cryst., m. 45-47°, IR spectrum	1671-2
$CH_2:CHSnCl_2(C_5H_5)Co(CO)SnCl_3$	IVA	--	Yellow cryst., IR spectrum (d)	1373
$BuSn[Co(CO)_4]_3$	IVA	47	m. 62-64°, IR, UV and NMR spectra	236a
$BuSnCl[Co(CO)_4]_2$	IIIA	60	Red cryst., m. 60-62°, IR spectrum	1671
$BuSnCl_2Co(CO)_4$	IVB	70	m. 29-30°, IR spectrum	1672
$BuSnCl_2Co(CO)_3PPh_3$	IVA	--	Orange liq., white cryst. at -80°, IR spectrum (h)	1673
$Ph_3SnR'Co(CO)_2$	VB (h)	--	Yellow cryst., m. 161-64° (dec.), IR spectrum R' = $CH_2:CHCH:CH_2$ (i)	1673
$Ph_3SnPhCoCO$	--	--	(i)	--
$(Ph_3SnC:CPh)_2CoC_5H_5$ (j)	IB	53	IR, UV, NMR and mass spectra, struct. (k)	1174
$Ph_3SnCo(D_2H_2)Py$	III (1)	--	Red orange cryst., dec. 194°, soly., stable in air	486
$Ph_3SnCo(D_2H_2)PBu_3$	III (1)	--	Orange cryst., m. 145-47°, soly., stable in air	486
$Ph_3SnCo(CO)_3PBu_3$	VA	--	IR spectrum and assignment	770
$Ph_3SnCo(CO)_3PPh_3$	VA	--	IR spectrum and assignment	770
$Ph_3SnCo(CO)_3P(OPh)_3$	VA	--	IR spectrum and assignment	770

Compound	Group		Description	Ref.
$Ph_2Sn[Co(CO)_3PBu_3]_2$	VA	--	Brownish yellow cryst., m. 121-22° (g)	759
$Ph_2Sn[Co(CO)_3PPh_3]_2$	V (m)	--	IR and UV spectra	2330
$Ph_2SnClCo(CO)_4$	IVA	--	Yellow cryst., m. 66-68°, IR spectrum (n, o)	1673
$Ph_2SnBrCo(CO)_4$	IVA	--	Yellow cryst., m. 55.5-57°, IR spectrum	1673
$Ph_2SnICo(CO)_4$	(o)	--	Yellow cryst., m. 69-70°, IR spectrum	1673
$Ph_2Sn[Co(CO)_4]_3$	IIIA	--	Red, m. 90-92°	1596
	IIIA	67	Red, m. 89-91° (sealed tube) (p)	1671
$PhSnCl[Co(CO)_4]_2$	IVB	--	Yellow cryst., m. 71-73°, IR spectrum	1672
$PhSnCl[Co(CO)_3PBu_3]_2$	IIIB	76	Canary yellow cryst., m. 105-106° (g)	759
$PhSnBr[Co(CO)_3PBu_3]_2$	IIIB	--	Straw yellow cryst., m. 110-12° (g)	759
$PhSnCl_2Co(CO)_4$	IVA	--	Yellow cryst., m. 63-64°, IR spectrum (o, q)	1673
$PhSnCl_2Co(CO)_3PPh_3$	VB (q)	--	Yellow, m. 158-62° (dec.), IR spectrum	1673
$PhSnCl_2(C_5H_5)CoCOSnCl_3$	IVA	--	Yellow cryst., IR spectrum (d)	1373
$PhSnBr_2Co(CO)_4$	IVA	--	Yellow needles, m. 73.5-75°, IR spectrum	1673
$PhSnBr_2(C_5H_5)CoCOBr$	IVA	--	Black cryst., IR spectrum (e)	1373
$PhSnI_2Co(CO)_4$	(o)	--	Yellow cryst., IR spectrum	1673
$PhSnI_2(C_5H_5)CoCOI$	IVA	--	Black cryst., IR spectrum, conductance (e)	1373
$Me_3SnRh(CO)_2(PPh_3)_2$	IIIA	59	Yellow solid, IR and NMR spectra, struct.	891

* Numbers in parenthesis refer to pages in main volume.

** The characters used here correspond to the synthetic scheme in the introduction to Chapter 6.5.7.
(a) IR spectrum (1672). (b) NMR spectrum (1672, 1674, 2034). (c) UV irradiation → $Co_2(CO)_8$ + $Co_4(CO)_{12}$ (1671). (d) Stable to air and light, soly. (1373). (e) Stable in air, soly. (1373). (f) IR (2198, 2338), assignment (1272), far IR spectra and assignment (1270), dipole moment (1270, 1272, 2057), force const. (2198, 2330). (g) IR spectrum, nonconducting (759). (h) Rxn. of $BuSnCl_2Co(CO)_4$ with PPh_3 → $BuSnCl_2Co(CO)_3PPh_3$ + CO (1673). (i) 1,3-Bistriphenyltin diphenylcyclobutadienyl-cyclopentadienyl cobalt. (j) Use as metal plating agent (603). (k) Stable to air and H_2O (1174), rxn. with HCl → $CPh:CHCPh:CHCoC_5H_5$ (1174). (l) DH_2 = diacetyldioxime, synth. from Ph_3SnX with $HCo(D_2H_2)B$; B = Py, Bu_3P (486). (m) Synth. from $(Ph_2Sn \cdot PhMgBr)_2$ with $Co_2(CO)_8$, followed by PPh_3 (2330). (n) Moessbauer resonance spectrum (1282). (o) Rxn. of $Ph_nSnCl_{2-n}Co(CO)_4$ with NaI in Me_2CO → $Ph_nSnI_{2-n}Co(CO)_4$; n = 1, 2 (1673). (p) IR, mass (1671) and Moessbauer resonance (2389) spectra. (q) Rxn. of $PhSnCl_2Co(CO)_4$ with Ph_3P at 100° in vacuo → $PhSnCl_2Co(CO)_3PPh_3$ + CO (1673).

Table 260 Continued

$R_nSnM(CO)_mL_o$

Formula*	Synth. Method**	Yield	Properties	Ref.
$Me_3SnIr(CO)_3PPh_3$	IIIA	61	Pale yellow, IR and NMR spectra, struct.	891
$Me_3Sn(MePh_2P)_2IrCOHCl$	II (r)	--	White, struct. (s)	2530
$Me_3Sn(Ph_3P)_2IrCOHCl$	II (t)	--	Yellow, NMR spectrum, struct. (s)	2530
$Me_3Sn(Ph_3P)_2IrCOHCl$	II (r)	--	White, struct. (s)	2530
$Me_3Sn(Ph_3P)_2IrCODCl$	II (t)	--	Yellow, struct. (s)	2530
	II (r)	--	White, struct. (s)	2530
$Me_3Sn(Ph_3P)_2IrHBr$	II (t)	--	Yellow, NMR spectrum, struct. (s)	2530
$Me_2Sn[Ir(CO)_3PPh_3]_2$	IIIA	>50	White solid, IR and NMR spectra, struct.	891
$Ph_3SnIr(CO)_3PPh_3$	IIIA	76	Pale yellow, IR spectrum, struct. (u)	891
$Ph_3Sn(MePh_2P)_2IrCOHCl$	II	--	White, NMR spectrum, struct. (s)	2530
$Ph_3Sn(MePh_2P)_2IrCODCl$	II	--	White, struct. (s)	2530
$Ph_3Sn(Ph_3P)_2IrCOHCl$	II	--	White, struct. (s)	2530
$Ph_3Sn(Ph_3P)_2IrCODCl$	II	--	White, struct. (s)	2530
$Ph_3Sn(Ph_3P)_2IrCOHBr$	II	--	White, struct. (s)	2530
$Ph_3Sn(Ph_3P)_2IrCODBr$	II	--	White, struct. (s)	2530
$Ph_3Sn(Ph_3P)_2IrCOHI$	II	--	White, struct. (s)	2530
$Ph_nSn[M(CO)_m]_o$	--	--	Moessbauer resonance spectrum	1115

(r) Synth. without solvent in excess Me_3SnH or Me_3SnD (2530). (s) (Far) IR spectra and assignment, dec. in air, mol. wt., soly. (2530). (t) Synth. in C_6H_6 (2530). (u) Rxn. with Br in C_6H_6 → $PhBr + Ph_3PIr(CO)_2Br_3$ (891), with I in THF → $Ph_3SnI + Ph_3PIr(CO)_2I_3 + CO$ (891).

6.5.8 ORGANOTIN COMPOUNDS WITH NICKEL AND PLATINUM

Triphenyltin Bistriphenylphosphine $SnPtC_{54}H_{45}ClP_2$ $Ph_3Sn(Ph_3P)_2PtCl$
  Platinum Chloride

Synth.: By rxn. of Ph_3SnCl with $(Ph_3P)_4Pt$ in C_6H_6 in N atm. (1397), with $(Ph_3P)_3Pt$ or $(Ph_3P)_2PtC_2H_4$ in C_6H_6 at 25° (2248).
By rxn. of Ph_3SnNO_3 with trans-$(Ph_3P)_2PtHCl$ in THF, in 55% yield (700).
By rxn. of Ph_3SnLi with $(Ph_3P)_2PtCl_2$ in THF (700).

Prop.: Colorless cryst., m. 278-82° (dec.) (700), white, m. 205° (1397), IR spectrum (700), conductance (1397), mol. wt., stable to air and light, struct. (700).

Rxn. with: HCl (anhyd.) → cis- and trans-$(Ph_3P)_2PtCl_2$ (700).
Me_2CO under reflux → $(Ph_2Sn)_x$ + $(Ph_3P)_2PhPtCl$ (700).

Table 261. Organotin Compounds with Nickel and Platinum* $R_nSnM(CO)_mL_o$

Formula	Properties	Ref.
$R_2SnNi(CO)_3$		
R = Me	Use as metal plating agent	603
Et	Use as metal plating agent	603
Pr	Use as metal plating agent	603
Me_2CH	Use as metal plating agent	603
Bu	Use as metal plating agent	603
Me_2CHCH_2	Use as metal plating agent	603
$C_{18}H_{37}$	Use as metal plating agent	603
$Ph_3SnC_5H_5NiCO$	Use as metal plating agent	603
$Me_3Sn(Ph_3P)_2PtCl$ (a)	Yellow brown solid	825
$Ph_nSn[M(CO)_m]_n$	Moessbauer resonance spectrum	1115

* Tris(triorganotin)phosphine and -arsine nickelcarbonyl derivatives are listed in Tables 246 and 247, respectively.
(a) Synth. from Me_3SnH and $(Ph_3P)_2Pt(NH_2)Cl$ in xylene (825).

6.6 POLYMETALLIC ORGANOTIN TRANSITION METAL COMPLEXES

This chapter comprises polynuclear organotin compounds. The derivatives contain at least three different metal atoms arranged in a metal-metal-metal-chain and contain one or more tin-carbon bonds. Tin is always linked to a transition metal. The third metal in the chain is a different transition metal or arsenic, antimony and silicon. All compounds are listed in Table 262. The following synthetic methods are reported.

I   Dealkylation reaction of polymetallic organotin metal complexes with chlorine, bromine or hydrochloric acid.

II Metathetic substitution reaction of organotin halide:
  (IIA) R_nSnCl_{4-n} with MX',
  (IIB) $R_2ClSnMn(CO)_5$ with MX',
  (IIC) Ph_3SnX with $HCo(D_2H_2)SbR_3$.

III Exchange reaction of arsine and stibine with $R_nSn[M'(CO)_5]_{4-n}$.

R = organic group, M = alkali metal, M' = manganese and ruthenium, n = 1-3, X = halogen, X' = $C_5H_5Mo(CO)_3$, $C_5H_5Mo(CO)_2SbR_3$, $C_5H_5W(CO)_3$, $Mn(CO)_4AsR_3$, $Mn(CO)_4SbR_3$, $Re(CO)_5$, $Ru(CO)_4SiMe_3$, $Co(CO)_4$.

Table 262. Polymetallic Organotin Transition Metal Complexes

Formula	Synth. Method*	Yield	Properties	$R_nSnM_mM'L_o$ Ref.
$Ph_3Sn(C_5H_5)Mo(CO)_2SbPh_3$	IIA	--	White cryst., m. 242-43° (dec.) (a)	1459
$C_5H_5Mo(CO)_3SnMe_2Mn(CO)_5$	IIB	--	m. 95-96°, IR and NMR spectra, monomeric	415
$C_5H_5Mo(CO)_3SnPh_2Mn(CO)_5$	IIB	31	Yellow, m. 147-48° (b)	387
$C_5H_5W(CO)_3SnMe_2Mn(CO)_5$	--	--	NMR spectrum	415
$C_5H_5W(CO)_3SnPh_2Mn(CO)_5$	IIB	35	Yellow, m. 157-58° (b)	387
$Me_3SnMn(CO)_4AsMe_3$	IIA	--	--	185
$Me_2PhSnMn(CO)_4As(C_5H_5)_3$	--	--	(c)	--
$Me_2ClSnMn(CO)_4As(C_5H_5)_3$	I (c)	--	--	187
$Et_2Sn[Mn(CO)_4AsPr_3]_2$	IIA	--	--	185
$PhSn[Mn(CO)_4AsR_3]_3$	IIA	--	R = mesityl	185
$(p-Me_2C_6H_4)_2Sn[Mn(CO)_4AsR_3]_2$	III	--	R = mesityl	185

PhCH$_2$Sn[Mn(CO)$_4$Sb(i-Bu)$_3$]$_3$	III	--	(d)	186
Pr$_3$SnMn(CO)$_4$Sb(C$_8$H$_{17}$)$_3$	III	--	--	186
CH$_2$:CHCH$_2$Sn[Mn(CO)$_4$SbPh$_3$]$_3$	IIA	--	--	185
(C$_5$H$_5$)$_3$SnMn(CO)$_4$SbR$_3$	--	--	R = m-Me$_2$CHC$_6$H$_4$ (e)	187
(C$_5$H$_5$)$_2$BrSnMn(CO)$_4$SbR$_3$	I (e)	--	R = m-Me$_2$CHC$_6$H$_4$	187
(CO)$_5$MnSnPh$_2$Re(CO)$_5$	IIB	38	White, m. 135-36° (f)	387
(CO)$_5$MnSnPh$_2$Co(CO)$_4$	IIB	59	Yellow, m. 134° (g, h)	387
(CO)$_5$MnSnPhClCo(CO)$_4$	I (h)	49	Yellow, m. 87-88°	387
Ph$_3$SnRe(CO)$_4$AsPh$_3$	III	--	White cryst., m. 213-15°, soly. (i)	1594
Ph$_3$SnRe(CO)$_4$SbPh$_3$	III	--	White cryst., m. 210-12°, soly. (i)	1594
Me$_3$SnRu(CO)$_4$SiMe$_3$	IIA	--	Mass spectrum	905
Ph$_3$SnCo(D$_2$H$_2$)SbMe$_3$ (j)	IIC	--	Red orange cryst., dec. 164-68° (k)	486

* The characters used here correspond to the synthetic scheme in the introduction to Chapter 6.6. (a) IR (1459), (far) IR, assignment (2560) and NMR (1459) spectra. (b) Rxn. with HCl → C$_5$H$_5$M(CO)$_3$SnCl$_2$Mn(CO)$_5$; M = Mo, W (387). (c) Rxn. of Me$_2$PhSnMn(CO)$_4$As(C$_5$H$_5$)$_3$ with Cl in (p-ClC$_6$H$_4$)$_2$O at 143-54° → Me$_2$ClSnMn(CO)$_4$As-(C$_5$H$_5$)$_3$ (187). (d) Use as antiknock agent (186). (e) Rxn. of (C$_5$H$_5$)$_3$SnMn(CO)$_4$SbR$_3$ with Br in (ClCH$_2$CH$_2$)$_2$O at 113-18° → (C$_5$H$_5$)$_2$BrSnMn(CO)$_4$SbR$_3$; R = m-Me$_2$CHC$_6$H$_4$ (187). (f) Rxn. with HCl → (CO)$_5$MnSnCl$_2$Re(CO)$_5$ (387). (g) Moessbauer resonance spectrum (1282), x-ray structural analysis (740, 2251). (h) Rxn. of (CO)$_5$MnSnPh$_2$Co(CO)$_4$ with HCl → (CO)$_5$MnSnPhClCo(CO)$_4$ (387). (i) IR spectrum (1594). (j) DH$_2$ = diacetyldioxime. (k) Soly., stable in air (486).

7. CYCLIC ORGANOTIN COMPOUNDS

This chapter comprises only such cyclic organotin compounds having the tin atom located between two carbon atoms in the heterocyclic system. Compounds containing hetero atoms linked through carbon atoms to tin are also included. Spiro compounds and monocyclic derivatives are known. Heterocyclic organotin compounds containing tin linked to oxygen, sulfur, nitrogen, metals and other heteroatoms in the ring system are located in the appropriate subsections of Chapters 4, 5 and 6.

7.1 CYCLIC TETRAORGANOTIN COMPOUNDS

The cyclic tetraorganotin compounds summarized in Tables 263 through 266 are prepared by methods from the following scheme.

I Hydrostannation with diorganotin dihydrides R_2SnH_2 :
   (IA) For diene $o\text{-}(CH_2:CH)_2C_6H_4$,
   (IB) For diynes $o\text{-}(HC\!:\!C)_2Y$.

II Synthesis of spiro compounds with $SnCl_4$ and :
   (IIA) $Y(MgBr)_2$,
   (IIB) YLi_2 ,
   (IIC) YNa_2 .

III Synthesis of monocyclic derivatives with diorganotin dihalides and organotin trihalides R_nSnX_{4-n} and :
   (IIIA) $Y(MgBr)_2$,
   (IIIB) YLi_2 ,
   (IIIC) YNa_2 ,
   (IIID) MeLi and MeMgBr for monocyclic organotin dihalide $YSnCl_2$

IV Arylation of metallic tin with $(o\text{-}IC_6F_4)_2$.

V Dihalocarbene addition to $\overline{CH_2(CH_2)_4SnMe_3}$ with $PhHgCCl_2Br$.

VI Cleavage reactions of cyclic organotin derivatives :
   (VIA) Y_2Sn with $SnCl_4$,
   (VIB) Y_2Sn with HBr ,
   (VIC) $YSnMe_2$ with $SnCl_4$.

R = organic group, n = 1 and 2, X = halogen, Y = hydrocarbon bridge.

7.1.1 UNSUBSTITUTED CYCLIC TETRAORGANOTIN COMPOUNDS

All compounds are listed in Table 268.

Table 263. Unsubstituted Cyclic Tetraorganotin Compounds ⬠SnR$_2$

Formula*	Synth. Method**	Yield	Properties	Ref.
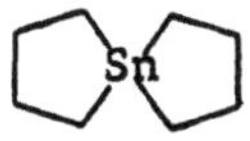 (cyclopentyl)$_2$Sn	IIA	low	Mass (850) and IR spectra	17
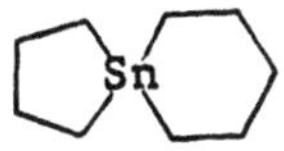 (cyclopentyl)(cyclohexyl)Sn	IIA	< 15	IR spectrum	17
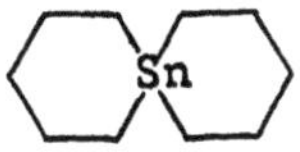 (cyclohexyl)$_2$Sn	IIA	<	b_{10} 119-20°, n_D^{25} 1.5362, IR spectrum (a)	17

⬠SnR$_2$

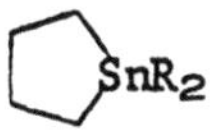				
R = Bu	IIA	52	$b_{0.04}$ 55°, n_D^{25} 1.4958, GLC, mol. wt. (b)	2888
Me$_3$CCH$_2$	IIA	32	$b_{0.3}$ 61°, n_D^{25} 1.4954, GLC, mol. wt. (c)	2888

⬡SnR$_2$

R = Me	--	--	(d)	--
Bu	IIIA	27	$b_{0.4}$ 87-88°, GLC (b)	2888
Me$_3$CCH$_2$	IIIA	27	$b_{0.08}$ 73-73.5°, n_D^{25} 1.4955, GLC (c)	2888
Ph (507)	IIIA	34	$b_{0.3}$ 143°, n_D^{25} 1.6042, GLC, mol. wt. (e)	2888

⬢SnR$_2$ (7-membered)

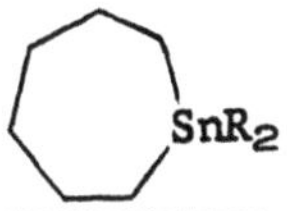				
R = Me	IIIA	19	$b_{4.6}$ 68°, n_D^{25} 1.5052, GLC (e)	2888
Et	IIIA	11	$b_{0.6}$ 61-63°, n_D^{25} 1.5085, GLC (e)	2888
Me$_3$CCH$_2$	IIIA	12	$b_{0.2}$ 99.5°, n_D^{25} 1.5014, GLC (e)	2888
Ph (507)	IIIA	26	$b_{0.07}$ 140°, n_D^{25} 1.6028 (e)	2888

* Numbers in parenthesis refer to pages in main volume.
** The characters used here correspond to the synthetic scheme in the introduction to Chapter 7.1.
(a) Rxn. with X → X(CH$_2$)$_5$SnX$_2$; X = Br, I (17), with HBr → $\overline{CH_2(CH_2)_4SnBr_2}$ (17).
(b) Dec. in air at r.t. (2888). (c) Stable in air < r.t. (2888). (d) Rxn. with PhHgCCl$_2$Br → $\overline{Me_2Sn(CH_2)_3CH(CHCl_2)CH_2}$ + PhHgBr (1889). (e) Stable in air (2888).

7.1.4 UNSATURATED CYCLIC TETRAORGANOTIN COMPOUNDS

1,1-Dibutyl-4,5-dihydrostannepin (M-507) $SnC_{14}H_{26}$

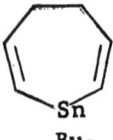

Synth.: By rxn. of Bu_2SnH_2 with $(CH_2C\vdots CH)_2$ in heptane and fast vacuum distn. at 200-300° (499).
By thermal depolymerization of $(Bu_2SnCH:CHCH_2CH_2CH:CH)_x$ in vacuo at 220-300°, in 80% crude yield (499).
Prop.: $b_{0.1}$ 93-94°, IR and NMR spectra, GLC, polymerizes > 90° (499).
Rxn. with: HCl → $(CH_2C\vdots CH)_2$ + Bu_2SnCl_2 (499).
BCl_3 → $CH:CHCH_2CH_2CH:CHBCl$ + Bu_2SnCl_2 (499).
$PhBBr_2$ → $CH:CHCH_2CH_2CH:CHBPh$ + Bu_2SnBr_2 (499).
AcOD → $(CH_2C\vdots CD)_2$ + $Bu_2Sn(OAc)_2$ (499).

Additional unsaturated cyclic tetraorganotin compounds are listed in Table 264.

7.1.6 FUNCTIONALLY SUBSTITUTED CYCLIC TETRAORGANOTIN COMPOUNDS

All compounds are listed in Table 265.

7.3 CYCLIC ORGANOTIN COMPOUNDS CONTAINING HETEROATOM TIN BONDS

All compounds are listed in Table 266.

Table 264. Unsaturated Cyclic Tetraorganotin Compounds

Formula*	Synth. Method**	Yield	Properties	Ref.
Ph₄Sn-cyclopentadienyl (506)	IIB	19	m. 281° (2533), 265-72° (a)	232
Bu,Ph,Ph,Ph,Bu-stannole	IIB	28	White cryst., m. 141.8-42.5°,	1736
	IIIB (b)	9	NMR spectrum, mol. wt.	1736
dibenzo-Sn structure	IIC	low	Yellow leaflets	1726
R"R'-SnR₂-R'R" (stannole)				
R = Me, R' = Ph, R" = H	IIIB	39	m. 120-21°	691
Me, Ph, Ph	--	--	Polarography and ESR spectrum (a, c) (506*)	964
Me, Ar, Ar (d)	IIIB	52	m. 189.5-91.5° (e)	2768
Ph, Ph, Ph	IIIB	--	m. 172-73° (506*)	232

* Numbers in parenthesis refer to pages in main volume.
** The characters used here correspond to the synthetic scheme in the introduction to Chapter 7.1.
(a) Use as antiknock agent and as insecticide (2533). (b) Synth. from $(CH_2:CH)_2SnCl_2$ (1736). (c) Forms stable blue radical anion (964). (d) Ar = Ph and/or p-C_6H_4D (2728). (e) Rxn. with Br → (d_2)-Me_2(BrCPh:CPhCPh:CPh)SnBr (2728), with AcO_2H → (d_2)-$PhĊ:CPhCPh:CPhȮ$ (2728).

Table 264 Continued

Formula*	Synth. Method**	Yield	Properties	Ref.
R''⟩⟨R' / R''⟩Sn R₂⟨R' (Contd.)				
R = Me, R' + R'' = (CH)₄	IIIB	--	m. 123-24°, soly. (f)	1254
CH₂:CH, (CH)₄	IIIB	--	(f)	1254
Ph, (CH)₄ (506)	IIIB	--	(f)	1254
Me CH:CMeCH:CH	--	--	(f)	--
[indanyl-SnR₂ bicyclic structure]				
R = Me	IIIC	--	m. 218-21°, sublimes (g)	1726
PhCH₂	IIIC	--	(g)	1726
CH₂:CH	IIIC	--	(g)	1726
Bu	IIIC	--	Yellow cryst., m. 225-35°, mol. wt. (g, h)	1726, 2172
C₁₈H₃₇	IIIA	--	Polymerizes > m. (g, h)	1727
Ph	IIIC	--	(g)	1726
	IIIC	--	m. 205-210°, soly. (g)	1726
[benzosuberane-SnR₂ structure]				
R = Et	IA	10	b₀.₀₈ 93-98°, n_D^{20} 1.5616, IR spectrum	310, 1410
Ph (507)	IA	16	m. 98-100°, IR spectrum	310, 1410

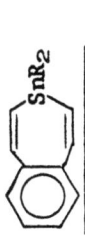

R = Me	IB	6	m. 41-42°, b$_{0.4}$ 88-90°, n$_D^{20}$ 1.6108 (i)	310, 1410
Et	IB	17	b$_{0.4}$ 95-96°, n$_D^{20}$ 1.5983 (i, j)	310, 1410
(Et + Ph)	IB	5	Crude (i)	310, 1410
Ph	IB	12	m. 282-85°	310
(o-CH$_2$CH$_2$C$_6$H$_4$CH$_2$CH$_2$SnPh$_2$)$_2$	IA	< 14	m. 277-82°, IR spectrum	310, 1410
(o-CH:CHC$_6$H$_4$CH:CHSnEtPh)$_2$	IB (k)	25	m. 184-86° (i)	310, 1410
(o-CH:CHC$_6$H$_4$CH:CHSnPh$_2$)$_2$	IB (k)	12	m. 282-85°, mol. wt.	1410

(f) Use as stabilizer and as insecticide (1254). (g) Use as stabilizer, fuel additive and fungicide (1726). (h) Use as antioxidant for mineral oil lubricants (2172). (i) IR spectrum, mol. wt. (1410). (j) Rxn. with I → Et$_2$SnI$_2$ + o-(ICH:CH)$_2$C$_6$H$_4$ (1410). (k) Synth. with EtPhSnH$_2$ (310, 1410).

Table 265. Functionally Substituted Cyclic Tetraorganotin Compounds

Formula*	Synth. Method**	Yield	Properties	Ref.
	IIB	17	m. 227-29°, IR spectrum, mol. wt.	883
	IV	--	Synth. at 230°	885

R	Y	Synth. Method**	Yield	Properties	Ref.
R = H	Y = O	IIB	7	m. 220-21°, (far) IR, UV and NMR spectra, mol. wt., soly. (b)	1378
H	SO$_2$	IIIB (a)	2		1378
H	NMe	IIB	15	White cryst., m. 315-20° (dec.), spectra (c)	1378
Me	NMe	IIB	49	m. 164-66°, spectra, mol. wt., soly. (d)	1376
Br	NMe	IIB	78	m. 209-11°, spectra, mol. wt., soly. (d)	1376
H	NMe	IIB	77	m. 259-60°, spectra, mol. wt., soly. (d, e)	1376
H	NEt (506)	--	--	(f)	--

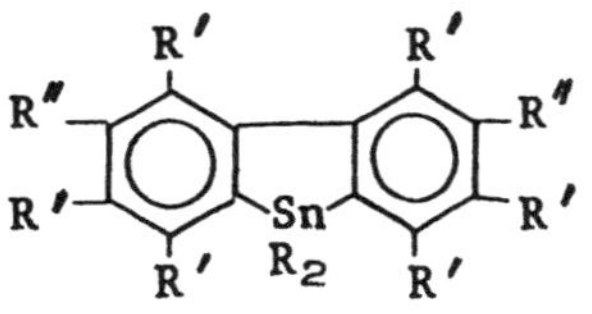

R = 2-C₄H₃S	R' = H	R'' = H	IIIB	--	1254
Me	H	MeO	--	--	--
Me	F	F	IIIB	Low	883
Ph	F	F	IIIB	24	883
Me₂Sn(CH₂)₃CH(CHCl₂)CH₂			V	50	1889

(g)
(g)
m. 131-33°, IR spectrum
n_D^{25} 1.5330, IR spectrum, mol. wt.

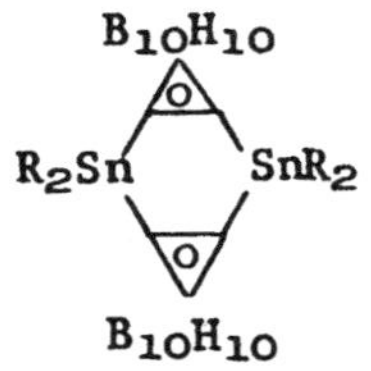

Derivative of 1,2-carborane

R = Me	IIIB	21	1835
Bu	IIIB	66	627

m. 352-54°
m. 186-86.5° (h)

---

* Numbers in parenthesis refer to pages in main volume.
** The characters used here correspond to the synthetic scheme in the introduction to Chapter 7.1.
(a) Synth. from Me₂SnCl₂ as by-prod. (1378). (b) Rxn. with SnCl₄ at 220° → o,o'-C₆H₄SnCl₂C₆H₄O (1378).
(c) (Far) IR, UV and NMR spectra (1378). (d) (Far) IR, UV and NMR spectra (1376). (d) Rxn. with SnCl₄ →
C₆H₃RSnCl₂C₆H₃RNMe (1376). (f) Use as stabilizer for mineral oil and synthetic lubricants (981). (g) Use as
stabilizer and insecticide (1254). (h) Moessbauer resonance spectrum (660).

Table 265 Continued

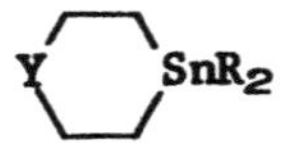

Formula*

R	R' = H	Y = O	Synth. Method**	Yield	Properties	Ref.
R = Me	H	O	IIIB	2.4	m. 259-60°, dimeric in MeEtCO	300, 1378
Et	H	O	IIID	35	Monomeric, spectra (c, g)	1378
Et	H	O	IIIB	0.5	m. 159-60°, spectra, mol. wt., soly.	1378
Bu	H	O	IIID	17	(c)	1378
Bu	H	O	IIIB	2	m. 118-19°, spectra, mol. wt., soly.	1378
Ph	H	O	IIID	7	(c)	4
Ph	H	O	IIID	29	m. 140-41.5°, IR spectrum, assignment	1378
Ph	H	O	IIID	28	m. 251-52°, spectra, mol. wt., soly. (c, i)	1378
Ar	H	O (j)	IIIB	2	m. 252-53°, spectra, mol. wt., soly. (b, c)	1378
Me	H	S	--	--	(g)	4
Ph	H	S	IIIB	--	m. 172-80°, crude, IR spectrum, assignment	1378
Me	H	SO$_2$	IIIB	18	m. 164-65° (c, g)	1378
Ph	H	SO$_2$	IIIB	13	m. 268.4-70.1°, IR spectrum, assignment	4
Me	H	NMe	IIIB	12	m. 164-65° (sic), spectra, mol. wt., soly. (c, i)	1378
Me	H	NMe	IIIB	51	m. 133-34°, spectra, mol. wt., soly. (d, k)	1376
Me	Me	NMe	IIIB	63	m. 119-20°, spectra, mol. wt., soly. (d, k)	1376
Me	Br	NMe	IIIB	73	m. 122-23°, spectra, mol. wt., soly. (d, k, l)	1376
Ph	H	NMe	IIIB	52	m. 148.5-50°, spectra, mol. wt., soly. (d)	1376
Ph	Me	NMe	IIIB	34	m. 146-48°, spectra, mol. wt., soly. (d)	1376
Ph	Br	NMe	IIIB	26	m. 191-93°, spectra, mol. wt., soly. (d)	1376
Ph	Br	NEt	IIIB	39	m. 177-78°	594a

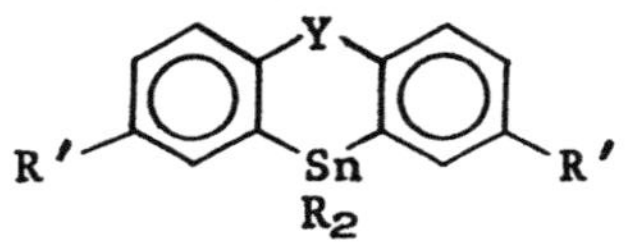

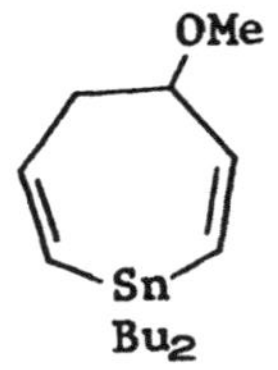 (structure: cycloheptadiene with OMe and SnBu₂)	IB	--	b₀.₂ 118-20°, IR and NMR spectra, dec. on activated alumina (m)	499, 499

(i) Purification by column chromatography (4). (j) Ar = 1/2 o-C₆H₄OC₆H₄ (compare structural formula I shown below) (1378). (k) Rxn. with PhPCl₂ → Me₂SnCl₂ + corresponding phosphazine (compare structural formula II shown below) (1376). (l) Rxn. with SnCl₄ → Me₂SnCl₂ + phenazastannine, R = Cl, R' = Br, Y = NMe (1376). (m) Rxn. with EtCO₂H → HC⋮CCH₂CHOMeC⋮CH (499), with BCl₃ → Bu₂SnCl₂ (499).

Formula I

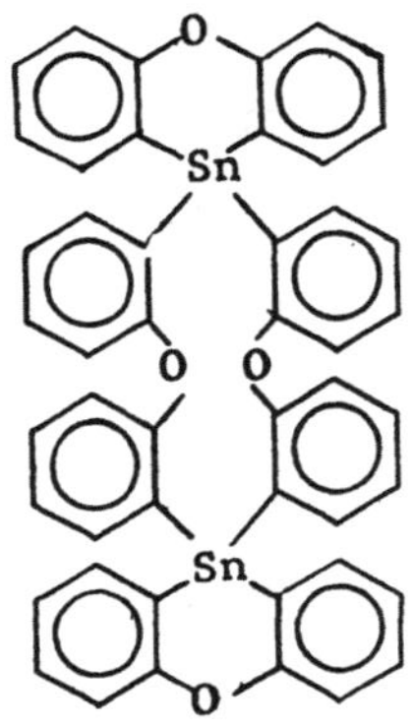

Formula II

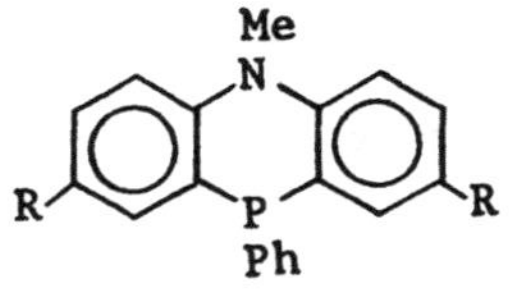

R = H, Me, Br

Table 266. Cyclic Organotin Hetero Compounds

Formula*	Synth. Method**	Yield	Properties	Ref.
$\overline{CH_2(CH_2)_n}SnRH$	--	--	(a)	--
[dibenzostannole (OEt)$_2$ structure]	IIIB	--	(b)	1254
[R, X stannole structure]				
R = Bu X = Cl	IIIC	--	(c, d)	1726
Bu OH	(c)	--	(d)	1726
R = X = Cl	IIC	--	Softening point 245-50°, m. > 300° (d, e)	1726
R + X = O	(e)	--	Fluoresces yellow white in hexane (d)	1726
R$_2$SnX$_2$ [cyclic]				
R = H X = Cl (508)	--	--	(f)	--
H Br (508)	VIB	--	(g)	17
H Br·BiPy	(g)	--	m. 207-210°	17
H Br·phen	(g)	--	m. 218-220°	17
Me Cl + R' (508)	--	--	(h)	--
Me Br + R' (508)	--	--	(h)	--
Me I + R' (508)	--	--	(h)	--

R				
Y = O	R = H	VIA	<63	m. 243-44°, (far) IR, UV and NMR spectra, 1378
		VIA (i)	74	mol. wt., soly. (j) 1378
NMe	Br	VIA	<68	m. 164.5-66.5°, (far) IR, UV and NMR spectra, 1376
		VIC	63	mol. wt., soly. 1376

* Numbers in parenthesis refer to pages in main volume.
** The characters used here correspond to the synthetic scheme in the introduction to Chapter 7.1.

(a) Rxn. with $[O_2CCH:CHCO_2(CH_2)(CH_2)_4]_x$ + AIBN → tin contg. polymer; n = 2-6 (599). (b) Use as stabilizer and insecticide (1254). (c) Rxn. of $C_{14}H_{10}SnBuCl$ with aq. NH_3 → $C_{14}H_{10}SnBuOH$ (1726). (d) Use as fuel additive and stabilizer (1726). (e) Rxn. of $C_{14}H_{10}SnCl_2$ with aq. NH_3 → $C_{14}H_{10}SnO$ (1726). (f) Rxn. with aromatic hydrocarbon + Na → organotin-contg. polynuclear hydrocarbon (2703). (g) $\overline{CH_2(CH_2)_4}SnBr_2$ forms (1:1) adducts with BiPy and 1,10-phenanthroline (17). (h) Biological activity as acaricide against Tetranychus species; R′ = $CH_2CH_2CMe_2CH_2CH_2X$ (433). (i) Synth. from $Sn_2C_{48}H_{32}O_2$, structure of formula I on page 885 (1378). (j) Rxn. with R′MgX → $SnC_{12}H_8OR'_2$; R′ = Me, Et, Ph (1378), with R′Li → $SnC_{12}H_8OR'_2$; R′ = Me, Et, Bu (1378).

8 MISCELLANEOUS ORGANOTIN COMPOUNDS

8.1 POLYMERIZATION REACTIONS WITH ORGANOTIN COMPOUNDS

Table 267 gives a short summary of polymerization reactions reported with unsaturated and functionally substituted organotin compounds. The following scheme serves as a key to Polymerization Systems.

I   Auto- or homopolymerization at:
    (IA)   room temperature,
    (IB)   below 90°,
    (IC)   above 100°.

II Catalytic polymerization with:
    (IIA) azobisisobutyronitrile (AIBN),
    (IIB) peroxide and/or mercaptan,
    (IIC) ^{60}Co irradiation,
    (IID) butyl lithium or potassium-naphthalene,
    (IIE) dialkylaluminum hydride,
    (IIF) triphenyl phosphite.

III Miscellaneous polymerization reactions:
    (IIIA) cohydrolysis,
    (IIIB) self-catalyzed esterification,
    (IIIC) condensation reactions (under cleavage of water, alcohol, metal halide, etc.).

Table 267. Polymerization Reactions with Organotin Compounds

Polymerization System	Method*	Remarks	Ref.
I OLEFIN POLYMERIZATION REACTIONS			
$Me_3SnCH:CH_2$ + (a) or (b)	II	Monomer reactivity ratios	1522
p-$Me_3SnC_6H_4CH:CH_2$ + (c)	IIA	Rubber-like polymer, prop.	1569
$Et_3SnCH:CH_2$	IIC	Rxn. mechanism	210
	IID	Prop., copolymer with (a, d)	1700
+ (e)	--	Thermal and electrostatic prop.	2514
$Et_2HSn(CH_2)_4CH:CH_2$	IIE	Mol. wt., prop.	1617
$Bu_3SnCH:CH_2$ + (a) or (b)	II	Monomer reactivity ratios	1522
$Bu_3SnC(:CH_2)CH:CH_2$	IIA	cis-struct., prop. (f)	14, 2933
$Bu_2Sn(CH_2CH:CH_2)_2$ + i-Bu_2SnH_2	IIE	Mol. wt., prop.	1617
+ $RSnH_3$	IIE	R = i-Bu, C_8H_{17}, crosslinked prod.	1617
+ (g)	IIE	Mol. wt., prop.	1617
p-$(c.C_6H_{11})_3SnC_6H_4CH:CH_2$ + (a)	IIA	Copolymerization const., prop.	284, 461
Ph_4Sn + (a)	IC	Prop., activity as scintillator	23
$Ph_3SnCH:CH_2$	IIC	Rxn. mechanism	210
p-$Ph_3SnC_6H_4CH:CH_2$ + (a)	IIA	Kinetic data, prop.	284
+ (h)	IB	Reactivity ratios (i)	461, 1784
p-$Ph_3Sn(C_6H_4)_2CH:CH_2$ + (a)	IIA	Copolymerization const., prop.	284
$Me_3SnO_2CCMe:CH_2$ + (a, j)	IIIB	Prop. (k)	1327
$Me_2Sn(O_2CCMe:CH_2)_2$ + (l)	IIIB	Activity as microbiocide, prop.	1328

* The characters used here refer to the system in the introduction to Chapter 8.1.
(a) $PhCH:CH_2$. (b) $CH_2:CMeCO_2Me$. (c) $(CH_2:CH)_2$. (d) $Et_3MCH:CH_2$; M = Si, Ge. (e) $CH_2:CHCl$. (f) Rxn. with Cl in CCl_4 at 0° → cis-$(CH_2CCl:CHCH_2)_x$ (14, 2933). (g) $(Et_2HSnCH_2CH_2CH_2)_2$. (h) $MeC_6H_4CH:CH_2$. (i) Use as scintillator (569, 1786). (j) $(:CHCO)_2O$. (k) Activity as bactericide against Bacillus subtilis, Escherichia coli, Brucella, Sarcina, Staphylococcus aureus (1327). (l) Styrene-maleic anhydride.

Table 267 Continued

Polymerization System	Method*	Remarks	Ref.
I OLEFIN POLYMERIZATION REACTIONS (Contd.)			
$(Et_3SnO_2CCH:)_2 + (a)$	IIB	m. 300° (dec.), prop.	1779
$Pr_3SnO_2CCH:CH_2$	IB	Prop., adhesive, laquer	2191
$Pr_3SnO_2CCMe:CH_2$	IIB	(m)	306
	IIB	Prop., struct. (m)	306, 1741
+ (b)	IIB	Prop., struct.	1741
+ (l)	IIIB	Activity as microbiocide, prop.	1328
$(Bu_3SnO_2CCHCH_2)_x$	--	(n)	1288
$Bu_3SnO_2CCMe:CH_2$	IIB	(m-306, 2302)	306
+ (b)	IIB	Prop., use as paint vehicle (o)	2949
+ (p)	IC	Resin (o)	2949
$Bu_3SnO_2CCH:CHCO_2C_3H_6O_2CCH:CH_2$	IIB	Prop. (o)	2974
+ (a) or (b)	IIB	Prop. (o)	2974
$p-Bu_3SnO_2CC_6H_4CH:CH_2$	IIB	(m-306, 2302)	306
$Ph_3SnO_2CCMe:CH_2$	IIB	(m, q)	306
$R_3SnO_2CCH:CH_2$	IIB	(m)	306
II ADDITION REACTIONS WITH ORGANOTIN HYDRIDES			
$Me_3SnH + (r)$	IIA	--	599
$Me_2SnH_2 + o-C_6H_4(C:CH)_2$	IB	IR spectrum	310, 1410
+ (r)	IIA	--	599
$Et_2SnH_2 + o-C_6H_4(CH:CH_2)_2$	IB	IR spectrum	310, 1410
+ $o-C_6H_4(C:CH)_2$	IB	IR spectrum	310, 1410
$Bu_3SnH + (r)$	IIB	Prop.	599
$Bu_2SnH_2 + 1,7$-octadiene	IIE	Mol. wt., prop. (s)	1617
+ 1,5-hexadiyne	IC	IR spectrum, dec. rxn.	499
+ $1-Bu_2Sn[(CH_2)_8CH:CH_2]_2$	IIE	Mol. wt., prop.	1617

+ p-C$_6$H$_4$(C!CH)$_2$	IC	Prop. thermogravimetric anal.	327
+ (r)	IIA	--	599
1-Bu$_2$SnH$_2$ + polybutadiene	IIB	Prop., copolymer with (a) - BuLi	2900
(c.C$_6$H$_{11}$)$_3$SnH + (r)	IIA	--	599
Ph$_3$SnH + (r)	IIA	--	599
Ph$_2$SnH$_2$ + o-C$_6$H$_4$(CH:CH$_2$)$_2$	IB	m. 60-70°	310, 1410
+ o-C$_6$H$_4$(C!CH)$_2$	IB	m. 60-80°, IR spectrum	310, 1410
+ (r)	IIA	--	599
R$_2$SnH$_2$ + (t)	I	(u)	1349

III POLYCONDENZATION REACTIONS

Me$_3$SnOR + (1); R = H, Ac, CH$_2$Cl	IIIB	Prop., activity as microbiocide	1328
Me$_2$Sn(CH$_2$SiMe$_2$)$_2$O + (v)	IIIA	Mol. wt., prop. (w)	159, 1066
Me$_2$SnCl$_2$ + (a) + Li	I	Glasslike polymer	383
+ (x)	IIIC	m. > 300°	1726
+ NaCH$_2$CO$_2$Na	IIIC	m. 232-33°, prop. (y)	124
+ (z) + (aa)	IIIA	Polyamide contg. 24% Sn (bb)	1245
+ (z) + (cc)	IIIA	Polyamide contg. 20% Sn (bb)	1245
+ (cc) + (dd)	IIIA	Prop. (bb, ee)	1246
+ R$_2$SiCl$_2$	IIIA	R = Me, CH$_2$:CH, Ph, prop.	1008
Me$_2$SnO + siloxanes	IIIB	Prop.	2185
(PhCH$_2$)$_3$SnOH + (1)	IIIB	Prop., activity as microbiocide	1328
(PhCH$_2$)$_2$SnCl$_2$ + (x)	IIIC	--	1726
(Et$_3$Sn)$_2$O + (a) + (j)	IIIB	Prop. (k)	1327

(m) Use as fungicide in paints, in marine coatings (306). (n) Use for SnO$_2$ plating (1288). (o) Use as microbiocide (2974). (p) AcCH$_2$CMe$_2$OH. (q) Activity as insecticide against Musca domestica (260). (r) Unsaturated polyester. (s) Effect of 1-BuSnH$_3$ as crosslinking agent (1617). (t) Acetylenic compounds. (u) Use as sensitizer in photosemiconductor (1349). (v) Me$_3$SiCl and Me$_2$SiCl$_2$. (w) Use as lubrication additive for polysiloxanes (159). (x) 9,10-Na$_2$C$_{14}$H$_{10}$. (y) Use as fungicide in paints, as stabilizer for PVC (124). (z) (CH$_2$CH$_2$CH$_2$NH$_2$)$_2$. (aa) p-C$_6$H$_4$(COCl)$_2$. (bb) Selfextinguishing after ignition (1245). (cc) (CH$_2$CH$_2$COCl)$_2$. (dd) (p-HOC$_6$H$_4$)$_2$CMe$_2$. (ee) Use as insect repellent and stabilizer for PVC (1246).

Table 267 Continued

Polymerization System	Method*		Ref.
	III POLYCONDENZATION REACTIONS (Contd.)		
Et$_3$SnOH + (a) + (j).	IIIB	White powder, dissocn. const.	868
+ (1)	IIIB	Prop. (k)	1327,2191
+ (1) + (j)	IIIB	--	1779
Et$_3$SnOMe + cellulose	IC	Prop. (ff)	1916
Et$_3$SnOSiEt$_3$ + (1)	IIIB	Prop., activity as microbiocide	1328
Et$_3$SnNMe$_2$ + CH$_2$:CHCHO	I	m. > 300°, struct.	2406
(CH$_2$:CH)$_2$SnCl$_2$ + (x)	IIIC	--	1726
Pr$_2$Sn(OAc)$_2$ + (1)	IIIB	Prop., activity as microbiocide	1328
(OCH$_2$CHCH$_2$O$_2$CCH$_2$CH$_2$)$_4$Sn + (gg)	IB	Elastomer, prop.	1948
(OCH$_2$CHCH$_2$O$_2$CCH$_2$CH$_2$)$_6$Sn$_2$ + (gg)	IC	Prop.	1948
Bu$_3$SnCl + cellulose	IB, c	Use as fungicide	1515
+ (hh)	IID	(ii)	2407
(Bu$_3$Sn)$_2$O + cellulose deriv.	IIIC	Use as fungicide	1515
+ PhHgOH + (j) + (jj)	IIIC	(kk)	2535
+ M(OAc)$_n$ + (j) + (jj)	IIIC	M = Cd, Hg, Pb, Sb, Bi (kk)	2535
+ Cl$_3$CCHO	I	Highly cryst., prop.	941
+ (1)	IIIB	Use as fungicide (1515)	1328,1515
+ (CH$_2$CHCO$_2$H)$_x$	IIIB	Use as fungicide	1515
+ (11)	IIIB	Use as fungicide	1515
+ (p) + (mm)	IIB	Prop. (nn)	2949
+ (MeHSiO)$_x$	IC		238,1239
Bu$_3$SnOH + (CH$_2$CMeCO$_2$H)$_x$	IIIB	Prop.	285
Bu$_3$SnOMe + Cl$_3$CCHO	I	Highly cryst., prop.	941
Bu$_3$SnOAc + (1)	IIIB	Prop., activity as microbiocide	1328
+ cellulose deriv.	IIIB	Use as fungicide	1515
Bu$_3$SnO$_2$CCH$_2$OH + PhHgOH + (j)	IIIB	(kk)	2535
+ M(OAc)$_n$ + (j)	IIIB	M = Cd, Hg, Sb, Bi (kk)	2535

Bu$_3$SnLi + PVC		Prop. and degradation	2598
	I	Struct. and IR spectrum	2700
Bu$_2$SnCl$_2$ + (x)	IIIC	White powder, m. > 300° (oo)	1726
+ R$_2$SiCl$_2$	IIIA	Elastomer prop.; R = Me, CH$_2$:CH, Ph	1008
+ RSiCl$_3$	IIIA	Elastomer prop.; R = Me, CH$_2$:CH, Ph	1008
Bu$_2$SnO + HSCH$_2$CH$_2$CO$_2$H	IIIB	Use as polymerization accelerator	1397
Bu$_2$Sn(OAc)$_2$ + (l)	IIIB	Prop., activity as microbiocide	1328
+ (pp)	IIIB	Prop.	2019
Bu$_2$Sn(ricinolate)$_2$ + (qq)	IIIB	Prop. (rr)	2959
Bu$_2$Sn(SCH$_2$CH$_2$OH)$_2$ + (qq)	IIIB	Prop. (rr)	2959
Bu$_2$Sn(NMe$_2$)$_2$ + C$_2$H$_2$	IB	Softening at 128°	250
[(C$_5$H$_{11}$)$_3$SnO$_2$CCMeCH$_2$]$_x$	--	Use in antifouling paints	2950
(C$_8$H$_{17}$)$_2$Sn(O$_2$CC$_6$H$_4$OH-p)$_2$ + (qq)	IIIB	Prop. (rr)	2959
[C$_{10}$F$_{21}$(CH$_2$)$_3$]$_2$Sn(OMe)$_2$ + MeSiCl$_3$	IIIA	Crosslinked polymer, prop. (ss)	745a
(C$_{18}$H$_{37}$)$_2$SnCl$_2$ + (x)	IIIC	--	1726
Ph$_3$SnCl + cellulose deriv.	IIIB	Use as fungicide	1515
Ph$_3$SnOH + (CH$_2$CMeCO$_2$H)$_x$	IIIB	Dec. 300°, prop.	285
+ (l)	IIIB	Use as fungicide	1515
Ph$_3$SnOAc + (l)	IIIB	Prop., activity as microbiocide	1328
Ph$_3$SnLi + PVC	IIIC	Mechanism of degradation	2598
+ PVC or (tt)	IIIC	IR spectrum, struct.	2700
Ph$_2$SnCl$_2$ + (x)	IIIC	m. > 300°	1726
+ HSCH$_2$CO$_2$Na	IIIC	m. > 300°	121
+ (z) + (cc)	IIIA	Dec. 265-75°, good adherence to Sn (bb)	1245
+ (aa) + (dd)	IIIA	White, m. 360° (bb, ee)	1246
Sn + (CH$_2$)$_n$Br$_2$ + (uu)	IIF	Use as stabilizer for PVC; n = 1-5	2980

(ff) Use as antioxidant, fungicide and stabilizer (1916). (gg) Anhydride of cyclohexane dicarboxylic acid. (hh) Polyvinyl alcohol. (ll) Activity as bactericide against Cryptococcus neoformans, as fungicide against Candida albicans (2407). (jj) HSCH$_2$CH$_2$OH. (kk) Activity as fungicide against Aspergillus niger, Penicillium funiculosum (2535). (ll) (CH$_2$CHC$_6$H$_4$SO$_3$H)$_x$. (mm) CH$_2$:CMeCO$_2$H. (nn) Rxn. with (MeHSiO)$_x$ → Bu$_3$SnH (1239). (oo) Rxn. with HCl → Bu$_2$SnCl$_2$ + 9,10-C$_{14}$H$_{12}$ (1726). (pp) (Bz$_2$CH)$_n$Ti(OCHMe$_2$)$_{4-n}$ or (1-PrO)$_4$Ti + Ac$_2$CH$_2$; n = 1,2. (qq) 1,2,4-MeC$_6$H$_3$(NCO)$_2$. (rr) Use as adhesive, elastomer, fungicide and stabilizer for PVC (2959). (ss) Use as dirt, oil and water repellent, release agent and antiadhesive (745a). (tt) Chlorinated polyethylene. (uu) BzSH and/or C$_{12}$H$_{25}$SH, HSCH$_2$CH$_2$CO$_2$H, C$_{11}$H$_{23}$CO$_2$Na, C$_8$H$_{17}$O$_2$CCH:CO$_2$H.

8.2 UNSPECIFIED ORGANOTIN COMPOUNDS

Table 268 comprises organotin entities that cannot be classified in any of the preceeding chapters.

Table 268. Unspecified Organotin Compounds

Me_3Sn-	Group electronegativity (234)
Et_3Sn-	Intoxication and effect on ^{32}P phospholipides and body temperature (5), effect on amino acid metabolism in brain cortex slices (99), effect on platelet shape and aggregation (407, 1647), effect on platelet morphology (2781), effect on oxidation of glucose and pyruvate in brain cortex slices (916), effect on brain glucose metabolism in vivo and vitro (918), effect on retention of potassium and amino acid in brain cortex slices (917).
$Et_2Sn=$	Spectrophotometric detn. with 4-(2-pyridylazo)resorcinol (PAR), (1697).
Bu_nSnAr_m	Organometallic polynuclear oil (2703).
Bu_3SnR	R = organic group contg. S and P, light resistance and fungicidal activity against Coriolellus palustris (1063).
Bu_3SnS-	Detn. and sepn. from $(C_8H_{17})_nSn(S-)_{4-n}$ by TLC; n = 2, 3 (215).
Bu_3Sn-	Use as bactericide and disinfectant (695), as marine preservative against Limnoria, Martensia and Teredine borers (226).
Bu_2SnS-	Detn. and sepn. from $(C_8H_{17})_nSn(S-)_{4-n}$ by TLC; n = 2, 3 (215), use in catalyst for polyurethan (979), for PVC foam (1786).
$(C_8H_{17})_3SnS-$	Detn. and sepn. from $Bu_nSn(S-)_m$ by TLC (215).
$(C_8H_{17})_2Sn(S-)_2$	Detn. and sepn. from $Bu_nSn(S-)_m$ by TLC (215).
$(C_8H_{17})_2Sn=$	Toxicity (1037).
Ph_3Sn-	Retention after oral and i.p. administration to rats and guinea pigs (1169).
Ph_nSnAr_m	Organometallic polynuclear oil (2703).
$(CH_2)_5SnAr_n$	Organometallic polynuclear oil (2703).
R_4Sn	Sn-C bond refractions (2688).
R_3SnAr	Ar = Ph, $PhCH_2$, IR and UV spectra (137).
$R_2Sn=$	Anal. detn. by coulometric titration with 4-(2-pyridylazo)-resorcinol (PAR) and 1-(2-pyridylazo-)-2-naphthal (PAN) (1696).
$RSn\equiv$	Anal. detn. in modified microanalytical train (594), influence of paint vehicle on mildewcide effect (449), use for deposition of tin films by electron beam treatment at 130-200° (2846).
$R_nSn(S-)_m$	Effect on thermogravimetric analysis of PVC (2563), microanalytical S- determination (1276).
Ar_xSn-	UV and IR spectra (1390).

9. COMMERCIAL APPLICATION OF ORGANOTIN COMPOUNDS

Organotin compounds are used as biocides and as catalysts for polymerization, esterification and condensation. The number of papers show their continuing importance as stabilizing agents for polymers. Aside from polyvinyl chloride and polyvinyl chloride copolymers increasing attention is given to polyurethan. Here, however, the main interest lies still in the activity of organotin compounds as blowing agents. There is also increasing activity for the organotin compounds in stabilizing systems for polyolefins, polyesters and polyamides.

A large number of papers deal with the biological activity of organotin derivatives as acaricides, bactericides, fungicides, insecticides, herbicides and molluscicides. Search for selective activity continues vigorously. One finds also several claims for bacteriostatic, fungistatic and chemosterilant properties. Several of the more reactive organotin compounds have registered trade marks, some with undisclosed compositions.

Commercial and legal considerations lead to publication of compositions and formulations with sometimes vague chemical characterization. Table 269 summarizes such publications containing a paucity of data on which a precise identification of the organotin entity can be based. The formulae given in the first column of the table frequently indicate only the general character of a product or give components of preparations.

Activity as stabilizer of defined organotin compounds have been given previously under the heading "use" at each separate compound and may be read in Chapters 1 through 7.

Table 269. Organotin Stabilizers

Compound or Reaction Components	Remarks	Ref.
$(Me_3Sn)_2O + 1\text{-}C_8H_{17}O_2CR$	(a) 9,10-epoxy stearate	1442
$Me_2SnO + (:CHCO)_2O + RO_2CCH:CHCO_2H$	(a) $R = MeOCH_2CH_2$, $BuOCH_2CH_2$	642
$+ OC_8H_{15}O_2CR$	(a) 9,10-epoxy stearate	1442
$+ B(OH)_3 + RSH + R'CO_2H$	(a)	2003
$Me_2Sn(SC_{12}H_{25})OC_6H_8O_2O_2CC_{17}H_{35}$	(b) Sorbide deriv.	2018
$Me_2Sn_2O_3 + RSH + R'CO_2H$	(a)	2372
$(PhCH_2)_2SnO + Me_3CO_2CR$	(a) 9,10-epoxy stearate	1442
$(PhCH_2)_2Sn(OR)O_2CR'$	(b) Sorbide deriv.	2018
$Et_2SnO + ClCH_2CH_2O_2CR$	(a) 9,10-epoxy stearate	1442
$(CH_2:CH)_2SnO + (p\text{-}HOC_6H_4)_2CHC_6H_{13}$	(a)	2123
$Pr_2SnO + ArCO_2H$	(a, c)	1289
$(Bu_3Sn)_2O + R'CO_2H$	(a) R' = epoxidized glycerin oleate	1442
$+ (CH_2CH_2OH)_2 + Y(CO)_2O$	(a) Y = 2-bicycloheptene	2020
$Bu_2SnCl_2 + p\text{-}OC_3H_5C_6H_4CONH_2 + NaOH$	(a)	1442
$Bu_2SnO + (p\text{-}HOC_6H_4)_2CR^2$	(a) Prop. for several deriv.	2123
$+ (CH_2OH)_2 + Y(CO)_2O$	(a) Y = 2-bicycloheptene	2020
$+ R'CO_2H + (:CHCO)_2O$	(a) R' = maleic acid halfester	642
$+ ArCO_2H$	(a, c) Prop. for several deriv.	1289
$+ R'CO_2H$	(a) epoxide deriv.	1442
$+ HO_2CCH_2CHSHCO_2H$	(a)	1708
$+ B(OH)_3 + RSH + R'CO_2H$	(a)	2003
$+ HO_2CCH_2CH_2SH$	(a, d)	1398
$+ 1\text{-}C_8H_{17}O_2CCH_2CH_2SH$	(a)	2861
$+ RSH + R'CO_2H$	(a)	1031, 2386
$+ BzSH + RSH$	(a)	1036
$+ RSH + R'SH$	(a)	445
$+ RSH$	(a)	1644, 2861
$+ RCHO + R'SH$	(a)	2122

Compound	Note	Reference
$Bu_2Sn(OMe)_2 + HSCH_2CO_2R$	(a)	1248
$Bu_2Sn(OR)O(CH_2CH_2O)_xR'$	(b) $R' = C_{17}H_{35}CO$	2017-8
	(b)	2817-8
$Bu_2Sn(SR)O(CH_2CH_2O)_xC_{18}H_{37}$	(b)	2817
$Bu_2Sn(SR)OC_6H_4C_8H_{17}$	(a, d, e) Benzthiazol deriv.	22
$Bu_2Sn(O_2CR)_2 + HSCH_2CO_2C_8H_{17}$	(a) Epoxide	1442
$Bu_2Sn(O_2CCH:CHCO_2R)_2$	(a)	635
$Bu_2Sn(O_2CC_{11}H_{23})O_2CR$	(f) Maleic acid halfester	1362
$Bu_2Sn(O_2CR')O_2CR$	(a) Maleic acid halfester	635
$[Bu_2Sn(O_2CR)O_2CCH:]_x$	(a)	635
$Bu_2Sn(O_2CR)O_2CR'$	(a) Thioether and maleate halfester	1208
$Bu_2Sn(SR)O_2CR$	(a) Maleic acid halfester	1208
	(b) Maleic acid halfester	2018
$Bu_2Sn(SR)_2$	(a)	2580
$Bu_2Sn(SR)_2$	(a) RS = cryptomercaptide	597
$Bu_2Sn(SNa)_2 + RX$	(a) R = alkyl, Bz	1640
$Bu_2Sn =$	(a) Detn. by TLC	215, 345
	(a) Detn. by photometry	1806, 2143
	(a) Detn. by paper chromatography	2077
	(a) Action of-, and mechanism	906
$Bu_2Sn_2O_3 + RCO_2H$	(a) $R = HOCH_2$, MeO_2CCH_2	1170a
$+ HSCH_2CO_2R$	(a)	2572
$+ HS(CH_2)_nCO_2R$	(a) $n = 1, 2$	1170a
$BuSnO_2H + RCO_2H$	(a) R = Me, $HOCH_2$	1170a
$+ HSCH_2CH_2CO_2R$	(a)	1170a
$[(\omega\text{-}ClC_4H_8)_3Sn]_2O + RCO_2Bu$	(a) 9,10-epoxy soyate	1442

(a) Use as stabilzer for PVC or similar resins. (b) Use as antistatic agent for thermoplastics. (c) Use as stabilizer for polyvinyl acetate. (d) Use as polymerization accelerator for vinyl monomers and unsaturated polyesters. (e) Use as vulcanization accelerator. (f) Use as stabilizer for dyeable polyolefins.

Table 269 Continued

Compound or Reaction Components	Remarks	Ref.
$(C_6H_{13})_4Sn + SnCl_4 + HCONMe_2$	(g)	742
$(C_6H_{13})_2SnO + diol + Y(CO)_2O$	(a) Y = 2-bicycloheptene	2020
$(c.C_6H_{11})_2SnO + RCO_2Ph$	(a) 9,10-epoxy stearate	1442
$[(C_8H_{17})_3Sn]_2O + RCO_2R'$	(a) Octyl 9,10-epoxy stearate	1442
$(C_8H_{17})_2SnO + RCO_2H + (:CHCO)_2O$	(a) Maleic acid halfester	642
$\quad + B(OH)_3 + RSH$	(a) R = thiol acid and alkyl	2003
$\quad + NCCH_2CO_2H + RCHO$	(a) R = aryl and alkyl	2565
$(C_8H_{17})_2Sn(OMe)_2 + RCONR_2$	(a) epoxide	1442
$(C_8H_{17})_2Sn(OR)OR'$	(b) R, R' = polyethoxide	2818
$(C_8H_{17})_2Sn(OR)O_2CR$	(b) R = polyethoxide	2017
	(b) R = polyethoxide	2817
$(C_8H_{17})_2Sn(OR)SC_{12}H_{25}$	(b) R = polyethoxide	2818
$(C_8H_{17})_2Sn =$	(a) Detn. by extraction, TLC	215, 345, 1172b
	(a) Detn. by photometry	1806, 2143
	(a) Detn. by paper chromatography	2077
$(C_8H_{17}Sn)_2O_3 + RCO_2H$	(a)	2372
$\quad + RSH$	(a)	2372
$C_8H_{17}SnO_2H + RCO_2H$	(a) Maleic acid halfester	1170a
$(C_{12}H_{25})_2SnO + RCO_2C_6H_{11}$	(a) 9,10-epoxy stearate	1442
$\quad + (p\text{-}HOC_6H_4)_2CMe_2$	(a) m. 165-67°	2123
$(C_{12}H_{25}Sn)_2O_3 + RCO_2H$	(a)	2372
$\quad + RSH$	(a)	2372
$(C_{18}H_{37})_2SnO + RCO_2Me$	(a) 9,10-epoxy stearate	1442
$(C_{18}H_{37}Sn)_2O_3 + RCO_2H$	(a)	2372
$\quad + RSH$	(a)	2372
$Ph_2SnO + RCO_2R$	(a) 9,10-epoxy stearate	1442
$\quad + B(OH)_3 + HSCH_2CH_2OH + (:CHCO)_2O$	(a)	2003
$Ph_2Sn(OR)O_2CC_{17}H_{35}$	(b) Sorbide deriv.	2018
$Ph_2Sn(OR)O_2CR'$	(b) R = polyethoxide	2818

$(C_{10}H_7)_2SnO + RCO_2Bu$	(a) 9,10-epoxy stearate	1442
$PhSnO_2H + HSCH_2CO_2R$	(a)	1170a
$(C_4H_3S)_2SnO + (C_4H_3S)CH_2O_2CR$	(a) 9,10-epoxy stearate	1442
R_3SnO_2CR'	(a) R = Me, $PhCH_2$, Bu, C_8H_{17}, $C_{18}H_{37}$, Ph; R' = thioether carboxylate	1170
$R_2SnO + HSCH_2CH_2CO_2H + R'_2SnCl_2$	(a)	178
$\{R_2Sn[O_2C(CH_2)_nCR'SR']_2\}$ $\{R_2Sn(O_2CC_6H_4S)_2CR'_2\}$	(a) R = Me, $PhCH_2$, Bu, Me_3CCH_2, C_6H_{13}, C_8H_{17}, $C_{12}H_{25}$, $C_{18}H_{37}$, Ph	1170
$(RSn)_n(O_2CR')_m$	(a) R = Me, $PhCH_2$, Bu, C_8H_{17}, $C_{12}H_{25}$, $C_{18}H_{37}$, Ph; R' = thioether carboxylate	1170
R_nSn	(a) Tin detn. by combustion and iodatometric titration	2655
	(a) Detn. in PVC by paper chromatography, oscillography and chem. methods	282
	(a) Method for rapid screening by tissue implanting	401
	(a) Effect on PVC dec. in solution	370
	(a) Liquid stabilizer (h)	2349

(g) Use as catalyst and stabilizer for polyurethan foam. (h) Activity as fungicide against Aspergillus niger, Fusarium culmorum, Nigrospora sphaerica, Penicillium funiculosum, Phoma suecica.

LEAD

1 SYMMETRIC TETRAORGANOLEAD COMPOUNDS

1.1 UNSUBSTITUTED DERIVATIVES

Tetramethyllead (M-519) $\qquad$ PbC_4H_{12} $\qquad$ Me_4Pb

<u>Synth.</u>: By rxn. of $Pb(OAc)_4$ with MeMgCl in THF at 5°, in 89% yield (2136).
By thermal dec. of Me_3PbCl at 112° (1194), of Me_6Pb_2 at r.t. (1936).
By rxn. of $PbCl_2$ with $Me_3Al_2Cl_3$ and alkali metal halide in nonpolar solvent or solid state, in good yield (1669).
By rxn. of Na-Pb alloy with MeCl in presence of Al, AlX_3 or R_3Al and $(MeO)_2CH_2$, in 78% yield (26), of R_3Al and $(MeOCH_2CH_2)_2O$ in 93% yield (273), of $Me_3Al_2Cl_3$ and R_3N, in 71% yield (27), and m-dioxane, in 91% yield (274), and R_2O (1324), of $Me_nAl_2X_{6-n}$ (715), of low amount water and MeOH or NH_3 (1203), in 95% yield (419), and $(MeOCH_2)_2$, in 95% yield (2729), and MeCN and NH_3 in MePh, in 91% yield (2676), and $MeOCH_2CH_2NH_2$, $MeNH_2$ or $(CH_2NH_2)_2$ at 25-90°, in 86% yield (1785), of MeOH, MeI and cat. $MeONH_2$, $H_2NOH \cdot HCl$ or $i-PrNH_2 \cdot HCl$, in 94% yield (2730).
By electrolysis of MeMgCl with MeCl on Pb anode (42), in presence of THF and Bu_2-carbitole (317), and glycol ethers and aromatic solvent (316), in 83-98% yield (774-78, 781), in $(BuOCH_2)_2$ (2271).
By electrolysis of MeBr in Et_4NBr and aq. MeCN on Pb cathode and Pt anode, in 84% yield (1946a).
By-prod. in electrolysis of $CH_2:CHMgX$ and Me_3M on Pb anode, M = B, Al (1325).
By electrical process from MeCl, Mg and Pb (195).
Electrolytical cell for manufacturing of - (46, 416, 779), Me_4Pb recovery after electrolysis (2273), MeCl recovery (2272).
Recovery of Me_4Pb from manufacturing process (2274, 2387, 2470), purification of Me_4Pb after manufacturing by treatment with alkane peroxyacid (89), with O_3 (888).
Stabilization of Me_4Pb during distillation by MeCN and MePh (1158).
<u>Prop.</u>: IR (546) and assignment (763), far IR and assignment (106-7), Raman (244, 1423), NMR (129, 518, 570, 985-6, 1936, 1949, 2296, 2583, 2723, 2875), ^{13}C NMR (2121) and mass (169, 1752, 1756, 2304) spectra, force const. (1067), self-diffusion and nuclear spin-lattice relaxion time (1307), thermodynamic data (106, 305, 2856), van der Waals volume and radius (762). Anal. detn. by GLC (432), comparison of detector systems (40), by coulometric titration (427, 1696), in air by colorimetric method after absorption with active carbon (1951), with ICl (1551), by rxn. with MeOH and I (315), by polarography (2197), in gasoline by GLC in presence of Me_nPbEt_{4-n}; n = 0-3 (1140, 1358), 0-4 (2790) and halogen scavengers (1957), by atomic absorption (1334), by colorimetric method (1552), removal from fuel with ICl (2614), soly. in AcOH and sepn. from Et_4Pb (418).
<u>Rxn. with</u>: Et_4Pb + catalyst → Me_nPbEt_{4-n}; cat.: $AlCl_3$ (93a, 432), $Me_3Al_2Cl_3$ or $BF_3 \cdot OEt_2$ (593), clay (88, 880), acidic ion exchange resin (2450).

$(CH_2:CH)_4Pb$ + activated clay → $Me_nPb(CH:CH_2)_{4-n}$; n = 0-4 (880, 2935), with $Me_3Al_2Cl_3$ (2935).
$Pb(OAc)_4$ + cat. $Hg(OAc)_2$ at -60° → $Me_3PbOAc + Me_2Pb(OAc)_2 + Pb(OAc)_2$ (2864).
B_2H_6 → $Me_3B + Me_2B_2H_4 + Me_3B_2H_3 + CH_4 + H + Pb$; Me_3PbH is discussed as rxn. intermediate (1195).
$Cl_2BCH:CHBCl_2$ → $Me_3PbCl + Me_3B + (MeBCH:CH)_x$ (1194).
$(Cl_2B)_2CHCH(BCl_2)_2$ → $Me_3PbCl + Me_3B + (Me_2B)_3C_2H_2BCl_3$ (1194).
Cl in MePh → Me_3PbCl at -78° and Me_2PbCl_2 at -20° (1056).
Br at -70° → Me_3PbBr (1936).
Br in MeOH, AcOH and Me_2SO → mechanism of cleavage rxn. (653).
I in MeOH → kinetic data for cleavage rxn. (1098), rate of cleavage, solvent effect (1698), mechanism (1750), in MeOH, AcOH and Me_2SO (653).
BF_3 → $Me_3PbF + Me_2BF + MeBF_2$ (1194).
BCl_3 → $Me_3PbCl + Me_3B$ (1194).
B_2Cl_4 → $Me_3PbCl + Me_3B + (BCl)_n$ (1194).
B_4Cl_4 → $Me_3B + (BMe)_x + PbCl_2$ (1194).
SO_2Cl_2 → $Me_3PbCl + Me_2PbCl_2$ (116).
SO_3 in CH_2Cl_2 at 0-5° → Me_3PbO_3SMe (1:1) + $Me_2Pb(O_3SMe)_2$ (1:2) (1087).
SO_3 in CH_2Cl_2 → Me_2PbSO_4 (1087).
SO_2 in dry C_6H_6 → $Me_2Pb(O_2SMe)_2$ (232a, 1217).
SO_2 in Et_2O → Me_3PbO_2SMe (1083).
SO_2 in aq. Et_2O → Me_2PbSO_3 (1217).
$Se(SeCN)_2$ → $Me_3PbSeCN + Se$ (697).
$HClO_4$ and/or AcOH → mechanism of cleavage rxn. (653).
ROH, + H_2O → Me_3PbOH; R = Me, Et, i-Pr, i-Bu (1678).
PhOH → $Me_2Pb(OPh)_2$ (342).
CF_3CO_2H → $Me_3PbO_2CCF_3$ (1:1) + $Me_2Pb(O_2CCF_3)_2$ (1:4) + CH_4 (1214).
Air in heptane → $PbO + PbO_2 + PbCO_3 + Pb(NO_3)_2$ (1085).
$H_2 + O_2$ → inhibition effect on flame speed (358).
Biol. Prop.: Toxicity (74, 75, 887), behavior of blood picture (76).
Use: As antiknock agent (369), in presence of ethers or tertiary esters of monocarboxylic acids (171), of organic phosphates (219, 410), of sulfur compounds (1513), of Me_nPbEt_{4-n} (880), of $Me_nPb(CH:CH_2)_{4-n}$ (880); n = 1-3 (1325), of $RMn(CO)_5$ (53), ferrocene and/or $Fe(CO)_5$ (576), of $(C_5H_5)_2Ni·C_2H_2$ (989), activity relative to Et_4Pb (197, 709), Me_nPbEt_{4-n} (877, 1104, 1157, 2938); n = 2 (877), $Me_nPb(CH:CH_2)_{4-n}$ (880, 1325, 2935), $Fe(CO)_5$ and ferrocene (197), rxn. mechanism (1783), in dual cycle diesel engine (326). Stabilized antiknock mixtures (93a, 560, 613, 899, 900, 1118-9, 1925, 2934).
As antidetonant (1148), in lubricants (1409).
As catalyst for polymerization of $(CH_2:CH)_2$ (381), of $CH_2:CHCl$ (764).
As chemiluminescent at altitudes > 90 Km (228).
As corrosion inhibitor for graphite in liq. CO_2 (1111).

Tetraethyllead (M-521) PbC_8H_{20} Et_4Pb
Synth.: By rxn. of Et_3PbOH with EtMgBr in Et_2O or THF, in 49% yield (1340).
By rxn. of $Pb(OAc)_4$ with EtMgCl in THF at 5°, in 90% yield (2136).
By rxn. of Et_2PbCl_2 with $(CH_2CH_2MgBr)_2$, in 10% yield (1260).

By thermal dec. of Et$_3$PbOMe at r.t., in 20% yield (951), of Et$_n$Pb(O$_2$SEt)$_{4-n}$; n = 2, 3, as by-prod. (1083).

By rxn. of Pb-Na alloy with EtCl; effect of work-up conditions in presence of polysulfide and salts of Fe and Al (1084), in presence of I, Me$_2$CO and Et$_3$PbI, in 91% yield (481, 1203), of sulfides of Fe, Co and Ni at 60-85°, in > 90% yield (2913), of Me$_2$NCN at 110°, in 40% yield (2496), of EtBr, effect of Pb content (1539), in presence of trace Ag and/or Cu at 40-150 ppm, in 90-94% yield (2788), of K, EtI, Et$_3$PbI and Me$_2$CO, in 90% yield (721).

By electrolysis of EtMgCl on Pb anode with excess EtCl (42), with ethers (777), with glycol ethers, THF (778, 781) and C$_6$H$_6$ (775) followed by extraction process (780).

By electrolysis on a Pb anode of NaAlEt$_4$ in presence of NaAlEt$_3$OR (616), of KAlEt$_4$ (638), of Et$_3$Al + EtONa (177), of Et$_3$Al or EtMg-compd. (2401).

By electrolysis of EtBr on electrodes of Pt or C in aq. alkaline suspension (2401).

By electrolysis on a Pb cathode with Et$_4$NBr in MeCN or Et$_3$SBr, in 12 and 35% yield, resp. (1927).

Electrolytic manufacturing process from Pb, Mg and EtCl (195), Pb, KAlEt$_4$ on Hg cathode (overall from C$_2$H$_4$ and Pb) (1401), Pb, NaBEt$_4$ on Hg cathode or with NaAlEt$_3$OBu and NaBEt$_4$, in 93% yield (2186).

Electrolytic cell for manufacturing (46, 416, 779, 1394a).

Methods of Et$_4$Pb recovery after electrolysis by solvent extraction (2272), of EtCl recovery (2272), of Et$_4$Pb purification by pressure steam distn. (2914), by two stage steam distn. (2936), by O$_3$ rxn. (889) of Pb reclamation (341), waste water treatment (1639).

Prop.: IR (244, 763, 1260) and assignment (673), Raman (244), NMR (1474, 2296) and mass (169, 850, 1756, 2268, 2294) spectra, absorption on active carbon (614), effect on molecular redistribution on refraction index (1175), spectroscopic measurements in shock discharge tube (797), thermodynamic data (305, 2856). Anal. detn. by GLC (431-2, 824, 1140, 1178, 1334, 1355, 1428, 1957, 2681, 2790), comparison of detector systems (48), by TLC (1260), by atomic absorption (1334, 1553, 2139), influence of ions and solvent (114), by β-ray reflection technique (687), by x-ray fluorescence (196, 208), by colorimetry (1977), by coulometric method (427, 1696), by polarographic method (641, 1769, 2197).

Anal. detn. in air (1945), with I in MeOH (315), with ICl and dithizone colorimetry (1551), with H$_2$SO$_4$ and colorimetry (960), by polarographic method (2197). Anal. detn. in gasoline by GLC technique (1178, 1344) in presence of Me$_n$Pb-Et$_{4-n}$ (1140, 1355, 1957, 2790), by atomic absorption (1334), by colorimetry after ICl rxn. (552), by conductometry or oscillometry after rxn. with Br (1258), by photometry (2413), by polarography (1765), by potentiometry (1699), by volumetry (236) after rxn. with ICl (2698), by gravimetry after combustion (2479) in presence of Mn (2527). Quantitative removal from fuel with ICl (2614), with impinger (2812). Solubility in AcOH and sepn. from Me$_4$Pb (418).

Rxn. with: UV irradiation in i-octane → stable Pb-dispersion (559).

Me$_4$Pb + catalyst → Me$_n$PbEt$_{4-n}$; n = 0-4, cat.: AlCl$_3$ (432), Me$_3$Al$_2$Cl$_3$ or BF$_3$·OEt$_2$ (593), clay (88, 880), acidic ion exchange resin (2450).

$Et_4{}^{210}Pb \to$ radioactive decay (992).

$(CH_2:CH)_4Pb$ + clay $\to Et_nPb(CH:CH_2)_{4-n}$ (880).

$Et_3PbO_2Et \to Et_3PbOEt$ (662).

Pr_4Sn at 140° + $AlCl_3 \to$ random redistribution (432).

Et_3SnH at 100-130° $\to Et_6Sn_2 + Et_4Sn + Pb + C_2H_6 + H$ + trace C_2H_4 (2850).

Et_3GeH at 165-70° $\to Et_6Ge_2 + Et_4Ge + Pb + C_2H_6$ + Ge-Pb-organic compd. (2850).

Et_3SiH at 165-70° $\to Et_4Si + Et_5Si_2H + Et_6Si_2 + C_2H_6 + Pb$ + Si-Pb-organic compd. (2850).

$HGeCl_3 \cdot OEt_2 \to EtGeHCl_2 + Et_3PbCl$ (366, 1471).

$GeCl_4 \to EtGeCl_3 + Et_3PbCl$ (366).

$GeCl_4$ (2:1) $\to Et_2GeCl_2 + Et_3PbCl$ (1364).

Br → mechanism of cleavage rxn., solvent effect (653), kinetic data in MeCN (1098).

I → mechanism of cleavage rxn. in polar solvent (653), kinetic data in MeCN (1098), rate const. in (non)polar solvent (1678, 1750).

$EtCl \to Et_3PbCl$ (1539).

$HCCl_3 + NaOEt \to Et_3PbCCl_2Et$ (2759).

$EtSOCl \to Et_3PbCl$ (166).

$SOCl_2 \to Et_3PbCl + Et_2PbCl_2 + PbCl_2 + EtSOCl$ (166).

$S_2Cl_2 \to Et_3PbCl + Et_2S_2$ (1441).

$MgBrOH \to Et_3PbBr$ (1340).

$HClO_4$ and/or AcOH → mechanism of cleavage rxn. (653).

AcOH in MePh at 60-100° $\to Et_3PbOAc + Et_2Pb(OAc)_2 + C_2H_6$; kinetic data and rxn. mechanism (1207).

$RCO_2H \to Et_3PbO_2CR$ (r.t.) + $Et_2Pb(O_2CR)_2$ (excess at 100°); R = CF_3, Me_3C (1214).

ROH, + $H_2O \to Et_3PbOH$; R = Me, Et, i-Pr, i-Bu (1678).

$PhOH \to Et_2Pb(OPh)_2$ (342).

Air in heptane $\to PbO + PbO_2 + PbCO_3 + Pb(NO_3)_2$ (1085).

O → kinetic data and oxidn. mechanism, effect of inhibitors (2214, 2220).

O + Et_3PbR → initial oxidn. rate; R = Me, n-, i-Pr, Et_3Pb, Cl, OH, Me_3CO_2 (1017).

$O_3 \to$ removal of Et_4Pb from air (2380).

O_3 at -78° $\to Et_3PbOH + Et_2PbO + Et_3PbOEt + AcH + EtOH$ (663).

O under UV irradiation $\to Et_3PbOH + Et_3PbO_2Et + Et_3PbOEt + PbO + AcH$; rxn. mechanism (662), influence of temp. (1017).

SO_2 in $Et_2O \to Et_3PbO_2SEt + Et_2Pb(O_2SEt)_2$ (1083).

    in aq. $Et_2O \to Et_2PbSO_3$ (1217).

    in dry $C_6H_6 \to Et_2Pb(O_2SEt)_2$ (232a, 1217).

SO_3 in CH_2Cl_2 at 0-5° $\to Et_3PbO_3SEt$ (1:1) + $Et_2Pb(O_3SEt)_2$ (1:2) (1087).

    in $CH_2Cl_2 \to Et_2PbSO_4$ (1087).

    in heptane $\to Et_2Pb(O_3SEt)_2$; method for "deleading" gasoline (1086).

$Se(SeCN)_2 \to Et_3PbSeCN + EtSeCN + Se$ (697).

$H_2 + O_2 \to$ effect of inhibiting flame speed (358).

Biol. Prop.: Toxicity (98, 2659, 2782) and absorption through skin (2026), relative to newborn and adult rats, carcinogenity (2381), acute lethal dose in gasoline (515), clinical features of chronical intoxication (170). Intoxication (460a) and effect on cerebral acetylcholinesterase in vivo and in vitro (1763), on human acetylcholinesterase in vitro (2402), on cholinesterase in sheep (1978), on 5-hydroxyindoleacetic acid excretion of rabbits (160), on monoamine oxidase in rabbits (1422) on porphyrin metabolism in vitro (1764), on protoporphyrin level in erythrocytes in humans and animals (919), on serotonin level in rabbit lungs (56), on tryptophan metabolism of rabbits (94) on human serum protein (351), tissue metabolism in mice (294). Contribution to Pb-content in human body (1154a).

Distribution and excretion in rats (2261), in rabbits (2262).

Effect on learning and memory of rats (811).

Use for contrast modification of tissue slices (1425).

U.S. threshhold limits (539).

Use: As antiknock agent (369) in two stroke engine fuel (2985, 3003), in C_3H_8 and C_7H_{10} (1255), activity in different gasolines (708), mechanism (1133, 1782-3, 2190) and flame propagation (2379), in presence of $(CH_2Cl)_2$ and $(CH_2Br)_2$ (808) of $(CH_2Br)_2$ and phosphates and thiophene (809), of $C_2H_2Cl_2F_2$, $C_2H_2Cl_3F$ and $CF_3CCl:CCl_2$ (1025), of ethers and tertiary esters of monocarboxylic acids (171), of esters or esters and phosphates (219, 410, 897, 1077, 1093), of aliphatic N-containing carboxylates (1075), of gem-diacylates (1176), of S-compounds (1146, 1513), of halogen scavengers and $Bu_3P:BH_3$ (29), of boron esters (2896), of $RMn(CO)_5$ (53), of $ArMn(CO)_3$ (83), of $C_5H_5Mn(CO)_3$ (1186), of $MeC_5H_4(CO)_3$ and $MePhNH$ (1636), of Mn-compd. and $AcOCH_2CH_2Cl$ (218) and Mn detn. (2526), of ferrocene and/or $Fe(CO)_5$ (197, 576), of $(RC_5H_4)_2Fe$ (975), of $(C_5H_5)_2Ni \cdot C_2H_2$ (989), of cyclopentadienyl metal deriv. (497, 2935), of Me_4Pb (197, 709), of Me_nPbEt_{4-n} (877, 880, 1104, 1157); n = 2 (877), of Cu, Pb and Zn naphthenates (988), of $Et_nPb(CH:CH_2)_{4-n}$ (880, 2935), effect on composition of automobile exhaust gas (1664), on particle size and distribution from exhaust (2054), on Pb-halide distribution from exhaust in Fairbanks, Alaska (2140), on Pb-concn. in air (79a), in nature (415a), on Pb-isomer distribution in Los Angeles basin (80), on slow combustion formed constituents in spark ignition engine (672), stabilized compositions (93, 93a, 149, 183, 560, 613, 834, 890, 892, 896, 898-9, 1118-9, 1833, 1925, 2937-8), computer controlled blending program (2390).

As additive for fire resistant polystyrene (1223), for propellant powders (561, 1759) for spontaneous ignition stabilized lubricants (1409).

As antidetonant (2177), influence of S-compounds (1148).

In catalyst for polymerization of butadiene (231, 381, 1584), C_2H_4 (1638), C_2H_4 copolymers with fluorovinyl monomers (763a), C_3H_6 (1638, 2995), $CH_2:CHF$ and vinyl copolymers (1923), $CH_2:CHCl$ (41a, 764, 1627, 1686), $CH_2:CHCN$ (765, 1250, 1686), $CH_2:CHCO_2Et$ (765), $CH_2:CMeCO_2Me$ (765, 1686), $CH_2:CHC(NO_2):CHMe$ (1353) of trioxane (2478).

Effect on black plague corrosion on high temperature turbine alloys (718), on graphite corrosion in liq. CO_2 (1111), on formation of frictional polymer (847), on reforming catalyst (577).

Lead vapor deposition by glow discharge on glass and bronze (2293).

Average annual U.S.A. production in 1962 (637a) in 1944-64 (1154a).

Tetrapropyllead (M-525) $PbC_{12}H_{28}$ Pr_4Pb

Synth.: By electrolytical manufacturing process with $NaBPr_3OMe$ and $NaBPr_4$

on Pb-anode (2186).

Prop.: (Far) IR, assignment (673) and NMR (1321) spectra, thermodynamic data (305). Anal. detn. by coulometric detn. after bromination (1696).

Rxn. with: UV irradiation in i-octane → stable Pb-dispersion (559).
Photochemical and thermal dec. in C_6H_6 → C_3H_6 + n-C_6H_{14} + PrPh + $Pr_2C_6H_4$ + C_3H_8, rxn. mechanism (2684).
I → rate of cleavage rxn., solvent effect (1698, 1750).
$GeCl_4$ → $PrGeCl_3$ + Pr_3PbCl (1364).
$SOCl_2$ → Pr_3PbCl + Pr_2PbCl_2 (166).
$HClO_4$ and/or AcOH → mechanism of cleavage rxn. (653).
Combustion in heptane → PbO + PbO_2 + $PbCO_3$ + $Pb(HO_3)_2$ (1085).

Use: As antiknock agent in presence of phosphates (219, 410), of organo-nickel compounds (989), stabilization by alkylcycloalkanes (560).
In catalyst for butadiene polymerization (381).

Tetrabutyllead (M-526) $PbC_{16}H_{36}$ Bu_4Pb

Synth.: By-prod. in rxn. of Bu_3PbOMe with CS_2 (951).
Prop.: Anal. detn. by coulometry after bromination (1696).
Rxn. with: $RGeCl_3$ → $BuRGeCl_2$ + Bu_3PbCl; R = Bu, Cl (1364).
$GeCl_4$ (2:1) → Bu_2GeCl_2 + Bu_3PbCl (1364).
$HClO_4$ and/or AcOH → mechanism of cleavage rxn. (653).
SO_2 in dry C_6H_6 → $Bu_2Pb(O_2SBu)_2$ (1217).
SO_2 in aq. Et_2O → Bu_2PbSO_3 (1217).
Biol. Prop.: Antiandrogenic activity (128).
Use: As antiknock agent in presence of phosphates (219).
In catalyst for polymerization of CH_2:CHCl, CH_2:CHCN and CH_2:$CMeCO_2Me$ (1685).

Tetraphenyllead (M-526) $PbC_{24}H_{20}$ Ph_4Pb

Synth.: By rxn. of PhMgBr with $Ph_nPb(O_2CCHMe_2)_{4-n}$; n = 0-2 in Et_2O-C_6H_6 under reflux in 27-66% yield (1424), with Ph_6Pb_2 in THF, in 78% yield (2865).
By rxn. of PhLi with Ph_6Pb_2 in THF, in 92% yield (2865), with Ph_nPb-$(O_2CCHMe_2)_{4-n}$; n = 0-4 in MePh < 30°, in 25-64% yield (1424).
By rxn. of $PhSiF_3$ with Ph_3PbF without solvent under reflux, in 90% yield (1554), in aq. $NH_4F \cdot HF$, in 74% yield (1555), with $Pb(OAc)_4$, in 71% yield (1554).
By rxn. of Ph_3PbLi with $(CH_2Br)_2$, in 42% yield (1928), in 70% yield (1127), with $ClCH_2\overline{CHCH_2O}$ (2:1) in THF, in 10% yield (607).
By rxn. of Ph_6Pb_2 with Ph_2S_2 in C_6H_6 in 7-17% yield (2865).
By polarographic reduction of Ph_3PbOAc in presence of Ph_6Pb_2 or Ph_6Sn_2 (967).
By thermal dec. of Ph_3PbN_3 or $Ph_2Pb(N_3)_2$ (2667), of $(Ph_3Pb)_2N_2O_2$ in high vacuo at 100-130° (716).

Prop.: IR (194, 207, 1260), far IR, assignment (51, 2784), Raman (2784), mass (850) and luminescence (2383) spectra, configuration by depolarized Raleigh diffusion (2059), dipole moment along Pb-C bond (194), effect of ^{60}Co irradiation (442), of neutron irradiation (84) on quenching scintillators in $C_{10}H_8$ (268), electrostriction in polar solvent (517), thermal dec. point 232° (547), thermodynamic data (305), x-ray structural analysis (819). Anal. detn.

and sepn. from Ph_nPbCl_{4-n} by TLC; n = 2, 3 (2635).

Rxn. with: $Ph_4{}^{210}Pb \rightarrow$ radioactive decay to $^{210}Bi^{3+}$, $^{210}Bi^0$, $^{210}BiPh_2{}^+$ + $^{210}BiPh_3$ (3).

UV or ^{60}Co irradiation $\rightarrow C_6H_6 + Ph_2$; quantum yield, luminescence quenching (1687).

^{60}Co irradiation $\rightarrow$ nature of radicals (693).

Ph_2PbCl_2 in $Me_2SO \rightarrow Ph_3PbCl + Ph_2PbCl_2 \cdot 2Me_2SO$ (1385).

$PhPb(OAc)_3$ + cat. $Hg(OAc)_2 \rightarrow Ph_2Pb(OAc)_2$ (2864).

$Pb(O_2CR)_4$ + cat. $Hg(OAc)_2 \rightarrow PhPb(O_2CR)_3$ (1:3) + $Ph_2Pb(OAc)_2$ (1:1); R = Me, i-Pr (2864).

$Ph_3SiLi \rightarrow Ph_4Si + Ph_3SiOH$ (1928).

$PdCl_2$ + CO in AcOH $\rightarrow$ BzOH (2439).

HCl at 50-60° $\rightarrow Ph_3PbCl + Ph_2PbCl_2$ (1421).

CCl_4 + cat. $BiCl_3 \rightarrow Ph_3PbCl + Ph_2PbCl_2 + PhCl$ (1716).

$PhSO_2Cl \rightarrow Ph_3PbCl + Ph_2PbCl_2$ (166).

$SOCl_2 \rightarrow Ph_2PbCl_2$ (166).

Cl_3CCO_2H at 140° $\rightarrow PbCl_2 + CO_2$ (1214).

$RCO_2H \rightarrow Ph_2Pb(O_2CR)_2$; R = CF_3 at 70°, CCl_3 in refluxing CCl_4, CMe_3 at 120° (1214).

SO_2 in dry $C_6H_6 \rightarrow Ph_2Pb(O_2SPh)_2$ (232a, 1217).

SO_2 in aq. $Et_2O \rightarrow Ph_2PbSO_3$ (1217).

SeO_2 in aq. dioxane $\rightarrow Ph_2PbSeO_3$ (1217).

$Se(CN)_2 \rightarrow Ph_3PbCN + PhSeCN$ (697).

$Se(SeCN)_2 \rightarrow Ph_3PbSeCN + PhSeCN + Se$ (697).

$PhCH:CH_2$ at 160° $\rightarrow$ hard colorless transparent polymer (23).

KNH_2 in liq. $NH_3 \rightarrow KPb(NH_2)_3 + N$ + black explosive Pb-N compd. (1826).

Biol. Prop.: Activity as fungicide against Aspergillus niger, Botrytis allii, Penicillium niger, Rhizopus nigricans (221).

Use: As additive to improve corona discharge in polyimide films (2585), for fire resistant polystyrene (1223).

As antifogging additive in photographic systems (591a).

As antiknock agent in presence of organonickel compounds (989).

In catalyst for polymerization of butadiene (381, 1584), for vinyl ether (2152).

As scintillator (575, 861, 2048).

As stabilizer for polyphenyl ether engine lubricants (209).

Tetraparatolyllead (M-529) $PbC_{28}H_{28}$ $(p\text{-}MeC_6H_4)_4Pb$

Prop.: m. 228-30° (166), NMR spectrum (1318).

Rxn. with: $SOCl_2$ in $C_6H_6 \rightarrow (p\text{-}MeC_6H_4)_2PbCl_2$ (166).

$(i\text{-}PrCO_2)_4Pb \rightarrow p\text{-}MeC_6H_4Pb(O_2CCHMe_2)_3$ (2864).

Use: In catalyst for butadiene polymerization (381).

As scintillator (2048).

Tetraorganolead Compounds (M-529) R_4Pb

Synth.: Electrolytical cell for manufacturing (416).

Process for recovery from rxn. mixtures by pressure steam distillation (2914), for purification (516).

Prop.: Magnetic rotation (1068). Anal. detn. by GLC (2986), by x-ray fluorescence (196, 208), in gasoline by emission spectroscopy (817), by chromate titration (1937), by polarography (1769).

Use: As antiknock agent (1176), distribution of Pb from automobile exhaust in atmosphere (2319).

Effect on burning of solid double based propellants (2780).

As stabilizer for Cl-contg. polymers and esterification catalyst (2973).

Listing of symmetric unsubstituted tetraorganolead compounds concludes in Table 270.

Table 270. Unsubstituted Symmetric Tetraorganolead Compounds R_4Pb

R in R_4Pb*	Properties	Ref.
$PhCH_2$ (528)	(a)	--
C_2D_5 (528)	IR and Raman spectra	224
Me_2CH (528)	Thermodynamic data (b, c)	2856
$\overline{CH_2CH_2CH}$ (528)	TLC	1260
Me_2CHCH_2 (528)	(d)	--
Me_3C	(a)	--
$Ph(CH_2)_4$	(a - 381, 1584)	--
$PhCMe_2CH_2$ (e)	m. 88-89°, NMR spectrum	2187
Me_3CCH_2 (f)	White cryst., m. 139-41°, subl.$_{0.5}$ 110°, IR and NMR spectra	1936
C_6H_{13} (528)	(a, g)	--
c.C_6H_{11} (529)	(a - 381, 1584, d, h)	--
c.4-MeC_6H_{10}	(a)	--
$C_{10}H_{21}$ (529)	(a)	--
$C_{13}H_{27}$	(a)	--
$C_{18}H_{37}$	(a)	--
2,4-$Et_2C_6H_3$	(a)	--
3,5-$(C_7H_{15})_2C_6H_3$	(a)	--
1-$C_{10}H_7$	(a)	--

* Numbers in parenthesis refer to pages in main volume.
(a) Use as catalyst for butadiene polymerization (381). (b) Effect of molecular redistribution on refractive index (1175). (c) Photochemical and thermal dec. → C_3H_6 + C_3H_8 + $(Me_2CH)_2$ + i-PrPh + i-$Pr_2C_6H_4$; rxn. mechanism (2684). (d) Rxn. with $GeCl_4$ → R_3PbCl + $RGeCl_3$; R = i-Bu, c.C_6H_{11} (1364). (e) Synth. from R_6Pb_2 with I in C_6H_6, followed by RMgCl in Et_2O, in 68% yield (2187). (f) Synth. from R_3PbBr with RMgCl in Et_2O-THF, in almost quant. yield (1936). (g) Rxn. with HCl in $HCCl_3$ → R_2PbCl_2 (2059). (h) UV irradiation → c.C_6H_{12} + c.C_6H_{10} + (c.$C_6H_{11})_2$ + PhC_6H_{11} (1281).

1.3 TETRAHETEROCYCLIC ORGANOLEAD COMPOUNDS

A reference to tetrathienyllead is listed at the bottom of Table 271.

1.4 SYMMETRIC UNSATURATED TETRAORGANOLEAD COMPOUNDS

Tetravinyllead (M-530) PbC_8H_{12} $(CH_2:CH)_4Pb$

Synth.: By rxn. of $PbCl_2$ with $CH_2:CHMgX$ in THF and steam distillation (2444). By-prod. in rxn. of $(CH_2:CH)_3PbCl$ with $LiAlH_4$ in Et_2O or $NaBH_4$ in NH_3 (1197).

Prop.: NMR (846) and ^{13}C NMR (328) spectra, TLC (1260), spontaneous ignition temp. > 180° (2444), thermodynamic data (305).

Rxn. with: Me_4Pb + cat. → $Me_nPb(CH:CH_2)_{4-n}$; n = 0-4, cat.: active clay (880), $Me_3Al_2Cl_3$ (2935).

PhLi → $CH_2:CHLi$ (1579).

$AgNO_3$ → $(CH_2:CH)_3PbNO_3$ + C_2H_4 + $(CH_2:CH)_2$ + Ag (1196a).

$SnCl_4$ → heat of redistribution rxn. (380).

Na in liq. NH_3 → Pb + $NaNH_2$ + C_2H_4 + H (1196).

Br in excess at r.t. → $PbBr_2$ + $CH_2BrCHBr_2$ (1196a).

Br (9:7) at r.t. → $CH_2:CHBr$ + $CH_2BrCHBr_2$ + $(CH_2:CH)_4Pb$ (1196a).

HCl → $(CH_2:CH)_3PbCl$ at -78° + $(CH_2:CH)_2PbCl_2$ at -20° (1196, 1196a).

BCl_3 at -78° → $(CH_2:CH)_3B$ + $(CH_2:CH)_2$ + C_2H_4 + Pb (1197).

B_2H_6 at -78° → EtB_2H_5 + $(CH_2:CH)_2$ + C_2H_4 + Pb + $1-C_4H_8$ + C_2H_6 (1197).

AcOH at 105° → $Pb(OAc)_2$ + C_2H_4 + $CH_2:CHOAc$ (1196a).

RCO_2H (1:1) → $(CH_2:CH)_3PbO_2CR$; R = Me, Et (1196a).

$EtCO_2H$ in excess at r.t. → C_2H_4 + $Pb(O_2CEt)_2$ (1196a).

Other unsaturated symmetric tetraorganolead compounds are listed in Table 271.

1.6 SYMMETRIC FUNCTIONALLY SUBSTITUTED TETRAORGANOLEAD COMPOUNDS

Several new functionally substituted tetraorganolead compounds are listed in this chapter. The derivatives summarized in Table 271 are prepared by methods from the following scheme.

  I Grignard synthesis from lead halide.

 II Synthesis with organolithium reagent from:
    (IIA) M_2PbCl_6,
    (IIB) $Pb(O_2CR)_4$.

III Dihalocarbene insertion by reaction of chloroform or trichloroacetate and strong base with:
    (IIIA) tetraalkyllead R_4Pb,
    (IIIB) olefinic organolead derivatives $(RCH:CH)_4Pb$.

IV By electrolysis of $NCCH_2CH_2I$ on lead cathode in basic solution.

R = organic group, M = potassium or rubidium.

Tetraperfluorophenyllead $PbC_{24}F_{20}$ $(C_6F_5)_4Pb$

<u>Synth.</u>: By rxn. of C_6F_5MgBr with $PbCl_2$ (146), in THF, in 3% yield (547), in $Et_2O-C_6H_6-THF$, in < 5% yield (1185), in $Et_2O-C_6H_6$, followed by Br in CH_2Cl_2 (2:1:2), in 54% yield (1185).
By rxn. of C_6F_5Li with $PbCl_2$ in Et_2O, in 30-50% yield (547), with $Pb(OAc)_4$ in Et_2O, in 16% yield (547).

<u>Prop.</u>: m. 204° (1185), 200-206° (547), 199-200° (146), subl. in vacuo at 140° (146), IR (146), assignment (547), NMR (146) and ^{19}F NMR (2475) spectra, stable in boiling water (146), thermal dec. temp. > 260° (547), dec. in 10% aq. NaOH at reflux (547).

<u>Rxn. with</u>: 6 N HCl → stable at reflux for 1 hour, dec. after 6 hours (547).

Listing of functionally substituted symmetric tetraorganolead derivatives continues in Table 271.

Table 271. Symmetric Substituted Tetraorganolead Compounds R'_4Pb

R' in R'_4Pb*	Synth. Method**	Yield	Properties	Ref.
$MeCCl_2$	IIIA	--	(a)	2759
$p-BrC_6H_4C⋮C$	IIA	22	m. 155°	204
$NCCH_2CH_2$	IV	13	n_D^{20} 1.5489, d_{20} 1.7480	566
$Me_3SiC⋮C$	I	--	m. 108°	1345
$Cl_2\overline{CCH_2CH}$	IIIB	--	(a)	2758
$BuCCl_2$	IIIA	--	(a)	2759
$Cl_2\overline{CCHEtCH}$	IIIB	--	(a)	2758
$Me_3CC⋮C$	IIA	85	m. 170-71° (dec.), mol. wt. (b, c)	2664
$c.C_6H_{11}C⋮C$	IIA	23	m. 94° (dec.)	206
$p-MeOC_6H_4$ (531)	--	--	(d)	--
$2-C_4H_3S$ (531)	IIB	54	m. 154.5°	1424

* Numbers in parenthesis refer to pages in main volume.
** The characters used here correspond to the synthetic scheme in the introduction to Chapter 1.6.
(a) Use as self-scavenging antiknock agent (2758 or 2759). (b) IR, NMR and mass spectra (2664). (c) Thermal dec. in decane at 161° (2664). (d) Rxn. with SO_2 in aq. Et_2O → $(R_3Pb)_2SO_3$ (1217).

2. UNSYMMETRIC TETRAORGANOLEAD COMPOUNDS

2.1 NONFUNCTIONAL DERIVATIVES

The nonfunctional unsymmetric tetraorganolead compounds listed in Table 272 are prepared by the following methods.

I Redistribution reactions in presence of $AlCl_3$:
$$R_4Pb + R'_4Pb \rightarrow R_nPbR'_{4-n} .$$

II Alkylation of organolead hydride R_3PbH with olefin:
$$R_3PbH + RCH:CH_2 \rightarrow R_3PbCH_2CH_2R .$$

III Grignard synthesis:
- (IIIA) $R_3PbX + R'MgX \rightarrow R_3PbR'$,
- (IIIB) $R_3PbOH + R'MgX \rightarrow R_3PbR'$,
- (IIIC) $R_3PbR'' + R'MgX \rightarrow R_3PbR'$,
- (IIID) $R_6Pb_2 + RMgX + R'Cl \rightarrow R_3PbR'$.

IV Synthesis with alkali metal compounds:
- (IVA) $R_3PbX + R'Li \rightarrow R_3PbR'$,
- (IVB) $R_3PbLi + R'Cl \rightarrow R_3PbR'$.

V Electrolysis of Grignard reagent on Pb anode:
$$Pb + RMgX + R'Cl \rightarrow R_nPbR'_{4-n} .$$

R, R', R'' = differing organic groups, X = halogen.

Ethyltrimethyllead* (M-533) PbC_5H_{14} Me_3PbEt
<u>Synth.</u>: By redistribution of Me_4Pb and Et_4Pb (1:1) on activated clay, in 21% yield (88) at 55-65°, in 20% yield (880), without cat. > 160° (432), in presence of $BF_3 \cdot OEt_2$ or $Me_3Al_2Cl_3$ (593), in presence of strongly acidic ion exchange resin in fraction of 24.2%, rxn. mechanism (2450).
By rxn. of Me_3PbOH with EtMgBr in Et_2O or THF in 50% yield (1340).
Electrolytical cell for manufacturing process (46, 779).
<u>Prop.</u>: Raman spectrum (244), GLC (432) in presence of Me_nPbEt_{4-n} (1140, 1355), thermodynamic data (305).
<u>Use</u>: As anitknock agent (410), stabilized by cycloalkanes (560).

Diethyldimethyllead* (M-533) PbC_6H_{16} Me_2PbEt_2
<u>Synth.</u>: By redistribution of Me_4Pb and Et_4Pb (1:1) on activated clay at 85°, in 38% yield (88) at 55-65°, in 42% yield (880), without cat > 160° (432), with $Me_3Al_2Cl_3$ or $BF_3 \cdot OEt_2$ (593), with strongly acidic ion exchange resin, in fraction of 38.1%; rxn. mechanism (2450).
Electrolytical cell for manufacturing process (46, 779).
<u>Prop.</u>: Raman spectrum (244), GLC (432), in presence of Me_nPbEt_{4-n} (1140, 1355), thermodynamic data (305).

---

* For additional references see Me_nPbEt_{4-n} .

Use: As antiknock agent (410, 989), stabilized by cycloalkanes (560), effect on automobile road performance in presence of Me$_4$Pb, Et$_4$Pb and mixtures (877). As catalyst for butadiene polymerization (381).

Methyltriethyllead* (M-534)   PbC$_7$H$_{18}$ MePbEt$_3$

Synth.: Synth. by redistribution of Me$_4$Pb and Et$_4$Pb (1:1) on activated clay at 85°, in 27% yield (88), at 55-65°, in 28% yield (880), without cat. at above 160° (432), with BF$_3$·OEt$_2$ or Me$_3$Al$_2$Cl$_3$ (593), with strongly acidic ion exchange resin in fraction of 25.9%; rxn. mechanism (2450).
Electrolytical cell for manufacturing process (46, 779).
Prop.: IR (244) and Raman (244) spectra, GLC (432), in presence of Me$_n$PbEt$_{4-n}$ (1140, 1355), thermodynamic data (305).
Use: As antiknock agent stabilized by cycloalkanes (560), effect on initial oxidn. rate of Et$_4$Pb (1017).

Methylethyllead (M-535)   Me$_n$PbEt$_{4-n}$

Synth.: By electrolysis of mixed Grignard reagent and excess RCl (775, 777), of MeMgCl and EtCl in (BuOCH$_2$)$_2$ on lead anode (2271).
By rxn. of Na-Pb with MeCl and EtCl in presence of Me$_3$Al (93a).
Prop.: Mass spectrum (2268), GLC (1140, 1355), composition of equil. mixt. of Me$_4$Pb and Et$_4$Pb in presence of strongly acidic ion exchange resin (2450), of AlCl$_3$ (93a). Anal. detn. in gasoline (1957, 2790).
Use: As antiknock agent (880), stabilized by alkanes or alkylbenzenes (93a), by halogen scavenger and MePh (2937) or C$_{10}$H$_8$ (2938), effect on automobile performance in presence of Me$_4$Pb and Et$_4$Pb (877).

Other nonfunctional unsymmetric tetraorganolead derivatives are listed in Table 272.

2.3 HETEROCYCLIC SUBSTITUTED TETRAORGANOLEAD COMPOUNDS

All compounds are listed in Table 273.

---

* For additional references see Me$_n$PbEt$_{4-n}$.

Table 272. Nonfunctional Unsymmetrical Tetraorganolead Compounds $R_nPbR'_{4-n}$

Formula*	Synth. Method**	Yield	Properties	Ref.
Me₃PbR'				
R' = Pr	--	--	Use as antiknock agent	710
Bu (533)	--	--	Mass spectrum (2268), use as antiknock agent	710
Me₂CHCH₂	--	--	Colorless liq., IR and NMR spectra (a)	475
Me₃C (533)	V	--	Mass spectrum (2268) (b)	775, 777
	V	--	Synth. with MeMgCl	2271
C₅H₁₁	IIIA	73	b. 63° (b)	710
Me₃CCH₂	IIIA	83	b.₂.₁ 32.5°, colorless liq., n_D^{25} 1.4990 (c)	1936
C₁₂H₂₅	--	--	Use as antiknock agent	710
Ph (533)	IIIC	57	m. 62-63°, use as antiknock agent (1755) (d)	608
	V	--	Synth. with PhCl	2270
Me₂PbR'₂				
R' = Me₃C	V	--	--	775, 777
	V	--	--	2271
Me₃CCH₂	IIIA	--	n_D^{25} 1.4973, mass spectrum GLC (c)	1936
Ph	V	--	Use as antiknock agent (989) (e)	2270
p-PhC₆H₄	IVA	~100	m. 178-80°	112
MeEtPbPh₂	--	--	(e)	--
MePb(CMe₃)₃	V	--	Synth. with t-BuCl	2271
	V	--	--	775, 777
MePb(CH₂CMe₃)₃	IIIA	92	n_D^{25} 1.4980, mass spectrum, GLC (c)	1936
MePbPh₃	V	--	Synth. with PhCl (f)	2270
PhCH₂PbPh₃ (534)	IVB	88	m. 93° (g)	2865
	IVB	98	m. 91-93°	1928
	IIID	21	Synth. with PhCl	2865

Et$_3$PbR'

R' = Pr				
Me$_2$CH	I	--	GLC (h)	432
C$_6$H$_{13}$	--	--	(h)	--
c.C$_6$H$_{11}$	IIIA	63	b. 115-17° (!), n_D^{18} 1.5047, d_{18} 1.415 (i)	2050
C$_7$H$_{15}$	V	--	Synth. with c.C$_6$H$_{11}$Cl	2271
C$_8$H$_{17}$	IIIA	52	b. 118-20° (!), n_D^{18} 1.4998, d_{18} 1.390 (i)	2050
C$_9$H$_{19}$	IIIA	56	b. 124-25° (!), n_D^{18} 1.4967, d_{18} 1.382 (i)	2050
C$_{10}$H$_{21}$	IIIA	52	b. 144-45° (!), n_D^{18} 1.4915, d_{18} 1.378 (i)	2050
Ph (534)	IIIA	52	b. 153-54° (!), n_D^{18} 1.4895, d_{18} 1.370 (i)	2050
	IIIB	50	Stabilization by cycloalkenes (560) (j)	1340

Et$_2$PbR'$_2$

R' = Pr (534)	I	--	GLC	432
Me$_3$CCH$_2$	IIIA	low	n_D^{25} 1.4994, IR spectrum, GLC	1936
c.C$_6$H$_{11}$	V	--	Synth. with c.C$_6$H$_{11}$Cl	2271
Ph (534)	--	--	Use as addition for propellent powders	561
EtPbPr$_3$	I	--	GLC	432
EtPb(CH$_2$CMe$_3$)$_3$	IIIA	90	n_D^{25} 1.5030, GLC (c)	1936
EtPb(c.C$_6$H$_{11}$)$_3$	V	--	Synth. with c.C$_6$H$_{11}$Cl	2271
PhCH$_2$CH$_2$PbBu$_3$	II	< 90	b$_{0.0001}$ 128-30° part. dec., IR spectrum	397
i-PrPb(CH$_2$CMe$_3$)$_3$	IIIA	84	n_D^{25} 1.5065, GLC (c)	1936
Bu$_3$PbPh	--	--	(e)	--

* Numbers in parenthesis refer to pages in main volume.

** The characters used here correspond to the synthetic scheme in the introduction to Chapter 2.1.

(a) Unpleasant odor, toxic (?), decrepitates in air > 100° (475). (b) Use as antiknock agent (710). (c) IR and NMR spectra (1936). (d) Use in catalyst for butadiene polymerization (231). (e) Use in catalyst for butadiene polymerization (381). (f) Activity as fungicide against Aspergillus niger, Botrytis allii, Penicillium niger, Rhiopus nigricans (221), activity as rodent repellent (221). (g) Use as antifogging additive for photographic systems (591a). (h) Effect on initial oxidation of Et$_4$Pb (1017). (i) Surface tension (2050). (j) Rxn. with HCl in dioxane → mechanism of Et$_3$Pb cleavage rxn., deuterium effect (43), use in catalyst for butadiene polymerization (1584).

Table 272 Continued $R_nPbR'_{4-n}$

Formula*	Synth. Method**	Yield	Properties	Ref.
BuPbPh$_3$	IVA	64	m. 47-48°	1928
Me$_3$CPb(CH$_2$CMe$_3$)$_3$	IIIA	75	m. 75-77°, chromatography on Al$_2$O$_3$ (c)	1936
C$_{10}$H$_{21}$PbPh$_3$	IIIA	52	m. 48-49°	2050
2,4-Me$_2$C$_6$H$_3$PbPh$_3$	--	--	Use as scintillator in polyvinyltoluene	936
p-EtC$_6$H$_4$PbPh$_3$ (535)	IIIA	--	m. 82-84°, soly., use as scintillator	2048
(p-EtC$_6$H$_4$)$_2$PbPh$_2$	--	--	Use as scintillator	2048
(MeC$_6$H$_4$)$_3$PbC$_6$H$_3$MeEt	--	--	Use as scintillator	2048
(Me$_2$C$_6$H$_3$)$_3$PbC$_6$H$_2$Me$_2$Et	--	--	Use as scintillator	2048
R$_3$PbAr	--	--	R = alkyl, Ar = aryl (k)	--

(k) Mechanism of protonolysis and halodemetallation (1379).

Table 273. Heterocyclic Substituted Tetraorganolead Compounds $R_nPbR'_{4-n}$

Formula	Synth. Method	Yield	Properties	Ref.
$Ph_3Pb(CH_2)_n$-N⟨triazole⟩-CMe_2OH 				
n = 3	(a)	56	m. 95-113°, IR spectrum, struct.	1127
4	(a)	65	m. 90-122°, IR spectrum, struct.	1127
5	(a)	--	m. 62-78°, IR spectrum, struct.	1127
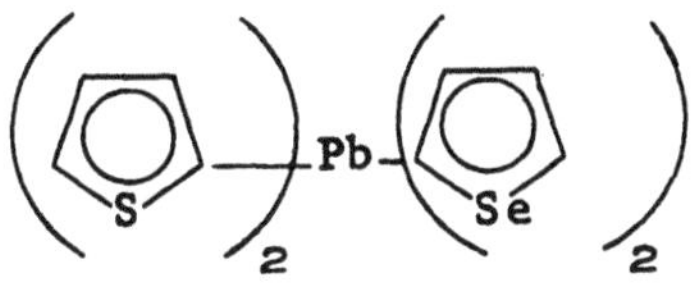 $((C_4H_3S)-Pb-(C_4H_3Se))_2$	(b)	6	m. 147° (c)	624

(a) Synth. from $Ph_3Pb(CH_2)_nN_3$ with $HC≡CCMe_2OH$ (1127). (b) Synth. from $(2-C_4H_3S)_2PbCl_2$ and rxn. prod. of $2-IC_4H_3Se$, Mg and $HgBr_2$ in Et_2O (624). (c) Rxn. with HCl → C_4H_4S + C_4H_4Se (624).

2.4 TETRAORGANOLEAD COMPOUNDS WITH OLEFINIC SUBSTITUENTS

This chapter comprises unsymmetric unsubstituted tetraorganolead compounds containing olefinic substituents, all listed in Table 274; derivatives with additional functions may be found in Tables 275-278. The tetraorganolead compounds listed in Tables 275-279 are prepared by methods from the following scheme.

I Grignard synthesis:
    (IA) $R_nPbX_{4-n} + R'MgX \rightarrow R_nPbR'_{4-n}$,
    (IB) $PbX_2 + RMgX + R'MgX$ and/or $R'X \rightarrow R_nPbR'_{4-n}$,
    (IC) $Pb + RMgX + R'Cl \rightarrow R_nPbR'_{4-n}$ by electrolysis on Pb anode.

II Synthesis with organic alkali metal derivatives:
    (IIA) $R_3PbX + R'Li \rightarrow R_3PbR'$,
    (IIB) $R_3PbX + R'Na \rightarrow R_3PbR'$,
    (IIC) $Ph_3PbLi + R'\overline{CH(CH_2)_nY} \rightarrow Ph_3Pb(CH_2)_nCHR'YH$; Y = O, S, NR; n = 1,2,
    (IID) $R_3PbLi + R'X \rightarrow R_3PbR'$.

III Addition reactions with triorganolead hydride:
    (IIIA) $R_3PbH + R'CH:CH_2 \rightarrow R_3PbCH_2CH_2R'$,
    (IIIB) $R_3PbH + R'C:CR'' \rightarrow R_3PbCR':CHR'' + R_3PbCR'':CHR'$; R'' = H, R'.

IV Reactions with organolead hydroxides, alkoxides and carboxylates:
  (IVA) $R_3PbOH + R'H \rightarrow R_3PbR'$,
  (IVB) $R_3PbOMe + R'H \rightarrow R_3PbR'$,
  (IVC) $R_3PbOMe + (R'C\vdots)_2 \rightarrow R_3PbCR'\!:\!CR'OMe$,
  (IVD) Decarboxylation of $Ph_3PbO_2CC\vdots CH$.

V Miscellaneous reactions:
  (VA) $R_4Pb + R'_4Pb + $ clay catalyst $\rightarrow R_nPbR'_{4-n}$,
  (VB) $R_4Pb + HCX_3$ or $CCl_3CO_2R + $ strong base $\rightarrow R_nPbCX_2R$,
  (VC) $R_nPb(CH\!:\!CHR)_{4-n} + HCX_3 + $ strong base $\rightarrow R_nPb(C_3H_2X_2R)_{4-n}$,
  (VD) $R_3Pb(CH_2)_nBr + NaN_3 \rightarrow R_3Pb(CH_2)_nN_3$.

R = organic group, R' = organic group, may be functionally substituted, n = 1-3, X = halogen.

Table 274. Tetraorganolead Compounds with Olefinic Substituents $R_nPbR'_{4-n}$

Formula*	Synth. Method**	Yield	Properties	Ref.
$Me_3PbCH\!:\!CH_2$ (537)	VA	--	(b)	880
	IC (a)	--	(b)	1325
	IC	--	--	1765a
	IB	--	(b)	2935
$Me_3PbCH\!:\!CHPh$	--	--	(c)	--
$Me_2Pb(CH\!:\!CH_2)_2$	VA	--	(b)	880
	IC (a)	--	(b)	1325
	IC	--	--	1765a
	IB	--	(b)	2935
$MePb(CH\!:\!CH_2)_3$	VA	--	(b)	880
	IC (a)	--	(b)	1325
	IC	--	--	1765a
	IB	--	(b)	2935
Et_3PbR'				
R' = $CH_2\!:\!CH$ (537)	VA	--	(b)	880
	IC	--	(d)	1765a
	IB	--	(b)	2935
$CH_2\!:\!CHCH_2$ (537)	IA	50	Synth. with Et_3PbOH	1340
	IA	62	m. > 250° (dec.), polymer	2050
C_5H_5 (537)	IVB	--	NMR spectrum	951
p-$CH_2\!:\!CHCH_2C_6H_4$	IA	45	Infusible powder	2050
$Et_2Pb(CH\!:\!CH_2)_2$ (538)	VA	--	(b)	880
	IC	--	--	1765a
	IB	--	(b)	2935

EtPb(CH:CH₂)₃ (538)	VA	--	(b)	880
	IB	--	(b)	2935
Bu₃PbCH:CHPh	IIIB	> 90	Dec. on distn. in vacuo, IR spectrum	397
Ph₃PbCH:CH₂ (538)	--	--	NMR spectrum	846
p-Ph₃PbC₆H₄CH:CH₂ (538)	--	--	m. 105-107° (e, f)	284

* Numbers in parenthesis refer to pages in main volume.
** The characters used here correspond to the synthetic scheme in the introduction to Chapter 2.4.
(a) Electrolysis with (CH₂:CH)₂Mg in presence of Me₃Al (1325). (b) Use as antiknock agent. (c) Accelerates radically or radiolytically induced polymerization of PhCH:CH₂ (935). (d) Rxn. with HCCl₃ and NaOEt → Et₃PhCHCH₂CCl₂ (2758). (e) Rxn. with PhCH:CH₂ + cat. AiBN → copolymer, relative reactivity (284, 461). (f) Rxn. with MeC₆H₄CH:CH₂ + AiBN → copolymer, relative reactivity (461).

2.5 TETRAORGANOLEAD COMPOUNDS WITH ACETYLENIC SUBSTITUENTS

This chapter includes unsymmetric unsubstituted tetraorganolead compounds containing acetylenic substituents; derivatives with additional functional groups may be found in Tables 276-279.

Triphenylleadphenylacetylene (M-540)   PbC₂₆H₂₀   Ph₃PbC⋮CPh
<u>Synth.</u>: By rxn. of Ph₃PbOH with PhC⋮CH in Et₂O in presence of CaH₂ in Ar atm., in 83% yield (2495).
By rxn. of Ph₃PbCl with NaOMe in alc. C₆H₆, followed by PhC⋮CH, in 65% yield (952).
By rxn. of Ph₃PbI with PhC⋮CNa in MePh, in 52% yield (1474).
By decarboxylation of Ph₃PbO₂CC⋮CPh in refluxing MePh, in 65% yield (952).
<u>Prop.</u>: m. 55-59° (952), IR (952, 1474, 2495) and UV (1474) spectra, anal. detn. by titration with I (2495).

Listing of tetraorganolead compounds with only acetylenic substituents concludes in Table 275.

2.6 FUNCTIONALLY SUBSTITUTED TETRAORGANOLEAD COMPOUNDS

2.6.1 TETRAORGANOLEAD COMPOUNDS CONTAINING HALOGEN

The compounds are listed in Table 276; derivatives with halogen and oxygen functions may be found in Table 277.

Table 275. Tetraorganolead Compounds with Acetylenic Substituents $R_nPbR'_{4-n}$

Formula*	Synth. Method**	Yield	Properties	Ref.
Me$_3$PbR'				
R' = HC⋮C	IVA	--	m. 82-84°, force const. (2144)	1917
PhC⋮C	IVA	--	b$_{1.5}$ 123°, n$_D^{20}$ 1.6133, d$_{20}$ 1.7798, IR spectrum	1917
MeC⋮C	IIA	--	b$_{16}$ 65.5°, IR spectrum (b)	20
EtC⋮C	IIB	11	b$_4$ 54° (b)	20
CH$_2$:CHC⋮C	IVA	--	b$_{13}$ 75-76°, n$_D^{20}$ 1.5618, d$_{20}$ 1.8747, IR spectrum	1917
Et$_3$PbR'				
R' = HC⋮C (540)	IVA	99	n$_D^{20}$ 1.5480, d$_{20}$ 1.7510, IR spectrum	1917
PhC⋮C (540)	IVB	86	b$_{0.05}$ 94-97°, IR and NMR spectra (c, d)	952
MeC⋮C (540)	--	--	NMR spectrum	1931
$\left\{\begin{array}{l}\text{HC⋮CCH}_2\\ \text{CH}_2\text{:C:CH}\end{array}\right\}$	IA	34	b$_{2.5}$ 75°	1473
	IA	62	b$_{0.6}$ 104°	1473
$\left\{\begin{array}{l}\text{PhC⋮CCH}_2\\ \text{PhCH:C:CH}_2\end{array}\right\}$	IVB	75	Colorless liq., b$_{0.05}$ 61-63°, IR spectrum (c)	952
BuC⋮C	IIB	30	b$_{1.5}$ 95°, IR and UV spectra	1474
1-c-C$_5$H$_7$	IVA	--	Violent dec. on distn., IR spectrum	2495
Bu$_3$PbC⋮CPh				
Ph$_3$PbR'				
R' = HC⋮C	IIB	50	Oil, IR (1474) and NMR (1474, 1931) spectra (e)	1474
MeC⋮C (540)	IIB	50	m. 65°, IR (1474) and NMR (1474, 1931) spectra	1474
$\left\{\begin{array}{l}\text{HC⋮CCH}_2\\ \text{CH}_2\text{:C:CH}_2\end{array}\right\}$ (540)	IA	70	m. 64°	1473
	IA	63	m. 74°	1473
$\left\{\begin{array}{l}\text{PhC⋮CCH}_2\\ \text{PhCH:C:CH}\end{array}\right\}$				1473

Compound			Ref.
EtC⋮C (540)	--	NMR spectrum	1474
{MeC⋮CCH₂ / MeCH:C:CH}	IA	54 / --	1473 / 1473
{HC⋮CCHMe / CH₂:C:CMe}	IA	65 / --	1473 / 1473
HC⋮CC⋮C (540)	IIB	30	UV (1474) and NMR (1474, 1931) spectra	1474
PhC⋮CC⋮C (540)	IIB	55	(e)	1474
MeC⋮CC⋮C (540)	IIB	42	NMR spectrum (1474, 1931)	1474
1-c-C₅H₇ (540)	IIB	41	m. 50°	1474
1-c-C₆H₉ (540)	IIB	53	m. 84°	1474

* Numbers in parenthesis refer to pages in main volume.
** The characters used here correspond to the synthetic scheme in the introduction to Chapter 2.4.
(a) IR spectrum (1074, 1917, 2144). (b) Use as antiknock agent (20). (c) Rxn. with AcCl → Et₃PbCl + RAc; R = PhC⋮C, BuC⋮C (952). (d) Rxn. with RNCO → PhCH:CNRCONRCO; R = Ph, 1-C₁₀H₇ (952). (e) Rxn. with HClO₄ or alc. NaOH → kinetic data and mechanism for cleavage rxn. (2562).

Table 276. Tetraorganolead Compounds Containing Halogen $R_nPbR'_{4-n}$

Formula*	Synth. Method**	Yield	Properties	Ref.
Me_3PbR'				
R' = $MeCCl_2$	VB	--	Use as antiknock agent	2759
ClC⋮C	IIA	--	$b_{50} \sim 102°$, exploded on distn. (a)	1980
$EtCCl_2$	VB	--	Use as antiknock agent	2759
EtCClBr	VB	--	Use as antiknock agent	2759
$EtCBr_2$	VB	--	Use as antiknock agent	2759
$Cl_2\overline{CCH_2CH}$	VC	--	Use as antiknock agent	2758
$ClBr\overline{CCH_2CH}$	VC	--	Use as antiknock agent	2758
$Br_2\overline{CCH_2CH}$	VC	--	Use as antiknock agent	2758
$PrCCl_2$	VB	--	Use as antiknock agent	2759
C_6F_5	IA	--	$b_{0.01}$ 40-45°, IR and NMR spectra (b, c)	146
$Me_2EtPbCCl_2Me$	VB	--	Use as antiknock agent	2759
$Me_2EtPbCCl_2Et$	VB	--	Use as antiknock agent	2759
$Me_2EtPbCHCH_2CCl_2$	VC	--	Use as antiknock agent	2758
$MeEtBuPbCCl_2Me$	VB	--	Use as antiknock agent	2759
$MeEtPb(CCl_2CH_2CH_2Br)_2$	VB	--	Use as antiknock agent	2759
$MeEtPb(\overline{CHCHBrCCl_2})_2$	VC	--	Use as antiknock agent	2758
$Me(CH_2{:}CH)_2Pb\overline{CHCH_2C}Br_2$	VC	--	Use as antiknock agent	2758
$MePr_2PbCBr_2Et$	VB	--	Use as antiknock agent	2759
$PhCH_2Pb(C_6F_5)_3$	IB	43	Pale yellow cryst., m. 108°, dec. slowly in air	1185
Et_3PbR'				
R' = $EtCCl_2$	VB	--	Use as antiknock agent	2759
$ClCH_2CH_2CCl_2$	VB	--	Use as antiknock agent	2759
$Cl_2\overline{CCH_2CH}$	VC	--	Use as antiknock agent	2758
$Cl_2\overline{CCHClCH}$	VC	--	Use as antiknock agent	2758
ClCH_2C⋮C	IVA	--	b_1 100°, n_D^{20} 1.5500, IR spectrum	1917

	VC		
(CH$_2$:CH)$_3$PbCHCH$_2$CCl$_2$	--	Use as antiknock agent	2758
Ph$_3$PbR'			
R' = CCl$_3$	IVB	Synth. with HCCl$_3$, m. 169-70°, TLC (d)	951
	IVB	Synth. with (CCl$_3$)$_2$CO, m. 171-71.5°, mol. wt.	951
CBr$_3$	IVB	Synth. with HCBr$_3$, m. 137-39°, TLC	951
	IVB	Synth. with CBr$_3$CHO, IR spectrum	951
BrC!C	IIB	m. 95°, IR spectrum	1474
Br(CH$_2$)$_3$	IID	m. 62-66°, mol. wt. (e)	1127
Br(CH$_2$)$_4$	IID	m. 61-63°, mol. wt. (e)	1127
Br(CH$_2$)$_5$	IID	m. 61-62°, mol. wt. (e)	1127
C$_6$F$_5$ (542)	IA	m. 85-87°, IR and NMR spectra	146
	IIB	(b, f)	146
Ph$_2$Pb(C$_6$F$_5$)$_2$	IA	White cryst., m. 91°	1185

* Numbers in parenthesis refer to pages in main volume.
** The characters used here correspond to the synthetic scheme in the introduction to Chapter 2.4.
(a) (Far) IR spectrum and assigment (1980). (b) Stable in vacuo at 100° (146), ^{19}F NMR spectra (2475). (c) Rxn. with HCl → Me$_3$PbCl + C$_6$HF$_5$ (146), with H$_2$O under reflux → 100% hydrolysis (146). (d) Rxn. with PhNH$_2$ + alkali → PhNCO (951). (e) Rxn. with NaN$_3$ in Me$_2$SO → Ph$_3$Pb(CH$_2$)$_n$N$_3$ + NaBr; n = 3, 4, 5 (1127). (f) Rxn. with HCl at r.t. → C$_6$H$_6$ + C$_6$HF$_5$ (146), with H$_2$O under reflux → 50% hydrolysis (146).

2.6.3 TETRAORGANOLEAD COMPOUNDS CONTAINING OXYGEN AND SULFUR

The compounds are listed in Table 277; derivatives containing additional functional groups may be found in Tables 273, 278 and 279.

Triphenylleadacetic Anhydride $Pb_2C_{40}H_{34}O_3$ $(Ph_3PbCH_2CO)_2O$
Synth.: By rxn. of Ph_3PbOH with $CH_2:C:O$ in dry Et_2O, in 84% yield (608).
Prop.: Dec. at 100-110°, monomeric in C_6H_6 (608).
Rxn. with: MeMgI → $MePbPh_3$ (608).
$LiAlH_4$ → Ph_6Pb_2 (608).
Br → Ph_2PbBr_2 (608).
HCl → Ph_3PbCl (608).

Listing of oxygen substituted unsymmetric tetraorganolead compounds concludes in Table 277.

2.6.5 TETRAORGANOLEAD COMPOUNDS CONTAINING NITROGEN AND PHOSPHORUS

The compounds are listed in Table 278; derivatives with additional functions may be found in Table 273.

2.7 POLYMETALLIC TETRAORGANOLEAD COMPOUNDS

All compounds are included in Table 279.

Table 277. Tetraorganolead Compounds Containing Oxygen and Sulfur $R_nPbR'_{4-n}$

Formula*	Synth. Method**	Yield	Properties	Ref.
Me₃PbR'				
R' = MeO₂CCH:CH	IIIB	85	IR and NMR spectra, cis:trans-ratio 16	1416
CH₂:C(CO₂Me)	IIIB	15	IR and NMR spectra	1416
BuOCHMeOCH₂C:C	IVA	50	b₀.₇ 105-106°, n₀²⁰ 1.5005, d₂₀ 1.5310, IR spectrum	2773
BuOCHMeO₂CC:C	IVA	--	b₀.₅ 135°, n₀²⁰ 1.5294, d₂₀ 1.6440, IR spectrum	2773
EtO₂CCH:C(CO₂Et)	IIIB	--	IR and NMR spectra, trans:cis-ratio 16	1416
MeO₂CC(OMe):C(CO₂Me)	IVC	63	Pale yellow liq., b₀.₁ 105-10° (a, b)	2346
Et₃PbC(CO₂Me):C(OMe)CO₂Me	IVC	75	b₀.₁ 100-110°, sensitive to air (a)	2346
	IVB	51-75	Synth. from MeO₂CCH:C(OMe)CO₂Me	2346
Bu₃PbCH₂CH₂CO₂Me	IIIA	>90	b₀.₀₀₀₁ 94-97°, IR spectrum	397
Bu₃PbCH:CHCO₂Me	IIIB	70	NMR spectrum	912
Ph₃PbR'				
R' = HOCH₂CH₂	IIC	91	Synth. from CH₂CH₂O, m. 72°	609
EtO₂CCH₂ (543)	IV	78	Synth. from Ph₃PbOAc and ketene, m. 61-62°	608
HSCH₂CH₂	IIC	90	Synth. from CH₂CH₂S, m. 104-106° (c)	607
HO(CH₂)₃	IIC	89	Synth. from CH₂CH₂CH₂O, m. 99°	607
ClCH₂CHOHCH₂	IIC	96	Synth. from ClCH₂CHCH₂O, m. 90-91° (d, e)	607
MeO₂CCH₂CH₂	IIC	95	Synth. from CH₂CH₂CO₂, + CH₂N₂, m. 52.5°	607

* Numbers in parenthesis refer to pages in main volume.
** The characters used here correspond to the synthetic scheme in the introduction to Chapter 2.4.
(a) IR and NMR spectra (2346). (b) Rxn. with MeOH → Me₃PbOMe + MeO₂CCH:C(OMe)CO₂Me; cis:trans-ratio 9 (2346).
(c) Monomeric, sensitive to light (607). (d) NMR spectrum (607). (e) Rxn. with HCl → Ph₂PbCl₂ + C₆H₆ + CH₂:CHCH₂Cl (607), with Cl → Ph(ClCH₂CHOHCH₂)PbCl₂ + PhCl (607).

Table 227 Continued $R_nPbR'_{4-n}$

Formula*	Synth. Method**	Yield	Properties	Ref.
Ph$_3$PbR' (Cont.)				
R' = p-HOCH$_2$C$_6$H$_4$ (543)	IIA	60	IR and NMR spectra, TLC (f, g)	1092
p-BzOCH$_2$C$_6$H$_4$	V (g)	--	Needles, m. 86-87°, IR and NMR spectra	1092
	V (g)	--	Prisms, m. 102-103°, TLC	1092
p-C$_{18}$H$_{21}$OCO$_2$CH$_2$C$_6$H$_4$ (h)	V (g)	--	m. 152.5-53.5°, TLC	1092
p-OHCC$_6$H$_4$	V (f)	50	White cryst., m. 110.5-112°, TLC (i, j)	1092
p-(C$_8$H$_{11}$O$_2$)$_2$CHC$_6$H$_4$ (k)	V (j)	95	m. 161.5-163°, TLC	1092

(f) Rxn. of p-Ph$_3$PbC$_6$H$_4$CH$_2$OH with MnO$_2$ in Et$_2$O → p-Ph$_3$PbC$_6$H$_4$CHO (1092).   (g) Rxn. of p-Ph$_3$PbC$_6$H$_4$CH$_2$OH with RCO$_2$Cl in Py → p-Ph$_3$PbC$_6$H$_4$CH$_2$O$_2$CR; RCO = Bz, (±)-cis-7-methylbisdehdroisynolyl C$_{18}$H$_{21}$OCO (I) (1092).   (h) See structural formula I shown below.   (i) IR and NMR spectra (1092).   (j) Rxn. of p-OHCC$_6$H$_4$PbPh$_3$ with dimedone → p-Ph$_3$PbC$_6$H$_4$CH(C$_8$H$_{11}$O$_2$)$_2$ (II) (1092).   (k) See structural formula II shown below.

Formula I   Formula II

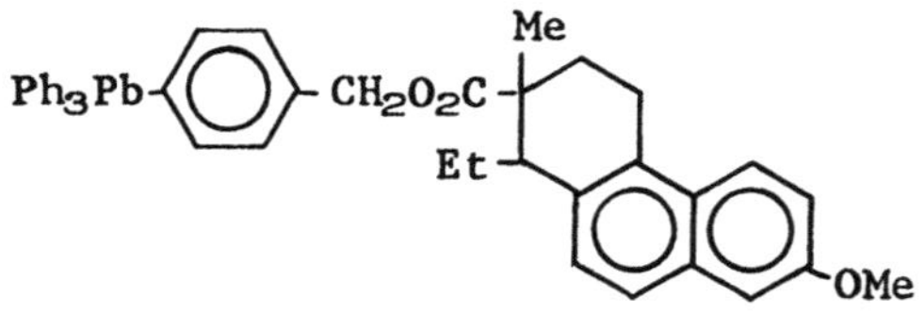

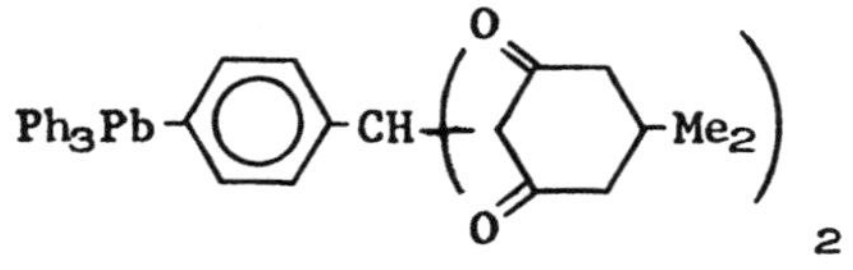

Table 278. Tetraorganolead Compounds Containing Nitrogen and Phosphorus $R_nPbR'_{4-n}$

Formula	Synth. Method*	Yield	Properties	Ref.
Me_3PbR'				
R′ = NCCH:CH	IIIB	98	IR and NMR spectra, cis:trans-ratio 15 (a)	1416
NCCD:CH	IIIB	5	Synth. from DC!CCN, IR and NMR spectra	1416
CH_2:C(CN)	IIIB	76	IR and NMR spectra	1416
CHD:C(CN)	IIIB	84	Synth. from DC!CCN, IR and NMR spectra	1416
Ph_2PC!C	IIA	43	Colorless cryst., m. 58-60°, IR and NMR spectra (b)	2778
Et_3PbCH:CHCN	IIIB (c)	15-37	NMR spectrum	912
$Et_3PbC(CN)$:CH_2	IIIB (c)	43-79	bo.1 71-72°, n_D^{20} 1.5337, NMR spectrum (d)	912
Bu_3PbR'				
R′ = PhN:C(NEt_2)	V	--	Synth. with Ph_3PbNR_2 + PhNC, IR spectrum (e)	1609
$NCCH_2CH_2$	IIIA	>90	bo.01 116-19°, IR spectrum	397
NCCH:CH	IIIB (c)	15	NMR spectrum	912
CH_2:C(CN)	IIIB (c)	45-57	bo.2 110-11°, n_D^{20} 1.5174, NMR spectrum (d)	912
$\{RO_2CCH(CN)CHPh\}$ $\{PhCH_2CCN(CO_2R)\}$	IIIA	Low	IR spectrum, struct. not decided	1607 1607
Ph_3PbR'				
R′ = $AcNHCH_2CH_2$	IIC	90	Synth. from CH_2CH_2NAc, m. 113°	607
$BzNHCH_2CH_2$	IIC	67	Synth. from CH_2CH_2NBz, m. 136-38°	607
Ph_2PC!C	IIA	64	Colorless cryst., m. 83-84.5°, IR spectrum, mol. wt.	2778
$N_3(CH_2)_3$	VD	73	m. 40.5-42°, IR spectrum, mol. wt. (f)	1127
$N_3(CH_2)_4$	VD	68	m. 42-43.5°, IR spectrum, mol. wt. (f)	1127
$N_3(CH_2)_5$	VD	--	Oil, IR spectrum, mol. wt. (f)	1127

* The characters used here correspond to the synthetic scheme in the introduction to Chapter 2.4.
(a) Rearrangement of cis- to trans- form in pentane (1416). (b) Dec. at 130-40°, slowly at r.t., mol. wt. (2778). (c) Yield and product ratio depending on mode of preparation of R_3PbH; R = Et, Bu (912). (d) Mixture not sepd. from preceding compd. (912). (e) Rxn. with anhyd. HCl → PhN:$CHNEt_2$ (1609). (f) Rxn. with HC!$CCMe_2OH$ → $Ph_3Pb(CH_2)_n$$\overline{NN:NC(CMe_2OH)}$:CH; n = 3, 4, 5 (1127).

Table 279. Metal Substituted Tetraorganolead Compounds $R_nPbR'_{4-n}$

Formula*	Synth. Method**	Yield	Properties	Ref.
(Me$_3$PbC!)$_2$	IVB	99	White, m. 99-101°, subl.o.1 80-100°, NMR spectrum (a)	952
(p-Me$_3$PbC$_6$H$_4$CHCH$_2$)$_x$	--	--	Study of thermal degradation	1410
(Et$_3$PbC!)$_2$ (548)	IIB	50	--	1474
	IVA	--	n_D^{20} 1.5666, d$_{20}$ 1.9026, IR spectrum	1917
(Ph$_3$Pb)$_2$CH$_2$	IID	40	Synth. with CH$_2$Cl$_2$, m. 97-97.5°	606
	IID	73	Synth. with CH$_2$Br$_2$, m. 98-99.5°, mol. wt.	1127
(Ph$_3$Pb)$_3$CH	IID	66	Synth. with HCCl$_3$, m. 166°	606
(Ph$_3$Pb)$_4$C	IID	66	Synth. with CCl$_4$, m. 292-94°, soly.	606
(Ph$_3$Pb)$_4$C·4HCCl$_3$	(b)	--	--	606
(Ph$_3$PbC!)$_2$ (548)	IVB	52	m. 134-36°	952
(Ph$_3$PbC!)$_2$ (548)	IIB	80	Synth. with (NaC!C)$_2$, stable at r.t.	1474
Me$_3$PbCH$_2$SiMe$_2$H	IA	86	Colorless liq., b$_1$ 26-27°, stable in air at r.t. (c)	475
Me$_3$PbC!CCHMeOSiMe$_3$	IVA	65	b$_{0.3}$ 61°, n_D^{20} 1.4940, d$_{20}$ 1.5151, IR spectrum	2773
Me$_3$PbC!CCHMeOSnEt$_3$	IVA	--	b$_{0.3}$ 120°, n_D^{20} 1.5400, d$_{20}$ 1.7387, IR spectrum	2773
Ph$_3$PbC!CSiPh$_3$	V (d)	59	m. 128-29°, IR spectrum, mol. wt.	1039
Ph$_3$PbC!CGePh$_3$	V (d)	58	m. 134-35°, mol. wt.	1039
Ph$_3$PbC!CSnPh$_3$	V (d)	56	m. 143-45°, mol. wt.	1039

* Numbers in parenthesis refer to pages in main volume.
** The characters used here correspond to the synthetic scheme in the introduction to Chapter 2.4.
(a) Rxn. with AcCl → Me$_3$PbCl + black tar (952). (b) Synth. by recrystallization of (Ph$_3$Pb)$_4$Pb in HCCl$_3$ (606).
(c) Unpleasant odor, toxic (?), decrepitates in air > 100° (475). (d) Synth. by rxn. of Ph$_3$PbCl with R$_2$NLi, followed by NaC!CH and Ph$_3$MCl; M = Si, Ge, Sn (1039).

3. ORGANOLEAD HYDRIDES

Trimethyllead Hydride (M-550) PbC_3H_{10} Me_3PbH

<u>Prop.</u>: NMR spectrum (246, 2361).
<u>Rxn. with</u>: R'C⋮CH at -70° → $Me_3PbCR:CH_2$ + $Me_3PbCH:CHR'$; R = CN, MeO_2C; rxn. mechanism and prod. distribution (1416).
$(EtO_2CC⋮)_2$ at -70° → $Me_3PbC(CO_2Et):CHCO_2Et$ (1416).
$Me_3PbCH:CHR$ in C_5H_{12} → rearrangement cis- to trans-form (1416).
<u>Addn. Compd.</u>: Me_3PbBH_4, (M-550), from Me_3PbOMe and B_2H_6 at -78°, dec. slowly at -78°, rxn. with MeOH at -78° → B_2H_6 + H (674) [$PbBC_3H_{13}$].

Triethyllead Hydride (M-551) PbC_6H_{16} Et_3PbH

<u>Synth.</u>: By rxn. of Et_3PbBH_4 at -78° with MeOH (674).
<u>Prop.</u>: Oil, dec. slowly at -20°, anal. detn. by thermal dec. and volumetric H detn. (674).
<u>Rxn. with</u>: Thermal dec. → Et_6Pb_2 + Pb + H (674).
<u>Addn. Compd.</u>: Et_3PbBH_4, from Et_3PbOMe and B_2H_6 at -78°, stable to -35°, rxn. with MeOH at -78° → B_2H_6 + H + Et_3PbH, thermal dec. → B_2H_6 + H + Pb + Et_6Pb_2 (674) [$PbBC_6H_{19}$].

Tripropyllead Hydride PbC_9H_{22} Pr_3PbH

<u>Synth.</u>: By rxn. of Pr_3PbCl with $LiAlH_4$ in $(MeOCH_2)_2$ at -60° (397, 1610).
By rxn. of Pr_3PbBH_4 with MeOH at -78° (674).
<u>Prop.</u>: IR spectrum, sensitive to air and light, stable at 0° in the dark for several days (397, 1610), dec. slowly at -20° (674).
<u>Rxn. with</u>: Thermal dec. → Pr_6Pb_2 + H + Pb (674).
<u>Addn. Compd.</u>: Pr_3PbBH_4, from Pr_3PbOMe and B_2H_6 at -78°, stable to -35°, anal. detn. by thermolysis and methanolysis (674), rxn. with MeOH at -78° → B_2H_6 + Pr_3PbH + H, thermal dec. → Pr_6Pb_2 + B_2H_6 + Pb + H (674) [$PbBC_9H_{25}$].

Tributyllead Hydride $PbC_{12}H_{28}$ Bu_3PbH

<u>Synth.</u>: By rxn. of Bu_3PbCl with $LiAlH_4$ in $(MeOCH_2)_2$ at -60° (397, 1610).
By rxn. of Bu_3PbBH_4 with MeOH at -78° (674).
<u>Prop.</u>: IR spectrum (397), sensitive to air and light, stable at 0° in the dark for several days (397), slow dec. at -20° (674), anal. detn. by rxn. with EtI (1610).
<u>Rxn. with</u>: $RCH:CH_2$ at 0° → $Bu_3PbCH_2CH_2R$; R = CN, CO_2Me, Ph (397, 1610).
$RCH:C(CN)_2$ → $Bu_3PbN:C:C(CN)CH_2R$; R = Me, Ph, $2-C_4H_3O$ (1607).
$PhCH:C(CN)CO_2R$ → $Bu_3PbOC(OR):C(CN)CH_2Ph$ + some 1,2-adduct (1607).
PhC⋮CH at -20° → trans-$Bu_3PbCH:CHPh$ (397, 1610).
PhNCO at 0° → [$Bu_3PbNPhCHO$] → Bu_6Pb_2 + $C_{27}H_{22}N_4O_2$ + $1,3,5-C_3(NPh)_3O_3$ (397).
RX → Bu_3PbX + RH; RX = $PhCH_2Cl$, $BzCH_2Br$, $CH_2:CHCH_2Br$, PhBr, PhI, Ph_3SnCl (397, 1610).
CCl_4 → Bu_3PbCl + $HCCl_3$ or H_2CCl_2 (397).
$(CH_2Br)_2$ → Bu_3PbBr + C_2H_4 (397, 1610).
<u>Addn. Compd.</u>: Bu_3PbBH_4, from Bu_3PbOMe and B_2H_6 at -78°, stable < -35°, thermal dec. → Bu_6Pb_2 + B_2H_6 + Pb + H, with MeOH at -78° → Bu_3PbH + B_2H_6 + H (674) [$PbBC_{12}H_{31}$].

Dibutyllead Dihydride   PbC$_8$H$_{20}$ Bu$_2$PbH$_2$

<u>Synth.</u>: By rxn. of Bu$_2$PbCl$_2$ with LiAlH$_4$ in (MeOCH$_2$)$_2$ at -60° (397, 1610).
<u>Prop.</u>: IR spectrum, sensitive to air and light, stable at 0° in the dark for several days (397).

Triisobutyllead Hydride   PbC$_{12}$H$_{28}$ i-Bu$_3$PbH

<u>Synth.</u>: By rxn. of i-Bu$_3$PbCl with LiAlH$_4$ in (MeOCH$_2$)$_2$ at -60° (397, 1610).
<u>Prop.</u>: IR spectrum, sensitive to air and light, stable at 0° in the dark for several days (397).

Tricyclohexyllead Hydride   PbC$_{18}$H$_{34}$ (c.C$_6$H$_{11}$)$_3$PbH

<u>Synth.</u>: By rxn. of (c.C$_6$H$_{11}$)$_3$PbCl with LiAlH$_4$ in (MeOCH$_2$)$_2$ at -60° (397, 1610).
<u>Prop.</u>: IR spectrum, sensitive to air and light, stable at 0° in the dark for several days (397).

4. ORGANOLEAD COMPOUNDS CONTAINING HETEROATOM-LEAD BONDS

4.1 ORGANOLEAD HALIDES

The organolead halides summarized in Tables 280-282 are prepared by methods from the ensuing scheme.

I Dealkylation reactions of tetraorganolead derivatives R$_4$Pb:
  (IA) R$_4$Pb + HCl → R$_2$PbCl$_2$,
  (IB) R$_4$Pb + SOCl$_2$ → R$_n$PbCl$_{4-n}$,
  (IC) R$_4$Pb + BF$_4$ → R$_3$PbF,
  (ID) Ph$_3$PbR + HCl → RPhPbCl$_2$.

II Dehydrogenation of triorganolead hydrides R$_3$PbH:
  Bu$_3$PbH + RX → Bu$_3$PbX.

III Alkylation of lead halide or acetate:
  (IIIA) PbCl$_2$ + RMgX → R$_3$PbX,
  (IIIB) PbCl$_2$ + Al$_4$C$_3$ + HCl → Me$_n$PbCl$_{4-n}$,
  (IIIC) Pb(OAc)$_4$ + RSiF$_3$ → R$_n$PbF$_{4-n}$,
  (IIID) Pb(OAc)$_4$ + (NH$_4$)$_2$PbRF$_5$ → RPbF$_3$.

IV Complex formation with diorganolead halides R$_2$PbX$_2$:
  (IVA) R$_2$PbX$_2$ + L → R$_2$PbX$_2$·L,
  (IVB) R$_2$PbX$_2$ + 8-HOC$_9$H$_6$N → R$_2$Pb(OC$_9$H$_6$N)X.

R = organic group, X = halogen, n = 3, 2, 1, L = 2,2'-bipyridyl, 1,10-phenanthroline.

4.1.1 TRIORGANOLEAD HALIDES

Trimethyllead Chloride (M-552)   PbC$_3$H$_9$Cl   Me$_3$PbCl

<u>Synth.</u>: By rxn. of Me$_4$Pb with Cl in EtOAc in presence of dry ice at -78° (1056), with BCl$_3$ or B$_2$Cl$_4$ at -78° to r.t. (1194), with SO$_2$Cl$_2$ in C$_6$H$_6$ (166). By rxn. of (Me$_3$PbCl)$_2$ with AcCl in CCl$_4$, in 84% yield (952).

By rxn. of PbCl$_2$ with Al$_4$C$_3$ and HCl in Me$_2$SO at 150-200° as by-prod. (2114).
Prop.: Subl.$_{0.1}$ 185-95° (952), IR, assignment (673, 2323), far IR, assignment (673, 1896, 2323) and NMR (1430, 1478, 1807, 1896) spectra polarography (1701).
Rxn. with: Thermal dec. at 112° → Me$_4$Pb + PbCl$_2$ + MeCl (1194).
RLi → Me$_3$PbR; R = MeC⋮C (20), ClC⋮C (1980), Ph$_2$PC⋮C (2778), Me$_3$SiN(CMe$_3$) (2749), Me$_3$GeMe (1855), (Me$_3$C)$_2$PNH (2739), B$_5$H$_8$ (2398).
p-PhC$_6$H$_4$Li → p-PhC$_6$H$_4$PbMe$_3$ + (p-PhC$_6$H$_4$)$_2$PbMe$_2$ (112).
Me$_3$SiNMeLi, + PhCN → Me$_3$PbN:CPhNMeSiMe$_3$ (2738).
RNa → Me$_3$PbR; R = EtC⋮C (20), polynuclear radical (2703), MeO (673, 1678), C$_5$H$_5$Mo(CO)$_3$ (1670).
YNa$_2$ → (Me$_3$Pb)$_2$Y; Y = S (18, 644), Ru(CO)$_4$ (905).
KSCN → Me$_3$PbNCS (2027).
AgX → Me$_3$PbX; X = NCO (2027), ClO$_4$·L; L = Py, (Me$_2$N)$_3$PO (1896).
RMgX → Me$_3$PbR; R = C$_5$H$_{11}$ (710), C$_6$F$_5$ (146), Me$_2$HSiCH$_2$ (475).
HN$_3$ → Me$_3$PbN$_3$ (556-7).
RSH + aq. Na$_2$CO$_3$ → Me$_3$PbSR; R = Me (644), Et (2), MeO$_2$CCH$_2$ (19).
YH$_3$ + Et$_3$N → (Me$_3$Pb)$_3$Y; Y = P, As (1839).
Addn. Compd.: Me$_3$PbCl·ONC$_5$H$_5$*, IR and NMR spectra (2340, 2494) [PbC$_8$H$_{14}$ClNO].
Me$_3$PbCl·ONC$_5$H$_4$Me-4*, from components in H$_2$CCl$_2$ or HCCl$_3$, white cryst., dec. 82-85°, NMR spectrum (2494) [PbC$_9$H$_{16}$ClNO].
Me$_3$PbCl·ONC$_5$H$_4$Cl-4*, from components in H$_2$CCl$_2$ or HCCl$_3$, white cryst., dec. 92-95°, NMR spectrum (2494) [PbC$_8$H$_{13}$Cl$_2$NO].
Me$_3$PbCl·ONC$_5$H$_4$OMe-4*, from components in H$_2$CCl$_2$ or HCCl$_3$, white cryst., dec. 98-101°, NMR spectrum (2494) [PbC$_9$H$_{16}$ClNO$_2$].
Me$_3$PbCl·OPPh$_3$, from components in HCCl$_3$ or H$_2$CCl$_2$, dec. 137-38°, IR and NMR spectra (2340) [PbC$_{21}$H$_{24}$ClOP].

Trimethyllead Bromide (M-522)   PbC$_3$H$_9$Br   Me$_3$PbBr
Synth.: By rxn. of Me$_4$Pb with Br in MePh-CH$_2$Cl$_2$-Et$_2$O-CCl$_4$ at -70°, in 95% yield (1936).
By-prod. in rxn. of Me$_3$PbOH with EtMgBr in Et$_2$O or THF (1340).
Prop.: m. 133°, subl.$_{1.0}$ 100° (1936), (far) IR, assignment (673, 2323) and NMR (1430, 1807) spectra.
Rxn. with: Me$_3$SiNMeLi → Me$_3$PbNMeSiMe$_3$ (472, 1814).
NaOMe → Me$_3$PbOMe (951).
t-BuCH$_2$MgCl → Me$_3$PbCH$_2$CMe$_3$ (1936).

Triethyllead Chloride (M-522)   PbC$_6$H$_{15}$Cl   Et$_3$PbCl
Synth.: By rxn. of Et$_4$Pb with SOCl$_2$ or EtSOCl in hexene, in 89% yield (166), with S$_2$Cl$_2$ in petr. ether at 30-40°, in > 90% yield (1441), with HGeCl$_3$·OEt$_2$, as by-prod. (366), with EtCl at 100-120° or in sunlight (1539).
By rxn. of Et$_3$PbC⋮CR with AcCl; R = Ph, Bu, in 96% yield (952).
As by-prod. in rxn. of PbCl$_2$ with Al, H and C$_2$H$_4$ and cat. Et$_3$Al at 200° in

---

* Pyridine oxides.

c.C_6H_{12} (1775), in rxn. of Na-Pb with EtCl (1539).
Prop.: Dec. 170-75° (166), subl.$_{0.01}$ 105-15° (673), IR (366, 589), far IR spectra and assignment (673), effect on initial oxidn. rate of Et_4Pb (1017), polarography (1701), soly. and partition between H_2O and CCl_4, complex formation with Cl^- (2008).
Rxn. with: $LiAl(C_6F_5)_2I_2 \rightarrow PbI_2$ + instable organolead compd. (977).
RC⋮CNa → $Et_3PbC⋮CR$; R = Me, Et, 1-c.C_5H_7, Ph (1474).
C_2H_2 + Na → $Et_3PbC⋮CH$ + $(Et_3PbC⋮)_2$ (1474).
NaX → Me_3PbX; X = MeO (673), $C_5H_5Fe(CO)_2$ (2057).
$Na_2Fe(CO)_4 \rightarrow (Et_3Pb)_2Fe(CO)_4$ (1268).
o-$C_6H_4(CO)_2NK \rightarrow$ o-$Et_3PbN(CO)_2C_6H_4$ (1214).
RMgBr → Et_3PbR; R = C_6H_{13}, C_7H_{15}, C_8H_{17}, C_9H_{19}, $C_{10}H_{21}$, $CH_2{:}CHCH_2$, p-$CH_2{:}CHCH_2C_6H_4$ (2050).
Biol. prop.: Effect on acetylcholinesterase activity (864).

Triethyllead Bromide (M-553) $PbC_6H_{15}Br$ Et_3PbBr
Synth.: By rxn. of Et_4Pb with BrMgOH (1340).
By-prod. in rxn. of Et_3PbOH with RMgBr in Et_2O or THF; R = Et, $CH_2{:}CHCH_2$, Ph (1340).
Rxn. with: NaOMe → Et_3PbOMe (951).
Stabilization by alkylcycloalkanes (560).

Trivinyllead Chloride (M-557) PbC_6H_9Cl $(CH_2{:}CH)_3PbCl$
Synth.: By rxn. of $(CH_2{:}CH)_4Pb$ with HCl in hexane at -78° (1196, 1196a).
Prop.: m. 119-21° (1196a).
Rxn. with: Na in liq. NH_3 → Pb + C_2H_4 + $NaNH_2$ + H (1196).
$NaBH_4$ in liq. $NH_3 \rightarrow [(CH_2{:}CH)_3PbBH_4 \cdot NH_3] \rightarrow (CH_2{:}CH)_4Pb$ + Pb + C_2H_4 (1197).
$LiAlH_4 \rightarrow (CH_2{:}CH)_4Pb + C_2H_4$ + H (1197).

Tripropyllead Chloride (M-556) $PbC_9H_{21}Cl$ Pr_3PbCl
Synth.: By rxn. of Pr_4Pb with $SOCl_2$ in hexane (166).
Prop.: Subl.$_{0.01}$ 100-110° (673), (far) IR spectrum and assignment (673).
Rxn. with: RLi → Pr_3PbR; R = $(PhCH_2)_3AsMn(CO)_4$ (185).
$LiAlH_4$ at -60° → Pr_3PbH (397).
NaOMe → Pr_3PbOMe (673).

Tributyllead Chloride (M-556) $PbC_{12}H_{27}Cl$ Bu_3PbCl
Synth.: By rxn. of Bu_3PbH with RCl at 0°; R = $PhCH_2$, Et_3Sn, CCl_3 (397).
Prop.: TLC (951).
Rxn. with: $LiAlH_4$ at -60° → Bu_3PbH (397).
NaOMe → Bu_3PbOMe (673, 951).
Biol. Prop.: Toxicity (1099).
Relative reactivity as bactericide (1099).
Activity as fungicide against Ceratostomella coerulescens, Fusarium culmorum, Penicillium funiculosum and Speira heptaspora (1099).
Activity as insecticide against Heliothis virescens and H. zea (2145).

Trineopentyllead Chloride $PbC_{15}H_{33}Br$ $(Me_3CCH_2)_3PbBr$

<u>Synth.</u>: By rxn. of $(Me_3CCH_2)_6Pb_2$ with Br in C_6H_6, in almost quant. yield (1935).

<u>Prop.</u>: m. 166°, NMR spectrum, soly. (1935).

<u>Rxn. with:</u> RMgCl → $(Me_3CCH_2)_3PbR$; R = Me, at -40 to -30°, Et at -40 to -30°, i-Pr, Me_3C, Me_3CCH_2 (1936).

RMgCl at 0° → $(Me_3CCH_2)_3PbR + (Me_3CCH_2)_2PbR_2$; R = Me, Et (1936).

Triphenyllead Chloride (M-554) $PbC_{18}H_{15}Cl$ Ph_3PbCl

<u>Synth.</u>: By rxn. of Ph_4Pb with Ph_2PbCl_2 in Me_2SO at 95°, in 15% yield (1385), with $PhSO_2Cl$ in C_6H_6 (166), with CCl_4 and cat. $BiCl_3$ in 27% yield (1716).
By rxn. of $(Ph_3PbCH_2CO)_2O$ with HCl in $HCCl_3$ at r.t., in 98% yield (608).
By photolysis of Ph_3PbN_3 in CH_2Cl_2 (2471).
By rxn. of Ph_3PbOH with Cl^- in aq. soln. and extraction with $HCCl_3$ (38).
By rxn. of Ph_3PbOMe with $(CCl_3)_2CO$, in 60% yield (951).
By-prod. in thermal dec. of $Ph_3PbOCH(OMe)CCl_3$ (951).
By rxn. of Ph_6Pb_2 with $KMnO_4$ in aq. Me_2CO (2865) or O_3 in $HCCl_3$, followed by HCl, in 71 and 76% yield, resp. (2865-6), with NaOCl and HCl, in aq. THF, in 95% yield (2865).

<u>Prop.</u>: m. 210° (2865), 206° (2866), 205-207° (951), IR spectrum (1385). Anal. detn. by differential thermal analysis (1385), by polarography (963), by TLC (951) in presence of Ph_nPbCl_{4-n}; n = 4, 2 (2635).

<u>Rxn. with:</u> ^{60}Co irradiation → evidence for Ph_3Pb-radical formation by ESR spectrum (693).

RLi → Ph_3PbR; R = C_6F_5 (146), Ph_2PC⋮C (2778), Ph_3GeY; Y = S, Se, Te (484), Ph_3PbY; Y = S, Se, Te (485), Ph_3SnPPh, $(Ph_3Sn)_2P$ (1850) $[Et_3SbMn(CO)_4]$ (185).

$Ph_2C:NLi$ → $Ph_3PbN:CPh_2 + (Ph_3Pb)_2O$ (852).

Et_2NLi, + NaC_2H, + Ph_3MX → Ph_3PbC⋮$CMPh_3$; M = Si, Ge, Sn (1039).

NaX → Ph_3PbX; X = MeO (951), PhO (791), R_2NCS_2; R = Me, Et, Ph, $PhCH_2$ + H (298), Ph_2As, Ph_2Sb (1847), $C_5H_5Mo(CO)_3$, $Mn(CO)_5$ (967), $C_5H_5Fe(CO)_2$ (2057).

NaOR, + PhC⋮CH → Ph_3PbC⋮CPh (952).

KSeCN in EtOH → $Ph_3PbSeCN$ (697).

RMgX → Ph_3PbR; R = $C_{10}H_{21}$ (2050), p-EtC_6H_4 (2048), C_6F_5 (146).

3-Cl-6-$MeC_6H_3CH:CH_2$ + Mg → 2-Ph_3Pb-5-Me-$C_6H_4CH:CH_2$ sic! (1784).

Hg → $Ph_2Hg + HgCl_2$ + Pb (963).

$Pb(SR)_2$ → Ph_3PbSR (121); R = Me, $PhCH_2$, Et, Pr, Bu, $C_{10}H_{21}$, Ac, Bz, 2-$C_{10}H_7$, MeO_2CCH_2, benzothiazolyl. benzooxazolyl (1179, 1685).

$Co[Co(CO)_4]_2$ in MeOH → $Ph_3PbCo(CO)_4$ (1672).

trans-$HPt(PPh_3)_2Cl$ → $PhPt(PPh_3)_2Cl$ (700).

C_6F_5SH + aq. alc. NaOH → $Ph_3PbSC_6F_5$ (2669).

$CS_2 + NH_3$ → $(Ph_3Pb)_2S$ (298).

$YH_3 + Et_3N$ → $(Ph_3Pb)_3Y$; Y = P (1839, 1849), As, Sb (1839).

$PhPH_2 + Et_3N$ → $(Ph_3Pb)_2PPh$ (1849).

$Ph_2AsO_2H + Et_3N$ → Ph_3PbO_2AsPh (1847).

<u>Addn. Compd.</u>: $Ph_3PbCl·Me_4NCl$, from components in EtOH (1218), in 81% yield (225), dec. at 298° (225), soly., stable in moist air, conductance (1218) $[PbC_{22}H_{27}Cl_2N]$.

Biol. Prop.: Toxicity (1099).
Relative reactivity as bactericide (1099).
Activity as fungicide against Aspergillus niger, Botrytis allii (221), Ceratostomella coerulescens, Fusarium culmorum, Penicillium funiculosum (1099), P. niger, Rhizopus nigricans (221), Speira heptaspora (1099).
Activity as rodent repellent (221).
Use: As stabilizer for polyphenyl ether lubricants (209).
As woodpreservative (2973).

Triphenyllead Bromide (M-554) $PbC_{18}H_{15}Br$ Ph_3PbBr
Synth.: By rxn. of Ph_6Pb_2 with $KMnO_4$ in aq. Me_2CO, followed by HBr, in 70% yield (2865).
By extraction of aq. Ph_3PbOH and Br^- with $HCCl_3$ (38).
By-prod. in rxn. of Ph_3PbOMe with Br_3CCHO (951).
Prop.: m. 166° (2865), TLC (951).
Rxn. with: $Ca[HFe(CO)_4]_2 \rightarrow (Ph_3Pb)_2Fe(CO)_4 + [Ph_2PbFe(CO)_4]_2$ (216).
Addn. Compd.: $Ph_3PbBr \cdot Me_4NBr$, from components in EtOH, m. 277° (dec.), stable in moist air, soly., conductance (1218) [$PbC_{22}H_{27}Br_2N$].
$Ph_3PbBr \cdot CsBr$, m. 209° (dec.) (1218) [$PbCsC_{18}H_{15}Br_2$].
Use: In x-ray sensitive films (592).
As stabilizer for polyphenyl ether lubricants (209).

Triphenyllead Iodide (M-555) $PbC_{18}H_{15}I$ Ph_3PbI
Synth.: By rxn. of Ph_6Pb_2 with KI_3 in alc. $HCCl_3$, in 92% yield (2865).
By extraction of aq. Ph_3PbOH and I^- with $HCCl_3$ (38).
Prop.: m. 143° (2865).
Rxn. with: $Ph_3{}^{210}PbI$ after β-desintegration $\rightarrow Ph_3{}^{210}Bi + Ph_2{}^{210}BiI + {}^{210}Bi^{+3}$; by paper chromatography (615).
$NaC \vdots CR \rightarrow Ph_3PbC \vdots CR$; R = Me, Et, 1-c.$C_5H_7$, 1-c.$C_6H_9$, $MeC \vdots C$, Br (1474).
$Na + YH_2$ in liq. $NH_3 \rightarrow Ph_3PbYH + (Ph_3Pb)_2Y$; Y = $C \vdots C$, $C \vdots CC \vdots C$ (1474).
$AgX \rightarrow Ph_3PbX$; X = $(CN)_2N$ (2501), $(CN)_2C:C:N$ (717).
$Ag_2N_2O_2 \rightarrow (Ph_3Pb)_2N_2O_2$ (716).
Use: As stabilizer for polyphenyl ether lubricants (209).

Listing of triorganolead halides continues in Table 280.

Table 280. Triorganolead Halides R₃PbX

Formula*	Synth. Method**	Yield	Properties	Ref.
Me₃PbF (555)	IC	--	(Far) IR spectrum and assignment (673, 2323) (a)	1194
Me₃PbI (555)	--	--	(Far) IR spectrum and assignment (a)	2323
Et₃PbI (555)	--	--	(b)	--
Bu₃PbBr (556)	II	--	--	397
Bu₃PbI	II	--	--	397
i-BuPbCl	--	--	+ LiAlH₄ at -60° → i-Bu₃PbH	397
(Me₃CCH₂)₃PbCl	IIIA	--	m. 207-208°, NMR spectrum, soly.	1935
(C₆H₁₃)₃PbCl	IIIA	42	--	2050
(c.C₆H₁₁)₃PbCl (556)	--	--	+ LiAlH₄ at -60° → (c.C₆H₁₁)₃PbH	397
Ph₃PbF	IIIC	75	White needles (c, d)	1555
(MeC₆H₄)₃PbCl	--	--	(e)	--
(Me₂C₆H₃)₃PbCl	--	--	(e)	--

* Numbers in parenthesis refer to pages in main volume.
** The characters used here correspond to the synthetic scheme in the introduction to Chapter 4.1.
(a) NMR spectrum (1430, 1807). (b) Use as catalyst in synthesis of Et₄Pb from EtCl and Na-Pb alloy (481), and Na-K-Pb alloy (721). (c) Rxn. with PhSiF₃ → Ph₄Pb + SiF₄ (1554). (d) Use as stabilizer for polyphenyl ether lubricants (209). (e) Rxn. with p-EtC₆H₄MgBr → p-R₃PbC₆H₄Et; R = MeC₆H₄, Me₂C₆H₃ (2048).

4.1.2 DIORGANOLEAD DIHALIDES

Dimethyllead Dichloride (M-560) $PbC_2H_6Cl_2$ Me_2PbCl_2
<u>Synth.</u>: By rxn. of Me_4Pb with Cl in MePh-EtOAc at $-78°$ to $-20°$ (1056).
By-prod. in rxn. of Me_4Pb with $SOCl_2$ in C_6H_6 (166).
By rxn. of $PbCl_2$ with Al_4C_3 and HCl in Me_2SO at 150-$200°$ (2114).
<u>Prop.</u>: Fine white powder, recryst. below $40°$ (1056), IR, assignment (2323),
far IR, assignment (1896, 2323) and NMR (1896) spectra, polarography (1701).
<u>Rxn. with</u>: $AgClO_4 + Me_2SO \rightarrow Me_2Pb(ClO_4)_2 \cdot 4OSMe_2 + AgCl$ (1896).
Aq. $Ag_2O \rightarrow Me_2Pb(OH)_2$ (1056).
CH_2Ac_2 + base $\rightarrow Me_2Pb(CHAc_2)_2$; with NaOMe (1294), with NH_3 (1649, 2055).
$HL \rightarrow Me_2PbL_2$; HL = 8-hydroxyquinoline, tropolone (1294).

Dimethyllead Dibromide (M-560) $PbC_2H_6Bu_2$ Me_2PbBr_2
<u>Prop.</u>: (Far) IR spectrum and assignment (2323).
<u>Rxn. with</u>: $Me_4NBr \rightarrow Me_4NPbBr_3 + C_2H_6$ (1218).
$9,10$-$Na_2C_{14}H_{10} \rightarrow 9,10$-$Me_2PbC_{14}H_{10}$ + $(9,10$-$Me_2PbC_{14}H_{10})_x$ (1726).

Diethyllead Dichloride (M-559) $PbC_4H_{10}Cl_2$ Et_2PbCl_2
<u>Synth.</u>: By rxn. of Et_4Pb with $SOCl_2$ in hexane or C_6H_6, in 4% yield (166).
By rxn. of $PbCl_2$ with Al, C_2H_4, H and cat. Et_3Al at $200°$ in c.C_6H_{12} (1775).
<u>Prop.</u>: Anal. detn. by polarography (1701), by spectrophotometry (1697).
<u>Rxn. with</u>: $(CH_2CH_2MgBr)_2 \rightarrow \overline{CH_2(CH_2)_3PbEt_2} + Et_4Pb$ (1260).
PVC $\rightarrow$ effect on thermal degradation (2277).
<u>Addn. Compd.</u>: $Et_2PbCl_2 \cdot BiPy$ (232a) [$PbC_{14}H_{18}Cl_2N_2$].
$Et_2PbCl_2 \cdot 1,10$-phenanthroline (232a) [$PbC_{16}H_{18}Cl_2N_2$].

Dibutyllead Dichloride (M-560) $PbC_8H_{18}Cl_2$ Bu_2PbCl_2
<u>Prop.</u>: TLC (951).
<u>Rxn. with</u>: $Bu_2SnO \rightarrow (Bu_2PbCl)O(Bu_2SnCl)$ (946).
$LiAlH_4$ at $-60° \rightarrow Bu_2PbH_2$ (397).
8-Hydroxyquinoline $\rightarrow Bu_2Pb(OC_9H_6N)Cl$ (232a).
<u>Addn. Compd.</u>: $Bu_2PbCl_2 \cdot BiPy$ (232a), from components in alc., m. $130°$ (1212) [$PbC_{18}H_{26}Cl_2N_2$].
$Bu_2PbCl_2 \cdot 1,10$-phenanthroline (232a), from components in alc., m. $132°$ (1212) [$PbC_{20}H_{26}Cl_2N_2$].

Diphenyllead Dichloride (M-559) $PbC_{12}H_{10}Cl_2$ Ph_2PbCl_2
<u>Synth.</u>: By rxn. of Ph_4Pb with CCl_4 and cat. $BiCl_3$ under reflux, in 68% yield (1716), with $SOCl_2$ in C_6H_6, in 25-98% yield (166), with $PhSO_2Cl$ in C_6H_6 (166).
By rxn. of $Ph_3PbCH_2CHOHCH_2Cl$ with 12 N HCl in Et_2O (607).
By rxn. of $(Ph_2PbOAc)_2O$ in alc. AcOH with dry HCl, in 81% yield (633).
<u>Prop.</u>: Polarography (963), x-ray struct. analysis (1455), TLC (951) and sepn. from Ph_nPbCl_{4-n}; n = 4, 3 (2635).
<u>Rxn. with</u>: ^{60}Co irradiation $\rightarrow$ evidence for Ph_2PbCl-radical formation by ESR spectrum (693).
Ph_4Pb in $Me_2SO \rightarrow Ph_2PbCl_2 \cdot 2Me_2SO + Ph_3PbCl$ (1385).

$YLi_2 \rightarrow Ph_2PbY$; $Y = (1,2,4,5-\overline{C_6H_2OCH_2O})_2$ (934).

$(SiPh_2)_4Li_2 \rightarrow Pb$ (1177).

$(PhNLi)_2 \rightarrow Pb + PhN:NPh$ (1089).

$C_6F_5MgBr \rightarrow Ph_2Pb(C_6F_5)_2$ (1185).

$(CH_2CH_2MgBr)_2 \rightarrow \overline{CH_2(CH_2)_3PbPh_2}$ (1260).

$CH_2Br_2 + Mg \rightarrow (Ph_2PbCH_2)_xPbPh_2(OH)_2$ (1352).

Hg in $(CH_2OMe)_2 \rightarrow Ph_2Hg + HgCl_2 + Pb$ (963).

Alc. $KOH \rightarrow Ph_2PbO$ (1421).

8-Hydroxyquinoline $\rightarrow Ph_2Pb(OC_9H_6N)Cl$ in MeOH (233), in aq. alc. (232a).

ArSH + aq. NaOH $\rightarrow Ph_3PbSAr + Pb(SAr)_2 + Ar_2S_2$; $Ar = C_6F_5$ (2669).

$R_2NCS_2Na \rightarrow Ph_2Pb(S_2CNR_2)_2$; R = Me, Et, Ph, PhCH$_2$ + H (298).

$Pb(SR)_2 \rightarrow Ph_2Pb(SR)_2$; R = Ac, benzothiazolyl (1685).

$Pb(SR)_2 \rightarrow Ph_3PbSR + R_2S_2$; R = Me, PhCH$_2$, Et, Ph (121).

$HSCH_2CO_2Na + Py \rightarrow (PbC_{14}H_{12}O_2S)_x$ (121).

$CS_2 + NH_3 \rightarrow (Ph_2PbS)_3$ (298).

Addn. Compd.: $Ph_2PbCl_2 \cdot Me_4NCl$, from components in EtOH, m. 301° (dec.) (1218) [$PbC_{16}H_{22}Cl_3N$].

$Ph_2PbCl_2 \cdot 2Py$, from components (225) in alc. (1212), m. > 300° (225) [$PbC_{22}H_{20}Cl_2N_2$].

$Ph_2PbCl_2 \cdot BiPy$ (232a), from components, in 55% yield (225) in Me$_2$CO (147), dec. 250° (225), > 300° (147) [$PbC_{22}H_{18}Cl_2N_2$].

$Ph_2PbCl_2 \cdot 1,10$-phenanthroline (232a), from components at 0.1 mm, in 30% yield (225), in alc. (1212), m. > 300° (225), 275° (1212) [$PbC_{24}H_{18}Cl_2N_2$].

$Ph_2PbCl_2 \cdot 2,2',2''$-terpyridyl, from components in alc., needles, m. 253-55°, conductance in Me$_2$SO (147) [$PbC_{27}H_{21}Cl_2N_3$].

$Ph_2PbCl_2 \cdot OCHNMe_2$, from components, dec. 76-80° (225) [$PbC_{15}H_{17}Cl_2NO$].

$Ph_2PbCl_2 \cdot OSMe_2$, from components (225), in 87% yield (1388), dec. 168-70° (225), by-prod. in rxn. of Ph$_4$Pb with Ph$_2$PbCl$_2$ in Me$_2$SO at 95° (1385), IR spectrum (1385, 1388), differential thermal analysis (1385) [$PbC_{14}H_{16}Cl_2OS$].

Biol. Prop.: Toxicity (1099).

Relative reactivity as bactericide (1099).

Activity as fungicide against Ceratostomella coerulescens, Fusarium culmorum, Penicillium funiculosum, Speira heptaspora (1099).

Activity as rodent repellent (221).

Use: As stabilizer for polyphenyl ether lubricants (209).

Diphenyllead Dibromide (M-560) $PbC_{12}H_{10}Br_2$ Ph_2PbBr_2

Synth.: By rxn. of $(Ph_3PbCH_2CO)_2O$ with Br in HCCl$_3$ at -40°, in 96% yield (608).

Rxn. with: $NaCo(CO)_4 \rightarrow Ph_2Pb[Co(CO)_4]_2$ (216).

$9,10-Na_2C_{14}H_{10} \rightarrow 9,10-Ph_2PbC_{14}H_{10} + (9,10-Ph_2PbC_{14}H_{10})_x$ (1726).

8-Hydroxyquinoline in MeOH $\rightarrow Ph_2Pb(OC_9H_6N)Br$ (233).

CsBr in EtOH $\rightarrow Cs_2PbBr_4$ (1218).

Addn. Compd.: $Ph_2PbBr_2 \cdot CsBr$, from components in dry Me$_2$CO, m. 223° (dec.) (1218) [$PbCsC_{12}H_{10}Br_3$].

$Ph_2PbBr_2 \cdot Me_4NBr$, from components in EtOH, m. 262° (dec.) (1218) [$PbC_{16}H_{22}Br_3N$].

Ph$_2$PbBr$_2$·BiPy (232a), from components in Me$_2$CO, m. 250° (dec.) (147) [PbC$_{22}$H$_{18}$Br$_2$N$_2$].

Ph$_2$PbBr$_2$·1,10-phenanthroline (232a) [PbC$_{24}$H$_{18}$Br$_2$N$_2$].

Ph$_2$PbBr$_2$·2,2',2"-terpyridyl, from components in alc., m. 233-34° (147) [PbC$_{27}$H$_{21}$Br$_2$N$_3$].

Ph$_2$PbBr$_2$·OSMe$_2$, from components, m. 165-70° (225) [PbC$_{14}$H$_{16}$Br$_2$OS].

Use: As stabilizer for polyphenyl ether lubricants (209).

Additional diorganolead dihalides are listed in Table 281.

Table 281. Diorganolead Dihalides   R$_2$PbX$_2$

Formula*	Synth. Method**	Yield	Properties	Ref.
Et$_2$PbBr$_2$ (560)	--	--	(a)	--
Et$_2$PbBr$_2$·BiPy	IVA	--	m. 118°	232a, 1212
Et$_2$PbBr$_2$·phen (b)	IVA	--	m. 106°	232a, 1212
(CH$_2$:CH)$_2$PbCl$_2$ (561)	IA	--	m. > 300° (c)	1196, 1196a
Pr$_2$PbCl$_2$ (560)	IB	Low	--	166
Bu$_2$PbBr$_2$	--	--	(d, e)	--
Bu$_2$PbBr$_2$·BiPy	--	--	--	232a
Bu$_2$PbBr$_2$·phen (b)	--	--	--	232a
(C$_6$H$_{13}$)$_2$PbCl$_2$	IA	71	Infusible solid, dec. 250°	2050
(C$_8$H$_{17}$)$_2$PbCl$_2$ (560)	--	--	(f)	--
Ph$_2$PbF$_2$ (560)	IIIC	--	(g)	1555
Ph$_2$PbI$_2$ (561)	--	--	(g)	--
Ph$_2$PbI$_2$·2Me$_4$NI	IVA	--	Yellow cryst., m. 333° (dec.)	1218
Ph$_2$PbI$_2$·BiPy	IVA	--	Pale yellow cryst., m. 180° (dec.)	147
Ph$_2$PbI$_2$·TerPy (h)	IVA	--	Yellow needles, m. 150°	147
(p-MeC$_6$H$_4$)$_2$PbCl$_2$ (561)	IB	--	Microcryst. powder	166
R$_2$PbX$_2$	--	--	Anal. detn. by coulometric titration	1696
(2-C$_4$H$_3$S)$_2$PbCl$_2$	--	--	(i)	--
Ph(ClCH$_2$CHOHCH$_2$)PbCl$_2$	ID	--	Dec. 135°	607
R$_2$Pb(OC$_9$H$_6$N)X				
R = Et X = Br	IVB (a)	--	--	232a
Bu Cl	IVB	--	Dec. > 122°	232a
Bu Br	IVB (d)	--	Dec. > 110°	233
Ph Cl	IVB	--	Dec. > 187°	232a, 233
Ph Br	IVB	--	Dec. > 195°	233

* Numbers in parenthesis refer to pages in main volume.
** The characters used here correspond to the synthetic scheme in the introduction to Chapter 4.1.
(a) Rxn. of Et_2PbBr_2 with alc. 8-hydroxyquinoline → $Et_2Pb(OC_9H_6N)Br$ (232a).
(b) 1,10-Phenanthroline. (c) Rxn. with Na in liq. NH_3 → $Pb + C_2H_4 + H + NaNH_2$ (1196). (d) Rxn. of Bu_2PbBr_2 with alc. 8-hydroxyquinoline → $Bu_2Pb(OC_9H_6N)Br$ (233). (e) Rxn. with $8-HOC_9H_6N + NaOAc$ → $Bu_2Pb(OC_9H_6N)_2$ (233). (f) Use in catalyst for polyurethan preparation (230). (g) Use as stabilizer for polyphenyl ether lubricants (209). (h) 2,2′,2″-Terpyridine. (i) Rxn. with RMgI + $HgBr_2$ → $(2-C_4H_3S)_2PbR_2$; R = $2-C_4H_3Se$ (624).

4.1.3 MONOORGANOLEAD TRIHALIDES

All compounds are listed in Table 282.

Table 282. Monoorganolead Trihalides $RPbX_3$

Formula*	Synth. Method**	Properties	Ref.
$MePbF_3$	IIID	--	1555
$MePbCl_3$	IIIB	--	2114
$EtPbF_3$	IIID	--	1555
$CH_2{:}CHPbF_3$	IIIC,D	--	1555
$PhPbF_3$	--	(a)	--
$PhPbCl_3$ (562)	--	(a)	--
$PhPbBr_3$ (562)	--	(a)	--
$PhPbI_3$	--	(a)	--

* Numbers in parenthesis refer to pages in main volume.
** The characters used here correspond to the synthetic scheme in the introduction to Chapter 4.1.
(a) Use as stabilizer for polyphenyl ether, lubricants (209).

4.2 ORGANOLEAD SALTS

4.2.1 ORGANOLEAD PSEUDOHALIDES

A number of new organolead pseudohalides are reported in the recent literature. These include derivatives of fulmic acid HCNO, dicyanamide $HN(CN)_2$ and tricyanomethane $HC(CN)_3$. The organolead pseudohalides listed in Table 283 are prepared by the following methods.

I Reaction of organolead halides R_3PbX with:
    (IA) silver pseudohalides,
    (IB) potassium pseudohalides,
    (IC) hydrazoic acid and extraction with organic solvent.

II Neutralization reaction of organolead oxide and hydroxide R_2PbO and R_3PbOH with:
  (IIA) hydropseudohalogenic acid,
  (IIB) sodium fulminate and dilute sulfuric acid.

III Cleavage reaction of tetraorganolead compounds R_4Pb with:
  (IIIA) selenodiselenocyanate $Se(SeCN)_2$,
  (IIIB) selenodicyanide $Se(CN)_2$.

IV Cleavage of triphenyllead selenocyanate with selenodiselenocyanate.

Triphenyllead Selenocyanate $PbC_{19}H_{15}NSe$ $Ph_3PbNCSe$
Synth.: By rxn. of $Se(SeCN)_2$ with Ph_4Pb in refluxing $HCCl_3$, in 81% yield (697), with Ph_6Pb_2 in $HCCl_3$ at r.t., in 58% yield (697).
By rxn. of Ph_3PbCl with $KSeCN$ in EtOH in presence of KOAc, in 75% yield (697).
Prop.: White cryst., IR spectrum, soly., stable in air, struct. (697).
Rxn. with: $Se(SeCN)_2 \rightarrow Ph_2Pb(NCSe)_2 + PhSeCN + Se$ (697).

Triphenyllead Azide (M-562) $PbC_{18}H_{15}N_3$ Ph_3PbN_3
Synth.: By rxn. of Ph_3PbOH with HN_3 in $HCCl_3$ below 35° (1421).
Prop.: IR spectrum (1127), exists in three polymorphic forms (903), differential thermal analysis (903, 2667), thermal gravimetric analysis (2667).
Rxn. with: UV irradiation $\rightarrow Ph_3PbCl$ (in H_2CCl_2) + Ph_3PbOH + $(Ph_3Pb)_2O$ (in THF) (2471).
$(!CCO_2Me)_2 \rightarrow Ph_3Pb\overline{NN:NC(CO_2Ne):C}CO_2Me$ (1127).

Listing of organolead pseudohalides concludes in Table 283.

Table 283. Organolead Pseudohalides $R_nPbX'_{4-n}$

Formula*	Synth. Method**	Yield	Properties	Ref.
Me_3PbCN	--	--	Use as stabilizer for PVC	1126
Me_3PbNCO	IA	10	Dec. 220° (far) IR spectrum and assignment	2027
Me_3PbNCS	IB	34	Dec. 145°, (far) IR spectrum and assignment	2027
$Me_3PbNCSe$	IIIA	91	White needles, IR spectrum, soly.	697
$(Me_3PbN:)_2C$	--	--	(a)	--
Me_3PbN_3	IC	38	Sinters at 220°, dec. 300°, IR and UV spectra	556-7
	--	--	Dec. 165°, (far) IR spectrum and assignment	2027
Et_3PbCN (563)	--	--	Use as stabilizer for PVC	1126
$Et_3PbNCSe$ (563)	IIIA	64	White, m. 30°, IR spectrum	697
Pr_3PbCN (563)	--	--	Use as stabilizer for PVC	1126
Pr_3PbCNO (b)	IIB	--	Yellow cryst., IR spectrum, explosive	28
$(C_5H_{11})_3PbCN$	--	--	Use as stabilizer for PVC	1126
$(C_6H_{13})_3PbCN$	--	--	Use as stabilizer for PVC	1126
Ph_3PbCN (563)	IIIB	91	--	697
Ph_3PbNCO	(c)	--	--	28
Ph_3PbNCS (563)	(d)	--	--	38
Ph_3PbCNO (b)	IIB	--	Colorless cryst., m. 174-75°, IR spectrum (c, e)	28
$Ph_3PbN(CN)_2$ (f)	IA	44	Colorless cryst., dec. > 240°, soly. (g)	2501
$Ph_3PbN:C:C(CN)_2$ (h)	IA	65	Colorless cryst., m. 196-98° (dec.), struct.	717
	IIA	53	(Far) IR spectrum and assignment, monomeric	717
$Ph_2Pb(N_3)_2$ (563)	IIA	--	Thermal dec. → Ph_4Pb (2667) (i)	1421
$Ph_2Pb(NCSe)_2$	IV	69	White cryst., IR spectrum	697

* Numbers in parenthesis refer to pages in main volume.
** The characters used here correspond to the synthetic scheme in the introduction to Chapter 4.2.1.

(a) Use as antiwear additive in lubrication oils (2239). (b) Fulminate. (c) Thermal rearrangement of Ph_3PbCNO at 155-165° → Ph_3PbOH (28). (d) Formation from Ph_3PbOH and distribution in H_2O-$HCCl_3$, pH dependence (38). (e) No conductance in Me_2CO, monomeric (28). (f) Dicyanamide. (g) (Far) IR spectrum and assignment, hydrolyzes easily (2501). (h) Tricyanomethanide. (i) Differential thermal and thermogravimetric analysis (2667).

4.2.4 ORGANOLEAD SALTS OR ESTERS OF OXYGEN ACIDS

This chapter comprises organolead derivatives of mineral acids. Included are compounds derived from acids having an oxo or hydroxy group substituted by fluor or an organic group. The compounds summarized in Table 284 are prepared by methods from the following scheme.

I Substitution reactions with organolead halides:
   (IA) $R_3PbX + AgX' \rightarrow R_3PbX'$,
   (IB) $R_3PbX + HX' + base \rightarrow R_3PbX'$.

II Neutralization or esterification of organolead hydroxides:
   $R_nPb(OH)_{4-n} + HX' \rightarrow R_nPbX'_{4-n}$.

III Dealkylation of tetraorganolead compounds:
   (IIIA) $R_4Pb + SO_3 \rightarrow R_nPb(O_3SR)_{4-n} + R_2PbSO_4$,
   (IIIB) $R_4Pb + SO_2 \rightarrow R_nPb(O_2SR)_{4-n} + R_2PbSO_3$,
   (IIIC) $R_4Pb + AgNO_3 \rightarrow R_3PbNO_3$,
   (IIID) $Me_4Pb + BF_3 \rightarrow Me_3PbBF_4$.

IV Oxidation of organolead arsenide:
   $R_3PbAsPh_2 + air\ or\ alkaline\ H_2O_2 \rightarrow R_3PbO_2AsPh_2$.

R = organic group, X = halogen, X' = mineral acid anion, n = 2,3 .

Dimethyllead Diperchlorate $PbC_2H_6Cl_2O_8$ $Me_2Pb(ClO_4)_2$
Synth.: By rxn. of Me_2PbO with dil. $HClO_4$ (1056).
Prop.: Raman (1056, 2596) and NMR (1056, 1896) spectra, EMF measurement (1056), normal-coordinate analysis for Me_2Pb^{+1} (2596). Anal. detn. (1056).
Rxn. with: $H_2O \rightarrow Me_2Pb(OH)_2 + Me_2Pb(OH)_3^- + [Me_2PbOH]^+ + [(Me_2Pb)_3(OH)_4]^{+2}$, hydrolysis const. (1056).
Addn. Compd.: $Me_2Pb(ClO_4)_2 \cdot 2Py$ from $Me_2Pb(ClO_4)_2 \cdot 4OSMe_2$ with Py in $HCCl_3$, prismatic cryst., m. 115°, far IR spectrum and assignment (1896) $[PbC_{12}H_{16}Cl_2N_2O_8]$.
$Me_2Pb(ClO_4)_2 \cdot 4OSMe_2$, from Me_2PbCl_2 and $AgClO_4$ in Me_2SO-i-PrOH, white, m. 66°, darkens on standing, rxn. with Py $\rightarrow Me_2Pb(ClO_4)_2 \cdot 2Py$ (1896) $[PbC_{10}H_{30}Cl_2O_{12}S_4]$.

Triethyllead Ethylsulfinate $PbC_8H_{20}O_2S$ Et_3PbO_2SEt
Synth.: By rxn. of Et_4Pb with SO_2 in Et_2O at 0°, in low yield (1083).
Prop.: Dec. > 130°, spectrophotometric detn. (1083).
Rxn. with: Thermal dec. on standing or at 100° $\rightarrow Et_4Pb + Et_2SO_2 + Pb(O_2SEt)_2$ (1083).

Diethyllead Diethylsulfinate $PbC_8H_{20}O_4S_2$ $Et_2Pb(O_2SEt)_2$
Synth.: By rxn. of Et_4Pb with SO_2 in heptane, in low yield (1083), in dry C_6H_6 (232a, 1217).
Prop.: Dec. 135° (1083), 230° (1217), IR spectrum (232a), spectrophotometric detn. (1083).
Rxn. with: Thermal dec. at 100° or on standing $\rightarrow Et_4Pb + Et_2SO + Pb(O_2SEt)_2$ (1083).
NaOH, + $(CH_2Br)_2 \rightarrow (EtSO_2CH_2)_2$ (1217).

Additional organolead salts or esters of mineral acids are summarized in Table 284.

Table 284. Organolead Salts or Esters of Oxygen Acids $R_nPbX'_{4-n}$

Formula*	Synth. Method**	Yield	Properties	Ref.
Me_3PbBF_4	IIID	--	NMR spectrum	1896
$(Me_3Pb)_2SO_4$	--	--	IR spectrum and assignment (a)	1974
Me_3PbO_3SMe	IIIA	86	m. 137-40° (dec.), IR spectrum and assignment (1974)	1087
Me_3PbO_2SMe	IIIB	92	Dec. 122-24°, IR spectrum and assignment (1974) (b)	1087
Me_3PbClO_4	--	--	NMR spectrum	1896
$Me_3PbClO_4 \cdot 2\ OP(NMe_2)_3$	IA	--	m. 86°, far IR, assignment and NMR spectra	1896
$Me_3PbClO_4 \cdot OP(NMe_2)_3$	IA	--	m. 96°	1896
$Me_3PbClO_4 \cdot Py \cdot H_2O$	IA	--	m. 135°, far IR, assignment and NMR spectra	1896
Me_2PbSO_4	IIIA	55	IR spectrum and assignment (1974) (a)	1087
$Me_2Pb(O_3SMe)_2$	IIIA	78	Dec. 165-68°, IR spectrum and assignment (1974)	1087
Me_2PbSO_3	IIIB	--	m. 220° (dec.)	1217
$Me_2Pb(O_2SMe)_2$	IIIB	35	IR spectrum, soly.	232a
$(Et_3Pb)_2SO_4$ (564)	--	--	(a)	--
Et_3PbO_3SEt	IIIA	58	--	1087
Et_2PbHPO_4	II	83	--	1214
Et_2PbSO_4 (565)	IIIA	36	Dec. 190° (a)	1087
$Et_2Pb(O_3SEt)_2$	IIIA	99	Dec. 172-74°	1086
Et_2PbSO_3 (565)	IIIA	63	(c)	1087
Et_2PbSO_3 (565)	IIIB	--	m. 230° (dec.)	1217
$Et_2Pb(ClO_4)_2$	--	--	Anal. detn. by spectrophotometry (d)	1695, 1697

* Numbers in parenthesis refer to pages in main volume.
** The characters used here correspond to the synthetic scheme in the introduction to Chapter 4.2.4.
(a) Refluxing in $H_2O \rightarrow PbSO_4$ (1087). (b) Oxidizes easily in air (1083). (c) Synthesis is useful method for deleading gasoline (1086). (d) Forms 1:1 chelate compounds with 1-(2-pyridylazo)-2-naphthole, PAN (1695) and 4-(2-pyridyl)resorcinol, PAR (1697).

Table 284 Continued

Formula*	Synth. Method**	Yield	Properties	Ref.
$(CH_2:CH)_3PbNO_3$	IIIC	--	Cream colored solid, dec. 80°, IR spectrum	1196a
Bu_2PbSO_3	IIIB	--	m. 250° (dec.)	1217
$Bu_2Pb(O_2SBu)_2$	IIIB	--	m. 190° (dec.)	1217
$(Ph_3Pb)_nY$				
$Y^{n-} = BO_2^-$	II	--	--	38
HCO_3^-	II	--	--	38
NO_3^- (565)	II	--	(e)	38
$N_2O_2^{2-}$ (f)	IA	44	m. 131° (dec.), (far) IR spectra, assignment (g)	716
PO_4^{3-}	II	--	--	38
HPO_4^{2-}	--	--	Toxicity, rel. reactivity as bactericide (h)	1099
AsO_4^{3-}	II	--	--	38
$Ph_2AsO_2^-$	IB	--	m. 280° (dec.)	1847
AsO_3^{3-}	IV	92	--	1847
SO_4^{2-}	II	--	Toxicity, rel. reactivity as bactericide (h)	1099
$S_2O_8^{2-}$	--	--	--	38
SeO_3^{2-}	II	--	--	38
TeO_3^{2-}	II	--	--	38
ClO_4^-	II	--	--	38
ReO_4^-	II	--	--	38
$Ph_2Pb(NO_3)_2$ (565)	--	--	(i)	--
$Ph_2Pb(OC_9H_6N)NO_3$	(i)	--	Yellow cryst., dec. > 228°	232a, 233
Ph_2PbSO_3	IIIB	--	m. 230° (dec.)	1217
$Ph_2Pb(O_2SPh)_2$	IIIB	--	m. 230° (dec.), soly.	232a, 1217
Ph_2PbSeO_3	IIIB	--	Synth. with SeO_2, white cryst.	1217
$[(p\text{-}MeC_6H_4)_3Pb]_2SO_3$	IIIB	--	m. 218-21° (dec.)	1217

$R_nPbX'_{4-n}$

(e) Rxn. with trans-HPt(PPh$_3$)$_2$Cl → Ph$_3$PbPt(PPh$_3$)$_2$Cl (700). (f) Hyponitrite. (g) Monomeric, dec. in high vacuo at 100-130° → Ph$_4$Pb + Ph$_2$ + N$_2$O + PbO (716). (h) Activity as fungicide against Ceratostomella coerulescens, Fusarium culmorum, Penicillium funiculosum, Speira heptaspora (1099). (i) Rxn. of Ph$_2$Pb(NO$_3$)$_2$ with 8-hydroxyquinoline → Ph$_2$Pb(OC$_9$H$_6$N)NO$_3$ in alc. (233), in aq. alc. (232a).

4.3 ORGANOLEAD OXYGEN COMPOUNDS

4.3.1 ORGANOLEAD OXIDES AND HYDROXIDES

The organolead oxides and hydroxides in Table 285 are prepared by the following methods.

I Dealkylation reactions:
   (IA) Et$_4$Pb + O$_3$ → Et$_2$PbO ,
   (IB) Et$_3$PbOH in vacuo → Et$_2$Pb(OH)$_2$.

II Hydrolysis reactions:
   (IIA) Me$_2$PbCl$_2$ + Ag$_2$O + H$_2$O → Me$_2$Pb(OH)$_2$,
   (IIB) RPb(OAc)$_3$ + aq. NH$_3$ → RPbO$_2$H .

III Interconversion reaction:
   R$_2$Pb(OH)$_2$ ⇌ R$_2$PbO + H$_2$O .

Trimethyllead Hydroxide   PbC$_3$H$_{10}$O   Me$_3$PbOH
(M-568)
<u>Synth.:</u> By rxn. of Me$_4$Pb with ROH, followed with cold H$_2$O; R = Me, Et, i-Pr, i-Bu (1678).
<u>Prop.:</u> NMR spectrum (1807).
<u>Rxn. with:</u> EtMgBr → Me$_3$PbEt (1340).
RC⋮CH → Me$_3$PbC⋮CR; R = H, CH$_2$:CH, Ph (1917),
BuOCHMeOCH$_2$, BuOCHMeO$_2$C , Me$_3$SiOCHMe, Et$_3$SnOCHMe (2773).
PhOH + molecular sieve in CCl$_4$ → Me$_3$PbOPh, rxn. suitable for anal. detn. (342).
ROH → Me$_3$PbOR; R = i-Pr, t-Bu, Me$_3$CCH$_2$ (951),
HC⋮CCH$_2$, HC⋮CCHMe, HC⋮CCMe$_2$ (2773).

Triethyllead Hydroxide (M-567)   PbC$_6$H$_{16}$O   Et$_3$PbOH
<u>Synth.:</u> By rxn. of Et$_4$Pb with ROH, followed by cold H$_2$O; R = Me, Et, i-Pr, i-Bu (1678), with air under UV irradiation, in 12% yield (662), with O$_3$ at -78° (663).
<u>Prop.:</u> (Far) IR spectrum and assignment (2499).
<u>Rxn. with:</u> Et$_4$Pb and air → effect on initial oxidation rate (1017).
Thermal dec. at 56° in vacuo → Et$_2$Pb(OH)$_2$ (342).
RMgBr → Et$_3$PbR + some Et$_3$PbBr; R = Et, CH$_2$:CHCH$_2$, Ph (1340).
HC⋮CCH$_2$Cl → Et$_3$PbC⋮CCH$_2$Cl (1917).
C$_2$H$_2$ → Et$_3$PbC⋮CH + (Et$_3$PbC⋮)$_2$ (1917).
Unsaturated polymers or copolymers → Pb-containing polymer (412).
HN:C(NHCN)NHX → Et$_3$PbN:C(NHCN)NHX; X = H, CN (2867).
ROH → Et$_3$PbOR; R = c.C$_6$H$_{11}$, MePhCH, Me$_3$CCH$_2$ (951).

$Me_2CO_3 \rightarrow (Et_3Pb)_2CO_3$ (951).
$C_9H_{19}CO_3H \rightarrow$ kinetic and thermodynamic data on dec. of peroxycaprate-intermediate, solvent effect and rxn. mechanism (2777).

Tributyllead Hydroxide (M-569)　　　　$PbC_{12}H_{28}O$　　　　Bu_3PbOH
Rxn. with: $PhC\vdots CH \rightarrow Bu_3PbC\vdots CPh$ (2495).
Unsaturated polymers or copolymers $\rightarrow$ Pb-containing polymer (412).
$HN\colon C(NHCN)NH_2 \rightarrow Bu_3PbN\colon C(NHCN)NH_2$ (2867).
Biol. prop.: Toxicity (1099).
Relative reactivity as bactericide (1099).
Activity as fungicide against Ceratostomella coerulescens, Fusarium culmorum, Penicillium funiculosum, Speira heptaspora (1099).

Bis(triphenyllead) Oxide　　　　$Pb_2C_{36}H_{30}O$　　　　$(Ph_3Pb)_2O$
Synth.: By photolysis of Ph_3PbH_3 in THF (2471).
By dehydration of Ph_3PbOH in $HCCl_3$ (38).
By precipitation of Ph_3Pb-salts in aq. $HCCl_3$ with dil. H_2SO_4 at pH < 3.5-4 (38).
Rxn. with: Mineral acids in aq. $HCCl_3$ $\rightarrow$ distribution of salts and anal. application (38).

Triphenyllead Hydroxide (M-568)　　　　$PbC_{18}H_{16}O$　　　　Ph_3PbOH
Synth.: By rxn. of Ph_6Pb_2 with O_3 in $HCCl_3$, followed by H_2O, in 94% yield (2865-6).
By hydrolysis of $Ph_3PbN(1-C_{10}H_7)CO_2Me$ (951), in presence of air and alc. KOH of $(Ph_3Pb)_2PPh$ and $(Ph_3Pb)_3P$ (1849), of $(Ph_3Sn)_2PPbPh_3$ and $Ph_3SnPPhPbPh_3$ (1850).
By rxn. of Ph_3PbOMe with HCX_3; X = Cl, and Br in 68 and 22% yield, resp. (951).
By photolysis of Ph_3PbN_3 in THF as by-prod. (2471).
By dec. of $Ph_3PbS_2CNPh_2$ at 150-200° and moisture (?), in 82% yield (298).
Prop.: m. 138-44° (2865-6), IR spectrum and assignment (342).
Rxn. with: $HCCl_3$ under reflux $\rightarrow (Ph_3P)_2O$ (38).
$Ph_3MOH \rightarrow Ph_3PbOMPh_3$; M = Si, Ge (2344).
$HX \rightarrow Ph_3PbX$; X = N_3 (1421), $(NC)_2C\colon C\colon N$ (717), $H_2N(NCNH)C\colon N$ (2867), several purine deriv. (1375).
Aq. $HX + HCCl_3 \rightarrow$ distribution of salts of several mineral acids between aq. and organic phase (38), rxn. followed by tracer method (487).
$HR \rightarrow Ph_3PbR$; R = $PhC\vdots C$ (2495), ONC (28).
$ROH \rightarrow Ph_3PbOR$; R = $2,6-Cl_2C_6H_3$ (2515), 2, or 4, $3,5-Cl_2C_5H_2N$ (2516), $C_{10}H_6NO$, $C_9H_5N(NO)$ and numerous substituted nitrosophenyl deriv. (2641).
$CH_2\colon CO \rightarrow (Ph_3PbCH_2CO)_2O + Ac_2O$ (608).
$RCO_2H \rightarrow Ph_3PbO_2CR$; R = C_5H_{11} (791), $HC\vdots C$, $PhC\vdots C$ (952).
$(\vdots CCO_2H)_2 \rightarrow (Ph_3PbO_2CC\vdots)_2$ (952).
$(\colon CHCO)_2O \rightarrow Ph_2PbO_2CCH\colon CHCO_2$ (1779).
Maleic anhydride-styrene-polymer $\rightarrow$ Pb-contg. polyester (1328).
$(CH_2CMeCO_2H)_x \rightarrow (Ph_3PbO_2CCMeCH_2)_x$ (285).
$C_9H_{19}CO_3H \rightarrow$ kinetic and thermodynamic data on dec. of intermediate peroxy-

caprate, solvent effect and rxn. mechanism (2777).
4-HSC$_5$H$_4$N → 4-Ph$_3$PbSC$_5$H$_4$N (2516).
6-HS-purine → 6-(N-Ph$_3$PbC$_5$H$_2$N$_4$)SPbPh$_3$·C$_5$H$_3$N$_4$PbPh$_3$ (1375).
PhCH$_2$NH$_2$ + CS$_2$ → Ph$_3$PbS$_2$CNHCH$_2$Ph (298).
PhCH$_2$NH$_3$S$_2$CNHCH$_2$Ph → (Ph$_3$Pb)$_2$S + (PhCH$_2$NH)$_2$CS (1377).
<u>Biol. rop.:</u> Activity as insecticide (1328).

Diphenyllead Oxide (M-569)   (PbC$_{12}$H$_{10}$O)$_x$ Ph$_2$PbO
<u>Synth.:</u> By thermal dec. of Ph$_2$Pb(S$_2$CNPh$_2$)$_2$ at 200° in air, in 70% yield (298).
<u>Rxn. with:</u> PhCH$_2$NH$_2$ + CS$_2$ → Ph$_2$Pb(S$_2$CNHCH$_2$Ph)$_2$ (298).
PhCH$_2$NH$_3$S$_2$CNHCH$_2$Ph → (Ph$_2$PbS)$_3$ + (PhCH$_2$NH)$_2$CS (1377).
HN$_3$ in HCCl$_3$ → Ph$_2$Pb(N$_3$)$_2$ (1421).

Listing of organolead oxides and hydroxides continues in Table 285.

Table 285. Organolead Oxides and Hydroxides

Formula*	Synth. Method**	Properties	Ref.
Me$_2$PbO (568)	III (a)	IR spectrum (a)	1056
Me$_2$Pb(OH)$_2$	IIA	(a)	1056
Et$_2$PbO (568)	IA	--	663
Et$_2$Pb(OH)$_2$ (567)	IB	(Far) IR spectrum and assignment (2499) (c)	342
Pr$_3$PbOH (568)	--	(d)	--
(c.C$_6$H$_{11}$)$_3$PbOH (569)	--	(e)	--
PhPbO$_2$H (569)	--	(f)	--
(p-MeC$_6$H$_4$)$_3$PbOH	--	(g)	--
p-MeC$_6$H$_4$PbO$_2$H (569)	--	(h)	--
p-IC$_6$H$_4$PbO$_2$H	IIB	--	319
2-C$_{10}$H$_7$PbO$_2$H (569)	--	(h)	--
HO(Ph$_2$PbCH$_2$)$_x$PbPh$_2$OH	(i)	Infusible solid	1352

* Numbers in parenthesis refer to pages in main volume.
** The characters used here correspond to the synthetic scheme in the introduction to Chapter 4.3.1.
(a) Dehydration of Me$_2$Pb(OH)$_2$ on storage over P$_2$O$_5$ → Me$_2$PbO (1056). (b) Rxn. with HClO$_4$ → Me$_2$Pb(ClO$_4$)$_2$ (1056). (c) Rxn. with aq. H$_3$PO$_4$ → Et$_2$PbHPO$_4$ (1214). (d) Rxn. with NaCNO with dil. H$_2$SO$_4$ → Pr$_3$PbCNO (28). (e) Rxn. with NaCo(CO)$_4$ → (c.C$_6$H$_{11}$)$_3$PbCo(CO)$_4$ (216). (f) Rxn. with alc. HCl → PbCl$_2$ (1424), with RCO$_2$H → PhPb(O$_2$CR)$_3$ + H$_2$O; R = ClCH$_2$, MeCHCl, MeCHBr, Cl(CH$_2$)$_3$, Br(CH$_2$)$_{10}$ (1661a). (g) Rxn. with CO$_2$ in C$_6$H$_6$ → [(p-MeC$_6$H$_4$)$_3$Pb]$_2$CO$_3$ (951), with C$_9$H$_{19}$CO$_3$H → kinetic and thermodynamic data on decomposition of intermediate peroxycaprate, solvent effect and rxn. mechanism (2777). (h) Rxn. with ClCH$_2$CH$_2$CO$_2$H → RPb(O$_2$CCH$_2$CH$_2$Cl)$_3$; R = p-MeC$_6$H$_4$, 2-C$_{10}$H$_7$ (1661a). (i) Synth. from Ph$_2$PbCl$_2$ with CH$_2$Br$_2$ and Mg in persence of cat. I and MeBr, followed by hydrolysis (1352).

4.3.2 ORGANOLEAD PEROXIDES

Triethyllead Ethylperoxide $PbC_8H_{20}O_2$ Et_3PbO_2Et
<u>Synth.</u>: By photolytic oxidation of Et_4Pb, in 19-24% yield (662).
<u>Prop.</u>: Instable, mechanism of oxidation and decomposition (662).
<u>Rxn. with</u>: $Et_4Pb \rightarrow Et_3PbOEt$ (662).

Triethyllead t-Butylperoxide (M-570) $PbC_{10}H_{24}O_2$ $Et_3PbO_2CMe_3$
<u>Prop.</u>: Effect on initial oxidation rate of Et_4Pb (1017).

Triphenyllead t-Butylperoxide (M-570) $PbC_{22}H_{24}O_2$ $Ph_3PbO_2CMe_3$
<u>Use</u>: For bonding atactic polypropylene to metal surfaces (2708).

4.3.3 ORGANOLEAD ALKOXIDES

The organolead alkoxides summarized in Tables 286 to 288 are prepared by methods from the following scheme.

  I Dealkylation of tetraorganolead compounds:
      $R_4Pb + R'OH \rightarrow R_2Pb(OR')_2$.

  II Metathetic substitution reactions with organolead halides:
      (IIA) $R_3PbX + NaOR' \rightarrow R_3PbOR'$,
      (IIB) $R_2PbX_2 + R'OH + NH_3 \rightarrow R_2Pb(OR')_2$.

  III Exchange reactions with organolead hydroxides and alkoxides:
      (IIIA) $R_3PbOH + R'OH \rightarrow R_3PbOR'$,
      (IIIB) $R_3PbOMe + R'OH \rightarrow R_3PbOR'$.

  IV Exchange reactions with organolead carboxylates:
      (IVA) $PhPb(OAc)_3 + R'OH \rightarrow PhPb(OR')_2OAc$,
      (IVB) $Ph_2Pb(O_2CR)_2 + R'OH \rightarrow Ph_2Pb(O_2CR)OR'$.

  V Insertion reaction with organolead amide:
      $Bu_3PbNEt_2 + RCHO \rightarrow Bu_3PbOCHRNEt_2$.

R = organic group, R' = organic group, may be functionally substituted, X = halogen.

4.3.3.1 UNSUBSTITUTED ORGANOLEAD ALKOXIDES

Trimethyllead Methoxide $PbC_4H_{12}O$ Me_3PbOMe
<u>Synth.</u>: By rxn. of Me_3PbCl with NaOMe in THF (1678), in Et_2O or MeOH in N atm., in 70% yield (673).
By rxn. of Me_3PbBr with NaOMe, in 67% yield (951).
By rxn. of $Me_3PbC(CO_2Me):C(OMe)CO_2Me$ with MeOH in CCl_4 at r.t. (2346).
<u>Prop.</u>: m. 250-60° (dec.) (673), 250° (dec.) (951), 90-92° (dec.) (1678), subl.$_{0.05}$ 70-100° (951), subl.$_{0.01}$ 50-80° (673), IR (951), (far) IR, assignment (673) and NMR (951) spectra, equivalent weight, easily hydrolyzed (951).
<u>Rxn. with</u>: B_2H_6 at -78° $\rightleftharpoons$ Me_3PbBH_4 + $(MeO)_2BH$ (674).

ROH → Me₃PbOR + MeOH; R = Et, Pr (951).
Cold H₂O → aq. soln. pH ~ 9 (951).
o(C₆H₄CO)₂O → o-Me₃PbO₂CC₆H₄CO₂Me (951).
PhNHCONPhCOC⋮CPh → PhCH:CNPhCONPhCO (952).
C₂H₂ → (Me₃PbC⋮)₂ + MeOH (952).
(MeO₂CC⋮)₂ → Me₃PbC(CO₂Me):C(OMe)CO₂Me (2346).
(MeO₂CC⋮)₂ + MeOH → MeO₂CCH:C(OMe)CO₂Me; cis-:trans-ratio, cat. effect of Me₃PbOMe (2346).

Triethyllead Methoxide (M-571)   PbC₇H₁₈O   Et₃PbOMe
<u>Synth.</u>: By rxn. of NaOMe in N atm. in Et₂O and THF with Et₃PbCl, in 59 and 93% yield, resp. (673), in Et₂O with Et₃PbBr, in 71% yield (951).
<u>Prop.</u>: m. 81-83° (dec.) (673), 73-75° (951), subl.₀.₀₁ 50-80° (673), 60-80° (951), IR (951), (far) IR, assignment (673) and NMR (951) spectra, mol. wt. (673), equivalent wt. (951), soly., easily hydrolyzed (951).
<u>Rxn. with</u>: B₂H₆ at -78° → Et₃PbBH₄ + (MeO)₂BH (674).
ROH → Et₃PbOR + MeOH; R = Et, n-, i-Pr, n-, i-, t-Bu (951).
RH → Et₃PbR + MeOH; R = CX₃ (?), c.C₅H₅ (951).
CO₂ in Et₂O → Et₃PbCO₃Me (951).
o-C₆H₄(CO)₂O → o-Et₃PbO₂CC₆H₄CO₂Me (951).
CH₂:CHCN → Et₃PbCH(CN)CH₂OMe (?) (951).
CH₂:CHCN, + MeOH → MeCH(OMe)CN (951).
CH₂:CHCN, + MeOH → MeOCH₂CH₂CN; cat. effect of Et₃PbOMe (2346).
CH₂:CHCN + CH₂(CO₂Et)₂ → NCCH₂CH₂CH(CO₂Et)₂ + (NCCH₂CH₂)₂C(CO₂Et)₂* (2346).
cis-(EtO₂CCH:)₂ → trans-(EtO₂CCH:)₂ (2346).
MeO₂CCH:C(OMe)CO₂Me ⇌ Et₃PbC(CO₂Me):C(OMe)CO₂Me + MeOH (2346).
RC⋮CH → Et₃PbC⋮CR; R = Bu, Ph (952).
(MeO₂CC⋮)₂ → Et₃PbC(CO₂Me):C(OMe)CO₂Me (2346).
Cl₃CCN + MeOH → Cl₃CC(:NH)OMe* (2346).
RNCO → (RNCO)₃; R = 1-C₁₀H₇* (2346).
RNCO + CH₂Ac₂ → RNHCOCHAc₂; R = Ph, 1-C₁₀H₇* (2346).
RNCO + AcCH₂CO₂Et → RNHCOCHAcCO₂Et; R = Ph, 1-C₁₀H₇* (2346).
PhNCO + CH₂(CO₂Et)₂ → PhNHCOCH(CO₂Et)₂* (2346).

Triethyllead Ethoxide (M-571)   PbC₈H₂₀O   Et₃PbOEt
<u>Synth.</u>: By rxn. of Et₄Pb with O under UV irradiation, in 78% yield (662), with Et₃PbO₂Et (662), with O₃ at -78° (663).
By rxn. of Et₃PbOMe with EtOH, in 86% yield (951).
<u>Prop.</u>: m. 37-39°, b₀.₀₄ 51-53°, IR and NMR spectra, easily hydrolyzed, equivalent weight (951).
<u>Rxn. with</u>: Dec. at r.t. in CCl₄ in sealed tube → Et₄Pb (951).

---

* Catalytic effect of Et₃PbOMe on addition reaction.

Tributyllead Methoxide $PbC_{13}H_{30}O$ Bu_3PbOMe

<u>Synth.</u>: By rxn. of Bu_3PbCl with NaOMe in Et_2O in N atm., in 91% yield (673), in 79% yield (951).

<u>Prop.</u>: m. 55-56° (673), 46-48° (951), subl.$_{0.01}$ 70-100°, mol. wt. (673), IR and NMR spectra, easily hydrolyzed, soly. (951).

<u>Rxn. with</u>: B_2H_6 at -78° → Bu_3PbBH_4 + $(MeO)_2BH$ (674).
RNCO → $Bu_3PbNRCO_2Me$ (951) or $(RNCO)_3$* (2346).
CS_2 → Bu_4Pb (951).
MeOH + BuNCO → $BuNHCO_2Me$* (2346).
MeOH + $(RN:)_2C$ → $RNHC(:NR)OMe$ (2346).

Triphenyllead Methoxide (M-572) $PbC_{19}H_{18}O$ Ph_3PbOMe

<u>Synth.</u>: By rxn. of Ph_3PbCl with NaOMe in C_6H_6, in 95% yield (951).

<u>Prop.</u>: m. 90-91°, IR and NMR spectra, equivalent wt., hydrolyzes easily, soly. (951).

<u>Rxn. with</u>: HCX_3, + H_2O → Ph_3PbCCl_3 + Ph_3PbOH; X = Cl, Br (951).
Br_3CCHO → Ph_3PbCBr_3 + Ph_3PbBr (951).
$(CCl_3)_2CO$ → Ph_3PbCCl_3 + Ph_3PbCl (951).
Cl_3CCHO → $Ph_3PbOCH(OMe)CCl_3$ (951).
Cl_3CCN → $Ph_3PbN:C(OMe)CCl_3$ (951).
RNCO → $Ph_3PbNRCO_2Me$; R = 1-$C_{10}H_7$ (951).
PhCH:CHCONPhCONHPh → PhCH:CHCONHPh (952).
$(RN:)_2C$ in C_6H_6 → $Ph_3PbNRC(OMe):NR$; R = 1-$C_{10}H_7$ (951).
PhNCS → $Ph_3PbSC(OMe):NPh$ (951).
CS_2 → Ph_3PbS_2COMe (951).

Listing of unsubstituted organolead alkoxides concludes in Table 286.

4.3.3.4 SUBSTITUTED ORGANOLEAD ALKOXIDES

This chapter comprises organolead alkoxides derived from unsaturated and functionally substituted alcohols. All compounds are listed in Table 287. Compounds with halogen and a functionally substituted alcohol linked to lead are included in Chapter 4.1.2.

4.3.3.8 ORGANOLEAD MIXED OXIDES

Organolead compounds having a germanium or silicon atom linked to lead via oxygen may be found at the top of Table 288.

4.3.3.9 MULTIFUNCTIONAL ORGANOLEAD ALKOXIDES

Organolead compounds with two different alkoxide moieties or an alkoxide and carboxylate group proposedly linked to one lead atom are summarized in Table 288.

---

* Catalytic effect of Bu_3PbOMe on reaction.

Table 286. Unsubstituted Organolead Alkoxides $R_nPb(OR)_{4-n}$

Formula	Synth. Method*	Yield	Properties	Ref.
Me_3PbOR				
R = Et	IIIB	38	m. 113-15°, subl.$_{0.1}$ 60-80°, soly. (a, b)	951
Pr	IIIB	57	m. 129-31°, subl.$_3$ 70-100°, soly. (a, b)	951
Me_2CH	IIIA	--	Solid, dec. on heating, NMR spectrum, soly. (b)	951
Me_3C	IIIA	--	Solid, dec. on distn., NMR spectrum, soly. (b)	951
Me_3CCH_2	IIIA	--	Liq., dec. on distn., NMR spectrum, soly. (b)	951
Ph	IIIA	100	IR, UV and NMR spectra (c)	342
$Me_2Pb(OPh)_2$	I	--	---	342
Et_3PbOR				
R = PhCMeH	IIIA	62	b$_{0.05}$ > 110°, soly. (a, b, d)	951
Pr	IIIB	57	b$_{0.2}$ 74-76°, IR and NMR spectra, soly. (b)	951
Me_2CH	IIIB	73	b$_{0.05}$ 48-50°, IR and NMR spectra, soly. (b)	951
Bu	IIIB	80	b$_{0.05}$ 78-79°, IR and NMR spectra, soly. (b)	951
Me_2CHCH_2	IIIB	85	b$_{0.1}$ 56-58°, IR and NMR spectra, soly. (b)	951
Me_3C	IIIB	84	b$_{0.2}$ 58-60°, IR spectrum, soly. (b)	951
Me_3CCH_2	IIIA	59	b$_{0.1}$ 107-110°, IR and NMR spectra, soly. (b)	951
$c.C_6H_{11}$	IIIA	82	b$_{0.1}$ > 110°, NMR spectrum, soly. (b)	951
Ph	IIIA	100	IR (985), UV and NMR spectra (c)	342
$Et_2Pb(OPh)_2$	I	--	---	342
Pr_3PbOMe	IIA	81	m. 51-53°, subl.$_{0.01}$ 45-65° (e, f)	673
Ph_3PbOPh	IIA	66	m. 113°, stable in soln., monomeric	791

* The characters used here correspond to the synthetic scheme in the introduction to Chapter 4.3.3.
(a) IR and NMR spectra (951). (b) Hydrolyzes easily, equivalent wt. (951). (c) Synth. rxn. suitable as anal. detn. for R_3PbOH (342). (d) Rxn. with PhNCO, followed by H_2O in $CCl_4 \rightarrow PhNHCO_2CHMePh$ (951). (e) (Far) IR spectra and assignment (673). (f) Rxn. with B_2H_6 at -78° $\rightarrow Pr_3PbBH_4 + (MeO)_2BH$ (674).

Table 287. Substituted Organolead Alkoxides

Formula*	Synth. Method**	Properties	$R_nPbOR'_{4-n}$ Ref.
Me₃PbOCH₂C⋮CH	IIIA	n_D^{20} 1.5725, d_{20} 2.0230, IR spectrum	2773
Me₃PbOCHMeC⋮CH	IIIA	n_D^{20} 1.5575, d_{20} 1.8940, IR spectrum (a)	2773
Me₃PbOCMe₂C⋮CH	IIIA	n_D^{20} 1.5510, d_{20} 1.8279, IR spectrum	2773
Me₂Pb(OC₉H₆N-8)₂ (b)	IIB	Yellow cryst., dec. > 135°	232a
Et₂Pb(OC₉H₆N-8)₂ (b)	IIB	Greenish yellow, dec. 147°, NMR spectrum	1294
Bu₃PbOCHEtNEt₂	IIB	Yellow cryst., dec. > 120°, soly.	232a
Bu₃PbOCH(C₆H₄Me-p)NEt₂	V	(c)	1609
Bu₂Pb(OC₉H₆N-8)₂ (b)	V	(c)	1609
	IIA (d)	Yellow cryst., dec. > 120°, soly.	232a, 233

Ph₃PbOR'

R' =			
2,6-Cl₂C₆H₃	IIIA	m. 92-93°, ³⁵Cl NQR spectrum, struct.	2515
m-ONC₆H₄	IIIA	m. 50° (dec.), visible and IR spectra (e)	2641
p-ONC₆H₄	IIIA	m. 175° (dec.), visible and IR spectra (e, f)	2641
p-O₂NC₆H₄	IIIA	m. 162° (dec.), IR spectrum	2641
2,4-Cl(ON)C₆H₃	IIIA	m. 153° (dec.), visible spectrum (e)	2641
2,4-Br₂(ON)C₆H₂	IIIA	m. 132° (dec.), visible and IR spectra (e)	2641
2,4-MeO(ON)C₆H₃	IIIA	m. 150° (dec.), visible and IR spectra (e)	2641
2,6,4-(MeO)₂ONC₆H₂	IIIA	m. 200° (dec.), visible and IR spectra (e)	2641
2,4-Me(ON)C₆H₃	IIIA	m. 185° (dec.), visible and IR spectra (e)	2641
3,4-Me(ON)C₆H₃	IIIA	m. 187° (dec.), visible and IR spectra (e)	2641
3,5,4-Me₂(ON)C₆H₂	IIIA	m. 210° (dec.), visible and IR spectra (e)	2641
p,p'-ONC₆H₄C₆H₄	IIIA	m. 125° (dec.), visible and IR spectra (e)	2641
4-ONC₁₀H₆	IIIA	m. 176° (dec.), IR spectrum (e)	2641
3,5-Cl₂C₅H₂N-2 (g)	IIIA	m. 118-20°, IR and UV spectra, soly., struct.	2516
3,5-Cl₂C₅H₂N-4 (g)	IIIA	m. 241-42°, soly., struct.	2516
8-C₉H₆N (b) (572)	IIA	Yellow, m. 108-11°, IR (1768) and UV spectra	142
5,8-(ON)C₉H₅N (b)	IIIA	m. 220° (dec.), IR spectrum (e)	2641

Ph$_2$Pb(OC$_9$H$_6$N)$_2$ (b) (572)	IIA	Yellow, m. 174-77°, UV spectrum 142
	IIA (d)	Yellow powder, dec. 167° 232a, 233
PhPb(OC$_9$H$_6$N)$_3$ (b)	IIA	In 98% yield, dec. > 120° 1213

---

* Numbers in parenthesis refer to pages in main volume.
** The characters used here correspond to the synthetic scheme in the introduction to Chapter 4.3.3.
(a) In 90% yield, anal. detn. by acidimetric titration (2773). (b) Derivative of 8-hydroxyquinoline. (c) Rxn. with H$_2$O → RCHO; R = Et, p-MeC$_6$H$_4$ (1609). (d) Synth. in presence of anhydrous NaOAc (232a, 233). (e) Compound in tautomeric equilibrium nitrosophenol ⇌ quinoneoxime, soly. (2641). (f) Elemental analysis by simultaneous detn. in one sample (1088), x-ray struct. analysis (1870). (g) Derivative of 2- and 4-hydroxypyridine.

Table 288. Organolead Mixed Oxides and Multifunctional Alkoxides

Formula*	Synth. Method**	Yield	Properties	Ref.
$Me_3PbOSiMe_3$ (572)	--	--	NMR (474), (far) IR spectra and assignment	673
$Ph_3PbOSiPh_3$ (572)	IIIA	100	m. 124-25°, IR spectrum	2344
$Ph_3PbOGePh_3$	IIIA	--	m. 127.5-28.5°, IR spectrum	2344
$Me_2Pb(O_2CCF_3)OC_9H_6N$ (a)	IVB	--	Dec. > 190°	232a
$Et_2Pb(O_2CCMe_3)OC_9H_6N$ (a)	IVB	60	Yellow cryst., m. 183° (dec.), dimeric	1214
$Et_2Pb(O_2CCF_3)OC_9H_6N$ (a)	IVB	--	Mol. wt.	1214
$Bu_2Pb(OAc)OC_9H_6N$ (a)	IVB	--	Dec. > 180°	232a, 233
$Ph_2Pb(OAc)OC_9H_6N$ (a)	IVB	--	Dec. > 200°	233
$Ph_2Pb(O_2CEt)OC_9H_6N$ (a)	IVB	--	Yellow cryst., dec. > 212°, soly.	232a, 233
$Ph_2Pb(O_2CCMe_3)OC_9H_6N$ (a)	IVB	79	Yellow cryst., m. 232° (dec.), mol. wt.	1214
$Ph_2Pb(O_2CCF_3)OC_9H_6N$ (a)	IVB	--	Yellow cryst., dec. > 240°, soly.	232a
	IVB	--	--	1214
$Ph_2Pb(O_2CCH_2Cl)OC_9H_6N$ (a)	IVB	--	Yellow cryst., dec. > 204°, soly.	232a, 233
$PhPb(OC_9H_6N)_2OMe$ (a)	(b)	--	Yellow, m. 148°	1213
$PhPb(OC_9H_6N)_2OEt$ (a)	(b)	--	Yellow, m. 152.5°	1213
	IVB	--	--	1213
$PhPb(OC_9H_6N)_2OAc$	IVA	83	Dec. > 158° (b)	1213

* Numbers in parenthesis refer to pages in main volume.
** The characters used here correspond to the synthetic scheme in the introduction of Chapter 4.3.3.
(a) Derivative of 8-hydroxyquinoline. (b) Rxn. of $PhPb(OC_9H_6N)_2Ac$ with ROH → $PhPb(OC_9H_6N)_2OR$; R = Me, Et (1213).

4.3.4 ORGANOLEAD HALF ACETALS AND DERIVATIVES OF DIKETONES

Dimethyllead Diacetylacetonate (M-572) $PbC_{12}H_{20}O_4$

Synth.: By rxn. of Me_2PbCl_2 with CH_2Ac_2 and NaOMe (1294), and NH_3 (1649) in EtOAc (2055).
Prop.: Pale blue cryst., m. 163-63.5° (2055), far IR (1301, 1649) and NMR (1294, 1298) spectra, soly. (2055), struct. (1649).

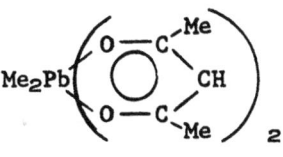

Dimethyllead Ditropolonate $PbC_{16}H_{16}O_4$

Synth.: By rxn. of Me_2PbCl_2 with tropolone in aq. NH_3 (1294).
Prop.: Pale yellow powder, dec. 191°, NMR spectrum (1294).

O-Tributyllead-keteneacetal $Bu_3PbOC(OR):C(CN)CH_2Ph$

Synth.: By-prod. in rxn. of Bu_3PbH with $PhCH:C(CN)CO_2R$ in C_6H_6 at 0° in absence of air, sole prod. by rxn. in Me_2CHCN (1607).
Prop.: IR spectrum (1607).

Triphenyllead Methyl Chloralacetal $PbC_{21}H_{19}Cl_3O_2$ $Ph_3PbOCH(OMe)CCl_3$

Synth.: By rxn. of Ph_3PbOMe with Cl_3CCHO in C_6H_6, in 100% yield (951).
Prop.: m. 96-100° (dec.), NMR spectrum (951).
Rxn. with: Thermal dec. → Ph_3PbCl (951).

Diphenyllead Dibenzoylacetonate $PbC_{32}H_{28}O_4$
Prop.: m. 160° (232a).

4.3.5 ORGANOLEAD CARBOXYLATES

4.3.5.1 TRIORGANOLEAD CARBOXYLATES

The organolead carboxylates compiled in Tables 290-293 are prepared by methods from the following scheme.

I Dealkylation of tetraorganolead compounds R_4Pb with:
  (IA) carboxylic acid,
  (IB) lead tetracarboxylate and,
  (IC) carboxylic acid from R_3PbO_2CR.

II Oxidation of hexaorganodilead compounds with ozone in presence of carboxylic acid.

III Reaction of triorganolead hydroxide R_3PbOH with:
  (IIIA) carboxylic acid,
  (IIIB) carboxylic acid anhydride or carbon dioxide,
  (IIIC) dimethyl carbonate.

IV Reaction of triorganolead methoxide R₃PbOMe with carbon dioxide or carboxylic acid anhydride.

V Reaction of triphenyllead lithium with β-propiolactone.

4.3.5.1.1 UNSUBSTITUTED TRIORGANOLEAD CARBOXYLATES

Triethyllead Acetate (M-574) PbC₈H₁₈O₂ Et₃PbOAc
Synth.: By rxn. of Et₄Pb with AcOH in MePh at 60-100° (1207).
Prop.: Spectrographic detn. (1358).
Rxn. with: Ph₃SnH → Ph₃SnOAc + Pb (912).
R₃SnH + HC⋮CCN → R₃SnOAc + Et₃PbCH:CHCN + Et₃PbC(CN):CH₂; R = Et, Ph (912).
AcOH at 60-100° in MePh → Et₂Pb(OAc)₂ + C₂H₆; kinetic data and rxn. mechanism (1207).
Biol. Prop.: Toxicity (843).
Activity as insecticide against Heliothis virescens, H. zea (2145), Musca domestica, LD₅₀ and ATP inhibition (425).
Use: In catalyst for polyurethan preparation (1460).

Tributyllead Acetate (M-575) PbC₁₄H₃₀O₂ Bu₃PbOAc
Synth.: By rxn. of Bu₆Pb₂ with O₃ in EtOH-AcOH, in 80-100% yield (2866).
Prop.: m. 85.5° (2866).
Rxn. with: R₃SnH + HC⋮CCN → R₃SnOAc + Bu₃PbCH:CHCN + Bu₃PbC(CN):CH₂; R = Et, Ph (912).
Et₃SnH + HC⋮CCO₂Me → Et₃SnOAc + Bu₃PbCH:CHCO₂Me (912).
Biol.Prop.: Toxicity (843).
Activity as insecticide against Heliothis virescens, H. zea (2145), Musca domestica, LD₅₀ and ATP inhibition (425).
Biological method for determination of 5 ppb in seawater (456).
Use: In antifouling paints (830), as antifouling agent and wood preservative (2973), as rot-proofing agent for fabrics (2359).

Triphenyllead Acetate (M-575) PbC₂₀H₁₈O₂ Ph₃PbOAc
Synth.: By rxn. of Ph₆Pb₂ with KMnO₄ in Me₂CO, followed by AcOH, in 76% yield (2865), with O₃ in AcOH-HCCl₃, in 97% yield (2865-6), with NaOEt in alc. C₆H₆, followed by aq. AcOH, in 78% yield (2865).
By-prod. in rxn. of Pb(OAc)₄ with PhSiF₃ (1554).
Prop.: m. 204-206° (2865-6), UV spectrum (967), polarography (963, 967).
Rxn. with: ⁶⁰Co irradiation → evidence for Ph₃Pb radical formation from ESR spectroscopy (693).
Polarography → Ph₂Hg (963).
  in presence of [C₅H₅Fe(CO)₂]₂ → Ph₃PbC₅H₅Fe(CO)₂ (967).
              Ph₆Sn₂ → Ph₄Pb (967).
              Ph₆Pb₂ → Ph₄Pb (967).
              [Mn(CO)₅]₂ → Ph₃PbMn(CO)₅ (967).
Pb(OAc)₄ (2:1) + cat. Hg(OAc)₂ → PhPb(OAc)₃ (2864).
CH₂:CO in EtOH → Ph₃PbCH₂CO₂Et + AcOH (608).

Biol. Prop.: Toxicity (1099), phytotoxicity to swamp rice (1202).
Activity as bactericide (1099).
Activity as fungicide against Ceratostomella coerulescens, Fusarium culmorum, Penicillium funiculosum, Speira heptaspora (1099).
Activity as insecticide against Heliothis virescens, H. zea (2145), Musca domestica, LD_{50} and ATP inhibition (425).
Activity as molluscicide against Biomphalaria glabrata (1202).
Use: In antifouling paints ( ), as rot-proofing agent for fabrics (2359).

Triphenyllead Laurate (M-575) $PbC_{30}H_{38}O_2$ $Ph_3PbO_2CC_{11}H_{23}$
Biol. Prop.: Toxicity (1099).
Activity as bactericide (1099).
Activity as fungicide against Ceratostomella coerulescens, Fusarium culmorum, Penicillium funiculosum, Speira heptaspora (1099).
Biological determination of 5 ppm in seawater (456).
Use: In antifouling paints (830), as rot-proofing agent for fabrics (2359).

Listing of unsubstituted triorganolead carboxylates continues in Table 289.

4.3.5.1.3 TRIORGANOLEAD CARBOXYLATES DERIVED FROM HETEROCYCLIC ACIDS

Triphenyllead orotate is listed at the bottom of Table 290.

4.3.5.1.4 UNSATURATED TRIORGANOLEAD CARBOXYLATES

All compounds are listed in Table 290.

4.3.5.1.6 FUNCTIONALLY SUBSTITUTED TRIORGANOLEAD CARBOXYLATES

All compounds are listed in Table 291.

Table 289. Unsubstituted Triorganolead Carboxylates

Formula*	Synth. Method**	Yield	Properties	R_3PbO_2CR Ref.
Me$_3$PbOAc (574)	IB	low	Toxicity (843)	2864
(Et$_3$Pb)$_2$CO$_3$ (574)	IIIC	51	m. >300°, IR spectrum, equiv. wt.	951
Et$_3$PbO$_2$CR				
R = MeO	IV	84	m. 60-62°, IR and NMR spectra, equiv. wt.	951
Et (574)	--	--	(a)	--
Me$_2$CH	--	--	(a)	--
C$_5$H$_{11}$ (574)	--	--	(a)	--
C$_7$H$_{15}$	--	--	(a)	--
C$_9$H$_{19}$	--	--	(a)	--
C$_{11}$H$_{23}$	--	--		--
Ph (574)	--	--	Spectrographic detn.	1358
Pr$_3$PbOAc (574)	II	85	m. 127-28°, toxicity (843)	2866
Bu$_3$PbO$_2$CC$_{11}$H$_{23}$	--	--	Biol. method for detn. of 5ppb in seawater (b)	456
Bu$_3$PbOBz	II	80-100	m. 76-77°	2866
(C$_7$H$_{15}$)$_3$PbOAc (575)	II	80-100	m. 25-30°	2866
(C$_{12}$H$_{25}$)$_3$PbOAc (575)	II	80-100	m. 64°	2866
Ph$_3$PbO$_2$CC$_5$H$_{11}$	IIIA	--	m. 116-18°	791
[(p-MeC$_6$H$_4$)$_3$Pb]$_2$CO$_3$	IIIB	--	m. 90-100°, NMR spectrum	951
R$_3$PbO$_2$CR'	--	--	Stabilizer for Cl-contg. polymers (c)	2973

* Numbers in parenthesis refer to pages in main volume.
** The characters used here correspond to the synthetic scheme in the introduction to Chapter 4.3.5.1.
(a) Use as catalyst for polyurethan synthesis (1460). (b) Use as fungicide in surface coatings (2973), in antifouling paints (830). (c) Use in catalyst for polyurethan synthesis (2561).

Table 290. Unsaturated Triorganolead Carboxylates R_3PbO_2CR'

Formula*	Synth. Method**	Yield	Properties	Ref.
$Me_3PbO_2CCMe:CH_2$	--	--	(Far) IR spectra and assignment (a)	2499
$Et_3PbO_2CCMe:CH_2$ (576)	--	--	(Far) IR spectra and assignment (a)	2499
$(CH_2:CH)_3PbOAc$ (574)	IB	--	IR spectrum	1196a
$(CH_2:CH)_3PbO_2CEt$	IB	--	IR spectrum	1196a
Ph_3PbO_2CR				
R = HC!C	IIIA	61	m. 189-90° (dec.), IR spectrum (b)	952
PhC!C	IIIA	81	m. 159-60° (dec.), IR spectrum, soly. (c)	952
$CH_2:CMe$ (576)	--	--	(Far) IR spectra and assignment (a, d)	2499
cis-$HO_2CCH:CH$ (576)	--	--	(e)	--
cis-$RO_2CCH:CH$	--	--	(e)	--
trans-$HO_2CCH:CH$	--	--	(e)	--
trans-$RO_2CCH:CH$	--	--	(e)	--
$(Ph_3PbO_2CC!)_2$	IIIA	62	m. 186-87°	952
$Ph_3PbO_2CC_4H_3H_2O_2$ (f) (576)	IIIA	--	(g)	--

* Numbers in parenthesis refer to pages in main volume.
** The characters used here correspond to the synthetic scheme in the introduction to Chapter 4.3.5.1.

(a) Thermo gravimetric analysis from r.t. to 800° (2499). (b) Refluxing in mesitylene → $(Ph_3PbC!)_2 + CO_2 + C_2H_2$ (952). (c) Refluxing in MePh → $Ph_3PbC!CPh + CO_2$ (952). (d) Rxn. with maleic anhydride-styrene-polymer → lead containing polymer (1328). (e) Biol. activity as microbiocide against Aspergillus niger, Bacillus subtilis, Candida albicans, Fusarium moriniformae, Rhizopus nigricans, Trichoderma viridae, Trichophyton mentagrophytes (2874). (f) Orotate. (g) Biol. properties resembling vitamin B_{13} (2500).

Table 291. Functionally Substituted Triorganolead Carboxylates R_3PbO_2CR'

Formula	Synth. Method*	Yield	Properties	Ref.
$Me_3PbO_2CCF_3$	IA	--	m. 153°, mol. wt., soly.	1214
$Et_3PbO_2CCF_3$	IA	--	m. 143°, mol. wt.	1214
$o\text{-}Et_3PbO_2CC_6H_4CO_2Me$	IV	84	m. 119-20°, IR spectrum, mol. wt.	951
$(Ph_3PbO_2CCMeCH_2)_x$	IIIA	--	Dec. 270°, thermal stability at 250°	285
$Ph_3PbO_2CCCl_3$	--	--	m. 174° (dec.)	2756
$(Ph_2PbCH_2CH_2CO_2)_3 \cdot H_2O$	V	76	m. > 240° (dec.), soly. (a)	607

* The characters used here correspond to the synthetic scheme in the introduction to Chapter 4.3.5.1.
(a) Dissolves in base, reprecipitated by acid (607).

4.3.5.2 DIORGANOLEAD DICARBOXYLATES

4.3.5.2.1 UNSUBSTITUTED DIORGANOLEAD DICARBOXYLATES

Dibutyllead Diacetate $\qquad$ $PbC_{12}H_{24}O_4$ $\qquad$ $Bu_2Pb(OAc)_2$
Prop.: TLC (951).
Rxn. with: 8-Hydroxyquinoline → $Bu_2Pb(OAc)OC_9H_6N$ (232a, 233).
Biol. Prop.: Toxicity (2421-2).
Activity as anthelmintic against Hymenolepsis fraterna (1134, 1138), H. cesticillus (2421), Raillietina cesticillus (1138).
Activity as insecticide against Heliothis virescens, H. zea (2145), Musca domestica, LD_{50}, ATP inhibition (425).
Use: Against coccidae parasites in poultry (2979).

Diphenyllead Diacetate (M-579) $\qquad$ $PbC_{16}H_{16}O_4$ $\qquad$ $Ph_2Pb(OAc)_2$
Synth.: By rxn. of Ph_4Pb with $Pb(OAc)_4$ (1:1) in $AcOH-Ac_2O$ and cat. $Hg(OAc)_2$ at 70°, in 93% yield (2864), with $PhPb(OAc)_3$ (1:2), in 100% yield (2864).
Prop.: m. 210° (2864), polarography (963).
Rxn. with: ^{60}Co irradiation → evidence for $MePh_2Pb$ radical formation by ESR spectroscopy (693).
$Pb(OAc)_4$ + cat. $Hg(OAc)_2$ → $PhPb(OAc)_3$ (2864).
8-Hydroxyquinoline → $Ph_2Pb(OAc)OC_9H_6N$ (233).
Addn. Compd.: $Ph_2Pb(OAc)_2 \cdot CsOAc$, from components in EtOH, m. 274° (dec.) (1218) $[PbCsC_{18}H_{19}O_6]$.
$Ph_2Pb(OAc)_2 \cdot Me_4NOAc$, from components in EtOH, m. 223° (dec.) (1218) $[PbC_{22}H_{31}NO_6]$.
$Ph_2Pb(OAc)_2 \cdot Py$, from components in EtOH, m. 185-93° (dec.) (1212) $[PbC_{21}H_{21}NO_4]$.
Biol. Prop.: Toxicity (1099, 2422).
Activity as bactericide (1099).
Activity as fungicide against Ceratostomella coerulescens, Fusarium culmorum, Penicillium funiculosum, Speira heptaspora (1099).

Diphenyllead Diisobutyrate (M-580) $\qquad$ $PbC_{20}H_{24}O_4$ $\qquad$ $Ph_2Pb(O_2CCHMe_2)_2$
Prop.: (Far) IR spectrum and assignment (41).
Rxn. with: PhLi → Ph_4Pb (1424).
PhMgBr → Ph_4Pb (1424).
Aq. Me_2CO → $Ph_2Pb(O_2CCHMe_2)OH$ (633).
Aq. Me_2CO + $RCHN_2$ → $(Ph_2PbO_2CCHMe_2)_2O$; R = H, Me Pr (633).

Additional unsubstituted diorganolead dicarboxylates are listed in Table 292.

Table 292. Unsubstituted Diorganolead Dicarboxylates

Formula*	Synth. Method**	Yield	Properties	Ref.
$Me_2Pb(OAc)_2$	IB	low	Toxicity (843)	2864
$Et_2Pb(O_2CR)_2$				
R = Me (580)	IA, C	--	Toxicity (2422) (a)	1207
Me_3C	IA	84	Dec. 149°, mol. wt. (b)	1214
C_7H_{15}	--	--	(c)	--
$C_{11}H_{23}$	--	--	(a, c)	--
$R_2Pb(OAc)_2$				
R = Pr (580)	--	--	Toxicity	2422
C_6H_{13}	--	--	Toxicity	2422
$c.C_6H_{11}$	--	--	(d)	--
C_7H_{15}	--	--	Toxicity	2422
C_8H_{17} (580)	--	--	Toxicity	2422
$Ph_2Pb(O_2CR)_2$				
R = Et (580)	--	--	(e)	--
Pr (580)	--	--	(Far) IR spectra, assignment (f)	41
Me_3C	IA	95	Dec. 210-31°, mol. wt. (b)	1214
Ph (580)	--	--	(Far) IR spectra, assignment (f, g)	41
$(2-C_{10}H_7)_2Pb(OAc)_2$ (581)	--	--	(Far) IR spectra, assignment (f)	41
$R_2Pb(O_2CR')_2$	--	--	R, R' = alkyl (h)	--
$Ar_2Pb(O_2CR)_2$	--	--	(Far) IR spectra, assignment	41

* Numbers in parenthesis refer to pages in main volume.

** The characters used here correspond to the synthetic scheme in the introduction to Chapter 4.3.5.1.

(a) Use as catalyst for polyurethan synthesis (1460). (b) Rxn. with 8-hydroxyquinoline → $R_2Pb(O_2CCMe_3)OC_9H_6N$; R = Et, Ph (1214). (c) Use as curing catalyst for silicone resins (636). (d) Anthelmintic activity against Hymenolepsis fraterna (1138). (e) Rxn. with 8-hydroxyquinoline → $Ph_2Pb(O_2CEt)OC_9H_6N$ (232a, 233). (f) Rxn. with $RCHN_2$ in aq. Me_2CO → $(Ar_2PhO_2CR)_2O$; Ar = Ph, $2-C_{10}H_7$, R = Pr, Ph (633). (g) Spectrographic determination (1358). (h) Use as catalyst for polyurethan synthesis (2561), as stabilizer for chlorine contg. resins (2973).

4.3.5.2.2 SUBSTITUTED DIORGANOLEAD DICARBOXYLATES

Substituted diorganolead dicarboxylates may contain olefinic unsaturated halogen and oxygen. Substitution occurs in the organic moiety linked to lead or in the carboxylate group. All compounds are listed in Table 293. Basic diorganolead carboxylates may be found in Table 295; derivatives containing carboxylic acid and alkoxide groups are included in Table 288.

Table 293. Substituted Diorganolead Dicarboxylates $\quad R_2Pb(O_2CR')_2$

Formula *	Synth. Method **	Properties	Ref.
$Me_2Pb(O_2CCF_3)_2$	IA	m. 180°	1214
$Me_2Pb(O_2CCF_3)_2 \cdot HOC_9H_6N$	--	--	1214
$Me_2PbO_2CCMe:CH_2$	--	(Far) IR spectra, assignment	2499
$Et_2Pb(O_2CCF_3)_2$	IA	m. 185-86° (a, b)	1214
$Et_2Pb(O_2CCMe:CH_2)_2$	--	(Far) IR spectra, assignment	2499
$Ph_2Pb(O_2CCF_3)_2$	IA	m. 307° (a, c)	1214
$Ph_2Pb(O_2CCH_2Cl)_2$ (582)	--	(c)	--
$Ph_2Pb(O_2CCCl_3)_2$	IA	In 90% yield, m. 213° (dec.)	1214
$Ph_2\overline{PbO_2CCH:CHCO_2}$	IIIB	In 70% yield, m. 300° (dec.)	1779
$(p\text{-}MeOC_6H_4)_2Pb(OAc)_2$	--	(Far) IR spectra, assignment (d)	41

* Numbers in parenthesis refer to pages in main volume.
** The characters used here correspond to the synthetic scheme in the introduction to Chapter 4.3.5.1.
(a) Rxn. with 8-hydroxyquinoline → $R_2Pb(O_2CCF_3)OC_9H_6N$; R = Et, Ph (1214).
(b) Detonates during combustion (1214). (c) Rxn. with 8-hydroxyquinoline → $Ph_2Pb(O_2CR)OC_9H_6N$; R = CF_3, $ClCH_2$ (232a, 233). (d) Rxn. with CH_2N_2 → $[(p\text{-}MeOC_6H_4)_2PbOAc]_2O$ (633).

4.3.5.3 MONOORGANOLEAD TRICARBOXYLATES

The organolead carboxylates summarized in Table 294 are prepared by the ensuing methods.

I Dealkylation of tetraorganolead compounds:
$R_4Pb + RCO_2H \rightarrow RPb(O_2CR)_3$.

II Esterification of organolead oxide:
$RPbO_2H + RCO_2H \rightarrow RPb(O_2CR)_3$.

III Transesterification of organolead carboxylate:
$RPb(OAc)_3 + RCO_2H \rightarrow RPb(OAc)_n(O_2CR)_{3-n}$.

IV Alkylation of lead tetracarboxylate:
(IVA) $Pb(O_2CR)_4 + R_2Hg \rightarrow RPb(O_2CR)_3$,
(IVB) $Pb(OAc)_4 + MeOPh \rightarrow p\text{-}MeOC_6H_4Pb(OAc)_3$.

Phenyllead Triacetate (M-583) $PbC_{12}H_{14}O_6$ $PhPb(OAc)_3$

Synth.: By rxn. of $Pb(OAc)_4$ with Ph_4Pb (3:1), Ph_3PbOAc (2:1) or $Ph_2Pb(OAc)_2$ (1:1) in $AcOH-Ac_2O$ and cat. $Hg(OAc)_3$ or Ph_2Hg at 70°, in 75-89% yield (2864).

Prop.: (Far) IR spectrum and assignment (41).

Rxn. with: ^{60}Co irradiation → evidence for $(AcO)_3Pb$ radical formation by ESR spectroscopy (693).

Ph_4Pb → $Ph_2Pb(OAc)_2$ (2864).

PhLi → Ph_4Pb (1424).

Me_4NI → Me_4NPbI_3 (1218).

$PhX + BF_3$ → o-, m- + p-PhC_6H_4X; X = Cl, MeO, NO_2, isomer distribution (696).

MeOPh at 80° → $MeOC_6H_4Ph$ (206a).

8-Hydroxyquinoline → $PhPb(OC_9H_6N)_2OAc + PhPb(OC_9H_6N)OEt$ in EtOH (1213).

8-Hydroxyquinoline + NH_3 → $PhPb(OC_9H_6N)_3$ (1213).

CO at 220 atm. and 100° in MeOH → $Pb(OAc)_2$ + BzOMe (2341).

RCO_2H in C_6H_6 at 100° → $PhPb(OAc)_2O_2CR + PhPb(O_2CR)_3$; R = MeCHBr (1661a).

Addn. Compd.: $PhPb(OAc)_3 \cdot H_2O$, m. 77-86° (2864) [$PbC_{12}H_{16}O_7$].

Use: As catalyst for polyurethan preparation (1661), relative reactivity as antioxidant (1661a).

Phenyllead Triisobutyrate (M-583) $PbC_{18}H_{26}O_6$ $PhPb(O_2CCHMe_2)_3$

Synth.: By rxn. of Ph_4Pb with $(i-PrCO_2)_4Pb$ in $i-PrCO_2H$ and cat. $Hg(O_2CPr-i)_2$ (2864).

Prop.: (Far) IR spectrum and assignment (41).

Rxn. with: PhMgBr → Ph_4Pb (1424).

$o-MeC_6H_4MgBr$ → $(o-MeC_6H_4)_6Pb_2$ (1424).

Alc. HCl → $PbCl_2$ + PhCl (1424).

Addn. Compd.: $PhPb(O_2CCHMe_2)_3 \cdot H_2O$, m. 75° (2864) [$PbC_{18}H_{28}O_7$].

Use: As catalyst for foaming polyurethan (1661).

Additional monoorganolead tricarboxylates are listed in Table 294.

Table 294. Monoorganolead Tricarboxylates $RPb(O_2CR)_3$

Formula*	Synth. Method**	Yield	Properties	Ref.
$PhPb(O_2CR)_3$				
R = $ClCH_2$	II	100	Light yellow oil, n_D^{20} 1.5958, IR spectrum (a)	1661a
MeCHCl	II	--	n_D^{20} 1.5825, IR spectrum (a)	1661a
MeCHBr	II	--	n_D^{20} 1.5815, IR spectrum	1661a
	III	100	n_D^{20} 1.5855 (a)	1661a
$Cl(CH_2)_4$	II	--	n_D^{20} 1.5638, IR spectrum (a)	1661a
BuEtCH	--	--	(b)	--
$Br(CH_2)_{10}$	II	--	n_D^{20} 1.5350, IR spectrum (a)	1661a
$C_{11}H_{23}$	--	--	(b, c)	--
2Me + MeCHBr	III	--	n_D^{20} 1.5878, IR spectrum (a)	1661a
$p\text{-}MeC_6H_4Pb(OAc)_3 \cdot H_2O$ (583)	I	--	m. 86-88° (b)	2864
$p\text{-}MeC_6H_4Pb(O_2CCH_2CH_2Cl)_3$	II	--	n_D^{20} 1.5734, IR spectrum (a)	1661a
$p\text{-}IC_6H_4Pb(O_2CR)_3$				
R = Me	IVA	--	m. 110-12°, (far) IR spectra, assignment (41) (d, e)	319
Et	II	50	m. 96°	319
	IVA	29	(Far) IR spectra, assignment (41)	319
Me_2CH	II	--	m. 78°	319
	IVA	--	(Far) IR spectra, assignment (41)	319
Bz	III (d)	26	m. 177-78°, (far) IR spectra, assignment (41)	319

* Numbers in parenthesis refer to pages in main volume.
** The characters used here correspond to the synthetic scheme in the introduction to Chapter 4.3.5.3.
(a) Use as stabilizer and catalyst for synthesis of polyurethan foam (1661a). (b) Potential catalyst for polyurethan foaming (413). (c) Use as catalyst for preparation of polyurethan foam (1661). (d) Rxn. of $p\text{-}IC_6H_4Pb(OAc)_3$ with BzOH → $p\text{-}IC_6H_4Pb(OBz)_3$ (319). (e) Rxn. with alc. NH_3 → $p\text{-}IC_6H_4PbO_2H$ (319).

Table 294 Continued

Formula*	Synth. Method**	Yield	Properties	RPb(O$_2$CR)$_3$ Ref.
p-MeOC$_6$H$_4$Pb(OAc)$_3$ (583)	IVB	24	(f)	206a
1-C$_{10}$H$_7$Pb(OAc)$_3$ (583)	--	--	(Far) IR spectra, assignment (g)	41
2-C$_{10}$H$_7$Pb(OAc)$_3$ (583)	--	--	(b, c)	--
2-C$_{10}$H$_7$Pb(O$_2$CCH$_2$CH$_2$Cl)$_3$	II	--	n_D^{20} 1.6068, IR spectrum (a)	1661a
ArPb(O$_2$CR)$_3$	--	--	(Far) IR spectra, assignment	41

(f) Rxn. with BF$_3$ in MePh → 2,2'- + 2,4'- + 4,4'-(MeOC$_6$H$_4$)$_2$ + p-MeOC$_6$H$_4$OAc (69%), thermal dec. at 80° → MeOPh + (p-MeOC$_6$H$_4$)$_2$ (206a). (g) Rxn. with alc. HCl → PbCl$_2$ (1424).

4.3.5.9 BASIC ORGANOLEAD CARBOXYLATES

Organolead carboxylates containing a lead oxide (Pb-O-Pb) and hydroxide (Pb-O-H) entity are compiled in Table 295. Organolead carboxylates supposedly having a carboxylate and alkoxide group linked to one lead atom are reported in Subchapter 4.3.3.9.

Table 295. Basic Organolead Carboxylates $(R_2PbO_2CR)_2O$

Formula*	Synth. Method	Remarks	Ref.
$(Ph_2PbOAc)_2O$ (569, 581)	---	(a)	--
$(Ph_2PbO_2CPr)_2O$	(b)	Dec. 225°	633
$(Ph_2PbO_2CPr-i)_2O$ (569)	(b)	m. 240°	633
$Ph_2Pb(OH)O_2CPr-i$	(c)	In 11% yield, dec. 200°	633
$(Ph_2PbOBz)_2O$	(b)	Dec. 260°	633
$[(p-MeOC_6H_4)_2PbOAc]_2O$	(b)	Solid	633
$[(2-C_{10}H_7)_2PbOAc]_2O$	(b)	Dec. 260°	633

* Numbers in parenthesis refer to pages in main volume.
(a) Rxn. with HCl-AcOH → Ph_2PbCl_2 (633). (b) Synth. from $R_2Pb(O_2CR)_2$ with $R'CHN_2$ in MeOH or Me_2CO, in 50-72% yield (633). (c) Synth. by hydrolysis of $Ph_2Pb(O_2CCHMe_2)_2$ in aq. Me_2CO (633).

4.4 ORGANOLEAD COMPOUNDS WITH SULFUR SELENIUM AND TELLURIUM

4.4.1 ORGANOLEAD SULFIDES, SELENIDES AND TELLURIDES

Bistrimethyllead Sulfide (M-584) $Pb_2C_6H_{18}S$ $(Me_3Pb)_2S$
Synth.: By rxn. of Me_3PbCl with aq. Na_2S, in 36% yield (18), with H_2S in aq. NaOH, in 66% yield (644).
Prop.: Brown cryst. (18), m. 95-100° (dec.) (644), IR (644) and NMR (646) spectra, relative basicity (644).
Use: As antiknock agent (18).

Bistriphenyllead Sulfide (M-585) $Pb_2C_{36}H_{30}S$ $(Ph_3Pb)_2S$
Synth.: By rxn. of Ph_3PbX with H_2S and organic base, in 100% yield (121). By rxn. of Ph_3PbCl with CS_2 and NH_3, in 90% yield (298), with Ph_3PbLi in THF, in 58% yield (485).
By rxn. of $(Ph_3Pb)_2O$ with $PhCH_2NH_3S_2CNHCH_2Ph$ in $HCCl_3$ at r.t., in 93% yield (1377).
By thermal dec. of $Ph_3PbS_2CNHCH_2Ph$ in C_6H_6 under reflux, in 64% yield (298).
Prop.: (Far) IR spectra and assignment (1848).
Biol. Prop.: Activity as fungicide against Aspergillus niger, Botrytis allii, Penicillium niger, Rhizopus nigricans (221).
Use: As rodent repellent (221).

Bistriphenyllead Selenide $Pb_2C_{36}H_{30}Se$ $(Ph_3Pb)_2Se$

<u>Synth.</u>: By rxn. of Ph_3PbCl with $Ph_3PbSeLi$ in THF in N atm., in 40% yield (485).

<u>Prop.</u>: m. 101° (485), (far) IR spectra and assignment (1848), monomeric (485), anal. detn. by x-ray fluorescence (1452).

<u>Rxn. with</u>: Thermal dec. at 118° in wet air → $(Ph_3Pb)_2O$ + Se (485).

Bistriphenyllead Telluride $Pb_2TeC_{36}H_{30}$ $(Ph_3Pb)_2Te$

<u>Synth.</u>: By rxn. of Ph_3PbCl with $Ph_3PbTeLi$ in THF in N atm., in 60% yield (485).

<u>Prop.</u>: Shiny yellow needles, m. 128-29° (485), (far) IR spectrum and assignment (1848), monomeric, stable in air at r.t. (485), anal. detn. by x-ray fluorescence (1452).

<u>Rxn. with</u>: Thermal dec. at 129° → Te (485).

Diphenyllead Sulfide (M-585) $Pb_3C_{36}H_{30}S_3$ $(Ph_2PbS)_3$

<u>Synth.</u>: By rxn. of Ph_2PbCl_2 with CS_2 and NH_3, in 60% yield (298).

By rxn. of Ph_2PbO with RNH_3S_2CNHR in $HCCl_3$ at r.t.; R = $PhCH_2$, in 36% yield (1377).

By thermal dec. of $Ph_2Pb(S_2CNHCH_2Ph)_2$ in Et_2O under reflux, in 60% yield (298).

4.4.2 ORGANOLEAD MERCAPTIDES, ORGANOSELENIDES AND ORGANOTELLURIDES

The organolead mercaptides and homologs summarized in Tables 296 and 297 are prepared by methods from the following scheme.

  I Substitution reactions with mercaptan:
     (IA) $R_3PbX + RSH$ + aq. NaOH or Na_2CO_3 → R_3PbSR,
     (IB) $R_nPbX_{4-n} + RSH$ + Py or R_3N → $R_nPb(SR)_{4-n}$.

  II Substitution reactions with lead mercaptide:
     (IIA) $R_3PbCl + Pb(SR)_2$ → R_3PbSR,
     (IIB) $R_2PbCl_2 + Pb(SR)_2$ → R_3PbSR.

  III Substitution reaction with organolead hydroxide:
     $R_3PbOH + RSH$ → R_3PbSR.

  IV Cleavage reaction with hexaorganodilead compounds:
     $R_6Pb_2 + R_2S_2$ → R_3PbSR.

R = organic group, may be functionally substituted, X = halogen, n = 2, 3.

4.4.2.1 UNSUBSTITUTED ORGANOLEAD MERCAPTIDES AND ORGANOTELLURIDES

Triphenyllead Benzylsulfide (M-585) $PbC_{25}H_{22}S$ Ph_3PbSCH_2Ph

<u>Synth.</u>: By rxn. of Ph_2PbCl_2 with $Pb(SCH_2Ph)_2$ in alc. C_6H_6 under reflux (121, 220-1).

By rxn. of Ph_6Pb_2 with O_3 in $HCCl_3$, followed by $PhCH_2SH$ (2866).

Prop.: m. 82-83° (1179, 1685, 2239).
Biol. Prop.: Antiinflammatory effect (220).
Activity as fungicide (220) against Aspergillus niger, Botrytis allii, Penicillium niger, Rhizopus nigricans (221).
Use: As antiwear agent in hydrocarbon lubricating oils (1685, 2239).

Triphenyllead Phenylsulfide (M-585) $PbC_{24}H_{20}S$ Ph_3PbSPh
Synth.: By rxn. of Ph_2PbCl_2 with $Pb(SPh)_2$ in alc. C_6H_6 under reflux (121).
By rxn. of Ph_6Pb_2 with O_3 in $HCCl_3$, followed by PhSH, in 66% yield (2865-6), with Ph_2S_2 in EtOH, in 76% yield (2865).
By-prod. in thermal dec. of $Ph_2Pb(SPh)_2$ in refluxing EtOH or C_6H_6 (121).
Prop.: m. 106-107° (1179), 105-106° (2865).
Biol. Prop.: Antiinflammatory effect (220).
Activity as fungicide (220).
Use: As antiwear additive in hydrocarbon lubrication oils (1685, 2239).
As rodent repellent (221, 1179).

Additional unsubstituted organolead mercaptides and organotellurides are listed in Table 296.

Table 296. Unsubstituted Organolead Mercaptides and Organotellurides $R_nPb(SR)_{4-n}$

Formula*	Synth. Method**	Yield	Properties	Ref.
Me_3PbSMe (584)	IA	24	$b_{0.01}$ 32°, n_D^{21} 1.6121 (a)	644
Me_3PbSEt	IA	53	$b_{0.05}$ 36°, n_D^{20} 1.5918 (a)	2
Ph_3PbSR				
R = Me (585)	IIB	--	(b-f)	121
Et (585)	IIB	--	(d, f)	121
Pr (585)	IIB	--	(b, d-f)	121
$CH_2{:}CHCH_2$	IIB	< 90	m. 45°	121
	IIB	--	--	121
Bu (585)	IIA	--	Liq., dec. on distn. (d)	1179
$C_{10}H_{21}$	IIA	--	Liq., dec. on distn. (d)	1179, 1685
$C_{10}H_7$	IIA	--	m. 73-75° (d)	1179, 1685
$Ph_2Pb(SPh)_2$	IB	--	Yellow cryst., m. 75° (g)	121
$Et_2Pb(TeMe)_2$	--	--	(h)	--

* Numbers in parenthesis refer to pages in main volume.
** The characters used here correspond to the synthetic scheme in the introduction to Chapter 4.4.2.
(a) Relative basicity from IR spectra in $DCCl_3$ (644), NMR spectrum (646).
(b) Activity as fungicide against Aspergillus niger, Botrytis allii, Penicillium niger, Rhizopus nigricans (221, 1179). (c) Use as antiwear lubrication additive (1179). (d) Use as antiwear additive in hydrocarbon lubrication additives (1685, 2239). (e) Use as rodent repellent (212, 1179). (f) Use as rotproofing agent for cotton fabrics (2358). (g) Thermal dec. in C_6H_6 → Ph_3PbSPh + Ph_2S_2 + $Pb(SPh)_2$ (121). (h) Use for metal plating on ceramics (603).

4.4.2.4 SUBSTITUTED ORGANOLEAD MERCAPTIDES

All compounds are compiled in Table 297.

Table 297. Substituted Organolead Mercaptides $R_nPb(SR')_{4-n}$

Formula*	Synth. Method**	Yield	Properties	Ref.
$Me_3PbSCH_2CO_2Me$ (584)	IA	--	Colorless liq., b_1 100° (dec.) (a)	19
Ph_3PbSR'				
$R' = HOCH_2CH_2$	IB	< 90	m. 78-79°	121, 220
MeO_2CCH_2	IB	< 90	m. 87-88°	121
	IIA	< 95	m. 85°	121, 1179
	IIB	--	m. 85° (b, c)	121, 1685
$i-C_9H_{19}O_2CCH_2$	--	--	Use as PVC stabilizer	2312
H_2NOCCH_2	IB	< 90	m. 124-25°	121
	IIA	< 95	--	121
	(c)	--	--	121
$H_2NCH_2CH_2$	IB	< 90	m. 40°	121, 220
C_6F_5	IA	93	Yellow, m. 93°, soly.	2669
	IB	< 90	m. 91°	121
	IIA	< 95	--	121
	IIB	--	IR and NMR spectra	2669
$p-ClC_6H_4$	IB	< 90	m. 102°	121
	IIA	< 95	--	121
	IV	80	m. 102.5°	2865
$p-H_2NC_6H_4$	IB	< 90	m. 135-36°	121
	IIA	< 95	--	121
$p-O_2NC_6H_4$	IB	< 90	m. 111°	121
	IIA	< 95	--	121
$17-\beta-C_{17}H_{23}O$ (d)	IB	< 90	m. 152° from C_6H_6	121, 220-1
	IIA	< 95	m. 125° from $HCCl_3$	121, 220-1
$4-C_5H_4N$	III	--	m. 134-36°, UV spectrum, struct., soly.	2516
C_7H_4NO (e)	IIA	--	m. 65° (b)	1179, 1685
C_7H_4NS (e)	IIA	51	m. 58°	121, 1179
	IB	< 90	m. 58°	121, 1179
	IIA	--	m. 58° (b)	1685
$6-C_5H_3N_4 \cdot L$ (f)	III	92	m. 197-98°, (far) IR and UV spectra	1375
$(Ph_3PbS)_2Y$				
$Y = CH_2CH_2$	IB	< 90	m. 141-42°	121
$3,5-C_2N_2S$ (g)	IB	< 90	m. 205° (dec.)	121

Ph$_2$Pb(SC$_6$H$_4$Cl-p)$_2$	IB	--	Yellow cryst., m. 71-72°	121,220
Ph$_2$Pb(SC$_7$H$_4$NS)$_2$ (e)	IB	--	m. 152-53°	121,1179
	IIB (h)	--	(b)	1685
	IB (i)	12	m. 148-50°	121
Ph$_2$$\overline{\text{PbSCH}_2\text{CH}_2\text{S}}$	IB	--	m. 132-33° (dec.), IR spectrum, monomeric	121

* Numbers in parenthesis refer to pages in main volume.
** The characters used here correspond to the synthetic scheme in the introduction to Chapter 4.4.2.
(a) Use as antiknock agent (19). (b) Use as antiwear additive for hydrocarbon lubricants (1685). (c) Rxn. of Ph$_3$PbSCH$_2$CO$_2$Me with NH$_3$ in anhyd. MeOH → Ph$_3$PbSCH$_2$CONH$_2$ (121). (d) Derivative of 17-β-mercaptotestosterone.
(e) Derivative of mercaptobenzoxazole and mercaptobenzothiazole, resp.
(f) L = 6-Ph$_3$PbS-9-Ph$_3$Pb-C$_5$H$_2$N$_4$, derivative of 6-mercaptopurine (1375).
(g) Derivative of 3,5-dimercapto-1,2,4-diazothiazole. (h) Synth. from Ph$_2$PbCl$_2$ (1685). (i) Synth. from Ph$_3$PbCl (121).

4.4.4 ORGANOLEAD DERIVATIVES OF ORGANIC THIOACIDS

A series of phenyllead derivatives of organic thioacids are listed in Table 298. The compounds comprise thiocarboxylates, xanthates, dithiocarbamates and isomeric thiocarbamate. The ensuing scheme summarizes synthetic routes.

I Reaction of organolead halide Ph$_n$PbCl$_{4-n}$ with:
  (IA) sodium and
  (IB) lead salts of the corresponding thioacid.

II Reaction of carbon disulfide and amine with:
  (IIA) triphenyllead hydroxide Ph$_3$PbOH,
  (IIB) diphenyllead oxide Ph$_2$PbO.

III Reaction of triphenyllead methoxide Ph$_3$PbOMe with carbon disulfide CS$_2$ or isophenylthiocyanate PhNCS.

IV Reaction of triphenyllead lithium sulfide Ph$_3$PbSLi with benzoyl chloride.

Organolead derivatives of thiocyanate and selenocyanate are listed in Chapter 4.2.1.

4.4.7 METAL SUBSTITUTED ORGANOLEAD SULFIDES, SELENIDES AND TELLURIDES

All compounds are listed in Table 299.

Table 298. Organolead Derivatives of Organic Thioacids $\qquad$ $R_nPb(SR')_{4-n}$

Formula*	Synth. Method**	Yield	Properties	Ref.
Ph_3PbSAc (585)	IB	--	m. 92°, antiandrogenic activity (220-1)	485
Ph_3PbSBz (585)	IB	--	m. 92-93° (a)	1179, 1685
	IB	--	m. 93-94°, use as repellent for mice	1179
	IB	--	m. 93-94°	1685
	IV	60	(a, b)	485
Ph_3PbS_2COMe	III	87	Colorless solid, NMR spectrum, mol. wt.	951
$Ph_3PbS_2CNHCH_2Ph$	IA	70	m. 107-109°, IR and UV spectra	298
	II	84	(c)	298
$Ph_3PbS_2CNMe_2$	IA	90	m. 163-65°, IR and UV spectra	298
$Ph_3PbSC(OMe):NPh$	III	72	Needles, m. 107-120° (dec.), mol. wt. (d)	951
$Ph_3PbS_2CNEt_2$	IA	90	m. 140-42°, IR and UV spectra	298
$Ph_3PbS_2CNPh_2$	IA	90	m. 163-65°, IR and UV spectra (e)	298
$Ph_2Pb(SAc)_2$ (585)	IB	--	m. 94-95°	1179
	IB	81	(a)	1685
$Ph_2Pb(S_2CNHCH_2Ph)_2$	IA	70	m. 102-103° (dec.)	298
	II	94	IR and UV spectra (f)	298
$Ph_2Pb(S_2CNMe_2)_2$	IA	88	m. 170° (dec.), IR and UV spectra	298
$Ph_2Pb(S_2CNEt_2)_2$	IA	91	m. 208-10° (dec.), IR and UV spectra	298
$Ph_2Pb(S_2CNPh_2)_3$	IA	63	m. 166-68° (dec.), IR and UV spectra (g)	298

* Numbers in parenthesis refer to pages in main volume.
** The characters used here correspond to the synthetic scheme in the introduction to Chapter 4.4.4.

(a) Use as antiwear additive for hydrocarbon lubricants (1685). (b) Activity as fungicide against Aspergillus niger, Botrytis allii, Penicillium niger, Rhizopus nigricans (221). (c) Thermal dec. in $C_6H_6 \rightarrow (Ph_3Pb)_2S$ (298). (d) IR and NMR spectra (951). (e) Thermal dec. at 150-200° $\rightarrow Ph_3PbOH + Ph_2NH$ (298). (f) Thermal dec. $\rightarrow (Ph_2PbS)_3$ (298). (g) Thermal dec. at 200° $\rightarrow (Ph_2PbO)_x + Ph_2NH$ (298).

Table 299. Metal Substituted Organolead Sulfides, Selenides and Tellurides   Ph₃PbYM

Formula*	Synth. Method	Yield	Properties	Ref.
Ph₃PbSLi	(a)	--	Orange soln., dec. on standing in N atm. (b, c)	485
Ph₃PbSGePh₃	(d)	45	Yellow needles, m. 128-29°, stable (e, f)	484
Ph₃PbSSnPh₃ (585)	(b)	47	m. 137° (e)	485
Ph₃PbSeLi	(a)	--	Dark green soln. dec. on standing in N atm. (b)	485
Ph₃PbSeGePh₃	(d)	45	Colorless cryst., m. 119°, monomeric (e, f)	484
Ph₃PbSeSnPh₃	(b)	35	Colorless cryst., m. 138° (e)	485
Ph₃PbTeLi	(a)	--	Black soln. dec. on standing in N atm. (b)	485
Ph₃PbTeGePh₃	(d)	42	Shiny yellow cryst., m. 115-17° (e, f)	484
Ph₃PbTeSnPh₃ (585)	(b)	--	(e)	485

* Numbers in parenthesis refer to pages in main volume.
(a) Synth. from Ph₃PbLi with S, Se and Te, resp. (485). (b) Rxn. of Ph₃PbYLi with Ph₃MCl → Ph₃PbYMPh₃ ; Y = S, Se, Te; M = Sn, Pb (485). (c) Rxn. with BzCl → Ph₃PbSBz (485). (d) Synth. from Ph₃PbCl with Ph₃GeYLi ; Y = S, Se and Te, resp. (484). (e) (Far) IR spectra and assignment (1848). (f) Rxn. with dil. acid → H₂S , Se and Te, resp. (484).

4.5 ORGANOLEAD DERIVATIVES OF NITROGEN, PHOSPHORUS, ARSENIC AND ANTIMONY

Most of the organolead derivatives of this chapter are very sensitive to air and moisture. Several derivatives synthesized by addition of organolead amides R_3PbNEt_2 to unsaturated entities are of uncertain structure and composition. N-Organolead derivatives of the "pseudohalogen-group" may be found in Chapter 4.2.1.

The organolead derivatives of Group V elements summarized in the ensuing tables are prepared by methods from the following scheme.

I Reactions with organolead halides:
  (IA) $R_nPbX_{4-n} + R'_2NLi \rightarrow R_nPb(NR'_2)_{4-n}$,
  (IB) $R_3PbX + R'_2NNa \rightarrow R_3PbNR'_2$,
  (IC) $R_3PbX + Ph_nY'H_{3-n} \rightarrow (R_3Pb)_{3-n}Y'Ph_n$; $Y' = P, As, Sb; n = 0, 1$.

II Reactions with organolead hydride and azide:
  (IIA) $R_3PbH + RCH:C(CN)_2 \rightarrow R_3PbN:C:C(CN)CH_2R$,
  (IIB) $R_3PbN_3 + (RC!)_2 \rightarrow R_3PbNN:NCR:CR$.

III Reactions with organolead oxide and hydroxide:
  $R_3PbOH + R'_2NH \rightarrow Ph_3PbNR'_2$.

IV Reactions with organolead alkoxides:
  (IVA) $R_3PbOMe + (RN:)_2C \rightarrow R_3PbNRC(:NR)OMe$,
  (IVB) $R_3PbOMe + RNCO \rightarrow R_3PbNRCO_2Me$,
  (IVC) $R_3PbOMe + RCN \rightarrow R_3PbN:CROMe$.

V Reactions with organolead amides:
  (VA) $R_3PbNEt_2 + (RN:)_2C \rightarrow R_3PbNRC(:NR)NEt_2$,
  (VB) $R_3PbNEt_2 + RNCY \rightarrow R_3PbNRC(Y)NEt_2$; $Y = 0, S$,
  (VC) $R_3PbNEt_2 + PhCN \rightarrow R_3PbN:CPhNEt_2$,
  (VD) $R_3PbNEt_2 + RCH:C(CN)_2 \rightarrow R_3PbN:C:C(CN)CHRNEt_2$.

VI Reactions with organolead phosphides:
  $(Ph_3Pb)_nPPh_{3-n} + PhN_3 \rightarrow (Ph_3PbNPh)_nPPh_{3-n}:NPh$; $n = 1, 2$.

R = organic group, R' = organic group, may be functionally substituted, R'_2NH = heterocyclic system, X = halogen.

4.5.1 UNSUBSTITUTED ORGANOLEAD AMIDES

Tributyllead Diethylamide $PbC_{16}H_{37}N$ Bu_3PbNEt_2
Synth.: By rxn. of Bu_3PbX with Et_2NLi in inert atm. (1608).
Prop.: Sensitive to air and moisture, dec. slowly at r.t. (1608).
Rxn. with: $Ph_3GeH \rightarrow Bu_3PbGePh_3 + Et_2NH$ (1608).
$PhN:C \rightarrow Bu_3PbC(:NPh)NEt_2$ (1609).
$RCHO \rightarrow Bu_3PbOCHRNEt_2$ (?); $R = Et, p-MeC_6H_4$ (1609).
$RNCO \rightarrow Bu_3PbNRCONEt_2$ (?) + $(RNCO)_3$; $R = c.C_6H_{11}$, Ph (1609).
$EtNCS \rightarrow Bu_3PbNEtC(:S)NEt_2$ (?) (1609).
$(RN:)_2C \rightarrow Bu_3PbNRC(:NR)NEt_2$; $R = c.C_6H_{11}$ (1609).
$PhCH:C(CN)_2 \rightarrow Bu_3PbN:C:C(CN)CHPhNEt_2$ (1609).

PhCN → Bu₃PbN:CPhNEt₂ (1609).

Listing of unsubstituted organolead amides continues in Table 300.

Table 300. Unsubstituted Organolead Amides and Phosphineimides $R_nPb(NR'_2)_{4-n}$

Formula	Synth. Method*	Yield	Properties	Ref.
Et₃PbNEt₂	IA	--	Sensitive to air and moisture (a, b)	1608
Pr₃PbNEt₂	IA	--	Sensitive to air and moisture (a)	1608
i-Bu₃PbNEt₂	IA	--	Sensitive to air and moisture (a)	1608
(c.C₆H₁₁)₃PbNEt₂	IA	--	Sensitive to air and moisture (a, b, c)	1608
Ph₃PbNEt₂	IA	--	Sensitive to air and moisture (a, b, c)	1608
Ph₂Pb(NEt₂)₂	IA	--	Sensitive to air and moisture (a, b, c)	1608
Me₃PbNHP(CMe₃)₂	IA	87	m. 33-35°, dec. 100°, very sensitive to air and moisture (e)	2739
Ph₃PbNPhPPh₂:NPh	VI	32	m. 173-78°, mol. wt., dec. slowly in N atm. (f)	1845
(Ph₃PbNPh)₂PPh:NPh	VI	18	m. 40° (dec.), dec. slowly in N atm. (f)	1845

* The characters used here correspond to the synthetic scheme in the introduction to Chapter 4.5.
(a) Dec. slowly at r.t. in inert atm. (1608). (b) Rxn. with R'NCO → R₃PbNR'-CONEt₂; or (R'NCO)₃; R = Et, c.C₆H₁₁, Ph; R' = c.C₆H₁₁, Ph (1609). (c) Rxn. with Ph₃GeH → R₃PbGePh₃ + Et₂NH; R = c.C₆H₁₁, Ph (1608). (d) Rxn. with Ph₂GeH₂ → (Ph₃Pb)₂GePh₂ + Pb + Et₂NH (1608). (e) IR and NMR spectra, mol. wt. (2739). (f) (Far) IR spectra and assignment, soly. (1845).

4.5.2 ORGANOLEAD COMPOUNDS WITH SUBSTITUTED AMINES

4.5.2.1 ORGANOLEAD DERIVATIVES OF HETEROCYCLIC AMINES

All compounds are listed in Table 301.

4.5.2.4 ORGANOLEAD DERIVATIVES OF AMIDINE, IMIDINE AND KETIMIDE

All compounds are summarized in Table 302.

4.5.3 ORGANOLEAD DERIVATIVES OF ACID AMIDES AND ACID IMIDES

All compounds are included in Table 303.

Table 301. Heterocyclic Triorganolead Amides $R_3PbNR'_2$

Formula	Synth. Method*	Yield	Properties	Ref.
$Et_3PbN_2C_3H_3$ (a)	--	--	(b)	912
$2\text{-}Et_3PbN_2C_3H_2R$ (a)	--	--	$R = C_{11}H_{23}$ (c)	2239
$Bu_3PbN_2C_3H_3$ (a)	--	--	(b-e)	912, 2239, 2357
$(C_7H_{15})_3PbN_2C_3H_3$ (a)	--	--	(c)	2239
$Ph_3PbN_2C_3H_3$ (a)	III (f)	--	m. 285° (dec.)	2866
$Ph_3PbN_3C_2(CO_2Me)_2$ (a)	IIB	--	White needles, m. 198-99.5°, IR spectrum	1127
$9\text{-}Ph_3Pb\text{-}6\text{-}X\text{-}C_5H_2N_4$ (a)				
X = H	III	80	m. 275-77°, stable in air and water (f)	1375
Cl	III	53	m. 230-32°, stable in air and water (f)	1375
MeS	III	50	m. 131-32°, stable in air and water (f)	1375

* The characters used here correspond to the synthetic scheme in the introduction to Chapter 4.5.
(a) $HN_2C_3H_3$ = imidazole, $HN_3C_2(CO_2Me)_2$ = 1,2,3-dimethyl triazoledicarboxylate, $9,6\text{-}HXC_5H_2N_4$ = 6-X-purine. (b) Rxn. with $R'_3SnH + HC\mathbin{!}CCN \rightarrow R_3PbCH\mathbin{:}CHCN + R_3PbC(CN)\mathbin{:}CH_2 + R'_3SnN_2C_3H_3$ (a); R = Et, Bu; R' = Et, Ph (912). (c) Use as antiwear additive in lubricating oils (2239). (d) Rxn. with $Ph_3SnH + HC\mathbin{!}CCO_2Me \rightarrow Bu_3PbCH\mathbin{:}CHCO_2Me + Ph_3SnN_2C_3H_3$ (a) (912). (e) Use as rot-proofing agent for textiles (2357). (f) UV, (far) IR spectra and assignment (1375).

Table 302. Organolead Derivatives of Amidine, Imidine and Ketimide

Formula	Synth. Method*	Properties	Ref.
$Et_3PbN:C(NH_2)NHCN$	III	White solid, dec. 260°, soly. (a)	2867
$Et_3PbN:C(NHCN)_2$	III	(a)	2867
$Bu_3PbNR'_2$			
$R'_2N = NCNHC(NH_2):N$	III	(a)	2867
$MeCH_2C(CN):C:N$	IIA	Light yellow oil, dec. < 100°, IR spectrum (b)	1607
$PhCH_2C(CN):C:N$	IIA	Light yellow oil, dec. < 100°, IR spectrum (b, c)	1607
$2\text{-}C_4H_3OCH_2C(CN):C:N$	IIA	Light yellow oil, dec. < 100°, IR spectrum (b)	1607
$Et_2NCHPhC(CN):C:N$	VD	IR spectrum, stable in ROH and H_2O	1609
$RN:C(NEt_2)NR$	VA	$R = c.C_6H_{11}$ (d)	1609
$Et_2NCPh:N$	VC	(e)	1609
$Ph_3PbN:CPh_2$	IA	Yellow cryst., UV spectrum	852
$Ph_3PbN:C(NH_2)NHCN$	III	(a)	2867

* The characters used here correspond to the synthetic scheme in the introduction to Chapter 4.5.
(a) Use as antiwear additive for lubricants (2867). (b) Sensitive to air (1607). (c) Rxn. with RBr → Bu_3PbBr + $PhCH_2C(CN)_2R$; R = $PhCH_2$, $CH_2:CHCH_2$, Br (1607), with $BzCl$ → Bu_3PbCl + $PhCH_2C(CN)_2Bz$ (1607), with ROH → $PhCH_2CH(CN)_2$; R = H, Et (1607). (d) Rxn. with H_2O → $c.C_6H_{11}N:C(NEt_2)NHc.C_6H_{11}$ (1609). (e) Rxn. with H_2O → $PhC(NEt_2):NH$ (1609).

Table 303. Triorganolead Derivatives of Acid Amides and Imides $R_3PbNR'_2$

Formula*	Synth. Method**	Yield	Properties	Ref.
o-Me$_3$PbN(OC)$_2$C$_6$H$_4$	--	--	Stabilizer for PVC	1125
Et$_3$PbNc.C$_6$H$_{11}$CONEt$_2$ (?)	VB	--	Spectra (a)	1609
Et$_3$PbNPhCONEt$_2$ (?)	VB	--	Spectra (a)	1609
o-Et$_3$PbN(OC)$_2$C$_6$H$_4$ (586)	--	--	m. 128°, mol. wt.	1214
Bu$_3$PbN(1-C$_{10}$H$_7$)CO$_2$Me	IVB	100	Colorless solid, m. 72-75°, dec. at r.t. (b, c)	951
Bu$_3$PbNRCONEt$_2$ (?)	VB	--	R = c.C$_6$H$_{11}$, IR spectrum (a)	1609
Bu$_3$PbNPhCONEt$_2$ (?)	VB	--	IR spectrum (a)	1609
Bu$_3$PbNEtCSNEt$_2$ (?)	VB	--	IR spectrum (d)	1609
(c.C$_6$H$_{11}$)$_3$PbNRCONEt$_2$ (?)	VB	--	R = c.C$_6$H$_{11}$, IR spectrum (a)	1609
(c.C$_6$H$_{11}$)$_3$PbNPhCONEt$_2$ (?)	VB	--	IR spectrum (a)	1609
Ph$_3$PbN(1-C$_{10}$H$_7$)CO$_2$Me	IVB	100	Colorless solid, m. 135-41° (dec.) (b, c)	951
Ph$_3$PbNRCONEt$_2$ (?)	VB	--	R = c.C$_6$H$_{11}$, spectra (a)	1609
Ph$_3$PbNPhCONEt$_2$ (?)	VB	--	Spectra (a)	1609
Ph$_3$PbNRC(OMe):NR	IVA	100	R = 1-C$_{10}$H$_7$, m. 120-40° (dec.), stable in air (b)	951
Ph$_3$PbN:C(OMe)CCl$_3$	IVC	100	m. 127-31°, IR and NMR spectra (e)	951

* Numbers in parenthesis refer to pages in main volume.
** The characters used here correspond to the synthetic scheme in the introduction to Chapter 4.5.
(a) Rxn. with H$_2$O → RNHCONEt$_2$; R = c.C$_6$H$_{11}$, Ph (1609). (b) IR and NMR spectra, mol. wt. (951). (c) Rxn. with H$_2$O → 1-C$_{10}$H$_7$NHCO$_2$Me (951). (d) Rxn. with H$_2$O → EtNHCSNEt$_2$ (1609). (e) Rxn. with H$_2$O → CCl$_3$C(OMe):NH (951).

4.5.4 ORGANOLEAD METAL IMIDES

The compounds may be found at the top of Table 304.

4.5.5 ORGANOLEAD PHOSPHORUS IMIDES

All compounds are listed at the bottom of Table 300.

4.5.6 ORGANOLEAD PHOSPHIDES ARSENIDES AND ANTIMONIDES

All compounds are reported at the bottom of Table 304.

5. ORGANO-POLYLEAD COMPOUNDS

5.1 HEXAORGANODILEAD COMPOUNDS

The hexeorganodilead compounds in Table 305 are prepared by methods from the following scheme.

I Grignard reaction with:
  (IA) $PbCl_2$,
  (IB) $PhPb(O_2CR)_3$.

II Reaction of triorganolead hydride (R_3PbH):
  (IIA) With PhNCO ,
  (IIB) Thermal decomposition also for R_3PbBH_4 .

Individual Hexaorganodilead Compounds may be found following Table 304.

Table 304. Organolead Metal Imides and Phosphorus, Arsenic and Antimony Derivatives R_3PbNM R_3PbYR_2

Formula*	Synth. Method**	Yield	Properties	Ref.
$Me_3PbNMeSiMe_3$	IA	52	b_3 42-43°, sensitive to air and moisture (a)	472, 1814
$Me_3PbNMeGeMe_3$ (587)	IA	48	b_2 49°, sensitive to light, NMR spectrum, soly.	1855
$Me_3PbNCMe_3SiMe_3$	IA	46	$b_{2.5}$ 63-65°, prepared in the dark, mol. wt. (b)	2749
$Me_3PbN:CPhNMeSiMe_3$	IA	53	Colorless oil, $b_{0.1}$ 121°, mol. wt. (c)	2738
$(Me_3Pb)_3P$	IC	40	m. 46-47°, dec. at 50°, in air, light, moisture (d-f)	1839
$(Me_3Pb)_3PNi(CO)_3$	(f)	100	Dec. 85°, stable in air (far) IR and NMR spectra	1851
$(Me_3Pb)_3As$	IC	40	m. 43-45°, dec. at 100°, in air, light, moisture (d, e)	1839
$(Ph_3Pb)_3P$	IC	65	Dec. at 100°, in light, air moisture (d)	1839
$(Ph_3Pb)_2PPh$	IC	--	Dec. 110°, (far) IR spectra, assigment (g)	1849
Ph_3PbPPh_2 (587)	IC	--	m. 110° (dec.), (far) IR spectra, assignment (h)	1849
$(Ph_3Pb)_3As$	--	--	(i)	--
$Ph_3PbAsPh_2$	IC	17	m. 158°, dec. at 160°, in air, light, moisture (d)	1839
$(Ph_3Pb)_3Sb$	IB	42	Dec. 115°, mol. wt. (j)	1847
$Ph_3PbSbPh_2$	IC	45	Dec. at 150°, in air, light, moisture	1839
	IB	11	Dec. 115°, mol. wt.	1847

* Numbers in parenthesis refer to pages in main volume.
** The characters used here correspond to the synthetic scheme in the introduction to Chapter 4.5.
(a) NMR spectrum (472). (b) (Far) IR, assignment and NMR spectra (2749). (c) IR and NMR spectra (2738). (d) (Far) IR spectra and assignment, struct. (1839). (e) NMR spectrum (1839). (f) Rxn. of $(M_3Pb)_3P$ with $Ni(CO)_4 \rightarrow (Me_3Pb)_3PN:(CO)_3$ (1851). (g) Rxn. with alc. KOH in air $\rightarrow Ph_3PbOH + H_3PO_4$ (1849). (h) Rxn. with PhN_3 (1:3) $\rightarrow (Ph_3PbNPh)_2PPh:NPh + N$ (1845), with alc. KOH in air $\rightarrow Ph_3PbOH + PhPO_3H_2$ (1849). (i) Rxn. with PhN_3 (1:2) $\rightarrow Ph_3PbNPhPPh_2:NPh + N$ (1845). (j) Rxn. with aq. alc. $H_2O_2 \rightarrow Ph_3PbO_2AsPh_2$ (1847).

Hexaneopentyldilead $Pb_2C_{30}H_{66}$ $(Me_3CCH_2)_6Pb_2$

Synth.: By rxn. of $PbCl_2$ with Me_3CCH_2MgCl at 0° as by-prod. (1935), in Et_2O at -10°, in 30% crude yield (2187).
By addn. of $Pb(OAc)_4$ to Me_3CCH_2MgCl in THF at 5°, in 50% yield (2136).
Prop.: Yellow solid (2136), m. 205-206° (dec.) (2187), 202-203° (dec.) (1935), IR (1935) and NMR (1935, 2136, 2187) spectra, soly. (1935, 2187).
Rxn. with: I, + Me_3CCH_2MgCl → $(Me_3CCH_2)_6Pb_2$ (2187).

Hexaphenyldilead (M-589) $Pb_2C_{36}H_{30}$ Ph_6Pb_2
Synth.: By rxn. of Ph_3PbLi with $SiCl_4$ in THF in N atm., in 37% yield (1928), with $ClCH_2\overline{CHCH_2O}$ in THF, in 79% yield (607).
By rxn. of $(Ph_3PbCH_2CO)_2O$ with $LiAlH_4$ in THF at r.t., in 50% yield (608).
By rxn. of $PbCl_2$ with $PhMgBr$ in Et_2O at 20-35°, in 72% yield (2685).
Prop.: m. 170° (dec.) (2865), UV spectrum (203, 967), electrochemical study, redox couple (2360), polarography (963, 965-7, 969), thermodynamic data (305).
Rxn. with: ^{60}Co irradiation → investigation by ESR spectroscopy (693).
γ-irradiation at -192° → evidence of radical formation by IR, visible and ESR spectra (2685).
Polarographic reduction followed with:
    $[C_5H_5Mo(CO)_3]_2$ and $Ph_3PbC_5H_5Mo(CO)_3$ → Ph_6Pb_2 (968),
    Ph_3PbOAc → Ph_4Pb (967),
    $C_5H_5Fe(CO)_2I$ → $Ph_3PbC_5H_5Fe(CO)_2$ (963, 967),
    $C_5H_5Mo(CO)_3C_5H_5Fe(CO)_2$ → $Ph_3PbC_5H_5Fe(CO)_2$ (968),
    Ph_3SnCl → unstable $[Ph_3PbSnPh_3]$ (963),
    $AgClO_4$ → unstable green $[Ph_3PbAg\ ?]$ (963).
PhLi in THF, + $PhCH_2Cl$ → Ph_4Pb + $PhCH_2PbPh_3$ (2865).
PhMgBr in THF, + $PhCH_2Cl$ → Ph_4Pb + $PhCH_2PbPh_3$ (2865).
Li in THF → Ph_3PbLi (2865).
KI_3 in aq. THF → Ph_3PbI + PbI_2 (2865).
NaOCl + HCl → Ph_3PbCl (2865).
NaOEt in alc. C_6H_6, + AcOH → Ph_3PbOAc + $Pb(OAc)_2$ + C_6H_6 (2865).
$KMnO_4$ in aq. Me_2CO, + HX → Ph_3PbX; X = Cl, Br, AcO (2865).
O_3 in $HCCl_3$ → Ph_3PbOH (2865-6).
O_3 in AcOH → Ph_3PbOAc (2865-6).
O_3, + HX → Ph_3PbX; X = Cl, PhS (2865-6), N-imidazolyl, $PhCH_2S$ (2866).
Ar_2S_2 → Ph_3PbSAr + Ph_4Pb; Ar = Ph, $p-ClC_6H_4$ (2865).
$Se(SeCN)_2$ → $Ph_3PbSeCN$ + Se + Pb (697).
Biol. Prop.: Activity as fungicide against Aspergillus niger, Botrytis allii, Penicillium niger, Rhizopus nigricans (221).
Biol. method of detg. 5 ppm in seawater (456).
Activity as rodent repellent (221).
Use: Effect as stabilizer for PVC (439).

Additional hexaorganodilead compounds are listed in Table 305.

Table 305.   Hexaorganodilead Compounds R_6Pb_2

R in R_6Pb_2*	Synth. Method**	Yield	Properties	Ref.
Me (587)	IA	--	Dec. at r.t. → Me_4Pb + Pb	1936
Et (588)	IIB	--	Use as additive for solid propellants (a)	1759
Pr (590)	IIB	--	(b)	674
Bu (590)	IIB	--	(b, c)	674
	IIA	--	---	397
$PhMe_2CCH_2$	IA	17	m. 132-33°, dec. in refluxing xylene, soly. (d, e)	2187
$c.C_6H_{11}$ (588)	--	--	(f)	--
C_7H_{15}	--	--	(b)	--
$C_{12}H_{25}$	--	--	(b)	--
$o-MeC_6H_4$ (590)	IB	79	m. 240-45°	1424
	IB (g)	48	---	1424
$p-MeC_6H_4$ (590)	--	--	NMR spectrum	1318

* Numbers in parenthesis refer to pages in main volume.
** The characters used here correspond to the synthetic scheme in the introduction to Chapter 5.1.
(a) Removal from Et_4Pb by O_3 treatment (889), effect on initial oxidation rate of Et_4Pb (1017), use in catalyst for C_3H_6-polymerization (2996).
(b) Rxn. with O_3 in $AcOH-Et_2O$ → R_3PbOAc; R = Pr, Bu, C_7H_{15}, $C_{12}H_{25}$ (2866).
(c) Rxn. with O_3 in $HCCl_3$, followed by BzOH → Bu_3PbOBz (2866). (d) NMR spectrum (2187). (e) Rxn. with I followed by $PhCMe_2CH_2MgCl$ → $(PhCMe_2CH_2)_4Pb$ (2187). (f) UV irradiation → $c.C_6H_{12}$ + $c.C_6H_{10}$ + $(c.C_6H_{11})_2$ + PhC_6H_{11} (1281). (g) Synth. from $PhPb(O_2CCHMe_2)_3$ (1424).

5.2 ORGANOLEAD COMPOUNDS WITH LESS THAN THREE ORGANIC GROUPS PER LEAD ATOM

Dicyclopentadienyllead (M-592) $PbC_{10}H_{10}$ $(C_5H_5)_2Pb$
Synth.: By rxn. of $PbCl_2$ with $(C_5H_5)_2Mg$ in xylene slurry at 160° in 85% yield (1743).
Of Pb-labeled prod. (1517).
Prop.: Yellow needles, brown-yellow prisms (1665), m. 138° (dec.), subl. 150-80° (1743), x-ray struct. analysis (758, 1665), exists in two modifications (758), struct. in vapor phase by electron scattering (671).
Use: In polymerization catalyst for olefins and dienes (2623).

Diphenyllead (M-592) $PbC_{12}H_{10}$ Ph_2Pb
Synth.: By rxn. of Ph_3PbLi, from Ph_3PbCl with Li in THF, with CO_2 (1928).
By thermal dec. of $Ph_3PbPt(PPh_3)_2Cl$ (700).
Prop.: m. 220° (turning black at 153°) (1928).

6 POLYMETALLIC ORGANOLEAD COMPOUNDS

6.1 ORGANOLEAD DERIVATIVES OF GROUP I-IV METALS

Organolead derivatives of germanium and tin may be found in Chapters 6.1.3 in sections GERMANIUM and TIN.

Trimethyllead Pentaborane(9) $PbB_5C_3H_{17}$ $Me_3PbB_5H_8$
Synth.: By rxn. of Me_3PbCl with LiB_5H_8 in Et_2O in inert atm. at -22° (2398).
Prop.: m. -5° (dec.), sublimes, ^{11}B NMR spectrum (2398).

Triethyllead Sodium (M-593) $PbNaC_6H_{15}$ Et_3PbNa
Use: As diene copolymerization catalyst (1639a).

Triphenyllead Lithium (M-591) $PbLiC_{18}H_{15}$ Ph_3PbLi
Synth.: By rxn. of Ph_6Pb_2 with PhLi in THF in inert atm. (2865).
Prop.: Dark green soln. (485).
Rxn. with: O-free H_2O, + AcOH → C_6H_6 + $Pb(OAc)_2$ + LiOAc; use for anal. detn. (2865).
$H_{4-n}CCl$ in THF → $(Ph_3Pb)_nCH_{4-n}$; n = 2, 3, 4 (606).
CH_2Br_2 → $(Ph_3Pb)_2CH_2$ (1127).
$(CH_2Br)_2$ → Ph_4Pb (1127, 1928).
$(CH_2)_nBr_2$ → $Ph_3Pb(CH_2)_nBr$; n = 3, 4, 5 (1127).
BuX → $BuPbPh_3$; X = Cl, Br (1928).
$SiCl_4$ → Ph_6Pb_2 (1928).
Y → Ph_3PbYLi; Y = S, Se, Te (485).
CO_2 → Ph_2Pb (1928).
$\overline{CH_2CH_2Y}$ → $Ph_3PbCH_2CH_2YH$; Y = O, S, NAc, NBz (607).
$ClCH_2\overline{CHCH_2O}$ (1:1) → $Ph_3PbCH_2CHOHCH_2Cl$ (607).
$ClCH_2\overline{CHCH_2O}$ (2:1) → Ph_6Pb_2 + Ph_4Pb (607).
$\overline{CH_2CH_2CH_2O}$ → $Ph_3Pb(CH_2)_3OH$ (607).
$\overline{OCH_2CH_2CO}$ → $(Ph_2PbCH_2CH_2CO_2)_3 \cdot H_2O$ (607).
$\overline{OCH_2CH_2CO}$, + CH_2N_2 → $Ph_3PbCH_2CH_2CO_2Me$ (607).

The following organolead derivatives of boron, aluminum and gallium are claimed as agents for metal plating (603). No other data are available for Me_3PbBEt_2, $Me_2EtPbBBu_2$, $MeEt_2PbBMe_2$, Et_3PbBEt_2, $MeEt_2PbAl(i-Bu)_2$, Et_3PbAlH_3, $Et_3PbAlMe_2$, $Et_3PbAlEt_2$, $(C_8H_{17})_3PbAl(i-C_5H_{11})_2$, $Me_3PbGaMe_2$, $Me_2EtPbGaMe_2$ and $Et_3PbGaPr_2$ (603).

6.5 POLYMETALLIC ORGANOLEAD DERIVATIVES OF TRANSITION METAL COORDINATION COMPOUNDS

6.5.1 BIMETALLIC ORGANOLEAD COMPOUNDS WITH GROUP I-VI METALS

The polymetallic organolead compounds of Tables 306 and 307 are prepared by the following methods.

I Reaction with organolead halide:
   $R_nPbX_{4-n} + MX' \rightarrow R_nPbX'_{4-n}$.

II Reaction with organolead metal derivative:
    $R_3PbM + X'Cl \rightarrow R_3PbX'$.

III Carbon monoxide exchange reaction:
     $PhPbX_2Mn(CO)_5 + L \rightarrow PhPbX_2Mn(CO)_4L$.

R = organic group, L = $(CH_2:CH)_3As$, Ph_3Sb, M = Li, Na, K, X = halogen, X' = $C_5H_5M'(CO)_3$, M' = Cr, Mo, W; $Mn(CO)_5$, $Re(CO)_5$, $Mn(CO)_4As(CH_2Ph)_3$, $Mn(CO)_4SbEt_3$, n = 2, 3.

Triphenyllead-π-cyclopentadienylmolybdenum-
   tricarbonyl (M-594) $PbMoC_{26}H_{20}O_3$ $Ph_3PbC_5H_5Mo(CO)_3$

<u>Synth.</u>: By rxn. of Ph_3PbX with $NaC_5H_5Mo(CO)_3$ in $(MeOCH_2)_2$ in N atm. (967), in THF, in 45% yield (1670).
<u>Prop.</u>: Pale yellow (1670) needles, dec. 201° (967), m. 200° (dec.) (1670), IR, UV and NMR spectra (1670), stable in air in solid state (1670), polarography (967-9), x-ray struct. anal. (1996).
<u>Rxn. with</u>: Polarographic reduction in presence of:
   $[C_5H_5Fe(CO)_2]_2 \rightarrow Ph_3PbC_5H_5Fe(CO)_2$ (968).

Listing of bimetallic organolead compounds with Group I-VI transition metals concludes in Table 306.

Table 306. Organolead Derivatives of Group I-VI Metals $R_nPbX'_{4-n}$

Formula*	Synth. Method**	Yield	Properties	Ref.
$Ph_3Pb(C_5H_5)_2TiCl$	II	--	Bright green soln., dec. < r.t.	907
$Me_2PbCr(CO)_5$	--	--	(a)	--
$Me_2Pb[C_5H_5Cr(CO)_3]_2$	--	--	(a)	--
$MeEtPbCr(CO)_5$	--	--	(a)	--
$Et_2PbCr(CO)_5$	--	--	(a)	--
$Pr_3Pb(C_5H_5)Cr(CO)_3$	--	--	(a)	--
$(C_7H_{15})_2PbCr(CO)_5$	--	--	(a)	--
$(C_{16}H_{33})_2PbCr(CO)_5$	--	--	(a)	--
$Ph_3PbC_5H_5Cr(CO)_3$	I	45	m. 195-97° (dec.), sensitive to air (b)	1670
$Me_3PbC_5H_5Mo(CO)_3$	I	6	Pale yellow, m. 93-95°, IR, UV and NMR spectra	1670
$Me_2Pb[C_5H_5Mo(CO)_3]_2$	--	--	(a)	--
$Pr_3Pb(C_5H_5)Mo(CO)_3$	--	--	(a)	--

Ph$_3$Pb(C$_5$H$_5$)W(CO)$_3$ (594)   I   65   Yellow, m. 214-15°, stable in air (c)   1670

* Numbers in parenthesis refer to pages in main volume.
** The characters used here correspond to the synthetic scheme in the introduction to Chapter 6.5.1.
(a) Use for metal plating on hot surfaces (603). (b) IR and NMR spectra, slow dec. in inert atm. (1670). (c) IR, UV and NMR spectra (1670), x-ray struct. anal. (1996).

6.5.5 POLYMETALLIC ORGANOLEAD COMPOUNDS WITH MANGANESE AND RHENIUM

Triphenyllead Manganesepentacarbonyl (M-594)   PbMnC$_{23}$H$_{15}$O$_5$ Ph$_3$PbMn(CO)$_5$
Synth.: By rxn. of Ph$_3$PbCl with NaMn(CO)$_5$ in THF in N atm. (967).
Prop.: Pale yellow (967) cryst., m. 146° (2057), 142-44° (1252), dec. 133° (967), IR (1252, 2057, 2337), assignment (770) and UV (967) spectra, dipole moment (2057), force const. (1132, 1252, 2337), polarography (967, 969).

Triphenyllead Rheniumpentacarbonyl (M-594)   PbReC$_{23}$H$_{15}$O$_5$ Ph$_3$PbRe(CO)$_5$
Synth.: By rxn. of Ph$_3$PbX with NaRe(CO)$_5$ in inert atm. (1252).
Prop.: Pale yellow, m. 133-35° (1252), IR spectrum (1252, 2337), force const. (1252, 2337), polarography (969), x-ray struct. anal. (1996).

Additional polymetallic organolead derivatives of manganese and rhenium are listed in Table 307.

Table 307. Organolead Derivatives of Group VII Metals   R$_n$PbX'$_{4-n}$

Formula*	Synth. Method**	Properties	Ref.
Me$_3$PbMn(CO)$_5$ (594)	--	NMR (1252), far IR spectra, assignment (a)	1271
Me$_2$Pb[Mn(CO)$_5$]$_2$ (594)	--	NMR spectrum (b)	1252
Me$_2$Pb[C$_4$H$_6$Mn(CO)$_3$]$_2$	--	C$_4$H$_6$ = butadiene (c)	--
Me$_2$Pb[C$_6$H$_6$Mn(CO)$_2$]$_2$	--	(c)	--
Et$_3$PbMn(CO)$_5$ (594)	--	Yellow liq., IR spectrum (d)	2057
Pr$_3$PbMn(CO)$_4$AsR$_3$	I	R = PhCH$_2$	185

* Numbers in parenthesis refer to pages in main volume.
** The characters used here correspond to the synthetic scheme in the introduction to Chapter 6.5.1.
(a) Use as additive for solid propellants (1759). (b) Use as metal plating agent on glass, cobalt (603). (c) Use as agent for metal plating on hot surfaces (603). (d) Polarography (969).

Table 307 Continued $R_nPbX'_{4-n}$

Formula*	Synth. Method**	Properties	Ref.
$Pr_3Pb(C_4H_6)Mn(CO)_2$	--	C_4H_6 = butadiene (c)	--
$Pr_3Pb(C_6H_6)Mn(CO)_3$	--	(c)	--
$Ph_3PbMn(CO)_4SbEt_3$	I	--	185
$Ph_2Pb[Mn(CO)_5]$	--	(e)	--
$PhPbCl_2Mn(CO)_5$	--	(f)	--
$PhPbCl_2Mn(CO)_4AsR_3$	III (f)	R = CH_2:CH (g)	186
$PhPbBr_2Mn(CO)_5$	--	(h)	--
$PhPbBr_2Mn(CO)_4SbPh_3$	III (h)	(g)	186

(e) Use as metal plating agent on glass (603). (f) Rxn. of $PhPbCl_2Mn(CO)_5$ with $(CH_2:CH)_3As$ at 200-208° in $PhCH_2OBu$ → $PhPbCl_2Mn(CO)_4As(CH:CH_2)_3$ (186). (g) Use as antiknock agent (186). (h) Rxn. of $PhPbBr_2Mn(CO)_5$ with Ph_3Sb in BuOPh under reflux → $PhPbBr_2Mn(CO)_4SbPh_3$ (186).

6.5.6 BIMETALLIC ORGANOLEAD COMPOUNDS WITH GROUP VIII METALS

The bimetallic organolead compounds with Group VIII metals listed in Table 308 are prepared by the following methods.

I Reactions with organolead halides:
  (IA) $R_3PbX + MX' → R_3PbX'$,
  (IB) $R_3PbX + CaX'_2 → R_3PbX' + R_2PbX'_2$,
  (IC) $R_3PbX + Co[Co(CO)_4]_2 → R_3PbCo(CO)_4$.

II Carbon monoxide exchange reaction:
  $Et_2PbFe(CO)_4 + PEt_3 → (Et_2Pb)_2Fe_2(CO)_7PEt_3$.

III Reaction with organolead nitrate:
  $Ph_3PbNO_3 + HX' → Ph_3PbX'$.

IV Reaction with organolead hydroxide:
  $R_3PbOH + NaX' → R_3PbX'$.

M = alkali metal, R = organic group, X = halogen, X' = $Fe(CO)_4$, $C_5H_5Fe(CO)_2$, $Ru(CO)_4$, $Co(CO)_4$.

A phosphinenickeltricarbonyl containing tris(trimethyllead)phosphine may be found in Table 304.

With the exception of one individual compound which appears on page 987, listing of bimetallic organolead compounds with Group VIII metals continues in Table 308.

Table 308. Bimetallic Organolead Derivatives of Group VIII Metals $R_nPbX'_{4-n}$

Formula*	Synth. Method**	Yield	Properties	Ref.
[Me$_2$PbFe(CO)$_4$]$_2$ (595)	--	--	IR spectrum (904) and force const.	2339
Me$_2$Pb[C$_5$H$_5$Fe(CO)$_2$]$_2$ (595)	--	--	X-ray struct. anal. (a)	2250
Me$_4$Pb$_3$Fe$_4$(CO)$_{18}$	--	--	IR spectrum (904) and force const.	2339
[MeEtPbFe(CO)$_4$]$_2$	--	--	(a)	--
(Et$_3$Pb)$_2$Fe(CO)$_4$	IA	--	IR (734, 1268, 1272) and Raman (734) spectra (b)	1268
Et$_3$Pb(C$_5$H$_5$)Fe(CO)$_2$	IA	--	Red liq., IR and NMR spectra (c)	2057
[Et$_2$PbFe(CO)$_4$]$_2$ (595)	--	--	IR spectrum (1272) and force const. (a, d-f)	2339
(Et$_2$Pb)$_2$Fe$_2$(CO)$_7$PEt$_3$	II (e)	--	IR spectrum, assignment, struct.	1269
Pr$_3$Pb(C$_5$H$_5$)Fe(CO)$_2$	--	--	(a)	--
[Bu$_2$PbFe(CO)$_4$]$_2$ (595)	--	--	(a)	--
(Ph$_3$Pb)$_2$Fe(CO)$_4$ (595)	IB	42	Dec. in air, magnetic moment	216
[Ph$_2$PbFe(CO)$_4$]$_2$ (595)	IB	11	Dec. in air, magnetic moment, soly.	216
(Me$_3$Pb)$_2$Ru(CO)$_4$	IA	Low	Pale yellow, subl., IR spectrum	905
Me$_3$PbCo(CO)$_4$	--	--	Far IR spectrum, assignment	1271
Me$_2$Pb[C$_4$H$_6$Co(CO)$_2$]$_2$	--	--	C$_4$H$_6$ = butadiene (a)	--
Me$_2$Pb[PhCo(CO)$_2$]$_2$	--	--	(a)	--
Et$_3$PbCo(CO)$_4$	--	--	IR (2338), Raman spectra and force const. (g)	734
Pr$_3$Pb(C$_4$H$_6$)Co(CO)$_2$	--	--	C$_4$H$_6$ = butadiene (a)	--

* Numbers in parenthesis refer to pages in main volume.
** The characters used here correspond to the synthetic scheme in the introduction to Chapter 6.5.6.
(a) Use for metal plating on hot surfaces (603). (b) Dipole moment (1272), force const. (734), struct. (1268, 1272). (c) Purified by distn. or column chromatography (2057). (d) (Far) IR spectrum and assignment (1269), dipole moment, struct. (1272). (e) Rxn. of [Et$_2$PbFe(CO)$_4$]$_2$ with PEt$_3$ → (Et$_2$Pb)$_2$Fe$_2$(CO)$_7$PEt$_3$ (1269). (f) Use for metal plating on mild steel (603). (g) Far (1270) IR spectra and assignment (1270, 1272), dipole moment (1270, 1272, 2057).

Table 308 Continued $R_nPbX'_{4-n}$

Formula*	Synth. Method**	Yield	Properties	Ref.
$Pr_3PbPhCo(CO)_2$	--	--	(a)	--
$(c\text{-}C_6H_{11})_3PbCo(CO)_4$	IV	76	Orange red, dec. in air, magnetic moment	216
$Ph_3PbCo(CO)_4$ (595)	IC	--	Yellow needles, m. 100-102° (h)	1672
$Ph_3PbCo(D_2H_2)\cdot Py$	I (i)	--	Dark brown cryst., dec. 190-95°, stable in air	486
$Ph_3PbCo(D_2H_2)\cdot PBu_3$	I (i)	--	Brown cryst., m. 167-68°, stable in air	486
$Ph_2Pb[Co(CO)_4]_2$	IA	76	Red, dec. at 103-106°, in air, magnetic moment	216
$Me_2PbNi(CO)_3$	--	--	(a)	--
$Me_2Pb(C_5H_5NiCO)_2$	--	--	(a)	--
$MeEtPbNi(CO)_3$	--	--	(a)	--
$Et_2PbNi(CO)_3$	--	--	(a)	--
$Pr_2PbNi(CO)_3$	--	--	(a)	--
$1\text{-}Bu_2PbNi(CO)_3$	--	--	(a)	--
$Ph_3PbPt(PPh_3)_2Cl$	III	25	White cryst., m. 205° (dec.), IR spectrum (j)	700

(h) IR spectrum (1672-3, 2338), force const. (2338). (i) DH_2 = Diacetyldioxime. (j) Struct., thermal dec. → $Ph(Cl)Pt(PPh_3)_2 + Ph_2Pb$ (700).

Triphenyllead-π-cyclopentadienyliron-
   dicarbonyl $PbFeC_{25}H_{20}O_2$ $Ph_3PbC_5H_5Fe(CO)_2$

Synth.: By rxn. of Ph_3PbCl with $NaC_5H_5Fe(CO)_2$ in THF in N atm. at r.t. (2057).
By polarographic reduction of Ph_6Pb_2 in $(MeOCH_2)_2$-Bu_4NClO_4, followed by
$C_5H_5Fe(CO)_2I$ (963, 967), by $(C_5H_5)Fe(CO)_2C_5H_5Mo(CO)_3$ (968).
By polarographic reduction of $[C_5H_5Fe(CO)_2]_2$, followed by $Ph_3PbC_5H_5Mo(CO)_3$
(968).

Prop.: Pale orange needles, m. 134-37° (967, 969), brown solid m. 130° (2057),
IR (2057), UV (967) and NMR (2057) spectra, polarography (967, 968).

Rxn. with: Polarographic reduction in presence of Ph_6Sn_2 → $Ph_3SnC_5H_5Fe(CO)_2$
(968).

7. CYCLIC ORGANOLEAD COMPOUNDS

Bis(9,10-Dihydro-9,10-anthranyl)lead $PbC_{28}H_{20}$

Synth.: By rxn. of $9,10-Na_2C_{14}H_{10}$ in THF in
inert atm. (1726).

Prop.: Exists in monomeric and polymeric form
(1726).

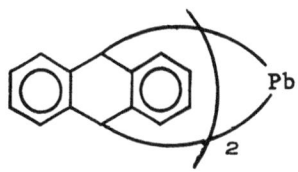

1,1-Diethyl-1-Plumbacyclopentane PbC_8H_{18}

Synth.: By rxn. of Et_3PbCl_2 with $(CH_2CH_2MgBr)_2$ in Et_2O in
inert atm., in 70% yield (1260).

Prop.: Liq., n_D^{29} 1.5452, IR spectrum, GLC, TLC (1260).

Rxn. with: Br in CCl_4 (1:3) → $PbBr_2$ (1260).
I at 0° → PbI_2 (1260).
RCO_2H → $\overline{CH_2(CH_2)_3Pb}(Et)O_2CR$; R = Me, CH_2Cl (1260).

1,1-Diphenyl-1-Plumbacyclopentane $PbC_{16}H_{18}$

Synth.: By rxn. of Ph_2PbCl_2 with $(CH_2CH_2MgBr)_2$ in THF in
inert atm., in 25% yield (1260).

Prop.: $b_{0.5}$ 145-65°, m. ~ 0°, IR spectrum and TLC (1260).

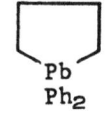

1,1-Ethylacetato-1-Plumbacyclopentane $PbC_8H_{16}O_2$

Synth.: By rxn. of $\overline{CH_2(CH_2)_3PbEt_2}$ with AcOH in Et_2O in
N atm., in 88% yield (1260).

Prop.: m. 152-53° (dec.) (1260).

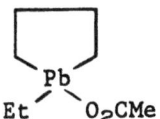

1,1-Ethylchloracetato-1-Plumbacyclopentane $PbC_8H_{15}ClO_2$

Synth.: By rxn. of $\overline{CH_2(CH_2)_3PbEt_2}$ with $ClCH_2CO_2H$ in Et_2O in
inert atm., in 68% yield (1260).

Prop.: m. 132-33° (dec.) (1260).

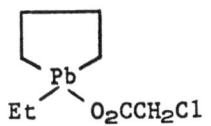

2,3,6,7-Bis(methylenedioxy)-9,9-Diphenyl-
   dibenzoplumbole $PbC_{26}H_{18}O_4$

<u>Synth.</u>: By rxn. of Ph_2PbCl_2 with (2-lithium-4,5-
methylenedioxyphenyl)$_2$ in Et_2O-C_6H_6 in inert atm.,
in 48% yield (934).
<u>Prop.</u>: m. 301-302° (934).

Dimethyllead-9,10-dihydro-9,10-anthracene $PbC_{16}H_{16}$

<u>Synth.</u>: By rxn. of Me_2PbBr_2 with 9,10-$Na_2C_{14}H_{10}$
in THF in inert atm. (1726).
<u>Use</u>: As fungicide, fuel additive and stabilizer
(1726).

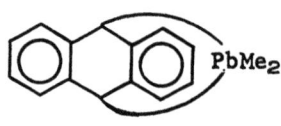

Diphenyllead-9,10-dihydro-9,10-anthracene $PbC_{26}H_{20}$

<u>Synth.</u>: By rxn. of Ph_2PbBr_2 with 9,10-$Na_2C_{14}H_{10}$
in THF in inert atm. (1726).
<u>Use</u>: As fungicide, fuel additive and stabilizer
(1726).

1,1-Diethyl-1-Plumbacyclohexane PbC_9H_{20}

<u>Prop.</u>: IR spectrum, TLC (1260).
Rxn. with: $RCO_2H \rightarrow Pb(O_2CR)_2$; R = Me, $ClCH_2$ (1260).

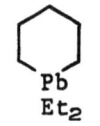

8   MISCELLANEOUS ORGANOLEAD COMPOUNDS

8.1   POLYMERIZATION REACTIONS WITH ORGANOLEAD COMPOUNDS

Table 309 yields a brief summary of polymerization and polycondensation
reactions with organolead compounds. The following scheme serves as a key
to the Polymerization systems.

  I   Autopolymerization at elevated temperature.

  II   Catalytic polymerization with azobisisobutyronitrile.

  III   Polycondensation and esterification reactions.

Table 309. Polymerization Reactions with Organolead Compounds

Polymerization System

Components	Method*	Remarks	Ref.
I OLEFIN POLYMERIZATION REACTIONS			
p-Ph$_3$PbC$_6$H$_4$CH:CH$_2$ + (a)	II	In bulk or in MePh, conversion < 25%, detn. of relative reactivity	284
+ (a)	II	--	461
+ (b)	II	--	461
+ (b)	I	Use in plastic scintillators	1784
+ (b)	I	Use in plastic scintillators	1784
+ (c)	I	Use in plastic scintillators	1784
2,5-Ph$_3$Pb(Me)C$_6$H$_3$CH:CH$_2$ + (b)	I	Use in plastic scintillators	1784
+ (c)	I	Use in plastic scintillators	1784
Ph$_4$Pb + (a ~ 70%)	I	Prop. of polymer, activity as γ-scintillator	23
Ph$_3$PbO$_2$CCMe:CH$_2$	I	Prop. of polymer, antimicrobial action	1328
II POLYCONDENSATION REACTIONS			
Me$_3$PbCl + RNa (d)	III	--	2703
Me$_2$PbBr$_2$ + 9,10-Na$_2$C$_{14}$H$_{10}$	III	--	1726
Et$_3$PbOH + (e)	III	Use as burning rate moderator for rocket fuel	412
Bu$_3$PbOH + (e)	III	Use as burning rate moderator for rocket fuel	412
Ph$_3$PbOH + (f)	III	Prop. of polymer, thermal stability	285
+ (g)	III	Prop. of polymer, antimicrobial action	1328
Ph$_2$PbBr$_2$ + 9,10-Na$_2$C$_{14}$H$_{10}$	III	--	1726
Ph$_2$PbCl$_2$ + HSCH$_2$CO$_2$Na	III	Synth. in Py, m. > 300°	121

* The numerals used here are explained in the introduction to Chapter 8.1.
(a) PhCH:CH$_2$. (b) MeC$_6$H$_4$CH:CH$_2$. (c) Me$_2$C$_6$H$_3$CH:CH$_2$. (d) R = polynuclear aromatic oil. (e) Butadiene-acrylic acid polymer. (f) (CH$_2$CMeCO$_2$H)$_x$. (g) Styrene-maleic anhydride polymer.

8.2 UNSPECIFIED ORGANOLEAD COMPOUNDS

A number of uncertain organolead derivatives are summarized in Table 310.

Table 310. Unspecified Organolead Compounds

Compound	Remarks	Ref.
$Et_3Pb\equiv$	Anal. detn. in blood and urine by spectrophotometry	756
	Comparative toxicity	98
	Effect on amino acid metabolism	99
	Effect on porphyrin metabolism in vitro	1764
Et_3Pb^+	Anal. detn. with $NaBPh_4$ in presence of Et_2Pb^{++} and Pb^{++}	1224
$RPb\equiv$	Anal. detn. of sulfur by micro method	1276
	Toxicity	2890

BIBLIOGRAPHY

1. Abel, E. W. et al., Trans. Faraday Soc. 60, 1214-9 (1964); CA 61, 3795.
2. --- and Brady, D. B., J. Chem. Soc. 1965, 1192-7; CA 62, 9164.
3. Adloff, M. and Adloff, J. P., Compt. Rend. 259, 141-3 (1964); CA 61, 9109.
4. Alderman, J. F., Ph. D. Thesis, Ohio Univ., 1965; Diss. Abstr. 25, 6952 (1965); CA 63, 13304.
5. Aldridge, W. N., Proc. Roy. Soc. Med. 58, 599 (1965); CA 63, 10566.
6. Aleksandrov, A. Yu. et al., Vysokomol. Soedin. 6, 2105-7 (1964); CA 62, 4161.
7. --- et al., Zh. Fiz. Khim. 38, 2190-7 (1964); CA 62, 144.
8. Aleksandrov, Yu. A. and Shushunov, V. A., Zh. Obshch. Khim. 35, 115-7 (1965); CA 62, 13167.
9. Aleksandrova, Yu. V. and Lakosina, T. A., Plast. Massy 1965 (7), 15-7; CA 63, 11305.
10. Alleston, D. L., Davies, A. G. and Hancock, M., J. Chem. Soc. 1964, 5744-8; CA 63, 1811.
11. Altrogge, K., Friker, H. H. and Kikuth, W., Belg. 628,032, Feb. 2, 1962; CA 61, 8142.
11a. Andrascheck, H. J., Ph. D. Thesis, Univ. Munich, 1962; see ref. 605a.
12. Andrianov, K. A. and Yakushkina, S. Ya., Zh. Obshch. Khim. 35, 330-3 (1965); CA 62, 13170.
13. Aufdermarsh, C. A., J. Org. Chem. 29, 1994-6 (1964); CA 61, 5675.
14. --- and Pariser, R., J. Polymer Sci. Pt. A2, 4727-33 (1964); CA 62, 2896.
15. Averbukh, B. S. et al., Zh. Fiz. Khim. 38, 2445-8 (1964); CA 62, 2396.
16. Bai, L. I., Yakubovich, A. Ya. and Muler, L. I., Zh. Obshch. Khim. 34, 3696-7 (1964); CA 62, 5293.
17. Bajer, F. J., Ph. D. Thesis, State Univ. New York, Buffalo, 1963; Diss. Abstr. 25, 1557 (1964); CA 62, 2492.
    --- and Post, H. W., J. Organometal. Chem. 11, 187-91 (1968); CA 68, 39751.
18. Ballinger, P., U.S. 3,143,399, Sept. 29, 1961; CA 61, 10520; cf. ref. M-1860.
19. ---, U.S. 3,143,400, Sept. 29, 1961; CA 61, 10520; cf. ref. M-1860.
20. ---, U.S. 3,185,533, Sept. 7, 1961; CA 63, 8403.
    ---, U.S. 3,206,490, July 30, 1962; CA 63, 18151.
21. Barbanson, E., U.S. 3,183,238, Jan. 3, 1962; CA 63, 1816.
22. Barnes, G. L., Plant Disease Reptr. 49, 285-7 (1965); CA 63, 1172.
23. Baroni, E. E. et al., At. Energ. (USSR) 17 (6), 497-500 (1964); CA 62, 10604.
24. Baukov, Yu. I. et al., Zh. Obshch. Khim. 35, 1092-4 (1965); CA 63, 9976.
24a. ---, Burlachenko, G. S. and Lutsenko, I. F., ibid. 35, 757-8 (1965); CA 63, 8395.
    ---, --- and ---, J. Organometal. Chem. 3, 478 (1965); CA 63, 4324.
24b. ---, --- and ---, Zh. Obshch. Khim. 35, 1173-7 (1965); CA 63, 11602.
25. --- and Lutsenko, I. F., ibid. 34, 3453-6 (1964); CA 62, 4046.

26. Beaird, F. J., Jr. and Kobetz, P., U.S. 3,188,333, June 6, 1963; CA 63, 13316.
    --- and ---, U.S. 3,391,086, Feb. 4, 1965; CA 69, 77505.
27. --- and ---, U.S. 3,188,334, Oct. 22, 1963; CA 63, 13316.
28. Beck, W. and Schuierer, E., Chem. Ber. 97, 3517-23 (1964); CA 62, 4047.
29. Bell, R. T., U.S. 3,158,451, Aug. 4, 1958; CA 62, 3870.
30. Bennett, R. F., Rubber Plastics Age 46, 260-1 (1965); CA 62, 16895.
31. van den Berghe, E. V. and van der Kelen, G. P., Ber. Bunsenges. Physik. Chem. 68, 652-6 (1964); CA 62, 1124.
32. Biritz, L. F., Am. Chem. Soc., Div. Petrol. Chem., Preprints 8, B57-62 (1963); CA 62, 644.
33. Biryukov, I. P. and Voronkov, M. G., Latvijas PSR Zinatnu Akad. Vestis, Khim. Ser. 1965, 115-6; CA 63, 14239.
34. Bitzer, D., Ger. 1,169,060, Oct. 26, 1963; CA 61, 5905.
    ---, Ger. 1,189,220, Jan. 23, 1964; CA 62, 16512.
35. Bloodworth, A. J. and Davies, A. G., Chem. Ind. (London) 1965, 900; CA 63, 5667.
36. --- and ---, J. Chem. Soc. 1965, 5238-44; CA 63, 16376.
37. Boboli, E. et al., Pol. 48,178, June 25, 1962; CA 62, 1688.
38. Bock, R. and Deister, H., Naturwissenschaften 50, 496 (1963); CA 61, 1241.
    --- and ---, Z. Anal. Chem. 230, 321-38 (1967); CA 67, 111830.
39. Boden, N. et al., Mol. Phys. 8, 133-49 (1964); CA 62, 2382.
40. Boettner, E. A. and Dallos, F. C., J. Gas Chromotog. 3, 190-1 (1965); CA 63, 7627.
41. Bogomolov, S. G., Veselkova, I. A. and Lodochnikova, V. I., Tr. Komis. po. Spektroskopii, Akad. Nauk, SSSR 1964, 475-82; CA 63, 13029.
41a. Borsini, G., Nicora, C. and Segalini, A., Belg. 651,260; Brit. 1,018,315; Fr. 1,403,414; Ital. 712,153; Neth. 64 8,302; U.S. 3,409,601, July 31, 1963; CA 63, 1900.
42. Bott, L. L., Hydrocarbon Process. Petrol. Refiner 44, 115-8 (1965); CA 62, 7788.
43. Bott, R. W., Eaborn, C. and Greasley, P. M., J. Chem. Soc. 1964, 4804-6; CA 62, 5170.
44. Boue, S., Gielen, M. and Nasielski, J., Bull. Soc. Chim. Belg. 73, 864-73 (1964); CA 62, 10316.
45. Bouissou, H. et al., Compt. Rend. 259, 3408-10 (1964); CA 62, 4504.
46. Braithwaite, D. G., Ger. 1,197,086, Aug. 1, 1962; U.S. 3,287,248-9, Aug. 31, 1962; CA 63, 11023.
47. Braun, J., Compt. Rend. 260, 218-20 (1965); CA 62, 14710.
48a. Brinckman, F. E. and Haiss, H. S., Chem. Ind. (London) 1963, 1124; CA 59, 10965.
48b. ---, --- and Robb, R. A., Inorg. Chem. 4, 936-42 (1965); CA 63, 3872.
49. Brook, A. G. et al., J. Organometal. Chem. 2, 491-3 (1964); CA 62, 4047.
50. Brooks, E. H., Glockling, F. and Hooton, K. A., J. Chem. Soc. 1965, 4283-8; CA 63, 11602.
51. Brown, D. H., Mohamed, A. and Sharp, D. A. W., Spectrochim. Acta 21, 659-62 (1965); CA 62, 12611.

52. Brown, D. H., Mohamed, A. and Sharp, D. A. W., ibid. 21, 1013-4 (1965); CA 63, 6473.
52a. Brown, G. C. et al., Chem. Ind. (London) 1965, 1634; CA 63, 14892.
53. Brown, J. E., U.S. 3,160,592, Oct. 16, 1959; CA 62, 3870.
54. Brown, L. H. and Hill, J. W., U.S. 3,211,769, Sept. 4, 1962; CA 63, 18152.
55. Brown, T. L. and Stark, K., J. Phys. Chem. 69, 2679-83 (1965); CA 63, 7798.
56. Brugnone, F., Galzigna, L. and Corsi, C., Med. Lav. 55, 411-3 (1964); CA 61, 16693.
57. Bruker, A. B., Balashova, L. D. and Soborovskii, L. Z., U.S.S.R. 170,976, Aug. 1, 1962; CA 63, 9985.
58. ---, --- and ---, U.S.S.R. 170,977, Aug. 1, 1962; CA 63, 9985.
59. Buchman, O. et al., Helv. Chim. Acta 47, 1688-95 (1964); CA 62, 7602.
60. --- et al., ibid. 47, 2037-41 (1964); CA 62, 7603.
61. ---, Grosjean, M. and Nasielski, J., ibid. 47, 1679-88 (1964); CA 62, 7602.
62. ---, --- and ---, ibid. 47, 1695-700 (1964); CA 62, 7603.
63. Burdon, J., Tetrahedron 21, 1101-8 (1965); CA 63, 2875.
64. ---, Coe, P. L. and Fulton, M., J. Chem. Soc. 1965, 2094-6; CA 62, 13067.
65. Burke, W. E. et al., Anal. Chem. 36, 2404-7 (1964); CA 62, 3856.
66. Bychkov, V. T. and Vyazankin, N. S., Zh. Obshch. Khim. 35, 687-9 (1965); CA 63, 4321.
67. Bykov, G. A., Ryasnyl, G. K. and Shpinel, V. S., Fiz. Tverd. Tela 7, 1657-62 (1965); CA 63, 7794.
68. Campbell, I. G. M., Fowles, G. W. A. and Nixon, L. A., J. Chem. Soc. 1964, 3026-9; CA 61, 12032.
68a. Cartledge, F. K., Ph. D. Thesis, Iowa State Univ., 1964; Diss. Abstr. 25, 6226 (1965); CA 63, 8387.
69. Cassol, A. and Barbieri, R., Ann. Chim. (Rome) 55, 606-14 (1965); CA 63, 14358.
70. --- and Magon, L., J. Inorg. Nucl. Chem. 27, 1297-303 (1965); CA 63, 3875.
71. ---, --- and Barbieri, R., J. Chromatog. 19, 57-63 (1965); CA 63, 12356.
72. ---, Portanova, R. and Barbieri, R., J. Inorg. Nucl. Chem. 27, 2275-6 (1965); CA 63, 17461.
73. Castel, P. et al., Trav. Soc. Pharm. Montpellier 23, 45-50 (1963); CA 61, 4901.
74. Castelino, N. et al., Arch. Maladies Profess. Med. Trav. Securite Sociale 25, 203-18 (1964); CA 61, 9950.
75. ---, Colicchio, G. and Piccoli, P., Folia Med. (Naples) 46, 825-34 (1963); CA 61, 1161.
76. ---, --- and Rossi, A., ibid. 46, 980-6 (1963); CA 61, 3599.
77. Chambers, R. D. and Chivers T. J., Chem. Soc. 1964, 4782-90; CA 62, 6501.
77a. --- and Cunningham, J., Tetrahedron Lett. 1965, 2389.
78. Chandler, R. H., Paint Manuf. 34, 35-7 (1964); CA 61, 13527.
79. Chernyshev, E. A., Petrov, A. D. and Krasnova, T. L., Sintez i Svoistva Monomerov, Akad. Nauk SSSR, ... Konf. po Vysokomol. Soedin. 1962, 103-8 (1964); CA 62, 6502.

79a. Cholak, J., Arch. Environ. Health 8, 314-23 (1964); CA 60, 15042.
80. Chow, T. J. and Johnstone, M. S., Science 147, 502-3 (1965); CA 62, 12362.
81. Christopher, P. M., J. Chem. Eng. Data 10, 44-5 (1965); CA 62, 7130.
82. Chumaevskii, N. A. and Borisov, A. E., Dokl. Akad. Nauk SSSR 161, 366-9 (1965); CA 63, 462.
83. Coffield, T. H., U.S. 3,100,212, Jan. 12, 1960; CA 60, 549.
84. Claridge, R. F. C., Merz, E. and Riedel, H. J., Nucleonik 7, 53-8 (1965); CA 62, 16293.
85. Clark, H. C. and Goel, R. G., Inorg. Chem. 4, 1428-32 (1965); CA 63, 16178.
86. ---, O'Brien, R. J. and Pickard, A. L., J. Organometal. Chem. 4, 43-9 (1965); CA 63, 4321.
87. --- and Tsai, J. H., Chem. Commun. 1965, 111; CA 62, 16020.
88. Closson, R. D., Brit. 997,883; Fr. 1,362,696; Ger. 1,195,751; Ger. 1,200,296-7; U.S. 3,231,510, May 1, 1962; CA 62, 4052.
89. Collier, H. E., Jr. and Hammond, G. S., U.S. 3,187,028, Oct. 9, 1963; CA 63, 17770.
90. Combarieu, J., Raitzyn, I. and Wetroff, G., Fr. 1,399,552, Sept. 19, 1958; CA 63, 9985.
91. Connolly, W. J., Brit. 1,017,670; Fr. 1,354,803, Sept. 6, 1962; CA 62, 2676.
92. Considine, W. J., Baum, G. A. and Jones, R. C., J. Organometal. Chem. 3, 308-13 (1965); CA 62, 11316.
93. Cook, S. E. and Sistrunk, T. A., Brit. 1,008,593; Fr. 1,367,570; Ger. 1,235,915, June 8, 1962; CA 62, 10276.
93a. --- and ---, Brit. 1,008,415; Fr. 1,412,943; Neth. 64 12,633; U.S. 3,221,038, Oct. 30, 1963; CA 63, 11223.
94. Corsi, G. C., Galzigna, F. and Brugnone, F., Med. Lav. 55, 665-78 (1964); CA 62, 16866.
95. Creemers, H. M. J. C. and Noltes, J. G., Rec. Trav. Chim. 84, 382-4 (1965); CA 62, 16288.
96. --- and ---, ibid. 84, 590-3 (1965); CA 63, 5464.
97. ---, --- and van der Kerk, G. J. M., ibid. 83, 1284-6 (1964); CA 62, 5291.
98. Cremer, J. E., Ann. Occupational Hyg. 3, 226-30 (1961); CA 61, 7583.
99. ---, J. Neurochem. 11, 165-85 (1964); CA 61, 1055.
100. --- and Aldridge, W. N., Brit. J. Ind. Med. 21, 214-7 (1964); CA 62, 2164.
101. Cross, R. J. and Glockling, F., Proc. Chem. Soc. 1964, 143; CA 61, 2705.
102. --- and ---, J. Organometal. Chem. 3, 253-4 (1965); CA 62, 13174.
103. --- and ---, J. Chem. Soc. 1965, 5422-32; CA 63, 15837.
104. --- and ---, ibid. 1964, 4125-33, CA 62, 1681.
105. --- and ---, J. Organometal. Chem. 3, 146-55 (1965); CA 62, 8527.
106. Crowder, G. A. et al., J. Mol. Spectrosc. 16, 115-21 (1965); CA 63, 1349.
107. --- and Scott, D. W., U. S. Bur. Mines, Rept. Invest. No. 6630, 37 pp. (1965); CA 63, 17324.

108. Cullen, W. R., Dawson, D. S. and Styan, G. E., J. Organometal. Chem. 3, 406-13 (1965); CA 63, 629.
109. --- and Styan, G. E., Inorg. Chem. 4, 1437-40 (1965); CA 63, 14353.
110. Cummins, R. A., Austr. J. Chem. 18, 98-101 (1963); CA 62, 8523.
111. --- and Evans, J. V., Spectrochim. Acta 21, 1016-8 (1965); CA 63, 7776.
112. Curtis, M. D., Ph. D. Thesis, Northwestern Univ., 1965; Diss. Abstr. 26, 3023 (1965); CA 64, 9207.
    --- and Allred, A. L., J. Am. Chem. Soc. 87, 2254-63 (1965); CA 63, 2545.
113. Curtis, R. J., Brit. 968,660; U.S. 3,252,944, Aug. 27, 1962; CA 61, 16321.
114. Dagnall, R. M. and West, T. S., Talanta 11, 1553-7 (1964); CA 61, 13863.
115. Dakli, I., Nigra, T. P. and Casiraghi, R., Belg. 645,883, March 29, 1963; CA 63, 11809.
116. Danek, O. and Willomitzer, J., Czech. 113,485, April 29, 1962; CA 63, 16140.
117. Danilov, S. N., Volozhin, A. I. and Kozmina, O. P., U.S.S.R. 163,167, July 10, 1963; CA 61, 13523.
118. Dannley, R. L. and Aue, W. A., J. Org. Chem. 30, 3845-8 (1965); CA 63, 17843.
    ---, NASA Accession No. N65-34348, Rept. No. AD468448, 7 pp. (1965); CA 67, 39704.
119. Dao-Huy-Giao, Compt. Rend. 260, 6937-8 (1965); CA 63, 7034.
119a. Dassese, P. and Dechenne, R., Belg. 642,208; Brit. 1,002,686; Fr. 1,379,988; U.S. 3,356,668, Jan. 10, 1963; CA 62, 16407; Neth. 287,653, Jan. 10, 1963; CA 61, 16185.
120. Dathe, C., Ger. (East) 36,060, June 23, 1962; CA 63, 15066.
121. Davidson, W. E., Hills, K. and Henry, M. C., J. Organometal. Chem. 3, 285-94 (1965); CA 62, 13167; cf. ref. 220.
122. Dawson, J. A., Belg. 645,019, March 11, 1964; CA 63, 10142.
122a. Delbouille, A. and Toussaint, A., Belg. 657,444; Brit. 1,042,649; Fr. 1,388-997; Neth. 64 14,879; U.S. 3,424,737, Jan. 3, 1964; CA 62, 16404.
123. Del Franco, G. L., Resnick, P. and Dillard, C. R., J. Organometal. Chem. 4, 57-66 (1965); CA 63, 4320.
124. Depree, D. O., U.S. 3,161,664, April 16, 1962; CA 62, 7890.
125. Deschiens, R., Floch, H. and Floch, T., Bull. Soc. Pathol. Exotique 57, 454-65 (1964) (Publ. 1965); CA 62, 16903.
126. Doetzer, R., Brit. 1,062,464; Fr. 1,429,264; Ger. 1,200,817, March 30, 1963; CA 63, 15896.
127. Dolgii, I. E., Meshcheryakov, A. P. and Shvedova, I. B., Izv. Akad. Nauk SSSR, Ser. Khim. 1965, 192-4; CA 62, 11843.
128. Dorfman, R. I., Proc. Soc. Exptl. Biol. Med. 116, 1055-7 (1964); CA 62, 3025.
129. Drago, R. S. and Matwiyoff, N. A., J. Organometal. Chem. 3, 62-9 (1965); CA 62, 7612.
130. Duffield, A. M., Budzikiewicz, H. and Djerassi, C., J. Am. Chem. Soc. 87, 2920-5 (1965); CA 63, 5500.

131. Duke, A. J. and Dobinson, B., Brit. 991,452, Aug. 18, 1961; CA 63, 5850.
132. Duyfjes, W. and de Lange, W., Brit. 981,238, May 27, 1960; CA 61, 9346.
--- and ---, U.S. 3,140,977, May 27, 1960; CA 61, 11273.
133. Dyer, E. and Pinkerton, R. B., J. Appl. Polymer Sci. 9, 1713-29 (1965); CA 63, 6967.
134. Eaborn, C. and Walton, D. R. M., J. Organometal. Chem. 4, 217-28 (1965); CA 63, 11308.
135. Egorov, Yu. P., Teor. i Eksperim. Khim., Akad. Nauk Ukr. SSR 1, 30-40 (1965); CA 63, 7773.
136. --- and Kirei, G. G., Zh. Obshch. Khim. 34, 3615-21 (1964); CA 62, 7252.
137. --- and Loktionova, R. A., Teor. i Eksperim. Khim., Akad. Nauk Ukr. SSR 1, 160-70 (1965); CA 63, 10852.
138. ---, Morozov, V. P. and Kovalenko, N. F., Ukr. Khim. Zh. 31, 123-32 (1965); CA 63, 3771.
139. Egorova, Z. S. et al., Tr. Komis. po Spektroskopii, Akad. Nauk SSSR 1964, 503-10; CA 63, 10119.
140. Eikelenboom, C., Chem. Weekblad 60 (16), 222-3 (1964); CA 61, 6261.
141. Epstein, L. M. and Straub, D. K., Inorg. Chem. 4, 1551-4 (1965); CA 63, 15839.
142. Faraglia, G., Roncucci, L. and Barbieri, R., Ric. Sci. Rend. Sez. A5, 205-12 (1965); CA 63, 12654.
143. Fekete, F., U.S. 3,133,892, April 26, 1960; CA 61, 12151.
144. Fentiman, A. F. et al., J. Organometal. Chem. 4, 302-7 (1965); CA 63, 11600.
--- et al., U. S. At. Energy Comm. BMI-1713, 14 pp. (1965); CA 63, 2549.
145. Fenton, D. E. and Massey, A. G., Chem. Ind. (London) 1964, 2100; CA 62, 5292.
146. --- and ---, J. Inorg. Nucl. Chem. 27, 329-33 (1965); CA 62, 13169.
147. Fergusson, J. E., Roper, W. R. and Wilkins, C. J., J. Chem. Soc. 1965, 3716-20; CA 63, 3873.
148. Fink, W., U.S. 3,127,431, April 28, 1960; CA 61, 3148.
149. Foldesi, I., Acta Chim. Acad. Sci. Hung. 45, 237-44 (1965); CA 63, 16376.
150. --- and Gomory, P., ibid. 45, 231-6 (1965); CA 63, 16377.
150a. Foschi, S. and Rapparini, G., Ric. Sci. Rend. Sez. B3, 341-4 (1963); CA 60, 15074.
151. Frankel, M. et al., J. Org. Chem. 30, 1596-9 (1965); CA 63, 669.
--- et al., Brit. 1,077,970; Fr. 1,514,934; Neth. 65 5,425, Oct. 20, 1964; CA 64, 2159.
152. Franzen, V., Kunststoffe 55 (5), 327-8 (1965); CA 63, 11792.
153. ---, Steiner, T. and Biese, V., Belg. 630,458, April 6, 1962; CA 60, 14686.
154. Frick, L. P. and De Jimenez, W. Q., Bull. World Health Organ. 31, 429 (1964); CA 62, 11053.
155. Fritz, G. and Scheer, H., Z. Anorg. Allg. Chem. 331, 151-3 (1964); CA 61, 12628.
156. --- and ---, ibid. 338, 1-8 (1965); CA 63, 11600.

157. Fritz, H. P. and Kreiter, C. G., J. Organometal. Chem. 4, 313-9 (1965); CA 63, 11287.
158. Frye, A. H., Horst, R. W. and Paliobagis, M. A., Am. Chem. Soc., Div. Polymer Chem., Preprints 4, 260-82 (1963); CA 62, 698.
159. Galashina, M. L., Sobolevskii, M. V. and Kaznina, G. V., U.S.S.R. 166,834, Nov. 4, 1963; CA 62, 10279.
160. Galzigna, L., Brugnone, F. and Corsi, G. C., Med. Lav. 55, 102-6 (1964); CA 61, 6249.
161. Gar, T. K. and Mironov, V. F., Izv. Akad. Nauk SSSR, Ser. Khim. 1965, 855-62; CA 63, 5666.
162. Gardner, B. G. and Poller, R. C., Bull. Entomol. Res. 55, 17-21 (1964); CA 63, 4888.
163. Gearhart, W. M. et al., Brit. 1,030,213; 1,030,215; Fr. 1,359,112; U.S. 3,211,561, Nov. 15, 1961; CA 61, 14536.
164. Gee, W., Shaw, R. A. and Smith, B. C., J. Chem. Soc. 1964, 2845-6; CA 61, 14705.
165. Geissler, H. and Kriegsmann, H., Z. Chem. 4, 354-5 (1964); CA 62, 30.
166. Gelius, R., Z. Anorg. Allg. Chem. 334, 72-80 (1964); CA 63, 1809.
167. George, T. A., Jones, K. and Lappert, M. F., J. Chem. Soc. 1965, 2157-65; CA 62, 14715.
168. Geyer, R. and Seidlitz, H. J., Z. Chem. 4, 468 (1964); CA 62, 9794.
169. Ghate, M. R. and Bhide, K. N., Indian J. Chem. 2, 243-6 (1964); CA 61, 9019.
170. Gherardi, M., Med. Lav. 55, 107-24 (1964); CA 61, 6249.
171. Giamaria, J. J. and Norris, H. D., U.S. 3,168,385, July 12, 1961; CA 62, 8917.
172. Gibbons, A. J., Brit. 1,055,995; Fr. 1,374,539; Ger. 1,196,197; Neth. 297,477, Sept. 4, 1962; CA 62, 6637.
172a. Gilbert, E. E., Belg. 654,853; Brit. 1,038,908; Fr. 1,412,782; Ger. 1,209,354; Neth. 64 12,559; U.S. 3,247,055, Oct. 28, 1963; CA 63, 15483.
173. Gilman, H. and Cartledge, F. K., J. Organometal. Chem. 2, 447-54 (1964); CA 62, 1084.
173a. See ref. 68a.
174. Gilman, H. and Cartledge, F. K., J. Organometal. Chem. 3, 255-7 (1965); CA 62, 14593.
175. --- and ---, ibid. 5, 48-56 (1966); CA 64, 2123; Chem. Ind. (London) 1964, 1231.
176. ---, --- and Sim, S.-Y., ibid. 4, 332-4 (1965); CA 63, 16172, cf. ref. 1928.
177. Giraitis, A. P., U.S. 3,177,130, July 11, 1960; CA 63, 1816.
178. Gloskey, C. R., Brit. 1,053,073; Fr. 1,386,988; Ger. 1,233,396; U.S. 3,293,273, April 9, 1963; CA 63, 10135.
179. Gmitter, G. T., U.S. 3,208,959, May 5, 1959; CA 63, 18394.
180. --- and Braidich, E. V., U.S. 3,148,162, April 27, 1960; CA 61, 14864.
181. Goldshtein, I. P. et al., Dokl. Akad. Nauk SSSR 163, 880-3 (1965); CA 63, 14357.

182. Goldshtein, I. P., Guryanova, E. N. and Kocheshkov, K. A., Sintez i Svoistva Monomerov, Akad. Nauk SSSR, ... Konf. po Vysokomol. Soedin. 1962, 109-12 (1964); CA 62, 6501.
183. Goppoldova, M. and Smrz, Z., Sb. Praci Vyzkumu Chem. Vyuziti Uhli, Dektu, Ropy 2, 141-55 (1962); CA 62, 5117.
184. Gordon, L. B. and Moote, T. P., U.S. 3,036,016, March 27, 1958; CA 57, 4877.
185. Gorsich, R. D., U.S. 3,030,396, March 29, 1961; CA 57, 7309.
186. ---, U.S. 3,030,397, March 31, 1961; CA 57, 7309.
187. ---, U.S. 3,033,885, April 8, 1961; CA 57, 13803.
188. Graber, M. and Gras, G., Rev. Elevage Med. Vet. Pays Trop. 16, 427-38 (1963); CA 61, 9986.
189. --- and ---, ibid. 17, 205-20 (1964); CA 63, 3372.
190. Graham, D. C., European Potato J. 7, 33-44 (1964); CA 61, 6299.
191. Grant, D. and VanWazer, J. R., J. Organometal. Chem. 4, 229-36 (1965); CA 63, 11599.
192. Graves, J. B., Bradley, J. R. and Bagent, J. L., J. Econ. Entomol. 58, 583-4 (1965); CA 63, 2339.
193. Greco, C. C., U.S. 3,201,408, June 5, 1963; CA 63, 13315.
194. Griffiths, V. S. and Derwish, G. A. W., J. Mol. Spectrosc. 13, 393-8 (1964); CA 61, 7843.
195. Guccione, E., Chem. Eng. 72 (13), 102-4 (1965); CA 63, 9454.
195a. Guinet, P. A. E. and Puthet, R. R., Belg. 659,271; Brit. 1,054,464; Fr. 1,392,648; Neth. 65 1432; U.S. 3,409,573, Feb. 5, 1964; CA 63, 1816.
196. Gunn, E. L., Appl. Spectrosc. 19 (3), 96-9 (1965); CA 63, 6765.
197. Gorsky, J. and Vesely, V., Freiberger Forschungsh. 340A, 303-21 (1964); CA 63, 2816.
--- and ---, Ropa Uhli 7, 53-7 (1965); CA 63, 2816.
198. Gverdtsiteli, I. M. and Buachidze, M. A., Dokl. Akad. Nauk SSSR 158, 147-50 (1964); CA 62, 1682.
199. --- and ---, Soobshch. Akad. Nauk Gruz. SSR 37, 59-64 (1965); CA 62, 14716.
200. --- and ---, ibid. 37, 323-30 (1965); CA 62, 14719.
201. ---, Guntsadse, T. P. and Petrov, A. D., ibid. 36, 579-84 (1964); CA 62, 11841.
202. Haglund, W. A., Plant Disease Reptr. 49, 793-6 (1965); CA 63, 18959.
203. Hague, D. N. and Prince, R. H., J. Chem. Soc. 1965, 4690-6; CA 63, 12995.
204. Hartmann, H. and El Assar, M. K., Naturwissenschaften 52, 304 (1965); CA 63, 5670.
205. ---, Karbstein, B. and Reiss, W., ibid. 52, 59 (1965); CA 62, 11852.
206. --- and Meyer, K., ibid. 52, 303 (1965); CA 63, 5670.
206a. Harvey, D. R. and Norman, R. O. C., J. Chem. Soc. 1964, 4860-8; CA 62, 5158.
207. Harvey, M. C. and Nebergall, W. H., Appl. Spectrosc. 16, 12-4 (1962); CA 57, 6764.
208. Hasegawa, K., Kajikawa, M. and Okamoto, N., Bunseki Kagaku 14, 717-20 (1965); CA 63, 11207.

209. Hatton, R. E. and Stark, L. R., Fr. 1,356,569; U.S. 3,423,469, Apr. 30, 1962; CA 62, 3870.
210. Hayakawa, K., Kawase, K. and Matsuda, T., Nature 206, 1038-9 (1965); CA 63, 7033.
211. Hayes, M. C., J. Inorg. Nucl. Chem. 26, 2306-8 (1964); CA 62, 8540.
212. Hechenbleikner, I., Bresser, R. E. and Homberg, O. A., Fr. 1,368,785, May 1, 1962; CA 62, 16460.
213. ---, --- and ---, Fr. 1,375,723, May 1, 1962; CA 62, 11975; U.S. 3,196,129, May 1, 1962; CA 64, 12906; cf. ref. 1170.
214. --- and Homberg, O. A., U.S. 3,209,017, Sept. 22, 1961; CA 63, 16382.
215. van der Heide, R. F., Z. Lebensm.-Untersuch.-Forsch. 124, 348-50 (1964); CA 61, 7708.
216. Hein, F. and Jehn, W., Ann. Chem. 684, 4-9 (1965); CA 63, 7042.
217. Henderson, A. and Holiday, A. K., J. Organometal. Chem. 4, 377-81 (1965); CA 63, 14688.
218. Henderson, H. T., U.S. 3,179,506, June 30, 1961; CA 62, 15971.
219. ---, Fr. 1,360,458; U.S. 3,188,187, June 4, 1962; CA 62, 8917.
219a. Henglein, F. A., Lang, R. and Schmack, L., Makromol. Chem. 22, 102 (1957); CA 51, 6506.
220. Henry, M. C., U.S. Dept. Comm., Office Tech. Serv., AD 611432, 469-76 (1964); CA 63, 14893.
221. ---, AD 612141, 9 pp. (1963); CA 63, 9975.
222. Herber, R. H. and Stoeckler, H. A., Trans. N.Y. Acad. Sci. 26, 929-33 (1964); CA 62, 2382; cf. ref. 1986.
223. --- and ---, Chem. Effects Nucl. Transformations, Proc. Symp., Vienna 1964, 2, 403-18 (1965); CA 63, 9326; cf. ref. 1986.
224. ---, --- and Reichle, W. T., J. Chem. Phys. 42, 2447-52 (1965); CA 62, 12628; cf. ref. 1986.
225. Hills, K. and Henry, M. C., J. Organometal. Chem. 3, 159-60 (1965); CA 62, 10456.
226. Hochman, H. and Roe, T. Jr., U.S. Dept. Comm., Office Tech. Serv., AD 401215, 52 pp. (1963); CA 62, 5827.
227. Hooten, K. A. and Allred, A. L., Inorg. Chem. 4, 671-8 (1965); CA 62, 16290.
228. Hord, R. A. and Tolefson, H. B., Virginia J. Sci. 16, 105-19 (1965); CA 63, 16086.
229. Hostettler, F., U.S. 3,194,773, Nov. 25, 1958; CA 63, 8577.
230. --- and Cox, E. F., U.S. 3,201,358, Dec. 8, 1958; CA 63, 11820.
231. Hsieh, H. L. and Strobel, C. W., Brit. 976,313; Fr. 1,363,464; U.S. 3,218,306, June 11, 1962; CA 62, 9331.
232. Hubel, K. W., Braye, E. H. and Caplier, I. H., U.S. 3,151,140, June 15, 1960; CA 61, 16097.
232a. Huber, F., Angew. Chem. 77, 1084-5 (1965).
233. --- and Enders, M., Z. Naturforsch. B20, 601 (1965); CA 63, 9976.
234. Huheey, J. E., J. Phys. Chem. 69, 3284-9 (1965); CA 63, 14058.
235. Hunter, D. N., Brit. 969,196, Jan. 4, 1960; CA 61, 13444.

236. Huartado De Mendoza Riquelme, J., Inform. Quim. Anal. (Madrid) 18, 27-30 (1964); CA 61, 2880.
236a. Ibekwe, S. D. and Newlands, M. J., Chem. Commun. 1965, 114-5; CA 62, 16285.
237. Issleib, K. and Krech, F., Z. Anorg. Allg. Chem. 328, 21-33 (1964); CA 62, 9169.
238. Itoi, K., Fr. 1,368,522, Sept. 5, 1962; CA 62, 2794.
239. --- and Kumano, S., Japan. 10,588 ('65), Feb. 9, 1963; CA 63, 16497.
240. --- and ---, Japan. 10,589 ('65), Feb. 9, 1963; CA 63, 16497.
241. --- and Yano, M., Fr. 1,375,867, Sept. 5, 1962; CA 62, 10541.
242. Iwamoto, H. and Kikuchi, M., Hakko Kyokaishi 22, 218-22 (1964); CA 63, 18706.
243. --- and ---, Kogyo Gijutsuin, Hakko Kenkyusho Kenkyu Hokoku No. 24, 1-11 (1963); CA 62, 15363.
244. Jackson, J. A. and Nielsen, J. R., J. Mol. Spectrosc. 14, 320-41 (1964); CA 62, 1209.
245. James, J. E., Brit. 994,167, Nov. 16, 1962; CA 63, 4090.
246. Jameson, C. J. and Gutovsky, H. S., J. Chem. Phys. 40, 1714-24 (1964); CA 60, 10085.
247. Jehring, H. and Mehner, H., Z. Chem. 4, 273-4 (1964); CA 61, 15358.
248. Johnson, F., Gohlke, R. S. and Nasutavicus, W. A., J. Organometal. Chem. 3, 233-44 (1965); CA 62, 11844.
249. Jones, J. P. and Everett, P. H., Plant Disease Reptr. 49, 29-32 (1965); CA 62, 9711.
250. Jones, K. and Lappert, M. F., J. Organometal. Chem. 3, 295-307 (1965); CA 62, 14715.
251. --- and ---, Proc. Chem. Soc. 1962, 358; J. Chem. Soc. 1965, 1944-51; CA 62, 13168.
252. Kafengauz, A. P., Kafengaus, I. M. and Murashova, V. I., Plast. Massy 1965 (9), 13-16; CA 63, 16547.
253. Karasev, A. N. et al., Kinet. Katal. 6, 710-6 (1965); CA 63, 15603.
    --- et al., ibid. 8, 232-5 (1967); CA 67, 6014.
    --- et al., Zh. Fiz. Khim. 39, 3117-8 (1965); CA 64, 8961.
254. Kartsev, G. N. et al., Zh. Strukt. Khim. 5, 492-3 (1964); CA 61, 8986.
255. --- et al., ibid. 5, 639 (1964); CA 61, 13988.
256. Kasai, N., Yasuda, K. and Okawara, R., J. Organometal. Chem. 3, 172-3 (1965); CA 62, 11239.
257. Katagiri, Y., Japan. 20,526 ('64), Dec. 12, 1962; CA 62, 16411.
258. Kawasaki, Y., Kawakami, K. and Tanaka, T., Bull. Chem. Soc. Japan 38, 1102-5 (1965); CA 63, 14236.
259. van der Kelen, G. P., Verdonck, L. and van de Vondel, D., Bull. Soc. Chim. Belges. 73, 733-40 (1964); CA 62, 3552.
260. Kenaga, E. E., J. Econ. Entomol. 58, 4-8 (1965); CA 62, 9715.
261. Kenton, J. R. and Adams, L. M., U.S. 3,168,590, Oct. 31, 1961; CA 62, 9006.
262. Khmelnitskii, R. A. et al., Zh. Obshch. Khim. 35, 773-6 (1965); CA 63, 6821.

263. Kimmel, E. E., U.S. 3,132,992, May 19, 1960; CA 61, 7642.
264. Kirkendall, R. K., Brit. 1,034,292; Fr. 1,373,716; Ger. 1,209,740, Aug. 8, 1962; CA 62, 11973.
265. Klimova, V. A. and Vitalina, M. D., Zh. Analit. Khim. 19, 1254-7 (1964); CA 62, 1070.
266. ---, Zabrodina, K. S. and Shitikova, N. L., Izv. Akad. Nauk SSSR, Ser. Khim. 1965, 178-80; CA 62, 11152.
267. Klose, G., Arch. Sci. (Geneva) 14, 427-9 (1961); CA 62, 6034.
268. Kobayashi, S. and Hayakawa, S., Japan. J. Appl. Phys. 4, 181-9 (1965); CA 63, 2598.
269. Kobayashi, Y., Fukuma, N. and Kitaoka, A., Japan. 4175 ('63), Jan. 14, 1961; CA 60, 10884.
270. ---, --- and ---, Japan. 1516 ('64), Oct. 30, 1961; CA 61, 2000.
271. ---, --- and ---, Japan. 1811 ('64), Oct. 30, 1961; CA 61, 2000.
272. Kobetz, P., U.S. 3,098,862, July 11, 1960; CA 60, 561.
273. --- and Beaird, F. M., Brit. 1,015,268; Fr. 1,372,724; Ger. 1,210,840; Neth. 297,327; U.S. 3,192,240, Aug. 31, 1962; CA 62, 586.
274. --- and ---, U.S. 3,188,332, May 29, 1963; CA 63, 9986.
275. Kochkin, D. A. et al., Zh. Obshch. Khim. 34, 4027-9 (1964); CA 62, 9042.
276. Koeller, H., U.S. At. Energy Comm. NP-13077, 209 pp. (1962); CA 61, 1422; cf. M-1781.
277. Koester-Pflugmacher, A. and Termin, E., Naturwissenschaften 51, 554-5 (1964); CA 62, 8637.
278. Kolesnikov, S. P. and Nefedov, O. M., Angew. Chem. 77, 345 (1965); CA 63, 1811; Zh. Vses. Khim. Obshchestva im. D.I. Mendeleeva 10, 478-9 (1965); CA 63, 14894.
279. Komarov, N. V., Shostakovskii, M. F. and Misyunas, V. K., U.S.S.R. 168,690, March 25, 1963; CA 63, 1815.
280. Komura, M. et al., J. Organometal. Chem. 4, 308-12 (1965); CA 63, 11600.
281. Koopmans, M. J., Ger. 1,155,630, Nov. 18, 1957; CA 62, 1032.
282. Korn, O. and Woggon, H., Ernaehrungsforschung 10, 57-65 (1965); CA 63, 4462.
283. Kostyanovskii, R. G. and Prokofev, A. K., Izv. Akad. Nauk SSSR, Ser. Khim. 1965, 175-8; CA 62, 11843.
284. Koton, M. M. and Dokukina, L. F., Vysokomol. Soedin. 6, 1791-4 (1964); CA 62, 2828.
285. ---, Kiseleva, T. M. and Arkhipova, I. L., ibid. 6, 1496-7 (1964); CA 61, 13441.
286. Kovacic, J., Powell, K. and Chadha, R. N., Abstr. 149th Nat. Meeting A.C.S. Detroit, Mich. 1965.
287. Kranz, J., Phytopathol. Z. 52, 59-72 (1965); CA 63, 3559.
288. ---, ibid. 52, 335-48 (1965); CA 63, 936.
289. Kriegsmann, H. and Pauly, S., Z. Anorg. Allg. Chem. 330, 275-89 (1964); CA 62, 3898.
290. Krizhanskii, L. M. et al., Dokl. Akad. Nauk SSSR 160, 1121-3 (1965); CA 63, 1373.

291. Kruglaya, O. A., Vyazankin, N. S. and Razuvaev, G. A., Zh. Obshch. Khim. 35, 394 (1965); CA 62, 14722.
292. Kubo, H., Agr. Biol. Chem. (Tokyo) 29, 43-55 (1965); CA 63, 7032.
293. Kubo, M., Hatakeyama, H. and Fukumoto, O., Japan. 12,375 ('65), June 21, 1962; CA 63, 18376.
294. Kubota, K., Igaku Kenkyu 29, 4223-47 (1959); CA 61, 6247.
295. Kuivila, H. G., Rahman, W. and Fish, R. H., J. Am. Chem. Soc. 87, 2835-40 (1965); CA 63, 6804; cf. ref. 1041.
296. Kula, M. R., Amberger, E. and Mayer, K. K., Chem. Ber. 98, 634-7 (1965); CA 62, 13166.
297. ---, --- and Rupprecht, H., ibid. 98, 629-33 (1965); CA 62, 13018.
298. Kupchik, E. J. and Calabretta, P. J., Inorg. Chem. 4, 973-8 (1965); CA 63, 4318.
299. --- and Kiesel, R. J., J. Org. Chem. 29, 3690-1 (1964); CA 62, 7597; cf. ref. 1310.
300. --- and Ursino, J. A., Chem. Ind. (London) 1965, 794-5; CA 63, 4321; cf. ref. 2841.
301. Labarre, J. F. and Mazerolles, P., Compt. Rend. 254, 3998-4000 (1962); CA 61, 11457.
302. Langer, H. G., Gohlke, R. S. and Smith, D. H., Anal. Chem. 37, 433-4 (1965); CA 62, 12586.
303. Lapkin, I. I. and Dumler, V. A., Zh. Obshch. Khim. 34, 3690-3 (1964); CA 62, 5291.
304. Lau, C. L., Rec. Trav. Chim. 84, 429-35 (1965); CA 63, 1345.
305. Lautsch, W. F. et al., Z. Chem. 4, 441-54 (1964); CA 62, 7177.
306. Leebrick, J. R., Belg. 637,090; Fr. 1,384,694; Neth. 296,988; U.S. 3,167,473, Jan. 3, 1963; CA 62, 6687.
307. Lefort, M., Brit. 1,016,813; Fr. 1,371,324; U.S. 3,365,479, July 24, 1963; CA 62, 4052.
308. Leites, L. A., Gar, T. K. and Mironov, V. F., Dokl. Akad. Nauk SSSR 158, 400-3 (1964); CA 62, 14064.
309. ---, Pavlova, I.D. and Egorov, Yu. P., Teor. i Experim. Khim., Akad. Nauk Ukr. SSR 1, 311-23 (1965); CA 63, 13024.
310. Leusink, A. J. et al., Rec. Trav. Chim. 83, 1036-8 (1964); CA 62, 1682.
311. --- et al., ibid. 84, 567-78 (1965); CA 63, 1809.
312. --- and Marsman, J. W., ibid. 84, 1123-8 (1965); CA 63, 12992.
313. ---, --- and Budding, H. A., ibid. 84, 689-703 (1965); CA 63, 11290.
314. --- and Noltes, J. G., ibid. 84, 585-9 (1965); CA 63, 1677.
315. Linch, A. L. et al., Am. Ind. Hyg. Assoc. J. 25, 69-93 (1964); CA 60, 15045.
316. Linsk, J., Carl, R. W. and Field, E., U.S. 3,164,537, Dec. 30, 1960; CA 62, 11429.
317. --- and Mayerle, E. A., U.S. 3,155,602, June 13, 1960; CA 62, 2794.
318. Litvyak, I. G. and Sumarokova, T. N., Zh. Obshch. Khim. 34, 3677-82 (1964); CA 62, 6500.
319. Lodochnikova, V. I., Panov, E. M. and Kocheshkov, K. A., ibid. 34, 4022-4 (1964); CA 62, 9164.

320. Loeb, W. E., U.S. 3,166,547, Sept. 15, 1959; CA 62, 9259.
321. Lohmann, D. H., J. Organometal. Chem. 4, 382-91 (1965); CA 63, 14228.
322. Lorberth, J., Chem. Ber. 98, 1201-4 (1965); CA 62, 16287.
323. --- and Kula, M. R., ibid. 97, 3444-51 (1964); CA 62, 6500.
324. --- and ---, ibid. 98, 520-5 (1965); CA 62, 13167.
325. --- and Noeth, H., ibid. 98, 969-76 (1965); CA 62, 14469.
326. Lovell, W. G., U.S. 3,202,141, Feb. 29, 1960; CA 63, 12957.
327. Luneva, L. K., Sladkov, A. M. and Korshak, V. V., Vysokomol. Soedin. 7, 427-31 (1965); CA 63, 1879.
328. Maciel, G. E., J. Phys. Chem. 69, 1947-51 (1965); CA 63, 2543.
329. Mack, G. P., Mod. Plastics 42 (4), 148, 150, 154, 158, 160, 194 (1964); CA 62, 6629.
330. ---, Fr. 1,369,815, Sept. 25, 1962; CA 62, 586.
331. Maddox, M. L., Flitcroft, N. and Kaesz, H. D., J. Organometal. Chem. 4, 50-6 (1965); CA 63, 4130; cf. ref. 1447.
332. Magie, R. O., Proc. Florida State Hort. Soc. 77, 549-51 (1964); CA 63, 13965.
333. Maire, J. C. et al., Compt. Rend. 260, 5290-6 (1965); CA 63, 6818.
334. Marchand, A., Mendelsohn, J. and Valade, J., ibid. 259, 1737-9 (1964); CA 62, 2369.
335. Marrot, J., Maire, J. C. and Cassan, J., ibid. 260, 3931-4 (1965); CA 63, 3784.
336. Mathia, R. and Mazerolles, P., Bull. Soc. Chim. Fr. 1962, 1913-4; CA 61, 2601.
337. Mathur, S. et al., J. Organometal. Chem. 4, 294-301 (1965); CA 63, 11601.
338. --- et al., ibid. 4, 371-6 (1965); CA 63, 14893.
339. Matsuda, S., Matsuda, H. and Yamane, Y., Kogyo Kagaku Zasshi 67, 467-9 (1964); CA 62, 16288.
340. Matsunami, Y., Ikebuchi, J. and Uchida, Y., Tokyo Nogyo Daigaku Nogaku Shuho 5, 90-8 (1959); CA 61, 1196.
341. Mattison, E. L. and Wolfe, R., Brit. 995,052; Fr. 1,361,089; U.S. 3,188,199, June 25, 1962; CA 62, 3709.
342. Matwiyoff, N. A. and Drago, R. S., J. Organometal. Chem. 3, 393-9 (1965); CA 63, 469.
343. Mazeroles, P., Bull. Soc. Chim. France 1965, 464-8; CA 63, 13308.
344. ---, Lesbre, M. and Dubac, J., Compt. Rend. 260, 2255-8 (1965); CA 62, 14715.
345. Mazur, H., Roczniki Panstwowego Zakladu Hig. 16, 275-80 (1965); CA 63, 11779.
345a. McCarthy, S. W., Ph. D. Thesis, Univ. of Tennessee, 1964; Diss. Abst. 25, 1552 (1964); CA 62, 4593; cf. ref. 487.
346. McEwen, F. L. and Davis, A. C., J. Econ. Entomol. 58, 369-70 (1965); CA 62, 15367.
347. McGrady, M. M. and Tobias, R. S., J. Am. Chem. Soc. 87, 1909-16 (1965); CA 63, 624.
---, Ph. D. Thesis, Univ. of Minesota, 1965; Diss. Abstr. 26, 76-7 (1965); CA 63, 16377.

348. McIntosh, A. H. and Eveling, D. W., Ann. Appl. Biol. 55, 397-407 (1965); CA 63, 13963.
349. Mehrotra, R. C. and Gupta, V. D., J. Organometal. Chem. 4, 145-50 (1965); CA 63, 16375.
350. --- and ---, ibid. 4, 237-40 (1965); CA 63, 16376.
351. Melnikova, E. A., Panasenko, Z. G. and Artamonova, T. A., Nauchn. Tr. Kubansk. gos. Med. Inst. 1962 (19), 77-83; CA 61, 6257.
352. Mendelsohn, J., Marchand, A. and Valade, J., Compt. Rend. 261, 135-8 (1965); CA 63, 12514.
353. Mendelsohn, J. C., Metras, F. and Valade, J., ibid. 261, 756-8 (1965); CA 63, 14354.
354. Meisner, J. and Ascher, K. R. S., Z. Pflanzenkrankh. Pflanzenschutz 72, 458-66 (1965); CA 63, 18964.
355. Merten, L. et al., Ger. 1,111,378, April 17, 1958; CA 56, 3664.
355a. ---, Loew, G. and Bayer, O., Ger. 1,111,377, Feb. 18, 1959; CA 56, 3664.
356. Merz, E., Radiochim. Acta 2, 172-9 (1964); CA 61, 14127.
--- and Riedel, H. J., ibid. 3, 35-45 (1964); CA 62, 15681.
--- and ---, Chem. Effects Nucl. Transf., Proc. Symp. Vienna 1964, 2, 179-94 (1965); CA 63, 9314.
357. Mikhailov, B. M., Bubnov, Yu. N. and Kiselev, V. G., Izv. Akad. Nauk SSSR, Ser. Khim. 1965, 68-72; CA 62, 11842.
358. Miller, D. R., Evers, R. L. and Skinner, G. B., Combust. Flame 7, 137-42 (1963); CA 60, 6693.
359. Miller, M. R., U.S. 3,175,982, April 5, 1960; CA 63, 1642.
360. Miller, S. M., Ind. Eng. Chem., Proc. Res. Develop. 3, 226-30 (1964); CA 61, 9986.
361. Minakami, S. et al., J. Biochem. (Tokyo) 57, 221-2 (1965); CA 63, 3489.
362. Mironov, V. F. and Fedotov, N. S., Zh. Obshch. Khim. 34, 4122 (1964); CA 62, 9163.
363. --- and Gar, T. K., Izv. Akad. Nauk SSSR, Ser. Khim. 1964, 1887-9; CA 62, 2787.
364. --- and ---, ibid. 1965, 291-300; CA 62, 14715.
---, Dzhurinskaya, N. G. and Gar, T. K., Sintez i Svoistva Monomerov, Akad. Nauk SSSR, ... Konf. po Vysokomol. Soedin. 1962, 150-2 (1964); CA 62, 5292.
365. --- and ---, ibid. 1965, 755-8; CA 63, 2993.
366. --- and Kravchenko, A. L., Dokl. Akad. Nauk SSSR 158, 656-9 (1964); CA 62, 580.
367. --- and ---, Izv. Akad. Nauk SSSR, Ser. Khim. 1965, 1026-35; CA 63, 8392.
367a. Mocotte, J., Belg. 639,250; Brit. 1,059,629; Fr. 1,352,843; Ger. 1,258,177; Neth. 299,805, Oct. 29, 1962; CA 61, 4903.
367b. Modic, F. J., Brit. 992,025; U.S. 3,205,283, Dec. 30, 1960; CA 63, 5855.
368. Moedritzer, K. and Van Wazer, J. R., J. Am. Chem. Soc. 87, 2360-5 (1965); CA 63, 2993.
369. Mori, T. et al., Bull. Japan. Petrol. Inst. 7, 7-16 (1965); CA 63, 17753.
370. Morikawa, T., Kagaku To Kogyo (Osaka) 38, 672-9 (1964); CA 62, 14892.

371. Morikawa, T. and Yoshida, K., ibid. **38**, 667-71 (1964); CA **62**, 14891.
372. Muetterties, E. L., Pure Appl. Chem. **10**, 53-9 (1965); CA **63**, 1454.
--- and Wright, C. M., J. Am. Chem. Soc. **86**, 5132-7 (1964); CA **62**, 3633.
373. Mufti, A. S. and Poller, R. C., J. Chem. Soc. **1965**, 5055-60; CA **63**, 13304.
--- and ---, J. Organometal. Chem. **3**, 99-100 (1965); CA **62**, 7789.
374. Nagasawa, S., Shinohara, H. and Shiba, M., Bochu-Kagaku **30**, 91-5 (1965); CA **63**, 18960.
375. Nagy, J., Borbely-Kuszmann, A. and Toronyi, M., Plaste Kaut. **10**, 402-4 (1963); CA **61**, 806.
376. Nakamura, K. and Fukunishi, S., Takamine Kenkyusho Nempo **13**, 245-9 (1961); CA **63**, 4880.
377. Nakane, T., Kubota, T. and Umezawa, Y., U.S. 3,108,010, July 25, 1960; CA **60**, 751.
378. Nakanishi, M. and Tsuda, A., Japan. 15,690 ('64), May 30, 1961; CA **62**, 6513.
379. Nakano, S. and Yamawaki, T., Japan. 7064 ('65), Nov. 16, 1961; CA **63**, 703.
380. Nash, G. A., Skinner, H. A. and Stack, W. F., Trans. Faraday Soc. **61**, 640-8 (1965); CA **62**, 15497.
381. Naylor, F. E. and Farrar, R. C., Jr., U.S. 3,177,183, Feb. 23, 1961; CA **63**, 4494.
381a. Nefedov, O. M. et al., Dokl. Akad. Nauk SSSR **162**, 589-92 (1965); CA **63**, 7795.
382. ---, Kolesnikov, S. P. and Novitskaya, N. N., Izv. Akad. Nauk SSSR, Ser. Khim. **1965**, 579-80; CA **63**, 624.
382a. ---, --- and Sheichenko, W. I., Angew. Chem. **76**, 498-9; Int. Ed. Engl. **3**, 226 (1964).
383. ---, Manakov, M. N. and Petrov, A. D., Sintez i Svoistva Monomerov, Akad. Nauk SSSR, ... Konf. po Vysokomol. Soedin. **1962**, 67-74 (1964); CA **62**, 6502.
384. Nefedov, V. D. et al., Radiokhimiya **6**, 112-3 (1964); CA **61**, 3721.
385. Nelson, W. H. and Martin, D. F., J. Inorg. Nucl. Chem. **27**, 89-93 (1965); CA **62**, 7788.
386. --- and ---, J. Organometal. Chem. **4**, 67-73 (1965); CA **63**, 4322.
387. Nesmeyanov, A. N. et al., Izv. Akad. Nauk SSSR, Ser. Khim. **1965**, 1122; CA **63**, 8393.
388. Neubert, G., Z. Anal. Chem. **203**, 265-72 (1964); CA **61**, 8900.
388a. Neumann, W. P., Angew. Chem. **75**, 679-80 (1963).
389. --- and Heymann, E., Ann. Chem. **683**, 11-23 (1965); CA **63**, 8390.
390. --- and ---, ibid. **683**, 24-29 (1965); CA **63**, 8391.
391. --- and Kleiner, F. G., Tetrahedron Lett. **1964**, 3779-82; CA **62**, 5291.
392. --- and Koenig, K., Ann. Chem. **677**, 1-11 (1964); CA **62**, 581.
393. --- and ---, ibid. **677**, 12-8 (1964); CA **62**, 582.
394. --- and ---, Angew. Chem. **76**, 892 (1964); CA **62**, 2787.
395. ---, --- and Burkhardt, G., Ann. Chem. **677**, 18-20 (1964); CA **62**, 583.
396. --- and Kuehlein, K., ibid. **683**, 1-11 (1965); CA **63**, 7031.

397. Neumann, W. P. and Kuehlein, K., Angew. Chem. 77, 808-9 (1965); Int. Ed. Engl. 4, 784; CA 63, 18138; cf. ref. 1610.
398. --- and Niermann, H., Belg. 638,642; Brit. 1,043,288; Fr. 1,381,395, Oct. 19, 1962; U.S. 3,347,891, Oct. 15, 1963; CA 62, 11455.
399. ---, Petersen, E. and Sommer, R., Angew. Chem. 77, 622 (1965); Int. Ed. Engl. 4, 599 (1965); CA 63, 11600.
400. --- and Schneider, B., ibid. 76, 891 (1964); Int. Ed. Engl. 3, 751 (1964); CA 62, 2789.
401. Nimni, M., J. Pharm. Sci. 53, 1262-4 (1964); CA 62, 990.
402. Noltes, J. G., Rec. Trav. Chim. 84, 799-805 (1965); CA 63, 8392.
403. Nosek, J., Collection Czech. Chem. Commun. 29, 3173-5 (1964); CA 62, 4046.
404. Oakes, V., Brit. 956,055, Oct. 2, 1961; CA 61, 2421.
405. ---, Brit. 985,721, Feb. 22, 1962; CA 62, 13334.
406. --- and Hutton, R. E., J. Organometal. Chem. 3, 472-7 (1965); CA 63, 4320.
407. O'Brien, J. R., J. Clin. Pathol. 17, 275-81 (1964); CA 62, 9641.
   ---, Nature 207, 306-7 (1965); CA 63, 10516.
408. Ohara, M. and Okawara, R., J. Organometal. Chem. 3, 484-5 (1965); CA 63, 4322.
409. ---, --- and Nakamura, Y., Bull. Chem. Soc. Japan 38, 1379-80 (1965); CA 63, 18137.
409a. Okawara, R. et al., J. Am. Chem. Soc. 83, 1342-4 (1961); CA 55, 18565.
410. Orloff, H. D. and Knapp, G. G., U.S. 3,145,177, May 2, 1960; CA 61, 13109.
411. Orzechowski, A. and MacKenzie, J. C., Brit. 1,038,882; Fr. 1,361,845; Neth. 295,137; U.S. 3,216,982, Mar. 28, 1963; U.S. 3,216,990, April 1, 1963; CA 62, 6588.
412. Osborn, S. W. and Yu, A. J., U.S. 3,201,376, June 27, 1961; CA 63, 15055.
   --- and ---, Brit. 973,318, July 16, 1962; CA 62, 1817.
413. Overmars, H. G. J. and van der Want, G. M., Chimia (Aaran) 19, 126-8 (1965); CA 62, 13326.
414. Ozawa, S. and Maekawa, K., Kobunski Kagaku 20, 357-62 (1963); CA 61, 14854.
415. Patil, H. R. H. and Graham, W. A. G., J. Am. Chem. Soc. 87, 673 (1965); CA 62, 10458.
415a. Patteson, C. C., Arch. Environ. Health 11, 344-60 (1965).
416. Pearce, F. G. et al., U.S. 3,180,810, July 31, 1961; CA 63, 3902.
417. Peddle, G. J. D., Ph. D. Thesis, Univ. of Toronto, 1963; Diss. Abstr. 25, 1553 (1964); CA 62, 3907.
418. Pedinelli, M. and Randi, M., Chim. Ind. (Milan) 46, 172 (1964); CA 61, 83.
419. Pedrotti, R. L. and Sandy, C. A., Brit. 1,015,227; CA 64, 8240; Fr. 1,406,132; U.S. 3,281,442, July 5, 1963; CA 63, 1490.
420. Penneck, R. J. and Pinner, S. H., Brit. 1,001,344, Feb. 13, 1962; CA 63, 13507.

421. Pereyre, M. and Valade, J., Compt. Rend. 260, 581-4 (1965); CA 62, 11841.
422. Peterson, D. B. et al., J. Phys. Chem. 69, 2880-6 (1965); CA 63, 13008.
423. Petrov, A. D., Cheltsova, M. A. and Komarova, S. D., Izv. Akad. Nauk SSSR, Ser. Khim. 1965, 550-2; CA 63, 625.
424. Petukhov, V. A., Mironov, V. F. and Shorygin, P. P., ibid. 1964, 2203-6; CA 62, 8973.
425. Pieper, G. R., Ph. D. Thesis, Univ. of Calif., 1965; Diss. Abstr. 26, 1897-8 (1965); CA 64, 4197.
--- and Casida, J. E., J. Econ. Entomol. 58 (3), 392-400 (1965); CA 63, 2334.
426. Pieters, H. and Buis, W. J., Microchem. J. 8, 383-4 (1964); CA 62, 11137.
427. Pilloni, G. and Plazzogna, G., Ric. Sci., Rend. Sez. A4, 27-33 (1964); CA 61, 1278.
428. Pohlmann, J. L. W. et al., Z. Naturforsch. 20b, 1-4 (1965); CA 62, 13168.
429. Pollard, F. H., Nickless, G. and Cooke, D. J., J. Chromatog. 17, 472-82 (1965); CA 63, 5666.
430. ---, --- and Dolan, D. N., Chem. Ind. (London) 1965, 1027; CA 63, 5667.
431. ---, --- and Uden, P. C., J. Chromatog. 14, 1-12 (1964); CA 60, 13855.
432. ---, --- and ---, ibid. 19, 28-56 (1965); CA 63, 14898.
433. Polster, R. and Adolphi, H., Ger. 1,181,977, March 9, 1963; CA 62, 4555.
434. Pommier, J. C., Pereyre, M. and Valade, J., Compt. Rend. 260, 6397-8 (1965); CA 63, 7032.
435. --- and Valade, J., ibid. 260, 4549-52 (1965); CA 63, 2890.
436. --- and ---, Bull. Soc. Chim. Fr. 1965, 975-80; CA 63, 4319.
437. Ponomarev, S. V. and Lutsenko, I. F., Zh. Obshch. Khim. 34, 3450-3 (1964); CA 62, 2787.
438. Pope, A. E. et al., NASA Accession No. N65-10358, Rept. No. AD442716, 12 pp. (1963); CA 62, 15497.
439. Popova, Z. V., Tikhova, N. V. and Vyazankin, N. S., Vysokomol. Soedin., Khim. Svoistva i Modifikatsiya Polimerov, Sb. Statei 1964, 175-8; CA 62, 14892.
440. Proskurnina, M. V., Novikova, Z. S. and Lutsenko, I. F., Dokl. Akad. Nauk SSSR 159, 619-21 (1964); CA 62, 6508.
441. Pudovik, A. N. and Muritova, A. A., ibid. 158, 419-22 (1964); CA 62, 11646.
442. Putnam, R. C. and Pu, H., J. Gas Chromatog. 3, 160-4 (1965); CA 63, 10725.
443. Quan, M. L. and Cadiot, P., Bull. Soc. Chim. Fr. 1965, 35-44; CA 62, 14716.
444. --- and ---, ibid. 1965, 45-7; CA 62, 14717.
445. Quattlebaum, W. M., Jr. and Hardwicke, J. E., U.S. 3,208,969, Feb. 15, 1962; CA 63, 18376.
446. Rand, L. et al., J. Appl. Polymer Sci. 9, 1787-95 (1965); CA 63, 453.
447. Reeves, L. W., J. Chem. Phys. 40, 2128-31 (1964); CA 60, 12796.
448. Reyerchov, R., Chim. Anal. (Paris) 47, 70-6 (1965); CA 62, 15428.

449. Reyne, J., Am. Paint J. **48** (21), 99-110 (1963); CA **62**, 6681.
450. Ricciardi, M. A. and Considine, W. J., U.S. 3,198,757, June 12, 1961; CA **63**, 11821.
451. Richardson, B. A., Tin Its Uses No. **64**, 5-9 (1964); Wood **1964** (June) 57-60; CA **62**, 3340.
452. Richardson, G. A. and Birum, G. H., U.S. 3,190,892, July 20, 1960; CA **63**, 11613.
453. Rieche, A., Kunststoffe **54**, 428-35 (1964); CA **62**, 2844.
454. Riedel, H. J. and Merz, E., Radiochim. Acta **4**, 48-51 (1965); CA **63**, 7828.
455. Ritchie, L. S. et al., Bull. World Health Organ. **31**, 147-9 (1964); CA **62**, 11093.
456. Rivett, P., J. Appl. Chem. (London) **15**, 469-73 (1965); CA **63**, 18461.
457. Robertson, J. and Hickman, J. B., Proc. West Va. Acad. Sci. **36**, 97-102 (1964); CA **61**, 15406.
458. Roeser, G. P., Fr. AD. 83,866, April 4, 1962; CA **62**, 10380.
459. Rosenthal, A. J., Belg. 632,815; Brit. 1,039,938; U.S. 3,220,866, March 25, 1962; CA **61**, 2784.
460. Ryan, M. T. and Lehn, W. L., AD 423,846 (OTS), 8 pp. 1963; CA **62**, 7276.
--- and ---, J. Organometal. Chem. **4**, 455-60 (1965); CA **64**, 169.
460a. Sanders, L. W., Arch. Environ. Health **8**, 270-7 (1964); CA **60**, 15042.
461. Sandler, S. R., Dannin, J. and Tsou, K. C., J. Polymer Sci. Pt. A**3**, 3199-207 (1965);
462. Sarankina, S. A. and Manulkin, Z. M., Usbeksk. Khim. Zh. **9** (3), 30-4 (1965); CA **63**, 14893.
463. --- and ---, Zh. Obshch. Khim. **35**, 845-8 (1965); CA **63**, 7032.
464. ---, --- and Kuchkarov, A. B., Dokl. Akad. Nauk Uz. SSR **22** (7), 35-7 (1965); CA **63**, 18137.
465. Satge, J. and Lesbre, M., Bull. Soc. Chim. Fr. **1965**, 2578-81; CA **63**, 18136.
466. ---, --- and Baudet, M., Compt. Rend. **259**, 4733 (1964); CA **62**, 11842.
467. --- and Massol, M., ibid. **261**, 170-3 (1965); CA **63**, 11601.
468. Sawyer, A. K., J. Am. Chem. Soc. **87**, 537-9 (1965); CA **62**, 9163; cf. ref. 1802.
469. ---, Brown, J. E. and Hanson, E. L., J. Organometal. Chem. **3**, 464-71 (1965); CA **63**, 4322.
470. Scherer, O. J. et al., Z. Naturforsch. B**20**, 183-4 (1965); CA **62**, 16288.
471. ---, Schmidt, J. F. and Schmidt, M., ibid. B**19**, 447 (1964); CA **61**, 5689.
472. --- and Schmidt, M., J. Organometal. Chem. **3**, 156-8 (1965); CA **62**, 11843.
473. Schloer, H. and Langheinrich, K., Ger. 1,161,076, Aug. 30, 1960; CA **61**, 2421.
474. Schmidbaur, H., J. Am. Chem. Soc. **85**, 2336-7 (1963); CA **61**, 1409.
474a. --- and Schmidt, M., ibid. **83**, 2963-4 (1961); CA **55**, 25734.
475. --- and Waldmann, S., Chem. Ber. **97**, 3381-91 (1964); CA **62**, 14050.
476. Schmidt, M. and Ruidisch, I., Ger. 1,179,550, Nov. 2, 1962; CA **62**, 1688.
477. --- and ---, Ger. 1,190,462, June 21, 1963; CA **63**, 631.
478. --- and ---, Ger. 1,194,858, Nov. 2, 1962; CA **63**, 13315.

479. Schmidt, T., Pflanzenschutz Ber. 28, 65-77 (1962); CA 60, 12603.
480. Schmitz-DuMont, O., Mueller, G. and Schaal, W., Z. Anorg. Allg. Chem. 332, 263-8 (1964); CA 62, 8654; cf. Z. Anorg. Allg. Chem. 357, 139-40 (1968).
481. Schuler, M. J., U.S. 3,197,491, Feb. 7, 1961; CA 63, 9985.
482. Schulz-Utermoehl, H. and Weissenstein, H., Gesundheitsw. Desinfekt. 57 (4), 41-5 (1965); CA 63, 6793.
483. Schumann, H. and Jutzi, P., Chem. Ber. 101, 24-8 (1968); CA 68, 49701.
---, --- and Schmidt, M., Angew. Chem. 77, 812; Int. Ed. Engl. 4, 787 (1965); CA 63, 18142.
483a. ---, Koepf, H. and Schmidt, M., Z. Anorg. Allg. Chem. 331, 200-5 (1964); CA 61, 12031.
484. ---, Thom, K. F. and Schmidt, M., J. Organometal. Chem. 4, 22-7 (1965); CA 63, 4321.
485. ---, --- and ---, ibid. 4, 28-33 (1965); CA 63, 4321.
486. Schrauzer, G. N. and Kratel, G., Angew. Chem. 77, 130 (1965); CA 62, 11406.
487. Schweitzer, G. K. and McCarthy, S. W., J. Inorg. Nucl. Chem. 27, 191-9 (1965); CA 62, 4683; cf. ref. 345a.
488. Scott, F. L., Brit. 1,001,369; U.S. 3,214,279, June 9, 1961; CA 63, 13569.
489. Seibel, A. D., Brit. 1,031,289; Fr. 1,377,380; Ger. 1,569,255, Sept. 6, 1962; CA 62, 11983.
490. Selivokhin, P. I., Vestn. Tekhn. i Ekon. Inform. Nauchn.-Issled. Gos. Kom. Khim. Prom. pri Gosplane SSSR 8, 27 (1964); CA 63, 17155.
491. Semlyen, J. A. et al., J. Chem. Soc. 1964, 4948-53; CA 62, 6502.
492. --- and Phillips, C. S. G., J. Chromatog. 18, 1-9 (1965); CA 63, 4947.
493. ---, Walker, G. R. and Phillips, C. S. G., J. Chem. Soc. 1965, 1197-1203; CA 62, 9163.
494. Seyferth, D. et al., J. Am. Chem. Soc. 87, 681-2 (1965); CA 62, 10347.
494a. --- and Cohen, H. M., J. Organometal. Chem. 1, 15-21 (1963); CA 60, 4165.
495. ---, Sato, Y. and Takamizawa, M., ibid. 2, 367-8 (1964); CA 61, 14707.
496. --- and Vaughan, L. G., J. Am. Chem. Soc. 88, 883-90 (1964); CA 60, 10703.
497. Shapiro, H., De Witt, E. G. and Brown, J. E., U.S. 3,130,017, Feb. 19, 1958; CA 61, 520.
498. Sharp, D. W. A. and Winfield, J. M., J. Chem. Soc. 1965, 2278-9; CA 63, 624.
499. Sheehan, D., Ph. D. Thesis, Yale Univ., 1964; Diss. Abstr. 25, 4417 (1965); CA 63, 2993.
500. Shikhiev, I. A. and Abdullaev, N. D., Zh. Obshch. Khim. 35, 1348-50 (1965); CA 63, 16377.
501. ---, Aslanov, I. A. and Mekhmandrova, N. T., ibid. 35, 459-61 (1965); CA 63, 624.
502. ---, Guseinzade, B. M. and Abdullaev, N. D., Dokl. Akad. Nauk. Azerb. SSSR 20 (11), 13-7 (1964); CA 62, 16289.

503. Shpinel, V. S. et al., Zh. Eksperim. i Teor. Fiz. 48, 69-71 (1965); CA 62, 14066.
504. Shostakovskii, M. F. et al., Dokl. Akad. Nauk SSSR 158, 918-21 (1964); CA 62, 2788.
505. --- et al., ibid. 161, 370-2 (1965); CA 63, 624.
506. --- et al., ibid. 163, 390-3 (1965); CA 63, 11601.
507. --- et al., Zh. Obshch. Khim. 34, 3178-80 (1964); CA 62, 4046.
508. --- et al., ibid. 35, 47-51 (1965); CA 62, 13168.
 --- et al., U.S.S.R. 165,454, Aug. 6, 1963; CA 64, 5138.
509. --- et al., Zh. Obshch. Khim. 35, 401-2 (1965); CA 62, 13167.
510. --- et al., ibid. 35, 751 (1965); CA 63, 5667.
511. ---, Vlasov, V. M. and Mirskov, R. G., Dokl. Akad. Nauk SSSR 159, 869-71 (1964); CA 62, 7788.
512. ---, --- and ---, Zh. Obshch. Khim. 35, 750 (1965); CA 62, 4322.
513. ---, --- and ---, ibid. 35, 1121 (1965); CA 63, 9976.
 --- et al., U.S.S.R. 175,505, April 4, 1964; CA 64, 5138.
514. Sijpesteijn, A. K. et al., Antonie van Leeuwenhoek, J. Microbiol. Serol. 30, 113-20 (1964); CA 62, 2005.
515. Sinitsyn, S. N., Farmakol. i Toksikol. 27, 619-20 (1964); CA 62, 3306.
516. Sirtl, E., Ger. 1,198,329, Feb. 23, 1961; CA 63, 14427.
517. Skinner, J. F. and Fuoss, R. M., J. Phys. Chem. 68, 2998-3003 (1964); CA 61, 13946.
517a. Skreckoski, G. R., Brit. 970,324; U.S. 3,224,988, Dec. 29, 1960; CA 62, 732.
518. Smith, G. W., J. Chem. Phys. 40, 2037-8 (1964); CA 61, 1408.
 ---, ibid. 42, 4229-43 (1965); CA 63, 5148.
519. Smith, J. A. S. and Wilkins, E. J., Chem. Commun. 1965, 381-2; CA 63, 14358.
520. Snegova, A. D., Markov, L. K. and Ponomarenko, V. A., Zh. Analit. Khim. 19, 610-4 (1964); CA 61, 6402.
521. Snegur, L. N. and Manulkin, Z. M., Zh. Obshch. Khim. 34, 4030-2 (1964); CA 62, 9164.
522. Soborovskii, L. Z., Gladshtein, B. M. and Kulylin, I. P., U.S.S.R. 168,693, March 25, 1958; CA 63, 1815.
523. Sokolovskaya, A. I., Tr. Fiz. Inst. Akad. Nauk SSSR 27, 63-110 (1964); CA 61, 7845.
524. Solodovnikov, S. P. and Chernyshev, E. A., Tr. Soveshch. po Fiz. Metodam Issled. Organ. Soedin. i Khim. Protsessov, Akad. Nauk Kirg. SSSR, Inst. Organ. Khim., Frunze 1962, 196-212 (1964); CA 62, 4803.
525. Sommer, R., Neumann, W. P. and Schneider, B., Tetrahedron Lett. 1964, 3875-8; CA 62, 6499.
526. Sone, N. and Hagihara, B., J. Biochem. (Tokyo) 56, 151-6 (1964); CA 61, 14922.
527. Sorokina, R. S., Panov, E. M. and Kocheshkov, K. A., Zh. Obshch. Khim. 35, 1625-8 (1965); CA 63, 18138.
528. Spialter, L., Buell, G. R. and Harris, C. W., J. Org. Chem. 30, 275-8 (1965); CA 62, 9165.

529. Srivastava, T. N. and Tandon, S. K., Indian J. Appl. Chem. 27, 116-8 (1964); CA 62, 4359.
530. Stadnichuk, M. D. and Petrov, A. A., Zh. Obshch. Khim. 35, 451-6 (1965); CA 63, 471.
531. --- and ---, ibid. 35, 700-4 (1965); CA 63, 4322.
532. Stamm, W., J. Org. Chem. 30, 693-5 (1965); CA 62, 18842.
532a. ---, Belg. 653,662; Brit. 1,021,200; Fr. 1,409,197; Ger. 1,212,976; Neth. 64 11,318; U.S. 3,326,906, Oct. 1, 1963; CA 63, 13316.
532b. ---, Belg. 653,823; Brit. 1,071,714; Fr. 1,458,934; Neth. 64 11,266; U.S. 3,311,647, Oct. 4, 1963; CA 63, 11613; Ger. 1,206,901, April 2, 1963; CA 64, 9767.
533. ---, Brit. 1,048,918; Fr. 1,405,428; Ger. 1,215,711; U.S. 3,409,653; CA 63, 14904.
534. --- and Lake, B. H., U.S. 3,206,489, July 3, 1962; CA 63, 18152.
535. Stegmann, H. B. and Scheffler, K., Tetrahedron Lett. 1964, 3387-92; CA 62, 6371.
536. Steinmeyer, R. D., Fentiman, A. F. and Kahler, E. J., Anal. Chem. 37, 520-3 (1965); CA 63, 2388.
537. Stern, A., Ph. D. Thesis, Polytech. Inst. of Brooklyn, N.Y., 1964; Diss. Abst. 25, 2768-9 (1964); CA 62, 6499; cf. ref. M-2559.
538. Stern, C. J., U.S. 3,179,676, March 28, 1961; CA 63, 2999.
539. Stokinger, H. E. et al., J. Occupational Med. 5, 491-8 (1963); CA 60, 6120.
539a. Strickland, A., Ger. 1,159,633, Feb. 27, 1959; CA 60, 9458.
540. Suenobu, Y., Japan. 2336 ('65), Jan. 23, 1962; CA 62, 14726.
541. Sulzbacher, M., Mfg. Chemist Aerosol News 35, 50, 52, 56 (1964); CA 61, 5469.
542. Swiger, E. D. and Graybeal, J. D., J. Am. Chem. Soc. 87, 1464-6 (1965); CA 62, 12630.
543. Taimsalu, P. and Wood, J. L., Spectrochim. Acta 20, 1357-68 (1964); CA 61, 14044.
544. Takami, Y., Tokyo Kogyo Shikensho Hokoku 57, 234-40 (1962); CA 62, 2826.
545. Takeda, Y. et al., Makromol. Chem. 83, 234-43 (1965); CA 63, 1874.
546. Takenaka, T. and Goto, R., Proc. Intern. Symp. Mol. Struct. Spectry., Tokyo 1962 (A211), 4 pp., CA 61, 1395.
547. Tamborski, C., Soloski, E. J. and Dec, S. M., J. Organometal. Chem. 4, 446-54 (1965); CA 63, 18136.
---, U.S. 3,392,178, March 31, 1964; CA 69, 87184.
548. Tamura, H., Nogyo Gijutsu Kenkyusho Hokoku, Byori Konchu 18, 135-204 (1965); CA 63, 18958.
549. ---, ibid. 19, 47-79 (1965); CA 63, 18958.
550. Tanaka, T. et al., Bull. Chem. Soc. Japan 37, 1554-5 (1964); CA 62, 2788.
551. Tanaka, Y. and Morikawa, T., Bunseki Kagaku 13, 753-9 (1964); CA 61, 11338.
552. Tappan, W. B., J. Econ. Entomol. 58, 730-2 (1965); CA 63, 7600.
553. Taylor, J. M. et al., Nature 204, 891-2 (1964); CA 62, 6768.

554. Tazewell, J. H. and Reid, R. J., U.S. 3,208,970, Oct. 7, 1960; CA <u>63</u>, 16554.
555. Telnoi, V. I. and Rabinovich, I. B., Zh. Fiz. Khim. <u>39</u>, 2076-7 (1965); CA <u>63</u>, 14146.
556. Thayer, J. S., Ph. D. Thesis, Univ. of Wisconsin 1964; Diss. Abstr. <u>25</u>, 2222 (1964); CA <u>62</u>, 6501.
557. --- and West, R., Inorg. Chem. <u>3</u>, 889-93 (1964); CA <u>61</u>, 1495.
558. --- and ---, ibid. <u>4</u>, 114-5 (1965); CA <u>62</u>, 6127.
559. Thomas, J. R., Brit. 964,014; Ger. 1,154,442; U.S. 3,167,525, March 21, 1960; CA <u>60</u>, 1143.
560. Thomas, W. H. and Cook, S. E., U.S. 3,197,492, June 8, 1962; CA <u>63</u>, 9730.
561. Thyssen-Bornemisza, I. v., Brit. 997,367, April 1, 1961; CA <u>63</u>, 11242.
562. Tillyaev, K. S. and Manulkin, Z. M., Dokl. Akad. Nauk Uz. SSR <u>22</u> (2), 45-7 (1965); CA <u>63</u>, 5668.
563. Tobias, R. S. and Freidline, C. E., Inorg. Chem. <u>4</u>, 215-20 (1965); CA <u>62</u>, 7171.
564. --- and Yasuda, M., J. Phys. Chem. <u>68</u>, 1820-8 (1964); CA <u>61</u>, 5009.
565. Tokunaga, H., Murayama, Yu. and Kijima, I., Japan. 24,958 ('64), Jan. 16, 1961; CA <u>62</u>, 14726.
566. Tomilov, A. P., Smirnov, Yu. D. and Varshavskii, S. L., Zh. Obshch. Khim. <u>35</u>, 391-3 (1965); CA <u>63</u>, 5238.
567. Treichel, P. M. and Goodrich, R. A., Inorg. Chem. <u>4</u>, 1424-8 (1965); CA <u>63</u>, 16380.
568. Tsai, T. T., Cutler, A. and Lehn, W. L., J. Org. Chem. <u>30</u>, 3049-52 (1965); CA <u>63</u>, 11323.
569. Tsou, K. C., IEEE, Trans. Nucl. Sci. <u>12</u>, 28-33 (1965); CA <u>63</u>, 2597.
570. Tzalmona, A., Mol. Phys. <u>7</u>, 497-8 (1963-4); CA <u>62</u>, 6035.
571. Uraneck, C. A. and Kahle, G. R., Belg. 644,681; Brit. 1,041,613, March 4, 1963; CA <u>63</u>, 13542; U.S. 3,383,377, March 4, 1963; CA <u>69</u>, 11250.
572. Updegraff, D. M. and Davis, H. R., U.S. 3,211,680, Sept. 18, 1961; CA <u>63</u>, 18476.
573. Verdonck, L. and van der Kelen, G. P., Ber. Bunsenges. Physik. Chem. <u>69</u>, 478-84 (1965); CA <u>63</u>, 7799.
574. Verhoeff, K., Neth. J. Plant Pathol. <u>69</u>, 265-78 (1963); CA <u>61</u>, 6300.
575. Vershinina, S. P. et al., Med. Radiol. <u>10</u> (4), 73-4 (1965); CA <u>63</u>, 9398.
576. Vesely, V. and Gursky, J., Ropa Uhli <u>7</u>, 215-20 (1965); CA <u>63</u>, 17753.
577. Vignes, S. and Cottin, A., Compt. Rend. Congr. Ind. Gaz <u>80</u>, 246-70 (1963); CA <u>62</u>, 14400.
578. Vilkov, L. V. et al., Zh. Fiz. Khim. <u>38</u>, 2674-5 (1964); CA <u>62</u>, 4760.
579. Vinogradova, L. M. et al., Plast. Massy <u>1964</u> (9), 18-20; CA <u>61</u>, 16235.
580. Vitalina, M. D. and Klimova, V. A., Zh. Analit. Khim. <u>17</u>, 1105-8 (1962); CA <u>58</u>, 11958.
581. Vladimiroff, T. and Malinowski, E. R., J. Chem. Phys. <u>42</u>, 440-2 (1965); CA <u>62</u>, 11316.
582. van der Vondel, D. F., J. Organometal. Chem. <u>3</u>, 400-5 (1965); CA <u>63</u>, 471.

583. Voronkov, M. G. and Biryukov, I. P., Teor. i Eksperim. Khim., Akad. Nauk Ukr. SSR 1, 124-6 (1965); CA 63, 3803.
583a. Vyazankin, N. A. et al., Dokl. Akad. Nauk SSSR 155, 1108-10 (1964).
584. Vyazankin, N. S. et al., ibid. 158, 884-7 (1964); CA 62, 2788.
585. --- and Bychkov, V. T., Zh. Obshch. Khim. 35, 684-7 (1965); CA 63, 4320.
586. ---, Razuvaev, G. A. and Brevnova, T. N., Dokl. Akad. Nauk SSSR 163, 1389-92 (1965); CA 63, 16379.
587. ---, --- and Bychkov, V. T., Izv. Akad. Nauk SSSR, Ser. Khim. 1965, 1665-7; CA 63, 18137.
587a. ---, --- and ---, Zh. Obshch. Khim. 35, 395-6 (1965).
588. Vyshinskii, N. N., Kozlova, T. V. and Rudnevskii, N. K., Tr. Komis. po Spektroskopii Akad. Nauk SSSR 1964, 451-9; CA 63, 13027.
589. --- and Rudnevskii, N. K., Spektroskopiya, Metody i Primenenie, Akad. Nauk SSSR, Sibirsk. Otd. 1964, 115-8; CA 62, 3533.
590. Wada, M., Kawakami, K. and Okawara, R., J. Organometal. Chem. 4, 159-60 (1965); CA 63, 16376.
591. ---, Nishino, M. and Okawara, R., ibid. 3, 70-5 (1965); CA 62, 10456.
591a. Wainer, E., Belg. 651,787; Fr. 1,406,796; Neth. 64 9367; U.S. 3,275,443, Aug. 14, 1963; CA 63, 181.
592. --- and Fotland, R. A., U.S. 3,147,117, May 26, 1961; CA 61, 12837.
593. Wall, H. H. Jr., U.S. 3,158,636, Oct. 29, 1963; CA 62, 3870.
594. Wang, S.-L., Hua Hsueh Hsueh Pao 30, 211-4 (1964); CA 61, 6378.
594a. Wasserman, D. et al., J. Org. Chem. 30, 3248-50 (1965); CA 63, 14895.
595. Watanabe, S. and Fujii, M., Japan. 5810 ('65), Sept. 8, 1962; CA 63, 1815.
596. Webbe, G. and Sturrock, R. F., Ann. Trop. Med. Parasitol. 58, 234-9 (1964); CA 61, 15286.
597. Weisfeld, L. B., Thacker, G. A. and Nass, L. I., S.P.E. J. 21, 649-58 (1965); CA 63, 13491.
598. Weissenberger, G., U.S. 3,188,331, July 21, 1961; CA 63, 5676.
599. ---, U.S. 3,208,978, Nov. 16, 1961; CA 63, 18372.
600. Wells, E. J. and Reeves, L. W., J. Chem. Phys. 40, 2036-7 (1964); CA 61, 1281.
601. West, R., J. Organometal. Chem. 3, 314-20 (1965); CA 62, 12609.
602. Westlake, A. H. and Martin, D. F., J. Inorg. Nucl. Chem. 27, 1579-89 (1965); CA 63, 5220.
603. Whaley, T. P. and Norman, V., U.S. 3,071,493, Nov. 15, 1961; CA 58, 4235.
604. Wheeler, P. P. and Richardson, A. C., Mikrochim. Acta 1964, 609-15; CA 61, 13874.
605. White, P., Electronics Reliability Microminiaturizn. 24, 161-6 (1963); CA 63, 12463.
605a. Wiberg, E. et al., Angew. Chem. Intern. Ed. Engl. 2, 507 (1963).
605b. Wilkins, C. J. and Haendler, H. M., J. Am. Chem. Soc. 1965, 3174-9; CA 63, 1464.
606. Willemsens, L. C. and van der Kerk, G. J. M., Rev. Trav. Chim. 84, 43-4 (1965); CA 62, 9163.

607. Willemsens, L. C. and van der Kerk, G. J. M., J. Organometal. Chem. 4, 34-42 (1965); CA 63, 4322.
608. --- and ---, ibid. 4, 241-4 (1965); CA 63, 11601.
609. Willomitzer, J., Ved. Prace Ustred. Vzykum. Ustavu Vet. Lekar. Brne 1962, 265-76; CA 61, 16675.
610. Wilson, G. R., U.S. 3,184,430, May 14, 1962; CA 63, 5774.
611. Winters, L. J. and Hill, D. T., Inorg. Chem. 4, 1433-6 (1965); CA 63, 16376.
612. Wissman, H. G., Rand, L. and Frisch, K. C., J. Appl. Polymer Sci. 8, 2971-8 (1964); CA 62, 696.
613. Wood, J. M. Jr., U.S. 3,197,414, June 26, 1961; CA 63, 9730.
614. Wronkowski, C., Gaz. Woda Tech. Sanit. 39 (4), 131-2 (1965); CA 63, 14082.
615. Wu, C.-L. et al., Yuan Tzu Neng 1, 27-31 (1965); CA 63, 14318.
616. Wunderlich, D. K. and Fussel, L. N., U.S. 3,159,557, June 1, 1961; CA 62, 11432.
616a. Wyant, R. E. et al., U.S. At. Energy Comm. BMI-1654, 12 pp., 1963; CA 60, 7603.
617. Yakubova, F. A. et al., Zh. Obshch. Khim. 35, 387-91 (1965); CA 62, 13169.
618. Yanagibashi, H., Tokyo Ika Daigaku Zasshi 20, 1-50 (1962); CA 61, 3600.
619. Yasuda, K. et al., Bull. Chem. Soc. Japan 38, 1216-8 (1965); CA 63, 17247.
620. --- and Okawara, R., J. Organometal. Chem. 3, 76-83 (1965); CA 62, 7789.
621. Yergey, A. L. and Lampe, F. W., J. Am. Chem. Soc. 87, 4204-5 (1965); CA 63, 17864; cf. ref. 2869.
622. Yoshikawa, K., Kurose, K. and Teramoto, S., Kogyo Kagaku Zasshi 67, 740-4 (1964); CA 61, 11032.
623. ---, --- and ---, ibid. 67, 1418-23 (1964); CA 62, 4359.
624. Yurev, Yu. K., Galbershtam, M. A. and Kandor, I. I., Zh. Obshch. Khim. 34, 4116 (1964); CA 62, 9163.
625. Zablotna, R., Bull. Acad. Polon. Sci., Ser. Chim. 12, 475-8 (1964); CA 62, 1311.
626. ---, Akerman, K. and Szuchnik, A., ibid. 12, 695-9 (1964); CA 62, 13167.
627. Zakharkin, L. I., Bregadze, V. I. and Okhlobystin, O. Yu., J. Organometal. Chem. 4, 211-6 (1965); CA 63, 11597.
628. Zavaglia, E. A., Mosher, W. A., and Billmeyer, F. W. Jr., Offic. Dig., J. Paint Technol. Eng. 37, 229-34 (1965); CA 63, 3054.
629. Zavgorodnii, V. S. and Petrov, A. A., Zh. Obshch. Khim. 35, 760 (1965); CA 63, 5666.
630. --- and ---, ibid. 35, 931-2 (1965); CA 63, 7033.
631. --- and ---, ibid. 35, 1313-4 (1965); CA 63, 11601.
632. Zemlyanskii, N. N. et al., Dokl. Akad. Nauk SSSR 156, 131-4 (1964); CA 61, 5046.
633. --- et al., Zh. Obshch. Khim. 35, 843-5 (1965); CA 63, 7031.
634. --- et al., ibid. 35, 1029-31 (1965); CA 63, 9978.

635. Zenbutsu, T., Ikeda, T. and Chizaki, E., Japan. 12,181 ('65), April 3, 1962; CA 63, 18293.
636. Zherdev, Yu. V., Korolev, A. Ya. and Leznov, N. S., Vysokomol. Soedin., Khim. Svoistva i Modifikatsiya Polimerov, Sb. Statei 1964, 260-4; CA 62, 2880.
637. Zhivukin, S. M. et al., U.S.S.R. 165,717, July 4, 1963; CA 62, 6513.
637a. Ziegfeld, R. L., Arch. Environ. Health 8, 202-12 (1964); CA 60, 15041.
638. Ziegler, K. and Lehmkuhl, H., Ger. 1,181,220, July 2, 1962; CA 62, 6156.
639. Zimmer, H. J., Brit. 999,805, Aug. 31, 1961; CA 63, 16522, 18338.
640. Zschitzschmann, H., Ger. (East) 39,200, Aug. 13, 1964; CA 63, 18480.
641. Zuliani, G., Metano (Padua) 14, 207-20 (1960); CA 62, 5113.
642. Abe, K., Uesugi, H. and Suzuki, S., Japan. 12,777 ('65), July 26, 1962; CA 64, 8410.
643. Abel, E. W., Armitage, D. A. and Brady, D. B., J. Organometal. Chem. 5, 130-5 (1966); CA 64, 6684.
644. ---, --- and ---, Trans. Faraday Soc. 62, 3459-62 (1962); CA 66, 37286.
645. ---, --- and Tyfield, S. P., J. Chem. Soc., A 1967, 554-7; CA 66, 115195.
646. --- and Brady, D. B., J. Organometal. Chem. 11, 145-9 (1968); CA 68, 44628.
647. ---, --- and Crosse, B. C., ibid. 5, 260-2 (1966); CA 64, 11245.
648. --- and Crosse, B. C., J. Chem. Soc., A 1966, 1377-8; CA 66, 18140.
649. --- and ---, ibid. A 1966, 1141-3; CA 65, 14816.
650. ---, --- and Brady, D. B., J. Am. Chem. Soc. 87, 4397-8 (1965); CA 64, 1609.
651. ---, --- and Hutson, G. V., Chem. Ind. (London) 1966, 238; J. Chem. Soc., A 1967, 2014-7; CA 68, 39789.
652. --- and Jenkins, C. R., J. Chem. Soc., A 1967, 1344-6; CA 67, 108726.
653. Abraham, M. H. and Hill, J. A., Chem. Ind. (London) 1965, 561-2; J. Organometal. Chem. 7, 11-21 (1967); CA 66, 45925.
654. --- and Spalding, T. R., Chem. Commun. 1968, 46; CA 68, 77383.
654a. Adachi, N., Brit. 1,077,190; Fr. 1,419,968; U.S. 3,418,162, Dec. 7, 1963; CA 66, 3788.
655. Affolder, J., Jacot-Guillarmod, A. and Bernauer, K., Helv. Chim. Acta 51, 293-300 (1968); CA 68, 81819.
656. Aftandilian, V. D., U.S. 3,285,891, Jan. 9, 1964; CA 66, 19013.
657. --- and MacKenzie, J. C., Fr. 1,399,045, May 31, 1963; CA 64, 5227.
658. Agolini, F. et al., Spectrochim. Acta, Pt. A 24, 169-86 (1968); CA 68, 91412.
659. Akerman, K. et al., Radioisotope Tracers Ind. Geophys., Proc. Symp., Prague 1966, 459-67 (1967); CA 68, 70854.
--- and Szuchnik, A., Intern. J. Appl. Radiation Isotopes 15, 319-24 (1964); CA 65, 16739.
660. Aleksandrov, A. Yu. et al., Dokl. Akad. Nauk SSSR 165, 593-6 (1965); CA 64, 9106.
661. Aleksandrov, Yu. A. and Radbil, B. A., Zh. Obshch. Khim. 36, 543-7 (1966); CA 65, 744.

662. Aleksandrov, Yu. A., Radbil, B. A. and Shushunov, V. A., Zh. Obshch. Khim. 37, 208-13 (1967); CA 66, 104491.
663. --- and Sheyanov, N. G., ibid. 36, 953 (1966); CA 65, 8955.
664. --- and ---, ibid. 37, 2136-7 (1967); CA 68, 29822.
665. --- and Suldin, B. V., ibid. 37, 2350-4 (1967); CA 68, 86622.
666. ---, --- and Kokurina, S. N., ibid. 36, 2198-202 (1966); CA 67, 32171.
667. ---, --- and ---, Tr. Khim. Khim. Tekhnol. 1965 (3), 228; CA 67, 3128.
668. Alexander, R. and Parker, A. J., J. Am. Chem. Soc. 89, 5549-51 (1967); CA 68, 6859.
669. Alexandri, A. V., Iosifescu, M. and Stoian, E., An. Inst. Cent. Cercet. Agr., Sect. Prot. Plant No. 3, 97-104 (1965); CA 68, 58673.
670. Ali-Sade, I. G. et al., Dokl. Akad. Nauk SSSR 173, 89-92 (1967); CA 67, 32723.
671. Almenningen, A., Haaland, A. and Motzfeld, T., J. Organometal. Chem. 7, 97-104 (1967); CA 66, 37323.
672. Alperstein, M. and Bradow, R. L., S.A.E. J. 75, 52-3 (1967); CA 67, 66298.
673. Amberger, E. and Hoenigschmid-Grossich, R., Chem. Ber. 98, 3795-803 (1965); CA 64, 6061.
674. --- and ---, ibid. 99, 1673-7 (1966); CA 65, 2292.
675. ---, Stoeger, W. and Hoenigschmid-Grossich, R., Angew. Chem. 78, 549 (1966); Int. Ed. Engl. 5, 522 (1966); CA 65, 8319.
676. --- and Salazar, R. W., Intern. Symp. Organosilicon Chem., Sci. Commun., Suppl., Prague 1965, 31-3; CA 65, 8955.
--- and ---, J. Organometal. Chem. 8, 111-4 (1967); CA 66, 85833.
677. Ameyama, M., Kami-pa Gikyoshi 20, 496-504 (1966); CA 65, 15657.
678. Anand, L. C., Deshpande, A. B. and Kapur, S. L., Indian J. Chem. 5, 188-90 (1967); CA 67, 82440.
679. Anderson, H. M. and Johnson, W. K., U.S. 3,225,022, Dec. 27, 1960; CA 64, 8340.
680. Anderson, H. H., J. Chem. Eng. Data 12 (3), 371-2 (1967); CA 67, 67864.
681. Ando, S. and Higuchi, H., Japan. 17,658 ('66), Dec. 18, 1963; CA 66, 66226.
682. Ando, T. et al., J. Am. Chem. Soc. 89, 5719-21 (1967); CA 68, 48736.
683. Armenskaya, L. V. et al., U.S.S.R. 172,785, Aug. 26, 1964; CA 64, 1662, cf. ref. 1348.
684. --- et al., U.S.S.R. 184,853, April 23, 1965; CA 66, 71949, cf. ref. 1348.
685. Armer, B. and Schmidbaur, H., Chem. Ber. 100, 1521-35 (1967); CA 67, 11552.
686. Ascher, K. R. S., Moscowitz, J. and Nissim, S., Tin Its Uses No. 73. 8-9 (1967); CA 67, 63281.
--- and ---, Int. Pest Contr. 10 (3), 10-13 (1968); CA 69, 85686.
687. Ashbel, F. B. et al., Zavodsk. Lab. 31, 1062-3 (1965); CA 64, 11.
688. Ashikara, N. and Nishioka, A., Japan. 21,435 ('65), March 30, 1963; CA 64, 2189.

689. Asso, M. and Carpeni, G., Omagiu Raluca Ripan. 1966, 81-96; CA 67, 76725.
690. Atavin, A. S., Dubova, R. I. and Vasilev, N. P., Zh. Obshch. Khim. 36, 1506-7 (1966); CA 66, 11017.
---, --- and ---, U.S.S.R. 184,854, May 17, 1965; CA 66, 95198.
691. Atwell, W. H., Weyenberg, D. R. and Gilman, H., J. Org. Chem. 32, 885-8 (1967); CA 66, 105013.
692. Avdeeva, V. I. et al., Zh. Obshch. Khim. 36, 1679-84 (1966); CA 66, 46463.
693. Ayers, C. L., Ph. D. Thesis, Univ. of Georgia 1966; Diss. Abstr. B27, 3476-7 (1967); CA 67, 86403.
694. Aylesworth, R. D., Huber, C. F. and Foulks, H. C., Brit. 1,087,674, April 12, 1965; CA 68, 22528.
695. Ayliffe, G. A. J., Collins, B. J. and Lowbury, E. J. L., Brit. Med. J. 1966-II, 442-5; CA 65, 12801.
696. Aylward, J. B., J. Chem. Soc., B 1967, 1268-70; CA 68, 38713.
697. Aynsley, E. E. et al., ibid. A 1966, 1344-7; CA 65, 18612.
698. Baekelmans, P., Gielen, M. and Nasielski, J., Tetrahedron Lett. 1967, 1149-51; CA 67, 11065.
699. Baeteman, N. and Baudet, J., C. R. Acad. Sci. Paris, Ser. C 265, 288-90 (1967); CA 67, 121120.
700. Baird, M. C., J. Inorg. Nucl. Chem. 29, 367-73 (1967); CA 67, 60495.
701. Baker, J. M. and Taylor, J. M., Ann. Appl. Biol. 60, 181-90 (1967); CA 68, 28714.
702. Balandin, A. A. and Gindin, L. G., Biofizik 10, 986-92 (1965); CA 64, 7299.
703. Balashova, L. D., Bruker, A. B. and Soborovskii, L. Z., Zh. Obshch. Khim. 35, 2207-9 (1965); CA 64, 12718.
704. Baldwin, F. P. et al., Brit. 1,128,105; Fr. 1,488,452, June 9, 1965; CA 68, 69892.
705. Banks, C. K., U.S. 3,297,732, July 25, 1963; CA 66, 115807.
706. Baranovskii, V. I. et al., Zh. Strukt. Khim. 7, 808-9 (1966); CA 66, 37244.
707. Barbanson, E., Ger. 1,264,768, Jan. 3, 1962; CA 68, 105794.
708. Barbatu, G. I., Craciun, G. and Raneti, R., Petrol Gaze (Bucharest) 17, 29-31 (1966); CA 64, 19267.
709. Barusch, M. R. et al., U.S. 3,316,071, Sept. 26, 1958; CA 67, 13586.
710. ---, Richardson, W. L. and Kautsky, G. J., U.S. 3,342,571, Sept. 17, 1964; CA 68, 51781.
711. Baukov, Yu. I. et al., Zh. Obshch. Khim. 36, 153-7 (1966); CA 64, 14213.
712. --- et al., Vestn. Mosk. Univ., Ser. II 22 (5), 118-27 (1967); CA 68, 39753; cf. ref. 1440.
713. Baum, G. A., Appl. Polym. Sym. No. 4, 189-204 (1967); CA 68, 30476.
---, Mod. Plast. 44, 148 (1967); CA 67, 3362.
714. --- and Considine, W. J., Brit. 1,037,030; Fr. 1,396,634; Ger. 1,219,631; U.S. 3,328,239, May 7, 1963; CA 64, 8678.

715. Beaird, F. M. Jr. and Kobetz, P., U.S. 3,226,408, June 6, 1963; CA 64, 9768.
--- and ---, U.S. 3,226,409, Oct. 22, 1963; CA 64, 9768.
--- and ---, U.S. 3,338,842, Feb. 17, 1965; CA 68, 49773.
716. Beck, W., Engelmann, H. and Smedal, H. S., Z. Anorg. Allg. Chem. 357, 134-8 (1968); CA 68, 74748.
717. ---, Smedal, H. S. and Koehler, H., ibid. 354, 69-73 (1967); CA 68, 13121.
718. Belcher, P. R., Bird, R. J. and Wilson, R. W., Amer. Soc. Test Mater., Spec. Tech. Publ. No. 421, 123-45 (1966); CA 68, 62222.
719. Bellegarde, B., Pereyre, M. and Valade, J., Bull. Soc. Chim. Fr. 1967, 746-7, 3082-3; CA 68, 39693.
719a. ---, --- and ---, C. R. Acad. Sci., Paris, Ser. C 264, 340-2 (1967); CA 67, 11527.
720. Bennett, R. F. and Zedler, R. J., J. Oil Colour Chemists' Assoc. 49, 928-53 (1966); CA 66, 10134.
721. Benning, A. F. and Sandy, C. A., U.S. 3,239,548, Jan. 11, 1963; CA 64, 15926.
722. Benton, J. L. B. and Marks, G. C., Brit. 1,071,397, Sept. 12, 1964; CA 67, 44424.
723. Berghe, E. V. van der and van der Kelen, G. P., Bull. Soc. Chim. Belges 74, 479-80 (1965); CA 64, 5968.
724. --- and ---, J. Organometal. Chem. 6, 515-21 (1966); CA 65, 19981.
725. --- and ---, ibid. 6, 522-7 (1966); CA 66, 1976.
726. --- and ---, ibid. 11, 479-85 (1966); CA 68, 91689.
727. ---, --- and Eckhaut, Z., Bull. Soc. Chem. Belges 76, 79-91 (1967); CA 67, 26226.
728. ---, van de Vondel, D. F. and van der Kelen, G. P., Inorg. Chim. Acta 1, 97-9 (1967); CA 68, 13111.
729. Berta, D. A., Ph. D. Thesis, West Va. Univ. 1967; Diss. Abstr. B28, 1441 (1967); CA 68, 54283.
729a. Besso, M. M., Brit. 1,084,344; U.S. 3,379,679, July 23, 1965; CA 68, 3565.
730. Beveridge, V. A. D. and Clark, H. C., Inorg. Nucl. Chem. Lett. 3, 95 (1967); J. Organometal. Chem. 11, 601-14 (1968); CA 68, 114730.
731. ---, --- and Kwon, J. T., Can. J. Chem. 44, 179-89 (1966); CA 64, 8232.
732. Bhalla, O. P. and Robinson, A. G., J. Econ. Entomol. 61, 552-5 (1968); CA 68, 104123.
733. Bichler, R. E. J., Booth, M. R. and Clark, H. C., Inorg. Nucl. Chem. Lett. 3, 71-4 (1967); CA 66, 95163.
734. Bigorgne, M. and Bernard, J., Rev. Chim. Miner. 3, 831-59 (1966); CA 67, 59211.
735. Birchall, T. and Dolly, W. L., Inorg. Chem. 5, 2177-80 (1966); CA 66, 15296.
736. Birnbaum, E. R. and Javora, P. H., J. Organometal. Chem. 9, 379-82 (1967); CA 68, 22026.

737. Biryukov, I. P. et al., Dokl. Akad. Nauk SSSR 173, 381-4 (1967); CA 67, 16519.
   ---, Voronkov, M. G. and Safin, I. A., Latv. P.S.R. Zinat Akad. Vestis, Khim. Ser. 1966, 638-48; CA 68, 73916.
738. Biryukov, B. P. et al., Chem. Commun. 1967, 749-50; CA 67, 120725.
739. --- et al., ibid. 1968, 159-60; CA 68, 82108.
740. --- et al., Zh. Struct. Khim. 8, 554-6 (1967); CA 67, 94822.
741. --- et al., ibid. 8, 556-7 (1967); CA 67, 94823.
742. Bitzer, D., Ger. 1,249,833, Feb. 3, 1964; CA 67, 100703.
743. Blears, D. J., Danyluk, S. S. and Cawley, S., J. Organometal. Chem. 6, 284-7 (1966); CA 65, 14679.
744. Bliznyk, N. K., Khokhlov, P. S. and Andrianov, Yu. A., U.S.S.R. 176,891, Nov. 21, 1964; CA 64, 12723.
745. ---, --- and ---, U.S.S.R. 181,103, April 16, 1965; CA 65, 8962.
745a. Bloechl, W., Belg. 667,480; Fr. 1,456,631; Ger. 1,469,252; Neth. 65 9564; U.S. 3,423,443, July 28, 1964; CA 65, 750.
746. Bloodworth, A. J. and Davis, A. G., Chem. Commun. 1965, 24-5; CA 64, 6653.
   --- and ---, J. Chem. Soc. 1965, 6858-63; CA 64, 5094.
747. --- and ---, Chem. Ind. (London) 1965, 1868-9; CA 64, 3592.
748. --- and ---, J. Chem. Soc. 1965, 6245-9; CA 64, 751.
749. --- and ---, ibid., C 1966, 299-303; CA 64, 8231.
750. ---, --- and Vasishtha, S. C., ibid., C 1967, 1309-13; CA 67, 63478.
751. Boboli, E., Kamionska, J. and Pazgan, A., Pol. 49,815, May 14, 1963; CA 64, 16082.
752. Boer, F. P. et al., J. Am. Chem. Soc. 89, 5068-9 (1967); CA 68, 54336.
753. Bogdanov, R. V. and Bondarevskii, S. I., Radiokhimiya 9, 521-2 (1967); CA 68, 74211.
754. Bokii, N. G. et al., Zh. Strukt. Khim. 6, 795-6 (1965); CA 64, 10502.
754a. --- and Struchkov, Yu. T., ibid. 7, 133-5 (1966); CA 65, 122.
755. --- and ---, ibid. 8, 122-37 (1967); CA 67, 6545.
   ---, Zakharova, G. N. and Struchkov, Yu. T., ibid. 8, 501-11 (1967); CA 67, 85831.
756. Bolanowska, W., Chem. Anal. (Warsaw) 12, 121-9 (1967); CA 67, 29769.
757. Bolles, T. F., Ph. D. Thesis, Univ. of Illinois, 1966; Diss. Abstr. B27, 1793-4; CA 66, 119500.
757a. --- and Drago, R. S., J. Am. Chem. Soc. 87, 5015-9 (1965); CA 64, 78.
757b. --- and ---, ibid. 88, 3921-5 (1966); CA 65, 13519.
757c. --- and ---, ibid. 88, 5730-4 (1966); CA 66, 37229.
758. Bombieri, G. and Panattoni, C., Acta Cryst. 20, 595-6 (1966); CA 65, 1504.
759. Bonati, F. et al., J. Chem. Soc., A 1966, 1052-5; CA 65, 8315.
760. ---, Cenini, S. and Ugo, R., J. Organometal. Chem. 9, 395-402 (1967); CA 67, 82251; Rend. Ist. Lombardo Sci. Lett., A 99, 825-8 (1965); CA 65, 1749.
761. --- and Ugo, R., J. Organometal. Chem. 10, 257-68 (1967); CA 68, 13128.
762. Bondi, A., J. Phys. Chem. 70, 3006-7 (1966); CA 65, 16068.

763. Borisov, A. E. et al., Dokl. Akad. Nauk SSSR 173, 855-8 (1967); CA 67, 37870; cf. ref. 867.
763a. Borsini, G. et al., Brit. 1,089,147; Fr. 1,493,391; Ital. 729,812; Neth. 66 13,478; U.S. 3,401,155, Oct. 4, 1965; CA 67, 44294.
764. --- and Nicora, C., J. Polym. Sci., Pt. A-1 6, 21-39 (1968); CA 68, 69404; Belg. 666,932; Brit. 1,043,026; Fr. 1,435,285; Ital. 718,588; Neth. 65 3797; U.S. 3,502,630, April 2, 1964; CA 64, 8340.
765. ---, --- and Cosmi, S., Ital. 729,811, Oct. 4, 1965; CA 68, 3352.
766. Bott, R. W., Eaborn, C. and Swaddle, T. W., J. Organometal. Chem. 5, 233-40 (1966); CA 64, 11039.
767. Boue, S., Gielen, M. and Nasielski, J., J. Organometal. Chem. 9, 443-60 (1967); CA 67, 116315.
768. ---, --- and ---, ibid. 9, 461-79 (1967); CA 67, 116316.
769. ---, --- and ---, ibid. 9, 481-94 (1967); CA 67, 116317.
770. Bower, L. M. and Stiddard, M. H. B., J. Chem. Soc., A 1968, 706-10; CA 68, 82673.
771. Bowman, M. C. and Beroza, M., J. Assoc. Offic. Agr. Chemists 48, 943-52 (1965); CA 64, 4192.
772. Braden, M. and Elliot, J. C., J. Deut. Res. 45, 1016-23 (1966); CA 66, 11594.
773. Brago, I. N., Kaabak, L. V. and Tomilov, A. P., Zh. Vses. Khim. Obshchest. 12, 472 (1967); CA 67, 104513.
774. Braithwaite, D. G., Ger. 1,202,790, Sept. 24, 1963; CA 64, 3064.
---, U.S. 3,256,161, March 6, 1961; CA 65, 6751.
775. ---, Ger. 1,226,100, Sept. 24, 1963; CA 65, 19690.
776. ---, Ger. 1,231,242, Sept. 24, 1963; CA 66, 95196.
777. ---, Ger. 1,231,700, Sept. 24, 1963; CA 66, 76155.
778. ---, U.S. 3,312,605, March 6, 1961; CA 67, 11589.
779. --- et al., U.S. 3,287,249, Aug. 31, 1962; CA 66, 51716.
---, d'Amico, J. S. and Hanzel, W., Ger. 1,221,223, Aug. 31, 1962; CA 66, 25471.
780. --- and Bott, L. L., U.S. 3,359,291, Oct. 5, 1964; CA 68, 95980.
781. ---, --- and Phillips, K. G., Belg. 671,841; Neth. 65 14,238; U.S. 3,380,900, Nov. 5, 1964; CA 65, 6751.
782. Braun, W. et al., Ger. (East) 42,962, Oct. 3, 1960; CA 64, 16092.
783. --- et al., Ger. (East) 55,657, Sept. 9, 1966; CA 68, 59725, cf. ref. 1380.
784. Bregadse, V. I. and Okhlobystin, O. Yu., Izv. Akad. Nauk SSSR, Ser. Khim. 1967, 2084-6; CA 68, 49702.
785. Bresadola, S., Rossetto, F. and Tagliavini, G., Chem. Commun. 1966, 623-4; CA 65, 17049; Eur. Polym. J. 4, 75-82 (1968); CA 68, 96185.
786. Breitschaft, S. and Basolo, F., J. Am. Chem. Soc. 88, 2702-6 (1966); CA 65, 4986.
787. Bridges, J. W., Davies, D. S. and Williams, R. T., Biochem. J. 98, 14P-15P (1966); CA 64, 14778.
---, --- and ---, ibid. 105, 1261-7 (1967); CA 68, 11264.
788. Brier, P. N. et al., J. Chem. Soc., A 1967, 1889-94; CA 68, 17225.

789. Brilkina, T. G., Safonova, M. K. and Shushunov, V. A., Tr. Khim. Tekhnol. 1965, 67-73; CA 66, 54769.
790. ---, --- and ---, ibid. 1965, 74-84; CA 66, 54770.
791. ---, --- and ---, Zh. Obshch. Khim. 36, 2202-6 (1966); CA 66, 76106.
792. Brook, A. G. et al., J. Am. Chem. Soc. 89, 431-4 (1967); CA 66, 46458.
792a. --- et al., ibid. 89, 704-6 (1967).
793. --- and Fieldhouse, S. A., J. Organometal. Chem. 10, 235-46 (1967); CA 68, 13090.
793a. ---, MacRae, D. M. and Limburg, W. W., J. Am. Chem. Soc. 89, 5493-5 (1967).
794. --- and Peddle, G. J. D., J. Organometal. Chem. 5, 106-7 (1966); CA 64, 5131.
---, Jones, P. F. and Peddle, G. J. D., Can. J. Chem. 46, 2119-27 (1968); CA 69, 36235.
795. Brooks, E. H. and Glockling, F., Chem. Commun. 1965, 510; CA 64, 5130.
--- and ---, J. Chem. Soc., A 1966, 1241-3; CA 65, 14817.
796. Brown, T. L. and Puckett, J. C., J. Chem. Phys. 44, 2238-43 (1966); CA 64, 13555.
797. Brown, W. A., Phys. Fluids 9, 1273-7 (1966); CA 65, 8215.
798. Bruck, S. D., J. Polym. Sci., Pt. B4, 933-7 (1966); CA 66, 46669.
799. ---, ibid., Pt. A-1 5, 2458-9 (1967); CA 67, 91125.
800. Bruker, A. B., Balashova, L. D. and Soborovskii, L. Z., Zh. Obshch. Khim. 36, 75-8 (1966); CA 64, 14211.
801. Bryan, R. F., Chem. Commun. 1967, 355-6; CA 67, 37115.
---, J. Chem. Soc., A 1968, 698-703; CA 68, 82057.
802. ---, ibid., A 1967, 172-81; CA 66, 41375.
803. ---, ibid., A 1967, 192-201, CA 66, 41377.
804. --- and Weber, H. P., ibid., A 1967, 843-53; CA 67, 6528.
805. Brydy, S., Ejmocki, Z. and Eckstein, Z., Bull. Acad. Pol. Sci., Ser. Chim. 13, 683-6 (1965); CA 64, 13324.
806. Bryskovskaya, A. V. and Albitskaya, V. M., Zh. Obshch. Khim. 37, 1553-8 (1967); CA 68, 29814.
807. ---, --- and Petrov, A. A., Zh. Org. Khim. 1, 1898-9 (1965); CA 64, 3446.
808. Buck, R. P. and Ryason, P. R., Am. Chem. Soc., Div. Petrol. Chem., Preprints 10, D5-20 (1965); CA 65, 18381.
809. --- and ---, ibid. 10, D21-39 (1965); CA 65, 18382.
810. Buerger, H. and Goetze, U., Angew. Chem. Int. Ed. Engl. 7, 212-3 (1968); CA 68, 95199.
811. Bullock, J. D. et al., Arch. Environ. Health 13, 21-2 (1966); CA 65, 9603.
812. Bulten, E. J. and Noltes, J. G., J. Organometal. Chem. 11, P19-20 (1968); CA 68, 90343.
813. --- and ---, Tetrahedron Lett. 1966, 3471-6; CA 65, 12232.
814. --- and ---, ibid. 1966, 4389-92; CA 65, 16996.
815. --- and ---, ibid. 1967, 1443-7; CA 67, 21988.
816. Buncak, P., Ropa Uhlie 8, 148-54 (1966); CA 65, 18381.

817. Burch, G. M. and Van Wazer, J. R., J. Chem. Soc., A 1966, 586-9; CA 64, 18941.
818. Burlachenko, G. S. et al., Zh. Obshch. Khim. 36, 512-8 (1966); CA 65, 742.
819. Busetti, V. et al., Inorg. Chim. Acta 1, 428-8 (1967); CA 68, 54311.
820. Caldwell, J. R. and Gilkey, R., Brit. 1,137,209; Fr. 1,474,377, April 5, 1965; CA 67, 54612.
--- and ---, Brit. 1,137,151; Fr. 1,474,508, April 5, 1965; CA 67, 55089.
--- and ---, Fr. 1,474,509; U.S. 3,408,334, April 5, 1965; CA 67, 55093.
821. ---, --- and Kuhfuss, H. F., Brit. 1,044,015, June 7, 1962; CA 66, 3017.
822. Calley, D., Guess, W. L. and Autian, J., J. Pharm. Sci. 56, 1267-72 (1967); CA 67, 115461.
823. Camey, T. and Paulini, E., Rev. Brasil. Malariol, Doencas Trop. 16, 487-91 (1964); CA 64, 20554.
824. Cantuti, V. and Cartoni, G. P., J. Chromatogr. 32, 641-7 (1968); CA 68, 71933.
825. Cardin, D. J. and Lappert, M. F., Chem. Commun. 1966, 506; CA 65, 12237.
826. --- and ---, ibid. 1967, 1034; CA 68, 13117.
827. Carey, N. A. D. and Clark, H. C., Can. J. Chem. 46, 643-7 (1968); CA 68, 84001.
828. --- and ---, Chem. Commun. 1967, 292-3; CA 68, 7783.
--- and ---, Inorg. Chem. 7, 94-9 (1968); CA 68, 34401.
829. Carlson, L. W., J. Am. Soc. Sugar Beet Techn. 14, 254-9 (1966); CA 67, 72727.
830. Carr, D. S., Paint Varn. Prod. 58 (2), 23-8 (1968); CA 68, 70208.
831. Carrick, A. and Glockling, F., J. Chem. Soc., A 1966, 623-9; CA 65, 2293.
832. --- and ---, ibid., A 1968, 913-20; CA 68, 101350.
833. Carrick, W. L., Macromol. Syn. 2, 33-8 (1966); CA 65, 18692.
834. Case, E. N., Patinkin, S. H. and Carlson, D. R., U.S. 3,222,146, Jan. 9, 1961; CA 64, 4841.
835. Cassol, A., Gazz. Chim. Italy 96, 1764-74 (1966); CA 67, 15436.
836. --- and Magon, L., ibid. 96, 1724-33 (1966); CA 67, 15433.
837. --- and ---, ibid. 96, 1752-63 (1966); CA 67, 15435.
838. ---, --- and Barbieri, R., Inorg. Nucl. Chem. Lett. 3, 25-9 (1967); CA 66, 89035.
839. --- and Portanova, R., Gazz. Chim. Ital. 96, 1734-51 (1966); CA 67, 15434.
840. ---, --- and Magon, L., Ric. Sci. 36, 1180-6 (1966); CA 66, 119421.
841. Cattanach, C. J. and Mooney, E. F., Spectrochim. Acta, Pt. A24, 407-15 (1968); CA 68, 100219.
842. Caujolle, D. et al., Ann. Biol. Clin. (Paris) 24, 479-85 (1966); CA 65, 2852.
843. --- and Voisin, M. C., Ann. Pharm. Tr. 24, 17-22 (1966); CA 65, 6165.
844. Caujolle, F. et al., ibid. 24, 23-8 (1966); CA 65, 4468.
--- et al., Bull. Trav. Soc. Pharm. Lyon 9, 221-35 (1965); CA 65, 19054.

845. --- et al., C. R. Acad. Sci. Paris, Ser. D. $\underline{262}$, 1302-4 (1966); CA $\underline{64}$, 18289.
846. Cawley, S., Ph. D. Thesis, Univ. of Toronto, 1964; Diss. Abstr. $\underline{26}$, 5057-8 (1966); CA $\underline{65}$, 1627.
--- and Danyluk, S. S., Can. J. Chem. $\underline{46}$, 2373-84 (1968); CA $\underline{69}$, 56092.
846a. Chadra, R. N., Brit. 1,119,577; Fr. 1,491,094; Neth. 66 12,421; U.S. 3,470,221, Sept. 2, 1965; CA $\underline{67}$, 44618.
847. Chaikin, S. W., Wear $\underline{10}$, 49-60 (1967); CA $\underline{68}$, 30313.
848. Chambers, D. B. et al., Chem. Commun. $\underline{1966}$, 281-2; CA $\underline{65}$, 3708.
849. --- and Glockling, F., J. Chem. Soc., A $\underline{1968}$, 735-41; CA $\underline{68}$, 82921.
850. ---, --- and Weston, M., ibid., A $\underline{1967}$, 1759-69; CA $\underline{68}$, 2488.
851. Chambers, R. D., Cunningham, J. A. and Pyke, D. A., Tetrahedron $\underline{24}$, 2783-7 (1968); CA $\underline{68}$, 78344.
852. Chan, L.-H. and Rochow, E. G., J. Organometal. Chem. $\underline{9}$, 231-50 (1967); CA $\underline{67}$, 90256.
853. Chan, S. S. and Willis, C. J., Can. J. Chem. $\underline{46}$, 1237-48 (1968); CA $\underline{68}$, 114717.
854. Chandra, G., George, T. A. and Lappert, M. F., Chem. Commun. $\underline{1967}$, 116-7; CA $\underline{66}$, 65594; cf. ref. 2406.
855. Chatt, J. et al., ibid. $\underline{1967}$, 869; CA $\underline{67}$, 104719.
856. ---, Eaborn, C. and Ibekwe, S., ibid. $\underline{1966}$, 700-1; CA $\underline{66}$, 16130.
857. Chauzov, V. A., Litvinova, O. V. and Baukov, Yu. I., Zh. Obshch. Khim. $\underline{36}$, 952 (1966); CA $\underline{65}$, 8955.
858. Cheng, P.-L. et al., Hua Kung Hsueh Pao $\underline{1965}$, 169-74; CA $\underline{65}$, 16997.
859. --- and Pai, Y.-L., ibid. $\underline{1965}$, 175-8; CA $\underline{65}$, 16997.
860. Cherayil, A., Kandera, J. and Lajtha, A., J. Neurochem. $\underline{14}$, 105-7 (1967); CA $\underline{66}$, 53561.
861. Chernobai, A. V. et al., Zh. Prikl. Spektrosk. $\underline{6}$, 90-4 (1967); CA $\underline{66}$, 116323.
862. --- and Kolesnikov, L. N., Pribory i Tekhn. Eksperim. $\underline{9}$, 120-1 (1964); CA $\underline{65}$, 11711.
863. Chernyshev, E. A., Zelenetskaya, A. A. and Krasnova, T. L., Izv. Akad. Nauk SSSR, Ser. Khim. $\underline{1966}$, 118-20; CA $\underline{65}$, 10616.
864. Chiesura, P., Brugnone, F. and Terribile, P. M., Med. Lav. $\underline{57}$, 641-6 (1966); CA $\underline{67}$, 1815.
865. Chivers, T. and David, B., J. Organometal. Chem. $\underline{10}$, P35-6 (1967); CA $\underline{68}$, 13118.
--- and ---, ibid. $\underline{13}$, 177-86 (1968); CA $\underline{69}$, 77399.
866. Cho, J.-H. et al., Hua Hsueh Hsueh Pao $\underline{32}$, 196-200 (1966); CA $\underline{65}$, 13753.
867. Chumaevskii, N. A., Tr. Komis. po Spektroskopii, Akad. Nauk SSSR $\underline{3}$, 84-91 (1964); CA $\underline{64}$, 10592; cf. ref. 763.
868. Chuveleva, E. A. et al., Zh. Fiz. Khim. $\underline{41}$, 1389-95 (1967); CA $\underline{68}$, 59917.
869. Clark, E. A., Ph. D. Thesis, Fordham. Univ. 1966; Diss. Abstr. B$\underline{27}$, 2600 (1966); CA $\underline{66}$, 120352.
--- and Weber, A., J. Chem. Phys. $\underline{45}$, 1759-66 (1966); CA $\underline{65}$, 13013.

870. Clark, H. C., Cyr, N. and Tsai, J. H., Can. J. Chem. 45, 1073-8 (1967); CA 67, 6940.
871. ---, Cotton, J. D. and Tsai, J. H., Can. J. Chem. 44, 903-16 (1966); CA 64, 15714.
872. ---, --- and ---, Inorg. Chem. 5, 1582-6 (1966); CA 65, 11746.
873. --- and Goel, R. C., J. Organometal. Chem. 7, 263-7 (1967); CA 66, 76117.
874. --- and Tsai, J. H., Inorg. Chem. 5, 1407-15 (1966); CA 65, 8957.
875. Clark, J. P., Langford, V. M. and Wilkins, C. J., J. Chem. Soc., A 1967, 792-4; CA 67, 21985.
876. --- and Wilkins, C. J., ibid., A 1966, 871-3; CA 65, 13020.
877. Clark, M. G., Rev. Ass. Fr. Tech. Petrole No. 183, 67-76 (1967); CA 67, 118803.
878. Clark, R. J. H., Record Chem. Progr. 26, 269-82 (1965); CA 64, 7537.
879. --- and Williams, C. S., Spectrochim. Acta 21, 1861-8 (1965); CA 64, 163.
880. Clossen, R. D., U.S. 3,231,511, Nov. 14, 1963; CA 64, 11251.
880a. Coates, H. and Hoye, P. A. T., Brit. 1,081,823; Fr. 1,420,048; Neth. 64 15,330; U.S. 3,387,011, Dec. 31, 1963; CA 64, 2129.
880b. --- and ---, Brit. 1,118,170; Fr. 1,455,706; Ger. 1,283,839; Neth. 65 15,697; U.S. 3,446,826, Dec. 2, 1964; CA 65, 13763.
881. Cochran, J. C., Ph. D. Thesis, Univ. of New Hampshire, 1967; Diss. Abstr. B28, 2334-5 (1967); CA 68, 95907; cf. ref. 1368.
882. Cohen, H. J., J. Organometal. Chem. 9, 177-9 (1967); CA 68, 39767.
883. Cohen, S. C. and Massey, A. G., ibid. 10, 471-81 (1967); CA 68, 22037.
884. --- and ---, Tetrahedron Lett. 1966, 4393-4; CA 66, 10668.
885. ---, Reddy, M. L. N. and Massey, A. G., Chem. Commun. 1967, 451-3; CA 67, 73655.
886. Coleman, D. J. and Skinner, H. A., Trans. Faraday Soc. 62, 1721-5 (1966); CA 65, 2106.
887. Colicchio, G., Rossi, A. and Grieco, B., Intern. Congr. Occupational Health, 14th, Madrid 1963, 915-6 (1964); CA 64, 20497.
888. Collier, H. E., Jr., U.S. 3,270,042, May 28, 1963; CA 65, 15428.
889. ---, Eberlein, J. W. and Hillman, W. S., U.S. 3,277,134, Nov. 21, 1963; CA 66, 95197.
890. --- and Sterling, J. D., U.S. 3,274,224, July 25, 1963; CA 65, 19910.
891. Collman, J. P., Vastine, F. D. and Roper, W. R., J. Am. Chem. Soc. 88, 5035-7 (1966); CA 66, 16104.
---, --- and ---, ibid. 90, 2282-7 (1968); CA 68, 110914.
892. Condo, A. C., Jr., U.S. 3,261,674, March 10, 1961; CA 65, 10402.
893. Conkey, J. H., Tappi 49 (8), 124A-5A (1966); CA 65, 19241.
894. --- and Carlson, J. A., ibid. 48 (11), Pt. 1, 109A (1965); CA 64, 7299.
895. Considine, W. J., J. Organometal. Chem. 5, 263-6 (1966); CA 64, 9760.
895a. --- and Reifenberg, G. H., Brit. 1,146,102; Fr. 1,510,254; Neth. 67 12; U.S. 3,417,116, Jan. 3, 1966; CA 68, 114745; cf. ref. 1747.
895b. --- and ---, Brit. 1,167,304; Fr. 1,510,255; Neth. 67 13; U.S. 3,471,538, Jan. 1, 1966; U.S. 3,475,474, Jan. 3, 1966; CA 67, 108761.

896. Cook, S. E., U.S. 3,312,726, March 11, 1963; CA 67, 4666; Belg. 645,044; Ger. 1,218,444; Neth. 64 2472; CA 62, 10277; Brit. 1,038,323; CA 66, 30785.
897. --- and Shapiro, H., U.S. 3,226,209, Nov. 11, 1960; CA 64, 7953.
898. --- and Sistrunk, T. O., U.S. 3,221,037, June 8, 1962; CA 64, 6378.
899. --- and ---, U.S. 3,221,039, Oct. 30, 1963; CA 64, 4841.
900. --- and Thomas, W. H., U.S. 3,340,284, April 13, 1964; CA 68, 39818.
901. Cooksley, M. V. and Parham, D. N., Surface Coatings 2, 280-2 (1966); CA 66, 76969.
902. Cordey-Hayes, M., Peacock, R. D. and Vucelic, M., J. Inorg. Nucl. Chem. 29, 1177-80 (1967); CA 67, 59336.
903. Cornell, J. H., Gorth, H. and Henry, M. C., ibid. 29, 1411-3 (1967); CA 67, 53354.
904. Cotton, J. D. et al., Chem. Commun. 1966, 253-4; CA 65, 3311.
--- et al., J. Chem. Soc., A 1967, 264-9; CA 66, 76139.
905. ---, Knox, S. A. R. and Stone, F. G. A., Chem. Commun. 1967, 965-6; CA 68, 39744
---, --- and ---, J. Chem. Soc., A 1968, 2758-62; CA 70, 25324.
906. Couanon, G. and Vinches, G., Ind. Plastiques Mod. Elastomers 16 (8), 71-3 (1964); CA 64, 9891.
907. Coutts, R. S. P. and Wailes, P. C., Chem. Commun. 1968, 260-1; CA 68, 83985.
908. Cowell, R. D., Ph. D. Thesis, Purdue Univ., 1965; Diss. Abstr. B27, 92 (1966); CA 65, 19981.
909. Coyle, T. D., Cooper, J. and Ritter, J. J., Inorg. Chem. 7, 1014-20 (1968); CA 68, 110939.
910. Cradock, S., Gibbon, G. A. and Van Dyke, C. H., ibid. 6, 1751 (1967); CA 67, 87284; cf. ref. 2409.
911. Cramer, C. R. and deBruijn, W. F. L., U.S. 3,242,201, April 6, 1959; CA 64, 17640.
912. Creemers, H. M. J. C. et al., Tetrahedron Lett. 1966, 3167-71; CA 65, 12231.
913. --- and Noltes, J. G., J. Organometal. Chem. 7, 237-47 (1967); CA 65, 95138.
914. --- and ---, Rec. Trav. Chim. 84, 1589-93 (1965); CA 64, 12718.
915. ---, Verbeck, F. and Noltes, J. G., J. Organometal. Chem. 8, 469-77 (1967); CA 67, 53186.
916. Cremer, J. E., Biochem. J. 104, 212-20 (1967); CA 67, 20079.
917. ---, ibid. 104, 223-8 (1967); CA 67, 19408.
918. ---, ibid. 106, 8P-9P (1968); CA 68, 85849.
919. Crepet, M. and Chiesura, P., Panminerva Med. (Engl. Ed.) 8, 295-301 (1966); CA 66, 1247.
920. Cullen, W. R., Deacon, G. B. and Green, J. H. S., Can. J. Chem. 43, 3193-200 (1965); CA 64, 2870.
921. --- and Leeder, W. R., Inorg. Chem. 5, 1004-8 (1966); CA 65, 2293.
922. --- and Styan, G. E., Can. J. Chem. 44, 1225-7 (1966); CA 65, 2292.
923. --- and ---, J. Organometal. Chem. 6, 117-25 (1966); CA 65, 10618.

924. Cullen, W. R. and Styan, G. E., J. Organometal. Chem. 6, 633-44 (1966); CA 66, 18749.
925. Cummins, R. A., Austr. J. Chem. 18, 985-92 (1965); CA 64, 162.
926. Cumper, C. W. N. et al., J. Chem. Soc., B 1966, 874-7; CA 65, 15197.
927. ---, Melnikoff, A. and Vogel, A. I., ibid., A 1966, 242-5; CA 64, 12502.
928. ---, --- and ---, ibid., A 1966, 246-9; CA 64, 12522.
929. ---, --- and ---, ibid., A 1966, 323-9; CA 64, 12522.
930. Curtis, M. D., Lee, R. L. and Allred, A. L., J. Am. Chem. Soc. 89, 5150-2 (1967); CA 68, 7841.
931. ---, ibid. 89, 4241-2 (1967); CA 67, 90873.
932. Cyr, N. and Cyr, T. J. R., J. Chem. Phys. 47, 3082-3 (1967); CA 68, 7937.
933. Czerwinska, E. et al., Bull. Acad. Pol. Sci., Ser. Chim. 15, 335-9 (1967); CA 68, 11909.
934. Dallacker, F. and Adolphen, G., Ann. Chem. 694, 110-6 (1966); CA 65, 12186.
935. Danielczyk, B. and Orszagh, A., Proc. Tihany Symp. Radiat. Chem., 2nd, Tihany, Hung. 1966, 479-82 (1967); CA 68, 30183.
936. Dannin, J., Sandler, S. R. and Baum, B., Intern. J. Appl. Radiation Isotopes 16, 589-97 (1965); CA 64, 2979.
937. Dannley, R. L. and Farrant, G., J. Am. Chem. Soc. 88, 627-8 (1966); CA 64, 12719.
938. Das, V. G. K. and Kitching, W., J. Organometal. Chem. 10, 59-69 (1967); CA 67, 100214.
939. Davidson, W. and Henry, M. C., ibid. 5, 29-34 (1966); CA 64, 3593.
940. Davies, A. G., U.S. 3,347,890, Oct. 26, 1964; CA 68, 49775.
941. ---, U.S. 3,376,328, Oct. 26, 1964; CA 68, 105364.
942. --- and Harrison, P. G., J. Chem. Soc., C 1967, 298-300; CA 66, 55563.
943. --- and ---, ibid., C 1967, 1313-7; CA 67, 63479.
944. --- and ---, J. Organometal. Chem. 7, P13-4 (1967); CA 66, 105034.
945. --- and ---, ibid. 8, P19-20 (1967); CA 67, 11563.
946. --- and ---, ibid. 10, P31-2 (1967); CA 68, 13120.
947. ---, --- and Palan, P. R., ibid. 10, P33-4 (1967); CA 68, 3239.
948. --- and Mitchell, T. N., ibid. 6, 568-9 (1966); CA 66, 11014.
949. ---, --- and Symes, W. R., J. Chem. Soc., C 1966, 1311-5; CA 65, 8955.
950. ---, Palan, P. R. and Vasishtha, S. C., Chem. Ind. (London) 1967, 229-30; CA 66, 85832.
951. --- and Puddephatt, R. J., J. Organometal. Chem. 5, 590-2 (1966); CA 65, 5482.
--- and ---, J. Chem. Soc., C 1967, 2663-9; CA 68, 29806.
952. --- and ---, Tetrahedron Lett. 1967, 2265-7; CA 67, 90763.
--- and ---, J. Chem. Soc., C 1968, 317-22; CA 68, 59670.
953. --- and Symes, W. R., Chem. Commun. 1965, 25-6; CA 64, 9761.
--- and ---, J. Chem. Soc., C 1967, 1009-16; CA 67, 11550.
954. --- and ---, J. Organometal. Chem. 5, 394-5 (1966); CA 64, 15918.
955. Dawson, J. A., Brit. 1,042,741, Nov. 7, 1961; CA 65, 17161.

956. Deacon, G. B. and Felder, P. W., Aust. J. Chem. 20, 1587-94 (1967); CA 67, 77567.
957. Dellepiane, G. and Zerbi, G., J. Mol. Spectrosc. 24, 62-86 (1967); CA 67, 112612.
958. Delman, A. D. et al., J. Polymer. Sci., Pt. A-1, 4, 2307-19 (1966); CA 65, 13753.
959. De Long, H. K. and Glesner, C. W., U.S. 3,252,215, Dec. 6, 1961; CA 65, 3506.
960. Demidowicz, A. et al., Pol. 49,789, May 6, 1964; CA 65, 4525.
961. Deschiens, R., Brottes, H. and Mvogo, L., Bull. Soc. Pathol. Exot. 59, 231-4 (1966); CA 66, 94208.
962. ---, Floch, H. and Floch, T., ibid. 58, 1058-67 (1965); CA 66, 94210.
963. Dessy, R. E., Kitching, W. and Chivers, T., J. Am. Chem. Soc. 88, 453-9 (1966); CA 64, 9571.
964. --- and Pohl, R. L., ibid. 90, 1995-2001 (1968); CA 68, 101187.
965. --- and ---, ibid. 90, 2005-8 (1968); CA 68, 101180.
966. ---, --- and King, R. B., ibid. 88, 5121-4 (1966); CA 66, 54850.
967. --- and Weissman, P. M., ibid. 88, 5124-9 (1966); CA 66, 54851.
968. --- and ---, ibid. 88, 5129-31 (1966); CA 66, 54852.
969. ---, --- and Pohl, R. L., ibid. 88, 5117-21 (1966); CA 66, 43138.
970. Devaud, M., C. R. Acad. Sci., Paris, Ser. C 262, 702-5 (1966); CA 64, 18967.
971. ---, ibid., C 263, 1269-72 (1966); CA 66, 61274.
---, J. Chim. Phys. 64, 791-8 (1967); CA 67, 78488.
972. ---, ibid. 63, 1335-45 (1966); CA 66, 61273.
973. --- and Souchay, P., ibid. 64, 1778-90 (1967); CA 68, 110811.
974. ---, --- and Person, M., ibid. 64, 646-57 (1967); CA 67, 39589.
975. DeWitt, E. G., Brown, J. E. and Shapiro, H., U.S. 3,285,946, March 22, 1954; CA 66, 20843.
976. Dickinson, L. A., Can. 769,798, Nov. 19, 1963; CA 68, 23255.
977. Dickson, R. S. and West, B. O., Aust. J. Chem. 19, 2073-8 (1966); CA 66, 37991.
978. Diener, U. L. and Garrett, F. E., Plant Dis. Reptr. 51, 185-7 (1967); CA 66, 104276.
979. Ditty, J. E., Brit. 1,086,069; Fr. 1,410,013, Sept. 25, 1963; CA 65, 5608.
980. Doerfelt, C., Ger. 1,227,658, Dec. 24, 1963; CA 66, 46483.
980a. --- and Lorz, W., Brit. 1,114,997; Fr. 1,480,655; Ger. 1,544,906; Neth. 66 6681; U.S. 3,442,852, May 21, 1965; CA 66, 86266.
981. Dolle, R. E., Jr., U.S. 3,322,671, Oct. 6, 1964; CA 67, 55952.
982. --- and Tamborski, C., U.S. 3,313,713, Jan. 7, 1966; CA 67, 3103.
983. Doretti, L. and Tagliavini, G., J. Organometal. Chem. 12, 203-8 (1968); CA 68, 83674.
984. Dorfman, Ya. G. et al., Zh. Strukt. Khim. 7, 200-4 (1966); CA 65, 3721.
985. Drago, R. S., Record Chem. Progr. 26 (3), 157-67 (1965); CA 64, 569.
986. Dreeskamp, H. and Stegmeier, G., Z. Naturforsch. A 22, 1458-64 (1967); CA 68, 118350.

987. Drenth, W., Rec. Trav. Chim. 85, 455-6 (1966); CA 65, 3714.
988. Drouven, E., Ger. 1,220,667, Sept. 26, 1958; CA 65, 18400.
989. Dubeck, M., U.S. 3,244,492, Nov. 9, 1962; CA 64, 17642; U.S. 3,097,225, Sept. 12, 1960; CA 59, 14025.
990. Dufermont, J. and Maire, J. C., J. Organometal. Chem. 7, 415-25 (1967); CA 66, 89961.
991. Duffield, A. M. et al., ibid. 12, 123-32 (1968); CA 68, 90974.
992. Duncan, J. F. and Thomas, F. G., J. Inorg. Nucl. Chem. 29, 869-90 (1967); CA 67, 7118.
993. Durig, J. R. and Sink, C. W., Spectrochim. Acta, Pt. A 24, 575-87 (1968); CA 68, 118130.
994. ---, --- and Bush, S. F., J. Chem. Phys. 45, 66-78 (1966); CA 65, 8214.
995. Dzurinskaya, N. G., Mikhailyants, S. A. and Evdakov, V. P., Zh. Obshch. Khim. 37, 2278-80 (1967); CA 68, 87361.
996. Eaborn, C. et al., J. Organometal. Chem. 9, 175-6 (1967); CA 67, 64499.
997. ---, Hornfeld, H. L. and Walton, D. R. M., J. Chem. Soc., B 1967, 1036-40; CA 67, 116312.
998. ---, --- and ---, J. Organometal. Chem. 10, 529-30 (1967); CA 68, 13125.
999. ---, Simpson, F. and Varma, I. D., J. Chem. Soc., A 1966, 1133-6; CA 65, 10616.
1000. ---, Skinner, G. A. and Walton, D. R. M., J. Organometal. Chem. 6, 438-41 (1966); CA 66, 11019.
1001. ---, Thompson, A. R. and Walton, D. R. M., J. Chem. Soc., C 1967, 1364-6; CA 67, 82234.
1002. ---, Treverton, J. A. and Walton, D. R. M., J. Organometal. Chem. 9, 259-62 (1967); CA 67, 99454.
1003. --- and Varma, I. D., ibid. 9, 377-8 (1967); CA 68, 13106.
1004. Eaton, D. R. and McClellan, W. R., Inorg. Chem. 6, 2134-8 (1967); CA 68, 33269.
1005. Eckstein, Z. and Ejmocki, Z., Pol. 51,771, Aug. 8, 1964; CA 68, 49776.
1006. Edmondson, R. C. and Newlands, M. J., Chem. Ind. (London) 1966, 1888-9; CA 66, 28873.
1007. Efer, J., Quaas, D. and Spichale, W., Z. Chem. 5, 390-1 (1965); CA 64, 7370.
1008. Eglin, S. B. and Landis, A. L., U.S. 3,287,285, Oct. 17, 1963; CA 66, 19577.
1009. Egorochkin, A. N. et al., Dokl. Akad. Nauk SSSR 170, 333-6 (1966); CA 66, 50604.
1010. --- et al., Zh. Obshch. Khim. 37, 1165-8 (1967); CA 68, 29076.
1011. --- et al., ibid. 37, 2308-11 (1967); CA 68, 91422.
1012. ---, Khidekel, M. S. and Razuvaev, G. A., Izv. Akad. Nauk SSSR, Ser. Khim. 1966, 437-43; CA 65, 596.
1013. Egorov, Yu. P., Khranovskii, V. A., Teor. i Eksperim. Khim., Akad. Nauk Ukr. SSR 2, 175-83 (1966); CA 65, 12092.
1014. ---, Morozov, V. P. and Kovalenko, N. F., Tr. Komis. po Spektroskopii, Akad. Nauk SSSR 2, 134-9 (1964); CA 64, 5941.

1015. Einstein, F. W. B. and Penfold, B. R., Chem. Commun. 1966, 780-1; CA 66, 23059.
1016. Eisch, J. J. and Foxton, M. W., J. Organometal. Chem. 11, P24-6 (1968); CA 68, 113900.
1017. Emelyanov, B. V. et al., Zh. Prikl. Khim. 40, 2501-6 (1967); CA 68, 69101.
1018. Engel, J. J., Guyer, V. L. and Schafer, R. J., U.S. 3,255,500, Feb. 1, 1965; CA 65, 6895.
1019. Engelhardt, G., Reich, P. and Schumann, H., Z. Naturforsch. B 22, 352-3 (1967); CA 67, 112528.
1019a. Englehart, O. D. and Michelotti, J. E., Belg. 657,345; Brit. 1,088,558; Fr. 1,420,024; Neth. 64 13,375; U.S. 3,411,934, Dec. 23, 1963; CA 64, 420.
1020. Entelis, S. G. and Nesterov, O. V., Khinetika i Kataliz 7, 464-9 (1966); CA 65, 10462.
---, --- and Zabrodin, V. B., ibid. 7, 627-31 (1966); CA 65, 18478.
1021. Evnin, A. B. and Seyferth, D., J. Am. Chem. Soc. 89, 952-9 (1967); CA 66, 95148.
1022. Faleschini, S. and Tagliavini, G., Gazz. Chim. Ital. 97, 1401-10 (1967); CA 68, 95053.
1023. Fahlstrom, G. B., U.S. 3,227,563, Oct. 19, 1961; CA 64, 8499.
1024. Farrer, H. N., McGrady, M. M. and Tobias, R. S., J. Am. Chem. Soc. 87, 5019-26 (1965); CA 64, 189.
1025. Fay, P. S. and Fox, F. J., U.S. 3,246,965, Dec. 24, 1958; CA 65, 2050.
1026. Fedin, E. I. et al., Dokl. Akad. Nauk SSSR 175, 879-81 (1967); CA 68, 73893.
1027. Fedorov, A. A. and Pokrovskii, L. I., Plast. Massy 1966, 59-61; CA 65, 7378.
1028. Fellows, O. N., J. Immunol. 99, 508-13 (1967); CA 67, 88445.
1029. Fenrick, H. W. and Willard, J. E., J. Am. Chem. Soc. 88, 412-16 (1966); CA 64, 9557.
1030. Fenster, A. N. and Becker, E. I., J. Organometal. Chem. 11, 549-55 (1968); CA 68, 114716; cf. ref. 2388.
1031. Fentiman, A. F. et al., ACE Accession No. 10870, Rept. No. BMI-1753 (1966); CA 65, 16293.
--- et al., J. Organometal. Chem. 6, 645-51 (1966); CA 66, 18748.
1032. Fenton, D. E. et al., Chem. Commun. 1967, 1097-8; CA 68, 44613.
1033. --- and Massey, A. G., Tetrahedron 21, 3009-18 (1965); CA 64, 6531.
1034. ---, --- and Urch, D. S., J. Organometal. Chem. 6, 352-8 (1966); CA 65, 18612.
1035. Fernley, A. M., Brit. 1,018,111, Apr. 24, 1961; CA 64, 12906.
1036. ---, Brit. 1,020,291, June 9, 1961; CA 64, 15926.
1037. Feuer, G., Goldberg, L. and LePelley, J. R., Food Cosmet. Toxicol. 3, 235-49 (1965); CA 64, 16431.
1038. Finch, A., Poller, R. C. and Steele, D., Trans. Faraday Soc. 61, 2682-4 (1965); CA 64, 10597.

1039. Findeiss, W., Davidsohn, W. and Henry, M. C., J. Organometal. Chem. $\underline{9}$, 435-41 (1967); CA $\underline{67}$, 82250.
1040. Finkner, R. E., Farus, D. E. and Calpouzos, L., J. Am. Soc. Sugar Beet Technol. $\underline{14}$, 232-7 (1966); CA $\underline{67}$, 72726.
1041. Fish, R. H., Ph. D. Thesis, Univ. New Hampshire, 1965; Diss. Abstr. $\underline{26}$, 7034 (1966); CA $\underline{65}$, 15420.
1042. --- and Kuivila, H. G., J. Org. Chem. $\underline{31}$, 2445-50 (1966); CA $\underline{65}$, 10617.
1043. ---, --- and Tyminski, I. J., J. Am. Chem. Soc. $\underline{89}$, 5861-8 (1967); CA $\underline{68}$, 49698.
1044. Fitzsimmons, B. W., Seeley, N. J. and Smith, A. W., Chem. Commun. $\underline{1968}$, 390-1; CA $\underline{68}$, 118317.
1045. Flitcroft, N. et al., J. Chem. Soc., A $\underline{1966}$, 1130-3; CA $\underline{65}$, 10104.
1046. Foldesi, I. and Straner, G., Acta Chim. Acad. Sci. Hung. $\underline{45}$, 313-22 (1965); CA $\underline{64}$, 3591.
1047. --- and ---, Ann. Univ. Sci. Budapest, Rolando Eotvos Nominatae, Sect. Chim. $\underline{7}$, 89-94 (1965); CA $\underline{65}$, 5437.
1048. Frankel, M. et al., Israel J. Chem. $\underline{4}$, 183-7 (1966); CA $\underline{66}$, 52043.
1049. --- et al., J. Appl. Polymer Sci. $\underline{9}$, 3383-8 (1965); CA $\underline{64}$, 2178.
--- et al., Brit. 1,134,891; Neth. 66 3742; U.S. 3,426,002, March 22, 1965; CA $\underline{66}$, 86538.
1050. --- et al., J. Organometal. Chem. $\underline{7}$, 518-20 (1967); CA $\underline{66}$, 94785.
1051. --- et al., ibid. $\underline{9}$, 83-8 (1967); CA $\underline{67}$, 54239.
1052. Franzen, V., Ernaehrungsforschung $\underline{11}$, 368-74 (1966); CA $\underline{66}$, 95774.
1053. Fraselle, J., Rev. Agr. (Brussels) $\underline{20}$, 1013-28 (1967); CA $\underline{68}$, 28709.
1054. Freiberg, A. H., U.S. 3,284,296, Dec. 9, 1964; CA $\underline{66}$, 45790.
1055. --- and Greco, C. C., U.S. 3,328,441, Dec. 10, 1963; CA $\underline{68}$, 49768.
1056. Freidline, C. E. and Tobias, R. S., Inorg. Chem. $\underline{5}$, 354-61 (1966); CA $\underline{64}$, 10456.
1057. Frensch, H. et al., U.S. 3,297,523, Sept. 1, 1964; CA $\underline{66}$, 54562; Belg. 669,056; Brit. 1,113,450; Fr. 1,445,986; Ger. 1,247,740; Neth. 65 11,285, Sept. 1, 1964; CA $\underline{65}$, 2944.
1058. Friese, K. and Lomb, M., Plaste Kaut. $\underline{14}$, 94-6 (1967); CA $\underline{66}$, 56110.
1059. Frisch, K. C. et al., J. Polym. Sci. Pt. A-1 $\underline{5}$, 35-42 (1967); CA $\underline{66}$, 64755.
1060. ---, Reegen, K. C. and Thir, B., ibid. Pt. C No. $\underline{16}$, 2191-2201 (1967); CA $\underline{67}$, 117319.
1061. Fry, J. L. and Wilson, H. R., Poultry Sci. $\underline{46}$, 319-22 (1967); CA $\underline{66}$, 92763.
1062. Fujimoto, T., Saiuchi, T. and Suwada, A., Japan. 25,767 ('67); CA $\underline{68}$, 87889.
1062a. Fukumoto, O., Kubo, M. and Hatakeyama, H., Brit. 1,055,299; Fr. 1,406,055; Ger. 1,261,317; U.S. 3,310,509, Feb. 5, 1964; CA $\underline{65}$, 5587.
1063. Fuse, G. and Nishimoto, K., Mokuzai Kenkyu No. $\underline{32}$, 15-25 (1964); CA $\underline{65}$, 9646.
1064. Fye, R. L., LaBrecque, G. C. and Gouck, H. K., J. Econ. Entomol. $\underline{59}$, 485-7 (1966); CA $\underline{64}$, 14895.

1065. Gager, H. M., Lewis, J. and Ware, M. L., Chem. Commun. 1966, 616-7; CA 66, 33340.
1066. Galashina, M. L., Kaznina, G. V. and Sobolevskii, M. V., Plast. Massy 1966, 26-7; CA 64, 16062.
1067. Galasso, V., Bigotto, A. and DeAlti, G., Z. Phys. Chem. (Frankfurt) 50, 38-45 (1966); CA 66, 6153.
1068. Gallais, F., Labarre, J. F. and deLoth, P., J. Chim. Phys. 64, 247-52 (1967); CA 67, 27408.
1069. Galley, D. J., Guess, W. L. and Autian, J., J. Pharm. Sci. 56, 240-3 (1967); CA 66, 63872.
1070. G'alpern, E. G. and Mayants, L. S., Zh. Strukt. Khim. 6, 785-7 (1965); CA 64, 5946.
1071. Gar, T. K. and Mironov, V. F., Zh. Obshch. Khim. 36, 1709-10 (1966); CA 66, 46461.
1072. Garzo, G., Fekete, J. and Blazso, M., Acta Chim. Acad. Sci. Hung. 51, 359-69 (1967); CA 67, 17613.
1073. Gashtold, N. S. et al., Vysokomol. Soedin., Ser. A9, 1489-93 (1967); CA 68, 30345.
1074. Gastilovich, E. A., Shigorin, D. N. and Komarov, N. V., Tr. Komis. po Spektroskopii, Akad. Nauk SSSR 3, 70-5 (1964); CA 65, 593.
1075. Gaston, G. E., U.S. 3,313,607, Aug. 12, 1964; CA 67, 4671.
1076. Gavrilenko, V. V., Iwanov, L. L. and Zakharkin, L. I., Zh. Obshch. Khim. 37, 550-3 (1967); CA 67, 21983.
1077. Gay, R. and Dupuy, G., Ger. 1,236,853, Sept. 29, 1960; CA 67, 4670.
1078. Geissler, H. and Kriegsmann, H., J. Organometal. Chem. 11, 85-95 (1968); CA 68, 34495.
1079. --- and ---, Z. Chem. 5, 423-4 (1965); CA 64, 11834.
1080. Gelfman, Ya. A. et al., Plast. Massy 1966, 10-11; CA 66, 11378.
1081. --- et al., Sb. Tr. Vses. Nauchn.-Issled. Inst. Novykh Stroit. Mater. 1960 (14), 58-61; CA 68, 79011.
1082. ---, Lauris, I. V. and Kuskova, V. P., ibid. 1966 (14), 54-8; CA 68, 79010.
1083. Gelius, R., Z. Anorg. Allg. Chem. 349, 22-32 (1966); CA 66, 76105.
1084. ---, Ger. 1,203,266, March 23, 1964; CA 64, 6693.
1085. --- and Franke, W., Brennstoff-Chem. 47, 280-5 (1966); CA 66, 12709.
1086. --- and Mueller, R., Chem. Tech. (Berlin) 18, 371 (1966); CA 65, 16743.
1087. --- and ---, Z. Anorg. Allg. Chem. 351, 42-7 (1967); CA 67, 3130.
1088. Gelman, N. E., Talanta 14, 1423-31 (1967); CA 68, 35646.
1089. George, M. V. et al., J. Organometal. Chem. 5, 397-404 (1966); CA 64, 19664.
1090. George, T. A. and Lappert, M. F., Chem. Commun. 1966, 463-4; CA 65, 12232.
1091. Georghiou, G. P., Metcalf, R. L. and Von Zboray, E. P., Bull. World Health Organ. 33, 479-84 (1965); CA 64, 5690.
1092. Gerrard, W. and Green, D. B., J. Organometal. Chem. 7, 91-6 (1967); CA 66, 28865.

1093. Giamaria, J. J. and Becker, M., U.S. 3,254,973, July 31, 1962; CA 65, 6979.
1094. Gibb, T. C. and Greenwood, N. N., J. Chem. Soc., A 1966, 43-6; CA 64, 9559.
1095. Gibbon, G. A., Wang, J. T. and van Dyke, C. H., Inorg. Chem. 6, 1989-94 (1967); CA 67, 116931; cf. ref. 2409.
1096. Gibbons, A. J., Jr., U.S. 3,307,973, Jan. 17, 1964; CA 66, 95179.
1097. --- and DeMarco, R. E., U.S. 3,361,775, Nov. 10, 1964; CA 68, 50618.
1098. Gielen, M. and Nasielski, J., J. Organometal. Chem. 7, 273-80 (1967); CA 66, 94535.
1099. Giesen, M., F.A.T.I.P.E.C., Congr. 8, 185-96 (1966); CA 65, 17625.
1100. Gillespie, R. J., Kapoor, R. and Robinson, E. A., Can. J. Chem. 44, 1197-202 (1966); CA 64, 18492.
1101. Gilman, H. and Sim, S.-Y., J. Organometal. Chem. 7, 249-58 (1967); CA 66, 65578; cf. ref. 1928.
1102. Giordano, J. J., U.S. 3,297,630, Sept. 25, 1963; CA 66, 56671.
1103. Gladshtein, B. M., Kulyulin, I. P. and Soborovskii, L. Z., Zh. Obshch. Khim. 36, 488-92 (1966); CA 65, 743.
1104. Glatte, W. et al., Chem. Tech. (Berlin) 19, 294-9 (1967); CA 67, 101626.
1105. Glockling, F. and Hooton, K. A., Inorg. Synth. 8, 31-4 (1966); CA 66, 38020.
1106. --- and ---, Chem. Commun. 1966, 218; CA 65, 1747.
   --- and ---, J. Chem. Soc., A 1967, 1066-75; CA 67, 73658.
1107. --- and ---, ibid., A 1968, 826-34; CA 68, 101357.
1108. --- and Light, J. R. C., ibid., A 1967, 623-7; CA 66, 115771.
1109. --- and ---, ibid., A 1968, 717-34; CA 68, 82290.
1110. Gloskey, C. R., Brit. 1,009,368, Dec. 31, 1962; CA 64, 8240; Belg. 661,479, March 22, 1965; CA 64, 19910.
1110a. ---, Brit. 1,074,407; Fr. 1,433,963; Neth. 65 6444; U.S. 3,340,283, May 20, 1964; CA 64, 11251.
1111. Goenvec, H., Fr. 1,461,586, Oct. 26, 1965; CA 67, 17108.
1112. Goldanskii, V. I. et al., Dokl. Akad. Nauk SSSR 169, 872-5 (1966); CA 65, 16095.
1113. --- et al., J. Organometal. Chem. 4, 160-2 (1965); CA 64, 2790.
1114. --- et al., Teor. Eksp. Khim. 3, 478-82 (1967); CA 68, 100418.
1115. ---, Makarov, E. F. and Stukan, R. A., J. Chem. Phys. 47, 4048-52 (1967); CA 68, 24620.
1116. Goldshtein, I. P. et al., Dokl. Akad. Nauk SSSR 175, 836-9 (1967); CA 68, 77577.
1117. --- et al., Izv. Akad. Nauk SSSR, Ser. Khim. 1967, 2201-7; CA 68, 25261.
1118. Goodacre, C. L. and Goodacre, U. M., Brit. 1,078,259, May 18, 1965; CA 67, 110323.
1119. --- and ---, Brit. 1,092,337, June 8, 1965; CA 68, 42033.
1120. Gormley, J. J. and Rees, R. G., J. Organometal. Chem. 5, 291-2 (1966); CA 64, 11245.
1121. Gorokhov, L. N., Zh. Strukt. Khim. 6, 766-8 (1965); CA 64, 129.

1122. Gorshkova, G. N. et al., Zh. Fiz. Khim. 39, 2695-700 (1965); CA 64, 7539.
1123. --- et al., ibid. 40, 1433-5 (1966); CA 65, 17900.
1124. Gorsich, R. D., J. Organometal. Chem. 5, 105-6 (1966); CA 64, 3594.
1125. ---, U.S. 3,261,806, May 31, 1963; CA 65, 17152.
1126. ---, U.S. 3,262,909, May 31, 1963; CA 65, 15614.
1127. Gorth, H. and Henry, M. C., J. Organometal. Chem. 9, 117-23 (1967); CA 67, 82249.
1127a. Gottlieb, J. B. and Mayo, W. E., Brit. 1,099,900, 1,110,709; Fr. 1,509,249; Neth. 67 14; U.S. 3,424,712, 3,424,717, Jan. 3, 1966; CA 67, 109344.
1128. deGotuzzo, E. A. and Docampo, D., Rev. Fac. Agron. Vet., Univ. Buenos Aires 16 (3), 3-26 (1966); CA 67, 52989.
1129. Graber, M. and Gras, G., Rev. Elevage Med. Vet. Pays Trop. 18, 404-5 (1965); CA 65, 4498.
1130. --- and ---, ibid. 18, 415-22 (1965); CA 65, 4498.
1131. --- and ---, ibid. 19, 7-14 (1966); CA 65, 4327.
1132. Graham, W. A. G., Inorg. Chem. 7, 315-21 (1968); CA 68, 53453.
1133. Graiff, L. B., S.A.E. J. 75 (8), 55-9 (1967); CA 67, 101625.
1134. Gras, G., Rev. Elevage Med. Vet. Pays Trop. 19, 15-20 (1966); CA 65, 4327.
1135. --- and Castel, J., Trav. Soc. Pharm. Montpellier 24, 116-9 (1964); CA 65, 9409.
1136. --- and ---, ibid. 25, 178-84 (1966); CA 65, 3672.
1137. --- and Rioux, I. A., Arch. Inst. Pasteur Tunis 42, 9-22 (1965); CA 65, 6222.
1138. --- and Un, S., ibid. 42, 337-49 (1965); CA 68, 20627.
1139. Gray, R. D. and Mayer, S. E., Belg. 673,045; Brit. 1,079,148; Fr. 1,456,268; U.S. 3,404,167, Nov. 30, 1964; CA 66, 115806.
1140. Green, L. E., Facts Methods 8 (4), 4-7 (1967); CA 68, 88740.
1141. Green, M., Mayne, N. and Stone, F. G. A., J. Chem. Soc., A 1968, 902-5; CA 68, 105329.
1142. Green, P. J., Ph. D. Thesis, West Va. Univ., 1967; Diss. Abstr. B28, 4897 (1968); CA 69, 82167.
--- and Graybeal, J. D., J. Am. Chem. Soc. 89, 4305-8 (1967); CA 67, 77741.
1143. Greene, F. D. and Lowry, N. N., J. Org. Chem. 32, 882-5 (1967); CA 66, 104457.
1143a. Greenwood, N. N. and Ruddick, J. N. R., J. Chem. Soc., A 1967, 1679-83; CA 67, 112662.
1143b. Grenoble, M. E., Brit. 1,091,650; Fr. 1,492,531; U.S. 3,419,508, Sept. 17, 1965; CA 68, 22572.
1144. Groagova, A. and Pribyl, M., Z. Anal. Chem. 234, 423-8 (1968); CA 68, 101663.
1145. Gruen, L., Gesundheitsw. Desinfek. 58, 81-4 (1966); CA 66, 610.
1146. Gureev, A. A. and Malyavinskii, L. V., Khim. Tekhnol. Topl. Masel 13, 45-8 (1968); CA 68, 97255.

1147. Gurevich, E. S. and Dolgopolskaya, M. A., Tr. Sevastopolsk. Biol. St. Akad. Nauk Ukr. SSR 15, 472-84 (1964); CA 64, 5293.
1148. Gursky, J. and Vesely, V., Scu. Azione 1967 (10), 5-24; CA 68, 51622.
1149. Gverdtsiteli, I. M. and Adamiya, S. V., Soobshch. Akad. Nauk Gruz. SSR 47, 55-9 (1967); CA 68, 29817.
1150. --- and Buachidse, M. A., ibid. 48, 571-4 (1967); CA 68, 105315.
1151. Haberman, S. et al., Tech. Pap. Reg. Tech. Conf., Soc. Plast. Eng., N.Y. Sect. 1967, 28-43; CA 67, 115358.
1152. Haefliger, H., Chem. Spec. Mfr. Ass. Proc. Mid-Year Meet. 53, 223-6 (1967); CA 68, 31340.
1153. Haggis, G. A. and Twitchett, H. J., Brit. 1,009,007, Nov. 23, 1963; CA 65, 20313.
1154. Hague, D. N. and Prince, R. H., J. Inorg. Nucl. Chem. 28, 1039-46 (1966); CA 64, 19669.
1154a. Haley, T. J., Arch. Environ. Health 12, 781 (1966).
1155. Halmi, G. and Advani, R., U.S. 3,236,814, Aug. 8, 1961; CA 63, 18338.
1156. --- and ---, U.S. 3,297,651, March 23, 1962; CA 66, 47232
1157. Hammerich, T., Erdoel Kohle, Erdgas, Petrochem. 20, 488-99 (1967); CA 67, 83608.
1158. Hannan, J. F., U.S. 3,362,889, Feb. 24, 1966; CA 68, 51783; Neth. 65 15,216, Oct. 27, 1966; CA 67, 73686.
1159. Harrendorf, K. and Klutts, R. E., J. Econ. Entomol. 60, 1471-2 (1967); CA 67, 107638.
1160. Harrison, R. W. and Trotter, J., J. Chem. Soc., A 1968, 258-66; CA 68, 63486.
1161. Harvey, A. B., J. Phys. Chem. 70, 3370-1 (1966); CA 66, 5852.
1162. Hashimoto, H. and Morimoto, Y., J. Organometal. Chem. 8, 271-5 (1967); CA 67, 11067.
1163. Hatakeyama, H., Kubo, M. and Fukumoto, O., Japan. 7687 ('67), Feb. 5, 1964; CA 68, 14003.
1164. Hayakawa, Y., Fueno, T. and Furukawa, J., J. Polym. Sci., Pt. A-1 5, 2099-109 (1967); CA 67, 82444.
1165. Hayashi, K., Iyoda, J. and Shiihara, I., J. Organometal. Chem. 10, 81-94 (1967); CA 67, 100215.
1166. Hayashi, T. et al., Kogyo Kagaku Zasshi 70, 714-8 (1967); CA 68, 13109.
1167. --- et al., ibid. 70, 2298-301 (1967); CA 68, 87363.
1168. ---, Kikkawa, S. and Matsuda, S., ibid. 70, 1389-93 (1967); CA 68, 59672.
1169. Heath, D. F., Radioisotop. Detection Pestic. Residues, Proc. Panel Vienna 1965, 18-26 (1966); CA 67, 1828.
1169a. Hechenbleikner, I., Fr. 1,482,907; Neth. 66 2980; U.S. 3,288,669, March 25, 1965; CA 66, 36822.
1170. ---, Bresser, R. E. and Homberg, O. A., U.S. 3,271,004, Dec. 30, 1963; CA 67, 90939; cf. ref. 213.
1170a. ---, Dalter, R. S. and Hussar, J. F., Brit. 1,111,398; Fr. 1,465,860; Neth. 66 846; U.S. 3,396,185, Jan. 22, 1965; CA 66, 3194.

1171. Heiss, H. L., Brit. 1,081,555; Fr. 1,438,111; Ger. 1,232,145; U.S. 3,480,656, July 8, 1964; CA 66, 38050.
1172. Helberg, D., Deut. Lebensm.-Rundsch. 63, 69-71 (1967); CA 68, 92801.
1173. Heldt, E., Hoeppner, K. and Krebs, K. H., Z. Anorg. Allg. Chem. 347, 95-100 (1966); CA 65, 16222.
1174. Helling, J. F. et al., J. Am. Chem. Soc. 89, 7140-1 (1967); CA 68, 39790.
1175. Hellwarth, R. W., Phys. Rev. 152, 156-65 (1966); CA 66, 23801.
1176. Henderson, H. T., Brit. 906,533; Ger. 1,221,488, Sept. 30, 1959; CA 65, 12044.
1177. Hengge, E. and Brychcy, V., Monatsh. Chem. 97, 1309-17 (1966); CA 66, 38002.
1178. Henneberg, D. and Schomburg, G., Z. Anal. Chem. 215, 424-30 (1965); CA 64, 13362.
1179. Henry, M. C. and Krebs, A. W., U.S. 3,322,779, April 1, 1963; CA 67, 73685.
1180. Herber, R. H. and Parisi, G. I., Inorg. Chem. 5, 769-74 (1966); CA 64, 18704.
1181. Herbstman, S., U.S. 3,355,469, April 9, 1964; CA 68, 49782.
1182. --- and Stamm, W. A., U.S. 3,311,648, Oct. 21, 1963; CA 67, 11590.
1183. Hess, F. G., U.S. 3,346,492, May 26, 1965; CA 68, 61469.
1184. Hester, R. E. and Jones, K., Chem. Commun. 1966, 317-8; CA 65, 7031.
1185. Hills, K. and Henry, M. C., J. Organometal. Chem. 9, 180-2 (1967); CA 68, 39760.
1185a. Hindersinn, R. R. and Marciniak, H. W., Belg. 672,846; Brit. 1,127,249; Fr. 1,456,239; Neth. 65 15,354; U.S. 3,418,263, Nov. 25, 1964; CA 64, 17154.
1186. Hirschler, D. A., Jr., Lyben, R. G. and Rifkin, E. B., Ger. 1,239,133, July 24, 1958; CA 67, 118894.
1186a. Hirshman, J. L. and Natoli, J. G., Brit. 1,084,076; Fr. 1,434,534, Neth. 65 4500; U.S. 3,355,468, April 8, 1964; CA 64, 8240.
1187. Hocking, D., Ann. Appl. Biol. 59, 363-73 (1967); CA 67, 72722.
1188. ---, Trop. Agr. (London) 44, 83-7 (1967); CA 66, 114874.
1189. ---, East Afr. Agr. Forest. J. 32, 356-8 (1967); CA 67, 81430.
   ---, ibid. 32, 359-62 (1967); CA 67, 81431.
1190. --- and White, P. J., ibid. 32, 380-2 (1967); CA 67, 63244.
1191. Hoeppner, K., Proc. Tihany Symp. Radiat. Chem., 2nd, Tihany, Hung. 1966, 33-6 (1967); CA 67, 59548; cf. ref. 2438.
1192. ---, Proesch, V. and Zoepfl, H. J., Abh. Deut. Akad. Wiss. Berlin, Kl. Chem., Geol., Biol. 1966, 393-406; CA 67, 65827.
1193. Hogg, W. H. and Watson, J. W., Brit. 1,026,594, Dec. 19, 1961; CA 64, 19993.
1194. Holliday, A. K. and Jessop, G. N., J. Chem. Soc., A 1967, 889-91; CA 68, 29808; CA 70, 20123.
1195. --- and ---, J. Organometal. Chem. 10, 291-3 (1967); CA 68, 13126.
1196. --- and Pendlebury, R. E., J. Chem. Soc. 1965, 6659-60; CA 64, 4572.
1196a. --- and ---, J. Organometal. Chem. 7, 281-4 (1967); CA 66, 65599.

1197. Holliday, A. K. and Pendlebury, R. E., J. Organometal. Chem. <u>10</u>, 295-300 (1967); CA <u>68</u>, 13028.
1198. Holmes, J. M., Peacock, R. D. and Tatlow, J. C., J. Chem. Soc., A <u>1966</u>, 150-3; CA <u>64</u>, 9760.
1199. Homberg, O. A., Ph. D. Thesis, Univ. of Cincinnati, 1965; Diss. Abstr. <u>26</u>, 5036 (1966); CA <u>65</u>, 2293; cf. ref. M-2611, 2187, 2188.
1200. Homrowski, S., Rocz. Panstw. Zakl. Hig. <u>18</u>, 283-91 (1967); CA <u>68</u>, 20583.
1201. Honda, M., Kawasaki, Y. and Tanaka, T., Tetrahedron Lett. <u>1967</u>, 3313-5; CA <u>68</u>, 48848.
1202. Hopf, H. S. et al., Bull. World Health Organ. <u>36</u>, 955-61 (1967); CA <u>68</u>, 21171.
1203. Hopkins, F. M., Brit. 1,111,949; Fr. 1,441,315; U.S. 3,384,559, Aug. 4, 1964; CA <u>66</u>, 30784.
1204. Horan, J. E. and Van Strien, R. E., U.S. 3,244,737, April 30, 1962; CA <u>65</u>, 4054.
1205. Horder, J. R. and Lappert, M. F., Chem. Commun. <u>1967</u>, 485-6; CA <u>67</u>, 54237.
1206. --- and ---, J. Chem. Soc., A <u>1968</u>, 1167-8; CA <u>68</u>, 118950.
1207. Horn, H. and Huber, F., Monatsh. Chem. <u>98</u>, 771-84 (1967); CA <u>67</u>, 90196.
1208. Horrocks, J. A., Brit. 1,061,747, Sept. 2, 1964; CA <u>67</u>, 12144.
1209. Hostettler, F. and Cox, E. F., U.S. 3,240,730, June 23, 1961; CA <u>64</u>, 16087.
1209a. Hoye, P. A. T., Brit. 1,115,646; Fr. 1,446,994; Ger. 1,277,255; Neth. 65 12,145; U.S. 3,415,857, Sept. 18, 1964; CA <u>65</u>, 8962.
1209b. --- and Sundeland, P., Brit. 1,079,641; Fr. 1,429,733; Neth. 65 4226, April 6, 1964; U.S. 3,400,141, March 23, 1965; CA <u>64</u>, 8240.
1210. Huang, H. H. and Hui, K. M., J. Organometal. Chem. <u>6</u>, 504-16 (1966); CA <u>66</u>, 10455.
1211. ---, --- and Chiu, K. K., ibid. <u>11</u>, 515-24 (1968); CA <u>68</u>, 113922.
1212. Huber, F., Enders, M. and Kaiser, R., Z. Naturforsch. B <u>21</u>, 83-4 (1966); CA <u>64</u>, 17631.
1213. --- and Haupt, H. J., ibid. B <u>21</u>, 808-9 (1966); CA <u>65</u>, 18611.
1214. ---, Horn, H. and Haupt, H. J., ibid. B <u>22</u>, 918-21 (1967); CA <u>68</u>, 49695.
1215. --- and Kaiser, R., ibid. B <u>20</u>, 1011-2 (1965); CA <u>64</u>, 6685.
1216. --- and ---, J. Organometal. Chem. <u>6</u>, 126-32 (1966); CA <u>65</u>, 10617.
1217. --- and Padberg, F. J., Z. Anorg. Allg. Chem. <u>351</u>, 1-8 (1967); CA <u>67</u>, 3131.
1218. --- and Schoenafinger, E., Angew. Chem. Int. Ed. Engl. <u>7</u>, 72-8 (1968); CA <u>68</u>, 59669.
1219. Hulea, A. et al., Probl. Agr. (Bucharest) <u>18</u>, 20-8 (1966); CA <u>67</u>, 2365.
1220. Hunter, B. K. and Reeves, L. W., Can. J. Chem. <u>46</u>, 1399-414 (1968); CA <u>68</u>, 118382.
1221. Ibekwe, S. D. and Newlands, M. J., J. Chem. Soc. <u>1965</u>, 4608-10; CA <u>64</u>, 3592.
1222. --- and ---, ibid., A <u>1967</u>, 1783-6; CA <u>68</u>, 26443; cf. ref. 236a.

1223. Ilgemann, R. and Rauschenbach, R. D., Fr. 1,411,363, Sept. 24, 1963; CA 65, 10765.
1224. Imura, S. and Fukutaka, K., Bunseki Kagaku 14, 1167 (1965); CA 64, 14952.
--- and Yamada, T., ibid. 16, 1351-4 (1967); CA 68, 111250.
1225. Inaba, T., Maeda, K. and Watabe, S., Japan. 12,788 ('65), Nov. 8, 1962; CA 64, 6694.
1226. --- and Wataribe, S., Japan. 10,102 ('66), Sept. 30, 1963; CA 65, 12240.
1227. Ingram, P. and Schindler, A., Makromol. Chem. 111, 267-70 (1968); CA 68, 59992.
1228. Ishii, Y. et al., Chem. Commun. 1967, 224; CA 66, 105035.
1229. Ishii, Y., Kawamura, H. and Yagi, S., Bunseki Kagaku 17, 3-7 (1968); CA 68, 96897.
1230. Ismail, R. M., J. Organometal. Chem. 6, 663-4 (1966); CA 66, 28864.
---, Fr. 1,475,896, Feb. 23, 1966; CA 68, 13178.
1231. ---, Brit. 1,169,011; Fr. 1,454,605, Oct. 11, 1965; CA 67, 32782.
1232. Issleib, K. and Walther, B., Angew. Chem. 79, 59-60 (1967); Int. Ed. Engl. 6, 88-9 (1967); CA 66, 76065.
1233. --- and ---, J. Organometal. Chem. 10, 177-80 (1967); CA 68, 22028.
1234. Ito, T. and Kawakami, Y., Japan. 13,259 ('65), Nov. 5, 1962; CA 65, 9116.
1235. Itoh, K. et al., J. Organometal. Chem. 10, 451-5 (1967); CA 68, 13119.
1236. --- et al., Kogyo Kagaku Zasshi 70, 935-8 (1967); CA 68, 28973.
1236a. --- et al., Tetrahedron Lett. 1967, 2667-70; CA 67, 90894.
1237. ---, Sakai, S. and Ishii, Y., Chem. Commun. 1967, 36-7; CA 66, 85218.
1238. ---, --- and ---, Tetrahedron Lett. 1966, 4941-5; CA 66, 11015.
1238a. Itoi, K., Brit. 1,107,188; Japan. 68 13,981; U.S. 3,501,450, July 14, 1964; CA 68, 105644.
1239. ---, Fr. 1,411,034, May 2, 1963; CA 64, 5136.
1240. ---, Brit. Amended 1,057,998; Fr. 1,453,036, Nov. 9, 1964; CA 70, 97391 and CA 66, 66047.
1241. ---, Japan. 29,025 ('65), Feb. 14, 1963; CA 64, 9838.
1242. --- and Kumano, S., Kogyo Kagaku Zasshi 70, 82-6 (1967); CA 67, 11556.
1243. Iwahara, S. et al., Eisei Shikensho Hokoku No. 82, 118-9 (1964); CA 65, 10771.
1244. Jackson, C. R., Plant Disease Reptr. 49, 928-31 (1965); CA 64, 4195.
1244a. Jasching, W. and Franzen, V., Belg. 669,340; Brit. 1,083,908; Fr. 1,446,611; Ger. 1,217,951; Neth. 65 11,702; U.S. 3,387,012, Sept. 12, 1964; CA 65, 5489.
1245. Jason, E. F. and Fields, E. K., U.S. 3,247,167, July 19, 1962; CA 65, 9053.
1246. --- and ---, U.S. 3,262,915, July 17, 1962; CA 65, 12359.
1247. Jaura, K. L., Churamani, L. K. and Sharma, K. K., Indian J. Chem. 4, 329-30 (1966); CA 65, 13751.
1247a. ---, Hundal, H. S. and Handa, R. D., ibid. 5, 211-2 (1967); CA 68, 13098.
1248. Jehn, W., Z. Anorg. Allg. Chem. 351, 260-7 (1967); CA 67, 78611.

1249. Jehring, H. and Mehner, H., Z. Anal. Chem. 224, 136-43 (1967); CA 66, 61266.
1250. Jenkins, L. T., U.S. 3,210,329, Oct. 24, 1960; CA 64, 2187.
1251. Jensen, F. R. and Patterson, D. B., Tetrahedron Lett. 1966, 3837-41; CA 65, 13570.
1252. Jetz, W. et al., Inorg. Chem. 5, 2217-22 (1966); CA 66, 76123.
1253. Jitsu, Y., Kudo, N. and Sugiyama, T., Noyaku Seisan Gijutsu 17, 17-22 (1967); CA 68, 56448.
1254. Johnson, F., U.S. 3,234,239, June 8, 1960; CA 64, 11251.
1255. Jones, H. F., J. Inst. Fuel 40, 569-79 (1967); CA 68, 51475.
1256. Jones, J. P. and Everett, P. H., Plant Disease Reptr. 50, 340-4 (1966); CA 65, 2930.
1257. Jones, M. T., Inorg. Chem. 6, 1249-51 (1967); CA 67, 38063.
1258. Jovanovic, M. S. et al., Chem. Anal. (Warsaw) 11, 479-81 (1966); CA 65, 15122.
1259. Juara, V. A. et al., Indian. J. Chem. 5, 211-2 (1967); CA 68, 13028.
1260. Juenge, E. C. and Gray, S., J. Organometal. Chem. 10, 465-70 (1967); CA 68, 29820.
1261. Kadina, M. A. et al., Zh. Obshch. Khim. 37, 1040-6 (1967); CA 68, 22019.
1262. Kado, M. and Maeda, T., Japan. 22,069 ('65), Nov. 14, 1962; CA 64, 3603.
1263. Kahana, L. and Sijpesteijn, A. K., Antomie van Leeuwenhoek, J. Microbiol. Serol. 33, 427-38 (1967); CA 67, 106167.
1264. Kahl, G. and Breuer, G., Ger. 1,262,590, July 30, 1964; CA 68, 87910.
1265. Kahle, G. R. and Buck, O. G., Brit. 1,071,535; Fr. 1,454,525, Sept. 8, 1964; CA 66, 96008.
1266. --- and Uraneck, C. A., U.S. 3,278,508, Jan. 14, 1963; CA 65, 20334; cf. ref. 571.
1267. Kahn, J. S., Biochim. Biophys. Acta 153, 203-10 (1968); CA 68, 46232.
1268. Kahn, O. and Biggorgne, M., Compt. Rend. 261, 2483-5 (1965); CA 64, 272.
1269. --- and ---, C.R., Acad. Sci., Paris, Ser. C 262, 906-9 (1966); CA 65, 1744.
1270. --- and ---, ibid., Ser. C 263, 973-6 (1966); CA 66, 24015.
1271. --- and ---, ibid., Ser. C 266, 792-5 (1968); CA 68, 118205.
1272. --- and ---, J. Organometal. Chem. 10, 137-67 (1967); CA 67, 121283.
1273. Kameyama, T. and Sekine, T., J. Biochem. (Tokyo) 58, 420-1 (1965); CA 64, 2603.
1274. Kaminaka, S., Seki, T. and Kawakami, Y., Japan. 19,176 ('67), April 14, 1964; CA 68, 69765.
1275. Kamitani, I. and Hasegawa, N., Kami-pa Gikyoshi 20, 55-63 (1966); CA 64, 9941.
1276. Kan, M., Tsukamoto, Y. and Suzuki, H., Takeda Kenkyusho Nempo 23, 76-85 (1964); CA 64, 18.
1277. Kanai, M. and Miyoshi, H., Japan. 17,657 ('66) Aug. 26, 1963; CA 66, 76567.

1278. Kandil, K. S. A. K., Ph. D. Thesis, Univ. of Cincinnati, 1966; Diss. Abstr. B28, 839-40 (1967); CA 68, 59634.
1279. Kankovskaya, E. N., Dmitrienko, S. S. and Pechennikova, T. I., U.S.S.R. 174,354, Feb. 17, 1964; CA 64, 879.
1280. Kaplan, L., J. Am. Chem. Soc. 88, 1833-4 (1966); CA 65, 3785.
    ---, ibid. 88, 4970-1 (1966); CA 65, 20039.
1281. Kaplin, Yu. A., Kudryavtsev, L. F. and Petukhov, G. G., Zh. Obshch. Khim. 36, 1061-3 (1966); CA 65, 10617.
1282. Karasev, A. N. et al., Teor. i Eksperim. Khim., Akad. Nauk Ukr. SSSR 2 (1), 126-30 (1966); CA 65, 6334.
1283. Kato, F., Tatsuhara, M. and Matsumoto, T., Japan. 864 ('67), Sept. 25, 1962; CA 67, 33316.
1284. ---, --- and ---, Japan. 865 ('67), Dec. 17, 1962; CA 67, 33317.
1285. --- and Yatsu, M., Japan. 2903 ('67), Jan. 24, 1963; CA 67, 33319.
1286. --- and ---, Japan. 2904 ('67), Jan. 24, 1963; CA 68, 40578.
1287. Katsumura, T., Kawakahi, Y. and Seki, T., Japan. 21,625 ('66), May 22, 1963; CA 67, 44433.
1288. Katsunuma, T. and Kano, K., Japan. 12,053 ('67), Oct. 24, 1963; CA 68, 96587.
1289. Kauder, O. S., Fr. 1,480,479, May 11, 1965; CA 67, 117764.
    ---, U.S. 3,384,649, May 11, 1965; CA 69, 19796.
1290. Kawakami, K., Kawasaki, Y. and Okawara, R., Bull. Chem. Soc. Jap. 40, 2693-5 (1967); CA 68, 49699.
1291. --- and Okawara, R., J. Organometal. Chem. 6, 249-58 (1966); CA 65, 12232.
1292. ---, Saito, T. and Okawara, R., ibid. 8, 377-81 (1967); CA 67, 38128.
1293. Kawasaki, Y., J. Inorg. Nucl. Chem. 29, 840-1 (1967); CA 67, 38089.
1294. ---, J. Organometal. Chem. 9, 549-52 (1967); CA 68, 73889.
1295. ---, Mol. Phys. 12, 287-8 (1967); CA 67, 69286.
1296. ---, Hori, M. and Uenaka, K., Bull. Chem. Soc. Jap. 40, 2463-7 (1967); CA 68, 53986.
1297. --- and Tanaka, T., J. Chem. Phys. 43, 3396-7 (1965); CA 64, 5970.
1298. ---, --- and Okawara, R., Bull. Soc. Chem. Jap. 40, 1562-5 (1967); CA 68, 7938.
1299. ---, --- and ---, Inorg. Nucl. Chem. Lett. 2, 9-12 (1966); CA 65, 1630.
1300. ---, --- and ---, J. Organometal. Chem. 6, 95-7 (1966); CA 65, 11552.
1301. ---, --- and ---, Spectrochim. Acta. 22, 1571-9 (1966); CA 65, 16262.
1302. Kawasumi, H., Japan. 26,853 ('67), Nov. 11, 1963; CA 68, 60052.
1303. Kazankova, M. A., Belkina, M. A. and Lutsenko, I. F., Zh. Obshch. Khim. 37, 1710-3 (1967); CA 68, 13133.
1304. Keblys, K. A. and Dubeck, M., U.S. 3,296,288, April 29, 1963; CA 68, 115804.
1305. Kelso, R. G., U.S. 3,232,905, Oct. 27, 1960; CA 64, 11399; cf. ref. M-1996.
1306. Kenaga, E. E., U.S. 3,264,177, Feb. 17, 1964; CA 65, 14364.
1307. Kessler, D., Weiss, A. and Witte, H., Ber. Bunsenges. Phys. Chem. 71, 3-19 (1967); CA 66, 70741.

1308. Khokhlov, P. S., Andrianov, Yu. A. and Bliznyuk, N. K., U.S.S.R. 180,592, April 16, 1965; CA 65, 13763.
1309. Khrapov, V. V. et al., Zh. Obshch. Khim. 37, 3-11 (1967); CA 66, 109968.
1310. Kiesel, R. J., Ph. D. Thesis, St. Johns Univ., 1966; Diss. Abstr. B27, 4311 (1967); CA 68, 2967; cf. ref. 299, 1374.
1311. Kilbourn, B. T., Blundell, T. L. and Powell, H. W., Chem. Commun. 1965, 444-5; CA 64, 95.
1312. --- and Powell, H. W., Chem. Ind. (London) 1964, 1578; CA 65, 14553.
1313. Kimmel, H. S., Ph. D. Thesis, City of New York Univ., 1967; Diss. Abstr. B28, 1884 (1967); CA 68, 73624.
--- and Dillard, C. R., Spectrochim. Acta, Pt. A 24, 909-19 (1968); CA 69, 31686.
1314. King, R. B., Inorg. Chem. 6, 25-9 (1967); CA 66, 34377.
1315. Kissam, J. B. and Hays, S. B., J. Econ. Entomol. 59, 748-9 (1966); CA 65, 2939.
1316. Kitching, W., J. Organometal. Chem. 6, 586-8 (1966); CA 66, 46465.
1317. ---, Tetrahedron Lett. 1966, 3689-93; CA 65, 13503.
1318. ---, Das, V. G. K. and Wells, P. R., Chem. Commun. 1967, 356-7; CA 67, 59337.
1319. Klotzer, D. et al., Brit. 1,081,969, May 31, 1966; CA 68, 2256; Fr. 1,497,486, Apr. 25, 1966; CA 69, 85724.
1320. --- et al., Ger. (East) 56,133, Jan. 4, 1966; CA 67, 81484.
1321. Klose, G., Ann. Phys. 10, 391-8 (1963); CA 64, 5971.
1322. Knaub, E. W., U.S. 3,338,951, July 28, 1965; CA 67, 82833.
1322a. Knusli, E. and Varsany, D., Belg. 666,691; Brit. 1,122,595; Fr. 1,455,942; Ger. 1,239,311; Neth. 65 8906; Swiss 417,216; U.S. 3,412,090, Jan. 10, 1966; CA 65, 17004.
1323. Kobayashi, Y., Fukuma, N. and Kitaoka, A., Brit. 1,019,916, Dec. 12, 1962; CA 64, 11413.
1324. Kobetz, P. and Beaird, F. M., U.S. 3,357,928, March 31, 1965; CA 68, 105365.
1325. --- and Thomas, W. H., U.S. 3,344,048, Oct. 26, 1964; CA 68, 45633.
1326. Kocheshkov, K. A. et al., Izv. Akad. Nauk SSSR, Ser. Khim. 1967, 1171; CA 67, 108723.
1327. Kochkin, D. A. et al., Vestn. Akad. Nauk Kaz. SSR 22 (10), 45-52 (1966); CA 66, 76330.
--- et al., Vysokomol. Soedin., Ser. A 9, 2208-13 (1967); CA 68, 3216.
1328. --- and Azerbaev, I. V., Vestn. Akad. Nauk Kaz. SSR 22 (12), 53-61 (1966); CA 67, 64467.
1329. Kocmid, V., Listy Cukrov. 82, 289-94 (1966); CA 66, 75221.
1330. Koehler, H. and Seifert, B., J. Organometal. Chem. 12, 253-5 (1968); CA 68, 105316.
1331. Koenig, K. and Neumann, W. P., Tetrahedron Lett. 1967, 495-8; CA 66, 105031.
1332. Koester-Pflugmacher, A. and Hirsch, A., Naturwissenschaften 54, 645 (1967); CA 68, 59680.

1333. Kokoreva, I. Yu. et al., Izv. Akad. Nauk SSSR, Ser. Khim. 1967, 411-2; CA 67, 15955.

1334. Kolb, B. et al., Z. Anal. Chem. 221, 166-75 (1966); CA 66, 12664.

1335. Kolesnikov, S. P. and Nefedov, O. M., Zh. Obshch. Khim. 37, 746 (1967); CA 67, 43894.

1336. ---, --- and Sheichenko, V. I., Izv. Akad. Nauk SSSR, Ser. Khim. 1966, 443-52; CA 64, 19668; cf. ref. 278.

1337. ---, Shiryaev, V. I. and Nefedov, O. M., ibid. 1966, 584; CA 65, 6705.

1338. Kolla, V. E. and Zalesov, V. S., Uch. Zap., Permskii Gos. Univ. No. 111, 196-202 (1964); CA 64, 1247.

1339. Komarov, N. V. et al., U.S.S.R. 173,758, March 25, 1963; CA 64, 3606.

1340. ---, Ermolova, T. I. and Chernov, N. F., Izv. Akad. Nauk SSSR, Ser. Khim. 1966, 1679; CA 66, 65605.

1341. --- and Guseva, I. S., Brit. 1,084,522, Nov. 10, 1964; CA 67, 108763; Fr. 1,415,111, Nov. 13, 1964; CA 64, 6694.

--- and ---, Ger. 1,227,021, Nov. 4, 1964; CA 66, 28893.

1342. ---, --- and Lvova, F. P., Izv. Akad. Nauk SSSR, Ser. Khim. 1966, 1479-81; CA 66, 65600.

---, --- and ---, "Khim. Atsetilena" p. 175-7, M. F. Shostakovskii, Ed., Nauka, Moscow, 1968; CA 71, 3442.

1343. --- and Misiunas, V., U.S.S.R. 180,591, Jan. 25, 1965; CA 65, 12240.

1344. --- and Yarosh, O. B., Zh. Obshch. Khim. 36, 101-3 (1966); CA 64, 14207.

1345. --- and ---, ibid. 37, 264-7 (1967); CA 66, 95126.

1346. Komura, M. et al., Inorg. Nucl. Chem. Lett. 3, 17-20 (1967); CA 66, 59446.

1347. --- and Okawara, R., ibid. 2 (4), 93-5 (1966); CA 65, 5480.

1348. Korotaevskii, K. N. et al., Zh. Obshch. Khim. 36, 167 (1966); CA 64, 13751.

1349. Korshak, V. V. et al., U.S.S.R. 200,033, March 20, 1965; CA 68, 54865.

1350. Kostyanovskii, R. G. and Prokofev, A. K., Dokl. Akad. Nauk SSSR 164, 1054-7 (1965); CA 64, 2123.

1351. --- and ---, Izv. Akad. Nauk SSSR, Ser. Khim. 1967, 473-4; CA 67, 21982.

1352. Koton, M. M. and Kiseleva, T. M., Zh. Obshch. Khim. 35, 2036-7 (1965); CA 64, 6684.

1353. Kozlov, L. M., Burmistrov, V. I. and Drabkina, L. S., Tr. Kazansk. Khim.-Tekhnol. Inst. No. 33, 227-31 (1964); CA 64, 19781.

1354. Kraihanzel, C. S. and Losee, M. L., J. Organometal. Chem. 10, 427-37 (1967); CA 68, 29758.

1355. Kramer, K., Erdoel Kohle 19, 182-5 (1966); CA 64, 15640.

1356. Kraus, W., Ger. 1,223,548, Nov. 8, 1963; CA 66, 56412.

1357. ---, Fr. 1,413,601, Nov. 9, 1963; CA 64, 5138; Ger. 1,229,088, Nov. 9, 1963; CA 66, 37422.

1358. Kreshkov, A. P. and Kuchkarev, E. A., Zavodsk. Lab. 32, 558-9 (1966); CA 65, 6292.

1359. Krexner, R., Pflanzenschutz Ber. 33 (3/4), 41-56 (1965); CA 65, 17626.

1360. Kriegsmann, H., Hoffmann, H. and Geissler, H., Z. Anorg. Allgem. Chem. <u>341</u>, 24-35 (1965); CA <u>64</u>, 9084.
1361. Kruglaya, O. A. et al., Dokl. Akad. Nauk SSSR <u>173</u>, 834-6 (1967); CA <u>67</u>, 43892.
1362. Kubo, M. and Fukumoto, O., Japan. 4275 ('67), Feb. 24, 1964; CA <u>67</u>, 44419.
1363. Kuchen, W., Jadat, A. and Metten, J., Chem. Ber. <u>98</u>, 3981-7 (1965); CA <u>64</u>, 6075.
1364. Kuehlein, K. and Neumann, W. P., Ann. Chem. <u>702</u>, 17-23 (1967); CA <u>66</u>, 105033.
1365. ---, --- and Becker, H. P., Angew. Chem. Int. Ed. Engl. <u>6</u>, 876 (1967); CA <u>68</u>, 2968.
1366. Kuiper, I. J., Aust. J. Exp. Agr. Anim. Hurb. <u>7</u>, 275-82 (1967); CA <u>67</u>, 90013.
1367. Kuivila, H. G., AEC Accession No. 6687, Rept. No. NYO-3453-1 (1965); CA <u>65</u>, 13473; cf. ref. 1457.
1368. --- and Cochran, J. C., J. Am. Chem. Soc. <u>89</u>, 7152-3 (1967); CA <u>68</u>, 77390; cf. ref. 881.
1369. --- and Sommer, R., ibid. <u>89</u>, 5616 (1967); CA <u>68</u>, 12134.
1370. ---, --- and Green, D. C., J. Org. Chem. <u>33</u>, 1119-22 (1968); CA <u>68</u>, 78391.
1371. --- and Walsh, E. J., Jr., J. Am. Chem. Soc. <u>88</u>, 571-6 (1966); CA <u>64</u>, 9539.
1372. Kummer, R. and Graham, W. A. G., Inorg. Chem. <u>7</u>, 310-5 (1968); CA <u>68</u>, 59674.
1373. --- and ---, ibid. <u>7</u>, 523-6 (1968); CA <u>68</u>, 69100.
1374. Kupchik, E. J. and Kiesel, R. J., J. Org. Chem. <u>31</u>, 456-61 (1966); CA <u>64</u>, 9761; cf. ref. 1310.
1375. --- and McInerney, E. F., J. Organometal. Chem. <u>11</u>, 291-8 (1968); CA <u>68</u>, 59675; cf. ref. 2584.
1376. --- and Perciaccante, V. A., ibid. <u>10</u>, 181-7 (1967); CA <u>68</u>, 13123; cf. ref. 2679.
1377. --- and Theisen, C. T., ibid. <u>11</u>, 627-30 (1968); CA <u>68</u>, 87362.
1378. ---, Ursino, J. A. and Boudjouk, P. R., ibid. <u>10</u>, 269-78 (1967); CA <u>68</u>, 22030; cf. ref. 2841.
1379. Kurts, A. L., Beletskaya, I. P. and Reutov, O. A., Izv. Akad. Nauk SSSR, Ser. Khim. <u>1967</u>, 2207-17; CA <u>68</u>, 154.
1380. Kuschk, R., Kaltwasser, H. and Braun, W., Chem. Tech. (Berlin) <u>17</u>, 749-51 (1965); CA <u>64</u>, 9761; cf. ref. 783.
1380a. Lach, A. A., Brit. 1,084,616; Fr. 1,458,039; Neth. 65 16,434; U.S. 3,349,109, Dec. 16, 1964; CA <u>65</u>, 13763.
1381. Laczko, M. and Szabo, A., Muanyag Gumi <u>4</u>, 29-31 (1967); CA <u>67</u>, 54708.
1382. Ladd, T. L., Jr., J. Econ. Entomol. <u>61</u>, 577-8 (1968); CA <u>68</u>, 104125.
1383. Laliberte, B. R., Davidsohn, W. and Henry, M. C., J. Organometal. Chem. <u>5</u>, 526-31 (1966); CA <u>65</u>, 2291.
1384. Lampe, W. R., Fr. 1,424,599, Dec. 16, 1963; CA <u>65</u>, 13928; U.S. 3,305,502, Dec. 16, 1963; CA <u>66</u>, 86473.

1385. Langer, H. G., Tetrahedron Lett. 1967, 43-7; CA 66, 75455.
1386. ---, U.S. 3,265,756, July 26, 1963; CA 65, 15223.
1387. ---, U.S. 3,345,391, Jan. 25, 1961; CA 68, 95972.
1388. --- and Blut, A. H., J. Organometal. Chem. 5, 288-91 (1966); CA 64, 14213.
1389. Langner, H., Plaste Kaut. 13, 76-8 (1966); CA 64, 14352.
1390. Lapkin, I. I. and Dummler, V. A., Uch. Zap., Permsk. Gos. Univ. No. 111, 185-9 (1964); CA 64, 12045.
1391. --- and ---, ibid. 111, 190-1 (1964); CA 64, 11068.
1392. Lappert, M. F. and Jones, K., Brit. 1,026,405, Sept. 26, 1962; CA 65, 2299.
1393. --- and Lorberth, J., Chem. Commun. 1967, 836-7; CA 68, 39752.
1394. Larrison, M. S., Fr. 1,455,466, Jan. 4, 1965; CA 66, 95813.
1394a. Laubach, J. E., Brit. 1,071,322; U.S. 3,368,961, April 20, 1964; CA 67, 60403.
1395. Lavell, H. H., Brit. 1,016,534, Aug. 29, 1962; CA 64, 8489.
1396. Lavigne, A. A. et al., Rec. Trav. Chim. 86, 746-8 (1967); CA 67, 100216.
1397. Layton, A. A. et al., Chem. Ind. (London) 1967, 465; CA 67, 28866.
1397a. Leebrick, J. R., Brit. 1,081,504; U.S. 3,395,039, Sept. 29, 1965; CA 67, 101119.
1398. --- and Kudisch, N., Ger. 1,150,814; U.S. 3,214,490, Aug. 26, 1960; CA 64, 8403.
1399. ---, Ross, A. and Zedler, R. J., U.S. 3,234,032, July 26, 1962; CA 64, 12966.
1400. Lees, J. K. and Flinn, P. A., J. Chem. Phys. 48, 882-9 (1968); CA 68, 100419.
1401. Lehmkuhl, H., U.S. 3,372,097, June 28, 1963; CA 68, 101279.
1402. Lehn, W. L., Inorg. Chem. 6, 1061-3 (1967); CA 67, 3129.
1403. Leistner, W. E. et al., Ger. 1,232,736, May 14, 1952; CA 66, 86267.
1404. --- et al., Ger. 1,232,740, May 14, 1952; CA 66, 82276.
1405. --- and Hecker, A. C., Ger. 1,232,737, Oct. 16, 1952; CA 66, 86273.
1406. --- and ---, Ger. 1,232,738, May 20, 1952; CA 66, 86275.
1407. --- and ---, Ger. 1,232,739, Sept. 8, 1952; CA 66, 86274.
1408. --- and Setzler, W. E., Ger. 1,233,589, May 23, 1952; CA 66, 76581.
1409. Leonardi, S. J. and Oberright, E. A., Brit. 1,100,001; Fr. 1,426,693; U.S. 3,303,132, March 19, 1964; CA 65, 8640.
1410. Leusink, A. J. et al., AD 629,554, 98 pp. (1965); CA 68, 13465.
1411. --- and Budding, H. A., J. Organometal. Chem. 11, 533-9 (1968); CA 68, 113749.
1412. ---, --- and Drenth, W., ibid. 9, 295-306 (1967); CA 67, 90195.
1413. ---, --- and ---, ibid. 11, 541-7 (1968); CA 68, 113750.
1414. ---, --- and Marsman, J. W., ibid. 9, 285-94 (1967); CA 67, 90194.
1415. ---, --- and Noltes, J. G., Rec. Trav. Chim. 85, 151-8 (1966); CA 65, 585.
1416. --- and van der Kerk, G. J. M., ibid. 84, 1617-20 (1965); CA 64, 11245.
1417. --- and Noltes, J. G., Tetrahedron Lett. 1966, 335-40; CA 64, 7795.
1418. --- and ---, ibid. 1966, 2221-5; CA 65, 2291.

1419. Lewis, J. et al., J. Chem. Soc., A 1966, 1663-70; CA 66, 25562.
1420. Lewis, J. T. and Frankland, P. G., Brit. 1,020,501, Nov. 13, 1961; CA 64, 19948.
1421. Lieber, E. and Keane, F. M., Inorg. Synth. 8, 56-63 (1966); CA 66, 34392.
1422. Lippi, U., Stefani, M. and Gritti, G., Riv. Anat. Pathol. Oncol. 27, 190-227 (1965); CA 65, 9583.
1423. Lippincott, E. R. and Nagarajan, G., Bull. Soc. Chim. Belges 74, 551-64 (1965); CA 64, 9101.
1424. Lodochnikova, V. I., Panov, E. M. and Kocheshkov, K. A., Zh. Obshch. Khim. 37, 547-9 (1967); CA 67, 21984.
1425. Lombardi, L. and Gautier, A., J. Microscopie 5, 255-8 (1966); CA 65, 14086.
1426. Lombardo, M. E., Brit. 1,118,600; Fr. 1,482,847, Dec. 29, 1964; CA 68, 49780.
1427. ---, U.S. 3,346,607, Dec. 29, 1964; CA 68, 69130.
---, Brit. 1,135,459; Fr. 1,526,906; Neth. 66 14,955; U.S. 3,443,010, Oct. 23, 1965; CA 67, 81485.
1428. Longi, P. and Mazzocchi, R., Chim. Ind. (Milan) 48, 718-20 (1966); CA 65, 14431.
1429. Lorberth, J., Krapf, H. and Noeth, H., Chem. Ber. 100, 3511-9 (1967); CA 68, 13110.
1430. --- and Vahrenkamp, H., J. Organometal. Chem. 11, 111-24 (1968); CA 68, 34537.
1431. Lovejoy, R. W. and Baker, D. R., J. Chem. Phys. 46, 658-65 (1967); CA 66, 50502.
1432. Lowe, J. P. and Parr, R. G., ibid. 44, 3001-9 (1966); CA 64, 16655.
1433. Lowes, F. J., Jr., U.S. 3,306,955, March 26, 1962; CA 66, 96178.
1434. Ludlum, D. B., U.S. 3,218,266, July 5, 1955; CA 64, 3714; U.S. 3,308,112, Sept. 15, 1965; CA 66, 116123.
1435. Luijten, J. G. A., Rec. Trav. Chim. 85, 873-8 (1966); CA 66, 28866.
1436. Lunazzi, L. and Taddei, F., Boll. Sci. Fac. Chim. Ind. Bologna 23, 359-69 (1965); CA 64, 18736.
1437. Luneva, L. K. et al., Izv. Akad. Nauk SSSR, Ser. Khim. 1967, 2095-6; CA 67, 117375.
1438. ---, Sladkov, A. M. and Korshak, V. V., ibid. 1968, 170-4; CA 68, 78657.
1439. ---, --- and ---, Vysokomol. Soedin. Ser. A 9, 910-4 (1967); CA 67, 54483.
1440. Lutsenko, I. F., Baukov, Yu. I. and Burlachenko, G. S., J. Organometal. Chem. 6, 496-503 (1966); CA 65, 20158; cf. ref. 712.
1441. Lutz, H. D., Z. Naturforsch. B20, 1011 (1965); CA 64, 8232.
1442. Mack, G. P., U.S. 3,264,256, May 1, 1964; CA 65, 10752; U.S. 3,147,285, July 20, 1956; CA 62, 11973.
1443. Mackay, K. M., Sowerby, D. B. and Young, W. C., Spectrochim. Acta Pt. A 24, 611-3 (1968); CA 68, 118182.
1444. --- and Watt, R., J. Organometal. Chem. 6, 336-51 (1966); CA 65, 18468.

1445. Mackay, K. M. and Watt, R. J., Spectrochim. Acta, Pt. A 23, 2761-78 (1967); CA 68, 64057.
1446. MacKenzie, J. C. and Orzechowski, A., U.S. 3,365,432, May 8, 1963; CA 68, 60033.
1447. Maddox, M. L., Ph. D. Thesis, Univ. of California, Los Angeles, 1966; Diss. Abstr. 26, 6361-2 (1966); CA 65, 8720; cf. ref. 331.
1448. Maeda, Y., Dillard, C. R. and Okawara, R., Inorg. Nucl. Chem. Lett. 2 (7), 197-9 (1966); CA 65, 13752.
1449. --- and Okawara, R., J. Organometal. Chem. 10, 247-56 (1967); CA 68, 13127.
1450. Magnuson, J. A. and Knaub, E. W., Anal. Chem. 37, 1607-9 (1965); CA 64, 2741.
1451. Magnuson, A. B. and Parker, J. A., U.S. 3,366,708, April 13, 1964; CA 68, 60236.
1452. Mahr, C. and Stork, G., Z. Anal. Chem. 221, 1-9 (1966); CA 65, 16062.
1453. Maire, J. C., J. Organometal. Chem. 9, 271-84 (1967); CA 67, 99470.
1454. --- and Dufermont, J., ibid. 10, 369-72 (1967); CA 68, 13104.
1455. Mammi, M., Busetti, V. and Del Pra, A., Inorg. Chim. Acta 1, 419-23 (1967); CA 68, 63469.
1456. Mandell, S. et al., J. Neurosurg. 24, 984-6 (1966); CA 65, 9573.
1457. Mangravite, J. A., Ph. D. Thesis, Univ. of New Hampshire, 1965; Diss. Abstr. 26, 7041 (1966); CA 65, 12073; cf. ref. 1367.
1458. Mannan, K. A. I. F. M., Acta Crystallogr., Sect. B 24, 603 (1968); CA 68, 99560.
1459. Manning, A. R., J. Chem. Soc., A 1968, 651-3; CA 68, 83983.
1460. Marchenko, G. N., Goldobin, S. F. and Smertin, G. Ya., Vysokomol. Soedin. 8, 2087-90 (1966); CA 66, 66095.
1461. Marchetti, A. P. and Kearns, D. R., J. Am. Chem. Soc. 89, 768-77 (1967); CA 66, 60940.
1462. Markova, S. V., Tr. Fiz. Inst. Akad. Nauk SSSR 35, 150-227 (1966); CA 66, 109813.
1463. --- and Zueva, G. Ya., Optika i Spektroskopiya 19, 716-20 (1965); CA 64, 10584.
1464. Marks, G. C., Benton, J. L. and Thomas, C. M., Soc. Chem. Ind. (London), Monogr. No. 25, 204-35 (1967); CA 68, 30349.
1465. Martin, D. F., Maybury, P. C. and Walton, R. D., J. Organometal. Chem. 7, 362-4 (1967); CA 67, 63554.
--- and Walton, R. D., ibid. 5, 57-62 (1966); CA 64, 1925.
1466. Martin, F. J. and Price, K. R., J. Appl. Polym. Sci. 12, 143-56 (1968); CA 68, 96441.
1467. Martin, J. C. and Gilkey, R., U.S. 3,254,061, Dec. 20, 1960; CA 65, 9054.
1468. Maruo, K. and Furutaka, Y., Japan. 21,021 ('66), April 17, 1964; CA 66, 46486.
1469. Masai, K. and Nakanishi, M., Japan. 5554 ('67), June 16, 1964; CA 68, 60275.

1470. Masai, Y. and Nakanishi, M., Japan. 19,171 ('67), April 27, 1964; CA 68, 60276.
1471. Massol, M. and Satge, J., Bull. Soc. Chim. Fr. 1966, 2737-43; CA 66, 95143.
1472. ---, --- and Lesbre, M., C. R., Akad. Sci., Paris, Ser. C 262, 1806-9 (1966); CA 65, 18611, cf. ref. 2150.
1473. Masson, J. C., Quan, M. L. and Cadiot, P., Bull. Soc. Chim. Fr. 1967, 777-80; CA 67, 32754.
1474. --- and Cadiot, P., ibid. 1965, 3518-24; CA 64, 11246.
1475. Mathiason, D. R. and Miller, N. E., Inorg. Chem. 7, 709-14 (1968); CA 68, 105292.
1476. Mathis, F., Wolf, R. and Mathis-Noel, R., Rev. Chim. (Bucharest) 17, 661-4 (1966); CA 66, 99872.
1477. Mathur, S. and Mehrotra, R. C., J. Organometal. Chem. 7, 227-31 (1967); CA 66, 65602.
1478. Matsubayashi, G. et al., Bull. Chem. Soc. Jap. 40, 1556-70 (1967); CA 67, 116454.
1479. --- et al., J. Inorg. Nucl. Chem. 28, 2937-43 (1966); CA 66, 46041.
1480. Matsubayashi, K. and Kawaguchi, T., Japan. 18,359 ('67), July 24, 1964; CA 68, 22682.
1481. Matsuda, S., Asahi Garasu Kogyo Gijutsu Shorei-Kai Kenkyu Hokoku 12, 329-38 (1966); CA 68, 39756.
1482. --- et al., Kogyo Kagaku Zasshi 70, 1747-50 (1967); CA 68, 87373.
1483. ---, Kikkawa, S. and Hayashi, T., ibid. 69, 256-9 (1966); CA 65, 5482; Technol. Rep. Osaka Univ. 17 (748-68), 193-203 (1967); CA 67, 82244.
1484. ---, --- and Kashiwa, N., ibid. 69, 1036-9 (1966); CA 65, 20160.
1485. ---, --- and Nomura, M., ibid. 69, 649-53 (1966); CA 65, 18612.
1486. ---, --- and Omae, I., ibid. 69, 646-9 (1966); CA 65, 18612.
1487. ---, Matsuda, H. and Mori, H., ibid. 70, 1751-5 (1967); CA 68, 78384.
1488. Maxfield, P. L., Ph. D. Thesis, Univ. of New Hampshire, 1965; Diss. Abstr. 26, 7041 (1966); CA 65, 12234.
1489. May, L. and Spijkerman, J. J., J. Chem. Phys. 46, 3272-3 (1967); CA 67, 16482.
1490. Mazerolles, P., Fr. 1,425,321, Oct. 1, 1964; CA 65, 20246.
1491. --- and Dubac, J., C. R. Acad. Sci., Paris, Ser. C 265, 403-6 (1967); CA 68, 22024.
---, --- and Lesbre, M., Tetrahedron Lett. 1967, 255-8; CA 66, 76107.
1492. ---, --- and ---, J. Organometal. Chem. 5, 35-47 (1966); CA 64, 2123.
1493. --- and Faucher, A., Bull. Soc. Chim. Fr. 1967, 2134-9; CA 68, 114713.
1494. ---, Lesbre, M. and Marre, S., Compt. Rend. 261, 4134-6 (1965); CA 64, 3592.
1495. --- and Manuel, G., Bull. Soc. Chim. Fr. 1966, 327-31; CA 64, 15918
1496. --- and ---, ibid. 1967, 2511-5; CA 68, 39759.
1497. McCord, R. S., U.S. 3,371,046, Dec. 30, 1963; CA 68, 80222.
1498. McFarlane, W., J. Chem. Soc., A 1967, 528-30; CA 66, 109972.
1499. McKusick, B. C. and Webster, O. W., U.S. 3,214,455, Oct. 16, 1961; CA 64, 1909.

1500. McWhinnie, W. R., Poller, R. C. and Thevarasa, M., J. Organometal. Chem. <u>11</u>, 499-502 (1968); CA <u>68</u>, 105319.
1501. Mehrotra, R. C. and Gupta, V. D., Indian J. Chem. <u>5</u>, 643-5 (1967); CA <u>68</u>, 101366.
1502. ---, --- and Sukhani, D., J. Organometal. Chem. <u>9</u>, 263-9 (1967); CA <u>67</u>, 100213.
1503. --- and Mathur, S., J. Indian Chem. Soc. <u>43</u>, 489-91 (1966); CA <u>65</u>, 20158.
1504. --- and ---, J. Organometal. Chem. <u>6</u>, 11-6 (1966); CA <u>65</u>, 5481.
1505. --- and ---, ibid. <u>6</u>, 425-7 (1966); CA <u>66</u>, 38022.
1506. --- and ---, ibid. <u>7</u>, 233-5 (1967); CA <u>66</u>, 65603.
1507. Meijer, H. J. de L., van den Hurk, J. W. G. and van der Kerk, G. J. M., Rec. Trav. Chim. <u>85</u>, 1018-24 (1966); CA <u>66</u>, 29150.
1508. ---, --- and ---, ibid. <u>85</u>, 1025-38 (1966); CA <u>66</u>, 29151.
1509. Mendelsohn, J., Marchand, A. and Valade, J., J. Organometal. Chem. <u>6</u>, 25-44 (1966); CA <u>65</u>, 8717.
1510. ---, Pommier, J. C. and Valade, J., C. R. Acad. Sci., Paris, Ser. C <u>263</u>, 921-4 (1966); CA <u>66</u>, 37212.
1511. Meyer, F. C., U.S. 3,252,929, Dec. 28, 1961; CA <u>65</u>, 20303.
1512. Meyling, A. H. and Pitchford, R. J., Bull. World Health Org. <u>34</u>, 141-6 (1966); CA <u>64</u>, 14899.
1513. Mieville, R. L. and Meguerian, G. H., Ind. Eng. Chem., Prod. Res. Develop. <u>6</u> (4), 253-7 (1967); CA <u>68</u>, 51624.
1514. Migdal, S., Gertner, D. and Zilkha, A., Can. J. Chem. <u>45</u>, 2987-92 (1967); CA <u>68</u>, 13115.
1515. ---, --- and ---, Israel J. Chem. <u>5</u>, 163-70 (1967); CA <u>67</u>, 100624.
1516. ---, --- and ---, J. Organometal. Chem. <u>11</u>, 441-5 (1968); CA <u>68</u>, 115043.
1517. Mikulaj, V., Macasek, F. and Kopunec, R., Radiochem. Conf., Abstr. Pap., Bratisheva <u>1966</u>, 76-7; CA <u>68</u>, 59689.
1518. Miller, J. J., Ph. D. Thesis, Univ. of Cincinnati, 1965; Diss. Abstr. <u>26</u>, 4243-4 (1966); CA <u>64</u>, 19669; cf. ref. 2189.
1519. Miller, J. M., J. Chem. Soc., A <u>1967</u>, 828-34; CA <u>67</u>, 6655.
1520. Miller, L. A., U.S. 3,257,194, Dec. 28, 1961; CA <u>65</u>, 12240.
1521. Miller, V. L. and Jerstad, A. C., J. Med. Chem. <u>9</u>, 208-10 (1966); CA <u>64</u>, 11588.
1522. Minoura, Y. et al., J. Polym. Sci., Pt. A-1 <u>4</u>, 2757-70 (1966); CA <u>66</u>, 2850.
--- et al., Kogyo Kagaku Zasshi <u>69</u>, 345-9 (1966); CA <u>66</u>, 95414.
1523. Minsker, et al., Plast. Massy <u>1966</u>, 56-9; CA <u>66</u>, 86188.
1524. --- et al., Vysokomol. Soedin. <u>8</u>, 1028-34 (1966); CA <u>65</u>, 9095; CA <u>66</u>, 38227.
1525. Miric, M., Zast. Bilja <u>17</u> (91-2), 267-76 (1966); CA <u>67</u>, 116085.
1526. Mironov, V. F. et al., Zh. Obshch. Khim. <u>37</u>, 2323-8 (1967); CA <u>68</u>, 87369.
1527. ---, Berliner, E. M. and Gar, T. K., ibid. <u>37</u>, 962 (1967); CA <u>67</u>, 90899.
1528. --- and Fedotov, N. S., ibid. <u>36</u>, 556-60 (1966); CA <u>65</u>, 743.

1529. Mironov, V. F. and Gar, T. K., Izv. Akad. Nauk SSSR, Ser. Khim. 1964, 1515-8; CA 64, 14213.
1530. --- and ---, ibid. 1966, 482-9; CA 65, 5481.
1531. ---, Kozyukov, V. P. and Sheludyakov, V. D., Zh. Obshch. Khim. 37, 1669-73 (1967); CA 68, 39746.
1532. ---, --- and ---, ibid. 37, 1915-9 (1967); CA 68, 29768.
1533. ---, Kravchenko, A. L. and Leites, L. A., Izv. Akad. Nauk SSSR, Ser. Khim. 1966, 1177-84; CA 65, 16997.
1534. ---, Sobolev, E. S. and Antipin, L. M., Zh. Obshch. Khim. 37, 1707-10 (1967); CA 68, 13112.
1535. ---, --- and ---, ibid. 37, 2573-6 (1967); CA 68, 87364.
1536. Mirskov, R. G. and Vlasov, V. M., ibid. 36, 166-7 (1966); CA 64, 14213.
1537. --- and ---, ibid. 36, 562 (1966); CA 65, 744.
1538. --- and ---, U.S.S.R. 196,839, April 30, 1965; CA 68, 49781.
1539. Mishima, S., Nippon Kagaku Zasshi 87, 162-6 (1966); CA 65, 15420.
1540. Moedritzer, K., J. Organometal. Chem. 6, 282-4 (1966); CA 65, 13754.
1541. --- and Van Wazer, J. R., Am. Chem. Soc. Org. Coatings Plastics Chem., Preprints 25 (2), 328-33 (1965); CA 66, 29273.
1542. --- and ---, Inorg. Chem. 4, 1753-60 (1965); CA 64, 3591.
1543. --- and ---, ibid. 5, 547-52 (1966); CA 64, 14062.
1544. --- and ---, J. Inorg. Nucl. Chem. 28, 957-70 (1966); CA 64, 19366.
1545. --- and ---, ibid. 29, 1571-6 (1967); CA 67, 76635.
1546. --- and ---, Inorg. Chim. Acta 1, 407-12 (1967); CA 68, 72901.
1547. --- and ---, J. Polym. Sci., Pt. A-1 6, 547-57 (1968); CA 68, 113834.
1548. Mohamed, A. and Satchell, D. P. N., Chem. Ind. (London) 1966, 2013-4; CA 66, 89857.
---, --- and Satchell, R. S., J. Chem. Soc., B 1967, 723-5; CA 67, 63613.
1548a. Molt, K. R. and Hechenbleikner, I., Belg. 672,867; Brit. 1,098,045-6; Ger. 1,235,914; Neth. 65 15,201; U.S. 3,311,649, Nov. 30, 1964; CA 65, 15428.
1549. Moore, M. and Lanning, F. C., Quart. J. Florida Acad. Sci. 29, 73-6 (1966); CA 66, 28483.
1550. Morris, M. D., Anal. Chem. 39 (4), 476-80 (1967); CA 66, 101077.
---, J. Electroanal. Chem. Interfacial Electrochem. 16, 569-73 (1968); CA 68, 65134.
1551. Moss, R. and Browsett, E. V., Analyst 91, 428-38 (1966); CA 65, 14323.
--- and ---, Technicon. Symp., 2nd, N.Y., London 1965, 285-90 (1966); CA 67, 57033.
1552. --- and Campbell, K., J. Inst. Petrol. 53 (518), 89-93 (1967); CA 66, 97169.
1553. Mostyn, R. A. and Cunningham, A. F., J. Inst. Petrol. 53, 101-11 (1967); CA 67, 4560.
1553a. Mueller, E. and Trense, V., Tetrahedron Lett. 1967, 4979-82; CA 68, 68106.
1554. Mueller, R., Reichel, S. and Dathe, C., Chem. Ber. 101, 783-6 (1968); CA 68, 95909.

1555. Mueller, R., Reichel, S. and Dathe, C., Inorg. Nucl. Chem. Lett. 3, 125-33 (1967); CA 67, 21989; Ger. (East) 62,833, Oct. 20, 1966; CA 70, 115334.
1556. Muenter, J. S. and Laurie, V. W., J. Chem. Phys. 45, 855-8 (1966); CA 65, 9872.
1557. Mufti, A. S. and Poller, R. C., J. Chem. Soc., C 1967, 1362-4; CA 67, 82248.
1558. --- and ---, ibid., C 1967, 1767-8; CA 68, 29804.
1559. Mullins, M. A. and Curran, C., Inorg. Chem. 6, 2017-9 (1967); CA 68, 7365.
1560. Mushina, E. A. et al., Dokl. Akad. Nauk SSSR 170, 344-6 (1966); CA 66, 95947.
1561. Musker, W. K. and Savitsky, G. B., J. Phys. Chem. 71, 431-4 (1967); CA 66, 50446.
1562. Mussell, D. R., U.S. 3,284,290, May 1, 1963; CA 66, 18296.
1563. Nadler, C., U.S. 3,364,161, Feb. 27, 1964; CA 68, 50828.
1564. Nagae, Y. and Wakamori, K., Japan. 4575 ('66), March 25, 1963; CA 65, 2298.
1565. --- and ---, Japan. 4576 ('66), March 25, 1963; CA 65, 2298.
1566. Nagarajan, G., Z. Naturforsch. A 21, 238-43 (1966); CA 65, 3018.
1567. Nagasawa, M., U.S. 3,268,347, Jan. 14, 1963; CA 65, 18861.
1568. Nagasawa, S., Shinohara, H. and Shiba, M., J. Stored Prod. Res. 3 (2), 177-84 (1967); CA 67, 116120.
1569. Nagibina, T. D. et al., Kauch. Rezina 26 (11), 2 (1966); CA 66, 38678.
1570. Nagy, J. et al., J. Organometal. Chem. 7, 393-404 (1967); CA 67, 37939.
--- et al., Intern. Symp. Organosilicon Chem., Sci. Commun., Prague 1965, 241-4; CA 66, 85308.
1571. --- and Borbely-Kuszmann, A., ibid. 1965, 201-3; CA 65, 7422.
--- and ---, Period. Polytech., Chem. Eng. (Budapest) 10 (2), 139-45 (1966); CA 67, 100805.
1572. ---, Ferenczi-Gresz, S. and Nefedov, O. M., ibid. 10 (3), 319-24 (1966); CA 67, 48245.
1573. Nagy, J., Mrs., Muanyag Gumi 3, 326-30 (1966); CA 66, 19119.
1574. Nakanishi, M. and Inamasu, S., Japan. 16,296 ('67), Oct. 12, 1964; CA 68, 105366.
1575. ---, Kuriyama, T. and Oe, T., Japan. 13,302 ('67), July 23, 1964; CA 68, 49767.
1576. --- and Tsuda, A., Japan. 15,290 ('66), April 6, 1962; CA 65, 20164.
1577. --- and ---, Japan. 17,141 ('66), Sept. 3, 1964; CA 66, 11049.
1578. --- and ---, Japan. 1977 ('67), Aug. 11, 1963; CA 66, 115803.
1579. Nametkin, N. S., Durgaryan, S. G. and Tikhonova, L. I., Dokl. Akad. Nauk SSSR 172, 615-7 (1967); CA 66, 76354.
1580. ---, --- and ---, ibid. 172, 867-9 (1967); CA 67, 11551.
1581. Narafu, T. et al., Shokuhin Eiseigaku Zasshi 7, 76-7 (1966); CA 65, 13937.
1582. --- et al., ibid. 7, 445-8 (1966); CA 66, 75003.

1583. Nasielski, J. et al., J. Organometal. Chem. $\underline{8}$, 97-104 (1967); CA $\underline{66}$, 89778.
1583a. Natoli, J. G., Brit. 1,069,964; Fr. 1,444,995; Neth. 65 7716; U.S. 3,432,531, June 16, 1964; CA $\underline{64}$, 17640.
1583b. ---, Brit. 1,082,904; Fr. 1,432,329; Neth. 65 5767; U.S. 3,355,470, May 6, 1964; CA $\underline{64}$, 12723.
1584. Naylor, F. E., U.S. 3,232,920, March 6, 1958; CA $\underline{64}$, 11423.
1585. Nefedov, O. M. et al., Dokl. Akad. Nauk SSSR $\underline{164}$, 822-5 (1965); CA $\underline{64}$, 2177.
1586. --- et al., Intern. Symp. Organosilicon Chem., Sci. Commun., Prague $\underline{1965}$, 65-8; CA $\underline{65}$, 12298.
1587. --- et al., U.S.S.R. 187,796, Aug. 12, 1965; CA $\underline{67}$, 43923.
1588. --- and Kolesnikov, S. P., Izv. Akad. Nauk SSSR, Ser. Khim. $\underline{1966}$, 201-11; CA $\underline{65}$, 743.
1589. --- and ---, Vysokomol. Soedin. $\underline{7}$, 1857-62 (1965); CA $\underline{64}$, 6770.
--- and ---, U.S.S.R. 173,951, Aug. 27, 1964; CA $\underline{64}$, 2190.
1590. ---, --- and Perlmutter, B. L., Angew. Chem. Int. Ed. Engl. $\underline{6}$, 628-9 (1967); CA $\underline{67}$, 82246; cf. ref. 2504.
1591. Nelson, W. H., Randall, W. J. and Martin, D. F., Inorg. Synth. $\underline{9}$, 52-5 (1967); CA $\underline{66}$, 121644.
1592. Nesmeyanov, A. N. et al., Dokl. Akad. Nauk SSSR $\underline{176}$, 844-7 (1967); CA $\underline{68}$, 69099.
1593. --- et al., Izv. Akad. Nauk SSSR, Ser. Khim. $\underline{1966}$, 160-2; CA $\underline{64}$, 12720.
1594. --- et al., ibid. $\underline{1966}$, 163-4; CA $\underline{64}$, 12718.
1595. --- et al., ibid. $\underline{1966}$, 1292; CA $\underline{65}$, 16999.
1596. --- et al., ibid. $\underline{1967}$, 1395; CA $\underline{68}$, 13113.
1597. --- et al., ibid. $\underline{1967}$, 2241-6; CA $\underline{68}$, 44413.
1598. --- and Borisov, A. E., Dokl. Akad. Nauk SSSR $\underline{174}$, 96-9 (1967); CA $\underline{67}$, 90903.
1599. --- and ---, Izv. Akad. Nauk SSSR, Ser. Khim. $\underline{1967}$, 226; CA $\underline{66}$, 95147.
1600. ---, --- and Novikova, N. V., Dokl. Akad. Nauk SSSR $\underline{165}$, 333-6 (1965); CA $\underline{64}$, 5129.
1601. ---, --- and ---, ibid. $\underline{172}$, 1329-32 (1967); CA $\underline{67}$, 3127.
1602. ---, --- and Wang, S.-H., Izv. Akad. Nauk SSSR, Ser. Khim. $\underline{1967}$, 1141-2; CA $\underline{68}$, 29807.
1603. Nesterov, O. V. et al., Kinet. Katal. $\underline{7}$, 805-8 (1966); CA $\underline{66}$, 37000.
1604. ---, Chirkov, Yu. N. and Entelis, S. G., ibid. $\underline{8}$, 1371-3 (1967); CA $\underline{68}$, 86516.
1605. Neumann, W. P., Albert, H. J. and Kaiser, W., Tetrahedron Lett. $\underline{1967}$, 2041-3; CA $\underline{67}$, 63484.
1606. --- and Kuehlein, K., Ann. Chem. $\underline{702}$, 13-16 (1967); CA $\underline{66}$, 105032.
1607. --- and ---, Tetrahedron Lett. $\underline{1966}$, 3415-8; CA $\underline{65}$, 12233.
1608. --- and ---, ibid. $\underline{1966}$, 3419-21; CA $\underline{65}$, 13754.
1609. --- and ---, ibid. $\underline{1966}$, 3423-5; CA $\underline{65}$, 13754.
1610. --- and ---, Nachrichten aus Chemie und Technik $\underline{14}$, 115-6 (1966); Chem. and Eng. News $\underline{43}$, 49-50 (1965).

1611. Neumann, W. P. and Lind, H., Angew. Chem. 79, 52 (1967); Int. Ed. Engl. 6, 76-7 (1967); CA 66, 75535.
1612. ---, Niermann, H. and Schneider, B., Ann. Chem. 707, 15-9 (1967); CA 67, 108725.
1613. --- and Pedain, J. A., Ger. 1,214,237, July 10, 1964; CA 65, 5490.
1614. ---, --- and Sommer, R., Ann. Chem. 694, 9-18 (1966); CA 65, 12233.
1615. --- and Ruebsamen, K., Chem. Ber. 100, 1621-6 (1967); CA 67, 21318.
1616. ---, --- and Sommer, R., ibid. 100, 1063-72 (1967); CA 66, 115029.
1617. --- and Schneider, B., Ann. Chem. 707, 20-5 (1967); CA 67, 117624.
1618. ---, Schneider, B. and Sommer, R., ibid. 692, 1-11 (1966); CA 64, 19665.
1619. --- and Sommer, R., ibid. 701, 28-39 (1967); CA 66, 105036.
1620. ---, --- and Lind, H., ibid. 688, 14-27 (1965); CA 64, 539.
1621. ---, --- and Mueller, E., Angew. Chem. 78, 545-6 (1966), Intern. Ed. Engl. 5, 514-5 (1966); CA 65, 5481.
1622. Newhall, A. G. and Diaz, F., Plant Disease Reptr. 50, 422-3 (1966); CA 65, 7929.
1623. Newsom, H. C. and Woods, W. G., Inorg. Chem. 7, 177-8 (1968); CA 68, 39678.
1624. Nicholson, D. A., Ph. D. Thesis, Northwestern Univ., 1965; Diss. Abstr. 26, 3026-7 (1965); CA 64, 9761.
1625. --- and Allred, A. L., Inorg. Chem. 4, 1747-50 (1965); CA 64, 3591.
1626. --- and ---, ibid. 4, 1751-3 (1965); CA 64, 3591.
1627. Nicora, C., Borsini, G. and Ratti, L., J. Polym. Sci. Pt. B 4, 151-4 (1966); CA 64, 14279.
--- and ---, Brit. 1,071,201; Fr. 1,469,833; Neth. 66 1847; U.S. 3,390,143, Feb. 22, 1965; CA 66, 19053.
1628. Nikiforova, T. I., Tr., Vses. Nauch.-Issled. Inst. Zashch. Rast. No. 24, 154-9 (1965); CA 67, 31884.
1629. Nishimoto, K., Fuse, G. and Nakanishi, M., Ger. 1,214,684, Nov. 28, 1962; CA 65, 3908.
1630. Nitzsche, S., Wick, M. and Schmidt, E., Ger. 1,224,139, May 1, 1960; CA 66, 4028.
1631. Noltes, J. G., Creemers, H. M. J. C. and van der Kerk, G. J. M., J. Organometal. Chem. 11, P21-3 (1968); CA 68, 114715.
1632. Nomura, M., Matsuda, S. and Kikkawa, S., Kogyo Kagaku Zasshi 70, 710-4 (1967); CA 68, 13108.
1633. Nomura, S. and Kara, Y., Japan. 2160 ('67), July 9, 1963; CA 67, 44411.
1634. Normant, H. et al., Bull. Soc. Chim. Fr. 1965, 3446-56; CA 64, 9759.
1635. --- and Cuvigny, T., ibid. 1966, 3344-51; CA 67, 11128.
1636. Nottes, G. and Wolf, W., Ger. 1,218,794, May 2, 1959; CA 65, 10402.
1637. Nowak, M., Intern. J. Appl. Radiation Isotopes 16, 649-53 (1965); CA 64, 6029.
1638. Nowlin, G. and Lyons, H. D., U.S. 3,378,539, July 28, 1965; CA 68, 115179.
1639. Nozaki, M. and Hatotani, H., Water Res. 1, 167-78 (1967); CA 66, 118631.

1639a. Nuetzel, K. and Holzrichter, H., Belg. 658,676; Brit. 1,025,118; Fr. 1,433,187; Neth. 65 606; U.S. 3,426,006, Jan. 22, 1964; CA 64, 5276.
1640. Oakes, V., Brit. 1,020,612, April 9, 1962; CA 64, 14219.
1641. ---, Brit. 1,043,609, May 6, 1964; CA 66, 28899; Ger. 1,246,735, Aug. 28, 1965; CA 68, 49766.
1642. ---, Brit. 1,064,178, Aug. 27, 1964; CA 67, 43924.
1643. ---, Brit. 1,070,942, Jan. 28, 1964; CA 67, 43925.
1644. --- and Hughes, B., Brit. 1,027,781, Feb. 14, 1963; CA 64, 19910.
1645. --- and Hutton, R. E., J. Organometal. Chem. 6, 133-40 (1966); CA 65, 10617.
1646. --- and Jankowski, C., Ger. 1,222,503, Dec. 13, 1963; CA 65, 13763.
1647. O'Brien, J. R., Thromb. Diath. Haemorrh., Suppl. 13, 307-10 (1965); CA 66, 114340.
1648. Occolowitz, J. L., Tetrahedron Lett. 1966, 5291-7; CA 66, 28273.
1649. Okawara, R., Kawasaki, Y. and Tanaka, T., Proc. Int. Conf. Coord. Chem., 8th, Vienna 1964, 432-4; CA 66, 120342.
1650. --- and Ohara, M., Japan. 6737 ('66), July 15, 1963; CA 65, 5490.
1650a. Okuya, E. et al., Brit. 1,100,933; U.S. 3,432,515, July 2, 1965; CA 68, 60365.
1651. O'Leary, R. K. et al., J. Pharm. Sci. 56, 494-8 (1967); CA 66, 116215.
1652. Omae, I. et al., Kogyo Kagaku Zasshi 70, 705-9 (1967); CA 68, 13107.
1653. ---, Matsuda, S. and Kikkawa, S., ibid. 70, 1759-61 (1967); CA 68, 78383.
1654. ---, Ohnishi, S. and Matsuda, S., ibid. 70, 1755-8 (1967); CA 68, 87371.
1655. Orlov, V. A. and Tarakanov, O. G., Plast. Massy 1966 (11), 46-8; CA 66, 38390.
--- and ---, Vysokomol. Soedin. 8, 1139 (1966); CA 65, 12352.
1656. Orzechowski, A. and MacKenzie, J. C., U.S. 3,326,877, Nov. 13, 1963; CA 67, 54576.
1657. Osugi, J., Hirayama, S. and Kusuhara, S., Nippon Kakaku Zasshi 88, 810-2 (1967); CA 68, 83080.
---, --- and ---, Rev. Phys. Chem. Jap. 36, 93-9 (1966); CA 67, 63520.
1658. Otera, J., Kawasaki, Y. and Tanaka, T., Inorg. Chim. Acta 1, 294-6 (1967); CA 68, 64398.
1659. Otto, P. P. H. L., Creemers, H. M. J. C. and Luijten, J. G. A., J. Label. Compd. 2, 339-48 (1967); CA 67, 11562.
1660. Overmars, H. G. J., Ger. 1,239,092, June 9, 1964; CA 67, 33314; Belg. 664,697; Brit. 1,077,948; Fr. 1,437,036; Neth. 64 6534; U.S. 3,427,263, June 9, 1964; CA 64, 16092.
1661. ---, Ger. 1,241,105, Oct. 17, 1963; CA 67, 109326; U.S. 3,324,054, Oct. 7, 1963; CA 67, 54798; Belg. 654,478; Brit. 1,009,965; Fr. 1,410,852, Oct. 7, 1963; CA 64, 6853.
1661a. ---, Belg. 660,428; Brit. 1,071,891; Fr. 1,429,538; Neth. 64 2097; U.S. 3,417,113, March 2, 1964; CA 64, 3792.
1662. Ozaki, M. et al., Japan. 11,493 ('67), Oct. 16, 1964; CA 68, 30594.
1663. Ozawa, K. et al., J. Biochem. (Tokyo) 61, 411-3 (1967); CA 66, 101815.

1664. Pahnke, A. J. and Squire, E. C., Oil Gas. J. 64 (50), 106-10 (1966); CA 66, 58638.
1665. Panattoni, C., Bombieri, G. and Croatto, U., Acta Crystallogr. 21, 823-6 (1966); CA 66, 23057.
1666. Parisi, G. I., Ph. D. Thesis, Rutgers State Univ., 1966; Diss. Abstr. B 27, 4351 (1967); CA 67, 112568.
1667. Parker, E. E. and Baker, J. G., U.S. 3,245,339, Dec. 20, 1963; CA 67, 109267.
1668. Parker, V. H., Biochem. J. 97, 658-62 (1965); CA 64, 930.
1669. Pasynkiewicz, S., Malinowski, S. and Bitter, J., Przemysl Chem. 44, 500-3 (1965); CA 64, 273.
--- et al., Pol. 54,578, May 7, 1965; CA 70, 29061.
1670. Patil, H. R. H. and Graham, W. A. G., Inorg. Chem. 5, 1401-5 (1966); CA 65, 8957.
1671. Patmore, D. J. and Graham, W. A. G., ibid. 5, 2222-6 (1966); CA 66, 76124.
1672. --- and ---, ibid. 6, 981-8 (1967); CA 67, 32189.
1673. --- and ---, ibid. 7, 771-6 (1968); CA 68, 92571.
1674. --- and ---, Inorg. Nucl. Chem. Lett. 2 (7), 179-82 (1966); CA 65, 13195.
1675. Pawlowska, Z., Prace Central. Inst. Ochrony Pracy 16 (49), 79-98 (1966); CA 65, 4019.
1676. Peach, M. E., Can. J. Chem. 46, 211-5 (1968); CA 68, 49697.
1677. Peddle, G. J. D., J. Organometal. Chem. 5, 486-8 (1966); CA 65, 2293.
1678. Pedinelli, M., Magri, R. and Randi, M., Chim. Ind. (Milan) 48, 144 (1966); CA 64, 12719.
1679. Pedinoff, M. E. and Seguin, H., Rev. Sci. Instrum. 38, 1342-4 (1967); CA 67, 85208.
1680. Pelissier, J. P., Papeterie 86, 1537 (1964); CA 66, 56754.
1681. Pereyre, M. et al., J. Organometal. Chem. 11, 97-110 (1968); CA 68, 78393.
1682. ---, Bellegarde, B. and Valade, J., C. R. Acad. Sci., Paris, Ser. C 265, 939-41 (1967); CA 68, 78379.
1683. ---, Colin, G. and Valade, J., Tetrahedron Lett. 1967, 4805-8; CA 68, 77680.
1684. --- and Valade, J., Bull. Soc. Chim. Fr. 1967, 1928-36; CA 67, 100212.
1685. Perilstein, W. L., U.S. 3,287,265, April 1, 1963; CA 66, 48102.
1686. Perry, E., U.S. 3,272,786, June 16, 1958 and April 21, 1961; CA 66, 11294.
1687. Petersen, D. B. et al., J. Chem. Phys. 71, 4506-8 (1967); CA 68, 29036.
1688. Petrov, A. A. et al., Teor. i Eksperim. Khim., Akad. Nauk Ukr. SSR 1, 697-700 (1965); CA 64, 13581.
1689. Petrovskaya, L. I. et al., Zh. Strukt. Khim. 6, 781-3 (1965); CA 64, 5970.
1690. Petukhov, G. G. and Guseva, T. V., Zavodsk. Lab. 32, 525-6 (1966); CA 65, 4662.

1691. Petukhov, G. G., Svirezheva, S. S. and Druzhkov, O. N., Zh. Obshch. Chim. 36, 914-6 (1966); CA 65, 10460.
1692. --- and Titov, V. A., Zh. Prikl. Khim. 39, 1200-3 (1966); CA 65, 5162.
1693. Petukhov, V. A., Mironov, V. F. and Kravchenko, A. L., Izv. Akad. Nauk SSSR, Ser. Khim. 1966, 156-8; CA 64, 15192.
1694. Picco, D., Notiz. Mal. Piante No. 72-73, 3-32 (1965); CA 65, 7929.
1695. Pilloni, G., Anal. Chim. Acta 37, 497-507 (1967); CA 67, 53526.
1696. ---, Farmaco, Ed. Prat. 22, 666-76 (1967); CA 68, 74927.
1697. --- and Plazzogna, G., Anal. Chim. Acta 35, 325-9 (1966); CA 65, 8006.
1698. --- and Tagliavini, G., J. Organometal. Chem. 11, 557-62 (1968); CA 68, 113747.
1699. Pison, R. E., Quim. Ind. (Bilbao) 14, 107-12 (1967); CA 68, 61249.
1700. Plate, N. A. et al., Vysokomol. Soedin. 8, 1890-4 (1966); CA 66, 18886.
--- et al., U.S.S.R. 176,408, Sept. 18, 1963; CA 64, 9838.
1701. Plazzogna, G. and Pilloni, G., Anal. Chim. Acta 37, 260-6 (1967); CA 66, 82188.
1701a. Plueddemann, E. P., Brit. 1,096,898; Fr. 1,459,412; Ger. 1,570,489; Neth. 65 16,388; U.S. 3,453,230, Dec. 17, 1964; CA 65, 15612.
1702. Plum, H. and Bokranz, A., Ger. 1,234,722, Aug. 5, 1965; CA 67, 22015.
1703. Poller, R. C., Spectrochim. Acta 22, 935-9 (1966); CA 65, 1600.
1704. --- and Spillman, J. A., J. Organometal. Chem. 6, 668-70 (1966); CA 66, 28861.
1705. --- and ---, ibid. 7, 259-62 (1967); CA 66, 65597.
1706. --- and Toley, D. L. B., J. Chem. Soc., A 1967, 1578-80; CA 67, 113311.
1707. --- and ---, ibid., A 1967, 2035-6; CA 68, 45819.
1708. Pollock, M. W., Belg. 658,003; Brit. 1,098,841; Fr. 1,440,654; U.S. 3,398,114, Jan. 10, 1965; CA 64, 3802.
1709. Ponomarev, S. V. et al., Zh. Obshch. Khim. 37, 2204-7 (1967); CA 68, 87333.
1710. ---, Lisina, Z. M. and Lutsenko, I. F., ibid. 36, 1818-22 (1966); CA 66, 55564.
1711. ---, Machigin, E. V. and Lutsenko, I. F., ibid. 36, 548-52 (1966); CA 65, 744.
1712. --- and Rogachev, B. G. and Lutsenko, I. F., ibid. 36, 1348 (1966); CA 65, 16998.
1713. Ponomarev, Yu. I., Kovalev, I. F. and Orlov, V. A., Opt. Spektrosk. 23, 483-5 (1967); CA 68, 25205.
1714. Prince, R. H. and Timms, R. E., Inorg. Chim. Acta 1, 129-33 (1967); CA 67, 116309.
1714a. Potter, G. H., Whitworth, C. J., Jr. and Zutty, N. L., Brit. 1,154,849; Fr. 1,484,581; Neth. 66 9051; U.S. 3,485,790, June 29, 1965; CA 67, 12242.
1715. Prokofev, A. K., Nechiporenko, V. P. and Kostyanovskii, R. G., Izv. Akad. Nauk SSSR, Ser. Khim. 1967, 794-801; CA 67, 73657.
1716. Puchinan, E. A. and Manulkin, Z. M., Tr. Tashkent. Farmatsev. Inst. 4, 354-60 (1966); CA 68, 78382.
1717. Putnam, R. C., Can. J. Chem. 44, 1343-50 (1966); CA 65, 2107.

1718. Putnam, R. C. and Pu, H., J. Gas Chromatog. 3, 289-93 (1965); CA 64, 8.
1719. Quan, L. M., C. R. Acad. Sci., Paris, Ser. C 266, 832-3 (1968); CA 68, 114712.
1720. Radbil, B. A., Glushakova, V. N. and Aleksandrov, Yu. A., Zh. Obshch. Khim. 37, 213-5 (1967); CA 66, 95145.
1721. Rafaeloff, R. et al., J. Inorg. Nucl. Chem. 28, 899-902 (1966); CA 65, 3303.
1722. Raicu, C. et al., An., Inst. Cent. Cercet. Agr., Sect. Prot. Plant No. 3, 53-64 (1965); CA 68, 11907.
1723. Ramaiah, K. and Martin, D. F., Chem. Commun. 1965, 130; CA 65, 13195.
1724. Ramos, T., U.S. 3,261,813, Feb. 13, 1963; CA 65, 18849.
1725. ---, U.S. 3,274,134, Sept. 14, 1961; CA 66, 3916.
1726. Ramsden, H. E., U.S. 3,240,795, July 2, 1962; CA 64, 14220; Fr. 1,467,549, Feb. 7, 1966; CA 68, 49769.
1727. ---, U.S. 3,354,190, Jan. 4, 1965; CA 68, 114744.
1728. --- and Engelhart, J. E., U.S. 3,351,646, Jan. 4, 1965; CA 68, 49723.
1729. Randall, E. W., Ellner, J. J. and Zuckerman, J. J., Inorg. Nucl. Chem. Lett. 1 (3), 109 (1965); CA 64, 10586.
1730. ---, Yoder, C. H. and Zuckerman, J. J., ibid. 1 (3), 105 (1965); CA 64, 11245.
1731. ---, --- and ---, Inorg. Chem. 6, 744-9 (1967); CA 66, 100020; cf. ref. 2160.
1732. ---, --- and ---, J. Am. Chem. Soc. 89, 3438-41 (1967); CA 67, 48852.
1733. --- and Zuckerman, J. J., Chem. Commun. 1966, 732-3; CA 66, 41900.
1734. Rao, D. V. R. A. and Rai, D. K., Curr. Sci. 37 (2), 41-2 (1968); CA 68, 108630.
1735. Rashkes, A. M., Manulkin, Z. M. and Kuchkarev, A. B., Zh. Obshch. Khim. 37, 1046-8 (1967); CA 68, 13129.
1736. Rausch, M. D. and Klemann, L. P., J. Am. Chem. Soc. 89, 5732-3 (1967); CA 68, 78368.
1737. Razuvaev, G. A., Dyachkovskaya, O. S. and Fionov, V. I., Dokl. Akad. Nauk SSSR 177, 1113-5 (1967); CA 68, 69096.
1738. Redfern, R. E., Walker, R. L. and Cantu, E., U. S. Dept. Agr. No. 33-122, 11 pp. (1967); CA 67, 99160.
1739. Reed, D. K., Crittenden, C. R. and Lyon, D. J., J. Econ. Entomol. 60, 668-71 (1967); CA 67, 20977.
1739a. Rees, D. M., Brit. 1,111,156; Fr. 1,443,657; Neth. 65 10,039; U.S. 3,436,251, Aug. 4, 1964; CA 65, 2471.
1740. Rees, R. G. and Webb, A. F., J. Organometal. Chem. 12, 239-40 (1968); CA 68, 105318.
1741. Rehage, G. and Schaefer, E. E., Z. Anal. Chem. 235, 137-43 (1968); CA 68, 96289.
1742. Reichle, W. T., Inorg. Chem. 5, 87-91 (1966); CA 64, 5131.
1743. Reid, A. F. and Wailes, P. C., Austr. J. Chem. 19, 309-12 (1966); CA 64, 15920.
1744. Reifenberg, G. H. and Considine, W. J., J. Organometal. Chem. 9, 495-504 (1967); CA 67, 90897.

1745. Reifenberg, G. H. and Considine, W. J., J. Organometal. Chem. 9, 505-9 (1967); CA 67, 90898.

--- and ---, Brit. 1,123,153, Jan. 3, 1966; CA 69, 87185.

1746. --- and ---, J. Organometal Chem. 10, 279-83 (1967); CA 68, 29815.

1747. --- and ---, ibid. 10, 285-9 (1967); CA 68, 29816; cf. ref. 895a.

1747a. --- and ---, Fr. 1,513,145; Neth. 67 3505; U.S. 3,507,893, March 7, 1966; CA 67, 116946.

1748. Reinhardt, H. F., U.S. 3,342,880, April 28, 1961; CA 68, 2636.

---, U.S. 3,378,587, Feb. 8, 1967; CA 69, 2583.

1749. Retzloff, J. B. and Kieninger, J. E., U.S. 3,251,661, Nov. 15, 1961; CA 65, 3649.

1750. Ricciboni, L. et al., J. Electroanal. Chem. 11, 340-9 (1966); CA 64, 15044.

1751. Rice, R. G., Faurote, P. D. and Geib, B. H., U.S. 3,313,774, Sept. 4, 1962; CA 67, 3406.

1752. Richards, J. R., Vacuum 16, 310-1 (1966); CA 65, 16047.

1753. Richardson, B. A., Int. Pest Contr. 10, 14-9 (1968); CA 68, 94834.

1754. ---, Brit. 1,026,692, April 27, 1961; CA 65, 919.

1755. Richardson, W. L., Barusch, M. R. and Kautsky, G. J., U.S. 3,356,472, Sept. 17, 1964; CA 68, 51780.

1756. de Ridder, J. J. and Dijkstra, G., Nature 216, 260-1 (1967); CA 68, 33390.

--- and ---, Rec. Trav. Chem. 86, 737-45 (1967); CA 67, 103569.

1757. ---, van Koten, G. and Dijkstra, G., ibid. 86, 1325-34 (1967); CA 68, 44111.

1758. Riethmayer, S. A., Swiss 417,950, Aug. 22, 1962; CA 67, 3420.

1759. Rifkin, E. B. and Closson, R. D., U.S. 3,336,751, May 29, 1961; CA 67, 101517.

1760. Rijkens, F. et al., Rec. Trav. Chim. 85, 1223-9 (1966); CA 66, 65595.

1761. ---, Janssen, M. J. and van der Kerk, G. J. M., ibid. 84, 1597-609 (1965); CA 64, 17631.

1762. Riviere, P. and Satge, J., Bull. Soc. Chim. Fr. 1967, 4039-46; CA 68, 87370.

1763. Rizzoli, A. and Galzigna, L., Biochem. Biol. Sper. 4 (3), 172-9 (1965); CA 65, 1281.

1764. --- and ---, Boll. Soc. Ital. Biol. Sper. 41, 1176-8 (1965); CA 64, 14843.

1765. Robins, J. and Updegraff, D. M., U.S. 3,236,793, Oct. 2, 1961; CA 64, 16135.

1765a. Robinson, G. C., Brit. 1,042,119; Ger. 1,250,441; Neth. 65 7727; U.S. 3,431,185, June 22, 1964; CA 64, 17640.

1766. Roeser, G. P., U.S. 3,245,959, Nov. 29, 1960; CA 64, 19830.

1767. Roesky, H. W., Chem. Ber. 100, 2147-50 (1967); CA 67, 90896.

1768. Roncucci, L., Faraglia, G. and Barbieri, R., J. Organometal. Chem. 6, 278-82 (1966); CA 65, 15420.

1769. Roschig, M. and Matschiner, H., Chem. Tech. (Berlin) 19, 103-4 (1967); CA 67, 17628.

1770. Rose, M. S. and Aldridge, W. N., Biochem. J. 106, 821-8 (1968); CA 68, 85779.
1771. --- and ---, J. Neurochem. 13, 103-8 (1966); CA 64, 14817.
1772. Rubinchik, G. F. and Manulkin, Z. M., Zh. Obshch. Khim. 36, 394-5 (1966); CA 64, 15918.
1773. --- and ---, ibid. 36, 748-50 (1966); CA 65, 8955.
1774. --- and ---, ibid. 36, 1301-4 (1966); CA 65, 16998.
1775. Rudner, B. and Moores, M. S., U.S. 3,314,980, April 10, 1962; CA 67, 22438.
1776. Ruff, J. K., Inorg. Chem. 5, 732-5 (1966); CA 64, 17023.
1777. Rush, J. J. and Hamilton, W. C., ibid. 5, 2238-9 (1966); CA 66, 98744.
1778. Ryabchenko, S. N. and Lisin, D. M., Mater. Soveshch. Rab. Lab. Geol. Organ., 9th, Kiev. 1965, 110-5; CA 66, 117696.
1779. Rzaev, Z. M., Kochkin, D. A. and Zubov, P. I., Dokl. Akad. Nauk SSSR 172, 364-7 (1967); CA 66, 95146.
1780. Sacher, R. E., Lemmon, D. H. and Miller, F. A., Spectrochim. Acta, Pt. A 23, 1169-76 (1967); CA 67, 38053.
1781. Sakurai, H. et al., Tetrahedron Lett. 1966, 5493-7; CA 66, 37955.
1782. Salooja, K. C., Combust. Flame 9 (3), 211-7 (1965); CA 64, 495.
1783. ---, J. Inst. Petrol. 53, 186-93 (1967); CA 67, 45736.
1784. Sandler, S. R., Tsou, K. C. and Dannin, J. E., U.S. 3,356,616, March 5, 1964; CA 68, 30797.
1785. Sandy, C. A., Pedrotti, R. L. and Tullio, V., Brit. 1,098,657; Fr. 1,480,011; U.S. 3,400,143, May 17, 1965; U.S. 3,408,375, Oct. 7, 1965; CA 68, 114743.
1786. Sano, K., Japan. 6264 ('67), July 24, 1963; CA 68, 40683.
1787. Sarankina, S. A. and Manulkin, Z. M., Zh. Obshch. Khim. 36, 1299-301 (1966); CA 65, 16998.
1788. ---, --- and Kuchkarov, A. B., ibid. 37, 217-9 (1967); CA 66, 95124.
1789. Sartori, P. and Weidenbruch, M., Angew. Chem. 77, 1138 (1965); Int. Ed. Engl. 4, 1097 (1965); CA 64, 12179.
       --- and ---, Chem. Ber. 100, 2049-63 (1967); CA 67, 32301.
1790. Saruto, K., Gono, T. and Suenobu, Y., Japan. 19,333 ('66), March 4, 1964; CA 66, 38049.
1791. ---, --- and ---, Japan. 19,416 ('66), April 28, 1964; CA 66, 46484.
1792. ---, --- and ---, Japan. 12,415 ('67), Feb. 19, 1964; CA 68, 49777.
1793. --- and Suenobu, Y., Japan. 3532 ('66), Feb. 22, 1963; CA 65, 2298.
1794. --- and ---, Japan. 6172 ('66), Nov. 28, 1961; CA 65, 5490.
1795. --- and ---, Japan. 8856 ('66), Feb. 23, 1963; CA 65, 12240.
1796. Satge, J. and Baudet, M., C. R. Acad. Sci., Paris, Ser. C 263, 435-8 (1966); CA 66, 11020.
       ---, --- and Lesbre, M., Bull. Soc. Chim. Fr. 1966, 2133.
1797. --- and Couret, C., C. R. Acad. Sci., Paris, Ser. C 264, 2169-72 (1967); CA 67, 108721.
1798. ---, Massol, M. and Lesbre, M., J. Organometal. Chem. 5, 241-53 (1966); CA 64, 11245.

1799. Satge, J. and Riviere, P., Bull. Soc. Chim. Fr. 1966, 1773-4; CA 65, 10616.
1800. Sato, H., Bull. Chem. Soc. Jap. 40, 410-1 (1967); CA 67, 11031.
1801. --- and Shimamine, M., Eisei Shikensho Hokoku No. 82, 39-42 (1964); CA 65, 8671.
1802. Sawyer, A. K., U.S. 3,322,801, July 31, 1963; CA 67, 22001; cf. ref. 468.
1803. ---, U.S. 3,347,889, July 31, 1963; CA 68, 49779.
1804. --- and Brown, J. E., J. Organometal. Chem. 5, 438-45 (1966); CA 64, 19668.
1805. ---, --- and May, G. S., ibid. 11, 192-4 (1968); CA 68, 95910.
1806. Sawyer, R., Analyst (London) 92, 569-74 (1967); CA 67, 91422.
1807. Schaefer, T., Hruska, F. and Hutton, H. M., Can. J. Chem. 45, 3143-51 (1967); CA 68, 34532.
1808. Schatz, M. and Heidingsfeld, V., Sb. Vys. Sk. Chem.-Technol. Praze, Org. Technol. 10, 37-50 (1966); CA 68, 50768.
1809. Scheinberg, L. C. et al., J. Neuropath. Exper. Neurol. 25, 202-13 (1966); CA 66, 9594.
1809a. Scherer, O, French, H. and Stenger, W., Belg. 664,034; Brit. 1,106,232; Fr. 1,439,363; Neth. 65 5966; U.S. 3,281,316, May 16, 1964; CA 64, 14901.
1810. Scherer, O. J. and Biller, D., Angew. Chem. Int. Ed. Engl. 6, 446 (1967); CA 67, 32755.
1811. ---, --- and Schmidt, M., Inorg. Nucl. Chem. Lett. 2, 103-5 (1966); CA 65, 16992.
1812. --- and Hornig, P., J. Organometal. Chem. 8, 465-8 (1967); CA 67, 53454.
1813. --- and Schieder, G., Angew. Chem. 80, 83 (1968); Int. Ed. Engl. 7, 75-6 (1968); CA 68, 59671.
1814. --- and Schmidt, M., Intern. Symp. Organosilicon Chem., Sci. Commun., Prague 1965, 315-6; CA 65, 10606.
1815. Schlemper, E. O., Inorg. Chem. 6, 2012-7 (1967); CA 68, 7248.
1816. --- and Britton, D., ibid. 5, 507-10 (1966); CA 64, 13488.
1817. --- and ---, ibid. 5, 511-4 (1966); CA 64, 13487.
1818. --- and Hamilton, W. C., ibid. 5, 995-8 (1966); CA 65, 1510.
1819. Schlimper, R. and Thinius, K., Ger. (East) 49,295, Feb. 24, 1960; CA 66, 29616.
1820. Schluenz, M. and Koester-Pflugmacher, A., Z. Anal. Chem. 232, 93-7 (1967); CA 68, 18356.
1821. Schmidbaur, H., Nucl. Magnetic Resonance Chem., Proc. Symp., Cagliari, Italy 1964, 185-7 (1965); CA 66, 6921.
1822. --- and Armer, B., Angew. Chem. 78, 305-6 (1966); Int. Ed. Engl. 5, 313 (1966); CA 64, 19668.
1823. --- and Findeiss, W., Chem. Ber. 99, 2187-96 (1966); CA 65, 8942.
1824. --- and Jonas, G., ibid. 100, 1120-8 (1967); CA 66, 105027.
1825. --- and Ruidisch, I., Proc. Int. Conf. Coord. Chem., 8th, Vienna 1964, 168-70; CA 67, 11530.
1826. --- and Tronich, W., Chem. Ber. 100, 1032-50 (1967); CA 66, 105026.

1827. Schmidt, U. et al., Chem. Ber. 98, 3827-30 (1965); CA 64, 5986.
1828. Schmitz-DuMont, O. and Jansen, W., Z. Anorg. Allg. Chem. 349, 189-201 (1967); CA 66, 76110.
1829. Schmutzler, G., Plaste Kaut. 13, 462-4 (1966); CA 67, 46950.
1830. Schneider, B. and Neumann, W. P., Ann. Chem. 707, 7-14 (1967); CA 67, 108724.
1831. Schoellkopf, U. and Rieber, N., Angew. Chem. Int. Ed. Engl. 6, 884 (1967); CA 68, 13002.
1832. --- and Traenckner, H. J., J. Organometal. Chem. 5, 300 (1966); CA 64, 14212.
1833. Schoen, W. F., U.S. 3,223,496, July 5, 1962; CA 64, 4841.
1833a. Schroeder, A. and Nieuwenhuis, P. G. J., Brit. 1,107,697; Fr. 1,518,397; Neth. 64 4827; U.S. 3,476,704, April 12, 1966; CA 68, 22547.
1834. Schroeder, E., Plaste Kaut. 13, 278-81 (1966); CA 65, 4041.
1835. Schroeder, H. et al., U. S., Clearing House Fed. Sci. Tech. Inform., AD 652,379, 18 pp. (1967); CA 68, 13463.
1836. Schroeder, H. D., Thomas, K. and Jerchel, D., Ger. 1,215,709, Aug. 20, 1964; CA 65, 5489; Neth. 65 10,858; CA 65, 17004.
1836a. Schroeder, L., Thomas, K. and Jerchel, D., Brit. 1,018,805; Fr. 1,437,384; Ger. 1,545,621; U.S. 3,321,480, Feb. 21, 1963; CA 64, 11251.
1837. ---, --- and ---, Ger. 1,204,226, Dec. 10, 1963; CA 64, 2128.
1838. ---, --- and ---, Ger. 1,216,300, Dec. 13, 1963; CA 65, 5490.
1838a. --- et al., Brit. 1,160,693; Fr. 1,495,647; Ger. 1,298,531; Neth. 66 12,312; U.S. 3,502,690, Sept. 3, 1965; CA 67, 73687.
1839. Schumann, H. et al., Inorg. Nucl. Chem. Lett. 2, 311-2 (1966); CA 66, 28863.
1840. --- et al., J. Organometal. Chem. 10, 71-9 (1967); CA 67, 100211.
1841. ---, Jutzi, P. and Schmidt, M., Angew. Chem. 77, 912 (1965); Int. Ed. Engl. 4, 869; CA 64, 3593.
1842. ---, Oestermann, T. and Schmidt, M., Chem. Ber. 99, 2057-62 (1966); CA 65, 7214.
1843. ---, --- and ---, J. Organometal. Chem. 8, 105-10 (1967); CA 66, 85834.
1844. --- and Ronecker, S., Z. Naturforsch., B 22, 452-3 (1967); CA 67, 68948.
1845. --- and Roth, A., J. Organometal. Chem. 11, 125-32 (1968); CA 68, 78381.
1846. ---, --- and Stelzer, O., Angew. Chem. 80, 240 (1968); Int. Ed. Engl. 7, 218-9 (1968); CA 68, 95908.
1847. --- and Schmidt, M., Inorg. Nucl. Chem. Lett. 1, 1-15 (1965); CA 64, 1603.
1848. --- and ---, J. Organometal. Chem. 3, 485-7 (1965); CA 64, 160.
1849. ---, Schwabe, P. and Schmidt, M., Inorg. Nucl. Chem. Lett. 2 (10), 309-10 (1966); CA 66, 28858.
1850. ---, --- and ---, ibid. 2 (10), 313-4 (1966); CA 66, 11011.
1851. --- and Stelzer, O., Angew. Chem. 79, 692 (1967); Int. Ed. Engl. 6, 701 (1967); CA 67, 90901.
1852. --- and ---, ibid. 80, 318 (1968); ibid. 7, 300 (1968); CA 68, 118920.
1853. Schumann-Ruidisch, I. and Blass, H., Z. Naturforsch. B 21, 1105 (1966); CA 66, 76113.

1854. Schumann-Ruidisch, I. and Blass, H., Z. Naturforsch. B 22, 1081-2 (1967); CA 68, 29819.
1855. --- and Jutzi-Mebert, B., J. Organometal. Chem. 11, 77-83 (1967); CA 68, 29818.
1856. ---, Lieb, V. and Jutzi-Mebert, B., Z. Anorg. Allg. Chem. 355, 64-72 (1967); CA 68, 13130.
1857. Schwartz, W. T., Jr., Ph. D. Thesis, State Univ. of New York, Buffalo, 1964; Diss. Abstr. B 27, 1429 (1966); CA 66, 76104.
1858. Scott, R. C., U.S. 3,324,058, July 6, 1965; CA 67, 44631.
1859. Secrist, D. R. and MacKenzie, J. D., Amer. Chem. Soc., Div. Fuel Chem. Preprints, Pt. 1, 11 (2), 203-10 (1967); CA 66, 119781.
1860. Seiffer, E. A. and Schoof, H. F., Publ. Health Rept. (U.S.) 82, 833-9 (1967); CA 68, 2250.
1861. Seil, C. A., U.S. 3,287,275, Dec. 30, 1963; CA 66, 20859.
1862. Seki, T. and Kawakami, Y., Japan. 11,495 ('67), Dec. 29, 1964; CA 68, 3561.
1863. ---, --- and Katsumura, T., Japan. 13,260 ('65), Dec. 30, 1962; CA 65, 9116.
1864. ---, --- and ---, Japan. 13,261 ('65), Dec. 30, 1962; CA 65, 9116.
1865. ---, --- and ---, Japan. 6755 ('66), April 4, 1963; CA 65, 13894.
1866. ---, --- and ---, Japan. 5345 ('67), April 26, 1964; CA 68, 40551.
1867. ---, --- and ---, Japan. 12,529 ('67), Jan. 5, 1963; CA 67, 109337.
1868. Selivokhin, P. I., Gig. Sanit. 31, 68-9 (1966); CA 65, 15973.
1869. Semenuk, N. S., Papetti, S. and Schroeder, H., U.S., Clearing House Fed. Sci. Tech. Inform., AD 652,378, 9 pp. (1967); CA 68, 13462.
1870. Semion, V. A. and Kravtsov, D. N., Zh. Strukt. Chem. 7, 814 (1966); CA 66, 23141.
1871. Sen, D. N. and Umapathy, P., Indian J. Chem. 5, 209-10 (1967); CA 67, 82436.
1871a. Senatore, P. J., Brit. 1,135,455; U.S. 3,449,451, Oct. 4, 1965; CA 70, 58025.
1872. Sergeev, N. M. et al., Teor. i Eksperim. Khim., Akad. Nauk Ukr. SSR 1, 695-7 (1965); CA 64, 15212.
1873. Seyferth, D., U.S. 3,347,888, May 27, 1963; CA 68, 59726.
1874. --- et al., J. Org. Chem. 32, 2980-4 (1967); CA 67, 108278; cf. ref. 494.
1875. --- et al., J. Organometal. Chem. 6, 573-6 (1966); CA 66, 11018; cf. ref. 2756.
1876. --- et al., ibid. 7, 405-13 (1967); CA 66, 94450.
1877. --- et al., ibid. 11, 63-76 (1968); CA 68, 29810.
1878. --- and Armbrecht, F. M., Jr., J. Am. Chem. Soc. 89, 2790-1 (1967); CA 67, 43895.
1879. ---, --- and Hanson, E. M., J. Organometal. Chem. 10, P25-7 (1967); CA 68, 13043.
1880. --- and Burlitch, J. M., Z. Naturforsch. B 22, 1358-9 (1967); CA 68, 69095.

1881. Seyferth, D., Cross, R. J. and Prokai, B., J. Organometal. Chem. 7, P19-20 (1967); CA 66, 105016.
1882. --- and Dertouzos, H., ibid. 11, 263-70 (1968); CA 68, 86517.
1883. --- and Evnin, A. B., J. Am. Chem. Soc. 89, 1468-75 (1967); CA 66, 104486.
1884. ---, --- and Blank, D. R., Inorg. Nucl. Chem. Lett. 3, 181-4 (1967); CA 67, 64459.
---, --- and ---, J. Organometal. Chem. 13, 25-36 (1968); CA 69, 106776.
1885. --- and Hetflejs, J., ibid. 11, 253-61 (1968); CA 68, 86535.
1886. --- and Jula, T. F., ibid. 8, P13-6 (1967); CA 67, 21279.
1887. ---, Singh, G. and Suzuki, R., Pure Appl. Chem. 13, 159-66 (1966); CA 67, 32709.
---, Suzuki, R. and Vaughan, L. G., J. Am. Chem. Soc. 88, 286-91 (1966); CA 64, 8231.
1888. --- and Vaughan, L. G., J. Organometal. Chem. 5, 580-2 (1966); CA 65, 5482.
1889. --- and Washburne, S. S., ibid. 5, 389-91 (1966); CA 64, 17633.
1890. Sezoko, M., Murayama, Yu. and Nakamura, M., Japan. 13,867 ('67), May 27, 1964; CA 68, 39817.
1891. Sharanina, L. G., Zavgorodnii, V. S. and Petrov, A. A., Zh. Obshch. Khim. 36, 1154 (1966); CA 65, 10617.
1892. Shaver, R. J., U.S. 3,226,292, April 18, 1962; CA 64, 6415.
1893. Shchembelov, G. A. and Ustynyuk, Yu. A., Dokl. Akad. Nauk SSSR 173, 847-50 (1967); CA 67, 43312.
--- and ---, ibid. 173, 1364-6 (1967); CA 68, 38929.
1894. Shergina, N. I. et al., Izv. Akad. Nauk SSSR, Ser. Khim. 1967, 1378-9; CA 68, 38836.
1895. Sheridan, J. E., Plant Pathol. 16 (2), 93-6 (1967); CA 67, 81438.
1896. Shier, G. D. and Drago, R. S., J. Organometal. Chem. 6, 359-63 (1966); CA 65, 18465.
1897. Shiihara, I. et al., Japan. 13,525 ('67), May 11, 1965; CA 68, 69983.
1898. Shikhiev, I. A. et al., Azerb. Khim. Zh. 1965 (4), 42-4; CA 64, 9760.
1899. ---, Abdullaev, N. D. and Aliev, M. I., Zh. Obshch. Khim. 36, 942-3 (1966); CA 65, 10616.
1900. ---, Aslanov, I. A. and Mekhmandarova, N. T., ibid. 36, 1295-7 (1966); CA 65, 20158.
1901. ---, --- and ---, ibid. 36, 1297-9 (1966); CA 65, 20159.
1902. Shindo, M., Matsumura, Y. and Okawara, R., J. Organometal. Chem. 11, 299-305 (1968); CA 68, 53979.
1903. --- and Okawara, R., Inorg. Nucl. Chem. Lett. 3, 75-7 (1967); CA 67, 54233.
1904. Shishido, K. and Takeda, Y., Japan. 1615 ('66), Oct. 4, 1961; CA 64, 12724.
1905. Shorygin, P. P. et al., Teor. i Eksperim. Khim., Akad. Nauk Ukr. SSR 2 (2), 190-5 (1966); CA 65, 14660.

1906. Shostakovskii, M. F. et al., Dokl. Akad. Nauk SSSR 176, 356-9 (1967); CA 68, 29824.
1907. --- et al., Izv. Akad. Nauk SSSR, Ser. Khim. 1967, 2118-20; CA 68, 29812.
1908. --- et al., Zh. Obshch. Khim. 35, 1768-70 (1965); CA 64, 1927.
1909. --- et al., ibid. 37, 567-70 (1967); CA 67, 21958.
1910. --- et al., ibid. 37, 1738-43 (1967); CA 68, 13103.
1911. --- et al., Zh. Prikl. Spektroskopii, Akad. Nauk Belorussk. SSR 4 (1), 46-51 (1966); CA 64, 18700.
1912. --- et al., U.S.S.R. 172,782, June 5, 1964; CA 64, 757.
1913. --- et al., U.S.S.R. 173,760, July 30, 1964; CA 64, 3602.
1914. --- et al., U.S.S.R. 176,582, Oct. 19, 1964; CA 64, 9766.
1915. --- et al., U.S.S.R. 180,593, April 5, 1965; CA 65, 12240.
1916. --- et al., U.S.S.R. 199,883, July 6, 1966; CA 68, 41235.
--- et al., Zh. Prikl. Khim. 41, 2796-8 (1968); CA 70, 88961.
1917. ---, Komarov, N. V. and Ermolova, T. I., Dokl. Akad. Nauk SSSR 175, 1079-81 (1967); CA 68, 2970; cf. ref. 2505.
1918. ---, --- and Misyunas, V. K., ibid. 173, 843-6 (1967); CA 67, 43857.
1919. ---, --- and ---, U.S.S.R. 172,777, July 2, 1963; CA 64, 758.
1920. ---, --- and Sklyanova, A. M., Ger. 1,235,917, March 30, 1965; CA 67, 22014; cf. ref. 2508.
---, --- and Stepanovich, A. M., Brit. 1,092,036, April 12, 1965; CA 68, 22059; Fr. 1,427,563, April 1, 1965; CA 65, 8962.
1921. ---, Vlasov, V. M. and Mirskov, R. G., U.S.S.R. 173,757, Nov. 5, 1963; CA 64, 5138.
1922. ---, --- and ---, U.S.S.R. 185,920, April 5, 1965; CA 66, 115305.
1923. Sianesi, D. and Caporiccio, G., Fr. 1,464,332, Jan. 11, 1965; CA 67, 54590.
1924. Sichel, J. M. and Whitehead, M. A., Theoret. Chim. Acta 5, 35-52 (1966); CA 65, 3173.
1925. Siegel, J. R., U.S. 3,265,474, Oct. 6, 1961; CA 65, 12044.
1926. Sijpesteijn, A. K., Antonie van Leeuwenhoek, J. Microbiol. Serol. 34, 85-92 (1968); CA 68, 36984.
1927. Silversmith, E. F. and Sloan, W. J., Ger. 1,246,734, Nov. 30, 1961; CA 67, 87255.
1928. Sim, S. Y., Ph. D. Thesis, Iowa State Univ., 1966; Diss. Abstr. B27, 1430 (1966); CA 66, 85836; cf. ref. 176, 1101.
1929. Simrock, R. A., Brit. 1,063,508, Oct. 8, 1965; CA 67, 22599.
1930. Simonnin, M. P., Bull. Soc. Chim. Fr. 1966, 1774-5; CA 65, 11576.
1931. ---, J. Organometal. Chem. 5, 155-65 (1966); CA 64, 6453.
1932. Simons, P. B. and Graham, W. A. G., ibid. 8, 479-90 (1967); CA 67, 54229.
1933. --- and ---, ibid. 10, 457-64 (1967); CA 68, 29078.
1934. Simpson, W. B., Chem. Ind. (London) 1966, 854; CA 65, 3904.
1935. Singh, G., J. Org. Chem. 31, 949-50 (1966); CA 64, 14064.
1936. ---, J. Organometal. Chem. 11, 133-43 (1968); CA 68, 29809.
---, Tetrahedron Lett. 1966, 4309-13; CA 65, 16826.

1937. Siniramed, C. and Renzanigo, F., Riv. Combust. <u>19</u>, 351-62 (1965); CA <u>64</u>, 4828.
1938. Sisido, K. et al., J. Organometal. Chem. <u>11</u>, 281-90 (1968); CA <u>68</u>, 59673.
1939. --- and Kozima, S., ibid. <u>11</u>, 503-13 (1968); CA <u>68</u>, 114714.
1940. ---, --- and Hanada, T., ibid. <u>9</u>, 99-107 (1967); CA <u>67</u>, 54230.
1941. ---, --- and Isibasi, T., ibid. <u>10</u>, 439-45 (1967); CA <u>68</u>, 29821.
1942. ---, --- and Takizawa, K., Tetrahedron Lett. <u>1967</u>, 33-6; CA <u>66</u>, 95144.
1943. ---, --- and Tuzi, T., J. Organometal. Chem. <u>9</u>, 109-15 (1967); CA <u>67</u>, 73654.
1944. Skachkova, I. N., Gig. Sanit. <u>32</u> (4), 11-7 (1967); CA <u>67</u>, 14750.
1945. Skalicka, B. and Cejka, M., Ropa Uhlie <u>8</u>, 246-8 (1966); CA <u>66</u>, 31814.
1946. Sladkov, A. M. and Luneva, L. K., Zh. Obshch. Khim. <u>36</u>, 553-6 (1966); CA <u>65</u>, 744.
1946a. Smeltz, K. C., Brit. 1,064,081; Fr. 1,450,613; Neth. 65 8049; U.S. 3,392,093, June 23, 1964; CA <u>64</u>, 17048.
1947. Smith, G., U.S. 3,332,970, Oct. 28, 1964; CA <u>68</u>, 13179.
1948. ---, U.S. 3,347,833, Dec. 8, 1964; CA <u>67</u>, 117696.
1949. Smith, G. W., Liquids, Struct., Prop. Solid Interactions, Proc. Symp., Warren, Mich. <u>1963</u>, 219-25 (1965); CA <u>64</u>, 16864.
1950. Smith, J. O. and McHugh, K. L., U.S. 3,244,629, Feb. 13, 1962; CA <u>64</u>, 19296.
1951. Snyder, L. J., Anal. Chem. <u>39</u>, 591-5 (1967); CA <u>66</u>, 118547.
1952. Sollott, G. P. and Peterson, W. R., Jr., J. Am. Chem. Soc. <u>89</u>, 6783-4 (1967); CA <u>68</u>, 59686.
1953. Sommer, R. and Kuivila, H. G., J. Org. Chem. <u>33</u>, 802-5 (1968); CA <u>68</u>, 68164.
1954. --- and Neumann, W. P., Angew. Chem. <u>78</u>, 546-7 (1966); Intern. Ed. Engl. <u>5</u>, 515 (1966); CA <u>65</u>, 5481.
1955. ---, Schneider, B. and Neumann, W. P., Ann. Chem. <u>692</u>, 12-21 (1966); CA <u>64</u>, 19666.
1956. Sorokin, G. V. et al., Dokl. Akad. Nauk SSSR <u>174</u>, 376-7 (1967); CA <u>67</u>, 117351.
1957. Soulage, N. L., Anal. Chem. <u>38</u>, 28-33 (1966); CA <u>65</u>, 2031.
    ---, ibid. <u>39</u>, 1340-1 (1967); CA <u>67</u>, 83606.
1958. Sowa, F. J., U.S. 3,222,158, May 26, 1960; CA <u>64</u>, 9767.
    ---, U.S. 3,287,103, Sept. 8, 1965; CA <u>67</u>, 52960.
    ---, U.S. 3,344,019, Sept. 8, 1965; CA <u>68</u>, 105368.
1959. Spichale, W., Kapitza, H. and Utschick, H., Z. Chem. <u>7</u>, 442 (1967); CA <u>68</u>, 33533.
1960. Spijkerman, J. J., Advan. Chem. Ser. No. <u>68</u>, 105-12 (1967); CA <u>68</u>, 24253.
1961. Sproat, A. D., U.S. 3,354,132, Sept. 2, 1964; CA <u>68</u>, 13701.
1962. Srivastava, T. S., J. Organometal. Chem. <u>10</u>, 373-4 (1967); CA <u>68</u>, 48833.
1963. ---, ibid. <u>10</u>, 375-6 (1967); CA <u>68</u>, 21407.

1964. Srivastava, T. S. and Bhattacharya, S. N., Indian J. Chem. 4, 474-5 (1966); CA 66, 95137.
1965. --- and ---, J. Inorg. Nucl. Chem. 28, 1480-2 (1966); CA 65, 7214.
1966. --- and ---, ibid. 28, 2445-7 (1966); CA 66, 38019.
1967. --- and ---, ibid. 29, 1873-6 (1967); CA 68, 22029.
1968. --- and ---, Z. Anorg. Allg. Chem. 344, 102-6 (1966); CA 64, 17630.
1969. --- and Tandon, S. K., Indian J. Chem. 3, 535-6 (1965); CA 64, 14211.
1970. --- and ---, Z. Anorg. Allg. Chem. 353, 87-92 (1967); CA 67, 90895.
1971. Stack, W. F., Nash, G. A. and Skinner, H. A., Trans. Faraday Soc. 61, 2122-5 (1965); CA 64, 10477.
1972. Stadnichuk, M. D., Zh. Obshch. Khim. 36, 937-41 (1966); CA 65, 10618.
1973. ---, Yakovleva, T. V. and Petrov, A. A., ibid. 37, 222-6 (1967); CA 66, 94566.
1974. Stahlberg, U., Gelius, R. and Mueller, R., Z. Anorg. Allg. Chem. 355, 230-7 (1967); CA 68, 44342.
1975. Stamm, W. and Breindel, A. W., U.S. 3,317,573, Jan. 22, 1964; CA 68, 39819.
1976. --- and Greco, C. C., U.S. 3,316,284, Nov. 14, 1963; CA 67, 64539.
1977. Stanilewicz, W., Farm. Pol. 23, 321-5 (1967); CA 68, 81159.
1978. Stankovic, M. and Mokranjac, M. S., Int. Congr. Occupational Health, 14th, Madrid, 1963 (2), 831-4 (1964); CA 65, 2883.
1979. Stark, L. R. and Hatton, R. E., U.S. 3,245,907, Aug. 28, 1961; CA 55, 5402.
1980. Steingross, W. and Zeil, W., J. Organometal. Chem. 6, 109-16 (1966); CA 65, 10617.
1981. --- and ---, ibid. 6, 464-73 (1966); CA 66, 2074.
1982. Stemniski, J. R., Belg. 653,242; Brit. 1,084,043-4; Fr. 1,422,900; U.S. 3,490,738, Sept. 19, 1963; CA 64, 9495.
1983. Sterman, S., Johnson, G. C. and Berger, S. E., Brit. 1,083,004; Fr. 1,419,054, Oct. 30, 1963; CA 65, 15663.
1984. Stern, C. J., Jr., U.S. 3,214,453, Aug. 25, 1960; CA 64, 9767.
1985. Stevens, D. R. and Sweet, R. L., U.S. 3,294,685, April 21, 1952; CA 66, 57619.
1986. Stoeckler, H. A., Ph. D. Thesis, Rutgers State Univ., 1967; Diss. Abstr. B28, 1890; CA 68, 73835.
1987. --- and Sano, H., Phys. Lett. A 25, 550-1 (1967); CA 68, 110009.
1988. --- and ---, Phys. Rev. 165, 406-8 (1968); CA 68, 55228.
1989. --- and ---, Trans. Faraday Soc. 64, 577-81 (1968); CA 68, 73833.
1990. ---, --- and Herber, R. H., J. Chem. Phys. 45, 1182-9 (1966); CA 65, 11536; cf. ref. 1986.
1991. ---, --- and ---, ibid. 47, 1567-71 (1967); CA 67, 86323; cf. ref. 1986.
1992. Stolzel, H., Klotzscher, I. and Hartmann, S., Brit. 1,089,243, May 23, 1966; CA 68, 78423.
---, --- and ---, Fr. 1,477,892, April 27, 1966; CA 68, 49778.
---, --- and ---, Ger. (East) 48,202, July 27, 1965; CA 65, 15428.
1993. Stoner, H. B., Brit. J. Ind. Med. 23, 222-9 (1966); CA 65, 11220.

1994. Stotskaya, L. L. et al., Plast. Massy 1967 (8), 14-6; CA 67, 82533.
1995. Stroganov, N. S. et al., Dokl. Akad. Nauk SSSR 170, 1189-91 (1966); CA 66, 36818.
1996. Struchkov, Yu. T. et al., ibid. 172, 107-10 (1967); CA 66, 119626.
1997. Sukhani, D., Gupta, V. D. and Mehrotra, R. C., J. Organometal. Chem. 7, 85-90 (1967); CA 66, 28860.
1998. Sulimov, I. G. and Stadnichuk, M. D., Zh. Obshch. Khim. 37, 2329-32 (1967); CA 68, 86511.
1999. Sumi, M., Kazama, S. and Nakai, K., Japan. 2947 ('68), May 20, 1964; CA 68, 105647.
2000. Sundermeyer, W. and Verbeek, W., Ger. 1,239,687, March 12, 1965; CA 68, 2989.
2001. --- and ---, Angew. Chem. Int. Ed. Engl. 5, 1-6 (1966); CA 64, 15918.
2002. Suzuki, K., Seki, T. and Kawakami, Y., Japan. 18,816 ('67), Nov. 5, 1964; CA 68, 60282.
2003. ---, --- and ---, Japan. 19,177 ('67), April 1, 1964; CA 68, 79165.
2004. Sweet, R. M., Fritchie, C. J., Jr. and Schunn, R. A., Inorg. Chem. 6, 749-54 (1967); CA 66, 99275.
2005. Syutkina, O. P. et al., Dokl. Akad. Nauk SSSR 177, 615-6 (1967); CA 68, 105317.
2006. Szalay, P. J., U.S. 3,363,028, March 1, 1965; CA 68, 50551.
2007. Tagliavini, G., Anal. Chim. Acta 34, 24-31 (1966); CA 64, 8927.
2008. ---, Ric. Sci., Rend., Sez. A 8, 1533-6 (1965); CA 65, 3071.
2009. --- et al., J. Organometal. Chem. 5, 136-46 (1966); CA 64, 6448.
2010. --- and Doretti, L., Chem. Commun. 1966, 562-3; CA 65, 19862.
2011. ---, Faleschini, S. and Genero, E., Ric. Sci. 36, 717-24 (1966); CA 67, 53417.
2012. ---, Pilloni, G. and Plazzogna, G., ibid. 36, 114-22 (1966); CA 65, 7023.
2013. --- and Zanella, P., Anal. Chim. Acta 40, 33-9 (1968); CA 68, 9172.
2014. --- and ---, J. Organometal. Chem. 5, 299 (1966); CA 64, 14213.
2015. ---, --- and Fiorani, M., Coord. Chem. Rev. 1, 249-54 (1966); CA 66, 38021.
2016. Takeda, T. et al., Japan. 7,695 ('67), July 26, 1963; CA 67, 100661.
2017. --- et al., Japan. 24,044 ('67), Oct. 29, 1964; CA 68, 87885.
2018. --- et al., Japan. 24,047 ('67), Oct. 29, 1964; CA 68, 87887.
2019. Takimoto, H. H. and Rust, J. B., U.S. 3,244,645, Sept. 28, 1960; CA 65, 822.
2020. Tamblyn, J. W. and Caldwell, J. R., U.S. 3,306,920, June 5, 1963; CA 66, 95815; U.S. 3,356,643, Sept. 26, 1966; CA 68, 96558.
2021. Tanaka, T., Inorg. Chim. Acta 1, 217-21 (1967); CA 67, 77622.
2022. Tareeva, A. I. and Borodina, G. M., Farmakol. Toksikol. 30, 207-9 (1967); CA 67, 20351.
2023. Taylor, J. L., U.S. 3,268,395, Feb. 12, 1965; CA 65, 14363.
2024. Telnoi, V. I. and Rabinovich, I. B., Zh. Fiz. Khim. 39, 2314-6 (1965); CA 64, 77.
2025. --- and ---, ibid. 40, 1556-63 (1966); CA 65, 15209.

2026. Tepper, L. B., U.S. Public Health Serv. Publ. No. 1440, 59-62 (1965); CA 65, 11224.
2027. Thayer, J. S. and Strommen, D. P., J. Organometal. Chem. 5, 383-7 (1966); CA 64, 18705.
2028. Thinius, K. and Schlimper, R., Ger. (East) 54,106, Sept. 17, 1963; CA 67, 74190.
2029. Thomas, E. C., Ph. D. Thesis, Stanford Univ., 1966; Diss. Abstr. B27, 2319 (1967); CA 66, 120530.
2030. --- and Laurie, V. W., J. Chem. Phys. 44, 2602-4 (1966); CA 64, 15204.
2031. Thomas, P. S. and Maguire, T. F., Fr. 1,446,021; U.S. 3,385,725, Sept. 1, 1964; CA 66, 66957.
2032. Thompson, C. J. et al., J. Gas Chromatogr. 5, 1-10 (1967); CA 66, 64715.
2033. Thompson, C. F., Brit. 1,078,870, June 23, 1966; CA 67, 91397.
2034. Thompson, J. A. J. and Graham, W. A. G., Inorg. Chem. 6, 1365-9 (1967); CA 67, 54245.
2035. --- and ---, ibid. 6, 1875-9 (1967); CA 67, 100219.
2036. Thompson, P. F., Fr. 1,400,314, July 11, 1963; CA 64, 4203.
2037. Tikhova, N. V., Popova, Z. V. and Razuvaev, G. A., Tr. Khim. Khim. Tekhnol. 1966 (2), 193-7; CA 67, 117502.
2038. Tillyaev, K. S. and Manulkin, Z. M., Tr. Tashkent. Farmatsevt. Inst. 4, 349-53 (1966); CA 68, 13116.
2039. Tirsell, J. B., Ph. D. Thesis, Indiana Univ., Bloomington, 1967; Diss. Abstr. B27, 3451 (1967); CA 67, 86224.
2040. Tobias, R. S. et al., Inorg. Chem. 5, 2052-5 (1966); CA 66, 22814.
2041. --- and Hutcheson, S., J. Organometal. Chem. 6, 535-41 (1966); CA 65, 19978.
2042. Toepsch, H., Belg. 651,846; Brit. 1,035,876; Fr. 1,411,819; Ger. 1,546,248, Aug. 16, 1963; CA 64, 2278.
2043. Tombe, F. J. A. des et al., Chem. Commun. 1966, 914-5; CA 66, 65607.
2044. Trepka, W. J., Belg. 663,321; Brit. 1,032,875; Fr. 1,441,715; U.S. 3,400,114, May 7, 1964; CA 65, 4079.
2044a. ---, Brit. 1,125,631; Fr. 1,466,025; Neth. 66 2265; U.S. 3,393,182, Feb. 23, 1965; CA 66, 3523.
2045. Trueblood, R. C. and Anderson, K., J. Paint Technol. 39, 650-4 (1967); CA 68, 22743.
2046. Tsai, T. T. and Lehn, W. L., J. Org. Chem. 31, 2981-5 (1966); CA 65, 13752.
2047. Tseng, C.-L., Cho, J.-H. and Ma, S.-C., K'o Hsueh T'ung Pao 17 (2), 77-8 (1966); CA 66, 28862.
2048. Tsou, K. C. and Sandler, S. R., U.S. 3,244,637, Oct. 1, 1962; CA 65, 1732.
2049. Tucker, W. P., Inorg. Nucl. Chem. Lett. 4, 83-6 (1968); CA 68, 78375.
2050. Tursunbaev, T. L. and Manulkin, Z. M., Tr. Tashkent. Farmatsevt. Inst. 4, 382-7 (1966); CA 68, 59679.
2051. --- and ---, Zh. Obshch. Khim. 37, 219-22 (1967); CA 66, 95141.

2052. Tyminski, I. J., Ph. D. Thesis, Univ. of New Hampshire, 1967; Diss. Abstr. B28, 2360 (1967); CA 68, 104377.

2053. Tzschach, A. and Reiss, E., J. Organometal. Chem. 8, 255-60 (1967); CA 67, 11555.

2054. Uchinuma, K., Kagaku Kogyo 18, 26-32 (1967); CA 68, 5989.

2055. Ueeda, R. et al., J. Organometal. Chem. 5, 194-7 (1966); CA 64, 9761.

2056. Ugo, R. II et al., Ric. Sci. 36, 253-6 (1966); CA 65, 6512.

2057. Ugo, R., Cenini, S. and Bonati, F., Inorg. Chim. Acta 1, 451-61 (1967); CA 68, 62852.

2058. Ulbricht, K. and Chvalovsky, V., J. Organometal. Chem. 12, 105-13 (1968); CA 68, 91379.

2059. Unanue, A. and Bothorel, P., Bull. Soc. Chim. Fr. 1965, 2827-32; CA 64, 11060.

2060. Van Wazer, J. R. and Groenweghe, L. C. D., N.M.R. Chem., Proc. Symp., Cagliari, Italy 1964, 283-98 (1965); CA 66, 6911.

2061. --- and Moedritzer, K., J. Am. Chem. Soc. 90, 47-52 (1968); CA 68, 95179.

2062. ---, --- and Groenweghe, L. C. D., J. Organometal. Chem. 5, 420-37 (1966); CA 65, 584.

2063. Varsany, D., Ger. 1,246,732, Aug. 12, 1964; CA 67, 90938; Belg. 668,247; Brit. 1,115,047; Fr. 1,449,940; Neth. 65 10,469, Aug. 12, 1964; CA 65, 7218.

2064. Vassiliou, E. and Rochow, E. G., U.S. Clearinghouse Fed. Sci. Tech. Inform., AD 661,201, 28 pp. (1967); CA 68, 78387.

2064a. Vegter, G. C. and Sinnema, F. H., Brit. 1,119,736; Fr. 1,512,005; Neth. 66 9502; U.S. 3,471,421, July 7, 1966; CA 68, 88279.

2065. Verburg, G. B. and Snowden, F. W., Text. Res. J. 37, 367-71 (1967); CA 67, 82894.

2066. Verdol, J. A. et al., Rubber Age 98 (7), 57-64 (1966); CA 65, 10772.

2067. Verdonck, L. and van der Kelen, G. P., Bull. Soc. Chim. Belges 74, 361-9 (1965); CA 64, 191.

2068. --- and ---, ibid. 76, 258-72 (1967); CA 67, 59345.

2069. --- and ---, J. Organometal. Chem. 5, 532-6 (1966); CA 65, 2095.

2070. --- and ---, ibid. 11, 491-7 (1968); CA 68, 73898.

2071. ---, --- and Eeckhaut, Z., ibid. 11, 487-90 (1968); CA 68, 91641.

2072. Vereshchinskii, I. V. et al., U.S.S.R. 181,104, April 23, 1965; CA 65, 8962.

2073. Verschuuren, H. G. et al., Food Cosmet. Toxicol. 4, 35-45 (1966); CA 65, 12764.

2074. Vilarem, M. and Maire, J. C., C. R. Acad. Sci. Paris, Ser. C 262, 480-3 (1966); CA 64, 18697.

2075. Vilkov, L. V. and Mastryukov, V. S., Zh. Strukt. Khim. 6, 811-6 (1965); CA 64, 12512.

2076. Vishnyakov, B. A., Osipov, K. A. and Otopkov, P. P., Izv. Akad. Nauk SSSR, Neorg. Mater. 2, 2234-6 (1966); CA 66, 69550.

2077. Visintin, B., Pepe, A. and Giuseppe, S. A., Ann. Ist. Super. Sanita 1, 767-9 (1965); CA 65, 12343.

2078. Vlasov, V. M., Mirskov, R. G. and Petrova, V. N., Zh. Obshch. Khim. 37, 954-7 (1967); CA 68, 13114.
2079. Volpin, M. E. et al., J. Organometal. Chem. 8, 87-91 (1967); CA 66, 85831.
2080. van der Vondel, D. F. and van der Kelen, G. P., Bull. Soc. Chim. Belges 74, 453-66 (1965); CA 64, 7537.
2081. --- and ---, ibid. 74, 467-78 (1965); CA 64, 7537.
2082. --- and ---, ibid. 74, 618-21 (1965); CA 64, 7537.
2083. Vondy, A. F. and Leuchten, W. E., U.S. 3,243,403, Feb. 27, 1961; CA 64, 19949.
2084. Voronkov, M. G. and Romadane, I., Khim. Geterotsikl. Soedin. 1966, 892-6; CA 67, 32751.
2085. ---, Zelchan, G. I. and Mironov, V. F., U.S.S.R. 190,897, June 16, 1965; CA 68, 69128.
2086. Vyazankin, N. S. et al., Dokl. Akad. Nauk SSSR 166, 99-102 (1966); CA 64, 11245.
--- et al., J. Organometal. Chem. 6, 474-83 (1966); CA 66, 11016.
2087. --- et al., ibid. 7, 353-7 (1967); CA 67, 11561.
--- et al., Zh. Obshch. Khim. 36, 952-3, 2025-6 (1966); CA 65, 8955 and CA 66, 76050, resp.
2088. --- et al., Izv. Akad. Nauk SSSR, Ser. Khim. 1966, 562-4; CA 65, 5483.
2089. --- et al., Zh. Obshch. Khim. 36, 160 (1966); CA 64, 14212.
2090. --- et al., ibid. 37, 2576-80 (1967); CA 68, 87365.
2091. ---, Bochkarev, M. N. and Sanina, L. P., ibid. 36, 166 (1966); CA 64, 14212.
---, --- and ---, ibid. 36, 1961-4 (1966); CA 66, 76114.
2092. ---, --- and ---, ibid. 36, 1154-5 (1966); CA 65, 10617.
---, --- and ---, ibid. 37, 1037-40 (1967); CA 68, 13099.
2093. ---, --- and ---, ibid. 37, 1545-8 (1967); CA 68, 29813.
2094. ---, Brevnova, T. N. and Razuvaev, G. A., ibid. 37, 204-7 (1967); CA 66, 94533.
2095. ---, --- and ---, ibid. 37, 2334-8 (1967); CA 68, 87372.
2096. --- and Bychkov, V. T., ibid. 36, 1684-7 (1966); CA 66, 46460.
2097. ---, Gladyshev, E. N. and Korneva, S. P., ibid. 37, 1736-8 (1967); CA 68, 13101.
2098. ---, Razuvaev, G. A. and Brevnova, T. N., ibid. 35, 2033-6 (1965); CA 64, 6684.
2099. Waack, R., U.S. 3,235,626, March 31, 1961; CA 64, 17746.
2100. ---, U.S. 3,242,105, June 13, 1962; CA 64, 17825.
2101. ---, U.S. 3,305,388, Aug. 28, 1963; CA 66, 97827.
---, U.S. 3,415,627, Aug. 15, 1966; CA 70, 70733.
2102. Wada, M. and Okawara, R., J. Organometal. Chem. 4, 487-8 (1965); CA 64, 751.
2103. --- and ---, ibid. 8, 261-70 (1967); CA 66, 120318.
2104. Wagner, P. J., J. Am. Chem. Soc. 89, 2503-5 (1967); CA 67, 27564.
2105. Wahren, M. et al., Isotopenpraxis 1 (2), 65-8 (1965); CA 65, 743.
2106. Walling, C. et al., J. Am. Chem. Soc. 88, 5361-3 (1966); CA 66, 94482.

2107. Walsh, E. J., Jr. and Kuivila, H. G., J. Am. Chem. Soc. 88, 576-81 (1966); CA 64, 9539.
2108. Walsh, E. N. and Kopacki, A. F., U.S. 3,296,193, Aug. 28, 1962; CA 66, 55981.
--- and ---, U.S. 3,358,006, Aug. 28, 1962; CA 68, 30705.
2109. Walton, D. R. M., J. Chem. Soc., C 1966, 1706-7; CA 65, 16992.
2110. Wardell, J. L., J. Organometal. Chem. 9, 89-98 (1967); CA 67, 47740.
2111. ---, ibid. 10, 53-8 (1967); CA 67, 103238.
2112. Warkentin, J. and Sanford, E., J. Am. Chem. Soc. 90, 1667-8 (1968); CA 68, 104375.
2113. Warner, C. R., Strunk, R. J. and Kuivila, H. G., J. Org. Chem. 31, 3381-4 (1966); CA 65, 18507.
2114. Wartik, T. and Barnes, R. L., U.S. 3,288,828, Feb. 26, 1963; CA 66, 28882.
2115. Washburn, R. M., U.S. 3,232,958, Nov. 3, 1961; CA 64, 17640.
2116. --- and Baldwin, R. A., U.S. 3,311,646, Jan. 11, 1963; CA 67, 12061.
--- and ---, U.S. 3,341,477, Aug. 1, 1966; CA 67, 100586.
--- and ---, U.S. 3,341,478, Aug. 1, 1966; CA 67, 100585.
2117. Watanabe, M. et al., Japan. 22,531 ('67), April 16, 1964; CA 68, 87849.
2118. Watson, J. W., Drakeford, G. E. and Pounder, D. W., Brit. 1,018,035, Feb. 5, 1963; CA 64, 12933.
2119. Weber, H. P. and Bryan, R. F., Chem. Commun. 1966, 443-4; CA 65, 12945.
--- and ---, Acta Crystallogr. 22, 822-36 (1967); CA 67, 26806.
2120. Weichert, D., Plaste Kaut. 14, 798-801 (1967); CA 68, 69648.
2121. Weigert, F. J., Winokur, M. and Roberts, J. D., J. Am. Chem. Soc. 90, 1566-9 (1966); CA 68, 91631.
2122. Weinberg, E. L., Banks, C. K. and Wyluda, B. J., Ger. 1,217,609, Aug. 26, 1954; CA 65, 10755.
2122a. Weisfeld, L. B., Brit. 1,099,106; Fr. 1,475,428; Ger. 1,277,853; Neth. 66 4233; U.S. 3,392,179, April 15, 1965; CA 66, 76156.
2123. Weissenberger, G., U.S. 3,221,036, Jan. 28, 1960; CA 64, 5138.
2124a. ---, U.S. 3,232,735, Dec. 21, 1960; CA 64, 11250.
2124b. ---, U.S. 3,236,727, Dec. 31, 1960; CA 64, 12724.
2125. ---, U.S. 3,275,659, April 28, 1961; CA 65, 20164.
---, U.S. 3,282,672, April 28, 1961; CA 66, 28891.
---, U.S. 3,419,662, April 28, 1961; CA 70, 78148.
2126. ---, U.S. 3,312,725, May 25, 1962; CA 67, 10686.
---, U.S. 3,361,554, May 25, 1962; CA 68, 78424.
2127. Wells, W. L. and Brown, T. L., J. Organometal. Chem. 11, 271-80 (1968); CA 68, 59678.
2128. Wiberg, E., Amberger, E. and Cambesi, H., Z. Anorg. Allg. Chem. 351, 164-79 (1967); CA 67, 54234.
2129. Wieber, M. and Frohning, C. D., Angew. Chem. 78, 1022 (1966); Int. Ed. Engl. 5, 966 (1966); CA 66, 18742.
2130. --- and ---, J. Organometal. Chem. 8, 459-63 (1967); CA 67, 54232.
2131. --- and ---, Z. Naturforsch., B 21, 492 (1966); CA 65, 5482.

2132. Wieber, E., Frohning, C. D. and Schmidt, M., J. Organometal. Chem. 6, 427-8 (1966); CA 66, 18751.
2133. ---, --- and Schwarzmann, G., Z. Anorg. Allg. Chem. 355, 79-82 (1967); CA 68, 13131.
2134. --- and Schwarzmann, G., Monatsh. Chem. 99, 255-60 (1968); CA 68, 78388.
2135. Wild, J. H. and Williams, D., Brit. 1,001,458, Dec. 3, 1962; CA 64, 17804.
2136. Williams, K. C., J. Org. Chem. 32, 4062-3 (1967); CA 68, 22023.
2137. Williamson, W. I., Brit. 1,077,390; Fr. 1,501,220; S. Afr. 66 06,709, Nov. 9, 1965; CA 68, 31196.
2138. Wilson, G. R., U.S. 3,213,119, May 14, 1962; CA 64, 3602.
2139. Wilson, H. W., Anal. Chem. 38, 920-1 (1966); CA 65, 13440.
2140. Winchester, J. W. et al., Atmos. Environ. 1, 105-19 (1967); CA 66, 118546.
2141. Winkler, J., U.S. 3,355,269, June 20, 1966; CA 68, 14665.
2142. Winkler, K. and Korsch, I., Z. Chem. 7 (3), 112-3 (1967); CA 66, 119357.
2143. Woggon, H., Koehler, U. and Uhde, W. J., Nahrung 11, 809-17 (1967); CA 68, 69831.
2144. Wojtkowiak, B. and Queignec, R., C. R. Acad. Sci. Paris, Ser. A, B 262, 811-4 (1966); CA 65, 1594.
2145. Wolfenbarger, D. A., Guerra, A. A. and Lowry, W. L., J. Econ. Entomol. 61, 78-81 (1968); CA 68, 58702.
2146. Woodward, P. et al., J. Am. Chem. Soc. 87, 5251-3 (1965); CA 64, 9217.
2146a. Wowk, A., Brit. 1,105,761; Fr. 1,468,233; Ger. 1,272,292; Neth. 66 2002; U.S. 3,391,174, Feb. 16, 1965; CA 65, 20164.
2147. --- and Digiovanni, S., Anal. Chem. 38 (6), 742-4 (1966); CA 65, 2985.
2148. Wright, C. M. and Muetterties, E. L., Inorg. Synth. 10, 137-9 (1967); CA 68, 78373.
2149. Wright, W. E., Benstead, J. C. and Shimmin, J. D., U.S. 3,324,160, April 12, 1963; CA 67, 73136.
2150. Wylde, J. et al., Bull. Soc. Chim. Fr. 1966, 3134-5; cf. ref. 1472.
2151. Yamada, N. et al., Japan. 617 ('68), April 24, 1964; CA 68, 115183.
2152. --- et al., Japan. 619 ('68), May 7, 1964; CA 68, 105683.
2153. --- and Shimada, K., Japan. 622 ('68), Aug. 13, 1964; CA 68, 115184.
2154. ---, --- and Takemura, T., Japan. 22,054 ('67), March 6, 1964; CA 68, 78743.
2155. ---, --- and ---, Japan. 22,056 ('67), April 24, 1964; CA 68, 78744.
2156. ---, --- and Tsutada, C., Japan. 615 ('68), Dec. 16, 1963; CA 68, 115182.
2156a. Yamamoto, T. et al., Belg. 657,488; Brit. 1,052,969; Fr. 1,418,632; Ger. 1,273,125; U.S. 3,437,620, Dec. 23, 1963; CA 64, 17779.
2157. Yastrebov, V. V. and Chernyshev, A. I., Zh. Obshch. Khim. 37, 2140-1 (1967); CA 68, 29823.
2158. Yasuda, K., Matsumoto, H. and Okawara, R., J. Organometal. Chem. 6, 528-34 (1966); CA 65, 20158.

2159. Yates, K. and Agolini, F., Can. J. Chem. 44, 2229-31 (1966); CA 65, 19987.
2160. Yoder, C. H., Ph. D. Thesis, Cornell Univ., 1966; Diss. Abstr. B27, 1801-2 (1966); CA 66, 115769; cf. ref. 1731, 2161-3.
2161. --- and Zuckerman, J. J., Inorg. Chem. 5, 2055-7 (1966); CA 66, 23965; cf. ref. 2160.
2162. --- and ---, J. Am. Chem. Soc. 88, 2170-4 (1966); CA 65, 737; cf. ref. 2160.
2163. --- and ---, ibid. 88, 4831-9 (1966); CA 65, 20222; cf. ref. 2160.
2164. Yokoo, A. and Wada, M., Japan. 8015 ('67), March 28, 1964; CA 67, 108759.
2165. --- and ---, Japan. 8366 ('67), March 28, 1964; CA 67, 108760.
2166. Yokoo, M. et al., Brit. 1,078,897, Sept. 18, 1964; CA 67, 100705.
2167. ---, Ogura, J. and Kanzawa, T., J. Polym. Sci., Pt. B 5, 57-63 (1967); CA 66, 66100.
2168. Yokoo, T. and Wada, M., Japan. 19,414 ('66), March 28, 1964; CA 66, 46485.
2169. --- and Yasuda, K., Japan. 10,888 ('66), Nov. 6, 1963; CA 65, 12240.
2170. Yoneda, H., Rezende, L. O. C. and Piedade, J. C., Biologico 32, 275-6 (1966); CA 66, 94012.
2171. Yoshida, Z. and Miyoshi, H., Kogyo Kagaku Zasshi 68, 580-2 (1965); CA 64, 9890.
2172. Younghouse, E. C. and Detweiler, W. K., U.S. 3,236,772, July 2, 1962; CA 64, 15658.
2173. Zablotna, R., Bull. Acad. Pol. Sci., Ser. Chim. 14, 835-41 (1966); CA 67, 25936.
2174. ---, Akerman, K. and Szuchnik, A., ibid. 13, 527-38 (1965); CA 64, 5131.
2175. ---, --- and ---, ibid. 14, 731-5 (1966); CA 66, 76108.
2176. ---, --- and ---, Pol. 51,899, Aug. 10, 1964; CA 67, 108762.
2177. Zabryanskii, E. I., Khim. Tekhnol. Topl. Masel 12, 44-7 (1967); CA 67, 83607.
2178. Zahnradnicek, J., Schmidt, L. and Havranek, A., Listy Cukrov. 83, 193-7 (1967); CA 68, 14258.
2179. Zavgorodnii, V. S., Ionin, B. I. and Petrov, A. A., Zh. Obshch. Khim. 37, 949-53 (1967); CA 68, 39754.
2180. --- and Petrov, A. A., ibid. 36, 1480-3 (1966); CA 66, 2624.
2181. ---, Sharanina, L. G. and Petrov, A. A., ibid. 37, 1548-53 (1967); CA 68, 39755.
2182. Zedler, R. J., Trav. Centre Rech. Etud. Oceanogr. (Paris) 6 (1-2-3-4) 401-5 (1965); CA 66, 105934.
2183. ---, Brit. 1,022,025, Nov. 6, 1961; CA 64, 20554.
2184. Zemlyanskii, N. N. et al., Izv. Akad. Nauk SSSR, Ser. Khim. 1967, 728-35; CA 67, 54235.
2185. Zhdanov, A. A. et al., Vysokomol. Soedin., Ser. B 9, 373-6 (1967); CA 68, 3641.

2186. Ziegler, K. and Lehmkuhl, H., Ger. 1,212,085, May 12, 1961; CA 64, 19675; cf. ref. 1401.
2187. Zimmer, H. and Homberg, O. A., J. Org. Chem. 31, 947-9 (1966); CA 64, 14212; cf. ref. 1199.
2188. ---, --- and Jayawant, M., ibid. 31, 3857-60 (1966); CA 66, 28867; cf. ref. 1199.
2189. --- and Miller, J. J., Naturwissenschaften 53, 38 (1966); CA 64, 12719; cf. ref. 1518.
2190. Zimpel, C. F. and Graiff, L. B., Symp. Combust. 11, 1015-25 (1966); CA 68, 70957.
2191. Zubov, P. I. et al., U.S.S.R. 181,289, May 15, 1964; CA 65, 13843.
2192. Zueva, G. Ya. et al., Izv. Akad. Nauk SSSR, Neorgan. Materialy 2, 229-38 (1966); CA 64, 17631.
2193. --- et al., ibid. 2, 1359-66 (1966); CA 66, 2625.
2194. --- et al., Izv. Akad. Nauk SSSR, Ser. Khim. 1966, 1843-5; CA 66, 65601.
2195. ---, Lukyankina, N. V. and Ponomarenko, V. A., ibid. 1967, 192-4; CA 66, 95140.
2196. --- and Ponomarenko, V. A., Izv. Akad. Nauk SSSR, Neorgan. Materialy 2, 472-7 (1966); CA 65, 16996.
2197. Zuliani, G., Perin, G. and Rausa, G., Med. Lab. 57, 771-80 (1966); CA 67, 25195.
2198. Abel, E. W. et al., J. Chem. Soc., A 1968, 1203-8; CA 69, 14333.
2199. --- and Crow, J. P., ibid. A 1968, 1361-3; CA 69, 15509.
2200. ---, --- and Illingworth, S. M., Chem. Commun. 1968, 817; CA 69, 55910.
2201. ---, Hoenigschmid-Grossich, R. and Illingworth, S. M., J. Chem. Soc., A 1968, 2623-5; CA 70, 11774.
2202. --- and Hutson, G. V., J. Inorg. Nucl. Chem. 30, 2339-44 (1968); CA 69, 113003.
2203. ---, Walker, D. J. and Wingfield, J. N., J. Chem. Soc., A 1968, 1814-6; CA 69, 67463.
2204. ---, Ison, R. R. and Newbold, G. T., S. African 67 7,058, Dec. 10, 1966; CA 70, 96958.
2205. Abraham, M. H. and Spalding, T. R., J. Chem. Soc., A 1968, 2530-5; CA 69, 105522.
2206. Adicoff, A. and Yukelson, A. A., J. Appl. Polym. Sci. 12 (8), 1959-66 (1968); CA 69, 67777.
2207. Agafonov, I. L. et al., Izv. Akad. Nauk SSSR, Ser. Khim. 1968, 1289-93; CA 69, 62794.
2208. Akhtar, M. and Clark, H. C., Can. J. Chem. 46, 633-42 (1968); CA 68, 78376.
2209. --- and ---, ibid. 46, 2165-73 (1968); CA 69, 36236.
2210. Alcock, N. W. and Timms, R. E., J. Chem. Soc., A 1968, 1873-6; CA 69, 62442.
2211. --- and ---, ibid. A 1968, 1876-8; CA 69, 62443.
2212. Aleksandrov, A. Yu. et al., Khim. Vys. Energ. 2, 331-7 (1968); CA 69, 59783.
2213. --- et al., Vysokomol. Soedin., Ser. B 10, 209-12 (1968); CA 69, 3295.

2214. Aleksandrov, Yu. A., Fomin, V. M. and Spiridonova, M. N., Zh. Obshch. Khim. 38, 1410 (1968); CA 69, 76213.
2215. ---, Glushakova, V. N. and Radbil, B. A., Tr. Khim. Khim. Tekhnol. 1967, 69-74; CA 69, 27506.
2216. --- and Radbil, B. A., Zh. Obshch. Khim. 37, 2345-50 (1967); CA 69, 2276.
2217. --- and ---, ibid. 38, 499-504 (1968); CA 69, 66686.
2218. --- and ---, ibid. 38, 1356-60 (1968); CA 69, 85916.
2219. ---, Sheyanov, N. G. and Shushunov, V. A., ibid. 38, 1352-6 (1968); CA 69, 77388.
2220. --- and Spiridonova, M. N., Dokl. Akad. Nauk SSSR 182, 1319-21 (1968); CA 70, 23361.
2221. Aleksandrova, Yu. V., Tarakanov, O. G. and Kryuchkov, F. A., Plast. Massy 1968 (10), 26-8; CA 70, 20609.
2222. Allison, W. E. et al., J. Econ. Entomol. 61, 1254-7 (1968); CA 69, 95401.
2223. Allred, A. L. and Bush, L. W., J. Am. Chem. Soc. 90, 3352-60 (1968); CA 69, 31927.
2224. Amberger, E. and Muehlhofer, E., J. Organometal. Chem. 12, 55-62 (1968); CA 69, 19273.
2225. ---, Roemer, R. and Layer, A., ibid. 12, 417-23 (1968); CA 69, 7913.
2226. Anisimov, K. N. et al., Izv. Akad. Nauk SSSR, Ser. Khim. 1968, 1024-30; CA 69, 47717.
2227. ---, Kolobova, N. E. and Antonova, A. B., ibid. 1968, 2664; CA 70, 53526.
2228. Antipin, L. M., Mironov, V. F. and Sobolev, E. S., U.S.S.R. 215,995, Dec. 1, 1966; CA 69, 59373; cf. ref. 2603.
2229. Armenskaya, L. V. et al., Khim. Prom. (Moscow) 44, 665-7 (1968); CA 69, 102461.
2230. Asher, K. S. R., Meisner, J. and Nissim, S., World Rev. Pest Contr. 7 (2), 84-96 (1968); CA 70, 46403.
2231. Asquith, D., J. Econ. Entomol. 61, 1044-6 (1968); CA 69, 58584.
2232. Asso, M., Rev. Chim. Miner. 5, 1051-84 (1968); CA 70, 81532.
--- and Carpeni, G., Can. J. Chem. 46, 1795-1802 (1968); CA 69, 22520.
2233. Auel, T. and Amma, E. L., J. Am. Chem. Soc. 90, 5941-2 (1968); CA 70, 23899.
2234. Baekelmans, P. et al., Bull. Soc. Chim. Belges 77, 85-97 (1968); CA 69, 58655.
2235. Balabanov, N. P. et al., Nauch. Tr. Vissh. Pedagog. Inst., Plodiv, Mat., Fiz., Khim., Biol. 5 (2), 49-58 (1967); CA 69, 48278.
2236. Bamberg, P., Ekstrom, B. and Sjoberg, B., Acta Chem. Scand. 22, 367-9 (1968); CA 69, 27318.
---, --- and ---, Fr. 1,507,866, Jan. 13, 1966; CA 70, 47434.
2237. Barnes, G. L., Okla. Agr. Exp. Sta. Processed Ser. 1968, No. P-595, 13 pp.; CA 70, 36614.
2238. Baukov, Yu. I. et al., Zh. Obshch. Khim. 38, 1899-900 (1968); CA 70, 4252.

2239. Beatty, H. A., Chem. Ind. (London) 1968, 733-6; CA 69, 37620.
2240. Belikin, A. V. and Grinshtein, S. A., U.S.S.R. 227,326, June 26, 1967; CA 70, 106656.
2241. Bennett, S. W. et al., J. Organometal. Chem. 15, P17 (1968); CA 70, 106617.
2242. Berger, A., U.S. 3,401,183, Dec. 23, 1965; CA 70, 29058.
2243. van den Berghe, E. V., van der Kelen, G. P. and Albrecht, J., Inorg. Chim. Acta 2, 89-92 (1968); CA 69, 67499.
2244. Berkowitz, G., Ph. D. Thesis, Univ. of Pennsylvania, 1968; Diss. Abstr. B29, 2334-5 (1969); CA 70, 106623.
2245. Berrios-Duran, L. A., Ritchie, L. S. and Wessel, H. B., Bull. WHO 39, 316-20 (1968); CA 70, 2700.
2246. Besolova, E. A., Foss, V. L. and Lutsenko, I. F., Zh. Obshch. Khim. 38, 1574-8 (1968); CA 69, 106849.
2247. Bickmore, J. T., U.S. 3,394,002, Oct. 21, 1964; CA 69, 63570.
2248. Birk, J. P., Halpern, J. and Pickard, A. L., Inorg. Chem. 7, 2672-3 (1968); CA 70, 23343.
2249. Biryukov, B. P. et al., Chem. Commun. 19, 1193-4 (1968); CA 69, 111040.
2250. --- et al., Zh. Strukt. Khim. 9, 922-5 (1968); CA 70, 15111.
2251. ---, Solodova, O. P. and Struchkov, Yu. T., ibid. 9, 228-30 (1968); CA 69, 46843.
2252. --- and Struchkov, Yu. T., ibid. 9, 488-502 (1968); CA 69, 62421.
2253. Blackmore, T. et al., J. Chem. Soc., A 1968, 2931-6; CA 70, 25320.
2254. Bleidelis, J. et al., Khim. Geterotsikl. Soedin. 1968, 184-5; CA 69, 100650.
2255. Bliznyuk, N. K., Streltsov, R. V. and Kiricina, L. E., U.S.S.R. 212,259, Nov. 23, 1966; CA 69, 44023.
2256. Bloodworth, A. J., J. Chem. Soc., C 1968, 2380-5; CA 69, 87134.
2257. ---, Davies, A. G. and Graham, I. F., J. Organometal. Chem. 13, 351-6 (1968); CA 69, 77382.
2258. ---, --- and Vasishtha, S. C., J. Chem. Soc., C 1968, 2640-6; CA 70, 4249.
2259. Boehland, H. and Niemann, E., Z. Chem. 8, 191-2 (1968); CA 69, 36238.
2260. Bokii, N. G. and Struchkov, Yu. T., Zh. Strukt. Khim. 9, 838-44 (1968); CA 70, 51781.
2261. Bolanowska, W., Brit. J. Ind. Med. 25, 203-8 (1968); CA 69, 75271.
2262. --- and Garczynski, H., Med. Pracy 19, 235-43 (1968); CA 70, 2298.
2263. Borisova, A. E. et al., Ukr. Fiz. Zh. 13, 75-82 (1968); CA 69, 14351; cf. ref. 763.
2264. ---, Borisova, A. I. and Kudryatseva, L. V., Izv. Akad. Nauk SSSR, Ser. Khim. 1968, 2287-90; CA 70, 29023.
2265. Bork, V. A. and Selivokhin, P. I., Plast. Massy 1968 (4), 56-7; CA 69, 10519.
2266. Boue, S. et al., J. Organometal. Chem. 15, 267-8 (1968); CA 70, 10854.
2267. ---, Gielen, M. and Nasielski, J., Bull. Soc. Chim. Belges 76, 559-65 (1967); CA 69, 14689.
2268. ---, --- and ---, ibid. 77, 43-58 (1968); CA 69, 35234.

2269. Boue, S., Gielen, M. and Nasielski, J., Tetrahedron Lett. 1968, 1047-8; CA 69, 43996.
2270. Braithwaite, D. G., U.S. 3,391,066, March 8, 1961; CA 69, 67537.
2271. ---, U.S. 3,391,067, March 6, 1961; CA 69, 64148.
2272. ---, U.S. 3,409,518, Jan. 6, 1966; CA 70, 37180.
2273. --- et al., U.S. 3,408,273, March 11, 1964; CA 70, 96952.
2274. --- and Bott, L. L., U.S. 3,380,889, Oct. 16, 1964; CA 69, 15414.
2275. Braun, D. B., U.S. 3,393,164, Sept. 23, 1964; CA 69, 59955.
2276. Braun, D. and Heimes, H. T., Z. Anal. Chem. 239, 6-14 (1968); CA 69, 56842.
2277. ---, Thallmaier, M. and Hepp, D., Angew. Makromol. Chem. 2, 71-85 (1968); CA 69, 10904.
2278. Brazhnikov, V. V. and Sakodynskii, K. I., J. Chromatogr. 38, 244-9 (1968); CA 70, 23107.
2279. Bresadola, S., Rossetto, F. and Tagliavini, G., Ann. Chim. (Rome) 58, 597-602 (1968); CA 69, 77385.
2280. Bretschneider, H., Veith, F. and Sextl, H., Ger. 1,271,113, Dec. 3, 1966; CA 69, 77504.
2281. Brook, A. G. and Anderson, D. G., Can. J. Chem. 46, 2115-8 (1968); CA 69, 36233.
2282. ---, Harrison, A. G. and Jones, P. F., ibid. 46, 2862-4 (1968); CA 69, 86108.
2283. ---, Pannell, K. H. and Anderson, D. G., J. Am. Chem. Soc. 90, 4374-7 (1968); CA 69, 87068.
2284. Brooks, E. H. et al., ibid. 90, 3587-8 (1968); CA 69, 77433.
2285. ---, Cross, R. J. and Glockling, F., Inorg. Chim. Acta 2, 17-21 (1968); CA 69, 24122.
2286. Brotherton, J., Brit. J. Dermatol. 80, 749-52 (1968); CA 70, 76698.
2287. Brotherton, T. K. and Smith, J., Jr., U.S. 3,422,165, Dec. 8, 1964; CA 70, 69009.
2288. Brown, D. S., Drago, R. S. and Bolles, T. F., J. Am. Chem. Soc. 90, 5706-12 (1968); CA 69, 101582.
2289. Bryan, R. F. and Manning, A. R., Chem. Commun. 1968, 1220-1; CA 70, 15106.
2290. Bryuchova, E. V. et al., ibid. 1968, 491-3; CA 69, 14709.
2291. Bullock, R. C. and Johnson, R. B., Fla. Entomol. 51, 223-7 (1968); CA 70, 76718.
2292. Bulten, E. J. and Noltes, J. G., J. Organometal. Chem. 15, P18-20 (1968); CA 71, 13187.
2293. Bushin, V. V., Litvishkov, Z. V. and Ivanova, A. A., Ukr. Fiz. Zh. (Ukr. Ed.) 12, 1038-40 (1967); CA 69, 62171.
2294. Butzert, H., Brennst.-Chem. 49, 283-6 (1968); CA 70, 82116.
--- and Beckey, H. D., Z. Phys. Chem. (Frankfurt) 62, 83-102 (1968); CA 70, 77074.
2295. Bychkov, V. T. et al., Izv. Akad. Nauk SSSR, Ser. Khim. 1968, 2141-3; CA 70, 20180.
2296. Bykov, G. V., ibid. 1968, 1773-9; CA 69, 111908.

2297. Byrdy, S., Rocz. Nauk Roln., Ser. A 93, 789-93 (1968); CA 69, 51184.
——, Ejmocki, Z. and Eckstein, Z., Meded. Rijksfac. Landbouwwetensch., Gent 31, 876-88 (1966); CA 69, 66441.
2298. Caldwell, J. R. and Gilkey, R., U.S. 3,399,073, July 22, 1964; CA 69, 87711.
2299. Callear, A. B. and Oldman, R. J., Spectrosc. Lett. 1, 149-51 (1968); CA 69, 47950.
2300. Cardarelli, N. F., U.S. 3,417,181, Feb. 21, 1966; CA 70, 38700.
2301. ——, Fr. 1,506,704, Dec. 20, 1965; CA 70, 12741.
2302. —— and Caprette, S. J., Jr., U.S. 3,426,473, March 31, 1966; CA 70, 69352.
2303. Cardin, D. J., Keppie, S. A. and Lappert, M. F., Inorg. Nucl. Chem. Lett. 4, 365-9 (1968); CA 69, 48798.
2304. Carlson, T. A. and White, R. M., J. Chem. Phys. 48, 5191-4 (1968); CA 69, 55467.
2305. Carlsson, D. J. and Ingold, K. U., J. Am. Chem. Soc. 90, 1055-6 (1968); CA 69, 35136.
2306. —— and ——, ibid. 90, 7047-55 (1968); CA 70, 19359.
2307. Carre, F. H., Corriu, R. J. P. and Thomassin, R. B., Chem. Commun. 1968, 560; CA 69, 43218.
2308. Case, L. C., U.S. 3,382,217, Jan. 6, 1964; CA 69, 10919.
2309. Challen, S. B. and Kucera, M., J. Chromatogr. 31, 345-53 (1967); CA 69, 9876.
2310. Chan, J. K. and Marcus, E., Chem. Ind. (London) 1968, 1767-8; CA 71, 13183.
2311. Chandra, G. and Lappert, M. F., J. Chem. Soc., A 1968, 1940-5; CA 69, 67510.
2312. Chang, S. B., Kungnip Kongop Yonguso Pogo 16, 197-201 (1966); CA 69, 19939.
2313. Channon, A. G. and Hampson, R. J., Ann. Appl. Biol. 62, 23-33 (1968); CA 69, 105287.
2314. Chatalic, A., Deschamps, P. and Pannetier, G., C. R. Acad. Sci., Paris, Ser. C 267, 948-50 (1968); CA 70, 24410.
2315. Cherkasov, L. N. and Zavgorodnii, V. S., Zh. Obshch. Khim. 38, 2812-3 (1968); CA 70, 78099.
2316. Chernyshev, E. A. et al., ibid. 38, 504-8 (1968); CA 69, 51331.
2317. Chiddix, M. E., Tracy, D. J. and Shah, V., U.S. 3,407,140, Aug. 5, 1966; CA 70, 12295.
2318. Childers, S., U.S. 3,391,094, Sept. 9, 1964; CA 69, 36813.
2319. Cholak, J., Schafer, L. J. and Yeager, D., Am. Ind. Hyg. Ass. J. 29, 562-8 (1968); CA 70, 70877.
2320. Chromy, L. and Uhacz, K., J. Oil Colour Chem. Ass. 51, 494-8 (1968); CA 69, 11445.
2321. Chromy, V. and Srp, L., Chem. Listy 61, 1509-12 (1967); CA 69, 16067.
2322. Clark, R. J. H., Davies, A. G. and Puddephatt, R. J., J. Am. Chem. Soc. 90, 6923-7 (1968); CA 70, 15519.
2323. ——, —— and ——, J. Chem. Soc., A 1968, 1824-34; CA 69, 64229.

2324. Cohen, J. H., U.S. 3,334,119, Jan. 17, 1964; CA 69, 77511.
2325. Cohen, S. C. and Massey, A. G., J. Organometal. Chem. 12, 341-7 (1968); CA 69, 52253.
2326. Considine, W. J. and Reifenberg, G. H., U.S. 3,412,120, Jan. 3, 1966; CA 70, 96951.
2327. --- and ---, U.S. 3,412,122, Jan. 3, 1966; CA 70, 58027.
2328. Cotton, J. D., Bruce, M. I. and Stone, F. G. A., J. Chem. Soc., A 1968, 2162-5; CA 69, 82974.
2329. Coyle, T. D. and Ritter, J. J., J. Organometal. Chem. 12, 269-80 (1968); CA 69, 19240.
2330. Creemers, H. M. J. C., Noltes, J. G. and van der Kerk, G. J. M., ibid. 14, 217-21 (1968); CA 70, 11772.
2331. ---, Verbeck, F. and Noltes, J. G., ibid. 15, 125-30 (1968); CA 70, 11784.
2332. Cullen, W. R. and Waldman, M. C., Inorg. Nucl. Chem. Lett. 4, 205-7 (1968); CA 69, 10523.
2333. Czekanski, J. and Smith, J., Brit. 1,108,182, Apr. 23, 1964; CA 69, 2090.
2334. Czerwinska, E. et al., Polimery 13, 355-60 (1968); CA 70, 78707.
2335. Dacus, A. M., U.S. 3,395,228, Oct. 10, 1966; CA 69, 66478.
2336. Dall'Asta, G., J. Polym. Sci., Pt. A-1 6, 2379-404 (1968); CA 69, 67786.
2337. Dalton, J. et al., J. Chem. Soc., A 1968, 1195-9; CA 69, 14331.
2338. --- et al., ibid., A 1968, 1199-202; CA 69, 14332.
2339. ---, Paul, I. and Stone, F. G. A., ibid., A 1968, 1215-17; CA 69, 14336.
2340. Das, V. G. K. and Kitching, W., J. Organometal. Chem. 13, 523-8 (1968); CA 69, 77381.
2341. Davidson, J. M. and Dyer, G., J. Chem. Soc., A 1968, 1616-7; CA 69, 52254.
2342. Davies, A. G., U.S. 3,396,167, Sept. 10, 1965; CA 69, 96784.
2343. ---, U.S. 3,417,117, Oct. 26, 1964; CA 70, 78149.
2344. --- et al., Chem. Ind. (London) 1968, 949-50; CA 69, 77392.
2345. --- and Kennedy, J. D., J. Chem. Soc., C 1968, 2630-40; CA 70, 4248.
2346. --- and Puddephatt, R. J., ibid., C 1968, 1479-83; CA 69, 27507.
2347. Debye, N. W. G., Rosenberg, E. and Zuckerman, J. J., J. Am. Chem. Soc. 90, 3234-6 (1968); CA 69, 23462.
2348. Delmond, B. and Pommier, J. C., Tetrahedron Lett. 1968, 6147-8; CA 70, 57967.
2349. Demmer, F., Mater. Organismen 3, 19-58 (1968); CA 69, 67971.
2350. Deschiens, R., C. R. Acad. Sci., Paris, Ser. D 266, 1860-1 (1968); CA 69, 42995.
2351. Devaud, M., Rev. Chim. Miner. 4, 921-35 (1967); CA 69, 87122.
2352. --- and Laviron, E., ibid. 5, 427-58 (1968); CA 69, 112852.
2353. Diekman, J., Thomson, J. B. and Djerassi, C., J. Org. Chem. 33, 2271-85 (1968); CA 69, 26607.
2354. Doerfelt, C. et al., J. Organometal. Chem. 14, P22-4 (1968); CA 69, 91003.

2355. Dolan, D. N. and Nickless, G., J. Chromatogr. 37, 1-13 (1968); CA 70, 20184.
2356. Donadille, M. et al., J. Organometal. Chem. 15, 244-6 (1968); CA 70, 7965.
2357. Donaldson, D. et al., Brit. 1,129,889, April 29, 1966; CA 70, 30015.
2358. --- et al., Brit. 1,129,900; U.S. 3,420,700, April 29, 1966; CA 70, 30014 and 50873, resp.
2359. --- et al., U.S. 3,420,701, April 29, 1966; CA 70, 58871.
2360. Doretti, L. and Tagliavini, G., J. Organometal. Chem. 13, 195-8 (1968); CA 69, 56529.
2361. Dreeskamp, H. and Schumann, C., Chem. Phys. Lett. 1, 555-6 (1968); CA 69, 14698.
2362. Dubac, J. and Mazerolles, P., C. R. Acad. Sci., Paris, Ser. C 267, 411-3 (1968); CA 69, 106844.
2363. Durig, J. R., Gibson, B. M. and Sink, C. W., J. Mol. Struct. 2, 1-17 (1968); CA 69, 47829.
2364. ---, Sink, D. W. and Turner, J. B., J. Chem. Phys. 49, 3422-41 (1968); CA 70, 7664.
2365. Dutton, W. A. and Onyszchuk, M., Inorg. Chem. 7, 1735-9 (1968); CA 69, 67455.
2366. Dzhurinskaya, N. G., Mikhailyants, S. A. and Evdakov, V. P., Zh. Obshch. Khim. 38, 1267-9 (1968); CA 69, 77380.
2367. Eaborn, C. et al., J. Organometal. Chem. 15, 241-3 (1968); CA 70, 19437.
2368. ---, Hill, R. E. E. and Simpson, P., Chem. Commun. 1968, 1077-8; CA 69, 105715.
2369. ---, --- and ---, J. Organometal. Chem. 15, P1-3 (1968).
2370. ---, Thompson, A. R. and Walton, D. R. M., Chem. Commun. 1968, 1051-2; CA 69, 92497
2371. Edmondson, R. C. and Newlands, M. J., ibid. 1968, 1219-20; CA 70, 11798.
2372. Eggensperger, H. et al., Fr. 1,529,957, May 24, 1966; CA 70, 115906.
2373. Egorochkin, A. N. et al., Dokl. Akad. Nauk SSSR 180, 861-4 (1968); CA 69, 72660.
2374. --- et al., Zh. Obshch. Khim. 38, 276-9 (1968); CA 69, 6600.
2375. --- et al., ibid. 38, 396-400 (1968); CA 69, 76371.
2376. Einstein, F. W. B. and Penfold, B. R., J. Chem. Soc., A 1968, 3019-24; CA 70, 23904.
2377. Elder, M. et al., J. Am. Chem. Soc. 90, 2189-90 (1968); CA 69, 43297.
2378. Elguero, J., Riviere-Baudet, M. and Satge, J., C. R. Acad. Sci., Paris, Ser. C 266, 44-7 (1968); CA 69, 19277.
2379. Ellison, R. J., Harrow, G. A. and Hayward, B. M., J. Inst. Petrol. 54, 243-50 (1968); CA 70, 5677.
2380. Emelyanov, B. V., Shemyakina, Z. N. and Shvarov, V. N., Khim. Prom. 44, 498-500 (1968); CA 69, 79964.
2381. Epstein, S. S. and Mantel, N., Experimentia 24, 580-1 (1968); CA 69, 17544.
2382. Ermanson, L. V. et al., Izv. Akad. Nauk SSSR 1968, 2664; CA 70, 57969.

2383. Fadeeva, M. S., Kuznetsov, V. A. and Kaplin, Yu. A., Zh. Prikl. Spektrosk. 9, 489-91 (1968); CA 70, 24470.

2384. Fantazier, R. M. and Poutsma, M. L., J. Am. Chem. Soc. 90, 5490-8 (1968); CA 69, 105576.

2385. Faraglia, G., Cassol, A. and Barbieri, R., Chromatogr. Methods Immed. Separ., Proc. Meet., Athens 1965, 2, 271-6; CA 69, 46300.

2386. Fath, J. and Deardorff, D. L., Brit. 1,129,725, July 26, 1966; CA 70, 12306.

2387. Feehs, R. H., U.S. 3,413,328, June 1, 1966; CA 70, 58024.

2388. Fenster, A. N., Ph. D. Thesis, Univ. Massachusetts, 1968; Diss. Abstr. B29, 103-4 (1968); CA 70, 4246; cf. ref. 1030.

2389. Fenton, D. E. and Zuckerman, J. J., J. Am. Chem. Soc. 90, 6226-8 (1968); CA 70, 7945.

2390. Finelonov, V. P. et al., Neftepererab. Neftekhim. (Moscow) 1968 (6), 8-10; CA 69, 68740.

2391. Fish, R. H. and Le Fevre, C. W., Fr. 1,499,737; U.S. 3,422,127, Nov. 15, 1965; CA 70, 37917.

2392. Fitzsimmons, B. W., Chem. Commun. 1968, 1485; CA 70, 36938.

2393. Flerov, V. N. and Tyurin, Yu. M., Zh. Obshch. Khim. 38, 1669-76 (1968); CA 69, 102484.

2394. Ford, B. F. E., Liengme, B. V. and Sams, J. P., Chem. Commun. 1968, 1333-4; CA 70, 19433.

2395. Frankel, M. et al., Israel J. Chem. 6, 817-21 (1968); CA 70, 96881.

2396. Friker, H. H., Altrogge, K. and Heuse, O., U.S. 3,415,935, Dec. 14, 1963; CA 70, 48645; Belg. 657,068; Fr. 1,419,988; Ger. 1,288,232; Neth. 64 14,475, Dec. 14, 1964; CA 65, 20024.

2397. Gaddy, R. H., Jr., U. S. At. Energy Comm. DP-1150, 12 pp. (1968); CA 70, 74108.

2398. Gaines, D. F. and Iorns, T. V., J. Am. Chem. Soc. 90, 6617-21 (1968); CA 70, 8522.

2399. Gaines, T. B. and Kimbrough, R. D., Toxicol. Appl. Pharmacol. 12, 397-403 (1968); CA 69, 43017.

2400. Galasso, V., De Alti, G. and Bigotto, A., Z. Phys. Chem. (Frankfurt) 57, 132-7 (1968); CA 69, 22618.

2401. Galli, R., Chim. Ind. (Milan) 50, 977-82 (1968); CA 70, 16599.

2402. Galzigna, L., Corsi, G. C. and Terribile, P. M., Boll. Soc. Ital. Biol. Sper. 64, 659-60 (1968); CA 69, 58084.

2403. Gar, T. K., Leites, L. A. and Mironov, V. F., Izv. Akad. Nauk SSSR, Ser. Khim. 1968, 1336-42; CA 69, 96849.

2404. Gaunt, I. F. et al., Food Cosmet. Toxicol. 6, 599-608 (1968); CA 70, 66361.

2405. Geissler, H., Radegia, R. and Kriegsmann, H., J. Organometal. Chem. 15, 349-57 (1968); CA 70, 23560.

2406. George, T. A. and Lappert, M. F., ibid. 14, 327-37 (1968); CA 70, 4244.

2407. Gertner, D., Migdal, S. and Zilkha, A., Brit. 1,107,929, Nov. 14, 1965; CA 69, 3461.

2408. Gevorkyan, A. A. and Sarksyan, Zh. G., Arm. Khim. Zh. $\underline{21}$, 269-70 (1968); CA $\underline{69}$, 106832.

2409. Gibbon, G. A., Ph. D. Thesis, Carnegie-Mellon Univ., 1968; Diss. Abstr. B$\underline{29}$, 2338 (1969); CA $\underline{70}$, 92714; cf. ref. 910, 1095.

2410. Gielen, M. and Mayence, G., J. Organometal. Chem. $\underline{12}$, 363-8 (1968); CA $\underline{69}$, 18428.

2411. --- and Nasielski, J., Bull. Soc. Chim. Belg. $\underline{77}$, 5-14 (1968); CA $\underline{69}$, 35233.

2412. ---, --- and Topart, J., Rec. Trav. Chim. $\underline{87}$, 1051-2 (1968); CA $\underline{70}$, 11779.

2413. Girko, I. P. and Shurupova, O. V., Khim. Tekhnol. Topl. Masel $\underline{13}$ (9), 56 (1968); CA $\underline{69}$, 108334.

2414. Gladyshev, E. N. et al., Dokl. Akad. Nauk SSSR $\underline{179}$, 1333-5 (1968); CA $\underline{69}$, 67518.

2415. --- et al., ibid. $\underline{183}$, 338-40 (1968); CA $\underline{70}$, 57962.

2416. --- et al., Zh. Obshch. Khim. $\underline{38}$, 662-3 (1968); CA $\underline{69}$, 10527.

2417. Glass, G. E., Ph. D. Thesis, Univ. of Minnesota, 1968; Diss. Abstr. B $\underline{29}$, 919-20 (1968); CA $\underline{70}$, 8520.

---, Schwabacher, W. B. and Tobias, R. S., Inorg. Chem. $\underline{7}$, 2471-8 (1968); CA $\underline{70}$, 33144.

2418. --- and Tobias, R. S., J. Organometal. Chem. $\underline{15}$, 481-90 (1968); CA $\underline{70}$, 23319.

2419. Glockling, F., Light, J. R. C. and Walker, J., Chem. Commun. $\underline{1968}$, 1052-3; CA $\underline{69}$, 91642.

2420. --- and Wilbey, M. D., J. Chem. Soc., A $\underline{1968}$, 2168-71; CA $\underline{69}$, 83036.

2421. Gras, G., S. African 67 01,320, March 24, 1966; CA $\underline{70}$, 80841.

2422. --- and Boucard, M., C. R. Soc. Biol. $\underline{162}$, 1456-8 (1968); CA $\underline{70}$, 104724.

2423. Green, M., Taunton-Rigby, A. and Stone, F. G. A., J. Chem. Soc., A $\underline{1968}$, 2762-5; CA $\underline{70}$, 25325.

2424. Greenwood, N. N., Perkins, P. G. and Wall, D. H., Symp. Faraday Soc. No. 1, 51-9 (1967); CA $\underline{69}$, 101526.

2425. Gverdtsiteli, I. M. and Baramidze, L. V., Zh. Obshch. Khim. $\underline{38}$, 1598-601 (1968); CA $\underline{69}$, 87076.

2426. ---, --- and Chelidze, M. V., ibid. $\underline{37}$, 2654-6 (1967); CA $\underline{69}$, 43994.

2427. --- and Gelashvili, E. S., Soobshch. Akad. Nauk Gruz. SSR $\underline{52}$, 69-74 (1968); CA $\underline{70}$, 78108.

2428. ---, Guntsadze, T. P. and Gudavadze, M. I., ibid. $\underline{50}$, 609-12 (1968); CA $\underline{69}$, 87124.

2429. Hagemeyer, H. J., Jr. and Robinson, A. G., U.S. 3,414,609, Dec. 22, 1961; CA $\underline{70}$, 38467.

2430. Hambling, J. K. and Jones, J. R., Brit. 1,123,474, April 25, 1966; CA $\underline{69}$, 105858.

2431. Harada, T., Bull. Chem. Soc. Jap. $\underline{41}$, 737-41 (1968); CA $\underline{69}$, 36231.

2432. Hardy, A., Carpenter, A. T. and Agger, R. T., Brit. 1,112,408, Dec. 5, 1963; CA $\underline{69}$, 11089.

2433. Hartmann, H., Ann. Chem. $\underline{714}$, 1-7 (1968); CA $\underline{69}$, 52229.

2434. Hayashi, T. et al., Technol. Rep. Osaka Univ. 18, 233-45 (1968); CA 70, 57966.
2435. ---, Kikkawa, S. and Matsuda, S., Kogyo Kagaku Zasshi 71, 710-5 (1968); CA 69, 87127.
2436. Hays, S. B., J. Econ. Entomol. 61, 1154-7 (1968); CA 69, 95390.
2437. Heaton, B. T. and Pidcock, A., J. Organometal. Chem. 14, 235-7 (1968); CA 69, 101597.
2438. Heldt, E., Hoeppner, K. and Krebs, K. H., Z. Anorg. Allg. Chem. 348, 113-6 (1966); CA 70, 15974; cf. ref. 1191.
2439. Henry, P. M., Tetrahedron Lett. 1968, 2285-7; CA 69, 58918.
2440. Herber, R. H. and Goscinny, Y., Inorg. Chem. 7, 1293-8 (1968); CA 69, 47794.
2441. Hirano, A., Zimmerman, H. M. and Levine, S., J. Neuropathol. Exp. Neurol. 27, 571-80 (1968); CA 70, 104730.
2442. Hocking, D. and Freeman, G. H., Trop. Agr. (London) 45, 141-5 (1968); CA 69, 26251.
2443. Hoeppner, K. and Lassmann, G., Z. Naturforsch. A23, 1758-62 (1968); CA 70, 52917.
2444. Holliday, A. K. et al., Chem. Ind. (London) 1968, 1699; CA 70, 47560.
2445. Honda, M. et al., J. Inorg. Nucl. Chem. 30, 3231-7 (1968); CA 70, 42470.
2446. Hota, N. K. and Willis, C. J., Can. J. Chem. 46, 3921-4 (1968); CA 70, 43465.
2447. --- and ---, J. Organometal. Chem. 15, 89-96 (1968); CA 70, 4250.
2448. Hoyer, R. F. and Plapp, F. W., Jr., J. Econ. Entomol. 61, 1269-76 (1968); CA 69, 95403.
2449. Hwang, S. Y., Ph. D. Thesis, Purdue Univ., 1967; Diss. Abstr. B28, 4528 (1968); CA 69, 47743.
2450. Imura, S. and Tamai, Y., U.S. 3,400,142, Feb. 11, 1964; CA 70, 20226.
2451. Isaacs, N. W., Kennard, C. H. L. and Kitching, W., Chem. Commun. 1968, 820-1; CA 69, 55139.
2452. Ishii, Y., Asahi Garasu, Kogyo Gijutsu Shorei-kai Kenkyu Hokoku 13, 479-96 (1967); CA 69, 96842.
2453. Issleib, K., Matschiner, H. and Naumann, S., Talanta 15, 379-84 (1968); CA 69, 8314.
2454. Itoh, K., Fukui, M. and Ishii, Y., Tetrahedron Lett. 1968, 3867-70; CA 69, 87133.
2455. ---, Fukumoto, Y. and Ishii, Y., ibid. 1968, 3199-202; CA 69, 96846.
2456. ---, Matsuzaki, K. and Ishii, Y., J. Chem. Soc., C 1968, 2709-12; CA 70, 11777.
2457. Itoi, K., Japan. 68 10,133, Jan. 23, 1965; CA 69, 106879.
2458. ---, Japan. 68 10,134, Feb. 2, 1965; CA 69, 106880.
2459. ---, Japan. 68 12,132, Feb. 18, 1965; CA 70, 37922.
2460. ---, Japan. 68 12,335, April 21, 1965; CA 70, 37921.
2461. ---, Japan. 68 12,336, April 22, 1965; CA 70, 37923.
2462. ---, Japan. 68 12,812, March 30, 1964; CA 69, 87600.

2463. Itoi, K., Japan. 68 12,813, April 10, 1964; CA <u>69</u>, 87585.
---, Japan. 68 19,015, Aug. 27, 1964; CA <u>70</u>, 58380.
---, Japan. 68 26,180, Oct. 12, 1964; CA <u>70</u>, 88429.
2464. ---, Japan. 68 13,981, July 14, 1964; CA <u>69</u>, 107267.
2465. ---, Japan. 68 14,928, Nov. 5, 1965; CA <u>70</u>, 20223.
2466. ---, Japan. 68 26,301, May 31, 1965; CA <u>70</u>, 78515.
2467. ---, Japan. 68 26,508, Oct. 18, 1965; CA <u>70</u>, 78524.
---, Japan. 68 28,833, Oct. 18, 1965; CA <u>70</u>, 78510.
2468. --- and Nishida, T., Japan. 68 26,181, Oct. 14, 1964; CA <u>70</u>, 97388.
2469. --- and ---, Japan. 68 30,316, March 18, 1966; CA <u>70</u>, 97384.
2470. Jaasma, W. C., U.S. 3,403,495, Oct. 24, 1967; CA <u>70</u>, 20222.
2471. Jappy, J. and Preston, P. N., Inorg. Nucl. Chem. Lett. <u>4</u>, 503-6 (1968); CA <u>70</u>, 8102.
2472. Jarvis, R. H., Short, J. L. and Shotton, F. E., Plant Pathol. <u>16</u>, 49-53 (1967); CA <u>70</u>, 27922.
2473. Jenks, G. J., Aust. Commonwealth, Dep. Supply, Def. Stand. Lab., Tech. Note No. 87, 8 pp. (1966); CA <u>69</u>, 6989.
2474. Jeppson, L. R., Jesser, M. J. and Complin, J. O., J. Econ. Entomol. <u>61</u>, 1502-5 (1968); CA <u>70</u>, 27942.
2475. Jolley, K. W. and Sutcliffe, L. H., Spectrochim. Acta, Pt. A <u>24</u>, 1191-203 (1968); CA <u>69</u>, 56104.
2476. Jula, T. F. and Seyferth, D., Inorg. Chem. <u>7</u>, 1245-6 (1968); CA <u>69</u>, 23495.
2477. Jurkat, P. A., Gas-Wasserfach <u>109</u>, 678-80 (1968); CA <u>69</u>, 45049.
2478. Kabaivanov, V., Glavchev, I. and Natov, M., Khim. Ind. (Sofia) <u>1968</u>, 302-5; CA <u>70</u>, 38180.
2479. Kaplan, G. and Necas, J., Czech. 128,294, March 1, 1966; CA <u>71</u>, 3484.
2480. Karasev, A. N., Polak, L. S. and Shlikhter, E. B., Probl. Kinet. Katal., Akad. Nauk SSSR <u>12</u>, 297-301 (1968); CA <u>70</u>, 23260.
2481. Katsumura, T., Kataoka, H. and Mizuno, Y., U.S. 3,390,159, Sept. 19, 1963; CA <u>69</u>, 77506.
2482. Kawasaki, Y., J. Inorg. Nucl. Chem. <u>30</u>, 3377-80 (1968); CA <u>70</u>, 72726.
2483. Kazankova, M. A., Protsenko, N. P. and Lutsenko, I. F., Zh. Obshch. Khim. <u>38</u>, 106-8 (1968); CA <u>69</u>, 67501.
2484. Kenaga, E. E., U.S. 3,389,048, Aug. 22, 1966; CA <u>69</u>, 43042.
2485. Keshi, A. et al., Brit. 1,124,459, July 29, 1966; CA <u>69</u>, 78168.
2486. Kester, K. B., Ph. D. Thesis, Harvard Univ., 1968; Diss. Abstr. B<u>29</u>, 920 (1968); CA <u>70</u>, 6879.
2487. Khokhlov, P. S., Bliznyuk, N. K. and Zhemchuzhin, S. G., U.S.S.R. 221,702, May 10, 1967; CA <u>70</u>, 4278.
2488. Khoo, L. E. and Lee, H. H., Tetrahedron Lett. <u>1968</u>, 4351-4 (1968); CA <u>69</u>, 95652.
2489. Khrapov, V. V. et al., Izv. Akad. Nauk SSSR, Ser. Khim. <u>1968</u>, 1261-7; CA <u>69</u>, 96850.
2490. King, R. B. and Pannell, K. H., Inorg. Chem. <u>7</u>, 1510-3 (1968); CA <u>69</u>, 56628.
2491. --- and ---, ibid. <u>7</u>, 2356-61 (1968); CA <u>70</u>, 4261.

2492. Kingston, B. M. and Lappert, M. F., Inorg. Nucl. Chem. Lett. 4, 371-3 (1968); CA 69, 48849.
2493. Kircher, J. G. et al., U.S. 3,397,131, Nov. 13, 1964; CA 69, 77503.
2494. Kitching, W. and Das, V. G. K., Aust. J. Chem. 21, 2401-9 (1968); CA 69, 102660.
2495. Kleiner, F. G. and Neumann, W. P., Ann. Chem. 716, 19-28 (1968); CA 70, 11780.
2496. Klinke, D. J., U.S. 3,412,123, April 27, 1966; CA 70, 58021.
2497. Kljajic, R. P. et al., Hrana Ishrana 8, 776-9 (1967); CA 69, 66211.
2498. Kobetz, P. and Shapiro, H., U.S. 3,376,329, March 31, 1965; CA 69, 52301.
2499. Kochkin, D. A. et al., Zh. Fiz. Khim. 42, 2345-9 (1968); CA 70, 46646.
2500. --- and Azerbaev, I. N., Izv. Akad. Nauk Kaz. SSR, Ser. Khim. 18 (4), 46-9 (1968); CA 70, 20010.
2501. Koehler, H. and Beck, W., Z. Anorg. Allg. Chem. 359, 241-5 (1968); CA 69, 52245.
2502. Koester-Pflugmacher, A. and Hirsch, A., J. Organometal. Chem. 12, 349-54 (1968); CA 69, 19279.
2503. Kolesnikov, G. S. et al., U.S.S.R. 218,438, Jan. 20, 1967; CA 69, 87753.
2504. Kolesnikov, S. P., Perlmutter, B. L. and Nefedov, O. M., Dokl. Akad. Nauk SSSR 180, 112-5 (1968); CA 69, 77383; cf. ref. 1590.
2505. Komarov, N. V. and Ermolova, T. I., U.S.S.R. 213,874, Jan. 12, 1967; CA 69, 59372; cf. ref. 1917.
2506. ---, Misiunas, V. and Sklyanova, A. M., "Khim. Atsetilena," pp. 178-81, M. F. Shostakovskii, Ed. Nauka, Moscow, 1968; CA 71, 3443.
2507. ---, Shostakovskii, M. F. and Burnashova, T. D., Zh. Obshch. Khim. 38, 1398-401 (1968); CA 69, 87132.
2508. ---, --- and Sklyanova, A. M., U.S.S.R. 228,687, June 23, 1962; CA 70, 87957; cf. ref. 1920.
2509. Komissarova, B. A. and Sorokin, A. A., Zh. Eksp. Teor. Fiz. 54, 423-31 (1968); CA 69, 14640.
2510. Komura, M., Tanaka, T. and Okawara, R., Inorg. Chim. Acta 2, 321-4 (1968); CA 69, 102598.
2511. Kostyanovskii, R. G., Izv. Akad. Nauk SSSR, Ser. Khim. 1967, 2784; CA 69, 6183.
2512. --- and Prokofev, A. K., ibid. 1968, 274-9; CA 69, 10518.
2513. Kostyuk, A. S. et al., Zh. Obshch. Khim. 38, 413-4 (1968); CA 69, 96851.
2514. Koyanagi, S. and Kitamura, H., Enka Biniiru To Porima 8, 24-33 (1968); CA 69, 107085.
2515. Kravtsov, D. N. et al., Izv. Akad. Nauk SSSR, Ser. Khim. 1968, 1703-8; CA 69, 111968.
2516. ---, Rokhlina, E. M. and Nesmeyanov, A. N., ibid. 1968, 1035-45; CA 70, 4254.
2517. Kriegsmann, H., Hoffmann, H. and Geissler, H., Z. Anorg. Allg. Chem. 359, 58-66 (1968); CA 69, 39959.
2518. Krueger, W., Phytopathol. Z. 62, 174-89 (1968); CA 69, 34913.

2519. Kryuchkov, F. A. and Shoshtaeva, M. V., Sin. Fiz. Khim. Polim. No. 5, 147-55 (1968); CA 70, 2974.
2520. Kuehlein, K. and Neumann, W. P., J. Organometal. Chem. 14, 317-25 (1968); CA 69, 100946.
2521. ---, --- and Mohring, H., Angew. Chem. 80, 438-9 (1968); Int. Ed. Engl. 7, 455-6 (1968); CA 69, 43141.
2522. Kuivila, H. G., U. S. Clearinghouse Fed. Sci. Tech. Inform., AD 664,792, 3 pp. (1967); CA 69, 87130.
2523. Kummer, R. and Graham, W. A. G., Inorg. Chem. 7, 1208-14 (1968); CA 69, 10514.
2524. Kushlefsky, B. G., U.S. 3,389,158, Aug. 5, 1963; CA 69, 77499.
2525. Kuznesof, P. M. and Jolly, W. L., Inorg. Chem. 7, 2574-7 (1968); CA 70, 25268.
2526. Kyriakopoulos, G. B., J. Inst. Petrol. 54, 369-75 (1968); CA 70, 98450.
2527. ---, ibid. 54, 376-9 (1968); CA 70, 98451.
2528. Lampe, F. W. and Niehaus, A., J. Chem. Phys. 49, 2949-53 (1968); CA 70, 8084.
2529. Lapitskii, G. A. et al., Zh. Obshch. Khim. 38, 2787-8 (1968); CA 70, 78100.
   --- et al., U.S.S.R. 228,027, June 26, 1967; CA 70, 58022.
2530. Lappert, M. F. and Travers, N. F., Chem. Commun. 1968, 1569-70; CA 70, 63683.
2531. Lassmann, G. and Hoeppner, K., Z. Naturforsch., A 23, 622-3 (1968); CA 69, 14723.
2532. Lavigne, A. A. et al., J. Organometal. Chem. 15, 57-64 (1968); CA 70, 4247.
2533. Leavitt, F. C. and Johnson, F., U.S. 3,412,119, July 22, 1963; CA 70, 106658.
2534. Lebedeff, M., Marchand, A. and Valade, J., C. R. Acad. Sci., Paris, Ser. C 267, 813-6 (1968); CA 69, 111589.
2535. Leebrick, J. R., U.S. 3,382,264, Feb. 11, 1965; CA 69, 59375.
2536. ---, U.S. 3,395,164, June 29, 1965; CA 69, 68052.
2537. Leusink, A. J., Budding, H. A. and Drenth, W., J. Organometal. Chem. 13, 163-8 (1968); CA 69, 51318.
2538. ---, --- and Marsman, J. W., ibid. 13, 155-62 (1968); CA 69, 67503.
2539. Levin, B. V., Rumyantseva, Z. G. and Mironova, V. V., Izv. Akad. Nauk SSSR, Neorg. Mater. 4, 1947-51 (1968); CA 70, 43499.
2540. Licht, K., Z. Chem. 8, 351-2 (1968); CA 70, 15768.
2541. --- and Kriegsmann, H., ibid. 8, 267-8 (1968); CA 69, 72615.
2542. Lipatova, T. E. et al., Vysokomol. Soedin., Ser. A 10, 859-66 (1968); CA 69, 19559.
2543. ---, Bakalo, L. A. and Loktionova, R. A., ibid. A 10, 1554-60 (1968); CA 69, 77842.
2544. Long, L. H. and Pulford, C. I., J. Inorg. Nucl. Chem. 30, 2071-5 (1968); CA 69, 106835.
2545. Lorbert, J., J. Organometal. Chem. 15, 251-3 (1968); CA 70, 20177.

2546. Luneva, L. K., Sladkov, A. M. and Korshak, V. V., Vysokomol. Soedin., Ser. B 10, 263-6 (1968); CA 69, 11019.
2547. Lutsenko, I. F. et al., J. Organometal. Chem. 14, 229-30 (1968); CA 69, 106840.
2548. Mack, G. P. and Parker, E., Ger. 1,270,799, Jan. 2, 1953; CA 69, 59838.
2549. --- and ---, Ger. 1,271,389, Jan. 2, 1953; CA 69, 59836.
2550. Mackay, K. M. et al., J. Chem. Soc., A 1968, 1920-3; CA 69, 67502.
2551. --- and Watt, R., J. Organometal. Chem. 14, 123-9 (1968); CA 69, 87126.
2552. Maebutsu, T. and Nakada, S., Japan. 68 04,868, Oct. 16, 1963; CA 69, 28085.
2553. Maggio, F. et al., Inorg. Nucl. Chem. Lett. 4, 389-92 (1968); CA 69, 56634.
2554. Magon, L. et al., Ric. Sci. 38, 782-6 (1968); CA 70, 92599.
2555. Maire, J. C. and Ouaki, R., Helv. Chim. Acta 51, 1151-5 (1968); CA 69, 47801.
2556. Malhotra, S. L., Deshpande, A. P. and Kapur, S. L., Indian J. Chem. 6, 587-91 (1968); CA 70, 20369.
2557. Malnar, M., Acta Pharm. Jugoslav. 18, 65-9 (1968); CA 69, 106847.
2558. Maltseva, E. N. et al., Zh. Obshch. Khim. 38, 203-4 (1968); CA 69, 52240.
2559. Mamedov, S. M., Shiekhieva, M. I. and Shiekhiev, I. A., Azerb. Khim. Zh. 1968, 85-91; CA 69, 106834.
2560. Manning, A. R., J. Chem. Soc., A 1968, 1670-3; CA 69, 47735.
2561. Marchenko, G. N., Golobin, S. F. and Smertin, G. Ya., U.S.S.R. 203,221, June 1, 1965; CA 69, 11078.
2562. Masson, J. C., Quan, Le M. and Cadiot, P., Bull. Soc. Chim. Fr. 1968, 1085-8; CA 69, 77398.
2563. Matlack, J. D. and Metzger, A. P., J. Appl. Polym. Sci. 12, 1745-5 (1968); CA 69, 52702.
2564. Matsubayashi, G., Tanaka, T. and Okawara, R., J. Inorg. Nucl. Chem. 30, 1831-6 (1968); CA 69, 67498.
2565. Matsuda, H. et al., Japan. 68 21,290, Dec. 11, 1965; CA 70, 58628.
2566. --- et al., Japan. 68 21,291, Dec. 22, 1965; CA 70, 58625.
2567. Matsuda, N. et al., Japan. 68 22,577, Jan. 24, 1966; CA 70, 96955.
2568. --- et al., Japan. 68 22,578, Jan. 27, 1966; CA 70, 96957.
2569. --- et al., Japan. 68 22,579, March 2, 1966; CA 70, 78154.
2570. --- et al., Japan. 68 22,580, March 3, 1966; CA 70, 78151.
2571. --- et al., Japan. 68 22,581, March 8, 1966; CA 70, 78153.
2572. Matsuda, S. et al., Kogyo Kagaku Zasshi 71, 2054-9 (1968); CA 70, 47920.
2573. Matsui, Y. and Goto, J., Japan. 68 09,211, Jan. 9, 1965; CA 69, 97780.
2574. Mazaev, V. T., Losev, N. I. and Voinov, V. A., Byull. Eksp. Biol. Med. 66, 72-4 (1968); CA 70, 18500.
2575. Mazerolles, P., Dubac, J. and Lesbre, M., C. R. Acad. Sci., Paris, Ser. C 266, 1794-6 (1968); CA 69, 87120.
2576. ---, --- and ---, J. Organometal. Chem. 12, 143-8 (1968); CA 69, 19278.
2577. ---, Lesbre, M. and Lavergne, J. P., C. R. Acad. Sci., Paris, Ser. C 266, 639-41 (1968); CA 69, 59354.

2578. Mazerolles, P. and Manuel, G., C. R. Acad. Sci., Paris, Ser. C 267, 1158-61 (1968); CA 70, 57961.
2579. --- and ---, Bull. Soc. Chim. Fr. 1965, 2447.
---, --- and Thoumas, F., C. R. Acad. Sci., Paris, Ser. C 267, 619-22 (1968); CA 69, 106841.
2580. Mazzolini, C. and Denti, F., U.S. 3,388,201, April 6, 1962; CA 69, 37032.
2581. McCauley, J. A., Ph. D. Thesis, Fordham Univ., 1968; Diss. Abstr. B29, 570-1 (1968); CA 69, 110553.
2582. McFarlane, W., J. Chem. Soc., A 1968, 1630-4; CA 69, 48078.
2583. ---, Mol. Phys. 13, 587-8 (1967); CA 69, 6978.
2584. McInerney, E. F., Ph. D. Thesis, St. John's Univ., 1967; Diss. Abstr. B28, 3652 (1968); CA 69, 36232; cf. ref. 1375.
2585. McKeown, J. J. and Wright, C. D., U.S. 3,389,111, Jan. 4, 1965; CA 69, 36706.
2586. Mednick, S. A. and Sklar, G., Brit. 1,136,688, April 28, 1966; CA 70, 37490.
2587. Megna, I. S. and Koroscil, A., J. Polym. Sci. Pt. B 6 (9), 653-8 (1968); CA 69, 87575.
2588. Mehner, H., Jehring, H. and Kriegsmann, H., J. Organometal. Chem. 15, 97-105 (1968); CA 69, 112875.
2589. ---, --- and ---, ibid. 15, 107-15 (1968); CA 70, 25183.
2590. Melnikov, N. N. et al., U.S.S.R. 211,537, Dec. 22, 1966; CA 69, 52303.
2591. Mendelsohn, J. C. et al., J. Organometal. Chem. 12, 327-40 (1968); CA 69, 52241.
2592. Meyer, H., ibid. 11, 525-32 (1968); CA 69, 2992.
2593. Meyer, J. M., Ph. D. Thesis, Northwestern Univ., 1968; Diss. Abstr. B29, 2340 (1969); CA 70, 92017.
--- and Allred, A. L., J. Phys. Chem. 72, 3043-5 (1968); CA 69, 81965.
2594. Migdal, S., Gertner, D. and Zilkha, A., Can. J. Chem. 46, 2409-13 (1968); CA 69, 59356.
2595. ---, --- and ---, Eur. Polym. J. 4, 465-72 (1968); CA 70, 29415.
2596. Miles, M. G. et al., Inorg. Chem. 7, 1721-9 (1968); CA 69, 72610.
2597. Mingaleva, K. S. et al., Zh. Obshch. Khim. 38, 606-8 (1968); CA 69, 76480.
2598. Minsker, K. S. et al., Vysokomol. Soedin., Ser. A 10, 1336-42 (1968); CA 69, 36562.
2599. --- et al., ibid., Ser. B 10, 454-7 (1968); CA 69, 59786.
2600. Miraldi, N. L., U.S. 3,385,345, Mar. 4, 1966; CA 69, 45711.
2601. Miretskii, V. Yu. et al., Khim. Prom. 44, 258-61 (1968); CA 69, 77397.
2602. Mironov, V. F. et al., Zh. Obshch. Khim. 38, 2292-300 (1968); CA 70, 29017.
2603. ---, Antipin, L. M. and Sobolev, E. S., ibid. 38, 251-5 (1968); CA 69, 96841; cf. ref. 2228.
2604. ---, Berliner, E. M. and Gar, T. K., ibid. 38, 1900-1 (1968); CA 70, 29016.

2605. Moedritzer, K., Groenweghe, L. C. D. and Van Wazer, J. R., J. Phys. Chem. 72, 4380-6 (1968); CA 70, 32016.

2606. --- and Van Wazer, J. R., Inorg. Chem. Acta 2, 111-5 (1968); CA 69, 54673.

--- and ---, J. Organometal. Chem. 13, 145-53 (1968); CA 69, 22544.

2607. --- and ---, U.S. 3,393,215, Dec. 28, 1964; CA 69, 52645.

2608. ---, --- and Miller, R. E., Inorg. Chem. 7, 1638-42 (1968); CA 69, 54689.

2609. Moore, M. and Lanning, F. C., Trans. Kans. Acad. Sci. 70, 426-31 (1967); CA 70, 47563.

2610. --- and ---, Quart. J. Fla. Acad. Sci. 29, 243-7 (1966); CA 69, 19275.

2611. Mori, H. et al., Kogyo Kagaku Zasshi 71, 399-403 (1968); CA 69, 77400.

2612. Morikawa, T., Kagaku To Kogyo (Osaka) 42, 465-90 (1968); CA 70, 38418.

2613. Mosby, W. L. and Klingsberg, E., U.S. 3,397,217, Aug. 3, 1965; CA 69, 77502.

2614. Moss, R., Campbell, K. and Griffiths, S. T., Brit. 1,126,630, Aug. 17, 1967; CA 69, 98222.

2615. Mueller, E., Sommer, R. and Neumann, W. P., Ann. Chem. 718, 1-10 (1968); CA 70, 37802.

2616. Mulla, M. S., Hilgardia 39, 297-324 (1968); CA 70, 2703.

2617. Mullins, M. A. and Curran, C., Inorg. Chem. 7, 2584-8 (1968); CA 70, 19473, 101458.

2618. Murayama, W., Nakamura, M. and Yokokoshi, K., Japan. 68 1,102, Dec. 14, 1965; CA 69, 3667.

2619. Murayama, Yu., Niino, H. and Taki, K., Japan. 68 26,858, Apr. 11, 1966; CA 70, 87953.

2620. --- and Shinno, H., Japan. 68 29,370, Sept. 16, 1966; CA 70, 87951.

2621. Murin, A. N. et al., Fiz. Tverd. Tela 10, 2803-6 (1968); CA 69, 102298.

2622. Myuller, B. E., Panova, N. V. and Apukhtina, N. P., Sin. Fiz.-Khim. Polim. No. 5, 89-95 (1968); CA 70, 5070.

2623. Naarmann, H., Ger. 1,266,506, Aug. 19, 1966; CA 69, 3307.

2624. Nagasawa, S. and Nakayama, I., Bochu-Kagaku 33, 146-52 (1968); CA 70, 95723.

2625. Nagornaya, L. L. et al., Monokris. Stsintill. Org. Lyuminofory No. 1, 72-5 (1967); CA 69, 40579.

2626. Nakanishi, M. and Inamasu, S., Japan. 67 21,334, March 26, 1965; CA 69, 27527.

2627. --- and ---, Japan. 67 24,573, Dec. 26, 1964; CA 69, 44022.

2628. --- and ---, Japan. 67 26,296, Aug. 4, 1964; CA 69, 52304.

2629. Nametkin, N. S. et al., Dokl. Akad. Nauk SSSR 181, 1138-41 (1968); CA 70, 47557.

2630. --- et al., Izv. Akad. Nauk SSSR, Ser. Khim. 1968, 2139-41; CA 70, 20181.

2631. Nasielski, J. et al., Bull. Soc. Chim. Belg. 77, 15-19 (1968); CA 69, 51294.

2632. --- et al., ibid. 77, 349-53 (1968); CA 69, 105662.

2633. Natoli, J. G., U.S. 3,402,189, March 29, 1966; CA 69, 96877.

2634. Necas, J. et al., Czech. 126,750, Oct. 14, 1965; CA 70, 37919.
2635. Nefedov, V. D. et al., Zh. Obshch. Khim. 38, 1219-21 (1968); CA 69, 87125.
2636. Nekhorosheva, E. V. and Albitskaya, V. M., ibid. 38, 1511-7 (1968); CA 70, 87926.
2637. Nesmeyanov, A. N. et al., Dokl. Akad. Nauk SSSR 181, 921-4 (1968); CA 69, 105684.
2638. --- et al., ibid. 183, 118-21 (1968); CA 70, 46660.
--- et al., J. Organometal. Chem. 15, 279-85 (1968); CA 70, 36992.
2639. --- et al., Dokl. Akad. Nauk SSSR 183, 1098-101 (1968); CA 70, 77095.
2640. --- et al., Izv. Akad. Nauk SSSR, Ser. Khim. 1968, 142-5; CA 69, 10526.
2641. --- et al., ibid. 1968, 296-306; CA 69, 52238.
2642. --- et al., ibid. 1968, 793-801; CA 69, 76380.
2643. --- et al., ibid. 1968, 1419-20; CA 69, 77428.
2644. Neumann, W. P. and Kleiner, F. G., Ann. Chem. 716, 29-36 (1968); CA 70, 11781.
2645. --- and Lind, H., Chem. Ber. 101, 2837-44 (1968); CA 69, 66694.
2646. ---, --- and Alester, G., ibid. 101, 2845-54 (1968); CA 69, 66695.
2647. Newton, D. W. and Hays, R. L., J. Econ. Entomol. 61, 1668-9 (1968); CA 70, 27958.
2648. Nguyen, D. H. et al., Zh. Obshch. Khim. 38, 191-2 (1968); CA 69, 19274.
2649. Nomura, M. et al., Kogyo Kagaku Zasshi 71, 1021-4 (1968); CA 70, 11785.
2650. ---, Ando, S. and Matsuda, S., ibid. 71, 394-8 (1968); CA 69, 77384.
2651. ---, Matsui, N. and Matsuda, S., ibid. 71, 1526-9 (1968); CA 70, 47572.
2652. Norman, A. D., Chem. Commun. 1968, 812-3; CA 69, 96840.
2653. Novikova, Z. S., Efimova, E. A. and Lutsenko, I. F., Zh. Obshch. Khim. 38, 2345 (1968); CA 70, 29014.
2654. Obtemperanskaya, S. I., Dudova, I. V. and Dikaya, G. F., Zh. Anal. Khim. 23, 784-6 (1968); CA 69, 41031.
2655. Ochsenbein, P., Kunststoffe 58 (5), 366-7 (1968); CA 69, 59787.
2656. Ohara, M., Okada, T. and Okawara, R., Tetrahedron Lett. 1968, 3489-90; CA 69, 96848.
2657. Orzechowski, A., U.S. 3,392,160, Aug. 29, 1963; CA 69, 44277.
2658. Osugi, J., Kusuhara, S. and Satoshi, H., Rev. Phys. Chem. Jap. 37, 94-104 (1968); CA 69, 51345.
2659. Oustrin, J., Clavel, M. J. and Pitet, G., C. R. Soc. Biol. 162, 1226-9 (1968); CA 70, 113458.
2660. Pande, K. C., J. Organometal. Chem. 13, 187-94 (1968).
2661. Pande, K. C., Fr. 1,506,186, Dec. 27, 1965; CA 69, 106878.
2662. --- and Lengnick, G. F., Fr. 1,506,842, Dec. 27, 1965; CA 70, 12439.
--- and ---, Fr. 1,506,843, Dec. 27, 1965; CA 70, 12452.
2663. Pankratova, V. N. and Stepovik, L. P., Zh. Obshch. Khim. 38, 844-6 (1968); CA 69, 77395.
2664. Pant, B. C. and Reiff, H. F., J. Organometal. Chem. 15, 65-8 (1968); CA 70, 11782.

2665. Parish, R. V. and Platt, R. H., Chem. Commun. 1968, 1118-20; CA 69, 111981.

2666. Pasquale, R. J. De and Tamborski, C., J. Organometal. Chem. 13, 273-82 (1968); CA 69, 77295.

2667. Patil, K. C. and Rao, C. N. R., Indian J. Chem. 6, 220-2 (1968); CA 69, 86031.

2668. Pavlinova, R. M., Nikolaeva, N. S. and Tuleuova, E. T., U.S.S.R. 220,389, July 9, 1966; CA 69, 88142.

2669. Peach, M. E., Can. J. Chem. 46, 2699-706 (1968); CA 69, 83092.

2670. Peddle, G. J. D., J. Organometal. Chem. 14, 115-21 (1968); CA 69, 87056.

2671. ---, ibid. 14, 139-47 (1968); CA 69, 87119.

2672. --- and Redl, G., Chem. Commun. 1968, 626-7; CA 69, 76419.

2673. --- and Roark, D. N., Can. J. Chem. 46, 2507-10 (1968); CA 69, 59320.

2674. ---, Shafir, J. M. and McGeachin, S. G., J. Organometal. Chem. 15, 505-7 (1968); CA 70, 36839.

2675. --- and Ward, J. E. H., ibid. 14, 131-7 (1968); CA 69, 87136.

2676. Pedrotti, R. L., U.S. 3,401,187, June 15, 1965; CA 70, 20224.

2677. Pelczar, F. A., Ph. D. Thesis, Univ. of New Hampshire, 1968; Diss. Abstr. B29, 2363 (1969); CA 70, 106622.

2678. Pelikan, Z. and Cerny, E., Arch. Tosikol. 23, 283-92 (1968); CA 69, 104751.

2679. Perciaccante, V. A., Ph. D. Thesis, St. John's Univ., 1967; Diss. Abstr. B28, 3654 (1968); CA 69, 27502; cf. ref. 1376.

2680. Pereyre, M., Colin, G. and Valade, J., Bull. Soc. Chim. Fr. 1968, 3358-70; CA 70, 11778.

2681. Perin, G., Med. Lav. 58, 624-31 (1967); CA 69, 21739.

2682. Perkins, F., Milne, A. and Hearson, A., S. African 67 4,668, Aug. 10, 1966; CA 70, 38965.

2683. Petrov, A. A. et al., Zh. Obshch. Khim. 38, 1196-7 (1968); CA 69, 56134.

2684. Petukhov, G. G. et al., Tr. Khim. Khim. Tekhnol. 1967, 152-6; CA 70, 77044.

2685. Plato, M., Schnabel, W. and Wendenburg, J., Z. Naturforsch. B 23, 1260-1 (1968); CA 70, 24579.

2686. Plazzogna, G., Bresadola, S. and Tagliavini, G., Inorg. Chim. Acta 1968, 333-6; CA 70, 6925.

2687. Pogorzhelskaya, N. A., Zavgorodnii, V. S. and Maretina, I. A., Zh. Obshch. Khim. 38, 2678-82 (1968); CA 70, 96882.

2688. Pohl, J. J., Monatsh. Chem. 99, 1705-12 (1968); CA 70, 6702.

2689. Poller, R. C. and Toley, D. L. B., J. Organometal. Chem. 14, 453-6 (1968); CA 70, 20207.

2690. Pommier, J. C., Delmont, B. and Valade, J., Tetrahedron Lett. 1967, 5289-91; CA 69, 2987.

2691. ---, Duchene, A. and Valade, J., Bull. Soc. Chim. Fr. 1968, 4677-8; CA 70, 47556.

2692. --- and Valade, J., J. Organometal. Chem. 12, 433-42 (1968); CA 69, 52247.

2693. Popham, E. J., McGlown, D. J. and Graham, T., Brit. 1,124,120, Sept. 13, 1966; CA 69, 107727.
2694. Poskozim, P. S., J. Organometal. Chem. 12, 115-21 (1968); CA 69, 2991.
2695. Preston, P. N., Rice, P. J. and Weir, N. A., Int. J. Mass Spectrom. Ion Phys. 1, 303-8 (1968); CA 69, 91220.
2696. --- and Weir, N. A., Inorg. Nucl. Chem. Lett. 4, 279-81 (1968); CA 69, 31409.
2697. Prince, R. H. and Timms, R. E., Inorg. Chim. Acta 2, 257-9 (1968); CA 70, 6892.
2698. Prohaska, G., Naphta (Zagreb) 19, 461-4 (1968); CA 70, 30576.
2699. Protanova, R. et al., Gazz. Chim. Ital. 98, 1290-300 (1968); CA 70, 61738.
2700. Purinson, Yu. A. et al., Vysokomol. Soedin., Ser. B 10, 257-62 (1968); CA 69, 11020.
2701. Quane, D. and Hunt, G. W., J. Organometal. Chem. 13, P16-20 (1968); CA 69, 77390.
2702. Raichle, K. et al., Ger. 1,266,497, Sept. 10, 1965; CA 69, 3448.
2703. Ramsden, H. E., U.S. 3,389,157, July 3, 1962; CA 69, 77500.
2704. Ramson, A. and Burth, U., Nachrichtenbl. Deut. Pflanzenschutzdienst (Berlin) 22, 146-50 (1968); CA 70, 67076.
2705. Randall, E. W. and Zuckerman, J. J., J. Am. Chem. Soc. 90, 3167-72 (1968); CA 69, 23537.
2706. Razuvaev, G. A. et al., Dokl. Akad. Nauk SSSR 180, 1119-21 (1968); CA 69, 77389.
--- et al., J. Organometal. Chem. 14, 339-47 (1968); CA 69, 105607.
2707. ---, Shcherbakov, V. I. and Zhiltsov, S. F., Izv. Akad. Nauk SSSR, Ser. Khim. 1968, 2803-5; CA 70, 78110.
2708. Reicherdt, W. and Wunsch, K., Ger. (East) 61,866, April 7, 1967; CA 70, 78893.
2709. Reiff, H. F. et al., J. Organometal. Chem. 15, 247-50 (1968); CA 71, 3448.
2710. Remtova, K. and Chvalovsky, V., Collect. Czech. Chem. Commun. 33, 3899-902 (1968); CA 70, 34118.
2711. Revukas, A. J., U.S. 3,401,184, March 6, 1964; CA 69, 95944.
2712. Richart, D. S., Fr. 1,511,876, Feb. 18, 1966; CA 70, 69368.
2713. Ridder, J. J., De and Dijkstra, G., Org. Mass Spectrom. 1, 647-57 (1968); CA 70, 36935.
2714. Riviere, P. and Satge, J., C. R. Acad. Sci., Paris, Ser. C 267, 267-9 (1968); CA 69, 77401.
2715. Roberts, R. M. G., J. Organometal. Chem. 12, 97-103 (1968); CA 69, 18411.
2716. --- and El Kaissi, F., ibid. 12, 79-88 (1968); CA 69, 18341.
2717. Robins, J. and Schafer, R. J., U.S. 3,403,721, June 13, 1966; CA 69, 109365.
2718. Rogozev, B. I. et al., Zh. Obshch. Khim. 38, 2064-9 (1968); CA 70, 7923.
2719. Rosenberg, D., De Haan, J. W. and Drenth, W., Rec. Trav. Chim. 87, 1387-93 (1968); CA 70, 82834.

2720. Roteman, J. and Arnowich, B., U.S. 3,411,936, March 1, 1965; CA 70, 33252.
2721. Rubinchik, G. F. and Manulkin, Z. M., Zh. Obshch. Khim. 38, 841-4 (1968); CA 69, 77394.
2722. Rudner, B., Achorn, G. S. and Hergenrother, P. M., U.S. 3,403,169, Aug. 23, 1963; CA 70, 11804.
2723. Rummens, F. H. A., Raynes, W. T. and Bernstein, H. J., J. Phys. Chem. 72, 2111-9 (1968); CA 69, 31903.
2724. Rustad, D. S., Birchall, T. and Jolly, W. L., Inorg. Synth. 11, 128-30 (1968); CA 69, 77393.
2725. Saegusa, T. et al., J. Am. Chem. Soc. 90, 4182 (1968); CA 69, 87137.
2726. Sage, S. H. and Tobias, R. S., Inorg. Nucl. Chem. Lett. 4, 459-61 (1968); CA 69, 102659.
2727. Sakai, S., Ito, H. and Ishii, Y., Kogyo Kagaku Zassni 71, 186-7 (1968); CA 69, 19568.
2728. Sandel, V. R. and Freedman, H. H., J. Am. Chem. Soc. 90, 2059-69 (1968); CA 69, 35194.
2729. Sandy, C. A., U.S. 3,401,188, Aug. 5, 1965; CA 69, 96875.
2730. --- and Tullio, V., U.S. 3,401,189, Feb. 7, 1966; CA 69, 96876.
2731. Saruto, K. et al., Japan. 68 16,369, March 5, 1965; CA 70, 68522.
2732. ---, Gono, T. and Suenobu, Yu., Japan. 68 7,941, Aug. 25, 1964; CA 69, 87186.
2733. --- and Nakajima, Y., Japan. 68 21,289, Sept. 30, 1965; CA 70, 58029.
2734. Sasaki, S. and Iyama, T., Japan. 68 12,130, Sept. 26, 1964; CA 70, 37924.
2735. Satge, J. and Couret, C., C. R. Acad. Sci., Paris, Ser. C 267, 173-5 (1968); CA 69, 77402.
2736. --- and Riviere-Baudet, M., Bull. Soc. Chim. Fr. 1968, 4093-6; CA 70, 47558.
2737. Sawyer, A. K., May, G. S. and Scofield, R. E., J. Organometal. Chem. 14, 213-6 (1968); CA 69, 106843.
2738. Scherer, O. J. and Hornig, P., Chem. Ber. 101, 2533-47 (1968); CA 69, 52246.
2739. --- and Schieder, G., ibid. 101, 4184-98 (1968); CA 70, 29020.
2740. --- and Schmitt, R., ibid. 101, 3302-12 (1968); CA 69, 95874.
2741. Schicke, P., Appel, K. R. and Schroeder, L., Pflanzenschutzberichte 38, 189-202 (1968); CA 70, 27921.
2742. Schmidbaur, H., Sechser, L. and Schmidt, M., J. Organometal. Chem. 15, 77-87 (1968); CA 70, 4251.
2743. --- and Wolfsberger, W., Chem. Ber. 101, 1664-9 (1968); CA 69, 2988.
2744. Schmitz-DuMont, O. and Jansen, W., Z. Anorg. Allg. Chem. 363, 140-4 (1968); CA 70, 37890.
2745. Scholer, F. R., Frederick, R. and Todd, L. J., J. Organometal. Chem. 14, 261-6 (1968); CA 70, 4178.
2746. Schumann, H. and Benda, H., Angew. Chem. 80, 845 (1968); Int. Ed. Engl. 7, 812-3 (1968); CA 70, 11776.
2747. --- and ---, ibid. 80, 846 (1968); ibid. 7, 813 (1968); CA 70, 11773.

2748. Schumann, H. and Stelzer, O., J. Organometal. Chem. 13, P25-7 (1968); CA 69, 77434.
2749. Schumann-Ruidisch, I., Kalk, W. and Bruening, R., Z. Naturforsch B 23, 307-12 (1968); CA 69, 43965.
2750. Seki, T. et al., Japan. 68 9,370, June 17, 1965; CA 69, 52843.
2751. Seltzer, R., J. Org. Chem. 33, 3896-900 (1968); CA 69, 106845.
2752. Semin, G. K. et al., Izv. Akad. Nauk SSSR, Ser. Khim. 1968, 1401-4; CA 69, 63419.
2753. --- et al., Radiospektrosk. Kvantovokhim. Metody Strukt. Issled. 1967, 225-30; CA 69, 111980, cf. ref. 737.
2754. --- and Bryuchova, E. V., Chem. Commun. 1968, 605-6; CA 69, 86181.
2755. Seyferth, D. and Jula, T. F., J. Am. Chem. Soc. 90, 2938-43 (1968); CA 69, 67505.
2756. ---, Prokai, B. and Cross, R. J., J. Organometal. Chem. 13, 169-75 (1968); CA 69, 96046; cf. ref. 1875.
2757. Sezoko, M., Japan. 68 1,505, Sept. 3, 1964; CA 69, 20337.
2758. Shapiro, H. and Hudson, R. L., U.S. 3,382,265, Feb. 23, 1965; CA 69, 59374.
2759. --- and ---, U.S. 3,393,216, Feb. 23, 1965; CA 69, 77501.
2760. Sharanina, L. G., Zavgorodnii, V. S. and Petrov, A. A., Zh. Obshch. Khim. 38, 1146-50 (1968); CA 69, 77374.
2761. Shelakova, I. D., Ivin, S. Z. and Promonenkov, V. K., U.S.S.R. 232,254, Aug. 31, 1967; CA 70, 96956.
2762. Sheridan, J. E. and Vachrabhorn, V., N. Z. J. Agr. Res. 11, 812-15 (1968); CA 70, 76740.
2763. Shergina, N. I. et al., Primen. Mol. Spektrosk. Khim., Sb. Dokl. Sib. Soveshch., 3rd, Krasnoyarsk. U.S.S.R. 1964, 93-6 (publ. 1966); CA 69, 14227.
2764. Shiina, K., Brennan, T. and Gilman, H., J. Organometal. Chem. 11, 471-7 (1968); CA 69, 2990.
2765. Shikhiev, I. A., Kulibekov, M. R. and Abdullaev, N. D., Azerb. Khim. Zh. 1967, 82-6; CA 69, 77396.
2766. ---, Mustafaev, R. M. and Abdullaev, N. D., Uch. Zap. Azerb. Gos. Univ., Ser. Khim. Nauk No. 2, 55-9 (1967); CA 70, 106624.
2767. --- and Nasirova, M. M., Azerb. Khim. Zh. 1968, 23-5; CA 70, 47561.
2768. Shimono, K., Katsuyama, A. and Katsuji, M., Shizuokaken Seishi Kogyo Shikensko Shiken Hokoku 19, 1-13 (1967); CA 69, 53016.
2769. Shimoi, M., Koda, Y. and Hattori, H., Japan. 68 3,012, Jan. 25, 1965; CA 69, 44396.
2770. Shinkawa, H. and Morikawa, T., Kobunshi Kagaku 25, 666-72 (1968); CA 70, 48077.
2771. Shorygin, P. P. et al., Zh. Fiz. Khim. 42, 1057-62 (1968); CA 69, 47866.
2772. Shostakovskii, M. F. et al., Izv. Akad. Nauk SSSR, Ser. Khim. 1968, 625-9; CA 69, 67500.
2773. --- et al., ibid. 1968, 2604-8; CA 70, 68478.
2774. --- et al., U.S.S.R. 208,947, Jan. 12, 1966; CA 69, 11110.

2775. Shostakovskii, M. F., Komarov, N. V. and Yarosh, O. G., Izv. Akad. Nauk SSSR, Ser. Khim. 1967, 2575-7; CA 69, 43959.
2776. ---, Mirskov, R. G. and Ivanova, N. P., ibid. 1968, 2842; CA 70, 78109.
2777. Shushunov, V. A., Brilkina, T. G. and Safonova, M. K., Dokl. Akad. Nauk SSSR 177, 621-4 (1967); CA 69, 18398.
2778. Siebert, W., Davidsohn, W. E. and Henry, M. C., J. Organometal. Chem. 15, 69-75 (1968); CA 70, 4242.
2779. Simonova, N. I., Babkin, B. M. and Shmidt, Ya. A., Sin. Fiz.-Khim. Polim. 1968, 52-8; CA 70, 4694.
2780. Sinha, S. K. and Patwardhan, W. D., Explosivstoffe 16, 223-5 (1968); CA 70, 49144.
2781. Sirigu, F. et al., Rass. Med. Sarda 71, 163-76 (1968); CA 69, 50734.
2782. Skachkova, I. N., Nauch. Tr. Aspir. Ordinatorov, 1 Mosk. Med. Inst. 1967, 204-6; CA 70, 104739.
2783. Slagan, P. M., Ph. D. Thesis, Univ. of North Carolina, 1967; Diss. Abstr. B28, 3675 (1968); CA 69, 36230.
2784. Smith, A. L., Spectrochim. Acta Pt. A 24, 695-706 (1968); CA 69, 14311.
2785. Smith, H. Q., U.S. 3,397,052, Aug. 29, 1966; CA 69, 66382.
2786. --- and Ivy, E. E., U.S. 3,400,202, March 31, 1966; CA 69, 105341.
2787. Snegur, L. N. and Manulkin, Z. M., Zh. Obshch. Khim. 38, 102-5 (1968); CA 69, 77378.
2788. Snuffer, R. K. and Wiley, D. E., U.S. 3,403,173, Sept. 21, 1965; CA 70, 87952.
2789. Sommer, R., Mueller, E. and Neumann, W. P., Ann. Chem. 718, 11-23 (1968); CA 70, 37414.
2790. Soulages, N. L., J. Gas Chromatogr. 6, 356-60 (1968); CA 69, 53351.
2791. Spirin, Yu. L., Getmanchuk, Yu. P. and Dryagileva, R. I., Vysokomol. Soedin., Ser. B 10, 378-81 (1968); CA 69, 27866.
2792. Splinter, R. C., Ph. D. Thesis, Univ. of Minnesota, 1968; Diss. Abstr. B29, 925-6 (1968); CA 70, 14940.
2793. Srivastava, T. N. and Bhattacharya, S. N., J. Indian Chem. Soc. 45, 764-8 (1968); CA 70, 29021.
2794. --- and Tandon, S. K., ibid. 45, 732-3 (1968); CA 70, 4245.
2795. --- and ---, J. Inorg. Nucl. Chem. 30, 1399-403 (1968); CA 69, 87117.
2796. --- and ---, Naturwissenschaften 55, 391 (1968); CA 69, 74709.
2797. Stadnichuk, M., Kaptyug, O. I. and Petrov, A. A., Zh. Obshch. Khim. 38, 1587-92 (1968); CA 69, 96845.
2798. Stallknecht, G. F. and Calpouzos, L., Phytopathology 58, 788-90 (1968); CA 69, 34910.
2799. Stamm, W., U.S. 3,417,115, Feb. 3, 1965; CA 70, 37925.
2800. Steward, O. W. et al., J. Chem. Soc., A 1968, 3119-22; CA 70, 23577.
2801. Stoeckler, H. A. and Sano, H., Chem. Phys. Lett. 2, 448-50 (1968); CA 70, 42686.
2802. Stoelzel, H., Kloetzscher, I. and Hartmann, S., Ger. 1,276,643, March 4, 1966; CA 70, 29060.
2803. Strong, R. G. and Sbur, D. E., J. Econ. Entomol. 61, 1034-41 (1968); CA 69, 58563.

2804. Struble, D. L., Beckwith, A. L. and Gream, G. E., Tetrahedron Lett. 1968, 3701-4; CA 69, 76727.
2805. Sukhani, D., Gupta, V. D. and Mehrotra, R. C., Aust. J. Chem. 21, 1175-9 (1968); CA 69, 77386.
2806. Sundaramurthi, V. T. and Abdul Kareem, A., Madras Agr. J. 55, 296-300 (1968); CA 70, 76234.
2807. Suzuki, H., Fukui, T. and Musashi, A., Japan. 68 26,319, June 24, 1965; CA 70, 69050.
2808. Suzuki, R. et al., Japan. 68 29,369, Aug. 22, 1966; CA 70, 78152.
2809. ---, Shioyama, H. and Takubo, K., Japan. 68 29,371, Sept. 20, 1966; CA 70, 78156.
2810. Suzuki, S., Kogyo Yosui No. 108, 19-21 (1967); CA 70, 48827.
2811. Tabara, T., Takubo, K. and Hachiya, I., Japan. 68 29,372, Oct. 11, 1966; CA 70, 78155.
2812. Tada, O., Rep. Inst. Sci. Labour No. 69, 10-22 (1968); CA 70, 70892.
2813. Takami, Y., Japan. 68 18,764, March 11, 1965; CA 70, 68524.
2814. --- and Kimura, T., Japan. 68 19,533, March 26, 1965; CA 70, 58030.
2815. --- and ---, Japan. 68 19,534, March 29, 1965; CA 70, 58028.
2816. Takaoka, K. and Igarashi, T., Sankyo Kenkyusho Nempo 19, 127-30 (1967); CA 69, 18258.
2817. Takeda, T. et al., Japan. 68 3,005, Oct. 29, 1964; CA 69, 20022.
2818. --- et al., Japan. 68 6,521, Oct. 29, 1964; CA 69, 52734.
2819. Tanaka, T. et al., Bull. Chem. Soc. Jap. 41, 1497-501 (1968); CA 69, 86068.
2820. --- and Abe, T., Inorg. Nucl. Chem. Lett. 4, 569-72 (1968); CA 70, 20183.
2821. --- and Kamitani, T., Inorg. Chim. Acta 2, 175-8 (1968); CA 69, 56632.
2822. Tarakanov, O. G., Orlov, V. A. and Belyakov, V. K., J. Polym. Sci. Pt. C No. 23 (Pt. 1), 117-25 (1966); CA 69, 52622.
2822a. Tarno, L. J., Brit. 1,110,429; Fr. 1,507,531; U.S. 3,419,516, Jan. 7, 1966; CA 69, 11260.
2823. Tatsuno, T. et al., Shokuhin Eiseigaku Zasshi 9, 488-94 (1968); CA 70, 97453.
2824. Tenny, K. S. and Tenny, A. M., J. Label. Compd. 4, 54-8 (1968); CA 69, 106839.
2825. Thayer, J. S., Inorg. Chem. 7, 2599-602 (1968); CA 70, 16737.
2826. Thomann, G. and Kuhnen, F., Swiss 456,232, Dec. 4, 1963; CA 70, 4277.
2827. Thorburn, J. A. and Bosman, C. J., S. African 67 61, Jan. 4, 1967; CA 70, 60807.
2828. Tishchenko, A. F., Tseshkovskaya, D. L. and Martynyuk, A. V., U.S.S.R. 212,106, Sept. 19, 1966; CA 69, 11579.
2829. Tokizawa, M., Japan. 68 8,480, March 19, 1963; CA 69, 28325.
2830. Tombe, F. J. A. des, van der Kerk, G. J. M. and Noltes, J. G., J. Organometal. Chem. 13, P9-12 (1968); CA 69, 106846.
2831. Tooke, J. W., U.S. 3,397,969, April 25, 1966; CA 69, 79047.
2832. Troitskaya, L. S. and Troitskii, B. B., Plast. Massy 1968, 12-15; CA 70, 4940.

2833. Trotter, W. and Testa, A. C., J. Am. Chem. Soc. 90, 7044-6 (1968); CA 70, 28193.
2834. Tursunbaev, T. L., Manulkin, Z. M. and Tatarenko, A. N., Zh. Obshch. Khim. 38, 2300-3 (1968); CA 70, 29015.
2835. Tzschach, A. et al., Ger. (East) 63,490, May 30, 1967; CA 70, 68521.
2836. Ulbricht, K. et al., J. Organometal. Chem. 13, 343-9 (1968); CA 69, 77379.
2837. ---, Jakoubkova, M. and Chvalovsky, V., Collect. Czech. Commun. 33, 1693-9 (1968); CA 69, 43286.
2838. Umilin, V. A. and Tsinovoi, Yu. N., Izv. Akad. Nauk SSSR, Ser. Khim. 1968, 1409-12; CA 69, 87138.
2839. ---, --- and Devyatykh, G. G., Zh. Fiz. Khim. 42, 2320-2 (1968); CA 70, 14659.
2840. Unland, M. L. and Letcher, J. H., J. Chem. Phys. 49, 2706-2 (1968); CA 69, 111736.
2841. Ursino, J. A., Ph. D. Thesis, St. John's Univ., 1967; Diss. Abstr. B28, 3662 (1968); CA 69, 27505; cf. ref. 300, 1378.
2842. Vanachayangkul, A. and Morris, M. D., Anal. Lett. 1, 885-90 (1968); CA 70, 43387.
2843. Vidhyasekaran, P. and Kothandaraman, R., Indian J. Agr. Sci. 38, 373-7 (1968); CA 69, 85658.
2844. Vinokurova, G. M. and Fattakhov, S. G., U.S.S.R. 228,686, July 6, 1967; CA 70, 96954.
2845. --- and ---, U.S.S.R. 233,668, Oct. 26, 1967; CA 70, 115336.
2846. Vishnyakov, B. A. and Osipov, K. A., Fiz.-Tekhnol. Vop. Kibern. Seminar No. 2, 90-108 (1967); CA 70, 52008.
2847. Voronkov, M. G. et al., Khim. Geterotsikl. Soedin. 1968, 227-9; CA 69, 87129.
2848. Vucelic, M., Croat. Chem. Acta 40, 255-6 (1968); CA 70, 72710.
2849. Vyazankin, N. S. et al., Izv. Akad. Nauk SSSR, Ser. Khim. 1968, 2081-5; CA 70, 29022.
2850. --- et al., Zh. Obshch. Khim. 38, 906-11 (1968); CA 69, 77387.
2851. --- et al., ibid. 38, 1800-3 (1968); CA 70, 78105.
2852. --- et al., ibid. 38, 1803-9 (1968); CA 70, 78104.
2853. ---, Bochkarev, M. N. and Sanina, L. P., ibid. 38, 414-5 (1968); CA 69, 96844.
2854. ---, Bychkov, V. T. and Vostokov, I. A., ibid. 38, 1345-8 (1968); CA 69, 77391.
2855. ---, --- and ---, ibid. 38, 2485-9 (1968); CA 70, 57963.
2856. Wagman, D. D. et al., Nat. Bur. Stand. (U.S.) Tech. Notes No. 270-3, 264 pp. (1968); CA 69, 30787.
2857. Wagner, D., Gertner, D. and Zilkha, A., Can. J. Chem. 46, 3612-4 (1968); CA 70, 16698.
2858. Walker, R. W. and Ramsden, H. E., U.S. 3,410,797, Nov. 13, 1964; CA 70, 21592.
2859. Watanabe, S. and Kawai, S., Japan. 68 19,928, Sept. 21, 1965; CA 70, 68520.

2860. Wegner, G., Nakabash, N. and Cassidi, H. G., J. Polym. Sci., Pt. A-1 6, 3151-6 (1968); CA 69, 107156.
2861. Weisfeld, L. B., Fr. 1,527,274, June 3, 1966; CA 70, 115337.
2862. ---, Thacker, G. A. and Giamundo, L., Advan. Chem. Ser. 85, 38-44 (1968); CA 70, 38428.
---, --- and Hampson, D. G., Annu. Tech. Conf. Soc. Plast. Eng., Tech. Pap. 26th, 1968, 211-5; CA 70, 78641.
2863. Weltzien, H. C., Zucker 21, 241-6 (1968); CA 69, 34917.
2864. Willemsens, L. C. and van der Kerk, G. M. J., J. Organometal. Chem. 13, 357-61 (1968); CA 69, 77376.
--- and ---, Fr. 1,558,723; Neth. 67 4,140, March 20, 1967; CA 70, 47595.
2865. --- and ---, J. Organometal. Chem. 15, 117-24 (1968); CA 70, 4243.
2866. ---, Brit. 1,168,360; Fr. 1,560,383; Neth. 66 18,311, Dec. 29, 1966; CA 70, 4279.
2867. Worrel, C. J., U.S. 3,436,414, Dec. 13, 1966; CA 70, 106659.
2868. Yamada, N. et al., Japan. 68 618, March 24, 1964; CA 69, 10941.
2869. Yergey, A. L., Ph. D. Thesis, Penn. State Univ., 1967; Diss. Abstr. B29, 908-9 (1969); CA 70, 15466; cf. ref. 621.
--- and Lampe, F. W., J. Organometal. Chem. 15, 339-48 (1968); CA 70, 36956.
2870. Yoda, K. and Kimoto, K., Bull. Chem. Soc. Jap. 41, 1687-9 (1968); CA 69, 77829.
2871. Yokoo, T., Hatakenaka, Y. and Okada, K., Japan. 67 18,818, Oct. 15, 1964; CA 69, 11133.
2872. ---, --- and ---, Japan. 68 2,329, Oct. 9, 1964; CA 69, 27528.
2873. Yoshida, M. et al., Bull. Chem. Soc. Jap. 41, 1113-9 (1968); CA 69, 31111.
2874. Yuasa, K. and Takeichi, K., Hakko Kogaku Zasshi 45, 743-9 (1967); CA 69, 9073.
2875. Yue, C. P., Can. J. Chem. 46, 2675-7 (1968); CA 69, 82115.
2876. Yurimoto, J. and Kojima, Y., S. African 68 342, Jan. 20, 1967; CA 70, 69049.
2877. Zablotna, R., Akerman, K. and Szuchnik, A., Bull. Acad. Pol. Sci. Ser. Chem. 16, 405-11 (1968); CA 70, 47559.
2878. Zakharkin, L. I., Kalinin, V. N. and Podvisotskaya, L. S., Izv. Akad. Nauk SSSR, Ser. Khim. 1968, 679-80; CA 69, 87147.
2879. Zambelli, N., Komblas, K. N. and Kovacs, A., Down Earth 24 (3), 25-31 (1968); CA 70, 86528.
2880. Zanella, P. and Tagliavini, G., J. Organometal. Chem. 12, 355-62 (1968); CA 69, 22583.
2881. Zavgorodnii, V. S., Sharanina, L. G. and Petrov, A. A., Zh. Obshch. Khim. 38, 1150-4 (1968); CA 69, 59355.
2882. Zavistoski, J. G., Ph. D. Thesis, Cornell Univ., 1968; Diss. Abstr. B29, 526 (1968); CA 70, 20179.
2883. --- and Zuckerman, J. J., J. Am. Chem. Soc. 90, 6612-6 (1968); CA 70, 11783.

2884. Zbirovska, E. and Hudecek, J., Plast. Hmoty Kauc. 5, 230-4 (1968); CA 70, 20811.

2885. Zeil, W. and Haas, B., Z. Naturforsch. A 22, 2011-4 (1967); CA 69, 7036.

2886. Zhitinkina, A. K. and Shoshtaeva, M. V., Sin. Fiz.-Khim. Polim. 1968, No. 5, 129-32; CA 70, 2976.

2887. Zimmer, H., Bayless, A. and Christopfel, W., J. Organometal. Chem. 14, 222-4 (1968); CA 69, 106842.

2888. ---, Blewett, C. W. and Brakas, A., Tetrahedron Lett. 1968, 1615-8; CA 69, 43995.

2889. Zinkov, Z. E. and Bogatyrev, P. M., Zh. Prikl. Khim. 41, 2257-61 (1968); CA 70, 29690.

2890. Zlochevskaya, I. V. and Rukhadze, E. G., Mikrobiologiya 37, 1116-21 (1968); CA 70, 55186.

2891. Albright and Wilson Ltd., Brit. 1,040,297, 1,042,138; Ger. 1,227,904; Neth. 296,687, Aug. 15, 1962; CA 63, 18152.

2892. --- and ---, Brit. 1,117,652; Fr. 1,472,990; Ger. 1,544,629; Neth. 65 12,796, Oct. 1, 1964; CA 65, 7393.

2893. --- and ---, Brit. 1,146,435; Fr. 1,504,635; Neth. 66 14,326, Oct. 11, 1965; CA 67, 100247.

2894. Allied Chemical Corp., Brit. 1,141,708; Fr. 1,469,395; Ger. 1,595,256; Neth. 66 2,164, Feb. 19, 1965; CA 66, 29583.

2895. Applicazioni Chimiche S.p.A., Belg. 630,561; Fr. 1,388,639; Ital. 693,188-89; Neth. 291,058, April 6, 1962; CA 61, 769.

2896. Aral A.-G., Neth. 65 16,432, Dec. 16, 1964; CA 65, 18399.

2897. Argus Chemical Corp., Brit. 1,038,723, March 26, 1962; CA 65, 20304.

2898. Asahi Chemical Industry Co., Ltd., Brit. 1,126,519; Fr. 1,535,278; Japan. 69 29,743, Aug. 30, 1966; CA 69, 87595.

2899. ---, Brit. 1,138,261; Fr. 1,535,197, Aug. 27, 1966; CA 70, 48223.

2900. Badische Anilin- & Soda-Fabrik A.-G., Brit. 1,118,135; Fr. 1,455,503; Ger. 1,495,208; Neth. 65 14,261, Nov. 5, 1964; CA 65, 15538.

2901. Bakelite Xylonite Ltd., Belg. 665,078; Brit. 1,081,462; Fr. 1,437,825; Ger. 1,519,042, June 9, 1964; CA 64, 20000.

2902. P. Beiersdorf & Co. A.-G., Ger. 1,569,897; Neth. 66 6239, May 14, 1965; CA 66, 77154.

2903. Billiton-Metal & Thermit Chemische Industrie N.V., Neth. 113,311, Oct. 23, 1958; CA 66, 28890.

2904. ---, Brit. 1,109,855; Fr. 1,437,980; Neth. 65 8201, June 26, 1964; CA 64, 17640.

2905. ---, Neth. 66 6272, May 9, 1966; CA 68, 95981.

2906. ---, Neth. 67 1522, Feb. 1, 1966; CA 67, 109314.

2907. C. H. Boehringer Sohn, Fr. 1,389,821, March 25, 1963; CA 63, 1816.

2908. ---, Neth. 66 14,757, Dec. 18, 1963; CA 64, 757.

2909. Carlisle Chemical Works, Inc., Neth. 65 13,659, Oct. 23, 1964; CA 65, 7218.

2910. Charlton, Weddle & Co., Ltd., Neth. 65 11,311, Aug. 30, 1965; CA 67, 53032.

2911. Chemische Werke Huels A.-G., Belg. 671,001; Brit. 1,116,633; Fr. 1,449,499; Ger. 1,258,082, Oct. 16, 1964; CA 65, 20315.
2912. ---, Fr. 1,389,997; Ger. 1,225,526, April 20, 1963; CA 64, 2288.
2913. Compagnia Italiana Petrolio S.p.A., Ital. 671,836, July 26, 1962; CA 65, 5489.
2914. ---, Ital. 674,927, Aug. 29, 1962; CA 63, 12717.
2915. Consortium fuer Elektrochemische Industrie G.m.b.H., Belg. 632,432; Brit. 1,037,085; Fr. 1,357,974; Ger. 1,495,233, May 23, 1962; CA 62, 7943.
2916. ---, Fr. 1,364,456; Neth. 295,379, July 27, 1962; CA 62, 6632.
2917. Deutsche Advance Produktion G.m.b.H., Brit. 1,004,663, April 17, 1961; CA 63, 16555.
2918. Deutsche Solvay-Werke G.m.b.H., Belg. 666,631, July 10, 1964; CA 65, 6206.
2919. Dow Chemical Co., Belg. 653,701; Brit. 1,058,385; Fr. 1,406,767; Ger. 1,495,543; Neth. 64 7,761, Sept. 30, 1963; CA 63, 10132.
2920. ---, Belg. 655,969; Fr. 1,412,642; Neth. 64 13,537, Nov. 20, 1963; CA 63, 18345.
2921. ---, Fr. 1,484,379, June 23, 1966; CA 68, 3512.
2922. Dow Corning Corp., Brit. 1,097,379; Fr. 1,506,971, Dec. 30, 1965; CA 68, 40776.
2923 - 2932 --- No entries.
2933. E. I. Du Pont de Nemour & Co., Brit. 1,044,659, Oct. 31, 1962; CA 66, 3516.
2934. Ethyl Corp., Belg. 635,357; Brit. 976,972; Ger. 1,420,937, July 11, 1960; CA 62, 6326.
2935. ---, Brit. 1,112,921; Fr. 1,441,803; Ger. 1,518,759; Neth. 65 5907, May 11, 1964; CA 64, 14005.
2936. ---, Fr. 1,400,715, June 3, 1964; CA 64, 5138.
2937. ---, Fr. 1,446,652; Ger. 1,545,496; Neth. 65 11,869, Sept. 10, 1964; CA 65, 10401.
2938. ---, Neth. 64 3049, March 20, 1964; CA 64, 6694.
2939. Farbwerke Hoechst A.-G., Brit. 1,078,386; Fr. 1,476,137; Ger. 1,544,902; Neth. 66 4743, April 14, 1965; CA 66, 66219.
2940. Federal Register, 29, 18055-6, Dec. 19, 1964; CA 62, 7028.
2941. ---, 30, 9575, July 13, 1965; CA 63, 13929.
2942. ---, 33, 731-2, Jan. 20, 1968; CA 68, 76971.
2943. ---, 33, 6089, April 20, 1968; CA 68, 113394.
2944. General Electric Co., Brit. 1,079,207; Fr. 1,424,599, Dec. 16, 1963; CA 67, 82838.
2945. Gevaert-Agfa N.V., Brit. 1,108,096; Neth. 67 1700, March 17, 1966; CA 68, 50515.
2946. Imperial Chemical Industries Ltd., Brit. 1,047,074; Neth. 65 11,664, Sept. 7, 1964; CA 65, 2340.
2947. ---, Brit. 1,073,312; Fr. 1,395,901; Ger. 1,224,489, March 25, 1963; CA 63, 18393.
2948. ---, Fr. 1,411,658; Ger. 1,419,482, April 12, 1960; CA 65, 7363.

2949. International Paints Ltd., Brit. 1,124,297; Neth. 65 13,952, Oct. 29, 1964; CA 65, 13955.
2950. Japan Oils & Fats Co., Ltd. and Nitto Chemical Industry Co., Ltd., Belg. 646,406; Brit. 1,062,324; Fr. 1,388,533, April 15, 1963; CA 63, 13568.
2951. Johnson & Johnson, Brit. 1,127,625; Fr. 1,456,964; Ger. 1,570,814; Neth. 65 12,579, Sept. 28, 1964; CA 65, 9111.
2952. August Kettenbach, Brit. 1,033,903, Feb. 17, 1962; CA 65, 10757.
---, Brit. 1,041,851, March 10, 1962; CA 65, 20337.
2953. Koninklijke Industrieele Maatschappij Noury & van der Lande N.V., Brit. 1,169,770; Fr. 1,542,351; Neth. 66 15,781, Nov. 9, 1966; CA 69, 44398.
2954. ---, Neth. 109,491, April 11, 1962; CA 62, 9173.
2955. Kyodo Chemical Co., Ltd., Brit. 1,129,986; Japan. 68 12,131; CA 70, 20225.
2956. Metal & Thermite Chemicals Inc., Belg. 661,478, March 22, 1965; CA 64, 19904.
2957. ---, Belg. 661,480, March 22, 1965; CA 64, 19904.
2958. ---, Brit. 979,995, Aug. 26, 1960; CA 64, 9901.
2959. ---, Brit. 1,016,529; Fr. 1,400,617; Ger. 1,520,387; Neth. 64 5136, May 10, 1963; CA 62, 11988.
2960. ---, Brit. 1,073,079; Fr. 1,451,853; Neth. 65 14,097, Oct. 30, 1964; CA 65, 12371.
2961. ---, Brit. 1,085,002; Fr. 1,466,771; Neth. 66 1352, Feb. 2, 1965; CA 66, 28897.
2962. ---, Brit. 1,096,922; Fr. 1,430,845; Ger. 1,279,018; Neth. 65 5520, May 5, 1964; CA 64, 9766.
2963. ---, Brit. 1,110,709; Neth. 67 14, March 2, 1966; CA 69, 11139.
2964. ---, Fr. 1,447,914; Neth. 297,476, Oct. 4, 1962; CA 64, 14365.
2965. ---, Neth. 64 13,688, Nov. 25, 1964; CA 65, 20303.
2966. Micro-Biological Laboratories, Inc., Fr. 1,386,350, Nov. 28, 1963; CA 62, 16296; Neth. 301,027, Nov. 27, 1963; CA 64, 5139.
2967. Minnesota Mining and Manufacturing Co., Brit. 947,586-7, Feb. 27, 1959; CA 60, 12215.
2968. Monsanto Co., Brit. 1,094,980, 1,097,557; Ger. 1,280,855-6; Neth. 65 16,374, 65 16,414, Dec. 17, 1964; CA 65, 20303.
2969. Montecatini Edison S.p.A., Ital. 792,997, March 24, 1967; CA 70, 20472.
2970. Nitto Chemical Industry Co., Ltd., Belg. 646,676; Brit. 1,053,996; Fr. 1,393,779; Ger. 1,240,081, 1,274,580, April 18, 1963; CA 63, 9985.
2971. ---, Brit. 1,080,178; Ger. 1,569,222; Neth. 65 4150, April 1, 1964; CA 64, 8406.
2972. ---, Brit. 1,158,548; Fr. 1,487,752; Japan. 69 8725, 69 30,382-3, July 26, 1965; CA 68, 60272.
2973. Paint Manuf. 35, 36-8 (1965); CA 62, 13386.
2974. Chas. Pfizer & Co., Inc., Brit. 1,089,428; Ger. 1,543,413, May 12, 1965; CA 68, 13889.

2975. N. V. Philips' Gloeilampenfabrieken, Fr. 1,556,456; Neth. 67 1062; S. Afr. 68 306, Jan. 24, 1967; CA 70, 27901.
2976. ---, Fr. 1,580,189; Neth. 67 4811; S. Afr. 68 1729, April 5, 1967; CA 70, 76767.
2977. ---, Neth. 103,105, May 17, 1958; CA 63, 1175.
2978. ---, Neth. 65 404, Jan. 15, 1965; CA 66, 1861.
2979. ---, Neth. 67 4285, March 23, 1967; CA 70, 55301.
2980. Pure Chemicals Ltd., Fr. 1,425,963, March 5, 1965; CA 65, 17153.
2981. Raytheon Co., Belg. 650,868; Brit. 1,073,254; Fr. 1,401,870; Ger. 1,519,324; Neth. 64 8317-8, July 23, 1963; CA 63, 800.
2982. Reichhold-Beckacite, Brit. 1,122,996; Fr. 1,477,822; Ger. 1,543,126; Swiss 474,551, April 9, 1965; CA 68, 3541.
2983. Rhone-Poulenc S.A., Brit. 1,144,129; Fr. 1,448,021; Neth. 66 2873, March 12, 1965; CA 66, 37608.
2984. Rohm & Haas Co., Brit. 1,119,804; Fr. 1,457,711; Neth. 65 13,108, Oct. 9, 1964; CA 65, 12359.
2985. Shell Internationale Research Maatschappij N.V., Brit. 1,159,295; Fr. 1,548,433; Neth. 67 12,456, Sept. 12, 1966; CA 69, 37712.
2986. ---, Neth. 65 11,077, Aug. 25, 1965; CA 68, 31816.
2987. Siemens-Schuckertwerke A.-G., Brit. 1,001,482, Aug. 9, 1960; CA 63, 12730.
2988. Societe Marles-Kuhlmann-Wyandotte, Fr. A.D. 90,598; Neth. 66 11,206, Aug. 19, 1965; CA 67, 55234.
2989. Societe de la Viscose Suisse, Belg. 666,138, July 18, 1964; CA 64, 19910.
2990. Solvay & Cie., Belg. 628,272, Feb. 11, 1963; CA 61, 7134.
2991. ---, Belg. 659,147; Fr. 1,381,947, Feb. 7, 1964; CA 62, 10542.
2992. ---, Belg. 709,095; Ger. 1,900,467, Jan. 9, 1968; CA 70, 48268.
2993. ---, Brit. 1,137,830; Fr. A.D. 90,940, Oct. 21, 1966; CA 70, 48030.
---, Fr. 1,516,800, Oct. 21, 1966; CA 70, 88427.
2994. ---, Fr. 1,449,872, July 8, 1965; CA 66, 95200.
2995. Stamicarbon N.V., Belg. 641,076; Neth. 286,789, Dec. 14, 1962; CA 63, 701.
2996. ---, Neth. 288,760, Feb. 8, 1963; CA 63, 4418.
2997. Sumitomo Chemical Co., Ltd., Brit. 1,123,725; Fr. 1,485,235; Japan. 68 17,983, June 30, 1967; CA 68, 30404.
2998. Takeda Chemical Industries, Ltd., Belg. 653,569; Brit. 1,078,896; Fr. 1,411,370; Ger. 1,248,931; Neth. 64 11,204, Sept. 27, 1963; CA 65, 4083.
2999. ---, Belg. 654,888; Brit. 1,032,059; Fr. 1,419,914; Ger. 1,272,532; Neth. 64 12,532; Japan. 69 18,877, Oct. 29, 1963; CA 64, 5267.
3000. ---, Brit. 983,814; Japan. 68 4520, June 2, 1961; CA 62, 10540.
3001. Thiem Products, Inc., Brit. 1,005,248, Jan. 9, 1961; CA 63, 15065.
3002. United States Steel Corp., Brit. 1,113,268; Neth. 65 10,567, Aug. 12, 1964; CA 65, 13962.
3003. VEB Leuna-Werke "Walter Ulbricht", Fr. 1,457,496, Aug. 21, 1965; CA 67, 4675.

3004. Wacker-Chemie G.m.b.H., Brit. 991,797, April 5, 1961; CA 63, 5904.
3005. Wyandotte Chemicals Corp., Brit. 994,348, March 10, 1961; CA 63, 5904.

REVIEW PUBLICATIONS AND MONOGRAPHS

Abel, E. W. and Armitage, D. A., "Organosulfur derivatives of silicon, germanium, tin and lead," Advan. Organometal. Chem. 5, 1-92 (1967); CA 67, 54183.

Adam, D. M., "Metal-Ligand and Related Vibrations: A critical survey of the infrared and Raman spectra of metallic and organometallic compounds," Arnold, London, 1967.

American Society for Testing and Materials, "Proposed methods of test for knock characteristics of motor fuel," Philadelphia, 1964; CA 62, 14404.

Andrianov, K. A., "Metalorganic Polymers" (Polymer Reviews, Vol. 8), Interscience, New York, 1965.

Andrianov, K. A., Eaborn, C., Fischer, E. O., Normant, H. and Seyferth, D., "Organometal. Chem. Rev., Vol. I," Elsevier, Amsterdam, 1966.

Ascher, K. R. S. and Nissim, S., "Organotin compounds and their potential use in insect control," World Rev. Pest Control 3, 188-211 (1964); CA 65, 9658.

Aylett, B. J., "Silicon hydrides and their derivatives," Advan. Inorg. Chem. Radiochem. 11, 249-307 (1968).

Bajer, F. J., "Organometallic spectra of silicon, germanium, tin and lead," Progr. Infrared Spectry (H. A. Syzmanski, ed., Plenum Press, New York) 2, 151-76.

Barnes, J. M. and Magos, L., "Toxicology of organometallic compounds," Organometal. Chem. Rev. 3, 137-50 (1968); CA 69, 1454.

Bass, K. C. I., "Homolytic decompositions of organometallic compounds," Lab. Pract. 14, 47-51 (1965); CA 62, 13163.

Beiter, C. B., "Comments on control of microorganisms in floor polish materials," Chem. Spec. Mfr. Ass., Proc. Mid-Year Meet. 53, 216-22 (1967); CA 68, 31145.

Boboli, E., "Industrial methods for preparation of organotin compounds for production of thermal stabilizers for poly(vinyl chloride) and other chlorinated polymers," Przem. Chem. 46, 377-8 (1967); CA 68, 87769.

Bodiot, D., "Infrared absorption spectra of organometallic compounds of tin and lead," Rev. Chim. Miner. 4, 957-75 (1967); CA 69, 31567.

Bokii, N. G. and Struchkov, Yu. T., "Structural chemistry of organic compounds of Group IV nontransition elements (silicon, germanium, tin, lead)," Zh. Strukt. Khim. 9, 722-65 (1968); CA 70, 23710.

Bolanowska, W., "Toxicology of Tetraethyllead," Med. Pracy 16, 476-83 (1965); CA 64, 14841.

Bonati, F., "Organometallic derivatives of β-diketones, "Organometal. Chem. Rev. 1 (2) (1966).

Brilkina, T. G. and Shushunov, V. A., "Oxidation of organometallic compounds," Usp. Khim. 35, 1430-47 (1966); CA 65, 15411.

Brook, A. G., "Keto derivatives of Group IV organometalloids," Advan. Organometal. Chem. 7 (1968).

Bruce, M. I., "Mass spectra of organometal. compounds," Advan. Organometal. Chem. 6 (1968).

Bulten, E. J., "Germanium. Annual survey covering the year 1967," Organometal. Chem. Rev. Sect. B 4, 339-58 (1968); CA 70, 20102.

Bursian, W., "Lead stabilization of hard poly(vinyl chloride) for the production of injection moldings," Plastverarbeiter 14, 581-5 (1963); CA 61, 5853.

Centre National, "Les Derives Organo-Metalliques," Paris, Centre National de la Recherche Scientifique 1963; CA 64, 12721.

Chalmers, L., "Chemistry and applications of organotin compounds," Mfg. Chem. Aerosol News 38, 37-41 (1967); CA 67, 54182.

Chambers, R. D. and Chivers, T., "Pentafluorophenyl-metal compounds," Organometal. Chem. Rev. 1, 279-304 (1966); Usp. Khim. 36, 1117-39 (1967); CA 67, 90861.

Chambers, D. B., Glockling, F. and Light, J. R. C., "Mass spectra of organometallic compounds," Quart. Rev. 22, 317-37 (1968); CA 69, 85831.

Chevassus, F. and Broutelles, R. de, "The Stabilization of Polyvinyl Chloride," Arnold, London, 1963.

Clark, H. C., "Perfluoroalkyl derivatives of the elements," Advan. Fluorine Chem. 3, 45-48, Butterworth, London.

Clark, H. C., "Organometallic cations," New Pathways Inorg. Chem. 1-13, E. A. V. Ebsworth, ed., Univ. Press, Cambridge, England, 1968.

Clark, H. C. and Goel, R. G., "Five coordination in tin compounds," Proc. Int. Conf. Coord. Chem., 8th, Vienna, 120-21 (1964); CA 67, 26797.

Coates, G. E. and Wade, K., "Organometallic Compounds, Vol. 1: The Main Group Elements," Methuen, London, 1967.

Croatt, O. and Barbieri, R., "Some new complexes of organometallic cations with halide and pseudohalide ligands," Ric. Sci. Rend. Sez. A 8, 441-8 (1965); CA 64, 4563.

Cross, R. J., "σ-Complexes of platinum(II) with hydrogen, carbon and other elements of Group IV," in Organometal. Chem. Rev. 2, 97-141 (1967).

Davidson, W. E. and Henry, M. C., "Organometallic acetylenes of the main groups III-V," Chem. Rev. 67, 73-106 (1967).

Davies, A. G., "Organometallic compounds," Annu. Rep. Progr. Chem. 62, 280-89 (1965), London.

Davies, A. G., "Organometallic compounds," Annu. Rep. Progr. Chem. 64, Pt. B, 219-41 (1967), London.

Davies, A. G., "Organotin chemistry," Chem. Brit. 4, 403-7 (1968); CA 69, 105697.

Davydov, V. I., "Germanium," engl. transl., Gordan and Breach, Sci. Publ., New York, London, Paris, 1966.

Dessy, R. E. and Kitching, W., "Organometallic reaction mechanism," Advan. Organometal. Chem. 4 (1966).

Dessy, R. E., Psarras, T. and Green, S., "Redistribution reactions," Ann. N. Y. Acad. Sci. 125, 43-56 (1965); CA 63, 4123.

Dorfmann, R. I., "Experimental and Biological Test Methods," Academic Press, 1962.

Downs, A. J., Ebsworth, E. A. V. and Turner, J. J., "The typical elements," Annu. Rep. Progr. Chem. 63, 138-85 (1966), London.

Eisch, J. J., "The Chemistry of Organometallic Compounds; the Main Group Elements," Macmillan, New York, 1967.

Ekelund, I. B., "Materials science and materials technology," Tidsskr. Kjemi, Bergvesen Met. 26, 93-100 (1966); CA 65, 16186.

Farkas, A. and Mills, G. A., "Tin compounds as catalysts for polyurethan,"

Advan. Catalysis 13, 363 (1962).

Firsoff, V. A., "Possible alternative chemistries of life," Spaceflight I (July) 132-6 (1965); CA 64, 2298.

Fowler, D. G., "Facts about lead and industrial hygiene," J. Occupational Med. 7, 324-9 (1965); CA 63, 17030.

Fowler, D. G., "Lead compounds: control of industrial hazards," Kirk-Othmer Encycl. Chem. Technol., 2nd Ed., 12, 301-3 (1967).

Friswell, N. J. and Gowenlock, B. G., "Inorganic H- and alkyl-containing free radicals," Advan. Free-Radical Chem. 1, 39-75 (1965); CA 66, 65545.

Furukawa, J., "Organometallic compounds and polymer chemistry," Kobunshi 16, 520-30 (1967); CA 69, 77751.

Gielen, M., "Mechanism of Cleavage of Carbon-Metal Bonds," Free University of Brussels, 1967.

Gielen, M. and Nasielski, J., "Struct. and reactivity in organometallic chemistry," Ind. Chim. Belge 29, 767-77 (1964); CA 61, 16083.

Gielen, M. and Sprecher, R., "Coordination of Group IV metals. Interaction of d-orbitals and reactivity," Organometal. Chem. Rev. 1, 455-89 (1966).

Gilman, H., "Some personal notes on more than one-half century of organometallic chemistry," Advan. Organometal. Chem. 7 (1968).

Gilman, H., Atwell, W. H. and Cartledge, F. K., "Catenated organic compounds of Si, Ge, Sn and Pb," Advan. Organometal. Chem. 4, 1-94 (1966); CA 65, 18611.

Glockling, F., "Organogermanium chemistry," Quart. Rev. 20, 45-65 (1966).

Glockling, F., "The Chemistry of Germanium," Academic Press, London, New York, 1969.

Goldanskii, V. I., "The Moessbauer Effect and Its Application in Chemistry," Izd. Akad. Nauk, Moscow 1965.

Goldanskii, V. I. and Herber, R. H., "Chemical Application of Moessbauer Spectroscopy," Academic Press, New York, London, 1968.

Goldanskii, V. I. and Stukan, R. A., "Use of gamma-resonance spectroscopy in structural chemistry," Zh. Strukt. Khim. 8, 875-90 (1967); CA 68, 25168.

Greenwood, N. N., "Some recent applications of Moessbauer spectroscopy,"

Rec. Chem. Progr. 28, 181-94 (1967), CA 68, 109423.

Hagihara, N., Kumada, M. and Okawara, R., "Handbook of Organometallic Compounds," Benjamin, New York, 1968.

Harner, H. R. and Laubengayer, A. W., "Germanium and germanium compounds," Kirk-Othmer Encycl. Chem. Technol., 2nd. Ed. 10, 519-27 (1966).

Harwood, J. H., "Development in organometallics," Chem. Process Eng. 47 (9), 65-69 (1966); CA 65, 18605.

Hashimoto, H., "Some reactions of σ-bonded organometallic compounds," Yuki Gosei Kagaku Kyokai Shi 24, 1156-69 (1966); CA 66, 76047.

Hayes, S., "Nuclear magnetic resonance spectra of organometallic compounds," Bull. Soc. Chim. France 1964, 2715-18; CA 62, 6033.

Hedges, E. S., "New Avenues for Tin Consumption," Intern. Tin Council, London, 1967.

Henry, M. C. and Davidson, W. E., "Organometallic polymers," Ann. N. Y. Acad. Sci. 125, 172-82 (1965); CA 63, 5745.

Herber, R. H., "Moessbauer parameters for metal-organic (Fe, Sn) compounds," Tech. Rept. Ser., Intern. At. Energy Agency No. 50, 121-33 (1966); CA 65, 9971.

Hess, G. G., Lampe, F. W. and Yergey, A. L., "Energetic properties of Group IV hydrides and their alkyl derivatives by electron impact," Ann. N. Y. Acad. Sci. 136, 106-21 (1966); CA 65, 17697.

Heuston, J. P., "Chemical additives in petroleum fuels," S. African Ind. Chemist 20, 65-70 (1966); CA 65, 16743.

Hoggarth, E., "Organometallic chemistry and its growing significance in industry," Advan. Sci. 23, 652-7 (1967).

Hooton, K. A., "Organogermanium compounds," Prep. Inorg. React. 4, 85-176 (1968); CA 69, 77291.

Hutchins, K. G., "Liquid lead stabilizers," Plastics 31 (347), 1130, 1132 (1966).

Inaba, T., "Applications of organotin compounds," Yuki Gosei Kagaku Kyokai Shi 23, 806-15 (1965); CA 63, 13304.

Inoue, S., "Mechanism of polymerization by organometallic compounds," Kagaku

No Ryoiki 22, 874-83 (1968); CA 70, 38116.

Itoh, K., Sakai, S. and Ishii, Y., "Some addition reactions of Group IVA organometallics to unsaturated compounds," Yuki Gosei Kagaku Kyokai Shi 24, 729-40 (1966); CA 65, 16998.

Johnson, C. A., "Analytical chemistry," Mfg. Chemist Aerosol News 36 (1), 72-4 (1965); CA 62, 11150.

Jones, K. and Lappert, M. F., "Organic tin-nitrogen compounds," Organometal. Chem. Rev. 1, 67-92 (1966); CA 65, 16997.

Jones, R. A. Y., "Proton magnetic resonance," Annu. Rev. NMR Spectrosc. 1, 1-42 (1968); CA 70, 62521.

Kerk, G. J. M. van der, "New developments in organolead chemistry," Ind. Eng. Chem. 58, 29-35 (1966); CA 65, 20158.

Kerk, G. J. M. van der, "Organolead catalysts for polyurethane foams," Aust. Chem. Process Eng. 21 (7), 20-3 (1968); CA 69, 97369.

King, R. B., "Reactions of alkali metal derivatives of metal carbonyls and related compounds," Advan. Organometal. Chem. 2, 157-256 (1964).

King, R. B., "Organometallic compounds containing two different transition metals. Annual survey covering the year 1967," Organometal. Chem. Rev., Sect. B. 4, 158-9 (1968).

Klamann, D. and Fischer, N., "Effect of fuel components and additives on knocking tendency of gasoline motors," Chem. Ing. Tech. 38, 925-40 (1966); CA 66, 4620.

Klema, F., "Chemistry of Organolead Compounds I," Mitt. Chem. Forschungsinst. Wirtsch. Oesterr. 20 (1), 1-6 (1966).

Klema, F., "Organolead compounds II," Mitt. Chem. Forschungsinst. Wirtsch. Oesterr. 20 (2), 43-6 (1966); CA 65, 15420.

Klema, F., "Organolead compounds," Chemiker Ztg. 90, 106-7 (1966).

Klemchuk, P. P., "Poly(vinyl chloride) stabilization mechanism," Advan. Chem. Ser. No. 85, 1-17 (1968).

Klimmer, O. R., "Applications of organotin fungicides in agriculture with regard to toxicity," Pflanzenschutzberichte 37, 57-66 (1966); CA 69, 18227.

Kochkin, D. A. and Azerbaev, I. N., "Tin and Lead Organic Monomers and Polymers," Alma-Ata, 1968; CA 70, 20111.

Kollonitsch, J., "Industrial use of organometallics in organic synthesis," Ann. N. Y. Acad. Sci. 125, 161-71 (1965).

Korshak, V. V., "Introduction (to organometallic polymers)," Usp. v. Obl. Sinteza Elementoorgan. Polimerov, Akad. Nauk SSSR, Inst. Elementoorgan. Soedin. 1966, 5-58; CA 65, 12285.

Korshak, V. V., "Progress in production of polymer materials and polymer science," Itogi Nauki Khim. Nauki, Akad. Nauk SSSR No. 8, 9-153 (1966); CA 67, 22154.

Krause, E. and van Grosse, A., "Die Chemie der Metall-Organischen Verbindungen," Saendig, Wiesbaden, 1965.

Krongauz, E. S. "Polymers containing germanium, tin and lead," Usp. v. Obl. Sinteza Elementoorgan. Polimerov, Akad. Nauk SSSR, Inst. Elementoorgan. Soedin. 1966, 129-46; CA 65, 12285.

Kuivila, H. G., "Reactions of organotin hydrides with organic compounds," Advan. Organometal. Chem. 1, 47-87 (1964).

Kuivila, H. G., "Organotin hydrides and organic free radicals," Accounts Chem. Res. 1 (10), 229-305 (1968); CA 70, 11727.

Lappert, M. F. and Prokai, B., "Insertion reactions of compounds of metals and metalloids involving unsaturated substrates," Advan. Organometal. Chem. 5, 225-319 (1967); CA 67, 54184.

Lautsch, W. F., Troeber, A., Koerner, H., Wagner, K., Kaden, R. and Blase, S., "Energy data on organic compounds," Z. Chem. 6 (5), 171-81 (1966); CA 65, 7037.

Lavigne, A. A., Tancrede, J. M. and Pike, R. M., "Coordination compounds of germanium," Coord. Chem. Rev. 3, 497-508 (1968); CA 70, 43440.

Ledwith, A., "The chemistry of carbenes," Roy. Inst. Chem. (London), Lecture Ser. 1964 (5), 1-66 (1965); CA 63, 14361.

Leusink, A. J., Noltes, J. G., Budding, H. A. and van der Kerk, G. J. M., "Synthesis of Group IV organometallic polymers and related compounds," AD 629554, 99 pp. (1965); CA 68, 13465.

Linke, W. F. and Seidell, A., "Solubilities of Inorganic and Metal Organic Compounds, Vol. II," 4th ed., Am. Chem. Soc., Washington, D.C., 1966.

Louis, R., "X-ray analysis in the petroleum industry," Erdoel Kohle 18, 187-90 (1965); CA 62, 15958.

Luijten, J. G. A., "Recent researches in the organotin field," Tin Its Uses 67, 1-3 (1965); CA 63, 14893.

Luijten, J. G. A., "Tin. Annual survey covering the year 1967," Organometal. Rev., Sect. B 4, 359-93 (1968); CA 70, 20104.

Luijten, J. G. A., Rijkens, F. and van der Kerk, G. J. M., "Organometallic nitrogen compounds of germanium, tin and lead," Advan. Organometal. Chem. 3, 397-446 (1965); CA 65, 3904.

Lukevics, E. and Voronkov, M. G., "Addition Products from Organic and Inorganic Hydrides of Silicon, Germanium, Tin and Lead," Izd. Akad. Nauk Latv. SSSR, Riga, 1964.

Luneva, L. K., "Ethynyl derivatives of silicon, germanium, tin and lead," Usp. Khim. 36, 1140-57 (1967); CA 68, 21975.

Ma, T. S. and Gutterson, M., "Organic microchemistry," Anal. Chem. 40 (5), 147R-58R (1968); CA 68, 111179.

MacDiarmid, A. G., ed., "Organometallic Compounds of the Group IV Elements, Vol. 1. The Bond to Carbon, Pt. 1 and 2., Dekker, New York, 1968.

Maddox, M. L., Stafford, S. L. and Kaesz, H. B., "Applications of NMR to the study of organometallic compounds," Advan. Organometal. Chem. 3 (1966).

Makarova, L. G. and Nesmeyanov, A. N., "Methods of Organometallic Chemistry; Mercury, Nauka Moscow 1965.

Marlett, E. M., "Electrochemical synthesis of organometallics," Ann. N. Y. Acad. Sci, 125, 12-24 (1965).

Martin, H., ed., "Insecticide and Fungicide Handbook for Crop Protection," F. A. Davis, Philadelphia, Pa., 1963.

Matsuda, S. and Kikkawa, S., "Organotin compounds having functional groups," Yuki Gosei Kagaku Kyokai Shi 24, 281-92 (1966).

McGinnety, J. A. and Mays, M. J., "Transition-metal carbonyls and organometallic complexes," Annu. Rep. Progr. Chem. (1967) 64, 319-63, Chem. Soc. London, 1968.

Mehrotra, R. C., Gupta, V. D. and Sukhani, D., "Thiol and thio-β-diketone derivatives of elements," Inorg. Chim. Acta 2, 111-21 (1968).

Mironov, V. F. and Gar, T. K., "Organic Compounds of Germanium," Nauka Moscow, 1967.

Mironov, V. F. and Gar, T. K., "Trichlorogermane chemistry," Organometal. Chem. Rev., Sec. A 3, 311-21 (1968); CA 69, 106761.

Misonov, A. and Ogata, I., "Hydrogenation with organometallic complex catalysts," Yuki Gosei Kagaku Kyokai Shi 22, 975-88 (1964); CA 62, 9844.

Moedritzer, K., "Redistribution reactions of organometallic compounds of silicon, germanium, tin and lead," Organometal. Chem. Rev. 1, 179-278 (1966); CA 65, 16997.

Moedritzer, K., "Redistribution equilibria of organometallic compounds," Advan. Organometal. Chem. 6, 171-271 (1968); CA 69, 10488.

Mooney, E. F. and Winson, P. H., "Fluorine-19 nuclear magnetic resonance spectroscopy," Annu. Rev. NMR Spectrosc. 1, 243-311 (1968); CA 70, 62526.

Morton, M., "The nature of organometallic polymerization," Advan. Chem. Ser. No. 52, 1-5 (1966); CA 64, 17713.

Muller, J. C., "Coupling constants between nuclei in nuclear magnetic resonance," Bull. Soc. Chim., France 1964, 2027-32.

National Bureau of Standards, "Selected Values of Chemical Thermodynamic Properties Pt. 1 (1965), Pt. 2 (1966), Pt. 3 (1968), N. B. S., Washington, D. C.

Nefedov, O. M. and Manakov, M. N., "Inorganic organometallic and organic analogs of carbenes," Angew. Chem. 78, 1039-56 (1966); Intern. Ed. Engl. 5, 1021-38 (1966); CA 66, 46445.

Nesmeyanov, A. N. and Kocheshkov, K. A., "Methods of Organometallic Chemistry," Nauka, Moscow, 1965.

Neumann, W. P., "The hydrostannation of unsaturated compounds," Angew. Chem. 76, 849-59 (1964); CA 62, 580.

Neumann, W. P., "Die Organische Chemie des Zinns," Enke Verlag Stuttgart, 1967.

Neumann, W. P., "Increasing significance of organometallic chemistry," Naturwissenschaften 55, 553-7 (1968); CA 70, 57925.

Neumann, W. P. and Kuehlein, K., "Recent developments in the organic chemistry of lead," Advan. Organometal. Chem. 7 (1968).

Nijesen, F. B., "Organotin compounds in sea vegetation - resistant paints," Ind. Vernice 22 (4), 3-7 (1968); CA 69, 37156.

Nilsson, M., "Organometallic chemistry," Svensk. Kem. Tidskr. 80 (6-7), 192-6, 202 (1968); CA 69, 87034.

Noeth, H. and Schmid, G., "Coordination compounds with metal-boron bonding," Allg. Prakt. Chem. 17, 610-12, 615-18 (1966); CA 66, 25503.

Noltes, J. G. and van der Kerk, G. J. M., "Functionally Substituted Organotin Compounds," Tin Research Inst. London, 1958.

Normant, H., "Industrial aspects of the chemistry of some organometallic compounds," Bull. Assoc. Franc. Techniciens Petrole 168, 673-89 (1964).

Ochiai, E., "Catalytic functions of metal ions and their complexes," Coord. Chem. Rev. 3, 49-89 (1968); CA 69, 13253.

Ogata, Y., "Synthesis by use of organic metal compounds," Kagaku No Ryoiki 19, 169-85 (1965); CA 63, 16145.

Okawara, R., "Chemical structure of organotin compounds," Nippon Kagaku Zasshi 86, 543-59 (1965); CA 64, 750.

Okawara, R. et al., "Organic tin compounds," Kagaku No Ryoiki 18, 1033-68 (1964); CA 63, 18137.

Okawara, R. and Wada, M., "Structural aspects of organotin compounds," Advan. Organometal. Chem. 5, 137-67 (1967); CA 67, 53177.

Oki, Y., "Application of organotin compounds in plastics," Yuki Gosei Kagaku Kyokai Shi 26, 688-98 (1968); CA 69, 78001.

Pauson, P. L., "Organometallic Chemistry," Arnold, London, 1967.

Perlman, D., "Microbial production of metal-organic compounds and complexes," Advan. Appl. Microbiol. 7, 103-33 (1965); CA 64, 7326.

Podall, H. E. and Mitchell, M. M., Jr., "The use of organometallic compounds in chemical vapor deposition," Ann. N. Y. Acad. Sci. 125, 218-28 (1965).

Poller, R. C., "Coordination in organotin chemistry," J. Organometal. Chem. 3, 321-9 (1965); CA 66, 8008.

Pommier, C., "Application of gas chromatography to the separation of mineral compounds and organometallics," Rev. Chim. Minerale 3, 401-37 (1966); CA 66, 8008.

Powell, H. M., Prout, C. K. and Wallwork, S. C., "Inorganic and organometallic structures," Annu. Rep. Progr. Chem. 61, 569-93 (1964), London.

Rabek, J. F., "The newest achievements in the research and synthesis of heat stable polymers. III.," Przem. Chem. 46 (3), 130-4 (1967); CA 67, 54433.

Rausch, M. D., "Cyclopentadienyl compounds of metals and metalloids," J. Chem. Educ. 37 (11), 568-78 (1960); CA 60, 5529.

Reutov, O. A. and Beletskaya, I. P., "Reaction Mechanisms of Organometallic Compounds," engl. transl., Interscience, New York, 1968.

Rijkens, F. and van der Kerk, G. J. M., "Organogermanium Chemistry," Germanium Research Committee, Utrecht, 1964.

Roberts, H. L., "Mercury chemistry," Advan. Inorg. Chem. Radiochem. 11, 309-39 (1968); CA 70, 53434.

Rochow, E. G., "Organometallic Chemistry," Reinhold, New York, 1964.

Rochow, E. G., "Direct synthesis of organometallic compounds," J. Chem. Educ. 43 (2), 58-62 (1966); CA 64, 10375.

Ross, A., "Industrial applications of organotin compounds," Ann. N. Y. Acad. Sci. 125, 107-23 (1965).

Ruidisch, I., Schmidbaur, H. and Schumann, H., "Organometallic halides of germanium, tin and lead," Halogen Chemistry 2, 233-349; V. Gutman, ed., Academic Press, New York, London, 1967.

Ruiz, J. L., "New applications of lead," Rev. Met. (Madrid) 2, 370-2 (1966).

Sabin, W. W. and Scales, R. K., "Evaluation of use versus nonuse of tetraethyllead in gasoline," U. S. Public Health Serv., Publ. No. 1440, 95-108 (1965).

Sakurai, H. and Kira, M., "Acyl-substituted Group IVB compounds," Kagaku No Ryoiki 22, 897-905 (1968); CA 70, 37855.

Sano, H., "Nuclear γ-ray resonance. Moessbauer effect," Kagaku No Ryoiki 19, 809-20 (1965); CA 68, 91248.

Saunders, J. H. and Frisch, K. C., "Polyurethans. Chemistry and Technology. Pt. I Chemistry," Interscience, New York, 1962.

Scherer, O. J., "Cleavage of organosilicon, germanium and tin nitrogen compounds by halides of Group III-VI," Organometal. Chem. Rev., Sect. A 3, 281-

309 (1968).

Schmidbaur, H., "Advances in heterosiloxane chemistry," Angew. Chem. 77, 206-16, Intern. Ed. Engl. 4, 201-11 (1965).

Schmidt. M., "Similarities and differences among organometallic compounds of silicon, germanium and tin," Pure Appl. Chem. 13, 15-33 (1966).

Schmitz-DuMont, O., "Ammonolysis of amine complexes and organometallic compounds in liquid ammonia," Rec. Chem. Progr. 29, 13-23 (1968); CA 69, 59301.

Schumann, H. and Schmidt, M., "New products of reaction of organometallic compounds with sulfur, selenium, tellurium and phosphorus," Angew. Chem. 77, 1049-55; Int. Ed. Engl. 4, 1007-13 (1965).

Schwanecke, R., "Accidents and health hazards caused by antiknock agents based on organolead compounds," Zentralbl. Arbeitsmed. Arbeitsschutz 18 (3), 69-78 (1968); CA 69, 5009.

Sehra, K. B. and Trivedi, P. K., "Fungitoxicity and chemical constitution," Labev No. 3 (4), 223-33 (1965); CA 64, 8863.

Sevestre, J., "Action of organostannic monohydrides on double and triple carbon-carbon bonds," Peintures, Pigments, Vernice 41, 278-84, 361-7 (1965); CA 63, 11273, 14896.

Seyferth, D., "Propenyl-metal compounds," Rec. Chem. Progr. 26, 87-100 (1965); CA 63, 9973.

Seyferth, D. and King, B., ed. "Annual Survey of Organometallic Chemistry, Vol. 1, Covering the Year 1964, Vol. 2, 1965, Vol. 3, 1966," Elsevier, New York, 1965, 1966 and 1967, resp.

Seyferth, D. et al., "The transmetallation reaction as a source of new organolithium reagents," Bull. Soc. Chim. Fr. 1963, 1364-67.

Shapiro, H. and Frey, F. W., "Lead compounds: organolead compounds," Kirk-Othmer Encycl. Chem. Technol., 2nd Ed. 12, 282-99 (1967).

Shapiro, H. and Frey, F. W., "The Organic Compounds of Lead," John Wiley Sons, New York, 1968.

Shikhiev, I. A., "The Chemistry of Element-Organic Compounds," Maarif. Baku, 1965.

Sisdo, K. and Kozima, S., "Direct synthesis of organometallic compounds," Kyoto Daigaku Nippon Kagakuseni Kenkyusho Koenshu 24, 29-33 (1967); CA 68, 59632.

Sittig, M., "Organometallics, 1966 (Chemical Process Monograph, No. 20),"
Noyes Development Corp., Park Ridge, N. J., 1966.

Skinner, H. A., "The strength of metal-to-carbon bonds," Advan. Organometal. Chem. 2, 1-48 (1964).

Smith, J. J., Carrick, W. L. and Ingberman, A. K., "Organometallic compounds in olefin polymerization," Ann. N. Y. Acad. Sci. 125, 183-8 (1965).

Soc. Ital. di Fitoiatria, "Index of Agricultural Antiparasitics Registered in Italy," 3rd ed., Pavia (1964); CA 64, 20544.

Sosnovsky, G. and Brown, J. H., "The chemistry of organometallic and organometalloid peroxides," Chem. Rev. 66, 529-55 (1966).

Stanford, E. G., Fearson, J. H. and McConnagle, W. J., ed., "Inorganic and organometallic structures," Progr. Appl. Mater. Res. p. 571-94, Gordon and Breach Sci. Publ. Inc., New York, 1964; CA 62, 5900.

Sterba, M. J., "Refining to produce gasolines of reduced lead content," U. S. Publ. Health Serv., Publ. No. 1440, 113-30 (1965).

Stone, F. G. A., "Organometallic compounds in the development of coordination chemistry," Pure Appl. Chem. 10, 37-51 (1965).

Stone, F. G. A. and West, R., ed., "Advances in Organometallic Chemistry," Vol. 1-7 (1964-68), Academic Press, New York, 1964-68.

Sugino, K., "Organic electrode processes and their application to organic synthesis," Yuki Gosei Kagaku Kyokai Shi 24, 1170-82 (1966).

Sundermeyer, W., "Chemical reactions in molten salts," Chem. Unserer Zeit 1 (5), 150-7 (1967); CA 68, 87320.

Tamborski, C., "Perfluoro organometallic compounds," Trans. N. Y. Acad. Sci. 28, 601-10 (1966); CA 65, 10601.

Tananaeva, I. V. and Shpirt, M. Ya., "Chemistry of Germanium," Khimiya, Moscow, 1967.

Thayer, J. S., "Azide derivatives of organometallic compounds," Organometal. Chem. Rev. 1, 157-78 (1966).

Thayer, J. S. and West, R., "Organometallic pseudohalides," Advan. Organometal. Chem. Vol. 5.

Thinius, K., "Stabilizer and antioxidant toxicity," Plaste Kantschuk 11,

324-8 (1964); CA 63, 10124.

Thompson, A. P., "Lead compounds: health and safety factors," Kirk-Othmer Encycl. Chem. Technol., 2nd Ed. 12, 299-301 (1967).

Tobias, R. S., "σ-Bonded organometallic cations in aqueous solution and crystals," Organometal. Chem. Rev. 1, 93-129 (1966).

Treichel, P. M. and Stone, F. G. A., "Fluorocarbon derivatives of metals," Advan. Organometal. Chem. 1, 143-220 (1964).

Uchinumi, K., "Problems of tetraethyllead in gasoline concerning air pollution," Kagaku Kogyo 18, 26-32 (1967).

U. S. Dept. of Health, Education and Welfare, "Survey of Lead in the Atmosphere of Three Urban Communities," H.E.W., Div. of Air Pollution, Cincinnati, Ohio, 1965.

Vereshchinski, I. V., Lebedev, L. D., Miretskii, V. Yu. and Podkhalyuzin, A. T., "Review of radiation-chemical synthesis reactions with alkyl halides," Advan. Chem. Ser. No. 82, 475-87 (1968); CA 69, 82222.

Volkenau, N. A. et al., "Chemistry of heteroorganic compounds," Razv. Org. Khim. SSSR, 1917-67, Akad. Nauk SSSR, Inst. Istor. Estestvozn. Tekh. 1967, 100-84; CA 68, 105270.

Vyazankin, N. S. and Kruglaya, O. A., "Covalent dimetalloorganic compounds," Usp. Khim. 35, 1388-403 (1966).

Vyazankin, N. S., Razuvaev, G. A. and Kruglaya, O. A., "Organometallic compounds with metal-metal bonds between different metals," Organometal. Chem. Rev., Sect. A 3, 323-423 (1968).

Wada, M. and Okawara, R., "Organic tin sulfides and related compounds," Kagaku No Ryoiki 20, 19-24 (1966).

Walsh, A. D., "Mechanism of the effect of antiknock agents," Kem. Kozlem. 26, 309-34 (1966); CA 66, 97104.

Want, G. M. van der, "Various aspects of organometallic research at the Organic Chemical Institute TNO," T.N.O. Nieuws 23, 98-103 (1968); CA 69, 26227.

Willemsens, L. C., "Lead Annual Survey Covering the Year 1967," Organometal. Chem. Rev., Sect. B 4, 394-404 (1968).

Willemsens, L. C. and van der Kerk, G. J. M., "Investigations in the Field of

Organolead Chemistry," Intern. Lead Zinc Res. Organ., Inc., New York, 1965.

Wittig, G., "The role of ate complexes as reaction-determining intermediates," Quart. Rev. (London) 20, 191-210 (1966).

Wolski, E., "Chemistry of Organometallic Compounds, Pt. A. Organosilicon and Organogermanium Chemistry," surveys of Communist World scientific and technical literature, AD-635221; CA 67, 3135.

Yamakawa, H. and Tanigawa, K., "Thin-layer chromatography of organic metal compounds," Kagaku No Ryoiki Zokan No. 64, 209-19 (1964); CA 62, 10800.

Yamamoto, A., "Oligomerization catalysts," Kobunshi 16 (181), 538-44 (1967); CA 70, 19484.

Young, J. F., "Transition metal complexes with Group IVB elements," Advan. Inorg. Chem. Radiochem. 11, 91-152 (1968).

Yuki, H., "Vinyl polymerization," Kobunshi 16 (181), 545-51 (1967); CA 69, 77758.

Zablotna, R., "Alkylhalogermanes," Wiad. Chem. 22, 861-71 (1968); CA 70, 106586.

Zhdanov, A. A., "Organoelemental polymers," Itogi Nauki, Khim. Nauki, Akad. Nauk SSSR No. 8, 512-41 (1966); CA 67, 22153.

Ziegler, K., "Metal-alkyls: their success and future prospects in industrial chemistry," Chim. Ind. (Paris) 92, 631-44 (1964); CA 62, 9841.

Ziegler, K., "Organometallic compounds in macromolecular chemistry," Plastics Inst. (London), Trans. J. 33 (103), 1-9 (1965); CA 62, 16385.

Ziegler, K., "Forty years' stroll through the realms of organometallic chemistry," Advan. Organometal. Chem. 6, 1-17 (1968).

Zolotareva, K. A. et al., "Stabilizers for polymer materials," Sintez i Issled. Effektivn. Stabilizatorov dlya Polimern. Materialov, Sb., Voronesh 1964, 5-33; CA 65, 18767.

MIX
Papier aus verantwortungsvollen Quellen
Paper from responsible sources
FSC® C105338

If you have any concerns about our products,
you can contact us on
ProductSafety@springernature.com

In case Publisher is established outside the EU,
the EU authorized representative is:
Springer Nature Customer Service Center GmbH
Europaplatz 3, 69115 Heidelberg, Germany

Printed by Libri Plureos GmbH
in Hamburg, Germany